DISTANCE AND MIDPOINT FORMULAS

Distance between $P_1(x_1, y_1)$ and $P_2(x_2, y_2)$:

$$d = \sqrt{(x_2 - x_1)^2 + (y_2 - y_1)^2}$$

Midpoint of P_1P_2: $\left(\dfrac{x_1 + x_2}{2}, \dfrac{y_1 + y_2}{2}\right)$

LINES

Slope of line through $P_1(x_1, y_1)$ and $P_2(x_2, y_2)$ $\qquad m = \dfrac{y_2 - y_1}{x_2 - x_1}$

Point-slope equation of line through $P_1(x_1, y_1)$ with slope m $\qquad y - y_1 = m(x - x_1)$

Slope-intercept equation of line with slope m and y-intercept b $\qquad y = mx + b$

Two-intercept equation of line with x-intercept a and y-intercept b $\qquad \dfrac{x}{a} + \dfrac{y}{b} = 1$

LOGARITHMS

$y = \log_a x$ means $a^y = x$

$\log_a a^x = x$	$a^{\log_a x} = x$
$\log_a 1 = 0$	$\log_a a = 1$
$\log x = \log_{10} x$	$\ln x = \log_e x$
$\log_a xy = \log_a x + \log_a y$	$\log_a\left(\dfrac{x}{y}\right) = \log_a x - \log_a y$
$\log_a x^b = b \log_a x$	$\log_b x = \dfrac{\log_a x}{\log_a b}$

EXPONENTIAL AND LOGARITHMIC FUNCTIONS

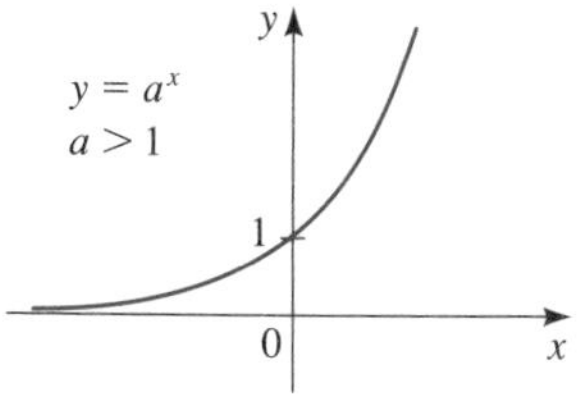

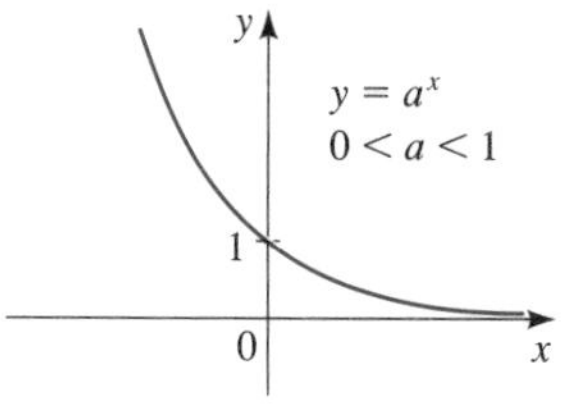

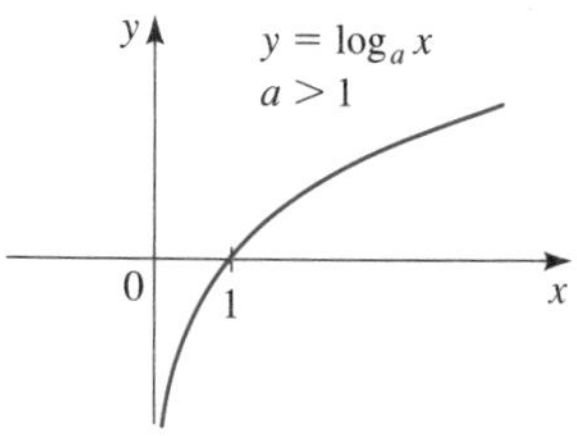

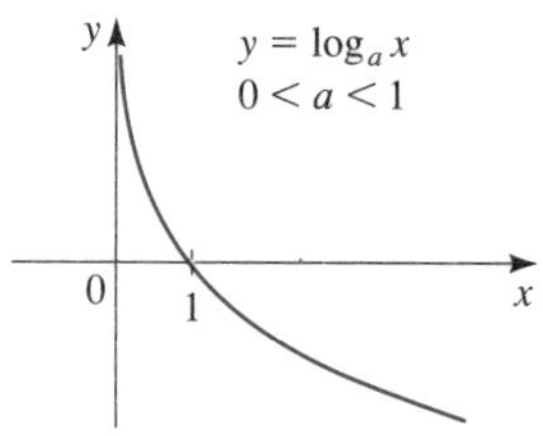

GRAPHS OF FUNCT

Linear functions:

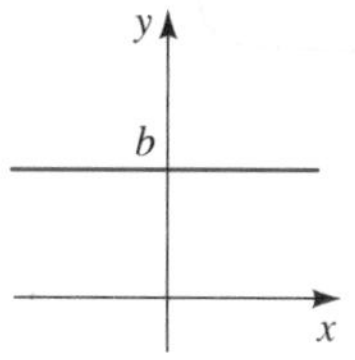

$f(x) = b$

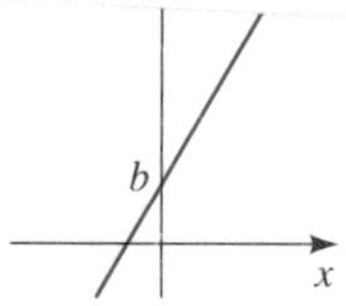

$f(x) = mx + b$

Power functions: $f(x) = x^n$

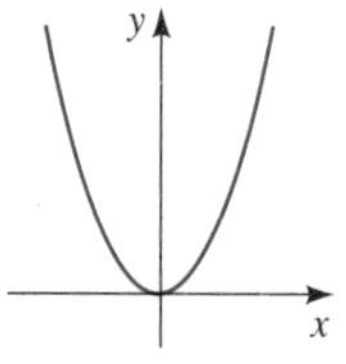

$f(x) = x^2$

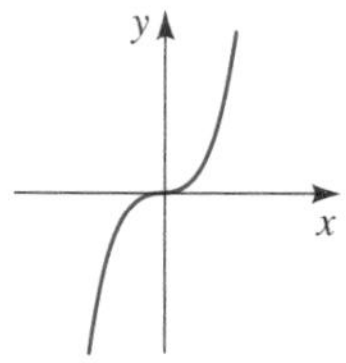

$f(x) = x^3$

Root functions: $f(x) = \sqrt[n]{x}$

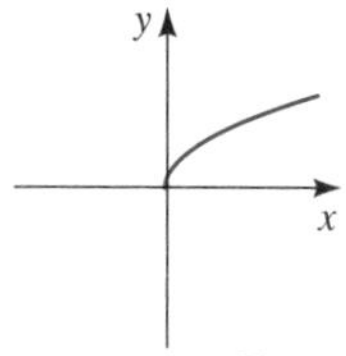

$f(x) = \sqrt{x}$

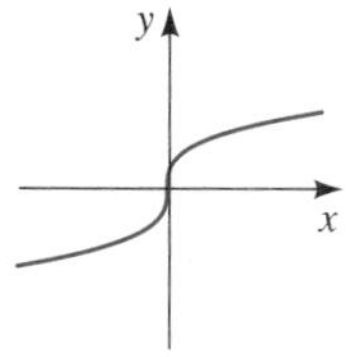

$f(x) = \sqrt[3]{x}$

Reciprocal functions: $f(x) = 1/x^n$

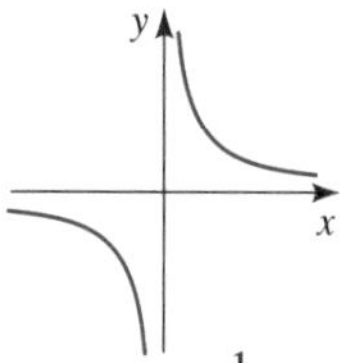

$f(x) = \dfrac{1}{x}$

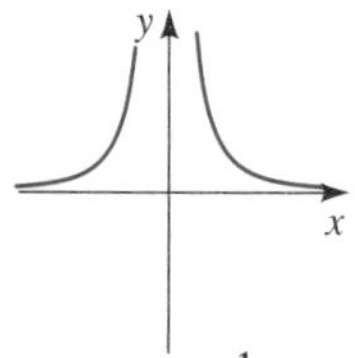

$f(x) = \dfrac{1}{x^2}$

Absolute value function

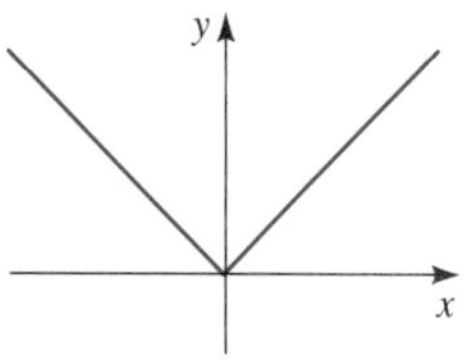

$f(x) = |x|$

Greatest integer function

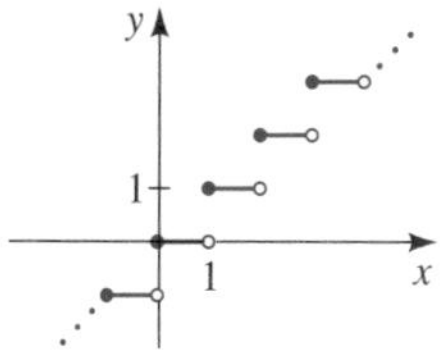

$f(x) = [\![x]\!]$

COMPLEX NUMBERS

For the complex number $z = a + bi$

the **conjugate** is $\bar{z} = a - bi$

the **modulus** is $|z| = \sqrt{a^2 + b^2}$

the **argument** is θ, where $\tan\theta = b/a$

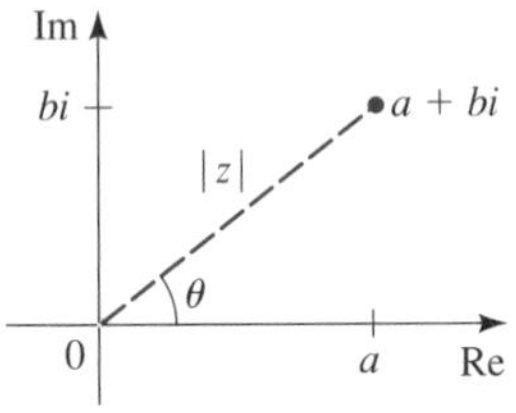

Polar form of a complex number

For $z = a + bi$, the **polar form** is

$$z = r(\cos\theta + i\sin\theta)$$

where $r = |z|$ is the modulus of z and θ is the argument of z

De Moivre's Theorem

$$z^n = [r(\cos\theta + i\sin\theta)]^n = r^n(\cos n\theta + i\sin n\theta)$$

$$\sqrt[n]{z} = [r(\cos\theta + i\sin\theta)]^{1/n}$$

$$= r^{1/n}\left(\cos\frac{\theta + 2k\pi}{n} + i\sin\frac{\theta + 2k\pi}{n}\right)$$

where $k = 0, 1, 2, \ldots, n - 1$

ROTATION OF AXES

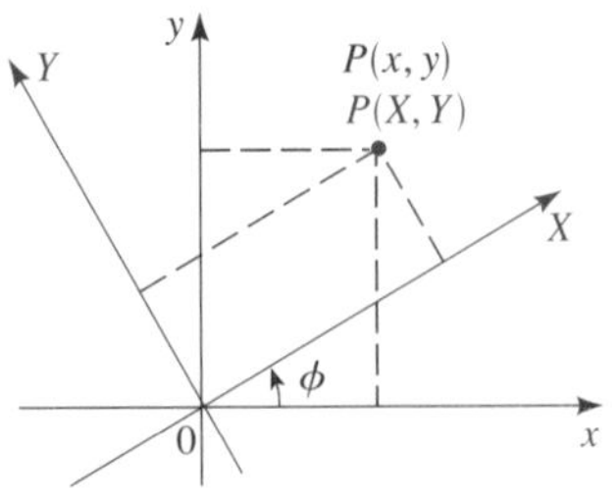

Rotation of axes formulas

$$x = X\cos\phi - Y\sin\phi$$

$$y = X\sin\phi + Y\cos\phi$$

Angle-of-rotation formula for conic sections

To eliminate the xy-term in the equation

$$Ax^2 + Bxy + Cy^2 + Dx + Ey + F = 0$$

rotate the axis by the angle ϕ that satisfies

$$\cot 2\phi = \frac{A - C}{B}$$

POLAR COORDINATES

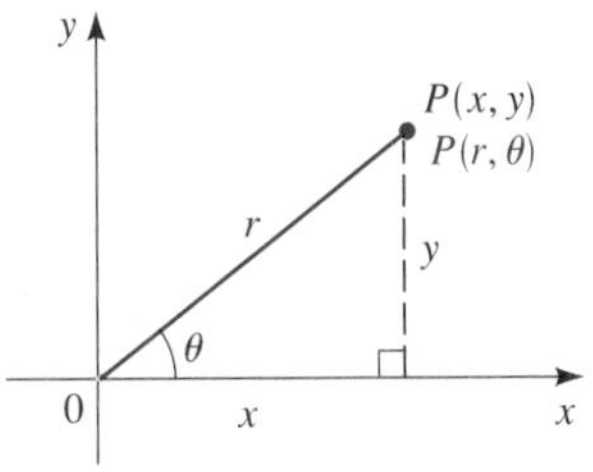

$$x = r\cos\theta$$

$$y = r\sin\theta$$

$$r^2 = x^2 + y^2$$

$$\tan\theta = \frac{y}{x}$$

SUMS OF POWERS OF INTEGERS

$$\sum_{k=1}^{n} 1 = n \qquad \sum_{k=1}^{n} k = \frac{n(n+1)}{2}$$

$$\sum_{k=1}^{n} k^2 = \frac{n(n+1)(2n+1)}{6} \qquad \sum_{k=1}^{n} k^3 = \frac{n^2(n+1)^2}{4}$$

THE DERIVATIVE

The **average rate of change** of f between a and b is

$$\frac{f(b) - f(a)}{b - a}$$

The **derivative** of f at a is

$$f'(a) = \lim_{x \to a} \frac{f(x) - f(a)}{x - a}$$

$$f'(a) = \lim_{h \to 0} \frac{f(a + h) - f(a)}{h}$$

AREA UNDER THE GRAPH OF f

The **area under the graph of f** on the interval $[a, b]$ is the limit of the sum of the areas of approximating rectangles

$$A = \lim_{n \to \infty} \sum_{k=1}^{n} f(x_k)\Delta x$$

where

$$\Delta x = \frac{b - a}{n}$$

$$x_k = a + k\,\Delta x$$

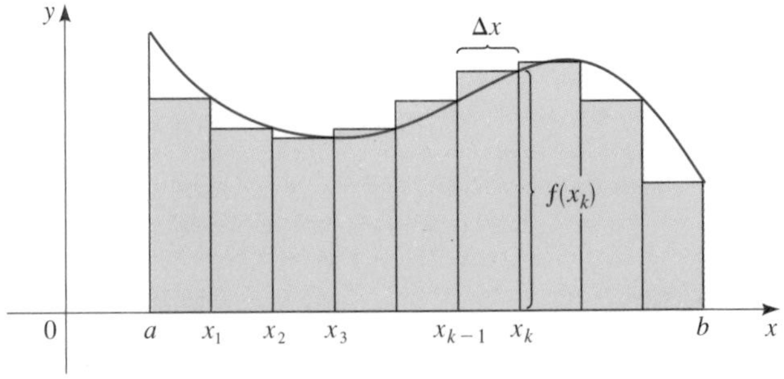

Precalculus: Mathematics for Calculus

University of Michigan - Dearborn

James Stewart

CENGAGE Learning™

Australia • Brazil • Japan • Korea • Mexico • Singapore • Spain • United Kingdom • United States

CENGAGE Learning™

Precalculus: Mathematics for Calculus: University of Michigan - Dearborn

Executive Editors:
Maureen Staudt
Michael Stranz

Senior Project Development Manager:
Linda deStefano

Marketing Specialist:
Courtney Sheldon

Senior Production/Manufacturing Manager:
Donna M. Brown

PreMedia Manager:
Joel Brennecke

Sr. Rights Acquisition Account Manager:
Todd Osborne

Cover Image:
Getty Images*

*Unless otherwise noted, all cover images used by Custom Solutions, a part of Cengage Learning, have been supplied courtesy of Getty Images with the exception of the Earthview cover image, which has been supplied by the National Aeronautics and Space Administration (NASA).

Printed in the United States of America

Precalculus: Mathematics for Calculus, University of Michigan - Dearborn
Stewart

ISBN-13: 978-1-133-35929-6

ISBN-10: 1-133-35929-9

Cengage Learning
5191 Natorp Boulevard
Mason, Ohio 45040
USA

Cengage Learning is a leading provider of customized learning solutions with office locations around the globe, including Singapore, the United Kingdom, Australia, Mexico, Brazil, and Japan. Locate your local office at: **international.cengage.com/region.**

Cengage Learning products are represented in Canada by Nelson Education, Ltd.

For your lifelong learning solutions, visit **www.cengage.com/custom.**
Visit our corporate website at **www.cengage.com.**

CONTENTS

PREFACE

What do students really need to know to be prepared for calculus? What tools do instructors really need to assist their students in preparing for calculus? These two questions have motivated the writing of this book.

To be prepared for calculus a student needs not only technical skill but also a clear understanding of concepts. Indeed, *conceptual understanding* and *technical skill* go hand in hand, each reinforcing the other. A student also needs to gain an appreciation for the power and utility of mathematics in *modeling* the real world. Every feature of this textbook is devoted to fostering these goals.

In writing this Sixth Edition our purpose is to further enhance the utility of the book as an instructional tool for teachers and as a learning tool for students. There are several major changes in this edition including a restructuring of each exercise set to better align the exercises with the examples of each section. In this edition each exercise set begins with *Concepts Exercises,* which encourage students to work with basic concepts and to use mathematical vocabulary appropriately. Several chapters have been reorganized and rewritten (as described below) to further focus the exposition on the main concepts; we have added a new chapter on vectors in two and three dimensions. In all these changes and numerous others (small and large) we have retained the main features that have contributed to the success of this book.

New to the Sixth Edition

- **Exercises** More than 20% of the exercises are new. This includes new *Concept* Exercises and new *Cumulative Review* exercises. Key exercises are now linked to examples in the text.
- **Book Companion Website** A new website **www.stewartmath.com** contains *Discovery Projects* for each chapter and *Focus on Problem Solving* sections that highlight different problem-solving principles outlined in the Prologue.
- **CHAPTER 2 Functions** This chapter has been completely rewritten to focus more sharply on the fundamental and crucial concept of *function.* The material on quadratic functions, formerly in this chapter, is now part of the chapter on polynomial functions.
- **CHAPTER 3 Polynomial and Rational Functions** This chapter now begins with a section on quadratic functions, leading to higher degree polynomial functions.
- **CHAPTER 4 Exponential and Logarithmic Functions** The material on the natural exponential function is now in a separate section.
- **CHAPTER 5 Trigonometric Functions: Unit Circle Approach** This chapter includes a new section on inverse trigonometric functions and their graphs. Introducing this topic here reinforces the function concept in the context of trigonometry.

- **CHAPTER 6 Trigonometric Functions: Right Triangle Approach** This chapter includes a new section on inverse trigonometric functions and right triangles (Section 6.4) which is needed in applying the Laws of Sines and Cosines in the following section, as well as for solving trigonometric equations in Chapter 7.
- **CHAPTER 7 Analytic Trigonometry** This chapter has been completely revised. There are two new sections on trigonometric equations (Sections 7.4 and 7.5). The material on this topic (formerly in Section 7.5) has been expanded and revised.
- **CHAPTER 8 Polar Coordinates and Parametric Equations** This chapter is now more sharply focused on the concept of a coordinate system. The section on parametric equations is new to this chapter. The material on vectors is now in its own chapter.
- **CHAPTER 9 Vectors in Two and Three Dimensions** This is a new chapter with a new *Focus on Modeling* section.
- **CHAPTER 10 Systems of Equations and Inequalities** The material on systems of nonlinear equations is now in a separate section.
- **CHAPTER 11 Conic Sections** This chapter is now more closely devoted to the topic of analytic geometry, especially the conic sections; the section on parametric equations has been moved to Chapter 8.

Teaching with the Help of This Book

We are keenly aware that good teaching comes in many forms, and that there are many different approaches to teaching the concepts and skills of precalculus. The organization of the topics in this book is designed to accommodate different teaching styles. For example, the trigonometry chapters have been organized so that either the unit circle approach or the right triangle approach can be taught first. Here are other special features that can be used to complement different teaching styles:

EXERCISE SETS The most important way to foster conceptual understanding and hone technical skill is through the problems that the instructor assigns. To that end we have provided a wide selection of exercises.

- **Concept Exercises** These exercises ask students to use mathematical language to state fundamental facts about the topics of each section.
- **Skills Exercises** Each exercise set is carefully graded, progressing from basic skill-development exercises to more challenging problems requiring synthesis of previously learned material with new concepts.
- **Applications Exercises** We have included substantial applied problems that we believe will capture the interest of students.
- **Discovery, Writing, and Group Learning** Each exercise set ends with a block of exercises labeled *Discovery* ■ *Discussion* ■ *Writing.* These exercises are designed to encourage students to experiment, preferably in groups, with the concepts developed in the section, and then to write about what they have learned, rather than simply look for the answer.
- **Now Try Exercise . . .** At the end of each example in the text the student is directed to a similar exercise in the section that helps reinforce the concepts and skills developed in that example (see, for instance, page 4).
- **Check Your Answer** Students are encouraged to check whether an answer they obtained is reasonable. This is emphasized throughout the text in numerous *Check Your Answer* sidebars that accompany the examples. (See, for instance, page 52).

A COMPLETE REVIEW CHAPTER We have included an extensive review chapter primarily as a handy reference for the basic concepts that are preliminary to this course.

- **Chapter 1** This is the review chapter; it contains the fundamental concepts from algebra and analytic geometry that a student needs in order to begin a precalculus course. As much or as little of this chapter can be covered in class as needed, depending on the background of the students.

- **Chapter 1 Test** The test at the end of Chapter 1 is designed as a diagnostic test for determining what parts of this review chapter need to be taught. It also serves to help students gauge exactly what topics they need to review.

Flexible Approach to Trigonometry The trigonometry chapters of this text have been written so that either the right triangle approach or the unit circle approach may be taught first. Putting these two approaches in different chapters, each with its relevant applications, helps to clarify the purpose of each approach. The chapters introducing trigonometry are as follows:

- **Chapter 5 Trigonometric Functions: Unit Circle Approach** This chapter introduces trigonometry through the unit circle approach. This approach emphasizes that the trigonometric functions are functions of real numbers, just like the polynomial and exponential functions with which students are already familiar.
- **Chapter 6 Trigonometric Functions: Right Triangle Approach** This chapter introduces trigonometry through the right triangle approach. This approach builds on the foundation of a conventional high-school course in trigonometry.

Another way to teach trigonometry is to intertwine the two approaches. Some instructors teach this material in the following order: Sections 5.1, 5.2, 6.1, 6.2, 6.3, 5.3, 5.4, 5.5, 5.6, 6.4, 6.5, and 6.6. Our organization makes it easy to do this without obscuring the fact that the two approaches involve distinct representations of the same functions.

Graphing Calculators and Computers We make use of graphing calculators and computers in examples and exercises throughout the book. Our calculator-oriented examples are always preceded by examples in which students must graph or calculate by hand, so that they can understand precisely what the calculator is doing when they later use it to simplify the routine, mechanical part of their work. The graphing calculator sections, subsections, examples, and exercises, all marked with the special symbol, are optional and may be omitted without loss of continuity. We use the following capabilities of the calculator.

- **Graphing, Regression, Matrix Algebra** The capabilities of the graphing calculator are used throughout the text to graph and analyze functions, families of functions, and sequences; to calculate and graph regression curves; to perform matrix algebra; to graph linear inequalities; and other powerful uses.
- **Simple Programs** We exploit the programming capabilities of a graphing calculator to simulate real-life situations, to sum series, or to compute the terms of a recursive sequence. (See, for instance, pages 787 and 791.)

Focus on Modeling The "modeling" theme has been used throughout to unify and clarify the many applications of precalculus. We have made a special effort to clarify the essential process of translating problems from English into the language of mathematics (see pages 214 and 636).

- **Constructing Models** There are numerous applied problems throughout the book where students are given a model to analyze (see, for instance, page 228). But the material on modeling, in which students are required to *construct* mathematical models, has been organized into clearly defined sections and subsections (see for example, pages 213, 340, and 427).
- **Focus on Modeling** Each chapter concludes with a *Focus on Modeling* section. The first such section, after Chapter 1, introduces the basic idea of modeling a real-life situation by fitting lines to data (linear regression). Other sections present ways in which polynomial, exponential, logarithmic, and trigonometric functions, and systems of inequalities can all be used to model familiar phenomena from the sciences and from everyday life (see for example pages 296, 357, and 427).

BOOK COMPANION WEBSITE A website that accompanies this book is located at **www. stewartmath.com**. The site includes many useful resources for teaching precalculus, including the following:

- **Discovery Projects** *Discovery Projects* for each chapter are available on the website. Each project provides a challenging but accessible set of activities that enable students (perhaps working in groups) to explore in greater depth an interesting aspect of the topic they have just learned. (See for instance the Discovery Projects *Visualizing a Formula, Relations and Functions, Will the Species Survive?,* and *Computer Graphics I* and *II.*)
- **Focus on Problem Solving** Several *Focus on Problem Solving* sections are available on the website. Each such section highlights one of the problem-solving principles introduced in the Prologue and includes several challenging problems. (See for instance *Recognizing Patterns, Using Analogy, Introducing Something Extra, Taking Cases,* and *Working Backward.*)

MATHEMATICAL VIGNETTES Throughout the book we make use of the margins to provide historical notes, key insights, or applications of mathematics in the modern world. These serve to enliven the material and show that mathematics is an important, vital activity, and that even at this elementary level it is fundamental to everyday life.

- **Mathematical Vignettes** These vignettes include biographies of interesting mathematicians and often include a key insight that the mathematician discovered and which is relevant to precalculus. (See, for instance, the vignettes on Viète, page 49; Salt Lake City, page 84; and radiocarbon dating, page 333).
- **Mathematics in the Modern World** This is a series of vignettes that emphasizes the central role of mathematics in current advances in technology and the sciences (see pages 283, 700, and 759, for example).

REVIEW SECTIONS AND CHAPTER TESTS Each chapter ends with an extensive review section including the following.

- **Concept Check** The *Concept Check* at the end of each chapter is designed to get the students to think about and explain in their own words the ideas presented in the chapter. These can be used as writing exercises, in a classroom discussion setting, or for personal study.
- **Review Exercises** The *Review Exercises* at the end of each chapter recapitulate the basic concepts and skills of the chapter and include exercises that combine the different ideas learned in the chapter.
- **Chapter Test** The review sections conclude with a *Chapter Test* designed to help students gauge their progress.
- **Cumulative Review Tests** The *Cumulative Review Tests* following Chapters 4, 7, 9, 11, and 13 combine skills and concepts from the preceding chapters and are designed to highlight the connections between the topics in these related chapters.
- **Answers** Brief answers to odd-numbered exercises in each section (including the review exercises), and to all questions in the Concepts Exercises and Chapter Tests, are given in the back of the book.

Acknowledgments

We thank the following reviewers for their thoughtful and constructive comments.

REVIEWERS FOR THE FIFTH EDITION Kenneth Berg, University of Maryland; Elizabeth Bowman, University of Alabama at Huntsville; William Cherry, University of North Texas; Barbara Cortzen, DePaul University; Gerry Fitch, Louisiana State University; Lana Grishchenko, Cal Poly State University, San Luis Obispo; Bryce Jenkins, Cal Poly State University, San Luis Obispo; Margaret Mary Jones, Rutgers University; Victoria

Kauffman, University of New Mexico; Sharon Keener, Georgia Perimeter College; YongHee Kim-Park, California State University Long Beach; Mangala Kothari, Rutgers University; Andre Mathurin, Bellarmine College Prep; Donald Robertson, Olympic College; Jude Socrates, Pasadena City College; Enefiok Umana, Georgia Perimeter College; Michele Wallace, Washington State University; and Linda Waymire, Daytona Beach Community College.

REVIEWERS FOR THE SIXTH EDITION Raji Baradwaj, UMBC; Chris Herman, Lorain County Community College; Irina Kloumova, Sacramento City College; Jim McCleery, Skagit Valley College, Whidbey Island Campus; Sally S. Shao, Cleveland State University; David Slutzky, Gainesville State College; Edward Stumpf, Central Carolina Community College; Ricardo Teixeira, University of Texas at Austin; Taixi Xu, Southern Polytechnic State University; and Anna Wlodarczyk, Florida International University.

We are grateful to our colleagues who continually share with us their insights into teaching mathematics. We especially thank Andrew Bulman-Fleming for writing the Study Guide and the Solutions Manual and Doug Shaw at the University of Northern Iowa for writing the Instructor Guide.

We thank Martha Emry, our production service and art editor; her energy, devotion, experience, and intelligence were essential components in the creation of this book. We thank Barbara Willette, our copy editor, for her attention to every detail in the manuscript. We thank Jade Myers and his staff at Matrix Art Services for their attractive and accurate graphs and Precision Graphics for bringing many of our illustrations to life. We thank our designer Lisa Henry for the elegant and appropriate design for the interior of the book.

At Brooks/Cole we especially thank Stacy Green, developmental editor, for guiding and facilitating every aspect of the production of this book. Of the many Brooks/Cole staff involved in this project we particularly thank the following: Jennifer Risden, content project manager, Cynthia Ashton, assistant editor; Lynh Pham, media editor; Vernon Boes, art director; and Myriah Fitzgibbon, marketing manager. They have all done an outstanding job.

Numerous other people were involved in the production of this book—including permissions editors, photo researchers, text designers, typesetters, compositors, proof readers, printers, and many more. We thank them all.

Above all, we thank our editor Gary Whalen. His vast editorial experience, his extensive knowledge of current issues in the teaching of mathematics, and especially his deep interest in mathematics textbooks, have been invaluable resources in the writing of this book.

INSTRUCTOR RESOURCES

Printed

Complete Solution Manual

ISBN-10: 0-8400-6880-8; ISBN-13: 978-0-8400-6880-4

The complete solutions manual provides worked-out solutions to all of the problems in the text.

Instructor's Guide ISBN-10: 0-8400-6883-2; ISBN-13: 978-0-8400-6883-5

Doug Shaw, author of the Instructor Guides for the widely used Stewart calculus texts, wrote this helpful teaching companion. It contains points to stress, suggested time to allot, text discussion topics, core materials for lectures, workshop/discussion suggestions, group work exercises in a form suitable for handout, solutions to group work exercises, and suggested homework problems.

Media

Enhanced WebAssign ISBN-10: 0-538-73810-3; ISBN-13: 978-0-538-73810-1

Exclusively from Cengage Learning, Enhanced WebAssign® offers an extensive online program for Precalculus to encourage the practice that's so critical for concept mastery. The meticulously crafted pedagogy and exercises in this text become even more effective in Enhanced WebAssign, supplemented by multimedia tutorial support and immediate feedback as students complete their assignments. Algorithmic problems allow you to assign unique versions to each student. The Practice Another Version feature (activated at your discretion) allows students to attempt the questions with new sets of values until they feel confident enough to work the original problem. Students benefit from a new Premium eBook with highlighting and search features; Personal Study Plans (based on diagnostic quizzing) that identify chapter topics they still need to master; and links to video solutions, interactive tutorials, and even live online help.

ExamView Computerized Testing

ExamView® testing software allows instructors to quickly create, deliver, and customize tests for class in print and online formats, and features automatic grading. Includes a test bank with hundreds of questions customized directly to the text. ExamView is available within the PowerLecture CD-ROM.

Solution Builder **www.cengage.com/solutionbuilder**

This online instructor database offers complete worked solutions to all exercises in the text, allowing you to create customized, secure solutions printouts (in PDF format) matched exactly to the problems you assign in class.

PowerLecture with ExamView

ISBN-10: 0-8400-6901-4; ISBN-13: 978-0-8400-6901-6

This CD-ROM provides the instructor with dynamic media tools for teaching. Create, deliver, and customize tests (both print and online) in minutes with ExamView® Computerized Testing Featuring Algorithmic Equations. Easily build solution sets for homework or exams using Solution Builder's online solutions manual. Microsoft® PowerPoint® lecture slides and figures from the book are also included on this CD-ROM.

STUDENT RESOURCES

Printed

Student Solution Manual

ISBN-10: 0-8400-6879-4; ISBN-13: 978-0-8400-6879-8

Contains fully worked-out solutions to all of the odd-numbered exercises in the text, giving students a way to check their answers and ensure that they took the correct steps to arrive at an answer.

Study Guide ISBN-10: 0-8400-6917-0; ISBN-13: 978-0-8400-6917-7

This carefully crafted learning resource helps students develop their problem-solving skills while reinforcing their understanding with detailed explanations, worked-out examples, and practice problems. Students will also find listings of key ideas to master. Each section of the main text has a corresponding section in the Study Guide.

Media

Enhanced WebAssign ISBN-10: 0-538-73810-3; ISBN-13: 978-0-538-73810-1

Exclusively from Cengage Learning, Enhanced WebAssign® offers an extensive online program for Precalculus to encourage the practice that's so critical for concept mastery. You'll receive multimedia tutorial support as you complete your assignments. You'll also benefit from a new Premium eBook with highlighting and search features; Personal Study Plans (based on diagnostic quizzing) that identify chapter topics you still need to master; and links to video solutions, interactive tutorials, and even live online help.

Book Companion Website

A new website **www.stewartmath.com** contains *Discovery Projects* for each chapter and *Focus on Problem Solving* sections that highlight different problem-solving principles outlined in the Prologue.

CengageBrain.com

Visit **www.cengagebrain.com** to access additional course materials and companion resources. At the CengageBrain.com home page, search for the ISBN of your title (from the back cover of your book) using the search box at the top of the page. This will take you to the product page where free companion resources can be found.

Text-Specific DVDs ISBN-10: 0-8400-6882-4; ISBN-13: 978-0-8400-6882-8

The Text-Specific DVDs include new learning objective based lecture videos. These DVDs provide comprehensive coverage of the course—along with additional explanations of concepts, sample problems, and applications—to help students review essential topics.

TO THE STUDENT

This textbook was written for you to use as a guide to mastering precalculus mathematics. Here are some suggestions to help you get the most out of your course.

First of all, you should read the appropriate section of text *before* you attempt your homework problems. Reading a mathematics text is quite different from reading a novel, a newspaper, or even another textbook. You may find that you have to reread a passage several times before you understand it. Pay special attention to the examples, and work them out yourself with pencil and paper as you read. Then do the linked exercises referred to in *"Now Try Exercise . . ."* at the end of each example. With this kind of preparation you will be able to do your homework much more quickly and with more understanding.

Don't make the mistake of trying to memorize every single rule or fact you may come across. Mathematics doesn't consist simply of memorization. Mathematics is a *problem-solving art*, not just a collection of facts. To master the subject you must solve problems—lots of problems. Do as many of the exercises as you can. Be sure to write your solutions in a logical, step-by-step fashion. Don't give up on a problem if you can't solve it right away. Try to understand the problem more clearly—reread it thoughtfully and relate it to what you have learned from your teacher and from the examples in the text. Struggle with it until you solve it. Once you have done this a few times you will begin to understand what mathematics is really all about.

Answers to the odd-numbered exercises, as well as all the answers to each chapter test, appear at the back of the book. If your answer differs from the one given, don't immediately assume that you are wrong. There may be a calculation that connects the two answers and makes both correct. For example, if you get $1/(\sqrt{2} - 1)$ but the answer given is $1 + \sqrt{2}$, your answer *is* correct, because you can multiply both numerator and denominator of your answer by $\sqrt{2} + 1$ to change it to the given answer. In rounding approximate answers, follow the guidelines in the Appendix: *Calculations and Significant Figures.*

The symbol ⊘ is used to warn against committing an error. We have placed this symbol in the margin to point out situations where we have found that many of our students make the same mistake.

ABBREVIATIONS

cm	centimeter	**mg**	milligram
dB	decibel	**MHz**	megahertz
F	farad	**mi**	mile
ft	foot	**min**	minute
g	gram	**mL**	milliliter
gal	gallon	**mm**	millimeter
h	hour	**N**	Newton
H	henry	**qt**	quart
Hz	Hertz	**oz**	ounce
in.	inch	**s**	second
J	Joule	**Ω**	ohm
kcal	kilocalorie	**V**	volt
kg	kilogram	**W**	watt
km	kilometer	**yd**	yard
kPa	kilopascal	**yr**	year
L	liter	**°C**	degree Celsius
lb	pound	**°F**	degree Fahrenheit
lm	lumen	**K**	Kelvin
M	mole of solute per liter of solution	⇒	implies
m	meter	⇔	is equivalent to

MATHEMATICAL VIGNETTES

MATHEMATICS IN THE MODERN WORLD

PROLOGUE PRINCIPLES OF PROBLEM SOLVING

Chuck Painter/Stanford News Service

GEORGE POLYA (1887–1985) is famous among mathematicians for his ideas on problem solving. His lectures on problem solving at Stanford University attracted overflow crowds whom he held on the edges of their seats, leading them to discover solutions for themselves. He was able to do this because of his deep insight into the psychology of problem solving. His well-known book *How To Solve It* has been translated into 15 languages. He said that Euler (see page 266) was unique among great mathematicians because he explained how he found his results. Polya often said to his students and colleagues, "Yes, I see that your proof is correct, but how did you discover it?" In the preface to *How To Solve It,* Polya writes, "A great discovery solves a great problem but there is a grain of discovery in the solution of any problem. Your problem may be modest; but if it challenges your curiosity and brings into play your inventive faculties, and if you solve it by your own means, you may experience the tension and enjoy the triumph of discovery."

The ability to solve problems is a highly prized skill in many aspects of our lives; it is certainly an important part of any mathematics course. There are no hard and fast rules that will ensure success in solving problems. However, in this Prologue we outline some general steps in the problem-solving process and we give principles that are useful in solving certain problems. These steps and principles are just common sense made explicit. They have been adapted from George Polya's insightful book *How To Solve It.*

1. Understand the Problem

The first step is to read the problem and make sure that you understand it. Ask yourself the following questions:

What is the unknown?
What are the given quantities?
What are the given conditions?

For many problems it is useful to

draw a diagram

and identify the given and required quantities on the diagram. Usually, it is necessary to

introduce suitable notation

In choosing symbols for the unknown quantities, we often use letters such as a, b, c, m, n, x, and y, but in some cases it helps to use initials as suggestive symbols, for instance, V for volume or t for time.

2. Think of a Plan

Find a connection between the given information and the unknown that enables you to calculate the unknown. It often helps to ask yourself explicitly: "How can I relate the given to the unknown?" If you don't see a connection immediately, the following ideas may be helpful in devising a plan.

▶ Try to Recognize Something Familiar

Relate the given situation to previous knowledge. Look at the unknown and try to recall a more familiar problem that has a similar unknown.

▶ Try to Recognize Patterns

Certain problems are solved by recognizing that some kind of pattern is occurring. The pattern could be geometric, numerical, or algebraic. If you can see regularity or repetition in a problem, then you might be able to guess what the pattern is and then prove it.

▶ Use Analogy

Try to think of an analogous problem, that is, a similar or related problem but one that is easier than the original. If you can solve the similar, simpler problem, then it might give you the clues you need to solve the original, more difficult one. For instance, if a problem involves very large numbers, you could first try a similar problem with smaller numbers. Or if the problem is in three-dimensional geometry, you could look for something similar in two-dimensional geometry. Or if the problem you start with is a general one, you could first try a special case.

▶ Introduce Something Extra

You might sometimes need to introduce something new—an auxiliary aid—to make the connection between the given and the unknown. For instance, in a problem for which a diagram is useful, the auxiliary aid could be a new line drawn in the diagram. In a more algebraic problem the aid could be a new unknown that relates to the original unknown.

▶ Take Cases

You might sometimes have to split a problem into several cases and give a different argument for each case. For instance, we often have to use this strategy in dealing with absolute value.

▶ Work Backward

Sometimes it is useful to imagine that your problem is solved and work backward, step by step, until you arrive at the given data. Then you might be able to reverse your steps and thereby construct a solution to the original problem. This procedure is commonly used in solving equations. For instance, in solving the equation $3x - 5 = 7$, we suppose that x is a number that satisfies $3x - 5 = 7$ and work backward. We add 5 to each side of the equation and then divide each side by 3 to get $x = 4$. Since each of these steps can be reversed, we have solved the problem.

▶ Establish Subgoals

In a complex problem it is often useful to set subgoals (in which the desired situation is only partially fulfilled). If you can attain or accomplish these subgoals, then you might be able to build on them to reach your final goal.

▶ Indirect Reasoning

Sometimes it is appropriate to attack a problem indirectly. In using **proof by contradiction** to prove that P implies Q, we assume that P is true and Q is false and try to see why this cannot happen. Somehow we have to use this information and arrive at a contradiction to what we absolutely know is true.

▶ Mathematical Induction

In proving statements that involve a positive integer n, it is frequently helpful to use the Principle of Mathematical Induction, which is discussed in Section 12.5.

3. Carry Out the Plan

In Step 2, a plan was devised. In carrying out that plan, you must check each stage of the plan and write the details that prove that each stage is correct.

4. Look Back

Having completed your solution, it is wise to look back over it, partly to see whether any errors have been made and partly to see whether you can discover an easier way to solve the problem. Looking back also familiarizes you with the method of solution, which may be useful for solving a future problem. Descartes said, "Every problem that I solved became a rule which served afterwards to solve other problems."

We illustrate some of these principles of problem solving with an example.

PROBLEM | Average Speed

A driver sets out on a journey. For the first half of the distance, she drives at the leisurely pace of 30 mi/h; during the second half she drives 60 mi/h. What is her average speed on this trip?

THINKING ABOUT THE PROBLEM

It is tempting to take the average of the speeds and say that the average speed for the entire trip is

$$\frac{30 + 60}{2} = 45 \text{ mi/h}$$

But is this simple-minded approach really correct?

Try a special case ▶

Let's look at an easily calculated special case. Suppose that the total distance traveled is 120 mi. Since the first 60 mi is traveled at 30 mi/h, it takes 2 h. The second 60 mi is traveled at 60 mi/h, so it takes one hour. Thus, the total time is $2 + 1 = 3$ hours and the average speed is

$$\frac{120}{3} = 40 \text{ mi/h}$$

So our guess of 45 mi/h was wrong.

SOLUTION

Understand the problem ▶

We need to look more carefully at the meaning of average speed. It is defined as

$$\text{average speed} = \frac{\text{distance traveled}}{\text{time elapsed}}$$

Introduce notation ▶

State what is given ▶

Let d be the distance traveled on each half of the trip. Let t_1 and t_2 be the times taken for the first and second halves of the trip. Now we can write down the information we have been given. For the first half of the trip we have

$$30 = \frac{d}{t_1}$$

and for the second half we have

$$60 = \frac{d}{t_2}$$

Identify the unknown ▶

Now we identify the quantity that we are asked to find:

$$\text{average speed for entire trip} = \frac{\text{total distance}}{\text{total time}} = \frac{2d}{t_1 + t_2}$$

Connect the given with the unknown ▶

To calculate this quantity, we need to know t_1 and t_2, so we solve the above equations for these times:

$$t_1 = \frac{d}{30} \qquad t_2 = \frac{d}{60}$$

Don't feel bad if you can't solve these problems right away. Problems 1 and 4 were sent to Albert Einstein by his friend Wertheimer. Einstein (and his friend Bucky) enjoyed the problems and wrote back to Wertheimer. Here is part of his reply:

> Your letter gave us a lot of amusement. The first intelligence test fooled both of us (Bucky and me). Only on working it out did I notice that no time is available for the downhill run! Mr. Bucky was also taken in by the second example, but I was not. Such drolleries show us how stupid we are!

(See *Mathematical Intelligencer*, Spring 1990, page 41.)

Now we have the ingredients needed to calculate the desired quantity:

$$\text{average speed} = \frac{2d}{t_1 + t_1} = \frac{2d}{\dfrac{d}{30} + \dfrac{d}{60}}$$

$$= \frac{60(2d)}{60\left(\dfrac{d}{30} + \dfrac{d}{60}\right)} \qquad \text{Multiply numerator and denominator by 60}$$

$$= \frac{120d}{2d + d} = \frac{120d}{3d} = 40$$

So the average speed for the entire trip is 40 mi/h. ■

PROBLEMS

1. **Distance, Time, and Speed** An old car has to travel a 2-mile route, uphill and down. Because it is so old, the car can climb the first mile—the ascent—no faster than an average speed of 15 mi/h. How fast does the car have to travel the second mile—on the descent it can go faster, of course—to achieve an average speed of 30 mi/h for the trip?

2. **Comparing Discounts** Which price is better for the buyer, a 40% discount or two successive discounts of 20%?

3. **Cutting up a Wire** A piece of wire is bent as shown in the figure. You can see that one cut through the wire produces four pieces and two parallel cuts produce seven pieces. How many pieces will be produced by 142 parallel cuts? Write a formula for the number of pieces produced by n parallel cuts.

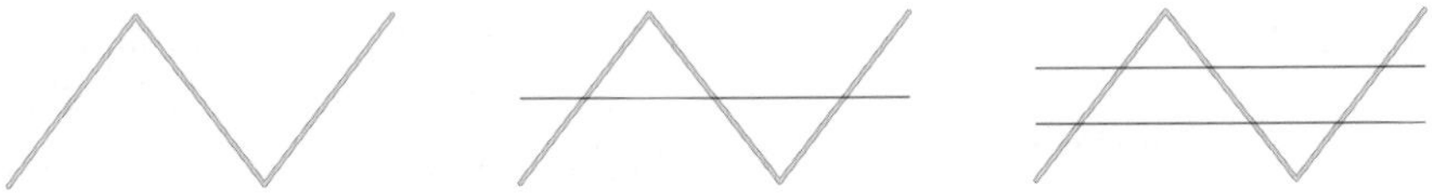

4. **Amoeba Propagation** An amoeba propagates by simple division; each split takes 3 minutes to complete. When such an amoeba is put into a glass container with a nutrient fluid, the container is full of amoebas in one hour. How long would it take for the container to be filled if we start with not one amoeba, but two?

5. **Batting Averages** Player A has a higher batting average than player B for the first half of the baseball season. Player A also has a higher batting average than player B for the second half of the season. Is it necessarily true that player A has a higher batting average than player B for the entire season?

6. **Coffee and Cream** A spoonful of cream is taken from a pitcher of cream and put into a cup of coffee. The coffee is stirred. Then a spoonful of this mixture is put into the pitcher of cream. Is there now more cream in the coffee cup or more coffee in the pitcher of cream?

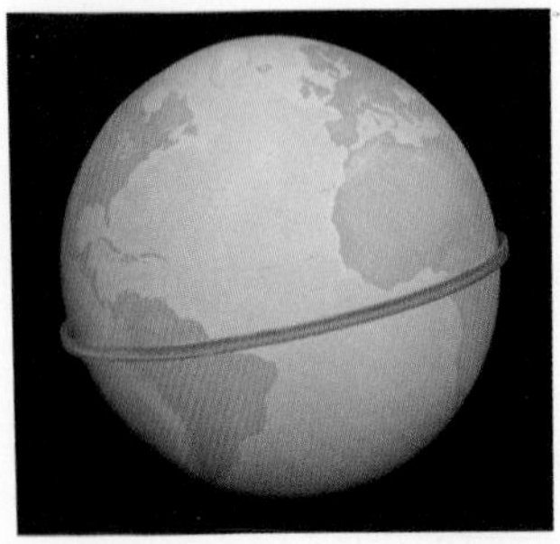

7. **Wrapping the World** A ribbon is tied tightly around the earth at the equator. How much more ribbon would you need if you raised the ribbon 1 ft above the equator everywhere? (You don't need to know the radius of the earth to solve this problem.)

8. **Ending Up Where You Started** A woman starts at a point P on the earth's surface and walks 1 mi south, then 1 mi east, then 1 mi north, and finds herself back at P, the starting point. Describe all points P for which this is possible. [*Hint:* There are infinitely many such points, all but one of which lie in Antarctica.]

Many more problems and examples that highlight different problem-solving principles are available at the book companion website: **www.stewartmath.com**. You can try them as you progress through the book.

CHAPTER 1

Image copyright Monkey Business Images 2010. Used under license from Shutterstock.com

FUNDAMENTALS

In this first chapter we review the real numbers, equations, and the coordinate plane. You are probably already familiar with these concepts, but it is helpful to get a fresh look at how these ideas work together to solve problems and model (or describe) real-world situations.

Let's see how all these ideas are used in a real-life situation: Suppose you get paid \$9 an hour at your part-time job. We can *model* your pay y for working x hours by the *equation* $y = 9x$. To find out how many hours you need to work to get paid 200 dollars, we solve the equation $200 = 9x$. Graphing the equation $y = 9x$ in a *coordinate plane* helps us "see" how pay increases with hours worked.

1.1 Real Numbers

Properties of Real Numbers ▶ Addition and Subtraction ▶ Multiplication and Division ▶ The Real Line ▶ Sets and Intervals ▶ Absolute Value and Distance

Let's review the types of numbers that make up the real number system. We start with the **natural numbers**:

$$1, 2, 3, 4, \ldots$$

The **integers** consist of the natural numbers together with their negatives and 0:

$$\ldots, -3, -2, -1, 0, 1, 2, 3, 4, \ldots$$

The different types of real numbers were invented to meet specific needs. For example, natural numbers are needed for counting, negative numbers for describing debt or below-zero temperatures, rational numbers for concepts like "half a gallon of milk," and irrational numbers for measuring certain distances, like the diagonal of a square.

We construct the **rational numbers** by taking ratios of integers. Thus, any rational number r can be expressed as

$$r = \frac{m}{n}$$

where m and n are integers and $n \neq 0$. Examples are

$$\tfrac{1}{2} \qquad -\tfrac{3}{7} \qquad 46 = \tfrac{46}{1} \qquad 0.17 = \tfrac{17}{100}$$

(Recall that division by 0 is always ruled out, so expressions like $\frac{3}{0}$ and $\frac{0}{0}$ are undefined.) There are also real numbers, such as $\sqrt{2}$, that cannot be expressed as a ratio of integers and are therefore called **irrational numbers**. It can be shown, with varying degrees of difficulty, that these numbers are also irrational:

$$\sqrt{3} \qquad \sqrt{5} \qquad \sqrt[3]{2} \qquad \pi \qquad \frac{3}{\pi^2}$$

The set of all real numbers is usually denoted by the symbol $\mathbb{R}$. When we use the word *number* without qualification, we will mean "real number." Figure 1 is a diagram of the types of real numbers that we work with in this book.

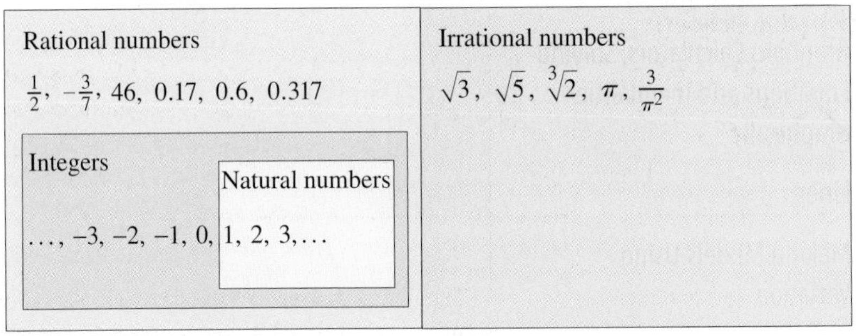

FIGURE 1 The real number system

Every real number has a decimal representation. If the number is rational, then its corresponding decimal is repeating. For example,

$$\tfrac{1}{2} = 0.5000\ldots = 0.5\overline{0} \qquad \tfrac{2}{3} = 0.66666\ldots = 0.\overline{6}$$

$$\tfrac{157}{495} = 0.3171717\ldots = 0.3\overline{17} \qquad \tfrac{9}{7} = 1.285714285714\ldots = 1.\overline{285714}$$

(The bar indicates that the sequence of digits repeats forever.) If the number is irrational, the decimal representation is nonrepeating:

$$\sqrt{2} = 1.414213562373095\ldots \qquad \pi = 3.141592653589793\ldots$$

A repeating decimal such as

$$x = 3.5474747\ldots$$

is a rational number. To convert it to a ratio of two integers, we write

$$\begin{aligned} 1000x &= 3547.47474747\ldots \\ 10x &= 35.47474747\ldots \\ \hline 990x &= 3512.0 \end{aligned}$$

Thus, $x = \frac{3512}{990}$. (The idea is to multiply x by appropriate powers of 10 and then subtract to eliminate the repeating part.)

If we stop the decimal expansion of any number at a certain place, we get an approximation to the number. For instance, we can write

$$\pi \approx 3.14159265$$

where the symbol $\approx$ is read "is approximately equal to." The more decimal places we retain, the better our approximation.

▼ Properties of Real Numbers

We all know that $2 + 3 = 3 + 2$, and $5 + 7 = 7 + 5$, and $513 + 87 = 87 + 513$, and so on. In algebra, we express all these (infinitely many) facts by writing

$$a + b = b + a$$

where a and b stand for any two numbers. In other words, "$a + b = b + a$" is a concise way of saying that "when we add two numbers, the order of addition doesn't matter." This fact is called the *Commutative Property* for addition. From our experience with numbers we know that the properties in the following box are also valid.

PROPERTIES OF REAL NUMBERS

Property	Example	Description
Commutative Properties		
$a + b = b + a$	$7 + 3 = 3 + 7$	When we add two numbers, order doesn't matter.
$ab = ba$	$3 \cdot 5 = 5 \cdot 3$	When we multiply two numbers, order doesn't matter.
Associative Properties		
$(a + b) + c = a + (b + c)$	$(2 + 4) + 7 = 2 + (4 + 7)$	When we add three numbers, it doesn't matter which two we add first.
$(ab)c = a(bc)$	$(3 \cdot 7) \cdot 5 = 3 \cdot (7 \cdot 5)$	When we multiply three numbers, it doesn't matter which two we multiply first.
Distributive Property		
$a(b + c) = ab + ac$ $(b + c)a = ab + ac$	$2 \cdot (3 + 5) = 2 \cdot 3 + 2 \cdot 5$ $(3 + 5) \cdot 2 = 2 \cdot 3 + 2 \cdot 5$	When we multiply a number by a sum of two numbers, we get the same result as we would get if we multiply the number by each of the terms and then add the results.

The Distributive Property applies whenever we multiply a number by a sum. Figure 2 explains why this property works for the case in which all the numbers are positive integers, but the property is true for any real numbers a, b, and c.

The Distributive Property is crucial because it describes the way addition and multiplication interact with each other.

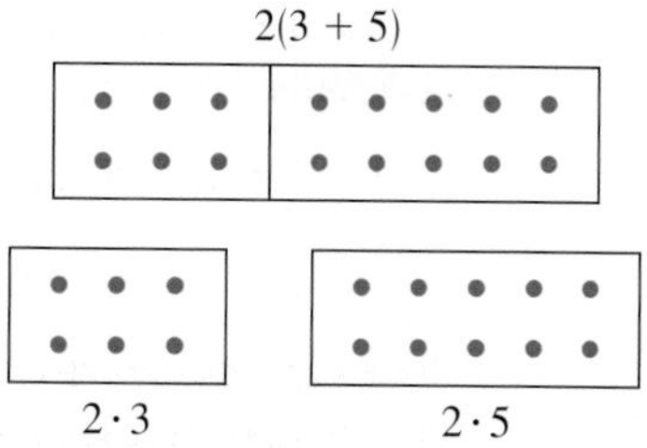

FIGURE 2 The Distributive Property

EXAMPLE 1 | Using the Distributive Property

(a) $2(x + 3) = 2 \cdot x + 2 \cdot 3$ — Distributive Property

$= 2x + 6$ — Simplify

(b) $(a + b)(x + y) = (a + b)x + (a + b)y$ — Distributive Property

$= (ax + bx) + (ay + by)$ — Distributive Property

$= ax + bx + ay + by$ — Associative Property of Addition

In the last step we removed the parentheses because, according to the Associative Property, the order of addition doesn't matter.

NOW TRY EXERCISE **11** ■

Addition and Subtraction

⊘ Don't assume that $-a$ is a negative number. Whether $-a$ is negative or positive depends on the value of a. For example, if $a = 5$, then $-a = -5$, a negative number, but if $a = -5$, then $-a = -(-5) = 5$ (Property 2), a positive number.

The number 0 is special for addition; it is called the **additive identity** because $a + 0 = a$ for any real number a. Every real number a has a **negative**, $-a$, that satisfies $a + (-a) = 0$. **Subtraction** is the operation that undoes addition; to subtract a number from another, we simply add the negative of that number. By definition

$$a - b = a + (-b)$$

To combine real numbers involving negatives, we use the following properties.

PROPERTIES OF NEGATIVES

Property	Example
1. $(-1)a = -a$	$(-1)5 = -5$
2. $-(-a) = a$	$-(-5) = 5$
3. $(-a)b = a(-b) = -(ab)$	$(-5)7 = 5(-7) = -(5 \cdot 7)$
4. $(-a)(-b) = ab$	$(-4)(-3) = 4 \cdot 3$
5. $-(a + b) = -a - b$	$-(3 + 5) = -3 - 5$
6. $-(a - b) = b - a$	$-(5 - 8) = 8 - 5$

Property 6 states the intuitive fact that $a - b$ and $b - a$ are negatives of each other. Property 5 is often used with more than two terms:

$$-(a + b + c) = -a - b - c$$

EXAMPLE 2 | Using Properties of Negatives

Let x, y, and z be real numbers.

(a) $-(x + 2) = -x - 2$ — Property 5: $-(a + b) = -a - b$

(b) $-(x + y - z) = -x - y - (-z)$ — Property 5: $-(a + b) = -a - b$

$= -x - y + z$ — Property 2: $-(-a) = a$

NOW TRY EXERCISE **23** ■

Multiplication and Division

The number 1 is special for multiplication; it is called the **multiplicative identity** because $a \cdot 1 = a$ for any real number a. Every nonzero real number a has an **inverse**, $1/a$, that satisfies $a \cdot (1/a) = 1$. **Division** is the operation that undoes multiplication; to divide by a number, we multiply by the inverse of that number. If $b \neq 0$, then, by definition,

$$a \div b = a \cdot \frac{1}{b}$$

We write $a \cdot (1/b)$ as simply a/b. We refer to a/b as the **quotient** of a and b or as the **fraction** a over b; a is the **numerator** and b is the **denominator** (or **divisor**). To combine real numbers using the operation of division, we use the following properties.

PROPERTIES OF FRACTIONS

Property	Example	Description
1. $\frac{a}{b} \cdot \frac{c}{d} = \frac{ac}{bd}$	$\frac{2}{3} \cdot \frac{5}{7} = \frac{2 \cdot 5}{3 \cdot 7} = \frac{10}{21}$	When **multiplying fractions**, multiply numerators and denominators.
2. $\frac{a}{b} \div \frac{c}{d} = \frac{a}{b} \cdot \frac{d}{c}$	$\frac{2}{3} \div \frac{5}{7} = \frac{2}{3} \cdot \frac{7}{5} = \frac{14}{15}$	When **dividing fractions**, invert the divisor and multiply.
3. $\frac{a}{c} + \frac{b}{c} = \frac{a+b}{c}$	$\frac{2}{5} + \frac{7}{5} = \frac{2+7}{5} = \frac{9}{5}$	When **adding fractions** with the **same denominator**, add the numerators.
4. $\frac{a}{b} + \frac{c}{d} = \frac{ad+bc}{bd}$	$\frac{2}{5} + \frac{3}{7} = \frac{2 \cdot 7 + 3 \cdot 5}{35} = \frac{29}{35}$	When **adding fractions** with **different denominators**, find a common denominator. Then add the numerators.
5. $\frac{ac}{bc} = \frac{a}{b}$	$\frac{2 \cdot 5}{3 \cdot 5} = \frac{2}{3}$	**Cancel** numbers that are **common factors** in numerator and denominator.
6. If $\frac{a}{b} = \frac{c}{d}$, then $ad = bc$	$\frac{2}{3} = \frac{6}{9}$, so $2 \cdot 9 = 3 \cdot 6$	**Cross-multiply.**

When adding fractions with different denominators, we don't usually use Property 4. Instead we rewrite the fractions so that they have the smallest possible common denominator (often smaller than the product of the denominators), and then we use Property 3. This denominator is the **L**east **C**ommon **D**enominator (LCD) described in the next example.

EXAMPLE 3 | Using the LCD to Add Fractions

Evaluate: $\frac{5}{36} + \frac{7}{120}$

SOLUTION Factoring each denominator into prime factors gives

$$36 = 2^2 \cdot 3^2 \qquad \text{and} \qquad 120 = 2^3 \cdot 3 \cdot 5$$

We find the least common denominator (LCD) by forming the product of all the factors that occur in these factorizations, using the highest power of each factor.

Thus the LCD is $2^3 \cdot 3^2 \cdot 5 = 360$. So

$$\frac{5}{36} + \frac{7}{120} = \frac{5 \cdot 10}{36 \cdot 10} + \frac{7 \cdot 3}{120 \cdot 3} \qquad \text{Use common denominator}$$

$$= \frac{50}{360} + \frac{21}{360} = \frac{71}{360} \qquad \text{Property 3: Adding fractions with the same denominator}$$

NOW TRY EXERCISE **25** ■

The Real Line

The real numbers can be represented by points on a line, as shown in Figure 3. The positive direction (toward the right) is indicated by an arrow. We choose an arbitrary reference point O, called the **origin**, which corresponds to the real number 0. Given any convenient unit of measurement, each positive number x is represented by the point on the line a distance of x units to the right of the origin, and each negative number $-x$ is represented by the point x units to the left of the origin. The number associated with the point P is called the coordinate of P, and the line is then called a **coordinate line**, or a **real number line**, or simply a **real line**. Often we identify the point with its coordinate and think of a number as being a point on the real line.

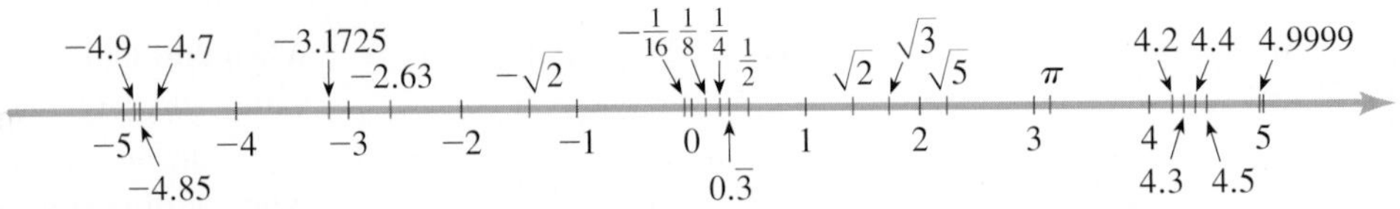

FIGURE 3 The real line

The real numbers are *ordered*. We say that ***a* is less than *b*** and write $a < b$ if $b - a$ is a positive number. Geometrically, this means that a lies to the left of b on the number line. Equivalently, we can say that ***b* is greater than *a*** and write $b > a$. The symbol $a \le b$ (or $b \ge a$) means that either $a < b$ or $a = b$ and is read "a is less than or equal to b." For instance, the following are true inequalities (see Figure 4):

$$7 < 7.4 < 7.5 \qquad -\pi < -3 \qquad \sqrt{2} < 2 \qquad 2 \le 2$$

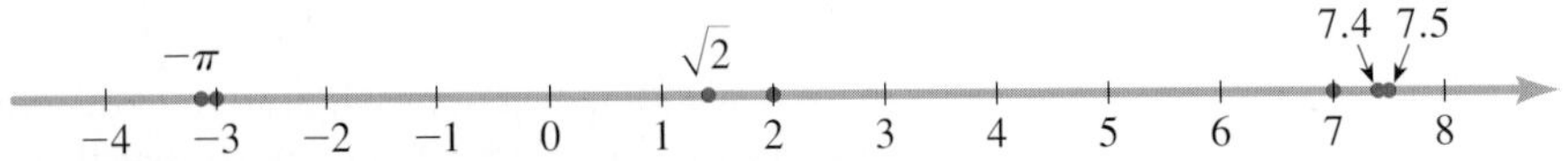

FIGURE 4

Sets and Intervals

A **set** is a collection of objects, and these objects are called the **elements** of the set. If S is a set, the notation $a \in S$ means that a is an element of S, and $b \notin S$ means that b is not an element of S. For example, if Z represents the set of integers, then $-3 \in Z$ but $\pi \notin Z$.

Some sets can be described by listing their elements within braces. For instance, the set A that consists of all positive integers less than 7 can be written as

$$A = \{1, 2, 3, 4, 5, 6\}$$

We could also write A in **set-builder notation** as

$$A = \{x \mid x \text{ is an integer and } 0 < x < 7\}$$

which is read "A is the set of all x such that x is an integer and $0 < x < 7$."

If S and T are sets, then their **union** $S \cup T$ is the set that consists of all elements that are in S *or* T (or in both). The **intersection** of S and T is the set $S \cap T$ consisting of all elements that are in both S *and* T. In other words, $S \cap T$ is the common part of S and T. The **empty set**, denoted by $\varnothing$, is the set that contains no element.

EXAMPLE 4 | Union and Intersection of Sets

If $S = \{1, 2, 3, 4, 5\}$, $T = \{4, 5, 6, 7\}$, and $V = \{6, 7, 8\}$, find the sets $S \cup T$, $S \cap T$, and $S \cap V$.

SOLUTION

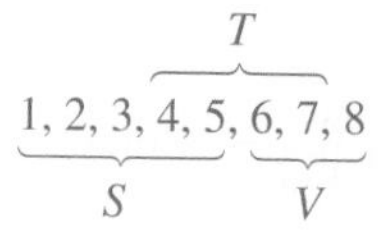

$S \cup T = \{1, 2, 3, 4, 5, 6, 7\}$ All elements in S or T

$S \cap T = \{4, 5\}$ Elements common to both S and T

$S \cap V = \varnothing$ S and V have no element in common

NOW TRY EXERCISE **39** ■

Certain sets of real numbers, called **intervals**, occur frequently in calculus and correspond geometrically to line segments. If $a < b$, then the **open interval** from a to b consists of all numbers between a and b and is denoted (a, b). The **closed interval** from a to b includes the endpoints and is denoted $[a, b]$. Using set-builder notation, we can write

$$(a, b) = \{x \mid a < x < b\} \qquad [a, b] = \{x \mid a \le x \le b\}$$

FIGURE 5 The open interval (a, b)

FIGURE 6 The closed interval $[a, b]$

Note that parentheses ( ) in the interval notation and open circles on the graph in Figure 5 indicate that endpoints are *excluded* from the interval, whereas square brackets [ ] and solid circles in Figure 6 indicate that the endpoints are *included*. Intervals may also include one endpoint but not the other, or they may extend infinitely far in one direction or both. The following table lists the possible types of intervals.

Notation	Set description	Graph
(a, b)	$\{x \mid a < x < b\}$	
$[a, b]$	$\{x \mid a \le x \le b\}$	
$[a, b)$	$\{x \mid a \le x < b\}$	
$(a, b]$	$\{x \mid a < x \le b\}$	
(a, ∞)	$\{x \mid a < x\}$	
$[a, \infty)$	$\{x \mid a \le x\}$	
$(-\infty, b)$	$\{x \mid x < b\}$	
$(-\infty, b]$	$\{x \mid x \le b\}$	
$(-\infty, \infty)$	$\mathbb{R}$ (set of all real numbers)	

The symbol ∞ ("infinity") does not stand for a number. The notation (a, ∞), for instance, simply indicates that the interval has no endpoint on the right but extends infinitely far in the positive direction.

EXAMPLE 5 | Graphing Intervals

Express each interval in terms of inequalities, and then graph the interval.

(a) $[-1, 2) = \{x \mid -1 \le x < 2\}$

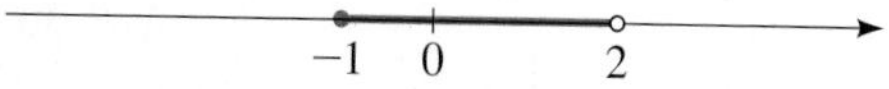

(b) $[1.5, 4] = \{x \mid 1.5 \le x \le 4\}$

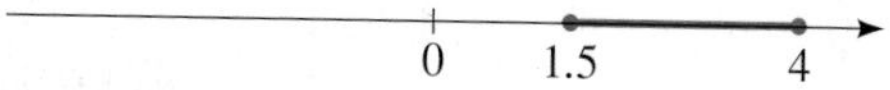

(c) $(-3, \infty) = \{x \mid -3 < x\}$

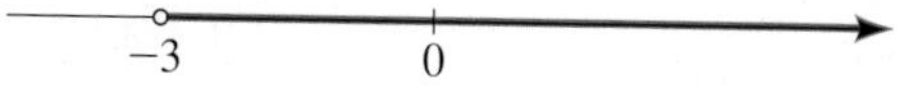

NOW TRY EXERCISE **45** ■

No Smallest or Largest Number in an Open Interval
Any interval contains infinitely many numbers—every point on the graph of an interval corresponds to a real number. In the closed interval $[0, 1]$, the smallest number is 0 and the largest is 1, but the open interval $(0, 1)$ contains no smallest or largest number. To see this, note that 0.01 is close to zero, but 0.001 is closer, 0.0001 is closer yet, and so on. We can always find a number in the interval $(0, 1)$ closer to zero than any given number. Since 0 itself is not in the interval, the interval contains no smallest number. Similarly, 0.99 is close to 1, but 0.999 is closer, 0.9999 closer yet, and so on. Since 1 itself is not in the interval, the interval has no largest number.

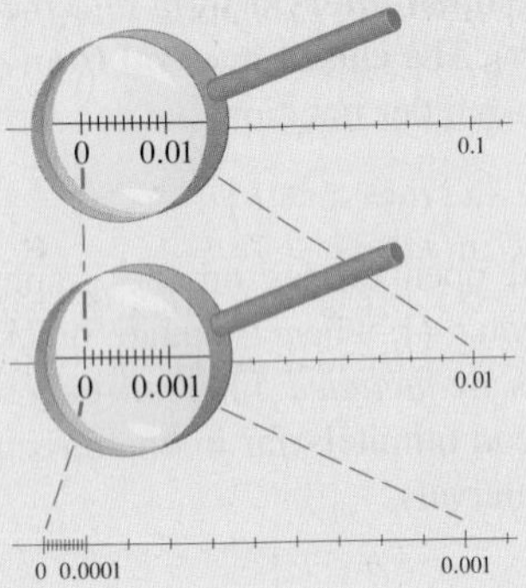

EXAMPLE 6 | Finding Unions and Intersections of Intervals

Graph each set.

(a) $(1, 3) \cap [2, 7]$ **(b)** $(1, 3) \cup [2, 7]$

SOLUTION

(a) The intersection of two intervals consists of the numbers that are in both intervals. Therefore

$$(1, 3) \cap [2, 7] = \{x \mid 1 < x < 3 \text{ and } 2 \le x \le 7\}$$
$$= \{x \mid 2 \le x < 3\} = [2, 3)$$

This set is illustrated in Figure 7.

(b) The union of two intervals consists of the numbers that are in either one interval or the other (or both). Therefore

$$(1, 3) \cup [2, 7] = \{x \mid 1 < x < 3 \text{ or } 2 \le x \le 7\}$$
$$= \{x \mid 1 < x \le 7\} = (1, 7]$$

This set is illustrated in Figure 8.

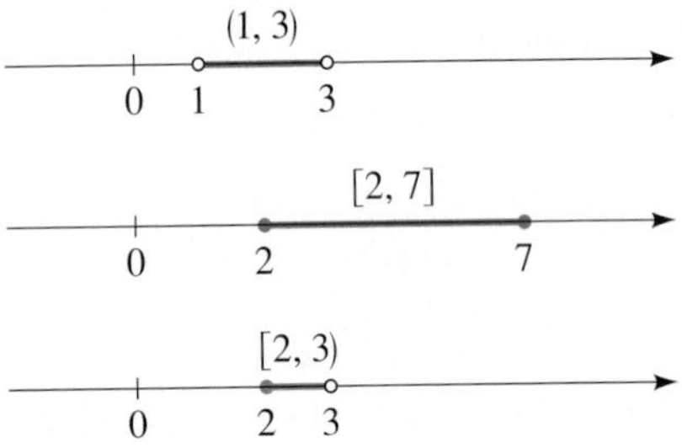

FIGURE 7 $(1, 3) \cap [2, 7] = [2, 3)$

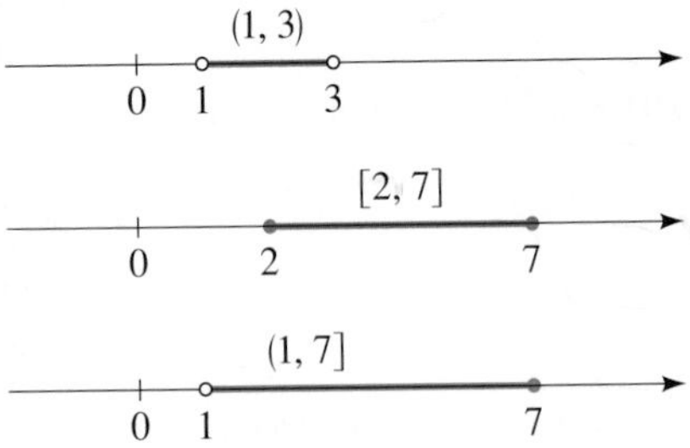

FIGURE 8 $(1, 3) \cup [2, 7] = (1, 7]$

NOW TRY EXERCISE **59**

Absolute Value and Distance

The **absolute value** of a number a, denoted by $|a|$, is the distance from a to 0 on the real number line (see Figure 9). Distance is always positive or zero, so we have $|a| \ge 0$ for every number a. Remembering that $-a$ is positive when a is negative, we have the following definition.

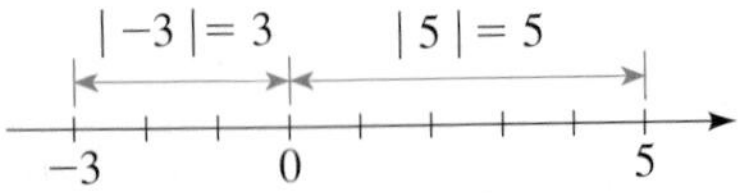

FIGURE 9

DEFINITION OF ABSOLUTE VALUE

If a is a real number, then the **absolute value** of a is

$$|a| = \begin{cases} a & \text{if } a \ge 0 \\ -a & \text{if } a < 0 \end{cases}$$

EXAMPLE 7 | Evaluating Absolute Values of Numbers

(a) $|3| = 3$

(b) $|-3| = -(-3) = 3$

(c) $|0| = 0$

(d) $|3 - \pi| = -(3 - \pi) = \pi - 3$ (since $3 < \pi \Rightarrow 3 - \pi < 0$)

NOW TRY EXERCISE **65**

When working with absolute values, we use the following properties.

PROPERTIES OF ABSOLUTE VALUE

Property	Example	Description
1. $\lvert a \rvert \geq 0$	$\lvert -3 \rvert = 3 \geq 0$	The absolute value of a number is always positive or zero.
2. $\lvert a \rvert = \lvert -a \rvert$	$\lvert 5 \rvert = \lvert -5 \rvert$	A number and its negative have the same absolute value.
3. $\lvert ab \rvert = \lvert a \rvert \lvert b \rvert$	$\lvert -2 \cdot 5 \rvert = \lvert -2 \rvert \lvert 5 \rvert$	The absolute value of a product is the product of the absolute values.
4. $\left\lvert \dfrac{a}{b} \right\rvert = \dfrac{\lvert a \rvert}{\lvert b \rvert}$	$\left\lvert \dfrac{12}{-3} \right\rvert = \dfrac{\lvert 12 \rvert}{\lvert -3 \rvert}$	The absolute value of a quotient is the quotient of the absolute values.

What is the distance on the real line between the numbers -2 and 11? From Figure 10 we see that the distance is 13. We arrive at this by finding either $\lvert 11 - (-2) \rvert = 13$ or $\lvert (-2) - 11 \rvert = 13$. From this observation we make the following definition (see Figure 11).

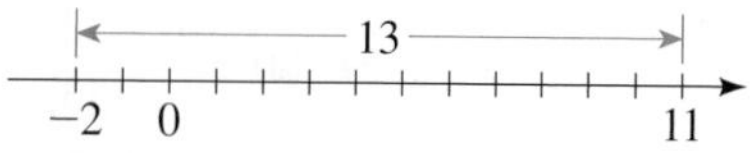

FIGURE 10

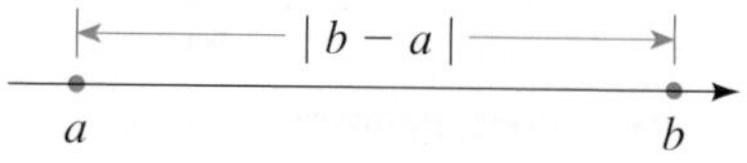

FIGURE 11 Length of a line segment is $\lvert b - a \rvert$

DISTANCE BETWEEN POINTS ON THE REAL LINE

If a and b are real numbers, then the **distance** between the points a and b on the real line is

$$d(a, b) = \lvert b - a \rvert$$

From Property 6 of negatives it follows that

$$\lvert b - a \rvert = \lvert a - b \rvert$$

This confirms that, as we would expect, the distance from a to b is the same as the distance from b to a.

EXAMPLE 8 | Distance Between Points on the Real Line

The distance between the numbers -8 and 2 is

$$d(a, b) = \lvert -8 - 2 \rvert = \lvert -10 \rvert = 10$$

We can check this calculation geometrically, as shown in Figure 12.

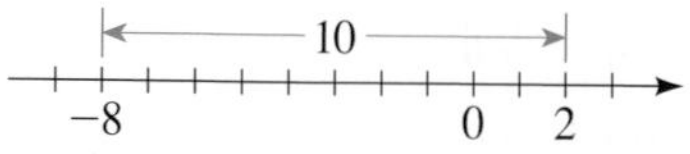

FIGURE 12

NOW TRY EXERCISE **73** ■

1.1 EXERCISES

CONCEPTS

1. Give an example of each of the following:

(a) A natural number

(b) An integer that is not a natural number

(c) A rational number that is not an integer

(d) An irrational number

2. Complete each statement and name the property of real numbers you have used.

(a) $ab =$ ______; ______ Property

(b) $a + (b + c) =$ ______; ______ Property

(c) $a(b + c) =$ ______; ______ Property

3. The set of numbers between but not including 2 and 7 can be written as follows:

______ in set-builder notation and

______ in interval notation.

4. The symbol $|x|$ stands for the ______ of the number x. If x is not 0, then the sign of $|x|$ is always ______.

SKILLS

5–6 ■ List the elements of the given set that are

(a) natural numbers

(b) integers

(c) rational numbers

(d) irrational numbers

5. $\{0, -10, 50, \frac{22}{7}, 0.538, \sqrt{7}, 1.2\overline{3}, -\frac{1}{3}, \sqrt[3]{2}\}$

6. $\{1.001, 0.333\ldots, -\pi, -11, 11, \frac{13}{15}, \sqrt{16}, 3.14, \frac{15}{3}\}$

7–14 ■ State the property of real numbers being used.

7. $7 + 10 = 10 + 7$

8. $2(3 + 5) = (3 + 5)2$

9. $(x + 2y) + 3z = x + (2y + 3z)$

10. $2(A + B) = 2A + 2B$

11. $(5x + 1)3 = 15x + 3$

12. $(x + a)(x + b) = (x + a)x + (x + a)b$

13. $2x(3 + y) = (3 + y)2x$

14. $7(a + b + c) = 7(a + b) + 7c$

15–18 ■ Rewrite the expression using the given property of real numbers.

15. Commutative Property of addition, $x + 3 =$ ______

16. Associative Property of multiplication, $7(3x) =$ ______

17. Distributive Property, $4(A + B) =$ ______

18. Distributive Property, $5x + 5y =$ ______

19–24 ■ Use properties of real numbers to write the expression without parentheses.

19. $3(x + y)$ **20.** $(a - b)8$

21. $4(2m)$ **22.** $\frac{4}{3}(-6y)$

23. $-\frac{5}{2}(2x - 4y)$ **24.** $(3a)(b + c - 2d)$

25–30 ■ Perform the indicated operations.

25. **(a)** $\frac{3}{10} + \frac{4}{15}$ **(b)** $\frac{1}{4} + \frac{1}{5}$

26. **(a)** $\frac{2}{3} - \frac{3}{5}$ **(b)** $1 + \frac{5}{8} - \frac{1}{6}$

27. **(a)** $\frac{2}{3}(6 - \frac{3}{2})$ **(b)** $0.25(\frac{8}{9} + \frac{1}{2})$

28. **(a)** $(3 + \frac{1}{4})(1 - \frac{4}{5})$ **(b)** $(\frac{1}{2} - \frac{1}{3})(\frac{1}{2} + \frac{1}{3})$

29. **(a)** $\dfrac{2}{\frac{2}{3}} - \dfrac{\frac{2}{3}}{2}$ **(b)** $\dfrac{\frac{1}{12}}{\frac{1}{8} - \frac{1}{9}}$

30. **(a)** $\dfrac{2 - \frac{3}{4}}{\frac{1}{2} - \frac{1}{3}}$ **(b)** $\dfrac{\frac{2}{5} + \frac{1}{2}}{\frac{1}{10} + \frac{3}{15}}$

31–32 ■ Place the correct symbol ($<$, $>$, or $=$) in the space.

31. **(a)** 3 ___ $\frac{7}{2}$ **(b)** -3 ___ $-\frac{7}{2}$ **(c)** 3.5 ___ $\frac{7}{2}$

32. **(a)** $\frac{2}{3}$ ___ 0.67 **(b)** $\frac{2}{3}$ ___ -0.67 **(c)** $|0.67|$ ___ $|-0.67|$

33–36 ■ State whether each inequality is true or false.

33. **(a)** $-6 < -10$ **(b)** $\sqrt{2} > 1.41$

34. **(a)** $\dfrac{10}{11} < \dfrac{12}{13}$ **(b)** $-\dfrac{1}{2} < -1$

35. **(a)** $-\pi > -3$ **(b)** $8 \le 9$

36. **(a)** $1.1 > 1.\overline{1}$ **(b)** $8 \le 8$

37–38 ■ Write each statement in terms of inequalities.

37. **(a)** x is positive

(b) t is less than 4

(c) a is greater than or equal to π

(d) x is less than $\frac{1}{3}$ and is greater than -5

(e) The distance from p to 3 is at most 5

38. **(a)** y is negative

(b) z is greater than 1

(c) b is at most 8

(d) w is positive and is less than or equal to 17

(e) y is at least 2 units from π

39–42 ■ Find the indicated set if

$$A = \{1, 2, 3, 4, 5, 6, 7\} \qquad B = \{2, 4, 6, 8\}$$
$$C = \{7, 8, 9, 10\}$$

39. **(a)** $A \cup B$ **(b)** $A \cap B$

40. **(a)** $B \cup C$ **(b)** $B \cap C$

41. **(a)** $A \cup C$ **(b)** $A \cap C$

42. **(a)** $A \cup B \cup C$ **(b)** $A \cap B \cap C$

43–44 ■ Find the indicated set if

$$A = \{x \mid x \geq -2\} \qquad B = \{x \mid x < 4\}$$
$$C = \{x \mid -1 < x \leq 5\}$$

43. **(a)** $B \cup C$ **(b)** $B \cap C$

44. **(a)** $A \cap C$ **(b)** $A \cap B$

45–50 ■ Express the interval in terms of inequalities, and then graph the interval.

45. $(-3, 0)$ **46.** $(2, 8]$

47. $[2, 8)$ **48.** $[-6, -\frac{1}{2}]$

49. $[2, \infty)$ **50.** $(-\infty, 1)$

51–56 ■ Express the inequality in interval notation, and then graph the corresponding interval.

51. $x \leq 1$ **52.** $1 \leq x \leq 2$

53. $-2 < x \leq 1$ **54.** $x \geq -5$

55. $x > -1$ **56.** $-5 < x < 2$

57–58 ■ Express each set in interval notation.

57. **(a)** 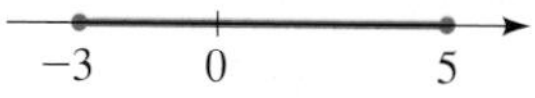

(b) 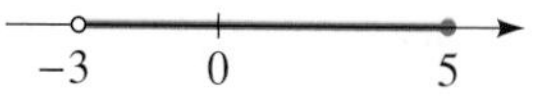

58. **(a)** 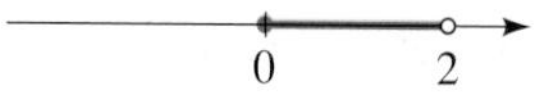

(b) 

59–64 ■ Graph the set.

59. $(-2, 0) \cup (-1, 1)$ **60.** $(-2, 0) \cap (-1, 1)$

61. $[-4, 6] \cap [0, 8)$ **62.** $[-4, 6) \cup [0, 8)$

63. $(-\infty, -4) \cup (4, \infty)$ **64.** $(-\infty, 6] \cap (2, 10)$

65–70 ■ Evaluate each expression.

65. **(a)** $|100|$ **(b)** $|-73|$

66. **(a)** $|\sqrt{5} - 5|$ **(b)** $|10 - \pi|$

67. **(a)** $||-6| - |-4||$ **(b)** $\dfrac{-1}{|-1|}$

68. **(a)** $|2 - |-12||$ **(b)** $-1 - |1 - |-1||$

69. **(a)** $|(-2) \cdot 6|$ **(b)** $|(-\frac{1}{3})(-15)|$

70. **(a)** $\left|\dfrac{-6}{24}\right|$ **(b)** $\left|\dfrac{7 - 12}{12 - 7}\right|$

71–74 ■ Find the distance between the given numbers.

71. 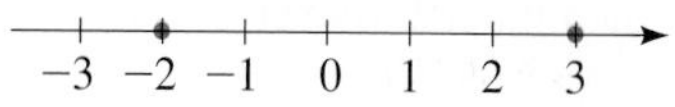

72. 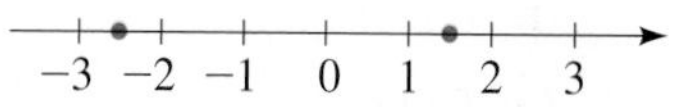

73. **(a)** 2 and 17 **(b)** -3 and 21 **(c)** $\frac{11}{8}$ and $-\frac{3}{10}$

74. **(a)** $\frac{7}{15}$ and $-\frac{1}{21}$ **(b)** -38 and -57 **(c)** -2.6 and -1.8

75–76 ■ Express each repeating decimal as a fraction. (See the margin note on page 2.)

75. **(a)** $0.\overline{7}$ **(b)** $0.2\overline{8}$ **(c)** $0.\overline{57}$

76. **(a)** $5.\overline{23}$ **(b)** $1.3\overline{7}$ **(c)** $2.1\overline{35}$

APPLICATIONS

77. Area of a Garden Mary's backyard vegetable garden measures 20 ft by 30 ft, so its area is $20 \times 30 = 600$ ft^2. She decides to make it longer, as shown in the figure, so that the area increases to $A = 20(30 + x)$. Which property of real numbers tells us that the new area can also be written $A = 600 + 20x$?

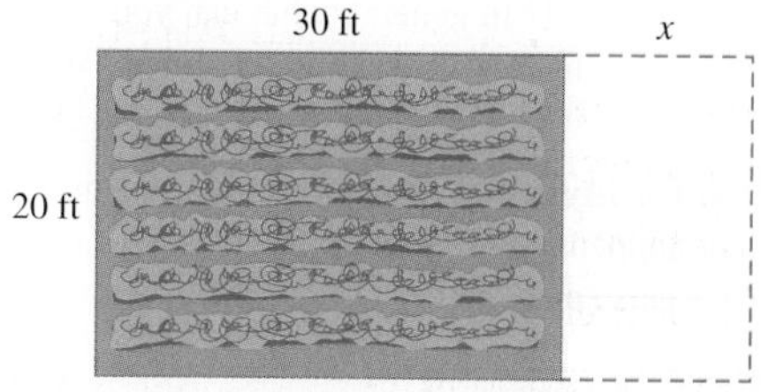

78. Temperature Variation The bar graph shows the daily high temperatures for Omak, Washington, and Geneseo, New York, during a certain week in June. Let T_O represent the temperature in Omak and T_G the temperature in Geneseo. Calculate $T_O - T_G$ and $|T_O - T_G|$ for each day shown. Which of these two values gives more information?

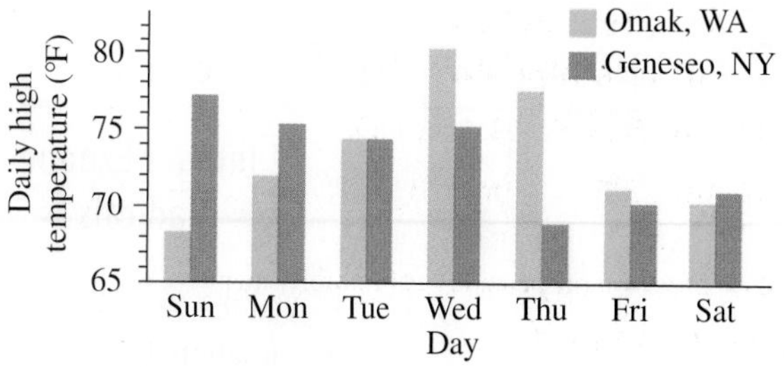

79. Mailing a Package The post office will only accept packages for which the length plus the "girth" (distance around) is no more than 108 inches. Thus, for the package in the figure, we must have

$$L + 2(x + y) \leq 108$$

(a) Will the post office accept a package that is 6 in. wide, 8 in. deep, and 5 ft long? What about a package that measures 2 ft by 2 ft by 4 ft?

(b) What is the greatest acceptable length for a package that has a square base measuring 9 in. by 9 in?

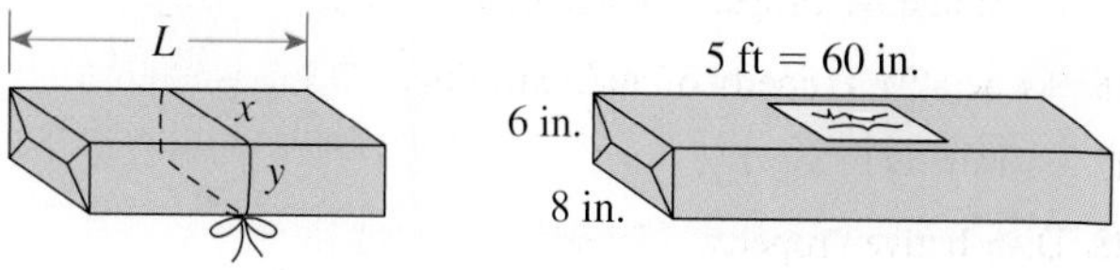

DISCOVERY ▪ DISCUSSION ▪ WRITING

80. Signs of Numbers Let a, b, and c be real numbers such that $a > 0$, $b < 0$, and $c < 0$. Find the sign of each expression.

(a) $-a$ (b) $-b$ (c) bc

(d) $a - b$ (e) $c - a$ (f) $a + bc$

(g) $ab + ac$ (h) $-abc$ (i) ab^2

81. Sums and Products of Rational and Irrational Numbers Explain why the sum, the difference, and the product of two rational numbers are rational numbers. Is the product of two irrational numbers necessarily irrational? What about the sum?

82. Combining Rational Numbers with Irrational Numbers Is $\frac{1}{2} + \sqrt{2}$ rational or irrational? Is $\frac{1}{2} \cdot \sqrt{2}$ rational or irrational? In general, what can you say about the sum of a rational and an irrational number? What about the product?

83. Limiting Behavior of Reciprocals Complete the tables. What happens to the size of the fraction $1/x$ as x gets large? As x gets small?

x	$1/x$
1	
2	
10	
100	
1000	

x	$1/x$
1.0	
0.5	
0.1	
0.01	
0.001	

84. Irrational Numbers and Geometry Using the following figure, explain how to locate the point $\sqrt{2}$ on a number line. Can you locate $\sqrt{5}$ by a similar method? What about $\sqrt{6}$? List some other irrational numbers that can be located this way.

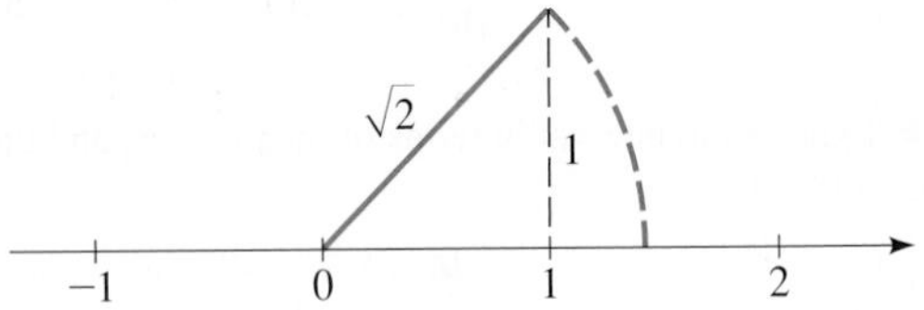

85. Commutative and Noncommutative Operations We have seen that addition and multiplication are both commutative operations.

(a) Is subtraction commutative?

(b) Is division of nonzero real numbers commutative?

1.2 Exponents and Radicals

Integer Exponents ▶ Rules for Working with Exponents ▶ Scientific Notation ▶ Radicals ▶ Rational Exponents ▶ Rationalizing the Denominator

In this section we give meaning to expressions such as $a^{m/n}$ in which the exponent m/n is a rational number. To do this, we need to recall some facts about integer exponents, radicals, and nth roots.

▼ Integer Exponents

A product of identical numbers is usually written in exponential notation. For example, $5 \cdot 5 \cdot 5$ is written as 5^3. In general, we have the following definition.

EXPONENTIAL NOTATION

If a is any real number and n is a positive integer, then the **nth power** of a is

$$a^n = \underbrace{a \cdot a \cdot \cdots \cdot a}_{n \text{ factors}}$$

The number a is called the **base**, and n is called the **exponent**.

EXAMPLE 1 | Exponential Notation

Note the distinction between $(-3)^4$ and -3^4. In $(-3)^4$ the exponent applies to -3, but in -3^4 the exponent applies only to 3.

(a) $\left(\frac{1}{2}\right)^5 = \left(\frac{1}{2}\right)\left(\frac{1}{2}\right)\left(\frac{1}{2}\right)\left(\frac{1}{2}\right)\left(\frac{1}{2}\right) = \frac{1}{32}$

(b) $(-3)^4 = (-3)\cdot(-3)\cdot(-3)\cdot(-3) = 81$

(c) $-3^4 = -(3\cdot 3\cdot 3\cdot 3) = -81$

NOW TRY EXERCISE **15**

We can state several useful rules for working with exponential notation. To discover the rule for multiplication, we multiply 5^4 by 5^2:

$$5^4\cdot 5^2 = \underbrace{(5\cdot5\cdot5\cdot5)}_{\text{4 factors}}\underbrace{(5\cdot5)}_{\text{2 factors}} = \underbrace{5\cdot5\cdot5\cdot5\cdot5\cdot5}_{\text{6 factors}} = 5^6 = 5^{4+2}$$

It appears that *to multiply two powers of the same base, we add their exponents.* In general, for any real number a and any positive integers m and n, we have

$$a^m a^n = \underbrace{(a\cdot a\cdot\cdots\cdot a)}_{m\text{ factors}}\underbrace{(a\cdot a\cdot\cdots\cdot a)}_{n\text{ factors}} = \underbrace{a\cdot a\cdot a\cdot\cdots\cdot a}_{m+n\text{ factors}} = a^{m+n}$$

Thus $a^m a^n = a^{m+n}$.

We would like this rule to be true even when m and n are 0 or negative integers. For instance, we must have

$$2^0\cdot 2^3 = 2^{0+3} = 2^3$$

But this can happen only if $2^0 = 1$. Likewise, we want to have

$$5^4\cdot 5^{-4} = 5^{4+(-4)} = 5^{4-4} = 5^0 = 1$$

and this will be true if $5^{-4} = 1/5^4$. These observations lead to the following definition.

ZERO AND NEGATIVE EXPONENTS

If $a \neq 0$ is any real number and n is a positive integer, then

$$a^0 = 1 \qquad \text{and} \qquad a^{-n} = \frac{1}{a^n}$$

EXAMPLE 2 | Zero and Negative Exponents

(a) $\left(\frac{4}{7}\right)^0 = 1$

(b) $x^{-1} = \dfrac{1}{x^1} = \dfrac{1}{x}$

(c) $(-2)^{-3} = \dfrac{1}{(-2)^3} = \dfrac{1}{-8} = -\dfrac{1}{8}$

NOW TRY EXERCISE **17**

▼ Rules for Working with Exponents

Familiarity with the following rules is essential for our work with exponents and bases. In the table the bases a and b are real numbers, and the exponents m and n are integers.

LAWS OF EXPONENTS

Law	Example	Description
1. $a^m a^n = a^{m+n}$	$3^2 \cdot 3^5 = 3^{2+5} = 3^7$	To multiply two powers of the same number, add the exponents.
2. $\dfrac{a^m}{a^n} = a^{m-n}$	$\dfrac{3^5}{3^2} = 3^{5-2} = 3^3$	To divide two powers of the same number, subtract the exponents.
3. $(a^m)^n = a^{mn}$	$(3^2)^5 = 3^{2 \cdot 5} = 3^{10}$	To raise a power to a new power, multiply the exponents.
4. $(ab)^n = a^n b^n$	$(3 \cdot 4)^2 = 3^2 \cdot 4^2$	To raise a product to a power, raise each factor to the power.
5. $\left(\dfrac{a}{b}\right)^n = \dfrac{a^n}{b^n}$	$\left(\dfrac{3}{4}\right)^2 = \dfrac{3^2}{4^2}$	To raise a quotient to a power, raise both numerator and denominator to the power.

PROOF OF LAW 3 If m and n are positive integers, we have

$$
\begin{aligned}
(a^m)^n &= (\underbrace{a \cdot a \cdot \cdots \cdot a}_{m \text{ factors}})^n \\
&= \underbrace{(\underbrace{a \cdot a \cdot \cdots \cdot a}_{m \text{ factors}})(\underbrace{a \cdot a \cdot \cdots \cdot a}_{m \text{ factors}}) \cdots (\underbrace{a \cdot a \cdot \cdots \cdot a}_{m \text{ factors}})}_{n \text{ groups of factors}} \\
&= \underbrace{a \cdot a \cdot \cdots \cdot a}_{mn \text{ factors}} = a^{mn}
\end{aligned}
$$

The cases for which $m \le 0$ or $n \le 0$ can be proved using the definition of negative exponents. ■

PROOF OF LAW 4 If n is a positive integer, we have

$$(ab)^n = \underbrace{(ab)(ab) \cdots (ab)}_{n \text{ factors}} = \underbrace{(a \cdot a \cdot \cdots \cdot a)}_{n \text{ factors}} \cdot \underbrace{(b \cdot b \cdot \cdots \cdot b)}_{n \text{ factors}} = a^n b^n$$

Here we have used the Commutative and Associative Properties repeatedly. If $n \le 0$, Law 4 can be proved by using the definition of negative exponents. ■

You are asked to prove Laws 2 and 5 in Exercise 94.

EXAMPLE 3 | Using Laws of Exponents

(a) $x^4 x^7 = x^{4+7} = x^{11}$ Law 1: $a^m a^n = a^{m+n}$

(b) $y^4 y^{-7} = y^{4-7} = y^{-3} = \dfrac{1}{y^3}$ Law 1: $a^m a^n = a^{m+n}$

(c) $\dfrac{c^9}{c^5} = c^{9-5} = c^4$ Law 2: $\dfrac{a^m}{a^n} = a^{m-n}$

(d) $(b^4)^5 = b^{4 \cdot 5} = b^{20}$ Law 3: $(a^m)^n = a^{mn}$

(e) $(3x)^3 = 3^3 x^3 = 27x^3$ Law 4: $(ab)^n = a^n b^n$

(f) $\left(\dfrac{x}{2}\right)^5 = \dfrac{x^5}{2^5} = \dfrac{x^5}{32}$ Law 5: $\left(\dfrac{a}{b}\right)^n = \dfrac{a^n}{b^n}$

NOW TRY EXERCISES **35, 37,** AND **39** ■

EXAMPLE 4 | Simplifying Expressions with Exponents

Simplify:

(a) $(2a^3b^2)(3ab^4)^3$ **(b)** $\left(\frac{x}{y}\right)^3\left(\frac{y^2x}{z}\right)^4$

SOLUTION

(a)
$$\begin{aligned}(2a^3b^2)(3ab^4)^3 &= (2a^3b^2)[3^3a^3(b^4)^3] && \text{Law 4: } (ab)^n = a^nb^n\\ &= (2a^3b^2)(27a^3b^{12}) && \text{Law 3: } (a^m)^n = a^{mn}\\ &= (2)(27)a^3a^3b^2b^{12} && \text{Group factors with the same base}\\ &= 54a^6b^{14} && \text{Law 1: } a^ma^n = a^{m+n}\end{aligned}$$

(b)
$$\begin{aligned}\left(\frac{x}{y}\right)^3\left(\frac{y^2x}{z}\right)^4 &= \frac{x^3}{y^3}\frac{(y^2)^4x^4}{z^4} && \text{Laws 5 and 4}\\ &= \frac{x^3}{y^3}\frac{y^8x^4}{z^4} && \text{Law 3}\\ &= (x^3x^4)\left(\frac{y^8}{y^3}\right)\frac{1}{z^4} && \text{Group factors with the same base}\\ &= \frac{x^7y^5}{z^4} && \text{Laws 1 and 2}\end{aligned}$$

NOW TRY EXERCISES **43** AND **47** ■

When simplifying an expression, you will find that many different methods will lead to the same result; you should feel free to use any of the rules of exponents to arrive at your own method. We now give two additional laws that are useful in simplifying expressions with negative exponents.

LAWS OF EXPONENTS

Law	Example	Description
6. $\left(\frac{a}{b}\right)^{-n} = \left(\frac{b}{a}\right)^n$	$\left(\frac{3}{4}\right)^{-2} = \left(\frac{4}{3}\right)^2$	To raise a fraction to a negative power, invert the fraction and change the sign of the exponent.
7. $\frac{a^{-n}}{b^{-m}} = \frac{b^m}{a^n}$	$\frac{3^{-2}}{4^{-5}} = \frac{4^5}{3^2}$	To move a number raised to a power from numerator to denominator or from denominator to numerator, change the sign of the exponent.

PROOF OF LAW 7 Using the definition of negative exponents and then Property 2 of fractions (page 5), we have

$$\frac{a^{-n}}{b^{-m}} = \frac{1/a^n}{1/b^m} = \frac{1}{a^n}\cdot\frac{b^m}{1} = \frac{b^m}{a^n}$$ ■

You are asked to prove Law 6 in Exercise 94.

EXAMPLE 5 | Simplifying Expressions with Negative Exponents

Eliminate negative exponents and simplify each expression.

(a) $\frac{6st^{-4}}{2s^{-2}t^2}$ **(b)** $\left(\frac{y}{3z^3}\right)^{-2}$

MATHEMATICS IN THE MODERN WORLD

Although we are often unaware of its presence, mathematics permeates nearly every aspect of life in the modern world. With the advent of modern technology, mathematics plays an ever greater role in our lives. Today you were probably awakened by a digital alarm clock, made a phone call that used digital transmission, sent an e-mail message over the Internet, drove a car with digitally controlled fuel injection, listened to music on a CD player or MP3 player, perhaps watched a DVD or digital television, then fell asleep in a room whose temperature is controlled by a digital thermostat. In each of these activities mathematics is crucially involved. In general, a property such as the intensity or frequency of sound, the oxygen level in the exhaust emission from a car, the colors in an image, or the temperature in your bedroom is transformed into sequences of numbers by sophisticated mathematical algorithms. These numerical data, which usually consist of many millions of bits (the digits 0 and 1), are then transmitted and reinterpreted. Dealing with such huge amounts of data was not feasible until the invention of computers, machines whose logical processes were invented by mathematicians.

The contributions of mathematics in the modern world are not limited to technological advances. The logical processes of mathematics are now used to analyze complex problems in the social, political, and life sciences in new and surprising ways. Advances in mathematics continue to be made, some of the most exciting of these just within the past decade.

In other *Mathematics in the Modern World,* we will describe in more detail how mathematics affects all of us in our everyday activities.

SOLUTION

(a) We use Law 7, which allows us to move a number raised to a power from the numerator to the denominator (or vice versa) by changing the sign of the exponent.

t^{-4} moves to denominator and becomes t^4

s^{-2} moves to numerator and becomes s^2

$$\frac{6st^{-4}}{2s^{-2}t^2} = \frac{6ss^2}{2t^2t^4} \qquad \text{Law 7}$$

$$= \frac{3s^3}{t^6} \qquad \text{Law 1}$$

(b) We use Law 6, which allows us to change the sign of the exponent of a fraction by inverting the fraction.

$$\left(\frac{y}{3z^3}\right)^{-2} = \left(\frac{3z^3}{y}\right)^2 \qquad \text{Law 6}$$

$$= \frac{9z^6}{y^2} \qquad \text{Laws 5 and 4}$$

NOW TRY EXERCISE **49** ■

▼ Scientific Notation

Scientists use exponential notation as a compact way of writing very large numbers and very small numbers. For example, the nearest star beyond the sun, Proxima Centauri, is approximately 40,000,000,000,000 km away. The mass of a hydrogen atom is about 0.00000000000000000000000166 g. Such numbers are difficult to read and to write, so scientists usually express them in *scientific notation*.

SCIENTIFIC NOTATION

A positive number x is said to be written in **scientific notation** if it is expressed as follows:

$$x = a \times 10^n \qquad \text{where } 1 \le a < 10 \text{ and } n \text{ is an integer}$$

For instance, when we state that the distance to the star Proxima Centauri is 4×10^{13} km, the positive exponent 13 indicates that the decimal point should be moved 13 places to the *right*:

$$4 \times 10^{13} = 40{,}000{,}000{,}000{,}000$$

Move decimal point 13 places to the right

When we state that the mass of a hydrogen atom is 1.66×10^{-24} g, the exponent -24 indicates that the decimal point should be moved 24 places to the *left*:

$$1.66 \times 10^{-24} = 0.00000000000000000000000166$$

Move decimal point 24 places to the left

EXAMPLE 6 | Changing from Decimal to Scientific Notation

Write each number in scientific notation.

(a) 56,920 **(b)** 0.000093

SOLUTION

(a) $\underbrace{56{,}920}_{\text{4 places}} = 5.692 \times 10^4$ **(b)** $\underbrace{0.000093}_{\text{5 places}} = 9.3 \times 10^{-5}$

NOW TRY EXERCISES **77** AND **79**

To use scientific notation on a calculator, press the key labeled EE or EXP or EEX to enter the exponent. For example, to enter the number 3.629×10^{15} on a TI-83 calculator, we enter

3.629 2ND EE 15

and the display reads

3.629E15

Scientific notation is often used on a calculator to display a very large or very small number. For instance, if we use a calculator to square the number 1,111,111, the display panel may show (depending on the calculator model) the approximation

1.234568 12 or 1.23468 E12

Here the final digits indicate the power of 10, and we interpret the result as

$$1.234568 \times 10^{12}$$

EXAMPLE 7 | Calculating with Scientific Notation

If $a \approx 0.00046$, $b \approx 1.697 \times 10^{22}$, and $c \approx 2.91 \times 10^{-18}$, use a calculator to approximate the quotient ab/c.

SOLUTION We could enter the data using scientific notation, or we could use laws of exponents as follows:

$$\begin{aligned}\frac{ab}{c} &\approx \frac{(4.6 \times 10^{-4})(1.697 \times 10^{22})}{2.91 \times 10^{-18}} \\ &= \frac{(4.6)(1.697)}{2.91} \times 10^{-4+22+18} \\ &\approx 2.7 \times 10^{36}\end{aligned}$$

For guidelines on working with significant figures, see the Appendix, *Calculations and Significant Figures.*

We state the answer rounded to two significant figures because the least accurate of the given numbers is stated to two significant figures.

NOW TRY EXERCISES **83** AND **85**

Radicals

We know what 2^n means whenever n is an integer. To give meaning to a power, such as $2^{4/5}$, whose exponent is a rational number, we need to discuss radicals.

The symbol $\sqrt{\ }$ means "the positive square root of." Thus

$$\sqrt{a} = b \quad \text{means} \quad b^2 = a \quad \text{and} \quad b \geq 0$$

It is true that the number 9 has two square roots, 3 and -3, but the notation $\sqrt{9}$ is reserved for the *positive* square root of 9 (sometimes called the *principal square root* of 9). If we want the negative root, we must write $-\sqrt{9}$, which is -3.

Since $a = b^2 \geq 0$, the symbol $\sqrt{a}$ makes sense only when $a \geq 0$. For instance,

$$\sqrt{9} = 3 \quad \text{because} \quad 3^2 = 9 \quad \text{and} \quad 3 \geq 0$$

Square roots are special cases of nth roots. The nth root of x is the number that, when raised to the nth power, gives x.

DEFINITION OF nth ROOT

If n is any positive integer, then the **principal nth root** of a is defined as follows:

$$\sqrt[n]{a} = b \qquad \text{means} \qquad b^n = a$$

If n is even, we must have $a \geq 0$ and $b \geq 0$.

Thus

$$\sqrt[4]{81} = 3 \qquad \text{because} \qquad 3^4 = 81 \qquad \text{and} \qquad 3 \geq 0$$

$$\sqrt[3]{-8} = -2 \qquad \text{because} \qquad (-2)^3 = -8$$

But $\sqrt{-8}$, $\sqrt[4]{-8}$, and $\sqrt[6]{-8}$ are not defined. (For instance, $\sqrt{-8}$ is not defined because the square of every real number is nonnegative.)

Notice that

$$\sqrt{4^2} = \sqrt{16} = 4 \qquad \text{but} \qquad \sqrt{(-4)^2} = \sqrt{16} = 4 = |-4|$$

So the equation $\sqrt{a^2} = a$ is not always true; it is true only when $a \geq 0$. However, we can always write $\sqrt{a^2} = |a|$. This last equation is true not only for square roots, but for any even root. This and other rules used in working with nth roots are listed in the following box. In each property we assume that all the given roots exist.

PROPERTIES OF nth ROOTS

Property	Example
1. $\sqrt[n]{ab} = \sqrt[n]{a}\sqrt[n]{b}$	$\sqrt[3]{-8 \cdot 27} = \sqrt[3]{-8}\sqrt[3]{27} = (-2)(3) = -6$
2. $\sqrt[n]{\dfrac{a}{b}} = \dfrac{\sqrt[n]{a}}{\sqrt[n]{b}}$	$\sqrt[4]{\dfrac{16}{81}} = \dfrac{\sqrt[4]{16}}{\sqrt[4]{81}} = \dfrac{2}{3}$
3. $\sqrt[m]{\sqrt[n]{a}} = \sqrt[mn]{a}$	$\sqrt{\sqrt[3]{729}} = \sqrt[6]{729} = 3$
4. $\sqrt[n]{a^n} = a$ if n is odd	$\sqrt[3]{(-5)^3} = -5, \quad \sqrt[5]{2^5} = 2$
5. $\sqrt[n]{a^n} = \|a\|$ if n is even	$\sqrt[4]{(-3)^4} = \|-3\| = 3$

EXAMPLE 8 | Simplifying Expressions Involving nth Roots

(a) $\sqrt[3]{x^4} = \sqrt[3]{x^3 x}$ Factor out the largest cube

$= \sqrt[3]{x^3}\sqrt[3]{x}$ Property 1: $\sqrt[3]{ab} = \sqrt[3]{a}\sqrt[3]{b}$

$= x\sqrt[3]{x}$ Property 4: $\sqrt[3]{a^3} = a$

(b) $\sqrt[4]{81x^8y^4} = \sqrt[4]{81}\sqrt[4]{x^8}\sqrt[4]{y^4}$ Property 1: $\sqrt[4]{abc} = \sqrt[4]{a}\sqrt[4]{b}\sqrt[4]{c}$

$= 3\sqrt[4]{(x^2)^4}|y|$ Property 5: $\sqrt[4]{a^4} = |a|$

$= 3x^2|y|$ Property 5: $\sqrt[4]{a^4} = |a|$, $|x^2| = x^2$

NOW TRY EXERCISES **55** AND **57** ■

It is frequently useful to combine like radicals in an expression such as $2\sqrt{3} + 5\sqrt{3}$. This can be done by using the Distributive Property. Thus

$$2\sqrt{3} + 5\sqrt{3} = (2 + 5)\sqrt{3} = 7\sqrt{3}$$

The next example further illustrates this process.

Avoid making the following error:

$$\sqrt{a+b} \neq \sqrt{a} + \sqrt{b}$$

For instance, if we let $a = 9$ and $b = 16$, then we see the error:

$$\sqrt{9+16} \stackrel{?}{=} \sqrt{9} + \sqrt{16}$$
$$\sqrt{25} \stackrel{?}{=} 3 + 4$$
$$5 \stackrel{?}{=} 7 \quad \text{Wrong!}$$

EXAMPLE 9 | Combining Radicals

(a) $\sqrt{32} + \sqrt{200} = \sqrt{16 \cdot 2} + \sqrt{100 \cdot 2}$ — Factor out the largest squares

$= \sqrt{16}\sqrt{2} + \sqrt{100}\sqrt{2}$ — Property 1: $\sqrt{ab} = \sqrt{a}\sqrt{b}$

$= 4\sqrt{2} + 10\sqrt{2} = 14\sqrt{2}$ — Distributive Property

(b) If $b > 0$, then

$\sqrt{25b} - \sqrt{b^3} = \sqrt{25}\sqrt{b} - \sqrt{b^2}\sqrt{b}$ — Property 1: $\sqrt{ab} = \sqrt{a}\sqrt{b}$

$= 5\sqrt{b} - b\sqrt{b}$ — Property 5, $b > 0$

$= (5 - b)\sqrt{b}$ — Distributive Property

NOW TRY EXERCISES **29** AND **33** ■

▼ Rational Exponents

To define what is meant by a *rational exponent* or, equivalently, a *fractional exponent* such as $a^{1/3}$, we need to use radicals. To give meaning to the symbol $a^{1/n}$ in a way that is consistent with the Laws of Exponents, we would have to have

$$(a^{1/n})^n = a^{(1/n)n} = a^1 = a$$

So by the definition of nth root,

$$a^{1/n} = \sqrt[n]{a}$$

In general, we define rational exponents as follows.

DEFINITION OF RATIONAL EXPONENTS

For any rational exponent m/n in lowest terms, where m and n are integers and $n > 0$, we define

$$a^{m/n} = (\sqrt[n]{a})^m \quad \text{or equivalently} \quad a^{m/n} = \sqrt[n]{a^m}$$

If n is even, then we require that $a \geq 0$.

With this definition it can be proved that *the Laws of Exponents also hold for rational exponents*.

EXAMPLE 10 | Using the Definition of Rational Exponents

(a) $4^{1/2} = \sqrt{4} = 2$

(b) $8^{2/3} = (\sqrt[3]{8})^2 = 2^2 = 4$ — Alternative solution: $8^{2/3} = \sqrt[3]{8^2} = \sqrt[3]{64} = 4$

(c) $125^{-1/3} = \dfrac{1}{125^{1/3}} = \dfrac{1}{\sqrt[3]{125}} = \dfrac{1}{5}$ **(d)** $\dfrac{1}{\sqrt[3]{x^4}} = \dfrac{1}{x^{4/3}} = x^{-4/3}$

NOW TRY EXERCISES **21** AND **23** ■

DIOPHANTUS lived in Alexandria about 250 A.D. His book *Arithmetica* is considered the first book on algebra. In it he gives methods for finding integer solutions of algebraic equations. *Arithmetica* was read and studied for more than a thousand years. Fermat (see page 99) made some of his most important discoveries while studying this book. Diophantus' major contribution is the use of symbols to stand for the unknowns in a problem. Although his symbolism is not as simple as what we use today, it was a major advance over writing everything in words. In Diophantus' notation the equation

$$x^5 - 7x^2 + 8x - 5 = 24$$

is written

$$\Delta K^{\gamma}\alpha\varsigma\eta \pitchfork \Delta^{\gamma}\zeta\mathring{M}\varepsilon\iota^{\sigma}\kappa\delta$$

Our modern algebraic notation did not come into common use until the 17th century.

EXAMPLE 11 | Using the Laws of Exponents with Rational Exponents

(a) $a^{1/3}a^{7/3} = a^{8/3}$ Law 1: $a^m a^n = a^{m+n}$

(b) $\dfrac{a^{2/5}a^{7/5}}{a^{3/5}} = a^{2/5+7/5-3/5} = a^{6/5}$ Law 1, Law 2: $\dfrac{a^m}{a^n} = a^{m-n}$

(c) $(2a^3b^4)^{3/2} = 2^{3/2}(a^3)^{3/2}(b^4)^{3/2}$ Law 4: $(abc)^n = a^n b^n c^n$

$= (\sqrt{2})^3 a^{3(3/2)} b^{4(3/2)}$ Law 3: $(a^m)^n = a^{mn}$

$= 2\sqrt{2}a^{9/2}b^6$

(d) $\left(\dfrac{2x^{3/4}}{y^{1/3}}\right)^3\left(\dfrac{y^4}{x^{-1/2}}\right) = \dfrac{2^3(x^{3/4})^3}{(y^{1/3})^3}\cdot(y^4x^{1/2})$ Laws 5, 4, and 7

$= \dfrac{8x^{9/4}}{y}\cdot y^4x^{1/2}$ Law 3

$= 8x^{11/4}y^3$ Laws 1 and 2

NOW TRY EXERCISES **61**, **63**, **67**, AND **69**

EXAMPLE 12 | Simplifying by Writing Radicals as Rational Exponents

(a) $(2\sqrt{x})(3\sqrt[3]{x}) = (2x^{1/2})(3x^{1/3})$ Definition of rational exponents

$= 6x^{1/2+1/3} = 6x^{5/6}$ Law 1

(b) $\sqrt{x\sqrt{x}} = (xx^{1/2})^{1/2}$ Definition of rational exponents

$= (x^{3/2})^{1/2}$ Law 1

$= x^{3/4}$ Law 3

NOW TRY EXERCISES **71** AND **75**

▼ Rationalizing the Denominator

It is often useful to eliminate the radical in a denominator by multiplying both numerator and denominator by an appropriate expression. This procedure is called **rationalizing the denominator**. If the denominator is of the form $\sqrt{a}$, we multiply numerator and denominator by $\sqrt{a}$. In doing this we multiply the given quantity by 1, so we do not change its value. For instance,

$$\frac{1}{\sqrt{a}} = \frac{1}{\sqrt{a}}\cdot 1 = \frac{1}{\sqrt{a}}\cdot\frac{\sqrt{a}}{\sqrt{a}} = \frac{\sqrt{a}}{a}$$

Note that the denominator in the last fraction contains no radical. In general, if the denominator is of the form $\sqrt[n]{a^m}$ with $m < n$, then multiplying the numerator and denominator by $\sqrt[n]{a^{n-m}}$ will rationalize the denominator, because (for $a > 0$)

$$\sqrt[n]{a^m}\sqrt[n]{a^{n-m}} = \sqrt[n]{a^{m+n-m}} = \sqrt[n]{a^n} = a$$

EXAMPLE 13 | Rationalizing Denominators

This equals 1

(a) $\dfrac{2}{\sqrt{3}} = \dfrac{2}{\sqrt{3}}\cdot\dfrac{\sqrt{3}}{\sqrt{3}} = \dfrac{2\sqrt{3}}{3}$

(b) $\dfrac{1}{\sqrt[3]{x^2}} = \dfrac{1}{\sqrt[3]{x^2}}\dfrac{\sqrt[3]{x}}{\sqrt[3]{x}} = \dfrac{\sqrt[3]{x}}{\sqrt[3]{x^3}} = \dfrac{\sqrt[3]{x}}{x}$

(c) $\sqrt[7]{\dfrac{1}{a^2}} = \dfrac{1}{\sqrt[7]{a^2}} = \dfrac{1}{\sqrt[7]{a^2}}\dfrac{\sqrt[7]{a^5}}{\sqrt[7]{a^5}} = \dfrac{\sqrt[7]{a^5}}{\sqrt[7]{a^7}} = \dfrac{\sqrt[7]{a^5}}{a}$

NOW TRY EXERCISES **89** AND **91** ■

1.2 EXERCISES

CONCEPTS

1. **(a)** Using exponential notation, we can write the product $5 \cdot 5 \cdot 5 \cdot 5 \cdot 5 \cdot 5$ as ______.

(b) In the expression 3^4, the number 3 is called the ______, and the number 4 is called the ______.

2. **(a)** When we multiply two powers with the same base, we ______ the exponents. So $3^4 \cdot 3^5 =$ ______.

(b) When we divide two powers with the same base, we ______ the exponents. So $\dfrac{3^5}{3^2} =$ ______.

3. **(a)** Using exponential notation, we can write $\sqrt[3]{5}$ as ______.

(b) Using radicals, we can write $5^{1/2}$ as ______.

(c) Is there a difference between $\sqrt{5^2}$ and $(\sqrt{5})^2$? Explain.

4. Explain what $4^{3/2}$ means, then calculate $4^{3/2}$ in two different ways:

$(4^{1/2})^{\square} =$ ______ or $(4^3)^{\square} =$ ______

5. Explain how we rationalize a denominator, then complete the following steps to rationalize $\dfrac{1}{\sqrt{3}}$:

$\dfrac{1}{\sqrt{3}} = \dfrac{1}{\sqrt{3}} \cdot \dfrac{\square}{\square} = \dfrac{\square}{\square}$

6. Find the missing power in the following calculation: $5^{1/3} \cdot 5^{\square} = 5$.

SKILLS

7–14 ■ Write each radical expression using exponents, and each exponential expression using radicals.

	Radical expression	Exponential expression
7.	$\dfrac{1}{\sqrt{5}}$	
8.	$\sqrt[3]{7^2}$	
9.		$4^{2/3}$
10.		$11^{-3/2}$
11.	$\sqrt[5]{5^3}$	
12.		$2^{-1.5}$
13.		$a^{2/5}$
14.	$\dfrac{1}{\sqrt{x^5}}$	

15–24 ■ Evaluate each expression.

15. **(a)** -3^2 **(b)** $(-3)^2$ **(c)** $(\frac{1}{3})^4(-3)^2$

16. **(a)** $5^4 \cdot 5^{-2}$ **(b)** $\dfrac{10^7}{10^4}$ **(c)** $\dfrac{3}{3^{-2}}$

17. **(a)** $(\frac{5}{3})^0 2^{-1}$ **(b)** $\dfrac{2^{-3}}{3^0}$ **(c)** $(\frac{1}{4})^{-2}$

18. **(a)** $(-\frac{2}{3})^{-3}$ **(b)** $(\frac{3}{2})^{-2} \cdot \frac{9}{16}$ **(c)** $(\frac{1}{2})^4 \cdot (\frac{5}{2})^{-2}$

19. **(a)** $\sqrt{16}$ **(b)** $\sqrt[4]{16}$ **(c)** $\sqrt[4]{\frac{1}{16}}$

20. **(a)** $\sqrt{64}$ **(b)** $\sqrt[3]{-64}$ **(c)** $\sqrt[5]{-32}$

21. **(a)** $\sqrt{\frac{4}{9}}$ **(b)** $\sqrt[4]{256}$ **(c)** $\sqrt[6]{\frac{1}{64}}$

22. **(a)** $\sqrt{7}\sqrt{28}$ **(b)** $\dfrac{\sqrt{48}}{\sqrt{3}}$ **(c)** $\sqrt[4]{24}\sqrt[4]{54}$

23. **(a)** $(\frac{4}{9})^{-1/2}$ **(b)** $(-32)^{2/5}$ **(c)** $-32^{2/5}$

24. **(a)** $1024^{-0.1}$ **(b)** $(-\frac{27}{8})^{2/3}$ **(c)** $(\frac{25}{64})^{-3/2}$

25–28 ■ Evaluate the expression using $x = 3$, $y = 4$, and $z = -1$.

25. $\sqrt{x^2 + y^2}$ **26.** $\sqrt[4]{x^3 + 14y + 2z}$

27. $(9x)^{2/3} + (2y)^{2/3} + z^{2/3}$ **28.** $(xy)^{2z}$

29–34 ■ Simplify the expression.

29. $\sqrt{32} + \sqrt{18}$ **30.** $\sqrt{75} + \sqrt{48}$

31. $\sqrt[5]{96} + \sqrt[5]{3}$ **32.** $\sqrt[4]{48} - \sqrt[4]{3}$

33. $\sqrt{16x} + \sqrt{x^5}$ **34.** $\sqrt[3]{2y^4} - \sqrt[3]{y}$

35–40 ■ Simplify each expression.

35. **(a)** x^8x^2 **(b)** $(3y^2)(4y^5)$ **(c)** x^2x^{-6}

36. **(a)** $x^{-5}x^3$ **(b)** $w^{-2}w^{-4}w^6$ **(c)** $z^5z^{-3}z^{-4}$

37. **(a)** $\dfrac{y^{10}y^0}{y^7}$ **(b)** $\dfrac{x^6}{x^{10}}$ **(c)** $\dfrac{a^9a^{-2}}{a}$

38. **(a)** $\dfrac{z^2z^4}{z^3z^{-1}}$ **(b)** $(2y^2)^3$ **(c)** $(8x)^2$

39. **(a)** $(a^2a^4)^3$ **(b)** $\left(\frac{a^2}{4}\right)^3$ **(c)** $(3z)^2(6z^2)^{-3}$

40. **(a)** $(2z^2)^{-5}z^{10}$ **(b)** $(2a^3a^2)^4$ **(c)** $\left(\frac{3x^4}{4x^2}\right)^2$

41–52 ■ Simplify the expression and eliminate any negative exponents(s).

41. **(a)** $(4x^2y^4)(2x^5y)$ **(b)** $(8a^2z)(\frac{1}{2}a^3z^4)$

42. **(a)** $b^4(3ab^3)(2a^2b^{-5})$ **(b)** $(2s^3t^{-2})(\frac{1}{4}s^7t)(16t^4)$

43. **(a)** $(5x^2y^3)(3x^2y^5)^4$ **(b)** $(2a^3b^2)^2(5a^2b^5)^3$

44. **(a)** $(s^{-2}t^2)^2(s^2t)^3$ **(b)** $(2u^2v^3)^3(3u^{-3}v)^2$

45. **(a)** $\frac{6y^3z}{2yz^2}$ **(b)** $\frac{(xy^2z^3)^4}{(x^2y^2z)^3}$

46. **(a)** $\frac{2x^3y^4}{x^5y^3}$ **(b)** $\frac{(2v^3w)^2}{v^3w^2}$

47. **(a)** $\left(\frac{a^2}{b}\right)^5\left(\frac{a^3b^2}{c^3}\right)^3$ **(b)** $\frac{(u^{-1}v^2)^2}{(u^3v^{-2})^3}$

48. **(a)** $\left(\frac{x^4z^2}{4y^5}\right)\left(\frac{2x^3y^2}{z^3}\right)^2$ **(b)** $\frac{(rs^2)^3}{(r^{-3}s^2)^2}$

49. **(a)** $\frac{8a^3b^{-4}}{2a^{-5}b^5}$ **(b)** $\left(\frac{y}{5x^{-2}}\right)^{-3}$

50. **(a)** $\frac{5xy^{-2}}{x^{-1}y^{-3}}$ **(b)** $\left(\frac{2a^{-1}b}{a^2b^{-3}}\right)^{-3}$

51. **(a)** $\left(\frac{3a}{b^3}\right)^{-1}$ **(b)** $\left(\frac{q^{-1}r^{-1}s^{-2}}{r^{-5}sq^{-8}}\right)^{-1}$

52. **(a)** $\left(\frac{s^2t^{-4}}{5s^{-1}t}\right)$ **(b)** $\left(\frac{xy^{-2}z^{-3}}{x^2y^3z^{-4}}\right)^{-3}$

53–60 ■ Simplify the expression. Assume that the letters denote any real numbers.

53. $\sqrt[4]{x^4}$

54. $\sqrt[5]{x^{10}}$

55. $\sqrt[4]{16x^8}$

56. $\sqrt[3]{x^3y^6}$

57. $\sqrt[6]{64a^6b^7}$

58. $\sqrt[3]{a^2b}\sqrt[3]{64a^4b}$

59. $\sqrt[3]{\sqrt{64x^6}}$

60. $\sqrt[4]{x^4y^2z^2}$

61–70 ■ Simplify the expression and eliminate any negative exponent(s). Assume that all letters denote positive numbers.

61. **(a)** $x^{3/4}x^{5/4}$ **(b)** $y^{2/3}y^{4/3}$

62. **(a)** $(4b)^{1/2}(8b^{1/4})$ **(b)** $(3a^{3/4})^2(5a^{1/2})$

63. **(a)** $\frac{w^{4/3}w^{2/3}}{w^{1/3}}$ **(b)** $\frac{s^{5/2}(2s^{5/4})^2}{s^{1/2}}$

64. **(a)** $(8y^3)^{-2/3}$ **(b)** $(u^4v^6)^{-1/3}$

65. **(a)** $(8a^6b^{3/2})^{2/3}$ **(b)** $(4a^6b^8)^{3/2}$

66. **(a)** $(x^{-5}y^{1/3})^{-3/5}$ **(b)** $(2x^3y^{-1/4})^2(8y^{-3/2})^{-1/3}$

67. **(a)** $\frac{(8s^3t^3)^{2/3}}{(s^4t^{-8})^{1/4}}$ **(b)** $\frac{(32y^{-5}z^{10})^{1/5}}{(64y^6z^{-12})^{-1/6}}$

68. **(a)** $\left(\frac{x^8y^{-4}}{16y^{4/3}}\right)^{-1/4}$ **(b)** $\left(\frac{-8y^{3/4}}{y^3z^6}\right)^{-1/3}$

69. **(a)** $\left(\frac{x^{-2/3}}{y^{1/2}}\right)\left(\frac{x^{-2}}{y^{-3}}\right)^{1/6}$ **(b)** $\left(\frac{4y^3z^{2/3}}{x^{1/2}}\right)^2\left(\frac{x^{-3}y^6}{8z^4}\right)^{1/3}$

70. **(a)** $\left(\frac{a^{1/6}b^{-3}}{x^{-1}y}\right)^3\left(\frac{x^{-2}b^{-1}}{a^{3/2}y^{1/3}}\right)$ **(b)** $\frac{(9st)^{3/2}}{(27s^3t^{-4})^{2/3}}\left(\frac{3s^{-2}}{4t^{1/3}}\right)^{-1}$

71–76 ■ Simplify the expression and eliminate any negative exponents(s). Assume that all letters denote positive numbers.

71. **(a)** $\sqrt[6]{y^5}\sqrt[3]{y^2}$ **(b)** $(5\sqrt[3]{x})(2\sqrt[4]{x})$

72. **(a)** $\sqrt[4]{b^3}\sqrt{b}$ **(b)** $(2\sqrt{a})(\sqrt[3]{a^2})$

73. **(a)** $\sqrt{4st^3}\sqrt[6]{s^3t^2}$ **(b)** $\frac{\sqrt[4]{x^7}}{\sqrt[4]{x^3}}$

74. **(a)** $\sqrt[5]{x^3y^2}\sqrt[10]{x^4y^{16}}$ **(b)** $\frac{\sqrt[3]{8x^2}}{\sqrt{x}}$

75. **(a)** $\sqrt[3]{y\sqrt{y}}$ **(b)** $\sqrt{\frac{16u^3v}{uv^5}}$

76. **(a)** $\sqrt{s\sqrt{s^3}}$ **(b)** $\sqrt[3]{\frac{54x^2y^4}{2x^5y}}$

77–78 ■ Write each number in scientific notation.

77. **(a)** 69,300,000 **(b)** 7,200,000,000,000
(c) 0.000028536 **(d)** 0.0001213

78. **(a)** 129,540,000 **(b)** 7,259,000,000
(c) 0.0000000014 **(d)** 0.0007029

79–80 ■ Write each number in decimal notation.

79. **(a)** 3.19×10^5 **(b)** 2.721×10^8
(c) 2.670×10^{-8} **(d)** 9.999×10^{-9}

80. **(a)** 7.1×10^{14} **(b)** 6×10^{12}
(c) 8.55×10^{-3} **(d)** 6.257×10^{-10}

81–82 ■ Write the number indicated in each statement in scientific notation.

81. **(a)** A light-year, the distance that light travels in one year, is about 5,900,000,000,000 mi.
(b) The diameter of an electron is about 0.0000000000004 cm.
(c) A drop of water contains more than 33 billion billion molecules.

82. **(a)** The distance from the earth to the sun is about 93 million miles.
(b) The mass of an oxygen molecule is about 0.0000000000000000000000053 g.
(c) The mass of the earth is about 5,970,000,000,000,000,000,000,000 kg.

83–88 ■ Use scientific notation, the Laws of Exponents, and a calculator to perform the indicated operations. State your answer rounded to the number of significant digits indicated by the given data.

83. $(7.2 \times 10^{-9})(1.806 \times 10^{-12})$

84. $(1.062 \times 10^{24})(8.61 \times 10^{19})$

85. $\dfrac{1.295643 \times 10^{9}}{(3.610 \times 10^{-17})(2.511 \times 10^{6})}$

86. $\dfrac{(73.1)(1.6341 \times 10^{28})}{0.0000000019}$

87. $\dfrac{(0.0000162)(0.01582)}{(594{,}621{,}000)(0.0058)}$

88. $\dfrac{(3.542 \times 10^{-6})^{9}}{(5.05 \times 10^{4})^{12}}$

89–92 ■ Rationalize the denominator.

89. **(a)** $\dfrac{1}{\sqrt{10}}$ **(b)** $\sqrt{\dfrac{2}{x}}$ **(c)** $\sqrt{\dfrac{x}{3}}$

90. **(a)** $\sqrt{\dfrac{5}{12}}$ **(b)** $\sqrt{\dfrac{x}{6}}$ **(c)** $\sqrt{\dfrac{y}{2z}}$

91. **(a)** $\dfrac{2}{\sqrt[3]{x}}$ **(b)** $\dfrac{1}{\sqrt[4]{y^3}}$ **(c)** $\dfrac{x}{y^{2/5}}$

92. **(a)** $\dfrac{1}{\sqrt[4]{a}}$ **(b)** $\dfrac{a}{\sqrt[3]{b^2}}$ **(c)** $\dfrac{1}{c^{3/7}}$

93. Let a, b, and c be real numbers with $a > 0$, $b < 0$, and $c < 0$. Determine the sign of each expression.

(a) b^5 **(b)** b^{10} **(c)** ab^2c^3

(d) $(b - a)^3$ **(e)** $(b - a)^4$ **(f)** $\dfrac{a^3c^3}{b^6c^6}$

94. Prove the given Laws of Exponents for the case in which m and n are positive integers and $m > n$.

(a) Law 2 **(b)** Law 5 **(c)** Law 6

APPLICATIONS

95. Distance to the Nearest Star Proxima Centauri, the star nearest to our solar system, is 4.3 light-years away. Use the information in Exercise 81(a) to express this distance in miles.

96. Speed of Light The speed of light is about 186,000 mi/s. Use the information in Exercise 82(a) to find how long it takes for a light ray from the sun to reach the earth.

97. Volume of the Oceans The average ocean depth is 3.7×10^3 m, and the area of the oceans is 3.6×10^{14} m^2. What is the total volume of the ocean in liters? (One cubic meter contains 1000 liters.)

98. National Debt As of July 2010, the population of the United States was 3.070×10^8, and the national debt was 1.320×10^{13} dollars. How much was each person's share of the debt?

99. Number of Molecules A sealed room in a hospital, measuring 5 m wide, 10 m long, and 3 m high, is filled with pure oxygen. One cubic meter contains 1000 L, and 22.4 L of any gas contains 6.02×10^{23} molecules (Avogadro's number). How many molecules of oxygen are there in the room?

100. How Far Can You See? Because of the curvature of the earth, the maximum distance D that you can see from the top of a tall building of height h is estimated by the formula

$$D = \sqrt{2rh + h^2}$$

where $r = 3960$ mi is the radius of the earth and D and h are also measured in miles. How far can you see from the observation deck of the Toronto CN Tower, 1135 ft above the ground?

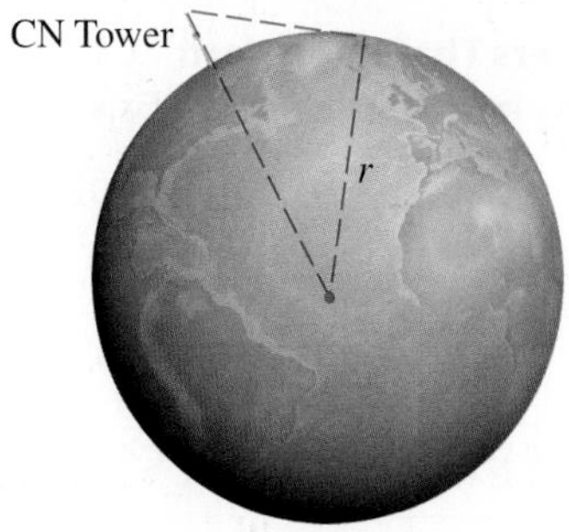

101. Speed of a Skidding Car Police use the formula $s = \sqrt{30fd}$ to estimate the speed s (in mi/h) at which a car is traveling if it skids d feet after the brakes are applied suddenly. The number f is the coefficient of friction of the road, which is a measure of the "slipperiness" of the road. The table gives some typical estimates for f.

	Tar	Concrete	Gravel
Dry	1.0	0.8	0.2
Wet	0.5	0.4	0.1

(a) If a car skids 65 ft on wet concrete, how fast was it moving when the brakes were applied?

(b) If a car is traveling at 50 mi/h, how far will it skid on wet tar?

102. Distance from the Earth to the Sun It follows from **Kepler's Third Law** of planetary motion that the average distance from a planet to the sun (in meters) is

$$d = \left(\frac{GM}{4\pi^2}\right)^{1/3} T^{2/3}$$

where $M = 1.99 \times 10^{30}$ kg is the mass of the sun, $G = 6.67 \times 10^{-11}\ \text{N} \cdot \text{m}^2/\text{kg}^2$ is the gravitational constant, and T is the period of the planet's orbit (in seconds). Use the fact that the period of the earth's orbit is about 365.25 days to find the distance from the earth to the sun.

DISCOVERY ■ DISCUSSION ■ WRITING

103. How Big Is a Billion? If you had a million (10^6) dollars in a suitcase, and you spent a thousand (10^3) dollars each day, how many years would it take you to use all the money? Spending at the same rate, how many years would it take you to empty a suitcase filled with a *billion* (10^9) dollars?

104. Easy Powers That Look Hard Calculate these expressions in your head. Use the Laws of Exponents to help you.

(a) $\dfrac{18^5}{9^5}$ **(b)** $20^6 \cdot (0.5)^6$

105. Limiting Behavior of Powers Complete the following tables. What happens to the nth root of 2 as n gets large? What about the nth root of $\frac{1}{2}$?

n	$2^{1/n}$
1	
2	
5	
10	
100	

n	$\left(\frac{1}{2}\right)^{1/n}$
1	
2	
5	
10	
100	

Construct a similar table for $n^{1/n}$. What happens to the nth root of n as n gets large?

106. Comparing Roots Without using a calculator, determine which number is larger in each pair.

(a) $2^{1/2}$ or $2^{1/3}$ **(b)** $\left(\frac{1}{2}\right)^{1/2}$ or $\left(\frac{1}{2}\right)^{1/3}$

(c) $7^{1/4}$ or $4^{1/3}$ **(d)** $\sqrt[3]{5}$ or $\sqrt{3}$

1.3 Algebraic Expressions

Adding and Subtracting Polynomials ▶ Multiplying Algebraic Expressions ▶ Special Product Formulas ▶ Factoring Common Factors ▶ Factoring Trinomials ▶ Special Factoring Formulas ▶ Factoring by Grouping Terms

A **variable** is a letter that can represent any number from a given set of numbers. If we start with variables, such as x, y, and z and some real numbers, and combine them using addition, subtraction, multiplication, division, powers, and roots, we obtain an **algebraic expression**. Here are some examples:

$$2x^2 - 3x + 4 \qquad \sqrt{x} + 10 \qquad \frac{y - 2z}{y^2 + 4}$$

A **monomial** is an expression of the form ax^k, where a is a real number and k is a nonnegative integer. A **binomial** is a sum of two monomials and a **trinomial** is a sum of three monomials. In general, a sum of monomials is called a *polynomial*. For example, the first expression listed above is a polynomial, but the other two are not.

POLYNOMIALS

A **polynomial** in the variable x is an expression of the form

$$a_n x^n + a_{n-1}x^{n-1} + \cdots + a_1 x + a_0$$

where $a_0, a_1, \ldots, a_n$ are real numbers, and n is a nonnegative integer. If $a_n \neq 0$, then the polynomial has **degree n**. The monomials $a_k x^k$ that make up the polynomial are called the **terms** of the polynomial.

Note that the degree of a polynomial is the highest power of the variable that appears in the polynomial.

Polynomial	Type	Terms	Degree
$2x^2 - 3x + 4$	trinomial	$2x^2, -3x, 4$	2
$x^8 + 5x$	binomial	$x^8, 5x$	8
$3 - x + x^2 - \frac{1}{2}x^3$	four terms	$-\frac{1}{2}x^3, x^2, -x, 3$	3
$5x + 1$	binomial	$5x, 1$	1
$9x^5$	monomial	$9x^5$	5
6	monomial	6	0

▼ Adding and Subtracting Polynomials

Distributive Property

$ac + bc = (a + b)c$

We **add** and **subtract** polynomials using the properties of real numbers that were discussed in Section 1.1. The idea is to combine **like terms** (that is, terms with the same variables raised to the same powers) using the Distributive Property. For instance,

$$5x^7 + 3x^7 = (5 + 3)x^7 = 8x^7$$

In subtracting polynomials, we have to remember that if a minus sign precedes an expression in parentheses, then the sign of every term within the parentheses is changed when we remove the parentheses:

$$-(b + c) = -b - c$$

[This is simply a case of the Distributive Property, $a(b + c) = ab + ac$, with $a = -1$.]

EXAMPLE 1 | Adding and Subtracting Polynomials

(a) Find the sum $(x^3 - 6x^2 + 2x + 4) + (x^3 + 5x^2 - 7x)$.

(b) Find the difference $(x^3 - 6x^2 + 2x + 4) - (x^3 + 5x^2 - 7x)$.

SOLUTION

(a) $(x^3 - 6x^2 + 2x + 4) + (x^3 + 5x^2 - 7x)$

$= (x^3 + x^3) + (-6x^2 + 5x^2) + (2x - 7x) + 4$ Group like terms

$= 2x^3 - x^2 - 5x + 4$ Combine like terms

(b) $(x^3 - 6x^2 + 2x + 4) - (x^3 + 5x^2 - 7x)$

$= x^3 - 6x^2 + 2x + 4 - x^3 - 5x^2 + 7x$ Distributive Property

$= (x^3 - x^3) + (-6x^2 - 5x^2) + (2x + 7x) + 4$ Group like terms

$= -11x^2 + 9x + 4$ Combine like terms

NOW TRY EXERCISES **15** AND **17** ■

▼ Multiplying Algebraic Expressions

To find the **product** of polynomials or other algebraic expressions, we need to use the Distributive Property repeatedly. In particular, using it three times on the product of two binomials, we get

$$(a + b)(c + d) = a(c + d) + b(c + d) = ac + ad + bc + bd$$

This says that we multiply the two factors by multiplying each term in one factor by each term in the other factor and adding these products. Schematically, we have

The acronym **FOIL** helps us remember that the product of two binomials is the sum of the products of the **F**irst terms, the **O**uter terms, the **I**nner terms, and the **L**ast terms.

$$(a + b)(c + d) = \underset{F}{ac} + \underset{O}{ad} + \underset{I}{bc} + \underset{L}{bd}$$

In general, we can multiply two algebraic expressions by using the Distributive Property and the Laws of Exponents.

EXAMPLE 2 | Multiplying Binomials Using FOIL

$$\begin{aligned} (2x+1)(3x-5) &= \underset{\text{F}}{6x^2} - \underset{\text{O}}{10x} + \underset{\text{I}}{3x} - \underset{\text{L}}{5} && \text{Distributive Property} \\ &= 6x^2 - 7x - 5 && \text{Combine like terms} \end{aligned}$$

NOW TRY EXERCISE 23 ■

When we multiply trinomials or other polynomials with more terms, we use the Distributive Property. It is also helpful to arrange our work in table form. The next example illustrates both methods.

EXAMPLE 3 | Multiplying Polynomials

Find the product: $(2x+3)(x^2-5x+4)$

SOLUTION 1: Using the Distributive Property

$$\begin{aligned} (2x+3)(x^2-5x+4) &= 2x(x^2-5x+4) + 3(x^2-5x+4) && \text{Distributive Property} \\ &= (2x \cdot x^2 - 2x \cdot 5x + 2x \cdot 4) + (3 \cdot x^2 - 3 \cdot 5x + 3 \cdot 4) && \text{Distributive Property} \\ &= (2x^3 - 10x^2 + 8x) + (3x^2 - 15x + 12) && \text{Laws of Exponents} \\ &= 2x^3 - 7x^2 - 7x + 12 && \text{Combine like terms} \end{aligned}$$

SOLUTION 2: Using Table Form

$$\begin{array}{rrrrl} & x^2 & -\ 5x & +\ 4 & \\ & & 2x & +\ 3 & \\ \hline & 3x^2 & -\ 15x & +\ 12 & \text{Multiply } x^2-5x+4 \text{ by } 3 \\ 2x^3 & -\ 10x^2 & +\ 8x & & \text{Multiply } x^2-5x+4 \text{ by } 2x \\ \hline 2x^3 & -\ 7x^2 & -\ 7x & +\ 12 & \text{Add like terms} \end{array}$$

NOW TRY EXERCISE 45 ■

▼ Special Product Formulas

Certain types of products occur so frequently that you should memorize them. You can verify the following formulas by performing the multiplications.

See the *Discovery Project* referenced on page 34 for a geometric interpretation of some of these formulas.

SPECIAL PRODUCT FORMULAS

If A and B are any real numbers or algebraic expressions, then

1. $(A+B)(A-B) = A^2 - B^2$ — Sum and product of same terms

2. $(A+B)^2 = A^2 + 2AB + B^2$ — Square of a sum

3. $(A-B)^2 = A^2 - 2AB + B^2$ — Square of a difference

4. $(A+B)^3 = A^3 + 3A^2B + 3AB^2 + B^3$ — Cube of a sum

5. $(A-B)^3 = A^3 - 3A^2B + 3AB^2 - B^3$ — Cube of a difference

The key idea in using these formulas (or any other formula in algebra) is the **Principle of Substitution**: We may substitute any algebraic expression for any letter in a formula. For example, to find $(x^2 + y^3)^2$ we use Product Formula 2, substituting x^2 for A and y^3 for B, to get

$$(x^2 + y^3)^2 = (x^2)^2 + 2(x^2)(y^3) + (y^3)^2$$

$$(A + B)^2 = A^2 + 2AB + B^2$$

EXAMPLE 4 | Using the Special Product Formulas

Use the Special Product Formulas to find each product.

(a) $(3x + 5)^2$ **(b)** $(x^2 - 2)^3$

SOLUTION

(a) Substituting $A = 3x$ and $B = 5$ in Product Formula 2, we get

$$(3x + 5)^2 = (3x)^2 + 2(3x)(5) + 5^2 = 9x^2 + 30x + 25$$

(b) Substituting $A = x^2$ and $B = 2$ in Product Formula 5, we get

$$\begin{aligned}(x^2 - 2)^3 &= (x^2)^3 - 3(x^2)^2(2) + 3(x^2)(2)^2 - 2^3 \\ &= x^6 - 6x^4 + 12x^2 - 8\end{aligned}$$

NOW TRY EXERCISES **29** AND **41**

EXAMPLE 5 | Using the Special Product Formulas

Find each product.

(a) $(2x - \sqrt{y})(2x + \sqrt{y})$ **(b)** $(x + y - 1)(x + y + 1)$

SOLUTION

(a) Substituting $A = 2x$ and $B = \sqrt{y}$ in Product Formula 1, we get

$$(2x - \sqrt{y})(2x + \sqrt{y}) = (2x)^2 - (\sqrt{y})^2 = 4x^2 - y$$

(b) If we group $x + y$ together and think of this as one algebraic expression, we can use Product Formula 1 with $A = x + y$ and $B = 1$.

$$\begin{aligned}(x + y - 1)(x + y + 1) &= [(x + y) - 1][(x + y) + 1] \\ &= (x + y)^2 - 1^2 && \text{Product Formula 1} \\ &= x^2 + 2xy + y^2 - 1 && \text{Product Formula 2}\end{aligned}$$

NOW TRY EXERCISES **55** AND **59**

Factoring Common Factors

We use the Distributive Property to expand algebraic expressions. We sometimes need to reverse this process (again using the Distributive Property) by **factoring** an expression as a product of simpler ones. For example, we can write

FACTORING →

$$x^2 - 4 = (x - 2)(x + 2)$$

← EXPANDING

We say that $x - 2$ and $x + 2$ are **factors** of $x^2 - 4$.

The easiest type of factoring occurs when the terms have a common factor.

EXAMPLE 6 | Factoring Out Common Factors

Factor each expression.

(a) $3x^2 - 6x$ **(b)** $8x^4y^2 + 6x^3y^3 - 2xy^4$

(c) $(2x + 4)(x - 3) - 5(x - 3)$

SOLUTION

(a) The greatest common factor of the terms $3x^2$ and $-6x$ is $3x$, so we have

$$3x^2 - 6x = 3x(x - 2)$$

CHECK YOUR ANSWER

Multiplying gives

$$3x(x - 2) = 3x^2 - 6x \checkmark$$

(b) We note that

8, 6, and -2 have the greatest common factor 2

x^4, x^3, and x have the greatest common factor x

y^2, y^3, and y^4 have the greatest common factor y^2

So the greatest common factor of the three terms in the polynomial is $2xy^2$, and we have

$$\begin{aligned} 8x^4y^2 + 6x^3y^3 - 2xy^4 &= (2xy^2)(4x^3) + (2xy^2)(3x^2y) + (2xy^2)(-y^2) \\ &= 2xy^2(4x^3 + 3x^2y - y^2) \end{aligned}$$

CHECK YOUR ANSWER

Multiplying gives

$$\begin{aligned} &2xy^2(4x^3 + 3x^2y - y^2) \\ &= 8x^4y^2 + 6x^3y^3 - 2xy^4 \checkmark \end{aligned}$$

(c) The two terms have the common factor $x - 3$.

$$\begin{aligned} (2x + 4)(x - 3) - 5(x - 3) &= [(2x + 4) - 5](x - 3) && \text{Distributive Property} \\ &= (2x - 1)(x - 3) && \text{Simplify} \end{aligned}$$

NOW TRY EXERCISES **61**, **63**, AND **65** ■

Factoring Trinomials

To factor a trinomial of the form $x^2 + bx + c$, we note that

$$(x + r)(x + s) = x^2 + (r + s)x + rs$$

so we need to choose numbers r and s so that $r + s = b$ and $rs = c$.

EXAMPLE 7 | Factoring $x^2 + bx + c$ by Trial and Error

Factor: $x^2 + 7x + 12$

SOLUTION We need to find two integers whose product is 12 and whose sum is 7. By trial and error we find that the two integers are 3 and 4. Thus, the factorization is

$$x^2 + 7x + 12 = (x + 3)(x + 4)$$

(3 and 4: factors of 12)

CHECK YOUR ANSWER

Multiplying gives

$$(x + 3)(x + 4) = x^2 + 7x + 12 \checkmark$$

NOW TRY EXERCISE **67** ■

To factor a trinomial of the form $ax^2 + bx + c$ with $a \neq 1$, we look for factors of the form $px + r$ and $qx + s$:

$$ax^2 + bx + c = (px + r)(qx + s) = pqx^2 + (ps + qr)x + rs$$

$$ax^2 + bx + c = (px + r)(qx + s)$$

(p, q: factors of a; r, s: factors of c)

Therefore, we try to find numbers p, q, r, and s such that $pq = a$, $rs = c$, $ps + qr = b$. If these numbers are all integers, then we will have a limited number of possibilities to try for p, q, r, and s.

EXAMPLE 8 | Factoring $ax^2 + bx + c$ by Trial and Error

Factor: $6x^2 + 7x - 5$

SOLUTION We can factor 6 as $6 \cdot 1$ or $3 \cdot 2$, and -5 as $-5 \cdot 1$ or $5 \cdot (-1)$. By trying these possibilities, we arrive at the factorization

factors of 6

$$6x^2 + 7x - 5 = (3x + 5)(2x - 1)$$

factors of -5

CHECK YOUR ANSWER

Multiplying gives

$(3x + 5)(2x - 1) = 6x^2 + 7x - 5$ ✔

NOW TRY EXERCISE **69** ■

EXAMPLE 9 | Recognizing the Form of an Expression

Factor each expression.

(a) $x^2 - 2x - 3$ **(b)** $(5a + 1)^2 - 2(5a + 1) - 3$

SOLUTION

(a) $x^2 - 2x - 3 = (x - 3)(x + 1)$ Trial and error

(b) This expression is of the form

$$\square^2 - 2\square - 3$$

where $\square$ represents $5a + 1$. This is the same form as the expression in part (a), so it will factor as $(\square - 3)(\square + 1)$.

$$(5a + 1)^2 - 2(5a + 1) - 3 = [(5a + 1) - 3][(5a + 1) + 1]$$
$$= (5a - 2)(5a + 2)$$

NOW TRY EXERCISE **71** ■

Special Factoring Formulas

Some special algebraic expressions can be factored using the following formulas. The first three are simply Special Product Formulas written backward.

SPECIAL FACTORING FORMULAS

Formula	Name
1. $A^2 - B^2 = (A - B)(A + B)$	Difference of squares
2. $A^2 + 2AB + B^2 = (A + B)^2$	Perfect square
3. $A^2 - 2AB + B^2 = (A - B)^2$	Perfect square
4. $A^3 - B^3 = (A - B)(A^2 + AB + B^2)$	Difference of cubes
5. $A^3 + B^3 = (A + B)(A^2 - AB + B^2)$	Sum of cubes

EXAMPLE 10 | Factoring Differences of Squares

Factor each expression.

(a) $4x^2 - 25$ **(b)** $(x + y)^2 - z^2$

SOLUTION

(a) Using the Difference of Squares Formula with $A = 2x$ and $B = 5$, we have

$$4x^2 - 25 = (2x)^2 - 5^2 = (2x - 5)(2x + 5)$$

$$A^2 - B^2 = (A - B)(A + B)$$

(b) We use the Difference of Squares Formula with $A = x + y$ and $B = z$.

$$(x + y)^2 - z^2 = (x + y - z)(x + y + z)$$

NOW TRY EXERCISES **75** AND **109** ■

EXAMPLE 11 | Factoring Differences and Sums of Cubes

Factor each polynomial.

(a) $27x^3 - 1$ **(b)** $x^6 + 8$

SOLUTION

(a) Using the Difference of Cubes Formula with $A = 3x$ and $B = 1$, we get

$$27x^3 - 1 = (3x)^3 - 1^3 = (3x - 1)[(3x)^2 + (3x)(1) + 1^2]$$
$$= (3x - 1)(9x^2 + 3x + 1)$$

(b) Using the Sum of Cubes Formula with $A = x^2$ and $B = 2$, we have

$$x^6 + 8 = (x^2)^3 + 2^3 = (x^2 + 2)(x^4 - 2x^2 + 4)$$

NOW TRY EXERCISES **77** AND **79** ■

A trinomial is a perfect square if it is of the form

$$A^2 + 2AB + B^2 \qquad \text{or} \qquad A^2 - 2AB + B^2$$

So we **recognize a perfect square** if the middle term ($2AB$ or $-2AB$) is plus or minus twice the product of the square roots of the outer two terms.

EXAMPLE 12 | Recognizing Perfect Squares

Factor each trinomial.

(a) $x^2 + 6x + 9$ **(b)** $4x^2 - 4xy + y^2$

SOLUTION

(a) Here $A = x$ and $B = 3$, so $2AB = 2 \cdot x \cdot 3 = 6x$. Since the middle term is $6x$, the trinomial is a perfect square. By the Perfect Square Formula we have

$$x^2 + 6x + 9 = (x + 3)^2$$

(b) Here $A = 2x$ and $B = y$, so $2AB = 2 \cdot 2x \cdot y = 4xy$. Since the middle term is $-4xy$, the trinomial is a perfect square. By the Perfect Square Formula we have

$$4x^2 - 4xy + y^2 = (2x - y)^2$$

NOW TRY EXERCISES **105** AND **107** ■

When we factor an expression, the result can sometimes be factored further. In general, *we first factor out common factors*, then inspect the result to see whether it can be factored by any of the other methods of this section. We repeat this process until we have factored the expression completely.

MATHEMATICS IN THE MODERN WORLD

Changing Words, Sound, and Pictures into Numbers

Pictures, sound, and text are routinely transmitted from one place to another via the Internet, fax machines, or modems. How can such things be transmitted through telephone wires? The key to doing this is to change them into numbers or bits (the digits 0 or 1). It's easy to see how to change text to numbers. For example, we could use the correspondence A = 00000001, B = 00000010, C = 00000011, D = 00000100, E = 00000101, and so on. The word "BED" then becomes 000000100000010100000100. By reading the digits in groups of eight, it is possible to translate this number back to the word "BED."

Changing sound to bits is more complicated. A sound wave can be graphed on an oscilloscope or a computer. The graph is then broken down mathematically into simpler components corresponding to the different frequencies of the original sound. (A branch of mathematics called Fourier analysis is used here.) The intensity of each component is a number, and the original sound can be reconstructed from these numbers. For example, music is stored on a CD as a sequence of bits; it may look like 101010001010010100101010 10000010 11110101000101011.... (One second of music requires 1.5 million bits!) The CD player reconstructs the music from the numbers on the CD.

Changing pictures into numbers involves expressing the color and brightness of each dot (or pixel) into a number. This is done very efficiently using a branch of mathematics called wavelet theory. The FBI uses wavelets as a compact way to store the millions of fingerprints they need on file.

EXAMPLE 13 | Factoring an Expression Completely

Factor each expression completely.

(a) $2x^4 - 8x^2$ **(b)** $x^5y^2 - xy^6$

SOLUTION

(a) We first factor out the power of x with the smallest exponent.

$$2x^4 - 8x^2 = 2x^2(x^2 - 4) \quad \text{Common factor is } 2x^2$$
$$= 2x^2(x - 2)(x + 2) \quad \text{Factor } x^2 - 4 \text{ as a difference of squares}$$

(b) We first factor out the powers of x and y with the smallest exponents.

$$x^5y^2 - xy^6 = xy^2(x^4 - y^4) \quad \text{Common factor is } xy^2$$
$$= xy^2(x^2 + y^2)(x^2 - y^2) \quad \text{Factor } x^4 - y^4 \text{ as a difference of squares}$$
$$= xy^2(x^2 + y^2)(x + y)(x - y) \quad \text{Factor } x^2 - y^2 \text{ as a difference of squares}$$

NOW TRY EXERCISES **115** AND **117** ■

In the next example we factor out variables with fractional exponents. This type of factoring occurs in calculus.

EXAMPLE 14 | Factoring Expressions with Fractional Exponents

Factor each expression.

(a) $3x^{3/2} - 9x^{1/2} + 6x^{-1/2}$ **(b)** $(2 + x)^{-2/3}x + (2 + x)^{1/3}$

SOLUTION

(a) Factor out the power of x with the *smallest exponent*, that is, $x^{-1/2}$.

$$3x^{3/2} - 9x^{1/2} + 6x^{-1/2} = 3x^{-1/2}(x^2 - 3x + 2) \quad \text{Factor out } 3x^{-1/2}$$
$$= 3x^{-1/2}(x - 1)(x - 2) \quad \text{Factor the quadratic } x^2 - 3x + 2$$

To factor out $x^{-1/2}$ from $x^{3/2}$, we *subtract* exponents:

$$x^{3/2} = x^{-1/2}(x^{3/2-(-1/2)})$$
$$= x^{-1/2}(x^{3/2+1/2})$$
$$= x^{-1/2}(x^2)$$

(b) Factor out the power of $2 + x$ with the *smallest exponent*, that is, $(2 + x)^{-2/3}$.

$$(2 + x)^{-2/3}x + (2 + x)^{1/3} = (2 + x)^{-2/3}[x + (2 + x)] \quad \text{Factor out } (2 + x)^{-2/3}$$
$$= (2 + x)^{-2/3}(2 + 2x) \quad \text{Simplify}$$
$$= 2(2 + x)^{-2/3}(1 + x) \quad \text{Factor out 2}$$

CHECK YOUR ANSWERS

To see that you have factored correctly, multiply using the Laws of Exponents.

(a) $3x^{-1/2}(x^2 - 3x + 2)$
$= 3x^{3/2} - 9x^{1/2} + 6x^{-1/2}$ ✔

(b) $(2 + x)^{-2/3}[x + (2 + x)]$
$= (2 + x)^{-2/3}x + (2 + x)^{1/3}$ ✔

NOW TRY EXERCISES **91** AND **93** ■

▼ Factoring by Grouping Terms

Polynomials with at least four terms can sometimes be factored by grouping terms. The following example illustrates the idea.

EXAMPLE 15 | Factoring by Grouping

Factor each polynomial.

(a) $x^3 + x^2 + 4x + 4$ **(b)** $x^3 - 2x^2 - 3x + 6$

SOLUTION

(a) $x^3 + x^2 + 4x + 4 = (x^3 + x^2) + (4x + 4)$ Group terms

$= x^2(x + 1) + 4(x + 1)$ Factor out common factors

$= (x^2 + 4)(x + 1)$ Factor out $x + 1$ from each term

(b) $x^3 - 2x^2 - 3x + 6 = (x^3 - 2x^2) - (3x - 6)$ Group terms

$= x^2(x - 2) - 3(x - 2)$ Factor out common factors

$= (x^2 - 3)(x - 2)$ Factor out $x - 2$ from each term

NOW TRY EXERCISE **83**

1.3 EXERCISES

CONCEPTS

1. Consider the polynomial $2x^5 + 6x^4 + 4x^3$.

How many terms does this polynomial have? ________

List the terms: ________

What factor is common to each term? ________

Factor the polynomial: $2x^5 + 6x^4 + 4x^3 =$ ________.

2. To factor the trinomial $x^2 + 7x + 10$, we look for two integers whose product is ________ and whose sum is ________. These integers are ________ and ________, so the trinomial factors as ________.

3. The Special Product Formula for the "square of a sum" is $(A + B)^2 =$ ________________.

So $(2x + 3)^2 =$ ________________.

4. The Special Product Formula for the "sum and difference of the same terms" is $(A + B)(A - B) =$ ________________.

So $(5 + x)(5 - x) =$ ________________.

5. The Special Factoring Formula for the "difference of squares" is $A^2 - B^2 =$ ________. So $4x^2 - 25$ factors as ________.

6. The Special Factoring Formula for a "perfect square" is $A^2 + 2AB + B^2 =$ ________. So $x^2 + 10x + 25$ factors as ________.

SKILLS

7–12 ■ Complete the following table by stating whether the polynomial is a monomial, binomial, or trinomial; then list its terms and state its degree.

Polynomial	Type	Terms	Degree
7. $x^2 - 3x + 7$			
8. $2x^5 + 4x^2$			
9. -8			
10. $\frac{1}{2}x^7$			
11. $x - x^2 + x^3 - x^4$			
12. $\sqrt{2}x - \sqrt{3}$			

13–22 ■ Find the sum, difference, or product.

13. $(12x - 7) - (5x - 12)$ **14.** $(5 - 3x) + (2x - 8)$

15. $(3x^2 + x + 1) + (2x^2 - 3x - 5)$

16. $(3x^2 + x + 1) - (2x^2 - 3x - 5)$

17. $(x^3 + 6x^2 - 4x + 7) - (3x^2 + 2x - 4)$

18. $3(x - 1) + 4(x + 2)$

19. $8(2x + 5) - 7(x - 9)$

20. $4(x^2 - 3x + 5) - 3(x^2 - 2x + 1)$

21. $2(2 - 5t) + t^2(t - 1) - (t^4 - 1)$

22. $5(3t - 4) - (t^2 + 2) - 2t(t - 3)$

23–28 ■ Multiply the algebraic expressions using the FOIL method and simplify.

23. $(3t - 2)(7t - 4)$ **24.** $(4s - 1)(2s + 5)$

25. $(3x + 5)(2x - 1)$ **26.** $(7y - 3)(2y - 1)$

27. $(x + 3y)(2x - y)$ **28.** $(4x - 5y)(3x - y)$

29–44 ■ Multiply the algebraic expressions using a Special Product Formula and simplify.

29. $(3x + 4)^2$ **30.** $(1 - 2y)^2$

31. $(2u + v)^2$ **32.** $(x - 3y)^2$

33. $(2x + 3y)^2$ **34.** $(r - 2s)^2$

35. $(x + 5)(x - 5)$ **36.** $(y - 3)(y + 3)$

37. $(3x - 4)(3x + 4)$ **38.** $(2y + 5)(2y - 5)$

39. $(\sqrt{x} + 2)(\sqrt{x} - 2)$ **40.** $(\sqrt{y} + \sqrt{2})(\sqrt{y} - \sqrt{2})$

41. $(y + 2)^3$
42. $(x - 3)^3$
43. $(1 - 2r)^3$
44. $(3 + 2y)^3$

45–60 ■ Perform the indicated operations and simplify.

45. $(x + 2)(x^2 + 2x + 3)$
46. $(x + 1)(2x^2 - x + 1)$
47. $(2x - 5)(x^2 - x + 1)$
48. $(1 + 2x)(x^2 - 3x + 1)$
49. $\sqrt{x}(x - \sqrt{x})$
50. $x^{3/2}(\sqrt{x} - 1/\sqrt{x})$
51. $y^{1/3}(y^{2/3} + y^{5/3})$
52. $x^{1/4}(2x^{3/4} - x^{1/4})$
53. $(x^2 - a^2)(x^2 + a^2)$
54. $(x^{1/2} + y^{1/2})(x^{1/2} - y^{1/2})$
55. $(\sqrt{a} - b)(\sqrt{a} + b)$
56. $(\sqrt{h^2 + 1} + 1)(\sqrt{h^2 + 1} - 1)$
57. $((x - 1) + x^2)((x - 1) - x^2)$
58. $(x + (2 + x^2))(x - (2 + x^2))$
59. $(2x + y - 3)(2x + y + 3)$
60. $(x + y + z)(x - y - z)$

61–66 ■ Factor out the common factor.

61. $-2x^3 + 16x$
62. $2x^4 + 4x^3 - 14x^2$
63. $y(y - 6) + 9(y - 6)$
64. $(z + 2)^2 - 5(z + 2)$
65. $2x^2y - 6xy^2 + 3xy$
66. $-7x^4y^2 + 14xy^3 + 21xy^4$

67–74 ■ Factor the trinomial.

67. $x^2 + 2x - 3$
68. $x^2 - 6x + 5$
69. $8x^2 - 14x - 15$
70. $6y^2 + 11y - 21$
71. $3x^2 - 16x + 5$
72. $5x^2 - 7x - 6$
73. $(3x + 2)^2 + 8(3x + 2) + 12$
74. $2(a + b)^2 + 5(a + b) - 3$

75–82 ■ Use a Special Factoring Formula to factor the expression.

75. $9a^2 - 16$
76. $(x + 3)^2 - 4$
77. $27x^3 + y^3$
78. $a^3 - b^6$
79. $8s^3 - 125t^3$
80. $1 + 1000y^3$
81. $x^2 + 12x + 36$
82. $16z^2 - 24z + 9$

83–88 ■ Factor the expression by grouping terms.

83. $x^3 + 4x^2 + x + 4$
84. $3x^3 - x^2 + 6x - 2$
85. $2x^3 + x^2 - 6x - 3$
86. $-9x^3 - 3x^2 + 3x + 1$
87. $x^3 + x^2 + x + 1$
88. $x^5 + x^4 + x + 1$

89–94 ■ Factor the expression completely. Begin by factoring out the lowest power of each common factor.

89. $x^{5/2} - x^{1/2}$
90. $3x^{-1/2} + 4x^{1/2} + x^{3/2}$
91. $x^{-3/2} + 2x^{-1/2} + x^{1/2}$
92. $(x - 1)^{7/2} - (x - 1)^{3/2}$
93. $(x^2 + 1)^{1/2} + 2(x^2 + 1)^{-1/2}$
94. $x^{-1/2}(x + 1)^{1/2} + x^{1/2}(x + 1)^{-1/2}$

95–124 ■ Factor the expression completely.

95. $12x^3 + 18x$
96. $30x^3 + 15x^4$
97. $x^2 - 2x - 8$
98. $x^2 - 14x + 48$
99. $2x^2 + 5x + 3$
100. $2x^2 + 7x - 4$
101. $9x^2 - 36x - 45$
102. $8x^2 + 10x + 3$
103. $49 - 4y^2$
104. $4t^2 - 9s^2$
105. $t^2 - 6t + 9$
106. $x^2 + 10x + 25$
107. $4x^2 + 4xy + y^2$
108. $r^2 - 6rs + 9s^2$
109. $(a + b)^2 - (a - b)^2$
110. $\left(1 + \frac{1}{x}\right)^2 - \left(1 - \frac{1}{x}\right)^2$
111. $x^2(x^2 - 1) - 9(x^2 - 1)$
112. $(a^2 - 1)b^2 - 4(a^2 - 1)$
113. $8x^3 - 125$
114. $x^6 + 64$
115. $x^3 + 2x^2 + x$
116. $3x^3 - 27x$
117. $x^4y^3 - x^2y^5$
118. $18y^3x^2 - 2xy^4$
119. $2x^3 + 4x^2 + x + 2$
120. $3x^3 + 5x^2 - 6x - 10$
121. $(x - 1)(x + 2)^2 - (x - 1)^2(x + 2)$
122. $y^4(y + 2)^3 + y^5(y + 2)^4$
123. $(a^2 + 1)^2 - 7(a^2 + 1) + 10$
124. $(a^2 + 2a)^2 - 2(a^2 + 2a) - 3$

125–128 ■ Factor the expression completely. (This type of expression arises in calculus when using the "Product Rule.")

125. $5(x^2 + 4)^4(2x)(x - 2)^4 + (x^2 + 4)^5(4)(x - 2)^3$
126. $3(2x - 1)^2(2)(x + 3)^{1/2} + (2x - 1)^3(\frac{1}{2})(x + 3)^{-1/2}$
127. $(x^2 + 3)^{-1/3} - \frac{2}{3}x^2(x^2 + 3)^{-4/3}$
128. $\frac{1}{2}x^{-1/2}(3x + 4)^{1/2} - \frac{3}{2}x^{1/2}(3x + 4)^{-1/2}$

129. **(a)** Show that $ab = \frac{1}{2}[(a + b)^2 - (a^2 + b^2)]$.
(b) Show that $(a^2 + b^2)^2 - (a^2 - b^2)^2 = 4a^2b^2$.
(c) Show that

$$(a^2 + b^2)(c^2 + d^2) = (ac + bd)^2 + (ad - bc)^2$$

(d) Factor completely: $4a^2c^2 - (a^2 - b^2 + c^2)^2$.

130. Verify Special Factoring Formulas 4 and 5 by expanding their right-hand sides.

APPLICATIONS

131. Volume of Concrete A culvert is constructed out of large cylindrical shells cast in concrete, as shown in the figure. Using the formula for the volume of a cylinder given on the inside front cover of this book, explain why the volume of the cylindrical shell is

$$V = \pi R^2h - \pi r^2h$$

Factor to show that

$$V = 2\pi \cdot \text{average radius} \cdot \text{height} \cdot \text{thickness}$$

Use the "unrolled" diagram to explain why this makes sense geometrically.

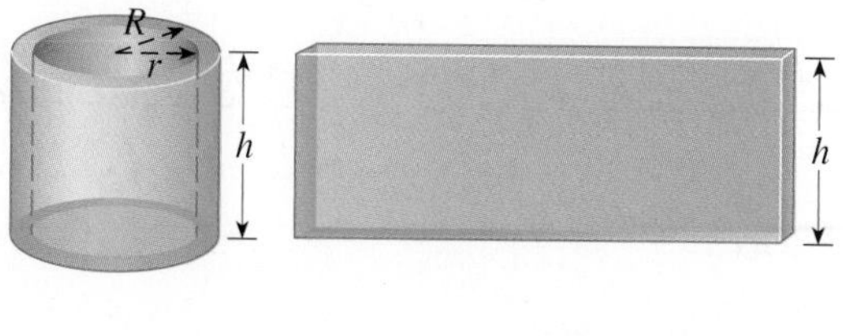

132. Mowing a Field A square field in a certain state park is mowed around the edges every week. The rest of the field is kept unmowed to serve as a habitat for birds and small animals (see the figure). The field measures b feet by b feet, and the mowed strip is x feet wide.

(a) Explain why the area of the mowed portion is $b^2 - (b - 2x)^2$.

(b) Factor the expression in part (a) to show that the area of the mowed portion is also $4x(b - x)$.

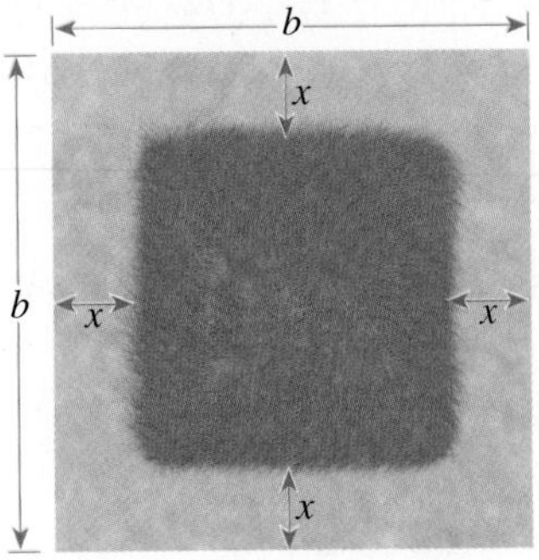

DISCOVERY ■ DISCUSSION ■ WRITING

133. Degrees of Sums and Products of Polynomials Make up several pairs of polynomials, then calculate the sum and product of each pair. On the basis of your experiments and observations, answer the following questions.

(a) How is the degree of the product related to the degrees of the original polynomials?

(b) How is the degree of the sum related to the degrees of the original polynomials?

134. The Power of Algebraic Formulas Use the Difference of Squares Formula to factor $17^2 - 16^2$. Notice that it is easy to calculate the factored form in your head but not so easy to calculate the original form in this way. Evaluate each expression in your head:

(a) $528^2 - 527^2$

(b) $122^2 - 120^2$

(c) $1020^2 - 1010^2$

Now use the Special Product Formula

$$(A + B)(A - B) = A^2 - B^2$$

to evaluate these products in your head:

(d) $79 \cdot 51$

(e) $998 \cdot 1002$

135. Differences of Even Powers

(a) Factor the expressions completely: $A^4 - B^4$ and $A^6 - B^6$.

(b) Verify that $18{,}335 = 12^4 - 7^4$ and that $2{,}868{,}335 = 12^6 - 7^6$.

(c) Use the results of parts (a) and (b) to factor the integers 18,335 and 2,868,335. Then show that in both of these factorizations, all the factors are prime numbers.

136. Factoring $A^n - 1$ Verify these formulas by expanding and simplifying the right-hand side.

$$A^2 - 1 = (A - 1)(A + 1)$$

$$A^3 - 1 = (A - 1)(A^2 + A + 1)$$

$$A^4 - 1 = (A - 1)(A^3 + A^2 + A + 1)$$

On the basis of the pattern displayed in this list, how do you think $A^5 - 1$ would factor? Verify your conjecture. Now generalize the pattern you have observed to obtain a factoring formula for $A^n - 1$, where n is a positive integer.

137. Factoring $x^4 + ax^2 + b$ A trinomial of the form $x^4 + ax^2 + b$ can sometimes be factored easily. For example,

$$x^4 + 3x^2 - 4 = (x^2 + 4)(x^2 - 1)$$

But $x^4 + 3x^2 + 4$ cannot be factored in this way. Instead, we can use the following method.

$x^4 + 3x^2 + 4 = (x^4 + 4x^2 + 4) - x^2$	Add and subtract x^2
$= (x^2 + 2)^2 - x^2$	Factor perfect square
$= [(x^2 + 2) - x][(x^2 + 2) + x]$	Difference of squares
$= (x^2 - x + 2)(x^2 + x + 2)$	

Factor the following, using whichever method is appropriate.

(a) $x^4 + x^2 - 2$

(b) $x^4 + 2x^2 + 9$

(c) $x^4 + 4x^2 + 16$

(d) $x^4 + 2x^2 + 1$

DISCOVERY PROJECT **Visualizing a Formula**

In this project we discover geometric interpretations of some of the Special Product Formulas. You can find the project at the book companion website: **www.stewartmath.com**

1.4 Rational Expressions

The Domain of an Algebraic Expression ▶ Simplifying Rational Expressions ▶ Multiplying and Dividing Rational Expressions ▶ Adding and Subtracting Rational Expressions ▶ Compound Fractions ▶ Rationalizing the Denominator or the Numerator ▶ Avoiding Common Errors

A quotient of two algebraic expressions is called a **fractional expression**. Here are some examples:

$$\frac{2x}{x-1} \qquad \frac{\sqrt{x}+3}{x+1} \qquad \frac{y-2}{y^2+4}$$

A **rational expression** is a fractional expression where both the numerator and denominator are polynomials. For example, the following are rational expressions:

$$\frac{2x}{x-1} \qquad \frac{x}{x^2+1} \qquad \frac{x^3-x}{x^2-5x+6}$$

In this section we learn how to perform algebraic operations on rational expressions.

▼ The Domain of an Algebraic Expression

Expression	Domain
$\frac{1}{x}$	$\{x \mid x \neq 0\}$
$\sqrt{x}$	$\{x \mid x \geq 0\}$
$\frac{1}{\sqrt{x}}$	$\{x \mid x > 0\}$

In general, an algebraic expression may not be defined for all values of the variable. The **domain** of an algebraic expression is the set of real numbers that the variable is permitted to have. The table in the margin gives some basic expressions and their domains.

EXAMPLE 1 | Finding the Domain of an Expression

Find the domains of the following expressions.

(a) $2x^2 + 3x - 1$ **(b)** $\dfrac{x}{x^2-5x+6}$ **(c)** $\dfrac{\sqrt{x}}{x-5}$

SOLUTION

(a) This polynomial is defined for every x. Thus, the domain is the set $\mathbb{R}$ of real numbers.

(b) We first factor the denominator.

$$\frac{x}{x^2-5x+6} = \frac{x}{(x-2)(x-3)}$$

Denominator would be 0 if $x = 2$ or $x = 3$

Since the denominator is zero when $x = 2$ or 3, the expression is not defined for these numbers. The domain is $\{x \mid x \neq 2 \text{ and } x \neq 3\}$.

(c) For the numerator to be defined, we must have $x \geq 0$. Also, we cannot divide by zero, so $x \neq 5$.

Must have $x \geq 0$ to take square root

$$\frac{\sqrt{x}}{x-5}$$

Denominator would be 0 if $x = 5$

Thus, the domain is $\{x \mid x \geq 0 \text{ and } x \neq 5\}$.

NOW TRY EXERCISE **11** ■

▼ Simplifying Rational Expressions

To **simplify rational expressions**, we factor both numerator and denominator and use the following property of fractions:

$$\frac{AC}{BC} = \frac{A}{B}$$

This allows us to **cancel** common factors from the numerator and denominator.

EXAMPLE 2 | Simplifying Rational Expressions by Cancellation

Simplify: $\dfrac{x^2 - 1}{x^2 + x - 2}$

SOLUTION

$$\frac{x^2 - 1}{x^2 + x - 2} = \frac{(x - 1)(x + 1)}{(x - 1)(x + 2)} \qquad \text{Factor}$$

$$= \frac{x + 1}{x + 2} \qquad \text{Cancel common factors}$$

We can't cancel the x^2's in $\dfrac{x^2 - 1}{x^2 + x - 2}$ because x^2 is not a factor.

NOW TRY EXERCISE **17** ■

▼ Multiplying and Dividing Rational Expressions

To **multiply rational expressions**, we use the following property of fractions:

$$\frac{A}{B} \cdot \frac{C}{D} = \frac{AC}{BD}$$

This says that to multiply two fractions we multiply their numerators and multiply their denominators.

EXAMPLE 3 | Multiplying Rational Expressions

Perform the indicated multiplication and simplify: $\dfrac{x^2 + 2x - 3}{x^2 + 8x + 16} \cdot \dfrac{3x + 12}{x - 1}$

SOLUTION We first factor.

$$\frac{x^2 + 2x - 3}{x^2 + 8x + 16} \cdot \frac{3x + 12}{x - 1} = \frac{(x - 1)(x + 3)}{(x + 4)^2} \cdot \frac{3(x + 4)}{x - 1} \qquad \text{Factor}$$

$$= \frac{3(x - 1)(x + 3)(x + 4)}{(x - 1)(x + 4)^2} \qquad \text{Property of fractions}$$

$$= \frac{3(x + 3)}{x + 4} \qquad \text{Cancel common factors}$$

NOW TRY EXERCISE **25** ■

To **divide rational expressions**, we use the following property of fractions:

$$\frac{A}{B} \div \frac{C}{D} = \frac{A}{B} \cdot \frac{D}{C}$$

This says that to divide a fraction by another fraction, we invert the divisor and multiply.

EXAMPLE 4 | Dividing Rational Expressions

Perform the indicated division and simplify:

$$\frac{x-4}{x^2-4} \div \frac{x^2-3x-4}{x^2+5x+6}$$

SOLUTION

$$\frac{x-4}{x^2-4} \div \frac{x^2-3x-4}{x^2+5x+6} = \frac{x-4}{x^2-4} \cdot \frac{x^2+5x+6}{x^2-3x-4}$$ Invert and multiply

$$= \frac{(x-4)(x+2)(x+3)}{(x-2)(x+2)(x-4)(x+1)}$$ Factor

$$= \frac{x+3}{(x-2)(x+1)}$$ Cancel common factors

NOW TRY EXERCISE 31 ■

Adding and Subtracting Rational Expressions

Avoid making the following error:

$$\frac{A}{B+C} \neq \frac{A}{B} + \frac{A}{C}$$

For instance, if we let $A = 2$, $B = 1$, and $C = 1$, then we see the error:

$$\frac{2}{1+1} \stackrel{?}{=} \frac{2}{1} + \frac{2}{1}$$

$$\frac{2}{2} \stackrel{?}{=} 2 + 2$$

$$1 \stackrel{?}{=} 4 \quad \text{Wrong!}$$

To **add or subtract rational expressions**, we first find a common denominator and then use the following property of fractions:

$$\frac{A}{C} + \frac{B}{C} = \frac{A+B}{C}$$

Although any common denominator will work, it is best to use the **least common denominator** (LCD) as explained in Section 1.1. The LCD is found by factoring each denominator and taking the product of the distinct factors, using the highest power that appears in any of the factors.

EXAMPLE 5 | Adding and Subtracting Rational Expressions

Perform the indicated operations and simplify:

(a) $\frac{3}{x-1} + \frac{x}{x+2}$ **(b)** $\frac{1}{x^2-1} - \frac{2}{(x+1)^2}$

SOLUTION

(a) Here the LCD is simply the product $(x-1)(x+2)$.

$$\frac{3}{x-1} + \frac{x}{x+2} = \frac{3(x+2)}{(x-1)(x+2)} + \frac{x(x-1)}{(x-1)(x+2)}$$ Write fractions using LCD

$$= \frac{3x+6+x^2-x}{(x-1)(x+2)}$$ Add fractions

$$= \frac{x^2+2x+6}{(x-1)(x+2)}$$ Combine terms in numerator

(b) The LCD of $x^2 - 1 = (x-1)(x+1)$ and $(x+1)^2$ is $(x-1)(x+1)^2$.

$$\frac{1}{x^2-1} - \frac{2}{(x+1)^2} = \frac{1}{(x-1)(x+1)} - \frac{2}{(x+1)^2} \qquad \text{Factor}$$

$$= \frac{(x+1) - 2(x-1)}{(x-1)(x+1)^2} \qquad \text{Combine fractions using LCD}$$

$$= \frac{x + 1 - 2x + 2}{(x-1)(x+1)^2} \qquad \text{Distributive Property}$$

$$= \frac{3 - x}{(x-1)(x+1)^2} \qquad \text{Combine terms in numerator}$$

NOW TRY EXERCISES **43** AND **45** ■

▼ Compound Fractions

A **compound fraction** is a fraction in which the numerator, the denominator, or both, are themselves fractional expressions.

EXAMPLE 6 | Simplifying a Compound Fraction

Simplify: $\dfrac{\dfrac{x}{y} + 1}{1 - \dfrac{y}{x}}$

SOLUTION 1 We combine the terms in the numerator into a single fraction. We do the same in the denominator. Then we invert and multiply.

$$\frac{\dfrac{x}{y} + 1}{1 - \dfrac{y}{x}} = \frac{\dfrac{x+y}{y}}{\dfrac{x-y}{x}} = \frac{x+y}{y} \cdot \frac{x}{x-y}$$

$$= \frac{x(x+y)}{y(x-y)}$$

MATHEMATICS IN THE MODERN WORLD

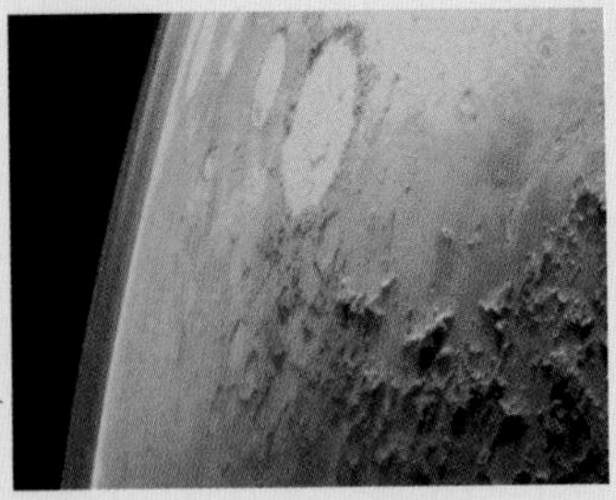

Courtesy of NASA

Error-Correcting Codes

The pictures sent back by the *Pathfinder* spacecraft from the surface of Mars on July 4, 1997, were astoundingly clear. But few viewing these pictures were aware of the complex mathematics used to accomplish that feat. The distance to Mars is enormous, and the background noise (or static) is many times stronger than the original signal emitted by the spacecraft. So, when scientists receive the signal, it is full of errors. To get a clear picture, the errors must be found and corrected. This same problem of errors is routinely encountered in transmitting bank records when you use an ATM machine or voice when you are talking on the telephone.

To understand how errors are found and corrected, we must first understand that to transmit pictures, sound, or text we transform them into bits (the digits 0 or 1; see page 30). To help the receiver recognize errors, the message is "coded" by inserting additional bits. For example, suppose you want to transmit the message "10100." A very simple-minded code is as follows: Send each digit a million times. The person receiving the message reads it in blocks of a million digits. If the first block is mostly 1's, he concludes that you are probably trying to transmit a 1, and so on. To say that this code is not efficient is a bit of an understatement; it requires sending a million times more data than the original message. Another method inserts "check digits." For example, for each block of eight digits insert a ninth digit; the inserted digit is 0 if there is an even number of 1's in the block and 1 if there is an odd number. So, if a single digit is wrong (a 0 changed to a 1, or vice versa), the check digits allow us to recognize that an error has occurred. This method does not tell us where the error is, so we can't correct it. Modern error-correcting codes use interesting mathematical algorithms that require inserting relatively few digits but that allow the receiver to not only recognize, but also correct, errors. The first error-correcting code was developed in the 1940s by Richard Hamming at MIT. It is interesting to note that the English language has a built-in error correcting mechanism; to test it, try reading this error-laden sentence: Gve mo libty ox giv ne deth.

SOLUTION 2 We find the LCD of all the fractions in the expression, then multiply numerator and denominator by it. In this example the LCD of all the fractions is xy. Thus

$$\frac{\frac{x}{y}+1}{1-\frac{y}{x}} = \frac{\frac{x}{y}+1}{1-\frac{y}{x}} \cdot \frac{xy}{xy} \qquad \text{Multiply numerator and denominator by } xy$$

$$= \frac{x^2+xy}{xy-y^2} \qquad \text{Simplify}$$

$$= \frac{x(x+y)}{y(x-y)} \qquad \text{Factor}$$

NOW TRY EXERCISES **59** AND **61** ■

The next two examples show situations in calculus that require the ability to work with fractional expressions.

EXAMPLE 7 | Simplifying a Compound Fraction

Simplify: $\dfrac{\frac{1}{a+h}-\frac{1}{a}}{h}$

SOLUTION We begin by combining the fractions in the numerator using a common denominator.

$$\frac{\frac{1}{a+h}-\frac{1}{a}}{h} = \frac{\frac{a-(a+h)}{a(a+h)}}{h} \qquad \text{Combine fractions in the numerator}$$

$$= \frac{a-(a+h)}{a(a+h)} \cdot \frac{1}{h} \qquad \text{Property 2 of fractions (invert divisor and multiply)}$$

$$= \frac{a-a-h}{a(a+h)} \cdot \frac{1}{h} \qquad \text{Distributive Property}$$

$$= \frac{-h}{a(a+h)} \cdot \frac{1}{h} \qquad \text{Simplify}$$

$$= \frac{-1}{a(a+h)} \qquad \text{Property 5 of fractions (cancel common factors)}$$

NOW TRY EXERCISE **69** ■

EXAMPLE 8 | Simplifying a Compound Fraction

Simplify: $\dfrac{(1+x^2)^{1/2}-x^2(1+x^2)^{-1/2}}{1+x^2}$

SOLUTION 1 Factor $(1+x^2)^{-1/2}$ from the numerator.

Factor out the power of $1+x^2$ with the *smallest* exponent, in this case $(1+x^2)^{-1/2}$.

$$\frac{(1+x^2)^{1/2}-x^2(1+x^2)^{-1/2}}{1+x^2} = \frac{(1+x^2)^{-1/2}[(1+x^2)-x^2]}{1+x^2}$$

$$= \frac{(1+x^2)^{-1/2}}{1+x^2} = \frac{1}{(1+x^2)^{3/2}}$$

SOLUTION 2 Since $(1 + x^2)^{-1/2} = 1/(1 + x^2)^{1/2}$ is a fraction, we can clear all fractions by multiplying numerator and denominator by $(1 + x^2)^{1/2}$.

$$\frac{(1 + x^2)^{1/2} - x^2(1 + x^2)^{-1/2}}{1 + x^2} = \frac{(1 + x^2)^{1/2} - x^2(1 + x^2)^{-1/2}}{1 + x^2} \cdot \frac{(1 + x^2)^{1/2}}{(1 + x^2)^{1/2}}$$

$$= \frac{(1 + x^2) - x^2}{(1 + x^2)^{3/2}} = \frac{1}{(1 + x^2)^{3/2}}$$

NOW TRY EXERCISE **77** ■

Rationalizing the Denominator or the Numerator

If a fraction has a denominator of the form $A + B\sqrt{C}$, we may rationalize the denominator by multiplying numerator and denominator by the **conjugate radical** $A - B\sqrt{C}$. This works because, by Special Product Formula 1 in Section 1.3, the product of the denominator and its conjugate radical does not contain a radical:

$$(A + B\sqrt{C})(A - B\sqrt{C}) = A^2 - B^2C$$

EXAMPLE 9 | Rationalizing the Denominator

Rationalize the denominator: $\dfrac{1}{1 + \sqrt{2}}$

SOLUTION We multiply both the numerator and the denominator by the conjugate radical of $1 + \sqrt{2}$, which is $1 - \sqrt{2}$.

Special Product Formula 1
$(A + B)(A - B) = A^2 - B^2$

$$\frac{1}{1 + \sqrt{2}} = \frac{1}{1 + \sqrt{2}} \cdot \frac{1 - \sqrt{2}}{1 - \sqrt{2}}$$ Multiply numerator and denominator by the conjugate radical

$$= \frac{1 - \sqrt{2}}{1^2 - (\sqrt{2})^2}$$ Special Product Formula 1

$$= \frac{1 - \sqrt{2}}{1 - 2} = \frac{1 - \sqrt{2}}{-1} = \sqrt{2} - 1$$

NOW TRY EXERCISE **81** ■

EXAMPLE 10 | Rationalizing the Numerator

Rationalize the numerator: $\dfrac{\sqrt{4 + h} - 2}{h}$

SOLUTION We multiply numerator and denominator by the conjugate radical $\sqrt{4 + h} + 2$.

Special Product Formula 1
$(A + B)(A - B) = A^2 - B^2$

$$\frac{\sqrt{4 + h} - 2}{h} = \frac{\sqrt{4 + h} - 2}{h} \cdot \frac{\sqrt{4 + h} + 2}{\sqrt{4 + h} + 2}$$ Multiply numerator and denominator by the conjugate radical

$$= \frac{(\sqrt{4 + h})^2 - 2^2}{h(\sqrt{4 + h} + 2)}$$ Special Product Formula 1

$$= \frac{4 + h - 4}{h(\sqrt{4 + h} + 2)}$$

$$= \frac{h}{h(\sqrt{4 + h} + 2)} = \frac{1}{\sqrt{4 + h} + 2}$$ Property 5 of fractions (cancel common factors)

NOW TRY EXERCISE **87** ■

Avoiding Common Errors

Don't make the mistake of applying properties of multiplication to the operation of addition. Many of the common errors in algebra involve doing just that. The following table states several properties of multiplication and illustrates the error in applying them to addition.

Correct multiplication property	Common error with addition
$(a \cdot b)^2 = a^2 \cdot b^2$	$(a + b)^2 \neq a^2 + b^2$
$\sqrt{a \cdot b} = \sqrt{a}\sqrt{b} \quad (a, b \geq 0)$	$\sqrt{a + b} \neq \sqrt{a} + \sqrt{b}$
$\sqrt{a^2 \cdot b^2} = a \cdot b \quad (a, b \geq 0)$	$\sqrt{a^2 + b^2} \neq a + b$
$\dfrac{1}{a} \cdot \dfrac{1}{b} = \dfrac{1}{a \cdot b}$	$\dfrac{1}{a} + \dfrac{1}{b} \neq \dfrac{1}{a + b}$
$\dfrac{ab}{a} = b$	$\dfrac{a + b}{a} \neq b$
$a^{-1} \cdot b^{-1} = (a \cdot b)^{-1}$	$a^{-1} + b^{-1} \neq (a + b)^{-1}$

To verify that the equations in the right-hand column are wrong, simply substitute numbers for a and b and calculate each side. For example, if we take $a = 2$ and $b = 2$ in the fourth error, we find that the left-hand side is

$$\frac{1}{a} + \frac{1}{b} = \frac{1}{2} + \frac{1}{2} = 1$$

whereas the right-hand side is

$$\frac{1}{a + b} = \frac{1}{2 + 2} = \frac{1}{4}$$

Since $1 \neq \frac{1}{4}$, the stated equation is wrong. You should similarly convince yourself of the error in each of the other equations. (See Exercise 105.)

1.4 EXERCISES

CONCEPTS

1. Which of the following are rational expressions?

   (a) $\dfrac{3x}{x^2 - 1}$ (b) $\dfrac{\sqrt{x + 1}}{2x + 3}$ (c) $\dfrac{x(x^2 - 1)}{x + 3}$

2. To simplify a rational expression, we cancel *factors* that are common to the ______ and ______. So the expression

   $$\frac{(x + 1)(x + 2)}{(x + 3)(x + 2)}$$

   simplifies to ______.

3. To multiply two rational expressions, we multiply their ______ together and multiply their ______ together.

   So $\dfrac{2}{x + 1} \cdot \dfrac{x}{x + 3}$ is the same as ______.

4. Consider the expression $\dfrac{1}{x} - \dfrac{2}{x + 1} - \dfrac{x}{(x + 1)^2}$.

   (a) How many terms does this expression have?

   (b) Find the least common denominator of all the terms.

   (c) Perform the addition and simplify.

SKILLS

5–12 ■ Find the domain of the expression.

5. $4x^2 - 10x + 3$

6. $-x^4 + x^3 + 9x$

7. $\dfrac{2x + 1}{x - 4}$

8. $\dfrac{2t^2 - 5}{3t + 6}$

9. $\sqrt{x + 3}$

10. $\dfrac{1}{\sqrt{x - 1}}$

11. $\dfrac{x^2 + 1}{x^2 - x - 2}$

12. $\dfrac{\sqrt{2x}}{x + 1}$

13–22 ■ Simplify the rational expression.

13. $\dfrac{3(x+2)(x-1)}{6(x-1)^2}$

14. $\dfrac{4(x^2-1)}{12(x+2)(x-1)}$

15. $\dfrac{x-2}{x^2-4}$

16. $\dfrac{x^2-x-2}{x^2-1}$

17. $\dfrac{x^2+6x+8}{x^2+5x+4}$

18. $\dfrac{x^2-x-12}{x^2+5x+6}$

19. $\dfrac{y^2+y}{y^2-1}$

20. $\dfrac{y^2-3y-18}{2y^2+5y+3}$

21. $\dfrac{2x^3-x^2-6x}{2x^2-7x+6}$

22. $\dfrac{1-x^2}{x^3-1}$

23–38 ■ Perform the multiplication or division and simplify.

23. $\dfrac{4x}{x^2-4}\cdot\dfrac{x+2}{16x}$

24. $\dfrac{x^2-25}{x^2-16}\cdot\dfrac{x+4}{x+5}$

25. $\dfrac{x^2-2x-15}{x^2-9}\cdot\dfrac{x+3}{x-5}$

26. $\dfrac{x^2+2x-3}{x^2-2x-3}\cdot\dfrac{3-x}{3+x}$

27. $\dfrac{t-3}{t^2+9}\cdot\dfrac{t+3}{t^2-9}$

28. $\dfrac{x^2-x-6}{x^2+2x}\cdot\dfrac{x^3+x^2}{x^2-2x-3}$

29. $\dfrac{x^2+7x+12}{x^2+3x+2}\cdot\dfrac{x^2+5x+6}{x^2+6x+9}$

30. $\dfrac{x^2+2xy+y^2}{x^2-y^2}\cdot\dfrac{2x^2-xy-y^2}{x^2-xy-2y^2}$

31. $\dfrac{x+3}{4x^2-9}\div\dfrac{x^2+7x+12}{2x^2+7x-15}$

32. $\dfrac{2x+1}{2x^2+x-15}\div\dfrac{6x^2-x-2}{x+3}$

33. $\dfrac{2x^2+3x+1}{x^2+2x-15}\div\dfrac{x^2+6x+5}{2x^2-7x+3}$

34. $\dfrac{4y^2-9}{2y^2+9y-18}\div\dfrac{2y^2+y-3}{y^2+5y-6}$

35. $\dfrac{\dfrac{x^3}{x+1}}{\dfrac{x}{x^2+2x+1}}$

36. $\dfrac{\dfrac{2x^2-3x-2}{x^2-1}}{\dfrac{2x^2+5x+2}{x^2+x-2}}$

37. $\dfrac{x/y}{z}$

38. $\dfrac{x}{y/z}$

39–58 ■ Perform the addition or subtraction and simplify.

39. $2+\dfrac{x}{x+3}$

40. $\dfrac{2x-1}{x+4}-1$

41. $\dfrac{1}{x+5}+\dfrac{2}{x-3}$

42. $\dfrac{1}{x+1}+\dfrac{1}{x-1}$

43. $\dfrac{1}{x+1}-\dfrac{1}{x+2}$

44. $\dfrac{x}{x-4}-\dfrac{3}{x+6}$

45. $\dfrac{x}{(x+1)^2}+\dfrac{2}{x+1}$

46. $\dfrac{5}{2x-3}-\dfrac{3}{(2x-3)^2}$

47. $u+1+\dfrac{u}{u+1}$

48. $\dfrac{2}{a^2}-\dfrac{3}{ab}+\dfrac{4}{b^2}$

49. $\dfrac{1}{x^2}+\dfrac{1}{x^2+x}$

50. $\dfrac{1}{x}+\dfrac{1}{x^2}+\dfrac{1}{x^3}$

51. $\dfrac{2}{x+3}-\dfrac{1}{x^2+7x+12}$

52. $\dfrac{x}{x^2-4}+\dfrac{1}{x-2}$

53. $\dfrac{1}{x+3}+\dfrac{1}{x^2-9}$

54. $\dfrac{x}{x^2+x-2}-\dfrac{2}{x^2-5x+4}$

55. $\dfrac{2}{x}+\dfrac{3}{x-1}-\dfrac{4}{x^2-x}$

56. $\dfrac{x}{x^2-x-6}-\dfrac{1}{x+2}-\dfrac{2}{x-3}$

57. $\dfrac{1}{x^2+3x+2}-\dfrac{1}{x^2-2x-3}$

58. $\dfrac{1}{x+1}-\dfrac{2}{(x+1)^2}+\dfrac{3}{x^2-1}$

59–68 ■ Simplify the compound fractional expression.

59. $\dfrac{x+\dfrac{1}{x+2}}{x-\dfrac{1}{x+2}}$

60. $\dfrac{1+\dfrac{1}{c-1}}{1-\dfrac{1}{c-1}}$

61. $\dfrac{\dfrac{x+2}{x-1}-\dfrac{x-3}{x-2}}{x+2}$

62. $\dfrac{\dfrac{x-3}{x-4}-\dfrac{x+2}{x+1}}{x+3}$

63. $\dfrac{\dfrac{x}{y}-\dfrac{y}{x}}{\dfrac{1}{x^2}-\dfrac{1}{y^2}}$

64. $x-\dfrac{y}{\dfrac{x}{y}+\dfrac{y}{x}}$

65. $\dfrac{x^{-2}-y^{-2}}{x^{-1}+y^{-1}}$

66. $\dfrac{x^{-1}+y^{-1}}{(x+y)^{-1}}$

67. $1-\dfrac{1}{1-\dfrac{1}{x}}$

68. $1+\dfrac{1}{1+\dfrac{1}{1+x}}$

69–74 ■ Simplify the fractional expression. (Expressions like these arise in calculus.)

69. $\dfrac{\dfrac{1}{1+x+h}-\dfrac{1}{1+x}}{h}$

70. $\dfrac{\dfrac{1}{\sqrt{x+h}}-\dfrac{1}{\sqrt{x}}}{h}$

71. $\dfrac{\dfrac{1}{(x+h)^2}-\dfrac{1}{x^2}}{h}$

72. $\dfrac{(x+h)^3-7(x+h)-(x^3-7x)}{h}$

73. $\sqrt{1+\left(\dfrac{x}{\sqrt{1-x^2}}\right)^2}$

74. $\sqrt{1+\left(x^3-\dfrac{1}{4x^3}\right)^2}$

75–80 ■ Simplify the expression. (This type of expression arises in calculus when using the "quotient rule.")

75. $\dfrac{3(x+2)^2(x-3)^2-(x+2)^3(2)(x-3)}{(x-3)^4}$

76. $\dfrac{2x(x+6)^4-x^2(4)(x+6)^3}{(x+6)^8}$

77. $\dfrac{2(1+x)^{1/2} - x(1+x)^{-1/2}}{x+1}$

78. $\dfrac{(1-x^2)^{1/2} + x^2(1-x^2)^{-1/2}}{1-x^2}$

79. $\dfrac{3(1+x)^{1/3} - x(1+x)^{-2/3}}{(1+x)^{2/3}}$

80. $\dfrac{(7-3x)^{1/2} + \frac{3}{2}x(7-3x)^{-1/2}}{7-3x}$

81–86 ■ Rationalize the denominator.

81. $\dfrac{1}{2-\sqrt{3}}$

82. $\dfrac{2}{3-\sqrt{5}}$

83. $\dfrac{2}{\sqrt{2}+\sqrt{7}}$

84. $\dfrac{1}{\sqrt{x}+1}$

85. $\dfrac{y}{\sqrt{3}+\sqrt{y}}$

86. $\dfrac{2(x-y)}{\sqrt{x}-\sqrt{y}}$

87–92 ■ Rationalize the numerator.

87. $\dfrac{1-\sqrt{5}}{3}$

88. $\dfrac{\sqrt{3}+\sqrt{5}}{2}$

89. $\dfrac{\sqrt{r}+\sqrt{2}}{5}$

90. $\dfrac{\sqrt{x}-\sqrt{x+h}}{h\sqrt{x}\sqrt{x+h}}$

91. $\sqrt{x^2+1} - x$

92. $\sqrt{x+1} - \sqrt{x}$

93–100 ■ State whether the given equation is true for all values of the variables. (Disregard any value that makes a denominator zero.)

93. $\dfrac{16+a}{16} = 1 + \dfrac{a}{16}$

94. $\dfrac{b}{b-c} = 1 - \dfrac{b}{c}$

95. $\dfrac{2}{4+x} = \dfrac{1}{2} + \dfrac{2}{x}$

96. $\dfrac{x+1}{y+1} = \dfrac{x}{y}$

97. $\dfrac{x}{x+y} = \dfrac{1}{1+y}$

98. $2\left(\dfrac{a}{b}\right) = \dfrac{2a}{2b}$

99. $\dfrac{-a}{b} = -\dfrac{a}{b}$

100. $\dfrac{1+x+x^2}{x} = \dfrac{1}{x} + 1 + x$

APPLICATIONS

101. Electrical Resistance If two electrical resistors with resistances R_1 and R_2 are connected in parallel (see the figure), then the total resistance R is given by

$$R = \frac{1}{\dfrac{1}{R_1} + \dfrac{1}{R_2}}$$

(a) Simplify the expression for R.

(b) If $R_1 = 10$ ohms and $R_2 = 20$ ohms, what is the total resistance R?

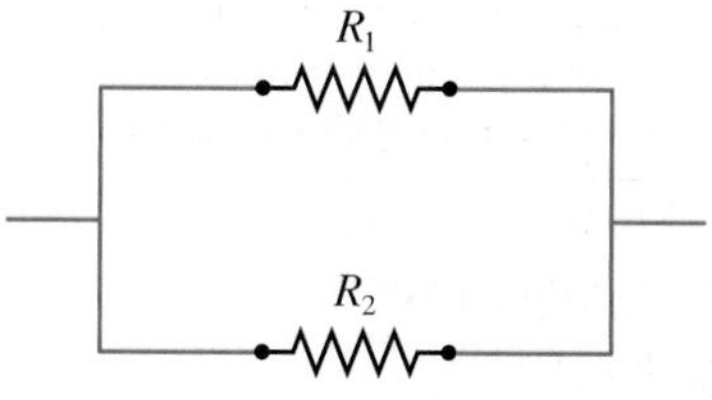

102. Average Cost A clothing manufacturer finds that the cost of producing x shirts is $500 + 6x + 0.01x^2$ dollars.

(a) Explain why the average cost per shirt is given by the rational expression

$$A = \frac{500 + 6x + 0.01x^2}{x}$$

(b) Complete the table by calculating the average cost per shirt for the given values of x.

x	Average cost
10	
20	
50	
100	
200	
500	
1000	

DISCOVERY ■ DISCUSSION ■ WRITING

103. Limiting Behavior of a Rational Expression The rational expression

$$\frac{x^2 - 9}{x - 3}$$

is not defined for $x = 3$. Complete the tables and determine what value the expression approaches as x gets closer and closer to 3. Why is this reasonable? Factor the numerator of the expression and simplify to see why.

x	$\dfrac{x^2-9}{x-3}$
2.80	
2.90	
2.95	
2.99	
2.999	

x	$\dfrac{x^2-9}{x-3}$
3.20	
3.10	
3.05	
3.01	
3.001	

104. Is This Rationalization? In the expression $2/\sqrt{x}$ we would eliminate the radical if we were to square both numerator and denominator. Is this the same thing as rationalizing the denominator?

105. Algebraic Errors The left-hand column in the table on the following page lists some common algebraic errors. In each case, give an example using numbers that show that the formula is not valid. An example of this type, which shows that a statement is false, is called a *counterexample*.

Algebraic error	Counterexample
$\frac{1}{a}+\frac{1}{b} \neq \frac{1}{a+b}$	$\frac{1}{2}+\frac{1}{2} \neq \frac{1}{2+2}$
$(a+b)^2 \neq a^2+b^2$	
$\sqrt{a^2+b^2} \neq a+b$	
$\frac{a+b}{a} \neq b$	
$(a^3+b^3)^{1/3} \neq a+b$	
$a^m/a^n \neq a^{m/n}$	
$a^{-1/n} \neq \frac{1}{a^n}$	

106. The Form of an Algebraic Expression An algebraic expression may look complicated, but its "form" is always simple; it must be a sum, a product, a quotient, or a power. For example, consider the following expressions:

$$(1+x^2)^2+\left(\frac{x+2}{x+1}\right)^3 \qquad (1+x)\left(1+\frac{x+5}{1+x^4}\right)$$

$$\frac{5-x^3}{1+\sqrt{1+x^2}} \qquad \sqrt{\frac{1+x}{1-x}}$$

With appropriate choices for A and B, the first has the form $A+B$, the second AB, the third A/B, and the fourth $A^{1/2}$. Recognizing the form of an expression helps us expand, simplify, or factor it correctly. Find the form of the following algebraic expressions.

(a) $x+\sqrt{1+\frac{1}{x}}$ **(b)** $(1+x^2)(1+x)^3$

(c) $\sqrt[3]{x^4(4x^2+1)}$ **(d)** $\frac{1-2\sqrt{1+x}}{1+\sqrt{1+x^2}}$

1.5 Equations

Solving Linear Equations ▶ Solving Quadratic Equations ▶ Other Types of Equations

An equation is a statement that two mathematical expressions are equal. For example,

$$3+5=8$$

is an equation. Most equations that we study in algebra contain variables, which are symbols (usually letters) that stand for numbers. In the equation

$$4x+7=19$$

the letter x is the variable. We think of x as the "unknown" in the equation, and our goal is to find the value of x that makes the equation true. The values of the unknown that make the equation true are called the **solutions** or **roots** of the equation, and the process of finding the solutions is called **solving the equation**.

$x=3$ is a solution of the equation $4x+7=19$, because substituting $x=3$ makes the equation true:

$x = 3$

$4(3)+7=19$ ✔

Two equations with exactly the same solutions are called **equivalent equations**. To solve an equation, we try to find a simpler, equivalent equation in which the variable stands alone on one side of the "equal" sign. Here are the properties that we use to solve an equation. (In these properties, A, B, and C stand for any algebraic expressions, and the symbol $\Leftrightarrow$ means "is equivalent to.")

PROPERTIES OF EQUALITY

Property	Description
1. $A=B \quad \Leftrightarrow \quad A+C=B+C$	Adding the same quantity to both sides of an equation gives an equivalent equation.
2. $A=B \quad \Leftrightarrow \quad CA=CB \quad (C \neq 0)$	Multiplying both sides of an equation by the same nonzero quantity gives an equivalent equation.

These properties require that you *perform the same operation on both sides of an equation* when solving it. Thus, if we say "*add* -7" when solving an equation, that is just a short way of saying "*add* -7 to each side of the equation."

Solving Linear Equations

The simplest type of equation is a *linear equation*, or first-degree equation, which is an equation in which each term is either a constant or a nonzero multiple of the variable.

> **LINEAR EQUATIONS**
>
> A **linear equation** in one variable is an equation equivalent to one of the form
>
> $$ax + b = 0$$
>
> where a and b are real numbers and x is the variable.

Here are some examples that illustrate the difference between linear and nonlinear equations.

Linear equations	**Nonlinear equations**	
$4x - 5 = 3$	$x^2 + 2x = 8$	Not linear; contains the square of the variable
$2x = \frac{1}{2}x - 7$	$\sqrt{x} - 6x = 0$	Not linear; contains the square root of the variable
$x - 6 = \dfrac{x}{3}$	$\dfrac{3}{x} - 2x = 1$	Not linear; contains the reciprocal of the variable

EXAMPLE 1 | Solving a Linear Equation

Solve the equation $7x - 4 = 3x + 8$.

SOLUTION We solve this by changing it to an equivalent equation with all terms that have the variable x on one side and all constant terms on the other.

$7x - 4 = 3x + 8$	Given equation
$(7x - 4) + 4 = (3x + 8) + 4$	Add 4
$7x = 3x + 12$	Simplify
$7x - 3x = (3x + 12) - 3x$	Subtract $3x$
$4x = 12$	Simplify
$\frac{1}{4} \cdot 4x = \frac{1}{4} \cdot 12$	Multiply by $\frac{1}{4}$
$x = 3$	Simplify

CHECK YOUR ANSWER

$x = 3$:

$x = 3$: LHS $= 7(3) - 4 = 17$

$x = 3$: RHS $= 3(3) + 8 = 17$

LHS = RHS ✔

NOW TRY EXERCISE 15 ■

Because it is important to CHECK YOUR ANSWER, we do this in many of our examples. In these checks, LHS stands for "left-hand side" and RHS stands for "right-hand side" of the original equation.

Many formulas in the sciences involve several variables, and it is often necessary to express one of the variables in terms of the others. In the next example we solve for a variable in Newton's Law of Gravity.

EXAMPLE 2 | Solving for One Variable in Terms of Others

This is Newton's Law of Gravity. It gives the gravitational force F between two masses m and M that are a distance r apart. The constant G is the universal gravitational constant.

Solve for the variable M in the equation

$$F = G\frac{mM}{r^2}$$

SOLUTION Although this equation involves more than one variable, we solve it as usual by isolating M on one side and treating the other variables as we would numbers.

$$F = \left(\frac{Gm}{r^2}\right)M \qquad \text{Factor } M \text{ from RHS}$$

$$\left(\frac{r^2}{Gm}\right)F = \left(\frac{r^2}{Gm}\right)\left(\frac{Gm}{r^2}\right)M \qquad \text{Multiply by reciprocal of } \frac{Gm}{r^2}$$

$$\frac{r^2F}{Gm} = M \qquad \text{Simplify}$$

The solution is $M = \dfrac{r^2F}{Gm}$.

NOW TRY EXERCISE 29

EXAMPLE 3 | Solving for One Variable in Terms of Others

The surface area A of the closed rectangular box shown in Figure 1 can be calculated from the length l, the width w, and the height h according to the formula

$$A = 2lw + 2wh + 2lh$$

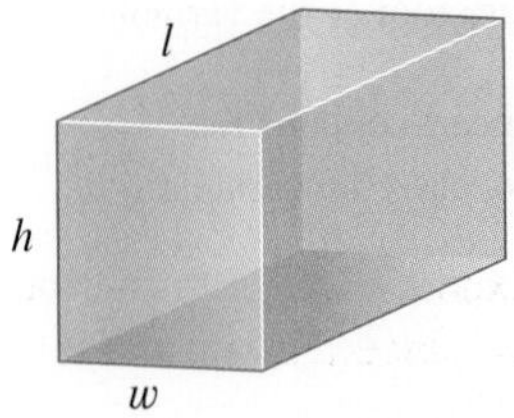

FIGURE 1 A closed rectangular box

Solve for w in terms of the other variables in this equation.

SOLUTION Although this equation involves more than one variable, we solve it as usual by isolating w on one side, treating the other variables as we would numbers.

$$A = (2lw + 2wh) + 2lh \qquad \text{Collect terms involving } w$$

$$A - 2lh = 2lw + 2wh \qquad \text{Subtract } 2lh$$

$$A - 2lh = (2l + 2h)w \qquad \text{Factor } w \text{ from RHS}$$

$$\frac{A - 2lh}{2l + 2h} = w \qquad \text{Divide by } 2l + 2h$$

The solution is $w = \dfrac{A - 2lh}{2l + 2h}$.

NOW TRY EXERCISE 31

▼ Solving Quadratic Equations

Linear equations are first-degree equations like $2x + 1 = 5$ or $4 - 3x = 2$. Quadratic equations are second-degree equations like $x^2 + 2x - 3 = 0$ or $2x^2 + 3 = 5x$.

Quadratic Equations

$x^2 - 2x - 8 = 0$

$3x + 10 = 4x^2$

$\frac{1}{2}x^2 + \frac{1}{3}x - \frac{1}{6} = 0$

QUADRATIC EQUATIONS

A **quadratic equation** is an equation of the form

$$ax^2 + bx + c = 0$$

where a, b, and c are real numbers with $a \neq 0$.

Some quadratic equations can be solved by factoring and using the following basic property of real numbers.

ZERO-PRODUCT PROPERTY

$AB = 0$ if and only if $A = 0$ or $B = 0$

This means that if we can factor the left-hand side of a quadratic (or other) equation, then we can solve it by setting each factor equal to 0 in turn. This method works only when the right-hand side of the equation is 0.

EXAMPLE 4 | Solving a Quadratic Equation by Factoring

Solve the equation $x^2 + 5x = 24$.

SOLUTION We must first rewrite the equation so that the right-hand side is 0.

$$x^2 + 5x = 24$$

$$x^2 + 5x - 24 = 0 \qquad \text{Subtract 24}$$

$$(x - 3)(x + 8) = 0 \qquad \text{Factor}$$

$$x - 3 = 0 \quad \text{or} \quad x + 8 = 0 \qquad \text{Zero-Product Property}$$

$$x = 3 \qquad x = -8 \qquad \text{Solve}$$

The solutions are $x = 3$ and $x = -8$.

CHECK YOUR ANSWERS

$x = 3$:

$(3)^2 + 5(3) = 9 + 15 = 24$ ✔

$x = -8$:

$(-8)^2 + 5(-8) = 64 - 40 = 24$ ✔

NOW TRY EXERCISE **43** ■

Do you see why one side of the equation must be 0 in Example 4? Factoring the equation as $x(x + 5) = 24$ does not help us find the solutions, since 24 can be factored in infinitely many ways, such as $6 \cdot 4$, $\frac{1}{2} \cdot 48$, $\left(-\frac{2}{5}\right) \cdot (-60)$, and so on.

A quadratic equation of the form $x^2 - c = 0$, where c is a positive constant, factors as $(x - \sqrt{c})(x + \sqrt{c}) = 0$, so the solutions are $x = \sqrt{c}$ and $x = -\sqrt{c}$. We often abbreviate this as $x = \pm\sqrt{c}$.

SOLVING A SIMPLE QUADRATIC EQUATION

The solutions of the equation $x^2 = c$ are $x = \sqrt{c}$ and $x = -\sqrt{c}$.

EXAMPLE 5 | Solving Simple Quadratics

Solve each equation.

(a) $x^2 = 5$ **(b)** $(x - 4)^2 = 5$

SOLUTION

(a) From the principle in the preceding box, we get $x = \pm\sqrt{5}$.

(b) We can take the square root of each side of this equation as well.

$$(x - 4)^2 = 5$$

$$x - 4 = \pm\sqrt{5} \qquad \text{Take the square root}$$

$$x = 4 \pm \sqrt{5} \qquad \text{Add 4}$$

The solutions are $x = 4 + \sqrt{5}$ and $x = 4 - \sqrt{5}$.

NOW TRY EXERCISES **51** AND **53** ■

See page 30 for how to recognize when a quadratic expression is a perfect square.

As we saw in Example 5, if a quadratic equation is of the form $(x \pm a)^2 = c$, then we can solve it by taking the square root of each side. In an equation of this form the left-hand side is a *perfect square*: the square of a linear expression in x. So if a quadratic equation does not factor readily, then we can solve it using the technique of **completing the square**. This means that we add a constant to an expression to make it a perfect square. For example, to make $x^2 - 6x$ a perfect square, we must add 9, since $x^2 - 6x + 9 = (x - 3)^2$.

Completing the Square

The area of the blue region is

$$x^2 + 2\left(\frac{b}{2}\right)x = x^2 + bx$$

Add a small square of area $(b/2)^2$ to "complete" the square.

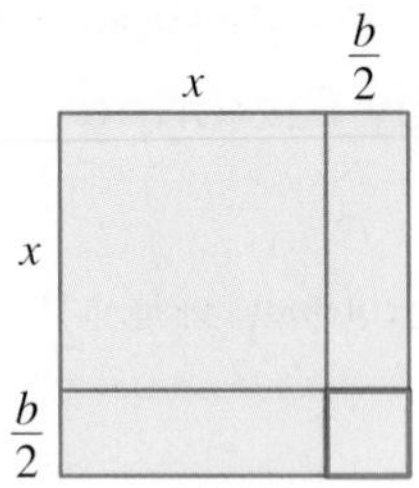

> **COMPLETING THE SQUARE**
>
> To make $x^2 + bx$ a perfect square, add $\left(\frac{b}{2}\right)^2$, the square of half the coefficient of x. This gives the perfect square
>
> $$x^2 + bx + \left(\frac{b}{2}\right)^2 = \left(x + \frac{b}{2}\right)^2$$

EXAMPLE 6 | Solving Quadratic Equations by Completing the Square

Solve each equation.

(a) $x^2 - 8x + 13 = 0$ **(b)** $3x^2 - 12x + 6 = 0$

SOLUTION

(a)

$x^2 - 8x + 13 = 0$	Given equation
$x^2 - 8x = -13$	Subtract 13
$x^2 - 8x + 16 = -13 + 16$	Complete the square: add $\left(\frac{-8}{2}\right)^2 = 16$
$(x - 4)^2 = 3$	Perfect square
$x - 4 = \pm\sqrt{3}$	Take square root
$x = 4 \pm \sqrt{3}$	Add 4

When completing the square, make sure the coefficient of x^2 is 1. If it isn't, you must factor this coefficient from both terms that contain x:

$$ax^2 + bx = a\left(x^2 + \frac{b}{a}x\right)$$

Then complete the square inside the parentheses. Remember that the term added inside the parentheses is multiplied by a.

(b) After subtracting 6 from each side of the equation, we must factor the coefficient of x^2 (the 3) from the left side to put the equation in the correct form for completing the square.

$3x^2 - 12x + 6 = 0$	Given equation
$3x^2 - 12x = -6$	Subtract 6
$3(x^2 - 4x) = -6$	Factor 3 from LHS

Now we complete the square by adding $(-2)^2 = 4$ *inside* the parentheses. Since everything inside the parentheses is multiplied by 3, this means that we are actually adding $3 \cdot 4 = 12$ to the left side of the equation. Thus, we must add 12 to the right side as well.

$3(x^2 - 4x + 4) = -6 + 3 \cdot 4$	Complete the square: add 4
$3(x - 2)^2 = 6$	Perfect square
$(x - 2)^2 = 2$	Divide by 3
$x - 2 = \pm\sqrt{2}$	Take square root
$x = 2 \pm \sqrt{2}$	Add 2

NOW TRY EXERCISES **55** AND **59** ■

We can use the technique of completing the square to derive a formula for the roots of the general quadratic equation $ax^2 + bx + c = 0$.

> **THE QUADRATIC FORMULA**
>
> The roots of the quadratic equation $ax^2 + bx + c = 0$, where $a \neq 0$, are
>
> $$x = \frac{-b \pm \sqrt{b^2 - 4ac}}{2a}$$

PROOF First, we divide each side of the equation by a and move the constant to the right side, giving

$$x^2 + \frac{b}{a}x = -\frac{c}{a} \qquad \text{Divide by } a$$

We now complete the square by adding $(b/2a)^2$ to each side of the equation:

$$x^2 + \frac{b}{a}x + \left(\frac{b}{2a}\right)^2 = -\frac{c}{a} + \left(\frac{b}{2a}\right)^2 \qquad \text{Complete the square: Add } \left(\frac{b}{2a}\right)^2$$

$$\left(x + \frac{b}{2a}\right)^2 = \frac{-4ac + b^2}{4a^2} \qquad \text{Perfect square}$$

$$x + \frac{b}{2a} = \pm\frac{\sqrt{b^2 - 4ac}}{2a} \qquad \text{Take square root}$$

$$x = \frac{-b \pm \sqrt{b^2 - 4ac}}{2a} \qquad \text{Subtract } \frac{b}{2a}$$

■

Library of Congress

FRANÇOIS VIÈTE (1540–1603) had a successful political career before taking up mathematics late in life. He became one of the most famous French mathematicians of the 16th century. Viète introduced a new level of abstraction in algebra by using letters to stand for *known* quantities in an equation. Before Viète's time, each equation had to be solved on its own. For instance, the quadratic equations

$$3x^2 + 2x + 8 = 0$$

$$5x^2 - 6x + 4 = 0$$

had to be solved separately by completing the square. Viète's idea was to consider all quadratic equations at once by writing

$$ax^2 + bx + c = 0$$

where a, b, and c are known quantities. Thus, he made it possible to write a *formula* (in this case, the quadratic formula) involving a, b, and c that can be used to solve all such equations in one fell swoop.

Viète's mathematical genius proved quite valuable during a war between France and Spain. To communicate with their troops, the Spaniards used a complicated code that Viète managed to decipher. Unaware of Viète's accomplishment, the Spanish king, Philip II, protested to the Pope, claiming that the French were using witchcraft to read his messages.

The Quadratic Formula could be used to solve the equations in Examples 4 and 6. You should carry out the details of these calculations.

EXAMPLE 7 | Using the Quadratic Formula

Find all solutions of each equation.

(a) $3x^2 - 5x - 1 = 0$ **(b)** $4x^2 + 12x + 9 = 0$ **(c)** $x^2 + 2x + 2 = 0$

SOLUTION

(a) In this quadratic equation $a = 3$, $b = -5$, and $c = -1$.

$b = -5$

$$3x^2 - 5x - 1 = 0$$

$a = 3$ $c = -1$

By the Quadratic Formula,

$$x = \frac{-(-5) \pm \sqrt{(-5)^2 - 4(3)(-1)}}{2(3)} = \frac{5 \pm \sqrt{37}}{6}$$

If approximations are desired, we can use a calculator to obtain

$$x = \frac{5 + \sqrt{37}}{6} \approx 1.8471 \quad \text{and} \quad x = \frac{5 - \sqrt{37}}{6} \approx -0.1805$$

(b) Using the Quadratic Formula with $a = 4$, $b = 12$, and $c = 9$ gives

$$x = \frac{-12 \pm \sqrt{(12)^2 - 4 \cdot 4 \cdot 9}}{2 \cdot 4} = \frac{-12 \pm 0}{8} = -\frac{3}{2}$$

This equation has only one solution, $x = -\frac{3}{2}$.

Another Method

$$4x^2 + 12x + 9 = 0$$
$$(2x + 3)^2 = 0$$
$$2x + 3 = 0$$
$$x = -\tfrac{3}{2}$$

(c) Using the Quadratic Formula with $a = 1$, $b = 2$, and $c = 2$ gives

$$x = \frac{-2 \pm \sqrt{2^2 - 4 \cdot 2}}{2} = \frac{-2 \pm \sqrt{-4}}{2} = \frac{-2 \pm 2\sqrt{-1}}{2} = -1 \pm \sqrt{-1}$$

Since the square of any real number is nonnegative, $\sqrt{-1}$ is undefined in the real number system. The equation has no real solution.

NOW TRY EXERCISES **65**, **69**, AND **75** ■

In Section 3.5 we study the complex number system, in which the square roots of negative numbers do exist. The equation in Example 7(c) does have solutions in the complex number system.

The quantity $b^2 - 4ac$ that appears under the square root sign in the quadratic formula is called the *discriminant* of the equation $ax^2 + bx + c = 0$ and is given the symbol D. If $D < 0$, then $\sqrt{b^2 - 4ac}$ is undefined, and the quadratic equation has no real solution, as in Example 7(c). If $D = 0$, then the equation has only one real solution, as in Example 7(b). Finally, if $D > 0$, then the equation has two distinct real solutions, as in Example 7(a). The following box summarizes these observations.

THE DISCRIMINANT

The **discriminant** of the general quadratic $ax^2 + bx + c = 0$ $(a \neq 0)$ is $D = b^2 - 4ac$.

1. If $D > 0$, then the equation has two distinct real solutions.
2. If $D = 0$, then the equation has exactly one real solution.
3. If $D < 0$, then the equation has no real solution.

EXAMPLE 8 | Using the Discriminant

Use the discriminant to determine how many real solutions each equation has.

(a) $x^2 + 4x - 1 = 0$ **(b)** $4x^2 - 12x + 9 = 0$ **(c)** $\frac{1}{3}x^2 - 2x + 4 = 0$

SOLUTION

(a) The discriminant is $D = 4^2 - 4(1)(-1) = 20 > 0$, so the equation has two distinct real solutions.

(b) The discriminant is $D = (-12)^2 - 4 \cdot 4 \cdot 9 = 0$, so the equation has exactly one real solution.

(c) The discriminant is $D = (-2)^2 - 4(\frac{1}{3})4 = -\frac{4}{3} < 0$, so the equation has no real solution.

NOW TRY EXERCISES **79**, **81**, AND **83** ■

Now let's consider a real-life situation that can be modeled by a quadratic equation.

EXAMPLE 9 | The Path of a Projectile

An object thrown or fired straight upward at an initial speed of v_0 ft/s will reach a height of h feet after t seconds, where h and t are related by the formula

$$h = -16t^2 + v_0 t$$

This formula depends on the fact that acceleration due to gravity is constant near the earth's surface. Here we neglect the effect of air resistance.

Suppose that a bullet is shot straight upward with an initial speed of 800 ft/s. Its path is shown in Figure 2.

(a) When does the bullet fall back to ground level?

(b) When does it reach a height of 6400 ft?

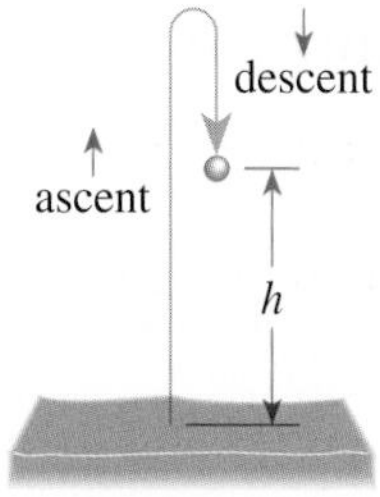

FIGURE 2

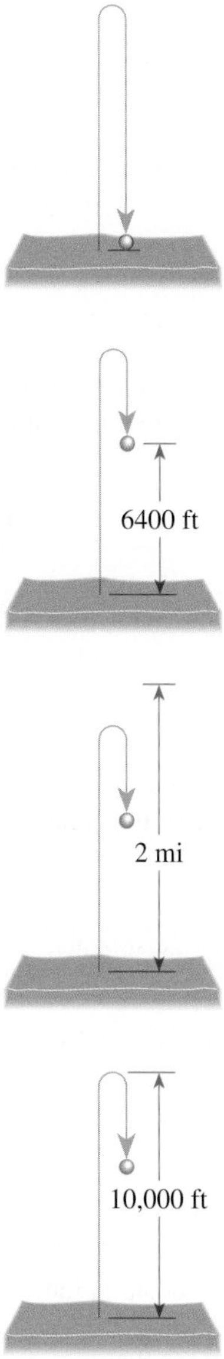

(c) When does it reach a height of 2 mi?

(d) How high is the highest point the bullet reaches?

SOLUTION Since the initial speed in this case is $v_0 = 800$ ft/s, the formula is

$$h = -16t^2 + 800t$$

(a) Ground level corresponds to $h = 0$, so we must solve the equation

$$0 = -16t^2 + 800t \qquad \text{Set } h = 0$$

$$0 = -16t(t - 50) \qquad \text{Factor}$$

Thus, $t = 0$ or $t = 50$. This means the bullet starts $(t = 0)$ at ground level and returns to ground level after 50 s.

(b) Setting $h = 6400$ gives the equation

$$6400 = -16t^2 + 800t \qquad \text{Set } h = 6400$$

$$16t^2 - 800t + 6400 = 0 \qquad \text{All terms to LHS}$$

$$t^2 - 50t + 400 = 0 \qquad \text{Divide by 16}$$

$$(t - 10)(t - 40) = 0 \qquad \text{Factor}$$

$$t = 10 \quad \text{or} \quad t = 40 \qquad \text{Solve}$$

The bullet reaches 6400 ft after 10 s (on its ascent) and again after 40 s (on its descent to earth).

(c) Two miles is $2 \times 5280 = 10{,}560$ ft.

$$10{,}560 = -16t^2 + 800t \qquad \text{Set } h = 10{,}560$$

$$16t^2 - 800t + 10{,}560 = 0 \qquad \text{All terms to LHS}$$

$$t^2 - 50t + 660 = 0 \qquad \text{Divide by 16}$$

The discriminant of this equation is $D = (-50)^2 - 4(660) = -140$, which is negative. Thus, the equation has no real solution. The bullet never reaches a height of 2 mi.

(d) Each height the bullet reaches is attained twice, once on its ascent and once on its descent. The only exception is the highest point of its path, which is reached only once. This means that for the highest value of h, the following equation has only one solution for t:

$$h = -16t^2 + 800t$$

$$16t^2 - 800t + h = 0 \qquad \text{All terms to LHS}$$

This in turn means that the discriminant D of the equation is 0, so

$$D = (-800)^2 - 4(16)h = 0$$

$$640{,}000 - 64h = 0$$

$$h = 10{,}000$$

The maximum height reached is 10,000 ft.

NOW TRY EXERCISE **111** ■

▼ Other Types of Equations

So far we have learned how to solve linear and quadratic equations. Now we study other types of equations, including those that involve higher powers, fractional expressions, and radicals.

EXAMPLE 10 | An Equation Involving Fractional Expressions

Solve the equation $\frac{3}{x} + \frac{5}{x+2} = 2$.

SOLUTION We eliminate the denominators by multiplying each side by the lowest common denominator.

$$\left(\frac{3}{x} + \frac{5}{x+2}\right)x(x+2) = 2x(x+2) \qquad \text{Multiply by LCD } x(x+2)$$

$$3(x+2) + 5x = 2x^2 + 4x \qquad \text{Expand}$$

$$8x + 6 = 2x^2 + 4x \qquad \text{Expand LHS}$$

$$0 = 2x^2 - 4x - 6 \qquad \text{Subtract } 8x + 6$$

$$0 = x^2 - 2x - 3 \qquad \text{Divide both sides by 2}$$

$$0 = (x-3)(x+1) \qquad \text{Factor}$$

$$x - 3 = 0 \quad \text{or} \quad x + 1 = 0 \qquad \text{Zero-Product Property}$$

$$x = 3 \qquad x = -1 \qquad \text{Solve}$$

We must check our answers because multiplying by an expression that contains the variable can introduce extraneous solutions. From *Check Your Answers* we see that the solutions are $x = 3$ and -1.

CHECK YOUR ANSWERS

$x = 3$:

$$\text{LHS} = \frac{3}{3} + \frac{5}{3+2} = 1 + 1 = 2$$

$$\text{RHS} = 2$$

LHS = RHS ✔

$x = -1$:

$$\text{LHS} = \frac{3}{-1} + \frac{5}{-1+2} = -3 + 5 = 2$$

$$\text{RHS} = 2$$

LHS = RHS ✔

NOW TRY EXERCISE **85** ■

When you solve an equation that involves radicals, you must be especially careful to check your final answers. The next example demonstrates why.

EXAMPLE 11 | An Equation Involving a Radical

Solve the equation $2x = 1 - \sqrt{2 - x}$.

SOLUTION To eliminate the square root, we first isolate it on one side of the equal sign, then square:

$$2x - 1 = -\sqrt{2 - x} \qquad \text{Subtract 1}$$

$$(2x - 1)^2 = 2 - x \qquad \text{Square each side}$$

$$4x^2 - 4x + 1 = 2 - x \qquad \text{Expand LHS}$$

$$4x^2 - 3x - 1 = 0 \qquad \text{Add } -2 + x$$

$$(4x + 1)(x - 1) = 0 \qquad \text{Factor}$$

$$4x + 1 = 0 \quad \text{or} \quad x - 1 = 0 \qquad \text{Zero-Product Property}$$

$$x = -\tfrac{1}{4} \qquad x = 1 \qquad \text{Solve}$$

The values $x = -\frac{1}{4}$ and $x = 1$ are only potential solutions. We must check them to see if they satisfy the original equation. From *Check Your Answers* we see that $x = -\frac{1}{4}$ is a solution but $x = 1$ is not. The only solution is $x = -\frac{1}{4}$.

CHECK YOUR ANSWERS

$x = -\frac{1}{4}$:

$$\text{LHS} = 2\left(-\tfrac{1}{4}\right) = -\tfrac{1}{2}$$

$$\text{RHS} = 1 - \sqrt{2 - \left(-\tfrac{1}{4}\right)} = 1 - \sqrt{\tfrac{9}{4}} = 1 - \tfrac{3}{2} = -\tfrac{1}{2}$$

LHS = RHS ✔

$x = 1$:

$$\text{LHS} = 2(1) = 2$$

$$\text{RHS} = 1 - \sqrt{2 - 1} = 1 - 1 = 0$$

LHS ≠ RHS ✕

NOW TRY EXERCISE **91** ■

When we solve an equation, we may end up with one or more **extraneous solutions**, that is, potential solutions that do not satisfy the original equation. In Example 11 the value $x = 1$ is an extraneous solution. Extraneous solutions may be introduced when we square each side of an equation because the operation of squaring can turn a false equation into a true one. For example, $-1 \neq 1$, but $(-1)^2 = 1^2$. Thus, the squared equation may be true for more values of the variable than the original equation. That is why you must always check your answers to make sure that each satisfies the original equation.

An equation of the form $aW^2 + bW + c = 0$, where W is an algebraic expression, is an equation of **quadratic type**. We solve equations of quadratic type by substituting for the algebraic expression, as we see in the next two examples.

EXAMPLE 12 | A Fourth-Degree Equation of Quadratic Type

Find all solutions of the equation $x^4 - 8x^2 + 8 = 0$.

SOLUTION If we set $W = x^2$, then we get a quadratic equation in the new variable W:

$$(x^2)^2 - 8x^2 + 8 = 0 \qquad \text{Write } x^4 \text{ as } (x^2)^2$$

$$W^2 - 8W + 8 = 0 \qquad \text{Let } W = x^2$$

$$W = \frac{-(-8) \pm \sqrt{(-8)^2 - 4 \cdot 8}}{2} = 4 \pm 2\sqrt{2} \qquad \text{Quadratic Formula}$$

$$x^2 = 4 \pm 2\sqrt{2} \qquad W = x^2$$

$$x = \pm\sqrt{4 \pm 2\sqrt{2}} \qquad \text{Take square roots}$$

So, there are four solutions:

$$\sqrt{4 + 2\sqrt{2}}, \quad \sqrt{4 - 2\sqrt{2}}, \quad -\sqrt{4 + 2\sqrt{2}}, \quad -\sqrt{4 - 2\sqrt{2}}$$

Using a calculator, we obtain the approximations $x \approx 2.61, 1.08, -2.61, -1.08$.

NOW TRY EXERCISE **95**

EXAMPLE 13 | An Equation Involving Fractional Powers

Find all solutions of the equation $x^{1/3} + x^{1/6} - 2 = 0$.

SOLUTION This equation is of quadratic type because if we let $W = x^{1/6}$, then $W^2 = (x^{1/6})^2 = x^{1/3}$.

$$x^{1/3} + x^{1/6} - 2 = 0$$

$$W^2 + W - 2 = 0 \qquad \text{Let } W = x^{1/6}$$

$$(W - 1)(W + 2) = 0 \qquad \text{Factor}$$

$$W - 1 = 0 \quad \text{or} \quad W + 2 = 0 \qquad \text{Zero-Product Property}$$

$$W = 1 \qquad W = -2 \qquad \text{Solve}$$

$$x^{1/6} = 1 \qquad x^{1/6} = -2 \qquad W = x^{1/6}$$

$$x = 1^6 = 1 \qquad x = (-2)^6 = 64 \qquad \text{Take the 6th power}$$

From *Check Your Answers* we see that $x = 1$ is a solution but $x = 64$ is not. The only solution is $x = 1$.

CHECK YOUR ANSWERS

$x = 1$:

$\text{LHS} = 1^{1/3} + 1^{1/6} - 2 = 0$

$\text{RHS} = 0$

$\text{LHS} = \text{RHS}$ ✔

$x = 64$:

$\text{LHS} = 64^{1/3} + 64^{1/6} - 2$

$= 4 + 2 - 2 = 4$

$\text{RHS} = 0$

$\text{LHS} \neq \text{RHS}$ ✗

NOW TRY EXERCISE **99**

When solving equations that involve absolute values, we usually take cases.

EXAMPLE 14 | An Absolute Value Equation

Solve the equation $|2x - 5| = 3$.

SOLUTION By the definition of absolute value, $|2x - 5| = 3$ is equivalent to

$$\begin{aligned} 2x - 5 &= 3 &\quad\text{or}\quad 2x - 5 &= -3 \\ 2x &= 8 & 2x &= 2 \\ x &= 4 & x &= 1 \end{aligned}$$

The solutions are $x = 1$, $x = 4$.

NOW TRY EXERCISE **105**

1.5 EXERCISES

CONCEPTS

1. *True or false?*

(a) Adding the same number to each side of an equation always gives an equivalent equation.

(b) Multiplying each side of an equation by the same number always gives an equivalent equation.

(c) Squaring each side of an equation always gives an equivalent equation.

2. Explain how you would use each method to solve the equation $x^2 - 4x - 5 = 0$.

(a) By factoring: ________

(b) By completing the square: ________

(c) By using the Quadratic Formula: ________

3. (a) The solutions of the equation $x^2(x - 4) = 0$ are ________.

(b) To solve the equation $x^3 - 4x^2 = 0$, we ________ the left-hand side.

4. Solve the equation $\sqrt{2x} + x = 0$ by doing the following steps.

(a) Isolate the radical: ________.

(b) Square both sides: ________.

(c) The solutions of the resulting quadratic equation are ________.

(d) The solution(s) that satisfy the original equation are ________.

5. The equation $(x + 1)^2 - 5(x + 1) + 6 = 0$ is of ________ type. To solve the equation, we set $W =$ ________. The resulting quadratic equation is ________.

6. The equation $x^6 + 7x^3 - 8 = 0$ is of ________ type. To solve the equation, we set $W =$ ________. The resulting quadratic equation is ________.

SKILLS

7–10 ■ Determine whether the given value is a solution of the equation.

7. $4x + 7 = 9x - 3$

(a) $x = -2$ **(b)** $x = 2$

8. $1 - [2 - (3 - x)] = 4x - (6 + x)$

(a) $x = 2$ **(b)** $x = 4$

9. $\dfrac{1}{x} - \dfrac{1}{x - 4} = 1$

(a) $x = 2$ **(b)** $x = 4$

10. $\dfrac{x^{3/2}}{x - 6} = x - 8$

(a) $x = 4$ **(b)** $x = 8$

11–28 ■ The given equation is either linear or equivalent to a linear equation. Solve the equation.

11. $2x + 7 = 31$

12. $5x - 3 = 4$

13. $\frac{1}{2}x - 8 = 1$

14. $3 + \frac{1}{3}x = 5$

15. $-7w = 15 - 2w$

16. $5t - 13 = 12 - 5t$

17. $\frac{1}{2}y - 2 = \frac{1}{3}y$

18. $\dfrac{z}{5} = \dfrac{3}{10}z + 7$

19. $2(1 - x) = 3(1 + 2x) + 5$

20. $\dfrac{2}{3}y + \dfrac{1}{2}(y - 3) = \dfrac{y + 1}{4}$

21. $x - \frac{1}{3}x - \frac{1}{2}x - 5 = 0$

22. $2x - \dfrac{x}{2} + \dfrac{x + 1}{4} = 6x$

23. $\dfrac{1}{x} = \dfrac{4}{3x} + 1$

24. $\dfrac{2x - 1}{x + 2} = \dfrac{4}{5}$

25. $\dfrac{3}{x + 1} - \dfrac{1}{2} = \dfrac{1}{3x + 3}$

26. $\dfrac{4}{x - 1} + \dfrac{2}{x + 1} = \dfrac{35}{x^2 - 1}$

27. $(t - 4)^2 = (t + 4)^2 + 32$

28. $\sqrt{3}x + \sqrt{12} = \dfrac{x + 5}{\sqrt{3}}$

29–42 ■ Solve the equation for the indicated variable.

29. $PV = nRT$; for R

30. $F = G\dfrac{mM}{r^2}$; for m

31. $P = 2l + 2w$; for w

32. $\frac{1}{R} = \frac{1}{R_1} + \frac{1}{R_2}$; for R_1

33. $\frac{ax + b}{cx + d} = 2$; for x

34. $a - 2[b - 3(c - x)] = 6$; for x

35. $a^2x + (a - 1) = (a + 1)x$; for x

36. $\frac{a + 1}{b} = \frac{a - 1}{b} + \frac{b + 1}{a}$; for a

37. $V = \frac{1}{3}\pi r^2 h$; for r

38. $F = G\frac{mM}{r^2}$; for r

39. $a^2 + b^2 = c^2$; for b

40. $A = P\left(1 + \frac{i}{100}\right)^2$; for i

41. $h = \frac{1}{2}gt^2 + v_0 t$; for t

42. $S = \frac{n(n + 1)}{2}$; for n

43–54 ■ Solve the equation by factoring.

43. $x^2 + x - 12 = 0$

44. $x^2 + 3x - 4 = 0$

45. $x^2 - 7x + 12 = 0$

46. $x^2 + 8x + 12 = 0$

47. $4x^2 - 4x - 15 = 0$

48. $2y^2 + 7y + 3 = 0$

49. $3x^2 + 5x = 2$

50. $6x(x - 1) = 21 - x$

51. $2x^2 = 8$

52. $3x^2 - 27 = 0$

53. $(3x + 2)^2 = 10$

54. $(2x - 1)^2 = 8$

55–62 ■ Solve the equation by completing the square.

55. $x^2 + 2x - 5 = 0$

56. $x^2 - 4x + 2 = 0$

57. $x^2 - 6x - 11 = 0$

58. $x^2 + 3x - \frac{7}{4} = 0$

59. $2x^2 + 8x + 1 = 0$

60. $3x^2 - 6x - 1 = 0$

61. $4x^2 - x = 0$

62. $x^2 = \frac{3}{4}x - \frac{1}{8}$

63–78 ■ Find all real solutions of the quadratic equation.

63. $x^2 - 2x - 15 = 0$

64. $x^2 + 5x - 6 = 0$

65. $x^2 - 7x + 10 = 0$

66. $x^2 + 30x + 200 = 0$

67. $2x^2 + x - 3 = 0$

68. $3x^2 + 7x + 4 = 0$

69. $3x^2 + 6x - 5 = 0$

70. $x^2 - 6x + 1 = 0$

71. $z^2 - \frac{3}{2}z + \frac{9}{16} = 0$

72. $2y^2 - y - \frac{1}{2} = 0$

73. $4x^2 + 16x - 9 = 0$

74. $0 = x^2 - 4x + 1$

75. $w^2 = 3(w - 1)$

76. $3 + 5z + z^2 = 0$

77. $10y^2 - 16y + 5 = 0$

78. $25x^2 + 70x + 49 = 0$

79–84 ■ Use the discriminant to determine the number of real solutions of the equation. Do not solve the equation.

79. $x^2 - 6x + 1 = 0$

80. $3x^2 = 6x - 9$

81. $x^2 + 2.20x + 1.21 = 0$

82. $x^2 + 2.21x + 1.21 = 0$

83. $4x^2 + 5x + \frac{13}{8} = 0$

84. $x^2 + rx - s = 0 \quad (s > 0)$

85–108 ■ Find all real solutions of the equation.

85. $\frac{1}{x - 1} + \frac{1}{x + 2} = \frac{5}{4}$

86. $\frac{10}{x} - \frac{12}{x - 3} + 4 = 0$

87. $\frac{x^2}{x + 100} = 50$

88. $\frac{1}{x - 1} - \frac{2}{x^2} = 0$

89. $\frac{x + 5}{x - 2} = \frac{5}{x + 2} + \frac{28}{x^2 - 4}$

90. $\frac{x}{2x + 7} - \frac{x + 1}{x + 3} = 1$

91. $\sqrt{2x + 1} + 1 = x$

92. $\sqrt{5 - x} + 1 = x - 2$

93. $2x + \sqrt{x + 1} = 8$

94. $\sqrt{\sqrt{x - 5} + x} = 5$

95. $x^4 - 13x^2 + 40 = 0$

96. $x^4 - 5x^2 + 4 = 0$

97. $2x^4 + 4x^2 + 1 = 0$

98. $x^6 - 2x^3 - 3 = 0$

99. $x^{4/3} - 5x^{2/3} + 6 = 0$

100. $\sqrt{x} - 3\sqrt[4]{x} - 4 = 0$

101. $4(x + 1)^{1/2} - 5(x + 1)^{3/2} + (x + 1)^{5/2} = 0$

102. $x^{1/2} + 3x^{-1/2} = 10x^{-3/2}$

103. $x^{1/2} - 3x^{1/3} = 3x^{1/6} - 9$

104. $x - 5\sqrt{x} + 6 = 0$

105. $|3x + 5| = 1$

106. $|2x| = 3$

107. $|x - 4| = 0.01$

108. $|x - 6| = -1$

APPLICATIONS

109–110 ■ **Falling-Body Problems** Suppose an object is dropped from a height h_0 above the ground. Then its height after t seconds is given by $h = -16t^2 + h_0$, where h is measured in feet. Use this information to solve the problem.

109. If a ball is dropped from 288 ft above the ground, how long does it take to reach ground level?

110. A ball is dropped from the top of a building 96 ft tall.
(a) How long will it take to fall half the distance to ground level?
(b) How long will it take to fall to ground level?

111–112 ■ **Falling-Body Problems** Use the formula $h = -16t^2 + v_0 t$ discussed in Example 9.

111. A ball is thrown straight upward at an initial speed of $v_0 = 40$ ft/s.
(a) When does the ball reach a height of 24 ft?
(b) When does it reach a height of 48 ft?
(c) What is the greatest height reached by the ball?
(d) When does the ball reach the highest point of its path?
(e) When does the ball hit the ground?

112. How fast would a ball have to be thrown upward to reach a maximum height of 100 ft? [*Hint:* Use the discriminant of the equation $16t^2 - v_0 t + h = 0$.]

113. **Shrinkage in Concrete Beams** As concrete dries, it shrinks—the higher the water content, the greater the shrinkage. If a concrete beam has a water content of w kg/m^3, then it will shrink by a factor

$$S = \frac{0.032w - 2.5}{10{,}000}$$

where S is the fraction of the original beam length that disappears due to shrinkage.
(a) A beam 12.025 m long is cast in concrete that contains 250 kg/m^3 water. What is the shrinkage factor S? How long will the beam be when it has dried?

(b) A beam is 10.014 m long when wet. We want it to shrink to 10.009 m, so the shrinkage factor should be $S = 0.00050$. What water content will provide this amount of shrinkage?

114. The Lens Equation If F is the focal length of a convex lens and an object is placed at a distance x from the lens, then its image will be at a distance y from the lens, where F, x, and y are related by the *lens equation*

$$\frac{1}{F} = \frac{1}{x} + \frac{1}{y}$$

Suppose that a lens has a focal length of 4.8 cm and that the image of an object is 4 cm closer to the lens than the object itself. How far from the lens is the object?

115. Fish Population The fish population in a certain lake rises and falls according to the formula

$$F = 1000(30 + 17t - t^2)$$

Here F is the number of fish at time t, where t is measured in years since January 1, 2002, when the fish population was first estimated.

(a) On what date will the fish population again be the same as it was on January 1, 2002?

(b) By what date will all the fish in the lake have died?

116. Fish Population A large pond is stocked with fish. The fish population P is modeled by the formula $P = 3t + 10\sqrt{t} + 140$, where t is the number of days since the fish were first introduced into the pond. How many days will it take for the fish population to reach 500?

117. Profit A small-appliance manufacturer finds that the profit P (in dollars) generated by producing x microwave ovens per week is given by the formula $P = \frac{1}{10}x(300 - x)$ provided that $0 \le x \le 200$. How many ovens must be manufactured in a given week to generate a profit of \$1250?

118. Gravity If an imaginary line segment is drawn between the centers of the earth and the moon, then the net gravitational force F acting on an object situated on this line segment is

$$F = \frac{-K}{x^2} + \frac{0.012K}{(239 - x)^2}$$

where $K > 0$ is a constant and x is the distance of the object from the center of the earth, measured in thousands of miles. How far from the center of the earth is the "dead spot" where no net gravitational force acts upon the object? (Express your answer to the nearest thousand miles.)

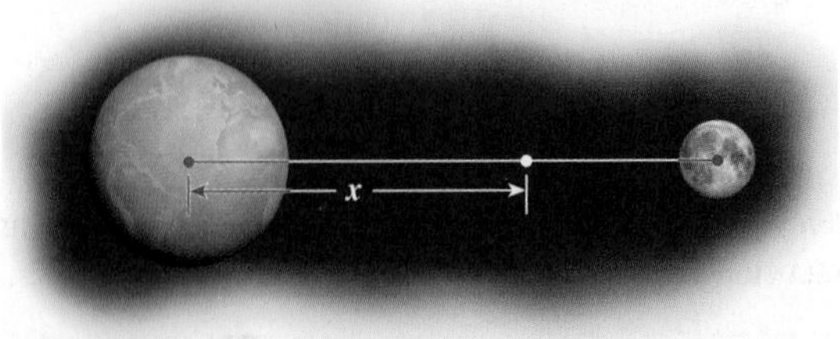

119. Depth of a Well One method for determining the depth of a well is to drop a stone into it and then measure the time it takes until the splash is heard. If d is the depth of the well (in feet) and t_1 the time (in seconds) it takes for the stone to fall, then $d = 16t_1^2$, so $t_1 = \sqrt{d}/4$. Now if t_2 is the time it takes for the sound to travel back up, then $d = 1090t_2$ because the speed of sound is 1090 ft/s. So $t_2 = d/1090$. Thus, the total time elapsed between dropping the stone and hearing the splash is

$$t_1 + t_2 = \frac{\sqrt{d}}{4} + \frac{d}{1090}$$

How deep is the well if this total time is 3 s?

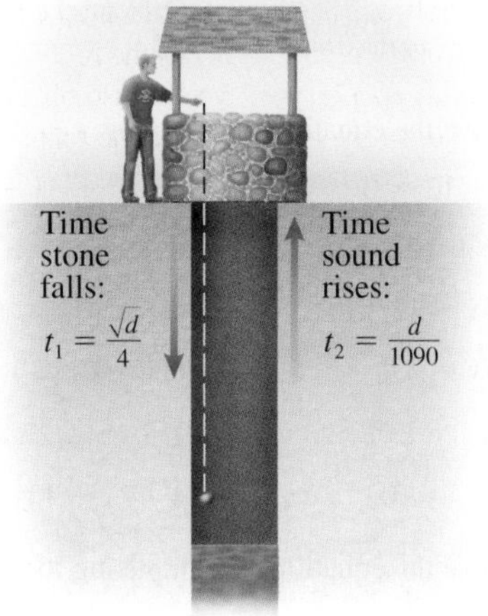

DISCOVERY ■ DISCUSSION ■ WRITING

120. A Family of Equations The equation

$$3x + k - 5 = kx - k + 1$$

is really a **family of equations**, because for each value of k, we get a different equation with the unknown x. The letter k is called a **parameter** for this family. What value should we pick for k to make the given value of x a solution of the resulting equation?

(a) $x = 0$ **(b)** $x = 1$ **(c)** $x = 2$

121. Proof That 0 = 1? The following steps appear to give equivalent equations, which seem to prove that $1 = 0$. Find the error.

$x = 1$	Given
$x^2 = x$	Multiply by x
$x^2 - x = 0$	Subtract x
$x(x - 1) = 0$	Factor
$\frac{x(x-1)}{x-1} = \frac{0}{x-1}$	Divide by $x - 1$
$x = 0$	Simplify
$1 = 0$	Given $x = 1$

122. Volumes of Solids The sphere, cylinder, and cone shown here all have the same radius r and the same volume V.

(a) Use the volume formulas given on the inside front cover of this book, to show that

$$\tfrac{4}{3}\pi r^3 = \pi r^2 h_1 \quad \text{and} \quad \tfrac{4}{3}\pi r^3 = \tfrac{1}{3}\pi r^2 h_2$$

(b) Solve these equations for h_1 and h_2.

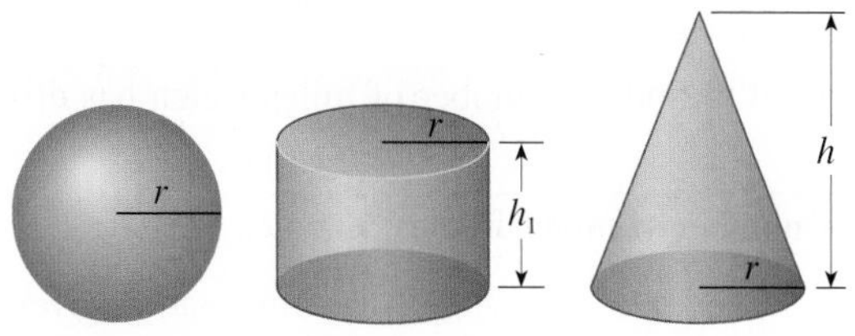

123. Relationship Between Roots and Coefficients The Quadratic Formula gives us the roots of a quadratic equation from its coefficients. We can also obtain the coefficients from the roots. For example, find the roots of the equation $x^2 - 9x + 20 = 0$ and show that the product of the roots is the constant term 20 and the sum of the roots is 9, the negative of the coefficient of x. Show that the same relationship between roots and coefficients holds for the following equations:

$$x^2 - 2x - 8 = 0$$
$$x^2 + 4x + 2 = 0$$

Use the Quadratic Formula to prove that in general, if the equation $x^2 + bx + c = 0$ has roots r_1 and r_2, then $c = r_1 r_2$ and $b = -(r_1 + r_2)$.

124. Solving an Equation in Different Ways We have learned several different ways to solve an equation in this section. Some equations can be tackled by more than one method. For example, the equation $x - \sqrt{x} - 2 = 0$ is of quadratic type. We can solve it by letting $\sqrt{x} = u$ and $x = u^2$, and factoring. Or we could solve for $\sqrt{x}$, square each side, and then solve the resulting quadratic equation. Solve the following equations using both methods indicated, and show that you get the same final answers.

(a) $x - \sqrt{x} - 2 = 0$ quadratic type; solve for the radical, and square

(b) $\dfrac{12}{(x-3)^2} + \dfrac{10}{x-3} + 1 = 0$ quadratic type; multiply by LCD

1.6 Modeling with Equations

Making and Using Models ▶ Problems About Interest ▶ Problems About Area or Length ▶ Problems About Mixtures ▶ Problems About the Time Needed to Do a Job ▶ Problems About Distance, Rate, and Time

Many problems in the sciences, economics, finance, medicine, and numerous other fields can be translated into algebra problems; this is one reason that algebra is so useful. In this section we use equations as mathematical models to solve real-life problems.

▼ Making and Using Models

We will use the following guidelines to help us set up equations that model situations described in words. To show how the guidelines can help you to set up equations, we note them as we work each example in this section.

GUIDELINES FOR MODELING WITH EQUATIONS

1. **Identify the Variable.** Identify the quantity that the problem asks you to find. This quantity can usually be determined by a careful reading of the question that is posed at the end of the problem. Then **introduce notation** for the variable (call it x or some other letter).
2. **Translate from Words to Algebra.** Read each sentence in the problem again, and express all the quantities mentioned in the problem in terms of the variable you defined in Step 1. To organize this information, it is sometimes helpful to **draw a diagram** or **make a table**.
3. **Set Up the Model.** Find the crucial fact in the problem that gives a relationship between the expressions you listed in Step 2. **Set up an equation** (or **model**) that expresses this relationship.
4. **Solve the Equation and Check Your Answer.** Solve the equation, check your answer, and express it as a sentence that answers the question posed in the problem.

The following example illustrates how these guidelines are used to translate a "word problem" into the language of algebra.

EXAMPLE 1 | Renting a Car

A car rental company charges \$30 a day and 15¢ a mile for renting a car. Helen rents a car for two days, and her bill comes to \$108. How many miles did she drive?

SOLUTION

Identify the variable. We are asked to find the number of miles Helen has driven. So we let

$$x = \text{number of miles driven}$$

Translate from words to algebra. Now we translate all the information given in the problem into the language of algebra.

In Words	In Algebra
Number of miles driven	x
Mileage cost (at \$0.15 per mile)	$0.15x$
Daily cost (at \$30 per day)	$2(30)$

Set up the model. Now we set up the model.

$$\text{mileage cost} + \text{daily cost} = \text{total cost}$$

$$0.15x + 2(30) = 108$$

Solve. Now we solve for x.

$$0.15x = 48 \qquad \text{Subtract 60}$$

$$x = \frac{48}{0.15} \qquad \text{Divide by 0.15}$$

$$x = 320 \qquad \text{Calculator}$$

CHECK YOUR ANSWER

total cost = mileage cost + daily cost

$= 0.15(320) + 2(30)$

$= 108$ ✔

Helen drove her rental car 320 miles.

NOW TRY EXERCISE 19 ■

In the examples and exercises that follow, we construct equations that model problems in many different real-life situations.

▼ Problems About Interest

When you borrow money from a bank or when a bank "borrows" your money by keeping it for you in a savings account, the borrower in each case must pay for the privilege of using the money. The fee that is paid is called **interest**. The most basic type of interest is **simple interest**, which is just an annual percentage of the total amount borrowed or deposited. The amount of a loan or deposit is called the **principal** P. The annual percentage paid for the use of this money is the **interest rate** r. We will use the variable t to stand for the number of years that the money is on deposit and the variable I to stand for the total interest earned. The following **simple interest formula** gives the amount of interest I earned when a principal P is deposited for t years at an interest rate r.

$$I = Prt$$

When using this formula, remember to convert r from a percentage to a decimal. For example, in decimal form, 5% is 0.05. So at an interest rate of 5%, the interest paid on a \$1000 deposit over a 3-year period is $I = Prt = 1000(0.05)(3) = \$150$.

EXAMPLE 2 | Interest on an Investment

Mary inherits \$100,000 and invests it in two certificates of deposit. One certificate pays 6% and the other pays $4\frac{1}{2}\%$ simple interest annually. If Mary's total interest is \$5025 per year, how much money is invested at each rate?

SOLUTION

Identify the variable. The problem asks for the amount she has invested at each rate. So we let

$$x = \text{the amount invested at } 6\%$$

Translate from words to algebra. Since Mary's total inheritance is \$100,000, it follows that she invested $100{,}000 - x$ at $4\frac{1}{2}\%$. We translate all the information given into the language of algebra.

In Words	In Algebra
Amount invested at 6%	x
Amount invested at $4\frac{1}{2}\%$	$100{,}000 - x$
Interest earned at 6%	$0.06x$
Interest earned at $4\frac{1}{2}\%$	$0.045(100{,}000 - x)$

Set up the model. We use the fact that Mary's total interest is \$5025 to set up the model.

$$\boxed{\text{interest at } 6\%} + \boxed{\text{interest at } 4\tfrac{1}{2}\%} = \boxed{\text{total interest}}$$

$$0.06x + 0.045(100{,}000 - x) = 5025$$

Solve. Now we solve for x.

$$0.06x + 4500 - 0.045x = 5025 \qquad \text{Distributive Property}$$

$$0.015x + 4500 = 5025 \qquad \text{Combine the } x\text{-terms}$$

$$0.015x = 525 \qquad \text{Subtract 4500}$$

$$x = \frac{525}{0.015} = 35{,}000 \qquad \text{Divide by 0.015}$$

So Mary has invested \$35,000 at 6% and the remaining \$65,000 at $4\frac{1}{2}\%$.

CHECK YOUR ANSWER

$$\text{total interest} = 6\% \text{ of } \$35{,}000 + 4\tfrac{1}{2}\% \text{ of } \$65{,}000$$
$$= \$2100 + \$2925 = \$5025 \quad ✔$$

NOW TRY EXERCISE **21** ■

▼ Problems About Area or Length

When we use algebra to model a physical situation, we must sometimes use basic formulas from geometry. For example, we may need a formula for an area or a perimeter, or the formula that relates the sides of similar triangles, or the Pythagorean Theorem. Most of these formulas are listed in the front endpapers of this book. The next two examples use these geometric formulas to solve some real-world problems.

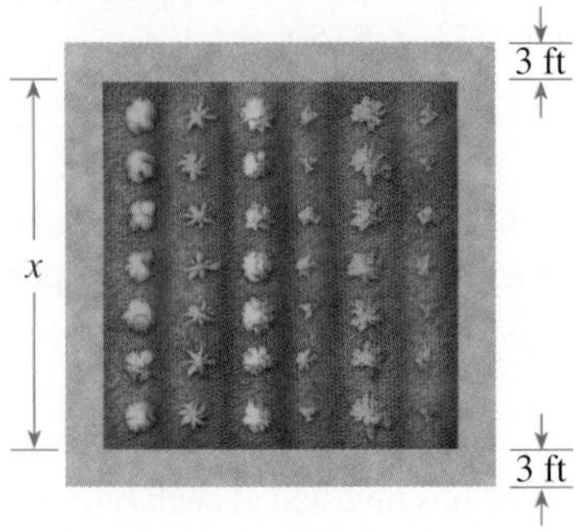

FIGURE 1

EXAMPLE 3 | Dimensions of a Garden

A square garden has a walkway 3 ft wide around its outer edge, as shown in Figure 1. If the area of the entire garden, including the walkway, is 18,000 ft^2, what are the dimensions of the planted area?

SOLUTION

Identify the variable. We are asked to find the length and width of the planted area. So we let

$$x = \text{the length of the planted area}$$

Translate from words to algebra. Next, translate the information from Figure 1 into the language of algebra.

In Words	In Algebra
Length of planted area	x
Length of entire garden	$x + 6$
Area of entire garden	$(x + 6)^2$

Set up the model. We now set up the model.

$$\text{area of entire garden} = 18{,}000\ \text{ft}^2$$

$$(x + 6)^2 = 18{,}000$$

Solve. Now we solve for x.

$$x + 6 = \sqrt{18{,}000} \qquad \text{Take square roots}$$

$$x = \sqrt{18{,}000} - 6 \qquad \text{Subtract 6}$$

$$x \approx 128$$

The planted area of the garden is about 128 ft by 128 ft.

NOW TRY EXERCISE **47**

EXAMPLE 4 | Dimensions of a Building Lot

A rectangular building lot is 8 ft longer than it is wide and has an area of 2900 ft^2. Find the dimensions of the lot.

SOLUTION

Identify the variable. We are asked to find the width and length of the lot. So let

$$w = \text{width of lot}$$

Translate from words to algebra. Then we translate the information given in the problem into the language of algebra (see Figure 2).

In Words	In Algebra
Width of lot	w
Length of Lot	$w + 8$

Set up the model. Now we set up the model.

$$\text{width of lot} \cdot \text{length of lot} = \text{area of lot}$$

$$w(w + 8) = 2900$$

Solve. Now we solve for w.

$$w^2 + 8w = 2900 \qquad \text{Expand}$$

$$w^2 + 8w - 2900 = 0 \qquad \text{Subtract 2900}$$

$$(w - 50)(w + 58) = 0 \qquad \text{Factor}$$

$$w = 50 \quad \text{or} \quad w = -58 \qquad \text{Zero-Product Property}$$

Since the width of the lot must be a positive number, we conclude that $w = 50$ ft. The length of the lot is $w + 8 = 50 + 8 = 58$ ft.

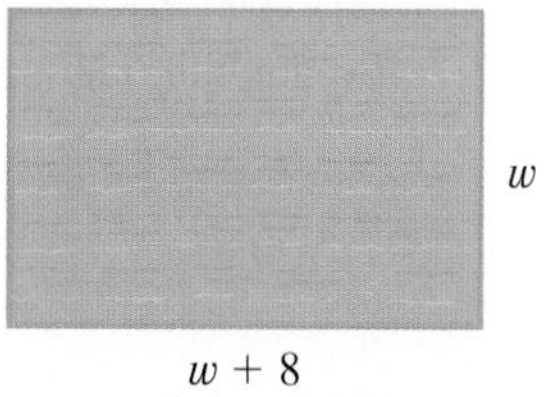

FIGURE 2

NOW TRY EXERCISE **39**

EXAMPLE 5 | Determining the Height of a Building Using Similar Triangles

A man who is 6 ft tall wishes to find the height of a certain four-story building. He measures its shadow and finds it to be 28 ft long, while his own shadow is $3\frac{1}{2}$ ft long. How tall is the building?

SOLUTION

Identify the variable. The problem asks for the height of the building. So let

$$h = \text{the height of the building}$$

Translate from words to algebra. We use the fact that the triangles in Figure 3 are similar. Recall that for any pair of similar triangles the ratios of corresponding sides are equal. Now we translate these observations into the language of algebra.

In Words	In Algebra
Height of building	h
Ratio of height to base in large triangle	$\frac{h}{28}$
Ratio of height to base in small triangle	$\frac{6}{3.5}$

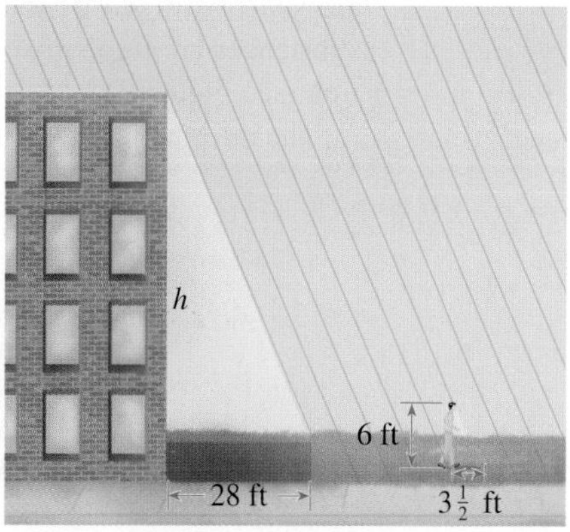

FIGURE 3

Set up the model. Since the large and small triangles are similar, we get the equation

$$\text{ratio of height to base in large triangle} = \text{ratio of height to base in small triangle}$$

$$\frac{h}{28} = \frac{6}{3.5}$$

Solve. Now we solve for h.

$$h = \frac{6 \cdot 28}{3.5} = 48 \qquad \text{Multiply by 28}$$

So the building is 48 ft tall.

NOW TRY EXERCISE **51**

Problems About Mixtures

Many real-world problems involve mixing different types of substances. For example, construction workers may mix cement, gravel, and sand; fruit juice from concentrate may involve mixing different types of juices. Problems involving mixtures and concentrations make use of the fact that if an amount x of a substance is dissolved in a solution with volume V, then the concentration C of the substance is given by

$$C = \frac{x}{V}$$

So if 10 g of sugar is dissolved in 5 L of water, then the sugar concentration is $C = 10/5 = 2$ g/L. Solving a mixture problem usually requires us to analyze the amount x of the substance that is in the solution. When we solve for x in this equation, we see that $x = CV$. Note that in many mixture problems the concentration C is expressed as a percentage, as in the next example.

EXAMPLE 6 | Mixtures and Concentration

A manufacturer of soft drinks advertises their orange soda as "naturally flavored," although it contains only 5% orange juice. A new federal regulation stipulates that to be called "natural," a drink must contain at least 10% fruit juice. How much pure orange juice must this manufacturer add to 900 gal of orange soda to conform to the new regulation?

SOLUTION

Identify the variable. The problem asks for the amount of pure orange juice to be added. So let

x = the amount (in gallons) of pure orange juice to be added

Translate from words to algebra. In any problem of this type—in which two different substances are to be mixed—drawing a diagram helps us to organize the given information (see Figure 4).

The information in the figure can be translated into the language of algebra, as follows:

In Words	In Algebra
Amount of orange juice to be added	x
Amount of the mixture	$900 + x$
Amount of orange juice in the first vat	$0.05(900) = 45$
Amount of orange juice in the second vat	$1 \cdot x = x$
Amount of orange juice in the mixture	$0.10(900 + x)$

Volume	900 gallons	x gallons	$900 + x$ gallons
Amount of orange juice	5% of 900 gallons = 45 gallons	100% of x gallons = x gallons	10% of $900 + x$ gallons = $0.1(900 + x)$ gallons

FIGURE 4

Set up the model. To set up the model, we use the fact that the total amount of orange juice in the mixture is equal to the orange juice in the first two vats.

$$\boxed{\begin{array}{c}\text{amount of}\\ \text{orange juice}\\ \text{in first vat}\end{array}} + \boxed{\begin{array}{c}\text{amount of}\\ \text{orange juice}\\ \text{in second vat}\end{array}} = \boxed{\begin{array}{c}\text{amount of}\\ \text{orange juice}\\ \text{in mixture}\end{array}}$$

$$45 + x = 0.1(900 + x) \qquad \text{From Figure 4}$$

Solve. Now we solve for x.

$$45 + x = 90 + 0.1x \qquad \text{Distributive Property}$$

$$0.9x = 45 \qquad \text{Subtract } 0.1x \text{ and } 45$$

$$x = \frac{45}{0.9} = 50 \qquad \text{Divide by 0.9}$$

The manufacturer should add 50 gal of pure orange juice to the soda.

CHECK YOUR ANSWER

amount of juice before mixing = 5% of 900 gal + 50 gal pure juice

= 45 gal + 50 gal = 95 gal

amount of juice after mixing = 10% of 950 gal = 95 gal

Amounts are equal. ✔

NOW TRY EXERCISE **53** ■

▼ Problems About the Time Needed to Do a Job

When solving a problem that involves determining how long it takes several workers to complete a job, we use the fact that if a person or machine takes H time units to complete the task, then in one time unit the fraction of the task that has been completed is $1/H$. For example, if a worker takes 5 hours to mow a lawn, then in 1 hour the worker will mow $1/5$ of the lawn.

EXAMPLE 7 | Time Needed to Do a Job

Because of an anticipated heavy rainstorm, the water level in a reservoir must be lowered by 1 ft. Opening spillway A lowers the level by this amount in 4 hours, whereas opening the smaller spillway B does the job in 6 hours. How long will it take to lower the water level by 1 ft if both spillways are opened?

SOLUTION Identify the variable. We are asked to find the time needed to lower the level by 1 ft if both spillways are open. So let

x = the time (in hours) it takes to lower the water level by 1 ft if both spillways are open

Translate from words to algebra. Finding an equation relating x to the other quantities in this problem is not easy. Certainly x is not simply $4 + 6$, because that would mean that together the two spillways require longer to lower the water level than either spillway alone. Instead, *we look at the fraction of the job that can be done in 1 hour by each spillway.*

In Words	In Algebra
Time it takes to lower level 1 ft with A and B together	x h
Distance A lowers level in 1 h	$\frac{1}{4}$ ft
Distance B lowers level in 1 h	$\frac{1}{6}$ ft
Distance A and B together lower levels in 1 h	$\frac{1}{x}$ ft

Set up the model. Now we set up the model.

fraction done by A + fraction done by B = fraction done by both

$$\frac{1}{4} + \frac{1}{6} = \frac{1}{x}$$

Solve. Now we solve for x.

$$3x + 2x = 12 \qquad \text{Multiply by the LCD, } 12x$$

$$5x = 12 \qquad \text{Add}$$

$$x = \frac{12}{5} \qquad \text{Divide by 5}$$

It will take $2\frac{2}{5}$ hours, or 2 h 24 min, to lower the water level by 1 ft if both spillways are open.

NOW TRY EXERCISE **61**

Problems About Distance, Rate, and Time

The next example deals with distance, rate (speed), and time. The formula to keep in mind here is

distance = rate × time

where the rate is either the constant speed or average speed of a moving object. For example, driving at 60 mi/h for 4 hours takes you a distance of $60 \cdot 4 = 240$ mi.

EXAMPLE 8 | A Distance-Speed-Time Problem

A jet flew from New York to Los Angeles, a distance of 4200 km. The speed for the return trip was 100 km/h faster than the outbound speed. If the total trip took 13 hours, what was the jet's speed from New York to Los Angeles?

SOLUTION Identify the variable. We are asked for the speed of the jet from New York to Los Angeles. So let

s = speed from New York to Los Angeles

Then $s + 100$ = speed from Los Angeles to New York

Translate from words to algebra. Now we organize the information in a table. We fill in the "Distance" column first, since we know that the cities are 4200 km apart. Then we fill in the "Speed" column, since we have expressed both speeds (rates) in terms of the variable s. Finally, we calculate the entries for the "Time" column, using

$$\text{time} = \frac{\text{distance}}{\text{rate}}$$

	Distance (km)	Speed (km/h)	Time (h)
N.Y. to L.A.	4200	s	$\frac{4200}{s}$
L.A. to N.Y.	4200	$s + 100$	$\frac{4200}{s + 100}$

Set up the model. The total trip took 13 hours, so we have the model

$$\boxed{\text{time from N.Y. to L.A.}} + \boxed{\text{time from L.A. to N.Y.}} = \boxed{\text{total time}}$$

$$\frac{4200}{s} + \frac{4200}{s + 100} = 13$$

Solve. Multiplying by the common denominator, $s(s + 100)$, we get

$$\begin{aligned} 4200(s + 100) + 4200s &= 13s(s + 100) \\ 8400s + 420{,}000 &= 13s^2 + 1300s \\ 0 &= 13s^2 - 7100s - 420{,}000 \end{aligned}$$

Although this equation does factor, with numbers this large it is probably quicker to use the Quadratic Formula and a calculator.

$$\begin{aligned} s &= \frac{7100 \pm \sqrt{(-7100)^2 - 4(13)(-420{,}000)}}{2(13)} \\ &= \frac{7100 \pm 8500}{26} \end{aligned}$$

$$s = 600 \qquad \text{or} \qquad s = \frac{-1400}{26} \approx -53.8$$

Since s represents speed, we reject the negative answer and conclude that the jet's speed from New York to Los Angeles was 600 km/h.

NOW TRY EXERCISE 67 ■

EXAMPLE 9 | Energy Expended in Bird Flight

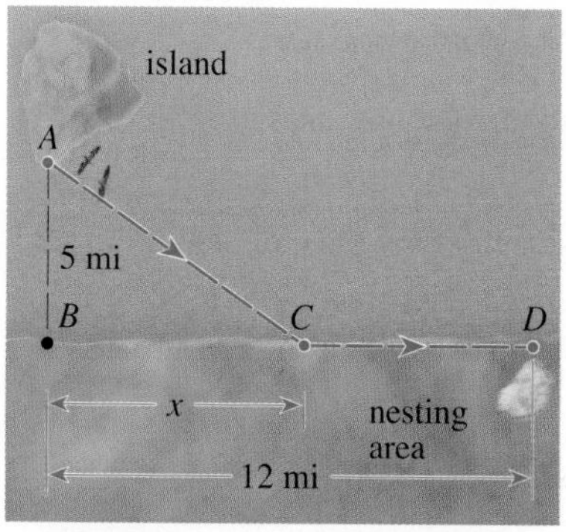

FIGURE 5

Ornithologists have determined that some species of birds tend to avoid flights over large bodies of water during daylight hours, because air generally rises over land and falls over water in the daytime, so flying over water requires more energy. A bird is released from point A on an island, 5 mi from B, the nearest point on a straight shoreline. The bird flies to a point C on the shoreline and then flies along the shoreline to its nesting area D, as shown in Figure 5. Suppose the bird has 170 kcal of energy reserves. It uses 10 kcal/mi flying over land and 14 kcal/mi flying over water.

(a) Where should the point C be located so that the bird uses exactly 170 kcal of energy during its flight?

(b) Does the bird have enough energy reserves to fly directly from A to D?

Bhaskara (born 1114) was an Indian mathematician, astronomer, and astrologer. Among his many accomplishments was an ingenious proof of the Pythagorean Theorem. (See *Focus on Problem Solving* 5, Problem 12, at the book companion website **www.stewartmath.com**.) His important mathematical book *Lilavati* [*The Beautiful*] consists of algebra problems posed in the form of stories to his daughter Lilavati. Many of the problems begin "Oh beautiful maiden, suppose . . ." The story is told that using astrology, Bhaskara had determined that great misfortune would befall his daughter if she married at any time other than at a certain hour of a certain day. On her wedding day, as she was anxiously watching the water clock, a pearl fell unnoticed from her headdress. It stopped the flow of water in the clock, causing her to miss the opportune moment for marriage. Bhaskara's *Lilavati* was written to console her.

SOLUTION

(a) Identify the variable. We are asked to find the location of C. So let

$$x = \text{ distance from } B \text{ to } C$$

Translate from words to algebra. From the figure, and from the fact that

$$\text{energy used} = \text{energy per mile} \times \text{miles flown}$$

we determine the following:

In Words	In Algebra	
Distance from B to C	x	
Distance flown over water (from A to C)	$\sqrt{x^2 + 25}$	Pythagorean Theorem
Distance flown over land (from C to D)	$12 - x$	
Energy used over water	$14\sqrt{x^2 + 25}$	
Energy used over land	$10(12 - x)$	

Set up the model. Now we set up the model.

$$\text{total energy used} = \text{energy used over water} + \text{energy used over land}$$

$$170 = 14\sqrt{x^2 + 25} + 10(12 - x)$$

Solve. To solve this equation, we eliminate the square root by first bringing all other terms to the left of the equal sign and then squaring each side.

$$170 - 10(12 - x) = 14\sqrt{x^2 + 25} \qquad \text{Isolate square-root term on RHS}$$

$$50 + 10x = 14\sqrt{x^2 + 25} \qquad \text{Simplify LHS}$$

$$(50 + 10x)^2 = (14)^2(x^2 + 25) \qquad \text{Square each side}$$

$$2500 + 1000x + 100x^2 = 196x^2 + 4900 \qquad \text{Expand}$$

$$0 = 96x^2 - 1000x + 2400 \qquad \text{All terms to RHS}$$

This equation could be factored, but because the numbers are so large it is easier to use the Quadratic Formula and a calculator:

$$x = \frac{1000 \pm \sqrt{(-1000)^2 - 4(96)(2400)}}{2(96)}$$

$$= \frac{1000 \pm 280}{192} = 6\tfrac{2}{3} \quad \text{or} \quad 3\tfrac{3}{4}$$

Point C should be either $6\frac{2}{3}$ mi or $3\frac{3}{4}$ mi from B so that the bird uses exactly 170 kcal of energy during its flight.

(b) By the Pythagorean Theorem (see page 219), the length of the route directly from A to D is $\sqrt{5^2 + 12^2} = 13$ mi, so the energy the bird requires for that route is $14 \times 13 = 182$ kcal. This is more energy than the bird has available, so it can't use this route.

NOW TRY EXERCISE **83** ■

1.6 EXERCISES

CONCEPTS

1. Explain in your own words what it means for an equation to model a real-world situation, and give an example.

2. In the formula $I = Prt$ for simple interest, P stands for ________, r for ________, and t for ________.

3. Give a formula for the area of the geometric figure.

   (a) A square of side x: $A =$ ________.

   (b) A rectangle of length l and width w: $A =$ ________.

   (c) A circle of radius r: $A =$ ________.

4. Balsamic vinegar contains 5% acetic acid, so a 32-oz bottle of balsamic vinegar contains ________ ounces of acetic acid.

5. A painter paints a wall in x hours, so the fraction of the wall that she paints in 1 hour is ________.

6. The formula $d = rt$ models the distance d traveled by an object moving at the constant rate r in time t. Find formulas for the following quantities.

   $r =$ ________ $t =$ ________.

SKILLS

7–18 ■ Express the given quantity in terms of the indicated variable.

7. The sum of three consecutive integers; $n =$ first integer of the three

8. The sum of three consecutive integers; $n =$ middle integer of the three

9. The average of three test scores if the first two scores are 78 and 82; $s =$ third test score

10. The average of four quiz scores if each of the first three scores is 8; $q =$ fourth quiz score

11. The interest obtained after one year on an investment at $2\frac{1}{2}\%$ simple interest per year; $x =$ number of dollars invested

12. The total rent paid for an apartment if the rent is \$795 a month; $n =$ number of months

13. The area (in ft^2) of a rectangle that is three times as long as it is wide; $w =$ width of the rectangle (in ft)

14. The perimeter (in cm) of a rectangle that is 5 cm longer than it is wide; $w =$ width of the rectangle (in cm)

15. The distance (in mi) that a car travels in 45 min; $s =$ speed of the car (in mi/h)

16. The time (in hours) it takes to travel a given distance at 55 mi/h; $d =$ given distance (in mi)

17. The concentration (in oz/gal) of salt in a mixture of 3 gal of brine containing 25 oz of salt to which some pure water has been added; $x =$ volume of pure water added (in gal)

18. The value (in cents) of the change in a purse that contains twice as many nickels as pennies, four more dimes than nickels, and as many quarters as dimes and nickels combined; $p =$ number of pennies

APPLICATIONS

19. **Renting a Truck** A rental company charges \$65 a day and 20 cents a mile for renting a truck. Michael rented a truck for 3 days, and his bill came to \$275. How many miles did he drive?

20. **Cell Phone Costs** A cell phone company charges a monthly fee of \$10 for the first 1000 text messages and 10 cents for each additional text message. Miriam's bill for text messages for the month of June is \$38.50. How many text messages did she send that month?

21. **Investments** Phyllis invested \$12,000, a portion earning a simple interest rate of $4\frac{1}{2}\%$ per year and the rest earning a rate of 4% per year. After 1 year the total interest earned on these investments was \$525. How much money did she invest at each rate?

22. **Investments** If Ben invests \$4000 at 4% interest per year, how much additional money must he invest at $5\frac{1}{2}\%$ annual interest to ensure that the interest he receives each year is $4\frac{1}{2}\%$ of the total amount invested?

23. **Investments** What annual rate of interest would you have to earn on an investment of \$3500 to ensure receiving \$262.50 interest after 1 year?

24. **Investments** Jack invests \$1000 at a certain annual interest rate, and he invests another \$2000 at an annual rate that is one-half percent higher. If he receives a total of \$190 interest in 1 year, at what rate is the \$1000 invested?

25. **Salaries** An executive in an engineering firm earns a monthly salary plus a Christmas bonus of \$8500. If she earns a total of \$97,300 per year, what is her monthly salary?

26. **Salaries** A woman earns 15% more than her husband. Together they make \$69,875 per year. What is the husband's annual salary?

27. **Inheritance** Craig is saving to buy a vacation home. He inherits some money from a wealthy uncle, then combines this with the \$22,000 he has already saved and doubles the total in a lucky investment. He ends up with \$134,000—just enough to buy a cabin on the lake. How much did he inherit?

28. **Overtime Pay** Helen earns \$7.50 an hour at her job, but if she works more than 35 hours in a week, she is paid $1\frac{1}{2}$ times her regular salary for the overtime hours worked. One week her gross pay was \$352.50. How many overtime hours did she work that week?

29. **Labor Costs** A plumber and his assistant work together to replace the pipes in an old house. The plumber charges \$45 an hour for his own labor and \$25 an hour for his assistant's labor. The plumber works twice as long as his assistant on this job, and the labor charge on the final bill is \$4025. How long did the plumber and his assistant work on this job?

30. A Riddle A father is four times as old as his daughter. In 6 years, he will be three times as old as she is. How old is the daughter now?

31. A Riddle A movie star, unwilling to give his age, posed the following riddle to a gossip columnist: "Seven years ago, I was eleven times as old as my daughter. Now I am four times as old as she is." How old is the movie star?

32. Career Home Runs During his major league career, Hank Aaron hit 41 more home runs than Babe Ruth hit during his career. Together they hit 1469 home runs. How many home runs did Babe Ruth hit?

33. Value of Coins A change purse contains an equal number of pennies, nickels, and dimes. The total value of the coins is \$1.44. How many coins of each type does the purse contain?

34. Value of Coins Mary has \$3.00 in nickels, dimes, and quarters. If she has twice as many dimes as quarters and five more nickels than dimes, how many coins of each type does she have?

35. Length of a Garden A rectangular garden is 25 ft wide. If its area is 1125 ft^2, what is the length of the garden?

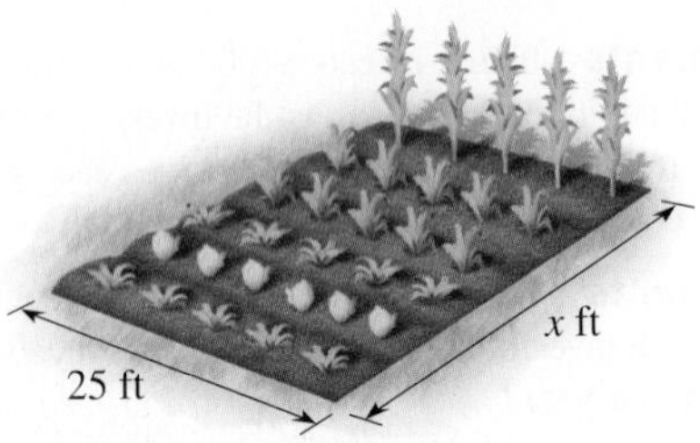

36. Width of a Pasture A pasture is twice as long as it is wide. Its area is 115,200 ft^2. How wide is the pasture?

37. Dimensions of a Lot A square plot of land has a building 60 ft long and 40 ft wide at one corner. The rest of the land outside the building forms a parking lot. If the parking lot has area 12,000 ft^2, what are the dimensions of the entire plot of land?

38. Dimensions of a Lot A half-acre building lot is five times as long as it is wide. What are its dimensions? [*Note:* 1 acre = 43,560 ft^2.]

39. Dimensions of a Garden A rectangular garden is 10 ft longer than it is wide. Its area is 875 ft^2. What are its dimensions?

40. Dimensions of a Room A rectangular bedroom is 7 ft longer than it is wide. Its area is 228 ft^2. What is the width of the room?

41. Dimensions of a Garden A farmer has a rectangular garden plot surrounded by 200 ft of fence. Find the length and width of the garden if its area is 2400 ft^2.

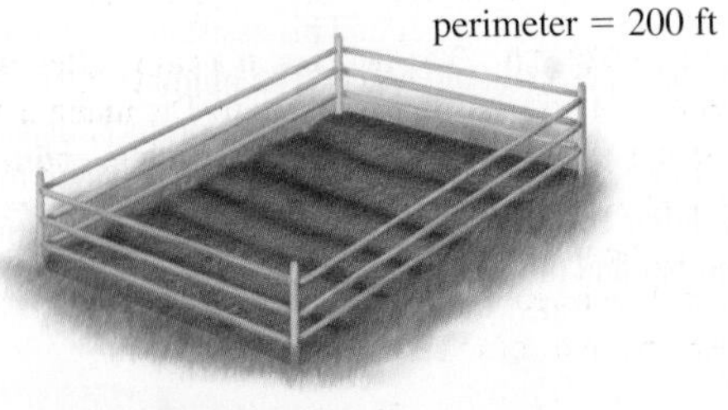

42. Dimensions of a Lot A parcel of land is 6 ft longer than it is wide. Each diagonal from one corner to the opposite corner is 174 ft long. What are the dimensions of the parcel?

43. Dimensions of a Lot A rectangular parcel of land is 50 ft wide. The length of a diagonal between opposite corners is 10 ft more than the length of the parcel. What is the length of the parcel?

44. Dimensions of a Track A running track has the shape shown in the figure, with straight sides and semicircular ends. If the length of the track is 440 yd and the two straight parts are each 110 yd long, what is the radius of the semicircular parts (to the nearest yard)?

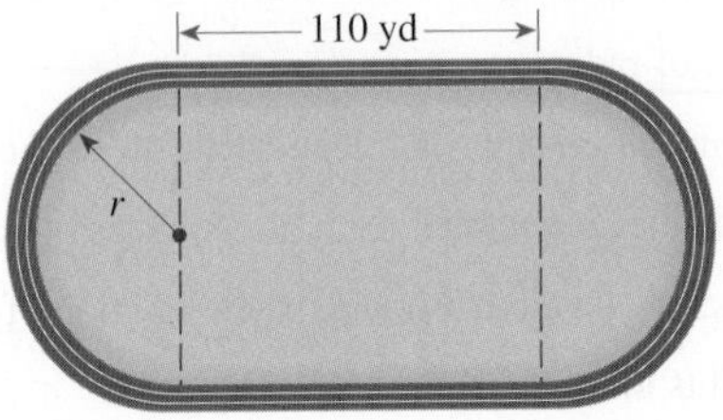

45. Length and Area Find the length x in the figure. The area of the shaded region is given.

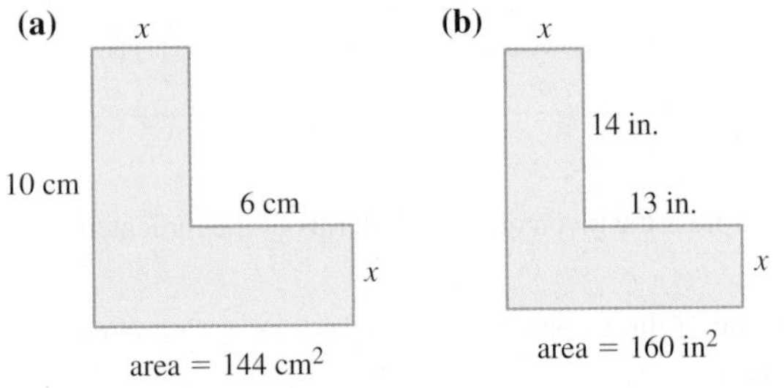

46. Length and Area Find the length y in the figure. The area of the shaded region is given.

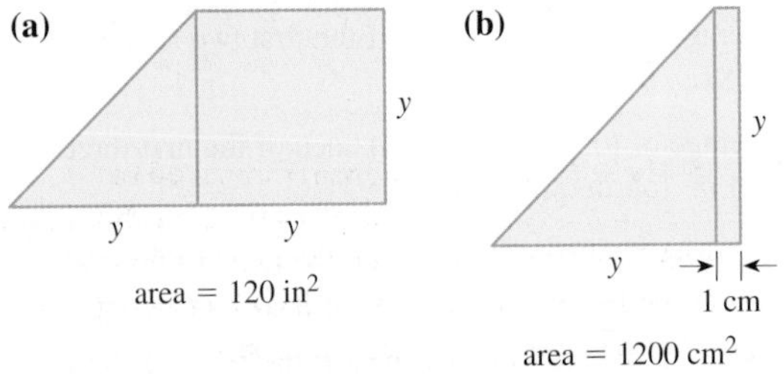

47. Framing a Painting Ali paints with watercolors on a sheet of paper 20 in. wide by 15 in. high. He then places this sheet on a mat so that a uniformly wide strip of the mat shows all around the picture. The perimeter of the mat is 102 in. How wide is the strip of the mat showing around the picture?

48. Dimensions of a Poster A poster has a rectangular printed area 100 cm by 140 cm and a blank strip of uniform width around the edges. The perimeter of the poster is $1\frac{1}{2}$ times the perimeter of the printed area. What is the width of the blank strip?

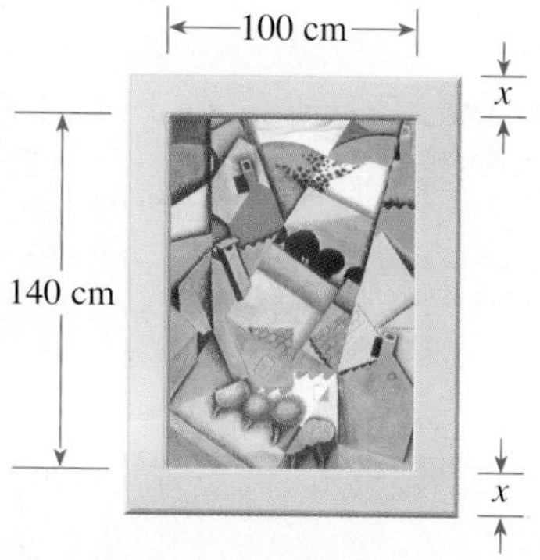

49. Reach of a Ladder A $19\frac{1}{2}$-foot ladder leans against a building. The base of the ladder is $7\frac{1}{2}$ ft from the building. How high up the building does the ladder reach?

50. Height of a Flagpole A flagpole is secured on opposite sides by two guy wires, each of which is 5 ft longer than the pole. The distance between the points where the wires are fixed to the ground is equal to the length of one guy wire. How tall is the flagpole (to the nearest inch)?

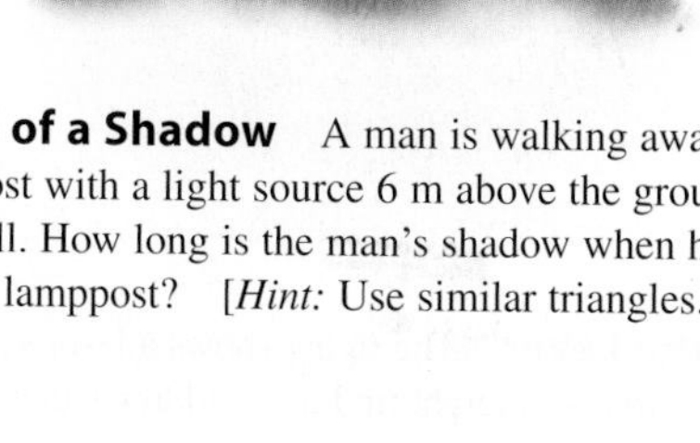

51. Length of a Shadow A man is walking away from a lamppost with a light source 6 m above the ground. The man is 2 m tall. How long is the man's shadow when he is 10 m from the lamppost? [*Hint:* Use similar triangles.]

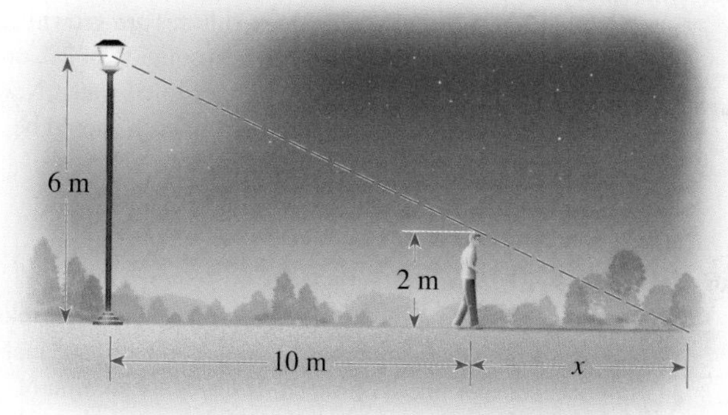

52. Height of a Tree A woodcutter determines the height of a tall tree by first measuring a smaller one 125 ft away, then moving so that his eyes are in the line of sight along the tops of the trees and measuring how far he is standing from the small tree (see the figure). Suppose the small tree is 20 ft tall, the man is 25 ft from the small tree, and his eye level is 5 ft above the ground. How tall is the taller tree?

53. Mixture Problem What quantity of a 60% acid solution must be mixed with a 30% solution to produce 300 mL of a 50% solution?

54. Mixture Problem What quantity of pure acid must be added to 300 mL of a 50% acid solution to produce a 60% acid solution?

55. Mixture Problem A jeweler has five rings, each weighing 18 g, made of an alloy of 10% silver and 90% gold. She decides to melt down the rings and add enough silver to reduce the gold content to 75%. How much silver should she add?

56. Mixture Problem A pot contains 6 L of brine at a concentration of 120 g/L. How much of the water should be boiled off to increase the concentration to 200 g/L?

57. Mixture Problem The radiator in a car is filled with a solution of 60% antifreeze and 40% water. The manufacturer of the antifreeze suggests that for summer driving, optimal cooling of the engine is obtained with only 50% antifreeze. If the capacity of the radiator is 3.6 L, how much coolant should be drained and replaced with water to reduce the antifreeze concentration to the recommended level?

58. **Mixture Problem** A health clinic uses a solution of bleach to sterilize petri dishes in which cultures are grown. The sterilization tank contains 100 gal of a solution of 2% ordinary household bleach mixed with pure distilled water. New research indicates that the concentration of bleach should be 5% for complete sterilization. How much of the solution should be drained and replaced with bleach to increase the bleach content to the recommended level?

59. **Mixture Problem** A bottle contains 750 mL of fruit punch with a concentration of 50% pure fruit juice. Jill drinks 100 mL of the punch and then refills the bottle with an equal amount of a cheaper brand of punch. If the concentration of juice in the bottle is now reduced to 48%, what was the concentration in the punch that Jill added?

60. **Mixture Problem** A merchant blends tea that sells for \$3.00 a pound with tea that sells for \$2.75 a pound to produce 80 lb of a mixture that sells for \$2.90 a pound. How many pounds of each type of tea does the merchant use in the blend?

61. **Sharing a Job** Candy and Tim share a paper route. It takes Candy 70 min to deliver all the papers, and it takes Tim 80 min. How long does it take the two when they work together?

62. **Sharing a Job** Stan and Hilda can mow the lawn in 40 min if they work together. If Hilda works twice as fast as Stan, how long does it take Stan to mow the lawn alone?

63. **Sharing a Job** Betty and Karen have been hired to paint the houses in a new development. Working together, the women can paint a house in two-thirds the time that it takes Karen working alone. Betty takes 6 h to paint a house alone. How long does it take Karen to paint a house working alone?

64. **Sharing a Job** Next-door neighbors Bob and Jim use hoses from both houses to fill Bob's swimming pool. They know that it takes 18 h using both hoses. They also know that Bob's hose, used alone, takes 20% less time than Jim's hose alone. How much time is required to fill the pool by each hose alone?

65. **Sharing a Job** Henry and Irene working together can wash all the windows of their house in 1 h 48 min. Working alone, it takes Henry $1\frac{1}{2}$ h more than Irene to do the job. How long does it take each person working alone to wash all the windows?

66. **Sharing a Job** Jack, Kay, and Lynn deliver advertising flyers in a small town. If each person works alone, it takes Jack 4 h to deliver all the flyers, and it takes Lynn 1 h longer than it takes Kay. Working together, they can deliver all the flyers in 40% of the time it takes Kay working alone. How long does it take Kay to deliver all the flyers alone?

67. **Distance, Speed, and Time** Wendy took a trip from Davenport to Omaha, a distance of 300 mi. She traveled part of the way by bus, which arrived at the train station just in time for Wendy to complete her journey by train. The bus averaged 40 mi/h, and the train averaged 60 mi/h. The entire trip took $5\frac{1}{2}$ h. How long did Wendy spend on the train?

68. **Distance, Speed, and Time** Two cyclists, 90 mi apart, start riding toward each other at the same time. One cycles twice as fast as the other. If they meet 2 h later, at what average speed is each cyclist traveling?

69. **Distance, Speed, and Time** A pilot flew a jet from Montreal to Los Angeles, a distance of 2500 mi. On the return trip, the average speed was 20% faster than the outbound speed. The round-trip took 9 h 10 min. What was the speed from Montreal to Los Angeles?

70. **Distance, Speed, and Time** A woman driving a car 14 ft long is passing a truck 30 ft long. The truck is traveling at 50 mi/h. How fast must the woman drive her car so that she can pass the truck completely in 6 s, from the position shown in figure (a) to the position shown in figure (b)? [*Hint:* Use feet and seconds instead of miles and hours.]

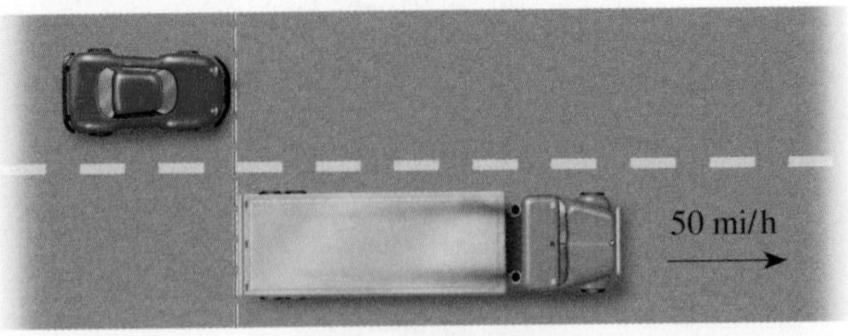

(a)

(b)

71. **Distance, Speed, and Time** A salesman drives from Ajax to Barrington, a distance of 120 mi, at a steady speed. He then increases his speed by 10 mi/h to drive the 150 mi from Barrington to Collins. If the second leg of his trip took 6 min more time than the first leg, how fast was he driving between Ajax and Barrington?

72. **Distance, Speed, and Time** Kiran drove from Tortula to Cactus, a distance of 250 mi. She increased her speed by 10 mi/h for the 360-mi trip from Cactus to Dry Junction. If the total trip took 11 h, what was her speed from Tortula to Cactus?

73. **Distance, Speed, and Time** It took a crew 2 h 40 min to row 6 km upstream and back again. If the rate of flow of the stream was 3 km/h, what was the rowing speed of the crew in still water?

74. **Speed of a Boat** Two fishing boats depart a harbor at the same time, one traveling east, the other south. The eastbound boat travels at a speed 3 mi/h faster than the southbound boat. After two hours the boats are 30 mi apart. Find the speed of the southbound boat.

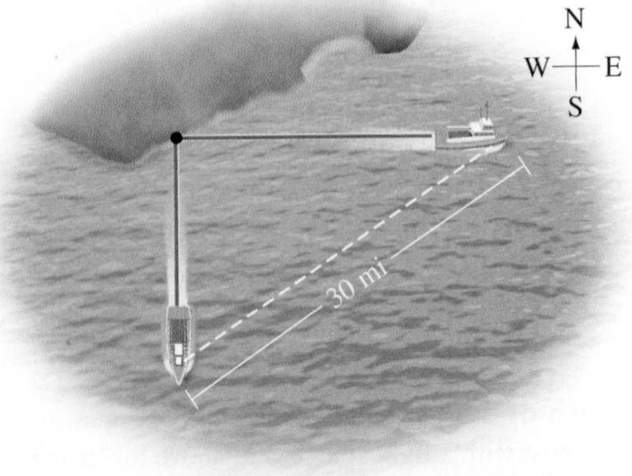

75. **Law of the Lever** The figure shows a lever system, similar to a seesaw that you might find in a children's playground. For

the system to balance, the product of the weight and its distance from the fulcrum must be the same on each side; that is,

$$w_1x_1 = w_2x_2$$

This equation is called the **law of the lever** and was first discovered by Archimedes (see page 729).

A woman and her son are playing on a seesaw. The boy is at one end, 8 ft from the fulcrum. If the son weighs 100 lb and the mother weighs 125 lb, where should the woman sit so that the seesaw is balanced?

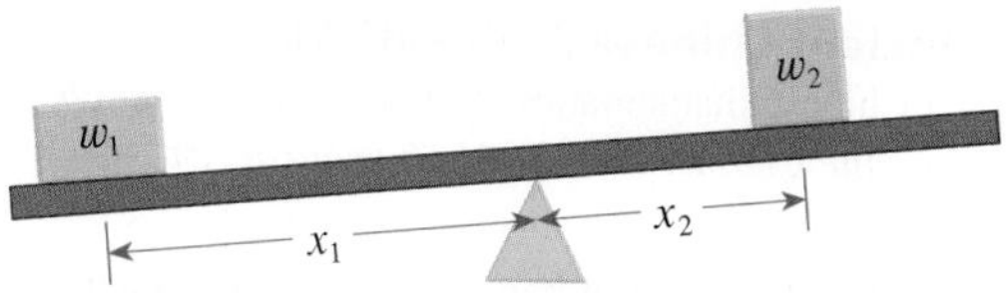

76. **Law of the Lever** A plank 30 ft long rests on top of a flat-roofed building, with 5 ft of the plank projecting over the edge, as shown in the figure. A worker weighing 240 lb sits on one end of the plank. What is the largest weight that can be hung on the projecting end of the plank if it is to remain in balance? (Use the law of the lever stated in Exercise 75.)

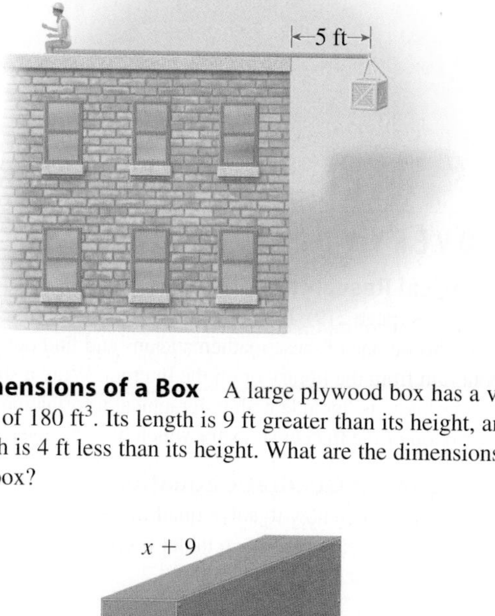

77. **Dimensions of a Box** A large plywood box has a volume of 180 ft^3. Its length is 9 ft greater than its height, and its width is 4 ft less than its height. What are the dimensions of the box?

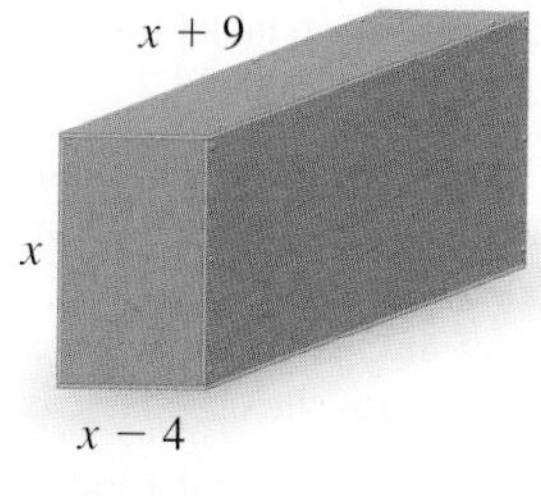

78. **Radius of a Sphere** A jeweler has three small solid spheres made of gold, of radius 2 mm, 3 mm, and 4 mm. He decides to melt these down and make just one sphere out of them. What will the radius of this larger sphere be?

79. **Dimensions of a Box** A box with a square base and no top is to be made from a square piece of cardboard by cutting 4-in. squares from each corner and folding up the sides, as shown in the figure. The box is to hold 100 in^3. How big a piece of cardboard is needed?

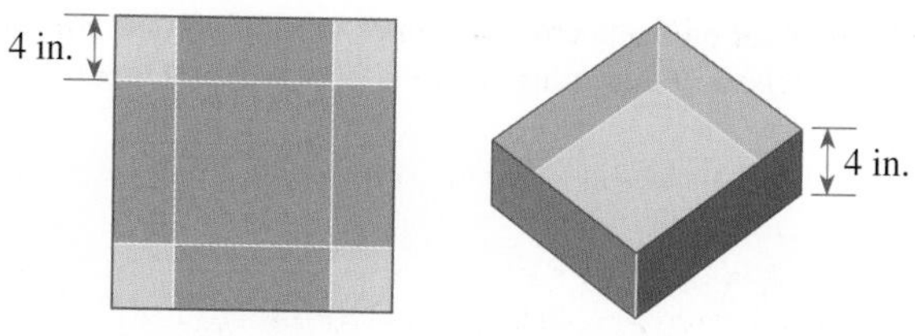

80. **Dimensions of a Can** A cylindrical can has a volume of 40π cm^3 and is 10 cm tall. What is its diameter? [*Hint:* Use the volume formula listed on the inside front cover of this book.]

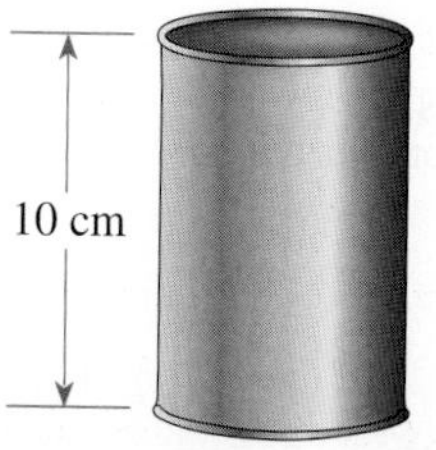

81. **Radius of a Tank** A spherical tank has a capacity of 750 gallons. Using the fact that one gallon is about 0.1337 ft^3, find the radius of the tank (to the nearest hundredth of a foot).

82. **Dimensions of a Lot** A city lot has the shape of a right triangle whose hypotenuse is 7 ft longer than one of the other sides. The perimeter of the lot is 392 ft. How long is each side of the lot?

83. **Construction Costs** The town of Foxton lies 10 mi north of an abandoned east-west road that runs through Grimley, as shown in the figure. The point on the abandoned road closest to Foxton is 40 mi from Grimley. County officials are about to build a new road connecting the two towns. They have determined that restoring the old road would cost \$100,000 per mile, whereas building a new road would cost \$200,000 per mile. How much of the abandoned road should be used (as indicated in the figure) if the officials intend to spend exactly \$6.8 million? Would it cost less than this amount to build a new road connecting the towns directly?

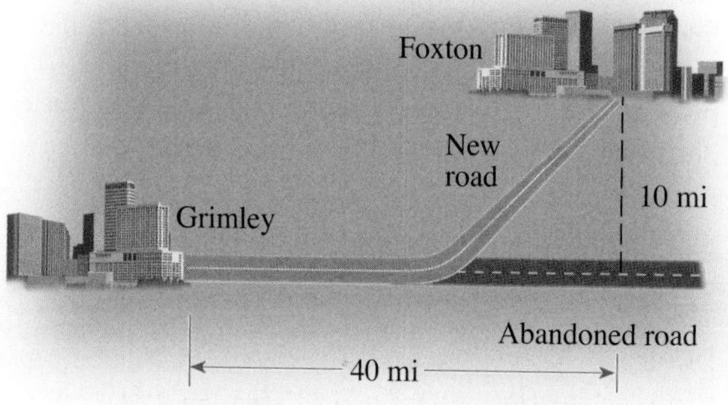

84. **Distance, Speed, and Time** A boardwalk is parallel to and 210 ft inland from a straight shoreline. A sandy beach lies between the boardwalk and the shoreline. A man is standing on the boardwalk, exactly 750 ft across the sand from his beach umbrella, which is right at the shoreline. The man walks 4 ft/s on the boardwalk and 2 ft/s on the sand. How far should

he walk on the boardwalk before veering off onto the sand if he wishes to reach his umbrella in exactly 4 min 45 s?

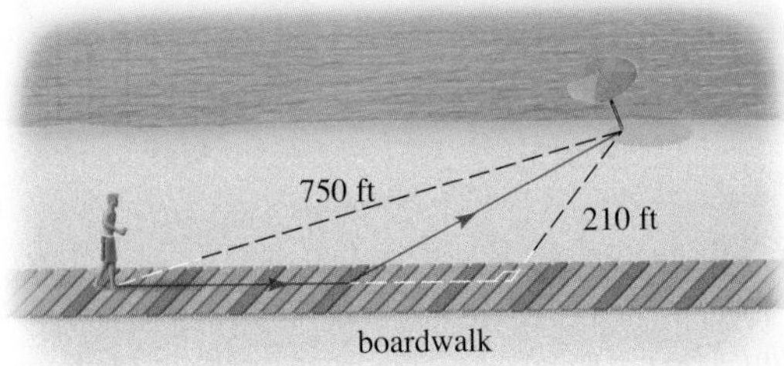

85. **Volume of Grain** Grain is falling from a chute onto the ground, forming a conical pile whose diameter is always three times its height. How high is the pile (to the nearest hundredth of a foot) when it contains 1000 ft^3 of grain?

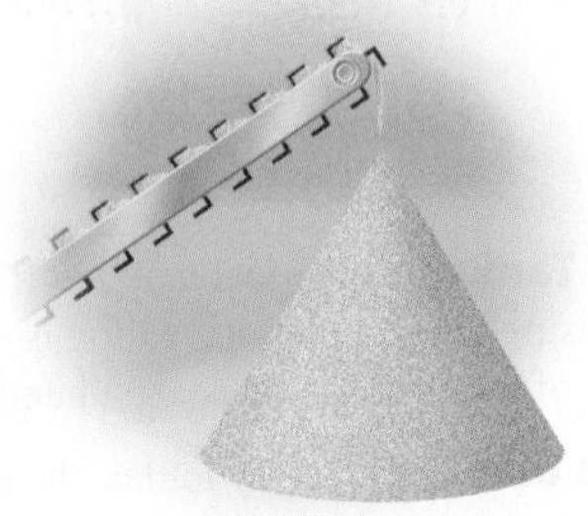

86. **TV Monitors** Two television monitors sitting side by side on a shelf in an appliance store have the same screen height. One has a conventional screen, which is 5 in. wider than it is high. The other has a wider, high-definition screen, which is 1.8 times as wide as it is high. The diagonal measure of the wider screen is 14 in. more than the diagonal measure of the smaller screen. What is the height of the screens, correct to the nearest 0.1 in.?

87. **Dimensions of a Structure** A storage bin for corn consists of a cylindrical section made of wire mesh, surmounted by a conical tin roof, as shown in the figure. The height of the roof is one-third the height of the entire structure. If the total volume of the structure is 1400π ft^3 and its radius is 10 ft, what is its height? [*Hint:* Use the volume formulas listed on the inside front cover of this book.]

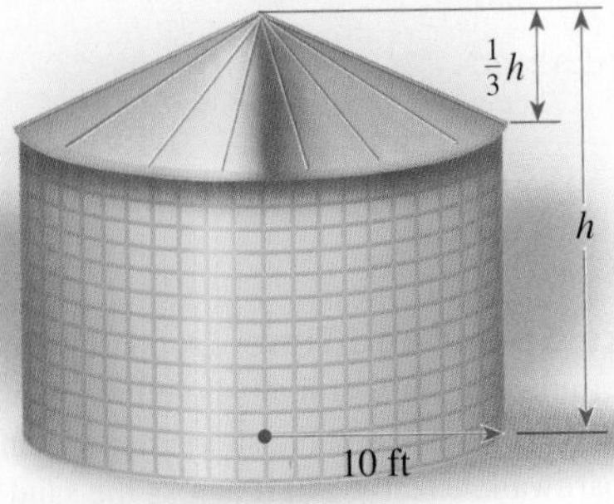

88. **Comparing Areas** A wire 360 in. long is cut into two pieces. One piece is formed into a square, and the other is formed into a circle. If the two figures have the same area, what are the lengths of the two pieces of wire (to the nearest tenth of an inch)?

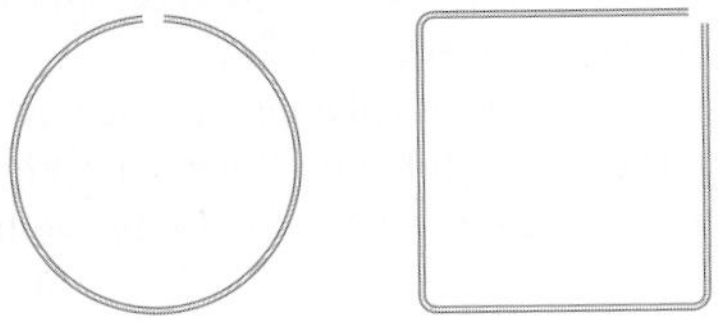

89. **An Ancient Chinese Problem** This problem is taken from a Chinese mathematics textbook called *Chui-chang suan-shu*, or *Nine Chapters on the Mathematical Art*, which was written about 250 B.C.

> A 10-ft-long stem of bamboo is broken in such a way that its tip touches the ground 3 ft from the base of the stem, as shown in the figure. What is the height of the break?

[*Hint:* Use the Pythagorean Theorem.]

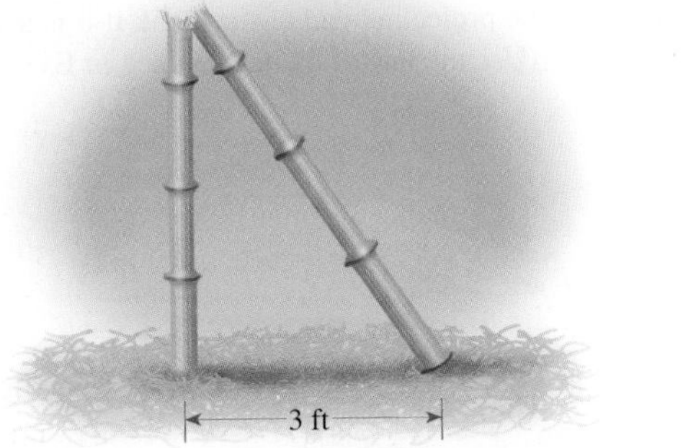

DISCOVERY ■ DISCUSSION ■ WRITING

90. **Historical Research** Read the biographical notes on Pythagoras (page 219), Euclid (page 497), and Archimedes (page 729). Choose one of these mathematicians and find out more about him from the library or on the Internet. Write a short essay on your findings. Include both biographical information and a description of the mathematics for which he is famous.

91. **A Babylonian Quadratic Equation** The ancient Babylonians knew how to solve quadratic equations. Here is a problem from a cuneiform tablet found in a Babylonian school dating back to about 2000 B.C.

> I have a reed, I know not its length. I broke from it one cubit, and it fit 60 times along the length of my field. I restored to the reed what I had broken off, and it fit 30 times along the width of my field. The area of my field is 375 square nindas. What was the original length of the reed?

Solve this problem. Use the fact that 1 ninda = 12 cubits.

DISCOVERY PROJECT **Equations Through the Ages**

In this project we study equations that were created and solved by the ancient peoples of Egypt, Babylon, India, and China. You can find the project at the book companion website: **www.stewartmath.com**

1.7 INEQUALITIES

Solving Linear Inequalities ▶ Solving Nonlinear Inequalities ▶ Absolute Value Inequalities ▶ Modeling with Inequalities

Some problems in algebra lead to **inequalities** instead of equations. An inequality looks just like an equation, except that in the place of the equal sign is one of the symbols, $<$, $>$, $\leq$, or $\geq$. Here is an example of an inequality:

$$4x + 7 \leq 19$$

x	$4x + 7 \leq 19$
1	$11 \leq 19$ ✔
2	$15 \leq 19$ ✔
3	$19 \leq 19$ ✔
4	$23 \leq 19$ ✕
5	$27 \leq 19$ ✕

The table in the margin shows that some numbers satisfy the inequality and some numbers don't.

To **solve** an inequality that contains a variable means to find all values of the variable that make the inequality true. Unlike an equation, an inequality generally has infinitely many solutions, which form an interval or a union of intervals on the real line. The following illustration shows how an inequality differs from its corresponding equation:

		Solution	**Graph**
Equation:	$4x + 7 = 19$	$x = 3$	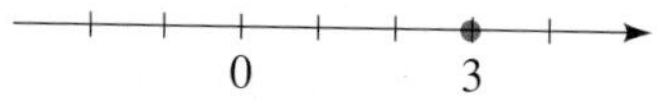
Inequality:	$4x + 7 \leq 19$	$x \leq 3$	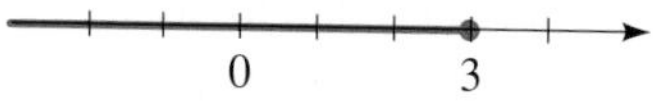

To solve inequalities, we use the following rules to isolate the variable on one side of the inequality sign. These rules tell us when two inequalities are *equivalent* (the symbol $\Leftrightarrow$ means "is equivalent to"). In these rules the symbols A, B, and C stand for real numbers or algebraic expressions. Here we state the rules for inequalities involving the symbol $\leq$, but they apply to all four inequality symbols.

RULES FOR INEQUALITIES

Rule	Description
1. $A \leq B \quad \Leftrightarrow \quad A + C \leq B + C$	**Adding** the same quantity to each side of an inequality gives an equivalent inequality.
2. $A \leq B \quad \Leftrightarrow \quad A - C \leq B - C$	**Subtracting** the same quantity from each side of an inequality gives an equivalent inequality.
3. If $C > 0$, then $A \leq B \quad \Leftrightarrow \quad CA \leq CB$	**Multiplying** each side of an inequality by the same *positive* quantity gives an equivalent inequality.
4. If $C < 0$, then $A \leq B \quad \Leftrightarrow \quad CA \geq CB$	**Multiplying** each side of an inequality by the same *negative* quantity *reverses the direction* of the inequality.
5. If $A > 0$ and $B > 0$, then $A \leq B \quad \Leftrightarrow \quad \dfrac{1}{A} \geq \dfrac{1}{B}$	**Taking reciprocals** of each side of an inequality involving *positive* quantities *reverses the direction* of the inequality.
6. If $A \leq B$ and $C \leq D$, then $A + C \leq B + D$	Inequalities can be added.

Pay special attention to Rules 3 and 4. Rule 3 says that we can multiply (or divide) each side of an inequality by a *positive* number, but Rule 4 says that if we multiply each side of an inequality by a *negative* number, then we reverse the direction of the inequality. For example, if we start with the inequality

$$3 < 5$$

and multiply by 2, we get

$$6 < 10$$

but if we multiply by -2, we get

$$-6 > -10$$

▼ Solving Linear Inequalities

An inequality is **linear** if each term is constant or a multiple of the variable. To solve a linear inequality, we isolate the variable on one side of the inequality sign.

EXAMPLE 1 | Solving a Linear Inequality

Solve the inequality $3x < 9x + 4$ and sketch the solution set.

SOLUTION

$3x < 9x + 4$	Given inequality
$3x - 9x < 9x + 4 - 9x$	Subtract $9x$
$-6x < 4$	Simplify
$(-\frac{1}{6})(-6x) > (-\frac{1}{6})(4)$	Multiply by $-\frac{1}{6}$ and reverse inequality
$x > -\frac{2}{3}$	Simplify

Multiplying by the negative number $-\frac{1}{6}$ *reverses* the direction of the inequality.

The solution set consists of all numbers greater than $-\frac{2}{3}$. In other words the solution of the inequality is the interval $\left(-\frac{2}{3}, \infty\right)$. It is graphed in Figure 1.

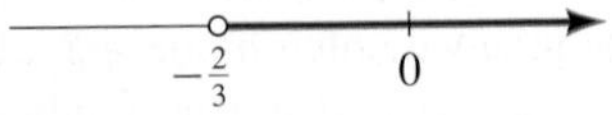

FIGURE 1

NOW TRY EXERCISE **21** ■

EXAMPLE 2 | Solving a Pair of Simultaneous Inequalities

Solve the inequalities $4 \le 3x - 2 < 13$.

SOLUTION The solution set consists of all values of x that satisfy both of the inequalities $4 \le 3x - 2$ and $3x - 2 < 13$. Using Rules 1 and 3, we see that the following inequalities are equivalent:

$4 \le 3x - 2 < 13$	Given inequality
$6 \le 3x < 15$	Add 2
$2 \le x < 5$	Divide by 3

FIGURE 2

Therefore, the solution set is $[2, 5)$, as shown in Figure 2.

NOW TRY EXERCISE **31** ■

▼ Solving Nonlinear Inequalities

To solve inequalities involving squares and other powers of the variable, we use factoring, together with the following principle.

> **THE SIGN OF A PRODUCT OR QUOTIENT**
>
> If a product or a quotient has an *even* number of *negative* factors, then its value is *positive*.
>
> If a product or a quotient has an *odd* number of *negative* factors, then its value is *negative*.

For example, to solve the inequality $x^2 - 5x \leq -6$, we first move all terms to the left-hand side and factor to get

$$(x - 2)(x - 3) \leq 0$$

This form of the inequality says that the product $(x - 2)(x - 3)$ must be negative or zero, so to solve the inequality, we must determine where each factor is negative or positive (because the sign of a product depends on the sign of the factors). The details are explained in Example 3, in which we use the following guidelines.

GUIDELINES FOR SOLVING NONLINEAR INEQUALITIES

1. **Move All Terms to One Side.** If necessary, rewrite the inequality so that all nonzero terms appear on one side of the inequality sign. If the nonzero side of the inequality involves quotients, bring them to a common denominator.
2. **Factor.** Factor the nonzero side of the inequality.
3. **Find the Intervals.** Determine the values for which each factor is zero. These numbers will divide the real line into intervals. List the intervals that are determined by these numbers.
4. **Make a Table or Diagram.** Use test values to make a table or diagram of the signs of each factor on each interval. In the last row of the table determine the sign of the product (or quotient) of these factors.
5. **Solve.** Determine the solution of the inequality from the last row of the sign table. Be sure to check whether the inequality is satisfied by some or all of the endpoints of the intervals. (This may happen if the inequality involves $\leq$ or $\geq$.)

The factoring technique that is described in these guidelines works only if all nonzero terms appear on one side of the inequality symbol. If the inequality is not written in this form, first rewrite it, as indicated in Step 1.

EXAMPLE 3 | Solving a Quadratic Inequality

Solve the inequality $x^2 \leq 5x - 6$.

SOLUTION We will follow the guidelines given above.

Move all terms to one side. We move all the terms to the left-hand side.

$$x^2 \leq 5x - 6 \qquad \text{Given inequality}$$

$$x^2 - 5x + 6 \leq 0 \qquad \text{Subtract } 5x\text{, add 6}$$

Factor. Factoring the left-hand side of the inequality, we get

$$(x - 2)(x - 3) \leq 0 \qquad \text{Factor}$$

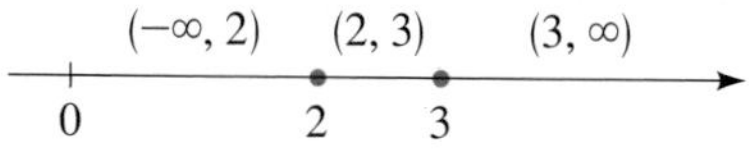

FIGURE 3

Find the intervals. The factors of the left-hand side are $x - 2$ and $x - 3$. These factors are zero when x is 2 and 3, respectively. As shown in Figure 3, the numbers 2 and 3 divide the real line into the three intervals

$$(-\infty, 2), (2, 3), (3, \infty)$$

The factors $x - 2$ and $x - 3$ change sign only at 2 and 3, respectively. So these factors maintain their sign on each of these three intervals.

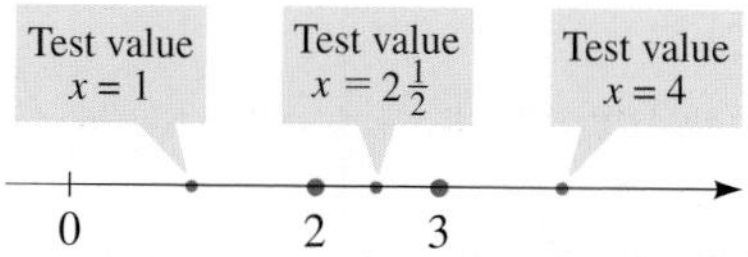

FIGURE 4

Make a table or diagram. To determine the sign of each factor on each of the intervals that we found, we use **test values**. We choose a number inside each interval and check the sign of the factors $x - 2$ and $x - 3$ at the number we chose. For the interval $(-\infty, 2)$, let's choose the test value 1 (see Figure 4). Substituting 1 for x in the factors $x - 2$ and $x - 3$, we get

$$x - 2 = 1 - 2 = -1 < 0$$

$$x - 3 = 1 - 3 = -2 < 0$$

So both factors are negative on this interval. Notice that we need to check only one test value for each interval because the factors $x - 2$ and $x - 3$ do not change sign on any of the three intervals we found.

Using the test values $x = 2\frac{1}{2}$ and $x = 4$ for the intervals $(2, 3)$ and $(3, \infty)$ (see Figure 4), respectively, we construct the following sign table. The final row of the table is obtained from the fact that the expression in the last row is the product of the two factors.

Interval	$(-\infty, 2)$	$(2, 3)$	$(3, \infty)$
Sign of $x - 2$	$-$	$+$	$+$
Sign of $x - 3$	$-$	$-$	$+$
Sign of $(x - 2)(x - 3)$	$+$	$-$	$+$

If you prefer, you can represent this information on a real line, as in the following sign diagram. The vertical lines indicate the points at which the real line is divided into intervals:

	(left of 2)	2 to 3	(right of 3)
Sign of $x - 2$	$-$	$+$	$+$
Sign of $x - 3$	$-$	$-$	$+$
Sign of $(x - 2)(x - 3)$	$+$	$-$	$+$

Solve. We read from the table or the diagram that $(x - 2)(x - 3)$ is negative on the interval $(2, 3)$. Thus, the solution of the inequality $(x - 2)(x - 3) \le 0$ is

$$\{x \mid 2 \le x \le 3\} = [2, 3]$$

We have included the endpoints 2 and 3 because we seek values of x such that the product is either less than *or equal to* zero. The solution is illustrated in Figure 5.

FIGURE 5

NOW TRY EXERCISE **41**

EXAMPLE 4 | Solving an Inequality with Repeated Factors

Solve the inequality $x(x - 1)^2(x - 3) < 0$.

SOLUTION All nonzero terms are already on one side of the inequality, and the nonzero side of the inequality is already factored. So we begin by finding the intervals for this inequality.

Find the intervals. The factors of the left-hand side are x, $(x - 1)^2$, and $x - 3$. These are zero when $x = 0, 1, 3$. These numbers divide the real line into the intervals

$$(-\infty, 0), (0, 1), (1, 3), (3, \infty)$$

Make a diagram. We make the following diagram, using test points to determine the sign of each factor in each interval.

	(left of 0)	0 to 1	1 to 3	(right of 3)
Sign of x	$-$	$+$	$+$	$+$
Sign of $(x - 1)^2$	$+$	$+$	$+$	$+$
Sign of $(x - 3)$	$-$	$-$	$-$	$+$
Sign of $x(x - 1)^2(x - 3)$	$+$	$-$	$-$	$+$

Solve. From the diagram we see that $x(x-1)^2(x-3) < 0$ for x in the interval $(0, 1)$ or for x in $(1, 3)$. So the solution set is the union of these two intervals:

$$(0, 1) \cup (1, 3)$$

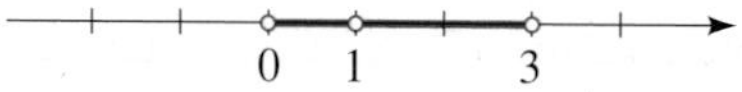

FIGURE 6

The solution set is graphed in Figure 6.

NOW TRY EXERCISE **53**

EXAMPLE 5 | Solving an Inequality Involving a Quotient

Solve the inequality $\dfrac{1+x}{1-x} \geq 1$.

SOLUTION

It is tempting to simply multiply both sides of the inequality by $1 - x$ (as you would if this were an *equation*). But this doesn't work because we don't know whether $1 - x$ is positive or negative, so we can't tell whether the inequality needs to be reversed. (See Exercise 123.)

Move all terms to one side. We move the terms to the left-hand side and simplify using a common denominator.

$$\frac{1+x}{1-x} \geq 1 \qquad \text{Given inequality}$$

$$\frac{1+x}{1-x} - 1 \geq 0 \qquad \text{Subtract 1}$$

$$\frac{1+x}{1-x} - \frac{1-x}{1-x} \geq 0 \qquad \text{Common denominator } 1-x$$

$$\frac{1+x-1+x}{1-x} \geq 0 \qquad \text{Combine the fractions}$$

$$\frac{2x}{1-x} \geq 0 \qquad \text{Simplify}$$

Find the intervals. The factors of the left-hand side are $2x$ and $1 - x$. These are zero when x is 0 and 1. These numbers divide the real line into the intervals

$$(-\infty, 0), (0, 1), (1, \infty)$$

Make a diagram. We make the following diagram using test points to determine the sign of each factor in each interval.

	$(-\infty, 0)$	$(0, 1)$	$(1, \infty)$
Sign of $2x$	−	+	+
Sign of $1 - x$	+	+	−
Sign of $\dfrac{2x}{1-x}$	−	+	−

Solve. From the diagram we see that $\dfrac{2x}{1-x} \geq 0$ for x in the interval $[0, 1)$. We include the endpoint 0 because the original inequality requires that the quotient be greater than *or equal to* 1. However, we do not include the other endpoint 1 because the quotient in the inequality is not defined at 1. So the solution set is the interval

$$[0, 1)$$

FIGURE 7

The solution set is graphed in Figure 7.

NOW TRY EXERCISE **59**

Example 5 shows that we should always check the endpoints of the solution set to see whether they satisfy the original inequality.

Absolute Value Inequalities

We use the following properties to solve inequalities that involve absolute value.

These properties hold when x is replaced by any algebraic expression. (In the figures we assume that $c > 0$.)

PROPERTIES OF ABSOLUTE VALUE INEQUALITIES

Inequality	Equivalent form	Graph
1. $\lvert x \rvert < c$	$-c < x < c$	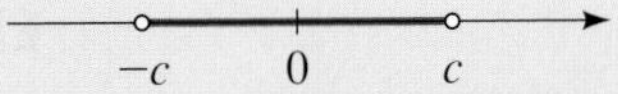
2. $\lvert x \rvert \le c$	$-c \le x \le c$	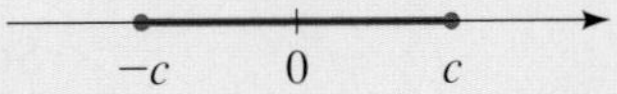
3. $\lvert x \rvert > c$	$x < -c$ or $c < x$	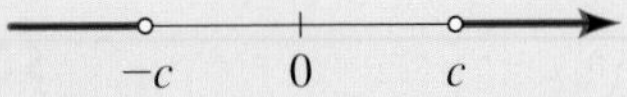
4. $\lvert x \rvert \ge c$	$x \le -c$ or $c \le x$	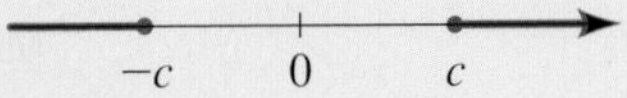

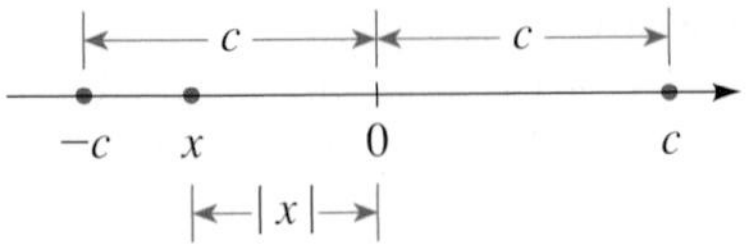

FIGURE 8

These properties can be proved using the definition of absolute value. To prove Property 1, for example, note that the inequality $\lvert x \rvert < c$ says that the distance from x to 0 is less than c, and from Figure 8 you can see that this is true if and only if x is between $-c$ and c.

EXAMPLE 6 | Solving an Absolute Value Inequality

Solve the inequality $\lvert x - 5 \rvert < 2$.

SOLUTION 1 The inequality $\lvert x - 5 \rvert < 2$ is equivalent to

$$-2 < x - 5 < 2 \qquad \text{Property 1}$$

$$3 < x < 7 \qquad \text{Add 5}$$

The solution set is the open interval $(3, 7)$.

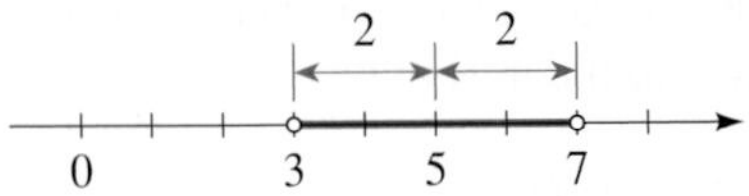

FIGURE 9

SOLUTION 2 Geometrically, the solution set consists of all numbers x whose distance from 5 is less than 2. From Figure 9 we see that this is the interval $(3,7)$.

NOW TRY EXERCISE **79** ■

EXAMPLE 7 | Solving an Absolute Value Inequality

Solve the inequality $\lvert 3x + 2 \rvert \ge 4$.

SOLUTION By Property 4 the inequality $\lvert 3x + 2 \rvert \ge 4$ is equivalent to

$$3x + 2 \ge 4 \quad \text{or} \quad 3x + 2 \le -4$$

$$3x \ge 2 \qquad 3x \le -6 \qquad \text{Subtract 2}$$

$$x \ge \tfrac{2}{3} \qquad x \le -2 \qquad \text{Divide by 3}$$

So the solution set is

$$\{x \mid x \le -2 \quad \text{or} \quad x \ge \tfrac{2}{3}\} = (-\infty, -2] \cup [\tfrac{2}{3}, \infty)$$

The set is graphed in Figure 10.

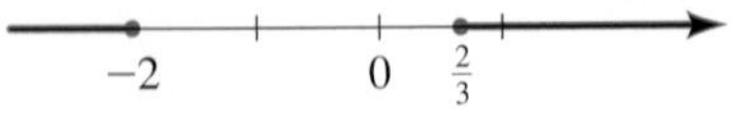

FIGURE 10

NOW TRY EXERCISE **83** ■

Modeling with Inequalities

Modeling real-life problems frequently leads to inequalities because we are often interested in determining when one quantity is more (or less) than another.

EXAMPLE 8 | Carnival Tickets

A carnival has two plans for tickets.

Plan A: $5 entrance fee and 25¢ each ride

Plan B: $2 entrance fee and 50¢ each ride

How many rides would you have to take for Plan A to be less expensive than Plan B?

SOLUTION Identify the variable. We are asked for the number of rides for which Plan A is less expensive than Plan B. So let

$$x = \text{number of rides}$$

Translate from words to algebra. The information in the problem may be organized as follows.

In Words	In Algebra
Number of rides	x
Cost with Plan A	$5 + 0.25x$
Cost with Plan B	$2 + 0.50x$

Set up the model. Now we set up the model.

cost with Plan A < cost with Plan B

$$5 + 0.25x < 2 + 0.50x$$

Solve. Now we solve for x.

$$3 + 0.25x < 0.50x \qquad \text{Subtract 2}$$

$$3 < 0.25x \qquad \text{Subtract } 0.25x$$

$$12 < x \qquad \text{Divide by 0.25}$$

So if you plan to take *more than* 12 rides, Plan A is less expensive.

NOW TRY EXERCISE **107**

EXAMPLE 9 | Relationship Between Fahrenheit and Celsius Scales

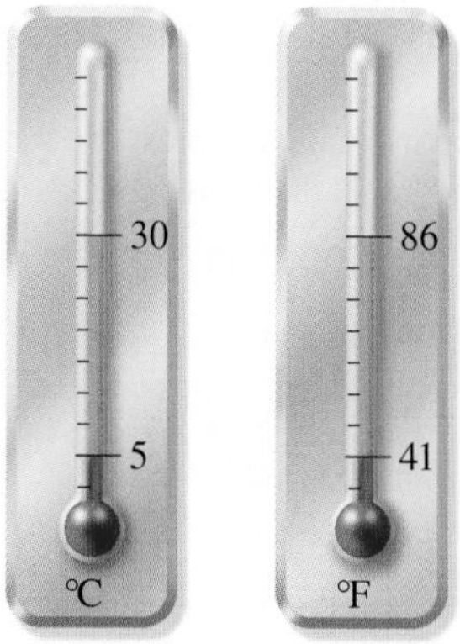

The instructions on a bottle of medicine indicate that the bottle should be stored at a temperature between 5 °C and 30 °C. What range of temperatures does this correspond to on the Fahrenheit scale?

SOLUTION The relationship between degrees Celsius (C) and degrees Fahrenheit (F) is given by the equation $C = \frac{5}{9}(F - 32)$. Expressing the statement on the bottle in terms of inequalities, we have

$$5 < C < 30$$

So the corresponding Fahrenheit temperatures satisfy the inequalities

$$5 < \tfrac{5}{9}(F - 32) < 30 \qquad \text{Substitute } C = \tfrac{5}{9}(F - 32)$$

$$\tfrac{9}{5} \cdot 5 < F - 32 < \tfrac{9}{5} \cdot 30 \qquad \text{Multiply by } \tfrac{9}{5}$$

$$9 < F - 32 < 54 \qquad \text{Simplify}$$

$$9 + 32 < F < 54 + 32 \qquad \text{Add 32}$$

$$41 < F < 86 \qquad \text{Simplify}$$

The medicine should be stored at a temperature between 41°F and 86°F.

NOW TRY EXERCISE **105**

1.7 EXERCISES

CONCEPTS

1. Fill in the blank with an appropriate inequality sign.

(a) If $x < 5$, then $x - 3$ _____ 2.

(b) If $x \le 5$, then $3x$ _____ 15.

(c) If $x \ge 2$, then $-3x$ _____ -6.

(d) If $x < -2$, then $-x$ _____ 2.

2. *True or false?*
(a) If $x(x + 1) > 0$, then x and $x + 1$ are either both positive or both negative.
(b) If $x(x + 1) > 5$, then x and $x + 1$ are each greater than 5.

3. (a) The solution of the inequality $|x| \le 3$ is the interval _____.

(b) The solution of the inequality $|x| \ge 3$ is a union of two intervals _____ ∪ _____.

4. (a) The set of all points on the real line whose distance from zero is less than 3 can be described by the absolute value inequality $|x|$_____.

(b) The set of all points on the real line whose distance from zero is greater than 3 can be described by the absolute value inequality $|x|$ _____.

SKILLS

5–10 ■ Let $S = \{-2, -1, 0, \frac{1}{2}, 1, \sqrt{2}, 2, 4\}$. Determine which elements of S satisfy the inequality.

5. $3 - 2x \le \frac{1}{2}$
6. $2x - 1 \ge x$
7. $1 < 2x - 4 \le 7$
8. $-2 \le 3 - x < 2$
9. $\frac{1}{x} \le \frac{1}{2}$
10. $x^2 + 2 < 4$

11–34 ■ Solve the linear inequality. Express the solution using interval notation and graph the solution set.

11. $2x \le 7$
12. $-4x \ge 10$
13. $2x - 5 > 3$
14. $3x + 11 < 5$
15. $7 - x \ge 5$
16. $5 - 3x \le -16$
17. $2x + 1 < 0$
18. $0 < 5 - 2x$
19. $3x + 11 \le 6x + 8$
20. $6 - x \ge 2x + 9$
21. $\frac{1}{2}x - \frac{2}{3} > 2$
22. $\frac{2}{5}x + 1 < \frac{1}{5} - 2x$
23. $\frac{1}{3}x + 2 < \frac{1}{6}x - 1$
24. $\frac{2}{3} - \frac{1}{2}x \ge \frac{1}{6} + x$
25. $4 - 3x \le -(1 + 8x)$
26. $2(7x - 3) \le 12x + 16$
27. $2 \le x + 5 < 4$
28. $5 \le 3x - 4 \le 14$
29. $-1 < 2x - 5 < 7$
30. $1 < 3x + 4 \le 16$
31. $-2 < 8 - 2x \le -1$
32. $-3 \le 3x + 7 \le \frac{1}{2}$
33. $\frac{1}{6} < \frac{2x - 13}{12} \le \frac{2}{3}$
34. $-\frac{1}{2} \le \frac{4 - 3x}{5} \le \frac{1}{4}$

35–72 ■ Solve the nonlinear inequality. Express the solution using interval notation and graph the solution set.

35. $(x + 2)(x - 3) < 0$
36. $(x - 5)(x + 4) \ge 0$
37. $x(2x + 7) \ge 0$
38. $x(2 - 3x) \le 0$
39. $x^2 - 3x - 18 \le 0$
40. $x^2 + 5x + 6 > 0$
41. $2x^2 + x \ge 1$
42. $x^2 < x + 2$
43. $3x^2 - 3x < 2x^2 + 4$
44. $5x^2 + 3x \ge 3x^2 + 2$
45. $x^2 > 3(x + 6)$
46. $x^2 + 2x > 3$
47. $x^2 < 4$
48. $x^2 \ge 9$
49. $(x + 2)(x - 1)(x - 3) \le 0$
50. $(x - 5)(x - 2)(x + 1) > 0$
51. $(x - 4)(x + 2)^2 < 0$
52. $(x + 3)^2(x + 1) > 0$
53. $(x - 2)^2(x - 3)(x + 1) \le 0$
54. $x^2(x^2 - 1) \ge 0$
55. $x^3 - 4x > 0$
56. $16x \le x^3$
57. $\frac{x - 3}{x + 1} \ge 0$
58. $\frac{2x + 6}{x - 2} < 0$
59. $\frac{4x}{2x + 3} > 2$
60. $-2 < \frac{x + 1}{x - 3}$
61. $\frac{2x + 1}{x - 5} \le 3$
62. $\frac{3 + x}{3 - x} \ge 1$
63. $\frac{4}{x} < x$
64. $\frac{x}{x + 1} > 3x$
65. $1 + \frac{2}{x + 1} \le \frac{2}{x}$
66. $\frac{3}{x - 1} - \frac{4}{x} \ge 1$
67. $\frac{6}{x - 1} - \frac{6}{x} \ge 1$
68. $\frac{x}{2} \ge \frac{5}{x + 1} + 4$
69. $\frac{x + 2}{x + 3} < \frac{x - 1}{x - 2}$
70. $\frac{1}{x + 1} + \frac{1}{x + 2} \le 0$
71. $x^4 > x^2$
72. $x^5 > x^2$

73–88 ■ Solve the absolute value inequality. Express the answer using interval notation and graph the solution set.

73. $|x| \le 4$
74. $|3x| < 15$
75. $|2x| > 7$
76. $\frac{1}{2}|x| \ge 1$
77. $|x - 5| \le 3$
78. $|x + 1| \ge 1$
79. $|2x - 3| \le 0.4$
80. $|5x - 2| < 6$
81. $|3x - 2| \ge 5$
82. $|8x + 3| > 12$
83. $\left|\frac{x - 2}{3}\right| < 2$
84. $\left|\frac{x + 1}{2}\right| \ge 4$
85. $|x + 6| < 0.001$
86. $3 - |2x + 4| \le 1$
87. $8 - |2x - 1| \ge 6$
88. $7|x + 2| + 5 > 4$

89–92 ■ A phrase describing a set of real numbers is given. Express the phrase as an inequality involving an absolute value.

89. All real numbers x less than 3 units from 0

90. All real numbers x more than 2 units from 0

91. All real numbers x at least 5 units from 7

92. All real numbers x at most 4 units from 2

93–98 ■ A set of real numbers is graphed. Find an inequality involving an absolute value that describes the set.

93. 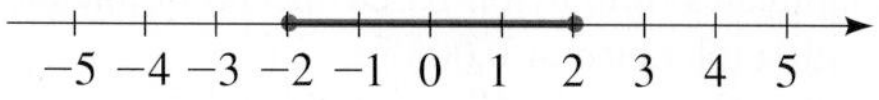

94. 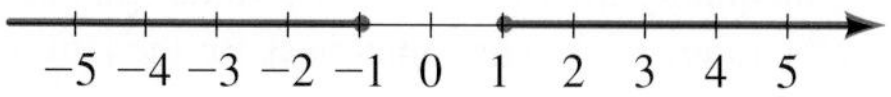

95. 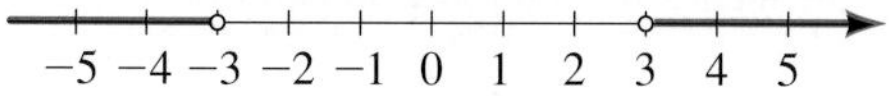

96. 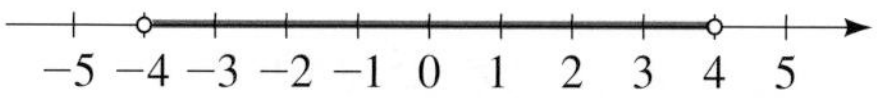

97. 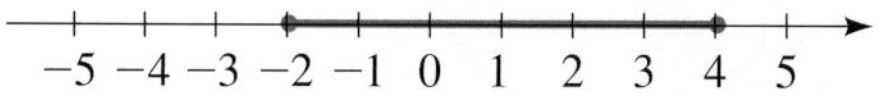

98. 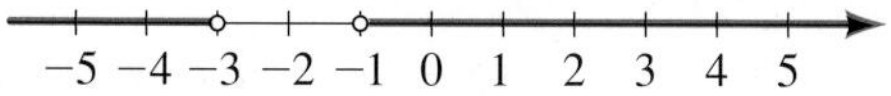

99–102 ■ Determine the values of the variable for which the expression is defined as a real number.

99. $\sqrt{16 - 9x^2}$ **100.** $\sqrt{3x^2 - 5x + 2}$

101. $\left(\dfrac{1}{x^2 - 5x - 14}\right)^{1/2}$ **102.** $\sqrt[4]{\dfrac{1 - x}{2 + x}}$

103. Solve the inequality for x, assuming that a, b, and c are positive constants.

(a) $a(bx - c) \geq bc$ (b) $a \leq bx + c < 2a$

104. Suppose that a, b, c, and d are positive numbers such that

$$\frac{a}{b} < \frac{c}{d}$$

Show that $\dfrac{a}{b} < \dfrac{a + c}{b + d} < \dfrac{c}{d}$.

APPLICATIONS

105. Temperature Scales Use the relationship between C and F given in Example 9 to find the interval on the Fahrenheit scale corresponding to the temperature range $20 \leq C \leq 30$.

106. Temperature Scales What interval on the Celsius scale corresponds to the temperature range $50 \leq F \leq 95$?

107. Car Rental Cost A car rental company offers two plans for renting a car.

Plan A: \$30 per day and 10¢ per mile

Plan B: \$50 per day with free unlimited mileage

For what range of miles will Plan B save you money?

108. Long-Distance Cost A telephone company offers two long-distance plans.

Plan A: \$25 per month and 5¢ per minute

Plan B: \$5 per month and 12¢ per minute

For how many minutes of long-distance calls would Plan B be financially advantageous?

109. Driving Cost It is estimated that the annual cost of driving a certain new car is given by the formula

$$C = 0.35m + 2200$$

where m represents the number of miles driven per year and C is the cost in dollars. Jane has purchased such a car and decides to budget between \$6400 and \$7100 for next year's driving costs. What is the corresponding range of miles that she can drive her new car?

110. Air Temperature As dry air moves upward, it expands and, in so doing, cools at a rate of about 1°C for each 100-meter rise, up to about 12 km.

(a) If the ground temperature is 20°C, write a formula for the temperature at height h.

(b) What range of temperatures can be expected if a plane takes off and reaches a maximum height of 5 km?

111. Airline Ticket Price A charter airline finds that on its Saturday flights from Philadelphia to London all 120 seats will be sold if the ticket price is \$200. However, for each \$3 increase in ticket price, the number of seats sold decreases by one.

(a) Find a formula for the number of seats sold if the ticket price is P dollars.

(b) Over a certain period the number of seats sold for this flight ranged between 90 and 115. What was the corresponding range of ticket prices?

112. Accuracy of a Scale A coffee merchant sells a customer 3 lb of Hawaiian Kona at \$6.50 per pound. The merchant's scale is accurate to within ±0.03 lb. By how much could the customer have been overcharged or undercharged because of possible inaccuracy in the scale?

113. Gravity The gravitational force F exerted by the earth on an object having a mass of 100 kg is given by the equation

$$F = \frac{4{,}000{,}000}{d^2}$$

where d is the distance (in km) of the object from the center of the earth, and the force F is measured in newtons (N). For what distances will the gravitational force exerted by the earth on this object be between 0.0004 N and 0.01 N?

114. Bonfire Temperature In the vicinity of a bonfire the temperature T in °C at a distance of x meters from the center of the fire was given by

$$T = \frac{600{,}000}{x^2 + 300}$$

At what range of distances from the fire's center was the temperature less than 500°C?

115. Falling Ball Using calculus, it can be shown that if a ball is thrown upward with an initial velocity of 16 ft/s from the top of a building 128 ft high, then its height h above the ground t seconds later will be

$$h = 128 + 16t - 16t^2$$

During what time interval will the ball be at least 32 ft above the ground?

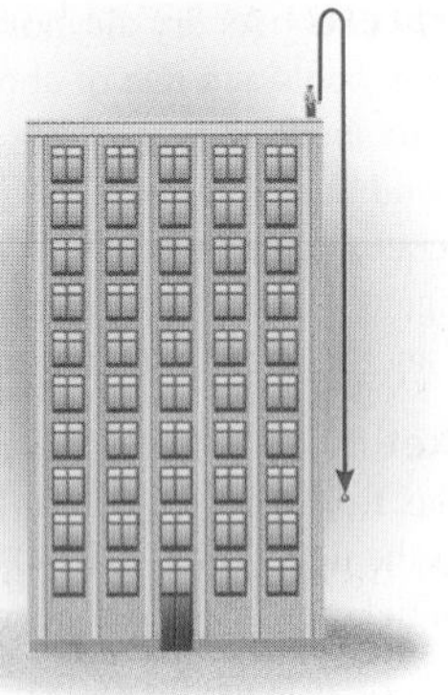

116. Gas Mileage The gas mileage g (measured in mi/gal) for a particular vehicle, driven at v mi/h, is given by the formula $g = 10 + 0.9v - 0.01v^2$, as long as v is between 10 mi/h and 75 mi/h. For what range of speeds is the vehicle's mileage 30 mi/gal or better?

117. Stopping Distance For a certain model of car the distance d required to stop the vehicle if it is traveling at v mi/h is given by the formula

$$d = v + \frac{v^2}{20}$$

where d is measured in feet. Kerry wants her stopping distance not to exceed 240 ft. At what range of speeds can she travel?

118. Manufacturer's Profit If a manufacturer sells x units of a certain product, revenue R and cost C (in dollars) are given by

$$R = 20x$$
$$C = 2000 + 8x + 0.0025x^2$$

Use the fact that

$$\text{profit} = \text{revenue} - \text{cost}$$

to determine how many units the manufacturer should sell to enjoy a profit of at least \$2400.

119. Fencing a Garden A determined gardener has 120 ft of deer-resistant fence. She wants to enclose a rectangular vegetable garden in her backyard, and she wants the area that is enclosed to be at least 800 ft^2. What range of values is possible for the length of her garden?

120. Thickness of a Laminate A company manufactures industrial laminates (thin nylon-based sheets) of thickness 0.020 in, with a tolerance of 0.003 in.

(a) Find an inequality involving absolute values that describes the range of possible thickness for the laminate.

(b) Solve the inequality you found in part (a).

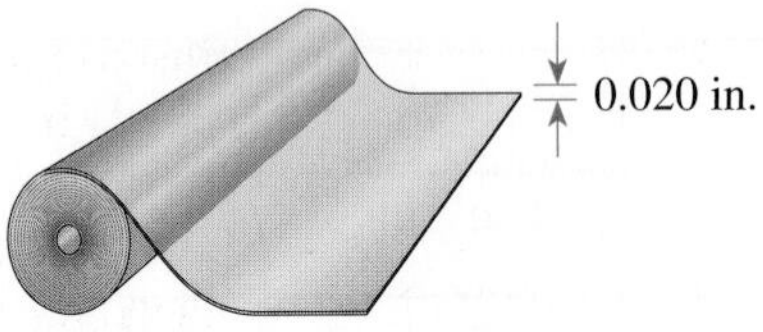

121. Range of Height The average height of adult males is 68.2 in, and 95% of adult males have height h that satisfies the inequality

$$\left| \frac{h - 68.2}{2.9} \right| \le 2$$

Solve the inequality to find the range of heights.

DISCOVERY ■ DISCUSSION ■ WRITING

122. Do Powers Preserve Order? If $a < b$, is $a^2 < b^2$? (Check both positive and negative values for a and b.) If $a < b$, is $a^3 < b^3$? On the basis of your observations, state a general rule about the relationship between a^n and b^n when $a < b$ and n is a positive integer.

123. What's Wrong Here? It is tempting to try to solve an inequality like an equation. For instance, we might try to solve $1 < 3/x$ by multiplying both sides by x, to get $x < 3$, so the solution would be $(-\infty, 3)$. But that's wrong; for example, $x = -1$ lies in this interval but does not satisfy the original inequality. Explain why this method doesn't work (think about the *sign* of x). Then solve the inequality correctly.

124. Using Distances to Solve Absolute Value Inequalities Recall that $|a - b|$ is the distance between a and b on the number line. For any number x, what do $|x - 1|$ and $|x - 3|$ represent? Use this interpretation to solve the inequality $|x - 1| < |x - 3|$ geometrically. In general, if $a < b$, what is the solution of the inequality $|x - a| < |x - b|$?

1.8 Coordinate Geometry

The Coordinate Plane ▶ The Distance and Midpoint Formulas ▶ Graphs of Equations in Two Variables ▶ Intercepts ▶ Circles ▶ Symmetry

The *coordinate plane* is the link between algebra and geometry. In the coordinate plane we can draw graphs of algebraic equations. The graphs, in turn, allow us to "see" the relationship between the variables in the equation. In this section we study the coordinate plane.

▼ The Coordinate Plane

The Cartesian plane is named in honor of the French mathematician René Descartes (1596–1650), although another Frenchman, Pierre Fermat (1601–1665), also invented the principles of coordinate geometry at the same time. (See their biographies on pages 181 and 99.)

Just as points on a line can be identified with real numbers to form the coordinate line, points in a plane can be identified with ordered pairs of numbers to form the **coordinate plane** or **Cartesian plane**. To do this, we draw two perpendicular real lines that intersect at 0 on each line. Usually, one line is horizontal with positive direction to the right and is called the ***x*-axis**; the other line is vertical with positive direction upward and is called the ***y*-axis**. The point of intersection of the x-axis and the y-axis is the **origin *O***, and the two axes divide the plane into four **quadrants**, labeled I, II, III, and IV in Figure 1. (The points *on* the coordinate axes are not assigned to any quadrant.)

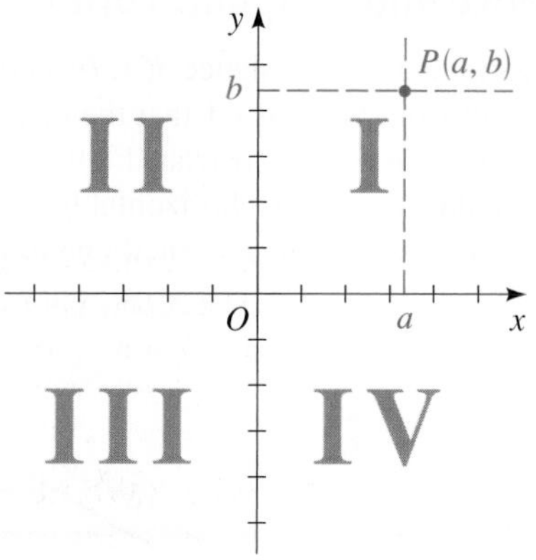

FIGURE 1

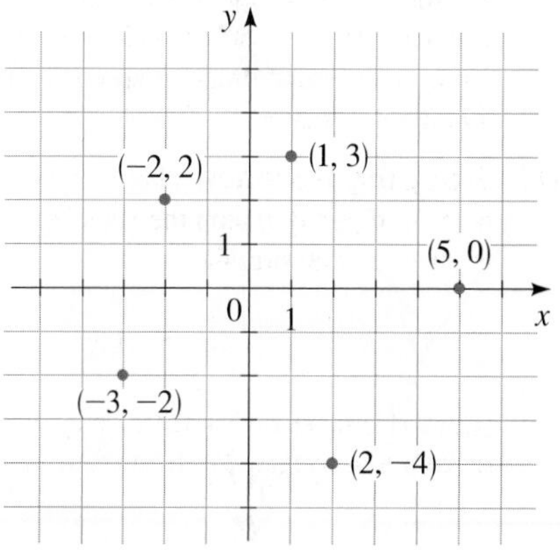

FIGURE 2

Although the notation for a point (a, b) is the same as the notation for an open interval (a, b), the context should make clear which meaning is intended.

Any point P in the coordinate plane can be located by a unique **ordered pair** of numbers (a, b), as shown in Figure 1. The first number a is called the ***x*-coordinate** of P; the second number b is called the ***y*-coordinate** of P. We can think of the coordinates of P as its "address," because they specify its location in the plane. Several points are labeled with their coordinates in Figure 2.

EXAMPLE 1 | Graphing Regions in the Coordinate Plane

Describe and sketch the regions given by each set.

(a) $\{(x, y) \mid x \geq 0\}$ **(b)** $\{(x, y) \mid y = 1\}$ **(c)** $\{(x, y) \mid |y| < 1\}$

SOLUTION

(a) The points whose x-coordinates are 0 or positive lie on the y-axis or to the right of it, as shown in Figure 3(a).

(b) The set of all points with y-coordinate 1 is a horizontal line one unit above the x-axis, as in Figure 3(b).

Coordinates as Addresses
The coordinates of a point in the xy-plane uniquely determine its location. We can think of the coordinates as the "address" of the point. In Salt Lake City, Utah, the addresses of most buildings are in fact expressed as coordinates. The city is divided into quadrants with Main Street as the vertical (North-South) axis and S. Temple Street as the horizontal (East-West) axis. An address such as

1760 W   2100 S

indicates a location 17.6 blocks west of Main Street and 21 blocks south of S. Temple Street. (This is the address of the main post office in Salt Lake City.) With this logical system it is possible for someone unfamiliar with the city to locate any address immediately, as easily as one locates a point in the coordinate plane.

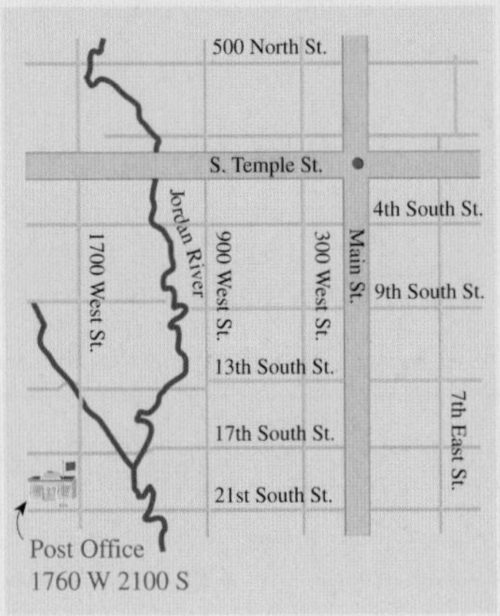

(c) Recall from Section 1.7 that

$$|y| < 1 \quad \text{if and only if} \quad -1 < y < 1$$

So the given region consists of those points in the plane whose y-coordinates lie between -1 and 1. Thus, the region consists of all points that lie between (but not on) the horizontal lines $y = 1$ and $y = -1$. These lines are shown as broken lines in Figure 3(c) to indicate that the points on these lines do not lie in the set.

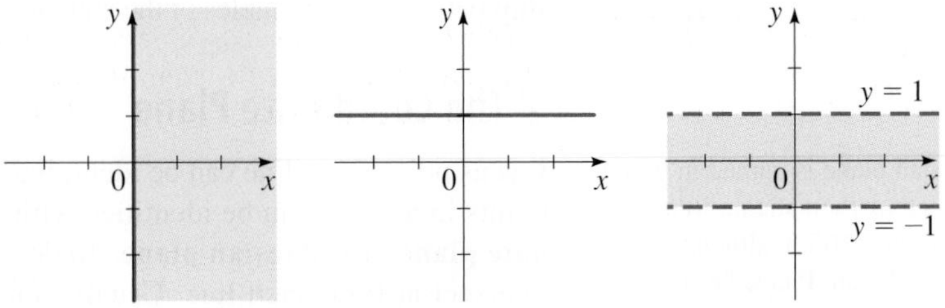

(a) $x \geq 0$ (b) $y = 1$ (c) $|y| < 1$

FIGURE 3

NOW TRY EXERCISES **23, 25,** AND **29**

▼ The Distance and Midpoint Formulas

We now find a formula for the distance $d(A, B)$ between two points $A(x_1, y_1)$ and $B(x_2, y_2)$ in the plane. Recall from Section 1.1 that the distance between points a and b on a number line is $d(a, b) = |b - a|$. So from Figure 4 we see that the distance between the points $A(x_1, y_1)$ and $C(x_2, y_1)$ on a horizontal line must be $|x_2 - x_1|$, and the distance between $B(x_2, y_2)$ and $C(x_2, y_1)$ on a vertical line must be $|y_2 - y_1|$.

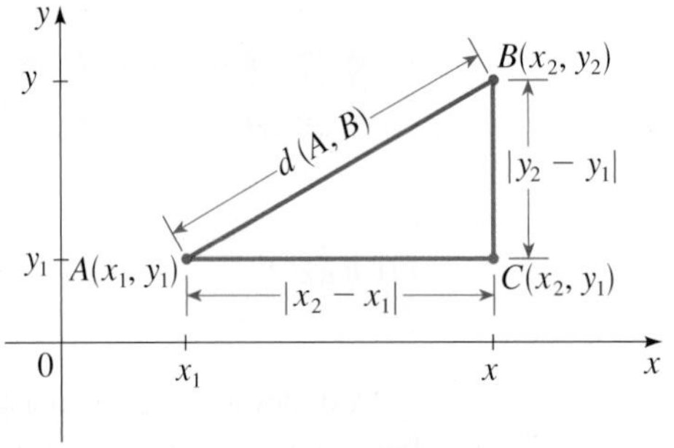

FIGURE 4

Since triangle ABC is a right triangle, the Pythagorean Theorem gives

$$d(A, B) = \sqrt{|x_2 - x_1|^2 + |y_2 - y_1|^2} = \sqrt{(x_2 - x_1)^2 + (y_2 - y_1)^2}$$

DISTANCE FORMULA

The distance between the points $A(x_1, y_1)$ and $B(x_2, y_2)$ in the plane is

$$d(A, B) = \sqrt{(x_2 - x_1)^2 + (y_2 - y_1)^2}$$

EXAMPLE 2 | Applying the Distance Formula

Which of the points $P(1, -2)$ or $Q(8, 9)$ is closer to the point $A(5, 3)$?

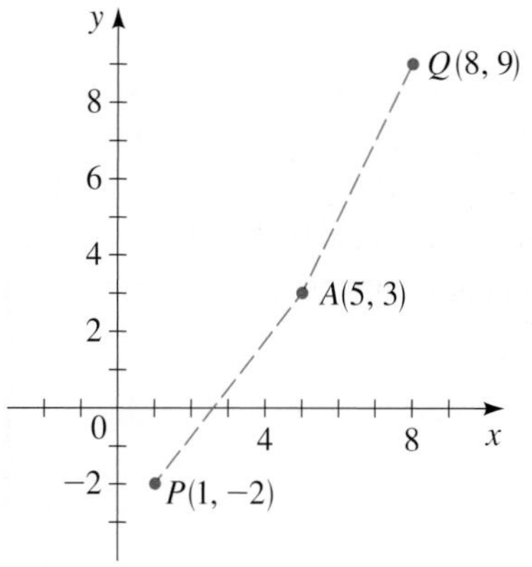

FIGURE 5

SOLUTION By the Distance Formula we have

$$d(P, A) = \sqrt{(5-1)^2 + [3-(-2)]^2} = \sqrt{4^2 + 5^2} = \sqrt{41}$$
$$d(Q, A) = \sqrt{(5-8)^2 + (3-9)^2} = \sqrt{(-3)^2 + (-6)^2} = \sqrt{45}$$

This shows that $d(P, A) < d(Q, A)$, so P is closer to A (see Figure 5).

NOW TRY EXERCISE **33** ■

Now let's find the coordinates (x, y) of the midpoint M of the line segment that joins the point $A(x_1, y_1)$ to the point $B(x_2, y_2)$. In Figure 6 notice that triangles APM and MQB are congruent because $d(A, M) = d(M, B)$ and the corresponding angles are equal.

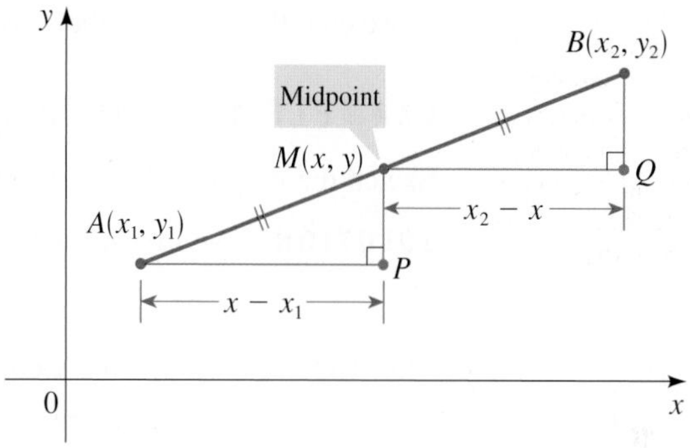

FIGURE 6

It follows that $d(A, P) = d(M, Q)$, so

$$x - x_1 = x_2 - x$$

Solving this equation for x, we get $2x = x_1 + x_2$, so $x = \dfrac{x_1 + x_2}{2}$. Similarly, $y = \dfrac{y_1 + y_2}{2}$.

> **MIDPOINT FORMULA**
>
> The midpoint of the line segment from $A(x_1, y_1)$ to $B(x_2, y_2)$ is
>
> $$\left(\frac{x_1 + x_2}{2}, \frac{y_1 + y_2}{2}\right)$$

EXAMPLE 3 | Applying the Midpoint Formula

Show that the quadrilateral with vertices $P(1, 2)$, $Q(4, 4)$, $R(5, 9)$, and $S(2, 7)$ is a parallelogram by proving that its two diagonals bisect each other.

SOLUTION If the two diagonals have the same midpoint, then they must bisect each other. The midpoint of the diagonal PR is

$$\left(\frac{1+5}{2}, \frac{2+9}{2}\right) = \left(3, \frac{11}{2}\right)$$

and the midpoint of the diagonal QS is

$$\left(\frac{4+2}{2}, \frac{4+7}{2}\right) = \left(3, \frac{11}{2}\right)$$

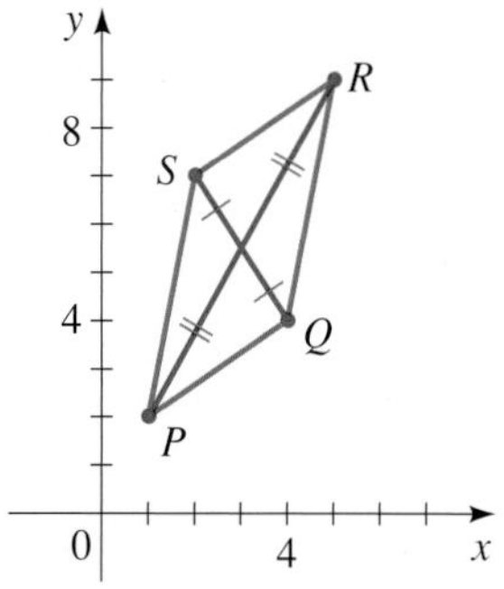

FIGURE 7

so each diagonal bisects the other, as shown in Figure 7. (A theorem from elementary geometry states that the quadrilateral is therefore a parallelogram.)

NOW TRY EXERCISE **37** ■

▼ Graphs of Equations in Two Variables

Fundamental Principle of Analytic Geometry

A point (x, y) lies on the graph of an equation if and only if its coordinates satisfy the equation.

An **equation in two variables**, such as $y = x^2 + 1$, expresses a relationship between two quantities. A point (x, y) **satisfies** the equation if it makes the equation true when the values for x and y are substituted into the equation. For example, the point $(3, 10)$ satisfies the equation $y = x^2 + 1$ because $10 = 3^2 + 1$, but the point $(1, 3)$ does not, because $3 \neq 1^2 + 1$.

> **THE GRAPH OF AN EQUATION**
>
> The **graph** of an equation in x and y is the set of all points (x, y) in the coordinate plane that satisfy the equation.

The graph of an equation is a curve, so to graph an equation, we plot as many points as we can, then connect them by a smooth curve.

EXAMPLE 4 | Sketching a Graph by Plotting Points

Sketch the graph of the equation $2x - y = 3$.

SOLUTION We first solve the given equation for y to get

$$y = 2x - 3$$

This helps us calculate the y-coordinates in the following table.

x	$y = 2x - 3$	(x, y)
-1	-5	$(-1, -5)$
0	-3	$(0, -3)$
1	-1	$(1, -1)$
2	1	$(2, 1)$
3	3	$(3, 3)$
4	5	$(4, 5)$

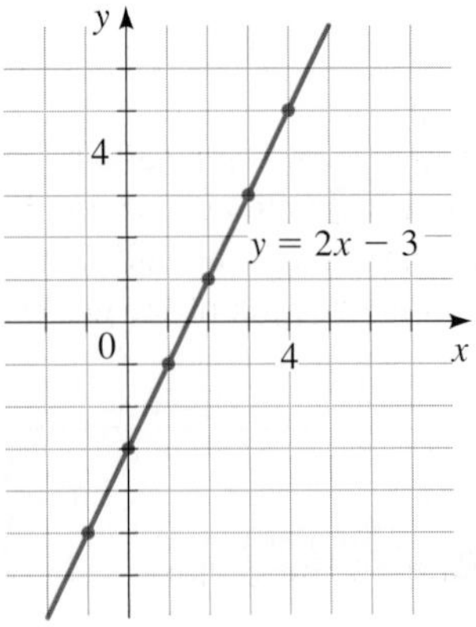

FIGURE 8

Of course, there are infinitely many points on the graph, and it is impossible to plot all of them. But the more points we plot, the better we can imagine what the graph represented by the equation looks like. We plot the points we found in Figure 8; they appear to lie on a line. So we complete the graph by joining the points by a line. (In Section 1.10 we verify that the graph of this equation is indeed a line.)

NOW TRY EXERCISE **59** ■

A detailed discussion of parabolas and their geometric properties is presented in Chapter 10.

EXAMPLE 5 | Sketching a Graph by Plotting Points

Sketch the graph of the equation $y = x^2 - 2$.

SOLUTION We find some of the points that satisfy the equation in the following table. In Figure 9 we plot these points and then connect them by a smooth curve. A curve with this shape is called a *parabola*.

x	$y = x^2 - 2$	(x, y)
-3	7	$(-3, 7)$
-2	2	$(-2, 2)$
-1	-1	$(-1, -1)$
0	-2	$(0, -2)$
1	-1	$(1, -1)$
2	2	$(2, 2)$
3	7	$(3, 7)$

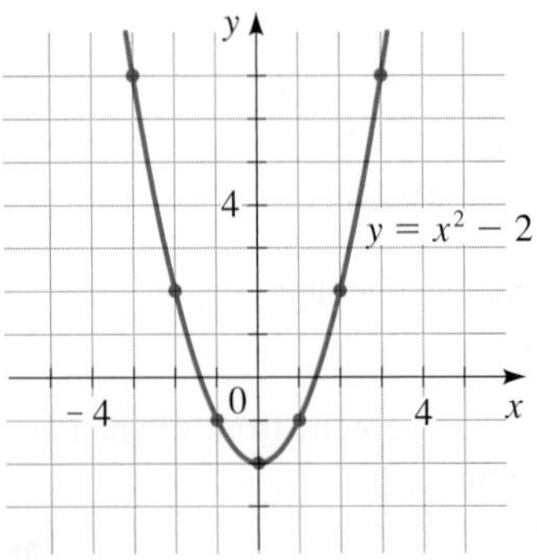

FIGURE 9

NOW TRY EXERCISE **63** ■

EXAMPLE 6 | Graphing an Absolute Value Equation

Sketch the graph of the equation $y = |x|$.

SOLUTION We make a table of values:

x	$y = \|x\|$	(x, y)
-3	3	$(-3, 3)$
-2	2	$(-2, 2)$
-1	1	$(-1, 1)$
0	0	$(0, 0)$
1	1	$(1, 1)$
2	2	$(2, 2)$
3	3	$(3, 3)$

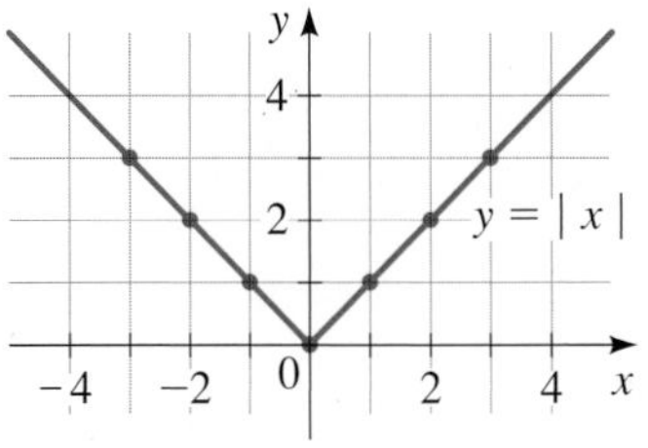

FIGURE 10

In Figure 10 we plot these points and use them to sketch the graph of the equation.

NOW TRY EXERCISE **75**

Intercepts

The x-coordinates of the points where a graph intersects the x-axis are called the **x-intercepts** of the graph and are obtained by setting $y = 0$ in the equation of the graph. The y-coordinates of the points where a graph intersects the y-axis are called the **y-intercepts** of the graph and are obtained by setting $x = 0$ in the equation of the graph.

DEFINITION OF INTERCEPTS

Intercepts	How to find them	Where they are on the graph
x-intercepts: The x-coordinates of points where the graph of an equation intersects the x-axis	Set $y = 0$ and solve for x	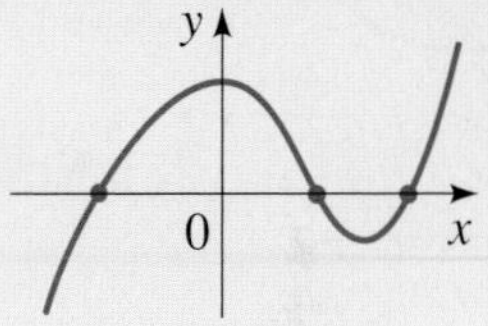
y-intercepts: The y-coordinates of points where the graph of an equation intersects the y-axis	Set $x = 0$ and solve for y	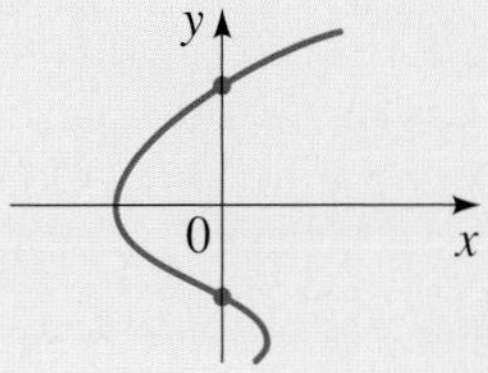

EXAMPLE 7 | Finding Intercepts

Find the x- and y-intercepts of the graph of the equation $y = x^2 - 2$.

SOLUTION To find the x-intercepts, we set $y = 0$ and solve for x. Thus

$$0 = x^2 - 2 \qquad \text{Set } y = 0$$

$$x^2 = 2 \qquad \text{Add 2 to each side}$$

$$x = \pm\sqrt{2} \qquad \text{Take the square root}$$

The x-intercepts are $\sqrt{2}$ and $-\sqrt{2}$.

To find the y-intercepts, we set $x = 0$ and solve for y. Thus

$$y = 0^2 - 2 \qquad \text{Set } x = 0$$
$$y = -2$$

The y-intercept is -2.

The graph of this equation was sketched in Example 5. It is repeated in Figure 11 with the x- and y-intercepts labeled.

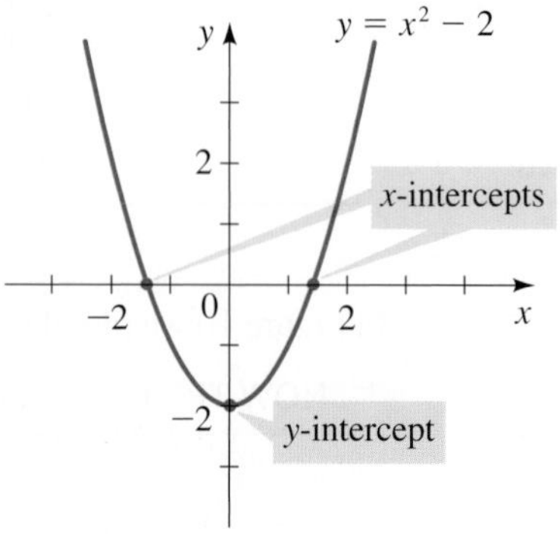

FIGURE 11

NOW TRY EXERCISE **65**

Circles

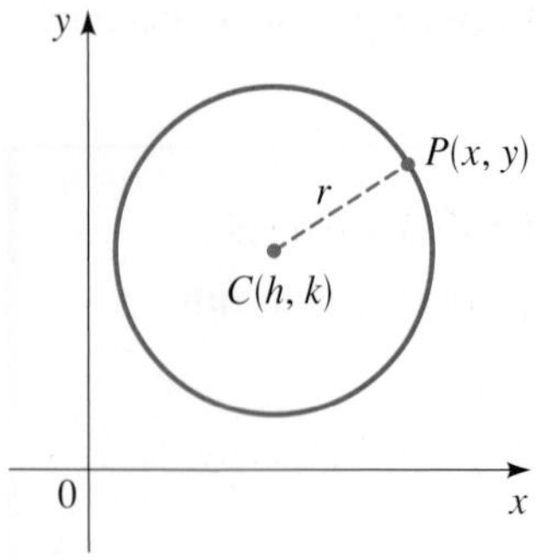

FIGURE 12

So far, we have discussed how to find the graph of an equation in x and y. The converse problem is to find an equation of a graph, that is, an equation that represents a given curve in the xy-plane. Such an equation is satisfied by the coordinates of the points on the curve and by no other point. This is the other half of the fundamental principle of analytic geometry as formulated by Descartes and Fermat. The idea is that if a geometric curve can be represented by an algebraic equation, then the rules of algebra can be used to analyze the curve.

As an example of this type of problem, let's find the equation of a circle with radius r and center (h, k). By definition the circle is the set of all points $P(x, y)$ whose distance from the center $C(h, k)$ is r (see Figure 12). Thus P is on the circle if and only if $d(P, C) = r$. From the distance formula we have

$$\sqrt{(x - h)^2 + (y - k)^2} = r$$
$$(x - h)^2 + (y - k)^2 = r^2 \qquad \text{Square each side}$$

This is the desired equation.

EQUATION OF A CIRCLE

An equation of the circle with center (h, k) and radius r is

$$(x - h)^2 + (y - k)^2 = r^2$$

This is called the **standard form** for the equation of the circle. If the center of the circle is the origin $(0, 0)$, then the equation is

$$x^2 + y^2 = r^2$$

EXAMPLE 8 | Graphing a Circle

Graph each equation.

(a) $x^2 + y^2 = 25$ **(b)** $(x - 2)^2 + (y + 1)^2 = 25$

SOLUTION

(a) Rewriting the equation as $x^2 + y^2 = 5^2$, we see that this is an equation of the circle of radius 5 centered at the origin. Its graph is shown in Figure 13.

(b) Rewriting the equation as $(x - 2)^2 + (y + 1)^2 = 5^2$, we see that this is an equation of the circle of radius 5 centered at $(2, -1)$. Its graph is shown in Figure 14.

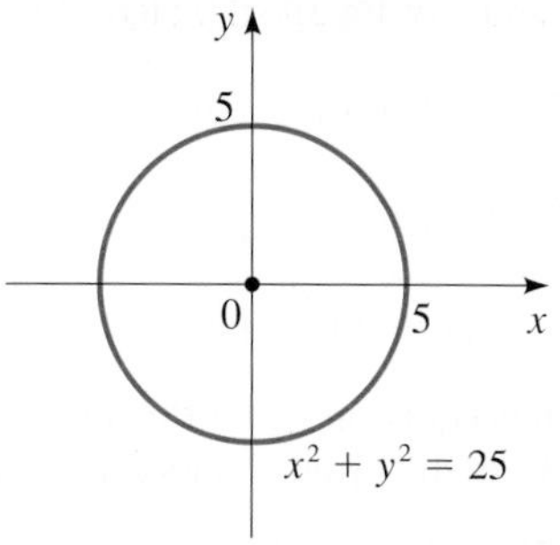

FIGURE 13

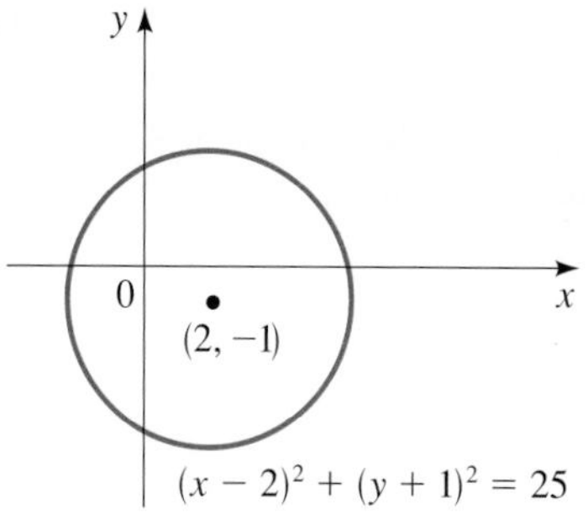

FIGURE 14

NOW TRY EXERCISES **87** AND **89**

EXAMPLE 9 | Finding an Equation of a Circle

(a) Find an equation of the circle with radius 3 and center $(2, -5)$.

(b) Find an equation of the circle that has the points $P(1, 8)$ and $Q(5, -6)$ as the endpoints of a diameter.

SOLUTION

(a) Using the equation of a circle with $r = 3$, $h = 2$, and $k = -5$, we obtain

$$(x - 2)^2 + (y + 5)^2 = 9$$

The graph is shown in Figure 15.

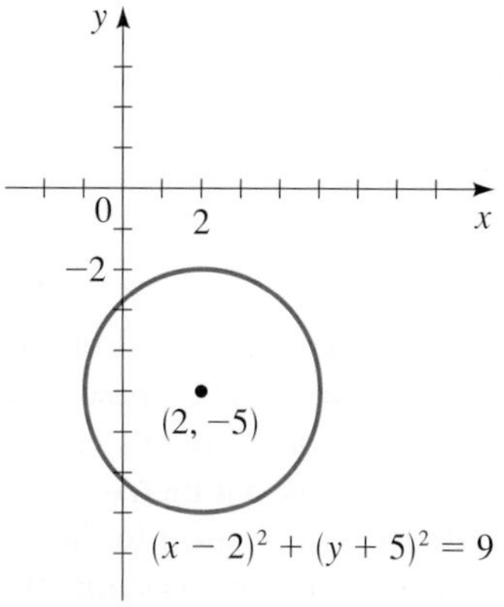

FIGURE 15

(b) We first observe that the center is the midpoint of the diameter PQ, so by the Midpoint Formula the center is

$$\left(\frac{1 + 5}{2}, \frac{8 - 6}{2}\right) = (3, 1)$$

The radius r is the distance from P to the center, so by the Distance Formula

$$r^2 = (3 - 1)^2 + (1 - 8)^2 = 2^2 + (-7)^2 = 53$$

Therefore, the equation of the circle is

$$(x - 3)^2 + (y - 1)^2 = 53$$

The graph is shown in Figure 16.

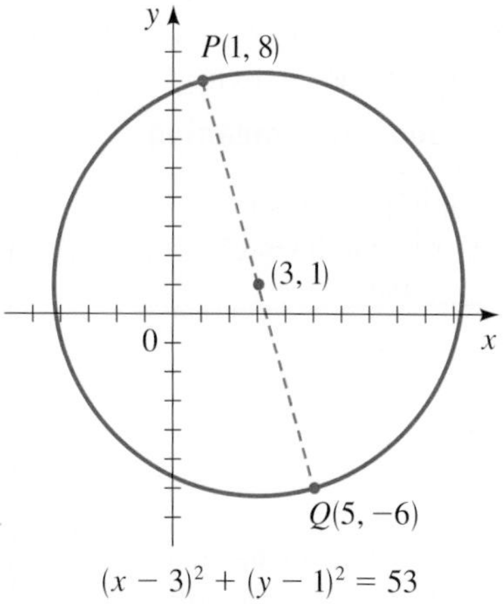

FIGURE 16

NOW TRY EXERCISES **93** AND **97**

Let's expand the equation of the circle in the preceding example.

$(x - 3)^2 + (y - 1)^2 = 53$	Standard form
$x^2 - 6x + 9 + y^2 - 2y + 1 = 53$	Expand the squares
$x^2 - 6x + y^2 - 2y = 43$	Subtract 10 to get expanded form

Suppose we are given the equation of a circle in expanded form. Then to find its center and radius, we must put the equation back in standard form. That means that we must reverse the steps in the preceding calculation, and to do that we need to know what to add to an expression like $x^2 - 6x$ to make it a perfect square—that is, we need to complete the square, as in the next example.

Completing the square is used in many contexts in algebra. In Section 1.5 we used completing the square to solve quadratic equations.

EXAMPLE 10 | Identifying an Equation of a Circle

Show that the equation $x^2 + y^2 + 2x - 6y + 7 = 0$ represents a circle, and find the center and radius of the circle.

SOLUTION We first group the x-terms and y-terms. Then we complete the square within each grouping. That is, we complete the square for $x^2 + 2x$ by adding $\left(\frac{1}{2}\cdot 2\right)^2 = 1$, and we complete the square for $y^2 - 6y$ by adding $\left[\frac{1}{2}\cdot(-6)\right]^2 = 9$.

We must add the same numbers to *each side* to maintain equality.

$$(x^2 + 2x \qquad) + (y^2 - 6y \qquad) = -7 \qquad \text{Group terms}$$

$$(x^2 + 2x + 1) + (y^2 - 6y + 9) = -7 + 1 + 9 \qquad \text{Complete the square by adding 1 and 9 to each side}$$

$$(x + 1)^2 + (y - 3)^2 = 3 \qquad \text{Factor and simplify}$$

Comparing this equation with the standard equation of a circle, we see that $h = -1$, $k = 3$, and $r = \sqrt{3}$, so the given equation represents a circle with center $(-1, 3)$ and radius $\sqrt{3}$.

NOW TRY EXERCISE **103**

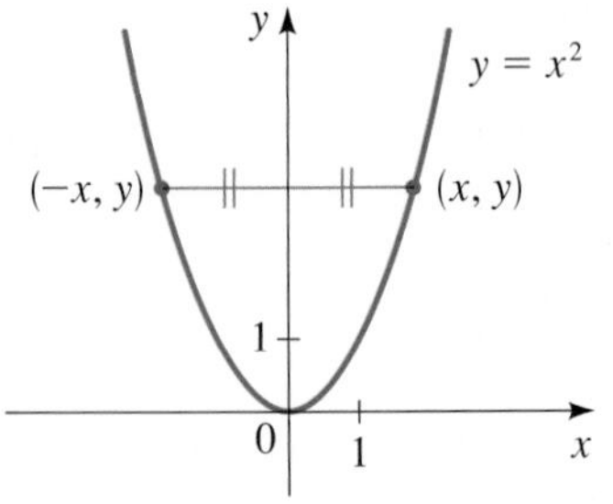

FIGURE 17

Symmetry

Figure 17 shows the graph of $y = x^2$. Notice that the part of the graph to the left of the y-axis is the mirror image of the part to the right of the y-axis. The reason is that if the point (x, y) is on the graph, then so is $(-x, y)$, and these points are reflections of each other about the y-axis. In this situation we say that the graph is **symmetric with respect to the *y*-axis**. Similarly, we say that a graph is **symmetric with respect to the *x*-axis** if whenever the point (x, y) is on the graph, then so is $(x, -y)$. A graph is **symmetric with respect to the origin** if whenever (x, y) is on the graph, so is $(-x, -y)$.

DEFINITION OF SYMMETRY

Type of symmetry	How to test for symmetry	What the graph looks like (figures in this section)	Geometric meaning
Symmetry with respect to the *x*-axis	The equation is unchanged when y is replaced by $-y$	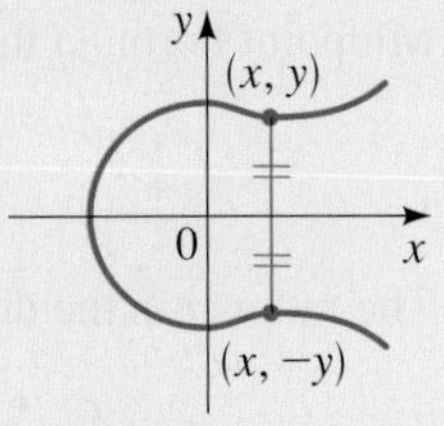 (Figures 13, 18)	Graph is unchanged when reflected in the x-axis
Symmetry with respect to the *y*-axis	The equation is unchanged when x is replaced by $-x$	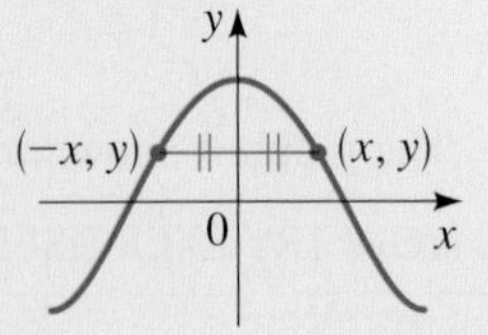 (Figures 9, 10, 11, 13, 17)	Graph is unchanged when reflected in the y-axis
Symmetry with respect to the origin	The equation is unchanged when x is replaced by $-x$ and y by $-y$	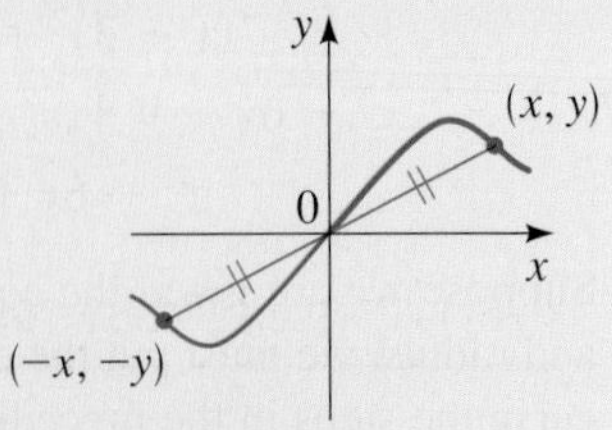 (Figures 13, 19)	Graph is unchanged when rotated 180° about the origin

The remaining examples in this section show how symmetry helps us sketch the graphs of equations.

EXAMPLE 11 | Using Symmetry to Sketch a Graph

Test the equation $x = y^2$ for symmetry and sketch the graph.

SOLUTION If y is replaced by $-y$ in the equation $x = y^2$, we get

$$x = (-y)^2 \qquad \text{Replace } y \text{ by } -y$$

$$x = y^2 \qquad \text{Simplify}$$

and so the equation is unchanged. Therefore, the graph is symmetric about the x-axis. But changing x to $-x$ gives the equation $-x = y^2$, which is not the same as the original equation, so the graph is not symmetric about the y-axis.

We use the symmetry about the x-axis to sketch the graph by first plotting points just for $y > 0$ and then reflecting the graph in the x-axis, as shown in Figure 18.

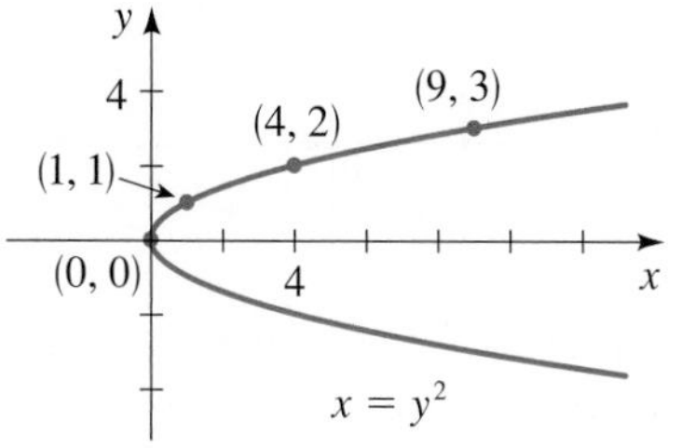

FIGURE 18

y	$x = y^2$	(x, y)
0	0	(0, 0)
1	1	(1, 1)
2	4	(4, 2)
3	9	(9, 3)

NOW TRY EXERCISE **77**

EXAMPLE 12 | Using Symmetry to Sketch a Graph

Test the equation $y = x^3 - 9x$ for symmetry and sketch its graph.

SOLUTION If we replace x by $-x$ and y by $-y$ in the equation, we get

$$-y = (-x)^3 - 9(-x) \qquad \text{Replace } x \text{ by } -x \text{ and } y \text{ by } -y$$

$$-y = -x^3 + 9x \qquad \text{Simplify}$$

$$y = x^3 - 9x \qquad \text{Multiply by } -1$$

and so the equation is unchanged. This means that the graph is symmetric with respect to the origin. We sketch it by first plotting points for $x > 0$ and then using symmetry about the origin (see Figure 19).

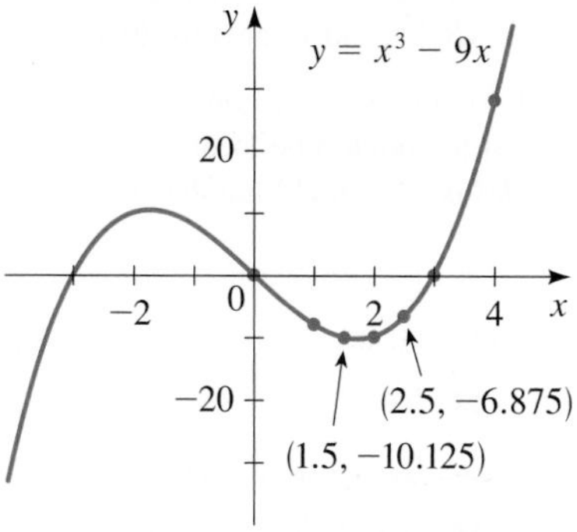

FIGURE 19

x	$y = x^3 - 9x$	(x, y)
0	0	(0, 0)
1	−8	(1, −8)
1.5	−10.125	(1.5, −10.125)
2	−10	(2, −10)
2.5	−6.875	(2.5, −6.875)
3	0	(3, 0)
4	28	(4, 28)

NOW TRY EXERCISE **79**

1.8 EXERCISES

CONCEPTS

1. The point that is 3 units to the right of the y-axis and 5 units below the x-axis has coordinates (____, ____).

2. The distance between the points (a, b) and (c, d) is ____________. So the distance between (1, 2) and (7, 10) is ________.

3. The point midway between (a, b) and (c, d) is ________. So the point midway between (1, 2) and (7, 10) is ________.

4. If the point (2, 3) is on the graph of an equation in x and y, then the equation is satisfied when we replace x by ________ and y by ________. Is the point (2, 3) on the graph of the equation $2y = x + 1$?

5. **(a)** To find the x-intercept(s) of the graph of an equation, we set ________ equal to 0 and solve for ________. So the x-intercept of $2y = x + 1$ is ________.

   (b) To find the y-intercept(s) of the graph of an equation, we set ________ equal to 0 and solve for ________. So the y-intercept of $2y = x + 1$ is ________.

6. The graph of the equation $(x - 1)^2 + (y - 2)^2 = 9$ is a circle with center (____, ____) and radius ________.

SKILLS

7. Plot the given points in a coordinate plane.

   $(2, 3), (-2, 3), (4, 5), (4, -5), (-4, 5), (-4, -5)$

8. Find the coordinates of the points shown in the figure.

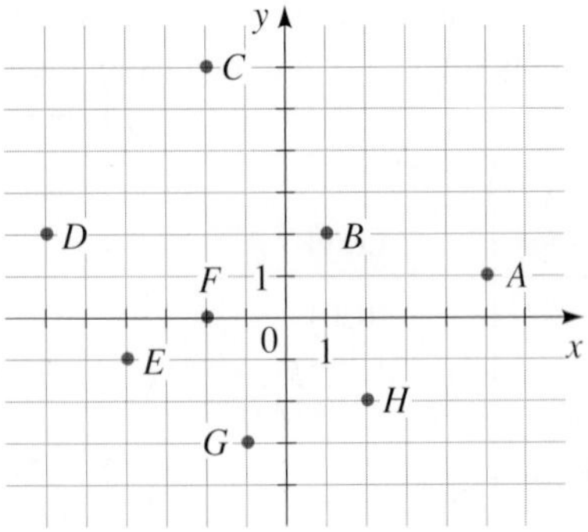

9–12 ■ A pair of points is graphed.
(a) Find the distance between them.
(b) Find the midpoint of the segment that joins them.

9.

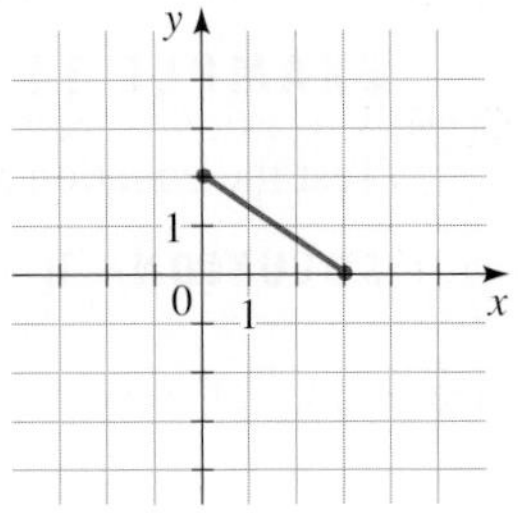

10.

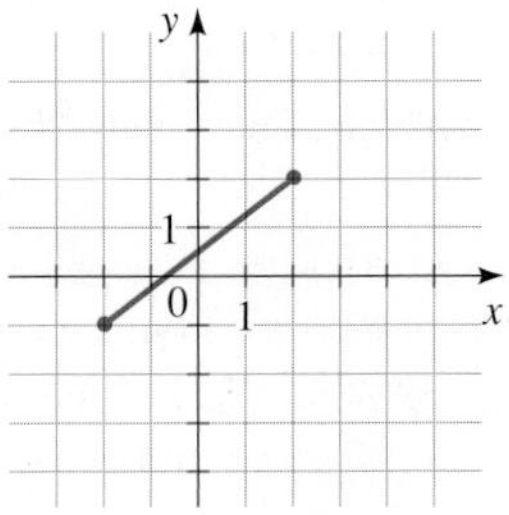

11.

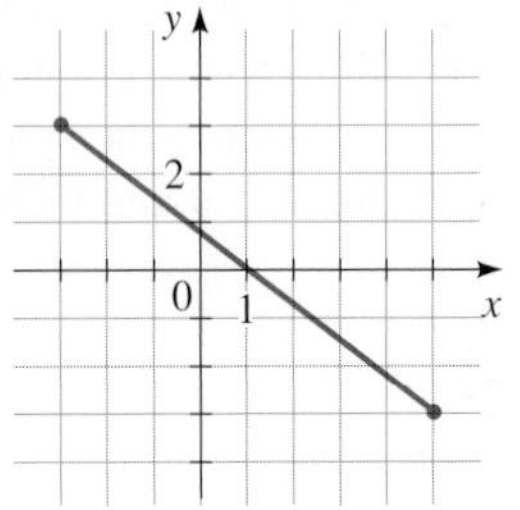

12.

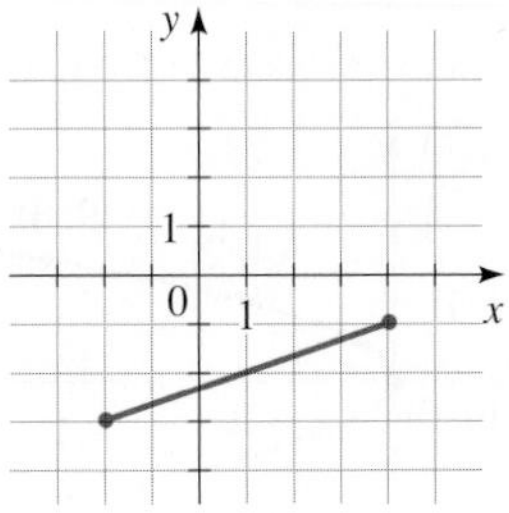

13–18 ■ A pair of points is graphed.
(a) Plot the points in a coordinate plane.
(b) Find the distance between them.
(c) Find the midpoint of the segment that joins them.

13. $(0, 8), (6, 16)$
14. $(-2, 5), (10, 0)$
15. $(-3, -6), (4, 18)$
16. $(-1, -1), (9, 9)$
17. $(6, -2), (-6, 2)$
18. $(0, -6), (5, 0)$

19. Draw the rectangle with vertices $A(1, 3)$, $B(5, 3)$, $C(1, -3)$, and $D(5, -3)$ on a coordinate plane. Find the area of the rectangle.

20. Draw the parallelogram with vertices $A(1, 2)$, $B(5, 2)$, $C(3, 6)$, and $D(7, 6)$ on a coordinate plane. Find the area of the parallelogram.

21. Plot the points $A(1, 0)$, $B(5, 0)$, $C(4, 3)$, and $D(2, 3)$ on a coordinate plane. Draw the segments AB, BC, CD, and DA. What kind of quadrilateral is $ABCD$, and what is its area?

22. Plot the points $P(5, 1)$, $Q(0, 6)$, and $R(-5, 1)$ on a coordinate plane. Where must the point S be located so that the quadrilateral $PQRS$ is a square? Find the area of this square.

23–32 ■ Sketch the region given by the set.

23. $\{(x, y) \mid x \geq 3\}$
24. $\{(x, y) \mid y < 3\}$
25. $\{(x, y) \mid y = 2\}$
26. $\{(x, y) \mid x = -1\}$
27. $\{(x, y) \mid 1 < x < 2\}$
28. $\{(x, y) \mid 0 \leq y \leq 4\}$
29. $\{(x, y) \mid |x| > 4\}$
30. $\{(x, y) \mid |y| \leq 2\}$
31. $\{(x, y) \mid x \geq 1 \text{ and } y < 3\}$
32. $\{(x, y) \mid |x| \leq 2 \text{ and } |y| \leq 3\}$

33. Which of the points $A(6, 7)$ or $B(-5, 8)$ is closer to the origin?

34. Which of the points $C(-6, 3)$ or $D(3, 0)$ is closer to the point $E(-2, 1)$?

35. Which of the points $P(3, 1)$ or $Q(-1, 3)$ is closer to the point $R(-1, -1)$?

36. **(a)** Show that the points $(7, 3)$ and $(3, 7)$ are the same distance from the origin.

(b) Show that the points (a, b) and (b, a) are the same distance from the origin.

37. Show that the triangle with vertices $A(0, 2)$, $B(-3, -1)$, and $C(-4, 3)$ is isosceles.

38. Find the area of the triangle shown in the figure.

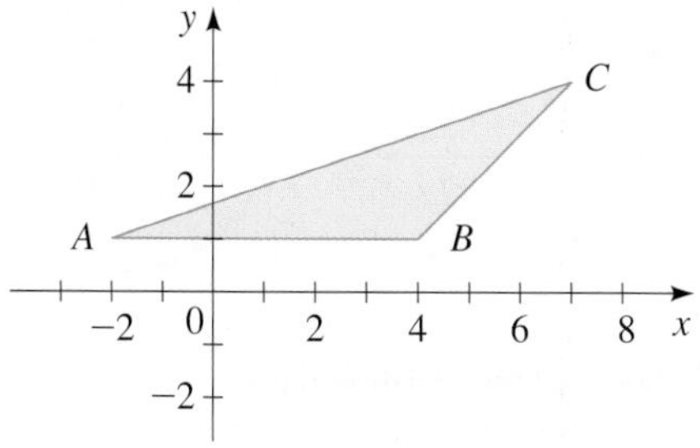

39. Refer to triangle ABC in the figure below.

(a) Show that triangle ABC is a right triangle by using the converse of the Pythagorean Theorem (see page 219).

(b) Find the area of triangle ABC.

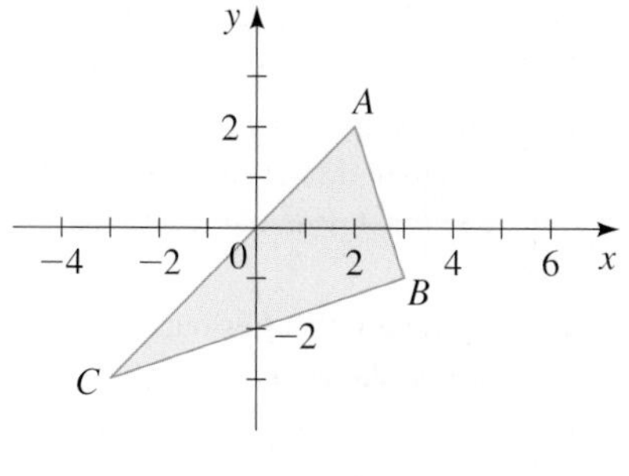

40. Show that the triangle with vertices $A(6, -7)$, $B(11, -3)$, and $C(2, -2)$ is a right triangle by using the converse of the Pythagorean Theorem. Find the area of the triangle.

41. Show that the points $A(-2, 9)$, $B(4, 6)$, $C(1, 0)$, and $D(-5, 3)$ are the vertices of a square.

42. Show that the points $A(-1, 3)$, $B(3, 11)$, and $C(5, 15)$ are collinear by showing that $d(A, B) + d(B, C) = d(A, C)$.

43. Find a point on the y-axis that is equidistant from the points $(5, -5)$ and $(1, 1)$.

44. Find the lengths of the medians of the triangle with vertices $A(1, 0)$, $B(3, 6)$, and $C(8, 2)$. (A *median* is a line segment from a vertex to the midpoint of the opposite side.)

45. Plot the points $P(-1, -4)$, $Q(1, 1)$, and $R(4, 2)$ on a coordinate plane. Where should the point S be located so that the figure $PQRS$ is a parallelogram?

46. If $M(6, 8)$ is the midpoint of the line segment AB and if A has coordinates $(2, 3)$, find the coordinates of B.

47. **(a)** Sketch the parallelogram with vertices $A(-2, -1)$, $B(4, 2)$, $C(7, 7)$, and $D(1, 4)$.

(b) Find the midpoints of the diagonals of this parallelogram.

(c) From part (b) show that the diagonals bisect each other.

48. The point M in the figure is the midpoint of the line segment AB. Show that M is equidistant from the vertices of triangle ABC.

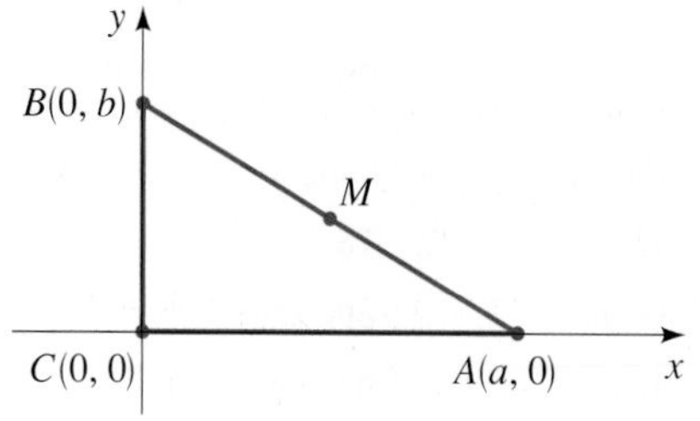

49–52 ■ Determine whether the given points are on the graph of the equation.

49. $x - 2y - 1 = 0$; $(0, 0), (1, 0), (-1, -1)$

50. $y(x^2 + 1) = 1$; $(1, 1), (1, \frac{1}{2}), (-1, \frac{1}{2})$

51. $x^2 + xy + y^2 = 4$; $(0, -2), (1, -2), (2, -2)$

52. $x^2 + y^2 = 1$; $(0, 1), \left(\frac{1}{\sqrt{2}}, \frac{1}{\sqrt{2}}\right), \left(\frac{\sqrt{3}}{2}, \frac{1}{2}\right)$

53–56 ■ An equation and its graph are given. Find the x- and y-intercepts.

53. $y = 4x - x^2$

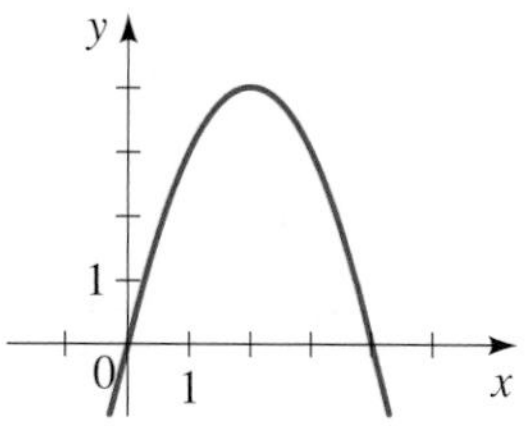

54. $\frac{x^2}{9} + \frac{y^2}{4} = 1$

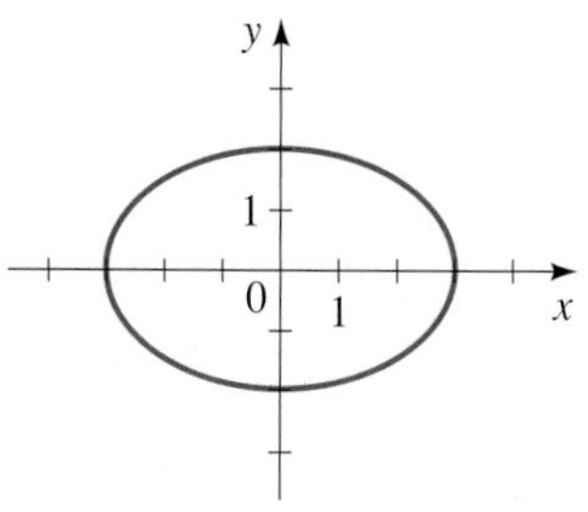

55. $x^4 + y^2 - xy = 16$

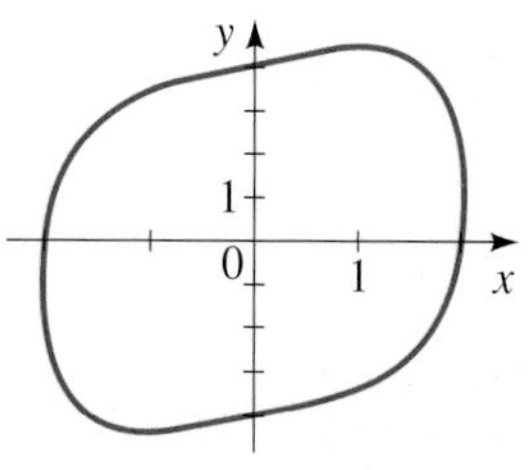

56. $x^2 + y^3 - x^2y^2 = 64$

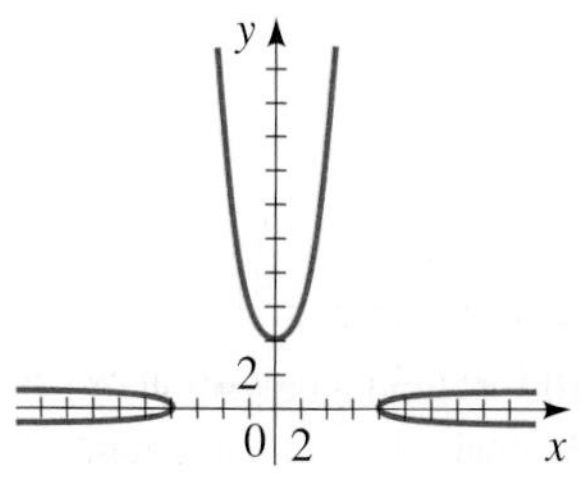

57–76 ■ Make a table of values and sketch the graph of the equation. Find the x- and y-intercepts and test for symmetry.

57. $y = -x + 4$ **58.** $y = 3x + 3$

59. $2x - y = 6$ **60.** $x + y = 3$

61. $y = 1 - x^2$ **62.** $y = x^2 + 2$

63. $4y = x^2$ **64.** $8y = x^3$

65. $y = x^2 - 9$ **66.** $y = 9 - x^2$

67. $xy = 2$ **68.** $y = \sqrt{x + 4}$

69. $y = \sqrt{4 - x^2}$ **70.** $y = -\sqrt{4 - x^2}$

71. $x + y^2 = 4$ **72.** $x = y^3$

73. $y = 16 - x^4$ **74.** $x = |y|$

75. $y = 4 - |x|$ **76.** $y = |4 - x|$

77–82 ■ Test the equation for symmetry.

77. $y = x^4 + x^2$ **78.** $x = y^4 - y^2$

79. $x^2y^2 + xy = 1$ **80.** $x^4y^4 + x^2y^2 = 1$

81. $y = x^3 + 10x$ **82.** $y = x^2 + |x|$

83–86 ■ Complete the graph using the given symmetry property.

83. Symmetric with respect to the y-axis

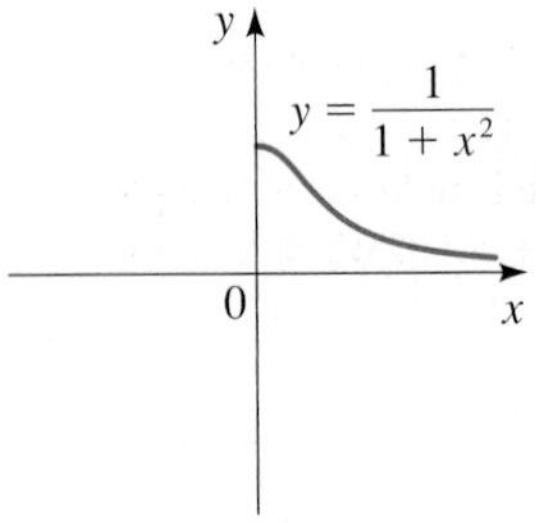

84. Symmetric with respect to the x-axis

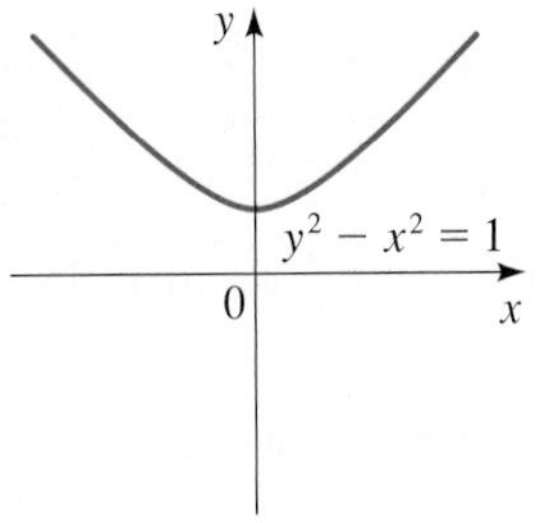

85. Symmetric with respect to the origin

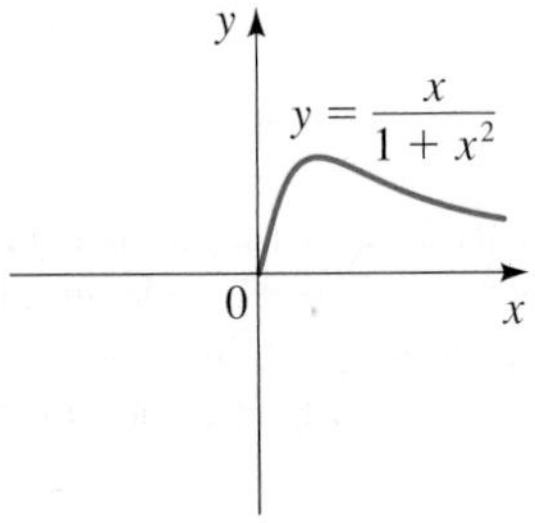

86. Symmetric with respect to the origin

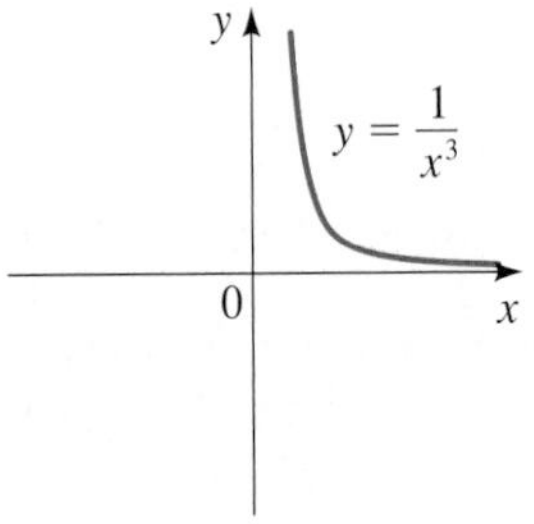

87–92 ■ Find the center and radius of the circle and sketch its graph

87. $x^2 + y^2 = 9$ **88.** $x^2 + y^2 = 5$

89. $(x - 3)^2 + y^2 = 16$ **90.** $x^2 + (y - 2)^2 = 4$

91. $(x + 3)^2 + (y - 4)^2 = 25$ **92.** $(x + 1)^2 + (y + 2)^2 = 36$

93–100 ■ Find an equation of the circle that satisfies the given conditions.

93. Center $(2, -1)$; radius 3

94. Center $(-1, -4)$; radius 8

95. Center at the origin; passes through $(4, 7)$

96. Center $(-1, 5)$; passes through $(-4, -6)$

97. Endpoints of a diameter are $P(-1, 1)$ and $Q(5, 9)$

98. Endpoints of a diameter are $P(-1, 3)$ and $Q(7, -5)$

99. Center $(7, -3)$; tangent to the x-axis

100. Circle lies in the first quadrant, tangent to both x-and y-axes; radius 5

101–102 ■ Find the equation of the circle shown in the figure.

101.

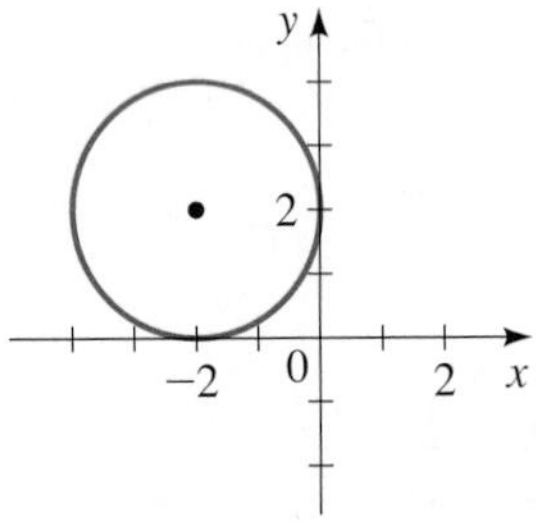

102.

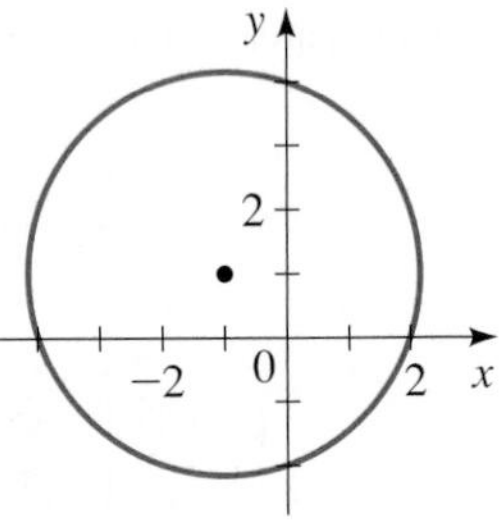

103–108 ■ Show that the equation represents a circle, and find the center and radius of the circle.

103. $x^2 + y^2 - 4x + 10y + 13 = 0$

104. $x^2 + y^2 + 6y + 2 = 0$

105. $x^2 + y^2 - \frac{1}{2}x + \frac{1}{2}y = \frac{1}{8}$

106. $x^2 + y^2 + \frac{1}{2}x + 2y + \frac{1}{16} = 0$

107. $2x^2 + 2y^2 - 3x = 0$

108. $3x^2 + 3y^2 + 6x - y = 0$

109–110 ■ Sketch the region given by the set.

109. $\{(x, y) \mid x^2 + y^2 \le 1\}$

110. $\{(x, y) \mid x^2 + y^2 > 4\}$

111. Find the area of the region that lies outside the circle $x^2 + y^2 = 4$ but inside the circle

$$x^2 + y^2 - 4y - 12 = 0$$

112. Sketch the region in the coordinate plane that satisfies both the inequalities $x^2 + y^2 \le 9$ and $y \ge |x|$. What is the area of this region?

APPLICATIONS

113. Distances in a City A city has streets that run north and south and avenues that run east and west, all equally spaced. Streets and avenues are numbered sequentially, as shown in the figure. The *walking* distance between points A and B is 7 blocks—that is, 3 blocks east and 4 blocks north. To find the *straight-line* distance d, we must use the Distance Formula.

(a) Find the straight-line distance (in blocks) between A and B.

(b) Find the walking distance and the straight-line distance between the corner of 4th St. and 2nd Ave. and the corner of 11th St. and 26th Ave.

(c) What must be true about the points P and Q if the walking distance between P and Q equals the straight-line distance between P and Q?

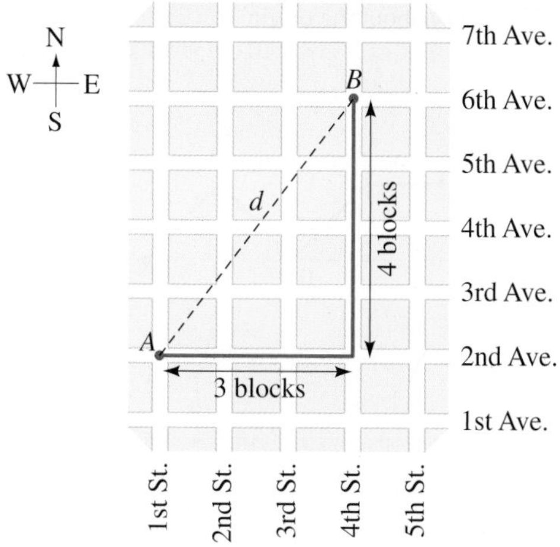

114. Halfway Point Two friends live in the city described in Exercise 113, one at the corner of 3rd St. and 7th Ave., the other at the corner of 27th St. and 17th Ave. They frequently meet at a coffee shop halfway between their homes.

(a) At what intersection is the coffee shop located?

(b) How far must each of them walk to get to the coffee shop?

115. Orbit of a Satellite A satellite is in orbit around the moon. A coordinate plane containing the orbit is set up with the center of the moon at the origin, as shown in the graph, with distances measured in megameters (Mm). The equation of the satellite's orbit is

$$\frac{(x-3)^2}{25} + \frac{y^2}{16} = 1$$

(a) From the graph, determine the closest and the farthest that the satellite gets to the center of the moon.

(b) There are two points in the orbit with y-coordinates 2. Find the x-coordinates of these points, and determine their distances to the center of the moon.

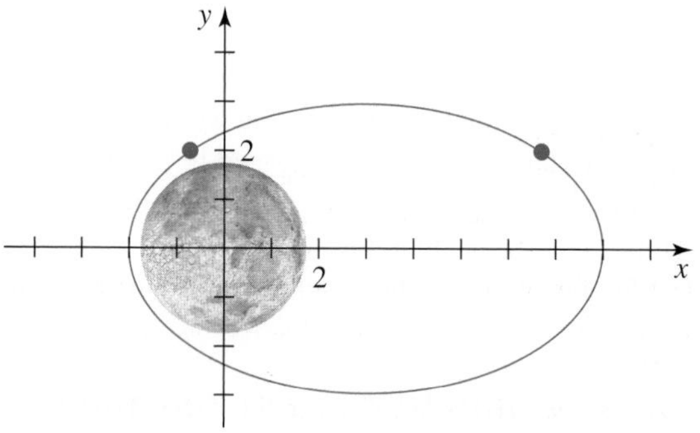

DISCOVERY ▪ DISCUSSION ▪ WRITING

116. Shifting the Coordinate Plane Suppose that each point in the coordinate plane is shifted 3 units to the right and 2 units upward.

(a) The point $(5, 3)$ is shifted to what new point?

(b) The point (a, b) is shifted to what new point?

(c) What point is shifted to $(3, 4)$?

(d) Triangle ABC in the figure has been shifted to triangle $A'B'C'$. Find the coordinates of the points A', B', and C'.

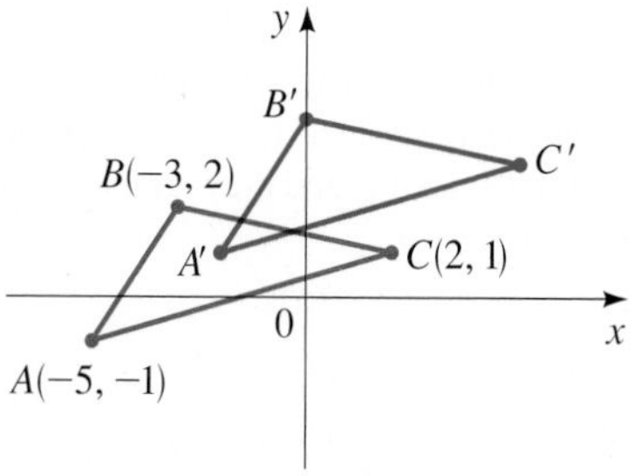

117. Reflecting in the Coordinate Plane Suppose that the y-axis acts as a mirror that reflects each point to the right of it into a point to the left of it.

(a) The point $(3, 7)$ is reflected to what point?

(b) The point (a, b) is reflected to what point?

(c) What point is reflected to $(-4, -1)$?

(d) Triangle ABC in the figure is reflected to triangle $A'B'C'$. Find the coordinates of the points A', B', and C'.

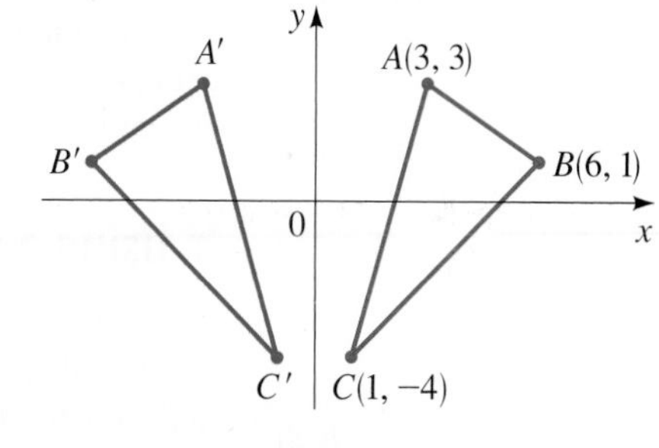

118. Completing a Line Segment Plot the points $M(6, 8)$ and $A(2, 3)$ on a coordinate plane. If M is the midpoint of the line segment AB, find the coordinates of B. Write a brief description of the steps you took to find B, and your reasons for taking them.

119. Completing a Parallelogram Plot the points $P(0, 3)$, $Q(2, 2)$, and $R(5, 3)$ on a coordinate plane. Where should the point S be located so that the figure $PQRS$ is a parallelogram? Write a brief description of the steps you took and your reasons for taking them.

120. Circle, Point, or Empty Set? Complete the squares in the general equation $x^2 + ax + y^2 + by + c = 0$ and simplify the result as much as possible. Under what conditions on the coefficients a, b, and c does this equation represent a circle? A single point? The empty set? In the case that the equation does represent a circle, find its center and radius.

121. Do the Circles Intersect?

(a) Find the radius of each circle in the pair and the distance between their centers; then use this information to determine whether the circles intersect.

(i) $(x - 2)^2 + (y - 1)^2 = 9$;
$(x - 6)^2 + (y - 4)^2 = 16$

(ii) $x^2 + (y - 2)^2 = 4$;
$(x - 5)^2 + (y - 14)^2 = 9$

(iii) $(x - 3)^2 + (y + 1)^2 = 1$;
$(x - 2)^2 + (y - 2)^2 = 25$

(b) How can you tell, just by knowing the radii of two circles and the distance between their centers, whether the circles intersect? Write a short paragraph describing how you would decide this and draw graphs to illustrate your answer.

122. Making a Graph Symmetric The graph shown in the figure is not symmetric about the x-axis, the y-axis, or the origin. Add more line segments to the graph so that it exhibits the indicated symmetry. In each case, add as little as possible.

(a) Symmetry about the x-axis

(b) Symmetry about the y-axis

(c) Symmetry about the origin

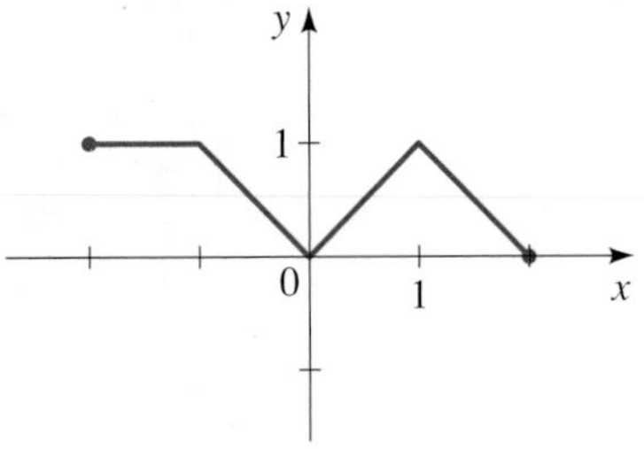

1.9 Graphing Calculators; Solving Equations and Inequalities Graphically

Using a Graphing Calculator ▶ Solving Equations Graphically ▶ Solving Inequalities Graphically

In Sections 1.5 and 1.7 we solved equations and inequalities algebraically. In Section 1.8 we learned how to sketch the graph of an equation in a coordinate plane. In this section we use graphs to solve equations and inequalities. To do this, we must first draw a graph using a graphing device. So we begin by giving a few guidelines to help us use graphing devices effectively.

▼ Using a Graphing Calculator

A graphing calculator or computer displays a rectangular portion of the graph of an equation in a display window or viewing screen, which we call a **viewing rectangle**. The default screen often gives an incomplete or misleading picture, so it is important to choose the viewing rectangle with care. If we choose the x-values to range from a minimum value of `Xmin` $= a$ to a maximum value of `Xmax` $= b$ and the y-values to range from a minimum value of `Ymin` $= c$ to a maximum value of `Ymax` $= d$, then the displayed portion of the graph lies in the rectangle

$$[a, b] \times [c, d] = \{(x, y) \mid a \leq x \leq b, c \leq y \leq d\}$$

as shown in Figure 1. We refer to this as the $[a, b]$ by $[c, d]$ viewing rectangle.

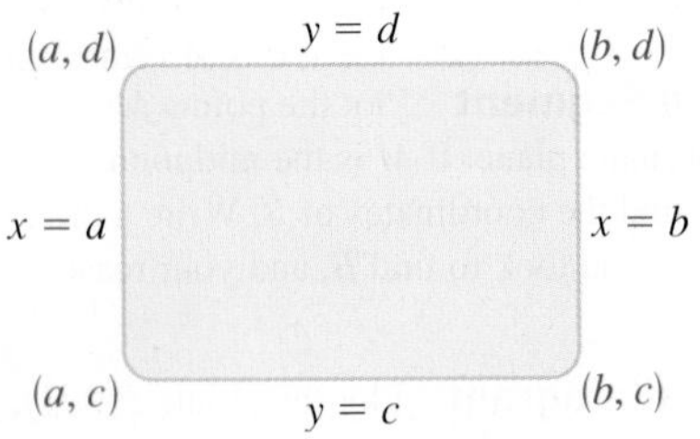

FIGURE 1 The viewing rectangle $[a, b]$ by $[c, d]$

The graphing device draws the graph of an equation much as you would. It plots points of the form (x, y) for a certain number of values of x, equally spaced between a and b. If the equation is not defined for an x-value or if the corresponding y-value lies outside the viewing rectangle, the device ignores this value and moves on to the next x-value. The machine connects each point to the preceding plotted point to form a representation of the graph of the equation.

EXAMPLE 1 | Choosing an Appropriate Viewing Rectangle

Graph the equation $y = x^2 + 3$ in an appropriate viewing rectangle.

SOLUTION Let's experiment with different viewing rectangles. We start with the viewing rectangle $[-2, 2]$ by $[-2, 2]$, so we set

```
Xmin = −2        Ymin = −2
Xmax = 2         Ymax = 2
```

The resulting graph in Figure 2(a) is blank! This is because $x^2 \geq 0$, so $x^2 + 3 \geq 3$ for all x. Thus, the graph lies entirely above the viewing rectangle, so this viewing rectangle is not appropriate. If we enlarge the viewing rectangle to $[-4, 4]$ by $[-4, 4]$, as in Figure 2(b), we begin to see a portion of the graph.

Now let's try the viewing rectangle $[-10, 10]$ by $[-5, 30]$. The graph in Figure 2(c) seems to give a more complete view of the graph. If we enlarge the viewing rectangle even further, as in Figure 2(d), the graph doesn't show clearly that the y-intercept is 3.

So the viewing rectangle $[-10, 10]$ by $[-5, 30]$ gives an appropriate representation of the graph.

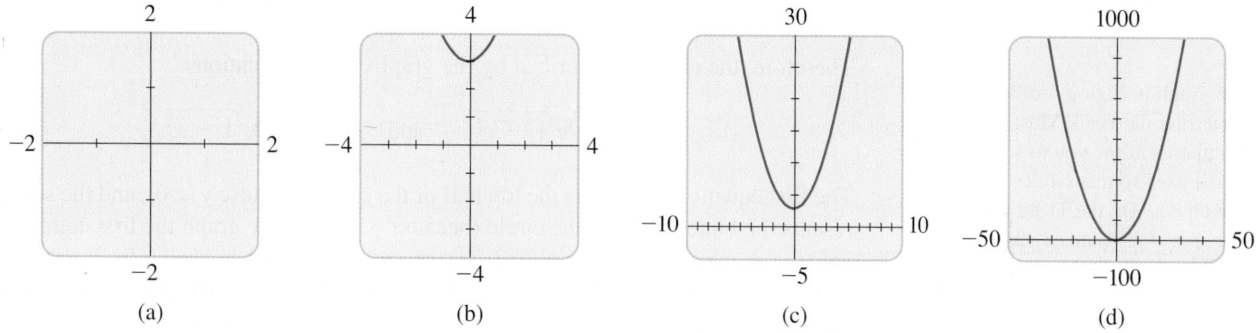

FIGURE 2 Graphs of $y = x^2 + 3$

NOW TRY EXERCISE **5**

EXAMPLE 2 | Two Graphs on the Same Screen

Graph the equations $y = 3x^2 - 6x + 1$ and $y = 0.23x - 2.25$ together in the viewing rectangle $[-1, 3]$ by $[-2.5, 1.5]$. Do the graphs intersect in this viewing rectangle?

SOLUTION Figure 3(a) shows the essential features of both graphs. One is a parabola, and the other is a line. It looks as if the graphs intersect near the point $(1, -2)$. However, if we zoom in on the area around this point as shown in Figure 3(b), we see that although the graphs almost touch, they do not actually intersect.

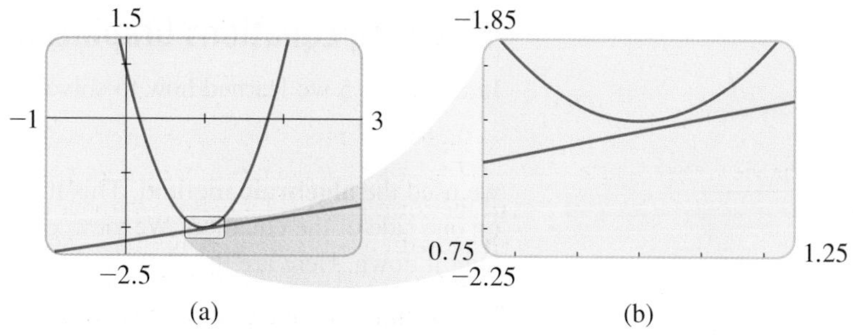

FIGURE 3

NOW TRY EXERCISE **23**

You can see from Examples 1 and 2 that the choice of a viewing rectangle makes a big difference in the appearance of a graph. If you want an overview of the essential features of a graph, you must choose a relatively large viewing rectangle to obtain a global view of the graph. If you want to investigate the details of a graph, you must zoom in to a small viewing rectangle that shows just the feature of interest.

Most graphing calculators can only graph equations in which y is isolated on one side of the equal sign. The next example shows how to graph equations that don't have this property.

EXAMPLE 3 | Graphing a Circle

Graph the circle $x^2 + y^2 = 1$.

SOLUTION We first solve for y, to isolate it on one side of the equal sign.

$$y^2 = 1 - x^2 \qquad \text{Subtract } x^2$$

$$y = \pm\sqrt{1 - x^2} \qquad \text{Take square roots}$$

Therefore, the circle is described by the graphs of *two* equations:

$$y = \sqrt{1 - x^2} \qquad \text{and} \qquad y = -\sqrt{1 - x^2}$$

The first equation represents the top half of the circle (because $y \geq 0$), and the second represents the bottom half of the circle (because $y \leq 0$). If we graph the first equation in the viewing rectangle $[-2, 2]$ by $[-2, 2]$, we get the semicircle shown in Figure 4(a). The graph of the second equation is the semicircle in Figure 4(b). Graphing these semicircles together on the same viewing screen, we get the full circle in Figure 4(c).

The graph in Figure 4(c) looks somewhat flattened. Most graphing calculators allow you to set the scales on the axes so that circles really look like circles. On the TI-82 and TI-83, from the ZOOM menu, choose ZSquare to set the scales appropriately. (On the TI-86 the command is Zsq.)

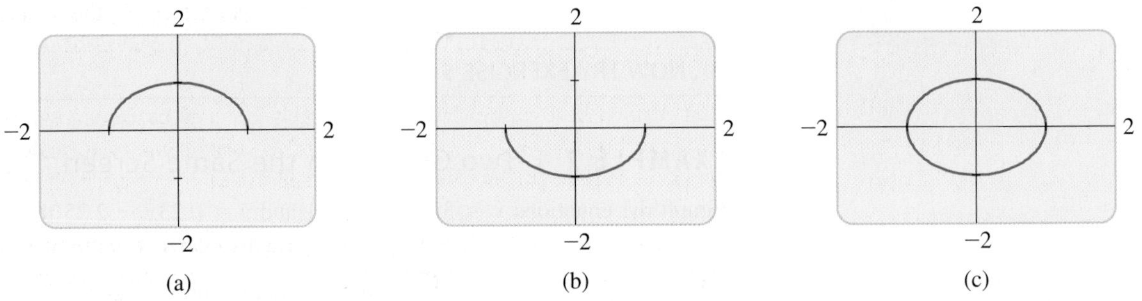

FIGURE 4 Graphing the equation $x^2 + y^2 = 1$

NOW TRY EXERCISE 27 ■

Solving Equations Graphically

In Section 1.5 we learned how to solve equations. To solve an equation such as

$$3x - 5 = 0$$

we used the **algebraic method**. This means that we used the rules of algebra to isolate x on one side of the equation. We view x as an *unknown*, and we use the rules of algebra to hunt it down. Here are the steps in the solution:

$$3x - 5 = 0$$

$$3x = 5 \qquad \text{Add 5}$$

$$x = \tfrac{5}{3} \qquad \text{Divide by 3}$$

So the solution is $x = \frac{5}{3}$.

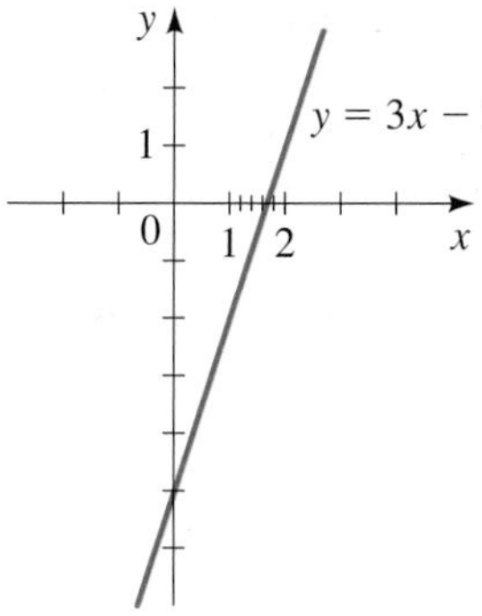

FIGURE 5

We can also solve this equation by the **graphical method**. In this method we view x as a *variable* and sketch the graph of the equation

$$y = 3x - 5$$

Different values for x give different values for y. Our goal is to find the value of x for which $y = 0$. From the graph in Figure 5 we see that $y = 0$ when $x \approx 1.7$. Thus, the solution is $x \approx 1.7$. Note that from the graph we obtain an approximate solution. We summarize these methods in the box below.

SOLVING AN EQUATION

Algebraic Method

Use the rules of algebra to isolate the unknown x on one side of the equation.

Example: $2x = 6 - x$

$3x = 6$ Add x

$x = 2$ Divide by 3

The solution is $x = 2$.

Graphical Method

Move all terms to one side and set equal to y. Sketch the graph to find the value of x where $y = 0$.

Example: $2x = 6 - x$

$0 = 6 - 3x$

Set $y = 6 - 3x$ and graph.

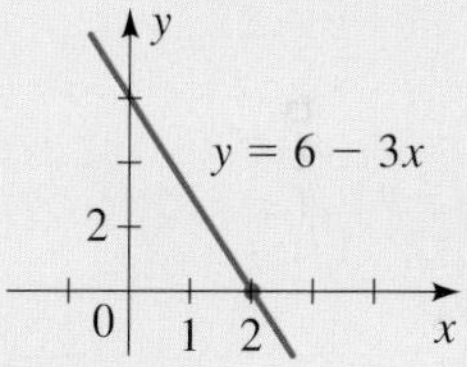

From the graph, the solution is $x \approx 2$.

The advantage of the algebraic method is that it gives exact answers. Also, the process of unraveling the equation to arrive at the answer helps us to understand the algebraic structure of the equation. On the other hand, for many equations it is difficult or impossible to isolate x.

The graphical method gives a numerical approximation to the answer. This is an advantage when a numerical answer is desired. (For example, an engineer might find an answer expressed as $x \approx 2.6$ more immediately useful than $x = \sqrt{7}$.) Also, graphing an equation helps us to visualize how the solution is related to other values of the variable.

The *Discovery Project* referenced on page 263 describes a numerical method for solving equations.

PIERRE DE FERMAT (1601–1665) was a French lawyer who became interested in mathematics at the age of 30. Because of his job as a magistrate, Fermat had little time to write complete proofs of his discoveries and often wrote them in the margin of whatever book he was reading at the time. After his death, his copy of Diophantus' *Arithmetica* (see page 20) was found to contain a particularly tantalizing comment. Where Diophantus discusses the solutions of $x^2 + y^2 = z^2$ (for example, $x = 3, y = 4$, and $z = 5$), Fermat states in the margin that for $n \geq 3$ there are no natural number solutions to the equation $x^n + y^n = z^n$. In other words, it's impossible for a cube to equal the sum of two cubes, a fourth power to equal the sum of two fourth powers, and so on. Fermat writes, "I have discovered a truly wonderful proof for this but the margin is too small to contain it." All the other margin comments in Fermat's copy of *Arithmetica* have been proved. This one, however, remained unproved, and it came to be known as "Fermat's Last Theorem."

In 1994, Andrew Wiles of Princeton University announced a proof of Fermat's Last Theorem, an astounding 350 years after it was conjectured. His proof is one of the most widely reported mathematical results in the popular press.

EXAMPLE 4 | Solving a Quadratic Equation Algebraically and Graphically

Solve the quadratic equations algebraically and graphically.

(a) $x^2 - 4x + 2 = 0$ **(b)** $x^2 - 4x + 4 = 0$ **(c)** $x^2 - 4x + 6 = 0$

SOLUTION 1: Algebraic

We use the Quadratic Formula to solve each equation.

The Quadratic Formula is discussed on page 49.

(a) $$x = \frac{-(-4) \pm \sqrt{(-4)^2 - 4 \cdot 1 \cdot 2}}{2} = \frac{4 \pm \sqrt{8}}{2} = 2 \pm \sqrt{2}$$

There are two solutions, $x = 2 + \sqrt{2}$ and $x = 2 - \sqrt{2}$.

(b) $$x = \frac{-(-4) \pm \sqrt{(-4)^2 - 4 \cdot 1 \cdot 4}}{2} = \frac{4 \pm \sqrt{0}}{2} = 2$$

There is just one solution, $x = 2$.

(c) $$x = \frac{-(-4) \pm \sqrt{(-4)^2 - 4 \cdot 1 \cdot 6}}{2} = \frac{4 \pm \sqrt{-8}}{2}$$

There is no real solution.

SOLUTION 2: Graphical

We graph the equations $y = x^2 - 4x + 2$, $y = x^2 - 4x + 4$, and $y = x^2 - 4x + 6$ in Figure 6. By determining the x-intercepts of the graphs, we find the following solutions.

(a) $x \approx 0.6$ and $x \approx 3.4$

(b) $x = 2$

(c) There is no x-intercept, so the equation has no solution.

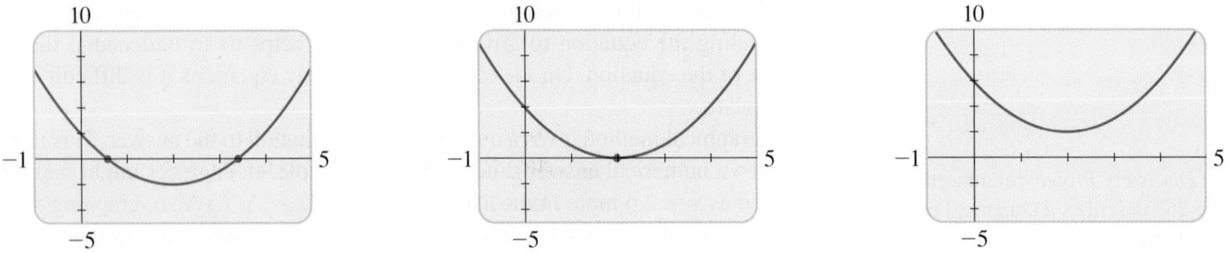

(a) $y = x^2 - 4x + 2$ (b) $y = x^2 - 4x + 4$ (c) $y = x^2 - 4x + 6$

FIGURE 6

NOW TRY EXERCISE **35**

ALAN TURING (1912–1954) was at the center of two pivotal events of the 20th century: World War II and the invention of computers. At the age of 23 Turing made his mark on mathematics by solving an important problem in the foundations of mathematics that had been posed by David Hilbert at the 1928 International Congress of Mathematicians (see page 683). In this research he invented a theoretical machine, now called a Turing machine, which was the inspiration for modern digital computers. During World War II Turing was in charge of the British effort to decipher secret German codes. His complete success in this endeavor played a decisive role in the Allies' victory. To carry out the numerous logical steps that are required to break a coded message, Turing developed decision procedures similar to modern computer programs. After the war he helped to develop the first electronic computers in Britain. He also did pioneering work on artificial intelligence and computer models of biological processes. At the age of 42 Turing died of poisoning after eating an apple that had mysteriously been laced with cyanide.

The graphs in Figure 6 show visually why a quadratic equation may have two solutions, one solution, or no real solution. We proved this fact algebraically in Section 1.5 when we studied the discriminant.

EXAMPLE 5 | Another Graphical Method

Solve the equation algebraically and graphically: $5 - 3x = 8x - 20$

SOLUTION 1: Algebraic

$$5 - 3x = 8x - 20$$

$$-3x = 8x - 25 \qquad \text{Subtract 5}$$

$$-11x = -25 \qquad \text{Subtract } 8x$$

$$x = \frac{-25}{-11} = 2\tfrac{3}{11} \qquad \text{Divide by } -11 \text{ and simplify}$$

SOLUTION 2: Graphical

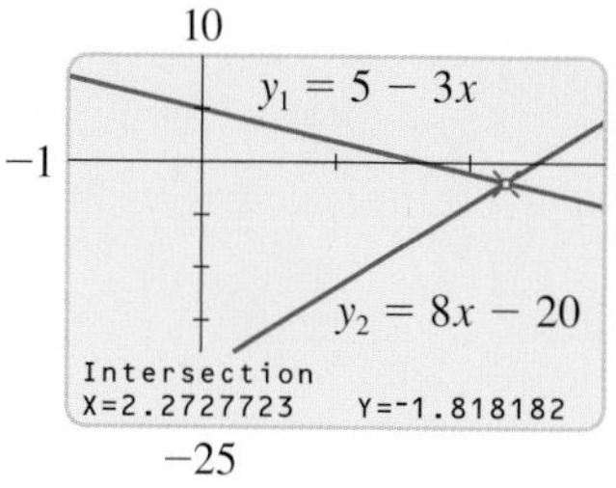

FIGURE 7

We could move all terms to one side of the equal sign, set the result equal to y, and graph the resulting equation. But to avoid all this algebra, we graph two equations instead:

$$y_1 = 5 - 3x \quad \text{and} \quad y_2 = 8x - 20$$

The solution of the original equation will be the value of x that makes y_1 equal to y_2; that is, the solution is the x-coordinate of the intersection point of the two graphs. Using the `TRACE` feature or the `intersect` command on a graphing calculator, we see from Figure 7 that the solution is $x \approx 2.27$.

NOW TRY EXERCISE **31** ■

In the next example we use the graphical method to solve an equation that is extremely difficult to solve algebraically.

EXAMPLE 6 | Solving an Equation in an Interval

Solve the equation

$$x^3 - 6x^2 + 9x = \sqrt{x}$$

in the interval $[1, 6]$.

SOLUTION We are asked to find all solutions x that satisfy $1 \le x \le 6$, so we will graph the equation in a viewing rectangle for which the x-values are restricted to this interval.

$$x^3 - 6x^2 + 9x = \sqrt{x}$$

$$x^3 - 6x^2 + 9x - \sqrt{x} = 0 \qquad \text{Subtract } \sqrt{x}$$

We can also use the `zero` command to find the solutions, as shown in Figures 8(a) and 8(b).

Figure 8 shows the graph of the equation $y = x^3 - 6x^2 + 9x - \sqrt{x}$ in the viewing rectangle $[1, 6]$ by $[-5, 5]$. There are two x-intercepts in this viewing rectangle; zooming in, we see that the solutions are $x \approx 2.18$ and $x \approx 3.72$.

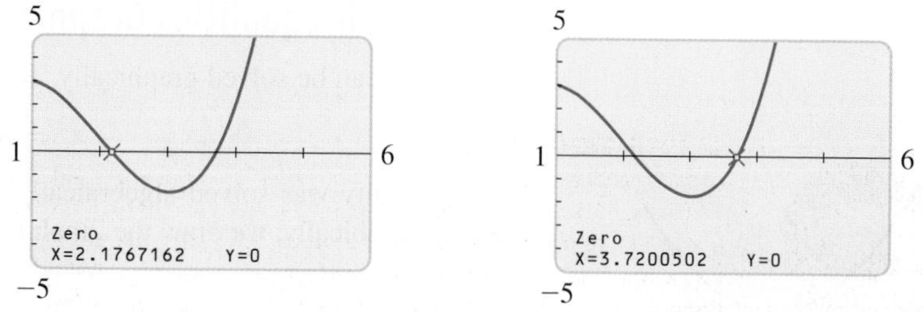

(a) (b)

FIGURE 8

NOW TRY EXERCISE **43** ■

The equation in Example 6 actually has four solutions. You are asked to find the other two in Exercise 71.

EXAMPLE 7 | Intensity of Light

Two light sources are 10 m apart. One is three times as intense as the other. The light intensity L (in lux) at a point x meters from the weaker source is given by

$$L = \frac{10}{x^2} + \frac{30}{(10 - x)^2}$$

(See Figure 9.) Find the points at which the light intensity is 4 lux.

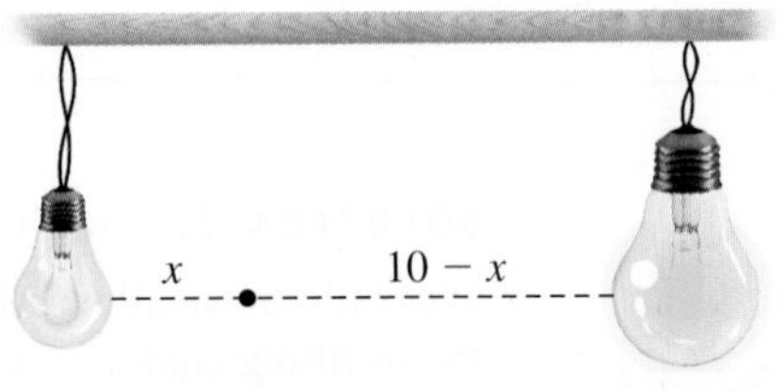

FIGURE 9

SOLUTION We need to solve the equation

$$4 = \frac{10}{x^2} + \frac{30}{(10 - x)^2}$$

The graphs of

$$y_1 = 4 \qquad \text{and} \qquad y_2 = \frac{10}{x^2} + \frac{30}{(10 - x)^2}$$

are shown in Figure 10. Zooming in (or using the `intersect` command), we find two solutions, $x \approx 1.67431$ and $x \approx 7.1927193$. So the light intensity is 4 lux at the points that are 1.67 m and 7.19 m from the weaker source.

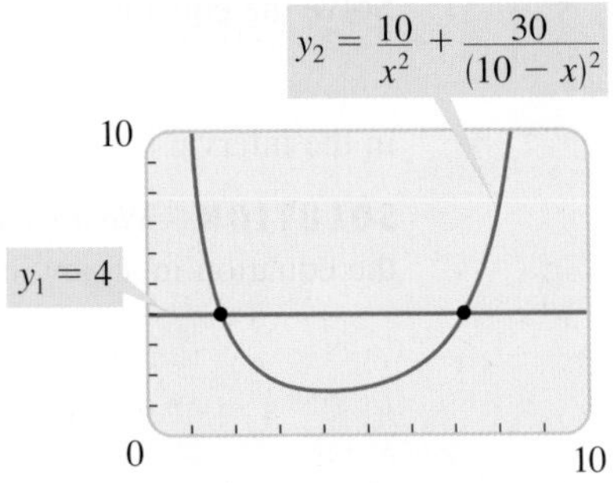

FIGURE 10

NOW TRY EXERCISE **73**

Solving Inequalities Graphically

Inequalities can be solved graphically. To describe the method, we solve

$$x^2 - 5x + 6 \le 0$$

This inequality was solved algebraically in Example 3 of Section 1.7. To solve the inequality graphically, we draw the graph of

$$y = x^2 - 5x + 6$$

Our goal is to find those values of x for which $y \le 0$. These are simply the x-values for which the graph lies below the x-axis. From Figure 11 we see that the solution of the inequality is the interval $[2, 3]$.

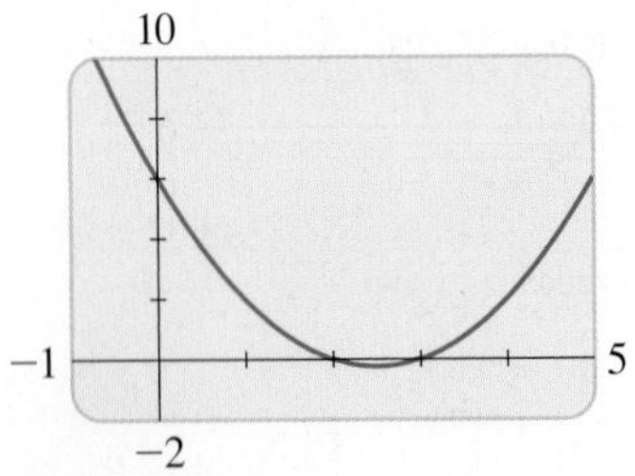

FIGURE 11 $x^2 - 5x + 6 \le 0$

EXAMPLE 8 | Solving an Inequality Graphically

Solve the inequality $3.7x^2 + 1.3x - 1.9 \le 2.0 - 1.4x$.

SOLUTION We graph the equations

$$y_1 = 3.7x^2 + 1.3x - 1.9 \qquad \text{and} \qquad y_2 = 2.0 - 1.4x$$

in the same viewing rectangle in Figure 12. We are interested in those values of x for which $y_1 \le y_2$; these are points for which the graph of y_2 lies on or above the graph of y_1. To determine the appropriate interval, we look for the x-coordinates of points where the graphs intersect. We conclude that the solution is (approximately) the interval $[-1.45, 0.72]$.

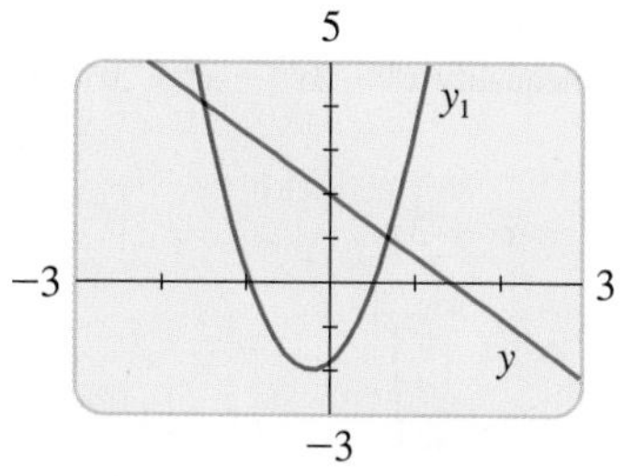

FIGURE 12
$y_1 = 3.7x^2 + 1.3x - 1.9$
$y_2 = 2.0 - 1.4x$

NOW TRY EXERCISE **59** ■

EXAMPLE 9 | Solving an Inequality Graphically

Solve the inequality $x^3 - 5x^2 \ge -8$.

SOLUTION We write the inequality as

$$x^3 - 5x^2 + 8 \ge 0$$

and then graph the equation

$$y = x^3 - 5x^2 + 8$$

in the viewing rectangle $[-6, 6]$ by $[-15, 15]$, as shown in Figure 13. The solution of the inequality consists of those intervals on which the graph lies on or above the x-axis. By moving the cursor to the x-intercepts, we find that, rounded to one decimal place, the solution is $[-1.1, 1.5] \cup [4.6, \infty)$.

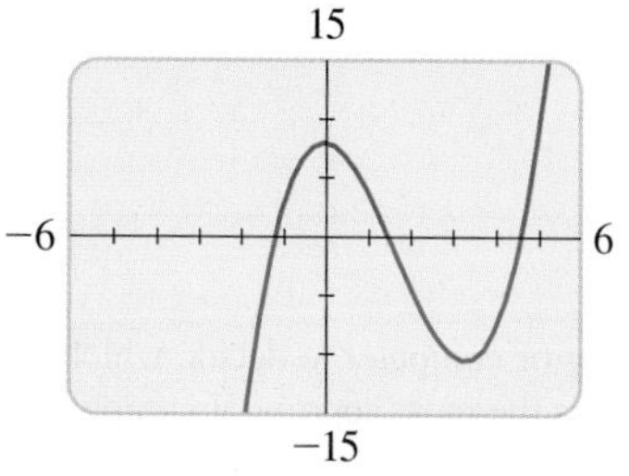

FIGURE 13 $x^3 - 5x^2 + 8 \ge 0$

NOW TRY EXERCISE **61** ■

1.9 EXERCISES

CONCEPTS

1. The solutions of the equation $x^2 - 2x - 3 = 0$ are the ____-intercepts of the graph of $y = x^2 - 2x - 3$.

2. The solutions of the inequality $x^2 - 2x - 3 > 0$ are the x-coordinates of the points on the graph of $y = x^2 - 2x - 3$ that lie ________ the x-axis.

3. The figure shows a graph of $y = x^4 - 3x^3 - x^2 + 3x$. Use the graph to do the following.
   (a) Find the solutions of the equation $x^4 - 3x^3 - x^2 + 3x = 0$.
   (b) Find the solutions of the inequality $x^4 - 3x^3 - x^2 + 3x \le 0$.

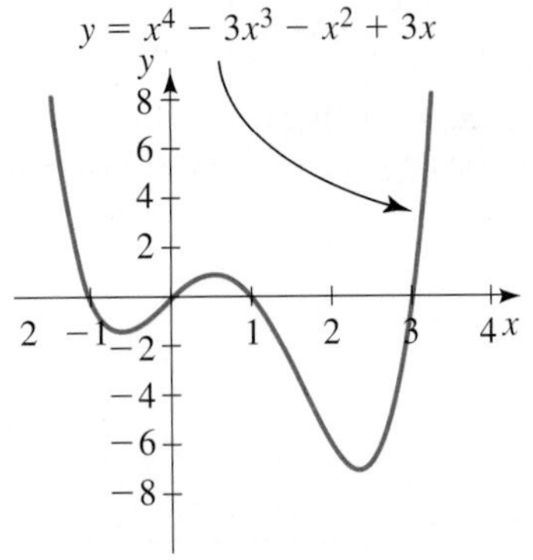

4. The figure shows the graphs of $y = 5x - x^2$ and $y = 4$. Use the graphs to do the following.
   (a) Find the solutions of the equation $5x - x^2 = 4$.
   (b) Find the solutions of the inequality $5x - x^2 > 4$.

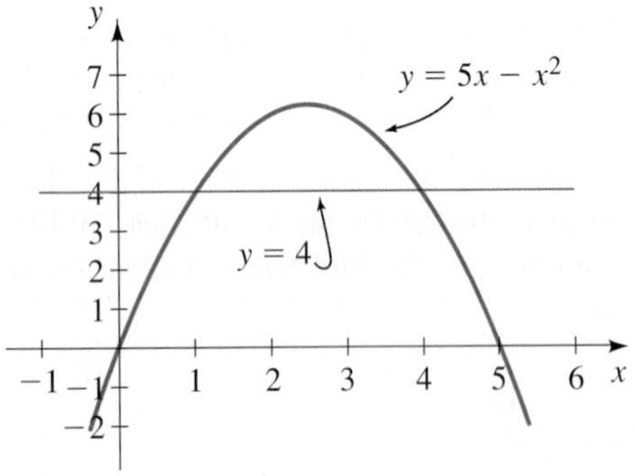

SKILLS

5–10 ■ Use a graphing calculator or computer to decide which viewing rectangle (a)–(d) produces the most appropriate graph of the equation.

5. $y = x^4 + 2$
   (a) $[-2, 2]$ by $[-2, 2]$
   (b) $[0, 4]$ by $[0, 4]$
   (c) $[-8, 8]$ by $[-4, 40]$
   (d) $[-40, 40]$ by $[-80, 800]$

6. $y = x^2 + 7x + 6$
   (a) $[-5, 5]$ by $[-5, 5]$
   (b) $[0, 10]$ by $[-20, 100]$
   (c) $[-15, 8]$ by $[-20, 100]$
   (d) $[-10, 3]$ by $[-100, 20]$

7. $y = 100 - x^2$
   (a) $[-4, 4]$ by $[-4, 4]$
   (b) $[-10, 10]$ by $[-10, 10]$
   (c) $[-15, 15]$ by $[-30, 110]$
   (d) $[-4, 4]$ by $[-30, 110]$

8. $y = 2x^2 - 1000$
   (a) $[-10, 10]$ by $[-10, 10]$
   (b) $[-10, 10]$ by $[-100, 100]$
   (c) $[-10, 10]$ by $[-1000, 1000]$
   (d) $[-25, 25]$ by $[-1200, 200]$

9. $y = 10 + 25x - x^3$
   (a) $[-4, 4]$ by $[-4, 4]$
   (b) $[-10, 10]$ by $[-10, 10]$
   (c) $[-20, 20]$ by $[-100, 100]$
   (d) $[-100, 100]$ by $[-200, 200]$

10. $y = \sqrt{8x - x^2}$
   (a) $[-4, 4]$ by $[-4, 4]$
   (b) $[-5, 5]$ by $[0, 100]$
   (c) $[-10, 10]$ by $[-10, 40]$
   (d) $[-2, 10]$ by $[-2, 6]$

11–22 ■ Determine an appropriate viewing rectangle for the equation, and use it to draw the graph.

11. $y = 100x^2$
12. $y = -100x^2$
13. $y = 4 + 6x - x^2$
14. $y = 0.3x^2 + 1.7x - 3$
15. $y = \sqrt[4]{256 - x^2}$
16. $y = \sqrt{12x - 17}$
17. $y = 0.01x^3 - x^2 + 5$
18. $y = x(x + 6)(x - 9)$
19. $y = x^4 - 4x^3$
20. $y = \dfrac{x}{x^2 + 25}$
21. $y = 1 + |x - 1|$
22. $y = 2x - |x^2 - 5|$

23–26 ■ Do the graphs intersect in the given viewing rectangle? If they do, how many points of intersection are there?

23. $y = -3x^2 + 6x - \frac{1}{2}$, $y = \sqrt{7 - \frac{7}{12}x^2}$; $[-4, 4]$ by $[-1, 3]$

24. $y = \sqrt{49 - x^2}$, $y = \frac{1}{5}(41 - 3x)$; $[-8, 8]$ by $[-1, 8]$

25. $y = 6 - 4x - x^2$, $y = 3x + 18$; $[-6, 2]$ by $[-5, 20]$

26. $y = x^3 - 4x$, $y = x + 5$; $[-4, 4]$ by $[-15, 15]$

27. Graph the circle $x^2 + y^2 = 9$ by solving for y and graphing two equations as in Example 3.

28. Graph the circle $(y - 1)^2 + x^2 = 1$ by solving for y and graphing two equations as in Example 3.

29. Graph the equation $4x^2 + 2y^2 = 1$ by solving for y and graphing two equations corresponding to the negative and positive square roots. (This graph is called an *ellipse*.)

30. Graph the equation $y^2 - 9x^2 = 1$ by solving for y and graphing the two equations corresponding to the positive and negative square roots. (This graph is called a *hyperbola.*)

31–42 ■ Solve the equation both algebraically and graphically.

31. $x - 4 = 5x + 12$ **32.** $\frac{1}{2}x - 3 = 6 + 2x$

33. $\dfrac{2}{x} + \dfrac{1}{2x} = 7$ **34.** $\dfrac{4}{x+2} - \dfrac{6}{2x} = \dfrac{5}{2x+4}$

35. $x^2 - 32 = 0$ **36.** $x^3 + 16 = 0$

37. $x^2 + 9 = 0$ **38.** $x^2 + 3 = 2x$

39. $16x^4 = 625$ **40.** $2x^5 - 243 = 0$

41. $(x - 5)^4 - 80 = 0$ **42.** $6(x + 2)^5 = 64$

43–50 ■ Solve the equation graphically in the given interval. State each answer rounded to two decimals.

43. $x^2 - 7x + 12 = 0$; $[0, 6]$

44. $x^2 - 0.75x + 0.125 = 0$; $[-2, 2]$

45. $x^3 - 6x^2 + 11x - 6 = 0$; $[-1, 4]$

46. $16x^3 + 16x^2 = x + 1$; $[-2, 2]$

47. $x - \sqrt{x + 1} = 0$; $[-1, 5]$

48. $1 + \sqrt{x} = \sqrt{1 + x^2}$; $[-1, 5]$

49. $x^{1/3} - x = 0$; $[-3, 3]$

50. $x^{1/2} + x^{1/3} - x = 0$; $[-1, 5]$

51–54 ■ Use the graphical method to solve the equation in the indicated exercise from Section 1.5.

51. Exercise 91 **52.** Exercise 92

53. Exercise 97 **54.** Exercise 98

55–58 ■ Find all real solutions of the equation, rounded to two decimals.

55. $x^3 - 2x^2 - x - 1 = 0$ **56.** $x^4 - 8x^2 + 2 = 0$

57. $x(x - 1)(x + 2) = \frac{1}{6}x$ **58.** $x^4 = 16 - x^3$

59–66 ■ Find the solutions of the inequality by drawing appropriate graphs. State each answer rounded to two decimals.

59. $x^2 \le 3x + 10$ **60.** $0.5x^2 + 0.875x \le 0.25$

61. $x^3 + 11x \le 6x^2 + 6$ **62.** $16x^3 + 24x^2 > -9x - 1$

63. $x^{1/3} < x$ **64.** $\sqrt{0.5x^2 + 1} \le 2|x|$

65. $(x + 1)^2 < (x - 1)^2$ **66.** $(x + 1)^2 \le x^3$

67–70 ■ Use the graphical method to solve the inequality in the indicated exercise from Section 1.7.

67. Exercise 43 **68.** Exercise 44

69. Exercise 53 **70.** Exercise 54

71. In Example 6 we found two solutions of the equation $x^3 - 6x^2 + 9x = \sqrt{x}$, the solutions that lie between 1 and 6. Find two more solutions, correct to two decimals.

APPLICATIONS

72. Estimating Profit An appliance manufacturer estimates that the profit y (in dollars) generated by producing x cooktops per month is given by the equation

$$y = 10x + 0.5x^2 - 0.001x^3 - 5000$$

where $0 \le x \le 450$.

(a) Graph the equation.

(b) How many cooktops must be produced to begin generating a profit?

(c) For what range of values of x is the company's profit greater than \$15,000?

73. How Far Can You See? If you stand on a ship in a calm sea, then your height x (in ft) above sea level is related to the farthest distance y (in mi) that you can see by the equation

$$y = \sqrt{1.5x + \left(\frac{x}{5280}\right)^2}$$

(a) Graph the equation for $0 \le x \le 100$.

(b) How high up do you have to be to be able to see 10 mi?

DISCOVERY ■ DISCUSSION ■ WRITING

74. Misleading Graphs Write a short essay describing different ways in which a graphing calculator might give a misleading graph of an equation.

75. Algebraic and Graphical Solution Methods Write a short essay comparing the algebraic and graphical methods for solving equations. Make up your own examples to illustrate the advantages and disadvantages of each method.

76. Equation Notation on Graphing Calculators When you enter the following equations into your calculator, how does what you see on the screen differ from the usual way of writing the equations? (Check your user's manual if you're not sure.)

(a) $y = |x|$ (b) $y = \sqrt[5]{x}$

(c) $y = \dfrac{x}{x - 1}$ (d) $y = x^3 + \sqrt[3]{x + 2}$

77. Enter Equations Carefully A student wishes to graph the equations

$$y = x^{1/3} \quad \text{and} \quad y = \frac{x}{x + 4}$$

on the same screen, so he enters the following information into his calculator:

```
Y1 = X^1/3        Y2 = X/X+4
```

The calculator graphs two lines instead of the equations he wanted. What went wrong?

1.10 Lines

The Slope of a Line ▶ Point-Slope Form of the Equation of a Line ▶ Slope-Intercept Form of the Equation of a Line ▶ Vertical and Horizontal Lines ▶ General Equation of a Line ▶ Parallel and Perpendicular Lines ▶ Modeling with Linear Equations: Slope as Rate of Change

In this section we find equations for straight lines lying in a coordinate plane. The equations will depend on how the line is inclined, so we begin by discussing the concept of slope.

▼ The Slope of a Line

We first need a way to measure the "steepness" of a line, or how quickly it rises (or falls) as we move from left to right. We define *run* to be the distance we move to the right and *rise* to be the corresponding distance that the line rises (or falls). The *slope* of a line is the ratio of rise to run:

$$\text{slope} = \frac{\text{rise}}{\text{run}}$$

Figure 1 shows situations in which slope is important. Carpenters use the term *pitch* for the slope of a roof or a staircase; the term *grade* is used for the slope of a road.

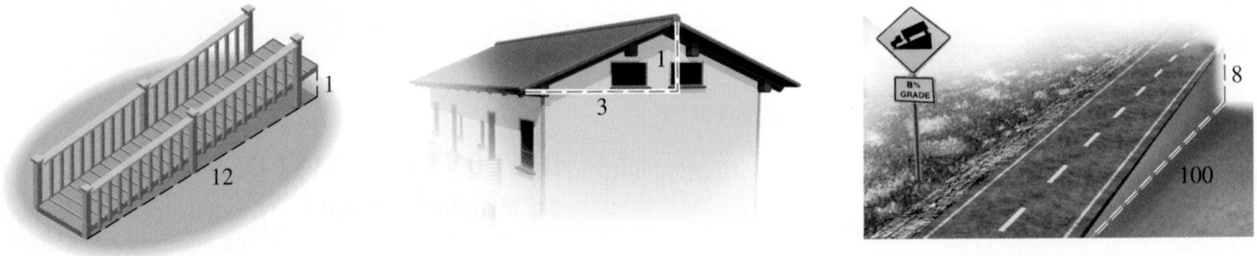

Slope of a ramp
Slope $= \frac{1}{12}$

Pitch of a roof
Slope $= \frac{1}{3}$

Grade of a road
Slope $= \frac{8}{100}$

FIGURE 1

If a line lies in a coordinate plane, then the **run** is the change in the x-coordinate and the **rise** is the corresponding change in the y-coordinate between any two points on the line (see Figure 2). This gives us the following definition of slope.

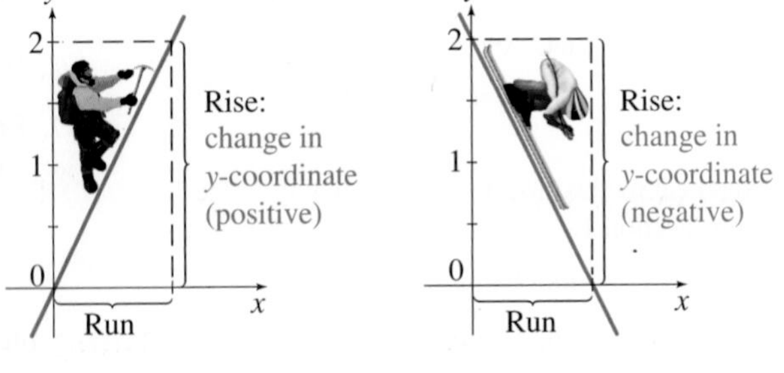

FIGURE 2

SLOPE OF A LINE

The **slope** m of a nonvertical line that passes through the points $A(x_1, y_1)$ and $B(x_2, y_2)$ is

$$m = \frac{\text{rise}}{\text{run}} = \frac{y_2 - y_1}{x_2 - x_1}$$

The slope of a vertical line is not defined.

The slope is independent of which two points are chosen on the line. We can see that this is true from the similar triangles in Figure 3:

$$\frac{y_2 - y_1}{x_2 - x_1} = \frac{y_2' - y_1'}{x_2' - x_1'}$$

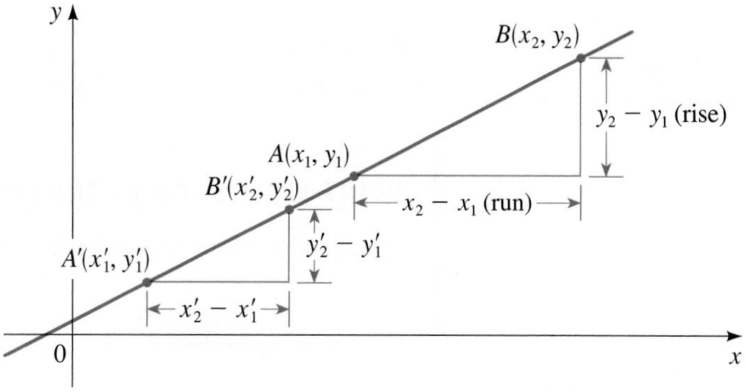

FIGURE 3

Figure 4 shows several lines labeled with their slopes. Notice that lines with positive slope slant upward to the right, whereas lines with negative slope slant downward to the right. The steepest lines are those for which the absolute value of the slope is the largest; a horizontal line has slope zero.

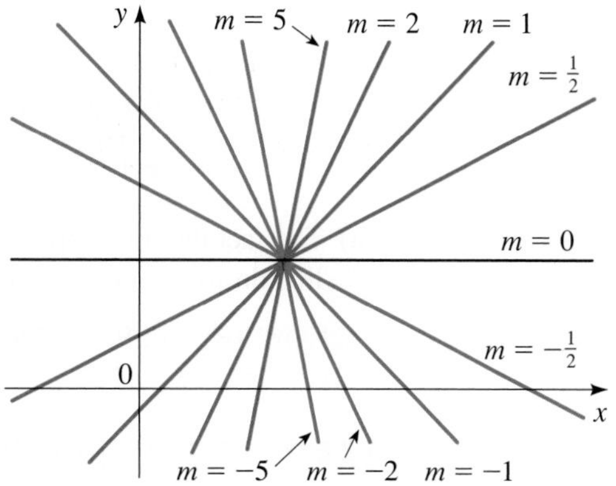

FIGURE 4 Lines with various slopes

EXAMPLE 1 | Finding the Slope of a Line Through Two Points

Find the slope of the line that passes through the points $P(2, 1)$ and $Q(8, 5)$.

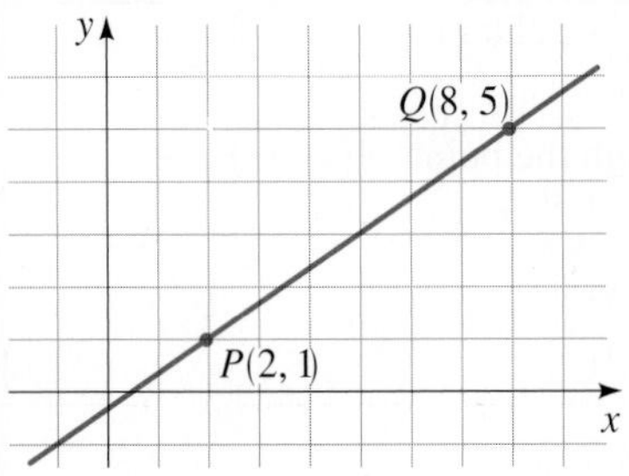

FIGURE 5

SOLUTION Since any two different points determine a line, only one line passes through these two points. From the definition the slope is

$$m = \frac{y_2 - y_1}{x_2 - x_1} = \frac{5 - 1}{8 - 2} = \frac{4}{6} = \frac{2}{3}$$

This says that for every 3 units we move to the right, the line rises 2 units. The line is drawn in Figure 5.

NOW TRY EXERCISE 5 ■

▼ Point-Slope Form of the Equation of a Line

Now let's find the equation of the line that passes through a given point $P(x_1, y_1)$ and has slope m. A point $P(x,y)$ with $x \neq x_1$ lies on this line if and only if the slope of the line through P_1 and P is equal to m (see Figure 6), that is,

$$\frac{y - y_1}{x - x_1} = m$$

This equation can be rewritten in the form $y - y_1 = m(x - x_1)$; note that the equation is also satisfied when $x = x_1$ and $y = y_1$. Therefore, it is an equation of the given line.

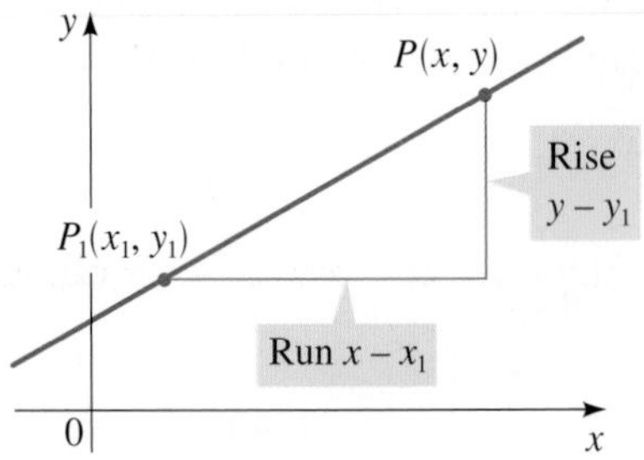

FIGURE 6

POINT-SLOPE FORM OF THE EQUATION OF A LINE

An equation of the line that passes through the point (x_1, y_1) and has slope m is

$$y - y_1 = m(x - x_1)$$

EXAMPLE 2 | Finding the Equation of a Line with Given Point and Slope

(a) Find an equation of the line through $(1, -3)$ with slope $-\frac{1}{2}$.

(b) Sketch the line.

SOLUTION

(a) Using the point-slope form with $m = -\frac{1}{2}$, $x_1 = 1$, and $y_1 = -3$, we obtain an equation of the line as

$y + 3 = -\frac{1}{2}(x - 1)$	Slope $m = -\frac{1}{2}$, point $(1, -3)$
$2y + 6 = -x + 1$	Multiply by 2
$x + 2y + 5 = 0$	Rearrange

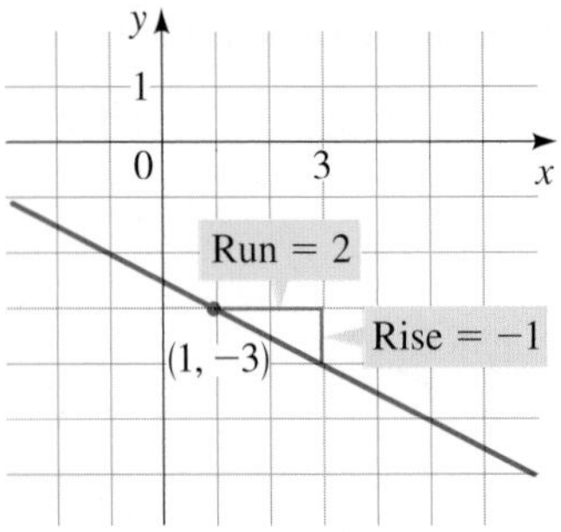

FIGURE 7

(b) The fact that the slope is $-\frac{1}{2}$ tells us that when we move to the right 2 units, the line drops 1 unit. This enables us to sketch the line in Figure 7.

NOW TRY EXERCISE 19 ■

EXAMPLE 3 | Finding the Equation of a Line Through Two Given Points

Find an equation of the line through the points $(-1, 2)$ and $(3, -4)$.

SOLUTION The slope of the line is

$$m = \frac{-4 - 2}{3 - (-1)} = -\frac{6}{4} = -\frac{3}{2}$$

We can use *either* point, $(-1, 2)$ *or* $(3, -4)$, in the point-slope equation. We will end up with the same final answer.

Using the point-slope form with $x_1 = -1$ and $y_1 = 2$, we obtain

$$y - 2 = -\tfrac{3}{2}(x + 1) \qquad \text{Slope } m = -\tfrac{3}{2}, \text{ point } (-1, 2)$$

$$2y - 4 = -3x - 3 \qquad \text{Multiply by 2}$$

$$3x + 2y - 1 = 0 \qquad \text{Rearrange}$$

NOW TRY EXERCISE **23**

Slope-Intercept Form of the Equation of a Line

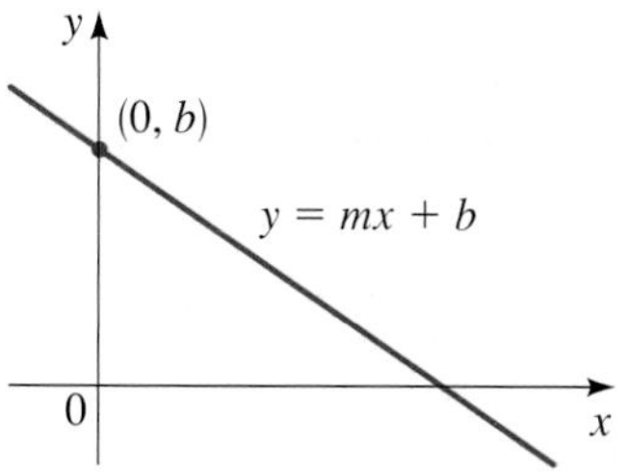

FIGURE 8

Suppose a nonvertical line has slope m and y-intercept b (see Figure 8). This means that the line intersects the y-axis at the point $(0, b)$, so the point-slope form of the equation of the line, with $x = 0$ and $y = b$, becomes

$$y - b = m(x - 0)$$

This simplifies to $y = mx + b$, which is called the **slope-intercept form** of the equation of a line.

SLOPE-INTERCEPT FORM OF THE EQUATION OF A LINE

An equation of the line that has slope m and y-intercept b is

$$y = mx + b$$

EXAMPLE 4 | Lines in Slope-Intercept Form

(a) Find the equation of the line with slope 3 and y-intercept -2.

(b) Find the slope and y-intercept of the line $3y - 2x = 1$.

SOLUTION

(a) Since $m = 3$ and $b = -2$, from the slope-intercept form of the equation of a line we get

$$y = 3x - 2$$

(b) We first write the equation in the form $y = mx + b$:

$$3y - 2x = 1$$

$$3y = 2x + 1 \qquad \text{Add } 2x$$

$$y = \tfrac{2}{3}x + \tfrac{1}{3} \qquad \text{Divide by 3}$$

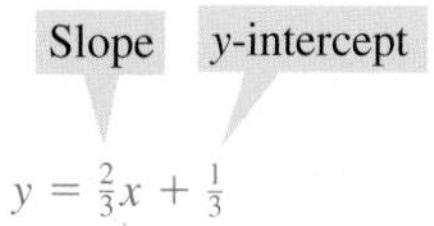

From the slope-intercept form of the equation of a line, we see that the slope is $m = \frac{2}{3}$ and the y-intercept is $b = \frac{1}{3}$.

NOW TRY EXERCISES **25** AND **47**

Vertical and Horizontal Lines

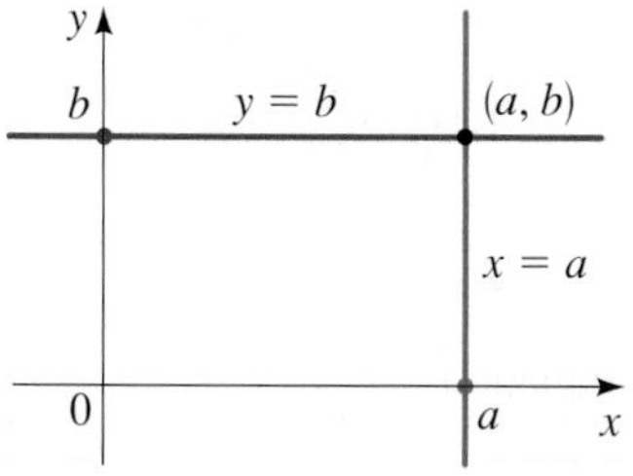

FIGURE 9

If a line is horizontal, its slope is $m = 0$, so its equation is $y = b$, where b is the y-intercept (see Figure 9). A vertical line does not have a slope, but we can write its equation as $x = a$, where a is the x-intercept, because the x-coordinate of every point on the line is a.

> **VERTICAL AND HORIZONTAL LINES**
>
> An equation of the vertical line through (a, b) is $x = a$.
>
> An equation of the horizontal line through (a, b) is $y = b$.

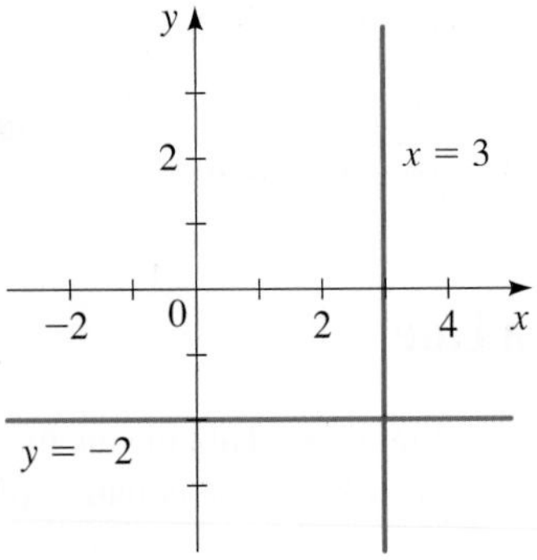

FIGURE 10

EXAMPLE 5 | Vertical and Horizontal Lines

(a) An equation for the vertical line through (3, 5) is $x = 3$.

(b) The graph of the equation $x = 3$ is a vertical line with x-intercept 3.

(c) An equation for the horizontal line through $(8, -2)$ is $y = -2$.

(d) The graph of the equation $y = -2$ is a horizontal line with y-intercept -2.

The lines are graphed in Figure 10.

NOW TRY EXERCISES **29** AND **33** ■

General Equation of a Line

A **linear equation** is an equation of the form

$$Ax + By + C = 0$$

where A, B, and C are constants and A and B are not both 0. The equation of a line is a linear equation:

- A nonvertical line has the equation $y = mx + b$ or $-mx + y - b = 0$, which is a linear equation with $A = -m$, $B = 1$, and $C = -b$.
- A vertical line has the equation $x = a$ or $x - a = 0$, which is a linear equation with $A = 1$, $B = 0$, and $C = -a$.

Conversely, the graph of a linear equation is a line.

- If $B \neq 0$, the equation becomes

$$y = -\frac{A}{B}x - \frac{C}{B} \qquad \text{Divide by } B$$

and this is the slope-intercept form of the equation of a line (with $m = -A/B$ and $b = -C/B$).

- If $B = 0$, the equation becomes

$$Ax + C = 0 \qquad \text{Set } B = 0$$

or $x = -C/A$, which represents a vertical line.

We have proved the following.

> **GENERAL EQUATION OF A LINE**
>
> The graph of every **linear equation**
>
> $$Ax + By + C = 0 \qquad (A, B \text{ not both zero})$$
>
> is a line. Conversely, every line is the graph of a linear equation.

EXAMPLE 6 | Graphing a Linear Equation

Sketch the graph of the equation $2x - 3y - 12 = 0$.

SOLUTION 1 Since the equation is linear, its graph is a line. To draw the graph, it is enough to find any two points on the line. The intercepts are the easiest points to find.

x-intercept: Substitute $y = 0$, to get $2x - 12 = 0$, so $x = 6$

y-intercept: Substitute $x = 0$, to get $-3y - 12 = 0$, so $y = -4$

With these points we can sketch the graph in Figure 11.

SOLUTION 2 We write the equation in slope-intercept form:

$$\begin{aligned} 2x - 3y - 12 &= 0 \\ 2x - 3y &= 12 && \text{Add 12} \\ -3y &= -2x + 12 && \text{Subtract } 2x \\ y &= \tfrac{2}{3}x - 4 && \text{Divide by } -3 \end{aligned}$$

This equation is in the form $y = mx + b$, so the slope is $m = \frac{2}{3}$ and the y-intercept is $b = -4$. To sketch the graph, we plot the y-intercept and then move 3 units to the right and 2 units up as shown in Figure 12.

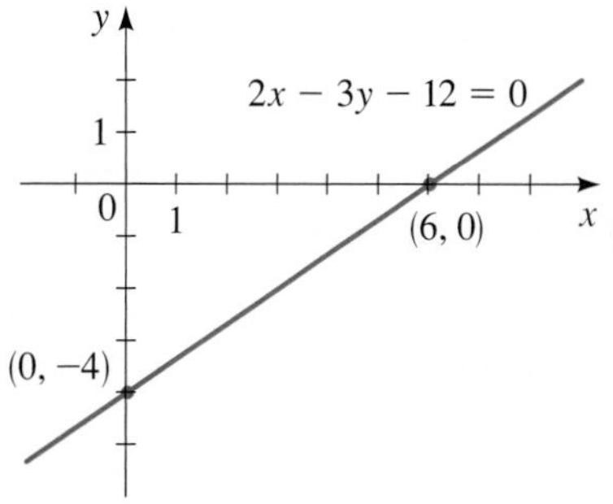

FIGURE 11

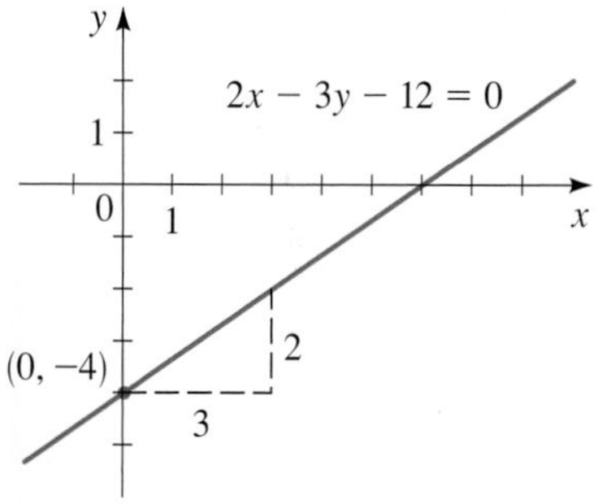

FIGURE 12

NOW TRY EXERCISE **53** ■

▼ Parallel and Perpendicular Lines

Since slope measures the steepness of a line, it seems reasonable that parallel lines should have the same slope. In fact, we can prove this.

PARALLEL LINES

Two nonvertical lines are parallel if and only if they have the same slope.

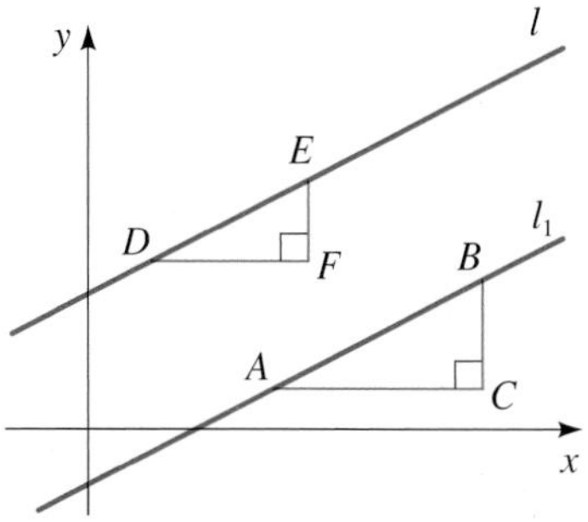

FIGURE 13

PROOF Let the lines l_1 and l_2 in Figure 13 have slopes m_1 and m_2. If the lines are parallel, then the right triangles ABC and DEF are similar, so

$$m_1 = \frac{d(B, C)}{d(A, C)} = \frac{d(E, F)}{d(D, F)} = m_2$$

Conversely, if the slopes are equal, then the triangles will be similar, so $\angle BAC = \angle EDF$ and the lines are parallel. ■

EXAMPLE 7 | Finding the Equation of a Line Parallel to a Given Line

Find an equation of the line through the point $(5, 2)$ that is parallel to the line $4x + 6y + 5 = 0$.

SOLUTION First we write the equation of the given line in slope-intercept form.

$$\begin{aligned} 4x + 6y + 5 &= 0 \\ 6y &= -4x - 5 && \text{Subtract } 4x + 5 \\ y &= -\tfrac{2}{3}x - \tfrac{5}{6} && \text{Divide by 6} \end{aligned}$$

So the line has slope $m = -\frac{2}{3}$. Since the required line is parallel to the given line, it also has slope $m = -\frac{2}{3}$. From the point-slope form of the equation of a line, we get

$$\begin{aligned} y - 2 &= -\tfrac{2}{3}(x - 5) && \text{Slope } m = -\tfrac{2}{3}, \text{ point } (5, 2) \\ 3y - 6 &= -2x + 10 && \text{Multiply by 3} \\ 2x + 3y - 16 &= 0 && \text{Rearrange} \end{aligned}$$

Thus, the equation of the required line is $2x + 3y - 16 = 0$.

NOW TRY EXERCISE **31** ■

The condition for perpendicular lines is not as obvious as that for parallel lines.

PERPENDICULAR LINES

Two lines with slopes m_1 and m_2 are perpendicular if and only if $m_1m_2 = -1$, that is, their slopes are negative reciprocals:

$$m_2 = -\frac{1}{m_1}$$

Also, a horizontal line (slope 0) is perpendicular to a vertical line (no slope).

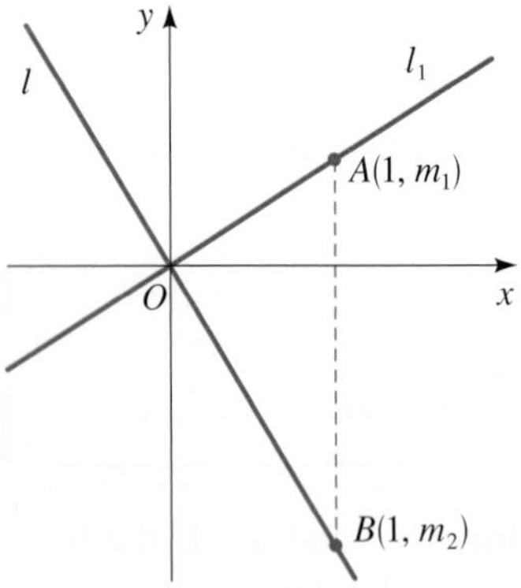

FIGURE 14

PROOF In Figure 14 we show two lines intersecting at the origin. (If the lines intersect at some other point, we consider lines parallel to these that intersect at the origin. These lines have the same slopes as the original lines.)

If the lines l_1 and l_2 have slopes m_1 and m_2, then their equations are $y = m_1x$ and $y = m_2x$. Notice that $A(1, m_1)$ lies on l_1 and $B(1, m_2)$ lies on l_2. By the Pythagorean Theorem and its converse (see page 219), $OA \perp OB$ if and only if

$$[d(O, A)]^2 + [d(O, B)]^2 = [d(A, B)]^2$$

By the Distance Formula this becomes

$$\begin{aligned} (1^2 + m_1^2) + (1^2 + m_2^2) &= (1 - 1)^2 + (m_2 - m_1)^2 \\ 2 + m_1^2 + m_2^2 &= m_2^2 - 2m_1m_2 + m_1^2 \\ 2 &= -2m_1m_2 \\ m_1m_2 &= -1 \end{aligned}$$

■

EXAMPLE 8 | Perpendicular Lines

Show that the points $P(3, 3)$, $Q(8, 17)$, and $R(11, 5)$ are the vertices of a right triangle.

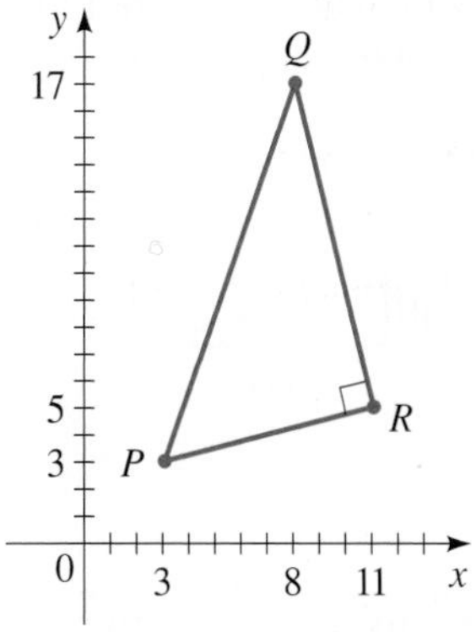

FIGURE 15

SOLUTION The slopes of the lines containing PR and QR are, respectively,

$$m_1 = \frac{5-3}{11-3} = \frac{1}{4} \quad \text{and} \quad m_2 = \frac{5-17}{11-8} = -4$$

Since $m_1 m_2 = -1$, these lines are perpendicular, so PQR is a right triangle. It is sketched in Figure 15.

NOW TRY EXERCISE **57** ■

EXAMPLE 9 | Finding an Equation of a Line Perpendicular to a Given Line

Find an equation of the line that is perpendicular to the line $4x + 6y + 5 = 0$ and passes through the origin.

SOLUTION In Example 7 we found that the slope of the line $4x + 6y + 5 = 0$ is $-\frac{2}{3}$. Thus, the slope of a perpendicular line is the negative reciprocal, that is, $\frac{3}{2}$. Since the required line passes through $(0, 0)$, the point-slope form gives

$$y - 0 = \tfrac{3}{2}(x - 0) \qquad \text{Slope } m = \tfrac{3}{2}, \text{ point } (0, 0)$$

$$y = \tfrac{3}{2}x \qquad \text{Simplify}$$

NOW TRY EXERCISE **35** ■

EXAMPLE 10 | Graphing a Family of Lines

Use a graphing calculator to graph the family of lines

$$y = 0.5x + b$$

for $b = -2, -1, 0, 1, 2$. What property do the lines share?

SOLUTION The lines are graphed in Figure 16 in the viewing rectangle $[-6, 6]$ by $[-6, 6]$. The lines all have the same slope, so they are parallel.

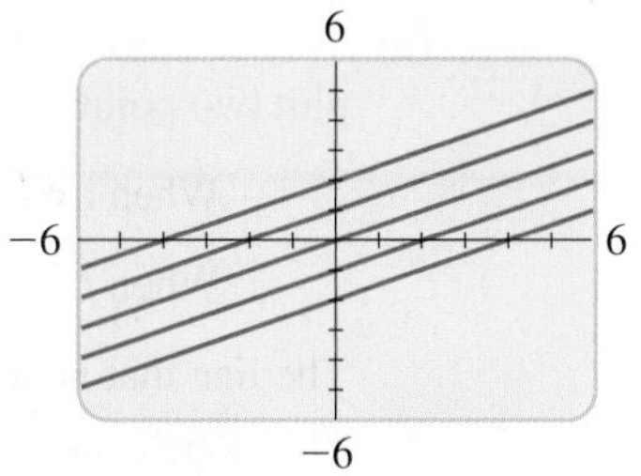

FIGURE 16 $y = 0.5x + b$

NOW TRY EXERCISE **41** ■

▼ Modeling with Linear Equations: Slope as Rate of Change

When a line is used to model the relationship between two quantities, the slope of the line is the **rate of change** of one quantity with respect to the other. For example, the graph in Figure 17(a) on the next page gives the amount of gas in a tank that is being filled. The slope between the indicated points is

$$m = \frac{6 \text{ gallons}}{3 \text{ minutes}} = 2 \text{ gal/min}$$

The slope is the *rate* at which the tank is being filled, 2 gallons per minute. In Figure 17(b) the tank is being drained at the *rate* of 0.03 gallon per minute, and the slope is −0.03.

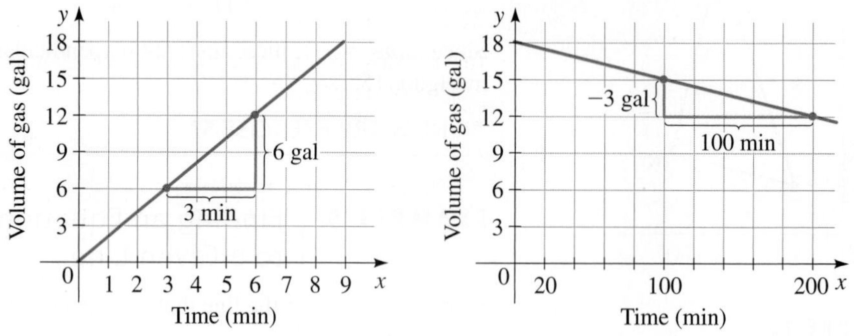

(a) Tank filled at 2 gal/min
Slope of line is 2

(b) Tank drained at 0.03 gal/min
Slope of line is −0.03

FIGURE 17

The next two examples give other situations in which the slope of a line is a rate of change.

EXAMPLE 11 | Slope as Rate of Change

A dam is built on a river to create a reservoir. The water level w in the reservoir is given by the equation

$$w = 4.5t + 28$$

where t is the number of years since the dam was constructed and w is measured in feet.

(a) Sketch a graph of this equation.

(b) What do the slope and w-intercept of this graph represent?

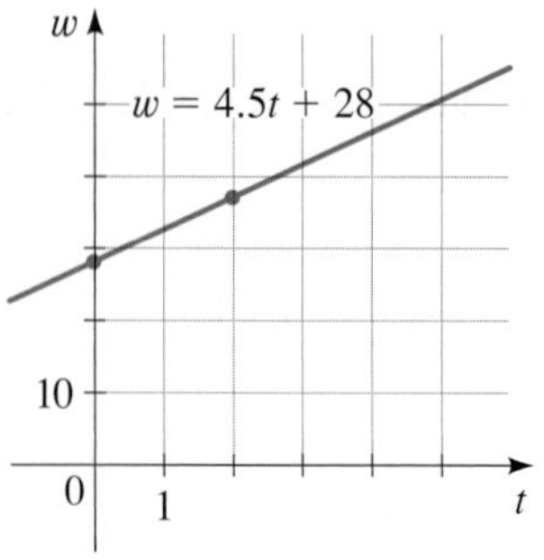

FIGURE 18

SOLUTION

(a) This equation is linear, so its graph is a line. Since two points determine a line, we plot two points that lie on the graph and draw a line through them.

When $t = 0$, then $w = 4.5(0) + 28 = 28$, so $(0, 28)$ is on the line.

When $t = 2$, then $w = 4.5(2) + 28 = 37$, so $(2, 37)$ is on the line.

The line that is determined by these points is shown in Figure 18.

(b) The slope is $m = 4.5$; it represents the rate of change of water level with respect to time. This means that the water level *increases* 4.5 ft per year. The w-intercept is 28 and occurs when $t = 0$, so it represents the water level when the dam was constructed.

NOW TRY EXERCISE **69** ■

EXAMPLE 12 | Linear Relationship Between Temperature and Elevation

(a) As dry air moves upward, it expands and cools. If the ground temperature is 20°C and the temperature at a height of 1 km is 10°C, express the temperature T (in °C) in terms of the height h (in km). (Assume that the relationship between T and h is linear.)

(b) Draw the graph of the linear equation. What does its slope represent?

(c) What is the temperature at a height of 2.5 km?

SOLUTION

(a) Because we are assuming a linear relationship between T and h, the equation must be of the form

$$T = mh + b$$

where m and b are constants. When $h = 0$, we are given that $T = 20$, so

$$20 = m(0) + b$$
$$b = 20$$

Thus, we have

$$T = mh + 20$$

When $h = 1$, we have $T = 10$, so

$$10 = m(1) + 20$$
$$m = 10 - 20 = -10$$

The required expression is

$$T = -10h + 20$$

(b) The graph is sketched in Figure 19. The slope is $m = -10°\text{C/km}$, and this represents the rate of change of temperature with respect to distance above the ground. So the temperature *decreases* 10°C per kilometer of height.

(c) At a height of $h = 2.5$ km the temperature is

$$T = -10(2.5) + 20 = -25 + 20 = -5°\text{C}$$

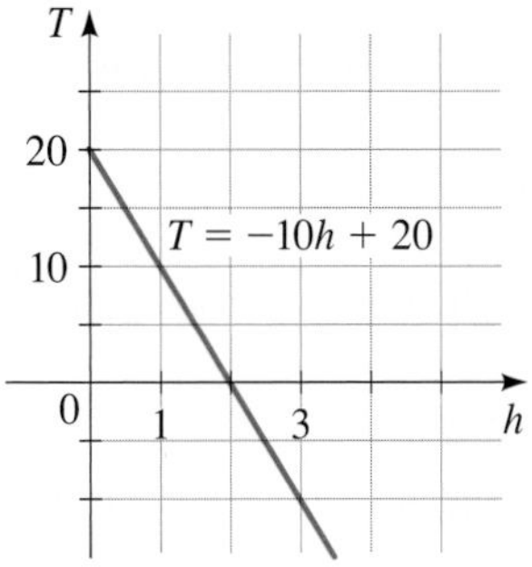

FIGURE 19

NOW TRY EXERCISE **73**

1.10 EXERCISES

CONCEPTS

1. We find the "steepness," or slope, of a line passing through two points by dividing the difference in the ____-coordinates of these points by the difference in the ____-coordinates. So the line passing through the points (0, 1) and (2, 5) has slope ________.

2. A line has the equation $y = 3x + 2$.

(a) This line has slope ________.

(b) Any line parallel to this line has slope ________.

(c) Any line perpendicular to this line has slope ________.

3. The point-slope form of the equation of the line with slope 3 passing through the point (1, 2) is ________.

4. (a) The slope of a horizontal line is ________. The equation of the horizontal line passing through (2, 3) is ________.

(b) The slope of a vertical line is ________. The equation of the vertical line passing through (2, 3) is ________.

SKILLS

5–12 ■ Find the slope of the line through P and Q.

5. $P(0, 0)$, $Q(4, 2)$

6. $P(0, 0)$, $Q(2, -6)$

7. $P(2, 2)$, $Q(-10, 0)$

8. $P(1, 2)$, $Q(3, 3)$

9. $P(2, 4)$, $Q(4, 3)$

10. $P(2, -5)$, $Q(-4, 3)$

11. $P(1, -3)$, $Q(-1, 6)$

12. $P(-1, -4)$, $Q(6, 0)$

13. Find the slopes of the lines l_1, l_2, l_3, and l_4 in the figure below.

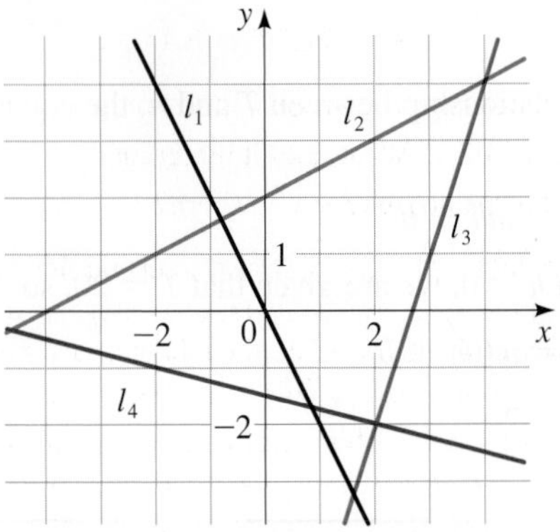

14. (a) Sketch lines through $(0, 0)$ with slopes $1, 0, \frac{1}{2}, 2$, and -1.

(b) Sketch lines through $(0, 0)$ with slopes $\frac{1}{3}, \frac{1}{2}, -\frac{1}{3}$, and 3.

15–18 ■ Find an equation for the line whose graph is sketched.

15.

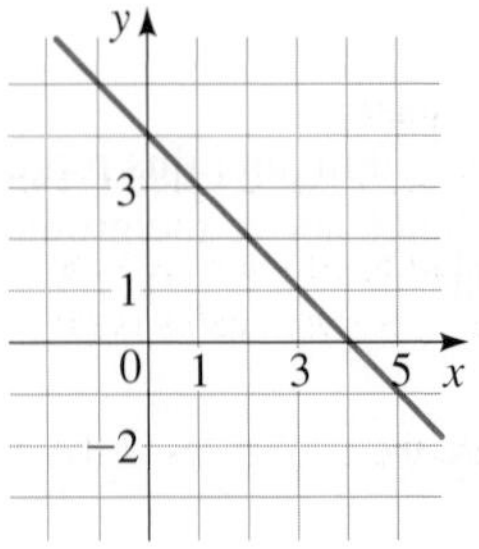

16.

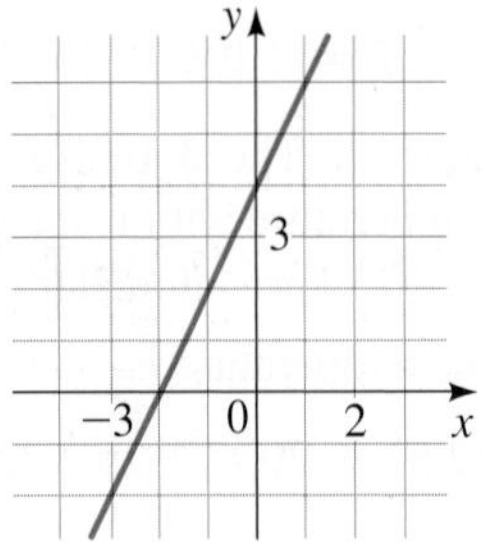

17.

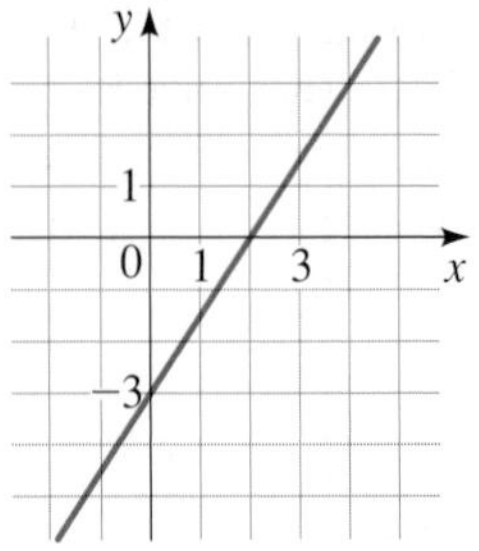

18.

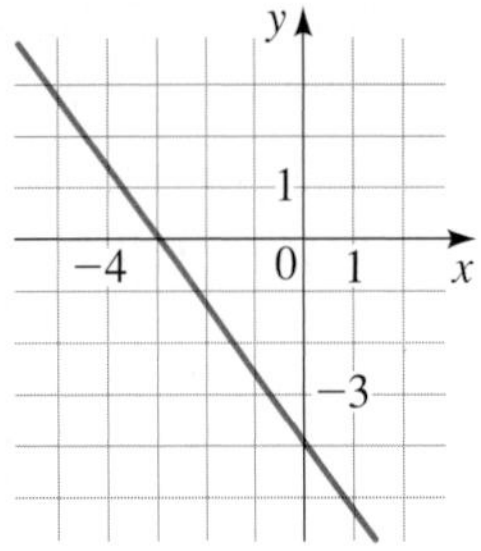

19–38 ■ Find an equation of the line that satisfies the given conditions.

19. Through $(2, 3)$; slope 5

20. Through $(-2, 4)$; slope -1

21. Through $(1, 7)$; slope $\frac{2}{3}$

22. Through $(-3, -5)$; slope $-\frac{7}{2}$

23. Through $(2, 1)$ and $(1, 6)$

24. Through $(-1, -2)$ and $(4, 3)$

25. Slope 3; y-intercept -2

26. Slope $\frac{2}{5}$; y-intercept 4

27. x-intercept 1; y-intercept -3

28. x-intercept -8; y-intercept 6

29. Through $(4, 5)$; parallel to the x-axis

30. Through $(4, 5)$; parallel to the y-axis

31. Through $(1, -6)$; parallel to the line $x + 2y = 6$

32. y-intercept 6; parallel to the line $2x + 3y + 4 = 0$

33. Through $(-1, 2)$; parallel to the line $x = 5$

34. Through $(2, 6)$; perpendicular to the line $y = 1$

35. Through $(-1, -2)$; perpendicular to the line $2x + 5y + 8 = 0$

36. Through $(\frac{1}{2}, -\frac{2}{3})$; perpendicular to the line $4x - 8y = 1$

37. Through $(1, 7)$; parallel to the line passing through $(2, 5)$ and $(-2, 1)$

38. Through $(-2, -11)$; perpendicular to the line passing through $(1, 1)$ and $(5, -1)$

39. (a) Sketch the line with slope $\frac{3}{2}$ that passes through the point $(-2, 1)$.

(b) Find an equation for this line.

40. (a) Sketch the line with slope -2 that passes through the point $(4, -1)$.

(b) Find an equation for this line.

41–44 ■ Use a graphing device to graph the given family of lines in the same viewing rectangle. What do the lines have in common?

41. $y = -2x + b$ for $b = 0, \pm 1, \pm 3, \pm 6$

42. $y = mx - 3$ for $m = 0, \pm 0.25, \pm 0.75, \pm 1.5$

43. $y = m(x - 3)$ for $m = 0, \pm 0.25, \pm 0.75, \pm 1.5$

44. $y = 2 + m(x + 3)$ for $m = 0, \pm 0.5, \pm 1, \pm 2, \pm 6$

45–56 ■ Find the slope and y-intercept of the line and draw its graph.

45. $x + y = 3$

46. $3x - 2y = 12$

47. $x + 3y = 0$

48. $2x - 5y = 0$

49. $\frac{1}{2}x - \frac{1}{3}y + 1 = 0$

50. $-3x - 5y + 30 = 0$

51. $y = 4$

52. $x = -5$

53. $3x - 4y = 12$

54. $4y + 8 = 0$

55. $3x + 4y - 1 = 0$

56. $4x + 5y = 10$

57. Use slopes to show that $A(1, 1)$, $B(7, 4)$, $C(5, 10)$, and $D(-1, 7)$ are vertices of a parallelogram.

58. Use slopes to show that $A(-3, -1)$, $B(3, 3)$, and $C(-9, 8)$ are vertices of a right triangle.

59. Use slopes to show that $A(1, 1)$, $B(11, 3)$, $C(10, 8)$, and $D(0, 6)$ are vertices of a rectangle.

60. Use slopes to determine whether the given points are collinear (lie on a line).

(a) $(1, 1), (3, 9), (6, 21)$

(b) $(-1, 3), (1, 7), (4, 15)$

61. Find an equation of the perpendicular bisector of the line segment joining the points $A(1, 4)$ and $B(7, -2)$.

62. Find the area of the triangle formed by the coordinate axes and the line

$$2y + 3x - 6 = 0$$

63. (a) Show that if the x- and y-intercepts of a line are nonzero numbers a and b, then the equation of the line can be written in the form

$$\frac{x}{a} + \frac{y}{b} = 1$$

This is called the **two-intercept form** of the equation of a line.

(b) Use part (a) to find an equation of the line whose x-intercept is 6 and whose y-intercept is -8.

64. (a) Find an equation for the line tangent to the circle $x^2 + y^2 = 25$ at the point $(3, -4)$. (See the figure.)

(b) At what other point on the circle will a tangent line be parallel to the tangent line in part (a)?

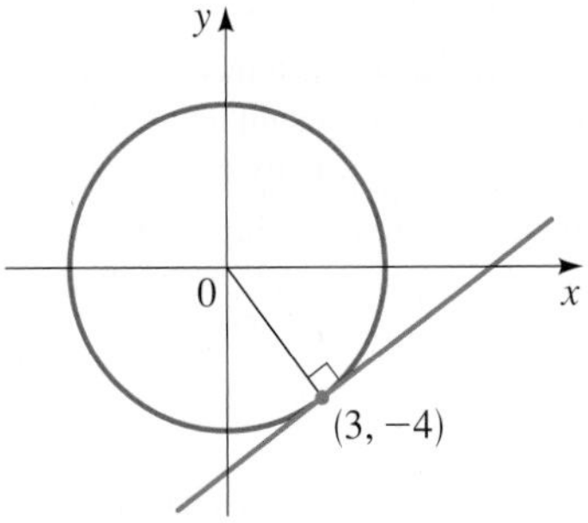

APPLICATIONS

65. Grade of a Road West of Albuquerque, New Mexico, Route 40 eastbound is straight and makes a steep descent toward the city. The highway has a 6% grade, which means that its slope is $-\frac{6}{100}$. Driving on this road, you notice from elevation signs that you have descended a distance of 1000 ft. What is the change in your horizontal distance?

66. Global Warming Some scientists believe that the average surface temperature of the world has been rising steadily. The average surface temperature can be modeled by

$$T = 0.02t + 15.0$$

where T is temperature in °C and t is years since 1950.

(a) What do the slope and T-intercept represent?

(b) Use the equation to predict the average global surface temperature in 2050.

67. Drug Dosages If the recommended adult dosage for a drug is D (in mg), then to determine the appropriate dosage c for a child of age a, pharmacists use the equation

$$c = 0.0417D(a + 1)$$

Suppose the dosage for an adult is 200 mg.

(a) Find the slope. What does it represent?

(b) What is the dosage for a newborn?

68. Flea Market The manager of a weekend flea market knows from past experience that if she charges x dollars for a rental space at the flea market, then the number y of spaces she can rent is given by the equation $y = 200 - 4x$.

(a) Sketch a graph of this linear equation. (Remember that the rental charge per space and the number of spaces rented must both be nonnegative quantities.)

(b) What do the slope, the y-intercept, and the x-intercept of the graph represent?

69. Production Cost A small-appliance manufacturer finds that if he produces x toaster ovens in a month, his production cost is given by the equation

$$y = 6x + 3000$$

(where y is measured in dollars).

(a) Sketch a graph of this linear equation.

(b) What do the slope and y-intercept of the graph represent?

70. Temperature Scales The relationship between the Fahrenheit (F) and Celsius (C) temperature scales is given by the equation $F = \frac{9}{5}C + 32$.

(a) Complete the table to compare the two scales at the given values.

(b) Find the temperature at which the scales agree. [*Hint:* Suppose that a is the temperature at which the scales agree. Set $F = a$ and $C = a$. Then solve for a.]

C	F
−30°	
−20°	
−10°	
0°	
	50°
	68°
	86°

71. Crickets and Temperature Biologists have observed that the chirping rate of crickets of a certain species is related to temperature, and the relationship appears to be very nearly linear. A cricket produces 120 chirps per minute at 70°F and 168 chirps per minute at 80°F.

(a) Find the linear equation that relates the temperature t and the number of chirps per minute n.

(b) If the crickets are chirping at 150 chirps per minute, estimate the temperature.

72. Depreciation A small business buys a computer for \$4000. After 4 years the value of the computer is expected to be \$200. For accounting purposes the business uses *linear depreciation* to assess the value of the computer at a given time.

This means that if V is the value of the computer at time t, then a linear equation is used to relate V and t.

(a) Find a linear equation that relates V and t.

(b) Sketch a graph of this linear equation.

(c) What do the slope and V-intercept of the graph represent?

(d) Find the depreciated value of the computer 3 years from the date of purchase.

73. Pressure and Depth At the surface of the ocean the water pressure is the same as the air pressure above the water, 15 lb/in^2. Below the surface the water pressure increases by 4.34 lb/in^2 for every 10 ft of descent.

(a) Find an equation for the relationship between pressure and depth below the ocean surface.

(b) Sketch a graph of this linear equation.

(c) What do the slope and y-intercept of the graph represent?

(d) At what depth is the pressure 100 lb/in^2?

74. Distance, Speed, and Time Jason and Debbie leave Detroit at 2:00 P.M. and drive at a constant speed, traveling west on I-90. They pass Ann Arbor, 40 mi from Detroit, at 2:50 P.M.

(a) Express the distance traveled in terms of the time elapsed.

(b) Draw the graph of the equation in part (a).

(c) What is the slope of this line? What does it represent?

75. Cost of Driving The monthly cost of driving a car depends on the number of miles driven. Lynn found that in May her driving cost was \$380 for 480 mi and in June her cost was \$460 for 800 mi. Assume that there is a linear relationship between the monthly cost C of driving a car and the distance driven d.

(a) Find a linear equation that relates C and d.

(b) Use part (a) to predict the cost of driving 1500 mi per month.

(c) Draw the graph of the linear equation. What does the slope of the line represent?

(d) What does the y-intercept of the graph represent?

(e) Why is a linear relationship a suitable model for this situation?

76. Manufacturing Cost The manager of a furniture factory finds that it costs \$2200 to manufacture 100 chairs in one day and \$4800 to produce 300 chairs in one day.

(a) Assuming that the relationship between cost and the number of chairs produced is linear, find an equation that expresses this relationship. Then graph the equation.

(b) What is the slope of the line in part (a), and what does it represent?

(c) What is the y-intercept of this line, and what does it represent?

DISCOVERY ■ DISCUSSION ■ WRITING

77. What Does the Slope Mean? Suppose that the graph of the outdoor temperature over a certain period of time is a line. How is the weather changing if the slope of the line is positive? If it is negative? If it is zero?

78. Collinear Points Suppose that you are given the coordinates of three points in the plane and you want to see whether they lie on the same line. How can you do this using slopes? Using the Distance Formula? Can you think of another method?

1.11 Making Models Using Variation

| Direct Variation ▶ Inverse Variation ▶ Joint Variation

When scientists talk about a mathematical model for a real-world phenomenon, they often mean an equation that describes the relationship between two quantities. For instance, the model might describe how the population of an animal species varies with time or how the pressure of a gas varies as its temperature changes. In this section we study a kind of modeling called *variation*.

Direct Variation

Two types of mathematical models occur so often that they are given special names. The first is called *direct variation* and occurs when one quantity is a constant multiple of the other, so we use an equation of the form $y = kx$ to model this dependence.

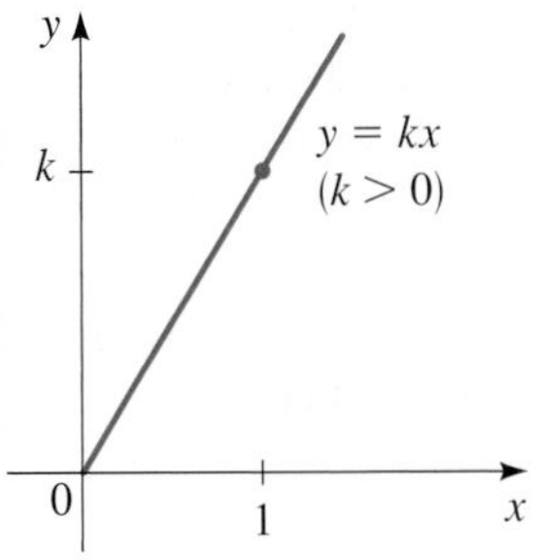

FIGURE 1

DIRECT VARIATION

If the quantities x and y are related by an equation

$$y = kx$$

for some constant $k \neq 0$, we say that y **varies directly as** x, or y is **directly proportional to** x, or simply y **is proportional to** x. The constant k is called the **constant of proportionality**.

Recall that the graph of an equation of the form $y = mx + b$ is a line with slope m and y-intercept b. So the graph of an equation $y = kx$ that describes direct variation is a line with slope k and y-intercept 0 (see Figure 1).

EXAMPLE 1 | Direct Variation

During a thunderstorm you see the lightning before you hear the thunder because light travels much faster than sound. The distance between you and the storm varies directly as the time interval between the lightning and the thunder.

(a) Suppose that the thunder from a storm 5400 ft away takes 5 s to reach you. Determine the constant of proportionality, and write the equation for the variation.

(b) Sketch the graph of this equation. What does the constant of proportionality represent?

(c) If the time interval between the lightning and thunder is now 8 s, how far away is the storm?

SOLUTION

(a) Let d be the distance from you to the storm, and let t be the length of the time interval. We are given that d varies directly as t, so

$$d = kt$$

where k is a constant. To find k, we use the fact that $t = 5$ when $d = 5400$. Substituting these values in the equation, we get

$$5400 = k(5) \qquad \text{Substitute}$$

$$k = \frac{5400}{5} = 1080 \qquad \text{Solve for } k$$

Substituting this value of k in the equation for d, we obtain

$$d = 1080t$$

as the equation for d as a function of t.

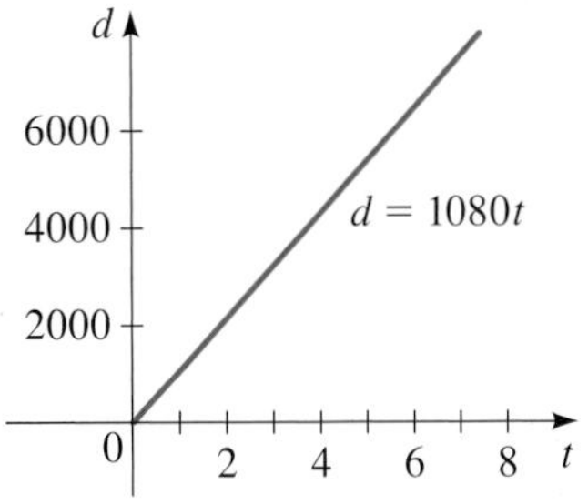

FIGURE 2

(b) The graph of the equation $d = 1080t$ is a line through the origin with slope 1080 and is shown in Figure 2. The constant $k = 1080$ is the approximate speed of sound (in ft/s).

(c) When $t = 8$, we have

$$d = 1080 \cdot 8 = 8640$$

So the storm is 8640 ft $\approx$ 1.6 mi away.

NOW TRY EXERCISES **17** AND **29**

▼ Inverse Variation

Another equation that is frequently used in mathematical modeling is $y = k/x$, where k is a constant.

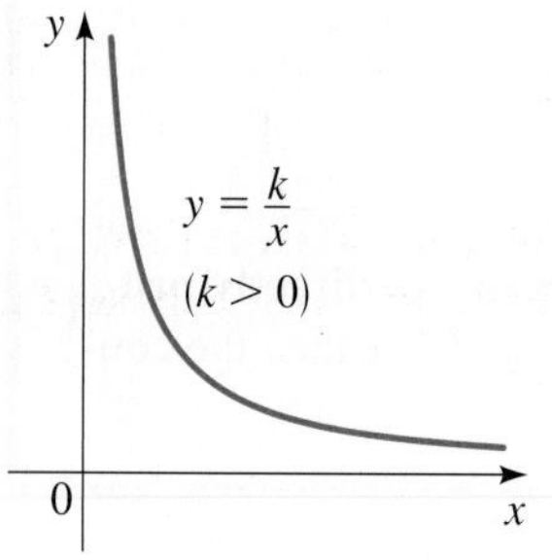

FIGURE 3 Inverse variation

> **INVERSE VARIATION**
>
> If the quantities x and y are related by the equation
>
> $$y = \frac{k}{x}$$
>
> for some constant $k \neq 0$, we say that y **is inversely proportional to** x or y **varies inversely as** x. The constant k is called the **constant of proportionality**.

The graph of $y = k/x$ for $x > 0$ is shown in Figure 3 for the case $k > 0$. It gives a picture of what happens when y is inversely proportional to x.

EXAMPLE 2 | Inverse Variation

Boyle's Law states that when a sample of gas is compressed at a constant temperature, the pressure of the gas is inversely proportional to the volume of the gas.

(a) Suppose the pressure of a sample of air that occupies 0.106 m^3 at 25°C is 50 kPa. Find the constant of proportionality, and write the equation that expresses the inverse proportionality.

(b) If the sample expands to a volume of 0.3 m^3, find the new pressure.

SOLUTION

(a) Let P be the pressure of the sample of gas, and let V be its volume. Then, by the definition of inverse proportionality, we have

$$P = \frac{k}{V}$$

where k is a constant. To find k, we use the fact that $P = 50$ when $V = 0.106$. Substituting these values in the equation, we get

$$50 = \frac{k}{0.106} \qquad \text{Substitute}$$

$$k = (50)(0.106) = 5.3 \qquad \text{Solve for } k$$

Putting this value of k in the equation for P, we have

$$P = \frac{5.3}{V}$$

(b) When $V = 0.3$, we have

$$P = \frac{5.3}{0.3} \approx 17.7$$

So the new pressure is about 17.7 kPa.

NOW TRY EXERCISES **19** AND **35** ■

▼ Joint Variation

A physical quantity often depends on more than one other quantity. If one quantity is proportional to two or more other quantities, we call this relationship *joint variation*.

JOINT VARIATION

If the quantities x, y, and z are related by the equation

$$z = kxy$$

where k is a nonzero constant, we say that z **varies jointly as** x and y or z **is jointly proportional to** x and y.

In the sciences, relationships between three or more variables are common, and any combination of the different types of proportionality that we have discussed is possible. For example, if

$$z = k\frac{x}{y}$$

we say that z **is proportional to** x and **inversely proportional to** y.

EXAMPLE 3 | Newton's Law of Gravitation

Newton's Law of Gravitation says that two objects with masses m_1 and m_2 attract each other with a force F that is jointly proportional to their masses and inversely proportional to the square of the distance r between the objects. Express Newton's Law of Gravitation as an equation.

SOLUTION Using the definitions of joint and inverse variation and the traditional notation G for the gravitational constant of proportionality, we have

$$F = G\frac{m_1m_2}{r^2}$$

NOW TRY EXERCISES **21** AND **41**

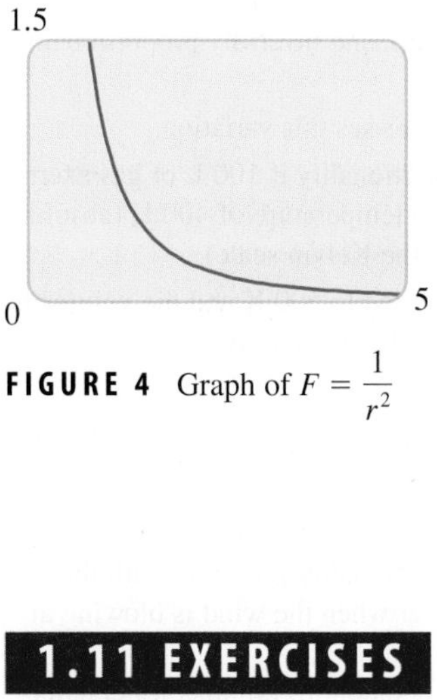

FIGURE 4 Graph of $F = \frac{1}{r^2}$

If m_1 and m_2 are fixed masses, then the gravitational force between them is $F = C/r^2$ (where $C = Gm_1m_2$ is a constant). Figure 4 shows the graph of this equation for $r > 0$ with $C = 1$. Observe how the gravitational attraction decreases with increasing distance.

1.11 EXERCISES

CONCEPTS

1. If the quantities x and y are related by the equation $y = 3x$, then we say that y is ________ ________ to x and the constant of ________ is 3.

2. If the quantities x and y are related by the equation $y = \frac{3}{x}$, then we say that y is ________ ________ to x and the constant of ________ is 3.

3. If the quantities x, y, and z are related by the equation $z = 3\frac{x}{y}$, then we say that z is ________ ________ to x and ________ ________ to y.

4. If z is jointly proportional to x and y and if z is 10 when x is 4 and y is 5, then x, y, and z are related by the equation $z =$ ________.

SKILLS

5–16 ■ Write an equation that expresses the statement.

5. T varies directly as x.

6. P is directly proportional to w.

7. v is inversely proportional to z.

8. w is jointly proportional to m and n.

9. y is proportional to s and inversely proportional to t.

10. P varies inversely as T.

11. z is proportional to the square root of y.

12. A is proportional to the square of t and inversely proportional to the cube of x.

13. V is jointly proportional to l, w, and h.

14. S is jointly proportional to the squares of r and θ.

15. R is proportional to i and inversely proportional to P and t.

16. A is jointly proportional to the square roots of x and y.

17–28 ■ Express the statement as an equation. Use the given information to find the constant of proportionality.

17. y is directly proportional to x. If $x = 6$, then $y = 42$.

18. z varies inversely as t. If $t = 3$, then $z = 5$.

19. R is inversely proportional to s. If $s = 4$, then $R = 3$.

20. P is directly proportional to T. If $T = 300$, then $P = 20$.

21. M varies directly as x and inversely as y. If $x = 2$ and $y = 6$, then $M = 5$.

22. S varies jointly as p and q. If $p = 4$ and $q = 5$, then $S = 180$.

23. W is inversely proportional to the square of r. If $r = 6$, then $W = 10$.

24. t is jointly proportional to x and y, and inversely proportional to r. If $x = 2$, $y = 3$, and $r = 12$, then $t = 25$.

25. C is jointly proportional to l, w, and h. If $l = w = h = 2$, then $C = 128$.

26. H is jointly proportional to the squares of l and w. If $l = 2$ and $w = \frac{1}{3}$, then $H = 36$.

27. s is inversely proportional to the square root of t. If $s = 100$, then $t = 25$.

28. M is jointly proportional to a, b, and c and inversely proportional to d. If a and d have the same value and if b and c are both 2, then $M = 128$.

APPLICATIONS

29. **Hooke's Law** Hooke's Law states that the force needed to keep a spring stretched x units beyond its natural length is directly proportional to x. Here the constant of proportionality is called the **spring constant**.
   (a) Write Hooke's Law as an equation.
   (b) If a spring has a natural length of 10 cm and a force of 40 N is required to maintain the spring stretched to a length of 15 cm, find the spring constant.
   (c) What force is needed to keep the spring stretched to a length of 14 cm?

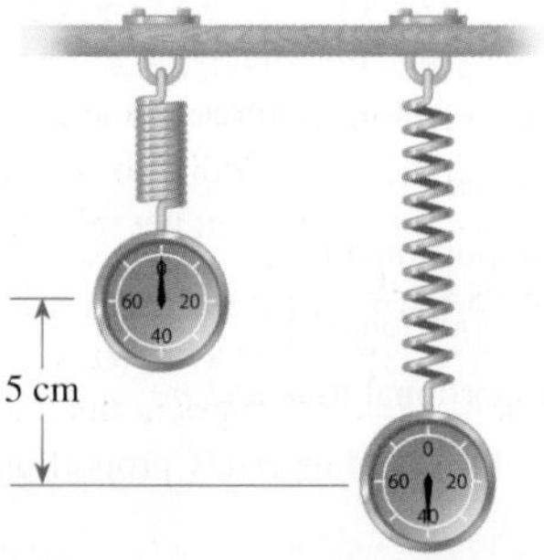

30. **Law of the Pendulum** The period of a pendulum (the time elapsed during one complete swing of the pendulum) varies directly with the square root of the length of the pendulum.
   (a) Express this relationship by writing an equation.
   (b) To double the period, how would we have to change the length l?

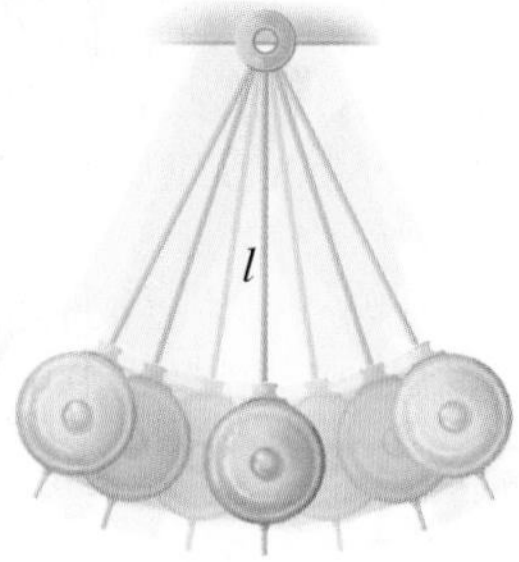

31. **Printing Costs** The cost C of printing a magazine is jointly proportional to the number of pages p in the magazine and the number of magazines printed m.
   (a) Write an equation that expresses this joint variation.
   (b) Find the constant of proportionality if the printing cost is \$60,000 for 4000 copies of a 120-page magazine.
   (c) How much would the printing cost be for 5000 copies of a 92-page magazine?

32. **Boyle's Law** The pressure P of a sample of gas is directly proportional to the temperature T and inversely proportional to the volume V.
   (a) Write an equation that expresses this variation.
   (b) Find the constant of proportionality if 100 L of gas exerts a pressure of 33.2 kPa at a temperature of 400 K (absolute temperature measured on the Kelvin scale).
   (c) If the temperature is increased to 500 K and the volume is decreased to 80 L, what is the pressure of the gas?

33. **Power from a Windmill** The power P that can be obtained from a windmill is directly proportional to the cube of the wind speed s.
   (a) Write an equation that expresses this variation.
   (b) Find the constant of proportionality for a windmill that produces 96 watts of power when the wind is blowing at 20 mi/h.
   (c) How much power will this windmill produce if the wind speed increases to 30 mi/h?

34. **Power Needed to Propel a Boat** The power P (measured in horse power, hp) needed to propel a boat is directly proportional to the cube of the speed s. An 80-hp engine is needed to propel a certain boat at 10 knots. Find the power needed to drive the boat at 15 knots.

35. Loudness of Sound The loudness L of a sound (measured in decibels, dB) is inversely proportional to the square of the distance d from the source of the sound. A person who is 10 ft from a lawn mower experiences a sound level of 70 dB. How loud is the lawn mower when the person is 100 ft away?

36. Stopping Distance The stopping distance D of a car after the brakes have been applied varies directly as the square of the speed s. A certain car traveling at 50 mi/h can stop in 240 ft. What is the maximum speed it can be traveling if it needs to stop in 160 ft?

37. A Jet of Water The power P of a jet of water is jointly proportional to the cross-sectional area A of the jet and to the cube of the velocity v. If the velocity is doubled and the cross-sectional area is halved, by what factor will the power increase?

38. Aerodynamic Lift The lift L on an airplane wing at takeoff varies jointly as the square of the speed s of the plane and the area A of its wings. A plane with a wing area of 500 ft^2 traveling at 50 mi/h experiences a lift of 1700 lb. How much lift would a plane with a wing area of 600 ft^2 traveling at 40 mi/h experience?

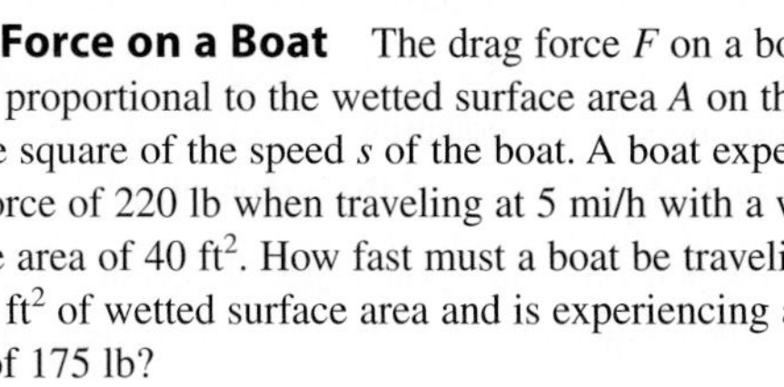

39. Drag Force on a Boat The drag force F on a boat is jointly proportional to the wetted surface area A on the hull and the square of the speed s of the boat. A boat experiences a drag force of 220 lb when traveling at 5 mi/h with a wetted surface area of 40 ft^2. How fast must a boat be traveling if it has 28 ft^2 of wetted surface area and is experiencing a drag force of 175 lb?

40. Skidding in a Curve A car is traveling on a curve that forms a circular arc. The force F needed to keep the car from skidding is jointly proportional to the weight w of the car and the square of its speed s, and is inversely proportional to the radius r of the curve.

(a) Write an equation that expresses this variation.

(b) A car weighing 1600 lb travels around a curve at 60 mi/h. The next car to round this curve weighs 2500 lb and requires the same force as the first car to keep from skidding. How fast is the second car traveling?

41. Electrical Resistance The resistance R of a wire varies directly as its length L and inversely as the square of its diameter d.

(a) Write an equation that expresses this joint variation.

(b) Find the constant of proportionality if a wire 1.2 m long and 0.005 m in diameter has a resistance of 140 ohms.

(c) Find the resistance of a wire made of the same material that is 3 m long and has a diameter of 0.008 m.

42. Kepler's Third Law Kepler's Third Law of planetary motion states that the square of the period T of a planet (the time it takes for the planet to make a complete revolution about the sun) is directly proportional to the cube of its average distance d from the sun.

(a) Express Kepler's Third Law as an equation.

(b) Find the constant of proportionality by using the fact that for our planet the period is about 365 days and the average distance is about 93 million miles.

(c) The planet Neptune is about 2.79×10^9 mi from the sun. Find the period of Neptune.

43. Radiation Energy The total radiation energy E emitted by a heated surface per unit area varies as the fourth power of its absolute temperature T. The temperature is 6000 K at the surface of the sun and 300 K at the surface of the earth.

(a) How many times more radiation energy per unit area is produced by the sun than by the earth?

(b) The radius of the earth is 3960 mi and the radius of the sun is 435,000 mi. How many times more total radiation does the sun emit than the earth?

44. Value of a Lot The value of a building lot on Galiano Island is jointly proportional to its area and the quantity of water produced by a well on the property. A 200 ft by 300 ft lot has a well producing 10 gallons of water per minute, and is valued at \$48,000. What is the value of a 400 ft by 400 ft lot if the well on the lot produces 4 gallons of water per minute?

45. Growing Cabbages In the short growing season of the Canadian arctic territory of Nunavut, some gardeners find it possible to grow gigantic cabbages in the midnight sun.

Assume that the final size of a cabbage is proportional to the amount of nutrients it receives and inversely proportional to the number of other cabbages surrounding it. A cabbage that received 20 oz of nutrients and had 12 other cabbages around it grew to 30 lb. What size would it grow to if it received 10 oz of nutrients and had only 5 cabbage "neighbors"?

46. **Heat of a Campfire** The heat experienced by a hiker at a campfire is proportional to the amount of wood on the fire and inversely proportional to the cube of his distance from the fire. If the hiker is 20 ft from the fire and someone doubles the amount of wood burning, how far from the fire would he have to be so that he feels the same heat as before?

47. **Frequency of Vibration** The frequency f of vibration of a violin string is inversely proportional to its length L. The constant of proportionality k is positive and depends on the tension and density of the string.
 (a) Write an equation that represents this variation.
 (b) What effect does doubling the length of the string have on the frequency of its vibration?

48. **Spread of a Disease** The rate r at which a disease spreads in a population of size P is jointly proportional to the number x of infected people and the number $P - x$ who are not infected. An infection erupts in a small town that has population $P = 5000$.
 (a) Write an equation that expresses r as a function of x.
 (b) Compare the rate of spread of this infection when 10 people are infected to the rate of spread when 1000 people are infected. Which rate is larger? By what factor?
 (c) Calculate the rate of spread when the entire population is infected. Why does this answer make intuitive sense?

DISCOVERY ■ DISCUSSION ■ WRITING

49. **Is Proportionality Everything?** A great many laws of physics and chemistry are expressible as proportionalities. Give at least one example of a function that occurs in the sciences that is *not* a proportionality.

CHAPTER 1 | REVIEW

■ CONCEPT CHECK

1. Define each term in your own words. (Check by referring to the definition in the text.)
 (a) An integer (b) A rational number
 (c) An irrational number (d) A real number
2. State each of these properties of real numbers.
 (a) Commutative Property
 (b) Associative Property
 (c) Distributive Property
3. What is an open interval? What is a closed interval? What notation is used for these intervals?
4. What is the absolute value of a number?
5. (a) In the expression a^x, which is the base and which is the exponent?
 (b) What does a^x mean if $x = n$, a positive integer?
 (c) What if $x = 0$?
 (d) What if x is a negative integer: $x = -n$, where n is a positive integer?
 (e) What if $x = m/n$, a rational number?
 (f) State the Laws of Exponents.
6. (a) What does $\sqrt[n]{a} = b$ mean?
 (b) Why is $\sqrt{a^2} = |a|$?
 (c) How many real nth roots does a positive real number have if n is odd? If n is even?
7. Explain how the procedure of rationalizing the denominator works.
8. State the Special Product Formulas for $(a + b)^2$, $(a - b)^2$, $(a + b)^3$, and $(a - b)^3$.
9. State each Special Factoring Formula.
 (a) Difference of squares (b) Difference of cubes
 (c) Sum of cubes
10. What is a solution of an equation?
11. How do you solve an equation involving radicals? Why is it important to check your answers when solving equations of this type?
12. How do you solve an equation
 (a) algebraically? (b) graphically?
13. Write the general form of each type of equation.
 (a) A linear equation (b) A quadratic equation
14. What are the three ways to solve a quadratic equation?
15. State the Zero-Product Property.
16. Describe the process of completing the square.
17. State the quadratic formula.
18. What is the discriminant of a quadratic equation?
19. State the rules for working with inequalities.

20. How do you solve
(a) a linear inequality?
(b) a nonlinear inequality?

21. **(a)** How do you solve an equation involving an absolute value?
(b) How do you solve an inequality involving an absolute value?

22. **(a)** Describe the coordinate plane.
(b) How do you locate points in the coordinate plane?

23. State each formula.
(a) The Distance Formula
(b) The Midpoint Formula

24. Given an equation, what is its graph?

25. How do you find the x-intercepts and y-intercepts of a graph?

26. Write an equation of the circle with center (h, k) and radius r.

27. Explain the meaning of each type of symmetry. How do you test for it?
(a) Symmetry with respect to the x-axis
(b) Symmetry with respect to the y-axis
(c) Symmetry with respect to the origin

28. Define the slope of a line.

29. Write each form of the equation of a line.
(a) The point-slope form
(b) The slope-intercept form

30. **(a)** What is the equation of a vertical line?
(b) What is the equation of a horizontal line?

31. What is the general equation of a line?

32. Given lines with slopes m_1 and m_2, explain how you can tell if the lines are
(a) parallel **(b)** perpendicular

33. Write an equation that expresses each relationship.
(a) y is directly proportional to x.
(b) y is inversely proportional to x.
(c) z is jointly proportional to x and y.

■ EXERCISES

1–4 ■ State the property of real numbers being used.

1. $3x + 2y = 2y + 3x$

2. $(a + b)(a - b) = (a - b)(a + b)$

3. $4(a + b) = 4a + 4b$

4. $(A + 1)(x + y) = (A + 1)x + (A + 1)y$

5–6 ■ Express the interval in terms of inequalities, and then graph the interval.

5. $[-2, 6)$ **6.** $(-\infty, 4]$

7–8 ■ Express the inequality in interval notation, and then graph the corresponding interval.

7. $x \geq 5$ **8.** $-1 < x \leq 5$

9–18 ■ Evaluate the expression.

9. $|3 - |-9||$ **10.** $1 - |1 - |-1||$

11. $2^{-3} - 3^{-2}$ **12.** $\sqrt[3]{-125}$

13. $216^{-1/3}$ **14.** $64^{2/3}$

15. $\dfrac{\sqrt{242}}{\sqrt{2}}$ **16.** $\sqrt[4]{4}\sqrt[4]{324}$

17. $2^{1/2}8^{1/2}$ **18.** $\sqrt{2}\sqrt{50}$

19–28 ■ Simplify the expression.

19. $\dfrac{x^2(2x)^4}{x^3}$ **20.** $(a^2)^{-3}(a^3b)^2(b^3)^4$

21. $(3xy^2)^3(\frac{2}{3}x^{-1}y)^2$ **22.** $\left(\dfrac{r^2s^{4/3}}{r^{1/3}s}\right)^6$

23. $\sqrt[3]{(x^3y)^2y^4}$ **24.** $\sqrt{x^2y^4}$

25. $\left(\dfrac{9x^3y}{y^{-3}}\right)^{1/2}$ **26.** $\left(\dfrac{x^{-2}y^3}{x^2y}\right)^{-1/2}\left(\dfrac{x^3y}{y^{1/2}}\right)^2$

27. $\dfrac{8r^{1/2}s^{-3}}{2r^{-2}s^4}$ **28.** $\left(\dfrac{ab^2c^{-3}}{2a^3b^{-4}}\right)^{-2}$

29. Write the number 78,250,000,000 in scientific notation.

30. Write the number 2.08×10^{-8} in ordinary decimal notation.

31. If $a \approx 0.00000293$, $b \approx 1.582 \times 10^{-14}$, and $c \approx 2.8064 \times 10^{12}$, use a calculator to approximate the number ab/c.

32. If your heart beats 80 times per minute and you live to be 90 years old, estimate the number of times your heart beats during your lifetime. State your answer in scientific notation.

33–48 ■ Factor the expression completely.

33. $12x^2y^4 - 3xy^5 + 9x^3y^2$ **34.** $x^2 - 9x + 18$

35. $x^2 + 3x - 10$ **36.** $6x^2 + x - 12$

37. $4t^2 - 13t - 12$ **38.** $x^4 - 2x^2 + 1$

39. $25 - 16t^2$ **40.** $2y^6 - 32y^2$

41. $x^6 - 1$ **42.** $y^3 - 2y^2 - y + 2$

43. $x^{-1/2} - 2x^{1/2} + x^{3/2}$ **44.** $a^4b^2 + ab^5$

45. $4x^3 - 8x^2 + 3x - 6$ **46.** $8x^3 + y^6$

47. $(x^2 + 2)^{5/2} + 2x(x^2 + 2)^{3/2} + x^2\sqrt{x^2 + 2}$

48. $3x^3 - 2x^2 + 18x - 12$

49–64 ■ Perform the indicated operations and simplify.

49. $(2x + 1)(3x - 2) - 5(4x - 1)$

50. $(2y - 7)(2y + 7)$

51. $(1 + x)(2 - x) - (3 - x)(3 + x)$

52. $\sqrt{x}(\sqrt{x} + 1)(2\sqrt{x} - 1)$

53. $x^2(x - 2) + x(x - 2)^2$ **54.** $\dfrac{x^2 - 2x - 3}{2x^2 + 5x + 3}$

55. $\dfrac{x^2 + 2x - 3}{x^2 + 8x + 16} \cdot \dfrac{3x + 12}{x - 1}$

56. $\dfrac{t^3 - 1}{t^2 - 1}$

57. $\dfrac{x^2 - 2x - 15}{x^2 - 6x + 5} \div \dfrac{x^2 - x - 12}{x^2 - 1}$

58. $\dfrac{2}{x} + \dfrac{1}{x - 2} + \dfrac{3}{(x - 2)^2}$

59. $\dfrac{1}{x - 1} - \dfrac{2}{x^2 - 1}$

60. $\dfrac{1}{x + 2} + \dfrac{1}{x^2 - 4} - \dfrac{2}{x^2 - x - 2}$

61. $\dfrac{\dfrac{1}{x} - \dfrac{1}{2}}{x - 2}$

62. $\dfrac{\dfrac{1}{x} - \dfrac{1}{x + 1}}{\dfrac{1}{x} + \dfrac{1}{x + 1}}$

63. $\dfrac{\sqrt{6}}{\sqrt{3} + \sqrt{2}}$ (rationalize the denominator)

64. $\dfrac{\sqrt{x + h} - \sqrt{x}}{h}$ (rationalize the numerator)

65–80 ■ Find all real solutions of the equation.

65. $7x - 6 = 4x + 9$

66. $8 - 2x = 14 + x$

67. $\dfrac{x + 1}{x - 1} = \dfrac{3x}{3x - 6}$

68. $(x + 2)^2 = (x - 4)^2$

69. $x^2 - 9x + 14 = 0$

70. $x^2 + 24x + 144 = 0$

71. $2x^2 + x = 1$

72. $3x^2 + 5x - 2 = 0$

73. $4x^3 - 25x = 0$

74. $x^3 - 2x^2 - 5x + 10 = 0$

75. $3x^2 + 4x - 1 = 0$

76. $\dfrac{1}{x} + \dfrac{2}{x - 1} = 3$

77. $\dfrac{x}{x - 2} + \dfrac{1}{x + 2} = \dfrac{8}{x^2 - 4}$

78. $x^4 - 8x^2 - 9 = 0$

79. $|x - 7| = 4$

80. $|2x - 5| = 9$

81. The owner of a store sells raisins for \$3.20 per pound and nuts for \$2.40 per pound. He decides to mix the raisins and nuts and sell 50 lb of the mixture for \$2.72 per pound. What quantities of raisins and nuts should he use?

82. Anthony leaves Kingstown at 2:00 P.M. and drives to Queensville, 160 mi distant, at 45 mi/h. At 2:15 P.M. Helen leaves Queensville and drives to Kingstown at 40 mi/h. At what time do they pass each other on the road?

83. A woman cycles 8 mi/h faster than she runs. Every morning she cycles 4 mi and runs $2\frac{1}{2}$ mi, for a total of one hour of exercise. How fast does she run?

84. The hypotenuse of a right triangle has length 20 cm. The sum of the lengths of the other two sides is 28 cm. Find the lengths of the other two sides of the triangle.

85. Abbie paints twice as fast as Beth and three times as fast as Cathie. If it takes them 60 min to paint a living room with all three working together, how long would it take Abbie if she works alone?

86. A homeowner wishes to fence in three adjoining garden plots, one for each of her children, as shown in the figure. If each plot is to be 80 ft^2 in area and she has 88 ft of fencing material at hand, what dimensions should each plot have?

87–94 ■ Solve the inequality. Express the solution using interval notation and graph the solution set on the real number line.

87. $3x - 2 > -11$

88. $-1 < 2x + 5 \leq 3$

89. $x^2 + 4x - 12 > 0$

90. $x^2 \leq 1$

91. $\dfrac{x - 4}{x^2 - 4} \leq 0$

92. $\dfrac{5}{x^3 - x^2 - 4x + 4} < 0$

93. $|x - 5| \leq 3$

94. $|x - 4| < 0.02$

95–98 ■ Solve the equation or inequality graphically.

95. $x^2 - 4x = 2x + 7$

96. $\sqrt{x + 4} = x^2 - 5$

97. $4x - 3 \geq x^2$

98. $x^3 - 4x^2 - 5x > 2$

99–100 ■ Two points P and Q are given.

(a) Plot P and Q on a coordinate plane.

(b) Find the distance from P to Q.

(c) Find the midpoint of the segment PQ.

(d) Sketch the line determined by P and Q, and find its equation in slope-intercept form.

(e) Sketch the circle that passes through Q and has center P, and find the equation of this circle.

99. $P(2, 0)$, $Q(-5, 12)$

100. $P(7, -1)$, $Q(2, -11)$

101–102 ■ Sketch the region given by the set.

101. $\{(x, y) \mid -4 < x < 4 \text{ and } -2 < y < 2\}$

102. $\{(x, y) \mid x \geq 4 \text{ or } y \geq 2\}$

103. Which of the points $A(4, 4)$ or $B(5, 3)$ is closer to the point $C(-1, -3)$?

104. Find an equation of the circle that has center $(2, -5)$ and radius $\sqrt{2}$.

105. Find an equation of the circle that has center $(-5, -1)$ and passes through the origin.

106. Find an equation of the circle that contains the points $P(2, 3)$ and $Q(-1, 8)$ and has the midpoint of the segment PQ as its center.

107–110 ■ Determine whether the equation represents a circle, represents a point, or has no graph. If the equation is that of a circle, find its center and radius.

107. $x^2 + y^2 + 2x - 6y + 9 = 0$

108. $2x^2 + 2y^2 - 2x + 8y = \frac{1}{2}$

109. $x^2 + y^2 + 72 = 12x$

110. $x^2 + y^2 - 6x - 10y + 34 = 0$

111–118 ■ Test the equation for symmetry, and sketch its graph.

111. $y = 2 - 3x$

112. $2x - y + 1 = 0$

113. $x + 3y = 21$

114. $x = 2y + 12$

115. $y = 16 - x^2$

116. $8x + y^2 = 0$

117. $x = \sqrt{y}$

118. $y = -\sqrt{1 - x^2}$

119–122 ■ Use a graphing device to graph the equation in an appropriate viewing rectangle.

119. $y = x^2 - 6x$

120. $y = \sqrt{5 - x}$

121. $y = x^3 - 4x^2 - 5x$

122. $\dfrac{x^2}{4} + y^2 = 1$

123. Find an equation for the line that passes through the points $(-1, -6)$ and $(2, -4)$.

124. Find an equation for the line that passes through the point $(6, -3)$ and has slope $-\frac{1}{2}$.

125. Find an equation for the line that has x-intercept 4 and y-intercept 12.

126. Find an equation for the line that passes through the point $(1, 7)$ and is perpendicular to the line $x - 3y + 16 = 0$.

127. Find an equation for the line that passes through the origin and is parallel to the line $3x + 15y = 22$.

128. Find an equation for the line that passes through the point $(5, 2)$ and is parallel to the line passing through $(-1, -3)$ and $(3, 2)$.

129–130 ■ Find equations for the circle and the line in the figure.

129.

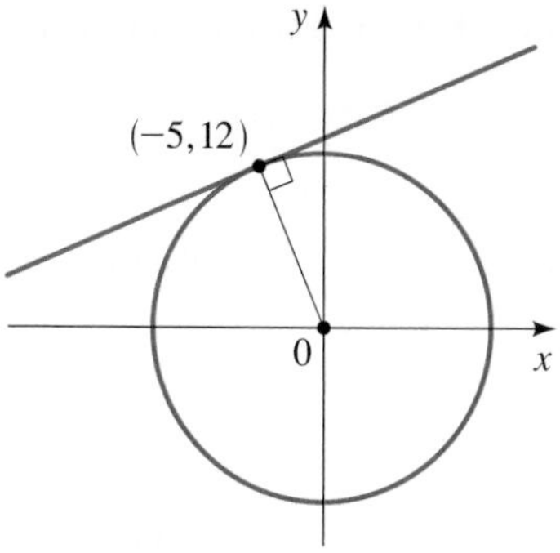

130.

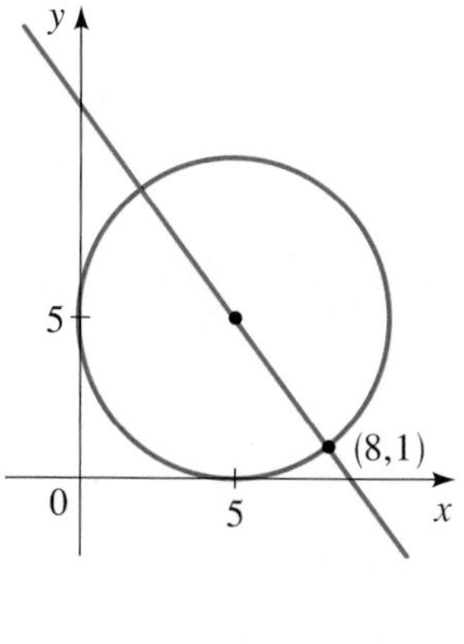

131. Hooke's Law states that if a weight w is attached to a hanging spring, then the stretched length s of the spring is linearly related to w. For a particular spring we have

$$s = 0.3w + 2.5$$

where s is measured in inches and w in pounds.

(a) What do the slope and s-intercept in this equation represent?

(b) How long is the spring when a 5-lb weight is attached?

132. Margarita is hired by an accounting firm at a salary of \$60,000 per year. Three years later her annual salary has increased to \$70,500. Assume that her salary increases linearly.

(a) Find an equation that relates her annual salary S and the number of years t that she has worked for the firm.

(b) What do the slope and S-intercept of her salary equation represent?

(c) What will her salary be after 12 years with the firm?

133. Suppose that M varies directly as z, and $M = 120$ when $z = 15$. Write an equation that expresses this variation.

134. Suppose that z is inversely proportional to y, and that $z = 12$ when $y = 16$. Write an equation that expresses z in terms of y.

135. The intensity of illumination I from a light varies inversely as the square of the distance d from the light.

(a) Write this statement as an equation.

(b) Determine the constant of proportionality if it is known that a lamp has an intensity of 1000 candles at a distance of 8 m.

(c) What is the intensity of this lamp at a distance of 20 m?

136. The frequency of a vibrating string under constant tension is inversely proportional to its length. If a violin string 12 inches long vibrates 440 times per second, to what length must it be shortened to vibrate 660 times per second?

137. The terminal velocity of a parachutist is directly proportional to the square root of his weight. A 160-lb parachutist attains a terminal velocity of 9 mi/h. What is the terminal velocity for a parachutist weighing 240 lb?

138. The maximum range of a projectile is directly proportional to the square of its velocity. A baseball pitcher throws a ball at 60 mi/h, with a maximum range of 242 ft. What is his maximum range if he throws the ball at 70 mi/h?

1. (a) Graph the intervals $(-5, 3]$ and $(2, \infty)$ on the real number line.
   (b) Express the inequalities $x \le 3$ and $-1 \le x < 4$ in interval notation.
   (c) Find the distance between -7 and 9 on the real number line.

2. Evaluate each expression.
   (a) $(-3)^4$ (b) -3^4 (c) 3^{-4} (d) $\dfrac{5^{23}}{5^{21}}$ (e) $\left(\dfrac{2}{3}\right)^{-2}$ (f) $16^{-3/4}$

3. Write each number in scientific notation.
   (a) 186,000,000,000 (b) 0.0000003965

4. Simplify each expression. Write your final answer without negative exponents.
   (a) $\sqrt{200} - \sqrt{32}$ (b) $(3a^3b^3)(4ab^2)^2$ (c) $\left(\dfrac{3x^{3/2}y^3}{x^2y^{-1/2}}\right)^{-2}$
   (d) $\dfrac{x^2 + 3x + 2}{x^2 - x - 2}$ (e) $\dfrac{x^2}{x^2 - 4} - \dfrac{x + 1}{x + 2}$ (f) $\dfrac{\dfrac{y}{x} - \dfrac{x}{y}}{\dfrac{1}{y} - \dfrac{1}{x}}$

5. Rationalize the denominator and simplify: $\dfrac{\sqrt{10}}{\sqrt{5} - 2}$

6. Perform the indicated operations and simplify.
   (a) $3(x + 6) + 4(2x - 5)$ (b) $(x + 3)(4x - 5)$ (c) $(\sqrt{a} + \sqrt{b})(\sqrt{a} - \sqrt{b})$
   (d) $(2x + 3)^2$ (e) $(x + 2)^3$

7. Factor each expression completely.
   (a) $4x^2 - 25$ (b) $2x^2 + 5x - 12$ (c) $x^3 - 3x^2 - 4x + 12$
   (d) $x^4 + 27x$ (e) $3x^{3/2} - 9x^{1/2} + 6x^{-1/2}$ (f) $x^3y - 4xy$

8. Find all real solutions.
   (a) $x + 5 = 14 - \frac{1}{2}x$ (b) $\dfrac{2x}{x + 1} = \dfrac{2x - 1}{x}$ (c) $x^2 - x - 12 = 0$
   (d) $2x^2 + 4x + 1 = 0$ (e) $\sqrt{3 - \sqrt{x + 5}} = 2$ (f) $x^4 - 3x^2 + 2 = 0$
   (g) $3|x - 4| = 10$

9. Mary drove from Amity to Belleville at a speed of 50 mi/h. On the way back, she drove at 60 mi/h. The total trip took $4\frac{2}{5}$ h of driving time. Find the distance between these two cities.

10. A rectangular parcel of land is 70 ft longer than it is wide. Each diagonal between opposite corners is 130 ft. What are the dimensions of the parcel?

11. Solve each inequality. Write the answer using interval notation, and sketch the solution on the real number line.
    (a) $-4 < 5 - 3x \le 17$ (b) $x(x - 1)(x + 2) > 0$
    (c) $|x - 4| < 3$ (d) $\dfrac{2x - 3}{x + 1} \le 1$

12. A bottle of medicine is to be stored at a temperature between 5°C and 10°C. What range does this correspond to on the Fahrenheit scale? [*Note:* Fahrenheit (F) and Celsius (C) temperatures satisfy the relation $C = \frac{5}{9}(F - 32)$.]

13. For what values of x is the expression $\sqrt{6x - x^2}$ defined as a real number?

14. Solve the equation and the inequality graphically.
    (a) $x^3 - 9x - 1 = 0$ (b) $x^2 - 1 \le |x + 1|$

15. (a) Plot the points $P(0, 3)$, $Q(3, 0)$, and $R(6, 3)$ in the coordinate plane. Where must the point S be located so that $PQRS$ is a square?
    (b) Find the area of $PQRS$.

16. (a) Sketch the graph of $y = x^2 - 4$.
    (b) Find the x- and y-intercepts of the graph.
    (c) Is the graph symmetric about the x-axis, the y-axis, or the origin?

17. Let $P(-3,1)$ and $Q(5,6)$ be two points in the coordinate plane.

(a) Plot P and Q in the coordinate plane.

(b) Find the distance between P and Q.

(c) Find the midpoint of the segment PQ.

(d) Find the slope of the line that contains P and Q.

(e) Find the perpendicular bisector of the line that contains P and Q.

(f) Find an equation for the circle for which the segment PQ is a diameter.

18. Find the center and radius of each circle and sketch its graph.

(a) $x^2 + y^2 = 25$ **(b)** $(x - 2)^2 + (y + 1)^2 = 9$ **(c)** $x^2 + 6x + y^2 - 2y + 6 = 0$

19. Write the linear equation $2x - 3y = 15$ in slope-intercept form, and sketch its graph. What are the slope and y-intercept?

20. Find an equation for the line with the given property.

(a) It passes through the point $(3, -6)$ and is parallel to the line $3x + y - 10 = 0$.

(b) It has x-intercept 6 and y-intercept 4.

21. A geologist uses a probe to measure the temperature T (in °C) of the soil at various depths below the surface, and finds that at a depth of x cm, the temperature is given by the linear equation $T = 0.08x - 4$.

(a) What is the temperature at a depth of one meter (100 cm)?

(b) Sketch a graph of the linear equation.

(c) What do the slope, the x-intercept, and T-intercept of the graph of this equation represent?

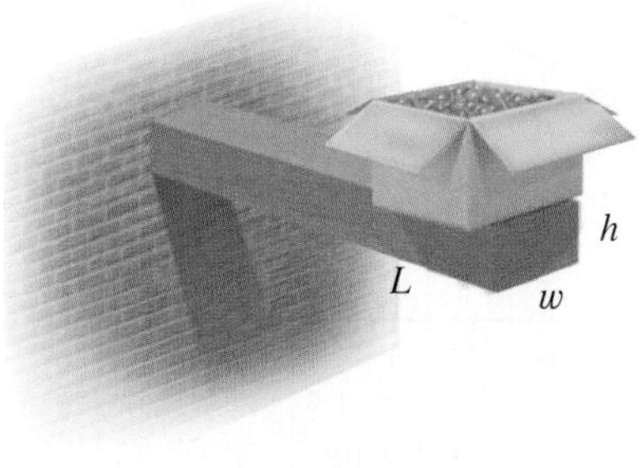

22. The maximum weight M that can be supported by a beam is jointly proportional to its width w and the square of its height h, and inversely proportional to its length L.

(a) Write an equation that expresses this proportionality.

(b) Determine the constant of proportionality if a beam 4 in. wide, 6 in. high, and 12 ft long can support a weight of 4800 lb.

(c) If a 10-ft beam made of the same material is 3 in. wide and 10 in. high, what is the maximum weight it can support?

If you had difficulty with any of these problems, you may wish to review the section of this chapter indicated below.

If you had trouble with this test problem	Review this section
1	Section 1.1
2, 3, 4(a), 4(b), 4(c)	Section 1.2
4(d), 4(e), 4(f), 5	Section 1.4
6, 7	Section 1.3
8	Section 1.5
9, 10	Section 1.6
11, 12, 13	Section 1.7
14	Section 1.9
15, 16, 17(a), 17(b)	Section 1.8
17(c), 17(d)	Section 1.10
17(e), 17(f), 18	Section 1.8
19, 20, 21	Section 1.10
22	Section 1.11

Fitting Lines to Data

A model is a representation of an object or process. For example, a toy Ferrari is a model of the actual car; a road map is a model of the streets in a city. A **mathematical model** is a mathematical representation (usually an equation) of an object or process. Once a mathematical model is made it can be used to obtain useful information or make predictions about the thing being modeled. In these *Focus on Modeling* sections we explore different ways in which mathematics is used to model real-world phenomena.

▼ The Line That Best Fits the Data

In Section 1.10 we used linear equations to model relationships between varying quantities. In practice, such relationships are discovered by collecting data. But real-world data seldom fall into a precise line. The **scatter plot** in Figure 1(a) shows the result of a study on childhood obesity. The graph plots the body mass index (BMI) versus the number of hours of television watched per day for 25 adolescent subjects. Of course, we would not expect the data to be exactly linear as in Figure 1(b). But there is a linear *trend* indicated by the blue line in Figure 1(a): The more hours a subject watches TV the higher the BMI. In this section we learn how to find the line that best fits the data.

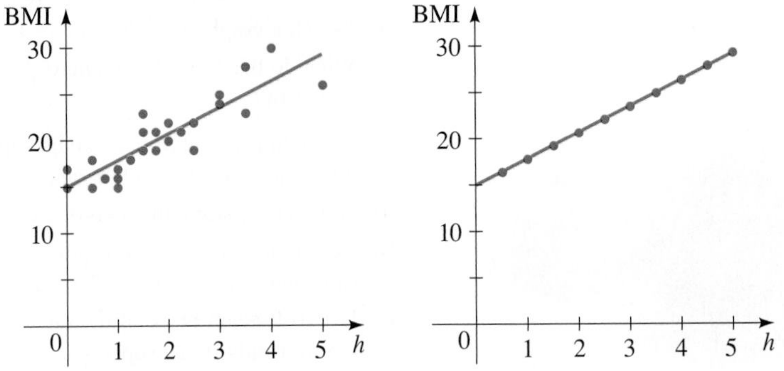

(a) Line of best fit

(b) Line fits data exactly

FIGURE 1

Table 1 gives the nationwide infant mortality rate for the period from 1950 to 2000. The *rate* is the number of infants who die before reaching their first birthday, out of every 1000 live births.

TABLE 1
U.S. Infant Mortality

Year	Rate
1950	29.2
1960	26.0
1970	20.0
1980	12.6
1990	9.2
2000	6.9

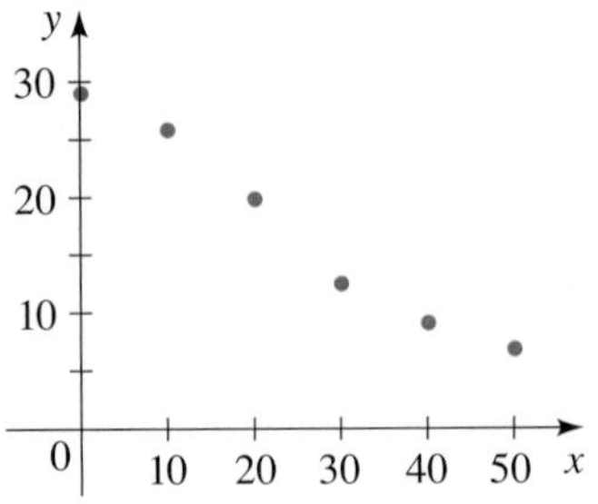

FIGURE 2 U.S. infant mortality rate

The scatter plot in Figure 2 shows that the data lie roughly on a straight line. We can try to fit a line visually to approximate the data points, but since the data aren't *exactly* linear, there are many lines that might seem to work. Figure 3 shows two attempts at "eyeballing" a line to fit the data.

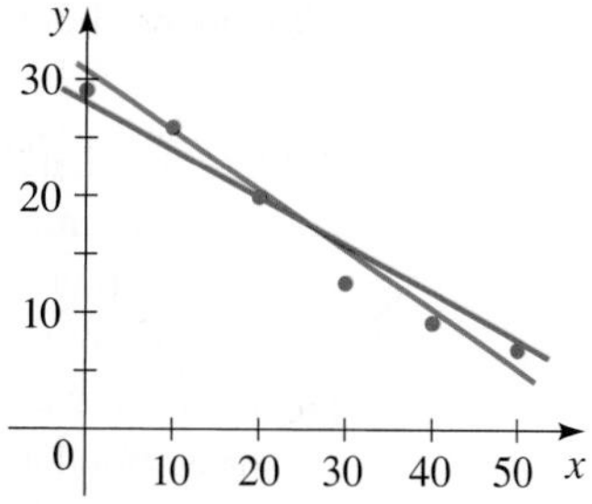

FIGURE 3 Visual attempts to fit line to data

Of all the lines that run through these data points, there is one that "best" fits the data, in the sense that it provides the most accurate linear model for the data. We now describe how to find this line.

It seems reasonable that the line of best fit is the line that is as close as possible to all the data points. This is the line for which the sum of the vertical distances from the data points to the line is as small as possible (see Figure 4). For technical reasons it is better to use the line where the sum of the squares of these distances is smallest. This is called the **regression line**. The formula for the regression line is found by using calculus, but fortunately, the formula is programmed into most graphing calculators. In Example 1 we see how to use a TI-83 calculator to find the regression line for the infant mortality data described above. (The process for other calculators is similar.)

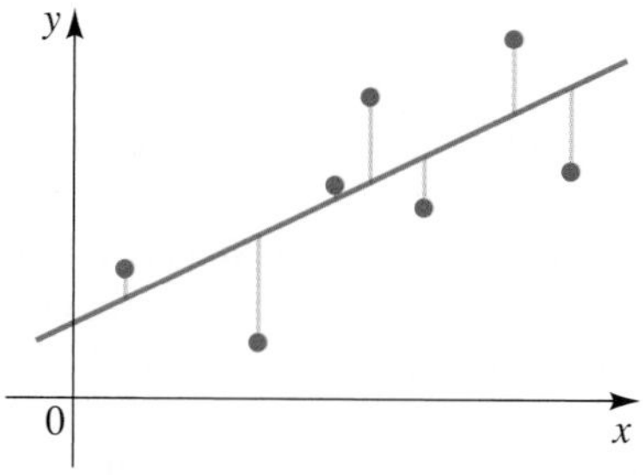

FIGURE 4 Distance from the data points to the line

EXAMPLE 1 | Regression Line for U.S. Infant Mortality Rates

(a) Find the regression line for the infant mortality data in Table 1.

(b) Graph the regression line on a scatter plot of the data.

(c) Use the regression line to estimate the infant mortality rates in 1995 and 2006.

SOLUTION

(a) To find the regression line using a TI-83 calculator, we must first enter the data into the lists L_1 and L_2, which are accessed by pressing the STAT key and selecting `Edit`. Figure 5 shows the calculator screen after the data have been entered. (Note that we are letting $x = 0$ correspond to the year 1950, so that $x = 50$ corresponds to 2000. This makes the equations easier to work with.) We then press the STAT key again and select `Calc`, then `4:LinReg(ax+b)`, which provides the output shown in Figure 6(a). This tells us that the regression line is

$$y = -0.48x + 29.4$$

Here x represents the number of years since 1950, and y represents the corresponding infant mortality rate.

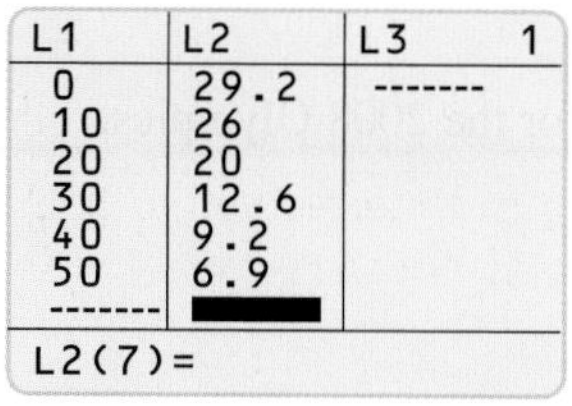

FIGURE 5 Entering the data

(b) The scatter plot and the regression line have been plotted on a graphing calculator screen in Figure 6(b).

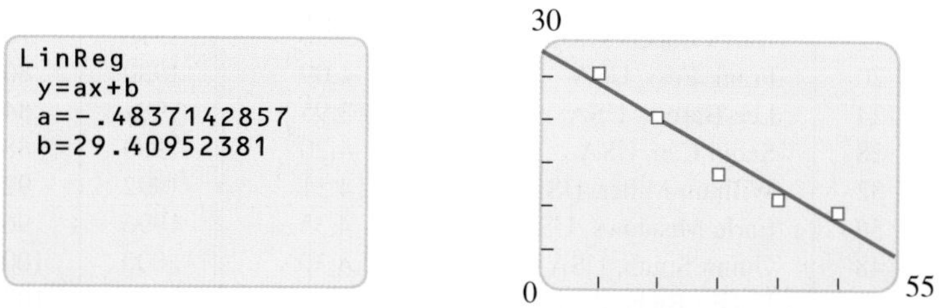

(a) Output of the `LinReg` command

(b) Scatter plot and regression line

FIGURE 6

(c) The year 1995 is 45 years after 1950, so substituting 45 for x, we find that $y = -0.48(45) + 29.4 = 7.8$. So the infant mortality rate in 1995 was about 7.8. Similarly, substituting 56 for x, we find that the infant mortality rate predicted for 2006 was about $-0.48(56) + 29.4 \approx 2.5$. ■

An Internet search shows that the actual infant mortality rate was 7.6 in 1995 and 6.4 in 2006. So the regression line is fairly accurate for 1995 (the actual rate was slightly lower than the predicted rate), but it is considerably off for 2006 (the actual rate was more than twice the predicted rate). The reason is that infant mortality in the United States stopped declining and actually started rising in 2002, for the first time in more than a century. This shows that we have to be very careful about extrapolating linear models outside the domain over which the data are spread.

AP Photo/Michael Probst

Steven Hooker, 2008 Olympic gold medal winner, men's pole vault

▼ Examples of Regression Analysis

Since the modern Olympic Games began in 1896, achievements in track and field events have been improving steadily. One example in which the winning records have shown an upward linear trend is the pole vault. Pole vaulting began in the northern Netherlands as a practical activity: When traveling from village to village, people would vault across the many canals that crisscrossed the area to avoid having to go out of their way to find a bridge. Households maintained a supply of wooden poles of lengths appropriate for each member of the family. Pole vaulting for height rather than distance became a collegiate track and field event in the mid-1800s and was one of the events in the first modern Olympics. In the next example we find a linear model for the gold-medal-winning records in the men's Olympic pole vault.

EXAMPLE 2 | Regression Line for Olympic Pole Vault Records

Table 2 gives the men's Olympic pole vault records up to 2004.

(a) Find the regression line for the data.

(b) Make a scatter plot of the data, and graph the regression line. Does the regression line appear to be a suitable model for the data?

(c) What does the slope of the regression line represent?

(d) Use the model to predict the winning pole vault height for the 2008 Olympics.

TABLE 2
Men's Olympic Pole Vault Records

Year	x	Gold medalist	Height (m)	Year	x	Gold medalist	Height (m)
1896	−4	William Hoyt, USA	3.30	1956	56	Robert Richards, USA	4.56
1900	0	Irving Baxter, USA	3.30	1960	60	Don Bragg, USA	4.70
1904	4	Charles Dvorak, USA	3.50	1964	64	Fred Hansen, USA	5.10
1906	6	Fernand Gonder, France	3.50	1968	68	Bob Seagren, USA	5.40
1908	8	A. Gilbert, E. Cook, USA	3.71	1972	72	W. Nordwig, E. Germany	5.64
1912	12	Harry Babcock, USA	3.95	1976	76	Tadeusz Slusarski, Poland	5.64
1920	20	Frank Foss, USA	4.09	1980	80	W. Kozakiewicz, Poland	5.78
1924	24	Lee Barnes, USA	3.95	1984	84	Pierre Quinon, France	5.75
1928	28	Sabin Can, USA	4.20	1988	88	Sergei Bubka, USSR	5.90
1932	32	William Miller, USA	4.31	1992	92	M. Tarassob, Unified Team	5.87
1936	36	Earle Meadows, USA	4.35	1996	96	Jean Jaffione, France	5.92
1948	48	Guinn Smith, USA	4.30	2000	100	Nick Hysong, USA	5.90
1952	52	Robert Richards, USA	4.55	2004	104	Timothy Mack, USA	5.95

```
LinReg
 y=ax+b
 a=.0265652857
 b=3.400989881
```

Output of the `LinReg` function on the TI-83

SOLUTION

(a) Let $x = \text{year} - 1900$, so 1896 corresponds to $x = -4$, 1900 to $x = 0$, and so on. Using a calculator, we find the following regression line:

$$y = 0.0266x + 3.40$$

(b) The scatter plot and the regression line are shown in Figure 7. The regression line appears to be a good model for the data.

(c) The slope is the average rate of increase in the pole vault record per year. So on average, the pole vault record increased by 0.0266 m/yr.

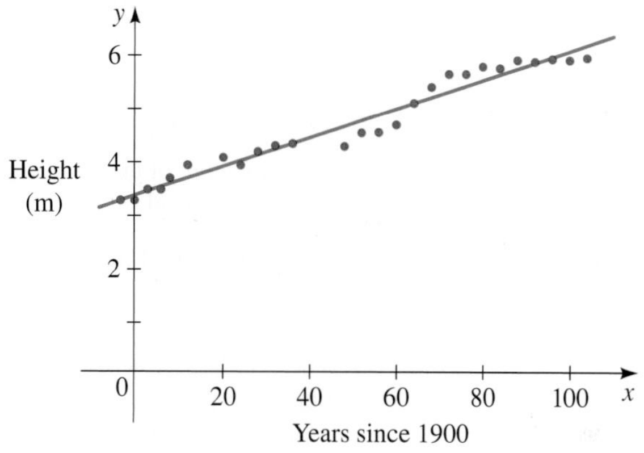

FIGURE 7 Scatter plot and regression line for pole vault data

(d) The year 2008 corresponds to $x = 108$ in our model. The model gives

$$y = 0.0266(108) + 3.40$$
$$\approx 6.27$$

So the model predicts that in 2008 the winning pole vault will be 6.27 m. ■

At the 2008 Olympics in Beijing, China, the men's Olympic gold medal in the pole vault was won by Steven Hooker of Australia, with a vault of 5.96 m. Although this height set an Olympic record, it was considerably lower than the 6.27 m predicted by the model of Example 2. In Problem 10 we find a regression line for the pole vault data from 1972 to 2004. Do the problem to see whether this restricted set of more recent data provides a better predictor for the 2008 record.

Is a linear model really appropriate for the data of Example 2? In subsequent *Focus on Modeling* sections, we study regression models that use other types of functions, and we learn how to choose the best model for a given set of data.

In the next example we see how linear regression is used in medical research to investigate potential causes of diseases such as cancer.

TABLE 3
Asbestos–Tumor Data

Asbestos exposure (fibers/mL)	Percent that develop lung tumors
50	2
400	6
500	5
900	10
1100	26
1600	42
1800	37
2000	28
3000	50

EXAMPLE 3 | Regression Line for Links Between Asbestos and Cancer

When laboratory rats are exposed to asbestos fibers, some of the rats develop lung tumors. Table 3 lists the results of several experiments by different scientists.

(a) Find the regression line for the data.

(b) Make a scatter plot and graph the regression line. Does the regression line appear to be a suitable model for the data?

(c) What does the y-intercept of the regression line represent?

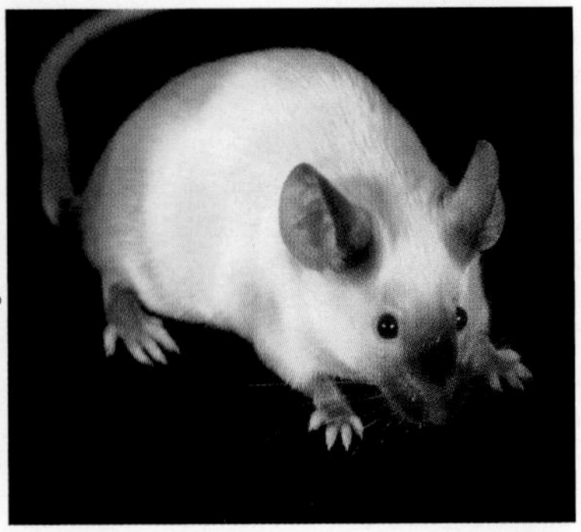

SOLUTION

(a) Using a calculator, we find the following regression line (see Figure 8(a)):

$$y = 0.0177x + 0.5405$$

(b) The scatter plot and regression line are graphed in Figure 8(b). The regression line appears to be a reasonable model for the data.

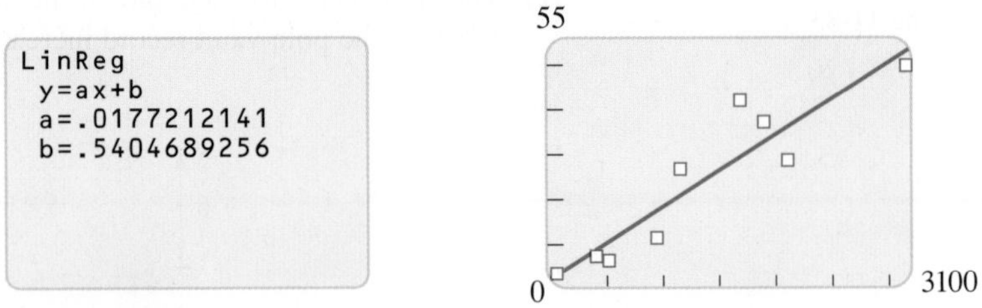

(a) Output of the `LinReg` command

(b) Scatter plot and regression line

FIGURE 8 Linear regression for the asbestos–tumor data

(c) The y-intercept is the percentage of rats that develop tumors when no asbestos fibers are present. In other words, this is the percentage that normally develop lung tumors (for reasons other than asbestos). ■

▼ How Good Is the Fit? The Correlation Coefficient

For any given set of two-variable data it is always possible to find a regression line, even if the data points do not tend to lie on a line and even if the variables don't seem to be related at all. Look at the three scatter plots in Figure 9. In the first scatter plot, the data points lie close to a line. In the second plot, there is still a linear trend but the points are more scattered. In the third plot there doesn't seem to be any trend at all, linear or otherwise.

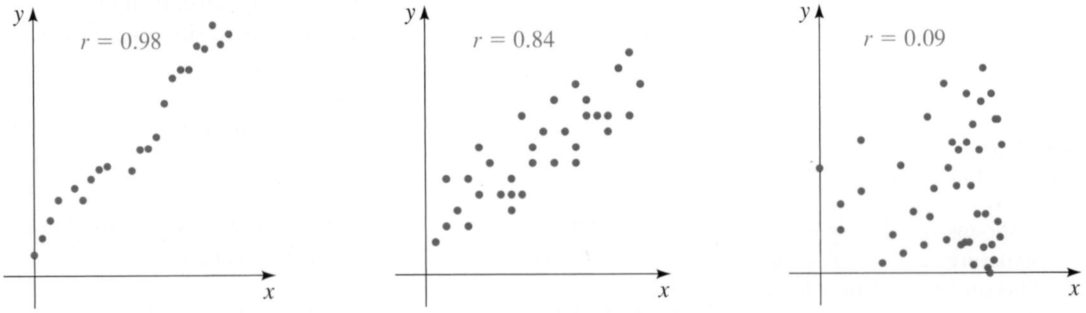

FIGURE 9

A graphing calculator can give us a regression line for each of these scatter plots. But how well do these lines represent or "fit" the data? To answer this question, statisticians have invented the **correlation coefficient**, usually denoted r. The correlation coefficient is a number between -1 and 1 that measures how closely the data follow the regression line—or, in other words, how strongly the variables are **correlated**. Many graphing calculators give the value of r when they compute a regression line. If r is close to -1

or 1, then the variables are strongly correlated—that is, the scatter plot follows the regression line closely. If r is close to 0, then the variables are weakly correlated or not correlated at all. (The sign of r depends on the slope of the regression line.) The correlation coefficients of the scatter plots in Figure 9 are indicated on the graphs. For the first plot, r is close to 1 because the data are very close to linear. The second plot also has a relatively large r, but it is not as large as the first, because the data, while fairly linear, are more diffuse. The third plot has an r close to 0, since there is virtually no linear trend in the data.

There are no hard and fast rules for deciding what values of r are sufficient for deciding that a linear correlation is "significant." The correlation coefficient is only a rough guide in helping us decide how much faith to put into a given regression line. In Example 1 the correlation coefficient is -0.99, indicating a very high level of correlation, so we can safely say that the drop in infant mortality rates from 1950 to 2000 was strongly linear. (The value of r is negative, since infant mortality trended *down* over this period.) In Example 3 the correlation coefficient is 0.92, which also indicates a strong correlation between the variables. So exposure to asbestos is clearly associated with the growth of lung tumors in rats. Does this mean that asbestos *causes* lung cancer?

If two variables are correlated, it does not necessarily mean that a change in one variable *causes* a change in the other. For example, the mathematician John Allen Paulos points out that shoe size is strongly correlated to mathematics scores among schoolchildren. Does this mean that big feet cause high math scores? Certainly not—both shoe size and math skills increase independently as children get older. So it is important not to jump to conclusions: Correlation and causation are not the same thing. Correlation is a useful tool in bringing important cause-and-effect relationships to light; but to prove causation, we must explain the mechanism by which one variable affects the other. For example, the link between smoking and lung cancer was observed as a correlation long before science found the mechanism through which smoking causes lung cancer.

PROBLEMS

1. **Femur Length and Height** Anthropologists use a linear model that relates femur length to height. The model allows an anthropologist to determine the height of an individual when only a partial skeleton (including the femur) is found. In this problem we find the model by analyzing the data on femur length and height for the eight males given in the table.

   (a) Make a scatter plot of the data.

   (b) Find and graph a linear function that models the data.

   (c) An anthropologist finds a femur of length 58 cm. How tall was the person?

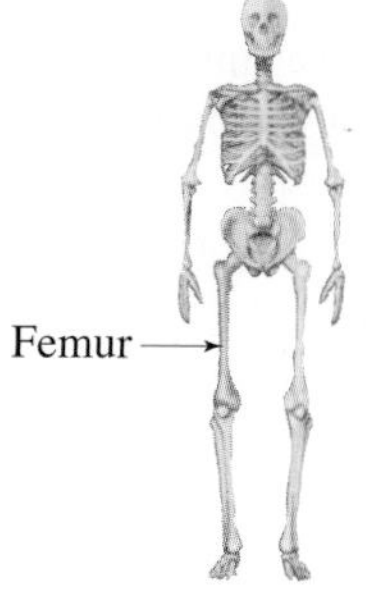

Femur length (cm)	Height (cm)
50.1	178.5
48.3	173.6
45.2	164.8
44.7	163.7
44.5	168.3
42.7	165.0
39.5	155.4
38.0	155.8

2. **Demand for Soft Drinks** A convenience store manager notices that sales of soft drinks are higher on hotter days, so he assembles the data in the table.

   (a) Make a scatter plot of the data.

   (b) Find and graph a linear function that models the data.

(c) Use the model to predict soft drink sales if the temperature is 95°F.

High temperature (°F)	Number of cans sold
55	340
58	335
64	410
68	460
70	450
75	610
80	735
84	780

3. Tree Diameter and Age To estimate ages of trees, forest rangers use a linear model that relates tree diameter to age. The model is useful because tree diameter is much easier to measure than tree age (which requires special tools for extracting a representative cross section of the tree and counting the rings). To find the model, use the data in the table, which were collected for a certain variety of oaks.

(a) Make a scatter plot of the data.

(b) Find and graph a linear function that models the data.

(c) Use the model to estimate the age of an oak whose diameter is 18 in.

Diameter (in.)	Age (years)
2.5	15
4.0	24
6.0	32
8.0	56
9.0	49
9.5	76
12.5	90
15.5	89

4. Carbon Dioxide Levels The Mauna Loa Observatory, located on the island of Hawaii, has been monitoring carbon dioxide (CO_2) levels in the atmosphere since 1958. The table lists the average annual CO_2 levels measured in parts per million (ppm) from 1984 to 2006.

(a) Make a scatter plot of the data.

(b) Find and graph the regression line.

(c) Use the linear model in part (b) to estimate the CO_2 level in the atmosphere in 2005. Compare your answer with the actual CO_2 level of 379.7 that was measured in 2005.

Year	CO_2 level (ppm)
1984	344.3
1986	347.0
1988	351.3
1990	354.0
1992	356.3
1994	358.9
1996	362.7
1998	366.5
2000	369.4
2002	372.0
2004	377.5
2006	380.9

5. Temperature and Chirping Crickets Biologists have observed that the chirping rate of crickets of a certain species appears to be related to temperature. The table shows the chirping rates for various temperatures.

(a) Make a scatter plot of the data.

(b) Find and graph the regression line.

(c) Use the linear model in part (b) to estimate the chirping rate at 100°F.

Temperature (°F)	Chirping rate (chirps/min)
50	20
55	46
60	79
65	91
70	113
75	140
80	173
85	198
90	211

6. Extent of Arctic Sea Ice The National Snow and Ice Data Center monitors the amount of ice in the Arctic year round. The table gives approximate values for the sea ice extent in millions of square kilometers from 1980 to 2006, in two-year intervals.

(a) Make a scatter plot of the data.

(b) Find and graph the regression line.

(c) Use the linear model in part (b) to estimate the ice extent in the year 2010.

Year	Ice extent (million km^2)	Year	Ice extent (million km^2)
1980	7.9	1994	7.1
1982	7.4	1996	7.9
1984	7.2	1998	6.6
1986	7.6	2000	6.3
1988	7.5	2002	6.0
1990	6.2	2004	6.1
1992	7.6	2006	5.7

7. Mosquito Prevalence The table lists the relative abundance of mosquitoes (as measured by the mosquito positive rate) versus the flow rate (measured as a percentage of maximum flow) of canal networks in Saga City, Japan.

(a) Make a scatter plot of the data.

(b) Find and graph the regression line.

(c) Use the linear model in part (b) to estimate the mosquito positive rate if the canal flow is 70% of maximum.

Flow rate (%)	Mosquito positive rate (%)
0	22
10	16
40	12
60	11
90	6
100	2

8. Noise and Intelligibility Audiologists study the intelligibility of spoken sentences under different noise levels. Intelligibility, the MRT score, is measured as the percent of a spoken sentence that the listener can decipher at a certain noise level in decibels (dB). The table shows the results of one such test.

(a) Make a scatter plot of the data.

(b) Find and graph the regression line.

(c) Find the correlation coefficient. Is a linear model appropriate?

(d) Use the linear model in part (b) to estimate the intelligibility of a sentence at a 94-dB noise level.

Noise level (dB)	MRT score (%)
80	99
84	91
88	84
92	70
96	47
100	23
104	11

9. Life Expectancy The average life expectancy in the United States has been rising steadily over the past few decades, as shown in the table.

(a) Make a scatter plot of the data.

(b) Find and graph the regression line.

(c) Use the linear model you found in part (b) to predict the life expectancy in the year 2006.

(d) Search the Internet or your campus library to find the actual 2006 average life expectancy. Compare to your answer in part (c).

Year	Life expectancy
1920	54.1
1930	59.7
1940	62.9
1950	68.2
1960	69.7
1970	70.8
1980	73.7
1990	75.4
2000	76.9

10. Olympic Pole Vault The graph in Figure 7 indicates that in recent years the winning Olympic men's pole vault height has fallen below the value predicted by the regression line in Example 2. This might have occurred because when the pole vault was a new event, there was much room for improvement in vaulters' performances, whereas now even the best training can produce only incremental advances. Let's see whether concentrating on more recent results gives a better predictor of future records.

(a) Use the data in Table 2 to complete the table of winning pole vault heights. (Note that we are using $x = 0$ to correspond to the year 1972, where this restricted data set begins.)

(b) Find the regression line for the data in part (a).

(c) Plot the data and the regression line on the same axes. Does the regression line seem to provide a good model for the data?

(d) What does the regression line predict as the winning pole vault height for the 2008 Olympics? Compare this predicted value to the actual 2008 winning height of 5.96 m, as described on page 133. Has this new regression line provided a better prediction than the line in Example 2?

Year	x	Height (m)
1972	0	5.64
1976	4	
1980	8	
1984		
1988		
1992		
1996		
2000		
2004		

11. Olympic Swimming Records The tables give the gold medal times in the men's and women's 100-m freestyle Olympic swimming event.

(a) Find the regression lines for the men's data and the women's data.

(b) Sketch both regression lines on the same graph. When do these lines predict that the women will overtake the men in the event? Does this conclusion seem reasonable?

MEN

Year	Gold medalist	Time (s)
1908	C. Daniels, USA	65.6
1912	D. Kahanamoku, USA	63.4
1920	D. Kahanamoku, USA	61.4
1924	J. Weissmuller, USA	59.0
1928	J. Weissmuller, USA	58.6
1932	Y. Miyazaki, Japan	58.2
1936	F. Csik, Hungary	57.6
1948	W. Ris, USA	57.3
1952	C. Scholes, USA	57.4
1956	J. Henricks, Australia	55.4
1960	J. Devitt, Australia	55.2
1964	D. Schollander, USA	53.4
1968	M. Wenden, Australia	52.2
1972	M. Spitz, USA	51.22
1976	J. Montgomery, USA	49.99
1980	J. Woithe, E. Germany	50.40
1984	R. Gaines, USA	49.80
1988	M. Biondi, USA	48.63
1992	A. Popov, Russia	49.02
1996	A. Popov, Russia	48.74
2000	P. van den Hoogenband, Netherlands	48.30
2004	P. van den Hoogenband, Netherlands	48.17
2008	A. Bernard, France	47.21

WOMEN

Year	Gold medalist	Time (s)
1912	F. Durack, Australia	82.2
1920	E. Bleibtrey, USA	73.6
1924	E. Lackie, USA	72.4
1928	A. Osipowich, USA	71.0
1932	H. Madison, USA	66.8
1936	H. Mastenbroek, Holland	65.9
1948	G. Andersen, Denmark	66.3
1952	K. Szoke, Hungary	66.8
1956	D. Fraser, Australia	62.0
1960	D. Fraser, Australia	61.2
1964	D. Fraser, Australia	59.5
1968	J. Henne, USA	60.0
1972	S. Nielson, USA	58.59
1976	K. Ender, E. Germany	55.65
1980	B. Krause, E. Germany	54.79
1984	(Tie) C. Steinseifer, USA	55.92
	N. Hogshead, USA	55.92
1988	K. Otto, E. Germany	54.93
1992	Z. Yong, China	54.64
1996	L. Jingyi, China	54.50
2000	I. DeBruijn, Netherlands	53.83
2004	J. Henry, Australia	53.84
2008	B. Steffen, Germany	53.12

12. Shoe Size and Height Do you think that shoe size and height are correlated? Find out by surveying the shoe sizes and heights of people in your class. (Of course, the data for men and women should be separate.) Find the correlation coefficient.

13. Demand for Candy Bars In this problem you will determine a linear demand equation that describes the demand for candy bars in your class. Survey your classmates to determine what price they would be willing to pay for a candy bar. Your survey form might look like the sample to the left.

Would you buy a candy bar from the vending machine in the hallway if the price is as indicated?

Price	Yes or No
30¢	
40¢	
50¢	
60¢	
70¢	
80¢	
90¢	
$1.00	
$1.10	
$1.20	

(a) Make a table of the number of respondents who answered "yes" at each price level.

(b) Make a scatter plot of your data.

(c) Find and graph the regression line $y = mp + b$, which gives the number of responents y who would buy a candy bar if the price were p cents. This is the *demand equation*. Why is the slope m negative?

(d) What is the p-intercept of the demand equation? What does this intercept tell you about pricing candy bars?

CHAPTER 2

FUNCTIONS

Perhaps the most useful idea for modeling the real world is the concept of *function.* Let's look at an example. If a rock climber drops a stone from a high cliff, we know that the stone will fall. But this general description doesn't help us figure out when the stone will hit the ground. To find out, we need a *rule* that relates the distance d the stone falls to the time it has been falling. Galileo was the first to discover the rule: In t seconds the stone falls $16t^2$ feet. This "rule" is called a *function;* we write this function as $d(t) = 16t^2$. Using this function model, we can *predict* when the stone will hit the ground. In this chapter we study properties of functions and how function models can help us to get precise information about the thing or process being modeled.

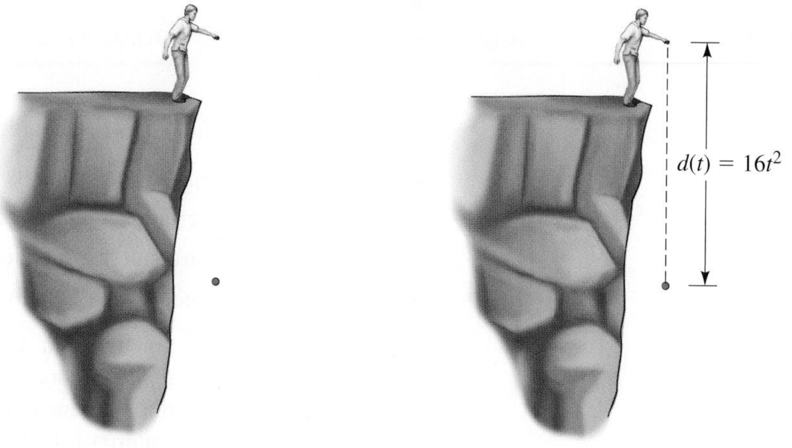

General description: The stone falls.

Function: In t seconds the stone falls $16t^2$ ft.

2.1 What Is a Function?

Functions All Around Us ▶ Definition of Function ▶ Evaluating a Function ▶ Domain of a Function ▶ Four Ways to Represent a Function

In this section we explore the idea of a function and then give the mathematical definition of function.

▼ Functions All Around Us

In nearly every physical phenomenon we observe that one quantity depends on another. For example, your height depends on your age, the temperature depends on the date, the cost of mailing a package depends on its weight (see Figure 1). We use the term *function* to describe this dependence of one quantity on another. That is, we say the following:

- Height is a function of age.
- Temperature is a function of date.
- Cost of mailing a package is a function of weight.

The U.S. Post Office uses a simple rule to determine the cost of mailing a first-class parcel on the basis of its weight. But it's not so easy to describe the rule that relates height to age or the rule that relates temperature to date.

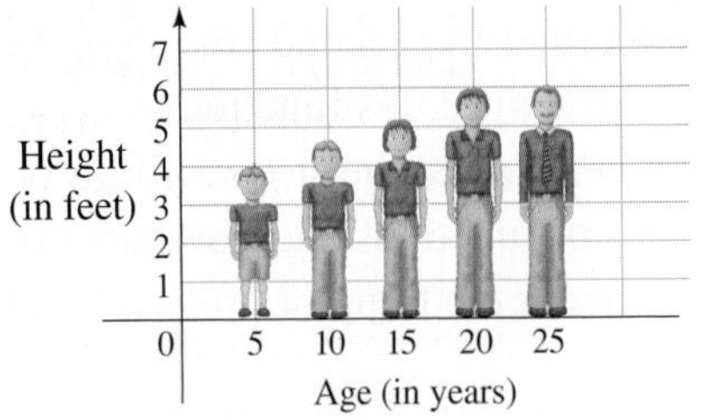

Height is a function of age.

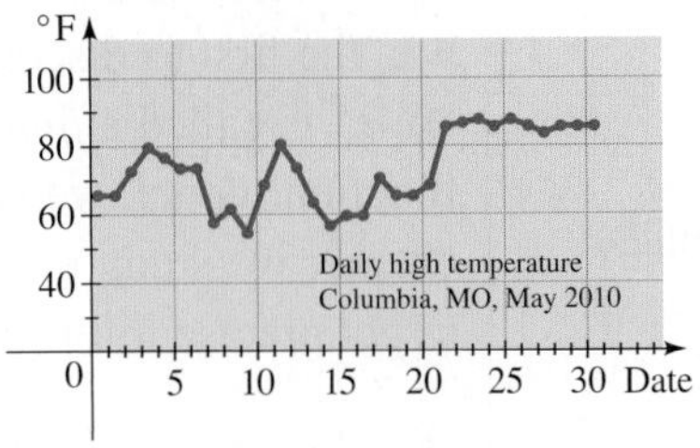

Temperature is a function of date.

w (ounces)	Postage (dollars)
$0 < w \leq 1$	1.22
$1 < w \leq 2$	1.39
$2 < w \leq 3$	1.56
$3 < w \leq 4$	1.73
$4 < w \leq 5$	1.90
$5 < w \leq 6$	2.07

Postage is a function of weight.

FIGURE 1

Can you think of other functions? Here are some more examples:

- The area of a circle is a function of its radius.
- The number of bacteria in a culture is a function of time.
- The weight of an astronaut is a function of her elevation.
- The price of a commodity is a function of the demand for that commodity.

The rule that describes how the area A of a circle depends on its radius r is given by the formula $A = \pi r^2$. Even when a precise rule or formula describing a function is not available, we can still describe the function by a graph. For example, when you turn on a hot water faucet, the temperature of the water depends on how long the water has been running. So we can say:

- The temperature of water from the faucet is a function of time.

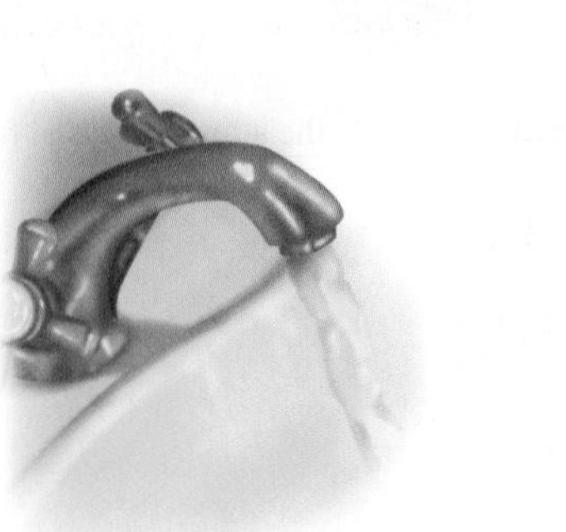

Figure 2 shows a rough graph of the temperature T of the water as a function of the time t that has elapsed since the faucet was turned on. The graph shows that the initial temperature of the water is close to room temperature. When the water from the hot water tank reaches the faucet, the water's temperature T increases quickly. In the next phase, T is constant at the temperature of the water in the tank. When the tank is drained, T decreases to the temperature of the cold water supply.

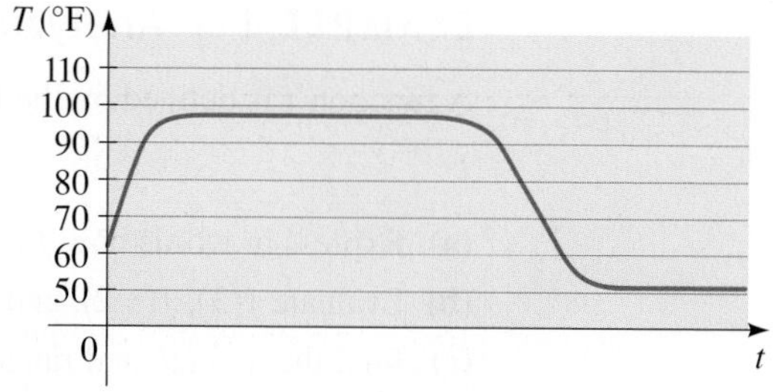

FIGURE 2 Graph of water temperature T as a function of time t

▼ Definition of Function

We have previously used letters to stand for numbers. Here we do something quite different: We use letters to represent *rules*.

A function is a rule. To talk about a function, we need to give it a name. We will use letters such as $f, g, h, \ldots$ to represent functions. For example, we can use the letter f to represent a rule as follows:

"f" is the rule "square the number"

When we write $f(2)$, we mean "apply the rule f to the number 2." Applying the rule gives $f(2) = 2^2 = 4$. Similarly, $f(3) = 3^2 = 9$, $f(4) = 4^2 = 16$, and in general $f(x) = x^2$.

DEFINITION OF A FUNCTION

A **function** f is a rule that assigns to each element x in a set A exactly one element, called $f(x)$, in a set B.

The $\boxed{\sqrt{\ }}$ key on your calculator is a good example of a function as a machine. First you input x into the display. Then you press the key labeled $\boxed{\sqrt{\ }}$. (On most *graphing* calculators the order of these operations is reversed.) If $x < 0$, then x is not in the domain of this function; that is, x is not an acceptable input, and the calculator will indicate an error. If $x \geq 0$, then an approximation to $\sqrt{x}$ appears in the display, correct to a certain number of decimal places. (Thus, the $\boxed{\sqrt{\ }}$ key on your calculator is not quite the same as the exact mathematical function f defined by $f(x) = \sqrt{x}$.)

We usually consider functions for which the sets A and B are sets of real numbers. The symbol $f(x)$ is read "f of x" or "f at x" and is called the **value of f at x**, or the **image of x under f**. The set A is called the **domain** of the function. The **range** of f is the set of all possible values of $f(x)$ as x varies throughout the domain, that is,

$$\text{range of } f = \{f(x) \mid x \in A\}$$

The symbol that represents an arbitrary number in the domain of a function f is called an **independent variable**. The symbol that represents a number in the range of f is called a **dependent variable**. So if we write $y = f(x)$, then x is the independent variable and y is the dependent variable.

It is helpful to think of a function as a **machine** (see Figure 3). If x is in the domain of the function f, then when x enters the machine, it is accepted as an **input** and the machine produces an **output** $f(x)$ according to the rule of the function. Thus, we can think of the domain as the set of all possible inputs and the range as the set of all possible outputs.

FIGURE 3 Machine diagram of f

Another way to picture a function is by an **arrow diagram** as in Figure 4. Each arrow connects an element of A to an element of B. The arrow indicates that $f(x)$ is associated with x, $f(a)$ is associated with a, and so on.

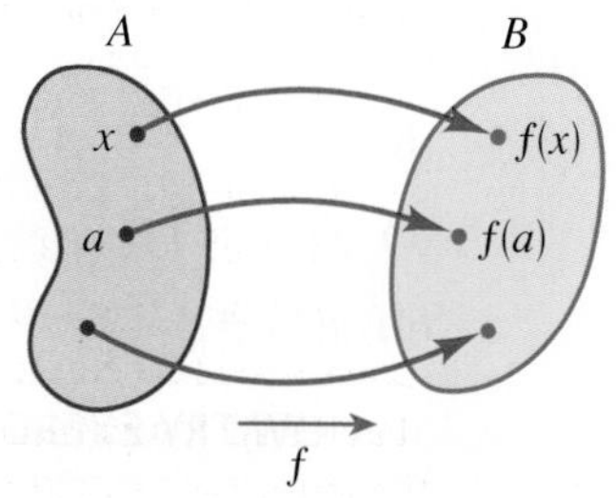

FIGURE 4 Arrow diagram of f

EXAMPLE 1 | Analyzing a Function

A function f is defined by the formula

$$f(x) = x^2 + 4$$

(a) Express in words how f acts on the input x to produce the output $f(x)$.

(b) Evaluate $f(3)$, $f(-2)$, and $f(\sqrt{5})$.

(c) Find the domain and range of f.

(d) Draw a machine diagram for f.

SOLUTION

(a) The formula tells us that f first squares the input x and then adds 4 to the result. So f is the function

"square, then add 4"

(b) The values of f are found by substituting for x in the formula $f(x) = x^2 + 4$.

$$f(3) = 3^2 + 4 = 13 \qquad \text{Replace } x \text{ by } 3$$

$$f(-2) = (-2)^2 + 4 = 8 \qquad \text{Replace } x \text{ by } -2$$

$$f(\sqrt{5}) = (\sqrt{5})^2 + 4 = 9 \qquad \text{Replace } x \text{ by } \sqrt{5}$$

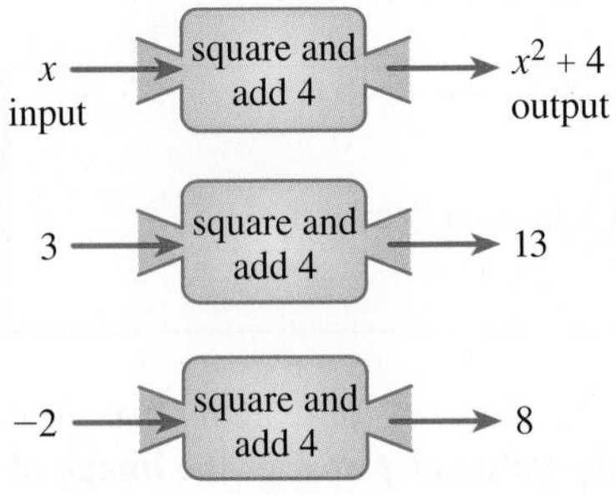

FIGURE 5 Machine diagram

(c) The domain of f consists of all possible inputs for f. Since we can evaluate the formula $f(x) = x^2 + 4$ for every real number x, the domain of f is the set $\mathbb{R}$ of all real numbers.

The range of f consists of all possible outputs of f. Because $x^2 \geq 0$ for all real numbers x, we have $x^2 + 4 \geq 4$, so for every output of f we have $f(x) \geq 4$. Thus, the range of f is $\{y \mid y \geq 4\} = [4, \infty)$.

(d) A machine diagram for f is shown in Figure 5.

NOW TRY EXERCISES **9**, **13**, **17**, AND **43**

▼ Evaluating a Function

In the definition of a function the independent variable x plays the role of a placeholder. For example, the function $f(x) = 3x^2 + x - 5$ can be thought of as

$$f(\square) = 3 \cdot \square^2 + \square - 5$$

To evaluate f at a number, we substitute the number for the placeholder.

EXAMPLE 2 | Evaluating a Function

Let $f(x) = 3x^2 + x - 5$. Evaluate each function value.

(a) $f(-2)$ **(b)** $f(0)$ **(c)** $f(4)$ **(d)** $f(\frac{1}{2})$

SOLUTION To evaluate f at a number, we substitute the number for x in the definition of f.

(a) $f(-2) = 3 \cdot (-2)^2 + (-2) - 5 = 5$

(b) $f(0) = 3 \cdot 0^2 + 0 - 5 = -5$

(c) $f(4) = 3 \cdot (4)^2 + 4 - 5 = 47$

(d) $f(\frac{1}{2}) = 3 \cdot (\frac{1}{2})^2 + \frac{1}{2} - 5 = -\frac{15}{4}$

NOW TRY EXERCISE **19**

EXAMPLE 3 | A Piecewise Defined Function

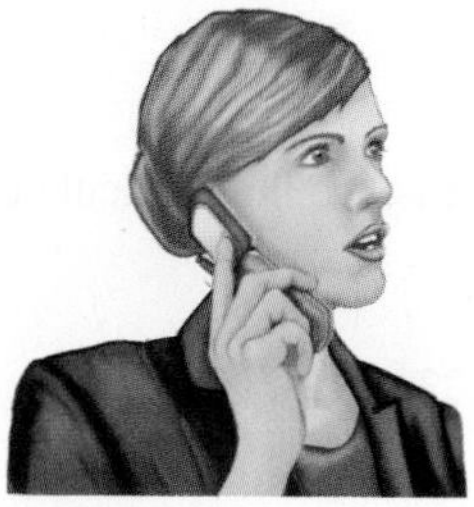

A cell phone plan costs \$39 a month. The plan includes 400 free minutes and charges 20¢ for each additional minute of usage. The monthly charges are a function of the number of minutes used, given by

$$C(x) = \begin{cases} 39 & \text{if } 0 \le x \le 400 \\ 39 + 0.20(x - 400) & \text{if } x > 400 \end{cases}$$

Find $C(100)$, $C(400)$, and $C(480)$.

A **piecewise defined function** is defined by different formulas on different parts of its domain. The function C of Example 3 is piecewise defined.

SOLUTION Remember that a function is a rule. Here is how we apply the rule for this function. First we look at the value of the input x. If $0 \le x \le 400$, then the value of $C(x)$ is 39. On the other hand, if $x > 400$, then the value of $C(x)$ is $39 + 0.20(x - 400)$.

Since $100 \le 400$, we have $C(100) = 39$.

Since $400 \le 400$, we have $C(400) = 39$.

Since $480 > 400$, we have $C(480) = 39 + 0.20(480 - 400) = 55$.

Thus, the plan charges \$39 for 100 minutes, \$39 for 400 minutes, and \$55 for 480 minutes.

NOW TRY EXERCISE **27** ■

EXAMPLE 4 | Evaluating a Function

Expressions like the one in part (d) of Example 4 occur frequently in calculus; they are called *difference quotients*, and they represent the average change in the value of f between $x = a$ and $x = a + h$.

If $f(x) = 2x^2 + 3x - 1$, evaluate the following.

(a) $f(a)$ **(b)** $f(-a)$

(c) $f(a + h)$ **(d)** $\dfrac{f(a + h) - f(a)}{h}$, $h \ne 0$

SOLUTION

(a) $f(a) = 2a^2 + 3a - 1$

(b) $f(-a) = 2(-a)^2 + 3(-a) - 1 = 2a^2 - 3a - 1$

(c)
$$\begin{aligned} f(a + h) &= 2(a + h)^2 + 3(a + h) - 1 \\ &= 2(a^2 + 2ah + h^2) + 3(a + h) - 1 \\ &= 2a^2 + 4ah + 2h^2 + 3a + 3h - 1 \end{aligned}$$

(d) Using the results from parts (c) and (a), we have

$$\begin{aligned} \frac{f(a + h) - f(a)}{h} &= \frac{(2a^2 + 4ah + 2h^2 + 3a + 3h - 1) - (2a^2 + 3a - 1)}{h} \\ &= \frac{4ah + 2h^2 + 3h}{h} = 4a + 2h + 3 \end{aligned}$$

NOW TRY EXERCISE **35** ■

EXAMPLE 5 | The Weight of an Astronaut

If an astronaut weighs 130 pounds on the surface of the earth, then her weight when she is h miles above the earth is given by the function

$$w(h) = 130\left(\frac{3960}{3960 + h}\right)^2$$

(a) What is her weight when she is 100 mi above the earth?

The weight of an object on or near the earth is the gravitational force that the earth exerts on it. When in orbit around the earth, an astronaut experiences the sensation of "weightlessness" because the centripetal force that keeps her in orbit is exactly the same as the gravitational pull of the earth.

(b) Construct a table of values for the function w that gives her weight at heights from 0 to 500 mi. What do you conclude from the table?

SOLUTION

(a) We want the value of the function w when $h = 100$; that is, we must calculate $w(100)$.

$$w(100) = 130\left(\frac{3960}{3960 + 100}\right)^2 \approx 123.67$$

So at a height of 100 mi she weighs about 124 lb.

(b) The table gives the astronaut's weight, rounded to the nearest pound, at 100-mile increments. The values in the table are calculated as in part (a).

h	$w(h)$
0	130
100	124
200	118
300	112
400	107
500	102

The table indicates that the higher the astronaut travels, the less she weighs.

NOW TRY EXERCISE **71**

The Domain of a Function

Recall that the *domain* of a function is the set of all inputs for the function. The domain of a function may be stated explicitly. For example, if we write

$$f(x) = x^2 \qquad 0 \le x \le 5$$

then the domain is the set of all real numbers x for which $0 \le x \le 5$. If the function is given by an algebraic expression and the domain is not stated explicitly, then by convention *the domain of the function is the domain of the algebraic expression—that is, the set of all real numbers for which the expression is defined as a real number.* For example, consider the functions

Domains of algebraic expressions are discussed on page 35.

$$f(x) = \frac{1}{x - 4} \qquad g(x) = \sqrt{x}$$

The function f is not defined at $x = 4$, so its domain is $\{x \mid x \ne 4\}$. The function g is not defined for negative x, so its domain is $\{x \mid x \ge 0\}$.

EXAMPLE 6 | Finding Domains of Functions

Find the domain of each function.

(a) $f(x) = \dfrac{1}{x^2 - x}$ **(b)** $g(x) = \sqrt{9 - x^2}$ **(c)** $h(t) = \dfrac{t}{\sqrt{t + 1}}$

SOLUTION

(a) A rational expression is not defined when the denominator is 0. Since

$$f(x) = \frac{1}{x^2 - x} = \frac{1}{x(x-1)}$$

we see that $f(x)$ is not defined when $x = 0$ or $x = 1$. Thus, the domain of f is

$$\{x \mid x \neq 0, x \neq 1\}$$

The domain may also be written in interval notation as

$$(\infty, 0) \cup (0, 1) \cup (1, \infty)$$

(b) We can't take the square root of a negative number, so we must have $9 - x^2 \geq 0$. Using the methods of Section 1.7, we can solve this inequality to find that $-3 \leq x \leq 3$. Thus, the domain of g is

$$\{x \mid -3 \leq x \leq 3\} = [-3, 3]$$

(c) We can't take the square root of a negative number, and we can't divide by 0, so we must have $t + 1 > 0$, that is, $t > -1$. So the domain of h is

$$\{t \mid t > -1\} = (-1, \infty)$$

NOW TRY EXERCISES **47** AND **51** ■

Four Ways to Represent a Function

To help us understand what a function is, we have used machine and arrow diagrams. We can describe a specific function in the following four ways:

- verbally (by a description in words)
- algebraically (by an explicit formula)
- visually (by a graph)
- numerically (by a table of values)

A single function may be represented in all four ways, and it is often useful to go from one representation to another to gain insight into the function. However, certain functions are described more naturally by one method than by the others. An example of a verbal description is the following rule for converting between temperature scales:

> "To find the Fahrenheit equivalent of a Celsius temperature, multiply the Celsius temperature by $\frac{9}{5}$, then add 32."

In Example 7 we see how to describe this verbal rule or function algebraically, graphically, and numerically. A useful representation of the area of a circle as a function of its radius is the algebraic formula

$$A(r) = \pi r^2$$

The graph produced by a seismograph (see the box on the next page) is a visual representation of the vertical acceleration function $a(t)$ of the ground during an earthquake. As a final example, consider the function $C(w)$, which is described verbally as "the cost of mailing a first-class letter with weight w." The most convenient way of describing this function is numerically—that is, using a table of values.

We will be using all four representations of functions throughout this book. We summarize them in the following box.

FOUR WAYS TO REPRESENT A FUNCTION

Verbal

Using words:

"To convert from Celsius to Fahrenheit, multiply the Celsius temperature by $\frac{9}{5}$, then add 32."

Relation between Celsius and Fahrenheit temperature scales

Algebraic

Using a formula:

$$A(r) = \pi r^2$$

Area of a circle

Visual

Using a graph:

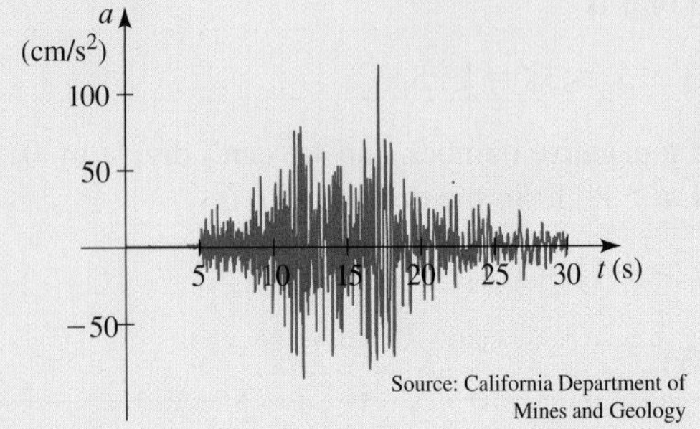

Vertical acceleration during an earthquake

Numerical

Using a table of values:

w (ounces)	*C*(*w*) (dollars)
$0 < w \le 1$	1.22
$1 < w \le 2$	1.39
$2 < w \le 3$	1.56
$3 < w \le 4$	1.73
$4 < w \le 5$	1.90
⋮	⋮

Cost of mailing a first-class parcel

EXAMPLE 7 | Representing a Function Verbally, Algebraically, Numerically, and Graphically

Let $F(C)$ be the Fahrenheit temperature corresponding to the Celsius temperature C. (Thus, F is the function that converts Celsius inputs to Fahrenheit outputs.) The box above gives a verbal description of this function. Find ways to represent this function

(a) Algebraically (using a formula)

(b) Numerically (using a table of values)

(c) Visually (using a graph)

SOLUTION

(a) The verbal description tells us that we should first multiply the input C by $\frac{9}{5}$ and then add 32 to the result. So we get

$$F(C) = \tfrac{9}{5}C + 32$$

(b) We use the algebraic formula for F that we found in part (a) to construct a table of values:

C (Celsius)	*F* (Fahrenheit)
−10	14
0	32
10	50
20	68
30	86
40	104

(c) We use the points tabulated in part (b) to help us draw the graph of this function in Figure 6.

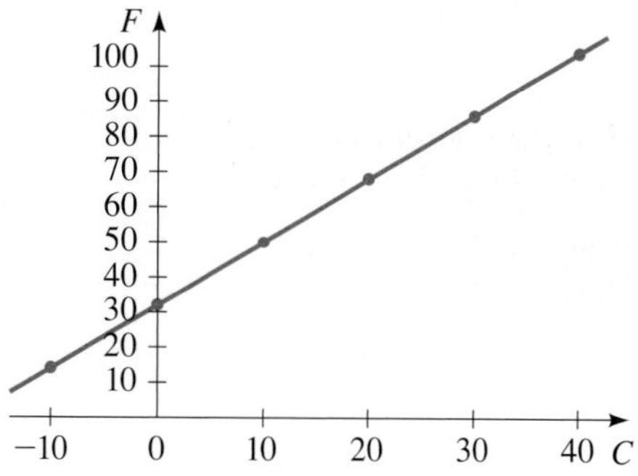

FIGURE 6 Celsius and Fahrenheit

NOW TRY EXERCISE **65**

2.1 EXERCISES

CONCEPTS

1. If a function f is given by the formula $y = f(x)$, then $f(a)$ is the ________ of f at $x = a$.

2. For a function f, the set of all possible inputs is called the ________ of f, and the set of all possible outputs is called the ________ of f.

3. (a) Which of the following functions have 5 in their domain?

$$f(x) = x^2 - 3x \qquad g(x) = \frac{x-5}{x} \qquad h(x) = \sqrt{x - 10}$$

(b) For the functions from part (a) that *do* have 5 in their domain, find the value of the function at 5.

4. A function is given algebraically by the formula $f(x) = (x-4)^2 + 3$. Complete these other ways to represent f:

(a) *Verbal:* "Subtract 4, then ________ and ________."

(b) *Numerical:*

x	$f(x)$
0	19
2	
4	
6	

SKILLS

5–8 ■ Express the rule in function notation. (For example, the rule "square, then subtract 5" is expressed as the function $f(x) = x^2 - 5$.)

5. Add 3, then multiply by 2

6. Divide by 7, then subtract 4

7. Subtract 5, then square

8. Take the square root, add 8, then multiply by $\frac{1}{3}$

9–12 ■ Express the function (or rule) in words.

9. $h(x) = x^2 + 2$

10. $k(x) = \sqrt{x+2}$

11. $f(x) = \dfrac{x-4}{3}$

12. $g(x) = \dfrac{x}{3} - 4$

13–14 ■ Draw a machine diagram for the function.

13. $f(x) = \sqrt{x-1}$

14. $f(x) = \dfrac{3}{x-2}$

15–16 ■ Complete the table.

15. $f(x) = 2(x-1)^2$

x	$f(x)$
−1	
0	
1	
2	
3	

16. $g(x) = |2x + 3|$

x	$g(x)$
−3	
−2	
0	
1	
3	

17–26 ■ Evaluate the function at the indicated values.

17. $f(x) = x^2 - 6;\quad f(-3), f(3), f(0), f(\frac{1}{2}), f(10)$

18. $f(x) = x^3 + 2x;\quad f(-2), f(1), f(0), f(\frac{1}{3}), f(0.2)$

19. $f(x) = 2x + 1;$

$f(1), f(-2), f(\frac{1}{2}), f(a), f(-a), f(a+b)$

20. $f(x) = x^2 + 2x;$

$f(0), f(3), f(-3), f(a), f(-x), f\left(\dfrac{1}{a}\right)$

21. $g(x) = \dfrac{1-x}{1+x};$

$g(2), g(-2), g(\frac{1}{2}), g(a), g(a-1), g(-1)$

22. $h(t) = t + \frac{1}{t}$;
$h(1), h(-1), h(2), h(\frac{1}{2}), h(x), h\left(\frac{1}{x}\right)$

23. $f(x) = 2x^2 + 3x - 4$;
$f(0), f(2), f(-2), f(\sqrt{2}), f(x+1), f(-x)$

24. $f(x) = x^3 - 4x^2$;
$f(0), f(1), f(-1), f(\frac{3}{2}), f\left(\frac{x}{2}\right), f(x^2)$

25. $f(x) = 2|x - 1|$;
$f(-2), f(0), f(\frac{1}{2}), f(2), f(x+1), f(x^2+2)$

26. $f(x) = \frac{|x|}{x}$;
$f(-2), f(-1), f(0), f(5), f(x^2), f\left(\frac{1}{x}\right)$

27–30 ■ Evaluate the piecewise defined function at the indicated values.

27. $f(x) = \begin{cases} x^2 & \text{if } x < 0 \\ x + 1 & \text{if } x \geq 0 \end{cases}$
$f(-2), f(-1), f(0), f(1), f(2)$

28. $f(x) = \begin{cases} 5 & \text{if } x \leq 2 \\ 2x - 3 & \text{if } x > 2 \end{cases}$
$f(-3), f(0), f(2), f(3), f(5)$

29. $f(x) = \begin{cases} x^2 + 2x & \text{if } x \leq -1 \\ x & \text{if } -1 < x \leq 1 \\ -1 & \text{if } x > 1 \end{cases}$
$f(-4), f(-\frac{3}{2}), f(-1), f(0), f(25)$

30. $f(x) = \begin{cases} 3x & \text{if } x < 0 \\ x + 1 & \text{if } 0 \leq x \leq 2 \\ (x-2)^2 & \text{if } x > 2 \end{cases}$
$f(-5), f(0), f(1), f(2), f(5)$

31–34 ■ Use the function to evaluate the indicated expressions and simplify.

31. $f(x) = x^2 + 1$; $f(x+2), f(x) + f(2)$

32. $f(x) = 3x - 1$; $f(2x), 2f(x)$

33. $f(x) = x + 4$; $f(x^2), (f(x))^2$

34. $f(x) = 6x - 18$; $f\left(\frac{x}{3}\right), \frac{f(x)}{3}$

35–42 ■ Find $f(a)$, $f(a+h)$, and the difference quotient $\frac{f(a+h) - f(a)}{h}$, where $h \neq 0$.

35. $f(x) = 3x + 2$

36. $f(x) = x^2 + 1$

37. $f(x) = 5$

38. $f(x) = \frac{1}{x+1}$

39. $f(x) = \frac{x}{x+1}$

40. $f(x) = \frac{2x}{x-1}$

41. $f(x) = 3 - 5x + 4x^2$

42. $f(x) = x^3$

43–64 ■ Find the domain of the function.

43. $f(x) = 2x$

44. $f(x) = x^2 + 1$

45. $f(x) = 2x, \quad -1 \leq x \leq 5$

46. $f(x) = x^2 + 1, \quad 0 \leq x \leq 5$

47. $f(x) = \frac{1}{x-3}$

48. $f(x) = \frac{1}{3x-6}$

49. $f(x) = \frac{x+2}{x^2-1}$

50. $f(x) = \frac{x^4}{x^2+x-6}$

51. $f(x) = \sqrt{x-5}$

52. $f(x) = \sqrt[4]{x+9}$

53. $f(t) = \sqrt[3]{t-1}$

54. $g(x) = \sqrt{7-3x}$

55. $h(x) = \sqrt{2x-5}$

56. $G(x) = \sqrt{x^2-9}$

57. $g(x) = \frac{\sqrt{2+x}}{3-x}$

58. $g(x) = \frac{\sqrt{x}}{2x^2+x-1}$

59. $g(x) = \sqrt[4]{x^2-6x}$

60. $g(x) = \sqrt{x^2-2x-8}$

61. $f(x) = \frac{3}{\sqrt{x-4}}$

62. $f(x) = \frac{x^2}{\sqrt{6-x}}$

63. $f(x) = \frac{(x+1)^2}{\sqrt{2x-1}}$

64. $f(x) = \frac{x}{\sqrt[4]{9-x^2}}$

65–68 ■ A verbal description of a function is given. Find **(a)** algebraic, **(b)** numerical, and **(c)** graphical representations for the function.

65. To evaluate $f(x)$, divide the input by 3 and add $\frac{2}{3}$ to the result.

66. To evaluate $g(x)$, subtract 4 from the input and multiply the result by $\frac{3}{4}$.

67. Let $T(x)$ be the amount of sales tax charged in Lemon County on a purchase of x dollars. To find the tax, take 8% of the purchase price.

68. Let $V(d)$ be the volume of a sphere of diameter d. To find the volume, take the cube of the diameter, then multiply by π and divide by 6.

APPLICATIONS

69. **Production Cost** The cost C in dollars of producing x yards of a certain fabric is given by the function

$$C(x) = 1500 + 3x + 0.02x^2 + 0.0001x^3$$

(a) Find $C(10)$ and $C(100)$.
(b) What do your answers in part (a) represent?
(c) Find $C(0)$. (This number represents the *fixed costs*.)

70. Area of a Sphere The surface area S of a sphere is a function of its radius r given by

$$S(r) = 4\pi r^2$$

(a) Find $S(2)$ and $S(3)$.
(b) What do your answers in part (a) represent?

71. Torricelli's Law A tank holds 50 gallons of water, which drains from a leak at the bottom, causing the tank to empty in 20 minutes. The tank drains faster when it is nearly full because the pressure on the leak is greater. **Torricelli's Law** gives the volume of water remaining in the tank after t minutes as

$$V(t) = 50\left(1 - \frac{t}{20}\right)^2 \qquad 0 \le t \le 20$$

(a) Find $V(0)$ and $V(20)$.
(b) What do your answers to part (a) represent?
(c) Make a table of values of $V(t)$ for $t = 0, 5, 10, 15, 20$.

72. How Far Can You See? Because of the curvature of the earth, the maximum distance D that you can see from the top of a tall building or from an airplane at height h is given by the function

$$D(h) = \sqrt{2rh + h^2}$$

where $r = 3960$ mi is the radius of the earth and D and h are measured in miles.
(a) Find $D(0.1)$ and $D(0.2)$.
(b) How far can you see from the observation deck of Toronto's CN Tower, 1135 ft above the ground?
(c) Commercial aircraft fly at an altitude of about 7 mi. How far can the pilot see?

73. Blood Flow As blood moves through a vein or an artery, its velocity v is greatest along the central axis and decreases as the distance r from the central axis increases (see the figure). The formula that gives v as a function of r is called the **law of laminar flow**. For an artery with radius 0.5 cm, the relationship between v (in cm/s) and r (in cm) is given by the function

$$v(r) = 18{,}500(0.25 - r^2) \qquad 0 \le r \le 0.5$$

(a) Find $v(0.1)$ and $v(0.4)$.
(b) What do your answers to part (a) tell you about the flow of blood in this artery?
(c) Make a table of values of $v(r)$ for $r = 0, 0.1, 0.2, 0.3, 0.4, 0.5$.

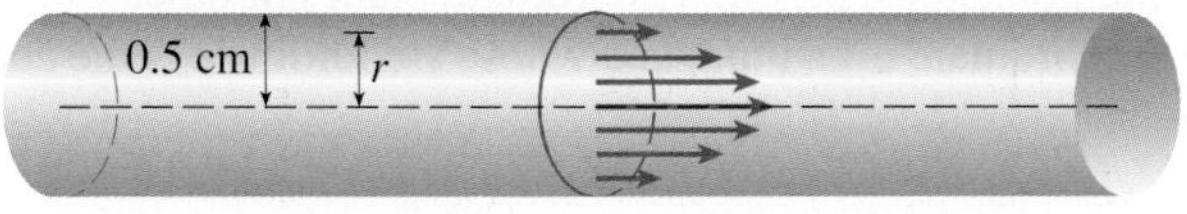

74. Pupil Size When the brightness x of a light source is increased, the eye reacts by decreasing the radius R of the pupil. The dependence of R on x is given by the function

$$R(x) = \sqrt{\frac{13 + 7x^{0.4}}{1 + 4x^{0.4}}}$$

where R is measured in millimeters and x is measured in appropriate units of brightness.
(a) Find $R(1)$, $R(10)$, and $R(100)$.
(b) Make a table of values of $R(x)$.

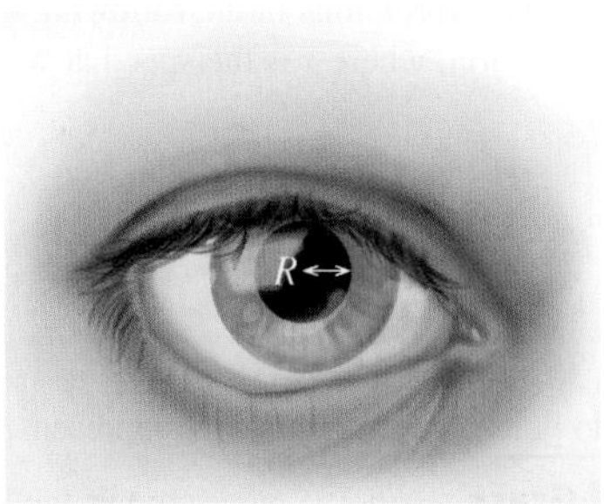

75. Relativity According to the Theory of Relativity, the length L of an object is a function of its velocity v with respect to an observer. For an object whose length at rest is 10 m, the function is given by

$$L(v) = 10\sqrt{1 - \frac{v^2}{c^2}}$$

where c is the speed of light (300,000 km/s).
(a) Find $L(0.5c)$, $L(0.75c)$, and $L(0.9c)$.
(b) How does the length of an object change as its velocity increases?

76. Income Tax In a certain country, income tax T is assessed according to the following function of income x:

$$T(x) = \begin{cases} 0 & \text{if } 0 \le x \le 10{,}000 \\ 0.08x & \text{if } 10{,}000 < x \le 20{,}000 \\ 1600 + 0.15x & \text{if } 20{,}000 < x \end{cases}$$

(a) Find $T(5{,}000)$, $T(12{,}000)$, and $T(25{,}000)$.
(b) What do your answers in part (a) represent?

77. Internet Purchases An Internet bookstore charges \$15 shipping for orders under \$100 but provides free shipping for orders of \$100 or more. The cost C of an order is a function of the total price x of the books purchased, given by

$$C(x) = \begin{cases} x + 15 & \text{if } x < 100 \\ x & \text{if } x \ge 100 \end{cases}$$

(a) Find $C(75)$, $C(90)$, $C(100)$, and $C(105)$.
(b) What do your answers in part (a) represent?

78. Cost of a Hotel Stay A hotel chain charges \$75 each night for the first two nights and \$50 for each additional night's stay. The total cost T is a function of the number of nights x that a guest stays.

(a) Complete the expressions in the following piecewise defined function.

$$T(x) = \begin{cases} \rule{2em}{1ex} & \text{if } 0 \le x \le 2 \\ \rule{2em}{1ex} & \text{if } x > 2 \end{cases}$$

(b) Find $T(2)$, $T(3)$, and $T(5)$.
(c) What do your answers in part (b) represent?

79. **Speeding Tickets** In a certain state the maximum speed permitted on freeways is 65 mi/h, and the minimum is 40. The fine F for violating these limits is \$15 for every mile above the maximum or below the minimum.
(a) Complete the expressions in the following piecewise defined function, where x is the speed at which you are driving.

$$F(x) = \begin{cases} \rule{2em}{1ex} & \text{if } 0 < x < 40 \\ \rule{2em}{1ex} & \text{if } 40 \le x \le 65 \\ \rule{2em}{1ex} & \text{if } x > 65 \end{cases}$$

(b) Find $F(30)$, $F(50)$, and $F(75)$.
(c) What do your answers in part (b) represent?

80. **Height of Grass** A home owner mows the lawn every Wednesday afternoon. Sketch a rough graph of the height of the grass as a function of time over the course of a four-week period beginning on a Sunday.

81. **Temperature Change** You place a frozen pie in an oven and bake it for an hour. Then you take the pie out and let it cool before eating it. Sketch a rough graph of the temperature of the pie as a function of time.

82. **Daily Temperature Change** Temperature readings T (in °F) were recorded every 2 hours from midnight to noon in Atlanta, Georgia, on March 18, 1996. The time t was measured in hours from midnight. Sketch a rough graph of T as a function of t.

t	0	2	4	6	8	10	12
T	58	57	53	50	51	57	61

83. **Population Growth** The population P (in thousands) of San Jose, California, from 1988 to 2000 is shown in the table. (Midyear estimates are given.) Draw a rough graph of P as a function of time t.

t	1988	1990	1992	1994	1996	1998	2000
P	733	782	800	817	838	861	895

DISCOVERY ■ DISCUSSION ■ WRITING

84. **Examples of Functions** At the beginning of this section we discussed three examples of everyday, ordinary functions: Height is a function of age, temperature is a function of date, and postage cost is a function of weight. Give three other examples of functions from everyday life.

85. **Four Ways to Represent a Function** In the box on page 148 we represented four different functions verbally, algebraically, visually, and numerically. Think of a function that can be represented in all four ways, and write the four representations.

2.2 GRAPHS OF FUNCTIONS

Graphing Functions by Plotting Points ▶ Graphing Functions with a Graphing Calculator ▶ Graphing Piecewise Defined Functions ▶ The Vertical Line Test ▶ Equations That Define Functions

The most important way to visualize a function is through its graph. In this section we investigate in more detail the concept of graphing functions.

▼ Graphing Functions by Plotting Points

To graph a function f, we plot the points $(x, f(x))$ in a coordinate plane. In other words, we plot the points (x, y) whose x-coordinate is an input and whose y-coordinate is the corresponding output of the function.

THE GRAPH OF A FUNCTION

If f is a function with domain A, then the **graph** of f is the set of ordered pairs

$$\{(x, f(x)) \mid x \in A\}$$

plotted in a coordinate plane. In other words, the graph of f is the set of all points (x, y) such that $y = f(x)$; that is, the graph of f is the graph of the equation $y = f(x)$.

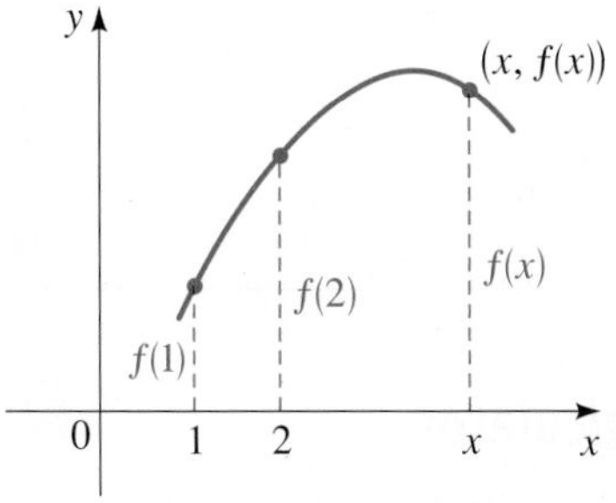

FIGURE 1 The height of the graph above the point x is the value of $f(x)$.

The graph of a function f gives a picture of the behavior or "life history" of the function. We can read the value of $f(x)$ from the graph as being the height of the graph above the point x (see Figure 1).

A function f of the form $f(x) = mx + b$ is called a **linear function** because its graph is the graph of the equation $y = mx + b$, which represents a line with slope m and y-intercept b. A special case of a linear function occurs when the slope is $m = 0$. The function $f(x) = b$, where b is a given number, is called a **constant function** because all its values are the same number, namely, b. Its graph is the horizontal line $y = b$. Figure 2 shows the graphs of the constant function $f(x) = 3$ and the linear function $f(x) = 2x + 1$.

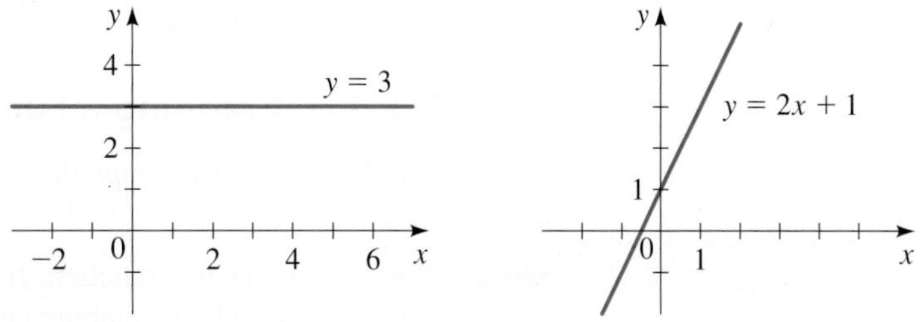

The constant function $f(x) = 3$ The linear function $f(x) = 2x + 1$

FIGURE 2

EXAMPLE 1 | Graphing Functions by Plotting Points

Sketch graphs of the following functions.

(a) $f(x) = x^2$ **(b)** $g(x) = x^3$ **(c)** $h(x) = \sqrt{x}$

SOLUTION We first make a table of values. Then we plot the points given by the table and join them by a smooth curve to obtain the graph. The graphs are sketched in Figure 3 on the next page.

x	$f(x) = x^2$
0	0
$\pm\frac{1}{2}$	$\frac{1}{4}$
±1	1
±2	4
±3	9

x	$g(x) = x^3$
0	0
$\frac{1}{2}$	$\frac{1}{8}$
1	1
2	8
$-\frac{1}{2}$	$-\frac{1}{8}$
-1	-1
-2	-8

x	$h(x) = \sqrt{x}$
0	0
1	1
2	$\sqrt{2}$
3	$\sqrt{3}$
4	2
5	$\sqrt{5}$

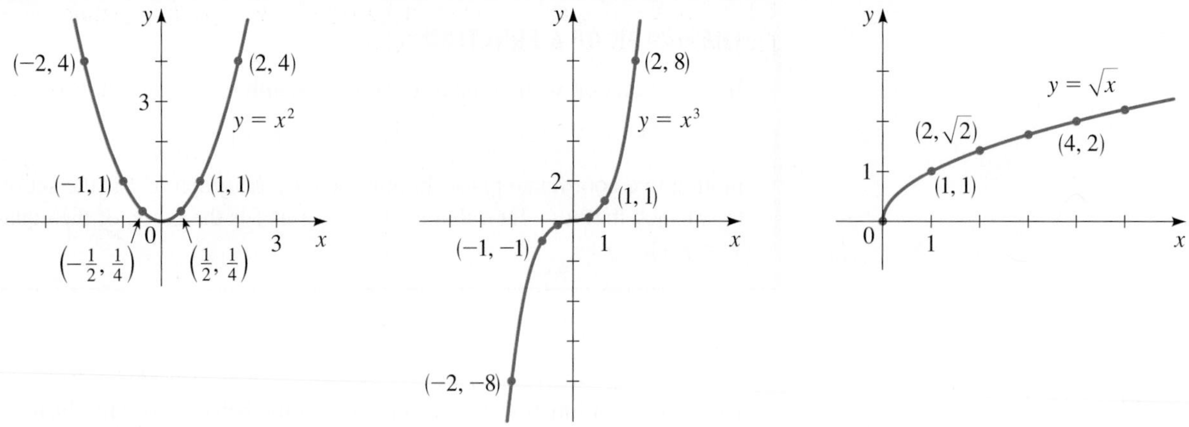

FIGURE 3 (a) $f(x) = x^2$ (b) $g(x) = x^3$ (c) $h(x) = \sqrt{x}$

NOW TRY EXERCISES **11, 15,** AND **19**

Graphing Functions with a Graphing Calculator

A convenient way to graph a function is to use a graphing calculator. Because the graph of a function f is the graph of the equation $y = f(x)$, we can use the methods of Section 1.9 to graph functions on a graphing calculator.

EXAMPLE 2 | Graphing a Function with a Graphing Calculator

Use a graphing calculator to graph the function $f(x) = x^3 - 8x^2$ in an appropriate viewing rectangle.

SOLUTION To graph the function $f(x) = x^3 - 8x^2$, we must graph the equation $y = x^3 - 8x^2$. On the TI-83 graphing calculator the default viewing rectangle gives the graph in Figure 4(a). But this graph appears to spill over the top and bottom of the screen. We need to expand the vertical axis to get a better representation of the graph. The viewing rectangle $[-4, 10]$ by $[-100, 100]$ gives a more complete picture of the graph, as shown in Figure 4(b).

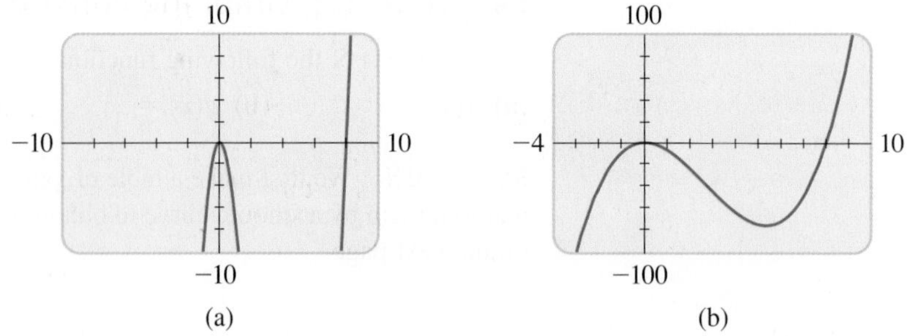

FIGURE 4 Graphing the function $f(x) = x^3 - 8x^2$

NOW TRY EXERCISE **29**

EXAMPLE 3 | A Family of Power Functions

(a) Graph the functions $f(x) = x^n$ for $n = 2, 4$, and 6 in the viewing rectangle $[-2, 2]$ by $[-1, 3]$.

(b) Graph the functions $f(x) = x^n$ for $n = 1, 3$, and 5 in the viewing rectangle $[-2, 2]$ by $[-2, 2]$.

(c) What conclusions can you draw from these graphs?

SOLUTION To graph the function $f(x) = x^n$, we graph the equation $y = x^n$. The graphs for parts (a) and (b) are shown in Figure 5.

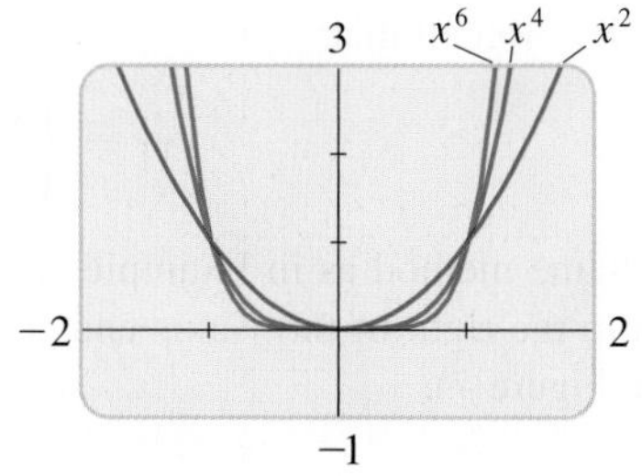

(a) Even powers of x

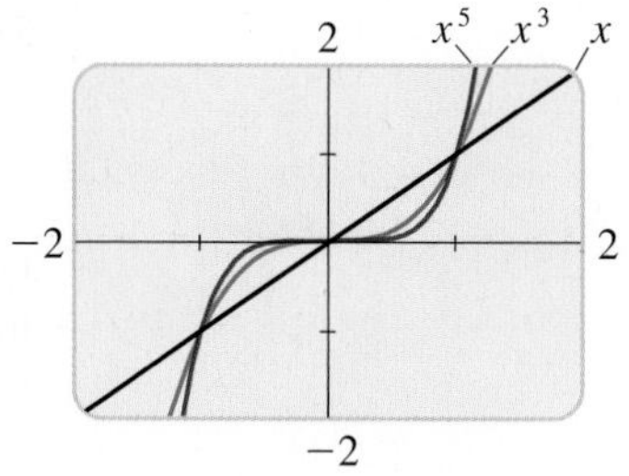

(b) Odd powers of x

FIGURE 5 A family of power functions $f(x) = x^n$

(c) We see that the general shape of the graph of $f(x) = x^n$ depends on whether n is even or odd.

If n is even, the graph of $f(x) = x^n$ is similar to the parabola $y = x^2$.

If n is odd, the graph of $f(x) = x^n$ is similar to that of $y = x^3$.

NOW TRY EXERCISE **69**

Notice from Figure 5 that as n increases, the graph of $y = x^n$ becomes flatter near 0 and steeper when $x > 1$. When $0 < x < 1$, the lower powers of x are the "bigger" functions. But when $x > 1$, the higher powers of x are the dominant functions.

Graphing Piecewise Defined Functions

A piecewise defined function is defined by different formulas on different parts of its domain. As you might expect, the graph of such a function consists of separate pieces.

EXAMPLE 4 | Graph of a Piecewise Defined Function

Sketch the graph of the function.

$$f(x) = \begin{cases} x^2 & \text{if } x \le 1 \\ 2x + 1 & \text{if } x > 1 \end{cases}$$

SOLUTION If $x \le 1$, then $f(x) = x^2$, so the part of the graph to the left of $x = 1$ coincides with the graph of $y = x^2$, which we sketched in Figure 3. If $x > 1$, then $f(x) = 2x + 1$, so the part of the graph to the right of $x = 1$ coincides with the line $y = 2x + 1$, which we graphed in Figure 2. This enables us to sketch the graph in Figure 6.

The solid dot at $(1, 1)$ indicates that this point is included in the graph; the open dot at $(1, 3)$ indicates that this point is excluded from the graph.

On many graphing calculators the graph in Figure 6 can be produced by using the logical functions in the calculator. For example, on the TI-83 the following equation gives the required graph:

`Y₁=(X≤1)X²+(X>1)(2X+1)`

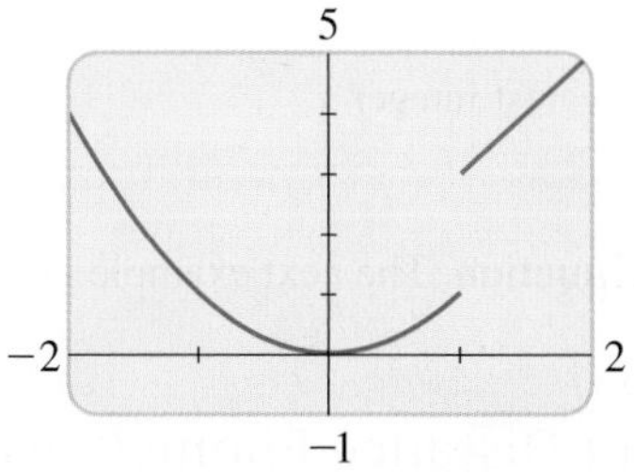

(To avoid the extraneous vertical line between the two parts of the graph, put the calculator in `Dot` mode.)

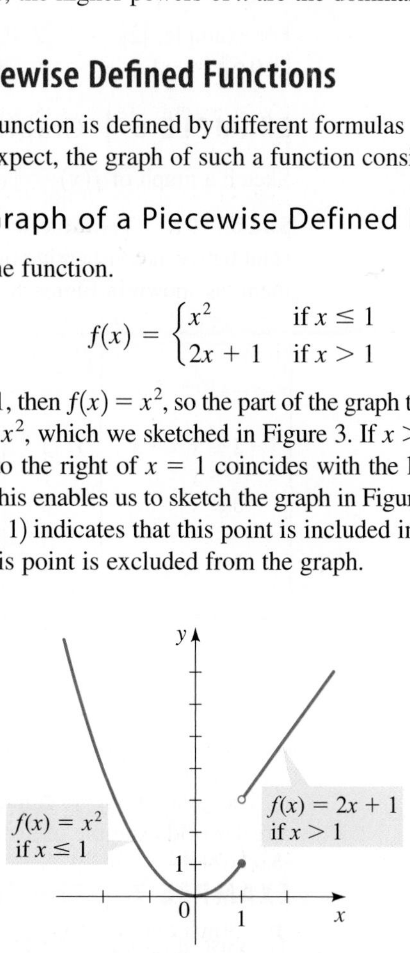

FIGURE 6 $f(x) = \begin{cases} x^2 & \text{if } x \le 1 \\ 2x + 1 & \text{if } x > 1 \end{cases}$

NOW TRY EXERCISE **35**

EXAMPLE 5 | Graph of the Absolute Value Function

Sketch a graph of the absolute value function $f(x) = |x|$.

SOLUTION Recall that

$$|x| = \begin{cases} x & \text{if } x \geq 0 \\ -x & \text{if } x < 0 \end{cases}$$

Using the same method as in Example 4, we note that the graph of f coincides with the line $y = x$ to the right of the y-axis and coincides with the line $y = -x$ to the left of the y-axis (see Figure 7).

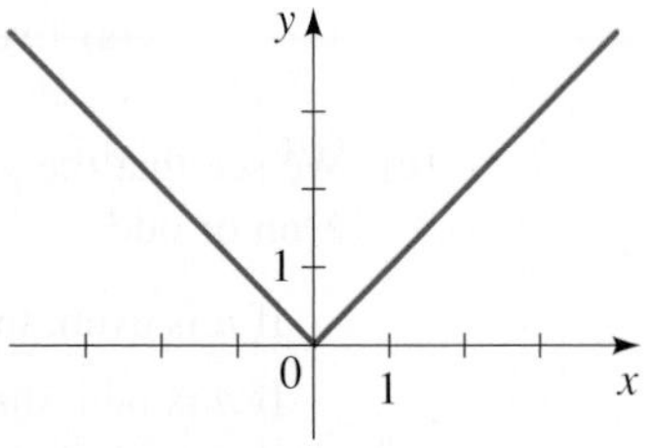

FIGURE 7 Graph of $f(x) = |x|$

NOW TRY EXERCISE **23** ■

The **greatest integer function** is defined by

$$[\![x]\!] = \text{greatest integer less than or equal to } x$$

For example, $[\![2]\!] = 2$, $[\![2.3]\!] = 2$, $[\![1.999]\!] = 1$, $[\![0.002]\!] = 0$, $[\![-3.5]\!] = -4$, and $[\![-0.5]\!] = -1$.

EXAMPLE 6 | Graph of the Greatest Integer Function

Sketch a graph of $f(x) = [\![x]\!]$.

SOLUTION The table shows the values of f for some values of x. Note that $f(x)$ is constant between consecutive integers, so the graph between integers is a horizontal line segment, as shown in Figure 8.

x	$[\![x]\!]$
⋮	⋮
$-2 \leq x < -1$	-2
$-1 \leq x < 0$	-1
$0 \leq x < 1$	0
$1 \leq x < 2$	1
$2 \leq x < 3$	2
⋮	⋮

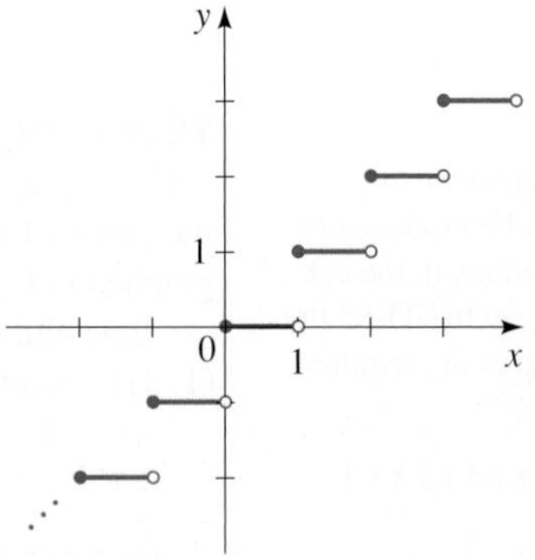

FIGURE 8 The greatest integer function, $y = [\![x]\!]$ ■

The greatest integer function is an example of a **step function**. The next example gives a real-world example of a step function.

EXAMPLE 7 | The Cost Function for Long-Distance Phone Calls

The cost of a long-distance daytime phone call from Toronto, Canada, to Mumbai, India, is 69 cents for the first minute and 58 cents for each additional minute (or part of a minute). Draw the graph of the cost C (in dollars) of the phone call as a function of time t (in minutes).

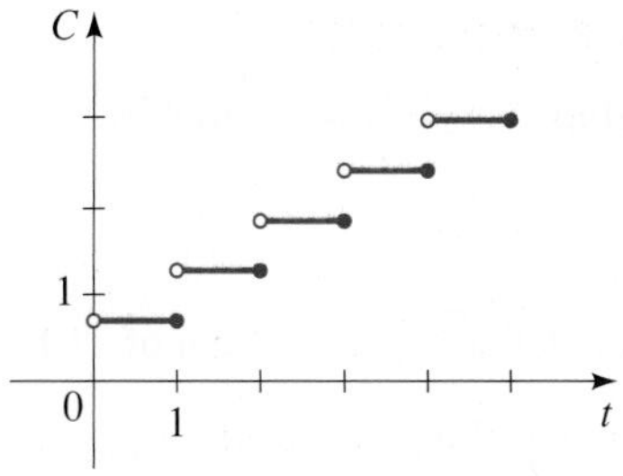

FIGURE 9 Cost of a long-distance call

SOLUTION Let $C(t)$ be the cost for t minutes. Since $t > 0$, the domain of the function is $(0, \infty)$. From the given information we have

$$C(t) = 0.69 \qquad \text{if } 0 < t \le 1$$
$$C(t) = 0.69 + 0.58 = 1.27 \qquad \text{if } 1 < t \le 2$$
$$C(t) = 0.69 + 2(0.58) = 1.85 \qquad \text{if } 2 < t \le 3$$
$$C(t) = 0.69 + 3(0.58) = 2.43 \qquad \text{if } 3 < t \le 4$$

and so on. The graph is shown in Figure 9.

NOW TRY EXERCISE **81** ■

Continuous functions are defined more precisely in Section 13.2, on page 851.

A function is called **continuous** if its graph has no "breaks" or "holes." The functions in Examples 1, 2, 3, and 5 are continuous; the functions in Examples 4, 6, and 7 are not continuous.

▼ The Vertical Line Test

The graph of a function is a curve in the xy-plane. But the question arises: Which curves in the xy-plane are graphs of functions? This is answered by the following test.

> **THE VERTICAL LINE TEST**
>
> A curve in the coordinate plane is the graph of a function if and only if no vertical line intersects the curve more than once.

We can see from Figure 10 why the Vertical Line Test is true. If each vertical line $x = a$ intersects a curve only once at (a, b), then exactly one functional value is defined by $f(a) = b$. But if a line $x = a$ intersects the curve twice, at (a, b) and at (a, c), then the curve cannot represent a function because a function cannot assign two different values to a.

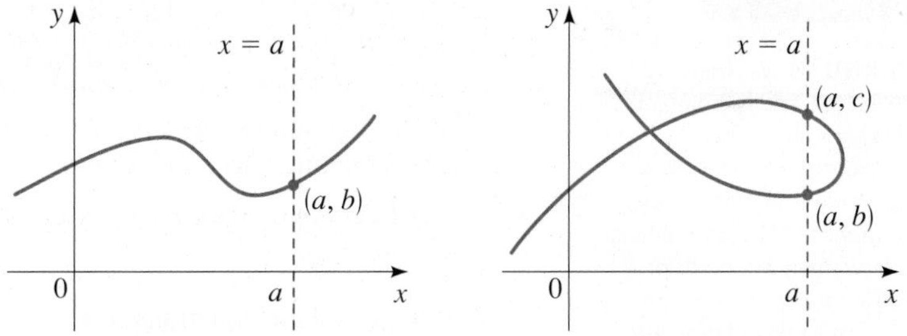

FIGURE 10 Vertical Line Test

Graph of a function

Not a graph of a function

EXAMPLE 8 | Using the Vertical Line Test

Using the Vertical Line Test, we see that the curves in parts (b) and (c) of Figure 11 represent functions, whereas those in parts (a) and (d) do not.

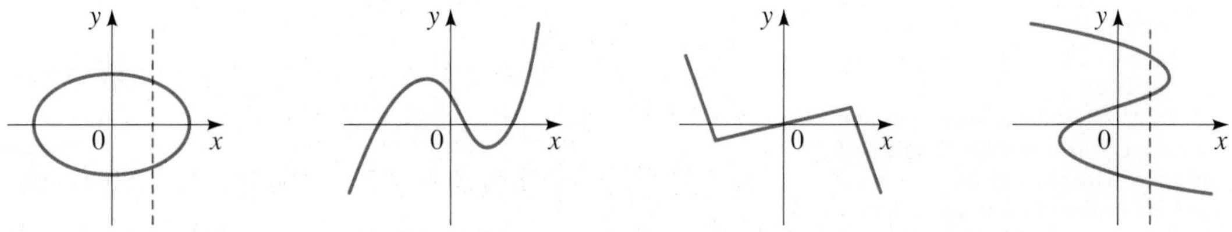

FIGURE 11 (a) (b) (c) (d)

NOW TRY EXERCISE **51** ■

▼ Equations That Define Functions

Any equation in the variables x and y defines a relationship between these variables. For example, the equation

$$y - x^2 = 0$$

defines a relationship between y and x. Does this equation define y as a *function* of x? To find out, we solve for y and get

$$y = x^2$$

We see that the equation defines a rule, or function, that gives one value of y for each value of x. We can express this rule in function notation as

$$f(x) = x^2$$

But not every equation defines y as a function of x, as the following example shows.

EXAMPLE 9 | Equations That Define Functions

Does the equation define y as a function of x?

(a) $y - x^2 = 2$ **(b)** $x^2 + y^2 = 4$

SOLUTION

(a) Solving for y in terms of x gives

$$y - x^2 = 2$$
$$y = x^2 + 2 \qquad \text{Add } x^2$$

The last equation is a rule that gives one value of y for each value of x, so it defines y as a function of x. We can write the function as $f(x) = x^2 + 2$.

(b) We try to solve for y in terms of x:

$$x^2 + y^2 = 4$$
$$y^2 = 4 - x^2 \qquad \text{Subtract } x^2$$
$$y = \pm\sqrt{4 - x^2} \qquad \text{Take square roots}$$

The last equation gives two values of y for a given value of x. Thus, the equation does not define y as a function of x.

NOW TRY EXERCISES **57** AND **61** ■

The graphs of the equations in Example 9 are shown in Figure 12. The Vertical Line Test shows graphically that the equation in Example 9(a) defines a function but the equation in Example 9(b) does not.

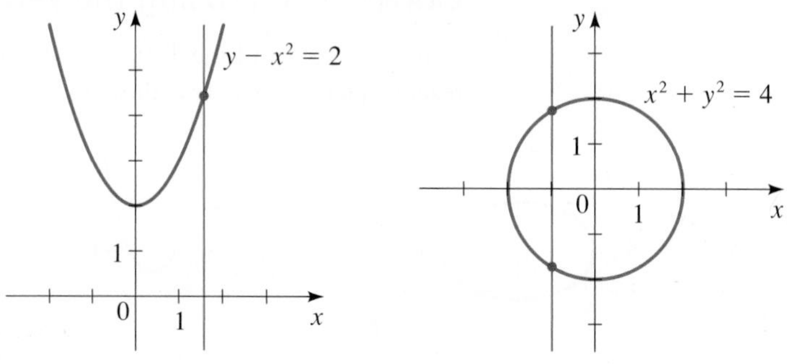

(a) (b)

FIGURE 12

Stanford University News Service

DONALD KNUTH was born in Milwaukee in 1938 and is Professor Emeritus of Computer Science at Stanford University. When Knuth was a high school student, he became fascinated with graphs of functions and laboriously drew many hundreds of them because he wanted to see the behavior of a great variety of functions. (Today, of course, it is far easier to use computers and graphing calculators to do this.) While still a graduate student at Caltech, he started writing a monumental series of books entitled *The Art of Computer Programming*.

Knuth is famous for his invention of TEX, a system of computer-assisted typesetting. This system was used in the preparation of the manuscript for this textbook.

Knuth has received numerous honors, among them election as an associate of the French Academy of Sciences, and as a Fellow of the Royal Society. President Carter awarded him the National Medal of Science in 1979.

The following table shows the graphs of some functions that you will see frequently in this book.

SOME FUNCTIONS AND THEIR GRAPHS

Linear functions
$f(x) = mx + b$

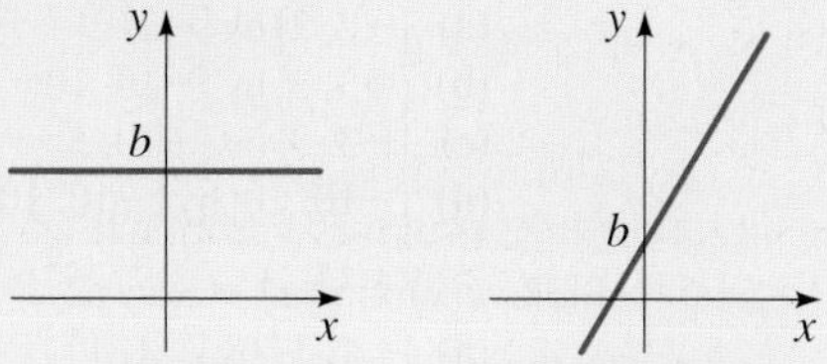

$f(x) = b$ $f(x) = mx + b$

Power functions
$f(x) = x^n$

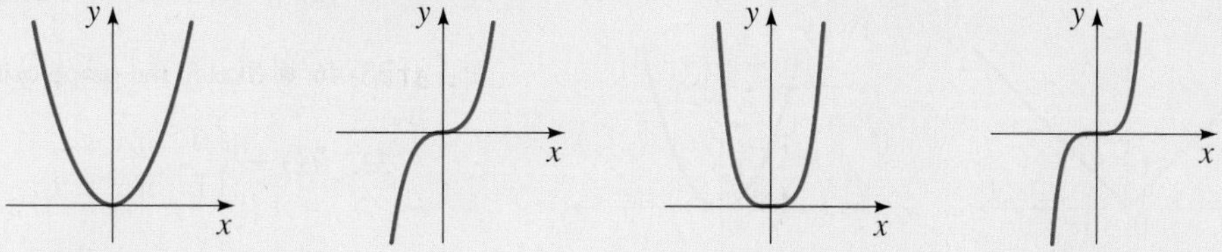

$f(x) = x^2$ $f(x) = x^3$ $f(x) = x^4$ $f(x) = x^5$

Root functions
$f(x) = \sqrt[n]{x}$

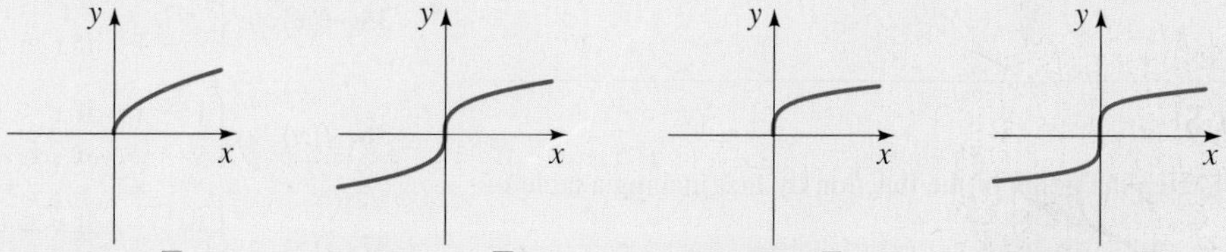

$f(x) = \sqrt{x}$ $f(x) = \sqrt[3]{x}$ $f(x) = \sqrt[4]{x}$ $f(x) = \sqrt[5]{x}$

Reciprocal functions
$f(x) = \dfrac{1}{x^n}$

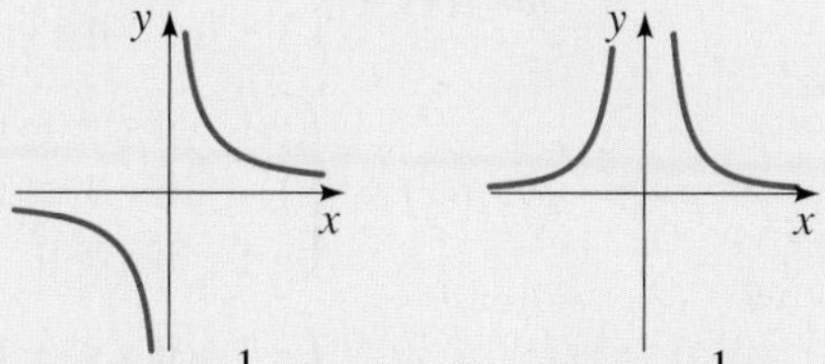

$f(x) = \dfrac{1}{x}$ $f(x) = \dfrac{1}{x^2}$

Absolute value function
$f(x) = |x|$

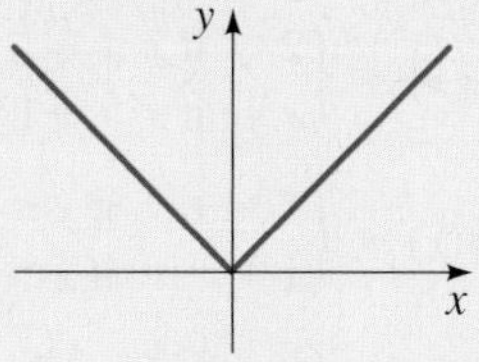

$f(x) = |x|$

Greatest integer function
$f(x) = [\![x]\!]$

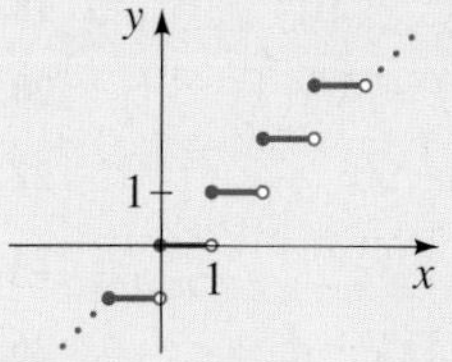

$f(x) = [\![x]\!]$

2.2 EXERCISES

CONCEPTS

1. To graph the function f, we plot the points (x, ________) in a coordinate plane. To graph $f(x) = x^3 + 2$, we plot the points (x, ________). So the point (2, ________) is on the graph of f. The height of the graph of f above the x-axis when $x = 2$ is ________.

2. If $f(2) = 3$, then the point (2, ________) is on the graph of f.

3. If the point (2, 3) is on the graph of f, then $f(2) =$ ________.

4. Match the function with its graph.

(a) $f(x) = x^2$ **(b)** $f(x) = x^3$

(c) $f(x) = \sqrt{x}$ **(d)** $f(x) = |x|$

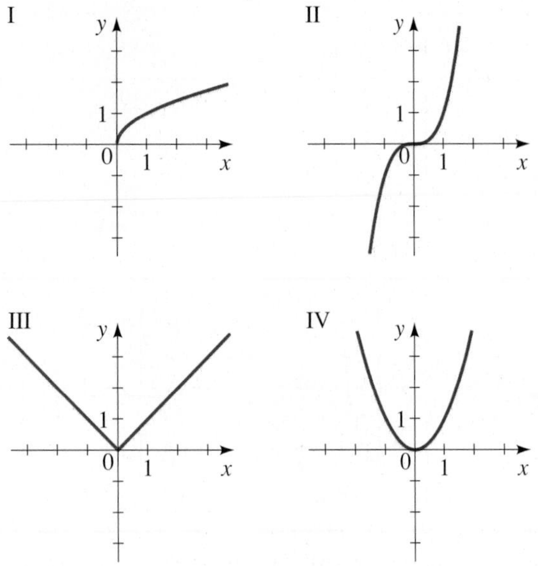

SKILLS

5–28 ■ Sketch the graph of the function by first making a table of values.

5. $f(x) = 2$

6. $f(x) = -3$

7. $f(x) = 2x - 4$

8. $f(x) = 6 - 3x$

9. $f(x) = -x + 3, \quad -3 \le x \le 3$

10. $f(x) = \dfrac{x - 3}{2}, \quad 0 \le x \le 5$

11. $f(x) = -x^2$

12. $f(x) = x^2 - 4$

13. $h(x) = 16 - x^2$

14. $g(x) = (x - 3)^2$

15. $g(x) = x^3 - 8$

16. $g(x) = (x + 2)^3$

17. $g(x) = x^2 - 2x$

18. $h(x) = 4x^2 - x^4$

19. $f(x) = 1 + \sqrt{x}$

20. $f(x) = \sqrt{x + 4}$

21. $g(x) = -\sqrt{x}$

22. $g(x) = \sqrt{-x}$

23. $H(x) = |2x|$

24. $H(x) = |x + 1|$

25. $G(x) = |x| + x$

26. $G(x) = |x| - x$

27. $f(x) = |2x - 2|$

28. $f(x) = \dfrac{x}{|x|}$

29–32 ■ Graph the function in each of the given viewing rectangles, and select the one that produces the most appropriate graph of the function.

29. $f(x) = 8x - x^2$

(a) $[-5, 5]$ by $[-5, 5]$

(b) $[-10, 10]$ by $[-10, 10]$

(c) $[-2, 10]$ by $[-5, 20]$

(d) $[-10, 10]$ by $[-100, 100]$

30. $g(x) = x^2 - x - 20$

(a) $[-2, 2]$ by $[-5, 5]$

(b) $[-10, 10]$ by $[-10, 10]$

(c) $[-7, 7]$ by $[-25, 20]$

(d) $[-10, 10]$ by $[-100, 100]$

31. $h(x) = x^3 - 5x - 4$

(a) $[-2, 2]$ by $[-2, 2]$

(b) $[-3, 3]$ by $[-10, 10]$

(c) $[-3, 3]$ by $[-10, 5]$

(d) $[-10, 10]$ by $[-10, 10]$

32. $k(x) = \frac{1}{32}x^4 - x^2 + 2$

(a) $[-1, 1]$ by $[-1, 1]$

(b) $[-2, 2]$ by $[-2, 2]$

(c) $[-5, 5]$ by $[-5, 5]$

(d) $[-10, 10]$ by $[-10, 10]$

33–46 ■ Sketch the graph of the piecewise defined function.

33. $f(x) = \begin{cases} 0 & \text{if } x < 2 \\ 1 & \text{if } x \ge 2 \end{cases}$

34. $f(x) = \begin{cases} 1 & \text{if } x \le 1 \\ x + 1 & \text{if } x > 1 \end{cases}$

35. $f(x) = \begin{cases} 3 & \text{if } x < 2 \\ x - 1 & \text{if } x \ge 2 \end{cases}$

36. $f(x) = \begin{cases} 1 - x & \text{if } x < -2 \\ 5 & \text{if } x \ge -2 \end{cases}$

37. $f(x) = \begin{cases} x & \text{if } x \le 0 \\ x + 1 & \text{if } x > 0 \end{cases}$

38. $f(x) = \begin{cases} 2x + 3 & \text{if } x < -1 \\ 3 - x & \text{if } x \ge -1 \end{cases}$

39. $f(x) = \begin{cases} -1 & \text{if } x < -1 \\ 1 & \text{if } -1 \le x \le 1 \\ -1 & \text{if } x > 1 \end{cases}$

40. $f(x) = \begin{cases} -1 & \text{if } x < -1 \\ x & \text{if } -1 \le x \le 1 \\ 1 & \text{if } x > 1 \end{cases}$

41. $f(x) = \begin{cases} 2 & \text{if } x \le -1 \\ x^2 & \text{if } x > -1 \end{cases}$

42. $f(x) = \begin{cases} 1 - x^2 & \text{if } x \le 2 \\ x & \text{if } x > 2 \end{cases}$

43. $f(x) = \begin{cases} 0 & \text{if } |x| \le 2 \\ 3 & \text{if } |x| > 2 \end{cases}$

44. $f(x) = \begin{cases} x^2 & \text{if } |x| \le 1 \\ 1 & \text{if } |x| > 1 \end{cases}$

45. $f(x) = \begin{cases} 4 & \text{if } x < -2 \\ x^2 & \text{if } -2 \le x \le 2 \\ -x + 6 & \text{if } x > 2 \end{cases}$

46. $f(x) = \begin{cases} -x & \text{if } x \le 0 \\ 9 - x^2 & \text{if } 0 < x \le 3 \\ x - 3 & \text{if } x > 3 \end{cases}$

47–48 ■ Use a graphing device to draw the graph of the piecewise defined function. (See the margin note on page 155.)

47. $f(x) = \begin{cases} x + 2 & \text{if } x \le -1 \\ x^2 & \text{if } x > -1 \end{cases}$

48. $f(x) = \begin{cases} 2x - x^2 & \text{if } x > 1 \\ (x - 1)^3 & \text{if } x \le 1 \end{cases}$

49–50 ■ The graph of a piecewise defined function is given. Find a formula for the function in the indicated form.

49.

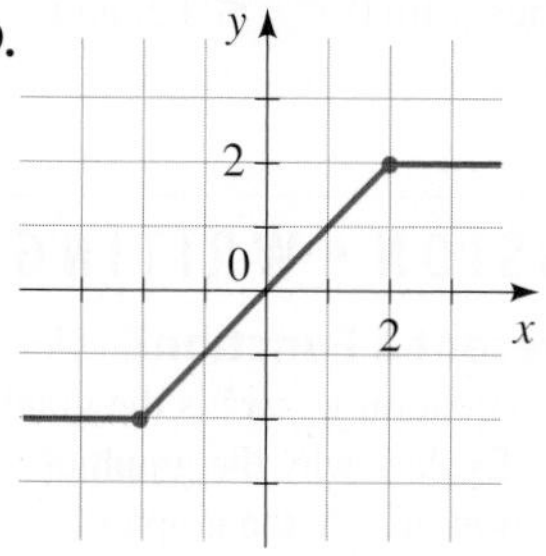

$f(x) = \begin{cases} \quad\quad & \text{if } x < -2 \\ \quad\quad & \text{if } -2 \le x \le 2 \\ \quad\quad & \text{if } x > 2 \end{cases}$

50.

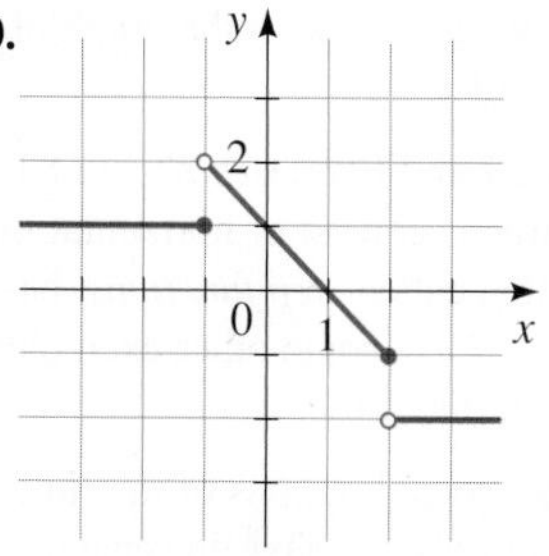

$f(x) = \begin{cases} \quad\quad & \text{if } x \le -1 \\ \quad\quad & \text{if } -1 < x \le 2 \\ \quad\quad & \text{if } x > 2 \end{cases}$

51–52 ■ Use the Vertical Line Test to determine whether the curve is the graph of a function of x.

51. (a)

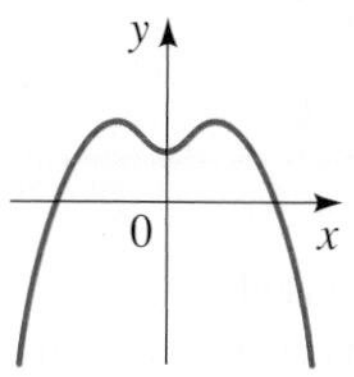

(b)

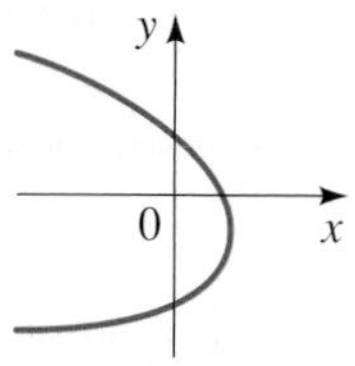

(c)

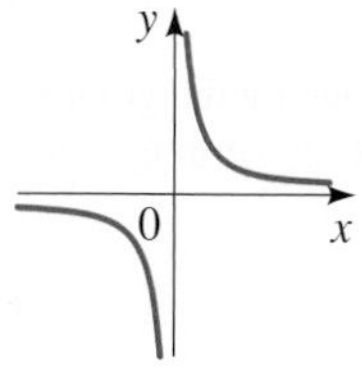

(d)

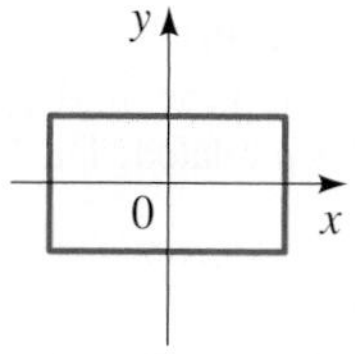

52. (a)

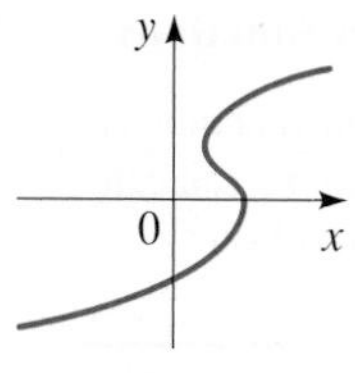

(b)

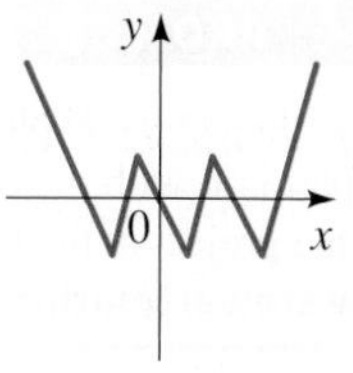

(c)

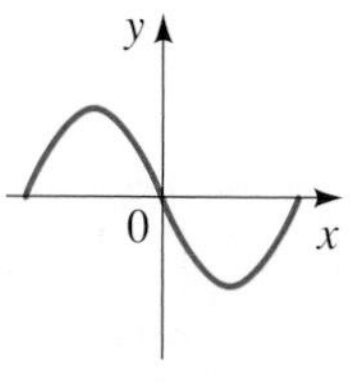

(d)

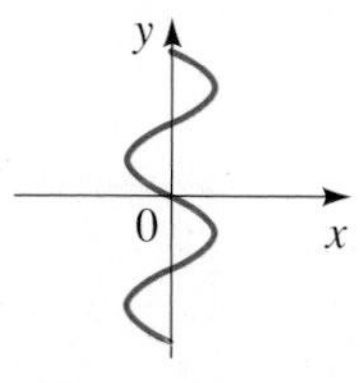

53–56 ■ Use the Vertical Line Test to determine whether the curve is the graph of a function of x. If it is, state the domain and range of the function.

53.

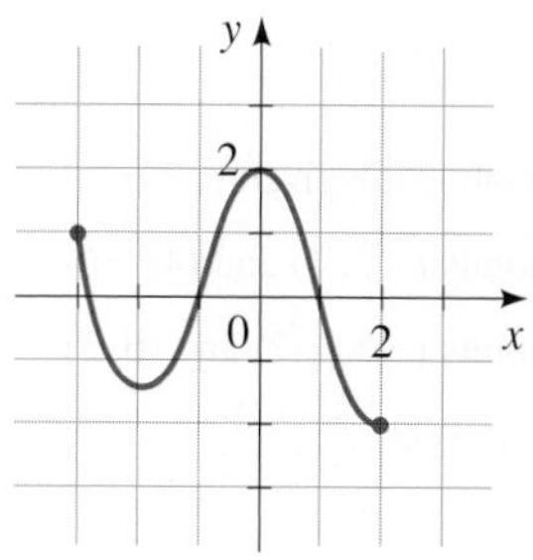

54.

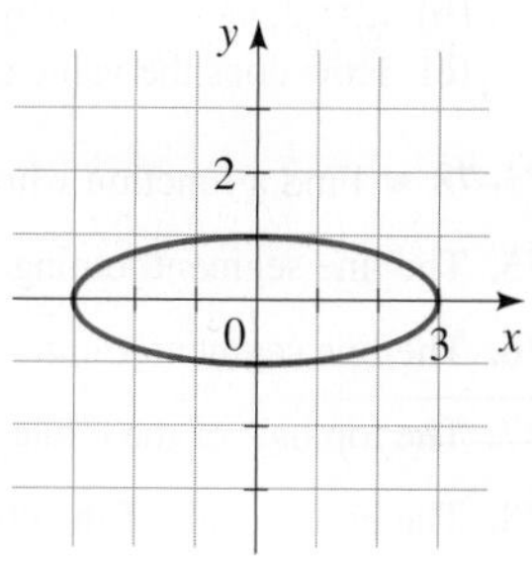

55.

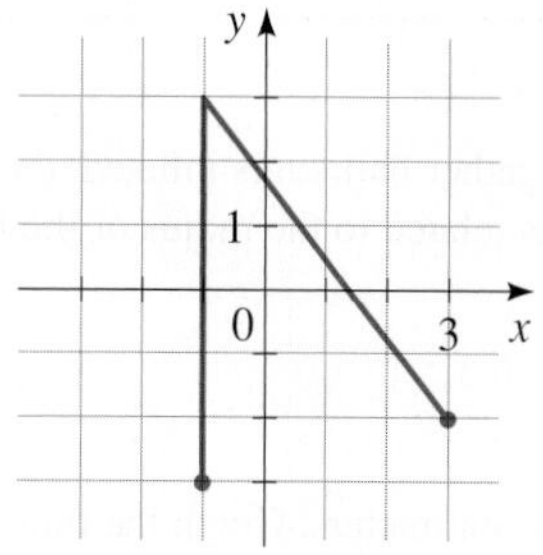

56.

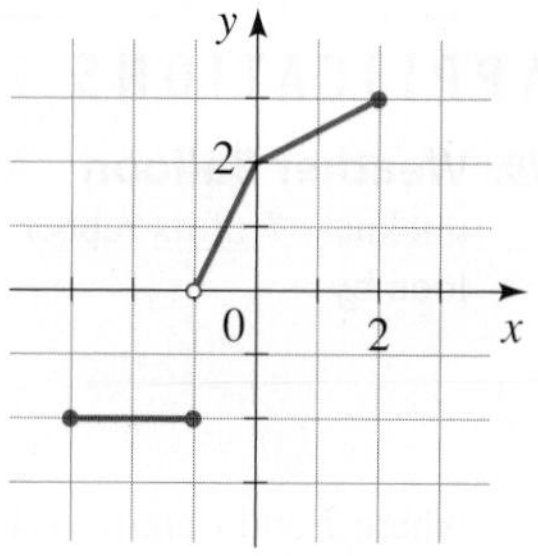

57–68 ■ Determine whether the equation defines y as a function of x. (See Example 9.)

57. $x^2 + 2y = 4$

58. $3x + 7y = 21$

59. $x = y^2$

60. $x^2 + (y - 1)^2 = 4$

61. $x + y^2 = 9$

62. $x^2 + y = 9$

63. $x^2y + y = 1$

64. $\sqrt{x} + y = 12$

65. $2|x| + y = 0$

66. $2x + |y| = 0$

67. $x = y^3$

68. $x = y^4$

69–74 ■ A family of functions is given. In parts (a) and (b) graph all the given members of the family in the viewing rectangle indicated. In part (c) state the conclusions that you can make from your graphs.

69. $f(x) = x^2 + c$

(a) $c = 0, 2, 4, 6$; $[-5, 5]$ by $[-10, 10]$

(b) $c = 0, -2, -4, -6$; $[-5, 5]$ by $[-10, 10]$

(c) How does the value of c affect the graph?

70. $f(x) = (x - c)^2$

(a) $c = 0, 1, 2, 3$; $[-5, 5]$ by $[-10, 10]$

(b) $c = 0, -1, -2, -3$; $[-5, 5]$ by $[-10, 10]$

(c) How does the value of c affect the graph?

71. $f(x) = (x - c)^3$

(a) $c = 0, 2, 4, 6$; $[-10, 10]$ by $[-10, 10]$

(b) $c = 0, -2, -4, -6$; $[-10, 10]$ by $[-10, 10]$

(c) How does the value of c affect the graph?

72. $f(x) = cx^2$

(a) $c = 1, \frac{1}{2}, 2, 4$; $[-5, 5]$ by $[-10, 10]$

(b) $c = 1, -1, -\frac{1}{2}, -2$; $[-5, 5]$ by $[-10, 10]$

(c) How does the value of c affect the graph?

73. $f(x) = x^c$

(a) $c = \frac{1}{2}, \frac{1}{4}, \frac{1}{6}$; $[-1, 4]$ by $[-1, 3]$

(b) $c = 1, \frac{1}{3}, \frac{1}{5}$; $[-3, 3]$ by $[-2, 2]$

(c) How does the value of c affect the graph?

74. $f(x) = \dfrac{1}{x^n}$

(a) $n = 1, 3;\ [-3, 3]$ by $[-3, 3]$
(b) $n = 2, 4;\ [-3, 3]$ by $[-3, 3]$
(c) How does the value of n affect the graph?

75–78 ■ Find a function whose graph is the given curve.

75. The line segment joining the points $(-2, 1)$ and $(4, -6)$

76. The line segment joining the points $(-3, -2)$ and $(6, 3)$

77. The top half of the circle $x^2 + y^2 = 9$

78. The bottom half of the circle $x^2 + y^2 = 9$

APPLICATIONS

79. **Weather Balloon** As a weather balloon is inflated, the thickness T of its rubber skin is related to the radius of the balloon by

$$T(r) = \frac{0.5}{r^2}$$

where T and r are measured in centimeters. Graph the function T for values of r between 10 and 100.

80. **Power from a Wind Turbine** The power produced by a wind turbine depends on the speed of the wind. If a windmill has blades 3 meters long, then the power P produced by the turbine is modeled by

$$P(v) = 14.1v^3$$

where P is measured in watts (W) and v is measured in meters per second (m/s). Graph the function P for wind speeds between 1 m/s and 10 m/s.

81. **Utility Rates** Westside Energy charges its electric customers a base rate of \$6.00 per month, plus 10¢ per kilowatt-hour (kWh) for the first 300 kWh used and 6¢ per kWh for all usage over 300 kWh. Suppose a customer uses x kWh of electricity in one month.

(a) Express the monthly cost E as a piecewise-defined function of x.
(b) Graph the function E for $0 \le x \le 600$.

82. **Taxicab Function** A taxi company charges \$2.00 for the first mile (or part of a mile) and 20 cents for each succeeding tenth of a mile (or part). Express the cost C (in dollars) of a ride as a piecewise-defined function of the distance x traveled (in miles) for $0 < x < 2$, and sketch the graph of this function.

83. **Postage Rates** The domestic postage rate for first-class letters weighing 3.5 oz or less is 44 cents for the first ounce (or less), plus 17 cents for each additional ounce (or part of an ounce). Express the postage P as a piecewise-defined function of the weight x of a letter, with $0 < x \le 3.5$, and sketch the graph of this function.

DISCOVERY ■ DISCUSSION ■ WRITING

84. **When Does a Graph Represent a Function?** For every integer n, the graph of the equation $y = x^n$ is the graph of a function, namely $f(x) = x^n$. Explain why the graph of $x = y^2$ is *not* the graph of a function of x. Is the graph of $x = y^3$ the graph of a function of x? If so, of what function of x is it the graph? Determine for what integers n the graph of $x = y^n$ is the graph of a function of x.

85. **Step Functions** In Example 7 and Exercises 82 and 83 we are given functions whose graphs consist of horizontal line segments. Such functions are often called *step functions*, because their graphs look like stairs. Give some other examples of step functions that arise in everyday life.

86. **Stretched Step Functions** Sketch graphs of the functions $f(x) = [\![x]\!]$, $g(x) = [\![2x]\!]$, and $h(x) = [\![3x]\!]$ on separate graphs. How are the graphs related? If n is a positive integer, what does the graph of $k(x) = [\![nx]\!]$ look like?

87. **Graph of the Absolute Value of a Function**

(a) Draw the graphs of the functions

$$f(x) = x^2 + x - 6$$

and

$$g(x) = |x^2 + x - 6|$$

How are the graphs of f and g related?

(b) Draw the graphs of the functions $f(x) = x^4 - 6x^2$ and $g(x) = |x^4 - 6x^2|$. How are the graphs of f and g related?

(c) In general, if $g(x) = |f(x)|$, how are the graphs of f and g related? Draw graphs to illustrate your answer.

DISCOVERY PROJECT **Relations and Functions**

In this project we explore the concept of function by comparing it with the concept of a *relation*. You can find the project at the book companion website: **www.stewartmath.com**

2.3 GETTING INFORMATION FROM THE GRAPH OF A FUNCTION

Values of a Function; Domain and Range ▶ Increasing and Decreasing Functions ▶ Local Maximum and Minimum Values of a Function

Many properties of a function are more easily obtained from a graph than from the rule that describes the function. We will see in this section how a graph tells us whether the values of a function are increasing or decreasing and also where the maximum and minimum values of a function are.

▼ Values of a Function; Domain and Range

A complete graph of a function contains all the information about a function, because the graph tells us which input values correspond to which output values. To analyze the graph of a function, we must keep in mind that *the height of the graph is the value of the function.* So we can read off the values of a function from its graph.

EXAMPLE 1 | Finding the Values of a Function from a Graph

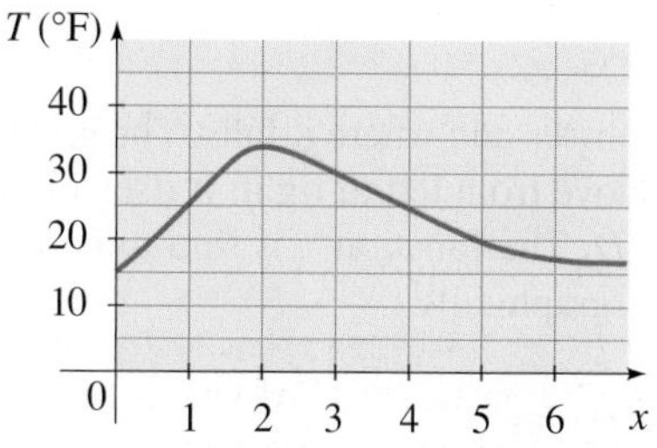

FIGURE 1 Temperature function

The function T graphed in Figure 1 gives the temperature between noon and 6:00 P.M. at a certain weather station.

(a) Find $T(1)$, $T(3)$, and $T(5)$.

(b) Which is larger, $T(2)$ or $T(4)$?

(c) Find the value(s) of x for which $T(x) = 25$.

(d) Find the value(s) of x for which $T(x) \geq 25$.

SOLUTION

(a) $T(1)$ is the temperature at 1:00 P.M. It is represented by the height of the graph above the x-axis at $x = 1$. Thus, $T(1) = 25$. Similarly, $T(3) = 30$ and $T(5) = 20$.

(b) Since the graph is higher at $x = 2$ than at $x = 4$, it follows that $T(2)$ is larger than $T(4)$.

(c) The height of the graph is 25 when x is 1 and when x is 4. In other words, the temperature is 25 at 1:00 P.M. and 4:00 P.M.

(d) The graph is higher than 25 for x between 1 and 4. In other words, the temperature was 25 or greater between 1:00 P.M. and 4:00 P.M.

NOW TRY EXERCISE **5** ■

The graph of a function helps us to picture the domain and range of the function on the x-axis and y-axis, as shown in Figure 2.

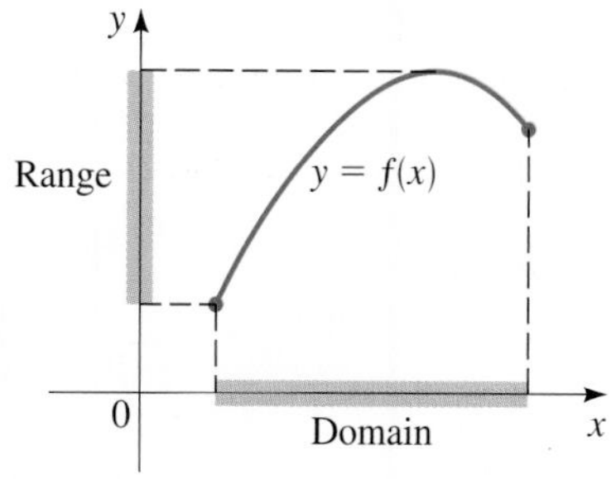

FIGURE 2 Domain and range of f

EXAMPLE 2 | Finding the Domain and Range from a Graph

(a) Use a graphing calculator to draw the graph of $f(x) = \sqrt{4 - x^2}$.

(b) Find the domain and range of f.

SOLUTION

(a) The graph is shown in Figure 3.

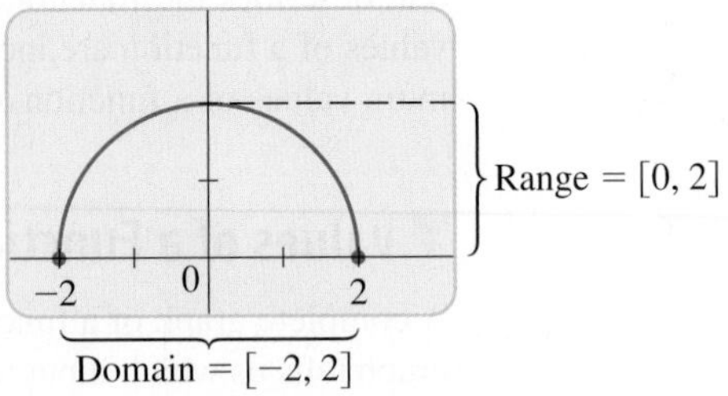

FIGURE 3 Graph of $f(x) = \sqrt{4 - x^2}$

(b) From the graph in Figure 3 we see that the domain is $[-2, 2]$ and the range is $[0, 2]$.

NOW TRY EXERCISE **15** ■

Increasing and Decreasing Functions

It is very useful to know where the graph of a function rises and where it falls. The graph shown in Figure 4 rises, falls, then rises again as we move from left to right: It rises from A to B, falls from B to C, and rises again from C to D. The function f is said to be *increasing* when its graph rises and *decreasing* when its graph falls.

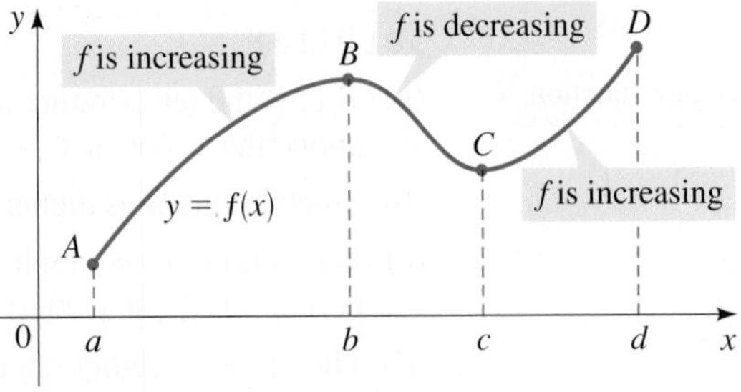

FIGURE 4 f is increasing on $[a, b]$ and $[c, d]$. f is decreasing on $[b, c]$.

We have the following definition.

DEFINITION OF INCREASING AND DECREASING FUNCTIONS

f is **increasing** on an interval I if $f(x_1) < f(x_2)$ whenever $x_1 < x_2$ in I.

f is **decreasing** on an interval I if $f(x_1) > f(x_2)$ whenever $x_1 < x_2$ in I.

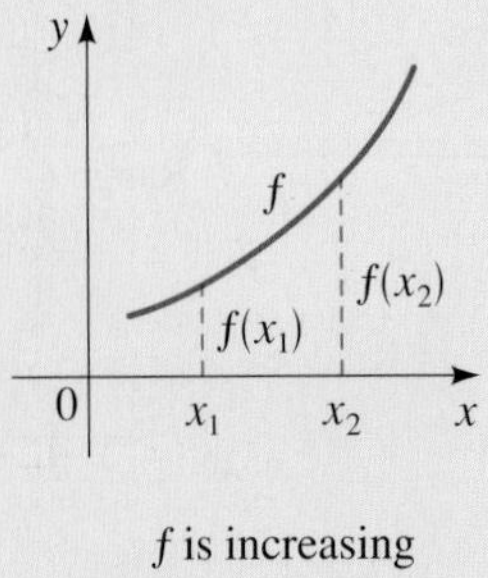

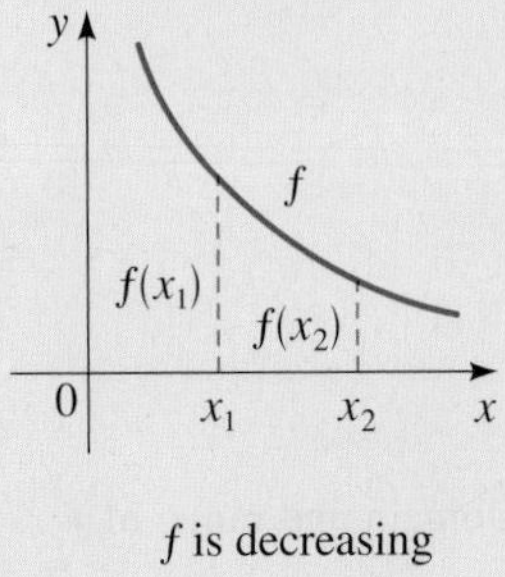

EXAMPLE 3 | Intervals on Which a Function Increases and Decreases

The graph in Figure 5 gives the weight W of a person at age x. Determine the intervals on which the function W is increasing and on which it is decreasing.

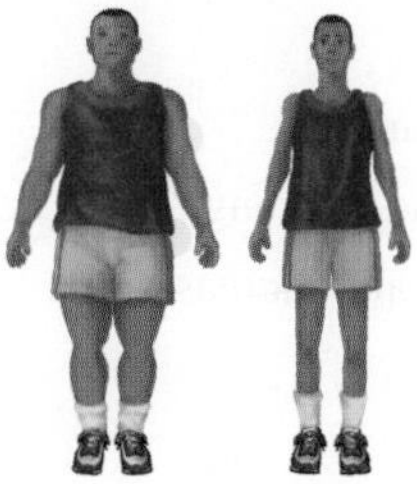

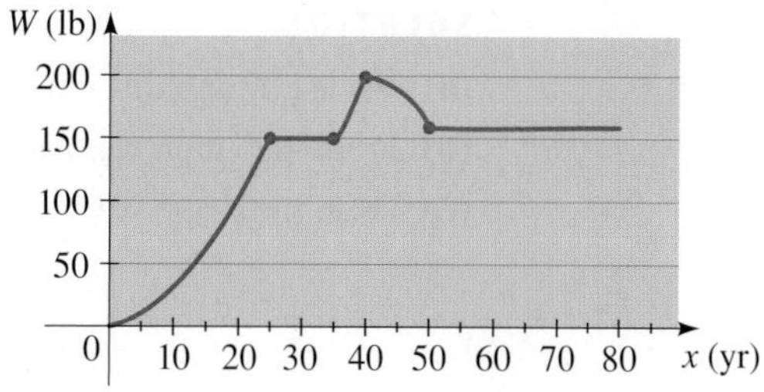

FIGURE 5 Weight as a function of age

SOLUTION The function W is increasing on $[0, 25]$ and $[35, 40]$. It is decreasing on $[40, 50]$. The function W is constant (neither increasing nor decreasing) on $[25, 30]$ and $[50, 80]$. This means that the person gained weight until age 25, then gained weight again between ages 35 and 40. He lost weight between ages 40 and 50.

NOW TRY EXERCISE **45** ■

EXAMPLE 4 | Finding Intervals Where a Function Increases and Decreases

(a) Sketch a graph of the function $f(x) = 12x^2 + 4x^3 - 3x^4$.

(b) Find the domain and range of f.

(c) Find the intervals on which f increases and decreases.

SOLUTION

(a) We use a graphing calculator to sketch the graph in Figure 6.

(b) The domain of f is $\mathbb{R}$ because f is defined for all real numbers. Using the TRACE feature on the calculator, we find that the highest value is $f(2) = 32$. So the range of f is $(-\infty, 32]$.

(c) From the graph we see that f is increasing on the intervals $(-\infty, -1]$ and $[0, 2]$ and is decreasing on $[-1, 0]$ and $[2, \infty)$.

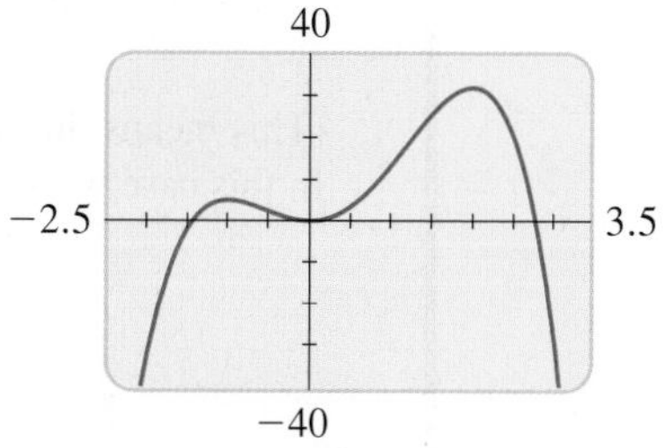

FIGURE 6 Graph of $f(x) = 12x^2 + 4x^3 - 3x^4$

NOW TRY EXERCISE **23** ■

EXAMPLE 5 | Finding Intervals Where a Function Increases and Decreases

(a) Sketch the graph of the function $f(x) = x^{2/3}$.

(b) Find the domain and range of the function.

(c) Find the intervals on which f increases and decreases.

SOLUTION

(a) We use a graphing calculator to sketch the graph in Figure 7.

(b) From the graph we observe that the domain of f is $\mathbb{R}$ and the range is $[0, \infty)$.

(c) From the graph we see that f is decreasing on $(-\infty, 0]$ and increasing on $[0, \infty)$.

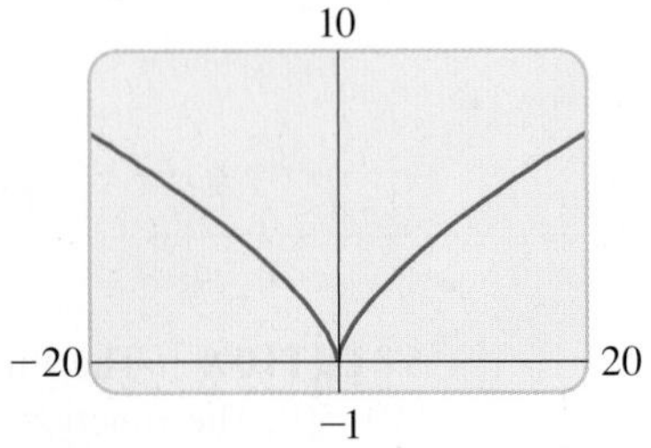

FIGURE 7 Graph of $f(x) = x^{2/3}$

NOW TRY EXERCISE **29** ■

▼ Local Maximum and Minimum Values of a Function

Finding the largest or smallest values of a function is important in many applications. For example, if a function represents revenue or profit, then we are interested in its maximum value. For a function that represents cost, we would want to find its minimum value. (See *Focus on Modeling: Modeling with Functions* on pages 213–222 for many such examples.) We can easily find these values from the graph of a function. We first define what we mean by a local maximum or minimum.

LOCAL MAXIMA AND MINIMA OF A FUNCTION

1. The function value $f(a)$ is a **local maximum value** of f if

$$f(a) \geq f(x) \quad \text{when } x \text{ is near } a$$

(This means that $f(a) \geq f(x)$ for all x in some open interval containing a.) In this case we say that f has a **local maximum** at $x = a$.

2. The function value $f(a)$ is a **local minimum** of f if

$$f(a) \leq f(x) \quad \text{when } x \text{ is near } a$$

(This means that $f(a) \leq f(x)$ for all x in some open interval containing a.) In this case we say that f has a **local minimum** at $x = a$.

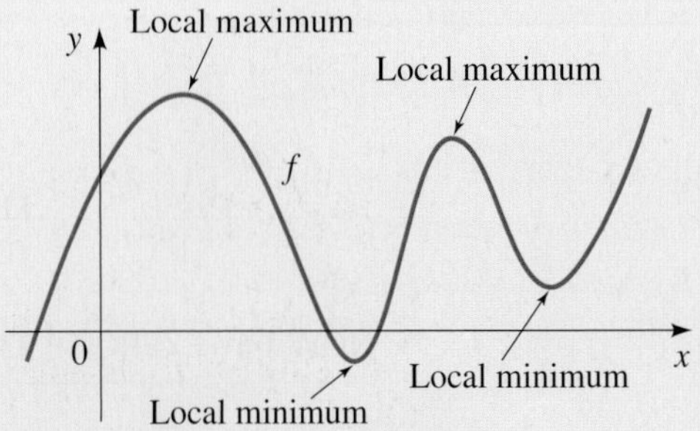

We can find the local maximum and minimum values of a function using a graphing calculator.

If there is a viewing rectangle such that the point $(a, f(a))$ is the highest point on the graph of f *within* the viewing rectangle (not on the edge), then the number $f(a)$ is a local maximum value of f (see Figure 8). Notice that $f(a) \geq f(x)$ for all numbers x that are close to a.

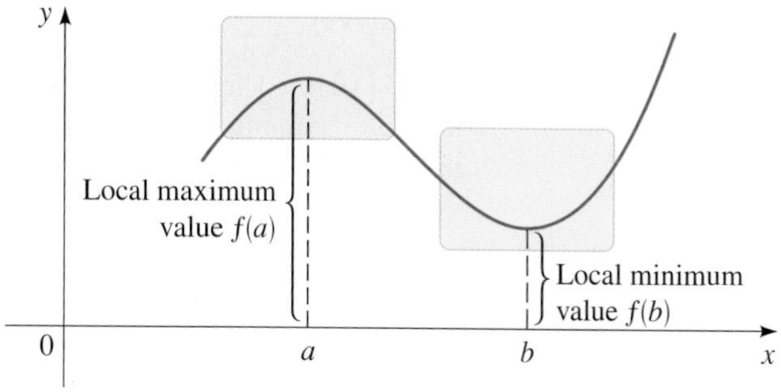

FIGURE 8

Similarly, if there is a viewing rectangle such that the point $(b, f(b))$ is the lowest point on the graph of f within the viewing rectangle, then the number $f(b)$ is a local minimum value of f. In this case, $f(b) \leq f(x)$ for all numbers x that are close to b.

EXAMPLE 6 | Finding Local Maxima and Minima from a Graph

Find the local maximum and minimum values of the function $f(x) = x^3 - 8x + 1$, correct to three decimal places.

SOLUTION The graph of f is shown in Figure 9. There appears to be one local maximum between $x = -2$ and $x = -1$, and one local minimum between $x = 1$ and $x = 2$.

Let's find the coordinates of the local maximum point first. We zoom in to enlarge the area near this point, as shown in Figure 10. Using the TRACE feature on the graphing device, we move the cursor along the curve and observe how the y-coordinates change. The local maximum value of y is 9.709, and this value occurs when x is -1.633, correct to three decimal places.

We locate the minimum value in a similar fashion. By zooming in to the viewing rectangle shown in Figure 11, we find that the local minimum value is about -7.709, and this value occurs when $x \approx 1.633$.

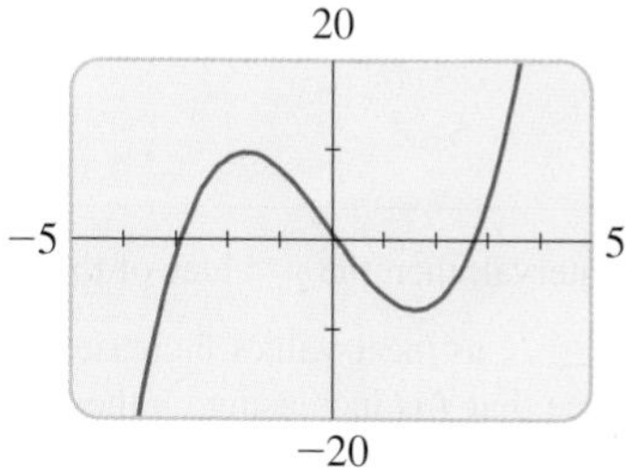

FIGURE 9 Graph of $f(x) = x^3 - 8x + 1$

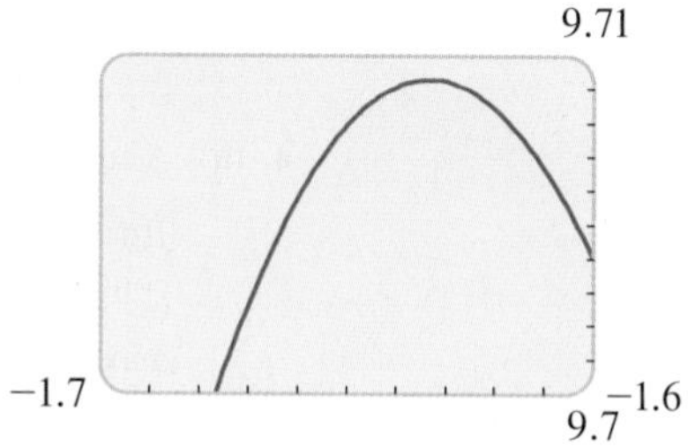

FIGURE 10

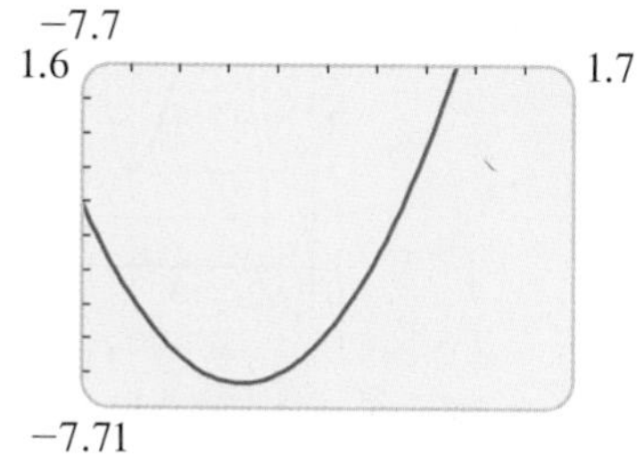

FIGURE 11

NOW TRY EXERCISE **35** ■

The `maximum` and `minimum` commands on a TI-83 or TI-84 calculator provide another method for finding extreme values of functions. We use this method in the next example.

EXAMPLE 7 | A Model for the Food Price Index

A model for the food price index (the price of a representative "basket" of foods) between 1990 and 2000 is given by the function

$$I(t) = -0.0113t^3 + 0.0681t^2 + 0.198t + 99.1$$

where t is measured in years since midyear 1990, so $0 \le t \le 10$, and $I(t)$ is scaled so that $I(3) = 100$. Estimate the time when food was most expensive during the period 1990–2000.

SOLUTION The graph of I as a function of t is shown in Figure 12(a). There appears to be a maximum between $t = 4$ and $t = 7$. Using the `maximum` command, as shown in Figure 12(b), we see that the maximum value of I is about 100.38, and it occurs when $t \approx 5.15$, which corresponds to August 1995.

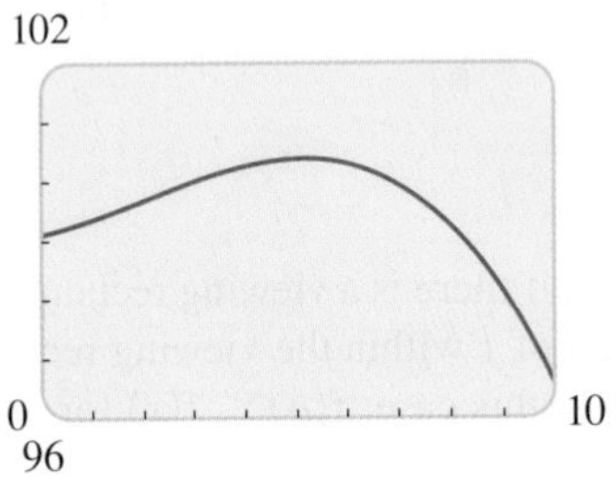

(a)

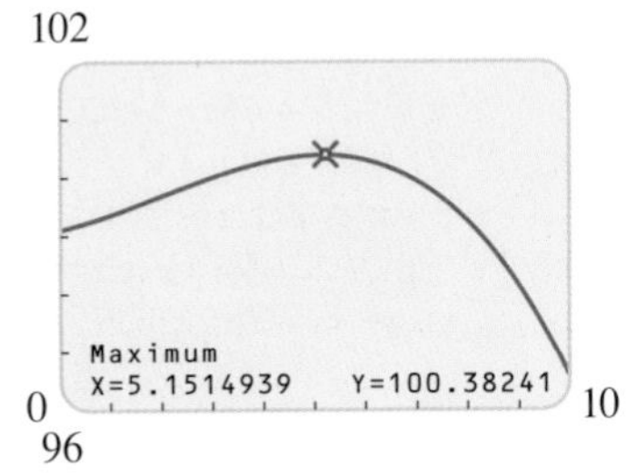

(b)

FIGURE 12

NOW TRY EXERCISE **53**

2.3 EXERCISES

CONCEPTS

1–4 ■ These exercises refer to the graph of the function f shown below.

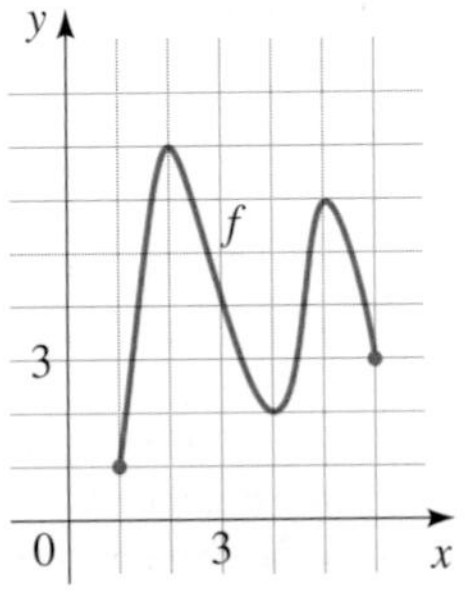

1. To find a function value $f(a)$ from the graph of f, we find the height of the graph above the x-axis at $x =$ ________. From the graph of f we see that $f(3) =$ ________.

2. The domain of the function f is all the ________-values of the points on the graph, and the range is all the corresponding ________-values. From the graph of f we see that the domain of f is the interval ________ and the range of f is the interval ________.

3. (a) If f is increasing on an interval, then the y-values of the points on the graph ________ as the x-values increase. From the graph of f we see that f is increasing on the intervals ________ and ________.

(b) If f is decreasing on an interval, then y-values of the points on the graph ________ as the x-values increase. From the graph of f we see that f is decreasing on the intervals ________ and ________.

4. (a) A function value $f(a)$ is a local maximum value of f if $f(a)$ is the ________ value of f on some interval containing a. From the graph of f we see that one local maximum value of f is ________ and that this value occurs when x is ________.

(b) The function value $f(a)$ is a local minimum value of f if $f(a)$ is the ________ value of f on some interval containing a. From the graph of f we see that one local minimum value of f is ________ and that this value occurs when x is ________.

SKILLS

5. The graph of a function h is given.
(a) Find $h(-2)$, $h(0)$, $h(2)$, and $h(3)$.
(b) Find the domain and range of h.
(c) Find the values of x for which $h(x) = 3$.
(d) Find the values of x for which $h(x) \leq 3$.

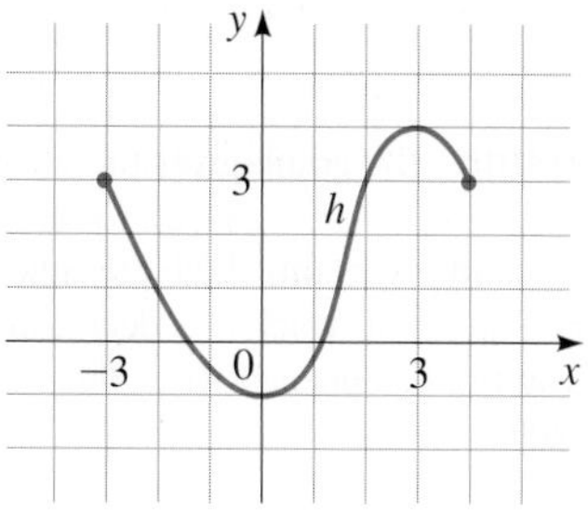

6. The graph of a function g is given.
(a) Find $g(-2)$, $g(0)$, and $g(7)$.
(b) Find the domain and range of g.
(c) Find the values of x for which $g(x) = 4$.
(d) Find the values of x for which $g(x) > 4$.

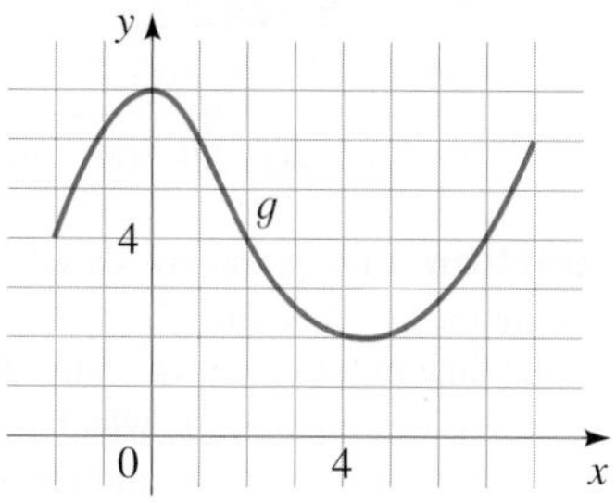

7. The graph of a function g is given.
(a) Find $g(-4)$, $g(-2)$, $g(0)$, $g(2)$, and $g(4)$.
(b) Find the domain and range of g.

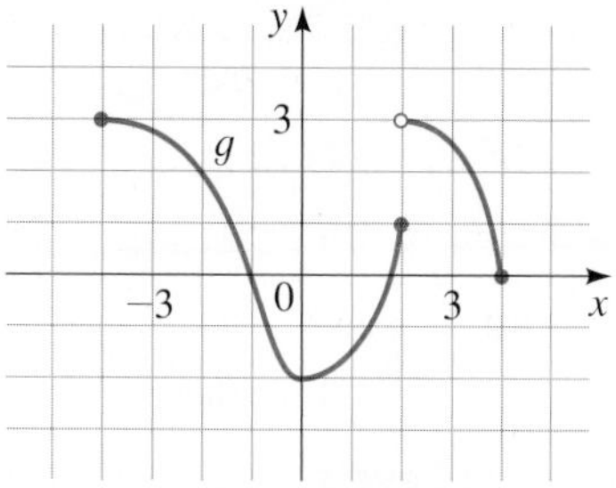

8. Graphs of the functions f and g are given.
(a) Which is larger, $f(0)$ or $g(0)$?
(b) Which is larger, $f(-3)$ or $g(-3)$?
(c) For which values of x is $f(x) = g(x)$?

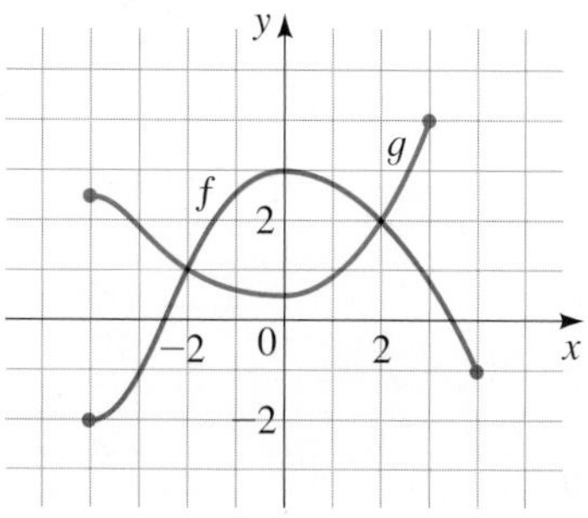

9–18 ■ A function f is given. **(a)** Use a graphing calculator to draw the graph of f. **(b)** Find the domain and range of f from the graph.

9. $f(x) = x - 1$
10. $f(x) = 2(x + 1)$
11. $f(x) = 4, \quad 1 \leq x \leq 3$
12. $f(x) = x^2, \quad -2 \leq x \leq 5$
13. $f(x) = 4 - x^2$
14. $f(x) = x^2 + 4$
15. $f(x) = \sqrt{16 - x^2}$
16. $f(x) = -\sqrt{25 - x^2}$
17. $f(x) = \sqrt{x - 1}$
18. $f(x) = \sqrt{x + 2}$

19–22 ■ The graph of a function is given. Determine the intervals on which the function is **(a)** increasing and **(b)** decreasing.

19.

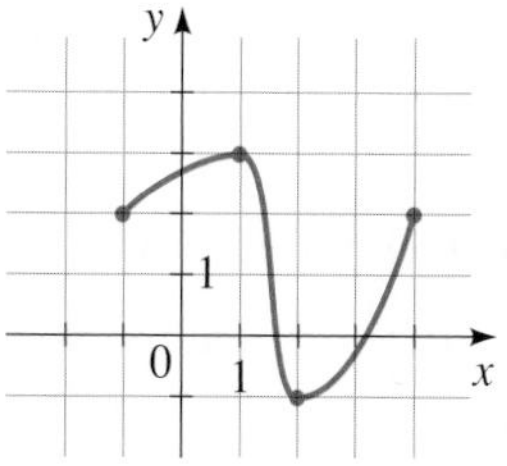

20.

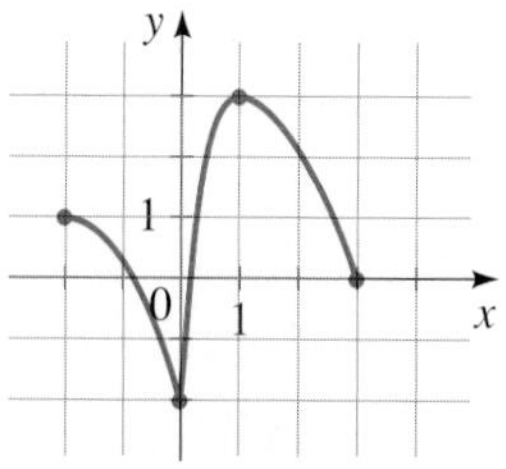

21.

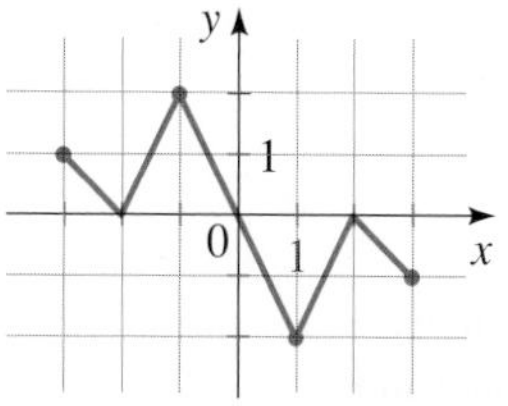

22.

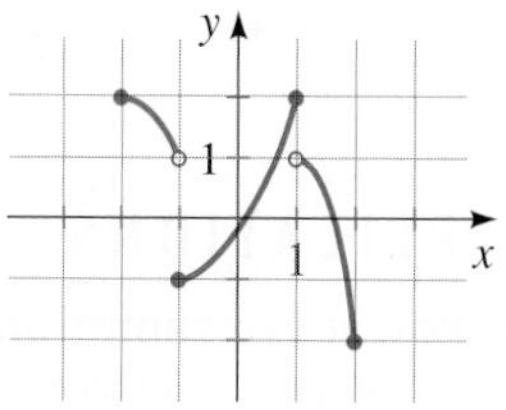

23–30 ■ A function f is given. **(a)** Use a graphing device to draw the graph of f. **(b)** State approximately the intervals on which f is increasing and on which f is decreasing.

23. $f(x) = x^2 - 5x$
24. $f(x) = x^3 - 4x$
25. $f(x) = 2x^3 - 3x^2 - 12x$
26. $f(x) = x^4 - 16x^2$
27. $f(x) = x^3 + 2x^2 - x - 2$
28. $f(x) = x^4 - 4x^3 + 2x^2 + 4x - 3$
29. $f(x) = x^{2/5}$
30. $f(x) = 4 - x^{2/3}$

31–34 ■ The graph of a function is given. **(a)** Find all the local maximum and minimum values of the function and the value of x at which each occurs. **(b)** Find the intervals on which the function is increasing and on which the function is decreasing.

31.

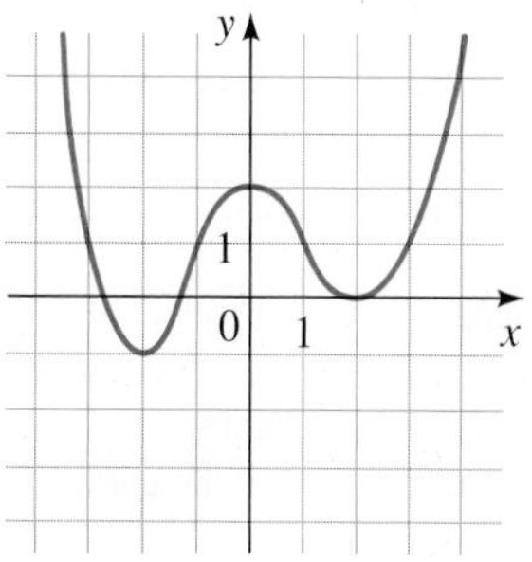

32.

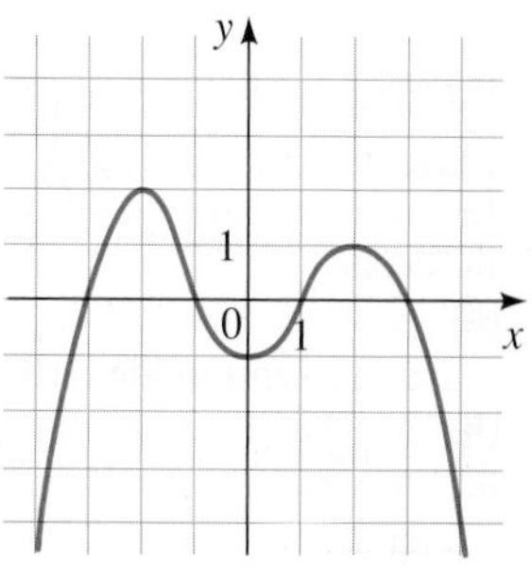

33.

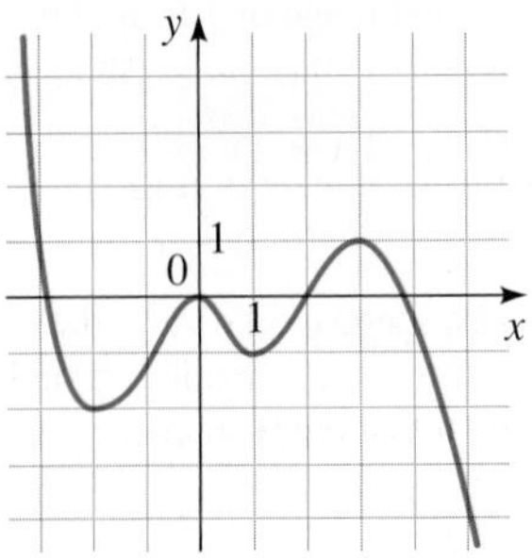

34.

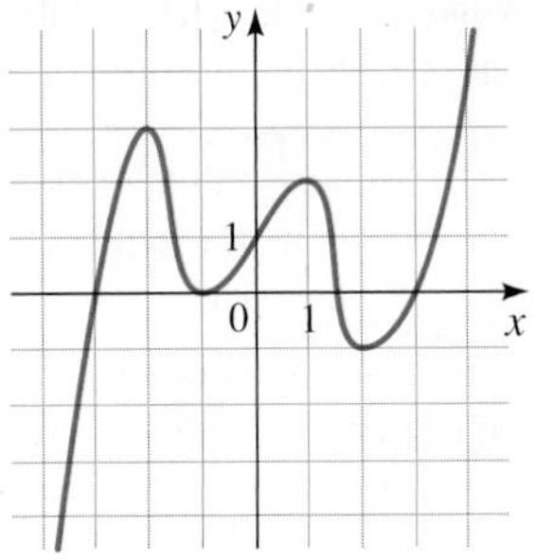

35–42 ■ A function is given. **(a)** Find all the local maximum and minimum values of the function and the value of x at which each occurs. State each answer correct to two decimal places. **(b)** Find the intervals on which the function is increasing and on which the function is decreasing. State each answer correct to two decimal places.

35. $f(x) = x^3 - x$

36. $f(x) = 3 + x + x^2 - x^3$

37. $g(x) = x^4 - 2x^3 - 11x^2$

38. $g(x) = x^5 - 8x^3 + 20x$

39. $U(x) = x\sqrt{6 - x}$

40. $U(x) = x\sqrt{x - x^2}$

41. $V(x) = \dfrac{1 - x^2}{x^3}$

42. $V(x) = \dfrac{1}{x^2 + x + 1}$

APPLICATIONS

43. **Power Consumption** The figure shows the power consumption in San Francisco for September 19, 1996 (P is measured in megawatts; t is measured in hours starting at midnight).

(a) What was the power consumption at 6:00 A.M.? At 6:00 P.M.?

(b) When was the power consumption the lowest?

(c) When was the power consumption the highest?

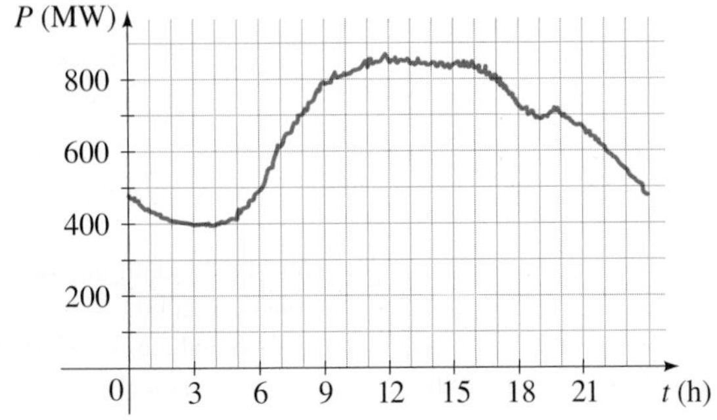

Source: Pacific Gas & Electric

44. **Earthquake** The graph shows the vertical acceleration of the ground from the 1994 Northridge earthquake in Los Angeles, as measured by a seismograph. (Here t represents the time in seconds.)

(a) At what time t did the earthquake first make noticeable movements of the earth?

(b) At what time t did the earthquake seem to end?

(c) At what time t was the maximum intensity of the earthquake reached?

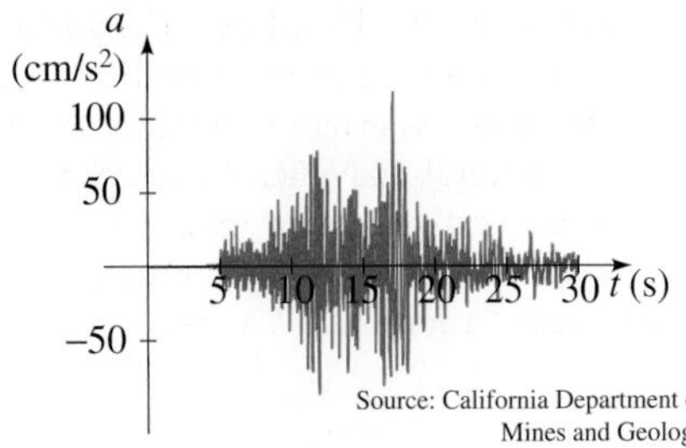

Source: California Department of Mines and Geology

45. **Weight Function** The graph gives the weight W of a person at age x.

(a) Determine the intervals on which the function W is increasing and those on which it is decreasing.

(b) What do you think happened when this person was 30 years old?

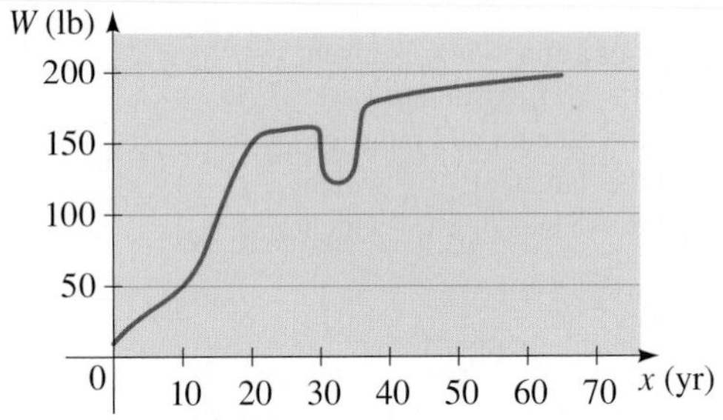

46. **Distance Function** The graph gives a sales representative's distance from his home as a function of time on a certain day.

(a) Determine the time intervals on which his distance from home was increasing and those on which it was decreasing.

(b) Describe in words what the graph indicates about his travels on this day.

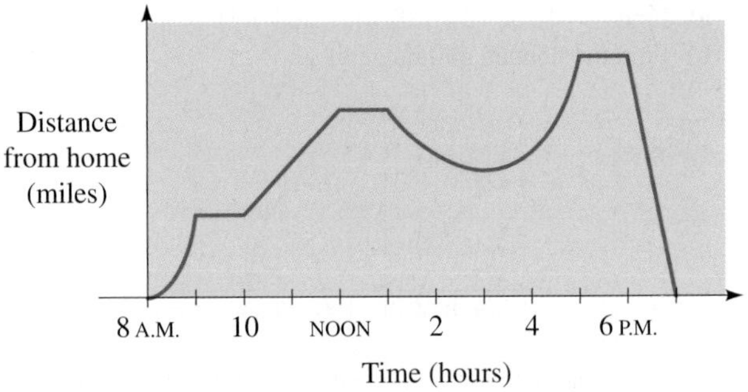

47. **Changing Water Levels** The graph shows the depth of water W in a reservoir over a one-year period as a function of the number of days x since the beginning of the year.

(a) Determine the intervals on which the function W is increasing and on which it is decreasing.

(b) At what value of x does W achieve a local maximum? A local minimum?

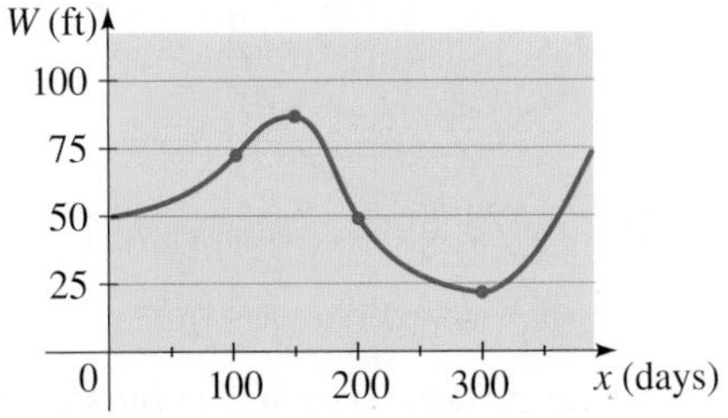

48. Population Growth and Decline The graph shows the population P in a small industrial city from 1950 to 2000. The variable x represents the number of years since 1950.

(a) Determine the intervals on which the function P is increasing and on which it is decreasing.

(b) What was the maximum population, and in what year was it attained?

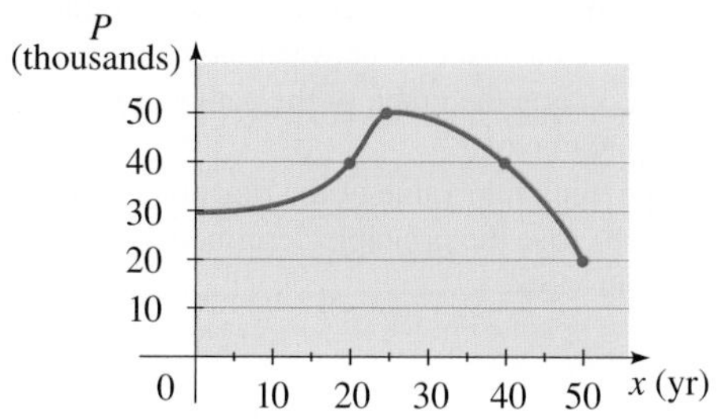

49. Hurdle Race Three runners compete in a 100-meter hurdle race. The graph depicts the distance run as a function of time for each runner. Describe in words what the graph tells you about this race. Who won the race? Did each runner finish the race? What do you think happened to runner B?

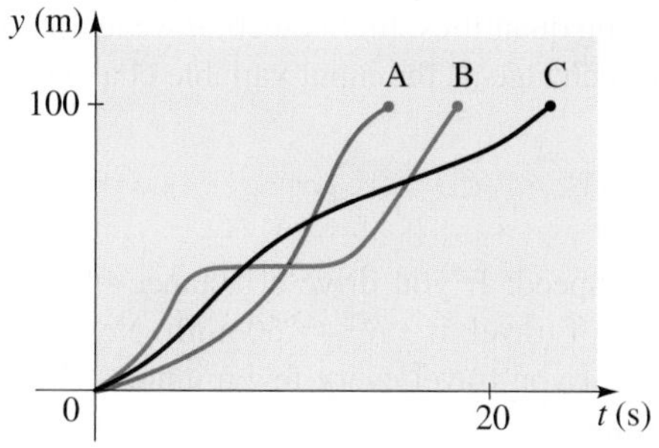

50. Gravity Near the Moon We can use Newton's Law of Gravity to measure the gravitational attraction between the moon and an algebra student in a space ship located a distance x above the moon's surface:

$$F(x) = \frac{350}{x^2}$$

Here F is measured in newtons (N), and x is measured in millions of meters.

(a) Graph the function F for values of x between 0 and 10.

(b) Use the graph to describe the behavior of the gravitational attraction F as the distance x increases.

51. Radii of Stars Astronomers infer the radii of stars using the Stefan Boltzmann Law:

$$E(T) = (5.67 \times 10^{-8})T^4$$

where E is the energy radiated per unit of surface area measured in watts (W) and T is the absolute temperature measured in kelvins (K).

(a) Graph the function E for temperatures T between 100 K and 300 K.

(b) Use the graph to describe the change in energy E as the temperature T increases.

52. Migrating Fish A fish swims at a speed v relative to the water, against a current of 5 mi/h. Using a mathematical model of energy expenditure, it can be shown that the total energy E required to swim a distance of 10 mi is given by

$$E(v) = 2.73v^3 \frac{10}{v - 5}$$

Biologists believe that migrating fish try to minimize the total energy required to swim a fixed distance. Find the value of v that minimizes energy required.

NOTE: This result has been verified; migrating fish swim against a current at a speed 50% greater than the speed of the current.

53. Highway Engineering A highway engineer wants to estimate the maximum number of cars that can safely travel a particular highway at a given speed. She assumes that each car is 17 ft long, travels at a speed s, and follows the car in front of it at the "safe following distance" for that speed. She finds that the number N of cars that can pass a given point per minute is modeled by the function

$$N(s) = \frac{88s}{17 + 17\left(\frac{s}{20}\right)^2}$$

At what speed can the greatest number of cars travel the highway safely?

54. Volume of Water Between 0°C and 30°C, the volume V (in cubic centimeters) of 1 kg of water at a temperature T is given by the formula

$$V = 999.87 - 0.06426T + 0.0085043T^2 - 0.0000679T^3$$

Find the temperature at which the volume of 1 kg of water is a minimum.

55. Coughing When a foreign object that is lodged in the trachea (windpipe) forces a person to cough, the diaphragm thrusts upward, causing an increase in pressure in the lungs. At the same time, the trachea contracts, causing the expelled air to move faster and increasing the pressure on the foreign object. According to a mathematical model of coughing, the velocity v of the airstream through an average-sized person's trachea is related to the radius r of the trachea (in centimeters) by the function

$$v(r) = 3.2(1 - r)r^2 \qquad \tfrac{1}{2} \le r \le 1$$

Determine the value of r for which v is a maximum.

DISCOVERY ▪ DISCUSSION ▪ WRITING

56. **Functions That Are Always Increasing or Decreasing** Sketch rough graphs of functions that are defined for all real numbers and that exhibit the indicated behavior (or explain why the behavior is impossible).
   (a) f is always increasing, and $f(x) > 0$ for all x
   (b) f is always decreasing, and $f(x) > 0$ for all x
   (c) f is always increasing, and $f(x) < 0$ for all x
   (d) f is always decreasing, and $f(x) < 0$ for all x

57. **Maxima and Minima** In Example 7 we saw a real-world situation in which the maximum value of a function is important. Name several other everyday situations in which a maximum or minimum value is important.

58. **Minimizing a Distance** When we seek a minimum or maximum value of a function, it is sometimes easier to work with a simpler function instead.
   (a) Suppose

$$g(x) = \sqrt{f(x)}$$

   where $f(x) \geq 0$ for all x. Explain why the local minima and maxima of f and g occur at the same values of x.
   (b) Let $g(x)$ be the distance between the point $(3, 0)$ and the point (x, x^2) on the graph of the parabola $y = x^2$. Express g as a function of x.
   (c) Find the minimum value of the function g that you found in part (b). Use the principle described in part (a) to simplify your work.

2.4 Average Rate of Change of a Function

Average Rate of Change ▶ Linear Functions Have Constant Rate of Change

Functions are often used to model changing quantities. In this section we learn how to find the rate at which the values of a function change as the input variable changes.

▼ Average Rate of Change

We are all familiar with the concept of speed: If you drive a distance of 120 miles in 2 hours, then your average speed, or rate of travel, is $\frac{120 \text{ mi}}{2 \text{ h}} = 60$ mi/h. Now suppose you take a car trip and record the distance that you travel every few minutes. The distance s you have traveled is a function of the time t:

$$s(t) = \text{total distance traveled at time } t$$

We graph the function s as shown in Figure 1. The graph shows that you have traveled a total of 50 miles after 1 hour, 75 miles after 2 hours, 140 miles after 3 hours, and so on. To find your *average* speed between any two points on the trip, we divide the distance traveled by the time elapsed.

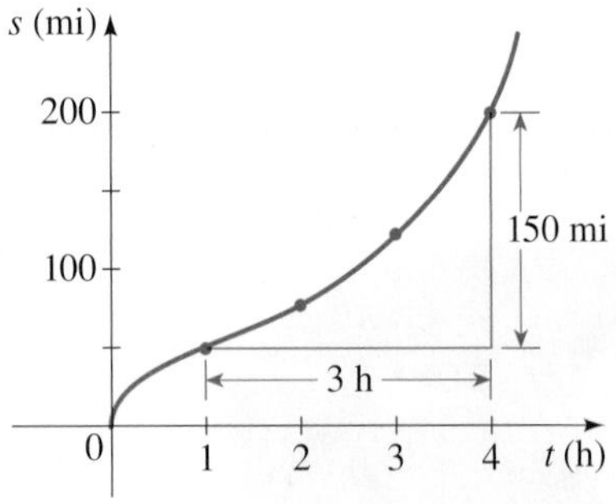

FIGURE 1
Average speed

Let's calculate your average speed between 1:00 P.M. and 4:00 P.M. The time elapsed is $4 - 1 = 3$ hours. To find the distance you traveled, we subtract the distance at 1:00 P.M. from the distance at 4:00 P.M., that is, $200 - 50 = 150$ mi. Thus, your average speed is

$$\text{average speed} = \frac{\text{distance traveled}}{\text{time elapsed}} = \frac{150 \text{ mi}}{3 \text{ h}} = 50 \text{ mi/h}$$

The average speed that we have just calculated can be expressed by using function notation:

$$\text{average speed} = \frac{s(4) - s(1)}{4 - 1} = \frac{200 - 50}{3} = 50 \text{ mi/h}$$

Note that the average speed is different over different time intervals. For example, between 2:00 P.M. and 3:00 P.M. we find that

$$\text{average speed} = \frac{s(3) - s(2)}{3 - 2} = \frac{140 - 75}{1} = 65 \text{ mi/h}$$

Finding average rates of change is important in many contexts. For instance, we might be interested in knowing how quickly the air temperature is dropping as a storm approaches or how fast revenues are increasing from the sale of a new product. So we need to know how to determine the average rate of change of the functions that model these quantities. In fact, the concept of average rate of change can be defined for any function.

AVERAGE RATE OF CHANGE

The **average rate of change** of the function $y = f(x)$ between $x = a$ and $x = b$ is

$$\text{average rate of change} = \frac{\text{change in } y}{\text{change in } x} = \frac{f(b) - f(a)}{b - a}$$

The average rate of change is the slope of the **secant line** between $x = a$ and $x = b$ on the graph of f, that is, the line that passes through $(a, f(a))$ and $(b, f(b))$.

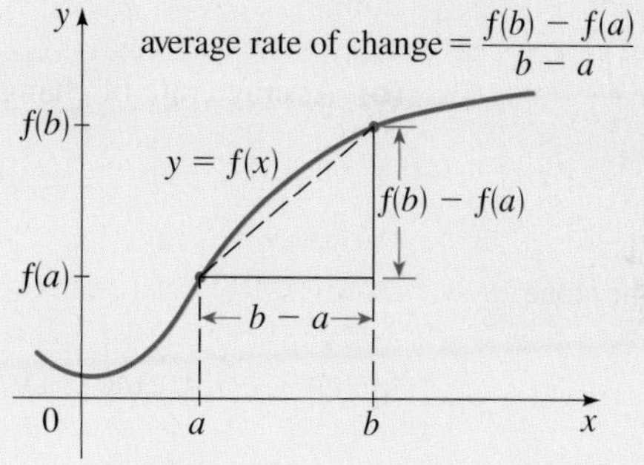

EXAMPLE 1 | Calculating the Average Rate of Change

For the function $f(x) = (x - 3)^2$, whose graph is shown in Figure 2, find the average rate of change between the following points:

(a) $x = 1$ and $x = 3$ **(b)** $x = 4$ and $x = 7$

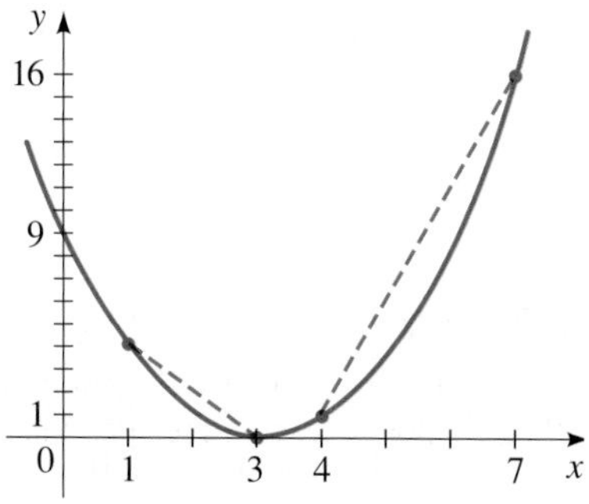

FIGURE 2 $f(x) = (x - 3)^2$

SOLUTION

(a)
$$\text{Average rate of change} = \frac{f(3) - f(1)}{3 - 1} \qquad \text{Definition}$$
$$= \frac{(3 - 3)^2 - (1 - 3)^2}{3 - 1} \qquad \text{Use } f(x) = (x - 3)^2$$
$$= \frac{0 - 4}{2} = -2$$

(b) Average rate of change $= \dfrac{f(7) - f(4)}{7 - 4}$ Definition

$= \dfrac{(7-3)^2 - (4-3)^2}{7-4}$ Use $f(x) = (x-3)^2$

$= \dfrac{16 - 1}{3} = 5$

NOW TRY EXERCISE **11**

EXAMPLE 2 | Average Speed of a Falling Object

If an object is dropped from a high cliff or a tall building, then the distance it has fallen after t seconds is given by the function $d(t) = 16t^2$. Find its average speed (average rate of change) over the following intervals:

(a) Between 1 s and 5 s   **(b)** Between $t = a$ and $t = a + h$

SOLUTION

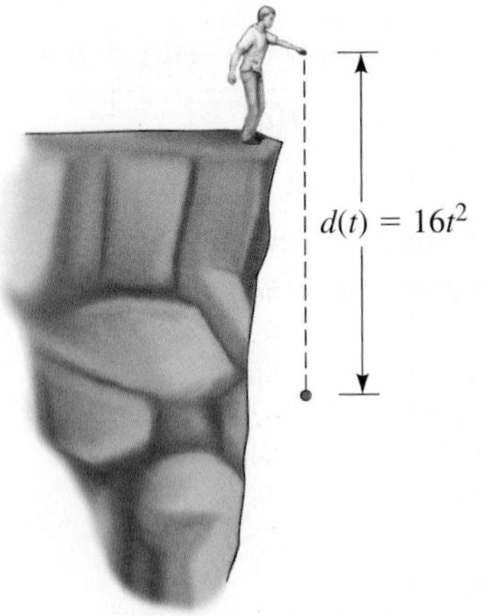

Function: In t seconds the stone falls $16t^2$ ft.

(a) Average rate of change $= \dfrac{d(5) - d(1)}{5 - 1}$ Definition

$= \dfrac{16(5)^2 - 16(1)^2}{5 - 1}$ Use $d(t) = 16t^2$

$= \dfrac{400 - 16}{4}$

$= 96$ ft/s

(b) Average rate of change $= \dfrac{d(a+h) - d(a)}{(a+h) - a}$ Definition

$= \dfrac{16(a+h)^2 - 16(a)^2}{(a+h) - a}$ Use $d(t) = 16t^2$

$= \dfrac{16(a^2 + 2ah + h^2 - a^2)}{h}$ Expand and factor 16

$= \dfrac{16(2ah + h^2)}{h}$ Simplify numerator

$= \dfrac{16h(2a + h)}{h}$ Factor h

$= 16(2a + h)$ Simplify

NOW TRY EXERCISE **15**

The average rate of change calculated in Example 2(b) is known as a *difference quotient*. In calculus we use difference quotients to calculate *instantaneous* rates of change. An example of an instantaneous rate of change is the speed shown on the speedometer of your car. This changes from one instant to the next as your car's speed changes.

The graphs in Figure 3 show that if a function is increasing on an interval, then the average rate of change between any two points is positive, whereas if a function is decreasing on an interval, then the average rate of change between any two points is negative.

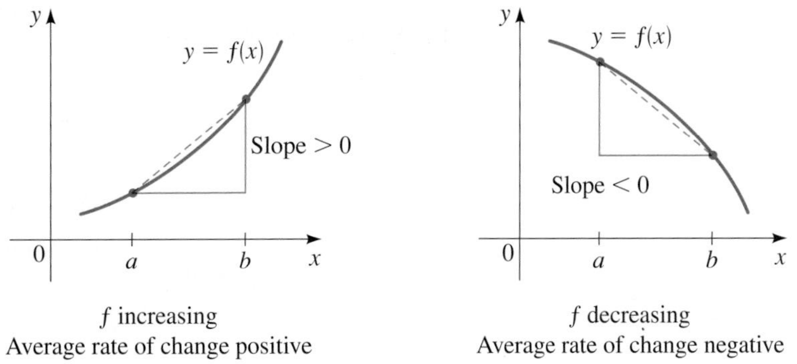

FIGURE 3

EXAMPLE 3 | Average Rate of Temperature Change

The table gives the outdoor temperatures observed by a science student on a spring day. Draw a graph of the data, and find the average rate of change of temperature between the following times:

(a) 8:00 A.M. and 9:00 A.M.

(b) 1:00 P.M. and 3:00 P.M.

(c) 4:00 P.M. and 7:00 P.M.

Time	Temperature (°F)
8:00 A.M.	38
9:00 A.M.	40
10:00 A.M.	44
11:00 A.M.	50
12:00 NOON	56
1:00 P.M.	62
2:00 P.M.	66
3:00 P.M.	67
4:00 P.M.	64
5:00 P.M.	58
6:00 P.M.	55
7:00 P.M.	51

SOLUTION A graph of the temperature data is shown in Figure 4. Let t represent time, measured in hours since midnight (so, for example, 2:00 P.M. corresponds to $t = 14$). Define the function F by

$$F(t) = \text{temperature at time } t$$

(a) $$\text{Average rate of change} = \frac{\text{temperature at 9 A.M.} - \text{temperature at 8 A.M.}}{9 - 8}$$

$$= \frac{F(9) - F(8)}{9 - 8} = \frac{40 - 38}{9 - 8} = 2$$

The average rate of change was 2°F per hour.

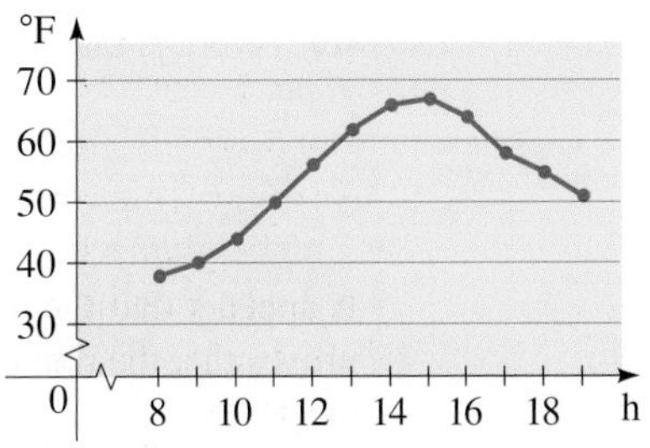

FIGURE 4

(b) $$\text{Average rate of change} = \frac{\text{temperature at 3 P.M.} - \text{temperature at 1 P.M.}}{15 - 13}$$

$$= \frac{F(15) - F(13)}{15 - 13} = \frac{67 - 62}{2} = 2.5$$

The average rate of change was 2.5°F per hour.

(c) $$\text{Average rate of change} = \frac{\text{temperature at 7 P.M.} - \text{temperature at 4 P.M.}}{19 - 16}$$

$$= \frac{F(19) - F(16)}{19 - 16} = \frac{51 - 64}{3} \approx -4.3$$

The average rate of change was about −4.3°F per hour during this time interval. The negative sign indicates that the temperature was dropping.

NOW TRY EXERCISE **25** ■

▼ Linear Functions Have Constant Rate of Change

For a linear function $f(x) = mx + b$ the average rate of change between any two points is the same constant m. This agrees with what we learned in Section 1.10: that the slope of a line $y = mx + b$ is the average rate of change of y with respect to x. On the other hand, if a function f has constant average rate of change, then it must be a linear function. You are asked to prove this fact in Exercise 33. In the next example we find the average rate of change for a particular linear function.

EXAMPLE 4 | Linear Functions Have Constant Rate of Change

Let $f(x) = 3x - 5$. Find the average rate of change of f between the following points.

(a) $x = 0$ and $x = 1$

(b) $x = 3$ and $x = 7$

(c) $x = a$ and $x = a + h$

What conclusion can you draw from your answers?

SOLUTION

(a) $$\text{Average rate of change} = \frac{f(1) - f(0)}{1 - 0} = \frac{(3 \cdot 1 - 5) - (3 \cdot 0 - 5)}{1}$$

$$= \frac{(-2) - (-5)}{1} = 3$$

(b) $$\text{Average rate of change} = \frac{f(7) - f(3)}{7 - 3} = \frac{(3 \cdot 7 - 5) - (3 \cdot 3 - 5)}{4}$$

$$= \frac{16 - 4}{4} = 3$$

(c) $$\text{Average rate of change} = \frac{f(a + h) - f(a)}{(a + h) - a} = \frac{[3(a + h) - 5] - [3a - 5]}{h}$$

$$= \frac{3a + 3h - 5 - 3a + 5}{h} = \frac{3h}{h} = 3$$

It appears that the average rate of change is always 3 for this function. In fact, part (c) proves that the rate of change between any two arbitrary points $x = a$ and $x = a + h$ is 3.

NOW TRY EXERCISE **21** ■

2.4 EXERCISES

CONCEPTS

1. If you travel 100 miles in two hours, then your average speed for the trip is

average speed = $\dfrac{\square}{\square}$ = ________

2. The average rate of change of a function f between $x = a$ and $x = b$ is

average rate of change = $\dfrac{\square}{\square}$

3. The average rate of change of the function $f(x) = x^2$ between $x = 1$ and $x = 5$ is

average rate of change = $\dfrac{\square}{\square}$ = ________

4. (a) The average rate of change of a function f between $x = a$ and $x = b$ is the slope of the ____________ line between $(a, f(a))$ and $(b, f(b))$.

(b) The average rate of change of the linear function $f(x) = 3x + 5$ between any two points is ________.

SKILLS

5–8 ■ The graph of a function is given. Determine the average rate of change of the function between the indicated points on the graph.

5.

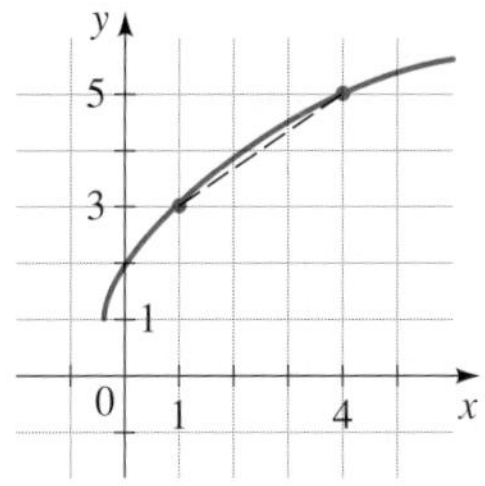

6.

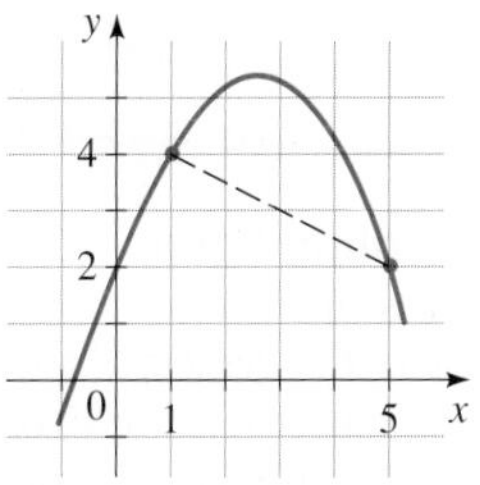

7.

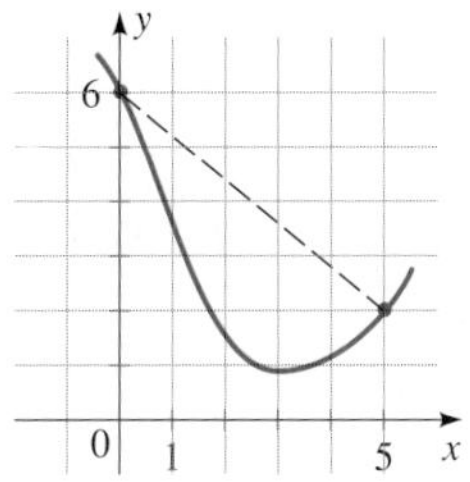

8.

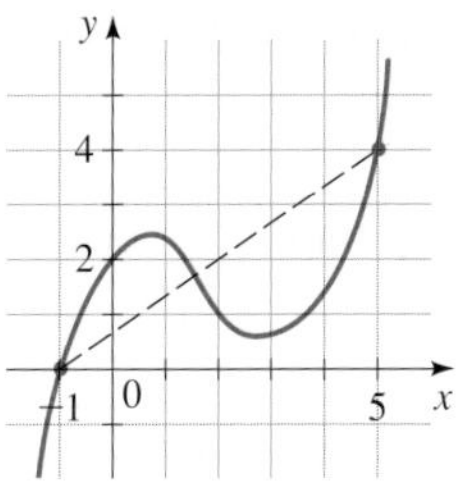

9–20 ■ A function is given. Determine the average rate of change of the function between the given values of the variable.

9. $f(x) = 3x - 2;\quad x = 2, x = 3$

10. $g(x) = 5 + \frac{1}{2}x;\quad x = 1, x = 5$

11. $h(t) = t^2 + 2t;\quad t = -1, t = 4$

12. $f(z) = 1 - 3z^2;\quad z = -2, z = 0$

13. $f(x) = x^3 - 4x^2;\quad x = 0, x = 10$

14. $f(x) = x + x^4;\quad x = -1, x = 3$

15. $f(x) = 3x^2;\quad x = 2, x = 2 + h$

16. $f(x) = 4 - x^2;\quad x = 1, x = 1 + h$

17. $g(x) = \dfrac{1}{x};\quad x = 1, x = a$

18. $g(x) = \dfrac{2}{x + 1};\quad x = 0, x = h$

19. $f(t) = \dfrac{2}{t};\quad t = a, t = a + h$

20. $f(t) = \sqrt{t};\quad t = a, t = a + h$

21–22 ■ A linear function is given. **(a)** Find the average rate of change of the function between $x = a$ and $x = a + h$. **(b)** Show that the average rate of change is the same as the slope of the line.

21. $f(x) = \frac{1}{2}x + 3$

22. $g(x) = -4x + 2$

APPLICATIONS

23. Changing Water Levels The graph shows the depth of water W in a reservoir over a one-year period as a function of the number of days x since the beginning of the year. What was the average rate of change of W between $x = 100$ and $x = 200$?

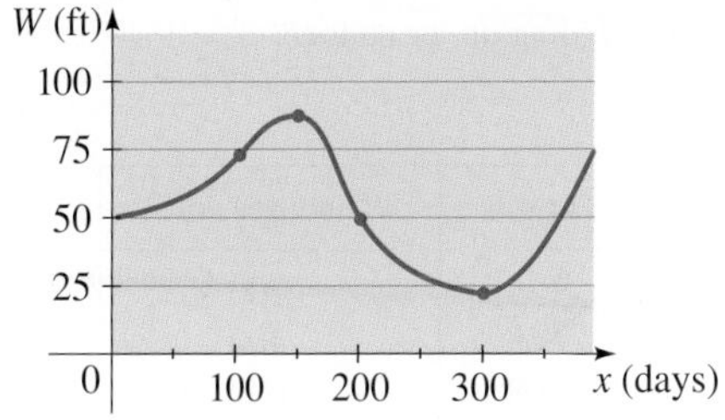

24. Population Growth and Decline The graph shows the population P in a small industrial city from 1950 to 2000. The variable x represents the number of years since 1950.

(a) What was the average rate of change of P between $x = 20$ and $x = 40$?

(b) Interpret the value of the average rate of change that you found in part (a).

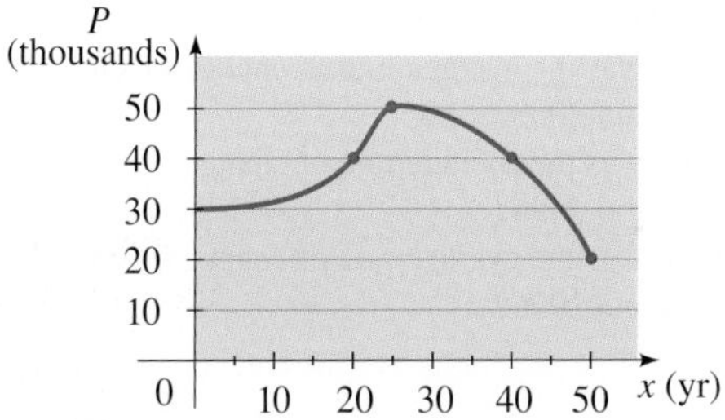

25. Population Growth and Decline The table gives the population in a small coastal community for the period 1997–2006. Figures shown are for January 1 in each year.

(a) What was the average rate of change of population between 1998 and 2001?

(b) What was the average rate of change of population between 2002 and 2004?

(c) For what period of time was the population increasing?

(d) For what period of time was the population decreasing?

Year	Population
1997	624
1998	856
1999	1,336
2000	1,578
2001	1,591
2002	1,483
2003	994
2004	826
2005	801
2006	745

26. Running Speed A man is running around a circular track that is 200 m in circumference. An observer uses a stopwatch to record the runner's time at the end of each lap, obtaining the data in the following table.

(a) What was the man's average speed (rate) between 68 s and 152 s?

(b) What was the man's average speed between 263 s and 412 s?

(c) Calculate the man's speed for each lap. Is he slowing down, speeding up, or neither?

Time (s)	Distance (m)
32	200
68	400
108	600
152	800
203	1000
263	1200
335	1400
412	1600

27. CD Player Sales The table shows the number of CD players sold in a small electronics store in the years 1993–2003.

(a) What was the average rate of change of sales between 1993 and 2003?

(b) What was the average rate of change of sales between 1993 and 1994?

(c) What was the average rate of change of sales between 1994 and 1996?

(d) Between which two successive years did CD player sales *increase* most quickly? *Decrease* most quickly?

Year	CD players sold
1993	512
1994	520
1995	413
1996	410
1997	468
1998	510
1999	590
2000	607
2001	732
2002	612
2003	584

28. Book Collection Between 1980 and 2000, a rare book collector purchased books for his collection at the rate of 40 books per year. Use this information to complete the following table. (Note that not every year is given in the table.)

Year	Number of books
1980	420
1981	460
1982	
1985	
1990	
1992	
1995	
1997	
1998	
1999	
2000	1220

29. Cooling Soup When a bowl of hot soup is left in a room, the soup eventually cools down to room temperature. The temperature T of the soup is a function of time t. The table below gives the temperature (in °F) of a bowl of soup t minutes after it was set on the table. Find the average rate of change of the temperature of the soup over the first 20 minutes and over the next 20 minutes. During which interval did the soup cool off more quickly?

t (min)	T (°F)	t (min)	T (°F)
0	200	35	94
5	172	40	89
10	150	50	81
15	133	60	77
20	119	90	72
25	108	120	70
30	100	150	70

30. Farms in the United States The graph gives the number of farms in the United States from 1850 to 2000.

(a) Estimate the average rate of change in the number of farms between (i) 1860 and 1890 and (ii) 1950 and 1970.

(b) In which decade did the number of farms experience the greatest average rate of decline?

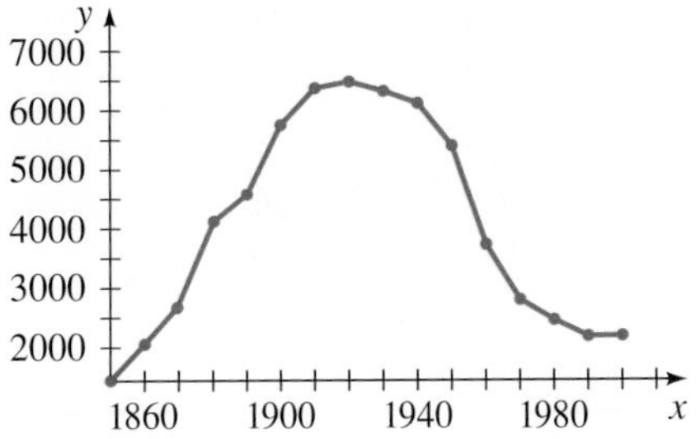

DISCOVERY ■ DISCUSSION ■ WRITING

31. 100-Meter Race A 100-m race ends in a three-way tie for first place. The graph at the top of the next column shows distance as a function of time for each of the three winners.

(a) Find the average speed for each winner.

(b) Describe the differences between the ways in which the three runners ran the race.

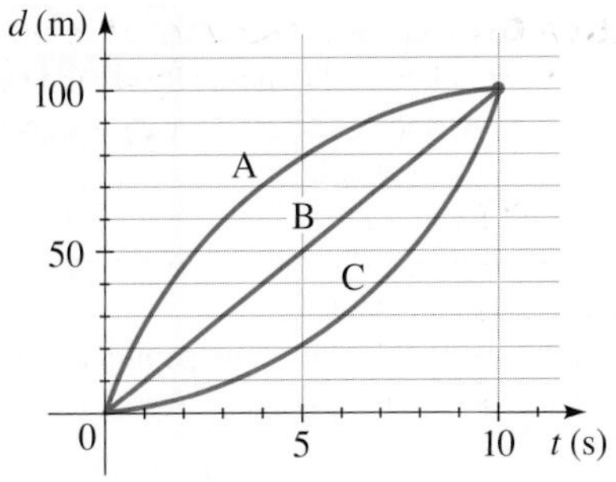

32. Linear Functions Have Constant Rate of Change If $f(x) = mx + b$ is a linear function, then the average rate of change of f between any two real numbers x_1 and x_2 is

$$\text{average rate of change} = \frac{f(x_2) - f(x_1)}{x_2 - x_1}$$

Calculate this average rate of change to show that it is the same as the slope m.

33. Functions with Constant Rate of Change Are Linear If the function f has the same average rate of change c between any two points, then for the points a and x we have

$$c = \frac{f(x) - f(a)}{x - a}$$

Rearrange this expression to show that

$$f(x) = cx + (f(a) - ca)$$

and conclude that f is a linear function.

2.5 Transformations of Functions

Vertical Shifting ▶ Horizontal Shifting ▶ Reflecting Graphs ▶ Vertical Stretching and Shrinking ▶ Horizontal Stretching and Shrinking ▶ Even and Odd Functions

In this section we study how certain transformations of a function affect its graph. This will give us a better understanding of how to graph functions. The transformations that we study are shifting, reflecting, and stretching.

▼ Vertical Shifting

Adding a constant to a function shifts its graph vertically: upward if the constant is positive and downward if it is negative.

In general, suppose we know the graph of $y = f(x)$. How do we obtain from it the graphs of

Recall that the graph of the function f is the same as the graph of the equation $y = f(x)$.

$$y = f(x) + c \quad \text{and} \quad y = f(x) - c \qquad (c > 0)$$

The y-coordinate of each point on the graph of $y = f(x) + c$ is c units above the y-coordinate of the corresponding point on the graph of $y = f(x)$. So we obtain the graph of $y = f(x) + c$ simply by shifting the graph of $y = f(x)$ upward c units. Similarly, we obtain the graph of $y = f(x) - c$ by shifting the graph of $y = f(x)$ downward c units.

VERTICAL SHIFTS OF GRAPHS

Suppose $c > 0$.

To graph $y = f(x) + c$, shift the graph of $y = f(x)$ upward c units.

To graph $y = f(x) - c$, shift the graph of $y = f(x)$ downward c units.

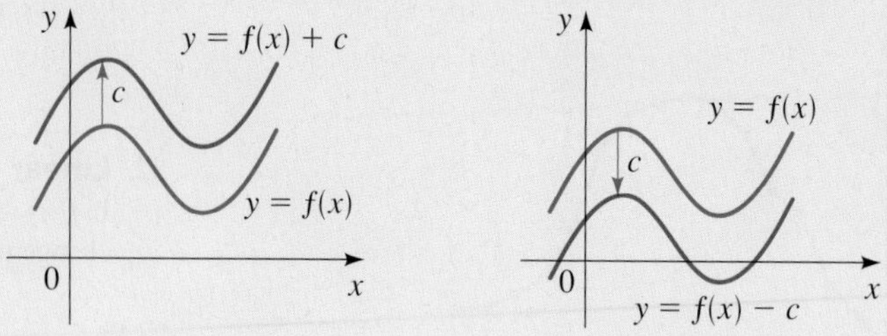

EXAMPLE 1 | Vertical Shifts of Graphs

Use the graph of $f(x) = x^2$ to sketch the graph of each function.

(a) $g(x) = x^2 + 3$ **(b)** $h(x) = x^2 - 2$

SOLUTION The function $f(x) = x^2$ was graphed in Example 1(a), Section 2.2. It is sketched again in Figure 1.

(a) Observe that

$$g(x) = x^2 + 3 = f(x) + 3$$

So the y-coordinate of each point on the graph of g is 3 units above the corresponding point on the graph of f. This means that to graph g, we shift the graph of f upward 3 units, as in Figure 1.

(b) Similarly, to graph h, we shift the graph of f downward 2 units, as shown in Figure 1.

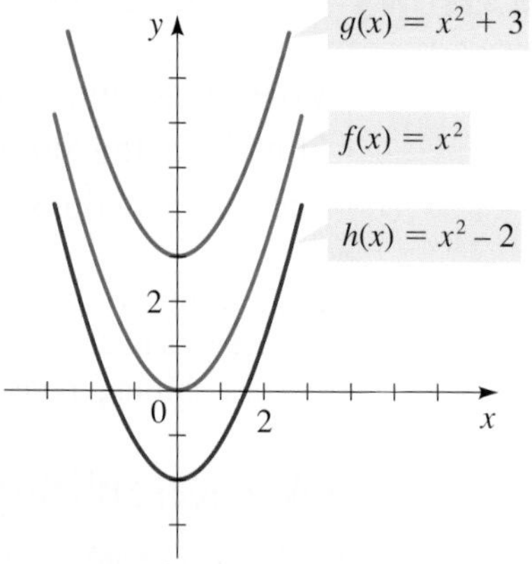

FIGURE 1

NOW TRY EXERCISES **21** AND **23**

▼ Horizontal Shifting

Suppose that we know the graph of $y = f(x)$. How do we use it to obtain the graphs of

$$y = f(x + c) \quad \text{and} \quad y = f(x - c) \quad (c > 0)$$

The value of $f(x - c)$ at x is the same as the value of $f(x)$ at $x - c$. Since $x - c$ is c units to the left of x, it follows that the graph of $y = f(x - c)$ is just the graph of $y = f(x)$

shifted to the right c units. Similar reasoning shows that the graph of $y = f(x + c)$ is the graph of $y = f(x)$ shifted to the left c units. The following box summarizes these facts.

HORIZONTAL SHIFTS OF GRAPHS

Suppose $c > 0$.

To graph $y = f(x - c)$, shift the graph of $y = f(x)$ to the right c units.

To graph $y = f(x + c)$, shift the graph of $y = f(x)$ to the left c units.

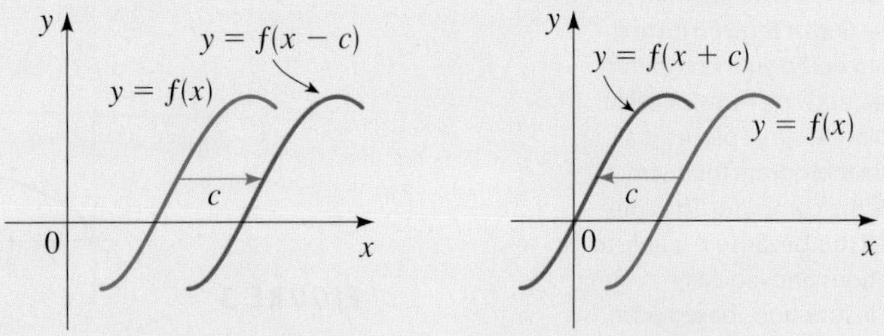

EXAMPLE 2 | Horizontal Shifts of Graphs

Use the graph of $f(x) = x^2$ to sketch the graph of each function.

(a) $g(x) = (x + 4)^2$ **(b)** $h(x) = (x - 2)^2$

SOLUTION

(a) To graph g, we shift the graph of f to the left 4 units.

(b) To graph h, we shift the graph of f to the right 2 units.

The graphs of g and h are sketched in Figure 2.

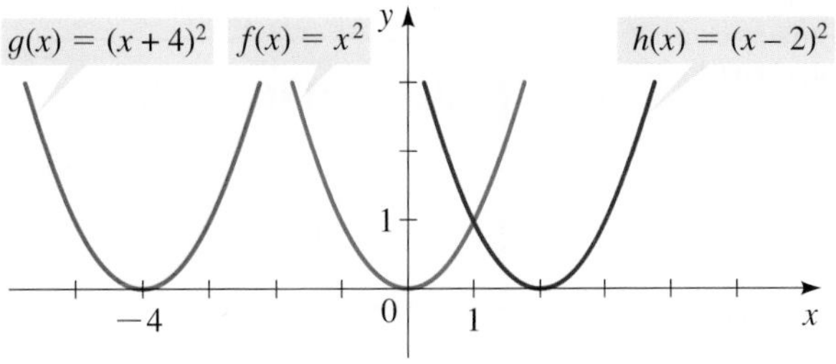

FIGURE 2

NOW TRY EXERCISE **25** AND **27** ■

RENÉ DESCARTES (1596–1650) was born in the town of La Haye in southern France. From an early age Descartes liked mathematics because of "the certainty of its results and the clarity of its reasoning." He believed that to arrive at truth, one must begin by doubting everything, including one's own existence; this led him to formulate perhaps the best-known sentence in all of philosophy: "I think, therefore I am." In his book *Discourse on Method* he described what is now called the Cartesian plane. This idea of combining algebra and geometry enabled mathematicians for the first time to graph functions and thus "see" the equations they were studying. The philosopher John Stuart Mill called this invention "the greatest single step ever made in the progress of the exact sciences." Descartes liked to get up late and spend the morning in bed thinking and writing. He invented the coordinate plane while lying in bed watching a fly crawl on the ceiling, reasoning that he could describe the exact location of the fly by knowing its distance from two perpendicular walls. In 1649 Descartes became the tutor of Queen Christina of Sweden. She liked her lessons at 5 o'clock in the morning, when, she said, her mind was sharpest. However, the change from his usual habits and the ice-cold library where they studied proved too much for Descartes. In February 1650, after just two months of this, he caught pneumonia and died.

MATHEMATICS IN THE MODERN WORLD

Computers

For centuries machines have been designed to perform specific tasks. For example, a washing machine washes clothes, a weaving machine weaves cloth, an adding machine adds numbers, and so on. The computer has changed all that.

The computer is a machine that does nothing—until it is given instructions on what to do. So your computer can play games, draw pictures, or calculate π to a million decimal places; it all depends on what program (or instructions) you give the computer. The computer can do all this because it is able to accept instructions and logically change those instructions based on incoming data. This versatility makes computers useful in nearly every aspect of human endeavor.

The idea of a computer was described theoretically in the 1940s by the mathematician Allan Turing (see page 100) in what he called a *universal machine*. In 1945 the mathematician John Von Neumann, extending Turing's ideas, built one of the first electronic computers.

Mathematicians continue to develop new theoretical bases for the design of computers. The heart of the computer is the "chip," which is capable of processing logical instructions. To get an idea of the chip's complexity, consider that the Pentium chip has over 3.5 million logic circuits!

EXAMPLE 3 | Combining Horizontal and Vertical Shifts

Sketch the graph of $f(x) = \sqrt{x-3} + 4$.

SOLUTION We start with the graph of $y = \sqrt{x}$ (Example 1(c), Section 2.2) and shift it to the right 3 units to obtain the graph of $y = \sqrt{x-3}$. Then we shift the resulting graph upward 4 units to obtain the graph of $f(x) = \sqrt{x-3} + 4$ shown in Figure 3.

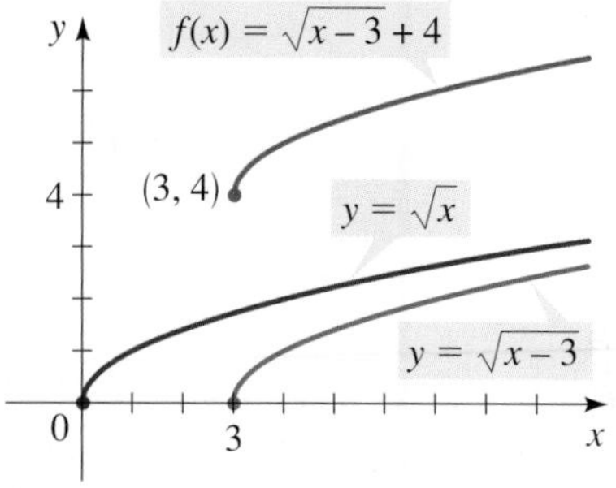

FIGURE 3

NOW TRY EXERCISE **37**

Reflecting Graphs

Suppose we know the graph of $y = f(x)$. How do we use it to obtain the graphs of $y = -f(x)$ and $y = f(-x)$? The y-coordinate of each point on the graph of $y = -f(x)$ is simply the negative of the y-coordinate of the corresponding point on the graph of $y = f(x)$. So the desired graph is the reflection of the graph of $y = f(x)$ in the x-axis. On the other hand, the value of $y = f(-x)$ at x is the same as the value of $y = f(x)$ at $-x$, so the desired graph here is the reflection of the graph of $y = f(x)$ in the y-axis. The following box summarizes these observations.

REFLECTING GRAPHS

To graph $y = -f(x)$, reflect the graph of $y = f(x)$ in the x-axis.

To graph $y = f(-x)$, reflect the graph of $y = f(x)$ in the y-axis.

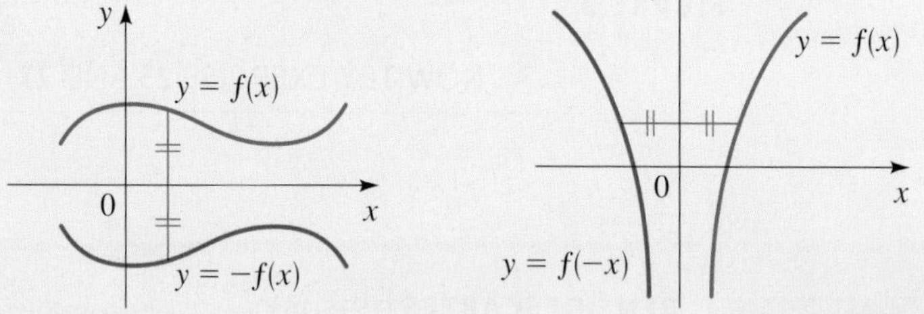

EXAMPLE 4 | Reflecting Graphs

Sketch the graph of each function.

(a) $f(x) = -x^2$ **(b)** $g(x) = \sqrt{-x}$

SOLUTION

(a) We start with the graph of $y = x^2$. The graph of $f(x) = -x^2$ is the graph of $y = x^2$ reflected in the x-axis (see Figure 4).

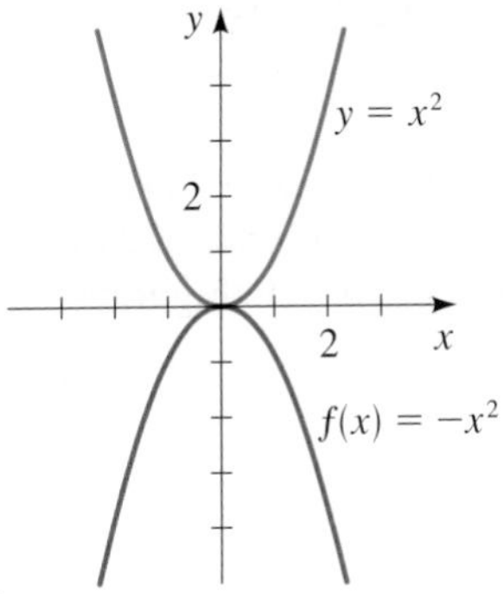

FIGURE 4

(b) We start with the graph of $y = \sqrt{x}$ (Example 1(c) in Section 2.2). The graph of $g(x) = \sqrt{-x}$ is the graph of $y = \sqrt{x}$ reflected in the y-axis (see Figure 5). Note that the domain of the function $g(x) = \sqrt{-x}$ is $\{x \mid x \le 0\}$.

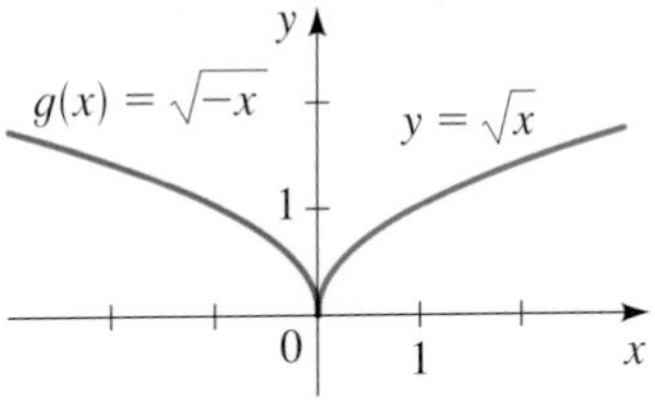

FIGURE 5

NOW TRY EXERCISES **29** AND **31** ■

Vertical Stretching and Shrinking

Suppose we know the graph of $y = f(x)$. How do we use it to obtain the graph of $y = cf(x)$? The y-coordinate of $y = cf(x)$ at x is the same as the corresponding y-coordinate of $y = f(x)$ multiplied by c. Multiplying the y-coordinates by c has the effect of vertically stretching or shrinking the graph by a factor of c.

VERTICAL STRETCHING AND SHRINKING OF GRAPHS

To graph $y = cf(x)$:

If $c > 1$, stretch the graph of $y = f(x)$ vertically by a factor of c.

If $0 < c < 1$, shrink the graph of $y = f(x)$ vertically by a factor of c.

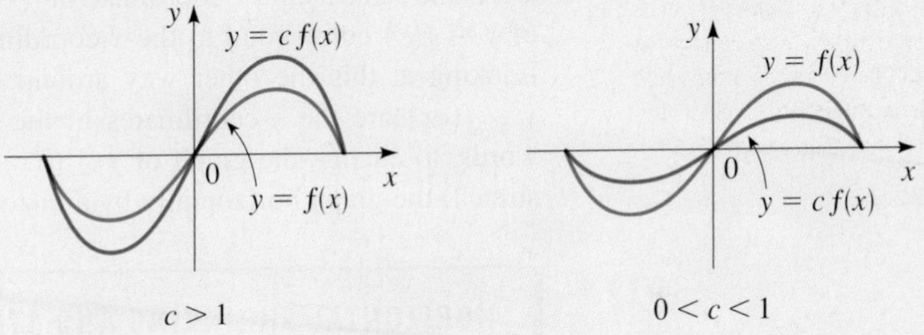

EXAMPLE 5 | Vertical Stretching and Shrinking of Graphs

Use the graph of $f(x) = x^2$ to sketch the graph of each function.

(a) $g(x) = 3x^2$ **(b)** $h(x) = \frac{1}{3}x^2$

SOLUTION

(a) The graph of g is obtained by multiplying the y-coordinate of each point on the graph of f by 3. That is, to obtain the graph of g, we stretch the graph of f vertically by a factor of 3. The result is the narrower parabola in Figure 6.

(b) The graph of h is obtained by multiplying the y-coordinate of each point on the graph of f by $\frac{1}{3}$. That is, to obtain the graph of h, we shrink the graph of f vertically by a factor of $\frac{1}{3}$. The result is the wider parabola in Figure 6.

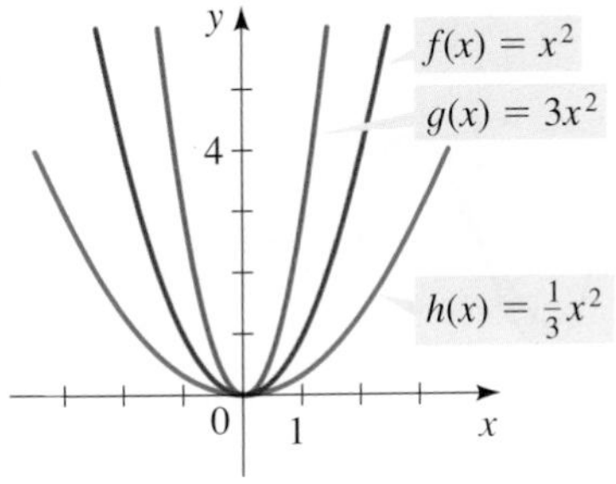

FIGURE 6

NOW TRY EXERCISES **33** AND **35** ■

We illustrate the effect of combining shifts, reflections, and stretching in the following example.

EXAMPLE 6 | Combining Shifting, Stretching, and Reflecting

Sketch the graph of the function $f(x) = 1 - 2(x - 3)^2$.

SOLUTION Starting with the graph of $y = x^2$, we first shift to the right 3 units to get the graph of $y = (x - 3)^2$. Then we reflect in the x-axis and stretch by a factor of 2 to get the graph of $y = -2(x - 3)^2$. Finally, we shift upward 1 unit to get the graph of $f(x) = 1 - 2(x - 3)^2$ shown in Figure 7.

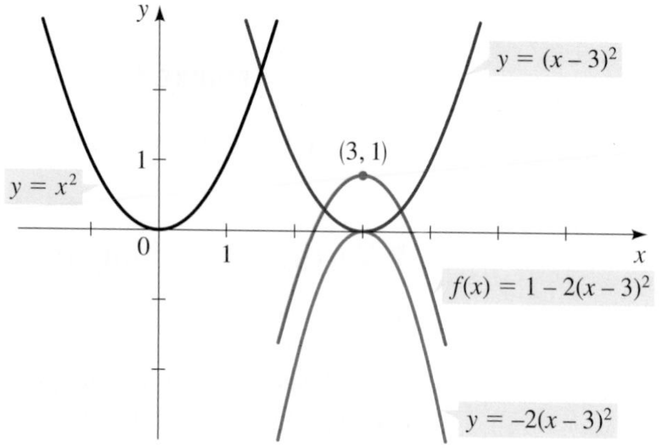

FIGURE 7

NOW TRY EXERCISE **39**

Horizontal Stretching and Shrinking

Now we consider horizontal shrinking and stretching of graphs. If we know the graph of $y = f(x)$, then how is the graph of $y = f(cx)$ related to it? The y-coordinate of $y = f(cx)$ at x is the same as the y-coordinate of $y = f(x)$ at cx. Thus, the x-coordinates in the graph of $y = f(x)$ correspond to the x-coordinates in the graph of $y = f(cx)$ multiplied by c. Looking at this the other way around, we see that the x-coordinates in the graph of $y = f(cx)$ are the x-coordinates in the graph of $y = f(x)$ multiplied by $1/c$. In other words, to change the graph of $y = f(x)$ to the graph of $y = f(cx)$, we must shrink (or stretch) the graph horizontally by a factor of $1/c$, as summarized in the following box.

HORIZONTAL SHRINKING AND STRETCHING OF GRAPHS

To graph $y = f(cx)$:

If $c > 1$, shrink the graph of $y = f(x)$ horizontally by a factor of $1/c$.

If $0 < c < 1$, stretch the graph of $y = f(x)$ horizontally by a factor of $1/c$.

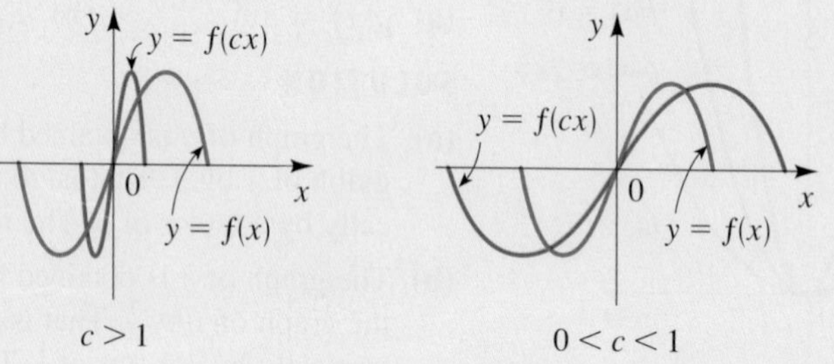

EXAMPLE 7 | Horizontal Stretching and Shrinking of Graphs

The graph of $y = f(x)$ is shown in Figure 8 on the next page. Sketch the graph of each function.

(a) $y = f(2x)$ **(b)** $y = f(\frac{1}{2}x)$

$-\frac{1}{4}(x-5)^2+4$

The Granger Collection, New York

SONYA KOVALEVSKY (1850–1891) is considered the most important woman mathematician of the 19th century. She was born in Moscow to an aristocratic family. While a child, she was exposed to the principles of calculus in a very unusual fashion: Her bedroom was temporarily wallpapered with the pages of a calculus book. She later wrote that she "spent many hours in front of that wall, trying to understand it." Since Russian law forbade women from studying in universities, she entered a marriage of convenience, which allowed her to travel to Germany and obtain a doctorate in mathematics from the University of Göttingen. She eventually was awarded a full professorship at the University of Stockholm, where she taught for eight years before dying in an influenza epidemic at the age of 41. Her research was instrumental in helping to put the ideas and applications of functions and calculus on a sound and logical foundation. She received many accolades and prizes for her research work.

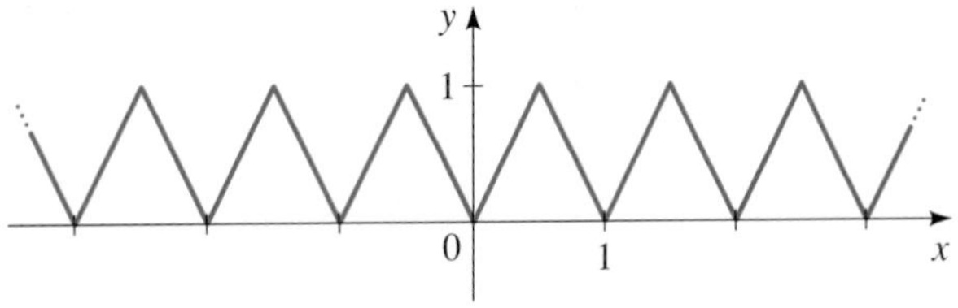

FIGURE 8 $y = f(x)$

SOLUTION Using the principles described in the preceding box, we obtain the graphs shown in Figures 9 and 10.

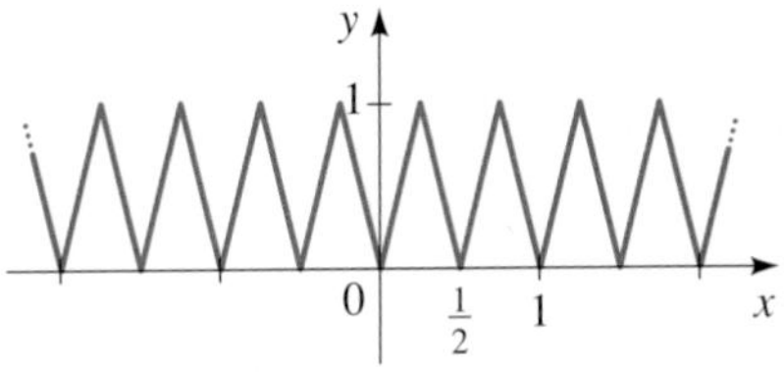

FIGURE 9 $y = f(2x)$

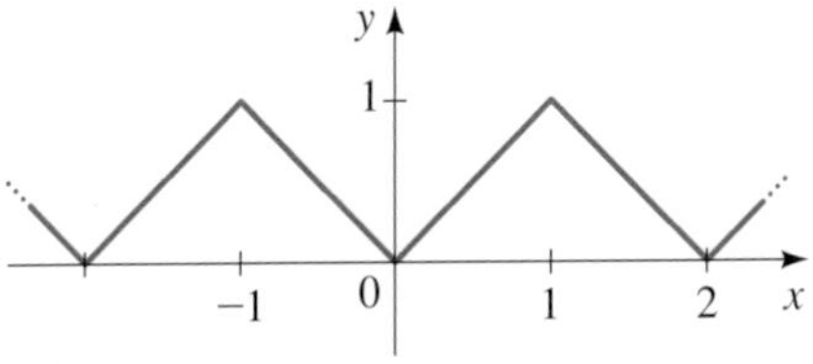

FIGURE 10 $y = f(\frac{1}{2}x)$

NOW TRY EXERCISE **63**

Even and Odd Functions

If a function f satisfies $f(-x) = f(x)$ for every number x in its domain, then f is called an **even function**. For instance, the function $f(x) = x^2$ is even because

$$f(-x) = (-x)^2 = (-1)^2x^2 = x^2 = f(x)$$

The graph of an even function is symmetric with respect to the y-axis (see Figure 11). This means that if we have plotted the graph of f for $x \geq 0$, then we can obtain the entire graph simply by reflecting this portion in the y-axis.

If f satisfies $f(-x) = -f(x)$ for every number x in its domain, then f is called an **odd function**. For example, the function $f(x) = x^3$ is odd because

$$f(-x) = (-x)^3 = (-1)^3x^3 = -x^3 = -f(x)$$

The graph of an odd function is symmetric about the origin (see Figure 12). If we have plotted the graph of f for $x \geq 0$, then we can obtain the entire graph by rotating this portion through 180° about the origin. (This is equivalent to reflecting first in the x-axis and then in the y-axis.)

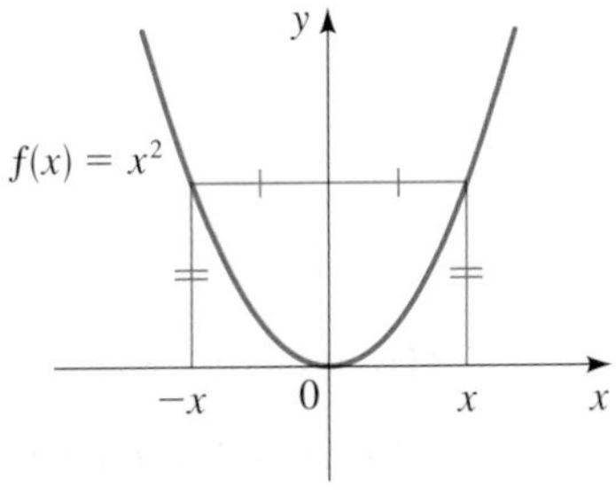

FIGURE 11 $f(x) = x^2$ is an even function.

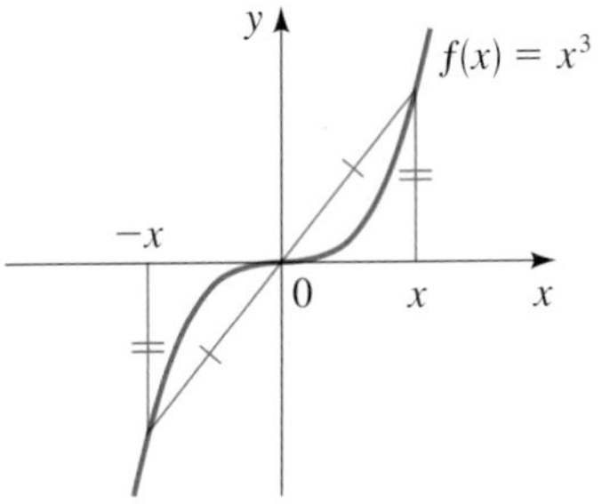

FIGURE 12 $f(x) = x^3$ is an odd function.

> **EVEN AND ODD FUNCTIONS**
>
> Let f be a function.
>
> f is **even** if $f(-x) = f(x)$ for all x in the domain of f.
>
> f is **odd** if $f(-x) = -f(x)$ for all x in the domain of f.
>
> 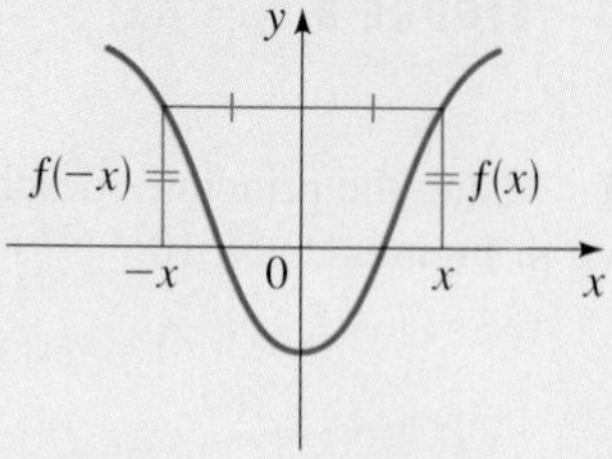
>
>
> The graph of an even function is symmetric with respect to the y-axis.
>
> 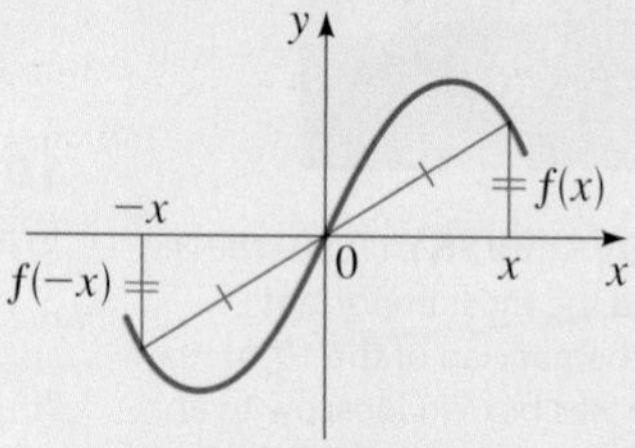
>
>
> The graph of an odd function is symmetric with respect to the origin.

EXAMPLE 8 | Even and Odd Functions

Determine whether the functions are even, odd, or neither even nor odd.

(a) $f(x) = x^5 + x$

(b) $g(x) = 1 - x^4$

(c) $h(x) = 2x - x^2$

SOLUTION

(a) $$\begin{aligned} f(-x) &= (-x)^5 + (-x) \\ &= -x^5 - x = -(x^5 + x) \\ &= -f(x) \end{aligned}$$

Therefore, f is an odd function.

(b) $g(-x) = 1 - (-x)^4 = 1 - x^4 = g(x)$

So g is even.

(c) $h(-x) = 2(-x) - (-x)^2 = -2x - x^2$

Since $h(-x) \neq h(x)$ and $h(-x) \neq -h(x)$, we conclude that h is neither even nor odd.

NOW TRY EXERCISES **75, 77,** AND **79** ■

The graphs of the functions in Example 8 are shown in Figure 13. The graph of f is symmetric about the origin, and the graph of g is symmetric about the y-axis. The graph of h is not symmetric either about the y-axis or the origin.

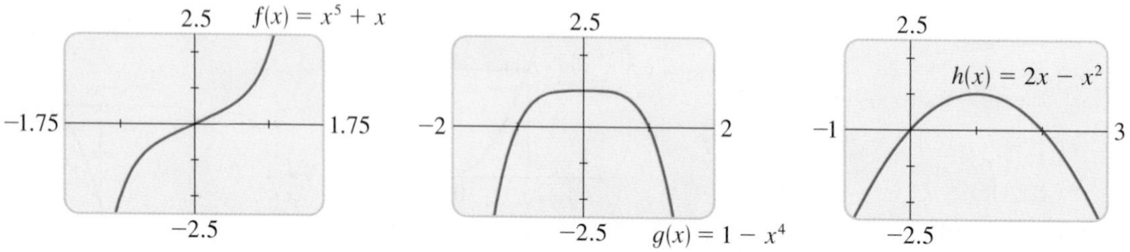

FIGURE 13 (a) (b) (c)

2.5 EXERCISES

CONCEPTS

1–2 ■ Fill in the blank with the appropriate direction (left, right, up, or down).

1. **(a)** The graph of $y = f(x) + 3$ is obtained from the graph of $y = f(x)$ by shifting ________ 3 units.
(b) The graph of $y = f(x + 3)$ is obtained from the graph of $y = f(x)$ by shifting ________ 3 units.

2. **(a)** The graph of $y = f(x) - 3$ is obtained from the graph of $y = f(x)$ by shifting ________ 3 units.
(b) The graph of $y = f(x - 3)$ is obtained from the graph of $y = f(x)$ by shifting ________ 3 units.

3. Fill in the blank with the appropriate axis (x-axis or y-axis).
(a) The graph of $y = -f(x)$ is obtained from the graph of $y = f(x)$ by reflecting in the ________.
(b) The graph of $y = f(-x)$ is obtained from the graph of $y = f(x)$ by reflecting in the ________.

4. Match the graph with the function.
(a) $y = |x + 1|$ **(b)** $y = |x - 1|$
(c) $y = |x| - 1$ **(d)** $y = -|x|$

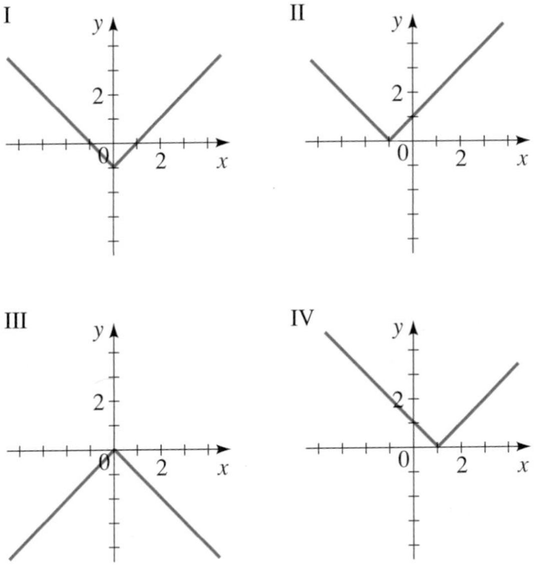

SKILLS

5–14 ■ Suppose the graph of f is given. Describe how the graph of each function can be obtained from the graph of f.

5. **(a)** $y = f(x) - 5$ **(b)** $y = f(x - 5)$

6. **(a)** $y = f(x + 7)$ **(b)** $y = f(x) + 7$

7. **(a)** $y = -f(x)$ **(b)** $y = f(-x)$

8. **(a)** $y = -2f(x)$ **(b)** $y = -\frac{1}{2}f(x)$

9. **(a)** $y = -f(x) + 5$ **(b)** $y = 3f(x) - 5$

10. **(a)** $y = f(x - 4) + \frac{3}{4}$ **(b)** $y = f(x + 4) - \frac{3}{4}$

11. **(a)** $y = 2f(x + 1) - 3$ **(b)** $y = 2f(x - 1) + 3$

12. **(a)** $y = 3 - 2f(x)$ **(b)** $y = 2 - f(-x)$

13. **(a)** $y = f(4x)$ **(b)** $y = f(\frac{1}{4}x)$

14. **(a)** $y = f(2x) - 1$ **(b)** $y = 2f(\frac{1}{2}x)$

15–18 ■ Explain how the graph of g is obtained from the graph of f.

15. **(a)** $f(x) = x^2, \quad g(x) = (x + 2)^2$
(b) $f(x) = x^2, \quad g(x) = x^2 + 2$

16. **(a)** $f(x) = x^3, \quad g(x) = (x - 4)^3$
(b) $f(x) = x^3, \quad g(x) = x^3 - 4$

17. **(a)** $f(x) = |x|, \quad g(x) = |x + 2| - 2$
(b) $f(x) = |x|, \quad g(x) = |x - 2| + 2$

18. **(a)** $f(x) = \sqrt{x}, \quad g(x) = -\sqrt{x} + 1$
(b) $f(x) = \sqrt{x}, \quad g(x) = \sqrt{-x} + 1$

19. Use the graph of $y = x^2$ in Figure 4 to graph the following.
(a) $g(x) = x^2 + 1$
(b) $g(x) = (x - 1)^2$
(c) $g(x) = -x^2$
(d) $g(x) = (x - 1)^2 + 3$

20. Use the graph of $y = \sqrt{x}$ in Figure 5 to graph the following.
(a) $g(x) = \sqrt{x - 2}$
(b) $g(x) = \sqrt{x} + 1$
(c) $g(x) = \sqrt{x + 2} + 2$
(d) $g(x) = -\sqrt{x} + 1$

21–44 ■ Sketch the graph of the function, not by plotting points, but by starting with the graph of a standard function and applying transformations.

21. $f(x) = x^2 - 1$

22. $f(x) = x^2 + 5$

23. $f(x) = \sqrt{x} + 1$

24. $f(x) = |x| - 1$

25. $f(x) = (x - 5)^2$

26. $f(x) = (x + 1)^2$

27. $f(x) = \sqrt{x + 4}$

28. $f(x) = |x - 3|$

29. $f(x) = -x^3$

30. $f(x) = -|x|$

31. $y = \sqrt[4]{-x}$

32. $y = \sqrt[3]{-x}$

33. $y = \frac{1}{4}x^2$

34. $y = -5\sqrt{x}$

35. $y = 3|x|$

36. $y = \frac{1}{2}|x|$

37. $y = (x - 3)^2 + 5$

38. $y = \sqrt{x + 4} - 3$

39. $y = 3 - \frac{1}{2}(x - 1)^2$

40. $y = 2 - \sqrt{x + 1}$

41. $y = |x + 2| + 2$

42. $y = 2 - |x|$

43. $y = \frac{1}{2}\sqrt{x + 4} - 3$

44. $y = 3 - 2(x - 1)^2$

45–54 ■ A function f is given, and the indicated transformations are applied to its graph (in the given order). Write the equation for the final transformed graph.

45. $f(x) = x^2$; shift upward 3 units

46. $f(x) = x^3$; shift downward 1 unit

47. $f(x) = \sqrt{x}$; shift 2 units to the left

48. $f(x) = \sqrt[3]{x}$; shift 1 unit to the right

49. $f(x) = |x|$; shift 3 units to the right and shift upward 1 unit

50. $f(x) = |x|$; shift 4 units to the left and shift downward 2 units

51. $f(x) = \sqrt[4]{x}$; reflect in the y-axis and shift upward 1 unit

52. $f(x) = x^2$; shift 2 units to the left and reflect in the x-axis

53. $f(x) = x^2$; stretch vertically by a factor of 2, shift downward 2 units, and shift 3 units to the right

54. $f(x) = |x|$; shrink vertically by a factor of $\frac{1}{2}$, shift to the left 1 unit, and shift upward 3 units

55–60 ■ The graphs of f and g are given. Find a formula for the function g.

55.

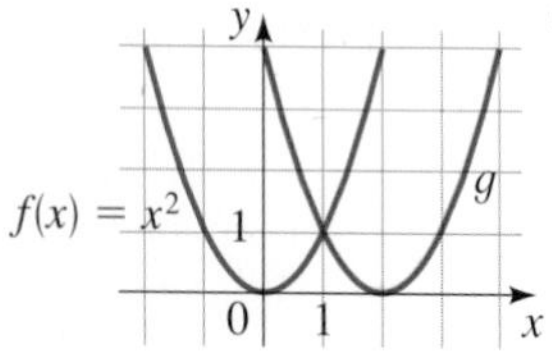

56.

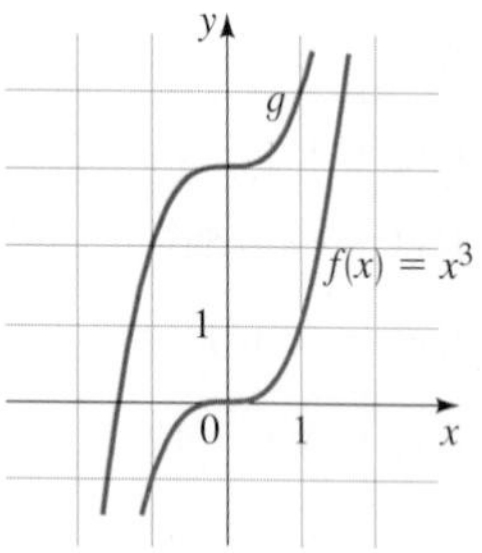

57.

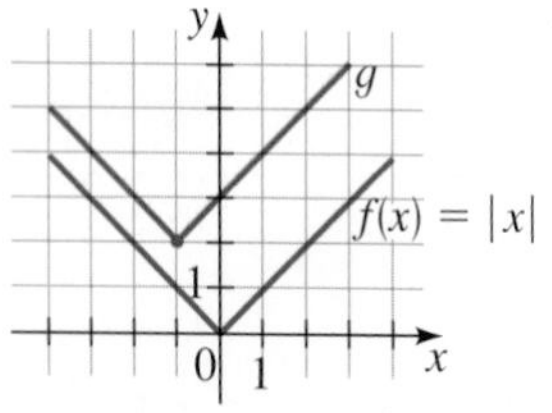

58.

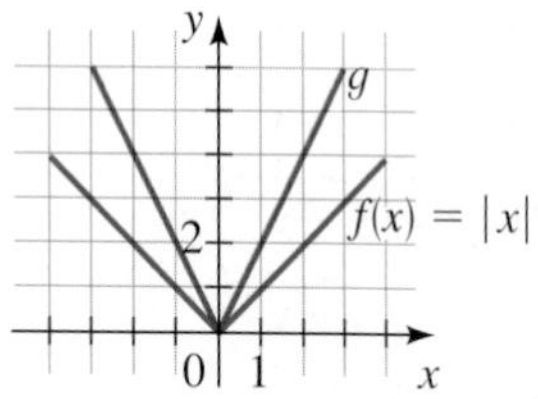

59.

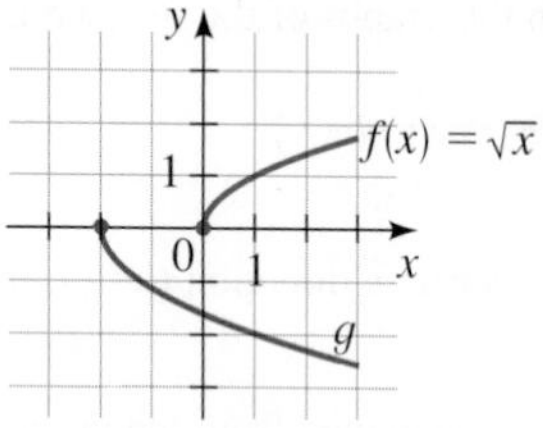

60.

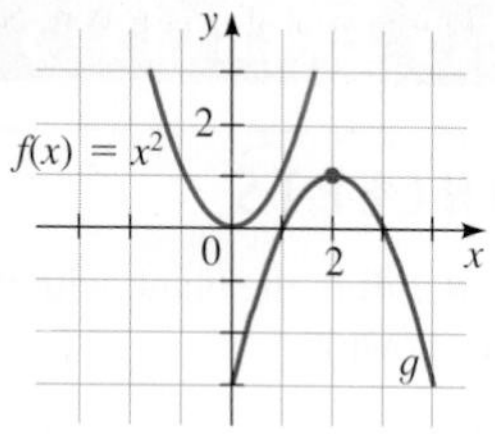

61–62 ■ The graph of $y = f(x)$ is given. Match each equation with its graph.

61. **(a)** $y = f(x - 4)$ **(b)** $y = f(x) + 3$
(c) $y = 2f(x + 6)$ **(d)** $y = -f(2x)$

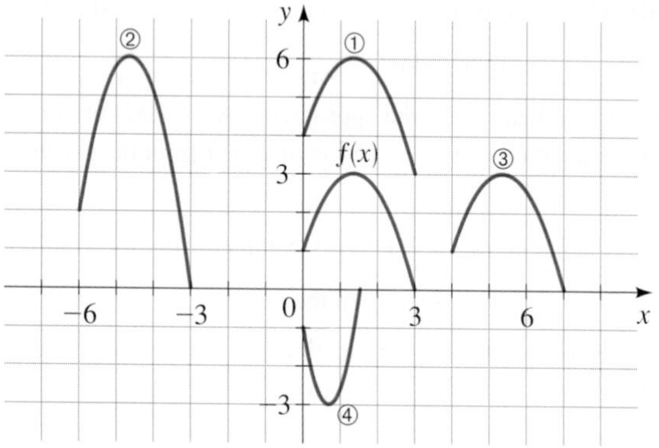

62. **(a)** $y = \frac{1}{3}f(x)$ **(b)** $y = -f(x + 4)$
(c) $y = f(x - 4) + 3$ **(d)** $y = f(-x)$

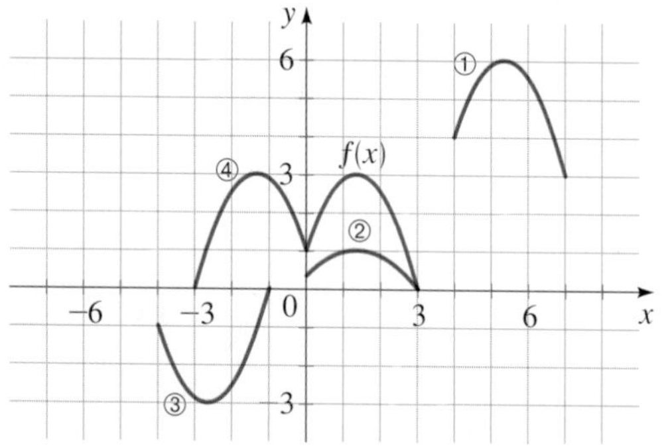

63. The graph of f is given. Sketch the graphs of the following functions.
(a) $y = f(x - 2)$ **(b)** $y = f(x) - 2$
(c) $y = 2f(x)$ **(d)** $y = -f(x) + 3$
(e) $y = f(-x)$ **(f)** $y = \frac{1}{2}f(x - 1)$

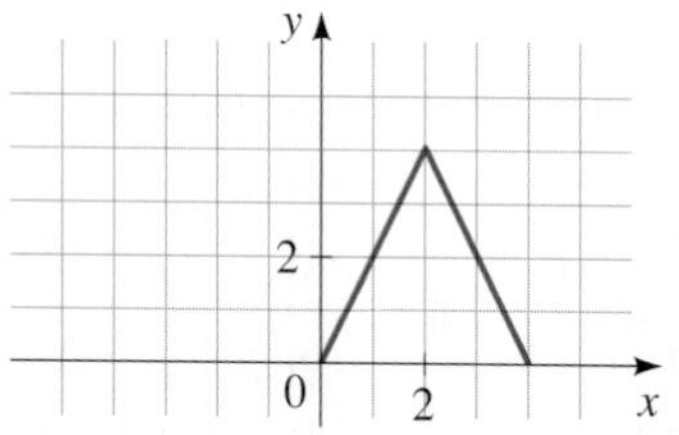

64. The graph of g is given. Sketch the graphs of the following functions.
(a) $y = g(x + 1)$ **(b)** $y = g(-x)$
(c) $y = g(x - 2)$ **(d)** $y = g(x) - 2$
(e) $y = -g(x)$ **(f)** $y = 2g(x)$

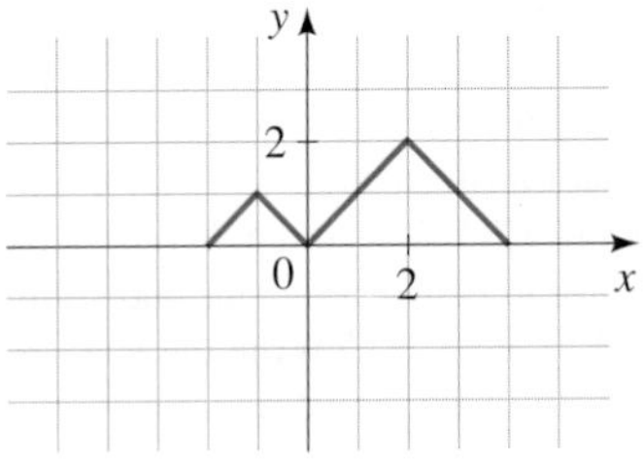

65. The graph of g is given. Use it to graph each of the following functions.
(a) $y = g(2x)$ **(b)** $y = g(\frac{1}{2}x)$

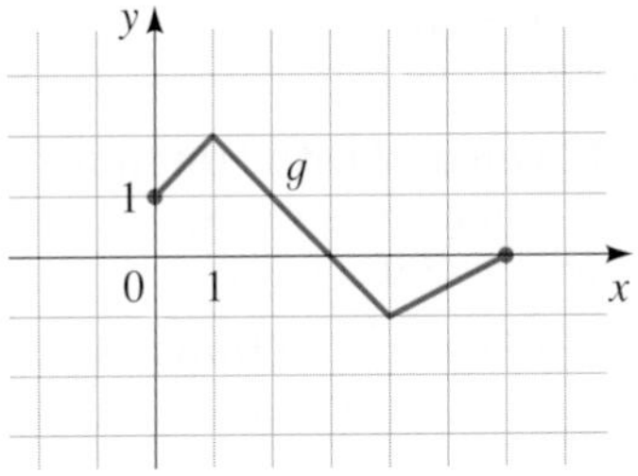

66. The graph of h is given. Use it to graph each of the following functions.
(a) $y = h(3x)$ **(b)** $y = h(\frac{1}{3}x)$

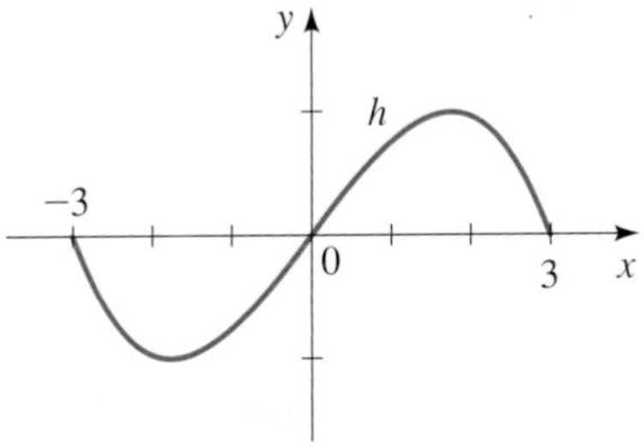

67–68 ■ Use the graph of $f(x) = [\![x]\!]$ described on page 156 to graph the indicated function.

67. $y = [\![2x]\!]$ **68.** $y = [\![\frac{1}{4}x]\!]$

69–72 ■ Graph the functions on the same screen using the given viewing rectangle. How is each graph related to the graph in part (a)?

69. Viewing rectangle $[-8, 8]$ by $[-2, 8]$
(a) $y = \sqrt[4]{x}$ **(b)** $y = \sqrt[4]{x + 5}$
(c) $y = 2\sqrt[4]{x + 5}$ **(d)** $y = 4 + 2\sqrt[4]{x + 5}$

70. Viewing rectangle $[-8, 8]$ by $[-6, 6]$
(a) $y = |x|$ **(b)** $y = -|x|$
(c) $y = -3|x|$ **(d)** $y = -3|x - 5|$

71. Viewing rectangle $[-4, 6]$ by $[-4, 4]$
(a) $y = x^6$ **(b)** $y = \frac{1}{3}x^6$
(c) $y = -\frac{1}{3}x^6$ **(d)** $y = -\frac{1}{3}(x - 4)^6$

72. Viewing rectangle $[-6, 6]$ by $[-4, 4]$
(a) $y = \dfrac{1}{\sqrt{x}}$ **(b)** $y = \dfrac{1}{\sqrt{x + 3}}$
(c) $y = \dfrac{1}{2\sqrt{x + 3}}$ **(d)** $y = \dfrac{1}{2\sqrt{x + 3}} - 3$

73. If $f(x) = \sqrt{2x - x^2}$, graph the following functions in the viewing rectangle $[-5, 5]$ by $[-4, 4]$. How is each graph related to the graph in part (a)?
(a) $y = f(x)$ **(b)** $y = f(2x)$ **(c)** $y = f(\frac{1}{2}x)$

74. If $f(x) = \sqrt{2x - x^2}$, graph the following functions in the viewing rectangle $[-5, 5]$ by $[-4, 4]$. How is each graph related to the graph in part (a)?
(a) $y = f(x)$ **(b)** $y = f(-x)$
(c) $y = -f(-x)$ **(d)** $y = f(-2x)$
(e) $y = f(-\frac{1}{2}x)$

75–82 ■ Determine whether the function f is even, odd, or neither. If f is even or odd, use symmetry to sketch its graph.

75. $f(x) = x^4$ **76.** $f(x) = x^3$

77. $f(x) = x^2 + x$ **78.** $f(x) = x^4 - 4x^2$

79. $f(x) = x^3 - x$ **80.** $f(x) = 3x^3 + 2x^2 + 1$

81. $f(x) = 1 - \sqrt[3]{x}$ **82.** $f(x) = x + \dfrac{1}{x}$

83–84 ■ The graph of a function defined for $x \geq 0$ is given. Complete the graph for $x < 0$ to make **(a)** an even function and **(b)** an odd function.

83.

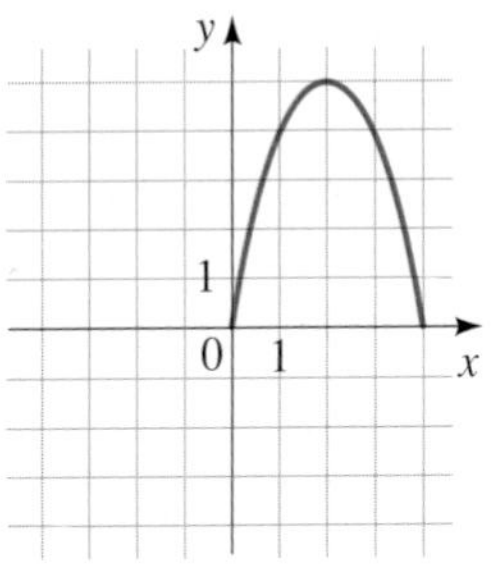

84.

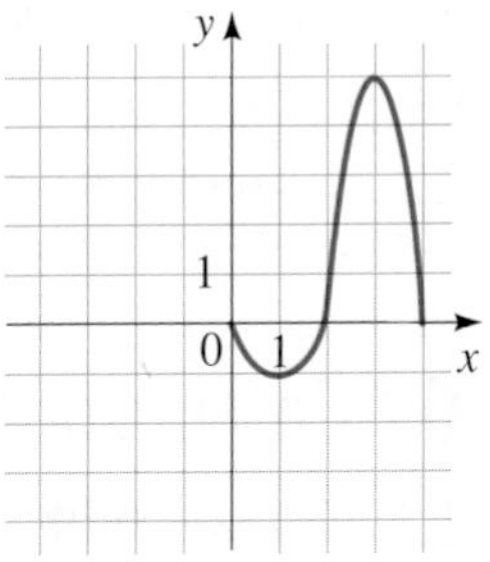

85–86 ■ These exercises show how the graph of $y = |f(x)|$ is obtained from the graph of $y = f(x)$.

85. The graphs of $f(x) = x^2 - 4$ and $g(x) = |x^2 - 4|$ are shown. Explain how the graph of g is obtained from the graph of f.

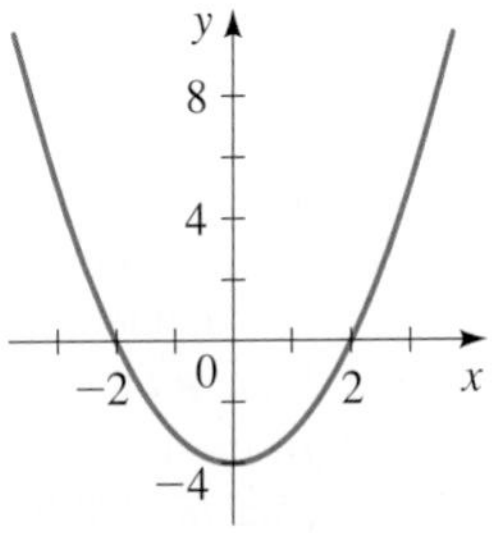

$f(x) = x^2 - 4$

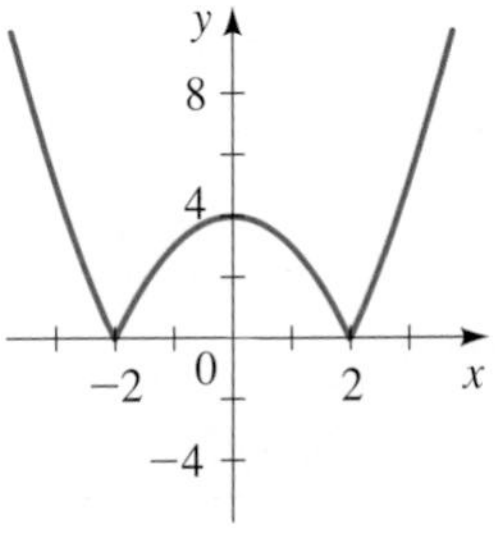

$g(x) = |x^2 - 4|$

86. The graph of $f(x) = x^4 - 4x^2$ is shown. Use this graph to sketch the graph of $g(x) = |x^4 - 4x^2|$.

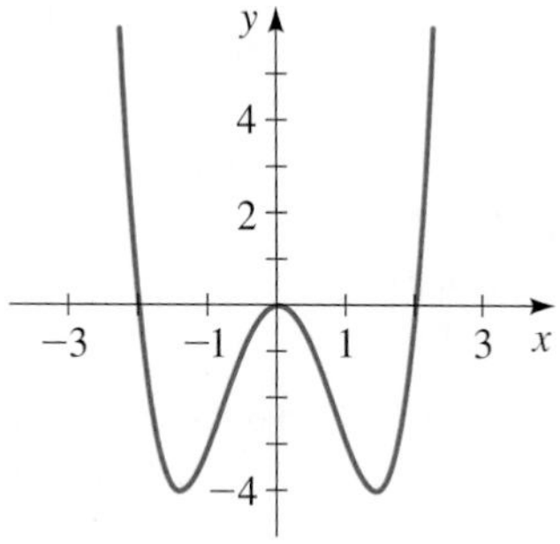

87–88 ■ Sketch the graph of each function.

87. (a) $f(x) = 4x - x^2$ **(b)** $g(x) = |4x - x^2|$

88. (a) $f(x) = x^3$ **(b)** $g(x) = |x^3|$

APPLICATIONS

89. Sales Growth The annual sales of a certain company can be modeled by the function $f(t) = 4 + 0.01t^2$, where t represents years since 1990 and $f(t)$ is measured in millions of dollars.

(a) What shifting and shrinking operations must be performed on the function $y = t^2$ to obtain the function $y = f(t)$?

(b) Suppose you want t to represent years since 2000 instead of 1990. What transformation would you have to apply to the function $y = f(t)$ to accomplish this? Write the new function $y = g(t)$ that results from this transformation.

90. Changing Temperature Scales The temperature on a certain afternoon is modeled by the function

$$C(t) = \tfrac{1}{2}t^2 + 2$$

where t represents hours after 12 noon $(0 \le t \le 6)$ and C is measured in °C.

(a) What shifting and shrinking operations must be performed on the function $y = t^2$ to obtain the function $y = C(t)$?

(b) Suppose you want to measure the temperature in °F instead. What transformation would you have to apply to the function $y = C(t)$ to accomplish this? (Use the fact that the relationship between Celsius and Fahrenheit degrees is given by $F = \tfrac{9}{5}C + 32$.) Write the new function $y = F(t)$ that results from this transformation.

DISCOVERY ■ DISCUSSION ■ WRITING

91. Sums of Even and Odd Functions If f and g are both even functions, is $f + g$ necessarily even? If both are odd, is their sum necessarily odd? What can you say about the sum if one is odd and one is even? In each case, prove your answer.

92. Products of Even and Odd Functions Answer the same questions as in Exercise 91, except this time consider the product of f and g instead of the sum.

93. Even and Odd Power Functions What must be true about the integer n if the function

$$f(x) = x^n$$

is an even function? If it is an odd function? Why do you think the names "even" and "odd" were chosen for these function properties?

2.6 Combining Functions

| Sums, Differences, Products, and Quotients ▶ Composition of Functions

In this section we study different ways to combine functions to make new functions.

▼ Sums, Differences, Products, and Quotients

The sum of f and g is defined by

$$(f + g)(x) = f(x) + g(x)$$

The name of the new function is "$f + g$." So this + sign stands for the operation of addition of *functions*. The + sign on the right side, however, stands for addition of the *numbers* $f(x)$ and $g(x)$.

Two functions f and g can be combined to form new functions $f + g$, $f - g$, fg, and f/g in a manner similar to the way we add, subtract, multiply, and divide real numbers. For example, we define the function $f + g$ by

$$(f + g)(x) = f(x) + g(x)$$

The new function $f + g$ is called the **sum** of the functions f and g; its value at x is $f(x) + g(x)$. Of course, the sum on the right-hand side makes sense only if both $f(x)$ and $g(x)$ are defined, that is, if x belongs to the domain of f and also to the domain of g. So if the domain of f is A and the domain of g is B, then the domain of $f + g$ is the intersection of these domains, that is, $A \cap B$. Similarly, we can define the **difference** $f - g$, the **product** fg, and the **quotient** f/g of the functions f and g. Their domains are $A \cap B$, but in the case of the quotient we must remember not to divide by 0.

ALGEBRA OF FUNCTIONS

Let f and g be functions with domains A and B. Then the functions $f + g$, $f - g$, fg, and f/g are defined as follows.

$$(f + g)(x) = f(x) + g(x) \qquad \text{Domain } A \cap B$$

$$(f - g)(x) = f(x) - g(x) \qquad \text{Domain } A \cap B$$

$$(fg)(x) = f(x)g(x) \qquad \text{Domain } A \cap B$$

$$\left(\frac{f}{g}\right)(x) = \frac{f(x)}{g(x)} \qquad \text{Domain } \{x \in A \cap B \mid g(x) \neq 0\}$$

EXAMPLE 1 | Combinations of Functions and Their Domains

Let $f(x) = \dfrac{1}{x - 2}$ and $g(x) = \sqrt{x}$.

(a) Find the functions $f + g$, $f - g$, fg, and f/g and their domains.

(b) Find $(f + g)(4)$, $(f - g)(4)$, $(fg)(4)$, and $(f/g)(4)$.

SOLUTION

(a) The domain of f is $\{x \mid x \neq 2\}$, and the domain of g is $\{x \mid x \geq 0\}$. The intersection of the domains of f and g is

$$\{x \mid x \geq 0 \text{ and } x \neq 2\} = [0, 2) \cup (2, \infty)$$

Thus, we have

$$(f + g)(x) = f(x) + g(x) = \frac{1}{x - 2} + \sqrt{x} \qquad \text{Domain } \{x \mid x \geq 0 \text{ and } x \neq 2\}$$

$$(f - g)(x) = f(x) - g(x) = \frac{1}{x - 2} - \sqrt{x} \qquad \text{Domain } \{x \mid x \geq 0 \text{ and } x \neq 2\}$$

$$(fg)(x) = f(x)g(x) = \frac{\sqrt{x}}{x - 2} \qquad \text{Domain } \{x \mid x \geq 0 \text{ and } x \neq 2\}$$

$$\left(\frac{f}{g}\right)(x) = \frac{f(x)}{g(x)} = \frac{1}{(x - 2)\sqrt{x}} \qquad \text{Domain } \{x \mid x > 0 \text{ and } x \neq 2\}$$

To divide fractions, invert the denominator and multiply:

$$\frac{1/(x - 2)}{\sqrt{x}} = \frac{1/(x - 2)}{\sqrt{x}/1} = \frac{1}{x - 2} \cdot \frac{1}{\sqrt{x}} = \frac{1}{(x - 2)\sqrt{x}}$$

Note that in the domain of f/g we exclude 0 because $g(0) = 0$.

(b) Each of these values exist because $x = 4$ is in the domain of each function.

$$(f + g)(4) = f(4) + g(4) = \frac{1}{4 - 2} + \sqrt{4} = \frac{5}{2}$$

$$(f - g)(4) = f(4) - g(4) = \frac{1}{4 - 2} - \sqrt{4} = -\frac{3}{2}$$

$$(fg)(4) = f(4)g(4) = \left(\frac{1}{4 - 2}\right)\sqrt{4} = 1$$

$$\left(\frac{f}{g}\right)(4) = \frac{f(4)}{g(4)} = \frac{1}{(4 - 2)\sqrt{4}} = \frac{1}{4}$$

NOW TRY EXERCISE 5 ■

The graph of the function $f + g$ can be obtained from the graphs of f and g by **graphical addition**. This means that we add corresponding y-coordinates, as illustrated in the next example.

EXAMPLE 2 | Using Graphical Addition

The graphs of f and g are shown in Figure 1. Use graphical addition to graph the function $f + g$.

SOLUTION We obtain the graph of $f + g$ by "graphically adding" the value of $f(x)$ to $g(x)$ as shown in Figure 2. This is implemented by copying the line segment PQ on top of PR to obtain the point S on the graph of $f + g$.

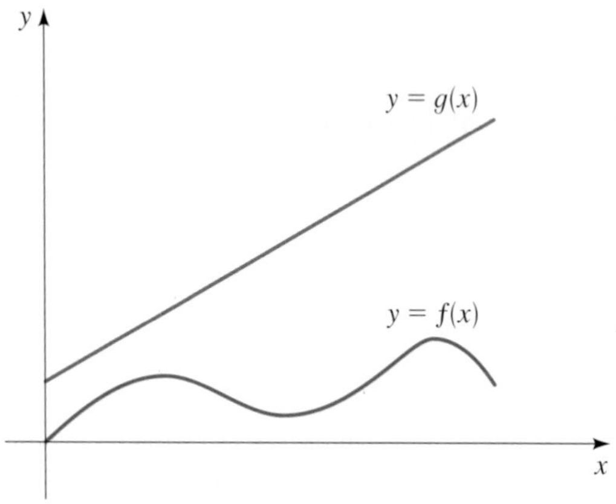

FIGURE 1

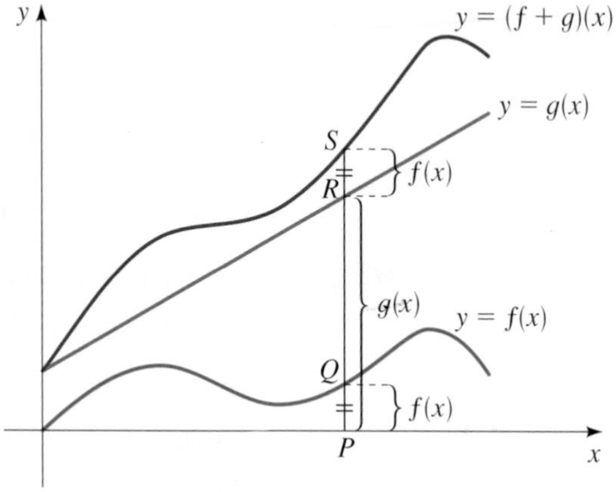

FIGURE 2 Graphical addition

NOW TRY EXERCISE **15** ■

Composition of Functions

Now let's consider a very important way of combining two functions to get a new function. Suppose $f(x) = \sqrt{x}$ and $g(x) = x^2 + 1$. We may define a new function h as

$$h(x) = f(g(x)) = f(x^2 + 1) = \sqrt{x^2 + 1}$$

The function h is made up of the functions f and g in an interesting way: Given a number x, we first apply the function g to it, then apply f to the result. In this case, f is the rule "take the square root," g is the rule "square, then add 1," and h is the rule "square, then add 1, then take the square root." In other words, we get the rule h by applying the rule g and then the rule f. Figure 3 shows a machine diagram for h.

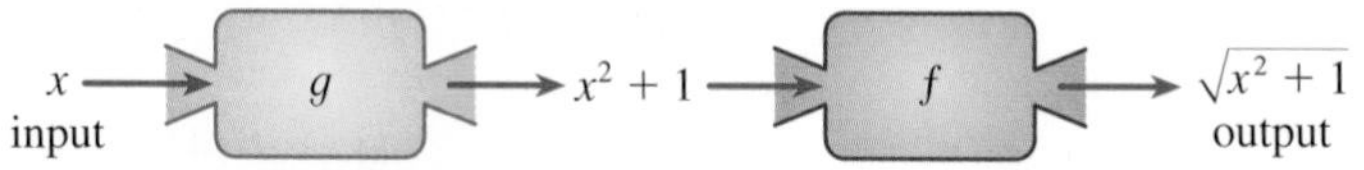

FIGURE 3 The h machine is composed of the g machine (first) and then the f machine.

In general, given any two functions f and g, we start with a number x in the domain of g and find its image $g(x)$. If this number $g(x)$ is in the domain of f, we can then calculate the value of $f(g(x))$. The result is a new function $h(x) = f(g(x))$ that is obtained by substituting g into f. It is called the *composition* (or *composite*) of f and g and is denoted by $f \circ g$ ("f composed with g").

COMPOSITION OF FUNCTIONS

Given two functions f and g, the **composite function** $f \circ g$ (also called the **composition** of f and g) is defined by

$$(f \circ g)(x) = f(g(x))$$

The domain of $f \circ g$ is the set of all x in the domain of g such that $g(x)$ is in the domain of f. In other words, $(f \circ g)(x)$ is defined whenever both $g(x)$ and $f(g(x))$ are defined. We can picture $f \circ g$ using an arrow diagram (Figure 4).

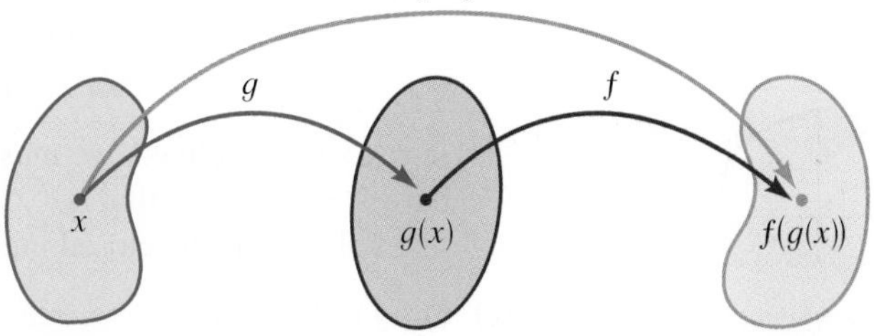

FIGURE 4 Arrow diagram for $f \circ g$

EXAMPLE 3 | Finding the Composition of Functions

Let $f(x) = x^2$ and $g(x) = x - 3$.

(a) Find the functions $f \circ g$ and $g \circ f$ and their domains.

(b) Find $(f \circ g)(5)$ and $(g \circ f)(7)$.

In Example 3, f is the rule "square" and g is the rule "subtract 3." The function $f \circ g$ *first* subtracts 3 and *then* squares; the function $g \circ f$ *first* squares and *then* subtracts 3.

SOLUTION

(a) We have

$$\begin{aligned}(f \circ g)(x) &= f(g(x)) && \text{Definition of } f \circ g\\ &= f(x - 3) && \text{Definition of } g\\ &= (x - 3)^2 && \text{Definition of } f\end{aligned}$$

and

$$\begin{aligned}(g \circ f)(x) &= g(f(x)) && \text{Definition of } g \circ f\\ &= g(x^2) && \text{Definition of } f\\ &= x^2 - 3 && \text{Definition of } g\end{aligned}$$

The domains of both $f \circ g$ and $g \circ f$ are $\mathbb{R}$.

(b) We have

$$(f \circ g)(5) = f(g(5)) = f(2) = 2^2 = 4$$

$$(g \circ f)(7) = g(f(7)) = g(49) = 49 - 3 = 46$$

NOW TRY EXERCISES **21** AND **35** ■

You can see from Example 3 that, in general, $f \circ g \neq g \circ f$. Remember that the notation $f \circ g$ means that the function g is applied first and then f is applied second.

The graphs of f and g of Example 4, as well as those of $f \circ g$, $g \circ f$, $f \circ f$, and $g \circ g$, are shown below. These graphs indicate that the operation of composition can produce functions that are quite different from the original functions.

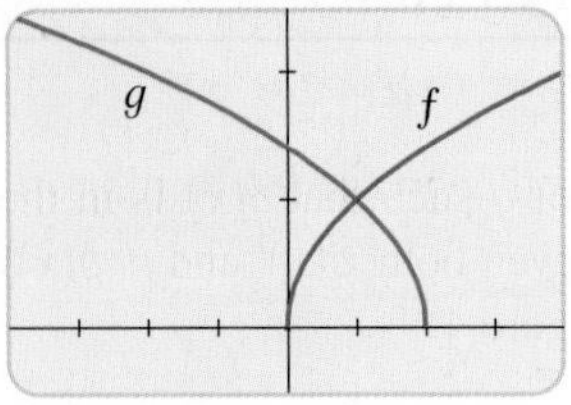

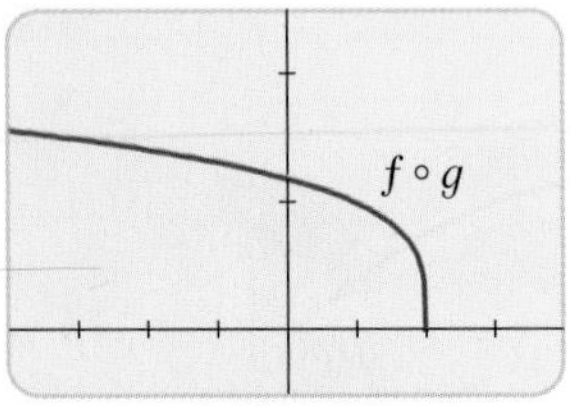

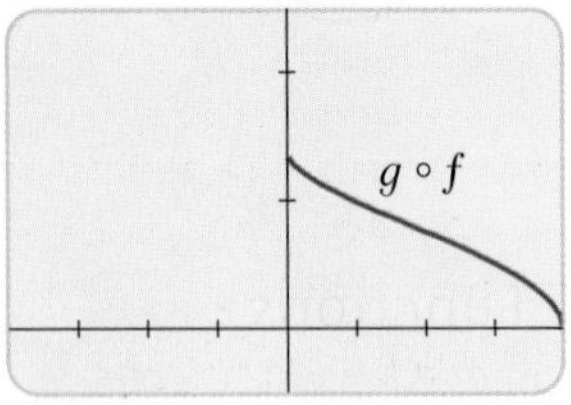

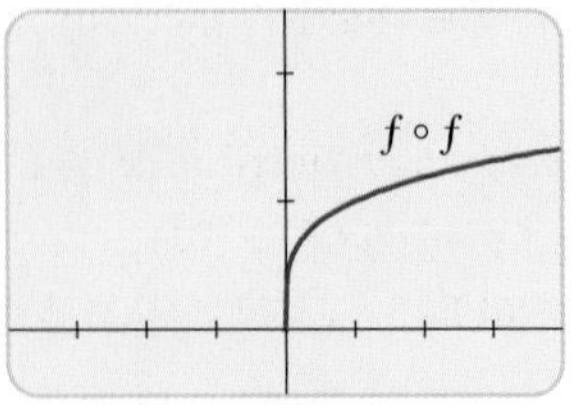

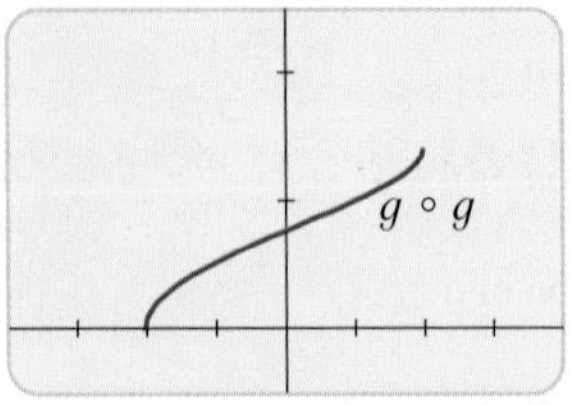

EXAMPLE 4 | Finding the Composition of Functions

If $f(x) = \sqrt{x}$ and $g(x) = \sqrt{2 - x}$, find the following functions and their domains.

(a) $f \circ g$ **(b)** $g \circ f$ **(c)** $f \circ f$ **(d)** $g \circ g$

SOLUTION

(a)

$$\begin{aligned}(f \circ g)(x) &= f(g(x)) && \text{Definition of } f \circ g\\ &= f(\sqrt{2-x}) && \text{Definition of } g\\ &= \sqrt{\sqrt{2-x}} && \text{Definition of } f\\ &= \sqrt[4]{2-x}\end{aligned}$$

The domain of $f \circ g$ is $\{x \mid 2 - x \geq 0\} = \{x \mid x \leq 2\} = (-\infty, 2]$.

(b)

$$\begin{aligned}(g \circ f)(x) &= g(f(x)) && \text{Definition of } g \circ f\\ &= g(\sqrt{x}) && \text{Definition of } f\\ &= \sqrt{2 - \sqrt{x}} && \text{Definition of } g\end{aligned}$$

For $\sqrt{x}$ to be defined, we must have $x \geq 0$. For $\sqrt{2 - \sqrt{x}}$ to be defined, we must have $2 - \sqrt{x} \geq 0$, that is, $\sqrt{x} \leq 2$, or $x \leq 4$. Thus, we have $0 \leq x \leq 4$, so the domain of $g \circ f$ is the closed interval $[0, 4]$.

(c)

$$\begin{aligned}(f \circ f)(x) &= f(f(x)) && \text{Definition of } f \circ f\\ &= f(\sqrt{x}) && \text{Definition of } f\\ &= \sqrt{\sqrt{x}} && \text{Definition of } f\\ &= \sqrt[4]{x}\end{aligned}$$

The domain of $f \circ f$ is $[0, \infty)$.

(d)

$$\begin{aligned}(g \circ g)(x) &= g(g(x)) && \text{Definition of } g \circ g\\ &= g(\sqrt{2-x}) && \text{Definition of } g\\ &= \sqrt{2 - \sqrt{2-x}} && \text{Definition of } g\end{aligned}$$

This expression is defined when both $2 - x \geq 0$ and $2 - \sqrt{2 - x} \geq 0$. The first inequality means $x \leq 2$, and the second is equivalent to $\sqrt{2 - x} \leq 2$, or $2 - x \leq 4$, or $x \geq -2$. Thus, $-2 \leq x \leq 2$, so the domain of $g \circ g$ is $[-2, 2]$.

NOW TRY EXERCISE **41** ■

It is possible to take the composition of three or more functions. For instance, the composite function $f \circ g \circ h$ is found by first applying h, then g, and then f as follows:

$$(f \circ g \circ h)(x) = f(g(h(x)))$$

EXAMPLE 5 | A Composition of Three Functions

Find $f \circ g \circ h$ if $f(x) = x/(x + 1)$, $g(x) = x^{10}$, and $h(x) = x + 3$.

SOLUTION

$$\begin{aligned}(f \circ g \circ h)(x) &= f(g(h(x))) && \text{Definition of } f \circ g \circ h\\ &= f(g(x+3)) && \text{Definition of } h\\ &= f((x+3)^{10}) && \text{Definition of } g\\ &= \frac{(x+3)^{10}}{(x+3)^{10} + 1} && \text{Definition of } f\end{aligned}$$

NOW TRY EXERCISE **45** ■

So far, we have used composition to build complicated functions from simpler ones. But in calculus it is useful to be able to "decompose" a complicated function into simpler ones, as shown in the following example.

EXAMPLE 6 | Recognizing a Composition of Functions

Given $F(x) = \sqrt[4]{x+9}$, find functions f and g such that $F = f \circ g$.

SOLUTION Since the formula for F says to first add 9 and then take the fourth root, we let

$$g(x) = x + 9 \qquad \text{and} \qquad f(x) = \sqrt[4]{x}$$

Then

$$\begin{aligned} (f \circ g)(x) &= f(g(x)) && \text{Definition of } f \circ g \\ &= f(x+9) && \text{Definition of } g \\ &= \sqrt[4]{x+9} && \text{Definition of } f \\ &= F(x) \end{aligned}$$

NOW TRY EXERCISE **49** ■

EXAMPLE 7 | An Application of Composition of Functions

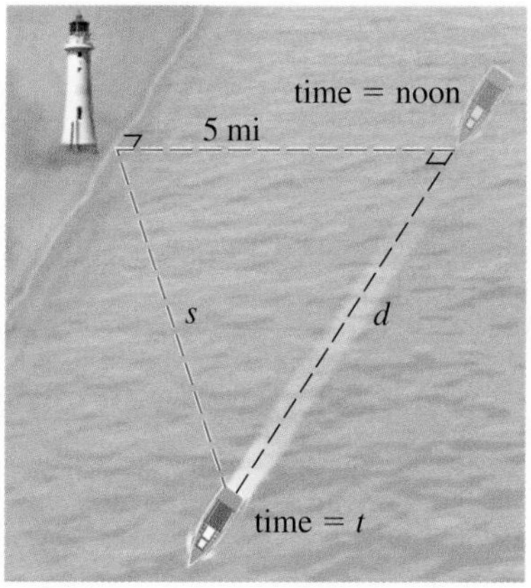

FIGURE 5

A ship is traveling at 20 mi/h parallel to a straight shoreline. The ship is 5 mi from shore. It passes a lighthouse at noon.

(a) Express the distance s between the lighthouse and the ship as a function of d, the distance the ship has traveled since noon; that is, find f so that $s = f(d)$.

(b) Express d as a function of t, the time elapsed since noon; that is, find g so that $d = g(t)$.

(c) Find $f \circ g$. What does this function represent?

SOLUTION We first draw a diagram as in Figure 5.

(a) We can relate the distances s and d by the Pythagorean Theorem. Thus, s can be expressed as a function of d by

$$s = f(d) = \sqrt{25 + d^2}$$

(b) Since the ship is traveling at 20 mi/h, the distance d it has traveled is a function of t as follows:

distance = rate × time

$$d = g(t) = 20t$$

(c) We have

$$\begin{aligned} (f \circ g)(t) &= f(g(t)) && \text{Definition of } f \circ g \\ &= f(20t) && \text{Definition of } g \\ &= \sqrt{25 + (20t)^2} && \text{Definition of } f \end{aligned}$$

The function $f \circ g$ gives the distance of the ship from the lighthouse as a function of time.

NOW TRY EXERCISE **63** ■

2.6 EXERCISES

CONCEPTS

1. From the graphs of f and g in the figure, we find

$(f+g)(2) =$ ______ $\quad (f-g)(2) =$ ______

$(fg)(2) =$ ______ $\quad \left(\dfrac{f}{g}\right)(2) =$ ______

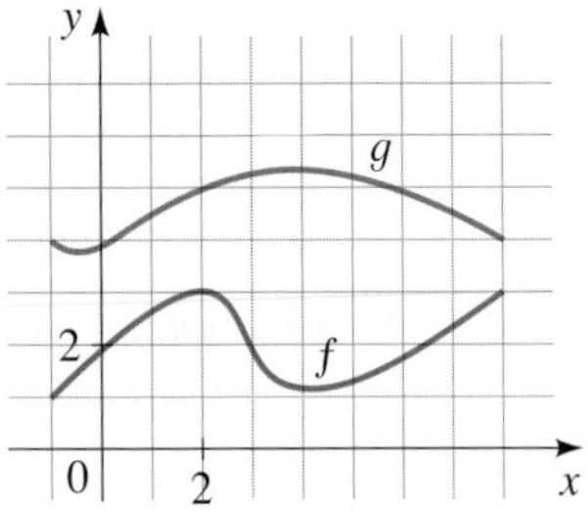

2. By definition, $f \circ g(x) =$ ______. So if $g(2) = 5$ and $f(5) = 12$, then $f \circ g(2) =$ ______.

3. If the rule of the function f is "add one" and the rule of the function g is "multiply by 2," then the rule of $f \circ g$ is

"________________________________"

and the rule of $g \circ f$ is

"________________________________."

4. We can express the functions in Exercise 3 algebraically as

$f(x) =$ ______ $\quad g(x) =$ ______

$f \circ g(x) =$ ______ $\quad g \circ f(x) =$ ______

SKILLS

5–10 ■ Find $f+g$, $f-g$, fg, and f/g and their domains.

5. $f(x) = x - 3, \quad g(x) = x^2$

6. $f(x) = x^2 + 2x, \quad g(x) = 3x^2 - 1$

7. $f(x) = \sqrt{4 - x^2}, \quad g(x) = \sqrt{1 + x}$

8. $f(x) = \sqrt{9 - x^2}, \quad g(x) = \sqrt{x^2 - 4}$

9. $f(x) = \dfrac{2}{x}, \quad g(x) = \dfrac{4}{x + 4}$

10. $f(x) = \dfrac{2}{x + 1}, \quad g(x) = \dfrac{x}{x + 1}$

11–14 ■ Find the domain of the function.

11. $f(x) = \sqrt{x} + \sqrt{1 - x}$

12. $g(x) = \sqrt{x + 1} - \dfrac{1}{x}$

13. $h(x) = (x - 3)^{-1/4}$

14. $k(x) = \dfrac{\sqrt{x + 3}}{x - 1}$

15–16 ■ Use graphical addition to sketch the graph of $f + g$.

15.

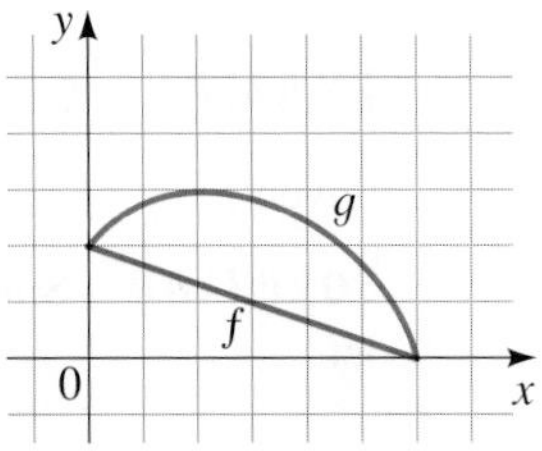

16.

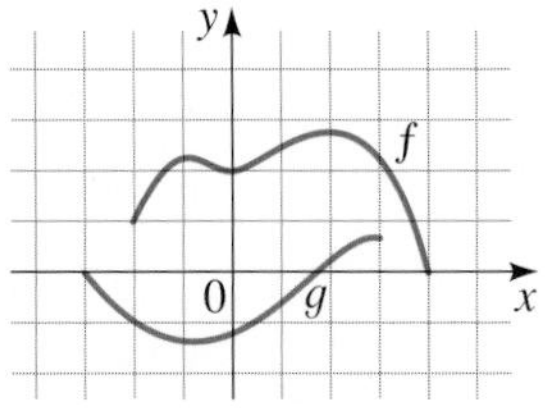

17–20 ■ Draw the graphs of f, g, and $f + g$ on a common screen to illustrate graphical addition.

17. $f(x) = \sqrt{1 + x}, \quad g(x) = \sqrt{1 - x}$

18. $f(x) = x^2, \quad g(x) = \sqrt{x}$

19. $f(x) = x^2, \quad g(x) = \frac{1}{3}x^3$

20. $f(x) = \sqrt[4]{1 - x}, \quad g(x) = \sqrt{1 - \dfrac{x^2}{9}}$

21–26 ■ Use $f(x) = 3x - 5$ and $g(x) = 2 - x^2$ to evaluate the expression.

21. (a) $f(g(0))$ **(b)** $g(f(0))$

22. (a) $f(f(4))$ **(b)** $g(g(3))$

23. (a) $(f \circ g)(-2)$ **(b)** $(g \circ f)(-2)$

24. (a) $(f \circ f)(-1)$ **(b)** $(g \circ g)(2)$

25. (a) $(f \circ g)(x)$ **(b)** $(g \circ f)(x)$

26. (a) $(f \circ f)(x)$ **(b)** $(g \circ g)(x)$

27–32 ■ Use the given graphs of f and g to evaluate the expression.

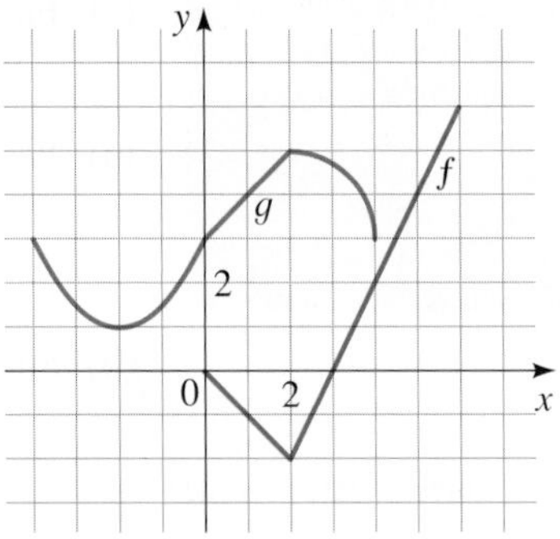

27. $f(g(2))$

28. $g(f(0))$

29. $(g \circ f)(4)$

30. $(f \circ g)(0)$

31. $(g \circ g)(-2)$

32. $(f \circ f)(4)$

33–44 ■ Find the functions $f \circ g$, $g \circ f$, $f \circ f$, and $g \circ g$ and their domains.

33. $f(x) = 2x + 3, \quad g(x) = 4x - 1$

34. $f(x) = 6x - 5, \quad g(x) = \dfrac{x}{2}$

35. $f(x) = x^2, \quad g(x) = x + 1$

36. $f(x) = x^3 + 2, \quad g(x) = \sqrt[3]{x}$

37. $f(x) = \dfrac{1}{x}, \quad g(x) = 2x + 4$

38. $f(x) = x^2, \quad g(x) = \sqrt{x - 3}$

39. $f(x) = |x|, \quad g(x) = 2x + 3$

40. $f(x) = x - 4, \quad g(x) = |x + 4|$

41. $f(x) = \dfrac{x}{x + 1}, \quad g(x) = 2x - 1$

42. $f(x) = \dfrac{1}{\sqrt{x}}, \quad g(x) = x^2 - 4x$

43. $f(x) = \dfrac{x}{x + 1}, \quad g(x) = \dfrac{1}{x}$

44. $f(x) = \dfrac{2}{x}, \quad g(x) = \dfrac{x}{x + 2}$

45–48 ■ Find $f \circ g \circ h$.

45. $f(x) = x - 1, \quad g(x) = \sqrt{x}, \quad h(x) = x - 1$

46. $f(x) = \dfrac{1}{x}, \quad g(x) = x^3, \quad h(x) = x^2 + 2$

47. $f(x) = x^4 + 1, \quad g(x) = x - 5, \quad h(x) = \sqrt{x}$

48. $f(x) = \sqrt{x}, \quad g(x) = \dfrac{x}{x - 1}, \quad h(x) = \sqrt[3]{x}$

49–54 ■ Express the function in the form $f \circ g$.

49. $F(x) = (x - 9)^5$

50. $F(x) = \sqrt{x} + 1$

51. $G(x) = \dfrac{x^2}{x^2 + 4}$

52. $G(x) = \dfrac{1}{x + 3}$

53. $H(x) = |1 - x^3|$

54. $H(x) = \sqrt{1 + \sqrt{x}}$

55–58 ■ Express the function in the form $f \circ g \circ h$.

55. $F(x) = \dfrac{1}{x^2 + 1}$

56. $F(x) = \sqrt[3]{\sqrt{x} - 1}$

57. $G(x) = (4 + \sqrt[3]{x})^9$

58. $G(x) = \dfrac{2}{(3 + \sqrt{x})^2}$

APPLICATIONS

59–60 ■ **Revenue, Cost, and Profit** A print shop makes bumper stickers for election campaigns. If x stickers are ordered (where $x < 10{,}000$), then the price per bumper sticker is $0.15 - 0.000002x$ dollars, and the total cost of producing the order is $0.095x - 0.0000005x^2$ dollars.

59. Use the fact that

revenue = price per item × number of items sold

to express $R(x)$, the revenue from an order of x stickers, as a product of two functions of x.

60. Use the fact that

profit = revenue − cost

to express $P(x)$, the profit on an order of x stickers, as a difference of two functions of x.

61. Area of a Ripple A stone is dropped in a lake, creating a circular ripple that travels outward at a speed of 60 cm/s.

(a) Find a function g that models the radius as a function of time.

(b) Find a function f that models the area of the circle as a function of the radius.

(c) Find $f \circ g$. What does this function represent?

62. Inflating a Balloon A spherical balloon is being inflated. The radius of the balloon is increasing at the rate of 1 cm/s.

(a) Find a function f that models the radius as a function of time.

(b) Find a function g that models the volume as a function of the radius.

(c) Find $g \circ f$. What does this function represent?

63. Area of a Balloon A spherical weather balloon is being inflated. The radius of the balloon is increasing at the rate of 2 cm/s. Express the surface area of the balloon as a function of time t (in seconds).

64. Multiple Discounts You have a \$50 coupon from the manufacturer good for the purchase of a cell phone. The store where you are purchasing your cell phone is offering a 20% discount on all cell phones. Let x represent the regular price of the cell phone.

(a) Suppose only the 20% discount applies. Find a function f that models the purchase price of the cell phone as a function of the regular price x.

(b) Suppose only the \$50 coupon applies. Find a function g that models the purchase price of the cell phone as a function of the sticker price x.

(c) If you can use the coupon and the discount, then the purchase price is either $f \circ g(x)$ or $g \circ f(x)$, depending on the order in which they are applied to the price. Find both $f \circ g(x)$ and $g \circ f(x)$. Which composition gives the lower price?

65. Multiple Discounts An appliance dealer advertises a 10% discount on all his washing machines. In addition, the manufacturer offers a \$100 rebate on the purchase of a washing machine. Let x represent the sticker price of the washing machine.

(a) Suppose only the 10% discount applies. Find a function f that models the purchase price of the washer as a function of the sticker price x.

(b) Suppose only the \$100 rebate applies. Find a function g that models the purchase price of the washer as a function of the sticker price x.

(c) Find $f \circ g$ and $g \circ f$. What do these functions represent? Which is the better deal?

66. Airplane Trajectory An airplane is flying at a speed of 350 mi/h at an altitude of one mile. The plane passes directly above a radar station at time $t = 0$.

(a) Express the distance s (in miles) between the plane and the radar station as a function of the horizontal distance d (in miles) that the plane has flown.

(b) Express d as a function of the time t (in hours) that the plane has flown.

(c) Use composition to express s as a function of t.

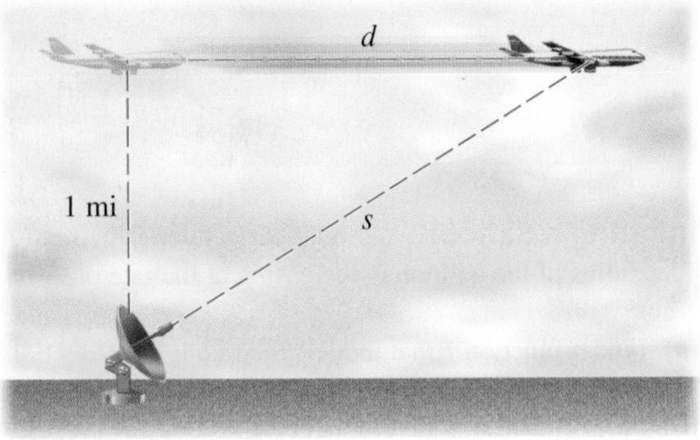

DISCOVERY ■ DISCUSSION ■ WRITING

67. Compound Interest A savings account earns 5% interest compounded annually. If you invest x dollars in such an account, then the amount $A(x)$ of the investment after one year is the initial investment plus 5%; that is,

$$A(x) = x + 0.05x = 1.05x$$

Find

$$A \circ A$$

$$A \circ A \circ A$$

$$A \circ A \circ A \circ A$$

What do these compositions represent? Find a formula for what you get when you compose n copies of A.

68. Composing Linear Functions The graphs of the functions

$$f(x) = m_1 x + b_1$$

$$g(x) = m_2 x + b_2$$

are lines with slopes m_1 and m_2, respectively. Is the graph of $f \circ g$ a line? If so, what is its slope?

69. Solving an Equation for an Unknown Function Suppose that

$$g(x) = 2x + 1$$

$$h(x) = 4x^2 + 4x + 7$$

Find a function f such that $f \circ g = h$. (Think about what operations you would have to perform on the formula for g to end up with the formula for h.) Now suppose that

$$f(x) = 3x + 5$$

$$h(x) = 3x^2 + 3x + 2$$

Use the same sort of reasoning to find a function g such that $f \circ g = h$.

70. Compositions of Odd and Even Functions Suppose that

$$h = f \circ g$$

If g is an even function, is h necessarily even? If g is odd, is h odd? What if g is odd and f is odd? What if g is odd and f is even?

DISCOVERY PROJECT **Iteration and Chaos**

In this project we explore the process of repeatedly composing a function with itself; the result can be regular or chaotic. You can find the project at the book companion website: **www.stewartmath.com**

2.7 One-to-One Functions and Their Inverses

One-to-One Functions ▶ The Inverse of a Function ▶ Graphing the Inverse of a Function

The *inverse* of a function is a rule that acts on the output of the function and produces the corresponding input. So the inverse "undoes" or reverses what the function has done. Not all functions have inverses; those that do are called *one-to-one*.

▼ One-to-One Functions

Let's compare the functions f and g whose arrow diagrams are shown in Figure 1. Note that f never takes on the same value twice (any two numbers in A have different images), whereas g does take on the same value twice (both 2 and 3 have the same image, 4). In symbols, $g(2) = g(3)$ but $f(x_1) \neq f(x_2)$ whenever $x_1 \neq x_2$. Functions that have this latter property are called *one-to-one*.

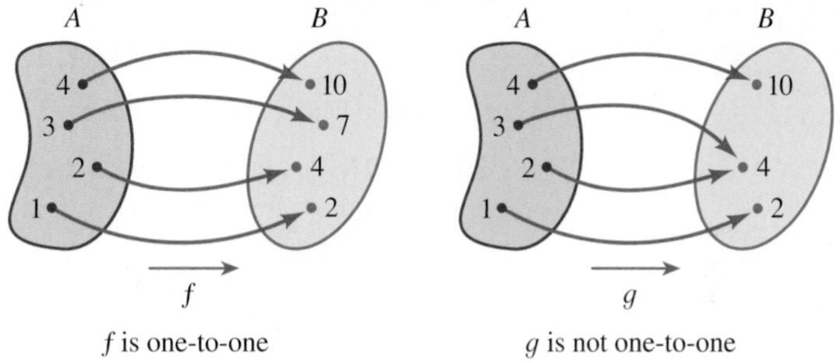

FIGURE 1

> **DEFINITION OF A ONE-TO-ONE FUNCTION**
>
> A function with domain A is called a **one-to-one function** if no two elements of A have the same image, that is,
>
> $$f(x_1) \neq f(x_2) \quad \text{whenever } x_1 \neq x_2$$

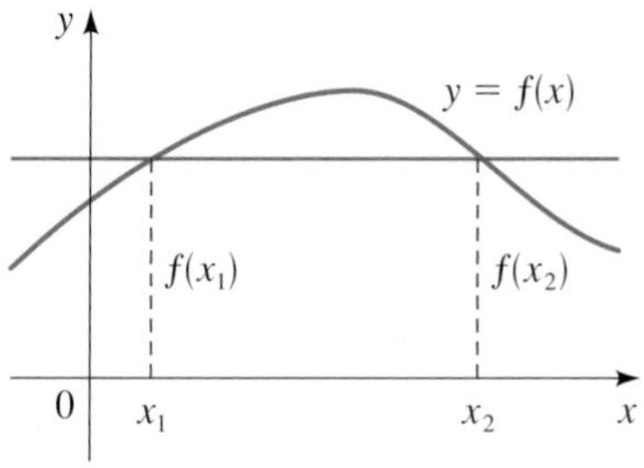

FIGURE 2 This function is not one-to-one because $f(x_1) = f(x_2)$.

An equivalent way of writing the condition for a one-to-one function is this:

$$\text{If } f(x_1) = f(x_2), \text{ then } x_1 = x_2.$$

If a horizontal line intersects the graph of f at more than one point, then we see from Figure 2 that there are numbers $x_1 \neq x_2$ such that $f(x_1) = f(x_2)$. This means that f is not one-to-one. Therefore, we have the following geometric method for determining whether a function is one-to-one.

> **HORIZONTAL LINE TEST**
>
> A function is one-to-one if and only if no horizontal line intersects its graph more than once.

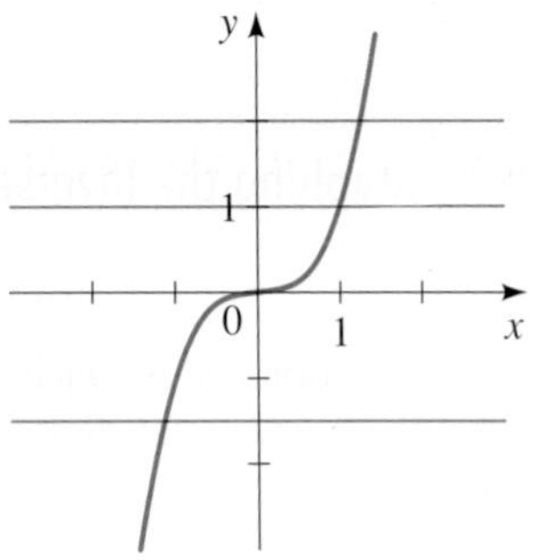

FIGURE 3 $f(x) = x^3$ is one-to-one.

EXAMPLE 1 | Deciding Whether a Function Is One-to-One

Is the function $f(x) = x^3$ one-to-one?

SOLUTION 1 If $x_1 \neq x_2$, then $x_1^3 \neq x_2^3$ (two different numbers cannot have the same cube). Therefore, $f(x) = x^3$ is one-to-one.

SOLUTION 2 From Figure 3 we see that no horizontal line intersects the graph of $f(x) = x^3$ more than once. Therefore, by the Horizontal Line Test, f is one-to-one.

NOW TRY EXERCISE **13** ■

Notice that the function f of Example 1 is increasing and is also one-to-one. In fact, it can be proved that *every increasing function and every decreasing function is one-to-one.*

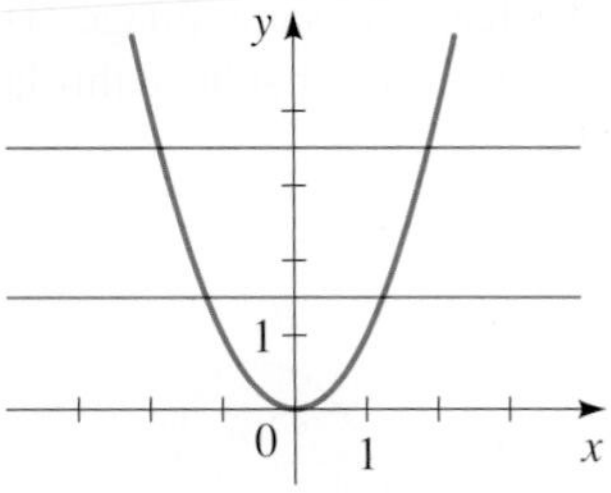

FIGURE 4 $f(x) = x^2$ is not one-to-one.

EXAMPLE 2 | Deciding Whether a Function Is One-to-One

Is the function $g(x) = x^2$ one-to-one?

SOLUTION 1 This function is not one-to-one because, for instance,

$$g(1) = 1 \quad \text{and} \quad g(-1) = 1$$

so 1 and -1 have the same image.

SOLUTION 2 From Figure 4 we see that there are horizontal lines that intersect the graph of g more than once. Therefore, by the Horizontal Line Test, g is not one-to-one.

NOW TRY EXERCISE **15** ■

Although the function g in Example 2 is not one-to-one, it is possible to restrict its domain so that the resulting function is one-to-one. In fact, if we define

$$h(x) = x^2 \qquad x \geq 0$$

then h is one-to-one, as you can see from Figure 5 and the Horizontal Line Test.

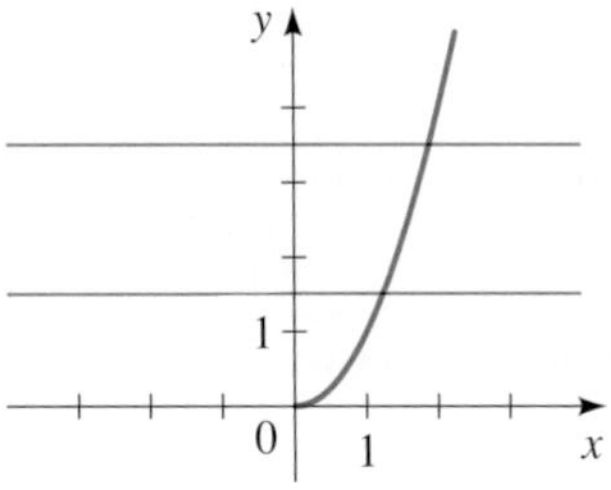

FIGURE 5 $f(x) = x^2\ (x \geq 0)$ is one-to-one.

EXAMPLE 3 | Showing That a Function Is One-to-One

Show that the function $f(x) = 3x + 4$ is one-to-one.

SOLUTION Suppose there are numbers x_1 and x_2 such that $f(x_1) = f(x_2)$. Then

$$\begin{aligned} 3x_1 + 4 &= 3x_2 + 4 && \text{Suppose } f(x_1) = f(x_2) \\ 3x_1 &= 3x_2 && \text{Subtract 4} \\ x_1 &= x_2 && \text{Divide by 3} \end{aligned}$$

Therefore, f is one-to-one.

NOW TRY EXERCISE **11** ■

The Inverse of a Function

One-to-one functions are important because they are precisely the functions that possess inverse functions according to the following definition.

DEFINITION OF THE INVERSE OF A FUNCTION

Let f be a one-to-one function with domain A and range B. Then its **inverse function** f^{-1} has domain B and range A and is defined by

$$f^{-1}(y) = x \quad \Leftrightarrow \quad f(x) = y$$

for any y in B.

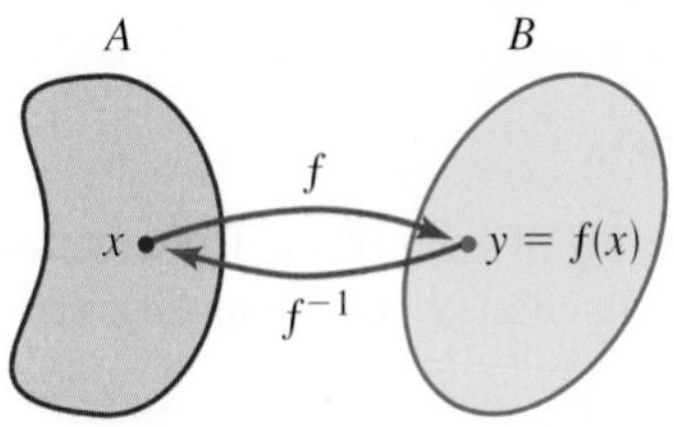

FIGURE 6

This definition says that if f takes x to y, then f^{-1} takes y back to x. (If f were not one-to-one, then f^{-1} would not be defined uniquely.) The arrow diagram in Figure 6 indicates that f^{-1} reverses the effect of f. From the definition we have

$$\text{domain of } f^{-1} = \text{range of } f$$

$$\text{range of } f^{-1} = \text{domain of } f$$

EXAMPLE 4 | Finding f^{-1} for Specific Values

Don't mistake the -1 in f^{-1} for an exponent.

$$f^{-1}(x) \quad \textit{does not mean} \quad \frac{1}{f(x)}$$

The reciprocal $1/f(x)$ is written as $(f(x))^{-1}$.

If $f(1) = 5$, $f(3) = 7$, and $f(8) = -10$, find $f^{-1}(5)$, $f^{-1}(7)$, and $f^{-1}(-10)$.

SOLUTION From the definition of f^{-1} we have

$$f^{-1}(5) = 1 \quad \text{because} \quad f(1) = 5$$

$$f^{-1}(7) = 3 \quad \text{because} \quad f(3) = 7$$

$$f^{-1}(-10) = 8 \quad \text{because} \quad f(8) = -10$$

Figure 7 shows how f^{-1} reverses the effect of f in this case.

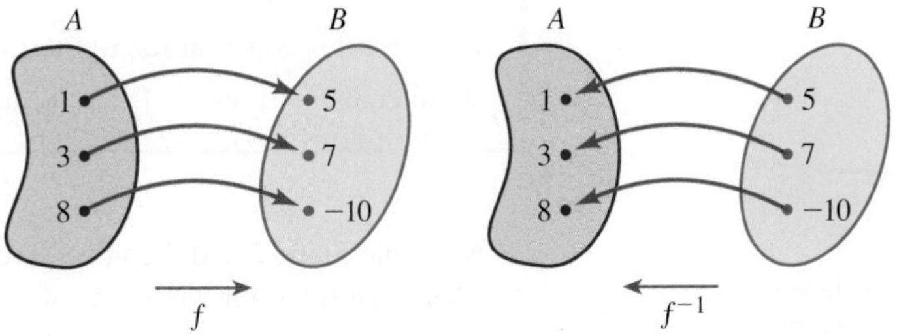

FIGURE 7

NOW TRY EXERCISE **21** ■

By definition the inverse function f^{-1} undoes what f does: If we start with x, apply f, and then apply f^{-1}, we arrive back at x, where we started. Similarly, f undoes what f^{-1} does. In general, any function that reverses the effect of f in this way must be the inverse of f. These observations are expressed precisely as follows.

INVERSE FUNCTION PROPERTY

Let f be a one-to-one function with domain A and range B. The inverse function f^{-1} satisfies the following cancellation properties:

$$f^{-1}(f(x)) = x \quad \text{for every } x \text{ in } A$$

$$f(f^{-1}(x)) = x \quad \text{for every } x \text{ in } B$$

Conversely, any function f^{-1} satisfying these equations is the inverse of f.

These properties indicate that f is the inverse function of f^{-1}, so we say that f and f^{-1} are *inverses of each other.*

EXAMPLE 5 | Verifying That Two Functions Are Inverses

Show that $f(x) = x^3$ and $g(x) = x^{1/3}$ are inverses of each other.

SOLUTION Note that the domain and range of both f and g is $\mathbb{R}$. We have

$$g(f(x)) = g(x^3) = (x^3)^{1/3} = x$$
$$f(g(x)) = f(x^{1/3}) = (x^{1/3})^3 = x$$

So by the Property of Inverse Functions, f and g are inverses of each other. These equations simply say that the cube function and the cube root function, when composed, cancel each other.

NOW TRY EXERCISE **27**

Now let's examine how we compute inverse functions. We first observe from the definition of f^{-1} that

$$y = f(x) \quad \Leftrightarrow \quad f^{-1}(y) = x$$

So if $y = f(x)$ and if we are able to solve this equation for x in terms of y, then we must have $x = f^{-1}(y)$. If we then interchange x and y, we have $y = f^{-1}(x)$, which is the desired equation.

HOW TO FIND THE INVERSE OF A ONE-TO-ONE FUNCTION

1. Write $y = f(x)$.
2. Solve this equation for x in terms of y (if possible).
3. Interchange x and y. The resulting equation is $y = f^{-1}(x)$.

Note that Steps 2 and 3 can be reversed. In other words, we can interchange x and y first and then solve for y in terms of x.

In Example 6 note how f^{-1} reverses the effect of f. The function f is the rule "Multiply by 3, then subtract 2," whereas f^{-1} is the rule "Add 2, then divide by 3."

EXAMPLE 6 | Finding the Inverse of a Function

Find the inverse of the function $f(x) = 3x - 2$.

SOLUTION First we write $y = f(x)$.

$$y = 3x - 2$$

Then we solve this equation for x.

$$3x = y + 2 \qquad \text{Add 2}$$
$$x = \frac{y + 2}{3} \qquad \text{Divide by 3}$$

Finally, we interchange x and y.

$$y = \frac{x + 2}{3}$$

Therefore, the inverse function is $f^{-1}(x) = \dfrac{x + 2}{3}$.

NOW TRY EXERCISE **37**

CHECK YOUR ANSWER

We use the Inverse Function Property.

$$f^{-1}(f(x)) = f^{-1}(3x - 2) = \frac{(3x - 2) + 2}{3} = \frac{3x}{3} = x$$

$$f(f^{-1}(x)) = f\left(\frac{x + 2}{3}\right) = 3\left(\frac{x + 2}{3}\right) - 2 = x + 2 - 2 = x \quad ✔$$

EXAMPLE 7 | Finding the Inverse of a Function

In Example 7 note how f^{-1} reverses the effect of f. The function f is the rule "Take the fifth power, subtract 3, then divide by 2," whereas f^{-1} is the rule "Multiply by 2, add 3, then take the fifth root."

Find the inverse of the function $f(x) = \dfrac{x^5 - 3}{2}$.

SOLUTION We first write $y = (x^5 - 3)/2$ and solve for x.

$y = \dfrac{x^5 - 3}{2}$	Equation defining function
$2y = x^5 - 3$	Multiply by 2
$x^5 = 2y + 3$	Add 3 (and switch sides)
$x = (2y + 3)^{1/5}$	Take fifth root of each side

Then we interchange x and y to get $y = (2x + 3)^{1/5}$. Therefore, the inverse function is $f^{-1}(x) = (2x + 3)^{1/5}$.

NOW TRY EXERCISE **53** ■

CHECK YOUR ANSWER

We use the Inverse Function Property.

$$f^{-1}(f(x)) = f^{-1}\left(\frac{x^5 - 3}{2}\right)$$
$$= \left[2\left(\frac{x^5 - 3}{2}\right) + 3\right]^{1/5}$$
$$= (x^5 - 3 + 3)^{1/5}$$
$$= (x^5)^{1/5} = x$$
$$f(f^{-1}(x)) = f((2x + 3)^{1/5})$$
$$= \frac{[(2x + 3)^{1/5}]^5 - 3}{2}$$
$$= \frac{2x + 3 - 3}{2}$$
$$= \frac{2x}{2} = x \ ✓$$

A **rational function** is a function defined by a rational expression. In the next example we find the inverse of a rational function.

Rational functions are studied in Section 3.7.

EXAMPLE 8 | Finding the Inverse of a Rational Function

Find the inverse of the function $f(x) = \dfrac{2x + 3}{x - 1}$.

SOLUTION We first write $y = (2x + 3)/(x - 1)$ and solve for x.

$y = \dfrac{2x + 3}{x - 1}$	Equation defining function
$y(x - 1) = 2x + 3$	Multiply by $x - 1$
$yx - y = 2x + 3$	Expand
$yx - 2x = y + 3$	Bring x-terms to LHS
$x(y - 2) = y + 3$	Factor x
$x = \dfrac{y + 3}{y - 2}$	Divide by $y - 2$

Therefore the inverse function is $f^{-1}(x) = \dfrac{x + 3}{x - 2}$.

NOW TRY EXERCISE **45** ■

▼ Graphing the Inverse of a Function

The principle of interchanging x and y to find the inverse function also gives us a method for obtaining the graph of f^{-1} from the graph of f. If $f(a) = b$, then $f^{-1}(b) = a$. Thus, the point (a, b) is on the graph of f if and only if the point (b, a) is on the graph of f^{-1}. But we get the point (b, a) from the point (a, b) by reflecting in the line $y = x$ (see Figure 8 on the next page). Therefore, as Figure 9 on the next page illustrates, the following is true.

> The graph of f^{-1} is obtained by reflecting the graph of f in the line $y = x$.

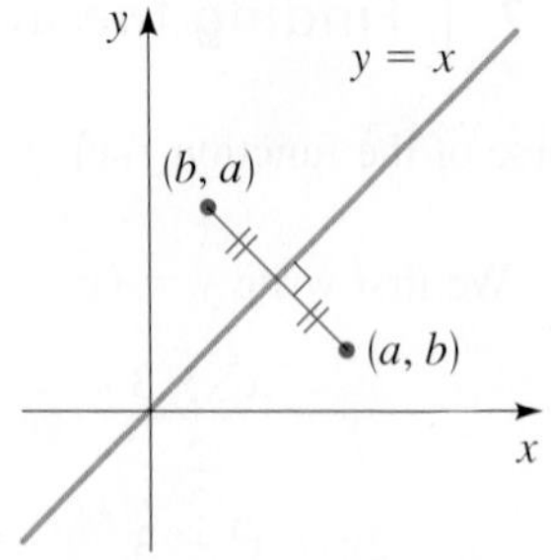

FIGURE 8

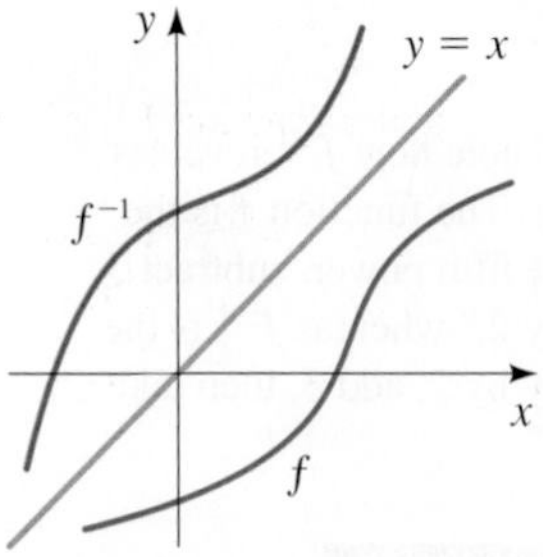

FIGURE 9

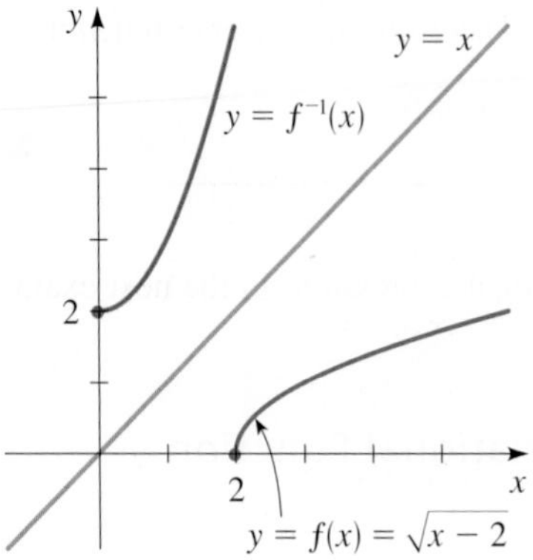

FIGURE 10

EXAMPLE 9 | Graphing the Inverse of a Function

(a) Sketch the graph of $f(x) = \sqrt{x-2}$.

(b) Use the graph of f to sketch the graph of f^{-1}.

(c) Find an equation for f^{-1}.

SOLUTION

(a) Using the transformations from Section 2.5, we sketch the graph of $y = \sqrt{x-2}$ by plotting the graph of the function $y = \sqrt{x}$ (Example 1(c) in Section 2.2) and moving it to the right 2 units.

(b) The graph of f^{-1} is obtained from the graph of f in part (a) by reflecting it in the line $y = x$, as shown in Figure 10.

(c) Solve $y = \sqrt{x-2}$ for x, noting that $y \geq 0$.

$$\sqrt{x-2} = y$$

$$x - 2 = y^2 \qquad \text{Square each side}$$

$$x = y^2 + 2 \quad y \geq 0 \qquad \text{Add 2}$$

Interchange x and y:

$$y = x^2 + 2 \quad x \geq 0$$

Thus $$f^{-1}(x) = x^2 + 2 \quad x \geq 0$$

This expression shows that the graph of f^{-1} is the right half of the parabola $y = x^2 + 2$, and from the graph shown in Figure 10, this seems reasonable.

NOW TRY EXERCISE **63**

In Example 9 note how f^{-1} reverses the effect of f. The function f is the rule "Subtract 2, then take the square root," whereas f^{-1} is the rule "Square, then add 2."

2.7 EXERCISES

CONCEPTS

1. A function f is one-to-one if different inputs produce ________ outputs. You can tell from the graph that a function is one-to-one by using the ________________ Test.

2. **(a)** For a function to have an inverse, it must be ________. So which one of the following functions has an inverse?

$$f(x) = x^2 \qquad g(x) = x^3$$

(b) What is the inverse of the function that you chose in part (a)?

3. A function f has the following verbal description: "Multiply by 3, add 5, and then take the third power of the result."

(a) Write a verbal description for f^{-1}.

(b) Find algebraic formulas that express f and f^{-1} in terms of the input x.

4. *True or false?*

(a) If f has an inverse, then $f^{-1}(x)$ is the same as $\dfrac{1}{f(x)}$.

(b) If f has an inverse, then $f^{-1}(f(x)) = x$.

SKILLS

5–10 ■ The graph of a function f is given. Determine whether f is one-to-one.

5.

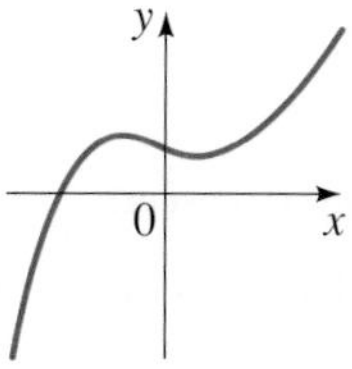

6.

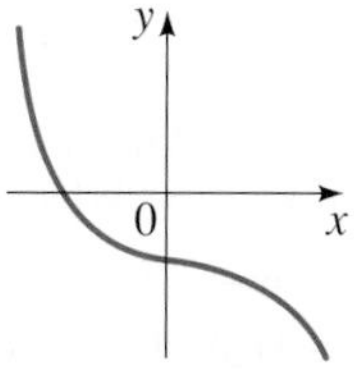

7.

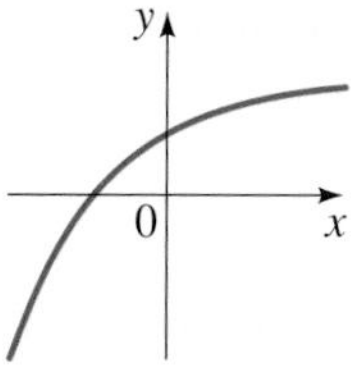

8.

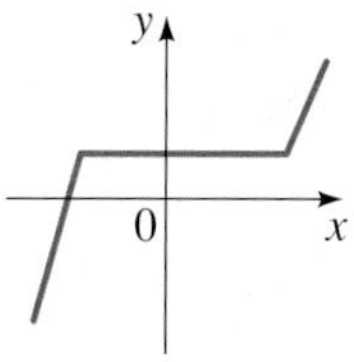

9.

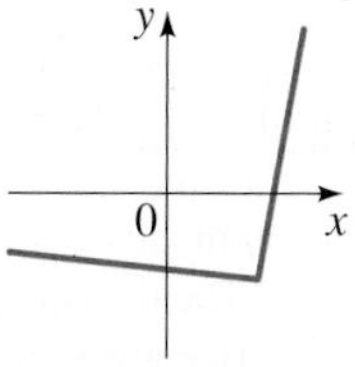

10.

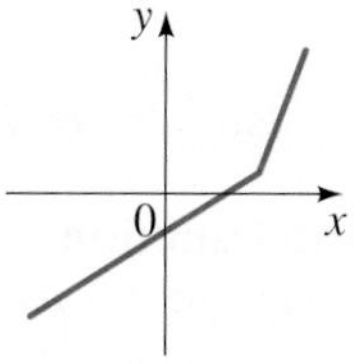

11–20 ■ Determine whether the function is one-to-one.

11. $f(x) = -2x + 4$

12. $f(x) = 3x - 2$

13. $g(x) = \sqrt{x}$

14. $g(x) = |x|$

15. $h(x) = x^2 - 2x$

16. $h(x) = x^3 + 8$

17. $f(x) = x^4 + 5$

18. $f(x) = x^4 + 5, \quad 0 \le x \le 2$

19. $f(x) = \dfrac{1}{x^2}$

20. $f(x) = \dfrac{1}{x}$

21–22 ■ Assume that f is a one-to-one function.

21. (a) If $f(2) = 7$, find $f^{-1}(7)$.
(b) If $f^{-1}(3) = -1$, find $f(-1)$.

22. (a) If $f(5) = 18$, find $f^{-1}(18)$.
(b) If $f^{-1}(4) = 2$, find $f(2)$.

23. If $f(x) = 5 - 2x$, find $f^{-1}(3)$.

24. If $g(x) = x^2 + 4x$ with $x \ge -2$, find $g^{-1}(5)$.

25–36 ■ Use the Inverse Function Property to show that f and g are inverses of each other.

25. $f(x) = x - 6; \quad g(x) = x + 6$

26. $f(x) = 3x; \quad g(x) = \dfrac{x}{3}$

27. $f(x) = 2x - 5; \quad g(x) = \dfrac{x + 5}{2}$

28. $f(x) = \dfrac{3 - x}{4}; \quad g(x) = 3 - 4x$

29. $f(x) = \dfrac{1}{x}; \quad g(x) = \dfrac{1}{x}$

30. $f(x) = x^5; \quad g(x) = \sqrt[5]{x}$

31. $f(x) = x^2 - 4, \quad x \ge 0;$
$g(x) = \sqrt{x + 4}, \quad x \ge -4$

32. $f(x) = x^3 + 1; \quad g(x) = (x - 1)^{1/3}$

33. $f(x) = \dfrac{1}{x - 1}, \quad x \ne 1; \quad g(x) = \dfrac{1}{x} + 1, \quad x \ne 0$

34. $f(x) = \sqrt{4 - x^2}, \quad 0 \le x \le 2;$
$g(x) = \sqrt{4 - x^2}, \quad 0 \le x \le 2$

35. $f(x) = \dfrac{x + 2}{x - 2}; \quad g(x) = \dfrac{2x + 2}{x - 1}$

36. $f(x) = \dfrac{x - 5}{3x + 4}; \quad g(x) = \dfrac{5 + 4x}{1 - 3x}$

37–60 ■ Find the inverse function of f.

37. $f(x) = 2x + 1$

38. $f(x) = 6 - x$

39. $f(x) = 4x + 7$

40. $f(x) = 3 - 5x$

41. $f(x) = 5 - 4x^3$

42. $f(x) = \dfrac{1}{x^2}, \quad x > 0$

43. $f(x) = \dfrac{1}{x + 2}$

44. $f(x) = \dfrac{x - 2}{x + 2}$

45. $f(x) = \dfrac{x}{x + 4}$

46. $f(x) = \dfrac{3x}{x - 2}$

47. $f(x) = \dfrac{2x + 5}{x - 7}$

48. $f(x) = \dfrac{4x - 2}{3x + 1}$

49. $f(x) = \dfrac{1 + 3x}{5 - 2x}$

50. $f(x) = \dfrac{2x - 1}{x - 3}$

51. $f(x) = \sqrt{2 + 5x}$

52. $f(x) = x^2 + x, \quad x \ge -\frac{1}{2}$

53. $f(x) = 4 - x^2, \quad x \ge 0$

54. $f(x) = \sqrt{2x - 1}$

55. $f(x) = 4 + \sqrt[3]{x}$

56. $f(x) = (2 - x^3)^5$

57. $f(x) = 1 + \sqrt{1 + x}$

58. $f(x) = \sqrt{9 - x^2}, \quad 0 \le x \le 3$

59. $f(x) = x^4, \quad x \ge 0$

60. $f(x) = 1 - x^3$

61–64 ■ A function f is given. **(a)** Sketch the graph of f. **(b)** Use the graph of f to sketch the graph of f^{-1}. **(c)** Find f^{-1}.

61. $f(x) = 3x - 6$

62. $f(x) = 16 - x^2, \quad x \ge 0$

63. $f(x) = \sqrt{x + 1}$

64. $f(x) = x^3 - 1$

65–70 ■ Draw the graph of f and use it to determine whether the function is one-to-one.

65. $f(x) = x^3 - x$

66. $f(x) = x^3 + x$

67. $f(x) = \dfrac{x + 12}{x - 6}$

68. $f(x) = \sqrt{x^3 - 4x + 1}$

69. $f(x) = |x| - |x - 6|$

70. $f(x) = x \cdot |x|$

71–74 ■ A one-to-one function is given. **(a)** Find the inverse of the function. **(b)** Graph both the function and its inverse on the same screen to verify that the graphs are reflections of each other in the line $y = x$.

71. $f(x) = 2 + x$

72. $f(x) = 2 - \frac{1}{2}x$

73. $g(x) = \sqrt{x + 3}$

74. $g(x) = x^2 + 1, \quad x \geq 0$

75–78 ■ The given function is not one-to-one. Restrict its domain so that the resulting function *is* one-to-one. Find the inverse of the function with the restricted domain. (There is more than one correct answer.)

75. $f(x) = 4 - x^2$

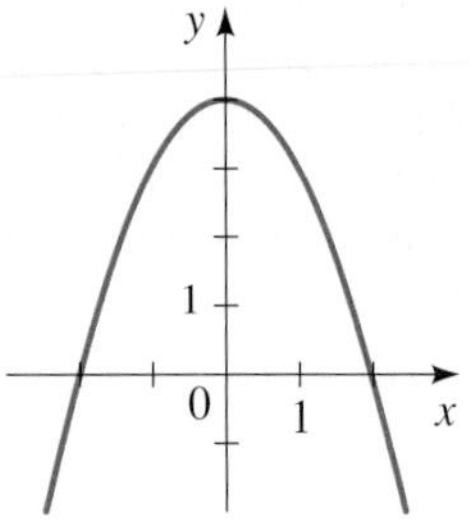

76. $g(x) = (x - 1)^2$

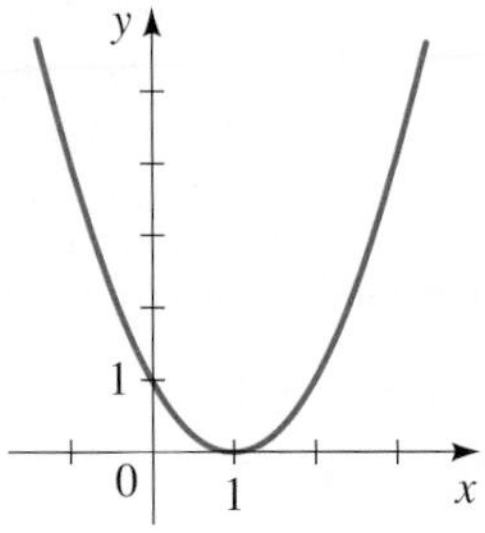

77. $h(x) = (x + 2)^2$

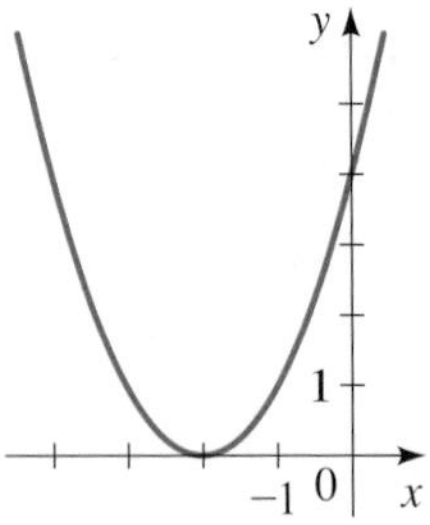

78. $k(x) = |x - 3|$

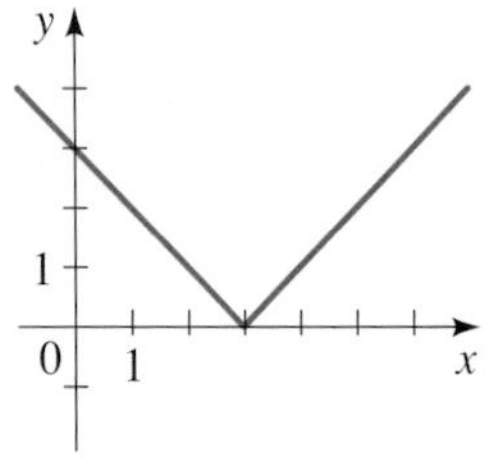

79–80 ■ Use the graph of f to sketch the graph of f^{-1}.

79.

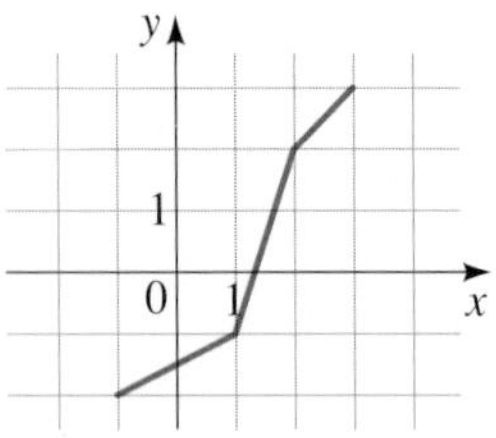

80.

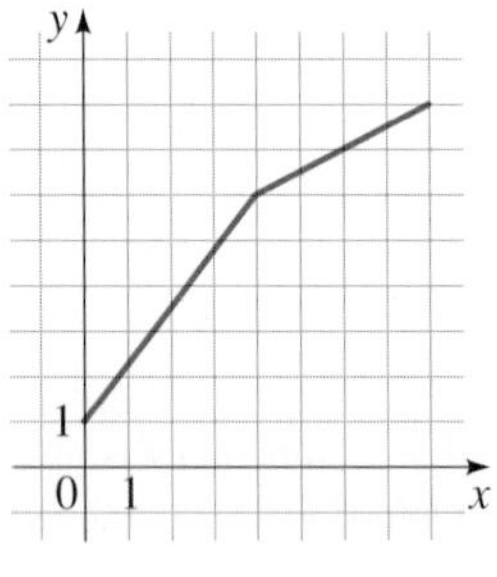

APPLICATIONS

81. Fee for Service For his services, a private investigator requires a \$500 retention fee plus \$80 per hour. Let x represent the number of hours the investigator spends working on a case.

(a) Find a function f that models the investigator's fee as a function of x.

(b) Find f^{-1}. What does f^{-1} represent?

(c) Find $f^{-1}(1220)$. What does your answer represent?

82. Toricelli's Law A tank holds 100 gallons of water, which drains from a leak at the bottom, causing the tank to empty in 40 minutes. Toricelli's Law gives the volume of water remaining in the tank after t minutes as

$$V(t) = 100\left(1 - \frac{t}{40}\right)^2$$

(a) Find V^{-1}. What does V^{-1} represent?

(b) Find $V^{-1}(15)$. What does your answer represent?

83. Blood Flow As blood moves through a vein or artery, its velocity v is greatest along the central axis and decreases as the distance r from the central axis increases (see the figure below). For an artery with radius 0.5 cm, v (in cm/s) is given as a function of r (in cm) by

$$v(r) = 18{,}500(0.25 - r^2)$$

(a) Find v^{-1}. What does v^{-1} represent?

(b) Find $v^{-1}(30)$. What does your answer represent?

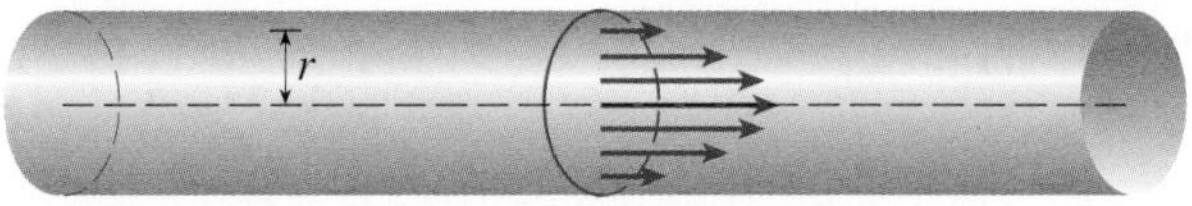

84. Demand Function The amount of a commodity that is sold is called the *demand* for the commodity. The demand D for a certain commodity is a function of the price given by

$$D(p) = -3p + 150$$

(a) Find D^{-1}. What does D^{-1} represent?

(b) Find $D^{-1}(30)$. What does your answer represent?

85. Temperature Scales The relationship between the Fahrenheit (F) and Celsius (C) scales is given by

$$F(C) = \tfrac{9}{5}C + 32$$

(a) Find F^{-1}. What does F^{-1} represent?

(b) Find $F^{-1}(86)$. What does your answer represent?

86. Exchange Rates The relative value of currencies fluctuates every day. When this problem was written, one Canadian dollar was worth 1.0573 U.S. dollar.

(a) Find a function f that gives the U.S. dollar value $f(x)$ of x Canadian dollars.

(b) Find f^{-1}. What does f^{-1} represent?

(c) How much Canadian money would \$12,250 in U.S. currency be worth?

87. Income Tax In a certain country, the tax on incomes less than or equal to €20,000 is 10%. For incomes that are more than €20,000, the tax is €2000 plus 20% of the amount over €20,000.

(a) Find a function f that gives the income tax on an income x. Express f as a piecewise defined function.

(b) Find f^{-1}. What does f^{-1} represent?

(c) How much income would require paying a tax of €10,000?

88. Multiple Discounts A car dealership advertises a 15% discount on all its new cars. In addition, the manufacturer offers a \$1000 rebate on the purchase of a new car. Let x represent the sticker price of the car.

(a) Suppose only the 15% discount applies. Find a function f that models the purchase price of the car as a function of the sticker price x.

(b) Suppose only the \$1000 rebate applies. Find a function g that models the purchase price of the car as a function of the sticker price x.

(c) Find a formula for $H = f \circ g$.

(d) Find H^{-1}. What does H^{-1} represent?

(e) Find $H^{-1}(13{,}000)$. What does your answer represent?

89. Pizza Cost Marcello's Pizza charges a base price of \$7 for a large pizza plus \$2 for each topping. Thus, if you order a large pizza with x toppings, the price of your pizza is given by the function $f(x) = 7 + 2x$. Find f^{-1}. What does the function f^{-1} represent?

DISCOVERY ■ DISCUSSION ■ WRITING

90. Determining When a Linear Function Has an Inverse For the linear function $f(x) = mx + b$ to be one-to-one, what must be true about its slope? If it is one-to-one, find its inverse. Is the inverse linear? If so, what is its slope?

91. Finding an Inverse "in Your Head" In the margin notes in this section we pointed out that the inverse of a function can be found by simply reversing the operations that make up the function. For instance, in Example 6 we saw that the inverse of

$$f(x) = 3x - 2 \quad \text{is} \quad f^{-1}(x) = \frac{x+2}{3}$$

because the "reverse" of "Multiply by 3 and subtract 2" is "Add 2 and divide by 3." Use the same procedure to find the inverse of the following functions.

(a) $f(x) = \dfrac{2x+1}{5}$ **(b)** $f(x) = 3 - \dfrac{1}{x}$

(c) $f(x) = \sqrt{x^3 + 2}$ **(d)** $f(x) = (2x - 5)^3$

Now consider another function:

$$f(x) = x^3 + 2x + 6$$

Is it possible to use the same sort of simple reversal of operations to find the inverse of this function? If so, do it. If not, explain what is different about this function that makes this task difficult.

92. The Identity Function The function $I(x) = x$ is called the **identity function**. Show that for any function f we have $f \circ I = f$, $I \circ f = f$, and $f \circ f^{-1} = f^{-1} \circ f = I$. (This means that the identity function I behaves for functions and composition just the way the number 1 behaves for real numbers and multiplication.)

93. Solving an Equation for an Unknown Function In Exercise 69 of Section 2.6 you were asked to solve equations in which the unknowns were functions. Now that we know about inverses and the identity function (see Exercise 92), we can use algebra to solve such equations. For instance, to solve $f \circ g = h$ for the unknown function f, we perform the following steps:

$f \circ g = h$	Problem: Solve for f
$f \circ g \circ g^{-1} = h \circ g^{-1}$	Compose with g^{-1} on the right
$f \circ I = h \circ g^{-1}$	Because $g \circ g^{-1} = I$
$f = h \circ g^{-1}$	Because $f \circ I = f$

So the solution is $f = h \circ g^{-1}$. Use this technique to solve the equation $f \circ g = h$ for the indicated unknown function.

(a) Solve for f, where $g(x) = 2x + 1$ and $h(x) = 4x^2 + 4x + 7$.

(b) Solve for g, where $f(x) = 3x + 5$ and $h(x) = 3x^2 + 3x + 2$.

CHAPTER 2 | REVIEW

■ CONCEPT CHECK

1. Define each concept in your own words. (Check by referring to the definition in the text.)

(a) Function

(b) Domain and range of a function

(c) Graph of a function

(d) Independent and dependent variables

2. Sketch by hand, on the same axes, the graphs of the following functions.

(a) $f(x) = x$ **(b)** $g(x) = x^2$

(c) $h(x) = x^3$ **(d)** $j(x) = x^4$

3. (a) State the Vertical Line Test.

(b) State the Horizontal Line Test.

4. How is the average rate of change of the function f between two points defined?

5. What can you say about the average rate of change of a linear function?

6. Define each concept in your own words.

(a) Increasing function

(b) Decreasing function

(c) Constant function

7. Suppose the graph of f is given. Write an equation for each graph that is obtained from the graph of f as follows.

(a) Shift 3 units upward.

(b) Shift 3 units downward.

(c) Shift 3 units to the right.

(d) Shift 3 units to the left.

(e) Reflect in the x-axis.

(f) Reflect in the y-axis.

(g) Stretch vertically by a factor of 3.

(h) Shrink vertically by a factor of $\frac{1}{3}$.

(i) Stretch horizontally by a factor of 2.

(j) Shrink horizontally by a factor of $\frac{1}{2}$.

8. **(a)** What is an even function? What symmetry does its graph possess? Give an example of an even function.

(b) What is an odd function? What symmetry does its graph possess? Give an example of an odd function.

9. What does it mean to say that $f(3)$ is a local maximum value of f?

10. Suppose that f has domain A and g has domain B.

(a) What is the domain of $f + g$?

(b) What is the domain of fg?

(c) What is the domain of f/g?

11. How is the composite function $f \circ g$ defined?

12. **(a)** What is a one-to-one function?

(b) How can you tell from the graph of a function whether it is one-to-one?

(c) Suppose f is a one-to-one function with domain A and range B. How is the inverse function f^{-1} defined? What is the domain of f^{-1}? What is the range of f^{-1}?

(d) If you are given a formula for f, how do you find a formula for f^{-1}?

(e) If you are given the graph of f, how do you find the graph of f^{-1}?

■ EXERCISES

1–2 ■ A verbal description of a function f is given. Find a formula that expresses f in function notation.

1. "Square, then subtract 5."

2. "Divide by 2, then add 9."

3–4 ■ A formula for a function f is given. Give a verbal description of the function.

3. $f(x) = 3(x + 10)$

4. $f(x) = \sqrt{6x - 10}$

5–6 ■ Complete the table of values for the given function.

5. $g(x) = x^2 - 4x$

x	$g(x)$
−1	
0	
1	
2	
3	

6. $h(x) = 3x^2 + 2x - 5$

x	$h(x)$
−2	
−1	
0	
1	
2	

7. A publisher estimates that the cost $C(x)$ of printing a run of x copies of a certain mathematics textbook is given by the function $C(x) = 5000 + 30x - 0.001x^2$.
(a) Find $C(1000)$ and $C(10{,}000)$.
(b) What do your answers in part (a) represent?
(c) Find $C(0)$. What does this number represent?

8. Reynalda works as a salesperson in the electronics division of a department store. She earns a base weekly salary plus a commission based on the retail price of the goods she has sold. If she sells x dollars worth of goods in a week, her earnings for that week are given by the function $E(x) = 400 + 0.03x$.
(a) Find $E(2000)$ and $E(15{,}000)$.
(b) What do your answers in part (a) represent?
(c) Find $E(0)$. What does this number represent?
(d) From the formula for E, determine what percentage Reynalda earns on the goods that she sells.

9. If $f(x) = x^2 - 4x + 6$, find $f(0)$, $f(2)$, $f(-2)$, $f(a)$, $f(-a)$, $f(x + 1)$, $f(2x)$, and $2f(x) - 2$.

10. If $f(x) = 4 - \sqrt{3x - 6}$, find $f(5)$, $f(9)$, $f(a + 2)$, $f(-x)$, $f(x^2)$, and $[f(x)]^2$.

11. Which of the following figures are graphs of functions? Which of the functions are one-to-one?

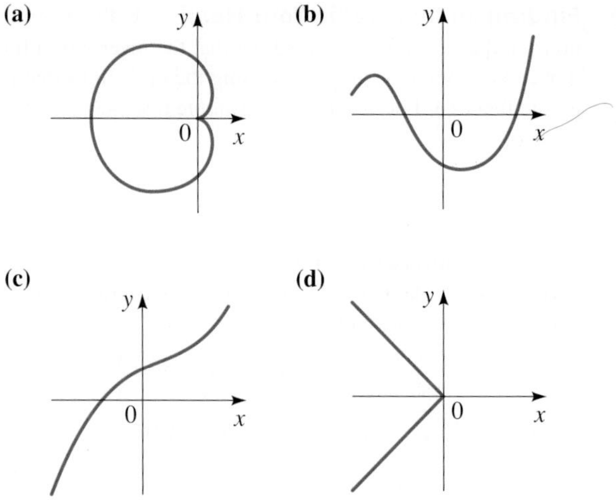

12. The graph of a function f is given.
(a) Find $f(-2)$ and $f(2)$.
(b) Find the domain of f.
(c) Find the range of f.
(d) On what intervals is f increasing? On what intervals is f decreasing?
(e) What are the local maximum values of f?
(f) Is f one-to-one?

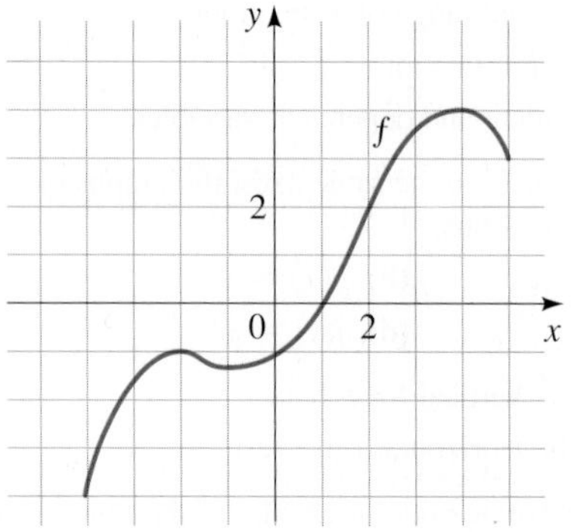

13–14 ■ Find the domain and range of the function.

13. $f(x) = \sqrt{x + 3}$

14. $F(t) = t^2 + 2t + 5$

15–22 ■ Find the domain of the function.

15. $f(x) = 7x + 15$

16. $f(x) = \dfrac{2x + 1}{2x - 1}$

17. $f(x) = \sqrt{x + 4}$

18. $f(x) = 3x - \dfrac{2}{\sqrt{x + 1}}$

19. $f(x) = \dfrac{1}{x} + \dfrac{1}{x + 1} + \dfrac{1}{x + 2}$

20. $g(x) = \dfrac{2x^2 + 5x + 3}{2x^2 - 5x - 3}$

21. $h(x) = \sqrt{4 - x} + \sqrt{x^2 - 1}$

22. $f(x) = \dfrac{\sqrt[3]{2x + 1}}{\sqrt[3]{2x + 2}}$

23–40 ■ Sketch the graph of the function.

23. $f(x) = 1 - 2x$

24. $f(x) = \frac{1}{3}(x - 5),\ 2 \le x \le 8$

25. $f(t) = 1 - \frac{1}{2}t^2$

26. $g(t) = t^2 - 2t$

27. $f(x) = x^2 - 6x + 6$

28. $f(x) = 3 - 8x - 2x^2$

29. $g(x) = 1 - \sqrt{x}$

30. $g(x) = -|x|$

31. $h(x) = \frac{1}{2}x^3$

32. $h(x) = \sqrt{x + 3}$

33. $h(x) = \sqrt[3]{x}$

34. $H(x) = x^3 - 3x^2$

35. $g(x) = \dfrac{1}{x^2}$

36. $G(x) = \dfrac{1}{(x - 3)^2}$

37. $f(x) = \begin{cases} 1 - x & \text{if } x < 0 \\ 1 & \text{if } x \ge 0 \end{cases}$

38. $f(x) = \begin{cases} 1 - 2x & \text{if } x \le 0 \\ 2x - 1 & \text{if } x > 0 \end{cases}$

39. $f(x) = \begin{cases} x + 6 & \text{if } x < -2 \\ x^2 & \text{if } x \ge -2 \end{cases}$

40. $f(x) = \begin{cases} -x & \text{if } x < 0 \\ x^2 & \text{if } 0 \le x < 2 \\ 1 & \text{if } x \ge 2 \end{cases}$

41–44 ■ Determine whether the equation defines y as a function of x.

41. $x + y^2 = 14$

42. $3x - \sqrt{y} = 8$

43. $x^3 - y^3 = 27$

44. $2x = y^4 - 16$

45. Determine which viewing rectangle produces the most appropriate graph of the function

$$f(x) = 6x^3 - 15x^2 + 4x - 1$$

(i) $[-2, 2]$ by $[-2, 2]$ (ii) $[-8, 8]$ by $[-8, 8]$
(iii) $[-4, 4]$ by $[-12, 12]$ (iv) $[-100, 100]$ by $[-100, 100]$

46. Determine which viewing rectangle produces the most appropriate graph of the function $f(x) = \sqrt{100 - x^3}$.
(i) $[-4, 4]$ by $[-4, 4]$
(ii) $[-10, 10]$ by $[-10, 10]$
(iii) $[-10, 10]$ by $[-10, 40]$
(iv) $[-100, 100]$ by $[-100, 100]$

47–50 ■ Draw the graph of the function in an appropriate viewing rectangle.

47. $f(x) = x^2 + 25x + 173$

48. $f(x) = 1.1x^3 - 9.6x^2 - 1.4x + 3.2$

49. $f(x) = \dfrac{x}{\sqrt{x^2 + 16}}$

50. $f(x) = |x(x + 2)(x + 4)|$

51. Find, approximately, the domain of the function

$$f(x) = \sqrt{x^3 - 4x + 1}$$

52. Find, approximately, the range of the function

$$f(x) = x^4 - x^3 + x^2 + 3x - 6$$

53–54 ■ Draw a graph of the function f, and determine the intervals on which f is increasing and on which f is decreasing.

53. $f(x) = x^3 - 4x^2$

54. $f(x) = |x^4 - 16|$

55–58 ■ Find the average rate of change of the function between the given points.

55. $f(x) = x^2 + 3x;\quad x = 0, x = 2$

56. $f(x) = \dfrac{1}{x - 2};\quad x = 4, x = 8$

57. $f(x) = \dfrac{1}{x};\quad x = 3, x = 3 + h$

58. $f(x) = (x + 1)^2;\quad x = a, x = a + h$

59. The population of a planned seaside community in Florida is given by the function $P(t) = 3000 + 200t + 0.1t^2$, where t represents the number of years since the community was incorporated in 1985.

(a) Find $P(10)$ and $P(20)$. What do these values represent?

(b) Find the average rate of change of P between $t = 10$ and $t = 20$. What does this number represent?

60. Ella is saving for her retirement by making regular deposits into a 401(k) plan. As her salary rises, she finds that she can deposit increasing amounts each year. Between 1995 and 2008, the annual amount (in dollars) that she deposited was given by the function $D(t) = 3500 + 15t^2$, where t represents the year of the deposit measured from the start of the plan (so 1995 corresponds to $t = 0$ and 1996 corresponds to $t = 1$, and so on).

(a) Find $D(0)$ and $D(15)$. What do these values represent?

(b) Assuming that her deposits continue to be modeled by the function D, in what year will she deposit \$17,000?

(c) Find the average rate of change of D between $t = 0$ and $t = 15$. What does this number represent?

61–62 ■ A function f is given. **(a)** Find the average rate of change of f between $x = 0$ and $x = 2$, and the average rate of change of f between $x = 15$ and $x = 50$. **(b)** Were the two average rates of change that you found in part (a) the same? Explain why or why not.

61. $f(x) = \frac{1}{2}x - 6$

62. $f(x) = 8 - 3x$

63. Suppose the graph of f is given. Describe how the graphs of the following functions can be obtained from the graph of f.

(a) $y = f(x) + 8$
(b) $y = f(x + 8)$
(c) $y = 1 + 2f(x)$
(d) $y = f(x - 2) - 2$
(e) $y = f(-x)$
(f) $y = -f(-x)$
(g) $y = -f(x)$
(h) $y = f^{-1}(x)$

64. The graph of f is given. Draw the graphs of the following functions.

(a) $y = f(x - 2)$ **(b)** $y = -f(x)$
(c) $y = 3 - f(x)$ **(d)** $y = \frac{1}{2}f(x) - 1$
(e) $y = f^{-1}(x)$ **(f)** $y = f(-x)$

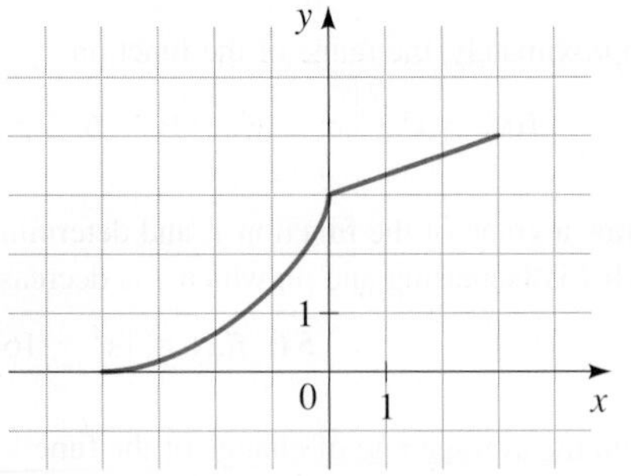

65. Determine whether f is even, odd, or neither.

(a) $f(x) = 2x^5 - 3x^2 + 2$ **(b)** $f(x) = x^3 - x^7$

(c) $f(x) = \dfrac{1 - x^2}{1 + x^2}$ **(d)** $f(x) = \dfrac{1}{x + 2}$

66. Determine whether the function in the figure is even, odd, or neither.

(a)

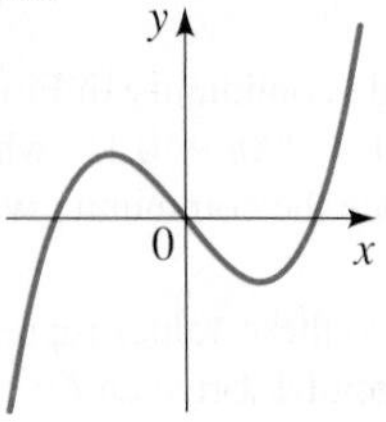

(b)

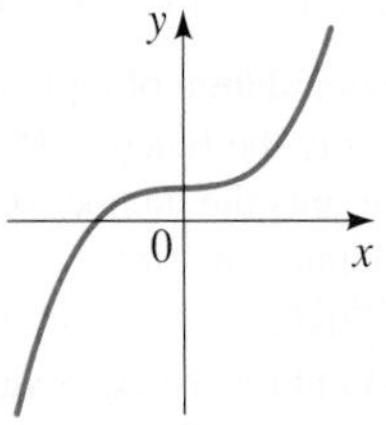

(c)

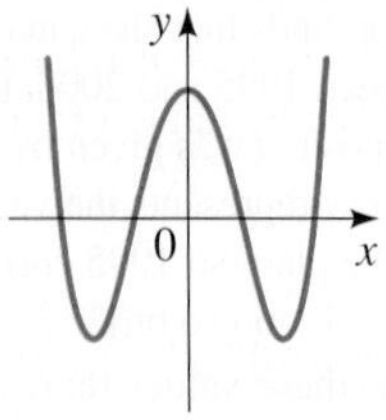

(d)

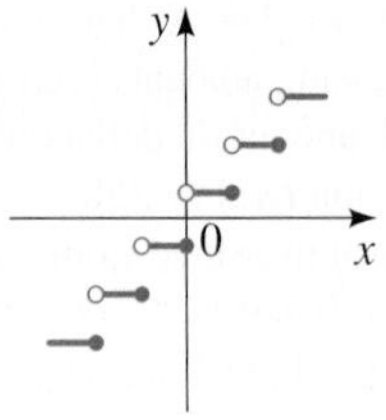

67. Find the minimum value of the function $g(x) = 2x^2 + 4x - 5$.

68. Find the maximum value of the function $f(x) = 1 - x - x^2$.

69. A stone is thrown upward from the top of a building. Its height (in feet) above the ground after t seconds is given by

$$h(t) = -16t^2 + 48t + 32$$

What maximum height does it reach?

70. The profit P (in dollars) generated by selling x units of a certain commodity is given by

$$P(x) = -1500 + 12x - 0.0004x^2$$

What is the maximum profit, and how many units must be sold to generate it?

71–72 ■ Find the local maximum and minimum values of the function and the values of x at which they occur. State each answer correct to two decimal places.

71. $f(x) = 3.3 + 1.6x - 2.5x^3$ **72.** $f(x) = x^{2/3}(6 - x)^{1/3}$

73–74 ■ Two functions, f and g, are given. Draw graphs of f, g, and $f + g$ on the same graphing calculator screen to illustrate the concept of graphical addition.

73. $f(x) = x + 2, \quad g(x) = x^2$

74. $f(x) = x^2 + 1, \quad g(x) = 3 - x^2$

75. If $f(x) = x^2 - 3x + 2$ and $g(x) = 4 - 3x$, find the following functions.

(a) $f + g$ **(b)** $f - g$ **(c)** fg
(d) f/g **(e)** $f \circ g$ **(f)** $g \circ f$

76. If $f(x) = 1 + x^2$ and $g(x) = \sqrt{x - 1}$, find the following.

(a) $f \circ g$ **(b)** $g \circ f$ **(c)** $(f \circ g)(2)$
(d) $(f \circ f)(2)$ **(e)** $f \circ g \circ f$ **(f)** $g \circ f \circ g$

77–78 ■ Find the functions $f \circ g$, $g \circ f$, $f \circ f$, and $g \circ g$ and their domains.

77. $f(x) = 3x - 1, \quad g(x) = 2x - x^2$

78. $f(x) = \sqrt{x}, \quad g(x) = \dfrac{2}{x - 4}$

79. Find $f \circ g \circ h$, where $f(x) = \sqrt{1 - x}$, $g(x) = 1 - x^2$, and $h(x) = 1 + \sqrt{x}$.

80. If $T(x) = \dfrac{1}{\sqrt{1 + \sqrt{x}}}$, find functions f, g, and h such that $f \circ g \circ h = T$.

81–86 ■ Determine whether the function is one-to-one.

81. $f(x) = 3 + x^3$

82. $g(x) = 2 - 2x + x^2$

83. $h(x) = \dfrac{1}{x^4}$

84. $r(x) = 2 + \sqrt{x + 3}$

85. $p(x) = 3.3 + 1.6x - 2.5x^3$

86. $q(x) = 3.3 + 1.6x + 2.5x^3$

87–90 ■ Find the inverse of the function.

87. $f(x) = 3x - 2$ **88.** $f(x) = \dfrac{2x + 1}{3}$

89. $f(x) = (x + 1)^3$ **90.** $f(x) = 1 + \sqrt[5]{x - 2}$

91. **(a)** Sketch the graph of the function

$$f(x) = x^2 - 4 \qquad x \geq 0$$

(b) Use part (a) to sketch the graph of f^{-1}.
(c) Find an equation for f^{-1}.

92. **(a)** Show that the function $f(x) = 1 + \sqrt[4]{x}$ is one-to-one.
(b) Sketch the graph of f.
(c) Use part (b) to sketch the graph of f^{-1}.
(d) Find an equation for f^{-1}.

1. Which of the following are graphs of functions? If the graph is that of a function, is it one-to-one?

(a)

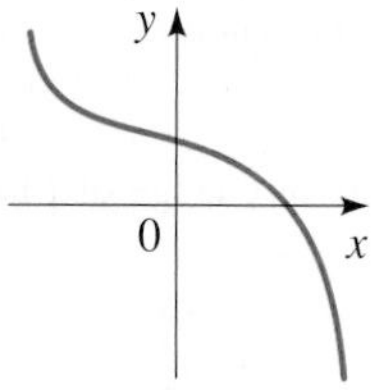

(b)

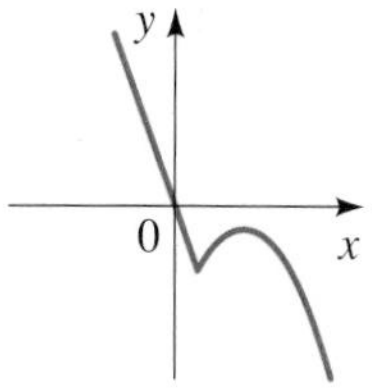

(c)

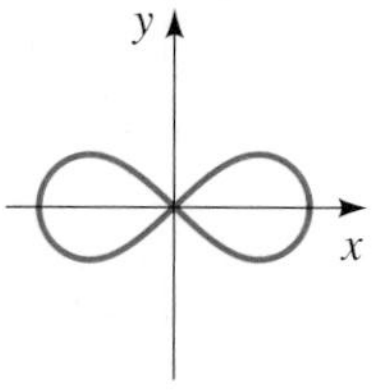

(d)

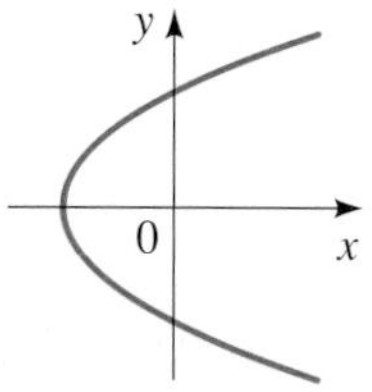

2. Let $f(x) = \dfrac{\sqrt{x+1}}{x}$.

(a) Evaluate $f(3)$, $f(5)$, and $f(a-1)$.

(b) Find the domain of f.

3. A function f has the following verbal description: "Subtract 2, then cube the result."

(a) Find a formula that expresses f algebraically.

(b) Make a table of values of f, for the inputs -1, 0, 1, 2, 3, and 4.

(c) Sketch a graph of f, using the table of values from part (b) to help you.

(d) How do we know that f has an inverse? Give a verbal description for f^{-1}.

(e) Find a formula that expresses f^{-1} algebraically.

4. A school fund-raising group sells chocolate bars to help finance a swimming pool for their physical education program. The group finds that when they set their price at x dollars per bar (where $0 < x \le 5$), their total sales revenue (in dollars) is given by the function $R(x) = -500x^2 + 3000x$.

(a) Evaluate $R(2)$ and $R(4)$. What do these values represent?

(b) Use a graphing calculator to draw a graph of R. What does the graph tell us about what happens to revenue as the price increases from 0 to 5 dollars?

(c) What is the maximum revenue, and at what price is it achieved?

5. Determine the average rate of change for the function $f(t) = t^2 - 2t$ between $t = 2$ and $t = 5$.

6. **(a)** Sketch the graph of the function $f(x) = x^3$.

(b) Use part (a) to graph the function $g(x) = (x-1)^3 - 2$.

7. **(a)** How is the graph of $y = f(x-3) + 2$ obtained from the graph of f?

(b) How is the graph of $y = f(-x)$ obtained from the graph of f?

8. Let $f(x) = \begin{cases} 1 - x & \text{if } x \le 1 \\ 2x + 1 & \text{if } x > 1 \end{cases}$

(a) Evaluate $f(-2)$ and $f(1)$.

(b) Sketch the graph of f.

9. If $f(x) = x^2 + 1$ and $g(x) = x - 3$, find the following.

(a) $f \circ g$

(b) $g \circ f$

(c) $f(g(2))$

(d) $g(f(2))$

(e) $g \circ g \circ g$

10. **(a)** If $f(x) = \sqrt{3 - x}$, find the inverse function f^{-1}.

(b) Sketch the graphs of f and f^{-1} on the same coordinate axes.

11. The graph of a function f is given.

(a) Find the domain and range of f.

(b) Sketch the graph of f^{-1}.

(c) Find the average rate of change of f between $x = 2$ and $x = 6$.

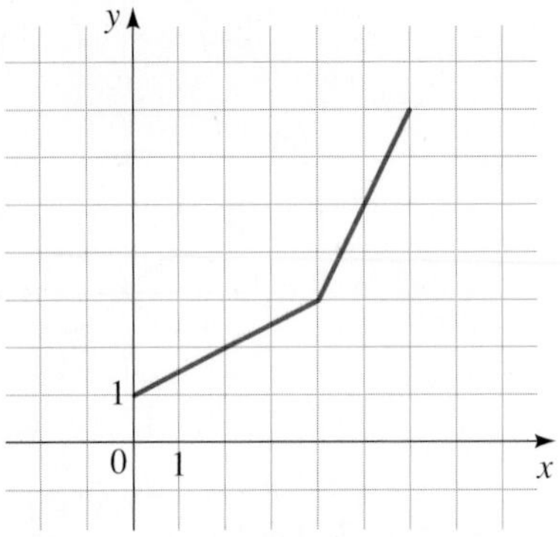

12. Let $f(x) = 3x^4 - 14x^2 + 5x - 3$.

(a) Draw the graph of f in an appropriate viewing rectangle.

(b) Is f one-to-one?

(c) Find the local maximum and minimum values of f and the values of x at which they occur. State each answer correct to two decimal places.

(d) Use the graph to determine the range of f.

(e) Find the intervals on which f is increasing and on which f is decreasing.

Modeling with Functions

Many of the processes that are studied in the physical and social sciences involve understanding how one quantity varies with respect to another. Finding a function that describes the dependence of one quantity on another is called *modeling*. For example, a biologist observes that the number of bacteria in a certain culture increases with time. He tries to model this phenomenon by finding the precise function (or rule) that relates the bacteria population to the elapsed time.

In this *Focus* we will learn how to find models that can be constructed using geometric or algebraic properties of the object under study. Once the model is found, we use it to analyze and predict properties of the object or process being studied.

▼ Modeling with Functions

We begin with a simple real-life situation that illustrates the modeling process.

EXAMPLE 1 | Modeling the Volume of a Box

A breakfast cereal company manufactures boxes to package their product. For aesthetic reasons, the box must have the following proportions: Its width is 3 times its depth, and its height is 5 times its depth.

(a) Find a function that models the volume of the box in terms of its depth.

(b) Find the volume of the box if the depth is 1.5 in.

(c) For what depth is the volume 90 in^3?

(d) For what depth is the volume greater than 60 in^3?

THINKING ABOUT THE PROBLEM

Let's experiment with the problem. If the depth is 1 in., then the width is 3 in. and the height is 5 in. So in this case, the volume is $V = 1 \times 3 \times 5 = 15$ in^3. The table gives other values. Notice that all the boxes have the same shape, and the greater the depth, the greater the volume.

Depth	Volume
1	$1 \times 3 \times 5 = 15$
2	$2 \times 6 \times 10 = 120$
3	$3 \times 9 \times 15 = 405$
4	$4 \times 12 \times 20 = 96$

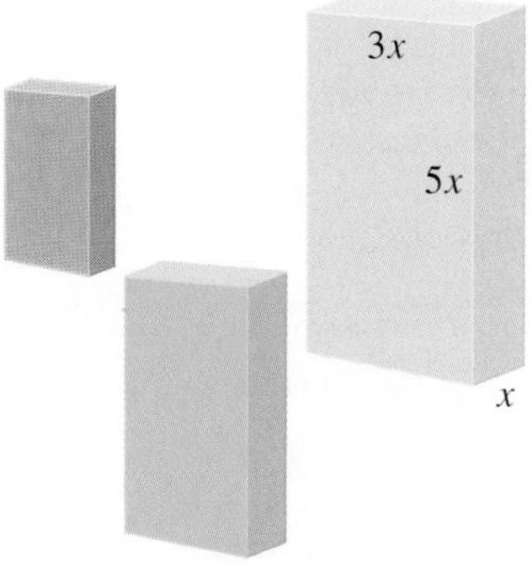

SOLUTION

(a) To find the function that models the volume of the box, we use the following steps.

▶ Express the Model in Words

We know that the volume of a rectangular box is

volume	=	depth	×	width	×	height

▶ Choose the Variable

There are three varying quantities: width, depth, and height. Because the function we want depends on the depth, we let

$$x = \text{depth of the box}$$

Then we express the other dimensions of the box in terms of x.

In Words	In Algebra
Depth	x
Width	$3x$
Height	$5x$

▶ Set Up the Model

The model is the function V that gives the volume of the box in terms of the depth x.

volume	=	depth	×	width	×	height

$$V(x) = x \cdot 3x \cdot 5x$$

$$V(x) = 15x^3$$

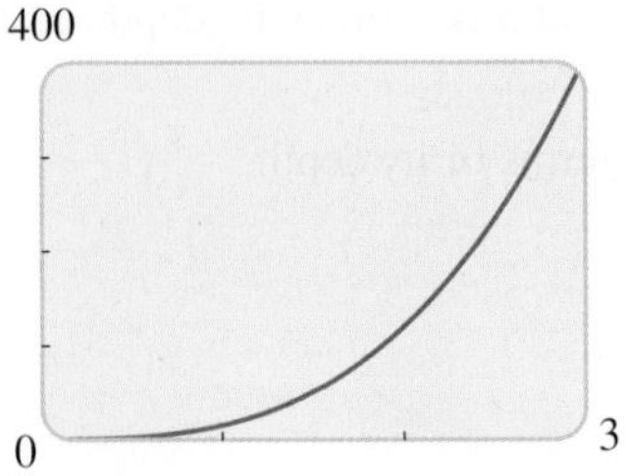

FIGURE 1

The volume of the box is modeled by the function $V(x) = 15x^3$. The function V is graphed in Figure 1.

▶ Use the Model

We use the model to answer the questions in parts (b), (c), and (d).

(b) If the depth is 1.5 in., the volume is $V(1.5) = 15(1.5)^3 = 50.625 \text{ in}^3$.

(c) We need to solve the equation $V(x) = 90$ or

$$15x^3 = 90$$

$$x^3 = 6$$

$$x = \sqrt[3]{6} \approx 1.82 \text{ in.}$$

The volume is 90 in^3 when the depth is about 1.82 in. (We can also solve this equation graphically, as shown in Figure 2.)

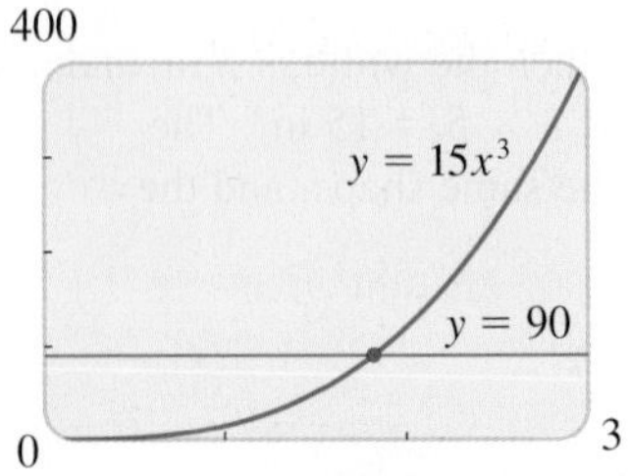

$15x^3 = 90$

FIGURE 2

(d) We need to solve the inequality $V(x) \geq 60$ or

$$15x^3 \geq 60$$

$$x^3 \geq 4$$

$$x \geq \sqrt[3]{4} \approx 1.59$$

The volume will be greater than 60 in^3 if the depth is greater than 1.59 in. (We can also solve this inequality graphically, as shown in Figure 3.) ■

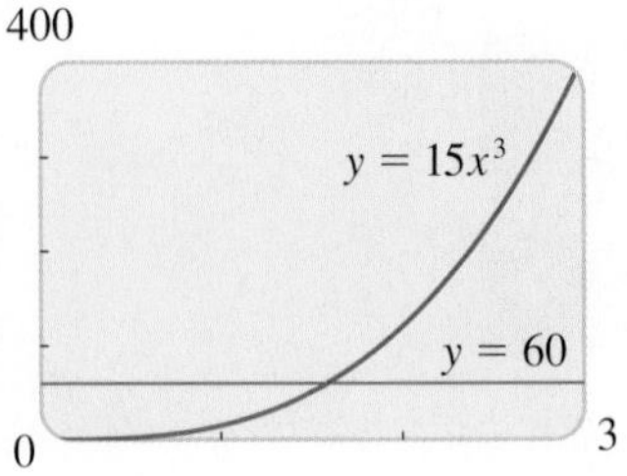

$15x^3 \geq 60$

FIGURE 3

The steps in Example 1 are typical of how we model with functions. They are summarized in the following box.

GUIDELINES FOR MODELING WITH FUNCTIONS

1. **Express the Model in Words.** Identify the quantity you want to model, and express it, in words, as a function of the other quantities in the problem.
2. **Choose the Variable.** Identify all the variables that are used to express the function in Step 1. Assign a symbol, such as x, to one variable, and express the other variables in terms of this symbol.
3. **Set up the Model.** Express the function in the language of algebra by writing it as a function of the single variable chosen in Step 2.
4. **Use the Model.** Use the function to answer the questions posed in the problem. (To find a maximum or a minimum, use the methods described in Section 3.3.)

EXAMPLE 2 | Fencing a Garden

A gardener has 140 feet of fencing to fence in a rectangular vegetable garden.

(a) Find a function that models the area of the garden she can fence.

(b) For what range of widths is the area greater than 825 ft^2?

(c) Can she fence a garden with area 1250 ft^2?

(d) Find the dimensions of the largest area she can fence.

THINKING ABOUT THE PROBLEM

If the gardener fences a plot with width 10 ft, then the length must be 60 ft, because $10 + 10 + 60 + 60 = 140$. So the area is

$$A = \text{width} \times \text{length} = 10 \cdot 60 = 600 \text{ ft}^2$$

The table shows various choices for fencing the garden. We see that as the width increases, the fenced area increases, then decreases.

Width	Length	Area
10	60	600
20	50	1000
30	40	1200
40	30	1200
50	20	1000
60	10	600

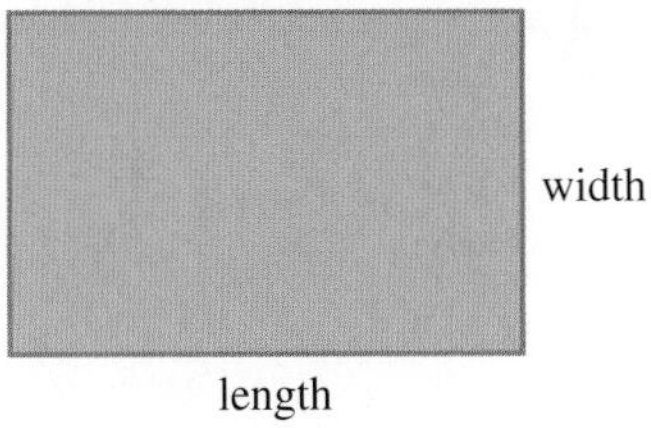

SOLUTION

(a) The model that we want is a function that gives the area she can fence.

▶ Express the Model in Words

We know that the area of a rectangular garden is

$$\text{area} = \text{width} \times \text{length}$$

▶ Choose the Variable

There are two varying quantities: width and length. Because the function we want depends on only one variable, we let

$$x = \text{width of the garden}$$

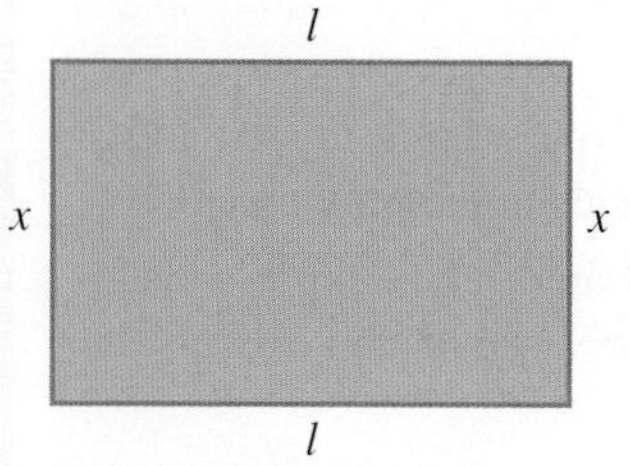

FIGURE 4

Then we must express the length in terms of x. The perimeter is fixed at 140 ft, so the length is determined once we choose the width. If we let the length be l, as in Figure 4, then $2x + 2l = 140$, so $l = 70 - x$. We summarize these facts.

In Words	In Algebra
Width	x
Length	$70 - x$

▶ Set Up the Model

The model is the function A that gives the area of the garden for any width x.

area = width × length

$$A(x) = x(70 - x)$$
$$A(x) = 70x - x^2$$

The area that she can fence is modeled by the function $A(x) = 70x - x^2$.

▶ Use the Model

We use the model to answer the questions in parts (b)–(d).

(b) We need to solve the inequality $A(x) \geq 825$. To solve graphically, we graph $y = 70x - x^2$ and $y = 825$ in the same viewing rectangle (see Figure 5). We see that $15 \leq x \leq 55$.

(c) From Figure 6 we see that the graph of $A(x)$ always lies below the line $y = 1250$, so an area of 1250 ft^2 is never attained.

Maximum values of functions are discussed on page 166.

(d) We need to find where the maximum value of the function $A(x) = 70x - x^2$ occurs. The function is graphed in Figure 7. Using the TRACE feature on a graphing calculator, we find that the function achieves its maximum value at $x = 35$. So the maximum area that she can fence is that when the garden's width is 35 ft and its length is $70 - 35 = 35$ ft. The maximum area then is $35 \times 35 = 1225$ ft^2.

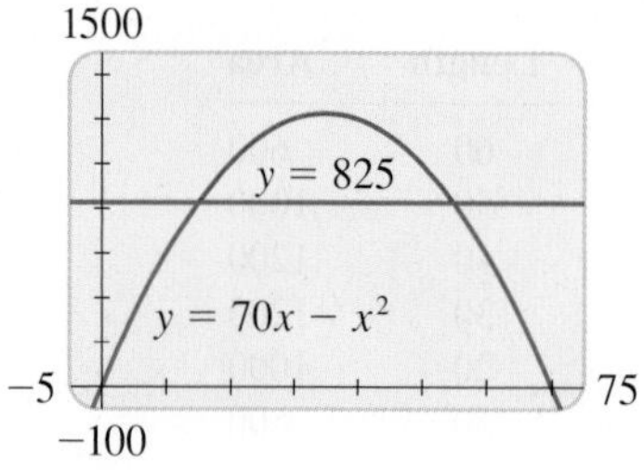

FIGURE 5

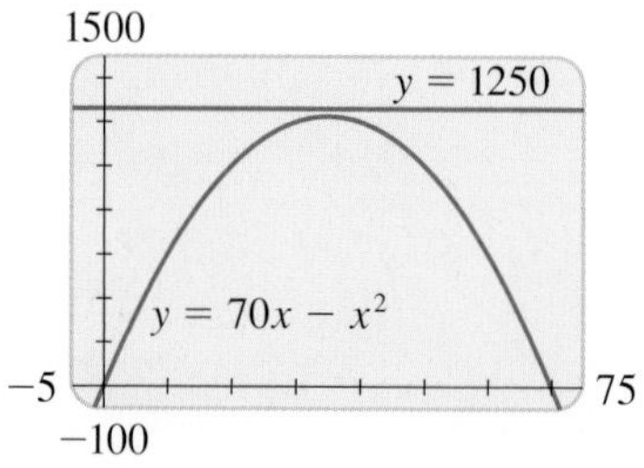

FIGURE 6

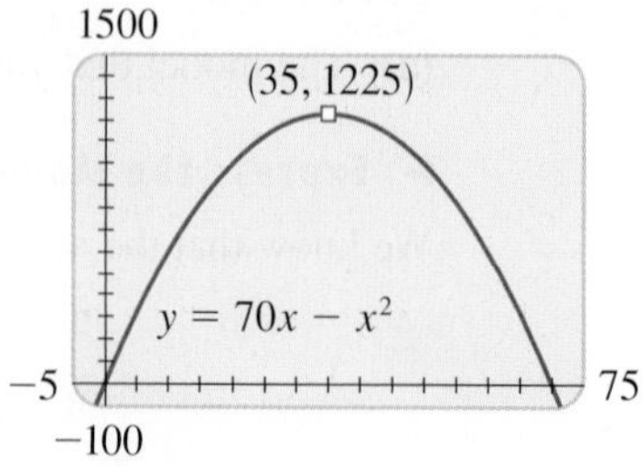

FIGURE 7

■

EXAMPLE 3 | Minimizing the Metal in a Can

A manufacturer makes a metal can that holds 1 L (liter) of oil. What radius minimizes the amount of metal in the can?

THINKING ABOUT THE PROBLEM

To use the least amount of metal, we must minimize the surface area of the can, that is, the area of the top, bottom, and the sides. The area of the top and bottom is $2\pi r^2$ and the area of the sides is $2\pi rh$ (see Figure 8), so the surface area of the can is

$$S = 2\pi r^2 + 2\pi rh$$

The radius and height of the can must be chosen so that the volume is exactly 1 L, or 1000 cm^3. If we want a small radius, say $r = 3$, then the height must be just tall enough to make the total volume 1000 cm^3. In other words, we must have

$$\pi(3)^2 h = 1000 \qquad \text{Volume of the can is } \pi r^2 h$$

$$h = \frac{1000}{9\pi} \approx 35.4 \text{ cm} \qquad \text{Solve for } h$$

Now that we know the radius and height, we can find the surface area of the can:

$$\text{surface area} = 2\pi(3)^2 + 2\pi(3)(35.4) \approx 723.8 \text{ cm}^3$$

If we want a different radius, we can find the corresponding height and surface area in a similar fashion.

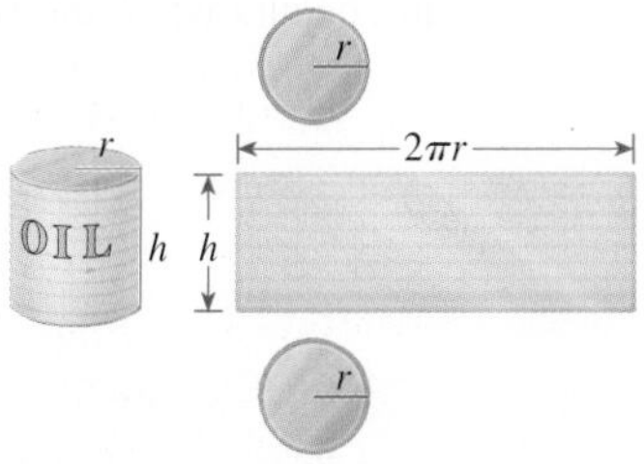

FIGURE 8

SOLUTION The model that we want is a function that gives the surface area of the can.

▶ Express the Model in Words

We know that for a cylindrical can

surface area = area of top and bottom + area of sides

▶ Choose the Variable

There are two varying quantities: radius and height. Because the function we want depends on the radius, we let

$$r = \text{radius of can}$$

Next, we must express the height in terms of the radius r. Because the volume of a cylindrical can is $V = \pi r^2 h$ and the volume must be 1000 cm^3, we have

$$\pi r^2 h = 1000 \qquad \text{Volume of can is 1000 cm}^3$$

$$h = \frac{1000}{\pi r^2} \qquad \text{Solve for } h$$

We can now express the areas of the top, bottom, and sides in terms of r only.

In Words	In Algebra
Radius of can	r
Height of can	$\dfrac{1000}{\pi r^2}$
Area of top and bottom	$2\pi r^2$
Area of sides ($2\pi rh$)	$2\pi r\left(\dfrac{1000}{\pi r^2}\right)$

▶ Set Up the Model

The model is the function S that gives the surface area of the can as a function of the radius r.

surface area = area of top and bottom + area of sides

$$S(r) = 2\pi r^2 + 2\pi r\left(\frac{1000}{\pi r^2}\right)$$

$$S(r) = 2\pi r^2 + \frac{2000}{r}$$

▶ Use the Model

We use the model to find the minimum surface area of the can. We graph S in Figure 9 and zoom in on the minimum point to find that the minimum value of S is about 554 cm^2 and occurs when the radius is about 5.4 cm.

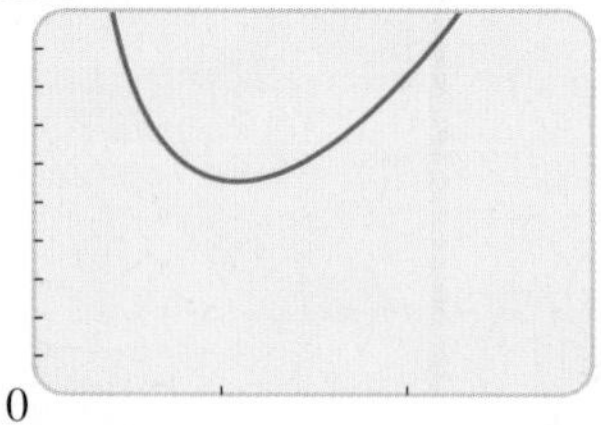

FIGURE 9 $S(r) = 2\pi r^2 + \dfrac{2000}{r}$

■

PROBLEMS

1–18 ■ In these problems you are asked to find a function that models a real-life situation. Use the principles of modeling described in this *Focus* to help you.

1. **Area** A rectangular building lot is three times as long as it is wide. Find a function that models its area A in terms of its width w.

2. **Area** A poster is 10 inches longer than it is wide. Find a function that models its area A in terms of its width w.

3. **Volume** A rectangular box has a square base. Its height is half the width of the base. Find a function that models its volume V in terms of its width w.

4. **Volume** The height of a cylinder is four times its radius. Find a function that models the volume V of the cylinder in terms of its radius r.

5. **Area** A rectangle has a perimeter of 20 ft. Find a function that models its area A in terms of the length x of one of its sides.

6. **Perimeter** A rectangle has an area of 16 m^2. Find a function that models its perimeter P in terms of the length x of one of its sides.

PYTHAGORAS (circa 580–500 B.C.) founded a school in Croton in southern Italy, devoted to the study of arithmetic, geometry, music, and astronomy. The Pythagoreans, as they were called, were a secret society with peculiar rules and initiation rites. They wrote nothing down and were not to reveal to anyone what they had learned from the Master. Although women were barred by law from attending public meetings, Pythagoras allowed women in his school, and his most famous student was Theano (whom he later married).

According to Aristotle, the Pythagoreans were convinced that "the principles of mathematics are the principles of all things." Their motto was "Everything is Number," by which they meant *whole* numbers. The outstanding contribution of Pythagoras is the theorem that bears his name: In a right triangle the area of the square on the hypotenuse is equal to the sum of the areas of the square on the other two sides.

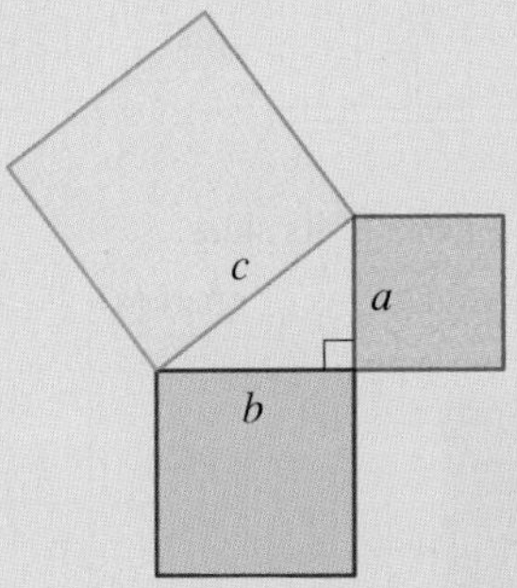

$$c^2 = a^2 + b^2$$

The converse of Pythagoras's Theorem is also true; that is, a triangle whose sides a, b, and c satisfy $a^2 + b^2 = c^2$ is a right triangle.

7. Area Find a function that models the area A of an equilateral triangle in terms of the length x of one of its sides.

8. Area Find a function that models the surface area S of a cube in terms of its volume V.

9. Radius Find a function that models the radius r of a circle in terms of its area A.

10. Area Find a function that models the area A of a circle in terms of its circumference C.

11. Area A rectangular box with a volume of 60 ft^3 has a square base. Find a function that models its surface area S in terms of the length x of one side of its base.

12. Length A woman 5 ft tall is standing near a street lamp that is 12 ft tall, as shown in the figure. Find a function that models the length L of her shadow in terms of her distance d from the base of the lamp.

13. Distance Two ships leave port at the same time. One sails south at 15 mi/h, and the other sails east at 20 mi/h. Find a function that models the distance D between the ships in terms of the time t (in hours) elapsed since their departure.

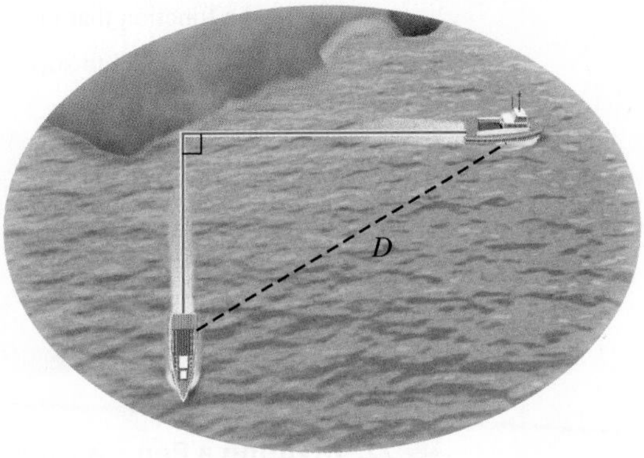

14. Product The sum of two positive numbers is 60. Find a function that models their product P in terms of x, one of the numbers.

15. Area An isosceles triangle has a perimeter of 8 cm. Find a function that models its area A in terms of the length of its base b.

16. Perimeter A right triangle has one leg twice as long as the other. Find a function that models its perimeter P in terms of the length x of the shorter leg.

17. Area A rectangle is inscribed in a semicircle of radius 10, as shown in the figure. Find a function that models the area A of the rectangle in terms of its height h.

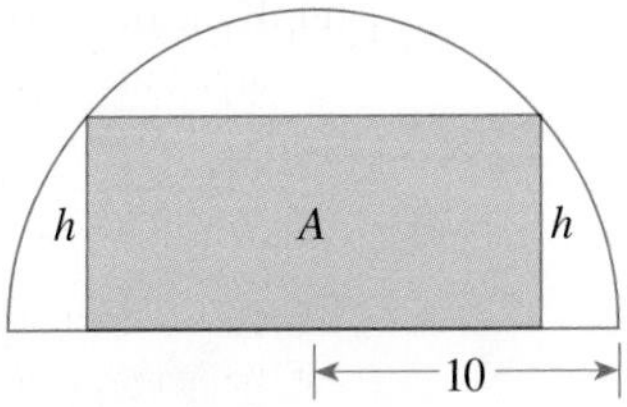

18. Height The volume of a cone is 100 in^3. Find a function that models the height h of the cone in terms of its radius r.

19–32 ■ In these problems you are asked to find a function that models a real-life situation, and then use the model to answer questions about the situation. Use the guidelines on page 215 to help you.

19. Maximizing a Product Consider the following problem: Find two numbers whose sum is 19 and whose product is as large as possible.

(a) Experiment with the problem by making a table like the one following, showing the product of different pairs of numbers that add up to 19. On the basis of the evidence in your table, estimate the answer to the problem.

(b) Find a function that models the product in terms of one of the two numbers.

(c) Use your model to solve the problem, and compare with your answer to part (a).

20. Minimizing a Sum Find two positive numbers whose sum is 100 and the sum of whose squares is a minimum.

21. Fencing a Field Consider the following problem: A farmer has 2400 ft of fencing and wants to fence off a rectangular field that borders a straight river. He does not need a fence along the river (see the figure). What are the dimensions of the field of largest area that he can fence?

(a) Experiment with the problem by drawing several diagrams illustrating the situation. Calculate the area of each configuration, and use your results to estimate the dimensions of the largest possible field.

First number	Second number	Product
1	18	18
2	17	34
3	16	48
⋮	⋮	⋮

(b) Find a function that models the area of the field in terms of one of its sides.

(c) Use your model to solve the problem, and compare with your answer to part (a).

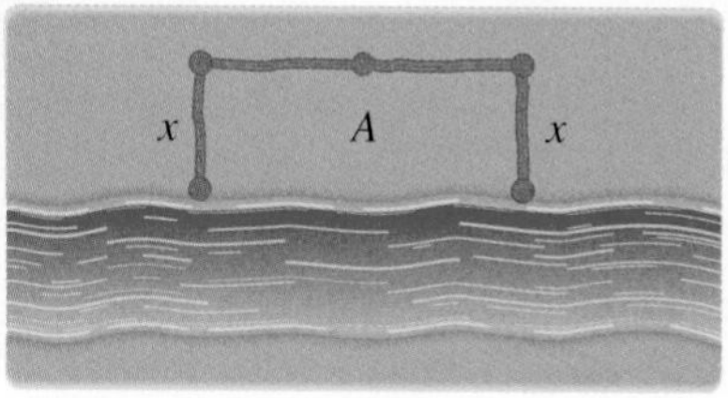

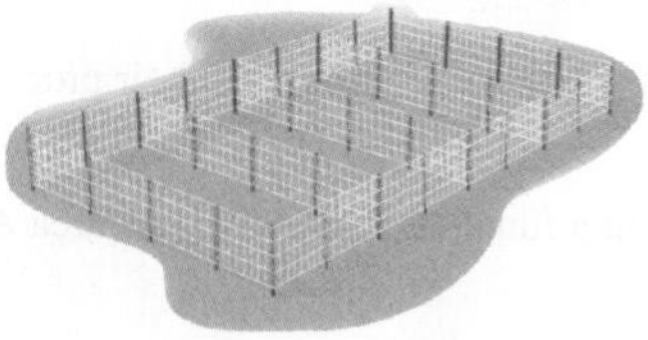

22. Dividing a Pen A rancher with 750 ft of fencing wants to enclose a rectangular area and then divide it into four pens with fencing parallel to one side of the rectangle (see the figure).

(a) Find a function that models the total area of the four pens.

(b) Find the largest possible total area of the four pens.

23. Fencing a Garden Plot A property owner wants to fence a garden plot adjacent to a road, as shown in the figure. The fencing next to the road must be sturdier and costs \$5 per foot, but the other fencing costs just \$3 per foot. The garden is to have an area of 1200 ft^2.

(a) Find a function that models the cost of fencing the garden.

(b) Find the garden dimensions that minimize the cost of fencing.

(c) If the owner has at most \$600 to spend on fencing, find the range of lengths he can fence along the road.

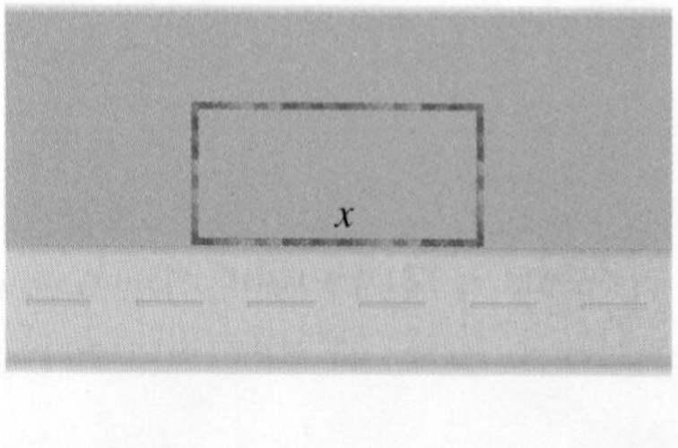

24. Maximizing Area A wire 10 cm long is cut into two pieces, one of length x and the other of length $10 - x$, as shown in the figure. Each piece is bent into the shape of a square.

(a) Find a function that models the total area enclosed by the two squares.

(b) Find the value of x that minimizes the total area of the two squares.

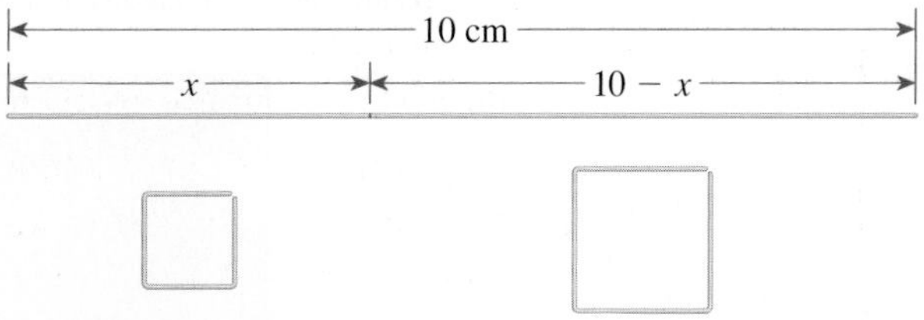

25. Light from a Window A Norman window has the shape of a rectangle surmounted by a semicircle, as shown in the figure to the left. A Norman window with perimeter 30 ft is to be constructed.

(a) Find a function that models the area of the window.

(b) Find the dimensions of the window that admits the greatest amount of light.

26. Volume of a Box A box with an open top is to be constructed from a rectangular piece of cardboard with dimensions 12 in. by 20 in. by cutting out equal squares of side x at each corner and then folding up the sides (see the figure).

(a) Find a function that models the volume of the box.

(b) Find the values of x for which the volume is greater than 200 in^3.

(c) Find the largest volume that such a box can have.

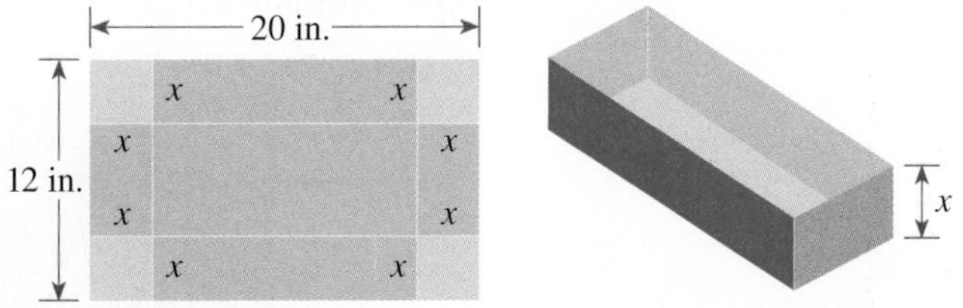

27. Area of a Box An open box with a square base is to have a volume of 12 ft^3.

(a) Find a function that models the surface area of the box.

(b) Find the box dimensions that minimize the amount of material used.

28. Inscribed Rectangle Find the dimensions that give the largest area for the rectangle shown in the figure. Its base is on the x-axis and its other two vertices are above the x-axis, lying on the parabola $y = 8 - x^2$.

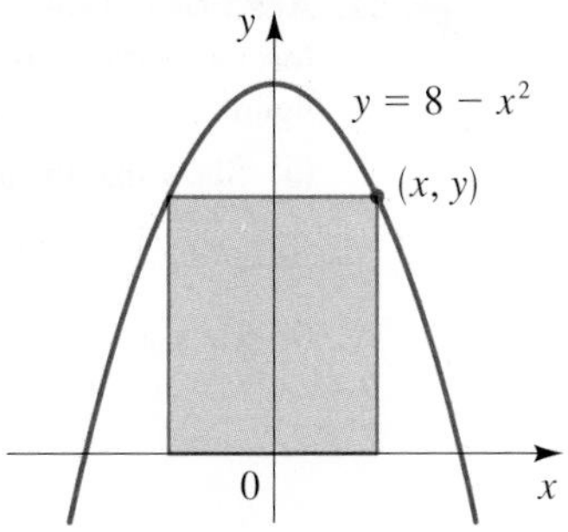

29. Minimizing Costs A rancher wants to build a rectangular pen with an area of 100 m^2.

(a) Find a function that models the length of fencing required.

(b) Find the pen dimensions that require the minimum amount of fencing.

30. Minimizing Time A man stands at a point A on the bank of a straight river, 2 mi wide. To reach point B, 7 mi downstream on the opposite bank, he first rows his boat to point P on the opposite bank and then walks the remaining distance x to B, as shown in the figure. He can row at a speed of 2 mi/h and walk at a speed of 5 mi/h.

(a) Find a function that models the time needed for the trip.

(b) Where should he land so that he reaches B as soon as possible?

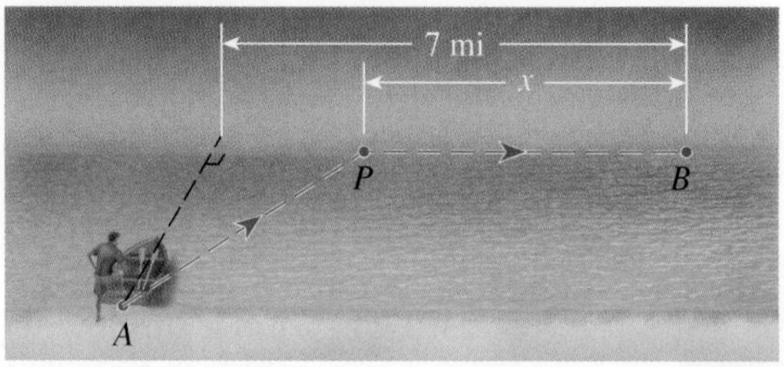

31. Bird Flight A bird is released from point A on an island, 5 mi from the nearest point B on a straight shoreline. The bird flies to a point C on the shoreline and then flies along the shoreline to its nesting area D (see the figure). Suppose the bird requires 10 kcal/mi of energy to fly over land and 14 kcal/mi to fly over water.

(a) Use the fact that

$$\text{energy used} = \text{energy per mile} \times \text{miles flown}$$

to show that the total energy used by the bird is modeled by the function

$$E(x) = 14\sqrt{x^2 + 25} + 10(12 - x)$$

(b) If the bird instinctively chooses a path that minimizes its energy expenditure, to what point does it fly?

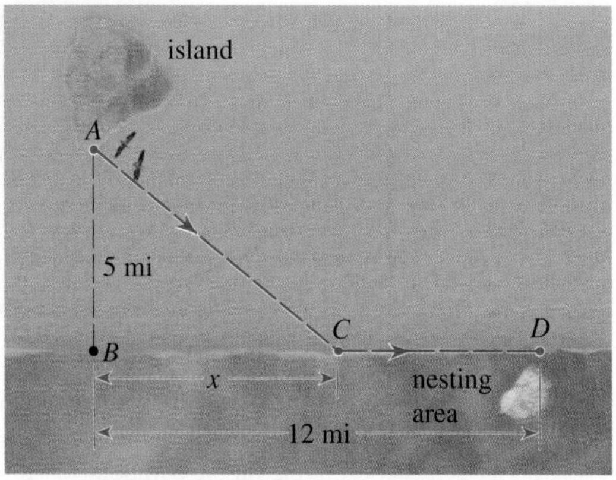

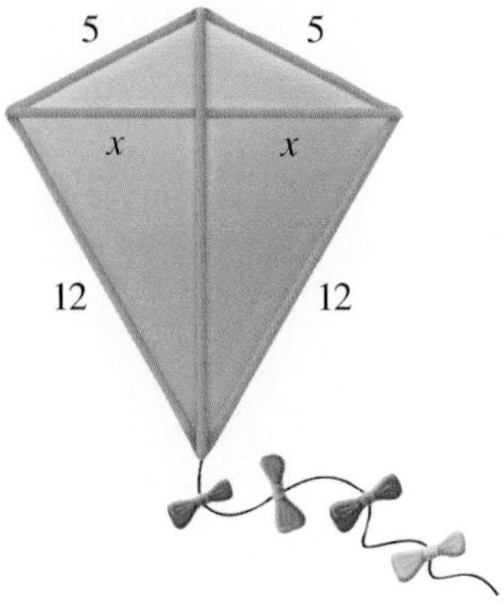

32. Area of a Kite A kite frame is to be made from six pieces of wood. The four pieces that form its border have been cut to the lengths indicated in the figure. Let x be as shown in the figure.

(a) Show that the area of the kite is given by the function

$$A(x) = x\left(\sqrt{25 - x^2} + \sqrt{144 - x^2}\right)$$

(b) How long should each of the two crosspieces be to maximize the area of the kite?

CHAPTER 3

POLYNOMIAL AND RATIONAL FUNCTIONS

Functions defined by polynomial expressions are called polynomial functions. The graphs of polynomial functions can have many peaks and valleys; this makes them suitable models for many real-world situations. For example, a factory owner notices that if she increases the number of workers, productivity increases, but if there are too many workers, productivity begins to decrease. This situation is modeled by a polynomial function of degree 2 (a quadratic function). As another example, when a volleyball is hit, it goes up and then down, following a path that is also modeled by a quadratic function.The graphs of polynomial functions are smooth curves that are used in designing many things. For example, sailboat designers put together portions of the graphs of different cubic functions (called cubic splines) to make the curves of the hull of a sailboat.

3.1 Quadratic Functions and Models

Graphing Quadratic Functions Using the Standard Form ▶ Maximum and Minimum Values of Quadratic Functions ▶ Modeling with Quadratic Functions

A polynomial function is a function that is defined by a polynomial expression. So a **polynomial function of degree *n*** is a function of the form

$$P(x) = a_n x^n + a_{n-1} x^{n-1} + \cdots + a_1 x + a_0$$

Polynomial expressions are defined in Section 1.3.

We have already studied polynomial functions of degree 0 and 1. These are functions of the form $P(x) = a_0$ and $P(x) = a_1 x + a_0$, respectively, whose graphs are lines. In this section we study polynomial functions of degree 2. These are called quadratic functions.

QUADRATIC FUNCTIONS

A **quadratic function** is a polynomial function of degree 2. So a quadratic function is a function of the form

$$f(x) = ax^2 + bx + c, \qquad a \neq 0$$

We see in this section how quadratic functions model many real-world phenomena. We begin by analyzing the graphs of quadratic functions.

▼ Graphing Quadratic Functions Using the Standard Form

For a geometric definition of parabolas, see Section 11.1.

If we take $a = 1$ and $b = c = 0$ in the quadratic function$f(x) = ax^2 + bx + c$, we get the quadratic function $f(x) = x^2$, whose graph is the parabola graphed in Example 1 of Section 2.2. In fact, the graph of any quadratic function is a **parabola**; it can be obtained from the graph of $f(x) = x^2$ by the transformations given in Section 2.5.

STANDARD FORM OF A QUADRATIC FUNCTION

A quadratic function $f(x) = ax^2 + bx + c$ can be expressed in the **standard form**

$$f(x) = a(x - h)^2 + k$$

by completing the square. The graph of f is a parabola with **vertex** (h, k); the parabola opens upward if $a > 0$ or downward if $a < 0$.

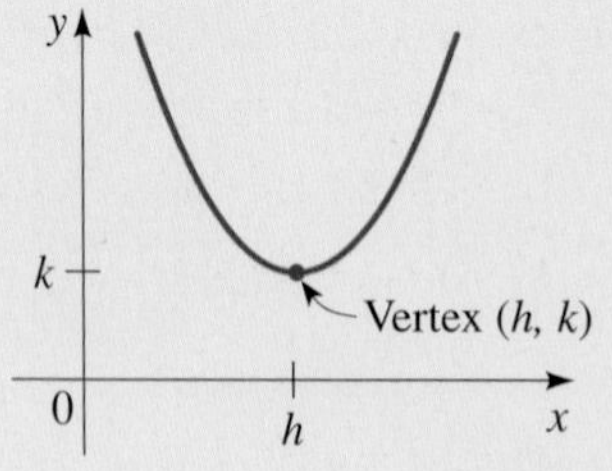

$f(x) = a(x - h)^2 + k,\ a > 0$

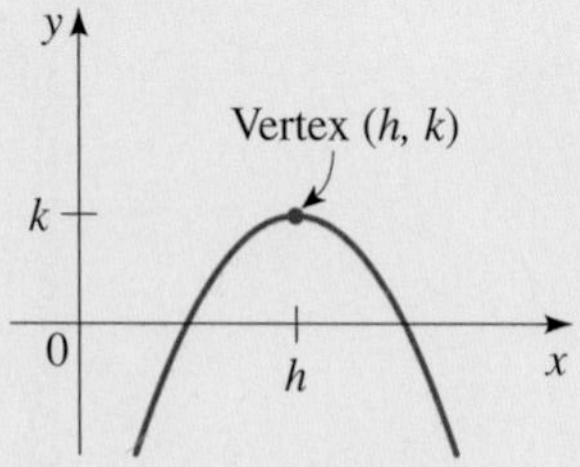

$f(x) = a(x - h)^2 + k,\ a < 0$

EXAMPLE 1 | Standard Form of a Quadratic Function

Let $f(x) = 2x^2 - 12x + 23$.

(a) Express f in standard form. **(b)** Sketch the graph of f.

Completing the square is discussed in Section 1.5.

$f(x) = 2(x - 3)^2 + 5$

Vertex is $(3, 5)$

SOLUTION

(a) Since the coefficient of x^2 is not 1, we must factor this coefficient from the terms involving x before we complete the square.

$$
\begin{aligned}
f(x) &= 2x^2 - 12x + 23 \\
&= 2(x^2 - 6x) + 23 && \text{Factor 2 from the } x\text{-terms} \\
&= 2(x^2 - 6x + 9) + 23 - 2 \cdot 9 && \text{Complete the square: Add 9 inside parentheses, subtract } 2 \cdot 9 \text{ outside} \\
&= 2(x - 3)^2 + 5 && \text{Factor and simplify}
\end{aligned}
$$

The standard form is $f(x) = 2(x - 3)^2 + 5$.

(b) The standard form tells us that we get the graph of f by taking the parabola $y = x^2$, shifting it to the right 3 units, stretching it by a factor of 2, and moving it upward 5 units. The vertex of the parabola is at $(3, 5)$, and the parabola opens upward. We sketch the graph in Figure 1 after noting that the y-intercept is $f(0) = 23$.

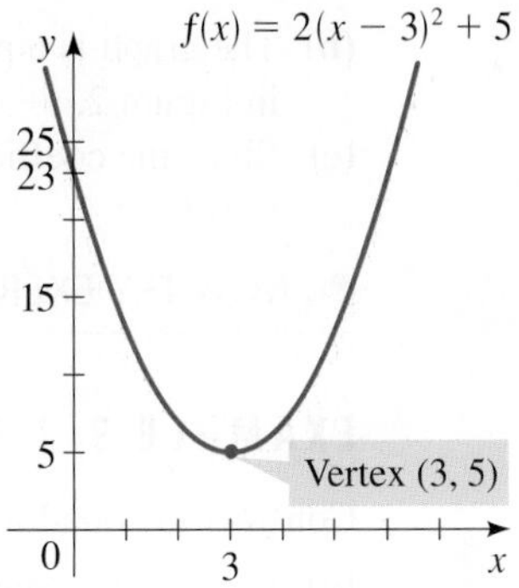

FIGURE 1

NOW TRY EXERCISE **13**

Maximum and Minimum Values of Quadratic Functions

If a quadratic function has vertex (h, k), then the function has a minimum value at the vertex if its graph opens upward and a maximum value at the vertex if its graph opens downward. For example, the function graphed in Figure 1 has minimum value 5 when $x = 3$, since the vertex $(3, 5)$ is the lowest point on the graph.

MAXIMUM OR MINIMUM VALUE OF A QUADRATIC FUNCTION

Let f be a quadratic function with standard form $f(x) = a(x - h)^2 + k$. The maximum or minimum value of f occurs at $x = h$.

If $a > 0$, then the **minimum value** of f is $f(h) = k$.

If $a < 0$, then the **maximum value** of f is $f(h) = k$.

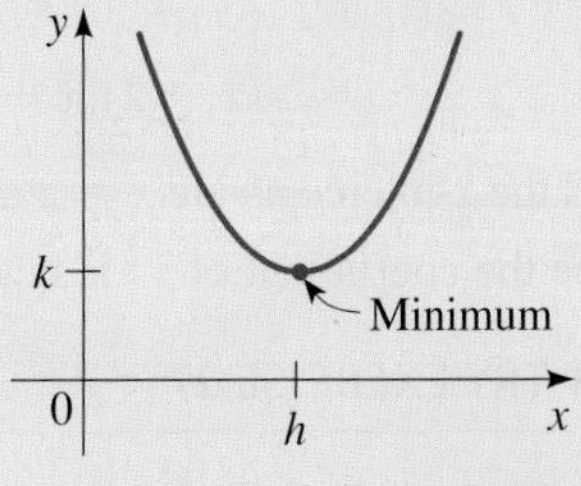

$f(x) = a(x - h)^2 + k,\ a > 0$

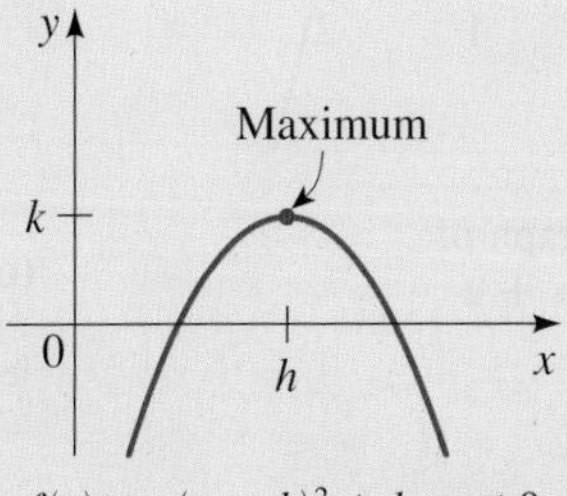

$f(x) = a(x - h)^2 + k,\ a < 0$

EXAMPLE 2 | Minimum Value of a Quadratic Function

Consider the quadratic function $f(x) = 5x^2 - 30x + 49$.

(a) Express f in standard form.

(b) Sketch the graph of f.

(c) Find the minimum value of f.

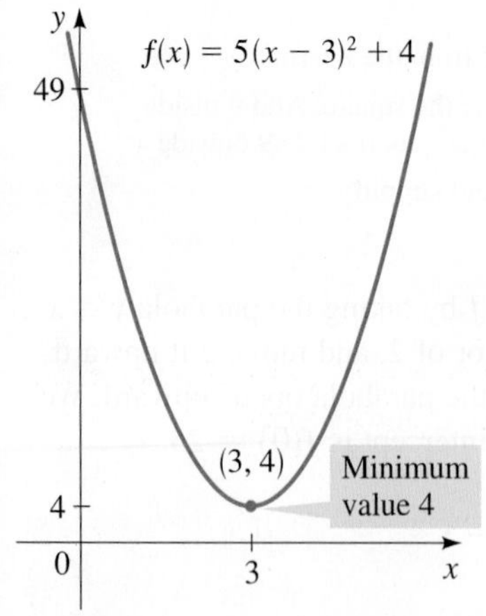

FIGURE 2

SOLUTION

(a) To express this quadratic function in standard form, we complete the square.

$$\begin{aligned} f(x) &= 5x^2 - 30x + 49 \\ &= 5(x^2 - 6x) + 49 && \text{Factor 5 from the } x\text{-terms} \\ &= 5(x^2 - 6x + 9) + 49 - 5 \cdot 9 && \text{Complete the square: Add 9 inside parentheses, subtract } 5 \cdot 9 \text{ outside} \\ &= 5(x - 3)^2 + 4 && \text{Factor and simplify} \end{aligned}$$

(b) The graph is a parabola that has its vertex at $(3, 4)$ and opens upward, as sketched in Figure 2.

(c) Since the coefficient of x^2 is positive, f has a minimum value. The minimum value is $f(3) = 4$.

NOW TRY EXERCISE **25**

EXAMPLE 3 | Maximum Value of a Quadratic Function

Consider the quadratic function $f(x) = -x^2 + x + 2$.

(a) Express f in standard form.

(b) Sketch the graph of f.

(c) Find the maximum value of f.

SOLUTION

(a) To express this quadratic function in standard form, we complete the square.

$$\begin{aligned} y &= -x^2 + x + 2 \\ &= -(x^2 - x) + 2 && \text{Factor } -1 \text{ from the } x\text{-terms} \\ &= -\left(x^2 - x + \tfrac{1}{4}\right) + 2 - (-1)\tfrac{1}{4} && \text{Complete the square: Add } \tfrac{1}{4} \text{ inside parentheses, subtract } (-1)\tfrac{1}{4} \text{ outside} \\ &= -\left(x - \tfrac{1}{2}\right)^2 + \tfrac{9}{4} && \text{Factor and simplify} \end{aligned}$$

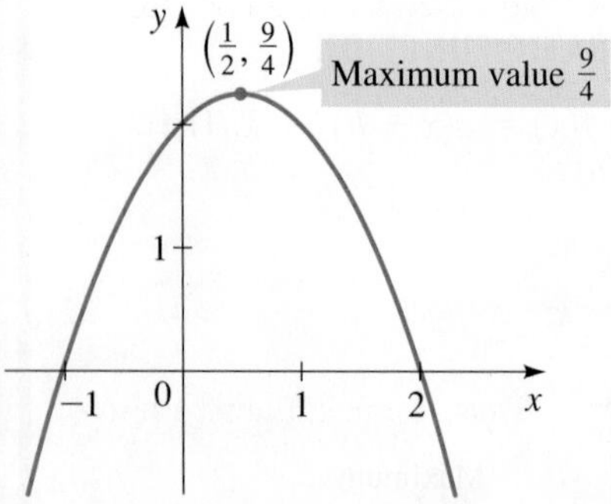

FIGURE 3 Graph of $f(x) = -x^2 + x + 2$

(b) From the standard form we see that the graph is a parabola that opens downward and has vertex $\left(\frac{1}{2}, \frac{9}{4}\right)$. As an aid to sketching the graph, we find the intercepts. The y-intercept is $f(0) = 2$. To find the x-intercepts, we set $f(x) = 0$ and factor the resulting equation.

$$\begin{aligned} -x^2 + x + 2 &= 0 && \text{Set } y = 0 \\ x^2 - x - 2 &= 0 && \text{Multiply by } -1 \\ (x - 2)(x + 1) &= 0 && \text{Factor} \end{aligned}$$

Thus the x-intercepts are $x = 2$ and $x = -1$. The graph of f is sketched in Figure 3.

(c) Since the coefficient of x^2 is negative, f has a maximum value, which is $f\left(\frac{1}{2}\right) = \frac{9}{4}$.

NOW TRY EXERCISE **27**

Expressing a quadratic function in standard form helps us to sketch its graph as well as to find its maximum or minimum value. If we are interested only in finding the maximum or minimum value, then a formula is available for doing so. This formula is obtained by completing the square for the general quadratic function as follows:

$$
\begin{aligned}
f(x) &= ax^2 + bx + c \\
&= a\left(x^2 + \frac{b}{a}x\right) + c && \text{Factor } a \text{ from the } x\text{-terms} \\
&= a\left(x^2 + \frac{b}{a}x + \frac{b^2}{4a^2}\right) + c - a\left(\frac{b^2}{4a^2}\right) && \text{Complete the square: Add } \frac{b^2}{4a^2} \text{ inside parentheses, subtract } a\left(\frac{b^2}{4a^2}\right) \text{ outside} \\
&= a\left(x + \frac{b}{2a}\right)^2 + c - \frac{b^2}{4a} && \text{Factor}
\end{aligned}
$$

This equation is in standard form with $h = -b/(2a)$ and $k = c - b^2/(4a)$. Since the maximum or minimum value occurs at $x = h$, we have the following result.

MAXIMUM OR MINIMUM VALUE OF A QUADRATIC FUNCTION

The maximum or minimum value of a quadratic function $f(x) = ax^2 + bx + c$ occurs at

$$x = -\frac{b}{2a}$$

If $a > 0$, then the **minimum value** is $f\left(-\frac{b}{2a}\right)$.

If $a < 0$, then the **maximum value** is $f\left(-\frac{b}{2a}\right)$.

EXAMPLE 4 | Finding Maximum and Minimum Values of Quadratic Functions

Find the maximum or minimum value of each quadratic function.

(a) $f(x) = x^2 + 4x$ **(b)** $g(x) = -2x^2 + 4x - 5$

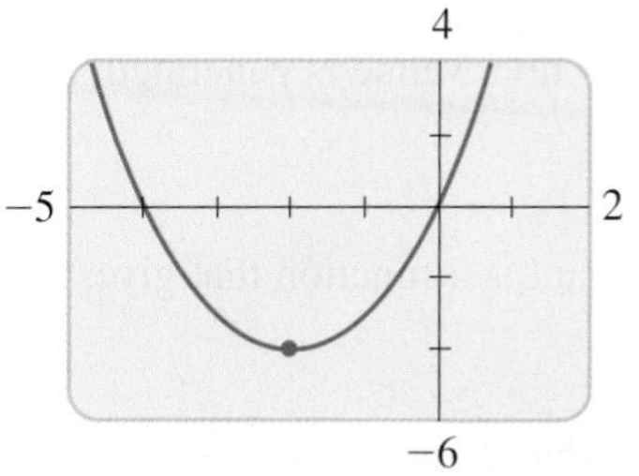

The minimum value occurs at $x = -2$.

SOLUTION

(a) This is a quadratic function with $a = 1$ and $b = 4$. Thus, the maximum or minimum value occurs at

$$x = -\frac{b}{2a} = -\frac{4}{2 \cdot 1} = -2$$

Since $a > 0$, the function has the *minimum* value

$$f(-2) = (-2)^2 + 4(-2) = -4$$

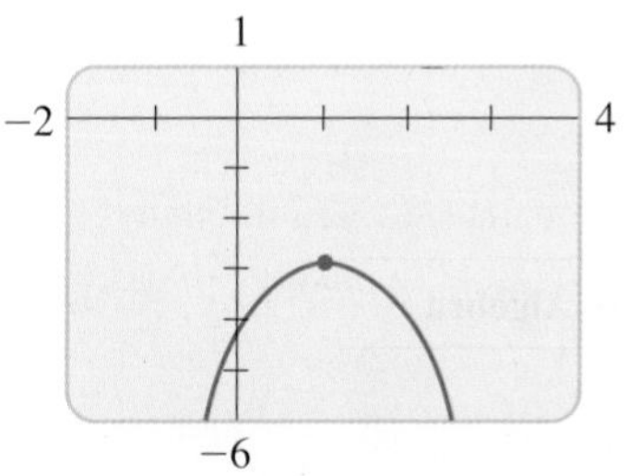

The maximum value occurs at $x = 1$.

(b) This is a quadratic function with $a = -2$ and $b = 4$. Thus, the maximum or minimum value occurs at

$$x = -\frac{b}{2a} = -\frac{4}{2 \cdot (-2)} = 1$$

Since $a < 0$, the function has the *maximum* value

$$f(1) = -2(1)^2 + 4(1) - 5 = -3$$

NOW TRY EXERCISES **33** AND **35** ■

Modeling with Quadratic Functions

We study some examples of real-world phenomena that are modeled by quadratic functions. These examples and the *Application* exercises for this section show some of the variety of situations that are naturally modeled by quadratic functions.

EXAMPLE 5 | Maximum Gas Mileage for a Car

Most cars get their best gas mileage when traveling at a relatively modest speed. The gas mileage M for a certain new car is modeled by the function

$$M(s) = -\frac{1}{28}s^2 + 3s - 31, \qquad 15 \le s \le 70$$

where s is the speed in mi/h and M is measured in mi/gal. What is the car's best gas mileage, and at what speed is it attained?

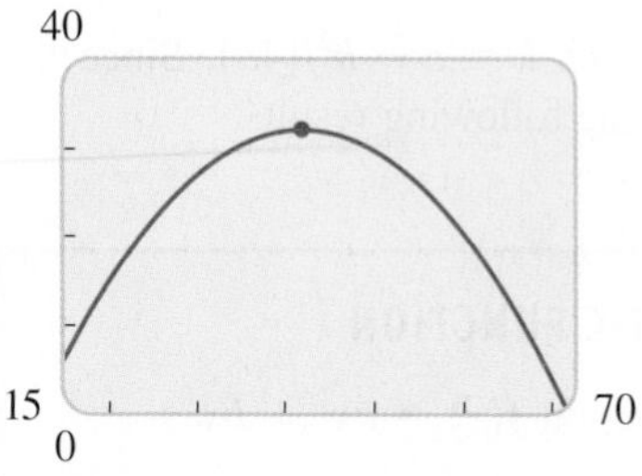

The maximum gas mileage occurs at 42 mi/h.

SOLUTION The function M is a quadratic function with $a = -\frac{1}{28}$ and $b = 3$. Thus, its maximum value occurs when

$$s = -\frac{b}{2a} = -\frac{3}{2\left(-\frac{1}{28}\right)} = 42$$

The maximum is $M(42) = -\frac{1}{28}(42)^2 + 3(42) - 31 = 32$. So the car's best gas mileage is 32 mi/gal, when it is traveling at 42 mi/h.

NOW TRY EXERCISE **67**

EXAMPLE 6 | Maximizing Revenue from Ticket Sales

A hockey team plays in an arena that has a seating capacity of 15,000 spectators. With the ticket price set at \$14, average attendance at recent games has been 9500. A market survey indicates that for each dollar the ticket price is lowered, the average attendance increases by 1000.

(a) Find a function that models the revenue in terms of ticket price.

(b) Find the price that maximizes revenue from ticket sales.

(c) What ticket price is so high that no one attends and so no revenue is generated?

SOLUTION

(a) Express the model in words. The model that we want is a function that gives the revenue for any ticket price.

revenue = ticket price × attendance

Choose the variable. There are two varying quantities: ticket price and attendance. Since the function we want depends on price, we let

$$x = \text{ticket price}$$

Next, we express attendance in terms of x.

In Words	In Algebra
Ticket price	x
Amount ticket price is lowered	$14 - x$
Increase in attendance	$1000(14 - x)$
Attendance	$9500 + 1000(14 - x)$

Set up the model. The model that we want is the function R that gives the revenue for a given ticket price x.

$$\text{revenue} = \text{ticket price} \times \text{attendance}$$

$$R(x) = x \times [9500 + 1000(14 - x)]$$
$$R(x) = x(23{,}500 - 1000x)$$
$$R(x) = 23{,}500x - 1000x^2$$

(b) **Use the model.** Since R is a quadratic function with $a = -1000$ and $b = 23{,}500$, the maximum occurs at

$$x = -\frac{b}{2a} = -\frac{23{,}500}{2(-1000)} = 11.75$$

So a ticket price of $11.75 gives the maximum revenue.

(c) **Use the model.** We want to find the ticket price for which $R(x) = 0$.

$23{,}500x - 1000x^2 = 0$	Set $R(x) = 0$
$23.5x - x^2 = 0$	Divide by 1000
$x(23.5 - x) = 0$	Factor
$x = 0$ or $x = 23.5$	Solve for x

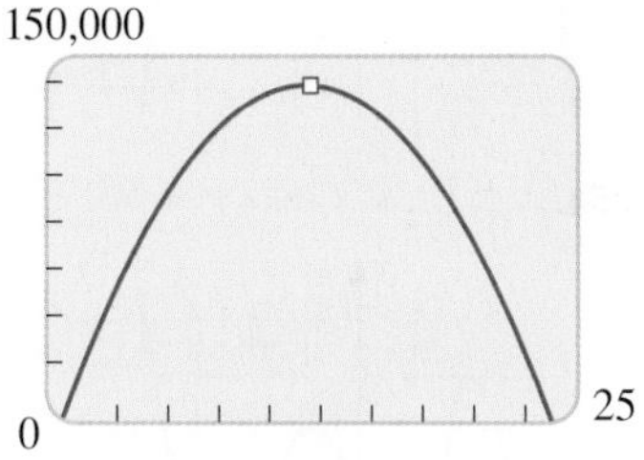

Maximum attendance occurs when ticket price is $11.75.

So according to this model, a ticket price of $23.50 is just too high; at that price, no one attends to watch this team play. (Of course, revenue is also zero if the ticket price is zero.)

NOW TRY EXERCISE **77** ■

3.1 EXERCISES

CONCEPTS

1. To put the quadratic function $f(x) = ax^2 + bx + c$ in standard form, we complete the ________.

2. The quadratic function $f(x) = a(x - h)^2 + k$ is in standard form.

(a) The graph of f is a parabola with vertex (____, ____).

(b) If $a > 0$, the graph of f opens ________. In this case $f(h) = k$ is the ________ value of f.

(c) If $a < 0$, the graph of f opens ________. In this case $f(h) = k$ is the ________ value of f.

3. The graph of $f(x) = 2(x - 3)^2 + 5$ is a parabola that opens ________, with its vertex at (____, ____), and $f(3) =$ ________ is the (minimum/maximum) ________ value of f.

4. The graph of $f(x) = -2(x - 3)^2 + 5$ is a parabola that opens ________, with its vertex at (____, ____), and $f(3) =$ ________ is the (minimum/maximum) ________ value of f.

SKILLS

5–8 ■ The graph of a quadratic function f is given. **(a)** Find the coordinates of the vertex. **(b)** Find the maximum or minimum value of f. **(c)** Find the domain and range of f.

5. $f(x) = -x^2 + 6x - 5$

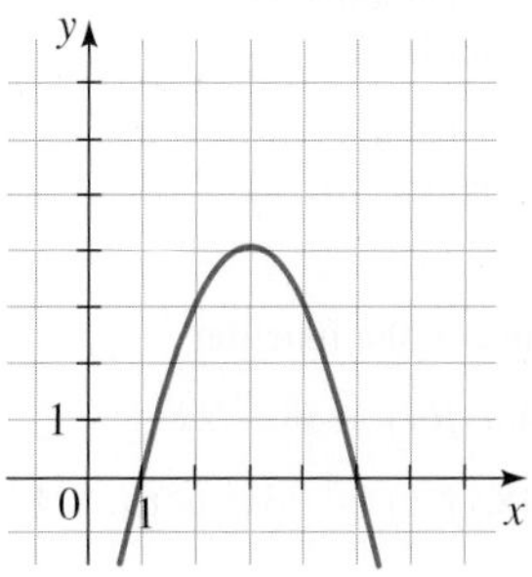

6. $f(x) = -\frac{1}{2}x^2 - 2x + 6$

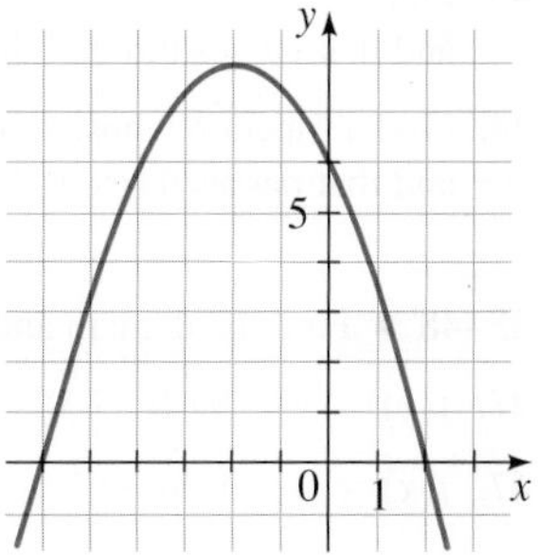

7. $f(x) = 2x^2 - 4x - 1$

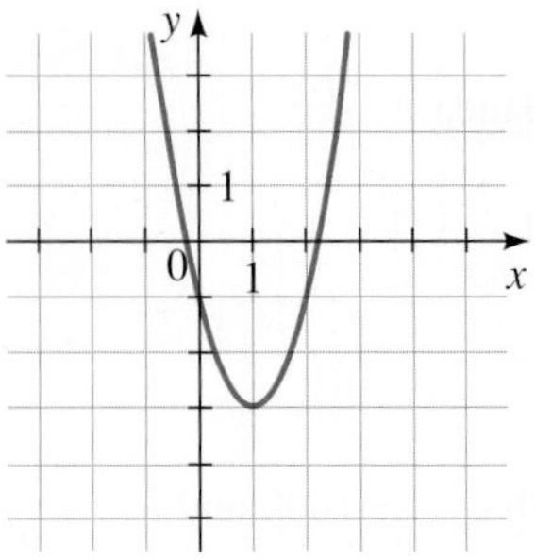

8. $f(x) = 3x^2 + 6x - 1$

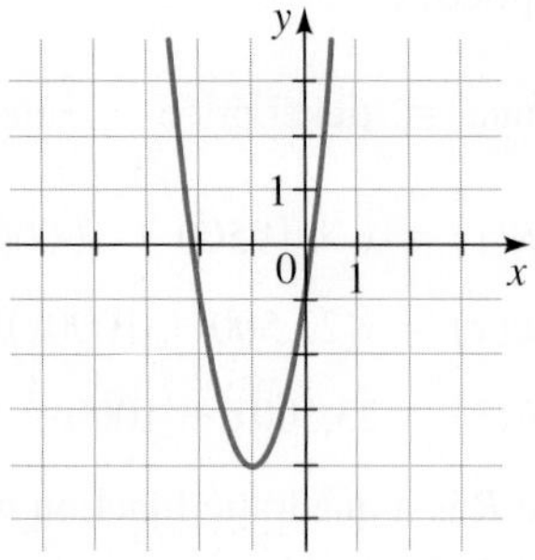

9–22 ■ A quadratic function is given. **(a)** Express the quadratic function in standard form. **(b)** Find its vertex and its x- and y-intercept(s). **(c)** Sketch its graph.

9. $f(x) = x^2 - 6x$

10. $f(x) = x^2 + 8x$

11. $f(x) = 2x^2 + 6x$

12. $f(x) = -x^2 + 10x$

13. $f(x) = x^2 + 4x + 3$

14. $f(x) = x^2 - 2x + 2$

15. $f(x) = -x^2 + 6x + 4$

16. $f(x) = -x^2 - 4x + 4$

17. $f(x) = 2x^2 + 4x + 3$

18. $f(x) = -3x^2 + 6x - 2$

19. $f(x) = 2x^2 - 20x + 57$

20. $f(x) = 2x^2 + x - 6$

21. $f(x) = -4x^2 - 16x + 3$

22. $f(x) = 6x^2 + 12x - 5$

23–32 ■ A quadratic function is given. **(a)** Express the quadratic function in standard form. **(b)** Sketch its graph. **(c)** Find its maximum or minimum value.

23. $f(x) = x^2 + 2x - 1$

24. $f(x) = x^2 - 8x + 8$

25. $f(x) = 3x^2 - 6x + 1$

26. $f(x) = 5x^2 + 30x + 4$

27. $f(x) = -x^2 - 3x + 3$

28. $f(x) = 1 - 6x - x^2$

29. $g(x) = 3x^2 - 12x + 13$

30. $g(x) = 2x^2 + 8x + 11$

31. $h(x) = 1 - x - x^2$

32. $h(x) = 3 - 4x - 4x^2$

33–42 ■ Find the maximum or minimum value of the function.

33. $f(x) = x^2 + x + 1$

34. $f(x) = 1 + 3x - x^2$

35. $f(t) = 100 - 49t - 7t^2$

36. $f(t) = 10t^2 + 40t + 113$

37. $f(s) = s^2 - 1.2s + 16$

38. $g(x) = 100x^2 - 1500x$

39. $h(x) = \frac{1}{2}x^2 + 2x - 6$

40. $f(x) = -\dfrac{x^2}{3} + 2x + 7$

41. $f(x) = 3 - x - \frac{1}{2}x^2$

42. $g(x) = 2x(x - 4) + 7$

43. Find a function whose graph is a parabola with vertex $(1, -2)$ and that passes through the point $(4, 16)$.

44. Find a function whose graph is a parabola with vertex $(3, 4)$ and that passes through the point $(1, -8)$.

45–48 ■ Find the domain and range of the function.

45. $f(x) = -x^2 + 4x - 3$

46. $f(x) = x^2 - 2x - 3$

47. $f(x) = 2x^2 + 6x - 7$

48. $f(x) = -3x^2 + 6x + 4$

49–50 ■ A quadratic function is given. **(a)** Use a graphing device to find the maximum or minimum value of the quadratic function f, correct to two decimal places. **(b)** Find the exact maximum or minimum value of f, and compare it with your answer to part (a).

49. $f(x) = x^2 + 1.79x - 3.21$

50. $f(x) = 1 + x - \sqrt{2}x^2$

51–54 ■ Find all local maximum and minimum values of the function whose graph is shown.

51.

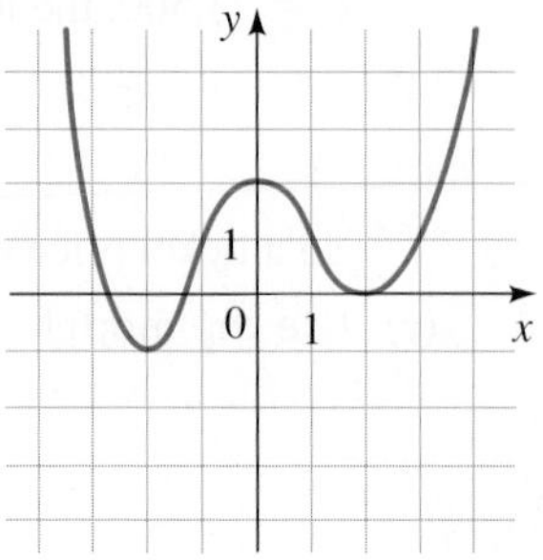

52.

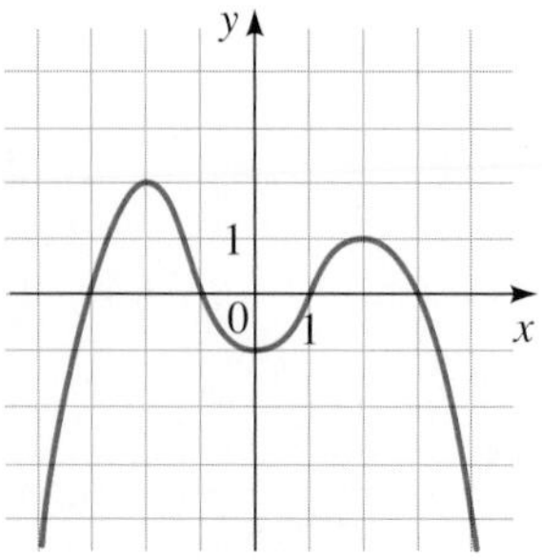

53.

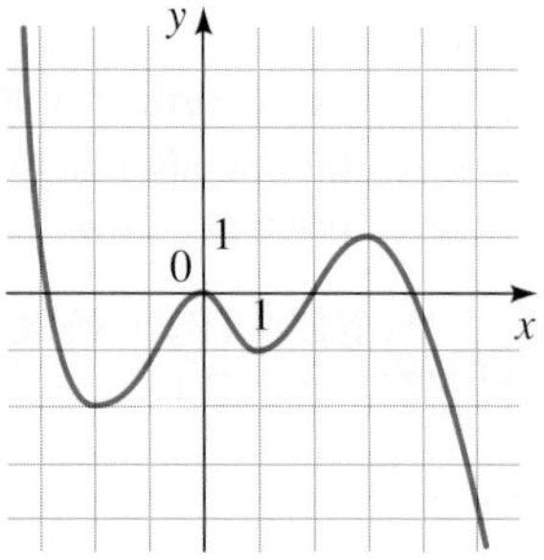

54.

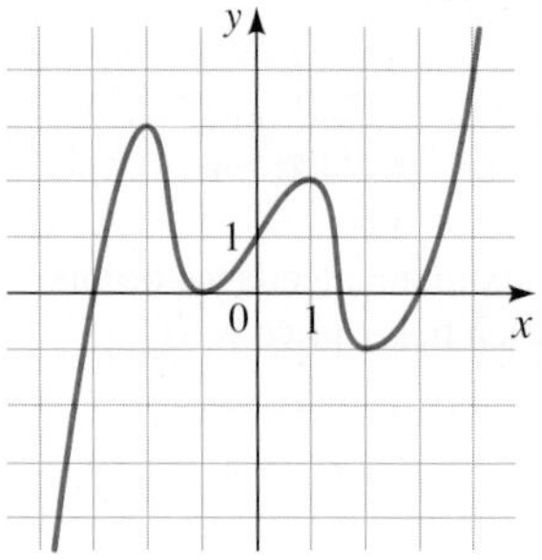

55–62 ■ Find the local maximum and minimum values of the function and the value of x at which each occurs. State each answer correct to two decimal places.

55. $f(x) = x^3 - x$

56. $f(x) = 3 + x + x^2 - x^3$

57. $g(x) = x^4 - 2x^3 - 11x^2$

58. $g(x) = x^5 - 8x^3 + 20x$

59. $U(x) = x\sqrt{6 - x}$

60. $U(x) = x\sqrt{x - x^2}$

61. $V(x) = \dfrac{1 - x^2}{x^3}$

62. $V(x) = \dfrac{1}{x^2 + x + 1}$

APPLICATIONS

63. Height of a Ball If a ball is thrown directly upward with a velocity of 40 ft/s, its height (in feet) after t seconds is given by $y = 40t - 16t^2$. What is the maximum height attained by the ball?

64. Path of a Ball A ball is thrown across a playing field from a height of 5 ft above the ground at an angle of 45° to the horizontal at a speed of 20 ft/s. It can be deduced from physical principles that the path of the ball is modeled by the function

$$y = -\frac{32}{(20)^2}x^2 + x + 5$$

where x is the distance in feet that the ball has traveled horizontally.

(a) Find the maximum height attained by the ball.

(b) Find the horizontal distance the ball has traveled when it hits the ground.

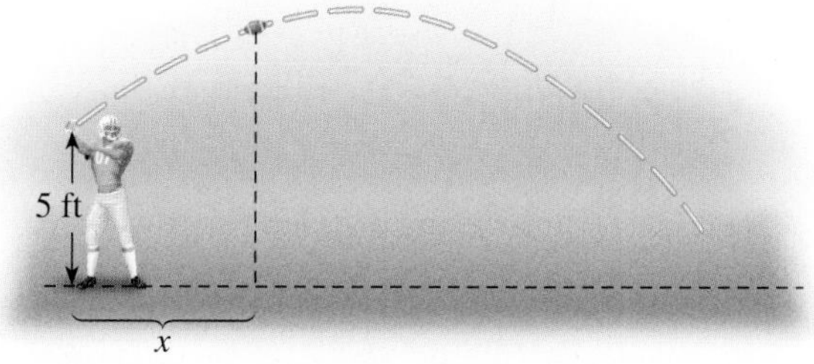

65. Revenue A manufacturer finds that the revenue generated by selling x units of a certain commodity is given by the function $R(x) = 80x - 0.4x^2$, where the revenue $R(x)$ is measured in dollars. What is the maximum revenue, and how many units should be manufactured to obtain this maximum?

66. Sales A soft-drink vendor at a popular beach analyzes his sales records and finds that if he sells x cans of soda pop in one day, his profit (in dollars) is given by

$$P(x) = -0.001x^2 + 3x - 1800$$

What is his maximum profit per day, and how many cans must he sell for maximum profit?

67. Advertising The effectiveness of a television commercial depends on how many times a viewer watches it. After some experiments an advertising agency found that if the effectiveness E is measured on a scale of 0 to 10, then

$$E(n) = \tfrac{2}{3}n - \tfrac{1}{90}n^2$$

where n is the number of times a viewer watches a given commercial. For a commercial to have maximum effectiveness, how many times should a viewer watch it?

68. Pharmaceuticals When a certain drug is taken orally, the concentration of the drug in the patient's bloodstream after t minutes is given by $C(t) = 0.06t - 0.0002t^2$, where $0 \le t \le 240$ and the concentration is measured in mg/L. When is the maximum serum concentration reached, and what is that maximum concentration?

69. Agriculture The number of apples produced by each tree in an apple orchard depends on how densely the trees are planted. If n trees are planted on an acre of land, then each tree produces $900 - 9n$ apples. So the number of apples produced per acre is

$$A(n) = n(900 - 9n)$$

How many trees should be planted per acre to obtain the maximum yield of apples?

70. Agriculture At a certain vineyard it is found that each grape vine produces about 10 pounds of grapes in a season when about 700 vines are planted per acre. For each additional vine that is planted, the production of each vine decreases by about 1 percent. So the number of pounds of grapes produced per acre is modeled by

$$A(n) = (700 + n)(10 - 0.01n)$$

where n is the number of additional vines planted. Find the number of vines that should be planted to maximize grape production.

71–74 ■ Use the formulas of this section to give an alternative solution to the indicated problem in *Focus on Modeling: Modeling with Functions* on pages 220–221.

71. Problem 21 **72.** Problem 22

73. Problem 25 **74.** Problem 24

75. Fencing a Horse Corral Carol has 2400 ft of fencing to fence in a rectangular horse corral.

(a) Find a function that models the area of the corral in terms of the width x of the corral.

(b) Find the dimensions of the rectangle that maximize the area of the corral.

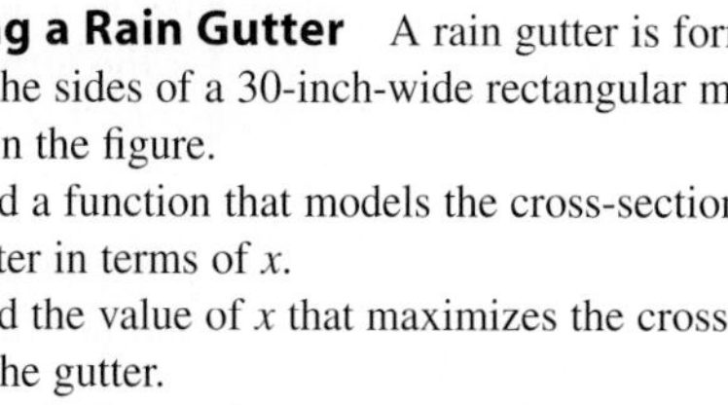

76. Making a Rain Gutter A rain gutter is formed by bending up the sides of a 30-inch-wide rectangular metal sheet as shown in the figure.

(a) Find a function that models the cross-sectional area of the gutter in terms of x.

(b) Find the value of x that maximizes the cross-sectional area of the gutter.

(c) What is the maximum cross-sectional area for the gutter?

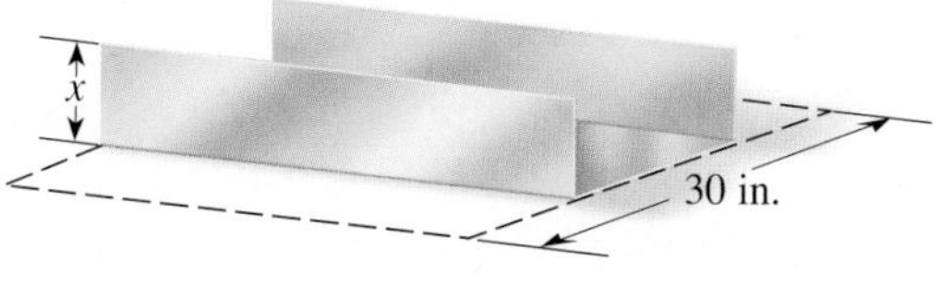

77. Stadium Revenue A baseball team plays in a stadium that holds 55,000 spectators. With the ticket price at \$10, the average attendance at recent games has been 27,000. A market survey indicates that for every dollar the ticket price is lowered, attendance increases by 3000.

(a) Find a function that models the revenue in terms of ticket price.

(b) Find the price that maximizes revenue from ticket sales.

(c) What ticket price is so high that no revenue is generated?

78. Maximizing Profit A community bird-watching society makes and sells simple bird feeders to raise money for its conservation activities. The materials for each feeder cost \$6, and the society sells an average of 20 per week at a price of \$10 each. The society has been considering raising the price, so it conducts a survey and finds that for every dollar increase, it loses 2 sales per week.

(a) Find a function that models weekly profit in terms of price per feeder.

(b) What price should the society charge for each feeder to maximize profits? What is the maximum weekly profit?

DISCOVERY ▪ DISCUSSION ▪ WRITING

79. Vertex and x-Intercepts We know that the graph of the quadratic function $f(x) = (x - m)(x - n)$ is a parabola. Sketch a rough graph of what such a parabola would look like. What are the x-intercepts of the graph of f? Can you tell from your graph the x-coordinate of the vertex in terms of m and n? (Use the symmetry of the parabola.) Confirm your answer by expanding and using the formulas of this section.

80. Maximum of a Fourth-Degree Polynomial Find the maximum value of the function

$$f(x) = 3 + 4x^2 - x^4$$

[*Hint:* Let $t = x^2$.]

3.2 Polynomial Functions and Their Graphs

Graphing Basic Polynomial Functions ▶ End Behavior and the Leading Term ▶ Using Zeros to Graph Polynomials ▶ Shape of the Graph Near a Zero ▶ Local Maxima and Minima of Polynomials

In this section we study polynomial functions of any degree. But before we work with polynomial functions, we must agree on some terminology.

POLYNOMIAL FUNCTIONS

A **polynomial function of degree n** is a function of the form

$$P(x) = a_n x^n + a_{n-1}x^{n-1} + \cdots + a_1 x + a_0$$

where n is a nonnegative integer and $a_n \neq 0$.

The numbers $a_0, a_1, a_2, \ldots, a_n$ are called the **coefficients** of the polynomial.

The number a_0 is the **constant coefficient** or **constant term**.

The number a_n, the coefficient of the highest power, is the **leading coefficient**, and the term $a_n x^n$ is the **leading term**.

We often refer to polynomial functions simply as *polynomials*. The following polynomial has degree 5, leading coefficient 3, and constant term -6.

Leading coefficient 3 | Degree 5 | Constant term -6

$$3x^5 + 6x^4 - 2x^3 + x^2 + 7x - 6$$

Leading term $3x^5$ | Coefficients 3, 6, -2, 1, 7, and -6

Here are some more examples of polynomials.

$P(x) = 3$	Degree 0
$Q(x) = 4x - 7$	Degree 1
$R(x) = x^2 + x$	Degree 2
$S(x) = 2x^3 - 6x^2 - 10$	Degree 3

If a polynomial consists of just a single term, then it is called a **monomial**. For example, $P(x) = x^3$ and $Q(x) = -6x^5$ are monomials.

▼ Graphing Basic Polynomial Functions

The graphs of polynomials of degree 0 or 1 are lines (Section 1.10), and the graphs of polynomials of degree 2 are parabolas (Section 3.1). The greater the degree of a polynomial, the more complicated its graph can be. However, the graph of a polynomial function is **continuous**. This means that the graph has no breaks or holes (see Figure 1). Moreover, the graph of a polynomial function is a smooth curve; that is, it has no corners or sharp points (cusps) as shown in Figure 1.

Continuous functions are studied in Section 13.2, page 851.

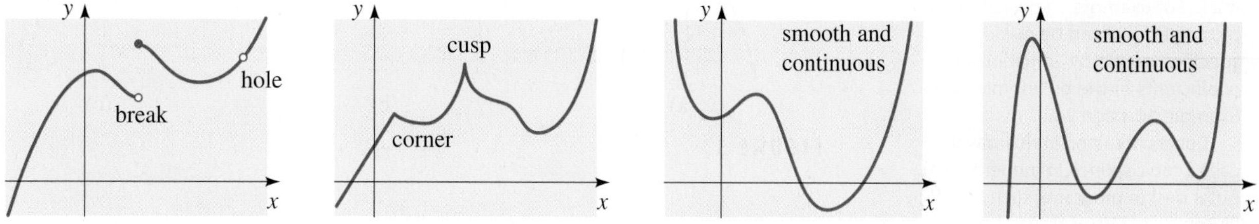

Not the graph of a polynomial function | Not the graph of a polynomial function | Graph of a polynomial function | Graph of a polynomial function

FIGURE 1

The simplest polynomial functions are the monomials $P(x) = x^n$, whose graphs are shown in Figure 2. As the figure suggests, the graph of $P(x) = x^n$ has the same general shape as the graph of $y = x^2$ when n is even and the same general shape as the graph of $y = x^3$ when n is odd. However, as the degree n becomes larger, the graphs become flatter around the origin and steeper elsewhere.

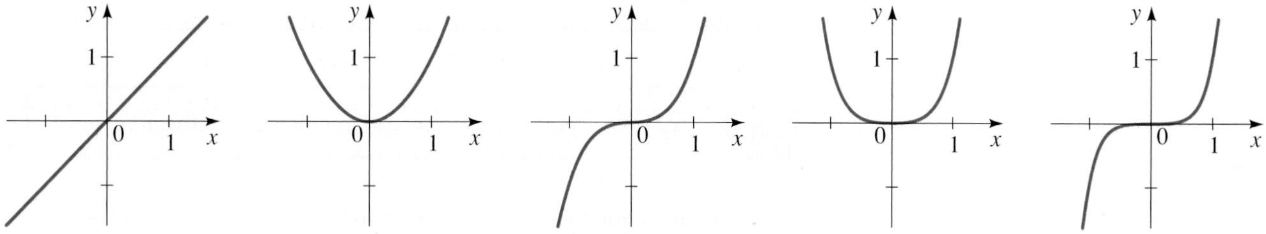

(a) $y = x$ (b) $y = x^2$ (c) $y = x^3$ (d) $y = x^4$ (e) $y = x^5$

FIGURE 2 Graphs of monomials

EXAMPLE 1 | Transformations of Monomials

Sketch the graphs of the following functions.

(a) $P(x) = -x^3$ **(b)** $Q(x) = (x - 2)^4$

(c) $R(x) = -2x^5 + 4$

SOLUTION We use the graphs in Figure 2 and transform them using the techniques of Section 2.5.

(a) The graph of $P(x) = -x^3$ is the reflection of the graph of $y = x^3$ in the x-axis, as shown in Figure 3(a) below.

(b) The graph of $Q(x) = (x - 2)^4$ is the graph of $y = x^4$ shifted to the right 2 units, as shown in Figure 3(b).

(c) We begin with the graph of $y = x^5$. The graph of $y = -2x^5$ is obtained by stretching the graph vertically and reflecting it in the x-axis (see the dashed blue graph in Figure 3(c)). Finally, the graph of $R(x) = -2x^5 + 4$ is obtained by shifting upward 4 units (see the red graph in Figure 3(c)).

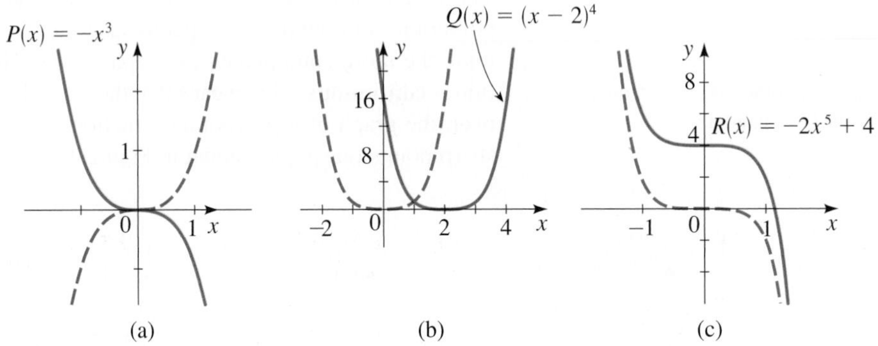

FIGURE 3

NOW TRY EXERCISE **5**

MATHEMATICS IN THE MODERN WORLD

Splines

A spline is a long strip of wood that is curved while held fixed at certain points. In the old days shipbuilders used splines to create the curved shape of a boat's hull. Splines are also used to make the curves of a piano, a violin, or the spout of a teapot.

Mathematicians discovered that the shapes of splines can be obtained by piecing together parts of polynomials. For example, the graph of a cubic polynomial can be made to fit specified points by adjusting the coefficients of the polynomial (see Example 10, page 242).

Curves obtained in this way are called cubic splines. In modern computer design programs, such as Adobe Illustrator or Microsoft Paint, a curve can be drawn by fixing two points, then using the mouse to drag one or more anchor points. Moving the anchor points amounts to adjusting the coefficients of a cubic polynomial.

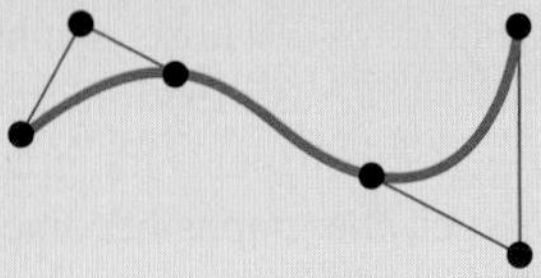

▼ End Behavior and the Leading Term

The **end behavior** of a polynomial is a description of what happens as x becomes large in the positive or negative direction. To describe end behavior, we use the following notation:

$x \to \infty$	means	"x becomes large in the positive direction"

$x \to -\infty$	means	"x becomes large in the negative direction"

For example, the monomial $y = x^2$ in Figure 2(b) has the following end behavior:

$$y \to \infty \quad \text{as} \quad x \to \infty \qquad \text{and} \qquad y \to \infty \quad \text{as} \quad x \to -\infty$$

The monomial $y = x^3$ in Figure 2(c) has the following end behavior:

$$y \to \infty \quad \text{as} \quad x \to \infty \qquad \text{and} \qquad y \to -\infty \quad \text{as} \quad x \to -\infty$$

For any polynomial *the end behavior is determined by the term that contains the highest power of* x, because when x is large, the other terms are relatively insignificant in size. The following box shows the four possible types of end behavior, based on the highest power and the sign of its coefficient.

END BEHAVIOR OF POLYNOMIALS

The end behavior of the polynomial $P(x) = a_n x^n + a_{n-1}x^{n-1} + \cdots + a_1 x + a_0$ is determined by the degree n and the sign of the leading coefficient a_n, as indicated in the following graphs.

***P* has odd degree** | ***P* has even degree**

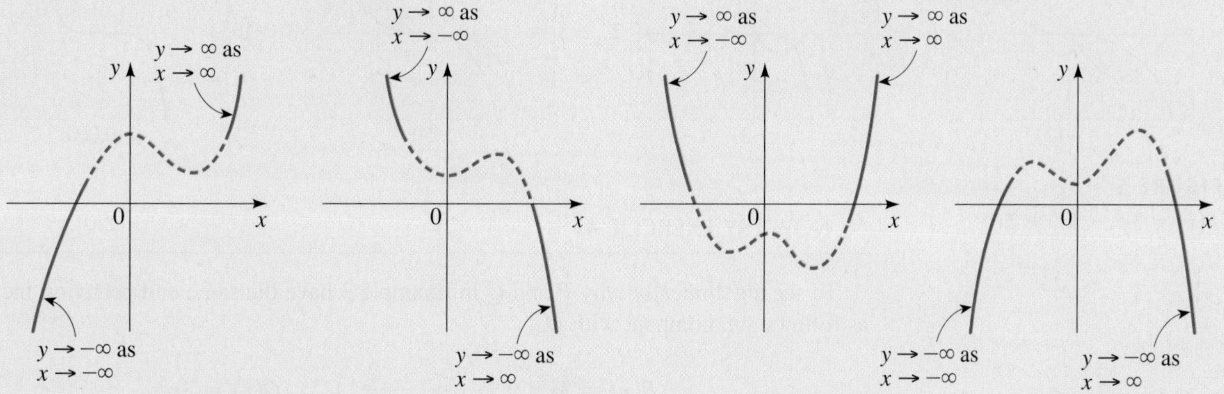

Leading coefficient positive | Leading coefficient negative | Leading coefficient positive | Leading coefficient negative

EXAMPLE 2 | End Behavior of a Polynomial

Determine the end behavior of the polynomial

$$P(x) = -2x^4 + 5x^3 + 4x - 7$$

SOLUTION The polynomial P has degree 4 and leading coefficient -2. Thus, P has *even* degree and *negative* leading coefficient, so it has the following end behavior:

$$y \to -\infty \quad \text{as} \quad x \to \infty \qquad \text{and} \qquad y \to -\infty \quad \text{as} \quad x \to -\infty$$

The graph in Figure 4 illustrates the end behavior of P.

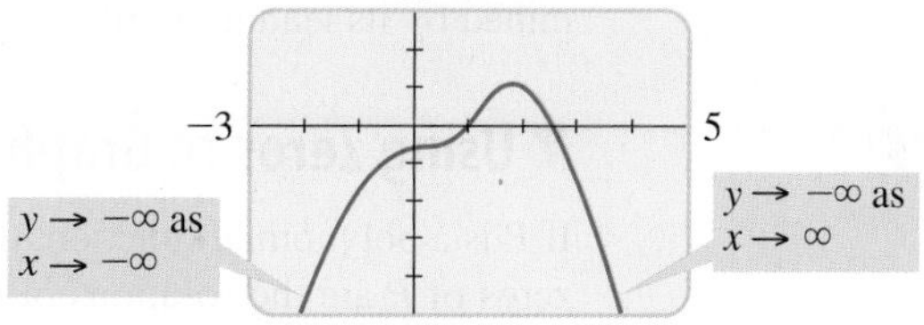

FIGURE 4
$P(x) = -2x^4 + 5x^3 + 4x - 7$

NOW TRY EXERCISE **11** ■

EXAMPLE 3 | End Behavior of a Polynomial

(a) Determine the end behavior of the polynomial $P(x) = 3x^5 - 5x^3 + 2x$.

(b) Confirm that P and its leading term $Q(x) = 3x^5$ have the same end behavior by graphing them together.

SOLUTION

(a) Since P has odd degree and positive leading coefficient, it has the following end behavior:

$$y \to \infty \quad \text{as} \quad x \to \infty \qquad \text{and} \qquad y \to -\infty \quad \text{as} \quad x \to -\infty$$

(b) Figure 5 shows the graphs of P and Q in progressively larger viewing rectangles. The larger the viewing rectangle, the more the graphs look alike. This confirms that they have the same end behavior.

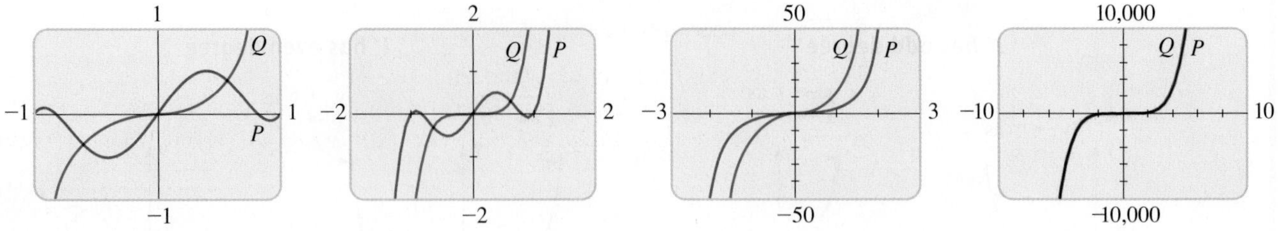

FIGURE 5
$P(x) = 3x^5 - 5x^3 + 2x$
$Q(x) = 3x^5$

NOW TRY EXERCISE **41** ■

To see algebraically why P and Q in Example 3 have the same end behavior, factor P as follows and compare with Q.

$$P(x) = 3x^5\left(1 - \frac{5}{3x^2} + \frac{2}{3x^4}\right) \qquad Q(x) = 3x^5$$

When x is large, the terms $5/3x^2$ and $2/3x^4$ are close to 0 (see Exercise 83 on page 12). So for large x, we have

$$P(x) \approx 3x^5(1 - 0 - 0) = 3x^5 = Q(x)$$

So when x is large, P and Q have approximately the same values. We can also see this numerically by making a table like the one shown below.

x	$P(x)$	$Q(x)$
15	2,261,280	2,278,125
30	72,765,060	72,900,000
50	936,875,100	937,500,000

By the same reasoning we can show that the end behavior of *any* polynomial is determined by its leading term.

▼ Using Zeros to Graph Polynomials

If P is a polynomial function, then c is called a **zero** of P if $P(c) = 0$. In other words, the zeros of P are the solutions of the polynomial equation $P(x) = 0$. Note that if $P(c) = 0$, then the graph of P has an x-intercept at $x = c$, so the x-intercepts of the graph are the zeros of the function.

REAL ZEROS OF POLYNOMIALS

If P is a polynomial and c is a real number, then the following are equivalent:

1. c is a zero of P.
2. $x = c$ is a solution of the equation $P(x) = 0$.
3. $x - c$ is a factor of $P(x)$.
4. c is an x-intercept of the graph of P.

To find the zeros of a polynomial P, we factor and then use the Zero-Product Property (see page 47). For example, to find the zeros of $P(x) = x^2 + x - 6$, we factor P to get

$$P(x) = (x - 2)(x + 3)$$

From this factored form we easily see that

1. 2 is a zero of P.
2. $x = 2$ is a solution of the equation $x^2 + x - 6 = 0$.
3. $x - 2$ is a factor of $x^2 + x - 6$.
4. 2 is an x-intercept of the graph of P.

The same facts are true for the other zero, -3.

The following theorem has many important consequences. (See, for instance, the *Discovery Project* referenced on page 263.) Here we use it to help us graph polynomial functions.

INTERMEDIATE VALUE THEOREM FOR POLYNOMIALS

If P is a polynomial function and $P(a)$ and $P(b)$ have opposite signs, then there exists at least one value c between a and b for which $P(c) = 0$.

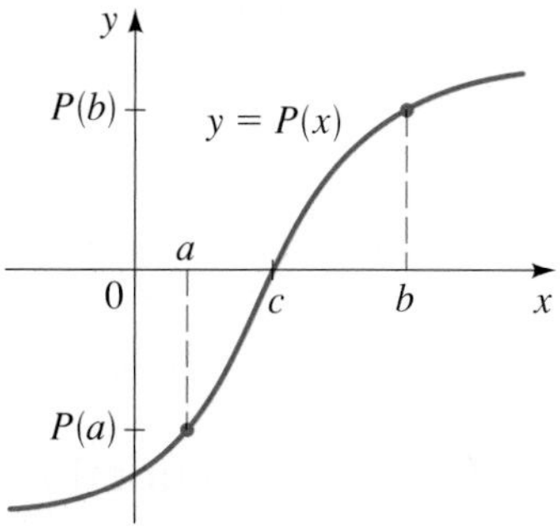

FIGURE 6

We will not prove this theorem, but Figure 6 shows why it is intuitively plausible.

One important consequence of this theorem is that between any two successive zeros the values of a polynomial are either all positive or all negative. That is, between two successive zeros the graph of a polynomial lies *entirely above* or *entirely below* the x-axis. To see why, suppose c_1 and c_2 are successive zeros of P. If P has both positive and negative values between c_1 and c_2, then by the Intermediate Value Theorem P must have another zero between c_1 and c_2. But that's not possible because c_1 and c_2 are successive zeros. This observation allows us to use the following guidelines to graph polynomial functions.

GUIDELINES FOR GRAPHING POLYNOMIAL FUNCTIONS

1. **Zeros.** Factor the polynomial to find all its real zeros; these are the x-intercepts of the graph.
2. **Test Points.** Make a table of values for the polynomial. Include test points to determine whether the graph of the polynomial lies above or below the x-axis on the intervals determined by the zeros. Include the y-intercept in the table.
3. **End Behavior.** Determine the end behavior of the polynomial.
4. **Graph.** Plot the intercepts and other points you found in the table. Sketch a smooth curve that passes through these points and exhibits the required end behavior.

EXAMPLE 4 | Using Zeros to Graph a Polynomial Function

Sketch the graph of the polynomial function $P(x) = (x + 2)(x - 1)(x - 3)$.

SOLUTION The zeros are $x = -2$, 1, and 3. These determine the intervals $(-\infty, -2)$, $(-2, 1)$, $(1, 3)$, and $(3, \infty)$. Using test points in these intervals, we get the information in the following sign diagram (see Section 1.7).

		-2		1		3	
	Test point $x = -3$ $P(-3) < 0$		Test point $x = -1$ $P(-1) > 0$		Test point $x = 2$ $P(2) < 0$		Test point $x = 4$ $P(3) > 0$
Sign of $P(x) = (x + 2)(x - 1)(x - 3)$	$-$		$+$		$-$		$+$
Graph of P	below x-axis		above x-axis		below x-axis		above x-axis

Plotting a few additional points and connecting them with a smooth curve helps us to complete the graph in Figure 7.

	x	$P(x)$
Test point →	-3	-24
	-2	0
Test point →	-1	8
	0	6
	1	0
Test point →	2	-4
	3	0
Test point →	4	18

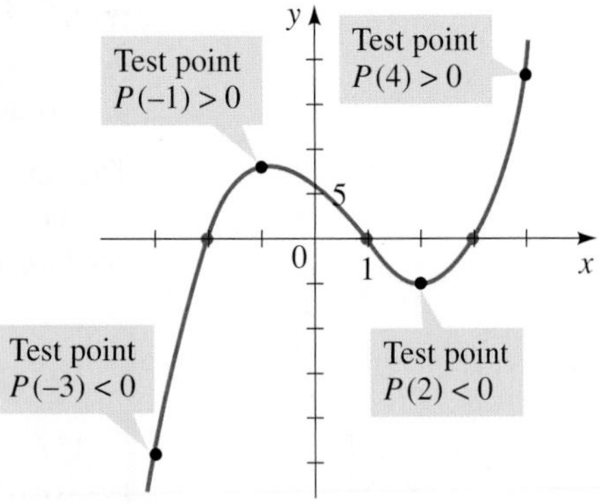

FIGURE 7 $P(x) = (x + 2)(x - 1)(x - 3)$

NOW TRY EXERCISE **17**

EXAMPLE 5 | Finding Zeros and Graphing a Polynomial Function

Let $P(x) = x^3 - 2x^2 - 3x$.

(a) Find the zeros of P.

(b) Sketch a graph of P.

SOLUTION

(a) To find the zeros, we factor completely.

$$P(x) = x^3 - 2x^2 - 3x$$

$$= x(x^2 - 2x - 3) \qquad \text{Factor } x$$

$$= x(x - 3)(x + 1) \qquad \text{Factor quadratic}$$

Thus, the zeros are $x = 0$, $x = 3$, and $x = -1$.

(b) The x-intercepts are $x = 0$, $x = 3$, and $x = -1$. The y-intercept is $P(0) = 0$. We make a table of values of $P(x)$, making sure that we choose test points between (and to the right and left of) successive zeros.

Since P is of odd degree and its leading coefficient is positive, it has the following end behavior:

$$y \to \infty \quad \text{as} \quad x \to \infty \qquad \text{and} \qquad y \to -\infty \quad \text{as} \quad x \to -\infty$$

We plot the points in the table and connect them by a smooth curve to complete the graph, as shown in Figure 8.

	x	$P(x)$
Test point →	-2	-10
	-1	0
Test point →	$-\frac{1}{2}$	$\frac{7}{8}$
	0	0
Test point →	1	-4
	2	-6
	3	0
Test point →	4	20

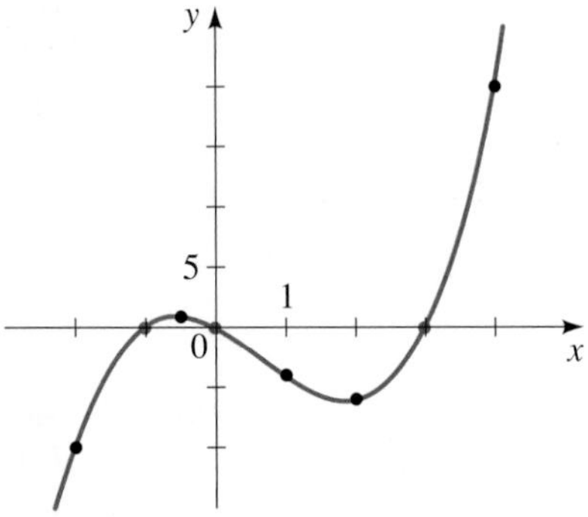

FIGURE 8 $P(x) = x^3 - 2x^2 - 3x$

NOW TRY EXERCISE **27**

MATHEMATICS IN THE MODERN WORLD

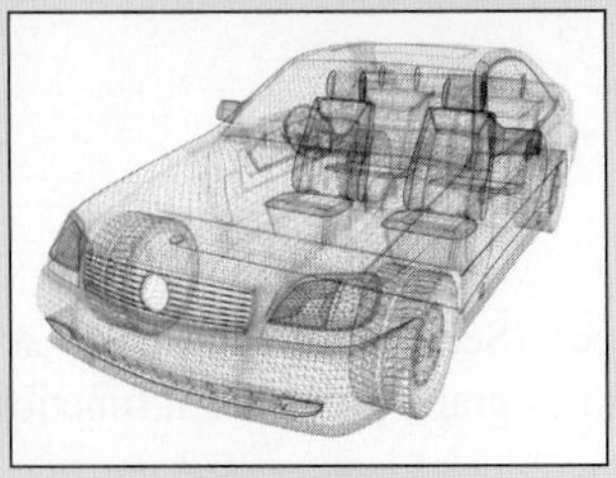

3DProfi/Shutterstock.com

Automotive Design

Computer-aided design (CAD) has completely changed the way in which car companies design and manufacture cars. Before the 1980s automotive engineers would build a full-scale "nuts and bolts" model of a proposed new car; this was really the only way to tell whether the design was feasible. Today automotive engineers build a mathematical model, one that exists only in the memory of a computer. The model incorporates all the main design features of the car. Certain polynomial curves, called *splines*, are used in shaping the body of the car. The resulting "mathematical car" can be tested for structural stability, handling, aerodynamics, suspension response, and more. All this testing is done before a prototype is built. As you can imagine, CAD saves car manufacturers millions of dollars each year. More importantly, CAD gives automotive engineers far more flexibility in design; desired changes can be created and tested within seconds. With the help of computer graphics, designers can see how good the "mathematical car" looks before they build the real one. Moreover, the mathematical car can be viewed from any perspective; it can be moved, rotated, or seen from the inside. These manipulations of the car on the computer monitor translate mathematically into solving large systems of linear equations.

EXAMPLE 6 | Finding Zeros and Graphing a Polynomial Function

Let $P(x) = -2x^4 - x^3 + 3x^2$.

(a) Find the zeros of P. **(b)** Sketch a graph of P.

SOLUTION

(a) To find the zeros, we factor completely.

$$\begin{aligned} P(x) &= -2x^4 - x^3 + 3x^2 \\ &= -x^2(2x^2 + x - 3) && \text{Factor } -x^2 \\ &= -x^2(2x + 3)(x - 1) && \text{Factor quadratic} \end{aligned}$$

Thus, the zeros are $x = 0$, $x = -\frac{3}{2}$, and $x = 1$.

(b) The x-intercepts are $x = 0$, $x = -\frac{3}{2}$, and $x = 1$. The y-intercept is $P(0) = 0$. We make a table of values of $P(x)$, making sure that we choose test points between (and to the right and left of) successive zeros.

Since P is of even degree and its leading coefficient is negative, it has the following end behavior:

$$y \to -\infty \quad \text{as} \quad x \to \infty \qquad \text{and} \qquad y \to -\infty \quad \text{as} \quad x \to -\infty$$

We plot the points from the table and connect the points by a smooth curve to complete the graph in Figure 9.

A table of values is most easily calculated by using a programmable calculator or a graphing calculator.

x	$P(x)$
−2	−12
−1.5	0
−1	2
−0.5	0.75
0	0
0.5	0.5
1	0
1.5	−6.75

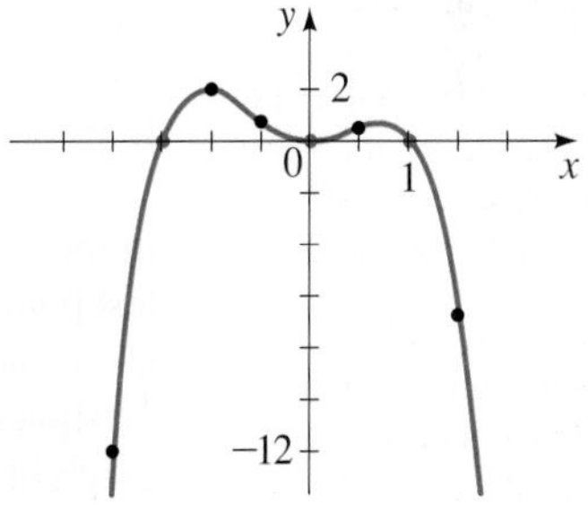

FIGURE 9 $P(x) = -2x^4 - x^3 + 3x^2$

NOW TRY EXERCISE **31**

EXAMPLE 7 | Finding Zeros and Graphing a Polynomial Function

Let $P(x) = x^3 - 2x^2 - 4x + 8$.

(a) Find the zeros of P. **(b)** Sketch a graph of P.

SOLUTION

(a) To find the zeros, we factor completely.

$$\begin{aligned} P(x) &= x^3 - 2x^2 - 4x + 8 \\ &= x^2(x - 2) - 4(x - 2) && \text{Group and factor} \\ &= (x^2 - 4)(x - 2) && \text{Factor } x - 2 \\ &= (x + 2)(x - 2)(x - 2) && \text{Difference of squares} \\ &= (x + 2)(x - 2)^2 && \text{Simplify} \end{aligned}$$

Thus the zeros are $x = -2$ and $x = 2$.

(b) The x-intercepts are $x = -2$ and $x = 2$. The y-intercept is $P(0) = 8$. The table gives additional values of $P(x)$.

Since P is of odd degree and its leading coefficient is positive, it has the following end behavior:

$$y \to \infty \quad \text{as} \quad x \to \infty \qquad \text{and} \qquad y \to -\infty \quad \text{as} \quad x \to -\infty$$

We connect the points by a smooth curve to complete the graph in Figure 10.

x	$P(x)$
-3	-25
-2	0
-1	9
0	8
1	3
2	0
3	5

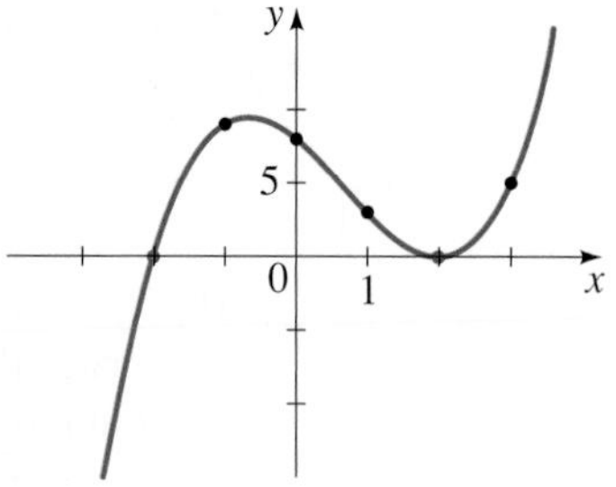

FIGURE 10 $P(x) = x^3 - 2x^2 - 4x + 8$

NOW TRY EXERCISE **33**

Shape of the Graph Near a Zero

Although $x = 2$ is a zero of the polynomial in Example 7, the graph does not cross the x-axis at the x-intercept 2. This is because the factor $(x - 2)^2$ corresponding to that zero is raised to an even power, so it doesn't change sign as we test points on either side of 2. In the same way the graph does not cross the x-axis at $x = 0$ in Example 6.

In general, if c is a zero of P, and the corresponding factor $x - c$ occurs exactly m times in the factorization of P, then we say that c is a **zero of multiplicity m**. By considering test points on either side of the x-intercept c, we conclude that the graph crosses the x-axis at c if the multiplicity m is odd and does not cross the x-axis if m is even. Moreover, it can be shown by using calculus that near $x = c$ the graph has the same general shape as the graph of $y = A(x - c)^m$.

SHAPE OF THE GRAPH NEAR A ZERO OF MULTIPLICITY m

If c is a zero of P of multiplicity m, then the shape of the graph of P near c is as follows.

Multiplicity of c	**Shape of the graph of P near the x-intercept c**
m odd, $m > 1$	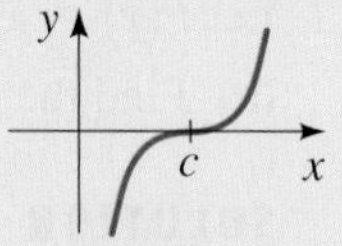 OR 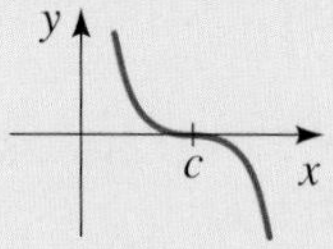
m even, $m > 1$	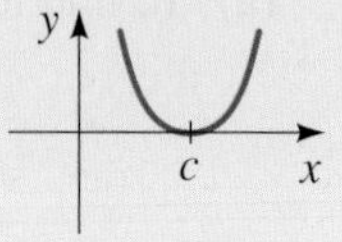 OR 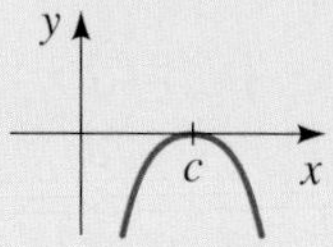

EXAMPLE 8 | Graphing a Polynomial Function Using Its Zeros

Graph the polynomial $P(x) = x^4(x - 2)^3(x + 1)^2$.

SOLUTION The zeros of P are -1, 0, and 2 with multiplicities 2, 4, and 3, respectively.

0 is a zero of multiplicity 4 — 2 is a zero of multiplicity 3 — -1 is a zero of multiplicity 2

$$P(x) = x^4(x-2)^3(x+1)^2$$

The zero 2 has *odd* multiplicity, so the graph crosses the x-axis at the x-intercept 2. But the zeros 0 and -1 have *even* multiplicity, so the graph does not cross the x-axis at the x-intercepts 0 and -1.

Since P is a polynomial of degree 9 and has positive leading coefficient, it has the following end behavior:

$$y \to \infty \quad \text{as} \quad x \to \infty \qquad \text{and} \qquad y \to -\infty \quad \text{as} \quad x \to -\infty$$

With this information and a table of values we sketch the graph in Figure 11.

x	$P(x)$
-1.3	-9.2
-1	0
-0.5	-3.9
0	0
1	-4
2	0
2.3	8.2

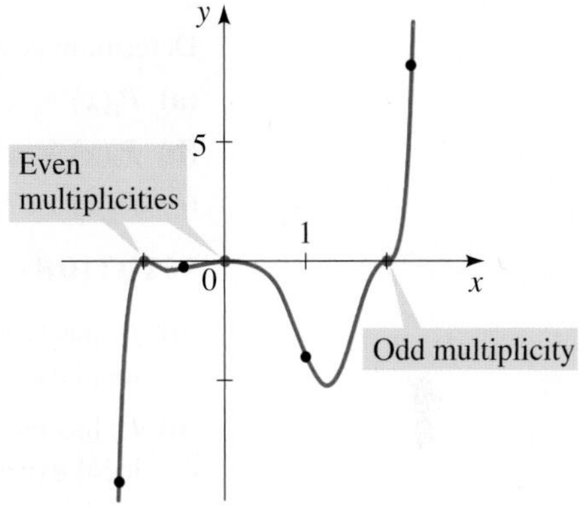

FIGURE 11 $P(x) = x^4(x-2)^3(x+1)^2$

NOW TRY EXERCISE **25**

Local Maxima and Minima of Polynomials

Recall from Section 2.3 that if the point $(a, f(a))$ is the highest point on the graph of f within some viewing rectangle, then $f(a)$ is a local maximum value of f, and if $(b, f(b))$ is the lowest point on the graph of f within a viewing rectangle, then $f(b)$ is a local minimum value (see Figure 12). We say that such a point $(a, f(a))$ is a **local maximum point** on the graph and that $(b, f(b))$ is a **local minimum point**. The local maximum and minimum points on the graph of a function are called its **local extrema**.

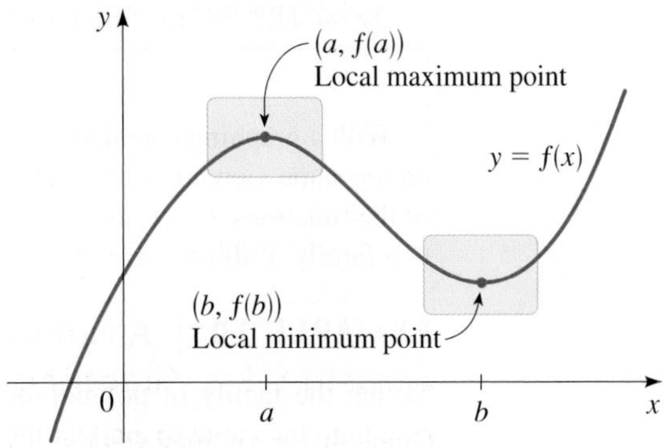

FIGURE 12

For a polynomial function the number of local extrema must be less than the degree, as the following principle indicates. (A proof of this principle requires calculus.)

LOCAL EXTREMA OF POLYNOMIALS

If $P(x) = a_n x^n + a_{n-1}x^{n-1} + \cdots + a_1 x + a_0$ is a polynomial of degree n, then the graph of P has at most $n - 1$ local extrema.

A polynomial of degree n may in fact have less than $n - 1$ local extrema. For example, $P(x) = x^5$ (graphed in Figure 2) has *no* local extrema, even though it is of degree 5. The preceding principle tells us only that a polynomial of degree n can have no more than $n - 1$ local extrema.

EXAMPLE 9 | The Number of Local Extrema

Determine how many local extrema each polynomial has.

(a) $P_1(x) = x^4 + x^3 - 16x^2 - 4x + 48$

(b) $P_2(x) = x^5 + 3x^4 - 5x^3 - 15x^2 + 4x - 15$

(c) $P_3(x) = 7x^4 + 3x^2 - 10x$

SOLUTION The graphs are shown in Figure 13.

(a) P_1 has two local minimum points and one local maximum point, for a total of three local extrema.

(b) P_2 has two local minimum points and two local maximum points, for a total of four local extrema.

(c) P_3 has just one local extremum, a local minimum.

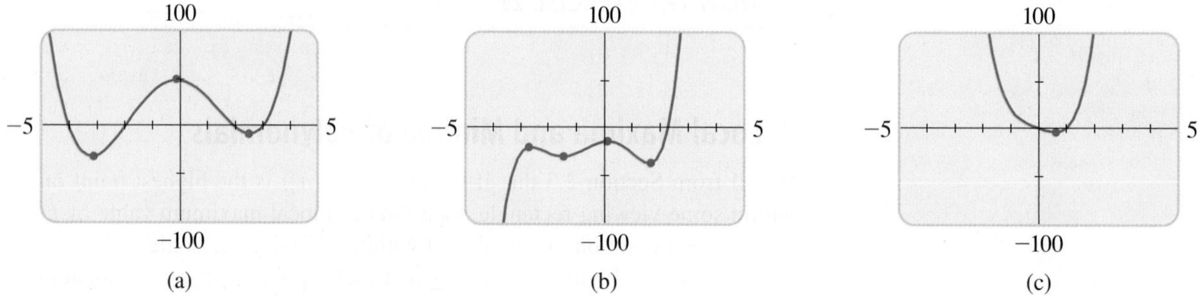

(a) $P_1(x) = x^4 + x^3 - 16x^2 - 4x + 48$

(b) $P_2(x) = x^5 + 3x^4 - 5x^3 - 15x^2 + 4x - 15$

(c) $P_3(x) = 7x^4 + 3x^2 - 10x$

FIGURE 13

NOW TRY EXERCISES **61** AND **63**

With a graphing calculator we can quickly draw the graphs of many functions at once, on the same viewing screen. This allows us to see how changing a value in the definition of the functions affects the shape of its graph. In the next example we apply this principle to a family of third-degree polynomials.

EXAMPLE 10 | A Family of Polynomials

Sketch the family of polynomials $P(x) = x^3 - cx^2$ for $c = 0, 1, 2,$ and 3. How does changing the value of c affect the graph?

SOLUTION The polynomials

$$P_0(x) = x^3 \qquad P_1(x) = x^3 - x^2$$
$$P_2(x) = x^3 - 2x^2 \qquad P_3(x) = x^3 - 3x^2$$

are graphed in Figure 14. We see that increasing the value of c causes the graph to develop an increasingly deep "valley" to the right of the y-axis, creating a local maximum at the origin and a local minimum at a point in Quadrant IV. This local minimum moves lower and farther to the right as c increases. To see why this happens, factor $P(x) = x^2(x - c)$. The polynomial P has zeros at 0 and c, and the larger c gets, the farther to the right the minimum between 0 and c will be.

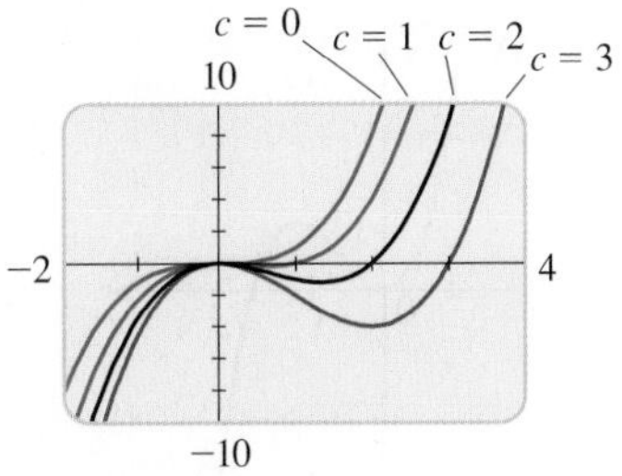

FIGURE 14 A family of polynomials $P(x) = x^3 - cx^2$

NOW TRY EXERCISE **71**

3.2 EXERCISES

CONCEPTS

1. Only one of the following graphs could be the graph of a polynomial function. Which one? Why are the others not graphs of polynomials?

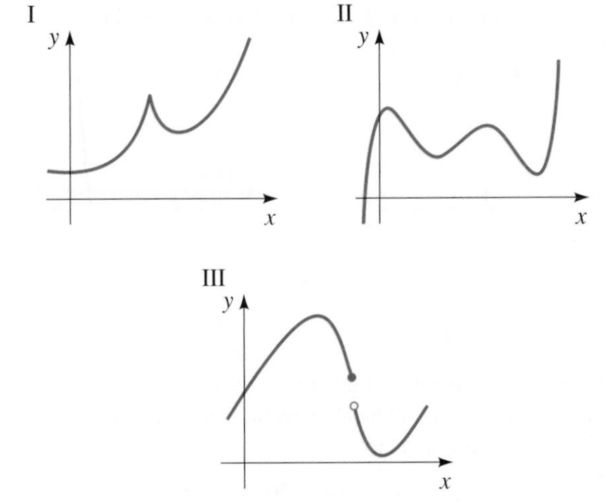

2. Every polynomial has one of the following behaviors:

(i) $y \to \infty$ as $x \to \infty$ and $y \to \infty$ as $x \to -\infty$
(ii) $y \to \infty$ as $x \to \infty$ and $y \to -\infty$ as $x \to -\infty$
(iii) $y \to -\infty$ as $x \to \infty$ and $y \to \infty$ as $x \to -\infty$
(iv) $y \to -\infty$ as $x \to \infty$ and $y \to -\infty$ as $x \to -\infty$

For each polynomial, choose the appropriate description of its end behavior from the list above.

(a) $y = x^3 - 8x^2 + 2x - 15$: end behavior ________.

(b) $y = -2x^4 + 12x + 100$: end behavior ________.

3. If c is a zero of the polynomial P, which of the following statements must be true?

(a) $P(c) = 0$. **(b)** $P(0) = c$.
(c) $x - c$ is a factor of $P(x)$.
(d) c is the y-intercept of the graph of P.

4. Which of the following statements couldn't possibly be true about the polynomial function P?

(a) P has degree 3, two local maxima, and two local minima.
(b) P has degree 3 and no local maxima or minima.
(c) P has degree 4, one local maximum, and no local minima.

SKILLS

5–8 ■ Sketch the graph of each function by transforming the graph of an appropriate function of the form $y = x^n$ from Figure 2. Indicate all x- and y-intercepts on each graph.

5. (a) $P(x) = x^2 - 4$ **(b)** $Q(x) = (x - 4)^2$
(c) $R(x) = 2x^2 - 2$ **(d)** $S(x) = 2(x - 2)^2$

6. (a) $P(x) = x^4 - 16$ **(b)** $Q(x) = (x + 2)^4$
(c) $R(x) = (x + 2)^4 - 16$ **(d)** $S(x) = -2(x + 2)^4$

7. (a) $P(x) = x^3 - 8$ **(b)** $Q(x) = -x^3 + 27$
(c) $R(x) = -(x + 2)^3$ **(d)** $S(x) = \frac{1}{2}(x - 1)^3 + 4$

8. (a) $P(x) = (x + 3)^5$ **(b)** $Q(x) = 2(x + 3)^5 - 64$
(c) $R(x) = -\frac{1}{2}(x - 2)^5$ **(d)** $S(x) = -\frac{1}{2}(x - 2)^5 + 16$

9–14 ■ Match the polynomial function with one of the graphs I–VI on the next page. Give reasons for your choice.

9. $P(x) = x(x^2 - 4)$ **10.** $Q(x) = -x^2(x^2 - 4)$

11. $R(x) = -x^5 + 5x^3 - 4x$ **12.** $S(x) = \frac{1}{2}x^6 - 2x^4$

13. $T(x) = x^4 + 2x^3$ **14.** $U(x) = -x^3 + 2x^2$

I
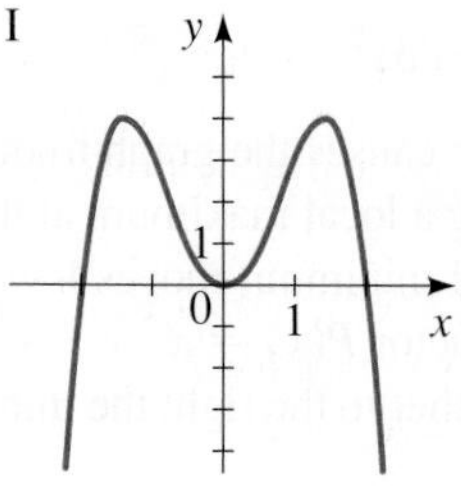

II
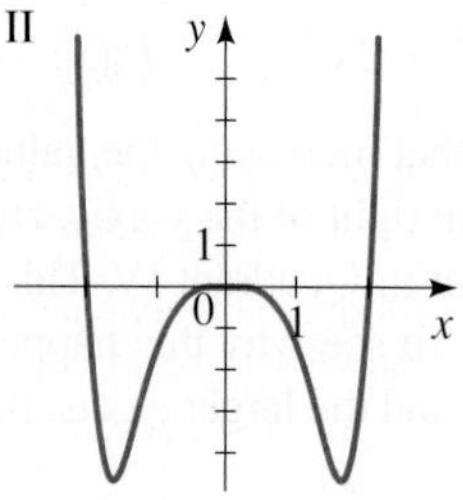

III
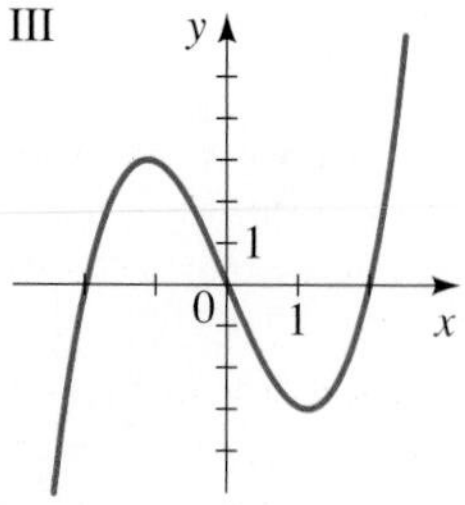

IV
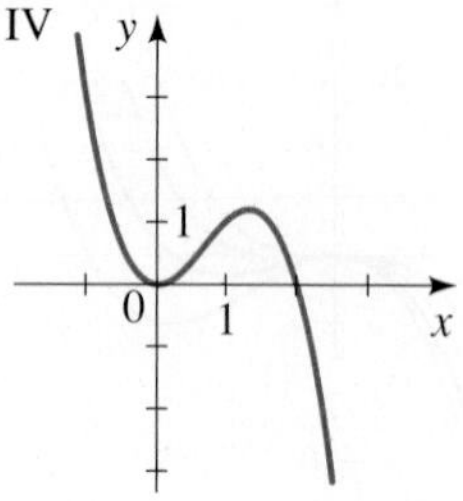

V
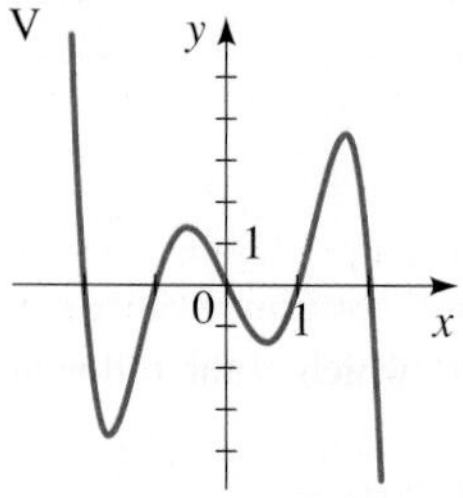

VI
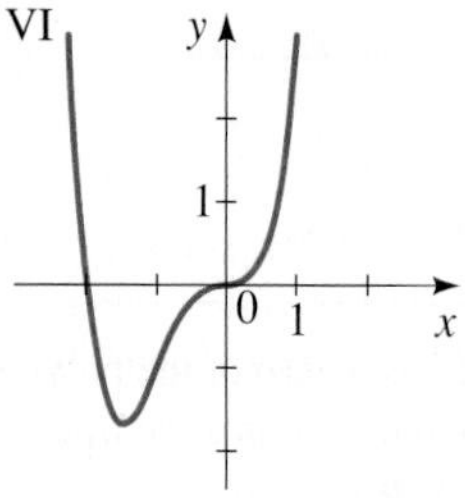

15–26 ■ Sketch the graph of the polynomial function. Make sure your graph shows all intercepts and exhibits the proper end behavior.

15. $P(x) = (x - 1)(x + 2)$

16. $P(x) = (x - 1)(x + 1)(x - 2)$

17. $P(x) = x(x - 3)(x + 2)$

18. $P(x) = (2x - 1)(x + 1)(x + 3)$

19. $P(x) = (x - 3)(x + 2)(3x - 2)$

20. $P(x) = \frac{1}{5}x(x - 5)^2$

21. $P(x) = (x - 1)^2(x - 3)$ **22.** $P(x) = \frac{1}{4}(x + 1)^3(x - 3)$

23. $P(x) = \frac{1}{12}(x + 2)^2(x - 3)^2$ **24.** $P(x) = (x - 1)^2(x + 2)^3$

25. $P(x) = x^3(x + 2)(x - 3)^2$ **26.** $P(x) = (x - 3)^2(x + 1)^2$

27–40 ■ Factor the polynomial and use the factored form to find the zeros. Then sketch the graph.

27. $P(x) = x^3 - x^2 - 6x$ **28.** $P(x) = x^3 + 2x^2 - 8x$

29. $P(x) = -x^3 + x^2 + 12x$ **30.** $P(x) = -2x^3 - x^2 + x$

31. $P(x) = x^4 - 3x^3 + 2x^2$ **32.** $P(x) = x^5 - 9x^3$

33. $P(x) = x^3 + x^2 - x - 1$ **34.** $P(x) = x^3 + 3x^2 - 4x - 12$

35. $P(x) = 2x^3 - x^2 - 18x + 9$

36. $P(x) = \frac{1}{8}(2x^4 + 3x^3 - 16x - 24)^2$

37. $P(x) = x^4 - 2x^3 - 8x + 16$

38. $P(x) = x^4 - 2x^3 + 8x - 16$

39. $P(x) = x^4 - 3x^2 - 4$ **40.** $P(x) = x^6 - 2x^3 + 1$

41–46 ■ Determine the end behavior of P. Compare the graphs of P and Q in large and small viewing rectangles, as in Example 3(b).

41. $P(x) = 3x^3 - x^2 + 5x + 1$; $Q(x) = 3x^3$

42. $P(x) = -\frac{1}{8}x^3 + \frac{1}{4}x^2 + 12x$; $Q(x) = -\frac{1}{8}x^3$

43. $P(x) = x^4 - 7x^2 + 5x + 5$; $Q(x) = x^4$

44. $P(x) = -x^5 + 2x^2 + x$; $Q(x) = -x^5$

45. $P(x) = x^{11} - 9x^9$; $Q(x) = x^{11}$

46. $P(x) = 2x^2 - x^{12}$; $Q(x) = -x^{12}$

47–50 ■ The graph of a polynomial function is given. From the graph, find **(a)** the x- and y-intercepts, and **(b)** the coordinates of all local extrema.

47. $P(x) = -x^2 + 4x$

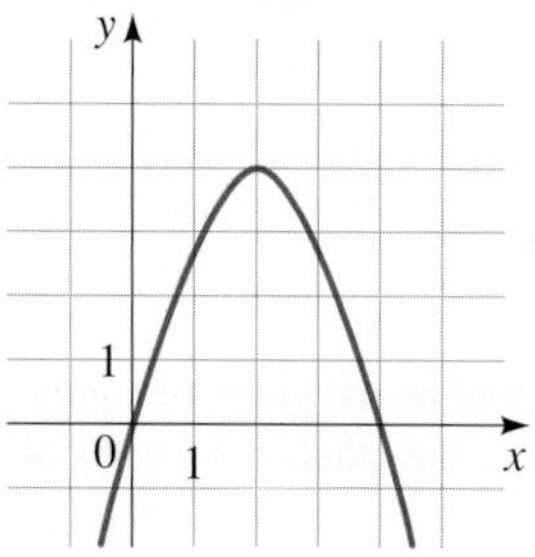

48. $P(x) = \frac{2}{9}x^3 - x^2$

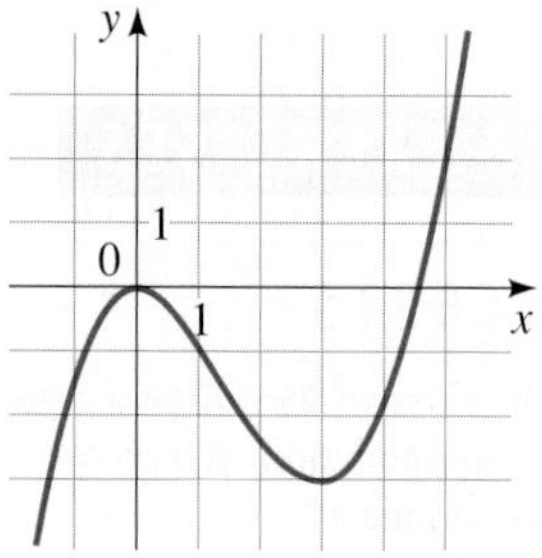

49. $P(x) = -\frac{1}{2}x^3 + \frac{3}{2}x - 1$

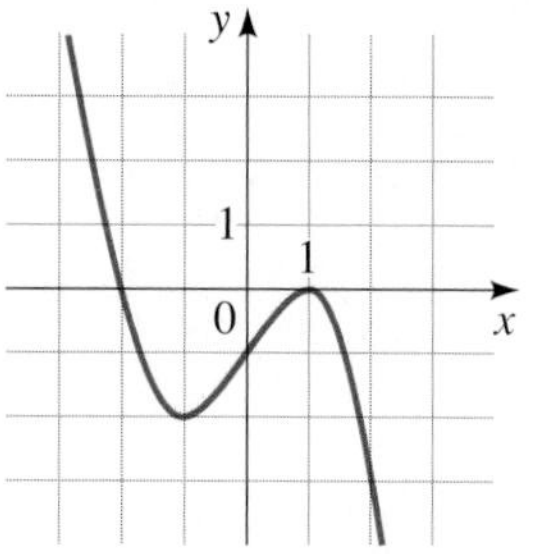

50. $P(x) = \frac{1}{9}x^4 - \frac{4}{9}x^3$

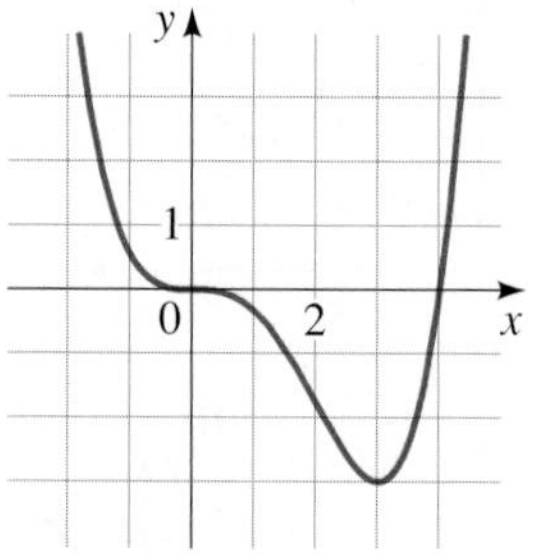

51–58 ■ Graph the polynomial in the given viewing rectangle. Find the coordinates of all local extrema. State each answer rounded to two decimal places.

51. $y = -x^2 + 8x$, $[-4, 12]$ by $[-50, 30]$

52. $y = x^3 - 3x^2$, $[-2, 5]$ by $[-10, 10]$

53. $y = x^3 - 12x + 9$, $[-5, 5]$ by $[-30, 30]$

54. $y = 2x^3 - 3x^2 - 12x - 32$, $[-5, 5]$ by $[-60, 30]$

55. $y = x^4 + 4x^3$, $[-5, 5]$ by $[-30, 30]$

56. $y = x^4 - 18x^2 + 32$, $[-5, 5]$ by $[-100, 100]$

57. $y = 3x^5 - 5x^3 + 3$, $[-3, 3]$ by $[-5, 10]$

58. $y = x^5 - 5x^2 + 6$, $[-3, 3]$ by $[-5, 10]$

59–68 ■ Graph the polynomial and determine how many local maxima and minima it has.

59. $y = -2x^2 + 3x + 5$

60. $y = x^3 + 12x$

61. $y = x^3 - x^2 - x$

62. $y = 6x^3 + 3x + 1$

63. $y = x^4 - 5x^2 + 4$

64. $y = 1.2x^5 + 3.75x^4 - 7x^3 - 15x^2 + 18x$

65. $y = (x - 2)^5 + 32$

66. $y = (x^2 - 2)^3$

67. $y = x^8 - 3x^4 + x$

68. $y = \frac{1}{3}x^7 - 17x^2 + 7$

69–74 ■ Graph the family of polynomials in the same viewing rectangle, using the given values of c. Explain how changing the value of c affects the graph.

69. $P(x) = cx^3$; $c = 1, 2, 5, \frac{1}{2}$

70. $P(x) = (x - c)^4$; $c = -1, 0, 1, 2$

71. $P(x) = x^4 + c$; $c = -1, 0, 1, 2$

72. $P(x) = x^3 + cx$; $c = 2, 0, -2, -4$

73. $P(x) = x^4 - cx$; $c = 0, 1, 8, 27$

74. $P(x) = x^c$; $c = 1, 3, 5, 7$

75. **(a)** On the same coordinate axes, sketch graphs (as accurately as possible) of the functions

$$y = x^3 - 2x^2 - x + 2 \quad \text{and} \quad y = -x^2 + 5x + 2$$

(b) On the basis of your sketch in part (a), at how many points do the two graphs appear to intersect?
(c) Find the coordinates of all intersection points.

76. Portions of the graphs of $y = x^2$, $y = x^3$, $y = x^4$, $y = x^5$, and $y = x^6$ are plotted in the figures. Determine which function belongs to each graph.

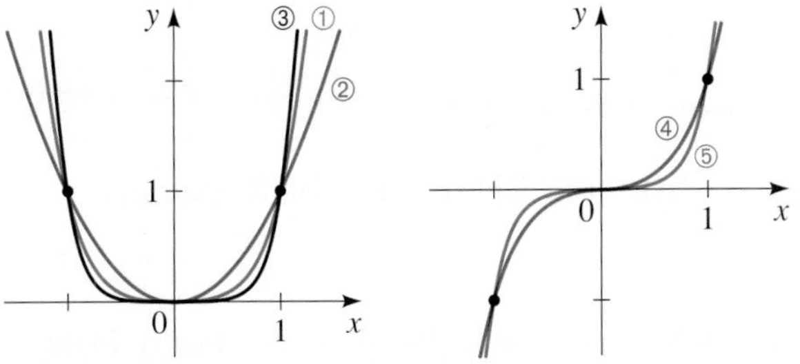

77. Recall that a function f is *odd* if $f(-x) = -f(x)$ or *even* if $f(-x) = f(x)$ for all real x.
(a) Show that a polynomial $P(x)$ that contains only odd powers of x is an odd function.
(b) Show that a polynomial $P(x)$ that contains only even powers of x is an even function.
(c) Show that if a polynomial $P(x)$ contains both odd and even powers of x, then it is neither an odd nor an even function.
(d) Express the function

$$P(x) = x^5 + 6x^3 - x^2 - 2x + 5$$

as the sum of an odd function and an even function.

78. **(a)** Graph the function $P(x) = (x - 1)(x - 3)(x - 4)$ and find all local extrema, correct to the nearest tenth.
(b) Graph the function

$$Q(x) = (x - 1)(x - 3)(x - 4) + 5$$

and use your answers to part (a) to find all local extrema, correct to the nearest tenth.

79. **(a)** Graph the function $P(x) = (x - 2)(x - 4)(x - 5)$ and determine how many local extrema it has.
(b) If $a < b < c$, explain why the function

$$P(x) = (x - a)(x - b)(x - c)$$

must have two local extrema.

80. **(a)** How many x-intercepts and how many local extrema does the polynomial $P(x) = x^3 - 4x$ have?
(b) How many x-intercepts and how many local extrema does the polynomial $Q(x) = x^3 + 4x$ have?
(c) If $a > 0$, how many x-intercepts and how many local extrema does each of the polynomials $P(x) = x^3 - ax$ and $Q(x) = x^3 + ax$ have? Explain your answer.

APPLICATIONS

81. **Market Research** A market analyst working for a small-appliance manufacturer finds that if the firm produces and sells x blenders annually, the total profit (in dollars) is

$$P(x) = 8x + 0.3x^2 - 0.0013x^3 - 372$$

Graph the function P in an appropriate viewing rectangle and use the graph to answer the following questions.
(a) When just a few blenders are manufactured, the firm loses money (profit is negative). (For example, $P(10) = -263.3$, so the firm loses \$263.30 if it produces and sells only 10 blenders.) How many blenders must the firm produce to break even?
(b) Does profit increase indefinitely as more blenders are produced and sold? If not, what is the largest possible profit the firm could have?

82. **Population Change** The rabbit population on a small island is observed to be given by the function

$$P(t) = 120t - 0.4t^4 + 1000$$

where t is the time (in months) since observations of the island began.
(a) When is the maximum population attained, and what is that maximum population?
(b) When does the rabbit population disappear from the island?

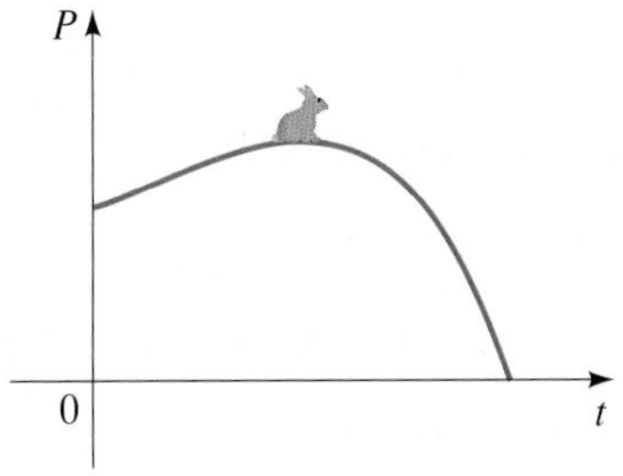

83. Volume of a Box An open box is to be constructed from a piece of cardboard 20 cm by 40 cm by cutting squares of side length x from each corner and folding up the sides, as shown in the figure.

(a) Express the volume V of the box as a function of x.

(b) What is the domain of V? (Use the fact that length and volume must be positive.)

(c) Draw a graph of the function V, and use it to estimate the maximum volume for such a box.

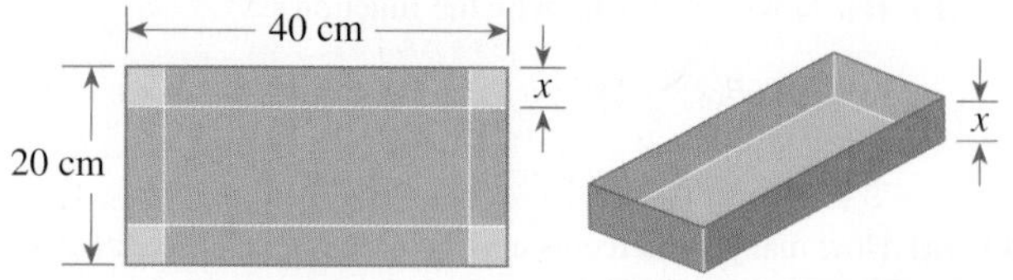

84. Volume of a Box A cardboard box has a square base, with each edge of the base having length x inches, as shown in the figure. The total length of all 12 edges of the box is 144 in.

(a) Show that the volume of the box is given by the function $V(x) = 2x^2(18 - x)$.

(b) What is the domain of V? (Use the fact that length and volume must be positive.)

(c) Draw a graph of the function V and use it to estimate the maximum volume for such a box.

DISCOVERY ■ DISCUSSION ■ WRITING

85. Graphs of Large Powers Graph the functions $y = x^2$, $y = x^3$, $y = x^4$, and $y = x^5$, for $-1 \le x \le 1$, on the same coordinate axes. What do you think the graph of $y = x^{100}$ would look like on this same interval? What about $y = x^{101}$? Make a table of values to confirm your answers.

86. Maximum Number of Local Extrema What is the smallest possible degree that the polynomial whose graph is shown can have? Explain.

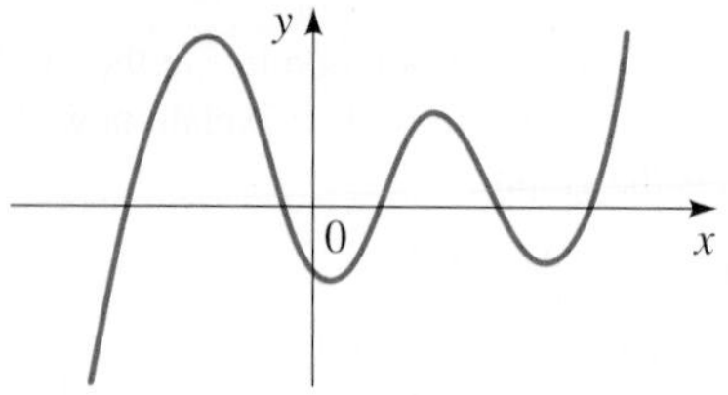

87. Possible Number of Local Extrema Is it possible for a third-degree polynomial to have exactly one local extremum? Can a fourth-degree polynomial have exactly two local extrema? How many local extrema can polynomials of third, fourth, fifth, and sixth degree have? (Think about the end behavior of such polynomials.) Now give an example of a polynomial that has six local extrema.

88. Impossible Situation? Is it possible for a polynomial to have two local maxima and no local minimum? Explain.

3.3 Dividing Polynomials

Long Division of Polynomials ▶ Synthetic Division ▶ The Remainder and Factor Theorems

So far in this chapter we have been studying polynomial functions *graphically*. In this section we begin to study polynomials *algebraically*. Most of our work will be concerned with factoring polynomials, and to factor, we need to know how to divide polynomials.

▼ Long Division of Polynomials

Dividing polynomials is much like the familiar process of dividing numbers. When we divide 38 by 7, the quotient is 5 and the remainder is 3. We write

Dividend, Remainder

$$\frac{38}{7} = 5 + \frac{3}{7}$$

Divisor, Quotient

To divide polynomials, we use long division, as follows.

To write the division algorithm another way, divide through by $D(x)$:

$$\frac{P(x)}{D(x)} = Q(x) + \frac{R(x)}{D(x)}$$

DIVISION ALGORITHM

If $P(x)$ and $D(x)$ are polynomials, with $D(x) \neq 0$, then there exist unique polynomials $Q(x)$ and $R(x)$, where $R(x)$ is either 0 or of degree less than the degree of $D(x)$, such that

$$P(x) = D(x) \cdot Q(x) + R(x)$$

Dividend · Divisor · Quotient · Remainder

The polynomials $P(x)$ and $D(x)$ are called the **dividend** and **divisor**, respectively, $Q(x)$ is the **quotient**, and $R(x)$ is the **remainder**.

EXAMPLE 1 | Long Division of Polynomials

Divide $6x^2 - 26x + 12$ by $x - 4$.

SOLUTION The *dividend* is $6x^2 - 26x + 12$ and the *divisor* is $x - 4$. We begin by arranging them as follows:

$$x - 4\overline{)6x^2 - 26x + 12}$$

Next we divide the leading term in the dividend by the leading term in the divisor to get the first term of the quotient: $6x^2/x = 6x$. Then we multiply the divisor by $6x$ and subtract the result from the dividend.

$$\begin{array}{rll}
 & 6x & \text{Divide leading terms: } \frac{6x^2}{x} = 6x \\
x - 4\big) & \overline{6x^2 - 26x + 12} & \\
 & \underline{6x^2 - 24x} & \text{Multiply: } 6x(x - 4) = 6x^2 - 24x \\
 & -2x + 12 & \text{Subtract and "bring down" 12}
\end{array}$$

We repeat the process using the last line $-2x + 12$ as the dividend.

$$\begin{array}{rll}
 & 6x - 2 & \text{Divide leading terms: } \frac{-2x}{x} = -2 \\
x - 4\big) & \overline{6x^2 - 26x + 12} & \\
 & \underline{6x^2 - 24x} & \\
 & -2x + 12 & \\
 & \underline{-2x + 8} & \text{Multiply: } -2(x - 4) = -2x + 8 \\
 & 4 & \text{Subtract}
\end{array}$$

The division process ends when the last line is of lesser degree than the divisor. The last line then contains the *remainder*, and the top line contains the *quotient*. The result of the division can be interpreted in either of two ways.

$$\frac{6x^2 - 26x + 12}{x - 4} = 6x - 2 + \frac{4}{x - 4}$$

Dividend · Divisor · Quotient · Remainder

or

$$6x^2 - 26x + 12 = (x - 4)(6x - 2) + 4$$

Dividend · Divisor · Quotient · Remainder

NOW TRY EXERCISE 3 ■

EXAMPLE 2 | Long Division of Polynomials

Let $P(x) = 8x^4 + 6x^2 - 3x + 1$ and $D(x) = 2x^2 - x + 2$. Find polynomials $Q(x)$ and $R(x)$ such that $P(x) = D(x) \cdot Q(x) + R(x)$.

SOLUTION We use long division after first inserting the term $0x^3$ into the dividend to ensure that the columns line up correctly.

$$
\begin{array}{rrl}
 & 4x^2 + 2x & \\
2x^2 - x + 2 \,\big) & \overline{8x^4 + 0x^3 + 6x^2 - 3x + 1} & \\
 & \underline{8x^4 - 4x^3 + 8x^2 } & \text{Multiply divisor by } 4x^2 \\
 & 4x^3 - 2x^2 - 3x & \text{Subtract} \\
 & \underline{4x^3 - 2x^2 + 4x } & \text{Multiply divisor by } 2x \\
 & -7x + 1 & \text{Subtract}
\end{array}
$$

The process is complete at this point because $-7x + 1$ is of lesser degree than the divisor $2x^2 - x + 2$. From the above long division we see that $Q(x) = 4x^2 + 2x$ and $R(x) = -7x + 1$, so

$$8x^4 + 6x^2 - 3x + 1 = (2x^2 - x + 2)(4x^2 + 2x) + (-7x + 1)$$

NOW TRY EXERCISE **19** ■

Synthetic Division

Synthetic division is a quick method of dividing polynomials; it can be used when the divisor is of the form $x - c$. In synthetic division we write only the essential parts of the long division. Compare the following long and synthetic divisions, in which we divide $2x^3 - 7x^2 + 5$ by $x - 3$. (We'll explain how to perform the synthetic division in Example 3.)

Long Division

$$
\begin{array}{rl}
 & 2x^2 - x - 3 \quad \leftarrow \text{Quotient} \\
x - 3 \,\big) & \overline{2x^3 - 7x^2 + 0x + 5} \\
 & \underline{2x^3 - 6x^2} \\
 & -x^2 + 0x \\
 & \underline{-x^2 + 3x} \\
 & -3x + 5 \\
 & \underline{-3x + 9} \\
 & -4 \quad \leftarrow \text{Remainder}
\end{array}
$$

Synthetic Division

$$
\begin{array}{r|rrrr}
3 & 2 & -7 & 0 & 5 \\
 & & 6 & -3 & -9 \\
\hline
 & 2 & -1 & -3 & -4
\end{array}
$$

Quotient: $2 \quad -1 \quad -3$ Remainder: -4

Note that in synthetic division we abbreviate $2x^3 - 7x^2 + 5$ by writing only the coefficients: $2 \quad -7 \quad 0 \quad 5$, and instead of $x - 3$, we simply write 3. (Writing 3 instead of -3 allows us to add instead of subtract, but this changes the sign of all the numbers that appear in the gold boxes.)

The next example shows how synthetic division is performed.

EXAMPLE 3 | Synthetic Division

Use synthetic division to divide $2x^3 - 7x^2 + 5$ by $x - 3$.

SOLUTION We begin by writing the appropriate coefficients to represent the divisor and the dividend.

$$
\text{Divisor } x - 3 \rightarrow \quad
\begin{array}{r|rrrr}
3 & 2 & -7 & 0 & 5 \\
\hline
\end{array}
\quad \leftarrow \text{Dividend } 2x^3 - 7x^2 + 0x + 5
$$

We bring down the 2, multiply $3 \cdot 2 = 6$, and write the result in the middle row. Then we add.

3	2	−7	0	5	
		6			Multiply: $3 \cdot 2 = 6$
	[2]	−1			Add: $-7 + 6 = -1$

We repeat this process of multiplying and then adding until the table is complete.

3	2	−7	0	5	
		6	−3		Multiply: $3(-1) = -3$
	2	[−1]	−3		Add: $0 + (-3) = -3$

3	2	−7	0	5	
		6	−3	−9	Multiply: $3(-3) = -9$
	2	−1	[−3]	−4	Add: $5 + (-9) = -4$

Quotient $2x^2 - x - 3$ — Remainder -4

From the last line of the synthetic division we see that the quotient is $2x^2 - x - 3$ and the remainder is -4. Thus

$$2x^3 - 7x^2 + 5 = (x - 3)(2x^2 - x - 3) - 4$$

NOW TRY EXERCISE **31**

The Remainder and Factor Theorems

The next theorem shows how synthetic division can be used to evaluate polynomials easily.

REMAINDER THEOREM

If the polynomial $P(x)$ is divided by $x - c$, then the remainder is the value $P(c)$.

PROOF If the divisor in the Division Algorithm is of the form $x - c$ for some real number c, then the remainder must be a constant (since the degree of the remainder is less than the degree of the divisor). If we call this constant r, then

$$P(x) = (x - c) \cdot Q(x) + r$$

Replacing x by c in this equation, we get $P(c) = (c - c) \cdot Q(x) + r = 0 + r = r$, that is, $P(c)$ is the remainder r.

EXAMPLE 4 | Using the Remainder Theorem to Find the Value of a Polynomial

Let $P(x) = 3x^5 + 5x^4 - 4x^3 + 7x + 3$.

(a) Find the quotient and remainder when $P(x)$ is divided by $x + 2$.

(b) Use the Remainder Theorem to find $P(-2)$.

SOLUTION

(a) Since $x + 2 = x - (-2)$, the synthetic division for this problem takes the following form.

-2	3	5	-4	0	7	3
		-6	2	4	-8	2
	3	-1	-2	4	-1	5

Remainder is 5, so $P(-2) = 5$

The quotient is $3x^4 - x^3 - 2x^2 + 4x - 1$, and the remainder is 5.

(b) By the Remainder Theorem, $P(-2)$ is the remainder when $P(x)$ is divided by $x - (-2) = x + 2$. From part (a) the remainder is 5, so $P(-2) = 5$.

NOW TRY EXERCISE **39**

The next theorem says that *zeros* of polynomials correspond to *factors*; we used this fact in Section 3.2 to graph polynomials.

FACTOR THEOREM

c is a zero of P if and only if $x - c$ is a factor of $P(x)$.

PROOF If $P(x)$ factors as $P(x) = (x - c) \cdot Q(x)$, then

$$P(c) = (c - c) \cdot Q(c) = 0 \cdot Q(c) = 0$$

Conversely, if $P(c) = 0$, then by the Remainder Theorem

$$P(x) = (x - c) \cdot Q(x) + 0 = (x - c) \cdot Q(x)$$

so $x - c$ is a factor of $P(x)$.

EXAMPLE 5 | Factoring a Polynomial Using the Factor Theorem

Let $P(x) = x^3 - 7x + 6$. Show that $P(1) = 0$, and use this fact to factor $P(x)$ completely.

1	1	0	-7	6
		1	1	-6
	1	1	-6	0

$$\begin{array}{r} x^2 + x - 6 \\ x - 1 \overline{)\, x^3 + 0x^2 - 7x + 6} \\ \underline{x^3 - x^2} \qquad\qquad \\ x^2 - 7x \qquad \\ \underline{x^2 - x} \qquad \\ -6x + 6 \\ \underline{-6x + 6} \\ 0 \end{array}$$

SOLUTION Substituting, we see that $P(1) = 1^3 - 7 \cdot 1 + 6 = 0$. By the Factor Theorem this means that $x - 1$ is a factor of $P(x)$. Using synthetic or long division (shown in the margin), we see that

$$\begin{aligned} P(x) &= x^3 - 7x + 6 && \text{Given polynomial} \\ &= (x - 1)(x^2 + x - 6) && \text{See margin} \\ &= (x - 1)(x - 2)(x + 3) && \text{Factor quadratic } x^2 + x - 6 \end{aligned}$$

NOW TRY EXERCISES **53** AND **57**

EXAMPLE 6 | Finding a Polynomial with Specified Zeros

Find a polynomial of degree 4 that has zeros -3, 0, 1, and 5.

SOLUTION By the Factor Theorem $x - (-3)$, $x - 0$, $x - 1$, and $x - 5$ must all be factors of the desired polynomial.

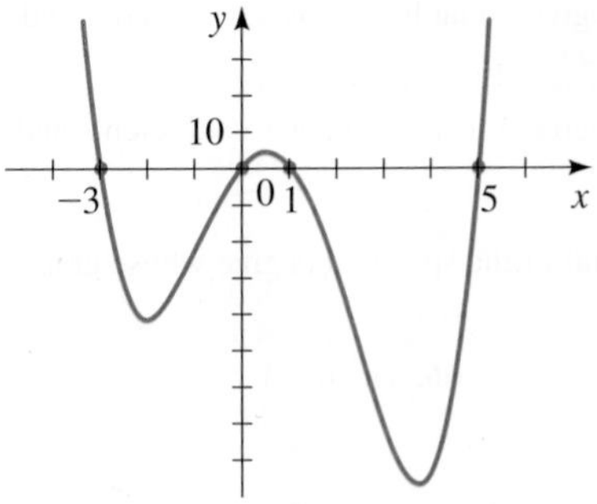

FIGURE 1
$P(x) = (x+3)x(x-1)(x-5)$ has zeros -3, 0, 1, and 5.

Let

$$\begin{aligned} P(x) &= (x+3)(x-0)(x-1)(x-5) \\ &= x^4 - 3x^3 - 13x^2 + 15x \end{aligned}$$

Since $P(x)$ is of degree 4, it is a solution of the problem. Any other solution of the problem must be a constant multiple of $P(x)$, since only multiplication by a constant does not change the degree.

NOW TRY EXERCISE **59** ■

The polynomial P of Example 6 is graphed in Figure 1. Note that the zeros of P correspond to the x-intercepts of the graph.

3.3 EXERCISES

CONCEPTS

1. If we divide the polynomial P by the factor $x - c$ and we obtain the equation $P(x) = (x-c)Q(x) + R(x)$, then we say that $x - c$ is the divisor, $Q(x)$ is the ______, and $R(x)$ is the ______.

2. (a) If we divide the polynomial $P(x)$ by the factor $x - c$ and we obtain a remainder of 0, then we know that c is a ______ of P.

(b) If we divide the polynomial $P(x)$ by the factor $x - c$ and we obtain a remainder of k, then we know that $P(c) =$ ______.

SKILLS

3–8 ■ Two polynomials P and D are given. Use either synthetic or long division to divide $P(x)$ by $D(x)$, and express P in the form $P(x) = D(x) \cdot Q(x) + R(x)$.

3. $P(x) = 3x^2 + 5x - 4$, $D(x) = x + 3$

4. $P(x) = x^3 + 4x^2 - 6x + 1$, $D(x) = x - 1$

5. $P(x) = 2x^3 - 3x^2 - 2x$, $D(x) = 2x - 3$

6. $P(x) = 4x^3 + 7x + 9$, $D(x) = 2x + 1$

7. $P(x) = x^4 - x^3 + 4x + 2$, $D(x) = x^2 + 3$

8. $P(x) = 2x^5 + 4x^4 - 4x^3 - x - 3$, $D(x) = x^2 - 2$

9–14 ■ Two polynomials P and D are given. Use either synthetic or long division to divide $P(x)$ by $D(x)$, and express the quotient $P(x)/D(x)$ in the form

$$\frac{P(x)}{D(x)} = Q(x) + \frac{R(x)}{D(x)}$$

9. $P(x) = x^2 + 4x - 8$, $D(x) = x + 3$

10. $P(x) = x^3 + 6x + 5$, $D(x) = x - 4$

11. $P(x) = 4x^2 - 3x - 7$, $D(x) = 2x - 1$

12. $P(x) = 6x^3 + x^2 - 12x + 5$, $D(x) = 3x - 4$

13. $P(x) = 2x^4 - x^3 + 9x^2$, $D(x) = x^2 + 4$

14. $P(x) = x^5 + x^4 - 2x^3 + x + 1$, $D(x) = x^2 + x - 1$

15–24 ■ Find the quotient and remainder using long division.

15. $\dfrac{x^2 - 6x - 8}{x - 4}$

16. $\dfrac{x^3 - x^2 - 2x + 6}{x - 2}$

17. $\dfrac{4x^3 + 2x^2 - 2x - 3}{2x + 1}$

18. $\dfrac{x^3 + 3x^2 + 4x + 3}{3x + 6}$

19. $\dfrac{x^3 + 6x + 3}{x^2 - 2x + 2}$

20. $\dfrac{3x^4 - 5x^3 - 20x - 5}{x^2 + x + 3}$

21. $\dfrac{6x^3 + 2x^2 + 22x}{2x^2 + 5}$

22. $\dfrac{9x^2 - x + 5}{3x^2 - 7x}$

23. $\dfrac{x^6 + x^4 + x^2 + 1}{x^2 + 1}$

24. $\dfrac{2x^5 - 7x^4 - 13}{4x^2 - 6x + 8}$

25–38 ■ Find the quotient and remainder using synthetic division.

25. $\dfrac{x^2 - 5x + 4}{x - 3}$

26. $\dfrac{x^2 - 5x + 4}{x - 1}$

27. $\dfrac{3x^2 + 5x}{x - 6}$

28. $\dfrac{4x^2 - 3}{x + 5}$

29. $\dfrac{x^3 + 2x^2 + 2x + 1}{x + 2}$

30. $\dfrac{3x^3 - 12x^2 - 9x + 1}{x - 5}$

31. $\dfrac{x^3 - 8x + 2}{x + 3}$

32. $\dfrac{x^4 - x^3 + x^2 - x + 2}{x - 2}$

33. $\dfrac{x^5 + 3x^3 - 6}{x - 1}$

34. $\dfrac{x^3 - 9x^2 + 27x - 27}{x - 3}$

35. $\dfrac{2x^3 + 3x^2 - 2x + 1}{x - \frac{1}{2}}$

36. $\dfrac{6x^4 + 10x^3 + 5x^2 + x + 1}{x + \frac{2}{3}}$

37. $\dfrac{x^3 - 27}{x - 3}$ 38. $\dfrac{x^4 - 16}{x + 2}$

39–51 ■ Use synthetic division and the Remainder Theorem to evaluate $P(c)$.

39. $P(x) = 4x^2 + 12x + 5, \quad c = -1$

40. $P(x) = 2x^2 + 9x + 1, \quad c = \frac{1}{2}$

41. $P(x) = x^3 + 3x^2 - 7x + 6, \quad c = 2$

42. $P(x) = x^3 - x^2 + x + 5, \quad c = -1$

43. $P(x) = x^3 + 2x^2 - 7, \quad c = -2$

44. $P(x) = 2x^3 - 21x^2 + 9x - 200, \quad c = 11$

45. $P(x) = 5x^4 + 30x^3 - 40x^2 + 36x + 14, \quad c = -7$

46. $P(x) = 6x^5 + 10x^3 + x + 1, \quad c = -2$

47. $P(x) = x^7 - 3x^2 - 1, \quad c = 3$

48. $P(x) = -2x^6 + 7x^5 + 40x^4 - 7x^2 + 10x + 112, \quad c = -3$

49. $P(x) = 3x^3 + 4x^2 - 2x + 1, \quad c = \frac{2}{3}$

50. $P(x) = x^3 - x + 1, \quad c = \frac{1}{4}$

51. $P(x) = x^3 + 2x^2 - 3x - 8, \quad c = 0.1$

52. Let

$$P(x) = 6x^7 - 40x^6 + 16x^5 - 200x^4 - 60x^3 - 69x^2 + 13x - 139$$

Calculate $P(7)$ by **(a)** using synthetic division and **(b)** substituting $x = 7$ into the polynomial and evaluating directly.

53–56 ■ Use the Factor Theorem to show that $x - c$ is a factor of $P(x)$ for the given value(s) of c.

53. $P(x) = x^3 - 3x^2 + 3x - 1, \quad c = 1$

54. $P(x) = x^3 + 2x^2 - 3x - 10, \quad c = 2$

55. $P(x) = 2x^3 + 7x^2 + 6x - 5, \quad c = \frac{1}{2}$

56. $P(x) = x^4 + 3x^3 - 16x^2 - 27x + 63, \quad c = 3, -3$

57–58 ■ Show that the given value(s) of c are zeros of $P(x)$, and find all other zeros of $P(x)$.

57. $P(x) = x^3 - x^2 - 11x + 15, \quad c = 3$

58. $P(x) = 3x^4 - x^3 - 21x^2 - 11x + 6, \quad c = \frac{1}{3}, -2$

59–62 ■ Find a polynomial of the specified degree that has the given zeros.

59. Degree 3; zeros $-1, 1, 3$

60. Degree 4; zeros $-2, 0, 2, 4$

61. Degree 4; zeros $-1, 1, 3, 5$

62. Degree 5; zeros $-2, -1, 0, 1, 2$

63. Find a polynomial of degree 3 that has zeros 1, -2, and 3 and in which the coefficient of x^2 is 3.

64. Find a polynomial of degree 4 that has integer coefficients and zeros 1, -1, 2, and $\frac{1}{2}$.

65–68 ■ Find the polynomial of the specified degree whose graph is shown.

65. Degree 3

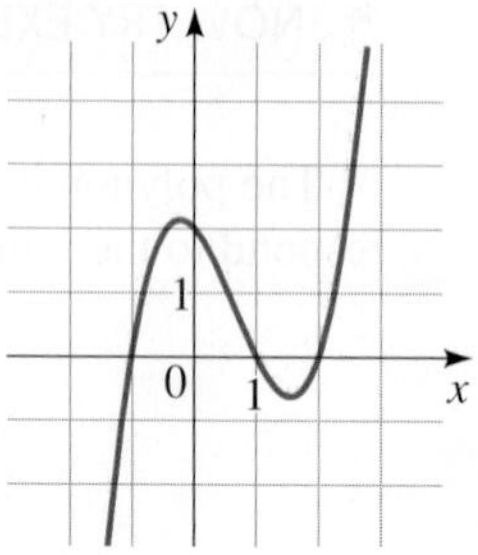

66. Degree 3

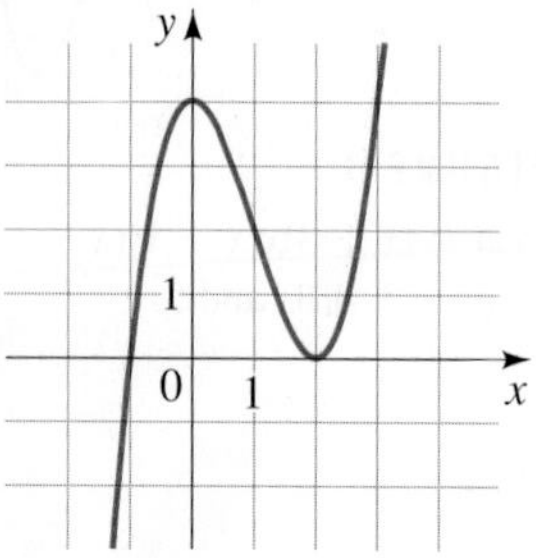

67. Degree 4

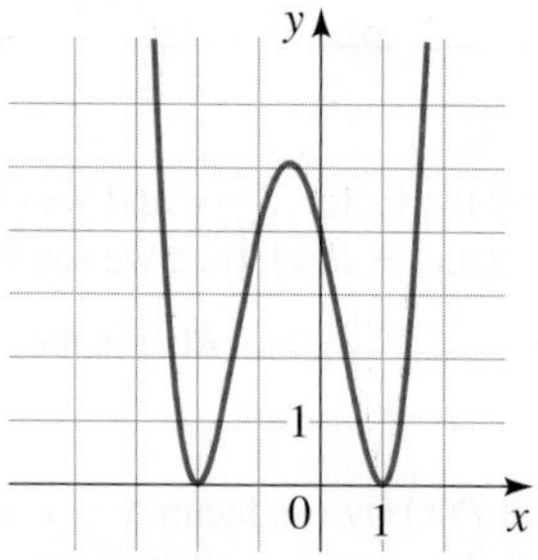

68. Degree 4

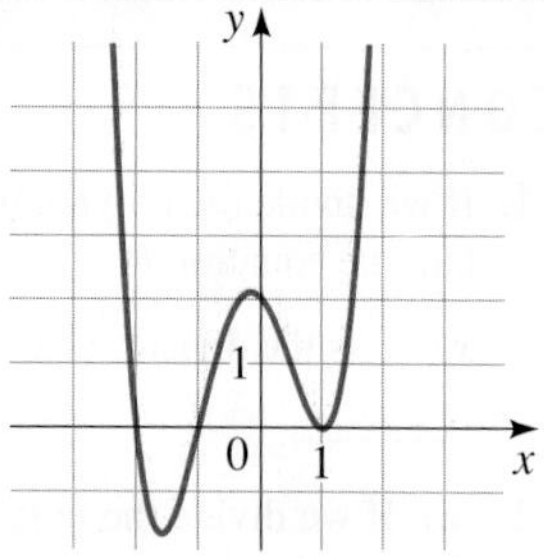

DISCOVERY ■ DISCUSSION ■ WRITING

69. **Impossible Division?** Suppose you were asked to solve the following two problems on a test:

A. Find the remainder when $6x^{1000} - 17x^{562} + 12x + 26$ is divided by $x + 1$.

B. Is $x - 1$ a factor of $x^{567} - 3x^{400} + x^9 + 2$?

Obviously, it's impossible to solve these problems by dividing, because the polynomials are of such large degree. Use one or more of the theorems in this section to solve these problems *without* actually dividing.

70. **Nested Form of a Polynomial** Expand Q to prove that the polynomials P and Q are the same.

$$P(x) = 3x^4 - 5x^3 + x^2 - 3x + 5$$

$$Q(x) = (((3x - 5)x + 1)x - 3)x + 5$$

Try to evaluate $P(2)$ and $Q(2)$ in your head, using the forms given. Which is easier? Now write the polynomial $R(x) = x^5 - 2x^4 + 3x^3 - 2x^2 + 3x + 4$ in "nested" form, like the polynomial Q. Use the nested form to find $R(3)$ in your head.

Do you see how calculating with the nested form follows the same arithmetic steps as calculating the value of a polynomial using synthetic division?

3.4 REAL ZEROS OF POLYNOMIALS

Rational Zeros of Polynomials ▶ Descartes' Rule of Signs and Upper and Lower Bounds for Roots ▶ Using Algebra and Graphing Devices to Solve Polynomial Equations

The Factor Theorem tells us that finding the zeros of a polynomial is really the same thing as factoring it into linear factors. In this section we study some algebraic methods that help us to find the real zeros of a polynomial and thereby factor the polynomial. We begin with the *rational* zeros of a polynomial.

▼ Rational Zeros of Polynomials

To help us understand the next theorem, let's consider the polynomial

$$P(x) = (x - 2)(x - 3)(x + 4) \qquad \text{Factored form}$$
$$= x^3 - x^2 - 14x + 24 \qquad \text{Expanded form}$$

From the factored form we see that the zeros of P are 2, 3, and -4. When the polynomial is expanded, the constant 24 is obtained by multiplying $(-2) \times (-3) \times 4$. This means that the zeros of the polynomial are all factors of the constant term. The following generalizes this observation.

RATIONAL ZEROS THEOREM

If the polynomial $P(x) = a_n x^n + a_{n-1}x^{n-1} + \cdots + a_1 x + a_0$ has integer coefficients, then every rational zero of P is of the form

$$\frac{p}{q}$$

where p is a factor of the constant coefficient a_0
and q is a factor of the leading coefficient a_n.

PROOF If p/q is a rational zero, in lowest terms, of the polynomial P, then we have

$$a_n\left(\frac{p}{q}\right)^n + a_{n-1}\left(\frac{p}{q}\right)^{n-1} + \cdots + a_1\left(\frac{p}{q}\right) + a_0 = 0$$

$$a_n p^n + a_{n-1}p^{n-1}q + \cdots + a_1 p q^{n-1} + a_0 q^n = 0 \qquad \text{Multiply by } q^n$$

$$p(a_n p^{n-1} + a_{n-1}p^{n-2}q + \cdots + a_1 q^{n-1}) = -a_0 q^n \qquad \text{Subtract } a_0q^n \text{ and factor LHS}$$

Now p is a factor of the left side, so it must be a factor of the right side as well. Since p/q is in lowest terms, p and q have no factor in common, so p must be a factor of a_0. A similar proof shows that q is a factor of a_n. ■

We see from the Rational Zeros Theorem that if the leading coefficient is 1 or -1, then the rational zeros must be factors of the constant term.

EXAMPLE 1 | Using the Rational Zeros Theorem

Find the rational zeros of $P(x) = x^3 - 3x + 2$.

Library of Congress

EVARISTE GALOIS (1811–1832) is one of the very few mathematicians to have an entire theory named in his honor. Not yet 21 when he died, he completely settled the central problem in the theory of equations by describing a criterion that reveals whether a polynomial equation can be solved by algebraic operations. Galois was one of the greatest mathematicians in the world at that time, although no one knew it but him. He repeatedly sent his work to the eminent mathematicians Cauchy and Poisson, who either lost his letters or did not understand his ideas. Galois wrote in a terse style and included few details, which probably played a role in his failure to pass the entrance exams at the Ecole Polytechnique in Paris. A political radical, Galois spent several months in prison for his revolutionary activities. His brief life came to a tragic end when he was killed in a duel over a love affair. The night before his duel, fearing that he would die, Galois wrote down the essence of his ideas and entrusted them to his friend Auguste Chevalier. He concluded by writing "there will, I hope, be people who will find it to their advantage to decipher all this mess." The mathematician Camille Jordan did just that, 14 years later.

SOLUTION Since the leading coefficient is 1, any rational zero must be a divisor of the constant term 2. So the possible rational zeros are ± 1 and ± 2. We test each of these possibilities.

$$P(1) = (1)^3 - 3(1) + 2 = 0$$

$$P(-1) = (-1)^3 - 3(-1) + 2 = 4$$

$$P(2) = (2)^3 - 3(2) + 2 = 4$$

$$P(-2) = (-2)^3 - 3(-2) + 2 = 0$$

The rational zeros of P are 1 and -2.

NOW TRY EXERCISE **15** ■

The following box explains how we use the Rational Zeros Theorem with synthetic division to factor a polynomial.

FINDING THE RATIONAL ZEROS OF A POLYNOMIAL

1. **List Possible Zeros.** List all possible rational zeros, using the Rational Zeros Theorem.
2. **Divide.** Use synthetic division to evaluate the polynomial at each of the candidates for the rational zeros that you found in Step 1. When the remainder is 0, note the quotient you have obtained.
3. **Repeat.** Repeat Steps 1 and 2 for the quotient. Stop when you reach a quotient that is quadratic or factors easily, and use the quadratic formula or factor to find the remaining zeros.

EXAMPLE 2 | Finding Rational Zeros

Factor the polynomial $P(x) = 2x^3 + x^2 - 13x + 6$, and find all its zeros.

SOLUTION By the Rational Zeros Theorem the rational zeros of P are of the form

$$\text{possible rational zero of } P = \frac{\text{factor of constant term}}{\text{factor of leading coefficient}}$$

The constant term is 6 and the leading coefficient is 2, so

$$\text{possible rational zero of } P = \frac{\text{factor of 6}}{\text{factor of 2}}$$

The factors of 6 are $\pm 1, \pm 2, \pm 3, \pm 6$, and the factors of 2 are $\pm 1, \pm 2$. Thus, the possible rational zeros of P are

$$\pm\frac{1}{1}, \quad \pm\frac{2}{1}, \quad \pm\frac{3}{1}, \quad \pm\frac{6}{1}, \quad \pm\frac{1}{2}, \quad \pm\frac{2}{2}, \quad \pm\frac{3}{2}, \quad \pm\frac{6}{2}$$

Simplifying the fractions and eliminating duplicates, we get the following list of possible rational zeros:

$$\pm 1, \quad \pm 2, \quad \pm 3, \quad \pm 6, \quad \pm\frac{1}{2}, \quad \pm\frac{3}{2}$$

To check which of these *possible* zeros actually *are* zeros, we need to evaluate P at each of these numbers. An efficient way to do this is to use synthetic division.

Test whether 1 is a zero

1	2	1	−13	6
		2	3	−10
	2	3	−10	−4

Remainder is *not* 0, so 1 is *not* a zero

Test whether 2 is a zero

2	2	1	−13	6
		4	10	−6
	2	5	−3	0

Remainder *is* 0, so 2 *is* a zero

From the last synthetic division we see that 2 is a zero of P and that P factors as

$$\begin{aligned} P(x) &= 2x^3 + x^2 - 13x + 6 && \text{Given polynomial} \\ &= (x-2)(2x^2 + 5x - 3) && \text{From synthetic division} \\ &= (x-2)(2x-1)(x+3) && \text{Factor } 2x^2 + 5x - 3 \end{aligned}$$

From the factored form we see that the zeros of P are 2, $\frac{1}{2}$, and –3.

NOW TRY EXERCISE **27**

EXAMPLE 3 | Using the Rational Zeros Theorem and the Quadratic Formula

Let $P(x) = x^4 - 5x^3 - 5x^2 + 23x + 10$.

(a) Find the zeros of P. **(b)** Sketch the graph of P.

SOLUTION

(a) The leading coefficient of P is 1, so all the rational zeros are integers: They are divisors of the constant term 10. Thus, the possible candidates are

$$\pm 1, \quad \pm 2, \quad \pm 5, \quad \pm 10$$

1	1	−5	−5	23	10
		1	−4	−9	14
	1	−4	−9	14	24

2	1	−5	−5	23	10
		2	−6	−22	2
	1	−3	−11	1	12

5	1	−5	−5	23	10
		5	0	−25	−10
	1	0	−5	−2	0

Using synthetic division (see the margin), we find that 1 and 2 are not zeros but that 5 is a zero and that P factors as

$$x^4 - 5x^3 - 5x^2 + 23x + 10 = (x-5)(x^3 - 5x - 2)$$

We now try to factor the quotient $x^3 - 5x - 2$. Its possible zeros are the divisors of -2, namely,

$$\pm 1, \quad \pm 2$$

Since we already know that 1 and 2 are not zeros of the original polynomial P, we don't need to try them again. Checking the remaining candidates, -1 and -2, we see that -2 is a zero (see the margin), and P factors as

−2	1	0	−5	−2
		−2	4	2
	1	−2	−1	0

$$\begin{aligned} x^4 - 5x^3 - 5x^2 + 23x + 10 &= (x-5)(x^3 - 5x - 2) \\ &= (x-5)(x+2)(x^2 - 2x - 1) \end{aligned}$$

Now we use the quadratic formula to obtain the two remaining zeros of P:

$$x = \frac{2 \pm \sqrt{(-2)^2 - 4(1)(-1)}}{2} = 1 \pm \sqrt{2}$$

The zeros of P are 5, -2, $1 + \sqrt{2}$, and $1 - \sqrt{2}$.

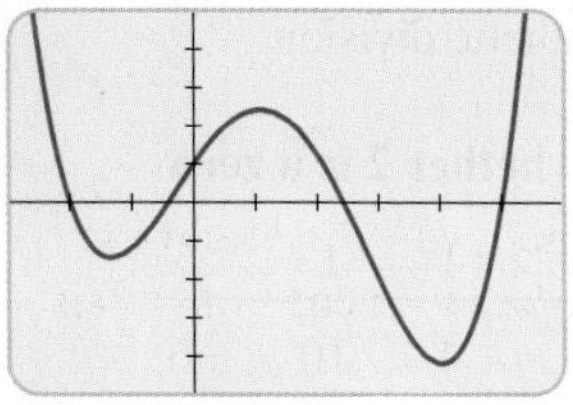

FIGURE 1
$P(x) = x^4 - 5x^3 - 5x^2 + 23x + 10$

(b) Now that we know the zeros of P, we can use the methods of Section 3.2 to sketch the graph. If we want to use a graphing calculator instead, knowing the zeros allows us to choose an appropriate viewing rectangle—one that is wide enough to contain all the x-intercepts of P. Numerical approximations to the zeros of P are

$$5, \quad -2, \quad 2.4, \quad \text{and} \quad -0.4$$

So in this case we choose the rectangle $[-3, 6]$ by $[-50, 50]$ and draw the graph shown in Figure 1.

NOW TRY EXERCISES **47** AND **57** ■

Descartes' Rule of Signs and Upper and Lower Bounds for Roots

In some cases, the following rule—discovered by the French philosopher and mathematician René Descartes around 1637 (see page 181)—is helpful in eliminating candidates from lengthy lists of possible rational roots. To describe this rule, we need the concept of *variation in sign*. If $P(x)$ is a polynomial with real coefficients, written with descending powers of x (and omitting powers with coefficient 0), then a **variation in sign** occurs whenever adjacent coefficients have opposite signs. For example,

$$P(x) = 5x^7 - 3x^5 - x^4 + 2x^2 + x - 3$$

has three variations in sign.

Polynomial	Variations in sign
$x^2 + 4x + 1$	0
$2x^3 + x - 6$	1
$x^4 - 3x^2 - x + 4$	2

DESCARTES' RULE OF SIGNS

Let P be a polynomial with real coefficients.

1. The number of positive real zeros of $P(x)$ either is equal to the number of variations in sign in $P(x)$ or is less than that by an even whole number.
2. The number of negative real zeros of $P(x)$ either is equal to the number of variations in sign in $P(-x)$ or is less than that by an even whole number.

EXAMPLE 4 | Using Descartes' Rule

Use Descartes' Rule of Signs to determine the possible number of positive and negative real zeros of the polynomial

$$P(x) = 3x^6 + 4x^5 + 3x^3 - x - 3$$

SOLUTION The polynomial has one variation in sign, so it has one positive zero. Now

$$\begin{aligned} P(-x) &= 3(-x)^6 + 4(-x)^5 + 3(-x)^3 - (-x) - 3 \\ &= 3x^6 - 4x^5 - 3x^3 + x - 3 \end{aligned}$$

So $P(-x)$ has three variations in sign. Thus, $P(x)$ has either three or one negative zero(s), making a total of either two or four real zeros.

NOW TRY EXERCISE **67** ■

We say that a is a **lower bound** and b is an **upper bound** for the zeros of a polynomial if every real zero c of the polynomial satisfies $a \le c \le b$. The next theorem helps us to find such bounds for the zeros of a polynomial.

THE UPPER AND LOWER BOUNDS THEOREM

Let P be a polynomial with real coefficients.

1. If we divide $P(x)$ by $x - b$ (with $b > 0$) using synthetic division and if the row that contains the quotient and remainder has no negative entry, then b is an upper bound for the real zeros of P.
2. If we divide $P(x)$ by $x - a$ (with $a < 0$) using synthetic division and if the row that contains the quotient and remainder has entries that are alternately nonpositive and nonnegative, then a is a lower bound for the real zeros of P.

A proof of this theorem is suggested in Exercise 97. The phrase "alternately nonpositive and nonnegative" simply means that the signs of the numbers alternate, with 0 considered to be positive or negative as required.

EXAMPLE 5 | Upper and Lower Bounds for Zeros of a Polynomial

Show that all the real zeros of the polynomial $P(x) = x^4 - 3x^2 + 2x - 5$ lie between -3 and 2.

SOLUTION We divide $P(x)$ by $x - 2$ and $x + 3$ using synthetic division.

2	1	0	−3	2	−5
		2	4	2	8
	1	2	1	4	3

All entries positive

−3	1	0	−3	2	−5
		−3	9	−18	48
	1	−3	6	−16	43

Entries alternate in sign

By the Upper and Lower Bounds Theorem, -3 is a lower bound and 2 is an upper bound for the zeros. Since neither -3 nor 2 is a zero (the remainders are not 0 in the division table), all the real zeros lie between these numbers.

NOW TRY EXERCISE **71** ■

EXAMPLE 6 | Factoring a Fifth-Degree Polynomial

Factor completely the polynomial

$$P(x) = 2x^5 + 5x^4 - 8x^3 - 14x^2 + 6x + 9$$

SOLUTION The possible rational zeros of P are $\pm\frac{1}{2}$, ± 1, $\pm\frac{3}{2}$, ± 3, $\pm\frac{9}{2}$, and ± 9. We check the positive candidates first, beginning with the smallest.

$\frac{1}{2}$	2	5	−8	−14	6	9
		1	3	$-\frac{5}{2}$	$-\frac{33}{4}$	$-\frac{9}{8}$
	2	6	−5	$-\frac{33}{2}$	$-\frac{9}{4}$	$\frac{63}{8}$

$\frac{1}{2}$ is not a zero

1	2	5	−8	−14	6	9
		2	7	−1	−15	−9
	2	7	−1	−15	−9	0

$P(1) = 0$

So 1 is a zero, and $P(x) = (x - 1)(2x^4 + 7x^3 - x^2 - 15x - 9)$. We continue by factoring the quotient. We still have the same list of possible zeros except that $\frac{1}{2}$ has been eliminated.

1	2	7	−1	−15	−9
		2	9	8	−7
	2	9	8	−7	−16

1 is not a zero

$\frac{3}{2}$	2	7	−1	−15	−9
		3	15	21	9
	2	10	14	6	0

$P(\frac{3}{2}) = 0$, all entries nonnegative

We see that $\frac{3}{2}$ is both a zero and an upper bound for the zeros of $P(x)$, so we do not need to check any further for positive zeros, because all the remaining candidates are greater than $\frac{3}{2}$.

$$P(x) = (x-1)(x-\tfrac{3}{2})(2x^3+10x^2+14x+6) \qquad \text{From synthetic division}$$

$$= (x-1)(2x-3)(x^3+5x^2+7x+3) \qquad \text{Factor 2 from last factor, multiply into second factor}$$

By Descartes' Rule of Signs, x^3+5x^2+7x+3 has no positive zero, so its only possible rational zeros are -1 and -3.

-1	1	5	7	3
		-1	-4	-3
	1	4	3	0

$P(-1) = 0$

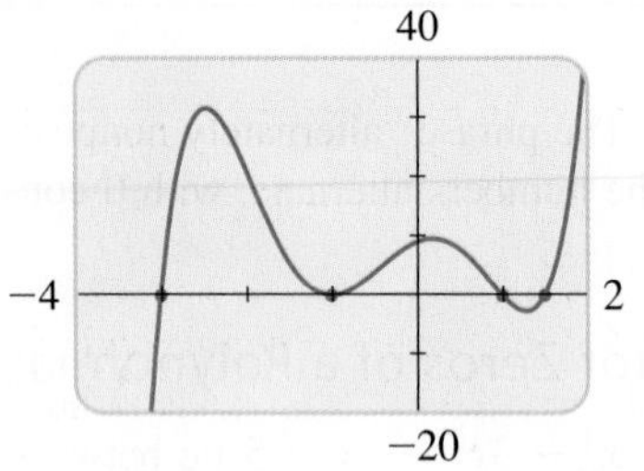

FIGURE 2
$P(x) = 2x^5+5x^4-8x^3-14x^2+6x+9$
$= (x-1)(2x-3)(x+1)^2(x+3)$

Therefore,

$$P(x) = (x-1)(2x-3)(x+1)(x^2+4x+3) \qquad \text{From synthetic division}$$

$$= (x-1)(2x-3)(x+1)^2(x+3) \qquad \text{Factor quadratic}$$

This means that the zeros of P are $1, \frac{3}{2}, -1$, and -3. The graph of the polynomial is shown in Figure 2.

NOW TRY EXERCISE **79**

Using Algebra and Graphing Devices to Solve Polynomial Equations

In Section 1.9 we used graphing devices to solve equations graphically. We can now use the algebraic techniques that we've learned to select an appropriate viewing rectangle when solving a polynomial equation graphically.

EXAMPLE 7 | Solving a Fourth-Degree Equation Graphically

Find all real solutions of the following equation, rounded to the nearest tenth.

$$3x^4+4x^3-7x^2-2x-3=0$$

SOLUTION To solve the equation graphically, we graph

$$P(x) = 3x^4+4x^3-7x^2-2x-3$$

We use the Upper and Lower Bounds Theorem to see where the solutions can be found.

First we use the Upper and Lower Bounds Theorem to find two numbers between which all the solutions must lie. This allows us to choose a viewing rectangle that is certain to contain all the x-intercepts of P. We use synthetic division and proceed by trial and error.

To find an upper bound, we try the whole numbers, $1, 2, 3, \ldots$, as potential candidates. We see that 2 is an upper bound for the solutions.

2	3	4	-7	-2	-3
		6	20	26	48
	3	10	13	24	45

All positive

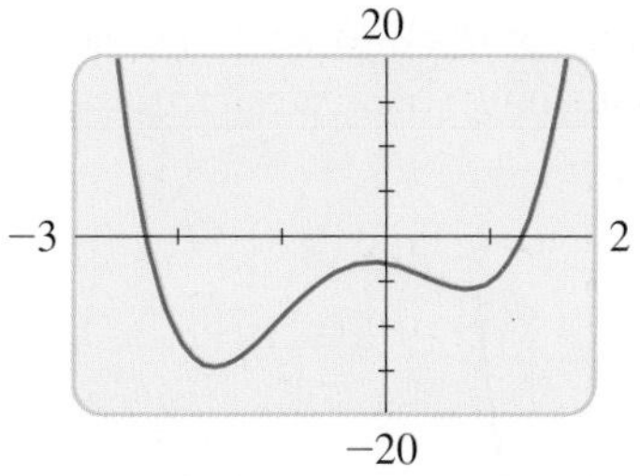

FIGURE 3
$y = 3x^4 + 4x^3 - 7x^2 - 2x - 3$

Now we look for a lower bound, trying the numbers -1, -2, and -3 as potential candidates. We see that -3 is a lower bound for the solutions.

-3	3	4	-7	-2	-3
		-9	15	-24	78
	3	-5	8	-26	75

Entries alternate in sign

Thus, all the solutions lie between -3 and 2. So the viewing rectangle $[-3, 2]$ by $[-20, 20]$ contains all the x-intercepts of P. The graph in Figure 3 has two x-intercepts, one between -3 and -2 and the other between 1 and 2. Zooming in, we find that the solutions of the equation, to the nearest tenth, are -2.3 and 1.3.

NOW TRY EXERCISE **93**

EXAMPLE 8 | Determining the Size of a Fuel Tank

A fuel tank consists of a cylindrical center section that is 4 ft long and two hemispherical end sections, as shown in Figure 4. If the tank has a volume of 100 ft^3, what is the radius r shown in the figure, rounded to the nearest hundredth of a foot?

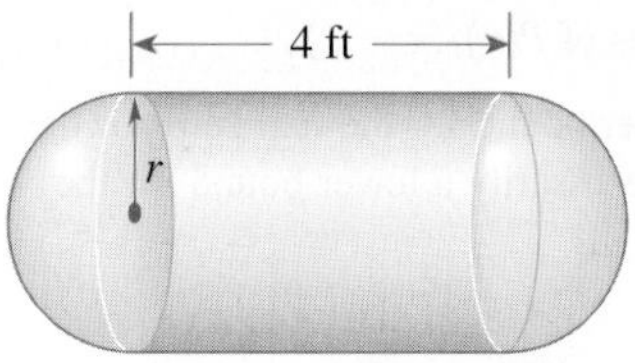

FIGURE 4

SOLUTION Using the volume formula listed on the inside back cover of this book, we see that the volume of the cylindrical section of the tank is

Volume of a cylinder: $V = \pi r^2 h$

$$\pi \cdot r^2 \cdot 4$$

The two hemispherical parts together form a complete sphere whose volume is

Volume of a sphere: $V = \frac{4}{3}\pi r^3$

$$\tfrac{4}{3}\pi r^3$$

Because the total volume of the tank is 100 ft^3, we get the following equation:

$$\tfrac{4}{3}\pi r^3 + 4\pi r^2 = 100$$

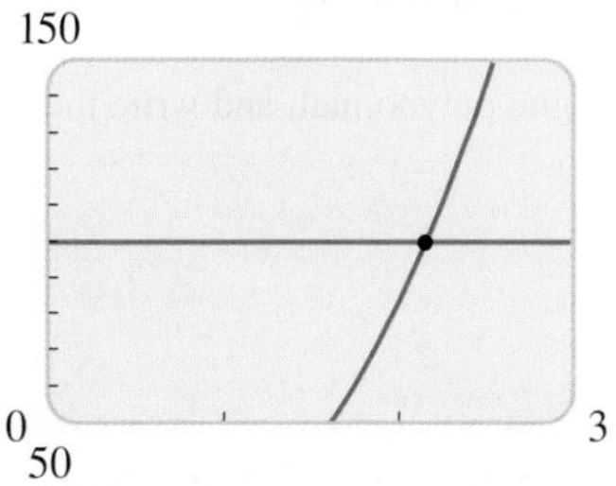

FIGURE 5
$y = \frac{4}{3}\pi x^3 + 4\pi x^2$ and $y = 100$

A negative solution for r would be meaningless in this physical situation, and by substitution we can verify that $r = 3$ leads to a tank that is over 226 ft^3 in volume, much larger than the required 100 ft^3. Thus, we know the correct radius lies somewhere between 0 and 3 ft, so we use a viewing rectangle of $[0, 3]$ by $[50, 150]$ to graph the function $y = \frac{4}{3}\pi x^3 + 4\pi x^2$, as shown in Figure 5. Since we want the value of this function to be 100, we also graph the horizontal line $y = 100$ in the same viewing rectangle. The correct radius will be the x-coordinate of the point of intersection of the curve and the line. Using the cursor and zooming in, we see that at the point of intersection $x \approx 2.15$, rounded to two decimal places. Thus the tank has a radius of about 2.15 ft.

NOW TRY EXERCISE **99**

Note that we also could have solved the equation in Example 8 by first writing it as

$$\tfrac{4}{3}\pi r^3 + 4\pi r^2 - 100 = 0$$

and then finding the x-intercept of the function $y = \frac{4}{3}\pi x^3 + 4\pi x^2 - 100$.

3.4 EXERCISES

CONCEPTS

1. If the polynomial function

$$P(x) = a_n x^n + a_{n-1}x^{n-1} + \cdots + a_1 x + a_0$$

has integer coefficients, then the only numbers that could possibly be rational zeros of P are all of the form $\frac{p}{q}$, where p is a factor of ________ and q is a factor of ________. The possible rational zeros of $P(x) = 6x^3 + 5x^2 - 19x - 10$ are

________________.

2. Using Descartes' Rule of Signs, we can tell that the polynomial $P(x) = x^5 - 3x^4 + 2x^3 - x^2 + 8x - 8$ has ________, ________, or ________ positive real zeros and ________ negative real zeros.

3. *True or false?* If c is a real zero of the polynomial P, then all the other zeros of P are zeros of $P(x)/(x - c)$.

4. *True or false?* If a is an upper bound for the real zeros of the polynomial P, then $-a$ is necessarily a lower bound for the real zeros of P.

SKILLS

5–10 ■ List all possible rational zeros given by the Rational Zeros Theorem (but don't check to see which actually are zeros).

5. $P(x) = x^3 - 4x^2 + 3$

6. $Q(x) = x^4 - 3x^3 - 6x + 8$

7. $R(x) = 2x^5 + 3x^3 + 4x^2 - 8$

8. $S(x) = 6x^4 - x^2 + 2x + 12$

9. $T(x) = 4x^4 - 2x^2 - 7$

10. $U(x) = 12x^5 + 6x^3 - 2x - 8$

11–14 ■ A polynomial function P and its graph are given. **(a)** List all possible rational zeros of P given by the Rational Zeros Theorem. **(b)** From the graph, determine which of the possible rational zeros actually turn out to be zeros.

11. $P(x) = 5x^3 - x^2 - 5x + 1$

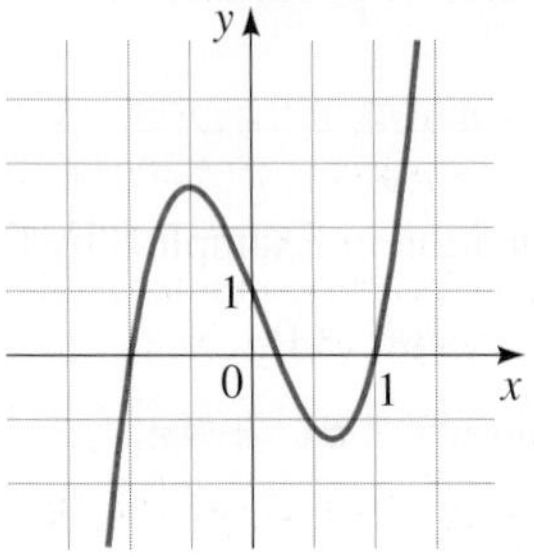

12. $P(x) = 3x^3 + 4x^2 - x - 2$

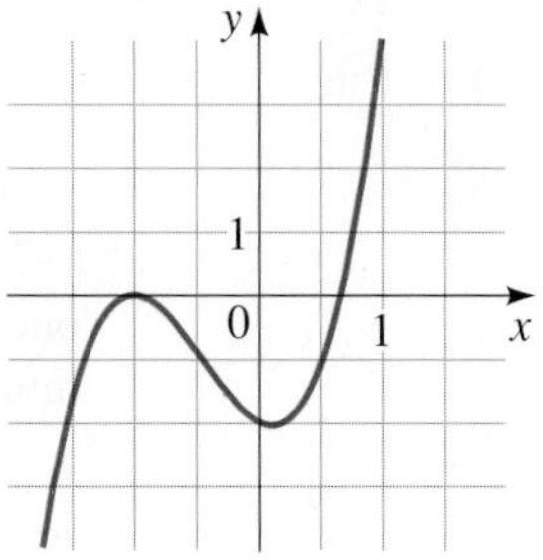

13. $P(x) = 2x^4 - 9x^3 + 9x^2 + x - 3$

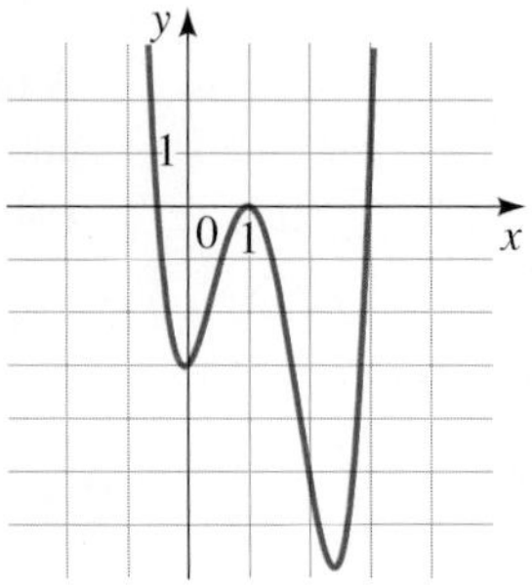

14. $P(x) = 4x^4 - x^3 - 4x + 1$

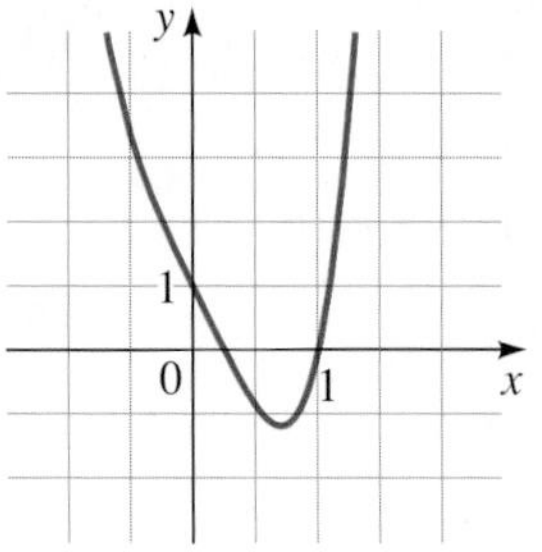

15–46 ■ Find all rational zeros of the polynomial, and write the polynomial in factored form.

15. $P(x) = x^3 + 3x^2 - 4$

16. $P(x) = x^3 - 7x^2 + 14x - 8$

17. $P(x) = x^3 - 3x - 2$

18. $P(x) = x^3 + 4x^2 - 3x - 18$

19. $P(x) = x^3 - 6x^2 + 12x - 8$

20. $P(x) = x^3 - x^2 - 8x + 12$

21. $P(x) = x^3 - 4x^2 + x + 6$

22. $P(x) = x^3 - 4x^2 - 7x + 10$

23. $P(x) = x^3 + 3x^2 - x - 3$

24. $P(x) = x^3 - 4x^2 - 11x + 30$

25. $P(x) = x^4 - 5x^2 + 4$

26. $P(x) = x^4 - 2x^3 - 3x^2 + 8x - 4$

27. $P(x) = x^4 + 6x^3 + 7x^2 - 6x - 8$

28. $P(x) = x^4 - x^3 - 23x^2 - 3x + 90$

29. $P(x) = 4x^4 - 25x^2 + 36$

30. $P(x) = 2x^4 - x^3 - 19x^2 + 9x + 9$

31. $P(x) = 3x^4 - 10x^3 - 9x^2 + 40x - 12$

32. $P(x) = 2x^3 + 7x^2 + 4x - 4$

33. $P(x) = 4x^3 + 4x^2 - x - 1$

34. $P(x) = 2x^3 - 3x^2 - 2x + 3$

35. $P(x) = 4x^3 - 7x + 3$

36. $P(x) = 8x^3 + 10x^2 - x - 3$

37. $P(x) = 4x^3 + 8x^2 - 11x - 15$

38. $P(x) = 6x^3 + 11x^2 - 3x - 2$

39. $P(x) = 20x^3 - 8x^2 - 5x + 2$

40. $P(x) = 12x^3 - 20x^2 + x + 3$

41. $P(x) = 2x^4 - 7x^3 + 3x^2 + 8x - 4$

42. $P(x) = 6x^4 - 7x^3 - 12x^2 + 3x + 2$

43. $P(x) = x^5 + 3x^4 - 9x^3 - 31x^2 + 36$

44. $P(x) = x^5 - 4x^4 - 3x^3 + 22x^2 - 4x - 24$

45. $P(x) = 3x^5 - 14x^4 - 14x^3 + 36x^2 + 43x + 10$

46. $P(x) = 2x^6 - 3x^5 - 13x^4 + 29x^3 - 27x^2 + 32x - 12$

47–56 ■ Find all the real zeros of the polynomial. Use the quadratic formula if necessary, as in Example 3(a).

47. $P(x) = x^3 + 4x^2 + 3x - 2$

48. $P(x) = x^3 - 5x^2 + 2x + 12$

49. $P(x) = x^4 - 6x^3 + 4x^2 + 15x + 4$

50. $P(x) = x^4 + 2x^3 - 2x^2 - 3x + 2$

51. $P(x) = x^4 - 7x^3 + 14x^2 - 3x - 9$

52. $P(x) = x^5 - 4x^4 - x^3 + 10x^2 + 2x - 4$

53. $P(x) = 4x^3 - 6x^2 + 1$

54. $P(x) = 3x^3 - 5x^2 - 8x - 2$

55. $P(x) = 2x^4 + 15x^3 + 17x^2 + 3x - 1$

56. $P(x) = 4x^5 - 18x^4 - 6x^3 + 91x^2 - 60x + 9$

57–64 ■ A polynomial P is given. **(a)** Find all the real zeros of P. **(b)** Sketch the graph of P.

57. $P(x) = x^3 - 3x^2 - 4x + 12$

58. $P(x) = -x^3 - 2x^2 + 5x + 6$

59. $P(x) = 2x^3 - 7x^2 + 4x + 4$

60. $P(x) = 3x^3 + 17x^2 + 21x - 9$

61. $P(x) = x^4 - 5x^3 + 6x^2 + 4x - 8$

62. $P(x) = -x^4 + 10x^2 + 8x - 8$

63. $P(x) = x^5 - x^4 - 5x^3 + x^2 + 8x + 4$

64. $P(x) = x^5 - x^4 - 6x^3 + 14x^2 - 11x + 3$

65–70 ■ Use Descartes' Rule of Signs to determine how many positive and how many negative real zeros the polynomial can have. Then determine the possible total number of real zeros.

65. $P(x) = x^3 - x^2 - x - 3$

66. $P(x) = 2x^3 - x^2 + 4x - 7$

67. $P(x) = 2x^6 + 5x^4 - x^3 - 5x - 1$

68. $P(x) = x^4 + x^3 + x^2 + x + 12$

69. $P(x) = x^5 + 4x^3 - x^2 + 6x$

70. $P(x) = x^8 - x^5 + x^4 - x^3 + x^2 - x + 1$

71–74 ■ Show that the given values for a and b are lower and upper bounds for the real zeros of the polynomial.

71. $P(x) = 2x^3 + 5x^2 + x - 2;\quad a = -3, b = 1$

72. $P(x) = x^4 - 2x^3 - 9x^2 + 2x + 8;\quad a = -3, b = 5$

73. $P(x) = 8x^3 + 10x^2 - 39x + 9;\quad a = -3, b = 2$

74. $P(x) = 3x^4 - 17x^3 + 24x^2 - 9x + 1;\quad a = 0, b = 6$

75–78 ■ Find integers that are upper and lower bounds for the real zeros of the polynomial.

75. $P(x) = x^3 - 3x^2 + 4$

76. $P(x) = 2x^3 - 3x^2 - 8x + 12$

77. $P(x) = x^4 - 2x^3 + x^2 - 9x + 2$

78. $P(x) = x^5 - x^4 + 1$

79–84 ■ Find all rational zeros of the polynomial, and then find the irrational zeros, if any. Whenever appropriate, use the Rational Zeros Theorem, the Upper and Lower Bounds Theorem, Descartes' Rule of Signs, the quadratic formula, or other factoring techniques.

79. $P(x) = 2x^4 + 3x^3 - 4x^2 - 3x + 2$

80. $P(x) = 2x^4 + 15x^3 + 31x^2 + 20x + 4$

81. $P(x) = 4x^4 - 21x^2 + 5$

82. $P(x) = 6x^4 - 7x^3 - 8x^2 + 5x$

83. $P(x) = x^5 - 7x^4 + 9x^3 + 23x^2 - 50x + 24$

84. $P(x) = 8x^5 - 14x^4 - 22x^3 + 57x^2 - 35x + 6$

85–88 ■ Show that the polynomial does not have any rational zeros.

85. $P(x) = x^3 - x - 2$

86. $P(x) = 2x^4 - x^3 + x + 2$

87. $P(x) = 3x^3 - x^2 - 6x + 12$

88. $P(x) = x^{50} - 5x^{25} + x^2 - 1$

89–92 ■ The real solutions of the given equation are rational. List all possible rational roots using the Rational Zeros Theorem, and then graph the polynomial in the given viewing rectangle to determine which values are actually solutions. (All solutions can be seen in the given viewing rectangle.)

89. $x^3 - 3x^2 - 4x + 12 = 0;\quad [-4, 4]$ by $[-15, 15]$

90. $x^4 - 5x^2 + 4 = 0;\quad [-4, 4]$ by $[-30, 30]$

91. $2x^4 - 5x^3 - 14x^2 + 5x + 12 = 0$; $[-2, 5]$ by $[-40, 40]$

92. $3x^3 + 8x^2 + 5x + 2 = 0$; $[-3, 3]$ by $[-10, 10]$

93–96 ■ Use a graphing device to find all real solutions of the equation, rounded to two decimal places.

93. $x^4 - x - 4 = 0$

94. $2x^3 - 8x^2 + 9x - 9 = 0$

95. $4.00x^4 + 4.00x^3 - 10.96x^2 - 5.88x + 9.09 = 0$

96. $x^5 + 2.00x^4 + 0.96x^3 + 5.00x^2 + 10.00x + 4.80 = 0$

97. Let $P(x)$ be a polynomial with real coefficients and let $b > 0$. Use the Division Algorithm to write

$$P(x) = (x - b) \cdot Q(x) + r$$

Suppose that $r \geq 0$ and that all the coefficients in $Q(x)$ are nonnegative. Let $z > b$.

(a) Show that $P(z) > 0$.

(b) Prove the first part of the Upper and Lower Bounds Theorem.

(c) Use the first part of the Upper and Lower Bounds Theorem to prove the second part. [*Hint:* Show that if $P(x)$ satisfies the second part of the theorem, then $P(-x)$ satisfies the first part.]

98. Show that the equation

$$x^5 - x^4 - x^3 - 5x^2 - 12x - 6 = 0$$

has exactly one rational root, and then prove that it must have either two or four irrational roots.

APPLICATIONS

99. Volume of a Silo A grain silo consists of a cylindrical main section and a hemispherical roof. If the total volume of the silo (including the part inside the roof section) is 15,000 ft^3 and the cylindrical part is 30 ft tall, what is the radius of the silo, rounded to the nearest tenth of a foot?

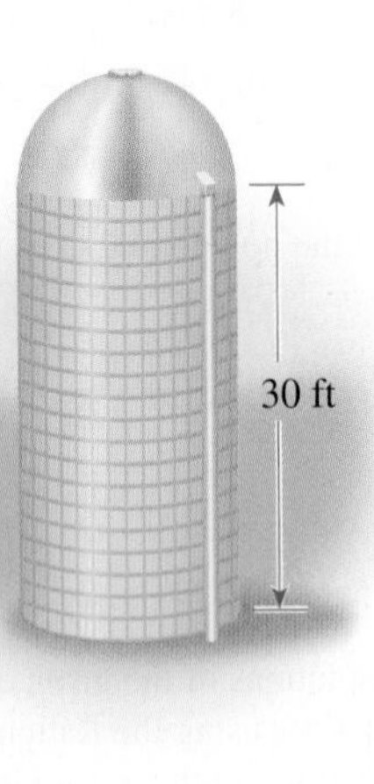

100. Dimensions of a Lot A rectangular parcel of land has an area of 5000 ft^2. A diagonal between opposite corners is measured to be 10 ft longer than one side of the parcel. What are the dimensions of the land, rounded to the nearest foot?

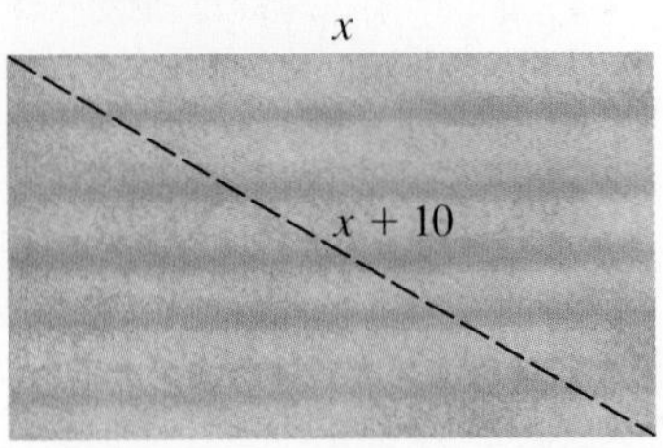

101. Depth of Snowfall Snow began falling at noon on Sunday. The amount of snow on the ground at a certain location at time t was given by the function

$$h(t) = 11.60t - 12.41t^2 + 6.20t^3 - 1.58t^4 + 0.20t^5 - 0.01t^6$$

where t is measured in days from the start of the snowfall and $h(t)$ is the depth of snow in inches. Draw a graph of this function, and use your graph to answer the following questions.

(a) What happened shortly after noon on Tuesday?

(b) Was there ever more than 5 in. of snow on the ground? If so, on what day(s)?

(c) On what day and at what time (to the nearest hour) did the snow disappear completely?

102. Volume of a Box An open box with a volume of 1500 cm^3 is to be constructed by taking a piece of cardboard 20 cm by 40 cm, cutting squares of side length x cm from each corner, and folding up the sides. Show that this can be done in two different ways, and find the exact dimensions of the box in each case.

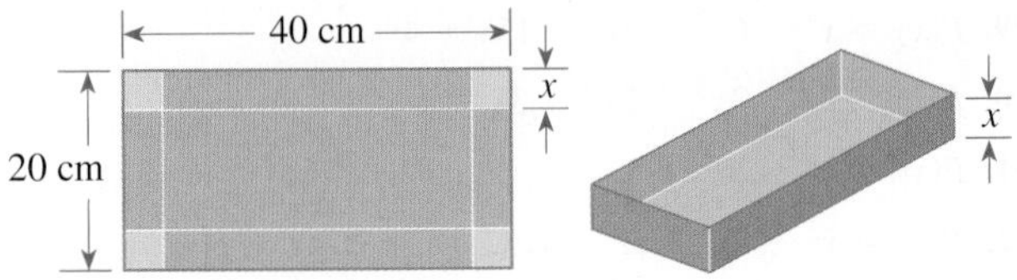

103. Volume of a Rocket A rocket consists of a right circular cylinder of height 20 m surmounted by a cone whose height and diameter are equal and whose radius is the same as that of the cylindrical section. What should this radius be (rounded to two decimal places) if the total volume is to be $500\pi/3$ m^3?

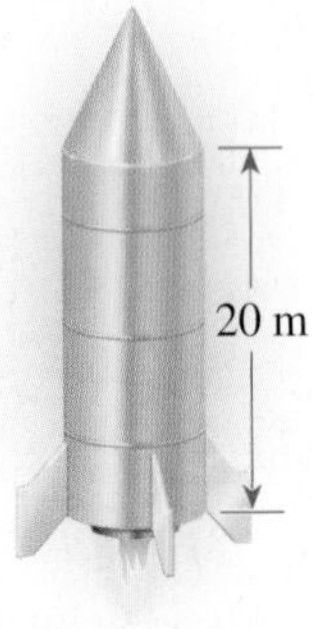

104. Volume of a Box A rectangular box with a volume of $2\sqrt{2}$ ft^3 has a square base as shown below. The diagonal of the box (between a pair of opposite corners) is 1 ft longer than each side of the base.

(a) If the base has sides of length x feet, show that

$$x^6 - 2x^5 - x^4 + 8 = 0$$

(b) Show that two different boxes satisfy the given conditions. Find the dimensions in each case, rounded to the nearest hundredth of a foot.

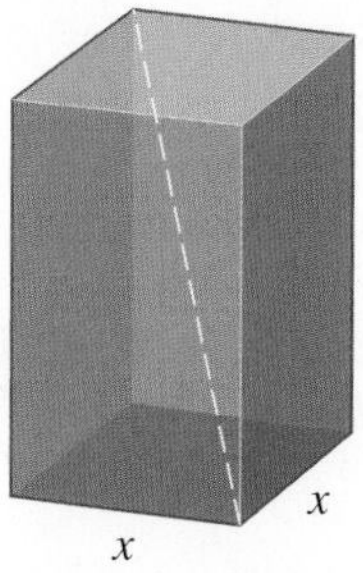

105. Girth of a Box A box with a square base has length plus girth of 108 in. (Girth is the distance "around" the box.) What is the length of the box if its volume is 2200 in^3?

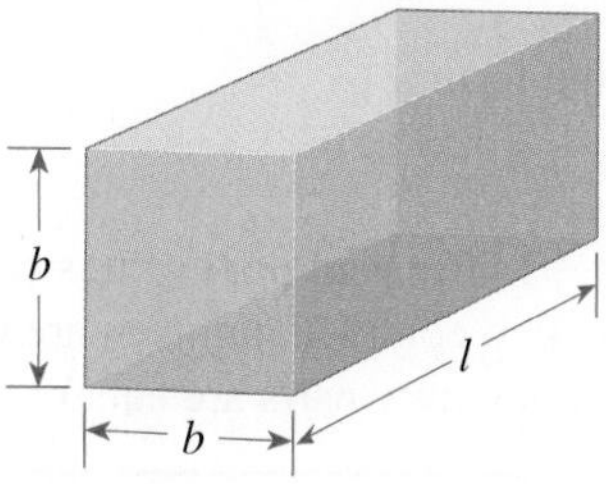

DISCOVERY ■ DISCUSSION ■ WRITING

106. How Many Real Zeros Can a Polynomial Have? Give examples of polynomials that have the following properties, or explain why it is impossible to find such a polynomial.

(a) A polynomial of degree 3 that has no real zeros

(b) A polynomial of degree 4 that has no real zeros

(c) A polynomial of degree 3 that has three real zeros, only one of which is rational

(d) A polynomial of degree 4 that has four real zeros, none of which is rational

What must be true about the degree of a polynomial with integer coefficients if it has no real zeros?

107. The Depressed Cubic The most general cubic (third-degree) equation with rational coefficients can be written as

$$x^3 + ax^2 + bx + c = 0$$

(a) Show that if we replace x by $X - a/3$ and simplify, we end up with an equation that doesn't have an X^2 term, that is, an equation of the form

$$X^3 + pX + q = 0$$

This is called a *depressed cubic*, because we have "depressed" the quadratic term.

(b) Use the procedure described in part (a) to depress the equation $x^3 + 6x^2 + 9x + 4 = 0$.

108. The Cubic Formula The quadratic formula can be used to solve any quadratic (or second-degree) equation. You might have wondered whether similar formulas exist for cubic (third-degree), quartic (fourth-degree), and higher-degree equations. For the depressed cubic $x^3 + px + q = 0$, Cardano (page 274) found the following formula for one solution:

$$x = \sqrt[3]{\frac{-q}{2} + \sqrt{\frac{q^2}{4} + \frac{p^3}{27}}} + \sqrt[3]{\frac{-q}{2} - \sqrt{\frac{q^2}{4} + \frac{p^3}{27}}}$$

A formula for quartic equations was discovered by the Italian mathematician Ferrari in 1540. In 1824 the Norwegian mathematician Niels Henrik Abel proved that it is impossible to write a quintic formula, that is, a formula for fifth-degree equations. Finally, Galois (page 254) gave a criterion for determining which equations can be solved by a formula involving radicals.

Use the cubic formula to find a solution for the following equations. Then solve the equations using the methods you learned in this section. Which method is easier?

(a) $x^3 - 3x + 2 = 0$

(b) $x^3 - 27x - 54 = 0$

(c) $x^3 + 3x + 4 = 0$

DISCOVERY PROJECT **Zeroing in on a Zero**

In this project we explore a numerical method for approximating the zeros of a polynomial function.
You can find the project at the book companion website:
www.stewartmath.com

3.5 Complex Numbers

Arithmetic Operations on Complex Numbers ▶ Square Roots of Negative Numbers ▶ Complex Solutions of Quadratic Equations

In Section 1.5 we saw that if the discriminant of a quadratic equation is negative, the equation has no real solution. For example, the equation

$$x^2 + 4 = 0$$

has no real solution. If we try to solve this equation, we get $x^2 = -4$, so

$$x = \pm\sqrt{-4}$$

But this is impossible, since the square of any real number is positive. [For example, $(-2)^2 = 4$, a positive number.] Thus, negative numbers don't have real square roots.

See the note on Cardano (page 274) for an example of how complex numbers are used to find real solutions of polynomial equations.

To make it possible to solve *all* quadratic equations, mathematicians invented an expanded number system, called the *complex number system*. First they defined the new number

$$i = \sqrt{-1}$$

This means that $i^2 = -1$. A complex number is then a number of the form $a + bi$, where a and b are real numbers.

DEFINITION OF COMPLEX NUMBERS

A **complex number** is an expression of the form

$$a + bi$$

where a and b are real numbers and $i^2 = -1$. The **real part** of this complex number is a and the **imaginary part** is b. Two complex numbers are **equal** if and only if their real parts are equal and their imaginary parts are equal.

Note that both the real and imaginary parts of a complex number are real numbers.

EXAMPLE 1 | Complex Numbers

The following are examples of complex numbers.

$3 + 4i$	Real part 3, imaginary part 4
$\frac{1}{2} - \frac{2}{3}i$	Real part $\frac{1}{2}$, imaginary part $-\frac{2}{3}$
$6i$	Real part 0, imaginary part 6
-7	Real part -7, imaginary part 0

NOW TRY EXERCISES **5** AND **9** ■

A number such as $6i$, which has real part 0, is called a **pure imaginary number**. A real number such as -7 can be thought of as a complex number with imaginary part 0.

In the complex number system every quadratic equation has solutions. The numbers $2i$ and $-2i$ are solutions of $x^2 = -4$ because

$$(2i)^2 = 2^2i^2 = 4(-1) = -4 \quad \text{and} \quad (-2i)^2 = (-2)^2i^2 = 4(-1) = -4$$

Although we use the term *imaginary* in this context, imaginary numbers should not be thought of as any less "real" (in the ordinary rather than the mathematical sense of that word) than negative numbers or irrational numbers. All numbers (except possibly the positive integers) are creations of the human mind—the numbers -1 and $\sqrt{2}$ as well as the number i. We study complex numbers because they complete, in a useful and elegant fashion, our study of the solutions of equations. In fact, imaginary numbers are useful not only in algebra and mathematics, but in the other sciences as well. To give just one example, in electrical theory the *reactance* of a circuit is a quantity whose measure is an imaginary number.

▼ Arithmetic Operations on Complex Numbers

Complex numbers are added, subtracted, multiplied, and divided just as we would any number of the form $a + b\sqrt{c}$. The only difference that we need to keep in mind is that $i^2 = -1$. Thus, the following calculations are valid.

$$\begin{aligned}(a + bi)(c + di) &= ac + (ad + bc)i + bdi^2 && \text{Multiply and collect like terms}\\ &= ac + (ad + bc)i + bd(-1) && i^2 = -1\\ &= (ac - bd) + (ad + bc)i && \text{Combine real and imaginary parts}\end{aligned}$$

We therefore define the sum, difference, and product of complex numbers as follows.

ADDING, SUBTRACTING, AND MULTIPLYING COMPLEX NUMBERS

Definition	Description
Addition	
$(a + bi) + (c + di) = (a + c) + (b + d)i$	To add complex numbers, add the real parts and the imaginary parts.
Subtraction	
$(a + bi) - (c + di) = (a - c) + (b - d)i$	To subtract complex numbers, subtract the real parts and the imaginary parts.
Multiplication	
$(a + bi) \cdot (c + di) = (ac - bd) + (ad + bc)i$	Multiply complex numbers like binomials, using $i^2 = -1$.

EXAMPLE 2 | Adding, Subtracting, and Multiplying Complex Numbers

Graphing calculators can perform arithmetic operations on complex numbers.

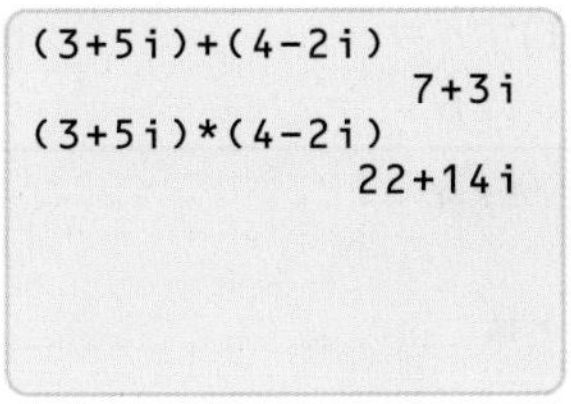

Express the following in the form $a + bi$.

(a) $(3 + 5i) + (4 - 2i)$ **(b)** $(3 + 5i) - (4 - 2i)$

(c) $(3 + 5i)(4 - 2i)$ **(d)** i^{23}

SOLUTION

(a) According to the definition, we add the real parts and we add the imaginary parts.

$$(3 + 5i) + (4 - 2i) = (3 + 4) + (5 - 2)i = 7 + 3i$$

(b) $(3 + 5i) - (4 - 2i) = (3 - 4) + [5 - (-2)]i = -1 + 7i$

(c) $(3 + 5i)(4 - 2i) = [3 \cdot 4 - 5(-2)] + [3(-2) + 5 \cdot 4]i = 22 + 14i$

(d) $i^{23} = i^{22+1} = (i^2)^{11}i = (-1)^{11}i = (-1)i = -i$

NOW TRY EXERCISES **15, 19, 25,** AND **33** ■

Complex Conjugates

Number	Conjugate
$3 + 2i$	$3 - 2i$
$1 - i$	$1 + i$
$4i$	$-4i$
5	5

Division of complex numbers is much like rationalizing the denominator of a radical expression, which we considered in Section 1.4. For the complex number $z = a + bi$ we define its **complex conjugate** to be $\overline{z} = a - bi$. Note that

$$z \cdot \overline{z} = (a + bi)(a - bi) = a^2 + b^2$$

Library of Congress

LEONHARD EULER (1707–1783) was born in Basel, Switzerland, the son of a pastor. When Euler was 13, his father sent him to the University at Basel to study theology, but Euler soon decided to devote himself to the sciences. Besides theology he studied mathematics, medicine, astronomy, physics, and Asian languages. It is said that Euler could calculate as effortlessly as "men breathe or as eagles fly." One hundred years before Euler, Fermat (see page 99) had conjectured that $2^{2^n} + 1$ is a prime number for all n. The first five of these numbers are 5, 17, 257, 65537, and 4,294,967,297. It is easy to show that the first four are prime. The fifth was also thought to be prime until Euler, with his phenomenal calculating ability, showed that it is the product $641 \times 6{,}700{,}417$ and so is not prime. Euler published more than any other mathematician in history. His collected works comprise 75 large volumes. Although he was blind for the last 17 years of his life, he continued to work and publish. In his writings he popularized the use of the symbols π, e, and i, which you will find in this textbook. One of Euler's most lasting contributions is his development of complex numbers.

So the product of a complex number and its conjugate is always a nonnegative real number. We use this property to divide complex numbers.

DIVIDING COMPLEX NUMBERS

To simplify the quotient $\dfrac{a+bi}{c+di}$, multiply the numerator and the denominator by the complex conjugate of the denominator:

$$\frac{a+bi}{c+di} = \left(\frac{a+bi}{c+di}\right)\left(\frac{c-di}{c-di}\right) = \frac{(ac+bd)+(bc-ad)i}{c^2+d^2}$$

Rather than memorizing this entire formula, it is easier to just remember the first step and then multiply out the numerator and the denominator as usual.

EXAMPLE 3 | Dividing Complex Numbers

Express the following in the form $a + bi$.

(a) $\dfrac{3+5i}{1-2i}$ **(b)** $\dfrac{7+3i}{4i}$

SOLUTION We multiply both the numerator and denominator by the complex conjugate of the denominator to make the new denominator a real number.

(a) The complex conjugate of $1 - 2i$ is $\overline{1-2i} = 1 + 2i$.

$$\frac{3+5i}{1-2i} = \left(\frac{3+5i}{1-2i}\right)\left(\frac{1+2i}{1+2i}\right) = \frac{-7+11i}{5} = -\frac{7}{5} + \frac{11}{5}i$$

(b) The complex conjugate of $4i$ is $-4i$. Therefore,

$$\frac{7+3i}{4i} = \left(\frac{7+3i}{4i}\right)\left(\frac{-4i}{-4i}\right) = \frac{12-28i}{16} = \frac{3}{4} - \frac{7}{4}i$$

NOW TRY EXERCISES **37** AND **43**

Square Roots of Negative Numbers

Just as every positive real number r has two square roots ($\sqrt{r}$ and $-\sqrt{r}$), every negative number has two square roots as well. If $-r$ is a negative number, then its square roots are $\pm i\sqrt{r}$, because $(i\sqrt{r})^2 = i^2 r = -r$ and $(-i\sqrt{r})^2 = (-1)^2 i^2 r = -r$.

SQUARE ROOTS OF NEGATIVE NUMBERS

If $-r$ is negative, then the **principal square root** of $-r$ is

$$\sqrt{-r} = i\sqrt{r}$$

The two square roots of $-r$ are $i\sqrt{r}$ and $-i\sqrt{r}$.

We usually write $i\sqrt{b}$ instead of $\sqrt{b}\,i$ to avoid confusion with $\sqrt{bi}$.

EXAMPLE 4 | Square Roots of Negative Numbers

(a) $\sqrt{-1} = i\sqrt{1} = i$ **(b)** $\sqrt{-16} = i\sqrt{16} = 4i$ **(c)** $\sqrt{-3} = i\sqrt{3}$

NOW TRY EXERCISES **47** AND **49**

Special care must be taken in performing calculations that involve square roots of negative numbers. Although $\sqrt{a} \cdot \sqrt{b} = \sqrt{ab}$ when a and b are positive, this is *not* true when both are negative. For example,

$$\sqrt{-2} \cdot \sqrt{-3} = i\sqrt{2} \cdot i\sqrt{3} = i^2\sqrt{6} = -\sqrt{6}$$

but $$\sqrt{(-2)(-3)} = \sqrt{6}$$

so $$\sqrt{-2} \cdot \sqrt{-3} \neq \sqrt{(-2)(-3)}$$

When multiplying radicals of negative numbers, express them first in the form $i\sqrt{r}$ (where $r > 0$) to avoid possible errors of this type.

EXAMPLE 5 | Using Square Roots of Negative Numbers

Evaluate $(\sqrt{12} - \sqrt{-3})(3 + \sqrt{-4})$ and express in the form $a + bi$.

SOLUTION

$$\begin{aligned}(\sqrt{12} - \sqrt{-3})(3 + \sqrt{-4}) &= (\sqrt{12} - i\sqrt{3})(3 + i\sqrt{4}) \\ &= (2\sqrt{3} - i\sqrt{3})(3 + 2i) \\ &= (6\sqrt{3} + 2\sqrt{3}) + i(2 \cdot 2\sqrt{3} - 3\sqrt{3}) \\ &= 8\sqrt{3} + i\sqrt{3}\end{aligned}$$

NOW TRY EXERCISE **51** ■

▼ Complex Solutions of Quadratic Equations

We have already seen that if $a \neq 0$, then the solutions of the quadratic equation $ax^2 + bx + c = 0$ are

$$x = \frac{-b \pm \sqrt{b^2 - 4ac}}{2a}$$

If $b^2 - 4ac < 0$, then the equation has no real solution. But in the complex number system, this equation will always have solutions, because negative numbers have square roots in this expanded setting.

EXAMPLE 6 | Quadratic Equations with Complex Solutions

Solve each equation.

(a) $x^2 + 9 = 0$ **(b)** $x^2 + 4x + 5 = 0$

SOLUTION

(a) The equation $x^2 + 9 = 0$ means $x^2 = -9$, so

$$x = \pm\sqrt{-9} = \pm i\sqrt{9} = \pm 3i$$

The solutions are therefore $3i$ and $-3i$.

(b) By the Quadratic Formula we have

$$\begin{aligned}x &= \frac{-4 \pm \sqrt{4^2 - 4 \cdot 5}}{2} \\ &= \frac{-4 \pm \sqrt{-4}}{2} \\ &= \frac{-4 \pm 2i}{2} = \frac{2(-2 \pm i)}{2} = -2 \pm i\end{aligned}$$

So the solutions are $-2 + i$ and $-2 - i$.

NOW TRY EXERCISES **57** AND **59** ■

We see from Example 6 that if a quadratic equation with real coefficients has complex solutions, then these solutions are complex conjugates of each other. So if $a + bi$ is a solution, then $a - bi$ is also a solution.

EXAMPLE 7 | Complex Conjugates as Solutions of a Quadratic

Show that the solutions of the equation

$$4x^2 - 24x + 37 = 0$$

are complex conjugates of each other.

SOLUTION We use the Quadratic Formula to get

$$x = \frac{24 \pm \sqrt{(24)^2 - 4(4)(37)}}{2(4)}$$

$$= \frac{24 \pm \sqrt{-16}}{8} = \frac{24 \pm 4i}{8} = 3 \pm \frac{1}{2}i$$

So the solutions are $3 + \frac{1}{2}i$ and $3 - \frac{1}{2}i$, and these are complex conjugates.

NOW TRY EXERCISE **65**

3.5 EXERCISES

CONCEPTS

1. The imaginary number i has the property that $i^2 =$ ______.
2. For the complex number $3 + 4i$ the real part is ______ and the imaginary part is ______.
3. (a) The complex conjugate of $3 + 4i$ is $\overline{3 + 4i} =$ ______.
   (b) $(3 + 4i)(\overline{3 + 4i}) =$ ______.
4. If $3 + 4i$ is a solution of a quadratic equation with real coefficients, then ______ is also a solution of the equation.

SKILLS

5–14 ■ Find the real and imaginary parts of the complex number.

5. $5 - 7i$
6. $-6 + 4i$
7. $\dfrac{-2 - 5i}{3}$
8. $\dfrac{4 + 7i}{2}$
9. 3
10. $-\frac{1}{2}$
11. $-\frac{2}{3}i$
12. $i\sqrt{3}$
13. $\sqrt{3} + \sqrt{-4}$
14. $2 - \sqrt{-5}$

15–46 ■ Evaluate the expression and write the result in the form $a + bi$.

15. $(2 - 5i) + (3 + 4i)$
16. $(2 + 5i) + (4 - 6i)$
17. $(-6 + 6i) + (9 - i)$
18. $(3 - 2i) + (-5 - \frac{1}{3}i)$
19. $(7 - \frac{1}{2}i) - (5 + \frac{3}{2}i)$
20. $(-4 + i) - (2 - 5i)$
21. $(-12 + 8i) - (7 + 4i)$
22. $6i - (4 - i)$
23. $4(-1 + 2i)$
24. $2i(\frac{1}{2} - i)$
25. $(7 - i)(4 + 2i)$
26. $(5 - 3i)(1 + i)$
27. $(3 - 4i)(5 - 12i)$
28. $(\frac{2}{3} + 12i)(\frac{1}{6} + 24i)$
29. $(6 + 5i)(2 - 3i)$
30. $(-2 + i)(3 - 7i)$
31. i^3
32. $(2i)^4$
33. i^{100}
34. i^{1002}
35. $\dfrac{1}{i}$
36. $\dfrac{1}{1 + i}$
37. $\dfrac{2 - 3i}{1 - 2i}$
38. $\dfrac{5 - i}{3 + 4i}$
39. $\dfrac{26 + 39i}{2 - 3i}$
40. $\dfrac{25}{4 - 3i}$
41. $\dfrac{10i}{1 - 2i}$
42. $(2 - 3i)^{-1}$
43. $\dfrac{4 + 6i}{3i}$
44. $\dfrac{-3 + 5i}{15i}$
45. $\dfrac{1}{1 + i} - \dfrac{1}{1 - i}$
46. $\dfrac{(1 + 2i)(3 - i)}{2 + i}$

47–56 ■ Evaluate the radical expression and express the result in the form $a + bi$.

47. $\sqrt{-25}$
48. $\sqrt{\dfrac{-9}{4}}$
49. $\sqrt{-3}\sqrt{-12}$
50. $\sqrt{\frac{1}{3}}\sqrt{-27}$

51. $(3 - \sqrt{-5})(1 + \sqrt{-1})$

52. $(\sqrt{3} - \sqrt{-4})(\sqrt{6} - \sqrt{-8})$

53. $\dfrac{2 + \sqrt{-8}}{1 + \sqrt{-2}}$ **54.** $\dfrac{1 - \sqrt{-1}}{1 + \sqrt{-1}}$

55. $\dfrac{\sqrt{-36}}{\sqrt{-2}\sqrt{-9}}$ **56.** $\dfrac{\sqrt{-7}\sqrt{-49}}{\sqrt{28}}$

57–72 ■ Find all solutions of the equation and express them in the form $a + bi$.

57. $x^2 + 49 = 0$ **58.** $9x^2 + 4 = 0$

59. $x^2 - 4x + 5 = 0$ **60.** $x^2 + 2x + 2 = 0$

61. $x^2 + 2x + 5 = 0$ **62.** $x^2 - 6x + 10 = 0$

63. $x^2 + x + 1 = 0$ **64.** $x^2 - 3x + 3 = 0$

65. $2x^2 - 2x + 1 = 0$ **66.** $2x^2 + 3 = 2x$

67. $t + 3 + \dfrac{3}{t} = 0$ **68.** $z + 4 + \dfrac{12}{z} = 0$

69. $6x^2 + 12x + 7 = 0$ **70.** $4x^2 - 16x + 19 = 0$

71. $\frac{1}{2}x^2 - x + 5 = 0$ **72.** $x^2 + \frac{1}{2}x + 1 = 0$

73–80 ■ Recall that the symbol $\bar{z}$ represents the complex conjugate of z. If $z = a + bi$ and $w = c + di$, prove each statement.

73. $\bar{z} + \bar{w} = \overline{z + w}$ **74.** $\overline{zw} = \bar{z} \cdot \bar{w}$

75. $(\bar{z})^2 = \overline{z^2}$ **76.** $\bar{\bar{z}} = z$

77. $z + \bar{z}$ is a real number.

78. $z - \bar{z}$ is a pure imaginary number.

79. $z \cdot \bar{z}$ is a real number.

80. $z = \bar{z}$ if and only if z is real.

DISCOVERY ■ DISCUSSION ■ WRITING

81. Complex Conjugate Roots Suppose that the equation $ax^2 + bx + c = 0$ has real coefficients and complex roots. Why must the roots be complex conjugates of each other? (Think about how you would find the roots using the Quadratic Formula.)

82. Powers of i Calculate the first 12 powers of i, that is, $i, i^2, i^3, \ldots, i^{12}$. Do you notice a pattern? Explain how you would calculate any whole number power of i, using the pattern that you have discovered. Use this procedure to calculate i^{4446}.

3.6 Complex Zeros and the Fundamental Theorem of Algebra

The Fundamental Theorem of Algebra and Complete Factorization ▶ Zeros and Their Multiplicities ▶ Complex Zeros Come in Conjugate Pairs ▶ Linear and Quadratic Factors

We have already seen that an nth-degree polynomial can have at most n real zeros. In the complex number system an nth-degree polynomial has exactly n zeros and so can be factored into exactly n linear factors. This fact is a consequence of the Fundamental Theorem of Algebra, which was proved by the German mathematician C. F. Gauss in 1799 (see page 272).

▼ The Fundamental Theorem of Algebra and Complete Factorization

The following theorem is the basis for much of our work in factoring polynomials and solving polynomial equations.

FUNDAMENTAL THEOREM OF ALGEBRA

Every polynomial

$$P(x) = a_n x^n + a_{n-1}x^{n-1} + \cdots + a_1 x + a_0 \qquad (n \geq 1, a_n \neq 0)$$

with complex coefficients has at least one complex zero.

Because any real number is also a complex number, the theorem applies to polynomials with real coefficients as well.

The Fundamental Theorem of Algebra and the Factor Theorem together show that a polynomial can be factored completely into linear factors, as we now prove.

COMPLETE FACTORIZATION THEOREM

If $P(x)$ is a polynomial of degree $n \geq 1$, then there exist complex numbers $a, c_1, c_2, \ldots, c_n$ (with $a \neq 0$) such that

$$P(x) = a(x - c_1)(x - c_2) \cdots (x - c_n)$$

PROOF By the Fundamental Theorem of Algebra, P has at least one zero. Let's call it c_1. By the Factor Theorem (see page 250), $P(x)$ can be factored as

$$P(x) = (x - c_1) \cdot Q_1(x)$$

where $Q_1(x)$ is of degree $n - 1$. Applying the Fundamental Theorem to the quotient $Q_1(x)$ gives us the factorization

$$P(x) = (x - c_1) \cdot (x - c_2) \cdot Q_2(x)$$

where $Q_2(x)$ is of degree $n - 2$ and c_2 is a zero of $Q_1(x)$. Continuing this process for n steps, we get a final quotient $Q_n(x)$ of degree 0, a nonzero constant that we will call a. This means that P has been factored as

$$P(x) = a(x - c_1)(x - c_2) \cdots (x - c_n)$$

■

To actually find the complex zeros of an nth-degree polynomial, we usually first factor as much as possible, then use the quadratic formula on parts that we can't factor further.

EXAMPLE 1 | Factoring a Polynomial Completely

Let $P(x) = x^3 - 3x^2 + x - 3$.

(a) Find all the zeros of P.

(b) Find the complete factorization of P.

SOLUTION

(a) We first factor P as follows.

$$\begin{aligned} P(x) &= x^3 - 3x^2 + x - 3 && \text{Given} \\ &= x^2(x - 3) + (x - 3) && \text{Group terms} \\ &= (x - 3)(x^2 + 1) && \text{Factor } x - 3 \end{aligned}$$

We find the zeros of P by setting each factor equal to 0:

$$P(x) = (x - 3)(x^2 + 1)$$

This factor is 0 when $x = 3$ | This factor is 0 when $x = i$ or $-i$

Setting $x - 3 = 0$, we see that $x = 3$ is a zero. Setting $x^2 + 1 = 0$, we get $x^2 = -1$, so $x = \pm i$. So the zeros of P are 3, i, and $-i$.

(b) Since the zeros are 3, i, and $-i$, by the Complete Factorization Theorem P factors as

$$\begin{aligned} P(x) &= (x - 3)(x - i)[x - (-i)] \\ &= (x - 3)(x - i)(x + i) \end{aligned}$$

NOW TRY EXERCISE 5 ■

EXAMPLE 2 | Factoring a Polynomial Completely

Let $P(x) = x^3 - 2x + 4$.

(a) Find all the zeros of P.

(b) Find the complete factorization of P.

SOLUTION

-2	1	-0	-2	-4
		-2	4	-4
	1	-2	-2	0

(a) The possible rational zeros are the factors of 4, which are $\pm 1, \pm 2, \pm 4$. Using synthetic division (see the margin), we find that -2 is a zero, and the polynomial factors as

$$P(x) = (x + 2)(x^2 - 2x + 2)$$

This factor is 0 when $x = -2$

Use the Quadratic Formula to find when this factor is 0

To find the zeros, we set each factor equal to 0. Of course, $x + 2 = 0$ means that $x = -2$. We use the quadratic formula to find when the other factor is 0.

$$x^2 - 2x + 2 = 0 \qquad \text{Set factor equal to 0}$$

$$x = \frac{2 \pm \sqrt{4 - 8}}{2} \qquad \text{Quadratic Formula}$$

$$x = \frac{2 \pm 2i}{2} \qquad \text{Take square root}$$

$$x = 1 \pm i \qquad \text{Simplify}$$

So the zeros of P are -2, $1 + i$, and $1 - i$.

(b) Since the zeros are -2, $1 + i$, and $1 - i$, by the Complete Factorization Theorem, P factors as

$$\begin{aligned} P(x) &= [x - (-2)][x - (1 + i)][x - (1 - i)] \\ &= (x + 2)(x - 1 - i)(x - 1 + i) \end{aligned}$$

NOW TRY EXERCISE **19** ■

▼ Zeros and Their Multiplicities

In the Complete Factorization Theorem the numbers $c_1, c_2, \ldots, c_n$ are the zeros of P. These zeros need not all be different. If the factor $x - c$ appears k times in the complete factorization of $P(x)$, then we say that c is a zero of **multiplicity k** (see page 240). For example, the polynomial

$$P(x) = (x - 1)^3(x + 2)^2(x + 3)^5$$

has the following zeros:

1 (multiplicity 3), $\qquad -2$ (multiplicity 2), $\qquad -3$ (multiplicity 5)

The polynomial P has the same number of zeros as its degree: It has degree 10 and has 10 zeros, provided that we count multiplicities. This is true for all polynomials, as we prove in the following theorem.

ZEROS THEOREM

Every polynomial of degree $n \geq 1$ has exactly n zeros, provided that a zero of multiplicity k is counted k times.

PROOF Let P be a polynomial of degree n. By the Complete Factorization Theorem

$$P(x) = a(x - c_1)(x - c_2) \cdots (x - c_n)$$

Now suppose that c is a zero of P other than $c_1, c_2, \ldots, c_n$. Then

$$P(c) = a(c - c_1)(c - c_2) \cdots (c - c_n) = 0$$

Thus, by the Zero-Product Property, one of the factors $c - c_i$ must be 0, so $c = c_i$ for some i. It follows that P has exactly the n zeros $c_1, c_2, \ldots, c_n$. ■

CARL FRIEDRICH GAUSS (1777–1855) is considered the greatest mathematician of modern times. His contemporaries called him the "Prince of Mathematics." He was born into a poor family; his father made a living as a mason. As a very small child, Gauss found a calculation error in his father's accounts, the first of many incidents that gave evidence of his mathematical precocity. (See also page 796.) At 19, Gauss demonstrated that the regular 17-sided polygon can be constructed with straight-edge and compass alone. This was remarkable because, since the time of Euclid, it had been thought that the only regular polygons constructible in this way were the triangle and pentagon. Because of this discovery Gauss decided to pursue a career in mathematics instead of languages, his other passion. In his doctoral dissertation, written at the age of 22, Gauss proved the Fundamental Theorem of Algebra: A polynomial of degree n with complex coefficients has n roots. His other accomplishments range over every branch of mathematics, as well as physics and astronomy.

EXAMPLE 3 | Factoring a Polynomial with Complex Zeros

Find the complete factorization and all five zeros of the polynomial

$$P(x) = 3x^5 + 24x^3 + 48x$$

SOLUTION Since $3x$ is a common factor, we have

$$\begin{aligned} P(x) &= 3x(x^4 + 8x^2 + 16) \\ &= 3x(x^2 + 4)^2 \end{aligned}$$

This factor is 0 when $x = 0$

This factor is 0 when $x = 2i$ or $x = -2i$

To factor $x^2 + 4$, note that $2i$ and $-2i$ are zeros of this polynomial. Thus, $x^2 + 4 = (x - 2i)(x + 2i)$, so

$$\begin{aligned} P(x) &= 3x[(x - 2i)(x + 2i)]^2 \\ &= 3x(x - 2i)^2(x + 2i)^2 \end{aligned}$$

0 is a zero of multiplicity 1

$2i$ is a zero of multiplicity 2

$-2i$ is a zero of multiplicity 2

The zeros of P are 0, $2i$, and $-2i$. Since the factors $x - 2i$ and $x + 2i$ each occur twice in the complete factorization of P, the zeros $2i$ and $-2i$ are of multiplicity 2 (or *double* zeros). Thus, we have found all five zeros.

NOW TRY EXERCISE **29** ■

The following table gives further examples of polynomials with their complete factorizations and zeros.

Degree	Polynomial	Zero(s)	Number of zeros
1	$P(x) = x - 4$	4	1
2	$P(x) = x^2 - 10x + 25$ $= (x - 5)(x - 5)$	5 (multiplicity 2)	2
3	$P(x) = x^3 + x$ $= x(x - i)(x + i)$	$0, i, -i$	3
4	$P(x) = x^4 + 18x^2 + 81$ $= (x - 3i)^2(x + 3i)^2$	$3i$ (multiplicity 2), $-3i$ (multiplicity 2)	4
5	$P(x) = x^5 - 2x^4 + x^3$ $= x^3(x - 1)^2$	0 (multiplicity 3), 1 (multiplicity 2)	5

EXAMPLE 4 | Finding Polynomials with Specified Zeros

(a) Find a polynomial $P(x)$ of degree 4, with zeros i, $-i$, 2, and -2, and with $P(3) = 25$.

(b) Find a polynomial $Q(x)$ of degree 4, with zeros -2 and 0, where -2 is a zero of multiplicity 3.

SOLUTION

(a) The required polynomial has the form

$$\begin{aligned} P(x) &= a(x - i)(x - (-i))(x - 2)(x - (-2)) \\ &= a(x^2 + 1)(x^2 - 4) && \text{Difference of squares} \\ &= a(x^4 - 3x^2 - 4) && \text{Multiply} \end{aligned}$$

We know that $P(3) = a(3^4 - 3 \cdot 3^2 - 4) = 50a = 25$, so $a = \frac{1}{2}$. Thus,

$$P(x) = \tfrac{1}{2}x^4 - \tfrac{3}{2}x^2 - 2$$

(b) We require

$$\begin{aligned} Q(x) &= a[x - (-2)]^3(x - 0) \\ &= a(x + 2)^3 x \\ &= a(x^3 + 6x^2 + 12x + 8)x && \text{Special Product Formula 4 (Section 1.3)} \\ &= a(x^4 + 6x^3 + 12x^2 + 8x) \end{aligned}$$

Since we are given no information about Q other than its zeros and their multiplicity, we can choose any number for a. If we use $a = 1$, we get

$$Q(x) = x^4 + 6x^3 + 12x^2 + 8x$$

NOW TRY EXERCISE **35** ■

EXAMPLE 5 | Finding All the Zeros of a Polynomial

Find all four zeros of $P(x) = 3x^4 - 2x^3 - x^2 - 12x - 4$.

SOLUTION Using the Rational Zeros Theorem from Section 3.4, we obtain the following list of possible rational zeros: $\pm1, \pm2, \pm4, \pm\frac{1}{3}, \pm\frac{2}{3}, \pm\frac{4}{3}$. Checking these using synthetic division, we find that 2 and $-\frac{1}{3}$ are zeros, and we get the following factorization:

$$\begin{aligned} P(x) &= 3x^4 - 2x^3 - x^2 - 12x - 4 \\ &= (x - 2)(3x^3 + 4x^2 + 7x + 2) && \text{Factor } x - 2 \\ &= (x - 2)\left(x + \tfrac{1}{3}\right)(3x^2 + 3x + 6) && \text{Factor } x + \tfrac{1}{3} \\ &= 3(x - 2)\left(x + \tfrac{1}{3}\right)(x^2 + x + 2) && \text{Factor 3} \end{aligned}$$

The zeros of the quadratic factor are

$$x = \frac{-1 \pm \sqrt{1 - 8}}{2} = -\frac{1}{2} \pm i\frac{\sqrt{7}}{2} \qquad \text{Quadratic Formula}$$

so the zeros of $P(x)$ are

$$2, \quad -\frac{1}{3}, \quad -\frac{1}{2} + i\frac{\sqrt{7}}{2}, \quad \text{and} \quad -\frac{1}{2} - i\frac{\sqrt{7}}{2}$$

NOW TRY EXERCISE **45** ■

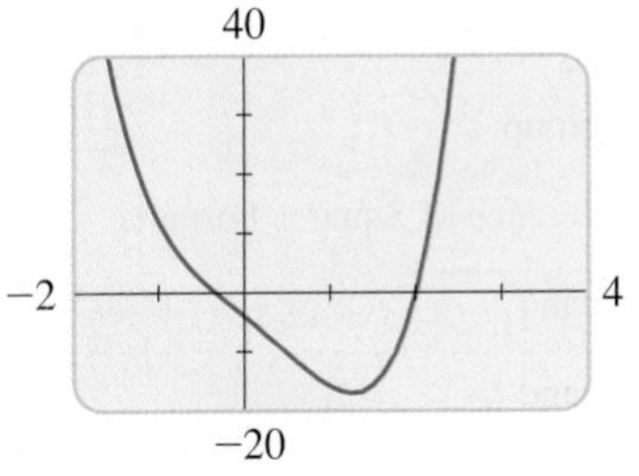

FIGURE 1
$P(x) = 3x^4 - 2x^3 - x^2 - 12x - 4$

Figure 1 shows the graph of the polynomial P in Example 5. The x-intercepts correspond to the real zeros of P. The imaginary zeros cannot be determined from the graph.

GEROLAMO CARDANO (1501–1576) is certainly one of the most colorful figures in the history of mathematics. He was the best-known physician in Europe in his day, yet throughout his life he was plagued by numerous maladies, including ruptures, hemorrhoids, and an irrational fear of encountering rabid dogs. He was a doting father, but his beloved sons broke his heart—his favorite was eventually beheaded for murdering his own wife. Cardano was also a compulsive gambler; indeed, this vice might have driven him to write the *Book on Games of Chance,* the first study of probability from a mathematical point of view.

In Cardano's major mathematical work, the *Ars Magna,* he detailed the solution of the general third- and fourth-degree polynomial equations. At the time of its publication, mathematicians were uncomfortable even with negative numbers, but Cardano's formulas paved the way for the acceptance not just of negative numbers, but also of imaginary numbers, because they occurred naturally in solving polynomial equations. For example, for the cubic equation

$$x^3 - 15x - 4 = 0$$

one of his formulas gives the solution

$$x = \sqrt[3]{2 + \sqrt{-121}} + \sqrt[3]{2 - \sqrt{-121}}$$

(See page263, Exercise 108). This value for x actually turns out to be the *integer* 4, yet to find it, Cardano had to use the imaginary number $\sqrt{-121} = 11i$.

▼ Complex Zeros Come in Conjugate Pairs

As you might have noticed from the examples so far, the complex zeros of polynomials with real coefficients come in pairs. Whenever $a + bi$ is a zero, its complex conjugate $a - bi$ is also a zero.

> **CONJUGATE ZEROS THEOREM**
>
> If the polynomial P has real coefficients and if the complex number z is a zero of P, then its complex conjugate $\overline{z}$ is also a zero of P.

PROOF Let

$$P(x) = a_n x^n + a_{n-1}x^{n-1} + \cdots + a_1 x + a_0$$

where each coefficient is real. Suppose that $P(z) = 0$. We must prove that $P(\overline{z}) = 0$. We use the facts that the complex conjugate of a sum of two complex numbers is the sum of the conjugates and that the conjugate of a product is the product of the conjugates.

$$\begin{aligned}
P(\overline{z}) &= a_n(\overline{z})^n + a_{n-1}(\overline{z})^{n-1} + \cdots + a_1\overline{z} + a_0 \\
&= \overline{a_n}\,\overline{z}^n + \overline{a_{n-1}}\,\overline{z}^{n-1} + \cdots + \overline{a_1}\,\overline{z} + \overline{a_0} && \text{Because the coefficients are real} \\
&= \overline{a_n z^n} + \overline{a_{n-1}z^{n-1}} + \cdots + \overline{a_1 z} + \overline{a_0} \\
&= \overline{a_n z^n + a_{n-1}z^{n-1} + \cdots + a_1 z + a_0} \\
&= \overline{P(z)} = \overline{0} = 0
\end{aligned}$$

This shows that $\overline{z}$ is also a zero of $P(x)$, which proves the theorem. ■

EXAMPLE 6 | A Polynomial with a Specified Complex Zero

Find a polynomial $P(x)$ of degree 3 that has integer coefficients and zeros $\frac{1}{2}$ and $3 - i$.

SOLUTION Since $3 - i$ is a zero, then so is $3 + i$ by the Conjugate Zeros Theorem. This means that $P(x)$ must have the following form.

$$\begin{aligned}
P(x) &= a\left(x - \tfrac{1}{2}\right)[x - (3 - i)][x - (3 + i)] \\
&= a\left(x - \tfrac{1}{2}\right)[(x - 3) + i][(x - 3) - i] && \text{Regroup} \\
&= a\left(x - \tfrac{1}{2}\right)[(x - 3)^2 - i^2] && \text{Difference of Squares Formula} \\
&= a\left(x - \tfrac{1}{2}\right)(x^2 - 6x + 10) && \text{Expand} \\
&= a\left(x^3 - \tfrac{13}{2}x^2 + 13x - 5\right) && \text{Expand}
\end{aligned}$$

To make all coefficients integers, we set $a = 2$ and get

$$P(x) = 2x^3 - 13x^2 + 26x - 10$$

Any other polynomial that satisfies the given requirements must be an integer multiple of this one.

NOW TRY EXERCISE **39** ■

▼ Linear and Quadratic Factors

We have seen that a polynomial factors completely into linear factors if we use complex numbers. If we don't use complex numbers, then a polynomial with real coefficients can always be factored into linear and quadratic factors. We use this property in Section 10.7 when we study partial fractions. A quadratic polynomial with no real zeros is called **irreducible** over the real numbers. Such a polynomial cannot be factored without using complex numbers.

LINEAR AND QUADRATIC FACTORS THEOREM

Every polynomial with real coefficients can be factored into a product of linear and irreducible quadratic factors with real coefficients.

PROOF We first observe that if $c = a + bi$ is a complex number, then

$$\begin{aligned}(x - c)(x - \bar{c}) &= [x - (a + bi)][x - (a - bi)] \\ &= [(x - a) - bi][(x - a) + bi] \\ &= (x - a)^2 - (bi)^2 \\ &= x^2 - 2ax + (a^2 + b^2)\end{aligned}$$

The last expression is a quadratic with *real* coefficients.

Now, if P is a polynomial with real coefficients, then by the Complete Factorization Theorem

$$P(x) = a(x - c_1)(x - c_2) \cdots (x - c_n)$$

Since the complex roots occur in conjugate pairs, we can multiply the factors corresponding to each such pair to get a quadratic factor with real coefficients. This results in P being factored into linear and irreducible quadratic factors. ■

EXAMPLE 7 | Factoring a Polynomial into Linear and Quadratic Factors

Let $P(x) = x^4 + 2x^2 - 8$.

(a) Factor P into linear and irreducible quadratic factors with real coefficients.

(b) Factor P completely into linear factors with complex coefficients.

SOLUTION

(a)

$$\begin{aligned}P(x) &= x^4 + 2x^2 - 8 \\ &= (x^2 - 2)(x^2 + 4) \\ &= (x - \sqrt{2})(x + \sqrt{2})(x^2 + 4)\end{aligned}$$

The factor $x^2 + 4$ is irreducible, since it has no real zeros.

(b) To get the complete factorization, we factor the remaining quadratic factor.

$$\begin{aligned}P(x) &= (x - \sqrt{2})(x + \sqrt{2})(x^2 + 4) \\ &= (x - \sqrt{2})(x + \sqrt{2})(x - 2i)(x + 2i)\end{aligned}$$

NOW TRY EXERCISE **65** ■

3.6 EXERCISES

CONCEPTS

1. The polynomial $P(x) = 3(x - 5)^3(x - 3)(x + 2)$ has degree ________. It has zeros 5, 3, and ________. The zero 5 has multiplicity ________, and the zero 3 has multiplicity ________.

2. (a) If a is a zero of the polynomial P, then ________ must be a factor of $P(x)$.
(b) If a is a zero of multiplicity m of the polynomial P, then ________ must be a factor of $P(x)$ when we factor P completely.

3. A polynomial of degree $n \geq 1$ has exactly ________ zeros if a zero of multiplicity m is counted m times.

4. If the polynomial function P has real coefficients and if $a + bi$ is a zero of P, then ________ is also a zero of P.

SKILLS

5–16 ■ A polynomial P is given. **(a)** Find all zeros of P, real and complex. **(b)** Factor P completely.

5. $P(x) = x^4 + 4x^2$
6. $P(x) = x^5 + 9x^3$
7. $P(x) = x^3 - 2x^2 + 2x$
8. $P(x) = x^3 + x^2 + x$
9. $P(x) = x^4 + 2x^2 + 1$
10. $P(x) = x^4 - x^2 - 2$
11. $P(x) = x^4 - 16$
12. $P(x) = x^4 + 6x^2 + 9$
13. $P(x) = x^3 + 8$
14. $P(x) = x^3 - 8$
15. $P(x) = x^6 - 1$
16. $P(x) = x^6 - 7x^3 - 8$

17–34 ■ Factor the polynomial completely, and find all its zeros. State the multiplicity of each zero.

17. $P(x) = x^2 + 25$
18. $P(x) = 4x^2 + 9$
19. $Q(x) = x^2 + 2x + 2$
20. $Q(x) = x^2 - 8x + 17$
21. $P(x) = x^3 + 4x$
22. $P(x) = x^3 - x^2 + x$
23. $Q(x) = x^4 - 1$
24. $Q(x) = x^4 - 625$
25. $P(x) = 16x^4 - 81$
26. $P(x) = x^3 - 64$
27. $P(x) = x^3 + x^2 + 9x + 9$
28. $P(x) = x^6 - 729$
29. $Q(x) = x^4 + 2x^2 + 1$
30. $Q(x) = x^4 + 10x^2 + 25$
31. $P(x) = x^4 + 3x^2 - 4$
32. $P(x) = x^5 + 7x^3$
33. $P(x) = x^5 + 6x^3 + 9x$
34. $P(x) = x^6 + 16x^3 + 64$

35–44 ■ Find a polynomial with integer coefficients that satisfies the given conditions.

35. P has degree 2 and zeros $1 + i$ and $1 - i$.

36. P has degree 2 and zeros $1 + i\sqrt{2}$ and $1 - i\sqrt{2}$.

37. Q has degree 3 and zeros 3, $2i$, and $-2i$.

38. Q has degree 3 and zeros 0 and i.

39. P has degree 3 and zeros 2 and i.

40. Q has degree 3 and zeros -3 and $1 + i$.

41. R has degree 4 and zeros $1 - 2i$ and 1, with 1 a zero of multiplicity 2.

42. S has degree 4 and zeros $2i$ and $3i$.

43. T has degree 4, zeros i and $1 + i$, and constant term 12.

44. U has degree 5, zeros $\frac{1}{2}$, -1, and $-i$, and leading coefficient 4; the zero -1 has multiplicity 2.

45–62 ■ Find all zeros of the polynomial.

45. $P(x) = x^3 + 2x^2 + 4x + 8$
46. $P(x) = x^3 - 7x^2 + 17x - 15$
47. $P(x) = x^3 - 2x^2 + 2x - 1$
48. $P(x) = x^3 + 7x^2 + 18x + 18$
49. $P(x) = x^3 - 3x^2 + 3x - 2$
50. $P(x) = x^3 - x - 6$
51. $P(x) = 2x^3 + 7x^2 + 12x + 9$
52. $P(x) = 2x^3 - 8x^2 + 9x - 9$
53. $P(x) = x^4 + x^3 + 7x^2 + 9x - 18$
54. $P(x) = x^4 - 2x^3 - 2x^2 - 2x - 3$
55. $P(x) = x^5 - x^4 + 7x^3 - 7x^2 + 12x - 12$
56. $P(x) = x^5 + x^3 + 8x^2 + 8$ [*Hint:* Factor by grouping.]
57. $P(x) = x^4 - 6x^3 + 13x^2 - 24x + 36$
58. $P(x) = x^4 - x^2 + 2x + 2$
59. $P(x) = 4x^4 + 4x^3 + 5x^2 + 4x + 1$
60. $P(x) = 4x^4 + 2x^3 - 2x^2 - 3x - 1$
61. $P(x) = x^5 - 3x^4 + 12x^3 - 28x^2 + 27x - 9$
62. $P(x) = x^5 - 2x^4 + 2x^3 - 4x^2 + x - 2$

63–68 ■ A polynomial P is given. **(a)** Factor P into linear and irreducible quadratic factors with real coefficients. **(b)** Factor P completely into linear factors with complex coefficients.

63. $P(x) = x^3 - 5x^2 + 4x - 20$
64. $P(x) = x^3 - 2x - 4$
65. $P(x) = x^4 + 8x^2 - 9$
66. $P(x) = x^4 + 8x^2 + 16$
67. $P(x) = x^6 - 64$
68. $P(x) = x^5 - 16x$

69. By the Zeros Theorem, every nth-degree polynomial equation has exactly n solutions (including possibly some that are repeated). Some of these may be real, and some may be imaginary. Use a graphing device to determine how many real and imaginary solutions each equation has.
(a) $x^4 - 2x^3 - 11x^2 + 12x = 0$
(b) $x^4 - 2x^3 - 11x^2 + 12x - 5 = 0$
(c) $x^4 - 2x^3 - 11x^2 + 12x + 40 = 0$

70–72 ■ So far, we have worked only with polynomials that have real coefficients. These exercises involve polynomials with real and imaginary coefficients.

70. Find all solutions of the equation.
(a) $2x + 4i = 1$
(b) $x^2 - ix = 0$
(c) $x^2 + 2ix - 1 = 0$
(d) $ix^2 - 2x + i = 0$

71. **(a)** Show that $2i$ and $1 - i$ are both solutions of the equation

$$x^2 - (1 + i)x + (2 + 2i) = 0$$

but that their complex conjugates $-2i$ and $1 + i$ are not.

(b) Explain why the result of part (a) does not violate the Conjugate Zeros Theorem.

72. **(a)** Find the polynomial with *real* coefficients of the smallest possible degree for which i and $1 + i$ are zeros and in which the coefficient of the highest power is 1.

(b) Find the polynomial with *complex* coefficients of the smallest possible degree for which i and $1 + i$ are zeros and in which the coefficient of the highest power is 1.

DISCOVERY ■ DISCUSSION ■ WRITING

73. Polynomials of Odd Degree The Conjugate Zeros Theorem says that the complex zeros of a polynomial with real coefficients occur in complex conjugate pairs. Explain how this fact proves that a polynomial with real coefficients and odd degree has at least one real zero.

74. Roots of Unity There are two square roots of 1, namely, 1 and -1. These are the solutions of $x^2 = 1$. The fourth roots of 1 are the solutions of the equation $x^4 = 1$ or $x^4 - 1 = 0$. How many fourth roots of 1 are there? Find them. The cube roots of 1 are the solutions of the equation $x^3 = 1$ or $x^3 - 1 = 0$. How many cube roots of 1 are there? Find them. How would you find the sixth roots of 1? How many are there? Make a conjecture about the number of nth roots of 1.

3.7 Rational Functions

Rational Functions and Asymptotes ▶ Transformations of $y = 1/x$ ▶ Asymptotes of Rational Functions ▶ Graphing Rational Functions ▶ Slant Asymptotes and End Behavior ▶ Applications

A rational function is a function of the form

$$r(x) = \frac{P(x)}{Q(x)}$$

where P and Q are polynomials. We assume that $P(x)$ and $Q(x)$ have no factor in common. Even though rational functions are constructed from polynomials, their graphs look quite different from the graphs of polynomial functions.

▼ Rational Functions and Asymptotes

Domains of rational expressions are discussed in Section 1.4.

The *domain* of a rational function consists of all real numbers x except those for which the denominator is zero. When graphing a rational function, we must pay special attention to the behavior of the graph near those x-values. We begin by graphing a very simple rational function.

EXAMPLE 1 | A Simple Rational Function

Graph the rational function $f(x) = \dfrac{1}{x}$, and state the domain and range.

SOLUTION The function f is not defined for $x = 0$. The following tables show that when x is close to zero, the value of $|f(x)|$ is large, and the closer x gets to zero, the larger $|f(x)|$ gets.

For positive real numbers,

$$\frac{1}{\text{BIG NUMBER}} = \text{small number}$$

$$\frac{1}{\text{small number}} = \text{BIG NUMBER}$$

x	$f(x)$
-0.1	-10
-0.01	-100
-0.00001	$-100{,}000$
Approaching 0^-	Approaching $-\infty$

x	$f(x)$
0.1	10
0.01	100
0.00001	100,000
Approaching 0^+	Approaching ∞

We describe this behavior in words and in symbols as follows. The first table shows that as x approaches 0 from the left, the values of $y = f(x)$ decrease without bound. In symbols,

$$f(x) \to -\infty \quad \text{as} \quad x \to 0^-$$ "y approaches negative infinity as x approaches 0 from the left"

The second table shows that as x approaches 0 from the right, the values of $f(x)$ increase without bound. In symbols,

$$f(x) \to \infty \quad \text{as} \quad x \to 0^+$$ "y approaches infinity as x approaches 0 from the right"

The next two tables show how $f(x)$ changes as $|x|$ becomes large.

x	$f(x)$
−10	−0.1
−100	−0.01
−100,000	−0.00001

Approaching $-\infty$ Approaching 0

x	$f(x)$
10	0.1
100	0.01
100,000	0.00001

Approaching ∞ Approaching 0

These tables show that as $|x|$ becomes large, the value of $f(x)$ gets closer and closer to zero. We describe this situation in symbols by writing

$$f(x) \to 0 \quad \text{as} \quad x \to -\infty \qquad \text{and} \qquad f(x) \to 0 \quad \text{as} \quad x \to \infty$$

Using the information in these tables and plotting a few additional points, we obtain the graph shown in Figure 1.

x	$f(x) = \frac{1}{x}$
−2	$-\frac{1}{2}$
−1	−1
$-\frac{1}{2}$	−2
$\frac{1}{2}$	2
1	1
2	$\frac{1}{2}$

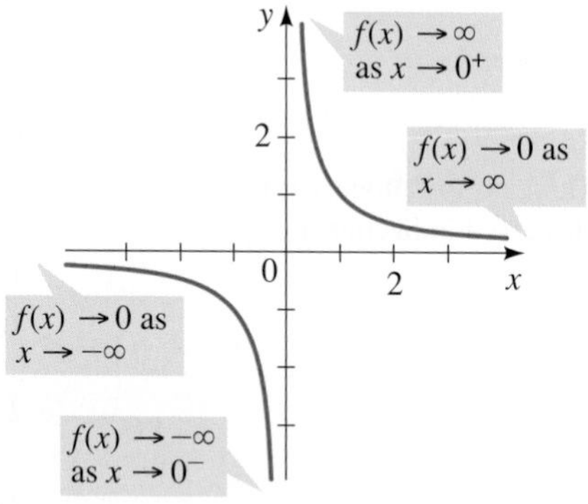

FIGURE 1
$f(x) = \frac{1}{x}$

The function f is defined for all values of x other than 0, so the domain is $\{x \mid x \neq 0\}$. From the graph we see that the range is $\{y \mid y \neq 0\}$.

NOW TRY EXERCISE 7 ■

In Example 1 we used the following arrow notation.

Symbol	Meaning
$x \to a^-$	x approaches a from the left
$x \to a^+$	x approaches a from the right
$x \to -\infty$	x goes to negative infinity; that is, x decreases without bound
$x \to \infty$	x goes to infinity; that is, x increases without bound

The line $x = 0$ is called a *vertical asymptote* of the graph in Figure 1, and the line $y = 0$ is a *horizontal asymptote*. Informally speaking, an asymptote of a function is a line to which the graph of the function gets closer and closer as one travels along that line.

DEFINITION OF VERTICAL AND HORIZONTAL ASYMPTOTES

1. The line $x = a$ is a **vertical asymptote** of the function $y = f(x)$ if y approaches $\pm\infty$ as x approaches a from the right or left.

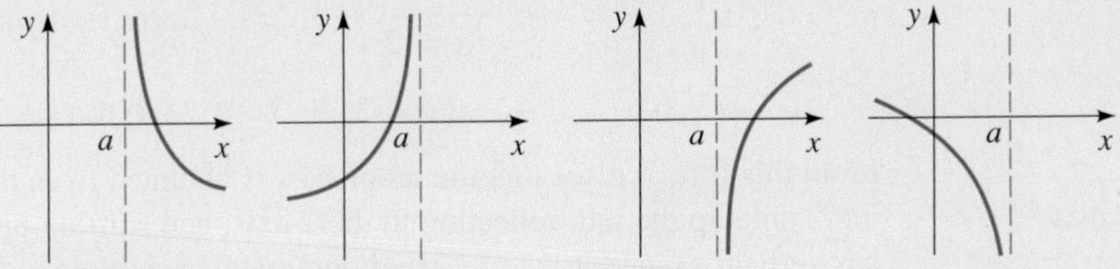

$y \to \infty$ as $x \to a^+$ $\quad$ $y \to \infty$ as $x \to a^-$ $\quad$ $y \to -\infty$ as $x \to a^+$ $\quad$ $y \to -\infty$ as $x \to a^-$

2. The line $y = b$ is a **horizontal asymptote** of the function $y = f(x)$ if y approaches b as x approaches $\pm\infty$.

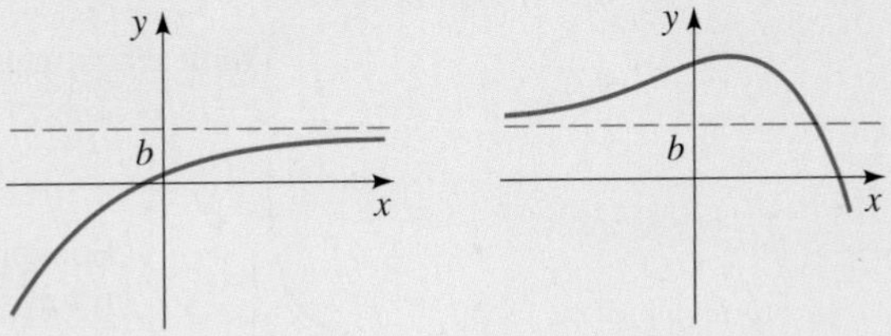

$y \to b$ as $x \to \infty$ $\quad$ $y \to b$ as $x \to -\infty$

A rational function has vertical asymptotes where the function is undefined, that is, where the denominator is zero.

▼ Transformations of $y = 1/x$

A rational function of the form

$$r(x) = \frac{ax + b}{cx + d}$$

can be graphed by shifting, stretching, and/or reflecting the graph of $f(x) = \frac{1}{x}$ shown in Figure 1, using the transformations studied in Section 2.5. (Such functions are called *linear fractional transformations*.)

EXAMPLE 2 | Using Transformations to Graph Rational Functions

Graph each rational function, and state the domain and range.

(a) $r(x) = \dfrac{2}{x - 3}$ $\qquad$ **(b)** $s(x) = \dfrac{3x + 5}{x + 2}$

SOLUTION

(a) Let $f(x) = \frac{1}{x}$. Then we can express r in terms of f as follows:

$$\begin{aligned} r(x) &= \frac{2}{x - 3} \\ &= 2\left(\frac{1}{x - 3}\right) && \text{Factor 2} \\ &= 2(f(x - 3)) && \text{Since } f(x) = \tfrac{1}{x} \end{aligned}$$

From this form we see that the graph of r is obtained from the graph of f by shifting 3 units to the right and stretching vertically by a factor of 2. Thus, r has vertical asymptote $x = 3$ and horizontal asymptote $y = 0$. The graph of r is shown in Figure 2.

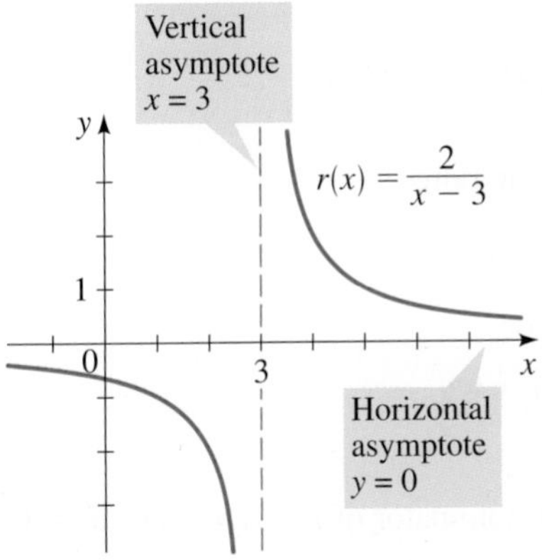

FIGURE 2

The function r is defined for all x other than 3, so the domain is $\{x \mid x \neq 3\}$. From the graph we see that the range is $\{y \mid y \neq 0\}$.

$$\begin{array}{r} 3 \\ x+2 \overline{)\,3x+5} \\ 3x+6 \\ \hline -1 \end{array}$$

(b) Using long division (see the margin), we get $s(x) = 3 - \frac{1}{x+2}$. Thus, we can express s in terms of f as follows:

$$s(x) = 3 - \frac{1}{x+2}$$

$$= -\frac{1}{x+2} + 3 \qquad \text{Rearrange terms}$$

$$= -f(x+2) + 3 \qquad \text{Since } f(x) = \tfrac{1}{x}$$

From this form we see that the graph of s is obtained from the graph of f by shifting 2 units to the left, reflecting in the x-axis, and shifting upward 3 units. Thus, s has vertical asymptote $x = -2$ and horizontal asymptote $y = 3$. The graph of s is shown in Figure 3.

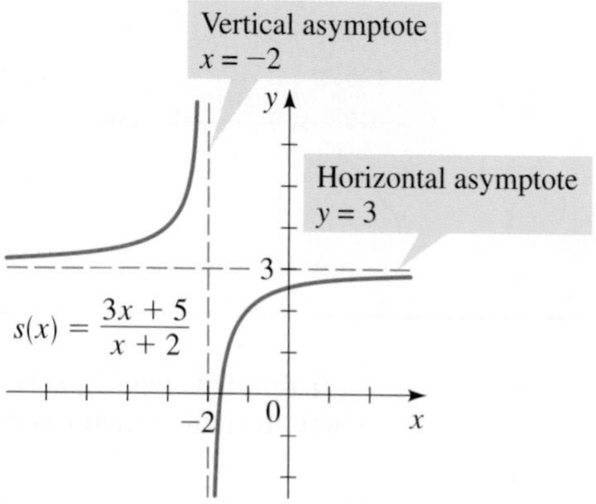

FIGURE 3

The function s is defined for all x other than -2, so the domain is $\{x \mid x \neq -2\}$. From the graph we see that the range is $\{y \mid y \neq 3\}$.

NOW TRY EXERCISES **35** AND **37**

Asymptotes of Rational Functions

The methods of Example 2 work only for simple rational functions. To graph more complicated ones, we need to take a closer look at the behavior of a rational function near its vertical and horizontal asymptotes.

EXAMPLE 3 | Asymptotes of a Rational Function

Graph $r(x) = \dfrac{2x^2 - 4x + 5}{x^2 - 2x + 1}$, and state the domain and range.

SOLUTION

Vertical asymptote: We first factor the denominator

$$r(x) = \frac{2x^2 - 4x + 5}{(x-1)^2}$$

The line $x = 1$ is a vertical asymptote because the denominator of r is zero when $x = 1$.

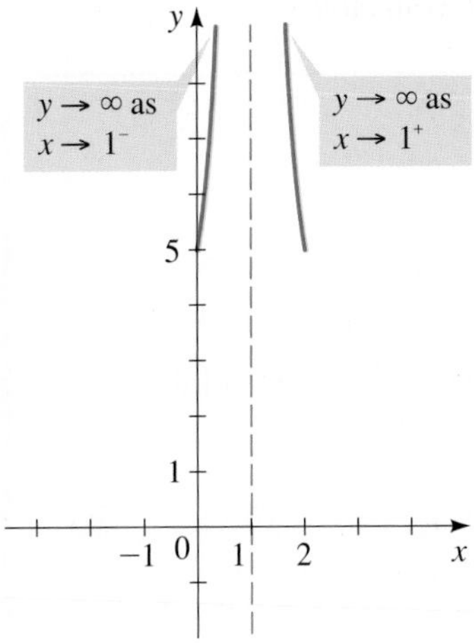

FIGURE 4

To see what the graph of r looks like near the vertical asymptote, we make tables of values for x-values to the left and to the right of 1. From the tables shown below we see that

$$y \to \infty \quad \text{as} \quad x \to 1^- \qquad \text{and} \qquad y \to \infty \quad \text{as} \quad x \to 1^+$$

$x \to 1^-$

x	y
0	5
0.5	14
0.9	302
0.99	30,002

Approaching 1^- — Approaching ∞

$x \to 1^+$

x	y
2	5
1.5	14
1.1	302
1.01	30,002

Approaching 1^+ — Approaching ∞

Thus, near the vertical asymptote $x = 1$, the graph of r has the shape shown in Figure 4.

Horizontal asymptote: The horizontal asymptote is the value that y approaches as $x \to \pm\infty$. To help us find this value, we divide both numerator and denominator by x^2, the highest power of x that appears in the expression:

$$y = \frac{2x^2 - 4x + 5}{x^2 - 2x + 1} \cdot \frac{\frac{1}{x^2}}{\frac{1}{x^2}} = \frac{2 - \frac{4}{x} + \frac{5}{x^2}}{1 - \frac{2}{x} + \frac{1}{x^2}}$$

The fractional expressions $\frac{4}{x}$, $\frac{5}{x^2}$, $\frac{2}{x}$, and $\frac{1}{x^2}$ all approach 0 as $x \to \pm\infty$ (see Exercise 83, page 12). So as $x \to \pm\infty$, we have

These terms approach 0

$$y = \frac{2 - \frac{4}{x} + \frac{5}{x^2}}{1 - \frac{2}{x} + \frac{1}{x^2}} \longrightarrow \frac{2 - 0 + 0}{1 - 0 + 0} = 2$$

These terms approach 0

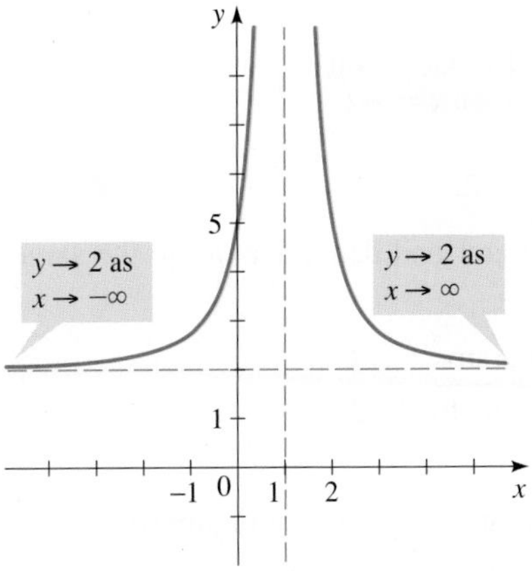

FIGURE 5

$$r(x) = \frac{2x^2 - 4x + 5}{x^2 - 2x + 1}$$

Thus, the horizontal asymptote is the line $y = 2$.

Since the graph must approach the horizontal asymptote, we can complete it as in Figure 5.

Domain and range: The function r is defined for all values of x other than 1, so the domain is $\{x \mid x \neq 1\}$. From the graph we see that the range is $\{y \mid y > 2\}$.

NOW TRY EXERCISE **45**

From Example 3 we see that the horizontal asymptote is determined by the leading coefficients of the numerator and denominator, since after dividing through by x^2 (the highest power of x), all other terms approach zero. In general, if $r(x) = P(x)/Q(x)$ and the degrees of P and Q are the same (both n, say), then dividing both numerator and denominator by x^n shows that the horizontal asymptote is

$$y = \frac{\text{leading coefficient of } P}{\text{leading coefficient of } Q}$$

The following box summarizes the procedure for finding asymptotes.

FINDING ASYMPTOTES OF RATIONAL FUNCTIONS

Let r be the rational function

$$r(x) = \frac{a_n x^n + a_{n-1}x^{n-1} + \cdots + a_1 x + a_0}{b_m x^m + b_{m-1}x^{m-1} + \cdots + b_1 x + b_0}$$

1. The vertical asymptotes of r are the lines $x = a$, where a is a zero of the denominator.
2. **(a)** If $n < m$, then r has horizontal asymptote $y = 0$.

   (b) If $n = m$, then r has horizontal asymptote $y = \dfrac{a_n}{b_m}$.

   (c) If $n > m$, then r has no horizontal asymptote.

EXAMPLE 4 | Asymptotes of a Rational Function

Find the vertical and horizontal asymptotes of $r(x) = \dfrac{3x^2 - 2x - 1}{2x^2 + 3x - 2}$.

SOLUTION

Vertical asymptotes: We first factor

$$r(x) = \frac{3x^2 - 2x - 1}{(2x - 1)(x + 2)}$$

This factor is 0 when $x = \frac{1}{2}$

This factor is 0 when $x = -2$

The vertical asymptotes are the lines $x = \frac{1}{2}$ and $x = -2$.

Horizontal asymptote: The degrees of the numerator and denominator are the same, and

$$\frac{\text{leading coefficient of numerator}}{\text{leading coefficient of denominator}} = \frac{3}{2}$$

Thus, the horizontal asymptote is the line $y = \frac{3}{2}$.

To confirm our results, we graph r using a graphing calculator (see Figure 6).

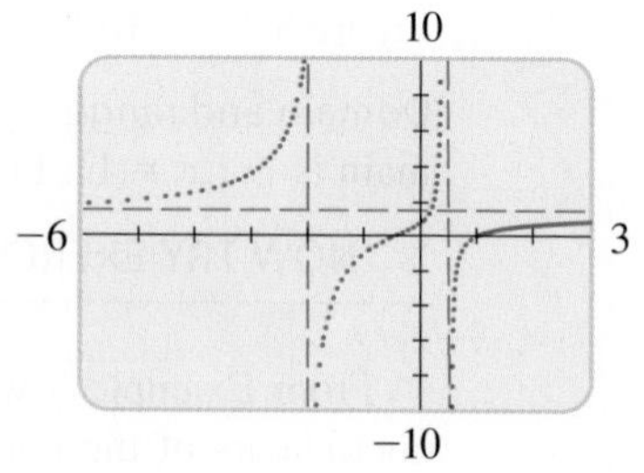

FIGURE 6
$r(x) = \dfrac{3x^2 - 2x - 1}{2x^2 + 3x - 2}$

Graph is drawn using dot mode to avoid extraneous lines.

NOW TRY EXERCISES **23** AND **25** ■

Graphing Rational Functions

We have seen that asymptotes are important when graphing rational functions. In general, we use the following guidelines to graph rational functions.

A fraction is 0 if and only if its numerator is 0.

SKETCHING GRAPHS OF RATIONAL FUNCTIONS

1. **Factor.** Factor the numerator and denominator.
2. **Intercepts.** Find the x-intercepts by determining the zeros of the numerator and the y-intercept from the value of the function at $x = 0$.
3. **Vertical Asymptotes.** Find the vertical asymptotes by determining the zeros of the denominator, and then see whether $y \to \infty$ or $y \to -\infty$ on each side of each vertical asymptote by using test values.
4. **Horizontal Asymptote.** Find the horizontal asymptote (if any), using the procedure described in the box on page 282.
5. **Sketch the Graph.** Graph the information provided by the first four steps. Then plot as many additional points as needed to fill in the rest of the graph of the function.

EXAMPLE 5 | Graphing a Rational Function

Graph $r(x) = \dfrac{2x^2 + 7x - 4}{x^2 + x - 2}$, and state the domain and range.

SOLUTION We factor the numerator and denominator, find the intercepts and asymptotes, and sketch the graph.

Factor: $$y = \frac{(2x - 1)(x + 4)}{(x - 1)(x + 2)}$$

x-Intercepts: The x-intercepts are the zeros of the numerator, $x = \frac{1}{2}$ and $x = -4$.

y-Intercept: To find the y-intercept, we substitute $x = 0$ into the original form of the function.

$$r(0) = \frac{2(0)^2 + 7(0) - 4}{(0)2 + (0) - 2} = \frac{-4}{-2} = 2$$

The y-intercept is 2.

Vertical asymptotes: The vertical asymptotes occur where the denominator is 0, that is, where the function is undefined. From the factored form we see that the vertical asymptotes are the lines $x = 1$ and $x = -2$.

When choosing test values, we must make sure that there is no x-intercept between the test point and the vertical asymptote.

Behavior near vertical asymptotes: We need to know whether $y \to \infty$ or $y \to -\infty$ on each side of each vertical asymptote. To determine the sign of y for x-values near the vertical asymptotes, we use test values. For instance, as $x \to 1^-$, we use a test value close to and to the left of 1 ($x = 0.9$, say) to check whether y is positive or negative to the left of $x = 1$.

$$y = \frac{(2(0.9) - 1)((0.9) + 4)}{((0.9) - 1)((0.9) + 2)} \quad \text{whose sign is} \quad \frac{(+)(+)}{(-)(+)} \quad \text{(negative)}$$

So $y \to -\infty$ as $x \to 1^-$. On the other hand, as $x \to 1^+$, we use a test value close to and to the right of 1 ($x = 1.1$, say), to get

$$y = \frac{(2(1.1) - 1)((1.1) + 4)}{((1.1) - 1)((1.1) + 2)} \quad \text{whose sign is} \quad \frac{(+)(+)}{(+)(+)} \quad \text{(positive)}$$

So $y \to \infty$ as $x \to 1^+$. The other entries in the following table are calculated similarly.

As $x \to$	-2^-	-2^+	1^-	1^+
the sign of $y = \dfrac{(2x-1)(x+4)}{(x-1)(x+2)}$ is	$\frac{(-)(+)}{(-)(-)}$	$\frac{(-)(+)}{(-)(+)}$	$\frac{(+)(+)}{(-)(+)}$	$\frac{(+)(+)}{(+)(+)}$
so $y \to$	$-\infty$	∞	$-\infty$	∞

MATHEMATICS IN THE MODERN WORLD

Unbreakable Codes

If you read spy novels, you know about secret codes and how the hero "breaks" the code. Today secret codes have a much more common use. Most of the information that is stored on computers is coded to prevent unauthorized use. For example, your banking records, medical records, and school records are coded. Many cellular and cordless phones code the signal carrying your voice so that no one can listen in. Fortunately, because of recent advances in mathematics, today's codes are "unbreakable."

Modern codes are based on a simple principle: Factoring is much harder than multiplying. For example, try multiplying 78 and 93; now try factoring 9991. It takes a long time to factor 9991 because it is a product of two primes 97 × 103, so to factor it, we have to find one of these primes. Now imagine trying to factor a number *N* that is the product of two primes *p* and *q*, each about 200 digits long. Even the fastest computers would take many millions of years to factor such a number! But the same computer would take less than a second to multiply two such numbers. This fact was used by Ron Rivest, Adi Shamir, and Leonard Adleman in the 1970s to devise the RSA code. Their code uses an extremely large number to encode a message but requires us to know its factors to decode it. As you can see, such a code is practically unbreakable.

The RSA code is an example of a "public key encryption" code. In such codes, anyone can code a message using a publicly known procedure based on *N*, but to decode the message, they must know *p* and *q*, the factors of *N*. When the RSA code was developed, it was thought that a carefully selected 80-digit number would provide an unbreakable code. But interestingly, recent advances in the study of factoring have made much larger numbers necessary.

Horizontal asymptote: The degrees of the numerator and denominator are the same, and

$$\frac{\text{leading coefficient of numerator}}{\text{leading coefficient of denominator}} = \frac{2}{1} = 2$$

Thus, the horizontal asymptote is the line $y = 2$.

Graph: We use the information we have found, together with some additional values, to sketch the graph in Figure 7.

x	y
−6	0.93
−3	−1.75
−1	4.50
1.5	6.29
2	4.50
3	3.50

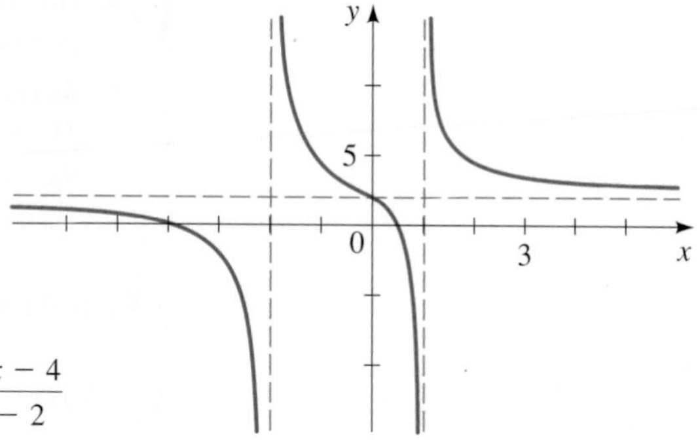

FIGURE 7

$r(x) = \dfrac{2x^2 + 7x - 4}{x^2 + x - 2}$

Domain and range: The domain is $\{x \mid x \neq 1, x \neq -2\}$. From the graph we see that the range is all real numbers.

NOW TRY EXERCISE **53**

EXAMPLE 6 | Graphing a Rational Function

Graph $r(x) = \dfrac{5x + 21}{x^2 + 10x + 25}$, and state the domain and range.

SOLUTION

Factor: $y = \dfrac{5x + 21}{(x + 5)^2}$

x-Intercept: $-\dfrac{21}{5}$, from $5x + 21 = 0$

y-Intercept: $\dfrac{21}{25}$, because $r(0) = \dfrac{5 \cdot 0 + 21}{0^2 + 10 \cdot 0 + 25} = \dfrac{21}{25}$

Vertical asymptote: $x = -5$, from the zeros of the denominator

Behavior near vertical asymptote:

As $x \to$	-5^-	-5^+
the sign of $y = \dfrac{5x + 21}{(x + 5)^2}$ is	$\dfrac{(-)}{(-)(-)}$	$\dfrac{(-)}{(+)(+)}$
so $y \to$	$-\infty$	$-\infty$

Horizontal asymptote: $y = 0$, because the degree of the numerator is less than the degree of the denominator

Graph: We use the information we have found, together with some additional values, to sketch the graph in Figure 8.

x	y
−15	−0.5
−10	−1.2
−3	1.5
−1	1.0
3	0.6
5	0.5
10	0.3

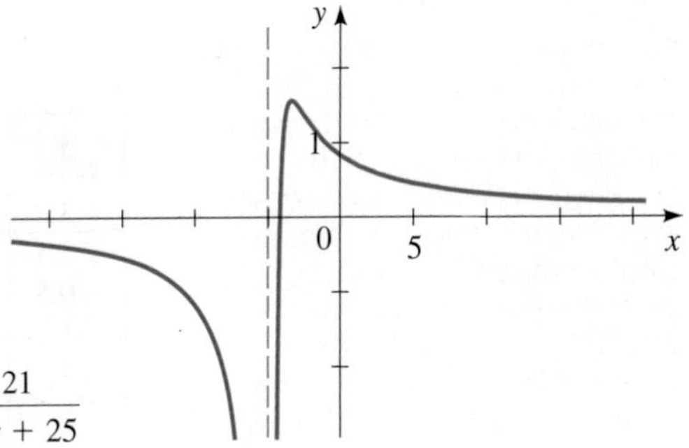

FIGURE 8

$$r(x) = \frac{5x + 21}{x^2 + 10x + 25}$$

Domain and range: The domain is $\{x \mid x \neq -5\}$. From the graph we see that the range is approximately the interval $(-\infty, 1.5]$.

NOW TRY EXERCISE 55

From the graph in Figure 8 we see that, contrary to common misconception, a graph may cross a horizontal asymptote. The graph in Figure 8 crosses the x-axis (the horizontal asymptote) from below, reaches a maximum value near $x = -3$, and then approaches the x-axis from above as $x \to \infty$.

EXAMPLE 7 | Graphing a Rational Function

Graph the rational function $r(x) = \dfrac{x^2 - 3x - 4}{2x^2 + 4x}$.

SOLUTION

Factor: $y = \dfrac{(x + 1)(x - 4)}{2x(x + 2)}$

x-Intercepts: -1 and 4, from $x + 1 = 0$ and $x - 4 = 0$

y-Intercept: None, because $r(0)$ is undefined

Vertical asymptotes: $x = 0$ and $x = -2$, from the zeros of the denominator

Behavior near vertical asymptotes:

As $x \to$	-2^-	-2^+	0^-	0^+
the sign of $y = \dfrac{(x+1)(x-4)}{2x(x+2)}$ is	$\dfrac{(-)(-)}{(-)(-)}$	$\dfrac{(-)(-)}{(-)(+)}$	$\dfrac{(+)(-)}{(-)(+)}$	$\dfrac{(+)(-)}{(+)(+)}$
so $y \to$	∞	$-\infty$	∞	$-\infty$

Horizontal asymptote: $y = \frac{1}{2}$, because the degree of the numerator and the degree of the denominator are the same and

$$\frac{\text{leading coefficient of numerator}}{\text{leading coefficient of denominator}} = \frac{1}{2}$$

Graph: We use the information we have found, together with some additional values, to sketch the graph in Figure 9.

x	y
−3	2.33
−2.5	3.90
−0.5	1.50
1	−1.00
3	−0.13
5	0.09

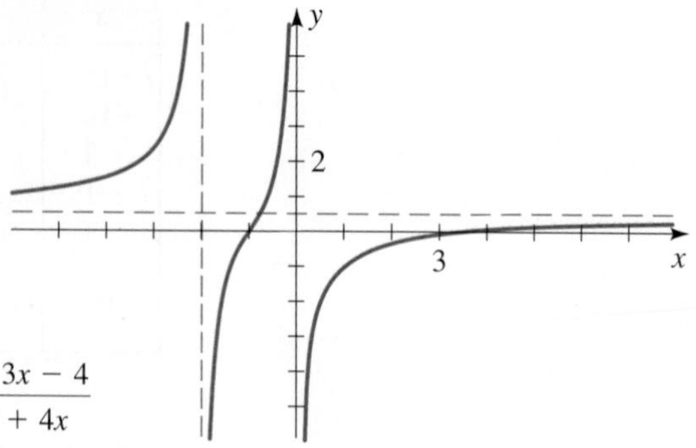

FIGURE 9

$$r(x) = \frac{x^2 - 3x - 4}{2x^2 + 4x}$$

Domain and range: The domain is $\{x \mid x \neq 0, x \neq -2\}$. From the graph we see that the range is all real numbers.

NOW TRY EXERCISE **57**

Slant Asymptotes and End Behavior

If $r(x) = P(x)/Q(x)$ is a rational function in which the degree of the numerator is one more than the degree of the denominator, we can use the Division Algorithm to express the function in the form

$$r(x) = ax + b + \frac{R(x)}{Q(x)}$$

where the degree of R is less than the degree of Q and $a \neq 0$. This means that as $x \to \pm\infty$, $R(x)/Q(x) \to 0$, so for large values of $|x|$ the graph of $y = r(x)$ approaches the graph of the line $y = ax + b$. In this situation we say that $y = ax + b$ is a **slant asymptote**, or an **oblique asymptote**.

EXAMPLE 8 | A Rational Function with a Slant Asymptote

Graph the rational function $r(x) = \dfrac{x^2 - 4x - 5}{x - 3}$.

SOLUTION

Factor: $y = \dfrac{(x+1)(x-5)}{x-3}$

x-Intercepts: -1 and 5, from $x + 1 = 0$ and $x - 5 = 0$

y-Intercepts: $\frac{5}{3}$, because $r(0) = \dfrac{0^2 - 4\cdot 0 - 5}{0 - 3} = \dfrac{5}{3}$

Horizontal asymptote: None, because the degree of the numerator is greater than the degree of the denominator

Vertical asymptote: $x = 3$, from the zero of the denominator

Behavior near vertical asymptote: $y \to \infty$ as $x \to 3^-$ and $y \to -\infty$ as $x \to 3^+$

$$\begin{array}{r} x - 1 \\ x - 3\overline{)x^2 - 4x - 5} \\ \underline{x^2 - 3x} \\ -x - 5 \\ \underline{-x + 3} \\ -8 \end{array}$$

Slant asymptote: Since the degree of the numerator is one more than the degree of the denominator, the function has a slant asymptote. Dividing (see the margin), we obtain

$$r(x) = x - 1 - \frac{8}{x - 3}$$

Thus, $y = x - 1$ is the slant asymptote.

Graph: We use the information we have found, together with some additional values, to sketch the graph in Figure 10.

x	y
−2	−1.4
1	4
2	9
4	−5
6	2.33

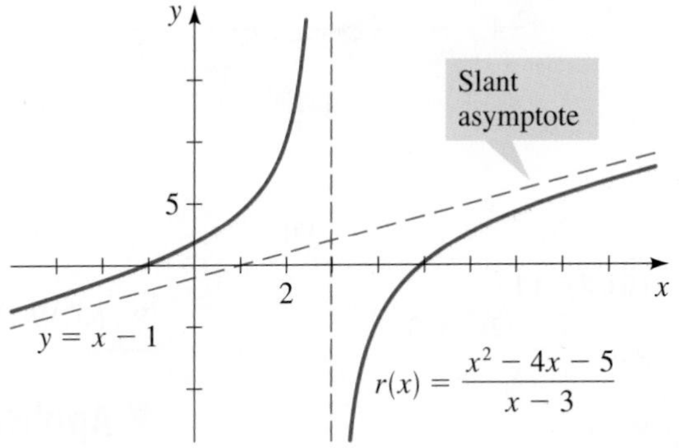

FIGURE 10

NOW TRY EXERCISE 65

So far, we have considered only horizontal and slant asymptotes as end behaviors for rational functions. In the next example we graph a function whose end behavior is like that of a parabola.

EXAMPLE 9 | End Behavior of a Rational Function

Graph the rational function

$$r(x) = \frac{x^3 - 2x^2 + 3}{x - 2}$$

and describe its end behavior.

SOLUTION

Factor: $y = \dfrac{(x + 1)(x^2 - 3x + 3)}{x - 2}$

x-Intercepts: -1, from $x + 1 = 0$ (The other factor in the numerator has no real zeros.)

y-Intercepts: $-\frac{3}{2}$, because $r(0) = \dfrac{0^3 - 2 \cdot 0^2 + 3}{0 - 2} = -\dfrac{3}{2}$

Vertical asymptote: $x = 2$, from the zero of the denominator

Behavior near vertical asymptote: $y \to -\infty$ as $x \to 2^-$ and $y \to \infty$ as $x \to 2^+$

Horizontal asymptote: None, because the degree of the numerator is greater than the degree of the denominator

$$\begin{array}{r} x^2 \\ x - 2\overline{)x^3 - 2x^2 + 0x + 3} \\ \underline{x^3 - 2x^2} \\ 3 \end{array}$$

End behavior: Dividing (see the margin), we get

$$r(x) = x^2 + \frac{3}{x - 2}$$

This shows that the end behavior of r is like that of the parabola $y = x^2$ because $3/(x - 2)$ is small when $|x|$ is large. That is, $3/(x - 2) \to 0$ as $x \to \pm\infty$. This means that the graph of r will be close to the graph of $y = x^2$ for large $|x|$.

Graph: In Figure 11(a) we graph r in a small viewing rectangle; we can see the intercepts, the vertical asymptotes, and the local minimum. In Figure 11(b) we graph r in a larger viewing rectangle; here the graph looks almost like the graph of a parabola. In Figure 11(c) we graph both $y = r(x)$ and $y = x^2$; these graphs are very close to each other except near the vertical asymptote.

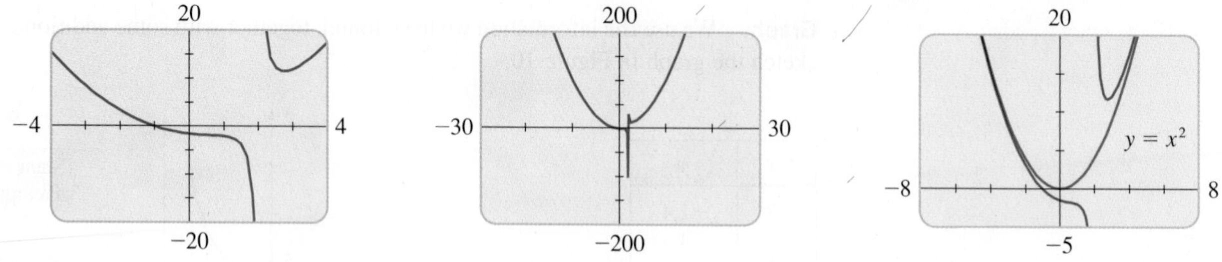

(a) (b) (c)

FIGURE 11
$r(x) = \dfrac{x^3 - 2x^2 + 3}{x - 2}$

NOW TRY EXERCISE **73**

Applications

Rational functions occur frequently in scientific applications of algebra. In the next example we analyze the graph of a function from the theory of electricity.

EXAMPLE 10 | Electrical Resistance

When two resistors with resistances R_1 and R_2 are connected in parallel, their combined resistance R is given by the formula

$$R = \frac{R_1 R_2}{R_1 + R_2}$$

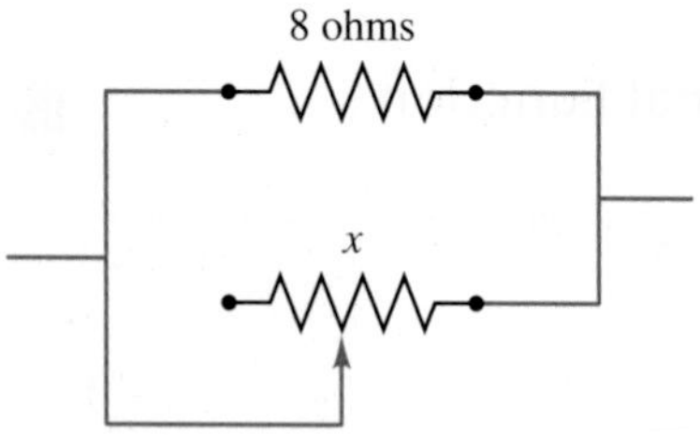

FIGURE 12

Suppose that a fixed 8-ohm resistor is connected in parallel with a variable resistor, as shown in Figure 12. If the resistance of the variable resistor is denoted by x, then the combined resistance R is a function of x. Graph R, and give a physical interpretation of the graph.

SOLUTION Substituting $R_1 = 8$ and $R_2 = x$ into the formula gives the function

$$R(x) = \frac{8x}{8 + x}$$

Since resistance cannot be negative, this function has physical meaning only when $x > 0$. The function is graphed in Figure 13(a) using the viewing rectangle $[0, 20]$ by $[0, 10]$. The function has no vertical asymptote when x is restricted to positive values. The combined resistance R increases as the variable resistance x increases. If we widen the viewing rectangle to $[0, 100]$ by $[0, 10]$, we obtain the graph in Figure 13(b). For large x the combined resistance R levels off, getting closer and closer to the horizontal asymptote $R = 8$. No matter how large the variable resistance x, the combined resistance is never greater than 8 ohms.

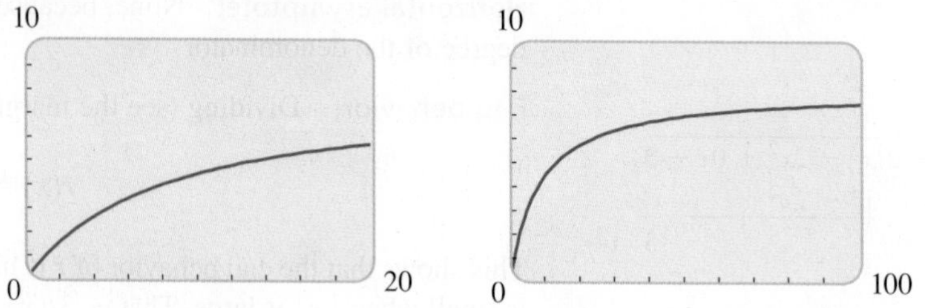

(a) (b)

FIGURE 13
$R(x) = \dfrac{8x}{8 + x}$

NOW TRY EXERCISE **83**

3.7 EXERCISES

CONCEPTS

1. If the rational function $y = r(x)$ has the vertical asymptote $x = 2$, then as $x \to 2^+$, either $y \to$ ______ or $y \to$ ______.

2. If the rational function $y = r(x)$ has the horizontal asymptote $y = 2$, then $y \to$ ______ as $x \to \pm\infty$.

3–6 ■ The following questions are about the rational function

$$r(x) = \frac{(x+1)(x-2)}{(x+2)(x-3)}$$

3. The function r has x-intercepts ______ and ______.

4. The function r has y-intercept ______.

5. The function r has vertical asymptotes $x =$ ______ and $x =$ ______.

6. The function r has horizontal asymptote $y =$ ______.

SKILLS

7–10 ■ A rational function is given. **(a)** Complete each table for the function. **(b)** Describe the behavior of the function near its vertical asymptote, based on Tables 1 and 2. **(c)** Determine the horizontal asymptote, based on Tables 3 and 4.

TABLE 1

x	$r(x)$
1.5	
1.9	
1.99	
1.999	

TABLE 2

x	$r(x)$
2.5	
2.1	
2.01	
2.001	

TABLE 3

x	$r(x)$
10	
50	
100	
1000	

TABLE 4

x	$r(x)$
−10	
−50	
−100	
−1000	

7. $r(x) = \dfrac{x}{x-2}$ **8.** $r(x) = \dfrac{4x+1}{x-2}$

9. $r(x) = \dfrac{3x-10}{(x-2)^2}$ **10.** $r(x) = \dfrac{3x^2+1}{(x-2)^2}$

11–16 ■ Find the x- and y-intercepts of the rational function.

11. $r(x) = \dfrac{x-1}{x+4}$ **12.** $s(x) = \dfrac{3x}{x-5}$

13. $t(x) = \dfrac{x^2-x-2}{x-6}$ **14.** $r(x) = \dfrac{2}{x^2+3x-4}$

15. $r(x) = \dfrac{x^2-9}{x^2}$ **16.** $r(x) = \dfrac{x^3+8}{x^2+4}$

17–20 ■ From the graph, determine the x- and y-intercepts and the vertical and horizontal asymptotes.

17.

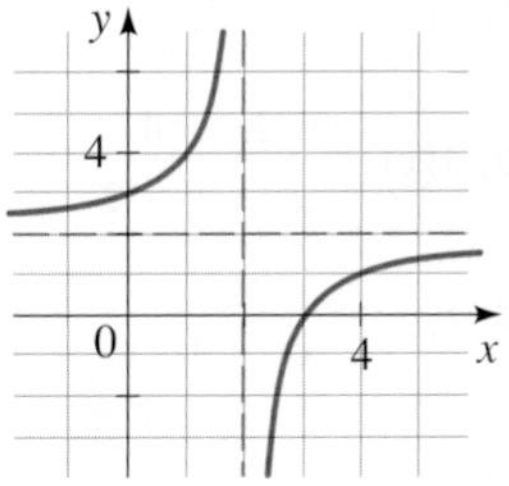

18.

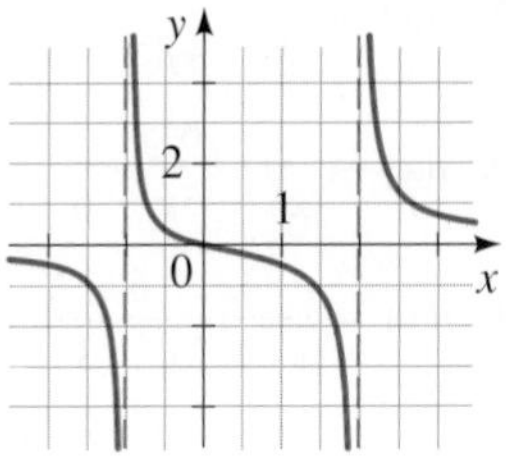

19.

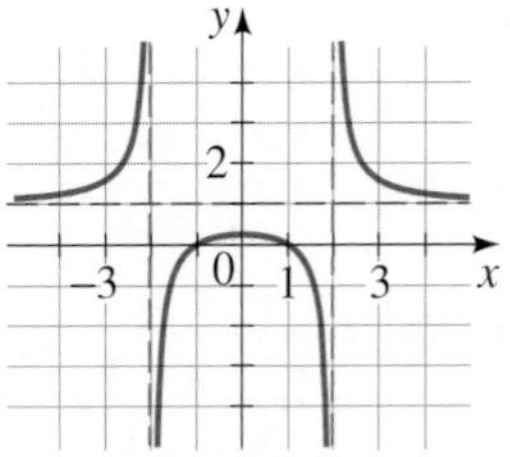

20.

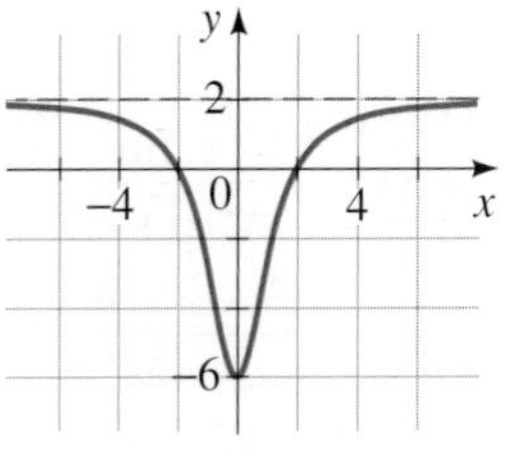

21–32 ■ Find all horizontal and vertical asymptotes (if any).

21. $r(x) = \dfrac{5}{x-2}$ **22.** $r(x) = \dfrac{2x-3}{x^2-1}$

23. $r(x) = \dfrac{6x}{x^2+2}$ **24.** $r(x) = \dfrac{2x-4}{x^2+x+1}$

25. $s(x) = \dfrac{6x^2+1}{2x^2+x-1}$ **26.** $s(x) = \dfrac{8x^2+1}{4x^2+2x-6}$

27. $s(x) = \dfrac{(5x-1)(x+1)}{(3x-1)(x+2)}$ **28.** $s(x) = \dfrac{(2x-1)(x+3)}{(3x-1)(x-4)}$

29. $r(x) = \dfrac{6x^3-2}{2x^3+5x^2+6x}$ **30.** $r(x) = \dfrac{5x^3}{x^3+2x^2+5x}$

31. $t(x) = \dfrac{x^2+2}{x-1}$ **32.** $r(x) = \dfrac{x^3+3x^2}{x^2-4}$

33–40 ■ Use transformations of the graph of $y = \frac{1}{x}$ to graph the rational function, as in Example 2.

33. $r(x) = \dfrac{1}{x-1}$ **34.** $r(x) = \dfrac{1}{x+4}$

35. $s(x) = \dfrac{3}{x+1}$ **36.** $s(x) = \dfrac{-2}{x-2}$

37. $t(x) = \dfrac{2x-3}{x-2}$ **38.** $t(x) = \dfrac{3x-3}{x+2}$

39. $r(x) = \dfrac{x+2}{x+3}$ **40.** $r(x) = \dfrac{2x-9}{x-4}$

41–64 ■ Find the intercepts and asymptotes, and then sketch a graph of the rational function and state the domain and range. Use a graphing device to confirm your answer.

41. $r(x) = \dfrac{4x-4}{x+2}$ **42.** $r(x) = \dfrac{2x+6}{-6x+3}$

43. $s(x) = \dfrac{4 - 3x}{x + 7}$

44. $s(x) = \dfrac{1 - 2x}{2x + 3}$

45. $r(x) = \dfrac{18}{(x - 3)^2}$

46. $r(x) = \dfrac{x - 2}{(x + 1)^2}$

47. $s(x) = \dfrac{4x - 8}{(x - 4)(x + 1)}$

48. $s(x) = \dfrac{x + 2}{(x + 3)(x - 1)}$

49. $s(x) = \dfrac{6}{x^2 - 5x - 6}$

50. $s(x) = \dfrac{2x - 4}{x^2 + x - 2}$

51. $t(x) = \dfrac{3x + 6}{x^2 + 2x - 8}$

52. $t(x) = \dfrac{x - 2}{x^2 - 4x}$

53. $r(x) = \dfrac{(x - 1)(x + 2)}{(x + 1)(x - 3)}$

54. $r(x) = \dfrac{2x(x + 2)}{(x - 1)(x - 4)}$

55. $r(x) = \dfrac{x^2 - 2x + 1}{x^2 + 2x + 1}$

56. $r(x) = \dfrac{4x^2}{x^2 - 2x - 3}$

57. $r(x) = \dfrac{2x^2 + 10x - 12}{x^2 + x - 6}$

58. $r(x) = \dfrac{2x^2 + 2x - 4}{x^2 + x}$

59. $r(x) = \dfrac{x^2 - x - 6}{x^2 + 3x}$

60. $r(x) = \dfrac{x^2 + 3x}{x^2 - x - 6}$

61. $r(x) = \dfrac{3x^2 + 6}{x^2 - 2x - 3}$

62. $r(x) = \dfrac{5x^2 + 5}{x^2 + 4x + 4}$

63. $s(x) = \dfrac{x^2 - 2x + 1}{x^3 - 3x^2}$

64. $t(x) = \dfrac{x^3 - x^2}{x^3 - 3x - 2}$

65–72 ■ Find the slant asymptote, the vertical asymptotes, and sketch a graph of the function.

65. $r(x) = \dfrac{x^2}{x - 2}$

66. $r(x) = \dfrac{x^2 + 2x}{x - 1}$

67. $r(x) = \dfrac{x^2 - 2x - 8}{x}$

68. $r(x) = \dfrac{3x - x^2}{2x - 2}$

69. $r(x) = \dfrac{x^2 + 5x + 4}{x - 3}$

70. $r(x) = \dfrac{x^3 + 4}{2x^2 + x - 1}$

71. $r(x) = \dfrac{x^3 + x^2}{x^2 - 4}$

72. $r(x) = \dfrac{2x^3 + 2x}{x^2 - 1}$

73–76 ■ Graph the rational function f, and determine all vertical asymptotes from your graph. Then graph f and g in a sufficiently large viewing rectangle to show that they have the same end behavior.

73. $f(x) = \dfrac{2x^2 + 6x + 6}{x + 3}$, $g(x) = 2x$

74. $f(x) = \dfrac{-x^3 + 6x^2 - 5}{x^2 - 2x}$, $g(x) = -x + 4$

75. $f(x) = \dfrac{x^3 - 2x^2 + 16}{x - 2}$, $g(x) = x^2$

76. $f(x) = \dfrac{-x^4 + 2x^3 - 2x}{(x - 1)^2}$, $g(x) = 1 - x^2$

77–82 ■ Graph the rational function, and find all vertical asymptotes, x- and y-intercepts, and local extrema, correct to the nearest decimal. Then use long division to find a polynomial that has the same end behavior as the rational function, and graph both functions in a sufficiently large viewing rectangle to verify that the end behaviors of the polynomial and the rational function are the same.

77. $y = \dfrac{2x^2 - 5x}{2x + 3}$

78. $y = \dfrac{x^4 - 3x^3 + x^2 - 3x + 3}{x^2 - 3x}$

79. $y = \dfrac{x^5}{x^3 - 1}$

80. $y = \dfrac{x^4}{x^2 - 2}$

81. $r(x) = \dfrac{x^4 - 3x^3 + 6}{x - 3}$

82. $r(x) = \dfrac{4 + x^2 - x^4}{x^2 - 1}$

APPLICATIONS

83. Population Growth Suppose that the rabbit population on Mr. Jenkins' farm follows the formula

$$p(t) = \frac{3000t}{t + 1}$$

where $t \geq 0$ is the time (in months) since the beginning of the year.

(a) Draw a graph of the rabbit population.
(b) What eventually happens to the rabbit population?

84. Drug Concentration After a certain drug is injected into a patient, the concentration c of the drug in the bloodstream is monitored. At time $t \geq 0$ (in minutes since the injection), the concentration (in mg/L) is given by

$$c(t) = \frac{30t}{t^2 + 2}$$

(a) Draw a graph of the drug concentration.
(b) What eventually happens to the concentration of drug in the bloodstream?

85. Drug Concentration A drug is administered to a patient, and the concentration of the drug in the bloodstream is monitored. At time $t \geq 0$ (in hours since giving the drug), the concentration (in mg/L) is given by

$$c(t) = \frac{5t}{t^2 + 1}$$

Graph the function c with a graphing device.
(a) What is the highest concentration of drug that is reached in the patient's bloodstream?
(b) What happens to the drug concentration after a long period of time?
(c) How long does it take for the concentration to drop below 0.3 mg/L?

86. **Flight of a Rocket** Suppose a rocket is fired upward from the surface of the earth with an initial velocity v (measured in meters per second). Then the maximum height h (in meters) reached by the rocket is given by the function

$$h(v) = \frac{Rv^2}{2gR - v^2}$$

where $R = 6.4 \times 10^6$ m is the radius of the earth and $g = 9.8$ m/s^2 is the acceleration due to gravity. Use a graphing device to draw a graph of the function h. (Note that h and v must both be positive, so the viewing rectangle need not contain negative values.) What does the vertical asymptote represent physically?

87. **The Doppler Effect** As a train moves toward an observer (see the figure), the pitch of its whistle sounds higher to the observer than it would if the train were at rest, because the crests of the sound waves are compressed closer together. This phenomenon is called the *Doppler effect*. The observed pitch P is a function of the speed v of the train and is given by

$$P(v) = P_0\left(\frac{s_0}{s_0 - v}\right)$$

where P_0 is the actual pitch of the whistle at the source and $s_0 = 332$ m/s is the speed of sound in air. Suppose that a train has a whistle pitched at $P_0 = 440$ Hz. Graph the function $y = P(v)$ using a graphing device. How can the vertical asymptote of this function be interpreted physically?

88. **Focusing Distance** For a camera with a lens of fixed focal length F to focus on an object located a distance x from the lens, the film must be placed a distance y behind the lens, where F, x, and y are related by

$$\frac{1}{x} + \frac{1}{y} = \frac{1}{F}$$

(See the figure.) Suppose the camera has a 55-mm lens ($F = 55$).

(a) Express y as a function of x and graph the function.

(b) What happens to the focusing distance y as the object moves far away from the lens?

(c) What happens to the focusing distance y as the object moves close to the lens?

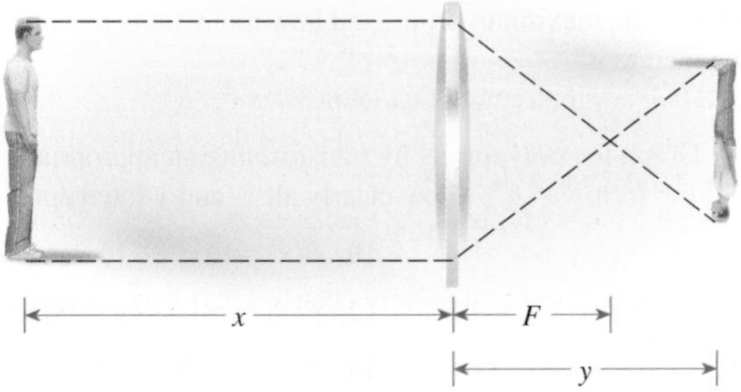

DISCOVERY ■ DISCUSSION ■ WRITING

89. **Constructing a Rational Function from Its Asymptotes** Give an example of a rational function that has vertical asymptote $x = 3$. Now give an example of one that has vertical asymptote $x = 3$ *and* horizontal asymptote $y = 2$. Now give an example of a rational function with vertical asymptotes $x = 1$ and $x = -1$, horizontal asymptote $y = 0$, and x-intercept 4.

90. **A Rational Function with No Asymptote** Explain how you can tell (without graphing it) that the function

$$r(x) = \frac{x^6 + 10}{x^4 + 8x^2 + 15}$$

has no x-intercept and no horizontal, vertical, or slant asymptote. What is its end behavior?'

91. **Graphs with Holes** In this chapter we adopted the convention that in rational functions, the numerator and denominator don't share a common factor. In this exercise we consider the graph of a rational function that does not satisfy this rule.

(a) Show that the graph of

$$r(x) = \frac{3x^2 - 3x - 6}{x - 2}$$

is the line $y = 3x + 3$ with the point $(2, 9)$ removed. [*Hint*: Factor. What is the domain of r?]

(b) Graph the rational functions:

$$s(x) = \frac{x^2 + x - 20}{x + 5}$$

$$t(x) = \frac{2x^2 - x - 1}{x - 1}$$

$$u(x) = \frac{x - 2}{x^2 - 2x}$$

92. **Transformations of $y = 1/x^2$** In Example 2 we saw that some simple rational functions can be graphed by shifting, stretching, or reflecting the graph of $y = 1/x$. In this exercise we consider rational functions that can be graphed by transforming the graph of $y = 1/x^2$, shown on the following page.

(a) Graph the function

$$r(x) = \frac{1}{(x - 2)^2}$$

by transforming the graph of $y = 1/x^2$.

(b) Use long division and factoring to show that the function

$$s(x) = \frac{2x^2 + 4x + 5}{x^2 + 2x + 1}$$

can be written as

$$s(x) = 2 + \frac{3}{(x + 1)^2}$$

Then graph s by transforming the graph of $y = 1/x^2$.

(c) One of the following functions can be graphed by transforming the graph of $y = 1/x^2$; the other cannot. Use transformations to graph the one that can be, and explain why this method doesn't work for the other one.

$$p(x) = \frac{2 - 3x^2}{x^2 - 4x + 4} \qquad q(x) = \frac{12x - 3x^2}{x^2 - 4x + 4}$$

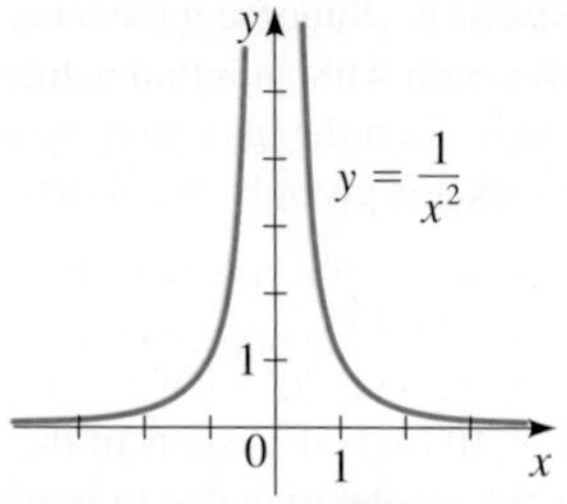

CHAPTER 3 | REVIEW

CONCEPT CHECK

1. (a) Write the defining equation for a polynomial P of degree n.
   (b) What does it mean to say that c is a zero of P?
2. Sketch graphs showing the possible end behaviors of polynomials of odd degree and of even degree.
3. What steps would you follow to graph a polynomial by hand?
4. (a) What is meant by a local maximum point or local minimum point of a polynomial?
   (b) How many local extrema can a polynomial of degree n have?
5. State the Division Algorithm and identify the dividend, divisor, quotient, and remainder.
6. How does synthetic division work?
7. (a) State the Remainder Theorem.
   (b) State the Factor Theorem.
8. (a) State the Rational Zeros Theorem.
   (b) What steps would you take to find the rational zeros of a polynomial?
9. State Descartes' Rule of Signs.
10. (a) What does it mean to say that a is a lower bound and b is an upper bound for the zeros of a polynomial?
    (b) State the Upper and Lower Bounds Theorem.
11. (a) What is a complex number?
    (b) What are the real and imaginary parts of a complex number?
    (c) What is the complex conjugate of a complex number?
    (d) How do you add, subtract, multiply, and divide complex numbers?
12. (a) State the Fundamental Theorem of Algebra.
    (b) State the Complete Factorization Theorem.
    (c) What does it mean to say that c is a zero of multiplicity k of a polynomial P?
    (d) State the Zeros Theorem.
    (e) State the Conjugate Zeros Theorem.
13. (a) What is a rational function?
    (b) What does it mean to say that $x = a$ is a vertical asymptote of $y = f(x)$?
    (c) How do you locate a vertical asymptote?
    (d) What does it mean to say that $y = b$ is a horizontal asymptote of $y = f(x)$?
    (e) How do you locate a horizontal asymptote?
    (f) What steps do you follow to sketch the graph of a rational function by hand?
    (g) Under what circumstances does a rational function have a slant asymptote? If one exists, how do you find it?
    (h) How do you determine the end behavior of a rational function?

EXERCISES

1–4 ■ A quadratic function is given. **(a)** Express the function in standard form. **(b)** Graph the function.

1. $f(x) = x^2 + 4x + 1$
2. $f(x) = -2x^2 + 12x + 12$
3. $g(x) = 1 + 8x - x^2$
4. $g(x) = 6x - 3x^2$

5–6 ■ Find the maximum or minimum value of the quadratic function.

5. $f(x) = 2x^2 + 4x - 5$
6. $g(x) = 1 - x - x^2$
7. A stone is thrown upward from the top of a building. Its height (in feet) above the ground after t seconds is given by the function $h(t) = -16t^2 + 48t + 32$. What maximum height does the stone reach?
8. The profit P (in dollars) generated by selling x units of a certain commodity is given by the function

$$P(x) = -1500 + 12x - 0.004x^2$$

What is the maximum profit, and how many units must be sold to generate it?

9–14 ■ Graph the polynomial by transforming an appropriate graph of the form $y = x^n$. Show clearly all x- and y-intercepts.

9. $P(x) = -x^3 + 64$
10. $P(x) = 2x^3 - 16$
11. $P(x) = 2(x + 1)^4 - 32$
12. $P(x) = 81 - (x - 3)^4$
13. $P(x) = 32 + (x - 1)^5$
14. $P(x) = -3(x + 2)^5 + 96$

15–16 ■ A polynomial function P is given. **(a)** Determine the multiplicity of each zero of P. **(b)** Sketch a graph of P.

15. $P(x) = x^3(x - 2)^2$ **16.** $P(x) = x(x + 1)^3(x - 1)^2$

17–20 ■ Use a graphing device to graph the polynomial. Find the x- and y-intercepts and the coordinates of all local extrema, correct to the nearest decimal. Describe the end behavior of the polynomial.

17. $P(x) = x^3 - 4x + 1$ **18.** $P(x) = -2x^3 + 6x^2 - 2$

19. $P(x) = 3x^4 - 4x^3 - 10x - 1$

20. $P(x) = x^5 + x^4 - 7x^3 - x^2 + 6x + 3$

21. The strength S of a wooden beam of width x and depth y is given by the formula $S = 13.8xy^2$. A beam is to be cut from a log of diameter 10 in., as shown in the figure.

(a) Express the strength S of this beam as a function of x only.

(b) What is the domain of the function S?

(c) Draw a graph of S.

(d) What width will make the beam the strongest?

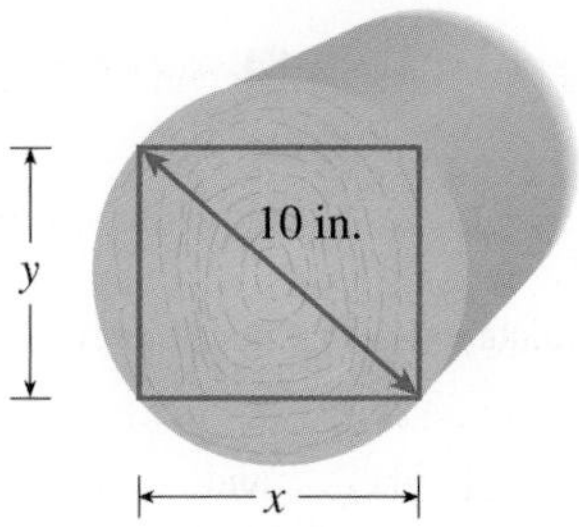

22. A small shelter for delicate plants is to be constructed of thin plastic material. It will have square ends and a rectangular top and back, with an open bottom and front, as shown in the figure. The total area of the four plastic sides is to be 1200 in^2.

(a) Express the volume V of the shelter as a function of the depth x.

(b) Draw a graph of V.

(c) What dimensions will maximize the volume of the shelter?

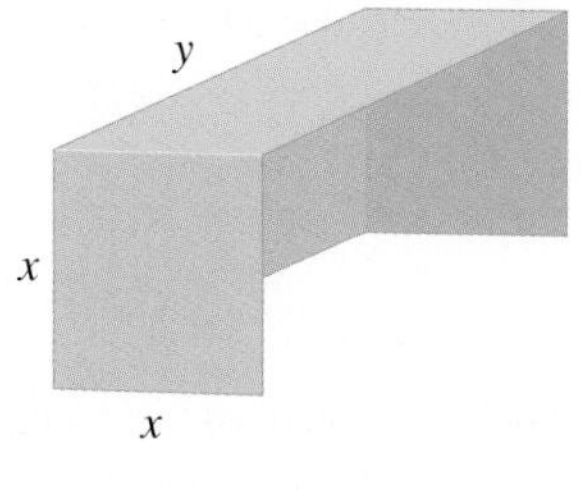

23–30 ■ Find the quotient and remainder.

23. $\dfrac{x^2 - 3x + 5}{x - 2}$ **24.** $\dfrac{x^2 + x - 12}{x - 3}$

25. $\dfrac{x^3 - x^2 + 11x + 2}{x - 4}$ **26.** $\dfrac{x^3 + 2x^2 - 10}{x + 3}$

27. $\dfrac{x^4 - 8x^2 + 2x + 7}{x + 5}$ **28.** $\dfrac{2x^4 + 3x^3 - 12}{x + 4}$

29. $\dfrac{2x^3 + x^2 - 8x + 15}{x^2 + 2x - 1}$ **30.** $\dfrac{x^4 - 2x^2 + 7x}{x^2 - x + 3}$

31–32 ■ Find the indicated value of the polynomial using the Remainder Theorem.

31. $P(x) = 2x^3 - 9x^2 - 7x + 13$; find $P(5)$

32. $Q(x) = x^4 + 4x^3 + 7x^2 + 10x + 15$; find $Q(-3)$

33. Show that $\frac{1}{2}$ is a zero of the polynomial

$$P(x) = 2x^4 + x^3 - 5x^2 + 10x - 4$$

34. Use the Factor Theorem to show that $x + 4$ is a factor of the polynomial

$$P(x) = x^5 + 4x^4 - 7x^3 - 23x^2 + 23x + 12$$

35. What is the remainder when the polynomial

$$P(x) = x^{500} + 6x^{201} - x^2 - 2x + 4$$

is divided by $x - 1$?

36. What is the remainder when $x^{101} - x^4 + 2$ is divided by $x + 1$?

37–38 ■ A polynomial P is given. **(a)** List all possible rational zeros (without testing to see whether they actually are zeros). **(b)** Determine the possible number of positive and negative real zeros using Descartes' Rule of Signs.

37. $P(x) = x^5 - 6x^3 - x^2 + 2x + 18$

38. $P(x) = 6x^4 + 3x^3 + x^2 + 3x + 4$

39–46 ■ A polynomial P is given. **(a)** Find all real zeros of P, and state their multiplicities. **(b)** Sketch the graph of P.

39. $P(x) = x^3 - 16x$

40. $P(x) = x^3 - 3x^2 - 4x$

41. $P(x) = x^4 + x^3 - 2x^2$

42. $P(x) = x^4 - 5x^2 + 4$

43. $P(x) = x^4 - 2x^3 - 7x^2 + 8x + 12$

44. $P(x) = x^4 - 2x^3 - 2x^2 + 8x - 8$

45. $P(x) = 2x^4 + x^3 + 2x^2 - 3x - 2$

46. $P(x) = 9x^5 - 21x^4 + 10x^3 + 6x^2 - 3x - 1$

47–56 ■ Evaluate the expression and write in the form $a + bi$.

47. $(2 - 3i) + (1 + 4i)$ **48.** $(3 - 6i) - (6 - 4i)$

49. $(2 + i)(3 - 2i)$ **50.** $4i(2 - \frac{1}{2}i)$

51. $\dfrac{4 + 2i}{2 - i}$ **52.** $\dfrac{8 + 3i}{4 + 3i}$

53. i^{25} **54.** $(1 + i)^3$

55. $(1 - \sqrt{-1})(1 + \sqrt{-1})$ **56.** $\sqrt{-10} \cdot \sqrt{-40}$

57. Find a polynomial of degree 3 with constant coefficient 12 and zeros $-\frac{1}{2}$, 2, and 3.

58. Find a polynomial of degree 4 that has integer coefficients and zeros $3i$ and 4, with 4 a double zero.

59. Does there exist a polynomial of degree 4 with integer coefficients that has zeros i, $2i$, $3i$, and $4i$? If so, find it. If not, explain why.

60. Prove that the equation $3x^4 + 5x^2 + 2 = 0$ has no real root.

61–70 ■ Find all rational, irrational, and complex zeros (and state their multiplicities). Use Descartes' Rule of Signs, the Upper and Lower Bounds Theorem, the Quadratic Formula, or other factoring techniques to help you whenever possible.

61. $P(x) = x^3 - 3x^2 - 13x + 15$

62. $P(x) = 2x^3 + 5x^2 - 6x - 9$

63. $P(x) = x^4 + 6x^3 + 17x^2 + 28x + 20$

64. $P(x) = x^4 + 7x^3 + 9x^2 - 17x - 20$

65. $P(x) = x^5 - 3x^4 - x^3 + 11x^2 - 12x + 4$

66. $P(x) = x^4 - 81$

67. $P(x) = x^6 - 64$

68. $P(x) = 18x^3 + 3x^2 - 4x - 1$

69. $P(x) = 6x^4 - 18x^3 + 6x^2 - 30x + 36$

70. $P(x) = x^4 + 15x^2 + 54$

71–74 ■ Use a graphing device to find all real solutions of the equation.

71. $2x^2 = 5x + 3$

72. $x^3 + x^2 - 14x - 24 = 0$

73. $x^4 - 3x^3 - 3x^2 - 9x - 2 = 0$

74. $x^5 = x + 3$

75–76 ■ A polynomial function P is given. Find all the real zeros of P, and factor P completely into linear and irreducible quadratic factors with real coefficients.

75. $P(x) = x^3 - 2x - 4$ **76.** $P(x) = x^4 + 3x^2 - 4$

77–82 ■ Graph the rational function. Show clearly all x- and y-intercepts and asymptotes.

77. $r(x) = \dfrac{3x - 12}{x + 1}$ **78.** $r(x) = \dfrac{1}{(x + 2)^2}$

79. $r(x) = \dfrac{x - 2}{x^2 - 2x - 8}$ **80.** $r(x) = \dfrac{2x^2 - 6x - 7}{x - 4}$

81. $r(x) = \dfrac{x^2 - 9}{2x^2 + 1}$ **82.** $r(x) = \dfrac{x^3 + 27}{x + 4}$

83–86 ■ Use a graphing device to analyze the graph of the rational function. Find all x- and y-intercepts and all vertical, horizontal, and slant asymptotes. If the function has no horizontal or slant asymptote, find a polynomial that has the same end behavior as the rational function.

83. $r(x) = \dfrac{x - 3}{2x + 6}$ **84.** $r(x) = \dfrac{2x - 7}{x^2 + 9}$

85. $r(x) = \dfrac{x^3 + 8}{x^2 - x - 2}$ **86.** $r(x) = \dfrac{2x^3 - x^2}{x + 1}$

87. Find the coordinates of all points of intersection of the graphs of

$$y = x^4 + x^2 + 24x \qquad \text{and} \qquad y = 6x^3 + 20$$

1. Express the quadratic function $f(x) = x^2 - x - 6$ in standard form, and sketch its graph.

2. Find the maximum or minimum value of the quadratic function $g(x) = 2x^2 + 6x + 3$.

3. A cannonball fired out to sea from a shore battery follows a parabolic trajectory given by the graph of the equation

$$h(x) = 10x - 0.01x^2$$

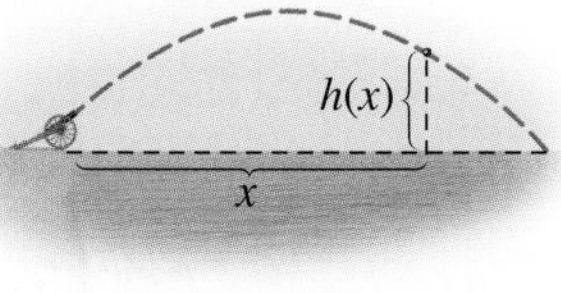

where $h(x)$ is the height of the cannonball above the water when it has traveled a horizontal distance of x feet.

(a) What is the maximum height that the cannonball reaches?

(b) How far does the cannonball travel horizontally before splashing into the water?

4. Graph the polynomial $P(x) = -(x + 2)^3 + 27$, showing clearly all x- and y-intercepts.

5. **(a)** Use synthetic division to find the quotient and remainder when $x^4 - 4x^2 + 2x + 5$ is divided by $x - 2$.

(b) Use long division to find the quotient and remainder when $2x^5 + 4x^4 - x^3 - x^2 + 7$ is divided by $2x^2 - 1$.

6. Let $P(x) = 2x^3 - 5x^2 - 4x + 3$.

(a) List all possible rational zeros of P.

(b) Find the complete factorization of P.

(c) Find the zeros of P.

(d) Sketch the graph of P.

7. Perform the indicated operation and write the result in the form $a + bi$.

(a) $(3 - 2i) + (4 + 3i)$ **(b)** $(3 - 2i) - (4 + 3i)$

(c) $(3 - 2i)(4 + 3i)$ **(d)** $\dfrac{3 - 2i}{4 + 3i}$

(e) i^{48} **(f)** $(\sqrt{2} - \sqrt{-2})(\sqrt{8} + \sqrt{-2})$

8. Find all real and complex zeros of $P(x) = x^3 - x^2 - 4x - 6$.

9. Find the complete factorization of $P(x) = x^4 - 2x^3 + 5x^2 - 8x + 4$.

10. Find a fourth-degree polynomial with integer coefficients that has zeros $3i$ and -1, with -1 a zero of multiplicity 2.

11. Let $P(x) = 2x^4 - 7x^3 + x^2 - 18x + 3$.

(a) Use Descartes' Rule of Signs to determine how many positive and how many negative real zeros P can have.

(b) Show that 4 is an upper bound and -1 is a lower bound for the real zeros of P.

(c) Draw a graph of P, and use it to estimate the real zeros of P, correct to two decimal places.

(d) Find the coordinates of all local extrema of P, correct to two decimals.

12. Consider the following rational functions:

$$r(x) = \frac{2x - 1}{x^2 - x - 2} \qquad s(x) = \frac{x^3 + 27}{x^2 + 4} \qquad t(x) = \frac{x^3 - 9x}{x + 2} \qquad u(x) = \frac{x^2 + x - 6}{x^2 - 25}$$

(a) Which of these rational functions has a horizontal asymptote?

(b) Which of these functions has a slant asymptote?

(c) Which of these functions has no vertical asymptote?

(d) Graph $y = u(x)$, showing clearly any asymptotes and x- and y-intercepts the function may have.

(e) Use long division to find a polynomial P that has the same end behavior as t. Graph both P and t on the same screen to verify that they have the same end behavior.

Fitting Polynomial Curves to Data

We have learned how to fit a line to data (see *Focus on Modeling*, page 130). The line models the increasing or decreasing trend in the data. If the data exhibit more variability, such as an increase followed by a decrease, then to model the data, we need to use a curve rather than a line. Figure 1 shows a scatter plot with three possible models that appear to fit the data. Which model fits the data best?

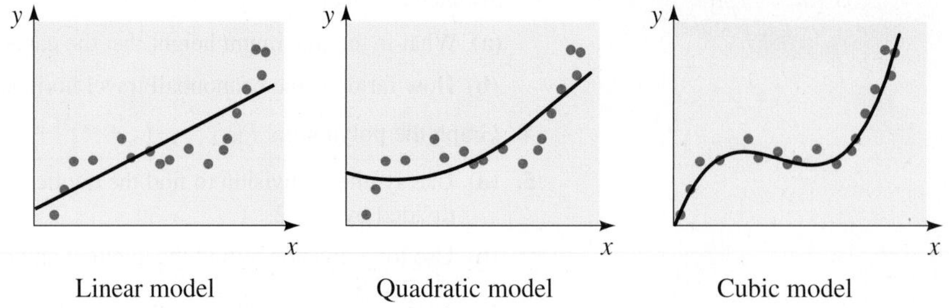

FIGURE 1

▼ Polynomial Functions as Models

Polynomial functions are ideal for modeling data for which the scatter plot has peaks or valleys (that is, local maxima or minima). For example, if the data have a single peak as in Figure 2(a), then it may be appropriate to use a quadratic polynomial to model the data. The more peaks or valleys the data exhibit, the higher the degree of the polynomial needed to model the data (see Figure 2).

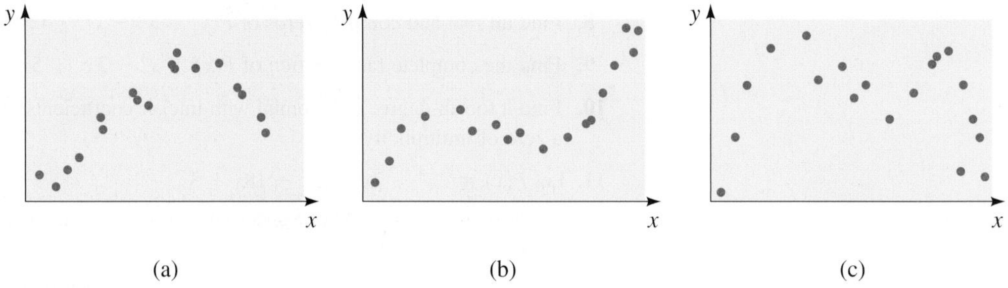

FIGURE 2

Graphing calculators are programmed to find the **polynomial of best fit** of a specified degree. As is the case for lines (see page 131), a polynomial of a given degree fits the data *best* if the sum of the squares of the distances between the graph of the polynomial and the data points is minimized.

EXAMPLE 1 | Rainfall and Crop Yield

Rain is essential for crops to grow, but too much rain can diminish crop yields. The data give rainfall and cotton yield per acre for several seasons in a certain county.

(a) Make a scatter plot of the data. What degree polynomial seems appropriate for modeling the data?

(b) Use a graphing calculator to find the polynomial of best fit. Graph the polynomial on the scatter plot.

(c) Use the model that you found to estimate the yield if there are 25 in. of rainfall.

Ted Wood/The Image Bank/Getty Images

Season	Rainfall (in.)	Yield (kg/acre)
1	23.3	5311
2	20.1	4382
3	18.1	3950
4	12.5	3137
5	30.9	5113
6	33.6	4814
7	35.8	3540
8	15.5	3850
9	27.6	5071
10	34.5	3881

SOLUTION

(a) The scatter plot is shown in Figure 3. The data appear to have a peak, so it is appropriate to model the data by a quadratic polynomial (degree 2).

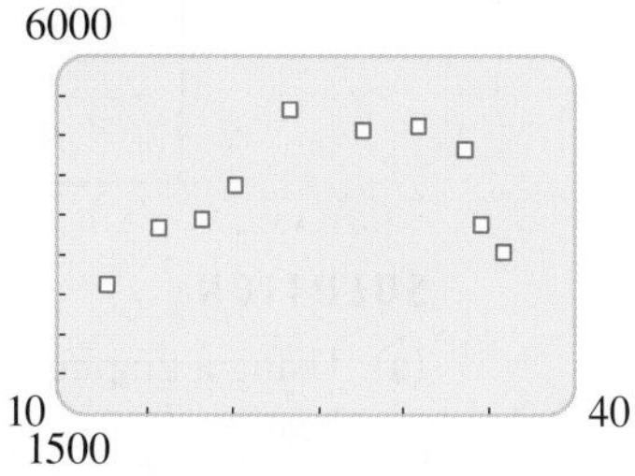

FIGURE 3 Scatter plot of yield vs. rainfall data

(b) Using a graphing calculator, we find that the quadratic polynomial of best fit is

$$y = -12.6x^2 + 651.5x - 3283.2$$

The calculator output and the scatter plot, together with the graph of the quadratic model, are shown in Figure 4.

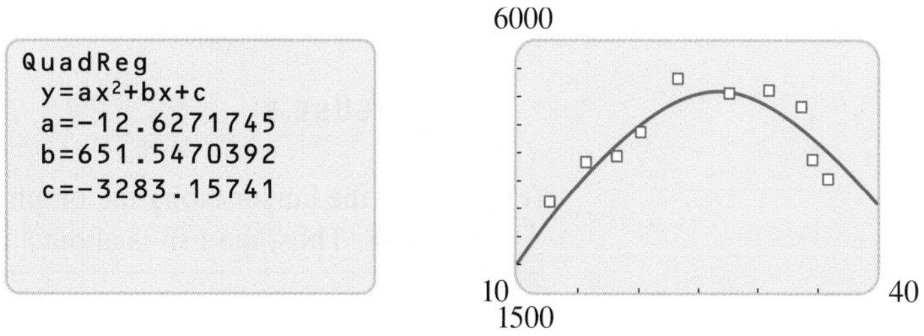

(a) (b)

FIGURE 4

(c) Using the model with $x = 25$, we get

$$y = -12.6(25)^2 + 651.5(25) - 3283.2 \approx 5129.3$$

We estimate the yield to be about 5130 kg/acre. ■

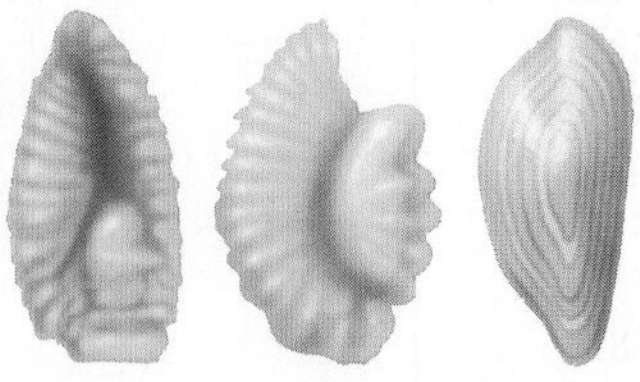

Otoliths for several fish species

EXAMPLE 2 | Length-at-Age Data for Fish

Otoliths ("earstones") are tiny structures that are found in the heads of fish. Microscopic growth rings on the otoliths, not unlike growth rings on a tree, record the age of a fish. The table gives the lengths of rock bass caught at different ages, as determined by the otoliths. Scientists have proposed a cubic polynomial to model this data.

(a) Use a graphing calculator to find the cubic polynomial of best fit for the data.

(b) Make a scatter plot of the data and graph the polynomial from part (a).

(c) A fisherman catches a rock bass 20 in. long. Use the model to estimate its age.

Age (yr)	Length (in.)	Age (yr)	Length (in.)
1	4.8	9	18.2
2	8.8	9	17.1
2	8.0	10	18.8
3	7.9	10	19.5
4	11.9	11	18.9
5	14.4	12	21.7
6	14.1	12	21.9
6	15.8	13	23.8
7	15.6	14	26.9
8	17.8	14	25.1

SOLUTION

(a) Using a graphing calculator (see Figure 5(a)), we find the cubic polynomial of best fit:

$$y = 0.0155x^3 - 0.372x^2 + 3.95x + 1.21$$

(b) The scatter plot of the data and the cubic polynomial are graphed in Figure 5(b).

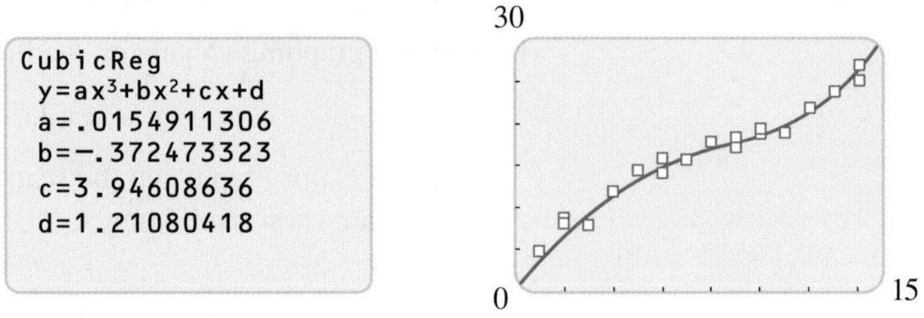

(a) (b)

FIGURE 5

(c) Moving the cursor along the graph of the polynomial, we find that $y = 20$ when $x \approx 10.8$. Thus, the fish is about 11 years old. ■

PROBLEMS

1. **Tire Inflation and Treadwear** Car tires need to be inflated properly. Overinflation or underinflation can cause premature treadwear. The data and scatter plot on the next page show tire life for different inflation values for a certain type of tire.

   (a) Find the quadratic polynomial that best fits the data.

   (b) Draw a graph of the polynomial from part (a) together with a scatter plot of the data.

   (c) Use your result from part (b) to estimate the pressure that gives the longest tire life.

Pressure (lb/in^2)	Tire life (mi)
26	50,000
28	66,000
31	78,000
35	81,000
38	74,000
42	70,000
45	59,000

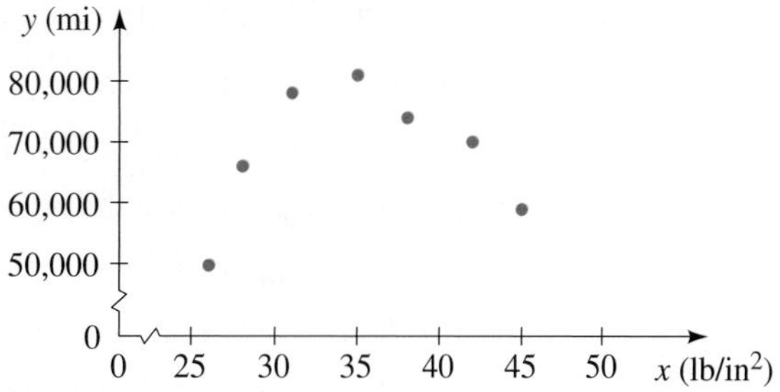

2. **Too Many Corn Plants per Acre?** The more corn a farmer plants per acre, the greater is the yield the farmer can expect, but only up to a point. Too many plants per acre can cause overcrowding and decrease yields. The data give crop yields per acre for various densities of corn plantings, as found by researchers at a university test farm.

   (a) Find the quadratic polynomial that best fits the data.

   (b) Draw a graph of the polynomial from part (a) together with a scatter plot of the data.

   (c) Use your result from part (b) to estimate the yield for 37,000 plants per acre.

Density (plants/acre)	Crop yield (bushels/acre)
15,000	43
20,000	98
25,000	118
30,000	140
35,000	142
40,000	122
45,000	93
50,000	67

3. **How Fast Can You List Your Favorite Things?** If you are asked to make a list of objects in a certain category, how fast you can list them follows a predictable pattern. For example, if you try to name as many vegetables as you can, you'll probably think of several right away—for example, carrots, peas, beans, corn, and so on. Then after a pause you might think of ones you eat less frequently—perhaps zucchini, eggplant, and asparagus. Finally, a few more exotic vegetables might come to mind—artichokes, jicama, bok choy, and the like. A psychologist performs this experiment on a number of subjects. The table below gives the average number of vegetables that the subjects named by a given number of seconds.

   (a) Find the cubic polynomial that best fits the data.

   (b) Draw a graph of the polynomial from part (a) together with a scatter plot of the data.

   (c) Use your result from part (b) to estimate the number of vegetables that subjects would be able to name in 40 seconds.

   (d) According to the model, how long (to the nearest 0.1 second) would it take a person to name five vegetables?

Seconds	Number of vegetables
1	2
2	6
5	10
10	12
15	14
20	15
25	18
30	21

4. Clothing Sales Are Seasonal Clothing sales tend to vary by season, with more clothes sold in spring and fall. The table gives sales figures for each month at a certain clothing store.

(a) Find the quartic (fourth-degree) polynomial that best fits the data.

(b) Draw a graph of the polynomial from part (a) together with a scatter plot of the data.

(c) Do you think that a quartic polynomial is a good model for these data? Explain.

Month	Sales ($)
January	8,000
February	18,000
March	22,000
April	31,000
May	29,000
June	21,000
July	22,000
August	26,000
September	38,000
October	40,000
November	27,000
December	15,000

5. Height of a Baseball A baseball is thrown upward, and its height measured at 0.5-second intervals using a strobe light. The resulting data are given in the table.

(a) Draw a scatter plot of the data. What degree polynomial is appropriate for modeling the data?

(b) Find a polynomial model that best fits the data, and graph it on the scatter plot.

(c) Find the times when the ball is 20 ft above the ground.

(d) What is the maximum height attained by the ball?

Time (s)	Height (ft)
0	4.2
0.5	26.1
1.0	40.1
1.5	46.0
2.0	43.9
2.5	33.7
3.0	15.8

6. Torricelli's Law Water in a tank will flow out of a small hole in the bottom faster when the tank is nearly full than when it is nearly empty. According to Torricelli's Law, the height $h(t)$ of water remaining at time t is a quadratic function of t.

A certain tank is filled with water and allowed to drain. The height of the water is measured at different times as shown in the table.

(a) Find the quadratic polynomial that best fits the data.

(b) Draw a graph of the polynomial from part (a) together with a scatter plot of the data.

(c) Use your graph from part (b) to estimate how long it takes for the tank to drain completely.

Time (min)	Height (ft)
0	5.0
4	3.1
8	1.9
12	0.8
16	0.2

CHAPTER 4

George Marks/Retrofile/Getty Images

EXPONENTIAL AND LOGARITHMIC FUNCTIONS

In this chapter we study a class of functions called *exponential functions.* These are functions, like $f(x) = 2^x$, where the independent variable is in the exponent. Exponential functions are used in modeling many real-world phenomena, such as the growth of a population or the growth of an investment that earns compound interest. Once an exponential model is obtained, we can use the model to predict population size or calculate the amount of an investment for any future date. To find out *when* a population will reach a certain level, we use the inverse functions of exponential functions, called *logarithmic functions.* So if we have an exponential model for population growth, we can answer questions like: When will my city be as crowded as the New York City street pictured above?

4.1 Exponential Functions

Exponential Functions ▶ Graphs of Exponential Functions ▶ Compound Interest

In this chapter we study a new class of functions called *exponential functions*. For example,

$$f(x) = 2^x$$

is an exponential function (with base 2). Notice how quickly the values of this function increase:

$$f(3) = 2^3 = 8$$

$$f(10) = 2^{10} = 1024$$

$$f(30) = 2^{30} = 1{,}073{,}741{,}824$$

Compare this with the function $g(x) = x^2$, where $g(30) = 30^2 = 900$. The point is that when the variable is in the exponent, even a small change in the variable can cause a dramatic change in the value of the function.

▼ Exponential Functions

To study exponential functions, we must first define what we mean by the exponential expression a^x when x is any real number. In Section 1.2 we defined a^x for $a > 0$ and x a rational number, but we have not yet defined irrational powers. So what is meant by $5^{\sqrt{3}}$ or 2^{π}? To define a^x when x is irrational, we approximate x by rational numbers.

For example, since

$$\sqrt{3} \approx 1.73205\ldots$$

is an irrational number, we successively approximate $a^{\sqrt{3}}$ by the following rational powers:

$$a^{1.7}, a^{1.73}, a^{1.732}, a^{1.7320}, a^{1.73205}, \ldots$$

Intuitively, we can see that these rational powers of a are getting closer and closer to $a^{\sqrt{3}}$. It can be shown by using advanced mathematics that there is exactly one number that these powers approach. We define $a^{\sqrt{3}}$ to be this number.

For example, using a calculator, we find

$$5^{\sqrt{3}} \approx 5^{1.732}$$

$$\approx 16.2411\ldots$$

The more decimal places of $\sqrt{3}$ we use in our calculation, the better our approximation of $5^{\sqrt{3}}$.

The Laws of Exponents are listed on page 14.

It can be proved that the *Laws of Exponents are still true when the exponents are real numbers.*

EXPONENTIAL FUNCTIONS

The **exponential function with base *a*** is defined for all real numbers x by

$$f(x) = a^x$$

where $a > 0$ and $a \neq 1$.

We assume that $a \neq 1$ because the function $f(x) = 1^x = 1$ is just a constant function. Here are some examples of exponential functions:

$$f(x) = 2^x \qquad g(x) = 3^x \qquad h(x) = 10^x$$

Base 2   Base 3   Base 10

EXAMPLE 1 | Evaluating Exponential Functions

Let $f(x) = 3^x$, and evaluate the following:

(a) $f(2)$ **(b)** $f(-\frac{2}{3})$

(c) $f(\pi)$ **(d)** $f(\sqrt{2})$

SOLUTION We use a calculator to obtain the values of f.

	Calculator keystrokes	Output
(a) $f(2) = 3^2 = 9$	3 ^ 2 ENTER	9
(b) $f(-\frac{2}{3}) = 3^{-2/3} \approx 0.4807$	3 ^ ((−) 2 ÷ 3) ENTER	0.4807498
(c) $f(\pi) = 3^\pi \approx 31.544$	3 ^ π ENTER	31.5442807
(d) $f(\sqrt{2}) = 3^{\sqrt{2}} \approx 4.7288$	3 ^ √ 2 ENTER	4.7288043

NOW TRY EXERCISE **5** ■

Graphs of Exponential Functions

We first graph exponential functions by plotting points. We will see that the graphs of such functions have an easily recognizable shape.

EXAMPLE 2 | Graphing Exponential Functions by Plotting Points

Draw the graph of each function.

(a) $f(x) = 3^x$ **(b)** $g(x) = \left(\frac{1}{3}\right)^x$

SOLUTION We calculate values of $f(x)$ and $g(x)$ and plot points to sketch the graphs in Figure 1.

x	$f(x) = 3^x$	$g(x) = (\frac{1}{3})^x$
-3	$\frac{1}{27}$	27
-2	$\frac{1}{9}$	9
-1	$\frac{1}{3}$	3
0	1	1
1	3	$\frac{1}{3}$
2	9	$\frac{1}{9}$
3	27	$\frac{1}{27}$

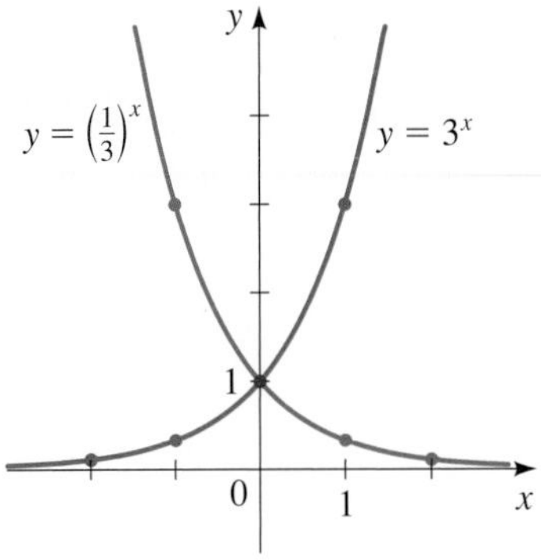

FIGURE 1

Notice that

$$g(x) = \left(\frac{1}{3}\right)^x = \frac{1}{3^x} = 3^{-x} = f(-x)$$

so we could have obtained the graph of g from the graph of f by reflecting in the y-axis.

Reflecting graphs is explained in Section 2.5.

NOW TRY EXERCISE **15** ■

Figure 2 shows the graphs of the family of exponential functions $f(x) = a^x$ for various values of the base a. All of these graphs pass through the point $(0, 1)$ because $a^0 = 1$ for $a \neq 0$. You can see from Figure 2 that there are two kinds of exponential functions: If $0 < a < 1$, the exponential function decreases rapidly. If $a > 1$, the function increases rapidly (see the margin note).

To see just how quickly $f(x) = 2^x$ increases, let's perform the following thought experiment. Suppose we start with a piece of paper that is a thousandth of an inch thick, and we fold it in half 50 times. Each time we fold the paper, the thickness of the paper stack doubles, so the thickness of the resulting stack would be $2^{50}/1000$ inches. How thick do you think that is? It works out to be more than 17 million miles!

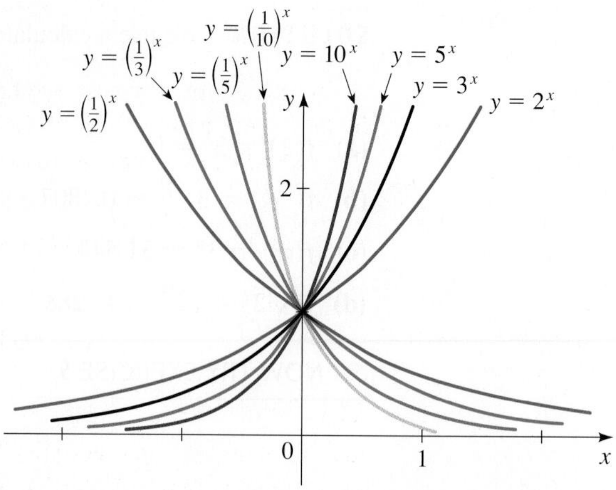

FIGURE 2 A family of exponential functions

The x-axis is a horizontal asymptote for the exponential function $f(x) = a^x$. This is because when $a > 1$, we have $a^x \to 0$ as $x \to -\infty$, and when $0 < a < 1$, we have $a^x \to 0$ as $x \to \infty$ (see Figure 2). Also, $a^x > 0$ for all $x \in \mathbb{R}$, so the function $f(x) = a^x$ has domain $\mathbb{R}$ and range $(0, \infty)$. These observations are summarized in the following box.

See Section 3.7, page 278, where the "arrow notation" used here is explained.

GRAPHS OF EXPONENTIAL FUNCTIONS

The exponential function

$$f(x) = a^x \qquad (a > 0, a \neq 1)$$

has domain $\mathbb{R}$ and range $(0, \infty)$. The line $y = 0$ (the x-axis) is a horizontal asymptote of f. The graph of f has one of the following shapes.

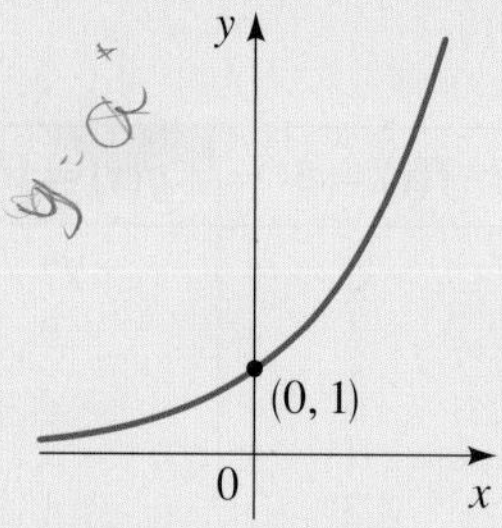

$f(x) = a^x$ for $a > 1$

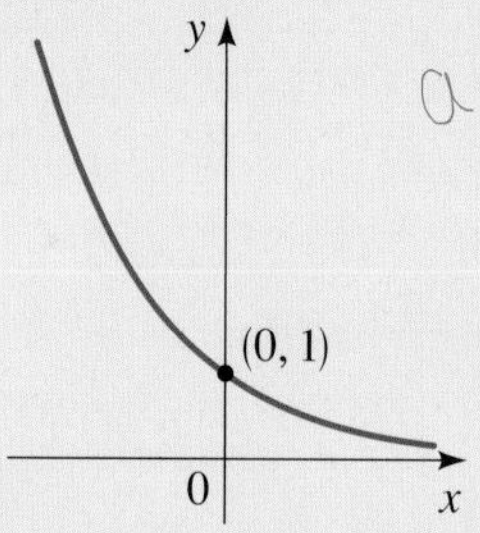

$f(x) = a^x$ for $0 < a < 1$

EXAMPLE 3 | Identifying Graphs of Exponential Functions

Find the exponential function $f(x) = a^x$ whose graph is given.

(a)

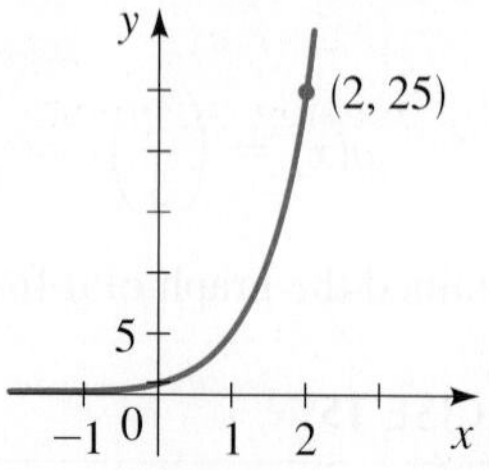

(b)

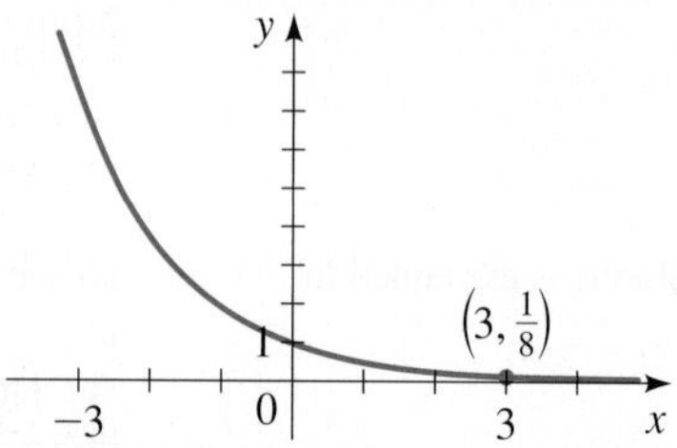

SOLUTION

(a) Since $f(2) = a^2 = 25$, we see that the base is $a = 5$. So $f(x) = 5^x$.

(b) Since $f(3) = a^3 = \frac{1}{8}$, we see that the base is $a = \frac{1}{2}$. So $f(x) = \left(\frac{1}{2}\right)^x$.

NOW TRY EXERCISE **19** ■

In the next example we see how to graph certain functions, not by plotting points, but by taking the basic graphs of the exponential functions in Figure 2 and applying the shifting and reflecting transformations of Section 2.5.

EXAMPLE 4 | Transformations of Exponential Functions

Use the graph of $f(x) = 2^x$ to sketch the graph of each function.

(a) $g(x) = 1 + 2^x$ **(b)** $h(x) = -2^x$ **(c)** $k(x) = 2^{x-1}$

SOLUTION

Shifting and reflecting of graphs is explained in Section 2.5.

(a) To obtain the graph of $g(x) = 1 + 2^x$, we start with the graph of $f(x) = 2^x$ and shift it upward 1 unit. Notice from Figure 3(a) that the line $y = 1$ is now a horizontal asymptote.

(b) Again we start with the graph of $f(x) = 2^x$, but here we reflect in the x-axis to get the graph of $h(x) = -2^x$ shown in Figure 3(b).

(c) This time we start with the graph of $f(x) = 2^x$ and shift it to the right by 1 unit to get the graph of $k(x) = 2^{x-1}$ shown in Figure 3(c).

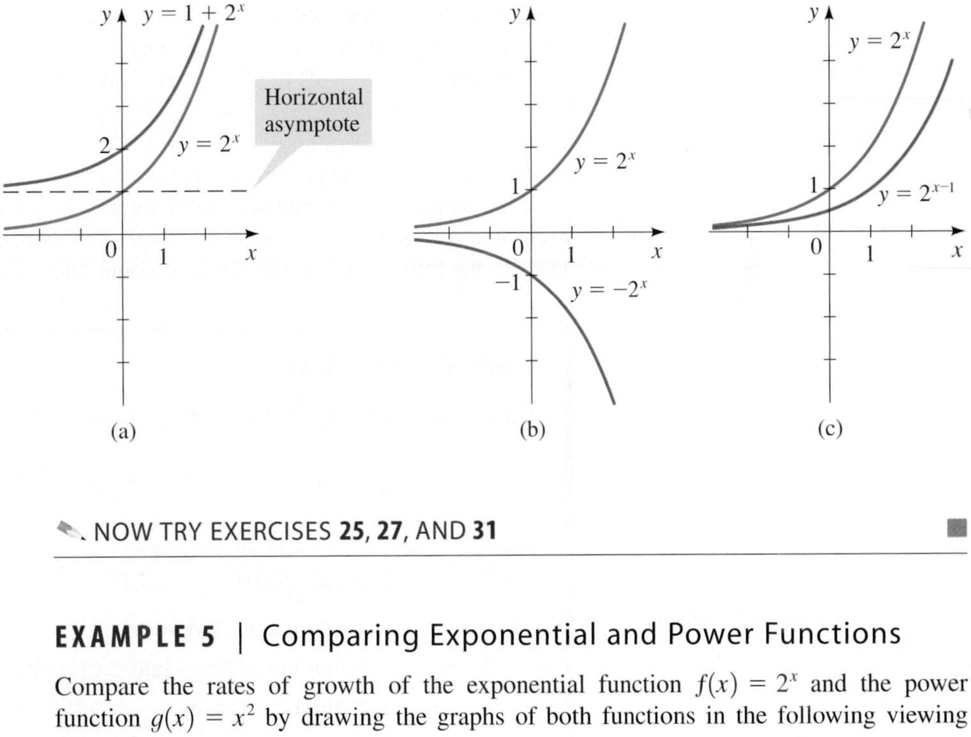

FIGURE 3

NOW TRY EXERCISES **25, 27,** AND **31** ■

EXAMPLE 5 | Comparing Exponential and Power Functions

Compare the rates of growth of the exponential function $f(x) = 2^x$ and the power function $g(x) = x^2$ by drawing the graphs of both functions in the following viewing rectangles.

(a) $[0, 3]$ by $[0, 8]$

(b) $[0, 6]$ by $[0, 25]$

(c) $[0, 20]$ by $[0, 1000]$

SOLUTION

(a) Figure 4(a) shows that the graph of $g(x) = x^2$ catches up with, and becomes higher than, the graph of $f(x) = 2^x$ at $x = 2$.

(b) The larger viewing rectangle in Figure 4(b) shows that the graph of $f(x) = 2^x$ overtakes that of $g(x) = x^2$ when $x = 4$.

(c) Figure 4(c) gives a more global view and shows that when x is large, $f(x) = 2^x$ is much larger than $g(x) = x^2$.

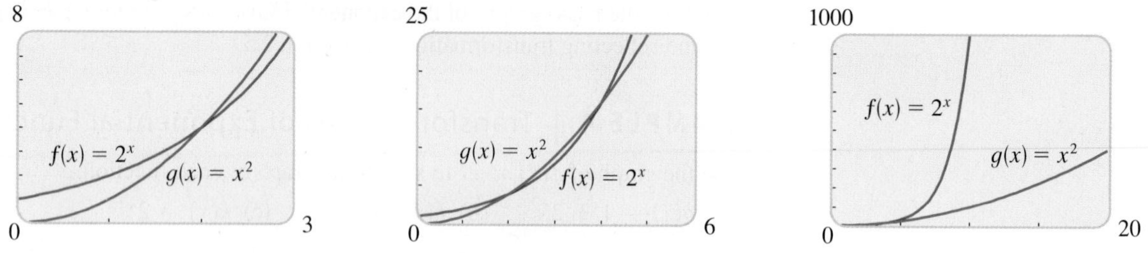

FIGURE 4 (a) (b) (c)

NOW TRY EXERCISE **41** ■

▼ Compound Interest

Exponential functions occur in calculating compound interest. If an amount of money P, called the **principal**, is invested at an interest rate i per time period, then after one time period the interest is Pi, and the amount A of money is

$$A = P + Pi = P(1 + i)$$

If the interest is reinvested, then the new principal is $P(1 + i)$, and the amount after another time period is $A = P(1 + i)(1 + i) = P(1 + i)^2$. Similarly, after a third time period the amount is $A = P(1 + i)^3$. In general, after k periods the amount is

$$A = P(1 + i)^k$$

Notice that this is an exponential function with base $1 + i$.

If the annual interest rate is r and if interest is compounded n times per year, then in each time period the interest rate is $i = r/n$, and there are nt time periods in t years. This leads to the following formula for the amount after t years.

COMPOUND INTEREST

Compound interest is calculated by the formula

$$A(t) = P\left(1 + \frac{r}{n}\right)^{nt}$$

where
$A(t)$ = amount after t years
P = principal
r = interest rate per year
n = number of times interest is compounded per year
t = number of years

r is often referred to as the nominal annual interest rate.

EXAMPLE 6 | Calculating Compound Interest

A sum of \$1000 is invested at an interest rate of 12% per year. Find the amounts in the account after 3 years if interest is compounded annually, semiannually, quarterly, monthly, and daily.

SOLUTION We use the compound interest formula with $P = \$1000$, $r = 0.12$, and $t = 3$.

Compounding	n	Amount after 3 years
Annual	1	$1000\left(1 + \frac{0.12}{1}\right)^{1(3)} = \$1404.93$
Semiannual	2	$1000\left(1 + \frac{0.12}{2}\right)^{2(3)} = \$1418.52$
Quarterly	4	$1000\left(1 + \frac{0.12}{4}\right)^{4(3)} = \$1425.76$
Monthly	12	$1000\left(1 + \frac{0.12}{12}\right)^{12(3)} = \$1430.77$
Daily	365	$1000\left(1 + \frac{0.12}{365}\right)^{365(3)} = \$1433.24$

NOW TRY EXERCISE **51** ■

If an investment earns compound interest, then the **annual percentage yield** (APY) is the *simple* interest rate that yields the same amount at the end of one year.

EXAMPLE 7 | Calculating the Annual Percentage Yield

Find the annual percentage yield for an investment that earns interest at a rate of 6% per year, compounded daily.

SOLUTION After one year, a principal P will grow to the amount

$$A = P\left(1 + \frac{0.06}{365}\right)^{365} = P(1.06183)$$

The formula for simple interest is

Simple interest is studied in Section 1.6.

$$A = P(1 + r)$$

Comparing, we see that $1 + r = 1.06183$, so $r = 0.06183$. Thus, the annual percentage yield is 6.183%.

NOW TRY EXERCISE **57** ■

4.1 EXERCISES

CONCEPTS

1. The function $f(x) = 5^x$ is an exponential function with base ________; $f(-2) =$ ________, $f(0) =$ ________, $f(2) =$ ________, and $f(6) =$ ________.

2. Match the exponential function with its graph.

(a) $f(x) = 2^x$

(b) $f(x) = 2^{-x}$

(c) $f(x) = -2^x$

(d) $f(x) = -2^{-x}$

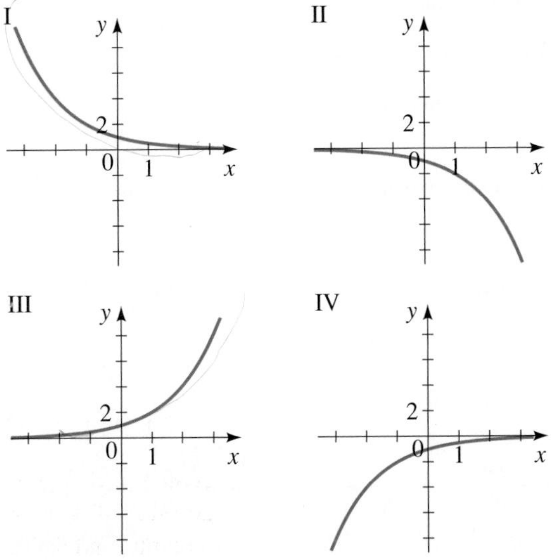

3. (a) To obtain the graph of $g(x) = 2^x - 1$, we start with the graph of $f(x) = 2^x$ and shift it __________ (upward/downward) 1 unit.

(b) To obtain the graph of $h(x) = 2^{x-1}$, we start with the graph of $f(x) = 2^x$ and shift it to the __________ (left/right) 1 unit.

4. In the formula $A(t) = P(1 + \frac{r}{n})^{nt}$ for compound interest the letters P, r, n, and t stand for __________, __________, __________, and __________, respectively, and $A(t)$ stands for __________. So if \$100 is invested at an interest rate of 6% compounded quarterly, then the amount after 2 years is __________.

SKILLS

5–10 ■ Use a calculator to evaluate the function at the indicated values. Round your answers to three decimals.

5. $f(x) = 4^x$; $f(0.5), f(\sqrt{2}), f(-\pi), f(\frac{1}{3})$

6. $f(x) = 3^{x+1}$; $f(-1.5), f(\sqrt{3}), f(e), f(-\frac{5}{4})$

7. $g(x) = (\frac{2}{3})^{x-1}$; $g(1.3), g(\sqrt{5}), g(2\pi), g(-\frac{1}{2})$

8. $g(x) = (\frac{3}{4})^{2x}$; $g(0.7), g(\sqrt{7}/2), g(1/\pi), g(\frac{2}{3})$

9–14 ■ Sketch the graph of the function by making a table of values. Use a calculator if necessary.

9. $f(x) = 2^x$

10. $g(x) = 8^x$

11. $f(x) = (\frac{1}{3})^x$

12. $h(x) = (1.1)^x$

13. $g(x) = 3(1.3)^x$

14. $h(x) = 2(\frac{1}{4})^x$

15–18 ■ Graph both functions on one set of axes.

15. $f(x) = 2^x$ and $g(x) = 2^{-x}$

16. $f(x) = 3^{-x}$ and $g(x) = (\frac{1}{3})^x$

17. $f(x) = 4^x$ and $g(x) = 7^x$

18. $f(x) = (\frac{2}{3})^x$ and $g(x) = (\frac{4}{3})^x$

19–22 ■ Find the exponential function $f(x) = a^x$ whose graph is given.

19.

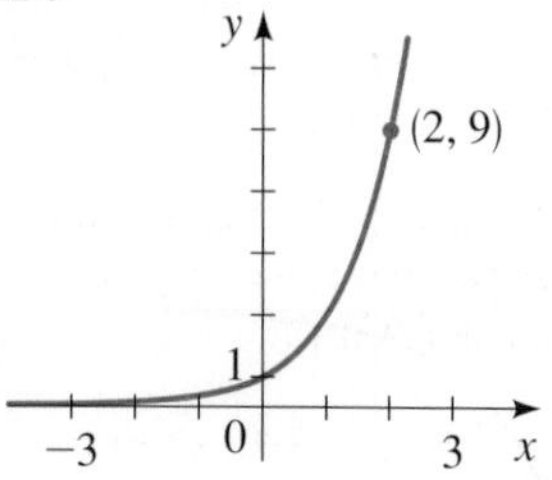

20.

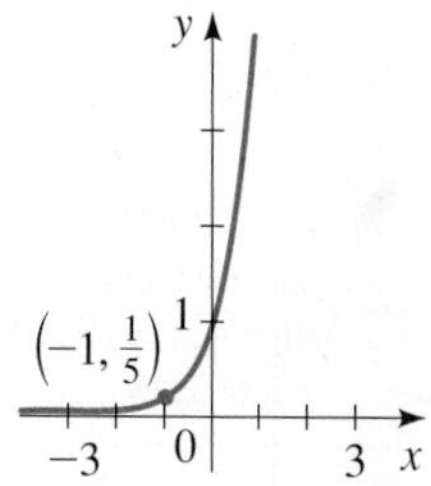

21.

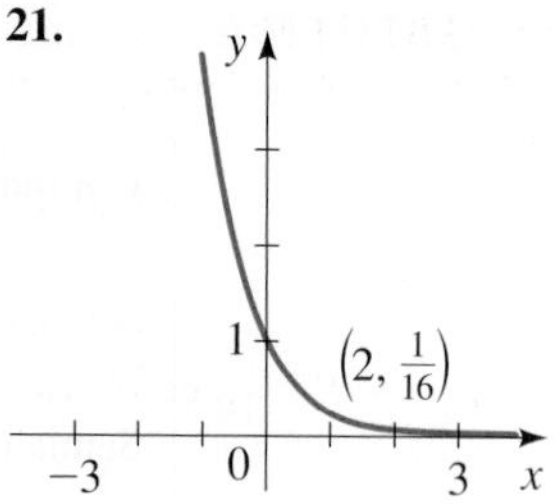

22.

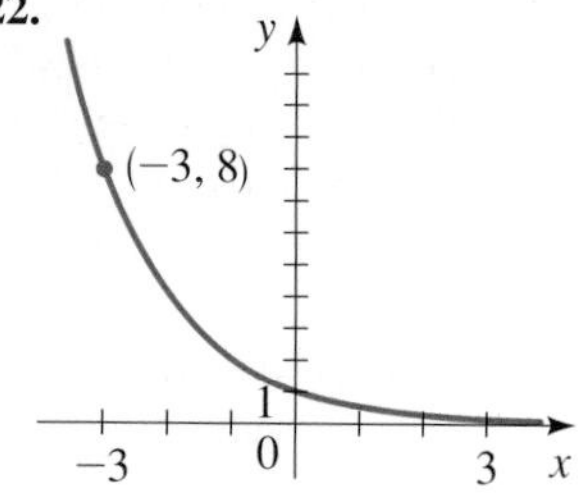

23–24 ■ Match the exponential function with one of the graphs labeled I or II.

23. $f(x) = 5^{x+1}$

24. $f(x) = 5^x + 1$

I

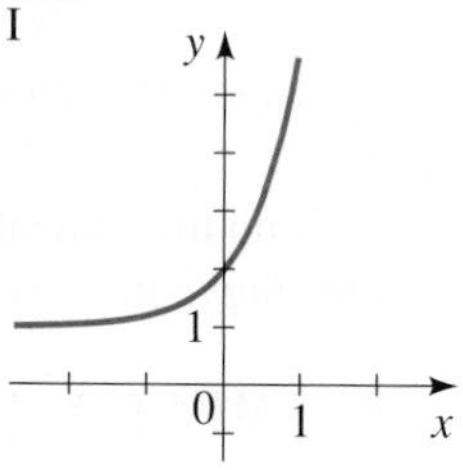

II

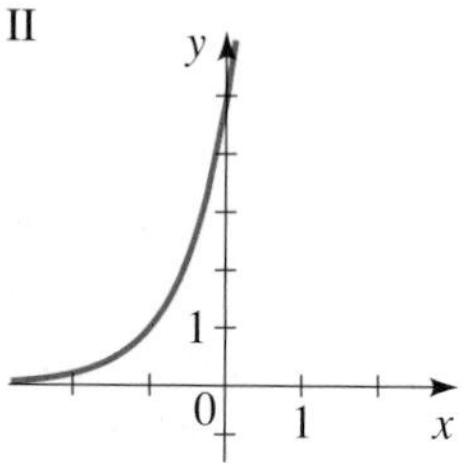

25–36 ■ Graph the function, not by plotting points, but by starting from the graphs in Figure 2. State the domain, range, and asymptote.

25. $f(x) = -3^x$

26. $f(x) = 10^{-x}$

27. $g(x) = 2^x - 3$

28. $g(x) = 2^{x-3}$

29. $h(x) = 4 + (\frac{1}{2})^x$

30. $h(x) = 6 - 3^x$

31. $f(x) = 10^{x+3}$

32. $f(x) = -(\frac{1}{5})^x$

33. $y = 5^{-x} + 1$

34. $g(x) = 1 - 3^{-x}$

35. $y = 3 - 10^{x-1}$

36. $h(x) = 2^{x-4} + 1$

37. (a) Sketch the graphs of $f(x) = 2^x$ and $g(x) = 3(2^x)$.
(b) How are the graphs related?

38. (a) Sketch the graphs of $f(x) = 9^{x/2}$ and $g(x) = 3^x$.
(b) Use the Laws of Exponents to explain the relationship between these graphs.

39. Compare the functions $f(x) = x^3$ and $g(x) = 3^x$ by evaluating both of them for $x = 0, 1, 2, 3, 4, 5, 6, 7, 8, 9, 10, 15$, and 20. Then draw the graphs of f and g on the same set of axes.

40. If $f(x) = 10^x$, show that $\dfrac{f(x + h) - f(x)}{h} = 10^x\left(\dfrac{10^h - 1}{h}\right)$.

41. (a) Compare the rates of growth of the functions $f(x) = 2^x$ and $g(x) = x^5$ by drawing the graphs of both functions in the following viewing rectangles.
(i) $[0, 5]$ by $[0, 20]$
(ii) $[0, 25]$ by $[0, 10^7]$
(iii) $[0, 50]$ by $[0, 10^8]$

(b) Find the solutions of the equation $2^x = x^5$, rounded to one decimal place.

42. **(a)** Compare the rates of growth of the functions $f(x) = 3^x$ and $g(x) = x^4$ by drawing the graphs of both functions in the following viewing rectangles:
(i) $[-4, 4]$ by $[0, 20]$
(ii) $[0, 10]$ by $[0, 5000]$
(iii) $[0, 20]$ by $[0, 10^5]$

(b) Find the solutions of the equation $3^x = x^4$, rounded to two decimal places.

43–44 ■ Draw graphs of the given family of functions for $c = 0.25, 0.5, 1, 2, 4$. How are the graphs related?

43. $f(x) = c2^x$ **44.** $f(x) = 2^{cx}$

45–46 ■ Find, rounded to two decimal places, **(a)** the intervals on which the function is increasing or decreasing and **(b)** the range of the function.

45. $y = 10^{x-x^2}$ **46.** $y = x2^x$

APPLICATIONS

47. Bacteria Growth A bacteria culture contains 1500 bacteria initially and doubles every hour.
(a) Find a function that models the number of bacteria after t hours.
(b) Find the number of bacteria after 24 hours.

48. Mouse Population A certain breed of mouse was introduced onto a small island with an initial population of 320 mice, and scientists estimate that the mouse population is doubling every year.
(a) Find a function that models the number of mice after t years.
(b) Estimate the mouse population after 8 years.

49–50 ■ **Compound Interest** An investment of \$5000 is deposited into an account in which interest is compounded monthly. Complete the table by filling in the amounts to which the investment grows at the indicated times or interest rates.

49. $r = 4\%$

Time (years)	Amount
1	
2	
3	
4	
5	
6	

50. $t = 5$ years

Rate per year	Amount
1%	
2%	
3%	
4%	
5%	
6%	

51. Compound Interest If \$10,000 is invested at an interest rate of 3% per year, compounded semiannually, find the value of the investment after the given number of years.
(a) 5 years **(b)** 10 years **(c)** 15 years

52. Compound Interest If \$2500 is invested at an interest rate of 2.5% per year, compounded daily, find the value of the investment after the given number of years.
(a) 2 years **(b)** 3 years **(c)** 6 years

53. Compound Interest If \$500 is invested at an interest rate of 3.75% per year, compounded quarterly, find the value of the investment after the given number of years.
(a) 1 year **(b)** 2 years **(c)** 10 years

54. Compound Interest If \$4000 is borrowed at a rate of 5.75% interest per year, compounded quarterly, find the amount due at the end of the given number of years.
(a) 4 years **(b)** 6 years **(c)** 8 years

55–56 ■ **Present Value** The **present value** of a sum of money is the amount that must be invested now, at a given rate of interest, to produce the desired sum at a later date.

55. Find the present value of \$10,000 if interest is paid at a rate of 9% per year, compounded semiannually, for 3 years.

56. Find the present value of \$100,000 if interest is paid at a rate of 8% per year, compounded monthly, for 5 years.

57. Annual Percentage Yield Find the annual percentage yield for an investment that earns 8% per year, compounded monthly.

58. Annual Percentage Yield Find the annual percentage yield for an investment that earns $5\frac{1}{2}\%$ per year, compounded quarterly.

DISCOVERY ■ DISCUSSION ■ WRITING

59. Growth of an Exponential Function Suppose you are offered a job that lasts one month, and you are to be very well paid. Which of the following methods of payment is more profitable for you?
(a) One million dollars at the end of the month
(b) Two cents on the first day of the month, 4 cents on the second day, 8 cents on the third day, and, in general, 2^n cents on the nth day

60. The Height of the Graph of an Exponential Function Your mathematics instructor asks you to sketch a graph of the exponential function

$$f(x) = 2^x$$

for x between 0 and 40, using a scale of 10 units to one inch. What are the dimensions of the sheet of paper you will need to sketch this graph?

DISCOVERY PROJECT **Exponential Explosion**

In this project we explore an example about collecting pennies that helps us experience how exponential growth works. You can find the project at the book companion website: **www.stewartmath.com**

4.2 The Natural Exponential Function

The Number e ▶ The Natural Exponential Function ▶ Continuously Compounded Interest

Any positive number can be used as a base for an exponential function. In this section we study the special base e, which is convenient for applications involving calculus.

▼ The Number e

The number e is defined as the value that $(1 + 1/n)^n$ approaches as n becomes large. (In calculus this idea is made more precise through the concept of a limit. See Chapter 13.) The table shows the values of the expression $(1 + 1/n)^n$ for increasingly large values of n.

n	$\left(1 + \frac{1}{n}\right)^n$
1	2.00000
5	2.48832
10	2.59374
100	2.70481
1000	2.71692
10,000	2.71815
100,000	2.71827
1,000,000	2.71828

The **Gateway Arch** in St. Louis, Missouri, is shaped in the form of the graph of a combination of exponential functions (*not* a parabola, as it might first appear). Specifically, it is a **catenary**, which is the graph of an equation of the form

$$y = a(e^{bx} + e^{-bx})$$

(see Exercise 17). This shape was chosen because it is optimal for distributing the internal structural forces of the arch. Chains and cables suspended between two points (for example, the stretches of cable between pairs of telephone poles) hang in the shape of a catenary.

It appears that, correct to five decimal places, $e \approx 2.71828$; in fact, the approximate value to 20 decimal places is

$$e \approx 2.71828182845904523536$$

It can be shown that e is an irrational number, so we cannot write its exact value in decimal form.

▼ The Natural Exponential Function

The notation e was chosen by Leonhard Euler (see page 266), probably because it is the first letter of the word *exponential*.

The number e is the base for the natural exponential function. Why use such a strange base for an exponential function? It might seem at first that a base such as 10 is easier to work with. We will see, however, that in certain applications the number e is the best possible base. In this section we study how e occurs in the description of compound interest.

THE NATURAL EXPONENTIAL FUNCTION

The **natural exponential function** is the exponential function

$$f(x) = e^x$$

with base e. It is often referred to as *the* exponential function.

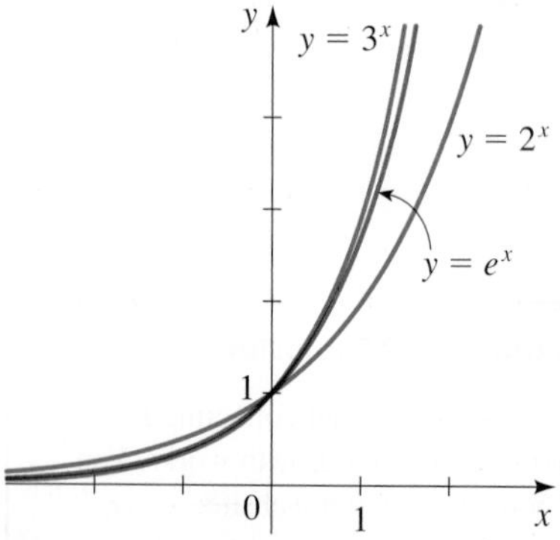

FIGURE 1 Graph of the natural exponential function

Since $2 < e < 3$, the graph of the natural exponential function lies between the graphs of $y = 2^x$ and $y = 3^x$, as shown in Figure 1.

Scientific calculators have a special key for the function $f(x) = e^x$. We use this key in the next example.

EXAMPLE 1 | Evaluating the Exponential Function

Evaluate each expression rounded to five decimal places.

(a) e^3 **(b)** $2e^{-0.53}$ **(c)** $e^{4.8}$

SOLUTION We use the $\boxed{e^x}$ key on a calculator to evaluate the exponential function.

(a) $e^3 \approx 20.08554$ **(b)** $2e^{-0.53} \approx 1.17721$ **(c)** $e^{4.8} \approx 121.51042$

NOW TRY EXERCISE **3**

EXAMPLE 2 | Transformations of the Exponential Function

Sketch the graph of each function.

(a) $f(x) = e^{-x}$ **(b)** $g(x) = 3e^{0.5x}$

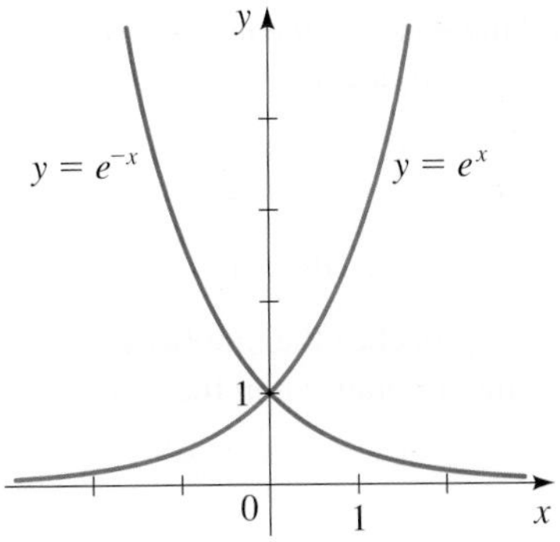

FIGURE 2

SOLUTION

(a) We start with the graph of $y = e^x$ and reflect in the y-axis to obtain the graph of $y = e^{-x}$ as in Figure 2.

(b) We calculate several values, plot the resulting points, then connect the points with a smooth curve. The graph is shown in Figure 3.

x	$f(x) = 3e^{0.5x}$
−3	0.67
−2	1.10
−1	1.82
0	3.00
1	4.95
2	8.15
3	13.45

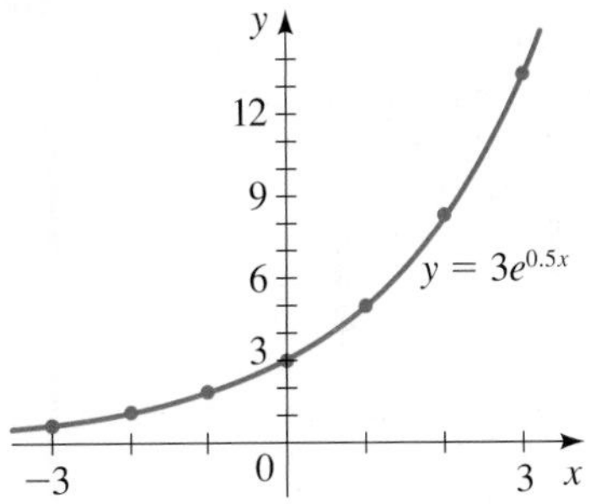

FIGURE 3

NOW TRY EXERCISES **5** AND **7**

EXAMPLE 3 | An Exponential Model for the Spread of a Virus

An infectious disease begins to spread in a small city of population 10,000. After t days, the number of people who have succumbed to the virus is modeled by the function

$$v(t) = \frac{10{,}000}{5 + 1245e^{-0.97t}}$$

(a) How many infected people are there initially (at time $t = 0$)?

(b) Find the number of infected people after one day, two days, and five days.

(c) Graph the function v, and describe its behavior.

SOLUTION

(a) Since $v(0) = 10{,}000/(5 + 1245e^0) = 10{,}000/1250 = 8$, we conclude that 8 people initially have the disease.

(b) Using a calculator, we evaluate $v(1)$, $v(2)$, and $v(5)$ and then round off to obtain the following values.

Days	Infected people
1	21
2	54
5	678

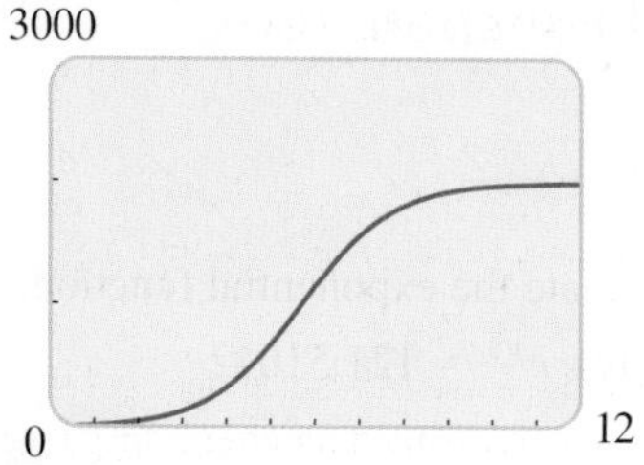

FIGURE 4

$$v(t) = \frac{10{,}000}{5 + 1245e^{-0.97t}}$$

(c) From the graph in Figure 4 we see that the number of infected people first rises slowly, then rises quickly between day 3 and day 8, and then levels off when about 2000 people are infected.

NOW TRY EXERCISE **25** ■

The graph in Figure 4 is called a *logistic curve* or a *logistic growth model*. Curves like it occur frequently in the study of population growth. (See Exercises 25–28.)

▼ Continuously Compounded Interest

In Example 6 of Section 4.1 we saw that the interest paid increases as the number of compounding periods n increases. Let's see what happens as n increases indefinitely. If we let $m = n/r$, then

$$A(t) = P\left(1 + \frac{r}{n}\right)^{nt} = P\left[\left(1 + \frac{r}{n}\right)^{n/r}\right]^{rt} = P\left[\left(1 + \frac{1}{m}\right)^{m}\right]^{rt}$$

Recall that as m becomes large, the quantity $(1 + 1/m)^m$ approaches the number e. Thus, the amount approaches $A = Pe^{rt}$. This expression gives the amount when the interest is compounded at "every instant."

CONTINUOUSLY COMPOUNDED INTEREST

Continuously compounded interest is calculated by the formula

$$A(t) = Pe^{rt}$$

where $A(t)$ = amount after t years
P = principal
r = interest rate per year
t = number of years

EXAMPLE 4 | Calculating Continuously Compounded Interest

Find the amount after 3 years if \$1000 is invested at an interest rate of 12% per year, compounded continuously.

SOLUTION We use the formula for continuously compounded interest with $P = \$1000$, $r = 0.12$, and $t = 3$ to get

$$A(3) = 1000e^{(0.12)3} = 1000e^{0.36} = \$1433.33$$

Compare this amount with the amounts in Example 6 of Section 4.1.

NOW TRY EXERCISE **31** ■

4.2 EXERCISES

CONCEPTS

1. The function $f(x) = e^x$ is called the ______ exponential function. The number e is approximately equal to ______.

2. In the formula $A(t) = Pe^{rt}$ for continuously compound interest, the letters P, r, and t stand for ______, ______, and ______, respectively, and $A(t)$ stands for ______. So if \$100 is invested at an interest rate of 6% compounded continuously, then the amount after 2 years is ______.

SKILLS

3–4 ■ Use a calculator to evaluate the function at the indicated values. Round your answers to three decimals.

3. $h(x) = e^x$; $h(3), h(0.23), h(1), h(-2)$

4. $h(x) = e^{-2x}$; $h(1), h(\sqrt{2}), h(-3), h(\frac{1}{2})$

5–6 ■ Complete the table of values, rounded to two decimal places, and sketch a graph of the function.

5.

x	$f(x) = 3e^x$
-2	
-1	
-0.5	
0	
0.5	
1	
2	

6.

x	$f(x) = 2e^{-0.5x}$
-3	
-2	
-1	
0	
1	
2	
3	

7–14 ■ Graph the function, not by plotting points, but by starting from the graph of $y = e^x$ in Figure 1. State the domain, range, and asymptote.

7. $f(x) = -e^x$

8. $y = 1 - e^x$

9. $y = e^{-x} - 1$

10. $f(x) = -e^{-x}$

11. $f(x) = e^{x-2}$

12. $y = e^{x-3} + 4$

13. $h(x) = e^{x+1} - 3$

14. $g(x) = -e^{x-1} - 2$

15. The *hyperbolic cosine function* is defined by

$$\cosh(x) = \frac{e^x + e^{-x}}{2}$$

(a) Sketch the graphs of the functions $y = \frac{1}{2}e^x$ and $y = \frac{1}{2}e^{-x}$ on the same axes, and use graphical addition (see Section 2.6) to sketch the graph of $y = \cosh(x)$.

(b) Use the definition to show that $\cosh(-x) = \cosh(x)$.

16. The *hyperbolic sine function* is defined by

$$\sinh(x) = \frac{e^x - e^{-x}}{2}$$

(a) Sketch the graph of this function using graphical addition as in Exercise 15.

(b) Use the definition to show that $\sinh(-x) = -\sinh(x)$

17. **(a)** Draw the graphs of the family of functions

$$f(x) = \frac{a}{2}(e^{x/a} + e^{-x/a})$$

for $a = 0.5$, 1, 1.5, and 2.

(b) How does a larger value of a affect the graph?

18–19 ■ Find the local maximum and minimum values of the function and the value of x at which each occurs. State each answer correct to two decimal places.

18. $g(x) = x^x \quad (x > 0)$

19. $g(x) = e^x + e^{-3x}$

APPLICATIONS

20. Medical Drugs When a certain medical drug is administered to a patient, the number of milligrams remaining in the patient's bloodstream after t hours is modeled by

$$D(t) = 50e^{-0.2t}$$

How many milligrams of the drug remain in the patient's bloodstream after 3 hours?

21. Radioactive Decay A radioactive substance decays in such a way that the amount of mass remaining after t days is given by the function

$$m(t) = 13e^{-0.015t}$$

where $m(t)$ is measured in kilograms.

(a) Find the mass at time $t = 0$.

(b) How much of the mass remains after 45 days?

22. Radioactive Decay Doctors use radioactive iodine as a tracer in diagnosing certain thyroid gland disorders. This type of iodine decays in such a way that the mass remaining after t days is given by the function

$$m(t) = 6e^{-0.087t}$$

where $m(t)$ is measured in grams.

(a) Find the mass at time $t = 0$.

(b) How much of the mass remains after 20 days?

23. Sky Diving A sky diver jumps from a reasonable height above the ground. The air resistance she experiences is proportional to her velocity, and the constant of proportionality is 0.2. It can be shown that the downward velocity of the sky diver at time t is given by

$$v(t) = 80(1 - e^{-0.2t})$$

where t is measured in seconds and $v(t)$ is measured in feet per second (ft/s).

(a) Find the initial velocity of the sky diver.

(b) Find the velocity after 5 s and after 10 s.

(c) Draw a graph of the velocity function $v(t)$.

(d) The maximum velocity of a falling object with wind resistance is called its *terminal velocity*. From the graph in part (c) find the terminal velocity of this sky diver.

24. Mixtures and Concentrations A 50-gallon barrel is filled completely with pure water. Salt water with a concentration of 0.3 lb/gal is then pumped into the barrel, and the resulting mixture overflows at the same rate. The amount of salt in the barrel at time t is given by

$$Q(t) = 15(1 - e^{-0.04t})$$

where t is measured in minutes and $Q(t)$ is measured in pounds.

(a) How much salt is in the barrel after 5 min?

(b) How much salt is in the barrel after 10 min?

(c) Draw a graph of the function $Q(t)$.

(d) Use the graph in part (c) to determine the value that the amount of salt in the barrel approaches as t becomes large. Is this what you would expect?

$$Q(t) = 15(1 - e^{-0.04t})$$

25. Logistic Growth Animal populations are not capable of unrestricted growth because of limited habitat and food supplies. Under such conditions the population follows a *logistic growth model*:

$$P(t) = \frac{d}{1 + ke^{-ct}}$$

where c, d, and k are positive constants. For a certain fish population in a small pond $d = 1200$, $k = 11$, $c = 0.2$, and t is measured in years. The fish were introduced into the pond at time $t = 0$.

(a) How many fish were originally put in the pond?
(b) Find the population after 10, 20, and 30 years.
(c) Evaluate $P(t)$ for large values of t. What value does the population approach as $t \to \infty$? Does the graph shown confirm your calculations?

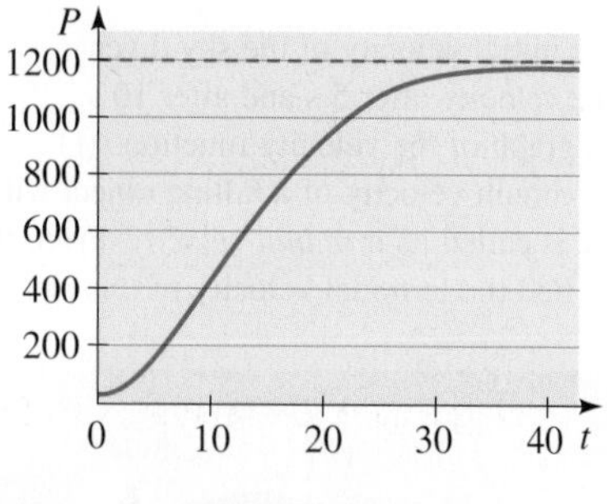

26. Bird Population The population of a certain species of bird is limited by the type of habitat required for nesting. The population behaves according to the logistic growth model

$$n(t) = \frac{5600}{0.5 + 27.5e^{-0.044t}}$$

where t is measured in years.

(a) Find the initial bird population.
(b) Draw a graph of the function $n(t)$.
(c) What size does the population approach as time goes on?

27. World Population The relative growth rate of world population has been decreasing steadily in recent years. On the basis of this, some population models predict that world population will eventually stabilize at a level that the planet can support. One such logistic model is

$$P(t) = \frac{73.2}{6.1 + 5.9e^{-0.02t}}$$

where $t = 0$ is the year 2000 and population is measured in billions.

(a) What world population does this model predict for the year 2200? For 2300?
(b) Sketch a graph of the function P for the years 2000 to 2500.
(c) According to this model, what size does the world population seem to approach as time goes on?

28. Tree Diameter For a certain type of tree the diameter D (in feet) depends on the tree's age t (in years) according to the logistic growth model

$$D(t) = \frac{5.4}{1 + 2.9e^{-0.01t}}$$

Find the diameter of a 20-year-old tree.

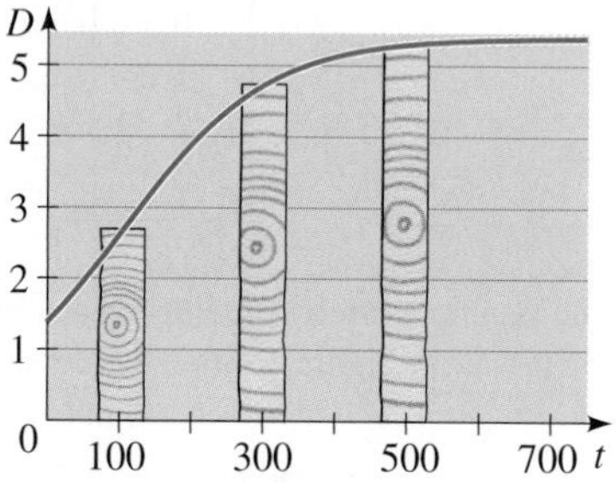

29–30 ■ Compound Interest An investment of \$7,000 is deposited into an account in which interest is compounded continuously. Complete the table by filling in the amounts to which the investment grows at the indicated times or interest rates.

29. $r = 3\%$

Time (years)	Amount
1	
2	
3	
4	
5	
6	

30. $t = 10$ years

Rate per year	Amount
1%	
2%	
3%	
4%	
5%	
6%	

31. Compound Interest If \$2000 is invested at an interest rate of 3.5% per year, compounded continuously, find the value of the investment after the given number of years.
(a) 2 years **(b)** 4 years **(c)** 12 years

32. Compound Interest If \$3500 is invested at an interest rate of 6.25% per year, compounded continuously, find the value of the investment after the given number of years.
(a) 3 years **(b)** 6 years **(c)** 9 years

33. Compound Interest If \$600 is invested at an interest rate of 2.5% per year, find the amount of the investment at the end of 10 years for the following compounding methods.
(a) Annually **(b)** Semiannually
(c) Quarterly **(d)** Continuously

34. Compound Interest If \$8000 is invested in an account for which interest is compounded continuously, find the amount of the investment at the end of 12 years for the following interest rates.
(a) 2% **(b)** 3% **(c)** 4.5% **(d)** 7%

35. Compound Interest Which of the given interest rates and compounding periods would provide the best investment?
(a) $2\frac{1}{2}\%$ per year, compounded semiannually
(b) $2\frac{1}{4}\%$ per year, compounded monthly
(c) 2% per year, compounded continuously

36. Compound Interest Which of the given interest rates and compounding periods would provide the better investment?
(a) $5\frac{1}{8}\%$ per year, compounded semiannually
(b) 5% per year, compounded continuously

37. Investment A sum of \$5000 is invested at an interest rate of 9% per year, compounded continuously.
(a) Find the value $A(t)$ of the investment after t years.
(b) Draw a graph of $A(t)$.
(c) Use the graph of $A(t)$ to determine when this investment will amount to \$25,000.

DISCOVERY ▪ DISCUSSION ▪ WRITING

38. The Definition of *e* Illustrate the definition of the number e by graphing the curve $y = (1 + 1/x)^x$ and the line $y = e$ on the same screen, using the viewing rectangle $[0, 40]$ by $[0, 4]$.

4.3 LOGARITHMIC FUNCTIONS

Logarithmic Functions ▶ Graphs of Logarithmic Functions ▶ Common Logarithms ▶ Natural Logarithms

In this section we study the inverses of exponential functions.

▼ Logarithmic Functions

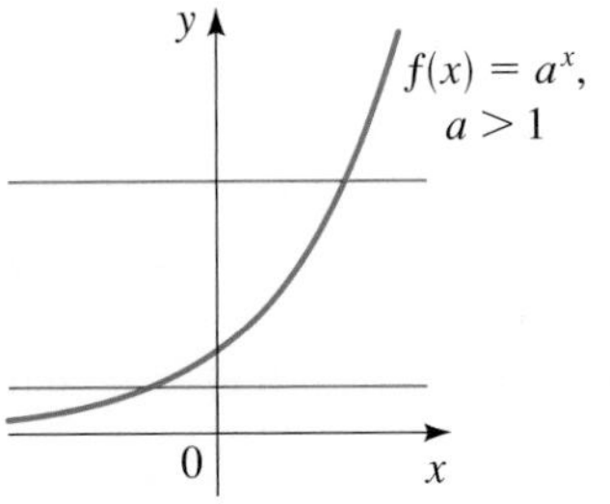

FIGURE 1 $f(x) = a^x$ is one-to-one.

Every exponential function $f(x) = a^x$, with $a > 0$ and $a \neq 1$, is a one-to-one function by the Horizontal Line Test (see Figure 1 for the case $a > 1$) and therefore has an inverse function. The inverse function f^{-1} is called the *logarithmic function with base a* and is denoted by $\log_a$. Recall from Section 2.6 that f^{-1} is defined by

$$f^{-1}(x) = y \quad \Leftrightarrow \quad f(y) = x$$

This leads to the following definition of the logarithmic function.

DEFINITION OF THE LOGARITHMIC FUNCTION

Let a be a positive number with $a \neq 1$. The **logarithmic function with base *a***, denoted by $\log_a$, is defined by

$$\log_a x = y \quad \Leftrightarrow \quad a^y = x$$

So $\log_a x$ is the *exponent* to which the base a must be raised to give x.

We read $\log_a x = y$ as "log base a of x is y."

When we use the definition of logarithms to switch back and forth between the **logarithmic form** $\log_a x = y$ and the **exponential form** $a^y = x$, it is helpful to notice that, in both forms, the base is the same:

By tradition the name of the logarithmic function is $\log_a$, not just a single letter. Also, we usually omit the parentheses in the function notation and write

$$\log_a(x) = \log_a x$$

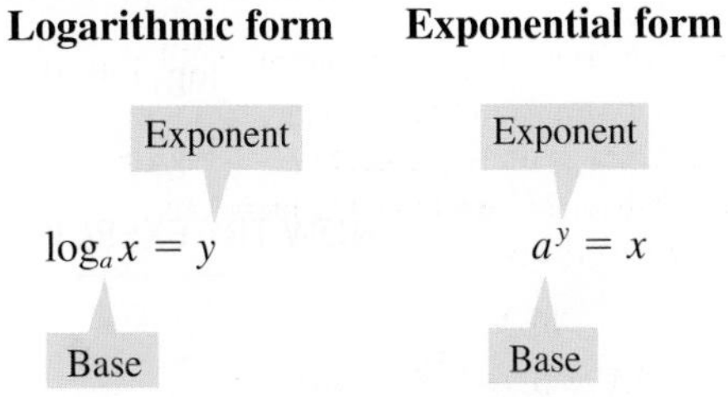

EXAMPLE 1 | Logarithmic and Exponential Forms

The logarithmic and exponential forms are equivalent equations: If one is true, then so is the other. So we can switch from one form to the other as in the following illustrations.

Logarithmic form	Exponential form
$\log_{10} 100{,}000 = 5$	$10^5 = 100{,}000$
$\log_2 8 = 3$	$2^3 = 8$
$\log_2\left(\frac{1}{8}\right) = -3$	$2^{-3} = \frac{1}{8}$
$\log_5 s = r$	$5^r = s$

NOW TRY EXERCISE **5**

x	$\log_{10} x$
10^4	4
10^3	3
10^2	2
10	1
1	0
10^{-1}	-1
10^{-2}	-2
10^{-3}	-3
10^{-4}	-4

It is important to understand that $\log_a x$ is an *exponent*. For example, the numbers in the right column of the table in the margin are the logarithms (base 10) of the numbers in the left column. This is the case for all bases, as the following example illustrates.

EXAMPLE 2 | Evaluating Logarithms

(a) $\log_{10} 1000 = 3$ because $10^3 = 1000$

(b) $\log_2 32 = 5$ because $2^5 = 32$

(c) $\log_{10} 0.1 = -1$ because $10^{-1} = 0.1$

(d) $\log_{16} 4 = \frac{1}{2}$ because $16^{1/2} = 4$

NOW TRY EXERCISES **7** AND **9**

Inverse Function Property:

$$f^{-1}(f(x)) = x$$

$$f(f^{-1}(x)) = x$$

When we apply the Inverse Function Property described on page 201 to $f(x) = a^x$ and $f^{-1}(x) = \log_a x$, we get

$$\log_a(a^x) = x, \qquad x \in \mathbb{R}$$

$$a^{\log_a x} = x, \qquad x > 0$$

We list these and other properties of logarithms discussed in this section.

PROPERTIES OF LOGARITHMS

Property	Reason
1. $\log_a 1 = 0$	We must raise a to the power 0 to get 1.
2. $\log_a a = 1$	We must raise a to the power 1 to get a.
3. $\log_a a^x = x$	We must raise a to the power x to get a^x.
4. $a^{\log_a x} = x$	$\log_a x$ is the power to which a must be raised to get x.

EXAMPLE 3 | Applying Properties of Logarithms

We illustrate the properties of logarithms when the base is 5.

$\log_5 1 = 0$	Property 1	$\log_5 5 = 1$	Property 2
$\log_5 5^8 = 8$	Property 3	$5^{\log_5 12} = 12$	Property 4

NOW TRY EXERCISES **19** AND **25**

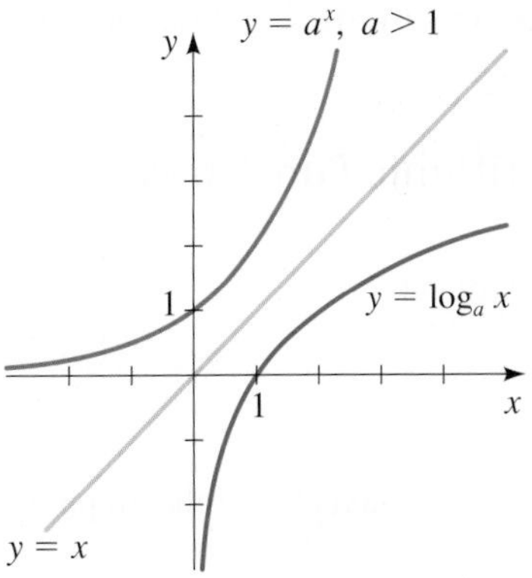

FIGURE 2 Graph of the logarithmic function $f(x) = \log_a x$

Graphs of Logarithmic Functions

Recall that if a one-to-one function f has domain A and range B, then its inverse function f^{-1} has domain B and range A. Since the exponential function $f(x) = a^x$ with $a \neq 1$ has domain $\mathbb{R}$ and range $(0, \infty)$, we conclude that its inverse function, $f^{-1}(x) = \log_a x$, has domain $(0, \infty)$ and range $\mathbb{R}$.

The graph of $f^{-1}(x) = \log_a x$ is obtained by reflecting the graph of $f(x) = a^x$ in the line $y = x$. Figure 2 shows the case $a > 1$. The fact that $y = a^x$ (for $a > 1$) is a very rapidly increasing function for $x > 0$ implies that $y = \log_a x$ is a very slowly increasing function for $x > 1$ (see Exercise 92).

Since $\log_a 1 = 0$, the x-intercept of the function $y = \log_a x$ is 1. The y-axis is a vertical asymptote of $y = \log_a x$ because $\log_a x \to -\infty$ as $x \to 0^+$.

EXAMPLE 4 | Graphing a Logarithmic Function by Plotting Points

Sketch the graph of $f(x) = \log_2 x$.

SOLUTION To make a table of values, we choose the x-values to be powers of 2 so that we can easily find their logarithms. We plot these points and connect them with a smooth curve as in Figure 3.

x	$\log_2 x$
2^3	3
2^2	2
2	1
1	0
2^{-1}	-1
2^{-2}	-2
2^{-3}	-3
2^{-4}	-4

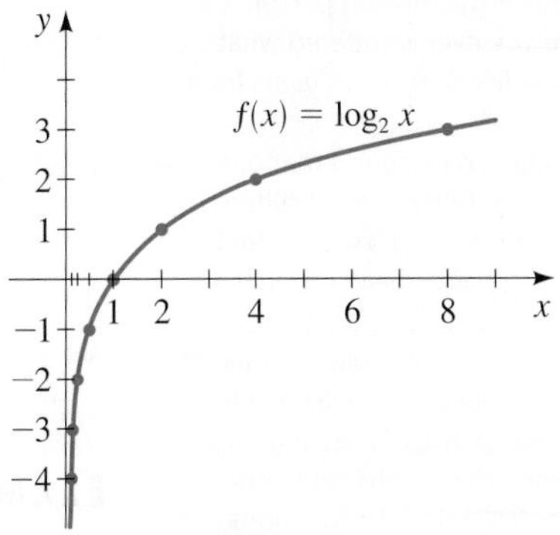

FIGURE 3

NOW TRY EXERCISE **41**

Figure 4 shows the graphs of the family of logarithmic functions with bases 2, 3, 5, and 10. These graphs are drawn by reflecting the graphs of $y = 2^x$, $y = 3^x$, $y = 5^x$, and $y = 10^x$ (see Figure 2 in Section 4.1) in the line $y = x$. We can also plot points as an aid to sketching these graphs, as illustrated in Example 4.

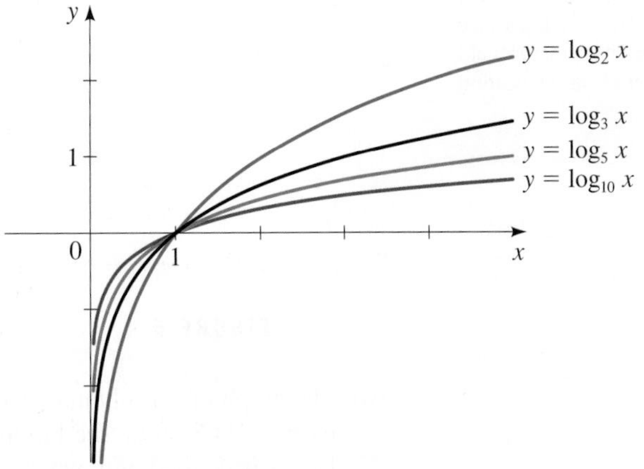

FIGURE 4 A family of logarithmic functions

MATHEMATICS IN THE MODERN WORLD

Law Enforcement

Mathematics aids law enforcement in numerous and surprising ways, from the reconstruction of bullet trajectories to determining the time of death to calculating the probability that a DNA sample is from a particular person. One interesting use is in the search for missing persons. A person who has been missing for several years might look quite different from his or her most recent available photograph. This is particularly true if the missing person is a child. Have you ever wondered what you will look like 5, 10, or 15 years from now?

Researchers have found that different parts of the body grow at different rates. For example, you have no doubt noticed that a baby's head is much larger relative to its body than an adult's. As another example, the ratio of arm length to height is $\frac{1}{3}$ in a child but about $\frac{2}{5}$ in an adult. By collecting data and analyzing the graphs, researchers are able to determine the functions that model growth. As in all growth phenomena, exponential and logarithmic functions play a crucial role. For instance, the formula that relates arm length l to height h is $l = ae^{kh}$ where a and k are constants. By studying various physical characteristics of a person, mathematical biologists model each characteristic by a function that describes how it changes over time. Models of facial characteristics can be programmed into a computer to give a picture of how a person's appearance changes over time. These pictures aid law enforcement agencies in locating missing persons.

In the next two examples we graph logarithmic functions by starting with the basic graphs in Figure 4 and using the transformations of Section 2.5.

EXAMPLE 5 | Reflecting Graphs of Logarithmic Functions

Sketch the graph of each function.

(a) $g(x) = -\log_2 x$

(b) $h(x) = \log_2(-x)$

SOLUTION

(a) We start with the graph of $f(x) = \log_2 x$ and reflect in the x-axis to get the graph of $g(x) = -\log_2 x$ in Figure 5(a).

(b) We start with the graph of $f(x) = \log_2 x$ and reflect in the y-axis to get the graph of $h(x) = \log_2(-x)$ in Figure 5(b).

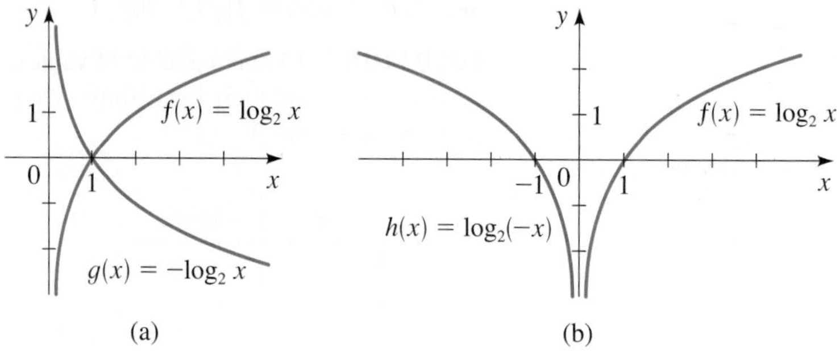

FIGURE 5

NOW TRY EXERCISE **55**

EXAMPLE 6 | Shifting Graphs of Logarithmic Functions

Find the domain of each function, and sketch the graph.

(a) $g(x) = 2 + \log_5 x$

(b) $h(x) = \log_{10}(x - 3)$

SOLUTION

(a) The graph of g is obtained from the graph of $f(x) = \log_5 x$ (Figure 4) by shifting upward 2 units (see Figure 6). The domain of f is $(0, \infty)$.

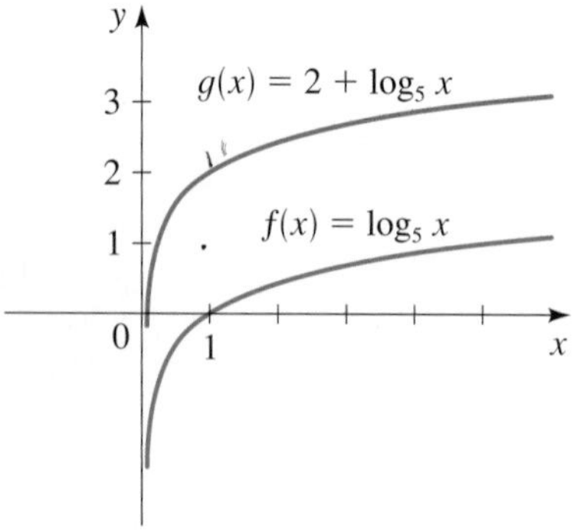

FIGURE 6

(b) The graph of h is obtained from the graph of $f(x) = \log_{10} x$ (Figure 4) by shifting to the right 3 units (see Figure 7). The line $x = 3$ is a vertical asymptote. Since $\log_{10} x$ is defined only when $x > 0$, the domain of $h(x) = \log_{10}(x - 3)$ is

$$\{x \mid x - 3 > 0\} = \{x \mid x > 3\} = (3, \infty)$$

Library of Congress

JOHN NAPIER (1550–1617) was a Scottish landowner for whom mathematics was a hobby. We know him today because of his key invention: logarithms, which he published in 1614 under the title *A Description of the Marvelous Rule of Logarithms*. In Napier's time, logarithms were used exclusively for simplifying complicated calculations. For example, to multiply two large numbers, we would write them as powers of 10. The exponents are simply the logarithms of the numbers. For instance,

$$\begin{aligned} &4532 \times 57783 \\ &\approx 10^{3.65629} \times 10^{4.76180} \\ &= 10^{8.41809} \\ &\approx 261{,}872{,}564 \end{aligned}$$

The idea is that multiplying powers of 10 is easy (we simply add their exponents). Napier produced extensive tables giving the logarithms (or exponents) of numbers. Since the advent of calculators and computers, logarithms are no longer used for this purpose. The logarithmic functions, however, have found many applications, some of which are described in this chapter.

Napier wrote on many topics. One of his most colorful works is a book entitled *A Plaine Discovery of the Whole Revelation of Saint John*, in which he predicted that the world would end in the year 1700.

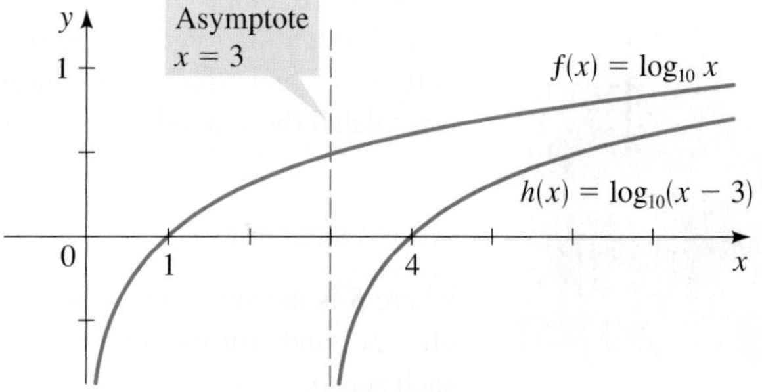

FIGURE 7

NOW TRY EXERCISES **53** AND **57** ■

▼ Common Logarithms

We now study logarithms with base 10.

> **COMMON LOGARITHM**
>
> The logarithm with base 10 is called the **common logarithm** and is denoted by omitting the base:
>
> $$\log x = \log_{10} x$$

From the definition of logarithms we can easily find that

$$\log 10 = 1 \qquad \text{and} \qquad \log 100 = 2$$

But how do we find log 50? We need to find the exponent y such that $10^y = 50$. Clearly, 1 is too small and 2 is too large. So

$$1 < \log 50 < 2$$

To get a better approximation, we can experiment to find a power of 10 closer to 50. Fortunately, scientific calculators are equipped with a LOG key that directly gives values of common logarithms.

EXAMPLE 7 | Evaluating Common Logarithms

Use a calculator to find appropriate values of $f(x) = \log x$ and use the values to sketch the graph.

SOLUTION We make a table of values, using a calculator to evaluate the function at those values of x that are not powers of 10. We plot those points and connect them by a smooth curve as in Figure 8.

x	$\log x$
0.01	−2
0.1	−1
0.5	−0.301
1	0
4	0.602
5	0.699
10	1

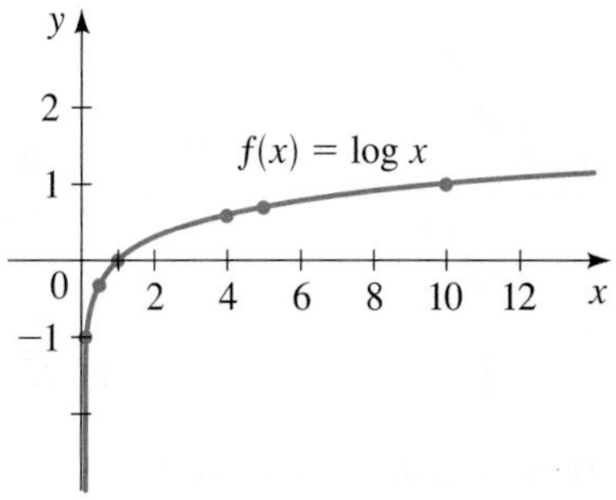

FIGURE 8

NOW TRY EXERCISE **43** ■

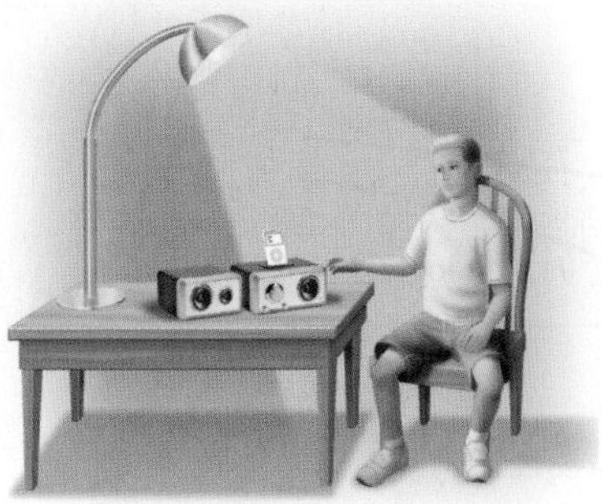

Human response to sound and light intensity is logarithmic.

Scientists model human response to stimuli (such as sound, light, or pressure) using logarithmic functions. For example, the intensity of a sound must be increased manyfold before we "feel" that the loudness has simply doubled. The psychologist Gustav Fechner formulated the law as

$$S = k \log\left(\frac{I}{I_0}\right)$$

where S is the subjective intensity of the stimulus, I is the physical intensity of the stimulus, I_0 stands for the threshold physical intensity, and k is a constant that is different for each sensory stimulus.

EXAMPLE 8 | Common Logarithms and Sound

We study the decibel scale in more detail in Section 4.6.

The perception of the loudness B (in decibels, dB) of a sound with physical intensity I (in W/m^2) is given by

$$B = 10 \log\left(\frac{I}{I_0}\right)$$

where I_0 is the physical intensity of a barely audible sound. Find the decibel level (loudness) of a sound whose physical intensity I is 100 times that of I_0.

SOLUTION We find the decibel level B by using the fact that $I = 100I_0$.

$$\begin{aligned} B &= 10 \log\left(\frac{I}{I_0}\right) && \text{Definition of } B \\ &= 10 \log\left(\frac{100I_0}{I_0}\right) && I = 100I_0 \\ &= 10 \log 100 && \text{Cancel } I_0 \\ &= 10 \cdot 2 = 20 && \text{Definition of log} \end{aligned}$$

The loudness of the sound is 20 dB.

NOW TRY EXERCISE **87**

Natural Logarithms

The notation ln is an abbreviation for the Latin name *logarithmus naturalis*.

Of all possible bases a for logarithms, it turns out that the most convenient choice for the purposes of calculus is the number e, which we defined in Section 4.2.

NATURAL LOGARITHM

The logarithm with base e is called the **natural logarithm** and is denoted by **ln**:

$$\ln x = \log_e x$$

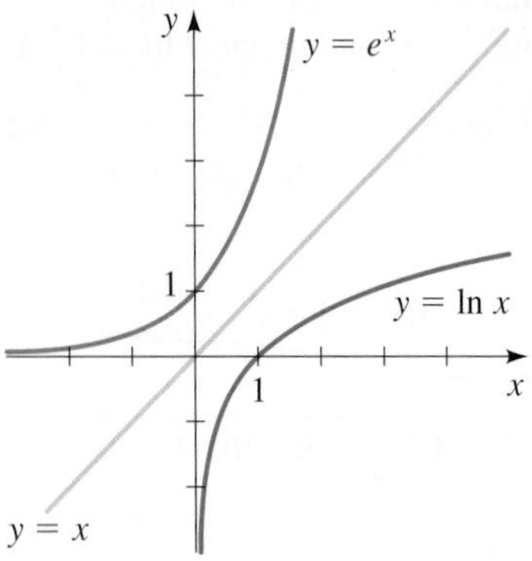

FIGURE 9 Graph of the natural logarithmic function

The natural logarithmic function $y = \ln x$ is the inverse function of the natural exponential function $y = e^x$. Both functions are graphed in Figure 9. By the definition of inverse functions we have

$$\ln x = y \quad \Leftrightarrow \quad e^y = x$$

If we substitute $a = e$ and write "ln" for "$\log_e$" in the properties of logarithms mentioned earlier, we obtain the following properties of natural logarithms.

PROPERTIES OF NATURAL LOGARITHMS

Property	Reason
1. $\ln 1 = 0$	We must raise e to the power 0 to get 1.
2. $\ln e = 1$	We must raise e to the power 1 to get e.
3. $\ln e^x = x$	We must raise e to the power x to get e^x.
4. $e^{\ln x} = x$	$\ln x$ is the power to which e must be raised to get x.

Calculators are equipped with an LN key that directly gives the values of natural logarithms.

EXAMPLE 9 | Evaluating the Natural Logarithm Function

(a) $\ln e^8 = 8$ — Definition of natural logarithm

(b) $\ln\left(\dfrac{1}{e^2}\right) = \ln e^{-2} = -2$ — Definition of natural logarithm

(c) $\ln 5 \approx 1.609$ — Use LN key on calculator

NOW TRY EXERCISE **39** ■

EXAMPLE 10 | Finding the Domain of a Logarithmic Function

Find the domain of the function $f(x) = \ln(4 - x^2)$.

SOLUTION As with any logarithmic function, $\ln x$ is defined when $x > 0$. Thus, the domain of f is

$$\{x \mid 4 - x^2 > 0\} = \{x \mid x^2 < 4\} = \{x \mid |x| < 2\}$$
$$= \{x \mid -2 < x < 2\} = (-2, 2)$$

NOW TRY EXERCISE **63** ■

EXAMPLE 11 | Drawing the Graph of a Logarithmic Function

Draw the graph of the function $y = x\ln(4 - x^2)$, and use it to find the asymptotes and local maximum and minimum values.

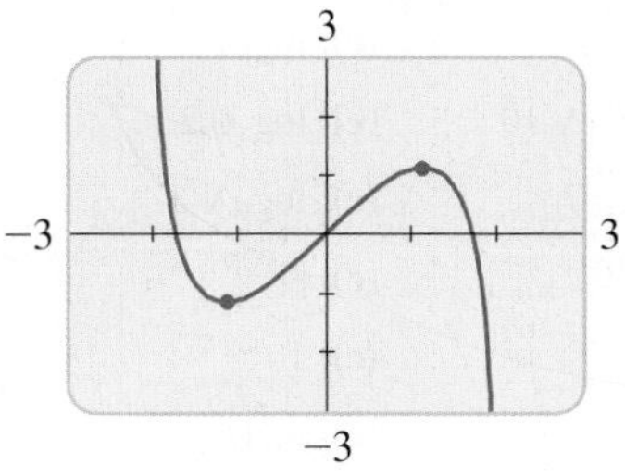

FIGURE 10
$y = x\ln(4 - x^2)$

SOLUTION As in Example 10 the domain of this function is the interval $(-2, 2)$, so we choose the viewing rectangle $[-3, 3]$ by $[-3, 3]$. The graph is shown in Figure 10, and from it we see that the lines $x = -2$ and $x = 2$ are vertical asymptotes.

The function has a local maximum point to the right of $x = 1$ and a local minimum point to the left of $x = -1$. By zooming in and tracing along the graph with the cursor, we find that the local maximum value is approximately 1.13 and this occurs when $x \approx 1.15$. Similarly (or by noticing that the function is odd), we find that the local minimum value is about -1.13, and it occurs when $x \approx -1.15$.

NOW TRY EXERCISE **69** ■

4.3 EXERCISES

CONCEPTS

1. $\log x$ is the exponent to which the base 10 must be raised to get ________. So we can complete the following table for $\log x$.

x	10^3	10^2	10^1	10^0	10^{-1}	10^{-2}	10^{-3}	$10^{1/2}$
$\log x$								

2. The function $f(x) = \log_9 x$ is the logarithm function with base ________. So $f(9) =$ ________, $f(1) =$ ________, $f(\frac{1}{9}) =$ ________, $f(81) =$ ________, and $f(3) =$ ________.

3. (a) $5^3 = 125$, so $\log_{\square} \square = \square$

(b) $\log_5 25 = 2$, so $\square^{\square} = \square$

4. Match the logarithmic function with its graph.
(a) $f(x) = \log_2 x$ **(b)** $f(x) = \log_2(-x)$
(c) $f(x) = -\log_2 x$ **(d)** $f(x) = -\log_2(-x)$

I

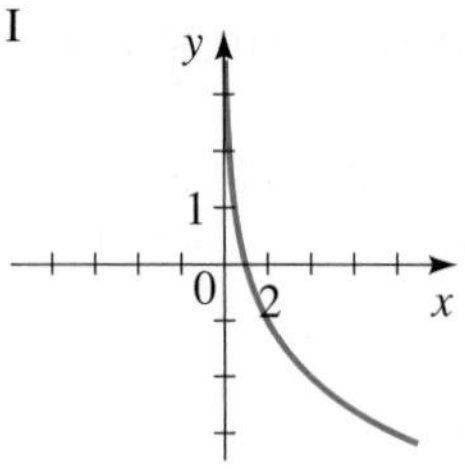

II

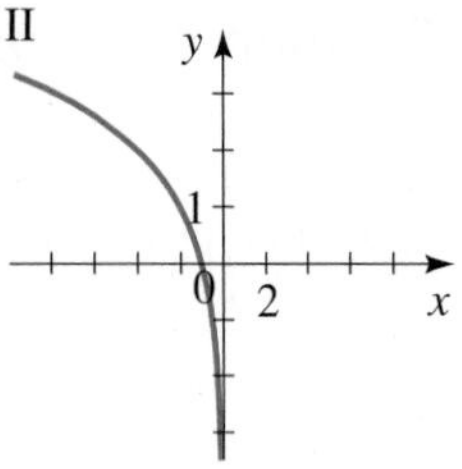

III

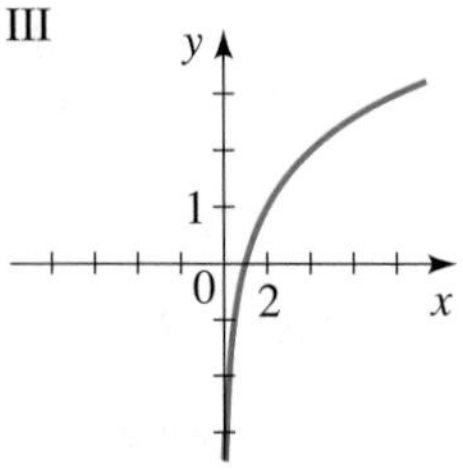

IV

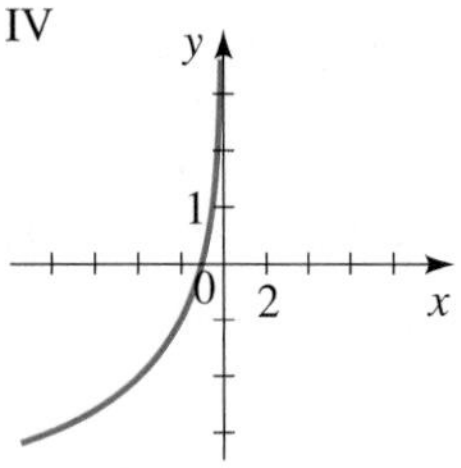

SKILLS

5–6 ■ Complete the table by finding the appropriate logarithmic or exponential form of the equation, as in Example 1.

5.

Logarithmic form	Exponential form
$\log_8 8 = 1$	
$\log_8 64 = 2$	
	$8^{2/3} = 4$
	$8^3 = 512$
$\log_8\left(\frac{1}{8}\right) = -1$	
	$8^{-2} = \frac{1}{64}$

6.

Logarithmic form	Exponential form
	$4^3 = 64$
$\log_4 2 = \frac{1}{2}$	
	$4^{3/2} = 8$
$\log_4\left(\frac{1}{16}\right) = -2$	
$\log_4\left(\frac{1}{2}\right) = -\frac{1}{2}$	
	$4^{-5/2} = \frac{1}{32}$

7–12 ■ Express the equation in exponential form.

7. (a) $\log_5 25 = 2$ **(b)** $\log_5 1 = 0$

8. (a) $\log_{10} 0.1 = -1$ **(b)** $\log_8 512 = 3$

9. (a) $\log_8 2 = \frac{1}{3}$ **(b)** $\log_2\left(\frac{1}{8}\right) = -3$

10. (a) $\log_3 81 = 4$ **(b)** $\log_8 4 = \frac{2}{3}$

11. (a) $\ln 5 = x$ **(b)** $\ln y = 5$

12. (a) $\ln(x + 1) = 2$ **(b)** $\ln(x - 1) = 4$

13–18 ■ Express the equation in logarithmic form.

13. (a) $5^3 = 125$ **(b)** $10^{-4} = 0.0001$

14. (a) $10^3 = 1000$ **(b)** $81^{1/2} = 9$

15. (a) $8^{-1} = \frac{1}{8}$ **(b)** $2^{-3} = \frac{1}{8}$

16. (a) $4^{-3/2} = 0.125$ **(b)** $7^3 = 343$

17. (a) $e^x = 2$ **(b)** $e^3 = y$

18. (a) $e^{x+1} = 0.5$ **(b)** $e^{0.5x} = t$

19–28 ■ Evaluate the expression.

19. (a) $\log_3 3$ **(b)** $\log_3 1$ **(c)** $\log_3 3^2$

20. (a) $\log_5 5^4$ **(b)** $\log_4 64$ **(c)** $\log_3 9$

21. (a) $\log_6 36$ **(b)** $\log_9 81$ **(c)** $\log_7 7^{10}$

22. (a) $\log_2 32$ **(b)** $\log_8 8^{17}$ **(c)** $\log_6 1$

23. (a) $\log_3\left(\frac{1}{27}\right)$ **(b)** $\log_{10}\sqrt{10}$ **(c)** $\log_5 0.2$

24. (a) $\log_5 125$ **(b)** $\log_{49} 7$ **(c)** $\log_9\sqrt{3}$

25. (a) $2^{\log_2 37}$ **(b)** $3^{\log_3 8}$ **(c)** $e^{\ln\sqrt{5}}$

26. (a) $e^{\ln \pi}$ **(b)** $10^{\log 5}$ **(c)** $10^{\log 87}$

27. (a) $\log_8 0.25$ **(b)** $\ln e^4$ **(c)** $\ln(1/e)$

28. (a) $\log_4\sqrt{2}$ **(b)** $\log_4\left(\frac{1}{2}\right)$ **(c)** $\log_4 8$

29–36 ■ Use the definition of the logarithmic function to find x.

29. (a) $\log_2 x = 5$ **(b)** $\log_2 16 = x$

30. (a) $\log_5 x = 4$ **(b)** $\log_{10} 0.1 = x$

31. (a) $\log_3 243 = x$ **(b)** $\log_3 x = 3$

32. (a) $\log_4 2 = x$ **(b)** $\log_4 x = 2$

33. (a) $\log_{10} x = 2$ **(b)** $\log_5 x = 2$

34. **(a)** $\log_x 1000 = 3$ **(b)** $\log_x 25 = 2$

35. **(a)** $\log_x 16 = 4$ **(b)** $\log_x 8 = \frac{3}{2}$

36. **(a)** $\log_x 6 = \frac{1}{2}$ **(b)** $\log_x 3 = \frac{1}{3}$

37–40 ■ Use a calculator to evaluate the expression, correct to four decimal places.

37. **(a)** $\log 2$ **(b)** $\log 35.2$ **(c)** $\log(\frac{2}{3})$

38. **(a)** $\log 50$ **(b)** $\log \sqrt{2}$ **(c)** $\log(3\sqrt{2})$

39. **(a)** $\ln 5$ **(b)** $\ln 25.3$ **(c)** $\ln(1 + \sqrt{3})$

40. **(a)** $\ln 27$ **(b)** $\ln 7.39$ **(c)** $\ln 54.6$

41–44 ■ Sketch the graph of the function by plotting points.

41. $f(x) = \log_3 x$ **42.** $g(x) = \log_4 x$

43. $f(x) = 2 \log x$ **44.** $g(x) = 1 + \log x$

45–48 ■ Find the function of the form $y = \log_a x$ whose graph is given.

45.

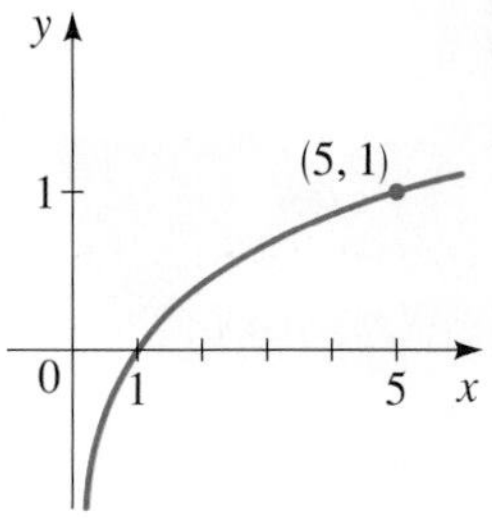

46.

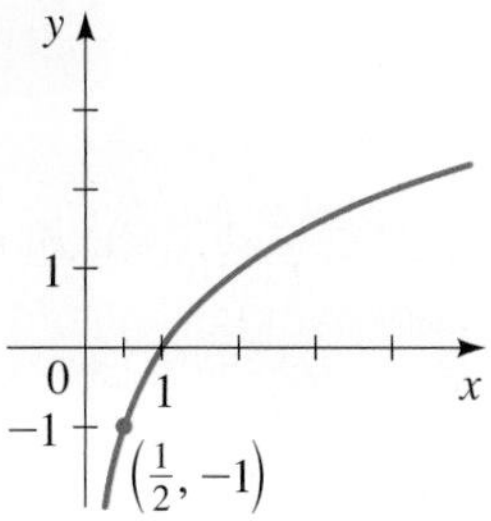

47.

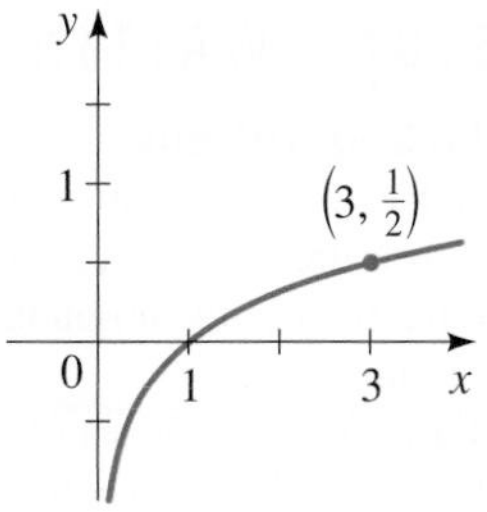

48.

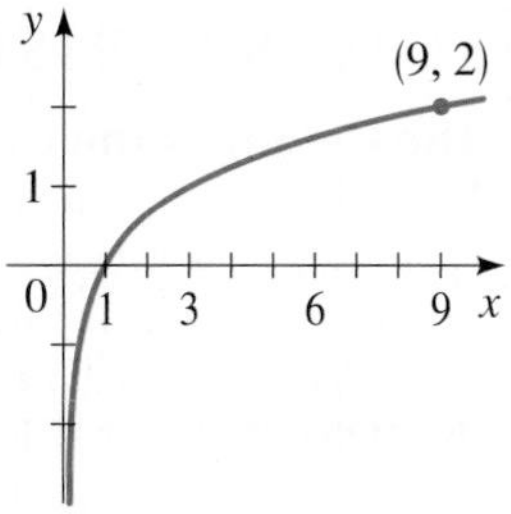

49–50 ■ Match the logarithmic function with one of the graphs labeled I or II.

49. $f(x) = 2 + \ln x$ **50.** $f(x) = \ln(x - 2)$

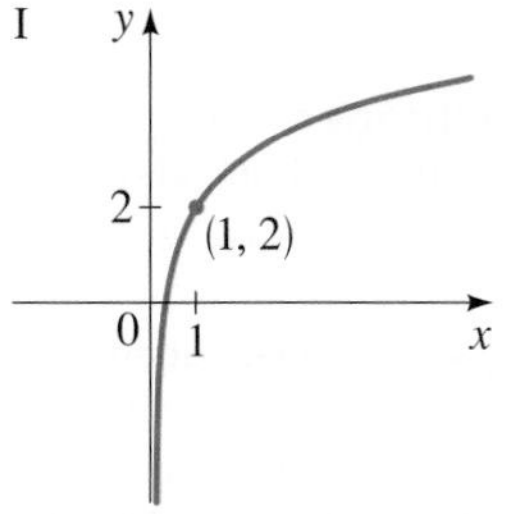

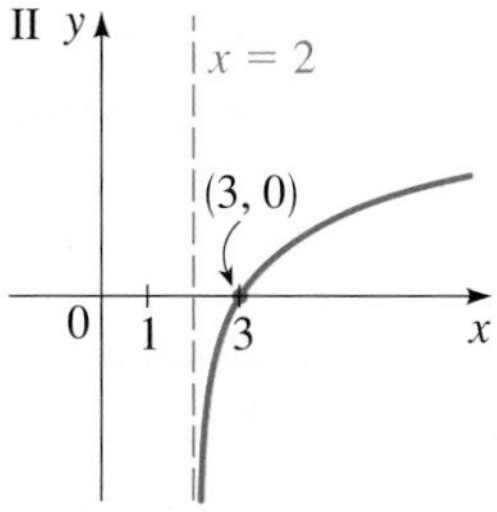

51. Draw the graph of $y = 4^x$, then use it to draw the graph of $y = \log_4 x$.

52. Draw the graph of $y = 3^x$, then use it to draw the graph of $y = \log_3 x$.

53–62 ■ Graph the function, not by plotting points, but by starting from the graphs in Figures 4 and 9. State the domain, range, and asymptote.

53. $f(x) = \log_2(x - 4)$ **54.** $f(x) = -\log_{10} x$

55. $g(x) = \log_5(-x)$ **56.** $g(x) = \ln(x + 2)$

57. $y = 2 + \log_3 x$ **58.** $y = \log_3(x - 1) - 2$

59. $y = 1 - \log_{10} x$ **60.** $y = 1 + \ln(-x)$

61. $y = |\ln x|$ **62.** $y = \ln |x|$

63–68 ■ Find the domain of the function.

63. $f(x) = \log_{10}(x + 3)$ **64.** $f(x) = \log_5(8 - 2x)$

65. $g(x) = \log_3(x^2 - 1)$ **66.** $g(x) = \ln(x - x^2)$

67. $h(x) = \ln x + \ln(2 - x)$

68. $h(x) = \sqrt{x - 2} - \log_5(10 - x)$

69–74 ■ Draw the graph of the function in a suitable viewing rectangle, and use it to find the domain, the asymptotes, and the local maximum and minimum values.

69. $y = \log_{10}(1 - x^2)$ **70.** $y = \ln(x^2 - x)$

71. $y = x + \ln x$ **72.** $y = x(\ln x)^2$

73. $y = \dfrac{\ln x}{x}$ **74.** $y = x \log_{10}(x + 10)$

75-78 ■ Find the functions $f \circ g$ and $g \circ f$ and their domains.

75. $f(x) = 2^x, \quad g(x) = x + 1$

76. $f(x) = 3^x, \quad g(x) = x^2 + 1$

77. $f(x) = \log_2 x, \quad g(x) = x - 2$

78. $f(x) = \log x, \quad g(x) = x^2$

79. Compare the rates of growth of the functions $f(x) = \ln x$ and $g(x) = \sqrt{x}$ by drawing their graphs on a common screen using the viewing rectangle $[-1, 30]$ by $[-1, 6]$.

80. **(a)** By drawing the graphs of the functions

$$f(x) = 1 + \ln(1 + x) \quad \text{and} \quad g(x) = \sqrt{x}$$

in a suitable viewing rectangle, show that even when a logarithmic function starts out higher than a root function, it is ultimately overtaken by the root function.

(b) Find, correct to two decimal places, the solutions of the equation $\sqrt{x} = 1 + \ln(1 + x)$.

81–82 ■ A family of functions is given. **(a)** Draw graphs of the family for $c = 1, 2, 3$, and 4. **(b)** How are the graphs in part (a) related?

81. $f(x) = \log(cx)$ **82.** $f(x) = c \log x$

83–84 ■ A function $f(x)$ is given. **(a)** Find the domain of the function f. **(b)** Find the inverse function of f.

83. $f(x) = \log_2(\log_{10} x)$ **84.** $f(x) = \ln(\ln(\ln x))$

85. **(a)** Find the inverse of the function $f(x) = \dfrac{2^x}{1 + 2^x}$.

(b) What is the domain of the inverse function?

APPLICATIONS

86. **Absorption of Light** A spectrophotometer measures the concentration of a sample dissolved in water by shining a light through it and recording the amount of light that emerges. In other words, if we know the amount of light that is absorbed, we can calculate the concentration of the sample. For a certain substance the concentration (in moles per liter) is found by using the formula

$$C = -2500 \ln\left(\frac{I}{I_0}\right)$$

where I_0 is the intensity of the incident light and I is the intensity of light that emerges. Find the concentration of the substance if the intensity I is 70% of I_0.

87. **Carbon Dating** The age of an ancient artifact can be determined by the amount of radioactive carbon-14 remaining in it. If D_0 is the original amount of carbon-14 and D is the amount remaining, then the artifact's age A (in years) is given by

$$A = -8267 \ln\left(\frac{D}{D_0}\right)$$

Find the age of an object if the amount D of carbon-14 that remains in the object is 73% of the original amount D_0.

88. **Bacteria Colony** A certain strain of bacteria divides every three hours. If a colony is started with 50 bacteria, then the time t (in hours) required for the colony to grow to N bacteria is given by

$$t = 3\frac{\log(N/50)}{\log 2}$$

Find the time required for the colony to grow to a million bacteria.

89. **Investment** The time required to double the amount of an investment at an interest rate r compounded continuously is given by

$$t = \frac{\ln 2}{r}$$

Find the time required to double an investment at 6%, 7%, and 8%.

90. **Charging a Battery** The rate at which a battery charges is slower the closer the battery is to its maximum charge C_0. The time (in hours) required to charge a fully discharged battery to a charge C is given by

$$t = -k \ln\left(1 - \frac{C}{C_0}\right)$$

where k is a positive constant that depends on the battery. For a certain battery, $k = 0.25$. If this battery is fully discharged, how long will it take to charge to 90% of its maximum charge C_0?

91. **Difficulty of a Task** The difficulty in "acquiring a target" (such as using your mouse to click on an icon on your computer screen) depends on the distance to the target and the size of the target. According to Fitts's Law, the index of difficulty (ID) is given by

$$\text{ID} = \frac{\log(2A/W)}{\log 2}$$

where W is the width of the target and A is the distance to the center of the target. Compare the difficulty of clicking on an icon that is 5 mm wide to clicking on one that is 10 mm wide. In each case, assume that the mouse is 100 mm from the icon.

DISCOVERY ▪ DISCUSSION ▪ WRITING

92. **The Height of the Graph of a Logarithmic Function** Suppose that the graph of $y = 2^x$ is drawn on a coordinate plane where the unit of measurement is an inch.
    (a) Show that at a distance 2 ft to the right of the origin the height of the graph is about 265 mi.
    (b) If the graph of $y = \log_2 x$ is drawn on the same set of axes, how far to the right of the origin do we have to go before the height of the curve reaches 2 ft?

93. **The Googolplex** A **googol** is 10^{100}, and a **googolplex** is 10^{googol}. Find

$$\log(\log(\text{googol})) \qquad \text{and} \qquad \log(\log(\log(\text{googolplex})))$$

94. **Comparing Logarithms** Which is larger, $\log_4 17$ or $\log_5 24$? Explain your reasoning.

95. **The Number of Digits in an Integer** Compare log 1000 to the number of digits in 1000. Do the same for 10,000. How many digits does any number between 1000 and 10,000 have? Between what two values must the common logarithm of such a number lie? Use your observations to explain why the number of digits in any positive integer x is $[\![\log x]\!] + 1$. (The symbol $[\![n]\!]$ is the greatest integer function defined in Section 2.2.) How many digits does the number 2^{100} have?

4.4 Laws of Logarithms

Laws of Logarithms ▶ Expanding and Combining Logarithmic Expressions ▶ Change of Base Formula

In this section we study properties of logarithms. These properties give logarithmic functions a wide range of applications, as we will see in Section 4.6.

▼ Laws of Logarithms

Since logarithms are exponents, the Laws of Exponents give rise to the Laws of Logarithms.

LAWS OF LOGARITHMS

Let a be a positive number, with $a \neq 1$. Let A, B, and C be any real numbers with $A > 0$ and $B > 0$.

Law	Description
1. $\log_a(AB) = \log_a A + \log_a B$	The logarithm of a product of numbers is the sum of the logarithms of the numbers.
2. $\log_a\left(\dfrac{A}{B}\right) = \log_a A - \log_a B$	The logarithm of a quotient of numbers is the difference of the logarithms of the numbers.
3. $\log_a(A^C) = C \log_a A$	The logarithm of a power of a number is the exponent times the logarithm of the number.

PROOF We make use of the property $\log_a a^x = x$ from Section 4.3.

Law 1 Let $\log_a A = u$ and $\log_a B = v$. When written in exponential form, these equations become

$$a^u = A \qquad \text{and} \qquad a^v = B$$

Thus

$$\log_a(AB) = \log_a(a^u a^v) = \log_a(a^{u+v})$$
$$= u + v = \log_a A + \log_a B$$

Law 2 Using Law 1, we have

$$\log_a A = \log_a\left[\left(\frac{A}{B}\right)B\right] = \log_a\left(\frac{A}{B}\right) + \log_a B$$

so

$$\log_a\left(\frac{A}{B}\right) = \log_a A - \log_a B$$

Law 3 Let $\log_a A = u$. Then $a^u = A$, so

$$\log_a(A^C) = \log_a(a^u)^C = \log_a(a^{uC}) = uC = C \log_a A$$ ■

EXAMPLE 1 | Using the Laws of Logarithms to Evaluate Expressions

Evaluate each expression.

(a) $\log_4 2 + \log_4 32$

(b) $\log_2 80 - \log_2 5$

(c) $-\frac{1}{3} \log 8$

SOLUTION

(a) $\log_4 2 + \log_4 32 = \log_4(2 \cdot 32)$ — Law 1

$= \log_4 64 = 3$ — Because $64 = 4^3$

(b) $\log_2 80 - \log_2 5 = \log_2\left(\frac{80}{5}\right)$ — Law 2

$= \log_2 16 = 4$ — Because $16 = 2^4$

(c) $-\frac{1}{3}\log 8 = \log 8^{-1/3}$ — Law 3

$= \log\left(\frac{1}{2}\right)$ — Property of negative exponents

≈ -0.301 — Calculator

NOW TRY EXERCISES **7**, **9**, AND **11**

Expanding and Combining Logarithmic Expressions

The Laws of Logarithms allow us to write the logarithm of a product or a quotient as the sum or difference of logarithms. This process, called *expanding* a logarithmic expression, is illustrated in the next example.

EXAMPLE 2 | Expanding Logarithmic Expressions

Use the Laws of Logarithms to expand each expression.

(a) $\log_2(6x)$ **(b)** $\log_5(x^3y^6)$ **(c)** $\ln\left(\dfrac{ab}{\sqrt[3]{c}}\right)$

SOLUTION

(a) $\log_2(6x) = \log_2 6 + \log_2 x$ — Law 1

(b) $\log_5(x^3y^6) = \log_5 x^3 + \log_5 y^6$ — Law 1

$= 3\log_5 x + 6\log_5 y$ — Law 3

(c) $\ln\left(\dfrac{ab}{\sqrt[3]{c}}\right) = \ln(ab) - \ln\sqrt[3]{c}$ — Law 2

$= \ln a + \ln b - \ln c^{1/3}$ — Law 1

$= \ln a + \ln b - \frac{1}{3}\ln c$ — Law 3

NOW TRY EXERCISES **19**, **21**, AND **33**

The Laws of Logarithms also allow us to reverse the process of expanding that was done in Example 2. That is, we can write sums and differences of logarithms as a single logarithm. This process, called *combining* logarithmic expressions, is illustrated in the next example.

EXAMPLE 3 | Combining Logarithmic Expressions

Combine $3\log x + \frac{1}{2}\log(x+1)$ into a single logarithm.

SOLUTION

$3\log x + \frac{1}{2}\log(x+1) = \log x^3 + \log(x+1)^{1/2}$ — Law 3

$= \log(x^3(x+1)^{1/2})$ — Law 1

NOW TRY EXERCISE **47**

EXAMPLE 4 | Combining Logarithmic Expressions

Combine $3\ln s + \frac{1}{2}\ln t - 4\ln(t^2+1)$ into a single logarithm.

SOLUTION

$$\begin{aligned} 3\ln s + \tfrac{1}{2}\ln t - 4\ln(t^2+1) &= \ln s^3 + \ln t^{1/2} - \ln(t^2+1)^4 && \text{Law 3} \\ &= \ln(s^3 t^{1/2}) - \ln(t^2+1)^4 && \text{Law 1} \\ &= \ln\left(\frac{s^3\sqrt{t}}{(t^2+1)^4}\right) && \text{Law 2} \end{aligned}$$

NOW TRY EXERCISE **49** ■

Warning Although the Laws of Logarithms tell us how to compute the logarithm of a product or a quotient, *there is no corresponding rule for the logarithm of a sum or a difference*. For instance,

$$\log_a(x+y) \neq \log_a x + \log_a y$$

In fact, we know that the right side is equal to $\log_a(xy)$. Also, don't improperly simplify quotients or powers of logarithms. For instance,

$$\frac{\log 6}{\log 2} \neq \log\left(\frac{6}{2}\right) \qquad \text{and} \qquad (\log_2 x)^3 \neq 3\log_2 x$$

Logarithmic functions are used to model a variety of situations involving human behavior. One such behavior is how quickly we forget things we have learned. For example, if you learn algebra at a certain performance level (say, 90% on a test) and then don't use algebra for a while, how much will you retain after a week, a month, or a year? Hermann Ebbinghaus (1850–1909) studied this phenomenon and formulated the law described in the next example.

EXAMPLE 5 | The Law of Forgetting

If a task is learned at a performance level P_0, then after a time interval t the performance level P satisfies

$$\log P = \log P_0 - c\log(t+1)$$

where c is a constant that depends on the type of task and t is measured in months.

(a) Solve for P.

(b) If your score on a history test is 90, what score would you expect to get on a similar test after two months? After a year? (Assume that $c = 0.2$.)

Forgetting what we've learned depends on how long ago we learned it.

SOLUTION

(a) We first combine the right-hand side.

$$\begin{aligned} \log P &= \log P_0 - c\log(t+1) && \text{Given equation} \\ \log P &= \log P_0 - \log(t+1)^c && \text{Law 3} \\ \log P &= \log\frac{P_0}{(t+1)^c} && \text{Law 2} \\ P &= \frac{P_0}{(t+1)^c} && \text{Because log is one-to-one} \end{aligned}$$

(b) Here $P_0 = 90$, $c = 0.2$, and t is measured in months.

In two months: $t = 2$ and $P = \dfrac{90}{(2+1)^{0.2}} \approx 72$

In one year: $t = 12$ and $P = \dfrac{90}{(12+1)^{0.2}} \approx 54$

Your expected scores after two months and one year are 72 and 54, respectively.

NOW TRY EXERCISE **69** ■

▼ Change of Base Formula

For some purposes we find it useful to change from logarithms in one base to logarithms in another base. Suppose we are given $\log_a x$ and want to find $\log_b x$. Let

$$y = \log_b x$$

We write this in exponential form and take the logarithm, with base a, of each side.

$$\begin{aligned} b^y &= x && \text{Exponential form} \\ \log_a(b^y) &= \log_a x && \text{Take } \log_a \text{ of each side} \\ y \log_a b &= \log_a x && \text{Law 3} \\ y &= \frac{\log_a x}{\log_a b} && \text{Divide by } \log_a b \end{aligned}$$

This proves the following formula.

We may write the Change of Base Formula as

$$\log_b x = \left(\frac{1}{\log_a b}\right)\log_a x$$

So $\log_b x$ is just a constant multiple of $\log_a x$; the constant is $\dfrac{1}{\log_a b}$.

CHANGE OF BASE FORMULA

$$\log_b x = \frac{\log_a x}{\log_a b}$$

In particular, if we put $x = a$, then $\log_a a = 1$, and this formula becomes

$$\log_b a = \frac{1}{\log_a b}$$

We can now evaluate a logarithm to *any* base by using the Change of Base Formula to express the logarithm in terms of common logarithms or natural logarithms and then using a calculator.

EXAMPLE 6 | Evaluating Logarithms with the Change of Base Formula

Use the Change of Base Formula and common or natural logarithms to evaluate each logarithm, correct to five decimal places.

(a) $\log_8 5$ **(b)** $\log_9 20$

SOLUTION

(a) We use the Change of Base Formula with $b = 8$ and $a = 10$:

$$\log_8 5 = \frac{\log_{10} 5}{\log_{10} 8} \approx 0.77398$$

(b) We use the Change of Base Formula with $b = 9$ and $a = e$:

$$\log_9 20 = \frac{\ln 20}{\ln 9} \approx 1.36342$$

NOW TRY EXERCISES **55** AND **57** ■

EXAMPLE 7 | Using the Change of Base Formula to Graph a Logarithmic Function

Use a graphing calculator to graph $f(x) = \log_6 x$.

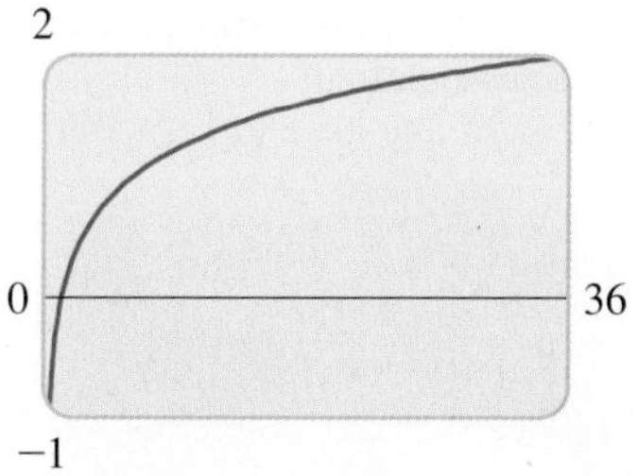

FIGURE 1 $f(x) = \log_6 x = \dfrac{\ln x}{\ln 6}$

SOLUTION Calculators don't have a key for $\log_6$, so we use the Change of Base Formula to write

$$f(x) = \log_6 x = \frac{\ln x}{\ln 6}$$

Since calculators do have an LN key, we can enter this new form of the function and graph it. The graph is shown in Figure 1.

NOW TRY EXERCISE **63** ■

4.4 EXERCISES

CONCEPTS

1. The logarithm of a product of two numbers is the same as the ________ of the logarithms of these numbers. So $\log_5(25 \cdot 125) =$ ________ + ________.

2. The logarithm of a quotient of two numbers is the same as the ________ of the logarithms of these numbers. So $\log_5(\frac{25}{125}) =$ ________ − ________.

3. The logarithm of a number raised to a power is the same as the power ________ the logarithm of the number. So $\log_5(25^{10}) =$ ________ · ________.

4. (a) We can expand $\log\left(\dfrac{x^2 y}{z}\right)$ to get ________________.

(b) We can combine $2\log x + \log y - \log z$ to get ________.

5. Most calculators can find logarithms with base ________ and base ________. To find logarithms with different bases, we use the ________________ Formula. To find $\log_7 12$, we write

$$\log_7 12 = \frac{\log \square}{\log \square} = \underline{\qquad}$$

6. *True or false?* We get the same answer if we do the calculation in Exercise 5 using ln in place of log.

SKILLS

7–18 ■ Evaluate the expression.

7. $\log_3 \sqrt{27}$

8. $\log_2 160 - \log_2 5$

9. $\log 4 + \log 25$

10. $\log \dfrac{1}{\sqrt{1000}}$

11. $\log_4 192 - \log_4 3$

12. $\log_{12} 9 + \log_{12} 16$

13. $\log_2 6 - \log_2 15 + \log_2 20$

14. $\log_3 100 - \log_3 18 - \log_3 50$

15. $\log_4 16^{100}$

16. $\log_2 8^{33}$

17. $\log(\log 10^{10{,}000})$

18. $\ln(\ln e^{e^{200}})$

19–44 ■ Use the Laws of Logarithms to expand the expression.

19. $\log_2(2x)$

20. $\log_3(5y)$

21. $\log_2(x(x-1))$

22. $\log_5 \dfrac{x}{2}$

23. $\log 6^{10}$

24. $\ln \sqrt{z}$

25. $\log_2(AB^2)$

26. $\log_6 \sqrt[4]{17}$

27. $\log_3(x\sqrt{y})$

28. $\log_2(xy)^{10}$

29. $\log_5 \sqrt[3]{x^2+1}$

30. $\log_a\left(\dfrac{x^2}{yz^3}\right)$

31. $\ln \sqrt{ab}$

32. $\ln \sqrt[3]{3r^2 s}$

33. $\log\left(\dfrac{x^3 y^4}{z^6}\right)$

34. $\log\left(\dfrac{a^2}{b^4\sqrt{c}}\right)$

35. $\log_2\left(\dfrac{x(x^2+1)}{\sqrt{x^2-1}}\right)$

36. $\log_5 \sqrt{\dfrac{x-1}{x+1}}$

37. $\ln\left(x\sqrt{\dfrac{y}{z}}\right)$

38. $\ln \dfrac{3x^2}{(x+1)^{10}}$

39. $\log \sqrt[4]{x^2+y^2}$

40. $\log\left(\dfrac{x}{\sqrt[3]{1-x}}\right)$

41. $\log \sqrt{\dfrac{x^2+4}{(x^2+1)(x^3-7)^2}}$

42. $\log \sqrt{x\sqrt{y\sqrt{z}}}$

43. $\ln\left(\dfrac{x^3\sqrt{x-1}}{3x+4}\right)$

44. $\log\left(\dfrac{10^x}{x(x^2+1)(x^4+2)}\right)$

45–54 ■ Use the Laws of Logarithms to combine the expression.

45. $\log_3 5 + 5\log_3 2$

46. $\log 12 + \frac{1}{2}\log 7 - \log 2$

47. $\log_2 A + \log_2 B - 2\log_2 C$

48. $\log_5(x^2-1) - \log_5(x-1)$

49. $4\log x - \frac{1}{3}\log(x^2+1) + 2\log(x-1)$

50. $\ln(a+b) + \ln(a-b) - 2\ln c$

51. $\ln 5 + 2\ln x + 3\ln(x^2+5)$

52. $2(\log_5 x + 2\log_5 y - 3\log_5 z)$

53. $\frac{1}{3}\log(x+2)^3 + \frac{1}{2}[\log x^4 - \log(x^2 - x - 6)^2]$

54. $\log_a b + c \log_a d - r \log_a s$

55–62 ■ Use the Change of Base Formula and a calculator to evaluate the logarithm, rounded to six decimal places. Use either natural or common logarithms.

55. $\log_2 5$

56. $\log_5 2$

57. $\log_3 16$

58. $\log_6 92$

59. $\log_7 2.61$

60. $\log_6 532$

61. $\log_4 125$

62. $\log_{12} 2.5$

63. Use the Change of Base Formula to show that

$$\log_3 x = \frac{\ln x}{\ln 3}$$

Then use this fact to draw the graph of the function $f(x) = \log_3 x$.

64. Draw graphs of the family of functions $y = \log_a x$ for $a = 2, e,$ 5, and 10 on the same screen, using the viewing rectangle $[0, 5]$ by $[-3, 3]$. How are these graphs related?

65. Use the Change of Base Formula to show that

$$\log e = \frac{1}{\ln 10}$$

66. Simplify: $(\log_2 5)(\log_5 7)$

67. Show that $-\ln(x - \sqrt{x^2 - 1}) = \ln(x + \sqrt{x^2 - 1})$.

APPLICATIONS

68. Forgetting Use the Law of Forgetting (Example 5) to estimate a student's score on a biology test two years after he got a score of 80 on a test covering the same material. Assume that $c = 0.3$ and t is measured in months.

69. Wealth Distribution Vilfredo Pareto (1848–1923) observed that most of the wealth of a country is owned by a few members of the population. **Pareto's Principle** is

$$\log P = \log c - k \log W$$

where W is the wealth level (how much money a person has) and P is the number of people in the population having that much money.

(a) Solve the equation for P.

(b) Assume that $k = 2.1$, $c = 8000$, and W is measured in millions of dollars. Use part (a) to find the number of people who have \$2 million or more. How many people have \$10 million or more?

70. Biodiversity Some biologists model the number of species S in a fixed area A (such as an island) by the species-area relationship

$$\log S = \log c + k \log A$$

where c and k are positive constants that depend on the type of species and habitat.

(a) Solve the equation for S.

(b) Use part (a) to show that if $k = 3$, then doubling the area increases the number of species eightfold.

71. Magnitude of Stars The magnitude M of a star is a measure of how bright a star appears to the human eye. It is defined by

$$M = -2.5 \log\left(\frac{B}{B_0}\right)$$

where B is the actual brightness of the star and B_0 is a constant.

(a) Expand the right-hand side of the equation.

(b) Use part (a) to show that the brighter a star, the less its magnitude.

(c) Betelgeuse is about 100 times brighter than Albiero. Use part (a) to show that Betelgeuse is 5 magnitudes less bright than Albiero.

DISCOVERY ■ DISCUSSION ■ WRITING

72. True or False? Discuss each equation and determine whether it is true for all possible values of the variables. (Ignore values of the variables for which any term is undefined.)

(a) $\log\left(\frac{x}{y}\right) = \frac{\log x}{\log y}$

(b) $\log_2(x - y) = \log_2 x - \log_2 y$

(c) $\log_5\left(\frac{a}{b^2}\right) = \log_5 a - 2\log_5 b$

(d) $\log 2^z = z \log 2$

(e) $(\log P)(\log Q) = \log P + \log Q$

(f) $\frac{\log a}{\log b} = \log a - \log b$

(g) $(\log_2 7)^x = x \log_2 7$

(h) $\log_a a^a = a$

(i) $\log(x - y) = \frac{\log x}{\log y}$

(j) $-\ln\left(\frac{1}{A}\right) = \ln A$

73. Find the Error What is wrong with the following argument?

$$\begin{aligned} \log 0.1 &< 2 \log 0.1 \\ &= \log(0.1)^2 \\ &= \log 0.01 \\ \log 0.1 &< \log 0.01 \\ 0.1 &< 0.01 \end{aligned}$$

74. Shifting, Shrinking, and Stretching Graphs of Functions Let $f(x) = x^2$. Show that $f(2x) = 4f(x)$, and explain how this shows that shrinking the graph of f horizontally has the same effect as stretching it vertically. Then use the identities $e^{2+x} = e^2 e^x$ and $\ln(2x) = \ln 2 + \ln x$ to show that for $g(x) = e^x$ a horizontal shift is the same as a vertical stretch and for $h(x) = \ln x$ a horizontal shrinking is the same as a vertical shift.

4.5 Exponential and Logarithmic Equations

Exponential Equations ▶ Logarithmic Equations ▶ Compound Interest

In this section we solve equations that involve exponential or logarithmic functions. The techniques that we develop here will be used in the next section for solving applied problems.

▼ Exponential Equations

An *exponential equation* is one in which the variable occurs in the exponent. For example,

$$2^x = 7$$

The variable x presents a difficulty because it is in the exponent. To deal with this difficulty, we take the logarithm of each side and then use the Laws of Logarithms to "bring down x" from the exponent.

$$\begin{aligned} 2^x &= 7 && \text{Given equation} \\ \ln 2^x &= \ln 7 && \text{Take ln of each side} \\ x \ln 2 &= \ln 7 && \text{Law 3 (bring down exponent)} \\ x &= \frac{\ln 7}{\ln 2} && \text{Solve for } x \\ &\approx 2.807 && \text{Calculator} \end{aligned}$$

Recall that Law 3 of the Laws of Logarithms says that $\log_a A^C = C \log_a A$.

The method that we used to solve $2^x = 7$ is typical of how we solve exponential equations in general.

GUIDELINES FOR SOLVING EXPONENTIAL EQUATIONS

1. Isolate the exponential expression on one side of the equation.
2. Take the logarithm of each side, then use the Laws of Logarithms to "bring down the exponent."
3. Solve for the variable.

EXAMPLE 1 | Solving an Exponential Equation

Find the solution of the equation $3^{x+2} = 7$, rounded to six decimal places.

SOLUTION We take the common logarithm of each side and use Law 3.

$3^{x+2} = 7$		Given equation
$\log(3^{x+2}) = \log 7$		Take log of each side
$(x + 2)\log 3 = \log 7$		Law 3 (bring down exponent)
$x + 2 = \dfrac{\log 7}{\log 3}$		Divide by log 3
$x = \dfrac{\log 7}{\log 3} - 2$		Subtract 2
≈ -0.228756		Calculator

We could have used natural logarithms instead of common logarithms. In fact, using the same steps, we get

$$x = \frac{\ln 7}{\ln 3} - 2 \approx -0.228756$$

CHECK YOUR ANSWER

Substituting $x = -0.228756$ into the original equation and using a calculator, we get

$$3^{(-0.228756)+2} \approx 7 \quad ✔$$

NOW TRY EXERCISE **7** ■

EXAMPLE 2 | Solving an Exponential Equation

Solve the equation $8e^{2x} = 20$.

SOLUTION We first divide by 8 to isolate the exponential term on one side of the equation.

$8e^{2x} = 20$	Given equation
$e^{2x} = \frac{20}{8}$	Divide by 8
$\ln e^{2x} = \ln 2.5$	Take ln of each side
$2x = \ln 2.5$	Property of ln
$x = \dfrac{\ln 2.5}{2}$	Divide by 2
≈ 0.458	Calculator

CHECK YOUR ANSWER

Substituting $x = 0.458$ into the original equation and using a calculator, we get

$$8e^{2(0.458)} \approx 20 \quad ✔$$

NOW TRY EXERCISE **9** ■

EXAMPLE 3 | Solving an Exponential Equation Algebraically and Graphically

Solve the equation $e^{3-2x} = 4$ algebraically and graphically.

SOLUTION 1: Algebraic

Since the base of the exponential term is e, we use natural logarithms to solve this equation.

$e^{3-2x} = 4$	Given equation
$\ln(e^{3-2x}) = \ln 4$	Take ln of each side
$3 - 2x = \ln 4$	Property of ln
$-2x = -3 + \ln 4$	Subtract 3
$x = \frac{1}{2}(3 - \ln 4) \approx 0.807$	Multiply by $-\frac{1}{2}$

You should check that this answer satisfies the original equation.

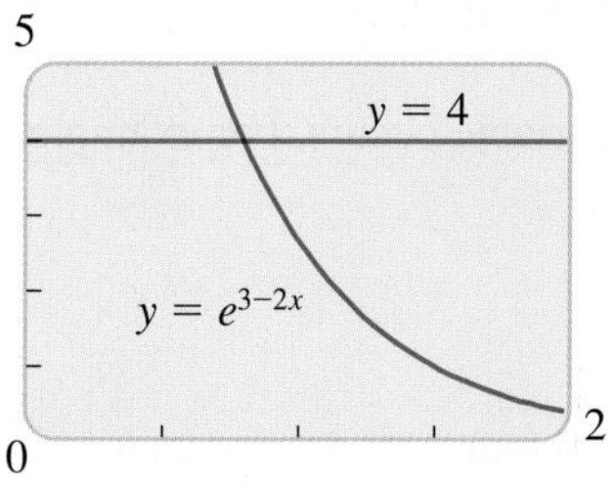

FIGURE 1

SOLUTION 2: Graphical

We graph the equations $y = e^{3-2x}$ and $y = 4$ in the same viewing rectangle as in Figure 1. The solutions occur where the graphs intersect. Zooming in on the point of intersection of the two graphs, we see that $x \approx 0.81$.

NOW TRY EXERCISE **11** ■

EXAMPLE 4 | An Exponential Equation of Quadratic Type

Solve the equation $e^{2x} - e^x - 6 = 0$.

SOLUTION To isolate the exponential term, we factor.

$$e^{2x} - e^x - 6 = 0 \qquad \text{Given equation}$$

$$(e^x)^2 - e^x - 6 = 0 \qquad \text{Law of Exponents}$$

$$(e^x - 3)(e^x + 2) = 0 \qquad \text{Factor (a quadratic in } e^x\text{)}$$

$$e^x - 3 = 0 \quad \text{or} \quad e^x + 2 = 0 \qquad \text{Zero-Product Property}$$

$$e^x = 3 \qquad\qquad e^x = -2$$

If we let $w = e^x$, we get the quadratic equation

$$w^2 - w - 6 = 0$$

which factors as

$$(w - 3)(w + 2) = 0$$

The equation $e^x = 3$ leads to $x = \ln 3$. But the equation $e^x = -2$ has no solution because $e^x > 0$ for all x. Thus, $x = \ln 3 \approx 1.0986$ is the only solution. You should check that this answer satisfies the original equation.

NOW TRY EXERCISE **29** ■

EXAMPLE 5 | Solving an Exponential Equation

Solve the equation $3xe^x + x^2e^x = 0$.

SOLUTION First we factor the left side of the equation.

$$3xe^x + x^2e^x = 0 \qquad \text{Given equation}$$

$$x(3 + x)e^x = 0 \qquad \text{Factor out common factors}$$

$$x(3 + x) = 0 \qquad \text{Divide by } e^x \text{ (because } e^x \neq 0\text{)}$$

$$x = 0 \quad \text{or} \quad 3 + x = 0 \qquad \text{Zero-Product Property}$$

Thus the solutions are $x = 0$ and $x = -3$.

CHECK YOUR ANSWER

$x = 0$:

$3(0)e^0 + 0^2e^0 = 0$ ✔

$x = -3$:

$3(-3)e^{-3} + (-3)^2e^{-3}$

$= -9e^{-3} + 9e^{-3} = 0$ ✔

NOW TRY EXERCISE **33** ■

Radiocarbon Dating is a method archeologists use to determine the age of ancient objects. The carbon dioxide in the atmosphere always contains a fixed fraction of radioactive carbon, carbon-14 (^{14}C), with a half-life of about 5730 years. Plants absorb carbon dioxide from the atmosphere, which then makes its way to animals through the food chain. Thus, all living creatures contain the same fixed proportions of ^{14}C to nonradioactive ^{12}C as the atmosphere.

After an organism dies, it stops assimilating ^{14}C, and the amount of ^{14}C in it begins to decay exponentially. We can then determine the time elapsed since the death of the organism by measuring the amount of ^{14}C left in it.

For example, if a donkey bone contains 73% as much ^{14}C as a living donkey and it died t years ago, then by the formula for radioactive decay (Section 4.6),

$$0.73 = (1.00)e^{-(t \ln 2)/5730}$$

We solve this exponential equation to find $t \approx 2600$, so the bone is about 2600 years old.

▼ Logarithmic Equations

A *logarithmic equation* is one in which a logarithm of the variable occurs. For example,

$$\log_2(x + 2) = 5$$

To solve for x, we write the equation in exponential form.

$$x + 2 = 2^5 \qquad \text{Exponential form}$$

$$x = 32 - 2 = 30 \qquad \text{Solve for } x$$

Another way of looking at the first step is to raise the base, 2, to each side of the equation.

$$2^{\log_2(x+2)} = 2^5 \qquad \text{Raise 2 to each side}$$

$$x + 2 = 2^5 \qquad \text{Property of logarithms}$$

$$x = 32 - 2 = 30 \qquad \text{Solve for } x$$

The method used to solve this simple problem is typical. We summarize the steps as follows.

GUIDELINES FOR SOLVING LOGARITHMIC EQUATIONS

1. Isolate the logarithmic term on one side of the equation; you might first need to combine the logarithmic terms.
2. Write the equation in exponential form (or raise the base to each side of the equation).
3. Solve for the variable.

EXAMPLE 6 | Solving Logarithmic Equations

Solve each equation for x.

(a) $\ln x = 8$ **(b)** $\log_2(25 - x) = 3$

SOLUTION

(a)

$$\ln x = 8 \qquad \text{Given equation}$$

$$x = e^8 \qquad \text{Exponential form}$$

Therefore, $x = e^8 \approx 2981$.

We can also solve this problem another way:

$$\ln x = 8 \qquad \text{Given equation}$$

$$e^{\ln x} = e^8 \qquad \text{Raise } e \text{ to each side}$$

$$x = e^8 \qquad \text{Property of ln}$$

(b) The first step is to rewrite the equation in exponential form.

$$\log_2(25 - x) = 3 \qquad \text{Given equation}$$

$$25 - x = 2^3 \qquad \text{Exponential form (or raise 2 to each side)}$$

$$25 - x = 8$$

$$x = 25 - 8 = 17$$

CHECK YOUR ANSWER

If $x = 17$, we get

$\log_2(25 - 17) = \log_2 8 = 3$ ✔

NOW TRY EXERCISES **37** AND **41** ■

EXAMPLE 7 | Solving a Logarithmic Equation

Solve the equation $4 + 3\log(2x) = 16$.

SOLUTION We first isolate the logarithmic term. This allows us to write the equation in exponential form.

$4 + 3\log(2x) = 16$	Given equation
$3\log(2x) = 12$	Subtract 4
$\log(2x) = 4$	Divide by 3
$2x = 10^4$	Exponential form (or raise 10 to each side)
$x = 5000$	Divide by 2

CHECK YOUR ANSWER

If $x = 5000$, we get

$$4 + 3\log 2(5000) = 4 + 3\log 10{,}000$$
$$= 4 + 3(4)$$
$$= 16 \quad ✔$$

NOW TRY EXERCISE **43**

EXAMPLE 8 | Solving a Logarithmic Equation Algebraically and Graphically

Solve the equation $\log(x + 2) + \log(x - 1) = 1$ algebraically and graphically.

SOLUTION 1: Algebraic

We first combine the logarithmic terms, using the Laws of Logarithms.

$\log[(x + 2)(x - 1)] = 1$	Law 1
$(x + 2)(x - 1) = 10$	Exponential form (or raise 10 to each side)
$x^2 + x - 2 = 10$	Expand left side
$x^2 + x - 12 = 0$	Subtract 10
$(x + 4)(x - 3) = 0$	Factor
$x = -4 \quad \text{or} \quad x = 3$	

We check these potential solutions in the original equation and find that $x = -4$ is not a solution (because logarithms of negative numbers are undefined), but $x = 3$ is a solution. (See *Check Your Answers.*)

CHECK YOUR ANSWER

$x = -4$:

$$\log(-4 + 2) + \log(-4 - 1)$$
$$= \log(-2) + \log(-5)$$
undefined ✗

$x = 3$:

$$\log(3 + 2) + \log(3 - 1)$$
$$= \log 5 + \log 2 = \log(5 \cdot 2)$$
$$= \log 10 = 1 \quad ✔$$

SOLUTION 2: Graphical

We first move all terms to one side of the equation:

$$\log(x + 2) + \log(x - 1) - 1 = 0$$

Then we graph

$$y = \log(x + 2) + \log(x - 1) - 1$$

as in Figure 2. The solutions are the x-intercepts of the graph. Thus, the only solution is $x \approx 3$.

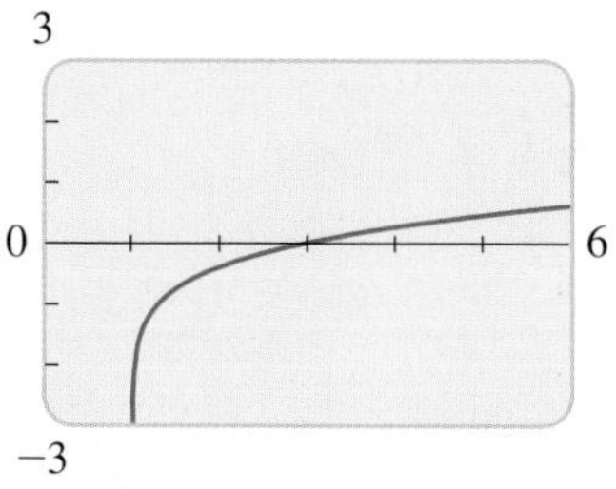

FIGURE 2

NOW TRY EXERCISE **49**

In Example 9 it's not possible to isolate x algebraically, so we must solve the equation graphically.

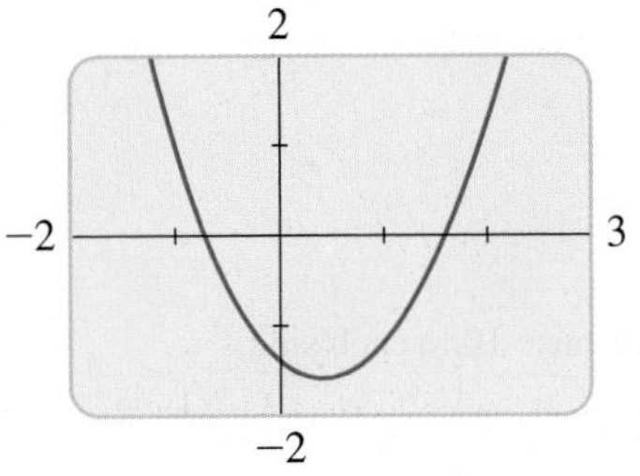

FIGURE 3

EXAMPLE 9 | Solving a Logarithmic Equation Graphically

Solve the equation $x^2 = 2\ln(x + 2)$.

SOLUTION We first move all terms to one side of the equation

$$x^2 - 2\ln(x + 2) = 0$$

Then we graph

$$y = x^2 - 2\ln(x + 2)$$

as in Figure 3. The solutions are the x-intercepts of the graph. Zooming in on the x-intercepts, we see that there are two solutions:

$$x \approx -0.71 \qquad \text{and} \qquad x \approx 1.60$$

NOW TRY EXERCISE **59** ■

Logarithmic equations are used in determining the amount of light that reaches various depths in a lake. (This information helps biologists to determine the types of life a lake can support.) As light passes through water (or other transparent materials such as glass or plastic), some of the light is absorbed. It's easy to see that the murkier the water, the more light is absorbed. The exact relationship between light absorption and the distance light travels in a material is described in the next example.

The intensity of light in a lake diminishes with depth.

EXAMPLE 10 | Transparency of a Lake

If I_0 and I denote the intensity of light before and after going through a material and x is the distance (in feet) the light travels in the material, then according to the **Beer-Lambert Law**,

$$-\frac{1}{k}\ln\left(\frac{I}{I_0}\right) = x$$

where k is a constant depending on the type of material.

(a) Solve the equation for I.

(b) For a certain lake $k = 0.025$, and the light intensity is $I_0 = 14$ lumens (lm). Find the light intensity at a depth of 20 ft.

SOLUTION

(a) We first isolate the logarithmic term.

$$-\frac{1}{k}\ln\left(\frac{I}{I_0}\right) = x \qquad \text{Given equation}$$

$$\ln\left(\frac{I}{I_0}\right) = -kx \qquad \text{Multiply by } -k$$

$$\frac{I}{I_0} = e^{-kx} \qquad \text{Exponential form}$$

$$I = I_0e^{-kx} \qquad \text{Multiply by } I_0$$

(b) We find I using the formula from part (a).

$$I = I_0e^{-kx} \qquad \text{From part (a)}$$

$$= 14e^{(-0.025)(20)} \qquad I_0 = 14,\ k = 0.025,\ x = 20$$

$$\approx 8.49 \qquad \text{Calculator}$$

The light intensity at a depth of 20 ft is about 8.5 lm.

NOW TRY EXERCISE **85** ■

▼ Compound Interest

Recall the formulas for interest that we found in Section 4.1. If a principal P is invested at an interest rate r for a period of t years, then the amount A of the investment is given by

$$A = P(1 + r) \qquad \text{Simple interest (for one year)}$$

$$A(t) = P\left(1 + \frac{r}{n}\right)^{nt} \qquad \text{Interest compounded } n \text{ times per year}$$

$$A(t) = Pe^{rt} \qquad \text{Interest compounded continuously}$$

We can use logarithms to determine the time it takes for the principal to increase to a given amount.

EXAMPLE 11 | Finding the Term for an Investment to Double

A sum of \$5000 is invested at an interest rate of 5% per year. Find the time required for the money to double if the interest is compounded according to the following method.

(a) Semiannually **(b)** Continuously

SOLUTION

(a) We use the formula for compound interest with $P = \$5000$, $A(t) = \$10{,}000$, $r = 0.05$, and $n = 2$ and solve the resulting exponential equation for t.

$$5000\left(1 + \frac{0.05}{2}\right)^{2t} = 10{,}000 \qquad P\left(1 + \frac{r}{n}\right)^{nt} = A$$

$$(1.025)^{2t} = 2 \qquad \text{Divide by 5000}$$

$$\log 1.025^{2t} = \log 2 \qquad \text{Take log of each side}$$

$$2t \log 1.025 = \log 2 \qquad \text{Law 3 (bring down the exponent)}$$

$$t = \frac{\log 2}{2 \log 1.025} \qquad \text{Divide by 2 log 1.025}$$

$$t \approx 14.04 \qquad \text{Calculator}$$

The money will double in 14.04 years.

(b) We use the formula for continuously compounded interest with $P = \$5000$, $A(t) = \$10{,}000$, and $r = 0.05$ and solve the resulting exponential equation for t.

$$5000e^{0.05t} = 10{,}000 \qquad Pe^{rt} = A$$

$$e^{0.05t} = 2 \qquad \text{Divide by 5000}$$

$$\ln e^{0.05t} = \ln 2 \qquad \text{Take ln of each side}$$

$$0.05t = \ln 2 \qquad \text{Property of ln}$$

$$t = \frac{\ln 2}{0.05} \qquad \text{Divide by 0.05}$$

$$t \approx 13.86 \qquad \text{Calculator}$$

The money will double in 13.86 years.

NOW TRY EXERCISE **75** ■

EXAMPLE 12 | Time Required to Grow an Investment

A sum of \$1000 is invested at an interest rate of 4% per year. Find the time required for the amount to grow to \$4000 if interest is compounded continuously.

SOLUTION We use the formula for continuously compounded interest with $P = \$1000$, $A(t) = \$4000$, and $r = 0.04$ and solve the resulting exponential equation for t.

$$
\begin{aligned}
1000e^{0.04t} &= 4000 && Pe^{rt} = A \\
e^{0.04t} &= 4 && \text{Divide by 1000} \\
0.04t &= \ln 4 && \text{Take ln of each side} \\
t &= \frac{\ln 4}{0.04} && \text{Divide by 0.04} \\
t &\approx 34.66 && \text{Calculator}
\end{aligned}
$$

The amount will be $4000 in about 34 years and 8 months.

NOW TRY EXERCISE **77**

4.5 EXERCISES

CONCEPTS

1. Let's solve the exponential equation $2e^x = 50$.

(a) First, we isolate e^x to get the equivalent equation ________.

(b) Next, we take ln of each side to get the equivalent equation ________.

(c) Now we use a calculator to find $x =$ ________.

2. Let's solve the logarithmic equation $\log 3 + \log(x - 2) = \log x$.

(a) First, we combine the logarithms to get the equivalent equation ________.

(b) Next, we write each side in exponential form to get the equivalent equation ________.

(c) Now we find $x =$ ________.

SKILLS

3–28 ■ Find the solution of the exponential equation, rounded to four decimal places.

3. $10^x = 25$

4. $10^{-x} = 4$

5. $e^{-2x} = 7$

6. $e^{3x} = 12$

7. $2^{1-x} = 3$

8. $3^{2x-1} = 5$

9. $3e^x = 10$

10. $2e^{12x} = 17$

11. $e^{1-4x} = 2$

12. $4(1 + 10^{5x}) = 9$

13. $4 + 3^{5x} = 8$

14. $2^{3x} = 34$

15. $8^{0.4x} = 5$

16. $3^{x/14} = 0.1$

17. $5^{-x/100} = 2$

18. $e^{3-5x} = 16$

19. $e^{2x+1} = 200$

20. $\left(\frac{1}{4}\right)^x = 75$

21. $5^x = 4^{x+1}$

22. $10^{1-x} = 6^x$

23. $2^{3x+1} = 3^{x-2}$

24. $7^{x/2} = 5^{1-x}$

25. $\dfrac{50}{1 + e^{-x}} = 4$

26. $\dfrac{10}{1 + e^{-x}} = 2$

27. $100(1.04)^{2t} = 300$

28. $(1.00625)^{12t} = 2$

29–36 ■ Solve the equation.

29. $e^{2x} - 3e^x + 2 = 0$

30. $e^{2x} - e^x - 6 = 0$

31. $e^{4x} + 4e^{2x} - 21 = 0$

32. $e^x - 12e^{-x} - 1 = 0$

33. $x^2 2^x - 2^x = 0$

34. $x^2 10^x - x10^x = 2(10^x)$

35. $4x^3e^{-3x} - 3x^4e^{-3x} = 0$

36. $x^2e^x + xe^x - e^x = 0$

37–54 ■ Solve the logarithmic equation for x.

37. $\ln x = 10$

38. $\ln(2 + x) = 1$

39. $\log x = -2$

40. $\log(x - 4) = 3$

41. $\log(3x + 5) = 2$

42. $\log_3(2 - x) = 3$

43. $4 - \log(3 - x) = 3$

44. $\log_2(x^2 - x - 2) = 2$

45. $\log_2 3 + \log_2 x = \log_2 5 + \log_2(x - 2)$

46. $2 \log x = \log 2 + \log(3x - 4)$

47. $\log x + \log(x - 1) = \log(4x)$

48. $\log_5 x + \log_5(x + 1) = \log_5 20$

49. $\log_5(x + 1) - \log_5(x - 1) = 2$

50. $\log_3(x + 15) - \log_3(x - 1) = 2$

51. $\log_2 x + \log_2(x - 3) = 2$

52. $\log x + \log(x - 3) = 1$

53. $\log_9(x - 5) + \log_9(x + 3) = 1$

54. $\ln(x - 1) + \ln(x + 2) = 1$

55. For what value of x is the following true?

$$\log(x + 3) = \log x + \log 3$$

56. For what value of x is it true that $(\log x)^3 = 3 \log x$?

57. Solve for x: $2^{2/\log_5 x} = \frac{1}{16}$

58. Solve for x: $\log_2(\log_3 x) = 4$

59–66 ■ Use a graphing device to find all solutions of the equation, rounded to two decimal places.

59. $\ln x = 3 - x$ **60.** $\log x = x^2 - 2$

61. $x^3 - x = \log(x + 1)$ **62.** $x = \ln(4 - x^2)$

63. $e^x = -x$ **64.** $2^{-x} = x - 1$

65. $4^{-x} = \sqrt{x}$ **66.** $e^{x^2} - 2 = x^3 - x$

67–70 ■ Solve the inequality.

67. $\log(x - 2) + \log(9 - x) < 1$

68. $3 \leq \log_2 x \leq 4$

69. $2 < 10^x < 5$ **70.** $x^2e^x - 2e^x < 0$

71-74 ■ Find the inverse function of f.

71. $f(x) = 2^{2x}$ **72.** $f(x) = 3^{x+1}$

73. $f(x) = \log_2(x - 1)$ **74.** $f(x) = \log 3x$

APPLICATIONS

75. Compound Interest A man invests \$5000 in an account that pays 8.5% interest per year, compounded quarterly.
(a) Find the amount after 3 years.
(b) How long will it take for the investment to double?

76. Compound Interest A woman invests \$6500 in an account that pays 6% interest per year, compounded continuously.
(a) What is the amount after 2 years?
(b) How long will it take for the amount to be \$8000?

77. Compound Interest Find the time required for an investment of \$5000 to grow to \$8000 at an interest rate of 7.5% per year, compounded quarterly.

78. Compound Interest Nancy wants to invest \$4000 in saving certificates that bear an interest rate of 9.75% per year, compounded semiannually. How long a time period should she choose to save an amount of \$5000?

79. Doubling an Investment How long will it take for an investment of \$1000 to double in value if the interest rate is 8.5% per year, compounded continuously?

80. Interest Rate A sum of \$1000 was invested for 4 years, and the interest was compounded semiannually. If this sum amounted to \$1435.77 in the given time, what was the interest rate?

81. Radioactive Decay A 15-g sample of radioactive iodine decays in such a way that the mass remaining after t days is given by $m(t) = 15e^{-0.087t}$, where $m(t)$ is measured in grams. After how many days is there only 5 g remaining?

82. Sky Diving The velocity of a sky diver t seconds after jumping is given by $v(t) = 80(1 - e^{-0.2t})$. After how many seconds is the velocity 70 ft/s?

83. Fish Population A small lake is stocked with a certain species of fish. The fish population is modeled by the function

$$P = \frac{10}{1 + 4e^{-0.8t}}$$

where P is the number of fish in thousands and t is measured in years since the lake was stocked.
(a) Find the fish population after 3 years.
(b) After how many years will the fish population reach 5000 fish?

84. Transparency of a Lake Environmental scientists measure the intensity of light at various depths in a lake to find the "transparency" of the water. Certain levels of transparency are required for the biodiversity of the submerged macrophyte population. In a certain lake the intensity of light at depth x is given by

$$I = 10e^{-0.008x}$$

where I is measured in lumens and x in feet.
(a) Find the intensity I at a depth of 30 ft.
(b) At what depth has the light intensity dropped to $I = 5$?

85. Atmospheric Pressure Atmospheric pressure P (in kilopascals, kPa) at altitude h (in kilometers, km) is governed by the formula

$$\ln\left(\frac{P}{P_0}\right) = -\frac{h}{k}$$

where $k = 7$ and $P_0 = 100$ kPa are constants.
(a) Solve the equation for P.
(b) Use part (a) to find the pressure P at an altitude of 4 km.

86. Cooling an Engine Suppose you're driving your car on a cold winter day (20°F outside) and the engine overheats (at about 220°F). When you park, the engine begins to cool down. The temperature T of the engine t minutes after you park satisfies the equation

$$\ln\left(\frac{T - 20}{200}\right) = -0.11t$$

(a) Solve the equation for T.
(b) Use part (a) to find the temperature of the engine after 20 min ($t = 20$).

87. Electric Circuits An electric circuit contains a battery that produces a voltage of 60 volts (V), a resistor with a resistance of 13 ohms (Ω), and an inductor with an inductance of 5 henrys (H), as shown in the figure. Using calculus, it can be shown that the current $I = I(t)$ (in amperes, A) t seconds after the switch is closed is $I = \frac{60}{13}(1 - e^{-13t/5})$.
(a) Use this equation to express the time t as a function of the current I.
(b) After how many seconds is the current 2 A?

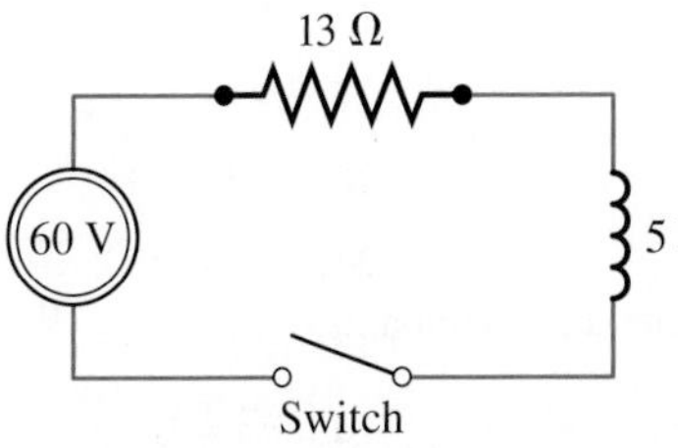

88. Learning Curve A *learning curve* is a graph of a function $P(t)$ that measures the performance of someone learning a skill as a function of the training time t. At first, the rate of learning is rapid. Then, as performance increases and approaches a maximal value M, the rate of learning decreases. It has been found that the function

$$P(t) = M - Ce^{-kt}$$

where k and C are positive constants and $C < M$ is a reasonable model for learning.

(a) Express the learning time t as a function of the performance level P.

(b) For a pole-vaulter in training, the learning curve is given by

$$P(t) = 20 - 14e^{-0.024t}$$

where $P(t)$ is the height he is able to pole-vault after t months. After how many months of training is he able to vault 12 ft?

(c) Draw a graph of the learning curve in part (b).

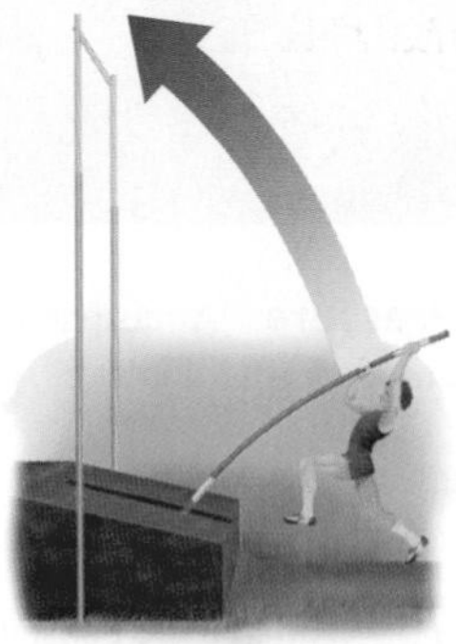

DISCOVERY ■ DISCUSSION ■ WRITING

89. Estimating a Solution Without actually solving the equation, find two whole numbers between which the solution of $9^x = 20$ must lie. Do the same for $9^x = 100$. Explain how you reached your conclusions.

90. A Surprising Equation Take logarithms to show that the equation

$$x^{1/\log x} = 5$$

has no solution. For what values of k does the equation

$$x^{1/\log x} = k$$

have a solution? What does this tell us about the graph of the function $f(x) = x^{1/\log x}$? Confirm your answer using a graphing device.

91. Disguised Equations Each of these equations can be transformed into an equation of linear or quadratic type by applying the hint. Solve each equation.

(a) $(x - 1)^{\log(x-1)} = 100(x - 1)$ [Take log of each side.]

(b) $\log_2 x + \log_4 x + \log_8 x = 11$ [Change all logs to base 2.]

(c) $4^x - 2^{x+1} = 3$ [Write as a quadratic in 2^x.]

4.6 Modeling with Exponential and Logarithmic Functions

Exponential Growth (Doubling Time) ▶ Exponential Growth (Relative Growth Rate) ▶ Radioactive Decay ▶ Newton's Law of Cooling ▶ Logarithmic Scales

Many processes that occur in nature, such as population growth, radioactive decay, heat diffusion, and numerous others, can be modeled by using exponential functions. Logarithmic functions are used in models for the loudness of sounds, the intensity of earthquakes, and many other phenomena. In this section we study exponential and logarithmic models.

▼ Exponential Growth (Doubling Time)

Suppose we start with a single bacterium, which divides every hour. After one hour we have 2 bacteria, after two hours we have 2^2 or 4 bacteria, after three hours we have 2^3 or 8 bacteria, and so on (see Figure 1). We see that we can model the bacteria population after t hours by $f(t) = 2^t$.

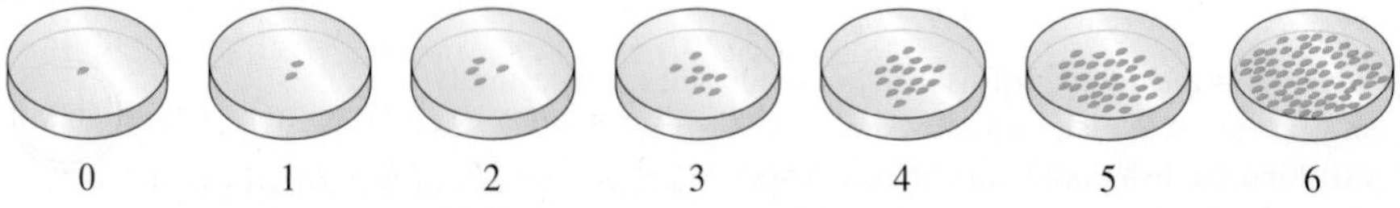

FIGURE 1 Bacteria population

If we start with 10 of these bacteria, then the population is modeled by $f(t) = 10 \cdot 2^t$. A slower-growing strain of bacteria doubles every 3 hours; in this case the population is modeled by $f(t) = 10 \cdot 2^{t/3}$. In general, we have the following.

EXPONENTIAL GROWTH (DOUBLING TIME)

If the intial size of a population is n_0 and the doubling time is a, then the size of the population at time t is

$$n(t) = n_0 2^{t/a}$$

where a and t are measured in the same time units (minutes, hours, days, years, and so on).

EXAMPLE 1 | Bacteria Population

Under ideal conditions a certain bacteria population doubles every three hours. Initially there are 1000 bacteria in a colony.

(a) Find a model for the bacteria population after t hours.

(b) How many bacteria are in the colony after 15 hours?

(c) When will the bacteria count reach 100,000?

SOLUTION

(a) The population at time t is modeled by

$$n(t) = 1000 \cdot 2^{t/3}$$

where t is measured in hours.

(b) After 15 hours the number of bacteria is

$$n(15) = 1000 \cdot 2^{15/3} = 32{,}000$$

(c) We set $n(t) = 100{,}000$ in the model that we found in part (a) and solve the resulting exponential equation for t.

$$100{,}000 = 1000 \cdot 2^{t/3} \qquad n(t) = 1000 \cdot 2^{t/3}$$

$$100 = 2^{t/3} \qquad \text{Divide by 1000}$$

$$\log 100 = \log 2^{t/3} \qquad \text{Take log of each side}$$

$$2 = \frac{t}{3} \log 2 \qquad \text{Properties of log}$$

$$t = \frac{6}{\log 2} \approx 19.93 \qquad \text{Solve for } t$$

For guidelines on working with significant figures, see the Appendix, *Calculations and Significant Figures.*

The bacteria level reaches 100,000 in about 20 hours.

NOW TRY EXERCISE **1** ■

EXAMPLE 2 | Rabbit Population

A certain breed of rabbit was introduced onto a small island 8 months ago. The current rabbit population on the island is estimated to be 4100 and doubling every 3 months.

(a) What was the initial size of the rabbit population?

(b) Estimate the population one year after the rabbits were introduced to the island.

(c) Sketch a graph of the rabbit population.

SOLUTION

(a) The doubling time is $a = 3$, so the population at time t is

$$n(t) = n_0 2^{t/3} \qquad \text{Model}$$

where n_0 is the initial population. Since the population is 4100 when t is 8 months, we have

$$n(8) = n_0 2^{8/3} \qquad \text{From model}$$

$$4100 = n_0 2^{8/3} \qquad \text{Because } n(8) = 4100$$

$$n_0 = \frac{4100}{2^{8/3}} \qquad \text{Divide by } 2^{8/3} \text{ and switch sides}$$

$$n_0 \approx 645 \qquad \text{Calculate}$$

Thus we estimate that 645 rabbits were introduced onto the island.

(b) From part (a) we know that the initial population is $n_0 = 645$, so we can model the population after t months by

$$n(t) = 645 \cdot 2^{t/3} \qquad \text{Model}$$

After one year $t = 12$, so

$$n(12) = 645 \cdot 2^{12/3} \approx 10{,}320$$

So after one year there would be about 10,000 rabbits.

(c) We first note that the domain is $t \geq 0$. The graph is shown in Figure 2.

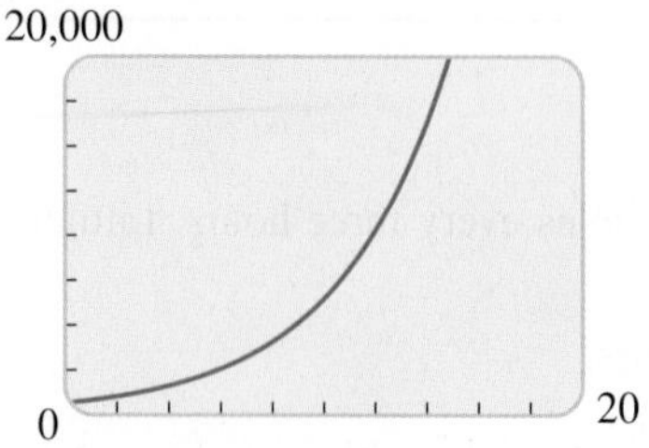

FIGURE 2 $n(t) = 645 \cdot 2^{t/3}$

For guidelines on working with significant figures, see the Appendix, *Calculations and Significant Figures.*

NOW TRY EXERCISE **3** ■

▼ Exponential Growth (Relative Growth Rate)

We have used an exponential function with base 2 to model population growth (in terms of the doubling time). We could also model the same population with an exponential function with base 3 (in terms of the tripling time). In fact, we can find an exponential model with any base. If we use the base e, we get the following model of a population in terms of the **relative growth rate** r: the rate of population growth expressed as a proportion of the population at any time. For instance, if $r = 0.02$, then at any time t the growth rate is 2% of the population at time t.

EXPONENTIAL GROWTH (RELATIVE GROWTH RATE)

A population that experiences **exponential growth** increases according to the model

$$n(t) = n_0 e^{rt}$$

where $n(t)$ = population at time t

n_0 = initial size of the population

r = relative rate of growth (expressed as a proportion of the population)

t = time

Notice that the formula for population growth is the same as that for continuously compounded interest. In fact, the same principle is at work in both cases: The growth of a population (or an investment) per time period is proportional to the size of the population (or the amount of the investment). A population of 1,000,000 will increase more in one year than a population of 1000; in exactly the same way, an investment of \$1,000,000 will increase more in one year than an investment of \$1000.

In the following examples we assume that the populations grow exponentially.

EXAMPLE 3 | Predicting the Size of a Population

The initial bacterium count in a culture is 500. A biologist later makes a sample count of bacteria in the culture and finds that the relative rate of growth is 40% per hour.

(a) Find a function that models the number of bacteria after t hours.

(b) What is the estimated count after 10 hours?

(c) When will the bacteria count reach 80,000?

(d) Sketch the graph of the function $n(t)$.

SOLUTION

(a) We use the exponential growth model with $n_0 = 500$ and $r = 0.4$ to get

$$n(t) = 500e^{0.4t}$$

where t is measured in hours.

(b) Using the function in part (a), we find that the bacterium count after 10 hours is

$$n(10) = 500e^{0.4(10)} = 500e^4 \approx 27{,}300$$

(c) We set $n(t) = 80{,}000$ and solve the resulting exponential equation for t:

$$80{,}000 = 500 \cdot e^{0.4t} \qquad n(t) = 500 \cdot e^{0.4t}$$

$$160 = e^{0.4t} \qquad \text{Divide by 500}$$

$$\ln 160 = 0.4t \qquad \text{Take ln of each side}$$

$$t = \frac{\ln 160}{0.4} \approx 12.68 \qquad \text{Solve for } t$$

The bacteria level reaches 80,000 in about 12.7 hours.

(d) The graph is shown in Figure 3.

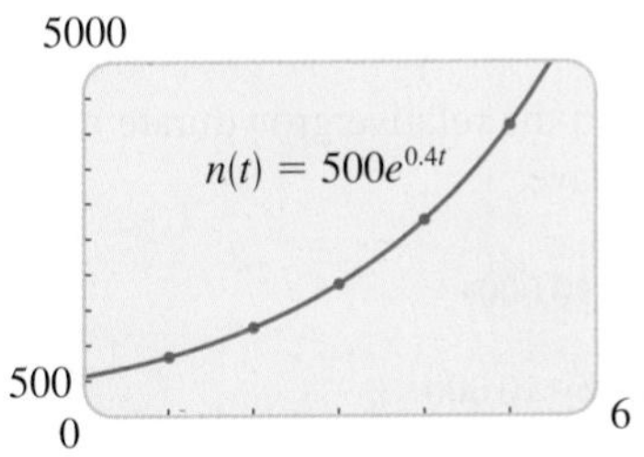

FIGURE 3

NOW TRY EXERCISE **5**

EXAMPLE 4 | Comparing Different Rates of Population Growth

In 2000 the population of the world was 6.1 billion, and the relative rate of growth was 1.4% per year. It is claimed that a rate of 1.0% per year would make a significant difference in the total population in just a few decades. Test this claim by estimating the population of the world in the year 2050 using a relative rate of growth of (a) 1.4% per year and (b) 1.0% per year.

Graph the population functions for the next 100 years for the two relative growth rates in the same viewing rectangle.

The relative growth of world population has been declining over the past few decades—from 2% in 1995 to 1.3% in 2006.

Standing Room Only
The population of the world was about 6.1 billion in 2000 and was increasing at 1.4% per year. Assuming that each person occupies an average of 4 ft^2 of the surface of the earth, the exponential model for population growth projects that by the year 2801 there will be standing room only! (The total land surface area of the world is about 1.8×10^{15} ft^2.)

SOLUTION

(a) By the exponential growth model we have

$$n(t) = 6.1e^{0.014t}$$

where $n(t)$ is measured in billions and t is measured in years since 2000. Because the year 2050 is 50 years after 2000, we find

$$n(50) = 6.1e^{0.014(50)} = 6.1e^{0.7} \approx 12.3$$

The estimated population in the year 2050 is about 12.3 billion.

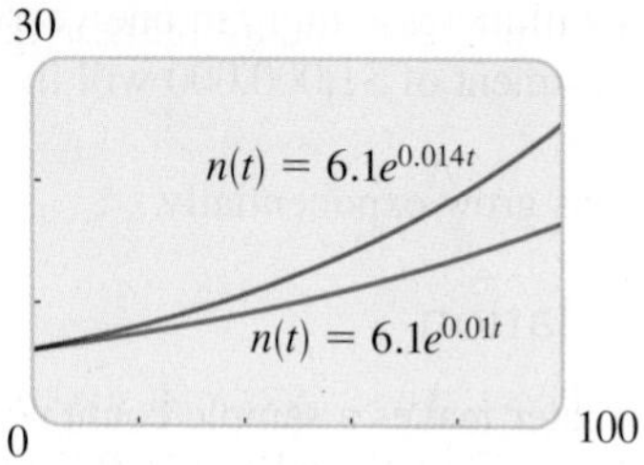

FIGURE 4

(b) We use the function

$$n(t) = 6.1e^{0.010t}$$

and find

$$n(50) = 6.1e^{0.010(50)} = 6.1e^{0.50} \approx 10.1$$

The estimated population in the year 2050 is about 10.1 billion.

The graphs in Figure 4 show that a small change in the relative rate of growth will, over time, make a large difference in population size.

NOW TRY EXERCISE **7** ■

EXAMPLE 5 | Expressing a Model in Terms of *e*

A culture starts with 10,000 bacteria, and the number doubles every 40 minutes.

(a) Find a function $n(t) = n_0 2^{t/a}$ that models the number of bacteria after t minutes.

(b) Find a function $n(t) = n_0 e^{rt}$ that models the number of bacteria after t minutes.

(c) Sketch a graph of the number of bacteria at time t.

SOLUTION

(a) The initial population is $n_0 = 10{,}000$. The doubling time is $a = 40 \text{ min} = 2/3$ h. Since $1/a = 3/2 = 1.5$, the model is

$$n(t) = 10{,}000 \cdot 2^{1.5t}$$

(b) The initial population is $n_0 = 10{,}000$. We need to find the relative growth rate r. Since there are 20,000 bacteria when $t = 2/3$ h, we have

$20{,}000 = 10{,}000e^{r(2/3)}$	$n(t) = 10{,}000e^{rt}$
$2 = e^{r(2/3)}$	Divide by 10,000
$\ln 2 = \ln e^{r(2/3)}$	Take ln of each side
$\ln 2 = r(2/3)$	Property of ln
$r = \dfrac{3 \ln 2}{2} \approx 1.0397$	Solve for r

Now that we know the relative growth rate r, we can find the model:

$$n(t) = 10{,}000e^{1.0397t}$$

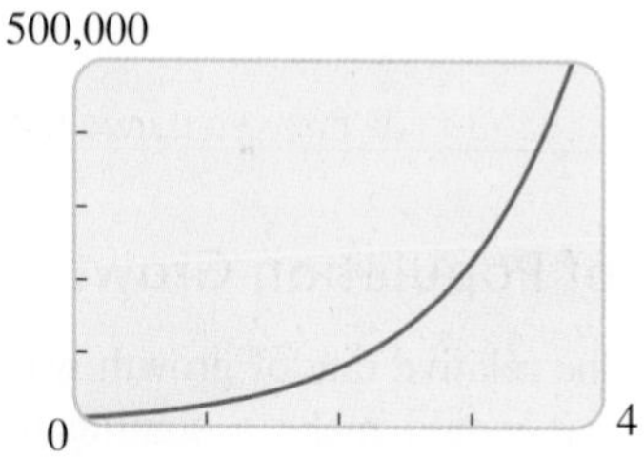

FIGURE 5
Graphs of $y = 10{,}000 \cdot 2^{1.5t}$ and $y = 10{,}000e^{1.0397t}$

(c) We can graph the model in part (a) or the one in part (b). The graphs are identical. See Figure 5.

NOW TRY EXERCISE **9** ■

▼ Radioactive Decay

Radioactive substances decay by spontaneously emitting radiation. The rate of decay is proportional to the mass of the substance. This is analogous to population growth except that the mass *decreases*. Physicists express the rate of decay in terms of **half-life**. For example, the half-life of radium-226 is 1600 years, so a 100-g sample decays to 50 g (or $\frac{1}{2} \times 100$ g) in 1600 years, then to 25 g (or $\frac{1}{2} \times \frac{1}{2} \times 100$ g) in 3200 years, and so on. In

The half-lives of **radioactive elements** vary from very long to very short. Here are some examples.

Element	Half-life
Thorium-232	14.5 billion years
Uranium-235	4.5 billion years
Thorium-230	80,000 years
Plutonium-239	24,360 years
Carbon-14	5,730 years
Radium-226	1,600 years
Cesium-137	30 years
Strontium-90	28 years
Polonium-210	140 days
Thorium-234	25 days
Iodine-135	8 days
Radon-222	3.8 days
Lead-211	3.6 minutes
Krypton-91	10 seconds

general, for a radioactive substance with mass m_0 and half-life h, the amount remaining at time t is modeled by

$$m(t) = m_0 2^{-t/h}$$

where h and t are measured in the same time units (minutes, hours, days, years, and so on).

To express this model in the form $m(t) = m_0 e^{rt}$, we need to find the relative decay rate r. Since h is the half-life, we have

$$m(t) = m_0 e^{-rt} \qquad \text{Model}$$

$$\frac{m_0}{2} = m_0 e^{-rh} \qquad h \text{ is the half-life}$$

$$\frac{1}{2} = e^{-rh} \qquad \text{Divide by } m_0$$

$$\ln\frac{1}{2} = -rh \qquad \text{Take ln of each side}$$

$$r = \frac{\ln 2}{h} \qquad \text{Solve for } r$$

This last equation allows us to find the rate r from the half-life h.

RADIOACTIVE DECAY MODEL

If m_0 is the initial mass of a radioactive substance with half-life h, then the mass remaining at time t is modeled by the function

$$m(t) = m_0 e^{-rt}$$

where $r = \dfrac{\ln 2}{h}$.

EXAMPLE 6 | Radioactive Decay

Polonium-210 (^{210}Po) has a half-life of 140 days. Suppose a sample of this substance has a mass of 300 mg.

(a) Find a function $m(t) = m_0 2^{-t/h}$ that models the mass remaining after t days.

(b) Find a function $m(t) = m_0 e^{-rt}$ that models the mass remaining after t days.

(c) Find the mass remaining after one year.

(d) How long will it take for the sample to decay to a mass of 200 mg?

(e) Draw a graph of the sample mass as a function of time.

SOLUTION

(a) We have $m_0 = 300$ and $h = 140$, so the amount remaining after t days is

$$m(t) = 300 \cdot 2^{-t/140}$$

(b) We have $m_0 = 300$ and $r = \ln 2/140 \approx -0.00495$, so the amount remaining after t days is

$$m(t) = 300 \cdot e^{-0.00495t}$$

(c) We use the function we found in part (a) with $t = 365$ (one year).

$$m(365) = 300e^{-0.00495(365)} \approx 49.256$$

Thus, approximately 49 mg of ^{210}Po remains after one year.

In parts (c) and (d) we can also use the model found in part (a). Check that the result is the same using either model.

Radioactive Waste

Harmful radioactive isotopes are produced whenever a nuclear reaction occurs, whether as the result of an atomic bomb test, a nuclear accident such as the one at Chernobyl in 1986, or the uneventful production of electricity at a nuclear power plant.

One radioactive material that is produced in atomic bombs is the isotope strontium-90 (^{90}Sr), with a half-life of 28 years. This is deposited like calcium in human bone tissue, where it can cause leukemia and other cancers. However, in the decades since atmospheric testing of nuclear weapons was halted, ^{90}Sr levels in the environment have fallen to a level that no longer poses a threat to health.

Nuclear power plants produce radioactive plutonium-239 (^{239}Pu), which has a half-life of 24,360 years. Because of its long half-life, ^{239}Pu could pose a threat to the environment for thousands of years. So great care must be taken to dispose of it properly. The difficulty of ensuring the safety of the disposed radioactive waste is one reason that nuclear power plants remain controversial.

(d) We use the function that we found in part (a) with $m(t) = 200$ and solve the resulting exponential equation for t.

$$300e^{-0.00495t} = 200 \qquad m(t) = m_0e^{-rt}$$

$$e^{-0.00495t} = \tfrac{2}{3} \qquad \text{Divided by 300}$$

$$\ln e^{-0.00495t} = \ln \tfrac{2}{3} \qquad \text{Take ln of each side}$$

$$-0.00495t = \ln \tfrac{2}{3} \qquad \text{Property of ln}$$

$$t = -\frac{\ln \frac{2}{3}}{0.00495} \qquad \text{Solve for } t$$

$$t \approx 81.9 \qquad \text{Calculator}$$

The time required for the sample to decay to 200 mg is about 82 days.

(e) We can graph the model in part (a) or the one in part (b). The graphs are identical. See Figure 6.

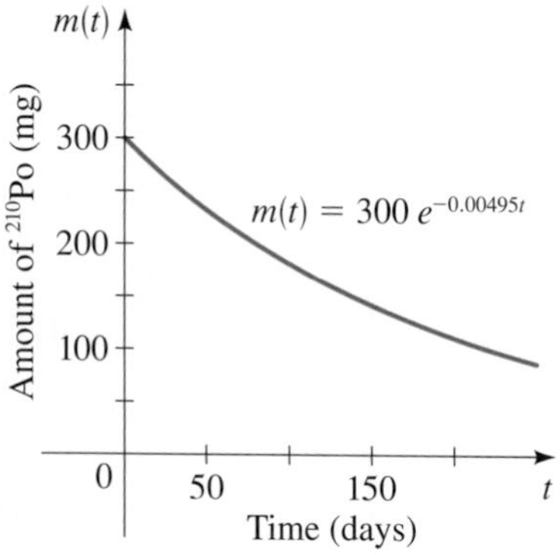

FIGURE 6

NOW TRY EXERCISE **17** ■

Newton's Law of Cooling

Newton's Law of Cooling states that the rate at which an object cools is proportional to the temperature difference between the object and its surroundings, provided that the temperature difference is not too large. By using calculus, the following model can be deduced from this law.

NEWTON'S LAW OF COOLING

If D_0 is the initial temperature difference between an object and its surroundings, and if its surroundings have temperature T_s, then the temperature of the object at time t is modeled by the function

$$T(t) = T_s + D_0e^{-kt}$$

where k is a positive constant that depends on the type of object.

EXAMPLE 7 | Newton's Law of Cooling

A cup of coffee has a temperature of 200°F and is placed in a room that has a temperature of 70°F. After 10 min the temperature of the coffee is 150°F.

(a) Find a function that models the temperature of the coffee at time t.

(b) Find the temperature of the coffee after 15 min.

(c) When will the coffee have cooled to 100°F?

(d) Illustrate by drawing a graph of the temperature function.

SOLUTION

(a) The temperature of the room is $T_s = 70$°F, and the initial temperature difference is

$$D_0 = 200 - 70 = 130°\text{F}$$

So by Newton's Law of Cooling, the temperature after t minutes is modeled by the function

$$T(t) = 70 + 130e^{-kt}$$

We need to find the constant k associated with this cup of coffee. To do this, we use the fact that when $t = 10$, the temperature is $T(10) = 150$. So we have

$$\begin{aligned} 70 + 130e^{-10k} &= 150 && T_s + D_0e^{-kt} = T(t) \\ 130e^{-10k} &= 80 && \text{Subtract 70} \\ e^{-10k} &= \tfrac{8}{13} && \text{Divide by 130} \\ -10k &= \ln \tfrac{8}{13} && \text{Take ln of each side} \\ k &= -\tfrac{1}{10} \ln \tfrac{8}{13} && \text{Solve for } k \\ k &\approx 0.04855 && \text{Calculator} \end{aligned}$$

Substituting this value of k into the expression for $T(t)$, we get

$$T(t) = 70 + 130e^{-0.04855t}$$

(b) We use the function that we found in part (a) with $t = 15$.

$$T(15) = 70 + 130e^{-0.04855(15)} \approx 133°\text{F}$$

(c) We use the function that we found in part (a) with $T(t) = 100$ and solve the resulting exponential equation for t.

$$\begin{aligned} 70 + 130e^{-0.04855t} &= 100 && T_s + D_0e^{-kt} = T(t) \\ 130e^{-0.04855t} &= 30 && \text{Subtract 70} \\ e^{-0.04855t} &= \tfrac{3}{13} && \text{Divide by 130} \\ -0.04855t &= \ln \tfrac{3}{13} && \text{Take ln of each side} \\ t &= \frac{\ln \frac{3}{13}}{-0.04855} && \text{Solve for } t \\ t &\approx 30.2 && \text{Calculator} \end{aligned}$$

The coffee will have cooled to 100°F after about half an hour.

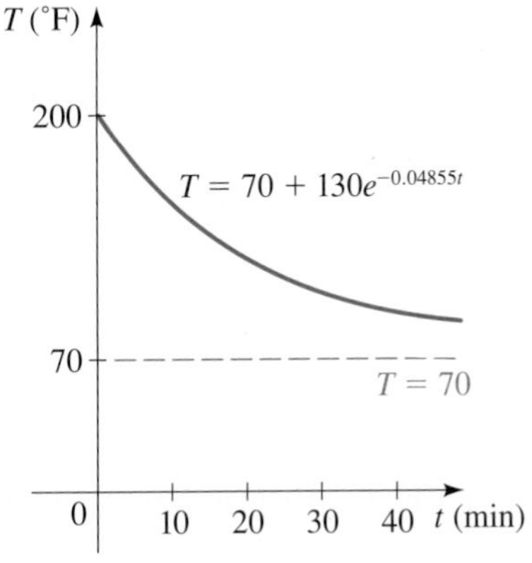

FIGURE 7 Temperature of coffee after t minutes

(d) The graph of the temperature function is sketched in Figure 7. Notice that the line $t = 70$ is a horizontal asymptote. (Why?)

NOW TRY EXERCISE **25** ■

▼ Logarithmic Scales

When a physical quantity varies over a very large range, it is often convenient to take its logarithm in order to have a more manageable set of numbers. We discuss three such situations: the pH scale, which measures acidity; the Richter scale, which measures the

pH for Some Common Substances

Substance	pH
Milk of magnesia	10.5
Seawater	8.0–8.4
Human blood	7.3–7.5
Crackers	7.0–8.5
Hominy	6.9–7.9
Cow's milk	6.4–6.8
Spinach	5.1–5.7
Tomatoes	4.1–4.4
Oranges	3.0–4.0
Apples	2.9–3.3
Limes	1.3–2.0
Battery acid	1.0

intensity of earthquakes; and the decibel scale, which measures the loudness of sounds. Other quantities that are measured on logarithmic scales are light intensity, information capacity, and radiation.

The pH Scale Chemists measured the acidity of a solution by giving its hydrogen ion concentration until Søren Peter Lauritz Sørensen, in 1909, proposed a more convenient measure. He defined

$$\text{pH} = -\log[\text{H}^+]$$

where $[\text{H}^+]$ is the concentration of hydrogen ions measured in moles per liter (M). He did this to avoid very small numbers and negative exponents. For instance,

$$\text{if} \quad [\text{H}^+] = 10^{-4}\text{ M}, \quad \text{then} \quad \text{pH} = -\log_{10}(10^{-4}) = -(-4) = 4$$

Solutions with a pH of 7 are defined as *neutral*, those with pH < 7 are *acidic*, and those with pH > 7 are *basic*. Notice that when the pH increases by one unit, $[\text{H}^+]$ decreases by a factor of 10.

EXAMPLE 8 | pH Scale and Hydrogen Ion Concentration

(a) The hydrogen ion concentration of a sample of human blood was measured to be $[\text{H}^+] = 3.16 \times 10^{-8}$ M. Find the pH and classify the blood as acidic or basic.

(b) The most acidic rainfall ever measured occurred in Scotland in 1974; its pH was 2.4. Find the hydrogen ion concentration.

SOLUTION

(a) A calculator gives

$$\text{pH} = -\log[\text{H}^+] = -\log(3.16 \times 10^{-8}) \approx 7.5$$

Since this is greater than 7, the blood is basic.

(b) To find the hydrogen ion concentration, we need to solve for $[\text{H}^+]$ in the logarithmic equation

$$\log[\text{H}^+] = -\text{pH}$$

So we write it in exponential form.

$$[\text{H}^+] = 10^{-\text{pH}}$$

In this case pH $= 2.4$, so

$$[\text{H}^+] = 10^{-2.4} \approx 4.0 \times 10^{-3}\text{ M}$$

NOW TRY EXERCISE **29** ■

Largest Earthquakes

Location	Date	Magnitude
Chile	1960	9.5
Alaska	1964	9.2
Sumatra	2004	9.1
Alaska	1957	9.1
Kamchatka	1952	9.0
Chile	2010	8.8
Ecuador	1906	8.8
Alaska	1965	8.7
Sumatra	2005	8.7
Tibet	1950	8.6
Kamchatka	1923	8.5
Indonesia	1938	8.5
Kuril Islands	1963	8.5

The Richter Scale In 1935 the American geologist Charles Richter (1900–1984) defined the magnitude M of an earthquake to be

$$M = \log \frac{I}{S}$$

where I is the intensity of the earthquake (measured by the amplitude of a seismograph reading taken 100 km from the epicenter of the earthquake) and S is the intensity of a "standard" earthquake (whose amplitude is 1 micron $= 10^{-4}$ cm). The magnitude of a standard earthquake is

$$M = \log \frac{S}{S} = \log 1 = 0$$

Richter studied many earthquakes that occurred between 1900 and 1950. The largest had magnitude 8.9 on the Richter scale, and the smallest had magnitude 0. This corresponds to a ratio of intensities of 800,000,000, so the Richter scale provides more manageable numbers to work with. For instance, an earthquake of magnitude 6 is ten times stronger than an earthquake of magnitude 5.

EXAMPLE 9 | Magnitude of Earthquakes

The 1906 earthquake in San Francisco had an estimated magnitude of 8.3 on the Richter scale. In the same year a powerful earthquake occurred on the Colombia-Ecuador border that was four times as intense. What was the magnitude of the Colombia-Ecuador earthquake on the Richter scale?

SOLUTION If I is the intensity of the San Francisco earthquake, then from the definition of magnitude we have

$$M = \log \frac{I}{S} = 8.3$$

The intensity of the Colombia-Ecuador earthquake was $4I$, so its magnitude was

$$M = \log \frac{4I}{S} = \log 4 + \log \frac{I}{S} = \log 4 + 8.3 \approx 8.9$$

NOW TRY EXERCISE **35** ■

EXAMPLE 10 | Intensity of Earthquakes

The 1989 Loma Prieta earthquake that shook San Francisco had a magnitude of 7.1 on the Richter scale. How many times more intense was the 1906 earthquake (see Example 9) than the 1989 event?

SOLUTION If I_1 and I_2 are the intensities of the 1906 and 1989 earthquakes, then we are required to find I_1/I_2. To relate this to the definition of magnitude, we divide the numerator and denominator by S.

$$\log \frac{I_1}{I_2} = \log \frac{I_1/S}{I_2/S} \qquad \text{Divide numerator and denominator by } S$$

$$= \log \frac{I_1}{S} - \log \frac{I_2}{S} \qquad \text{Law 2 of logarithms}$$

$$= 8.3 - 7.1 = 1.2 \qquad \text{Definition of earthquake magnitude}$$

Therefore,

$$\frac{I_1}{I_2} = 10^{\log(I_1/I_2)} = 10^{1.2} \approx 16$$

The 1906 earthquake was about 16 times as intense as the 1989 earthquake.

NOW TRY EXERCISE **37** ■

The Decibel Scale The ear is sensitive to an extremely wide range of sound intensities. We take as a reference intensity $I_0 = 10^{-12}\ \text{W/m}^2$ (watts per square meter) at a frequency of 1000 hertz, which measures a sound that is just barely audible (the threshold of hearing). The psychological sensation of loudness varies with the logarithm of the intensity (the Weber-Fechner Law), so the **intensity level** B, measured in decibels (dB), is defined as

$$B = 10 \log \frac{I}{I_0}$$

The intensity level of the barely audible reference sound is

$$B = 10 \log \frac{I_0}{I_0} = 10 \log 1 = 0 \text{ dB}$$

The **intensity levels of sounds** that we can hear vary from very loud to very soft. Here are some examples of the decibel levels of commonly heard sounds.

Source of sound	B (dB)
Jet takeoff	140
Jackhammer	130
Rock concert	120
Subway	100
Heavy traffic	80
Ordinary traffic	70
Normal conversation	50
Whisper	30
Rustling leaves	10–20
Threshold of hearing	0

EXAMPLE 11 | Sound Intensity of a Jet Takeoff

Find the decibel intensity level of a jet engine during takeoff if the intensity was measured at 100 W/m^2.

SOLUTION From the definition of intensity level we see that

$$B = 10 \log \frac{I}{I_0} = 10 \log \frac{10^2}{10^{-12}} = 10 \log 10^{14} = 140 \text{ dB}$$

Thus, the intensity level is 140 dB.

NOW TRY EXERCISE **41**

The table in the margin lists decibel intensity levels for some common sounds ranging from the threshold of human hearing to the jet takeoff of Example 11. The threshold of pain is about 120 dB.

4.6 EXERCISES

APPLICATIONS

1–16 ■ These exercises use the population growth model.

1. **Bacteria Culture** A certain culture of the bacterium *Streptococcus A* initially has 10 bacteria and is observed to double every 1.5 hours.
   (a) Find an exponential model $n(t) = n_0 2^{t/a}$ for the number of bacteria in the culture after t hours.
   (b) Estimate the number of bacteria after 35 hours.
   (c) When will the bacteria count reach 10,000?

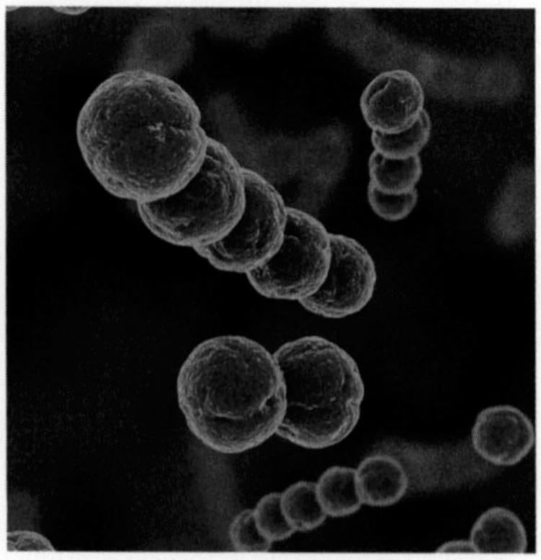

Streptococcus A
(12,000 × magnification)

2. **Bacteria Culture** A certain culture of the bacterium *Rhodobacter sphaeroides* initially has 25 bacteria and is observed to double every 5 hours.
   (a) Find an exponential model $n(t) = n_0 2^{t/a}$ for the number of bacteria in the culture after t hours.
   (b) Estimate the number of bacteria after 18 hours.
   (c) After how many hours will the bacteria count reach 1 million?

3. **Squirrel Population** A grey squirrel population was introduced in a certain county of Great Britain 30 years ago. Biologists observe that the population doubles every 6 years, and now the population is 100,000.
   (a) What was the initial size of the squirrel population?
   (b) Estimate the squirrel population 10 years from now.
   (c) Sketch a graph of the squirrel population.

4. **Bird Population** A certain species of bird was introduced in a certain county 25 years ago. Biologists observe that the population doubles every 10 years, and now the population is 13,000.
   (a) What was the initial size of the bird population?
   (b) Estimate the bird population 5 years from now.
   (c) Sketch a graph of the bird population.

5. **Fox Population** The fox population in a certain region has a relative growth rate of 8% per year. It is estimated that the population in 2005 was 18,000.
   (a) Find a function $n(t) = n_0 e^{rt}$ that models the population t years after 2005.
   (b) Use the function from part (a) to estimate the fox population in the year 2013.
   (c) Sketch a graph of the fox population function for the years 2005–2013.

6. **Fish Population** The population of a certain species of fish has a relative growth rate of 1.2% per year. It is estimated that the population in 2000 was 12 million.
   (a) Find an exponential model $n(t) = n_0 e^{rt}$ for the population t years after 2000.
   (b) Estimate the fish population in the year 2005.
   (c) Sketch a graph of the fish population.

7. **Population of a Country** The population of a country has a relative growth rate of 3% per year. The government is trying to reduce the growth rate to 2%. The population in 1995 was approximately 110 million. Find the projected population for the year 2020 for the following conditions.
   (a) The relative growth rate remains at 3% per year.
   (b) The relative growth rate is reduced to 2% per year.

8. **Bacteria Culture** It is observed that a certain bacteria culture has a relative growth rate of 12% per hour, but in the presence of an antibiotic the relative growth rate is reduced to 5% per hour. The initial number of bacteria in the culture is 22. Find the projected population after 24 hours for the following conditions.
   (a) No antibiotic is present, so the relative growth rate is 12%.
   (b) An antibiotic is present in the culture, so the relative growth rate is reduced to 5%.

9. **Population of a City** The population of a certain city was 112,000 in 2006, and the observed doubling time for the population is 18 years.
   (a) Find an exponential model $n(t) = n_0 2^{t/a}$ for the population t years after 2006.
   (b) Find an exponential model $n(t) = n_0 e^{rt}$ for the population t years after 2006.
   (c) Sketch a graph of the population at time t.
   (d) Estimate when the population will reach 500,000.

10. **Bat Population** The bat population in a certain Midwestern county was 350,000 in 2009, and the observed doubling time for the population is 25 years.
   (a) Find an exponential model $n(t) = n_0 2^{t/a}$ for the population t years after 2006.
   (b) Find an exponential model $n(t) = n_0 e^{rt}$ for the population t years after 2006.
   (c) Sketch a graph of the population at time t.
   (d) Estimate when the population will reach 2 million.

11. **Deer Population** The graph shows the deer population in a Pennsylvania county between 2003 and 2007. Assume that the population grows exponentially.
   (a) What was the deer population in 2003?
   (b) Find a function that models the deer population t years after 2003.
   (c) What is the projected deer population in 2011?
   (d) In what year will the deer population reach 100,000?

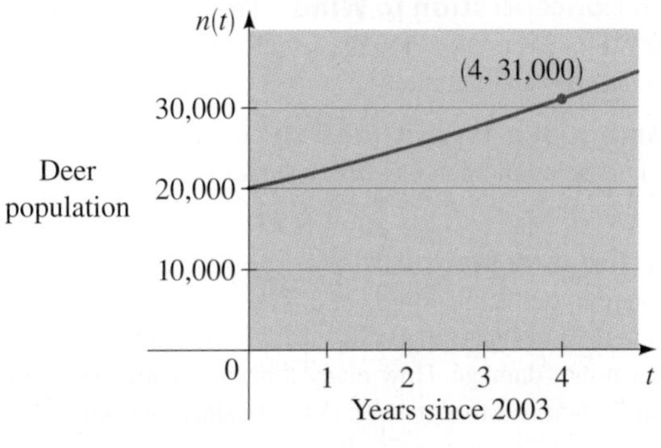

12. **Frog Population** Some bullfrogs were introduced into a small pond. The graph shows the bullfrog population for the next few years. Assume that the population grows exponentially.
   (a) What was the initial bullfrog population?
   (b) Find a function that models the bullfrog population t years since the bullfrogs were put into the pond.
   (c) What is the projected bullfrog population after 15 years?
   (d) Estimate how long it takes the population to reach 75,000.

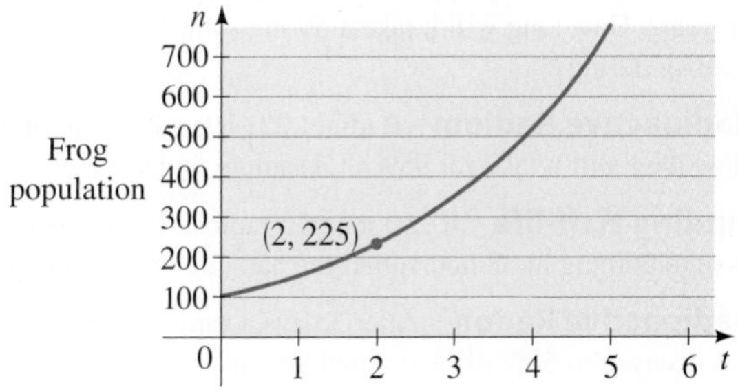

13. **Bacteria Culture** A culture starts with 8600 bacteria. After one hour the count is 10,000.
   (a) Find a function that models the number of bacteria $n(t)$ after t hours.
   (b) Find the number of bacteria after 2 hours.
   (c) After how many hours will the number of bacteria double?

14. **Bacteria Culture** The count in a culture of bacteria was 400 after 2 hours and 25,600 after 6 hours.
   (a) What is the relative rate of growth of the bacteria population? Express your answer as a percentage.
   (b) What was the initial size of the culture?
   (c) Find a function that models the number of bacteria $n(t)$ after t hours.
   (d) Find the number of bacteria after 4.5 hours.
   (e) When will the number of bacteria be 50,000?

15. **Population of California** The population of California was 29.76 million in 1990 and 33.87 million in 2000. Assume that the population grows exponentially.
   (a) Find a function that models the population t years after 1990.
   (b) Find the time required for the population to double.
   (c) Use the function from part (a) to predict the population of California in the year 2010. Look up California's actual population in 2010, and compare.

16. **World Population** The population of the world was 5.7 billion in 1995, and the observed relative growth rate was 2% per year.
   (a) By what year will the population have doubled?
   (b) By what year will the population have tripled?

17–24 ■ These exercises use the radioactive decay model.

17. **Radioactive Radium** The half-life of radium-226 is 1600 years. Suppose we have a 22-mg sample.
   (a) Find a function $m(t) = m_0 2^{-t/h}$ that models the mass remaining after t years.
   (b) Find a function $m(t) = m_0 e^{-rt}$ that models the mass remaining after t years.
   (c) How much of the sample will remain after 4000 years?
   (d) After how long will only 18 mg of the sample remain?

18. **Radioactive Cesium** The half-life of cesium-137 is 30 years. Suppose we have a 10-g sample.
   (a) Find a function $m(t) = m_0 2^{-t/h}$ that models the mass remaining after t years.

(b) Find a function $m(t) = m_0 e^{-rt}$ that models the mass remaining after t years.
(c) How much of the sample will remain after 80 years?
(d) After how long will only 2 g of the sample remain?

19. **Radioactive Strontium** The half-life of strontium-90 is 28 years. How long will it take a 50-mg sample to decay to a mass of 32 mg?

20. **Radioactive Radium** Radium-221 has a half-life of 30 s. How long will it take for 95% of a sample to decay?

21. **Finding Half-life** If 250 mg of a radioactive element decays to 200 mg in 48 hours, find the half-life of the element.

22. **Radioactive Radon** After 3 days a sample of radon-222 has decayed to 58% of its original amount.
(a) What is the half-life of radon-222?
(b) How long will it take the sample to decay to 20% of its original amount?

23. **Carbon-14 Dating** A wooden artifact from an ancient tomb contains 65% of the carbon-14 that is present in living trees. How long ago was the artifact made? (The half-life of carbon-14 is 5730 years.)

24. **Carbon-14 Dating** The burial cloth of an Egyptian mummy is estimated to contain 59% of the carbon-14 it contained originally. How long ago was the mummy buried? (The half-life of carbon-14 is 5730 years.)

25–28 ■ These exercises use Newton's Law of Cooling.

25. **Cooling Soup** A hot bowl of soup is served at a dinner party. It starts to cool according to Newton's Law of Cooling, so its temperature at time t is given by

$$T(t) = 65 + 145e^{-0.05t}$$

where t is measured in minutes and T is measured in °F.
(a) What is the initial temperature of the soup?
(b) What is the temperature after 10 min?
(c) After how long will the temperature be 100°F?

26. **Time of Death** Newton's Law of Cooling is used in homicide investigations to determine the time of death. The normal body temperature is 98.6°F. Immediately following death, the body begins to cool. It has been determined experimentally that the constant in Newton's Law of Cooling is approximately $k = 0.1947$, assuming that time is measured in hours. Suppose that the temperature of the surroundings is 60°F.
(a) Find a function $T(t)$ that models the temperature t hours after death.
(b) If the temperature of the body is now 72°F, how long ago was the time of death?

27. **Cooling Turkey** A roasted turkey is taken from an oven when its temperature has reached 185°F and is placed on a table in a room where the temperature is 75°F.
(a) If the temperature of the turkey is 150°F after half an hour, what is its temperature after 45 min?
(b) When will the turkey cool to 100°F?

28. **Boiling Water** A kettle full of water is brought to a boil in a room with temperature 20°C. After 15 min the temperature of the water has decreased from 100°C to 75°C. Find the temperature after another 10 min. Illustrate by graphing the temperature function.

29–43 ■ These exercises deal with logarithmic scales.

29. **Finding pH** The hydrogen ion concentration of a sample of each substance is given. Calculate the pH of the substance.
(a) Lemon juice: $[H^+] = 5.0 \times 10^{-3}$ M
(b) Tomato juice: $[H^+] = 3.2 \times 10^{-4}$ M
(c) Seawater: $[H^+] = 5.0 \times 10^{-9}$ M

30. **Finding pH** An unknown substance has a hydrogen ion concentration of $[H^+] = 3.1 \times 10^{-8}$ M. Find the pH and classify the substance as acidic or basic.

31. **Ion Concentration** The pH reading of a sample of each substance is given. Calculate the hydrogen ion concentration of the substance.
(a) Vinegar: pH = 3.0
(b) Milk: pH = 6.5

32. **Ion Concentration** The pH reading of a glass of liquid is given. Find the hydrogen ion concentration of the liquid.
(a) Beer: pH = 4.6
(b) Water: pH = 7.3

33. **Finding pH** The hydrogen ion concentrations in cheeses range from 4.0×10^{-7} M to 1.6×10^{-5} M. Find the corresponding range of pH readings.

34. **Ion Concentration in Wine** The pH readings for wines vary from 2.8 to 3.8. Find the corresponding range of hydrogen ion concentrations.

35. **Earthquake Magnitudes** If one earthquake is 20 times as intense as another, how much larger is its magnitude on the Richter scale?

36. **Earthquake Magnitudes** The 1906 earthquake in San Francisco had a magnitude of 8.3 on the Richter scale. At the same time in Japan an earthquake with magnitude 4.9 caused only minor damage. How many times more intense was the San Francisco earthquake than the Japanese earthquake?

37. **Earthquake Magnitudes** The Alaska earthquake of 1964 had a magnitude of 8.6 on the Richter scale. How many times more intense was this than the 1906 San Francisco earthquake? (See Exercise 36.)

38. **Earthquake Magnitudes** The Northridge, California, earthquake of 1994 had a magnitude of 6.8 on the Richter scale. A year later, a 7.2-magnitude earthquake struck Kobe, Japan. How many times more intense was the Kobe earthquake than the Northridge earthquake?

39. **Earthquake Magnitudes** The 1985 Mexico City earthquake had a magnitude of 8.1 on the Richter scale. The 1976 earthquake in Tangshan, China, was 1.26 times as intense. What was the magnitude of the Tangshan earthquake?

40. **Subway Noise** The intensity of the sound of a subway train was measured at 98 dB. Find the intensity in W/m^2.

41. **Traffic Noise** The intensity of the sound of traffic at a busy intersection was measured at 2.0×10^{-5} W/m^2. Find the intensity level in decibels.

42. **Comparing Decibel Levels** The noise from a power mower was measured at 106 dB. The noise level at a rock concert was measured at 120 dB. Find the ratio of the intensity of the rock music to that of the power mower.

43. **Inverse Square Law for Sound** A law of physics states that the intensity of sound is inversely proportional to the square of the distance d from the source: $I = k/d^2$.

(a) Use this model and the equation

$$B = 10 \log \frac{I}{I_0}$$

(described in this section) to show that the decibel levels B_1 and B_2 at distances d_1 and d_2 from a sound source are related by the equation

$$B_2 = B_1 + 20 \log \frac{d_1}{d_2}$$

(b) The intensity level at a rock concert is 120 dB at a distance 2 m from the speakers. Find the intensity level at a distance of 10 m.

CHAPTER 4 | REVIEW

CONCEPT CHECK

1. **(a)** Write an equation that defines the exponential function with base a.
   (b) What is the domain of this function?
   (c) What is the range of this function?
   (d) Sketch the general shape of the graph of the exponential function for each case.
   (i) $a > 1$ (ii) $0 < a < 1$

2. If x is large, which function grows faster, $y = 2^x$ or $y = x^2$?

3. **(a)** How is the number e defined?
   (b) What is the natural exponential function?

4. **(a)** How is the logarithmic function $y = \log_a x$ defined?
   (b) What is the domain of this function?
   (c) What is the range of this function?
   (d) Sketch the general shape of the graph of the function $y = \log_a x$ if $a > 1$.
   (e) What is the natural logarithm?
   (f) What is the common logarithm?

5. State the three Laws of Logarithms.

6. State the Change of Base Formula.

7. **(a)** How do you solve an exponential equation?
   (b) How do you solve a logarithmic equation?

8. Suppose an amount P is invested at an interest rate r and A is the amount after t years.
   (a) Write an expression for A if the interest is compounded n times per year.
   (b) Write an expression for A if the interest is compounded continuously.

9. The initial size of a population is n_0 and the population grows exponentially.
   (a) Write an expression for the population in terms of the doubling time a.
   (b) Write an expression for the population in terms of the relative growth rate r.

10. **(a)** What is the half-life of a radioactive substance?
    (b) If a radioactive substance has initial mass m_0 and half-life h, write an expression for the mass $m(t)$ remaining at time t.

11. What does Newton's Law of Cooling say?

12. What do the pH scale, the Richter scale, and the decibel scale have in common? What do they measure?

EXERCISES

1–4 ■ Use a calculator to find the indicated values of the exponential function, correct to three decimal places.

1. $f(x) = 5^x$; $f(-1.5), f(\sqrt{2}), f(2.5)$
2. $f(x) = 3 \cdot 2^x$; $f(-2.2), f(\sqrt{7}), f(5.5)$
3. $g(x) = 4 \cdot (\frac{2}{3})^{x-2}$; $g(-0.7), g(e), g(\pi)$
4. $g(x) = \frac{7}{4}e^{x+1}$; $g(-2), g(\sqrt{3}), g(3.6)$

5–16 ■ Sketch the graph of the function. State the domain, range, and asymptote.

5. $f(x) = 2^{-x+1}$
6. $f(x) = 3^{x-2}$
7. $g(x) = 3 + 2^x$
8. $g(x) = 5^{-x} - 5$

9. $f(x) = \log_3(x - 1)$ **10.** $g(x) = \log(-x)$

11. $f(x) = 2 - \log_2 x$ **12.** $f(x) = 3 + \log_5(x + 4)$

13. $F(x) = e^x - 1$ **14.** $G(x) = \frac{1}{2}e^{x-1}$

15. $g(x) = 2 \ln x$ **16.** $g(x) = \ln(x^2)$

17–20 ■ Find the domain of the function.

17. $f(x) = 10^{x^2} + \log(1 - 2x)$

18. $g(x) = \log(2 + x - x^2)$

19. $h(x) = \ln(x^2 - 4)$ **20.** $k(x) = \ln|x|$

21–24 ■ Write the equation in exponential form.

21. $\log_2 1024 = 10$ **22.** $\log_6 37 = x$

23. $\log x = y$ **24.** $\ln c = 17$

25–28 ■ Write the equation in logarithmic form.

25. $2^6 = 64$ **26.** $49^{-1/2} = \frac{1}{7}$

27. $10^x = 74$ **28.** $e^k = m$

29–44 ■ Evaluate the expression without using a calculator.

29. $\log_2 128$ **30.** $\log_8 1$

31. $10^{\log 45}$ **32.** $\log 0.000001$

33. $\ln(e^6)$ **34.** $\log_4 8$

35. $\log_3(\frac{1}{27})$ **36.** $2^{\log_2 13}$

37. $\log_5 \sqrt{5}$ **38.** $e^{2\ln 7}$

39. $\log 25 + \log 4$ **40.** $\log_3 \sqrt{243}$

41. $\log_2 16^{23}$ **42.** $\log_5 250 - \log_5 2$

43. $\log_8 6 - \log_8 3 + \log_8 2$ **44.** $\log \log 10^{100}$

45–50 ■ Expand the logarithmic expression.

45. $\log(AB^2C^3)$ **46.** $\log_2(x\sqrt{x^2 + 1})$

47. $\ln\sqrt{\dfrac{x^2 - 1}{x^2 + 1}}$ **48.** $\log\left(\dfrac{4x^3}{y^2(x - 1)^5}\right)$

49. $\log_5\left(\dfrac{x^2(1 - 5x)^{3/2}}{\sqrt{x^3 - x}}\right)$ **50.** $\ln\left(\dfrac{\sqrt[3]{x^4 + 12}}{(x + 16)\sqrt{x - 3}}\right)$

51–56 ■ Combine into a single logarithm.

51. $\log 6 + 4 \log 2$ **52.** $\log x + \log(x^2y) + 3 \log y$

53. $\frac{3}{2}\log_2(x - y) - 2 \log_2(x^2 + y^2)$

54. $\log_5 2 + \log_5(x + 1) - \frac{1}{3}\log_5(3x + 7)$

55. $\log(x - 2) + \log(x + 2) - \frac{1}{2}\log(x^2 + 4)$

56. $\frac{1}{2}[\ln(x - 4) + 5 \ln(x^2 + 4x)]$

57–68 ■ Solve the equation. Find the exact solution if possible; otherwise, use a calculator to approximate to two decimals.

57. $3^{2x-7} = 27$ **58.** $5^{4-x} = \frac{1}{125}$

59. $2^{3x-5} = 7$ **60.** $10^{6-3x} = 18$

61. $4^{1-x} = 3^{2x+5}$ **62.** $e^{3x/4} = 10$

63. $x^2e^{2x} + 2xe^{2x} = 8e^{2x}$ **64.** $3^{2x} - 3^x - 6 = 0$

65. $\log_2(1 - x) = 4$

66. $\log x + \log(x + 1) = \log 12$

67. $\log_8(x + 5) - \log_8(x - 2) = 1$

68. $\ln(2x - 3) + 1 = 0$

69–72 ■ Use a calculator to find the solution of the equation, rounded to six decimal places.

69. $5^{-2x/3} = 0.63$ **70.** $2^{3x-5} = 7$

71. $5^{2x+1} = 3^{4x-1}$ **72.** $e^{-15k} = 10{,}000$

73–76 ■ Draw a graph of the function and use it to determine the asymptotes and the local maximum and minimum values.

73. $y = e^{x/(x+2)}$ **74.** $y = 10^x - 5^x$

75. $y = \log(x^3 - x)$ **76.** $y = 2x^2 - \ln x$

77–78 ■ Find the solutions of the equation, rounded to two decimal places.

77. $3 \log x = 6 - 2x$ **78.** $4 - x^2 = e^{-2x}$

79–80 ■ Solve the inequality graphically.

79. $\ln x > x - 2$ **80.** $e^x < 4x^2$

81. Use a graph of $f(x) = e^x - 3e^{-x} - 4x$ to find, approximately, the intervals on which f is increasing and on which f is decreasing.

82. Find an equation of the line shown in the figure.

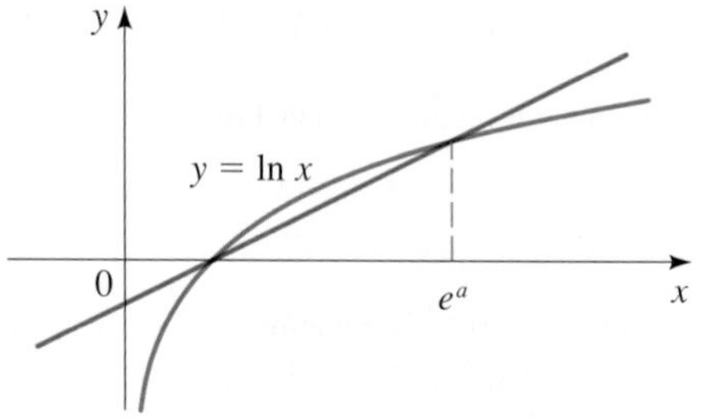

83–86 ■ Use the Change of Base Formula to evaluate the logarithm, rounded to six decimal places.

83. $\log_4 15$ **84.** $\log_7(\frac{3}{4})$

85. $\log_9 0.28$ **86.** $\log_{100} 250$

87. Which is larger, $\log_4 258$ or $\log_5 620$?

88. Find the inverse of the function $f(x) = 2^{3^x}$, and state its domain and range.

89. If \$12,000 is invested at an interest rate of 10% per year, find the amount of the investment at the end of 3 years for each compounding method.

(a) Semiannually **(b)** Monthly

(c) Daily **(d)** Continuously

90. A sum of \$5000 is invested at an interest rate of $8\frac{1}{2}\%$ per year, compounded semiannually.

(a) Find the amount of the investment after $1\frac{1}{2}$ years.

(b) After what period of time will the investment amount to \$7000?

(c) If interest were compounded continously instead of semi-annually, how long would it take for the amount to grow to \$7000?

91. A money market account pays 5.2% annual interest, compounded daily. If \$100,000 is invested in this account, how long will it take for the account to accumulate \$10,000 in interest?

92. A retirement savings plan pays 4.5% interest, compounded continuously. How long will it take for an investment in this plan to double?

93–94 ■ Determine the annual percentage yield (APY) for the given nominal annual interest rate and compounding frequency.

93. 4.25%; daily

94. 3.2%; monthly

95. The stray-cat population in a small town grows exponentially. In 1999 the town had 30 stray cats, and the relative growth rate was 15% per year.

(a) Find a function that models the stray-cat population $n(t)$ after t years.

(b) Find the projected population after 4 years.

(c) Find the number of years required for the stray-cat population to reach 500.

96. A culture contains 10,000 bacteria initially. After an hour the bacteria count is 25,000.

(a) Find the doubling period.

(b) Find the number of bacteria after 3 hours.

97. Uranium-234 has a half-life of 2.7×10^5 years.

(a) Find the amount remaining from a 10-mg sample after a thousand years.

(b) How long will it take this sample to decompose until its mass is 7 mg?

98. A sample of bismuth-210 decayed to 33% of its original mass after 8 days.

(a) Find the half-life of this element.

(b) Find the mass remaining after 12 days.

99. The half-life of radium-226 is 1590 years.

(a) If a sample has a mass of 150 mg, find a function that models the mass that remains after t years.

(b) Find the mass that will remain after 1000 years.

(c) After how many years will only 50 mg remain?

100. The half-life of palladium-100 is 4 days. After 20 days a sample has been reduced to a mass of 0.375 g.

(a) What was the initial mass of the sample?

(b) Find a function that models the mass remaining after t days.

(c) What is the mass after 3 days?

(d) After how many days will only 0.15 g remain?

101. The graph shows the population of a rare species of bird, where t represents years since 1999 and $n(t)$ is measured in thousands.

(a) Find a function that models the bird population at time t in the form $n(t) = n_0e^{rt}$.

(b) What is the bird population expected to be in the year 2010?

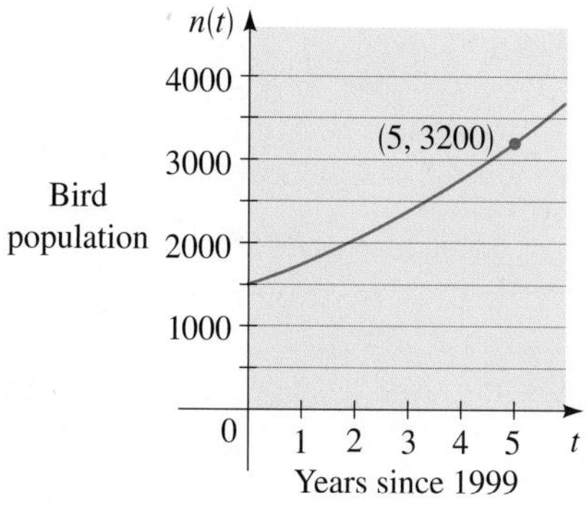

102. A car engine runs at a temperature of 190°F. When the engine is turned off, it cools according to Newton's Law of Cooling with constant $k = 0.0341$, where the time is measured in minutes. Find the time needed for the engine to cool to 90°F if the surrounding temperature is 60°F.

103. The hydrogen ion concentration of fresh egg whites was measured as

$$[H^+] = 1.3 \times 10^{-8}\ M$$

Find the pH, and classify the substance as acidic or basic.

104. The pH of lime juice is 1.9. Find the hydrogen ion concentration.

105. If one earthquake has magnitude 6.5 on the Richter scale, what is the magnitude of another quake that is 35 times as intense?

106. The drilling of a jackhammer was measured at 132 dB. The sound of whispering was measured at 28 dB. Find the ratio of the intensity of the drilling to that of the whispering.

1. Sketch the graph of each function, and state its domain, range, and asymptote. Show the x- and y-intercepts on the graph.

(a) $f(x) = 2^{-x} + 4$ **(b)** $g(x) = \log_3(x + 3)$

2. **(a)** Write the equation $6^{2x} = 25$ in logarithmic form.

(b) Write the equation $\ln A = 3$ in exponential form.

3. Find the exact value of each expression.

(a) $10^{\log 36}$ **(b)** $\ln e^3$

(c) $\log_3 \sqrt{27}$ **(d)** $\log_2 80 - \log_2 10$

(e) $\log_8 4$ **(f)** $\log_6 4 + \log_6 9$

4. Use the Laws of Logarithms to expand the expression:

$$\log \sqrt[3]{\frac{x + 2}{x^4(x^2 + 4)}}$$

5. Combine into a single logarithm: $\ln x - 2\ln(x^2 + 1) + \frac{1}{2}\ln(3 - x^4)$

6. Find the solution of the equation, correct to two decimal places.

(a) $2^{x-1} = 10$ **(b)** $5\ln(3 - x) = 4$

(c) $10^{x+3} = 6^{2x}$ **(d)** $\log_2(x + 2) + \log_2(x - 1) = 2$

7. The initial size of a culture of bacteria is 1000. After one hour the bacteria count is 8000.

(a) Find a function that models the population after t hours.

(b) Find the population after 1.5 hours.

(c) When will the population reach 15,000?

(d) Sketch the graph of the population function.

8. Suppose that \$12,000 is invested in a savings account paying 5.6% interest per year.

(a) Write the formula for the amount in the account after t years if interest is compounded monthly.

(b) Find the amount in the account after 3 years if interest is compounded daily.

(c) How long will it take for the amount in the account to grow to \$20,000 if interest is compounded semiannually?

9. The half-life of krypton-91 (^{91}Kr) is 10 seconds. At time $t = 0$ a heavy canister contains 3 g of this radioactive gas.

(a) Find a function that models the amount $A(t)$ of ^{91}Kr remaining in the canister after t seconds.

(b) How much ^{91}Kr remains after one minute?

(c) When will the amount of ^{91}Kr remaining be reduced to 1 μg (1 microgram, or 10^{-6} g)?

10. An earthquake measuring 6.4 on the Richter scale struck Japan in July 2007, causing extensive damage. Earlier that year, a minor earthquake measuring 3.1 on the Richter scale was felt in parts of Pennsylvania. How many times more intense was the Japanese earthquake than the Pennsylvania earthquake?

Fitting Exponential and Power Curves to Data

In a previous *Focus on Modeling* (page 296) we learned that the shape of a scatter plot helps us to choose the type of curve to use in modeling data. The first plot in Figure 1 fairly screams for a line to be fitted through it, and the second one points to a cubic polynomial. For the third plot it is tempting to fit a second-degree polynomial. But what if an exponential curve fits better? How do we decide this? In this section we learn how to fit exponential and power curves to data and how to decide which type of curve fits the data better. We also learn that for scatter plots like those in the last two plots in Figure 1, the data can be modeled by logarithmic or logistic functions.

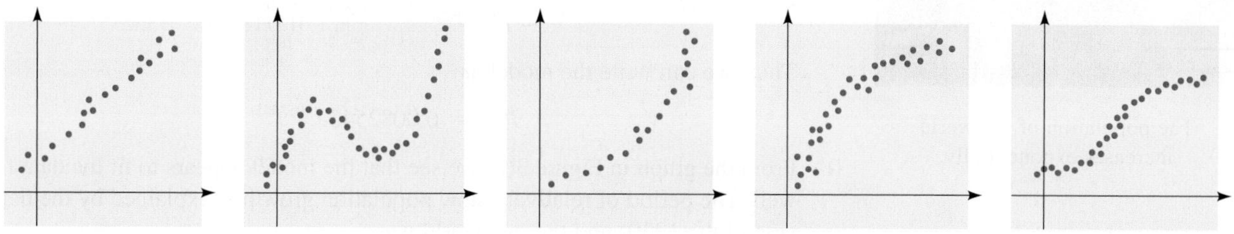

FIGURE 1

Modeling with Exponential Functions

If a scatter plot shows that the data increase rapidly, we might want to model the data using an *exponential model*, that is, a function of the form

$$f(x) = Ce^{kx}$$

where C and k are constants. In the first example we model world population by an exponential model. Recall from Section 4.6 that population tends to increase exponentially.

TABLE 1
World population

Year (t)	World population (P in millions)
1900	1650
1910	1750
1920	1860
1930	2070
1940	2300
1950	2520
1960	3020
1970	3700
1980	4450
1990	5300
2000	6060

EXAMPLE 1 | An Exponential Model for World Population

Table 1 gives the population of the world in the 20th century.

(a) Draw a scatter plot, and note that a linear model is not appropriate.

(b) Find an exponential function that models population growth.

(c) Draw a graph of the function that you found together with the scatter plot. How well does the model fit the data?

(d) Use the model that you found to predict world population in the year 2020.

SOLUTION

(a) The scatter plot is shown in Figure 2. The plotted points do not appear to lie along a straight line, so a linear model is not appropriate.

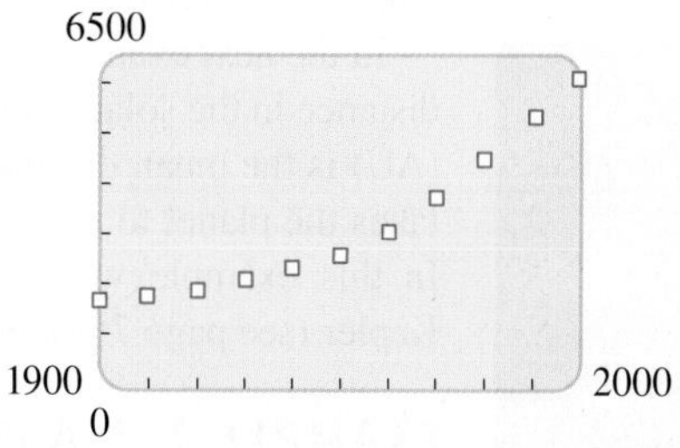

FIGURE 2 Scatter plot of world population

The population of the world increases exponentially.

(b) Using a graphing calculator and the ExpReg command (see Figure 3(a)), we get the exponential model

$$P(t) = (0.0082543) \cdot (1.0137186)^t$$

This is a model of the form $y = Cb^t$. To convert this to the form $y = Ce^{kt}$, we use the properties of exponentials and logarithms as follows:

$$1.0137186^t = e^{\ln 1.0137186^t} \qquad A = e^{\ln A}$$

$$= e^{t \ln 1.0137186} \qquad \ln A^B = B \ln A$$

$$= e^{0.013625t} \qquad \ln 1.0137186 \approx 0.013625$$

Thus, we can write the model as

$$P(t) = 0.0082543e^{0.013625t}$$

(c) From the graph in Figure 3(b) we see that the model appears to fit the data fairly well. The period of relatively slow population growth is explained by the depression of the 1930s and the two world wars.

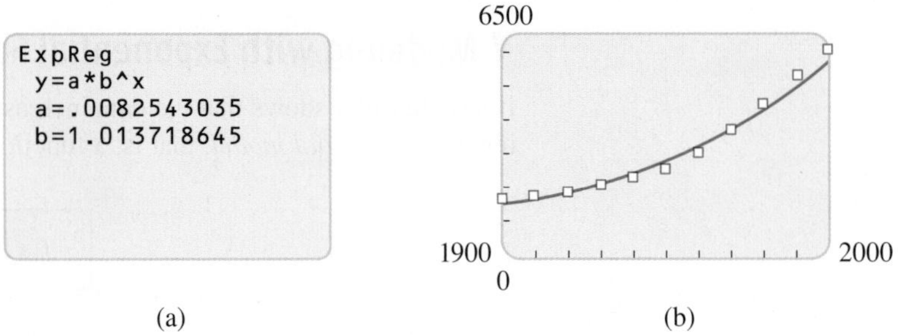

FIGURE 3 Exponential model for world population

(d) The model predicts that the world population in 2020 will be

$$P(2020) = 0.0082543e^{(0.013625)(2020)}$$

$$\approx 7{,}405{,}400{,}000$$

▼ Modeling with Power Functions

If the scatter plot of the data we are studying resembles the graph of $y = ax^2$, $y = ax^{1.32}$, or some other power function, then we seek a *power model*, that is, a function of the form

$$f(x) = ax^n$$

where a is a positive constant and n is any real number.

In the next example we seek a power model for some astronomical data. In astronomy, distance in the solar system is often measured in astronomical units. An *astronomical unit* (AU) is the mean distance from the earth to the sun. The *period* of a planet is the time it takes the planet to make a complete revolution around the sun (measured in earth years). In this example we derive the remarkable relationship, first discovered by Johannes Kepler (see page 754), between the mean distance of a planet from the sun and its period.

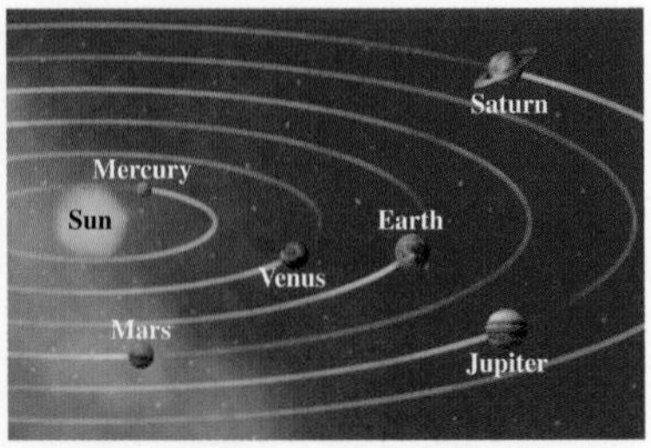

EXAMPLE 2 | A Power Model for Planetary Periods

Table 2 gives the mean distance d of each planet from the sun in astronomical units and its period T in years.

TABLE 2
Distances and periods of the planets

Planet	d	T
Mercury	0.387	0.241
Venus	0.723	0.615
Earth	1.000	1.000
Mars	1.523	1.881
Jupiter	5.203	11.861
Saturn	9.541	29.457
Uranus	19.190	84.008
Neptune	30.086	164.784
Pluto	39.507	248.350

(a) Sketch a scatter plot. Is a linear model appropriate?

(b) Find a power function that models the data.

(c) Draw a graph of the function you found and the scatter plot on the same graph. How well does the model fit the data?

(d) Use the model that you found to calculate the period of an asteroid whose mean distance from the sun is 5 AU.

SOLUTION

(a) The scatter plot shown in Figure 4 indicates that the plotted points do not lie along a straight line, so a linear model is not appropriate.

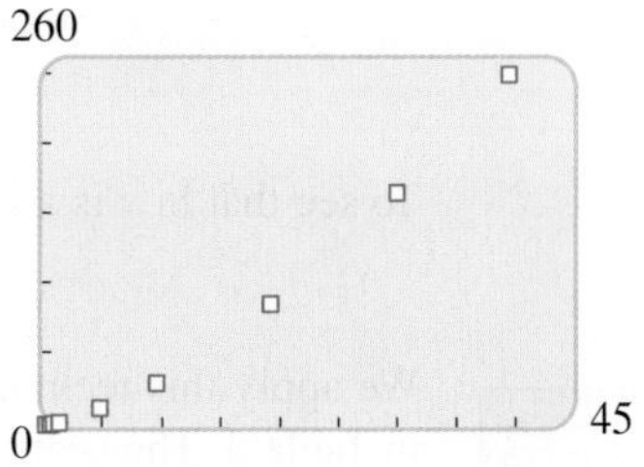

FIGURE 4 Scatter plot of planetary data

(b) Using a graphing calculator and the `PwrReg` command (see Figure 5(a)), we get the power model

$$T = 1.000396d^{1.49966}$$

If we round both the coefficient and the exponent to three significant figures, we can write the model as

$$T = d^{1.5}$$

This is the relationship discovered by Kepler (see page 754). Sir Isaac Newton (page 852) later used his Law of Gravity to derive this relationship theoretically, thereby providing strong scientific evidence that the Law of Gravity must be true.

(c) The graph is shown in Figure 5(b). The model appears to fit the data very well.

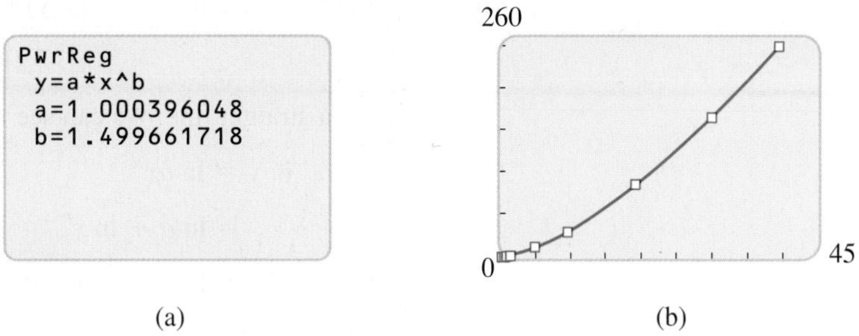

FIGURE 5 Power model for planetary data

(d) In this case $d = 5$ AU, so our model gives

$$T = 1.00039 \cdot 5^{1.49966} \approx 11.22$$

The period of the asteroid is about 11.2 years. ■

▼ Linearizing Data

We have used the shape of a scatter plot to decide which type of model to use: linear, exponential, or power. This works well if the data points lie on a straight line. But it's difficult to distinguish a scatter plot that is exponential from one that requires a power model. So to help decide which model to use, we can *linearize* the data, that is, apply a function that "straightens" the scatter plot. The inverse of the linearizing function is then an appropriate model. We now describe how to linearize data that can be modeled by exponential or power functions.

▶ Linearizing exponential data

If we suspect that the data points (x, y) lie on an exponential curve $y = Ce^{kx}$, then the points

$$(x, \ln y)$$

should lie on a straight line. We can see this from the following calculations:

$$\begin{aligned} \ln y &= \ln Ce^{kx} && \text{Assume that } y = Ce^{kx} \text{ and take ln} \\ &= \ln e^{kx} + \ln C && \text{Property of ln} \\ &= kx + \ln C && \text{Property of ln} \end{aligned}$$

To see that $\ln y$ is a linear function of x, let $Y = \ln y$ and $A = \ln C$; then

$$Y = kx + A$$

We apply this technique to the world population data (t, P) to obtain the points $(t, \ln P)$ in Table 3. The scatter plot of $(t, \ln P)$ in Figure 6, called a **semi-log plot**, shows that the linearized data lie approximately on a straight line, so an exponential model should be appropriate.

TABLE 3
World population data

t	Population P (in millions)	$\ln P$
1900	1650	21.224
1910	1750	21.283
1920	1860	21.344
1930	2070	21.451
1940	2300	21.556
1950	2520	21.648
1960	3020	21.829
1970	3700	22.032
1980	4450	22.216
1990	5300	22.391
2000	6060	22.525

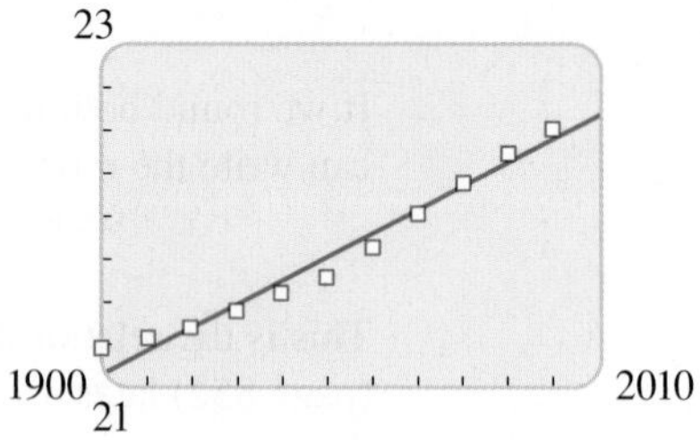

FIGURE 6 Semi-log plot of data in Table 3

▶ Linearizing power data

If we suspect that the data points (x, y) lie on a power curve $y = ax^n$, then the points

$$(\ln x, \ln y)$$

should be on a straight line. We can see this from the following calculations:

$$\begin{aligned} \ln y &= \ln ax^n && \text{Assume that } y = ax^n \text{ and take ln} \\ &= \ln a + \ln x^n && \text{Property of ln} \\ &= \ln a + n \ln x && \text{Property of ln} \end{aligned}$$

To see that $\ln y$ is a linear function of $\ln x$, let $Y = \ln y$, $X = \ln x$, and $A = \ln a$; then

$$Y = nX + A$$

We apply this technique to the planetary data (d, T) in Table 2 to obtain the points $(\ln d, \ln T)$ in Table 4. The scatter plot of $(\ln d, \ln T)$ in Figure 7, called a **log-log plot**, shows that the data lie on a straight line, so a power model seems appropriate.

TABLE 4
Log-log table

$\ln d$	$\ln T$
−0.94933	−1.4230
−0.32435	−0.48613
0	0
0.42068	0.6318
1.6492	2.4733
2.2556	3.3829
2.9544	4.4309
3.4041	5.1046
3.6765	5.5148

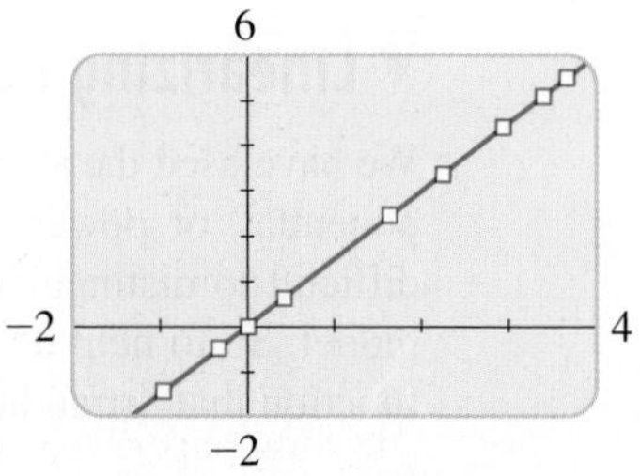

FIGURE 7 Log-log plot of data in Table 4

An Exponential or Power Model?

Suppose that a scatter plot of the data points (x, y) shows a rapid increase. Should we use an exponential function or a power function to model the data? To help us decide, we draw two scatter plots: one for the points $(x, \ln y)$ and the other for the points $(\ln x, \ln y)$. If the first scatter plot appears to lie along a line, then an exponential model is appropriate. If the second plot appears to lie along a line, then a power model is appropriate.

EXAMPLE 3 | An Exponential or Power Model?

Data points (x, y) are shown in Table 5.

(a) Draw a scatter plot of the data.

(b) Draw scatter plots of $(x, \ln y)$ and $(\ln x, \ln y)$.

(c) Is an exponential function or a power function appropriate for modeling this data?

(d) Find an appropriate function to model the data.

TABLE 5

x	y
1	2
2	6
3	14
4	22
5	34
6	46
7	64
8	80
9	102
10	130

SOLUTION

(a) The scatter plot of the data is shown in Figure 8.

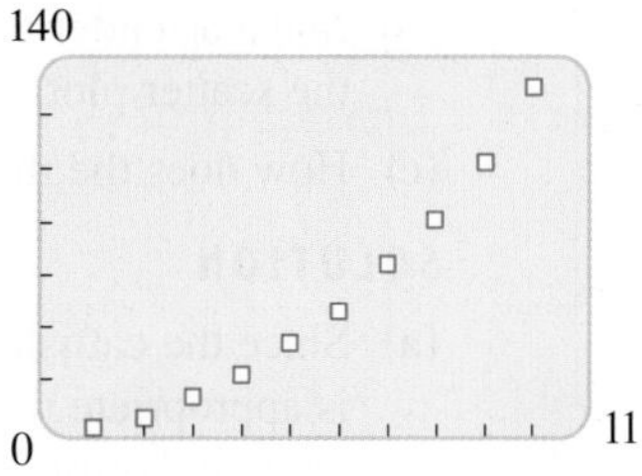

FIGURE 8

(b) We use the values from Table 6 to graph the scatter plots in Figures 9 and 10.

TABLE 6

x	$\ln x$	$\ln y$
1	0	0.7
2	0.7	1.8
3	1.1	2.6
4	1.4	3.1
5	1.6	3.5
6	1.8	3.8
7	1.9	4.2
8	2.1	4.4
9	2.2	4.6
10	2.3	4.9

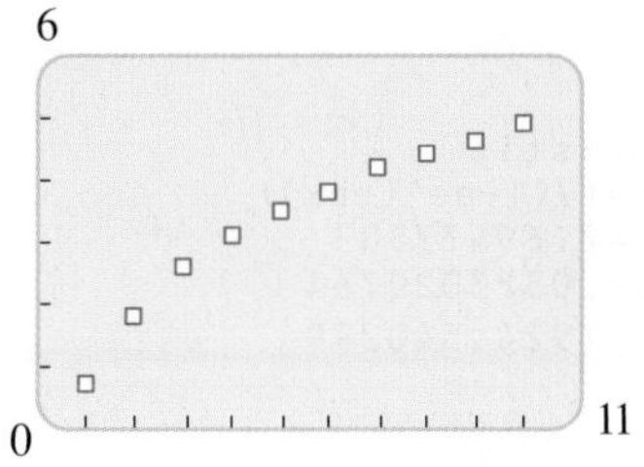

FIGURE 9 Semi-log plot

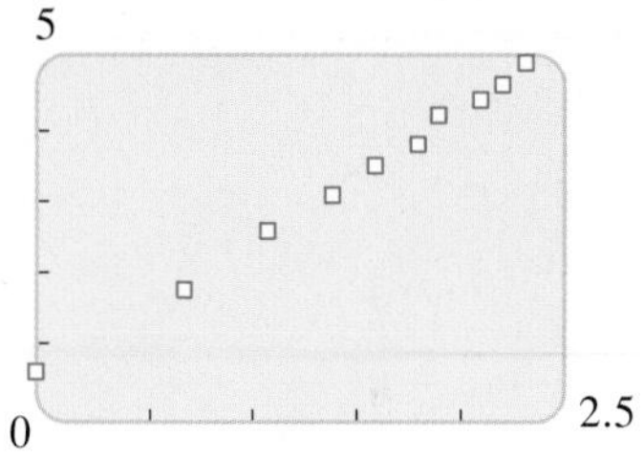

FIGURE 10 Log-log plot

(c) The scatter plot of $(x, \ln y)$ in Figure 9 does not appear to be linear, so an exponential model is not appropriate. On the other hand, the scatter plot of $(\ln x, \ln y)$ in Figure 10 is very nearly linear, so a power model is appropriate.

(d) Using the `PwrReg` command on a graphing calculator, we find that the power function that best fits the data point is

$$y = 1.85x^{1.82}$$

The graph of this function and the original data points are shown in Figure 11. ■

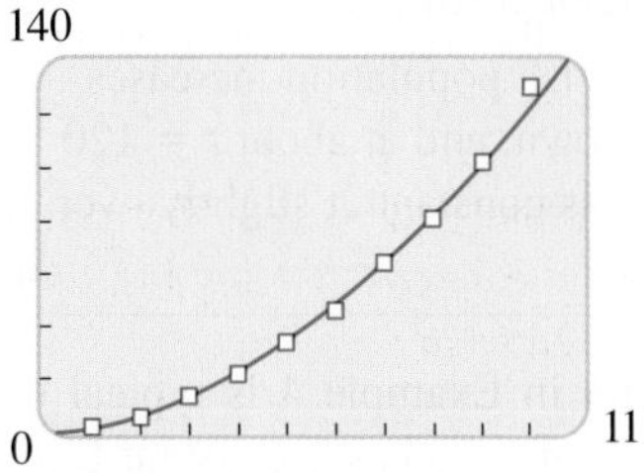

FIGURE 11

Before graphing calculators and statistical software became common, exponential and power models for data were often constructed by first finding a linear model for the linearized data. Then the model for the actual data was found by taking exponentials. For instance, if we find that $\ln y = A \ln x + B$, then by taking exponentials we get the model $y = e^B \cdot e^{A \ln x}$, or $y = Cx^A$ (where $C = e^B$). Special graphing paper called "log paper" or "log-log paper" was used to facilitate this process.

▼ Modeling with Logistic Functions

A logistic growth model is a function of the form

$$f(t) = \frac{c}{1 + ae^{-bt}}$$

where a, b, and c are positive constants. Logistic functions are used to model populations where the growth is constrained by available resources. (See Exercises 25–28 of Section 4.2.)

EXAMPLE 4 | Stocking a Pond with Catfish

TABLE 7

Week	Catfish
0	1000
15	1500
30	3300
45	4400
60	6100
75	6900
90	7100
105	7800
120	7900

Much of the fish that is sold in supermarkets today is raised on commercial fish farms, not caught in the wild. A pond on one such farm is initially stocked with 1000 catfish, and the fish population is then sampled at 15-week intervals to estimate its size. The population data are given in Table 7.

(a) Find an appropriate model for the data.

(b) Make a scatter plot of the data and graph the model that you found in part (a) on the scatter plot.

(c) How does the model predict that the fish population will change with time?

SOLUTION

(a) Since the catfish population is restricted by its habitat (the pond), a logistic model is appropriate. Using the `Logistic` command on a calculator (see Figure 12(a)), we find the following model for the catfish population $P(t)$:

$$P(t) = \frac{7925}{1 + 7.7e^{-0.052t}}$$

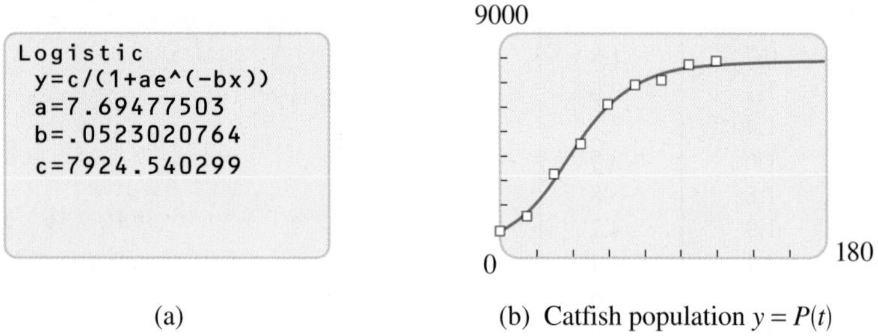

FIGURE 12 (a) (b) Catfish population $y = P(t)$

(b) The scatter plot and the logistic curve are shown in Figure 12(b).

(c) From the graph of P in Figure 12(b) we see that the catfish population increases rapidly until about $t = 80$ weeks. Then growth slows down, and at about $t = 120$ weeks the population levels off and remains more or less constant at slightly over 7900. ■

The behavior that is exhibited by the catfish population in Example 4 is typical of logistic growth. After a rapid growth phase, the population approaches a constant level called the **carrying capacity** of the environment. This occurs because as $t \to \infty$, we have $e^{-bt} \to 0$ (see Section 4.2), and so

$$P(t) = \frac{c}{1 + ae^{-bt}} \quad \longrightarrow \quad \frac{c}{1 + 0} = c$$

Thus, the carrying capacity is c.

PROBLEMS

1. U.S. Population The U.S. Constitution requires a census every 10 years. The census data for 1790–2000 are given in the table.

(a) Make a scatter plot of the data.

(b) Use a calculator to find an exponential model for the data.

(c) Use your model to predict the population at the 2010 census.

(d) Use your model to estimate the population in 1965.

(e) Compare your answers from parts (c) and (d) to the values in the table. Do you think an exponential model is appropriate for these data?

Year	Population (in millions)	Year	Population (in millions)	Year	Population (in millions)
1790	3.9	1870	38.6	1950	151.3
1800	5.3	1880	50.2	1960	179.3
1810	7.2	1890	63.0	1970	203.3
1820	9.6	1900	76.2	1980	226.5
1830	12.9	1910	92.2	1990	248.7
1840	17.1	1920	106.0	2000	281.4
1850	23.2	1930	123.2		
1860	31.4	1940	132.2		

2. A Falling Ball In a physics experiment a lead ball is dropped from a height of 5 m. The students record the distance the ball has fallen every one-tenth of a second. (This can be done by using a camera and a strobe light.)

(a) Make a scatter plot of the data.

(b) Use a calculator to find a power model.

(c) Use your model to predict how far a dropped ball would fall in 3 s.

Time (s)	Distance (m)
0.1	0.048
0.2	0.197
0.3	0.441
0.4	0.882
0.5	1.227
0.6	1.765
0.7	2.401
0.8	3.136
0.9	3.969
1.0	4.902

3. Health-Care Expenditures The U.S. health-care expenditures for 1970–2001 are given in the table below, and a scatter plot of the data is shown in the figure.

(a) Does the scatter plot shown suggest an exponential model?

(b) Make a table of the values $(t, \ln E)$ and a scatter plot. Does the scatter plot appear to be linear?

(c) Find the regression line for the data in part (b).

(d) Use the results of part (c) to find an exponential model for the growth of health-care expenditures.

(e) Use your model to predict the total health-care expenditures in 2009.

Year	Health-care expenditures (in billions of dollars)
1970	74.3
1980	251.1
1985	434.5
1987	506.2
1990	696.6
1992	820.3
1994	937.2
1996	1039.4
1998	1150.0
2000	1310.0
2001	1424.5

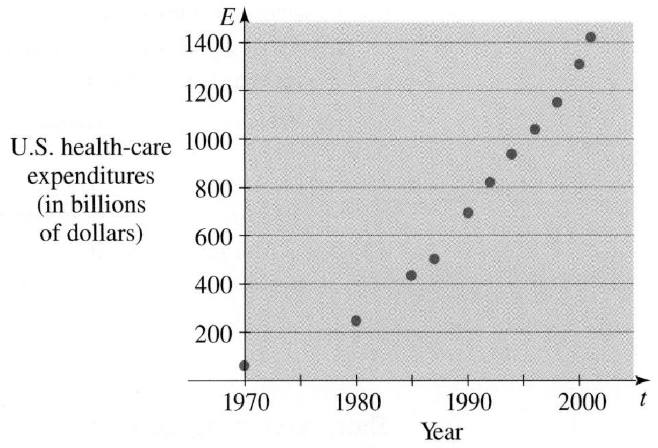

Time (h)	Amount of ^{131}I (g)
0	4.80
8	4.66
16	4.51
24	4.39
32	4.29
40	4.14
48	4.04

4. Half-Life of Radioactive Iodine A student is trying to determine the half-life of radioactive iodine-131. He measures the amount of iodine-131 in a sample solution every 8 hours. His data are shown in the table in the margin.

(a) Make a scatter plot of the data.

(b) Use a calculator to find an exponential model.

(c) Use your model to find the half-life of iodine-131.

5. The Beer-Lambert Law As sunlight passes through the waters of lakes and oceans, the light is absorbed, and the deeper it penetrates, the more its intensity diminishes. The light intensity I at depth x is given by the Beer-Lambert Law:

$$I = I_0 e^{-kx}$$

where I_0 is the light intensity at the surface and k is a constant that depends on the murkiness of the water (see page 336). A biologist uses a photometer to investigate light penetration in a northern lake, obtaining the data in the table.

(a) Use a graphing calculator to find an exponential function of the form given by the Beer-Lambert Law to model these data. What is the light intensity I_0 at the surface on this day, and what is the "murkiness" constant k for this lake? [*Hint:* If your calculator gives you a function of the form $I = ab^x$, convert this to the form you want using the identities $b^x = e^{\ln(b^x)} = e^{x \ln b}$. See Example 1(b).]

(b) Make a scatter plot of the data, and graph the function that you found in part (a) on your scatter plot.

(c) If the light intensity drops below 0.15 lumen (lm), a certain species of algae can't survive because photosynthesis is impossible. Use your model from part (a) to determine the depth below which there is insufficient light to support this algae.

Light intensity decreases exponentially with depth.

Depth (ft)	Light intensity (lm)	Depth (ft)	Light intensity (lm)
5	13.0	25	1.8
10	7.6	30	1.1
15	4.5	35	0.5
20	2.7	40	0.3

6. Experimenting with "Forgetting" Curves Every one of us is all too familiar with the phenomenon of forgetting. Facts that we clearly understood at the time we first learned them sometimes fade from our memory by the time the final exam rolls around. Psychologists have proposed several ways to model this process. One such model is Ebbinghaus' Law of Forgetting, described on page 327. Other models use exponential or logarithmic functions. To develop her own model, a psychologist performs an experiment on a group of volunteers by asking them to memorize a list of 100 related words. She then tests how many of these words they can recall after various periods of time. The average results for the group are shown in the table.

(a) Use a graphing calculator to find a *power* function of the form $y = at^b$ that models the average number of words y that the volunteers remember after t hours. Then find an *exponential* function of the form $y = ab^t$ to model the data.

(b) Make scatter plot of the data, and graph both the functions that you found in part (a) on your scatter plot.

(c) Which of the two functions seems to provide the better model?

Time	Words recalled
15 min	64.3
1 h	45.1
8 h	37.3
1 day	32.8
2 days	26.9
3 days	25.6
5 days	22.9

The number of different bat species in a cave is related to the size of the cave by a power function.

7. Modeling the Species-Area Relation The table gives the areas of several caves in central Mexico and the number of bat species that live in each cave.*

(a) Find a power function that models the data.

(b) Draw a graph of the function you found in part (a) and a scatter plot of the data on the same graph. Does the model fit the data well?

(c) The cave called El Sapo near Puebla, Mexico, has a surface area of $A = 205\ \text{m}^2$. Use the model to estimate the number of bat species you would expect to find in that cave.

Cave	Area (m^2)	Number of species
La Escondida	18	1
El Escorpion	19	1
El Tigre	58	1
Mision Imposible	60	2
San Martin	128	5
El Arenal	187	4
La Ciudad	344	6
Virgen	511	7

8. Auto Exhaust Emissions A study by the U.S. Office of Science and Technology in 1972 estimated the cost of reducing automobile emissions by certain percentages. Find an exponential model that captures the "diminishing returns" trend of these data shown in the table below.

Reduction in emissions (%)	Cost per car ($)
50	45
55	55
60	62
65	70
70	80
75	90
80	100
85	200
90	375
95	600

9. Exponential or Power Model? Data points (x, y) are shown in the table.

(a) Draw a scatter plot of the data.

(b) Draw scatter plots of $(x, \ln y)$ and $(\ln x, \ln y)$.

(c) Which is more appropriate for modeling this data: an exponential function or a power function?

(d) Find an appropriate function to model the data.

x	y
2	0.08
4	0.12
6	0.18
8	0.25
10	0.36
12	0.52
14	0.73
16	1.06

*A. K. Brunet and R. A. Medallin, "The Species-Area Relationship in Bat Assemblages of Tropical Caves." *Journal of Mammalogy,* 82(4):1114–1122, 2001.

x	y
10	29
20	82
30	151
40	235
50	330
60	430
70	546
80	669
90	797

10. Exponential or Power Model? Data points (x, y) are shown in the table in the margin.

(a) Draw a scatter plot of the data.

(b) Draw scatter plots of $(x, \ln y)$ and $(\ln x, \ln y)$.

(c) Which is more appropriate for modeling this data: an exponential function or a power function?

(d) Find an appropriate function to model the data.

11. Logistic Population Growth The table and scatter plot give the population of black flies in a closed laboratory container over an 18-day period.

(a) Use the `Logistic` command on your calculator to find a logistic model for these data.

(b) Use the model to estimate the time when there were 400 flies in the container.

Time (days)	Number of flies
0	10
2	25
4	66
6	144
8	262
10	374
12	446
16	492
18	498

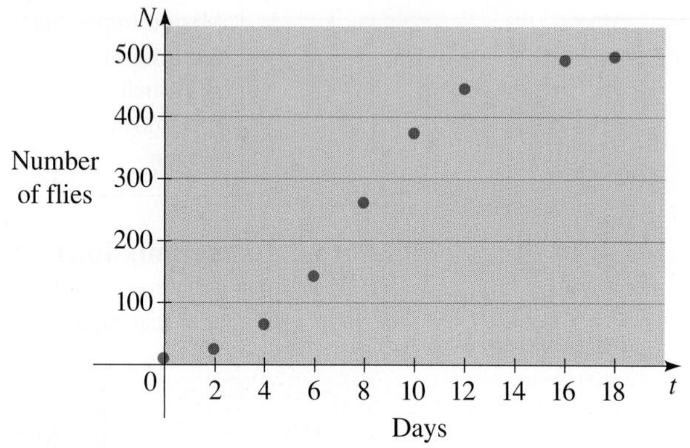

12. Logarithmic Models A **logarithmic model** is a function of the form

$$y = a + b \ln x$$

Many relationships between variables in the real world can be modeled by this type of function. The table and the scatter plot show the coal production (in metric tons) from a small mine in northern British Columbia.

(a) Use the `LnReg` command on your calculator to find a logarithmic model for these production figures.

(b) Use the model to predict coal production from this mine in 2010.

Year	Metric tons of coal
1950	882
1960	889
1970	894
1980	899
1990	905
2000	909

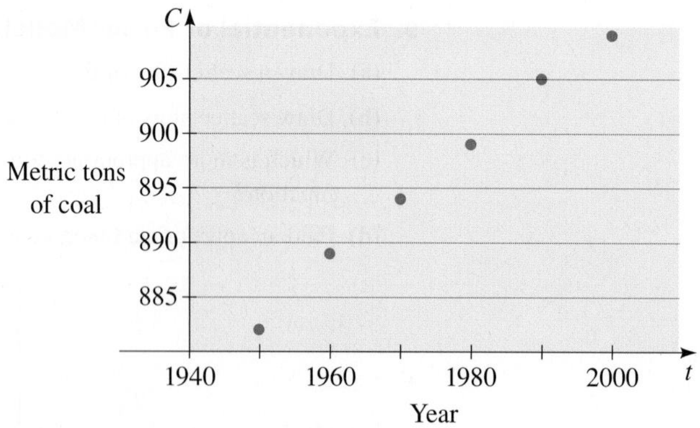

CUMULATIVE REVIEW TEST CHAPTERS 2, 3, and 4

1. Let $f(x) = x^2 - 4x$ and $g(x) = \sqrt{x + 4}$. Find each of the following:

(a) The domain of f

(b) The domain of g

(c) $f(-2), f(0), f(4), g(0), g(8), g(-6)$

(d) $f(x + 2), g(x + 2), f(2 + h)$

(e) The average rate of change of g between $x = 5$ and $x = 21$

(f) $f \circ g, g \circ f, f(g(12)), g(f(12))$

(g) The inverse of g

2. Let $f(x) = \begin{cases} 4 & \text{if } x \le 2 \\ x - 3 & \text{if } x > 2 \end{cases}$

(a) Evaluate $f(0), f(1), f(2), f(3)$, and $f(4)$.

(b) Sketch the graph of f.

3. Let f be the quadratic function $f(x) = -2x^2 + 8x + 5$.

(a) Express f in standard form.

(b) Find the maximum or minimum value of f.

(c) Sketch the graph of f.

(d) Find the interval on which f is increasing and the interval on which f is decreasing.

(e) How is the graph of $g(x) = -2x^2 + 8x + 10$ obtained from the graph of f?

(f) How is the graph of $h(x) = -2(x + 3)^2 + 8(x + 3) + 5$ obtained from the graph of f?

4. Without using a graphing calculator, match each of the following functions to the graphs below. Give reasons for your choices.

$$f(x) = x^3 - 8x \qquad g(x) = -x^4 + 8x^2 \qquad r(x) = \frac{2x + 3}{x^2 - 9}$$

$$s(x) = \frac{2x - 3}{x^2 + 9} \qquad h(x) = 2^x - 5 \qquad k(x) = 2^{-x} + 3$$

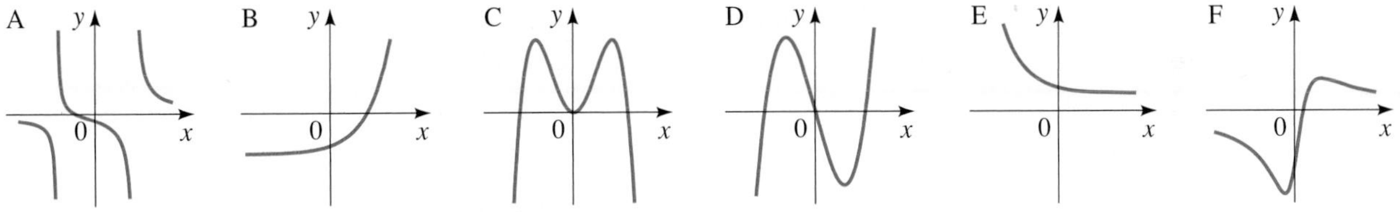

5. Let $P(x) = 2x^3 - 11x^2 + 10x + 8$.

(a) List all possible rational zeros of P.

(b) Determine which of the numbers you listed in part (a) actually are zeros of P.

(c) Factor P completely.

(d) Sketch a graph of P.

6. Let $Q(x) = x^5 - 3x^4 + 3x^3 + x^2 - 4x + 2$.

(a) Find all zeros of Q, real and complex, and state their multiplicities.

(b) Factor Q completely.

(c) Factor Q into linear and irreducible quadratic factors.

7. Let $r(x) = \dfrac{3x^2 + 6x}{x^2 - x - 2}$. Find the x- and y-intercepts and the horizontal and vertical asymptotes. Then sketch the graph of r.

8. Sketch graphs of the following functions on the same coordinate plane.

(a) $f(x) = 2 - e^x$ **(b)** $g(x) = \ln(x + 1)$

9. (a) Find the exact value of $\log_3 16 - 2 \log_3 36$.

(b) Use the Laws of Logarithms to expand the expression

$$\log\left(\frac{x^5\sqrt{x-1}}{2x-3}\right)$$

10. Solve the equations.

(a) $\log_2 x + \log_2(x - 2) = 3$

(b) $2e^{3x} - 11e^{2x} + 10e^x + 8 = 0$ [*Hint:* Compare to the polynomial in Problem 5.]

11. A sum of \$25,000 is deposited into an account paying 5.4% interest per year, compounded daily.

(a) What will the amount in the account be after 3 years?

(b) When will the account have grown to \$35,000?

(c) How long will it take for the initial deposit to double?

12. After a shipwreck, 120 rats manage to swim from the wreckage to a deserted island. The rat population on the island grows exponentially, and after 15 months there are 280 rats on the island.

(a) Find a function that models the population t months after the arrival of the rats.

(b) What will the population be 3 years after the shipwreck?

(c) When will the population reach 2000?

CHAPTER 5

TRIGONOMETRIC FUNCTIONS: UNIT CIRCLE APPROACH

If you've ever taken a ferris wheel ride, then you know about periodic motion—that is, motion that repeats over and over. Periodic motion is common in nature. Think about the daily rising and setting of the sun (day, night, day, night, . . .), the daily variation in tide levels (high, low, high, low, . . .), or the vibrations of a leaf in the wind (left, right, left, right, . . .). To model such motion, we need a function whose values increase, then decrease, then increase, and so on. To understand how to define such a function, let's look at a person riding on a ferris wheel. The graph shows the height of the person above the center of the wheel at time t. Notice that the graph goes up and down repeatedly.

The trigonometric function *sine* is defined in a similar way, using the unit circle (in place of the ferris wheel). The trigonometric functions can be defined in two different but equivalent ways: as functions of real numbers (Chapter 5) or as functions of angles (Chapter 6). The two approaches are independent of each other, so either Chapter 5 or Chapter 6 may be studied first. We study both approaches because the different approaches are required for different applications.

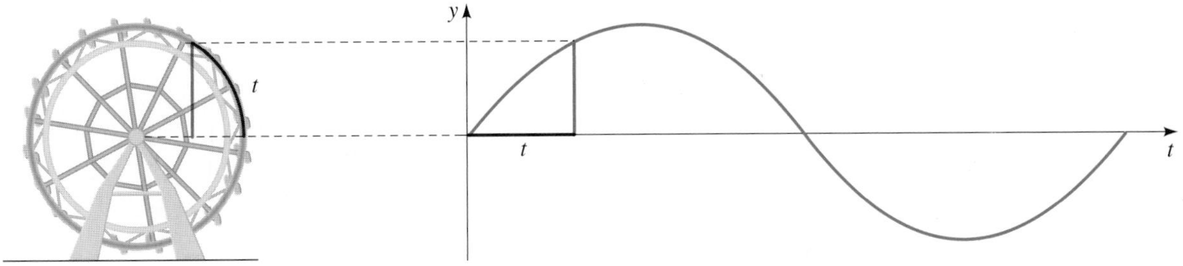

5.1 The Unit Circle

| The Unit Circle ▶ Terminal Points on the Unit Curcle ▶ The Reference Number

In this section we explore some properties of the circle of radius 1 centered at the origin. These properties are used in the next section to define the trigonometric functions.

▼ The Unit Circle

The set of points at a distance 1 from the origin is a circle of radius 1 (see Figure 1). In Section 1.8 we learned that the equation of this circle is $x^2 + y^2 = 1$.

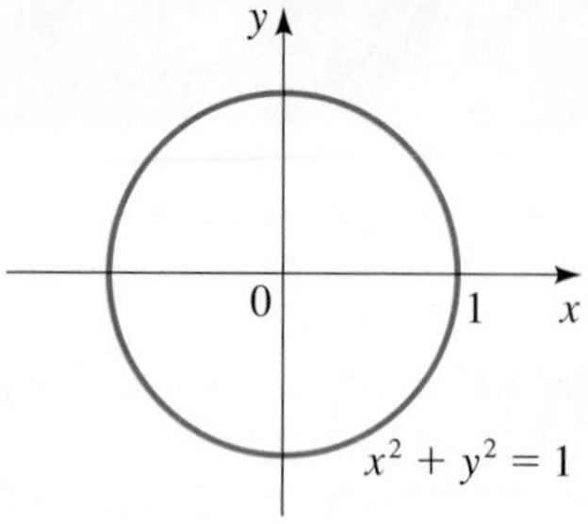

FIGURE 1 The unit circle

THE UNIT CIRCLE

The **unit circle** is the circle of radius 1 centered at the origin in the *xy*-plane. Its equation is

$$x^2 + y^2 = 1$$

EXAMPLE 1 | A Point on the Unit Circle

Show that the point $P\left(\frac{\sqrt{3}}{3}, \frac{\sqrt{6}}{3}\right)$ is on the unit circle.

SOLUTION We need to show that this point satisfies the equation of the unit circle, that is, $x^2 + y^2 = 1$. Since

$$\left(\frac{\sqrt{3}}{3}\right)^2 + \left(\frac{\sqrt{6}}{3}\right)^2 = \frac{3}{9} + \frac{6}{9} = 1$$

P is on the unit circle.

NOW TRY EXERCISE **3** ■

EXAMPLE 2 | Locating a Point on the Unit Circle

The point $P(\sqrt{3}/2, y)$ is on the unit circle in Quadrant IV. Find its y-coordinate.

SOLUTION Since the point is on the unit circle, we have

$$\left(\frac{\sqrt{3}}{2}\right)^2 + y^2 = 1$$

$$y^2 = 1 - \frac{3}{4} = \frac{1}{4}$$

$$y = \pm\frac{1}{2}$$

Since the point is in Quadrant IV, its y-coordinate must be negative, so $y = -\frac{1}{2}$.

NOW TRY EXERCISE **9** ■

▼ Terminal Points on the Unit Circle

Suppose t is a real number. Let's mark off a distance t along the unit circle, starting at the point $(1, 0)$ and moving in a counterclockwise direction if t is positive or in a clockwise direction if t is negative (Figure 2). In this way we arrive at a point $P(x, y)$ on the unit circle. The point $P(x, y)$ obtained in this way is called the **terminal point** determined by the real number t.

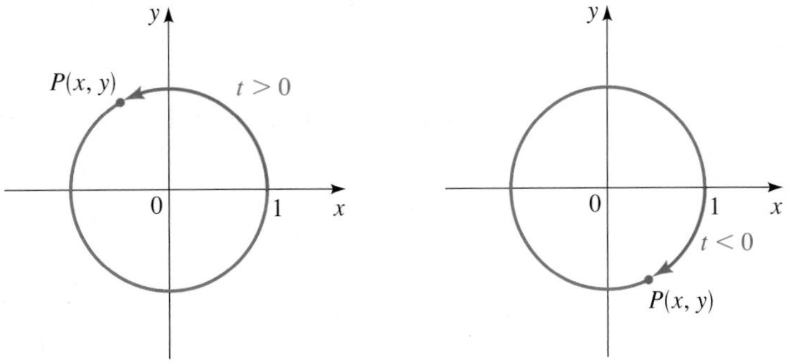

FIGURE 2

(a) Terminal point $P(x, y)$ determined by $t > 0$

(b) Terminal point $P(x, y)$ determined by $t < 0$

The circumference of the unit circle is $C = 2\pi(1) = 2\pi$. So if a point starts at $(1, 0)$ and moves counterclockwise all the way around the unit circle and returns to $(1, 0)$, it travels a distance of 2π. To move halfway around the circle, it travels a distance of $\frac{1}{2}(2\pi) = \pi$. To move a quarter of the distance around the circle, it travels a distance of $\frac{1}{4}(2\pi) = \pi/2$. Where does the point end up when it travels these distances along the circle? From Figure 3 we see, for example, that when it travels a distance of π starting at $(1, 0)$, its terminal point is $(-1, 0)$.

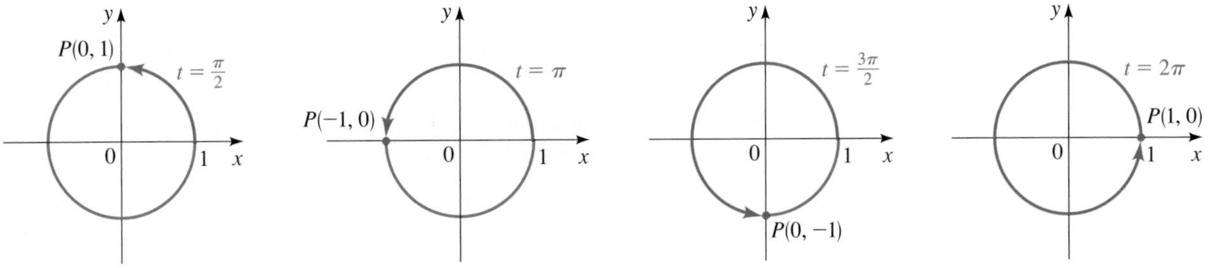

FIGURE 3 Terminal points determined by $t = \frac{\pi}{2}, \pi, \frac{3\pi}{2}$, and 2π

EXAMPLE 3 | Finding Terminal Points

Find the terminal point on the unit circle determined by each real number t.

(a) $t = 3\pi$ **(b)** $t = -\pi$ **(c)** $t = -\dfrac{\pi}{2}$

SOLUTION From Figure 4 we get the following:

(a) The terminal point determined by 3π is $(-1, 0)$.

(b) The terminal point determined by $-\pi$ is $(-1, 0)$.

(c) The terminal point determined by $-\pi/2$ is $(0, -1)$.

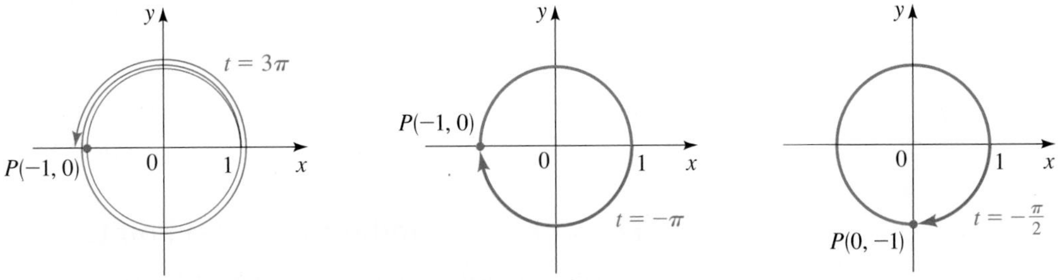

FIGURE 4

Notice that different values of t can determine the same terminal point.

NOW TRY EXERCISE **23** ■

The terminal point $P(x, y)$ determined by $t = \pi/4$ is the same distance from $(1, 0)$ as from $(0, 1)$ along the unit circle (see Figure 5).

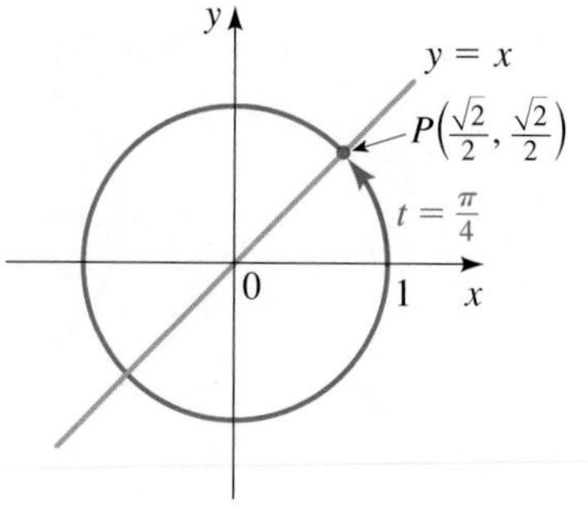

FIGURE 5

Since the unit circle is symmetric with respect to the line $y = x$, it follows that P lies on the line $y = x$. So P is the point of intersection (in the first quadrant) of the circle $x^2 + y^2 = 1$ and the line $y = x$. Substituting x for y in the equation of the circle, we get

$$x^2 + x^2 = 1$$

$$2x^2 = 1 \qquad \text{Combine like terms}$$

$$x^2 = \frac{1}{2} \qquad \text{Divide by 2}$$

$$x = \pm\frac{1}{\sqrt{2}} \qquad \text{Take square roots}$$

Since P is in the first quadrant, $x = 1/\sqrt{2}$ and since $y = x$, we have $y = 1/\sqrt{2}$ also. Thus, the terminal point determined by $\pi/4$ is

$$P\left(\frac{1}{\sqrt{2}}, \frac{1}{\sqrt{2}}\right) = P\left(\frac{\sqrt{2}}{2}, \frac{\sqrt{2}}{2}\right)$$

Similar methods can be used to find the terminal points determined by $t = \pi/6$ and $t = \pi/3$ (see Exercises 57 and 58). Table 1 and Figure 6 give the terminal points for some special values of t.

TABLE 1

t	Terminal point determined by t
0	$(1,0)$
$\frac{\pi}{6}$	$\left(\frac{\sqrt{3}}{2}, \frac{1}{2}\right)$
$\frac{\pi}{4}$	$\left(\frac{\sqrt{2}}{2}, \frac{\sqrt{2}}{2}\right)$
$\frac{\pi}{3}$	$\left(\frac{1}{2}, \frac{\sqrt{3}}{2}\right)$
$\frac{\pi}{2}$	$(0,1)$

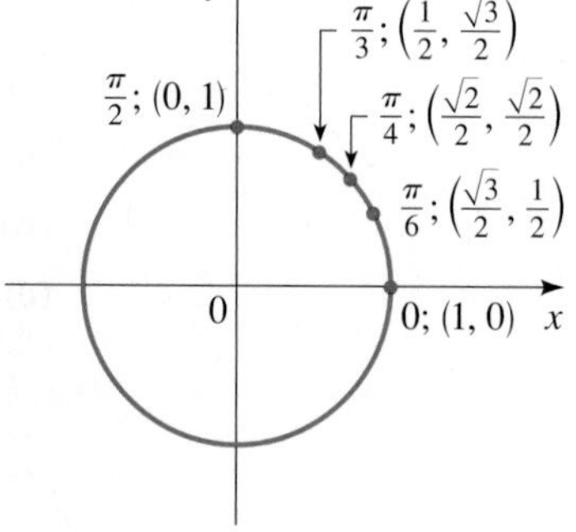

FIGURE 6

EXAMPLE 4 | Finding Terminal Points

Find the terminal point determined by each given real number t.

(a) $t = -\dfrac{\pi}{4}$ **(b)** $t = \dfrac{3\pi}{4}$ **(c)** $t = -\dfrac{5\pi}{6}$

SOLUTION

(a) Let P be the terminal point determined by $-\pi/4$, and let Q be the terminal point determined by $\pi/4$. From Figure 7(a) we see that the point P has the same coordinates as Q except for sign. Since P is in quadrant IV, its x-coordinate is positive and its y-coordinate is negative. Thus, the terminal point is $P\left(\sqrt{2}/2, -\sqrt{2}/2\right)$.

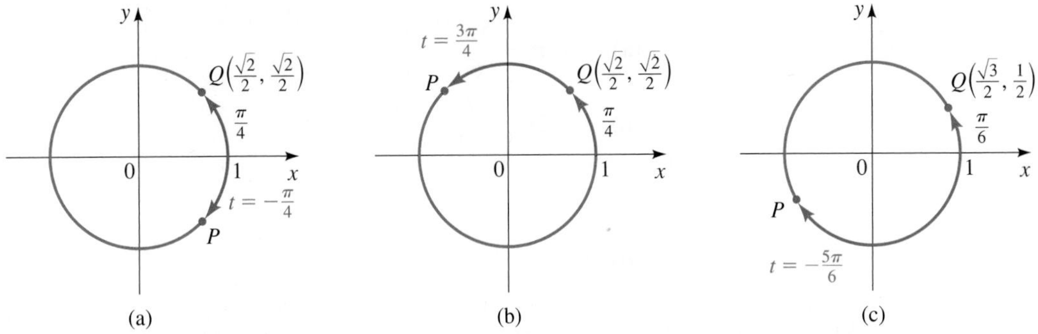

FIGURE 7

(b) Let P be the terminal point determined by $3\pi/4$, and let Q be the terminal point determined by $\pi/4$. From Figure 7(b) we see that the point P has the same coordinates as Q except for sign. Since P is in quadrant II, its x-coordinate is negative and its y-coordinate is positive. Thus, the terminal point is $P\left(-\sqrt{2}/2, \sqrt{2}/2\right)$.

(c) Let P be the terminal point determined by $-5\pi/6$, and let Q be the terminal point determined by $\pi/6$. From Figure 7(c) we see that the point P has the same coordinates as Q except for sign. Since P is in quadrant III, its coordinates are both negative. Thus, the terminal point is $P\left(-\sqrt{3}/2, -\frac{1}{2}\right)$.

NOW TRY EXERCISE **25** ■

The Reference Number

From Examples 3 and 4 we see that to find a terminal point in any quadrant we need only know the "corresponding" terminal point in the first quadrant. We use the idea of the *reference number* to help us find terminal points.

REFERENCE NUMBER

Let t be a real number. The **reference number** $\bar{t}$ associated with t is the shortest distance along the unit circle between the terminal point determined by t and the x-axis.

Figure 8 shows that to find the reference number $\bar{t}$, it's helpful to know the quadrant in which the terminal point determined by t lies. If the terminal point lies in quadrants I or IV, where x is positive, we find $\bar{t}$ by moving along the circle to the *positive* x-axis. If it lies in quadrants II or III, where x is negative, we find $\bar{t}$ by moving along the circle to the *negative* x-axis.

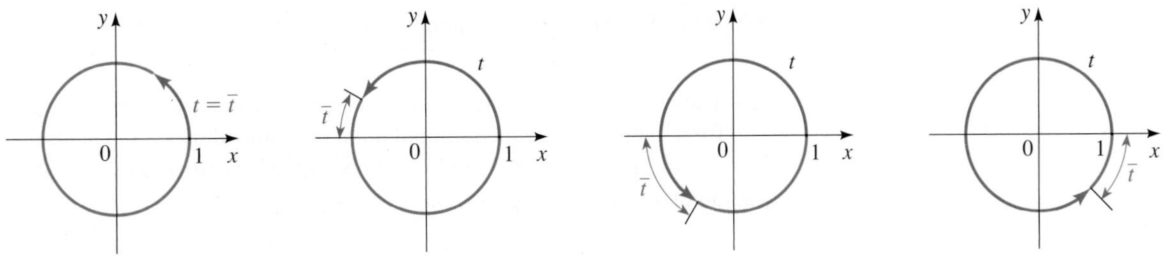

FIGURE 8 The reference number $\bar{t}$ for t

EXAMPLE 5 | Finding Reference Numbers

Find the reference number for each value of t.

(a) $t = \frac{5\pi}{6}$ **(b)** $t = \frac{7\pi}{4}$ **(c)** $t = -\frac{2\pi}{3}$ **(d)** $t = 5.80$

SOLUTION From Figure 9 we find the reference numbers as follows:

(a) $\bar{t} = \pi - \frac{5\pi}{6} = \frac{\pi}{6}$

(b) $\bar{t} = 2\pi - \frac{7\pi}{4} = \frac{\pi}{4}$

(c) $\bar{t} = \pi - \frac{2\pi}{3} = \frac{\pi}{3}$

(d) $\bar{t} = 2\pi - 5.80 \approx 0.48$

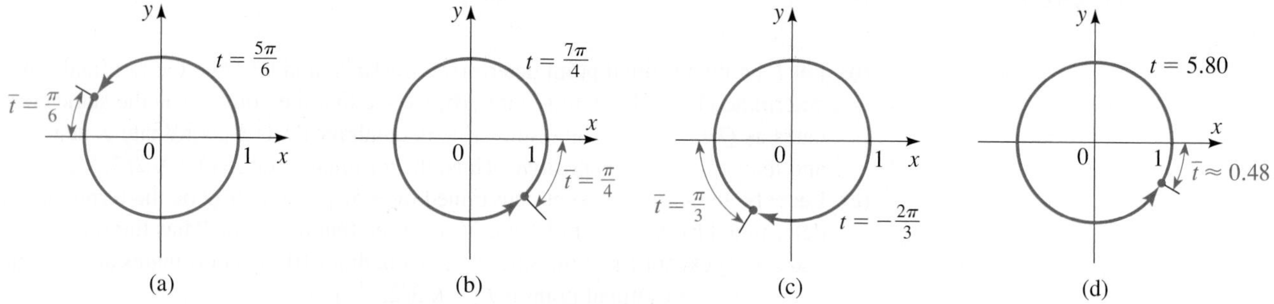

FIGURE 9

NOW TRY EXERCISE 35

USING REFERENCE NUMBERS TO FIND TERMINAL POINTS

To find the terminal point P determined by any value of t, we use the following steps:

1. Find the reference number $\bar{t}$.
2. Find the terminal point $Q(a, b)$ determined by $\bar{t}$.
3. The terminal point determined by t is $P(\pm a, \pm b)$, where the signs are chosen according to the quadrant in which this terminal point lies.

EXAMPLE 6 | Using Reference Numbers to Find Terminal Points

Find the terminal point determined by each given real number t.

(a) $t = \frac{5\pi}{6}$ **(b)** $t = \frac{7\pi}{4}$ **(c)** $t = -\frac{2\pi}{3}$

SOLUTION The reference numbers associated with these values of t were found in Example 5.

(a) The reference number is $\bar{t} = \pi/6$, which determines the terminal point $(\sqrt{3}/2, \frac{1}{2})$ from Table 1. Since the terminal point determined by t is in Quadrant II, its x-coordinate is negative and its y-coordinate is positive. Thus, the desired terminal point is

$$\left(-\frac{\sqrt{3}}{2}, \frac{1}{2}\right)$$

(b) The reference number is $\bar{t} = \pi/4$, which determines the terminal point $(\sqrt{2}/2, \sqrt{2}/2)$ from Table 1. Since the terminal point is in Quadrant IV, its x-coordinate is positive and its y-coordinate is negative. Thus, the desired terminal point is

$$\left(\frac{\sqrt{2}}{2}, -\frac{\sqrt{2}}{2}\right)$$

(c) The reference number is $\bar{t} = \pi/3$, which determines the terminal point $(\frac{1}{2}, \sqrt{3}/2)$ from Table 1. Since the terminal point determined by t is in Quadrant III, its coordinates are both negative. Thus, the desired terminal point is

$$\left(-\frac{1}{2}, -\frac{\sqrt{3}}{2}\right)$$

NOW TRY EXERCISE **39**

Since the circumference of the unit circle is 2π, the terminal point determined by t is the same as that determined by $t + 2\pi$ or $t - 2\pi$. In general, we can add or subtract 2π any number of times without changing the terminal point determined by t. We use this observation in the next example to find terminal points for large t.

EXAMPLE 7 | Finding the Terminal Point for Large t

Find the terminal point determined by $t = \dfrac{29\pi}{6}$.

SOLUTION Since

$$t = \frac{29\pi}{6} = 4\pi + \frac{5\pi}{6}$$

we see that the terminal point of t is the same as that of $5\pi/6$ (that is, we subtract 4π). So by Example 6(a) the terminal point is $(-\sqrt{3}/2, \frac{1}{2})$. (See Figure 10.)

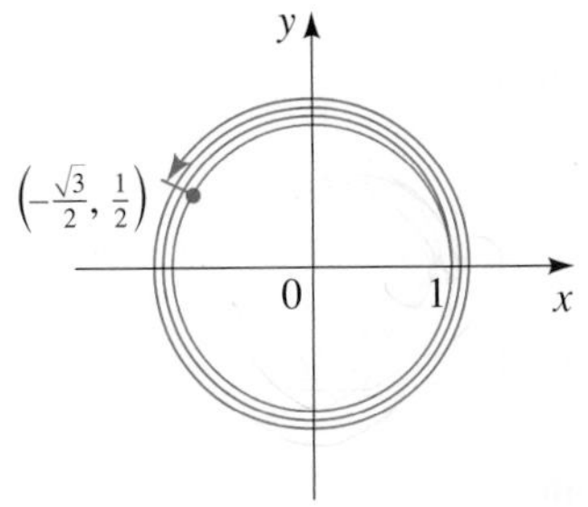

FIGURE 10

NOW TRY EXERCISE **45**

5.1 EXERCISES

CONCEPTS

1. **(a)** The unit circle is the circle centered at ______ with radius ______.
   (b) The equation of the unit circle is ______.
   (c) Suppose the point $P(x, y)$ is on the unit circle. Find the missing coordinate:
   (i) $P(1, \;\;)$ (ii) $P(\;\;, 1)$
   (iii) $P(-1, \;\;)$ (iv) $P(\;\;, -1)$

2. **(a)** If we mark off a distance t along the unit circle, starting at $(1, 0)$ and moving in a counterclockwise direction, we arrive at the ______ point determined by t.
   (b) The terminal points determined by $\pi/2$, π, $-\pi/2$, 2π are ______, ______, ______, and ______, respectively.

SKILLS

3–8 ■ Show that the point is on the unit circle.

3. $\left(\frac{4}{5}, -\frac{3}{5}\right)$
4. $\left(-\frac{5}{13}, \frac{12}{13}\right)$
5. $\left(\frac{7}{25}, \frac{24}{25}\right)$
6. $\left(-\frac{5}{7}, -\frac{2\sqrt{6}}{7}\right)$
7. $\left(-\frac{\sqrt{5}}{3}, \frac{2}{3}\right)$
8. $\left(\frac{\sqrt{11}}{6}, \frac{5}{6}\right)$

9–14 ■ Find the missing coordinate of P, using the fact that P lies on the unit circle in the given quadrant.

Coordinates	Quadrant
9. $P(-\frac{3}{5}, \;\;)$	III
10. $P(\;\;, -\frac{7}{25})$	IV
11. $P(\;\;, \frac{1}{3})$	II

Coordinates	Quadrant
12. $P(\frac{2}{5}, \quad)$	I
13. $P(\quad, -\frac{2}{7})$	IV
14. $P(-\frac{2}{3}, \quad)$	II

15–20 ■ The point P is on the unit circle. Find $P(x, y)$ from the given information.

15. The x-coordinate of P is $\frac{4}{5}$, and the y-coordinate is positive.

16. The y-coordinate of P is $-\frac{1}{3}$, and the x-coordinate is positive.

17. The y-coordinate of P is $\frac{2}{3}$, and the x-coordinate is negative.

18. The x-coordinate of P is positive, and the y-coordinate of P is $-\sqrt{5}/5$.

19. The x-coordinate of P is $-\sqrt{2}/3$, and P lies below the x-axis.

20. The x-coordinate of P is $-\frac{2}{5}$, and P lies above the x-axis.

21–22 ■ Find t and the terminal point determined by t for each point in the figure. In Exercise 21, t increases in increments of $\pi/4$; in Exercise 22, t increases in increments of $\pi/6$.

21.

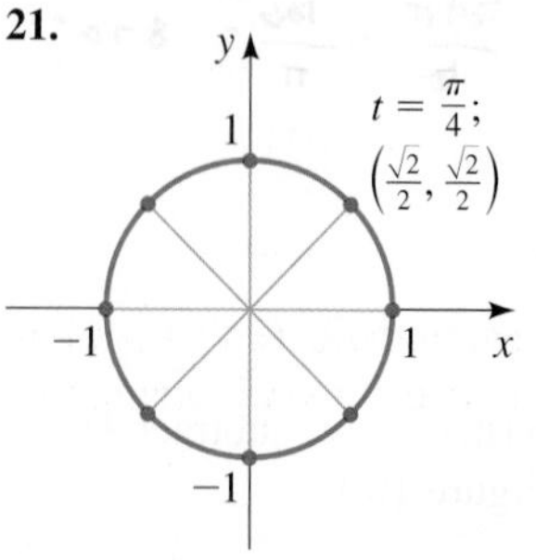

22.

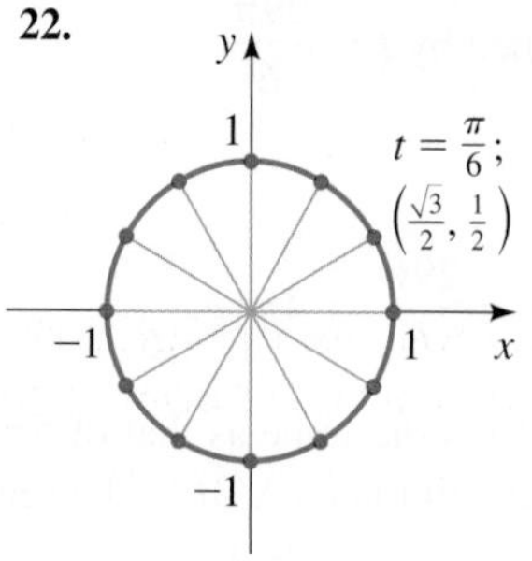

23–32 ■ Find the terminal point $P(x, y)$ on the unit circle determined by the given value of t.

23. $t = \dfrac{\pi}{2}$

24. $t = \dfrac{3\pi}{2}$

25. $t = \dfrac{5\pi}{6}$

26. $t = \dfrac{7\pi}{6}$

27. $t = -\dfrac{\pi}{3}$

28. $t = \dfrac{5\pi}{3}$

29. $t = \dfrac{2\pi}{3}$

30. $t = -\dfrac{\pi}{2}$

31. $t = -\dfrac{3\pi}{4}$

32. $t = \dfrac{11\pi}{6}$

33. Suppose that the terminal point determined by t is the point $\left(\frac{3}{5}, \frac{4}{5}\right)$ on the unit circle. Find the terminal point determined by each of the following.

(a) $\pi - t$ **(b)** $-t$
(c) $\pi + t$ **(d)** $2\pi + t$

34. Suppose that the terminal point determined by t is the point $\left(\frac{3}{4}, \sqrt{7}/4\right)$ on the unit circle. Find the terminal point determined by each of the following.

(a) $-t$ **(b)** $4\pi + t$
(c) $\pi - t$ **(d)** $t - \pi$

35–38 ■ Find the reference number for each value of t.

35. **(a)** $t = \dfrac{5\pi}{4}$ **(b)** $t = \dfrac{7\pi}{3}$
(c) $t = -\dfrac{4\pi}{3}$ **(d)** $t = \dfrac{\pi}{6}$

36. **(a)** $t = \dfrac{5\pi}{6}$ **(b)** $t = \dfrac{7\pi}{6}$
(c) $t = \dfrac{11\pi}{3}$ **(d)** $t = -\dfrac{7\pi}{4}$

37. **(a)** $t = \dfrac{5\pi}{7}$ **(b)** $t = -\dfrac{7\pi}{9}$
(c) $t = -3$ **(d)** $t = 5$

38. **(a)** $t = \dfrac{11\pi}{5}$ **(b)** $t = -\dfrac{9\pi}{7}$
(c) $t = 6$ **(d)** $t = -7$

39–52 ■ Find **(a)** the reference number for each value of t and **(b)** the terminal point determined by t.

39. $t = \dfrac{2\pi}{3}$

40. $t = \dfrac{4\pi}{3}$

41. $t = \dfrac{3\pi}{4}$

42. $t = \dfrac{7\pi}{3}$

43. $t = -\dfrac{2\pi}{3}$

44. $t = -\dfrac{7\pi}{6}$

45. $t = \dfrac{13\pi}{4}$

46. $t = \dfrac{13\pi}{6}$

47. $t = \dfrac{7\pi}{6}$

48. $t = \dfrac{17\pi}{4}$

49. $t = -\dfrac{11\pi}{3}$

50. $t = \dfrac{31\pi}{6}$

51. $t = \dfrac{16\pi}{3}$

52. $t = -\dfrac{41\pi}{4}$

53–56 ■ Use the figure to find the terminal point determined by the real number t, with coordinates rounded to one decimal place.

53. $t = 1$

54. $t = 2.5$

55. $t = -1.1$

56. $t = 4.2$

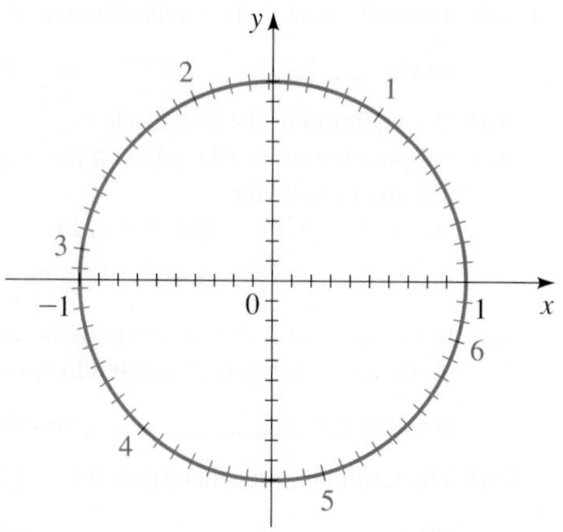

DISCOVERY ■ DISCUSSION ■ WRITING

57. Finding the Terminal Point for $\pi/6$ Suppose the terminal point determined by $t = \pi/6$ is $P(x, y)$ and the points Q and R are as shown in the figure. Why are the distances PQ and PR the same? Use this fact, together with the Distance Formula, to show that the coordinates of P satisfy the equation $2y = \sqrt{x^2 + (y - 1)^2}$. Simplify this equation using the fact that $x^2 + y^2 = 1$. Solve the simplified equation to find $P(x, y)$.

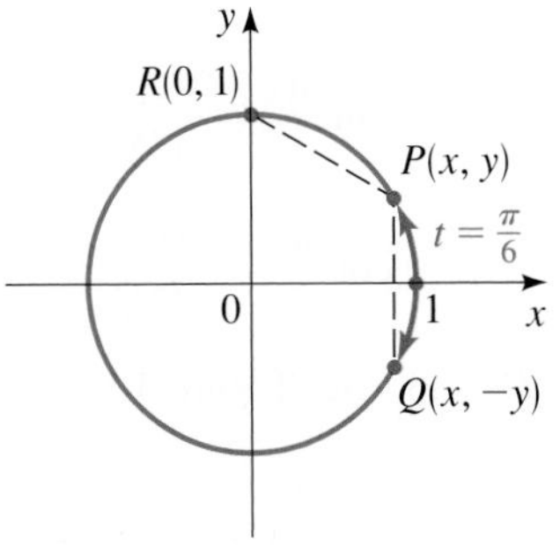

58. Finding the Terminal Point for $\pi/3$ Now that you know the terminal point determined by $t = \pi/6$, use symmetry to find the terminal point determined by $t = \pi/3$ (see the figure). Explain your reasoning.

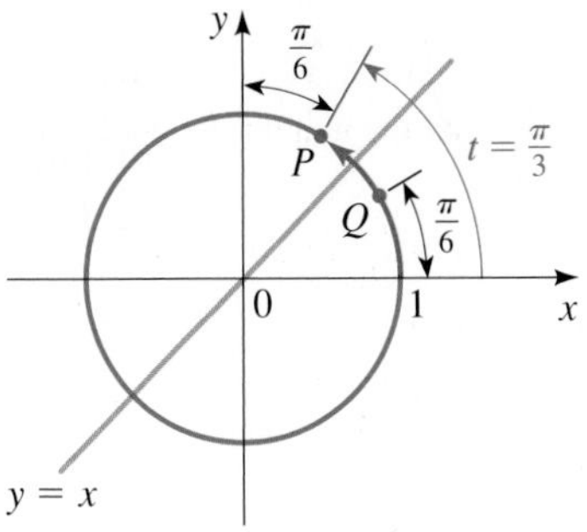

5.2 Trigonometric Functions of Real Numbers

The Trigonometric Functions ► Values of the Trigonometric Functions ► Fundamental Identities

A function is a rule that assigns to each real number another real number. In this section we use properties of the unit circle from the preceding section to define the trigonometric functions.

▼ The Trigonometric Functions

Recall that to find the terminal point $P(x, y)$ for a given real number t, we move a distance t along the unit circle, starting at the point $(1, 0)$. We move in a counterclockwise direction if t is positive and in a clockwise direction if t is negative (see Figure 1). We now use the x- and y-coordinates of the point $P(x, y)$ to define several functions. For instance, we define the function called *sine* by assigning to each real number t the y-coordinate of the terminal point $P(x, y)$ determined by t. The functions *cosine*, *tangent*, *cosecant*, *secant*, and *cotangent* are also defined by using the coordinates of $P(x, y)$.

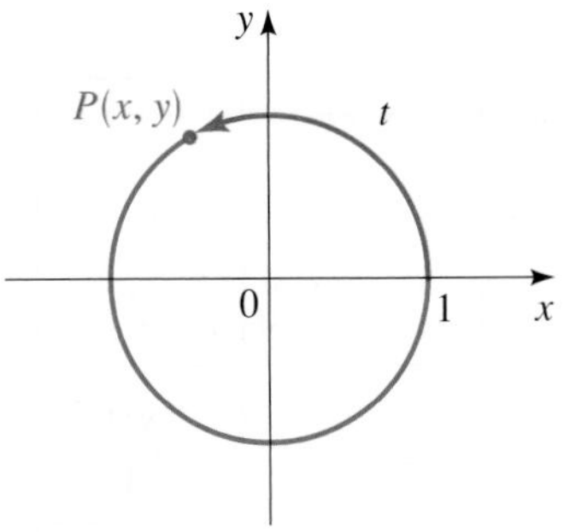

FIGURE 1

DEFINITION OF THE TRIGONOMETRIC FUNCTIONS

Let t be any real number and let $P(x, y)$ be the terminal point on the unit circle determined by t. We define

$$\sin t = y \qquad \cos t = x \qquad \tan t = \frac{y}{x} \quad (x \neq 0)$$

$$\csc t = \frac{1}{y} \quad (y \neq 0) \qquad \sec t = \frac{1}{x} \quad (x \neq 0) \qquad \cot t = \frac{x}{y} \quad (y \neq 0)$$

Because the trigonometric functions can be defined in terms of the unit circle, they are sometimes called the **circular functions**.

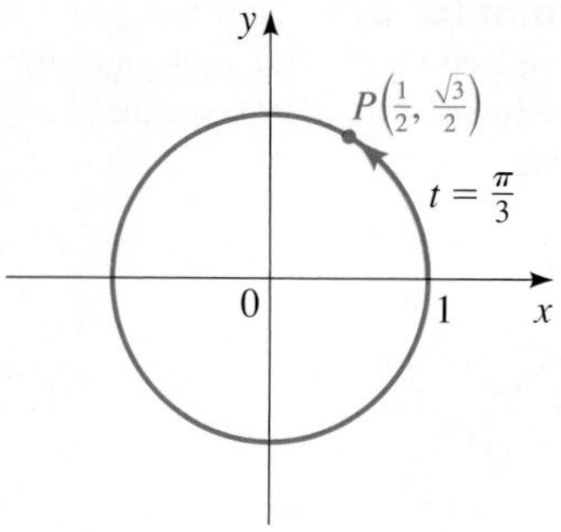

FIGURE 2

EXAMPLE 1 | Evaluating Trigonometric Functions

Find the six trigonometric functions of each given real number t.

(a) $t = \dfrac{\pi}{3}$ **(b)** $t = \dfrac{\pi}{2}$

SOLUTION

(a) From Table 1 on page 372, we see that the terminal point determined by $t = \pi/3$ is $P(\frac{1}{2}, \sqrt{3}/2)$. (See Figure 2.) Since the coordinates are $x = \frac{1}{2}$ and $y = \sqrt{3}/2$, we have

$$\sin\frac{\pi}{3} = \frac{\sqrt{3}}{2} \qquad \cos\frac{\pi}{3} = \frac{1}{2} \qquad \tan\frac{\pi}{3} = \frac{\sqrt{3}/2}{1/2} = \sqrt{3}$$

$$\csc\frac{\pi}{3} = \frac{2\sqrt{3}}{3} \qquad \sec\frac{\pi}{3} = 2 \qquad \cot\frac{\pi}{3} = \frac{1/2}{\sqrt{3}/2} = \frac{\sqrt{3}}{3}$$

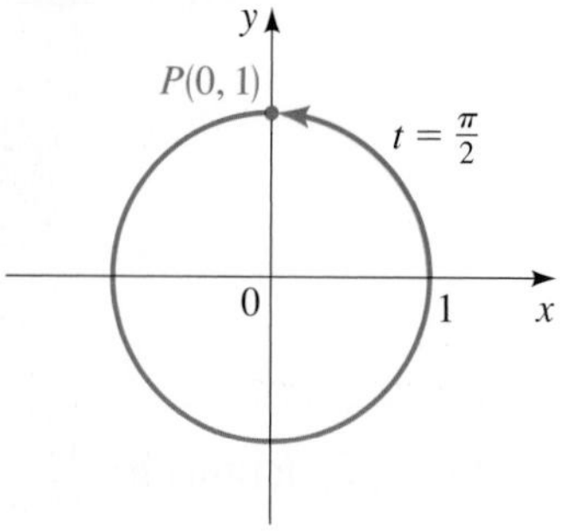

FIGURE 3

(b) The terminal point determined by $\pi/2$ is $P(0, 1)$. (See Figure 3.) So

$$\sin\frac{\pi}{2} = 1 \qquad \cos\frac{\pi}{2} = 0 \qquad \csc\frac{\pi}{2} = \frac{1}{1} = 1 \qquad \cot\frac{\pi}{2} = \frac{0}{1} = 0$$

But $\tan \pi/2$ and $\sec \pi/2$ are undefined because $x = 0$ appears in the denominator in each of their definitions.

NOW TRY EXERCISE **3**

Some special values of the trigonometric functions are listed in Table 1. This table is easily obtained from Table 1 of Section 5.1, together with the definitions of the trigonometric functions.

TABLE 1
Special values of the trigonometric functions

t	$\sin t$	$\cos t$	$\tan t$	$\csc t$	$\sec t$	$\cot t$
0	0	1	0	—	1	—
$\frac{\pi}{6}$	$\frac{1}{2}$	$\frac{\sqrt{3}}{2}$	$\frac{\sqrt{3}}{3}$	2	$\frac{2\sqrt{3}}{3}$	$\sqrt{3}$
$\frac{\pi}{4}$	$\frac{\sqrt{2}}{2}$	$\frac{\sqrt{2}}{2}$	1	$\sqrt{2}$	$\sqrt{2}$	1
$\frac{\pi}{3}$	$\frac{\sqrt{3}}{2}$	$\frac{1}{2}$	$\sqrt{3}$	$\frac{2\sqrt{3}}{3}$	2	$\frac{\sqrt{3}}{3}$
$\frac{\pi}{2}$	1	0	—	1	—	0

We can easily remember the sines and cosines of the basic angles by writing them in the form $\sqrt{\blacksquare}/2$:

t	$\sin t$	$\cos t$
0	$\sqrt{0}/2$	$\sqrt{4}/2$
$\pi/6$	$\sqrt{1}/2$	$\sqrt{3}/2$
$\pi/4$	$\sqrt{2}/2$	$\sqrt{2}/2$
$\pi/3$	$\sqrt{3}/2$	$\sqrt{1}/2$
$\pi/2$	$\sqrt{4}/2$	$\sqrt{0}/2$

Example 1 shows that some of the trigonometric functions fail to be defined for certain real numbers. So we need to determine their domains. The functions sine and cosine are defined for all values of t. Since the functions cotangent and cosecant have y in the denominator of their definitions, they are not defined whenever the y-coordinate of the terminal point $P(x, y)$ determined by t is 0. This happens when $t = n\pi$ for any integer n, so their domains do not include these points. The functions tangent and secant have x in the denominator in their definitions, so they are not defined whenever $x = 0$. This happens when $t = (\pi/2) + n\pi$ for any integer n.

Relationship to the Trigonometric Functions of Angles

If you have previously studied trigonometry of right triangles (Chapter 6), you are probably wondering how the sine and cosine of an *angle* relate to those of this section. To see how, let's start with a right triangle, ΔOPQ.

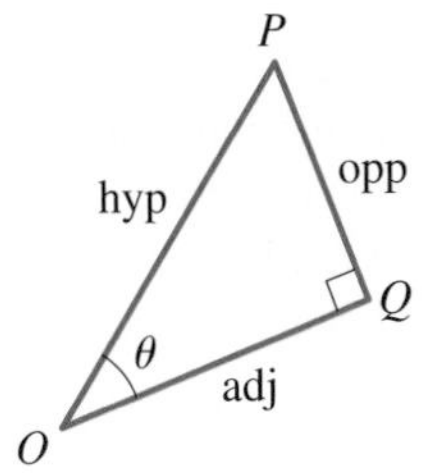

Right triangle *OPQ*

Place the triangle in the coordinate plane as shown, with angle θ in standard position.

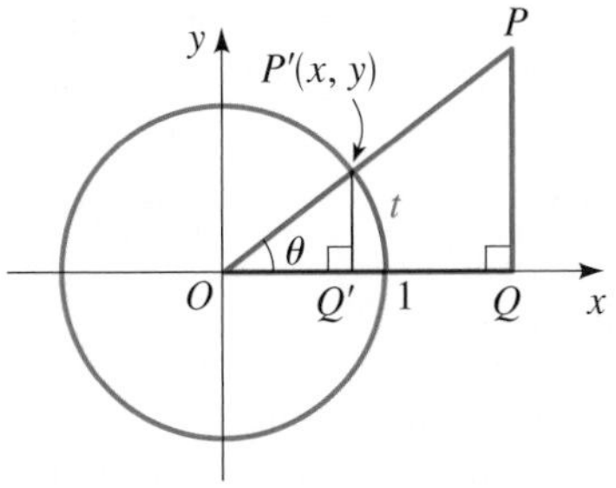

$P'(x, y)$ is the terminal point determined by t.

The point $P'(x, y)$ in the figure is the terminal point determined by the arc t. Note that triangle OPQ is similar to the small triangle $OP'Q'$ whose legs have lengths x and y.

Now, by the definition of the trigonometric functions of the *angle* θ we have

$$\sin\theta = \frac{\text{opp}}{\text{hyp}} = \frac{PQ}{OP} = \frac{P'Q'}{OP'}$$

$$= \frac{y}{1} = y$$

$$\cos\theta = \frac{\text{adj}}{\text{hyp}} = \frac{OQ}{OP} = \frac{OQ'}{OP'}$$

$$= \frac{x}{1} = x$$

By the definition of the trigonometric functions of the *real number t,* we have

$$\sin t = y \qquad \cos t = x$$

Now, if θ is measured in radians, then $\theta = t$ (see the figure). So the trigonometric functions of the angle with radian measure θ are exactly the same as the trigonometric functions defined in terms of the terminal point determined by the real number t.

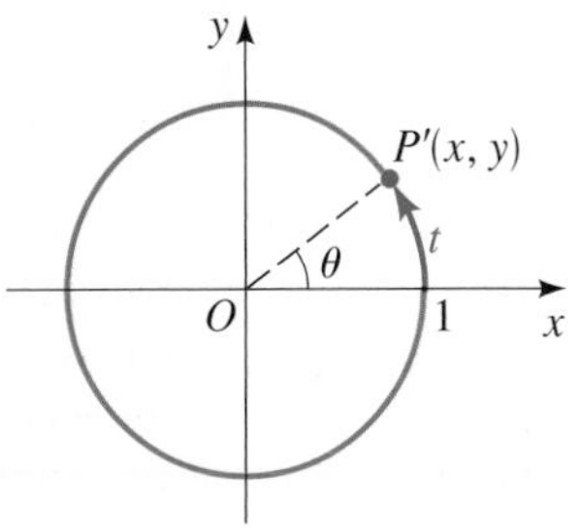

The radian measure of angle θ is t.

Why then study trigonometry in two different ways? Because different applications require that we view the trigonometric functions differently. (Compare Section 5.6 with Sections 6.2, 6.5, and 6.6.)

DOMAINS OF THE TRIGONOMETRIC FUNCTIONS

Function	Domain
sin, cos	All real numbers
tan, sec	All real numbers other than $\frac{\pi}{2} + n\pi$ for any integer n
cot, csc	All real numbers other than $n\pi$ for any integer, n

▼ Values of the Trigonometric Functions

To compute other values of the trigonometric functions, we first determine their signs. The signs of the trigonometric functions depend on the quadrant in which the terminal point of t lies. For example, if the terminal point $P(x, y)$ determined by t lies in Quadrant III, then its coordinates are both negative. So $\sin t$, $\cos t$, $\csc t$, and $\sec t$ are all negative, whereas $\tan t$ and $\cot t$ are positive. You can check the other entries in the following box.

The following mnemonic device will help you remember which trigonometric functions are positive in each quadrant: **A**ll of them, **S**ine, **T**angent, or **C**osine.

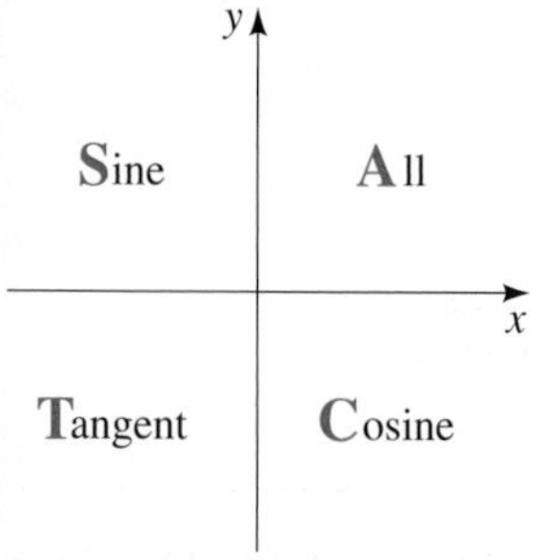

You can remember this as "All Students Take Calculus."

SIGNS OF THE TRIGONOMETRIC FUNCTIONS

Quadrant	Positive Functions	Negative Functions
I	all	none
II	sin, csc	cos, sec, tan, cot
III	tan, cot	sin, csc, cos, sec
IV	cos, sec	sin, csc, tan, cot

For example $\cos(2\pi/3) < 0$ because the terminal point of $t = 2\pi/3$ is in Quadrant II, whereas $\tan 4 > 0$ because the terminal point of $t = 4$ is in Quadrant III.

In Section 5.1 we used the reference number to find the terminal point determined by a real number t. Since the trigonometric functions are defined in terms of the coordinates of terminal points, we can use the reference number to find values of the trigonometric functions. Suppose that $\bar{t}$ is the reference number for t. Then the terminal point of $\bar{t}$ has the same coordinates, except possibly for sign, as the terminal point of t. So the values of the trigonometric functions at t are the same, except possibly for sign, as their values at $\bar{t}$. We illustrate this procedure in the next example.

EXAMPLE 2 | Evaluating Trigonometric Functions

Find each value.

(a) $\cos \frac{2\pi}{3}$ **(b)** $\tan\left(-\frac{\pi}{3}\right)$ **(c)** $\sin \frac{19\pi}{4}$

SOLUTION

(a) The reference number for $2\pi/3$ is $\pi/3$ (see Figure 4(a)). Since the terminal point of $2\pi/3$ is in Quadrant II, $\cos(2\pi/3)$ is negative. Thus,

$$\cos \frac{2\pi}{3} = -\cos \frac{\pi}{3} = -\frac{1}{2}$$

Sign | Reference number | From Table 1

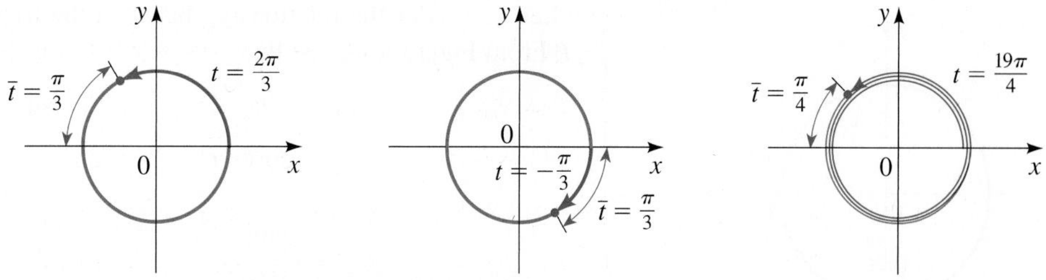

FIGURE 4 (a) (b) (c)

(b) The reference number for $-\pi/3$ is $\pi/3$ (see Figure 4(b)). Since the terminal point of $-\pi/3$ is in Quadrant IV, $\tan(-\pi/3)$ is negative. Thus,

$$\tan\left(-\frac{\pi}{3}\right) = -\tan\frac{\pi}{3} = -\sqrt{3}$$

Sign | Reference number | From Table 1

(c) Since $(19\pi/4) - 4\pi = 3\pi/4$, the terminal points determined by $19\pi/4$ and $3\pi/4$ are the same. The reference number for $3\pi/4$ is $\pi/4$ (see Figure 4(c)). Since the terminal point of $3\pi/4$ is in Quadrant II, $\sin(3\pi/4)$ is positive. Thus,

$$\sin\frac{19\pi}{4} = \sin\frac{3\pi}{4} = +\sin\frac{\pi}{4} = \frac{\sqrt{2}}{2}$$

Subtract 4π | Sign | Reference number | From Table 1

NOW TRY EXERCISE **7** ■

So far, we have been able to compute the values of the trigonometric functions only for certain values of t. In fact, we can compute the values of the trigonometric functions whenever t is a multiple of $\pi/6$, $\pi/4$, $\pi/3$, and $\pi/2$. How can we compute the trigonometric functions for other values of t? For example, how can we find $\sin 1.5$? One way is to carefully sketch a diagram and read the value (see Exercises 39–46); however, this method is not very accurate. Fortunately, programmed directly into scientific calculators are mathematical procedures (see the margin note on page 400) that find the values of *sine*, *cosine*, and *tangent* correct to the number of digits in the display. The calculator must be put in *radian mode* to evaluate these functions. To find values of cosecant, secant, and cotangent using a calculator, we need to use the following *reciprocal relations*:

$$\csc t = \frac{1}{\sin t} \qquad \sec t = \frac{1}{\cos t} \qquad \cot t = \frac{1}{\tan t}$$

These identities follow from the definitions of the trigonometric functions. For instance, since $\sin t = y$ and $\csc t = 1/y$, we have $\csc t = 1/y = 1/(\sin t)$. The others follow similarly.

EXAMPLE 3 | Using a Calculator to Evaluate Trigonometric Functions

Making sure our calculator is set to radian mode and rounding the results to six decimal places, we get

(a) $\sin 2.2 \approx 0.808496$

(b) $\cos 1.1 \approx 0.453596$

(c) $\cot 28 = \dfrac{1}{\tan 28} \approx -3.553286$

(d) $\csc 0.98 = \dfrac{1}{\sin 0.98} \approx 1.204098$

NOW TRY EXERCISES **41** AND **43** ■

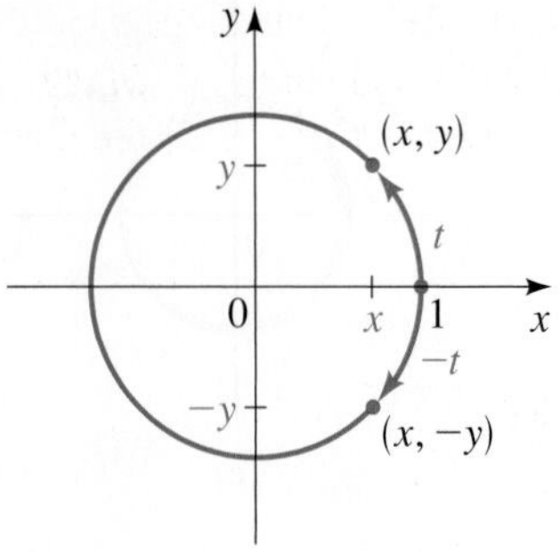

FIGURE 5

Let's consider the relationship between the trigonometric functions of t and those of $-t$. From Figure 5 we see that

$$\sin(-t) = -y = -\sin t$$
$$\cos(-t) = x = \cos t$$
$$\tan(-t) = \frac{-y}{x} = -\frac{y}{x} = -\tan t$$

These equations show that sine and tangent are odd functions, whereas cosine is an even function. It's easy to see that the reciprocal of an even function is even and the reciprocal of an odd function is odd. This fact, together with the reciprocal relations, completes our knowledge of the even-odd properties for all the trigonometric functions.

Even and odd functions are defined in Section 2.5.

EVEN-ODD PROPERTIES

Sine, cosecant, tangent, and cotangent are odd functions; cosine and secant are even functions.

$$\sin(-t) = -\sin t \qquad \cos(-t) = \cos t \qquad \tan(-t) = -\tan t$$
$$\csc(-t) = -\csc t \qquad \sec(-t) = \sec t \qquad \cot(-t) = -\cot t$$

EXAMPLE 4 | Even and Odd Trigonometric Functions

Use the even-odd properties of the trigonometric functions to determine each value.

(a) $\sin\left(-\frac{\pi}{6}\right)$ **(b)** $\cos\left(-\frac{\pi}{4}\right)$

SOLUTION By the even-odd properties and Table 1 we have

(a) $\sin\left(-\frac{\pi}{6}\right) = -\sin\frac{\pi}{6} = -\frac{1}{2}$ Sine is odd

(b) $\cos\left(-\frac{\pi}{4}\right) = \cos\frac{\pi}{4} = \frac{\sqrt{2}}{2}$ Cosine is even

NOW TRY EXERCISE **13**

Fundamental Identities

The trigonometric functions are related to each other through equations called **trigonometric identities**. We give the most important ones in the following box.*

FUNDAMENTAL IDENTITIES

Reciprocal Identities

$$\csc t = \frac{1}{\sin t} \qquad \sec t = \frac{1}{\cos t} \qquad \cot t = \frac{1}{\tan t} \qquad \tan t = \frac{\sin t}{\cos t} \qquad \cot t = \frac{\cos t}{\sin t}$$

Pythagorean Identities

$$\sin^2 t + \cos^2 t = 1 \qquad \tan^2 t + 1 = \sec^2 t \qquad 1 + \cot^2 t = \csc^2 t$$

*We follow the usual convention of writing $\sin^2 t$ for $(\sin t)^2$. In general, we write $\sin^n t$ for $(\sin t)^n$ for all integers n except $n = -1$. The exponent $n = -1$ will be assigned another meaning in Section 5.5. Of course, the same convention applies to the other five trigonometric functions.

The Value of π

The number π is the ratio of the circumference of a circle to its diameter. It has been known since ancient times that this ratio is the same for all circles. The first systematic effort to find a numerical approximation for π was made by Archimedes (ca. 240 B.C.), who proved that $\frac{22}{7} < \pi < \frac{223}{71}$ by finding the perimeters of regular polygons inscribed in and circumscribed about a circle.

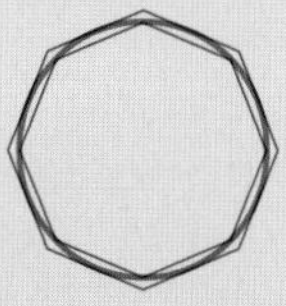

In about A.D. 480, the Chinese physicist Tsu Ch'ung-chih gave the approximation

$$\pi \approx \tfrac{355}{113} = 3.141592\ldots$$

which is correct to six decimals. This remained the most accurate estimation of π until the Dutch mathematician Adrianus Romanus (1593) used polygons with more than a billion sides to compute π correct to 15 decimals. In the 17th century, mathematicians began to use infinite series and trigonometric identities in the quest for π. The Englishman William Shanks spent 15 years (1858–1873) using these methods to compute π to 707 decimals, but in 1946 it was found that his figures were wrong beginning with the 528th decimal. Today, with the aid of computers, mathematicians routinely determine π correct to millions of decimals. The current record is that π has been computed to 2,576,980,370,000 (more than two trillion) decimal places by T. Daesuke and his team.

PROOF The reciprocal identities follow immediately from the definitions on page 377. We now prove the Pythagorean identities. By definition, $\cos t = x$ and $\sin t = y$, where x and y are the coordinates of a point $P(x, y)$ on the unit circle. Since $P(x, y)$ is on the unit circle, we have $x^2 + y^2 = 1$. Thus

$$\sin^2 t + \cos^2 t = 1$$

Dividing both sides by $\cos^2 t$ (provided that $\cos t \neq 0$), we get

$$\frac{\sin^2 t}{\cos^2 t} + \frac{\cos^2 t}{\cos^2 t} = \frac{1}{\cos^2 t}$$

$$\left(\frac{\sin t}{\cos t}\right)^2 + 1 = \left(\frac{1}{\cos t}\right)^2$$

$$\tan^2 t + 1 = \sec^2 t$$

We have used the reciprocal identities $\sin t/\cos t = \tan t$ and $1/\cos t = \sec t$. Similarly, dividing both sides of the first Pythagorean identity by $\sin^2 t$ (provided that $\sin t \neq 0$) gives us $1 + \cot^2 t = \csc^2 t$. ■

As their name indicates, the fundamental identities play a central role in trigonometry because we can use them to relate any trigonometric function to any other. So, if we know the value of any one of the trigonometric functions at t, then we can find the values of all the others at t.

EXAMPLE 5 | Finding All Trigonometric Functions from the Value of One

If $\cos t = \frac{3}{5}$ and t is in Quadrant IV, find the values of all the trigonometric functions at t.

SOLUTION From the Pythagorean identities we have

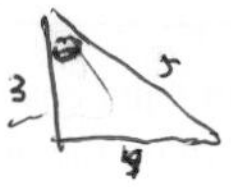

$$\sin^2 t + \cos^2 t = 1$$

$$\sin^2 t + \left(\tfrac{3}{5}\right)^2 = 1 \qquad \text{Substitute } \cos t = \tfrac{3}{5}$$

$$\sin^2 t = 1 - \tfrac{9}{25} = \tfrac{16}{25} \qquad \text{Solve for } \sin^2 t$$

$$\sin t = \pm\tfrac{4}{5} \qquad \text{Take square roots}$$

Since this point is in Quadrant IV, $\sin t$ is negative, so $\sin t = -\frac{4}{5}$. Now that we know both $\sin t$ and $\cos t$, we can find the values of the other trigonometric functions using the reciprocal identities:

$$\sin t = -\frac{4}{5} \qquad \cos t = \frac{3}{5} \qquad \tan t = \frac{\sin t}{\cos t} = \frac{-\frac{4}{5}}{\frac{3}{5}} = -\frac{4}{3}$$

$$\csc t = \frac{1}{\sin t} = -\frac{5}{4} \qquad \sec t = \frac{1}{\cos t} = \frac{5}{3} \qquad \cot t = \frac{1}{\tan t} = -\frac{3}{4}$$

NOW TRY EXERCISE **65** ■

EXAMPLE 6 | Writing One Trigonometric Function in Terms of Another

Write $\tan t$ in terms of $\cos t$, where t is in Quadrant III.

SOLUTION Since $\tan t = \sin t/\cos t$, we need to write $\sin t$ in terms of $\cos t$. By the Pythagorean identities we have

$$\sin^2 t + \cos^2 t = 1$$

$$\sin^2 t = 1 - \cos^2 t \qquad \text{Solve for } \sin^2 t$$

$$\sin t = \pm\sqrt{1 - \cos^2 t} \qquad \text{Take square roots}$$

Since $\sin t$ is negative in Quadrant III, the negative sign applies here. Thus,

$$\tan t = \frac{\sin t}{\cos t} = \frac{-\sqrt{1 - \cos^2 t}}{\cos t}$$

NOW TRY EXERCISE **55**

5.2 EXERCISES

CONCEPTS

1. Let $P(x, y)$ be the terminal point on the unit circle determined by t. Then $\sin t =$ ________, $\cos t =$ ________, and $\tan t =$ ________.

2. If $P(x, y)$ is on the unit circle, then $x^2 + y^2 =$ ________. So for all t we have $\sin^2 t + \cos^2 t =$ ________.

SKILLS

3–4 ■ Find $\sin t$ and $\cos t$ for the values of t whose terminal points are shown on the unit circle in the figure. In Exercise 3, t increases in increments of $\pi/4$; in Exercise 4, t increases in increments of $\pi/6$. (See Exercises 21 and 22 in Section 5.1.)

3.

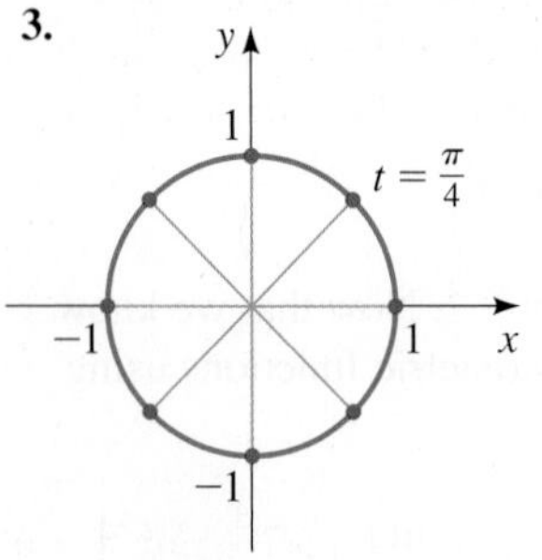

4.

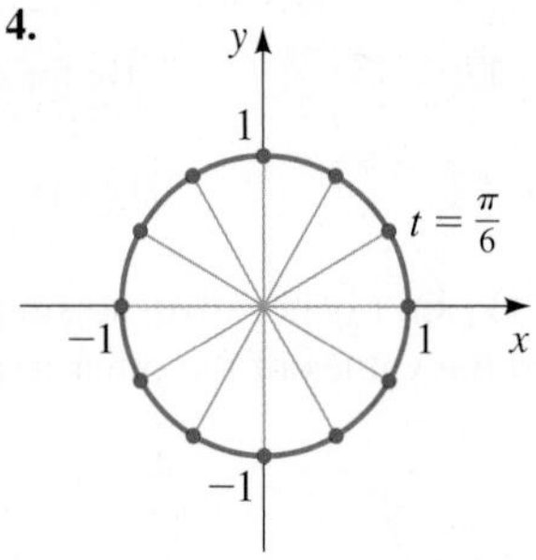

5–24 ■ Find the exact value of the trigonometric function at the given real number.

5. **(a)** $\sin \frac{2\pi}{3}$ **(b)** $\cos \frac{2\pi}{3}$ **(c)** $\tan \frac{2\pi}{3}$

6. **(a)** $\sin \frac{5\pi}{6}$ **(b)** $\cos \frac{5\pi}{6}$ **(c)** $\tan \frac{5\pi}{6}$

7. **(a)** $\sin \frac{7\pi}{6}$ **(b)** $\sin\left(-\frac{\pi}{6}\right)$ **(c)** $\sin \frac{11\pi}{6}$

8. **(a)** $\cos \frac{5\pi}{3}$ **(b)** $\cos\left(-\frac{5\pi}{3}\right)$ **(c)** $\cos \frac{7\pi}{3}$

9. **(a)** $\cos \frac{3\pi}{4}$ **(b)** $\cos \frac{5\pi}{4}$ **(c)** $\cos \frac{7\pi}{4}$

10. **(a)** $\sin \frac{3\pi}{4}$ **(b)** $\sin \frac{5\pi}{4}$ **(c)** $\sin \frac{7\pi}{4}$

11. **(a)** $\sin \frac{7\pi}{3}$ **(b)** $\csc \frac{7\pi}{3}$ **(c)** $\cot \frac{7\pi}{3}$

12. **(a)** $\cos\left(-\frac{\pi}{3}\right)$ **(b)** $\sec\left(-\frac{\pi}{3}\right)$ **(c)** $\tan\left(-\frac{\pi}{3}\right)$

13. **(a)** $\sin\left(-\frac{\pi}{2}\right)$ **(b)** $\cos\left(-\frac{\pi}{2}\right)$ **(c)** $\cot\left(-\frac{\pi}{2}\right)$

14. **(a)** $\sin\left(-\frac{3\pi}{2}\right)$ **(b)** $\cos\left(-\frac{3\pi}{2}\right)$ **(c)** $\cot\left(-\frac{3\pi}{2}\right)$

15. **(a)** $\sec \frac{11\pi}{3}$ **(b)** $\csc \frac{11\pi}{3}$ **(c)** $\sec\left(-\frac{\pi}{3}\right)$

16. **(a)** $\cos \frac{7\pi}{6}$ **(b)** $\sec \frac{7\pi}{6}$ **(c)** $\csc \frac{7\pi}{6}$

17. **(a)** $\tan \frac{5\pi}{6}$ **(b)** $\tan \frac{7\pi}{6}$ **(c)** $\tan \frac{11\pi}{6}$

18. **(a)** $\cot\left(-\frac{\pi}{3}\right)$ **(b)** $\cot \frac{2\pi}{3}$ **(c)** $\cot \frac{5\pi}{3}$

19. **(a)** $\cos\left(-\frac{\pi}{4}\right)$ **(b)** $\csc\left(-\frac{\pi}{4}\right)$ **(c)** $\cot\left(-\frac{\pi}{4}\right)$

20. **(a)** $\sin \frac{5\pi}{4}$ **(b)** $\sec \frac{5\pi}{4}$ **(c)** $\tan \frac{5\pi}{4}$

21. **(a)** $\csc\left(-\frac{\pi}{2}\right)$ **(b)** $\csc \frac{\pi}{2}$ **(c)** $\csc \frac{3\pi}{2}$

22. **(a)** $\sec(-\pi)$ **(b)** $\sec \pi$ **(c)** $\sec 4\pi$

23. **(a)** $\sin 13\pi$ **(b)** $\cos 14\pi$ **(c)** $\tan 15\pi$

24. **(a)** $\sin \frac{25\pi}{2}$ **(b)** $\cos \frac{25\pi}{2}$ **(c)** $\cot \frac{25\pi}{2}$

25–28 ■ Find the value of each of the six trigonometric functions (if it is defined) at the given real number t. Use your answers to complete the table.

25. $t = 0$ **26.** $t = \frac{\pi}{2}$ **27.** $t = \pi$ **28.** $t = \frac{3\pi}{2}$

t	$\sin t$	$\cos t$	$\tan t$	$\csc t$	$\sec t$	$\cot t$
0	0	1		undefined		
$\frac{\pi}{2}$						
π			0			undefined
$\frac{3\pi}{2}$						

29–38 ■ The terminal point $P(x, y)$ determined by a real number t is given. Find $\sin t$, $\cos t$, and $\tan t$.

29. $\left(\frac{3}{5}, \frac{4}{5}\right)$ **30.** $\left(-\frac{3}{5}, \frac{4}{5}\right)$

31. $\left(\frac{\sqrt{5}}{4}, -\frac{\sqrt{11}}{4}\right)$ **32.** $\left(-\frac{1}{3}, -\frac{2\sqrt{2}}{3}\right)$

33. $\left(-\frac{6}{7}, \frac{\sqrt{13}}{7}\right)$ **34.** $\left(\frac{40}{41}, \frac{9}{41}\right)$

35. $\left(-\frac{5}{13}, -\frac{12}{13}\right)$ **36.** $\left(\frac{\sqrt{5}}{5}, \frac{2\sqrt{5}}{5}\right)$

37. $\left(-\frac{20}{29}, \frac{21}{29}\right)$ **38.** $\left(\frac{24}{25}, -\frac{7}{25}\right)$

39–46 ■ Find an approximate value of the given trigonometric function by using **(a)** the figure and **(b)** a calculator. Compare the two values.

39. $\sin 1$

40. $\cos 0.8$

41. $\sin 1.2$

42. $\cos 5$

43. $\tan 0.8$

44. $\tan(-1.3)$

45. $\cos 4.1$

46. $\sin(-5.2)$

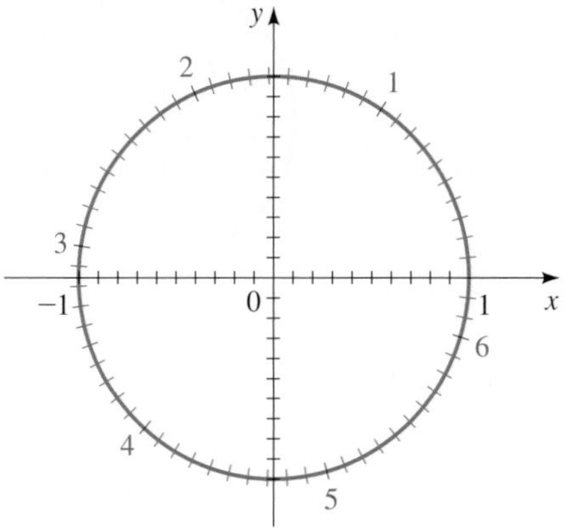

47–50 ■ Find the sign of the expression if the terminal point determined by t is in the given quadrant.

47. $\sin t \cos t$, Quadrant II **48.** $\tan t \sec t$, Quadrant IV

49. $\frac{\tan t \sin t}{\cot t}$, Quadrant III **50.** $\cos t \sec t$, any quadrant

51–54 ■ From the information given, find the quadrant in which the terminal point determined by t lies.

51. $\sin t > 0$ and $\cos t < 0$ **52.** $\tan t > 0$ and $\sin t < 0$

53. $\csc t > 0$ and $\sec t < 0$ **54.** $\cos t < 0$ and $\cot t < 0$

55–64 ■ Write the first expression in terms of the second if the terminal point determined by t is in the given quadrant.

55. $\sin t$, $\cos t$; Quadrant II **56.** $\cos t$, $\sin t$; Quadrant IV

57. $\tan t$, $\sin t$; Quadrant IV **58.** $\tan t$, $\cos t$; Quadrant III

59. $\sec t$, $\tan t$; Quadrant II **60.** $\csc t$, $\cot t$; Quadrant III

61. $\tan t$, $\sec t$; Quadrant III **62.** $\sin t$, $\sec t$; Quadrant IV

63. $\tan^2 t$, $\sin t$; any quadrant

64. $\sec^2 t \sin^2 t$, $\cos t$; any quadrant

65–72 ■ Find the values of the trigonometric functions of t from the given information.

65. $\sin t = \frac{3}{5}$, terminal point of t is in Quadrant II

66. $\cos t = -\frac{4}{5}$, terminal point of t is in Quadrant III

67. $\sec t = 3$, terminal point of t is in Quadrant IV

68. $\tan t = \frac{1}{4}$, terminal point of t is in Quadrant III

69. $\tan t = -\frac{3}{4}$, $\cos t > 0$

70. $\sec t = 2$, $\sin t < 0$

71. $\sin t = -\frac{1}{4}$, $\sec t < 0$

72. $\tan t = -4$, $\csc t > 0$

73–80 ■ Determine whether the function is even, odd, or neither.

73. $f(x) = x^2 \sin x$ **74.** $f(x) = x^2 \cos 2x$

75. $f(x) = \sin x \cos x$ **76.** $f(x) = \sin x + \cos x$

77. $f(x) = |x| \cos x$ **78.** $f(x) = x \sin^3 x$

79. $f(x) = x^3 + \cos x$ **80.** $f(x) = \cos(\sin x)$

APPLICATIONS

81. Harmonic Motion The displacement from equilibrium of an oscillating mass attached to a spring is given by $y(t) = 4 \cos 3\pi t$ where y is measured in inches and t in seconds. Find the displacement at the times indicated in the table.

t	$y(t)$
0	
0.25	
0.50	
0.75	
1.00	
1.25	

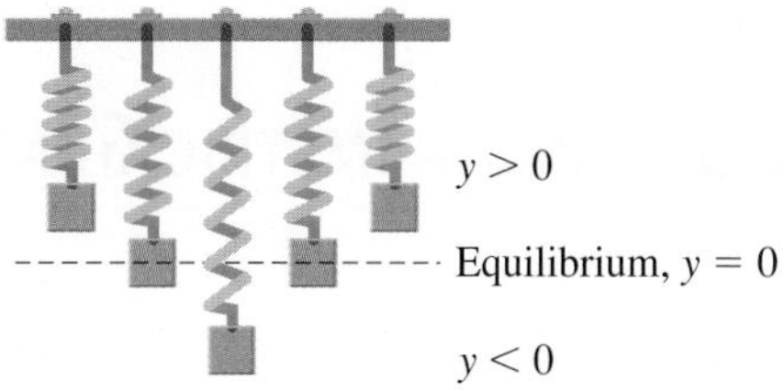

82. Circadian Rhythms Everybody's blood pressure varies over the course of the day. In a certain individual the resting diastolic blood pressure at time t is given by $B(t) = 80 + 7 \sin(\pi t/12)$, where t is measured in hours since midnight and $B(t)$ in mmHg (millimeters of mercury). Find this person's diastolic blood pressure at
(a) 6:00 A.M. **(b)** 10:30 A.M. **(c)** Noon **(d)** 8:00 P.M.

83. Electric Circuit After the switch is closed in the circuit shown, the current t seconds later is $I(t) = 0.8e^{-3t}\sin 10t$. Find the current at the times
(a) $t = 0.1$ s and **(b)** $t = 0.5$ s.

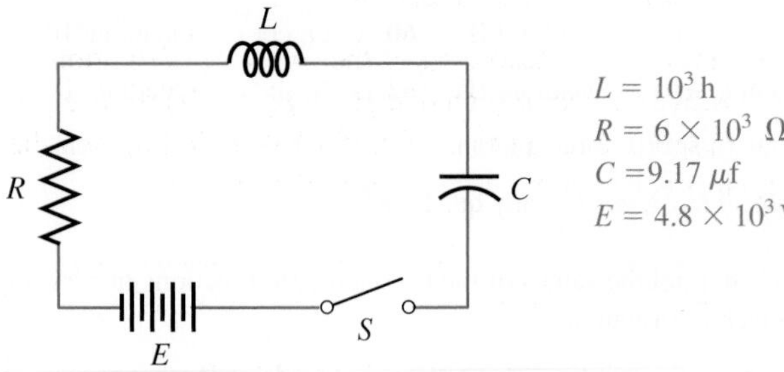

84. Bungee Jumping A bungee jumper plummets from a high bridge to the river below and then bounces back over and over again. At time t seconds after her jump, her height H (in meters) above the river is given by $H(t) = 100 + 75e^{-t/20}\cos(\frac{\pi}{4}t)$. Find her height at the times indicated in the table.

t	$H(t)$
0	
1	
2	
4	
6	
8	
12	

DISCOVERY ▪ DISCUSSION ▪ WRITING

85. Reduction Formulas A *reduction formula* is one that can be used to "reduce" the number of terms in the input for a trigonometric function. Explain how the figure shows that the following reduction formulas are valid:

$$\sin(t + \pi) = -\sin t \qquad \cos(t + \pi) = -\cos t$$

$$\tan(t + \pi) = \tan t$$

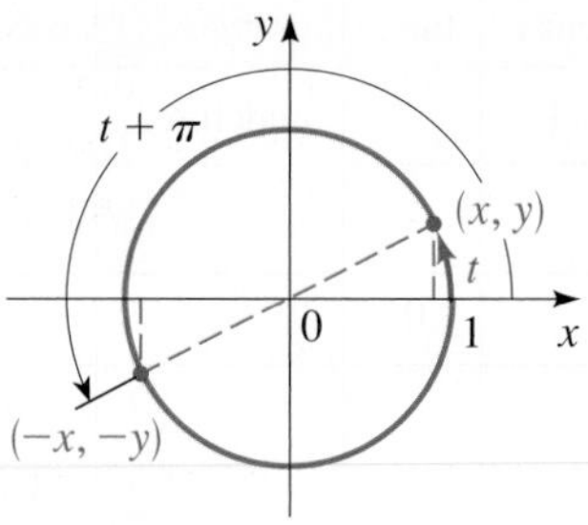

86. More Reduction Formulas By the "Angle-Side-Angle" theorem from elementary geometry, triangles CDO and AOB in the figure are congruent. Explain how this proves that if B has coordinates (x, y), then D has coordinates $(-y, x)$. Then explain how the figure shows that the following reduction formulas are valid:

$$\sin\left(t + \frac{\pi}{2}\right) = \cos t$$

$$\cos\left(t + \frac{\pi}{2}\right) = -\sin t$$

$$\tan\left(t + \frac{\pi}{2}\right) = -\cot t$$

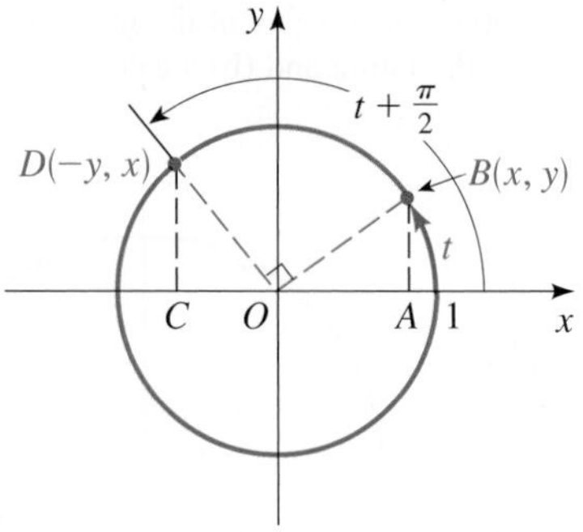

5.3 Trigonometric Graphs

Graphs of Sine and Cosine ▶ Graphs of Transformations of Sine and Cosine ▶ Using Graphing Devices to Graph Trigonometric Functions

The graph of a function gives us a better idea of its behavior. So, in this section we graph the sine and cosine functions and certain transformations of these functions. The other trigonometric functions are graphed in the next section.

▼ Graphs of Sine and Cosine

To help us graph the sine and cosine functions, we first observe that these functions repeat their values in a regular fashion. To see exactly how this happens, recall that the circumference of the unit circle is 2π. It follows that the terminal point $P(x, y)$ determined by the real number t is the same as that determined by $t + 2\pi$. Since the sine and cosine functions are defined in terms of the coordinates of $P(x, y)$, it follows that their values are unchanged by the addition of any integer multiple of 2π. In other words,

$$\sin(t + 2n\pi) = \sin t \qquad \text{for any integer } n$$
$$\cos(t + 2n\pi) = \cos t \qquad \text{for any integer } n$$

Thus, the sine and cosine functions are *periodic* according to the following definition: A function f is **periodic** if there is a positive number p such that $f(t + p) = f(t)$ for every t. The least such positive number (if it exists) is the **period** of f. If f has period p, then the graph of f on any interval of length p is called **one complete period** of f.

PERIODIC PROPERTIES OF SINE AND COSINE

The functions sine and cosine have period 2π:

$$\sin(t + 2\pi) = \sin t \qquad \cos(t + 2\pi) = \cos t$$

TABLE 1

t	$\sin t$	$\cos t$
$0 \to \frac{\pi}{2}$	$0 \to 1$	$1 \to 0$
$\frac{\pi}{2} \to \pi$	$1 \to 0$	$0 \to -1$
$\pi \to \frac{3\pi}{2}$	$0 \to -1$	$-1 \to 0$
$\frac{3\pi}{2} \to 2\pi$	$-1 \to 0$	$0 \to 1$

So the sine and cosine functions repeat their values in any interval of length 2π. To sketch their graphs, we first graph one period. To sketch the graphs on the interval $0 \le t \le 2\pi$, we could try to make a table of values and use those points to draw the graph. Since no such table can be complete, let's look more closely at the definitions of these functions.

Recall that $\sin t$ is the y-coordinate of the terminal point $P(x, y)$ on the unit circle determined by the real number t. How does the y-coordinate of this point vary as t increases? It's easy to see that the y-coordinate of $P(x, y)$ increases to 1, then decreases to -1 repeatedly as the point $P(x, y)$ travels around the unit circle. (See Figure 1.) In fact, as t increases from 0 to $\pi/2$, $y = \sin t$ increases from 0 to 1. As t increases from $\pi/2$ to π, the value of $y = \sin t$ decreases from 1 to 0. Table 1 shows the variation of the sine and cosine functions for t between 0 and 2π.

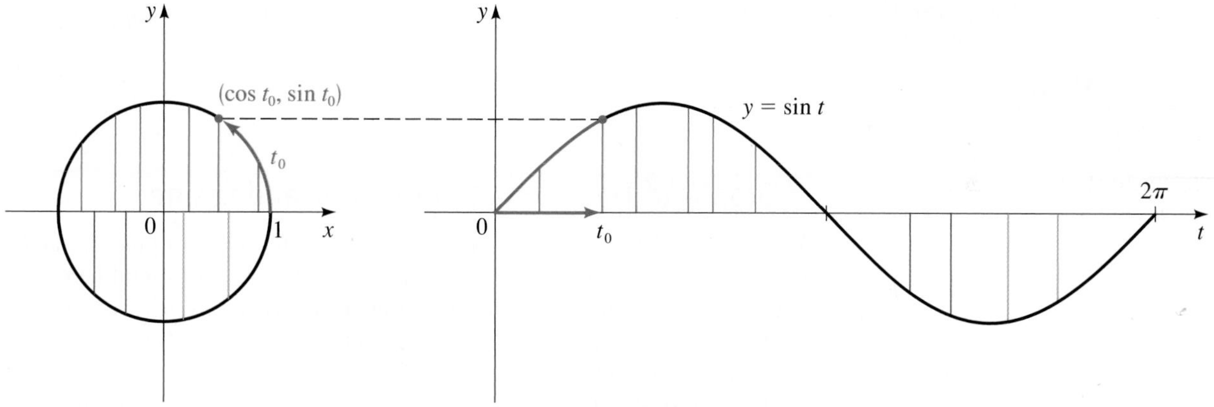

FIGURE 1

To draw the graphs more accurately, we find a few other values of $\sin t$ and $\cos t$ in Table 2. We could find still other values with the aid of a calculator.

TABLE 2

t	0	$\frac{\pi}{6}$	$\frac{\pi}{3}$	$\frac{\pi}{2}$	$\frac{2\pi}{3}$	$\frac{5\pi}{6}$	π	$\frac{7\pi}{6}$	$\frac{4\pi}{3}$	$\frac{3\pi}{2}$	$\frac{5\pi}{3}$	$\frac{11\pi}{6}$	2π
$\sin t$	0	$\frac{1}{2}$	$\frac{\sqrt{3}}{2}$	1	$\frac{\sqrt{3}}{2}$	$\frac{1}{2}$	0	$-\frac{1}{2}$	$-\frac{\sqrt{3}}{2}$	-1	$-\frac{\sqrt{3}}{2}$	$-\frac{1}{2}$	0
$\cos t$	1	$\frac{\sqrt{3}}{2}$	$\frac{1}{2}$	0	$-\frac{1}{2}$	$-\frac{\sqrt{3}}{2}$	-1	$-\frac{\sqrt{3}}{2}$	$-\frac{1}{2}$	0	$\frac{1}{2}$	$\frac{\sqrt{3}}{2}$	1

Now we use this information to graph the functions sin t and cos t for t between 0 and 2π in Figures 2 and 3. These are the graphs of one period. Using the fact that these functions are periodic with period 2π, we get their complete graphs by continuing the same pattern to the left and to the right in every successive interval of length 2π.

The graph of the sine function is symmetric with respect to the origin. This is as expected, since sine is an odd function. Since the cosine function is an even function, its graph is symmetric with respect to the y-axis.

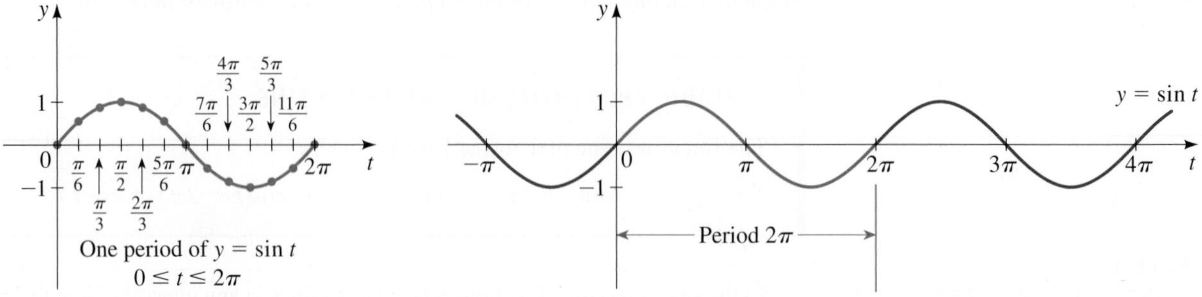

FIGURE 2 Graph of sin t

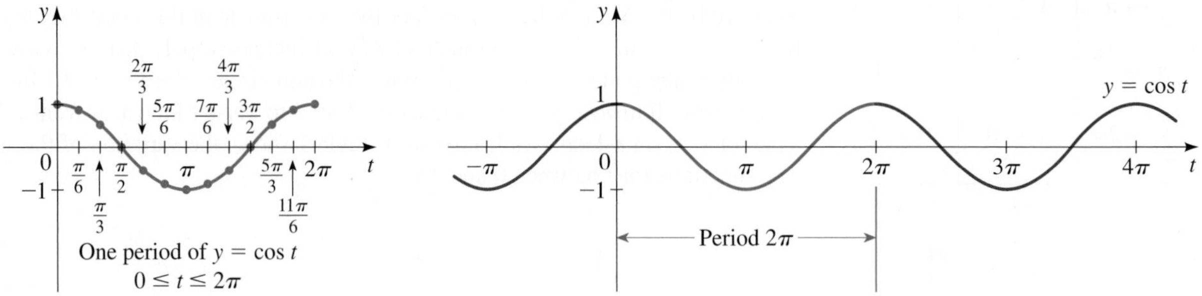

FIGURE 3 Graph of cos t

▼ Graphs of Transformations of Sine and Cosine

We now consider graphs of functions that are transformations of the sine and cosine functions. Thus, the graphing techniques of Section 2.5 are very useful here. The graphs we obtain are important for understanding applications to physical situations such as harmonic motion (see Section 5.6), but some of them are beautiful graphs that are interesting in their own right.

It's traditional to use the letter x to denote the variable in the domain of a function. So from here on we use the letter x and write $y = \sin x$, $y = \cos x$, $y = \tan x$, and so on to denote these functions.

EXAMPLE 1 | Cosine Curves

Sketch the graph of each function.

(a) $f(x) = 2 + \cos x$ **(b)** $g(x) = -\cos x$

SOLUTION

(a) The graph of $y = 2 + \cos x$ is the same as the graph of $y = \cos x$, but shifted up 2 units (see Figure 4(a)).

(b) The graph of $y = -\cos x$ in Figure 4(b) is the reflection of the graph of $y = \cos x$ in the x-axis.

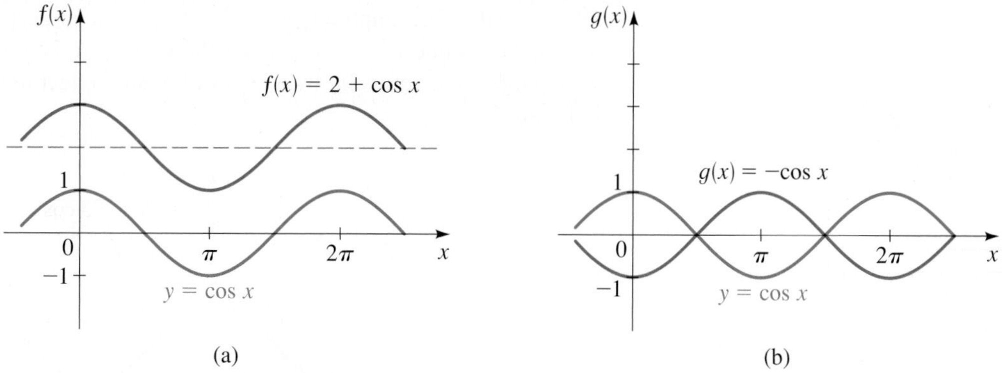

FIGURE 4

NOW TRY EXERCISES **3** AND **5**

Let's graph $y = 2 \sin x$. We start with the graph of $y = \sin x$ and multiply the y-coordinate of each point by 2. This has the effect of stretching the graph vertically by a factor of 2. To graph $y = \frac{1}{2} \sin x$, we start with the graph of $y = \sin x$ and multiply the y-coordinate of each point by $\frac{1}{2}$. This has the effect of shrinking the graph vertically by a factor of $\frac{1}{2}$ (see Figure 5).

Vertical stretching and shrinking of graphs is discussed in Section 2.5.

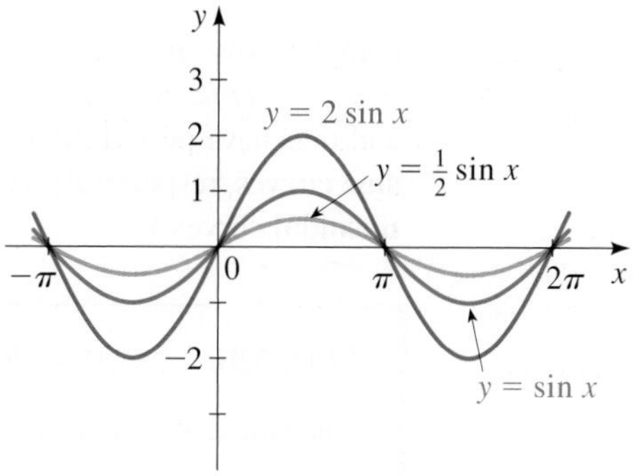

FIGURE 5

In general, for the functions

$$y = a \sin x \qquad \text{and} \qquad y = a \cos x$$

the number $|a|$ is called the **amplitude** and is the largest value these functions attain. Graphs of $y = a \sin x$ for several values of a are shown in Figure 6.

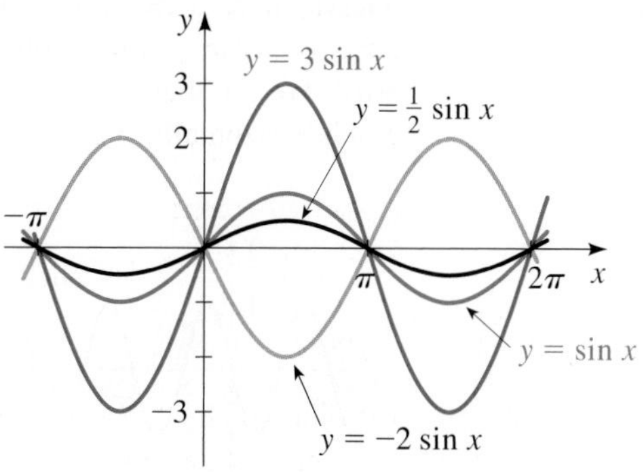

FIGURE 6

EXAMPLE 2 | Stretching a Cosine Curve

Find the amplitude of $y = -3 \cos x$, and sketch its graph.

SOLUTION The amplitude is $|-3| = 3$, so the largest value the graph attains is 3 and the smallest value is -3. To sketch the graph, we begin with the graph of $y = \cos x$, stretch the graph vertically by a factor of 3, and reflect in the x-axis, arriving at the graph in Figure 7.

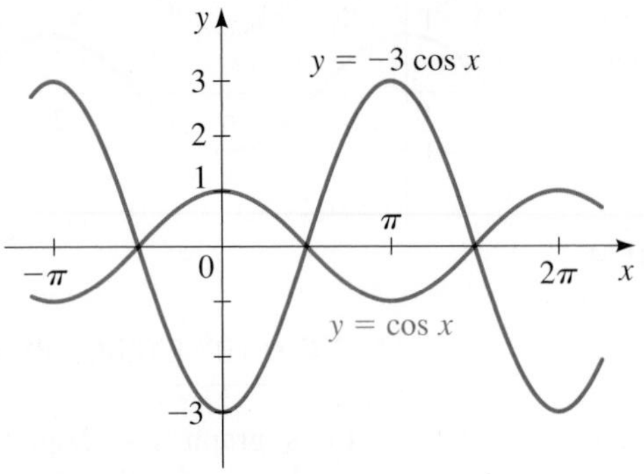

FIGURE 7

NOW TRY EXERCISE 9

Since the sine and cosine functions have period 2π, the functions

$$y = a \sin kx \qquad \text{and} \qquad y = a \cos kx \qquad (k > 0)$$

complete one period as kx varies from 0 to 2π, that is, for $0 \le kx \le 2\pi$ or for $0 \le x \le 2\pi/k$. So these functions complete one period as x varies between 0 and $2\pi/k$ and thus have period $2\pi/k$. The graphs of these functions are called **sine curves** and **cosine curves**, respectively. (Collectively, sine and cosine curves are often referred to as **sinusoidal** curves.)

SINE AND COSINE CURVES

The sine and cosine curves

$$y = a \sin kx \qquad \text{and} \qquad y = a \cos kx \qquad (k > 0)$$

have **amplitude** $|a|$ and **period** $2\pi/k$.

An appropriate interval on which to graph one complete period is $[0, 2\pi/k]$.

To see how the value of k affects the graph of $y = \sin kx$, let's graph the sine curve $y = \sin 2x$. Since the period is $2\pi/2 = \pi$, the graph completes one period in the interval $0 \le x \le \pi$ (see Figure 8(a)). For the sine curve $y = \sin \frac{1}{2}x$, the period is $2\pi \div \frac{1}{2} = 4\pi$, so the graph completes one period in the interval $0 \le x \le 4\pi$ (see Figure 8(b)). We see that the effect is to *shrink* the graph horizontally if $k > 1$ or to *stretch* the graph horizontally if $k < 1$.

Horizontal stretching and shrinking of graphs is discussed in Section 2.5.

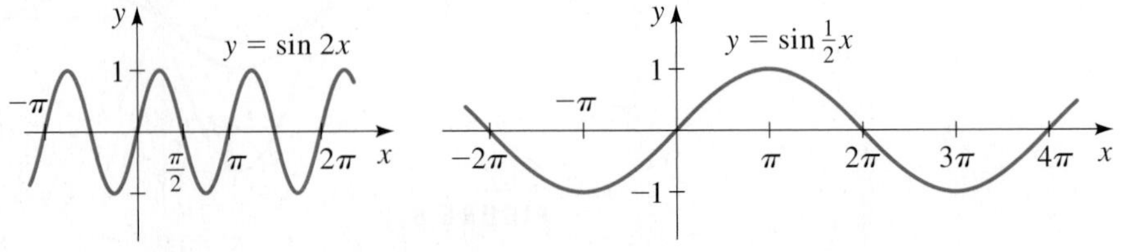

FIGURE 8

For comparison, in Figure 9 we show the graphs of one period of the sine curve $y = a \sin kx$ for several values of k.

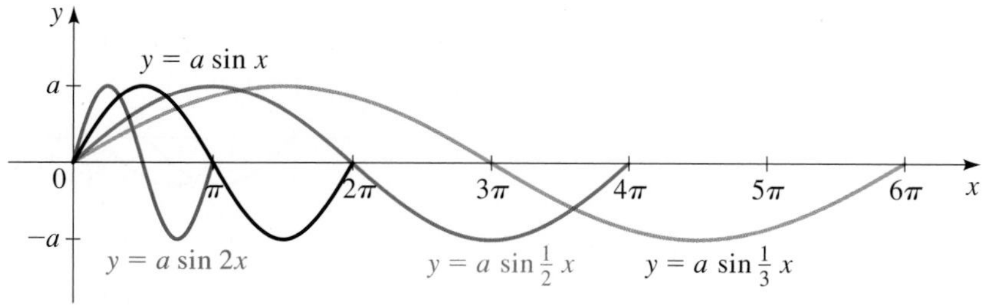

FIGURE 9

EXAMPLE 3 | Amplitude and Period

Find the amplitude and period of each function, and sketch its graph.

(a) $y = 4 \cos 3x$ **(b)** $y = -2 \sin \frac{1}{2}x$

SOLUTION

(a) We get the amplitude and period from the form of the function as follows:

$$\text{amplitude} = |a| = 4$$

$$y = 4 \cos 3x$$

$$\text{period} = \frac{2\pi}{k} = \frac{2\pi}{3}$$

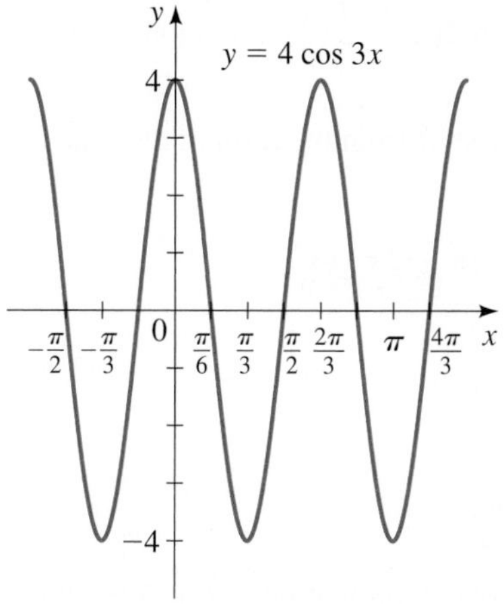

FIGURE 10

The amplitude is 4 and the period is $2\pi/3$. The graph is shown in Figure 10.

(b) For $y = -2 \sin \frac{1}{2}x$,

$$\text{amplitude} = |a| = |-2| = 2$$

$$\text{period} = \frac{2\pi}{\frac{1}{2}} = 4\pi$$

The graph is shown in Figure 11.

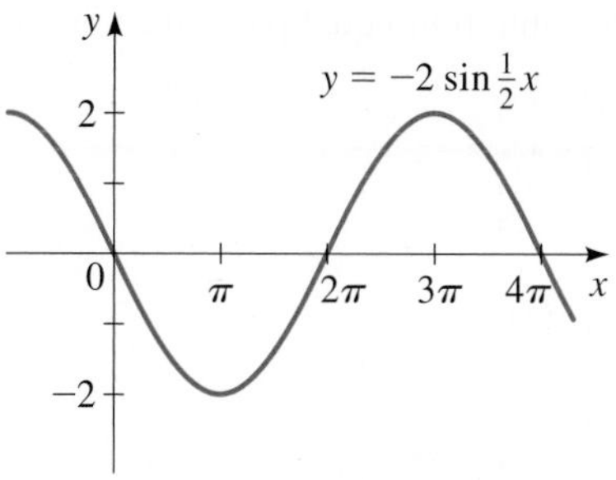

FIGURE 11

NOW TRY EXERCISES **19** AND **21** ■

The graphs of functions of the form $y = a \sin k(x - b)$ and $y = a \cos k(x - b)$ are simply sine and cosine curves shifted horizontally by an amount $|b|$. They are shifted to the right if $b > 0$ or to the left if $b < 0$. The number b is the *phase shift*. We summarize the properties of these functions in the following box.

SHIFTED SINE AND COSINE CURVES

The sine and cosine curves

$$y = a \sin k(x - b) \qquad \text{and} \qquad y = a \cos k(x - b) \qquad (k > 0)$$

have **amplitude** $|a|$, **period** $2\pi/k$, and **phase shift** b.

An appropriate interval on which to graph one complete period is $[b, b + (2\pi/k)]$.

The graphs of $y = \sin\left(x - \dfrac{\pi}{3}\right)$ and $y = \sin\left(x + \dfrac{\pi}{6}\right)$ are shown in Figure 12.

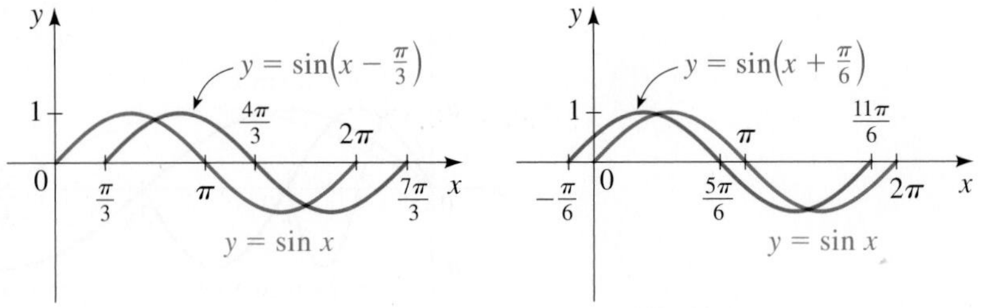

FIGURE 12

EXAMPLE 4 | A Shifted Sine Curve

Find the amplitude, period, and phase shift of $y = 3\sin 2\left(x - \dfrac{\pi}{4}\right)$, and graph one complete period.

SOLUTION We get the amplitude, period, and phase shift from the form of the function as follows:

amplitude $= |a| = 3$ period $= \dfrac{2\pi}{k} = \dfrac{2\pi}{2} = \pi$

$$y = 3\sin 2\left(x - \frac{\pi}{4}\right)$$

phase shift $= \dfrac{\pi}{4}$ (to the right)

Here is another way to find an appropriate interval on which to graph one complete period. Since the period of $y = \sin x$ is 2π, the function $y = 3\sin 2(x - \frac{\pi}{4})$ will go through one complete period as $2(x - \frac{\pi}{4})$ varies from 0 to 2π.

Start of period:	End of period:
$2(x - \frac{\pi}{4}) = 0$	$2(x - \frac{\pi}{4}) = 2\pi$
$x - \frac{\pi}{4} = 0$	$x - \frac{\pi}{4} = \pi$
$x = \frac{\pi}{4}$	$x = \frac{5\pi}{4}$

So we graph one period on the interval $[\frac{\pi}{4}, \frac{5\pi}{4}]$.

Since the phase shift is $\pi/4$ and the period is π, one complete period occurs on the interval

$$\left[\frac{\pi}{4}, \frac{\pi}{4} + \pi\right] = \left[\frac{\pi}{4}, \frac{5\pi}{4}\right]$$

As an aid in sketching the graph, we divide this interval into four equal parts, then graph a sine curve with amplitude 3 as in Figure 13.

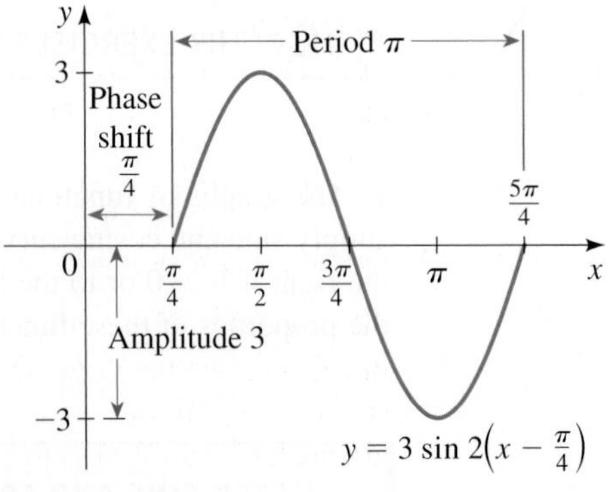

FIGURE 13

NOW TRY EXERCISE **33** ■

EXAMPLE 5 | A Shifted Cosine Curve

Find the amplitude, period, and phase shift of

$$y = \frac{3}{4}\cos\left(2x + \frac{2\pi}{3}\right)$$

and graph one complete period.

SOLUTION We first write this function in the form $y = a\cos k(x - b)$. To do this, we factor 2 from the expression $2x + \frac{2\pi}{3}$ to get

$$y = \frac{3}{4}\cos 2\left[x - \left(-\frac{\pi}{3}\right)\right]$$

Thus we have

$$\text{amplitude} = |a| = \frac{3}{4}$$

$$\text{period} = \frac{2\pi}{k} = \frac{2\pi}{2} = \pi$$

$$\text{phase shift} = b = -\frac{\pi}{3} \qquad \text{Shift } \frac{\pi}{3} \text{ to the } \textit{left}$$

We can also find one complete period as follows:

Start of period:	End of period:
$2x + \frac{2\pi}{3} = 0$	$2x + \frac{2\pi}{3} = 2\pi$
$2x = -\frac{2\pi}{3}$	$2x = \frac{4\pi}{3}$
$x = -\frac{\pi}{3}$	$x = \frac{2\pi}{3}$

So we graph one period on the interval $\left[-\frac{\pi}{3}, \frac{2\pi}{3}\right]$.

From this information it follows that one period of this cosine curve begins at $-\pi/3$ and ends at $(-\pi/3) + \pi = 2\pi/3$. To sketch the graph over the interval $[-\pi/3, 2\pi/3]$, we divide this interval into four equal parts and graph a cosine curve with amplitude $\frac{3}{4}$ as shown in Figure 14.

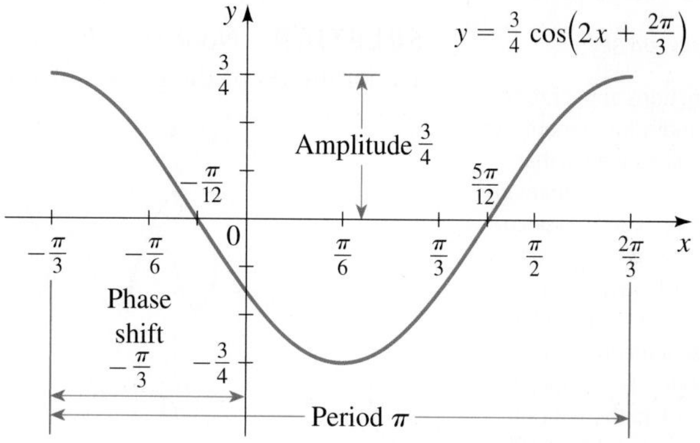

FIGURE 14

NOW TRY EXERCISE **35**

Using Graphing Devices to Graph Trigonometric Functions

When using a graphing calculator or a computer to graph a function, it is important to choose the viewing rectangle carefully in order to produce a reasonable graph of the function. This is especially true for trigonometric functions; the next example shows that, if care is not taken, it's easy to produce a very misleading graph of a trigonometric function.

See Section 1.9 for guidelines on choosing an appropriate viewing rectangle.

EXAMPLE 6 | Choosing the Viewing Rectangle

Graph the function $f(x) = \sin 50x$ in an appropriate viewing rectangle.

SOLUTION Figure 15(a) shows the graph of f produced by a graphing calculator using the viewing rectangle $[-12, 12]$ by $[-1.5, 1.5]$. At first glance the graph appears to be reasonable. But if we change the viewing rectangle to the ones shown in Figure 15, the graphs look very different. Something strange is happening.

The appearance of the graphs in Figure 15 depends on the machine used. The graphs you get with your own graphing device might not look like these figures, but they will also be quite inaccurate.

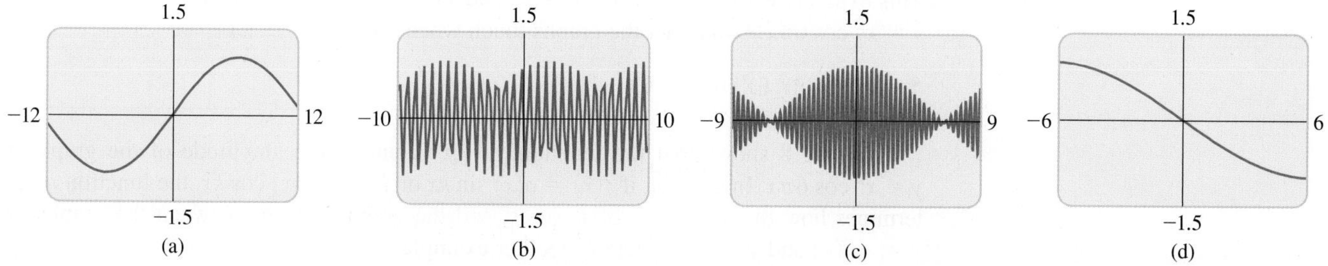

FIGURE 15 Graphs of $f(x) = \sin 50x$ in different viewing rectangles

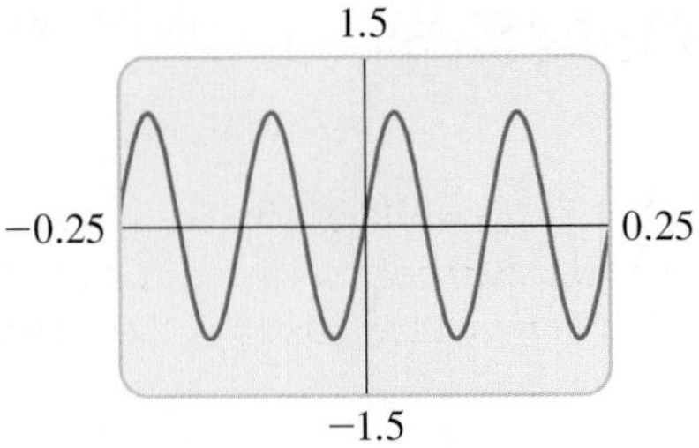

FIGURE 16 $f(x) = \sin 50x$

To explain the big differences in appearance of these graphs and to find an appropriate viewing rectangle, we need to find the period of the function $y = \sin 50x$:

$$\text{period} = \frac{2\pi}{50} = \frac{\pi}{25} \approx 0.126$$

This suggests that we should deal only with small values of x in order to show just a few oscillations of the graph. If we choose the viewing rectangle $[-0.25, 0.25]$ by $[-1.5, 1.5]$, we get the graph shown in Figure 16.

Now we see what went wrong in Figure 15. The oscillations of $y = \sin 50x$ are so rapid that when the calculator plots points and joins them, it misses most of the maximum and minimum points and therefore gives a very misleading impression of the graph.

NOW TRY EXERCISE **51** ■

The function h in Example 7 is **periodic** with period 2π. In general, functions that are sums of functions from the following list are periodic:

$$1, \cos kx, \cos 2kx, \cos 3kx, \dots$$

$$\sin kx, \sin 2kx, \sin 3kx, \dots$$

Although these functions appear to be special, they are actually fundamental to describing all periodic functions that arise in practice. The French mathematician J. B. J. Fourier (see page 501) discovered that nearly every periodic function can be written as a sum (usually an infinite sum) of these functions. This is remarkable because it means that any situation in which periodic variation occurs can be described mathematically using the functions sine and cosine. A modern application of Fourier's discovery is the digital encoding of sound on compact discs.

EXAMPLE 7 | A Sum of Sine and Cosine Curves

Graph $f(x) = 2\cos x$, $g(x) = \sin 2x$, and $h(x) = 2\cos x + \sin 2x$ on a common screen to illustrate the method of graphical addition.

SOLUTION Notice that $h = f + g$, so its graph is obtained by adding the corresponding y-coordinates of the graphs of f and g. The graphs of f, g, and h are shown in Figure 17.

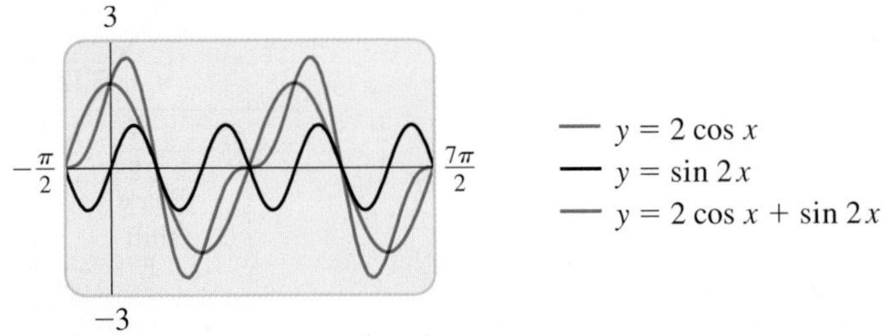

FIGURE 17

NOW TRY EXERCISE **59** ■

EXAMPLE 8 | A Cosine Curve with Variable Amplitude

Graph the functions $y = x^2$, $y = -x^2$, and $y = x^2 \cos 6\pi x$ on a common screen. Comment on and explain the relationship among the graphs.

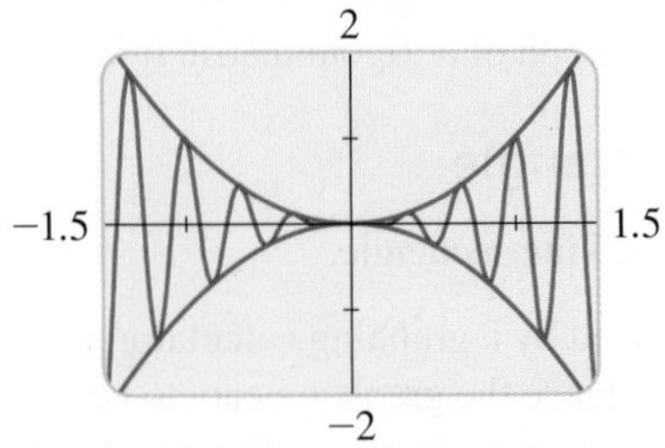

FIGURE 18 $y = x^2 \cos 6\pi x$

SOLUTION Figure 18 shows all three graphs in the viewing rectangle $[-1.5, 1.5]$ by $[-2, 2]$. It appears that the graph of $y = x^2 \cos 6\pi x$ lies between the graphs of the functions $y = x^2$ and $y = -x^2$.

To understand this, recall that the values of $\cos 6\pi x$ lie between -1 and 1, that is,

$$-1 \le \cos 6\pi x \le 1$$

for all values of x. Multiplying the inequalities by x^2 and noting that $x^2 \ge 0$, we get

$$-x^2 \le x^2 \cos 6\pi x \le x^2$$

This explains why the functions $y = x^2$ and $y = -x^2$ form a boundary for the graph of $y = x^2 \cos 6\pi x$. (Note that the graphs touch when $\cos 6\pi x = \pm 1$.)

NOW TRY EXERCISE **63** ■

Example 8 shows that the function $y = x^2$ controls the amplitude of the graph of $y = x^2 \cos 6\pi x$. In general, if $f(x) = a(x) \sin kx$ or $f(x) = a(x) \cos kx$, the function a determines how the amplitude of f varies, and the graph of f lies between the graphs of $y = -a(x)$ and $y = a(x)$. Here is another example.

AM and FM Radio

Radio transmissions consist of sound waves superimposed on a harmonic electromagnetic wave form called the **carrier signal.**

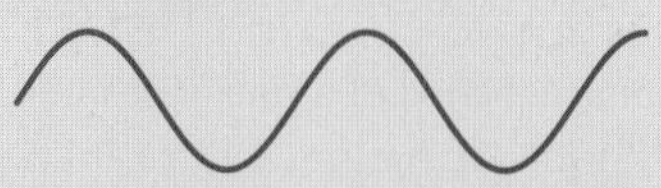

Sound wave

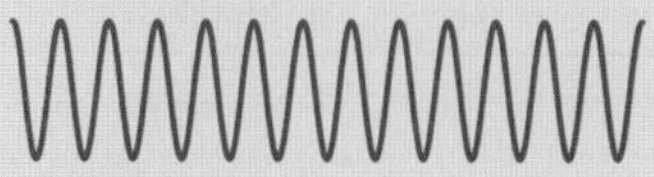

Carrier signal

There are two types of radio transmission, called **amplitude modulation (AM)** and **frequency modulation (FM).** In AM broadcasting, the sound wave changes, or **modulates,** the amplitude of the carrier, but the frequency remains unchanged.

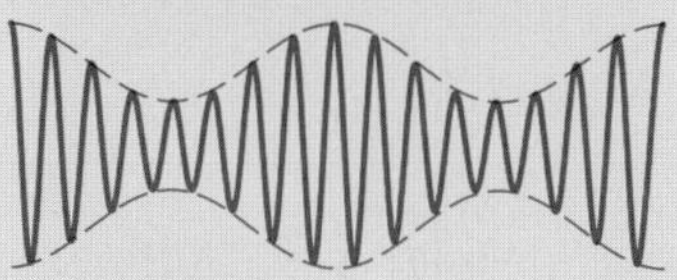

AM signal

In FM broadcasting, the sound wave modulates the frequency, but the amplitude remains the same.

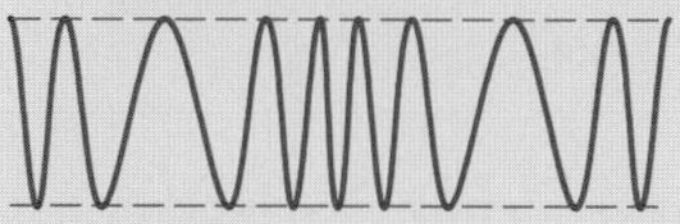

FM signal

EXAMPLE 9 | A Cosine Curve with Variable Amplitude

Graph the function $f(x) = \cos 2\pi x \cos 16\pi x$.

SOLUTION The graph is shown in Figure 19. Although it was drawn by a computer, we could have drawn it by hand, by first sketching the boundary curves $y = \cos 2\pi x$ and $y = -\cos 2\pi x$. The graph of f is a cosine curve that lies between the graphs of these two functions.

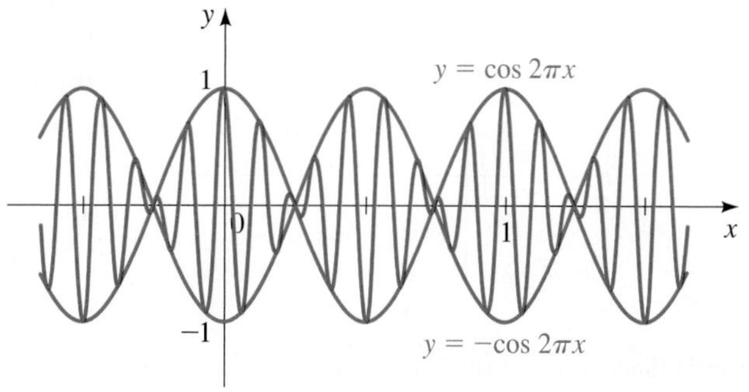

FIGURE 19 $f(x) = \cos 2\pi x \cos 16\pi x$

NOW TRY EXERCISE **65**

EXAMPLE 10 | A Sine Curve with Decaying Amplitude

The function $f(x) = \dfrac{\sin x}{x}$ is important in calculus. Graph this function and comment on its behavior when x is close to 0.

SOLUTION The viewing rectangle $[-15, 15]$ by $[-0.5, 1.5]$ shown in Figure 20(a) gives a good global view of the graph of f. The viewing rectangle $[-1, 1]$ by $[-0.5, 1.5]$ in Figure 20(b) focuses on the behavior of f when $x \approx 0$. Notice that although $f(x)$ is not defined when $x = 0$ (in other words, 0 is not in the domain of f), the values of f seem to approach 1 when x gets close to 0. This fact is crucial in calculus.

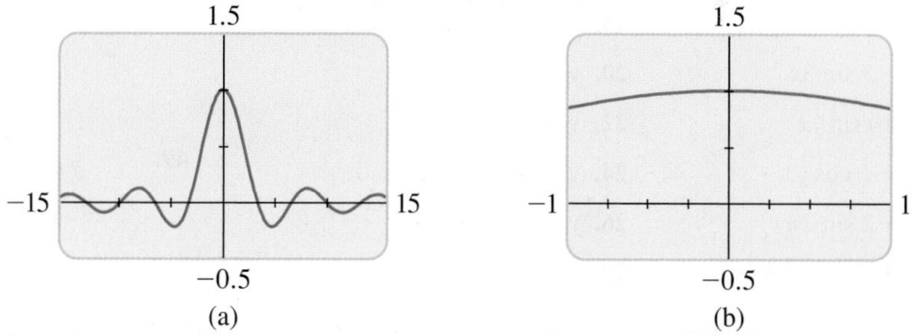

FIGURE 20 $f(x) = \dfrac{\sin x}{x}$

NOW TRY EXERCISE **75**

The function in Example 10 can be written as

$$f(x) = \frac{1}{x} \sin x$$

and may thus be viewed as a sine function whose amplitude is controlled by the function $a(x) = 1/x$.

5.3 EXERCISES

CONCEPTS

1. The trigonometric functions $y = \sin x$ and $y = \cos x$ have amplitude ________ and period ________. Sketch a graph of each function on the interval $[0, 2\pi]$.

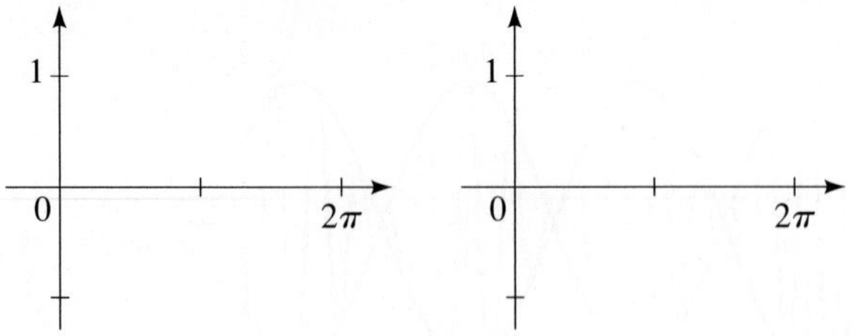

2. The trigonometric function $y = 3 \sin 2x$ has amplitude ________ and period ________.

SKILLS

3–16 ■ Graph the function.

3. $f(x) = 1 + \cos x$
4. $f(x) = 3 + \sin x$
5. $f(x) = -\sin x$
6. $f(x) = 2 - \cos x$
7. $f(x) = -2 + \sin x$
8. $f(x) = -1 + \cos x$
9. $g(x) = 3 \cos x$
10. $g(x) = 2 \sin x$
11. $g(x) = -\frac{1}{2} \sin x$
12. $g(x) = -\frac{2}{3} \cos x$
13. $g(x) = 3 + 3 \cos x$
14. $g(x) = 4 - 2 \sin x$
15. $h(x) = |\cos x|$
16. $h(x) = |\sin x|$

17–28 ■ Find the amplitude and period of the function, and sketch its graph.

17. $y = \cos 2x$
18. $y = -\sin 2x$
19. $y = -3 \sin 3x$
20. $y = \frac{1}{2} \cos 4x$
21. $y = 10 \sin \frac{1}{2}x$
22. $y = 5 \cos \frac{1}{4}x$
23. $y = -\frac{1}{3} \cos \frac{1}{3}x$
24. $y = 4 \sin(-2x)$
25. $y = -2 \sin 2\pi x$
26. $y = -3 \sin \pi x$
27. $y = 1 + \frac{1}{2} \cos \pi x$
28. $y = -2 + \cos 4\pi x$

29–42 ■ Find the amplitude, period, and phase shift of the function, and graph one complete period.

29. $y = \cos\left(x - \dfrac{\pi}{2}\right)$
30. $y = 2 \sin\left(x - \dfrac{\pi}{3}\right)$
31. $y = -2 \sin\left(x - \dfrac{\pi}{6}\right)$
32. $y = 3 \cos\left(x + \dfrac{\pi}{4}\right)$
33. $y = -4 \sin 2\left(x + \dfrac{\pi}{2}\right)$
34. $y = \sin \dfrac{1}{2}\left(x + \dfrac{\pi}{4}\right)$
35. $y = 5 \cos\left(3x - \dfrac{\pi}{4}\right)$
36. $y = 2 \sin\left(\dfrac{2}{3}x - \dfrac{\pi}{6}\right)$
37. $y = \dfrac{1}{2} - \dfrac{1}{2} \cos\left(2x - \dfrac{\pi}{3}\right)$
38. $y = 1 + \cos\left(3x + \dfrac{\pi}{2}\right)$
39. $y = 3 \cos \pi(x + \frac{1}{2})$
40. $y = 3 + 2 \sin 3(x + 1)$
41. $y = \sin(\pi + 3x)$
42. $y = \cos\left(\dfrac{\pi}{2} - x\right)$

43–50 ■ The graph of one complete period of a sine or cosine curve is given.

(a) Find the amplitude, period, and phase shift.

(b) Write an equation that represents the curve in the form

$$y = a \sin k(x - b) \quad \text{or} \quad y = a \cos k(x - b)$$

43.

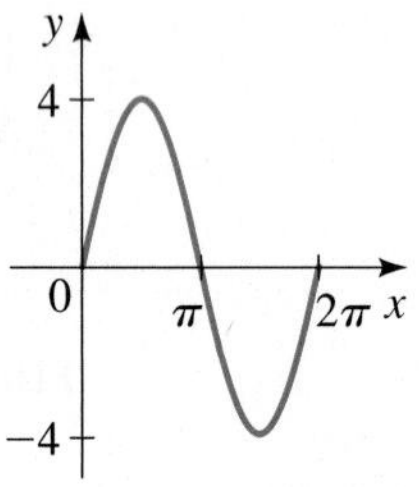

44.

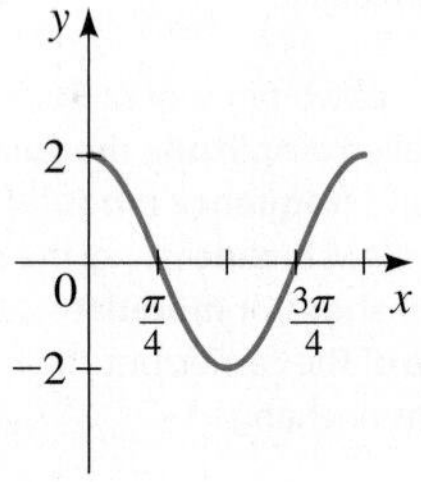

45.

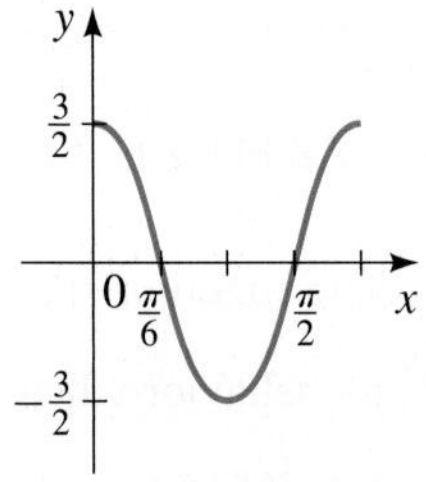

46.

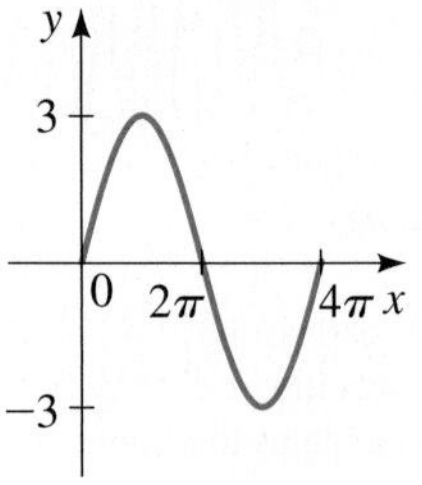

47.

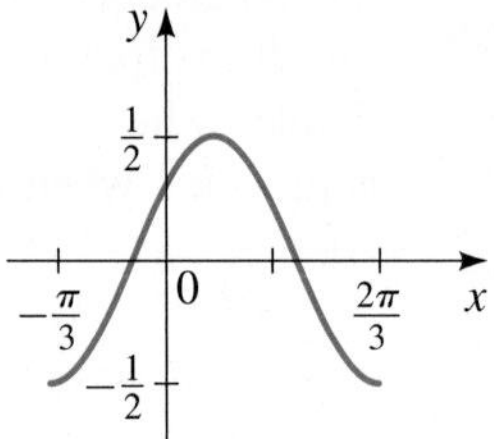

48.

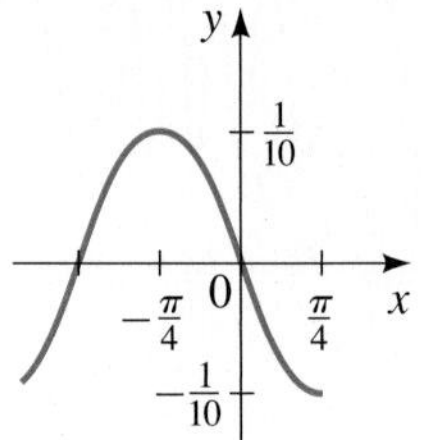

49.

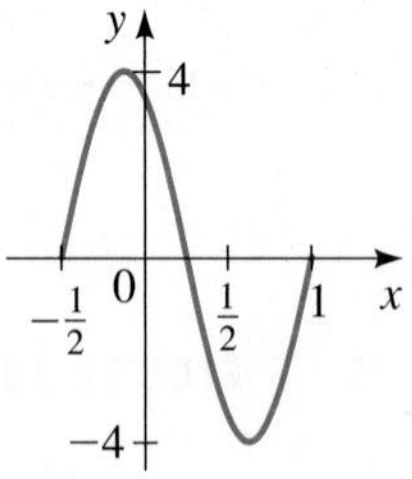

50.

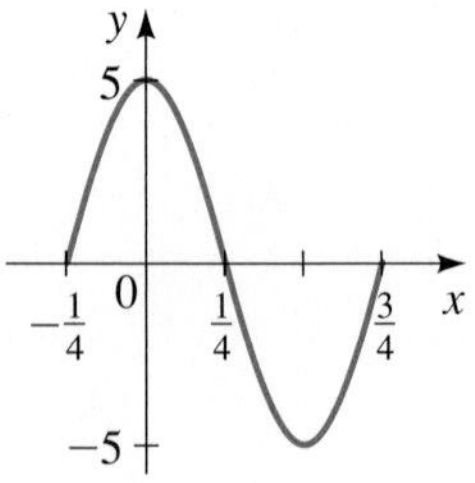

51–58 ■ Determine an appropriate viewing rectangle for each function, and use it to draw the graph.

51. $f(x) = \cos 100x$
52. $f(x) = 3 \sin 120x$
53. $f(x) = \sin(x/40)$
54. $f(x) = \cos(x/80)$
55. $y = \tan 25x$
56. $y = \csc 40x$
57. $y = \sin^2 20x$
58. $y = \sqrt{\tan 10\pi x}$

59–60 ■ Graph f, g, and $f + g$ on a common screen to illustrate graphical addition.

59. $f(x) = x, \quad g(x) = \sin x$

60. $f(x) = \sin x, \quad g(x) = \sin 2x$

61–66 ■ Graph the three functions on a common screen. How are the graphs related?

61. $y = x^2, \quad y = -x^2, \quad y = x^2 \sin x$

62. $y = x, \quad y = -x, \quad y = x \cos x$

63. $y = \sqrt{x}, \quad y = -\sqrt{x}, \quad y = \sqrt{x} \sin 5\pi x$

64. $y = \dfrac{1}{1 + x^2}, \quad y = -\dfrac{1}{1 + x^2}, \quad y = \dfrac{\cos 2\pi x}{1 + x^2}$

65. $y = \cos 3\pi x, \quad y = -\cos 3\pi x, \quad y = \cos 3\pi x \cos 21\pi x$

66. $y = \sin 2\pi x, \quad y = -\sin 2\pi x, \quad y = \sin 2\pi x \sin 10\pi x$

67–70 ■ Find the maximum and minimum values of the function.

67. $y = \sin x + \sin 2x$

68. $y = x - 2 \sin x, \; 0 \le x \le 2\pi$

69. $y = 2 \sin x + \sin^2 x$

70. $y = \dfrac{\cos x}{2 + \sin x}$

71–74 ■ Find all solutions of the equation that lie in the interval $[0, \pi]$. State each answer correct to two decimal places.

71. $\cos x = 0.4$

72. $\tan x = 2$

73. $\csc x = 3$

74. $\cos x = x$

75–76 ■ A function f is given.

(a) Is f even, odd, or neither?
(b) Find the x-intercepts of the graph of f.
(c) Graph f in an appropriate viewing rectangle.
(d) Describe the behavior of the function as $x \to \pm\infty$.
(e) Notice that $f(x)$ is not defined when $x = 0$. What happens as x approaches 0?

75. $f(x) = \dfrac{1 - \cos x}{x}$

76. $f(x) = \dfrac{\sin 4x}{2x}$

APPLICATIONS

77. Height of a Wave As a wave passes by an offshore piling, the height of the water is modeled by the function

$$h(t) = 3 \cos\left(\frac{\pi}{10} t\right)$$

where $h(t)$ is the height in feet above mean sea level at time t seconds.

(a) Find the period of the wave.
(b) Find the wave height, that is, the vertical distance between the trough and the crest of the wave.

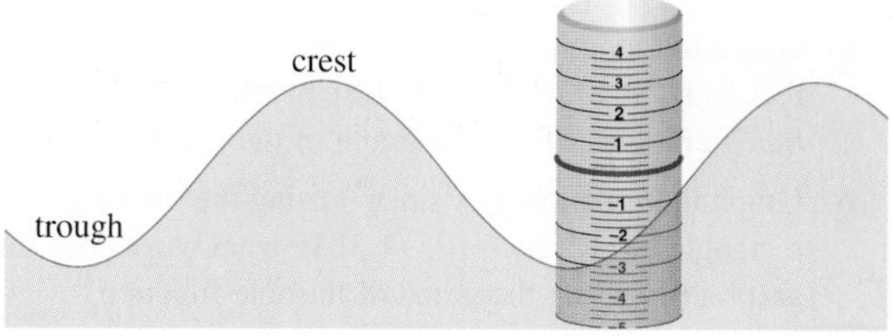

78. Sound Vibrations A tuning fork is struck, producing a pure tone as its tines vibrate. The vibrations are modeled by the function

$$v(t) = 0.7 \sin(880\pi t)$$

where $v(t)$ is the displacement of the tines in millimeters at time t seconds.

(a) Find the period of the vibration.
(b) Find the frequency of the vibration, that is, the number of times the fork vibrates per second.
(c) Graph the function v.

79. Blood Pressure Each time your heart beats, your blood pressure first increases and then decreases as the heart rests between beats. The maximum and minimum blood pressures are called the *systolic* and *diastolic* pressures, respectively. Your *blood pressure reading* is written as systolic/diastolic. A reading of 120/80 is considered normal.

A certain person's blood pressure is modeled by the function

$$p(t) = 115 + 25 \sin(160\pi t)$$

where $p(t)$ is the pressure in mmHg (millimeters of mercury), at time t measured in minutes.

(a) Find the period of p.
(b) Find the number of heartbeats per minute.
(c) Graph the function p.
(d) Find the blood pressure reading. How does this compare to normal blood pressure?

80. Variable Stars Variable stars are ones whose brightness varies periodically. One of the most visible is R Leonis; its brightness is modeled by the function

$$b(t) = 7.9 - 2.1 \cos\left(\frac{\pi}{156} t\right)$$

where t is measured in days.

(a) Find the period of R Leonis.
(b) Find the maximum and minimum brightness.
(c) Graph the function b.

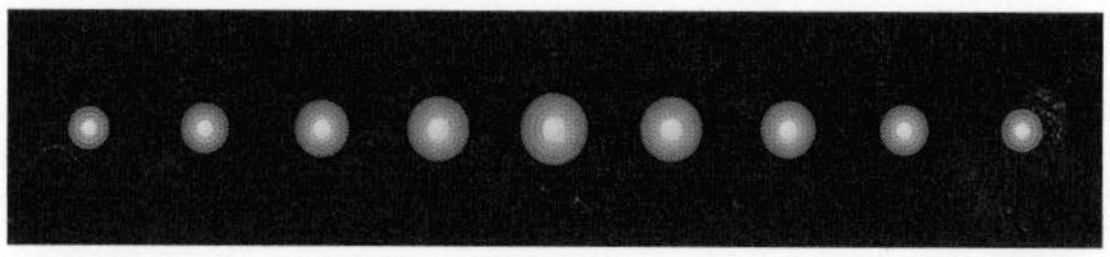

DISCOVERY ■ DISCUSSION ■ WRITING

81. Compositions Involving Trigonometric Functions This exercise explores the effect of the inner function g on a composite function $y = f(g(x))$.

(a) Graph the function $y = \sin\sqrt{x}$ using the viewing rectangle $[0, 400]$ by $[-1.5, 1.5]$. In what ways does this graph differ from the graph of the sine function?

(b) Graph the function $y = \sin(x^2)$ using the viewing rectangle $[-5, 5]$ by $[-1.5, 1.5]$. In what ways does this graph differ from the graph of the sine function?

82. Periodic Functions I Recall that a function f is *periodic* if there is a positive number p such that $f(t + p) = f(t)$ for every t, and the least such p (if it exists) is the *period* of f. The graph of a function of period p looks the same on each interval of length p, so we can easily determine the period from the graph. Determine whether the function whose graph is shown is periodic; if it is periodic, find the period.

(a)

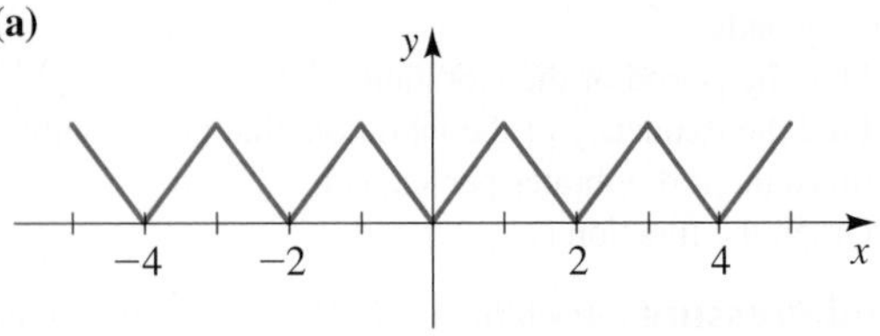

(b)

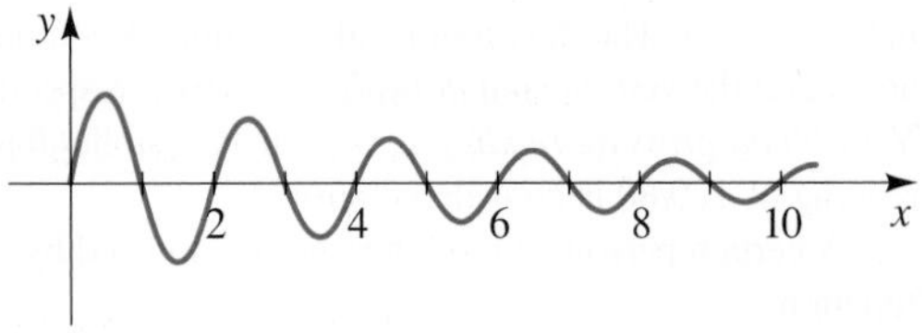

(c)

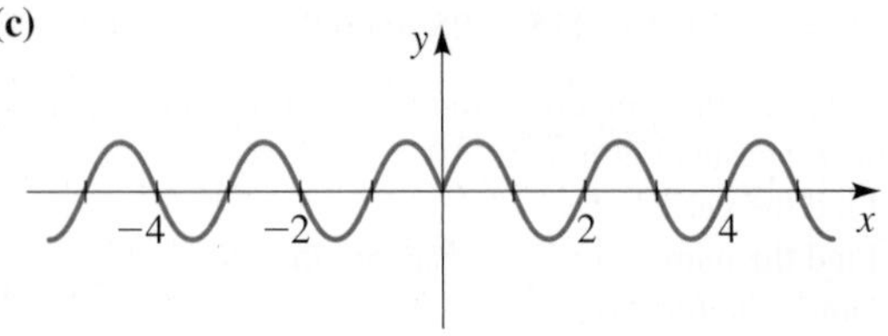

(d)

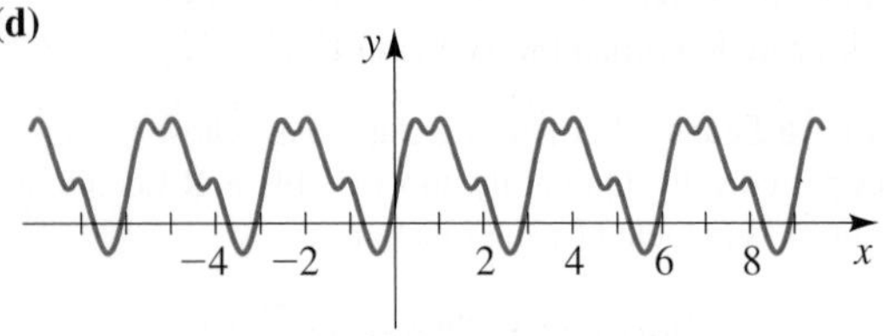

83. Periodic Functions II Use a graphing device to graph the following functions. From the graph, determine whether the function is periodic; if it is periodic, find the period. (See page 156 for the definition of $[\![x]\!]$.)

(a) $y = |\sin x|$

(b) $y = \sin|x|$

(c) $y = 2^{\cos x}$

(d) $y = x - [\![x]\!]$

(e) $y = \cos(\sin x)$

(f) $y = \cos(x^2)$

84. Sinusoidal Curves The graph of $y = \sin x$ is the same as the graph of $y = \cos x$ shifted to the right $\pi/2$ units. So the sine curve $y = \sin x$ is also at the same time a cosine curve: $y = \cos(x - \pi/2)$. In fact, any sine curve is also a cosine curve with a different phase shift, and any cosine curve is also a sine curve. Sine and cosine curves are collectively referred to as *sinusoidal*. For the curve whose graph is shown, find all possible ways of expressing it as a sine curve $y = a\sin(x - b)$ or as a cosine curve $y = a\cos(x - b)$. Explain why you think you have found all possible choices for a and b in each case.

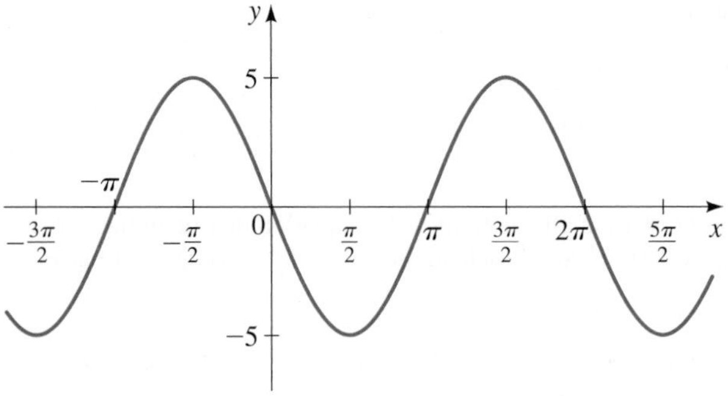

DISCOVERY PROJECT **Predator/Prey Models**

In this project we explore the use of sine functions in modeling the population of a predator and its prey. You can find the project at the book companion website: **www.stewartmath.com**

5.4 More Trigonometric Graphs

Graphs of Tangent, Cotangent, Secant, and Cosecant ▶ Graphs of Transformations of Tangent and Cotangent ▶ Graphs of Transformations of Cosecant and Secant

In this section we graph the tangent, cotangent, secant, and cosecant functions and transformations of these functions.

▼ Graphs of Tangent, Cotangent, Secant, and Cosecant

We begin by stating the periodic properties of these functions. Recall that sine and cosine have period 2π. Since cosecant and secant are the reciprocals of sine and cosine, respectively, they also have period 2π (see Exercise 55). Tangent and cotangent, however, have period π (see Exercise 85 of Section 5.2).

PERIODIC PROPERTIES

The functions tangent and cotangent have period π:

$$\tan(x + \pi) = \tan x \qquad \cot(x + \pi) = \cot x$$

The functions cosecant and secant have period 2π:

$$\csc(x + 2\pi) = \csc x \qquad \sec(x + 2\pi) = \sec x$$

x	$\tan x$
0	0
$\pi/6$	0.58
$\pi/4$	1.00
$\pi/3$	1.73
1.4	5.80
1.5	14.10
1.55	48.08
1.57	1,255.77
1.5707	10,381.33

We first sketch the graph of tangent. Since it has period π, we need only sketch the graph on any interval of length π and then repeat the pattern to the left and to the right. We sketch the graph on the interval $(-\pi/2, \pi/2)$. Since $\tan(\pi/2)$ and $\tan(-\pi/2)$ aren't defined, we need to be careful in sketching the graph at points near $\pi/2$ and $-\pi/2$. As x gets near $\pi/2$ through values less than $\pi/2$, the value of $\tan x$ becomes large. To see this, notice that as x gets close to $\pi/2$, $\cos x$ approaches 0 and $\sin x$ approaches 1 and so $\tan x = \sin x/\cos x$ is large. A table of values of $\tan x$ for x close to $\pi/2$ (≈ 1.570796) is shown in the margin.

Thus, by choosing x close enough to $\pi/2$ through values less than $\pi/2$, we can make the value of $\tan x$ larger than any given positive number. We express this by writing

$$\tan x \to \infty \qquad \text{as} \qquad x \to \frac{\pi}{2}^-$$

This is read "$\tan x$ approaches infinity as x approaches $\pi/2$ from the left."

In a similar way, by choosing x close to $-\pi/2$ through values greater than $-\pi/2$, we can make $\tan x$ smaller than any given negative number. We write this as

Arrow notation is discussed in Section 3.7.

$$\tan x \to -\infty \qquad \text{as} \qquad x \to -\frac{\pi}{2}^+$$

This is read "$\tan x$ approaches negative infinity as x approaches $-\pi/2$ from the right."

Asymptotes are discussed in Section 3.7.

Thus, the graph of $y = \tan x$ approaches the vertical lines $x = \pi/2$ and $x = -\pi/2$. So these lines are **vertical asymptotes**. With the information we have so far, we sketch the graph of

$y = \tan x$ for $-\pi/2 < x < \pi/2$ in Figure 1. The complete graph of tangent (see Figure 5(a) on the next page) is now obtained using the fact that tangent is periodic with period π.

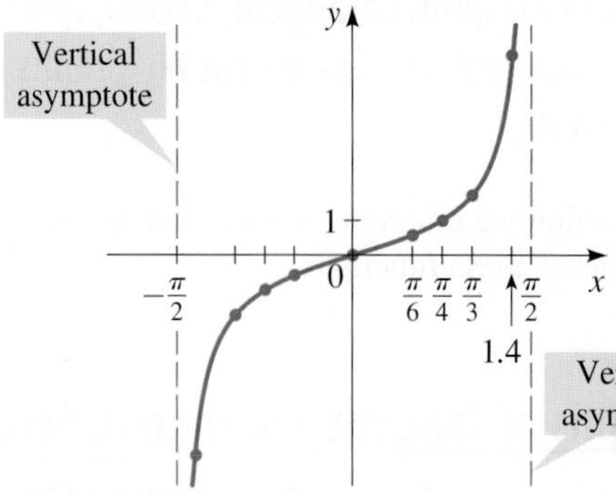

FIGURE 1
One period of $y = \tan x$

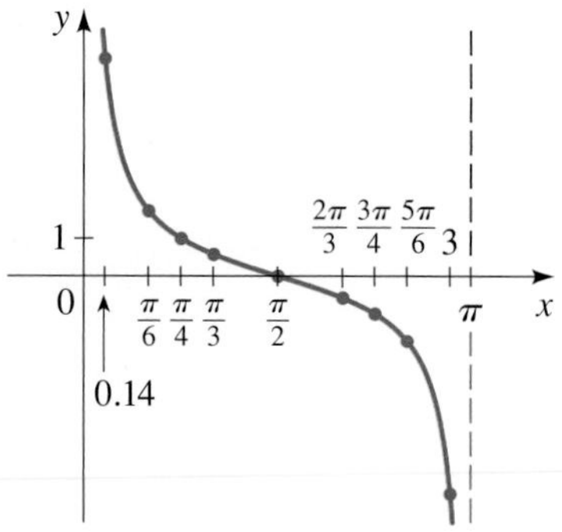

FIGURE 2
One period of $y = \cot x$

The function $y = \cot x$ is graphed on the interval $(0, \pi)$ by a similar analysis (see Figure 2). Since $\cot x$ is undefined for $x = n\pi$ with n an integer, its complete graph (in Figure 5(b) on the next page) has vertical asymptotes at these values.

To graph the cosecant and secant functions, we use the reciprocal identities

$$\csc x = \frac{1}{\sin x} \qquad \text{and} \qquad \sec x = \frac{1}{\cos x}$$

So to graph $y = \csc x$, we take the reciprocals of the y-coordinates of the points of the graph of $y = \sin x$. (See Figure 3.) Similarly, to graph $y = \sec x$, we take the reciprocals of the y-coordinates of the points of the graph of $y = \cos x$. (See Figure 4.)

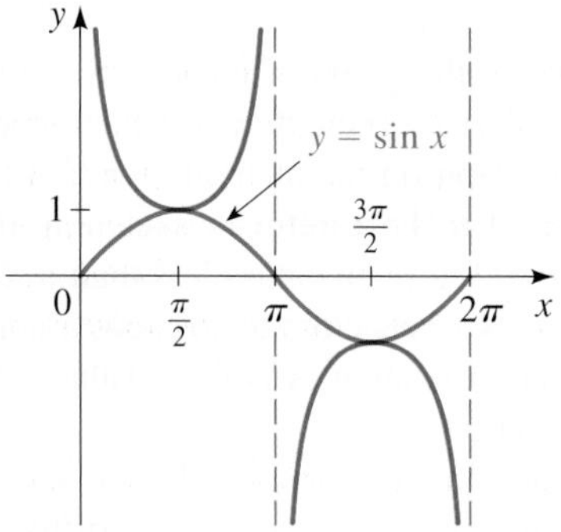

FIGURE 3
One period of $y = \csc x$

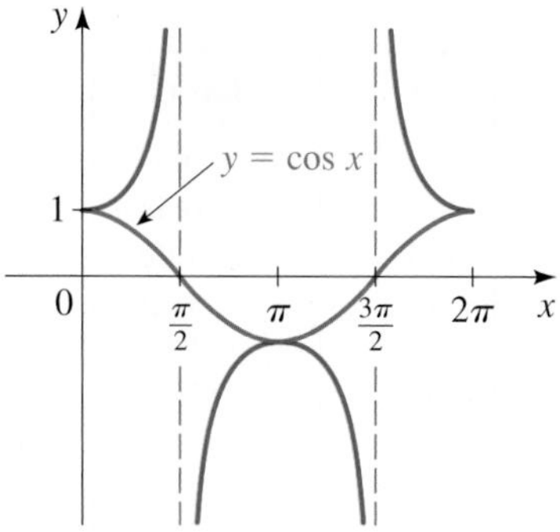

FIGURE 4
One period of $y = \sec x$

Let's consider more closely the graph of the function $y = \csc x$ on the interval $0 < x < \pi$. We need to examine the values of the function near 0 and π, since at these values $\sin x = 0$, and $\csc x$ is thus undefined. We see that

$$\csc x \to \infty \quad \text{as} \quad x \to 0^+$$

$$\csc x \to \infty \quad \text{as} \quad x \to \pi^-$$

Thus, the lines $x = 0$ and $x = \pi$ are vertical asymptotes. In the interval $\pi < x < 2\pi$ the graph is sketched in the same way. The values of $\csc x$ in that interval are the same as those in the interval $0 < x < \pi$ except for sign (see Figure 3). The complete graph in Figure 5(c) is now obtained from the fact that the function cosecant is periodic with period

MATHEMATICS IN THE MODERN WORLD

Evaluating Functions on a Calculator

How does your calculator evaluate $\sin t$, $\cos t$, e^t, $\ln t$, $\sqrt{t}$, and other such functions? One method is to approximate these functions by polynomials, because polynomials are easy to evaluate. For example,

$$\sin t = t - \frac{t^3}{3!} + \frac{t^5}{5!} - \frac{t^7}{7!} + \cdots$$

$$\cos t = 1 - \frac{t^2}{2!} + \frac{t^4}{4!} - \frac{t^6}{6!} + \cdots$$

where $n! = 1 \cdot 2 \cdot 3 \cdot \cdots \cdot n$. These remarkable formulas were found by the British mathematician Brook Taylor (1685–1731). For instance, if we use the first three terms of Taylor's series to find cos(0.4), we get

$$\cos 0.4 \approx 1 - \frac{(0.4)^2}{2!} + \frac{(0.4)^4}{4!}$$

$$\approx 0.92106667$$

(Compare this with the value you get from your calculator.) The graph shows that the more terms of the series we use, the more closely the polynomials approximate the function $\cos t$.

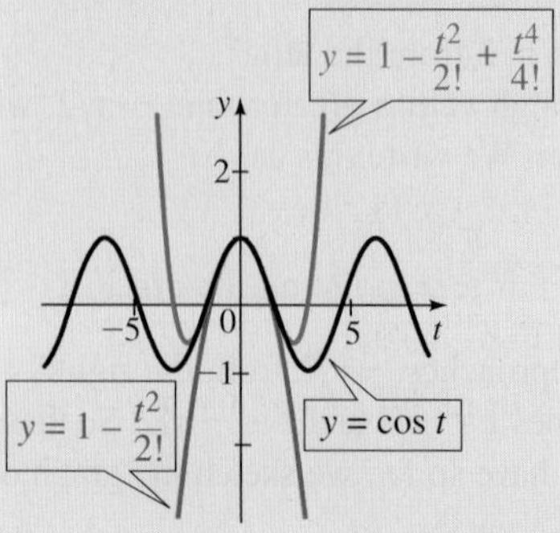

2π. Note that the graph has vertical asymptotes at the points where $\sin x = 0$, that is, at $x = n\pi$, for n an integer.

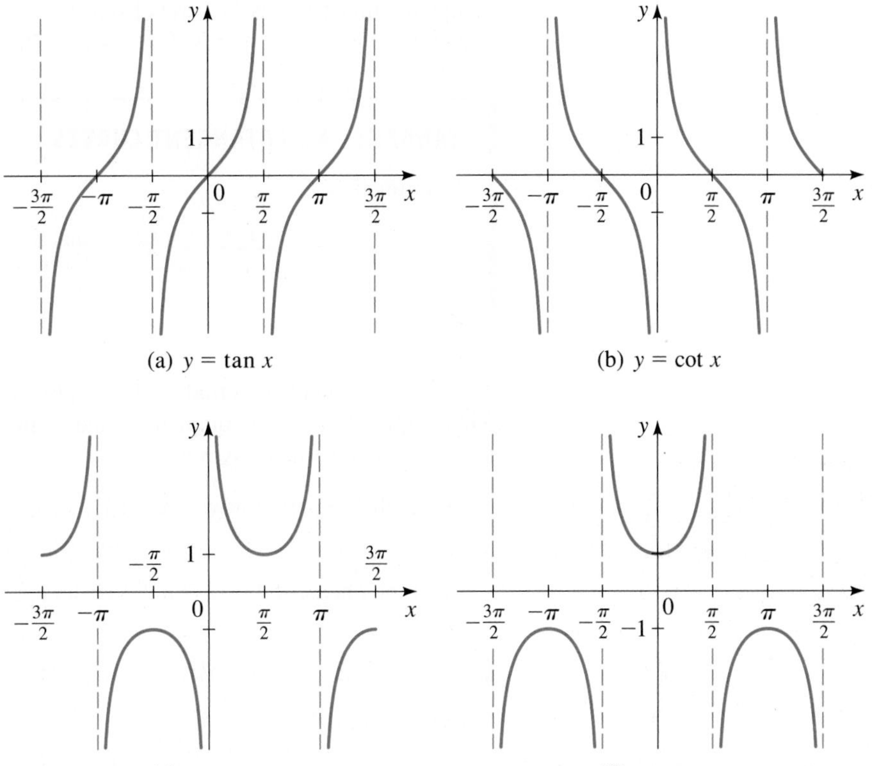

(a) $y = \tan x$ (b) $y = \cot x$

(c) $y = \csc x$ (d) $y = \sec x$

FIGURE 5

The graph of $y = \sec x$ is sketched in a similar manner. Observe that the domain of $\sec x$ is the set of all real numbers other than $x = (\pi/2) + n\pi$, for n an integer, so the graph has vertical asymptotes at those points. The complete graph is shown in Figure 5(d).

It is apparent that the graphs of $y = \tan x$, $y = \cot x$, and $y = \csc x$ are symmetric about the origin, whereas that of $y = \sec x$ is symmetric about the y-axis. This is because tangent, cotangent, and cosecant are odd functions, whereas secant is an even function.

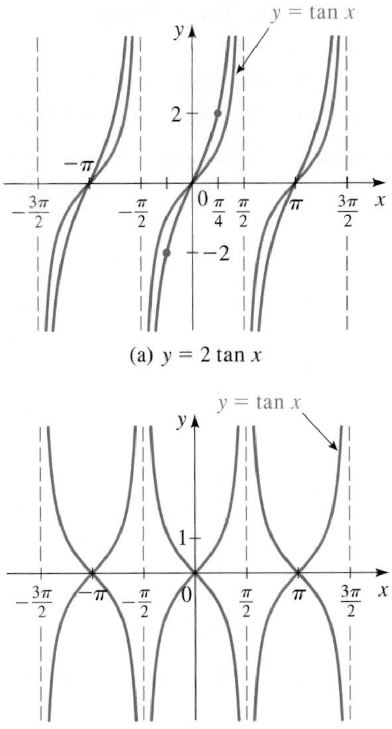

(a) $y = 2 \tan x$

(b) $y = -\tan x$

FIGURE 6

▼ Graphs of Transformations of Tangent and Cotangent

We now consider graphs of transformations of the tangent and cotangent functions.

EXAMPLE 1 | Graphing Tangent Curves

Graph each function.

(a) $y = 2 \tan x$ **(b)** $y = -\tan x$

SOLUTION We first graph $y = \tan x$ and then transform it as required.

(a) To graph $y = 2 \tan x$, we multiply the y-coordinate of each point on the graph of $y = \tan x$ by 2. The resulting graph is shown in Figure 6(a).

(b) The graph of $y = -\tan x$ in Figure 6(b) is obtained from that of $y = \tan x$ by reflecting in the x-axis.

NOW TRY EXERCISES **9** AND **11** ■

Since the tangent and cotangent functions have period π, the functions

$$y = a \tan kx \qquad \text{and} \qquad y = a \cot kx \qquad (k > 0)$$

complete one period as kx varies from 0 to π, that is, for $0 \le kx \le \pi$. Solving this inequality, we get $0 \le x \le \pi/k$. So they each have period π/k.

TANGENT AND COTANGENT CURVES

The functions

$$y = a \tan kx \qquad \text{and} \qquad y = a \cot kx \qquad (k > 0)$$

have period π/k.

Thus, one complete period of the graphs of these functions occurs on any interval of length π/k. To sketch a complete period of these graphs, it's convenient to select an interval between vertical asymptotes:

To graph one period of $y = a \tan kx$, an appropriate interval is $\left(-\dfrac{\pi}{2k}, \dfrac{\pi}{2k}\right)$.

To graph one period of $y = a \cot kx$, an appropriate interval is $\left(0, \dfrac{\pi}{k}\right)$.

EXAMPLE 2 | Graphing Tangent Curves

Graph each function.

(a) $y = \tan 2x$ **(b)** $y = \tan 2\left(x - \dfrac{\pi}{4}\right)$

SOLUTION

(a) The period is $\pi/2$ and an appropriate interval is $(-\pi/4, \pi/4)$. The endpoints $x = -\pi/4$ and $x = \pi/4$ are vertical asymptotes. Thus, we graph one complete period of the function on $(-\pi/4, \pi/4)$. The graph has the same shape as that of the tangent function, but is shrunk horizontally by a factor of $\frac{1}{2}$. We then repeat that portion of the graph to the left and to the right. See Figure 7(a).

(b) The graph is the same as that in part (a), but it is shifted to the right $\pi/4$, as shown in Figure 7(b).

Since $y = \tan x$ completes one period between $x = -\frac{\pi}{2}$ and $x = \frac{\pi}{2}$, the function $y = \tan 2(x - \frac{\pi}{4})$ completes one period as $2(x - \frac{\pi}{4})$ varies from $-\frac{\pi}{2}$ to $\frac{\pi}{2}$.

Start of period:	End of period:
$2(x - \frac{\pi}{4}) = -\frac{\pi}{2}$	$2(x - \frac{\pi}{4}) = \frac{\pi}{2}$
$x - \frac{\pi}{4} = -\frac{\pi}{4}$	$x - \frac{\pi}{4} = \frac{\pi}{4}$
$x = 0$	$x = \frac{\pi}{2}$

So we graph one period on the interval $(0, \frac{\pi}{2})$.

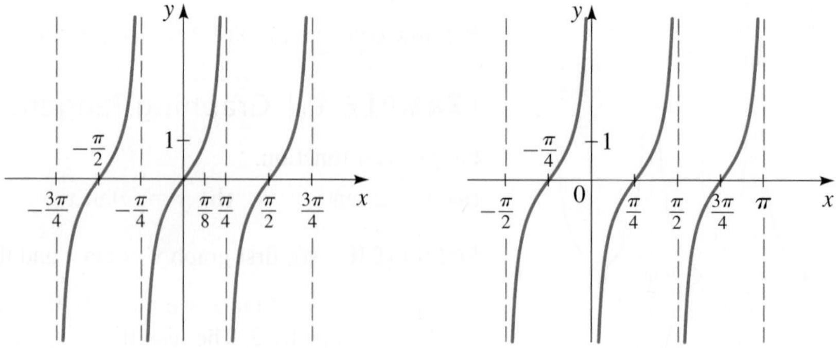

(a) $y = \tan 2x$ (b) $y = \tan 2\left(x - \frac{\pi}{4}\right)$

FIGURE 7

NOW TRY EXERCISES **27** AND **39**

EXAMPLE 3 | A Shifted Cotangent Curve

Graph $y = 2\cot\left(3x - \frac{\pi}{2}\right)$.

SOLUTION We first put this in the form $y = a\cot k(x - b)$ by factoring 3 from the expression $3x - \frac{\pi}{2}$:

$$y = 2\cot\left(3x - \frac{\pi}{2}\right) = 2\cot 3\left(x - \frac{\pi}{6}\right)$$

Since $y = \cot x$ completes one period between $x = 0$ and $x = \pi$, the function $y = 2\cot(3x - \frac{\pi}{2})$ completes one period as $3x - \frac{\pi}{2}$ varies from 0 to π.

Start of period:	End of period:
$3x - \frac{\pi}{2} = 0$	$3x - \frac{\pi}{2} = \pi$
$3x = \frac{\pi}{2}$	$3x = \frac{3\pi}{2}$
$x = \frac{\pi}{6}$	$x = \frac{\pi}{2}$

So we graph one period on the interval $(\frac{\pi}{6}, \frac{\pi}{2})$.

Thus the graph is the same as that of $y = 2\cot 3x$ but is shifted to the right $\pi/6$. The period of $y = 2\cot 3x$ is $\pi/3$, and an appropriate interval is $(0, \pi/3)$. To get the corresponding interval for the desired graph, we shift this interval to the right $\pi/6$. This gives

$$\left(0 + \frac{\pi}{6}, \frac{\pi}{3} + \frac{\pi}{6}\right) = \left(\frac{\pi}{6}, \frac{\pi}{2}\right)$$

Finally, we graph one period in the shape of cotangent on the interval $(\pi/6, \pi/2)$ and repeat that portion of the graph to the left and to the right. (See Figure 8.)

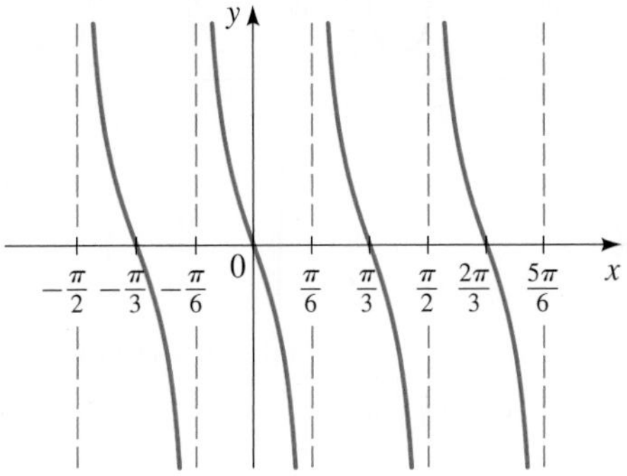

FIGURE 8
$y = 2\cot\left(3x - \frac{\pi}{2}\right)$

NOW TRY EXERCISE 43

Graphs of Transformations of Cosecant and Secant

We have already observed that the cosecant and secant functions are the reciprocals of the sine and cosine functions. Thus, the following result is the counterpart of the result for sine and cosine curves in Section 5.3.

COSECANT AND SECANT CURVES

The functions

$$y = a\csc kx \quad \text{and} \quad y = a\sec kx \quad (k > 0)$$

have period $2\pi/k$.

An appropriate interval on which to graph one complete period is $[0, 2\pi/k]$.

EXAMPLE 4 | Graphing Cosecant Curves

Graph each function.

(a) $y = \frac{1}{2}\csc 2x$ **(b)** $y = \frac{1}{2}\csc\left(2x + \frac{\pi}{2}\right)$

SOLUTION

(a) The period is $2\pi/2 = \pi$. An appropriate interval is $[0, \pi]$, and the asymptotes occur in this interval whenever $\sin 2x = 0$. So the asymptotes in this interval are $x = 0$, $x = \pi/2$, and $x = \pi$. With this information we sketch on the interval $[0, \pi]$ a graph with the same general shape as that of one period of the cosecant function. The complete graph in Figure 9(a) is obtained by repeating this portion of the graph to the left and to the right.

Since $y = \csc x$ completes one period between $x = 0$ and $x = 2\pi$, the function $y = \frac{1}{2}\csc(2x + \frac{\pi}{2})$ completes one period as $2x + \frac{\pi}{2}$ varies from 0 to 2π.

Start of period:	End of period:
$2x + \frac{\pi}{2} = 0$	$2x + \frac{\pi}{2} = 2\pi$
$2x = -\frac{\pi}{2}$	$2x = \frac{3\pi}{2}$
$x = -\frac{\pi}{4}$	$x = \frac{3\pi}{4}$

So we graph one period on the interval $(-\frac{\pi}{4}, \frac{3\pi}{4})$.

(b) We first write

$$y = \frac{1}{2}\csc\left(2x + \frac{\pi}{2}\right) = \frac{1}{2}\csc 2\left(x + \frac{\pi}{4}\right)$$

From this we see that the graph is the same as that in part (a) but shifted to the left $\pi/4$. The graph is shown in Figure 9(b).

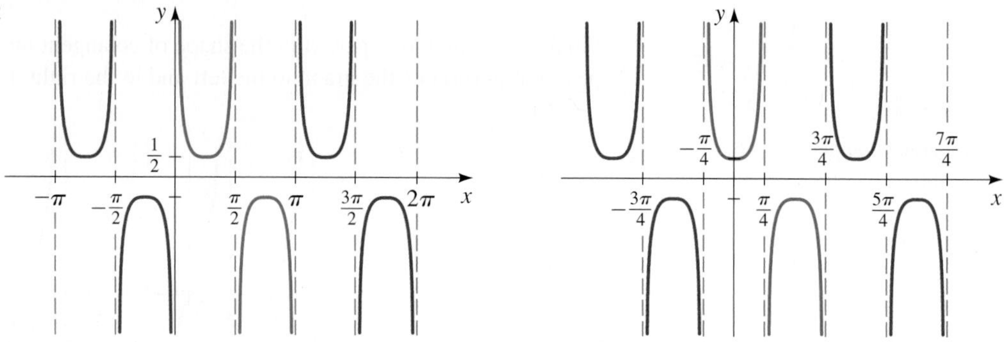

(a) $y = \frac{1}{2}\csc 2x$ (b) $y = \frac{1}{2}\csc\left(2x + \frac{\pi}{2}\right)$

FIGURE 9

NOW TRY EXERCISES **33** AND **45**

EXAMPLE 5 | Graphing a Secant Curve

Graph $y = 3\sec\frac{1}{2}x$.

SOLUTION The period is $2\pi \div \frac{1}{2} = 4\pi$. An appropriate interval is $[0, 4\pi]$, and the asymptotes occur in this interval wherever $\cos\frac{1}{2}x = 0$. Thus, the asymptotes in this interval are $x = \pi$, $x = 3\pi$. With this information we sketch on the interval $[0, 4\pi]$ a graph with the same general shape as that of one period of the secant function. The complete graph in Figure 10 is obtained by repeating this portion of the graph to the left and to the right.

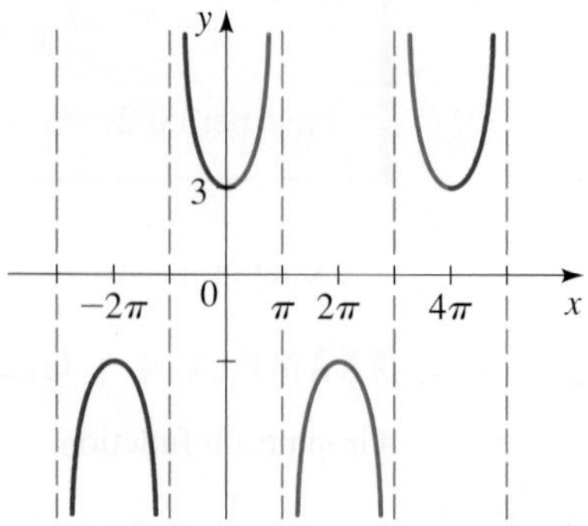

FIGURE 10
$y = 3\sec\frac{1}{2}x$

NOW TRY EXERCISE **31**

5.4 EXERCISES

CONCEPTS

1. The trigonometric function $y = \tan x$ has period ________ and asymptotes $x =$ ________. Sketch a graph of this function on the interval $(-\pi/2, \pi/2)$.

2. The trigonometric function $y = \csc x$ has period ________ and asymptotes $x =$ ________. Sketch a graph of this function on the interval $(-\pi, \pi)$.

SKILLS

3–8 ■ Match the trigonometric function with one of the graphs I–VI.

3. $f(x) = \tan\left(x + \dfrac{\pi}{4}\right)$
4. $f(x) = \sec 2x$
5. $f(x) = \cot 2x$
6. $f(x) = -\tan x$
7. $f(x) = 2 \sec x$
8. $f(x) = 1 + \csc x$

I
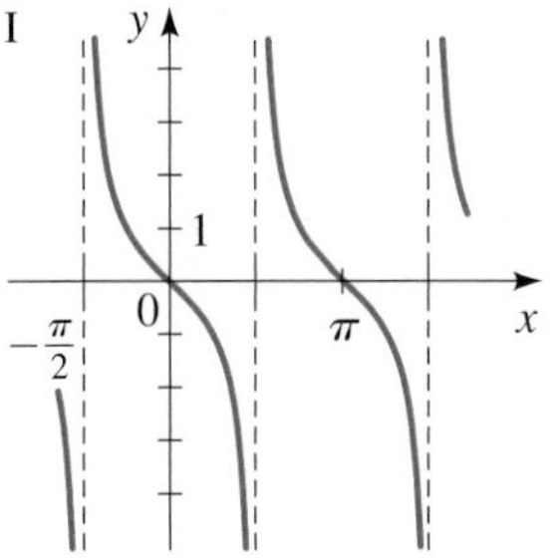

II
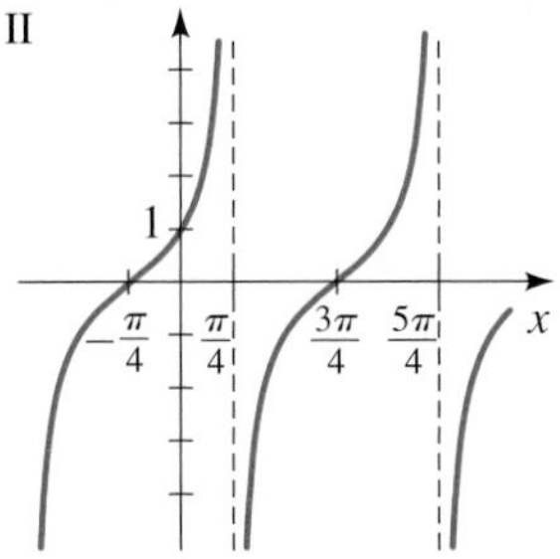

III
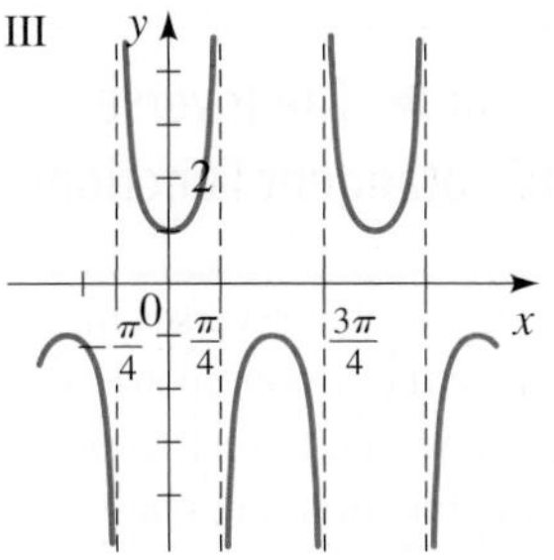

IV
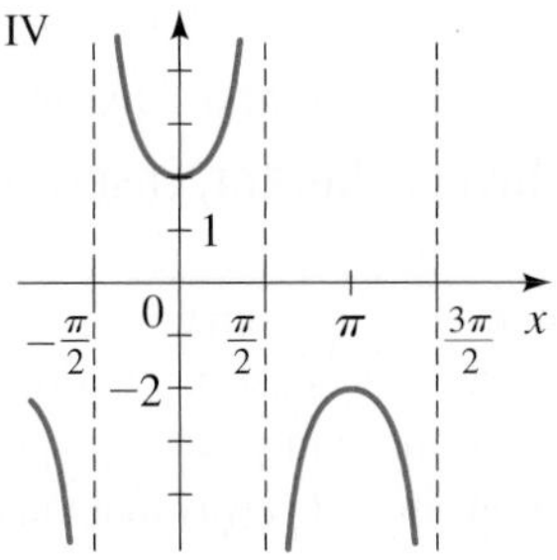

V
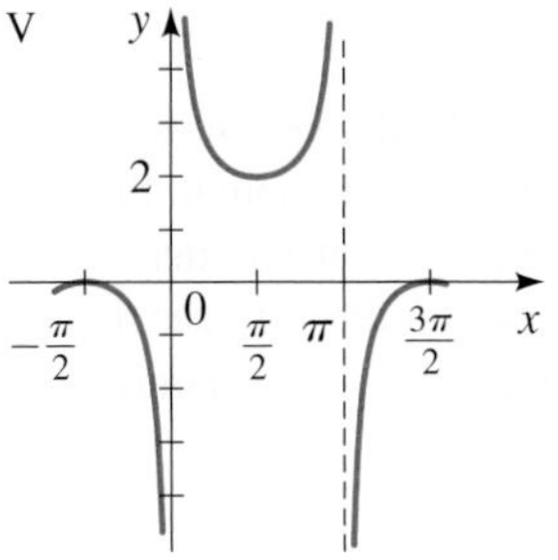

VI
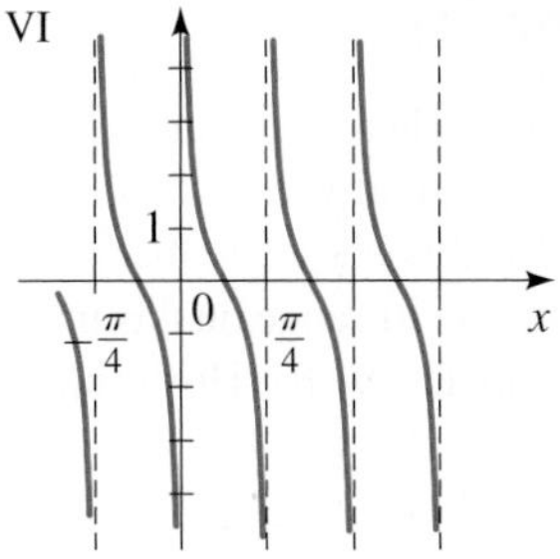

9–54 ■ Find the period and graph the function.

9. $y = 4 \tan x$
10. $y = -4 \tan x$
11. $y = -\frac{1}{2} \tan x$
12. $y = \frac{1}{2} \tan x$
13. $y = -\cot x$
14. $y = 2 \cot x$
15. $y = 2 \csc x$
16. $y = \frac{1}{2} \csc x$
17. $y = 3 \sec x$
18. $y = -3 \sec x$
19. $y = \tan\left(x + \dfrac{\pi}{2}\right)$
20. $y = \tan\left(x - \dfrac{\pi}{4}\right)$
21. $y = \csc\left(x - \dfrac{\pi}{2}\right)$
22. $y = \sec\left(x + \dfrac{\pi}{4}\right)$
23. $y = \cot\left(x + \dfrac{\pi}{4}\right)$
24. $y = 2\csc\left(x - \dfrac{\pi}{3}\right)$
25. $y = \dfrac{1}{2}\sec\left(x - \dfrac{\pi}{6}\right)$
26. $y = 3\csc\left(x + \dfrac{\pi}{2}\right)$
27. $y = \tan 4x$
28. $y = \tan \frac{1}{2}x$
29. $y = \tan \dfrac{\pi}{4}x$
30. $y = \cot \dfrac{\pi}{2}x$
31. $y = \sec 2x$
32. $y = 5 \csc 3x$
33. $y = \csc 4x$
34. $y = \csc \frac{1}{2}x$
35. $y = 2 \tan 3\pi x$
36. $y = 2 \tan \dfrac{\pi}{2}x$
37. $y = 5 \csc \dfrac{3\pi}{2}x$
38. $y = 5 \sec 2\pi x$
39. $y = \tan 2\left(x + \dfrac{\pi}{2}\right)$
40. $y = \csc 2\left(x + \dfrac{\pi}{2}\right)$
41. $y = \tan 2(x - \pi)$
42. $y = \sec 2\left(x - \dfrac{\pi}{2}\right)$
43. $y = \cot\left(2x - \dfrac{\pi}{2}\right)$
44. $y = \frac{1}{2}\tan(\pi x - \pi)$
45. $y = 2\csc\left(\pi x - \dfrac{\pi}{3}\right)$
46. $y = 2\sec\left(\dfrac{1}{2}x - \dfrac{\pi}{3}\right)$
47. $y = 5\sec\left(3x - \dfrac{\pi}{2}\right)$
48. $y = \frac{1}{2}\sec(2\pi x - \pi)$
49. $y = \tan\left(\dfrac{2}{3}x - \dfrac{\pi}{6}\right)$
50. $y = \tan \dfrac{1}{2}\left(x + \dfrac{\pi}{4}\right)$
51. $y = 3 \sec \pi\left(x + \dfrac{1}{2}\right)$
52. $y = \sec\left(3x + \dfrac{\pi}{2}\right)$
53. $y = -2\tan\left(2x - \dfrac{\pi}{3}\right)$
54. $y = 2\csc(3x + 3)$

55. **(a)** Prove that if f is periodic with period p, then $1/f$ is also periodic with period p.
(b) Prove that cosecant and secant each have period 2π.

56. Prove that if f and g are periodic with period p, then f/g is also periodic, but its period could be smaller than p.

APPLICATIONS

57. **Lighthouse** The beam from a lighthouse completes one rotation every two minutes. At time t, the distance d shown in the figure on the next page is

$$d(t) = 3 \tan \pi t$$

where t is measured in minutes and d in miles.
(a) Find $d(0.15)$, $d(0.25)$, and $d(0.45)$.

(b) Sketch a graph of the function d for $0 \le t < \frac{1}{2}$.
(c) What happens to the distance d as t approaches $\frac{1}{2}$?

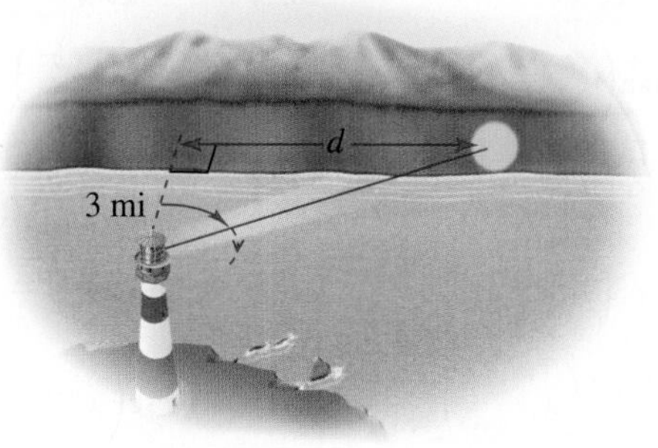

58. Length of a Shadow On a day when the sun passes directly overhead at noon, a six-foot-tall man casts a shadow of length

$$S(t) = 6\left|\cot \frac{\pi}{12} t\right|$$

where S is measured in feet and t is the number of hours since 6 A.M.

(a) Find the length of the shadow at 8:00 A.M., noon, 2:00 P.M., and 5:45 P.M.
(b) Sketch a graph of the function S for $0 < t < 12$.
(c) From the graph determine the values of t at which the length of the shadow equals the man's height. To what time of day does each of these values correspond?
(d) Explain what happens to the shadow as the time approaches 6 P.M. (that is, as $t \to 12^-$).

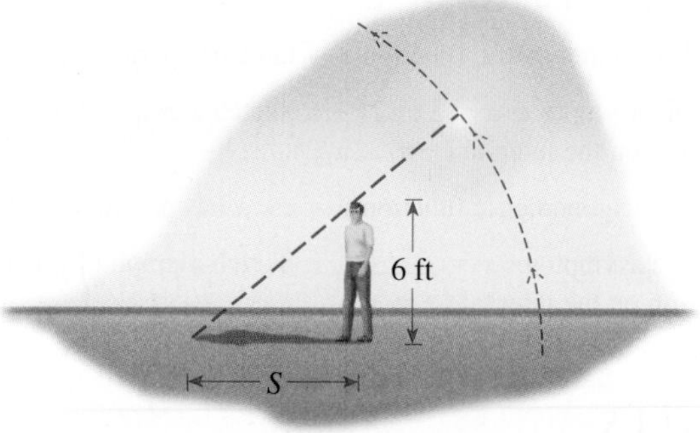

DISCOVERY ■ DISCUSSION ■ WRITING

59. Reduction Formulas Use the graphs in Figure 5 to explain why the following formulas are true.

$$\tan\left(x - \frac{\pi}{2}\right) = -\cot x$$

$$\sec\left(x - \frac{\pi}{2}\right) = \csc x$$

5.5 Inverse Trigonometric Functions and Their Graphs

The Inverse Sine Function ▶ The Inverse Cosine Function ▶ The Inverse Tangent Function ▶ The Inverse Secant, Cosecant, and Cotangent Functions

We study applications of inverse trigonometric functions to triangles in Sections 6.4–6.6.

Recall from Section 2.7 that the inverse of a function f is a function f^{-1} that reverses the rule of f. For a function to have an inverse, it must be one-to-one. Since the trigonometric functions are not one-to-one, they do not have inverses. It is possible, however, to restrict the domains of the trigonometric functions in such a way that the resulting functions are one-to-one.

▼ The Inverse Sine Function

Let's first consider the sine function. There are many ways to restrict the domain of sine so that the new function is one-to-one. A natural way to do this is to restrict the domain to the interval $[-\pi/2, \pi/2]$. The reason for this choice is that sine is one-to-one on this interval and moreover attains each of the values in its range on this interval. From Figure 1 we see that sine is one-to-one on this restricted domain (by the Horizontal Line Test) and so has an inverse.

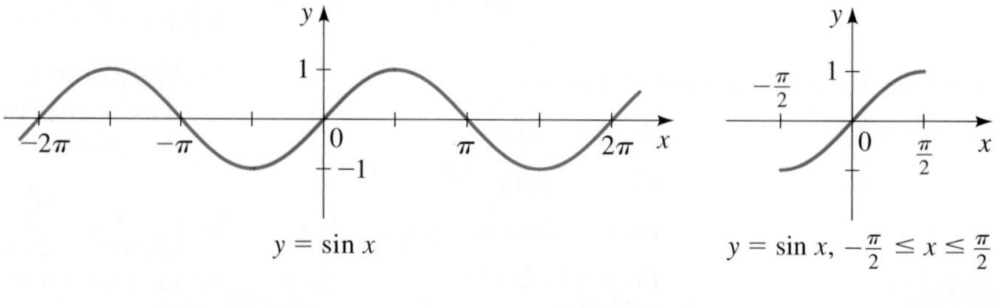

FIGURE 1 Graphs of the sine function and the restricted sine function

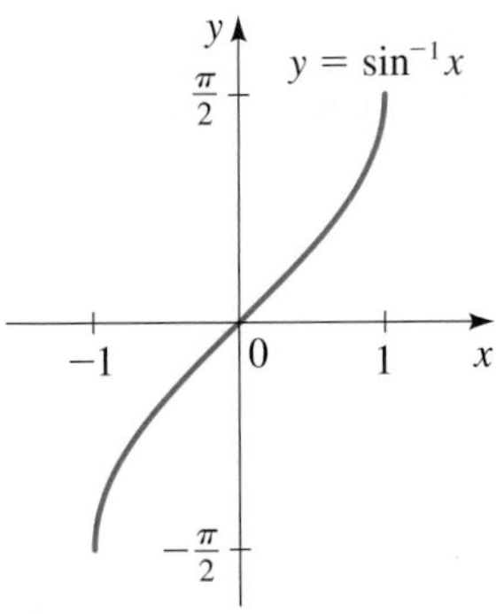

FIGURE 2 Graph of $y = \sin^{-1} x$

We can now define an inverse sine function on this restricted domain. The graph of $y = \sin^{-1} x$ is shown in Figure 2; it is obtained by reflecting the graph of $y = \sin x$, $-\pi/2 \le x \le \pi/2$, in the line $y = x$.

DEFINITION OF THE INVERSE SINE FUNCTION

The **inverse sine function** is the function $\sin^{-1}$ with domain $[-1, 1]$ and range $[-\pi/2, \pi/2]$ defined by

$$\sin^{-1} x = y \quad \Leftrightarrow \quad \sin y = x$$

The inverse sine function is also called **arcsine**, denoted by **arcsin**.

Thus, $y = \sin^{-1} x$ *is the number in the interval* $[-\pi/2, \pi/2]$ *whose sine is* x. In other words, $\sin(\sin^{-1} x) = x$. In fact, from the general properties of inverse functions studied in Section 2.7, we have the following **cancellation properties**.

$$\sin(\sin^{-1} x) = x \quad \text{for} \quad -1 \le x \le 1$$

$$\sin^{-1}(\sin x) = x \quad \text{for} \quad -\frac{\pi}{2} \le x \le \frac{\pi}{2}$$

EXAMPLE 1 | Evaluating the Inverse Sine Function

Find each value.

(a) $\sin^{-1} \frac{1}{2}$ **(b)** $\sin^{-1}\left(-\frac{1}{2}\right)$ **(c)** $\sin^{-1} \frac{3}{2}$

SOLUTION

(a) The number in the interval $[-\pi/2, \pi/2]$ whose sine is $\frac{1}{2}$ is $\pi/6$. Thus, $\sin^{-1}\frac{1}{2} = \pi/6$.

(b) The number in the interval $[-\pi/2, \pi/2]$ whose sine is $-\frac{1}{2}$ is $-\pi/6$. Thus, $\sin^{-1}(-\frac{1}{2}) = -\pi/6$.

(c) Since $\frac{3}{2} > 1$, it is not in the domain of $\sin^{-1} x$, so $\sin^{-1} \frac{3}{2}$ is not defined.

NOW TRY EXERCISE **3** ■

EXAMPLE 2 | Using a Calculator to Evaluate Inverse Sine

Find approximate values for **(a)** $\sin^{-1}(0.82)$ and **(b)** $\sin^{-1}\frac{1}{3}$.

SOLUTION

We use a calculator to approximate these values. Using the [SIN⁻¹], or [INV][SIN], or [ARC][SIN] key(s) on the calculator (with the calculator in radian mode), we get

(a) $\sin^{-1}(0.82) \approx 0.96141$ **(b)** $\sin^{-1}\frac{1}{3} \approx 0.33984$

NOW TRY EXERCISES **11** AND **21** ■

When evaluating expressions involving $\sin^{-1}$, we need to remember that the range of $\sin^{-1}$ is the interval $[-\pi/2, \pi/2]$.

EXAMPLE 3 | Evaluating Expressions with Inverse Sine

Find each value.

(a) $\sin^{-1}\left(\sin \frac{\pi}{3}\right)$ **(b)** $\sin^{-1}\left(\sin \frac{2\pi}{3}\right)$

SOLUTION

(a) Since $\pi/3$ is in the interval $[-\pi/2, \pi/2]$, we can use the above cancellation properties of inverse functions:

$$\sin^{-1}\left(\sin\frac{\pi}{3}\right) = \frac{\pi}{3} \qquad \text{Cancellation property: } -\frac{\pi}{2} \le \frac{\pi}{3} \le \frac{\pi}{2}$$

(b) We first evaluate the expression in the parentheses:

$$\sin^{-1}\left(\sin\frac{2\pi}{3}\right) = \sin^{-1}\left(\frac{\sqrt{3}}{2}\right) \qquad \text{Evaluate}$$

$$= \frac{\pi}{3} \qquad \text{Because } \sin\frac{\pi}{3} = \frac{\sqrt{3}}{2}$$

Note: $\sin^{-1}(\sin x) = x$ only if $-\frac{\pi}{2} \le x \le \frac{\pi}{2}$.

NOW TRY EXERCISES **31** AND **33**

The Inverse Cosine Function

If the domain of the cosine function is restricted to the interval $[0, \pi]$, the resulting function is one-to-one and so has an inverse. We choose this interval because on it, cosine attains each of its values exactly once (see Figure 3).

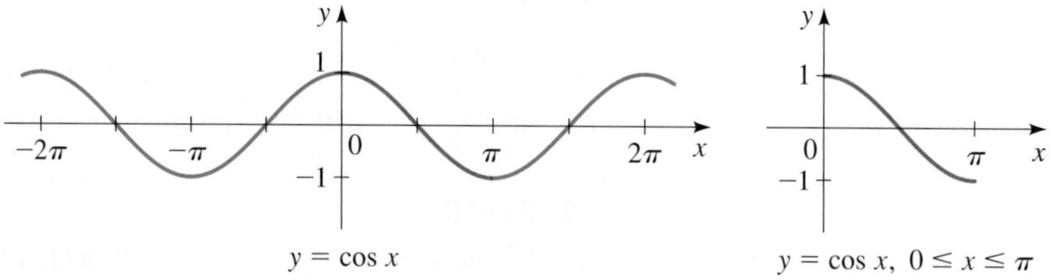

FIGURE 3 Graphs of the cosine function and the restricted cosine function

DEFINITION OF THE INVERSE COSINE FUNCTION

The **inverse cosine function** is the function $\cos^{-1}$ with domain $[-1, 1]$ and range $[0, \pi]$ defined by

$$\cos^{-1} x = y \quad \Leftrightarrow \quad \cos y = x$$

The inverse cosine function is also called **arccosine**, denoted by **arccos**.

Thus, $y = \cos^{-1} x$ *is the number in the interval* $[0, \pi]$ *whose cosine is* x. The following **cancellation properties** follow from the inverse function properties.

$$\cos(\cos^{-1} x) = x \quad \text{for} \quad -1 \le x \le 1$$
$$\cos^{-1}(\cos x) = x \quad \text{for} \quad 0 \le x \le \pi$$

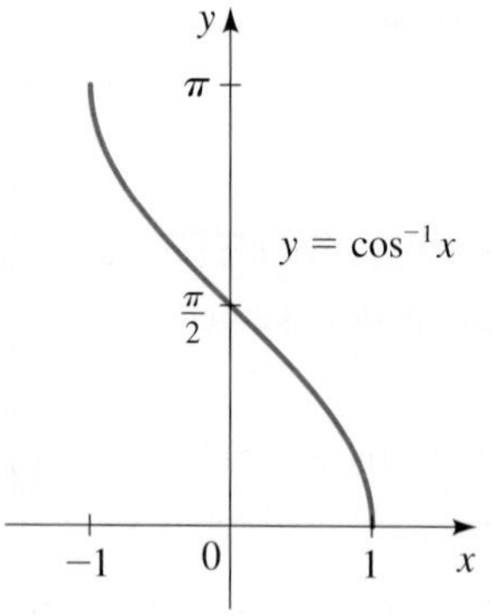

FIGURE 4 Graph of $y = \cos^{-1} x$

The graph of $y = \cos^{-1} x$ is shown in Figure 4; it is obtained by reflecting the graph of $y = \cos x$, $0 \le x \le \pi$, in the line $y = x$.

EXAMPLE 4 | Evaluating the Inverse Cosine Function

Find each value.

(a) $\cos^{-1}\frac{\sqrt{3}}{2}$ **(b)** $\cos^{-1} 0$ **(c)** $\cos^{-1}\frac{5}{7}$

SOLUTION

(a) The number in the interval $[0, \pi]$ whose cosine is $\sqrt{3}/2$ is $\pi/6$. Thus, $\cos^{-1}(\sqrt{3}/2) = \pi/6$.

(b) The number in the interval $[0, \pi]$ whose cosine is 0 is $\pi/2$. Thus, $\cos^{-1} 0 = \pi/2$.

(c) Since no rational multiple of π has cosine $\frac{5}{7}$, we use a calculator (in radian mode) to find this value approximately:

$$\cos^{-1}\frac{5}{7} \approx 0.77519$$

NOW TRY EXERCISES **5** AND **13** ■

EXAMPLE 5 | Evaluating Expressions with Inverse Cosine

Find each value.

(a) $\cos^{-1}\left(\cos\frac{2\pi}{3}\right)$ **(b)** $\cos^{-1}\left(\cos\frac{5\pi}{3}\right)$.

SOLUTION

(a) Since $2\pi/3$ is in the interval $[0, \pi]$ we can use the above cancellation properties:

$$\cos^{-1}\left(\cos\frac{2\pi}{3}\right) = \frac{2\pi}{3} \qquad \text{Cancellation property: } 0 \le \frac{2\pi}{3} \le \pi$$

(b) We first evaluate the expression in the parentheses:

$$\cos^{-1}\left(\cos\frac{5\pi}{3}\right) = \cos^{-1}\left(\tfrac{1}{2}\right) \qquad \text{Evaluate}$$

$$= \frac{\pi}{3} \qquad \text{Because } \cos\frac{\pi}{3} = \frac{1}{2}$$

Note: $\cos^{-1}(\cos x) = x$ only if $0 \le x \le \pi$.

NOW TRY EXERCISES **29** AND **35** ■

▼ The Inverse Tangent Function

We restrict the domain of the tangent function to the interval $(-\pi/2, \pi/2)$ in order to obtain a one-to-one function.

DEFINITION OF THE INVERSE TANGENT FUNCTION

The **inverse tangent function** is the function $\tan^{-1}$ with domain $\mathbb{R}$ and range $(-\pi/2, \pi/2)$ defined by

$$\tan^{-1} x = y \quad \Leftrightarrow \quad \tan y = x$$

The inverse tangent function is also called **arctangent**, denoted by **arctan**.

Thus, $y = \tan^{-1} x$ *is the number in the interval* $(-\pi/2, \pi/2)$ *whose tangent is* x. The following **cancellation properties** follow from the inverse function properties.

$$\tan(\tan^{-1} x) = x \quad \text{for} \quad x \in \mathbb{R}$$

$$\tan^{-1}(\tan x) = x \quad \text{for} \quad -\frac{\pi}{2} < x < \frac{\pi}{2}$$

Figure 5 shows the graph of $y = \tan x$ on the interval $(-\pi/2, \pi/2)$ and the graph of its inverse function, $y = \tan^{-1} x$.

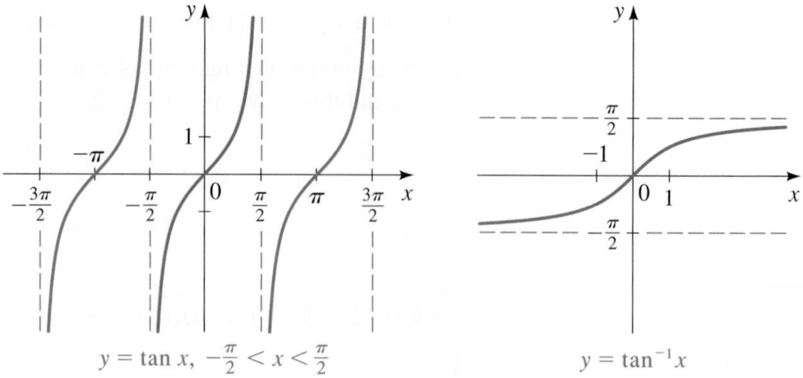

FIGURE 5 Graphs of the restricted tangent function and the inverse tangent function

EXAMPLE 6 | Evaluating the Inverse Tangent Function

Find each value.

(a) $\tan^{-1} 1$ **(b)** $\tan^{-1} \sqrt{3}$ **(c)** $\tan^{-1}(20)$

SOLUTION

(a) The number in the interval $(-\pi/2, \pi/2)$ with tangent 1 is $\pi/4$. Thus, $\tan^{-1} 1 = \pi/4$.

(b) The number in the interval $(-\pi/2, \pi/2)$ with tangent $\sqrt{3}$ is $\pi/3$. Thus, $\tan^{-1}\sqrt{3} = \pi/3$.

(c) We use a calculator (in radian mode) to find that $\tan^{-1}(20) \approx -1.52084$.

NOW TRY EXERCISES **7** AND **17**

The Inverse Secant, Cosecant, and Cotangent Functions

See Exercise 44 in Section 6.4 (page 469) for a way of finding the values of these inverse trigonometric functions on a calculator.

To define the inverse functions of the secant, cosecant, and cotangent functions, we restrict the domain of each function to a set on which it is one-to-one and on which it attains all its values. Although any interval satisfying these criteria is appropriate, we choose to restrict the domains in a way that simplifies the choice of sign in computations involving inverse trigonometric functions. The choices we make are also appropriate for calculus. This explains the seemingly strange restriction for the domains of the secant and cosecant functions. We end this section by displaying the graphs of the secant, cosecant, and cotangent functions with their restricted domains and the graphs of their inverse functions (Figures 6–8).

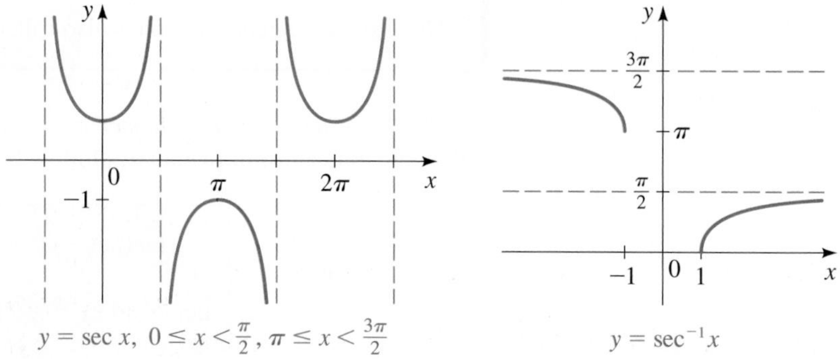

FIGURE 6 The inverse secant function

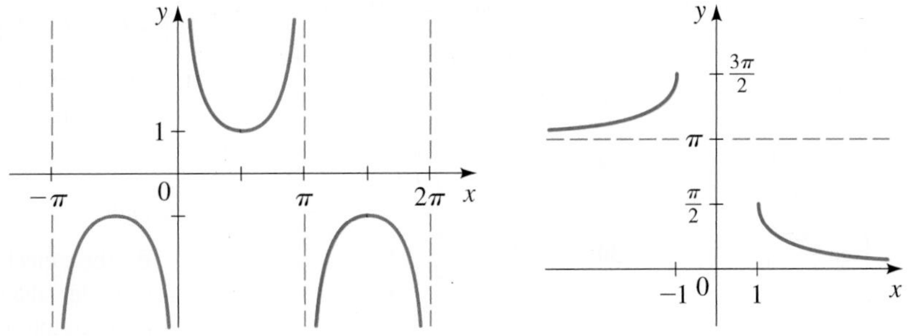

FIGURE 7 The inverse cosecant function

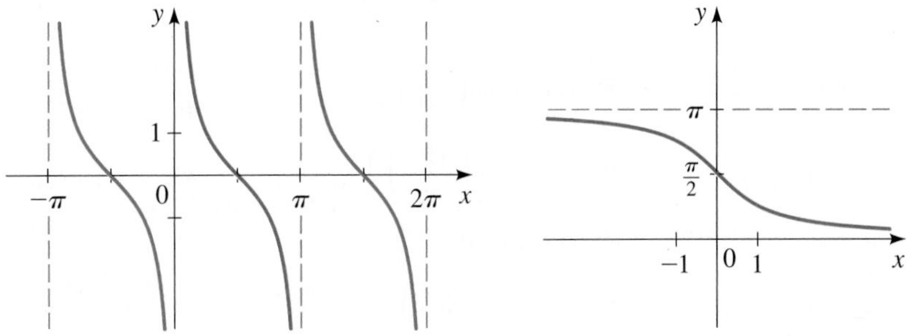

FIGURE 8 The inverse cotangent function

5.5 EXERCISES

CONCEPTS

1. (a) To define the inverse sine function, we restrict the domain of sine to the interval ____________. On this interval the sine function is one-to-one, and its inverse function $\sin^{-1}$ is defined by $\sin^{-1} x = y \Leftrightarrow \sin$ _____ = _____. For example, $\sin^{-1}\frac{1}{2} =$ _____ because sin _____ = _____.

(b) To define the inverse cosine function we restrict the domain of cosine to the interval ____________. On this interval the cosine function is one-to-one and its inverse function $\cos^{-1}$ is defined by $\cos^{-1} x = y \Leftrightarrow$ cos _____ = _____. For example, $\cos^{-1}\frac{1}{2} =$ _____ because cos _____ = _____.

2. The cancellation property $\sin^{-1}(\sin x) = x$ is valid for x in the interval ____________. Which of the following is not true?

(a) $\sin^{-1}\left(\sin \dfrac{\pi}{3}\right) = \dfrac{\pi}{3}$

(b) $\sin^{-1}\left(\sin \dfrac{10\pi}{3}\right) = \dfrac{10\pi}{3}$

SKILLS

3–10 ■ Find the exact value of each expression, if it is defined.

3. (a) $\sin^{-1} 1$ **(b)** $\sin^{-1} \dfrac{\sqrt{3}}{2}$ **(c)** $\sin^{-1} 2$

4. (a) $\sin^{-1}(-1)$ **(b)** $\sin^{-1} \dfrac{\sqrt{2}}{2}$ **(c)** $\sin^{-1}(-2)$

5. (a) $\cos^{-1}(-1)$ **(b)** $\cos^{-1} \frac{1}{2}$ **(c)** $\cos^{-1}\left(-\dfrac{\sqrt{3}}{2}\right)$

6. (a) $\cos^{-1}\left(\dfrac{\sqrt{2}}{2}\right)$ **(b)** $\cos^{-1} 1$ **(c)** $\cos^{-1}\left(-\dfrac{\sqrt{2}}{2}\right)$

7. (a) $\tan^{-1}(-1)$ **(b)** $\tan^{-1}\sqrt{3}$ **(c)** $\tan^{-1} \dfrac{\sqrt{3}}{3}$

8. (a) $\tan^{-1} 0$ **(b)** $\tan^{-1}(-\sqrt{3})$ **(c)** $\tan^{-1}\left(-\dfrac{\sqrt{3}}{3}\right)$

9. (a) $\cos^{-1}(-\frac{1}{2})$ **(b)** $\sin^{-1}\left(-\dfrac{\sqrt{2}}{2}\right)$ **(c)** $\tan^{-1} 1$

10. (a) $\cos^{-1} 0$ **(b)** $\sin^{-1} 0$ **(c)** $\sin^{-1}(-\frac{1}{2})$

11–22 ■ Use a calculator to find an approximate value of each expression correct to five decimal places, if it is defined.

11. $\sin^{-1} \frac{2}{3}$ **12.** $\sin^{-1}(-\frac{8}{9})$

13. $\cos^{-1}(-\frac{3}{7})$ **14.** $\cos^{-1}(\frac{4}{9})$

15. $\cos^{-1}(-0.92761)$ **16.** $\sin^{-1}(0.13844)$

17. $\tan^{-1} 10$ **18.** $\tan^{-1}(-26)$

19. $\tan^{-1}(1.23456)$ **20.** $\cos^{-1}(1.23456)$

21. $\sin^{-1}(-0.25713)$ **22.** $\tan^{-1}(-0.25713)$

23–44 ■ Find the exact value of the expression, if it is defined.

23. $\sin(\sin^{-1}\frac{1}{4})$

24. $\cos(\cos^{-1}\frac{2}{3})$

25. $\tan(\tan^{-1} 5)$

26. $\sin(\sin^{-1} 5)$

27. $\sin\left(\sin^{-1}\left(\frac{3}{2}\right)\right)$

28. $\tan\left(\tan^{-1}\left(\frac{3}{2}\right)\right)$

29. $\cos^{-1}\left(\cos\frac{5\pi}{6}\right)$

30. $\tan^{-1}\left(\tan\left(\frac{\pi}{4}\right)\right)$

31. $\sin^{-1}\left(\sin\left(-\frac{\pi}{6}\right)\right)$

32. $\tan^{-1}\left(\tan\left(-\frac{\pi}{4}\right)\right)$

33. $\sin^{-1}\left(\sin\left(\frac{5\pi}{6}\right)\right)$

34. $\cos^{-1}\left(\cos\left(-\frac{\pi}{6}\right)\right)$

35. $\cos^{-1}\left(\cos\left(\frac{17\pi}{6}\right)\right)$

36. $\tan^{-1}\left(\tan\left(\frac{4\pi}{3}\right)\right)$

37. $\tan^{-1}\left(\tan\left(\frac{2\pi}{3}\right)\right)$

38. $\sin^{-1}\left(\sin\left(\frac{11\pi}{4}\right)\right)$

39. $\tan(\sin^{-1}\frac{1}{2})$

40. $\cos(\sin^{-1} 0)$

41. $\cos\left(\sin^{-1}\frac{\sqrt{3}}{2}\right)$

42. $\tan\left(\sin^{-1}\frac{\sqrt{2}}{2}\right)$

43. $\sin(\tan^{-1}(-1))$

44. $\sin(\tan^{-1}(-\sqrt{3}))$

DISCOVERY ■ DISCUSSION ■ WRITING

45. Two Different Compositions Let f and g be the functions

$$f(x) = \sin(\sin^{-1} x)$$

and

$$g(x) = \sin^{-1}(\sin x)$$

By the cancellation properties, $f(x) = x$ and $g(x) = x$ for suitable values of x. But these functions are not the same for all x. Graph both f and g to show how the functions differ. (Think carefully about the domain and range of $\sin^{-1}$).

46–47 ■ **Graphing Inverse Trigonometric Functions**
(a) Graph the function and make a conjecture, and **(b)** prove that your conjecture is true.

46. $y = \sin^{-1} x + \cos^{-1} x$

47. $y = \tan^{-1} x + \tan^{-1}\frac{1}{x}$

5.6 Modeling Harmonic Motion

Simple Harmonic Motion ▶ Damped Harmonic Motion

Periodic behavior—behavior that repeats over and over again—is common in nature. Perhaps the most familiar example is the daily rising and setting of the sun, which results in the repetitive pattern of day, night, day, night, ... Another example is the daily variation of tide levels at the beach, which results in the repetitive pattern of high tide, low tide, high tide, low tide, ... Certain animal populations increase and decrease in a predictable periodic pattern: A large population exhausts the food supply, which causes the population to dwindle; this in turn results in a more plentiful food supply, which makes it possible for the population to increase; and the pattern then repeats over and over (see the Discovery Project *Predator/Prey Models* referenced on page 398).

Other common examples of periodic behavior involve motion that is caused by vibration or oscillation. A mass suspended from a spring that has been compressed and then allowed to vibrate vertically is a simple example. This "back and forth" motion also occurs in such diverse phenomena as sound waves, light waves, alternating electrical current, and pulsating stars, to name a few. In this section we consider the problem of modeling periodic behavior.

▼ Simple Harmonic Motion

The trigonometric functions are ideally suited for modeling periodic behavior. A glance at the graphs of the sine and cosine functions, for instance, tells us that these functions themselves exhibit periodic behavior. Figure 1 shows the graph of $y = \sin t$. If we think of

t as time, we see that as time goes on, $y = \sin t$ increases and decreases over and over again. Figure 2 shows that the motion of a vibrating mass on a spring is modeled very accurately by $y = \sin t$.

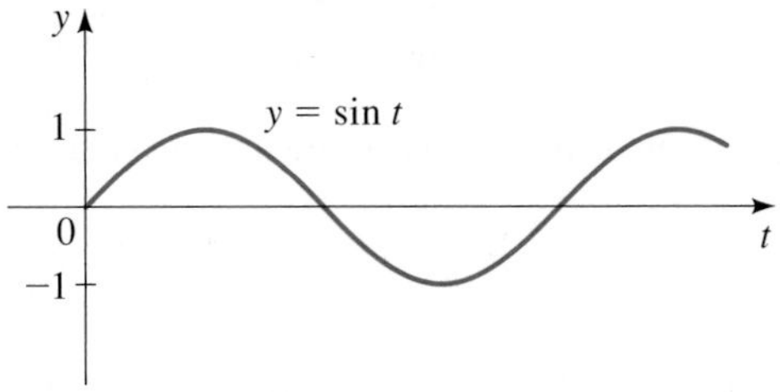

FIGURE 1 $y = \sin t$

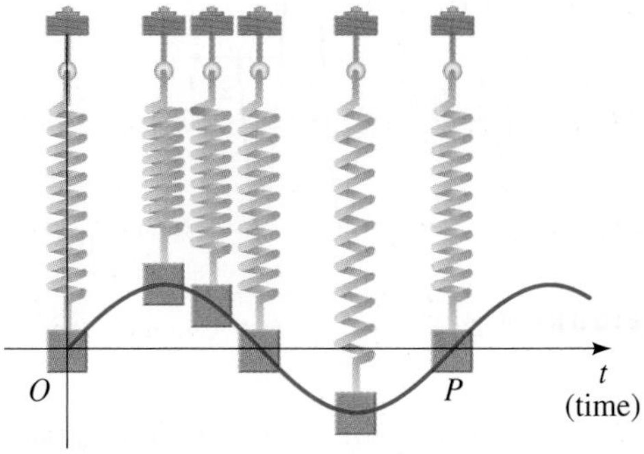

FIGURE 2 Motion of a vibrating spring is modeled by $y = \sin t$.

Notice that the mass returns to its original position over and over again. A **cycle** is one complete vibration of an object, so the mass in Figure 2 completes one cycle of its motion between O and P. Our observations about how the sine and cosine functions model periodic behavior are summarized in the following box.

The main difference between the two equations describing simple harmonic motion is the starting point. At $t = 0$ we get

$$y = a \sin \omega \cdot 0 = 0$$
$$y = a \cos \omega \cdot 0 = a$$

In the first case the motion "starts" with zero displacement, whereas in the second case the motion "starts" with the displacement at maximum (at the amplitude a).

SIMPLE HARMONIC MOTION

If the equation describing the displacement y of an object at time t is

$$y = a \sin \omega t \qquad \text{or} \qquad y = a \cos \omega t$$

then the object is in **simple harmonic motion**. In this case,

amplitude $= \lvert a \rvert$	Maximum displacement of the object
period $= \dfrac{2\pi}{\omega}$	Time required to complete one cycle
frequency $= \dfrac{\omega}{2\pi}$	Number of cycles per unit of time

The symbol ω is the lowercase Greek letter "omega," and ν is the letter "nu."

Notice that the functions

$$y = a \sin 2\pi\nu t \qquad \text{and} \qquad y = a \cos 2\pi\nu t$$

have frequency ν, because $2\pi\nu/(2\pi) = \nu$. Since we can immediately read the frequency from these equations, we often write equations of simple harmonic motion in this form.

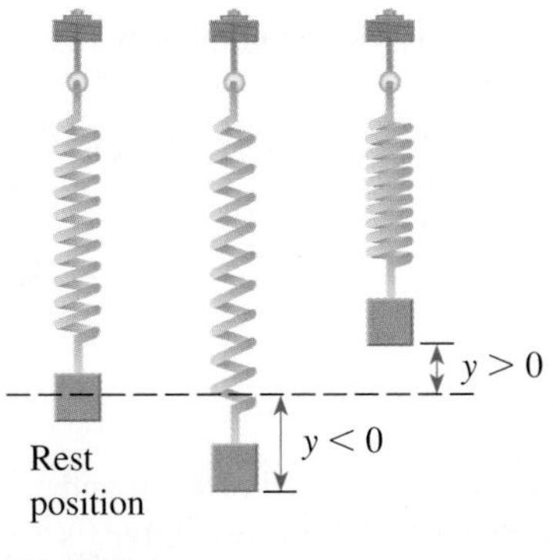

FIGURE 3

EXAMPLE 1 | A Vibrating Spring

The displacement of a mass suspended by a spring is modeled by the function

$$y = 10 \sin 4\pi t$$

where y is measured in inches and t in seconds (see Figure 3).

(a) Find the amplitude, period, and frequency of the motion of the mass.

(b) Sketch a graph of the displacement of the mass.

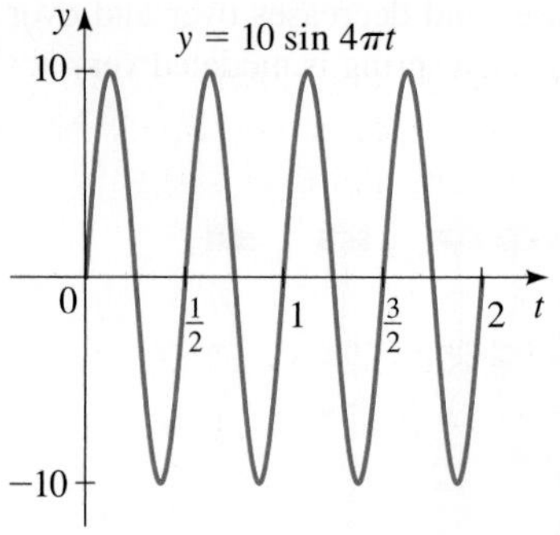

FIGURE 4

SOLUTION

(a) From the formulas for amplitude, period, and frequency we get

$$\text{amplitude} = |a| = 10 \text{ in.}$$

$$\text{period} = \frac{2\pi}{\omega} = \frac{2\pi}{4\pi} = \frac{1}{2} \text{ s}$$

$$\text{frequency} = \frac{\omega}{2\pi} = \frac{4\pi}{2\pi} = 2 \text{ cycles per second (Hz)}$$

(b) The graph of the displacement of the mass at time t is shown in Figure 4.

NOW TRY EXERCISE **3** ■

An important situation in which simple harmonic motion occurs is in the production of sound. Sound is produced by a regular variation in air pressure from the normal pressure. If the pressure varies in simple harmonic motion, then a pure sound is produced. The tone of the sound depends on the frequency, and the loudness depends on the amplitude.

EXAMPLE 2 | Vibrations of a Musical Note

A tuba player plays the note E and sustains the sound for some time. For a pure E the variation in pressure from normal air pressure is given by

$$V(t) = 0.2 \sin 80\pi t$$

where V is measured in pounds per square inch and t is measured in seconds.

(a) Find the amplitude, period, and frequency of V.

(b) Sketch a graph of V.

(c) If the tuba player increases the loudness of the note, how does the equation for V change?

(d) If the player is playing the note incorrectly and it is a little flat, how does the equation for V change?

SOLUTION

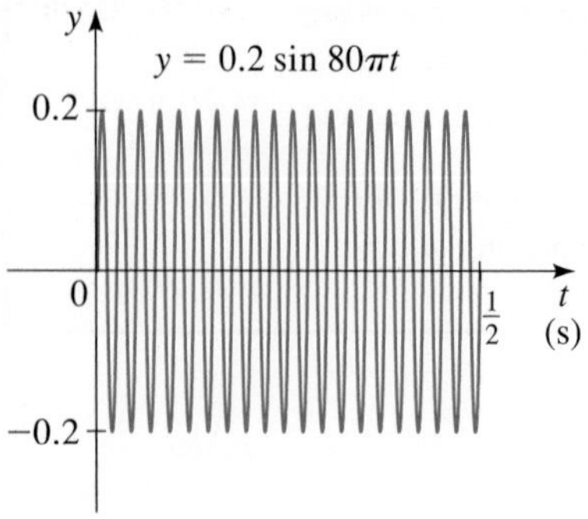

FIGURE 5

(a) From the formulas for amplitude, period, and frequency we get

$$\text{amplitude} = |0.2| = 0.2$$

$$\text{period} = \frac{2\pi}{80\pi} = \frac{1}{40}$$

$$\text{frequency} = \frac{80\pi}{2\pi} = 40$$

(b) The graph of V is shown in Figure 5.

(c) If the player increases the loudness the amplitude increases. So the number 0.2 is replaced by a larger number.

(d) If the note is flat, then the frequency is decreased. Thus, the coefficient of t is less than 80π.

NOW TRY EXERCISE **29** ■

EXAMPLE 3 | Modeling a Vibrating Spring

A mass is suspended from a spring. The spring is compressed a distance of 4 cm and then released. It is observed that the mass returns to the compressed position after $\frac{1}{3}$ s.

(a) Find a function that models the displacement of the mass.

(b) Sketch the graph of the displacement of the mass.

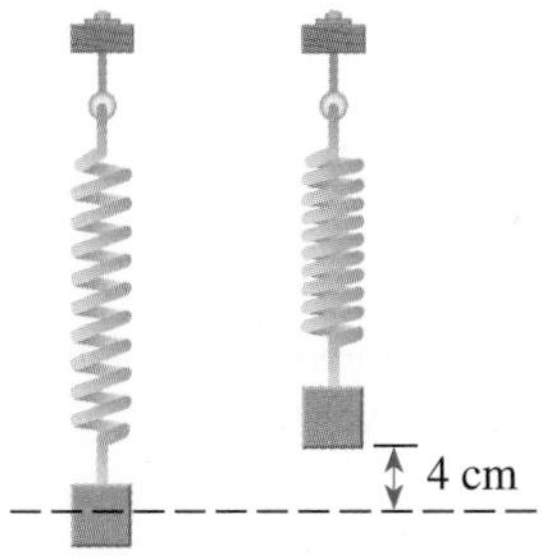

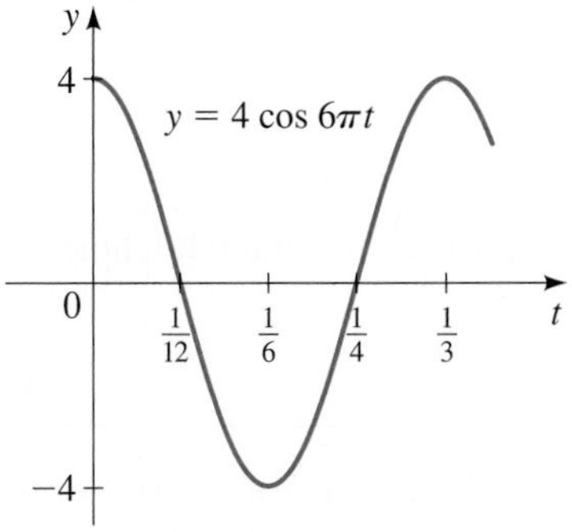

FIGURE 6

SOLUTION

(a) The motion of the mass is given by one of the equations for simple harmonic motion. The amplitude of the motion is 4 cm. Since this amplitude is reached at time $t = 0$, an appropriate function that models the displacement is of the form

$$y = a \cos \omega t$$

Since the period is $p = \frac{1}{3}$, we can find ω from the following equation:

$$\text{period} = \frac{2\pi}{\omega}$$

$$\frac{1}{3} = \frac{2\pi}{\omega} \qquad \text{Period} = \tfrac{1}{3}$$

$$\omega = 6\pi \qquad \text{Solve for } \omega$$

So the motion of the mass is modeled by the function

$$y = 4 \cos 6\pi t$$

where y is the displacement from the rest position at time t. Notice that when $t = 0$, the displacement is $y = 4$, as we expect.

(b) The graph of the displacement of the mass at time t is shown in Figure 6.

NOW TRY EXERCISES **15** AND **35** ■

In general, the sine or cosine functions representing harmonic motion may be shifted horizontally or vertically. In this case, the equations take the form

$$y = a \sin(\omega(t - c)) + b \qquad \text{or} \qquad y = a \cos(\omega(t - c)) + b$$

The vertical shift b indicates that the variation occurs around an average value b. The horizontal shift c indicates the position of the object at $t = 0$. (See Figure 7.)

FIGURE 7

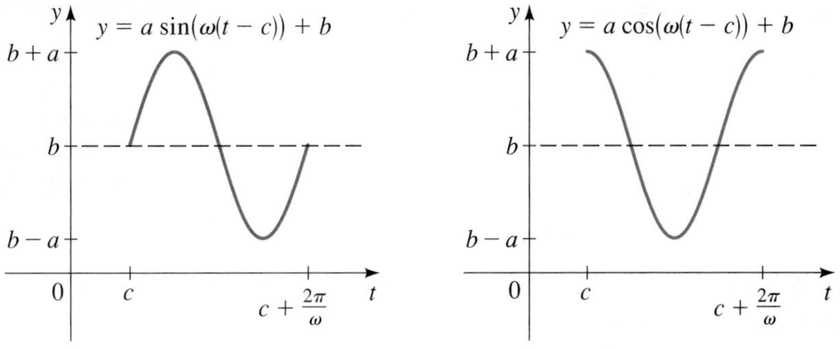

(a) (b)

EXAMPLE 4 | Modeling the Brightness of a Variable Star

A variable star is one whose brightness alternately increases and decreases. For the variable star Delta Cephei, the time between periods of maximum brightness is 5.4 days. The average brightness (or magnitude) of the star is 4.0, and its brightness varies by ± 0.35 magnitude.

(a) Find a function that models the brightness of Delta Cephei as a function of time.

(b) Sketch a graph of the brightness of Delta Cephei as a function of time.

SOLUTION

(a) Let's find a function in the form

$$y = a\cos(\omega(t - c)) + b$$

The amplitude is the maximum variation from average brightness, so the amplitude is $a = 0.35$ magnitude. We are given that the period is 5.4 days, so

$$\omega = \frac{2\pi}{5.4} \approx 1.164$$

Since the brightness varies from an average value of 4.0 magnitudes, the graph is shifted upward by $b = 4.0$. If we take $t = 0$ to be a time when the star is at maximum brightness, there is no horizontal shift, so $c = 0$ (because a cosine curve achieves its maximum at $t = 0$). Thus, the function we want is

$$y = 0.35\cos(1.16t) + 4.0$$

where t is the number of days from a time when the star is at maximum brightness.

(b) The graph is sketched in Figure 8.

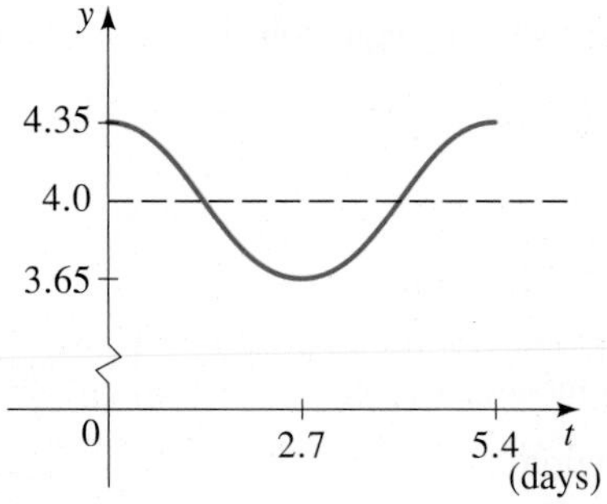

FIGURE 8

NOW TRY EXERCISE **39**

The number of hours of daylight varies throughout the course of a year. In the Northern Hemisphere, the longest day is June 21, and the shortest is December 21. The average length of daylight is 12 h, and the variation from this average depends on the latitude. (For example, Fairbanks, Alaska, experiences more than 20 h of daylight on the longest day and less than 4 h on the shortest day!) The graph in Figure 9 shows the number of hours of daylight at different times of the year for various latitudes. It's apparent from the graph that the variation in hours of daylight is simple harmonic.

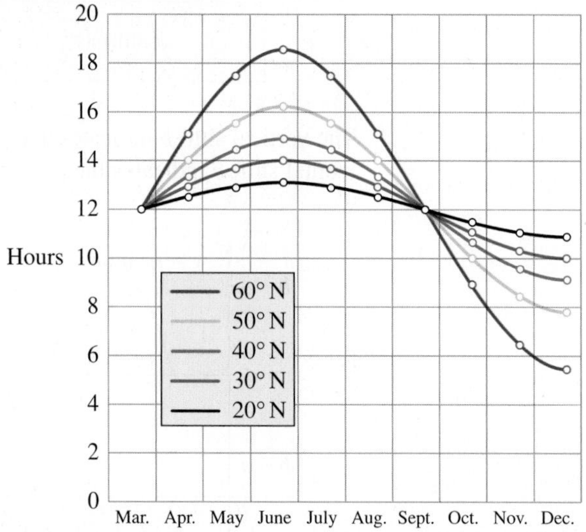

Source: Lucia C. Harrison, *Daylight, Twilight, Darkness and Time* (New York: Silver, Burdett, 1935), page 40.

FIGURE 9 Graph of the length of daylight from March 21 through December 21 at various latitudes

EXAMPLE 5 | Modeling the Number of Hours of Daylight

In Philadelphia (40° N latitude) the longest day of the year has 14 h 50 min of daylight, and the shortest day has 9 h 10 min of daylight.

(a) Find a function L that models the length of daylight as a function of t, the number of days from January 1.

(b) An astronomer needs at least 11 hours of darkness for a long exposure astronomical photograph. On what days of the year are such long exposures possible?

SOLUTION

(a) We need to find a function in the form

$$y = a \sin(\omega(t - c)) + b$$

whose graph is the 40° N latitude curve in Figure 9. From the information given, we see that the amplitude is

$$a = \tfrac{1}{2}\left(14\tfrac{5}{6} - 9\tfrac{1}{6}\right) \approx 2.83 \text{ h}$$

Since there are 365 days in a year, the period is 365, so

$$\omega = \frac{2\pi}{365} \approx 0.0172$$

Since the average length of daylight is 12 h, the graph is shifted upward by 12, so $b = 12$. Since the curve attains the average value (12) on March 21, the 80th day of the year, the curve is shifted 80 units to the right. Thus, $c = 80$. So a function that models the number of hours of daylight is

$$y = 2.83 \sin(0.0172(t - 80)) + 12$$

where t is the number of days from January 1.

(b) A day has 24 h, so 11 h of night correspond to 13 h of daylight. So we need to solve the inequality $y \le 13$. To solve this inequality graphically, we graph $y = 2.83 \sin 0.0172(t - 80) + 12$ and $y = 13$ on the same graph. From the graph in Figure 10 we see that there are fewer than 13 h of daylight between day 1 (January 1) and day 101 (April 11) and from day 241 (August 29) to day 365 (December 31).

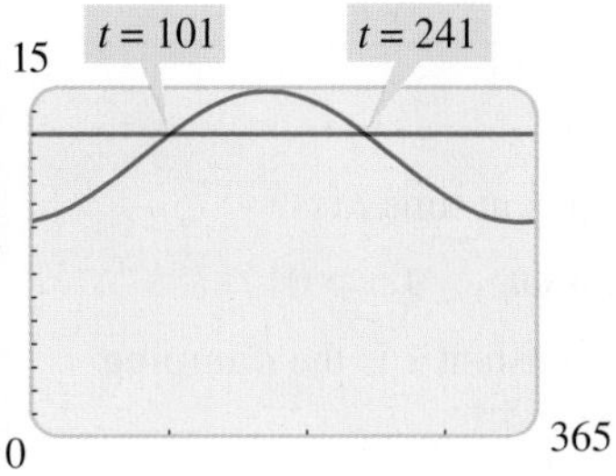

FIGURE 10

NOW TRY EXERCISE **41**

Another situation in which simple harmonic motion occurs is in alternating current (AC) generators. Alternating current is produced when an armature rotates about its axis in a magnetic field.

Figure 11 represents a simple version of such a generator. As the wire passes through the magnetic field, a voltage E is generated in the wire. It can be shown that the voltage generated is given by

$$E(t) = E_0 \cos \omega t$$

where E_0 is the maximum voltage produced (which depends on the strength of the magnetic field) and $\omega/(2\pi)$ is the number of revolutions per second of the armature (the frequency).

Why do we say that household current is 110 V when the maximum voltage produced is 155 V? From the symmetry of the cosine function we see that the average voltage produced is zero. This average value would be the same for all AC generators and so gives no information about the voltage generated. To obtain a more informative measure of voltage, engineers use the **root-mean-square** (rms) method. It can be shown that the rms voltage is $1/\sqrt{2}$ times the maximum voltage. So for household current the rms voltage is

$$155 \times \frac{1}{\sqrt{2}} \approx 110 \text{ V}$$

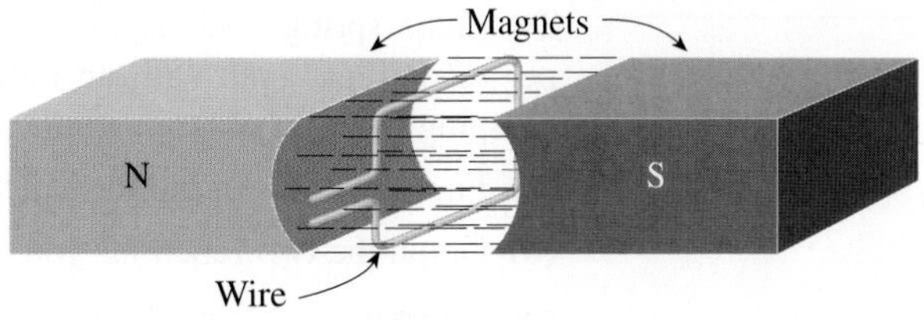

FIGURE 11

EXAMPLE 6 | Modeling Alternating Current

Ordinary 110-V household alternating current varies from +155 V to −155 V with a frequency of 60 Hz (cycles per second). Find an equation that describes this variation in voltage.

SOLUTION The variation in voltage is simple harmonic. Since the frequency is 60 cycles per second, we have

$$\frac{\omega}{2\pi} = 60 \qquad \text{or} \qquad \omega = 120\pi$$

Let's take $t = 0$ to be a time when the voltage is $+155$ V. Then

$$E(t) = a \cos \omega t = 155 \cos 120\pi t$$

NOW TRY EXERCISE **43**

Damped Harmonic Motion

The spring in Figure 2 on page 413 is assumed to oscillate in a frictionless environment. In this hypothetical case the amplitude of the oscillation will not change. In the presence of friction, however, the motion of the spring eventually "dies down"; that is, the amplitude of the motion decreases with time. Motion of this type is called *damped harmonic motion.*

DAMPED HARMONIC MOTION

If the equation describing the displacement y of an object at time t is

$$y = ke^{-ct} \sin \omega t \qquad \text{or} \qquad y = ke^{-ct} \cos \omega t \qquad (c > 0)$$

then the object is in **damped harmonic motion**. The constant c is the **damping constant**, k is the initial amplitude, and $2\pi/\omega$ is the period.*

Damped harmonic motion is simply harmonic motion for which the amplitude is governed by the function $a(t) = ke^{-ct}$. Figure 12 shows the difference between harmonic motion and damped harmonic motion.

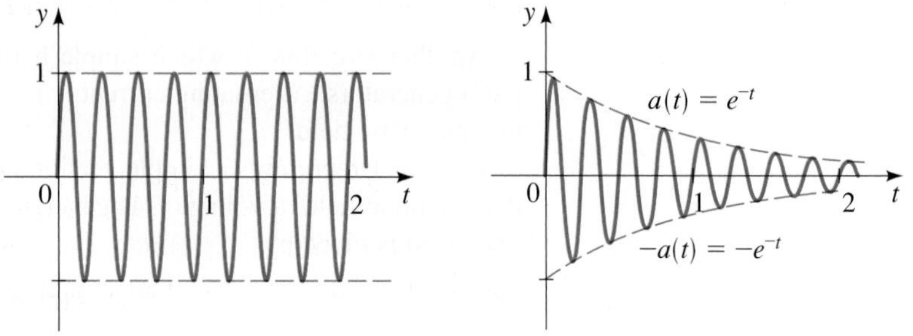

FIGURE 12 (a) Harmonic motion: $y = \sin 8\pi t$ (b) Damped harmonic motion: $y = e^{-t} \sin 8\pi t$

EXAMPLE 7 | Modeling Damped Harmonic Motion

Two mass-spring systems are experiencing damped harmonic motion, both at 0.5 cycles per second and both with an initial maximum displacement of 10 cm. The first has a damping constant of 0.5, and the second has a damping constant of 0.1.

(a) Find functions of the form $g(t) = ke^{-ct} \cos \omega t$ to model the motion in each case.

(b) Graph the two functions you found in part (a). How do they differ?

SOLUTION

Hz is the abbreviation for hertz. One hertz is one cycle per second.

(a) At time $t = 0$ the displacement is 10 cm. Thus, $g(0) = ke^{-c \cdot 0} \cos(\omega \cdot 0) = k$, so $k = 10$. Also, the frequency is $f = 0.5$ Hz, and since $\omega = 2\pi f$ (see page 413), we get $\omega = 2\pi(0.5) = \pi$. Using the given damping constants, we find that the motions of the two springs are given by the functions

$$g_1(t) = 10e^{-0.5t} \cos \pi t \qquad \text{and} \qquad g_2(t) = 10e^{-0.1t} \cos \pi t$$

*In the case of damped harmonic motion the term *quasi-period* is often used instead of *period* because the motion is not actually periodic—it diminishes with time. However, we will continue to use the term *period* to avoid confusion.

(b) The functions g_1 and g_2 are graphed in Figure 13. From the graphs we see that in the first case (where the damping constant is larger) the motion dies down quickly, whereas in the second case, perceptible motion continues much longer.

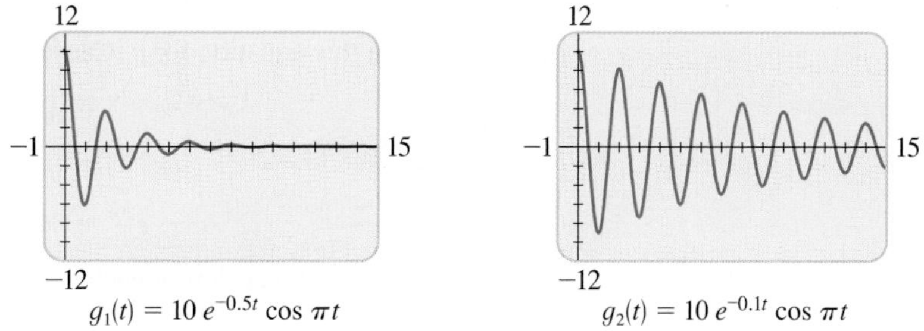

FIGURE 13

NOW TRY EXERCISE **19** ■

As the preceding example indicates, the larger the damping constant c, the quicker the oscillation dies down. When a guitar string is plucked and then allowed to vibrate freely, a point on that string undergoes damped harmonic motion. We hear the damping of the motion as the sound produced by the vibration of the string fades. How fast the damping of the string occurs (as measured by the size of the constant c) is a property of the size of the string and the material it is made of. Another example of damped harmonic motion is the motion that a shock absorber on a car undergoes when the car hits a bump in the road. In this case the shock absorber is engineered to damp the motion as quickly as possible (large c) and to have the frequency as small as possible (small ω). On the other hand, the sound produced by a tuba player playing a note is undamped as long as the player can maintain the loudness of the note. The electromagnetic waves that produce light move in simple harmonic motion that is not damped.

EXAMPLE 8 | A Vibrating Violin String

The G-string on a violin is pulled a distance of 0.5 cm above its rest position, then released and allowed to vibrate. The damping constant c for this string is determined to be 1.4. Suppose that the note produced is a pure G (frequency = 200 Hz). Find an equation that describes the motion of the point at which the string was plucked.

SOLUTION Let P be the point at which the string was plucked. We will find a function $f(t)$ that gives the distance at time t of the point P from its original rest position. Since the maximum displacement occurs at $t = 0$, we find an equation in the form

$$y = ke^{-ct}\cos \omega t$$

From this equation we see that $f(0) = k$. But we know that the original displacement of the string is 0.5 cm. Thus, $k = 0.5$. Since the frequency of the vibration is 200, we have $\omega = 2\pi f = 2\pi(200) = 400\pi$. Finally, since we know that the damping constant is 1.4, we get

$$f(t) = 0.5e^{-1.4t}\cos 400\pi t$$

NOW TRY EXERCISE **45** ■

EXAMPLE 9 | Ripples on a Pond

A stone is dropped in a calm lake, causing waves to form. The up-and-down motion of a point on the surface of the water is modeled by damped harmonic motion. At some time the amplitude of the wave is measured, and 20 s later it is found that the amplitude has dropped to $\frac{1}{10}$ of this value. Find the damping constant c.

SOLUTION The amplitude is governed by the coefficient ke^{-ct} in the equations for damped harmonic motion. Thus, the amplitude at time t is ke^{-ct}, and 20 s later, it is $ke^{-c(t+20)}$. So, because the later value is $\frac{1}{10}$ the earlier value, we have

$$ke^{-c(t+20)} = \tfrac{1}{10}ke^{-ct}$$

We now solve this equation for c. Canceling k and using the Laws of Exponents, we get

$$e^{-ct} \cdot e^{-20c} = \tfrac{1}{10}e^{-ct}$$

$$e^{-20c} = \tfrac{1}{10} \qquad \text{Cancel } e^{-ct}$$

$$e^{20c} = 10 \qquad \text{Take reciprocals}$$

Taking the natural logarithm of each side gives

$$20c = \ln(10)$$

$$c = \tfrac{1}{20}\ln(10) \approx \tfrac{1}{20}(2.30) \approx 0.12$$

Thus, the damping constant is $c \approx 0.12$.

NOW TRY EXERCISE **47** ■

5.6 EXERCISES

CONCEPTS

1. For an object in simple harmonic motion with amplitude a and period $2\pi/\omega$, find an equation that models the displacement y at time t if

(a) $y = 0$ at time $t = 0$: $y =$ __________.

(b) $y = a$ at time $t = 0$: $y =$ __________.

2. For an object in damped harmonic motion with initial amplitude k, period $2\pi/\omega$, and damping constant c, find an equation that models the displacement y at time t if

(a) $y = 0$ at time $t = 0$: $y =$ __________.

(b) $y = a$ at time $t = 0$: $y =$ __________.

SKILLS

3–10 ■ The given function models the displacement of an object moving in simple harmonic motion.
(a) Find the amplitude, period, and frequency of the motion.
(b) Sketch a graph of the displacement of the object over one complete period.

3. $y = 2\sin 3t$

4. $y = 3\cos\frac{1}{2}t$

5. $y = -\cos 0.3t$

6. $y = 2.4\sin 3.6t$

7. $y = -0.25\cos\left(1.5t - \dfrac{\pi}{3}\right)$

8. $y = -\frac{3}{2}\sin(0.2t + 1.4)$

9. $y = 5\cos(\frac{2}{3}t + \frac{3}{4})$

10. $y = 1.6\sin(t - 1.8)$

11–14 ■ Find a function that models the simple harmonic motion having the given properties. Assume that the displacement is zero at time $t = 0$.

11. amplitude 10 cm, period 3 s

12. amplitude 24 ft, period 2 min

13. amplitude 6 in., frequency $5/\pi$ Hz

14. amplitude 1.2 m, frequency 0.5 Hz

15–18 ■ Find a function that models the simple harmonic motion having the given properties. Assume that the displacement is at its maximum at time $t = 0$.

15. amplitude 60 ft, period 0.5 min

16. amplitude 35 cm, period 8 s

17. amplitude 2.4 m, frequency 750 Hz

18. amplitude 6.25 in., frequency 60 Hz

19–26 ■ An initial amplitude k, damping constant c, and frequency f or period p are given. (Recall that frequency and period are related by the equation $f = 1/p$.)
(a) Find a function that models the damped harmonic motion. Use a function of the form $y = ke^{-ct}\cos\omega t$ in Exercises 19–22, and of the form $y = ke^{-ct}\sin\omega t$ in Exercises 23–26.
(b) Graph the function.

19. $k = 2$, $c = 1.5$, $f = 3$

20. $k = 15$, $c = 0.25$, $f = 0.6$

21. $k = 100$, $c = 0.05$, $p = 4$

22. $k = 0.75$, $c = 3$, $p = 3\pi$

23. $k = 7$, $c = 10$, $p = \pi/6$

24. $k = 1$, $c = 1$, $p = 1$

25. $k = 0.3$, $c = 0.2$, $f = 20$

26. $k = 12$, $c = 0.01$, $f = 8$

APPLICATIONS

27. A Bobbing Cork A cork floating in a lake is bobbing in simple harmonic motion. Its displacement above the bottom of the lake is modeled by

$$y = 0.2 \cos 20\pi t + 8$$

where y is measured in meters and t is measured in minutes.
(a) Find the frequency of the motion of the cork.
(b) Sketch a graph of y.
(c) Find the maximum displacement of the cork above the lake bottom.

28. FM Radio Signals The carrier wave for an FM radio signal is modeled by the function

$$y = a \sin(2\pi(9.15 \times 10^7)t)$$

where t is measured in seconds. Find the period and frequency of the carrier wave.

29. Blood Pressure Each time your heart beats, your blood pressure increases, then decreases as the heart rests between beats. A certain person's blood pressure is modeled by the function

$$p(t) = 115 + 25 \sin(160\pi t)$$

where $p(t)$ is the pressure in mmHg at time t, measured in minutes.
(a) Find the amplitude, period, and frequency of p.
(b) Sketch a graph of p.
(c) If a person is exercising, his or her heart beats faster. How does this affect the period and frequency of p?

30. Predator Population Model In a predator/prey model the predator population is modeled by the function

$$y = 900 \cos 2t + 8000$$

where t is measured in years.
(a) What is the maximum population?
(b) Find the length of time between successive periods of maximum population.

31. Spring–Mass System A mass attached to a spring is moving up and down in simple harmonic motion. The graph gives its displacement $d(t)$ from equilibrium at time t. Express the function d in the form $d(t) = a \sin \omega t$.

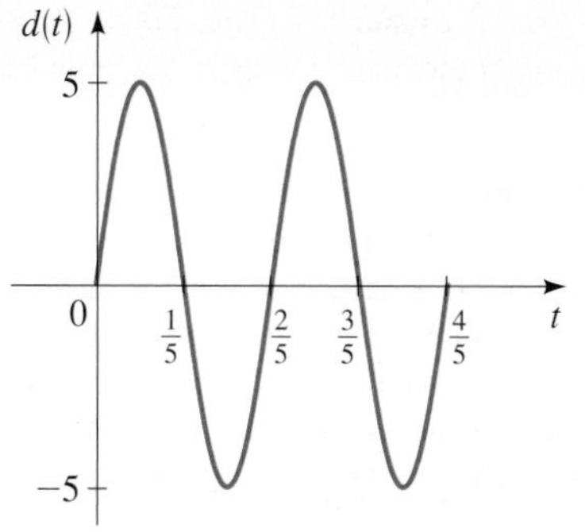

32. Tides The graph shows the variation of the water level relative to mean sea level in Commencement Bay at Tacoma, Washington, for a particular 24-hour period. Assuming that this variation is modeled by simple harmonic motion, find an equation of the form $y = a \sin \omega t$ that describes the variation in water level as a function of the number of hours after midnight.

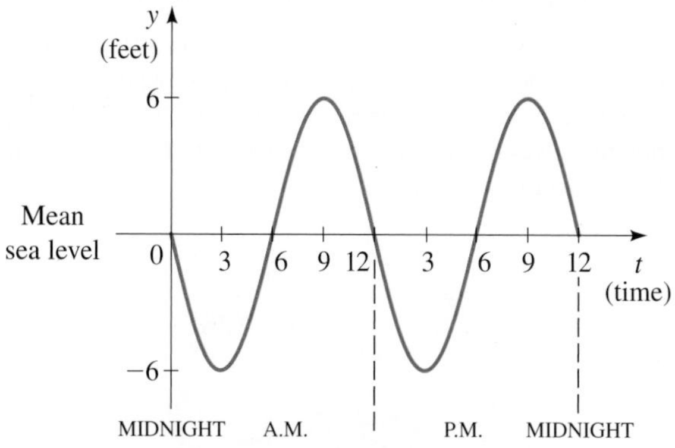

33. Tides The Bay of Fundy in Nova Scotia has the highest tides in the world. In one 12-hour period the water starts at mean sea level, rises to 21 ft above, drops to 21 ft below, then returns to mean sea level. Assuming that the motion of the tides is simple harmonic, find an equation that describes the height of the tide in the Bay of Fundy above mean sea level. Sketch a graph that shows the level of the tides over a 12-hour period.

34. Spring–Mass System A mass suspended from a spring is pulled down a distance of 2 ft from its rest position, as shown in the figure. The mass is released at time $t = 0$ and allowed to oscillate. If the mass returns to this position after 1 s, find an equation that describes its motion.

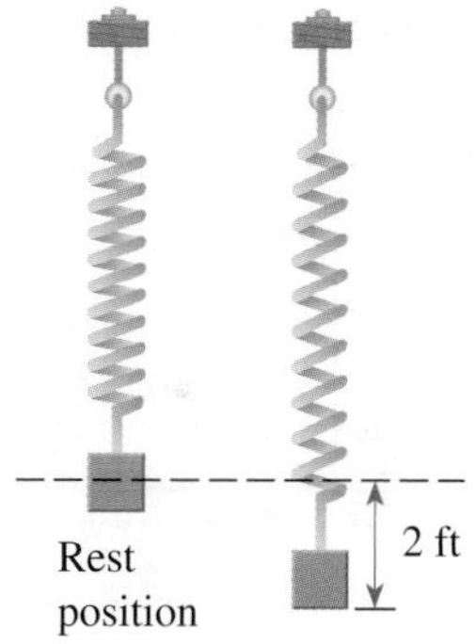

35. Spring–Mass System A mass is suspended on a spring. The spring is compressed so that the mass is located 5 cm above its rest position. The mass is released at time $t = 0$ and allowed to oscillate. It is observed that the mass reaches its lowest point $\frac{1}{2}$ s after it is released. Find an equation that describes the motion of the mass.

36. Spring–Mass System The frequency of oscillation of an object suspended on a spring depends on the stiffness k of the spring (called the *spring constant*) and the mass m of the object. If the spring is compressed a distance a and then allowed to oscillate, its displacement is given by

$$f(t) = a \cos \sqrt{k/m}\, t$$

(a) A 10-g mass is suspended from a spring with stiffness $k = 3$. If the spring is compressed a distance 5 cm and then released, find the equation that describes the oscillation of the spring.
(b) Find a general formula for the frequency (in terms of k and m).

(c) How is the frequency affected if the mass is increased? Is the oscillation faster or slower?

(d) How is the frequency affected if a stiffer spring is used (larger k)? Is the oscillation faster or slower?

37. Ferris Wheel A ferris wheel has a radius of 10 m, and the bottom of the wheel passes 1 m above the ground. If the ferris wheel makes one complete revolution every 20 s, find an equation that gives the height above the ground of a person on the ferris wheel as a function of time.

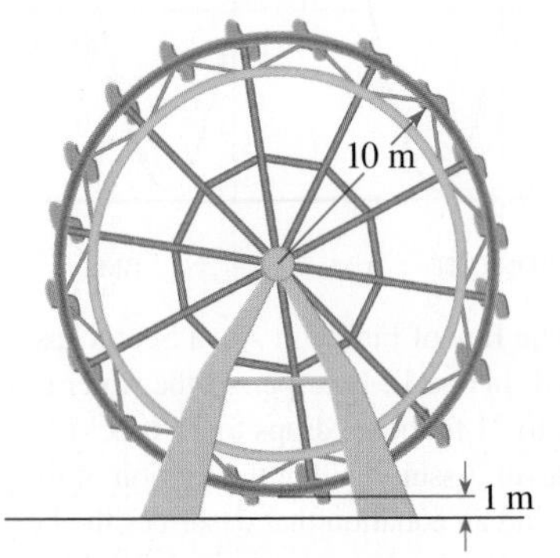

38. Clock Pendulum The pendulum in a grandfather clock makes one complete swing every 2 s. The maximum angle that the pendulum makes with respect to its rest position is 10°. We know from physical principles that the angle θ between the pendulum and its rest position changes in simple harmonic fashion. Find an equation that describes the size of the angle θ as a function of time. (Take $t = 0$ to be a time when the pendulum is vertical.)

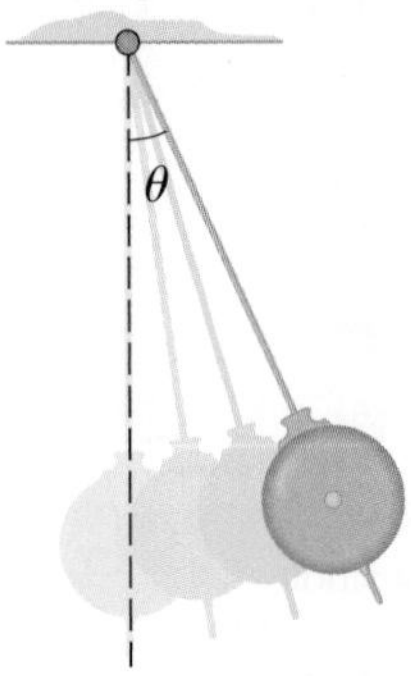

39. Variable Stars The variable star Zeta Gemini has a period of 10 days. The average brightness of the star is 3.8 magnitudes, and the maximum variation from the average is 0.2 magnitude. Assuming that the variation in brightness is simple harmonic, find an equation that gives the brightness of the star as a function of time.

40. Variable Stars Astronomers believe that the radius of a variable star increases and decreases with the brightness of the star. The variable star Delta Cephei (Example 4) has an average radius of 20 million miles and changes by a maximum of 1.5 million miles from this average during a single pulsation. Find an equation that describes the radius of this star as a function of time.

41. Biological Clocks *Circadian rhythms* are biological processes that oscillate with a period of approximately 24 hours. That is, a circadian rhythm is an internal daily biological clock. Blood pressure appears to follow such a rhythm. For a certain individual the average resting blood pressure varies from a maximum of 100 mmHg at 2:00 P.M. to a minimum of 80 mmHg at 2:00 A.M. Find a sine function of the form

$$f(t) = a \sin(\omega(t - c)) + b$$

that models the blood pressure at time t, measured in hours from midnight.

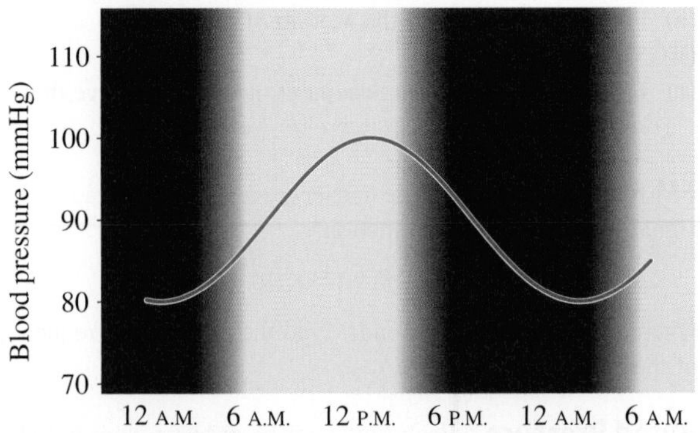

42. Electric Generator The armature in an electric generator is rotating at the rate of 100 revolutions per second (rps). If the maximum voltage produced is 310 V, find an equation that describes this variation in voltage. What is the rms voltage? (See Example 6 and the margin note adjacent to it.)

43. Electric Generator The graph shows an oscilloscope reading of the variation in voltage of an AC current produced by a simple generator.

(a) Find the maximum voltage produced.

(b) Find the frequency (cycles per second) of the generator.

(c) How many revolutions per second does the armature in the generator make?

(d) Find a formula that describes the variation in voltage as a function of time.

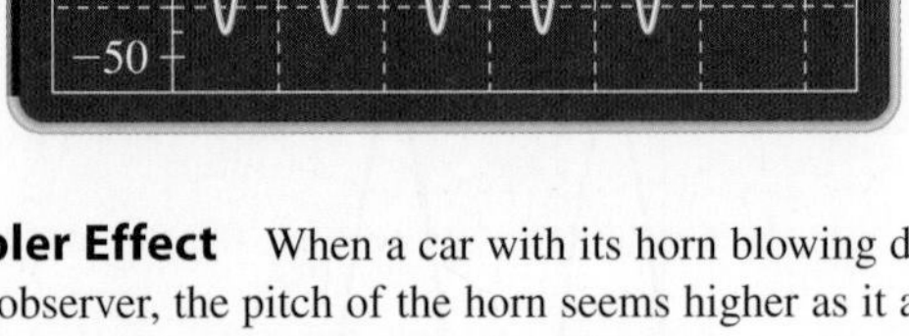

44. Doppler Effect When a car with its horn blowing drives by an observer, the pitch of the horn seems higher as it approaches and lower as it recedes (see the figure on the next page). This phenomenon is called the **Doppler effect**. If the sound source is moving at speed v relative to the observer and if the speed of sound is v_0, then the perceived frequency f is related to the actual frequency f_0 as follows:

$$f = f_0\left(\frac{v_0}{v_0 \pm v}\right)$$

We choose the minus sign if the source is moving toward the observer and the plus sign if it is moving away.

Suppose that a car drives at 110 ft/s past a woman standing on the shoulder of a highway, blowing its horn, which has a frequency of 500 Hz. Assume that the speed of sound is 1130 ft/s. (This is the speed in dry air at 70°F.)

(a) What are the frequencies of the sounds that the woman hears as the car approaches her and as it moves away from her?

(b) Let A be the amplitude of the sound. Find functions of the form

$$y = A \sin \omega t$$

that model the perceived sound as the car approaches the woman and as it recedes.

45. Motion of a Building A strong gust of wind strikes a tall building, causing it to sway back and forth in damped harmonic motion. The frequency of the oscillation is 0.5 cycle per second, and the damping constant is $c = 0.9$. Find an equation that describes the motion of the building. (Assume that $k = 1$, and take $t = 0$ to be the instant when the gust of wind strikes the building.)

46. Shock Absorber When a car hits a certain bump on the road, a shock absorber on the car is compressed a distance of 6 in., then released (see the figure). The shock absorber vibrates in damped harmonic motion with a frequency of 2 cycles per second. The damping constant for this particular shock absorber is 2.8.

(a) Find an equation that describes the displacement of the shock absorber from its rest position as a function of time. Take $t = 0$ to be the instant that the shock absorber is released.

(b) How long does it take for the amplitude of the vibration to decrease to 0.5 in?

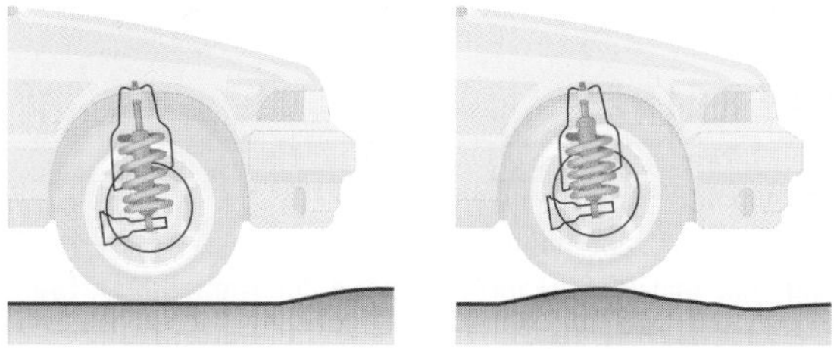

47. Tuning Fork A tuning fork is struck and oscillates in damped harmonic motion. The amplitude of the motion is measured, and 3 s later it is found that the amplitude has dropped to $\frac{1}{4}$ of this value. Find the damping constant c for this tuning fork.

48. Guitar String A guitar string is pulled at point P a distance of 3 cm above its rest position. It is then released and vibrates in damped harmonic motion with a frequency of 165 cycles per second. After 2 s, it is observed that the amplitude of the vibration at point P is 0.6 cm.

(a) Find the damping constant c.

(b) Find an equation that describes the position of point P above its rest position as a function of time. Take $t = 0$ to be the instant that the string is released.

CHAPTER 5 | REVIEW

CONCEPT CHECK

1. (a) What is the unit circle?

(b) Use a diagram to explain what is meant by the terminal point determined by a real number t.

(c) What is the reference number $\bar{t}$ associated with t?

(d) If t is a real number and $P(x, y)$ is the terminal point determined by t, write equations that define $\sin t$, $\cos t$, $\tan t$, $\cot t$, $\sec t$, and $\csc t$.

(e) What are the domains of the six functions that you defined in part (d)?

(f) Which trigonometric functions are positive in Quadrants I, II, III, and IV?

2. (a) What is an even function?

(b) Which trigonometric functions are even?

(c) What is an odd function?

(d) Which trigonometric functions are odd?

3. (a) State the reciprocal identities.

(b) State the Pythagorean identities.

4. (a) What is a periodic function?

(b) What are the periods of the six trigonometric functions?

5. Graph the sine and cosine functions. How is the graph of cosine related to the graph of sine?

6. Write expressions for the amplitude, period, and phase shift of the sine curve $y = a \sin k(x - b)$ and the cosine curve $y = a \cos k(x - b)$.

7. (a) Graph the tangent and cotangent functions.

(b) State the periods of the tangent curve $y = a \tan kx$ and the cotangent curve $y = a \cot kx$.

8. (a) Graph the secant and cosecant functions.

(b) State the periods of the secant curve $y = a \sec kx$ and the cosecant curve $y = a \csc kx$.

9. (a) Define the inverse sine function $\sin^{-1}$. What are its domain and range?

(b) For what values of x is the equation $\sin(\sin^{-1} x) = x$ true?

(c) For what values of x is the equation $\sin^{-1}(\sin x) = x$ true?

10. **(a)** Define the inverse cosine function $\cos^{-1}$. What are its domain and range?
(b) For what values of x is the equation $\cos(\cos^{-1} x) = x$ true?
(c) For what values of x is the equation $\cos^{-1}(\cos x) = x$ true?

11. **(a)** Define the inverse tangent function $\tan^{-1}$. What are its domain and range?
(b) For what values of x is the equation $\tan(\tan^{-1} x) = x$ true?
(c) For what values of x is the equation $\tan^{-1}(\tan x) = x$ true?

12. **(a)** What is simple harmonic motion?
(b) What is damped harmonic motion?
(c) Give three real-life examples of simple harmonic motion and of damped harmonic motion.

■ EXERCISES

1–2 ■ A point $P(x, y)$ is given.
(a) Show that P is on the unit circle.
(b) Suppose that P is the terminal point determined by t. Find $\sin t$, $\cos t$, and $\tan t$.

1. $P\left(-\dfrac{\sqrt{3}}{2}, \dfrac{1}{2}\right)$ **2.** $P\left(\dfrac{3}{5}, -\dfrac{4}{5}\right)$

3–6 ■ A real number t is given.
(a) Find the reference number for t.
(b) Find the terminal point $P(x, y)$ on the unit circle determined by t.
(c) Find the six trigonometric functions of t.

3. $t = \dfrac{2\pi}{3}$ **4.** $t = \dfrac{5\pi}{3}$

5. $t = -\dfrac{11\pi}{4}$ **6.** $t = -\dfrac{7\pi}{6}$

7–16 ■ Find the value of the trigonometric function. If possible, give the exact value; otherwise, use a calculator to find an approximate value correct to five decimal places.

7. **(a)** $\sin \dfrac{3\pi}{4}$ **(b)** $\cos \dfrac{3\pi}{4}$

8. **(a)** $\tan \dfrac{\pi}{3}$ **(b)** $\tan\left(-\dfrac{\pi}{3}\right)$

9. **(a)** $\sin 1.1$ **(b)** $\cos 1.1$

10. **(a)** $\cos \dfrac{\pi}{5}$ **(b)** $\cos\left(-\dfrac{\pi}{5}\right)$

11. **(a)** $\cos \dfrac{9\pi}{2}$ **(b)** $\sec \dfrac{9\pi}{2}$

12. **(a)** $\sin \dfrac{\pi}{7}$ **(b)** $\csc \dfrac{\pi}{7}$

13. **(a)** $\tan \dfrac{5\pi}{2}$ **(b)** $\cot \dfrac{5\pi}{2}$

14. **(a)** $\sin 2\pi$ **(b)** $\csc 2\pi$

15. **(a)** $\tan \dfrac{5\pi}{6}$ **(b)** $\cot \dfrac{5\pi}{6}$

16. **(a)** $\cos \dfrac{\pi}{3}$ **(b)** $\sin \dfrac{\pi}{6}$

17–20 ■ Use the fundamental identities to write the first expression in terms of the second.

17. $\dfrac{\tan t}{\cos t}$, $\sin t$ **18.** $\tan^2 t \sec t$, $\cos t$

19. $\tan t$, $\sin t$; t in Quadrant IV

20. $\sec t$, $\sin t$; t in Quadrant II

21–24 ■ Find the values of the remaining trigonometric functions at t from the given information.

21. $\sin t = \frac{5}{13}$, $\cos t = -\frac{12}{13}$

22. $\sin t = -\frac{1}{2}$, $\cos t > 0$

23. $\cot t = -\frac{1}{2}$, $\csc t = \sqrt{5}/2$

24. $\cos t = -\frac{3}{5}$, $\tan t < 0$

25. If $\tan t = \frac{1}{4}$ and the terminal point for t is in Quadrant III, find $\sec t + \cot t$.

26. If $\sin t = -\frac{8}{17}$ and the terminal point for t is in Quadrant IV, find $\csc t + \sec t$.

27. If $\cos t = \frac{3}{5}$ and the terminal point for t is in Quadrant I, find $\tan t + \sec t$.

28. If $\sec t = -5$ and the terminal point for t is in Quadrant II, find $\sin^2 t + \cos^2 t$.

29–36 ■ A trigonometric function is given.
(a) Find the amplitude, period, and phase shift of the function.
(b) Sketch the graph.

29. $y = 10 \cos \frac{1}{2}x$ **30.** $y = 4 \sin 2\pi x$

31. $y = -\sin \frac{1}{2}x$ **32.** $y = 2 \sin\left(x - \dfrac{\pi}{4}\right)$

33. $y = 3 \sin(2x - 2)$ **34.** $y = \cos 2\left(x - \dfrac{\pi}{2}\right)$

35. $y = -\cos\left(\dfrac{\pi}{2}x + \dfrac{\pi}{6}\right)$ **36.** $y = 10 \sin\left(2x - \dfrac{\pi}{2}\right)$

37–40 ■ The graph of one period of a function of the form $y = a \sin k(x - b)$ or $y = a \cos k(x - b)$ is shown. Determine the function.

37.

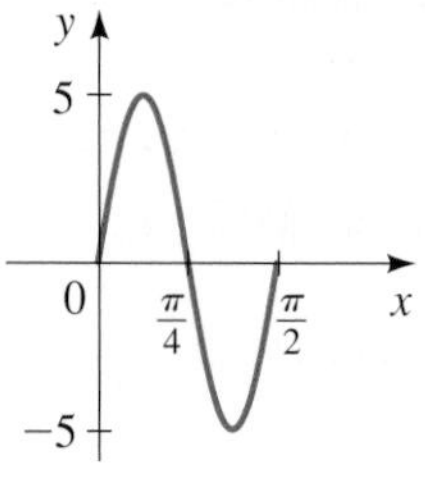

38.

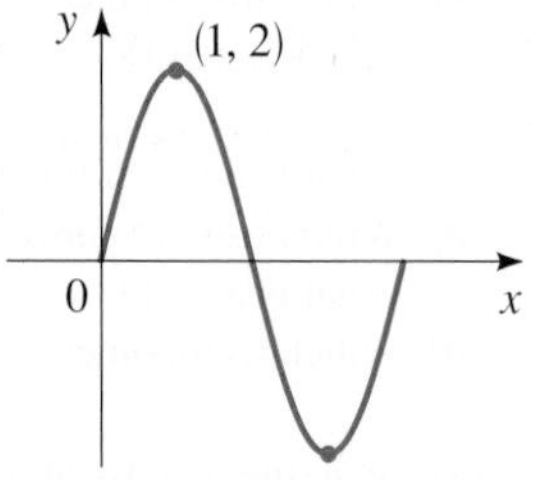

39.

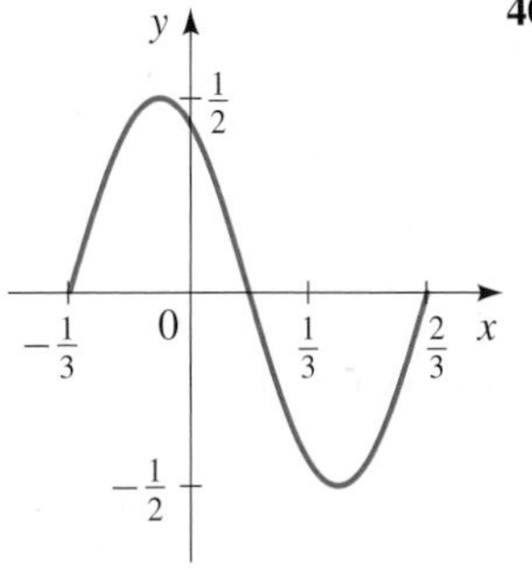

40.

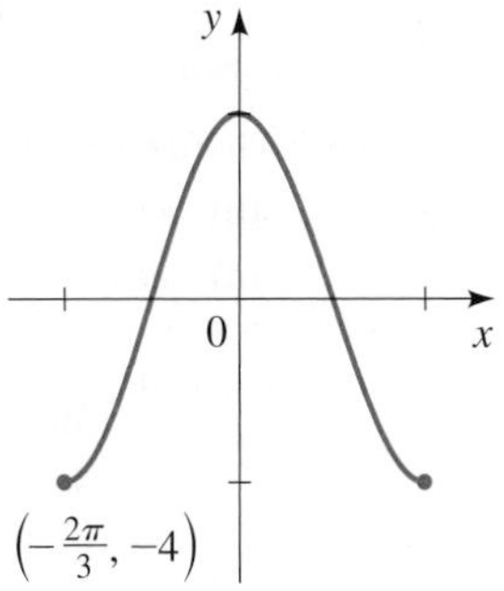

41–48 ■ Find the period, and sketch the graph.

41. $y = 3 \tan x$

42. $y = \tan \pi x$

43. $y = 2 \cot\left(x - \frac{\pi}{2}\right)$

44. $y = \sec\left(\frac{1}{2}x - \frac{\pi}{2}\right)$

45. $y = 4 \csc(2x + \pi)$

46. $y = \tan\left(x + \frac{\pi}{6}\right)$

47. $y = \tan\left(\frac{1}{2}x - \frac{\pi}{8}\right)$

48. $y = -4 \sec 4\pi x$

49–52 ■ Find the exact value of each expression, if it is defined.

49. $\sin^{-1} 1$

50. $\cos^{-1}\left(-\frac{1}{2}\right)$

51. $\sin^{-1}\left(\sin \frac{13\pi}{6}\right)$

52. $\tan\left(\cos^{-1}\left(\frac{1}{2}\right)\right)$

53–58 ■ A function is given.
(a) Use a graphing device to graph the function.
(b) Determine from the graph whether the function is periodic and, if so, determine the period.
(c) Determine from the graph whether the function is odd, even, or neither.

53. $y = |\cos x|$

54. $y = \sin(\cos x)$

55. $y = \cos(2^{0.1x})$

56. $y = 1 + 2^{\cos x}$

57. $y = |x| \cos 3x$

58. $y = \sqrt{x} \sin 3x \quad (x > 0)$

59–62 ■ Graph the three functions on a common screen. How are the graphs related?

59. $y = x, \quad y = -x, \quad y = x \sin x$

60. $y = 2^{-x}, \quad y = -2^{-x}, \quad y = 2^{-x} \cos 4\pi x$

61. $y = x, \quad y = \sin 4x, \quad y = x + \sin 4x$

62. $y = \sin^2 x, \quad y = \cos^2 x, \quad y = \sin^2 x + \cos^2 x$

63–64 ■ Find the maximum and minimum values of the function.

63. $y = \cos x + \sin 2x$

64. $y = \cos x + \sin^2 x$

65. Find the solutions of $\sin x = 0.3$ in the interval $[0, 2\pi]$.

66. Find the solutions of $\cos 3x = x$ in the interval $[0, \pi]$.

67. Let $f(x) = \dfrac{\sin^2 x}{x}$.
(a) Is the function f even, odd, or neither?
(b) Find the x-intercepts of the graph of f.
(c) Graph f in an appropriate viewing rectangle.
(d) Describe the behavior of the function as x becomes large.
(e) Notice that $f(x)$ is not defined when $x = 0$. What happens as x approaches 0?

68. Let $y_1 = \cos(\sin x)$ and $y_2 = \sin(\cos x)$.
(a) Graph y_1 and y_2 in the same viewing rectangle.
(b) Determine the period of each of these functions from its graph.
(c) Find an inequality between $\sin(\cos x)$ and $\cos(\sin x)$ that is valid for all x.

69. A point P moving in simple harmonic motion completes 8 cycles every second. If the amplitude of the motion is 50 cm, find an equation that describes the motion of P as a function of time. Assume that the point P is at its maximum displacement when $t = 0$.

70. A mass suspended from a spring oscillates in simple harmonic motion at a frequency of 4 cycles per second. The distance from the highest to the lowest point of the oscillation is 100 cm. Find an equation that describes the distance of the mass from its rest position as a function of time. Assume that the mass is at its lowest point when $t = 0$.

71. The graph shows the variation of the water level relative to mean sea level in the Long Beach harbor for a particular 24-hour period. Assuming that this variation is simple harmonic, find an equation of the form $y = a \cos \omega t$ that describes the variation in water level as a function of the number of hours after midnight.

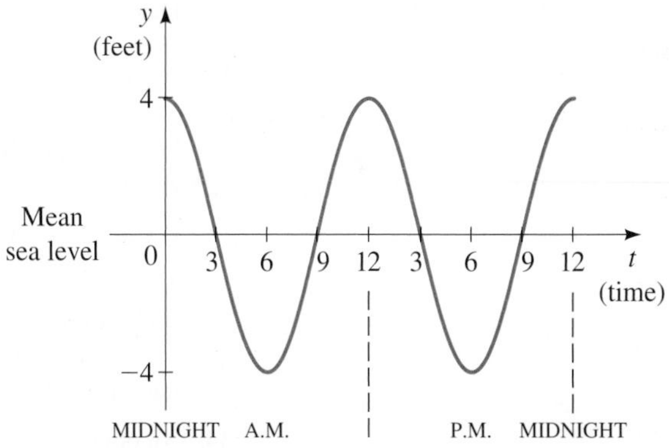

72. The top floor of a building undergoes damped harmonic motion after a sudden brief earthquake. At time $t = 0$ the displacement is at a maximum, 16 cm from the normal position. The damping constant is $c = 0.72$ and the building vibrates at 1.4 cycles per second.
(a) Find a function of the form $y = ke^{-ct} \cos \omega t$ to model the motion.
(b) Graph the function you found in part (a).
(c) What is the displacement at time $t = 10$ s?

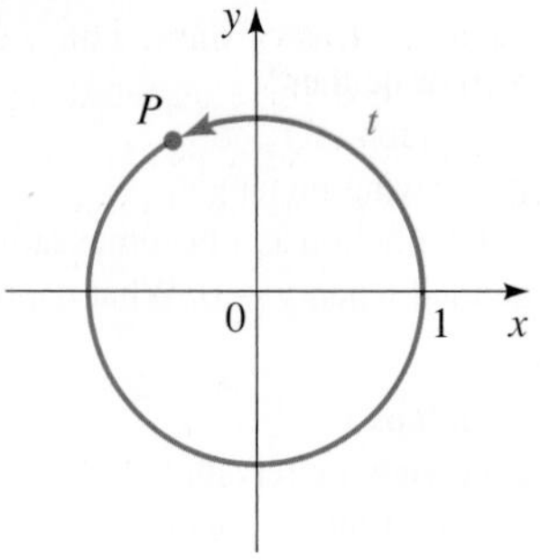

1. The point $P(x, y)$ is on the unit circle in Quadrant IV. If $x = \sqrt{11}/6$, find y.

2. The point P in the figure at the left has y-coordinate $\frac{4}{5}$. Find:
   (a) $\sin t$ **(b)** $\cos t$
   (c) $\tan t$ **(d)** $\sec t$

3. Find the exact value.
   (a) $\sin \dfrac{7\pi}{6}$ **(b)** $\cos \dfrac{13\pi}{4}$
   (c) $\tan\left(-\dfrac{5\pi}{3}\right)$ **(d)** $\csc \dfrac{3\pi}{2}$

4. Express $\tan t$ in terms of $\sin t$, if the terminal point determined by t is in Quadrant II.

5. If $\cos t = -\frac{8}{17}$ and if the terminal point determined by t is in Quadrant III, find $\tan t \cot t + \csc t$.

6–7 ■ A trigonometric function is given.
(a) Find the amplitude, period, and phase shift of the function.
(b) Sketch the graph.

6. $y = -5 \cos 4x$

7. $y = 2 \sin\left(\dfrac{1}{2}x - \dfrac{\pi}{6}\right)$

8–9 ■ Find the period, and graph the function.

8. $y = -\csc 2x$

9. $y = \tan\left(2x - \dfrac{\pi}{2}\right)$

10. Find the exact value of each expression, if it is defined.
   (a) $\tan^{-1} 1$ **(b)** $\cos^{-1}\left(-\dfrac{\sqrt{3}}{2}\right)$
   (c) $\tan^{-1}(\tan 3\pi)$ **(d)** $\cos(\tan^{-1}(-\sqrt{3}))$

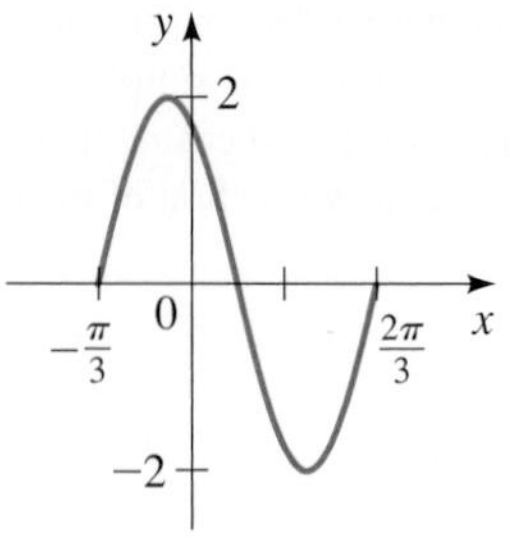

11. The graph shown at left is one period of a function of the form $y = a \sin k(x - b)$. Determine the function.

12. Let $f(x) = \dfrac{\cos x}{1 + x^2}$.
   (a) Use a graphing device to graph f in an appropriate viewing rectangle.
   (b) Determine from the graph if f is even, odd, or neither.
   (c) Find the minimum and maximum values of f.

13. A mass suspended from a spring oscillates in simple harmonic motion. The mass completes 2 cycles every second, and the distance between the highest point and the lowest point of the oscillation is 10 cm. Find an equation of the form $y = a \sin \omega t$ that gives the distance of the mass from its rest position as a function of time.

14. An object is moving up and down in damped harmonic motion. Its displacement at time $t = 0$ is 16 in; this is its maximum displacement. The damping constant is $c = 0.1$, and the frequency is 12 Hz.
   (a) Find a function that models this motion.
   (b) Graph the function.

Fitting Sinusoidal Curves to Data

In previous *Focus on Modeling* sections, we learned how to fit linear, polynomial, exponential, and power models to data. Figure 1 shows some scatter plots of data. The scatter plots can help guide us in choosing an appropriate model. (Try to determine what type of function would best model the data in each graph.) If the scatter plot indicates simple harmonic motion, then we might try to model the data with a sine or cosine function. The next example illustrates this process.

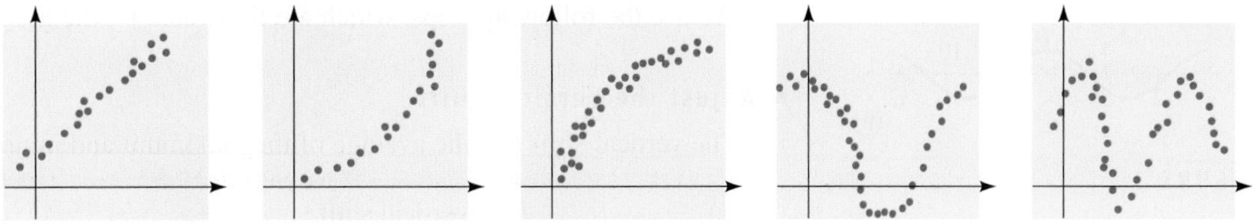

FIGURE 1

EXAMPLE 1 | Modeling the Height of a Tide

The water depth in a narrow channel varies with the tides. Table 1 shows the water depth over a 12-hour period.

(a) Make a scatter plot of the water depth data.

(b) Find a function that models the water depth with respect to time.

(c) If a boat needs at least 11 ft of water to cross the channel, during which times can it safely do so?

TABLE 1

Time	Depth (ft)
12:00 A.M.	9.8
1:00 A.M.	11.4
2:00 A.M.	11.6
3:00 A.M.	11.2
4:00 A.M.	9.6
5:00 A.M.	8.5
6:00 A.M.	6.5
7:00 A.M.	5.7
8:00 A.M.	5.4
9:00 A.M.	6.0
10:00 A.M.	7.0
11:00 A.M.	8.6
12:00 P.M.	10.0

SOLUTION

(a) A scatter plot of the data is shown in Figure 2.

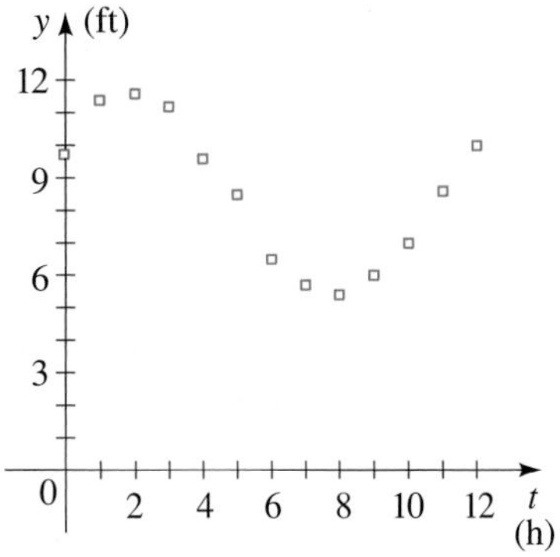

FIGURE 2

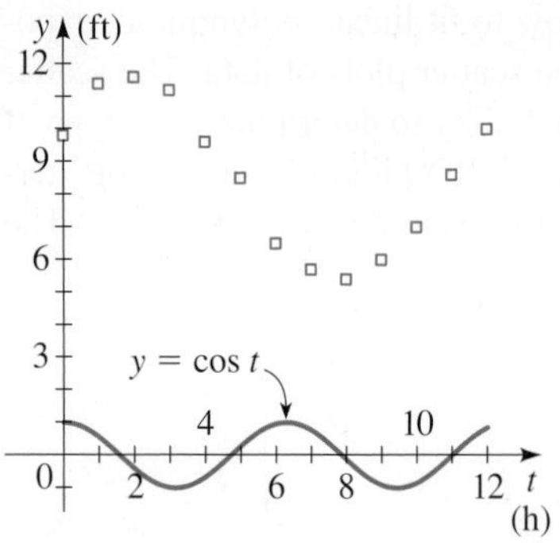

FIGURE 3

(b) The data appear to lie on a cosine (or sine) curve. But if we graph $y = \cos t$ on the same graph as the scatter plot, the result in Figure 3 is not even close to the data. To fit the data, we need to adjust the vertical shift, amplitude, period, and phase shift of the cosine curve. In other words, we need to find a function of the form

$$y = a\cos(\omega(t - c)) + b$$

We use the following steps, which are illustrated by the graphs in the margin.

▶ Adjust the Vertical Shift

The vertical shift b is the average of the maximum and minimum values:

$$\begin{aligned} b &= \text{vertical shift} \\ &= \frac{1}{2}\cdot(\text{maximum value} + \text{minimum value}) \\ &= \frac{1}{2}(11.6 + 5.4) = 8.5 \end{aligned}$$

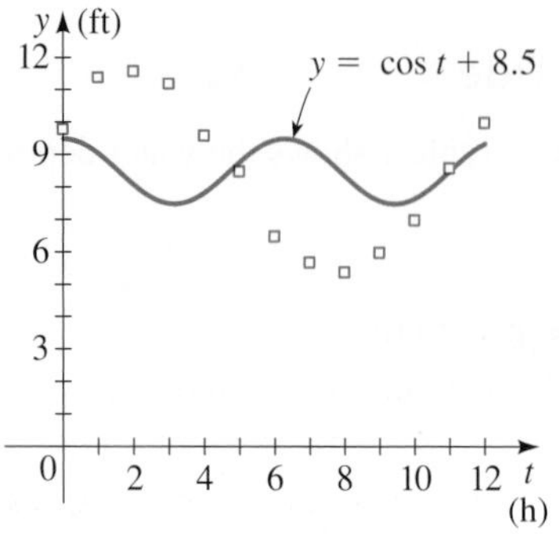

▶ Adjust the Amplitude

The amplitude a is half of the difference between the maximum and minimum values:

$$\begin{aligned} a &= \text{amplitude} \\ &= \frac{1}{2}\cdot(\text{maximum value} - \text{minimum value}) \\ &= \frac{1}{2}(11.6 - 5.4) = 3.1 \end{aligned}$$

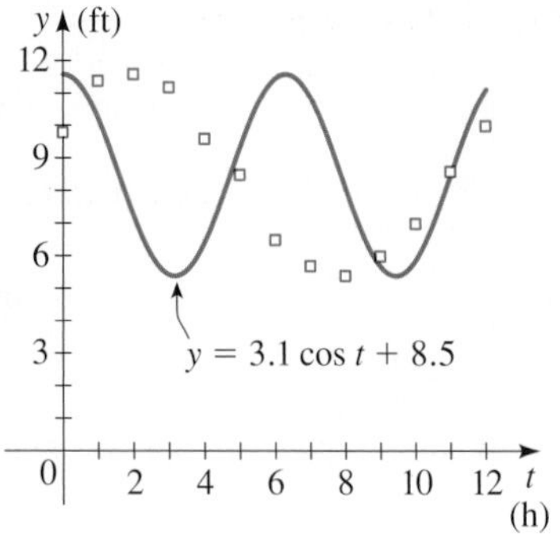

▶ Adjust the Period

The time between consecutive maximum and minimum values is half of one period. Thus

$$\begin{aligned} \frac{2\pi}{\omega} &= \text{period} \\ &= 2\cdot(\text{time of maximum value} - \text{time of minimum value}) \\ &= 2(8 - 2) = 12 \end{aligned}$$

Thus, $\omega = 2\pi/12 = 0.52$.

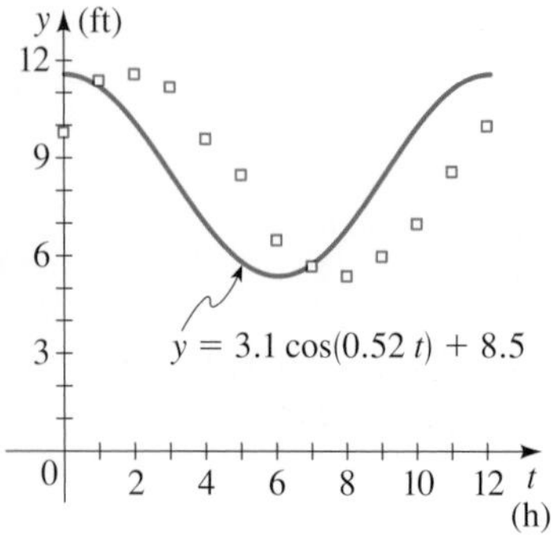

▶ Adjust the Horizontal Shift

Since the maximum value of the data occurs at approximately $t = 2.0$, it represents a cosine curve shifted 2 h to the right. So

$$\begin{aligned} c &= \text{phase shift} \\ &= \text{time of maximum value} \\ &= 2.0 \end{aligned}$$

▶ The Model

We have shown that a function that models the tides over the given time period is given by

$$y = 3.1\cos(0.52(t - 2.0)) + 8.5$$

A graph of the function and the scatter plot are shown in Figure 4. It appears that the model we found is a good approximation to the data.

(c) We need to solve the inequality $y \geq 11$. We solve this inequality graphically by graphing $y = 3.1 \cos 0.52(t - 2.0) + 8.5$ and $y = 11$ on the same graph. From the graph in Figure 5 we see the water depth is higher than 11 ft between $t \approx 0.8$ and $t \approx 3.2$. This corresponds to the times 12:48 A.M. to 3:12 A.M.

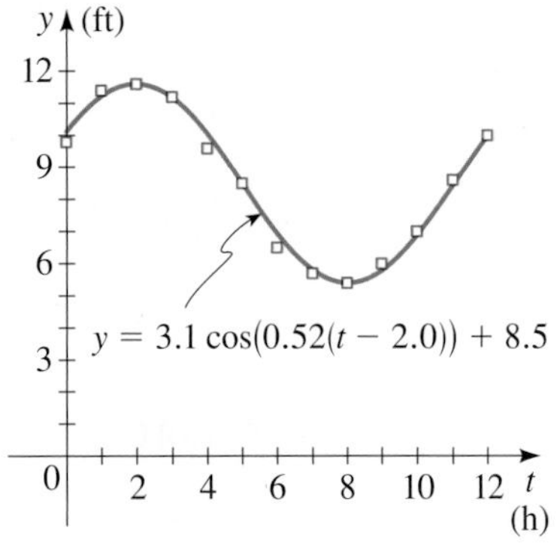

FIGURE 4

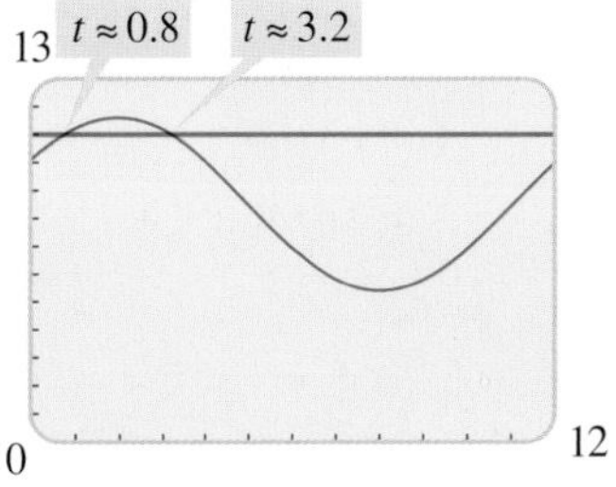

FIGURE 5

For the TI-83 and TI-86 the command `SinReg` (for sine regression) finds the sine curve that best fits the given data.

In Example 1 we used the scatter plot to guide us in finding a cosine curve that gives an approximate model of the data. Some graphing calculators are capable of finding a sine or cosine curve that best fits a given set of data points. The method these calculators use is similar to the method of finding a line of best fit, as explained on page 131.

EXAMPLE 2 | Fitting a Sine Curve to Data

(a) Use a graphing device to find the sine curve that best fits the depth of water data in Table 1 on page 427.

(b) Compare your result to the model found in Example 1.

SOLUTION

```
SinReg
 y=a*sin(bx+c)+d
 a=3.097877596
 b=.5268322697
 c=.5493035195
 d=8.424021899
```

Output of the `SinReg` function on the TI-83.

(a) Using the data in Table 1 and the `SinReg` command on the TI-83 calculator, we get a function of the form

$$y = a \sin(bt + c) + d$$

where

$$a = 3.1 \qquad b = 0.53$$

$$c = 0.55 \qquad d = 8.42$$

So the sine function that best fits the data is

$$y = 3.1 \sin(0.53t + 0.55) + 8.42$$

(b) To compare this with the function in Example 1, we change the sine function to a cosine function by using the reduction formula $\sin u = \cos(u - \pi/2)$.

$$y = 3.1 \sin(0.53t + 0.55) + 8.42$$

$$= 3.1 \cos\left(0.53t + 0.55 - \frac{\pi}{2}\right) + 8.42 \qquad \text{Reduction formula}$$

$$= 3.1 \cos(0.53t - 1.02) + 8.42$$

$$= 3.1 \cos(0.53(t - 1.92)) + 8.42 \qquad \text{Factor 0.53}$$

Comparing this with the function we obtained in Example 1, we see that there are small differences in the coefficients. In Figure 6 we graph a scatter plot of the data together with the sine function of best fit.

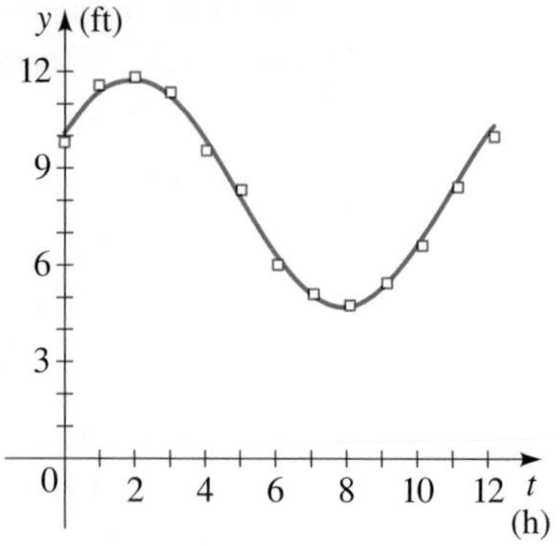

FIGURE 6

In Example 1 we estimated the values of the amplitude, period, and shifts from the data. In Example 2 the calculator computed the sine curve that best fits the data (that is, the curve that deviates least from the data as explained on page 131). The different ways of obtaining the model account for the differences in the functions.

PROBLEMS

1–4 ■ Modeling Periodic Data A set of data is given.

(a) Make a scatter plot of the data.

(b) Find a cosine function of the form $y = a\cos(\omega(t - c)) + b$ that models the data, as in Example 1.

(c) Graph the function you found in part (b) together with the scatter plot. How well does the curve fit the data?

(d) Use a graphing calculator to find the sine function that best fits the data, as in Example 2.

(e) Compare the functions you found in parts (b) and (d). [Use the reduction formula $\sin u = \cos(u - \pi/2)$.]

1.

t	y
0	2.1
2	1.1
4	−0.8
6	−2.1
8	−1.3
10	0.6
12	1.9
14	1.5

2.

t	y
0	190
25	175
50	155
75	125
100	110
125	95
150	105
175	120
200	140
225	165
250	185
275	200
300	195
325	185
350	165

3.

t	y
0.1	21.1
0.2	23.6
0.3	24.5
0.4	21.7
0.5	17.5
0.6	12.0
0.7	5.6
0.8	2.2
0.9	1.0
1.0	3.5
1.1	7.6
1.2	13.2
1.3	18.4
1.4	23.0
1.5	25.1

4.

t	y
0.0	0.56
0.5	0.45
1.0	0.29
1.5	0.13
2.0	0.05
2.5	−0.10
3.0	0.02
3.5	0.12
4.0	0.26
4.5	0.43
5.0	0.54
5.5	0.63
6.0	0.59

5. Annual Temperature Change The table gives the average monthly temperature in Montgomery County, Maryland.

(a) Make a scatter plot of the data.

(b) Find a cosine curve that models the data (as in Example 1).

(c) Graph the function you found in part (b) together with the scatter plot.

(d) Use a graphing calculator to find the sine curve that best fits the data (as in Example 2).

Month	Average temperature (°F)	Month	Average temperature (°F)
January	40.0	July	85.8
February	43.1	August	83.9
March	54.6	September	76.9
April	64.2	October	66.8
May	73.8	November	55.5
June	81.8	December	44.5

6. Circadian Rhythms Circadian rhythm (from the Latin *circa*—about, and *diem*—day) is the daily biological pattern by which body temperature, blood pressure, and other physiological variables change. The data in the table below show typical changes in human body temperature over a 24-hour period ($t = 0$ corresponds to midnight).

(a) Make a scatter plot of the data.

(b) Find a cosine curve that models the data (as in Example 1).

(c) Graph the function you found in part (b) together with the scatter plot.

(d) Use a graphing calculator to find the sine curve that best fits the data (as in Example 2).

Time	Body temperature (°C)	Time	Body temperature (°C)
0	36.8	14	37.3
2	36.7	16	37.4
4	36.6	18	37.3
6	36.7	20	37.2
8	36.8	22	37.0
10	37.0	24	36.8
12	37.2		

7. Predator Population When two species interact in a predator/prey relationship, the populations of both species tend to vary in a sinusoidal fashion. (See the Discovery Project *Predator/Prey Models* referenced on page 398). In a certain midwestern county, the main food source for barn owls consists of field mice and other small mammals. The table gives the population of barn owls in this county every July 1 over a 12-year period.

(a) Make a scatter plot of the data.

(b) Find a sine curve that models the data (as in Example 1).

(c) Graph the function you found in part (b) together with the scatter plot.

(d) Use a graphing calculator to find the sine curve that best fits the data (as in Example 2). Compare to your answer from part (b).

Year	Owl population
0	50
1	62
2	73
3	80
4	71
5	60
6	51
7	43
8	29
9	20
10	28
11	41
12	49

8. Salmon Survival For reasons not yet fully understood, the number of fingerling salmon that survive the trip from their riverbed spawning grounds to the open ocean varies approximately sinusoidally from year to year. The table shows the number of salmon that hatch in a certain British Columbia creek and then make their way to the Strait of Georgia. The data is given in thousands of fingerlings, over a period of 16 years.

(a) Make a scatter plot of the data.

(b) Find a sine curve that models the data (as in Example 1).

(c) Graph the function you found in part (b) together with the scatter plot.

(d) Use a graphing calculator to find the sine curve that best fits the data (as in Example 2). Compare to your answer from part (b).

Year	Salmon (× 1000)	Year	Salmon (× 1000)
1985	43	1993	56
1986	36	1994	63
1987	27	1995	57
1988	23	1996	50
1989	26	1997	44
1990	33	1998	38
1991	43	1999	30
1992	50	2000	22

9. Sunspot Activity Sunspots are relatively "cool" regions on the sun that appear as dark spots when observed through special solar filters. The number of sunspots varies in an 11-year cycle. The table gives the average daily sunspot count for the years 1975–2004.

(a) Make a scatter plot of the data.

(b) Find a cosine curve that models the data (as in Example 1).

(c) Graph the function you found in part (b) together with the scatter plot.

(d) Use a graphing calculator to find the sine curve that best fits the data (as in Example 2). Compare to your answer in part (b).

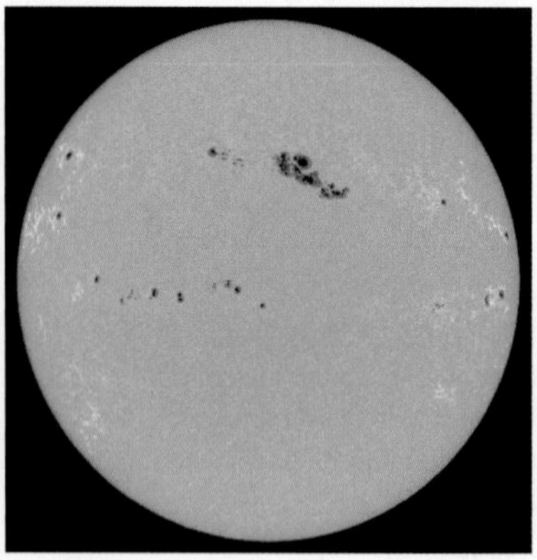

Solar & Heliospheric Observatory/Photo Researchers

Year	Sunspots	Year	Sunspots
1975	16	1990	143
1976	13	1991	146
1977	28	1992	94
1978	93	1993	55
1979	155	1994	30
1980	155	1995	18
1981	140	1996	9
1982	116	1997	21
1983	67	1998	64
1984	46	1999	93
1985	18	2000	119
1986	13	2001	111
1987	29	2002	104
1988	100	2003	64
1989	158	2004	40

© age fotostock/SuperStock

CHAPTER 6

TRIGONOMETRIC FUNCTIONS: RIGHT TRIANGLE APPROACH

Suppose we want to find the distance from the earth to the sun. Using a tape measure is obviously impractical, so we need something other than simple measurements to tackle this problem. Angles are easier to measure than distances. For example, we can find the angle formed by the sun, earth, and moon by simply pointing to the sun with one arm and to the moon with the other and estimating the angle between them. The key idea is to find relationships between angles and distances. So if we had a way of determining distances from angles, we would be able to find the distance to the sun without having to go there. The trigonometric functions provide us with just the tools we need.

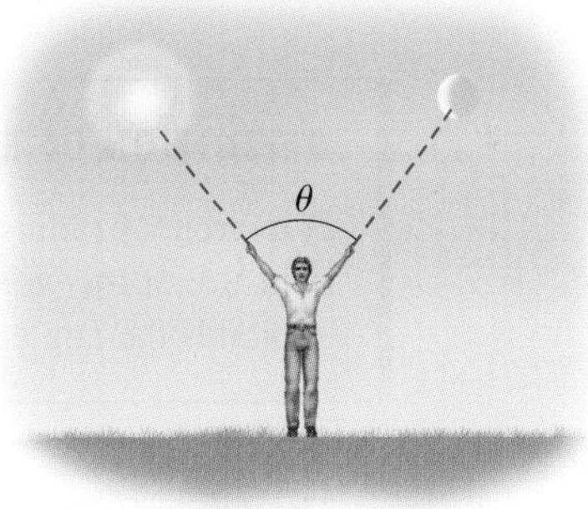

If θ is an angle in a right triangle, then the trigonometric ratio $\sin\theta$ is defined as the length of the side opposite θ divided by the length of the hypotenuse. This ratio is the same in *any* similar right triangle, including the huge triangle formed by the sun, earth, and moon! (See Section 6.2, Exercise 61.)

The trigonometric functions can be defined in two different but equivalent ways: as functions of real numbers (Chapter 5) or as functions of angles (Chapter 6). The two approaches are independent of each other, so either Chapter 5 or Chapter 6 may be studied first. We study both approaches because the different approaches are required for different applications.

6.1 Angle Measure

Angle Measure ▶ Angles in Standard Position ▶ Length of a Circular Arc ▶ Area of a Circular Sector ▶ Circular Motion

An **angle** AOB consists of two rays R_1 and R_2 with a common vertex O (see Figure 1). We often interpret an angle as a rotation of the ray R_1 onto R_2. In this case, R_1 is called the **initial side**, and R_2 is called the **terminal side** of the angle. If the rotation is counterclockwise, the angle is considered **positive**, and if the rotation is clockwise, the angle is considered **negative**.

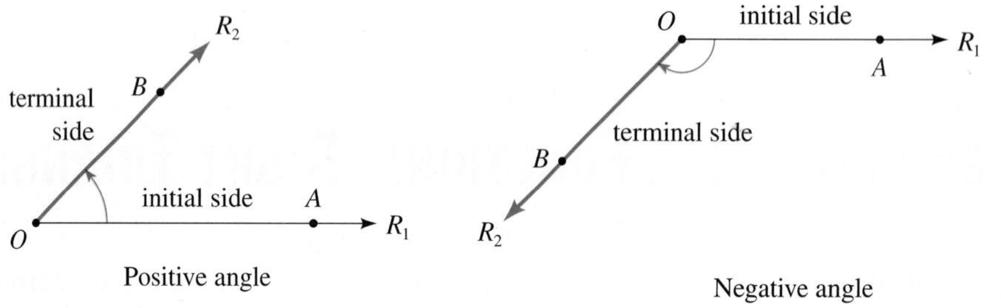

FIGURE 1

▼ Angle Measure

The **measure** of an angle is the amount of rotation about the vertex required to move R_1 onto R_2. Intuitively, this is how much the angle "opens." One unit of measurement for angles is the **degree**. An angle of measure 1 degree is formed by rotating the initial side $\frac{1}{360}$ of a complete revolution. In calculus and other branches of mathematics, a more natural method of measuring angles is used—*radian measure*. The amount an angle opens is measured along the arc of a circle of radius 1 with its center at the vertex of the angle.

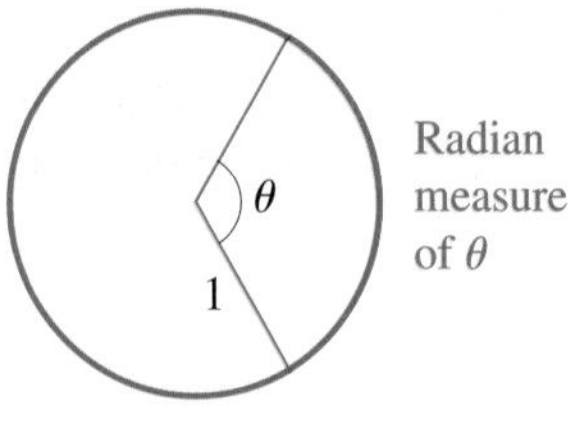

FIGURE 2

DEFINITION OF RADIAN MEASURE

If a circle of radius 1 is drawn with the vertex of an angle at its center, then the measure of this angle in **radians** (abbreviated **rad**) is the length of the arc that subtends the angle (see Figure 2).

The circumference of the circle of radius 1 is 2π and so a complete revolution has measure 2π rad, a straight angle has measure π rad, and a right angle has measure $\pi/2$ rad. An angle that is subtended by an arc of length 2 along the unit circle has radian measure 2 (see Figure 3).

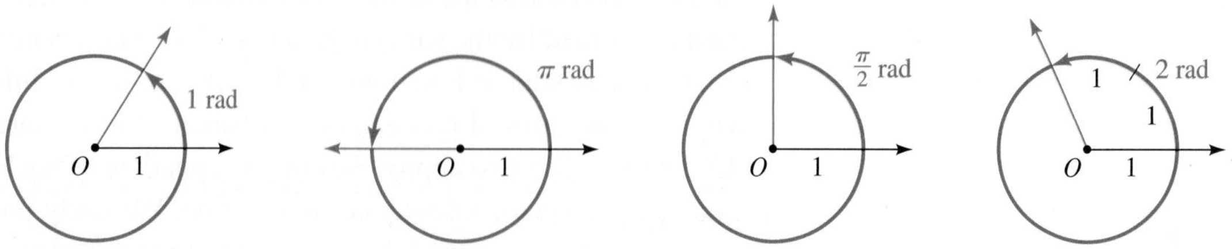

FIGURE 3 Radian measure

Since a complete revolution measured in degrees is 360° and measured in radians is 2π rad, we get the following simple relationship between these two methods of angle measurement.

> **RELATIONSHIP BETWEEN DEGREES AND RADIANS**
>
> $$180° = \pi \text{ rad} \qquad 1 \text{ rad} = \left(\frac{180}{\pi}\right)^{\circ} \qquad 1° = \frac{\pi}{180} \text{ rad}$$
>
> **1.** To convert degrees to radians, multiply by $\dfrac{\pi}{180}$.
>
> **2.** To convert radians to degrees, multiply by $\dfrac{180}{\pi}$.

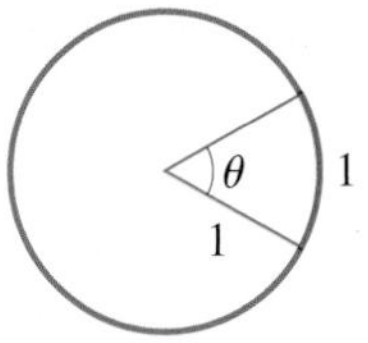

Measure of $\theta = 1$ rad
Measure of $\theta \approx 57.296°$

FIGURE 4

To get some idea of the size of a radian, notice that

$$1 \text{ rad} \approx 57.296° \qquad \text{and} \qquad 1° \approx 0.01745 \text{ rad}$$

An angle θ of measure 1 rad is shown in Figure 4.

EXAMPLE 1 | Converting Between Radians and Degrees

(a) Express 60° in radians. **(b)** Express $\dfrac{\pi}{6}$ rad in degrees.

SOLUTION The relationship between degrees and radians gives

(a) $60° = 60\left(\dfrac{\pi}{180}\right) \text{ rad} = \dfrac{\pi}{3} \text{ rad}$ **(b)** $\dfrac{\pi}{6} \text{ rad} = \left(\dfrac{\pi}{6}\right)\left(\dfrac{180}{\pi}\right) = 30°$

NOW TRY EXERCISES **3** AND **15** ■

A note on terminology: We often use a phrase such as "a 30° angle" to mean *an angle whose measure is* 30°. Also, for an angle θ, we write $\theta = 30°$ or $\theta = \pi/6$ to mean *the measure of* θ *is* 30° *or* $\pi/6$ *rad.* When no unit is given, the angle is assumed to be measured in radians.

▼ Angles in Standard Position

An angle is in **standard position** if it is drawn in the xy-plane with its vertex at the origin and its initial side on the positive x-axis. Figure 5 gives examples of angles in standard position.

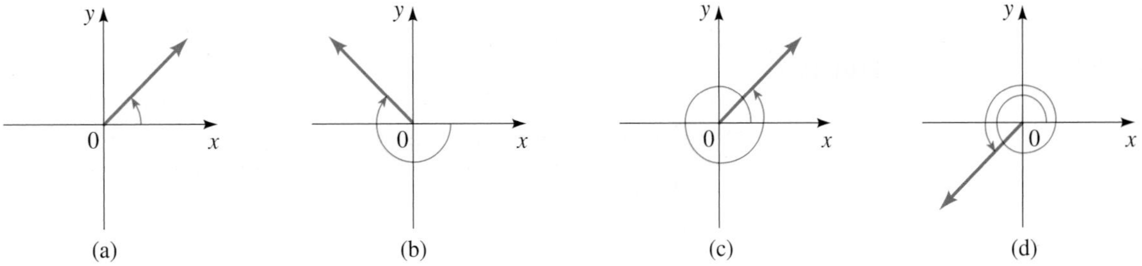

FIGURE 5 Angles in standard position

Two angles in standard position are **coterminal** if their sides coincide. In Figure 5 the angles in (a) and (c) are coterminal.

EXAMPLE 2 | Coterminal Angles

(a) Find angles that are coterminal with the angle $\theta = 30°$ in standard position.

(b) Find angles that are coterminal with the angle $\theta = \dfrac{\pi}{3}$ in standard position.

SOLUTION

(a) To find positive angles that are coterminal with θ, we add any multiple of 360°. Thus

$$30° + 360° = 390° \quad \text{and} \quad 30° + 720° = 750°$$

are coterminal with $\theta = 30°$. To find negative angles that are coterminal with θ, we subtract any multiple of 360°. Thus

$$30° - 360° = -330° \quad \text{and} \quad 30° - 720° = -690°$$

are coterminal with θ. (See Figure 6.)

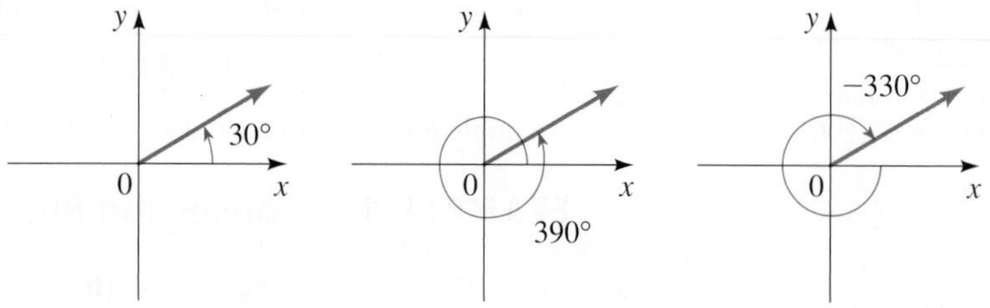

FIGURE 6

(b) To find positive angles that are coterminal with θ, we add any multiple of 2π. Thus

$$\frac{\pi}{3} + 2\pi = \frac{7\pi}{3} \quad \text{and} \quad \frac{\pi}{3} + 4\pi = \frac{13\pi}{3}$$

are coterminal with $\theta = \pi/3$. To find negative angles that are coterminal with θ, we subtract any multiple of 2π. Thus

$$\frac{\pi}{3} - 2\pi = -\frac{5\pi}{3} \quad \text{and} \quad \frac{\pi}{3} - 4\pi = -\frac{11\pi}{3}$$

are coterminal with θ. (See Figure 7.)

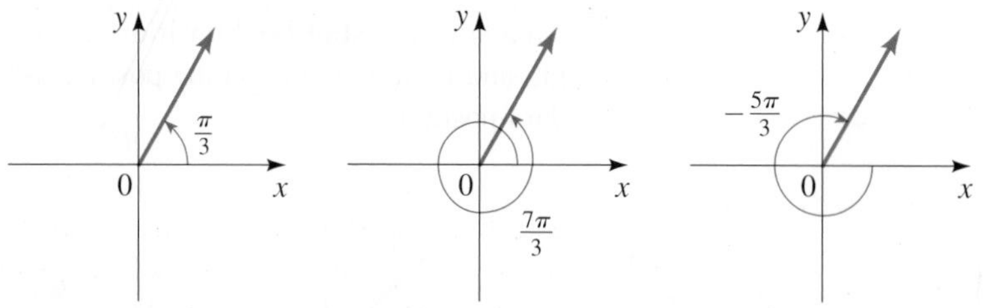

FIGURE 7

NOW TRY EXERCISES **27** AND **29**

EXAMPLE 3 | Coterminal Angles

Find an angle with measure between 0° and 360° that is coterminal with the angle of measure 1290° in standard position.

SOLUTION We can subtract 360° as many times as we wish from 1290°, and the resulting angle will be coterminal with 1290°. Thus, $1290° - 360° = 930°$ is coterminal with 1290°, and so is the angle $1290° - 2(360)° = 570°$.

To find the angle we want between 0° and 360°, we subtract 360° from 1290° as many times as necessary. An efficient way to do this is to determine how many times 360° goes into 1290°, that is, divide 1290 by 360, and the remainder will be the angle we are looking for. We see that 360 goes into 1290 three times with a remainder of 210. Thus, 210° is the desired angle (see Figure 8).

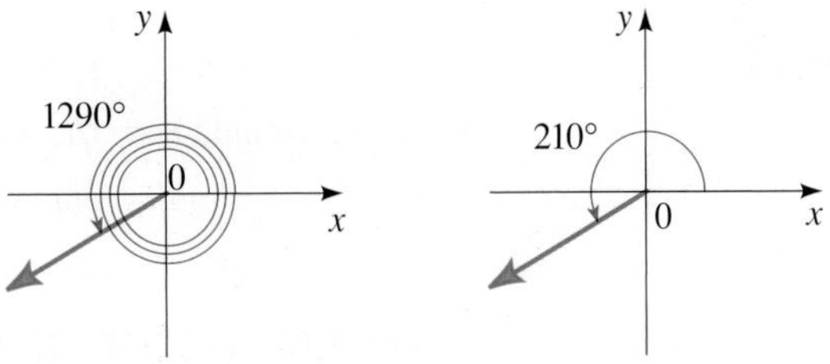

FIGURE 8

NOW TRY EXERCISE **39**

▼ Length of a Circular Arc

An angle whose radian measure is θ is subtended by an arc that is the fraction $\theta/(2\pi)$ of the circumference of a circle. Thus, in a circle of radius r, the length s of an arc that subtends the angle θ (see Figure 9) is

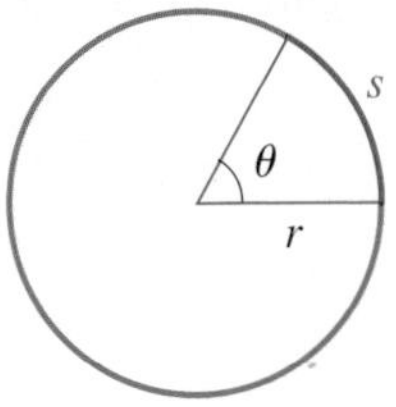

FIGURE 9 $s = \theta r$

$$s = \frac{\theta}{2\pi} \times \text{circumference of circle}$$

$$= \frac{\theta}{2\pi}(2\pi r) = \theta r$$

LENGTH OF A CIRCULAR ARC

In a circle of radius r, the length s of an arc that subtends a central angle of θ radians is

$$s = r\theta$$

Solving for θ, we get the important formula

$$\theta = \frac{s}{r}$$

This formula allows us to define radian measure using a circle of any radius r: The radian measure of an angle θ is s/r, where s is the length of the circular arc that subtends θ in a circle of radius r (see Figure 10).

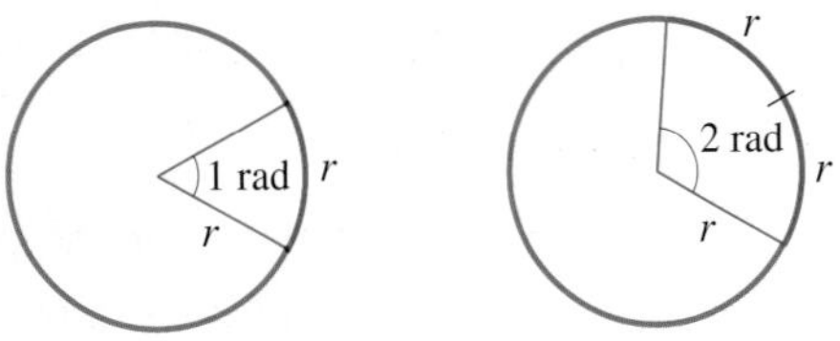

FIGURE 10 The radian measure of θ is the number of "radiuses" that can fit in the arc that subtends θ; hence the term *radian*.

EXAMPLE 4 | Arc Length and Angle Measure

(a) Find the length of an arc of a circle with radius 10 m that subtends a central angle of 30°.

(b) A central angle θ in a circle of radius 4 m is subtended by an arc of length 6 m. Find the measure of θ in radians.

SOLUTION

(a) From Example 1(b) we see that $30° = \pi/6$ rad. So the length of the arc is

The formula $s = r\theta$ is true only when θ is measured in radians.

$$s = r\theta = (10)\frac{\pi}{6} = \frac{5\pi}{3} \text{ m}$$

(b) By the formula $\theta = s/r$, we have

$$\theta = \frac{s}{r} = \frac{6}{4} = \frac{3}{2} \text{ rad}$$

NOW TRY EXERCISES **55** AND **57**

Area of a Circular Sector

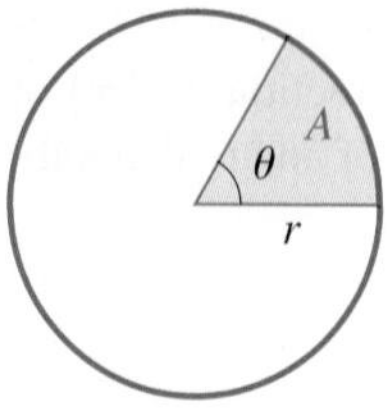

FIGURE 11
$A = \frac{1}{2}r^2\theta$

The area of a circle of radius r is $A = \pi r^2$. A sector of this circle with central angle θ has an area that is the fraction $\theta/(2\pi)$ of the area of the entire circle (see Figure 11). So the area of this sector is

$$A = \frac{\theta}{2\pi} \times \text{area of circle}$$

$$= \frac{\theta}{2\pi}(\pi r^2) = \frac{1}{2}r^2\theta$$

AREA OF A CIRCULAR SECTOR

In a circle of radius r, the area A of a sector with a central angle of θ radians is

$$A = \frac{1}{2}r^2\theta$$

EXAMPLE 5 | Area of a Sector

Find the area of a sector of a circle with central angle 60° if the radius of the circle is 3 m.

SOLUTION To use the formula for the area of a circular sector, we must find the central angle of the sector in radians: $60° = 60(\pi/180)$ rad $= \pi/3$ rad. Thus, the area of the sector is

The formula $A = \frac{1}{2}r^2\theta$ is true only when θ is measured in radians.

$$A = \frac{1}{2}r^2\theta = \frac{1}{2}(3)^2\left(\frac{\pi}{3}\right) = \frac{3\pi}{2} \text{ m}^2$$

NOW TRY EXERCISE **61**

Circular Motion

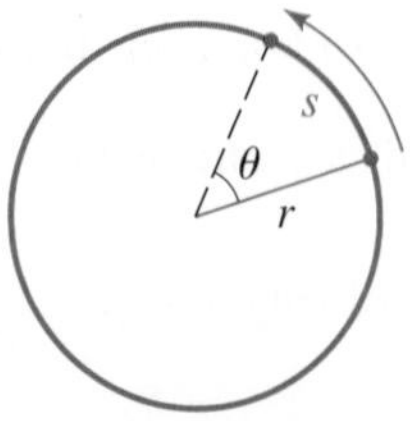

FIGURE 12

Suppose a point moves along a circle as shown in Figure 12. There are two ways to describe the motion of the point: linear speed and angular speed. **Linear speed** is the rate at which the distance traveled is changing, so linear speed is the distance traveled divided by the time elapsed. **Angular speed** is the rate at which the central angle θ is changing, so angular speed is the number of radians this angle changes divided by the time elapsed.

LINEAR SPEED AND ANGULAR SPEED

Suppose a point moves along a circle of radius r and the ray from the center of the circle to the point traverses θ radians in time t. Let $s = r\theta$ be the distance the point travels in time t. Then the speed of the object is given by

$$\textbf{Angular speed} \qquad \omega = \frac{\theta}{t}$$

$$\textbf{Linear speed} \qquad v = \frac{s}{t}$$

The symbol ω is the Greek letter "omega."

EXAMPLE 6 | Finding Linear and Angular Speed

A boy rotates a stone in a 3-ft-long sling at the rate of 15 revolutions every 10 seconds. Find the angular and linear velocities of the stone.

SOLUTION In 10 s, the angle θ changes by $15 \cdot 2\pi = 30\pi$ radians. So the *angular speed* of the stone is

$$\omega = \frac{\theta}{t} = \frac{30\pi \text{ rad}}{10 \text{ s}} = 3\pi \text{ rad/s}$$

The distance traveled by the stone in 10 s is $s = 15 \cdot 2\pi r = 15 \cdot 2\pi \cdot 3 = 90\pi$ ft. So the *linear speed* of the stone is

$$v = \frac{s}{t} = \frac{90\pi \text{ ft}}{10 \text{ s}} = 9\pi \text{ ft/s}$$

NOW TRY EXERCISE **79** ■

Notice that angular speed does *not* depend on the radius of the circle, but only on the angle θ. However, if we know the angular speed ω and the radius r, we can find linear speed as follows: $v = s/t = r\theta/t = r(\theta/t) = r\omega$.

RELATIONSHIP BETWEEN LINEAR AND ANGULAR SPEED

If a point moves along a circle of radius r with angular speed ω, then its linear speed v is given by

$$v = r\omega$$

EXAMPLE 7 | Finding Linear Speed from Angular Speed

A woman is riding a bicycle whose wheels are 26 inches in diameter. If the wheels rotate at 125 revolutions per minute (rpm), find the speed at which she is traveling, in mi/h.

SOLUTION The angular speed of the wheels is $2\pi \cdot 125 = 250\pi$ rad/min. Since the wheels have radius 13 in. (half the diameter), the linear speed is

$$v = r\omega = 13 \cdot 250\pi \approx 10{,}210.2 \text{ in./min}$$

Since there are 12 inches per foot, 5280 feet per mile, and 60 minutes per hour, her speed in miles per hour is

$$\frac{10{,}210.2 \text{ in./min} \times 60 \text{ min/h}}{12 \text{ in./ft} \times 5280 \text{ ft/mi}} = \frac{612{,}612 \text{ in./h}}{63{,}360 \text{ in./mi}}$$

$$\approx 9.7 \text{ mi/h}$$

NOW TRY EXERCISE **81** ■

6.1 EXERCISES

CONCEPTS

1. (a) The radian measure of an angle θ is the length of the ________ that subtends the angle in a circle of radius ________.

(b) To convert degrees to radians, we multiply by ________.

(c) To convert radians to degrees, we multiply by ________.

2. A central angle θ is drawn in a circle of radius r.

(a) The length of the arc subtended by θ is $s =$ ________.

(b) The area of the sector with central angle θ is $A =$ ________.

SKILLS

3–14 ■ Find the radian measure of the angle with the given degree measure.

3. 72° **4.** 54° **5.** −45°
6. −60° **7.** −75° **8.** −300°
9. 1080° **10.** 3960° **11.** 96°
12. 15° **13.** 7.5° **14.** 202.5°

15–26 ■ Find the degree measure of the angle with the given radian measure.

15. $\frac{7\pi}{6}$ **16.** $\frac{11\pi}{3}$ **17.** $-\frac{5\pi}{4}$
18. $-\frac{3\pi}{2}$ **19.** 3 **20.** −2
21. −1.2 **22.** 3.4 **23.** $\frac{\pi}{10}$
24. $\frac{5\pi}{18}$ **25.** $-\frac{2\pi}{15}$ **26.** $-\frac{13\pi}{12}$

27–32 ■ The measure of an angle in standard position is given. Find two positive angles and two negative angles that are coterminal with the given angle.

27. 50° **28.** 135° **29.** $\frac{3\pi}{4}$
30. $\frac{11\pi}{6}$ **31.** $-\frac{\pi}{4}$ **32.** −45°

33–38 ■ The measures of two angles in standard position are given. Determine whether the angles are coterminal.

33. 70°, 430° **34.** −30°, 330°
35. $\frac{5\pi}{6}, \frac{17\pi}{6}$ **36.** $\frac{32\pi}{3}, \frac{11\pi}{3}$
37. 155°, 875° **38.** 50°, 340°

39–44 ■ Find an angle between 0° and 360° that is coterminal with the given angle.

39. 733° **40.** 361° **41.** 1110°
42. −100° **43.** −800° **44.** 1270°

45–50 ■ Find an angle between 0 and 2π that is coterminal with the given angle.

45. $\frac{17\pi}{6}$ **46.** $-\frac{7\pi}{3}$ **47.** 87π
48. 10 **49.** $\frac{17\pi}{4}$ **50.** $\frac{51\pi}{2}$

51. Find the length of the arc s in the figure.

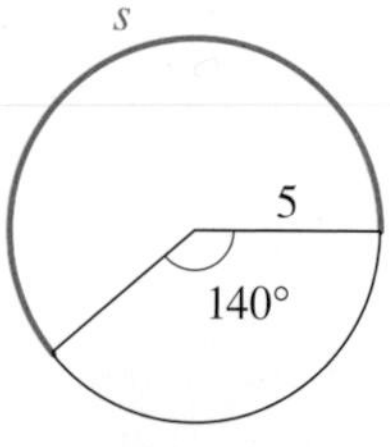

52. Find the angle θ in the figure.

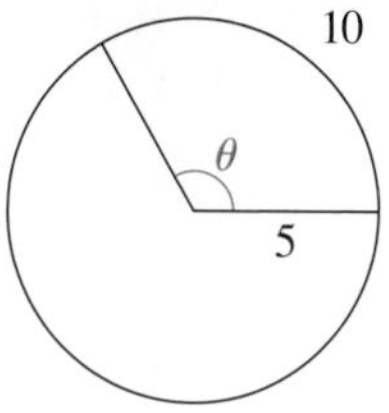

53. Find the radius r of the circle in the figure.

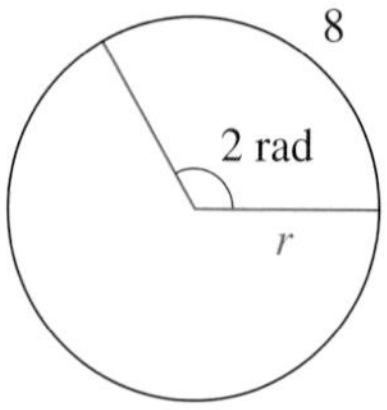

54. Find the length of an arc that subtends a central angle of 45° in a circle of radius 10 m.

55. Find the length of an arc that subtends a central angle of 2 rad in a circle of radius 2 mi.

56. A central angle θ in a circle of radius 5 m is subtended by an arc of length 6 m. Find the measure of θ in degrees and in radians.

57. An arc of length 100 m subtends a central angle θ in a circle of radius 50 m. Find the measure of θ in degrees and in radians.

58. A circular arc of length 3 ft subtends a central angle of 25°. Find the radius of the circle.

59. Find the radius of the circle if an arc of length 6 m on the circle subtends a central angle of $\pi/6$ rad.

60. Find the radius of the circle if an arc of length 4 ft on the circle subtends a central angle of 135°.

61. Find the area of the sector shown in each figure.

(a)

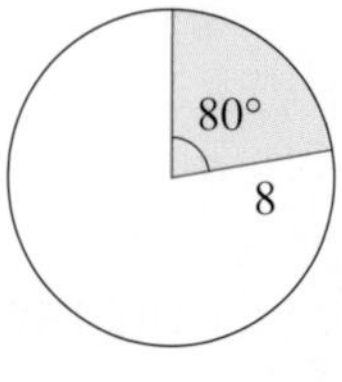

(b)

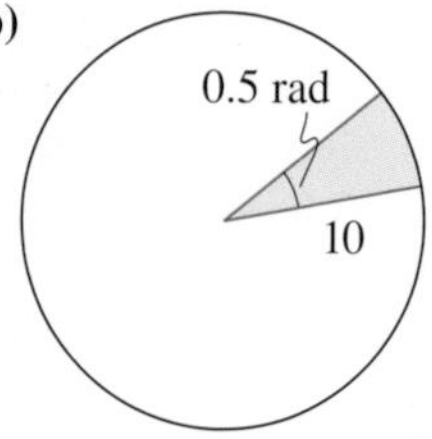

62. Find the radius of each circle if the area of the sector is 12.

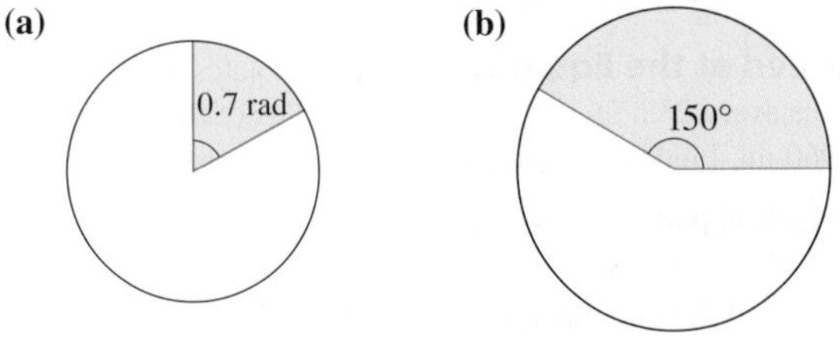

63. Find the area of a sector with central angle 1 rad in a circle of radius 10 m.

64. A sector of a circle has a central angle of 60°. Find the area of the sector if the radius of the circle is 3 mi.

65. The area of a sector of a circle with a central angle of 2 rad is 16 m^2. Find the radius of the circle.

66. A sector of a circle of radius 24 mi has an area of 288 mi^2. Find the central angle of the sector.

67. The area of a circle is 72 cm^2. Find the area of a sector of this circle that subtends a central angle of $\pi/6$ rad.

68. Three circles with radii 1, 2, and 3 ft are externally tangent to one another, as shown in the figure. Find the area of the sector of the circle of radius 1 that is cut off by the line segments joining the center of that circle to the centers of the other two circles.

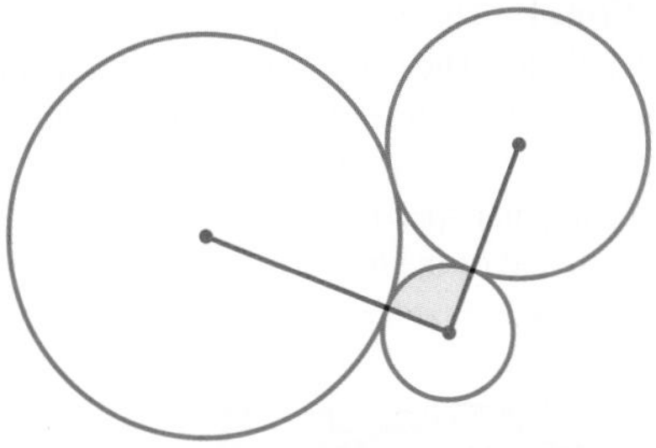

APPLICATIONS

69. **Travel Distance** A car's wheels are 28 in. in diameter. How far (in miles) will the car travel if its wheels revolve 10,000 times without slipping?

70. **Wheel Revolutions** How many revolutions will a car wheel of diameter 30 in. make as the car travels a distance of one mile?

71. **Latitudes** Pittsburgh, Pennsylvania, and Miami, Florida, lie approximately on the same meridian. Pittsburgh has a latitude of 40.5° N, and Miami has a latitude of 25.5° N. Find the distance between these two cities. (The radius of the earth is 3960 mi.)

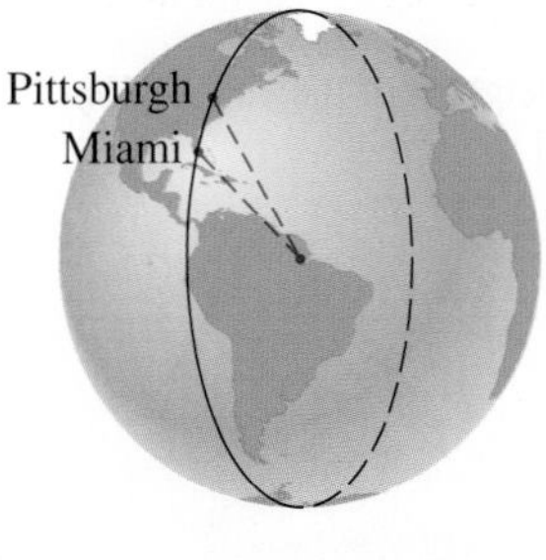

72. **Latitudes** Memphis, Tennessee, and New Orleans, Louisiana, lie approximately on the same meridian. Memphis has a latitude of 35° N, and New Orleans has a latitude of 30° N. Find the distance between these two cities. (The radius of the earth is 3960 mi.)

73. **Orbit of the Earth** Find the distance that the earth travels in one day in its path around the sun. Assume that a year has 365 days and that the path of the earth around the sun is a circle of radius 93 million miles. [The path of the earth around the sun is actually an *ellipse* with the sun at one focus (see Section 10.2). This ellipse, however, has very small eccentricity, so it is nearly circular.]

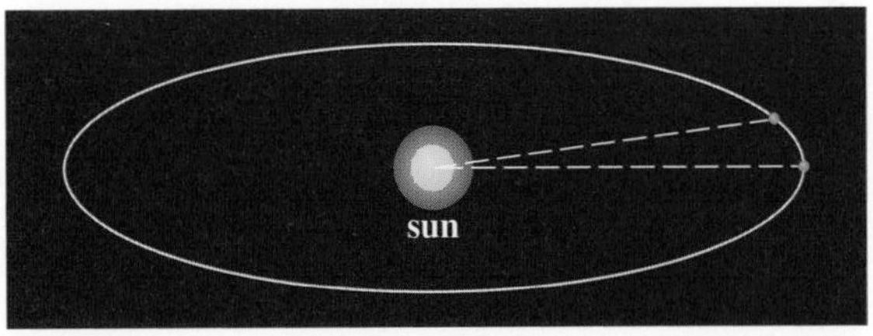

74. **Circumference of the Earth** The Greek mathematician Eratosthenes (ca. 276–195 B.C.) measured the circumference of the earth from the following observations. He noticed that on a certain day the sun shone directly down a deep well in Syene (modern Aswan). At the same time in Alexandria, 500 miles north (on the same meridian), the rays of the sun shone at an angle of 7.2° to the zenith. Use this information and the figure to find the radius and circumference of the earth.

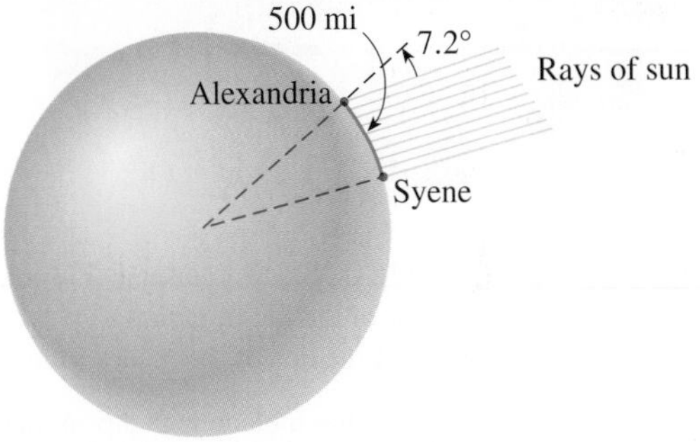

75. **Nautical Miles** Find the distance along an arc on the surface of the earth that subtends a central angle of 1 minute (1 minute = $\frac{1}{60}$ degree). This distance is called a *nautical mile*. (The radius of the earth is 3960 mi.)

76. **Irrigation** An irrigation system uses a straight sprinkler pipe 300 ft long that pivots around a central point as shown. Due to an obstacle the pipe is allowed to pivot through 280° only. Find the area irrigated by this system.

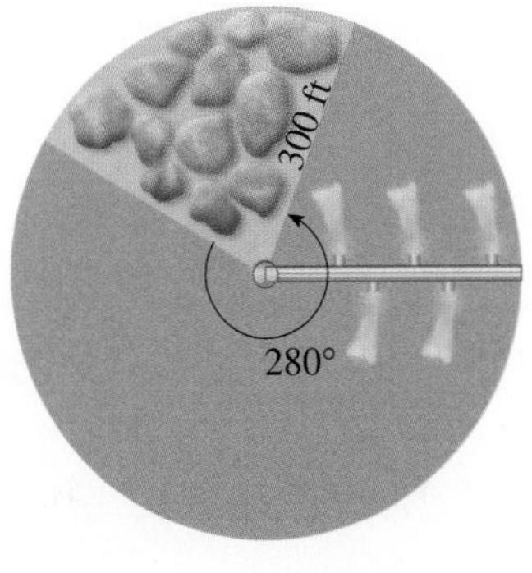

77. **Windshield Wipers** The top and bottom ends of a windshield wiper blade are 34 in. and 14 in., respectively, from the pivot point. While in operation, the wiper sweeps through 135°. Find the area swept by the blade.

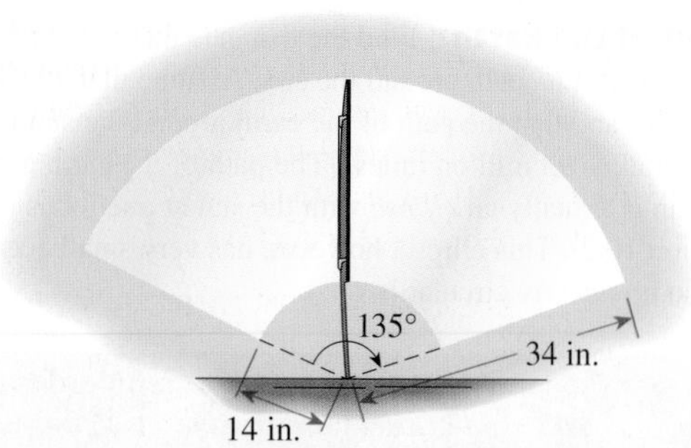

78. **The Tethered Cow** A cow is tethered by a 100-ft rope to the inside corner of an L-shaped building, as shown in the figure. Find the area that the cow can graze.

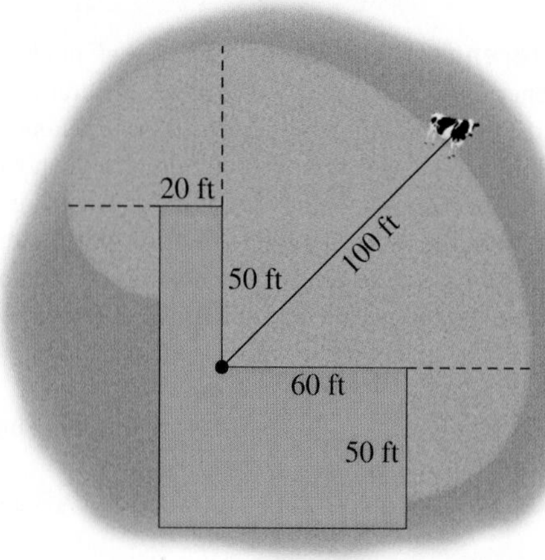

79. **Fan** A ceiling fan with 16-in. blades rotates at 45 rpm.
(a) Find the angular speed of the fan in rad/min.
(b) Find the linear speed of the tips of the blades in in./min.

80. **Radial Saw** A radial saw has a blade with a 6-in. radius. Suppose that the blade spins at 1000 rpm.
(a) Find the angular speed of the blade in rad/min.
(b) Find the linear speed of the sawteeth in ft/s.

81. **Winch** A winch of radius 2 ft is used to lift heavy loads. If the winch makes 8 revolutions every 15 s, find the speed at which the load is rising.

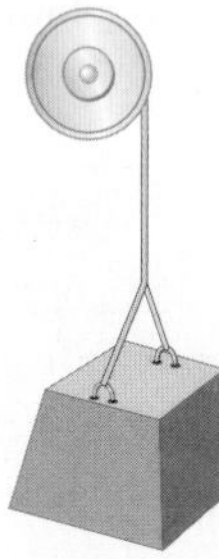

82. **Speed of a Car** The wheels of a car have radius 11 in. and are rotating at 600 rpm. Find the speed of the car in mi/h.

83. **Speed at the Equator** The earth rotates about its axis once every 23 h 56 min 4 s, and the radius of the earth is 3960 mi. Find the linear speed of a point on the equator in mi/h.

84. **Truck Wheels** A truck with 48-in.-diameter wheels is traveling at 50 mi/h.
(a) Find the angular speed of the wheels in rad/min.
(b) How many revolutions per minute do the wheels make?

85 **Speed of a Current** To measure the speed of a current, scientists place a paddle wheel in the stream and observe the rate at which it rotates. If the paddle wheel has radius 0.20 m and rotates at 100 rpm, find the speed of the current in m/s.

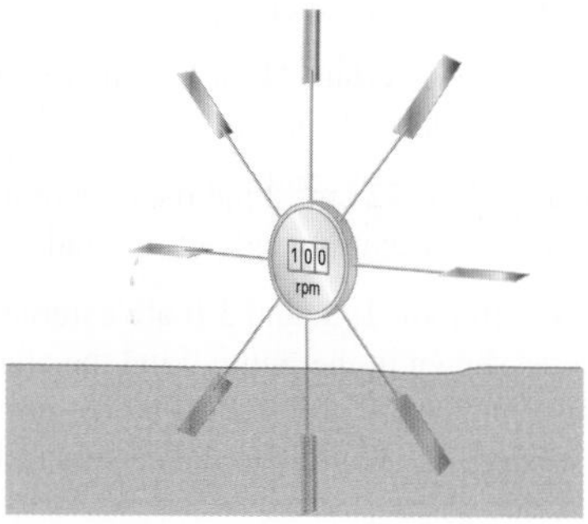

86. **Bicycle Wheel** The sprockets and chain of a bicycle are shown in the figure. The pedal sprocket has a radius of 4 in., the wheel sprocket a radius of 2 in., and the wheel a radius of 13 in. The cyclist pedals at 40 rpm.
(a) Find the angular speed of the wheel sprocket.
(b) Find the speed of the bicycle. (Assume that the wheel turns at the same rate as the wheel sprocket.)

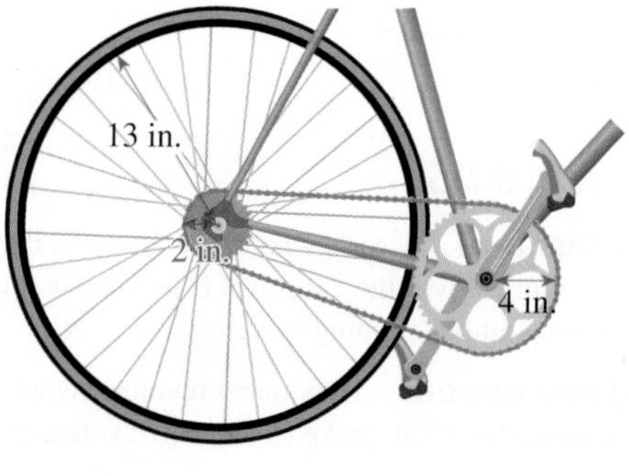

87. **Conical Cup** A conical cup is made from a circular piece of paper with radius 6 cm by cutting out a sector and joining the edges as shown on the next page. Suppose $\theta = 5\pi/3$.
(a) Find the circumference C of the opening of the cup.
(b) Find the radius r of the opening of the cup. [*Hint:* Use $C = 2\pi r$.]
(c) Find the height h of the cup. [*Hint:* Use the Pythagorean Theorem.]

(d) Find the volume of the cup.

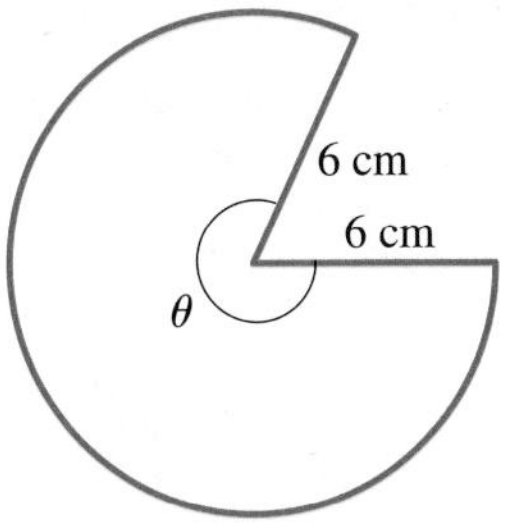

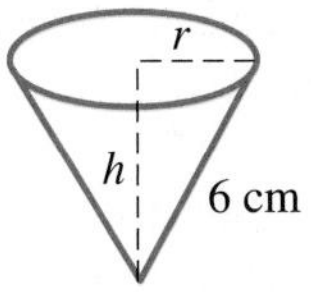

88. Conical Cup In this exercise we find the volume of the conical cup in Exercise 87 for any angle θ.

(a) Follow the steps in Exercise 87 to show that the volume of the cup as a function of θ is

$$V(\theta) = \frac{9}{\pi^2}\theta^2\sqrt{4\pi^2 - \theta^2}, \qquad 0 < \theta < 2\pi$$

(b) Graph the function V.

(c) For what angle θ is the volume of the cup a maximum?

DISCOVERY ■ DISCUSSION ■ WRITING

89. Different Ways of Measuring Angles The custom of measuring angles using degrees, with 360° in a circle, dates back to the ancient Babylonians, who used a number system based on groups of 60. Another system of measuring angles divides the circle into 400 units, called *grads*. In this system a right angle is 100 grad, so this fits in with our base 10 number system.

Write a short essay comparing the advantages and disadvantages of these two systems and the radian system of measuring angles. Which system do you prefer? Why?

90. Clocks and Angles In one hour, the minute hand on a clock moves through a complete circle, and the hour hand moves through $\frac{1}{12}$ of a circle. Through how many radians do the minute and the hour hand move between 1:00 P.M. and 6:45 P.M. (on the same day)?

6.2 TRIGONOMETRY OF RIGHT TRIANGLES

Trigonometric Ratios ▶ Special Triangles ▶ Applications of Trigonometry of Right Triangles

In this section we study certain ratios of the sides of right triangles, called trigonometric ratios, and give several applications.

▼ Trigonometric Ratios

Consider a right triangle with θ as one of its acute angles. The trigonometric ratios are defined as follows (see Figure 1).

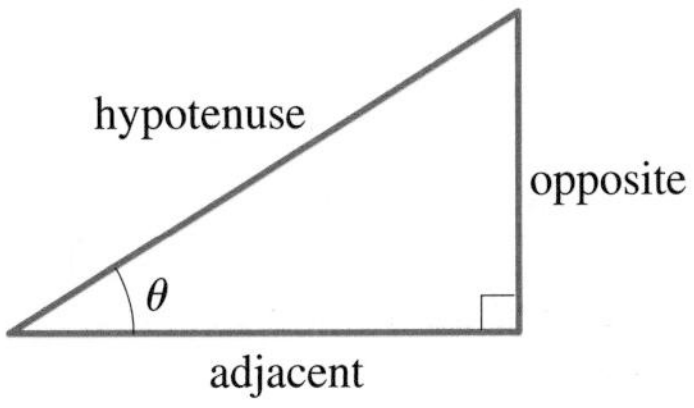

FIGURE 1

THE TRIGONOMETRIC RATIOS

$$\sin\theta = \frac{\text{opposite}}{\text{hypotenuse}} \qquad \cos\theta = \frac{\text{adjacent}}{\text{hypotenuse}} \qquad \tan\theta = \frac{\text{opposite}}{\text{adjacent}}$$

$$\csc\theta = \frac{\text{hypotenuse}}{\text{opposite}} \qquad \sec\theta = \frac{\text{hypotenuse}}{\text{adjacent}} \qquad \cot\theta = \frac{\text{adjacent}}{\text{opposite}}$$

The symbols we use for these ratios are abbreviations for their full names: **sine**, **cosine**, **tangent**, **cosecant**, **secant**, **cotangent**. Since any two right triangles with angle θ are

HIPPARCHUS (circa 140 B.C.) is considered the founder of trigonometry. He constructed tables for a function closely related to the modern sine function and evaluated for angles at half-degree intervals. These are considered the first trigonometric tables. He used his tables mainly to calculate the paths of the planets through the heavens.

similar, these ratios are the same, regardless of the size of the triangle; the trigonometric ratios depend only on the angle θ (see Figure 2).

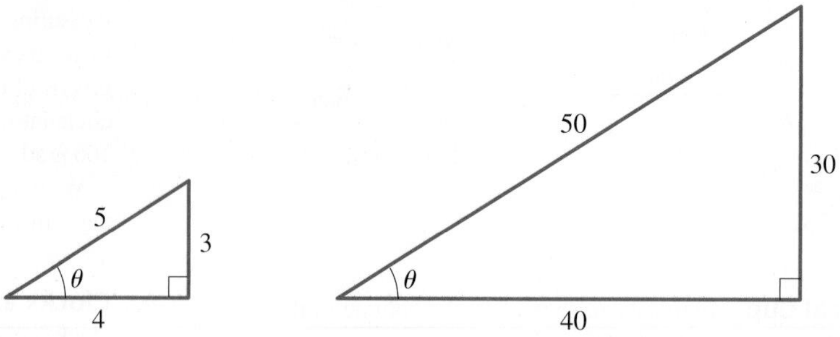

FIGURE 2 $\sin\theta = \frac{3}{5}$ $\quad \sin\theta = \frac{30}{50} = \frac{3}{5}$

EXAMPLE 1 | Finding Trigonometric Ratios

Find the six trigonometric ratios of the angle θ in Figure 3.

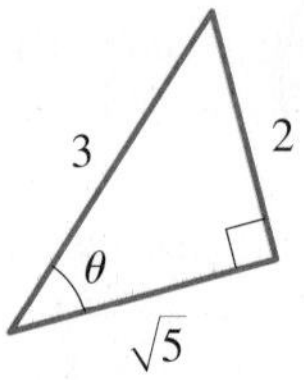

FIGURE 3

SOLUTION

$$\sin\theta = \frac{2}{3} \qquad \cos\theta = \frac{\sqrt{5}}{3} \qquad \tan\theta = \frac{2}{\sqrt{5}}$$

$$\csc\theta = \frac{3}{2} \qquad \sec\theta = \frac{3}{\sqrt{5}} \qquad \cot\theta = \frac{\sqrt{5}}{2}$$

NOW TRY EXERCISE **3** ■

EXAMPLE 2 | Finding Trigonometric Ratios

If $\cos\alpha = \frac{3}{4}$, sketch a right triangle with acute angle α, and find the other five trigonometric ratios of α.

SOLUTION Since $\cos\alpha$ is defined as the ratio of the adjacent side to the hypotenuse, we sketch a triangle with hypotenuse of length 4 and a side of length 3 adjacent to α. If the opposite side is x, then by the Pythagorean Theorem, $3^2 + x^2 = 4^2$ or $x^2 = 7$, so $x = \sqrt{7}$. We then use the triangle in Figure 4 to find the ratios.

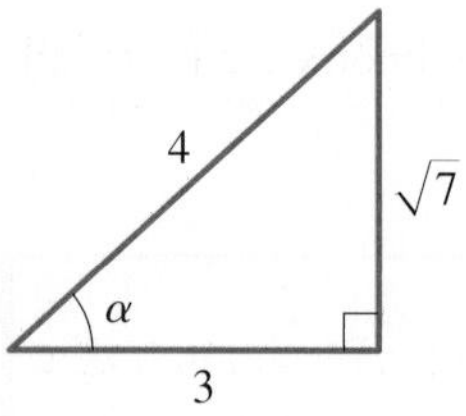

FIGURE 4

$$\sin\alpha = \frac{\sqrt{7}}{4} \qquad \cos\alpha = \frac{3}{4} \qquad \tan\alpha = \frac{\sqrt{7}}{3}$$

$$\csc\alpha = \frac{4}{\sqrt{7}} \qquad \sec\alpha = \frac{4}{3} \qquad \cot\alpha = \frac{3}{\sqrt{7}}$$

NOW TRY EXERCISE **19** ■

▼ Special Triangles

Certain right triangles have ratios that can be calculated easily from the Pythagorean Theorem. Since they are used frequently, we mention them here.

The first triangle is obtained by drawing a diagonal in a square of side 1 (see Figure 5). By the Pythagorean Theorem this diagonal has length $\sqrt{2}$. The resulting triangle has angles 45°, 45°, and 90° (or $\pi/4$, $\pi/4$, and $\pi/2$). To get the second triangle, we start with an equilateral triangle ABC of side 2 and draw the perpendicular bisector DB of the base, as in Figure 6. By the Pythagorean Theorem the length of DB is $\sqrt{3}$. Since DB bisects angle ABC, we obtain a triangle with angles 30°, 60°, and 90° (or $\pi/6$, $\pi/3$, and $\pi/2$).

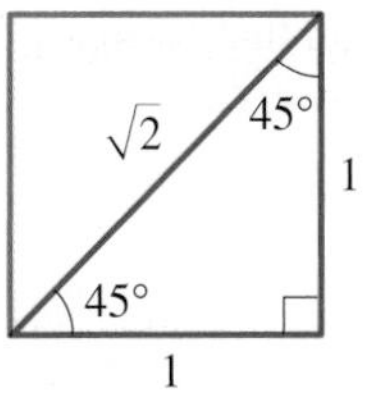

FIGURE 5

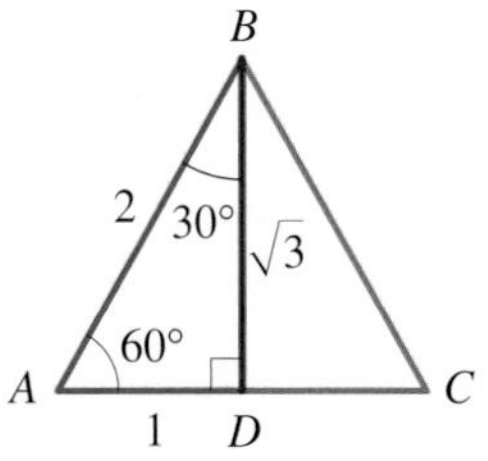

FIGURE 6

We can now use the special triangles in Figures 5 and 6 to calculate the trigonometric ratios for angles with measures 30°, 45°, and 60° (or $\pi/6$, $\pi/4$, and $\pi/3$). These are listed in Table 1.

TABLE 1
Values of the trigonometric ratios for special angles

θ in degrees	θ in radians	$\sin\theta$	$\cos\theta$	$\tan\theta$	$\csc\theta$	$\sec\theta$	$\cot\theta$
30°	$\frac{\pi}{6}$	$\frac{1}{2}$	$\frac{\sqrt{3}}{2}$	$\frac{\sqrt{3}}{3}$	2	$\frac{2\sqrt{3}}{3}$	$\sqrt{3}$
45°	$\frac{\pi}{4}$	$\frac{\sqrt{2}}{2}$	$\frac{\sqrt{2}}{2}$	1	$\sqrt{2}$	$\sqrt{2}$	1
60°	$\frac{\pi}{3}$	$\frac{\sqrt{3}}{2}$	$\frac{1}{2}$	$\sqrt{3}$	$\frac{2\sqrt{3}}{3}$	2	$\frac{\sqrt{3}}{3}$

It's useful to remember these special trigonometric ratios because they occur often. Of course, they can be recalled easily if we remember the triangles from which they are obtained.

For an explanation of numerical methods, see the margin note on page 400.

To find the values of the trigonometric ratios for other angles, we use a calculator. Mathematical methods (called *numerical methods*) used in finding the trigonometric ratios are programmed directly into scientific calculators. For instance, when the SIN key is pressed, the calculator computes an approximation to the value of the sine of the given angle. Calculators give the values of sine, cosine, and tangent; the other ratios can be easily calculated from these by using the following *reciprocal relations*:

$$\csc t = \frac{1}{\sin t} \qquad \sec t = \frac{1}{\cos t} \qquad \cot t = \frac{1}{\tan t}$$

You should check that these relations follow immediately from the definitions of the trigonometric ratios.

We follow the convention that when we write $\sin t$, *we mean the sine of the angle whose radian measure is* t. For instance, $\sin 1$ means the sine of the angle whose radian measure is 1. When using a calculator to find an approximate value for this number, set your calculator to radian mode; you will find that

$$\sin 1 \approx 0.841471$$

If you want to find the sine of the angle whose measure is 1°, set your calculator to degree mode; you will find that

$$\sin 1^\circ \approx 0.0174524$$

▼ Applications of Trigonometry of Right Triangles

A triangle has six parts: three angles and three sides. To **solve a triangle** means to determine all of its parts from the information known about the triangle, that is, to determine the lengths of the three sides and the measures of the three angles.

EXAMPLE 3 | Solving a Right Triangle

Solve triangle ABC, shown in Figure 7 on the next page.

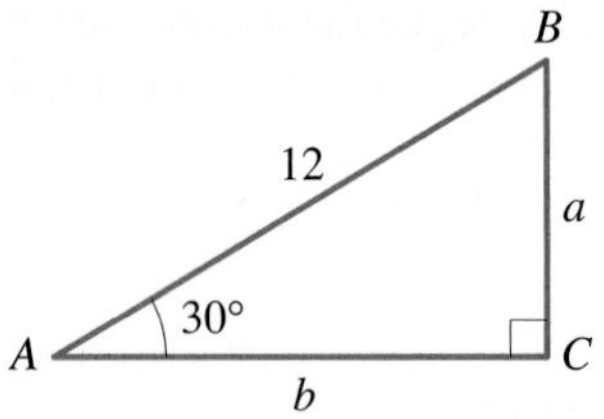

FIGURE 7

SOLUTION It's clear that $\angle B = 60°$. To find a, we look for an equation that relates a to the lengths and angles we already know. In this case, we have $\sin 30° = a/12$, so

$$a = 12 \sin 30° = 12(\tfrac{1}{2}) = 6$$

Similarly, $\cos 30° = b/12$, so

$$b = 12 \cos 30° = 12\left(\frac{\sqrt{3}}{2}\right) = 6\sqrt{3}$$

NOW TRY EXERCISE **31**

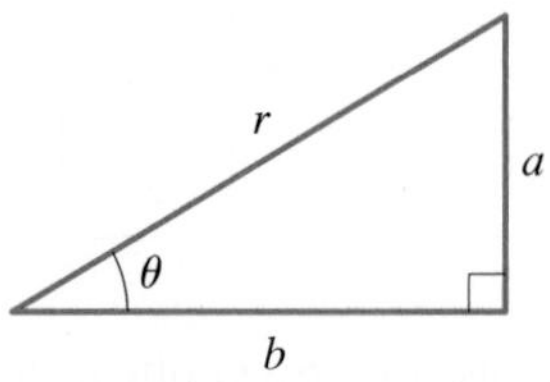

FIGURE 8
$a = r \sin \theta$
$b = r \cos \theta$

Figure 8 shows that if we know the hypotenuse r and an acute angle θ in a right triangle, then the legs a and b are given by

$$a = r \sin \theta \qquad \text{and} \qquad b = r \cos \theta$$

The ability to solve right triangles by using the trigonometric ratios is fundamental to many problems in navigation, surveying, astronomy, and the measurement of distances. The applications we consider in this section always involve right triangles, but as we will see in the next three sections, trigonometry is also useful in solving triangles that are not right triangles.

To discuss the next examples, we need some terminology. If an observer is looking at an object, then the line from the eye of the observer to the object is called the **line of sight** (Figure 9). If the object being observed is above the horizontal, then the angle between the line of sight and the horizontal is called the **angle of elevation**. If the object is below the horizontal, then the angle between the line of sight and the horizontal is called the **angle of depression**. In many of the examples and exercises in this chapter, angles of elevation and depression will be given for a hypothetical observer at ground level. If the line of sight follows a physical object, such as an inclined plane or a hillside, we use the term **angle of inclination**.

ARISTARCHUS OF SAMOS (310–230 B.C.) was a famous Greek scientist, musician, astronomer, and geometer. In his book *On the Sizes and Distances of the Sun and the Moon,* he estimated the distance to the sun by observing that when the moon is exactly half full, the triangle formed by the sun, moon, and the earth has a right angle at the moon. His method was similar to the one described in Exercise 61 in this section. Aristarchus was the first to advance the theory that the earth and planets move around the sun, an idea that did not gain full acceptance until after the time of Copernicus, 1800 years later. For this reason Aristarchus is often called "the Copernicus of antiquity."

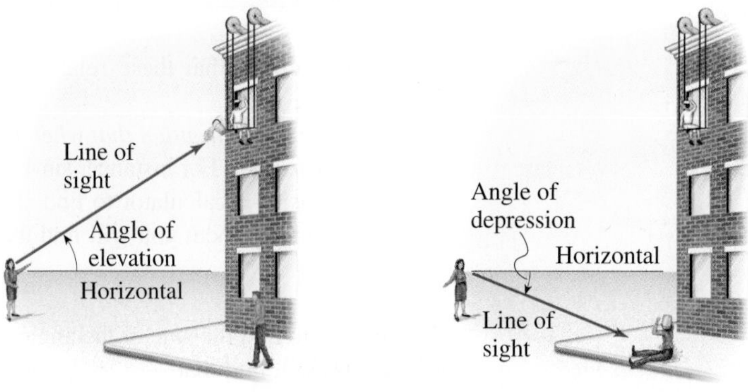

FIGURE 9

The next example gives an important application of trigonometry to the problem of measurement: We measure the height of a tall tree without having to climb it! Although the example is simple, the result is fundamental to understanding how the trigonometric ratios are applied to such problems.

EXAMPLE 4 | Finding the Height of a Tree

A giant redwood tree casts a shadow 532 ft long. Find the height of the tree if the angle of elevation of the sun is 25.7°.

THALES OF MILETUS (circa 625–547 B.C.) is the legendary founder of Greek geometry. It is said that he calculated the height of a Greek column by comparing the length of the shadow of his staff with that of the column. Using properties of similar triangles, he argued that the ratio of the height h of the column to the height h' of his staff was equal to the ratio of the length s of the column's shadow to the length s' of the staff's shadow:

$$\frac{h}{h'} = \frac{s}{s'}$$

Since three of these quantities are known, Thales was able to calculate the height of the column.

According to legend, Thales used a similar method to find the height of the Great Pyramid in Egypt, a feat that impressed Egypt's king. Plutarch wrote that "although he [the king of Egypt] admired you [Thales] for other things, yet he particularly liked the manner by which you measured the height of the pyramid without any trouble or instrument." The principle Thales used, the fact that ratios of corresponding sides of similar triangles are equal, is the foundation of the subject of trigonometry.

SOLUTION Let the height of the tree be h. From Figure 10 we see that

$$\frac{h}{532} = \tan 25.7° \qquad \text{Definition of tangent}$$

$$h = 532 \tan 25.7° \qquad \text{Multiply by 532}$$

$$\approx 532(0.48127) \approx 256 \qquad \text{Use a calculator}$$

Therefore, the height of the tree is about 256 ft.

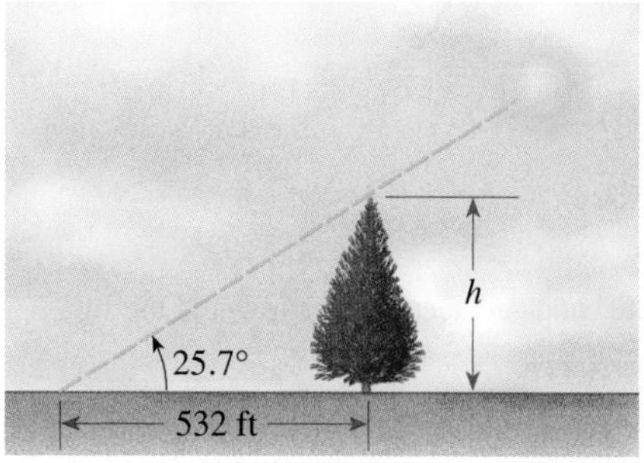

FIGURE 10

NOW TRY EXERCISE **47** ■

EXAMPLE 5 | A Problem Involving Right Triangles

From a point on the ground 500 ft from the base of a building, an observer finds that the angle of elevation to the top of the building is 24° and that the angle of elevation to the top of a flagpole atop the building is 27°. Find the height of the building and the length of the flagpole.

SOLUTION Figure 11 illustrates the situation. The height of the building is found in the same way that we found the height of the tree in Example 4.

$$\frac{h}{500} = \tan 24° \qquad \text{Definition of tangent}$$

$$h = 500 \tan 24° \qquad \text{Multiply by 500}$$

$$\approx 500(0.4452) \approx 223 \qquad \text{Use a calculator}$$

The height of the building is approximately 223 ft.

To find the length of the flagpole, let's first find the height from the ground to the top of the pole:

$$\frac{k}{500} = \tan 27°$$

$$k = 500 \tan 27°$$

$$\approx 500(0.5095)$$

$$\approx 255$$

To find the length of the flagpole, we subtract h from k. So the length of the pole is approximately $255 - 223 = 32$ ft.

NOW TRY EXERCISE **55** ■

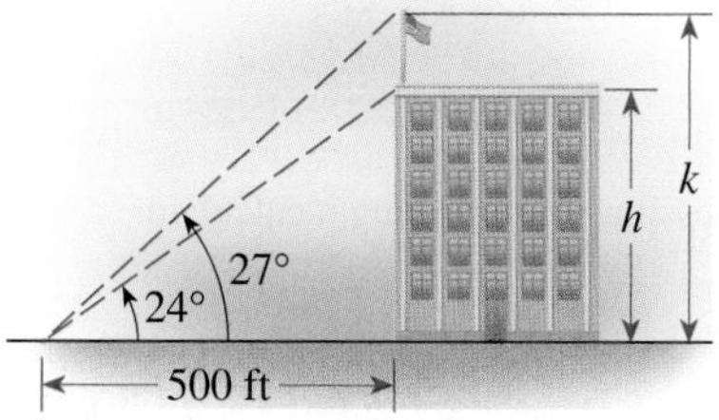

FIGURE 11

6.2 EXERCISES

CONCEPTS

1. A right triangle with an angle θ is shown in the figure.

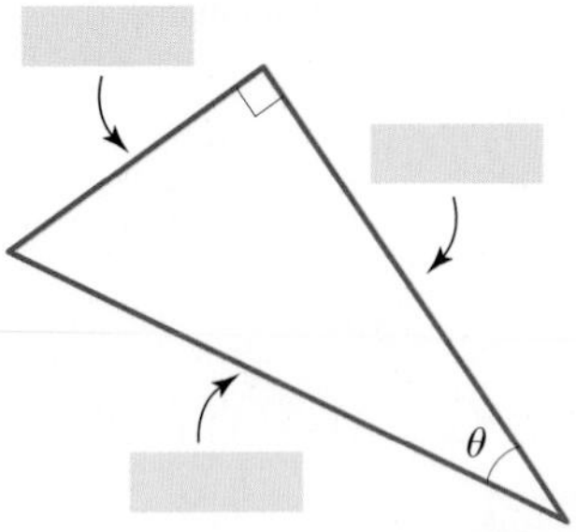

(a) Label the "opposite" and "adjacent" sides of θ and the hypotenuse of the triangle.

(b) The trigonometric functions of the angle θ are defined as follows:

$\sin\theta = \dfrac{\quad}{\quad}$ $\cos\theta = \dfrac{\quad}{\quad}$ $\tan\theta = \dfrac{\quad}{\quad}$

(c) The trigonometric ratios do not depend on the size of the triangle. This is because all right triangles with an acute angle θ are ________.

2. The reciprocal identities state that

$\csc\theta = \dfrac{1}{\quad}$ $\sec\theta = \dfrac{1}{\quad}$ $\cot\theta = \dfrac{1}{\quad}$

SKILLS

3–8 ■ Find the exact values of the six trigonometric ratios of the angle θ in the triangle.

3.

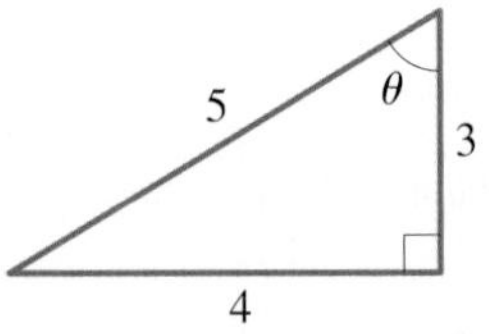

4.

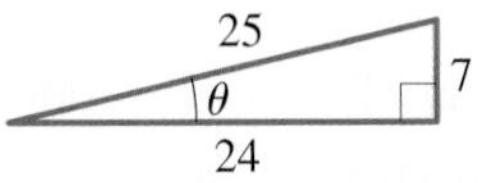

5.

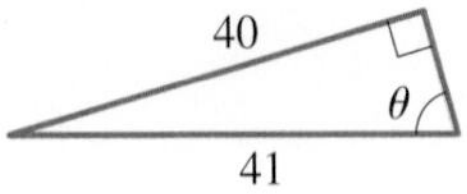

6.

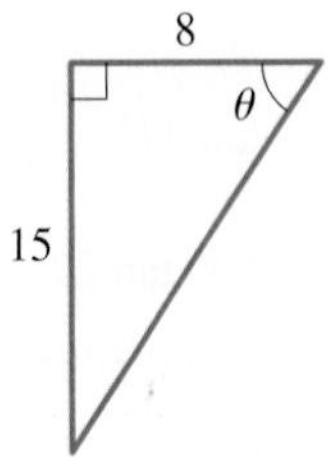

7.

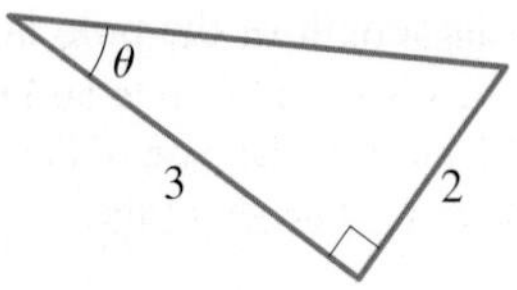

8.

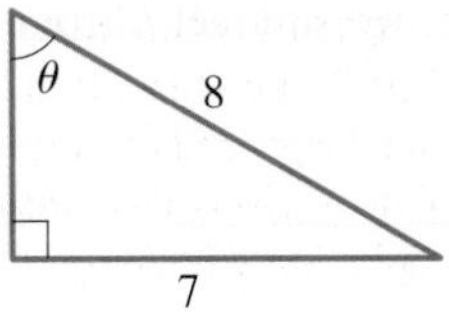

9–10 ■ Find **(a)** $\sin\alpha$ and $\cos\beta$, **(b)** $\tan\alpha$ and $\cot\beta$, and **(c)** $\sec\alpha$ and $\csc\beta$.

9.

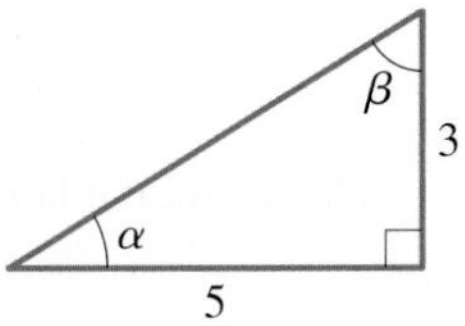

10.

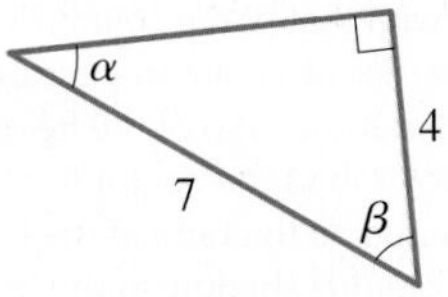

11–16 ■ Find the side labeled x. In Exercises 13 and 14 state your answer rounded to five decimal places.

11.

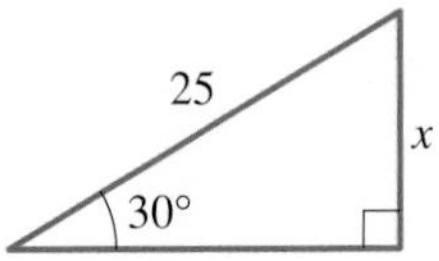

12.

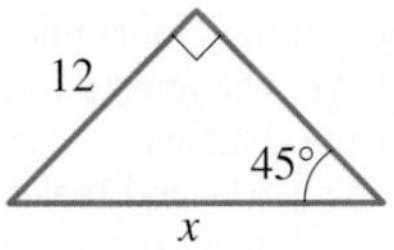

13.

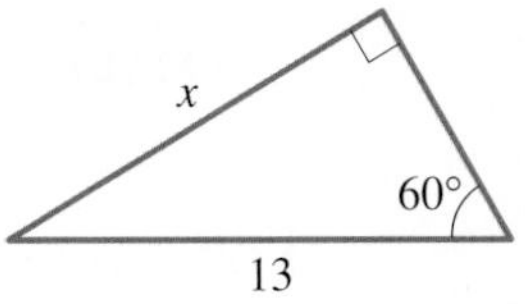

14.

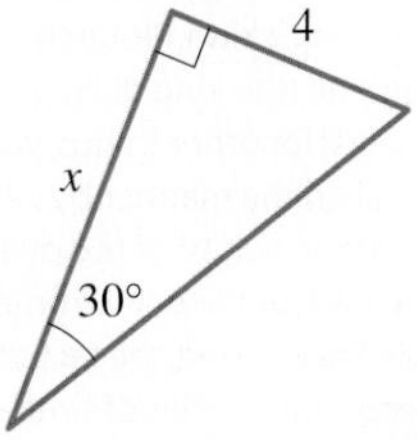

15.

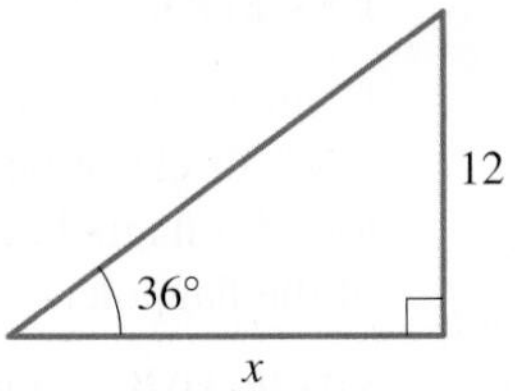

16.

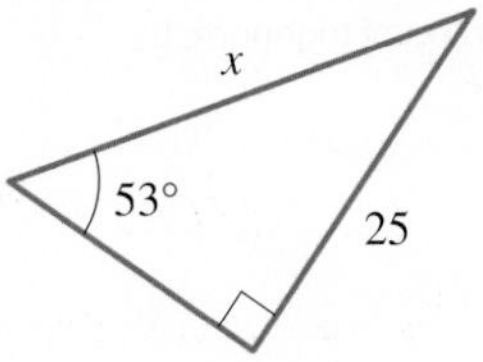

17–18 ■ Express x and y in terms of trigonometric ratios of θ.

17.

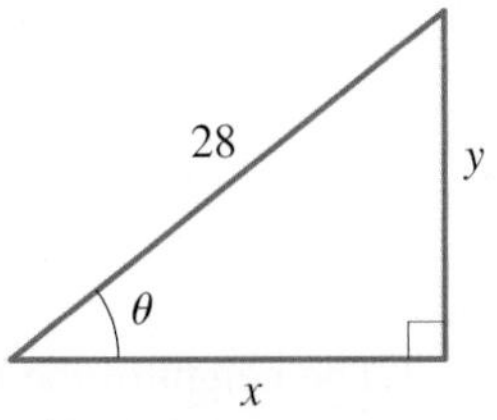

18.

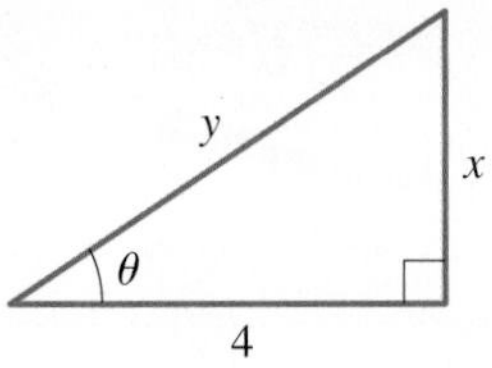

19–24 ■ Sketch a triangle that has acute angle θ, and find the other five trigonometric ratios of θ.

19. $\sin\theta = \frac{3}{5}$ **20.** $\cos\theta = \frac{9}{40}$

21. $\cot\theta = 1$ **22.** $\tan\theta = \sqrt{3}$

23. $\sec\theta = \frac{7}{2}$ **24.** $\csc\theta = \frac{13}{12}$

25–30 ■ Evaluate the expression without using a calculator.

25. $\sin\dfrac{\pi}{6} + \cos\dfrac{\pi}{6}$

26. $\sin 30° \csc 30°$

27. $\sin 30° \cos 60° + \sin 60° \cos 30°$

28. $(\sin 60°)^2 + (\cos 60°)^2$

29. $(\cos 30°)^2 - (\sin 30°)^2$

30. $\left(\sin \dfrac{\pi}{3} \cos \dfrac{\pi}{4} - \sin \dfrac{\pi}{4} \cos \dfrac{\pi}{3}\right)^2$

31–38 ■ Solve the right triangle.

31.

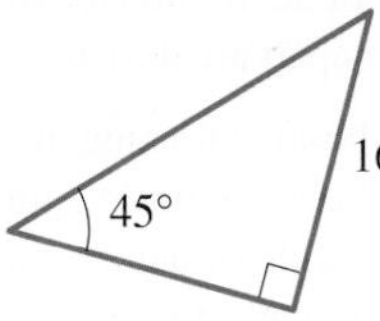

32.

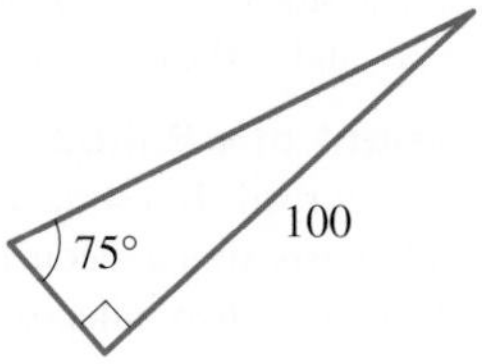

33.

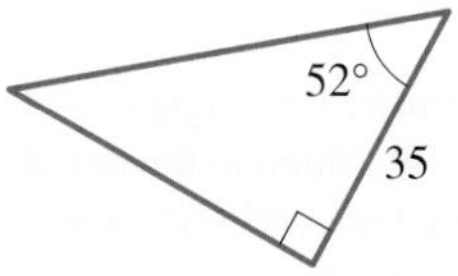

34.

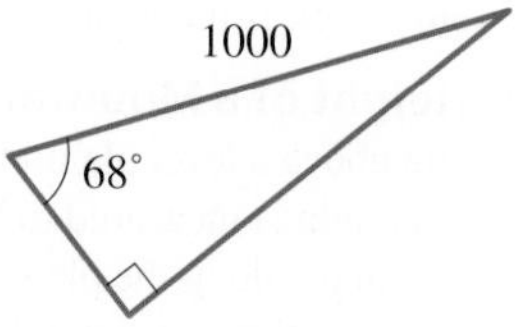

35.

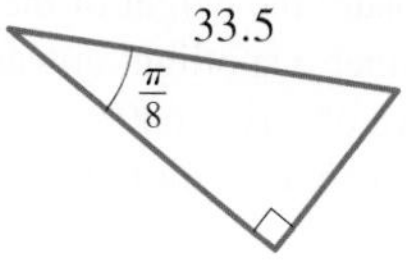

36.

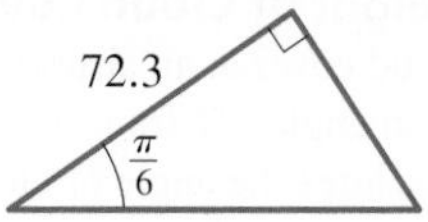

37.

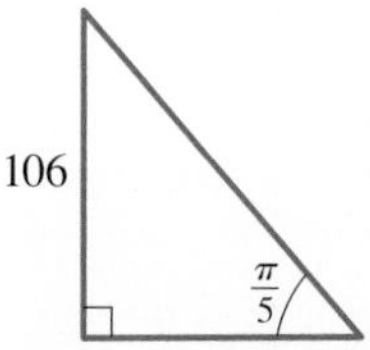

38.

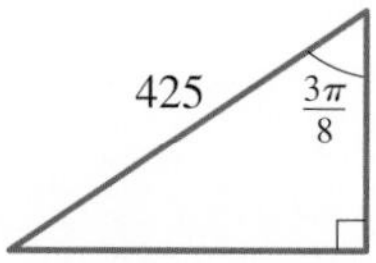

39. Use a ruler to carefully measure the sides of the triangle, and then use your measurements to estimate the six trigonometric ratios of θ.

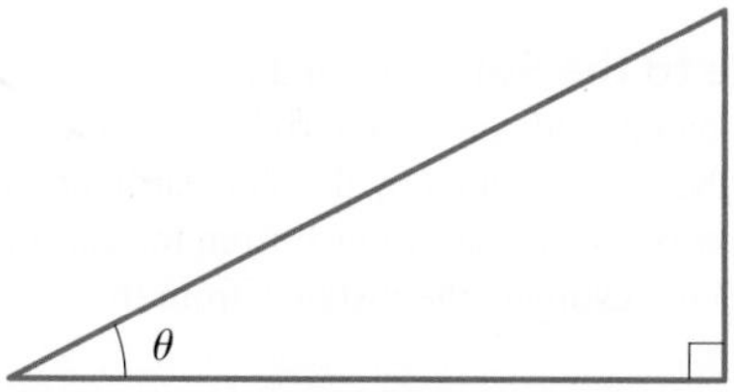

40. Using a protractor, sketch a right triangle that has the acute angle 40°. Measure the sides carefully, and use your results to estimate the six trigonometric ratios of 40°.

41–44 ■ Find x rounded to one decimal place.

41.

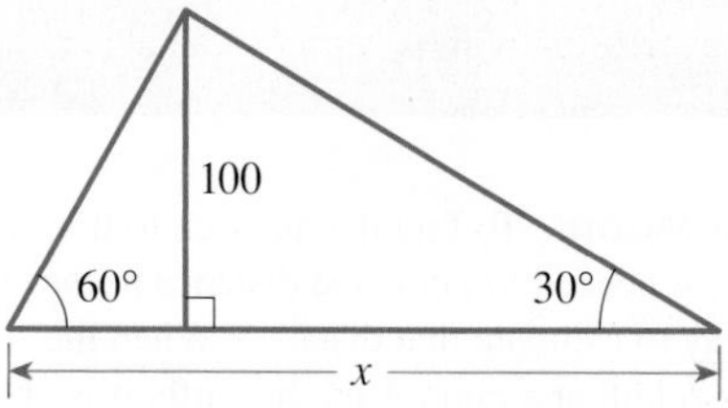

42.

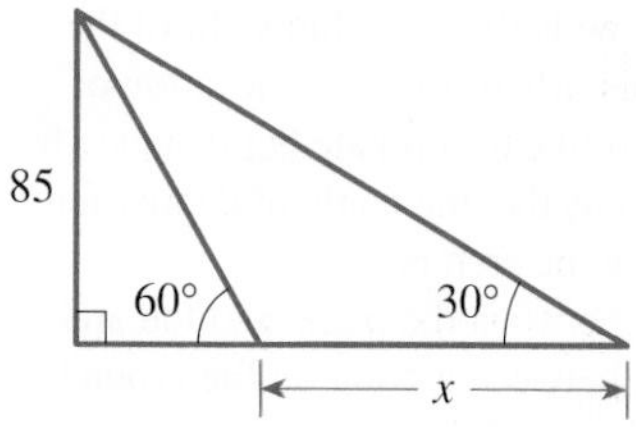

43.

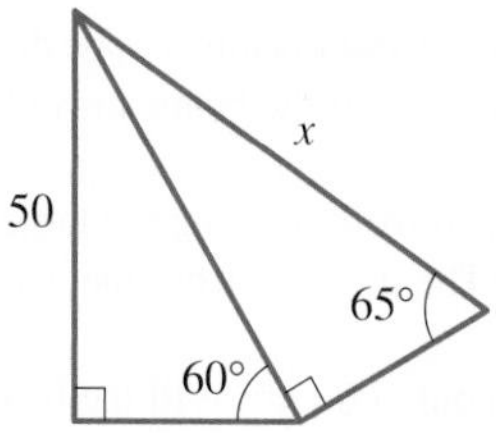

44.

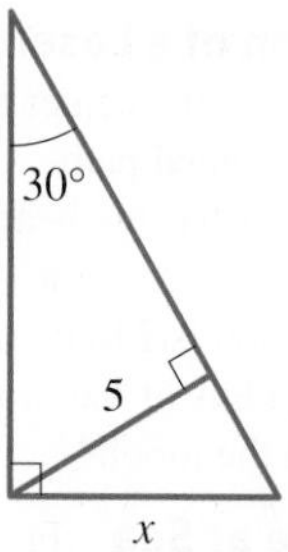

45. Express the length x in terms of the trigonometric ratios of θ.

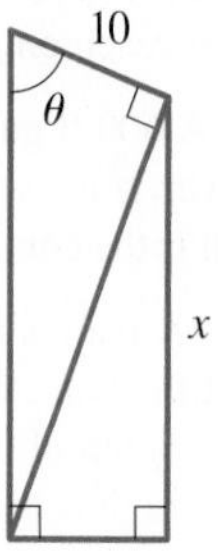

46. Express the length a, b, c, and d in the figure in terms of the trigonometric ratios of θ.

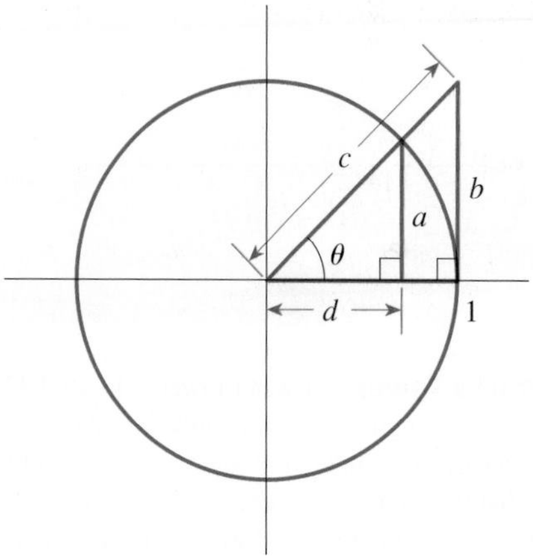

APPLICATIONS

47. Height of a Building The angle of elevation to the top of the Empire State Building in New York is found to be 11° from the ground at a distance of 1 mi from the base of the building. Using this information, find the height of the Empire State Building.

48. Gateway Arch A plane is flying within sight of the Gateway Arch in St. Louis, Missouri, at an elevation of 35,000 ft. The pilot would like to estimate her distance from the Gateway Arch. She finds that the angle of depression to a point on the ground below the arch is 22°.

(a) What is the distance between the plane and the arch?

(b) What is the distance between a point on the ground directly below the plane and the arch?

49. Deviation of a Laser Beam A laser beam is to be directed toward the center of the moon, but the beam strays 0.5° from its intended path.

(a) How far has the beam diverged from its assigned target when it reaches the moon? (The distance from the earth to the moon is 240,000 mi.)

(b) The radius of the moon is about 1000 mi. Will the beam strike the moon?

50. Distance at Sea From the top of a 200-ft lighthouse, the angle of depression to a ship in the ocean is 23°. How far is the ship from the base of the lighthouse?

51. Leaning Ladder A 20-ft ladder leans against a building so that the angle between the ground and the ladder is 72°. How high does the ladder reach on the building?

52. Height of a Tower A 600-ft guy wire is attached to the top of a communications tower. If the wire makes an angle of 65° with the ground, how tall is the communications tower?

53. Elevation of a Kite A man is lying on the beach, flying a kite. He holds the end of the kite string at ground level, and estimates the angle of elevation of the kite to be 50°. If the string is 450 ft long, how high is the kite above the ground?

54. Determining a Distance A woman standing on a hill sees a flagpole that she knows is 60 ft tall. The angle of depression to the bottom of the pole is 14°, and the angle of elevation to the top of the pole is 18°. Find her distance x from the pole.

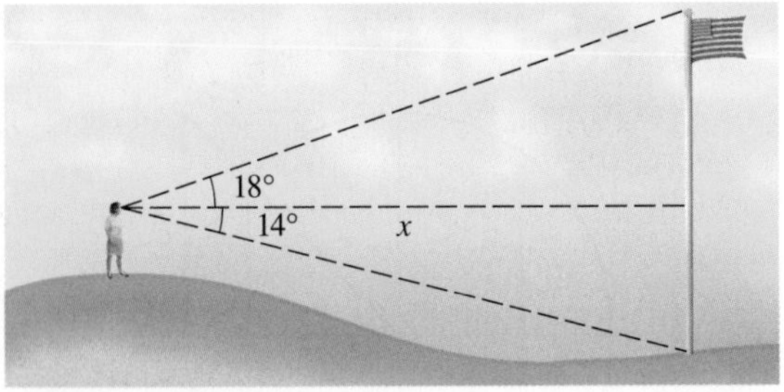

55. Height of a Tower A water tower is located 325 ft from a building (see the figure). From a window in the building, an observer notes that the angle of elevation to the top of the tower is 39° and that the angle of depression to the bottom of the tower is 25°. How tall is the tower? How high is the window?

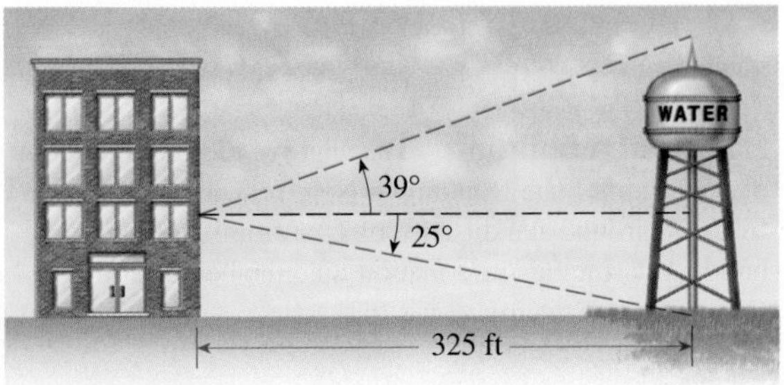

56. Determining a Distance An airplane is flying at an elevation of 5150 ft, directly above a straight highway. Two motorists are driving cars on the highway on opposite sides of the plane, and the angle of depression to one car is 35° and to the other is 52°. How far apart are the cars?

57. Determining a Distance If both cars in Exercise 56 are on one side of the plane and if the angle of depression to one car is 38° and to the other car is 52°, how far apart are the cars?

58. Height of a Balloon A hot-air balloon is floating above a straight road. To estimate their height above the ground, the balloonists simultaneously measure the angle of depression to two consecutive mileposts on the road on the same side of the balloon. The angles of depression are found to be 20° and 22°. How high is the balloon?

59. Height of a Mountain To estimate the height of a mountain above a level plain, the angle of elevation to the top of the mountain is measured to be 32°. One thousand feet closer to the mountain along the plain, it is found that the angle of elevation is 35°. Estimate the height of the mountain.

60. Height of Cloud Cover To measure the height of the cloud cover at an airport, a worker shines a spotlight upward at an angle 75° from the horizontal. An observer 600 m away measures the angle of elevation to the spot of light to be 45°. Find the height h of the cloud cover.

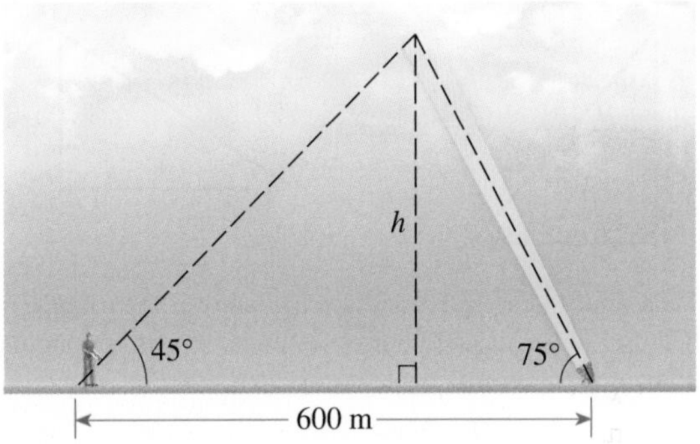

61. Distance to the Sun When the moon is exactly half full, the earth, moon, and sun form a right angle (see the figure). At that time the angle formed by the sun, earth, and moon is measured to be 89.85°. If the distance from the earth to the moon is 240,000 mi, estimate the distance from the earth to the sun.

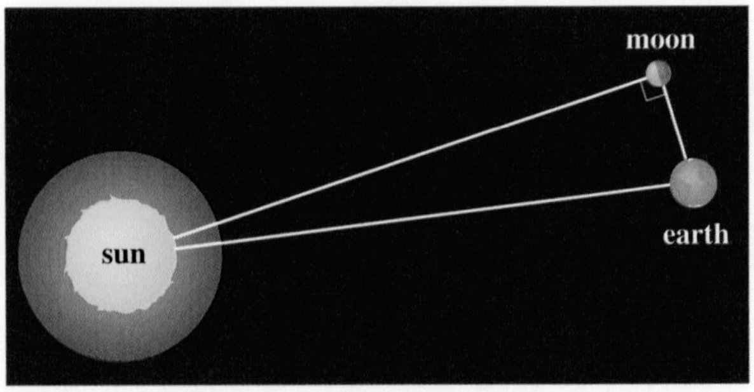

62. Distance to the Moon To find the distance to the sun as in Exercise 61, we needed to know the distance to the moon. Here is a way to estimate that distance: When the moon is seen at its zenith at a point A on the earth, it is observed to be at the horizon from point B (see the following figure). Points A and B are 6155 mi apart, and the radius of the earth is 3960 mi.

(a) Find the angle θ in degrees.

(b) Estimate the distance from point A to the moon.

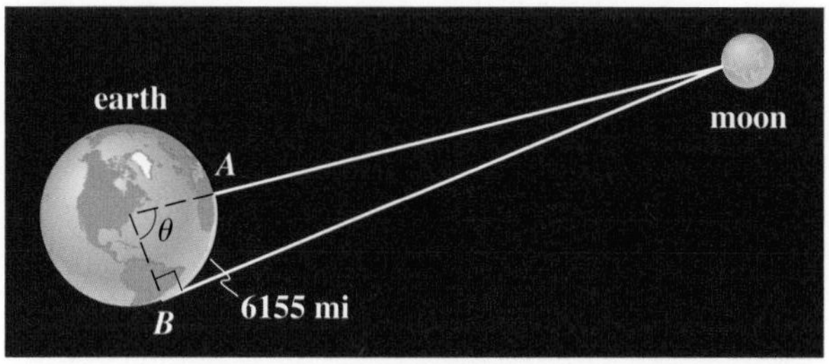

63. **Radius of the Earth** In Exercise 74 of Section 6.1 a method was given for finding the radius of the earth. Here is a more modern method: From a satellite 600 mi above the earth, it is observed that the angle formed by the vertical and the line of sight to the horizon is 60.276°. Use this information to find the radius of the earth.

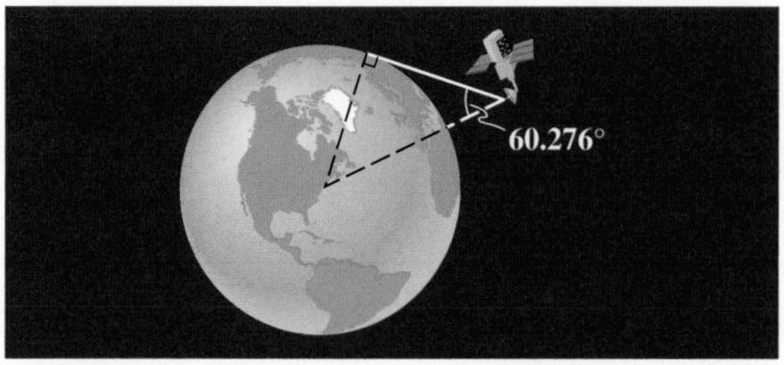

64. **Parallax** To find the distance to nearby stars, the method of parallax is used. The idea is to find a triangle with the star at one vertex and with a base as large as possible. To do this, the star is observed at two different times exactly 6 months apart, and its apparent change in position is recorded. From these two observations, $\angle E_1SE_2$ can be calculated. (The times are chosen so that $\angle E_1SE_2$ is as large as possible, which guarantees that $\angle E_1OS$ is 90°.) The angle E_1SO is called the *parallax* of the star. Alpha Centauri, the star nearest the earth, has a parallax of 0.000211°. Estimate the distance to this star. (Take the distance from the earth to the sun to be 9.3×10^7 mi.)

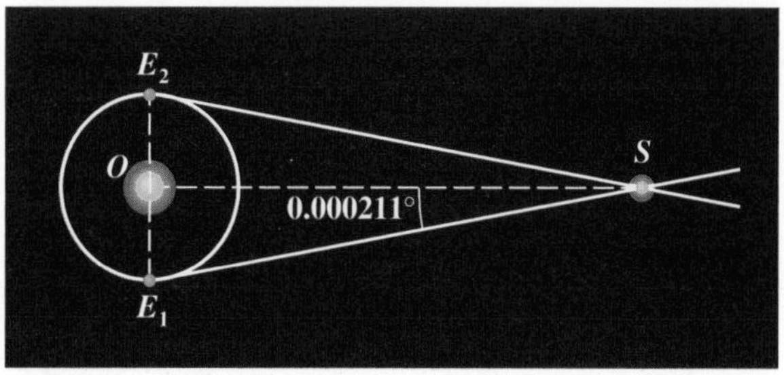

65. **Distance from Venus to the Sun** The **elongation** α of a planet is the angle formed by the planet, earth, and sun (see the figure). When Venus achieves its maximum elongation of 46.3°, the earth, Venus, and the sun form a triangle with a right angle at Venus. Find the distance between Venus and the sun in astronomical units (AU). (By definition the distance between the earth and the sun is 1 AU.)

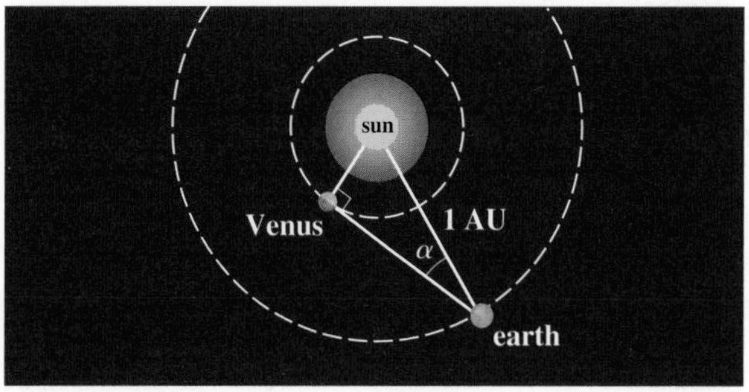

DISCOVERY ▪ DISCUSSION ▪ WRITING

66. **Similar Triangles** If two triangles are similar, what properties do they share? Explain how these properties make it possible to define the trigonometric ratios without regard to the size of the triangle.

6.3 Trigonometric Functions of Angles

Trigonometric Functions of Angles ▶ Evaluating Trigonometric Functions at Any Angle ▶ Trigonometric Identities ▶ Areas of Triangles

In the preceding section we defined the trigonometric ratios for acute angles. Here we extend the trigonometric ratios to all angles by defining the trigonometric functions of angles. With these functions we can solve practical problems that involve angles that are not necessarily acute.

Trigonometric Functions of Angles

Let POQ be a right triangle with acute angle θ as shown in Figure 1(a). Place θ in standard position as shown in Figure 1(b).

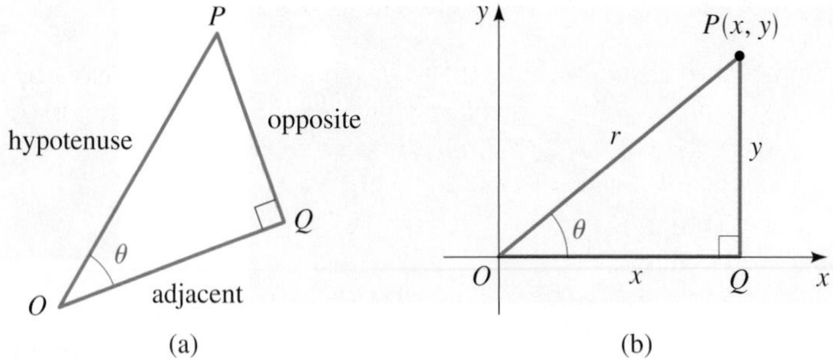

FIGURE 1

Then $P = P(x, y)$ is a point on the terminal side of θ. In triangle POQ, the opposite side has length y and the adjacent side has length x. Using the Pythagorean Theorem, we see that the hypotenuse has length $r = \sqrt{x^2 + y^2}$. So

$$\sin\theta = \frac{y}{r} \qquad \cos\theta = \frac{x}{r} \qquad \tan\theta = \frac{y}{x}$$

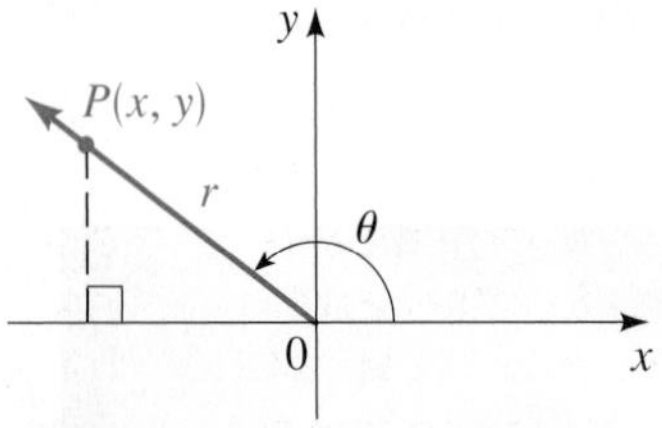

FIGURE 2

The other trigonometric ratios can be found in the same way.

These observations allow us to extend the trigonometric ratios to any angle. We define the trigonometric functions of angles as follows (see Figure 2).

DEFINITION OF THE TRIGONOMETRIC FUNCTIONS

Let θ be an angle in standard position and let $P(x, y)$ be a point on the terminal side. If $r = \sqrt{x^2 + y^2}$ is the distance from the origin to the point $P(x, y)$, then

$$\sin\theta = \frac{y}{r} \qquad \cos\theta = \frac{x}{r} \qquad \tan\theta = \frac{y}{x} \quad (x \neq 0)$$

$$\csc\theta = \frac{r}{y} \quad (y \neq 0) \qquad \sec\theta = \frac{r}{x} \quad (x \neq 0) \qquad \cot\theta = \frac{x}{y} \quad (y \neq 0)$$

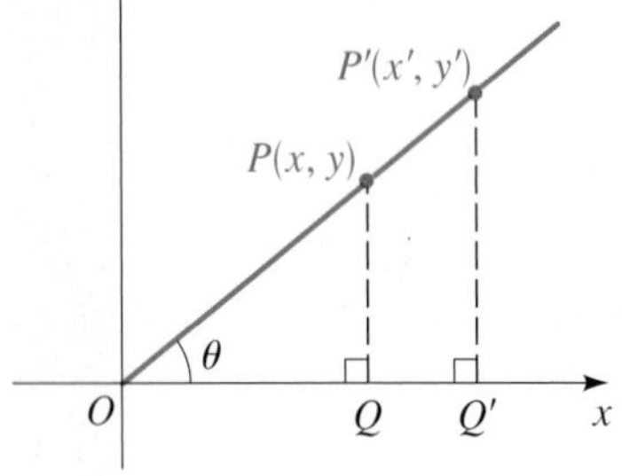

FIGURE 3

Since division by 0 is an undefined operation, certain trigonometric functions are not defined for certain angles. For example, $\tan 90° = y/x$ is undefined because $x = 0$. The angles for which the trigonometric functions may be undefined are the angles for which either the x- or y-coordinate of a point on the terminal side of the angle is 0. These are **quadrantal angles**—angles that are coterminal with the coordinate axes.

It is a crucial fact that the values of the trigonometric functions do *not* depend on the choice of the point $P(x, y)$. This is because if $P'(x', y')$ is any other point on the terminal side, as in Figure 3, then triangles POQ and $P'OQ'$ are similar.

Evaluating Trigonometric Functions at Any Angle

From the definition we see that the values of the trigonometric functions are all positive if the angle θ has its terminal side in Quadrant I. This is because x and y are positive in this quadrant. [Of course, r is always positive, since it is simply the distance from the origin to the point $P(x, y)$.] If the terminal side of θ is in Quadrant II, however, then x is

Relationship to the Trigonometric Functions of Real Numbers

You may have already studied the trigonometric functions defined using the unit circle (Chapter 5). To see how they relate to the trigonometric functions of an *angle*, let's start with the unit circle in the coordinate plane.

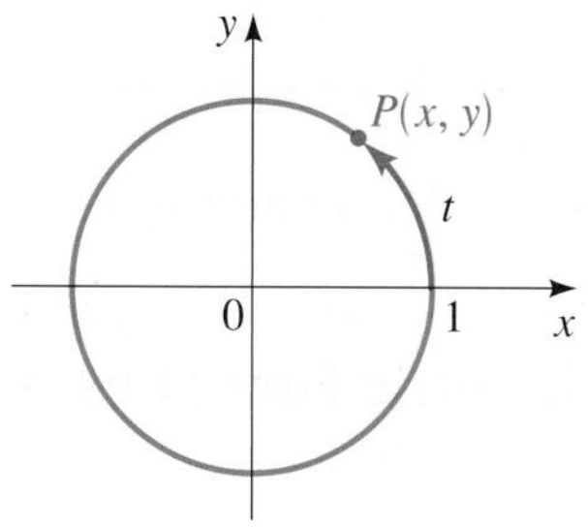

$P(x, y)$ is the terminal point determined by t.

Let $P(x, y)$ be the terminal point determined by an arc of length t on the unit circle. Then t subtends an angle θ at the center of the circle. If we drop a perpendicular from P onto the point Q on the x-axis, then triangle ΔOPQ is a right triangle with legs of length x and y, as shown in the figure.

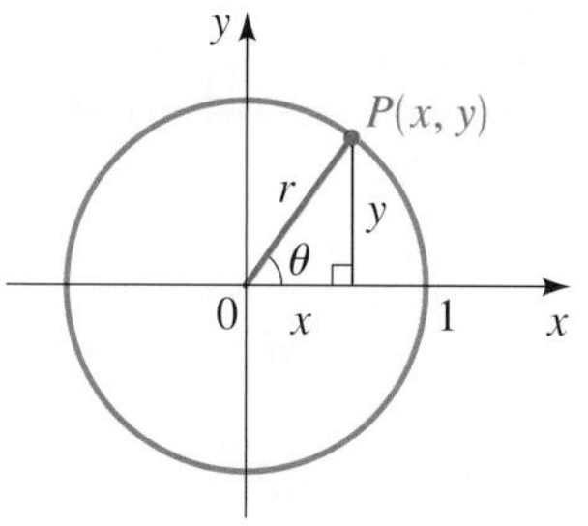

Triangle OPQ is a right triangle.

Now, by the definition of the trigonometric functions of the *real number* t we have

$$\sin t = y$$

$$\cos t = x$$

By the definition of the trigonometric functions of the *angle* θ we have

$$\sin \theta = \frac{\text{opp}}{\text{hyp}} = \frac{y}{1} = y$$

$$\cos \theta = \frac{\text{adj}}{\text{hyp}} = \frac{x}{1} = x$$

If θ is measured in radians, then $\theta = t$. (See the figure below.) Comparing the two ways of defining the trigonometric functions, we see that they are identical. In other words, as functions, they assign identical values to a given real number. (The real number is the radian measure of θ in one case or the length t of an arc in the other.)

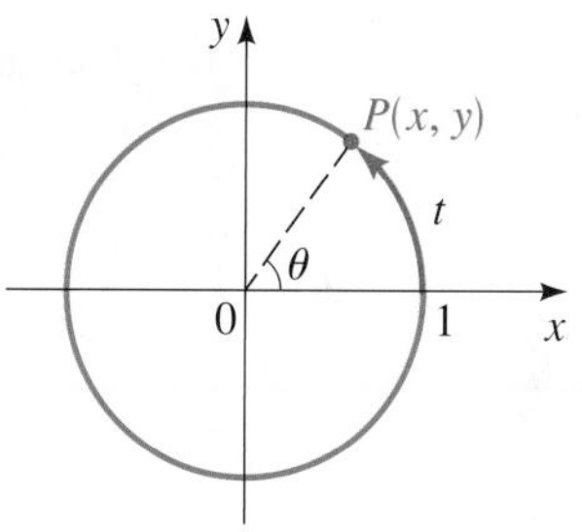

The radian measure of angle θ is t.

Why then do we study trigonometry in two different ways? Because different applications require that we view the trigonometric functions differently. (See *Focus on Modeling*, pages 427, 489, and 533, and Sections 6.2, 6.5, and 6.6.)

The following mnemonic device can be used to remember which trigonometric functions are positive in each quadrant: **A**ll of them, **S**ine, **T**angent, or **C**osine.

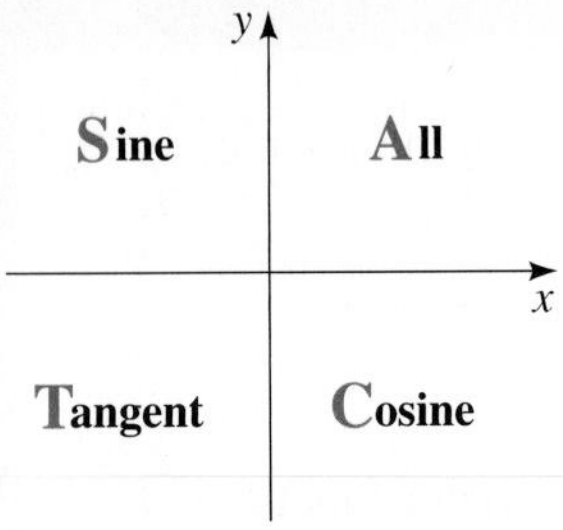

You can remember this as "All Students Take Calculus."

negative and y is positive. Thus, in Quadrant II the functions $\sin\theta$ and $\csc\theta$ are positive, and all the other trigonometric functions have negative values. You can check the other entries in the following table.

SIGNS OF THE TRIGONOMETRIC FUNCTIONS

Quadrant	Positive Functions	Negative Functions
I	all	none
II	sin, csc	cos, sec, tan, cot
III	tan, cot	sin, csc, cos, sec
IV	cos, sec	sin, csc, tan, cot

We now turn our attention to finding the values of the trigonometric functions for angles that are not acute.

EXAMPLE 1 | Finding Trigonometric Functions of Angles

Find **(a)** $\cos 135°$ and **(b)** $\tan 390°$.

SOLUTION

(a) From Figure 4 we see that $\cos 135° = -x/r$. But $\cos 45° = x/r$, and since $\cos 45° = \sqrt{2}/2$, we have

$$\cos 135° = -\frac{\sqrt{2}}{2}$$

(b) The angles 390° and 30° are coterminal. From Figure 5 it's clear that $\tan 390° = \tan 30°$ and, since $\tan 30° = \sqrt{3}/3$, we have

$$\tan 390° = \frac{\sqrt{3}}{3}$$

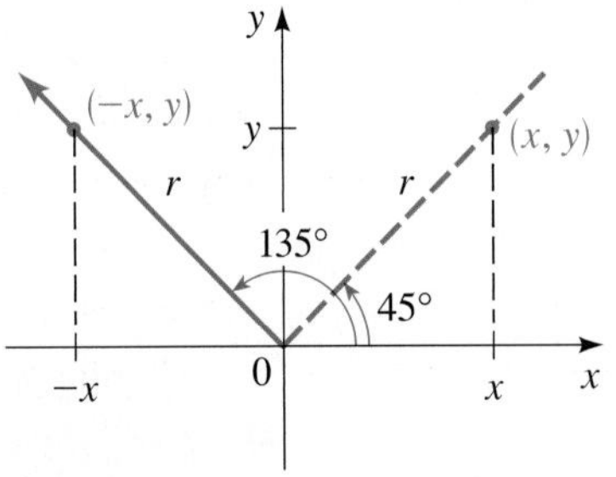

FIGURE 4

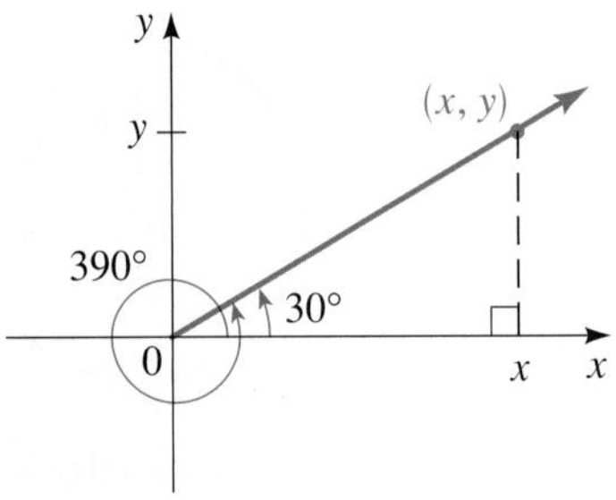

FIGURE 5

NOW TRY EXERCISES **11** AND **13** ■

From Example 1 we see that the trigonometric functions for angles that aren't acute have the same value, except possibly for sign, as the corresponding trigonometric functions of an acute angle. That acute angle will be called the *reference angle*.

REFERENCE ANGLE

Let θ be an angle in standard position. The **reference angle** $\bar{\theta}$ associated with θ is the acute angle formed by the terminal side of θ and the x-axis.

Figure 6 shows that to find a reference angle $\overline{\theta}$, it's useful to know the quadrant in which the terminal side of the angle θ lies.

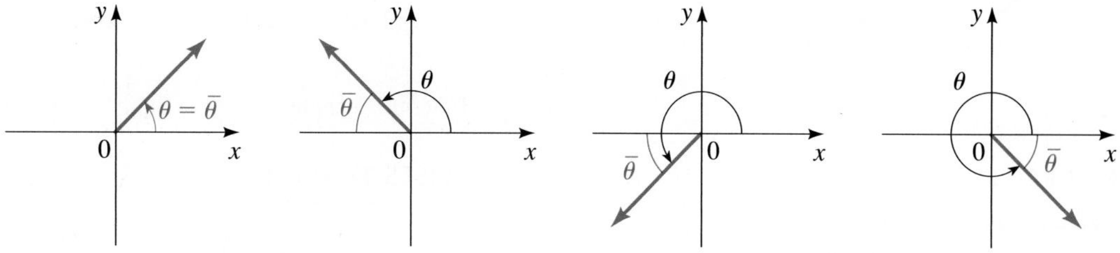

FIGURE 6 The reference angle $\overline{\theta}$ for an angle θ

EXAMPLE 2 | Finding Reference Angles

Find the reference angle for **(a)** $\theta = \dfrac{5\pi}{3}$ and **(b)** $\theta = 870°$.

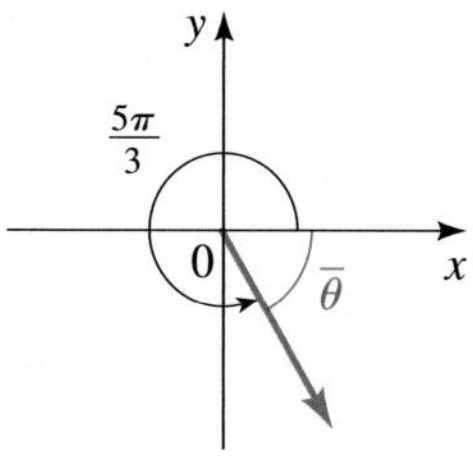

FIGURE 7

SOLUTION

(a) The reference angle is the acute angle formed by the terminal side of the angle $5\pi/3$ and the x-axis (see Figure 7). Since the terminal side of this angle is in Quadrant IV, the reference angle is

$$\overline{\theta} = 2\pi - \frac{5\pi}{3} = \frac{\pi}{3}$$

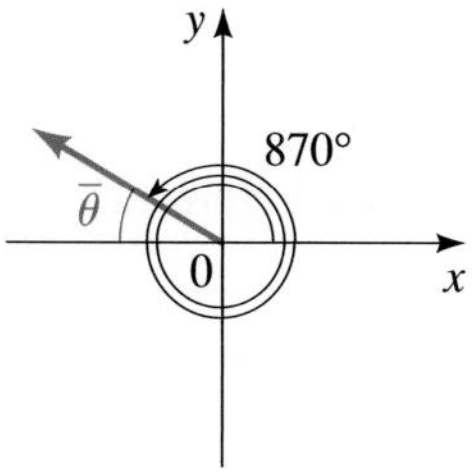

FIGURE 8

(b) The angles 870° and 150° are coterminal [because $870 - 2(360) = 150$]. Thus, the terminal side of this angle is in Quadrant II (see Figure 8). So the reference angle is

$$\overline{\theta} = 180° - 150° = 30°$$

NOW TRY EXERCISES **3** AND **7**

EVALUATING TRIGONOMETRIC FUNCTIONS FOR ANY ANGLE

To find the values of the trigonometric functions for any angle θ, we carry out the following steps.

1. Find the reference angle $\overline{\theta}$ associated with the angle θ.
2. Determine the sign of the trigonometric function of θ by noting the quadrant in which θ lies.
3. The value of the trigonometric function of θ is the same, except possibly for sign, as the value of the trigonometric function of $\overline{\theta}$.

EXAMPLE 3 | Using the Reference Angle to Evaluate Trigonometric Functions

Find **(a)** sin 240° and **(b)** cot 495°.

SOLUTION

(a) This angle has its terminal side in Quadrant III, as shown in Figure 9. The reference angle is therefore $240° - 180° = 60°$, and the value of sin 240° is negative. Thus

$$\sin 240° = -\sin 60° = -\frac{\sqrt{3}}{2}$$

Sign — Reference angle

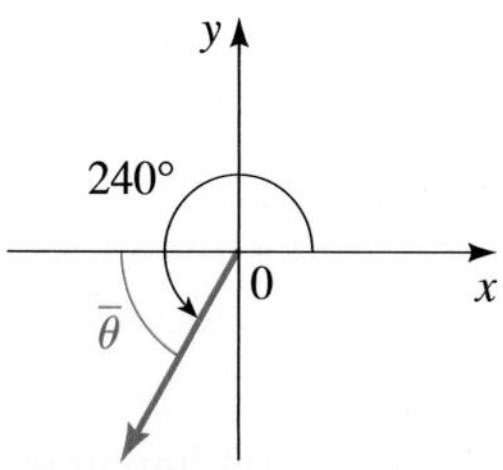

FIGURE 9

sin 240° is *negative*.

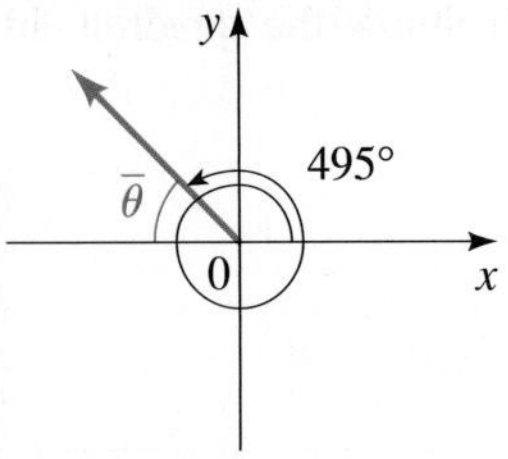

FIGURE 10

S | A
T | C

tan 495° is negative, so cot 495° is negative.

(b) The angle 495° is coterminal with the angle 135°, and the terminal side of this angle is in Quadrant II, as shown in Figure 10. So the reference angle is $180° - 135° = 45°$, and the value of cot 495° is negative. We have

$$\cot 495° = \cot 135° = -\cot 45° = -1$$

Coterminal angles | Sign | Reference angle

NOW TRY EXERCISES **17** AND **19** ■

EXAMPLE 4 | Using the Reference Angle to Evaluate Trigonometric Functions

Find **(a)** $\sin \dfrac{16\pi}{3}$ and **(b)** $\sec\left(-\dfrac{\pi}{4}\right)$.

SOLUTION

(a) The angle $16\pi/3$ is coterminal with $4\pi/3$, and these angles are in Quadrant III (see Figure 11). Thus, the reference angle is $(4\pi/3) - \pi = \pi/3$. Since the value of sine is negative in Quadrant III, we have

$$\sin \frac{16\pi}{3} = \sin \frac{4\pi}{3} = -\sin \frac{\pi}{3} = -\frac{\sqrt{3}}{2}$$

Coterminal angles | Sign | Reference angle

(b) The angle $-\pi/4$ is in Quadrant IV, and its reference angle is $\pi/4$ (see Figure 12). Since secant is positive in this quadrant, we get

$$\sec\left(-\frac{\pi}{4}\right) = +\sec \frac{\pi}{4} = \sqrt{2}$$

Sign | Reference angle

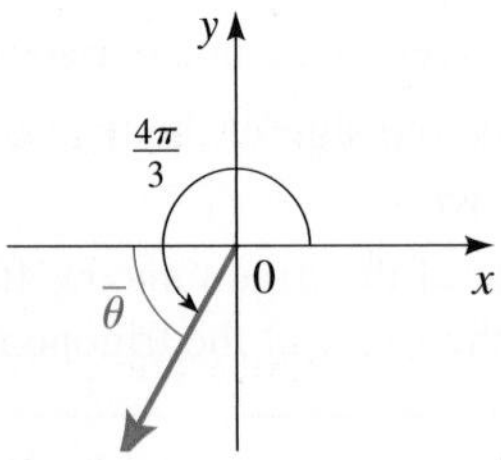

FIGURE 11

S | A
T | C

$\sin \frac{16\pi}{3}$ *is negative.*

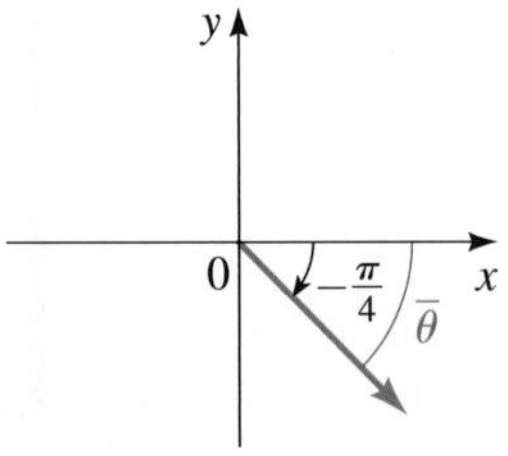

FIGURE 12

S | A
T | C

$\cos(-\frac{\pi}{4})$ *is positive, so* $\sec(-\frac{\pi}{4})$ *is positive.*

NOW TRY EXERCISES **23** AND **25** ■

▼ Trigonometric Identities

The trigonometric functions of angles are related to each other through several important equations called **trigonometric identities**. We've already encountered the reciprocal identities. These identities continue to hold for any angle θ, provided that both sides of the equation are defined. The Pythagorean identities are a consequence of the Pythagorean Theorem.*

FUNDAMENTAL IDENTITIES

Reciprocal Identities

$$\csc\theta = \frac{1}{\sin\theta} \qquad \sec\theta = \frac{1}{\cos\theta} \qquad \cot\theta = \frac{1}{\tan\theta}$$

$$\tan\theta = \frac{\sin\theta}{\cos\theta} \qquad \cot\theta = \frac{\cos\theta}{\sin\theta}$$

Pythagorean Identities

$$\sin^2\theta + \cos^2\theta = 1 \qquad \tan^2\theta + 1 = \sec^2\theta \qquad 1 + \cot^2\theta = \csc^2\theta$$

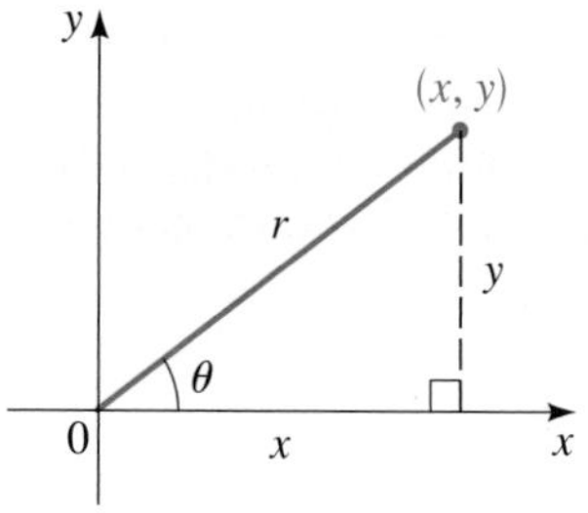

FIGURE 13

PROOF Let's prove the first Pythagorean identity. Using $x^2 + y^2 = r^2$ (the Pythagorean Theorem) in Figure 13, we have

$$\sin^2\theta + \cos^2\theta = \left(\frac{y}{r}\right)^2 + \left(\frac{x}{r}\right)^2 = \frac{x^2 + y^2}{r^2} = \frac{r^2}{r^2} = 1$$

Thus, $\sin^2\theta + \cos^2\theta = 1$. (Although the figure indicates an acute angle, you should check that the proof holds for all angles θ.) ■

See Exercises 61 and 62 for the proofs of the other two Pythagorean identities.

EXAMPLE 5 | Expressing One Trigonometric Function in Terms of Another

(a) Express $\sin\theta$ in terms of $\cos\theta$.

(b) Express $\tan\theta$ in terms of $\sin\theta$, where θ is in Quadrant II.

SOLUTION

(a) From the first Pythagorean identity we get

$$\sin\theta = \pm\sqrt{1 - \cos^2\theta}$$

where the sign depends on the quadrant. If θ is in Quadrant I or II, then $\sin\theta$ is positive, and so

$$\sin\theta = \sqrt{1 - \cos^2\theta}$$

whereas if θ is in Quadrant III or IV, $\sin\theta$ is negative, and so

$$\sin\theta = -\sqrt{1 - \cos^2\theta}$$

(b) Since $\tan\theta = \sin\theta/\cos\theta$, we need to write $\cos\theta$ in terms of $\sin\theta$. By part (a)

$$\cos\theta = \pm\sqrt{1 - \sin^2\theta}$$

and since $\cos\theta$ is negative in Quadrant II, the negative sign applies here. Thus

$$\tan\theta = \frac{\sin\theta}{\cos\theta} = \frac{\sin\theta}{-\sqrt{1 - \sin^2\theta}}$$

NOW TRY EXERCISE **39** ■

---

* We follow the usual convention of writing $\sin^2\theta$ for $(\sin\theta)^2$. In general, we write $\sin^n\theta$ for $(\sin\theta)^n$ for all integers n except $n = -1$. The exponent $n = -1$ will be assigned another meaning in Section 6.4. Of course, the same convention applies to the other five trigonometric functions.

EXAMPLE 6 | Evaluating a Trigonometric Function

If $\tan\theta = \frac{2}{3}$ and θ is in Quadrant III, find $\cos\theta$.

SOLUTION 1 We need to write $\cos\theta$ in terms of $\tan\theta$. From the identity $\tan^2\theta + 1 = \sec^2\theta$ we get $\sec\theta = \pm\sqrt{\tan^2\theta + 1}$. In Quadrant III, $\sec\theta$ is negative, so

$$\sec\theta = -\sqrt{\tan^2\theta + 1}$$

Thus,

$$\cos\theta = \frac{1}{\sec\theta} = \frac{1}{-\sqrt{\tan^2\theta + 1}}$$

$$= \frac{1}{-\sqrt{\left(\frac{2}{3}\right)^2 + 1}} = \frac{1}{-\sqrt{\frac{13}{9}}} = -\frac{3}{\sqrt{13}}$$

If you wish to rationalize the denominator, you can express $\cos\theta$ as

$$-\frac{3}{\sqrt{13}}\cdot\frac{\sqrt{13}}{\sqrt{13}} = -\frac{3\sqrt{3}}{13}$$

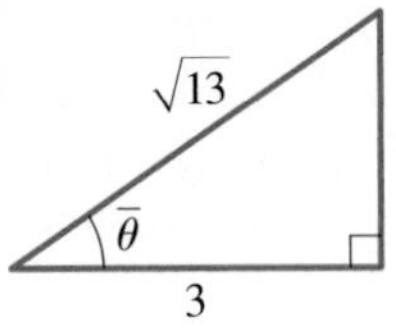

FIGURE 14

SOLUTION 2 This problem can be solved more easily by using the method of Example 2 of Section 6.2. Recall that, except for sign, the values of the trigonometric functions of any angle are the same as those of an acute angle (the reference angle). So, ignoring the sign for the moment, let's sketch a right triangle with an acute angle $\bar{\theta}$ satisfying $\tan\bar{\theta} = \frac{2}{3}$ (see Figure 14). By the Pythagorean Theorem the hypotenuse of this triangle has length $\sqrt{13}$. From the triangle in Figure 14 we immediately see that $\cos\bar{\theta} = 3/\sqrt{13}$. Since θ is in Quadrant III, $\cos\theta$ is negative, so

$$\cos\theta = -\frac{3}{\sqrt{13}}$$

NOW TRY EXERCISE **45** ■

EXAMPLE 7 | Evaluating Trigonometric Functions

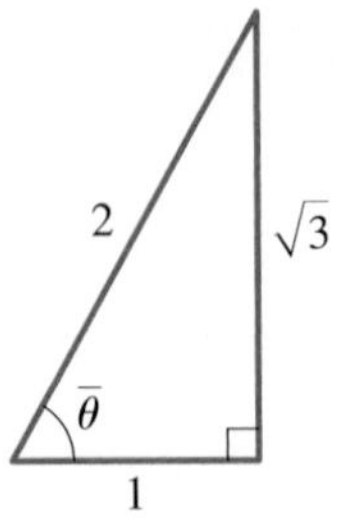

FIGURE 15

If $\sec\theta = 2$ and θ is in Quadrant IV, find the other five trigonometric functions of θ.

SOLUTION We sketch a triangle as in Figure 15 so that $\sec\bar{\theta} = 2$. Taking into account the fact that θ is in Quadrant IV, we get

$$\sin\theta = -\frac{\sqrt{3}}{2} \qquad \cos\theta = \frac{1}{2} \qquad \tan\theta = -\sqrt{3}$$

$$\csc\theta = -\frac{2}{\sqrt{3}} \qquad \sec\theta = 2 \qquad \cot\theta = -\frac{1}{\sqrt{3}}$$

NOW TRY EXERCISE **47** ■

▼ Areas of Triangles

We conclude this section with an application of the trigonometric functions that involves angles that are not necessarily acute. More extensive applications appear in the next two sections.

The area of a triangle is $\mathcal{A} = \frac{1}{2} \times \text{base} \times \text{height}$. If we know two sides and the included angle of a triangle, then we can find the height using the trigonometric functions, and from this we can find the area.

If θ is an acute angle, then the height of the triangle in Figure 16(a) is given by $h = b\sin\theta$. Thus the area is

$$\mathcal{A} = \tfrac{1}{2} \times \text{base} \times \text{height} = \tfrac{1}{2}ab\sin\theta$$

If the angle θ is not acute, then from Figure 16(b) we see that the height of the triangle is

$$h = b\sin(180° - \theta) = b\sin\theta$$

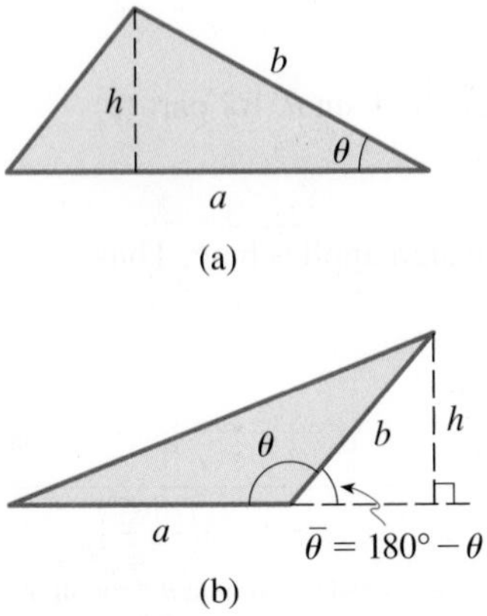

FIGURE 16

This is so because the reference angle of θ is the angle $180° - \theta$. Thus, in this case also, the area of the triangle is

$$\mathcal{A} = \tfrac{1}{2} \times \text{base} \times \text{height} = \tfrac{1}{2}ab\sin\theta$$

AREA OF A TRIANGLE

The area $\mathcal{A}$ of a triangle with sides of lengths a and b and with included angle θ is

$$\mathcal{A} = \tfrac{1}{2}ab\sin\theta$$

EXAMPLE 8 | Finding the Area of a Triangle

Find the area of triangle ABC shown in Figure 17.

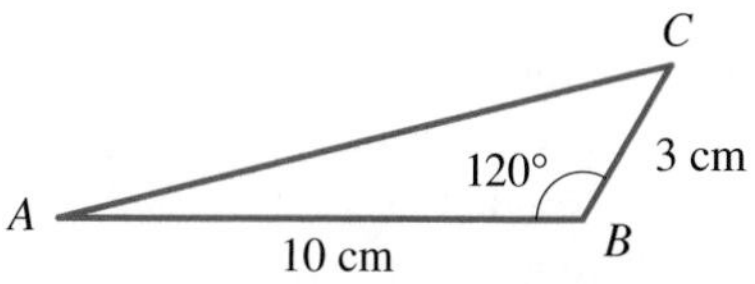

FIGURE 17

SOLUTION The triangle has sides of length 10 cm and 3 cm, with included angle 120°. Therefore

$$\begin{aligned}\mathcal{A} &= \tfrac{1}{2}ab\sin\theta \\ &= \tfrac{1}{2}(10)(3)\sin 120° \\ &= 15\sin 60° \qquad \text{Reference angle} \\ &= 15\frac{\sqrt{3}}{2} \approx 13 \text{ cm}^2\end{aligned}$$

NOW TRY EXERCISE **55**

6.3 EXERCISES

CONCEPTS

1. If the angle θ is in standard position and $P(x, y)$ is a point on the terminal side of θ, and r is the distance from the origin to P, then

$\sin\theta = \frac{\square}{\square}$ $\quad\cos\theta = \frac{\square}{\square}$ $\quad\tan\theta = \frac{\square}{\square}$

2. The sign of a trigonometric function of θ depends on the ________ in which the terminal side of the angle θ lies.

In Quadrant II, $\sin\theta$ is ________ (positive / negative).

In Quadrant III, $\cos\theta$ is ________ (positive / negative).

In Quadrant IV, $\sin\theta$ is ________ (positive / negative).

SKILLS

3–10 ■ Find the reference angle for the given angle.

3. (a) 150° **(b)** 330° **(c)** −30°

4. (a) 120° **(b)** −210° **(c)** 780°

5. (a) 225° **(b)** 810° **(c)** −105°

6. (a) 99° **(b)** −199° **(c)** 359°

7. (a) $\frac{11\pi}{4}$ **(b)** $-\frac{11\pi}{6}$ **(c)** $\frac{11\pi}{3}$

8. (a) $\frac{4\pi}{3}$ **(b)** $\frac{33\pi}{4}$ **(c)** $-\frac{23\pi}{6}$

9. (a) $\frac{5\pi}{7}$ **(b)** -1.4π **(c)** 1.4

10. (a) 2.3π **(b)** 2.3 **(c)** -10π

11–34 ■ Find the exact value of the trigonometric function.

11. sin 150° **12.** sin 225° **13.** cos 210°

14. cos(−60°) **15.** tan(−60°) **16.** sec 300°

17. csc(−630°) **18.** cot 210° **19.** cos 570°

20. sec 120° **21.** tan 750° **22.** cos 660°

23. $\sin\frac{2\pi}{3}$ **24.** $\sin\frac{5\pi}{3}$ **25.** $\sin\frac{3\pi}{2}$

26. $\cos\frac{7\pi}{3}$ **27.** $\cos\left(-\frac{7\pi}{3}\right)$ **28.** $\tan\frac{5\pi}{6}$

29. $\sec\frac{17\pi}{3}$ **30.** $\csc\frac{5\pi}{4}$ **31.** $\cot\left(-\frac{\pi}{4}\right)$

32. $\cos\frac{7\pi}{4}$ **33.** $\tan\frac{5\pi}{2}$ **34.** $\sin\frac{11\pi}{6}$

35–38 ■ Find the quadrant in which θ lies from the information given.

35. $\sin\theta < 0$ and $\cos\theta < 0$

36. $\tan\theta < 0$ and $\sin\theta < 0$

37. $\sec\theta > 0$ and $\tan\theta < 0$

38. $\csc\theta > 0$ and $\cos\theta < 0$

39–44 ■ Write the first trigonometric function in terms of the second for θ in the given quadrant.

39. $\tan\theta$, $\cos\theta$; θ in Quadrant III

40. $\cot\theta$, $\sin\theta$; θ in Quadrant II

41. $\cos\theta$, $\sin\theta$; θ in Quadrant IV

42. $\sec\theta$, $\sin\theta$; θ in Quadrant I

43. $\sec\theta$, $\tan\theta$; θ in Quadrant II

44. $\csc\theta$, $\cot\theta$; θ in Quadrant III

45–52 ■ Find the values of the trigonometric functions of θ from the information given.

45. $\sin\theta = \frac{3}{5}$, θ in Quadrant II

46. $\cos\theta = -\frac{7}{12}$, θ in Quadrant III

47. $\tan\theta = -\frac{3}{4}$, $\cos\theta > 0$

48. $\sec\theta = 5$, $\sin\theta < 0$

49. $\csc\theta = 2$, θ in Quadrant I

50. $\cot\theta = \frac{1}{4}$, $\sin\theta < 0$

51. $\cos\theta = -\frac{2}{7}$, $\tan\theta < 0$

52. $\tan\theta = -4$, $\sin\theta > 0$

53. If $\theta = \pi/3$, find the value of each expression.
(a) $\sin 2\theta$, $2\sin\theta$ **(b)** $\sin\frac{1}{2}\theta$, $\frac{1}{2}\sin\theta$
(c) $\sin^2\theta$, $\sin(\theta^2)$

54. Find the area of a triangle with sides of length 7 and 9 and included angle 72°.

55. Find the area of a triangle with sides of length 10 and 22 and included angle 10°.

56. Find the area of an equilateral triangle with side of length 10.

57. A triangle has an area of 16 in^2, and two of the sides of the triangle have lengths 5 in. and 7 in. Find the angle included by these two sides.

58. An isosceles triangle has an area of 24 cm^2, and the angle between the two equal sides is $5\pi/6$. What is the length of the two equal sides?

59–60 ■ Find the area of the shaded region in the figure.

59.

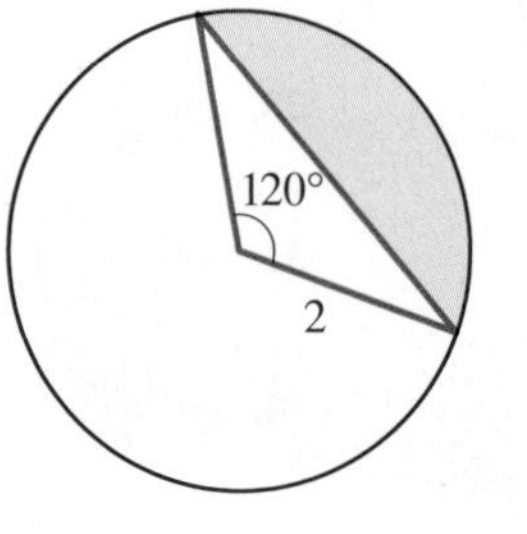

60.

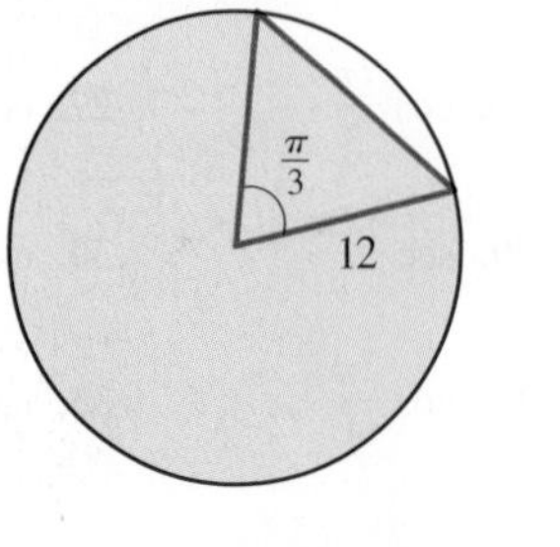

61. Use the first Pythagorean identity to prove the second. [*Hint:* Divide by $\cos^2\theta$.]

62. Use the first Pythagorean identity to prove the third.

APPLICATIONS

63. Height of a Rocket A rocket fired straight up is tracked by an observer on the ground a mile away.
(a) Show that when the angle of elevation is θ, the height of the rocket in feet is $h = 5280\tan\theta$.
(b) Complete the table to find the height of the rocket at the given angles of elevation.

θ	20°	60°	80°	85°
h				

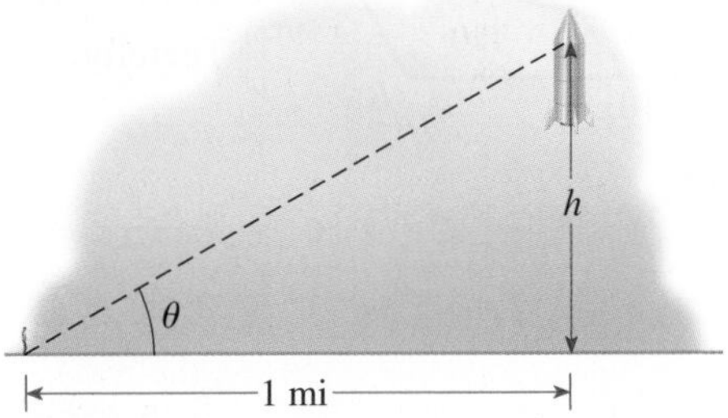

64. Rain Gutter A rain gutter is to be constructed from a metal sheet of width 30 cm by bending up one-third of the sheet on each side through an angle θ.
(a) Show that the cross-sectional area of the gutter is modeled by the function

$$A(\theta) = 100\sin\theta + 100\sin\theta\cos\theta$$

(b) Graph the function A for $0 \le \theta \le \pi/2$.
(c) For what angle θ is the largest cross-sectional area achieved?

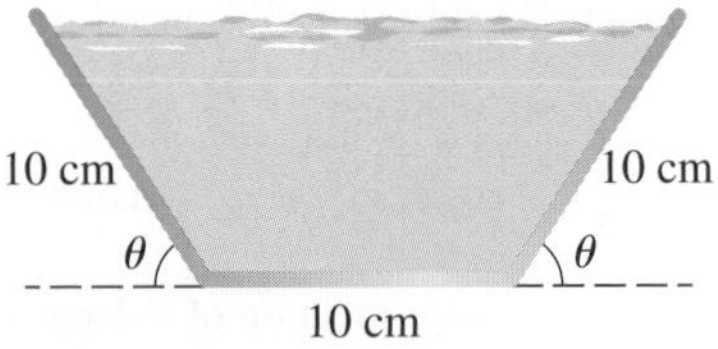

65. Wooden Beam A rectangular beam is to be cut from a cylindrical log of diameter 20 cm. The figures show different ways this can be done.
(a) Express the cross-sectional area of the beam as a function of the angle θ in the figures.
(b) Graph the function you found in part (a).
(c) Find the dimensions of the beam with largest cross-sectional area.

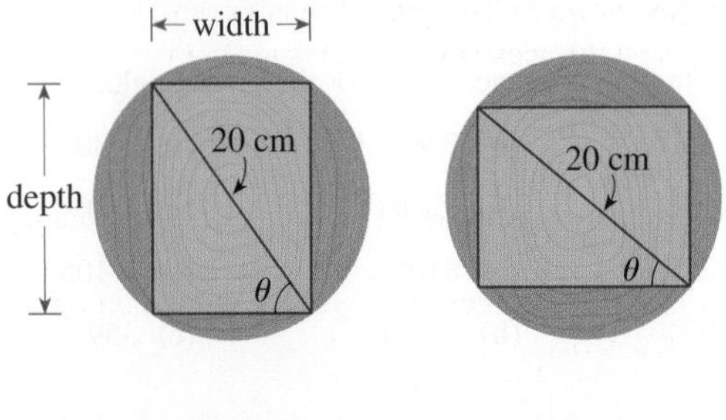

66. Strength of a Beam The strength of a beam is proportional to the width and the square of the depth. A beam is cut from a log as in Exercise 65. Express the strength of the beam as a function of the angle θ in the figures.

67. Throwing a Shot Put The range R and height H of a shot put thrown with an initial velocity of v_0 ft/s at an angle θ are given by

$$R = \frac{v_0^2 \sin(2\theta)}{g}$$

$$H = \frac{v_0^2 \sin^2\theta}{2g}$$

On the earth $g = 32$ ft/s^2 and on the moon $g = 5.2$ ft/s^2. Find the range and height of a shot put thrown under the given conditions.

(a) On the earth with $v_0 = 12$ ft/s and $\theta = \pi/6$
(b) On the moon with $v_0 = 12$ ft/s and $\theta = \pi/6$

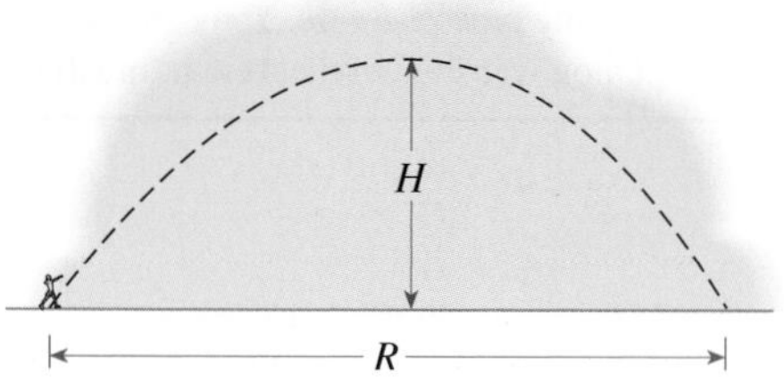

68. Sledding The time in seconds that it takes for a sled to slide down a hillside inclined at an angle θ is

$$t = \sqrt{\frac{d}{16 \sin \theta}}$$

where d is the length of the slope in feet. Find the time it takes to slide down a 2000-ft slope inclined at 30°.

69. Beehives In a beehive each cell is a regular hexagonal prism, as shown in the figure. The amount of wax W in the cell depends on the apex angle θ and is given by

$$W = 3.02 - 0.38 \cot \theta + 0.65 \csc \theta$$

Bees instinctively choose θ so as to use the least amount of wax possible.

(a) Use a graphing device to graph W as a function of θ for $0 < \theta < \pi$.
(b) For what value of θ does W have its minimum value? [*Note:* Biologists have discovered that bees rarely deviate from this value by more than a degree or two.]

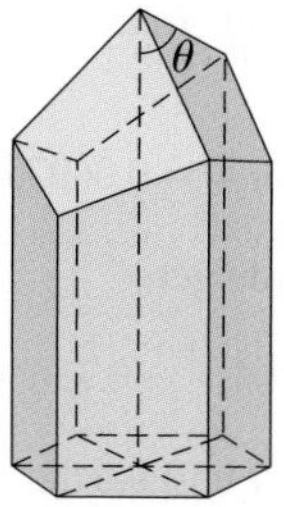

70. Turning a Corner A steel pipe is being carried down a hallway that is 9 ft wide. At the end of the hall there is a right-angled turn into a narrower hallway 6 ft wide.

(a) Show that the length of the pipe in the figure is modeled by the function

$$L(\theta) = 9 \csc \theta + 6 \sec \theta$$

(b) Graph the function L for $0 < \theta < \pi/2$.
(c) Find the minimum value of the function L.
(d) Explain why the value of L you found in part (c) is the length of the longest pipe that can be carried around the corner.

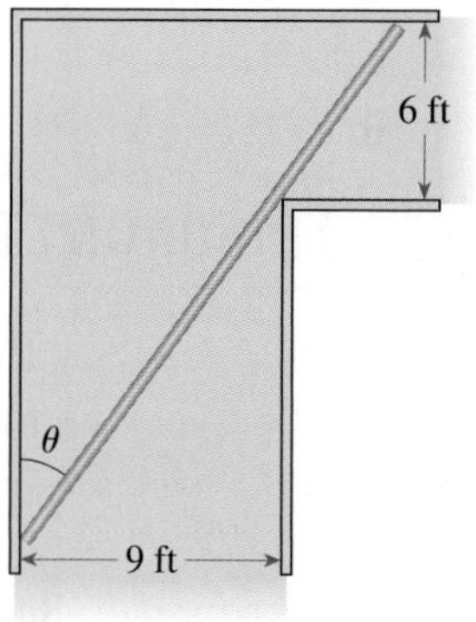

71. Rainbows Rainbows are created when sunlight of different wavelengths (colors) is refracted and reflected in raindrops. The angle of elevation θ of a rainbow is always the same. It can be shown that $\theta = 4\beta - 2\alpha$, where

$$\sin \alpha = k \sin \beta$$

and $\alpha = 59.4°$ and $k = 1.33$ is the index of refraction of water. Use the given information to find the angle of elevation θ of a rainbow. (For a mathematical explanation of rainbows see *Calculus Early Transcendentals,* 7th Edition, by James Stewart, page 282.)

DISCOVERY ■ DISCUSSION ■ WRITING

72. **Using a Calculator** To solve a certain problem, you need to find the sine of 4 rad. Your study partner uses his calculator and tells you that

$$\sin 4 = 0.0697564737$$

On your calculator you get

$$\sin 4 = -0.7568024953$$

What is wrong? What mistake did your partner make?

73. **Viète's Trigonometric Diagram** In the 16th century the French mathematician François Viète (see page 49) published the following remarkable diagram. Each of the six trigonometric functions of θ is equal to the length of a line segment in the figure. For instance, $\sin\theta = |PR|$, since from $\triangle OPR$ we see that

$$\sin\theta = \frac{\text{opp}}{\text{hyp}} = \frac{|PR|}{|OR|} = \frac{|PR|}{1} = |PR|$$

For each of the five other trigonometric functions, find a line segment in the figure whose length equals the value of the function at θ. (*Note:* The radius of the circle is 1, the center is O, segment QS is tangent to the circle at R, and $\angle SOQ$ is a right angle.)

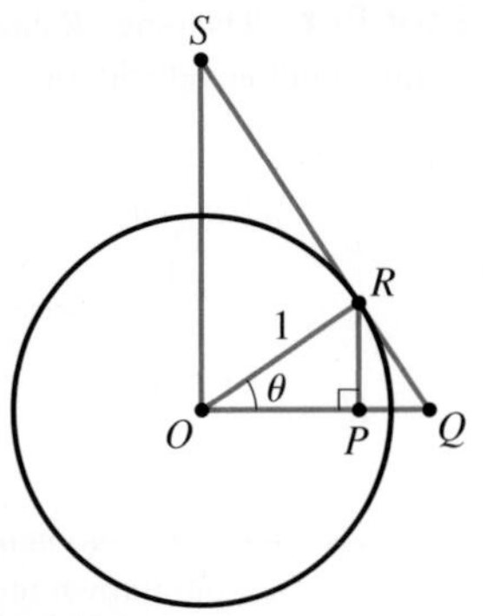

DISCOVERY PROJECT **Similarity**

In this project we explore the idea of similarity and some of its consequences for any type of figure. You can find the project at the book companion website: **www.stewartmath.com**

6.4 Inverse Trigonometric Functions and Right Triangles

The Inverse Sine, Inverse Cosine, and Inverse Tangent Functions ▶ Solving for Angles in Right Triangles ▶ Evaluating Expressions Involving Inverse Trigonometric Functions

The graphs of the inverse trigonometric functions are studied in Section 5.5.

Recall that for a function to have an inverse, it must be one-to-one. Since the trigonometric functions are not one-to-one, they do not have inverses. So we restrict the domain of each of the trigonometric functions to intervals on which they attain all their values and on which they are one-to-one. The resulting functions have the same range as the original functions but are one-to-one.

▼ The Inverse Sine, Inverse Cosine, and Inverse Tangent Functions

Let's first consider the sine function. We restrict the domain of the sine function to angles θ with $-\pi/2 \le \theta \le \pi/2$. From Figure 1 we see that on this domain the sine function attains each of the values in the interval $[-1, 1]$ exactly once and so is one-to-one. Similarly, we restrict the domains of cosine and tangent as shown in Figure 1.

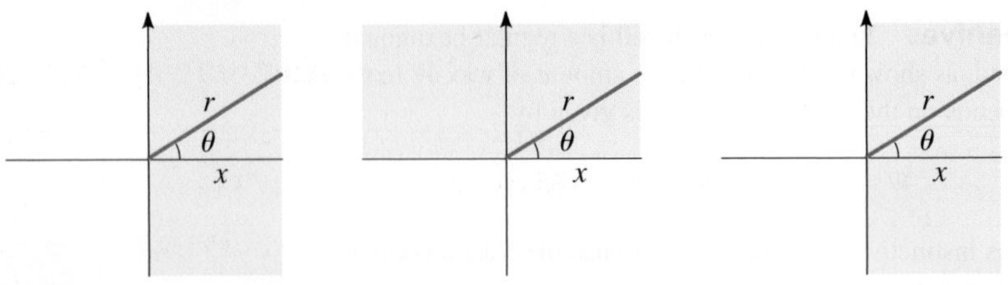

$\sin\theta = \frac{y}{r}$, $-\frac{\pi}{2} \le \theta \le \frac{\pi}{2}$

$\cos\theta = \frac{x}{r}$, $0 \le \theta \le \pi$

$\tan\theta = \frac{x}{y}$, $-\frac{\pi}{2} < \theta < \frac{\pi}{2}$

FIGURE 1 Restricted domains of the sine, cosine, and tangent functions

On these restricted domains we can define an inverse for each of these functions. By the definition of inverse function we have

$\sin^{-1}x = y$	$\Leftrightarrow$	$\sin y = x$
$\cos^{-1}x = y$	$\Leftrightarrow$	$\cos y = x$
$\tan^{-1}x = y$	$\Leftrightarrow$	$\tan y = x$

We summarize the domains and ranges of the inverse trigonometric functions in the following box.

THE INVERSE SINE, INVERSE COSINE, AND INVERSE TANGENT FUNCTIONS

The sine, cosine, and tangent functions on the restricted domains $[-\pi/2, \pi/2]$, $[0, \pi]$, and $(-\pi/2, \pi/2)$, respectively, are one-to one and so have inverses. The inverse functions have domain and range as follows.

Function	Domain	Range
$\sin^{-1}$	$[-1, 1]$	$[-\pi/2, \pi/2]$
$\cos^{-1}$	$[-1, 1]$	$[0, \pi]$
$\tan^{-1}$	$\mathbb{R}$	$(-\pi/2, \pi/2)$

The functions $\sin^{-1}$, $\cos^{-1}$, and $\tan^{-1}$ are sometimes called **arcsine**, **arccosine**, and **arctangent**, respectively.

Since these are inverse functions, they reverse the rule of the original function. For example, since $\sin \pi/6 = \frac{1}{2}$, it follows that $\sin^{-1}\frac{1}{2} = \pi/6$. The following example gives further illustrations.

EXAMPLE 1 | Evaluating Inverse Trigonometric Functions

Find the exact value.

(a) $\sin^{-1}\dfrac{\sqrt{3}}{2}$ **(b)** $\cos^{-1}(-\frac{1}{2})$ **(c)** $\tan^{-1} 1$

SOLUTION

(a) The angle in the interval $[-\pi/2, \pi/2]$ whose sine is $\sqrt{3}/2$ is $\pi/3$. Thus $\sin^{-1}(\sqrt{3}/2) = \pi/3$.

(b) The angle in the interval $[0, \pi]$ whose cosine is $-\frac{1}{2}$ is $2\pi/3$. Thus $\cos^{-1}(-\frac{1}{2}) = 2\pi/3$.

(c) The angle in the interval $[-\pi/2, \pi/2]$ whose tangent is 1 is $\pi/4$. Thus $\tan^{-1} 1 = \pi/4$.

NOW TRY EXERCISE **3** ■

EXAMPLE 2 | Evaluating Inverse Trigonometric Functions

Find approximate values for the given expression.

(a) $\sin^{-1}(0.71)$ **(b)** $\tan^{-1}(2)$ **(c)** $\cos^{-1} 2$

SOLUTION We use a calculator to approximate these values.

(a) Using the [INV] [SIN], or [SIN⁻¹], or [ARC] [SIN] key(s) on the calculator (with the calculator in radian mode), we get

$$\sin^{-1}(0.71) \approx 0.78950$$

(b) Using the INV TAN, or TAN^{-1}, or ARC TAN key(s) on the calculator (with the calculator in radian mode), we get

$$\tan^{-1} 2 \approx 1.10715$$

(c) Since $2 > 1$, it is not in the domain of $\cos^{-1}$, so $\cos^{-1} 2$ is not defined.

NOW TRY EXERCISES **7, 11,** AND **13**

Solving for Angles in Right Triangles

In Section 6.2 we solved triangles by using the trigonometric functions to find the unknown sides. We now use the inverse trigonometric functions to solve for *angles* in a right triangle.

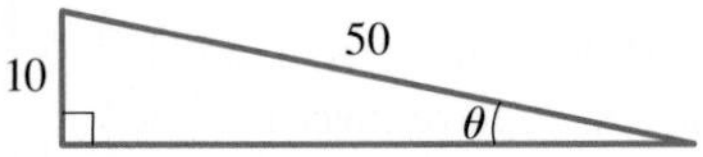

FIGURE 2

EXAMPLE 3 | Finding an Angle in a Right Triangle

Find the angle θ in the triangle shown in Figure 2.

SOLUTION Since θ is the angle opposite the side of length 10 and the hypotenuse has length 50, we have

$$\sin\theta = \frac{10}{50} = \frac{1}{5} \qquad \sin\theta = \frac{\text{opp}}{\text{hyp}}$$

Now we can use $\sin^{-1}$ to find θ:

$$\theta = \sin^{-1}\tfrac{1}{5} \qquad \text{Definition of } \sin^{-1}$$

$$\theta \approx 11.5° \qquad \text{Calculator (in degree mode)}$$

NOW TRY EXERCISE **15**

EXAMPLE 4 | Solving for an Angle in a Right Triangle

A 40-ft ladder leans against a building. If the base of the ladder is 6 ft from the base of the building, what is the angle formed by the ladder and the building?

SOLUTION First we sketch a diagram as in Figure 3. If θ is the angle between the ladder and the building, then

$$\sin\theta = \frac{6}{40} = 0.15 \qquad \sin\theta = \frac{\text{opp}}{\text{hyp}}$$

Now we use $\sin^{-1}$ to find θ:

$$\theta = \sin^{-1}(0.15) \qquad \text{Definition of } \sin^{-1}$$

$$\theta \approx 8.6° \qquad \text{Calculator (in degree mode)}$$

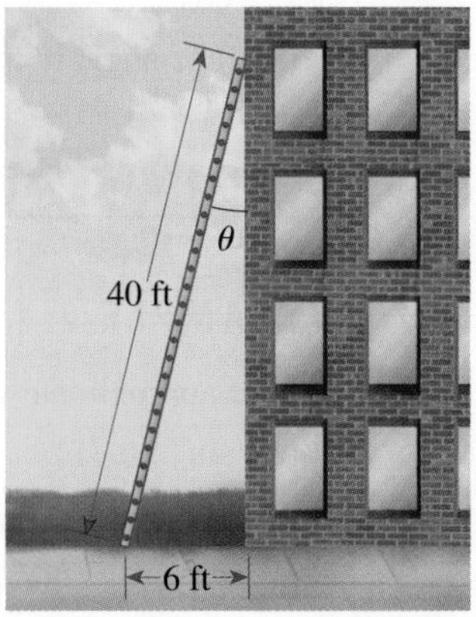

FIGURE 3

NOW TRY EXERCISE **37**

EXAMPLE 5 | The Angle of a Beam of Light

A lighthouse is located on an island that is 2 mi off a straight shoreline (see Figure 4). Express the angle formed by the beam of light and the shoreline in terms of the distance d in the figure.

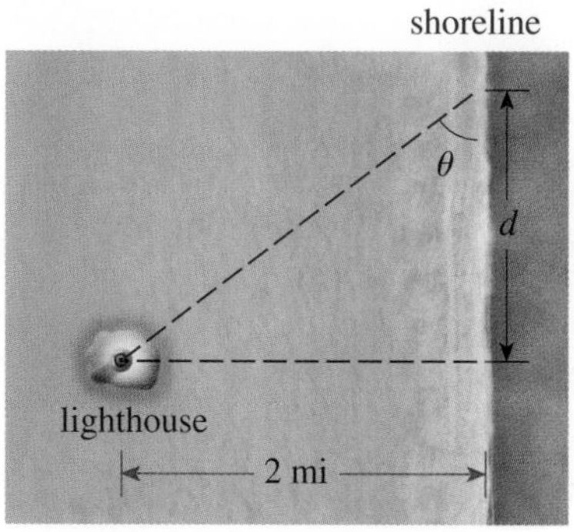

FIGURE 4

SOLUTION From the figure we see that

$$\tan\theta = \frac{2}{d} \qquad \tan\theta = \frac{\text{opp}}{\text{adj}}$$

Taking the inverse tangent of both sides, we get

$$\tan^{-1}(\tan\theta) = \tan^{-1}\left(\frac{2}{d}\right) \qquad \text{Take } \tan^{-1} \text{ of both sides}$$

$$\theta = \tan^{-1}\left(\frac{2}{d}\right) \qquad \text{Property of inverse functions: } \tan^{-1}(\tan\theta) = \theta$$

NOW TRY EXERCISE **39** ■

In Section 6.5 we learn how to solve any triangle (not necessarily a right triangle). The angles in a triangle are always in the interval $[0, \pi]$ (or between 0° and 180°). We'll see that to solve such triangles we need to find all angles in the interval $[0, \pi]$ that have a specified sine or cosine. We do this in the next example.

EXAMPLE 6 | Solving a Basic Trigonometric Equation on an Interval

Find all angles θ between 0° and 180° satisfying the given equation.

(a) $\sin\theta = 0.4$ **(b)** $\cos\theta = 0.4$

SOLUTION

(a) We use $\sin^{-1}$ to find one solution in the interval $[-\pi/2, \pi/2]$.

$$\sin\theta = 0.4 \qquad \text{Equation}$$
$$\theta = \sin^{-1}(0.4) \qquad \text{Take } \sin^{-1} \text{ of each side}$$
$$\theta \approx 23.6° \qquad \text{Calculator (in degree mode)}$$

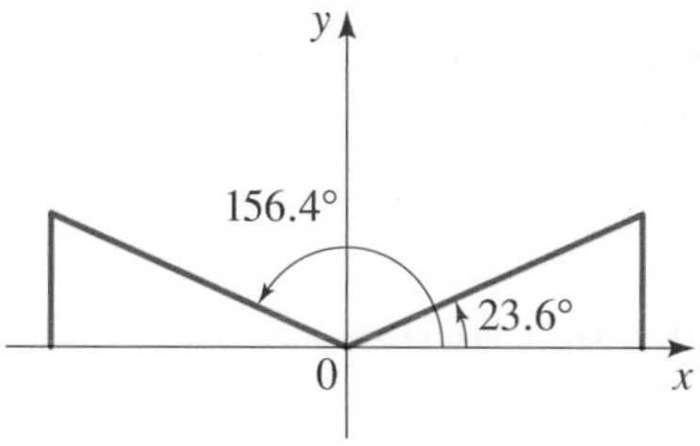

FIGURE 5

Another solution with θ between 0° and 180° is obtained by taking the supplement of the angle: $180° - 23.6° = 156.4°$ (see Figure 5). So the solutions of the equation with θ between 0° and 180° are

$$\theta \approx 23.6° \quad \text{and} \quad \theta \approx 156.4°$$

(b) The cosine function is one-to-one on the interval $[0, \pi]$, so there is only one solution of the equation with θ between 0° and 180°. We find that solution by taking $\cos^{-1}$ of each side.

$$\cos\theta = 0.4$$
$$\theta = \cos^{-1}(0.4) \qquad \text{Take } \cos^{-1} \text{ of each side}$$
$$\theta \approx 66.4° \qquad \text{Calculator (in degree mode)}$$

The solution is $\theta \approx 66.4°$

NOW TRY EXERCISES **23** AND **25** ■

▼ Evaluating Expressions Involving Inverse Trigonometric Functions

Expressions like $\cos(\sin^{-1}x)$ arise in calculus. We find exact values of such expressions using trigonometric identities or right triangles.

EXAMPLE 7 | Composing Trigonometric Functions and Their Inverses

Find $\cos(\sin^{-1}\frac{3}{5})$.

SOLUTION 1

Let $\theta = \sin^{-1}\frac{3}{5}$. Then θ is the number in the interval $[-\pi/2, \pi/2]$ whose sine is $\frac{3}{5}$. Let's interpret θ as an angle and draw a right triangle with θ as one of its acute angles, with opposite side 3 and hypotenuse 5 (see Figure 6). The remaining leg of the triangle is found by the Pythagorean Theorem to be 4. From the figure we get

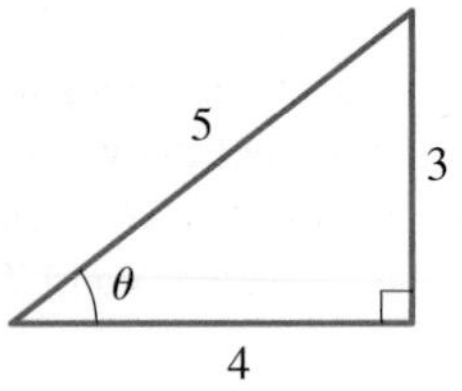

FIGURE 6 $\cos\theta = \frac{4}{5}$

$$\cos(\sin^{-1}\tfrac{3}{5}) = \cos\theta \qquad \theta = \sin^{-1}\tfrac{3}{5}$$
$$= \frac{4}{5} \qquad \cos\theta = \frac{\text{adj}}{\text{hyp}}$$

So $\cos(\sin^{-1}\frac{3}{5}) = \frac{4}{5}$.

SOLUTION 2

It's easy to find $\sin(\sin^{-1}\frac{3}{5})$. In fact, by the cancellation properties of inverse functions, this value is exactly $\frac{3}{5}$. To find $\cos(\sin^{-1}\frac{3}{5})$, we first write the cosine function in terms of the sine function. Let $u = \sin^{-1}\frac{3}{5}$. Since $-\pi/2 \le u \le \pi/2$, $\cos u$ is positive, and we can write the following:

$$\cos u = +\sqrt{1 - \sin^2 u} \qquad \cos^2 u + \sin^2 u = 1$$
$$= \sqrt{1 - \sin^2(\sin^{-1}\tfrac{3}{5})} \qquad u = \sin^{-1}\tfrac{3}{5}$$
$$= \sqrt{1 - (\tfrac{3}{5})^2} \qquad \text{Property of inverse functions: } \sin(\sin^{-1}\tfrac{3}{5}) = \tfrac{3}{5}$$
$$= \sqrt{1 - \tfrac{9}{25}} = \sqrt{\tfrac{16}{25}} = \tfrac{4}{5} \qquad \text{Calculate}$$

So $\cos(\sin^{-1}\frac{3}{5}) = \frac{4}{5}$.

NOW TRY EXERCISE 27

EXAMPLE 8 | Composing Trigonometric Functions and Their Inverses

Write $\sin(\cos^{-1}x)$ and $\tan(\cos^{-1}x)$ as algebraic expressions in x for $-1 \le x \le 1$.

SOLUTION 1

Let $\theta = \cos^{-1}x$; then $\cos\theta = x$. In Figure 7 we sketch a right triangle with an acute angle θ, adjacent side x, and hypotenuse 1. By the Pythagorean Theorem the remaining leg is $\sqrt{1 - x^2}$. From the figure we have

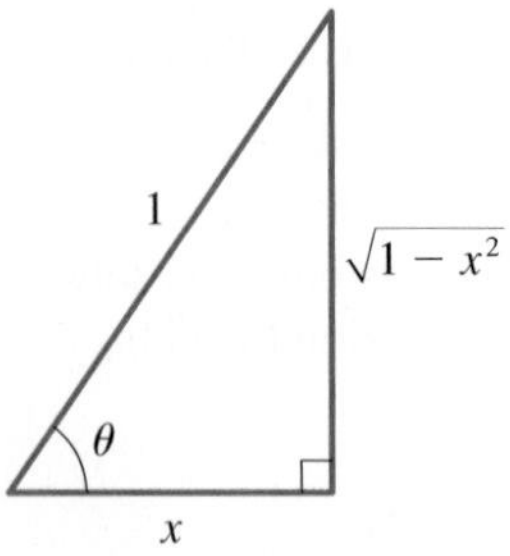

FIGURE 7 $\cos\theta = \frac{x}{1} = x$

$$\sin(\cos^{-1}x) = \sin\theta = \sqrt{1 - x^2} \qquad \text{and} \qquad \tan(\cos^{-1}x) = \tan\theta = \frac{\sqrt{1 - x^2}}{x}$$

SOLUTION 2

Let $u = \cos^{-1}x$. We need to find $\sin u$ and $\tan u$ in terms of x. As in Example 5 the idea here is to write sine and tangent in terms of cosine. Note that $0 \le u \le \pi$ because $u = \cos^{-1}x$. We have

$$\sin u = \pm\sqrt{1 - \cos^2 u} \qquad \text{and} \qquad \tan u = \frac{\sin u}{\cos u} = \frac{\pm\sqrt{1 - \cos^2 u}}{\cos u}$$

To choose the proper signs, note that u lies in the interval $[0, \pi]$ because $u = \cos^{-1} x$. Since $\sin u$ is positive on this interval, the $+$ sign is the correct choice. Substituting $u = \cos^{-1} x$ in the displayed equations and using the cancellation property $\cos(\cos^{-1} x) = x$, we get

$$\sin(\cos^{-1} x) = \sqrt{1 - x^2} \qquad \text{and} \qquad \tan(\cos^{-1} x) = \frac{\sqrt{1 - x^2}}{x}$$

NOW TRY EXERCISES **33** AND **35** ■

Note: In Solution 1 of Example 8 it might seem that because we are sketching a triangle, the angle $\theta = \cos^{-1} x$ must be acute. But it turns out that the triangle method works for any x. The domains and ranges of all six inverse trigonometric functions have been chosen in such a way that we can always use a triangle to find $S(T^{-1}(x))$, where S and T are any trigonometric functions.

6.4 EXERCISES

CONCEPTS

1. The inverse sine, inverse cosine, and inverse tangent functions have the followings domains and ranges.

(a) The function $\sin^{-1}$ has domain ________ and range ________.

(b) The function $\cos^{-1}$ has domain ________ and range ________.

(c) The function $\tan^{-1}$ has domain ________ and range ________.

2. In the triangle shown, we can find the angle θ as follows:

(a) $\theta = \sin^{-1} \frac{\square}{\square}$

(b) $\theta = \cos^{-1} \frac{\square}{\square}$

(c) $\theta = \tan^{-1} \frac{\square}{\square}$

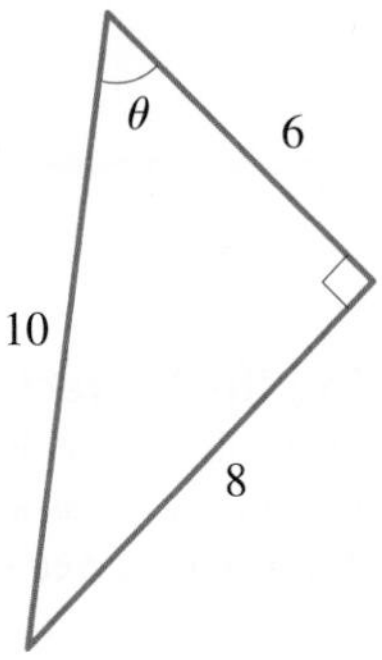

SKILLS

3–6 ■ Find the exact value of each expression, if it is defined.

3. (a) $\sin^{-1} \frac{1}{2}$ **(b)** $\cos^{-1}\left(-\frac{\sqrt{3}}{2}\right)$ **(c)** $\tan^{-1}(-1)$

4. (a) $\sin^{-1}\left(-\frac{\sqrt{3}}{2}\right)$ **(b)** $\cos^{-1}\left(-\frac{\sqrt{2}}{2}\right)$ **(c)** $\tan^{-1}(-\sqrt{3})$

5. (a) $\sin^{-1}\left(-\frac{1}{2}\right)$ **(b)** $\cos^{-1} \frac{1}{2}$ **(c)** $\tan^{-1}\left(\frac{\sqrt{3}}{3}\right)$

6. (a) $\sin^{-1}(-1)$ **(b)** $\cos^{-1} 1$ **(c)** $\tan^{-1} 0$

7–14 ■ Use a calculator to find an approximate value of each expression rounded to five decimal places, if it is defined.

7. $\sin^{-1}(0.45)$ **8.** $\cos^{-1}(-0.75)$

9. $\cos^{-1}(-\frac{1}{4})$ **10.** $\sin^{-1} \frac{1}{3}$

11. $\tan^{-1} 3$ **12.** $\tan^{-1}(-4)$

13. $\cos^{-1} 3$ **14.** $\sin^{-1}(-2)$

15–20 ■ Find the angle θ in degrees, rounded to one decimal.

15.

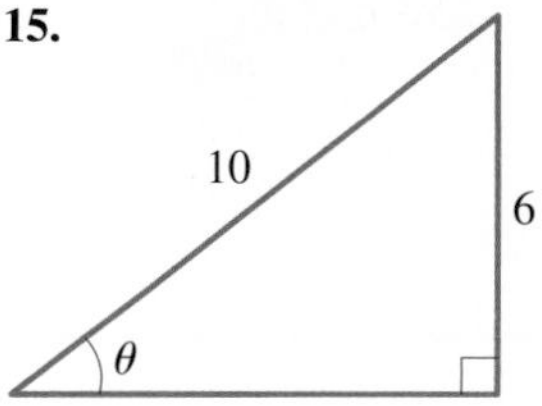

16.

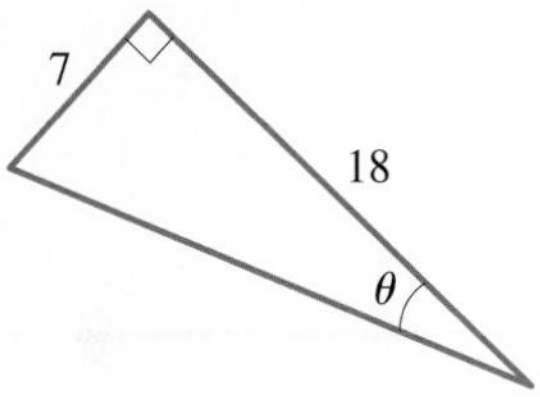

17.

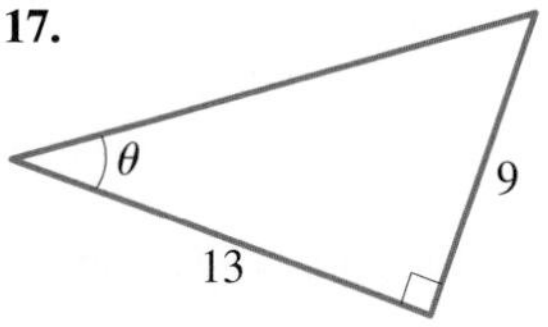

18.

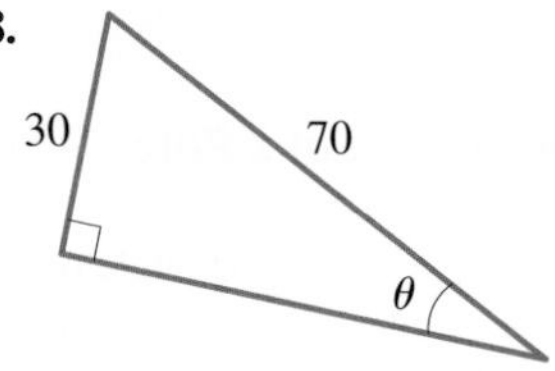

19.

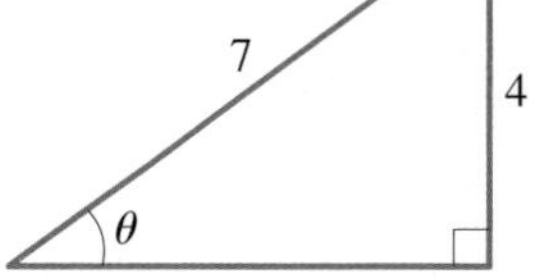

20.

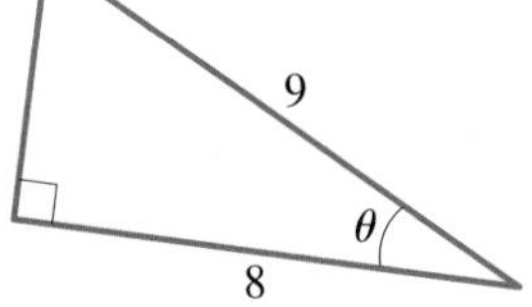

21–26 ■ Find all angles θ between 0° and 180° satisfying the given equation.

21. $\sin \theta = \frac{1}{2}$ **22.** $\sin \theta = \frac{\sqrt{3}}{2}$

23. $\sin\theta = 0.7$

24. $\sin\theta = \dfrac{1}{4}$

25. $\cos\theta = 0.7$

26. $\cos\theta = \dfrac{1}{9}$

27–32 ■ Find the exact value of the expression.

27. $\sin(\cos^{-1}\frac{3}{5})$

28. $\tan(\sin^{-1}\frac{4}{5})$

29. $\sec(\sin^{-1}\frac{12}{13})$

30. $\csc(\cos^{-1}\frac{7}{25})$

31. $\tan(\sin^{-1}\frac{12}{13})$

32. $\cot(\sin^{-1}\frac{2}{3})$

33–36 ■ Rewrite the expression as an algebraic expression in x.

33. $\cos(\sin^{-1} x)$

34. $\sin(\tan^{-1} x)$

35. $\tan(\sin^{-1} x)$

36. $\cos(\tan^{-1} x)$

APPLICATIONS

37. **Leaning Ladder** A 20-ft ladder is leaning against a building. If the base of the ladder is 6 ft from the base of the building, what is the angle of elevation of the ladder? How high does the ladder reach on the building?

38. **Angle of the Sun** A 96-ft tree casts a shadow that is 120 ft long. What is the angle of elevation of the sun?

39. **Height of the Space Shuttle** An observer views the space shuttle from a distance of 2 mi from the launch pad.
(a) Express the height of the space shuttle as a function of the angle of elevation θ.
(b) Express the angle of elevation θ as a function of the height h of the space shuttle.

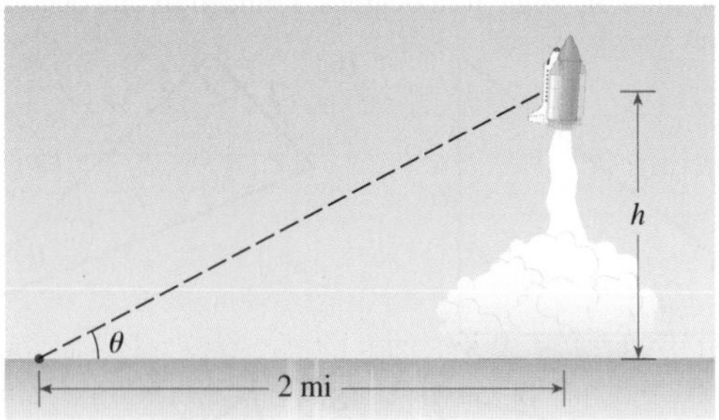

40. **Height of a Pole** A 50-ft pole casts a shadow as shown in the figure.
(a) Express the angle of elevation θ of the sun as a function of the length s of the shadow.
(b) Find the angle θ of elevation of the sun when the shadow is 20 ft long.

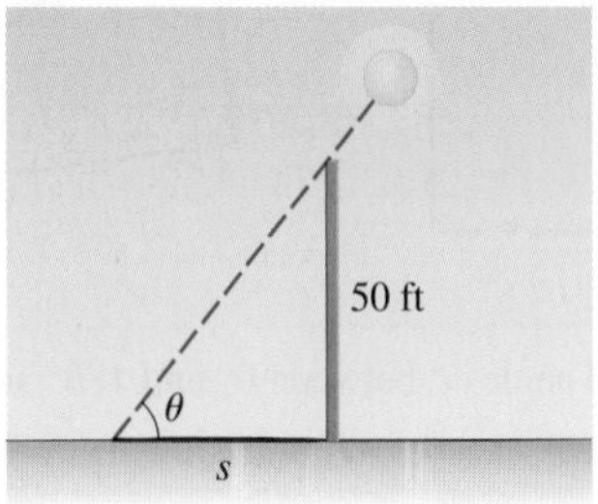

41. **Height of a Balloon** A 680-ft rope anchors a hot-air balloon as shown in the figure.
(a) Express the angle θ as a function of the height h of the balloon.
(b) Find the angle θ if the balloon is 500 ft high.

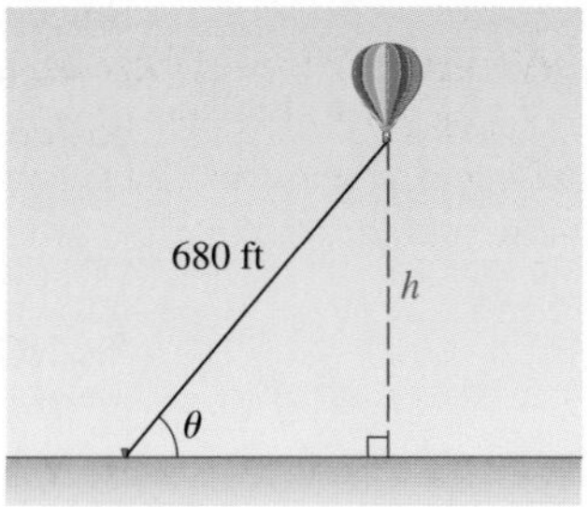

42. **View from a Satellite** The figures indicate that the higher the orbit of a satellite, the more of the earth the satellite can "see." Let θ, s, and h be as in the figure, and assume the earth is a sphere of radius 3960 mi.
(a) Express the angle θ as a function of h.
(b) Express the distance s as a function of θ.
(c) Express the distance s as a function of h. [*Hint:* Find the composition of the functions in parts (a) and (b).]
(d) If the satellite is 100 mi above the earth, what is the distance s that it can see?
(e) How high does the satellite have to be to see both Los Angeles and New York, 2450 mi apart?

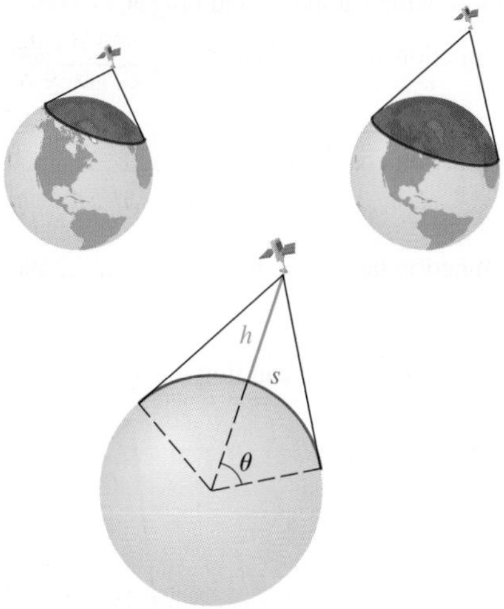

43. **Surfing the Perfect Wave** For a wave to be surfable, it can't break all at once. Robert Guza and Tony Bowen have shown that a wave has a surfable shoulder if it hits the shoreline at an angle θ given by

$$\theta = \sin^{-1}\left(\frac{1}{(2n+1)\tan\beta}\right)$$

where β is the angle at which the beach slopes down and where $n = 0, 1, 2, \ldots$.
(a) For $\beta = 10°$, find θ when $n = 3$.
(b) For $\beta = 15°$, find θ when $n = 2$, 3, and 4. Explain why the formula does not give a value for θ when $n = 0$ or 1.

DISCOVERY ■ DISCUSSION ■ WRITING

44. Inverse Trigonometric Functions on a Calculator Most calculators do not have keys for $\sec^{-1}$, $\csc^{-1}$, or $\cot^{-1}$. Prove the following identities, and then use these identities and a calculator to find $\sec^{-1} 2$, $\csc^{-1} 3$, and $\cot^{-1} 4$.

$$\sec^{-1} x = \cos^{-1}\left(\frac{1}{x}\right), \qquad x \geq 1$$

$$\csc^{-1} x = \sin^{-1}\left(\frac{1}{x}\right), \qquad x \geq 1$$

$$\cot^{-1} x = \tan^{-1}\left(\frac{1}{x}\right), \qquad x > 0$$

6.5 The Law of Sines

| The Law of Sines ▶ The Ambiguous Case

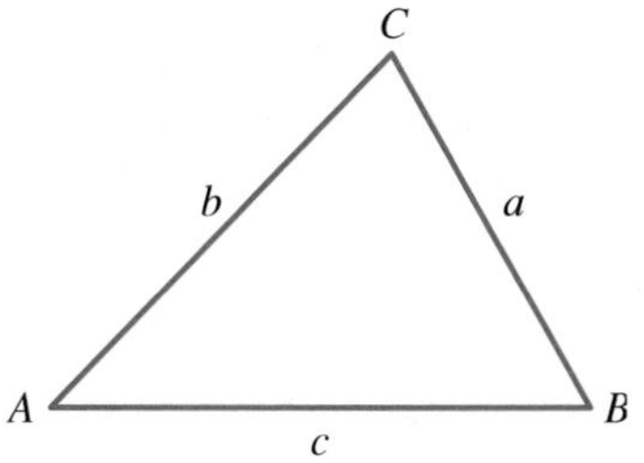

FIGURE 1

In Section 6.2 we used the trigonometric ratios to solve right triangles. The trigonometric functions can also be used to solve *oblique triangles*, that is, triangles with no right angles. To do this, we first study the Law of Sines here and then the Law of Cosines in the next section. To state these laws (or formulas) more easily, we follow the convention of labeling the angles of a triangle as A, B, C and the lengths of the corresponding opposite sides as a, b, c, as in Figure 1.

To solve a triangle, we need to know certain information about its sides and angles. To decide whether we have enough information, it's often helpful to make a sketch. For instance, if we are given two angles and the included side, then it's clear that one and only one triangle can be formed (see Figure 2(a)). Similarly, if two sides and the included angle are known, then a unique triangle is determined (Figure 2(c)). But if we know all three angles and no sides, we cannot uniquely determine the triangle because many triangles can have the same three angles. (All these triangles would be similar, of course.) So we won't consider this last case.

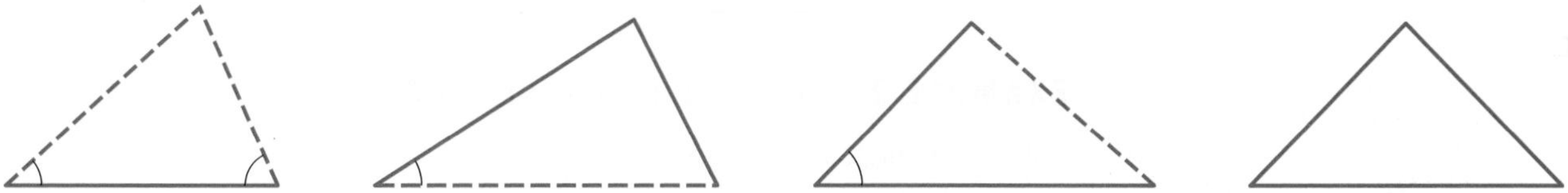

(a) ASA or SAA (b) SSA (c) SAS (d) SSS

FIGURE 2

In general, a triangle is determined by three of its six parts (angles and sides) as long as at least one of these three parts is a side. So the possibilities, illustrated in Figure 2, are as follows.

Case 1 One side and two angles (ASA or SAA)

Case 2 Two sides and the angle opposite one of those sides (SSA)

Case 3 Two sides and the included angle (SAS)

Case 4 Three sides (SSS)

Cases 1 and 2 are solved by using the Law of Sines; Cases 3 and 4 require the Law of Cosines.

▼ The Law of Sines

The **Law of Sines** says that in any triangle the lengths of the sides are proportional to the sines of the corresponding opposite angles.

THE LAW OF SINES

In triangle ABC we have

$$\frac{\sin A}{a} = \frac{\sin B}{b} = \frac{\sin C}{c}$$

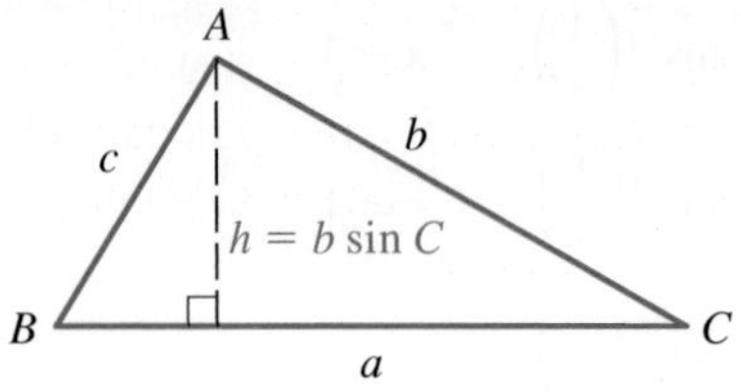

FIGURE 3

PROOF To see why the Law of Sines is true, refer to Figure 3. By the formula in Section 6.3 the area of triangle ABC is $\frac{1}{2}ab \sin C$. By the same formula the area of this triangle is also $\frac{1}{2}ac \sin B$ and $\frac{1}{2}bc \sin A$. Thus,

$$\tfrac{1}{2}bc \sin A = \tfrac{1}{2}ac \sin B = \tfrac{1}{2}ab \sin C$$

Multiplying by $2/(abc)$ gives the Law of Sines. ■

EXAMPLE 1 | Tracking a Satellite (ASA)

A satellite orbiting the earth passes directly overhead at observation stations in Phoenix and Los Angeles, 340 mi apart. At an instant when the satellite is between these two stations, its angle of elevation is simultaneously observed to be 60° at Phoenix and 75° at Los Angeles. How far is the satellite from Los Angeles?

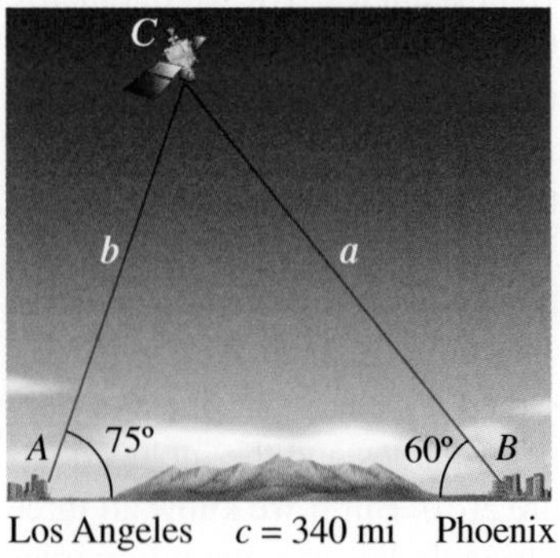

FIGURE 4

SOLUTION We need to find the distance b in Figure 4. Since the sum of the angles in any triangle is 180°, we see that $\angle C = 180° - (75° + 60°) = 45°$ (see Figure 4), so we have

$$\frac{\sin B}{b} = \frac{\sin C}{c} \qquad \text{Law of Sines}$$

$$\frac{\sin 60°}{b} = \frac{\sin 45°}{340} \qquad \text{Substitute}$$

$$b = \frac{340 \sin 60°}{\sin 45°} \approx 416 \qquad \text{Solve for } b$$

The distance of the satellite from Los Angeles is approximately 416 mi.

NOW TRY EXERCISES **5** AND **33** ■

EXAMPLE 2 | Solving a Triangle (SAA)

Solve the triangle in Figure 5.

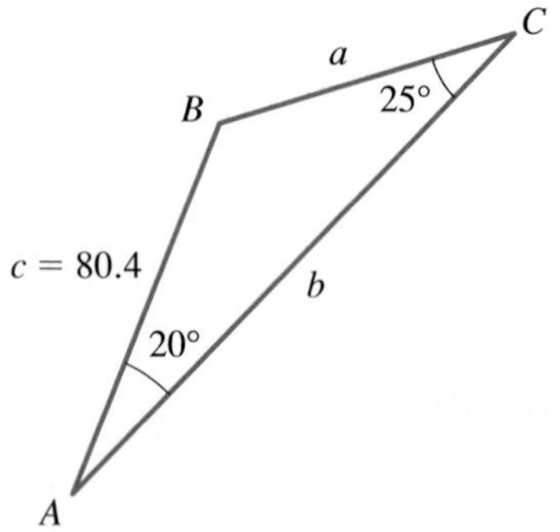

FIGURE 5

SOLUTION First, $\angle B = 180° - (20° + 25°) = 135°$. Since side c is known, to find side a we use the relation

$$\frac{\sin A}{a} = \frac{\sin C}{c} \qquad \text{Law of Sines}$$

$$a = \frac{c \sin A}{\sin C} = \frac{80.4 \sin 20°}{\sin 25°} \approx 65.1 \qquad \text{Solve for } a$$

Similarly, to find b, we use

$$\frac{\sin B}{b} = \frac{\sin C}{c} \qquad \text{Law of Sines}$$

$$b = \frac{c \sin B}{\sin C} = \frac{80.4 \sin 135°}{\sin 25°} \approx 134.5 \qquad \text{Solve for } b$$

NOW TRY EXERCISE **13** ■

▼ The Ambiguous Case

In Examples 1 and 2 a unique triangle was determined by the information given. This is always true of Case 1 (ASA or SAA). But in Case 2 (SSA) there may be two triangles, one triangle, or no triangle with the given properties. For this reason, Case 2 is sometimes called the **ambiguous case**. To see why this is so, we show in Figure 6 the possibilities when angle A and sides a and b are given. In part (a) no solution is possible, since side a is too short to complete the triangle. In part (b) the solution is a right triangle. In part (c) two solutions are possible, and in part (d) there is a unique triangle with the given properties. We illustrate the possibilities of Case 2 in the following examples.

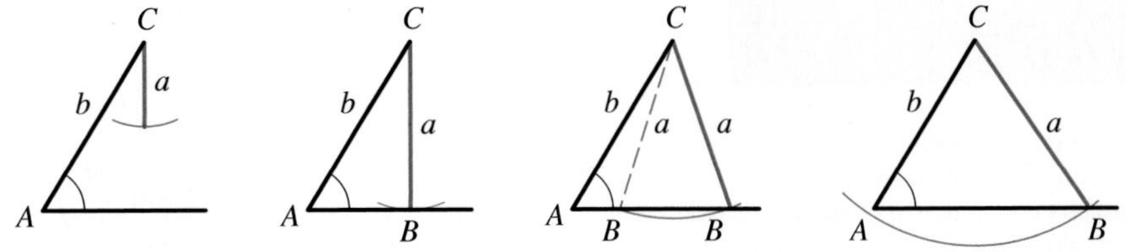

FIGURE 6 The ambiguous case

EXAMPLE 3 | SSA, the One-Solution Case

Solve triangle ABC, where $\angle A = 45°$, $a = 7\sqrt{2}$, and $b = 7$.

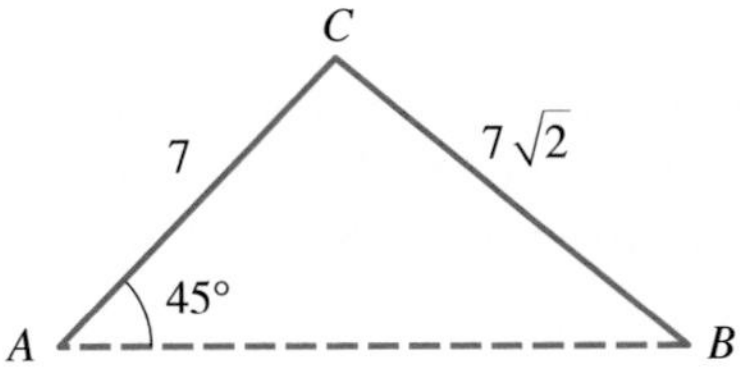

FIGURE 7

SOLUTION We first sketch the triangle with the information we have (see Figure 7). Our sketch is necessarily tentative, since we don't yet know the other angles. Nevertheless, we can now see the possibilities.

We first find $\angle B$.

$$\frac{\sin A}{a} = \frac{\sin B}{b} \qquad \text{Law of Sines}$$

$$\sin B = \frac{b \sin A}{a} = \frac{7}{7\sqrt{2}} \sin 45° = \left(\frac{1}{\sqrt{2}}\right)\left(\frac{\sqrt{2}}{2}\right) = \frac{1}{2} \qquad \text{Solve for } \sin B$$

Which angles B have $\sin B = \frac{1}{2}$? From the preceding section we know that there are two such angles smaller than 180° (they are 30° and 150°). Which of these angles is compatible with what we know about triangle ABC? Since $\angle A = 45°$, we cannot have $\angle B = 150°$, because $45° + 150° > 180°$. So $\angle B = 30°$, and the remaining angle is $\angle C = 180° - (30° + 45°) = 105°$.

We consider only angles smaller than 180°, since no triangle can contain an angle of 180° or larger.

Now we can find side c.

$$\frac{\sin B}{b} = \frac{\sin C}{c} \qquad \text{Law of Sines}$$

$$c = \frac{b \sin C}{\sin B} = \frac{7 \sin 105°}{\sin 30°} = \frac{7 \sin 105°}{\frac{1}{2}} \approx 13.5 \qquad \text{Solve for } c$$

NOW TRY EXERCISE **19** ■

In Example 3 there were two possibilities for angle B, and one of these was not compatible with the rest of the information. In general, if $\sin A < 1$, we must check the angle and its supplement as possibilities, because any angle smaller than 180° can be in the triangle. To decide whether either possibility works, we check to see whether the resulting sum of the angles exceeds 180°. It can happen, as in Figure 6(c), that both possibilities are compatible with the given information. In that case, two different triangles are solutions to the problem.

The *supplement* of an angle θ (where $0 \le \theta \le 180°$) is the angle $180° - \theta$.

EXAMPLE 4 | SSA, the Two-Solution Case

Solve triangle ABC if $\angle A = 43.1°$, $a = 186.2$, and $b = 248.6$.

Surveying is a method of land measurement used for mapmaking. Surveyors use a process called *triangulation* in which a network of thousands of interlocking triangles is created on the area to be mapped. The process is started by measuring the length of a *baseline* between two surveying stations. Then, with the use of an instrument called a *theodolite*, the angles between these two stations and a third station are measured. The Law of Sines is then used to calculate the two other sides of the triangle formed by the three stations. The calculated sides are used as baselines, and the process is repeated over and over to create a network of triangles. In this method the only distance measured is the initial baseline; all other distances are calculated from the Law of Sines. This method is practical because it is much easier to measure angles than distances.

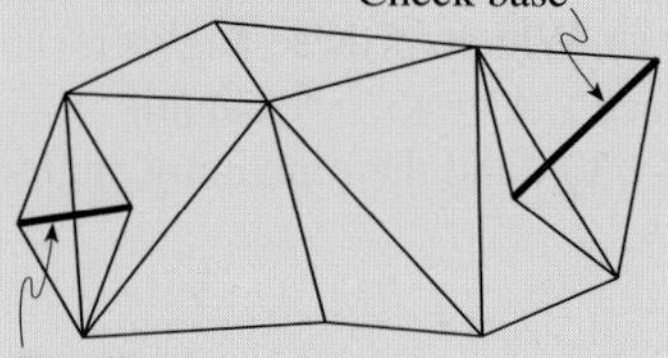

One of the most ambitious mapmaking efforts of all time was the Great Trigonometric Survey of India (see Problem 8, page 492) which required several expeditions and took over a century to complete. The famous expedition of 1823, led by **Sir George Everest**, lasted 20 years. Ranging over treacherous terrain and encountering the dreaded malaria-carrying mosquitoes, this expedition reached the foothills of the Himalayas. A later expedition, using triangulation, calculated the height of the highest peak of the Himalayas to be 29,002 ft. The peak was named in honor of Sir George Everest.

Today, with the use of satellites, the height of Mt. Everest is estimated to be 29,028 ft. The very close agreement of these two estimates shows the great accuracy of the trigonometric method.

SOLUTION From the given information we sketch the triangle shown in Figure 8. Note that side a may be drawn in two possible positions to complete the triangle. From the Law of Sines

$$\sin B = \frac{b \sin A}{a} = \frac{248.6 \sin 43.1°}{186.2} \approx 0.91225$$

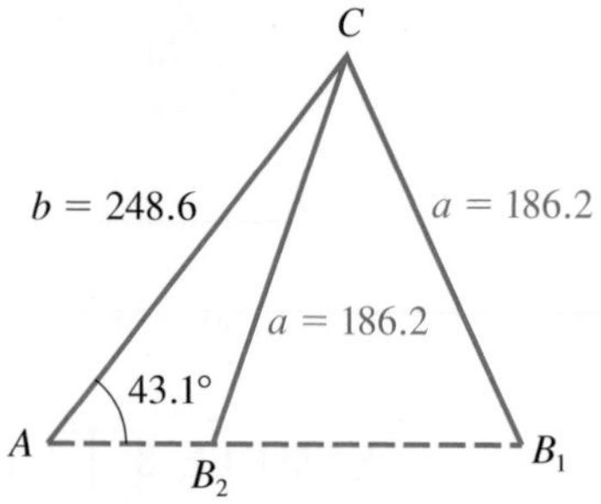

FIGURE 8

There are two possible angles B between 0° and 180° such that $\sin B = 0.91225$. Using a calculator, we find that one of the angles is $\sin^{-1}(0.91225) \approx 65.8°$. The other angle is approximately $180° - 65.8° = 114.2°$. We denote these two angles by B_1 and B_2 so that

$$\angle B_1 \approx 65.8° \qquad \text{and} \qquad \angle B_2 \approx 114.2°$$

Thus two triangles satisfy the given conditions: triangle $A_1B_1C_1$ and triangle $A_2B_2C_2$.

Solve triangle $A_1B_1C_1$:

$$\angle C_1 \approx 180° - (43.1° + 65.8°) = 71.1° \qquad \text{Find } \angle C_1$$

Thus $$c_1 = \frac{a_1 \sin C_1}{\sin A_1} \approx \frac{186.2 \sin 71.1°}{\sin 43.1°} \approx 257.8 \qquad \text{Law of Sines}$$

Solve triangle $A_2B_2C_2$:

$$\angle C_2 \approx 180° - (43.1° + 114.2°) = 22.7° \qquad \text{Find } \angle C_2$$

Thus $$c_2 = \frac{a_2 \sin C_2}{\sin A_2} \approx \frac{186.2 \sin 22.7°}{\sin 43.1°} \approx 105.2 \qquad \text{Law of Sines}$$

Triangles $A_1B_1C_1$ and $A_2B_2C_2$ are shown in Figure 9.

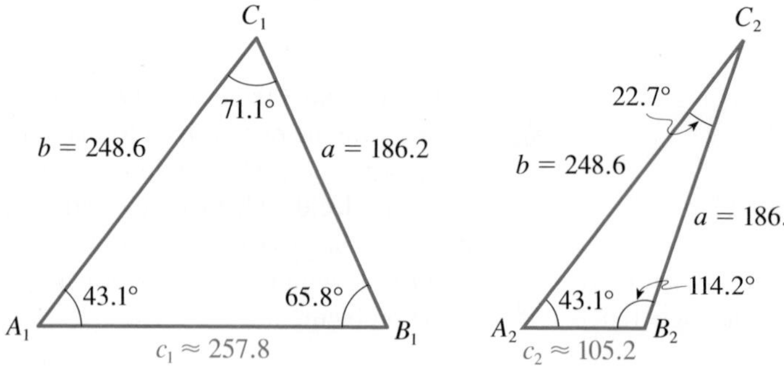

FIGURE 9

NOW TRY EXERCISE **23**

The next example presents a situation for which no triangle is compatible with the given data.

EXAMPLE 5 | SSA, the No-Solution Case

Solve triangle ABC, where $\angle A = 42°$, $a = 70$, and $b = 122$.

SOLUTION To organize the given information, we sketch the diagram in Figure 10. Let's try to find $\angle B$. We have

$$\frac{\sin A}{a} = \frac{\sin B}{b} \qquad \text{Law of Sines}$$

$$\sin B = \frac{b \sin A}{a} = \frac{122 \sin 42°}{70} \approx 1.17 \qquad \text{Solve for } \sin B$$

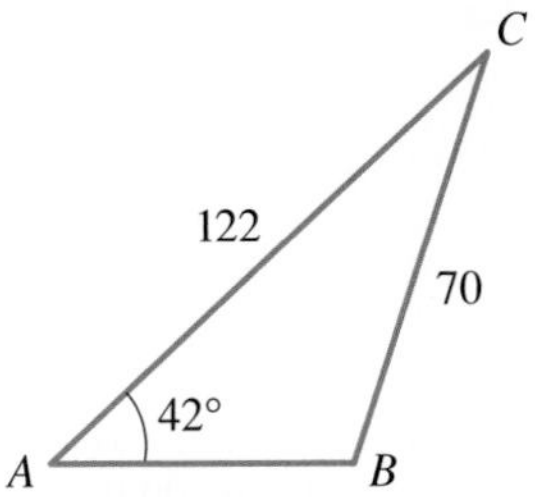

FIGURE 10

Since the sine of an angle is never greater than 1, we conclude that no triangle satisfies the conditions given in this problem.

NOW TRY EXERCISE 21 ■

6.5 EXERCISES

CONCEPTS

1. In triangle ABC with sides a, b, and c the Law of Sines states that

$$\frac{\square}{\square} = \frac{\square}{\square} = \frac{\square}{\square}$$

2. In which of the following cases can we use the Law of Sines to solve a triangle?

ASA SSS SAS SSA

SKILLS

3–8 ■ Use the Law of Sines to find the indicated side x or angle θ.

3.
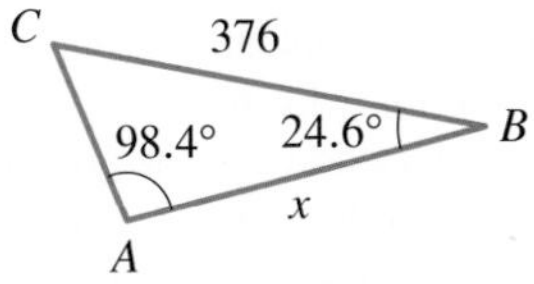

4.
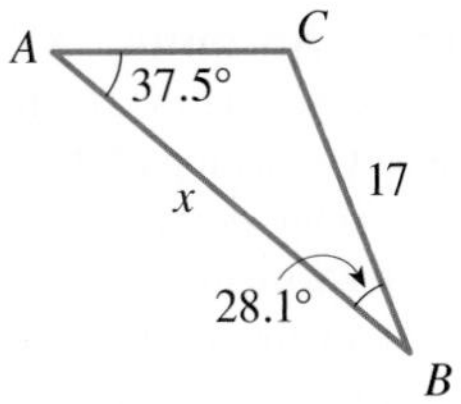

5.
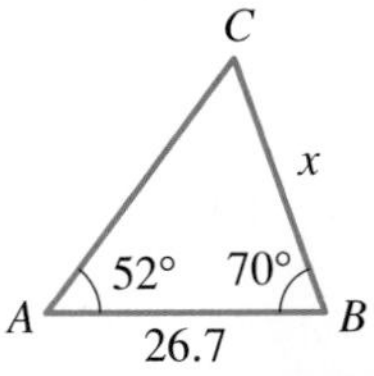

6.
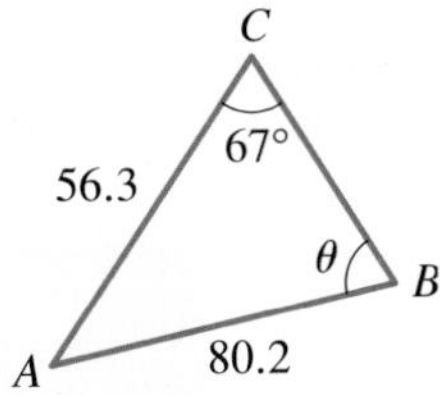

7.
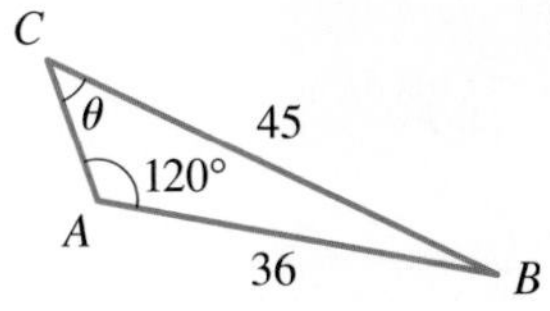

8.
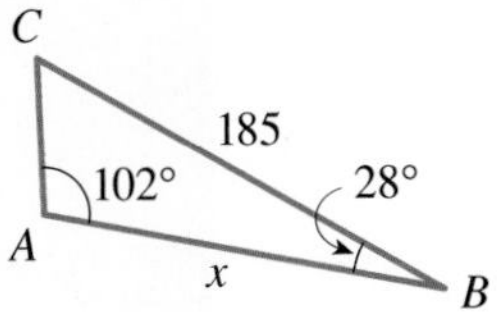

9–12 ■ Solve the triangle using the Law of Sines.

9.
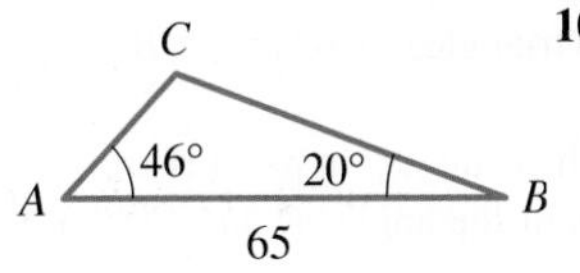

10.
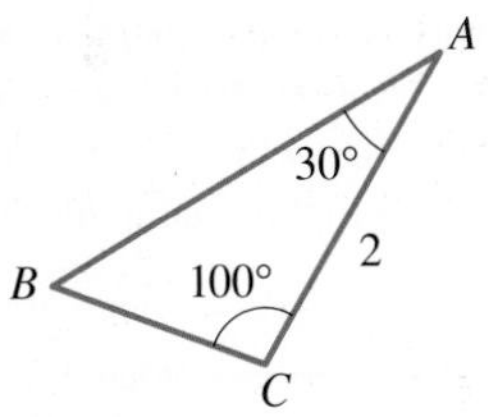

11.
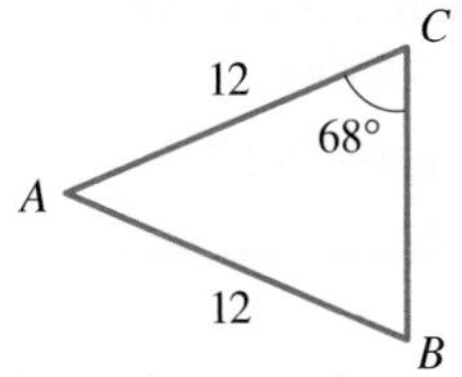

12.
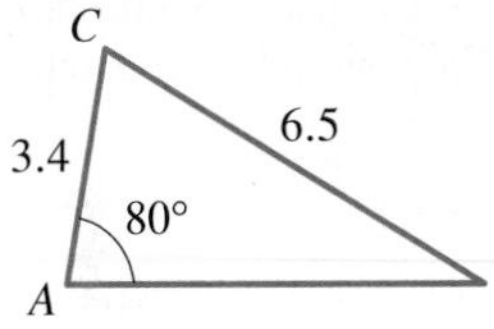

13–18 ■ Sketch each triangle, and then solve the triangle using the Law of Sines.

13. $\angle A = 50°$, $\angle B = 68°$, $c = 230$

14. $\angle A = 23°$, $\angle B = 110°$, $c = 50$

15. $\angle A = 30°$, $\angle C = 65°$, $b = 10$

16. $\angle A = 22°$, $\angle B = 95°$, $a = 420$

17. $\angle B = 29°$, $\angle C = 51°$, $b = 44$

18. $\angle B = 10°$, $\angle C = 100°$, $c = 115$

19–28 ■ Use the Law of Sines to solve for all possible triangles that satisfy the given conditions.

19. $a = 28$, $b = 15$, $\angle A = 110°$

20. $a = 30$, $c = 40$, $\angle A = 37°$

21. $a = 20$, $c = 45$, $\angle A = 125°$

22. $b = 45$, $c = 42$, $\angle C = 38°$

23. $b = 25, \quad c = 30, \quad \angle B = 25°$

24. $a = 75, \quad b = 100, \quad \angle A = 30°$

25. $a = 50, \quad b = 100, \quad \angle A = 50°$

26. $a = 100, \quad b = 80, \quad \angle A = 135°$

27. $a = 26, \quad c = 15, \quad \angle C = 29°$

28. $b = 73, \quad c = 82, \quad \angle B = 58°$

29. For the triangle shown, find

(a) $\angle BCD$ and

(b) $\angle DCA$.

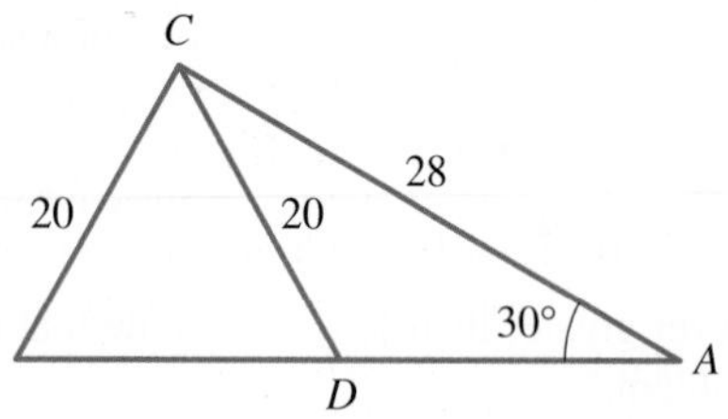

30. For the triangle shown, find the length AD.

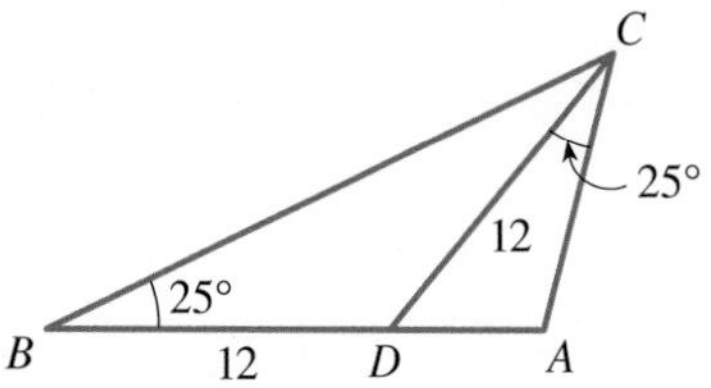

31. In triangle ABC, $\angle A = 40°$, $a = 15$, and $b = 20$.

(a) Show that there are two triangles, ABC and $A'B'C'$, that satisfy these conditions.

(b) Show that the areas of the triangles in part (a) are proportional to the sines of the angles C and C', that is,

$$\frac{\text{area of } \triangle ABC}{\text{area of } \triangle A'B'C'} = \frac{\sin C}{\sin C'}$$

32. Show that, given the three angles A, B, C of a triangle and one side, say a, the area of the triangle is

$$\text{area} = \frac{a^2 \sin B \sin C}{2 \sin A}$$

APPLICATIONS

33. Tracking a Satellite The path of a satellite orbiting the earth causes it to pass directly over two tracking stations A and B, which are 50 mi apart. When the satellite is on one side of the two stations, the angles of elevation at A and B are measured to be 87.0° and 84.2°, respectively.

(a) How far is the satellite from station A?

(b) How high is the satellite above the ground?

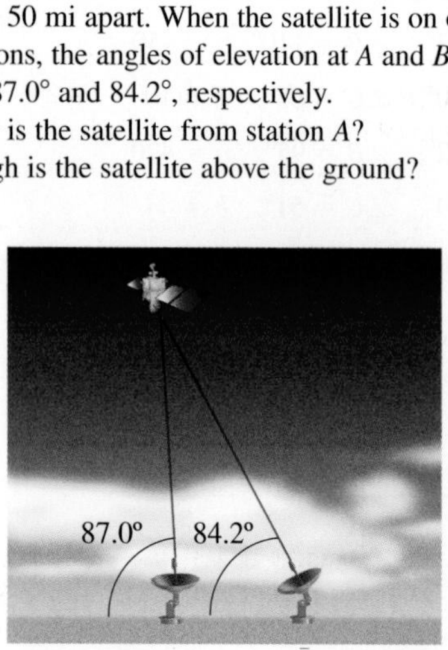

34. Flight of a Plane A pilot is flying over a straight highway. He determines the angles of depression to two mileposts, 5 mi apart, to be 32° and 48°, as shown in the figure.

(a) Find the distance of the plane from point A.

(b) Find the elevation of the plane.

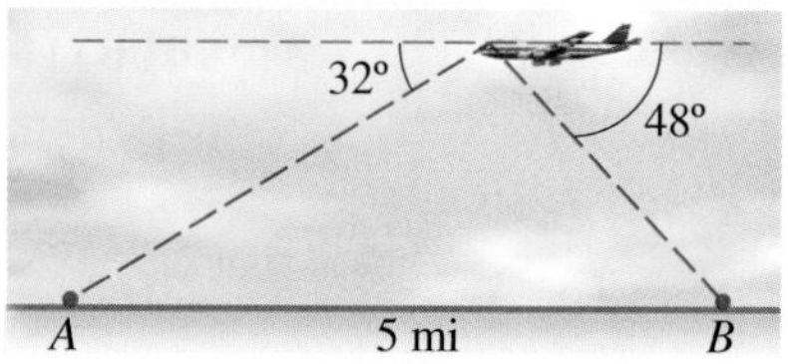

35. Distance Across a River To find the distance across a river, a surveyor chooses points A and B, which are 200 ft apart on one side of the river (see the figure). She then chooses a reference point C on the opposite side of the river and finds that $\angle BAC \approx 82°$ and $\angle ABC \approx 52°$. Approximate the distance from A to C.

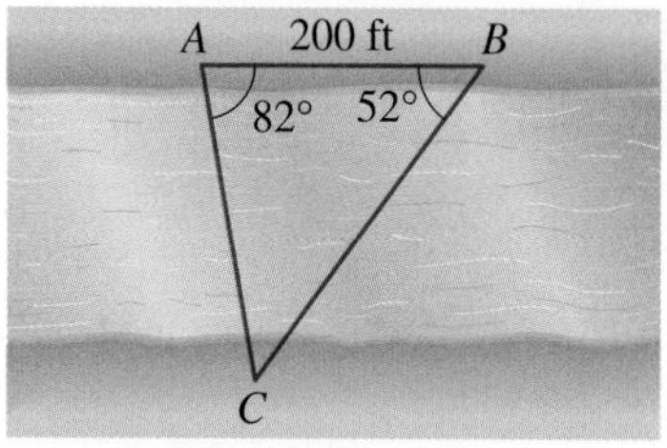

36. Distance Across a Lake Points A and B are separated by a lake. To find the distance between them, a surveyor locates a point C on land such that $\angle CAB = 48.6°$. He also measures CA as 312 ft and CB as 527 ft. Find the distance between A and B.

37. The Leaning Tower of Pisa The bell tower of the cathedral in Pisa, Italy, leans 5.6° from the vertical. A tourist stands 105 m from its base, with the tower leaning directly toward her. She measures the angle of elevation to the top of the tower to be 29.2°. Find the length of the tower to the nearest meter.

38. Radio Antenna A short-wave radio antenna is supported by two guy wires, 165 ft and 180 ft long. Each wire is attached to the top of the antenna and anchored to the ground, at two anchor points on opposite sides of the antenna. The shorter wire makes an angle of 67° with the ground. How far apart are the anchor points?

39. Height of a Tree A tree on a hillside casts a shadow 215 ft down the hill. If the angle of inclination of the hillside is 22° to the horizontal and the angle of elevation of the sun is 52°, find the height of the tree.

40. **Length of a Guy Wire** A communications tower is located at the top of a steep hill, as shown. The angle of inclination of the hill is 58°. A guy wire is to be attached to the top of the tower and to the ground, 100 m downhill from the base of the tower. The angle α in the figure is determined to be 12°. Find the length of cable required for the guy wire.

41. **Calculating a Distance** Observers at P and Q are located on the side of a hill that is inclined 32° to the horizontal, as shown. The observer at P determines the angle of elevation to a hot-air balloon to be 62°. At the same instant the observer at Q measures the angle of elevation to the balloon to be 71°. If P is 60 m down the hill from Q, find the distance from Q to the balloon.

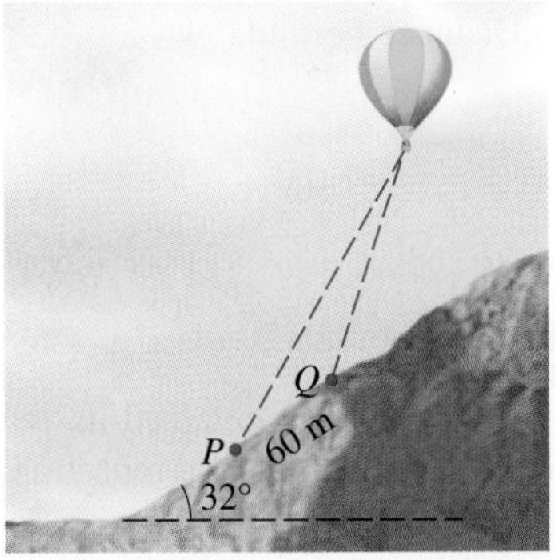

42. **Calculating an Angle** A water tower 30 m tall is located at the top of a hill. From a distance of 120 m down the hill, it is observed that the angle formed between the top and base of the tower is 8°. Find the angle of inclination of the hill.

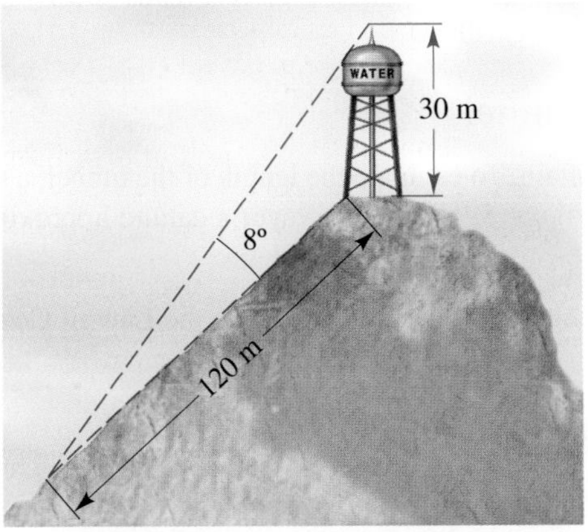

43. **Distances to Venus** The *elongation* α of a planet is the angle formed by the planet, earth, and sun (see the figure). It is known that the distance from the sun to Venus is 0.723 AU (see Exercise 65 in Section 6.2). At a certain time the elongation of Venus is found to be 39.4°. Find the possible distances from the earth to Venus at that time in astronomical units (AU).

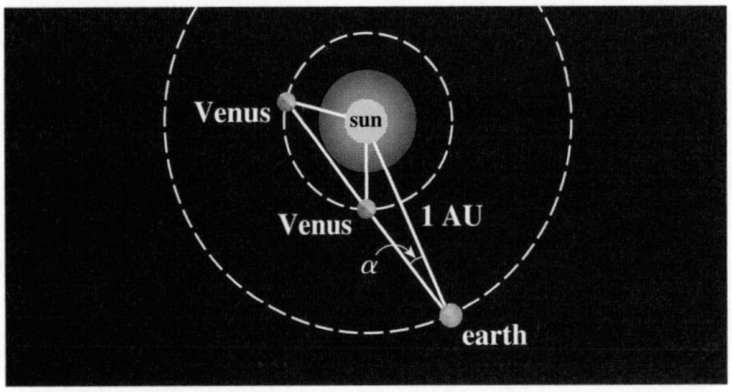

44. **Soap Bubbles** When two bubbles cling together in midair, their common surface is part of a sphere whose center D lies on the line passing throught the centers of the bubbles (see the figure). Also, angles ACB and ACD each have measure 60°.

(a) Show that the radius r of the common face is given by

$$r = \frac{ab}{a - b}$$

[*Hint:* Use the Law of Sines together with the fact that an angle θ and its supplement $180° - \theta$ have the same sine.]

(b) Find the radius of the common face if the radii of the bubbles are 4 cm and 3 cm.

(c) What shape does the common face take if the two bubbles have equal radii?

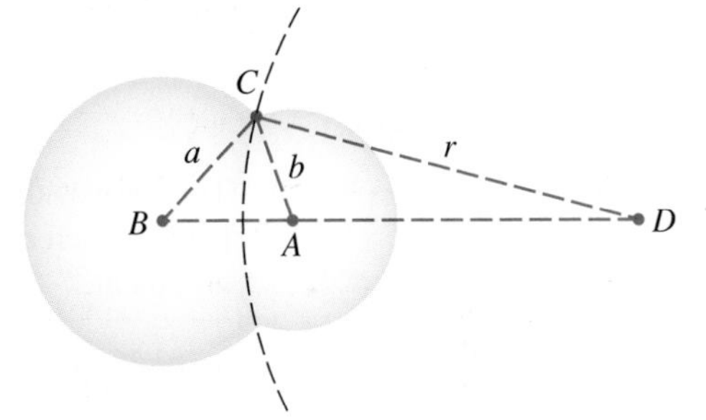

DISCOVERY ■ DISCUSSION ■ WRITING

45. **Number of Solutions in the Ambiguous Case** We have seen that when the Law of Sines is used to solve a triangle in the SSA case, there may be two, one, or no solution(s). Sketch triangles like those in Figure 6 to verify the criteria in the table for the number of solutions if you are given $\angle A$ and sides a and b.

Criterion	Number of solutions
$a \geq b$	1
$b > a > b \sin A$	2
$a = b \sin A$	1
$a < b \sin A$	0

If $\angle A = 30°$ and $b = 100$, use these criteria to find the range of values of a for which the triangle ABC has two solutions, one solution, or no solution.

6.6 The Law of Cosines

| The Law of Cosines ▶ Navigation: Heading and Bearing ▶ The Area of a Triangle

▼ The Law of Cosines

The Law of Sines cannot be used directly to solve triangles if we know two sides and the angle between them or if we know all three sides (these are Cases 3 and 4 of the preceding section). In these two cases the **Law of Cosines** applies.

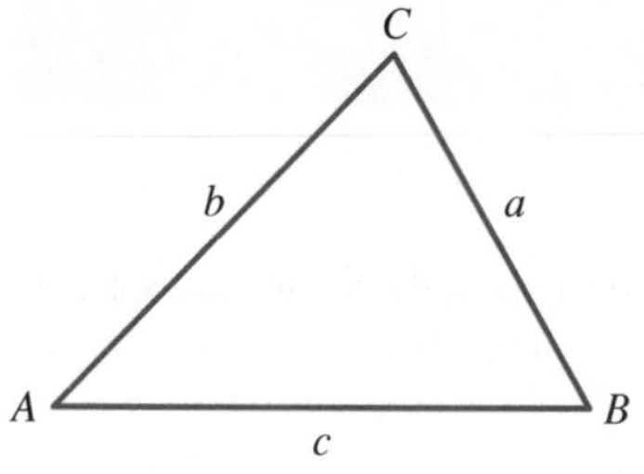

FIGURE 1

THE LAW OF COSINES

In any triangle ABC (see Figure 1), we have

$$a^2 = b^2 + c^2 - 2bc\cos A$$

$$b^2 = a^2 + c^2 - 2ac\cos B$$

$$c^2 = a^2 + b^2 - 2ab\cos C$$

PROOF To prove the Law of Cosines, place triangle ABC so that $\angle A$ is at the origin, as shown in Figure 2. The coordinates of vertices B and C are $(c, 0)$ and $(b\cos A, b\sin A)$, respectively. (You should check that the coordinates of these points will be the same if we draw angle A as an acute angle.) Using the Distance Formula, we get

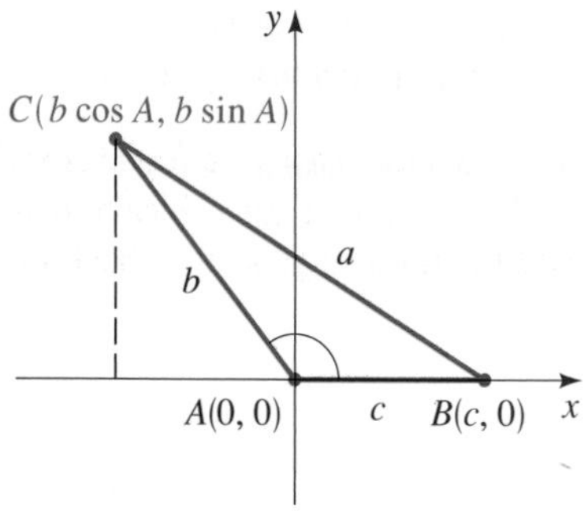

FIGURE 2

$$\begin{aligned} a^2 &= (b\cos A - c)^2 + (b\sin A - 0)^2 \\ &= b^2\cos^2 A - 2bc\cos A + c^2 + b^2\sin^2 A \\ &= b^2(\cos^2 A + \sin^2 A) - 2bc\cos A + c^2 \\ &= b^2 + c^2 - 2bc\cos A \qquad \text{Because } \sin^2 A + \cos^2 A = 1 \end{aligned}$$

This proves the first formula. The other two formulas are obtained in the same way by placing each of the other vertices of the triangle at the origin and repeating the preceding argument. ■

In words, the Law of Cosines says that the square of any side of a triangle is equal to the sum of the squares of the other two sides, minus twice the product of those two sides times the cosine of the included angle.

If one of the angles of a triangle, say $\angle C$, is a right angle, then $\cos C = 0$, and the Law of Cosines reduces to the Pythagorean Theorem, $c^2 = a^2 + b^2$. Thus the Pythagorean Theorem is a special case of the Law of Cosines.

EXAMPLE 1 | Length of a Tunnel

A tunnel is to be built through a mountain. To estimate the length of the tunnel, a surveyor makes the measurements shown in Figure 3. Use the surveyor's data to approximate the length of the tunnel.

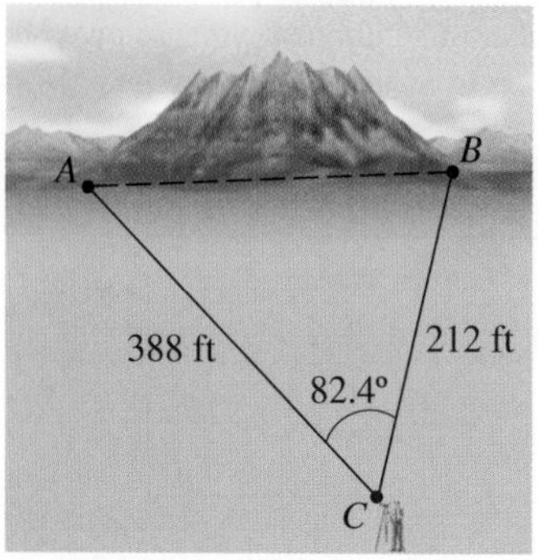

FIGURE 3

SOLUTION To approximate the length c of the tunnel, we use the Law of Cosines:

$$\begin{aligned} c^2 &= a^2 + b^2 - 2ab\cos C && \text{Law of Cosines} \\ &= 388^2 + 212^2 - 2(388)(212)\cos 82.4° && \text{Substitute} \\ &\approx 173730.2367 && \text{Use a calculator} \\ c &\approx \sqrt{173730.2367} \approx 416.8 && \text{Take square roots} \end{aligned}$$

Thus the tunnel will be approximately 417 ft long.

NOW TRY EXERCISES 3 AND 39 ■

EXAMPLE 2 | SSS, the Law of Cosines

The sides of a triangle are $a = 5$, $b = 8$, and $c = 12$ (see Figure 4). Find the angles of the triangle.

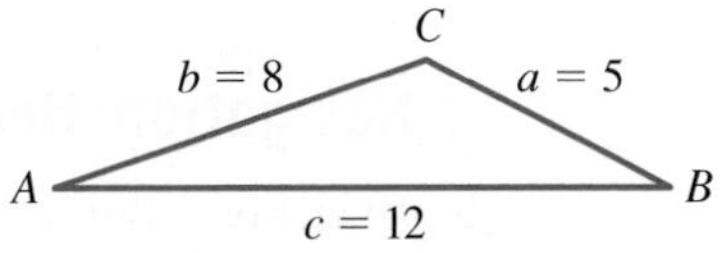

FIGURE 4

SOLUTION We first find $\angle A$. From the Law of Cosines, $a^2 = b^2 + c^2 - 2bc \cos A$. Solving for $\cos A$, we get

$$\cos A = \frac{b^2 + c^2 - a^2}{2bc} = \frac{8^2 + 12^2 - 5^2}{2(8)(12)} = \frac{183}{192} = 0.953125$$

COS⁻¹ OR INV COS OR ARC COS

Using a calculator, we find that $\angle A = \cos^{-1}(0.953125) \approx 18°$. In the same way we get

$$\cos B = \frac{a^2 + c^2 - b^2}{2ac} = \frac{5^2 + 12^2 - 8^2}{2(5)(12)} = 0.875$$

$$\cos C = \frac{a^2 + b^2 - c^2}{2ab} = \frac{5^2 + 8^2 - 12^2}{2(5)(8)} = -0.6875$$

Using a calculator, we find that

$$\angle B = \cos^{-1}(0.875) \approx 29° \qquad \text{and} \qquad \angle C = \cos^{-1}(-0.6875) \approx 133°$$

Of course, once two angles have been calculated, the third can more easily be found from the fact that the sum of the angles of a triangle is 180°. However, it's a good idea to calculate all three angles using the Law of Cosines and add the three angles as a check on your computations.

NOW TRY EXERCISE **7**

EXAMPLE 3 | SAS, the Law of Cosines

Solve triangle ABC, where $\angle A = 46.5°$, $b = 10.5$, and $c = 18.0$.

SOLUTION We can find a using the Law of Cosines.

$$a^2 = b^2 + c^2 - 2bc \cos A$$
$$= (10.5)^2 + (18.0)^2 - 2(10.5)(18.0)(\cos 46.5°) \approx 174.05$$

Thus, $a \approx \sqrt{174.05} \approx 13.2$. We also use the Law of Cosines to find $\angle B$ and $\angle C$, as in Example 2.

$$\cos B = \frac{a^2 + c^2 - b^2}{2ac} = \frac{13.2^2 + 18.0^2 - 10.5^2}{2(13.2)(18.0)} \approx 0.816477$$

$$\cos C = \frac{a^2 + b^2 - c^2}{2ab} = \frac{13.2^2 + 10.5^2 - 18.0^2}{2(13.2)(10.5)} \approx -0.142532$$

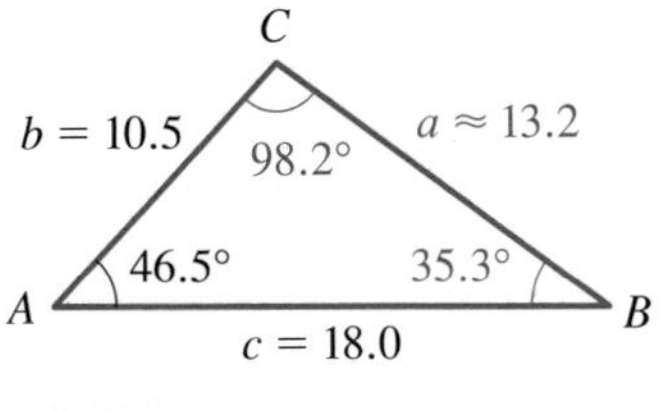

FIGURE 5

Using a calculator, we find that

$$\angle B = \cos^{-1}(0.816477) \approx 35.3° \qquad \text{and} \qquad \angle C = \cos^{-1}(-0.142532) \approx 98.2°$$

To summarize: $\angle B \approx 35.3°$, $\angle C \approx 98.2°$, and $a \approx 13.2$. (See Figure 5.)

NOW TRY EXERCISE **13**

We could have used the Law of Sines to find $\angle B$ and $\angle C$ in Example 3, since we knew all three sides and an angle in the triangle. But knowing the sine of an angle does not uniquely specify the angle, since an angle θ and its supplement $180° - \theta$ both have the same sine. Thus, we would need to decide which of the two angles is the correct choice. This ambiguity does not arise when we use the Law of Cosines, because every angle between 0° and 180° has a unique cosine. So using only the Law of Cosines is preferable in problems like Example 3.

▼ Navigation: Heading and Bearing

In navigation a direction is often given as a **bearing**, that is, as an acute angle measured from due north or due south. The bearing N 30° E, for example, indicates a direction that points 30° to the east of due north (see Figure 6).

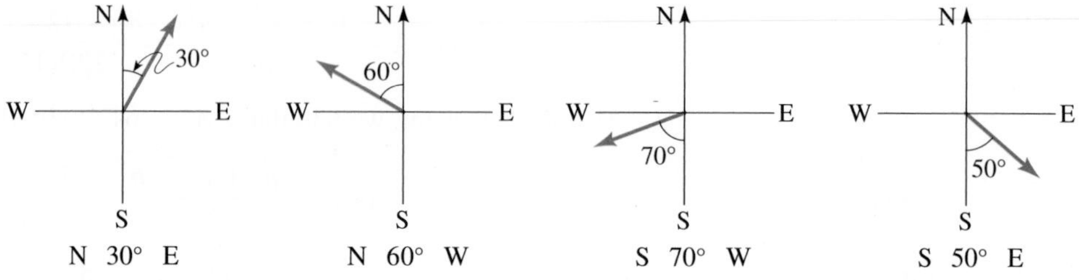

FIGURE 6

EXAMPLE 4 | Navigation

A pilot sets out from an airport and heads in the direction N 20° E, flying at 200 mi/h. After one hour, he makes a course correction and heads in the direction N 40° E. Half an hour after that, engine trouble forces him to make an emergency landing.

(a) Find the distance between the airport and his final landing point.

(b) Find the bearing from the airport to his final landing point.

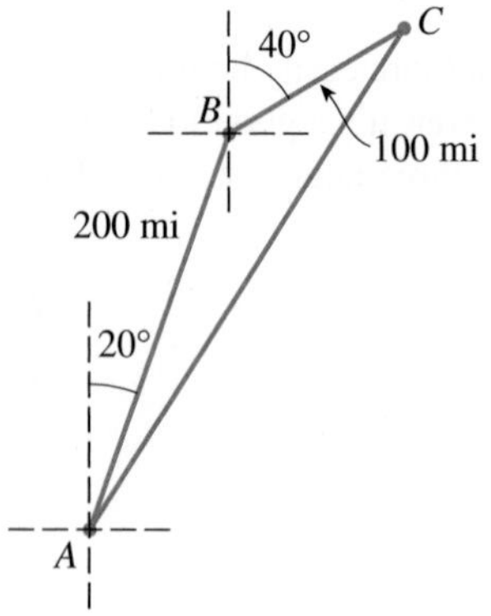

FIGURE 7

SOLUTION

(a) In one hour the plane travels 200 mi, and in half an hour it travels 100 mi, so we can plot the pilot's course as in Figure 7. When he makes his course correction, he turns 20° to the right, so the angle between the two legs of his trip is $180° - 20° = 160°$. So by the Law of Cosines we have

$$b^2 = 200^2 + 100^2 - 2 \cdot 200 \cdot 100 \cos 160°$$
$$\approx 87{,}587.70$$

Thus, $b \approx 295.95$. The pilot lands about 296 mi from his starting point.

(b) We first use the Law of Sines to find $\angle A$.

$$\frac{\sin A}{100} = \frac{\sin 160°}{295.95}$$
$$\sin A = 100 \cdot \frac{\sin 160°}{295.95}$$
$$\approx 0.11557$$

Another angle with sine 0.11557 is $180° - 6.636° = 173.364°$. But this is clearly too large to be $\angle A$ in $\angle ABC$.

Using the SIN⁻¹ key on a calculator, we find that $\angle A \approx 6.636°$. From Figure 7 we see that the line from the airport to the final landing site points in the direction $20° + 6.636° = 26.636°$ east of due north. Thus, the bearing is about N 26.6° E.

NOW TRY EXERCISE **45** ■

▼ The Area of a Triangle

An interesting application of the Law of Cosines involves a formula for finding the area of a triangle from the lengths of its three sides (see Figure 8).

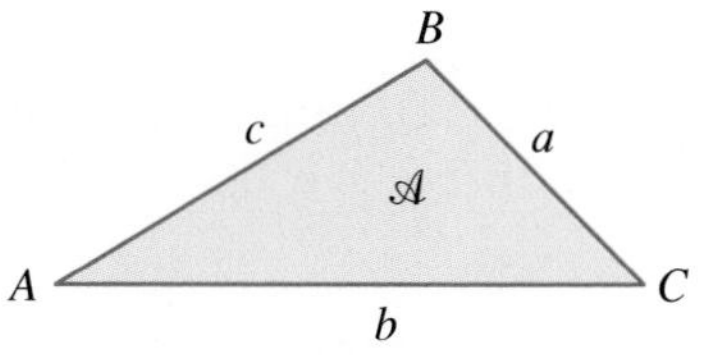

FIGURE 8

> **HERON'S FORMULA**
>
> The area $\mathscr{A}$ of triangle ABC is given by
>
> $$\mathscr{A} = \sqrt{s(s-a)(s-b)(s-c)}$$
>
> where $s = \frac{1}{2}(a + b + c)$ is the **semiperimeter** of the triangle; that is, s is half the perimeter.

PROOF We start with the formula $\mathscr{A} = \frac{1}{2}ab \sin C$ from Section 6.3. Thus

$$\begin{aligned} \mathscr{A}^2 &= \tfrac{1}{4}a^2b^2 \sin^2 C \\ &= \tfrac{1}{4}a^2b^2(1 - \cos^2 C) && \text{Pythagorean identity} \\ &= \tfrac{1}{4}a^2b^2(1 - \cos C)(1 + \cos C) && \text{Factor} \end{aligned}$$

Next, we write the expressions $1 - \cos C$ and $1 + \cos C$ in terms of a, b, and c. By the Law of Cosines we have

$$\begin{aligned} \cos C &= \frac{a^2 + b^2 - c^2}{2ab} && \text{Law of Cosines} \\ 1 + \cos C &= 1 + \frac{a^2 + b^2 - c^2}{2ab} && \text{Add 1} \\ &= \frac{2ab + a^2 + b^2 - c^2}{2ab} && \text{Common denominator} \\ &= \frac{(a + b)^2 - c^2}{2ab} && \text{Factor} \\ &= \frac{(a + b + c)(a + b - c)}{2ab} && \text{Difference of squares} \end{aligned}$$

Similarly

$$1 - \cos C = \frac{(c + a - b)(c - a + b)}{2ab}$$

Substituting these expressions in the formula we obtained for $\mathscr{A}^2$ gives

$$\begin{aligned} \mathscr{A}^2 &= \tfrac{1}{4}a^2b^2 \frac{(a + b + c)(a + b - c)}{2ab} \frac{(c + a - b)(c - a + b)}{2ab} \\ &= \frac{(a + b + c)}{2} \frac{(a + b - c)}{2} \frac{(c + a - b)}{2} \frac{(c - a + b)}{2} \\ &= s(s - c)(s - b)(s - a) \end{aligned}$$

Heron's Formula now follows by taking the square root of each side. ■

To see that the factors in the last two products are equal, note for example that

$$\frac{a + b - c}{2} = \frac{a + b + c}{2} - c = s - c$$

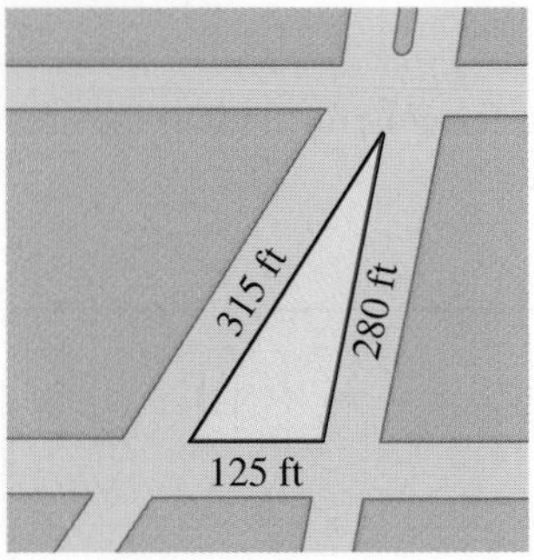

FIGURE 9

EXAMPLE 5 | Area of a Lot

A businessman wishes to buy a triangular lot in a busy downtown location (see Figure 9). The lot frontages on the three adjacent streets are 125, 280, and 315 ft. Find the area of the lot.

SOLUTION The semiperimeter of the lot is

$$s = \frac{125 + 280 + 315}{2} = 360$$

By Heron's Formula the area is

$$\mathcal{A} = \sqrt{360(360 - 125)(360 - 280)(360 - 315)} \approx 17{,}451.6$$

Thus, the area is approximately 17,452 ft^2.

NOW TRY EXERCISES **29** AND **53**

6.6 EXERCISES

CONCEPTS

1. For triangle ABC with sides a, b, and c the Law of Cosines states

$c^2 =$ ____________________

2. In which of the following cases must the Law of Cosines be used to solve a triangle?

ASA SSS SAS SSA

SKILLS

3–10 ■ Use the Law of Cosines to determine the indicated side x or angle θ.

3.
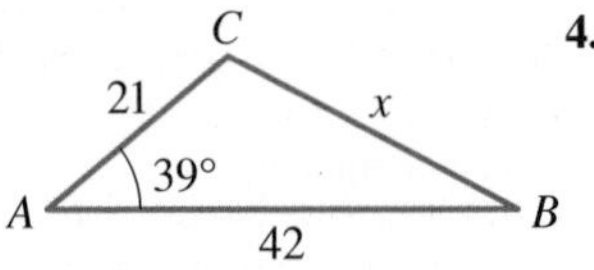

4.
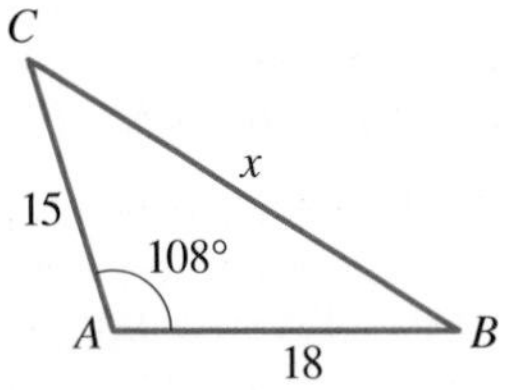

5.
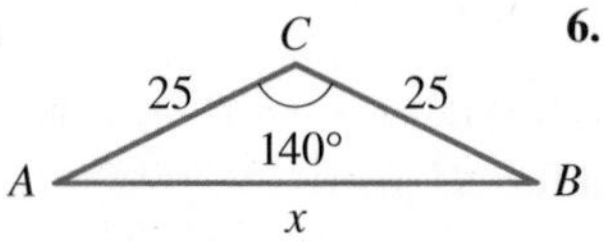

6.
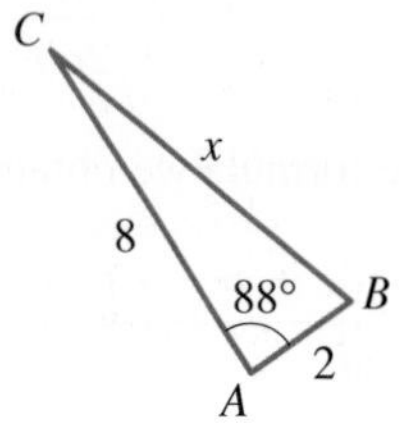

7.
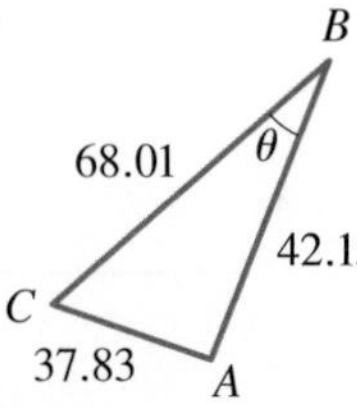

8.
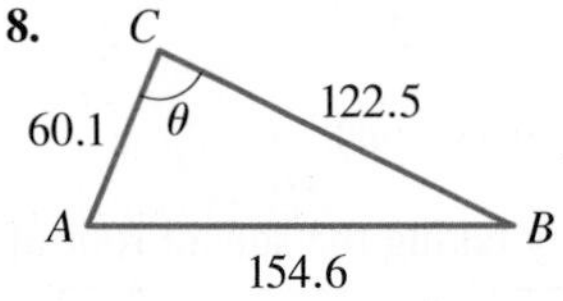

9.
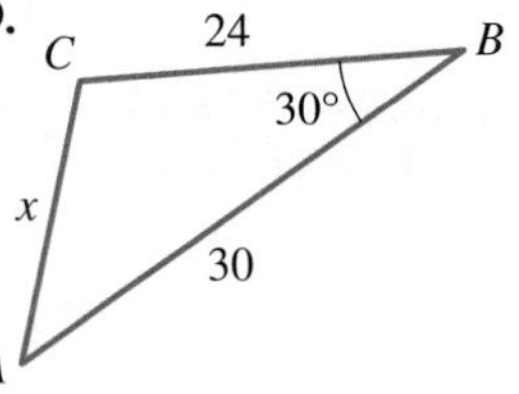

10.
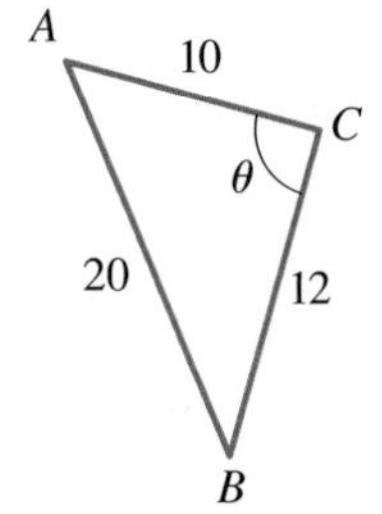

11–20 ■ Solve triangle ABC.

11.
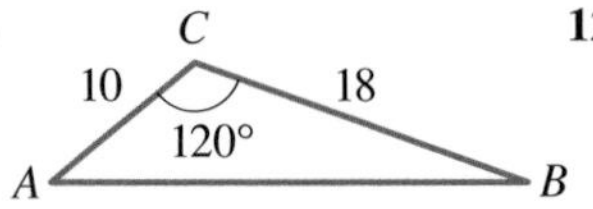

12.
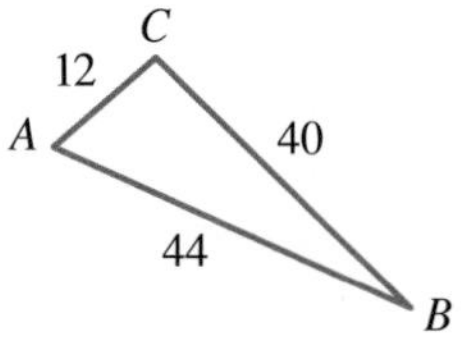

13. $a = 3.0$, $b = 4.0$, $\angle C = 53°$

14. $b = 60$, $c = 30$, $\angle A = 70°$

15. $a = 20$, $b = 25$, $c = 22$

16. $a = 10$, $b = 12$, $c = 16$

17. $b = 125$, $c = 162$, $\angle B = 40°$

18. $a = 65$, $c = 50$, $\angle C = 52°$

19. $a = 50$, $b = 65$, $\angle A = 55°$

20. $a = 73.5$, $\angle B = 61°$, $\angle C = 83°$

21–28 ■ Find the indicated side x or angle θ. (Use either the Law of Sines or the Law of Cosines, as appropriate.)

21.
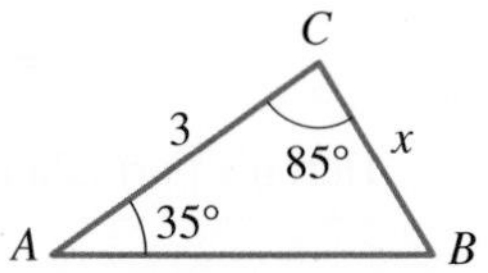

22.
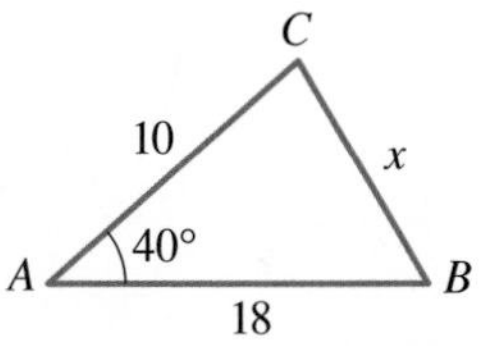

23.
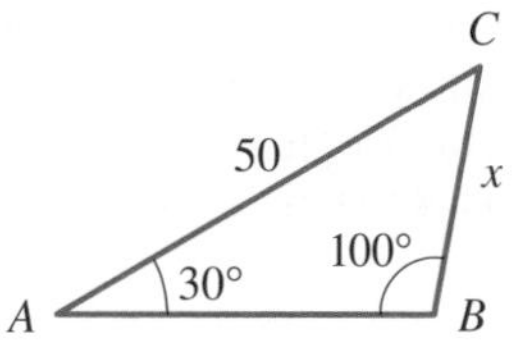

24.
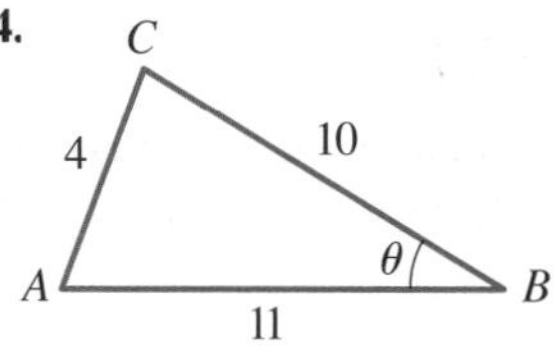

25.
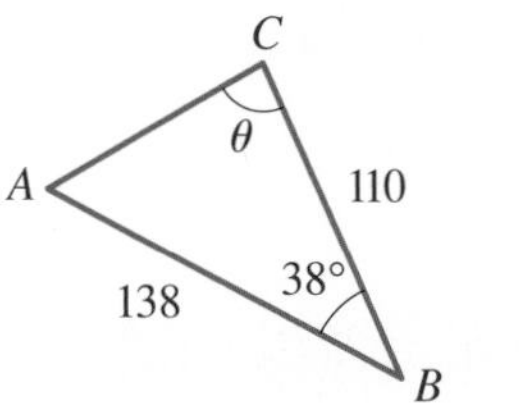

26.
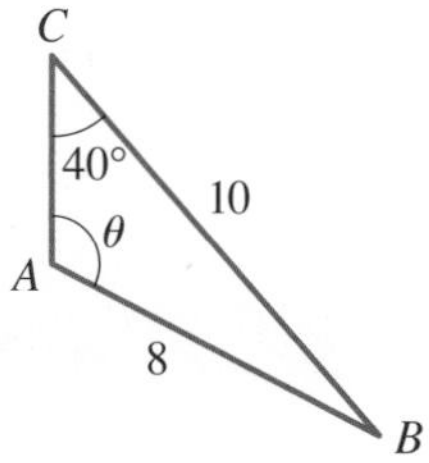

27.
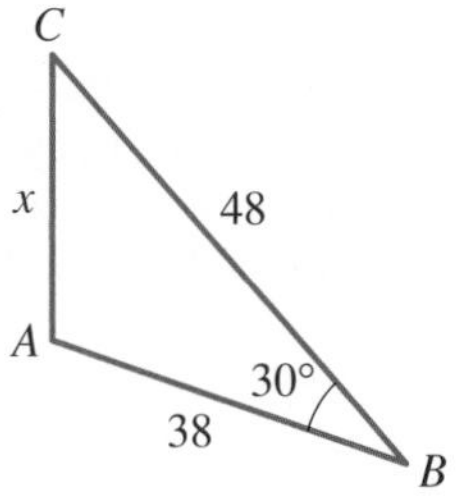

28.
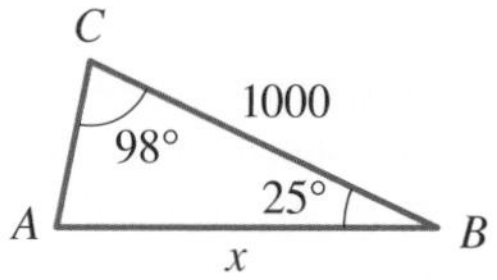

29–32 ■ Find the area of the triangle whose sides have the given lengths.

29. $a = 9, \quad b = 12, \quad c = 15$ 30. $a = 1, \quad b = 2, \quad c = 2$

31. $a = 7, \quad b = 8, \quad c = 9$

32. $a = 11, \quad b = 100, \quad c = 101$

33–36 ■ Find the area of the shaded figure, rounded to two decimals.

33.
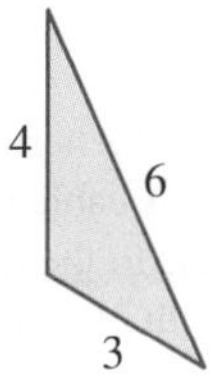

34.
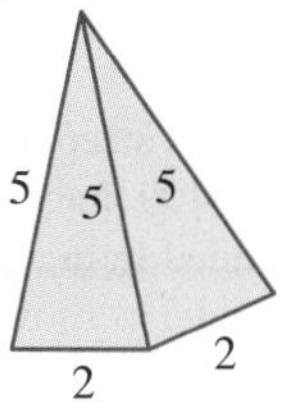

35.
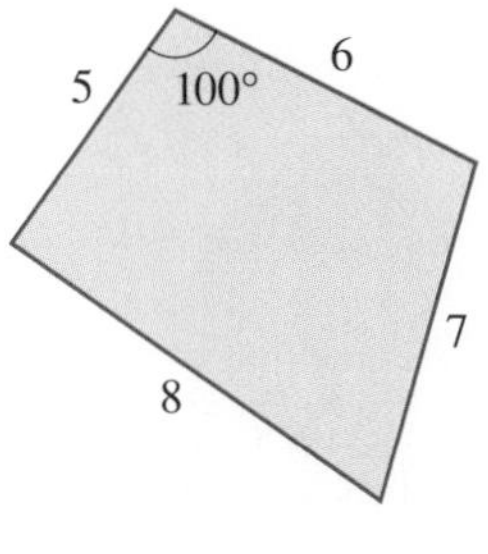

36.
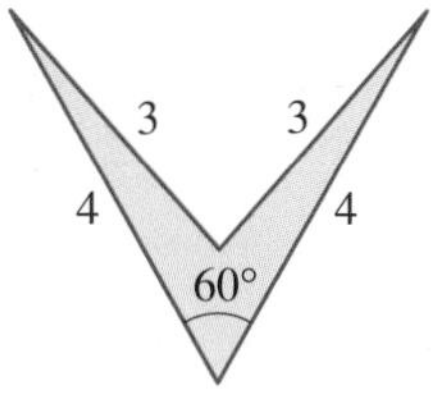

37. Three circles of radii 4, 5, and 6 cm are mutually tangent. Find the shaded area enclosed between the circles.

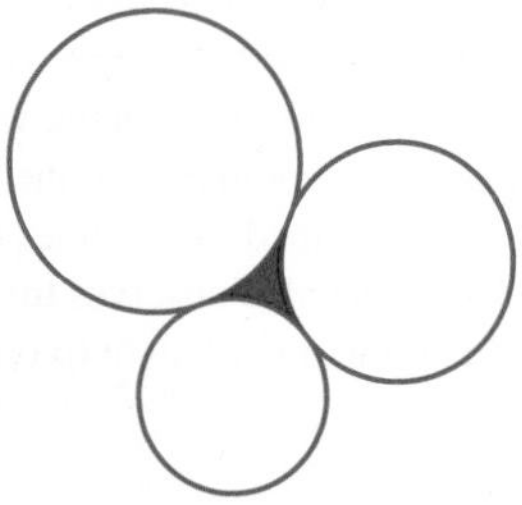

38. Prove that in triangle ABC

$$a = b\cos C + c\cos B$$
$$b = c\cos A + a\cos C$$
$$c = a\cos B + b\cos A$$

These are called the *Projection Laws*. [*Hint:* To get the first equation, add the second and third equations in the Law of Cosines and solve for a.]

APPLICATIONS

39. **Surveying** To find the distance across a small lake, a surveyor has taken the measurements shown. Find the distance across the lake using this information.

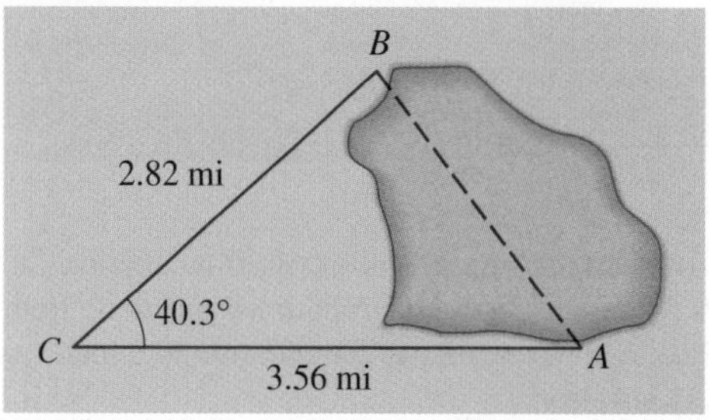

40. **Geometry** A parallelogram has sides of lengths 3 and 5, and one angle is 50°. Find the lengths of the diagonals.

41. **Calculating Distance** Two straight roads diverge at an angle of 65°. Two cars leave the intersection at 2:00 P.M., one traveling at 50 mi/h and the other at 30 mi/h. How far apart are the cars at 2:30 P.M.?

42. **Calculating Distance** A car travels along a straight road, heading east for 1 h, then traveling for 30 min on another road that leads northeast. If the car has maintained a constant speed of 40 mi/h, how far is it from its starting position?

43. **Dead Reckoning** A pilot flies in a straight path for 1 h 30 min. She then makes a course correction, heading 10° to the right of her original course, and flies 2 h in the new direction. If she maintains a constant speed of 625 mi/h, how far is she from her starting position?

44. **Navigation** Two boats leave the same port at the same time. One travels at a speed of 30 mi/h in the direction N 50° E and the other travels at a speed of 26 mi/h in a direction S 70° E (see the figure). How far apart are the two boats after one hour?

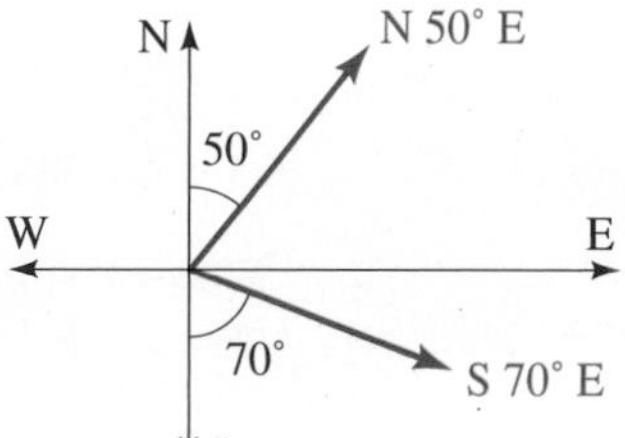

45. Navigation A fisherman leaves his home port and heads in the direction N 70° W. He travels 30 mi and reaches Egg Island. The next day he sails N 10° E for 50 mi, reaching Forrest Island.

(a) Find the distance between the fisherman's home port and Forrest Island.

(b) Find the bearing from Forrest Island back to his home port.

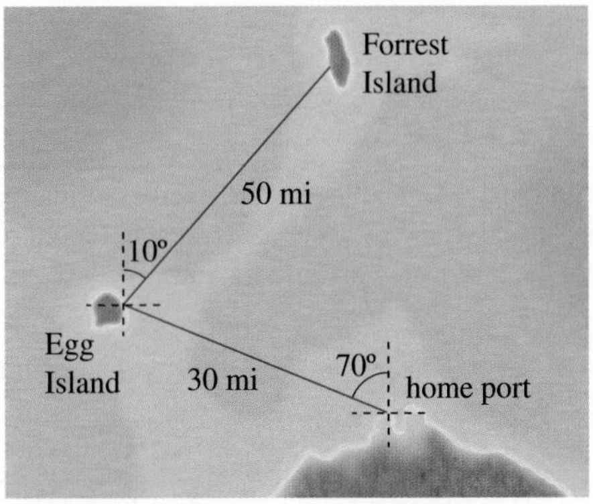

46. Navigation Airport B is 300 mi from airport A at a bearing N 50° E (see the figure). A pilot wishing to fly from A to B mistakenly flies due east at 200 mi/h for 30 minutes, when he notices his error.

(a) How far is the pilot from his destination at the time he notices the error?

(b) What bearing should he head his plane in order to arrive at airport B?

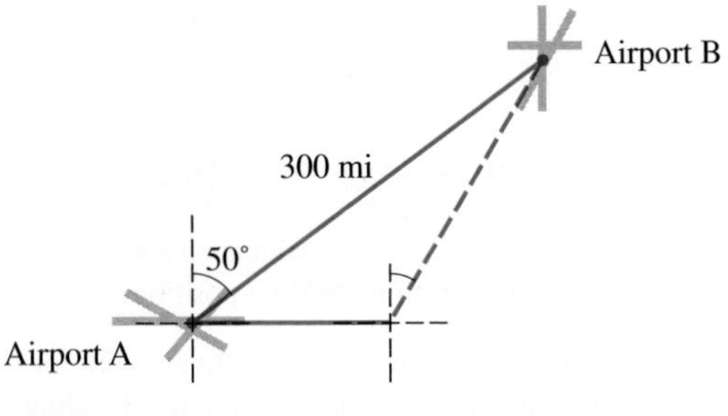

47. Triangular Field A triangular field has sides of lengths 22, 36, and 44 yd. Find the largest angle.

48. Towing a Barge Two tugboats that are 120 ft apart pull a barge, as shown. If the length of one cable is 212 ft and the length of the other is 230 ft, find the angle formed by the two cables.

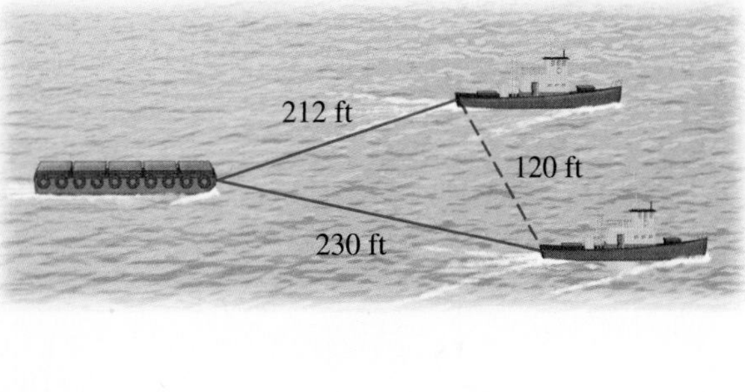

49. Flying Kites A boy is flying two kites at the same time. He has 380 ft of line out to one kite and 420 ft to the other. He estimates the angle between the two lines to be 30°. Approximate the distance between the kites.

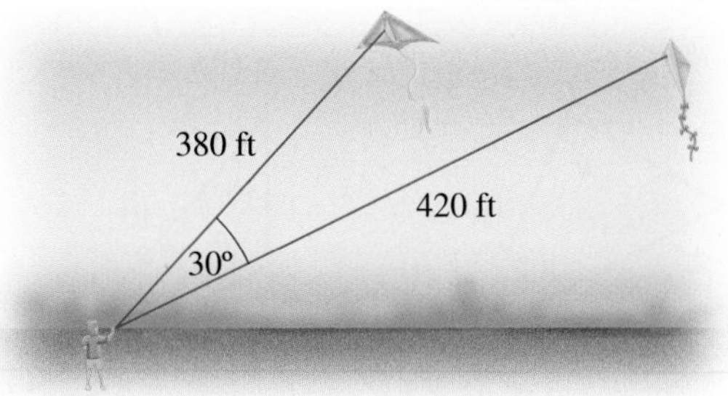

50. Securing a Tower A 125-ft tower is located on the side of a mountain that is inclined 32° to the horizontal. A guy wire is to be attached to the top of the tower and anchored at a point 55 ft downhill from the base of the tower. Find the shortest length of wire needed.

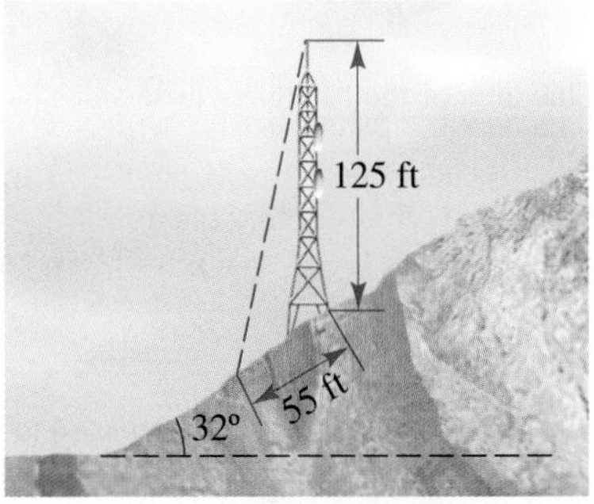

51. Cable Car A steep mountain is inclined 74° to the horizontal and rises 3400 ft above the surrounding plain. A cable car is to be installed from a point 800 ft from the base to the top of the mountain, as shown. Find the shortest length of cable needed.

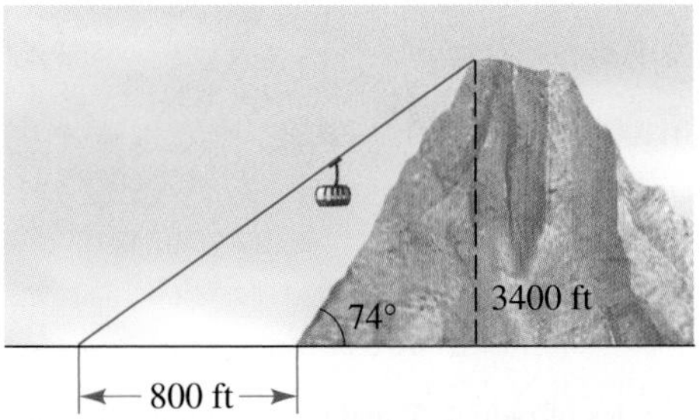

52. CN Tower The CN Tower in Toronto, Canada, is the tallest free-standing structure in North America. A woman on the observation deck, 1150 ft above the ground, wants to determine the distance between two landmarks on the ground below. She observes that the angle formed by the lines of sight to these two landmarks is 43°. She also observes that the angle between the vertical and the line of sight to one of the landmarks is 62° and to the other landmark is 54°. Find the distance between the two landmarks.

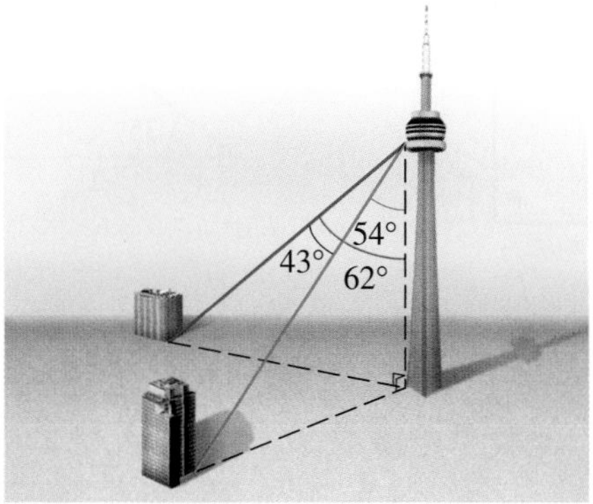

53. **Land Value** Land in downtown Columbia is valued at \$20 a square foot. What is the value of a triangular lot with sides of lengths 112, 148, and 190 ft?

DISCOVERY ■ DISCUSSION ■ WRITING

54. **Solving for the Angles in a Triangle** The paragraph that follows the solution of Example 3 on page 477 explains an alternative method for finding $\angle B$ and $\angle C$, using the Law of Sines. Use this method to solve the triangle in the example, finding $\angle B$ first and then $\angle C$. Explain how you chose the appropriate value for the measure of $\angle B$. Which method do you prefer for solving an SAS triangle problem, the one explained in Example 3 or the one you used in this exercise?

CHAPTER 6 | REVIEW

■ CONCEPT CHECK

1. (a) Explain the difference between a positive angle and a negative angle.
   (b) How is an angle of measure 1 degree formed?
   (c) How is an angle of measure 1 radian formed?
   (d) How is the radian measure of an angle θ defined?
   (e) How do you convert from degrees to radians?
   (f) How do you convert from radians to degrees?
2. (a) When is an angle in standard position?
   (b) When are two angles coterminal?
3. (a) What is the length s of an arc of a circle with radius r that subtends a central angle of θ radians?
   (b) What is the area A of a sector of a circle with radius r and central angle θ radians?
4. If θ is an acute angle in a right triangle, define the six trigonometric ratios in terms of the adjacent and opposite sides and the hypotenuse.
5. What does it mean to solve a triangle?
6. If θ is an angle in standard position, $P(x, y)$ is a point on the terminal side, and r is the distance from the origin to P, write expressions for the six trigonometric functions of θ.
7. Which trigonometric functions are positive in Quadrants I, II, III, and IV?
8. If θ is an angle in standard position, what is its reference angle $\overline{\theta}$?
9. (a) State the reciprocal identities.
   (b) State the Pythagorean identities.
10. (a) What is the area of a triangle with sides of length a and b and with included angle θ?
    (b) What is the area of a triangle with sides of length a, b, and c?
11. Define the inverse sine function $\sin^{-1}$. What are its domain and range?
12. Define the inverse cosine function $\cos^{-1}$. What are its domain and range?
13. Define the inverse tangent function $\tan^{-1}$. What are its domain and range?
14. (a) State the Law of Sines.
    (b) State the Law of Cosines.
15. Explain the ambiguous case in the Law of Sines.

■ EXERCISES

1–2 ■ Find the radian measure that corresponds to the given degree measure.

1. (a) 60° (b) 330° (c) −135° (d) −90°
2. (a) 24° (b) −330° (c) 750° (d) 5°

3–4 ■ Find the degree measure that corresponds to the given radian measure.

3. (a) $\dfrac{5\pi}{2}$ (b) $-\dfrac{\pi}{6}$ (c) $\dfrac{9\pi}{4}$ (d) 3.1
4. (a) 8 (b) $-\frac{5}{2}$ (c) $\dfrac{11\pi}{6}$ (d) $\dfrac{3\pi}{5}$
5. Find the length of an arc of a circle of radius 8 m if the arc subtends a central angle of 1 rad.
6. Find the measure of a central angle θ in a circle of radius 5 ft if the angle is subtended by an arc of length 7 ft.
7. A circular arc of length 100 ft subtends a central angle of 70°. Find the radius of the circle.
8. How many revolutions will a car wheel of diameter 28 in. make over a period of half an hour if the car is traveling at 60 mi/h?

9. New York and Los Angeles are 2450 mi apart. Find the angle that the arc between these two cities subtends at the center of the earth. (The radius of the earth is 3960 mi.)

10. Find the area of a sector with central angle 2 rad in a circle of radius 5 m.

11. Find the area of a sector with central angle 52° in a circle of radius 200 ft.

12. A sector in a circle of radius 25 ft has an area of 125 ft^2. Find the central angle of the sector.

13. A potter's wheel with radius 8 in. spins at 150 rpm. Find the angular and linear speeds of a point on the rim of the wheel.

14. In an automobile transmission a *gear ratio* g is the ratio

$$g = \frac{\text{angular speed of engine}}{\text{angular speed of wheels}}$$

The angular speed of the engine is shown on the tachometer (in rpm).

A certain sports car has wheels with radius 11 in. Its gear ratios are shown in the following table. Suppose the car is in fourth gear and the tachometer reads 3500 rpm.

(a) Find the angular speed of the engine.

(b) Find the angular speed of the wheels.

(c) How fast (in mi/h) is the car traveling?

Gear	Ratio
1st	4.1
2nd	3.0
3rd	1.6
4th	0.9
5th	0.7

15–16 ■ Find the values of the six trigonometric ratios of θ.

15.

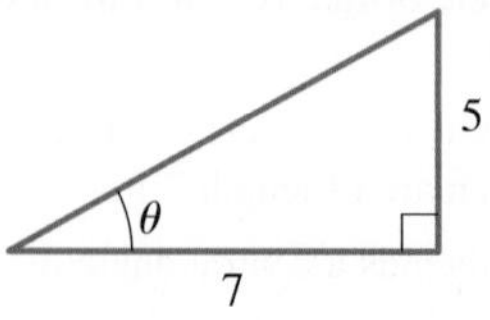

16.

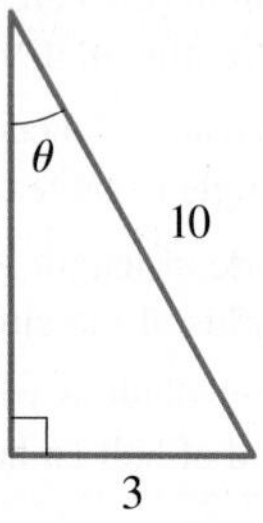

17–20 ■ Find the sides labeled x and y, rounded to two decimal places.

17.

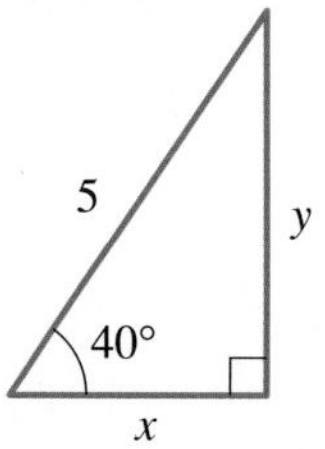

18.

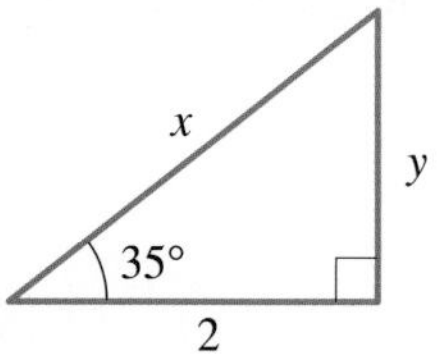

19.

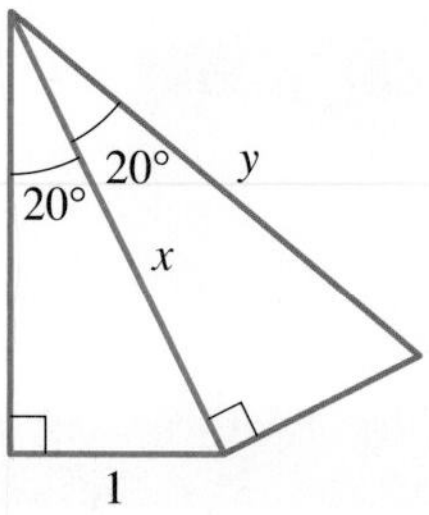

20.

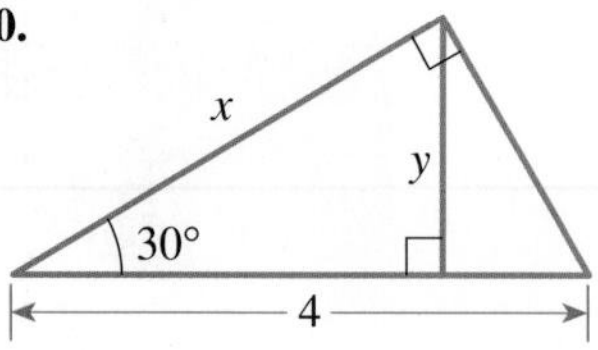

21–24 ■ Solve the triangle.

21.

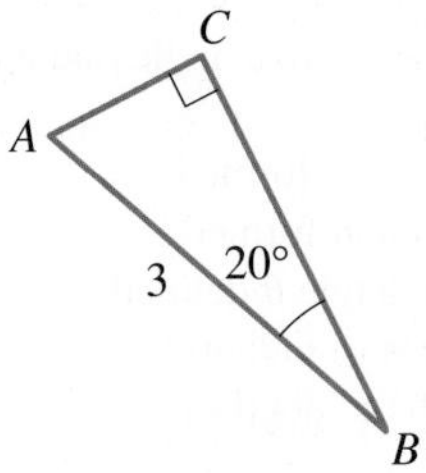

22.

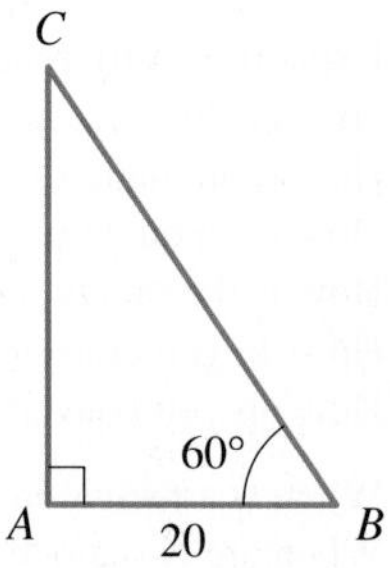

23.

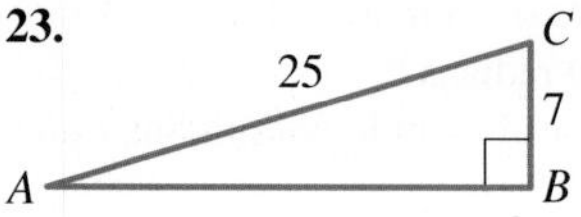

24.

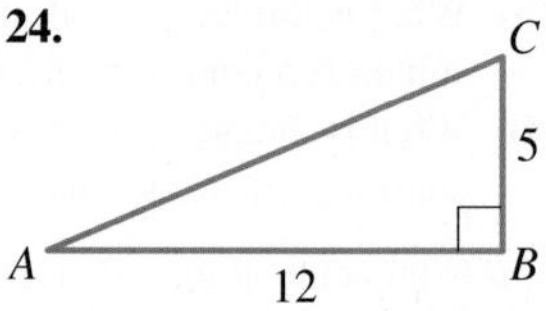

25. Express the lengths a and b in the figure in terms of the trigonometric ratios of θ.

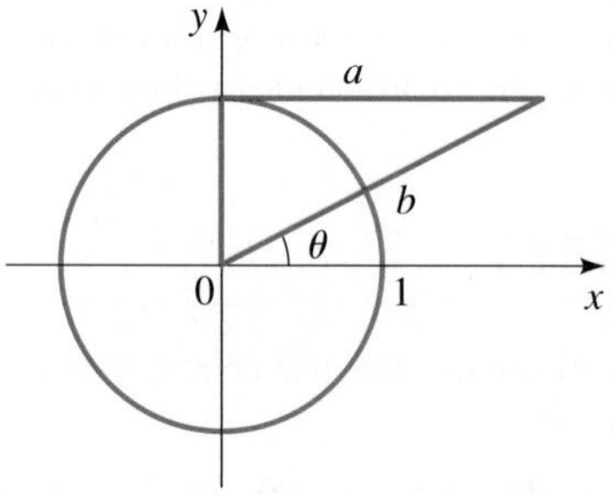

26. The highest free-standing tower in North America is the CN Tower in Toronto, Canada. From a distance of 1 km from its base, the angle of elevation to the top of the tower is 28.81°. Find the height of the tower.

27. Find the perimeter of a regular hexagon that is inscribed in a circle of radius 8 m.

28. The pistons in a car engine move up and down repeatedly to turn the crankshaft, as shown. Find the height of the point P above the center O of the crankshaft in terms of the angle θ.

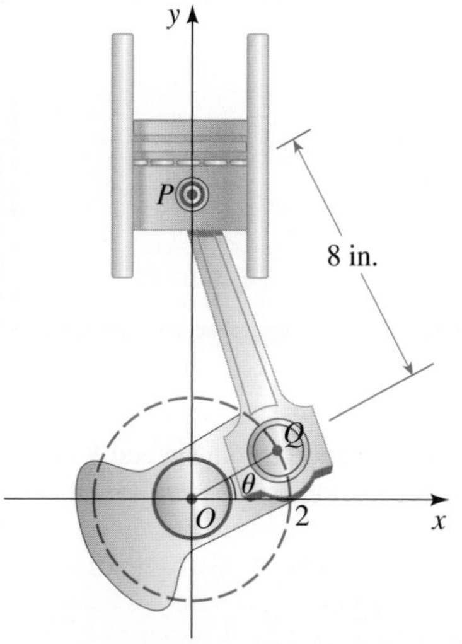

29. As viewed from the earth, the angle subtended by the full moon is 0.518°. Use this information and the fact that the distance AB from the earth to the moon is 236,900 mi to find the radius of the moon.

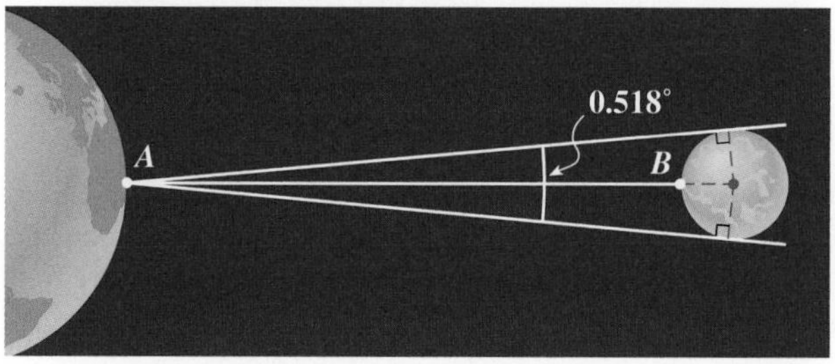

30. A pilot measures the angles of depression to two ships to be 40° and 52° (see the figure). If the pilot is flying at an elevation of 35,000 ft, find the distance between the two ships.

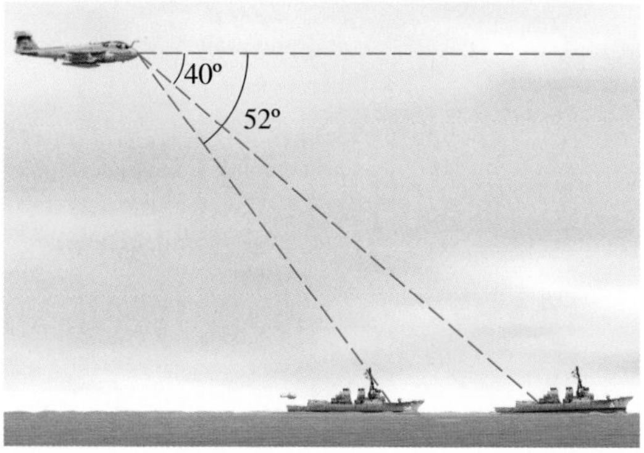

31–42 ■ Find the exact value.

31. $\sin 315°$ **32.** $\csc \dfrac{9\pi}{4}$

33. $\tan(-135°)$ **34.** $\cos \dfrac{5\pi}{6}$

35. $\cot\left(-\dfrac{22\pi}{3}\right)$ **36.** $\sin 405°$

37. $\cos 585°$ **38.** $\sec \dfrac{22\pi}{3}$

39. $\csc \dfrac{8\pi}{3}$ **40.** $\sec \dfrac{13\pi}{6}$

41. $\cot(-390°)$ **42.** $\tan \dfrac{23\pi}{4}$

43. Find the values of the six trigonometric ratios of the angle θ in standard position if the point $(-5, 12)$ is on the terminal side of θ.

44. Find $\sin \theta$ if θ is in standard position and its terminal side intersects the circle of radius 1 centered at the origin at the point $(-\sqrt{3}/2, \frac{1}{2})$.

45. Find the acute angle that is formed by the line $y - \sqrt{3}x + 1 = 0$ and the x-axis.

46. Find the six trigonometric ratios of the angle θ in standard position if its terminal side is in Quadrant III and is parallel to the line $4y - 2x - 1 = 0$.

47–50 ■ Write the first expression in terms of the second, for θ in the given quadrant.

47. $\tan \theta$, $\cos \theta$; θ in Quadrant II

48. $\sec \theta$, $\sin \theta$; θ in Quadrant III

49. $\tan^2 \theta$, $\sin \theta$; θ in any quadrant

50. $\csc^2 \theta \cos^2 \theta$, $\sin \theta$; θ in any quadrant

51–54 ■ Find the values of the six trigonometric functions of θ from the information given.

51. $\tan \theta = \sqrt{7}/3$, $\sec \theta = \frac{4}{3}$ **52.** $\sec \theta = \frac{41}{40}$, $\csc \theta = -\frac{41}{9}$

53. $\sin \theta = \frac{3}{5}$, $\cos \theta < 0$ **54.** $\sec \theta = -\frac{13}{5}$, $\tan \theta > 0$

55. If $\tan \theta = -\frac{1}{2}$ for θ in Quadrant II, find $\sin \theta + \cos \theta$.

56. If $\sin \theta = \frac{1}{2}$ for θ in Quadrant I, find $\tan \theta + \sec \theta$.

57. If $\tan \theta = -1$, find $\sin^2 \theta + \cos^2 \theta$.

58. If $\cos \theta = -\sqrt{3}/2$ and $\pi/2 < \theta < \pi$, find $\sin 2\theta$.

59–62 ■ Find the exact value of the expression.

59. $\sin^{-1}(\sqrt{3}/2)$ **60.** $\tan^{-1}(\sqrt{3}/3)$

61. $\tan(\sin^{-1} \frac{2}{5})$ **62.** $\sin(\cos^{-1} \frac{3}{8})$

63–64 ■ Rewrite the expression as an algebraic expression in x.

63. $\sin(\tan^{-1} x)$ **64.** $\sec(\sin^{-1} x)$

65–66 ■ Express θ in terms of x.

65.

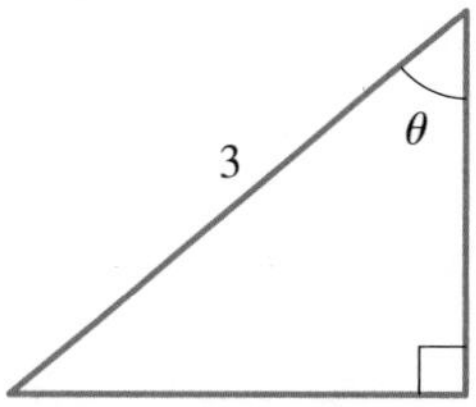

66.

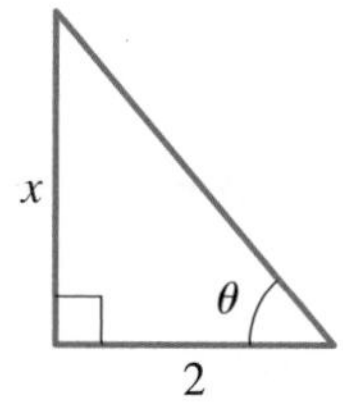

67–76 ■ Find the side labeled x or the angle labeled θ.

67.

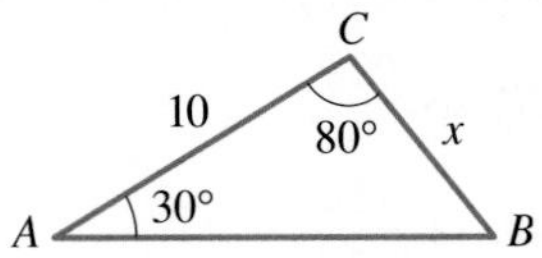

68.

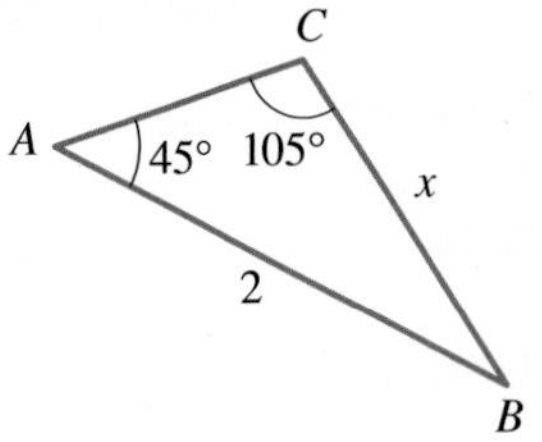

69.

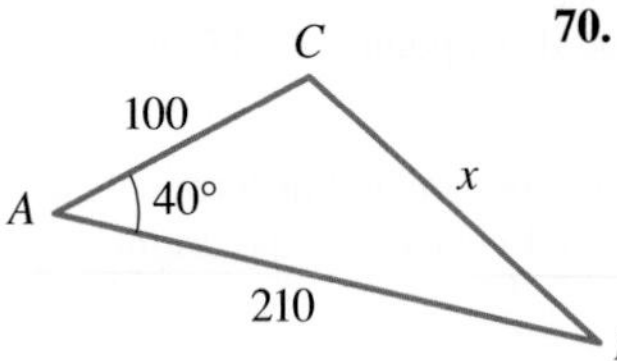

70.

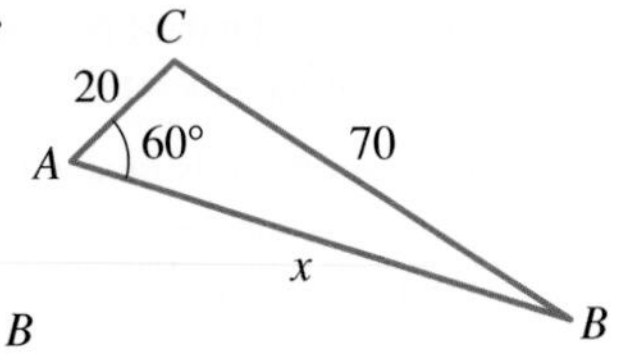

71.

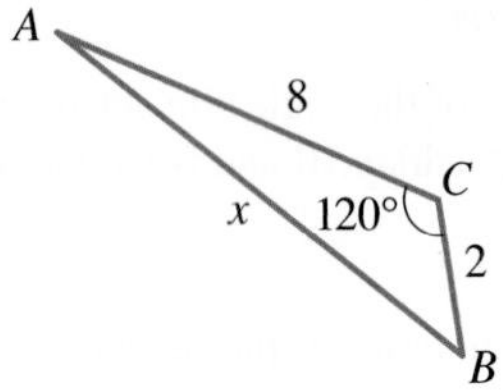

72.

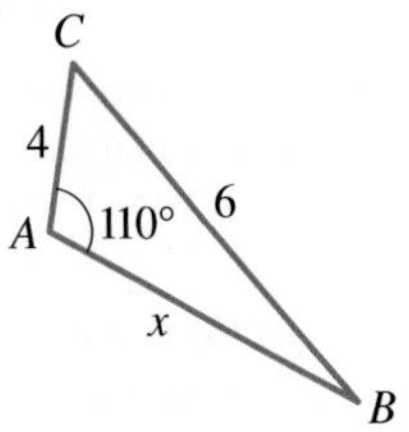

73.

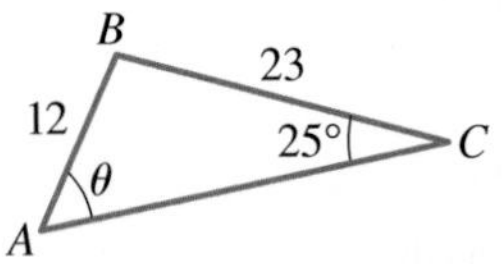

74.

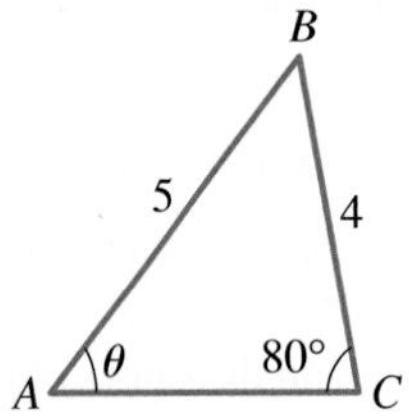

75.

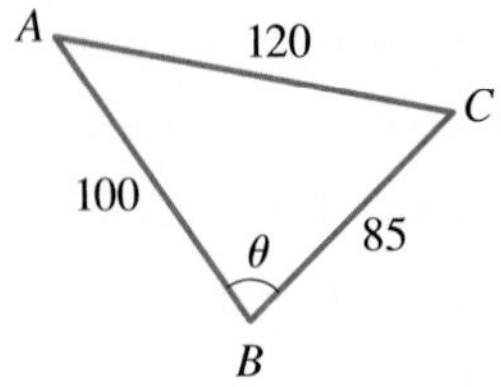

76.

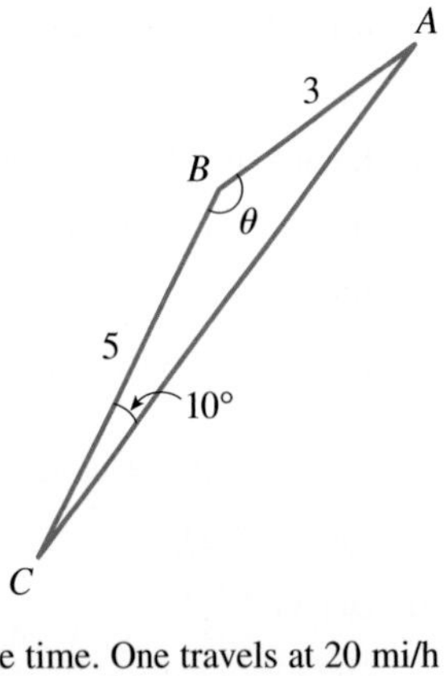

77. Two ships leave a port at the same time. One travels at 20 mi/h in a direction N 32° E, and the other travels at 28 mi/h in a direction S 42° E (see the figure). How far apart are the two ships after 2 h?

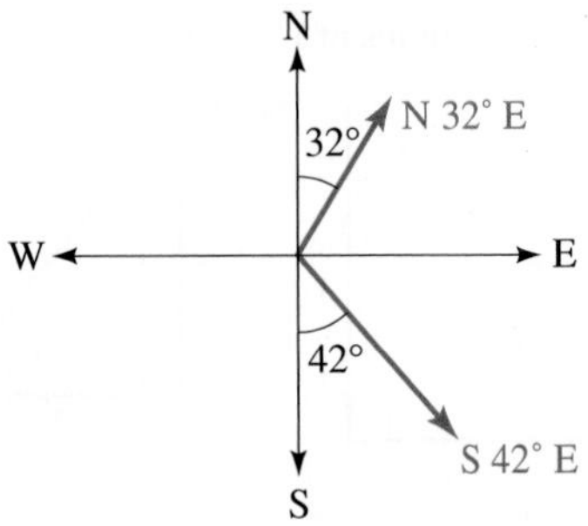

78. From a point A on the ground, the angle of elevation to the top of a tall building is 24.1°. From a point B, which is 600 ft closer to the building, the angle of elevation is measured to be 30.2°. Find the height of the building.

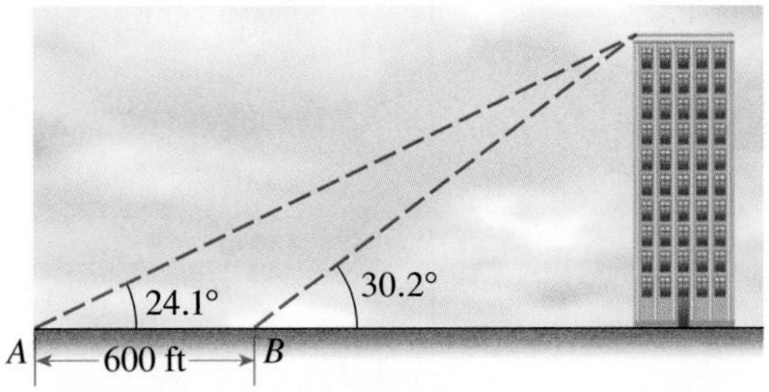

79. Find the distance between points A and B on opposite sides of a lake from the information shown.

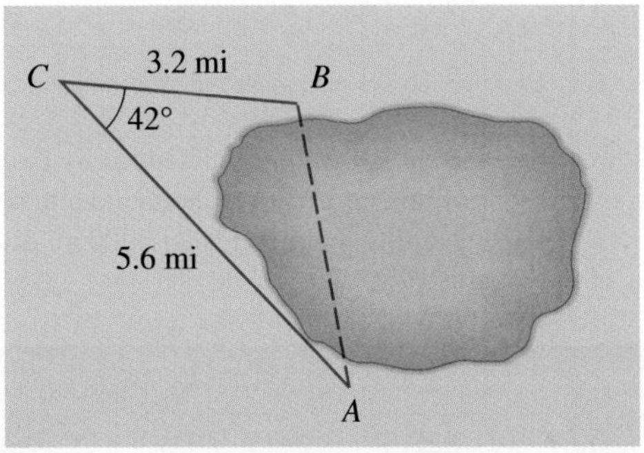

80. A boat is cruising the ocean off a straight shoreline. Points A and B are 120 mi apart on the shore, as shown. It is found that $\angle A = 42.3°$ and $\angle B = 68.9°$. Find the shortest distance from the boat to the shore.

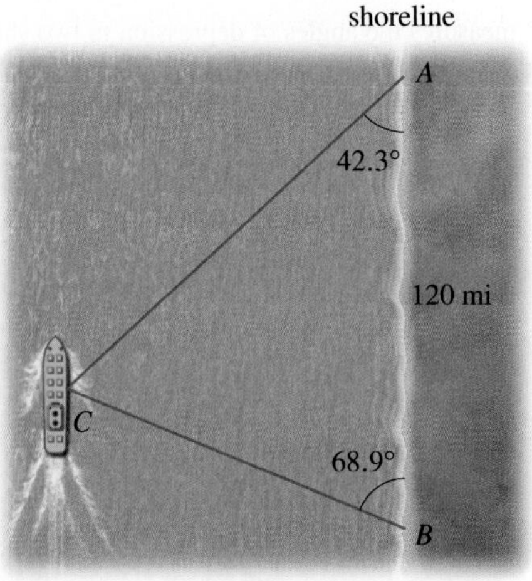

81. Find the area of a triangle with sides of length 8 and 14 and included angle 35°.

82. Find the area of a triangle with sides of length 5, 6, and 8.

1. Find the radian measures that correspond to the degree measures 330° and −135°.

2. Find the degree measures that correspond to the radian measures $\frac{4\pi}{3}$ and −1.3.

3. The rotor blades of a helicopter are 16 ft long and are rotating at 120 rpm.
   (a) Find the angular speed of the rotor.
   (b) Find the linear speed of a point on the tip of a blade.

4. Find the exact value of each of the following.

   (a) $\sin 405°$ **(b)** $\tan(-150°)$ **(c)** $\sec \frac{5\pi}{3}$ **(d)** $\csc \frac{5\pi}{2}$

5. Find $\tan\theta + \sin\theta$ for the angle θ shown.

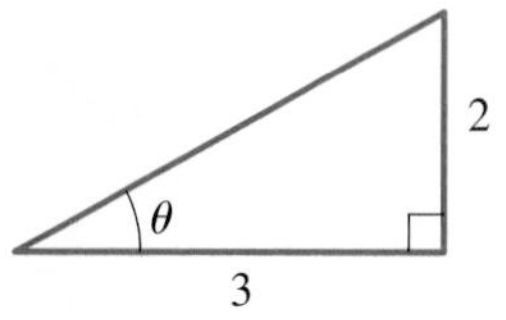

6. Express the lengths a and b shown in the figure in terms of θ.

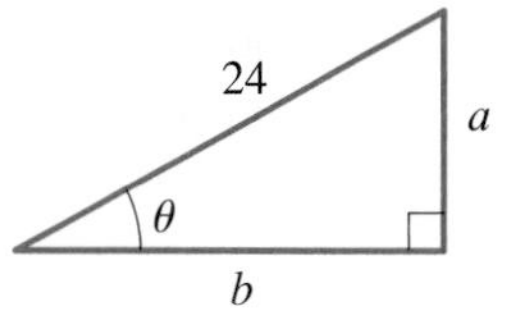

7. If $\cos\theta = -\frac{1}{3}$ and θ is in Quadrant III, find $\tan\theta \cot\theta + \csc\theta$.

8. If $\sin\theta = \frac{5}{13}$ and $\tan\theta = -\frac{5}{12}$, find $\sec\theta$.

9. Express $\tan\theta$ in terms of $\sec\theta$ for θ in Quadrant II.

10. The base of the ladder in the figure is 6 ft from the building, and the angle formed by the ladder and the ground is 73°. How high up the building does the ladder touch?

11. Express θ in each figure in terms of x.

(a)

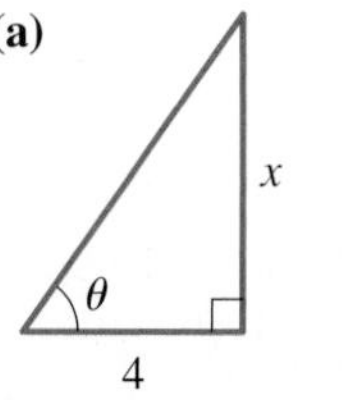

(b)

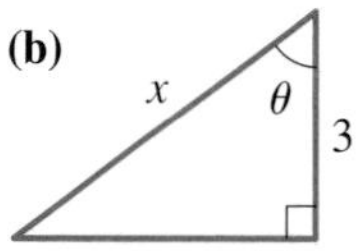

12. Find the exact value of $\cos\left(\tan^{-1}\frac{9}{40}\right)$.

13–18 ■ Find the side labeled x or the angle labeled θ.

13.

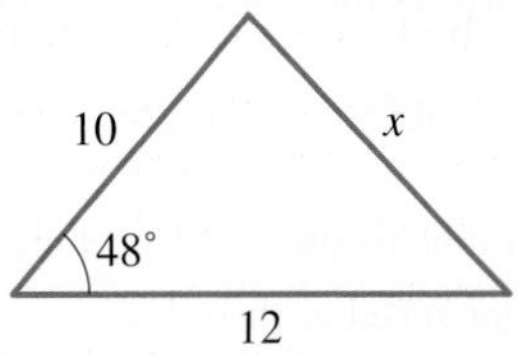

14.

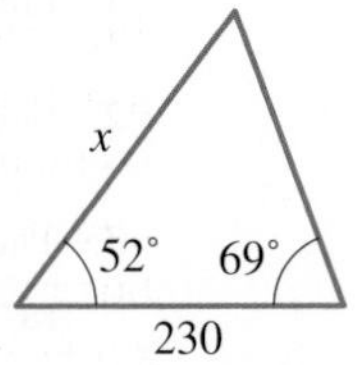

15.

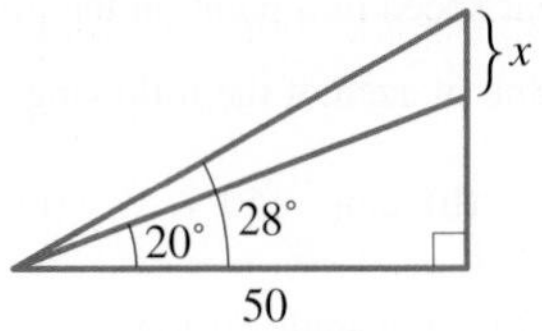

16.

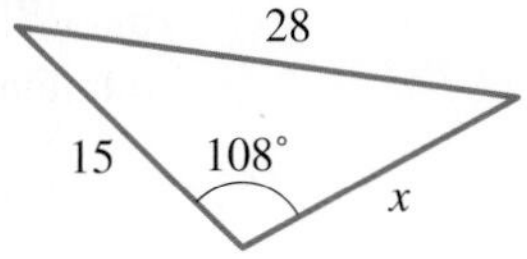

17.

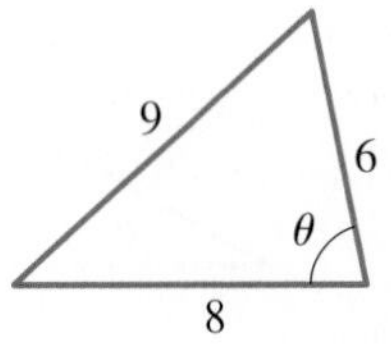

18.

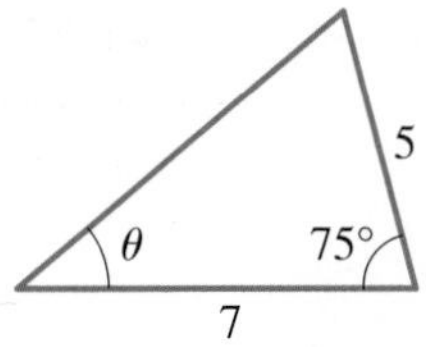

19. Refer to the figure below.

(a) Find the area of the shaded region.

(b) Find the perimeter of the shaded region.

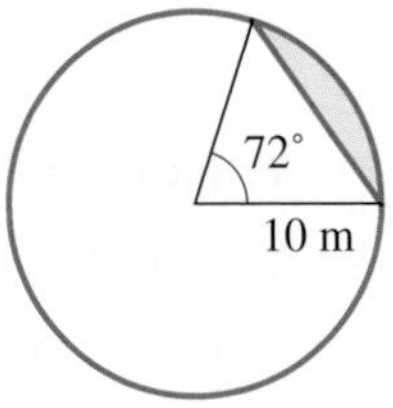

20. Refer to the figure below.

(a) Find the angle opposite the longest side.

(b) Find the area of the triangle.

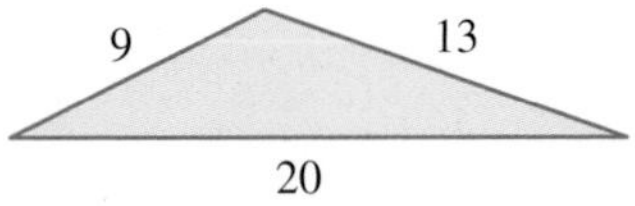

21. Two wires tether a balloon to the ground, as shown. How high is the balloon above the ground?

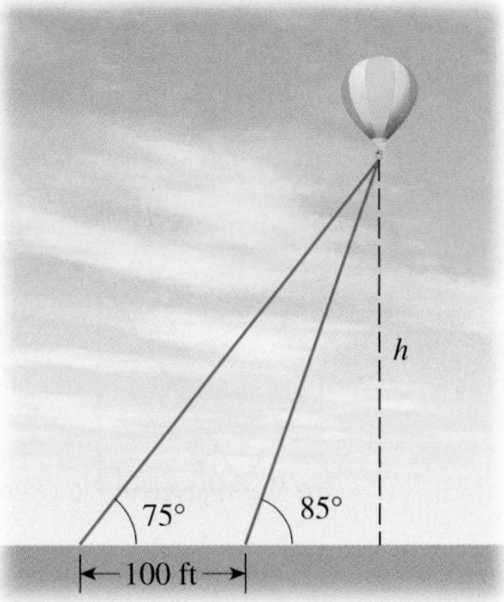

Surveying

How can we measure the height of a mountain or the distance across a lake? Obviously, it may be difficult, inconvenient, or impossible to measure these distances directly (that is, by using a tape measure or a yardstick). On the other hand, it is easy to measure *angles* involving distant objects. That's where trigonometry comes in: the trigonometric ratios relate angles to distances, so they can be used to *calculate* distances from the *measured* angles. In this *Focus* we examine how trigonometry is used to map a town. Modern mapmaking methods use satellites and the Global Positioning System, but mathematics remains at the core of the process.

▼ Mapping a Town

A student wants to draw a map of his hometown. To construct an accurate map (or scale model), he needs to find distances between various landmarks in the town. The student makes the measurements shown in Figure 1. Note that only one distance is measured, between City Hall and the first bridge. All other measurements are angles.

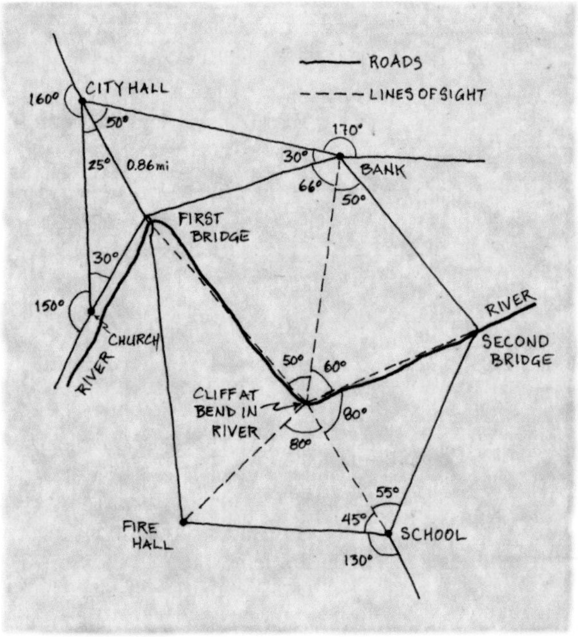

FIGURE 1

The distances between other landmarks can now be found by using the Law of Sines. For example, the distance x from the bank to the first bridge is calculated by applying the Law of Sines to the triangle with vertices at City Hall, the bank, and the first bridge:

$$\frac{x}{\sin 50°} = \frac{0.86}{\sin 30°} \qquad \text{Law of Sines}$$

$$x = \frac{0.86 \sin 50°}{\sin 30°} \qquad \text{Solve for } x$$

$$\approx 1.32 \text{ mi} \qquad \text{Calculator}$$

So the distance between the bank and the first bridge is 1.32 mi.

The distance we just found can now be used to find other distances. For instance, we find the distance y between the bank and the cliff as follows:

$$\frac{y}{\sin 64°} = \frac{1.32}{\sin 50°} \qquad \text{Law of Sines}$$

$$y = \frac{1.32 \sin 64°}{\sin 50°} \qquad \text{Solve for } y$$

$$\approx 1.55 \text{ mi} \qquad \text{Calculator}$$

Continuing in this fashion, we can calculate all the distances between the landmarks shown in the rough sketch in Figure 1. We can use this information to draw the map shown in Figure 2.

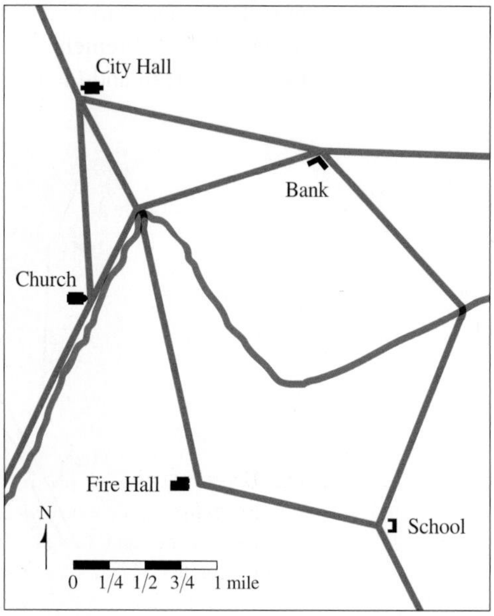

FIGURE 2

To make a topographic map, we need to measure elevation. This concept is explored in Problems 4–6.

PROBLEMS

1. **Completing the Map** Find the distance between the church and City Hall.
2. **Completing the Map** Find the distance between the fire hall and the school. (You will need to find other distances first.)
3. **Determining a Distance** A surveyor on one side of a river wishes to find the distance between points A and B on the opposite side of the river. On her side, she chooses points C and D, which are 20 m apart, and measures the angles shown in the figure below. Find the distance between A and B.

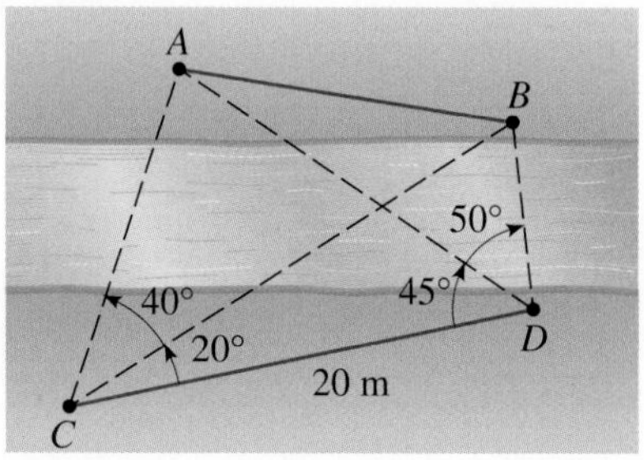

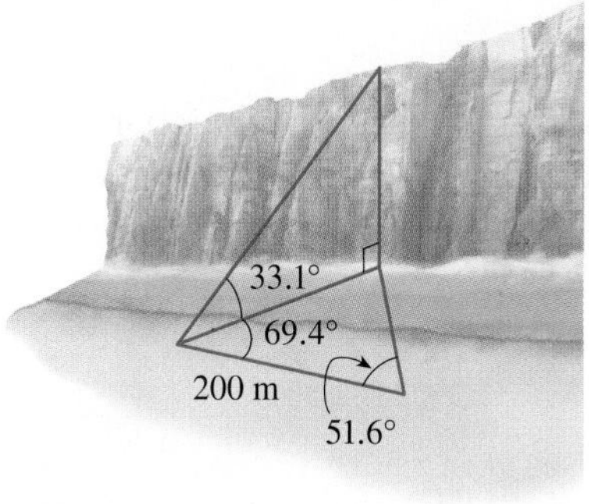

4. Height of a Cliff To measure the height of an inaccessible cliff on the opposite side of a river, a surveyor makes the measurements shown in the figure at the left. Find the height of the cliff.

5. Height of a Mountain To calculate the height h of a mountain, angle α, β, and distance d are measured, as shown in the figure below.

(a) Show that

$$h = \frac{d}{\cot \alpha - \cot \beta}$$

(b) Show that

$$h = d\frac{\sin \alpha \sin \beta}{\sin(\beta - \alpha)}$$

(c) Use the formulas from parts (a) and (b) to find the height of a mountain if $\alpha = 25°$, $\beta = 29°$, and $d = 800$ ft. Do you get the same answer from each formula?

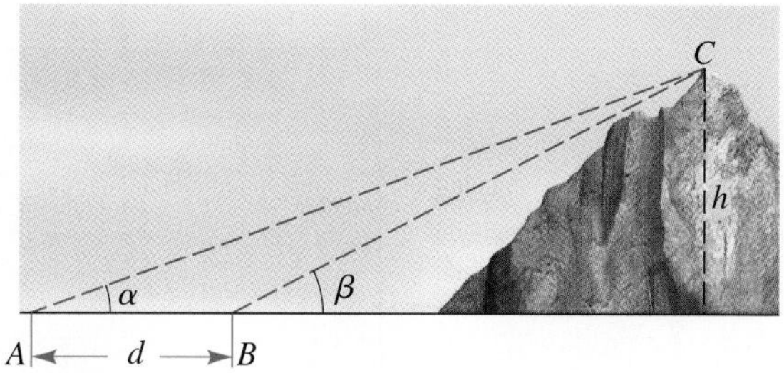

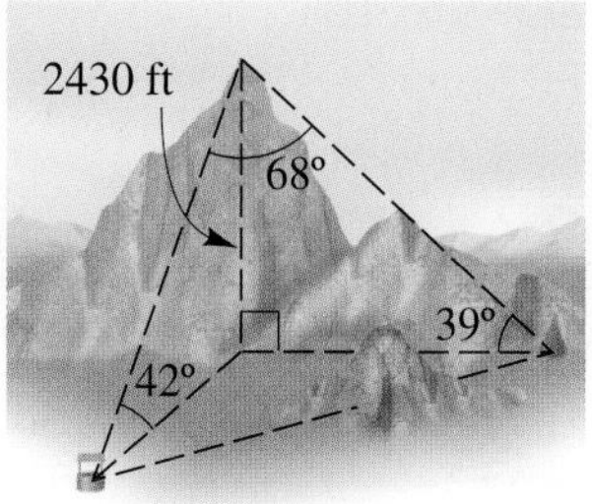

6. Determining a Distance A surveyor has determined that a mountain is 2430 ft high. From the top of the mountain he measures the angles of depression to two landmarks at the base of the mountain and finds them to be 42° and 39°. (Observe that these are the same as the angles of elevation from the landmarks as shown in the figure at the left.) The angle between the lines of sight to the landmarks is 68°. Calculate the distance between the two landmarks.

7. Surveying Building Lots A surveyor surveys two adjacent lots and makes the following rough sketch showing his measurements. Calculate all the distances shown in the figure, and use your result to draw an accurate map of the two lots.

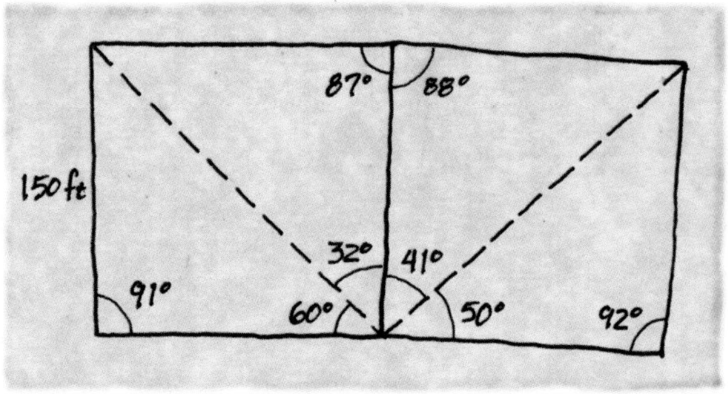

8. Great Survey of India The Great Trigonometric Survey of India was one of the most massive mapping projects ever undertaken (see the margin note on page 472). Do some research at your library or on the Internet to learn more about the Survey, and write a report on your findings.

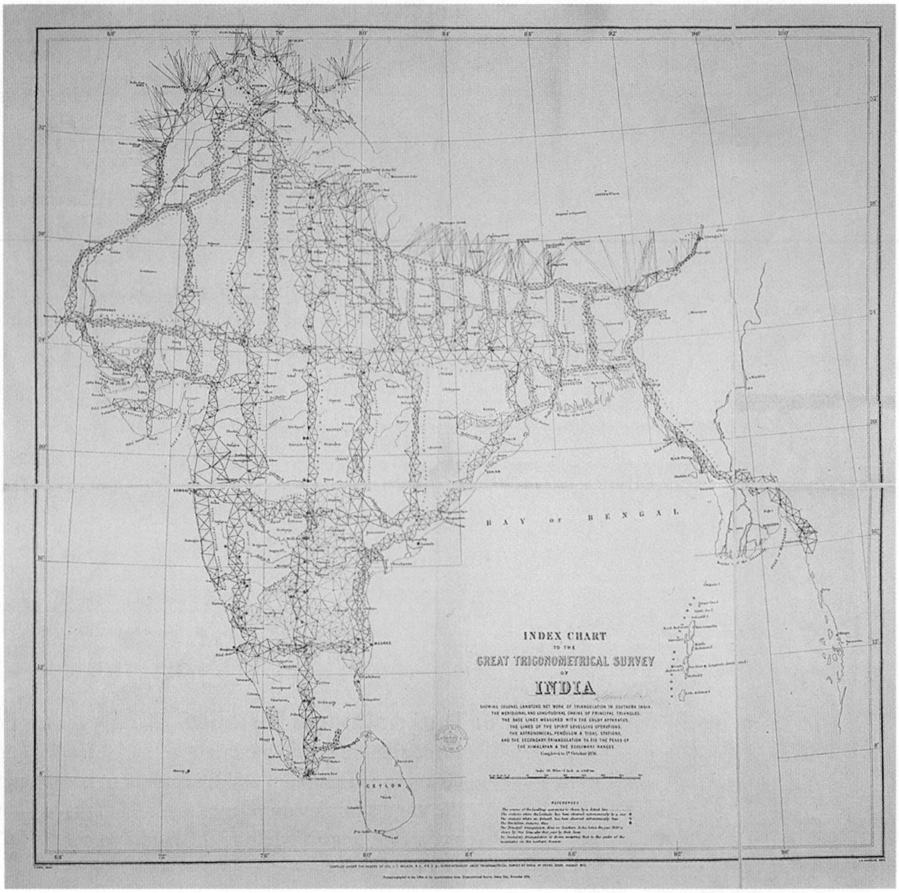

© Sony Pictures Classics/Courtesy of Everett Collection

CHAPTER 7

Analytic Trigonometry

In Chapters 5 and 6 we studied graphical and geometric properties of the trigonometric functions. In this chapter we study algebraic properties of these functions, that is, simplifying and factoring expressions and solving equations that involve trigonometric functions.

We have used the trigonometric functions to model different real-world phenomena, including periodic motion (such as the motion of an ocean wave). To obtain information from a model, we often need to solve equations. If the model involves trigonometric functions, we need to solve trigonometric equations. Solving trigonometric equations often involves using trigonometric identities. We've already encountered some basic trigonometric identities in the preceding chapters. We begin this chapter by finding many new identities.

7.1 Trigonometric Identities

| Simplifying Trigonometric Expressions ▶ Proving Trigonometric Identities

We begin by listing some of the basic trigonometric identities. We studied most of these in Chapters 5 and 6; you are asked to prove the cofunction identities in Exercise 102.

FUNDAMENTAL TRIGONOMETRIC IDENTITIES

Reciprocal Identities

$$\csc x = \frac{1}{\sin x} \qquad \sec x = \frac{1}{\cos x} \qquad \cot x = \frac{1}{\tan x}$$

$$\tan x = \frac{\sin x}{\cos x} \qquad \cot x = \frac{\cos x}{\sin x}$$

Pythagorean Identities

$$\sin^2 x + \cos^2 x = 1 \qquad \tan^2 x + 1 = \sec^2 x \qquad 1 + \cot^2 x = \csc^2 x$$

Even-Odd Identities

$$\sin(-x) = -\sin x \qquad \cos(-x) = \cos x \qquad \tan(-x) = -\tan x$$

Cofunction Identities

$$\sin\left(\frac{\pi}{2} - u\right) = \cos u \qquad \tan\left(\frac{\pi}{2} - u\right) = \cot u \qquad \sec\left(\frac{\pi}{2} - u\right) = \csc u$$

$$\cos\left(\frac{\pi}{2} - u\right) = \sin u \qquad \cot\left(\frac{\pi}{2} - u\right) = \tan u \qquad \csc\left(\frac{\pi}{2} - u\right) = \sec u$$

▼ Simplifying Trigonometric Expressions

Identities enable us to write the same expression in different ways. It is often possible to rewrite a complicated-looking expression as a much simpler one. To simplify algebraic expressions, we used factoring, common denominators, and the Special Product Formulas. To simplify trigonometric expressions, we use these same techniques together with the fundamental trigonometric identities.

EXAMPLE 1 | Simplifying a Trigonometric Expression

Simplify the expression $\cos t + \tan t \sin t$.

SOLUTION We start by rewriting the expression in terms of sine and cosine:

$$\begin{aligned}
\cos t + \tan t \sin t &= \cos t + \left(\frac{\sin t}{\cos t}\right)\sin t && \text{Reciprocal identity} \\
&= \frac{\cos^2 t + \sin^2 t}{\cos t} && \text{Common denominator} \\
&= \frac{1}{\cos t} && \text{Pythagorean identity} \\
&= \sec t && \text{Reciprocal identity}
\end{aligned}$$

NOW TRY EXERCISE 3 ■

EXAMPLE 2 | Simplifying by Combining Fractions

Simplify the expression $\dfrac{\sin\theta}{\cos\theta} + \dfrac{\cos\theta}{1+\sin\theta}$.

SOLUTION We combine the fractions by using a common denominator:

$$\frac{\sin\theta}{\cos\theta} + \frac{\cos\theta}{1+\sin\theta} = \frac{\sin\theta\,(1+\sin\theta)+\cos^2\theta}{\cos\theta\,(1+\sin\theta)} \quad \text{Common denominator}$$

$$= \frac{\sin\theta + \sin^2\theta + \cos^2\theta}{\cos\theta\,(1+\sin\theta)} \quad \text{Distribute } \sin\theta$$

$$= \frac{\sin\theta + 1}{\cos\theta\,(1+\sin\theta)} \quad \text{Pythagorean identity}$$

$$= \frac{1}{\cos\theta} = \sec\theta \quad \text{Cancel and use reciprocal identity}$$

NOW TRY EXERCISE **21** ■

Proving Trigonometric Identities

Many identities follow from the fundamental identities. In the examples that follow, we learn how to prove that a given trigonometric equation is an identity, and in the process we will see how to discover new identities.

First, it's easy to decide when a given equation is *not* an identity. All we need to do is show that the equation does not hold for some value of the variable (or variables). Thus the equation

$$\sin x + \cos x = 1$$

is not an identity, because when $x = \pi/4$, we have

$$\sin\frac{\pi}{4} + \cos\frac{\pi}{4} = \frac{\sqrt{2}}{2} + \frac{\sqrt{2}}{2} = \sqrt{2} \neq 1$$

To verify that a trigonometric equation is an identity, we transform one side of the equation into the other side by a series of steps, each of which is itself an identity.

GUIDELINES FOR PROVING TRIGONOMETRIC IDENTITIES

1. **Start with one side.** Pick one side of the equation and write it down. Your goal is to transform it into the other side. It's usually easier to start with the more complicated side.
2. **Use known identities.** Use algebra and the identities you know to change the side you started with. Bring fractional expressions to a common denominator, factor, and use the fundamental identities to simplify expressions.
3. **Convert to sines and cosines.** If you are stuck, you may find it helpful to rewrite all functions in terms of sines and cosines.

Warning: To prove an identity, we do *not* just perform the same operations on both sides of the equation. For example, if we start with an equation that is not an identity, such as

$$\sin x = -\sin x \quad (1)$$

and square both sides, we get the equation

$$\sin^2 x = \sin^2 x \quad (2)$$

which is clearly an identity. Does this mean that the original equation is an identity? Of course not. The problem here is that the operation of squaring is not **reversible** in the sense that we cannot arrive back at (1) from (2) by taking square roots (reversing the procedure). Only operations that are reversible will necessarily transform an identity into an identity.

EXAMPLE 3 | Proving an Identity by Rewriting in Terms of Sine and Cosine

Consider the equation $\cos\theta\,(\sec\theta - \cos\theta) = \sin^2\theta$.

(a) Verify algebraically that the equation is an identity.

(b) Confirm graphically that the equation is an identity.

SOLUTION

(a) The left-hand side looks more complicated, so we start with it and try to transform it into the right-hand side:

$$\begin{aligned} \text{LHS} &= \cos\theta\,(\sec\theta - \cos\theta) \\ &= \cos\theta\left(\frac{1}{\cos\theta} - \cos\theta\right) && \text{Reciprocal identity} \\ &= 1 - \cos^2\theta && \text{Expand} \\ &= \sin^2\theta = \text{RHS} && \text{Pythagorean identity} \end{aligned}$$

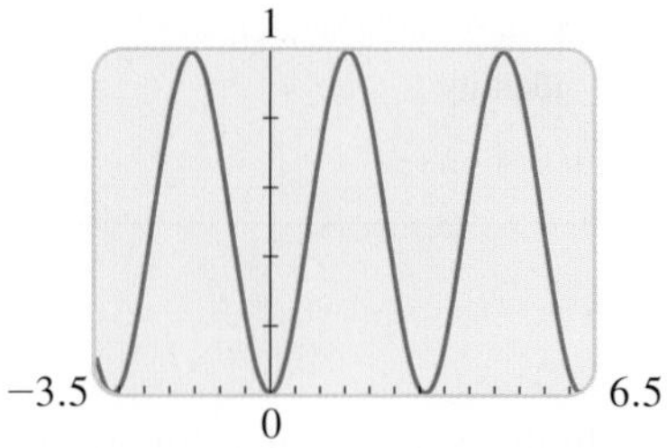

FIGURE 1

(b) We graph each side of the equation to see whether the graphs coincide. From Figure 1 we see that the graphs of $y = \cos\theta\,(\sec\theta - \cos\theta)$ and $y = \sin^2\theta$ are identical. This confirms that the equation is an identity.

NOW TRY EXERCISE **27** ■

In Example 3 it isn't easy to see how to change the right-hand side into the left-hand side, but it's definitely possible. Simply notice that each step is reversible. In other words, if we start with the last expression in the proof and work backward through the steps, the right-hand side is transformed into the left-hand side. You will probably agree, however, that it's more difficult to prove the identity this way. That's why it's often better to change the more complicated side of the identity into the simpler side.

EXAMPLE 4 | Proving an Identity by Combining Fractions

Verify the identity

$$2\tan x\sec x = \frac{1}{1 - \sin x} - \frac{1}{1 + \sin x}$$

SOLUTION Finding a common denominator and combining the fractions on the right-hand side of this equation, we get

$$\begin{aligned} \text{RHS} &= \frac{1}{1 - \sin x} - \frac{1}{1 + \sin x} \\ &= \frac{(1 + \sin x) - (1 - \sin x)}{(1 - \sin x)(1 + \sin x)} && \text{Common denominator} \\ &= \frac{2\sin x}{1 - \sin^2 x} && \text{Simplify} \\ &= \frac{2\sin x}{\cos^2 x} && \text{Pythagorean identity} \\ &= 2\,\frac{\sin x}{\cos x}\left(\frac{1}{\cos x}\right) && \text{Factor} \\ &= 2\tan x\sec x = \text{LHS} && \text{Reciprocal identities} \end{aligned}$$

NOW TRY EXERCISE **79** ■

See the Prologue: *Principles of Problem Solving,* pages P1–P4.

In Example 5 we introduce "something extra" to the problem by multiplying the numerator and the denominator by a trigonometric expression, chosen so that we can simplify the result.

EXAMPLE 5 | Proving an Identity by Introducing Something Extra

Verify the identity $\dfrac{\cos u}{1 - \sin u} = \sec u + \tan u$.

SOLUTION We start with the left-hand side and multiply the numerator and denominator by $1 + \sin u$:

We multiply by $1 + \sin u$ because we know by the difference of squares formula that $(1 - \sin u)(1 + \sin u) = 1 - \sin^2 u$, and this is just $\cos^2 u$, a simpler expression.

$$\text{LHS} = \frac{\cos u}{1 - \sin u}$$

$$= \frac{\cos u}{1 - \sin u} \cdot \frac{1 + \sin u}{1 + \sin u} \qquad \text{Multiply numerator and denominator by } 1 + \sin u$$

$$= \frac{\cos u\,(1 + \sin u)}{1 - \sin^2 u} \qquad \text{Expand denominator}$$

$$= \frac{\cos u\,(1 + \sin u)}{\cos^2 u} \qquad \text{Pythagorean identity}$$

$$= \frac{1 + \sin u}{\cos u} \qquad \text{Cancel common factor}$$

$$= \frac{1}{\cos u} + \frac{\sin u}{\cos u} \qquad \text{Separate into two fractions}$$

$$= \sec u + \tan u \qquad \text{Reciprocal identities}$$

NOW TRY EXERCISE **53** ■

EUCLID (circa 300 B.C.) taught in Alexandria. His *Elements* is the most widely influential scientific book in history. For 2000 years it was the standard introduction to geometry in the schools, and for many generations it was considered the best way to develop logical reasoning. Abraham Lincoln, for instance, studied the *Elements* as a way to sharpen his mind. The story is told that King Ptolemy once asked Euclid if there was a faster way to learn geometry than through the *Elements.* Euclid replied that there is "no royal road to geometry"—meaning by this that mathematics does not respect wealth or social status. Euclid was revered in his own time and was referred to by the title "The Geometer" or "The Writer of the *Elements.*" The greatness of the *Elements* stems from its precise, logical, and systematic treatment of geometry. For dealing with equality, Euclid lists the following rules, which he calls "common notions."

1. Things that are equal to the same thing are equal to each other.
2. If equals are added to equals, the sums are equal.
3. If equals are subtracted from equals, the remainders are equal.
4. Things that coincide with one another are equal.
5. The whole is greater than the part.

Here is another method for proving that an equation is an identity. If we can transform each side of the equation *separately*, by way of identities, to arrive at the same result, then the equation is an identity. Example 6 illustrates this procedure.

EXAMPLE 6 | Proving an Identity by Working with Both Sides Separately

Verify the identity $\dfrac{1 + \cos\theta}{\cos\theta} = \dfrac{\tan^2\theta}{\sec\theta - 1}$.

SOLUTION We prove the identity by changing each side separately into the same expression. Supply the reasons for each step:

$$\text{LHS} = \frac{1 + \cos\theta}{\cos\theta} = \frac{1}{\cos\theta} + \frac{\cos\theta}{\cos\theta} = \sec\theta + 1$$

$$\text{RHS} = \frac{\tan^2\theta}{\sec\theta - 1} = \frac{\sec^2\theta - 1}{\sec\theta - 1} = \frac{(\sec\theta - 1)(\sec\theta + 1)}{\sec\theta - 1} = \sec\theta + 1$$

It follows that LHS = RHS, so the equation is an identity.

NOW TRY EXERCISE **81** ■

We conclude this section by describing the technique of *trigonometric substitution*, which we use to convert algebraic expressions to trigonometric ones. This is often useful in calculus, for instance, in finding the area of a circle or an ellipse.

EXAMPLE 7 | Trigonometric Substitution

Substitute $\sin\theta$ for x in the expression $\sqrt{1 - x^2}$ and simplify. Assume that $0 \le \theta \le \pi/2$.

SOLUTION Setting $x = \sin\theta$, we have

$$\begin{aligned} \sqrt{1 - x^2} &= \sqrt{1 - \sin^2\theta} && \text{Substitute } x = \sin\theta \\ &= \sqrt{\cos^2\theta} && \text{Pythagorean identity} \\ &= \cos\theta && \text{Take square root} \end{aligned}$$

The last equality is true because $\cos\theta \ge 0$ for the values of θ in question.

NOW TRY EXERCISE **91**

7.1 EXERCISES

CONCEPTS

1. An equation is called an identity if it is valid for ________ values of the variable. The equation $2x = x + x$ is an algebraic identity, and the equation $\sin^2 x + \cos^2 x =$ ________ is a trigonometric identity.

2. For any x it is true that $\cos(-x)$ has the same value as $\cos x$. We express this fact as the identity ____________.

SKILLS

3–12 ■ Write the trigonometric expression in terms of sine and cosine, and then simplify.

3. $\cos t \tan t$

4. $\cos t \csc t$

5. $\sin\theta \sec\theta$

6. $\tan\theta \csc\theta$

7. $\tan^2 x - \sec^2 x$

8. $\dfrac{\sec x}{\csc x}$

9. $\sin u + \cot u \cos u$

10. $\cos^2\theta\,(1 + \tan^2\theta)$

11. $\dfrac{\sec\theta - \cos\theta}{\sin\theta}$

12. $\dfrac{\cot\theta}{\csc\theta - \sin\theta}$

13–26 ■ Simplify the trigonometric expression.

13. $\dfrac{\sin x \sec x}{\tan x}$

14. $\cos^3 x + \sin^2 x \cos x$

15. $\dfrac{1 + \cos y}{1 + \sec y}$

16. $\dfrac{\tan x}{\sec(-x)}$

17. $\dfrac{\sec^2 x - 1}{\sec^2 x}$

18. $\dfrac{\sec x - \cos x}{\tan x}$

19. $\dfrac{1 + \csc x}{\cos x + \cot x}$

20. $\dfrac{\sin x}{\csc x} + \dfrac{\cos x}{\sec x}$

21. $\dfrac{1 + \sin u}{\cos u} + \dfrac{\cos u}{1 + \sin u}$

22. $\tan x \cos x \csc x$

23. $\dfrac{2 + \tan^2 x}{\sec^2 x} - 1$

24. $\dfrac{1 + \cot A}{\csc A}$

25. $\tan\theta + \cos(-\theta) + \tan(-\theta)$

26. $\dfrac{\cos x}{\sec x + \tan x}$

27–28 ■ Consider the given equation. **(a)** Verify algebraically that the equation is an identity. **(b)** Confirm graphically that the equation is an identity.

27. $\dfrac{\cos x}{\sec x \sin x} = \csc x - \sin x$

28. $\dfrac{\tan y}{\csc y} = \sec y - \cos y$

29–90 ■ Verify the identity.

29. $\dfrac{\sin\theta}{\tan\theta} = \cos\theta$

30. $\dfrac{\tan x}{\sec x} = \sin x$

31. $\dfrac{\cos u \sec u}{\tan u} = \cot u$

32. $\dfrac{\cot x \sec x}{\csc x} = 1$

33. $\sin B + \cos B \cot B = \csc B$

34. $\cos(-x) - \sin(-x) = \cos x + \sin x$

35. $\cot(-\alpha)\cos(-\alpha) + \sin(-\alpha) = -\csc\alpha$

36. $\csc x\,[\csc x + \sin(-x)] = \cot^2 x$

37. $\tan\theta + \cot\theta = \sec\theta \csc\theta$

38. $(\sin x + \cos x)^2 = 1 + 2\sin x \cos x$

39. $(1 - \cos\beta)(1 + \cos\beta) = \dfrac{1}{\csc^2\beta}$

40. $\dfrac{\cos x}{\sec x} + \dfrac{\sin x}{\csc x} = 1$

41. $\dfrac{(\sin x + \cos x)^2}{\sin^2 x - \cos^2 x} = \dfrac{\sin^2 x - \cos^2 x}{(\sin x - \cos x)^2}$

42. $(\sin x + \cos x)^4 = (1 + 2\sin x \cos x)^2$

43. $\dfrac{\sec t - \cos t}{\sec t} = \sin^2 t$

44. $\dfrac{1 - \sin x}{1 + \sin x} = (\sec x - \tan x)^2$

45. $\dfrac{1}{1 - \sin^2 y} = 1 + \tan^2 y$

46. $\csc x - \sin x = \cos x \cot x$

47. $(\cot x - \csc x)(\cos x + 1) = -\sin x$

48. $\sin^4\theta - \cos^4\theta = \sin^2\theta - \cos^2\theta$

49. $(1 - \cos^2 x)(1 + \cot^2 x) = 1$

50. $\cos^2 x - \sin^2 x = 2\cos^2 x - 1$

51. $2\cos^2 x - 1 = 1 - 2\sin^2 x$

52. $(\tan y + \cot y)\sin y \cos y = 1$

53. $\dfrac{1 - \cos\alpha}{\sin\alpha} = \dfrac{\sin\alpha}{1 + \cos\alpha}$

54. $\sin^2\alpha + \cos^2\alpha + \tan^2\alpha = \sec^2\alpha$

55. $\tan^2\theta - \sin^2\theta = \tan^2\theta \sin^2\theta$

56. $\cot^2\theta \cos^2\theta = \cot^2\theta - \cos^2\theta$

57. $\dfrac{\sin x - 1}{\sin x + 1} = \dfrac{-\cos^2 x}{(\sin x + 1)^2}$

58. $\dfrac{\sin w}{\sin w + \cos w} = \dfrac{\tan w}{1 + \tan w}$

59. $\dfrac{(\sin t + \cos t)^2}{\sin t \cos t} = 2 + \sec t \csc t$

60. $\sec t \csc t(\tan t + \cot t) = \sec^2 t + \csc^2 t$

61. $\dfrac{1 + \tan^2 u}{1 - \tan^2 u} = \dfrac{1}{\cos^2 u - \sin^2 u}$

62. $\dfrac{1 + \sec^2 x}{1 + \tan^2 x} = 1 + \cos^2 x$

63. $\dfrac{\sec x}{\sec x - \tan x} = \sec x(\sec x + \tan x)$

64. $\dfrac{\sec x + \csc x}{\tan x + \cot x} = \sin x + \cos x$

65. $\sec v - \tan v = \dfrac{1}{\sec v + \tan v}$

66. $\dfrac{\sin A}{1 - \cos A} - \cot A = \csc A$

67. $\dfrac{\sin x + \cos x}{\sec x + \csc x} = \sin x \cos x$

68. $\dfrac{1 - \cos x}{\sin x} + \dfrac{\sin x}{1 - \cos x} = 2\csc x$

69. $\dfrac{\csc x - \cot x}{\sec x - 1} = \cot x$

70. $\dfrac{\csc^2 x - \cot^2 x}{\sec^2 x} = \cos^2 x$

71. $\tan^2 u - \sin^2 u = \tan^2 u \sin^2 u$

72. $\dfrac{\tan v \sin v}{\tan v + \sin v} = \dfrac{\tan v - \sin v}{\tan v \sin v}$

73. $\sec^4 x - \tan^4 x = \sec^2 x + \tan^2 x$

74. $\dfrac{\cos\theta}{1 - \sin\theta} = \sec\theta + \tan\theta$

75. $\dfrac{\cos\theta}{1 - \sin\theta} = \dfrac{\sin\theta - \csc\theta}{\cos\theta - \cot\theta}$

76. $\dfrac{1 + \tan x}{1 - \tan x} = \dfrac{\cos x + \sin x}{\cos x - \sin x}$

77. $\dfrac{\cos^2 t + \tan^2 t - 1}{\sin^2 t} = \tan^2 t$

78. $\dfrac{1}{1 - \sin x} - \dfrac{1}{1 + \sin x} = 2\sec x \tan x$

79. $\dfrac{1}{\sec x + \tan x} + \dfrac{1}{\sec x - \tan x} = 2\sec x$

80. $\dfrac{1 + \sin x}{1 - \sin x} - \dfrac{1 - \sin x}{1 + \sin x} = 4\tan x \sec x$

81. $(\tan x + \cot x)^2 = \sec^2 x + \csc^2 x$

82. $\tan^2 x - \cot^2 x = \sec^2 x - \csc^2 x$

83. $\dfrac{\sec u - 1}{\sec u + 1} = \dfrac{1 - \cos u}{1 + \cos u}$

84. $\dfrac{\cot x + 1}{\cot x - 1} = \dfrac{1 + \tan x}{1 - \tan x}$

85. $\dfrac{\sin^3 x + \cos^3 x}{\sin x + \cos x} = 1 - \sin x \cos x$

86. $\dfrac{\tan v - \cot v}{\tan^2 v - \cot^2 v} = \sin v \cos v$

87. $\dfrac{1 + \sin x}{1 - \sin x} = (\tan x + \sec x)^2$

88. $\dfrac{\tan x + \tan y}{\cot x + \cot y} = \tan x \tan y$

89. $(\tan x + \cot x)^4 = \csc^4 x \sec^4 x$

90. $(\sin\alpha - \tan\alpha)(\cos\alpha - \cot\alpha) = (\cos\alpha - 1)(\sin\alpha - 1)$

91–96 ■ Make the indicated trigonometric substitution in the given algebraic expression and simplify (see Example 7). Assume that $0 \le \theta < \pi/2$.

91. $\dfrac{x}{\sqrt{1 - x^2}}$, $x = \sin\theta$

92. $\sqrt{1 + x^2}$, $x = \tan\theta$

93. $\sqrt{x^2 - 1}$, $x = \sec\theta$

94. $\dfrac{1}{x^2\sqrt{4 + x^2}}$, $x = 2\tan\theta$

95. $\sqrt{9 - x^2}$, $x = 3\sin\theta$

96. $\dfrac{\sqrt{x^2 - 25}}{x}$, $x = 5\sec\theta$

97–100 ■ Graph f and g in the same viewing rectangle. Do the graphs suggest that the equation $f(x) = g(x)$ is an identity? Prove your answer.

97. $f(x) = \cos^2 x - \sin^2 x$, $g(x) = 1 - 2\sin^2 x$

98. $f(x) = \tan x(1 + \sin x)$, $g(x) = \dfrac{\sin x \cos x}{1 + \sin x}$

99. $f(x) = (\sin x + \cos x)^2$, $g(x) = 1$

100. $f(x) = \cos^4 x - \sin^4 x$, $g(x) = 2\cos^2 x - 1$

101. Show that the equation is not an identity.

(a) $\sin 2x = 2\sin x$ **(b)** $\sin(x + y) = \sin x + \sin y$

(c) $\sec^2 x + \csc^2 x = 1$

(d) $\dfrac{1}{\sin x + \cos x} = \csc x + \sec x$

DISCOVERY ■ DISCUSSION ■ WRITING

102. Cofunction Identities In the right triangle shown, explain why $v = (\pi/2) - u$. Explain how you can obtain all six cofunction identities from this triangle for $0 < u < \pi/2$.

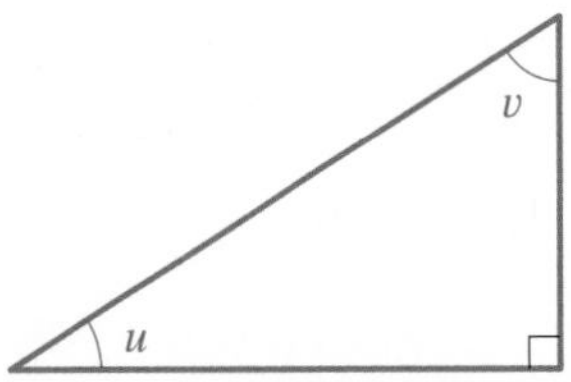

103. Graphs and Identities Suppose you graph two functions, f and g, on a graphing device and their graphs appear identical in the viewing rectangle. Does this prove that the equation $f(x) = g(x)$ is an identity? Explain.

104. Making Up Your Own Identity If you start with a trigonometric expression and rewrite it or simplify it, then setting the original expression equal to the rewritten expression yields a trigonometric identity. For instance, from Example 1 we get the identity

$$\cos t + \tan t \sin t = \sec t$$

Use this technique to make up your own identity, then give it to a classmate to verify.

7.2 Addition and Subtraction Formulas

Addition and Subtraction Formulas ▶ Evaluating Expressions Involving Inverse Trigonometric Functions ▶ Expressions of the form $A \sin x + B \cos x$

▼ Addition and Subtraction Formulas

We now derive identities for trigonometric functions of sums and differences.

ADDITION AND SUBTRACTION FORMULAS

Formulas for sine:
$$\sin(s + t) = \sin s \cos t + \cos s \sin t$$
$$\sin(s - t) = \sin s \cos t - \cos s \sin t$$

Formulas for cosine:
$$\cos(s + t) = \cos s \cos t - \sin s \sin t$$
$$\cos(s - t) = \cos s \cos t + \sin s \sin t$$

Formulas for tangent:
$$\tan(s + t) = \frac{\tan s + \tan t}{1 - \tan s \tan t}$$
$$\tan(s - t) = \frac{\tan s - \tan t}{1 + \tan s \tan t}$$

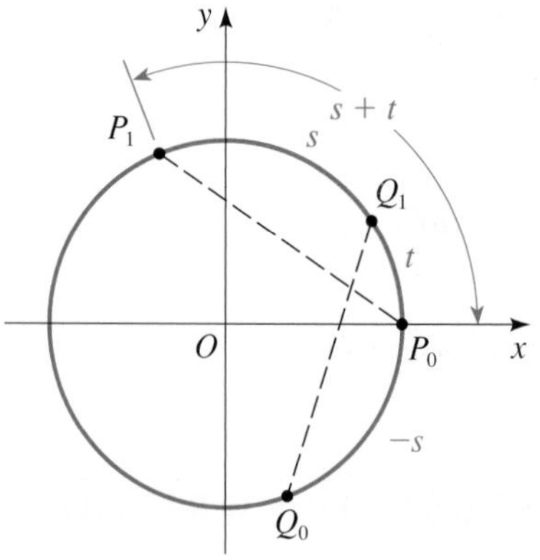

FIGURE 1

PROOF OF ADDITION FORMULA FOR COSINE To prove the formula $\cos(s + t) = \cos s \cos t - \sin s \sin t$, we use Figure 1. In the figure, the distances t, $s + t$, and $-s$ have been marked on the unit circle, starting at $P_0(1, 0)$ and terminating at Q_1, P_1, and Q_0, respectively. The coordinates of these points are

$$P_0(1, 0) \qquad Q_0(\cos(-s), \sin(-s))$$
$$P_1(\cos(s + t), \sin(s + t)) \qquad Q_1(\cos t, \sin t)$$

Since $\cos(-s) = \cos s$ and $\sin(-s) = -\sin s$, it follows that the point Q_0 has the coordinates $Q_0(\cos s, -\sin s)$. Notice that the distances between P_0 and P_1 and between Q_0 and Q_1 measured along the arc of the circle are equal. Since equal arcs are subtended by equal chords, it follows that $d(P_0, P_1) = d(Q_0, Q_1)$. Using the Distance Formula, we get

$$\sqrt{[\cos(s + t) - 1]^2 + [\sin(s + t) - 0]^2} = \sqrt{(\cos t - \cos s)^2 + (\sin t + \sin s)^2}$$

Squaring both sides and expanding, we have

$$\underbrace{\cos^2(s + t)}_{\text{These add to 1}} - 2\cos(s + t) + 1 + \underbrace{\sin^2(s + t)}_{\text{These add to 1}}$$
$$= \cos^2 t - 2\cos s \cos t + \cos^2 s + \sin^2 t + 2\sin s \sin t + \sin^2 s$$

These add to 1 ($\cos^2 t$ and $\sin^2 t$); These add to 1 ($\cos^2 s$ and $\sin^2 s$)

JEAN BAPTISTE JOSEPH FOURIER (1768–1830) is responsible for the most powerful application of the trigonometric functions (see the margin note on page 394). He used sums of these functions to describe such physical phenomena as the transmission of sound and the flow of heat.

Orphaned as a young boy, Fourier was educated in a military school, where he became a mathematics teacher at the age of 20. He was later appointed professor at the École Polytechnique but resigned this position to accompany Napoleon on his expedition to Egypt, where Fourier served as governor. After returning to France he began conducting experiments on heat. The French Academy refused to publish his early papers on this subject because of his lack of rigor. Fourier eventually became Secretary of the Academy and in this capacity had his papers published in their original form. Probably because of his study of heat and his years in the deserts of Egypt, Fourier became obsessed with keeping himself warm—he wore several layers of clothes, even in the summer, and kept his rooms at unbearably high temperatures. Evidently, these habits overburdened his heart and contributed to his death at the age of 62.

Using the Pythagorean identity $\sin^2\theta + \cos^2\theta = 1$ three times gives

$$2 - 2\cos(s + t) = 2 - 2\cos s \cos t + 2\sin s \sin t$$

Finally, subtracting 2 from each side and dividing both sides by -2, we get

$$\cos(s + t) = \cos s \cos t - \sin s \sin t$$

which proves the Addition Formula for Cosine. ■

PROOF OF SUBTRACTION FORMULA FOR COSINE Replacing t with $-t$ in the Addition Formula for Cosine, we get

$$\begin{aligned}\cos(s - t) &= \cos(s + (-t)) \\ &= \cos s \cos(-t) - \sin s \sin(-t) && \text{Addition Formula for Cosine} \\ &= \cos s \cos t + \sin s \sin t && \text{Even-odd identities}\end{aligned}$$

This proves the Subtraction Formula for Cosine. ■

See Exercises 70 and 71 for proofs of the other Addition Formulas.

EXAMPLE 1 | Using the Addition and Subtraction Formulas

Find the exact value of each expression.

(a) $\cos 75°$ **(b)** $\cos \dfrac{\pi}{12}$

SOLUTION

(a) Notice that $75° = 45° + 30°$. Since we know the exact values of sine and cosine at 45° and 30°, we use the Addition Formula for Cosine to get

$$\begin{aligned}\cos 75° &= \cos(45° + 30°) \\ &= \cos 45° \cos 30° - \sin 45° \sin 30° \\ &= \frac{\sqrt{2}}{2}\frac{\sqrt{3}}{2} - \frac{\sqrt{2}}{2}\frac{1}{2} = \frac{\sqrt{2}\sqrt{3} - \sqrt{2}}{4} = \frac{\sqrt{6} - \sqrt{2}}{4}\end{aligned}$$

(b) Since $\dfrac{\pi}{12} = \dfrac{\pi}{4} - \dfrac{\pi}{6}$, the Subtraction Formula for Cosine gives

$$\begin{aligned}\cos\frac{\pi}{12} &= \cos\left(\frac{\pi}{4} - \frac{\pi}{6}\right) \\ &= \cos\frac{\pi}{4}\cos\frac{\pi}{6} + \sin\frac{\pi}{4}\sin\frac{\pi}{6} \\ &= \frac{\sqrt{2}}{2}\frac{\sqrt{3}}{2} + \frac{\sqrt{2}}{2}\frac{1}{2} = \frac{\sqrt{6} + \sqrt{2}}{4}\end{aligned}$$

NOW TRY EXERCISES **3** AND **9** ■

EXAMPLE 2 | Using the Addition Formula for Sine

Find the exact value of the expression $\sin 20° \cos 40° + \cos 20° \sin 40°$.

SOLUTION We recognize the expression as the right-hand side of the Addition Formula for Sine with $s = 20°$ and $t = 40°$. So we have

$$\sin 20° \cos 40° + \cos 20° \sin 40° = \sin(20° + 40°) = \sin 60° = \frac{\sqrt{3}}{2}$$

NOW TRY EXERCISE **15** ■

EXAMPLE 3 | Proving a Cofunction Identity

Prove the cofunction identity $\cos\left(\frac{\pi}{2} - u\right) = \sin u$.

SOLUTION By the Subtraction Formula for Cosine, we have

$$\cos\left(\frac{\pi}{2} - u\right) = \cos\frac{\pi}{2}\cos u + \sin\frac{\pi}{2}\sin u$$
$$= 0 \cdot \cos u + 1 \cdot \sin u = \sin u$$

NOW TRY EXERCISE **21** ■

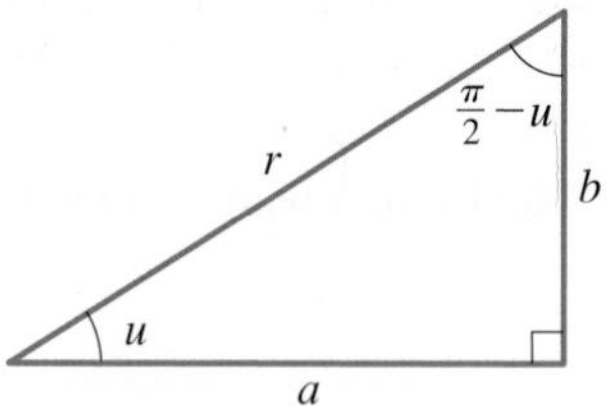

$\cos\left(\frac{\pi}{2} - u\right) = \frac{b}{r} = \sin u$

The cofunction identity in Example 3, as well as the other cofunction identities, can also be derived from the figure in the margin.

EXAMPLE 4 | Proving an Identity

Verify the identity $\dfrac{1 + \tan x}{1 - \tan x} = \tan\left(\dfrac{\pi}{4} + x\right)$.

SOLUTION Starting with the right-hand side and using the Addition Formula for Tangent, we get

$$\text{RHS} = \tan\left(\frac{\pi}{4} + x\right) = \frac{\tan\frac{\pi}{4} + \tan x}{1 - \tan\frac{\pi}{4}\tan x}$$
$$= \frac{1 + \tan x}{1 - \tan x} = \text{LHS}$$

NOW TRY EXERCISE **25** ■

The next example is a typical use of the Addition and Subtraction Formulas in calculus.

EXAMPLE 5 | An Identity from Calculus

If $f(x) = \sin x$, show that

$$\frac{f(x + h) - f(x)}{h} = \sin x\left(\frac{\cos h - 1}{h}\right) + \cos x\left(\frac{\sin h}{h}\right)$$

SOLUTION

$$\frac{f(x + h) - f(x)}{h} = \frac{\sin(x + h) - \sin x}{h} \qquad \text{Definition of } f$$

$$= \frac{\sin x\cos h + \cos x\sin h - \sin x}{h} \qquad \text{Addition Formula for Sine}$$

$$= \frac{\sin x(\cos h - 1) + \cos x\sin h}{h} \qquad \text{Factor}$$

$$= \sin x\left(\frac{\cos h - 1}{h}\right) + \cos x\left(\frac{\sin h}{h}\right) \qquad \text{Separate the fraction}$$

NOW TRY EXERCISE **61** ■

▼ Evaluating Expressions Involving Inverse Trigonometric Functions

Expressions involving trigonometric functions and their inverses arise in calculus. In the next examples we illustrate how to evaluate such expressions.

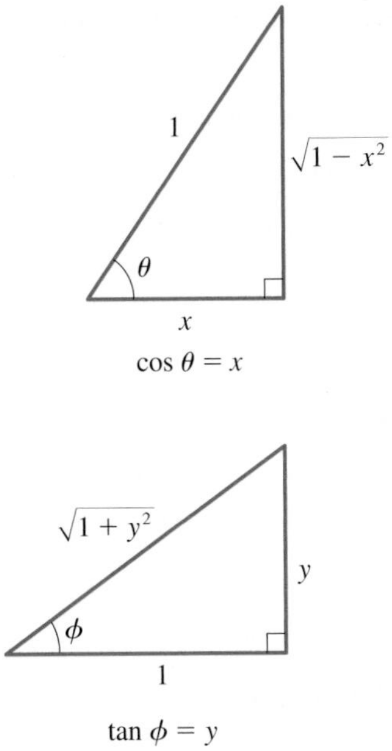

FIGURE 2

EXAMPLE 6 | Simplifying an Expression Involving Inverse Trigonometric Functions

Write $\sin(\cos^{-1}x + \tan^{-1}y)$ as an algebraic expression in x and y, where $-1 \le x \le 1$ and y is any real number.

SOLUTION Let $\theta = \cos^{-1}x$ and $\phi = \tan^{-1}y$. Using the methods of Section 6.4, we sketch triangles with angles θ and ϕ such that $\cos\theta = x$ and $\tan\phi = y$ (see Figure 2). From the triangles we have

$$\sin\theta = \sqrt{1-x^2} \qquad \cos\phi = \frac{1}{\sqrt{1+y^2}} \qquad \sin\phi = \frac{y}{\sqrt{1+y^2}}$$

From the Addition Formula for Sine we have

$$\begin{aligned}
\sin(\cos^{-1}x + \tan^{-1}y) &= \sin(\theta+\phi) \\
&= \sin\theta\cos\phi + \cos\theta\sin\phi && \text{Addition Formula for Sine} \\
&= \sqrt{1-x^2}\frac{1}{\sqrt{1+y^2}} + x\frac{y}{\sqrt{1+y^2}} && \text{From triangles} \\
&= \frac{1}{\sqrt{1+y^2}}\left(\sqrt{1-x^2} + xy\right) && \text{Factor } \frac{1}{\sqrt{1+y^2}}
\end{aligned}$$

NOW TRY EXERCISES **43** AND **47** ■

EXAMPLE 7 | Evaluating an Expression Involving Trigonometric Functions

Evaluate $\sin(\theta+\phi)$, where $\sin\theta = \frac{12}{13}$ with θ in Quadrant II and $\tan\phi = \frac{3}{4}$ with ϕ in Quadrant III.

SOLUTION We first sketch the angles θ and ϕ in standard position with terminal sides in the appropriate quadrants as in Figure 3. Since $\sin\theta = y/r = \frac{12}{13}$ we can label a side and the hypotenuse in the triangle in Figure 3(a). To find the remaining side, we use the Pythagorean Theorem:

$$\begin{aligned}
x^2 + y^2 &= r^2 && \text{Pythagorean Theorem} \\
x^2 + 12^2 &= 13^2 && y = 12, \quad r = 13 \\
x^2 &= 25 && \text{Solve for } x^2 \\
x &= -5 && \text{Because } x < 0
\end{aligned}$$

Similarly, since $\tan\phi = y/x = \frac{3}{4}$, we can label two sides of the triangle in Figure 3(b) and then use the Pythagorean Theorem to find the hypotenuse.

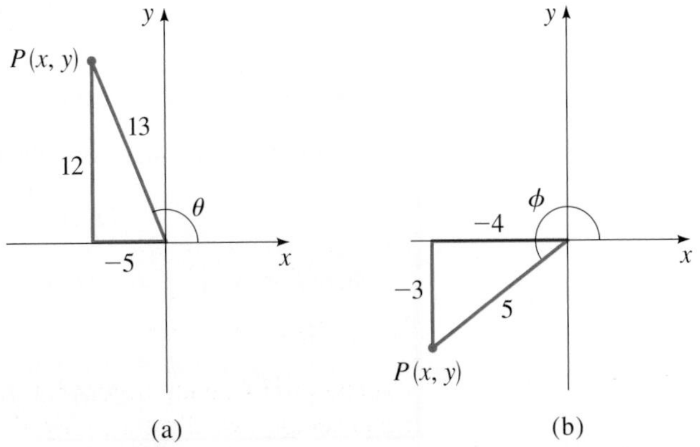

FIGURE 3

Now, to find $\sin(\theta + \phi)$, we use the Addition Formula for Sine and the triangles in Figure 3:

$$\begin{aligned} \sin(\theta + \phi) &= \sin\theta\cos\phi + \cos\theta\sin\phi && \text{Addition Formula} \\ &= \left(\tfrac{12}{13}\right)\left(-\tfrac{4}{5}\right) + \left(-\tfrac{5}{13}\right)\left(-\tfrac{3}{5}\right) && \text{From triangles} \\ &= -\tfrac{33}{65} && \text{Calculate} \end{aligned}$$

NOW TRY EXERCISE **51**

Expressions of the Form $A \sin x + B \cos x$

We can write expressions of the form $A \sin x + B \cos x$ in terms of a single trigonometric function using the Addition Formula for Sine. For example, consider the expression

$$\frac{1}{2}\sin x + \frac{\sqrt{3}}{2}\cos x$$

If we set $\phi = \pi/3$, then $\cos\phi = \frac{1}{2}$ and $\sin\phi = \sqrt{3}/2$, and we can write

$$\begin{aligned} \frac{1}{2}\sin x + \frac{\sqrt{3}}{2}\cos x &= \cos\phi\sin x + \sin\phi\cos x \\ &= \sin(x + \phi) = \sin\left(x + \frac{\pi}{3}\right) \end{aligned}$$

We are able to do this because the coefficients $\frac{1}{2}$ and $\sqrt{3}/2$ are precisely the cosine and sine of a particular number, in this case, $\pi/3$. We can use this same idea in general to write $A \sin x + B \cos x$ in the form $k \sin(x + \phi)$. We start by multiplying the numerator and denominator by $\sqrt{A^2 + B^2}$ to get

$$A\sin x + B\cos x = \sqrt{A^2 + B^2}\left(\frac{A}{\sqrt{A^2 + B^2}}\sin x + \frac{B}{\sqrt{A^2 + B^2}}\cos x\right)$$

We need a number ϕ with the property that

$$\cos\phi = \frac{A}{\sqrt{A^2 + B^2}} \quad \text{and} \quad \sin\phi = \frac{B}{\sqrt{A^2 + B^2}}$$

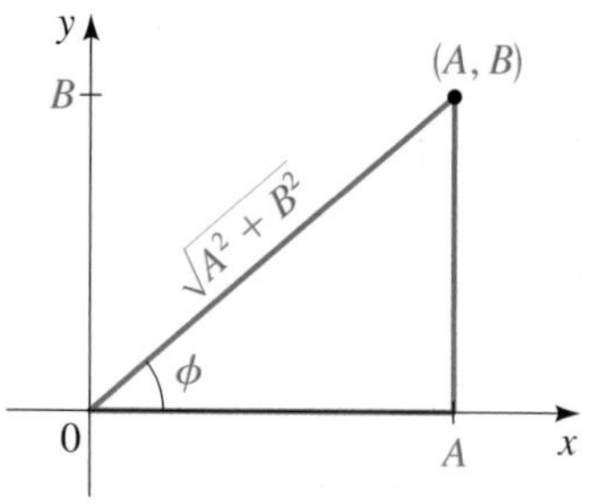

FIGURE 4

Figure 4 shows that the point (A, B) in the plane determines a number ϕ with precisely this property. With this ϕ we have

$$\begin{aligned} A\sin x + B\cos x &= \sqrt{A^2 + B^2}(\cos\phi\sin x + \sin\phi\cos x) \\ &= \sqrt{A^2 + B^2}\sin(x + \phi) \end{aligned}$$

We have proved the following theorem.

SUMS OF SINES AND COSINES

If A and B are real numbers, then

$$A\sin x + B\cos x = k\sin(x + \phi)$$

where $k = \sqrt{A^2 + B^2}$ and ϕ satisfies

$$\cos\phi = \frac{A}{\sqrt{A^2 + B^2}} \quad \text{and} \quad \sin\phi = \frac{B}{\sqrt{A^2 + B^2}}$$

EXAMPLE 8 | A Sum of Sine and Cosine Terms

Express $3 \sin x + 4 \cos x$ in the form $k \sin(x + \phi)$.

SOLUTION By the preceding theorem, $k = \sqrt{A^2 + B^2} = \sqrt{3^2 + 4^2} = 5$. The angle ϕ has the property that $\sin \phi = \frac{4}{5}$ and $\cos \phi = \frac{3}{5}$. Using a calculator, we find $\phi \approx 53.1°$. Thus

$$3 \sin x + 4 \cos x \approx 5 \sin(x + 53.1°)$$

NOW TRY EXERCISE **55** ■

EXAMPLE 9 | Graphing a Trigonometric Function

Write the function $f(x) = -\sin 2x + \sqrt{3} \cos 2x$ in the form $k \sin(2x + \phi)$, and use the new form to graph the function.

SOLUTION Since $A = -1$ and $B = \sqrt{3}$, we have $k = \sqrt{A^2 + B^2} = \sqrt{1 + 3} = 2$. The angle ϕ satisfies $\cos \phi = -\frac{1}{2}$ and $\sin \phi = \sqrt{3}/2$. From the signs of these quantities we conclude that ϕ is in Quadrant II. Thus $\phi = 2\pi/3$. By the preceding theorem we can write

$$f(x) = -\sin 2x + \sqrt{3} \cos 2x = 2 \sin\left(2x + \frac{2\pi}{3}\right)$$

Using the form

$$f(x) = 2 \sin 2\left(x + \frac{\pi}{3}\right)$$

we see that the graph is a sine curve with amplitude 2, period $2\pi/2 = \pi$, and phase shift $-\pi/3$. The graph is shown in Figure 5.

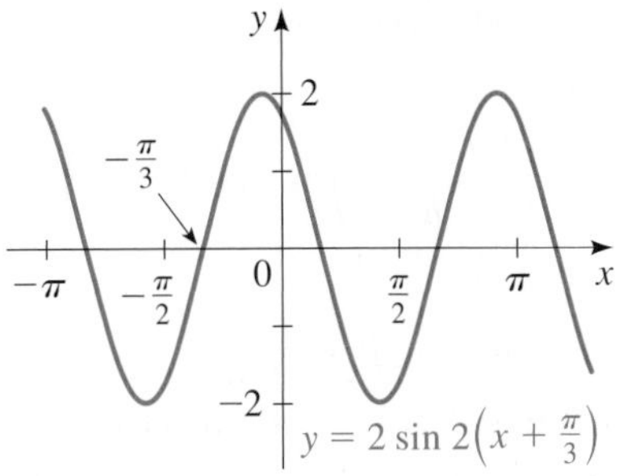

FIGURE 5

NOW TRY EXERCISE **59** ■

7.2 EXERCISES

CONCEPTS

1. If we know the values of the sine and cosine of x and y, we can find the value of $\sin(x + y)$ by using the ________ Formula for Sine. State the formula: $\sin(x + y) =$ ____________.

2. If we know the values of the sine and cosine of x and y, we can find the value of $\cos(x - y)$ by using the ________ Formula for Cosine. State the formula: $\cos(x - y) =$ ____________.

SKILLS

3–14 ■ Use an Addition or Subtraction Formula to find the exact value of the expression, as demonstrated in Example 1.

3. $\sin 75°$ **4.** $\sin 15°$

5. $\cos 105°$ **6.** $\cos 195°$

7. $\tan 15°$ **8.** $\tan 165°$

9. $\sin \dfrac{19\pi}{12}$ **10.** $\cos \dfrac{17\pi}{12}$

11. $\tan\left(-\dfrac{\pi}{12}\right)$ **12.** $\sin\left(-\dfrac{5\pi}{12}\right)$

13. $\cos \dfrac{11\pi}{12}$ **14.** $\tan \dfrac{7\pi}{12}$

15–20 ■ Use an Addition or Subtraction Formula to write the expression as a trigonometric function of one number, and then find its exact value.

15. $\sin 18° \cos 27° + \cos 18° \sin 27°$

16. $\cos 10° \cos 80° - \sin 10° \sin 80°$

17. $\cos \dfrac{3\pi}{7} \cos \dfrac{2\pi}{21} + \sin \dfrac{3\pi}{7} \sin \dfrac{2\pi}{21}$

18. $\dfrac{\tan \dfrac{\pi}{18} + \tan \dfrac{\pi}{9}}{1 - \tan \dfrac{\pi}{18} \tan \dfrac{\pi}{9}}$

19. $\dfrac{\tan 73° - \tan 13°}{1 + \tan 73° \tan 13°}$

20. $\cos \dfrac{13\pi}{15} \cos\left(-\dfrac{\pi}{5}\right) - \sin \dfrac{13\pi}{15} \sin\left(-\dfrac{\pi}{5}\right)$

21–24 ■ Prove the cofunction identity using the Addition and Subtraction Formulas.

21. $\tan\left(\dfrac{\pi}{2} - u\right) = \cot u$ **22.** $\cot\left(\dfrac{\pi}{2} - u\right) = \tan u$

23. $\sec\left(\dfrac{\pi}{2} - u\right) = \csc u$ **24.** $\csc\left(\dfrac{\pi}{2} - u\right) = \sec u$

25–42 ■ Prove the identity.

25. $\sin\left(x - \dfrac{\pi}{2}\right) = -\cos x$

26. $\cos\left(x - \dfrac{\pi}{2}\right) = \sin x$

27. $\sin(x - \pi) = -\sin x$ **28.** $\cos(x - \pi) = -\cos x$

29. $\tan(x - \pi) = \tan x$

30. $\sin\left(\dfrac{\pi}{2} - x\right) = \sin\left(\dfrac{\pi}{2} + x\right)$

31. $\cos\left(x + \dfrac{\pi}{6}\right) + \sin\left(x - \dfrac{\pi}{3}\right) = 0$

32. $\tan\left(x - \dfrac{\pi}{4}\right) = \dfrac{\tan x - 1}{\tan x + 1}$

33. $\sin(x + y) - \sin(x - y) = 2 \cos x \sin y$

34. $\cos(x + y) + \cos(x - y) = 2 \cos x \cos y$

35. $\cot(x - y) = \dfrac{\cot x \cot y + 1}{\cot y - \cot x}$

36. $\cot(x + y) = \dfrac{\cot x \cot y - 1}{\cot x + \cot y}$

37. $\tan x - \tan y = \dfrac{\sin(x - y)}{\cos x \cos y}$

38. $1 - \tan x \tan y = \dfrac{\cos(x + y)}{\cos x \cos y}$

39. $\dfrac{\sin(x + y) - \sin(x - y)}{\cos(x + y) + \cos(x - y)} = \tan y$

40. $\cos(x + y) \cos(x - y) = \cos^2 x - \sin^2 y$

41. $\sin(x + y + z) = \sin x \cos y \cos z + \cos x \sin y \cos z + \cos x \cos y \sin z - \sin x \sin y \sin z$

42. $\tan(x - y) + \tan(y - z) + \tan(z - x) = \tan(x - y) \tan(y - z) \tan(z - x)$

43–46 ■ Write the given expression in terms of x and y only.

43. $\cos(\sin^{-1} x - \tan^{-1} y)$ **44.** $\tan(\sin^{-1} x + \cos^{-1} y)$

45. $\sin(\tan^{-1} x - \tan^{-1} y)$ **46.** $\sin(\sin^{-1} x + \cos^{-1} y)$

47–50 ■ Find the exact value of the expression.

47. $\sin\left(\cos^{-1} \frac{1}{2} + \tan^{-1} 1\right)$ **48.** $\cos\left(\sin^{-1} \frac{\sqrt{3}}{2} + \cot^{-1} \sqrt{3}\right)$

49. $\tan\left(\sin^{-1} \frac{3}{4} - \cos^{-1} \frac{1}{3}\right)$ **50.** $\sin\left(\cos^{-1} \frac{2}{3} - \tan^{-1} \frac{1}{2}\right)$

51–54 ■ Evaluate each expression under the given conditions.

51. $\cos(\theta - \phi)$; $\cos \theta = \frac{3}{5}$, θ in Quadrant IV, $\tan \phi = -\sqrt{3}$, ϕ in Quadrant II.

52. $\sin(\theta - \phi)$; $\tan \theta = \frac{4}{3}$, θ in Quadrant III, $\sin \phi = -\sqrt{10}/10$, ϕ in Quadrant IV

53. $\sin(\theta + \phi)$; $\sin \theta = \frac{5}{13}$, θ in Quadrant I, $\cos \phi = -2\sqrt{5}/5$, ϕ in Quadrant II

54. $\tan(\theta + \phi)$; $\cos \theta = -\frac{1}{3}$, θ in Quadrant III, $\sin \phi = \frac{1}{4}$, ϕ in Quadrant II

55–58 ■ Write the expression in terms of sine only.

55. $-\sqrt{3} \sin x + \cos x$ **56.** $\sin x + \cos x$

57. $5(\sin 2x - \cos 2x)$ **58.** $3 \sin \pi x + 3\sqrt{3} \cos \pi x$

59–60 ■ **(a)** Express the function in terms of sine only. **(b)** Graph the function.

59. $g(x) = \cos 2x + \sqrt{3} \sin 2x$ **60.** $f(x) = \sin x + \cos x$

61. Let $g(x) = \cos x$. Show that

$$\frac{g(x + h) - g(x)}{h} = -\cos x \left(\frac{1 - \cos h}{h}\right) - \sin x \left(\frac{\sin h}{h}\right)$$

62. Show that if $\beta - \alpha = \pi/2$, then

$$\sin(x + \alpha) + \cos(x + \beta) = 0$$

63. Refer to the figure. Show that $\alpha + \beta = \gamma$, and find $\tan \gamma$.

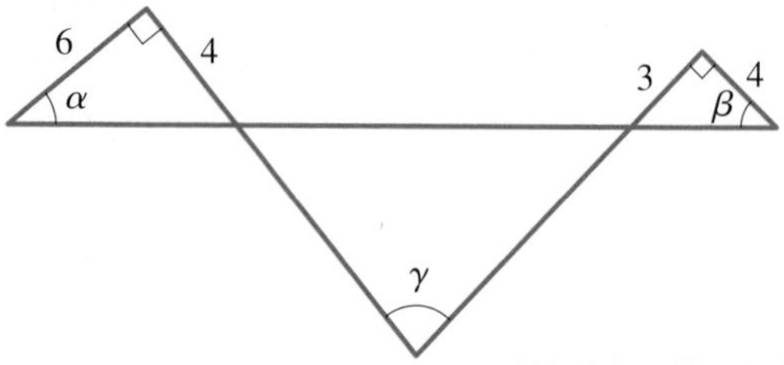

64. **(a)** If L is a line in the plane and θ is the angle formed by the line and the x-axis as shown in the figure, show that the slope m of the line is given by

$$m = \tan \theta$$

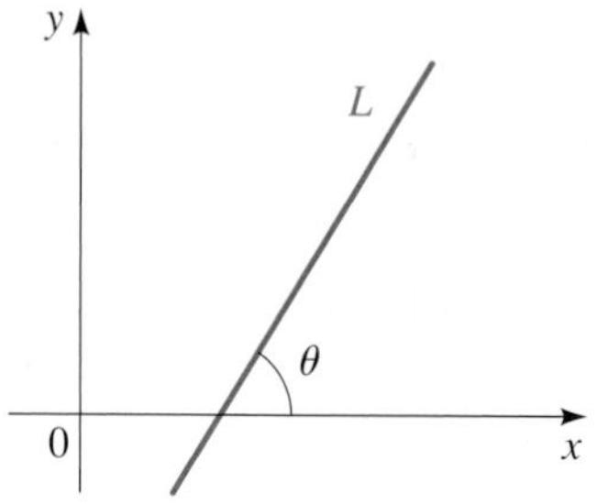

(b) Let L_1 and L_2 be two nonparallel lines in the plane with slopes m_1 and m_2, respectively. Let ψ be the acute angle formed by the two lines (see the following figure). Show that

$$\tan \psi = \frac{m_2 - m_1}{1 + m_1 m_2}$$

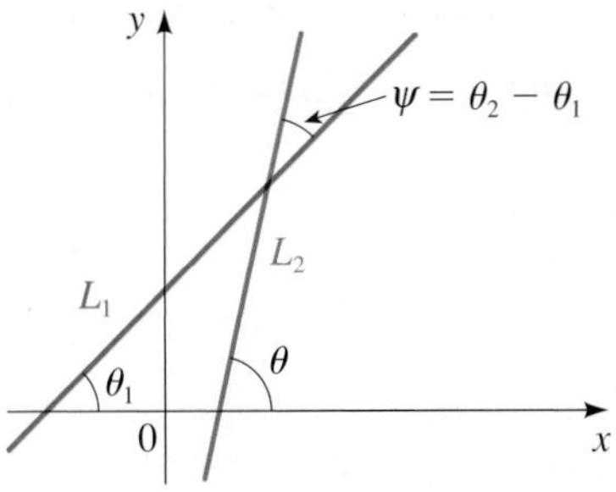

(c) Find the acute angle formed by the two lines

$$y = \tfrac{1}{3}x + 1 \quad \text{and} \quad y = -\tfrac{1}{2}x - 3$$

(d) Show that if two lines are perpendicular, then the slope of one is the negative reciprocal of the slope of the other. [*Hint:* First find an expression for cot ψ.]

65–66 ■ **(a)** Graph the function and make a conjecture, then **(b)** prove that your conjecture is true.

65. $y = \sin^2\left(x + \dfrac{\pi}{4}\right) + \sin^2\left(x - \dfrac{\pi}{4}\right)$

66. $y = -\frac{1}{2}[\cos(x + \pi) + \cos(x - \pi)]$

67. Find $\angle A + \angle B + \angle C$ in the figure. [*Hint:* First use an addition formula to find $\tan(A + B)$.]

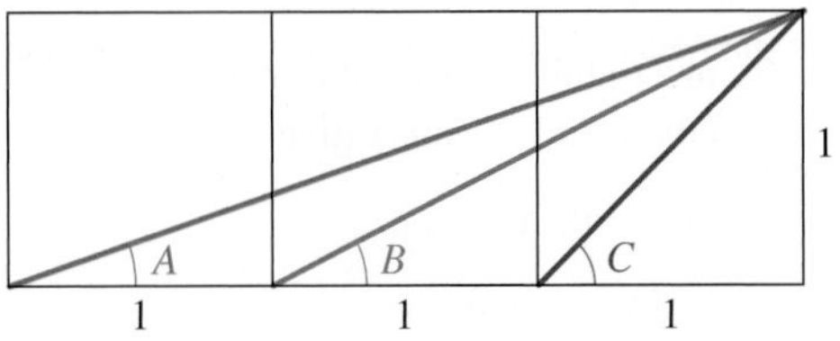

APPLICATIONS

68. Adding an Echo A digital delay device echoes an input signal by repeating it a fixed length of time after it is received. If such a device receives the pure note $f_1(t) = 5 \sin t$ and echoes the pure note $f_2(t) = 5 \cos t$, then the combined sound is $f(t) = f_1(t) + f_2(t)$.

(a) Graph $y = f(t)$ and observe that the graph has the form of a sine curve $y = k \sin(t + \phi)$.

(b) Find k and ϕ.

69. Interference Two identical tuning forks are struck, one a fraction of a second after the other. The sounds produced are modeled by $f_1(t) = C \sin \omega t$ and $f_2(t) = C \sin(\omega t + \alpha)$. The two sound waves interfere to produce a single sound modeled by the sum of these functions

$$f(t) = C \sin \omega t + C \sin(\omega t + \alpha)$$

(a) Use the Addition Formula for Sine to show that f can be written in the form $f(t) = A \sin \omega t + B \cos \omega t$, where A and B are constants that depend on α.

(b) Suppose that $C = 10$ and $\alpha = \pi/3$. Find constants k and ϕ so that $f(t) = k \sin(\omega t + \phi)$.

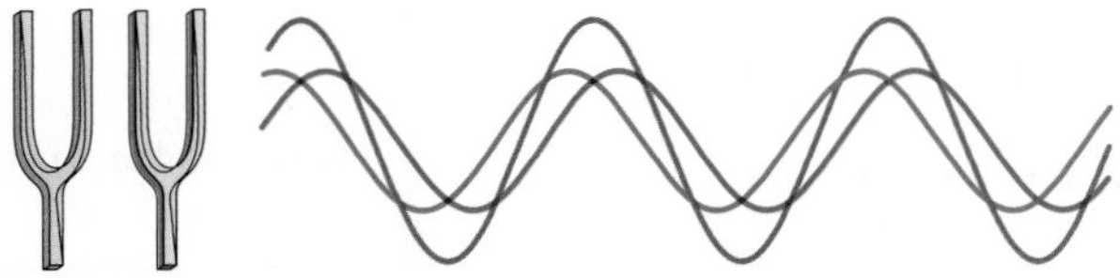

DISCOVERY ■ DISCUSSION ■ WRITING

70. Addition Formula for Sine In the text we proved only the Addition and Subtraction Formulas for Cosine. Use these formulas and the cofunction identities

$$\sin x = \cos\left(\frac{\pi}{2} - x\right)$$

$$\cos x = \sin\left(\frac{\pi}{2} - x\right)$$

to prove the Addition Formula for Sine. [*Hint:* To get started, use the first cofunction identity to write

$$\begin{aligned}\sin(s + t) &= \cos\left(\frac{\pi}{2} - (s + t)\right)\\ &= \cos\left(\left(\frac{\pi}{2} - s\right) - t\right)\end{aligned}$$

and use the Subtraction Formula for Cosine.]

71. Addition Formula for Tangent Use the Addition Formulas for Cosine and Sine to prove the Addition Formula for Tangent. [*Hint:* Use

$$\tan(s + t) = \frac{\sin(s + t)}{\cos(s + t)}$$

and divide the numerator and denominator by $\cos s \cos t$.]

7.3 Double-Angle, Half-Angle, and Product-Sum Formulas

Double-Angle Formulas ▶ Half-Angle Formulas ▶ Simplifying Expressions Involving Inverse Trigonometric Functions ▶ Product-Sum Formulas

The identities we consider in this section are consequences of the addition formulas. The **Double-Angle Formulas** allow us to find the values of the trigonometric functions at $2x$ from their values at x. The **Half-Angle Formulas** relate the values of the trigonometric functions at $\frac{1}{2}x$ to their values at x. The **Product-Sum Formulas** relate products of sines and cosines to sums of sines and cosines.

▼ Double-Angle Formulas

The formulas in the following box are immediate consequences of the addition formulas, which we proved in the preceding section.

DOUBLE-ANGLE FORMULAS

Formula for sine: $\sin 2x = 2 \sin x \cos x$

Formulas for cosine:

$$\cos 2x = \cos^2 x - \sin^2 x$$
$$= 1 - 2\sin^2 x$$
$$= 2\cos^2 x - 1$$

Formula for tangent: $\tan 2x = \dfrac{2 \tan x}{1 - \tan^2 x}$

The proofs for the formulas for cosine are given here. You are asked to prove the remaining formulas in Exercises 35 and 36.

PROOF OF DOUBLE-ANGLE FORMULAS FOR COSINE

$$\begin{aligned} \cos 2x &= \cos(x + x) \\ &= \cos x \cos x - \sin x \sin x \\ &= \cos^2 x - \sin^2 x \end{aligned}$$

The second and third formulas for $\cos 2x$ are obtained from the formula we just proved and the Pythagorean identity. Substituting $\cos^2 x = 1 - \sin^2 x$ gives

$$\begin{aligned} \cos 2x &= \cos^2 x - \sin^2 x \\ &= (1 - \sin^2 x) - \sin^2 x \\ &= 1 - 2\sin^2 x \end{aligned}$$

The third formula is obtained in the same way, by substituting $\sin^2 x = 1 - \cos^2 x$. ■

EXAMPLE 1 | Using the Double-Angle Formulas

If $\cos x = -\frac{2}{3}$ and x is in Quadrant II, find $\cos 2x$ and $\sin 2x$.

SOLUTION Using one of the Double-Angle Formulas for Cosine, we get

$$\begin{aligned} \cos 2x &= 2\cos^2 x - 1 \\ &= 2\left(-\frac{2}{3}\right)^2 - 1 = \frac{8}{9} - 1 = -\frac{1}{9} \end{aligned}$$

To use the formula $\sin 2x = 2 \sin x \cos x$, we need to find $\sin x$ first. We have

$$\sin x = \sqrt{1 - \cos^2 x} = \sqrt{1 - \left(-\tfrac{2}{3}\right)^2} = \frac{\sqrt{5}}{3}$$

where we have used the positive square root because $\sin x$ is positive in Quadrant II. Thus

$$\begin{aligned} \sin 2x &= 2 \sin x \cos x \\ &= 2\left(\frac{\sqrt{5}}{3}\right)\left(-\frac{2}{3}\right) = -\frac{4\sqrt{5}}{9} \end{aligned}$$

NOW TRY EXERCISE 3 ■

EXAMPLE 2 | A Triple-Angle Formula

Write $\cos 3x$ in terms of $\cos x$.

SOLUTION

$$\begin{aligned}
\cos 3x &= \cos(2x + x) \\
&= \cos 2x \cos x - \sin 2x \sin x && \text{Addition formula} \\
&= (2\cos^2 x - 1)\cos x - (2\sin x \cos x)\sin x && \text{Double-Angle Formulas} \\
&= 2\cos^3 x - \cos x - 2\sin^2 x \cos x && \text{Expand} \\
&= 2\cos^3 x - \cos x - 2\cos x\,(1 - \cos^2 x) && \text{Pythagorean identity} \\
&= 2\cos^3 x - \cos x - 2\cos x + 2\cos^3 x && \text{Expand} \\
&= 4\cos^3 x - 3\cos x && \text{Simplify}
\end{aligned}$$

NOW TRY EXERCISE **101** ■

Example 2 shows that $\cos 3x$ can be written as a polynomial of degree 3 in $\cos x$. The identity $\cos 2x = 2\cos^2 x - 1$ shows that $\cos 2x$ is a polynomial of degree 2 in $\cos x$. In fact, for any natural number n, we can write $\cos nx$ as a polynomial in $\cos x$ of degree n (see the note following Exercise 101). The analogous result for $\sin nx$ is not true in general.

EXAMPLE 3 | Proving an Identity

Prove the identity $\dfrac{\sin 3x}{\sin x \cos x} = 4\cos x - \sec x$.

SOLUTION We start with the left-hand side:

$$\begin{aligned}
\frac{\sin 3x}{\sin x \cos x} &= \frac{\sin(x + 2x)}{\sin x \cos x} \\
&= \frac{\sin x \cos 2x + \cos x \sin 2x}{\sin x \cos x} && \text{Addition Formula} \\
&= \frac{\sin x\,(2\cos^2 x - 1) + \cos x\,(2\sin x \cos x)}{\sin x \cos x} && \text{Double-Angle Formulas} \\
&= \frac{\sin x\,(2\cos^2 x - 1)}{\sin x \cos x} + \frac{\cos x\,(2\sin x \cos x)}{\sin x \cos x} && \text{Separate fraction} \\
&= \frac{2\cos^2 x - 1}{\cos x} + 2\cos x && \text{Cancel} \\
&= 2\cos x - \frac{1}{\cos x} + 2\cos x && \text{Separate fraction} \\
&= 4\cos x - \sec x && \text{Reciprocal identity}
\end{aligned}$$

NOW TRY EXERCISE **81** ■

▼ Half-Angle Formulas

The following formulas allow us to write any trigonometric expression involving even powers of sine and cosine in terms of the first power of cosine only. This technique is important in calculus. The Half-Angle Formulas are immediate consequences of these formulas.

FORMULAS FOR LOWERING POWERS

$$\sin^2 x = \frac{1 - \cos 2x}{2} \qquad \cos^2 x = \frac{1 + \cos 2x}{2}$$

$$\tan^2 x = \frac{1 - \cos 2x}{1 + \cos 2x}$$

PROOF The first formula is obtained by solving for $\sin^2 x$ in the double-angle formula $\cos 2x = 1 - 2\sin^2 x$. Similarly, the second formula is obtained by solving for $\cos^2 x$ in the Double-Angle Formula $\cos 2x = 2\cos^2 x - 1$.

The last formula follows from the first two and the reciprocal identities:

$$\tan^2 x = \frac{\sin^2 x}{\cos^2 x} = \frac{\dfrac{1 - \cos 2x}{2}}{\dfrac{1 + \cos 2x}{2}} = \frac{1 - \cos 2x}{1 + \cos 2x}$$

■

EXAMPLE 4 | Lowering Powers in a Trigonometric Expression

Express $\sin^2 x \cos^2 x$ in terms of the first power of cosine.

SOLUTION We use the formulas for lowering powers repeatedly:

$$\begin{aligned}
\sin^2 x \cos^2 x &= \left(\frac{1 - \cos 2x}{2}\right)\left(\frac{1 + \cos 2x}{2}\right) \\
&= \frac{1 - \cos^2 2x}{4} = \frac{1}{4} - \frac{1}{4}\cos^2 2x \\
&= \frac{1}{4} - \frac{1}{4}\left(\frac{1 + \cos 4x}{2}\right) = \frac{1}{4} - \frac{1}{8} - \frac{\cos 4x}{8} \\
&= \frac{1}{8} - \frac{1}{8}\cos 4x = \frac{1}{8}(1 - \cos 4x)
\end{aligned}$$

Another way to obtain this identity is to use the Double-Angle Formula for Sine in the form $\sin x \cos x = \frac{1}{2}\sin 2x$. Thus

$$\begin{aligned}
\sin^2 x \cos^2 x &= \frac{1}{4}\sin^2 2x = \frac{1}{4}\left(\frac{1 - \cos 4x}{2}\right) \\
&= \frac{1}{8}(1 - \cos 4x)
\end{aligned}$$

NOW TRY EXERCISE **11** ■

HALF-ANGLE FORMULAS

$$\sin\frac{u}{2} = \pm\sqrt{\frac{1 - \cos u}{2}} \qquad \cos\frac{u}{2} = \pm\sqrt{\frac{1 + \cos u}{2}}$$

$$\tan\frac{u}{2} = \frac{1 - \cos u}{\sin u} = \frac{\sin u}{1 + \cos u}$$

The choice of the + or − sign depends on the quadrant in which $u/2$ lies.

PROOF We substitute $x = u/2$ in the formulas for lowering powers and take the square root of each side. This gives the first two Half-Angle Formulas. In the case of the Half-Angle Formula for Tangent we get

$$\tan\frac{u}{2} = \pm\sqrt{\frac{1-\cos u}{1+\cos u}}$$

$$= \pm\sqrt{\left(\frac{1-\cos u}{1+\cos u}\right)\left(\frac{1-\cos u}{1-\cos u}\right)} \qquad \text{Multiply numerator and denominator by } 1-\cos u$$

$$= \pm\sqrt{\frac{(1-\cos u)^2}{1-\cos^2 u}} \qquad \text{Simplify}$$

$$= \pm\frac{|1-\cos u|}{|\sin u|} \qquad \sqrt{A^2} = |A| \text{ and } 1-\cos^2 u = \sin^2 u$$

Now, $1 - \cos u$ is nonnegative for all values of u. It is also true that $\sin u$ and $\tan(u/2)$ always have the same sign. (Verify this.) It follows that

$$\tan\frac{u}{2} = \frac{1-\cos u}{\sin u}$$

The other Half-Angle Formula for Tangent is derived from this by multiplying the numerator and denominator by $1 + \cos u$. ■

EXAMPLE 5 | Using a Half-Angle Formula

Find the exact value of $\sin 22.5°$.

SOLUTION Since 22.5° is half of 45°, we use the Half-Angle Formula for Sine with $u = 45°$. We choose the + sign because 22.5° is in the first quadrant:

$$\sin\frac{45°}{2} = \sqrt{\frac{1-\cos 45°}{2}} \qquad \text{Half-Angle Formula}$$

$$= \sqrt{\frac{1-\sqrt{2}/2}{2}} \qquad \cos 45° = \sqrt{2}/2$$

$$= \sqrt{\frac{2-\sqrt{2}}{4}} \qquad \text{Common denominator}$$

$$= \tfrac{1}{2}\sqrt{2-\sqrt{2}} \qquad \text{Simplify}$$

NOW TRY EXERCISE **17** ■

EXAMPLE 6 | Using a Half-Angle Formula

Find $\tan(u/2)$ if $\sin u = \frac{2}{5}$ and u is in Quadrant II.

SOLUTION To use the Half-Angle Formula for Tangent, we first need to find $\cos u$. Since cosine is negative in Quadrant II, we have

$$\cos u = -\sqrt{1-\sin^2 u}$$

$$= -\sqrt{1-\left(\tfrac{2}{5}\right)^2} = -\frac{\sqrt{21}}{5}$$

Thus

$$\tan\frac{u}{2} = \frac{1-\cos u}{\sin u}$$

$$= \frac{1+\sqrt{21}/5}{\frac{2}{5}} = \frac{5+\sqrt{21}}{2}$$

NOW TRY EXERCISE **37** ■

▼ Evaluating Expressions Involving Inverse Trigonometric Functions

Expressions involving trigonometric functions and their inverses arise in calculus. In the next examples we illustrate how to evaluate such expressions.

EXAMPLE 7 | Simplifying an Expression Involving an Inverse Trigonometric Function

Write $\sin(2\cos^{-1}x)$ as an algebraic expression in x only, where $-1 \le x \le 1$.

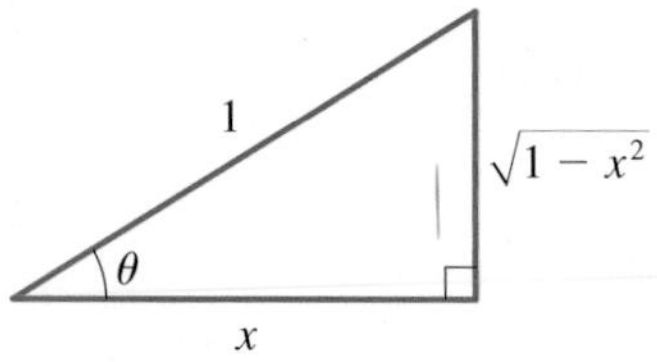

FIGURE 1

SOLUTION Let $\theta = \cos^{-1}x$, and sketch a triangle as in Figure 1. We need to find $\sin 2\theta$, but from the triangle we can find trigonometric functions of θ only, not 2θ. So we use the Double-Angle Formula for Sine.

$$\begin{aligned} \sin(2\cos^{-1}x) &= \sin 2\theta && \cos^{-1}x = \theta \\ &= 2\sin\theta\cos\theta && \text{Double-Angle Formula} \\ &= 2x\sqrt{1-x^2} && \text{From the triangle} \end{aligned}$$

NOW TRY EXERCISES **43** AND **47** ■

EXAMPLE 8 | Evaluating an Expression Involving Inverse Trigonometric Functions

Evaluate $\sin 2\theta$, where $\cos\theta = -\frac{2}{5}$ with θ in Quadrant II.

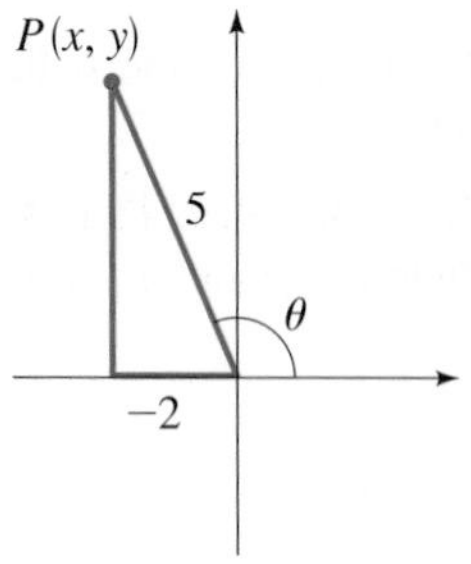

FIGURE 2

SOLUTION We first sketch the angle θ in standard position with terminal side in Quadrant II as in Figure 2. Since $\cos\theta = x/r = -\frac{2}{5}$, we can label a side and the hypotenuse of the triangle in Figure 2. To find the remaining side, we use the Pythagorean Theorem:

$$\begin{aligned} x^2 + y^2 &= r^2 && \text{Pythagorean Theorem} \\ (-2)^2 + y^2 &= 5^2 && x = -2, \quad r = 5 \\ y &= \pm\sqrt{21} && \text{Solve for } y^2 \\ y &= +\sqrt{21} && \text{Because } y > 0 \end{aligned}$$

We can now use the Double-Angle Formula for Sine:

$$\begin{aligned} \sin 2\theta &= 2\sin\theta\cos\theta && \text{Double-Angle Formula} \\ &= 2\left(\frac{\sqrt{21}}{5}\right)\left(-\frac{2}{5}\right) && \text{From the triangle} \\ &= -\frac{4\sqrt{21}}{25} && \text{Simplify} \end{aligned}$$

NOW TRY EXERCISE **51** ■

▼ Product-Sum Formulas

It is possible to write the product $\sin u \cos v$ as a sum of trigonometric functions. To see this, consider the addition and subtraction formulas for the sine function:

$$\sin(u+v) = \sin u\cos v + \cos u \sin v$$

$$\sin(u-v) = \sin u\cos v - \cos u \sin v$$

Adding the left- and right-hand sides of these formulas gives

$$\sin(u+v) + \sin(u-v) = 2\sin u\cos v$$

Dividing by 2 gives the formula

$$\sin u \cos v = \tfrac{1}{2}[\sin(u + v) + \sin(u - v)]$$

The other three **Product-to-Sum Formulas** follow from the addition formulas in a similar way.

PRODUCT-TO-SUM FORMULAS

$$\sin u \cos v = \tfrac{1}{2}[\sin(u + v) + \sin(u - v)]$$
$$\cos u \sin v = \tfrac{1}{2}[\sin(u + v) - \sin(u - v)]$$
$$\cos u \cos v = \tfrac{1}{2}[\cos(u + v) + \cos(u - v)]$$
$$\sin u \sin v = \tfrac{1}{2}[\cos(u - v) - \cos(u + v)]$$

EXAMPLE 9 | Expressing a Trigonometric Product as a Sum

Express $\sin 3x \sin 5x$ as a sum of trigonometric functions.

SOLUTION Using the fourth Product-to-Sum Formula with $u = 3x$ and $v = 5x$ and the fact that cosine is an even function, we get

$$\begin{aligned}\sin 3x \sin 5x &= \tfrac{1}{2}[\cos(3x - 5x) - \cos(3x + 5x)]\\ &= \tfrac{1}{2}\cos(-2x) - \tfrac{1}{2}\cos 8x\\ &= \tfrac{1}{2}\cos 2x - \tfrac{1}{2}\cos 8x\end{aligned}$$

NOW TRY EXERCISE **55** ■

The Product-to-Sum Formulas can also be used as Sum-to-Product Formulas. This is possible because the right-hand side of each Product-to-Sum Formula is a sum and the left side is a product. For example, if we let

$$u = \frac{x + y}{2} \qquad \text{and} \qquad v = \frac{x - y}{2}$$

in the first Product-to-Sum Formula, we get

$$\sin\frac{x + y}{2}\cos\frac{x - y}{2} = \tfrac{1}{2}(\sin x + \sin y)$$

so

$$\sin x + \sin y = 2\sin\frac{x + y}{2}\cos\frac{x - y}{2}$$

The remaining three of the following **Sum-to-Product Formulas** are obtained in a similar manner.

SUM-TO-PRODUCT FORMULAS

$$\sin x + \sin y = 2\sin\frac{x + y}{2}\cos\frac{x - y}{2}$$

$$\sin x - \sin y = 2\cos\frac{x + y}{2}\sin\frac{x - y}{2}$$

$$\cos x + \cos y = 2\cos\frac{x + y}{2}\cos\frac{x - y}{2}$$

$$\cos x - \cos y = -2\sin\frac{x + y}{2}\sin\frac{x - y}{2}$$

EXAMPLE 10 | Expressing a Trigonometric Sum as a Product

Write $\sin 7x + \sin 3x$ as a product.

SOLUTION The first Sum-to-Product Formula gives

$$\sin 7x + \sin 3x = 2 \sin \frac{7x + 3x}{2} \cos \frac{7x - 3x}{2}$$

$$= 2 \sin 5x \cos 2x$$

NOW TRY EXERCISE **61** ■

EXAMPLE 11 | Proving an Identity

Verify the identity $\dfrac{\sin 3x - \sin x}{\cos 3x + \cos x} = \tan x$.

SOLUTION We apply the second Sum-to-Product Formula to the numerator and the third formula to the denominator:

$$\text{LHS} = \frac{\sin 3x - \sin x}{\cos 3x + \cos x} = \frac{2 \cos \frac{3x + x}{2} \sin \frac{3x - x}{2}}{2 \cos \frac{3x + x}{2} \cos \frac{3x - x}{2}} \qquad \text{Sum-to-Product Formulas}$$

$$= \frac{2 \cos 2x \sin x}{2 \cos 2x \cos x} \qquad \text{Simplify}$$

$$= \frac{\sin x}{\cos x} = \tan x = \text{RHS} \qquad \text{Cancel}$$

NOW TRY EXERCISE **89** ■

7.3 EXERCISES

CONCEPTS

1. If we know the values of $\sin x$ and $\cos x$, we can find the value of $\sin 2x$ by using the ________ Formula for Sine. State the formula: $\sin 2x =$ ________________.

2. If we know the value of $\cos x$ and the quadrant in which $x/2$ lies, we can find the value of $\sin(x/2)$ by using the ________ Formula for Sine. State the formula: $\sin(x/2) =$ ________________.

SKILLS

3–10 ■ Find $\sin 2x$, $\cos 2x$, and $\tan 2x$ from the given information.

3. $\sin x = \frac{5}{13}$, x in Quadrant I

4. $\tan x = -\frac{4}{3}$, x in Quadrant II

5. $\cos x = \frac{4}{5}$, $\csc x < 0$

6. $\csc x = 4$, $\tan x < 0$

7. $\sin x = -\frac{3}{5}$, x in Quadrant III

8. $\sec x = 2$, x in Quadrant IV

9. $\tan x = -\frac{1}{3}$, $\cos x > 0$

10. $\cot x = \frac{2}{3}$, $\sin x > 0$

11–16 ■ Use the formulas for lowering powers to rewrite the expression in terms of the first power of cosine, as in Example 4.

11. $\sin^4 x$

12. $\cos^4 x$

13. $\cos^2 x \sin^4 x$

14. $\cos^4 x \sin^2 x$

15. $\cos^4 x \sin^4 x$

16. $\cos^6 x$

17–28 ■ Use an appropriate Half-Angle Formula to find the exact value of the expression.

17. $\sin 15°$

18. $\tan 15°$

19. $\tan 22.5°$

20. $\sin 75°$

21. $\cos 165°$

22. $\cos 112.5°$

23. $\tan \dfrac{\pi}{8}$ **24.** $\cos \dfrac{3\pi}{8}$

25. $\cos \dfrac{\pi}{12}$ **26.** $\tan \dfrac{5\pi}{12}$

27. $\sin \dfrac{9\pi}{8}$ **28.** $\sin \dfrac{11\pi}{12}$

29–34 ■ Simplify the expression by using a Double-Angle Formula or a Half-Angle Formula.

29. **(a)** $2 \sin 18° \cos 18°$ **(b)** $2 \sin 3\theta \cos 3\theta$

30. **(a)** $\dfrac{2 \tan 7°}{1 - \tan^2 7°}$ **(b)** $\dfrac{2 \tan 7\theta}{1 - \tan^2 7\theta}$

31. **(a)** $\cos^2 34° - \sin^2 34°$ **(b)** $\cos^2 5\theta - \sin^2 5\theta$

32. **(a)** $\cos^2 \dfrac{\theta}{2} - \sin^2 \dfrac{\theta}{2}$ **(b)** $2 \sin \dfrac{\theta}{2} \cos \dfrac{\theta}{2}$

33. **(a)** $\dfrac{\sin 8°}{1 + \cos 8°}$ **(b)** $\dfrac{1 - \cos 4\theta}{\sin 4\theta}$

34. **(a)** $\sqrt{\dfrac{1 - \cos 30°}{2}}$ **(b)** $\sqrt{\dfrac{1 - \cos 8\theta}{2}}$

35. Use the Addition Formula for Sine to prove the Double-Angle Formula for Sine.

36. Use the Addition Formula for Tangent to prove the Double-Angle Formula for Tangent.

37–42 ■ Find $\sin \dfrac{x}{2}$, $\cos \dfrac{x}{2}$, and $\tan \dfrac{x}{2}$ from the given information.

37. $\sin x = \frac{3}{5}$, $0° < x < 90°$

38. $\cos x = -\frac{4}{5}$, $180° < x < 270°$

39. $\csc x = 3$, $90° < x < 180°$

40. $\tan x = 1$, $0° < x < 90°$

41. $\sec x = \frac{3}{2}$, $270° < x < 360°$

42. $\cot x = 5$, $180° < x < 270°$

43–46 ■ Write the given expression as an algebraic expression in x.

43. $\sin(2 \tan^{-1} x)$ **44.** $\tan(2 \cos^{-1} x)$

45. $\sin(\frac{1}{2} \cos^{-1} x)$ **46.** $\cos(2 \sin^{-1} x)$

47–50 ■ Find the exact value of the given expression.

47. $\sin(2 \cos^{-1} \frac{7}{25})$ **48.** $\cos(2 \tan^{-1} \frac{12}{5})$

49. $\sec(2 \sin^{-1} \frac{1}{4})$ **50.** $\tan(\frac{1}{2} \cos^{-1} \frac{2}{3})$

51–54 ■ Evaluate each expression under the given conditions.

51. $\cos 2\theta$; $\sin \theta = -\frac{3}{5}$, θ in Quadrant III

52. $\sin(\theta/2)$; $\tan \theta = -\frac{5}{12}$, θ in Quadrant IV

53. $\sin 2\theta$; $\sin \theta = \frac{1}{7}$, θ in Quadrant II

54. $\tan 2\theta$; $\cos \theta = \frac{3}{5}$, θ in Quadrant I

55–60 ■ Write the product as a sum.

55. $\sin 2x \cos 3x$ **56.** $\sin x \sin 5x$

57. $\cos x \sin 4x$ **58.** $\cos 5x \cos 3x$

59. $3 \cos 4x \cos 7x$ **60.** $11 \sin \dfrac{x}{2} \cos \dfrac{x}{4}$

61–66 ■ Write the sum as a product.

61. $\sin 5x + \sin 3x$ **62.** $\sin x - \sin 4x$

63. $\cos 4x - \cos 6x$ **64.** $\cos 9x + \cos 2x$

65. $\sin 2x - \sin 7x$ **66.** $\sin 3x + \sin 4x$

67–72 ■ Find the value of the product or sum.

67. $2 \sin 52.5° \sin 97.5°$ **68.** $3 \cos 37.5° \cos 7.5°$

69. $\cos 37.5° \sin 7.5°$ **70.** $\sin 75° + \sin 15°$

71. $\cos 255° - \cos 195°$

72. $\cos \dfrac{\pi}{12} + \cos \dfrac{5\pi}{12}$

73–90 ■ Prove the identity.

73. $\cos^2 5x - \sin^2 5x = \cos 10x$

74. $\sin 8x = 2 \sin 4x \cos 4x$

75. $(\sin x + \cos x)^2 = 1 + \sin 2x$

76. $\dfrac{2 \tan x}{1 + \tan^2 x} = \sin 2x$

77. $\dfrac{\sin 4x}{\sin x} = 4 \cos x \cos 2x$

78. $\dfrac{1 + \sin 2x}{\sin 2x} = 1 + \frac{1}{2} \sec x \csc x$

79. $\dfrac{2(\tan x - \cot x)}{\tan^2 x - \cot^2 x} = \sin 2x$

80. $\cot 2x = \dfrac{1 - \tan^2 x}{2 \tan x}$

81. $\tan 3x = \dfrac{3 \tan x - \tan^3 x}{1 - 3 \tan^2 x}$

82. $4(\sin^6 x + \cos^6 x) = 4 - 3 \sin^2 2x$

83. $\cos^4 x - \sin^4 x = \cos 2x$

84. $\tan^2\left(\dfrac{x}{2} + \dfrac{\pi}{4}\right) = \dfrac{1 + \sin x}{1 - \sin x}$

85. $\dfrac{\sin x + \sin 5x}{\cos x + \cos 5x} = \tan 3x$

86. $\dfrac{\sin 3x + \sin 7x}{\cos 3x - \cos 7x} = \cot 2x$

87. $\dfrac{\sin 10x}{\sin 9x + \sin x} = \dfrac{\cos 5x}{\cos 4x}$

88. $\dfrac{\sin x + \sin 3x + \sin 5x}{\cos x + \cos 3x + \cos 5x} = \tan 3x$

89. $\dfrac{\sin x + \sin y}{\cos x + \cos y} = \tan\left(\dfrac{x + y}{2}\right)$

90. $\tan y = \dfrac{\sin(x + y) - \sin(x - y)}{\cos(x + y) + \cos(x - y)}$

91. Show that $\sin 130° - \sin 110° = -\sin 10°$.

92. Show that $\cos 100° - \cos 200° = \sin 50°$.

93. Show that $\sin 45° + \sin 15° = \sin 75°$.

94. Show that $\cos 87° + \cos 33° = \sin 63°$.

95. Prove the identity

$$\frac{\sin x + \sin 2x + \sin 3x + \sin 4x + \sin 5x}{\cos x + \cos 2x + \cos 3x + \cos 4x + \cos 5x} = \tan 3x$$

96. Use the identity

$$\sin 2x = 2 \sin x \cos x$$

n times to show that

$$\sin(2^n x) = 2^n \sin x \cos x \cos 2x \cos 4x \cdots \cos 2^{n-1} x$$

97. (a) Graph $f(x) = \dfrac{\sin 3x}{\sin x} - \dfrac{\cos 3x}{\cos x}$ and make a conjecture.

(b) Prove the conjecture you made in part (a).

98. (a) Graph $f(x) = \cos 2x + 2 \sin^2 x$ and make a conjecture.

(b) Prove the conjecture you made in part (a).

99. Let $f(x) = \sin 6x + \sin 7x$.

(a) Graph $y = f(x)$.

(b) Verify that $f(x) = 2 \cos \frac{1}{2}x \sin \frac{13}{2}x$.

(c) Graph $y = 2 \cos \frac{1}{2}x$ and $y = -2 \cos \frac{1}{2}x$, together with the graph in part (a), in the same viewing rectangle. How are these graphs related to the graph of f?

100. Let $3x = \pi/3$ and let $y = \cos x$. Use the result of Example 2 to show that y satisfies the equation

$$8y^3 - 6y - 1 = 0$$

NOTE This equation has roots of a certain kind that are used to show that the angle $\pi/3$ cannot be trisected by using a ruler and compass only.

101. (a) Show that there is a polynomial $P(t)$ of degree 4 such that $\cos 4x = P(\cos x)$ (see Example 2).

(b) Show that there is a polynomial $Q(t)$ of degree 5 such that $\cos 5x = Q(\cos x)$.

NOTE In general, there is a polynomial $P_n(t)$ of degree n such that $\cos nx = P_n(\cos x)$. These polynomials are called *Tchebycheff polynomials*, after the Russian mathematician P. L. Tchebycheff (1821–1894).

102. In triangle ABC (see the figure) the line segment s bisects angle C. Show that the length of s is given by

$$s = \frac{2ab \cos x}{a + b}$$

[*Hint:* Use the Law of Sines.]

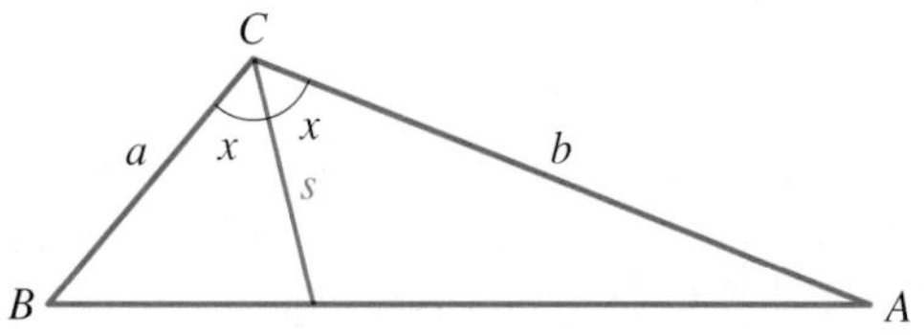

103. If A, B, and C are the angles in a triangle, show that

$$\sin 2A + \sin 2B + \sin 2C = 4 \sin A \sin B \sin C$$

104. A rectangle is to be inscribed in a semicircle of radius 5 cm as shown in the following figure.

(a) Show that the area of the rectangle is modeled by the function

$$A(\theta) = 25 \sin 2\theta$$

(b) Find the largest possible area for such an inscribed rectangle.

(c) Find the dimensions of the inscribed rectangle with the largest possible area.

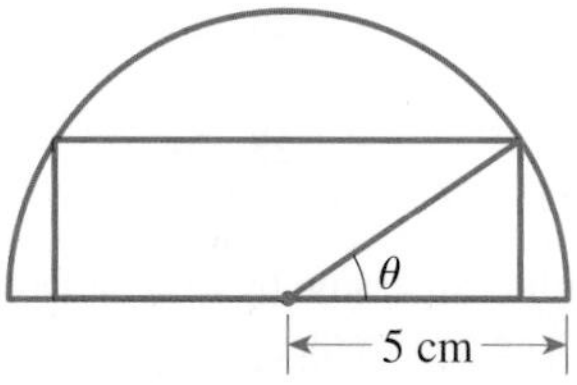

APPLICATIONS

105. Sawing a Wooden Beam A rectangular beam is to be cut from a cylindrical log of diameter 20 in.

(a) Show that the cross-sectional area of the beam is modeled by the function

$$A(\theta) = 200 \sin 2\theta$$

where θ is as shown in the figure.

(b) Show that the maximum cross-sectional area of such a beam is 200 in^2. [*Hint:* Use the fact that $\sin u$ achieves its maximum value at $u = \pi/2$.]

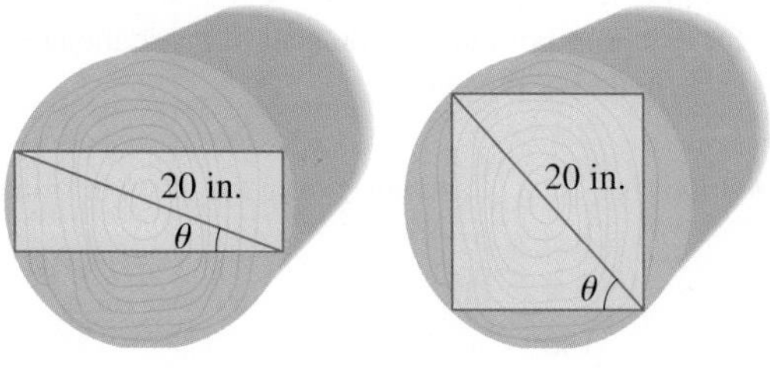

106. Length of a Fold The lower right-hand corner of a long piece of paper 6 in. wide is folded over to the left-hand edge as shown. The length L of the fold depends on the angle θ. Show that

$$L = \frac{3}{\sin \theta \cos^2 \theta}$$

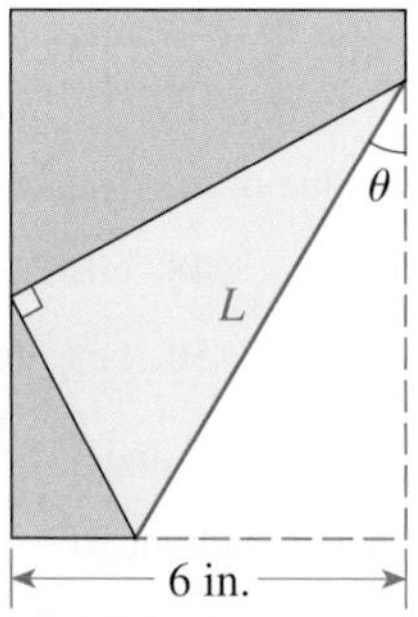

107. Sound Beats When two pure notes that are close in frequency are played together, their sounds interfere to produce *beats*; that is, the loudness (or amplitude) of the sound alternately increases and decreases. If the two notes are given by

$$f_1(t) = \cos 11t \qquad \text{and} \qquad f_2(t) = \cos 13t$$

the resulting sound is $f(t) = f_1(t) + f_2(t)$.

(a) Graph the function $y = f(t)$.

(b) Verify that $f(t) = 2 \cos t \cos 12t$.

(c) Graph $y = 2 \cos t$ and $y = -2 \cos t$, together with the graph in part (a), in the same viewing rectangle. How do these graphs describe the variation in the loudness of the sound?

108. Touch-Tone Telephones When a key is pressed on a touch-tone telephone, the keypad generates two pure tones, which combine to produce a sound that uniquely identifies the key. The figure shows the low frequency f_1 and the high frequency f_2 associated with each key. Pressing a key produces the sound wave $y = \sin(2\pi f_1 t) + \sin(2\pi f_2 t)$.

(a) Find the function that models the sound produced when the 4 key is pressed.

(b) Use a Sum-to-Product Formula to express the sound generated by the 4 key as a product of a sine and a cosine function.

(c) Graph the sound wave generated by the 4 key, from $t = 0$ to $t = 0.006$ s.

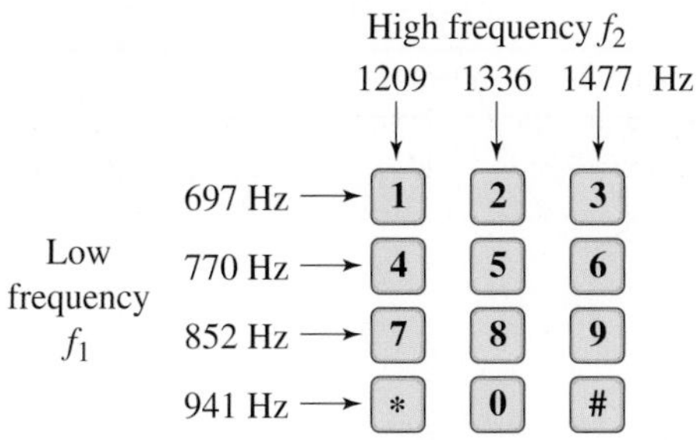

DISCOVERY ■ DISCUSSION ■ WRITING

109. Geometric Proof of a Double-Angle Formula Use the figure to prove that $\sin 2\theta = 2 \sin \theta \cos \theta$.

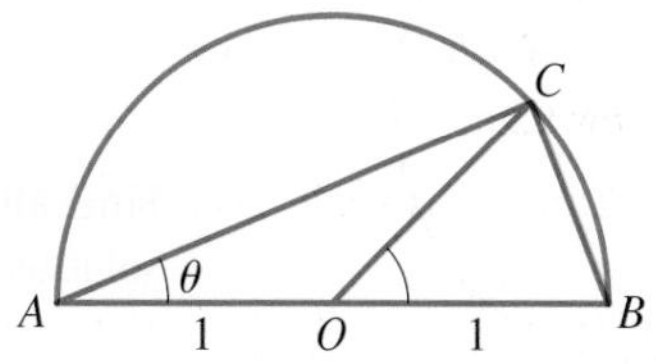

[*Hint:* Find the area of triangle ABC in two different ways. You will need the following facts from geometry:

An angle inscribed in a semicircle is a right angle, so $\angle ACB$ is a right angle.

The central angle subtended by the chord of a circle is twice the angle subtended by the chord on the circle, so $\angle BOC$ is 2θ.]

DISCOVERY PROJECT **Where to Sit at the Movies**

In this project we use trigonometry to find the best location to observe such things as a painting or a movie. You can find the project at the book companion website: **www.stewartmath.com**

7.4 Basic Trigonometric Equations

Basic Trigonometric Equations ▶ Solving Trigonometric Equations by Factoring

An equation that contains trigonometric functions is called a **trigonometric equation**. For example, the following are trigonometric equations:

$$\sin^2\theta + \cos^2\theta = 1 \qquad 2 \sin \theta - 1 = 0 \qquad \tan 2\theta - 1 = 0$$

The first equation is an *identity*—that is, it is true for every value of the variable θ. The other two equations are true only for certain values of θ. To solve a trigonometric equation, we find all the values of the variable that make the equation true.

▼ Basic Trigonometric Equations

Solving any trigonometric equation always reduces to solving a **basic trigonometric equation**—an equation of the form $T(\theta) = c$, where T is a trigonometric function and c is a constant. In the next three examples we solve such basic equations.

EXAMPLE 1 | Solving a Basic Trigonometric Equation

Solve the equation $\sin \theta = \dfrac{1}{2}$.

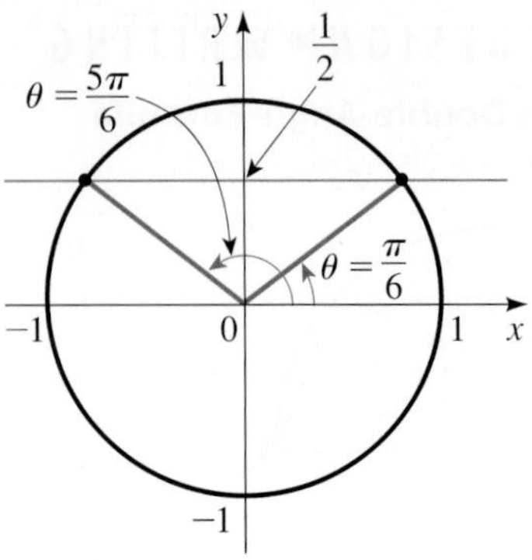

FIGURE 1

SOLUTION

Find the solutions in one period. Because sine has period 2π, we first find the solutions in any interval of length 2π. To find these solutions, we look at the unit circle in Figure 1. We see that $\sin\theta = \frac{1}{2}$ in Quadrants I and II, so the solutions in the interval $[0, 2\pi)$ are

$$\theta = \frac{\pi}{6} \qquad \theta = \frac{5\pi}{6}$$

Find all solutions. Because the sine function repeats its values every 2π units, we get all solutions of the equation by adding integer multiples of 2π to these solutions:

$$\theta = \frac{\pi}{6} + 2k\pi \qquad \theta = \frac{5\pi}{6} + 2k\pi$$

where k is any integer. Figure 2 gives a graphical representation of the solutions.

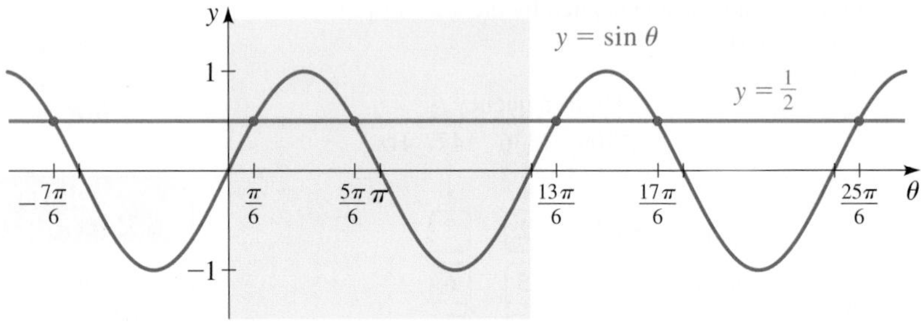

FIGURE 2

NOW TRY EXERCISE **5**

EXAMPLE 2 | Solving a Basic Trigonometric Equation

Solve the equation $\cos\theta = -\frac{\sqrt{2}}{2}$, and list eight specific solutions.

SOLUTION

Find the solutions in one period. Because cosine has period 2π, we first find the solutions in any interval of length 2π. From the unit circle in Figure 3 we see that $\cos\theta = -\sqrt{2}/2$ in Quadrants II and III, so the solutions in the interval $[0, 2\pi)$ are

$$\theta = \frac{3\pi}{4} \qquad \theta = \frac{5\pi}{4}$$

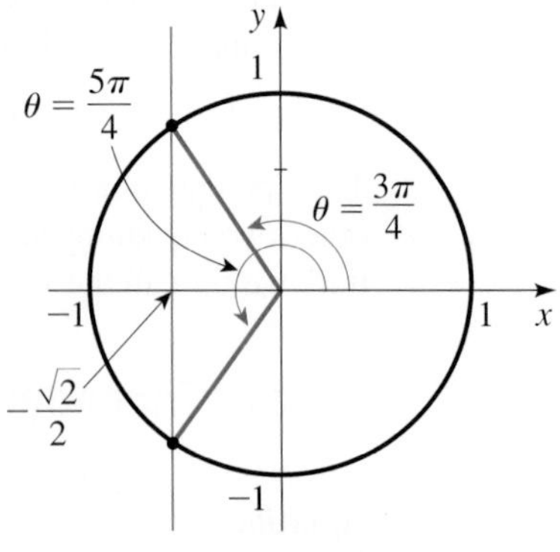

FIGURE 3

Find all solutions. Because the cosine function repeats its values every 2π units, we get all solutions of the equation by adding integer multiples of 2π to these solutions:

$$\theta = \frac{3\pi}{4} + 2k\pi \qquad \theta = \frac{5\pi}{4} + 2k\pi$$

where k is any integer. You can check that for $k = -1, 0, 1, 2$ we get the following specific solutions:

$$\theta = \underbrace{-\frac{5\pi}{4}, -\frac{3\pi}{4}}_{k=-1}, \underbrace{\frac{3\pi}{4}, \frac{5\pi}{4}}_{k=0}, \underbrace{\frac{11\pi}{4}, \frac{13\pi}{4}}_{k=1}, \underbrace{\frac{19\pi}{4}, \frac{21\pi}{4}}_{k=2}$$

Figure 4 gives a graphical representation of the solutions.

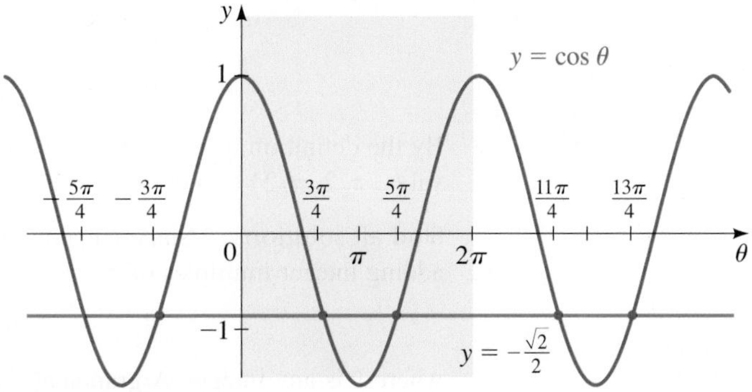

FIGURE 4

NOW TRY EXERCISE **17**

EXAMPLE 3 | Solving a Basic Trigonometric Equation

Solve the equation $\cos\theta = 0.65$.

SOLUTION

Find the solutions in one period. We first find one solution by taking $\cos^{-1}$ of each side of the equation.

$\cos\theta = 0.65$	Given equation
$\theta = \cos^{-1}(0.65)$	Take $\cos^{-1}$ of each side
$\theta \approx 0.86$	Calculator (in radian mode)

Because cosine has period 2π, we next find the solutions in any interval of length 2π. To find these solutions, we look at the unit circle in Figure 5. We see that $\cos\theta = 0.85$ in Quadrants I and IV, so the solutions are

$$\theta \approx 0.86 \qquad \theta \approx 2\pi - 0.86 \approx 5.42$$

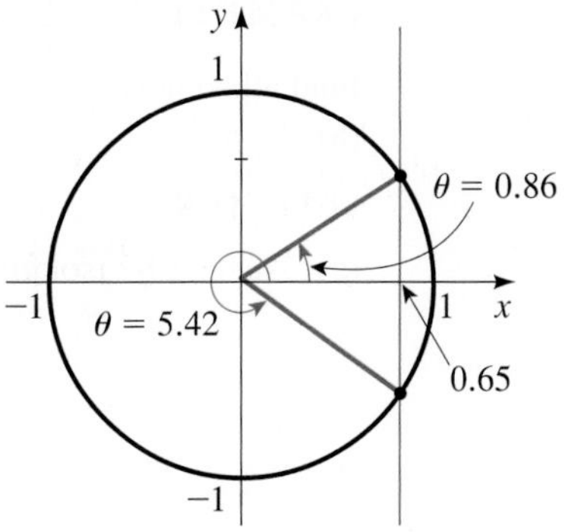

FIGURE 5

Find all solutions. To get all solutions of the equation, we add integer multiples of 2π to these solutions:

$$\theta \approx 0.86 + 2k\pi \qquad \theta \approx 5.42 + 2k\pi$$

where k is any integer.

NOW TRY EXERCISE **21**

EXAMPLE 4 | Solving a Basic Trigonometric Equation

Solve the equation $\tan\theta = 2$.

SOLUTION

Find the solutions in one period. We first find one solution by taking $\tan^{-1}$ of each side of the equation:

$$\begin{aligned} \tan\theta &= 2 && \text{Given equation} \\ \theta &= \tan^{-1}(2) && \text{Take } \tan^{-1} \text{ of each side} \\ \theta &\approx 1.12 && \text{Calculator (in radian mode)} \end{aligned}$$

By the definition of $\tan^{-1}$, the solution that we obtained is the only solution in the interval $(-\pi/2, \pi/2)$ (which is an interval of length π).

Find all solutions. Since tangent has period π, we get all solutions of the equation by adding integer multiples of π:

$$\theta \approx 1.12 + k\pi$$

where k is any integer. A graphical representation of the solutions is shown in Figure 6. You can check that the solutions shown in the graph correspond to $k = -1, 0, 1, 2, 3$.

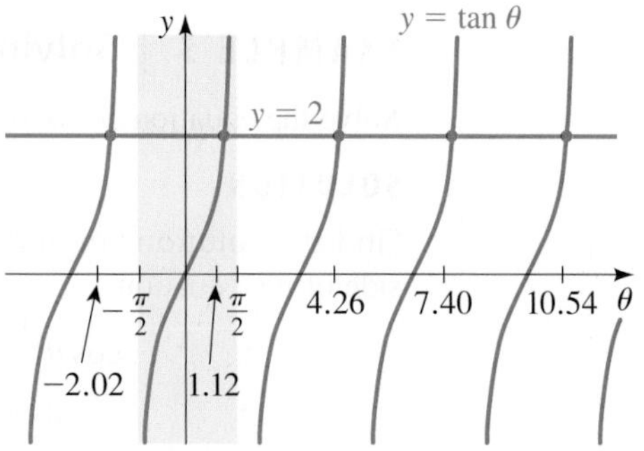

FIGURE 6

NOW TRY EXERCISE **23**

In the next example we solve trigonometric equations that are algebraically equivalent to basic trigonometric equations.

EXAMPLE 5 | Solving Trigonometric Equations

Find all solutions of the equation.

(a) $2\sin\theta - 1 = 0$ **(b)** $\tan^2\theta - 3 = 0$

SOLUTION

(a) We start by isolating $\sin\theta$:

$$\begin{aligned} 2\sin\theta - 1 &= 0 && \text{Given equation} \\ 2\sin\theta &= 1 && \text{Add 1} \\ \sin\theta &= \frac{1}{2} && \text{Divide by 2} \end{aligned}$$

This last equation is the same as that in Example 1. The solutions are

$$\theta = \frac{\pi}{6} + 2k\pi \qquad \theta = \frac{5\pi}{6} + 2k\pi$$

where k is any integer.

(b) We start by isolating $\tan\theta$:

$$\tan^2\theta - 3 = 0 \qquad \text{Given equation}$$

$$\tan^2\theta = 3 \qquad \text{Add 3}$$

$$\tan\theta = \pm\sqrt{3} \qquad \text{Take the square root}$$

Because tangent has period π, we first find the solutions in any interval of length π. In the interval $(-\pi/2, \pi/2)$ the solutions are $\theta = \pi/3$ and $\theta = -\pi/3$. To get all solutions, we add integer multiples of π to these solutions:

$$\theta = \frac{\pi}{3} + k\pi \qquad \theta = -\frac{\pi}{3} + k\pi$$

where k is any integer.

NOW TRY EXERCISES **27** AND **33**

Solving Trigonometric Equations by Factoring

Zero-Product Property
If $AB = 0$, then $A = 0$ or $B = 0$.

Factoring is one of the most useful techniques for solving equations, including trigonometric equations. The idea is to move all terms to one side of the equation, factor, and then use the Zero-Product Property (see Section 1.5).

EXAMPLE 6 | A Trigonometric Equation of Quadratic Type

Solve the equation $2\cos^2\theta - 7\cos\theta + 3 = 0$.

Equation of Quadratic Type

$$2C^2 - 7C + 3 = 0$$

$$(2C - 1)(C - 3) = 0$$

SOLUTION We factor the left-hand side of the equation.

$$2\cos^2\theta - 7\cos\theta + 3 = 0 \qquad \text{Given equation}$$

$$(2\cos\theta - 1)(\cos\theta - 3) = 0 \qquad \text{Factor}$$

$$2\cos\theta - 1 = 0 \quad \text{or} \quad \cos\theta - 3 = 0 \qquad \text{Set each factor equal to 0}$$

$$\cos\theta = \frac{1}{2} \quad \text{or} \quad \cos\theta = 3 \qquad \text{Solve for } \cos\theta$$

Because cosine has period 2π, we first find the solutions in the interval $[0, 2\pi)$. For the first equation the solutions are $\theta = \pi/3$ and $\theta = 5\pi/3$ (see Figure 7). The second equation has no solution because $\cos\theta$ is never greater than 1. Thus the solutions are

$$\theta = \frac{\pi}{3} + 2k\pi \qquad \theta = \frac{5\pi}{3} + 2k\pi$$

where k is any integer.

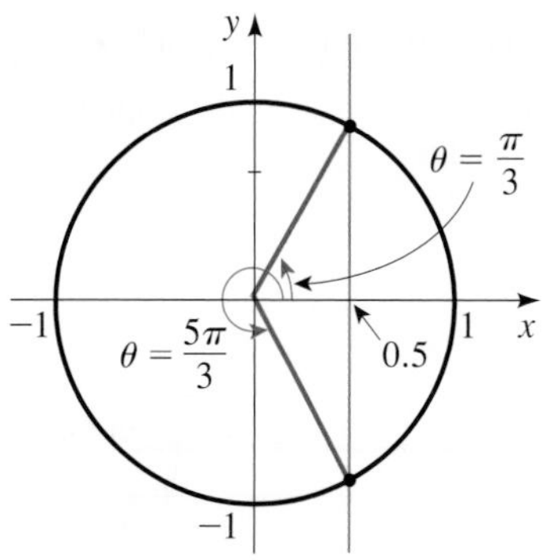

FIGURE 7

NOW TRY EXERCISE **41**

EXAMPLE 7 | Solving a Trigonometric Equation by Factoring

Solve the equation $5 \sin \theta \cos \theta + 4 \cos \theta = 0$.

SOLUTION We factor the left-hand side of the equation:

$$5 \sin \theta \cos \theta + 2 \cos \theta = 0 \qquad \text{Given equation}$$

$$\cos \theta(5 \sin \theta + 2) = 0 \qquad \text{Factor}$$

$$\cos \theta = 0 \quad \text{or} \quad 5 \sin \theta + 4 = 0 \qquad \text{Set each factor equal to 0}$$

$$\sin \theta = -0.8 \qquad \text{Solve for } \sin \theta$$

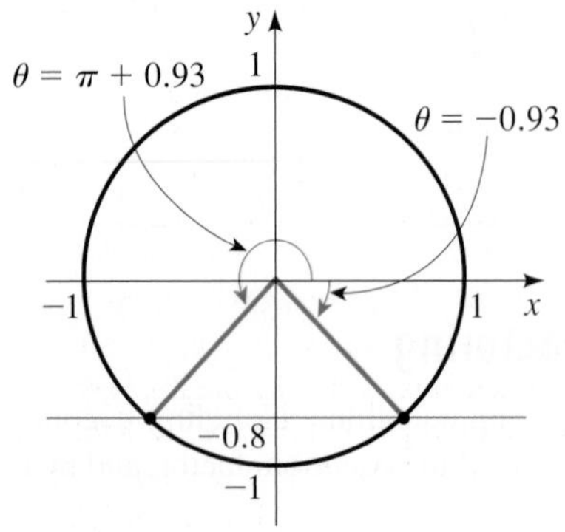

FIGURE 8

Because sine and cosine have period 2π, we first find the solutions of these equations in an interval of length 2π. For the first equation the solutions in the interval $[0, 2\pi)$ are $\theta = \pi/2$ and $\theta = 3\pi/2$. To solve the second equation, we take $\sin^{-1}$ of each side:

$$\sin \theta = -0.80 \qquad \text{Second equation}$$

$$\theta = \sin^{-1}(-0.80) \qquad \text{Take } \sin^{-1} \text{ of each side}$$

$$\theta \approx -0.93 \qquad \text{Calculator (in radian mode)}$$

So the solutions in an interval of length 2π are $\theta = -0.93$ and $\theta = \pi + 0.93 \approx 4.07$ (see Figure 8). We get all the solutions of the equation by adding integer multiples of 2π to these solutions.

$$\theta = \frac{\pi}{2} + 2k\pi, \qquad \theta = \frac{3\pi}{2} + 2k\pi, \qquad \theta \approx -0.93 + 2k\pi, \qquad \theta \approx 4.07 + 2k\pi$$

where k is any integer.

NOW TRY EXERCISE **53**

7.4 EXERCISES

CONCEPTS

1. Because the trigonometric functions are periodic, if a basic trigonometric equation has one solution, it has ________ (several/infinitely many) solutions.

2. The basic equation $\sin x = 2$ has ________ (no/one/infinitely many) solutions, whereas the basic equation $\sin x = 0.3$ has ________ (no/one/infinitely many) solutions.

3. We can find some of the solutions of $\sin x = 0.3$ graphically by graphing $y = \sin x$ and $y =$ ________. Use the graph below to estimate some of the solutions.

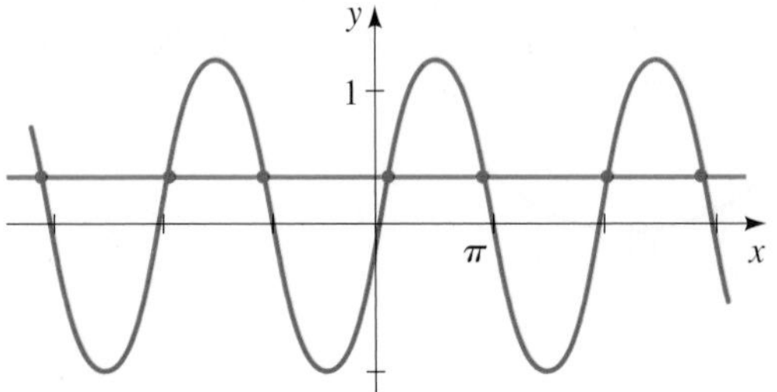

4. We can find the solutions of $\sin x = 0.3$ algebraically.
   (a) First we find the solutions in the interval $[-\pi, \pi]$. We get one such solution by taking $\sin^{-1}$ to get $x =$ ________. The other solution in this interval is $x =$ ________.
   (b) We find all solutions by adding multiples of ________ to the solutions in $[-\pi, \pi]$. The solutions are $x =$ ________ and $x =$ ________.

SKILLS

5–16 ■ Solve the given equation.

5. $\sin \theta = \frac{\sqrt{3}}{2}$

6. $\sin \theta = -\frac{\sqrt{2}}{2}$

7. $\cos \theta = -1$

8. $\cos \theta = \frac{\sqrt{3}}{2}$

9. $\cos \theta = \frac{1}{4}$

10. $\sin \theta = -0.3$

11. $\sin \theta = -0.45$

12. $\cos \theta = 0.32$

13. $\tan \theta = -\sqrt{3}$

14. $\tan \theta = 1$

15. $\tan \theta = 5$

16. $\tan \theta = -\frac{1}{3}$

17–24 ■ Solve the given equation, and list six specific solutions.

17. $\cos\theta = -\frac{\sqrt{3}}{2}$ **18.** $\cos\theta = \frac{1}{2}$

19. $\sin\theta = \frac{\sqrt{2}}{2}$ **20.** $\sin\theta = -\frac{\sqrt{3}}{2}$

21. $\cos\theta = 0.28$ **22.** $\tan\theta = 2.5$

23. $\tan\theta = -10$ **24.** $\sin\theta = -0.9$

25–38 ■ Find all solutions of the given equation.

25. $\cos\theta + 1 = 0$ **26.** $\sin\theta + 1 = 0$

27. $\sqrt{2}\sin\theta + 1 = 0$ **28.** $\sqrt{2}\cos\theta - 1 = 0$

29. $5\sin\theta - 1 = 0$ **30.** $4\cos\theta + 1 = 0$

31. $3\tan^2\theta - 1 = 0$ **32.** $\cot\theta + 1 = 0$

33. $2\cos^2\theta - 1 = 0$ **34.** $4\sin^2\theta - 3 = 0$

35. $\tan^2\theta - 4 = 0$ **36.** $9\sin^2\theta - 1 = 0$

37. $\sec^2\theta - 2 = 0$ **38.** $\csc^2\theta - 4 = 0$

39–56 ■ Solve the given equation.

39. $(\tan^2\theta - 4)(2\cos\theta + 1) = 0$

40. $(\tan\theta - 2)(16\sin^2\theta - 1) = 0$

41. $4\cos^2\theta - 4\cos\theta + 1 = 0$

42. $2\sin^2\theta - \sin\theta - 1 = 0$

43. $3\sin^2\theta - 7\sin\theta + 2 = 0$

44. $\tan^4\theta - 13\tan^2\theta + 36 = 0$

45. $2\cos^2\theta - 7\cos\theta + 3 = 0$

46. $\sin^2\theta - \sin\theta - 2 = 0$

47. $\cos^2\theta - \cos\theta - 6 = 0$ **48.** $2\sin^2\theta + 5\sin\theta - 12 = 0$

49. $\sin^2\theta = 2\sin\theta + 3$ **50.** $3\tan^3\theta = \tan\theta$

51. $\cos\theta(2\sin\theta + 1) = 0$ **52.** $\sec\theta(2\cos\theta - \sqrt{2}) = 0$

53. $\cos\theta\sin\theta - 2\cos\theta = 0$ **54.** $\tan\theta\sin\theta + \sin\theta = 0$

55. $3\tan\theta\sin\theta - 2\tan\theta = 0$ **56.** $4\cos\theta\sin\theta + 3\cos\theta = 0$

APPLICATIONS

57. Refraction of Light It has been observed since ancient times that light refracts or "bends" as it travels from one medium to another (from air to water, for example). If v_1 is the speed of light in one medium and v_2 its speed in another medium, then according to **Snell's Law**,

$$\frac{\sin\theta_1}{\sin\theta_2} = \frac{v_1}{v_2}$$

where θ_1 is the *angle of incidence* and θ_2 is the *angle of refraction* (see the figure). The number v_1/v_2 is called the *index of refraction*. The index of refraction for several substances is given in the table.

If a ray of light passes through the surface of a lake at an angle of incidence of 70°, what is the angle of refraction?

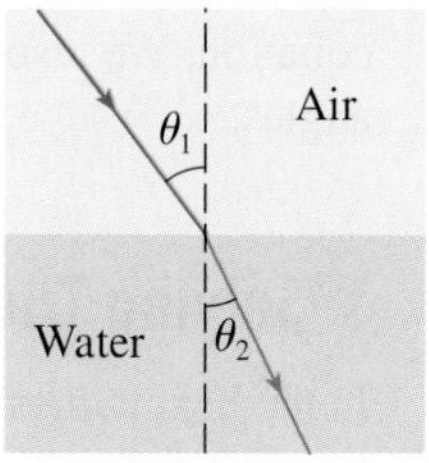

Substance	Refraction from air to substance
Water	1.33
Alcohol	1.36
Glass	1.52
Diamond	2.41

58. Total Internal Reflection When light passes from a more-dense to a less-dense medium—from glass to air, for example—the angle of refraction predicted by Snell's Law (see Exercise 57) can be 90° or larger. In this case the light beam is actually reflected back into the denser medium. This phenomenon, called *total internal reflection*, is the principle behind fiber optics. Set $\theta_2 = 90°$ in Snell's Law, and solve for θ_1 to determine the critical angle of incidence at which total internal reflection begins to occur when light passes from glass to air. (Note that the index of refraction from glass to air is the reciprocal of the index from air to glass.)

59. Phases of the Moon As the moon revolves around the earth, the side that faces the earth is usually just partially illuminated by the sun. The phases of the moon describe how much of the surface appears to be in sunlight. An astronomical measure of phase is given by the fraction F of the lunar disc that is lit. When the angle between the sun, earth, and moon is θ $(0 \le \theta \le 360°)$, then

$$F = \frac{1}{2}(1 - \cos\theta)$$

Determine the angles θ that correspond to the following phases:

(a) $F = 0$ (new moon)
(b) $F = 0.25$ (a crescent moon)
(c) $F = 0.5$ (first or last quarter)
(d) $F = 1$ (full moon)

DISCOVERY ■ DISCUSSION ■ WRITING

60. Equations and Identities Which of the following statements is true?

A. Every identity is an equation.

B. Every equation is an identity.

Give examples to illustrate your answer. Write a short paragraph to explain the difference between an equation and an identity.

7.5 More Trigonometric Equations

Solving Trigonometric Equations by Using Identities ▶ Equations with Trigonometric Functions of Multiples of Angles

In this section we solve trigonometric equations by first using identities to simplify the equation. We also solve trigonometric equations in which the terms contain multiples of angles.

▼ Solving Trigonometric Equations by Using Identities

In the next two examples we use trigonometric identities to express a trigonometric equation in a form in which it can be factored.

EXAMPLE 1 | Using a Trigonometric Identity

Solve the equation $1 + \sin\theta = 2\cos^2\theta$.

SOLUTION We first need to rewrite this equation so that it contains only one trigonometric function. To do this, we use a trigonometric identity:

$$1 + \sin\theta = 2\cos^2\theta \qquad \text{Given equation}$$

$$1 + \sin\theta = 2(1 - \sin^2\theta) \qquad \text{Pythagorean identity}$$

$$2\sin^2\theta + \sin\theta - 1 = 0 \qquad \text{Put all terms on one side}$$

$$(2\sin\theta - 1)(\sin\theta + 1) = 0 \qquad \text{Factor}$$

$$2\sin\theta - 1 = 0 \quad \text{or} \quad \sin\theta + 1 = 0 \qquad \text{Set each factor equal to 0}$$

$$\sin\theta = \frac{1}{2} \quad \text{or} \quad \sin\theta = -1 \qquad \text{Solve for } \sin\theta$$

$$\theta = \frac{\pi}{6}, \frac{5\pi}{6} \quad \text{or} \quad \theta = \frac{3\pi}{2} \qquad \text{Solve for } \theta \text{ in the interval } [0, 2\pi)$$

Because sine has period 2π, we get all the solutions of the equation by adding integer multiples of 2π to these solutions. Thus the solutions are

$$\theta = \frac{\pi}{6} + 2k\pi \qquad \theta = \frac{5\pi}{6} + 2k\pi \qquad \theta = \frac{3\pi}{2} + 2k\pi$$

where k is any integer.

NOW TRY EXERCISE 3 ■

EXAMPLE 2 | Using a Trigonometric Identity

Solve the equation $\sin 2\theta - \cos\theta = 0$.

SOLUTION The first term is a function of 2θ, and the second is a function of θ, so we begin by using a trigonometric identity to rewrite the first term as a function of θ only:

$$\sin 2\theta - \cos\theta = 0 \qquad \text{Given equation}$$

$$2\sin\theta\cos\theta - \cos\theta = 0 \qquad \text{Double-Angle Formula}$$

$$\cos\theta\,(2\sin\theta - 1) = 0 \qquad \text{Factor}$$

$$\cos\theta = 0 \quad \text{or} \quad 2\sin\theta - 1 = 0 \qquad \text{Set each factor equal to 0}$$

$$\sin\theta = \frac{1}{2} \qquad \text{Solve for } \sin\theta$$

$$\theta = \frac{\pi}{2}, \frac{3\pi}{2} \quad \text{or} \quad \theta = \frac{\pi}{6}, \frac{5\pi}{6} \qquad \text{Solve for } \theta \text{ in } [0, 2\pi)$$

Both sine and cosine have period 2π, so we get all the solutions of the equation by adding integer multiples of 2π to these solutions. Thus the solutions are

$$\theta = \frac{\pi}{2} + 2k\pi \qquad \theta = \frac{3\pi}{2} + 2k\pi \qquad \theta = \frac{\pi}{6} + 2k\pi \qquad \theta = \frac{5\pi}{6} + 2k\pi$$

where k is any integer.

NOW TRY EXERCISES **7** AND **11**

EXAMPLE 3 | Squaring and Using an Identity

Solve the equation $\cos\theta + 1 = \sin\theta$ in the interval $[0, 2\pi)$.

SOLUTION To get an equation that involves either sine only or cosine only, we square both sides and use a Pythagorean identity:

$$\cos\theta + 1 = \sin\theta \qquad \text{Given equation}$$

$$\cos^2\theta + 2\cos\theta + 1 = \sin^2\theta \qquad \text{Square both sides}$$

$$\cos^2\theta + 2\cos\theta + 1 = 1 - \cos^2\theta \qquad \text{Pythagorean identity}$$

$$2\cos^2\theta + 2\cos\theta = 0 \qquad \text{Simplify}$$

$$2\cos\theta(\cos\theta + 1) = 0 \qquad \text{Factor}$$

$$2\cos\theta = 0 \quad \text{or} \quad \cos\theta + 1 = 0 \qquad \text{Set each factor equal to 0}$$

$$\cos\theta = 0 \quad \text{or} \quad \cos\theta = -1 \qquad \text{Solve for } \cos\theta$$

$$\theta = \frac{\pi}{2}, \frac{3\pi}{2} \quad \text{or} \quad \theta = \pi \qquad \text{Solve for } \theta \text{ in } [0, 2\pi)$$

Because we squared both sides, we need to check for extraneous solutions. From *Check Your Answers* we see that the solutions of the given equation are $\pi/2$ and π.

CHECK YOUR ANSWER

$\theta = \frac{\pi}{2}$	$\theta = \frac{3\pi}{2}$	$\theta = \pi$
$\cos\frac{\pi}{2} + 1 = \sin\frac{\pi}{2}$	$\cos\frac{3\pi}{2} + 1 = \sin\frac{3\pi}{2}$	$\cos\pi + 1 = \sin\pi$
$0 + 1 = 1$ ✔	$0 + 1 \stackrel{?}{=} -1$ ✗	$-1 + 1 = 0$ ✔

NOW TRY EXERCISE **13**

EXAMPLE 4 | Finding Intersection Points

Find the values of x for which the graphs of $f(x) = \sin x$ and $g(x) = \cos x$ intersect.

SOLUTION 1: Graphical

The graphs intersect where $f(x) = g(x)$. In Figure 1 we graph $y_1 = \sin x$ and $y_2 = \cos x$ on the same screen, for x between 0 and 2π. Using TRACE or the `intersect` command on the graphing calculator, we see that the two points of intersection in this interval occur where $x \approx 0.785$ and $x \approx 3.927$. Since sine and cosine are periodic with period 2π, the intersection points occur where

$$x \approx 0.785 + 2k\pi \qquad \text{and} \qquad x \approx 3.927 + 2k\pi$$

where k is any integer.

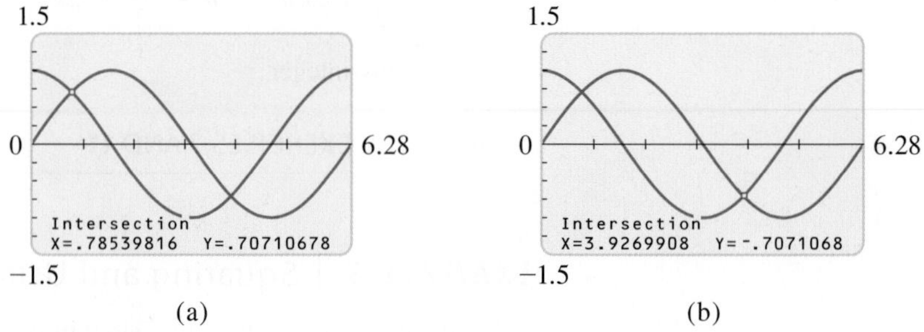

FIGURE 1

SOLUTION 2: Algebraic

To find the exact solution, we set $f(x) = g(x)$ and solve the resulting equation algebraically:

$$\sin x = \cos x \qquad \text{Equate functions}$$

Since the numbers x for which $\cos x = 0$ are not solutions of the equation, we can divide both sides by $\cos x$:

$$\frac{\sin x}{\cos x} = 1 \qquad \text{Divide by } \cos x$$

$$\tan x = 1 \qquad \text{Reciprocal identity}$$

The only solution of this equation in the interval $(-\pi/2, \pi/2)$ is $x = \pi/4$. Since tangent has period π, we get all solutions of the equation by adding integer multiples of π:

$$x = \frac{\pi}{4} + k\pi$$

where k is any integer. The graphs intersect for these values of x. You should use your calculator to check that, rounded to three decimals, these are the same values that we obtained in Solution 1.

NOW TRY EXERCISE **35** ■

Equations with Trigonometric Functions of Multiples of Angles

When solving trigonometric equations that involve functions of multiples of angles, we first solve for the multiple of the angle, then divide to solve for the angle.

EXAMPLE 5 | A Trigonometric Equation Involving a Multiple of an Angle

Consider the equation $2 \sin 3\theta - 1 = 0$.

(a) Find all solutions of the equation.

(b) Find the solutions in the interval $[0, 2\pi)$.

SOLUTION

(a) We first isolate $\sin 3\theta$ and then solve for the angle 3θ.

$$2\sin 3\theta - 1 = 0 \qquad \text{Given equation}$$

$$2\sin 3\theta = 1 \qquad \text{Add 1}$$

$$\sin 3\theta = \frac{1}{2} \qquad \text{Divide by 2}$$

$$3\theta = \frac{\pi}{6}, \frac{5\pi}{6} \qquad \text{Solve for } 3\theta \text{ in the interval } [0, 2\pi) \text{ (see Figure 2)}$$

To get all solutions, we add integer multiples of 2π to these solutions. So the solutions are of the form

$$3\theta = \frac{\pi}{6} + 2k\pi \qquad 3\theta = \frac{5\pi}{6} + 2k\pi$$

To solve for θ, we divide by 3 to get the solutions

$$\theta = \frac{\pi}{18} + \frac{2k\pi}{3} \qquad \theta = \frac{5\pi}{18} + \frac{2k\pi}{3}$$

where k is any integer.

(b) The solutions from part (a) that are in the interval $[0, 2\pi)$ correspond to $k = 0, 1$, and 2. For all other values of k the corresponding values of θ lie outside this interval. So the solutions in the interval $[0, 2\pi)$ are

$$\theta = \underbrace{\frac{\pi}{18}, \frac{5\pi}{18}}_{k=0}, \underbrace{\frac{13\pi}{18}, \frac{17\pi}{18}}_{k=1}, \underbrace{\frac{25\pi}{18}, \frac{29\pi}{18}}_{k=2}$$

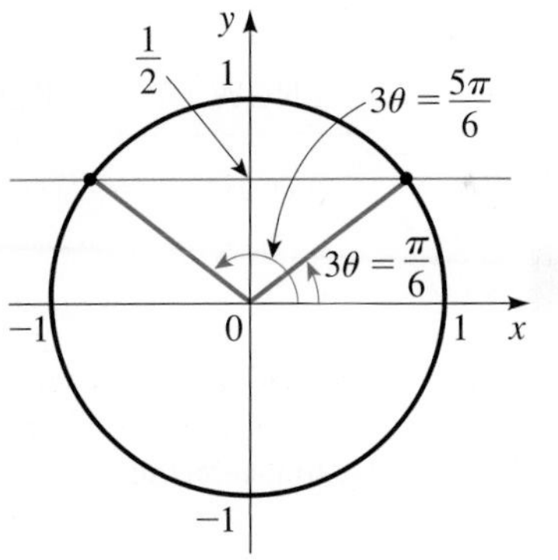

FIGURE 2

NOW TRY EXERCISE **17**

EXAMPLE 6 | A Trigonometric Equation Involving a Half Angle

Consider the equation $\sqrt{3}\tan\frac{\theta}{2} - 1 = 0$.

(a) Find all solutions of the equation.

(b) Find the solutions in the interval $[0, 4\pi)$.

SOLUTION

(a) We start by isolating $\tan \frac{\theta}{2}$:

$$\sqrt{3} \tan \frac{\theta}{2} - 1 = 0 \qquad \text{Given equation}$$

$$\sqrt{3} \tan \frac{\theta}{2} = 1 \qquad \text{Add 1}$$

$$\tan \frac{\theta}{2} = \frac{1}{\sqrt{3}} \qquad \text{Divide by } \sqrt{3}$$

$$\frac{\theta}{2} = \frac{\pi}{6} \qquad \text{Solve for } \frac{\theta}{2} \text{ in the interval } \left(-\frac{\pi}{2}, \frac{\pi}{2}\right)$$

Since tangent has period π, to get all solutions we add integer multiples of π to this solution. So the solutions are of the form

$$\frac{\theta}{2} = \frac{\pi}{6} + k\pi$$

Multiplying by 2, we get the solutions

$$\theta = \frac{\pi}{3} + 2k\pi$$

where k is any integer.

(b) The solutions from part (a) that are in the interval $[0, 4\pi)$ correspond to $k = 0$ and $k = 1$. For all other values of k the corresponding values of x lie outside this interval. Thus the solutions in the interval $[0, 4\pi)$ are

$$x = \frac{\pi}{3}, \frac{7\pi}{3}$$

NOW TRY EXERCISE **23** ■

7.5 EXERCISES

CONCEPTS

1–2 ■ We can use identities to help us solve trigonometric equations.

1. Using a Pythagorean identity we see that the equation $\sin x + \sin^2 x + \cos^2 x = 1$ is equivalent to the basic equation ________________ whose solutions are $x =$ ________.

2. Using a Double-Angle Formula we see that the equation $\sin x + \sin 2x = 0$ is equivalent to the equation ________. Factoring we see that solving this equation is equivalent to solving the two basic equations ________ and ________.

SKILLS

3–16 ■ Solve the given equation.

3. $2\cos^2\theta + \sin\theta = 1$

4. $\sin^2\theta = 4 - 2\cos^2\theta$

5. $\tan^2\theta - 2\sec\theta = 2$

6. $\csc^2\theta = \cot\theta + 3$

7. $2\sin 2\theta - 3\sin\theta = 0$

8. $3\sin 2\theta - 2\sin\theta = 0$

9. $\cos 2\theta = 3\sin\theta - 1$

10. $\cos 2\theta = \cos^2\theta - \frac{1}{2}$

11. $2\sin^2\theta - \cos\theta = 1$

12. $\tan\theta - 3\cot\theta = 0$

13. $\sin\theta - 1 = \cos\theta$

14. $\cos\theta - \sin\theta = 1$

15. $\tan\theta + 1 = \sec\theta$

16. $2\tan\theta + \sec^2\theta = 4$

17–34 ■ An equation is given. **(a)** Find all solutions of the equation. **(b)** Find the solutions in the interval $[0, 2\pi)$.

17. $2\cos 3\theta = 1$

18. $3\csc^2\theta = 4$

19. $2\cos 2\theta + 1 = 0$

20. $2\sin 3\theta + 1 = 0$

21. $\sqrt{3}\tan 3\theta + 1 = 0$

22. $\sec 4\theta - 2 = 0$

23. $\cos\frac{\theta}{2} - 1 = 0$

24. $\tan\frac{\theta}{4} + \sqrt{3} = 0$

25. $2\sin\frac{\theta}{3} + \sqrt{3} = 0$

26. $\sec\frac{\theta}{2} = \cos\frac{\theta}{2}$

27. $\sin 2\theta = 3 \cos 2\theta$

28. $\csc 3\theta = 5 \sin 3\theta$

29. $\sec \theta - \tan \theta = \cos \theta$

30. $\tan 3\theta + 1 = \sec 3\theta$

31. $3 \tan^3\theta - 3 \tan^2\theta - \tan \theta + 1 = 0$

32. $4 \sin \theta \cos \theta + 2 \sin \theta - 2 \cos \theta - 1 = 0$

33. $2 \sin \theta \tan \theta - \tan \theta = 1 - 2 \sin \theta$

34. $\sec \theta \tan \theta - \cos \theta \cot \theta = \sin \theta$

35–38 ■ **(a)** Graph f and g in the given viewing rectangle and find the intersection points graphically, rounded to two decimal places. **(b)** Find the intersection points of f and g algebraically. Give exact answers.

35. $f(x) = 3 \cos x + 1,\ g(x) = \cos x - 1$; $[-2\pi, 2\pi]$ by $[-2.5, 4.5]$

36. $f(x) = \sin 2x + 1,\ g(x) = 2 \sin 2x + 1$; $[-2\pi, 2\pi]$ by $[-1.5, 3.5]$

37. $f(x) = \tan x,\ g(x) = \sqrt{3}$; $\left[-\frac{\pi}{2}, \frac{\pi}{2}\right]$ by $[-10, 10]$

38. $f(x) = \sin x - 1,\ g(x) = \cos x$; $[-2\pi, 2\pi]$ by $[-2.5, 1.5]$

39–42 ■ Use an Addition or Subtraction Formula to simplify the equation. Then find all solutions in the interval $[0, 2\pi)$.

39. $\cos \theta \cos 3\theta - \sin \theta \sin 3\theta = 0$

40. $\cos \theta \cos 2\theta + \sin \theta \sin 2\theta = \frac{1}{2}$

41. $\sin 2\theta \cos \theta - \cos 2\theta \sin \theta = \sqrt{3}/2$

42. $\sin 3\theta \cos \theta - \cos 3\theta \sin \theta = 0$

43–52 ■ Use a Double- or Half-Angle Formula to solve the equation in the interval $[0, 2\pi)$.

43. $\sin 2\theta + \cos \theta = 0$

44. $\tan \frac{\theta}{2} - \sin \theta = 0$

45. $\cos 2\theta + \cos \theta = 2$

46. $\tan \theta + \cot \theta = 4 \sin 2\theta$

47. $\cos 2\theta - \cos^2\theta = 0$

48. $2 \sin^2\theta = 2 + \cos 2\theta$

49. $\cos 2\theta - \cos 4\theta = 0$

50. $\sin 3\theta - \sin 6\theta = 0$

51. $\cos \theta - \sin \theta = \sqrt{2} \sin \frac{\theta}{2}$

52. $\sin \theta - \cos \theta = \frac{1}{2}$

53–56 ■ Solve the equation by first using a Sum-to-Product Formula.

53. $\sin \theta + \sin 3\theta = 0$

54. $\cos 5\theta - \cos 7\theta = 0$

55. $\cos 4\theta + \cos 2\theta = \cos \theta$

56. $\sin 5\theta - \sin 3\theta = \cos 4\theta$

57–62 ■ Use a graphing device to find the solutions of the equation, correct to two decimal places.

57. $\sin 2x = x$

58. $\cos x = \frac{x}{3}$

59. $2^{\sin x} = x$

60. $\sin x = x^3$

61. $\frac{\cos x}{1 + x^2} = x^2$

62. $\cos x = \frac{1}{2}(e^x + e^{-x})$

APPLICATIONS

63. **Range of a Projectile** If a projectile is fired with velocity v_0 at an angle θ, then its *range*, the horizontal distance it travels (in feet), is modeled by the function

$$R(\theta) = \frac{v_0^2 \sin 2\theta}{32}$$

(See page 576.) If $v_0 = 2200$ ft/s, what angle (in degrees) should be chosen for the projectile to hit a target on the ground 5000 ft away?

64. **Damped Vibrations** The displacement of a spring vibrating in damped harmonic motion is given by

$$y = 4e^{-3t} \sin 2\pi t$$

Find the times when the spring is at its equilibrium position ($y = 0$).

65. **Hours of Daylight** In Philadelphia the number of hours of daylight on day t (where t is the number of days after January 1) is modeled by the function

$$L(t) = 12 + 2.83 \sin\left(\frac{2\pi}{365}(t - 80)\right)$$

(a) Which days of the year have about 10 hours of daylight?
(b) How many days of the year have more than 10 hours of daylight?

66. **Belts and Pulleys** A thin belt of length L surrounds two pulleys of radii R and r, as shown in the figure.
(a) Show that the angle θ (in radians) where the belt crosses itself satisfies the equation

$$\theta + 2 \cot \frac{\theta}{2} = \frac{L}{R + r} - \pi$$

[*Hint:* Express L in terms of R, r, and θ by adding up the lengths of the curved and straight parts of the belt.]
(b) Suppose that $R = 2.42$ ft, $r = 1.21$ ft, and $L = 27.78$ ft. Find θ by solving the equation in part (a) graphically. Express your answer both in radians and in degrees.

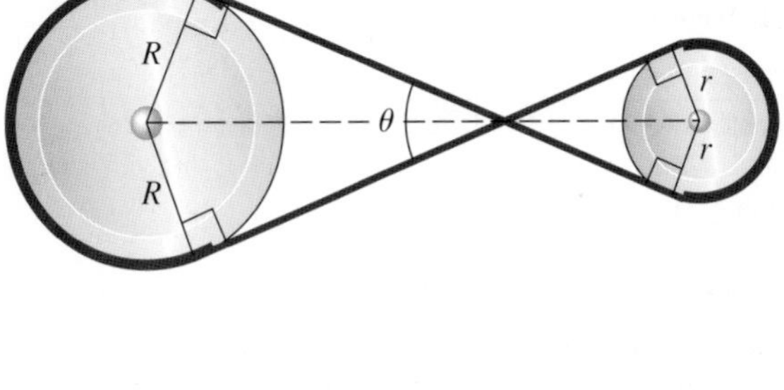

DISCOVERY ■ DISCUSSION ■ WRITING

67. **A Special Trigonometric Equation** What makes the equation $\sin(\cos x) = 0$ different from all the other equations we've looked at in this section? Find all solutions of this equation.

CHAPTER 7 | REVIEW

CONCEPT CHECK

1. **(a)** State the reciprocal identities.
   (b) State the Pythagorean identities.
   (c) State the even-odd identities.
   (d) State the cofunction identities.

2. Explain the difference between an equation and an identity.

3. How do you prove a trigonometric identity?

4. **(a)** State the Addition Formulas for Sine, Cosine, and Tangent.
   (b) State the Subtraction Formulas for Sine, Cosine, and Tangent.

5. **(a)** State the Double-Angle Formulas for Sine, Cosine, and Tangent.
   (b) State the formulas for lowering powers.
   (c) State the Half-Angle Formulas.

6. **(a)** State the Product-to-Sum Formulas.
   (b) State the Sum-to-Product Formulas.

7. Explain how you solve a trigonometric equation by factoring.

8. What identity would you use to solve the equation $\cos x - \sin 2x = 0$?

EXERCISES

1–24 ■ Verify the identity.

1. $\sin\theta\,(\cot\theta + \tan\theta) = \sec\theta$

2. $(\sec\theta - 1)(\sec\theta + 1) = \tan^2\theta$

3. $\cos^2 x \csc x - \csc x = -\sin x$

4. $\dfrac{1}{1 - \sin^2 x} = 1 + \tan^2 x$

5. $\dfrac{\cos^2 x - \tan^2 x}{\sin^2 x} = \cot^2 x - \sec^2 x$

6. $\dfrac{1 + \sec x}{\sec x} = \dfrac{\sin^2 x}{1 - \cos x}$

7. $\dfrac{\cos^2 x}{1 - \sin x} = \dfrac{\cos x}{\sec x - \tan x}$

8. $(1 - \tan x)(1 - \cot x) = 2 - \sec x \csc x$

9. $\sin^2 x \cot^2 x + \cos^2 x \tan^2 x = 1$

10. $(\tan x + \cot x)^2 = \csc^2 x \sec^2 x$

11. $\dfrac{\sin 2x}{1 + \cos 2x} = \tan x$

12. $\dfrac{\cos(x + y)}{\cos x \sin y} = \cot y - \tan x$

13. $\tan\dfrac{x}{2} = \csc x - \cot x$

14. $\dfrac{\sin(x + y) + \sin(x - y)}{\cos(x + y) + \cos(x - y)} = \tan x$

15. $\sin(x + y)\sin(x - y) = \sin^2 x - \sin^2 y$

16. $\csc x - \tan\dfrac{x}{2} = \cot x$

17. $1 + \tan x \tan\dfrac{x}{2} = \sec x$

18. $\dfrac{\sin 3x + \cos 3x}{\cos x - \sin x} = 1 + 2\sin 2x$

19. $\left(\cos\dfrac{x}{2} - \sin\dfrac{x}{2}\right)^2 = 1 - \sin x$

20. $\dfrac{\cos 3x - \cos 7x}{\sin 3x + \sin 7x} = \tan 2x$

21. $\dfrac{\sin 2x}{\sin x} - \dfrac{\cos 2x}{\cos x} = \sec x$

22. $(\cos x + \cos y)^2 + (\sin x - \sin y)^2 = 2 + 2\cos(x + y)$

23. $\tan\left(x + \dfrac{\pi}{4}\right) = \dfrac{1 + \tan x}{1 - \tan x}$

24. $\dfrac{\sec x - 1}{\sin x \sec x} = \tan\dfrac{x}{2}$

25–28 ■ **(a)** Graph f and g. **(b)** Do the graphs suggest that the equation $f(x) = g(x)$ is an identity? Prove your answer.

25. $f(x) = 1 - \left(\cos\dfrac{x}{2} - \sin\dfrac{x}{2}\right)^2, \quad g(x) = \sin x$

26. $f(x) = \sin x + \cos x, \quad g(x) = \sqrt{\sin^2 x + \cos^2 x}$

27. $f(x) = \tan x \tan\dfrac{x}{2}, \quad g(x) = \dfrac{1}{\cos x}$

28. $f(x) = 1 - 8\sin^2 x + 8\sin^4 x, \quad g(x) = \cos 4x$

29–30 ■ **(a)** Graph the function(s) and make a conjecture, and **(b)** prove your conjecture.

29. $f(x) = 2\sin^2 3x + \cos 6x$

30. $f(x) = \sin x \cot\dfrac{x}{2}, \quad g(x) = \cos x$

31–48 ■ Solve the equation in the interval $[0, 2\pi)$.

31. $4\sin\theta - 3 = 0$

32. $5\cos\theta + 3 = 0$

33. $\cos x \sin x - \sin x = 0$

34. $\sin x - 2\sin^2 x = 0$

35. $2\sin^2 x - 5\sin x + 2 = 0$

36. $\sin x - \cos x - \tan x = -1$

37. $2\cos^2 x - 7\cos x + 3 = 0$

38. $4\sin^2 x + 2\cos^2 x = 3$

39. $\dfrac{1-\cos x}{1+\cos x} = 3$

40. $\sin x = \cos 2x$

41. $\tan^3 x + \tan^2 x - 3\tan x - 3 = 0$

42. $\cos 2x \csc^2 x = 2\cos 2x$

43. $\tan \frac{1}{2}x + 2\sin 2x = \csc x$

44. $\cos 3x + \cos 2x + \cos x = 0$

45. $\tan x + \sec x = \sqrt{3}$

46. $2\cos x - 3\tan x = 0$

47. $\cos x = x^2 - 1$

48. $e^{\sin x} = x$

49. If a projectile is fired with velocity v_0 at an angle θ, then the maximum height it reaches (in feet) is modeled by the function

$$M(\theta) = \frac{v_0^2 \sin^2\theta}{64}$$

Suppose $v_0 = 400$ ft/s.

(a) At what angle θ should the projectile be fired so that the maximum height it reaches is 2000 ft?

(b) Is it possible for the projectile to reach a height of 3000 ft?

(c) Find the angle θ for which the projectile will travel highest.

50. The displacement of an automobile shock absorber is modeled by the function

$$f(t) = 2^{-0.2t}\sin 4\pi t$$

Find the times when the shock absorber is at its equilibrium position (that is, when $f(t) = 0$). [*Hint:* $2^x > 0$ for all real x.]

51–60 ■ Find the exact value of the expression.

51. $\cos 15°$

52. $\sin \dfrac{5\pi}{12}$

53. $\tan \dfrac{\pi}{8}$

54. $2\sin \dfrac{\pi}{12}\cos \dfrac{\pi}{12}$

55. $\sin 5° \cos 40° + \cos 5° \sin 40°$

56. $\dfrac{\tan 66° - \tan 6°}{1 + \tan 66° \tan 6°}$

57. $\cos^2 \dfrac{\pi}{8} - \sin^2 \dfrac{\pi}{8}$

58. $\dfrac{1}{2}\cos \dfrac{\pi}{12} + \dfrac{\sqrt{3}}{2}\sin \dfrac{\pi}{12}$

59. $\cos 37.5° \cos 7.5°$

60. $\cos 67.5° + \cos 22.5°$

61–66 ■ Find the exact value of the expression given that $\sec x = \frac{3}{2}$, $\csc y = 3$, and x and y are in Quadrant I.

61. $\sin(x+y)$

62. $\cos(x-y)$

63. $\tan(x+y)$

64. $\sin 2x$

65. $\cos \dfrac{y}{2}$

66. $\tan \dfrac{y}{2}$

67–68 ■ Find the exact value of the expression.

67. $\tan\left(2\cos^{-1}\frac{3}{7}\right)$

68. $\sin\left(\tan^{-1}\frac{3}{4} + \cos^{-1}\frac{5}{13}\right)$

69–70 ■ Write the expression as an algebraic expression in the variable(s).

69. $\tan(2\tan^{-1}x)$

70. $\cos(\sin^{-1}x + \cos^{-1}y)$

71. A 10-ft-wide highway sign is adjacent to a roadway, as shown in the figure. As a driver approaches the sign, the viewing angle θ changes.

(a) Express viewing angle θ as a function of the distance x between the driver and the sign.

(b) The sign is legible when the viewing angle is 2° or greater. At what distance x does the sign first become legible?

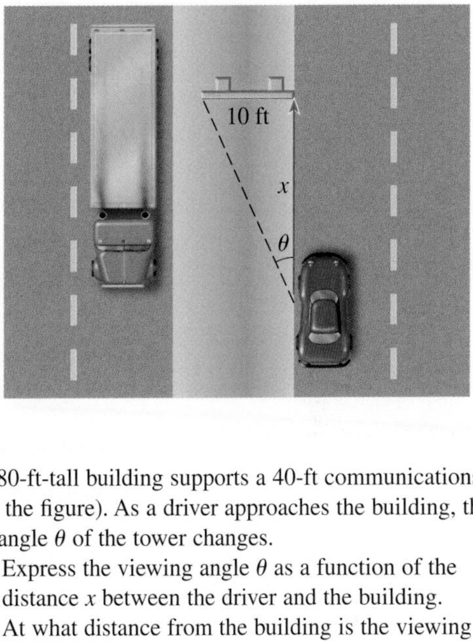

72. A 380-ft-tall building supports a 40-ft communications tower (see the figure). As a driver approaches the building, the viewing angle θ of the tower changes.

(a) Express the viewing angle θ as a function of the distance x between the driver and the building.

(b) At what distance from the building is the viewing angle θ as large as possible?

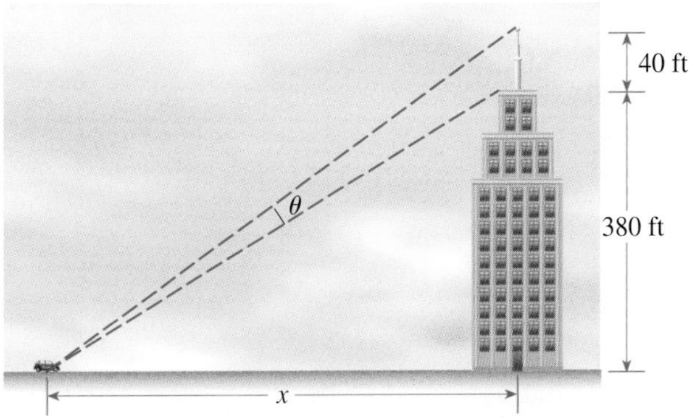

1. Verify each identity.

(a) $\tan\theta \sin\theta + \cos\theta = \sec\theta$

(b) $\dfrac{\tan x}{1 - \cos x} = \csc x\,(1 + \sec x)$

(c) $\dfrac{2\tan x}{1 + \tan^2 x} = \sin 2x$

2. Let $x = 2\sin\theta$, $-\pi/2 < \theta < \pi/2$. Simplify the expression

$$\frac{x}{\sqrt{4 - x^2}}$$

3. Find the exact value of each expression.

(a) $\sin 8^\circ \cos 22^\circ + \cos 8^\circ \sin 22^\circ$ **(b)** $\sin 75^\circ$ **(c)** $\sin \dfrac{\pi}{12}$

4. For the angles α and β in the figures, find $\cos(\alpha + \beta)$.

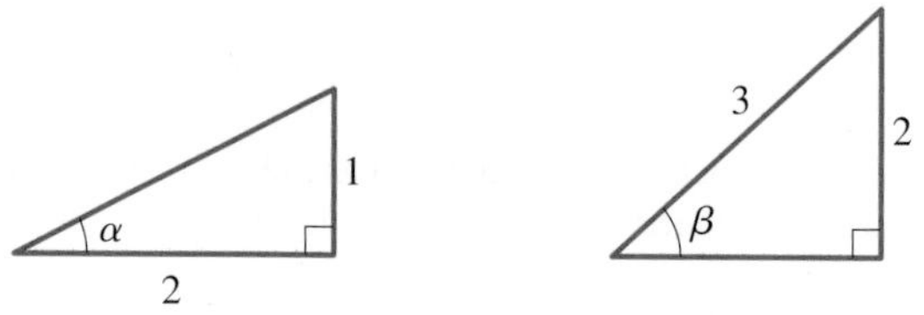

5. **(a)** Write $\sin 3x \cos 5x$ as a sum of trigonometric functions.

(b) Write $\sin 2x - \sin 5x$ as a product of trigonometric functions.

6. If $\sin\theta = -\frac{4}{5}$ and θ is in Quadrant III, find $\tan(\theta/2)$.

7. Solve each trigonometric equation in the interval $[0, 2\pi)$, rounded to two decimal places.

(a) $3\sin\theta - 1 = 0$

(b) $(2\cos\theta - 1)(\sin\theta - 1) = 0$

(c) $2\cos^2\theta + 5\cos\theta + 2 = 0$

(d) $\sin 2\theta - \cos\theta = 0$

8. Find all solutions in the interval $[0, 2\pi)$, rounded to five decimal places:

$$5\cos 2\theta = 2$$

9. Find the exact value of $\cos\left(2\tan^{-1}\frac{9}{40}\right)$.

10. Rewrite the expression as an algebraic function of x and y:

$$\sin(\cos^{-1}x - \tan^{-1}y)$$

Traveling and Standing Waves

We've learned that the position of a particle in simple harmonic motion is described by a function of the form $y = A \sin \omega t$ (see Section 5.6). For example, if a string is moved up and down as in Figure 1, then the red dot on the string moves up and down in simple harmonic motion. Of course, the same holds true for each point on the string.

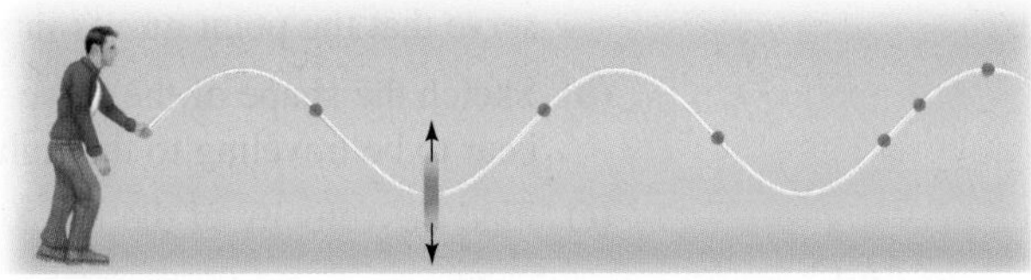

FIGURE 1

What function describes the shape of the whole string? If we fix an instant in time ($t = 0$) and snap a photograph of the string, we get the shape in Figure 2, which is modeled by

$$y = A \sin kx$$

where y is the height of the string above the x-axis at the point x.

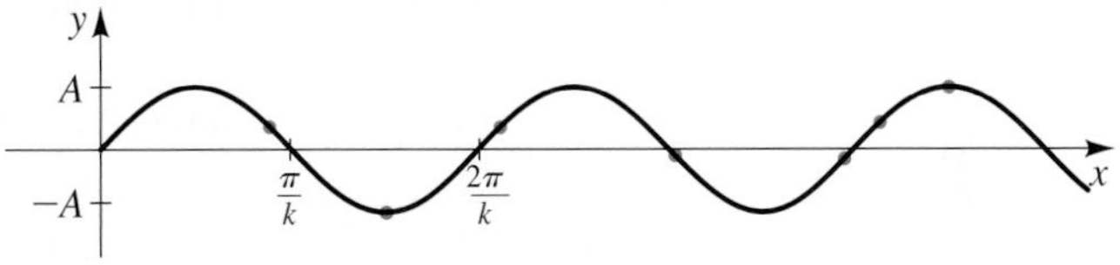

FIGURE 2 $y = A \sin kx$

Traveling Waves

If we snap photographs of the string at other instants, as in Figure 3, it appears that the waves in the string "travel" or shift to the right.

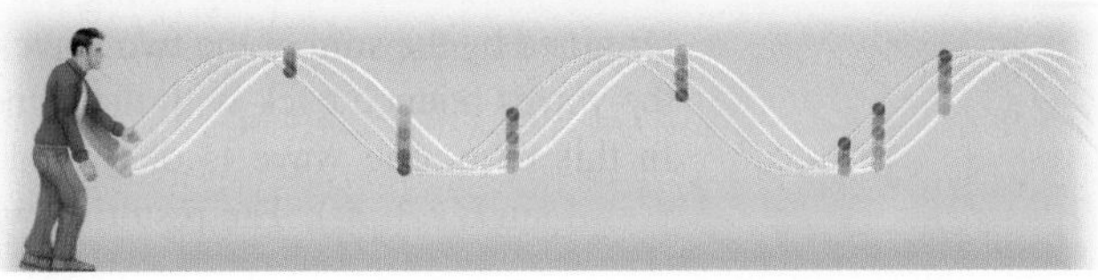

FIGURE 3

The **velocity** of the wave is the rate at which it moves to the right. If the wave has velocity v, then it moves to the right a distance vt in time t. So the graph of the shifted wave at time t is

$$y(x, t) = A \sin k(x - vt)$$

This function models the position of any point x on the string at any time t. We use the notation $y(x, t)$ to indicate that the function depends on the *two* variables x and t. Here is how this function models the motion of the string.

- **If we fix x**, then $y(x, t)$ is a function of t only, which gives the position of the fixed point x at time t.
- **If we fix t**, then $y(x, t)$ is a function of x only, whose graph is the shape of the string at the fixed time t.

EXAMPLE 1 | A Traveling Wave

A traveling wave is described by the function

$$y(x, t) = 3 \sin\left(2x - \frac{\pi}{2}t\right), \qquad x \geq 0$$

(a) Find the function that models the position of the point $x = \pi/6$ at any time t. Observe that the point moves in simple harmonic motion.

(b) Sketch the shape of the wave when $t = 0, 0.5, 1.0, 1.5$, and 2.0. Does the wave appear to be traveling to the right?

(c) Find the velocity of the wave.

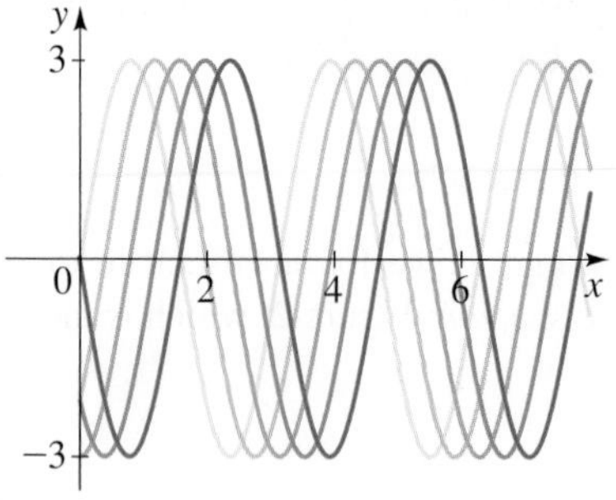

FIGURE 4 Traveling wave

SOLUTION

(a) Substituting $x = \pi/6$ we get

$$y\left(\frac{\pi}{6}, t\right) = 3 \sin\left(2 \cdot \frac{\pi}{6} - \frac{\pi}{2}t\right) = 3 \sin\left(\frac{\pi}{3} - \frac{\pi}{2}t\right)$$

The function $y = 3 \sin(\frac{\pi}{3} - \frac{\pi}{2}t)$ describes simple harmonic motion with amplitude 3 and period $2\pi/(\pi/2) = 4$.

(b) The graphs are shown in Figure 4. As t increases, the wave moves to the right.

(c) We express the given function in the standard form $y(x, t) = A \sin k(x - vt)$:

$$y(x, t) = 3 \sin\left(2x - \frac{\pi}{2}t\right) \qquad \text{Given}$$

$$= 3 \sin 2\left(x - \frac{\pi}{4}t\right) \qquad \text{Factor 2}$$

Comparing this to the standard form, we see that the wave is moving with velocity $v = \pi/4$. ■

▼ Standing Waves

If two waves are traveling along the same string, then the movement of the string is determined by the sum of the two waves. For example, if the string is attached to a wall, then the waves bounce back with the same amplitude and speed but in the opposite direction. In this case, one wave is described by $y = A \sin k(x - vt)$ and the reflected wave by $y = A \sin k(x + vt)$. The resulting wave is

$$y(x, t) = A \sin k(x - vt) + A \sin k(x + vt) \qquad \text{Add the two waves}$$

$$= 2A \sin kx \cos kvt \qquad \text{Sum-to-Product Formula}$$

The points where kx is a multiple of 2π are special, because at these points $y = 0$ for any time t. In other words, these points never move. Such points are called **nodes**. Figure 5 shows the graph of the wave for several values of t. We see that the wave does not travel, but simply vibrates up and down. Such a wave is called a **standing wave**.

FIGURE 5 A standing wave

EXAMPLE 2 | A Standing Wave

Traveling waves are generated at each end of a wave tank 30 ft long, with equations

$$y = 1.5\sin\left(\frac{\pi}{5}x - 3t\right)$$

and

$$y = 1.5\sin\left(\frac{\pi}{5}x + 3t\right)$$

(a) Find the equation of the combined wave, and find the nodes.

(b) Sketch the graph for $t = 0, 0.17, 0.34, 0.51, 0.68, 0.85$, and 1.02. Is this a standing wave?

SOLUTION

(a) The combined wave is obtained by adding the two equations:

$$y = 1.5\sin\left(\frac{\pi}{5}x - 3t\right) + 1.5\sin\left(\frac{\pi}{5}x + 3t\right) \qquad \text{Add the two waves}$$

$$= 3\sin\frac{\pi}{5}x\cos 3t \qquad \text{Sum-to-Product Formula}$$

The nodes occur at the values of x for which $\sin\frac{\pi}{5}x = 0$, that is, where $\frac{\pi}{5}x = k\pi$ (k an integer). Solving for x, we get $x = 5k$. So the nodes occur at

$$x = 0, 5, 10, 15, 20, 25, 30$$

(b) The graphs are shown in Figure 6. From the graphs we see that this is a standing wave.

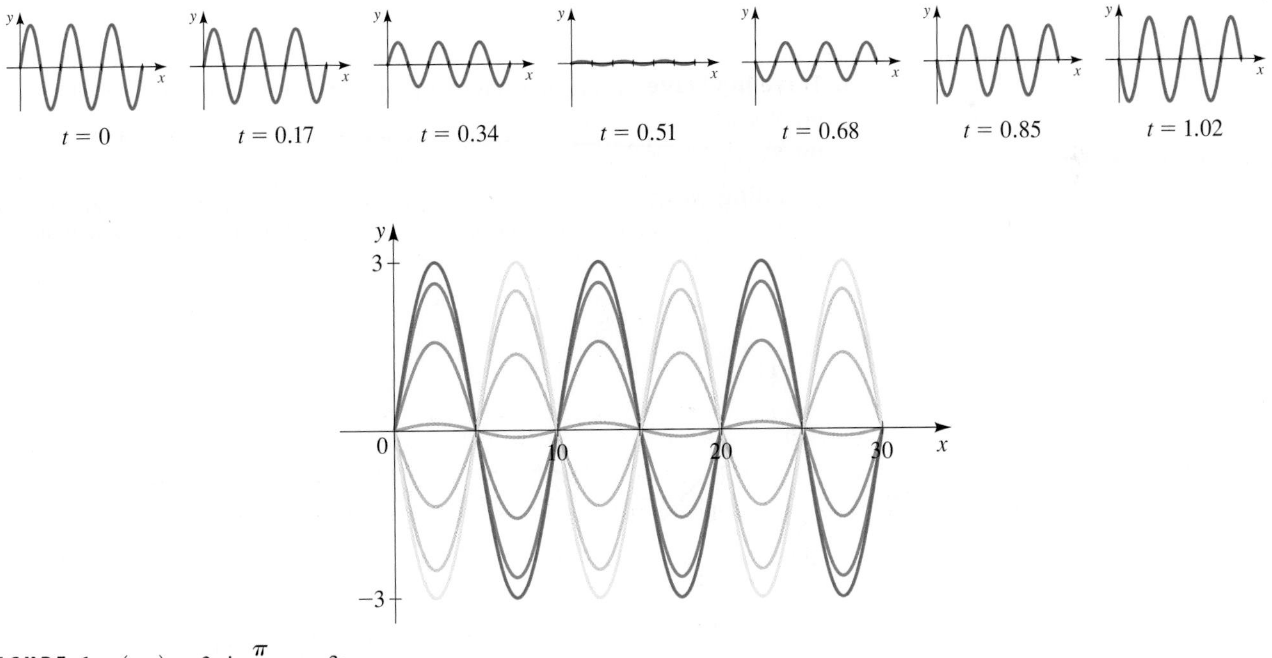

FIGURE 6 $y(x, t) = 3\sin\frac{\pi}{5}x\cos 3t$

■

PROBLEMS

1. Wave on a Canal A wave on the surface of a long canal is described by the function

$$y(x, t) = 5 \sin\left(2x - \frac{\pi}{2}t\right), \quad x \geq 0$$

(a) Find the function that models the position of the point $x = 0$ at any time t.

(b) Sketch the shape of the wave when $t = 0, 0.4, 0.8, 1.2$, and 1.6. Is this a traveling wave?

(c) Find the velocity of the wave.

2. Wave in a Rope Traveling waves are generated at each end of a tightly stretched rope 24 ft long, with equations

$$y = 0.2 \sin(1.047x - 0.524t) \qquad \text{and} \qquad y = 0.2 \sin(1.047x + 0.524t)$$

(a) Find the equation of the combined wave, and find the nodes.

(b) Sketch the graph for $t = 0, 1, 2, 3, 4, 5$, and 6. Is this a standing wave?

3. Traveling Wave A traveling wave is graphed at the instant $t = 0$. If it is moving to the right with velocity 6, find an equation of the form $y(x, t) = A \sin(kx - kvt)$ for this wave.

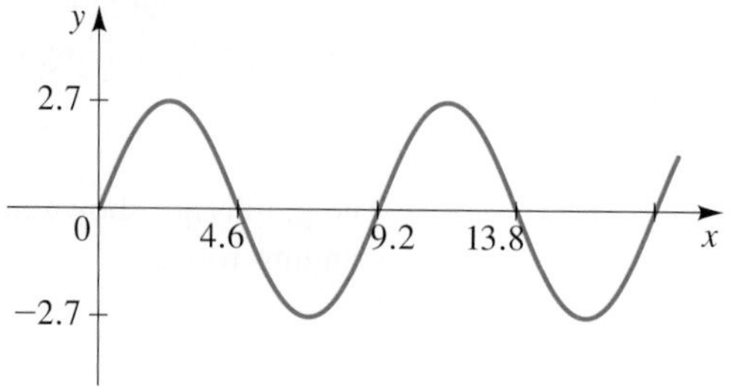

4. Traveling Wave A traveling wave has period $2\pi/3$, amplitude 5, and velocity 0.5.

(a) Find the equation of the wave.

(b) Sketch the graph for $t = 0, 0.5, 1, 1.5$, and 2.

5. Standing Wave A standing wave with amplitude 0.6 is graphed at several times t as shown in the figure. If the vibration has a frequency of 20 Hz, find an equation of the form $y(x, t) = A \sin \alpha x \cos \beta t$ that models this wave.

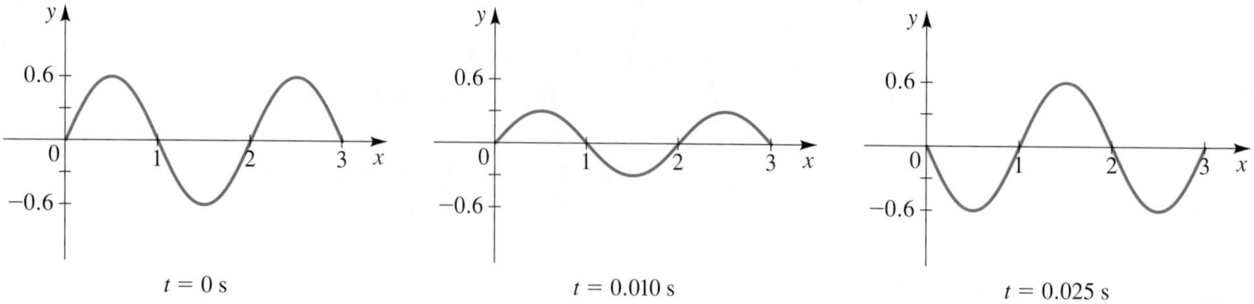

6. **Standing Wave** A standing wave has maximum amplitude 7 and nodes at 0, $\pi/2$, π, $3\pi/2$, 2π, as shown in the figure. Each point that is not a node moves up and down with period 4π. Find a function of the form $y(x, t) = A \sin \alpha x \cos \beta t$ that models this wave.

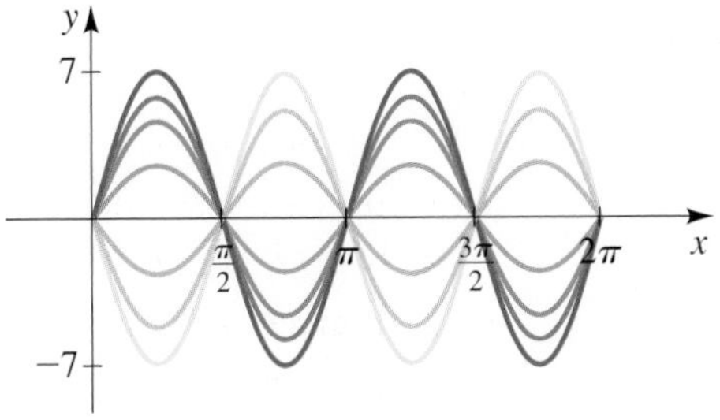

7. **Vibrating String** When a violin string vibrates, the sound produced results from a combination of standing waves that have evenly placed nodes. The figure illustrates some of the possible standing waves. Let's assume that the string has length π.
   (a) For fixed t, the string has the shape of a sine curve $y = A \sin \alpha x$. Find the appropriate value of α for each of the illustrated standing waves.
   (b) Do you notice a pattern in the values of α that you found in part (a)? What would the next two values of α be? Sketch rough graphs of the standing waves associated with these new values of α.
   (c) Suppose that for fixed t, each point on the string that is not a node vibrates with frequency 440 Hz. Find the value of β for which an equation of the form $y = A \cos \beta t$ would model this motion.
   (d) Combine your answers for parts (a) and (c) to find functions of the form $y(x, t) = A \sin \alpha x \cos \beta t$ that model each of the standing waves in the figure. (Assume that $A = 1$.)

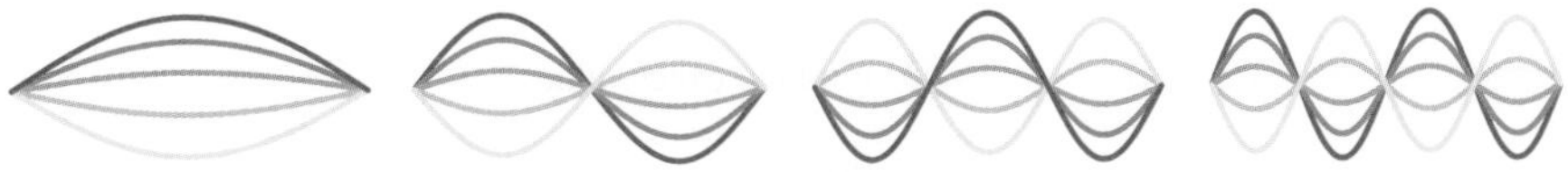

8. **Waves in a Tube** Standing waves in a violin string must have nodes at the ends of the string because the string is fixed at its endpoints. But this need not be the case with sound waves in a tube (such as a flute or an organ pipe). The figure shows some possible standing waves in a tube.

   Suppose that a standing wave in a tube 37.7 ft long is modeled by the function

$$y(x, t) = 0.3 \cos \tfrac{1}{2}x \cos 50\pi t$$

   Here $y(x, t)$ represents the variation from normal air pressure at the point x feet from the end of the tube, at time t seconds.
   (a) At what points x are the nodes located? Are the endpoints of the tube nodes?
   (b) At what frequency does the air vibrate at points that are not nodes?

CUMULATIVE REVIEW TEST CHAPTERS 5, 6, and 7

1. The point $P(x, y)$ shown in the figure has y-coordinate $\sqrt{5}/3$. Find:

(a) $\sin t$ **(b)** $\cos t$ **(c)** $\tan t$ **(d)** $\csc t$

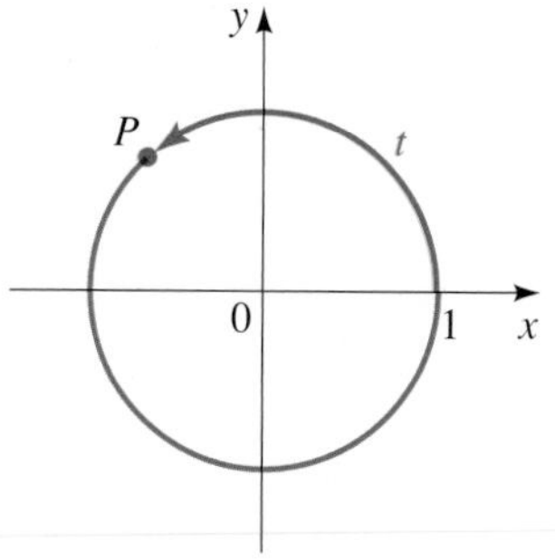

2. For the angle θ shown in the figure, find:

(a) $\sin \theta$ **(b)** $\sec \theta$ **(c)** $\cot \theta$

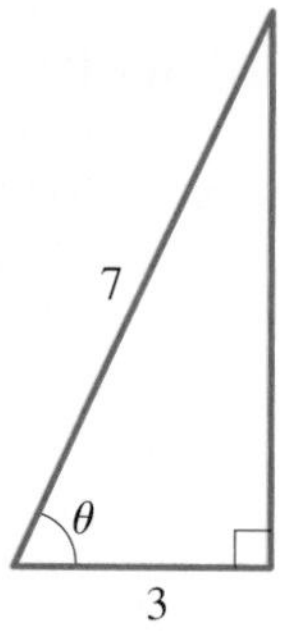

3. Find the exact value:

(a) $\cos \dfrac{7\pi}{6}$ **(b)** $\tan 135°$ **(c)** $\csc 240°$ **(d)** $\sin\left(-\dfrac{9\pi}{2}\right)$

4. Suppose that $\cos t = \dfrac{7}{25}$ and $\tan t < 0$. Find the values of $\sin t$, $\tan t$, $\cot t$, $\sec t$, and $\csc t$.

5. Let $f(x) = -2\sin\left(2x - \dfrac{\pi}{2}\right)$.

(a) Find the amplitude, period, and phase shift of f.

(b) Sketch the graph of one complete period of f.

6. One period of a function of the form $y = a\cos k(x - b)$ is shown in the figure. Determine the function.

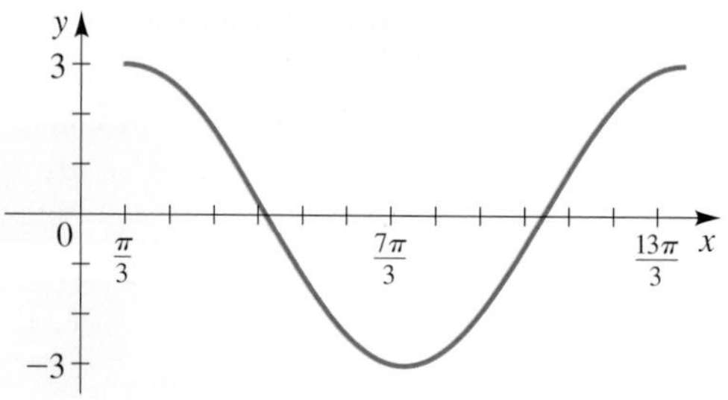

7. The figure below shows a model ferris wheel that a child has constructed using a toy building kit. The wheel has a radius of 40 cm, and the center of the wheel is 45 cm above the floor. An electric motor turns the wheel at 4 rotations per minute.

(a) Let $h(t)$ be the vertical distance between the point P and the floor at the time t. Express the function h in the form $h(t) = a + b \cos kt$. (Assume that at $t = 0$ the point P is at the lowest point of its travel.)

(b) The support struts AB and AC are each 50 cm long. Find the distance between B and C.

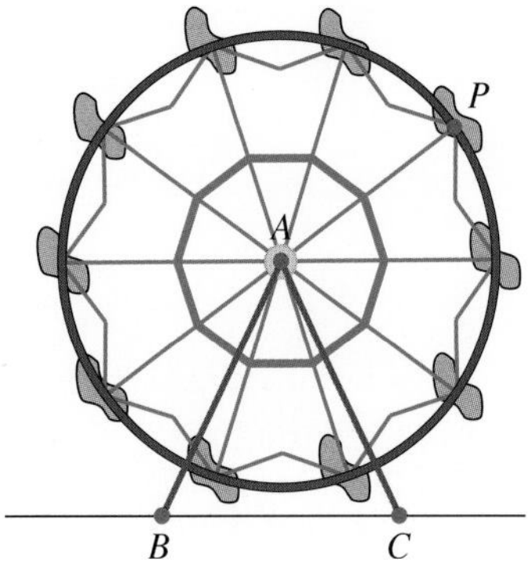

8. Find the side or angle labeled x:

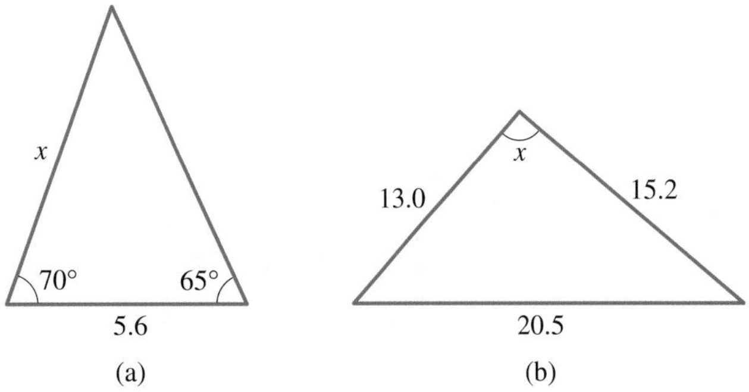

9. Verify each identity:

(a) $\dfrac{\sec \theta - 1}{\tan \theta} = \dfrac{\tan \theta}{\sec \theta + 1}$ **(b)** $8 \sin^2\theta \cos^2\theta = 1 - \cos 4\theta$

10. Write $\cos 3x + \cos 4x$ as a product of trigonometric functions.

11. **(a)** What are the domain and range of the function $f(x) = \cos^{-1}x$? Sketch a graph of this function.

(b) Find the exact value of $\cos^{-1}(\cos(7\pi/6))$.

(c) Express $\tan(\cos^{-1}x)$ as an algebraic function of x.

12. Find all solutions of the equation $\cos 2x - \sin x = 0$ in the interval $[0, 2\pi)$.

AP Photo/Mark Duncan, File

CHAPTER 10

SYSTEMS OF EQUATIONS AND INEQUALITIES

Throughout the preceding chapters we modeled real-world situations by equations. But many real-world situations involve too many variables to be modeled by a single equation. For example, weather depends on the relationship among many variables, including temperature, wind speed, air pressure, and humidity. So to model (and forecast) the weather, scientists use many equations, each having many variables. Such collections of equations, called systems of equations, *work together* to describe the weather. Systems of equations with hundreds of variables are used by airlines to establish consistent flight schedules and by telecommunications companies to find efficient routings for telephone calls. In this chapter we learn how to solve systems of equations that consist of several equations in several variables.

10.1 Systems of Linear Equations in Two Variables

Systems of Linear Equations and Their Solutions ▶ Substitution Method ▶ Elimination Method ▶ Graphical Method ▶ The Number of Solutions of a Linear System in Two Variables ▶ Modeling with Linear Systems

▼ Systems of Linear Equations and Their Solutions

A linear equation in two variables is an equation of the form

$$ax + by = c$$

The graph of a linear equation is a line (see Section 1.10).

A **system of equations** is a set of equations that involve the same variables. A **system of linear equations** is a system of equations in which each equation is linear. A **solution** of a system is an assignment of values for the variables that makes *each* equation in the system true. To **solve** a system means to find all solutions of the system.

Here is an example of a system of linear equations in two variables:

$$\begin{cases} 2x - y = 5 & \text{Equation 1} \\ x + 4y = 7 & \text{Equation 2} \end{cases}$$

We can check that $x = 3$ and $y = 1$ is a solution of this system.

Equation 1	**Equation 2**
$2x - y = 5$	$x + 4y = 7$
$2(3) - 1 = 5$ ✔	$3 + 4(1) = 7$ ✔

The solution can also be written as the ordered pair $(3, 1)$.

Note that the graphs of Equations 1 and 2 are lines (see Figure 1). Since the solution $(3, 1)$ satisfies each equation, the point $(3, 1)$ lies on each line. So it is the point of intersection of the two lines.

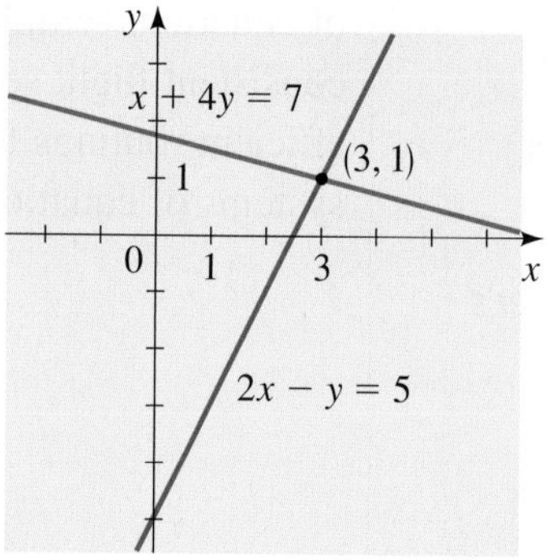

FIGURE 1

▼ Substitution Method

In the **substitution method** we start with one equation in the system and solve for one variable in terms of the other variable. The following box describes the procedure.

> **SUBSTITUTION METHOD**
>
> 1. **Solve for One Variable.** Choose one equation, and solve for one variable in terms of the other variable.
> 2. **Substitute.** Substitute the expression you found in Step 1 into the other equation to get an equation in one variable, then solve for that variable.
> 3. **Back-Substitute.** Substitute the value you found in Step 2 back into the expression found in Step 1 to solve for the remaining variable.

EXAMPLE 1 | Substitution Method

Find all solutions of the system.

$$\begin{cases} 2x + y = 1 & \text{Equation 1} \\ 3x + 4y = 14 & \text{Equation 2} \end{cases}$$

SOLUTION Solve for one variable. We solve for y in the first equation.

$$y = 1 - 2x \qquad \text{Solve for } y \text{ in Equation 1}$$

Substitute. Now we substitute for y in the second equation and solve for x.

$$\begin{aligned} 3x + 4(1 - 2x) &= 14 && \text{Substitute } y = 1 - 2x \text{ into Equation 2} \\ 3x + 4 - 8x &= 14 && \text{Expand} \\ -5x + 4 &= 14 && \text{Simplify} \\ -5x &= 10 && \text{Subtract 4} \\ x &= -2 && \text{Solve for } x \end{aligned}$$

Back-substitute. Next we back-substitute $x = -2$ into the equation $y = 1 - 2x$.

$$y = 1 - 2(-2) = 5 \qquad \text{Back-substitute}$$

Thus $x = -2$ and $y = 5$, so the solution is the ordered pair $(-2, 5)$. Figure 2 shows that the graphs of the two equations intersect at the point $(-2, 5)$.

CHECK YOUR ANSWER

$x = -2, y = 5$:

$$\begin{cases} 2(-2) + 5 = 1 \\ 3(-2) + 4(5) = 14 \end{cases} \checkmark$$

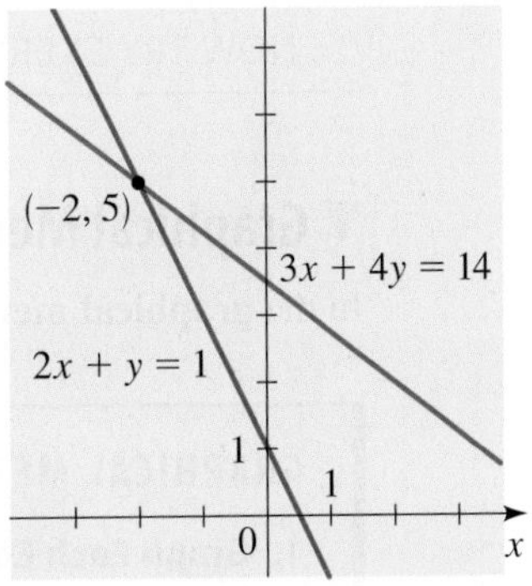

FIGURE 2

NOW TRY EXERCISE **5**

Elimination Method

To solve a system using the **elimination method**, we try to combine the equations using sums or differences so as to eliminate one of the variables.

ELIMINATION METHOD

1. **Adjust the Coefficients.** Multiply one or more of the equations by appropriate numbers so that the coefficient of one variable in one equation is the negative of its coefficient in the other equation.
2. **Add the Equations.** Add the two equations to eliminate one variable, then solve for the remaining variable.
3. **Back-Substitute.** Substitute the value that you found in Step 2 back into one of the original equations, and solve for the remaining variable.

EXAMPLE 2 | Elimination Method

Find all solutions of the system.

$$\begin{cases} 3x + 2y = 14 & \text{Equation 1} \\ x - 2y = 2 & \text{Equation 2} \end{cases}$$

SOLUTION Since the coefficients of the y-terms are negatives of each other, we can add the equations to eliminate y.

$$\begin{cases} 3x + 2y = 14 \\ x - 2y = 2 \end{cases} \qquad \text{System}$$

$$\begin{aligned} 4x \qquad &= 16 && \text{Add} \\ x &= 4 && \text{Solve for } x \end{aligned}$$

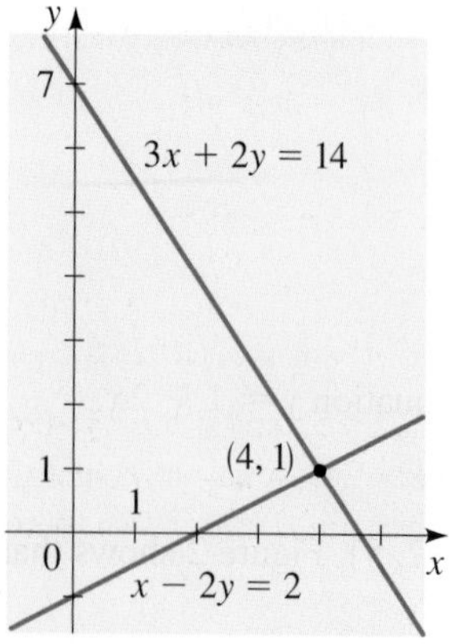

FIGURE 3

Now we back-substitute $x = 4$ into one of the original equations and solve for y. Let's choose the second equation because it looks simpler.

$$\begin{aligned} x - 2y &= 2 && \text{Equation 2} \\ 4 - 2y &= 2 && \text{Back-substitute } x = 4 \text{ into Equation 2} \\ -2y &= -2 && \text{Subtract 4} \\ y &= 1 && \text{Solve for } y \end{aligned}$$

The solution is $(4, 1)$. Figure 3 shows that the graphs of the equations in the system intersect at the point $(4, 1)$.

NOW TRY EXERCISE **9**

▼ Graphical Method

In the **graphical method** we use a graphing device to solve the system of equations.

GRAPHICAL METHOD

1. **Graph Each Equation.** Express each equation in a form suitable for the graphing calculator by solving for y as a function of x. Graph the equations on the same screen.
2. **Find the Intersection Points.** The solutions are the x- and y-coordinates of the points of intersection.

MATHEMATICS IN THE MODERN WORLD

Weather Prediction

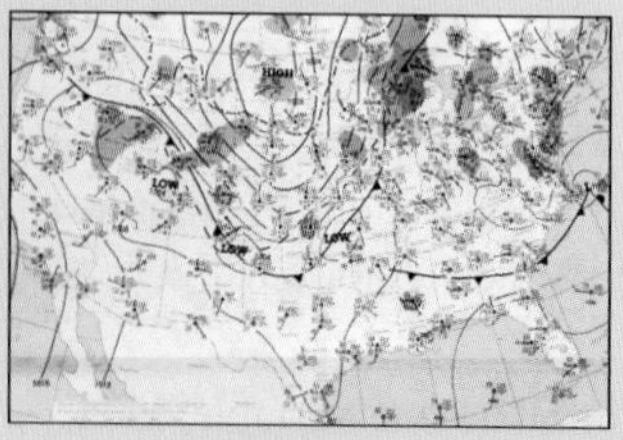

© Rachel Epstein/Photo Edit

Modern meteorologists do much more than predict tomorrow's weather. They research long-term weather patterns, depletion of the ozone layer, global warming, and other effects of human activity on the weather. But daily weather prediction is still a major part of meteorology; its value is measured by the innumerable human lives that are saved each year through accurate prediction of hurricanes, blizzards, and other catastrophic weather phenomena. Early in the 20th century mathematicians proposed to model weather with equations that used the current values of hundreds of atmospheric variables. Although this model worked in principle, it was impossible to predict future weather patterns with it because of the difficulty of measuring all the variables accurately and solving all the equations. Today, new mathematical models combined with high-speed computer simulations and better data have vastly improved weather prediction. As a result, many human as well as economic disasters have been averted. Mathematicians at the National Oceanographic and Atmospheric Administration (NOAA) are continually researching better methods of weather prediction.

EXAMPLE 3 | Graphical Method

Find all solutions of the system

$$\begin{cases} 1.35x - 2.13y = -2.36 \\ 2.16x + 0.32y = 1.06 \end{cases}$$

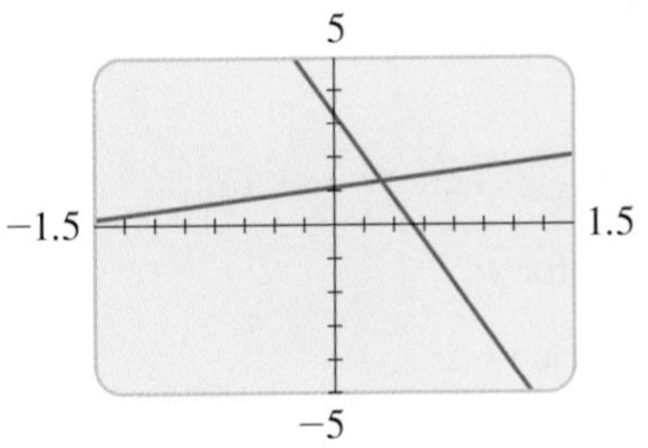

FIGURE 4

SOLUTION Solving for y in terms of x, we get the equivalent system

$$\begin{cases} y = 0.63x + 1.11 \\ y = -6.75x + 3.31 \end{cases}$$

where we have rounded the coefficients to two decimals. Figure 4 shows that the two lines intersect. Zooming in, we see that the solution is approximately $(0.30, 1.3)$.

NOW TRY EXERCISES **13** AND **49**

The Number of Solutions of a Linear System in Two Variables

The graph of a linear system in two variables is a pair of lines, so to solve the system graphically, we must find the intersection point(s) of the lines. Two lines may intersect in a single point, they may be parallel, or they may coincide, as shown in Figure 5. So there are three possible outcomes in solving such a system.

NUMBER OF SOLUTIONS OF A LINEAR SYSTEM IN TWO VARIABLES

For a system of linear equations in two variables, exactly one of the following is true. (See Figure 5.)

1. The system has exactly one solution.
2. The system has no solution.
3. The system has infinitely many solutions.

A system that has no solution is said to be **inconsistent**. A system with infinitely many solutions is called **dependent**.

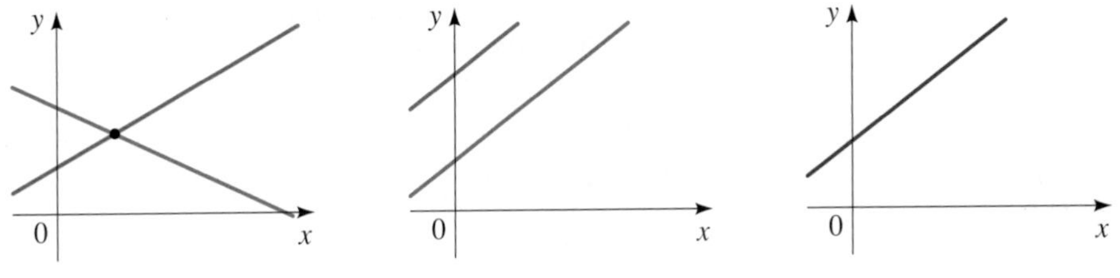

FIGURE 5 (a) Lines intersect at a single point. The system has one solution. (b) Lines are parallel and do not intersect. The system has no solution. (c) Lines coincide—equations are for the same line. The system has infinitely many solutions.

EXAMPLE 4 | A Linear System with One Solution

Solve the system and graph the lines.

$$\begin{cases} 3x - y = 0 & \text{Equation 1} \\ 5x + 2y = 22 & \text{Equation 2} \end{cases}$$

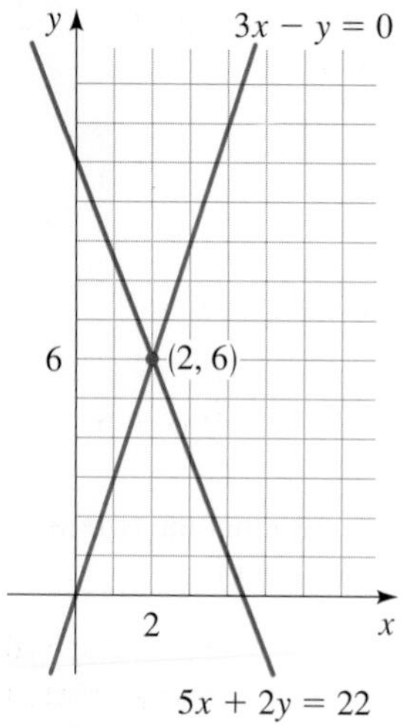

FIGURE 6

CHECK YOUR ANSWER

$x = 2, \quad y = 6$:

$$\begin{cases} 3(2) - (6) = 0 \\ 5(2) + 2(6) = 22 \end{cases} \checkmark$$

SOLUTION We eliminate y from the equations and solve for x.

$$\begin{cases} 6x - 2y = 0 & 2 \times \text{Equation 1} \\ 5x + 2y = 22 & \end{cases}$$

$$11x \qquad = 22 \qquad \text{Add}$$

$$x = 2 \qquad \text{Solve for } x$$

Now we back-substitute into the first equation and solve for y:

$$6(2) - 2y = 0 \qquad \text{Back-substitute } x = 2$$

$$-2y = -12 \qquad \text{Subtract } 6 \times 2 = 12$$

$$y = 6 \qquad \text{Solve for } y$$

The solution of the system is the ordered pair (2, 6), that is,

$$x = 2, \qquad y = 6$$

The graph in Figure 6 shows that the lines in the system intersect at the point (2, 6).

NOW TRY EXERCISE **23** ■

EXAMPLE 5 | A Linear System with No Solution

Solve the system.

$$\begin{cases} 8x - 2y = 5 & \text{Equation 1} \\ -12x + 3y = 7 & \text{Equation 2} \end{cases}$$

SOLUTION This time we try to find a suitable combination of the two equations to eliminate the variable y. Multiplying the first equation by 3 and the second equation by 2 gives

$$\begin{cases} 24x - 6y = 15 & 3 \times \text{Equation 1} \\ -24x + 6y = 14 & 2 \times \text{Equation 2} \end{cases}$$

$$0 = 29 \qquad \text{Add}$$

Adding the two equations eliminates *both x and y* in this case, and we end up with $0 = 29$, which is obviously false. No matter what values we assign to x and y, we cannot make this statement true, so the system has *no solution*. Figure 7 shows that the lines in the system are parallel and do not intersect. The system is inconsistent.

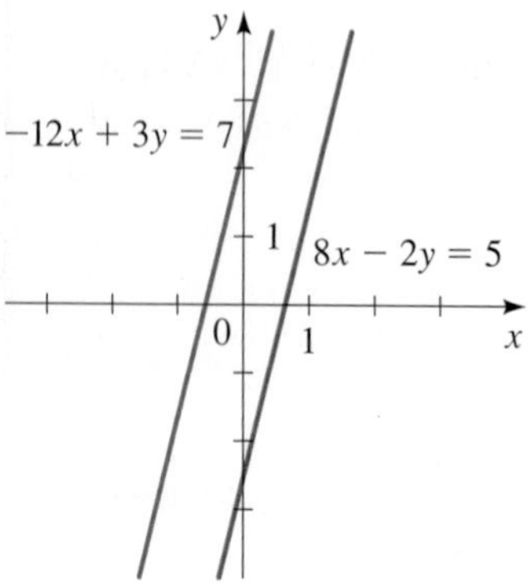

FIGURE 7

NOW TRY EXERCISE **35** ■

EXAMPLE 6 | A Linear System with Infinitely Many Solutions

Solve the system.

$$\begin{cases} 3x - 6y = 12 & \text{Equation 1} \\ 4x - 8y = 16 & \text{Equation 2} \end{cases}$$

SOLUTION We multiply the first equation by 4 and the second by 3 to prepare for subtracting the equations to eliminate x. The new equations are

$$\begin{cases} 12x - 24y = 48 & 4 \times \text{Equation 1} \\ 12x - 24y = 48 & 3 \times \text{Equation 2} \end{cases}$$

We see that the two equations in the original system are simply different ways of expressing the equation of one single line. The coordinates of any point on this line give a

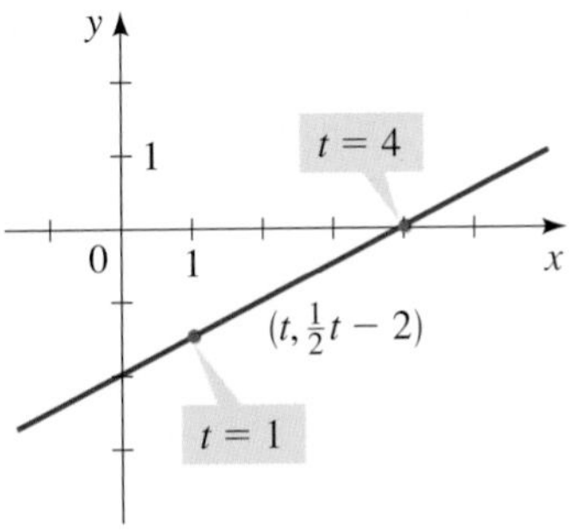

FIGURE 8

solution of the system. Writing the equation in slope-intercept form, we have $y = \frac{1}{2}x - 2$. So if we let t represent any real number, we can write the solution as

$$x = t$$
$$y = \tfrac{1}{2}t - 2$$

We can also write the solution in ordered-pair form as

$$\left(t, \tfrac{1}{2}t - 2\right)$$

where t is any real number. The system has infinitely many solutions (see Figure 8).

NOW TRY EXERCISE 37 ■

In Example 3, to get specific solutions we have to assign values to t. For instance, if $t = 1$, we get the solution $\left(1, -\frac{3}{2}\right)$. If $t = 4$, we get the solution $(4, 0)$. For every value of t we get a different solution. (See Figure 8.)

▼ Modeling with Linear Systems

Frequently, when we use equations to solve problems in the sciences or in other areas, we obtain systems like the ones we've been considering. When modeling with systems of equations, we use the following guidelines, which are similar to those in Section 1.6.

GUIDELINES FOR MODELING WITH SYSTEMS OF EQUATIONS

1. **Identify the Variables.** Identify the quantities that the problem asks you to find. These are usually determined by a careful reading of the question posed at the end of the problem. Introduce notation for the variables (call them x and y or some other letters).
2. **Express All Unknown Quantities in Terms of the Variables.** Read the problem again, and express all the quantities mentioned in the problem in terms of the variables you defined in Step 1.
3. **Set Up a System of Equations.** Find the crucial facts in the problem that give the relationships between the expressions you found in Step 2. Set up a system of equations (or a model) that expresses these relationships.
4. **Solve the System and Interpret the Results.** Solve the system you found in Step 3, check your solutions, and state your final answer as a sentence that answers the question posed in the problem.

The next two examples illustrate how to model with systems of equations.

EXAMPLE 7 | A Distance-Speed-Time Problem

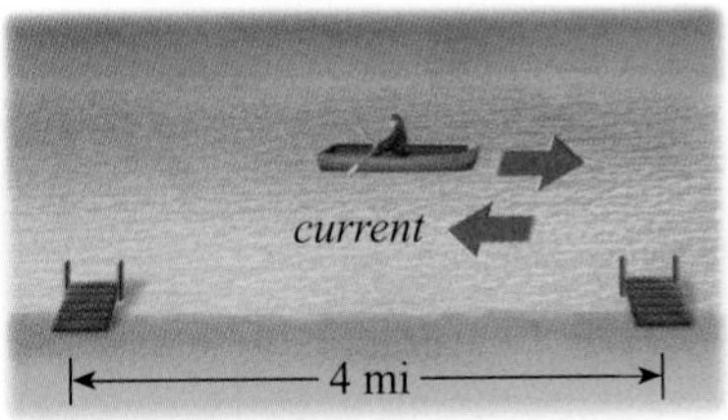

A woman rows a boat upstream from one point on a river to another point 4 mi away in $1\frac{1}{2}$ hours. The return trip, traveling with the current, takes only 45 min. How fast does she row relative to the water, and at what speed is the current flowing?

SOLUTION Identify the variables. We are asked to find the rowing speed and the speed of the current, so we let

$$x = \text{rowing speed (mi/h)}$$
$$y = \text{current speed (mi/h)}$$

Express unknown quantites in terms of the variable. The woman's speed when she rows upstream is her rowing speed minus the speed of the current; her speed downstream is her rowing speed plus the speed of the current. Now we translate this information into the language of algebra.

In Words	In Algebra
Rowing speed	x
Current speed	y
Speed upstream	$x - y$
Speed downstream	$x + y$

Set up a system of equations. The distance upstream and downstream is 4 mi, so using the fact that speed × time = distance for both legs of the trip, we get

speed upstream × time upstream = distance traveled

speed downstream × time downstream = distance traveled

In algebraic notation this translates into the following equations.

$$(x - y)\tfrac{3}{2} = 4 \qquad \text{Equation 1}$$

$$(x + y)\tfrac{3}{4} = 4 \qquad \text{Equation 2}$$

(The times have been converted to hours, since we are expressing the speeds in miles per *hour*.)

Solve the system. We multiply the equations by 2 and 4, respectively, to clear the denominators.

$$\begin{cases} 3x - 3y = 8 & 2 \times \text{Equation 1} \\ 3x + 3y = 16 & 4 \times \text{Equation 2} \end{cases}$$

$$6x = 24 \qquad \text{Add}$$

$$x = 4 \qquad \text{Solve for } x$$

Back-substituting this value of x into the first equation (the second works just as well) and solving for y gives

$$3(4) - 3y = 8 \qquad \text{Back-substitute } x = 4$$

$$-3y = 8 - 12 \qquad \text{Subtract 12}$$

$$y = \tfrac{4}{3} \qquad \text{Solve for } y$$

The woman rows at 4 mi/h, and the current flows at $1\frac{1}{3}$ mi/h.

CHECK YOUR ANSWER

Speed upstream is

$$\frac{\text{distance}}{\text{time}} = \frac{4\text{ mi}}{1\frac{1}{2}\text{ h}} = 2\tfrac{2}{3}\text{ mi/h}$$

and this should equal

rowing speed − current flow

$$= 4\text{ mi/h} - \tfrac{4}{3}\text{ mi/h} = 2\tfrac{2}{3}\text{ mi/h}$$

Speed downstream is

$$\frac{\text{distance}}{\text{time}} = \frac{4\text{ mi}}{\frac{3}{4}\text{ h}} = 5\tfrac{1}{3}\text{ mi/h}$$

and this should equal

rowing speed + current flow

$$= 4\text{ mi/h} + \tfrac{4}{3}\text{ mi/h} = 5\tfrac{1}{3}\text{ mi/h}$$ ✔

NOW TRY EXERCISE **63** ■

EXAMPLE 8 | A Mixture Problem

A vintner fortifies wine that contains 10% alcohol by adding a 70% alcohol solution to it. The resulting mixture has an alcoholic strength of 16% and fills 1000 one-liter bottles. How many liters (L) of the wine and of the alcohol solution does the vintner use?

SOLUTION Identify the variables. Since we are asked for the amounts of wine and alcohol, we let

$$x = \text{amount of wine used (L)}$$

$$y = \text{amount of alcohol solution used (L)}$$

Express all unknown quantities in terms of the variable. From the fact that the wine contains 10% alcohol and the solution contains 70% alcohol, we get the following.

In Words	In Algebra
Amount of wine used (L)	x
Amount of alcohol solution used (L)	y
Amount of alcohol in wine (L)	$0.10x$
Amount of alcohol in solution (L)	$0.70y$

Set up a system of equations. The volume of the mixture must be the total of the two volumes the vintner is adding together, so

$$x + y = 1000$$

Also, the amount of alcohol in the mixture must be the total of the alcohol contributed by the wine and by the alcohol solution, that is,

$$0.10x + 0.70y = (0.16)1000$$

$$0.10x + 0.70y = 160 \qquad \text{Simplify}$$

$$x + 7y = 1600 \qquad \text{Multiply by 10 to clear decimals}$$

Thus we get the system

$$\begin{cases} x + y = 1000 & \text{Equation 1} \\ x + 7y = 1600 & \text{Equation 2} \end{cases}$$

Solve the system. Subtracting the first equation from the second eliminates the variable x, and we get

$$6y = 600 \qquad \text{Subtract Equation 1 from Equation 2}$$

$$y = 100 \qquad \text{Solve for } y$$

We now back-substitute $y = 100$ into the first equation and solve for x.

$$x + 100 = 1000 \qquad \text{Back-substitute } y = 100$$

$$x = 900 \qquad \text{Solve for } x$$

The vintner uses 900 L of wine and 100 L of the alcohol solution.

NOW TRY EXERCISE **65** ■

10.1 EXERCISES

CONCEPTS

1. The system of equations

$$\begin{cases} 2x + 3y = 7 \\ 5x - y = 9 \end{cases}$$

is a system of two equations in the two variables ________ and ________. To determine whether $(5, -1)$ is a solution of this system, we check whether $x = 5$ and $y = -1$ satisfy each ________ in the system. Which of the following are solutions of this system?

$$(5, -1), \quad (-1, 3), \quad (2, 1)$$

2. A system of equations in two variables can be solved by the ____________ method, the ____________ method, or the ____________ method.

3. A system of two linear equations in two variables can have one solution, ________ solution, or ________ ________ solutions.

4. The following is a system of two linear equations in two variables.

$$\begin{cases} x + y = 1 \\ 2x + 2y = 2 \end{cases}$$

The graph of the first equation is the same as the graph of the second equation, so the system has ________ ________ solutions. We express these solutions by writing

$$x = t$$
$$y = \_\_\_\_\_\_$$

where t is any real number. Some of the solutions of this system are $(1, \_\_)$, $(-3, \_\_)$, and $(5, \_\_)$.

SKILLS

5–8 ■ Use the substitution method to find all solutions of the system of equations.

5. $\begin{cases} x - y = 1 \\ 4x + 3y = 18 \end{cases}$ **6.** $\begin{cases} 3x + y = 1 \\ 5x + 2y = 1 \end{cases}$

7. $\begin{cases} x - y = 2 \\ 2x + 3y = 9 \end{cases}$ **8.** $\begin{cases} 2x + y = 7 \\ x + 2y = 2 \end{cases}$

9–12 ■ Use the elimination method to find all solutions of the system of equations.

9. $\begin{cases} 3x + 4y = 10 \\ x - 4y = -2 \end{cases}$ **10.** $\begin{cases} 2x + 5y = 15 \\ 4x + y = 21 \end{cases}$

11. $\begin{cases} x + 2y = 5 \\ 2x + 3y = 8 \end{cases}$ **12.** $\begin{cases} 4x - 3y = 11 \\ 8x + 4y = 12 \end{cases}$

13–14 ■ Two equations and their graphs are given. Find the intersection point(s) of the graphs by solving the system.

13. $\begin{cases} 2x + y = -1 \\ x - 2y = -8 \end{cases}$ **14.** $\begin{cases} x + y = 2 \\ 2x + y = 5 \end{cases}$

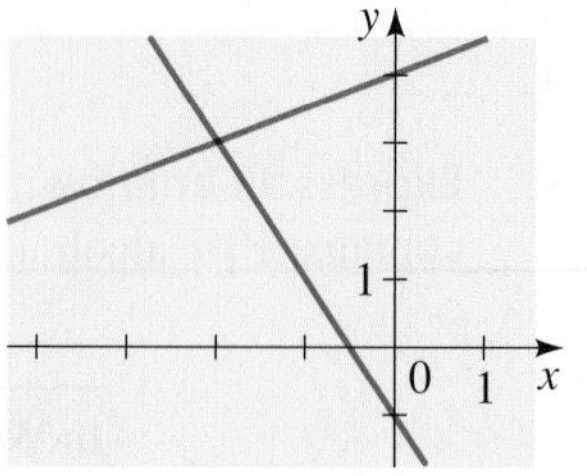

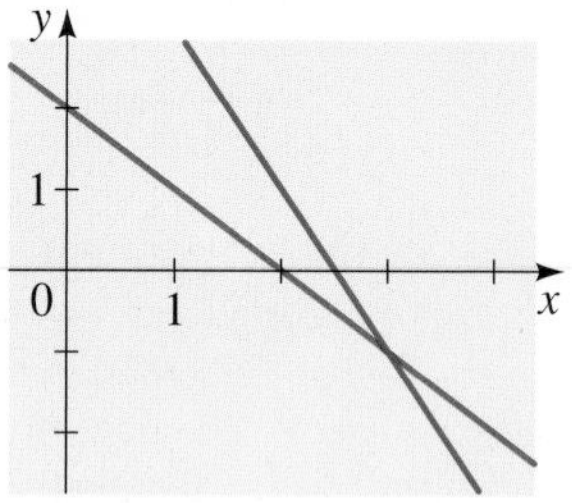

15–20 ■ Graph each linear system, either by hand or using a graphing device. Use the graph to determine whether the system has one solution, no solution, or infinitely many solutions. If there is exactly one solution, use the graph to find it.

15. $\begin{cases} x - y = 4 \\ 2x + y = 2 \end{cases}$ **16.** $\begin{cases} 2x - y = 4 \\ 3x + y = 6 \end{cases}$

17. $\begin{cases} 2x - 3y = 12 \\ -x + \frac{3}{2}y = 4 \end{cases}$ **18.** $\begin{cases} 2x + 6y = 0 \\ -3x - 9y = 18 \end{cases}$

19. $\begin{cases} -x + \frac{1}{2}y = -5 \\ 2x - y = 10 \end{cases}$ **20.** $\begin{cases} 12x + 15y = -18 \\ 2x + \frac{5}{2}y = -3 \end{cases}$

21–48 ■ Solve the system, or show that it has no solution. If the system has infinitely many solutions, express them in the ordered-pair form given in Example 6.

21. $\begin{cases} x + y = 4 \\ -x + y = 0 \end{cases}$ **22.** $\begin{cases} x - y = 3 \\ x + 3y = 7 \end{cases}$

23. $\begin{cases} 2x - 3y = 9 \\ 4x + 3y = 9 \end{cases}$ **24.** $\begin{cases} 3x + 2y = 0 \\ -x - 2y = 8 \end{cases}$

25. $\begin{cases} x + 3y = 5 \\ 2x - y = 3 \end{cases}$ **26.** $\begin{cases} x + y = 7 \\ 2x - 3y = -1 \end{cases}$

27. $\begin{cases} -x + y = 2 \\ 4x - 3y = -3 \end{cases}$ **28.** $\begin{cases} 4x - 3y = 28 \\ 9x - y = -6 \end{cases}$

29. $\begin{cases} x + 2y = 7 \\ 5x - y = 2 \end{cases}$ **30.** $\begin{cases} -4x + 12y = 0 \\ 12x + 4y = 160 \end{cases}$

31. $\begin{cases} \frac{1}{2}x + \frac{1}{3}y = 2 \\ \frac{1}{5}x - \frac{2}{3}y = 8 \end{cases}$ **32.** $\begin{cases} 0.2x - 0.2y = -1.8 \\ -0.3x + 0.5y = 3.3 \end{cases}$

33. $\begin{cases} 3x + 2y = 8 \\ x - 2y = 0 \end{cases}$ **34.** $\begin{cases} 4x + 2y = 16 \\ x - 5y = 70 \end{cases}$

35. $\begin{cases} x + 4y = 8 \\ 3x + 12y = 2 \end{cases}$ **36.** $\begin{cases} -3x + 5y = 2 \\ 9x - 15y = 6 \end{cases}$

37. $\begin{cases} 2x - 6y = 10 \\ -3x + 9y = -15 \end{cases}$ **38.** $\begin{cases} 2x - 3y = -8 \\ 14x - 21y = 3 \end{cases}$

39. $\begin{cases} 6x + 4y = 12 \\ 9x + 6y = 18 \end{cases}$

40. $\begin{cases} 25x - 75y = 100 \\ -10x + 30y = -40 \end{cases}$

41. $\begin{cases} 8s - 3t = -3 \\ 5s - 2t = -1 \end{cases}$

42. $\begin{cases} u - 30v = -5 \\ -3u + 80v = 5 \end{cases}$

43. $\begin{cases} \frac{1}{2}x + \frac{3}{5}y = 3 \\ \frac{5}{3}x + 2y = 10 \end{cases}$

44. $\begin{cases} \frac{3}{2}x - \frac{1}{3}y = \frac{1}{2} \\ 2x - \frac{1}{2}y = -\frac{1}{2} \end{cases}$

45. $\begin{cases} 0.4x + 1.2y = 14 \\ 12x - 5y = 10 \end{cases}$

46. $\begin{cases} 26x - 10y = -4 \\ -0.6x + 1.2y = 3 \end{cases}$

47. $\begin{cases} \frac{1}{3}x - \frac{1}{4}y = 2 \\ -8x + 6y = 10 \end{cases}$

48. $\begin{cases} -\frac{1}{10}x + \frac{1}{2}y = 4 \\ 2x - 10y = -80 \end{cases}$

49–52 ■ Use a graphing device to graph both lines in the same viewing rectangle. (Note that you must solve for y in terms of x before graphing if you are using a graphing calculator.) Solve the system rounded to two decimal places, either by zooming in and using TRACE or by using Intersect.

49. $\begin{cases} 0.21x + 3.17y = 9.51 \\ 2.35x - 1.17y = 5.89 \end{cases}$

50. $\begin{cases} 18.72x - 14.91y = 12.33 \\ 6.21x - 12.92y = 17.82 \end{cases}$

51. $\begin{cases} 2371x - 6552y = 13{,}591 \\ 9815x + 992y = 618{,}555 \end{cases}$

52. $\begin{cases} -435x + 912y = 0 \\ 132x + 455y = 994 \end{cases}$

53–56 ■ Find x and y in terms of a and b.

53. $\begin{cases} x + y = 0 \\ x + ay = 1 \end{cases} \quad (a \neq 1)$

54. $\begin{cases} ax + by = 0 \\ x + y = 1 \end{cases} \quad (a \neq b)$

55. $\begin{cases} ax + by = 1 \\ bx + ay = 1 \end{cases} \quad (a^2 - b^2 \neq 0)$

56. $\begin{cases} ax + by = 0 \\ a^2x + b^2y = 1 \end{cases} \quad (a \neq 0, b \neq 0, a \neq b)$

APPLICATIONS

57. **Number Problem** Find two numbers whose sum is 34 and whose difference is 10.

58. **Number Problem** The sum of two numbers is twice their difference. The larger number is 6 more than twice the smaller. Find the numbers.

59. **Value of Coins** A man has 14 coins in his pocket, all of which are dimes and quarters. If the total value of his change is \$2.75, how many dimes and how many quarters does he have?

60. **Admission Fees** The admission fee at an amusement park is \$1.50 for children and \$4.00 for adults. On a certain day, 2200 people entered the park, and the admission fees that were collected totaled \$5050. How many children and how many adults were admitted?

61. **Gas Station** A gas station sells regular gas for \$2.20 per gallon and premium gas for \$3.00 a gallon. At the end of a business day 280 gallons of gas were sold, and receipts totaled \$680. How many gallons of each type of gas were sold?

62. **Fruit Stand** A fruit stand sells two varieties of strawberries: standard and deluxe. A box of standard strawberries sells for \$7, and a box of deluxe strawberries sells for \$10. In one day the stand sells 135 boxes of strawberries for a total of \$1110. How many boxes of each type were sold?

63. **Airplane Speed** A man flies a small airplane from Fargo to Bismarck, North Dakota—a distance of 180 mi. Because he is flying into a head wind, the trip takes him 2 hours. On the way back, the wind is still blowing at the same speed, so the return trip takes only 1 h 12 min. What is his speed in still air, and how fast is the wind blowing?

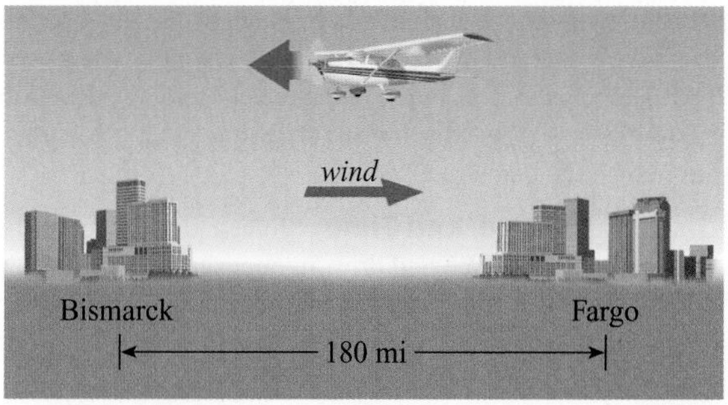

64. **Boat Speed** A boat on a river travels downstream between two points, 20 mi apart, in one hour. The return trip against the current takes $2\frac{1}{2}$ hours. What is the boat's speed, and how fast does the current in the river flow?

65. **Nutrition** A researcher performs an experiment to test a hypothesis that involves the nutrients niacin and retinol. She feeds one group of laboratory rats a daily diet of precisely 32 units of niacin and 22,000 units of retinol. She uses two types of commercial pellet foods. Food A contains 0.12 unit of niacin and 100 units of retinol per gram. Food B contains 0.20 unit of niacin and 50 units of retinol per gram. How many grams of each food does she feed this group of rats each day?

66. **Coffee Blends** A customer in a coffee shop purchases a blend of two coffees: Kenyan, costing \$3.50 a pound, and Sri Lankan, costing \$5.60 a pound. He buys 3 lb of the blend, which costs him \$11.55. How many pounds of each kind went into the mixture?

67. **Mixture Problem** A chemist has two large containers of sulfuric acid solution, with different concentrations of acid in each container. Blending 300 mL of the first solution and 600 mL of the second gives a mixture that is 15% acid, whereas blending 100 mL of the first with 500 mL of the second gives a $12\frac{1}{2}\%$ acid mixture. What are the concentrations of sulfuric acid in the original containers?

68. **Mixture Problem** A biologist has two brine solutions, one containing 5% salt and another containing 20% salt. How many milliliters of each solution should she mix to obtain 1 L of a solution that contains 14% salt?

69. **Investments** A woman invests a total of \$20,000 in two accounts, one paying 5% and the other paying 8% simple interest per year. Her annual interest is \$1180. How much did she invest at each rate?

70. **Investments** A man invests his savings in two accounts, one paying 6% and the other paying 10% simple interest per year. He puts twice as much in the lower-yielding account because it is less risky. His annual interest is \$3520. How much did he invest at each rate?

71. **Distance, Speed, and Time** John and Mary leave their house at the same time and drive in opposite directions. John drives at 60 mi/h and travels 35 mi farther than Mary, who drives at 40 mi/h. Mary's trip takes 15 min longer than John's. For what length of time does each of them drive?

72. **Aerobic Exercise** A woman keeps fit by bicycling and running every day. On Monday she spends $\frac{1}{2}$ hour at each activity, covering a total of $12\frac{1}{2}$ mi. On Tuesday she runs for 12 min and cycles for 45 min, covering a total of 16 mi. Assuming that her running and cycling speeds don't change from day to day, find these speeds.

73. **Number Problem** The sum of the digits of a two-digit number is 7. When the digits are reversed, the number is increased by 27. Find the number.

74. **Area of a Triangle** Find the area of the triangle that lies in the first quadrant (with its base on the x-axis) and that is bounded by the lines $y = 2x - 4$ and $y = -4x + 20$.

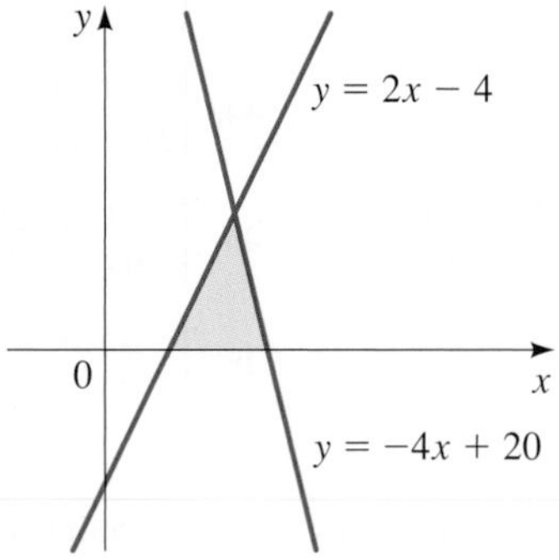

DISCOVERY ■ DISCUSSION ■ WRITING

75. **The Least Squares Line** The *least squares* line or *regression* line is the line that best fits a set of points in the plane. We studied this line in the *Focus on Modeling* that follows Chapter 1 (see page 130). By using calculus, it can be shown that the line that best fits the n data points $(x_1, y_1), (x_2, y_2), \ldots, (x_n, y_n)$ is the line $y = ax + b$, where the coefficients a and b satisfy the following pair of linear equations. (The notation $\sum_{k=1}^{n} x_k$ stands for the sum of all the x's. See Section 12.1 for a complete description of sigma (Σ) notation.)

$$\left(\sum_{k=1}^{n} x_k\right)a + nb = \sum_{k=1}^{n} y_k$$

$$\left(\sum_{k=1}^{n} x_k^2\right)a + \left(\sum_{k=1}^{n} x_k\right)b = \sum_{k=1}^{n} x_k y_k$$

Use these equations to find the least squares line for the following data points.

$$(1, 3),\quad (2, 5),\quad (3, 6),\quad (5, 6),\quad (7, 9)$$

Sketch the points and your line to confirm that the line fits these points well. If your calculator computes regression lines, see whether it gives you the same line as the formulas.

10.2 Systems of Linear Equations in Several Variables

Solving a Linear System ▶ The Number of Solutions of a Linear System ▶ Modeling Using Linear Systems

A **linear equation in n variables** is an equation that can be put in the form

$$a_1x_1 + a_2x_2 + \cdots + a_nx_n = c$$

where $a_1, a_2, \ldots, a_n$ and c are real numbers, and $x_1, x_2, \ldots, x_n$ are the variables. If we have only three or four variables, we generally use x, y, z, and w instead of x_1, x_2, x_3, and x_4. Such equations are called *linear* because if we have just two variables, the equation is $a_1x + a_2y = c$, which is the equation of a line. Here are some examples of equations in three variables that illustrate the difference between linear and nonlinear equations.

Linear equations	Nonlinear equations	
$6x_1 - 3x_2 + \sqrt{5}x_3 = 10$	$x^2 + 3y - \sqrt{z} = 5$	Not linear because it contains the square and the square root of a variable
$x + y + z = 2w - \frac{1}{2}$	$x_1x_2 + 6x_3 = -6$	Not linear because it contains a product of variables

In this section we study systems of linear equations in three or more variables.

▼ Solving a Linear System

The following are two examples of systems of linear equations in three variables. The second system is in **triangular form**; that is, the variable x doesn't appear in the second equation, and the variables x and y do not appear in the third equation.

A system of linear equations

$$\begin{cases} x - 2y - z = 1 \\ -x + 3y + 3z = 4 \\ 2x - 3y + z = 10 \end{cases}$$

A system in triangular form

$$\begin{cases} x - 2y - z = 1 \\ y + 2z = 5 \\ z = 3 \end{cases}$$

It's easy to solve a system that is in triangular form by using back-substitution. So our goal in this section is to start with a system of linear equations and change it to a system in triangular form that has the same solutions as the original system. We begin by showing how to use back-substitution to solve a system that is already in triangular form.

EXAMPLE 1 | Solving a Triangular System Using Back-Substitution

Solve the system using back-substitution:

$$\begin{cases} x - 2y - z = 1 & \text{Equation 1} \\ y + 2z = 5 & \text{Equation 2} \\ z = 3 & \text{Equation 3} \end{cases}$$

SOLUTION From the last equation we know that $z = 3$. We back-substitute this into the second equation and solve for y.

$$y + 2(3) = 5 \qquad \text{Back-substitute } z = 3 \text{ into Equation 2}$$

$$y = -1 \qquad \text{Solve for } y$$

Then we back-substitute $y = -1$ and $z = 3$ into the first equation and solve for x.

$$x - 2(-1) - (3) = 1 \qquad \text{Back-substitute } y = -1 \text{ and } z = 3 \text{ into Equation 1}$$

$$x = 2 \qquad \text{Solve for } x$$

The solution of the system is $x = 2$, $y = -1$, $z = 3$. We can also write the solution as the ordered triple $(2, -1, 3)$.

NOW TRY EXERCISE 7 ■

To change a system of linear equations to an **equivalent system** (that is, a system with the same solutions as the original system), we use the elimination method. This means that we can use the following operations.

OPERATIONS THAT YIELD AN EQUIVALENT SYSTEM

1. Add a nonzero multiple of one equation to another.
2. Multiply an equation by a nonzero constant.
3. Interchange the positions of two equations.

To solve a linear system, we use these operations to change the system to an equivalent triangular system. Then we use back-substitution as in Example 1. This process is called **Gaussian elimination**.

EXAMPLE 2 | Solving a System of Three Equations in Three Variables

Solve the system using Gaussian elimination.

$$\begin{cases} x - 2y + 3z = 1 & \text{Equation 1} \\ x + 2y - z = 13 & \text{Equation 2} \\ 3x + 2y - 5z = 3 & \text{Equation 3} \end{cases}$$

SOLUTION We need to change this to a triangular system, so we begin by eliminating the x-term from the second equation.

$$\begin{array}{rl} x + 2y - z = 13 & \text{Equation 2} \\ x - 2y + 3z = 1 & \text{Equation 1} \\ \hline 4y - 4z = 12 & \text{Equation 2} + (-1) \times \text{Equation 1} = \text{new Equation 2} \end{array}$$

This gives us a new, equivalent system that is one step closer to triangular form.

$$\begin{cases} x - 2y + 3z = 1 & \text{Equation 1} \\ 4y - 4z = 12 & \text{Equation 2} \\ 3x + 2y - 5z = 3 & \text{Equation 3} \end{cases}$$

Now we eliminate the x-term from the third equation.

$$\begin{array}{r} 3x + 2y - 5z = 3 \\ -3x + 6y - 9z = -3 \\ \hline 8y - 14z = 0 \end{array}$$

$$\begin{cases} x - 2y + 3z = 1 \\ 4y - 4z = 12 \\ 8y - 14z = 0 & \text{Equation 3} + (-3) \times \text{Equation 1} = \text{new Equation 3} \end{cases}$$

Then we eliminate the y-term from the third equation.

$$\begin{array}{r} 8y - 14z = 0 \\ -8y + 8z = -24 \\ \hline -6z = -24 \end{array}$$

$$\begin{cases} x - 2y + 3z = 1 \\ 4y - 4z = 12 \\ -6z = -24 & \text{Equation 3} + (-2) \times \text{Equation 2} = \text{new Equation 3} \end{cases}$$

The system is now in triangular form, but it will be easier to work with if we divide the second and third equations by the common factors of each term.

$$\begin{cases} x - 2y + 3z = 1 \\ y - z = 3 & \frac{1}{4} \times \text{Equation 2} = \text{new Equation 2} \\ z = 4 & -\frac{1}{6} \times \text{Equation 3} = \text{new Equation 3} \end{cases}$$

Now we use back-substitution to solve the system. From the third equation we get $z = 4$. We back-substitute this into the second equation and solve for y.

$$y - (4) = 3 \qquad \text{Back-substitute } z = 4 \text{ into Equation 2}$$

$$y = 7 \qquad \text{Solve for } y$$

CHECK YOUR ANSWER

$x = 3, y = 7, z = 4$:

$$\begin{aligned} (3) - 2(7) + 3(4) &= 1 \\ (3) + 2(7) - (4) &= 13 \\ 3(3) + 2(7) - 5(4) &= 3 \end{aligned} \quad ✔$$

Now we back-substitute $y = 7$ and $z = 4$ into the first equation and solve for x.

$$x - 2(7) + 3(4) = 1 \qquad \text{Back-substitute } y = 7 \text{ and } z = 4 \text{ into Equation 1}$$

$$x = 3 \qquad \text{Solve for } x$$

The solution of the system is $x = 3$, $y = 7$, $z = 4$, which we can write as the ordered triple $(3, 7, 4)$.

NOW TRY EXERCISE 17

The Number of Solutions of a Linear System

The graph of a linear equation in three variables is a plane in three dimensional space (see Section 9.6). A system of three equations in three variables represents three planes in space. The solutions of the system are the points where all three planes intersect. Three planes may intersect in a point, a line, not at all, or all three planes may coincide. Figure 1 illustrates some of these possibilities. Checking these possibilities we see that there are three possible outcomes when solving such a system.

NUMBER OF SOLUTIONS OF A LINEAR SYSTEM

For a system of linear equations, exactly one of the following is true.

1. The system has exactly one solution.
2. The system has no solution.
3. The system has infinitely many solutions.

A system with no solutions is said to be **inconsistent**, and a system with infinitely many solutions is said to be **dependent**. As we see in the next example, a linear system has no solution if we end up with a *false equation* after applying Gaussian elimination to the system.

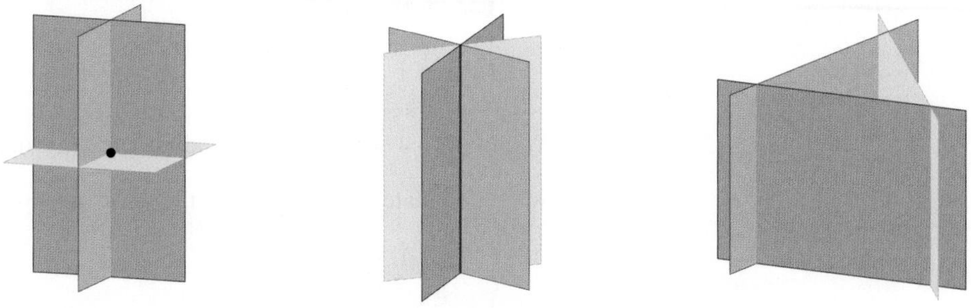

FIGURE 1

(a) The three planes intersect at a single point. The system has one solution.

(b) The three planes intersect at more than one point. The system has infinitely many solutions.

(c) The three planes have no point in common. The system has no solution.

EXAMPLE 3 | A System with No Solution

Solve the following system.

$$\begin{cases} x + 2y - 2z = 1 & \text{Equation 1} \\ 2x + 2y - z = 6 & \text{Equation 2} \\ 3x + 4y - 3z = 5 & \text{Equation 3} \end{cases}$$

SOLUTION To put this in triangular form, we begin by eliminating the x-terms from the second equation and the third equation.

$$\begin{cases} x + 2y - 2z = 1 \\ \quad -2y + 3z = 4 \\ 3x + 4y + 3z = 5 \end{cases}$$ Equation 2 + (–2) × Equation 1 = new Equation 2

$$\begin{cases} x + 2y - 2z = 1 \\ \quad -2y + 3z = 4 \\ \quad -2y + 3z = 2 \end{cases}$$ Equation 3 + (–3) × Equation 1 = new Equation 3

Now we eliminate the y-term from the third equation.

$$\begin{cases} x + 2y - 2z = 1 \\ \quad -2y + 3z = 4 \\ \qquad\qquad 0 = 2 \end{cases}$$ Equation 3 + (–1) × Equation 2 = new Equation 3

The system is now in triangular form, but the third equation says $0 = 2$, which is false. No matter what values we assign to x, y, and z, the third equation will never be true. This means that the system has *no solution*.

NOW TRY EXERCISE **27** ■

EXAMPLE 4 | A System with Infinitely Many Solutions

Solve the following system.

$$\begin{cases} x - y + 5z = -2 & \text{Equation 1} \\ 2x + y + 4z = 2 & \text{Equation 2} \\ 2x + 4y - 2z = 8 & \text{Equation 3} \end{cases}$$

SOLUTION To put this in triangular form, we begin by eliminating the x-terms from the second equation and the third equation.

$$\begin{cases} x - y + 5z = -2 \\ \quad 3y - 6z = 6 \\ 2x + 4y - 2z = 8 \end{cases}$$ Equation 2 + (–2) × Equation 1 = new Equation 2

$$\begin{cases} x - y + 5z = -2 \\ \quad 3y - 6z = 6 \\ \quad 6y - 12z = 12 \end{cases}$$ Equation 3 + (–2) × Equation 1 = new Equation 3

Now we eliminate the y-term from the third equation.

$$\begin{cases} x - y + 5z = -2 \\ \quad 3y - 6z = 6 \\ \qquad\quad 0 = 0 \end{cases}$$ Equation 3 + (–2) × Equation 2 = new Equation 3

The new third equation is true, but it gives us no new information, so we can drop it from the system. Only two equations are left. We can use them to solve for x and y in terms of z, but z can take on any value, so there are infinitely many solutions.

To find the complete solution of the system, we begin by solving for y in terms of z, using the new second equation.

$$3y - 6z = 6 \qquad \text{Equation 2}$$

$$y - 2z = 2 \qquad \text{Multiply by } \tfrac{1}{3}$$

$$y = 2z + 2 \qquad \text{Solve for } y$$

Then we solve for x in terms of z, using the first equation.

$$x - (2z + 2) + 5z = -2 \qquad \text{Substitute } y = 2z + 2 \text{ into Equation 1}$$

$$x + 3z - 2 = -2 \qquad \text{Simplify}$$

$$x = -3z \qquad \text{Solve for } x$$

To describe the complete solution, we let t represent any real number. The solution is

$$x = -3t$$
$$y = 2t + 2$$
$$z = t$$

We can also write this as the ordered triple $(-3t, 2t + 2, t)$.

NOW TRY EXERCISE **25** ■

In the solution of Example 4 the variable t is called a **parameter**. To get a specific solution, we give a specific value to the parameter t. For instance, if we set $t = 2$, we get

$$x = -3(2) = -6$$
$$y = 2(2) + 2 = 6$$
$$z = 2$$

Thus $(-6, 6, 2)$ is a solution of the system. Here are some other solutions of the system obtained by substituting other values for the parameter t.

Parameter t	Solution $(-3t, 2t + 2, t)$
-1	$(3, 0, -1)$
0	$(0, 2, 0)$
3	$(-9, 8, 3)$
10	$(-30, 22, 10)$

You should check that these points satisfy the original equations. There are infinitely many choices for the parameter t, so the system has infinitely many solutions.

▼ Modeling Using Linear Systems

Linear systems are used to model situations that involve several varying quantities. In the next example we consider an application of linear systems to finance.

EXAMPLE 5 | Modeling a Financial Problem Using a Linear System

Jason receives an inheritance of \$50,000. His financial advisor suggests that he invest this in three mutual funds: a money-market fund, a blue-chip stock fund, and a high-tech stock fund. The advisor estimates that the money-market fund will return 5% over the next year, the blue-chip fund 9%, and the high-tech fund 16%. Jason wants a total first-year return of \$4000. To avoid excessive risk, he decides to invest three times as much in the money-market fund as in the high-tech stock fund. How much should he invest in each fund?

SOLUTION

Let x = amount invested in the money-market fund

y = amount invested in the blue-chip stock fund

z = amount invested in the high-tech stock fund

We convert each fact given in the problem into an equation.

$$x + y + z = 50{,}000 \qquad \text{Total amount invested is \$50,000}$$

$$0.05x + 0.09y + 0.16z = 4000 \qquad \text{Total investment return is \$4000}$$

$$x = 3z \qquad \text{Money-market amount is } 3 \times \text{high-tech amount}$$

Multiplying the second equation by 100 and rewriting the third gives the following system, which we solve using Gaussian elimination.

$$\begin{cases} x + y + z = 50{,}000 \\ 5x + 9y + 16z = 400{,}000 \\ x - 3z = 0 \end{cases} \qquad \begin{matrix} \\ 100 \times \text{Equation 2} \\ \text{Subtract } 3z \end{matrix}$$

$$\begin{cases} x + y + z = 50{,}000 \\ 4y + 11z = 150{,}000 \\ -y - 4z = -50{,}000 \end{cases} \qquad \begin{matrix} \\ \text{Equation 2} + (-5) \times \text{Equation 1} = \text{new Equation 2} \\ \text{Equation 3} + (-1) \times \text{Equation 1} = \text{new Equation 3} \end{matrix}$$

$$\begin{cases} x + y + z = 50{,}000 \\ -5z = -50{,}000 \\ -y - 4z = -50{,}000 \end{cases} \qquad \text{Equation 2} + 4 \times \text{Equation 3} = \text{new Equation 3}$$

$$\begin{cases} x + y + z = 50{,}000 \\ z = 10{,}000 \\ y + 4z = 50{,}000 \end{cases} \qquad \begin{matrix} \\ \left(-\frac{1}{5}\right) \times \text{Equation 2} \\ (-1) \times \text{Equation 3} \end{matrix}$$

$$\begin{cases} x + y + z = 50{,}000 \\ y + 4z = 50{,}000 \\ z = 10{,}000 \end{cases} \qquad \text{Interchange Equations 2 and 3}$$

Now that the system is in triangular form, we use back-substitution to find that $x = 30{,}000$, $y = 10{,}000$, and $z = 10{,}000$. This means that Jason should invest

$30,000 in the money-market fund

$10,000 in the blue-chip stock fund

$10,000 in the high-tech stock fund

NOW TRY EXERCISE **37** ■

10.2 EXERCISES

CONCEPTS

1–2 ■ These exercises refer to the following system.

$$\begin{cases} x - y + z = 2 \\ -x + 2y + z = -3 \\ 3x + y - 2z = 2 \end{cases}$$

1. If we add 2 times the first equation to the second equation, the second equation becomes ________________ = _____.

2. To eliminate x from the third equation, we add ________ times the first equation to the third equation. The third equation becomes ________________ = _____.

SKILLS

3–6 ■ State whether the equation or system of equations is linear.

3. $6x - \sqrt{3}y + \frac{1}{2}z = 0$

4. $x^2 + y^2 + z^2 = 4$

5. $\begin{cases} xy - 3y + z = 5 \\ x - y^2 + 5z = 0 \\ 2x + yz = 3 \end{cases}$

6. $\begin{cases} x - 2y + 3z = 10 \\ 2x + 5y = 2 \\ y + 2z = 4 \end{cases}$

7–12 ■ Use back-substitution to solve the triangular system.

7. $\begin{cases} x - 2y + 4z = 3 \\ y + 2z = 7 \\ z = 2 \end{cases}$

8. $\begin{cases} x + y - 3z = 8 \\ y - 3z = 5 \\ z = -1 \end{cases}$

9. $\begin{cases} x + 2y + z = 7 \\ -y + 3z = 9 \\ 2z = 6 \end{cases}$

10. $\begin{cases} x - 2y + 3z = 10 \\ 2y - z = 2 \\ 3z = 12 \end{cases}$

11. $\begin{cases} 2x - y + 6z = 5 \\ y + 4z = 0 \\ -2z = 1 \end{cases}$

12. $\begin{cases} 4x + 3z = 10 \\ 2y - z = -6 \\ \frac{1}{2}z = 4 \end{cases}$

13–16 ■ Perform an operation on the given system that eliminates the indicated variable. Write the new equivalent system.

13. $\begin{cases} x - 2y - z = 4 \\ x - y + 3z = 0 \\ 2x + y + z = 0 \end{cases}$

Eliminate the x-term from the second equation.

14. $\begin{cases} x + y - 3z = 3 \\ -2x + 3y + z = 2 \\ x - y + 2z = 0 \end{cases}$

Eliminate the x-term from the second equation.

15. $\begin{cases} 2x - y + 3z = 2 \\ x + 2y - z = 4 \\ -4x + 5y + z = 10 \end{cases}$

Eliminate the x-term from the third equation.

16. $\begin{cases} x - 4y + z = 3 \\ y - 3z = 10 \\ 3y - 8z = 24 \end{cases}$

Eliminate the y-term from the third equation.

17–36 ■ Find the complete solution of the linear system, or show that it is inconsistent.

17. $\begin{cases} x - y - z = 4 \\ 2y + z = -1 \\ -x + y - 2z = 5 \end{cases}$

18. $\begin{cases} x - y + z = 0 \\ y + 2z = -2 \\ x + y - z = 2 \end{cases}$

19. $\begin{cases} x + y + z = 4 \\ x + 3y + 3z = 10 \\ 2x + y - z = 3 \end{cases}$

20. $\begin{cases} x + y + z = 0 \\ -x + 2y + 5z = 3 \\ 3x - y = 6 \end{cases}$

21. $\begin{cases} x - 4z = 1 \\ 2x - y - 6z = 4 \\ 2x + 3y - 2z = 8 \end{cases}$

22. $\begin{cases} x - y + 2z = 2 \\ 3x + y + 5z = 8 \\ 2x - y - 2z = -7 \end{cases}$

23. $\begin{cases} 2x + 4y - z = 2 \\ x + 2y - 3z = -4 \\ 3x - y + z = 1 \end{cases}$

24. $\begin{cases} 2x + y - z = -8 \\ -x + y + z = 3 \\ -2x + 4z = 18 \end{cases}$

25. $\begin{cases} y - 2z = 0 \\ 2x + 3y = 2 \\ -x - 2y + z = -1 \end{cases}$

26. $\begin{cases} 2y + z = 3 \\ 5x + 4y + 3z = -1 \\ x - 3y = -2 \end{cases}$

27. $\begin{cases} x + 2y - z = 1 \\ 2x + 3y - 4z = -3 \\ 3x + 6y - 3z = 4 \end{cases}$

28. $\begin{cases} -x + 2y + 5z = 4 \\ x - 2z = 0 \\ 4x - 2y - 11z = 2 \end{cases}$

29. $\begin{cases} 2x + 3y - z = 1 \\ x + 2y = 3 \\ x + 3y + z = 4 \end{cases}$

30. $\begin{cases} x - 2y - 3z = 5 \\ 2x + y - z = 5 \\ 4x - 3y - 7z = 5 \end{cases}$

31. $\begin{cases} x + y - z = 0 \\ x + 2y - 3z = -3 \\ 2x + 3y - 4z = -3 \end{cases}$

32. $\begin{cases} x - 2y + z = 3 \\ 2x - 5y + 6z = 7 \\ 2x - 3y - 2z = 5 \end{cases}$

33. $\begin{cases} x + 3y - 2z = 0 \\ 2x + 4z = 4 \\ 4x + 6y = 4 \end{cases}$

34. $\begin{cases} 2x + 4y - z = 3 \\ x + 2y + 4z = 6 \\ x + 2y - 2z = 0 \end{cases}$

35. $\begin{cases} x + z + 2w = 6 \\ y - 2z = -3 \\ x + 2y - z = -2 \\ 2x + y + 3z - 2w = 0 \end{cases}$

36. $\begin{cases} x + y + z + w = 0 \\ x + y + 2z + 2w = 0 \\ 2x + 2y + 3z + 4w = 1 \\ 2x + 3y + 4z + 5w = 2 \end{cases}$

APPLICATIONS

37–38 ■ **Finance** An investor has \$100,000 to invest in three types of bonds: short-term, intermediate-term, and long-term. How much should she invest in each type to satisfy the given conditions?

37. Short-term bonds pay 4% annually, intermediate-term bonds pay 5%, and long-term bonds pay 6%. The investor wishes to realize a total annual income of 5.1%, with equal amounts invested in short- and intermediate-term bonds.

38. Short-term bonds pay 4% annually, intermediate-term bonds pay 6%, and long-term bonds pay 8%. The investor wishes to have a total annual return of \$6700 on her investment, with equal amounts invested in intermediate- and long-term bonds.

39. Agriculture A farmer has 1200 acres of land on which he grows corn, wheat, and soybeans. It costs \$45 per acre to grow corn, \$60 to grow wheat, and \$50 to grow soybeans. Because of market demand, the farmer will grow twice as many acres of wheat as of corn. He has allocated \$63,750 for the cost of growing his crops. How many acres of each crop should he plant?

40. Gas Station A gas station sells three types of gas: Regular for \$3.00 a gallon, Performance Plus for \$3.20 a gallon, and Premium for \$3.30 a gallon. On a particular day 6500 gallons of gas were sold for a total of \$20,050. Three times as many gallons of Regular as Premium gas were sold. How many gallons of each type of gas were sold that day?

41. Nutrition A biologist is performing an experiment on the effects of various combinations of vitamins. She wishes to feed each of her laboratory rabbits a diet that contains exactly 9 mg of niacin, 14 mg of thiamin, and 32 mg of riboflavin. She has available three different types of commercial rabbit pellets; their vitamin content (per ounce) is given in the table. How many ounces of each type of food should each rabbit be given daily to satisfy the experiment requirements?

	Type A	Type B	Type C
Niacin (mg)	2	3	1
Thiamin (mg)	3	1	3
Riboflavin (mg)	8	5	7

42. Diet Program Nicole started a new diet that requires each meal to have 460 calories, 6 grams of fiber, and 11 grams of fat. The table shows the fiber, fat, and calorie content of one serving of each of three breakfast foods. How many servings of each food should Nicole eat to follow her diet?

Food	Fiber	Fat	Calories
Toast	2	1	100
Cottage cheese	0	5	120
Fruit	2	0	60

43. Juice Blends The Juice Company offers three kinds of smoothies: Midnight Mango, Tropical Torrent, and Pineapple Power. Each smoothie contains the amounts of juices shown in the table.

Smoothie	Mango juice (oz)	Pineapple juice (oz)	Orange juice (oz)
Midnight Mango	8	3	3
Tropical Torrent	6	5	3
Pineapple Power	2	8	4

On a particular day the Juice Company used 820 oz of mango juice, 690 oz of pineapple juice, and 450 oz of orange juice. How many smoothies of each kind were sold that day?

44. Appliance Manufacturing Kitchen Korner produces refrigerators, dishwashers, and stoves at three different factories. The table gives the number of each product produced at each factory per day. Kitchen Korner receives an order for 110 refrigerators, 150 dishwashers, and 114 ovens. How many days should each plant be scheduled to fill this order?

Appliance	Factory A	Factory B	Factory C
Refrigerators	8	10	14
Dishwashers	16	12	10
Stoves	10	18	6

45. Stock Portfolio An investor owns three stocks: A, B, and C. The closing prices of the stocks on three successive trading days are given in the table.

	Stock A	Stock B	Stock C
Monday	\$10	\$25	\$29
Tuesday	\$12	\$20	\$32
Wednesday	\$16	\$15	\$32

Despite the volatility in the stock prices, the total value of the investor's stocks remained unchanged at \$74,000 at the end of each of these three days. How many shares of each stock does the investor own?

46. Elecricity By using Kirchhoff's Laws, it can be shown that the currents I_1, I_2, and I_3 that pass through the three branches of the circuit in the figure satisfy the given linear system. Solve the system to find I_1, I_2, and I_3.

$$\begin{cases} I_1 + I_2 - I_3 = 0 \\ 16I_1 - 8I_2 = 4 \\ 8I_2 + 4I_3 = 5 \end{cases}$$

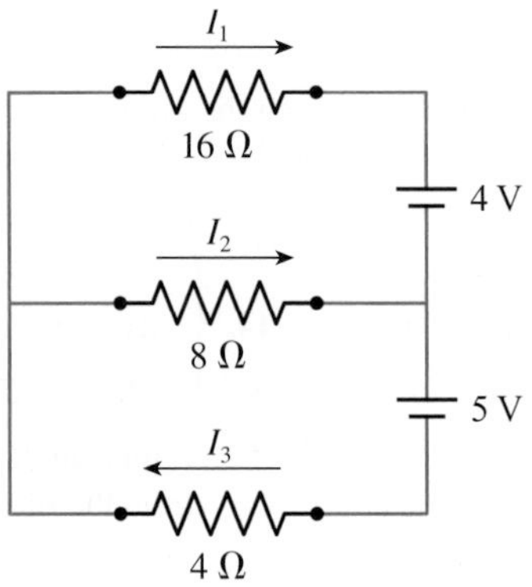

DISCOVERY ▪ DISCUSSION ▪ WRITING

47. Can a Linear System Have Exactly Two Solutions?

(a) Suppose that (x_0, y_0, z_0) and (x_1, y_1, z_1) are solutions of the system

$$\begin{cases} a_1x + b_1y + c_1z = d_1 \\ a_2x + b_2y + c_2z = d_2 \\ a_3x + b_3y + c_3z = d_3 \end{cases}$$

Show that $\left(\dfrac{x_0 + x_1}{2}, \dfrac{y_0 + y_1}{2}, \dfrac{z_0 + z_1}{2}\right)$ is also a solution.

(b) Use the result of part (a) to prove that if the system has two different solutions, then it has infinitely many solutions.

DISCOVERY PROJECT **Best Fit Versus Exact Fit**

In this project we use linear systems to find quadratic functions whose graphs pass through a set of given points. You can find the project at the book companion website: **www.stewartmath.com**

10.3 Matrices and Systems of Linear Equations

Matrices ▶ The Augmented Matrix of a Linear System ▶ Elementary Row Operations ▶ Gaussian Elimination ▶ Gauss-Jordan Elimination ▶ Inconsistent and Dependent Systems ▶ Modeling with Linear Systems

A *matrix* is simply a rectangular array of numbers. Matrices* are used to organize information into categories that correspond to the rows and columns of the matrix. For example, a scientist might organize information on a population of endangered whales as follows:

	Immature	**Juvenile**	**Adult**
Male	12	52	18
Female	15	42	11

This is a compact way of saying that there are 12 immature males, 15 immature females, 18 adult males, and so on.

In this section we represent a linear system by a matrix, called the *augmented matrix* of the system:

Linear system — **Augmented matrix**

$$\begin{cases} 2x - y = 5 \\ x + 4y = 7 \end{cases} \qquad \begin{bmatrix} 2 & -1 & 5 \\ 1 & 4 & 7 \end{bmatrix}$$

Equation 1

Equation 2

x y

The augmented matrix contains the same information as the system, but in a simpler form. The operations we learned for solving systems of equations can now be performed on the augmented matrix.

▼ Matrices

We begin by defining the various elements that make up a matrix.

DEFINITION OF MATRIX

An $m \times n$ **matrix** is a rectangular array of numbers with m **rows** and n **columns**.

$$\begin{bmatrix} a_{11} & a_{12} & a_{13} & \cdots & a_{1n} \\ a_{21} & a_{22} & a_{23} & \cdots & a_{2n} \\ a_{31} & a_{32} & a_{33} & \cdots & a_{3n} \\ \vdots & \vdots & \vdots & \ddots & \vdots \\ a_{m1} & a_{m2} & a_{m3} & \cdots & a_{mn} \end{bmatrix}$$

m rows

n columns

We say that the matrix has **dimension** $m \times n$. The numbers a_{ij} are the **entries** of the matrix. The subscript on the entry a_{ij} indicates that it is in the ith row and the jth column.

*The plural of *matrix* is *matrices*.

Here are some examples of matrices.

Matrix	Dimension	
$\begin{bmatrix} 1 & 3 & 0 \\ 2 & 4 & -1 \end{bmatrix}$	2×3	2 rows by 3 columns
$\begin{bmatrix} 6 & -5 & 0 & 1 \end{bmatrix}$	1×4	1 row by 4 columns

The Augmented Matrix of a Linear System

We can write a system of linear equations as a matrix, called the **augmented matrix** of the system, by writing only the coefficients and constants that appear in the equations. Here is an example.

Linear system

$$\begin{cases} 3x - 2y + z = 5 \\ x + 3y - z = 0 \\ -x + 4z = 11 \end{cases}$$

Augmented matrix

$$\begin{bmatrix} 3 & -2 & 1 & 5 \\ 1 & 3 & -1 & 0 \\ -1 & 0 & 4 & 11 \end{bmatrix}$$

Notice that a missing variable in an equation corresponds to a 0 entry in the augmented matrix.

EXAMPLE 1 | Finding the Augmented Matrix of a Linear System

Write the augmented matrix of the system of equations.

$$\begin{cases} 6x - 2y - z = 4 \\ x + 3z = 1 \\ 7y + z = 5 \end{cases}$$

SOLUTION First we write the linear system with the variables lined up in columns.

$$\begin{cases} 6x - 2y - z = 4 \\ x + 3z = 1 \\ 7y + z = 5 \end{cases}$$

The augmented matrix is the matrix whose entries are the coefficients and the constants in this system.

$$\begin{bmatrix} 6 & -2 & -1 & 4 \\ 1 & 0 & 3 & 1 \\ 0 & 7 & 1 & 5 \end{bmatrix}$$

NOW TRY EXERCISE **2** ■

Elementary Row Operations

The operations that we used in Section 10.2 to solve linear systems correspond to operations on the rows of the augmented matrix of the system. For example, adding a multiple of one equation to another corresponds to adding a multiple of one row to another.

ELEMENTARY ROW OPERATIONS

1. Add a multiple of one row to another.
2. Multiply a row by a nonzero constant.
3. Interchange two rows.

Note that performing any of these operations on the augmented matrix of a system does not change its solution. We use the following notation to describe the elementary row operations:

Symbol	Description
$R_i + kR_j \rightarrow R_i$	Change the *i*th row by adding *k* times row *j* to it, and then put the result back in row *i*.
kR_i	Multiply the *i*th row by *k*.
$R_i \leftrightarrow R_j$	Interchange the *i*th and *j*th rows.

In the next example we compare the two ways of writing systems of linear equations.

EXAMPLE 2 | Using Elementary Row Operations to Solve a Linear System

Solve the system of linear equations.

$$\begin{cases} x - y + 3z = 4 \\ x + 2y - 2z = 10 \\ 3x - y + 5z = 14 \end{cases}$$

SOLUTION Our goal is to eliminate the *x*-term from the second equation and the *x*- and *y*-terms from the third equation. For comparison, we write both the system of equations and its augmented matrix.

	System		Augmented matrix
	$\begin{cases} x - y + 3z = 4 \\ x + 2y - 2z = 10 \\ 3x - y + 5z = 14 \end{cases}$		$\begin{bmatrix} 1 & -1 & 3 & 4 \\ 1 & 2 & -2 & 10 \\ 3 & -1 & 5 & 14 \end{bmatrix}$
Add $(-1) \times$ Equation 1 to Equation 2. Add $(-3) \times$ Equation 1 to Equation 3.	$\begin{cases} x - y + 3z = 4 \\ 3y - 5z = 6 \\ 2y - 4z = 2 \end{cases}$	$\xrightarrow[R_3 - 3R_1 \rightarrow R_3]{R_2 - R_1 \rightarrow R_2}$	$\begin{bmatrix} 1 & -1 & 3 & 4 \\ 0 & 3 & -5 & 6 \\ 0 & 2 & -4 & 2 \end{bmatrix}$
Multiply Equation 3 by $\frac{1}{2}$.	$\begin{cases} x - y + 3z = 4 \\ 3y - 5z = 6 \\ y - 2z = 1 \end{cases}$	$\xrightarrow{\frac{1}{2}R_3}$	$\begin{bmatrix} 1 & -1 & 3 & 4 \\ 0 & 3 & -5 & 6 \\ 0 & 1 & -2 & 1 \end{bmatrix}$
Add $(-3) \times$ Equation 3 to Equation 2 (to eliminate *y* from Equation 2).	$\begin{cases} x - y + 3z = 4 \\ z = 3 \\ y - 2z = 1 \end{cases}$	$\xrightarrow{R_2 - 3R_3 \rightarrow R_2}$	$\begin{bmatrix} 1 & -1 & 3 & 4 \\ 0 & 0 & 1 & 3 \\ 0 & 1 & -2 & 1 \end{bmatrix}$
Interchange Equations 2 and 3.	$\begin{cases} x - y + 3z = 4 \\ y - 2z = 1 \\ z = 3 \end{cases}$	$\xrightarrow{R_2 \leftrightarrow R_3}$	$\begin{bmatrix} 1 & -1 & 3 & 4 \\ 0 & 1 & -2 & 1 \\ 0 & 0 & 1 & 3 \end{bmatrix}$

Now we use back-substitution to find that $x = 2$, $y = 7$, and $z = 3$. The solution is $(2, 7, 3)$.

NOW TRY EXERCISE **19** ■

▼ Gaussian Elimination

In general, to solve a system of linear equations using its augmented matrix, we use elementary row operations to arrive at a matrix in a certain form. This form is described in the following box.

ROW-ECHELON FORM AND REDUCED ROW-ECHELON FORM OF A MATRIX

A matrix is in **row-echelon form** if it satisfies the following conditions.

1. The first nonzero number in each row (reading from left to right) is 1. This is called the **leading entry**.
2. The leading entry in each row is to the right of the leading entry in the row immediately above it.
3. All rows consisting entirely of zeros are at the bottom of the matrix.

A matrix is in **reduced row-echelon** form if it is in row-echelon form and also satisfies the following condition.

4. Every number above and below each leading entry is a 0.

In the following matrices, the first one is not in row-echelon form. The second one *is* in row-echelon form, and the third one is in reduced row-echelon form. The entries in red are the leading entries.

Not in row-echelon form

$$\begin{bmatrix} 0 & 1 & -\frac{1}{2} & 0 & 6 \\ 1 & 0 & 3 & 4 & -5 \\ 0 & 0 & 0 & 1 & 0.4 \\ 0 & 1 & 1 & 0 & 0 \end{bmatrix}$$

Leading 1's do *not* shift to the right in successive rows

Row-echelon form

$$\begin{bmatrix} 1 & 3 & -6 & 10 & 0 \\ 0 & 0 & 1 & 4 & -3 \\ 0 & 0 & 0 & 1 & \frac{1}{2} \\ 0 & 0 & 0 & 0 & 0 \end{bmatrix}$$

Leading 1's shift to the right in successive rows

Reduced row-echelon form

$$\begin{bmatrix} 1 & 3 & 0 & 0 & 0 \\ 0 & 0 & 1 & 0 & -3 \\ 0 & 0 & 0 & 1 & \frac{1}{2} \\ 0 & 0 & 0 & 0 & 0 \end{bmatrix}$$

Leading 1's have 0's above and below them

Here is a systematic way to put a matrix in row-echelon form using elementary row operations:

- Start by obtaining 1 in the top left corner. Then obtain zeros below that 1 by adding appropriate multiples of the first row to the rows below it.
- Next, obtain a leading 1 in the next row, and then obtain zeros below that 1.
- At each stage make sure that every leading entry is to the right of the leading entry in the row above it—rearrange the rows if necessary.
- Continue this process until you arrive at a matrix in row-echelon form.

This is how the process might work for a 3×4 matrix:

$$\begin{bmatrix} 1 & \blacksquare & \blacksquare & \blacksquare \\ 0 & \blacksquare & \blacksquare & \blacksquare \\ 0 & \blacksquare & \blacksquare & \blacksquare \end{bmatrix} \quad \begin{bmatrix} 1 & \blacksquare & \blacksquare & \blacksquare \\ 0 & 1 & \blacksquare & \blacksquare \\ 0 & 0 & \blacksquare & \blacksquare \end{bmatrix} \quad \begin{bmatrix} 1 & \blacksquare & \blacksquare & \blacksquare \\ 0 & 1 & \blacksquare & \blacksquare \\ 0 & 0 & 1 & \blacksquare \end{bmatrix}$$

Once an augmented matrix is in row-echelon form, we can solve the corresponding linear system using back-substitution. This technique is called **Gaussian elimination**, in honor of its inventor, the German mathematician C. F. Gauss (see page 272).

SOLVING A SYSTEM USING GAUSSIAN ELIMINATION

1. **Augmented Matrix.** Write the augmented matrix of the system.
2. **Row-Echelon Form.** Use elementary row operations to change the augmented matrix to row-echelon form.
3. **Back-Substitution.** Write the new system of equations that corresponds to the row-echelon form of the augmented matrix and solve by back-substitution.

EXAMPLE 3 | Solving a System Using Row-Echelon Form

Solve the system of linear equations using Gaussian elimination.

$$\begin{cases} 4x + 8y - 4z = 4 \\ 3x + 8y + 5z = -11 \\ -2x + y + 12z = -17 \end{cases}$$

SOLUTION We first write the augmented matrix of the system, and then we use elementary row operations to put it in row-echelon form.

Need a 1 here

Augmented matrix:

$$\begin{bmatrix} \textcircled{4} & 8 & -4 & 4 \\ 3 & 8 & 5 & -11 \\ -2 & 1 & 12 & -17 \end{bmatrix}$$

$$\xrightarrow{\frac{1}{4}R_1} \begin{bmatrix} 1 & 2 & -1 & 1 \\ \textcircled{3} & 8 & 5 & -11 \\ \textcircled{-2} & 1 & 12 & -17 \end{bmatrix}$$

Need 0's here

$$\xrightarrow[R_3 + 2R_1 \to R_3]{R_2 - 3R_1 \to R_2} \begin{bmatrix} 1 & 2 & -1 & 1 \\ 0 & \textcircled{2} & 8 & -14 \\ 0 & 5 & 10 & -15 \end{bmatrix}$$

Need a 1 here

$$\xrightarrow{\frac{1}{2}R_2} \begin{bmatrix} 1 & 2 & -1 & 1 \\ 0 & 1 & 4 & -7 \\ 0 & \textcircled{5} & 10 & -15 \end{bmatrix}$$

Need a 0 here

$$\xrightarrow{R_3 - 5R_2 \to R_3} \begin{bmatrix} 1 & 2 & -1 & 1 \\ 0 & 1 & 4 & -7 \\ 0 & 0 & \textcircled{-10} & 20 \end{bmatrix}$$

Need a 1 here

Row-echelon form:

$$\xrightarrow{-\frac{1}{10}R_3} \begin{bmatrix} 1 & 2 & -1 & 1 \\ 0 & 1 & 4 & -7 \\ 0 & 0 & 1 & -2 \end{bmatrix}$$

We now have an equivalent matrix in row-echelon form, and the corresponding system of equations is

$$\begin{cases} x + 2y - z = 1 \\ y + 4z = -7 \\ z = -2 \end{cases}$$

Back-substitute: We use back-substitution to solve the system.

$$y + 2(-2) = -3 \qquad \text{Back-substitute } z = -2 \text{ into Equation 2}$$

$$y = 1 \qquad \text{Solve for } y$$

$$x + 2(1) - (-2) = 1 \qquad \text{Back-substitute } y = 1 \text{ and } z = -2 \text{ into Equation 1}$$

$$x = -3 \qquad \text{Solve for } x$$

So the solution of the system is $(-3, 1, -2)$.

NOW TRY EXERCISE 21 ■

```
ref([A])
  [[1 2 -1 1 ]
   [0 1 2  -3]
   [0 0 1  -2]]
```

FIGURE 1

Graphing calculators have a "row-echelon form" command that puts a matrix in row-echelon form. (On the TI-83 this command is `ref`.) For the augmented matrix in Example 3 the `ref` command gives the output shown in Figure 1. Notice that the row-echelon form that is obtained by the calculator differs from the one we got in Example 3. This is because the calculator used different row operations than we did. You should check that your calculator's row-echelon form leads to the same solution as ours.

▼ Gauss-Jordan Elimination

If we put the augmented matrix of a linear system in *reduced* row-echelon form, then we don't need to back-substitute to solve the system. To put a matrix in reduced row-echelon form, we use the following steps.

- Use the elementary row operations to put the matrix in row-echelon form.
- Obtain zeros above each leading entry by adding multiples of the row containing that entry to the rows above it. Begin with the last leading entry and work up.

Here is how the process works for a 3×4 matrix:

$$\begin{bmatrix} 1 & \blacksquare & \blacksquare & \blacksquare \\ 0 & 1 & \blacksquare & \blacksquare \\ 0 & 0 & 1 & \blacksquare \end{bmatrix} \rightarrow \begin{bmatrix} 1 & \blacksquare & 0 & \blacksquare \\ 0 & 1 & 0 & \blacksquare \\ 0 & 0 & 1 & \blacksquare \end{bmatrix} \rightarrow \begin{bmatrix} 1 & 0 & 0 & \blacksquare \\ 0 & 1 & 0 & \blacksquare \\ 0 & 0 & 1 & \blacksquare \end{bmatrix}$$

Using the reduced row-echelon form to solve a system is called **Gauss-Jordan elimination**. The process is illustrated in the next example.

EXAMPLE 4 | Solving a System Using Reduced Row-Echelon Form

Solve the system of linear equations, using Gauss-Jordan elimination.

$$\begin{cases} 4x + 8y - 4z = 4 \\ 3x + 8y + 5z = -11 \\ -2x + y + 12z = -17 \end{cases}$$

SOLUTION In Example 3 we used Gaussian elimination on the augmented matrix of this system to arrive at an equivalent matrix in row-echelon form. We continue using elementary row operations on the last matrix in Example 3 to arrive at an equivalent matrix in reduced row-echelon form.

Need 0's here

$$\begin{bmatrix} 1 & 2 & \textcircled{-1} & 1 \\ 0 & 1 & \textcircled{4} & -7 \\ 0 & 0 & 1 & -2 \end{bmatrix}$$

Need a 0 here

$$\xrightarrow[R_1 + R_3 \rightarrow R_1]{R_2 - 4R_3 \rightarrow R_2} \begin{bmatrix} 1 & \textcircled{2} & 0 & -1 \\ 0 & 1 & 0 & 1 \\ 0 & 0 & 1 & -2 \end{bmatrix}$$

$$\xrightarrow{R_1 - 2R_2 \rightarrow R_1} \begin{bmatrix} 1 & 0 & 0 & -3 \\ 0 & 1 & 0 & 1 \\ 0 & 0 & 1 & -2 \end{bmatrix}$$

We now have an equivalent matrix in reduced row-echelon form, and the corresponding system of equations is

$$\begin{cases} x = -3 \\ y = 1 \\ z = -2 \end{cases}$$

Since the system is in reduced row-echelon form, back-subsitution is not required to get the solution.

Hence we immediately arrive at the solution $(-3, 1, -2)$.

NOW TRY EXERCISE 23 ■

```
rref([A])
   [[1 0 0 -3]
    [0 1 0 1 ]
    [0 0 1 -2]]
```

FIGURE 2

Graphing calculators also have a command that puts a matrix in reduced row-echelon form. (On the TI-83 this command is `rref`.) For the augmented matrix in Example 4, the `rref` command gives the output shown in Figure 2. The calculator gives the same reduced row-echelon form as the one we got in Example 4. This is because every matrix has a *unique* reduced row-echelon form.

▼ Inconsistent and Dependent Systems

The systems of linear equations that we considered in Examples 1–4 had exactly one solution. But as we know from Section 10.2, a linear system may have one solution, no solution, or infinitely many solutions. Fortunately, the row-echelon form of a system allows us to determine which of these cases applies, as described in the following box.

First we need some terminology. A **leading variable** in a linear system is one that corresponds to a leading entry in the row-echelon form of the augmented matrix of the system.

THE SOLUTIONS OF A LINEAR SYSTEM IN ROW-ECHELON FORM

Suppose the augmented matrix of a system of linear equations has been transformed by Gaussian elimination into row-echelon form. Then exactly one of the following is true.

1. **No solution.** If the row-echelon form contains a row that represents the equation $0 = c$, where c is not zero, then the system has no solution. A system with no solution is called **inconsistent**.
2. **One solution.** If each variable in the row-echelon form is a leading variable, then the system has exactly one solution, which we find using back-substitution or Gauss-Jordan elimination.
3. **Infinitely many solutions.** If the variables in the row-echelon form are not all leading variables and if the system is not inconsistent, then it has infinitely many solutions. In this case the system is called **dependent**. We solve the system by putting the matrix in reduced row-echelon form and then expressing the leading variables in terms of the nonleading variables. The nonleading variables may take on any real numbers as their values.

The matrices below, all in row-echelon form, illustrate the three cases described above.

No solution

$$\begin{bmatrix} 1 & 2 & 5 & 7 \\ 0 & 1 & 3 & 4 \\ 0 & 0 & 0 & 1 \end{bmatrix}$$

Last equation says $0 = 1$

One solution

$$\begin{bmatrix} 1 & 6 & -1 & 3 \\ 0 & 1 & 2 & -2 \\ 0 & 0 & 1 & 8 \end{bmatrix}$$

Each variable is a leading variable

Infinitely many solutions

$$\begin{bmatrix} 1 & 2 & -3 & 1 \\ 0 & 1 & 5 & -2 \\ 0 & 0 & 0 & 0 \end{bmatrix}$$

z is not a leading variable

EXAMPLE 5 | A System with No Solution

Solve the system.

$$\begin{cases} x - 3y + 2z = 12 \\ 2x - 5y + 5z = 14 \\ x - 2y + 3z = 20 \end{cases}$$

SOLUTION We transform the system into row-echelon form.

$$\left[\begin{array}{ccc|c} 1 & -3 & 2 & 12 \\ 2 & -5 & 5 & 14 \\ 1 & -2 & 3 & 20 \end{array}\right] \xrightarrow[R_3 - R_1 \to R_3]{R_2 - 2R_1 \to R_2} \left[\begin{array}{ccc|c} 1 & -3 & 2 & 12 \\ 0 & 1 & 1 & -10 \\ 0 & 1 & 1 & 8 \end{array}\right]$$

$$\xrightarrow{R_3 - R_2 \to R_3} \left[\begin{array}{ccc|c} 1 & -3 & 2 & 12 \\ 0 & 1 & 1 & -10 \\ 0 & 0 & 0 & 18 \end{array}\right] \xrightarrow{\frac{1}{18}R_3} \left[\begin{array}{ccc|c} 1 & -3 & 2 & 12 \\ 0 & 1 & 1 & -10 \\ 0 & 0 & 0 & 1 \end{array}\right]$$

This last matrix is in row-echelon form, so we can stop the Gaussian elimination process. Now if we translate the last row back into equation form, we get $0x + 0y + 0z = 1$, or $0 = 1$, which is false. No matter what values we pick for x, y, and z, the last equation will never be a true statement. This means that the system *has no solution.*

NOW TRY EXERCISE **29**

```
ref([A])
[[1  -2.5 2.5 7   ]
 [0  1    1    -10]
 [0  0    0    1  ]]
```

FIGURE 3

Figure 3 shows the row-echelon form produced by a TI-83 calculator for the augmented matrix in Example 5. You should check that this gives the same solution.

EXAMPLE 6 | A System with Infinitely Many Solutions

Find the complete solution of the system.

$$\begin{cases} -3x - 5y + 36z = 10 \\ -x + 7z = 5 \\ x + y - 10z = -4 \end{cases}$$

SOLUTION We transform the system into reduced row-echelon form.

$$\left[\begin{array}{ccc|c} -3 & -5 & 36 & 10 \\ -1 & 0 & 7 & 5 \\ 1 & 1 & -10 & -4 \end{array}\right] \xrightarrow{R_1 \leftrightarrow R_3} \left[\begin{array}{ccc|c} 1 & 1 & -10 & -4 \\ -1 & 0 & 7 & 5 \\ -3 & -5 & 36 & 10 \end{array}\right]$$

$$\xrightarrow[R_3 + 3R_1 \to R_3]{R_2 + R_1 \to R_2} \left[\begin{array}{ccc|c} 1 & 1 & -10 & -4 \\ 0 & 1 & -3 & 1 \\ 0 & -2 & 6 & -2 \end{array}\right] \xrightarrow{R_3 + 2R_2 \to R_3} \left[\begin{array}{ccc|c} 1 & 1 & -10 & -4 \\ 0 & 1 & -3 & 1 \\ 0 & 0 & 0 & 0 \end{array}\right]$$

$$\xrightarrow{R_1 - R_2 \to R_1} \left[\begin{array}{ccc|c} 1 & 0 & -7 & -5 \\ 0 & 1 & -3 & 1 \\ 0 & 0 & 0 & 0 \end{array}\right]$$

Reduced row-echelon form on the TI-83 calculator:

```
rref([A])
  [[1 0 -7 -5 ]
   [0 1 -3 1  ]
   [0 0 0  0  ]]
```

The third row corresponds to the equation $0 = 0$. This equation is always true, no matter what values are used for x, y, and z. Since the equation adds no new information about the variables, we can drop it from the system. So the last matrix corresponds to the system

$$\begin{cases} x - 7z = -5 & \text{Equation 1} \\ y - 3z = 1 & \text{Equation 2} \end{cases}$$

Leading variables

Now we solve for the leading variables x and y in terms of the nonleading variable z:

$$x = 7z - 5 \qquad \text{Solve for } x \text{ in Equation 1}$$

$$y = 3z + 1 \qquad \text{Solve for } y \text{ in Equation 2}$$

To obtain the complete solution, we let t represent any real number, and we express x, y, and z in terms of t:

$$x = 7t - 5$$
$$y = 3t + 1$$
$$z = t$$

We can also write the solution as the ordered triple $(7t - 5, 3t + 1, t)$, where t is any real number.

NOW TRY EXERCISE **31** ■

In Example 6, to get specific solutions, we give a specific value to t. For example, if $t = 1$, then

$$x = 7(1) - 5 = 2$$
$$y = 3(1) + 1 = 4$$
$$z = 1$$

Here are some other solutions of the system obtained by substituting other values for the parameter t.

Parameter t	**Solution $(7t - 5, 3t + 1, t)$**
-1	$(-12, -2, -1)$
0	$(-5, 1, 0)$
2	$(9, 7, 2)$
5	$(30, 16, 5)$

EXAMPLE 7 | A System with Infinitely Many Solutions

Find the complete solution of the system.

$$\begin{cases} x + 2y - 3z - 4w = 10 \\ x + 3y - 3z - 4w = 15 \\ 2x + 2y - 6z - 8w = 10 \end{cases}$$

SOLUTION We transform the system into reduced row-echelon form.

$$\begin{bmatrix} 1 & 2 & -3 & -4 & 10 \\ 1 & 3 & -3 & -4 & 15 \\ 2 & 2 & -6 & -8 & 10 \end{bmatrix} \xrightarrow[R_3 - 2R_1 \to R_3]{R_2 - R_1 \to R_2} \begin{bmatrix} 1 & 2 & -3 & -4 & 10 \\ 0 & 1 & 0 & 0 & 5 \\ 0 & -2 & 0 & 0 & -10 \end{bmatrix}$$

$$\xrightarrow{R_3 + 2R_2 \to R_3} \begin{bmatrix} 1 & 2 & -3 & -4 & 10 \\ 0 & 1 & 0 & 0 & 5 \\ 0 & 0 & 0 & 0 & 0 \end{bmatrix} \xrightarrow{R_1 - 2R_2 \to R_1} \begin{bmatrix} 1 & 0 & -3 & -4 & 0 \\ 0 & 1 & 0 & 0 & 5 \\ 0 & 0 & 0 & 0 & 0 \end{bmatrix}$$

This is in reduced row-echelon form. Since the last row represents the equation $0 = 0$, we may discard it. So the last matrix corresponds to the system

$$\begin{cases} x \qquad -3z - 4w = 0 \\ \quad y \qquad\qquad\quad = 5 \end{cases}$$

Leading variables

To obtain the complete solution, we solve for the leading variables x and y in terms of the nonleading variables z and w, and we let z and w be any real numbers. Thus the complete solution is

$$\begin{aligned} x &= 3s + 4t \\ y &= 5 \\ z &= s \\ w &= t \end{aligned}$$

where s and t are any real numbers.

NOW TRY EXERCISE **51**

Note that s and t do *not* have to be the *same* real number in the solution for Example 7. We can choose arbitrary values for each if we wish to construct a specific solution to the system. For example, if we let $s = 1$ and $t = 2$, then we get the solution $(11, 5, 1, 2)$. You should check that this does indeed satisfy all three of the original equations in Example 7.

Examples 6 and 7 illustrate this general fact: If a system in row-echelon form has n nonzero equations in m variables $(m > n)$, then the complete solution will have $m - n$ nonleading variables. For instance, in Example 6 we arrived at *two* nonzero equations in the *three* variables x, y, and z, which gave us $3 - 2 = 1$ nonleading variable.

Modeling with Linear Systems

Linear equations, often containing hundreds or even thousands of variables, occur frequently in the applications of algebra to the sciences and to other fields. For now, let's consider an example that involves only three variables.

EXAMPLE 8 | Nutritional Analysis Using a System of Linear Equations

A nutritionist is performing an experiment on student volunteers. He wishes to feed one of his subjects a daily diet that consists of a combination of three commercial diet foods: MiniCal, LiquiFast, and SlimQuick. For the experiment it is important that the subject consume exactly 500 mg of potassium, 75 g of protein, and 1150 units of vitamin D every day. The amounts of these nutrients in one ounce of each food are given in the table. How many ounces of each food should the subject eat every day to satisfy the nutrient requirements exactly?

	MiniCal	LiquiFast	SlimQuick
Potassium (mg)	50	75	10
Protein (g)	5	10	3
Vitamin D (units)	90	100	50

SOLUTION Let x, y, and z represent the number of ounces of MiniCal, LiquiFast, and SlimQuick, respectively, that the subject should eat every day. This means that he will get $50x$ mg of potassium from MiniCal, $75y$ mg from LiquiFast, and $10z$ mg from SlimQuick, for a total of $50x + 75y + 10z$ mg potassium in all. Since the potassium requirement is 500 mg, we get the first equation below. Similar reasoning for the protein and vitamin D requirements leads to the system

$$\begin{cases} 50x + 75y + 10z = 500 & \text{Potassium} \\ 5x + 10y + 3z = 75 & \text{Protein} \\ 90x + 100y + 50z = 1150 & \text{Vitamin D} \end{cases}$$

```
rref([A])
  [[1 0 0 5 ]
   [0 1 0 2 ]
   [0 0 1 10]]
```

FIGURE 4

Dividing the first equation by 5 and the third one by 10 gives the system

$$\begin{cases} 10x + 15y + 2z = 100 \\ 5x + 10y + 3z = 75 \\ 9x + 10y + 5z = 115 \end{cases}$$

We can solve this system using Gaussian elimination, or we can use a graphing calculator to find the reduced row-echelon form of the augmented matrix of the system. Using the `rref` command on the TI-83, we get the output in Figure 4. From the reduced row-echelon form we see that $x = 5$, $y = 2$, $z = 10$. The subject should be fed 5 oz of Mini-Cal, 2 oz of LiquiFast, and 10 oz of SlimQuick every day.

CHECK YOUR ANSWER

$x = 5, y = 2, z = 10$:

$$\begin{cases} 10(5) + 15(2) + 2(10) = 100 \\ 5(5) + 10(2) + 3(10) = 75 \\ 9(5) + 10(2) + 5(10) = 115 \end{cases} \checkmark$$

NOW TRY EXERCISE **55**

A more practical application might involve dozens of foods and nutrients rather than just three. Such problems lead to systems with large numbers of variables and equations. Computers or graphing calculators are essential for solving such large systems.

10.3 EXERCISES

CONCEPTS

1. If a system of linear equations has infinitely many solutions, then the system is called ________. If a system of linear equations has no solution, then the system is called ________.

2. Write the augmented matrix of the following system of equations.

System

$$\begin{cases} x + y - z = 1 \\ x + 2z = -3 \\ 2y - z = 3 \end{cases}$$

Augmented matrix

$$\begin{bmatrix} \blacksquare & \blacksquare & \blacksquare & \blacksquare \\ \blacksquare & \blacksquare & \blacksquare & \blacksquare \\ \blacksquare & \blacksquare & \blacksquare & \blacksquare \end{bmatrix}$$

3. The following matrix is the augmented matrix of a system of linear equations in the variables x, y, and z. (It is given in reduced row-echelon form.)

$$\begin{bmatrix} 1 & 0 & -1 & 3 \\ 0 & 1 & 2 & 5 \\ 0 & 0 & 0 & 0 \end{bmatrix}$$

(a) The leading variables are ________.

(b) Is the system inconsistent or dependent?________

(c) The solution of the system is:

$x =$ _____, $y =$ _____, $z =$ _____

4. The augmented matrix of a system of linear equations is given in reduced row-echelon form. Find the solution of the system.

(a) $\begin{bmatrix} 1 & 0 & 0 & 2 \\ 0 & 1 & 0 & 1 \\ 0 & 0 & 1 & 3 \end{bmatrix}$ **(b)** $\begin{bmatrix} 1 & 0 & 1 & 2 \\ 0 & 1 & 1 & 1 \\ 0 & 0 & 0 & 0 \end{bmatrix}$ **(c)** $\begin{bmatrix} 1 & 0 & 0 & 2 \\ 0 & 1 & 0 & 1 \\ 0 & 0 & 0 & 3 \end{bmatrix}$

(a) $x =$ _____ $y =$ _____ $z =$ _____

(b) $x =$ _____ $y =$ _____ $z =$ _____

(c) $x =$ _____ $y =$ _____ $z =$ _____

SKILLS

5–10 ■ State the dimension of the matrix.

5. $\begin{bmatrix} 2 & 7 \\ 0 & -1 \\ 5 & -3 \end{bmatrix}$ **6.** $\begin{bmatrix} -1 & 5 & 4 & 0 \\ 0 & 2 & 11 & 3 \end{bmatrix}$ **7.** $\begin{bmatrix} 12 \\ 35 \end{bmatrix}$

8. $\begin{bmatrix} -3 \\ 0 \\ 1 \end{bmatrix}$ **9.** $\begin{bmatrix} 1 & 4 & 7 \end{bmatrix}$ **10.** $\begin{bmatrix} 1 & 0 \\ 0 & 1 \end{bmatrix}$

11–18 ■ A matrix is given. **(a)** Determine whether the matrix is in row-echelon form. **(b)** Determine whether the matrix is in reduced row-echelon form. **(c)** Write the system of equations for which the given matrix is the augmented matrix.

11. $\begin{bmatrix} 1 & 0 & -3 \\ 0 & 1 & 5 \end{bmatrix}$ **12.** $\begin{bmatrix} 1 & 3 & -3 \\ 0 & 1 & 5 \end{bmatrix}$

13. $\begin{bmatrix} 1 & 2 & 8 & 0 \\ 0 & 1 & 3 & 2 \\ 0 & 0 & 0 & 0 \end{bmatrix}$ **14.** $\begin{bmatrix} 1 & 0 & -7 & 0 \\ 0 & 1 & 3 & 0 \\ 0 & 0 & 0 & 1 \end{bmatrix}$

15. $\begin{bmatrix} 1 & 0 & 0 & 0 \\ 0 & 0 & 0 & 0 \\ 0 & 1 & 5 & 1 \end{bmatrix}$ **16.** $\begin{bmatrix} 1 & 0 & 0 & 1 \\ 0 & 1 & 0 & 2 \\ 0 & 0 & 1 & 3 \end{bmatrix}$

17. $\begin{bmatrix} 1 & 3 & 0 & -1 & 0 \\ 0 & 0 & 1 & 2 & 0 \\ 0 & 0 & 0 & 0 & 1 \\ 0 & 0 & 0 & 0 & 0 \end{bmatrix}$ **18.** $\begin{bmatrix} 1 & 3 & 0 & 1 & 0 & 0 \\ 0 & 1 & 0 & 4 & 0 & 0 \\ 0 & 0 & 0 & 1 & 1 & 2 \\ 0 & 0 & 0 & 1 & 0 & 0 \end{bmatrix}$

19–28 ■ The system of linear equations has a unique solution. Find the solution using Gaussian elimination or Gauss-Jordan elimination.

19. $\begin{cases} x - 2y + z = 1 \\ y + 2z = 5 \\ x + y + 3z = 8 \end{cases}$ **20.** $\begin{cases} x + y + 6z = 3 \\ x + y + 3z = 3 \\ x + 2y + 4z = 7 \end{cases}$

21. $\begin{cases} x + y + z = 2 \\ 2x - 3y + 2z = 4 \\ 4x + y - 3z = 1 \end{cases}$

22. $\begin{cases} x + y + z = 4 \\ -x + 2y + 3z = 17 \\ 2x - y = -7 \end{cases}$

23. $\begin{cases} x + 2y - z = -2 \\ x + z = 0 \\ 2x - y - z = -3 \end{cases}$

24. $\begin{cases} 2y + z = 4 \\ x + y = 4 \\ 3x + 3y - z = 10 \end{cases}$

25. $\begin{cases} x_1 + 2x_2 - x_3 = 9 \\ 2x_1 - x_3 = -2 \\ 3x_1 + 5x_2 + 2x_3 = 22 \end{cases}$

26. $\begin{cases} 2x_1 + x_2 = 7 \\ 2x_1 - x_2 + x_3 = 6 \\ 3x_1 - 2x_2 + 4x_3 = 11 \end{cases}$

27. $\begin{cases} 2x - 3y - z = 13 \\ -x + 2y - 5z = 6 \\ 5x - y - z = 49 \end{cases}$

28. $\begin{cases} 10x + 10y - 20z = 60 \\ 15x + 20y + 30z = -25 \\ -5x + 30y - 10z = 45 \end{cases}$

29–38 ■ Determine whether the system of linear equations is inconsistent or dependent. If it is dependent, find the complete solution.

29. $\begin{cases} x + y + z = 2 \\ y - 3z = 1 \\ 2x + y + 5z = 0 \end{cases}$

30. $\begin{cases} x + 3z = 3 \\ 2x + y - 2z = 5 \\ -y + 8z = 8 \end{cases}$

31. $\begin{cases} 2x - 3y - 9z = -5 \\ x + 3z = 2 \\ -3x + y - 4z = -3 \end{cases}$

32. $\begin{cases} x - 2y + 5z = 3 \\ -2x + 6y - 11z = 1 \\ 3x - 16y - 20z = -26 \end{cases}$

33. $\begin{cases} x - y + 3z = 3 \\ 4x - 8y + 32z = 24 \\ 2x - 3y + 11z = 4 \end{cases}$

34. $\begin{cases} -2x + 6y - 2z = -12 \\ x - 3y + 2z = 10 \\ -x + 3y + 2z = 6 \end{cases}$

35. $\begin{cases} x + 4y - 2z = -3 \\ 2x - y + 5z = 12 \\ 8x + 5y + 11z = 30 \end{cases}$

36. $\begin{cases} 3r + 2s - 3t = 10 \\ r - s - t = -5 \\ r + 4s - t = 20 \end{cases}$

37. $\begin{cases} 2x + y - 2z = 12 \\ -x - \frac{1}{2}y + z = -6 \\ 3x + \frac{3}{2}y - 3z = 18 \end{cases}$

38. $\begin{cases} y - 5z = 7 \\ 3x + 2y = 12 \\ 3x + 10z = 80 \end{cases}$

39–54 ■ Solve the system of linear equations.

39. $\begin{cases} 4x - 3y + z = -8 \\ -2x + y - 3z = -4 \\ x - y + 2z = 3 \end{cases}$

40. $\begin{cases} 2x - 3y + 5z = 14 \\ 4x - y - 2z = -17 \\ -x - y + z = 3 \end{cases}$

41. $\begin{cases} 2x + y + 3z = 9 \\ -x - 7z = 10 \\ 3x + 2y - z = 4 \end{cases}$

42. $\begin{cases} -4x - y + 36z = 24 \\ x - 2y + 9z = 3 \\ -2x + y + 6z = 6 \end{cases}$

43. $\begin{cases} x + 2y - 3z = -5 \\ -2x - 4y - 6z = 10 \\ 3x + 7y - 2z = -13 \end{cases}$

44. $\begin{cases} 3x + y = 2 \\ -4x + 3y + z = 4 \\ 2x + 5y + z = 0 \end{cases}$

45. $\begin{cases} x - y + 6z = 8 \\ x + z = 5 \\ x + 3y - 14z = -4 \end{cases}$

46. $\begin{cases} 3x - y + 2z = -1 \\ 4x - 2y + z = -7 \\ -x + 3y - 2z = -1 \end{cases}$

47. $\begin{cases} -x + 2y + z - 3w = 3 \\ 3x - 4y + z + w = 9 \\ -x - y + z + w = 0 \\ 2x + y + 4z - 2w = 3 \end{cases}$

48. $\begin{cases} x + y - z - w = 6 \\ 2x + z - 3w = 8 \\ x - y + 4w = -10 \\ 3x + 5y - z - w = 20 \end{cases}$

49. $\begin{cases} x + y + 2z - w = -2 \\ 3y + z + 2w = 2 \\ x + y + 3w = 2 \\ -3x + z + 2w = 5 \end{cases}$

50. $\begin{cases} x - 3y + 2z + w = -2 \\ x - 2y - 2w = -10 \\ z + 5w = 15 \\ 3x + 2z + w = -3 \end{cases}$

51. $\begin{cases} x - y + w = 0 \\ 3x - z + 2w = 0 \\ x - 4y + z + 2w = 0 \end{cases}$

52. $\begin{cases} 2x - y + 2z + w = 5 \\ -x + y + 4z - w = 3 \\ 3x - 2y - z = 0 \end{cases}$

53. $\begin{cases} x + z + w = 4 \\ y - z = -4 \\ x - 2y + 3z + w = 12 \\ 2x - 2z + 5w = -1 \end{cases}$

54. $\begin{cases} y - z + 2w = 0 \\ 3x + 2y + w = 0 \\ 2x + 4w = 12 \\ -2x - 2z + 5w = 6 \end{cases}$

APPLICATIONS

55. Nutrition A doctor recommends that a patient take 50 mg each of niacin, riboflavin, and thiamin daily to alleviate a vitamin deficiency. In his medicine chest at home the patient finds three brands of vitamin pills. The amounts of the relevant vitamins per pill are given in the table. How many pills of each type should he take every day to get 50 mg of each vitamin?

	VitaMax	Vitron	VitaPlus
Niacin (mg)	5	10	15
Riboflavin (mg)	15	20	0
Thiamin (mg)	10	10	10

56. Mixtures A chemist has three acid solutions at various concentrations. The first is 10% acid, the second is 20%, and the third is 40%. How many milliliters of each should she use to make 100 mL of 18% solution, if she has to use four times as much of the 10% solution as the 40% solution?

57. Distance, Speed, and Time Amanda, Bryce, and Corey enter a race in which they have to run, swim, and cycle over a marked course. Their average speeds are given in the table. Corey finishes first with a total time of 1 h 45 min. Amanda comes in second with a time of 2 h 30 min. Bryce finishes last with a time of 3 h. Find the distance (in miles) for each part of the race.

	Average speed (mi/h)		
	Running	**Swimming**	**Cycling**
Amanda	10	4	20
Bryce	$7\frac{1}{2}$	6	15
Corey	15	3	40

58. Classroom Use A small school has 100 students who occupy three classrooms: A, B, and C. After the first period of the school day, half the students in room A move to room B, one-fifth of the students in room B move to room C, and one-third of the students in room C move to room A. Nevertheless, the total number of students in each room is the same for both periods. How many students occupy each room?

59. Manufacturing Furniture A furniture factory makes wooden tables, chairs, and armoires. Each piece of furniture requires three operations: cutting the wood, assembling, and finishing. Each operation requires the number of hours (h) given in the table. The workers in the factory can provide 300 hours of cutting, 400 hours of assembling, and 590 hours of finishing each work week. How many tables, chairs, and armoires should be produced so that all available labor-hours are used? Or is this impossible?

	Table	**Chair**	**Armoire**
Cutting (h)	$\frac{1}{2}$	1	1
Assembling (h)	$\frac{1}{2}$	$1\frac{1}{2}$	1
Finishing (h)	1	$1\frac{1}{2}$	2

60. Traffic Flow A section of a city's street network is shown in the figure. The arrows indicate one-way streets, and the numbers show how many cars enter or leave this section of the city via the indicated street in a certain one-hour period. The variables x, y, z, and w represent the number of cars that travel along the portions of First, Second, Avocado, and Birch Streets during this period. Find x, y, z, and w, assuming that none of the cars stop or park on any of the streets shown.

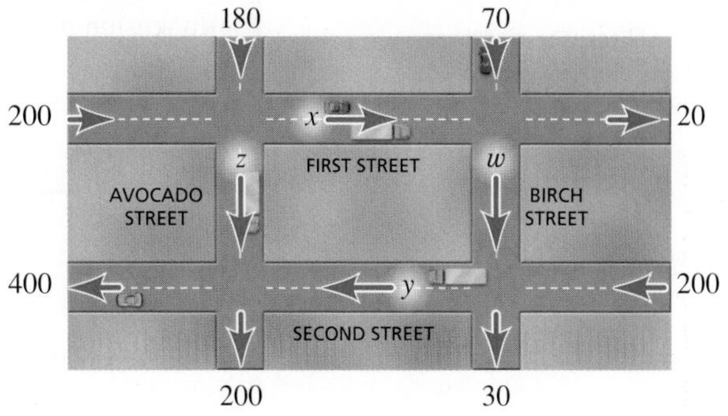

DISCOVERY ▪ DISCUSSION ▪ WRITING

61. Polynomials Determined by a Set of Points We all know that two points uniquely determine a line $y = ax + b$ in the coordinate plane. Similarly, three points uniquely determine a quadratic (second-degree) polynomial

$$y = ax^2 + bx + c$$

four points uniquely determine a cubic (third-degree) polynomial

$$y = ax^3 + bx^2 + cx + d$$

and so on. (Some exceptions to this rule are if the three points actually lie on a line, or the four points lie on a quadratic or line, and so on.) For the following set of five points, find the line that contains the first two points, the quadratic that contains the first three points, the cubic that contains the first four points, and the fourth-degree polynomial that contains all five points.

$$(0, 0), \quad (1, 12), \quad (2, 40), \quad (3, 6), \quad (-1, -14)$$

Graph the points and functions in the same viewing rectangle using a graphing device.

10.4 The Algebra of Matrices

Equality of Matrices ▶ Addition, Subtraction, and Scalar Multiplication of Matrices ▶ Multiplication of Matrices ▶ Properties of Matrix Multiplication ▶ Applications of Matrix Multiplication ▶ Computer Graphics

Thus far, we have used matrices simply for notational convenience when solving linear systems. Matrices have many other uses in mathematics and the sciences, and for most of these applications a knowledge of matrix algebra is essential. Like numbers, matrices can be added, subtracted, multiplied, and divided. In this section we learn how to perform these algebraic operations on matrices.

▼ Equality of Matrices

Two matrices are equal if they have the same entries in the same positions.

Equal matrices

$$\begin{bmatrix} \sqrt{4} & 2^2 & e^0 \\ 0.5 & 1 & 1-1 \end{bmatrix} = \begin{bmatrix} 2 & 4 & 1 \\ \frac{1}{2} & \frac{2}{2} & 0 \end{bmatrix}$$

Unequal matrices

$$\begin{bmatrix} 1 & 2 \\ 3 & 4 \\ 5 & 6 \end{bmatrix} \neq \begin{bmatrix} 1 & 3 & 5 \\ 2 & 4 & 6 \end{bmatrix}$$

EQUALITY OF MATRICES

The matrices $A = [a_{ij}]$ and $B = [b_{ij}]$ are **equal** if and only if they have the same dimension $m \times n$, and corresponding entries are equal, that is,

$$a_{ij} = b_{ij}$$

for $i = 1, 2, \ldots, m$ and $j = 1, 2, \ldots, n$.

EXAMPLE 1 | Equal Matrices

Find a, b, c, and d, if

$$\begin{bmatrix} a & b \\ c & d \end{bmatrix} = \begin{bmatrix} 1 & 3 \\ 5 & 2 \end{bmatrix}$$

SOLUTION Since the two matrices are equal, corresponding entries must be the same. So we must have $a = 1$, $b = 3$, $c = 5$, and $d = 2$.

NOW TRY EXERCISE 5 ■

▼ Addition, Subtraction, and Scalar Multiplication of Matrices

Two matrices can be added or subtracted if they have the same dimension. (Otherwise, their sum or difference is undefined.) We add or subtract the matrices by adding or subtracting corresponding entries. To multiply a matrix by a number, we multiply every element of the matrix by that number. This is called the *scalar product*.

SUM, DIFFERENCE, AND SCALAR PRODUCT OF MATRICES

Let $A = [a_{ij}]$ and $B = [b_{ij}]$ be matrices of the same dimension $m \times n$, and let c be any real number.

1. The **sum** $A + B$ is the $m \times n$ matrix obtained by adding corresponding entries of A and B.

$$A + B = [a_{ij} + b_{ij}]$$

2. The **difference** $A - B$ is the $m \times n$ matrix obtained by subtracting corresponding entries of A and B.

$$A - B = [a_{ij} - b_{ij}]$$

3. The **scalar product** cA is the $m \times n$ matrix obtained by multiplying each entry of A by c.

$$cA = [ca_{ij}]$$

EXAMPLE 2 | Performing Algebraic Operations on Matrices

Let $A = \begin{bmatrix} 2 & -3 \\ 0 & 5 \\ 7 & -\frac{1}{2} \end{bmatrix}$ $\quad B = \begin{bmatrix} 1 & 0 \\ -3 & 1 \\ 2 & 2 \end{bmatrix}$

$$C = \begin{bmatrix} 7 & -3 & 0 \\ 0 & 1 & 5 \end{bmatrix} \qquad D = \begin{bmatrix} 6 & 0 & -6 \\ 8 & 1 & 9 \end{bmatrix}$$

JULIA ROBINSON (1919–1985) was born in St. Louis, Missouri, and grew up at Point Loma, California. Because of an illness, Robinson missed two years of school, but later, with the aid of a tutor, she completed fifth, sixth, seventh, and eighth grades, all in one year. Later, at San Diego State University, reading biographies of mathematicians in E. T. Bell's *Men of Mathematics* awakened in her what became a lifelong passion for mathematics. She said, "I cannot overemphasize the importance of such books . . . in the intellectual life of a student." Robinson is famous for her work on Hilbert's tenth problem (page 683), which asks for a general procedure for determining whether an equation has integer solutions. Her ideas led to a complete answer to the problem. Interestingly, the answer involved certain properties of the Fibonacci numbers (page 787) discovered by the then 22-year-old Russian mathematician Yuri Matijasevič. As a result of her brilliant work on Hilbert's tenth problem, Robinson was offered a professorship at the University of California, Berkeley, and became the first woman mathematician elected to the National Academy of Sciences. She also served as president of the American Mathematical Society.

Carry out each indicated operation, or explain why it cannot be performed.

(a) $A + B$ **(b)** $C - D$ **(c)** $C + A$ **(d)** $5A$

SOLUTION

(a) $$A + B = \begin{bmatrix} 2 & -3 \\ 0 & 5 \\ 7 & -\frac{1}{2} \end{bmatrix} + \begin{bmatrix} 1 & 0 \\ -3 & 1 \\ 2 & 2 \end{bmatrix} = \begin{bmatrix} 3 & -3 \\ -3 & 6 \\ 9 & \frac{3}{2} \end{bmatrix}$$

(b) $$C - D = \begin{bmatrix} 7 & -3 & 0 \\ 0 & 1 & 5 \end{bmatrix} - \begin{bmatrix} 6 & 0 & -6 \\ 8 & 1 & 9 \end{bmatrix}$$

$$= \begin{bmatrix} 1 & -3 & 6 \\ -8 & 0 & -4 \end{bmatrix}$$

(c) $C + A$ is undefined because we can't add matrices of different dimensions.

(d) $$5A = 5\begin{bmatrix} 2 & -3 \\ 0 & 5 \\ 7 & -\frac{1}{2} \end{bmatrix} = \begin{bmatrix} 10 & -15 \\ 0 & 25 \\ 35 & -\frac{5}{2} \end{bmatrix}$$

NOW TRY EXERCISES **21** AND **23**

The properties in the box follow from the definitions of matrix addition and scalar multiplication and the corresponding properties of real numbers.

PROPERTIES OF ADDITION AND SCALAR MULTIPLICATION OF MATRICES

Let A, B, and C be $m \times n$ matrices and let c and d be scalars.

$A + B = B + A$	Commutative Property of Matrix Addition
$(A + B) + C = A + (B + C)$	Associative Property of Matrix Addition
$c(dA) = cdA$	Associative Property of Scalar Multiplication
$(c + d)A = cA + dA$ $c(A + B) = cA + cB$	Distributive Properties of Scalar Multiplication

EXAMPLE 3 | Solving a Matrix Equation

Solve the matrix equation

$$2X - A = B$$

for the unknown matrix X, where

$$A = \begin{bmatrix} 2 & 3 \\ -5 & 1 \end{bmatrix} \qquad B = \begin{bmatrix} 4 & -1 \\ 1 & 3 \end{bmatrix}$$

SOLUTION We use the properties of matrices to solve for X.

$2X - A = B$	Given equation
$2X = B + A$	Add the matrix A to each side
$X = \frac{1}{2}(B + A)$	Multiply each side by the scalar $\frac{1}{2}$

So

$$X = \frac{1}{2}\left(\begin{bmatrix} 4 & -1 \\ 1 & 3 \end{bmatrix} + \begin{bmatrix} 2 & 3 \\ -5 & 1 \end{bmatrix}\right) \qquad \text{Substitute the matrices } A \text{ and } B$$

$$= \frac{1}{2}\begin{bmatrix} 6 & 2 \\ -4 & 4 \end{bmatrix} \qquad \text{Add matrices}$$

$$= \begin{bmatrix} 3 & 1 \\ -2 & 2 \end{bmatrix} \qquad \text{Multiply by the scalar } \tfrac{1}{2}$$

NOW TRY EXERCISE **15** ■

▼ Multiplication of Matrices

Multiplying two matrices is more difficult to describe than other matrix operations. In later examples we will see why multiplying matrices involves a rather complex procedure, which we now describe.

First, the product AB (or $A \cdot B$) of two matrices A and B is defined only when the number of columns in A is equal to the number of rows in B. This means that if we write their dimensions side by side, the two inner numbers must match:

Matrices	A	B
Dimensions	$m \times n$	$n \times k$
	Columns in A	Rows in B

If we think of the row of A and the column of B as vectors, then their inner product is the same as their dot product (see Sections 9.2 and 9.4).

If the dimensions of A and B match in this fashion, then the product AB is a matrix of dimension $m \times k$. Before describing the procedure for obtaining the elements of AB, we define the *inner product* of a row of A and a column of B.

If $[a_1 \quad a_2 \quad \cdots \quad a_n]$ is a row of A, and if $\begin{bmatrix} b_1 \\ b_2 \\ \vdots \\ b_n \end{bmatrix}$ is a column of B, then their **inner product**

is the number $a_1b_1 + a_2b_2 + \cdots + a_nb_n$. For example, taking the inner product of

$$[2 \quad -1 \quad 0 \quad 4] \text{ and } \begin{bmatrix} 5 \\ 4 \\ -3 \\ \frac{1}{2} \end{bmatrix} \text{ gives}$$

$$2 \cdot 5 + (-1) \cdot 4 + 0 \cdot (-3) + 4 \cdot \tfrac{1}{2} = 8$$

We now define the **product** AB of two matrices.

MATRIX MULTIPLICATION

If $A = [a_{ij}]$ is an $m \times n$ matrix and $B = [b_{ij}]$ an $n \times k$ matrix, then their product is the $m \times k$ matrix

$$C = [c_{ij}]$$

where c_{ij} is the inner product of the ith row of A and the jth column of B. We write the product as

$$C = AB$$

This definition of matrix product says that each entry in the matrix AB is obtained from a *row* of A and a *column* of B as follows: The entry c_{ij} in the ith row and jth column of the matrix AB is obtained by multiplying the entries in the ith row of A with the corresponding entries in the jth column of B and adding the results.

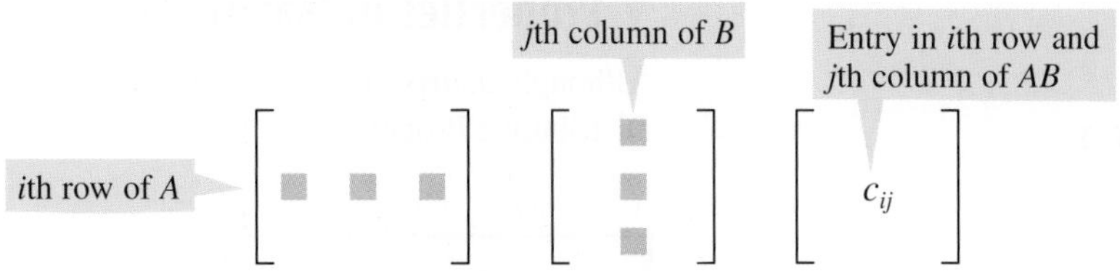

EXAMPLE 4 | Multiplying Matrices

Let

$$A = \begin{bmatrix} 1 & 3 \\ -1 & 0 \end{bmatrix} \quad \text{and} \quad B = \begin{bmatrix} -1 & 5 & 2 \\ 0 & 4 & 7 \end{bmatrix}$$

Calculate, if possible, the products AB and BA.

Inner numbers match, so product is defined

$2 \times 2 \quad 2 \times 3$

Outer numbers give dimension of product: 2×3

SOLUTION Since A has dimension 2×2 and B has dimension 2×3, the product AB is defined and has dimension 2×3. We can therefore write

$$AB = \begin{bmatrix} 1 & 3 \\ -1 & 0 \end{bmatrix}\begin{bmatrix} -1 & 5 & 2 \\ 0 & 4 & 7 \end{bmatrix} = \begin{bmatrix} ? & ? & ? \\ ? & ? & ? \end{bmatrix}$$

where the question marks must be filled in using the rule defining the product of two matrices. If we define $C = AB = [c_{ij}]$, then the entry c_{11} is the inner product of the first row of A and the first column of B:

$$\begin{bmatrix} 1 & 3 \\ -1 & 0 \end{bmatrix}\begin{bmatrix} -1 & 5 & 2 \\ 0 & 4 & 7 \end{bmatrix} \qquad 1 \cdot (-1) + 3 \cdot 0 = -1$$

Similarly, we calculate the remaining entries of the product as follows.

Entry	Inner product of:	Value	Product matrix
c_{12}	$\begin{bmatrix} 1 & 3 \\ -1 & 0 \end{bmatrix}\begin{bmatrix} -1 & 5 & 2 \\ 0 & 4 & 7 \end{bmatrix}$	$1 \cdot 5 + 3 \cdot 4 = 17$	$\begin{bmatrix} -1 & 17 & \\ & & \end{bmatrix}$
c_{13}	$\begin{bmatrix} 1 & 3 \\ -1 & 0 \end{bmatrix}\begin{bmatrix} -1 & 5 & 2 \\ 0 & 4 & 7 \end{bmatrix}$	$1 \cdot 2 + 3 \cdot 7 = 23$	$\begin{bmatrix} -1 & 17 & 23 \\ & & \end{bmatrix}$
c_{21}	$\begin{bmatrix} 1 & 3 \\ -1 & 0 \end{bmatrix}\begin{bmatrix} -1 & 5 & 2 \\ 0 & 4 & 7 \end{bmatrix}$	$(-1) \cdot (-1) + 0 \cdot 0 = 1$	$\begin{bmatrix} -1 & 17 & 23 \\ 1 & & \end{bmatrix}$
c_{22}	$\begin{bmatrix} 1 & 3 \\ -1 & 0 \end{bmatrix}\begin{bmatrix} -1 & 5 & 2 \\ 0 & 4 & 7 \end{bmatrix}$	$(-1) \cdot 5 + 0 \cdot 4 = -5$	$\begin{bmatrix} -1 & 17 & 23 \\ 1 & -5 & \end{bmatrix}$
c_{23}	$\begin{bmatrix} 1 & 3 \\ -1 & 0 \end{bmatrix}\begin{bmatrix} -1 & 5 & 2 \\ 0 & 4 & 7 \end{bmatrix}$	$(-1) \cdot 2 + 0 \cdot 7 = -2$	$\begin{bmatrix} -1 & 17 & 23 \\ 1 & -5 & -2 \end{bmatrix}$

Thus, we have

$$AB = \begin{bmatrix} -1 & 17 & 23 \\ 1 & -5 & -2 \end{bmatrix}$$

Not equal, so product not defined

$2 \times 3 \quad 2 \times 2$

The product BA is not defined, however, because the dimensions of B and A are

$$2 \times 3 \quad \text{and} \quad 2 \times 2$$

The inner two numbers are not the same, so the rows and columns won't match up when we try to calculate the product.

NOW TRY EXERCISE **25** ■

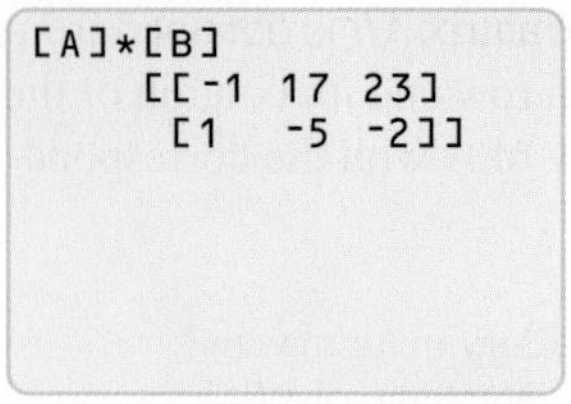

FIGURE 1

Graphing calculators and computers are capable of performing matrix algebra. For instance, if we enter the matrices in Example 4 into the matrix variables [A] and [B] on a TI-83 calculator, then the calculator finds their product as shown in Figure 1.

Properties of Matrix Multiplication

Although matrix multiplication is not commutative, it does obey the Associative and Distributive Properties.

PROPERTIES OF MATRIX MULTIPLICATION

Let A, B, and C be matrices for which the following products are defined. Then

$$A(BC) = (AB)C \qquad \text{Associative Property}$$

$$\begin{aligned} A(B + C) &= AB + AC \\ (B + C)A &= BA + CA \end{aligned} \qquad \text{Distributive Property}$$

The next example shows that even when both AB and BA are defined, they aren't necessarily equal. This result proves that matrix multiplication is *not* commutative.

EXAMPLE 5 | Matrix Multiplication Is Not Commutative

Let
$$A = \begin{bmatrix} 5 & 7 \\ -3 & 0 \end{bmatrix} \quad \text{and} \quad B = \begin{bmatrix} 1 & 2 \\ 9 & -1 \end{bmatrix}$$

Calculate the products AB and BA.

SOLUTION Since both matrices A and B have dimension 2×2, both products AB and BA are defined, and each product is also a 2×2 matrix.

$$\begin{aligned} AB &= \begin{bmatrix} 5 & 7 \\ -3 & 0 \end{bmatrix}\begin{bmatrix} 1 & 2 \\ 9 & -1 \end{bmatrix} = \begin{bmatrix} 5\cdot 1 + 7\cdot 9 & 5\cdot 2 + 7\cdot(-1) \\ (-3)\cdot 1 + 0\cdot 9 & (-3)\cdot 2 + 0\cdot(-1) \end{bmatrix} \\ &= \begin{bmatrix} 68 & 3 \\ -3 & -6 \end{bmatrix} \\ BA &= \begin{bmatrix} 1 & 2 \\ 9 & -1 \end{bmatrix}\begin{bmatrix} 5 & 7 \\ -3 & 0 \end{bmatrix} = \begin{bmatrix} 1\cdot 5 + 2\cdot(-3) & 1\cdot 7 + 2\cdot 0 \\ 9\cdot 5 + (-1)\cdot(-3) & 9\cdot 7 + (-1)\cdot 0 \end{bmatrix} \\ &= \begin{bmatrix} -1 & 7 \\ 48 & 63 \end{bmatrix} \end{aligned}$$

This shows that, in general, $AB \neq BA$. In fact, in this example AB and BA don't even have an entry in common.

NOW TRY EXERCISE **27** ■

Applications of Matrix Multiplication

We now consider some applied examples that give some indication of why mathematicians chose to define the matrix product in such an apparently bizarre fashion. Example 6 shows how our definition of matrix product allows us to express a system of linear equations as a single matrix equation.

EXAMPLE 6 | Writing a Linear System as a Matrix Equation

Show that the following matrix equation is equivalent to the system of equations in Example 2 of Section 10.3.

Matrix equations like this one are described in more detail on page 677.

$$\begin{bmatrix} 1 & -1 & 3 \\ 1 & 2 & -2 \\ 3 & -1 & 5 \end{bmatrix}\begin{bmatrix} x \\ y \\ z \end{bmatrix} = \begin{bmatrix} 4 \\ 10 \\ 14 \end{bmatrix}$$

SOLUTION If we perform matrix multiplication on the left side of the equation, we get

$$\begin{bmatrix} x - y + 3z \\ x + 2y - 2z \\ 3x - y + 5z \end{bmatrix} = \begin{bmatrix} 4 \\ 10 \\ 14 \end{bmatrix}$$

Because two matrices are equal only if their corresponding entries are equal, we equate entries to get

$$\begin{cases} x - y + 3z = 4 \\ x + 2y - 2z = 10 \\ 3x - y + 5z = 14 \end{cases}$$

This is exactly the system of equations in Example 2 of Section 10.3.

NOW TRY EXERCISE **39** ■

EXAMPLE 7 | Representing Demographic Data by Matrices

In a certain city the proportions of voters in each age group who are registered as Democrats, Republicans, or Independents are given by the following matrix.

	Age 18–30	Age 31–50	Age Over 50
Democrat	0.30	0.60	0.50
Republican	0.50	0.35	0.25
Independent	0.20	0.05	0.25

$$\begin{bmatrix} 0.30 & 0.60 & 0.50 \\ 0.50 & 0.35 & 0.25 \\ 0.20 & 0.05 & 0.25 \end{bmatrix} = A$$

The next matrix gives the distribution, by age and sex, of the voting population of this city.

Age	Male	Female
18–30	5,000	6,000
31–50	10,000	12,000
Over 50	12,000	15,000

$$\begin{bmatrix} 5{,}000 & 6{,}000 \\ 10{,}000 & 12{,}000 \\ 12{,}000 & 15{,}000 \end{bmatrix} = B$$

For this problem, let's make the (highly unrealistic) assumption that within each age group, political preference is not related to gender. That is, the percentage of Democrat males in the 18–30 group, for example, is the same as the percentage of Democrat females in this group.

(a) Calculate the product AB.

(b) How many males are registered as Democrats in this city?

(c) How many females are registered as Republicans?

OLGA TAUSSKY-TODD (1906–1995) was instrumental in developing applications of matrix theory. Described as "in love with anything matrices can do," she successfully applied matrices to aerodynamics, a field used in the design of airplanes and rockets. Taussky-Todd was also famous for her work in number theory, which deals with prime numbers and divisibility. Although number theory has often been called the least applicable branch of mathematics, it is now used in significant ways throughout the computer industry.

Taussky-Todd studied mathematics at a time when young women rarely aspired to be mathematicians. She said, "When I entered university I had no idea what it meant to study mathematics." One of the most respected mathematicians of her day, she was for many years a professor of mathematics at Caltech in Pasadena.

SOLUTION

(a) $$AB = \begin{bmatrix} 0.30 & 0.60 & 0.50 \\ 0.50 & 0.35 & 0.25 \\ 0.20 & 0.05 & 0.25 \end{bmatrix} \begin{bmatrix} 5{,}000 & 6{,}000 \\ 10{,}000 & 12{,}000 \\ 12{,}000 & 15{,}000 \end{bmatrix} = \begin{bmatrix} 13{,}500 & 16{,}500 \\ 9{,}000 & 10{,}950 \\ 4{,}500 & 5{,}550 \end{bmatrix}$$

(b) When we take the inner product of a row in A with a column in B, we are adding the number of people in each age group who belong to the category in question. For example, the entry c_{21} of AB (the 9000) is obtained by taking the inner product of the Republican row in A with the Male column in B. This number is therefore the total number of male Republicans in this city. We can label the rows and columns of AB as follows.

	Male	Female
Democrat	13,500	16,500
Republican	9,000	10,950
Independent	4,500	5,550

$= AB$

Thus 13,500 males are registered as Democrats in this city.

(c) There are 10,950 females registered as Republicans.

NOW TRY EXERCISE **45**

In Example 7 the entries in each column of A add up to 1. (Can you see why this has to be true, given what the matrix describes?) A matrix with this property is called **stochastic**. Stochastic matrices are used extensively in statistics, where they arise frequently in situations like the one described here.

Computer Graphics

One important use of matrices is in the digital representation of images. A digital camera or a scanner converts an image into a matrix by dividing the image into a rectangular array of elements called pixels. Each pixel is assigned a value that represents the color, brightness, or some other feature of that location. For example, in a 256-level gray-scale image each pixel is assigned a value between 0 and 255, where 0 represents white, 255 represents black, and the numbers in between represent increasing gradations of gray. The gradations of a much simpler 8-level gray scale are shown in Figure 2. We use this eight-level gray scale to illustrate the process.

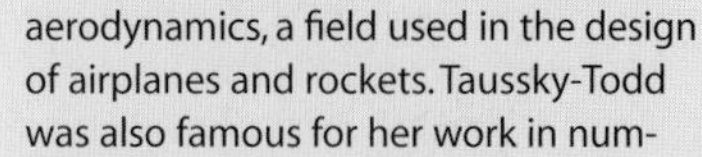

FIGURE 2

To digitize the black and white image in Figure 3(a), we place a grid over the picture as shown in Figure 3(b). Each cell in the grid is compared to the gray scale and then assigned a value between 0 and 7 depending on which gray square in the scale most closely matches

(a) Original image

(b) 10×10 grid

$$\begin{bmatrix} 1 & 1 & 1 & 1 & 1 & 1 & 1 & 2 & 2 & 1 \\ 1 & 1 & 1 & 1 & 1 & 1 & 4 & 6 & 5 & 2 \\ 1 & 1 & 1 & 1 & 2 & 3 & 3 & 5 & 5 & 3 \\ 1 & 1 & 1 & 1 & 3 & 5 & 4 & 6 & 3 & 2 \\ 1 & 1 & 1 & 1 & 1 & 2 & 3 & 2 & 2 & 1 \\ 1 & 1 & 1 & 1 & 1 & 3 & 3 & 2 & 1 & 1 \\ 1 & 1 & 1 & 1 & 1 & 1 & 4 & 1 & 1 & 1 \\ 1 & 1 & 1 & 1 & 2 & 2 & 4 & 2 & 2 & 2 \\ 2 & 2 & 3 & 5 & 5 & 2 & 2 & 3 & 4 & 4 \\ 3 & 3 & 3 & 4 & 3 & 2 & 3 & 3 & 3 & 4 \end{bmatrix}$$

(c) Matrix representation

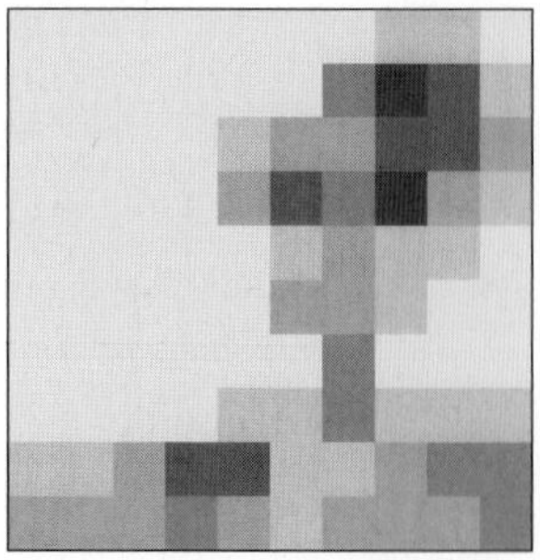

(d) Digital image

FIGURE 3

the "darkness" of the cell. (If the cell is not uniformly gray, an average value is assigned.) The values are stored in the matrix shown in Figure 3(c). The digital image corresponding to this matrix is shown in Figure 3(d). Obviously, the grid that we have used is far too coarse to provide good image resolution. In practice, currently available high-resolution digital cameras use matrices with dimension as large as 2048×2048.

Once the image is stored as a matrix, it can be manipulated by using matrix operations. For example, to darken the image, we add a constant to each entry in the matrix; to lighten the image, we subtract a constant. To increase the contrast, we darken the darker areas and lighten the lighter areas, so we could add 1 to each entry that is 4, 5, or 6 and subtract 1 from each entry that is 1, 2, or 3. (Note that we cannot darken an entry of 7 or lighten a 0.) Applying this process to the matrix in Figure 3(c) produces the new matrix in Figure 4(a). This generates the high-contrast image shown in Figure 4(b).

$$\begin{bmatrix} 0&0&0&0&0&0&0&1&1&0 \\ 0&0&0&0&0&0&5&7&6&1 \\ 0&0&0&0&1&2&2&6&6&2 \\ 0&0&0&0&2&6&5&7&2&1 \\ 0&0&0&0&0&1&2&1&1&0 \\ 0&0&0&0&0&2&2&1&0&0 \\ 0&0&0&0&0&0&5&0&0&0 \\ 0&0&0&0&1&1&5&1&1&1 \\ 1&1&2&6&6&1&1&2&5&5 \\ 2&2&2&5&2&1&2&2&2&5 \end{bmatrix}$$

(a) Matrix modified to increase contrast

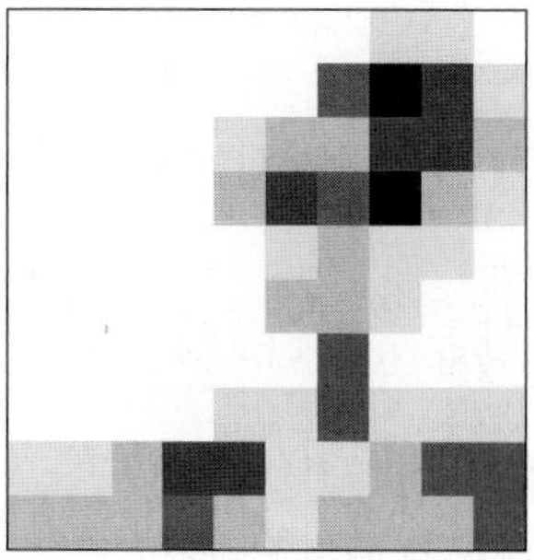

(b) High contrast image

FIGURE 4

Other ways of representing and manipulating images using matrices are discussed in the Discovery Projects *Computer Graphics I* and *II* at the book companion website: **www.stewartmath.com.**

10.4 EXERCISES

CONCEPTS

1. We can add (or subtract) two matrices only if they have the same ______________.

2. (a) We can multiply two matrices only if the number of ________ in the first matrix is the same as the number of ________ in the second matrix.

(b) If A is a 3×3 matrix and B is a 4×3 matrix, which of the following matrix multiplications are possible?

(i) AB (ii) BA (iii) AA (iv) BB

3. Which of the following operations can we perform for a matrix A of any dimension?

(i) $A + A$ (ii) $2A$ (iii) $A \cdot A$

4. Fill in the missing entries in the product matrix.

$$\begin{bmatrix} 3 & 1 & 2 \\ -1 & 2 & 0 \\ 1 & 3 & -2 \end{bmatrix} \begin{bmatrix} -1 & 3 & -2 \\ 3 & -2 & -1 \\ 2 & 1 & 0 \end{bmatrix} = \begin{bmatrix} 4 & \blacksquare & -7 \\ 7 & -7 & \blacksquare \\ \blacksquare & -5 & -5 \end{bmatrix}$$

SKILLS

5–6 ■ Determine whether the matrices A and B are equal.

5. $A = \begin{bmatrix} 1 & -2 & 0 \\ \frac{1}{2} & 6 & 0 \end{bmatrix} \quad B = \begin{bmatrix} 1 & -2 \\ \frac{1}{2} & 6 \end{bmatrix}$

6. $A = \begin{bmatrix} \frac{1}{4} & \ln 1 \\ 2 & 3 \end{bmatrix} \quad B = \begin{bmatrix} 0.25 & 0 \\ \sqrt{4} & \frac{6}{2} \end{bmatrix}$

7–14 ■ Perform the matrix operation, or if it is impossible, explain why.

7. $\begin{bmatrix} 2 & 6 \\ -5 & 3 \end{bmatrix} + \begin{bmatrix} -1 & -3 \\ 6 & 2 \end{bmatrix}$

8. $\begin{bmatrix} 0 & 1 & 1 \\ 1 & 1 & 0 \end{bmatrix} - \begin{bmatrix} 2 & 1 & -1 \\ 1 & 3 & -2 \end{bmatrix}$

9. $3\begin{bmatrix} 1 & 2 \\ 4 & -1 \\ 1 & 0 \end{bmatrix}$

10. $2\begin{bmatrix} 1 & 1 & 0 \\ 1 & 0 & 1 \\ 0 & 1 & 1 \end{bmatrix} + \begin{bmatrix} 1 & 1 \\ 2 & 1 \\ 3 & 1 \end{bmatrix}$

11. $\begin{bmatrix} 2 & 6 \\ 1 & 3 \\ 2 & 4 \end{bmatrix} \begin{bmatrix} 1 & -2 \\ 3 & 6 \\ -2 & 0 \end{bmatrix}$

12. $\begin{bmatrix} 2 & 1 & 2 \\ 6 & 3 & 4 \end{bmatrix} \begin{bmatrix} 1 & -2 \\ 3 & 6 \\ -2 & 0 \end{bmatrix}$

13. $\begin{bmatrix} 1 & 2 \\ -1 & 4 \end{bmatrix}\begin{bmatrix} 1 & -2 & 3 \\ 2 & 2 & -1 \end{bmatrix}$ **14.** $\begin{bmatrix} 2 & -3 \\ 0 & 1 \\ 1 & 2 \end{bmatrix}\begin{bmatrix} 5 \\ 1 \end{bmatrix}$

15–20 ■ Solve the matrix equation for the unknown matrix X, or explain why no solution exists.

$$A = \begin{bmatrix} 4 & 6 \\ 1 & 3 \end{bmatrix} \qquad B = \begin{bmatrix} 2 & 5 \\ 3 & 7 \end{bmatrix}$$

$$C = \begin{bmatrix} 2 & 3 \\ 1 & 0 \\ 0 & 2 \end{bmatrix} \qquad D = \begin{bmatrix} 10 & 20 \\ 30 & 20 \\ 10 & 0 \end{bmatrix}$$

15. $2X + A = B$ **16.** $3X - B = C$

17. $2(B - X) = D$ **18.** $5(X - C) = D$

19. $\frac{1}{5}(X + D) = C$ **20.** $2A = B - 3X$

21–34 ■ The matrices A, B, C, D, E, F, G and H are defined as follows.

$$A = \begin{bmatrix} 2 & -5 \\ 0 & 7 \end{bmatrix} \qquad B = \begin{bmatrix} 3 & \frac{1}{2} & 5 \\ 1 & -1 & 3 \end{bmatrix} \qquad C = \begin{bmatrix} 2 & -\frac{5}{2} & 0 \\ 0 & 2 & -3 \end{bmatrix}$$

$$D = \begin{bmatrix} 7 & 3 \end{bmatrix} \qquad E = \begin{bmatrix} 1 \\ 2 \\ 0 \end{bmatrix} \qquad F = \begin{bmatrix} 1 & 0 & 0 \\ 0 & 1 & 0 \\ 0 & 0 & 1 \end{bmatrix}$$

$$G = \begin{bmatrix} 5 & -3 & 10 \\ 6 & 1 & 0 \\ -5 & 2 & 2 \end{bmatrix} \qquad H = \begin{bmatrix} 3 & 1 \\ 2 & -1 \end{bmatrix}$$

Carry out the indicated algebraic operation, or explain why it cannot be performed.

21. **(a)** $B + C$ **(b)** $B + F$

22. **(a)** $C - B$ **(b)** $2C - 6B$

23. **(a)** $5A$ **(b)** $C - 5A$

24. **(a)** $3B + 2C$ **(b)** $2H + D$

25. **(a)** AD **(b)** DA

26. **(a)** DH **(b)** HD

27. **(a)** AH **(b)** HA

28. **(a)** BC **(b)** BF

29. **(a)** GF **(b)** GE

30. **(a)** B^2 **(b)** F^2

31. **(a)** A^2 **(b)** A^3

32. **(a)** $(DA)B$ **(b)** $D(AB)$

33. **(a)** ABE **(b)** AHE

34. **(a)** $DB + DC$ **(b)** $BF + FE$

35–38 ■ Solve for x and y.

35. $\begin{bmatrix} x & 2y \\ 4 & 6 \end{bmatrix} = \begin{bmatrix} 2 & -2 \\ 2x & -6y \end{bmatrix}$ **36.** $3\begin{bmatrix} x & y \\ y & x \end{bmatrix} = \begin{bmatrix} 6 & -9 \\ -9 & 6 \end{bmatrix}$

37. $2\begin{bmatrix} x & y \\ x + y & x - y \end{bmatrix} = \begin{bmatrix} 2 & -4 \\ -2 & 6 \end{bmatrix}$

38. $\begin{bmatrix} x & y \\ -y & x \end{bmatrix} - \begin{bmatrix} y & x \\ x & -y \end{bmatrix} = \begin{bmatrix} 4 & -4 \\ -6 & 6 \end{bmatrix}$

39–42 ■ Write the system of equations as a matrix equation (see Example 6).

39. $\begin{cases} 2x - 5y = 7 \\ 3x + 2y = 4 \end{cases}$ **40.** $\begin{cases} 6x - y + z = 12 \\ 2x + z = 7 \\ y - 2z = 4 \end{cases}$

41. $\begin{cases} 3x_1 + 2x_2 - x_3 + x_4 = 0 \\ x_1 - x_3 = 5 \\ 3x_2 + x_3 - x_4 = 4 \end{cases}$

42. $\begin{cases} x - y + z = 2 \\ 4x - 2y - z = 2 \\ x + y + 5z = 2 \\ -x - y - z = 2 \end{cases}$

43. Let

$$A = \begin{bmatrix} 1 & 0 & 6 & -1 \\ 2 & \frac{1}{2} & 4 & 0 \end{bmatrix} \qquad C = \begin{bmatrix} 1 \\ 0 \\ -1 \\ -2 \end{bmatrix}$$

$$B = \begin{bmatrix} 1 & 7 & -9 & 2 \end{bmatrix}$$

Determine which of the following products are defined, and calculate the ones that are.

$ABC \quad ACB \quad BAC$

$BCA \quad CAB \quad CBA$

44. **(a)** Prove that if A and B are 2×2 matrices, then

$$(A + B)^2 = A^2 + AB + BA + B^2$$

(b) If A and B are 2×2 matrices, is it necessarily true that

$$(A + B)^2 \overset{?}{=} A^2 + 2AB + B^2$$

APPLICATIONS

45. Fast-Food Sales A small fast-food chain with restaurants in Santa Monica, Long Beach, and Anaheim sells only hamburgers, hot dogs, and milk shakes. On a certain day, sales were distributed according to the following matrix.

	Number of items sold		
	Santa Monica	**Long Beach**	**Anaheim**
Hamburgers	4000	1000	3500
Hot dogs	400	300	200
Milk shakes	700	500	9000

$= A$

The price of each item is given by the following matrix.

Hamburger	**Hot dog**	**Milk shake**
[\$0.90	\$0.80	\$1.10]

$= B$

(a) Calculate the product BA.
(b) Interpret the entries in the product matrix BA.

46. Car-Manufacturing Profits A specialty-car manufacturer has plants in Auburn, Biloxi, and Chattanooga. Three models are produced, with daily production given in the following matrix.

Cars produced each day

	Model K	Model R	Model W	
Auburn	12	10	0	
Biloxi	4	4	20	$= A$
Chattanooga	8	9	12	

Because of a wage increase, February profits are lower than January profits. The profit per car is tabulated by model in the following matrix.

	January	February	
Model K	\$1000	\$500	
Model R	\$2000	\$1200	$= B$
Model W	\$1500	\$1000	

(a) Calculate AB.

(b) Assuming that all cars produced were sold, what was the daily profit in January from the Biloxi plant?

(c) What was the total daily profit (from all three plants) in February?

47. Canning Tomato Products Jaeger Foods produces tomato sauce and tomato paste, canned in small, medium, large, and giant sized cans. The matrix A gives the size (in ounces) of each container.

	Small	Medium	Large	Giant	
Ounces	6	10	14	28	$= A$

The matrix B tabulates one day's production of tomato sauce and tomato paste.

	Cans of sauce	Cans of paste	
Small	2000	2500	
Medium	3000	1500	$= B$
Large	2500	1000	
Giant	1000	500	

(a) Calculate the product of AB.

(b) Interpret the entries in the product matrix AB.

48. Produce Sales A farmer's three children, Amy, Beth, and Chad, run three roadside produce stands during the summer months. One weekend they all sell watermelons, yellow squash, and tomatoes. The matrices A and B tabulate the number of pounds of each product sold by each sibling on Saturday and Sunday.

Saturday

	Melons	Squash	Tomatoes	
Amy	120	50	60	
Beth	40	25	30	$= A$
Chad	60	30	20	

Sunday

	Melons	Squash	Tomatoes	
Amy	100	60	30	
Beth	35	20	20	$= B$
Chad	60	25	30	

The matrix C gives the price per pound (in dollars) for each type of produce that they sell.

	Price per pound	
Melons	0.10	
Squash	0.50	$= C$
Tomatoes	1.00	

Perform each of the following matrix operations, and interpret the entries in each result.

(a) AC **(b)** BC **(c)** $A + B$ **(d)** $(A + B)C$

49. Digital Images A four-level gray scale is shown below.

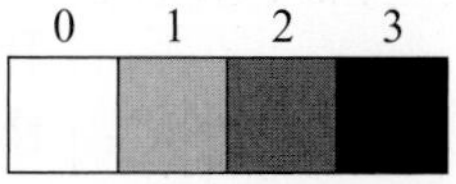

(a) Use the gray scale to find a 6×6 matrix that digitally represents the image in the figure.

(b) Find a matrix that represents a darker version of the image in the figure.

(c) The **negative** of an image is obtained by reversing light and dark, as in the negative of a photograph. Find the matrix that represents the negative of the image in the figure. How do you change the elements of the matrix to create the negative?

(d) Increase the contrast of the image by changing each 1 to a 0 and each 2 to a 3 in the matrix you found in part (b). Draw the image represented by the resulting matrix. Does this clarify the image?

(e) Draw the image represented by the matrix I. Can you recognize what this is? If you don't, try increasing the contrast.

$$I = \begin{bmatrix} 1 & 2 & 3 & 3 & 2 & 0 \\ 0 & 3 & 0 & 1 & 0 & 1 \\ 1 & 3 & 2 & 3 & 0 & 0 \\ 0 & 3 & 0 & 1 & 0 & 1 \\ 1 & 3 & 3 & 2 & 3 & 0 \\ 0 & 1 & 0 & 1 & 0 & 1 \end{bmatrix}$$

DISCOVERY ■ DISCUSSION ■ WRITING

50. When Are Both Products Defined? What must be true about the dimensions of the matrices A and B if both products AB and BA are defined?

51. Powers of a Matrix Let

$$A = \begin{bmatrix} 1 & 1 \\ 0 & 1 \end{bmatrix}$$

Calculate $A^2, A^3, A^4, \ldots$ until you detect a pattern. Write a general formula for A^n.

52. Powers of a Matrix Let $A = \begin{bmatrix} 1 & 1 \\ 1 & 1 \end{bmatrix}$. Calculate A^2, $A^3, A^4, \ldots$ until you detect a pattern. Write a general formula for A^n.

53. Square Roots of Matrices A **square root** of a matrix B is a matrix A with the property that $A^2 = B$. (This is the same definition as for a square root of a number.) Find as many square roots as you can of each matrix:

$$\begin{bmatrix} 4 & 0 \\ 0 & 9 \end{bmatrix} \qquad \begin{bmatrix} 1 & 5 \\ 0 & 9 \end{bmatrix}$$

[*Hint:* If $A = \begin{bmatrix} a & b \\ c & d \end{bmatrix}$, write the equations that a, b, c, and d would have to satisfy if A is the square root of the given matrix.]

DISCOVERY PROJECT **Will the Species Survive?**

In this project we investigate matrix models for species populations and how multiplication by a transition matrix can predict future population trends. You can find the project at the book companion website: **www.stewartmath.com**

10.5 Inverses of Matrices and Matrix Equations

The Inverse of a Matrix ▶ Finding the Inverse of a 2 × 2 Matrix ▶ Finding the Inverse of an $n \times n$ Matrix ▶ Matrix Equations ▶ Modeling with Matrix Equations

In the preceding section we saw that when the dimensions are appropriate, matrices can be added, subtracted, and multiplied. In this section we investigate division of matrices. With this operation we can solve equations that involve matrices.

▼ The Inverse of a Matrix

First, we define *identity matrices*, which play the same role for matrix multiplication as the number 1 does for ordinary multiplication of numbers; that is, $1 \cdot a = a \cdot 1 = a$ for all numbers a. A **square matrix** is one that has the same number of rows as columns. The **main diagonal** of a square matrix consists of the entries whose row and column numbers are the same. These entries stretch diagonally down the matrix, from top left to bottom right.

IDENTITY MATRIX

The **identity matrix** I_n is the $n \times n$ matrix for which each main diagonal entry is a 1 and for which all other entries are 0.

Thus the 2×2, 3×3, and 4×4 identity matrices are

$$I_2 = \begin{bmatrix} 1 & 0 \\ 0 & 1 \end{bmatrix} \qquad I_3 = \begin{bmatrix} 1 & 0 & 0 \\ 0 & 1 & 0 \\ 0 & 0 & 1 \end{bmatrix} \qquad I_4 = \begin{bmatrix} 1 & 0 & 0 & 0 \\ 0 & 1 & 0 & 0 \\ 0 & 0 & 1 & 0 \\ 0 & 0 & 0 & 1 \end{bmatrix}$$

Identity matrices behave like the number 1 in the sense that

$$A \cdot I_n = A \qquad \text{and} \qquad I_n \cdot B = B$$

whenever these products are defined.

EXAMPLE 1 | Identity Matrices

The following matrix products show how multiplying a matrix by an identity matrix of the appropriate dimension leaves the matrix unchanged.

$$\begin{bmatrix} 1 & 0 \\ 0 & 1 \end{bmatrix}\begin{bmatrix} 3 & 5 & 6 \\ -1 & 2 & 7 \end{bmatrix} = \begin{bmatrix} 3 & 5 & 6 \\ -1 & 2 & 7 \end{bmatrix}$$

$$\begin{bmatrix} -1 & 7 & \frac{1}{2} \\ 12 & 1 & 3 \\ -2 & 0 & 7 \end{bmatrix}\begin{bmatrix} 1 & 0 & 0 \\ 0 & 1 & 0 \\ 0 & 0 & 1 \end{bmatrix} = \begin{bmatrix} -1 & 7 & \frac{1}{2} \\ 12 & 1 & 3 \\ -2 & 0 & 7 \end{bmatrix}$$

NOW TRY EXERCISE **1(a), (b)** ■

If A and B are $n \times n$ matrices, and if $AB = BA = I_n$, then we say that B is the *inverse* of A, and we write $B = A^{-1}$. The concept of the inverse of a matrix is analogous to that of the reciprocal of a real number.

> **INVERSE OF A MATRIX**
>
> Let A be a square $n \times n$ matrix. If there exists an $n \times n$ matrix A^{-1} with the property that
>
> $$AA^{-1} = A^{-1}A = I_n$$
>
> then we say that A^{-1} is the **inverse** of A.

EXAMPLE 2 | Verifying That a Matrix Is an Inverse

Verify that B is the inverse of A, where

$$A = \begin{bmatrix} 2 & 1 \\ 5 & 3 \end{bmatrix} \qquad \text{and} \qquad B = \begin{bmatrix} 3 & -1 \\ -5 & 2 \end{bmatrix}$$

SOLUTION We perform the matrix multiplications to show that $AB = I$ and $BA = I$.

$$\begin{bmatrix} 2 & 1 \\ 5 & 3 \end{bmatrix}\begin{bmatrix} 3 & -1 \\ -5 & 2 \end{bmatrix} = \begin{bmatrix} 2 \cdot 3 + 1(-5) & 2(-1) + 1 \cdot 2 \\ 5 \cdot 3 + 3(-5) & 5(-1) + 3 \cdot 2 \end{bmatrix} = \begin{bmatrix} 1 & 0 \\ 0 & 1 \end{bmatrix}$$

$$\begin{bmatrix} 3 & -1 \\ -5 & 2 \end{bmatrix}\begin{bmatrix} 2 & 1 \\ 5 & 3 \end{bmatrix} = \begin{bmatrix} 3 \cdot 2 + (-1)5 & 3 \cdot 1 + (-1)3 \\ (-5)2 + 2 \cdot 5 & (-5)1 + 2 \cdot 3 \end{bmatrix} = \begin{bmatrix} 1 & 0 \\ 0 & 1 \end{bmatrix}$$

NOW TRY EXERCISE **3** ■

▼ Finding the Inverse of a 2 × 2 Matrix

The following rule provides a simple way for finding the inverse of a 2×2 matrix, when it exists. For larger matrices there is a more general procedure for finding inverses, which we consider later in this section.

The Granger Collection, New York

ARTHUR CAYLEY (1821–1895) was an English mathematician who was instrumental in developing the theory of matrices. He was the first to use a single symbol such as A to represent a matrix, thereby introducing the idea that a matrix is a single entity rather than just a collection of numbers. Cayley practiced law until the age of 42, but his primary interest from adolescence was mathematics, and he published almost 200 articles on the subject in his spare time. In 1863 he accepted a professorship in mathematics at Cambridge, where he taught until his death. Cayley's work on matrices was of purely theoretical interest in his day, but in the 20th century many of his results found application in physics, the social sciences, business, and other fields. One of the most common uses of matrices today is in computers, where matrices are employed for data storage, error correction, image manipulation, and many other purposes. These applications have made matrix algebra more useful than ever.

INVERSE OF A 2 × 2 MATRIX

If $A = \begin{bmatrix} a & b \\ c & d \end{bmatrix}$, then

$$A^{-1} = \frac{1}{ad - bc}\begin{bmatrix} d & -b \\ -c & a \end{bmatrix}$$

If $ad - bc = 0$, then A has no inverse.

EXAMPLE 3 | Finding the Inverse of a 2 × 2 Matrix

Let

$$A = \begin{bmatrix} 4 & 5 \\ 2 & 3 \end{bmatrix}$$

Find A^{-1}, and verify that $AA^{-1} = A^{-1}A = I_2$.

SOLUTION Using the rule for the inverse of a 2×2 matrix, we get

$$A^{-1} = \frac{1}{4 \cdot 3 - 5 \cdot 2}\begin{bmatrix} 3 & -5 \\ -2 & 4 \end{bmatrix} = \frac{1}{2}\begin{bmatrix} 3 & -5 \\ -2 & 4 \end{bmatrix} = \begin{bmatrix} \frac{3}{2} & -\frac{5}{2} \\ -1 & 2 \end{bmatrix}$$

To verify that this is indeed the inverse of A, we calculate AA^{-1} and $A^{-1}A$:

$$AA^{-1} = \begin{bmatrix} 4 & 5 \\ 2 & 3 \end{bmatrix}\begin{bmatrix} \frac{3}{2} & -\frac{5}{2} \\ -1 & 2 \end{bmatrix} = \begin{bmatrix} 4 \cdot \frac{3}{2} + 5(-1) & 4(-\frac{5}{2}) + 5 \cdot 2 \\ 2 \cdot \frac{3}{2} + 3(-1) & 2(-\frac{5}{2}) + 3 \cdot 2 \end{bmatrix} = \begin{bmatrix} 1 & 0 \\ 0 & 1 \end{bmatrix}$$

$$A^{-1}A = \begin{bmatrix} \frac{3}{2} & -\frac{5}{2} \\ -1 & 2 \end{bmatrix}\begin{bmatrix} 4 & 5 \\ 2 & 3 \end{bmatrix} = \begin{bmatrix} \frac{3}{2} \cdot 4 + (-\frac{5}{2})2 & \frac{3}{2} \cdot 5 + (-\frac{5}{2})3 \\ (-1)4 + 2 \cdot 2 & (-1)5 + 2 \cdot 3 \end{bmatrix} = \begin{bmatrix} 1 & 0 \\ 0 & 1 \end{bmatrix}$$

NOW TRY EXERCISE **9**

The quantity $ad - bc$ that appears in the rule for calculating the inverse of a 2×2 matrix is called the **determinant** of the matrix. If the determinant is 0, then the matrix does not have an inverse (since we cannot divide by 0).

Finding the Inverse of an $n \times n$ Matrix

For 3×3 and larger square matrices the following technique provides the most efficient way to calculate their inverses. If A is an $n \times n$ matrix, we first construct the $n \times 2n$ matrix that has the entries of A on the left and of the identity matrix I_n on the right:

$$\left[\begin{array}{cccc|cccc} a_{11} & a_{12} & \cdots & a_{1n} & 1 & 0 & \cdots & 0 \\ a_{21} & a_{22} & \cdots & a_{2n} & 0 & 1 & \cdots & 0 \\ \vdots & \vdots & \ddots & \vdots & \vdots & \vdots & \ddots & \vdots \\ a_{n1} & a_{n2} & \cdots & a_{nn} & 0 & 0 & \cdots & 1 \end{array}\right]$$

We then use the elementary row operations on this new large matrix to change the left side into the identity matrix. (This means that we are changing the large matrix to reduced row-echelon form.) The right side is transformed automatically into A^{-1}. (We omit the proof of this fact.)

EXAMPLE 4 | Finding the Inverse of a 3 × 3 Matrix

Let A be the matrix

$$A = \begin{bmatrix} 1 & -2 & -4 \\ 2 & -3 & -6 \\ -3 & 6 & 15 \end{bmatrix}$$

(a) Find A^{-1}.

(b) Verify that $AA^{-1} = A^{-1}A = I_3$.

SOLUTION

(a) We begin with the 3×6 matrix whose left half is A and whose right half is the identity matrix.

$$\left[\begin{array}{ccc|ccc} 1 & -2 & -4 & 1 & 0 & 0 \\ 2 & -3 & -6 & 0 & 1 & 0 \\ -3 & 6 & 15 & 0 & 0 & 1 \end{array}\right]$$

We then transform the left half of this new matrix into the identity matrix by performing the following sequence of elementary row operations on the *entire* new matrix.

$$\xrightarrow[\mathrm{R}_3 + 3\mathrm{R}_1 \to \mathrm{R}_3]{\mathrm{R}_2 - 2\mathrm{R}_1 \to \mathrm{R}_2} \left[\begin{array}{ccc|ccc} 1 & -2 & -4 & 1 & 0 & 0 \\ 0 & 1 & 2 & -2 & 1 & 0 \\ 0 & 0 & 3 & 3 & 0 & 1 \end{array}\right]$$

$$\xrightarrow{\frac{1}{3}\mathrm{R}_3} \left[\begin{array}{ccc|ccc} 1 & -2 & -4 & 1 & 0 & 0 \\ 0 & 1 & 2 & -2 & 1 & 0 \\ 0 & 0 & 1 & 1 & 0 & \frac{1}{3} \end{array}\right]$$

$$\xrightarrow{\mathrm{R}_1 + 2\mathrm{R}_2 \to \mathrm{R}_1} \left[\begin{array}{ccc|ccc} 1 & 0 & 0 & -3 & 2 & 0 \\ 0 & 1 & 2 & -2 & 1 & 0 \\ 0 & 0 & 1 & 1 & 0 & \frac{1}{3} \end{array}\right]$$

$$\xrightarrow{\mathrm{R}_2 - 2\mathrm{R}_3 \to \mathrm{R}_2} \left[\begin{array}{ccc|ccc} 1 & 0 & 0 & -3 & 2 & 0 \\ 0 & 1 & 0 & -4 & 1 & -\frac{2}{3} \\ 0 & 0 & 1 & 1 & 0 & \frac{1}{3} \end{array}\right]$$

We have now transformed the left half of this matrix into an identity matrix. (This means that we have put the entire matrix in reduced row-echelon form.) Note that to do this in as systematic a fashion as possible, we first changed the elements below the main diagonal to zeros, just as we would if we were using Gaussian elimination. We then changed each main diagonal element to a 1 by multiplying by the appropriate constant(s). Finally, we completed the process by changing the remaining entries on the left side to zeros.

The right half is now A^{-1}.

$$A^{-1} = \begin{bmatrix} -3 & 2 & 0 \\ -4 & 1 & -\frac{2}{3} \\ 1 & 0 & \frac{1}{3} \end{bmatrix}$$

(b) We calculate AA^{-1} and $A^{-1}A$ and verify that both products give the identity matrix I_3.

$$AA^{-1} = \begin{bmatrix} 1 & -2 & -4 \\ 2 & -3 & -6 \\ -3 & 6 & 15 \end{bmatrix} \begin{bmatrix} -3 & 2 & 0 \\ -4 & 1 & -\frac{2}{3} \\ 1 & 0 & \frac{1}{3} \end{bmatrix} = \begin{bmatrix} 1 & 0 & 0 \\ 0 & 1 & 0 \\ 0 & 0 & 1 \end{bmatrix}$$

$$A^{-1}A = \begin{bmatrix} -3 & 2 & 0 \\ -4 & 1 & -\frac{2}{3} \\ 1 & 0 & \frac{1}{3} \end{bmatrix} \begin{bmatrix} 1 & -2 & -4 \\ 2 & -3 & -6 \\ -3 & 6 & 15 \end{bmatrix} = \begin{bmatrix} 1 & 0 & 0 \\ 0 & 1 & 0 \\ 0 & 0 & 1 \end{bmatrix}$$

NOW TRY EXERCISE **17** ■

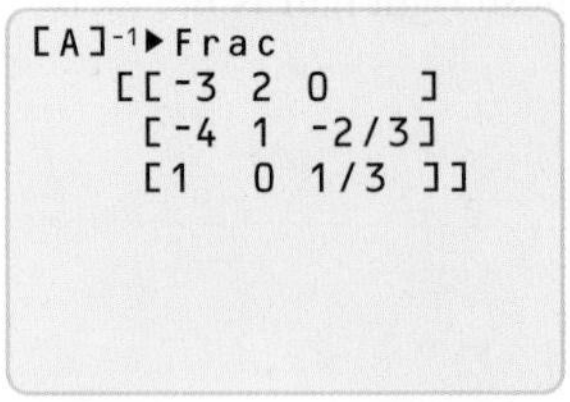

FIGURE 1

Graphing calculators are also able to calculate matrix inverses. On the TI-83 and TI-84 calculators, matrices are stored in memory using names such as [A], [B], [C], ... To find the inverse of [A], we key in

[A] x^{-1} ENTER

For the matrix of Example 4 this results in the output shown in Figure 1 (where we have also used the ▶Frac command to display the output in fraction form rather than in decimal form).

The next example shows that not every square matrix has an inverse.

EXAMPLE 5 | A Matrix That Does Not Have an Inverse

Find the inverse of the matrix.

$$\begin{bmatrix} 2 & -3 & -7 \\ 1 & 2 & 7 \\ 1 & 1 & 4 \end{bmatrix}$$

SOLUTION We proceed as follows.

$$\left[\begin{array}{ccc|ccc} 2 & -3 & -7 & 1 & 0 & 0 \\ 1 & 2 & 7 & 0 & 1 & 0 \\ 1 & 1 & 4 & 0 & 0 & 1 \end{array}\right] \xrightarrow{R_1 \leftrightarrow R_2} \left[\begin{array}{ccc|ccc} 1 & 2 & 7 & 0 & 1 & 0 \\ 2 & -3 & -7 & 1 & 0 & 0 \\ 1 & 1 & 4 & 0 & 0 & 1 \end{array}\right]$$

$$\xrightarrow[R_3 - R_1 \to R_3]{R_2 - 2R_1 \to R_2} \left[\begin{array}{ccc|ccc} 1 & 2 & 7 & 0 & 1 & 0 \\ 0 & -7 & -21 & 1 & -2 & 0 \\ 0 & -1 & -3 & 0 & -1 & 1 \end{array}\right]$$

$$\xrightarrow{-\frac{1}{7}R_2} \left[\begin{array}{ccc|ccc} 1 & 2 & 7 & 0 & 1 & 0 \\ 0 & 1 & 3 & -\frac{1}{7} & \frac{2}{7} & 0 \\ 0 & -1 & -3 & 0 & -1 & 1 \end{array}\right]$$

$$\xrightarrow[R_1 - 2R_2 \to R_1]{R_3 + R_2 \to R_3} \left[\begin{array}{ccc|ccc} 1 & 0 & 1 & \frac{2}{7} & \frac{3}{7} & 0 \\ 0 & 1 & 3 & -\frac{1}{7} & \frac{2}{7} & 0 \\ 0 & 0 & 0 & -\frac{1}{7} & -\frac{5}{7} & 1 \end{array}\right]$$

At this point we would like to change the 0 in the (3, 3) position of this matrix to a 1 without changing the zeros in the (3, 1) and (3, 2) positions. But there is no way to accomplish this, because no matter what multiple of rows 1 and/or 2 we add to row 3, we can't change the third zero in row 3 without changing the first or second zero as well. Thus we cannot change the left half to the identity matrix, so the original matrix doesn't have an inverse.

NOW TRY EXERCISE **19** ■

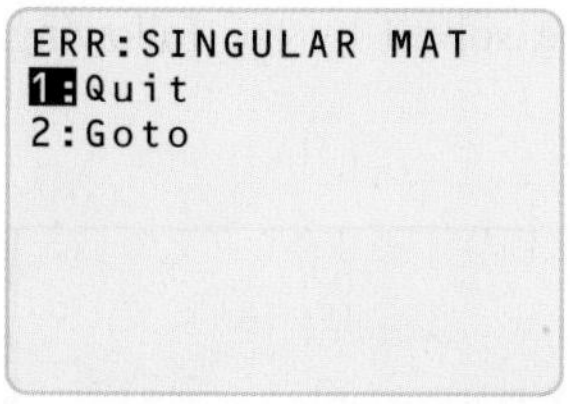

FIGURE 2

If we encounter a row of zeros on the left when trying to find an inverse, as in Example 5, then the original matrix does not have an inverse. If we try to calculate the inverse of the matrix from Example 5 on a TI-83 calculator, we get the error message shown in Figure 2. (A matrix that has no inverse is called *singular*.)

▼ Matrix Equations

We saw in Example 6 in Section 10.4 that a system of linear equations can be written as a single matrix equation. For example, the system

$$\begin{cases} x - 2y - 4z = 7 \\ 2x - 3y - 6z = 5 \\ -3x + 6y + 15z = 0 \end{cases}$$

is equivalent to the matrix equation

$$\underbrace{\begin{bmatrix} 1 & -2 & -4 \\ 2 & -3 & -6 \\ -3 & 6 & 15 \end{bmatrix}}_{A} \underbrace{\begin{bmatrix} x \\ y \\ z \end{bmatrix}}_{X} = \underbrace{\begin{bmatrix} 7 \\ 5 \\ 0 \end{bmatrix}}_{B}$$

If we let

$$A = \begin{bmatrix} 1 & -2 & -4 \\ 2 & -3 & -6 \\ -3 & 6 & 15 \end{bmatrix} \qquad X = \begin{bmatrix} x \\ y \\ z \end{bmatrix} \qquad B = \begin{bmatrix} 7 \\ 5 \\ 0 \end{bmatrix}$$

then this matrix equation can be written as

$$AX = B$$

The matrix A is called the **coefficient matrix**.

We solve this matrix equation by multiplying each side by the inverse of A (provided that this inverse exists):

Solving the matrix equation $AX = B$ is very similar to solving the simple real-number equation

$$3x = 12$$

which we do by multiplying each side by the reciprocal (or inverse) of 3.

$$\tfrac{1}{3}(3x) = \tfrac{1}{3}(12)$$

$$x = 4$$

$AX = B$	
$A^{-1}(AX) = A^{-1}B$	Multiply on left by A^{-1}
$(A^{-1}A)X = A^{-1}B$	Associative Property
$I_3X = A^{-1}B$	Property of inverses
$X = A^{-1}B$	Property of identity matrix

In Example 4 we showed that

$$A^{-1} = \begin{bmatrix} -3 & 2 & 0 \\ -4 & 1 & -\frac{2}{3} \\ 1 & 0 & \frac{1}{3} \end{bmatrix}$$

So from $X = A^{-1}B$ we have

$$\underbrace{\begin{bmatrix} x \\ y \\ z \end{bmatrix}}_{X} = \underbrace{\begin{bmatrix} -3 & 2 & 0 \\ -4 & 1 & -\frac{2}{3} \\ 1 & 0 & \frac{1}{3} \end{bmatrix}}_{A^{-1}} \underbrace{\begin{bmatrix} 7 \\ 5 \\ 0 \end{bmatrix}}_{B} = \begin{bmatrix} -11 \\ -23 \\ 7 \end{bmatrix}$$

Thus $x = -11$, $y = -23$, $z = 7$ is the solution of the original system.

We have proved that the matrix equation $AX = B$ can be solved by the following method.

SOLVING A MATRIX EQUATION

If A is a square $n \times n$ matrix that has an inverse A^{-1} and if X is a variable matrix and B a known matrix, both with n rows, then the solution of the matrix equation

$$AX = B$$

is given by

$$X = A^{-1}B$$

EXAMPLE 6 | Solving a System Using a Matrix Inverse

A system of equations is given.

(a) Write the system of equations as a matrix equation.

(b) Solve the system by solving the matrix equation.

$$\begin{cases} 2x - 5y = 15 \\ 3x - 6y = 36 \end{cases}$$

SOLUTION

(a) We write the system as a matrix equation of the form $AX = B$.

$$\underbrace{\begin{bmatrix} 2 & -5 \\ 3 & -6 \end{bmatrix}}_{A} \underbrace{\begin{bmatrix} x \\ y \end{bmatrix}}_{X} = \underbrace{\begin{bmatrix} 15 \\ 36 \end{bmatrix}}_{B}$$

(b) Using the rule for finding the inverse of a 2×2 matrix, we get

$$A^{-1} = \begin{bmatrix} 2 & -5 \\ 3 & -6 \end{bmatrix}^{-1} = \frac{1}{2(-6) - (-5)3} \begin{bmatrix} -6 & -(-5) \\ -3 & 2 \end{bmatrix} = \frac{1}{3} \begin{bmatrix} -6 & 5 \\ -3 & 2 \end{bmatrix}$$

Multiplying each side of the matrix equation by this inverse matrix, we get

$$\underbrace{\begin{bmatrix} x \\ y \end{bmatrix}}_{X} = \underbrace{\frac{1}{3} \begin{bmatrix} -6 & 5 \\ -3 & 2 \end{bmatrix}}_{A^{-1}} \underbrace{\begin{bmatrix} 15 \\ 36 \end{bmatrix}}_{B} = \begin{bmatrix} 30 \\ 9 \end{bmatrix}$$

So $x = 30$ and $y = 9$.

NOW TRY EXERCISE 25

Modeling with Matrix Equations

Suppose we need to solve several systems of equations with the same coefficient matrix. Then converting the systems to matrix equations provides an efficient way to obtain the solutions, because we need to find the inverse of the coefficient matrix only once. This procedure is particularly convenient if we use a graphing calculator to perform the matrix operations, as in the next example.

EXAMPLE 7 | Modeling Nutritional Requirements Using Matrix Equations

A pet-store owner feeds his hamsters and gerbils different mixtures of three types of rodent food: KayDee Food, Pet Pellets, and Rodent Chow. He wishes to feed his animals the correct amount of each brand to satisfy their daily requirements for protein, fat, and carbohydrates exactly. Suppose that hamsters require 340 mg of protein, 280 mg of fat, and 440 mg of carbohydrates, and gerbils need 480 mg of protein, 360 mg of fat, and 680 mg of carbohydrates each day. The amount of each nutrient (in mg) in one gram of each brand is given in the following table. How many grams of each food should the storekeeper feed his hamsters and gerbils daily to satisfy their nutrient requirements?

	KayDee Food	Pet Pellets	Rodent Chow
Protein (mg)	10	0	20
Fat (mg)	10	20	10
Carbohydrates (mg)	5	10	30

SOLUTION We let x_1, x_2, and x_3 be the respective amounts (in grams) of KayDee Food, Pet Pellets, and Rodent Chow that the hamsters should eat and y_1, y_2, and y_3 be the corresponding amounts for the gerbils. Then we want to solve the matrix equations

$$\begin{bmatrix} 10 & 0 & 20 \\ 10 & 20 & 10 \\ 5 & 10 & 30 \end{bmatrix} \begin{bmatrix} x_1 \\ x_2 \\ x_3 \end{bmatrix} = \begin{bmatrix} 340 \\ 280 \\ 440 \end{bmatrix} \qquad \text{Hamster equation}$$

$$\begin{bmatrix} 10 & 0 & 20 \\ 10 & 20 & 10 \\ 5 & 10 & 30 \end{bmatrix} \begin{bmatrix} y_1 \\ y_2 \\ y_3 \end{bmatrix} = \begin{bmatrix} 480 \\ 360 \\ 680 \end{bmatrix} \qquad \text{Gerbil equation}$$

Let

$$A = \begin{bmatrix} 10 & 0 & 20 \\ 10 & 20 & 10 \\ 5 & 10 & 30 \end{bmatrix} \quad B = \begin{bmatrix} 340 \\ 280 \\ 440 \end{bmatrix} \quad C = \begin{bmatrix} 480 \\ 360 \\ 680 \end{bmatrix} \quad X = \begin{bmatrix} x_1 \\ x_2 \\ x_3 \end{bmatrix} \quad Y = \begin{bmatrix} y_1 \\ y_2 \\ y_3 \end{bmatrix}$$

Then we can write these matrix equations as

$$AX = B \qquad \text{Hamster equation}$$

$$AY = C \qquad \text{Gerbil equation}$$

We want to solve for X and Y, so we multiply both sides of each equation by A^{-1}, the inverse of the coefficient matrix. We could find A^{-1} by hand, but it is more convenient to use a graphing calculator as shown in Figure 3.

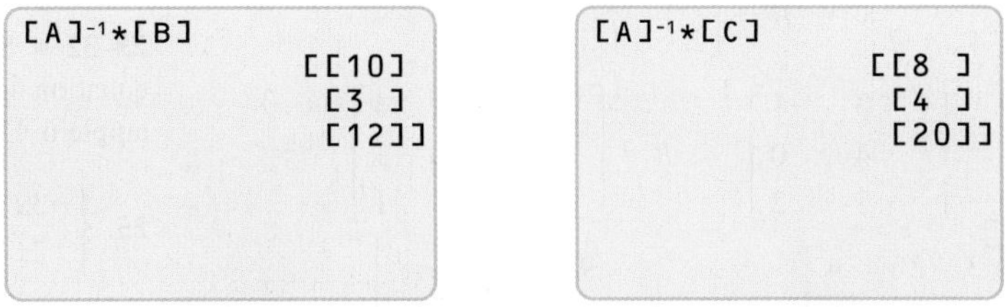

FIGURE 3 (a) (b)

MATHEMATICS IN THE MODERN WORLD

Art Wolfe/Stone/Getty Images

Mathematical Ecology

In the 1970s humpback whales became a center of controversy. Environmentalists believed that whaling threatened the whales with imminent extinction; whalers saw their livelihood threatened by any attempt to stop whaling. Are whales really threatened to extinction by whaling? What level of whaling is safe to guarantee survival of the whales? These questions motivated mathematicians to study population patterns of whales and other species more closely.

As early as the 1920s Lotka and Volterra had founded the field of mathematical biology by creating predator-prey models. Their models, which draw on a branch of mathematics called differential equations, take into account the rates at which predator eats prey and the rates of growth of each population. Note that as predator eats prey, the prey population decreases; this means less food supply for the predators, so their population begins to decrease; with fewer predators the prey population begins to increase, and so on. Normally, a state of equilibrium develops, and the two populations alternate between a minimum and a maximum. Notice that if the predators eat the prey too fast, they will be left without food and will thus ensure their own extinction.

Since Lotka and Volterra's time, more detailed mathematical models of animal populations have been developed. For many species the population is divided into several stages: immature, juvenile, adult, and so on. The proportion of each stage that survives or reproduces in a given time period is entered into a matrix (called a transition matrix); matrix multiplication is then used to predict the population in succeeding time periods. (See the Discovery Project *Will the Species Survive?* at the book companion website: www.stewartmath.com.)

As you can see, the power of mathematics to model and predict is an invaluable tool in the ongoing debate over the environment.

So

$$X = A^{-1}B = \begin{bmatrix} 10 \\ 3 \\ 12 \end{bmatrix} \qquad Y = A^{-1}C = \begin{bmatrix} 8 \\ 4 \\ 20 \end{bmatrix}$$

Thus each hamster should be fed 10 g of KayDee Food, 3 g of Pet Pellets, and 12 g of Rodent Chow; and each gerbil should be fed 8 g of KayDee Food, 4 g of Pet Pellets, and 20 g of Rodent Chow daily.

NOW TRY EXERCISE **47** ■

10.5 EXERCISES

CONCEPTS

1. **(a)** The matrix $I = \begin{bmatrix} 1 & 0 \\ 0 & 1 \end{bmatrix}$ is called an ________ matrix.

   (b) If A is a 2×2 matrix, then $A \times I =$ ________ and $I \times A =$ ________.

   (c) If A and B are 2×2 matrices with $AB = I$, then B is the ________ of A.

2. **(a)** Write the following system as a matrix equation $AX = B$.

System	**Matrix equation**
	$A \cdot X = B$
$5x + 3y = 4$ $3x + 2y = 3$	$\begin{bmatrix} \blacksquare & \blacksquare \\ \blacksquare & \blacksquare \end{bmatrix}\begin{bmatrix} \blacksquare \\ \blacksquare \end{bmatrix} = \begin{bmatrix} \blacksquare \\ \blacksquare \end{bmatrix}$

   (b) The inverse of A is $A^{-1} = \begin{bmatrix} \blacksquare & \blacksquare \\ \blacksquare & \blacksquare \end{bmatrix}$.

   (c) The solution of the matrix equation is $X = A^{-1}B$.

$$\begin{matrix} X & = & A^{-1} & B \end{matrix}$$

$$\begin{bmatrix} x \\ y \end{bmatrix} = \begin{bmatrix} \blacksquare & \blacksquare \\ \blacksquare & \blacksquare \end{bmatrix}\begin{bmatrix} \blacksquare \\ \blacksquare \end{bmatrix} = \begin{bmatrix} \blacksquare \\ \blacksquare \end{bmatrix}$$

   (d) The solution of the system is $x =$ ________, $y =$ ________.

SKILLS

3–6 ■ Calculate the products AB and BA to verify that B is the inverse of A.

3. $A = \begin{bmatrix} 4 & 1 \\ 7 & 2 \end{bmatrix} \quad B = \begin{bmatrix} 2 & -1 \\ -7 & 4 \end{bmatrix}$

4. $A = \begin{bmatrix} 2 & -3 \\ 4 & -7 \end{bmatrix} \quad B = \begin{bmatrix} \frac{7}{2} & -\frac{3}{2} \\ 2 & -1 \end{bmatrix}$

5. $A = \begin{bmatrix} 1 & 3 & -1 \\ 1 & 4 & 0 \\ -1 & -3 & 2 \end{bmatrix} \quad B = \begin{bmatrix} 8 & -3 & 4 \\ -2 & 1 & -1 \\ 1 & 0 & 1 \end{bmatrix}$

6. $A = \begin{bmatrix} 3 & 2 & 4 \\ 1 & 1 & -6 \\ 2 & 1 & 12 \end{bmatrix} \quad B = \begin{bmatrix} 9 & -10 & -8 \\ -12 & 14 & 11 \\ -\frac{1}{2} & \frac{1}{2} & \frac{1}{2} \end{bmatrix}$

7–8 ■ Find the inverse of the matrix and verify that $A^{-1}A = AA^{-1} = I_2$ and $B^{-1}B = BB^{-1} = I_3$.

7. $A = \begin{bmatrix} 7 & 4 \\ 3 & 2 \end{bmatrix}$

8. $B = \begin{bmatrix} 1 & 3 & 2 \\ 0 & 2 & 2 \\ -2 & -1 & 0 \end{bmatrix}$

9–24 ■ Find the inverse of the matrix if it exists.

9. $\begin{bmatrix} -3 & -5 \\ 2 & 3 \end{bmatrix}$

10. $\begin{bmatrix} 3 & 4 \\ 7 & 9 \end{bmatrix}$

11. $\begin{bmatrix} 2 & 5 \\ -5 & -13 \end{bmatrix}$

12. $\begin{bmatrix} -7 & 4 \\ 8 & -5 \end{bmatrix}$

13. $\begin{bmatrix} 6 & -3 \\ -8 & 4 \end{bmatrix}$

14. $\begin{bmatrix} \frac{1}{2} & \frac{1}{3} \\ 5 & 4 \end{bmatrix}$

15. $\begin{bmatrix} 0.4 & -1.2 \\ 0.3 & 0.6 \end{bmatrix}$

16. $\begin{bmatrix} 4 & 2 & 3 \\ 3 & 3 & 2 \\ 1 & 0 & 1 \end{bmatrix}$

17. $\begin{bmatrix} 2 & 4 & 1 \\ -1 & 1 & -1 \\ 1 & 4 & 0 \end{bmatrix}$

18. $\begin{bmatrix} 5 & 7 & 4 \\ 3 & -1 & 3 \\ 6 & 7 & 5 \end{bmatrix}$

19. $\begin{bmatrix} 1 & 2 & 3 \\ 4 & 5 & -1 \\ 1 & -1 & -10 \end{bmatrix}$

20. $\begin{bmatrix} 2 & 1 & 0 \\ 1 & 1 & 4 \\ 2 & 1 & 2 \end{bmatrix}$

21. $\begin{bmatrix} 0 & -2 & 2 \\ 3 & 1 & 3 \\ 1 & -2 & 3 \end{bmatrix}$

22. $\begin{bmatrix} 3 & -2 & 0 \\ 5 & 1 & 1 \\ 2 & -2 & 0 \end{bmatrix}$

23. $\begin{bmatrix} 1 & 2 & 0 & 3 \\ 0 & 1 & 1 & 1 \\ 0 & 1 & 0 & 1 \\ 1 & 2 & 0 & 2 \end{bmatrix}$

24. $\begin{bmatrix} 1 & 0 & 1 & 0 \\ 0 & 1 & 0 & 1 \\ 1 & 1 & 1 & 0 \\ 1 & 1 & 1 & 1 \end{bmatrix}$

25–32 ■ Solve the system of equations by converting to a matrix equation and using the inverse of the coefficient matrix, as in Example 6. Use the inverses from Exercises 9–12, 17, 18, 21, and 23.

25. $\begin{cases} -3x - 5y = 4 \\ 2x + 3y = 0 \end{cases}$

26. $\begin{cases} 3x + 4y = 10 \\ 7x + 9y = 20 \end{cases}$

27. $\begin{cases} 2x + 5y = 2 \\ -5x - 13y = 20 \end{cases}$

28. $\begin{cases} -7x + 4y = 0 \\ 8x - 5y = 100 \end{cases}$

29. $\begin{cases} 2x + 4y + z = 7 \\ -x + y - z = 0 \\ x + 4y = -2 \end{cases}$

30. $\begin{cases} 5x + 7y + 4z = 1 \\ 3x - y + 3z = 1 \\ 6x + 7y + 5z = 1 \end{cases}$

31. $\begin{cases} -2y + 2z = 12 \\ 3x + y + 3z = -2 \\ x - 2y + 3z = 8 \end{cases}$

32. $\begin{cases} x + 2y + 3w = 0 \\ y + z + w = 1 \\ y + w = 2 \\ x + 2y + 2w = 3 \end{cases}$

33–38 ■ Use a calculator that can perform matrix operations to solve the system, as in Example 7.

33. $\begin{cases} x + y - 2z = 3 \\ 2x + 5z = 11 \\ 2x + 3y = 12 \end{cases}$

34. $\begin{cases} 3x + 4y - z = 2 \\ 2x - 3y + z = -5 \\ 5x - 2y + 2z = -3 \end{cases}$

35. $\begin{cases} 12x + \frac{1}{2}y - 7z = 21 \\ 11x - 2y + 3z = 43 \\ 13x + y - 4z = 29 \end{cases}$

36. $\begin{cases} x + \frac{1}{2}y - \frac{1}{3}z = 4 \\ x - \frac{1}{4}y + \frac{1}{6}z = 7 \\ x + y - z = -6 \end{cases}$

37. $\begin{cases} x + y - 3w = 0 \\ x - 2z = 8 \\ 2y - z + w = 5 \\ 2x + 3y - 2w = 13 \end{cases}$

38. $\begin{cases} x + y + z + w = 15 \\ x - y + z - w = 5 \\ x + 2y + 3z + 4w = 26 \\ x - 2y + 3z - 4w = 2 \end{cases}$

39–40 ■ Solve the matrix equation by multiplying each side by the appropriate inverse matrix.

39. $\begin{bmatrix} 3 & -2 \\ -4 & 3 \end{bmatrix}\begin{bmatrix} x & y & z \\ u & v & w \end{bmatrix} = \begin{bmatrix} 1 & 0 & -1 \\ 2 & 1 & 3 \end{bmatrix}$

40. $\begin{bmatrix} 0 & -2 & 2 \\ 3 & 1 & 3 \\ 1 & -2 & 3 \end{bmatrix}\begin{bmatrix} x & u \\ y & v \\ z & w \end{bmatrix} = \begin{bmatrix} 3 & 6 \\ 6 & 12 \\ 0 & 0 \end{bmatrix}$

41–42 ■ Find the inverse of the matrix.

41. $\begin{bmatrix} a & -a \\ a & a \end{bmatrix}$ $(a \neq 0)$

42. $\begin{bmatrix} a & 0 & 0 & 0 \\ 0 & b & 0 & 0 \\ 0 & 0 & c & 0 \\ 0 & 0 & 0 & d \end{bmatrix}$ $(abcd \neq 0)$

43–46 ■ Find the inverse of the matrix. For what value(s) of x, if any, does the matrix have no inverse?

43. $\begin{bmatrix} 2 & x \\ x & x^2 \end{bmatrix}$

44. $\begin{bmatrix} e^x & -e^{2x} \\ e^{2x} & e^{3x} \end{bmatrix}$

45. $\begin{bmatrix} 1 & e^x & 0 \\ e^x & -e^{2x} & 0 \\ 0 & 0 & 2 \end{bmatrix}$

46. $\begin{bmatrix} x & 1 \\ -x & \dfrac{1}{x-1} \end{bmatrix}$

APPLICATIONS

47. Nutrition A nutritionist is studying the effects of the nutrients folic acid, choline, and inositol. He has three types of food available, and each type contains the following amounts of these nutrients per ounce.

	Type A	Type B	Type C
Folic acid (mg)	3	1	3
Choline (mg)	4	2	4
Inositol (mg)	3	2	4

(a) Find the inverse of the matrix

$$\begin{bmatrix} 3 & 1 & 3 \\ 4 & 2 & 4 \\ 3 & 2 & 4 \end{bmatrix}$$

and use it to solve the remaining parts of this problem.

(b) How many ounces of each food should the nutritionist feed his laboratory rats if he wants their daily diet to contain 10 mg of folic acid, 14 mg of choline, and 13 mg of inositol?

(c) How much of each food is needed to supply 9 mg of folic acid, 12 mg of choline, and 10 mg of inositol?

(d) Will any combination of these foods supply 2 mg of folic acid, 4 mg of choline, and 11 mg of inositol?

48. Nutrition Refer to Exercise 47. Suppose food type C has been improperly labeled, and it actually contains 4 mg of folic acid, 6 mg of choline, and 5 mg of inositol per ounce. Would it still be possible to use matrix inversion to solve parts (b), (c), and (d) of Exercise 47? Why or why not?

49. Sales Commissions An encyclopedia saleswoman works for a company that offers three different grades of bindings for its encyclopedias: standard, deluxe, and leather. For each set that she sells, she earns a commission based on the set's binding grade. One week she sells one standard, one deluxe, and two leather sets and makes $675 in commission. The next week she sells two standard, one deluxe, and one leather set for a $600 commission. The third week she sells one standard, two deluxe, and one leather set, earning $625 in commission.

(a) Let x, y, and z represent the commission she earns on standard, deluxe, and leather sets, respectively. Translate the given information into a system of equations in x, y, and z.

(b) Express the system of equations you found in part (a) as a matrix equation of the form $AX = B$.

(c) Find the inverse of the coefficient matrix A and use it to solve the matrix equation in part (b). How much commission does the saleswoman earn on a set of encyclopedias in each grade of binding?

DISCOVERY ■ DISCUSSION ■ WRITING

50. No Zero-Product Property for Matrices We have used the Zero-Product Property to solve algebraic equations. Matrices do *not* have this property. Let O represent the **2 × 2 zero matrix**

$$O = \begin{bmatrix} 0 & 0 \\ 0 & 0 \end{bmatrix}$$

Find 2×2 matrices $A \neq O$ and $B \neq O$ such that $AB = O$. Can you find a matrix $A \neq O$ such that $A^2 = O$?

10.6 Determinants and Cramer's Rule

Determinant of a 2 × 2 Matrix ▶ Determinant of an $n \times n$ Matrix ▶ Row and Column Transformations ▶ Cramer's Rule ▶ Areas of Triangles Using Determinants

If a matrix is **square** (that is, if it has the same number of rows as columns), then we can assign to it a number called its *determinant*. Determinants can be used to solve systems of linear equations, as we will see later in this section. They are also useful in determining whether a matrix has an inverse.

Determinant of a 2 × 2 Matrix

We denote the determinant of a square matrix A by the symbol $\det(A)$ or $|A|$. We first define $\det(A)$ for the simplest cases. If $A = [a]$ is a 1×1 matrix, then $\det(A) = a$. The following box gives the definition of a 2×2 determinant.

We will use both notations, $\det(A)$ and $|A|$, for the determinant of A. Although the symbol $|A|$ looks like the absolute value symbol, it will be clear from the context which meaning is intended.

DETERMINANT OF A 2 × 2 MATRIX

The **determinant** of the 2×2 matrix $A = \begin{bmatrix} a & b \\ c & d \end{bmatrix}$ is

$$\det(A) = |A| = \begin{vmatrix} a & b \\ c & d \end{vmatrix} = ad - bc$$

EXAMPLE 1 | Determinant of a 2 × 2 Matrix

Evaluate $|A|$ for $A = \begin{bmatrix} 6 & -3 \\ 2 & 3 \end{bmatrix}$.

SOLUTION

To evaluate a 2×2 determinant, we take the product of the diagonal from top left to bottom right and subtract the product from top right to bottom left, as indicated by the arrows.

$$\begin{vmatrix} 6 & -3 \\ 2 & 3 \end{vmatrix} = 6 \cdot 3 - (-3)2 = 18 - (-6) = 24$$

NOW TRY EXERCISE 5 ■

Determinant of an $n \times n$ Matrix

To define the concept of determinant for an arbitrary $n \times n$ matrix, we need the following terminology.

MINORS AND COFACTORS

Let A be an $n \times n$ matrix.

1. The **minor** M_{ij} of the element a_{ij} is the determinant of the matrix obtained by deleting the ith row and jth column of A.
2. The **cofactor** A_{ij} of the element a_{ij} is

$$A_{ij} = (-1)^{i+j} M_{ij}$$

DAVID HILBERT (1862–1943) was born in Königsberg, Germany, and became a professor at Göttingen University. He is considered by many to be the greatest mathematician of the 20th century. At the International Congress of Mathematicians held in Paris in 1900, Hilbert set the direction of mathematics for the about-to-dawn 20th century by posing 23 problems that he believed to be of crucial importance. He said that "these are problems whose solutions we expect from the future." Most of Hilbert's problems have now been solved (see Julia Robinson, page 663, and Alan Turing, page 100), and their solutions have led to important new areas of mathematical research. Yet as we proceed into the new millennium, some of his problems remain unsolved. In his work, Hilbert emphasized structure, logic, and the foundations of mathematics. Part of his genius lay in his ability to see the most general possible statement of a problem. For instance, Euler proved that every whole number is the sum of four squares; Hilbert proved a similar statement for all powers of positive integers.

For example, if A is the matrix

$$\begin{bmatrix} 2 & 3 & -1 \\ 0 & 2 & 4 \\ -2 & 5 & 6 \end{bmatrix}$$

then the minor M_{12} is the determinant of the matrix obtained by deleting the first row and second column from A. Thus

$$M_{12} = \begin{vmatrix} 2 & 3 & -1 \\ 0 & 2 & 4 \\ -2 & 5 & 6 \end{vmatrix} = \begin{vmatrix} 0 & 4 \\ -2 & 6 \end{vmatrix} = 0(6) - 4(-2) = 8$$

So the cofactor $A_{12} = (-1)^{1+2}M_{12} = -8$. Similarly,

$$M_{33} = \begin{vmatrix} 2 & 3 & -1 \\ 0 & 2 & 4 \\ 2 & 5 & 6 \end{vmatrix} = \begin{vmatrix} 2 & 3 \\ 0 & 2 \end{vmatrix} = 2 \cdot 2 - 3 \cdot 0 = 4$$

So $A_{33} = (-1)^{3+3}M_{33} = 4$.

Note that the cofactor of a_{ij} is simply the minor of a_{ij} multiplied by either 1 or -1, depending on whether $i + j$ is even or odd. Thus in a 3×3 matrix we obtain the cofactor of any element by prefixing its minor with the sign obtained from the following checkerboard pattern.

$$\begin{bmatrix} + & - & + \\ - & + & - \\ + & - & + \end{bmatrix}$$

We are now ready to define the determinant of any square matrix.

THE DETERMINANT OF A SQUARE MATRIX

If A is an $n \times n$ matrix, then the **determinant** of A is obtained by multiplying each element of the first row by its cofactor and then adding the results. In symbols,

$$\det(A) = |A| = \begin{vmatrix} a_{11} & a_{12} & \cdots & a_{1n} \\ a_{21} & a_{22} & \cdots & a_{2n} \\ \vdots & \vdots & \ddots & \vdots \\ a_{n1} & a_{n2} & \cdots & a_{nn} \end{vmatrix} = a_{11}A_{11} + a_{12}A_{12} + \cdots + a_{1n}A_{1n}$$

EXAMPLE 2 | Determinant of a 3×3 Matrix

Evaluate the determinant of the matrix.

$$A = \begin{bmatrix} 2 & 3 & -1 \\ 0 & 2 & 4 \\ -2 & 5 & 6 \end{bmatrix}$$

SOLUTION

$$\det(A) = \begin{vmatrix} 2 & 3 & -1 \\ 0 & 2 & 4 \\ -2 & 5 & 6 \end{vmatrix} = 2\begin{vmatrix} 2 & 4 \\ 5 & 6 \end{vmatrix} - 3\begin{vmatrix} 0 & 4 \\ -2 & 6 \end{vmatrix} + (-1)\begin{vmatrix} 0 & 2 \\ -2 & 5 \end{vmatrix}$$

$$= 2(2\cdot 6 - 4\cdot 5) - 3[0\cdot 6 - 4(-2)] - [0\cdot 5 - 2(-2)]$$

$$= -16 - 24 - 4$$

$$= -44$$

NOW TRY EXERCISE **19** ■

In our definition of the determinant we used the cofactors of elements in the first row only. This is called **expanding the determinant by the first row**. In fact, *we can expand the determinant by any row or column in the same way and obtain the same result in each case* (although we won't prove this). The next example illustrates this principle.

EXAMPLE 3 | Expanding a Determinant About a Row and a Column

Let A be the matrix of Example 2. Evaluate the determinant of A by expanding

(a) by the second row

(b) by the third column

Verify that each expansion gives the same value.

SOLUTION

(a) Expanding by the second row, we get

$$\det(A) = \begin{vmatrix} 2 & 3 & -1 \\ 0 & 2 & 4 \\ -2 & 5 & 6 \end{vmatrix} = -0\begin{vmatrix} 3 & -1 \\ 5 & 6 \end{vmatrix} + 2\begin{vmatrix} 2 & -1 \\ -2 & 6 \end{vmatrix} - 4\begin{vmatrix} 2 & 3 \\ -2 & 5 \end{vmatrix}$$

$$= 0 + 2[2\cdot 6 - (-1)(-2)] - 4[2\cdot 5 - 3(-2)]$$

$$= 0 + 20 - 64 = -44$$

(b) Expanding by the third column gives

$$\det(A) = \begin{vmatrix} 2 & 3 & -1 \\ 0 & 2 & 4 \\ -2 & 5 & 6 \end{vmatrix}$$

$$= -1\begin{vmatrix} 0 & 2 \\ -2 & 5 \end{vmatrix} - 4\begin{vmatrix} 2 & 3 \\ -2 & 5 \end{vmatrix} + 6\begin{vmatrix} 2 & 3 \\ 0 & 2 \end{vmatrix}$$

$$= -[0\cdot 5 - 2(-2)] - 4[2\cdot 5 - 3(-2)] + 6(2\cdot 2 - 3\cdot 0)$$

$$= -4 - 64 + 24 = -44$$

In both cases we obtain the same value for the determinant as when we expanded by the first row in Example 2.

NOW TRY EXERCISE **31** ■

Graphing calculators are capable of computing determinants. Here is the output when the TI-83 is used to calculate the determinant in Example 3:

```
[A]
        [[2   3  -1]
         [0   2  4 ]
         [-2  5  6 ]]
det([A])
                 -44
```

The following criterion allows us to determine whether a square matrix has an inverse without actually calculating the inverse. This is one of the most important uses of the determinant in matrix algebra, and it is the reason for the name *determinant*.

INVERTIBILITY CRITERION

If A is a square matrix, then A has an inverse if and only if $\det(A) \neq 0$.

We will not prove this fact, but from the formula for the inverse of a 2×2 matrix (page 674) you can see why it is true in the 2×2 case.

EXAMPLE 4 | Using the Determinant to Show That a Matrix Is Not Invertible

Show that the matrix A has no inverse.

$$A = \begin{bmatrix} 1 & 2 & 0 & 4 \\ 0 & 0 & 0 & 3 \\ 5 & 6 & 2 & 6 \\ 2 & 4 & 0 & 9 \end{bmatrix}$$

SOLUTION We begin by calculating the determinant of A. Since all but one of the elements of the second row is zero, we expand the determinant by the second row. If we do this, we see from the following equation that only the cofactor A_{24} will have to be calculated.

$$\det(A) = \begin{vmatrix} 1 & 2 & 0 & 4 \\ 0 & 0 & 0 & 3 \\ 5 & 6 & 2 & 6 \\ 2 & 4 & 0 & 9 \end{vmatrix}$$

$$= -0 \cdot A_{21} + 0 \cdot A_{22} - 0 \cdot A_{23} + 3 \cdot A_{24} = 3A_{24}$$

$$= 3\begin{vmatrix} 1 & 2 & 0 \\ 5 & 6 & 2 \\ 2 & 4 & 0 \end{vmatrix} \qquad \text{Expand this by column 3}$$

$$= 3(-2)\begin{vmatrix} 1 & 2 \\ 2 & 4 \end{vmatrix}$$

$$= 3(-2)(1 \cdot 4 - 2 \cdot 2) = 0$$

Since the determinant of A is zero, A cannot have an inverse, by the Invertibility Criterion.

NOW TRY EXERCISE **23** ■

Row and Column Transformations

The preceding example shows that if we expand a determinant about a row or column that contains many zeros, our work is reduced considerably because we don't have to evaluate the cofactors of the elements that are zero. The following principle often simplifies the process of finding a determinant by introducing zeros into the matrix without changing the value of the determinant.

The Granger Collection, New York

EMMY NOETHER (1882–1935) was one of the foremost mathematicians of the early 20th century. Her groundbreaking work in abstract algebra provided much of the foundation for this field, and her work in invariant theory was essential in the development of Einstein's theory of general relativity. Although women weren't allowed to study at German universities at that time, she audited courses unofficially and went on to receive a doctorate at Erlangen *summa cum laude,* despite the opposition of the academic senate, which declared that women students would "overthrow all academic order." She subsequently taught mathematics at Göttingen, Moscow, and Frankfurt. In 1933 she left Germany to escape Nazi persecution, accepting a position at Bryn Mawr College in suburban Philadelphia. She lectured there and at the Institute for Advanced Study in Princeton, New Jersey, until her untimely death in 1935.

ROW AND COLUMN TRANSFORMATIONS OF A DETERMINANT

If A is a square matrix and if the matrix B is obtained from A by adding a multiple of one row to another or a multiple of one column to another, then $\det(A) = \det(B)$.

EXAMPLE 5 | Using Row and Column Transformations to Calculate a Determinant

Find the determinant of the matrix A. Does it have an inverse?

$$A = \begin{bmatrix} 8 & 2 & -1 & -4 \\ 3 & 5 & -3 & 11 \\ 24 & 6 & 1 & -12 \\ 2 & 2 & 7 & -1 \end{bmatrix}$$

SOLUTION If we add -3 times row 1 to row 3, we change all but one element of row 3 to zeros:

$$\begin{bmatrix} 8 & 2 & -1 & -4 \\ 3 & 5 & -3 & 11 \\ 0 & 0 & 4 & 0 \\ 2 & 2 & 7 & -1 \end{bmatrix}$$

This new matrix has the same determinant as A, and if we expand its determinant by the third row, we get

$$\det(A) = 4\begin{vmatrix} 8 & 2 & -4 \\ 3 & 5 & 11 \\ 2 & 2 & -1 \end{vmatrix}$$

Now, adding 2 times column 3 to column 1 in this determinant gives us

$$\det(A) = 4\begin{vmatrix} 0 & 2 & -4 \\ 25 & 5 & 11 \\ 0 & 2 & -1 \end{vmatrix} \qquad \text{Expand this by column 1}$$

$$= 4(-25)\begin{vmatrix} 2 & -4 \\ 2 & -1 \end{vmatrix}$$

$$= 4(-25)[2(-1) - (-4)2] = -600$$

Since the determinant of A is not zero, A does have an inverse.

NOW TRY EXERCISE **27**

Cramer's Rule

The solutions of linear equations can sometimes be expressed by using determinants. To illustrate, let's solve the following pair of linear equations for the variable x.

$$\begin{cases} ax + by = r \\ cx + dy = s \end{cases}$$

To eliminate the variable y, we multiply the first equation by d and the second by b and subtract.

$$\begin{aligned} adx + bdy &= rd \\ bcx + bdy &= bs \\ \hline adx - bcx &= rd - bs \end{aligned}$$

Factoring the left-hand side, we get $(ad - bc)x = rd - bs$. Assuming that $ad - bc \neq 0$, we can now solve this equation for x:

$$x = \frac{rd - bs}{ad - bc}$$

Similarly, we find

$$y = \frac{as - cr}{ad - bc}$$

The numerator and denominator of the fractions for x and y are determinants of 2×2 matrices. So we can express the solution of the system using determinants as follows.

CRAMER'S RULE FOR SYSTEMS IN TWO VARIABLES

The linear system

$$\begin{cases} ax + by = r \\ cx + dy = s \end{cases}$$

has the solution

$$x = \frac{\begin{vmatrix} r & b \\ s & d \end{vmatrix}}{\begin{vmatrix} a & b \\ c & d \end{vmatrix}} \qquad y = \frac{\begin{vmatrix} a & r \\ c & s \end{vmatrix}}{\begin{vmatrix} a & b \\ c & d \end{vmatrix}}$$

provided that $\begin{vmatrix} a & b \\ c & d \end{vmatrix} \neq 0$.

Using the notation

$$D = \begin{bmatrix} a & b \\ c & d \end{bmatrix} \qquad D_x = \begin{bmatrix} r & b \\ s & d \end{bmatrix} \qquad D_y = \begin{bmatrix} a & r \\ c & s \end{bmatrix}$$

Coefficient matrix

Replace first column of D by r and s

Replace second column of D by r and s

we can write the solution of the system as

$$x = \frac{|D_x|}{|D|} \quad \text{and} \quad y = \frac{|D_y|}{|D|}$$

EXAMPLE 6 | Using Cramer's Rule to Solve a System with Two Variables

Use Cramer's Rule to solve the system.

$$\begin{cases} 2x + 6y = -1 \\ x + 8y = 2 \end{cases}$$

SOLUTION For this system we have

$$|D| = \begin{vmatrix} 2 & 6 \\ 1 & 8 \end{vmatrix} = 2 \cdot 8 - 6 \cdot 1 = 10$$

$$|D_x| = \begin{vmatrix} -1 & 6 \\ 2 & 8 \end{vmatrix} = (-1)8 - 6 \cdot 2 = -20$$

$$|D_y| = \begin{vmatrix} 2 & -1 \\ 1 & 2 \end{vmatrix} = 2 \cdot 2 - (-1)1 = 5$$

The solution is

$$x = \frac{|D_x|}{|D|} = \frac{-20}{10} = -2$$

$$y = \frac{|D_y|}{|D|} = \frac{5}{10} = \frac{1}{2}$$

NOW TRY EXERCISE **33**

Cramer's Rule can be extended to apply to any system of n linear equations in n variables in which the determinant of the coefficient matrix is not zero. As we saw in the preceding section, any such system can be written in matrix form as

$$\begin{bmatrix} a_{11} & a_{12} & \cdots & a_{1n} \\ a_{21} & a_{22} & \cdots & a_{2n} \\ \vdots & \vdots & \ddots & \vdots \\ a_{n1} & a_{n2} & \cdots & a_{nn} \end{bmatrix} \begin{bmatrix} x_1 \\ x_2 \\ \vdots \\ x_n \end{bmatrix} = \begin{bmatrix} b_1 \\ b_2 \\ \vdots \\ b_n \end{bmatrix}$$

By analogy with our derivation of Cramer's Rule in the case of two equations in two unknowns, we let D be the coefficient matrix in this system, and D_{x_i} be the matrix obtained by replacing the ith column of D by the numbers $b_1, b_2, \ldots, b_n$ that appear to the right of the equal sign. The solution of the system is then given by the following rule.

CRAMER'S RULE

If a system of n linear equations in the n variables $x_1, x_2, \ldots, x_n$ is equivalent to the matrix equation $DX = B$, and if $|D| \neq 0$, then its solutions are

$$x_1 = \frac{|D_{x_1}|}{|D|} \qquad x_2 = \frac{|D_{x_2}|}{|D|} \qquad \cdots \qquad x_n = \frac{|D_{x_n}|}{|D|}$$

where D_{x_i} is the matrix obtained by replacing the ith column of D by the $n \times 1$ matrix B.

EXAMPLE 7 | Using Cramer's Rule to Solve a System with Three Variables

Use Cramer's Rule to solve the system.

$$\begin{cases} 2x - 3y + 4z = 1 \\ x \quad\quad\; + 6z = 0 \\ 3x - 2y \quad\quad\; = 5 \end{cases}$$

SOLUTION First, we evaluate the determinants that appear in Cramer's Rule. Note that D is the coefficient matrix and that D_x, D_y, and D_z are obtained by replacing the first, second, and third columns of D by the constant terms.

$$|D| = \begin{vmatrix} 2 & -3 & 4 \\ 1 & 0 & 6 \\ 3 & -2 & 0 \end{vmatrix} = -38 \qquad |D_x| = \begin{vmatrix} 1 & -3 & 4 \\ 0 & 0 & 6 \\ 5 & -2 & 0 \end{vmatrix} = -78$$

$$|D_y| = \begin{vmatrix} 2 & 1 & 4 \\ 1 & 0 & 6 \\ 3 & 5 & 0 \end{vmatrix} = -22 \qquad |D_z| = \begin{vmatrix} 2 & -3 & 1 \\ 1 & 0 & 0 \\ 3 & -2 & 5 \end{vmatrix} = 13$$

Now we use Cramer's Rule to get the solution:

$$x = \frac{|D_x|}{|D|} = \frac{-78}{-38} = \frac{39}{19} \qquad y = \frac{|D_y|}{|D|} = \frac{-22}{-38} = \frac{11}{19}$$

$$z = \frac{|D_z|}{|D|} = \frac{13}{-38} = -\frac{13}{38}$$

NOW TRY EXERCISE **39**

Solving the system in Example 7 using Gaussian elimination would involve matrices whose elements are fractions with fairly large denominators. Thus in cases like Examples 6 and 7, Cramer's Rule gives us an efficient way to solve systems of linear equations. But in systems with more than three equations, evaluating the various determinants that are involved is usually a long and tedious task (unless you are using a graphing calculator). Moreover, the rule doesn't apply if $|D| = 0$ or if D is not a square matrix. So Cramer's Rule is a useful alternative to Gaussian elimination, but only in some situations.

Areas of Triangles Using Determinants

Determinants provide a simple way to calculate the area of a triangle in the coordinate plane.

AREA OF A TRIANGLE

If a triangle in the coordinate plane has vertices (a_1, b_1), (a_2, b_2), and (a_3, b_3), then its area is

$$\mathcal{A} = \pm\frac{1}{2}\begin{vmatrix} a_1 & b_1 & 1 \\ a_2 & b_2 & 1 \\ a_3 & b_3 & 1 \end{vmatrix}$$

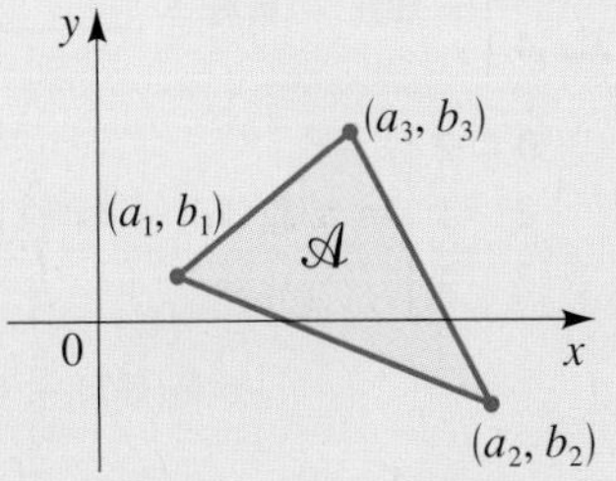

where the sign is chosen to make the area positive.

You are asked to prove this formula in Exercise 63.

EXAMPLE 8 | Area of a Triangle

Find the area of the triangle shown in Figure 1.

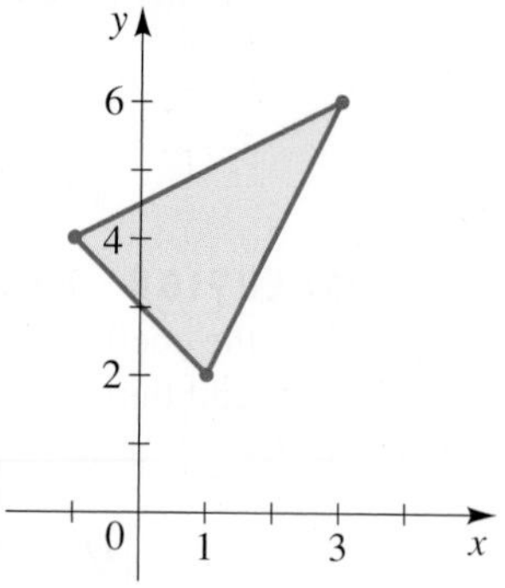

FIGURE 1

We can calculate the determinant by hand or by using a graphing calculator.

```
[A]
        [[-1 4 1]
         [3  6 1]
         [1  2 1]]
det([A])
                -12
```

SOLUTION The vertices are (1, 2), (3, 6), and (−1, 4). Using the formula in the preceding box, we get

$$\mathscr{A} = \pm\tfrac{1}{2}\begin{vmatrix} -1 & 4 & 1 \\ 3 & 6 & 1 \\ 1 & 2 & 1 \end{vmatrix} = \pm\tfrac{1}{2}(-12)$$

To make the area positive, we choose the negative sign in the formula. Thus, the area of the triangle is

$$\mathscr{A} = -\tfrac{1}{2}(-12) = 6$$

NOW TRY EXERCISE **55** ■

10.6 EXERCISES

CONCEPTS

1. *True or false*? det(A) is defined only for a square matrix A.

2. *True or false*? det(A) is a number, not a matrix.

3. *True or false*? If det(A) = 0, then A is not invertible.

4. Fill in the blanks with appropriate numbers to calculate the determinant. Where there is "±", choose the appropriate sign (+ or −).

(a) $\begin{vmatrix} 2 & 1 \\ -3 & 4 \end{vmatrix} = ■ - ■ = $ ______

(b) $\begin{vmatrix} 1 & 0 & 2 \\ 3 & 2 & 1 \\ 0 & -3 & 4 \end{vmatrix} = \pm ■(■ - ■) \pm ■(■ - ■)$

$\pm ■(■ - ■) = $ ______

SKILLS

5–12 ■ Find the determinant of the matrix, if it exists.

5. $\begin{bmatrix} 2 & 0 \\ 0 & 3 \end{bmatrix}$

6. $\begin{bmatrix} 0 & -1 \\ 2 & 0 \end{bmatrix}$

7. $\begin{bmatrix} 4 & 5 \\ 0 & -1 \end{bmatrix}$

8. $\begin{bmatrix} -2 & 1 \\ 3 & -2 \end{bmatrix}$

9. $\begin{bmatrix} 2 & 5 \end{bmatrix}$

10. $\begin{bmatrix} 3 \\ 0 \end{bmatrix}$

11. $\begin{bmatrix} \frac{1}{2} & \frac{1}{8} \\ 1 & \frac{1}{2} \end{bmatrix}$

12. $\begin{bmatrix} 2.2 & -1.4 \\ 0.5 & 1.0 \end{bmatrix}$

13–18 ■ Evaluate the minor and cofactor using the matrix A.

$$A = \begin{bmatrix} 1 & 0 & \frac{1}{2} \\ -3 & 5 & 2 \\ 0 & 0 & 4 \end{bmatrix}$$

13. M_{11}, A_{11} **14.** M_{33}, A_{33} **15.** M_{12}, A_{12}

16. M_{13}, A_{13} **17.** M_{23}, A_{23} **18.** M_{32}, A_{32}

19–26 ■ Find the determinant of the matrix. Determine whether the matrix has an inverse, but don't calculate the inverse.

19. $\begin{bmatrix} 2 & 1 & 0 \\ 0 & -2 & 4 \\ 0 & 1 & -3 \end{bmatrix}$

20. $\begin{bmatrix} 1 & 2 & 5 \\ -2 & -3 & 2 \\ 3 & 5 & 3 \end{bmatrix}$

21. $\begin{bmatrix} 30 & 0 & 20 \\ 0 & -10 & -20 \\ 40 & 0 & 10 \end{bmatrix}$

22. $\begin{bmatrix} -2 & -\frac{3}{2} & \frac{1}{2} \\ 2 & 4 & 0 \\ \frac{1}{2} & 2 & 1 \end{bmatrix}$

23. $\begin{bmatrix} 1 & 3 & 7 \\ 2 & 0 & 8 \\ 0 & 2 & 2 \end{bmatrix}$

24. $\begin{bmatrix} 0 & -1 & 0 \\ 2 & 6 & 4 \\ 1 & 0 & 3 \end{bmatrix}$

25. $\begin{bmatrix} 1 & 3 & 3 & 0 \\ 0 & 2 & 0 & 1 \\ -1 & 0 & 0 & 2 \\ 1 & 6 & 4 & 1 \end{bmatrix}$

26. $\begin{bmatrix} 1 & 2 & 0 & 2 \\ 3 & -4 & 0 & 4 \\ 0 & 1 & 6 & 0 \\ 1 & 0 & 2 & 0 \end{bmatrix}$

27–30 ■ Evaluate the determinant, using row or column operations whenever possible to simplify your work.

27. $\begin{vmatrix} 0 & 0 & 4 & 6 \\ 2 & 1 & 1 & 3 \\ 2 & 1 & 2 & 3 \\ 3 & 0 & 1 & 7 \end{vmatrix}$

28. $\begin{vmatrix} -2 & 3 & -1 & 7 \\ 4 & 6 & -2 & 3 \\ 7 & 7 & 0 & 5 \\ 3 & -12 & 4 & 0 \end{vmatrix}$

29. $\begin{vmatrix} 1 & 2 & 3 & 4 & 5 \\ 0 & 2 & 4 & 6 & 8 \\ 0 & 0 & 3 & 6 & 9 \\ 0 & 0 & 0 & 4 & 8 \\ 0 & 0 & 0 & 0 & 5 \end{vmatrix}$

30. $\begin{vmatrix} 2 & -1 & 6 & 4 \\ 7 & 2 & -2 & 5 \\ 4 & -2 & 10 & 8 \\ 6 & 1 & 1 & 4 \end{vmatrix}$

31. Let

$$B = \begin{bmatrix} 4 & 1 & 0 \\ -2 & -1 & 1 \\ 4 & 0 & 3 \end{bmatrix}$$

(a) Evaluate det(B) by expanding by the second row.
(b) Evaluate det(B) by expanding by the third column.
(c) Do your results in parts (a) and (b) agree?

32. Consider the system

$$\begin{cases} x + 2y + 6z = 5 \\ -3x - 6y + 5z = 8 \\ 2x + 6y + 9z = 7 \end{cases}$$

(a) Verify that $x = -1$, $y = 0$, $z = 1$ is a solution of the system.
(b) Find the determinant of the coefficient matrix.
(c) Without solving the system, determine whether there are any other solutions.
(d) Can Cramer's Rule be used to solve this system? Why or why not?

33–48 ■ Use Cramer's Rule to solve the system.

33. $\begin{cases} 2x - y = -9 \\ x + 2y = 8 \end{cases}$

34. $\begin{cases} 6x + 12y = 33 \\ 4x + 7y = 20 \end{cases}$

35. $\begin{cases} x - 6y = 3 \\ 3x + 2y = 1 \end{cases}$

36. $\begin{cases} \frac{1}{2}x + \frac{1}{3}y = 1 \\ \frac{1}{4}x - \frac{1}{6}y = -\frac{3}{2} \end{cases}$

37. $\begin{cases} 0.4x + 1.2y = 0.4 \\ 1.2x + 1.6y = 3.2 \end{cases}$

38. $\begin{cases} 10x - 17y = 21 \\ 20x - 31y = 39 \end{cases}$

39. $\begin{cases} x - y + 2z = 0 \\ 3x + z = 11 \\ -x + 2y = 0 \end{cases}$

40. $\begin{cases} 5x - 3y + z = 6 \\ 4y - 6z = 22 \\ 7x + 10y = -13 \end{cases}$

41. $\begin{cases} 2x_1 + 3x_2 - 5x_3 = 1 \\ x_1 + x_2 - x_3 = 2 \\ 2x_2 + x_3 = 8 \end{cases}$

42. $\begin{cases} -2a + c = 2 \\ a + 2b - c = 9 \\ 3a + 5b + 2c = 22 \end{cases}$

43. $\begin{cases} \frac{1}{3}x - \frac{1}{5}y + \frac{1}{2}z = \frac{7}{10} \\ -\frac{2}{3}x + \frac{2}{5}y + \frac{3}{2}z = \frac{11}{10} \\ x - \frac{4}{5}y + z = \frac{9}{5} \end{cases}$

44. $\begin{cases} 2x - y = 5 \\ 5x + 3z = 19 \\ 4y + 7z = 17 \end{cases}$

45. $\begin{cases} 3y + 5z = 4 \\ 2x - z = 10 \\ 4x + 7y = 0 \end{cases}$

46. $\begin{cases} 2x - 5y = 4 \\ x + y - z = 8 \\ 3x + 5z = 0 \end{cases}$

47. $\begin{cases} x + y + z + w = 0 \\ 2x + w = 0 \\ y - z = 0 \\ x + 2z = 1 \end{cases}$

48. $\begin{cases} x + y = 1 \\ y + z = 2 \\ z + w = 3 \\ w - x = 4 \end{cases}$

49–50 ■ Evaluate the determinants.

49. $\begin{vmatrix} a & 0 & 0 & 0 & 0 \\ 0 & b & 0 & 0 & 0 \\ 0 & 0 & c & 0 & 0 \\ 0 & 0 & 0 & d & 0 \\ 0 & 0 & 0 & 0 & e \end{vmatrix}$

50. $\begin{vmatrix} a & a & a & a & a \\ 0 & a & a & a & a \\ 0 & 0 & a & a & a \\ 0 & 0 & 0 & a & a \\ 0 & 0 & 0 & 0 & a \end{vmatrix}$

51–54 ■ Solve for x.

51. $\begin{vmatrix} x & 12 & 13 \\ 0 & x - 1 & 23 \\ 0 & 0 & x - 2 \end{vmatrix} = 0$

52. $\begin{vmatrix} x & 1 & 1 \\ 1 & 1 & x \\ x & 1 & x \end{vmatrix} = 0$

53. $\begin{vmatrix} 1 & 0 & x \\ x^2 & 1 & 0 \\ x & 0 & 1 \end{vmatrix} = 0$

54. $\begin{vmatrix} a & b & x - a \\ x & x + b & x \\ 0 & 1 & 1 \end{vmatrix} = 0$

55–58 ■ Sketch the triangle with the given vertices, and use a determinant to find its area.

55. $(0, 0), (6, 2), (3, 8)$

56. $(1, 0), (3, 5), (-2, 2)$

57. $(-1, 3), (2, 9), (5, -6)$

58. $(-2, 5), (7, 2), (3, -4)$

59. Show that $\begin{vmatrix} 1 & x & x^2 \\ 1 & y & y^2 \\ 1 & z & z^2 \end{vmatrix} = (x - y)(y - z)(z - x)$

APPLICATIONS

60. **Buying Fruit** A roadside fruit stand sells apples at 75¢ a pound, peaches at 90¢ a pound, and pears at 60¢ a pound. Muriel buys 18 pounds of fruit at a total cost of $13.80. Her peaches and pears together cost $1.80 more than her apples.
   (a) Set up a linear system for the number of pounds of apples, peaches, and pears that she bought.
   (b) Solve the system using Cramer's Rule.

61. **The Arch of a Bridge** The opening of a railway bridge over a roadway is in the shape of a parabola. A surveyor measures the heights of three points on the bridge, as shown in the figure. He wishes to find an equation of the form

$$y = ax^2 + bx + c$$

to model the shape of the arch.
   (a) Use the surveyed points to set up a system of linear equations for the unknown coefficients a, b, and c.
   (b) Solve the system using Cramer's Rule.

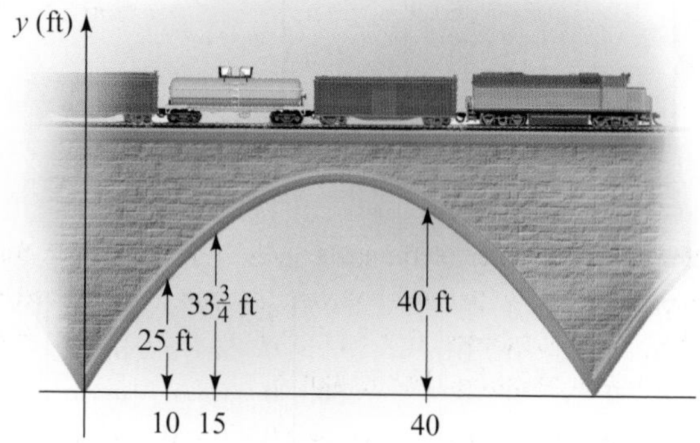

62. **A Triangular Plot of Land** An outdoors club is purchasing land to set up a conservation area. The last remaining piece they need to buy is the triangular plot shown in the figure. Use the determinant formula for the area of a triangle to find the area of the plot.

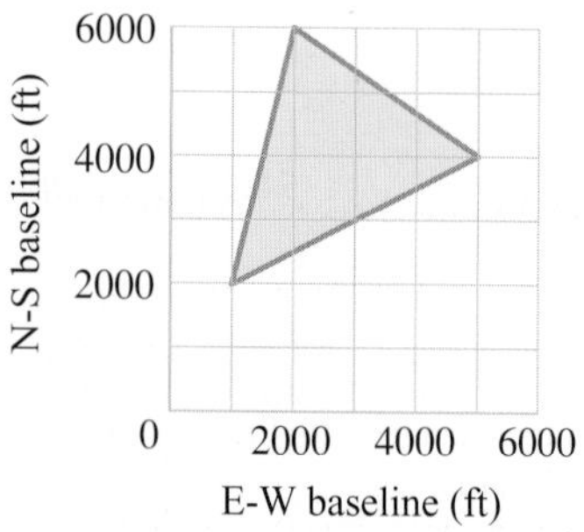

DISCOVERY ▪ DISCUSSION ▪ WRITING

63. **Determinant Formula for the Area of a Triangle** The figure shows a triangle in the plane with vertices (a_1, b_1), (a_2, b_2), and (a_3, b_3).
   (a) Find the coordinates of the vertices of the surrounding rectangle, and find its area.
   (b) Find the area of the red triangle by subtracting the areas of the three blue triangles from the area of the rectangle.
   (c) Use your answer to part (b) to show that the area of the red triangle is given by

$$\text{area} = \pm\frac{1}{2}\begin{vmatrix} a_1 & b_1 & 1 \\ a_2 & b_2 & 1 \\ a_3 & b_3 & 1 \end{vmatrix}$$

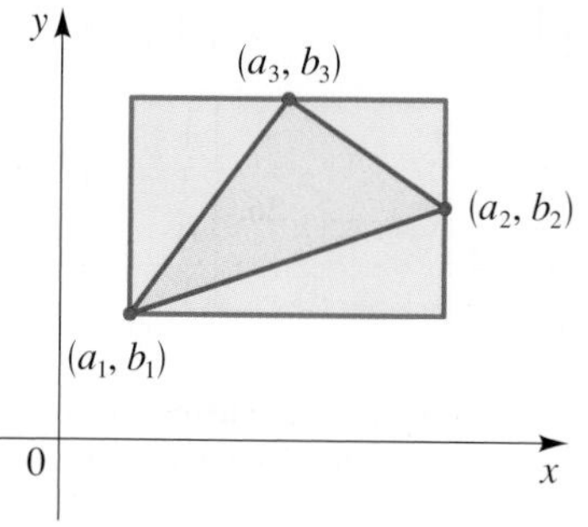

64. **Collinear Points and Determinants**
   (a) If three points lie on a line, what is the area of the "triangle" that they determine? Use the answer to this question, together with the determinant formula for the area of a triangle, to explain why the points (a_1, b_1), (a_2, b_2), and (a_3, b_3) are collinear if and only if

$$\begin{vmatrix} a_1 & b_1 & 1 \\ a_2 & b_2 & 1 \\ a_3 & b_3 & 1 \end{vmatrix} = 0$$

   (b) Use a determinant to check whether each set of points is collinear. Graph them to verify your answer.
   (i) $(-6, 4), (2, 10), (6, 13)$
   (ii) $(-5, 10), (2, 6), (15, -2)$

65. **Determinant Form for the Equation of a Line**
   (a) Use the result of Exercise 64(a) to show that the equation of the line containing the points (x_1, y_1) and (x_2, y_2) is

$$\begin{vmatrix} x & y & 1 \\ x_1 & y_1 & 1 \\ x_2 & y_2 & 1 \end{vmatrix} = 0$$

   (b) Use the result of part (a) to find an equation for the line containing the points $(20, 50)$ and $(-10, 25)$.

66. **Matrices with Determinant Zero** Use the definition of determinant and the elementary row and column operations to explain why matrices of the following types have determinant 0.
   (a) A matrix with a row or column consisting entirely of zeros
   (b) A matrix with two rows the same or two columns the same
   (c) A matrix in which one row is a multiple of another row, or one column is a multiple of another column

67. Solving Linear Systems Suppose you have to solve a linear system with five equations and five variables without the assistance of a calculator or computer. Which method would you prefer: Cramer's Rule or Gaussian elimination? Write a short paragraph explaining the reasons for your answer.

DISCOVERY PROJECT **Computer Graphics I**

In this project we investigate how matrices are used to manipulate images on a computer screen by compressing, stretching, reflecting, and shearing. You can find the project at the book companion website: **www.stewartmath.com**

10.7 Partial Fractions

Distinct Linear Factors ▶ Repeated Linear Factors ▶ Irreducible Quadratic Factors ▶ Repeated Irreducible Quadratic Factors

Common denominator →

$$\frac{1}{x-1} + \frac{1}{2x+1} = \frac{3x}{2x^2 - x - 1}$$

← Partial fractions

To write a sum or difference of fractional expressions as a single fraction, we bring them to a common denominator. For example,

$$\frac{1}{x-1} + \frac{1}{2x+1} = \frac{(2x+1)+(x-1)}{(x-1)(2x+1)} = \frac{3x}{2x^2 - x - 1}$$

But for some applications of algebra to calculus we must reverse this process—that is, we must express a fraction such as $3x/(2x^2 - x - 1)$ as the sum of the simpler fractions $1/(x-1)$ and $1/(2x+1)$. These simpler fractions are called *partial fractions*; we learn how to find them in this section.

Let r be the rational function

$$r(x) = \frac{P(x)}{Q(x)}$$

where the degree of P is less than the degree of Q. By the Linear and Quadratic Factors Theorem in Section 3.6, every polynomial with real coefficients can be factored completely into linear and irreducible quadratic factors, that is, factors of the form $ax + b$ and $ax^2 + bx + c$, where a, b, and c are real numbers. For instance,

$$x^4 - 1 = (x^2 - 1)(x^2 + 1) = (x - 1)(x + 1)(x^2 + 1)$$

After we have completely factored the denominator Q of r, we can express $r(x)$ as a sum of **partial fractions** of the form

$$\frac{A}{(ax+b)^i} \quad \text{and} \quad \frac{Ax+B}{(ax^2+bx+c)^j}$$

This sum is called the **partial fraction decomposition** of r. Let's examine the details of the four possible cases.

▼ Distinct Linear Factors

We first consider the case in which the denominator factors into distinct linear factors.

CASE 1: THE DENOMINATOR IS A PRODUCT OF DISTINCT LINEAR FACTORS

Suppose that we can factor $Q(x)$ as

$$Q(x) = (a_1x + b_1)(a_2x + b_2)\cdots(a_nx + b_n)$$

with no factor repeated. In this case the partial fraction decomposition of $P(x)/Q(x)$ takes the form

$$\frac{P(x)}{Q(x)} = \frac{A}{a_1x + b_1} + \frac{A_2}{a_2x + b_2} + \cdots + \frac{A_n}{a_nx + b_n}$$

The Rhind Papyrus is the oldest known mathematical document. It is an Egyptian scroll written in 1650 B.C. by the scribe Ahmes, who explains that it is an exact copy of a scroll written 200 years earlier. Ahmes claims that his papyrus contains "a thorough study of all things, insight into all that exists, knowledge of all obscure secrets." Actually, the document contains rules for doing arithmetic, including multiplication and division of fractions and several exercises with solutions. The exercise shown below reads: "A heap and its seventh make nineteen; how large is the heap?" In solving problems of this sort, the Egyptians used partial fractions because their number system required all fractions to be written as sums of reciprocals of whole numbers. For example, $\frac{7}{12}$ would be written as $\frac{1}{3} + \frac{1}{4}$.

The papyrus gives a correct formula for the volume of a truncated pyramid, which the ancient Egyptians used when building the pyramids at Giza. It also gives the formula $A = (\frac{8}{9}d)^2$ for the area of a circle with diameter d. How close is this to the actual area?

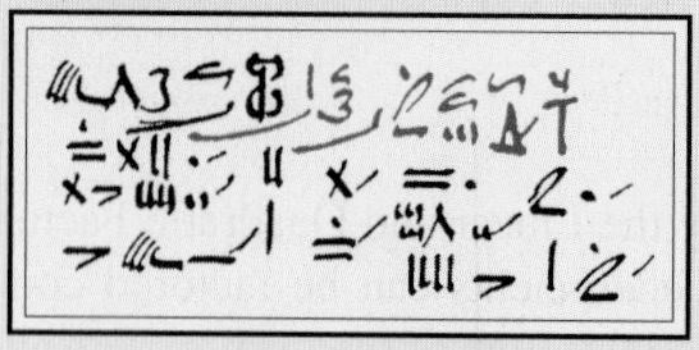

The constants $A_1, A_2, \ldots, A_n$ are determined as in the following example.

EXAMPLE 1 | Distinct Linear Factors

Find the partial fraction decomposition of $\dfrac{5x + 7}{x^3 + 2x^2 - x - 2}$.

SOLUTION The denominator factors as follows.

$$\begin{aligned} x^3 + 2x^2 - x - 2 &= x^2(x + 2) - (x + 2) = (x^2 - 1)(x + 2) \\ &= (x - 1)(x + 1)(x + 2) \end{aligned}$$

This gives us the partial fraction decomposition

$$\frac{5x + 7}{x^3 + 2x^2 - x - 2} = \frac{A}{x - 1} + \frac{B}{x + 1} + \frac{C}{x + 2}$$

Multiplying each side by the common denominator, $(x - 1)(x + 1)(x + 2)$, we get

$$\begin{aligned} 5x + 7 &= A(x + 1)(x + 2) + B(x - 1)(x + 2) + C(x - 1)(x + 1) \\ &= A(x^2 + 3x + 2) + B(x^2 + x - 2) + C(x^2 - 1) && \text{Expand} \\ &= (A + B + C)x^2 + (3A + B)x + (2A - 2B - C) && \text{Combine like terms} \end{aligned}$$

If two polynomials are equal, then their coefficients are equal. Thus since $5x + 7$ has no x^2-term, we have $A + B + C = 0$. Similarly, by comparing the coefficients of x, we see that $3A + B = 5$, and by comparing constant terms, we get $2A - 2B - C = 7$. This leads to the following system of linear equations for A, B, and C.

$$\begin{cases} A + B + C = 0 & \text{Equation 1: Coefficients of } x^2 \\ 3A + B = 5 & \text{Equation 2: Coefficients of } x \\ 2A - 2B - C = 7 & \text{Equation 3: Constant coefficients} \end{cases}$$

We use Gaussian elimination to solve this system.

$$\begin{cases} A + B + C = 0 & \\ \quad - 2B - 3C = 5 & \text{Equation 2} + (-3) \times \text{Equation 1} \\ \quad - 4B - 3C = 7 & \text{Equation 3} + (-2) \times \text{Equation 1} \end{cases}$$

$$\begin{cases} A + B + C = 0 & \\ \quad - 2B - 3C = 5 & \\ \quad\quad 3C = -3 & \text{Equation 3} + (-2) \times \text{Equation 2} \end{cases}$$

From the third equation we get $C = -1$. Back-substituting, we find that $B = -1$ and $A = 2$. So the partial fraction decomposition is

$$\frac{5x + 7}{x^3 + 2x^2 - x - 2} = \frac{2}{x - 1} + \frac{-1}{x + 1} + \frac{-1}{x + 2}$$

NOW TRY EXERCISES **3** AND **13**

The same approach works in the remaining cases. We set up the partial fraction decomposition with the unknown constants $A, B, C, \ldots$. Then we multiply each side of the resulting equation by the common denominator, simplify the right-hand side of the equation, and equate coefficients. This gives a set of linear equations that will always have a unique solution (provided that the partial fraction decomposition has been set up correctly).

▼ Repeated Linear Factors

We now consider the case in which the denominator factors into linear factors, some of which are repeated.

CASE 2: THE DENOMINATOR IS A PRODUCT OF LINEAR FACTORS, SOME OF WHICH ARE REPEATED

Suppose the complete factorization of $Q(x)$ contains the linear factor $ax + b$ repeated k times; that is, $(ax + b)^k$ is a factor of $Q(x)$. Then, corresponding to each such factor, the partial fraction decomposition for $P(x)/Q(x)$ contains

$$\frac{A_1}{ax + b} + \frac{A_2}{(ax + b)^2} + \cdots + \frac{A_k}{(ax + b)^k}$$

EXAMPLE 2 | Repeated Linear Factors

Find the partial fraction decomposition of $\dfrac{x^2 + 1}{x(x - 1)^3}$.

SOLUTION Because the factor $x - 1$ is repeated three times in the denominator, the partial fraction decomposition has the form

$$\frac{x^2 + 1}{x(x - 1)^3} = \frac{A}{x} + \frac{B}{x - 1} + \frac{C}{(x - 1)^2} + \frac{D}{(x - 1)^3}$$

Multiplying each side by the common denominator, $x(x - 1)^3$, gives

$$\begin{aligned} x^2 + 1 &= A(x - 1)^3 + Bx(x - 1)^2 + Cx(x - 1) + Dx \\ &= A(x^3 - 3x^2 + 3x - 1) + B(x^3 - 2x^2 + x) + C(x^2 - x) + Dx && \text{Expand} \\ &= (A + B)x^3 + (-3A - 2B + C)x^2 + (3A + B - C + D)x - A && \text{Combine like terms} \end{aligned}$$

Equating coefficients, we get the following equations.

$$\begin{cases} A + B = 0 & \text{Coefficients of } x^3 \\ -3A - 2B + C = 1 & \text{Coefficients of } x^2 \\ 3A + B - C + D = 0 & \text{Coefficients of } x \\ -A = 1 & \text{Constant coefficients} \end{cases}$$

If we rearrange these equations by putting the last one in the first position, we can easily see (using substitution) that the solution to the system is $A = -1$, $B = 1$, $C = 0$, $D = 2$, so the partial fraction decomposition is

$$\frac{x^2 + 1}{x(x - 1)^3} = \frac{-1}{x} + \frac{1}{x - 1} + \frac{2}{(x - 1)^3}$$

NOW TRY EXERCISES **5** AND **29** ■

Irreducible Quadratic Factors

We now consider the case in which the denominator has distinct irreducible quadratic factors.

CASE 3: THE DENOMINATOR HAS IRREDUCIBLE QUADRATIC FACTORS, NONE OF WHICH IS REPEATED

Suppose the complete factorization of $Q(x)$ contains the quadratic factor $ax^2 + bx + c$ (which can't be factored further). Then, corresponding to this, the partial fraction decomposition of $P(x)/Q(x)$ will have a term of the form

$$\frac{Ax + B}{ax^2 + bx + c}$$

EXAMPLE 3 | Distinct Quadratic Factors

Find the partial fraction decomposition of $\dfrac{2x^2 - x + 4}{x^3 + 4x}$.

SOLUTION Since $x^3 + 4x = x(x^2 + 4)$, which can't be factored further, we write

$$\frac{2x^2 - x + 4}{x^3 + 4x} = \frac{A}{x} + \frac{Bx + C}{x^2 + 4}$$

Multiplying by $x(x^2 + 4)$, we get

$$\begin{aligned} 2x^2 - x + 4 &= A(x^2 + 4) + (Bx + C)x \\ &= (A + B)x^2 + Cx + 4A \end{aligned}$$

Equating coefficients gives us the equations

$$\begin{cases} A + B = 2 & \text{Coefficients of } x^2 \\ C = -1 & \text{Coefficients of } x \\ 4A = 4 & \text{Constant coefficients} \end{cases}$$

so $A = 1$, $B = 1$, and $C = -1$. The required partial fraction decomposition is

$$\frac{2x^2 - x + 4}{x^3 + 4x} = \frac{1}{x} + \frac{x - 1}{x^2 + 4}$$

NOW TRY EXERCISES **7** AND **37** ■

▼ Repeated Irreducible Quadratic Factors

We now consider the case in which the denominator has irreducible quadratic factors, some of which are repeated.

CASE 4: THE DENOMINATOR HAS A REPEATED IRREDUCIBLE QUADRATIC FACTOR

Suppose the complete factorization of $Q(x)$ contains the factor $(ax^2 + bx + c)^k$, where $ax^2 + bx + c$ can't be factored further. Then the partial fraction decomposition of $P(x)/Q(x)$ will have the terms

$$\frac{A_1x + B_1}{ax^2 + bx + c} + \frac{A_2x + B_2}{(ax^2 + bx + c)^2} + \cdots + \frac{A_kx + B_k}{(ax^2 + bx + c)^k}$$

EXAMPLE 4 | Repeated Quadratic Factors

Write the form of the partial fraction decomposition of

$$\frac{x^5 - 3x^2 + 12x - 1}{x^3(x^2 + x + 1)(x^2 + 2)^3}$$

SOLUTION

$$\frac{x^5 - 3x^2 + 12x - 1}{x^3(x^2 + x + 1)(x^2 + 2)^3}$$

$$= \frac{A}{x} + \frac{B}{x^2} + \frac{C}{x^3} + \frac{Dx + E}{x^2 + x + 1} + \frac{Fx + G}{x^2 + 2} + \frac{Hx + I}{(x^2 + 2)^2} + \frac{Jx + K}{(x^2 + 2)^3}$$

NOW TRY EXERCISES **11** AND **41** ■

To find the values of $A, B, C, D, E, F, G, H, I, J$, and K in Example 4, we would have to solve a system of 11 linear equations. Although possible, this would certainly involve a great deal of work!

The techniques that we have described in this section apply only to rational functions $P(x)/Q(x)$ in which the degree of P is less than the degree of Q. If this isn't the case, we must first use long division to divide Q into P.

EXAMPLE 5 | Using Long Division to Prepare for Partial Fractions

Find the partial fraction decomposition of

$$\frac{2x^4 + 4x^3 - 2x^2 + x + 7}{x^3 + 2x^2 - x - 2}$$

SOLUTION Since the degree of the numerator is larger than the degree of the denominator, we use long division to obtain

$$\begin{array}{r} 2x \\ x^3 + 2x^2 - x - 2 \overline{\smash{)}\,2x^4 + 4x^3 - 2x^2 + x + 7} \\ \underline{2x^4 + 4x^3 - 2x^2 - 4x} \\ 5x + 7 \end{array}$$

$$\frac{2x^4 + 4x^3 - 2x^2 + x + 7}{x^3 + 2x^2 - x - 2} = 2x + \frac{5x + 7}{x^3 + 2x^2 - x - 2}$$

The remainder term now satisfies the requirement that the degree of the numerator is less than the degree of the denominator. At this point we proceed as in Example 1 to obtain the decomposition

$$\frac{2x^4 + 4x^3 - 2x^2 + x + 7}{x^3 + 2x^2 - x - 2} = 2x + \frac{2}{x - 1} + \frac{-1}{x + 1} + \frac{-1}{x + 2}$$

NOW TRY EXERCISE **43**

10.7 EXERCISES

CONCEPTS

1–2 ■ For each rational function r, choose from (i)–(iv) the appropriate form for its partial fraction decomposition.

1. $r(x) = \dfrac{4}{x(x - 2)^2}$

(i) $\dfrac{A}{x} + \dfrac{B}{x - 2}$ (ii) $\dfrac{A}{x} + \dfrac{B}{(x - 2)^2}$

(iii) $\dfrac{A}{x} + \dfrac{B}{x - 2} + \dfrac{C}{(x - 2)^2}$ (iv) $\dfrac{A}{x} + \dfrac{B}{x - 2} + \dfrac{Cx + D}{(x - 2)^2}$

2. $r(x) = \dfrac{2x + 8}{(x - 1)(x^2 + 4)}$

(i) $\dfrac{A}{x - 1} + \dfrac{B}{x^2 + 4}$

(ii) $\dfrac{A}{x - 1} + \dfrac{Bx + C}{x^2 + 4}$

(iii) $\dfrac{A}{x - 1} + \dfrac{B}{x + 2} + \dfrac{C}{x^2 + 4}$

(iv) $\dfrac{Ax + B}{x - 1} + \dfrac{Cx + D}{x^2 + 4}$

SKILLS

3–12 ■ Write the form of the partial fraction decomposition of the function (as in Example 4). Do not determine the numerical values of the coefficients.

3. $\dfrac{1}{(x - 1)(x + 2)}$ **4.** $\dfrac{x}{x^2 + 3x - 4}$

5. $\dfrac{x^2 - 3x + 5}{(x - 2)^2(x + 4)}$ **5.** $\dfrac{1}{x^4 - x^3}$

7. $\dfrac{x^2}{(x - 3)(x^2 + 4)}$ **8.** $\dfrac{1}{x^4 - 1}$

9. $\dfrac{x^3 - 4x^2 + 2}{(x^2 + 1)(x^2 + 2)}$ **10.** $\dfrac{x^4 + x^2 + 1}{x^2(x^2 + 4)^2}$

11. $\dfrac{x^3 + x + 1}{x(2x - 5)^3(x^2 + 2x + 5)^2}$ **12.** $\dfrac{1}{(x^3 - 1)(x^2 - 1)}$

13–44 ■ Find the partial fraction decomposition of the rational function.

13. $\dfrac{2}{(x - 1)(x + 1)}$ **14.** $\dfrac{2x}{(x - 1)(x + 1)}$

15. $\dfrac{5}{(x-1)(x+4)}$

16. $\dfrac{x+6}{x(x+3)}$

17. $\dfrac{12}{x^2-9}$

18. $\dfrac{x-12}{x^2-4x}$

19. $\dfrac{4}{x^2-4}$

20. $\dfrac{2x+1}{x^2+x-2}$

21. $\dfrac{x+14}{x^2-2x-8}$

22. $\dfrac{8x-3}{2x^2-x}$

23. $\dfrac{x}{8x^2-10x+3}$

24. $\dfrac{7x-3}{x^3+2x^2-3x}$

25. $\dfrac{9x^2-9x+6}{2x^3-x^2-8x+4}$

26. $\dfrac{-3x^2-3x+27}{(x+2)(2x^2+3x-9)}$

27. $\dfrac{x^2+1}{x^3+x^2}$

28. $\dfrac{3x^2+5x-13}{(3x+2)(x^2-4x+4)}$

29. $\dfrac{2x}{4x^2+12x+9}$

30. $\dfrac{x-4}{(2x-5)^2}$

31. $\dfrac{4x^2-x-2}{x^4+2x^3}$

32. $\dfrac{x^3-2x^2-4x+3}{x^4}$

33. $\dfrac{-10x^2+27x-14}{(x-1)^3(x+2)}$

34. $\dfrac{-2x^2+5x-1}{x^4-2x^3+2x-1}$

35. $\dfrac{3x^3+22x^2+53x+41}{(x+2)^2(x+3)^2}$

36. $\dfrac{3x^2+12x-20}{x^4-8x^2+16}$

37. $\dfrac{x-3}{x^3+3x}$

38. $\dfrac{3x^2-2x+8}{x^3-x^2+2x-2}$

39. $\dfrac{2x^3+7x+5}{(x^2+x+2)(x^2+1)}$

40. $\dfrac{x^2+x+1}{2x^4+3x^2+1}$

41. $\dfrac{x^4+x^3+x^2-x+1}{x(x^2+1)^2}$

42. $\dfrac{2x^2-x+8}{(x^2+4)^2}$

43. $\dfrac{x^5-2x^4+x^3+x+5}{x^3-2x^2+x-2}$

44. $\dfrac{x^5-3x^4+3x^3-4x^2+4x+12}{(x-2)^2(x^2+2)}$

45. Determine A and B in terms of a and b.

$$\frac{ax+b}{x^2-1}=\frac{A}{x-1}+\frac{B}{x+1}$$

46. Determine A, B, C, and D in terms of a and b.

$$\frac{ax^3+bx^2}{(x^2+1)^2}=\frac{Ax+B}{x^2+1}+\frac{Cx+D}{(x^2+1)^2}$$

DISCOVERY ▪ DISCUSSION ▪ WRITING

47. Recognizing Partial Fraction Decompositions For each expression, determine whether it is already a partial fraction decomposition or whether it can be decomposed further.

(a) $\dfrac{x}{x^2+1}+\dfrac{1}{x+1}$ **(b)** $\dfrac{x}{(x+1)^2}$

(c) $\dfrac{1}{x+1}+\dfrac{2}{(x+1)^2}$ **(d)** $\dfrac{x+2}{(x^2+1)^2}$

48. Assembling and Disassembling Partial Fractions The following expression is a partial fraction decomposition.

$$\frac{2}{x-1}+\frac{1}{(x-1)^2}+\frac{1}{x+1}$$

Use a common denominator to combine the terms into one fraction. Then use the techniques of this section to find its partial fraction decomposition. Did you get back the original expression?

10.8 Systems of Nonlinear Equations

| Substitution and Elimination Methods ▶ Graphical Method

In this section we solve systems of equations in which the equations are not all linear. The methods we learned in Section 10.1 can also be used to solve nonlinear systems.

▼ Substitution and Elimination Methods

To solve a system of nonlinear equations, we can use the substitution or elimination method, as illustrated in the next examples.

EXAMPLE 1 | Substitution Method

Find all solutions of the system.

$$\begin{cases} x^2+y^2=100 & \text{Equation 1} \\ 3x-y=10 & \text{Equation 2} \end{cases}$$

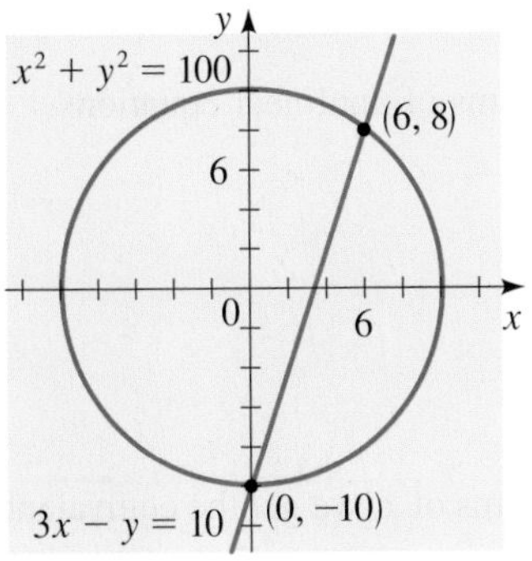

FIGURE 1

CHECK YOUR ANSWERS

$x = 0, y = -10$:

$$\begin{cases} (0)^2 + (-10)^2 = 100 \\ 3(0) - (-10) = 10 \quad \checkmark \end{cases}$$

$x = 6, y = 8$:

$$\begin{cases} (6)^2 + (8)^2 = 36 + 64 = 100 \\ 3(6) - (8) = 18 - 8 = 10 \quad \checkmark \end{cases}$$

SOLUTION Solve for one variable. We start by solving for y in the second equation:

$$y = 3x - 10 \qquad \text{Solve for } y \text{ in Equation 2}$$

Substitute. Next we substitute for y in the first equation and solve for x:

$$\begin{aligned} x^2 + (3x - 10)^2 &= 100 && \text{Substitute } y = 3x - 10 \text{ into Equation 1} \\ x^2 + (9x^2 - 60x + 100) &= 100 && \text{Expand} \\ 10x^2 - 60x &= 0 && \text{Simplify} \\ 10x(x - 6) &= 0 && \text{Factor} \\ x = 0 \quad \text{or} \quad x &= 6 && \text{Solve for } x \end{aligned}$$

Back-subsitute. Now we back-substitute these values of x into the equation $y = 3x - 10$:

$$\begin{aligned} &\text{For } x = 0: && y = 3(0) - 10 = -10 && \text{Back-substitute} \\ &\text{For } x = 6: && y = 3(6) - 10 = 8 && \text{Back-substitute} \end{aligned}$$

So we have two solutions: $(0, -10)$ and $(6, 8)$.

The graph of the first equation is a circle, and the graph of the second equation is a line; Figure 1 shows that the graphs intersect at the two points $(0, -10)$ and $(6, 8)$.

NOW TRY EXERCISE **5** ■

EXAMPLE 2 | Elimination Method

Find all solutions of the system.

$$\begin{cases} 3x^2 + 2y = 26 & \text{Equation 1} \\ 5x^2 + 7y = 3 & \text{Equation 2} \end{cases}$$

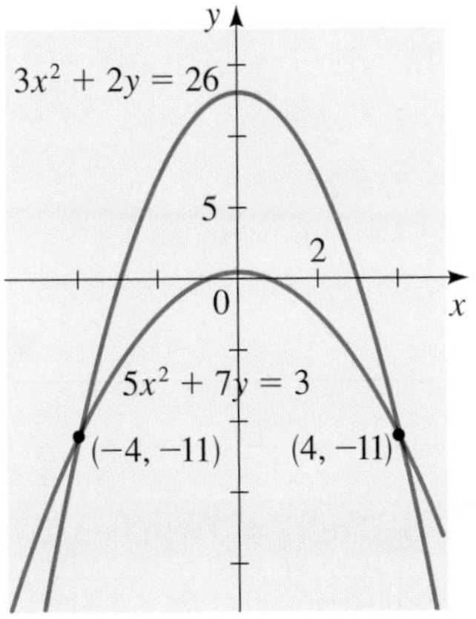

FIGURE 2

CHECK YOUR ANSWERS

$x = -4, y = -11$:

$$\begin{cases} 3(-4)^2 + 2(-11) = 26 \\ 5(-4)^2 + 7(-11) = 3 \quad \checkmark \end{cases}$$

$x = 4, y = -11$:

$$\begin{cases} 3(4)^2 + 2(-11) = 26 \\ 5(4)^2 + 7(-11) = 3 \quad \checkmark \end{cases}$$

SOLUTION We choose to eliminate the x-term, so we multiply the first equation by 5 and the second equation by -3. Then we add the two equations and solve for y.

$$\begin{aligned} &\begin{cases} 15x^2 + 10y = 130 \\ -15x^2 - 21y = -9 \end{cases} && \begin{aligned} &5 \times \text{Equation 1} \\ &(-3) \times \text{Equation 2} \end{aligned} \\ &\quad -11y = 121 && \text{Add} \\ &\quad y = -11 && \text{Solve for } y \end{aligned}$$

Now we back-substitute $y = -11$ into one of the original equations, say $3x^2 + 2y = 26$, and solve for x.

$$\begin{aligned} 3x^2 + 2(-11) &= 26 && \text{Back-substitute } y = -11 \text{ into Equation 1} \\ 3x^2 &= 48 && \text{Add 22} \\ x^2 &= 16 && \text{Divide by 3} \\ x = -4 \quad \text{or} \quad x &= 4 && \text{Solve for } x \end{aligned}$$

So we have two solutions: $(-4, -11)$ and $(4, -11)$.

The graphs of both equations are parabolas (see Section 3.1). Figure 2 shows that the graphs intersect at the two points $(-4, -11)$ and $(4, -11)$.

NOW TRY EXERCISE **11** ■

▼ Graphical Method

The graphical method is particularly useful in solving systems of nonlinear equations.

EXAMPLE 3 | Graphical Method

Find all solutions of the system

$$\begin{cases} x^2 - y = 2 \\ 2x - y = -1 \end{cases}$$

SOLUTION Graph each equation. Solving for y in terms of x, we get the equivalent system

$$\begin{cases} y = x^2 - 2 \\ y = 2x + 1 \end{cases}$$

Find intersection points. Figure 3 shows that the graphs of these equations intersect at two points. Zooming in, we see that the solutions are

$$(-1, -1) \quad \text{and} \quad (3, 7)$$

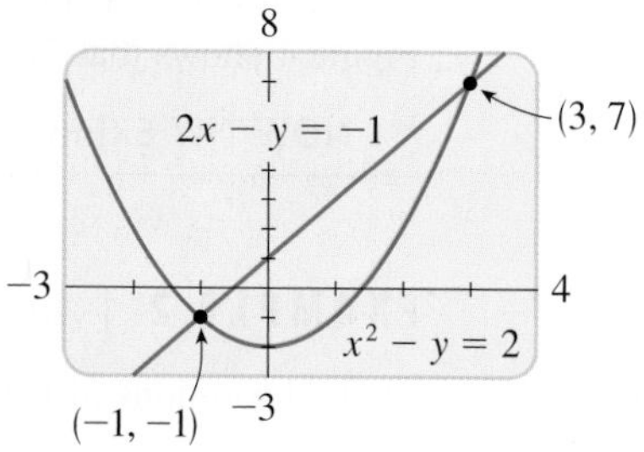

FIGURE 3

CHECK YOUR ANSWERS

$x = -1, y = -1$:

$$\begin{cases} (-1)^2 - (-1) = 2 \\ 2(-1) - (-1) = -1 \end{cases} \checkmark$$

$x = 3, y = 7$:

$$\begin{cases} 3^2 - 7 = 2 \\ 2(3) - 7 = -1 \end{cases} \checkmark$$

NOW TRY EXERCISE **33** ■

MATHEMATICS IN THE MODERN WORLD

Global Positioning System (GPS)

Courtesy of NASA

On a cold, foggy day in 1707 a British naval fleet was sailing home at a fast clip. The fleet's navigators didn't know it, but the fleet was only a few yards from the rocky shores of En-gland. In the ensuing disaster the fleet was totally destroyed. This tragedy could have been avoided had the navigators known their positions. In those days latitude was determined by the position of the North Star (and this could only be done at night in good weather), and longitude by the position of the sun relative to where it would be in England *at that same time*. So navigation required an accurate method of telling time on ships. (The invention of the spring-loaded clock brought about the eventual solution.)

Since then, several different methods have been developed to determine position, and all rely heavily on mathematics (see LORAN, page 747). The latest method, called the Global Positioning System (GPS), uses triangulation. In this system, 24 satellites are strategically located above the surface of the earth. A handheld GPS device measures distance from a satellite, using the travel time of radio signals emitted from the satellite. Knowing the distances to three different satellites tells us that we are at the point of intersection of three different spheres. This uniquely determines our position (see Exercise 47, page 703).

EXAMPLE 4 | Solving a System of Equations Graphically

Find all solutions of the system, correct to one decimal place.

$$\begin{cases} x^2 + y^2 = 12 & \text{Equation 1} \\ y = 2x^2 - 5x & \text{Equation 2} \end{cases}$$

SOLUTION The graph of the first equation is a circle, and the graph of the second is a parabola. To graph the circle on a graphing calculator, we must first solve for y in terms of x (see Section 1.9).

$$x^2 + y^2 = 12$$

$$y^2 = 12 - x^2 \qquad \text{Isolate } y^2 \text{ on LHS}$$

$$y = \pm\sqrt{12 - x^2} \qquad \text{Take square roots}$$

To graph the circle, we must graph both functions.

$$y = \sqrt{12 - x^2} \quad \text{and} \quad y = -\sqrt{12 - x^2}$$

In Figure 4 the graph of the circle is shown in red, and the parabola is shown in blue. The graphs intersect in Quadrants I and II. Zooming in, or using the `Intersect` command, we see that the intersection points are $(-0.559, 3.419)$ and $(2.847, 1.974)$. There also appears to be an intersection point in Quadrant IV. However, when we zoom in, we see that the curves come close to each other but don't intersect (see Figure 5). Thus the system has two solutions; rounded to the nearest tenth, they are

$$(-0.6, 3.4) \quad \text{and} \quad (2.8, 2.0)$$

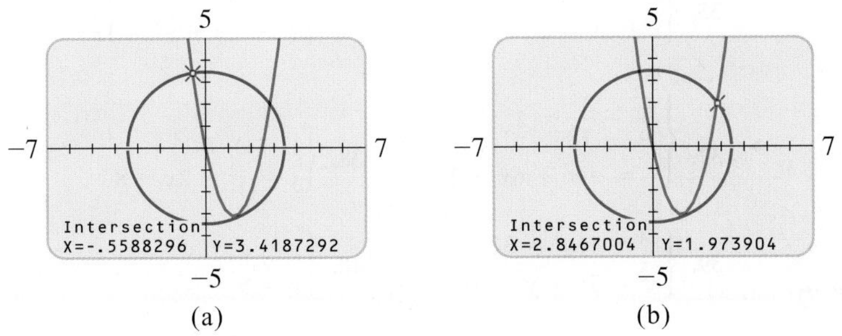

FIGURE 4 $x^2 + y^2 = 12,\ y = 2x^2 - 5x$

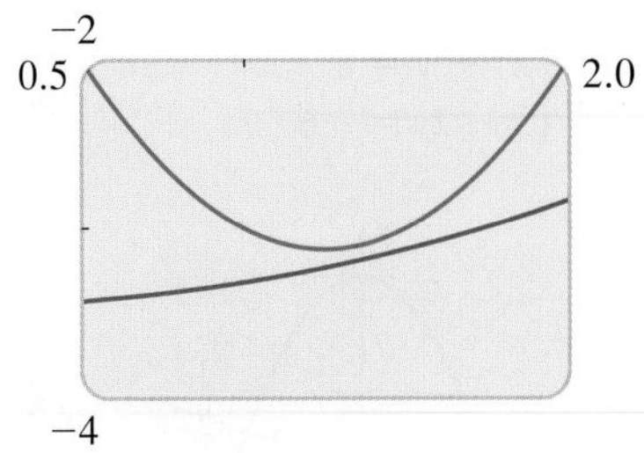

FIGURE 5 Zooming in

NOW TRY EXERCISE 37

10.8 EXERCISES

CONCEPTS

1–2 ■ The system of equations

$$\begin{cases} 2y - x^2 = 0 \\ y - x = 4 \end{cases}$$

is graphed to the right.

1. Use the graph to find the solution(s) of the system.

2. Check that the solutions you found in Exercise 1 satisfy the system.

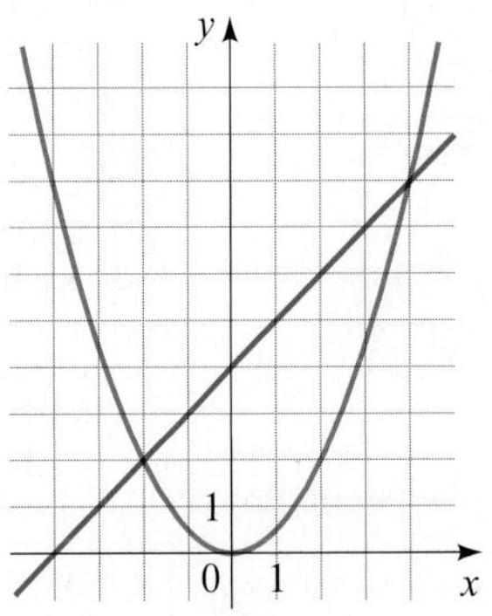

SKILLS

3–8 ■ Use the substitution method to find all solutions of the system of equations.

3. $\begin{cases} y = x^2 \\ y = x + 12 \end{cases}$

4. $\begin{cases} x^2 + y^2 = 25 \\ y = 2x \end{cases}$

5. $\begin{cases} x^2 + y^2 = 8 \\ x + y = 0 \end{cases}$

6. $\begin{cases} x^2 + y = 9 \\ x - y + 3 = 0 \end{cases}$

7. $\begin{cases} x + y^2 = 0 \\ 2x + 5y^2 = 75 \end{cases}$

8. $\begin{cases} x^2 - y = 1 \\ 2x^2 + 3y = 17 \end{cases}$

9–14 ■ Use the elimination method to find all solutions of the system of equations.

9. $\begin{cases} x^2 - 2y = 1 \\ x^2 + 5y = 29 \end{cases}$

10. $\begin{cases} 3x^2 + 4y = 17 \\ 2x^2 + 5y = 2 \end{cases}$

11. $\begin{cases} 3x^2 - y^2 = 11 \\ x^2 + 4y^2 = 8 \end{cases}$

12. $\begin{cases} 2x^2 + 4y = 13 \\ x^2 - y^2 = \frac{7}{2} \end{cases}$

13. $\begin{cases} x - y^2 + 3 = 0 \\ 2x^2 + y^2 - 4 = 0 \end{cases}$

14. $\begin{cases} x^2 - y^2 = 1 \\ 2x^2 - y^2 = x + 3 \end{cases}$

15–18 ■ Two equations and their graphs are given. Find the intersection point(s) of the graphs by solving the system.

15. $\begin{cases} x^2 + y = 8 \\ x - 2y = -6 \end{cases}$

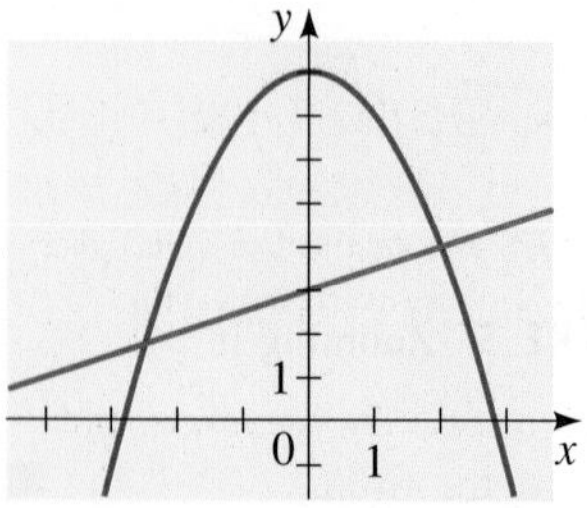

16. $\begin{cases} x - y^2 = -4 \\ x - y = 2 \end{cases}$

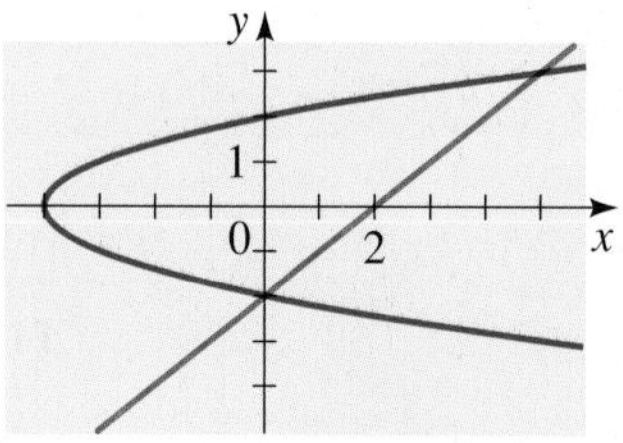

17. $\begin{cases} x^2 + y = 0 \\ x^3 - 2x - y = 0 \end{cases}$

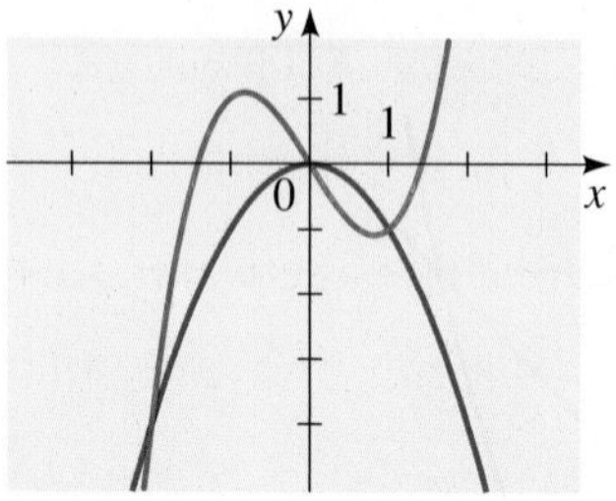

18. $\begin{cases} x^2 + y^2 = 4x \\ x = y^2 \end{cases}$

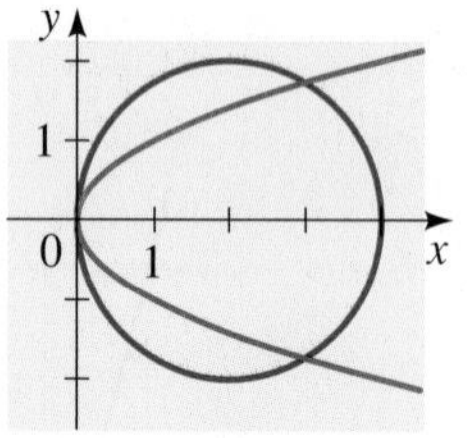

19–32 ■ Find all solutions of the system of equations.

19. $\begin{cases} y + x^2 = 4x \\ y + 4x = 16 \end{cases}$

20. $\begin{cases} x - y^2 = 0 \\ y - x^2 = 0 \end{cases}$

21. $\begin{cases} x - 2y = 2 \\ y^2 - x^2 = 2x + 4 \end{cases}$

22. $\begin{cases} y = 4 - x^2 \\ y = x^2 - 4 \end{cases}$

23. $\begin{cases} x - y = 4 \\ xy = 12 \end{cases}$

24. $\begin{cases} xy = 24 \\ 2x^2 - y^2 + 4 = 0 \end{cases}$

25. $\begin{cases} x^2y = 16 \\ x^2 + 4y + 16 = 0 \end{cases}$

26. $\begin{cases} x + \sqrt{y} = 0 \\ y^2 - 4x^2 = 12 \end{cases}$

27. $\begin{cases} x^2 + y^2 = 9 \\ x^2 - y^2 = 1 \end{cases}$

28. $\begin{cases} x^2 + 2y^2 = 2 \\ 2x^2 - 3y = 15 \end{cases}$

29. $\begin{cases} 2x^2 - 8y^3 = 19 \\ 4x^2 + 16y^3 = 34 \end{cases}$

30. $\begin{cases} x^4 + y^3 = 17 \\ 3x^4 + 5y^3 = 53 \end{cases}$

31. $\begin{cases} \dfrac{2}{x} - \dfrac{3}{y} = 1 \\ -\dfrac{4}{x} + \dfrac{7}{y} = 1 \end{cases}$

32. $\begin{cases} \dfrac{4}{x^2} + \dfrac{6}{y^4} = \dfrac{7}{2} \\ \dfrac{1}{x^2} - \dfrac{2}{y^4} = 0 \end{cases}$

33–40 ■ Use the graphical method to find all solutions of the system of equations, rounded to two decimal places.

33. $\begin{cases} y = x^2 + 8x \\ y = 2x + 16 \end{cases}$

34. $\begin{cases} y = x^2 - 4x \\ 2x - y = 2 \end{cases}$

35. $\begin{cases} x^2 + y^2 = 25 \\ x + 3y = 2 \end{cases}$

36. $\begin{cases} x^2 + y^2 = 17 \\ x^2 - 2x + y^2 = 13 \end{cases}$

37. $\begin{cases} \dfrac{x^2}{9} + \dfrac{y^2}{18} = 1 \\ y = -x^2 + 6x - 2 \end{cases}$

38. $\begin{cases} x^2 - y^2 = 3 \\ y = x^2 - 2x - 8 \end{cases}$

39. $\begin{cases} x^4 + 16y^4 = 32 \\ x^2 + 2x + y = 0 \end{cases}$

40. $\begin{cases} y = e^x + e^{-x} \\ y = 5 - x^2 \end{cases}$

APPLICATIONS

41. Dimensions of a Rectangle A rectangle has an area of 180 cm^2 and a perimeter of 54 cm. What are its dimensions?

42. Legs of a Right Triangle A right triangle has an area of 84 ft^2 and a hypotenuse 25 ft long. What are the lengths of its other two sides?

43. Dimensions of a Rectangle The perimeter of a rectangle is 70, and its diagonal is 25. Find its length and width.

44. Dimensions of a Rectangle A circular piece of sheet metal has a diameter of 20 in. The edges are to be cut off to form a rectangle of area 160 in^2 (see the figure). What are the dimensions of the rectangle?

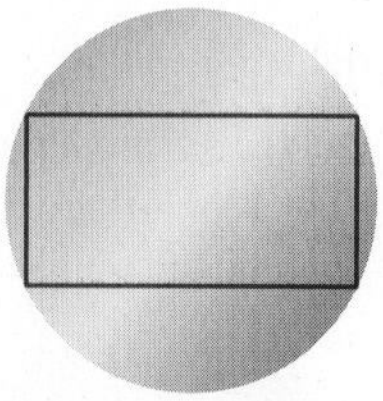

45. Flight of a Rocket A hill is inclined so that its "slope" is $\frac{1}{2}$, as shown in the figure. We introduce a coordinate system with the origin at the base of the hill and with the scales on the axes measured in meters. A rocket is fired from the base of the hill in such a way that its trajectory is the parabola $y = -x^2 + 401x$. At what point does the rocket strike the hillside? How far is this point from the base of the hill (to the nearest centimeter)?

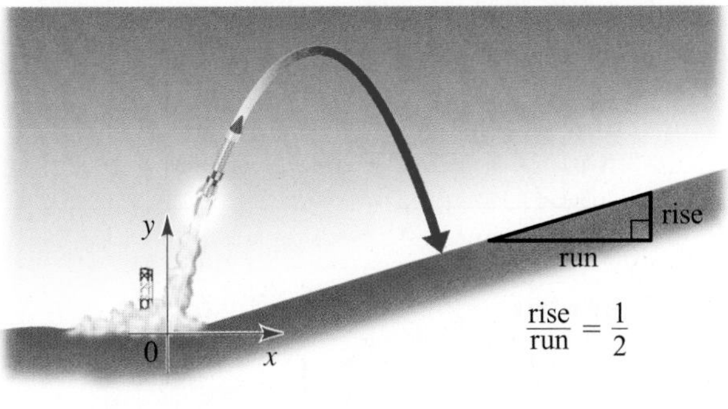

46. Making a Stovepipe A rectangular piece of sheet metal with an area of 1200 in^2 is to be bent into a cylindrical length of stovepipe having a volume of 600 in^3. What are the dimensions of the sheet metal?

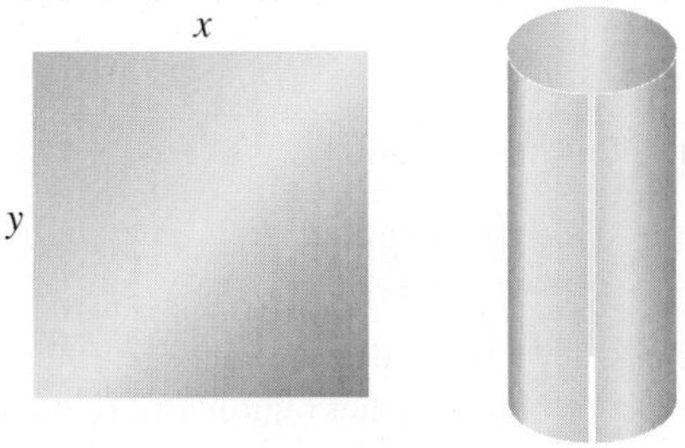

47. Global Positioning System (GPS) The Global Positioning System determines the location of an object from its distances to satellites in orbit around the earth. In the simplified, two-dimensional situation shown in the following figure, determine the coordinates of P from the fact that P is 26 units from satellite A and 20 units from satellite B.

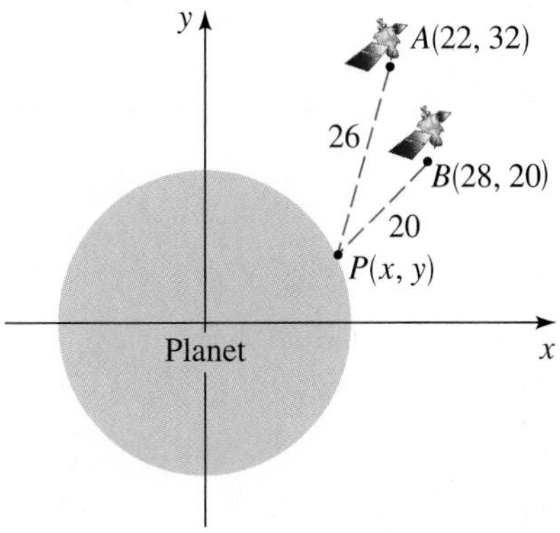

DISCOVERY ■ DISCUSSION ■ WRITING

48. Intersection of a Parabola and a Line On a sheet of graph paper or using a graphing calculator, draw the parabola $y = x^2$. Then draw the graphs of the linear equation $y = x + k$ on the same coordinate plane for various values of k. Try to choose values of k so that the line and the parabola intersect at two points for some of your k's and not for others. For what value of k is there exactly one intersection point? Use the results of your experiment to make a conjecture about the values of k for which the following system has two solutions, one solution, and no solution. Prove your conjecture.

$$\begin{cases} y = x^2 \\ y = x + k \end{cases}$$

49. Some Trickier Systems Follow the hints and solve the systems.

(a) $\begin{cases} \log x + \log y = \frac{3}{2} \\ 2 \log x - \log y = 0 \end{cases}$ [*Hint:* Add the equations.]

(b) $\begin{cases} 2^x + 2^y = 10 \\ 4^x + 4^y = 68 \end{cases}$ [*Hint:* Note that $4^x = 2^{2x} = (2^x)^2$.]

(c) $\begin{cases} x - y = 3 \\ x^3 - y^3 = 387 \end{cases}$ [*Hint:* Factor the left-hand side of the second equation.]

(d) $\begin{cases} x^2 + xy = 1 \\ xy + y^2 = 3 \end{cases}$ [*Hint:* Add the equations, and factor the result.]

10.9 SYSTEMS OF INEQUALITIES

Graphing an Inequality ▶ Systems of Inequalities ▶ Systems of Linear Inequalities ▶ Application: Feasible Regions

In this section we study systems of inequalities in two variables from a graphical point of view.

▼ Graphing an Inequality

We begin by considering the graph of a single inequality. We already know that the graph of $y = x^2$, for example, is the *parabola* in Figure 1. If we replace the equal sign by the symbol $\geq$, we obtain the *inequality*

$$y \geq x^2$$

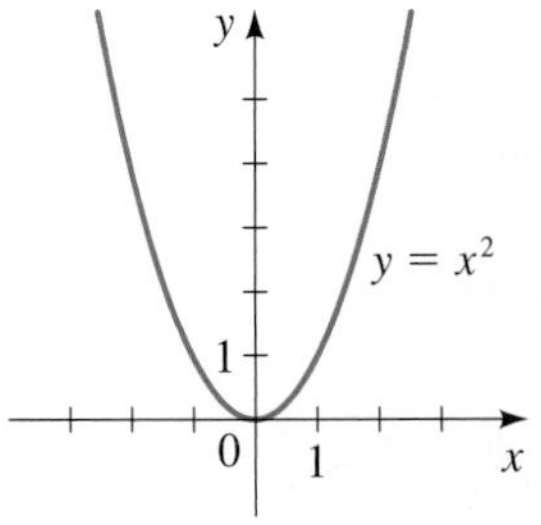

FIGURE 1

Its graph consists of not just the parabola in Figure 1, but also every point whose y-coordinate is *larger* than x^2. We indicate the solution in Figure 2(a) by shading the points *above* the parabola.

Similarly, the graph of $y \le x^2$ in Figure 2(b) consists of all points on and *below* the parabola. However, the graphs of $y > x^2$ and $y < x^2$ do not include the points on the parabola itself, as indicated by the dashed curves in Figures 2(c) and 2(d).

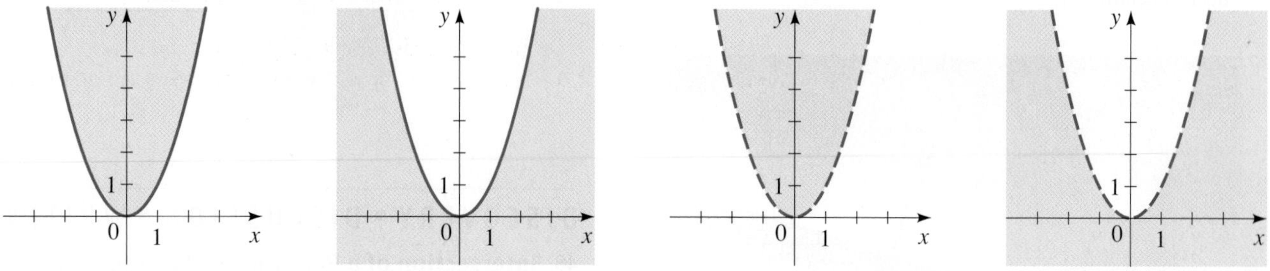

(a) $y \ge x^2$ (b) $y \le x^2$ (c) $y > x^2$ (d) $y < x^2$

FIGURE 2

The graph of an inequality, in general, consists of a region in the plane whose boundary is the graph of the equation obtained by replacing the inequality sign ($\ge$, $\le$, $>$, or $<$) with an equal sign. To determine which side of the graph gives the solution set of the inequality, we need only check **test points**.

GRAPHING INEQUALITIES

To graph an inequality, we carry out the following steps.

1. **Graph Equation.** Graph the equation corresponding to the inequality. Use a dashed curve for $>$ or $<$ and a solid curve for $\le$ or $\ge$.
2. **Test Points.** Test one point in each region formed by the graph in Step 1. If the point satisfies the inequality, then all the points in that region satisfy the inequality. (In that case, shade the region to indicate that it is part of the graph.) If the test point does not satisfy the inequality, then the region isn't part of the graph.

EXAMPLE 1 | Graphs of Inequalities

Graph each inequality.

(a) $x^2 + y^2 < 25$ **(b)** $x + 2y \ge 5$

SOLUTION

(a) The graph of $x^2 + y^2 = 25$ is a circle of radius 5 centered at the origin. The points on the circle itself do not satisfy the inequality because it is of the form $<$, so we graph the circle with a dashed curve, as shown in Figure 3.

To determine whether the inside or the outside of the circle satisfies the inequality, we use the test points (0, 0) on the inside and (6, 0) on the outside. To do this, we substitute the coordinates of each point into the inequality and check whether the result satisfies the inequality. (Note that *any* point inside or outside the circle can serve as a test point. We have chosen these points for simplicity.)

Test point	$x^2 + y^2 < 25$	Conclusion
(0, 0)	$0^2 + 0^2 = 0 < 25$	Part of graph
(6, 0)	$6^2 + 0^2 = 36 \not< 25$	Not part of graph

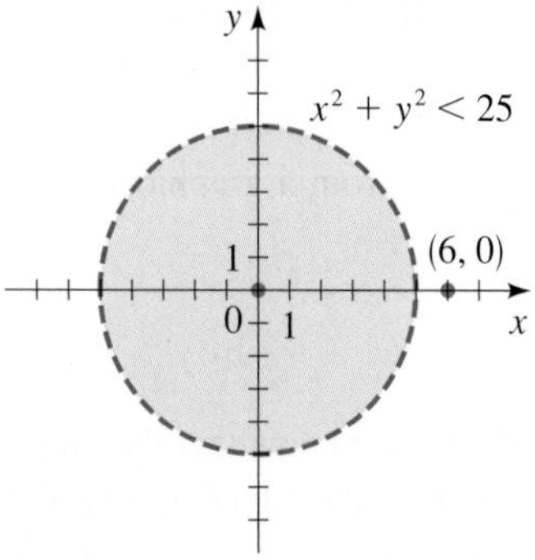

FIGURE 3

Thus the graph of $x^2 + y^2 < 25$ is the set of all points *inside* the circle (see Figure 3).

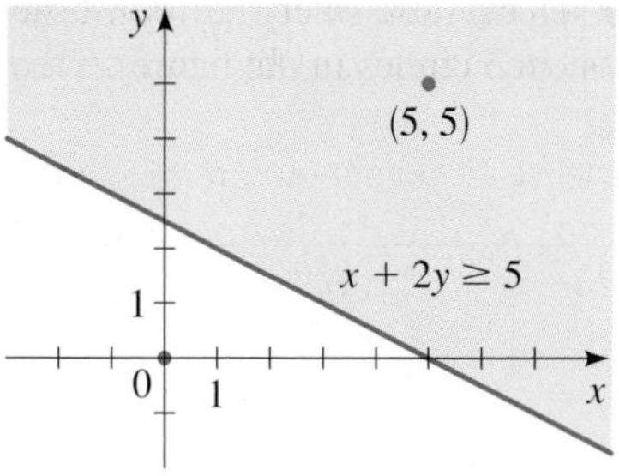

FIGURE 4

(b) The graph of $x + 2y = 5$ is the line shown in Figure 4. We use the test points (0, 0) and (5, 5) on opposite sides of the line.

Test point	$x + 2y \geq 5$	Conclusion
(0, 0)	$0 + 2(0) = 0 \ngeq 5$	Not part of graph
(5, 5)	$5 + 2(5) = 15 \geq 5$	Part of graph

Our check shows that the points *above* the line satisfy the inequality.

Alternatively, we could put the inequality into slope-intercept form and graph it directly:

$$x + 2y \geq 5$$
$$2y \geq -x + 5$$
$$y \geq -\tfrac{1}{2}x + \tfrac{5}{2}$$

From this form we see that the graph includes all points whose y-coordinates are *greater* than those on the line $y = -\frac{1}{2}x + \frac{5}{2}$; that is, the graph consists of the points *on or above* this line, as shown in Figure 4.

NOW TRY EXERCISES **7** AND **15** ■

Systems of Inequalities

We now consider *systems* of inequalities. The solution of such a system is the set of all points in the coordinate plane that satisfy every inequality in the system.

EXAMPLE 2 | A System of Two Inequalities

Graph the solution of the system of inequalities, and label its vertices.

$$\begin{cases} x^2 + y^2 < 25 \\ x + 2y \geq 5 \end{cases}$$

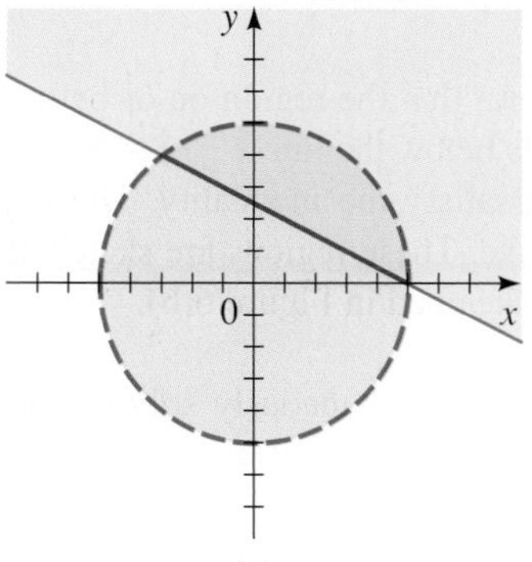

(a)

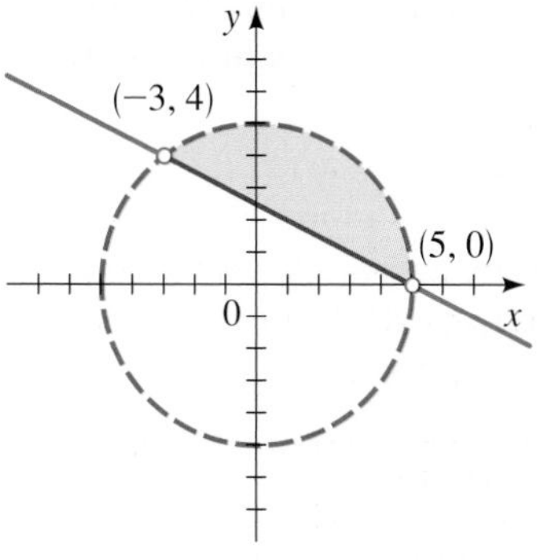

(b)

FIGURE 5 $\begin{cases} x^2 + y^2 < 25 \\ x + 2y \geq 5 \end{cases}$

SOLUTION These are the two inequalities of Example 1. In this example we wish to graph only those points that simultaneously satisfy both inequalities. The solution consists of the intersection of the graphs in Example 1. In Figure 5(a) we show the two regions on the same coordinate plane (in different colors), and in Figure 5(b) we show their intersection.

Vertices The points $(-3, 4)$ and $(5, 0)$ in Figure 5(b) are the **vertices** of the solution set. They are obtained by solving the system of *equations*

$$\begin{cases} x^2 + y^2 = 25 \\ x + 2y = 5 \end{cases}$$

We solve this system of equations by substitution. Solving for x in the second equation gives $x = 5 - 2y$, and substituting this into the first equation gives

$$(5 - 2y)^2 + y^2 = 25 \qquad \text{Substitute } x = 5 - 2y$$
$$(25 - 20y + 4y^2) + y^2 = 25 \qquad \text{Expand}$$
$$-20y + 5y^2 = 0 \qquad \text{Simplify}$$
$$-5y(4 - y) = 0 \qquad \text{Factor}$$

Thus $y = 0$ or $y = 4$. When $y = 0$, we have $x = 5 - 2(0) = 5$, and when $y = 4$, we have $x = 5 - 2(4) = -3$. So the points of intersection of these curves are $(5, 0)$ and $(-3, 4)$.

Note that in this case the vertices are not part of the solution set, since they don't satisfy the inequality $x^2 + y^2 < 25$ (so they are graphed as open circles in the figure). They simply show where the "corners" of the solution set lie.

NOW TRY EXERCISE **33** ■

Systems of Linear Inequalities

An inequality is **linear** if it can be put into one of the following forms:

$$ax + by \geq c \qquad ax + by \leq c \qquad ax + by > c \qquad ax + by < c$$

In the next example we graph the solution set of a system of linear inequalities.

EXAMPLE 3 | A System of Four Linear Inequalities

Graph the solution set of the system, and label its vertices.

$$\begin{cases} x + 3y \leq 12 \\ x + y \leq 8 \\ x \geq 0 \\ y \geq 0 \end{cases}$$

SOLUTION In Figure 6 we first graph the lines given by the equations that correspond to each inequality. To determine the graphs of the linear inequalities, we need to check only one test point. For simplicity let's use the point (0, 0).

Inequality	Test point (0, 0)	Conclusion
$x + 3y \leq 12$	$0 + 3(0) = 0 \leq 12$	Satisfies inequality
$x + y \leq 8$	$0 + 0 = 0 \leq 8$	Satisfies inequality

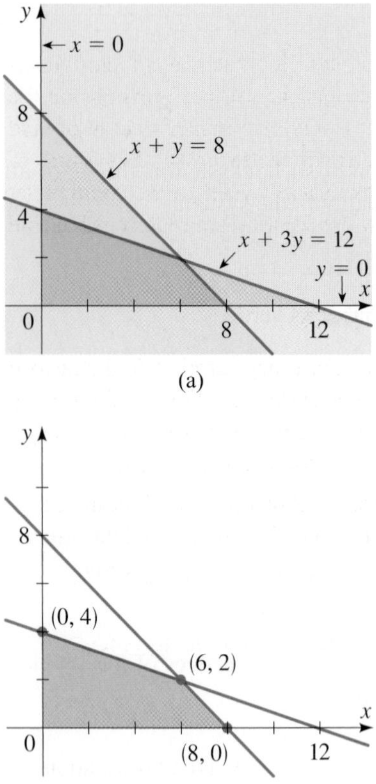

(a)

(b)

FIGURE 6

Since (0, 0) is below the line $x + 3y = 12$, our check shows that the region on or below the line must satisfy the inequality. Likewise, since (0, 0) is below the line $x + y = 8$, our check shows that the region on or below this line must satisfy the inequality. The inequalities $x \geq 0$ and $y \geq 0$ say that x and y are nonnegative. These regions are sketched in Figure 6(a), and the intersection—the solution set—is sketched in Figure 6(b).

Vertices The coordinates of each vertex are obtained by simultaneously solving the equations of the lines that intersect at that vertex. From the system

$$\begin{cases} x + 3y = 12 \\ x + y = 8 \end{cases}$$

we get the vertex (6, 2). The origin (0, 0) is also clearly a vertex. The other two vertices are at the x- and y-intercepts of the corresponding lines: (8, 0) and (0, 4). In this case all the vertices *are* part of the solution set.

NOW TRY EXERCISE **39** ■

EXAMPLE 4 | A System of Linear Inequalities

Graph the solution set of the system.

$$\begin{cases} x + 2y \geq 8 \\ -x + 2y \leq 4 \\ 3x - 2y \leq 8 \end{cases}$$

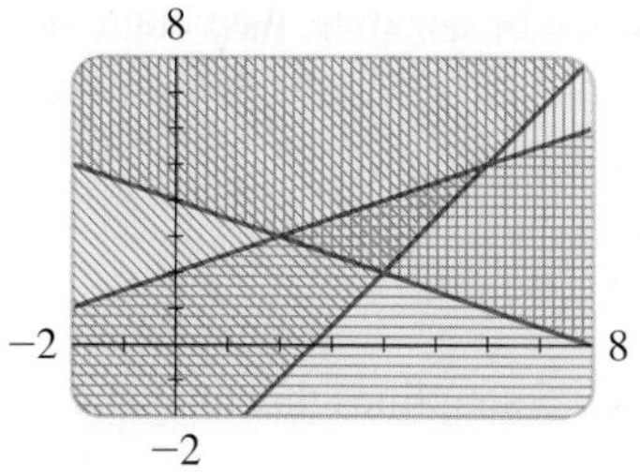

FIGURE 7

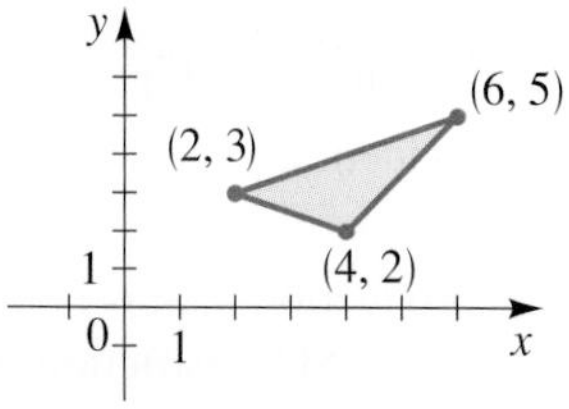

FIGURE 8

SOLUTION We must graph the lines that correspond to these inequalities and then shade the appropriate regions, as in Example 3. We will use a graphing calculator, so we must first isolate y on the left-hand side of each inequality.

$$\begin{cases} y \geq -\frac{1}{2}x + 4 \\ y \leq \frac{1}{2}x + 2 \\ y \geq \frac{3}{2}x - 4 \end{cases}$$

Using the shading feature of the calculator, we obtain the graph in Figure 7. The solution set is the triangular region that is shaded in all three patterns. We then use TRACE or the `Intersect` command to find the vertices of the region. The solution set is graphed in Figure 8.

NOW TRY EXERCISE **47** ■

When a region in the plane can be covered by a (sufficiently large) circle, it is said to be **bounded**. A region that is not bounded is called **unbounded**. For example, the regions graphed in Figures 3, 5(b), 6(b), and 8 are bounded, whereas those in Figures 2 and 4 are unbounded. An unbounded region cannot be "fenced in"—it extends infinitely far in at least one direction.

Application: Feasible Regions

Many applied problems involve *constraints* on the variables. For instance, a factory manager has only a certain number of workers who can be assigned to perform jobs on the factory floor. A farmer deciding what crops to cultivate has only a certain amount of land that can be seeded. Such constraints or limitations can usually be expressed as systems of inequalities. When dealing with applied inequalities, we usually refer to the solution set of a system as a *feasible region*, because the points in the solution set represent feasible (or possible) values for the quantities being studied.

EXAMPLE 5 | Restricting Pollutant Outputs

A factory produces two agricultural pesticides, A and B. For every barrel of A, the factory emits 0.25 kg of carbon monoxide (CO) and 0.60 kg of sulfur dioxide (SO_2); and for every barrel of B, it emits 0.50 kg of CO and 0.20 kg of SO_2. Pollution laws restrict the factory's output of CO to a maximum of 75 kg and SO_2 to a maximum of 90 kg per day.

(a) Find a system of inequalities that describes the number of barrels of each pesticide the factory can produce and still satisfy the pollution laws. Graph the feasible region.

(b) Would it be legal for the factory to produce 100 barrels of A and 80 barrels of B per day?

(c) Would it be legal for the factory to produce 60 barrels of A and 160 barrels of B per day?

SOLUTION

(a) To set up the required inequalities, it is helpful to organize the given information into a table.

	A	B	Maximum
CO (kg)	0.25	0.50	75
SO_2 (kg)	0.60	0.20	90

We let

x = number of barrels of A produced per day

y = number of barrels of B produced per day

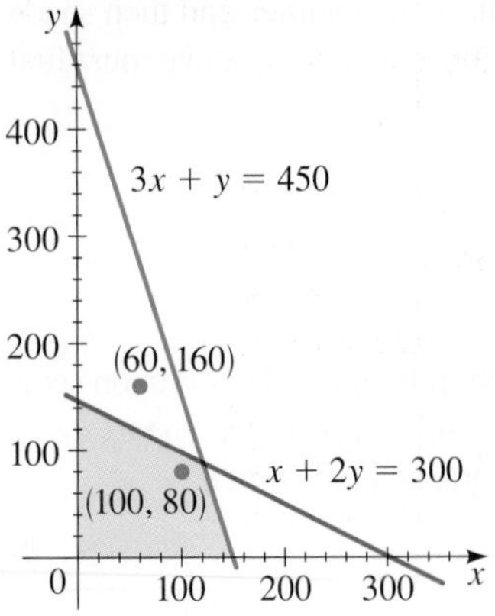

FIGURE 9

From the data in the table and the fact that x and y can't be negative, we obtain the following inequalities.

$$\begin{cases} 0.25x + 0.50y \le 75 & \text{CO inequality} \\ 0.60x + 0.20y \le 90 & \text{SO}_2 \text{ inequality} \\ x \ge 0, \quad y \ge 0 \end{cases}$$

Multiplying the first inequality by 4 and the second by 5 simplifies this to

$$\begin{cases} x + 2y \le 300 \\ 3x + y \le 450 \\ x \ge 0, \quad y \ge 0 \end{cases}$$

The feasible region is the solution of this system of inequalities, shown in Figure 9.

(b) Since the point (100, 80) lies inside the feasible region, this production plan is legal (see Figure 9).

(c) Since the point (60, 160) lies outside the feasible region, this production plan is not legal. It violates the CO restriction, although it does not violate the SO_2 restriction (see Figure 9).

NOW TRY EXERCISE **51** ■

10.9 EXERCISES

CONCEPTS

1. To graph an inequality, we first graph the corresponding ________. So to graph $y \le x + 1$, we first graph the equation ________. To decide which side of the graph of the equation is the graph of the inequality, we use ________ points. Using (0, 0) as such a point, graph the inequality by shading the appropriate region.

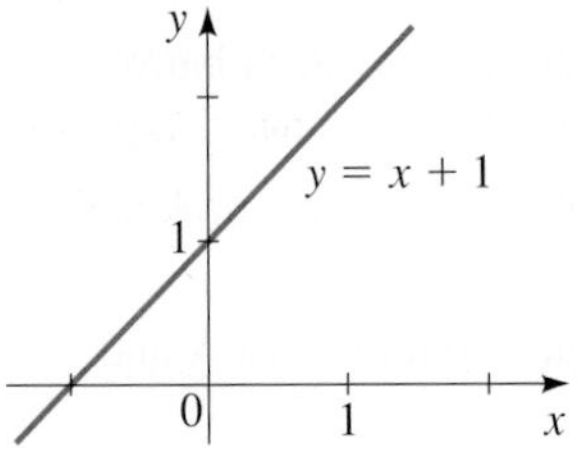

2. Shade the solution of each system of inequalities on the given graph.

(a) $\begin{cases} x - y \ge 0 \\ x + y \ge 2 \end{cases}$

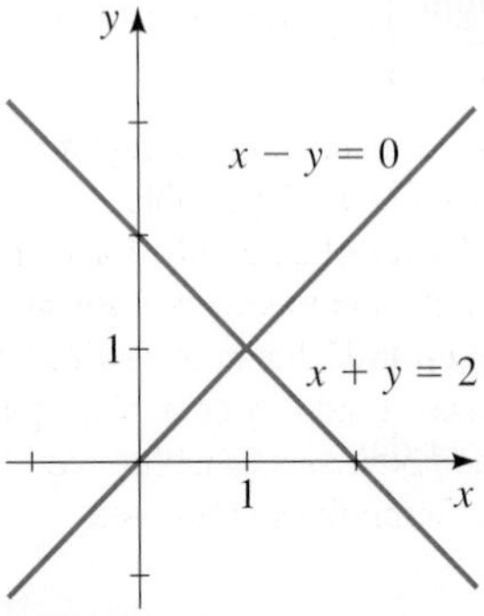

(b) $\begin{cases} x - y \le 0 \\ x + y \le 2 \end{cases}$

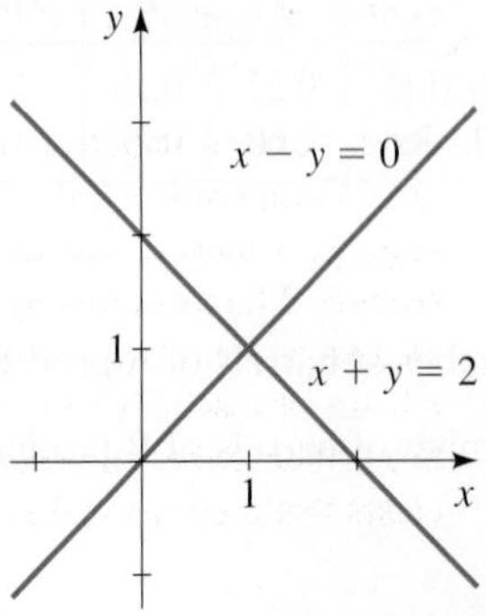

(c) $\begin{cases} x - y \ge 0 \\ x + y \le 2 \end{cases}$

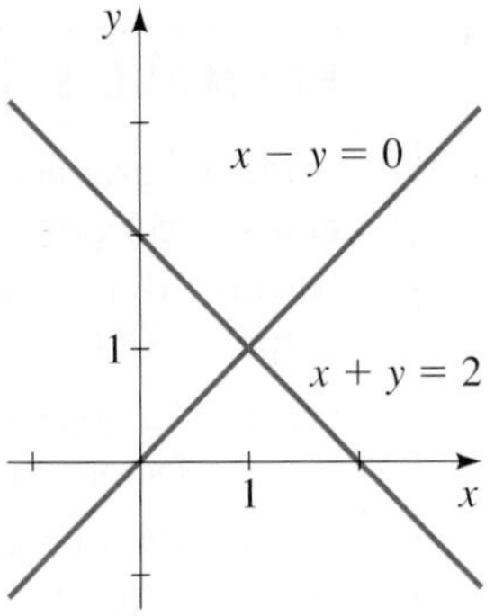

(d) $\begin{cases} x - y \le 0 \\ x + y \ge 2 \end{cases}$

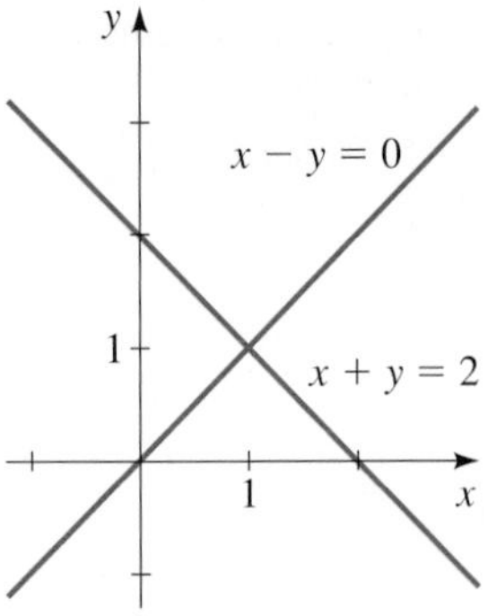

SKILLS

3–16 ■ Graph the inequality.

3. $x < 3$

4. $y \ge -2$

5. $y > x$

6. $y < x + 2$

7. $y \le 2x + 2$

8. $y < -x + 5$

9. $2x - y \le 8$

10. $3x + 4y + 12 > 0$

11. $4x + 5y < 20$

12. $-x^2 + y \ge 10$

13. $y > x^2 + 1$

14. $x^2 + y^2 \ge 9$

15. $x^2 + y^2 \le 25$

16. $x^2 + (y - 1)^2 \le 1$

17–20 ■ An equation and its graph are given. Find an inequality whose solution is the shaded region.

17. $y = \frac{1}{2}x - 1$

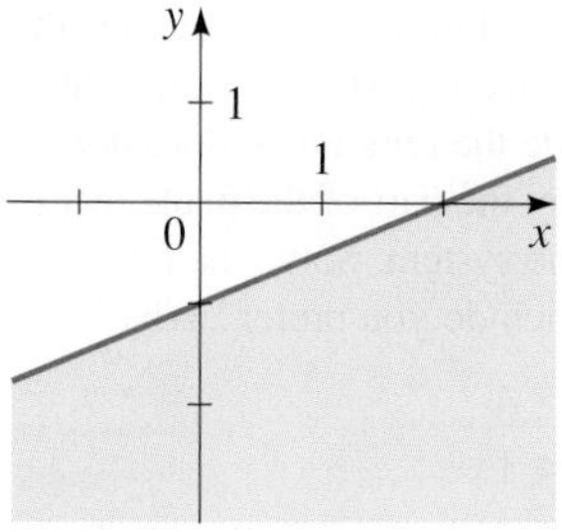

18. $y = x^2 + 2$

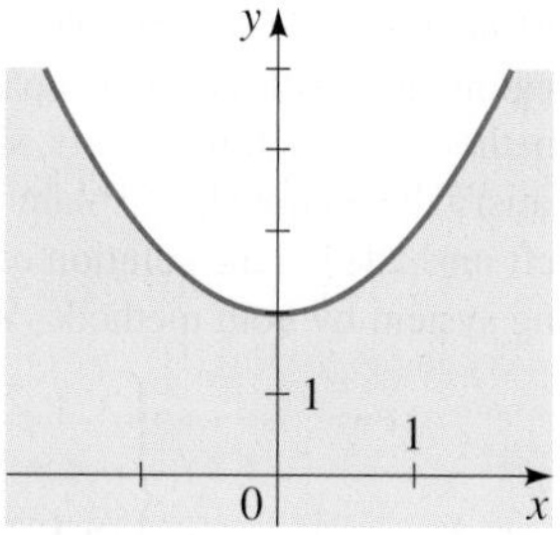

19. $x^2 + y^2 = 4$

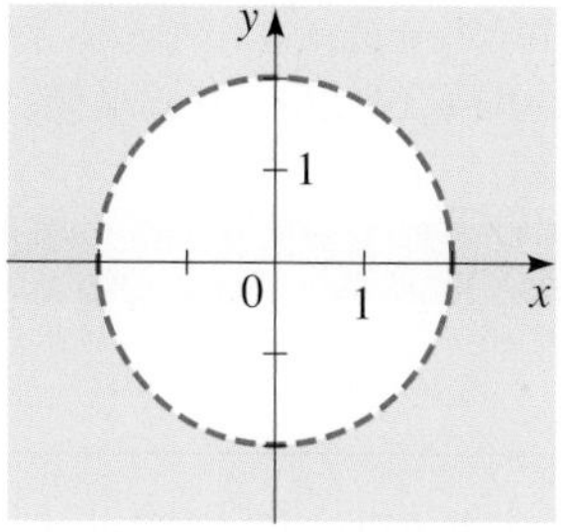

20. $y = x^3 - 4x$

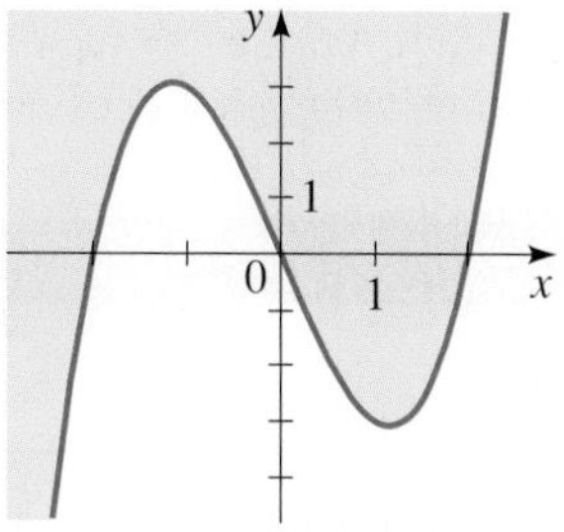

21–46 ■ Graph the solution of the system of inequalities. Find the coordinates of all vertices, and determine whether the solution set is bounded.

21. $\begin{cases} x + y \le 4 \\ y \ge x \end{cases}$

22. $\begin{cases} 2x + 3y > 12 \\ 3x - y < 21 \end{cases}$

23. $\begin{cases} y < \frac{1}{4}x + 2 \\ y \ge 2x - 5 \end{cases}$

24. $\begin{cases} x - y > 0 \\ 4 + y \le 2x \end{cases}$

25. $\begin{cases} y \le -2x + 8 \\ y \le -\frac{1}{2}x + 5 \\ x \ge 0, \quad y \ge 0 \end{cases}$

26. $\begin{cases} 4x + 3y \le 18 \\ 2x + y \le 8 \\ x \ge 0, \quad y \ge 0 \end{cases}$

27. $\begin{cases} x \ge 0 \\ y \ge 0 \\ 3x + 5y \le 15 \\ 3x + 2y \le 9 \end{cases}$

28. $\begin{cases} x > 2 \\ y < 12 \\ 2x - 4y > 8 \end{cases}$

29. $\begin{cases} y \le 9 - x^2 \\ x \ge 0, \quad y \ge 0 \end{cases}$

30. $\begin{cases} y \ge x^2 \\ y \le 4 \\ x \ge 0 \end{cases}$

31. $\begin{cases} y < 9 - x^2 \\ y \ge x + 3 \end{cases}$

32. $\begin{cases} y \ge x^2 \\ x + y \ge 6 \end{cases}$

33. $\begin{cases} x^2 + y^2 \le 4 \\ x - y > 0 \end{cases}$

34. $\begin{cases} x > 0 \\ y > 0 \\ x + y < 10 \\ x^2 + y^2 > 9 \end{cases}$

35. $\begin{cases} x^2 - y \le 0 \\ 2x^2 + y \le 12 \end{cases}$

36. $\begin{cases} x^2 + y^2 < 9 \\ 2x + y^2 \ge 1 \end{cases}$

37. $\begin{cases} x + 2y \le 14 \\ 3x - y \ge 0 \\ x - y \ge 2 \end{cases}$

38. $\begin{cases} y < x + 6 \\ 3x + 2y \ge 12 \\ x - 2y \le 2 \end{cases}$

39. $\begin{cases} x \ge 0 \\ y \ge 0 \\ x \le 5 \\ x + y \le 7 \end{cases}$

40. $\begin{cases} x \ge 0 \\ y \ge 0 \\ y \le 4 \\ 2x + y \le 8 \end{cases}$

41. $\begin{cases} y > x + 1 \\ x + 2y \le 12 \\ x + 1 > 0 \end{cases}$

42. $\begin{cases} x + y > 12 \\ y < \frac{1}{2}x - 6 \\ 3x + y < 6 \end{cases}$

43. $\begin{cases} x^2 + y^2 \le 8 \\ x \ge 2 \\ y \ge 0 \end{cases}$

44. $\begin{cases} x^2 - y \ge 0 \\ x + y < 6 \\ x - y < 6 \end{cases}$

45. $\begin{cases} x^2 + y^2 < 9 \\ x + y > 0 \\ x \le 0 \end{cases}$

46. $\begin{cases} y \ge x^3 \\ y \le 2x + 4 \\ x + y \ge 0 \end{cases}$

47–50 ■ Use a graphing calculator to graph the solution of the system of inequalities. Find the coordinates of all vertices, rounded to one decimal place.

47. $\begin{cases} y \ge x - 3 \\ y \ge -2x + 6 \\ y \le 8 \end{cases}$

48. $\begin{cases} x + y \ge 12 \\ 2x + y \le 24 \\ x - y \ge -6 \end{cases}$

49. $\begin{cases} y \le 6x - x^2 \\ x + y \ge 4 \end{cases}$

50. $\begin{cases} y \ge x^3 \\ 2x + y \ge 0 \\ y \le 2x + 6 \end{cases}$

APPLICATIONS

51. Publishing Books A publishing company publishes a total of no more than 100 books every year. At least 20 of these are nonfiction, but the company always publishes at least as much fiction as nonfiction. Find a system of inequalities that describes the possible numbers of fiction and nonfiction books that the company can produce each year consistent with these policies. Graph the solution set.

52. Furniture Manufacturing A man and his daughter manufacture unfinished tables and chairs. Each table requires 3 hours of sawing and 1 hour of assembly. Each chair requires 2 hours of sawing and 2 hours of assembly. Between the two of them, they can put in up to 12 hours of sawing and 8 hours of assembly work each day. Find a system of inequalities that describes all possible combinations of tables and chairs that they can make daily. Graph the solution set.

53. **Coffee Blends** A coffee merchant sells two different coffee blends. The Standard blend uses 4 oz of arabica and 12 oz of robusta beans per package; the Deluxe blend uses 10 oz of arabica and 6 oz of robusta beans per package. The merchant has 80 lb of arabica and 90 lb of robusta beans available. Find a system of inequalities that describes the possible number of Standard and Deluxe packages the merchant can make. Graph the solution set.

54. **Nutrition** A cat food manufacturer uses fish and beef byproducts. The fish contains 12 g of protein and 3 g of fat per ounce. The beef contains 6 g of protein and 9 g of fat per ounce. Each can of cat food must contain at least 60 g of protein and 45 g of fat. Find a system of inequalities that describes the possible number of ounces of fish and beef that can be used in each can to satisfy these minimum requirements. Graph the solution set.

DISCOVERY ■ DISCUSSION ■ WRITING

55. **Shading Unwanted Regions** To graph the solution of a system of inequalities, we have shaded the solution of each inequality in a different color; the solution of the system is the region where all the shaded parts overlap. Here is a different method: For each inequality, shade the region that does *not* satisfy the inequality. Explain why the part of the plane that is left unshaded is the solution of the system. Solve the following system by both methods. Which do you prefer? Why?

$$\begin{cases} x + 2y > 4 \\ -x + y < 1 \\ x + 3y < 9 \\ x < 3 \end{cases}$$

CHAPTER 10 | REVIEW

■ CONCEPT CHECK

1. Suppose you are asked to solve a system of two equations in two variables. Explain how you would solve the system
   (a) by the substitution method
   (b) by the elimination method
   (c) graphically

2. Suppose you are asked to solve a system of two *linear* equations in two variables.
   (a) Would you prefer to use the substitution method or the elimination method?
   (b) How many solutions are possible? Draw diagrams to illustrate the possibilities.

3. What operations can be performed on a linear system that result in an equivalent system?

4. Explain how Gaussian elimination works. Your explanation should include a discussion of the steps used to obtain a system in triangular form, and back-substitution.

5. What does it mean to say that A is a matrix with dimension $m \times n$?

6. What is the augmented matrix of a system? Describe the role of elementary row operations, row-echelon form, back-substitution, and leading variables when solving a system in matrix form.

7. (a) What is meant by an inconsistent system?
   (b) What is meant by a dependent system?

8. Suppose you have used Gaussian elimination to transform the augmented matrix of a linear system into row-echelon form. How can you tell whether the system has
   (a) exactly one solution?
   (b) no solution?
   (c) infinitely many solutions?

9. How can you tell whether a matrix is in reduced row-echelon form?

10. How do Gaussian elimination and Gauss-Jordan elimination differ? What advantage does Gauss-Jordan elimination have?

11. If A and B are matrices with the same dimension and k is a real number, how do you find $A + B$, $A - B$, and kA?

12. (a) What must be true of the dimensions of A and B for the product AB to be defined?
    (b) If the product AB is defined, how do you calculate it?

13. (a) What is the identity matrix I_n?
    (b) If A is a square $n \times n$ matrix, what is its inverse matrix?
    (c) Write a formula for the inverse of a 2×2 matrix.
    (d) Explain how you would find the inverse of a 3×3 matrix.

14. (a) Explain how to express a linear system as a matrix equation of the form $AX = B$.
    (b) If A has an inverse, how would you solve the matrix equation $AX = B$?

15. Suppose A is an $n \times n$ matrix.
    (a) What is the minor M_{ij} of the element a_{ij}?
    (b) What is the cofactor A_{ij}?
    (c) How do you find the determinant of A?
    (d) How can you tell if A has an inverse?

16. State Cramer's Rule for solving a system of linear equations in terms of determinants. Do you prefer to use Cramer's Rule or Gaussian elimination? Explain.

17. Explain how to find the partial fraction decomposition of a rational expression. Include in your explanation a discussion of each of the four cases that arise.

18. How do you graph an inequality in two variables?

19. How do you graph the solution set of a system of inequalities?

EXERCISES

1–6 ■ Solve the system of equations, and graph the lines.

1. $\begin{cases} 3x - y = 5 \\ 2x + y = 5 \end{cases}$ **2.** $\begin{cases} y = 2x + 6 \\ y = -x + 3 \end{cases}$

3. $\begin{cases} 2x - 7y = 28 \\ y = \frac{2}{7}x - 4 \end{cases}$ **4.** $\begin{cases} 6x - 8y = 15 \\ -\frac{3}{2}x + 2y = -4 \end{cases}$

5. $\begin{cases} 2x - y = 1 \\ x + 3y = 10 \\ 3x + 4y = 15 \end{cases}$ **6.** $\begin{cases} 2x + 5y = 9 \\ -x + 3y = 1 \\ 7x - 2y = 14 \end{cases}$

7–10 ■ Solve the system of equations.

7. $\begin{cases} y = x^2 + 2x \\ y = 6 + x \end{cases}$ **8.** $\begin{cases} x^2 + y^2 = 8 \\ y = x + 2 \end{cases}$

9. $\begin{cases} 3x + \dfrac{4}{y} = 6 \\ x - \dfrac{8}{y} = 4 \end{cases}$ **10.** $\begin{cases} x^2 + y^2 = 10 \\ x^2 + 2y^2 - 7y = 0 \end{cases}$

11–14 ■ Use a graphing device to solve the system, rounded to the nearest hundredth.

11. $\begin{cases} 0.32x + 0.43y = 0 \\ 7x - 12y = 341 \end{cases}$ **12.** $\begin{cases} \sqrt{12}x - 3\sqrt{2}y = 660 \\ 7137x + 3931y = 20{,}000 \end{cases}$

13. $\begin{cases} x - y^2 = 10 \\ x = \frac{1}{22}y + 12 \end{cases}$ **14.** $\begin{cases} y = 5^x + x \\ y = x^5 + 5 \end{cases}$

15–20 ■ A matrix is given.

(a) State the dimension of the matrix.
(b) Is the matrix in row-echelon form?
(c) Is the matrix in reduced row-echelon form?
(d) Write the system of equations for which the given matrix is the augmented matrix.

15. $\begin{bmatrix} 1 & 2 & -5 \\ 0 & 1 & 3 \end{bmatrix}$ **16.** $\begin{bmatrix} 1 & 0 & 6 \\ 0 & 1 & 0 \end{bmatrix}$

17. $\begin{bmatrix} 1 & 0 & 8 & 0 \\ 0 & 1 & 5 & -1 \\ 0 & 0 & 0 & 0 \end{bmatrix}$ **18.** $\begin{bmatrix} 1 & 3 & 6 & 2 \\ 2 & 1 & 0 & 5 \\ 0 & 0 & 1 & 0 \end{bmatrix}$

19. $\begin{bmatrix} 0 & 1 & -3 & 4 \\ 1 & 1 & 0 & 7 \\ 1 & 2 & 1 & 2 \end{bmatrix}$ **20.** $\begin{bmatrix} 1 & 8 & 6 & -4 \\ 0 & 1 & -3 & 5 \\ 0 & 0 & 2 & -7 \\ 1 & 1 & 1 & 0 \end{bmatrix}$

21–42 ■ Find the complete solution of the system, or show that the system has no solution.

21. $\begin{cases} x + y + 2z = 6 \\ 2x + 5z = 12 \\ x + 2y + 3z = 9 \end{cases}$ **22.** $\begin{cases} x - 2y + 3z = 1 \\ x - 3y - z = 0 \\ 2x - 6z = 6 \end{cases}$

23. $\begin{cases} x - 2y + 3z = 1 \\ 2x - y + z = 3 \\ 2x - 7y + 11z = 2 \end{cases}$ **24.** $\begin{cases} x + y + z + w = 2 \\ 2x - 3z = 5 \\ x - 2y + 4w = 9 \\ x + y + 2z + 3w = 5 \end{cases}$

25. $\begin{cases} x + 2y + 2z = 6 \\ x - y = -1 \\ 2x + y + 3z = 7 \end{cases}$ **26.** $\begin{cases} x - y + z = 2 \\ x + y + 3z = 6 \\ 2y + 3z = 5 \end{cases}$

27. $\begin{cases} x - 2y + 3z = -2 \\ 2x - y + z = 2 \\ 2x - 7y + 11z = -9 \end{cases}$ **28.** $\begin{cases} x - y + z = 2 \\ x + y + 3z = 6 \\ 3x - y + 5z = 10 \end{cases}$

29. $\begin{cases} x + y + z + w = 0 \\ x - y - 4z - w = -1 \\ x - 2y + 4w = -7 \\ 2x + 2y + 3z + 4w = -3 \end{cases}$

30. $\begin{cases} x + 3z = -1 \\ y - 4w = 5 \\ 2y + z + w = 0 \\ 2x + y + 5z - 4w = 4 \end{cases}$

31. $\begin{cases} x - 3y + z = 4 \\ 4x - y + 15z = 5 \end{cases}$ **32.** $\begin{cases} 2x - 3y + 4z = 3 \\ 4x - 5y + 9z = 13 \\ 2x + 7z = 0 \end{cases}$

33. $\begin{cases} -x + 4y + z = 8 \\ 2x - 6y + z = -9 \\ x - 6y - 4z = -15 \end{cases}$ **34.** $\begin{cases} x - z + w = 2 \\ 2x + y - 2w = 12 \\ 3y + z + w = 4 \\ x + y - z = 10 \end{cases}$

35. $\begin{cases} x - y + 3z = 2 \\ 2x + y + z = 2 \\ 3x + 4z = 4 \end{cases}$ **36.** $\begin{cases} x - y = 1 \\ x + y + 2z = 3 \\ x - 3y - 2z = -1 \end{cases}$

37. $\begin{cases} x - y + z - w = 0 \\ 3x - y - z - w = 2 \end{cases}$ **38.** $\begin{cases} x - y = 3 \\ 2x + y = 6 \\ x - 2y = 9 \end{cases}$

39. $\begin{cases} x - y + z = 0 \\ 3x + 2y - z = 6 \\ x + 4y - 3z = 3 \end{cases}$ **40.** $\begin{cases} x + 2y + 3z = 2 \\ 2x - y - 5z = 1 \\ 4x + 3y + z = 6 \end{cases}$

41. $\begin{cases} x + y - z - w = 2 \\ x - y + z - w = 0 \\ 2x + 2w = 2 \\ 2x + 4y - 4z - 2w = 6 \end{cases}$

42. $\begin{cases} x - y - 2z + 3w = 0 \\ y - z + w = 1 \\ 3x - 2y - 7z + 10w = 2 \end{cases}$

43. A man invests his savings in two accounts, one paying 6% interest per year and the other paying 7%. He has twice as much invested in the 7% account as in the 6% account, and his annual interest income is $600. How much is invested in each account?

44. A piggy bank contains 50 coins, all of them nickels, dimes, or quarters. The total value of the coins is $5.60, and the value of the dimes is five times the value of the nickels. How many coins of each type are there?

45. Clarisse invests $60,000 in money-market accounts at three different banks. Bank A pays 2% interest per year, bank B pays 2.5%, and bank C pays 3%. She decides to invest twice as much in bank B as in the other two banks. After one year, Clarisse has earned \$1575 in interest. How much did she invest in each bank?

46. A commercial fisherman fishes for haddock, sea bass, and red snapper. He is paid \$1.25 a pound for haddock, \$0.75 a pound for sea bass, and \$2.00 a pound for red snapper. Yesterday he caught 560 lb of fish worth \$575. The haddock and red snapper together are worth \$320. How many pounds of each fish did he catch?

47–58 ■ Let

$$A = \begin{bmatrix} 2 & 0 & -1 \end{bmatrix} \qquad B = \begin{bmatrix} 1 & 2 & 4 \\ -2 & 1 & 0 \end{bmatrix}$$

$$C = \begin{bmatrix} \frac{1}{2} & 3 \\ 2 & \frac{3}{2} \\ -2 & 1 \end{bmatrix} \qquad D = \begin{bmatrix} 1 & 4 \\ 0 & -1 \\ 2 & 0 \end{bmatrix}$$

$$E = \begin{bmatrix} 2 & -1 \\ -\frac{1}{2} & 1 \end{bmatrix} \qquad F = \begin{bmatrix} 4 & 0 & 2 \\ -1 & 1 & 0 \\ 7 & 5 & 0 \end{bmatrix}$$

$$G = \begin{bmatrix} 5 \end{bmatrix}$$

Carry out the indicated operation, or explain why it cannot be performed.

47. $A + B$ **48.** $C - D$ **49.** $2C + 3D$

50. $5B - 2C$ **51.** GA **52.** AG

53. BC **54.** CB **55.** BF

56. FC **57.** $(C + D)E$ **58.** $F(2C - D)$

59–60 ■ Verify that the matrices A and B are inverses of each other by calculating the products AB and BA.

59. $A = \begin{bmatrix} 2 & -5 \\ -2 & 6 \end{bmatrix}, \quad B = \begin{bmatrix} 3 & \frac{5}{2} \\ 1 & 1 \end{bmatrix}$

60. $A = \begin{bmatrix} 2 & -1 & 3 \\ 2 & -2 & 1 \\ 0 & 1 & 1 \end{bmatrix}, \quad B = \begin{bmatrix} -\frac{3}{2} & 2 & \frac{5}{2} \\ -1 & 1 & 2 \\ 1 & -1 & -1 \end{bmatrix}$

61–66 ■ Solve the matrix equation for the unknown matrix, X, or show that no solution exists, where

$$A = \begin{bmatrix} 2 & 1 \\ 3 & 2 \end{bmatrix}, \quad B = \begin{bmatrix} 1 & -2 \\ -2 & 4 \end{bmatrix}, \quad C = \begin{bmatrix} 0 & 1 & 3 \\ -2 & 4 & 0 \end{bmatrix}$$

61. $A + 3X = B$ **62.** $\frac{1}{2}(X - 2B) = A$

63. $2(X - A) = 3B$ **64.** $2X + C = 5A$

65. $AX = C$ **66.** $AX = B$

67–74 ■ Find the determinant and, if possible, the inverse of the matrix.

67. $\begin{bmatrix} 1 & 4 \\ 2 & 9 \end{bmatrix}$ **68.** $\begin{bmatrix} 2 & 2 \\ 1 & -3 \end{bmatrix}$

69. $\begin{bmatrix} 4 & -12 \\ -2 & 6 \end{bmatrix}$ **70.** $\begin{bmatrix} 2 & 4 & 0 \\ -1 & 1 & 2 \\ 0 & 3 & 2 \end{bmatrix}$

71. $\begin{bmatrix} 3 & 0 & 1 \\ 2 & -3 & 0 \\ 4 & -2 & 1 \end{bmatrix}$ **72.** $\begin{bmatrix} 1 & 2 & 3 \\ 2 & 4 & 5 \\ 2 & 5 & 6 \end{bmatrix}$

73. $\begin{bmatrix} 1 & 0 & 0 & 1 \\ 0 & 2 & 0 & 2 \\ 0 & 0 & 3 & 3 \\ 0 & 0 & 0 & 4 \end{bmatrix}$ **74.** $\begin{bmatrix} 1 & 0 & 1 & 0 \\ 0 & 1 & 0 & 1 \\ 1 & 1 & 1 & 2 \\ 1 & 2 & 1 & 2 \end{bmatrix}$

75–78 ■ Express the system of linear equations as a matrix equation. Then solve the matrix equation by multiplying each side by the inverse of the coefficient matrix.

75. $\begin{cases} 12x - 5y = 10 \\ 5x - 2y = 17 \end{cases}$ **76.** $\begin{cases} 6x - 5y = 1 \\ 8x - 7y = -1 \end{cases}$

77. $\begin{cases} 2x + y + 5z = \frac{1}{3} \\ x + 2y + 2z = \frac{1}{4} \\ x + 3z = \frac{1}{6} \end{cases}$ **78.** $\begin{cases} 2x + 3z = 5 \\ x + y + 6z = 0 \\ 3x - y + z = 5 \end{cases}$

79–82 ■ Solve the system using Cramer's Rule.

79. $\begin{cases} 2x + 7y = 13 \\ 6x + 16y = 30 \end{cases}$

80. $\begin{cases} 12x - 11y = 140 \\ 7x + 9y = 20 \end{cases}$

81. $\begin{cases} 2x - y + 5z = 0 \\ -x + 7y = 9 \\ 5x + 4y + 3z = -9 \end{cases}$

82. $\begin{cases} 3x + 4y - z = 10 \\ x - 4z = 20 \\ 2x + y + 5z = 30 \end{cases}$

83–84 ■ Use the determinant formula for the area of a triangle to find the area of the triangle in the figure.

83.

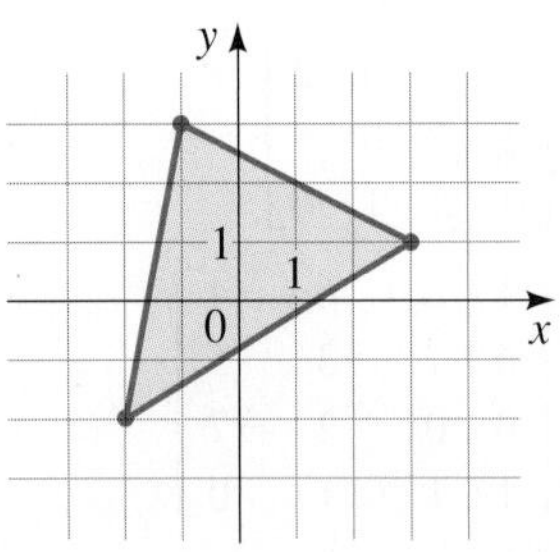

84.

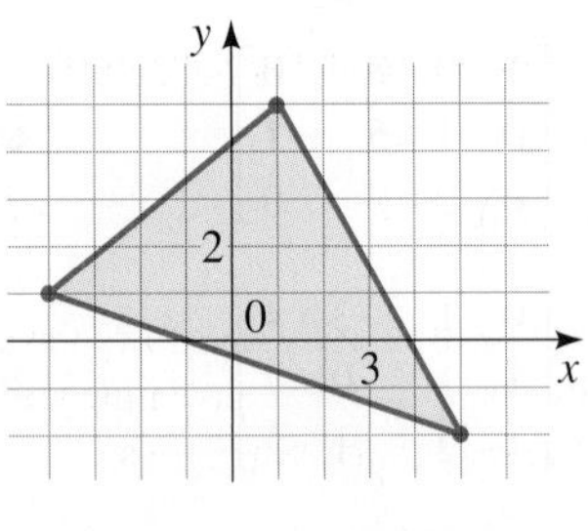

85–90 ■ Find the partial fraction decomposition of the rational function.

85. $\dfrac{3x + 1}{x^2 - 2x - 15}$ **86.** $\dfrac{8}{x^3 - 4x}$

87. $\dfrac{2x - 4}{x(x - 1)^2}$ **88.** $\dfrac{x + 6}{x^3 - 2x^2 + 4x - 8}$

89. $\dfrac{2x - 1}{x^3 + x}$ **90.** $\dfrac{5x^2 - 3x + 10}{x^4 + x^2 - 2}$

91–94 ■ Two equations and their graphs are given. Find the intersection point(s) of the graphs by solving the system.

91. $\begin{cases} 2x + 3y = 7 \\ x - 2y = 0 \end{cases}$

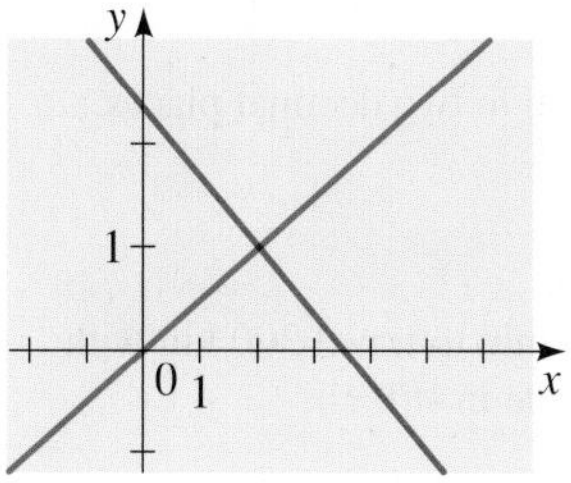

92. $\begin{cases} 3x + y = 8 \\ y = x^2 - 5x \end{cases}$

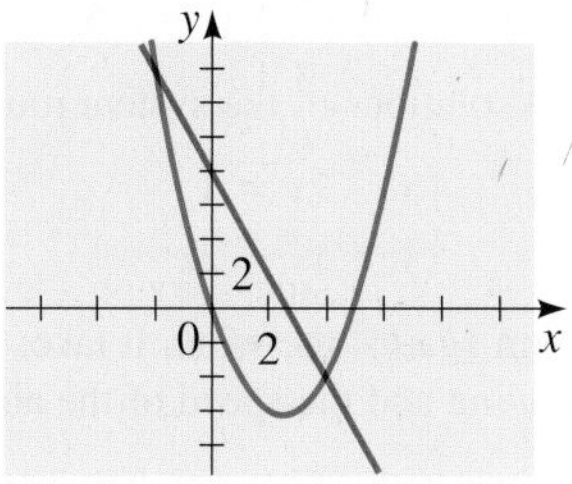

93. $\begin{cases} x^2 + y = 2 \\ x^2 - 3x - y = 0 \end{cases}$

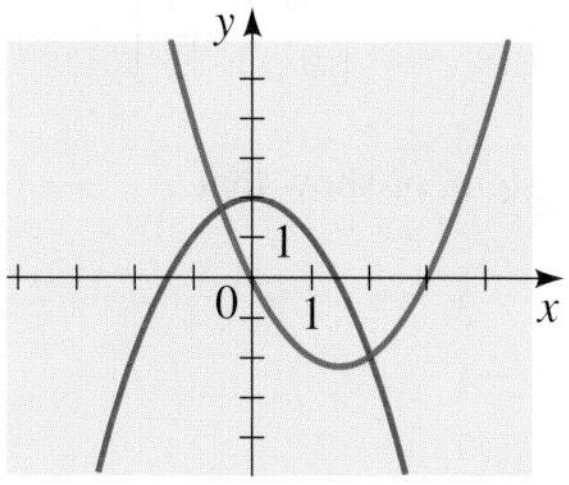

94. $\begin{cases} x - y = -2 \\ x^2 + y^2 - 4y = 4 \end{cases}$

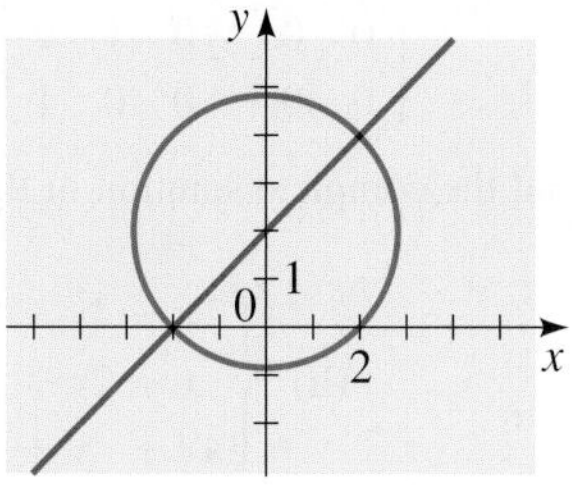

95–96 ■ An equation and its graph are given. Find an inequality whose solution is the shaded region.

95. $x + y^2 = 4$

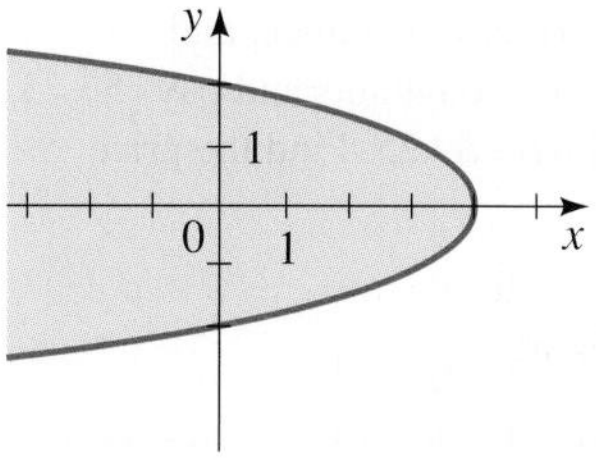

96. $x^2 + y^2 = 8$

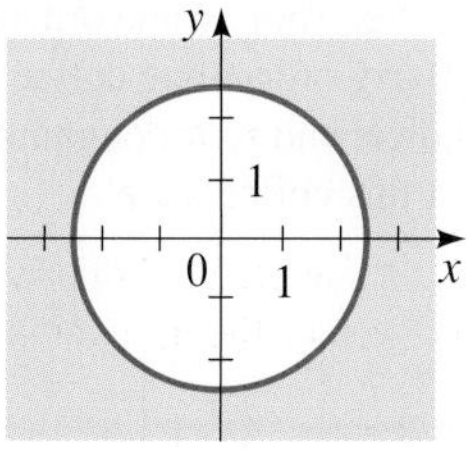

97–100 ■ Graph the inequality.

97. $3x + y \le 6$

98. $y \ge x^2 - 3$

99. $x^2 + y^2 > 9$

100. $x - y^2 < 4$

101–104 ■ The figure shows the graphs of the equations corresponding to the given inequalities. Shade the solution set of the system of inequalities.

101. $\begin{cases} y \ge x^2 - 3x \\ y \le \frac{1}{3}x - 1 \end{cases}$

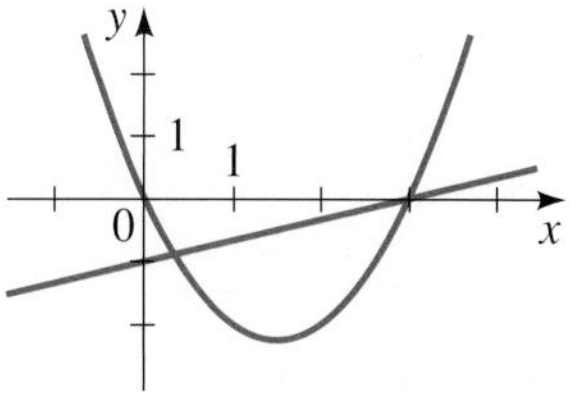

102. $\begin{cases} y \ge x - 1 \\ x^2 + y^2 \le 1 \end{cases}$

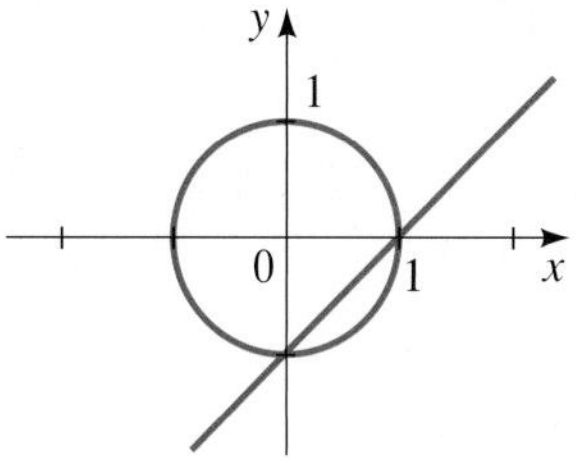

103. $\begin{cases} x + y \ge 2 \\ y - x \le 2 \\ x \le 3 \end{cases}$

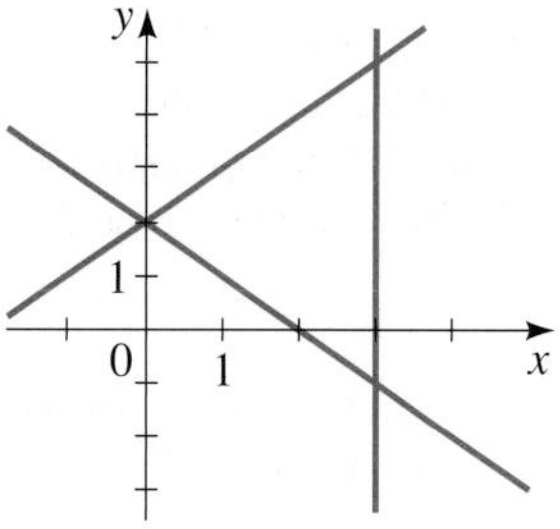

104. $\begin{cases} y \ge -2x \\ y \le 2x \\ y \le -\frac{1}{2}x + 2 \end{cases}$

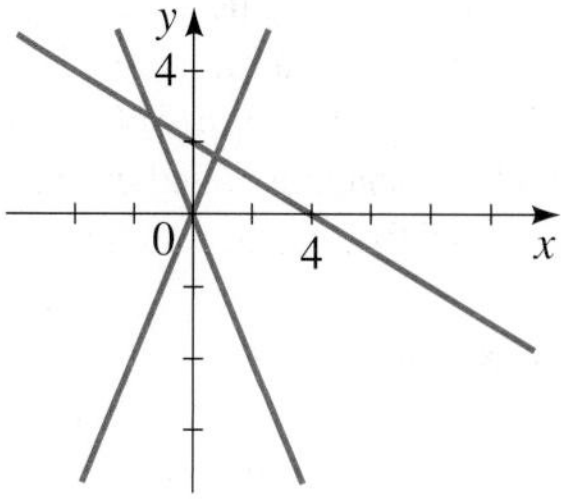

105–108 ■ Graph the solution set of the system of inequalities. Find the coordinates of all vertices, and determine whether the solution set is bounded or unbounded.

105. $\begin{cases} x^2 + y^2 < 9 \\ x + y < 0 \end{cases}$

106. $\begin{cases} y - x^2 \ge 4 \\ y < 20 \end{cases}$

107. $\begin{cases} x \ge 0, \quad y \ge 0 \\ x + 2y \le 12 \\ y \le x + 4 \end{cases}$

108. $\begin{cases} x \ge 4 \\ x + y \ge 24 \\ x \le 2y + 12 \end{cases}$

109–110 ■ Solve for x, y, and z in terms of a, b, and c.

109. $\begin{cases} -x + y + z = a \\ x - y + z = b \\ x + y - z = c \end{cases}$

110. $\begin{cases} ax + by + cz = a - b + c \\ bx + by + cz = c \\ cx + cy + cz = c \end{cases} \qquad (a \ne b, b \ne c, c \ne 0)$

111. For what values of k do the following three lines have a common point of intersection?

$$x + y = 12$$
$$kx - y = 0$$
$$y - x = 2k$$

112. For what value of k does the following system have infinitely many solutions?

$$\begin{cases} kx + y + z = 0 \\ x + 2y + kz = 0 \\ -x + 3z = 0 \end{cases}$$

1–2 ■ A system of equations is given. **(a)** Determine whether the system is linear or nonlinear. **(b)** Find all solutions of the system.

1. $\begin{cases} x + 3y = 7 \\ 5x + 2y = -4 \end{cases}$ **2.** $\begin{cases} 6x + y^2 = 10 \\ 3x - y = 5 \end{cases}$

3. Use a graphing device to find all solutions of the system rounded to two decimal places.

$$\begin{cases} x - 2y = 1 \\ y = x^3 - 2x^2 \end{cases}$$

4. In $2\frac{1}{2}$ h an airplane travels 600 km against the wind. It takes 50 min to travel 300 km with the wind. Find the speed of the wind and the speed of the airplane in still air.

5. Determine whether each matrix is in reduced row-echelon form, row-echelon form, or neither.

(a) $\begin{bmatrix} 1 & 2 & 4 & -6 \\ 0 & 1 & -3 & 0 \end{bmatrix}$ **(b)** $\begin{bmatrix} 1 & 0 & -1 & 0 & 0 \\ 0 & 1 & 3 & 0 & 0 \\ 0 & 0 & 0 & 1 & 0 \\ 0 & 0 & 0 & 0 & 1 \end{bmatrix}$ **(c)** $\begin{bmatrix} 1 & 1 & 0 \\ 0 & 0 & 1 \\ 0 & 1 & 3 \end{bmatrix}$

6. Use Gaussian elimination to find the complete solution of the system, or show that no solution exists.

(a) $\begin{cases} x - y + 2z = 0 \\ 2x - 4y + 5z = -5 \\ 2y - 3z = 5 \end{cases}$ **(b)** $\begin{cases} 2x - 3y + z = 3 \\ x + 2y + 2z = -1 \\ 4x + y + 5z = 4 \end{cases}$

7. Use Gauss-Jordan elimination to find the complete solution of the system.

$$\begin{cases} x + 3y - z = 0 \\ 3x + 4y - 2z = -1 \\ -x + 2y = 1 \end{cases}$$

8. Anne, Barry, and Cathy enter a coffee shop. Anne orders two coffees, one juice, and two doughnuts and pays \$6.25. Barry orders one coffee and three doughnuts and pays \$3.75. Cathy orders three coffees, one juice, and four doughnuts and pays \$9.25. Find the price of coffee, juice, and doughnuts at this coffee shop.

9. Let

$$A = \begin{bmatrix} 2 & 3 \\ 2 & 4 \end{bmatrix} \qquad B = \begin{bmatrix} 2 & 4 \\ -1 & 1 \\ 3 & 0 \end{bmatrix} \qquad C = \begin{bmatrix} 1 & 0 & 4 \\ -1 & 1 & 2 \\ 0 & 1 & 3 \end{bmatrix}$$

Carry out the indicated operation, or explain why it cannot be performed.

(a) $A + B$ **(b)** AB **(c)** $BA - 3B$ **(d)** CBA

(e) A^{-1} **(f)** B^{-1} **(g)** $\det(B)$ **(h)** $\det(C)$

10. **(a)** Write a matrix equation equivalent to the following system.

$$\begin{cases} 4x - 3y = 10 \\ 3x - 2y = 30 \end{cases}$$

(b) Find the inverse of the coefficient matrix, and use it to solve the system.

11. Only one of the following matrices has an inverse. Find the determinant of each matrix, and use the determinants to identify the one that has an inverse. Then find the inverse.

$$A = \begin{bmatrix} 1 & 4 & 1 \\ 0 & 2 & 0 \\ 1 & 0 & 1 \end{bmatrix} \qquad B = \begin{bmatrix} 1 & 4 & 0 \\ 0 & 2 & 0 \\ -3 & 0 & 1 \end{bmatrix}$$

12. Solve using Cramer's Rule:

$$\begin{cases} 2x \qquad\quad - z = 14 \\ 3x - y + 5z = 0 \\ 4x + 2y + 3z = -2 \end{cases}$$

13. Find the partial fraction decomposition of the rational function.

(a) $\dfrac{4x - 1}{(x - 1)^2(x + 2)}$ **(b)** $\dfrac{2x - 3}{x^3 + 3x}$

14. Graph the solution set of the system of inequalities. Label the vertices with their coordinates.

(a) $\begin{cases} 2x + y \le 8 \\ x - y \ge -2 \\ x + 2y \ge 4 \end{cases}$ **(b)** $\begin{cases} x^2 + y \le 5 \\ y \le 2x + 5 \end{cases}$

Linear Programming

Linear programming is a modeling technique that is used to determine the optimal allocation of resources in business, the military, and other areas of human endeavor. For example, a manufacturer who makes several different products from the same raw materials can use linear programming to determine how much of each product should be produced to maximize the profit. This modeling technique is probably the most important practical application of systems of linear inequalities. In 1975 Leonid Kantorovich and T. C. Koopmans won the Nobel Prize in economics for their work in the development of this technique.

Although linear programming can be applied to very complex problems with hundreds or even thousands of variables, we consider only a few simple examples to which the graphical methods of Section 10.9 can be applied. (For large numbers of variables a linear programming method based on matrices is used.) Let's examine a typical problem.

EXAMPLE 1 | Manufacturing for Maximum Profit

A small shoe manufacturer makes two styles of shoes: oxfords and loafers. Two machines are used in the process: a cutting machine and a sewing machine. Each type of shoe requires 15 min per pair on the cutting machine. Oxfords require 10 min of sewing per pair, and loafers require 20 min of sewing per pair. Because the manufacturer can hire only one operator for each machine, each process is available for just 8 hours per day. If the profit is \$15 on each pair of oxfords and \$20 on each pair of loafers, how many pairs of each type should be produced per day for maximum profit?

Because loafers produce more profit, it would seem best to manufacture only loafers. Surprisingly, this does not turn out to be the most profitable solution.

SOLUTION First we organize the given information into a table. To be consistent, let's convert all times to hours.

	Oxfords	Loafers	Time available
Time on cutting machine (h)	$\frac{1}{4}$	$\frac{1}{4}$	8
Time on sewing machine (h)	$\frac{1}{6}$	$\frac{1}{3}$	8
Profit	\$15	\$20	

We describe the model and solve the problem in four steps.

▶ **Choose the Variables.** To make a mathematical model, we first give names to the variable quantities. For this problem we let

$$x = \text{number of pairs of oxfords made daily}$$

$$y = \text{number of pairs of loafers made daily}$$

▶ **Find the Objective Function.** Our goal is to determine which values for x and y give maximum profit. Since each pair of oxfords provides \$15 profit and each pair of loafers \$20, the total profit is given by

$$P = 15x + 20y$$

This function is called the *objective function*.

▶ **Graph the Feasible Region.** The larger x and y are, the greater is the profit. But we cannot choose arbitrarily large values for these variables because of the restrictions, or *constraints*, in the problem. Each restriction is an inequality in the variables.

In this problem the total number of cutting hours needed is $\frac{1}{4}x + \frac{1}{4}y$. Since only 8 hours are available on the cutting machine, we have

$$\tfrac{1}{4}x + \tfrac{1}{4}y \le 8$$

Similarly, by considering the amount of time needed and available on the sewing machine, we get

$$\tfrac{1}{6}x + \tfrac{1}{3}y \le 8$$

We cannot produce a negative number of shoes, so we also have

$$x \ge 0 \qquad \text{and} \qquad y \ge 0$$

Thus, x and y must satisfy the constraints

$$\begin{cases} \frac{1}{4}x + \frac{1}{4}y \le 8 \\ \frac{1}{6}x + \frac{1}{3}y \le 8 \\ x \ge 0 \\ y \ge 0 \end{cases}$$

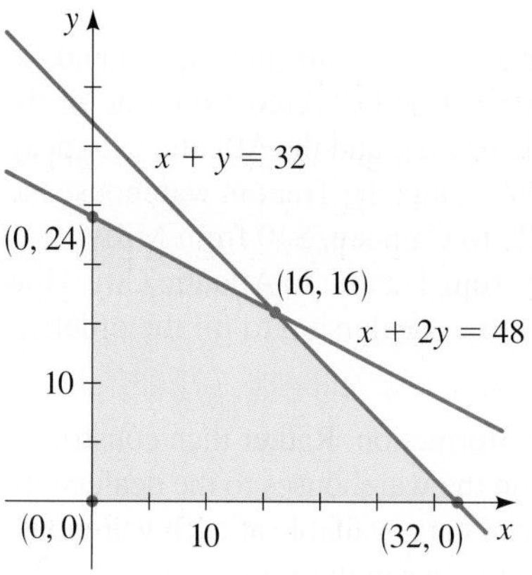

FIGURE 1

If we multiply the first inequality by 4 and the second by 6, we obtain the simplified system

$$\begin{cases} x + \ \ y \le 32 \\ x + 2y \le 48 \\ x \le 0 \\ y \le 0 \end{cases}$$

The solution of this system (with vertices labeled) is sketched in Figure 1. The only values that satisfy the restrictions of the problem are the ones that correspond to points of the shaded region in Figure 1. This is called the *feasible region* for the problem.

▶ **Find the Maximum Profit.** As x or y increases, profit increases as well. Thus it seems reasonable that the maximum profit will occur at a point on one of the outside edges of the feasible region, where it is impossible to increase x or y without going outside the region. In fact, it can be shown that the maximum value occurs at a vertex. This means that we need to check the profit only at the vertices. The largest value of P occurs at the point (16, 16), where $P = \$560$. Thus the manufacturer should make 16 pairs of oxfords and 16 pairs of loafers, for a maximum daily profit of \$560.

Linear Programming helps the telephone industry to determine the most efficient way to route telephone calls. The computerized routing decisions must be made very rapidly so that callers are not kept waiting for connections. Since the database of customers and routes is huge, an extremely fast method for solving linear programming problems is essential. In 1984 the 28-year-old mathematician **Narendra Karmarkar**, working at Bell Labs in Murray Hill, New Jersey, discovered just such a method. His idea is so ingenious and his method so fast that the discovery caused a sensation in the mathematical world. Although mathematical discoveries rarely make the news, this one was reported in *Time*, on December 3, 1984. Today airlines routinely use Karmarkar's technique to minimize costs in scheduling passengers, flight personnel, fuel, baggage, and maintenance workers.

Vertex	$P = 15x + 20y$	
(0, 0)	0	
(0, 24)	$15(0) + 20(24) = \$480$	
(16, 16)	$15(16) + 20(16) = \$560$	Maximum profit
(32, 0)	$15(32) + 20(0) = \$480$	

The linear programming problems that we consider all follow the pattern of Example 1. Each problem involves two variables. The problem describes restrictions, called **constraints**, that lead to a system of linear inequalities whose solution is called the **feasible region**. The function that we wish to maximize or minimize is called the **objective function**. This function always attains its largest and smallest values at the **vertices** of the feasible region. This modeling technique involves four steps, summarized in the following box.

GUIDELINES FOR LINEAR PROGRAMMING

1. **Choose the Variables.** Decide what variable quantities in the problem should be named x and y.
2. **Find the Objective Function.** Write an expression for the function we want to maximize or minimize.
3. **Graph the Feasible Region.** Express the constraints as a system of inequalities, and graph the solution of this system (the feasible region).
4. **Find the Maximum or Minimum.** Evaluate the objective function at the vertices of the feasible region to determine its maximum or minimum value.

EXAMPLE 2 | A Shipping Problem

A car dealer has warehouses in Millville and Trenton and dealerships in Camden and Atlantic City. Every car that is sold at the dealerships must be delivered from one of the warehouses. On a certain day the Camden dealers sell 10 cars, and the Atlantic City dealers sell 12. The Millville warehouse has 15 cars available, and the Trenton warehouse has 10. The cost of shipping one car is \$50 from Millville to Camden, \$40 from Millville to Atlantic City, \$60 from Trenton to Camden, and \$55 from Trenton to Atlantic City. How many cars should be moved from each warehouse to each dealership to fill the orders at minimum cost?

SOLUTION Our first step is to organize the given information. Rather than construct a table, we draw a diagram to show the flow of cars from the warehouses to the dealerships (see Figure 2 below). The diagram shows the number of cars available at each warehouse or required at each dealership and the cost of shipping between these locations.

▶ **Choose the Variables.** The arrows in Figure 2 indicate four possible routes, so the problem seems to involve four variables. But we let

$$x = \text{number of cars to be shipped from Millville to Camden}$$

$$y = \text{number of cars to be shipped from Millville to Atlantic City}$$

To fill the orders, we must have

$$10 - x = \text{number of cars shipped from Trenton to Camden}$$

$$12 - y = \text{number of cars shipped from Trenton to Atlantic City}$$

So the only variables in the problem are x and y.

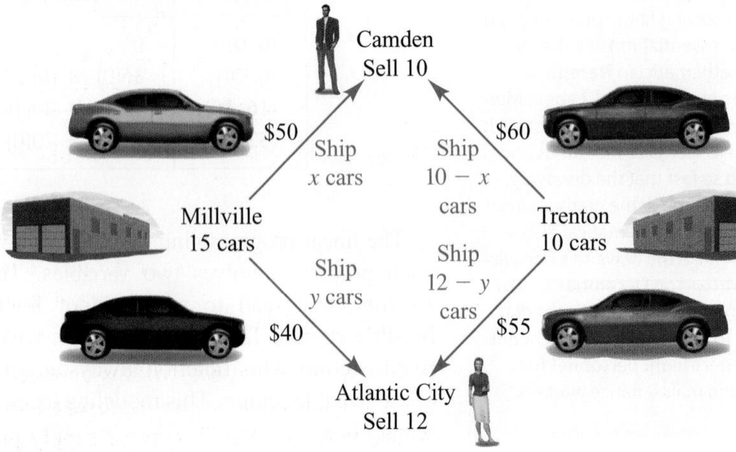

FIGURE 2

▶ **Find the Objective Function.** The objective of this problem is to minimize cost. From Figure 2 we see that the total cost C of shipping the cars is

$$\begin{aligned} C &= 50x + 40y + 60(10 - x) + 55(12 - y) \\ &= 50x + 40y + 600 - 60x + 660 - 55y \\ &= 1260 - 10x - 15y \end{aligned}$$

This is the objective function.

▶ **Graph the Feasible Region.** Now we derive the constraint inequalities that define the feasible region. First, the number of cars shipped on each route can't be negative, so we have

$$\begin{aligned} x &\geq 0 & y &\geq 0 \\ 10 - x &\geq 0 & 12 - y &\geq 0 \end{aligned}$$

Second, the total number of cars shipped from each warehouse can't exceed the number of cars available there, so

$$\begin{aligned} x + y &\leq 15 \\ (10 - x) + (12 - y) &\leq 10 \end{aligned}$$

Simplifying the latter inequality, we get

$$\begin{aligned} 22 - x - y &\leq 10 \\ -x - y &\leq -12 \\ x + y &\geq 12 \end{aligned}$$

The inequalities $10 - x \geq 0$ and $12 - y \geq 0$ can be rewritten as $x \leq 10$ and $y \leq 12$. Thus the feasible region is described by the constraints

$$\begin{cases} x + y \leq 15 \\ x + y \geq 12 \\ 0 \leq x \leq 10 \\ 0 \leq y \leq 12 \end{cases}$$

The feasible region is graphed in Figure 3.

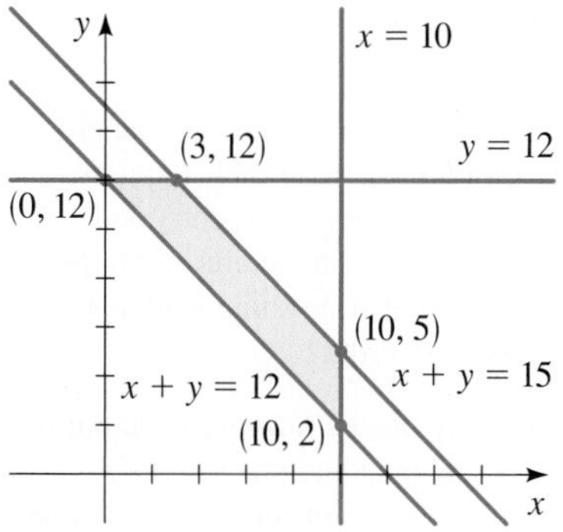

FIGURE 3

▶ **Find the Minimum Cost.** We check the value of the objective function at each vertex of the feasible region.

Vertex	$C = 1260 - 10x - 15y$	
(0, 12)	$1260 - 10(0) - 15(12) = \$1080$	
(3, 12)	$1260 - 10(3) - 15(12) = \$1050$	Minimum cost
(10, 5)	$1260 - 10(10) - 15(5) = \$1085$	
(10, 2)	$1260 - 10(10) - 15(2) = \$1130$	

The lowest cost is incurred at the point (3, 12). Thus, the dealer should ship

3 cars from Millville to Camden
12 cars from Millville to Atlantic City
7 cars from Trenton to Camden
0 cars from Trenton to Atlantic City ■

In the 1940s mathematicians developed matrix methods for solving linear programming problems that involve more than two variables. These methods were first used by the Allies in World War II to solve supply problems similar to (but, of course, much more complicated than) Example 2. Improving such matrix methods is an active and exciting area of current mathematical research.

PROBLEMS

1–4 ■ Find the maximum and minimum values of the given objective function on the indicated feasible region.

1. $M = 200 - x - y$

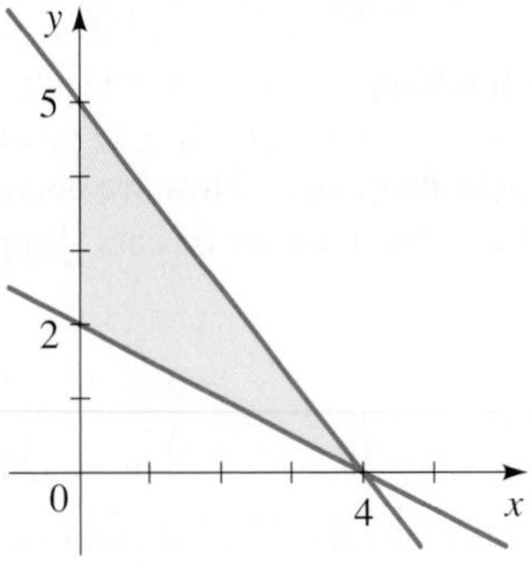

2. $N = \frac{1}{2}x + \frac{1}{4}y + 40$

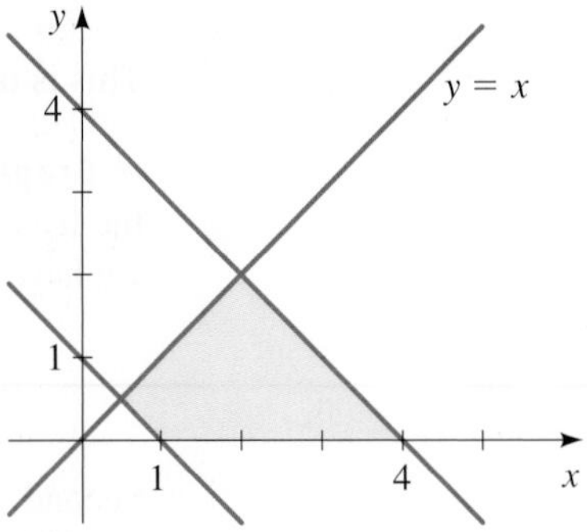

3. $P = 140 - x + 3y$

$$\begin{cases} x \geq 0,\ y \geq 0 \\ 2x + y \leq 10 \\ 2x + 4y \leq 28 \end{cases}$$

4. $Q = 70x + 82y$

$$\begin{cases} x \geq 0,\ y \geq 0 \\ x \leq 10,\ y \leq 20 \\ x + y \geq 5 \\ x + 2y \leq 18 \end{cases}$$

5. Making Furniture A furniture manufacturer makes wooden tables and chairs. The production process involves two basic types of labor: carpentry and finishing. A table requires 2 hours of carpentry and 1 hour of finishing, and a chair requires 3 hours of carpentry and $\frac{1}{2}$ hour of finishing. The profit is \$35 per table and \$20 per chair. The manufacturer's employees can supply a maximum of 108 hours of carpentry work and 20 hours of finishing work per day. How many tables and chairs should be made each day to maximize profit?

6. A Housing Development A housing contractor has subdivided a farm into 100 building lots. She has designed two types of homes for these lots: colonial and ranch style. A colonial requires \$30,000 of capital and produces a profit of \$4000 when sold. A ranch-style house requires \$40,000 of capital and provides an \$8000 profit. If the contractor has \$3.6 million of capital on hand, how many houses of each type should she build for maximum profit? Will any of the lots be left vacant?

7. Hauling Fruit A trucker hauls citrus fruit from Florida to Montreal. Each crate of oranges is 4 ft^3 in volume and weighs 80 lb. Each crate of grapefruit has a volume of 6 ft^3 and weighs 100 lb. His truck has a maximum capacity of 300 ft^3 and can carry no more than 5600 lb. Moreover, he is not permitted to carry more crates of grapefruit than crates of oranges. If his profit is \$2.50 on each crate of oranges and \$4 on each crate of grapefruit, how many crates of each fruit should he carry for maximum profit?

8. Manufacturing Calculators A manufacturer of calculators produces two models: standard and scientific. Long-term demand for the two models mandates that the company manufacture at least 100 standard and 80 scientific calculators each day. However, because of limitations on production capacity, no more than 200 standard and 170 scientific calculators can be made daily. To satisfy a shipping contract, a total of at least 200 calculators must be shipped every day.

(a) If the production cost is \$5 for a standard calculator and \$7 for a scientific one, how many of each model should be produced daily to minimize this cost?

(b) If each standard calculator results in a \$2 loss but each scientific one produces a \$5 profit, how many of each model should be made daily to maximize profit?

9. Shipping Stereos An electronics discount chain has a sale on a certain brand of stereo. The chain has stores in Santa Monica and El Toro and warehouses in Long Beach and Pasadena. To satisfy rush orders, 15 sets must be shipped from the warehouses to the Santa Monica store, and 19 must be shipped to the El Toro store. The cost of shipping a set is \$5 from Long Beach to Santa Monica, \$6 from Long Beach to El Toro, \$4 from Pasadena to Santa Monica, and \$5.50 from Pasadena to El Toro. If the Long Beach warehouse has 24 sets and the Pasadena warehouse has 18 sets in stock, how many sets should be shipped from each warehouse to each store to fill the orders at a minimum shipping cost?

10. Delivering Plywood A man owns two building supply stores, one on the east side and one on the west side of a city. Two customers order some $\frac{1}{2}$-inch plywood. Customer A needs 50 sheets, and customer B needs 70 sheets. The east-side store has 80 sheets, and the west-side store has 45 sheets of this plywood in stock. The east-side store's delivery costs per sheet are \$0.50 to customer A and \$0.60 to customer B. The west-side store's delivery costs per sheet are \$0.40 to customer A and \$0.55 to customer B. How many sheets should be shipped from each store to each customer to minimize delivery costs?

11. Packaging Nuts A confectioner sells two types of nut mixtures. The standard-mixture package contains 100 g of cashews and 200 g of peanuts and sells for \$1.95. The deluxe-mixture package contains 150 g of cashews and 50 g of peanuts and sells for \$2.25. The confectioner has 15 kg of cashews and 20 kg of peanuts available. On the basis of past sales, the confectioner needs to have at least as many standard as deluxe packages available. How many bags of each mixture should he package to maximize his revenue?

12. Feeding Lab Rabbits A biologist wishes to feed laboratory rabbits a mixture of two types of foods. Type I contains 8 g of fat, 12 g of carbohydrate, and 2 g of protein per ounce. Type II contains 12 g of fat, 12 g of carbohydrate, and 1 g of protein per ounce. Type I costs \$0.20 per ounce and type II costs \$0.30 per ounce. The rabbits each receive a daily minimum of 24 g of fat, 36 g of carbohydrate, and 4 g of protein, but get no more than 5 oz of food per day. How many ounces of each food type should be fed to each rabbit daily to satisfy the dietary requirements at minimum cost?

13. Investing in Bonds A woman wishes to invest \$12,000 in three types of bonds: municipal bonds paying 7% interest per year, bank investment certificates paying 8%, and high-risk bonds paying 12%. For tax reasons she wants the amount invested in municipal bonds to be at least three times the amount invested in bank certificates. To keep her level of risk manageable, she will invest no more than \$2000 in high-risk bonds. How much should she invest in each type of bond to maximize her annual interest yield? [*Hint:* Let x = amount in municipal bonds and y = amount in bank certificates. Then the amount in high-risk bonds will be $12{,}000 - x - y$.]

14. Annual Interest Yield Refer to Problem 13. Suppose the investor decides to increase the maximum invested in high-risk bonds to \$3000 but leaves the other conditions unchanged. By how much will her maximum possible interest yield increase?

15. Business Strategy A small software company publishes computer games and educational and utility software. Their business strategy is to market a total of 36 new programs each year, at least four of these being games. The number of utility programs published is never more than twice the number of educational programs. On average, the company makes an annual profit of \$5000 on each computer game, \$8000 on each educational program, and \$6000 on each utility program. How many of each type of software should the company publish annually for maximum profit?

16. Feasible Region All parts of this problem refer to the following feasible region and objective function.

$$\begin{cases} x \geq 0 \\ x \geq y \\ x + 2y \leq 12 \\ x + y \leq 10 \end{cases}$$

$$P = x + 4y$$

(a) Graph the feasible region.

(b) On your graph from part (a), sketch the graphs of the linear equations obtained by setting P equal to 40, 36, 32, and 28.

(c) If you continue to decrease the value of P, at which vertex of the feasible region will these lines first touch the feasible region?

(d) Verify that the maximum value of P on the feasible region occurs at the vertex you chose in part (c).

APPENDIX Calculations and Significant Figures

Most of the applied examples and exercises in this book involve approximate values. For example, one exercise states that the moon has a radius of 1074 miles. This does not mean that the moon's radius is exactly 1074 miles but simply that this is the radius rounded to the nearest mile.

One simple method for specifying the accuracy of a number is to state how many **significant digits** it has. The significant digits in a number are the ones from the first nonzero digit to the last nonzero digit (reading from left to right). Thus, 1074 has four significant digits, 1070 has three, 1100 has two, and 1000 has one significant digit. This rule may sometimes lead to ambiguities. For example, if a distance is 200 km to the nearest kilometer, then the number 200 really has three significant digits, not just one. This ambiguity is avoided if we use scientific notation—that is, if we express the number as a multiple of a power of 10:

$$2.00 \times 10^2$$

When working with approximate values, students often make the mistake of giving a final answer with *more* significant digits than the original data. This is incorrect because you cannot "create" precision by using a calculator. The final result can be no more accurate than the measurements given in the problem. For example, suppose we are told that the two shorter sides of a right triangle are measured to be 1.25 and 2.33 inches long. By the Pythagorean Theorem, we find, using a calculator, that the hypotenuse has length

$$\sqrt{1.25^2 + 2.33^2} \approx 2.644125564 \text{ in.}$$

But since the given lengths were expressed to three significant digits, the answer cannot be any more accurate. We can therefore say only that the hypotenuse is 2.64 in. long, rounding to the nearest hundredth.

In general, the final answer should be expressed with the same accuracy as the *least-accurate* measurement given in the statement of the problem. The following rules make this principle more precise.

RULES FOR WORKING WITH APPROXIMATE DATA

1. When multiplying or dividing, round off the final result so that it has as many *significant digits* as the given value with the fewest number of significant digits.
2. When adding or subtracting, round off the final result so that it has its last significant digit in the *decimal place* in which the least-accurate given value has its last significant digit.
3. When taking powers or roots, round off the final result so that it has the same number of *significant digits* as the given value.

As an example, suppose that a rectangular table top is measured to be 122.64 in. by 37.3 in. We express its area and perimeter as follows:

$$\text{Area} = \text{length} \times \text{width} = 122.64 \times 37.3 \approx 4570 \text{ in}^2 \qquad \text{Three significant digits}$$

$$\text{Perimeter} = 2(\text{length} + \text{width}) = 2(122.64 + 37.3) \approx 319.9 \text{ in.} \qquad \text{Tenths digit}$$

Note that in the formula for the perimeter, the value 2 is an exact value, not an approximate measurement. It therefore does not affect the accuracy of the final result. In general, if a problem involves only exact values, we may express the final answer with as many significant digits as we wish.

Note also that to make the final result as accurate as possible, *you should wait until the last step to round off your answer*. If necessary, use the memory feature of your calculator to retain the results of intermediate calculations.

69. 500 mi/h **71.** 50 mi/h (or 240 mi/h) **73.** 6 km/h **75.** 6.4 ft from the fulcrum **77.** 2 ft by 6 ft by 15 ft **79.** 13 in. by 13 in. **81.** 2.88 ft **83.** 16 mi; no **85.** 7.52 ft **87.** 18 ft **89.** 4.55 ft

SECTION 1.7 ■ PAGE 80

1. (a) $<$ **(b)** $\le$ **(c)** $\le$ **(d)** $>$ **2. (a)** True **(b)** False **3. (a)** $[-3, 3]$ **(b)** $(-\infty, -3], [3, \infty)$ **4. (a)** < 3 **(b)** > 3 **5.** $\{\sqrt{2}, 2, 4\}$ **7.** $\{4\}$ **9.** $\{-2, -1, 2, 4\}$
11. $(-\infty, \frac{7}{2}]$ **13.** $(4, \infty)$ **15.** $(-\infty, 2]$ **17.** $(-\infty, -\frac{1}{2})$ **19.** $[1, \infty)$ **21.** $(\frac{16}{3}, \infty)$ **23.** $(-\infty, -18)$ **25.** $(-\infty, -1]$ **27.** $[-3, -1)$ **29.** $(2, 6)$ **31.** $[\frac{9}{2}, 5)$ **33.** $(\frac{15}{2}, \frac{21}{2}]$ **35.** $(-2, 3)$ **37.** $(-\infty, -\frac{7}{2}] \cup [0, \infty)$ **39.** $[-3, 6]$ **41.** $(-\infty, -1] \cup [\frac{1}{2}, \infty)$ **43.** $(-1, 4)$ **45.** $(-\infty, -3) \cup (6, \infty)$ **47.** $(-2, 2)$ **49.** $(-\infty, -2] \cup [1, 3]$ **51.** $(-\infty, -2] \cup (-2, 4)$ **53.** $[-1, 3]$ **55.** $(-2, 0) \cup (2, \infty)$ **57.** $(-\infty, -1) \cup [3, \infty)$ **59.** $(-\infty, -\frac{3}{2})$ **61.** $(-\infty, 5) \cup [16, \infty)$ **63.** $(-2, 0) \cup (2, \infty)$ **65.** $[-2, -1) \cup (0, 1]$ **67.** $[-2, 0) \cup (1, 3]$ **69.** $(-3, -\frac{1}{2}) \cup (2, \infty)$ **71.** $(-\infty, -1) \cup (1, \infty)$ **73.** $[-4, 4]$ **75.** $(-\infty, -\frac{7}{2}) \cup (\frac{7}{2}, \infty)$ **77.** $[2, 8]$ **79.** $[1.3, 1.7]$ **81.** $(-\infty, -1] \cup [\frac{7}{3}, \infty)$ **83.** $(-4, 8)$ **85.** $(-6.001, -5.999)$ **87.** $[-\frac{1}{2}, \frac{3}{2}]$

89. $|x| < 3$ **91.** $|x - 7| \ge 5$ **93.** $|x| \le 2$ **95.** $|x| > 3$ **97.** $|x - 1| \le 3$ **99.** $-\frac{4}{3} \le x \le \frac{4}{3}$ **101.** $x < -2$ or $x > 7$
103. (a) $x \ge \frac{c}{a} + \frac{c}{b}$ **(b)** $\frac{a - c}{b} \le x < \frac{2a - c}{b}$
105. $68 \le F \le 86$ **107.** More than 200 mi **109.** Between 12,000 mi and 14,000 mi **111. (a)** $-\frac{1}{3}P + \frac{560}{3}$ **(b)** From \$215 to \$290 **113.** Distances between 20,000 km and 100,000 km **115.** From 0 s to 3 s **117.** Between 0 and 60 mi/h **119.** Between 20 and 40 ft **121.** Between 62.4 and 74.0 in.

SECTION 1.8 ■ PAGE 92

1. $(3, -5)$ **2.** $\sqrt{(c - a)^2 + (d - b)^2}$; 10
3. $\left(\frac{a + c}{2}, \frac{b + d}{2}\right)$; $(4, 6)$ **4.** 2; 3; No **5. (a)** y; x; -1 **(b)** x; y; $\frac{1}{2}$ **6.** $(1, 2)$; 3

7.

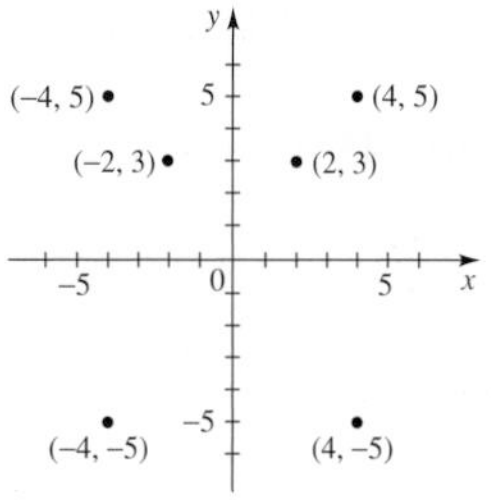

9. (a) $\sqrt{13}$ **(b)** $(\frac{3}{2}, 1)$ **11. (a)** 10 **(b)** $(1, 0)$
13. (a)

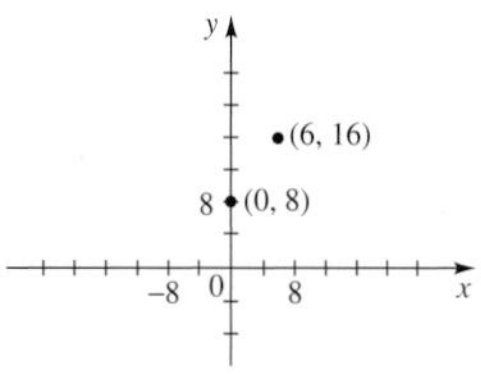

(b) 10 **(c)** $(3, 12)$

15. (a)

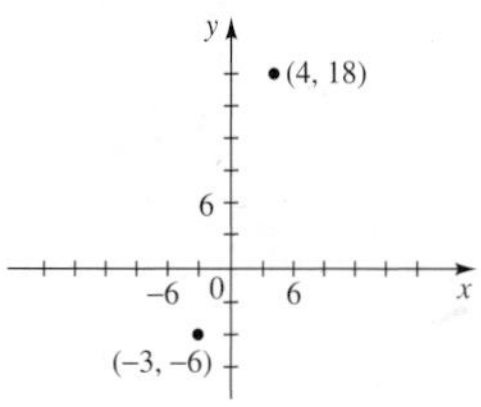

(b) 25 **(c)** $(\frac{1}{2}, 6)$

17. (a)

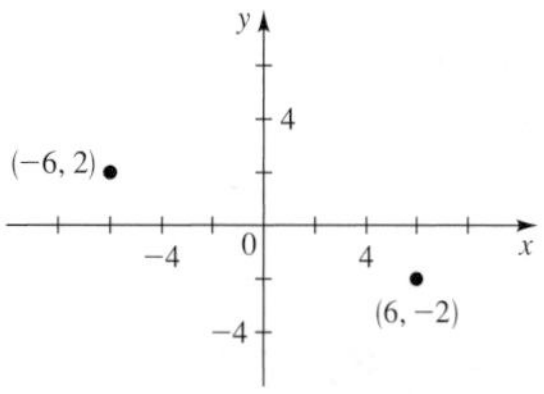

(b) $4\sqrt{10}$ **(c)** $(0, 0)$

19. 24

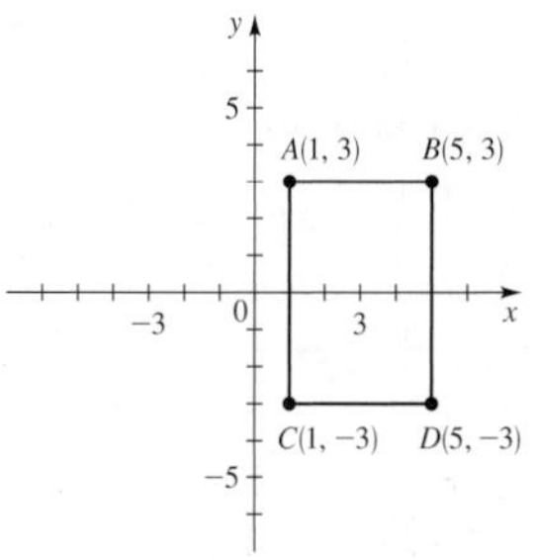

21. Trapezoid, area = 9

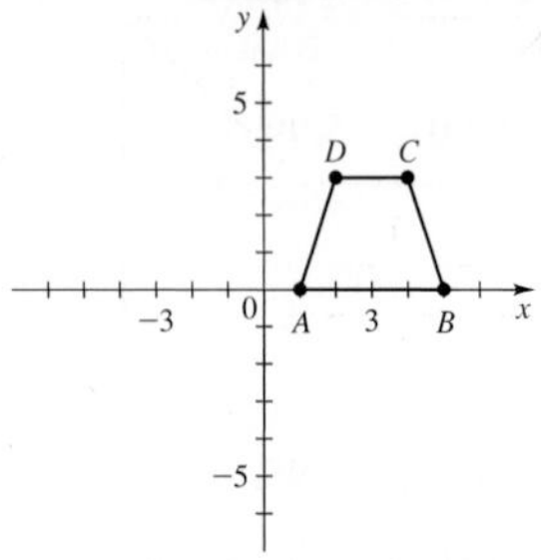

23.

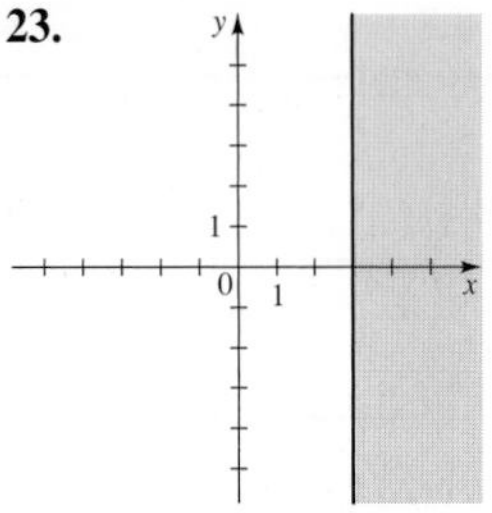

25.

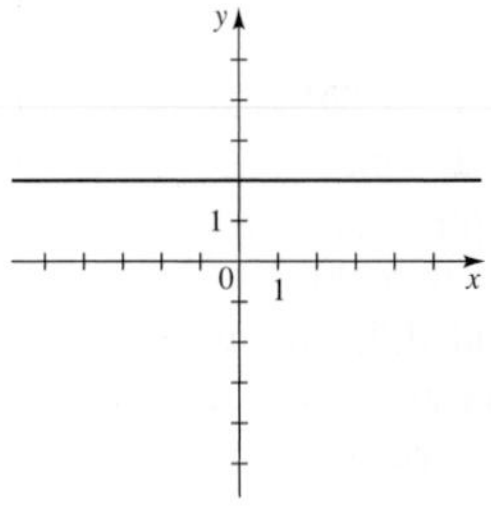

27.

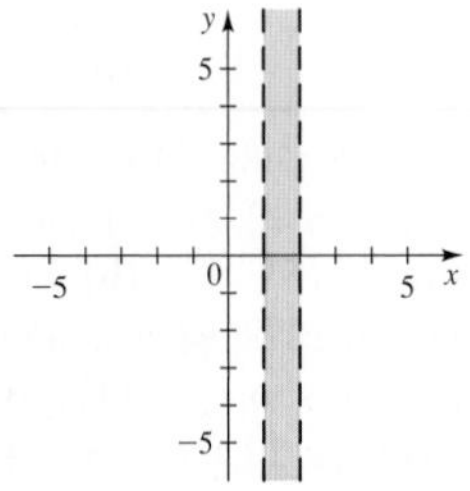

29.

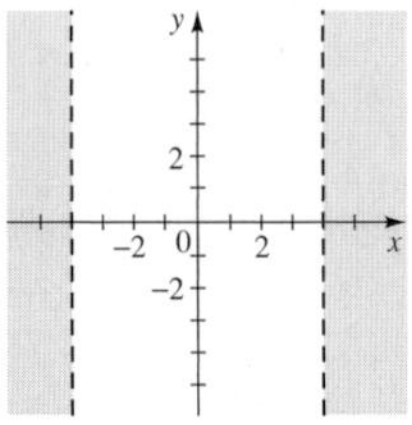

31.

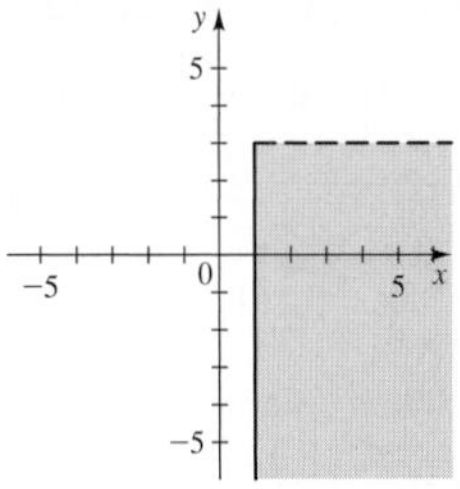

33. $A(6, 7)$ **35.** $Q(-1, 3)$ **39. (b)** 10 **43.** $(0, -4)$

45. $(2, -3)$

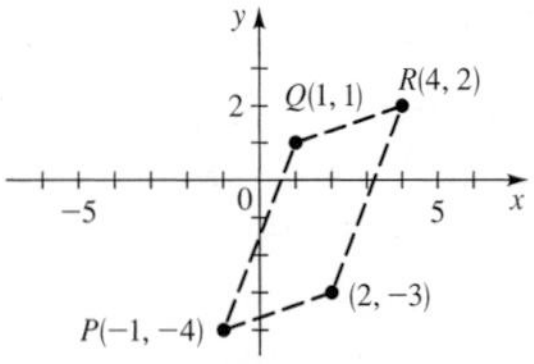

47. (a)

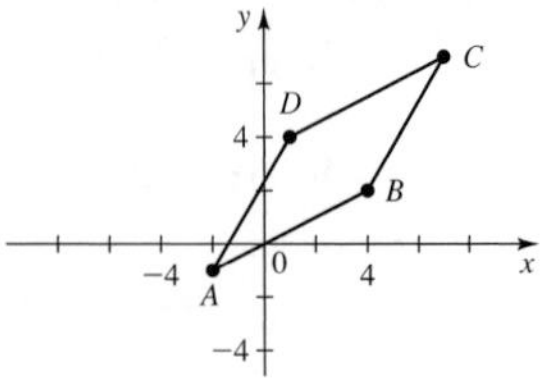

(b) $(\frac{5}{2}, 3), (\frac{5}{2}, 3)$

49. No, yes, yes **51.** Yes, no, yes

53. x-intercepts 0, 4; y-intercept 0

55. x-intercepts -2, 2; y-intercepts -4, 4

57. x-intercept 4, y-intercept 4, no symmetry

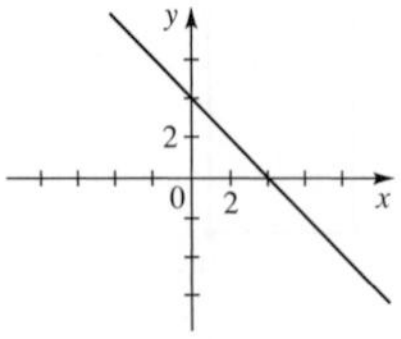

59. x-intercept 3, y-intercept -6, no symmetry

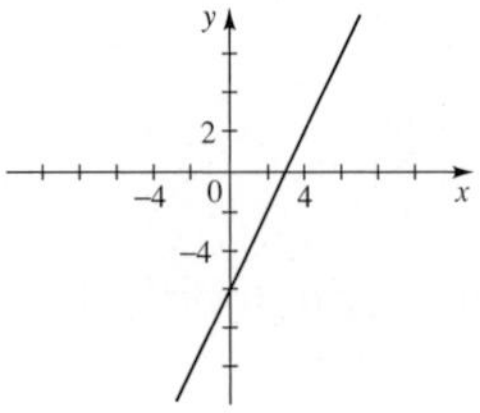

61. x-intercepts ± 1, y-intercept 1, symmetry about y-axis

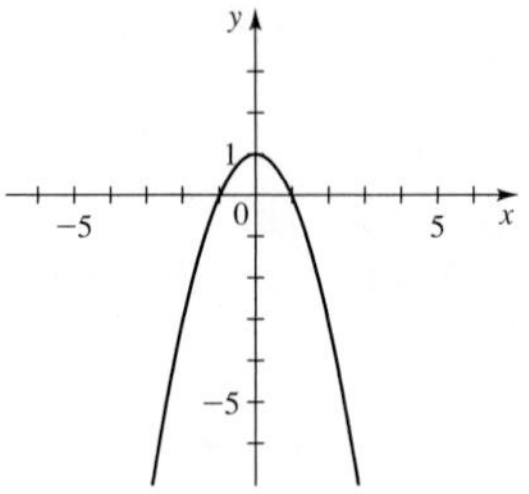

63. x-intercept 0, y-intercept 0, symmetry about y-axis

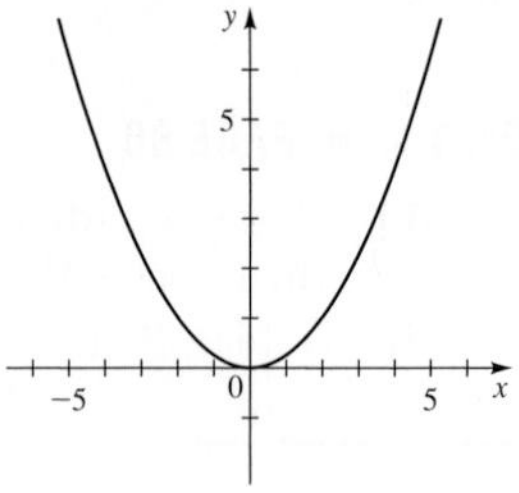

65. x-intercepts ± 3, y-intercept -9, symmetry about y-axis

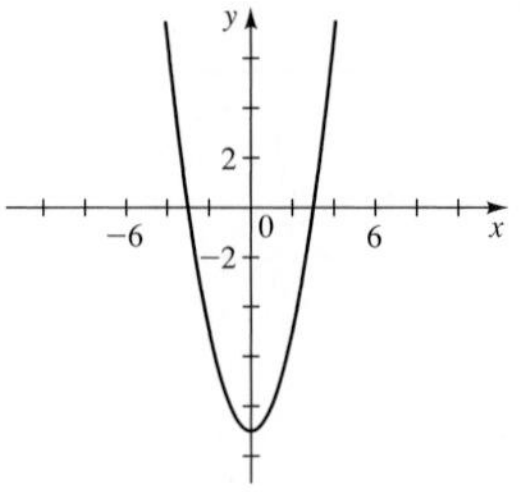

67. No intercepts, symmetry about origin

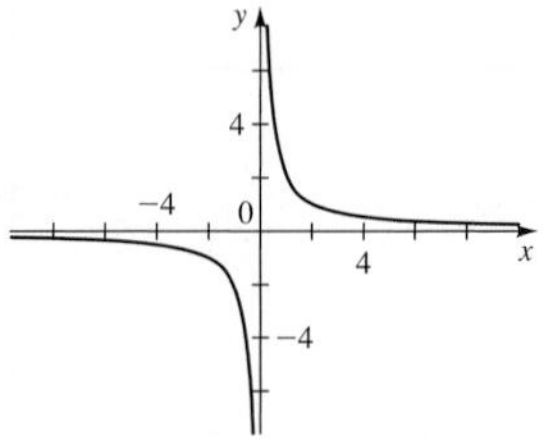

69. x-intercepts ± 2, y-intercept 2, symmetry about y-axis

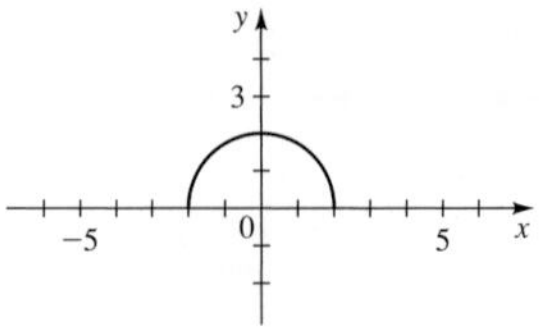

71. x-intercept 4, y-intercepts -2, 2, symmetry about x-axis

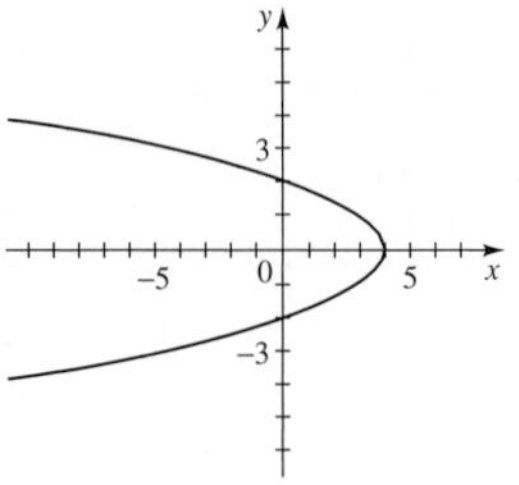

73. x-intercepts ± 2, y-intercept 16, symmetry about y-axis

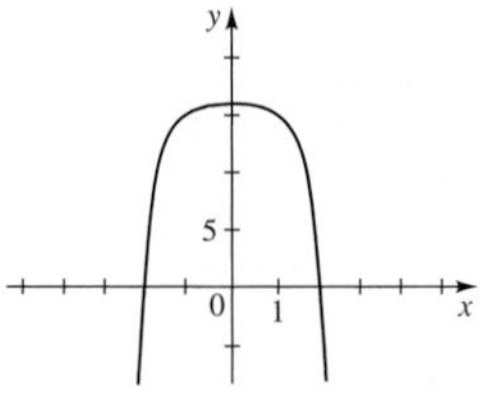

75. x-intercepts ± 4, y-intercept 4, symmetry about y-axis

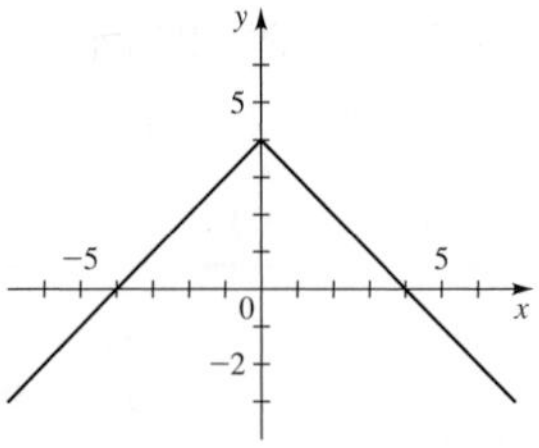

77. Symmetry about y-axis

79. Symmetry about origin

81. Symmetry about origin

83.

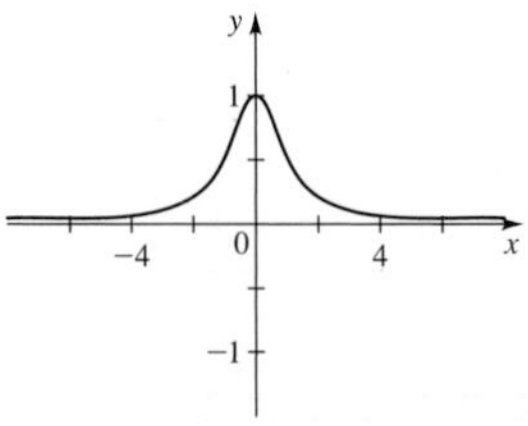

85.

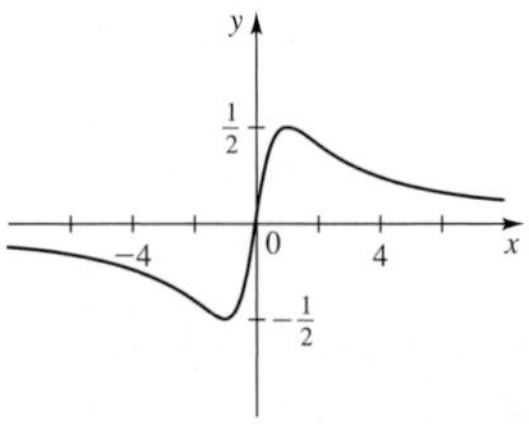

87. $(0, 0), 3$

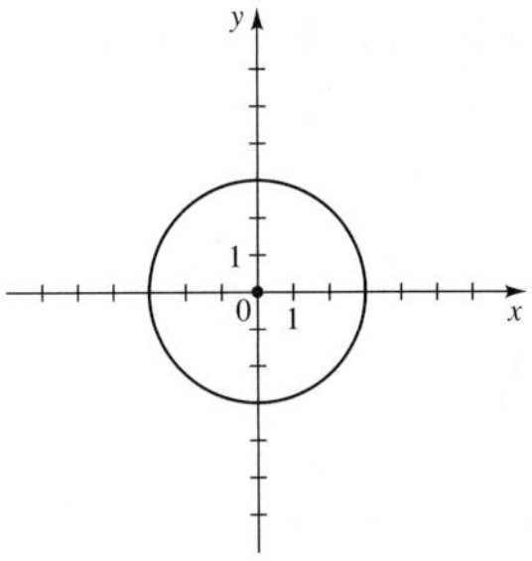

89. $(3, 0), 4$

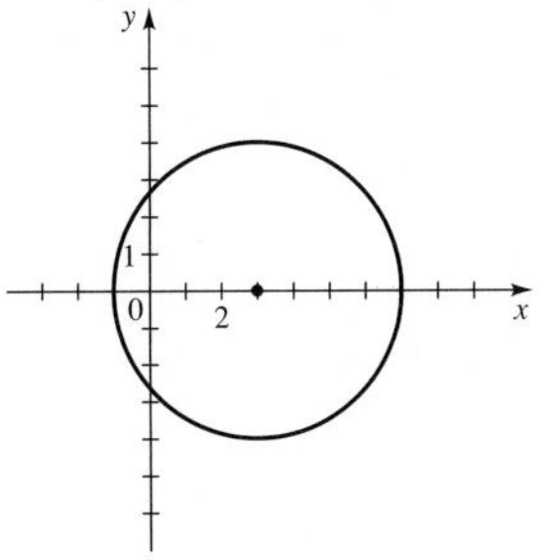

91. $(-3, 4), 5$

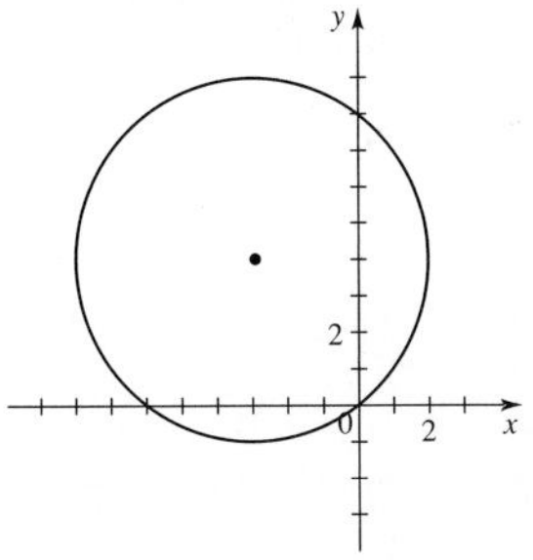

93. $(x - 2)^2 + (y + 1)^2 = 9$ **95.** $x^2 + y^2 = 65$
97. $(x - 2)^2 + (y - 5)^2 = 25$ **99.** $(x - 7)^2 + (y + 3)^2 = 9$
101. $(x + 2)^2 + (y - 2)^2 = 4$ **103.** $(2, -5), 4$ **105.** $(\frac{1}{4}, -\frac{1}{4}), \frac{1}{2}$
107. $(\frac{3}{4}, 0), \frac{3}{4}$
109.

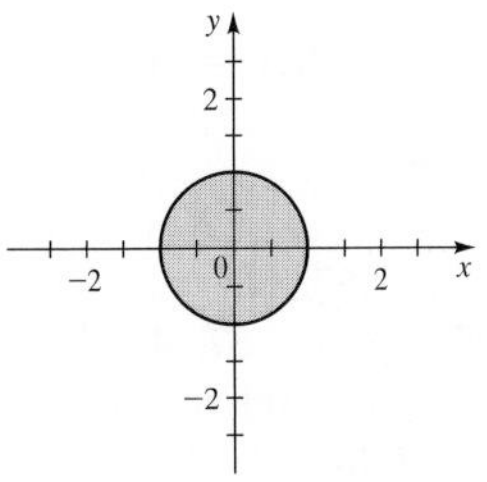

111. 12π **113. (a)** 5 **(b)** 31; 25 **(c)** Points P and Q must either be on the same street or the same avenue.
115. (a) 2 Mm, 8 Mm **(b)** −1.33, 7.33; 2.40 Mm, 7.60 Mm

SECTION 1.9 ■ PAGE 104

1. x **2.** above **3. (a)** $x = -1, 0, 1, 3$ **(b)** $[-1, 0] \cup [1, 3]$
4. (a) $x = 1, 4$ **(b)** $(1, 4)$ **5.** (c) **7.** (c) **9.** (c)
11.

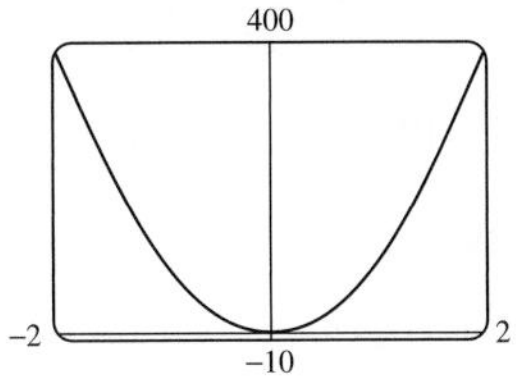

13.

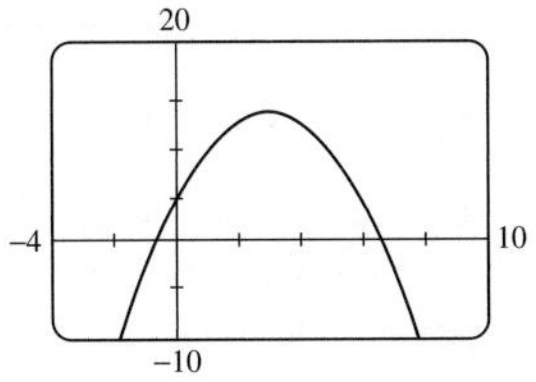

15.

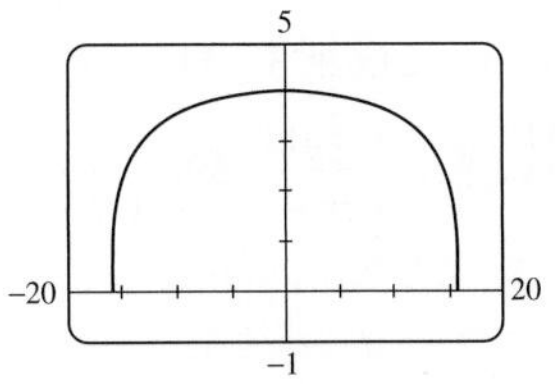

17.

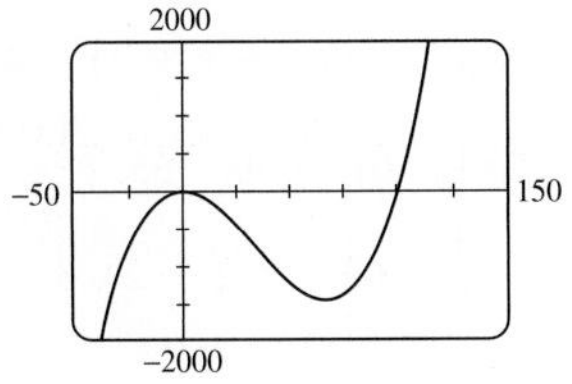

19.

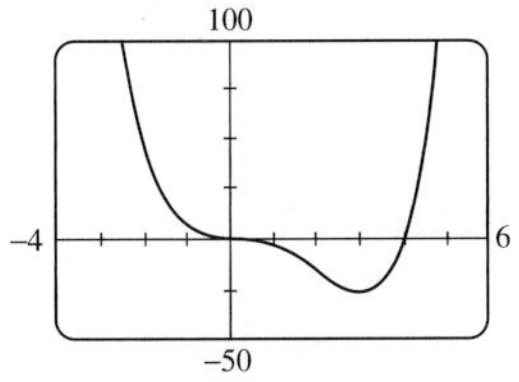

21.

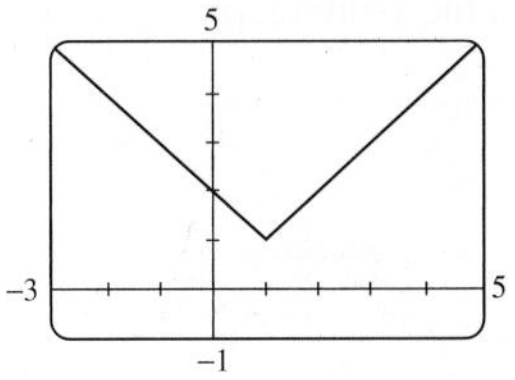

23. No **25.** Yes, 2
27.

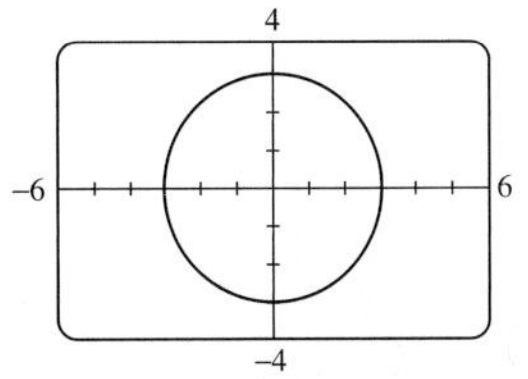

29.

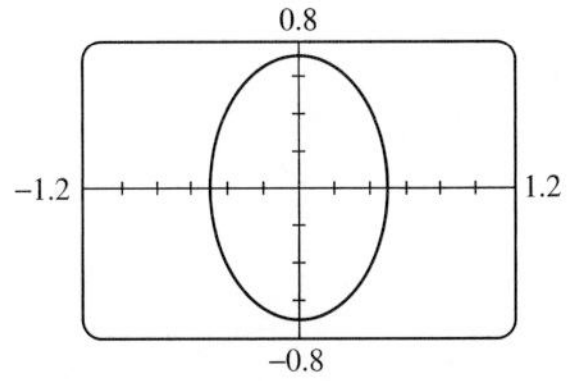

31. -4 **33.** $\frac{5}{14}$ **35.** $\pm 4\sqrt{2} \approx \pm 5.7$ **37.** No solution
39. 2.5, −2.5 **41.** $5 + 2\sqrt[4]{5} \approx 7.99, 5 - 2\sqrt[4]{5} \approx 2.01$
43. 3.00, 4.00 **45.** 1.00, 2.00, 3.00 **47.** 1.62
49. −1.00, 0.00, 1.00 **51.** 4 **53.** No solution
55. 2.55 **57.** −2.05, 0, 1.05 **59.** $[-2.00, 5.00]$
61. $(-\infty, 1.00] \cup [2.00, 3.00]$ **63.** $(-1.00, 0) \cup (1.00, \infty)$
65. $(-\infty, 0)$ **67.** $(-1, 4)$ **69.** $[-1, 3]$ **71.** 0, 0.01
73. (a)

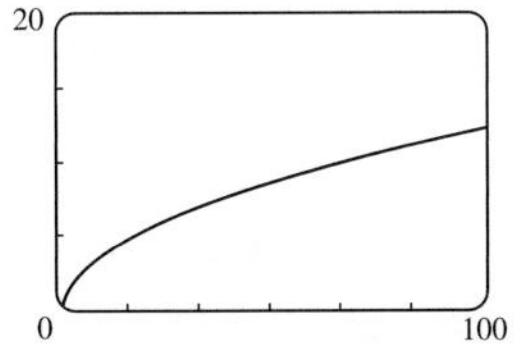

(b) 67 mi

SECTION 1.10 ■ PAGE 115

1. y; x; 2 **2. (a)** 3 **(b)** 3 **(c)** $-\frac{1}{3}$ **3.** $y - 2 = 3(x - 1)$
4. (a) 0; $y = 3$ **(b)** Undefined; $x = 2$ **5.** $\frac{1}{2}$ **7.** $\frac{1}{6}$
9. $-\frac{1}{2}$ **11.** $-\frac{9}{2}$ **13.** $-2, \frac{1}{2}, 3, -\frac{1}{4}$ **15.** $x + y - 4 = 0$
17. $3x - 2y - 6 = 0$ **19.** $5x - y - 7 = 0$
21. $2x - 3y + 19 = 0$ **23.** $5x + y - 11 = 0$
25. $3x - y - 2 = 0$ **27.** $3x - y - 3 = 0$ **29.** $y = 5$
31. $x + 2y + 11 = 0$ **33.** $x = -1$ **35.** $5x - 2y + 1 = 0$
37. $x - y + 6 = 0$
39. (a)

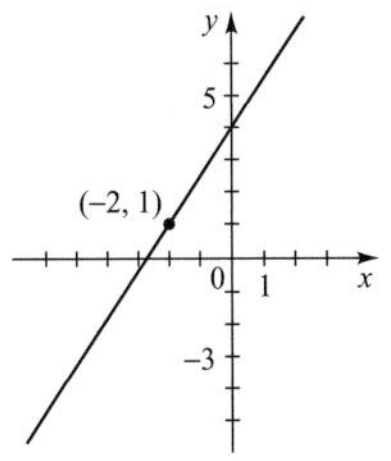

(b) $3x - 2y + 8 = 0$

41. They all have the same slope.

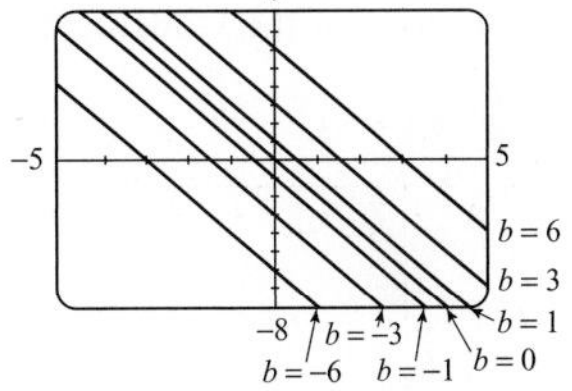

43. They all have the same x-intercept.

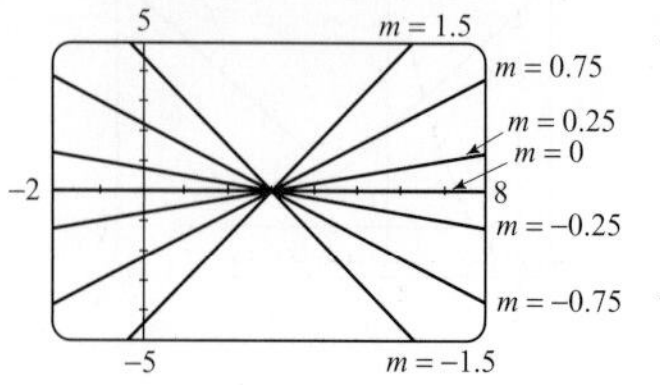

45. $-1, 3$

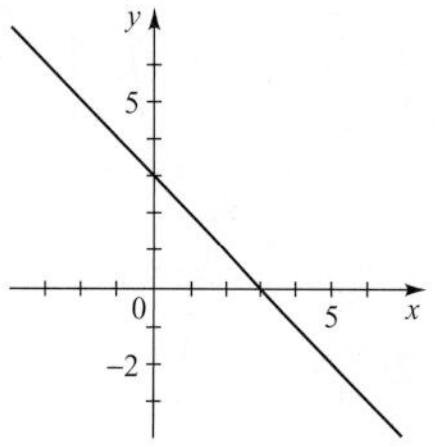

47. $-\frac{1}{3}, 0$

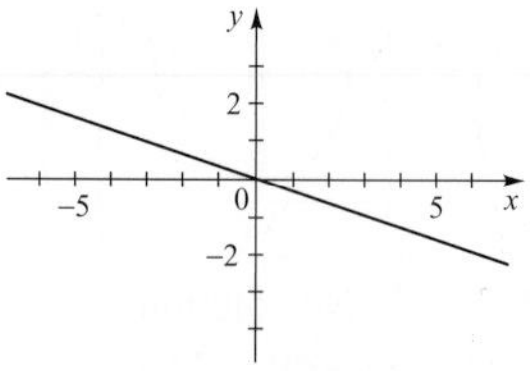

49. $\frac{3}{2}, 3$

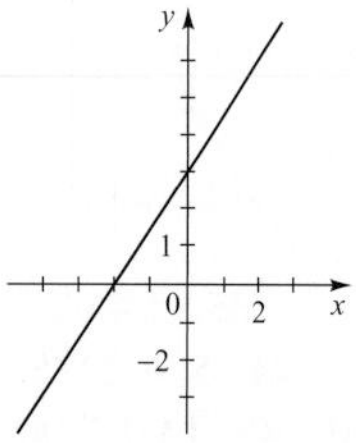

51. $0, 4$

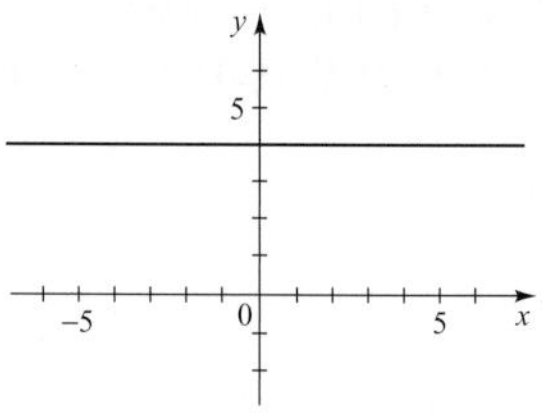

53. $\frac{3}{4}, -3$

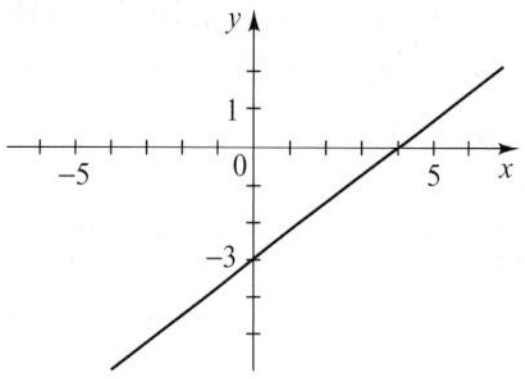

55. $-\frac{3}{4}, \frac{1}{4}$

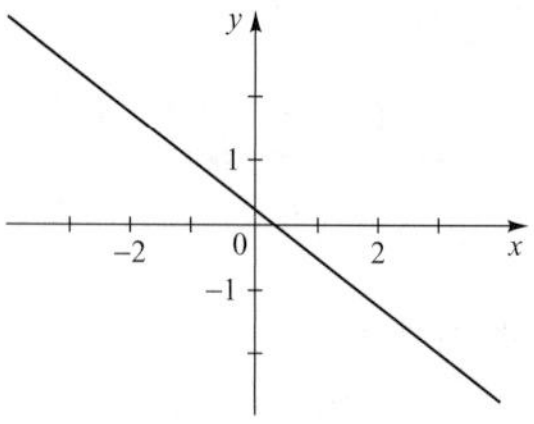

61. $x - y - 3 = 0$ **63.** **(b)** $4x - 3y - 24 = 0$ **65.** 16,667 ft
67. **(a)** 8.34; the slope represents the increase in dosage for a one-year increase in age. **(b)** 8.34 mg
69. **(a)**

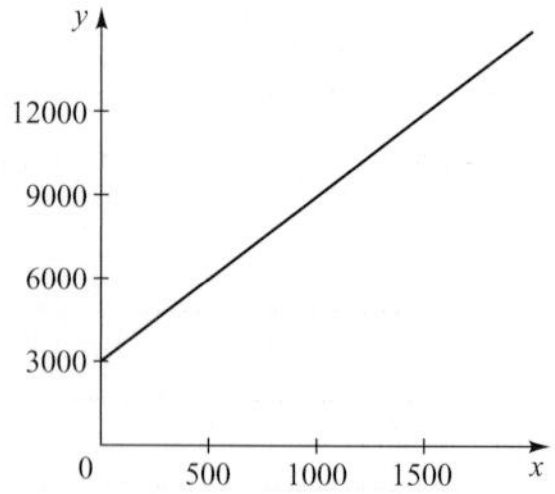

(b) The slope represents production cost per toaster; the y-intercept represents monthly fixed cost.

71. **(a)** $t = \frac{5}{24}n + 45$ **(b)** 76°F
73. **(a)** $P = 0.434d + 15$, where P is pressure in lb/in^2 and d is depth in feet
(b)

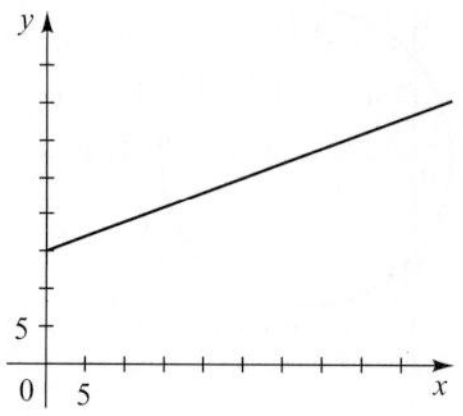

(c) The slope is the rate of increase in water pressure, and the y-intercept is the air pressure at the surface. **(d)** 196 ft
75. **(a)** $C = \frac{1}{4}d + 260$
(b) \$635
(c) The slope represents cost per mile.
(d) The y-intercept represents monthly fixed cost.

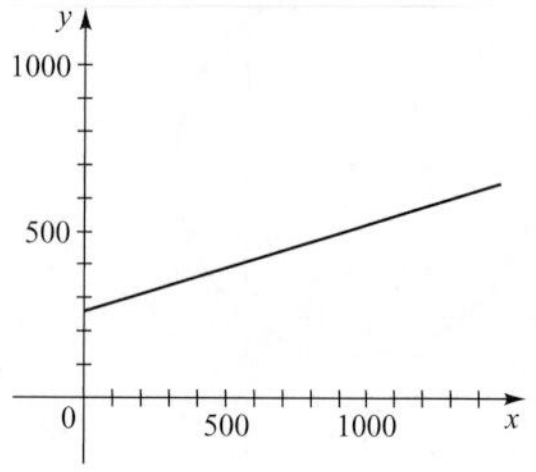

SECTION 1.11 ▪ PAGE 121

1. directly proportional; proportionality **2.** inversely proportional; proportionality **3.** directly proportional; inversely proportional **4.** $\frac{1}{2}xy$ **5.** $T = kx$ **7.** $v = k/z$ **9.** $y = ks/t$

11. $z = k\sqrt{y}$ **13.** $V = klwh$ **15.** $R = k\dfrac{i}{Pt}$ **17.** $y = 7x$

19. $R = 12/s$ **21.** $M = 15x/y$ **23.** $W = 360/r^2$
25. $C = 16lwh$ **27.** $s = 500/\sqrt{t}$ **29.** **(a)** $F = kx$ **(b)** 8
(c) 32 N **31.** **(a)** $C = kpm$ **(b)** 0.125 **(c)** \$57,500
33. **(a)** $P = ks^3$ **(b)** 0.012 **(c)** 324 **35.** 0.7 dB **37.** 4
39. 5.3 mi/h **41.** **(a)** $R = kL/d^2$ **(b)** $0.00291\overline{6}$ **(c)** $R \approx 137\ \Omega$
43. **(a)** 160,000 **(b)** 1,930,670,340 **45.** 36 lb

47. **(a)** $f = \dfrac{k}{L}$ **(b)** Halves it

CHAPTER 1 REVIEW ▪ PAGE 125

1. Commutative Property for addition
3. Distributive Property

5. $-2 \le x < 6$

7. $[5, \infty)$

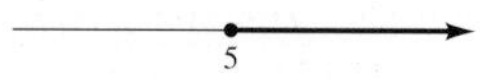

9. 6 **11.** $\frac{1}{72}$ **13.** $\frac{1}{6}$ **15.** 11 **17.** 4 **19.** $16x^3$ **21.** $12xy^8$

23. x^2y^2 **25.** $3x^{3/2}y^2$ **27.** $\dfrac{4r^{5/2}}{s^7}$ **29.** 7.825×10^{10}

31. 1.65×10^{-32} **33.** $3xy^2(4xy^2 - y^3 + 3x^2)$
35. $(x - 2)(x + 5)$ **37.** $(4t + 3)(t - 4)$ **39.** $(5 - 4t)(5 + 4t)$
41. $(x - 1)(x^2 + x + 1)(x + 1)(x^2 - x + 1)$
43. $x^{-1/2}(x - 1)^2$ **45.** $(x - 2)(4x^2 + 3)$
47. $\sqrt{x^2 + 2}(x^2 + x + 2)^2$ **49.** $6x^2 - 21x + 3$ **51.** $-7 + x$

53. $2x^3 - 6x^2 + 4x$ **55.** $\dfrac{3(x + 3)}{x + 4}$ **57.** $\dfrac{x + 1}{x - 4}$ **59.** $\dfrac{1}{x + 1}$

61. $-\dfrac{1}{2x}$ **63.** $3\sqrt{2} - 2\sqrt{3}$ **65.** 5 **67.** No solution

69. 2, 7 **71.** $-1, \frac{1}{2}$ **73.** $0, \pm\frac{5}{2}$ **75.** $\dfrac{-2 \pm \sqrt{7}}{3}$ **77.** -5

79. 3, 11 **81.** 20 lb raisins, 30 lb nuts

83. $\frac{1}{4}(\sqrt{329} - 3) \approx 3.78$ mi/h **85.** 1 h 50 min

87. $(-3, \infty)$

89. $(-\infty, -6) \cup (2, \infty)$

91. $(-\infty, -2) \cup (2, 4]$

93. $[2, 8]$

95. $-1, 7$ **97.** $[1, 3]$

99. (a)

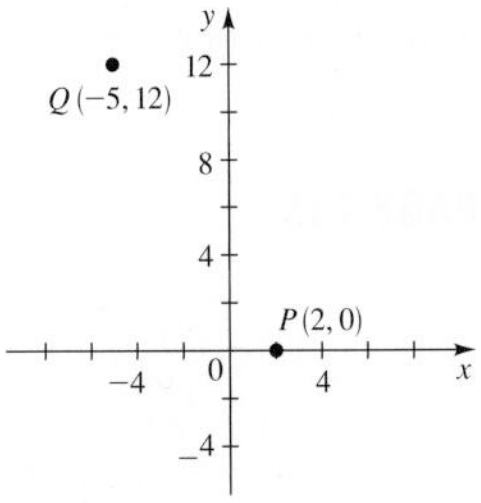

(b) $\sqrt{193}$ **(c)** $\left(-\frac{3}{2}, 6\right)$

(d) $y = -\frac{12}{7}x + \frac{24}{7}$ **(e)** $(x - 2)^2 + y^2 = 193$

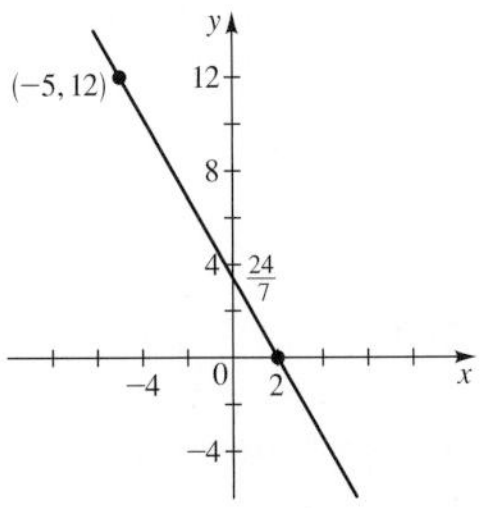

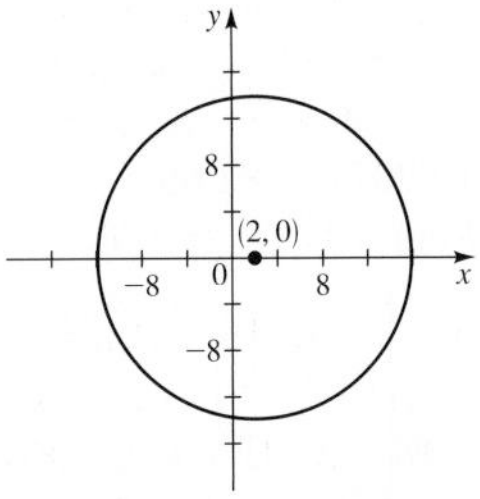

101.

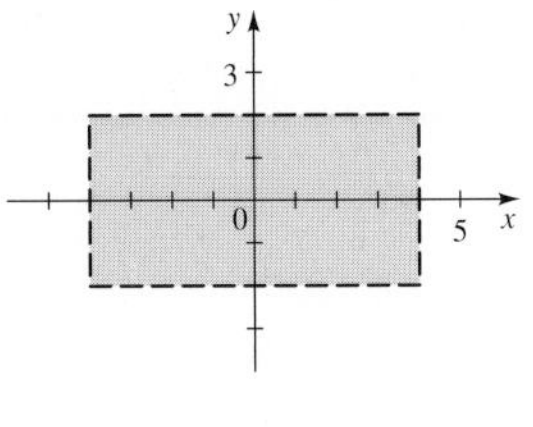

103. B **105.** $(x + 5)^2 + (y + 1)^2 = 26$

107. Circle, center $(-1, 3)$, radius 1 **109.** No graph

111. No symmetry **113.** No symmetry

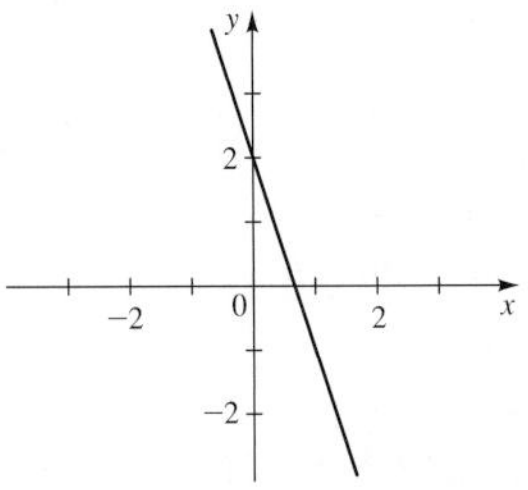

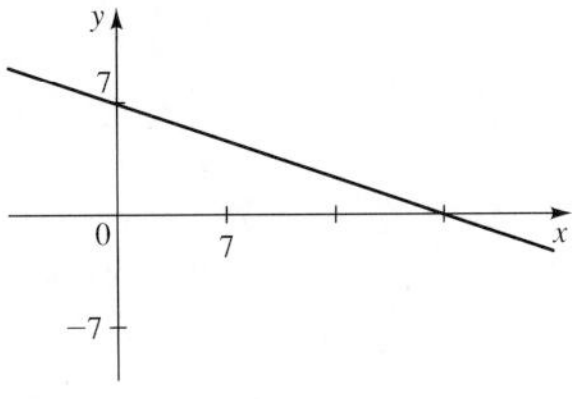

115. Symmetry about y-axis **117.** No symmetry

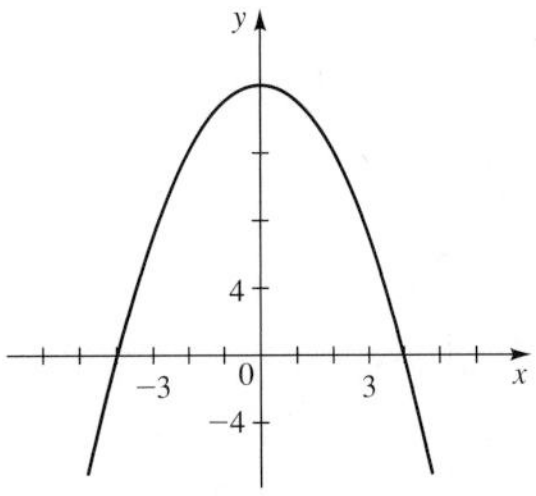

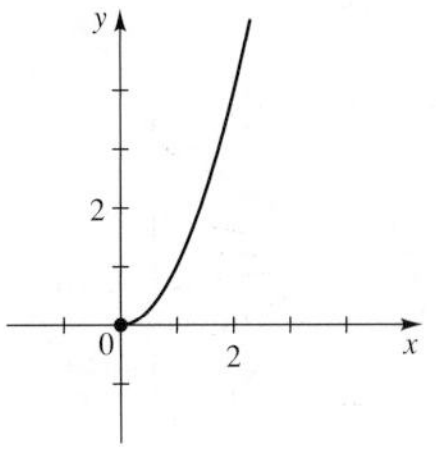

119.

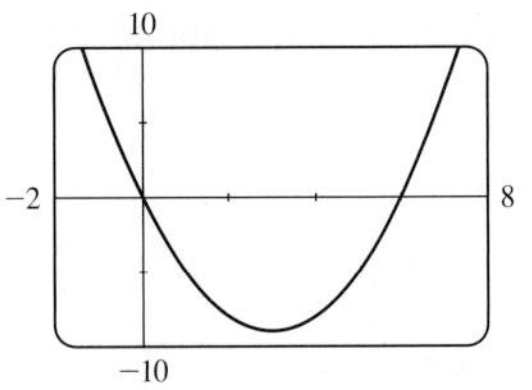

121.

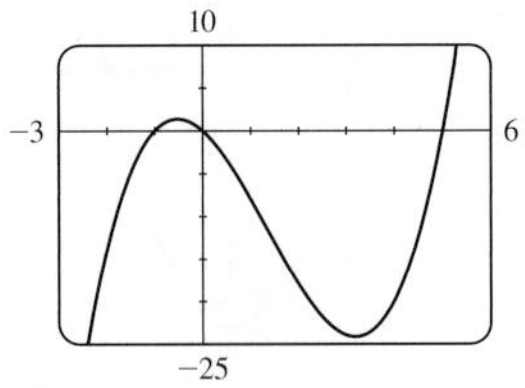

123. $2x - 3y - 16 = 0$ **125.** $3x + y - 12 = 0$

127. $x + 5y = 0$ **129.** $x^2 + y^2 = 169$, $5x - 12y + 169 = 0$

131. (a) The slope represents the amount the spring lengthens for a one-pound increase in weight. The S-intercept represents the unstretched length of the spring. **(b)** 4 in.

133. $M = 8z$ **135. (a)** $I = k/d^2$ **(b)** 64,000 **(c)** 160 candles

137. 11.0 mi/h

CHAPTER 1 TEST ■ PAGE 128

1. (a)

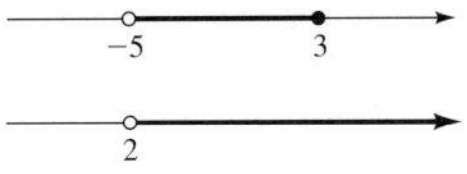

(b) $(-\infty, 3], [-1, 4)$ **(c)** 16

2. (a) 81 **(b)** -81 **(c)** $\frac{1}{81}$ **(d)** 25 **(e)** $\frac{9}{4}$ **(f)** $\frac{1}{8}$

3. (a) 1.86×10^{11} **(b)** 3.965×10^{-7}

4. (a) $6\sqrt{2}$ **(b)** $48a^5b^7$ **(c)** $\dfrac{x}{9y^7}$ **(d)** $\dfrac{x + 2}{x - 2}$ **(e)** $\dfrac{1}{x - 2}$

(f) $-(x + y)$ **5.** $5\sqrt{2} + 2\sqrt{10}$

6. (a) $11x - 2$ **(b)** $4x^2 + 7x - 15$ **(c)** $a - b$

(d) $4x^2 + 12x + 9$ **(e)** $x^3 + 6x^2 + 12x + 8$

7. (a) $(2x - 5)(2x + 5)$ **(b)** $(2x - 3)(x + 4)$

(c) $(x - 3)(x - 2)(x + 2)$

(d) $x(x + 3)(x^2 - 3x + 9)$

(e) $3x^{-1/2}(x - 1)(x - 2)$ **(f)** $xy(x - 2)(x + 2)$

8. (a) 6 **(b)** 1 **(c)** $-3, 4$ **(d)** $-1 \pm \dfrac{\sqrt{2}}{2}$

(e) No real solution **(f)** $\pm 1, \pm\sqrt{2}$ **(g)** $\frac{2}{3}, \frac{22}{3}$ **9.** 120 mi

10. 50 ft by 120 ft

11. (a) $[-4, 3)$

(b) $(-2, 0) \cup (1, \infty)$

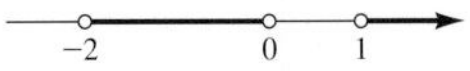

(c) $(1, 7)$

(d) $(-1, 4]$

12. Between 41°F and 50°F **13.** $0 \le x \le 6$

14. (a) $-2.94, -0.11, 3.05$ **(b)** $[-1, 2]$

15. (a) $S(3, 6)$

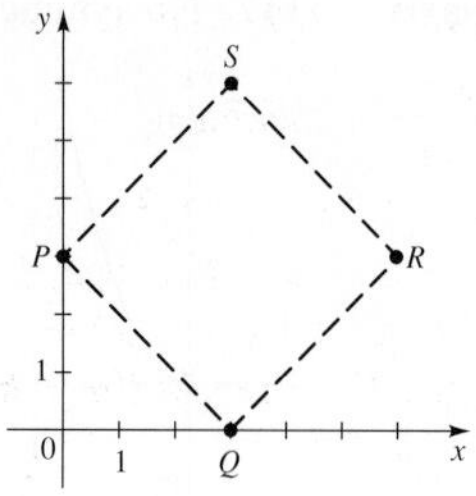

(b) 18

16. (a)

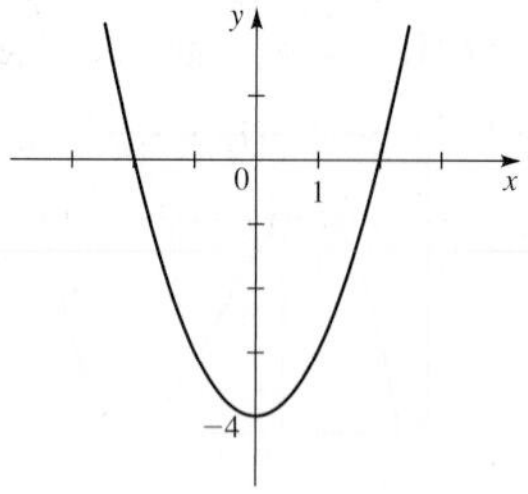

(b) x-intercepts $-2, 2$
y-intercept -4
(c) Symmetric about y-axis

17. (a)

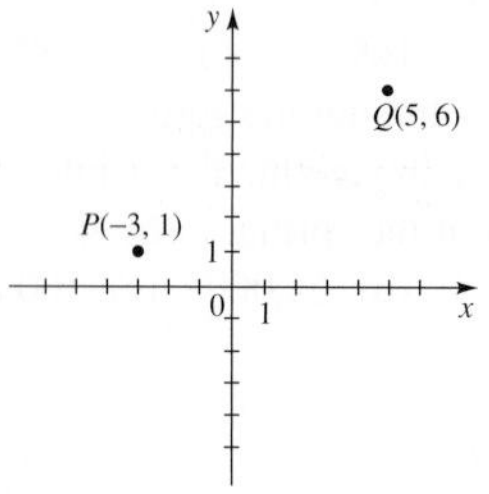

(b) $\sqrt{89}$ **(c)** $\left(1, \frac{7}{2}\right)$ **(d)** $\frac{5}{8}$ **(e)** $y = -\frac{8}{5}x + \frac{51}{10}$
(f) $(x - 1)^2 + \left(y - \frac{7}{2}\right)^2 = \frac{89}{4}$

18. (a) $(0, 0), 5$

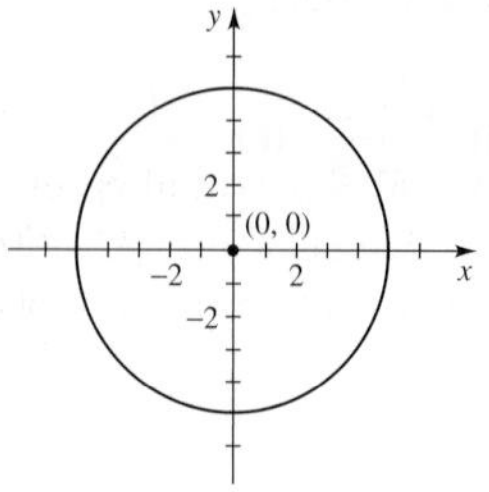

(b) $(2, -1), 3$

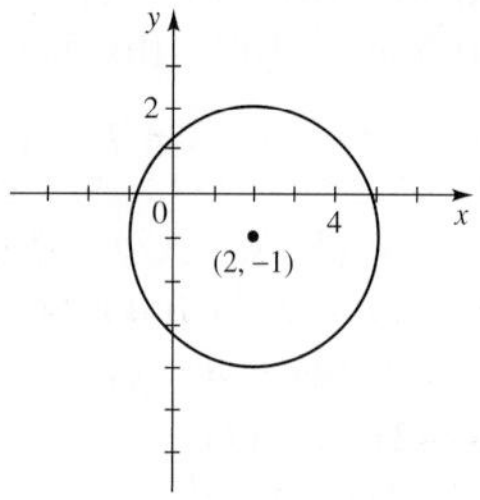

(c) $(-3, 1), 2$

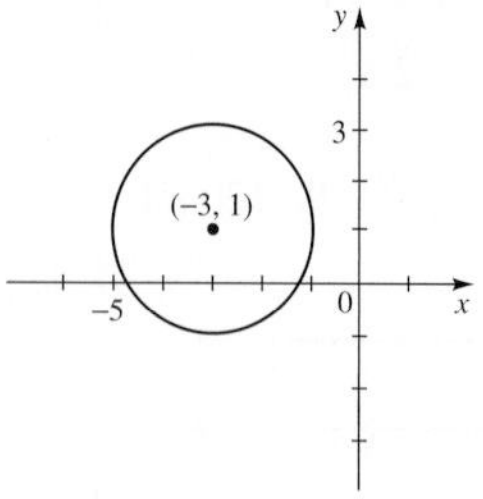

19. $y = \frac{2}{3}x - 5$

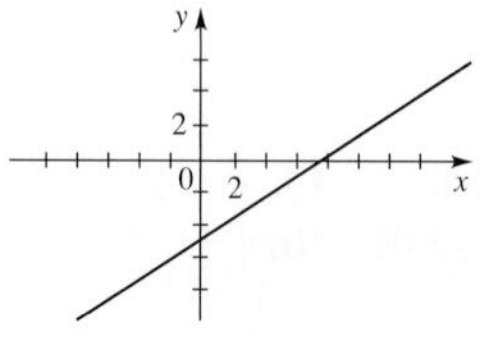

slope $\frac{2}{3}$; y-intercept -5

20. (a) $3x + y - 3 = 0$ **(b)** $2x + 3y - 12 = 0$

21. (a) 4°C **(b)**

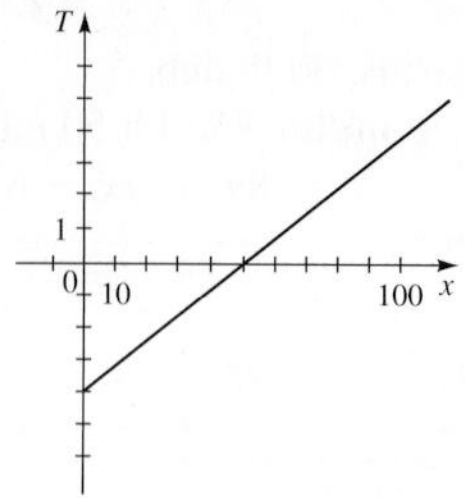

(c) The slope is the rate of change in temperature, the x-intercept is the depth at which the temperature is 0°C, and the T-intercept is the temperature at ground level.

22. (a) $M = kwh^2/L$ **(b)** 400 **(c)** 12,000 lb

FOCUS ON MODELING ▪ PAGE 135

1. (a)

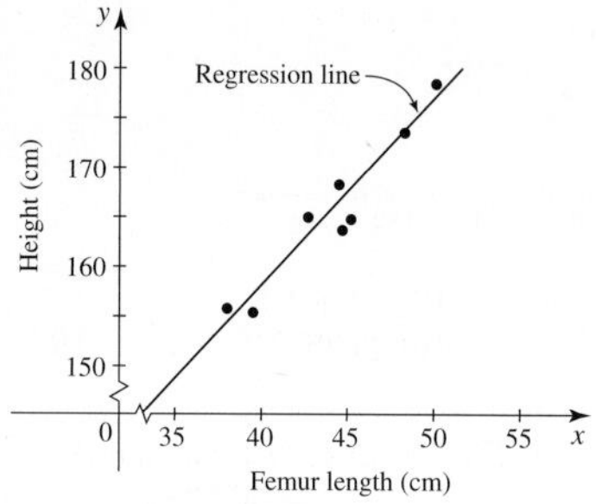

(b) $y = 1.8807x + 82.65$
(c) 191.7 cm

3. (a)

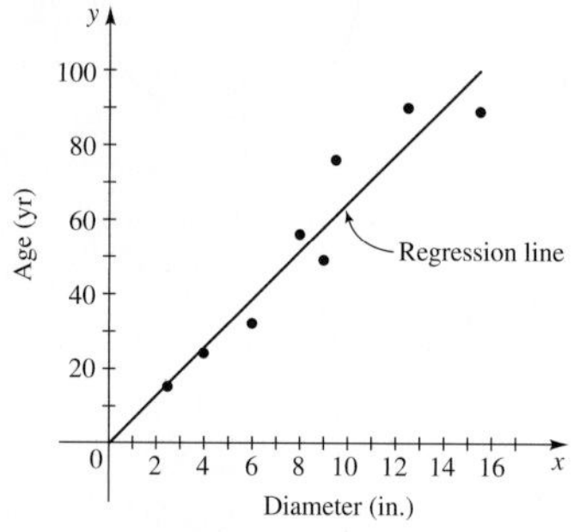

(b) $y = 6.451x - 0.1523$
(c) 116 years

5. (a)

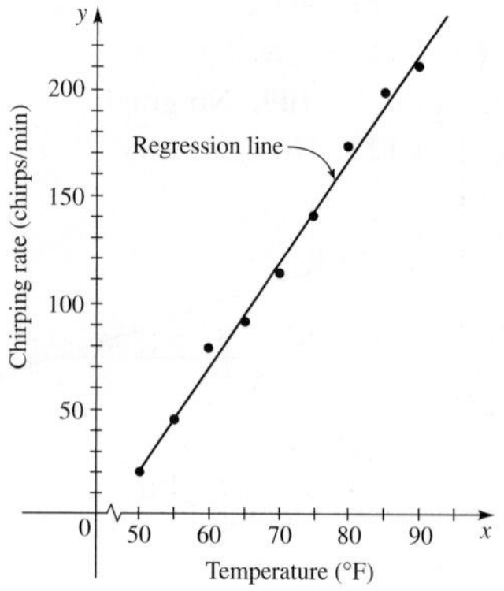

(b) $y = 4.857x - 220.97$
(c) 265 chirps/min

7. (a)

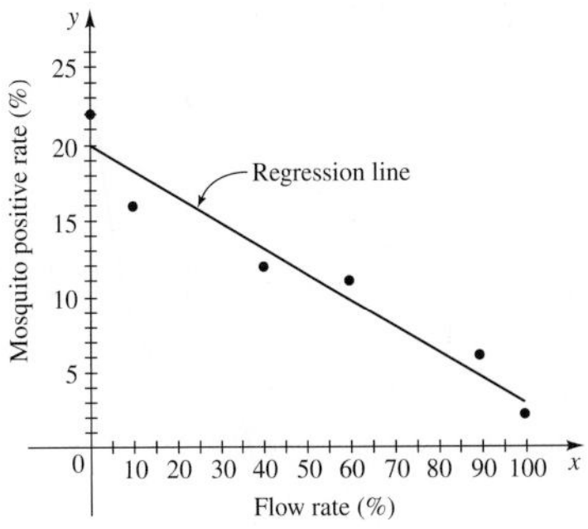

(b) $y = -0.168x + 19.89$ **(c)** 8.13%

9. (a)

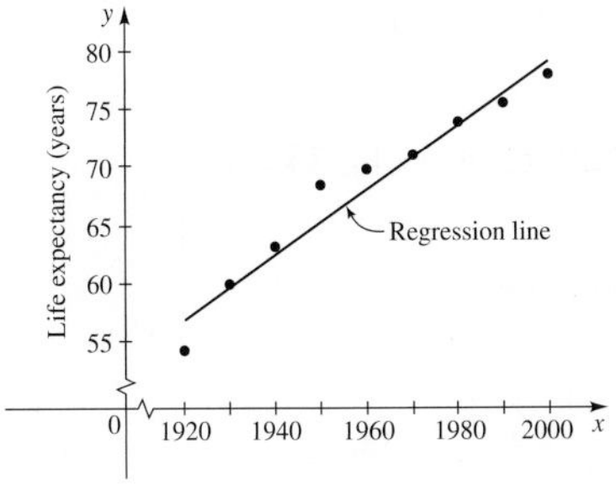

(b) $y = 0.2708x - 462.9$ **(c)** 80.3 years

11. (a) Men: $y = -0.1703x + 64.61$, women $y = -0.2603x + 78.27$; x represents years since 1900 **(b)** 2052

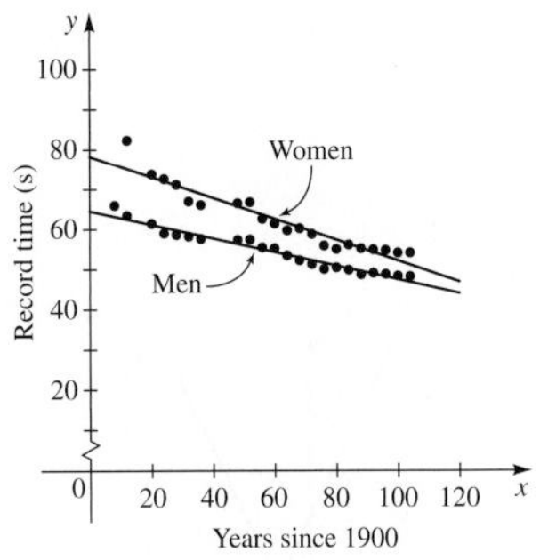

CHAPTER 2

SECTION 2.1 ■ PAGE 149

1. value **2.** domain, range **3. (a)** f and g **(b)** $f(5) = 10, g(5) = 0$ **4. (a)** square, add 3

(b)

x	0	2	4	6
$f(x)$	19	7	3	7

5. $f(x) = 2(x + 3)$ **7.** $f(x) = (x - 5)^2$ **9.** Square, then add 2 **11.** Subtract 4, then divide by 3

13.

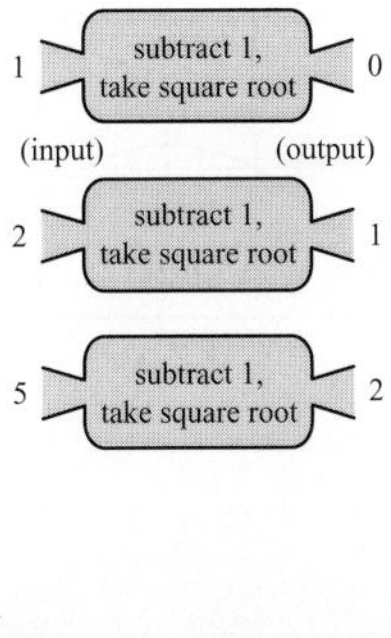

15.

x	$f(x)$
−1	8
0	2
1	0
2	2
3	8

17. $3, 3, -6, -\frac{23}{4}, 94$ **19.** $3, -3, 2, 2a + 1, -2a + 1, 2a + 2b + 1$

21. $-\dfrac{1}{3}, -3, \dfrac{1}{3}, \dfrac{1 - a}{1 + a}, \dfrac{2 - a}{a}$, undefined

23. $-4, 10, -2, 3\sqrt{2}, 2x^2 + 7x + 1, 2x^2 - 3x - 4$

25. $6, 2, 1, 2, 2|x|, 2(x^2 + 1)$ **27.** $4, 1, 1, 2, 3$

29. $8, -\frac{3}{4}, -1, 0, -1$ **31.** $x^2 + 4x + 5, x^2 + 6$

33. $x^2 + 4, x^2 + 8x + 16$ **35.** $3a + 2, 3(a + h) + 2, 3$

37. $5, 5, 0$ **39.** $\dfrac{a}{a + 1}, \dfrac{a + h}{a + h + 1}, \dfrac{1}{(a + h + 1)(a + 1)}$

41. $3 - 5a + 4a^2, 3 - 5a - 5h + 4a^2 + 8ah + 4h^2, -5 + 8a + 4h$

43. $(-\infty, \infty)$ **45.** $[-1, 5]$ **47.** $\{x \mid x \neq 3\}$ **49.** $\{x \mid x \neq \pm 1\}$

51. $[5, \infty)$ **53.** $(-\infty, \infty)$ **55.** $[\frac{5}{2}, \infty)$ **57.** $[-2, 3) \cup (3, \infty)$

59. $(-\infty, 0] \cup [6, \infty)$ **61.** $(4, \infty)$ **63.** $(\frac{1}{2}, \infty)$

65. (a) $f(x) = \dfrac{x}{3} + \dfrac{2}{3}$

(b)

x	$f(x)$
2	$\frac{4}{3}$
4	2
6	$\frac{8}{3}$
8	$\frac{10}{3}$

(c)

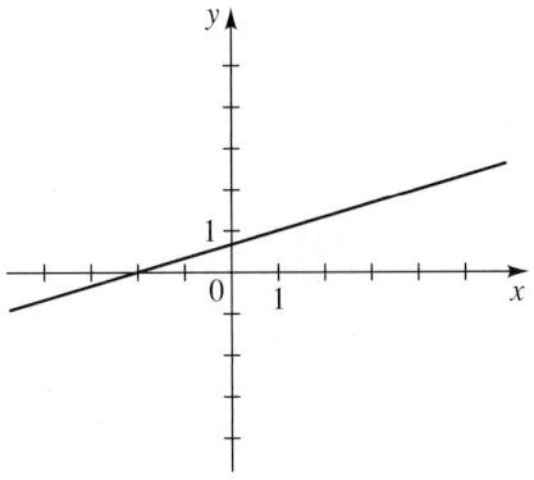

67. (a) $T(x) = 0.08x$

(b)

x	$T(x)$
2	0.16
4	0.32
6	0.48
8	0.64

(c)

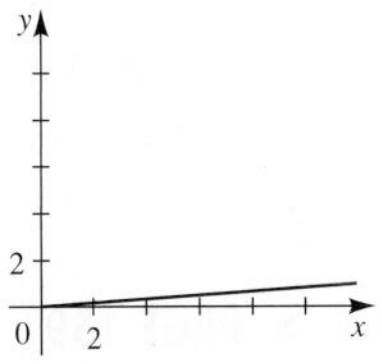

69. (a) $C(10) = 1532.1, C(100) = 2100$ **(b)** The cost of producing 10 yd and 100 yd **(c)** $C(0) = 1500$ **71. (a)** 50, 0 **(b)** $V(0)$ is the volume of the full tank, and $V(20)$ is the volume of the empty tank, 20 minutes later.

(c)

x	$V(x)$
0	50
5	28.125
10	12.5
15	3.125
20	0

73. (a) $v(0.1) = 4440, v(0.4) = 1665$ **(b)** Flow is faster near central axis.

(c)

r	$v(r)$
0	4625
0.1	4440
0.2	3885
0.3	2960
0.4	1665
0.5	0

75. **(a)** 8.66 m, 6.61 m, 4.36 m
(b) It will appear to get shorter.
77. **(a)** \$90, \$105, \$100, \$105
(b) Total cost of an order, including shipping

79. **(a)** $F(x) = \begin{cases} 15(40 - x) & \text{if } 0 < x < 40 \\ 0 & \text{if } 40 \le x \le 65 \\ 15(x - 65) & \text{if } x > 65 \end{cases}$

(b) \$150, \$0, \$150 **(c)** Fines for violating the speed limits

81.

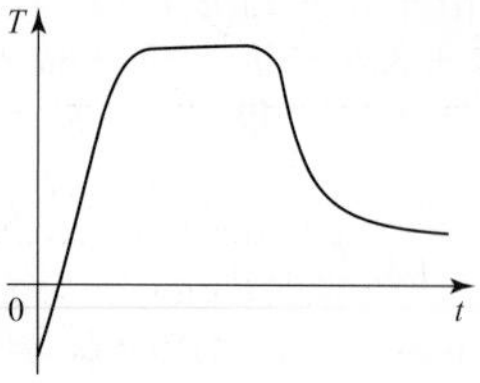

83.

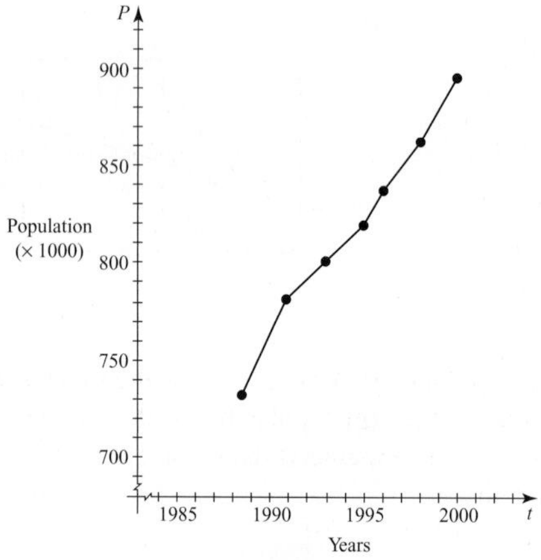

SECTION 2.2 ■ PAGE 159

1. $f(x)$, $x^3 + 2$, 10, 10 **2.** 3 **3.** 3
4. **(a)** IV **(b)** II **(c)** I **(d)** III

5.

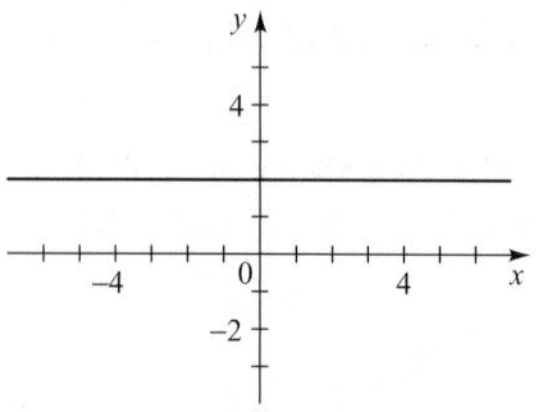

7.

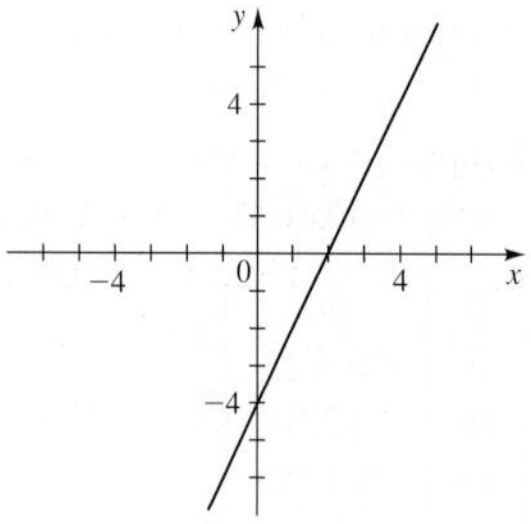

9.

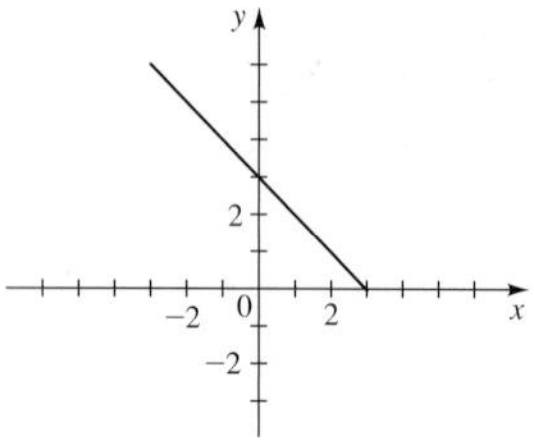

11.

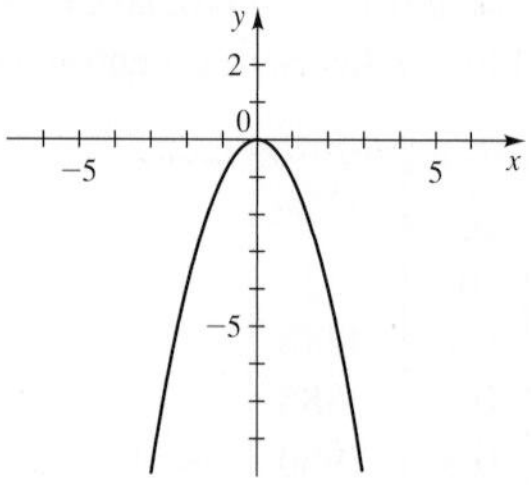

13.

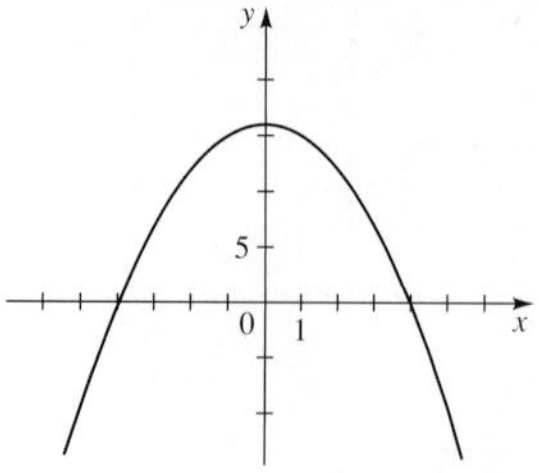

15.

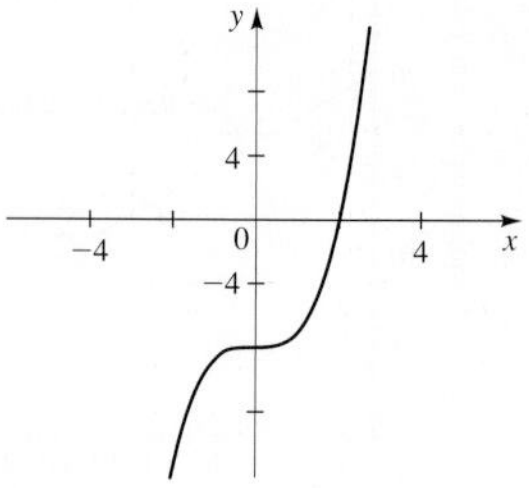

17.

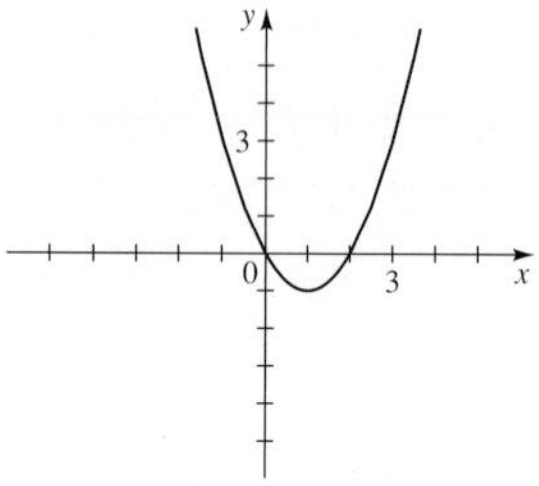

19.

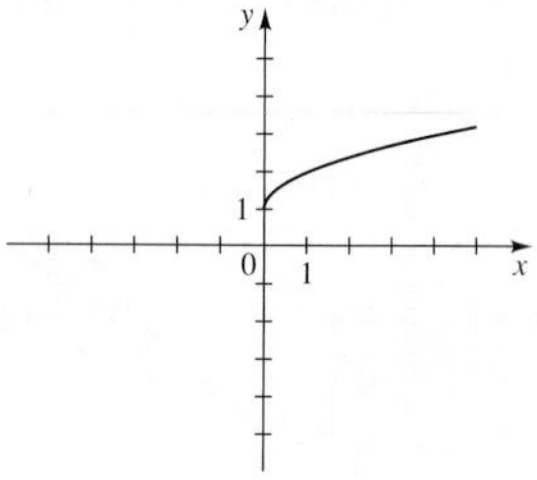

21.

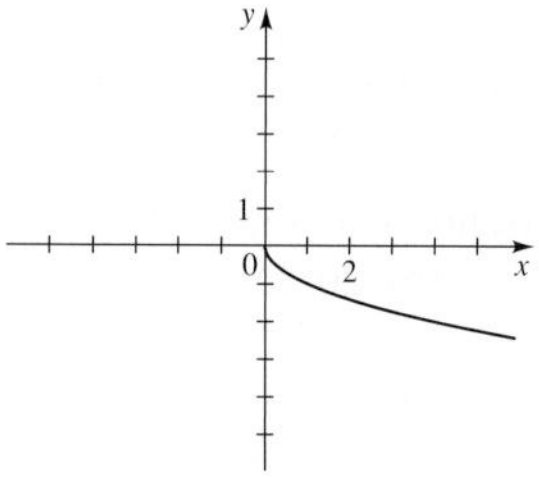

23.

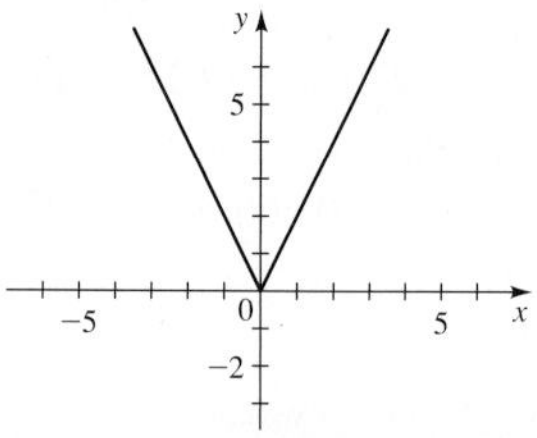

25.

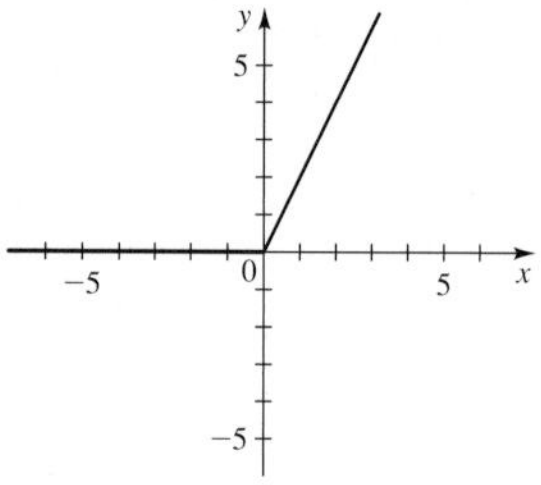

27.

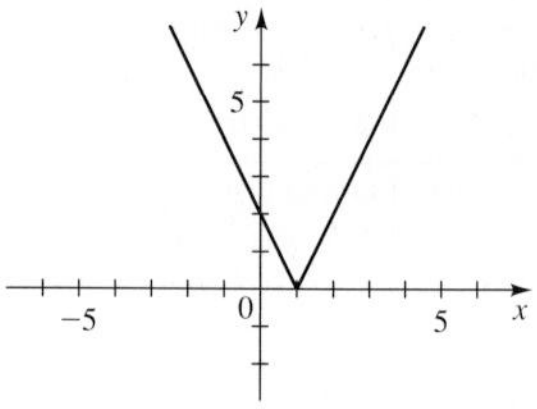

29. **(a)**

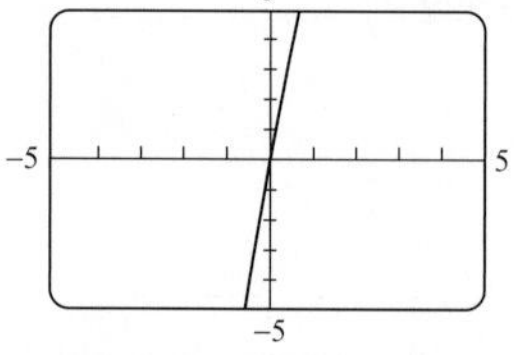

(b)

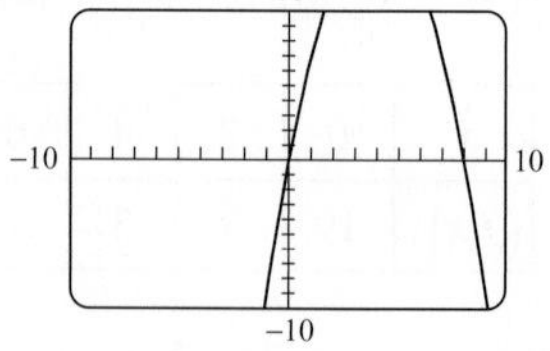

(c)

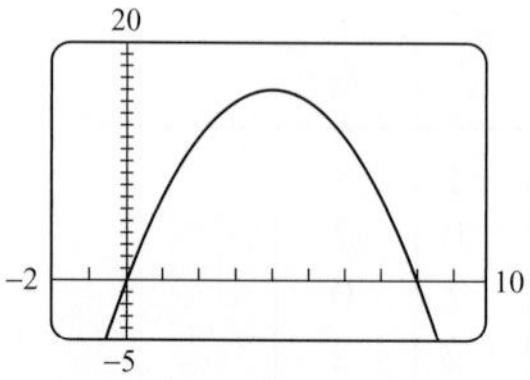

(d)

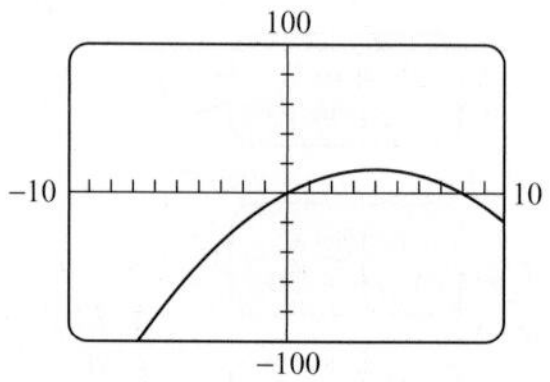

Graph (c) is the most appropriate.

31. (a)

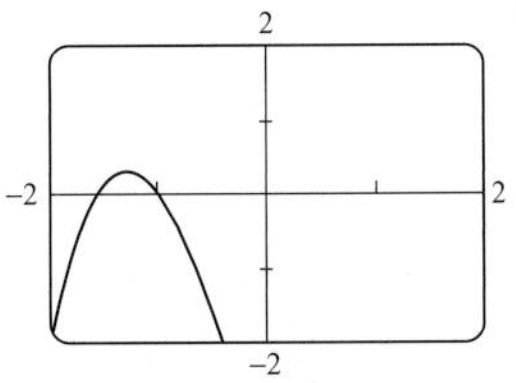

(b)

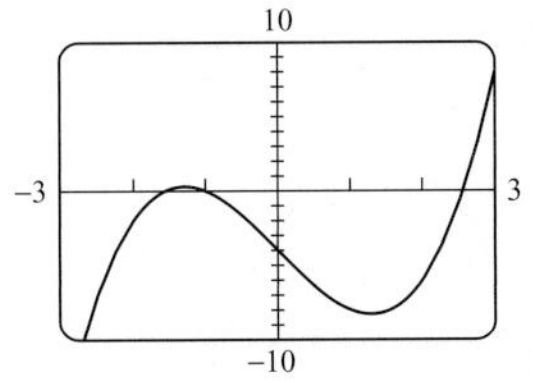

(c)

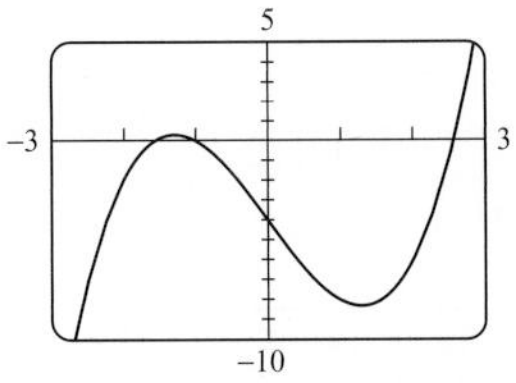

(d)

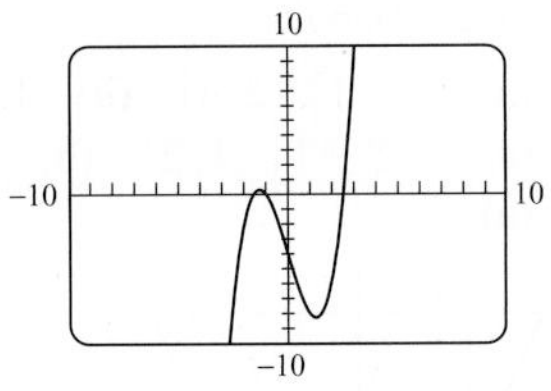

Graph (c) is the most appropriate.

33.

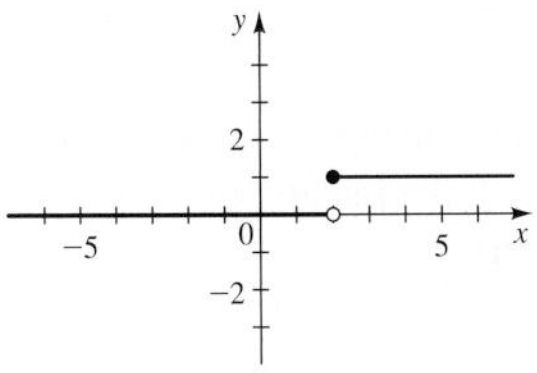

35.

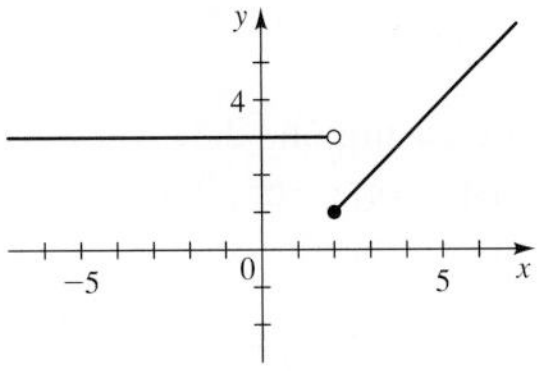

37.

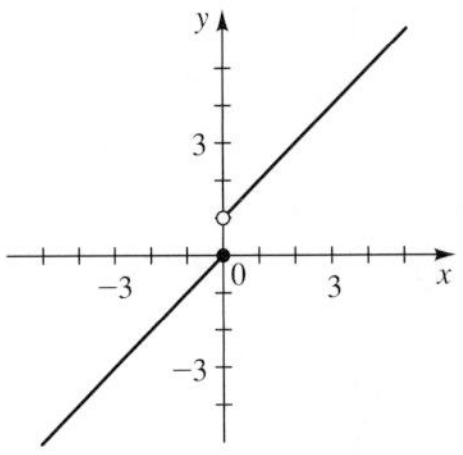

39.

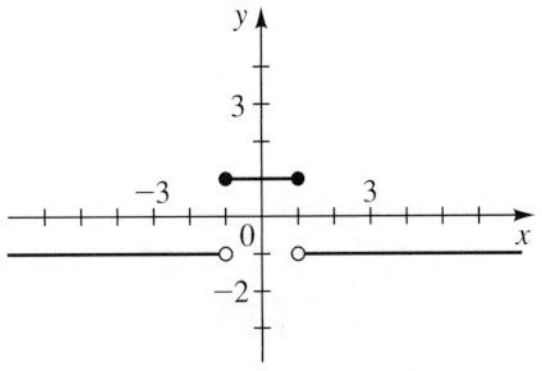

41.

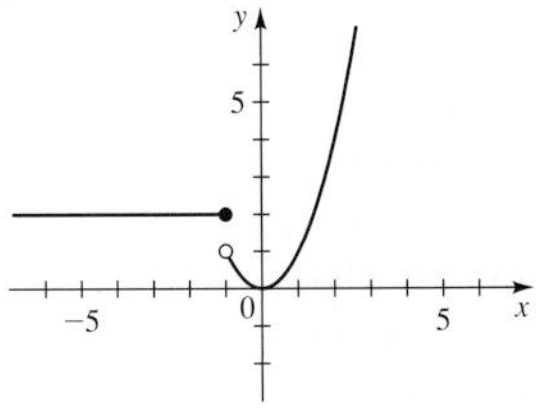

43.

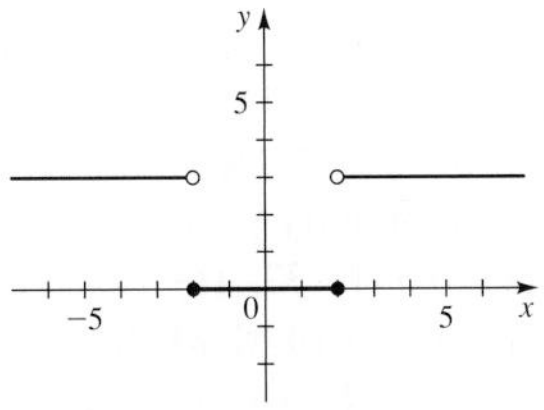

45.

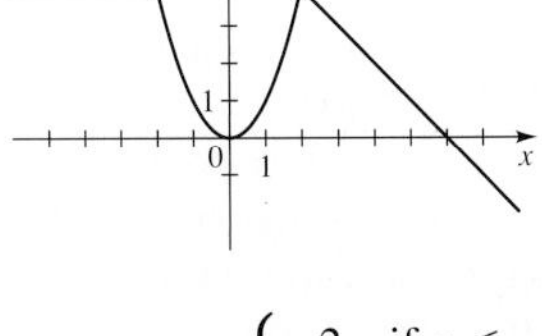

47.

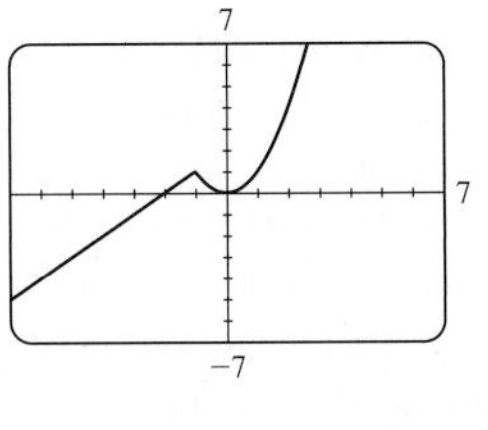

49. $f(x) = \begin{cases} -2 & \text{if } x < -2 \\ x & \text{if } -2 \le x \le 2 \\ 2 & \text{if } x > 2 \end{cases}$

51. (a) Yes **(b)** No **(c)** Yes **(d)** No

53. Function, domain $[-3, 2]$, range $[-2, 2]$ **55.** Not a function

57. Yes **59.** No **61.** No **63.** Yes **65.** Yes **67.** Yes

69. (a)

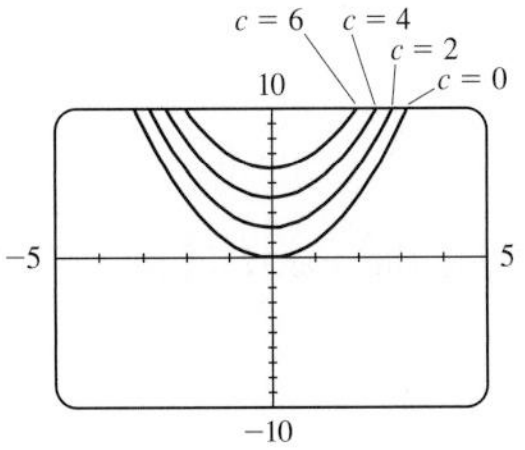

(b)

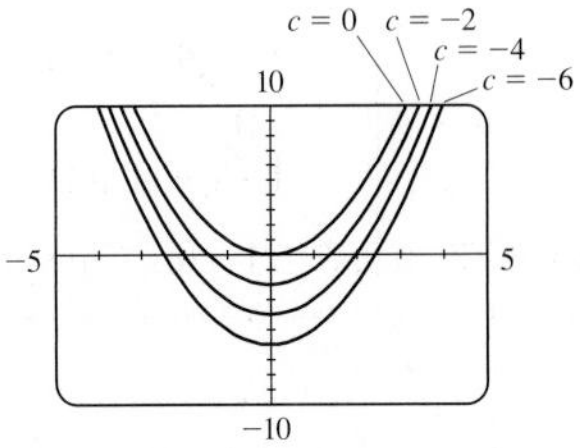

(c) If $c > 0$, then the graph of $f(x) = x^2 + c$ is the same as the graph of $y = x^2$ shifted upward c units. If $c < 0$, then the graph of $f(x) = x^2 + c$ is the same as the graph of $y = x^2$ shifted downward c units.

71. (a)

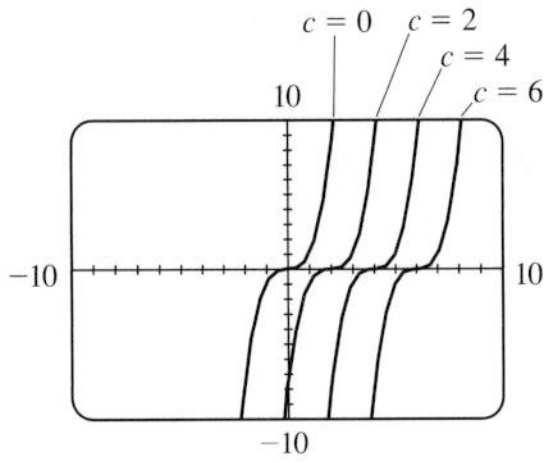

(b)

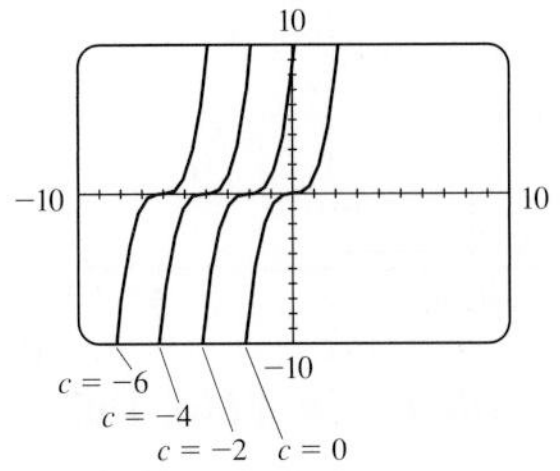

(c) If $c > 0$, then the graph of $f(x) = (x - c)^3$ is the same as the graph of $y = x^3$ shifted to the right c units. If $c < 0$, then the graph of $f(x) = (x - c)^3$ is the same as the graph of $y = x^3$ shifted to the left c units.

73. (a)

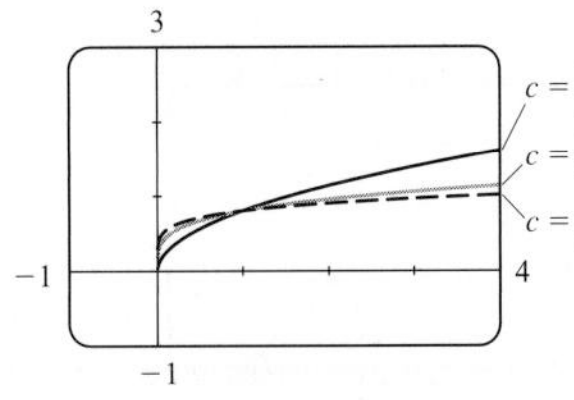

(b)

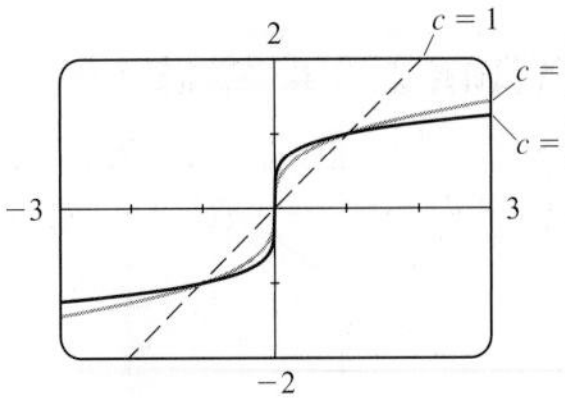

(c) Graphs of even roots are similar to $\sqrt{x}$; graphs of odd roots are similar to $\sqrt[3]{x}$. As c increases, the graph of $y = \sqrt[c]{x}$ becomes steeper near 0 and flatter when $x > 1$.

75. $f(x) = -\frac{7}{6}x - \frac{4}{3}$, $-2 \le x \le 4$

77. $f(x) = \sqrt{9 - x^2}$, $-3 \le x \le 3$

79.

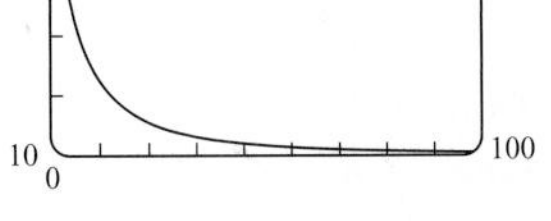

81. **(a)** $E(x) = \begin{cases} 6 + 0.10x & 0 \le x \le 300 \\ 36 + 0.06(x - 300), & x > 300 \end{cases}$

(b)

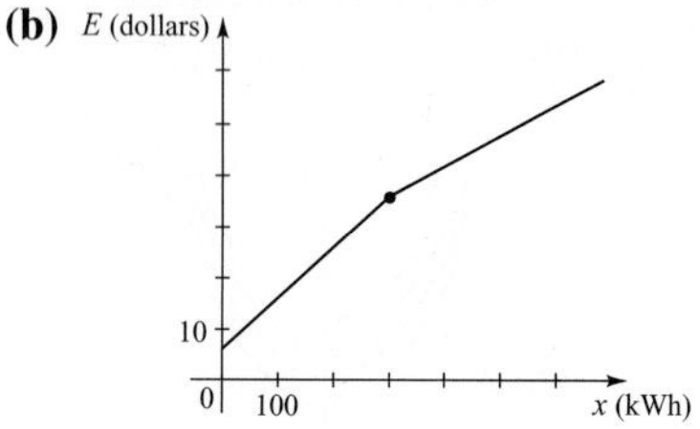

83. $P(x) = \begin{cases} 0.44 & \text{if } 0 < x \le 1 \\ 0.61 & \text{if } 1 < x \le 2 \\ 0.78 & \text{if } 2 < x \le 3 \\ 0.95 & \text{if } 3 < x \le 3.5 \end{cases}$

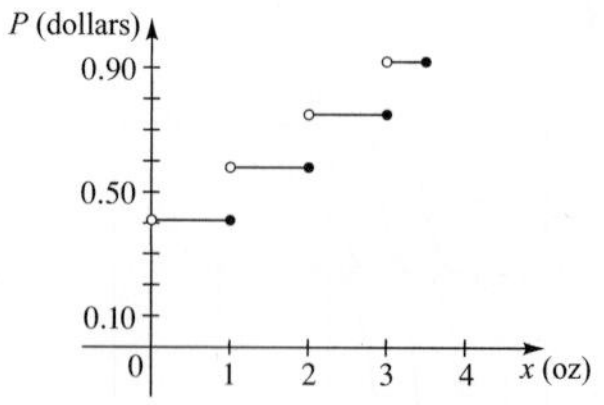

SECTION 2.3 ■ PAGE 168

1. a, 4 **2.** x, y, $[1, 6]$, $[1, 7]$
3. **(a)** increase, $[1, 2]$, $[4, 5]$ **(b)** decrease, $[2, 4]$, $[5, 6]$
4. **(a)** largest, 7, 2 **(b)** smallest, 2, 4
5. **(a)** 1, −1, 3, 4 **(b)** Domain $[-3, 4]$, range $[-1, 4]$
(c) −3, 2, 4 **(d)** $-3 \le x \le 2$ and $x = 4$
7. **(a)** 3, 2, −2, 1, 0 **(b)** Domain $[-4, 4]$, range $[-2, 3]$

9. **(a)**

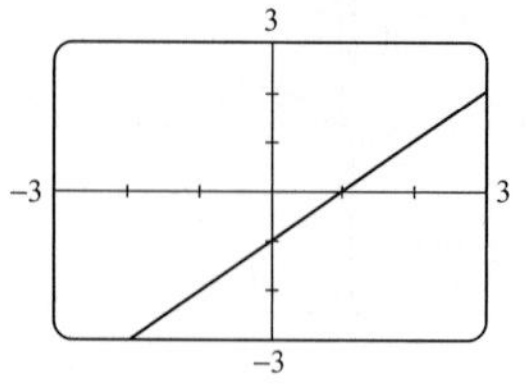

(b) Domain $(-\infty, \infty)$, range $(-\infty, \infty)$

11. **(a)**

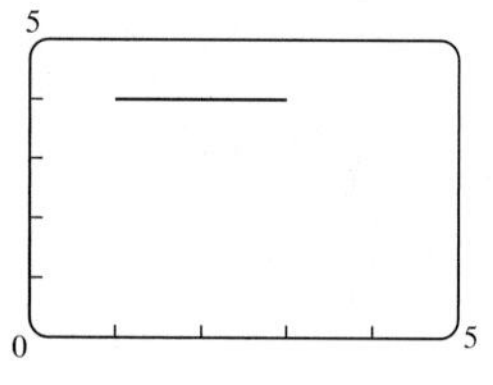

(b) Domain $[1, 3]$, range $\{4\}$

13. **(a)**

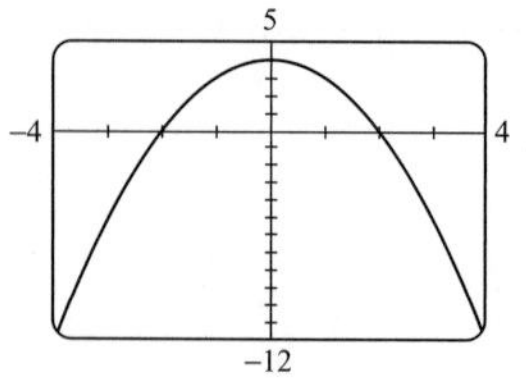

(b) Domain $(-\infty, \infty)$, range $(-\infty, 4]$

15. **(a)**

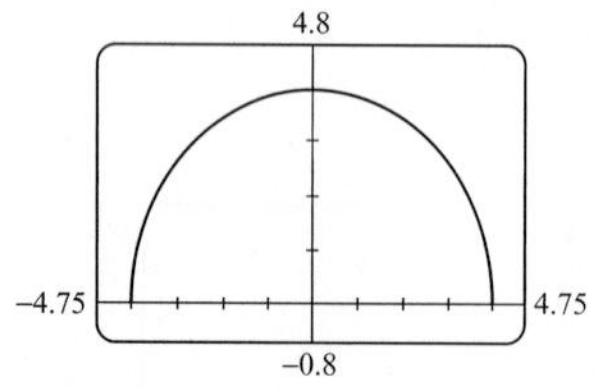

(b) Domain $[-4, 4]$, range $[0, 4]$

17. **(a)**

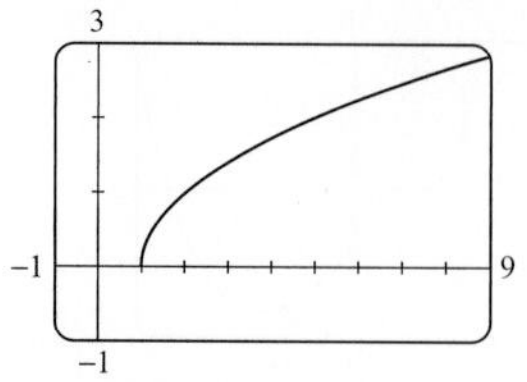

(b) Domain $[1, \infty)$, range $[0, \infty)$

19. **(a)** $[-1, 1]$, $[2, 4]$ **(b)** $[1, 2]$
21. **(a)** $[-2, -1]$, $[1, 2]$ **(b)** $[-3, -2]$, $[-1, 1]$, $[2, 3]$

23. **(a)**

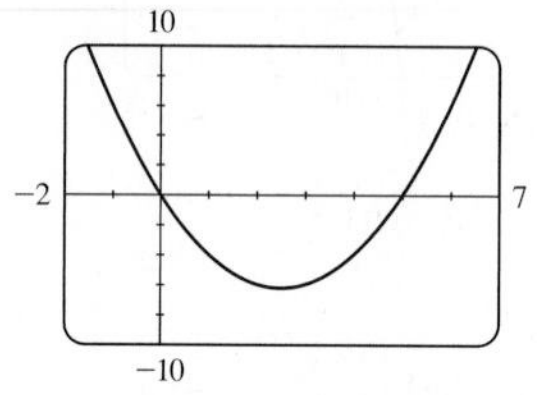

(b) Increasing on $[2.5, \infty)$; decreasing on $(-\infty, 2.5]$

25. **(a)**

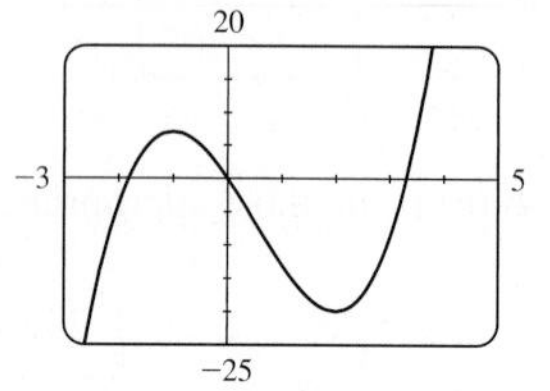

(b) Increasing on $(-\infty, -1]$, $[2, \infty)$; decreasing on $[-1, 2]$

27. **(a)**

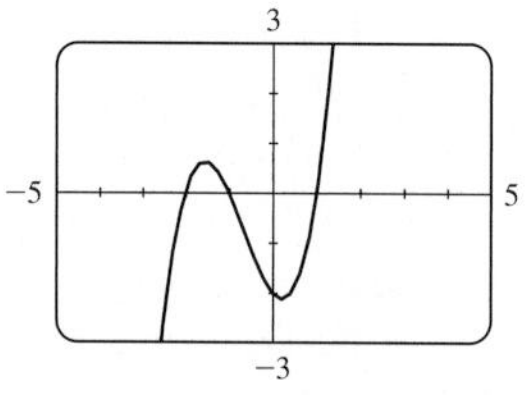

(b) Increasing on $(-\infty, -1.55]$, $[0.22, \infty)$; decreasing on $[-1.55, 0.22]$

29. **(a)**

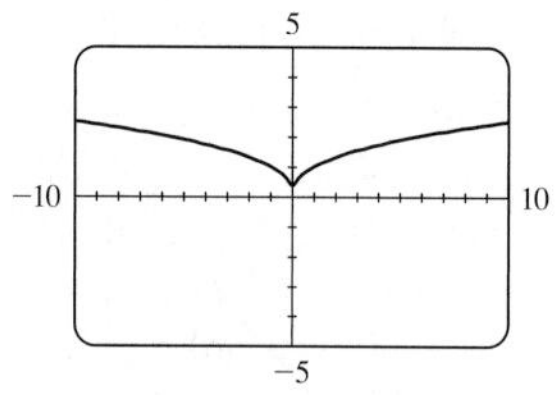

(b) Increasing on $[0, \infty)$; decreasing on $(-\infty, 0]$

31. **(a)** Local maximum 2 when $x = 0$; local minimum −1 when $x = -2$, local minimum 0 when $x = 2$ **(b)** Increasing on $[-2, 0] \cup [2, \infty)$; decreasing on $(-\infty, -2] \cup [0, 2]$
33. **(a)** Local maximum 0 when $x = 0$; local maximum 1 when $x = 3$, local minimum −2 when $x = -2$, local minimum −1 when $x = 1$ **(b)** Increasing on $[-2, 0] \cup [1, 3]$; decreasing on $(-\infty, -2] \cup [0, 1] \cup [3, \infty)$ **35.** **(a)** Local maximum ≈ 0.38 when $x \approx -0.58$; local minimum ≈ -0.38 when $x \approx 0.58$
(b) Increasing on $(-\infty, -0.58] \cup [0.58, \infty)$; decreasing on $[-0.58, 0.58]$ **37.** **(a)** Local maximum ≈ 0 when $x = 0$; local minimum ≈ -13.61 when $x \approx -1.71$, local minimum ≈ -73.32 when $x \approx 3.21$ **(b)** Increasing on $[-1.71, 0] \cup [3.21, \infty)$; decreasing on $(-\infty, -1.71] \cup [0, 3.21]$ **39.** **(a)** Local maximum ≈ 5.66 when $x \approx 4.00$ **(b)** Increasing on $(-\infty, 4.00]$; decreasing on $[4.00, 6.00]$ **41.** **(a)** Local maximum ≈ 0.38 when $x \approx -1.73$; local minimum ≈ -0.38 when $x \approx 1.73$
(b) Increasing on $(-\infty, -1.73] \cup [1.73, \infty]$; decreasing on $[-1.73, 0) \cup (0, 1.73]$ **43.** **(a)** 500 MW, 725 MW
(b) Between 3:00 A.M. and 4:00 A.M. **(c)** Just before noon
45. **(a)** Increasing on $[0, 30] \cup [32, 68]$; decreasing on $[30, 32]$
(b) He went on a crash diet and lost weight, only to regain it again later. **47.** **(a)** Increasing on $[0, 150] \cup [300, \infty)$; decreasing on $[150, 300]$ **(b)** Local maximum when $x = 150$; local minimum when $x = 300$

49. Runner A won the race. All runners finished. Runner B fell but got up again to finish second.

51. (a)

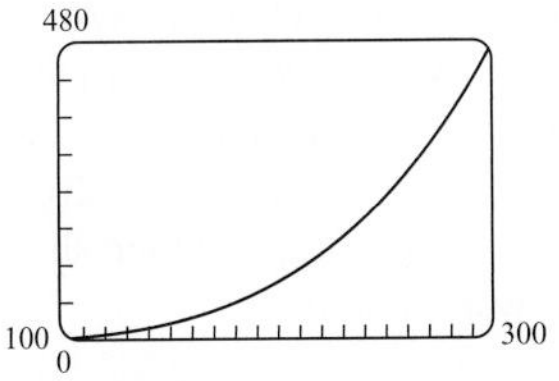

(b) Increases **53.** 20 mi/h **55.** $r \approx 0.67$ cm

SECTION 2.4 ■ PAGE 177

1. $\dfrac{100 \text{ miles}}{2 \text{ hours}} = 50$ mi/h **2.** $\dfrac{f(b) - f(a)}{b - a}$ **3.** $\dfrac{25 - 1}{5 - 1} = 6$
4. (a) secant **(b)** 3 **5.** $\frac{2}{3}$ **7.** $-\frac{4}{5}$ **9.** 3 **11.** 5 **13.** 60
15. $12 + 3h$ **17.** $-\dfrac{1}{a}$ **19.** $\dfrac{-2}{a(a + h)}$ **21. (a)** $\frac{1}{2}$
23. -0.25 ft/day **25. (a)** 245 persons/yr
(b) -328.5 persons/yr **(c)** 1997–2001 **(d)** 2001–2006
27. (a) 7.2 units/yr **(b)** 8 units/yr **(c)** -55 units/yr
(d) 2000–2001, 2001–2002 **29.** First 20 minutes: 4.05°F/min, next 20 minutes: 1.5°F/min; first interval

SECTION 2.5 ■ PAGE 187

1. (a) up **(b)** left **2. (a)** down **(b)** right **3. (a)** x-axis **(b)** y-axis **4. (a)** II **(b)** IV **(c)** I **(d)** III **5. (a)** Shift downward 5 units **(b)** Shift to the right 5 units **7. (a)** Reflect in the x-axis **(b)** Reflect in the y-axis **9. (a)** Reflect in the x-axis, then shift upward 5 units **(b)** Stretch vertically by a factor of 3, then shift downward 5 units **11. (a)** Shift to the left 1 unit, stretch vertically by a factor of 2, then shift downward 3 units **(b)** Shift to the right 1 unit, stretch vertically by a factor of 2, then shift upward 3 units **13. (a)** Shrink horizontally by a factor of $\frac{1}{4}$ **(b)** Stretch horizontally by a factor of 4 **15. (a)** Shift to the left 2 units **(b)** Shift upward 2 units **17. (a)** Shift to the left 2 units, then shift downward 2 units **(b)** Shift to the right 2 units, then shift upward 2 units

19. (a)

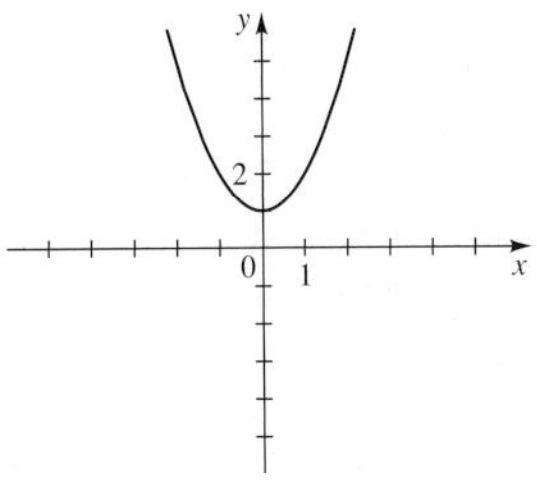

(b)

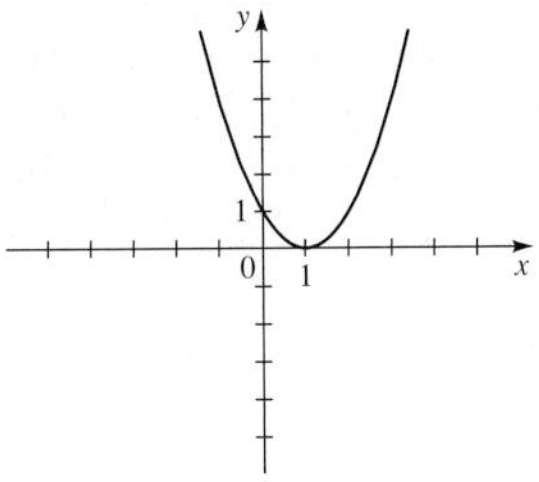

(c)

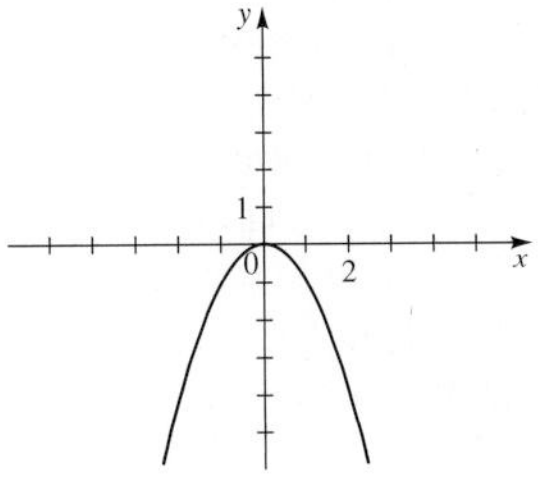

(d)

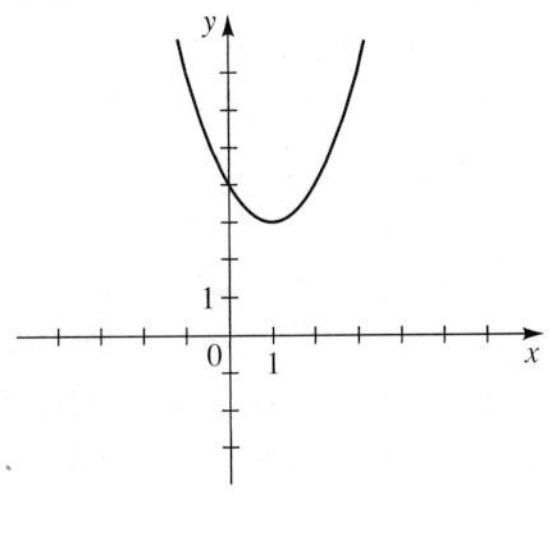

21.

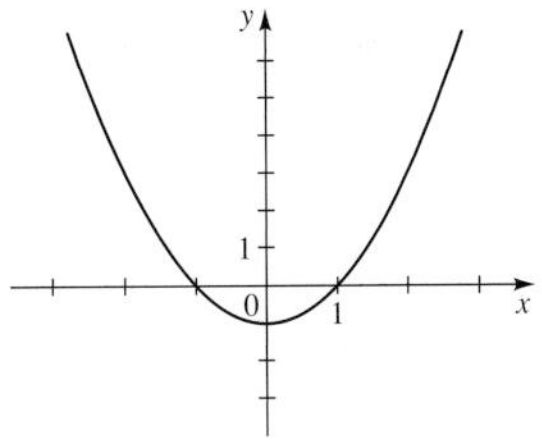

23.

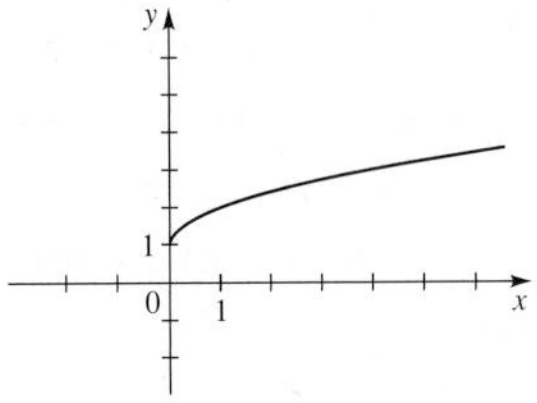

25.

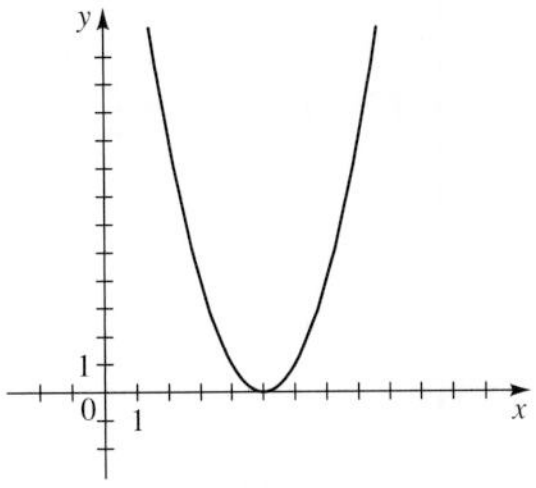

27.

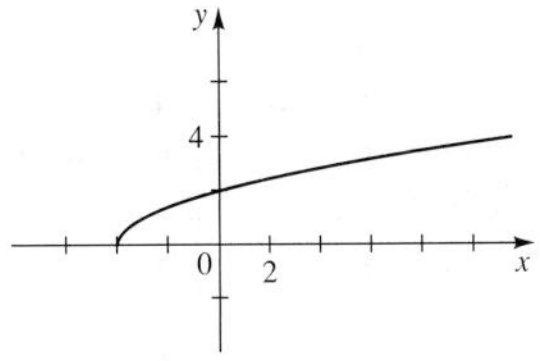

29.

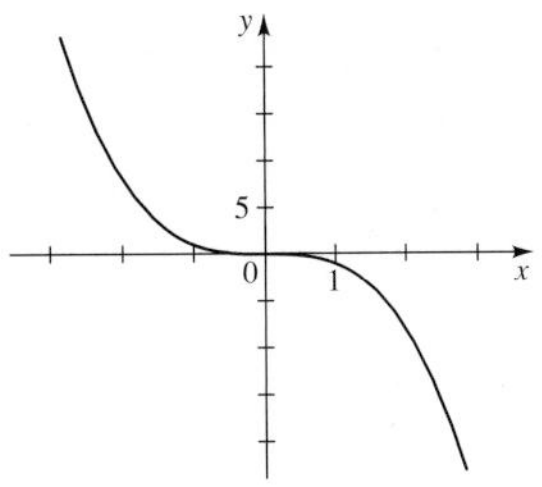

31.

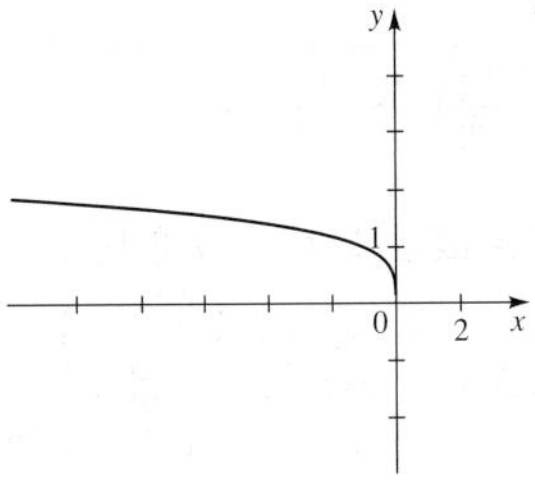

33.

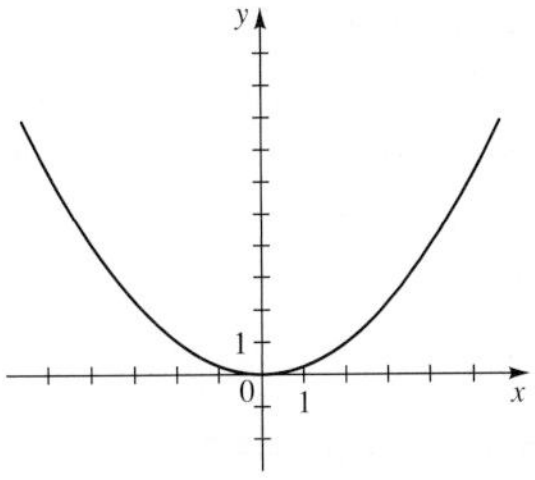

35.

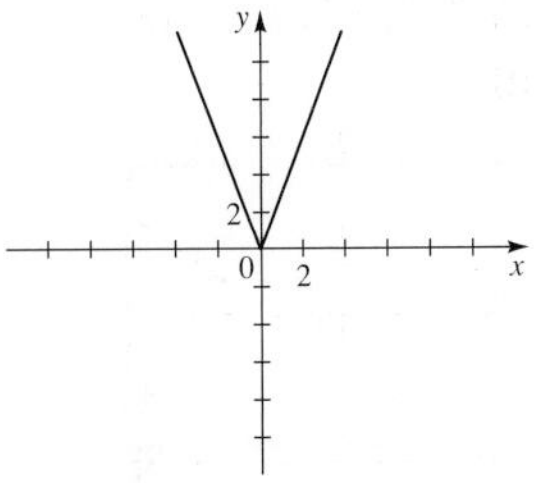

37.

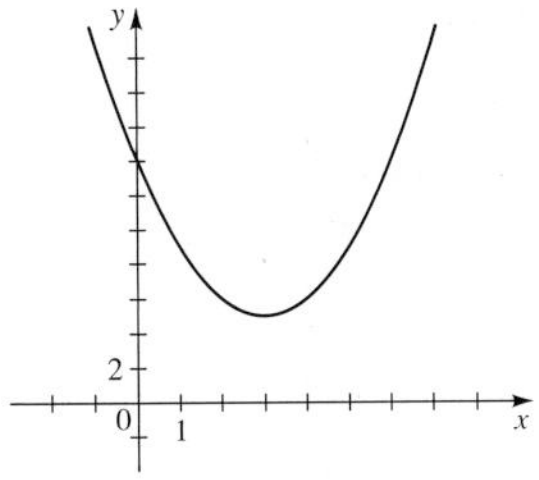

39.

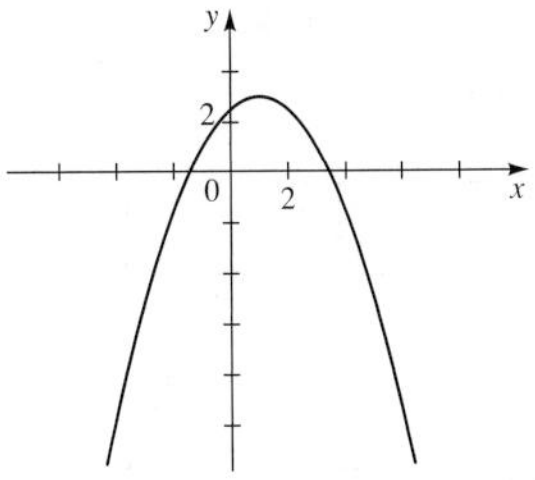

41.

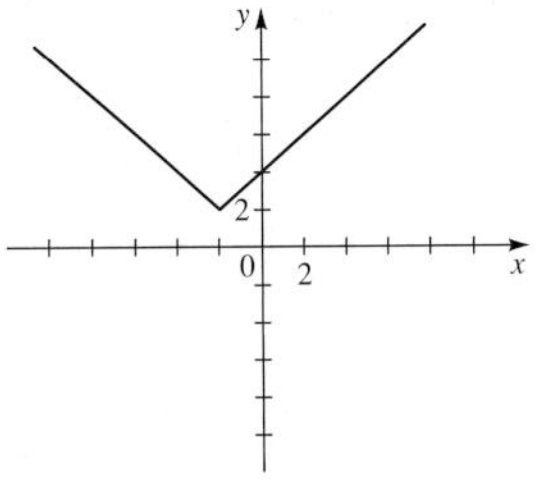

43.

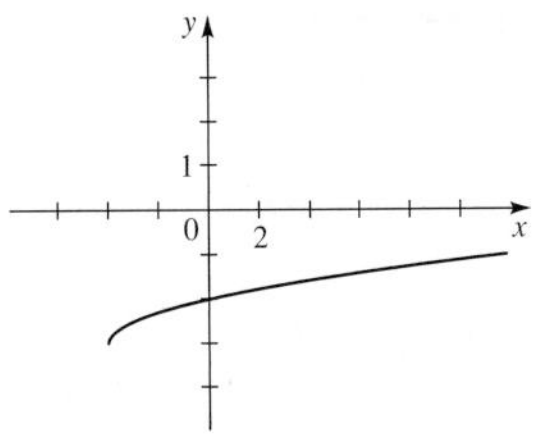

45. $f(x) = x^2 + 3$ **47.** $f(x) = \sqrt{x + 2}$
49. $f(x) = |x - 3| + 1$ **51.** $f(x) = \sqrt[4]{-x} + 1$
53. $f(x) = 2(x - 3)^2 - 2$ **55.** $g(x) = (x - 2)^2$
57. $g(x) = |x + 1| + 2$ **59.** $g(x) = -\sqrt{x + 2}$
61. (a) 3 **(b)** 1 **(c)** 2 **(d)** 4

63. (a)

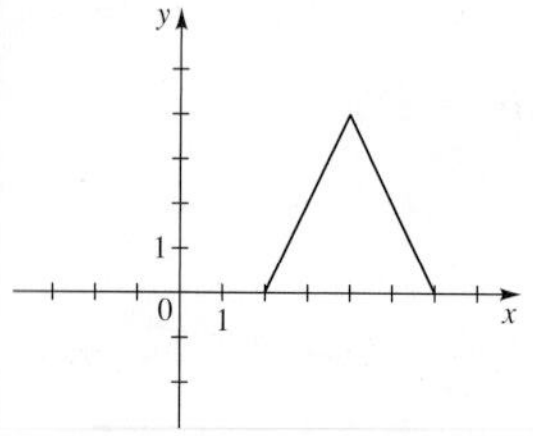

(b)

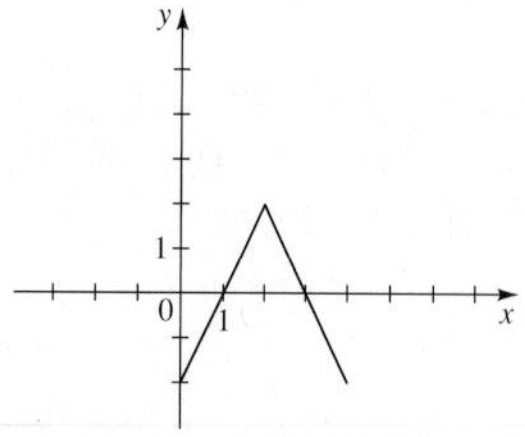

(c)

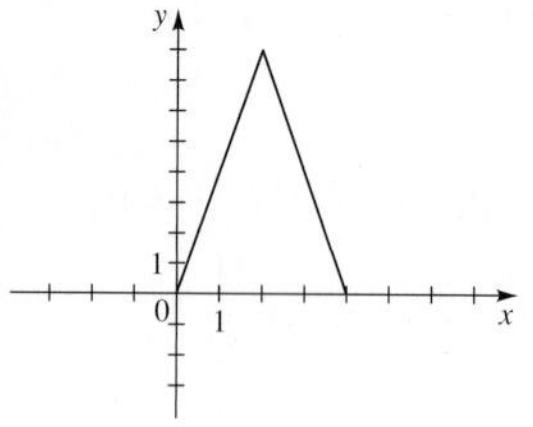

(d)

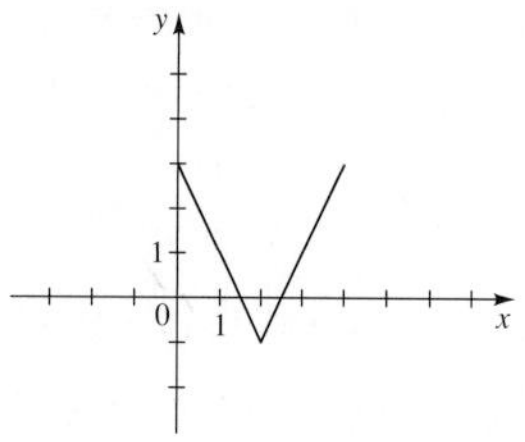

(e)

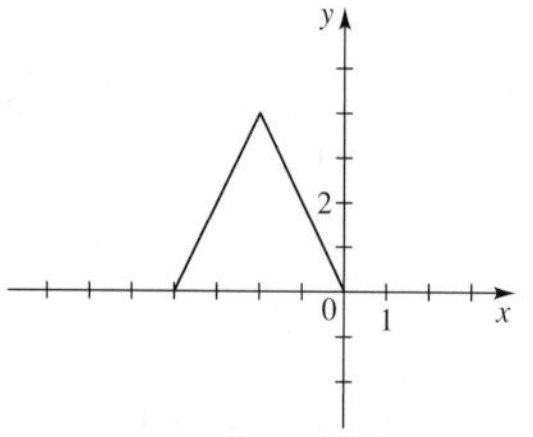

(f)

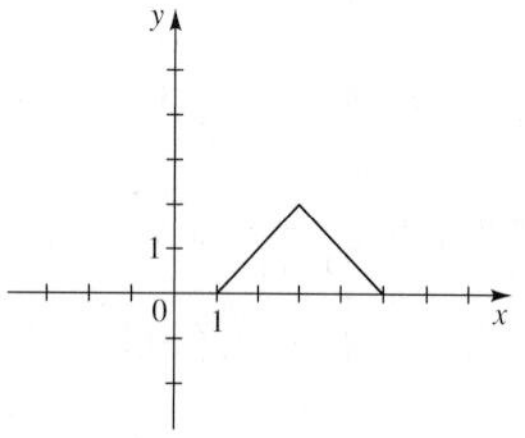

65. (a)

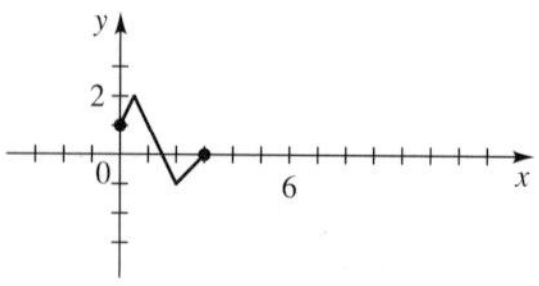

(b)

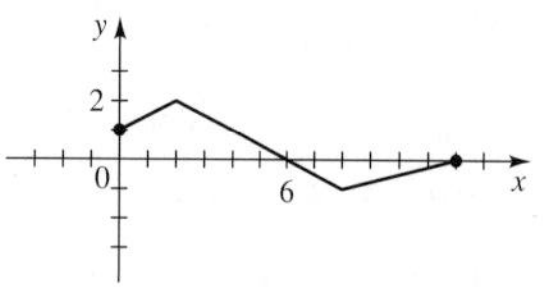

67.

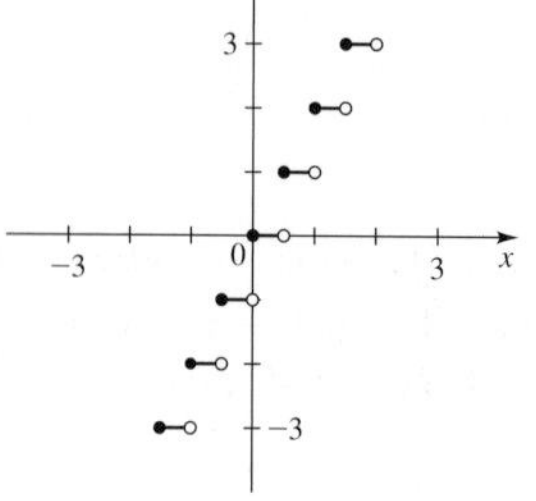

69.

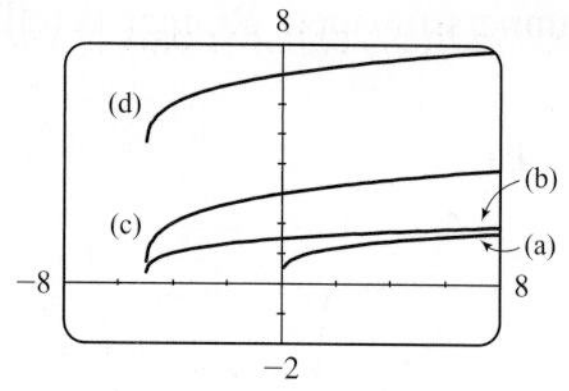

For part (b) shift the graph in (a) to the left 5 units; for part (c) shift the graph in (a) to the left 5 units and stretch vertically by a factor of 2; for part (d) shift the graph in (a) to the left 5 units, stretch vertically by a factor of 2, and then shift upward 4 units.

71.

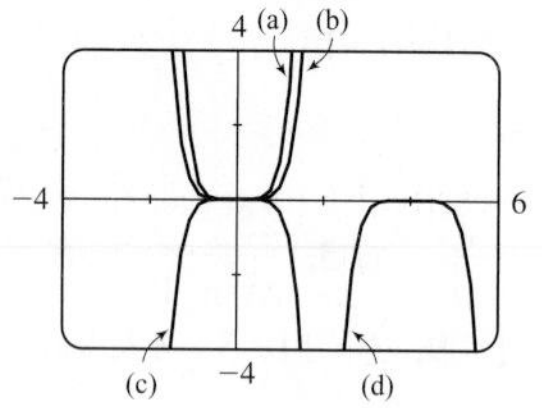

For part (b) shrink the graph in (a) vertically by a factor of $\frac{1}{3}$; for part (c) shrink the graph in (a) vertically by a factor of $\frac{1}{3}$ and reflect in the x-axis; for part (d) shift the graph in (a) to the right 4 units, shrink vertically by a factor of $\frac{1}{3}$, and then reflect in the x-axis.

73.

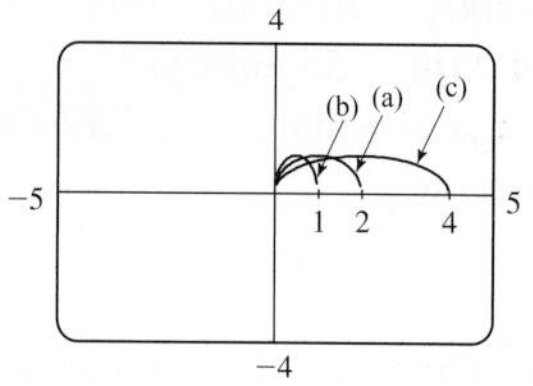

The graph in part (b) is shrunk horizontally by a factor of $\frac{1}{2}$ and the graph in part (c) is stretched by a factor of 2.

75. Even

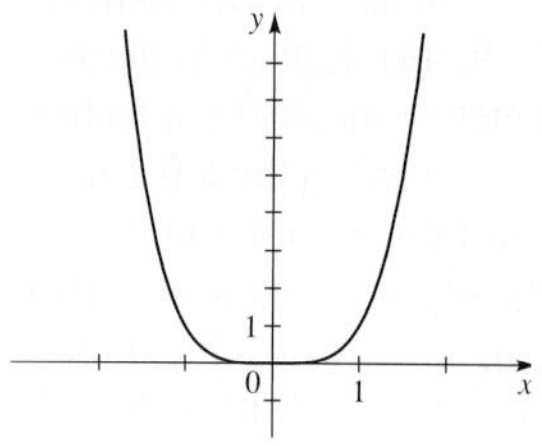

77. Neither

79. Odd

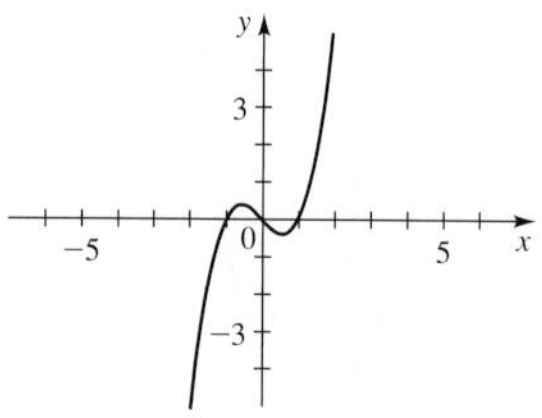

81. Neither

83. (a)

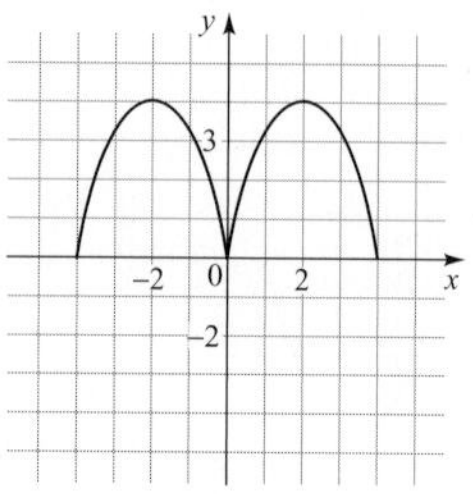

(b)

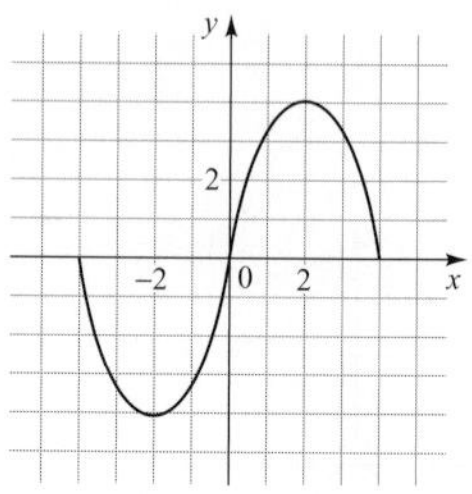

85. To obtain the graph of g, reflect in the x-axis the part of the graph of f that is below the x-axis.

87. (a)

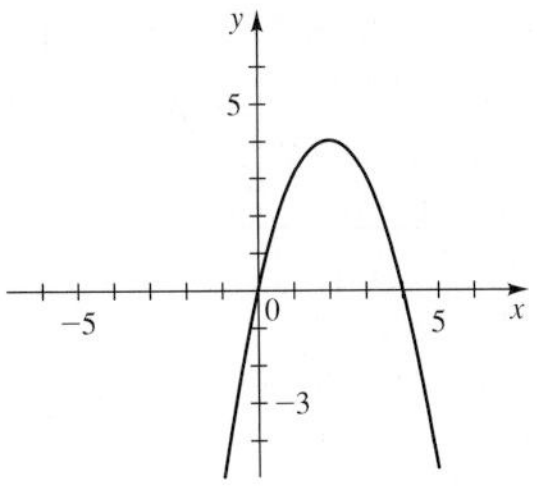

(b)

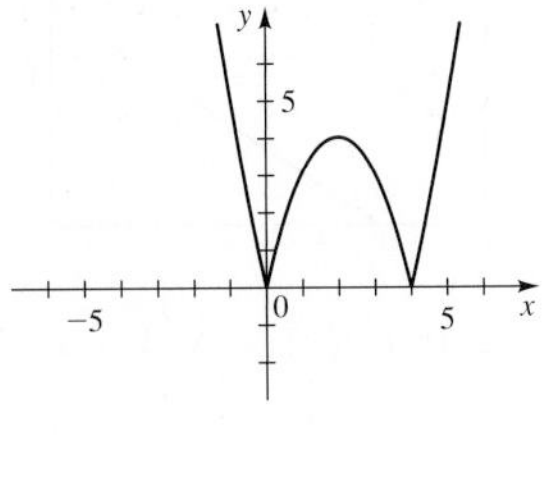

89. (a) Shift upward 4 units, shrink vertically by a factor of 0.01 **(b)** Shift to the right 10 units; $g(t) = 4 + 0.01(t - 10)^2$

SECTION 2.6 ■ PAGE 196

1. $8, -2, 15, \frac{3}{5}$ **2.** $f(g(x))$, 12 **3.** Multiply by 2, then add 1; Add 1, then multiply by 2 **4.** $x + 1, 2x, 2x + 1, 2(x + 1)$

5. $(f + g)(x) = x^2 + x - 3, (-\infty, \infty)$;
$(f - g)(x) = -x^2 + x - 3, (-\infty, \infty)$;
$(fg)(x) = x^3 - 3x^2, (-\infty, \infty)$;
$\left(\frac{f}{g}\right)(x) = \frac{x - 3}{x^2}, (-\infty, 0) \cup (0, \infty)$

7. $(f + g)(x) = \sqrt{4 - x^2} + \sqrt{1 + x}, [-1, 2]$;
$(f - g)(x) = \sqrt{4 - x^2} - \sqrt{1 + x}, [-1, 2]$;
$(fg)(x) = \sqrt{-x^3 - x^2 + 4x + 4}, [-1, 2]$;
$\left(\frac{f}{g}\right)(x) = \sqrt{\frac{4 - x^2}{1 + x}}, (-1, 2]$

9. $(f + g)(x) = \frac{6x + 8}{x^2 + 4x}, x \neq -4, x \neq 0$;
$(f - g)(x) = \frac{-2x + 8}{x^2 + 4x}, x \neq -4, x \neq 0$;
$(fg)(x) = \frac{8}{x^2 + 4x}, x \neq -4, x \neq 0$;
$\left(\frac{f}{g}\right)(x) = \frac{x + 4}{2x}, x \neq -4, x \neq 0$

11. $[0, 1]$ **13.** $(3, \infty)$

15.

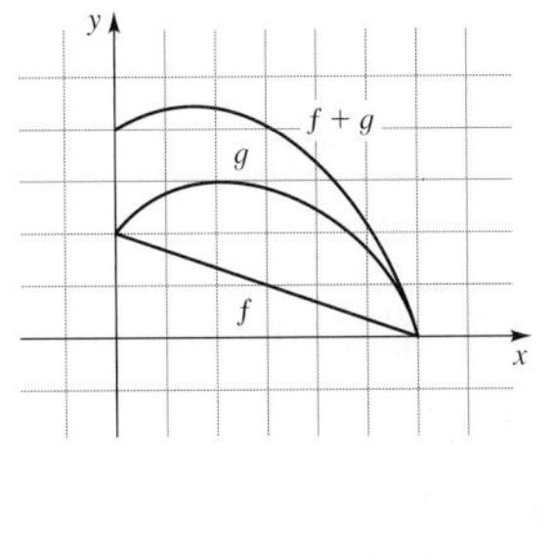

17.

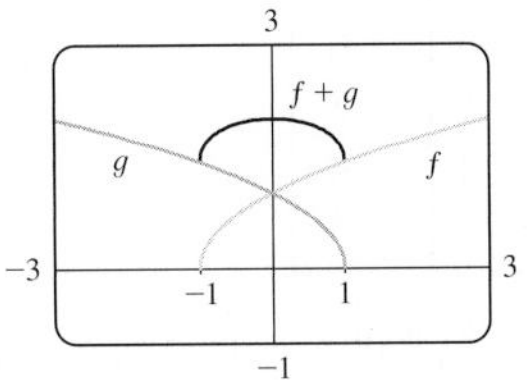

19.

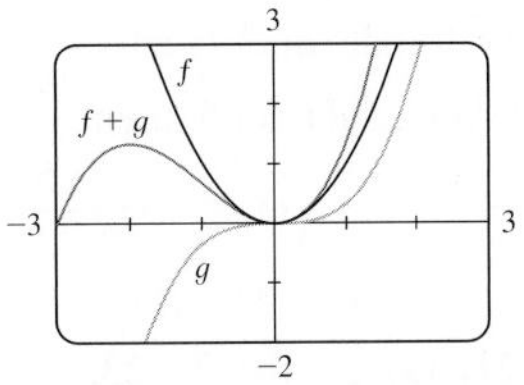

21. (a) 1 **(b)** −23 **23. (a)** −11 **(b)** −119
25. (a) $-3x^2 + 1$ **(b)** $-9x^2 + 30x - 23$
27. 4 **29.** 5 **31.** 4

33. $(f \circ g)(x) = 8x + 1, (-\infty, \infty)$;
$(g \circ f)(x) = 8x + 11, (-\infty, \infty)$; $(f \circ f)(x) = 4x + 9, (-\infty, \infty)$;
$(g \circ g)(x) = 16x - 5, (-\infty, \infty)$

35. $(f \circ g)(x) = (x + 1)^2, (-\infty, \infty)$;
$(g \circ f)(x) = x^2 + 1, (-\infty, \infty)$; $(f \circ f)(x) = x^4, (-\infty, \infty)$;
$(g \circ g)(x) = x + 2, (-\infty, \infty)$

37. $(f \circ g)(x) = \frac{1}{2x + 4}, x \neq -2$; $(g \circ f)(x) = \frac{2}{x} + 4, x \neq 0$;
$(f \circ f)(x) = x, x \neq 0$, $(g \circ g)(x) = 4x + 12, (-\infty, \infty)$

39. $(f \circ g)(x) = |2x + 3|, (-\infty, \infty)$;
$(g \circ f)(x) = 2|x| + 3, (-\infty, \infty)$; $(f \circ f)(x) = |x|, (-\infty, \infty)$;
$(g \circ g)(x) = 4x + 9, (-\infty, \infty)$

41. $(f \circ g)(x) = \frac{2x - 1}{2x}, x \neq 0$; $(g \circ f)(x) = \frac{2x}{x + 1} - 1, x \neq -1$;
$(f \circ f)(x) = \frac{x}{2x + 1}, x \neq -1, x \neq -\frac{1}{2}$;
$(g \circ g)(x) = 4x - 3, (-\infty, \infty)$

43. $(f \circ g)(x) = \frac{1}{x + 1}, x \neq -1, x \neq 0$; $(g \circ f)(x) = \frac{x + 1}{x}$,
$x \neq -1, x \neq 0$; $(f \circ f)(x) = \frac{x}{2x + 1}, x \neq -1, x \neq -\frac{1}{2}$;
$(g \circ g)(x) = x, x \neq 0$

45. $(f \circ g \circ h)(x) = \sqrt{x - 1} - 1$
47. $(f \circ g \circ h)(x) = (\sqrt{x} - 5)^4 + 1$
49. $g(x) = x - 9, f(x) = x^5$
51. $g(x) = x^2, f(x) = x/(x + 4)$
53. $g(x) = 1 - x^3, f(x) = |x|$
55. $h(x) = x^2, g(x) = x + 1, f(x) = 1/x$
57. $h(x) = \sqrt[3]{x}, g(x) = 4 + x, f(x) = x^9$
59. $R(x) = 0.15x - 0.000002x^2$
61. (a) $g(t) = 60t$ **(b)** $f(r) = \pi r^2$ **(c)** $(f \circ g)(t) = 3600\pi t^2$
63. $A(t) = 16\pi t^2$ **65. (a)** $f(x) = 0.9x$ **(b)** $g(x) = x - 100$
(c) $(f \circ g)(x) = 0.9x - 90$, $(g \circ f)(x) = 0.9x - 100$, $f \circ g$: first rebate, then discount, $g \circ f$: first discount, then rebate, $g \circ f$ is the better deal

SECTION 2.7 ■ PAGE 204

1. different, Horizontal Line **2. (a)** one-to-one, $g(x) = x^3$ **(b)** $g^{-1}(x) = x^{1/3}$ **3. (a)** Take the cube root, subtract 5, then divide the result by 3. **(b)** $f(x) = (3x + 5)^3, f^{-1}(x) = \dfrac{x^{1/3} - 5}{3}$
4. (a) False **(b)** True **5.** No **7.** Yes **9.** No **11.** Yes
13. Yes **15.** No **17.** No **19.** No **21. (a)** 2 **(b)** 3 **23.** 1
37. $f^{-1}(x) = \frac{1}{2}(x - 1)$ **39.** $f^{-1}(x) = \frac{1}{4}(x - 7)$
41. $f^{-1}(x) = \sqrt[3]{\frac{1}{4}(5 - x)}$ **43.** $f^{-1}(x) = (1/x) - 2$
45. $f^{-1}(x) = \dfrac{4x}{1 - x}$ **47.** $f^{-1}(x) = \dfrac{7x + 5}{x - 2}$
49. $f^{-1}(x) = (5x - 1)/(2x + 3)$
51. $f^{-1}(x) = \frac{1}{5}(x^2 - 2), x \ge 0$
53. $f^{-1}(x) = \sqrt{4 - x}, x \le 4$
55. $f^{-1}(x) = (x - 4)^3$
57. $f^{-1}(x) = x^2 - 2x, x \ge 1$
59. $f^{-1}(x) = \sqrt[4]{x}$
61. (a)

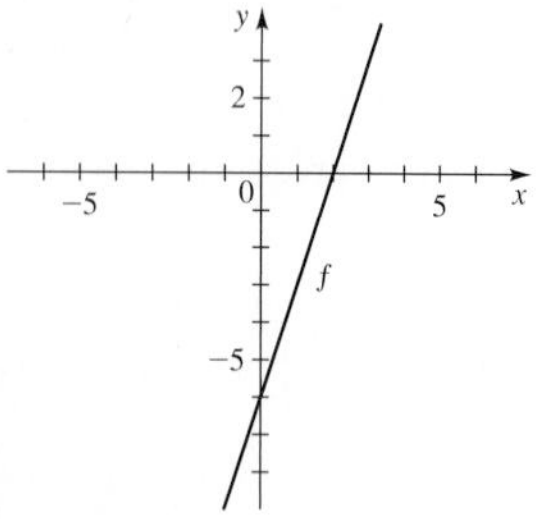

(b)

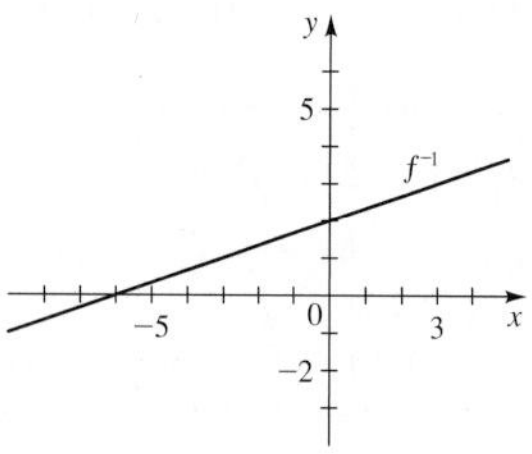

(c) $f^{-1}(x) = \frac{1}{3}(x + 6)$
63. (a)

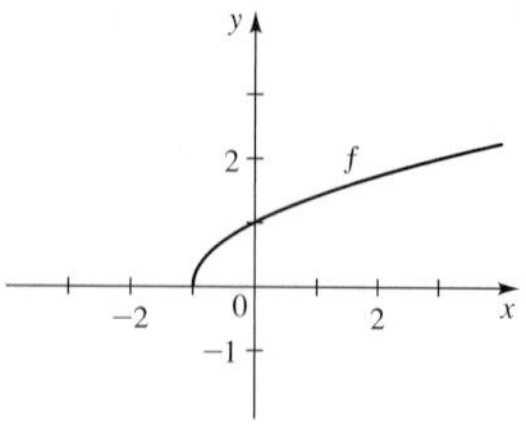

(b)

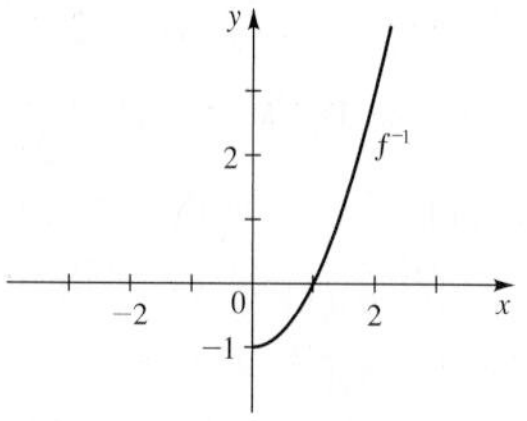

(c) $f^{-1}(x) = x^2 - 1, x \ge 0$
65. Not one-to-one

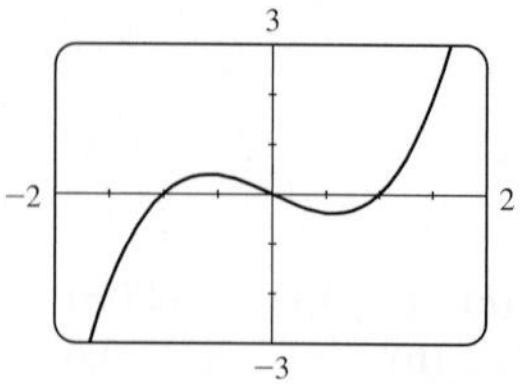

67. One-to-one

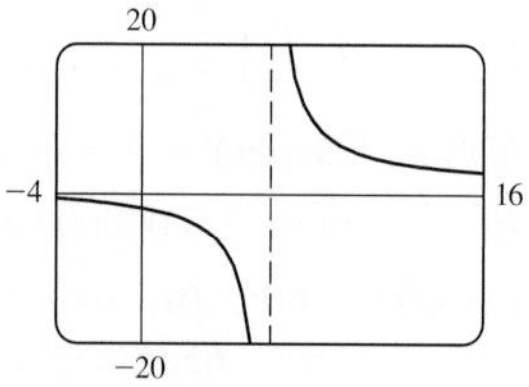

69. Not one-to-one

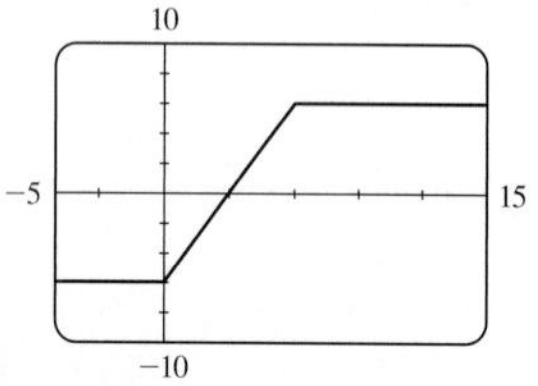

71. (a) $f^{-1}(x) = x - 2$
(b)

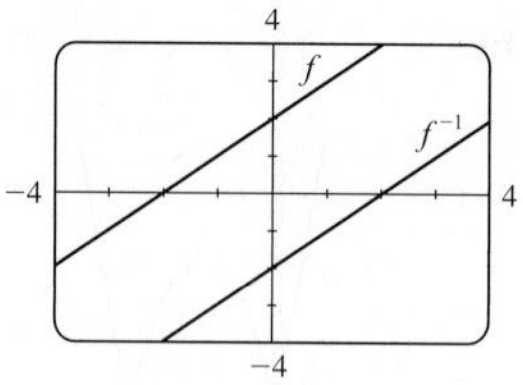

73. (a) $g^{-1}(x) = x^2 - 3, x \ge 0$
(b)

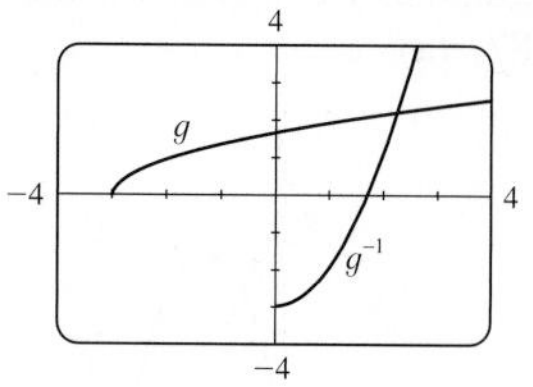

75. $x \ge 0, f^{-1}(x) = \sqrt{4 - x}$ **77.** $x \ge -2, h^{-1}(x) = \sqrt{x} - 2$
79.

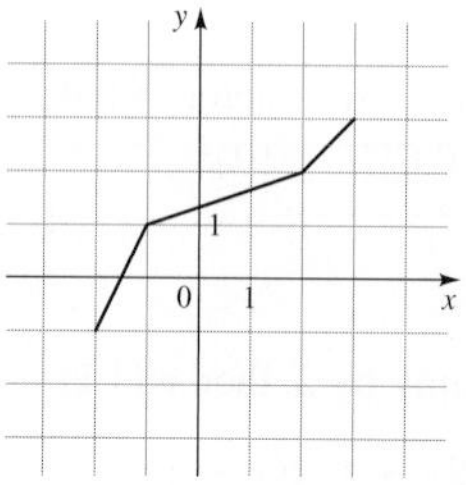

81. (a) $f(x) = 500 + 80x$ **(b)** $f^{-1}(x) = \frac{1}{80}(x - 500)$, the number of hours worked as a function of the fee
(c) 9; if he charges $1220, he worked 9 h
83. (a) $v^{-1}(t) = \sqrt{0.25 - \dfrac{t}{18{,}500}}$ **(b)** 0.498; at a distance 0.498 from the central axis the velocity is 30
85. (a) $F^{-1}(x) = \frac{5}{9}(x - 32)$; the Celsius temperature when the Fahrenheit temperature is x **(b)** $F^{-1}(86) = 30$; when the temperature is 86°F, it is 30°C
87. (a) $f(x) = \begin{cases} 0.1x & \text{if } 0 \le x \le 20{,}000 \\ 2000 + 0.2(x - 20{,}000) & \text{if } x > 20{,}000 \end{cases}$
(b) $f^{-1}(x) = \begin{cases} 10x & \text{if } 0 \le x \le 2000 \\ 10{,}000 + 5x & \text{if } x > 2000 \end{cases}$
If you pay x euros (€) in taxes, your income is $f^{-1}(x)$.
(c) $f^{-1}(10{,}000) =$ € 60,000
89. $f^{-1}(x) = \frac{1}{2}(x - 7)$. A pizza costing x dollars has $f^{-1}(x)$ toppings.

CHAPTER 2 REVIEW ■ PAGE 208

1. $f(x) = x^2 - 5$ **3.** Add 10, then multiply the result by 3.
5.

x	$g(x)$
−1	5
0	0
1	−3
2	−4
3	−3

7. (a) $C(1000) = 34{,}000$, $C(10{,}000) = 205{,}000$ **(b)** The costs of printing 1000 and 10,000 copies of the book **(c)** $C(0) = 5000$; fixed costs **9.** $6, 2, 18, a^2 - 4a + 6, a^2 + 4a + 6, x^2 - 2x + 3, 4x^2 - 8x + 6, 2x^2 - 8x + 10$ **11. (a)** Not a function **(b)** Function **(c)** Function, one-to-one **(d)** Not a function
13. Domain $[-3, \infty)$, range $[0, \infty)$ **15.** $(-\infty, \infty)$

17. $[-4, \infty)$ **19.** $\{x \mid x \neq -2, -1, 0\}$ **21.** $(-\infty, -1] \cup [1, 4]$

23.

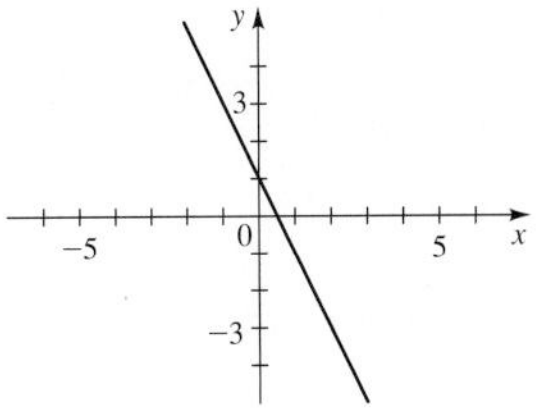

25.

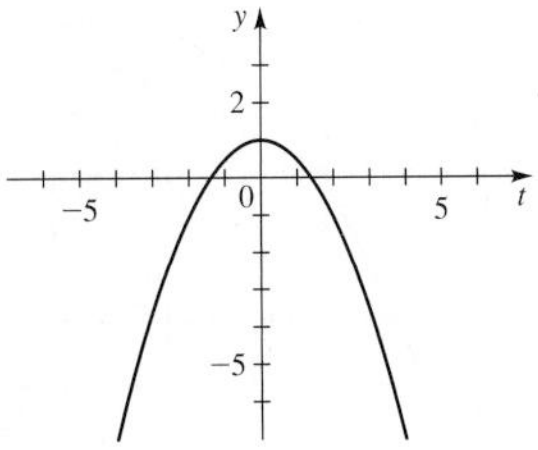

27.

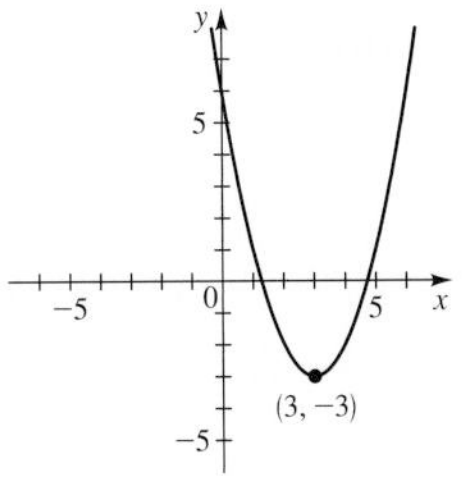

29.

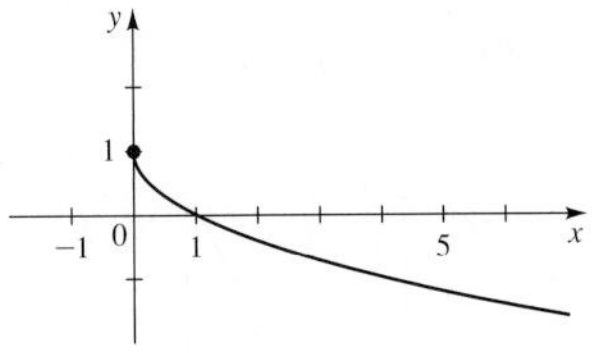

31.

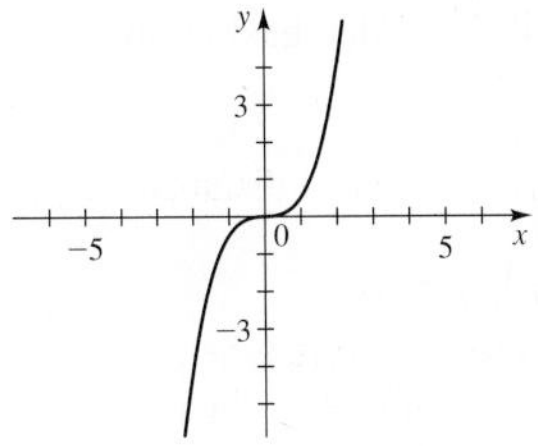

33.

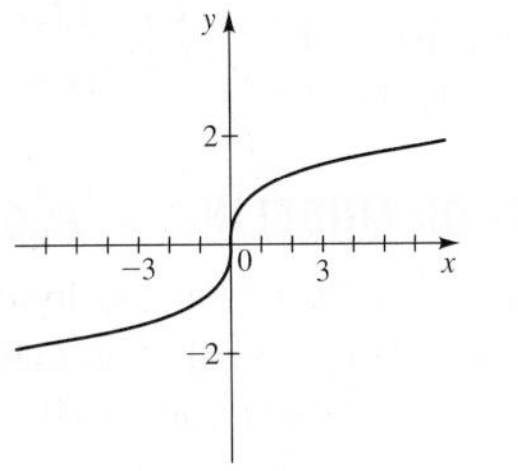

35.

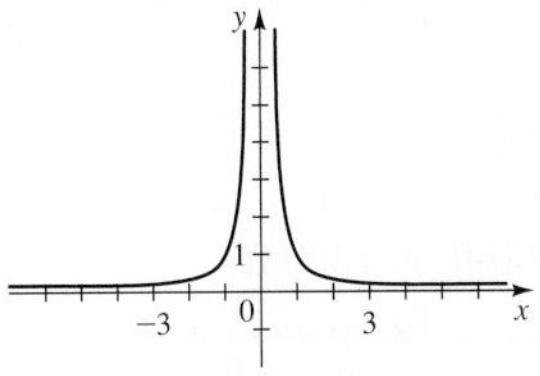

37.

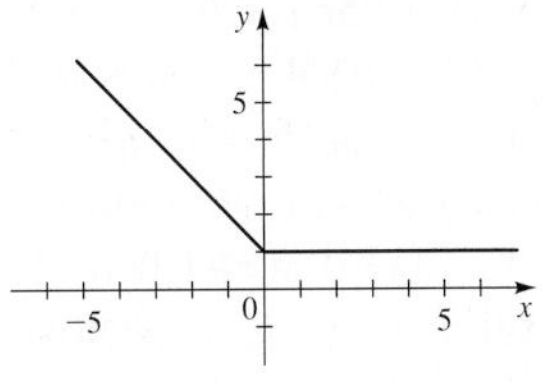

39.

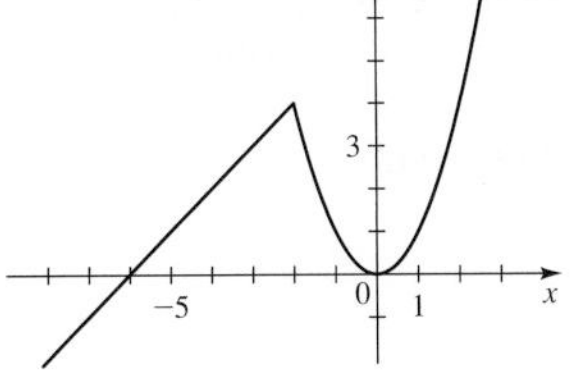

41. No **43.** Yes **45.** (iii)

47.

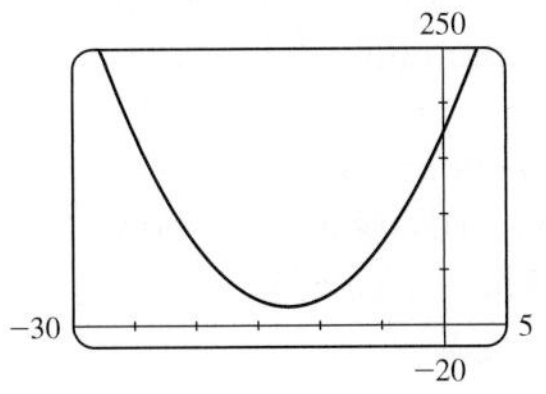

49.

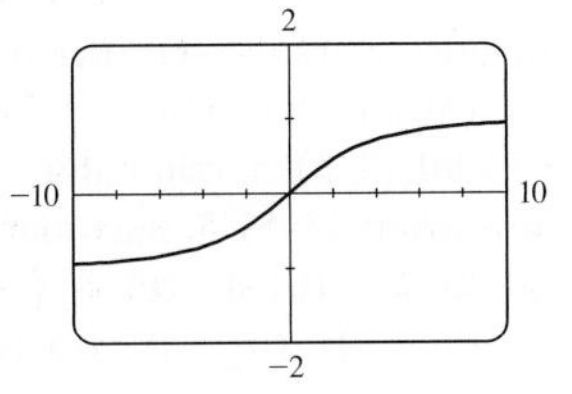

51. $[-2.1, 0.2] \cup [1.9, \infty)$

53.

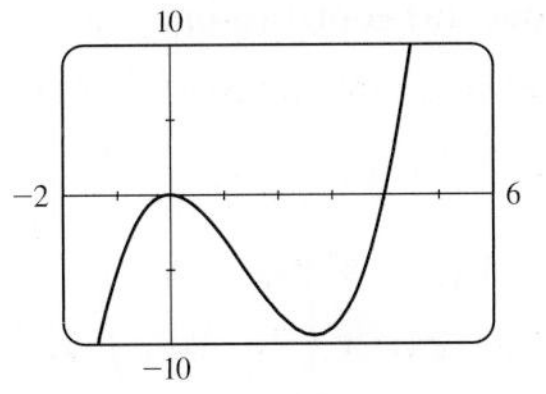

Increasing on $(-\infty, 0]$, $[2.67, \infty)$; decreasing on $[0, 2.67]$

55. 5 **57.** $\dfrac{-1}{3(3+h)}$ **59.** **(a)** $P(10) = 5010$, $P(20) = 7040$; the populations in 1995 and 2005 **(b)** 203 people/yr; average annual population increase **61.** **(a)** $\frac{1}{2}, \frac{1}{2}$ **(b)** Yes, because it is a linear function **63.** **(a)** Shift upward 8 units **(b)** Shift to the left 8 units **(c)** Stretch vertically by a factor of 2, then shift upward 1 unit **(d)** Shift to the right 2 units and downward 2 units **(e)** Reflect in y-axis **(f)** Reflect in y-axis, then in x-axis **(g)** Reflect in x-axis **(h)** Reflect in line $y = x$

65. **(a)** Neither **(b)** Odd **(c)** Even **(d)** Neither

67. $g(-1) = -7$ **69.** 68 ft **71.** Local maximum ≈ 3.79 when $x \approx 0.46$; local minimum ≈ 2.81 when $x \approx -0.46$

73.

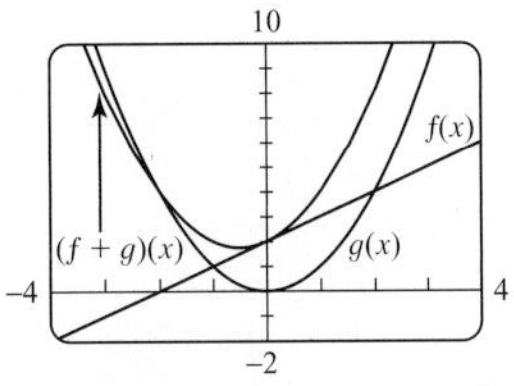

75. **(a)** $(f + g)(x) = x^2 - 6x + 6$

(b) $(f - g)(x) = x^2 - 2$

(c) $(fg)(x) = -3x^3 + 13x^2 - 18x + 8$

(d) $(f/g)(x) = (x^2 - 3x + 2)/(4 - 3x)$

(e) $(f \circ g)(x) = 9x^2 - 15x + 6$

(f) $(g \circ f)(x) = -3x^2 + 9x - 2$

77. $(f \circ g)(x) = -3x^2 + 6x - 1, (-\infty, \infty)$;

$(g \circ f)(x) = -9x^2 + 12x - 3, (-\infty, \infty)$;

$(f \circ f)(x) = 9x - 4, (-\infty, \infty)$;

$(g \circ g)(x) = -x^4 + 4x^3 - 6x^2 + 4x, (-\infty, \infty)$

79. $(f \circ g \circ h)(x) = 1 + \sqrt{x}$ **81.** Yes **83.** No

85. No **87.** $f^{-1}(x) = \dfrac{x+2}{3}$

89. $f^{-1}(x) = \sqrt[3]{x} - 1$

91. **(a), (b)**

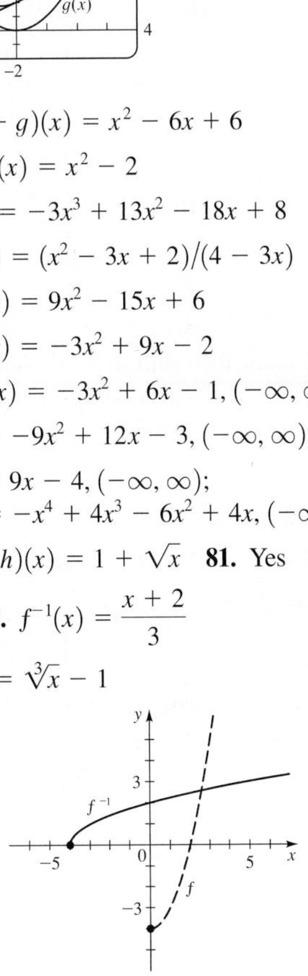

(c) $f^{-1}(x) = \sqrt{x + 4}$

CHAPTER 2 TEST ■ PAGE 211

1. (a) and (b) are graphs of functions, (a) is one-to-one
2. **(a)** $2/3$, $\sqrt{6}/5$, $\sqrt{a}/(a-1)$ **(b)** $[-1, 0) \cup (0, \infty)$
3. **(a)** $f(x) = (x-2)^3$

(b)

x	$f(x)$
-1	-27
0	-8
1	-1
2	0
3	1
4	8

(c)

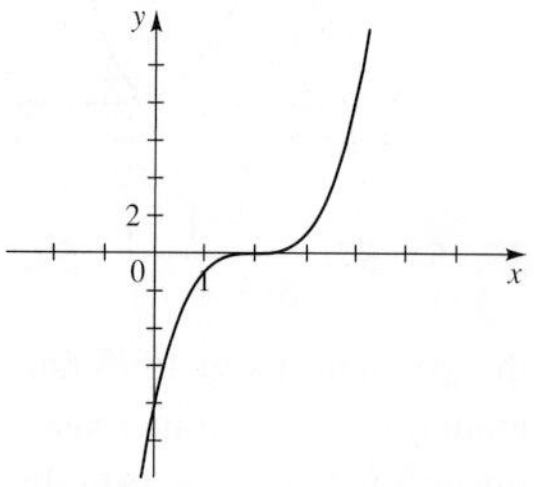

(d) By the Horizontal Line Test; take the cube root, then add 2
(e) $f^{-1}(x) = x^{1/3} + 2$ **4.** **(a)** $R(2) = 4000$, $R(4) = 4000$; total sales revenue with prices of \$2 and \$4
(b)

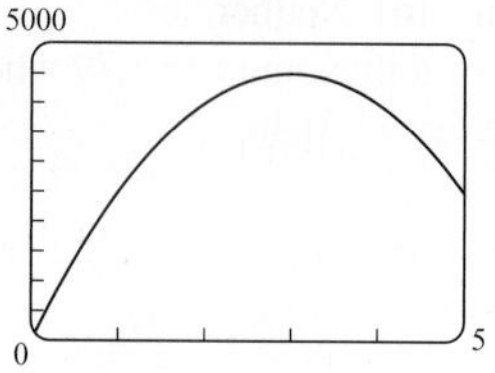

Revenue increases until price reaches \$3, then decreases

(c) \$4500; \$3 **5.** 5
6. **(a)**

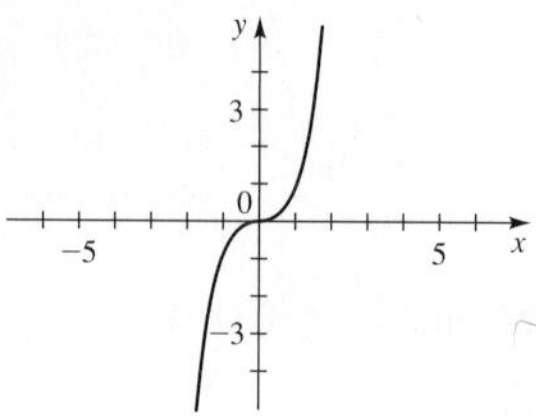

(b)

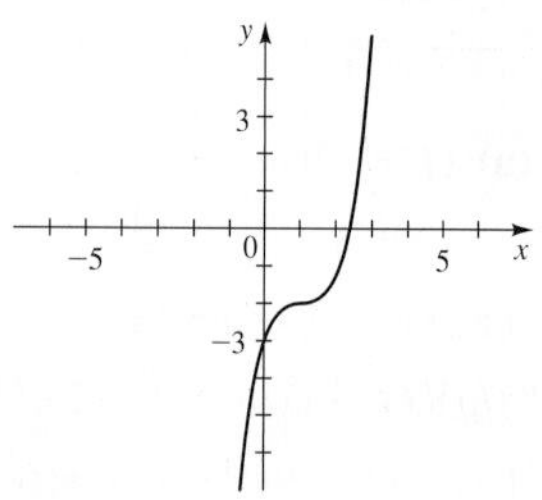

7. **(a)** Shift to the right 3 units, then shift upward 2 units
(b) Reflect in y-axis
8. **(a)** 3, 0 **(b)**

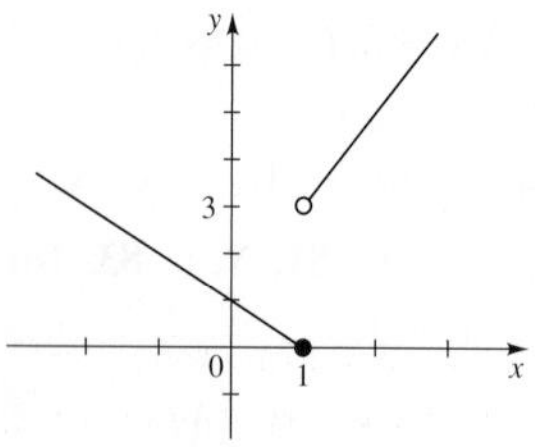

9. **(a)** $(f \circ g)(x) = (x-3)^2 + 1$ **(b)** $(g \circ f)(x) = x^2 - 2$
(c) 2 **(d)** 2 **(e)** $(g \circ g \circ g)(x) = x - 9$
10. **(a)** $f^{-1}(x) = 3 - x^2, x \geq 0$
(b)

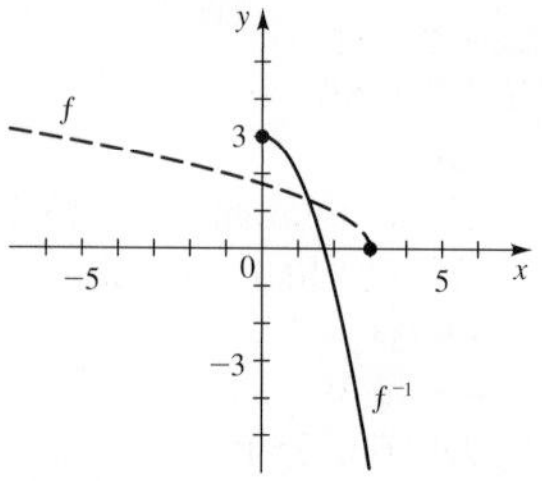

11. **(a)** Domain $[0, 6]$, range $[1, 7]$
(b)

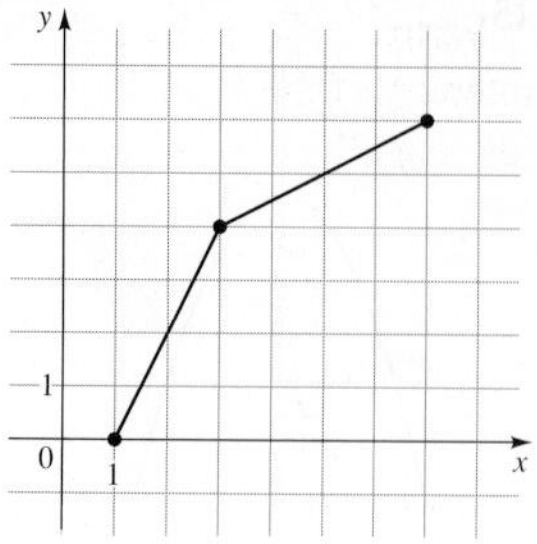

(c) $\frac{5}{4}$

12. **(a)**

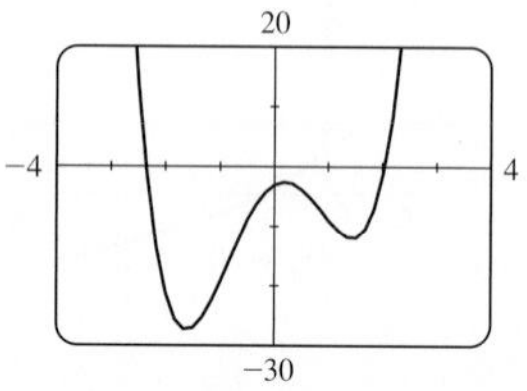

(b) No

(c) Local minimum ≈ -27.18 when $x \approx -1.61$; local maximum ≈ -2.55 when $x \approx 0.18$; local minimum ≈ -11.93 when $x \approx 1.43$ **(d)** $[-27.18, \infty)$
(e) Increasing on $[-1.61, 0.18] \cup [1.43, \infty)$; decreasing on $(-\infty, -1.61] \cup [0.18, 1.43]$

FOCUS ON MODELING ■ PAGE 218

1. $A(w) = 3w^2, w > 0$ **3.** $V(w) = \frac{1}{2}w^3, w > 0$
5. $A(x) = 10x - x^2, 0 < x < 10$
7. $A(x) = (\sqrt{3}/4)x^2, x > 0$
9. $r(A) = \sqrt{A/\pi}, A > 0$
11. $S(x) = 2x^2 + 240/x, x > 0$
13. $D(t) = 25t, t \geq 0$
15. $A(b) = b\sqrt{4-b}, 0 < b < 4$
17. $A(h) = 2h\sqrt{100 - h^2}, 0 < h < 10$
19. **(b)** $p(x) = x(19 - x)$ **(c)** 9.5, 9.5
21. **(b)** $A(x) = x(2400 - 2x)$ **(c)** 600 ft by 1200 ft
23. **(a)** $f(w) = 8w + 7200/w$
(b) Width along road is 30 ft, length is 40 ft **(c)** 15 ft to 60 ft
25. **(a)** $A(x) = 15x - \left(\dfrac{\pi + 4}{8}\right)x^2$
(b) Width $\approx$ 8.40 ft, height of rectangular part $\approx$ 4.20 ft
27. **(a)** $A(x) = x^2 + 48/x$ **(b)** Height $\approx$ 1.44 ft, width $\approx$ 2.88 ft
29. **(a)** $A(x) = 2x + \dfrac{200}{x}$ **(b)** 10 m by 10 m
31. **(b)** To point C, 5.1 mi from B

CHAPTER 3

SECTION 3.1 ■ PAGE 229

1. square **2.** **(a)** (h, k) **(b)** upward, minimum
(c) downward, maximum
3. upward, (3, 5), 5, minimum
4. downward, (3, 5), 5, maximum
5. **(a)** (3, 4) **(b)** 4 **(c)** $\mathbb{R}$, $(-\infty, 4]$
7. **(a)** $(1, -3)$ **(b)** -3 **(c)** $\mathbb{R}$, $[-3, \infty)$

9. (a) $f(x) = (x-3)^2 - 9$

(b) Vertex $(3, -9)$
x-intercepts 0, 6
y-intercept 0

(c)

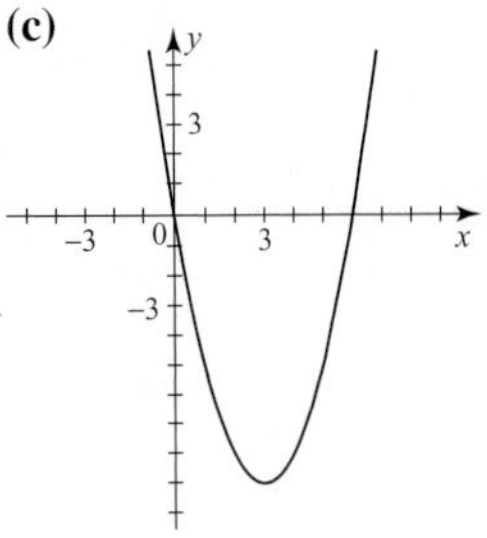

11. (a) $f(x) = 2\left(x + \frac{3}{2}\right)^2 - \frac{9}{2}$

(b) Vertex $\left(-\frac{3}{2}, -\frac{9}{2}\right)$
x-intercepts 0, −3,
y-intercept 0

(c)

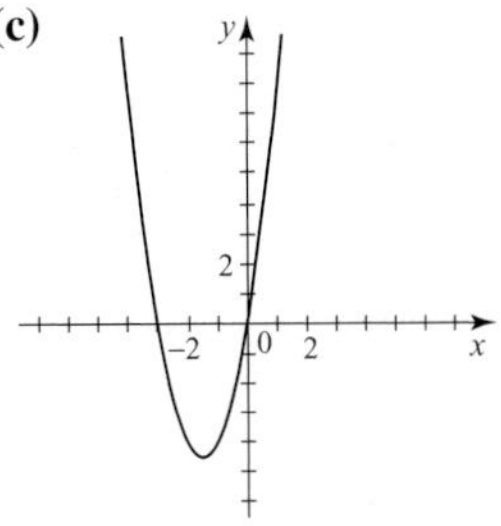

13. (a) $f(x) = (x+2)^2 - 1$

(b) Vertex $(-2, -1)$, x-intercepts $-1, -3$, y-intercept 3

(c)

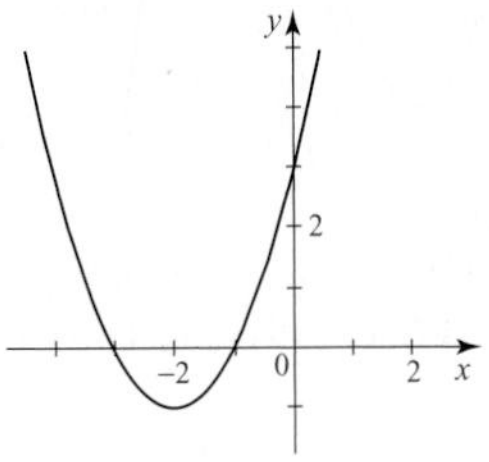

15. (a) $f(x) = -(x-3)^2 + 13$

(b) Vertex $(3, 13)$; x-intercepts $3 \pm \sqrt{13}$; y-intercept 4

(c)

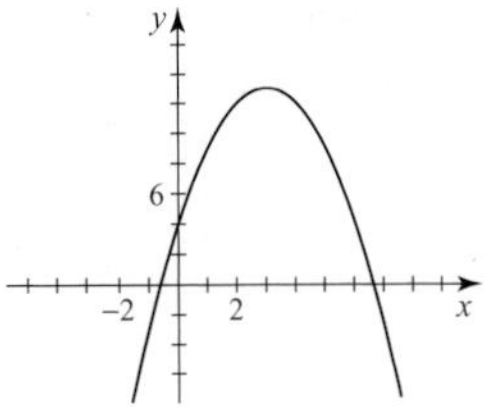

17. (a) $f(x) = 2(x+1)^2 + 1$

(b) Vertex $(-1, 1)$; no x-intercept; y-intercept 3

(c)

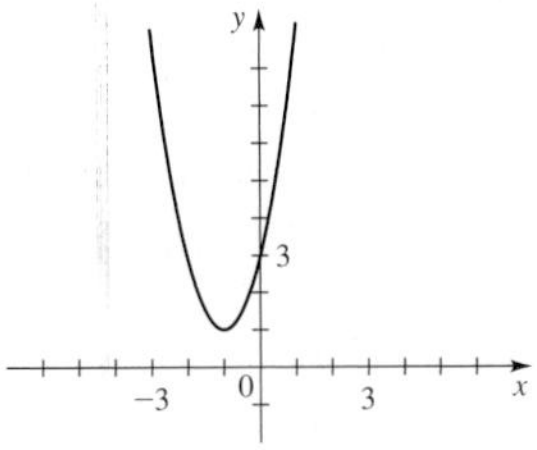

19. (a) $f(x) = 2(x-5)^2 + 7$

(b) Vertex $(5, 7)$; no x-intercept; y-intercept 57

(c)

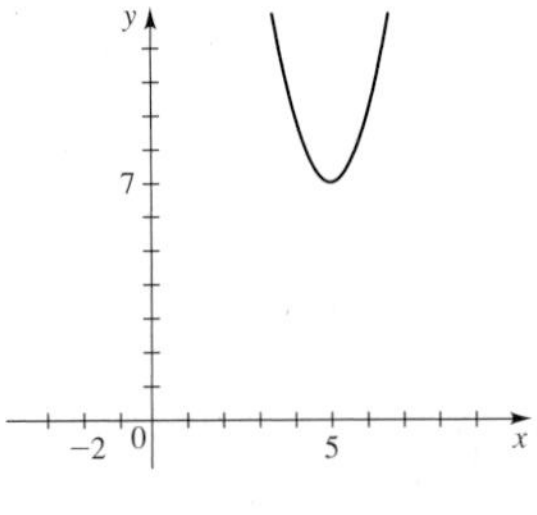

21. (a) $f(x) = -4(x+2)^2 + 19$

(b) Vertex $(-2, 19)$; x-intercepts $-2 \pm \frac{1}{2}\sqrt{19}$; y-intercept 3

(c)

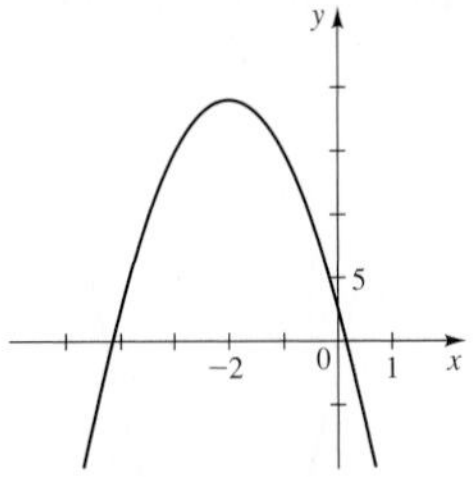

23. (a) $f(x) = (x+1)^2 - 2$

(b)

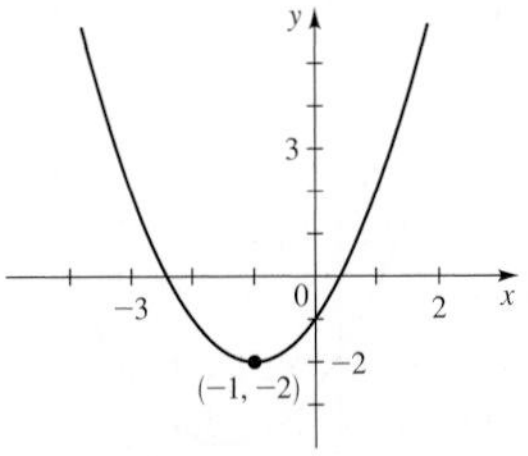

(c) Minimum $f(-1) = -2$

25. (a) $f(x) = 3(x-1)^2 - 2$

(b)

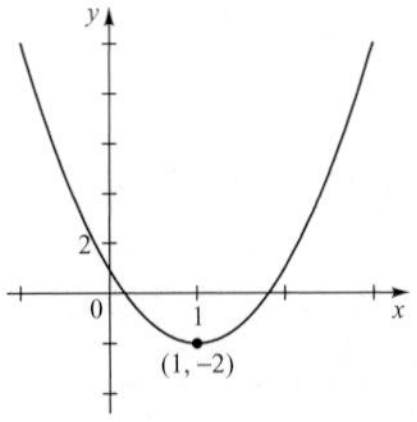

(c) Minimum $f(1) = -2$

27. (a) $f(x) = -\left(x + \frac{3}{2}\right)^2 + \frac{21}{4}$

(b)

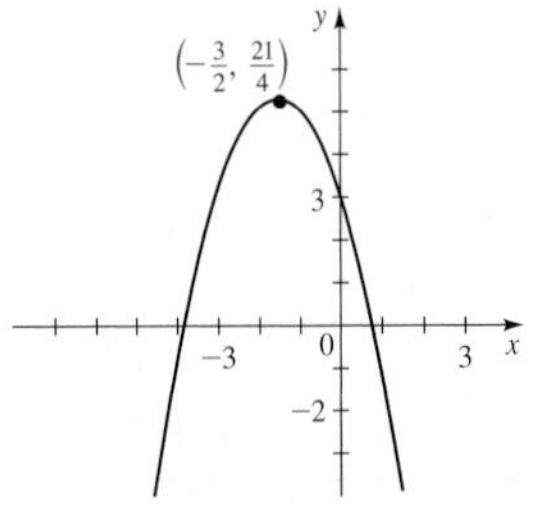

(c) Maximum $f\left(-\frac{3}{2}\right) = \frac{21}{4}$

29. (a) $g(x) = 3(x-2)^2 + 1$

(b)

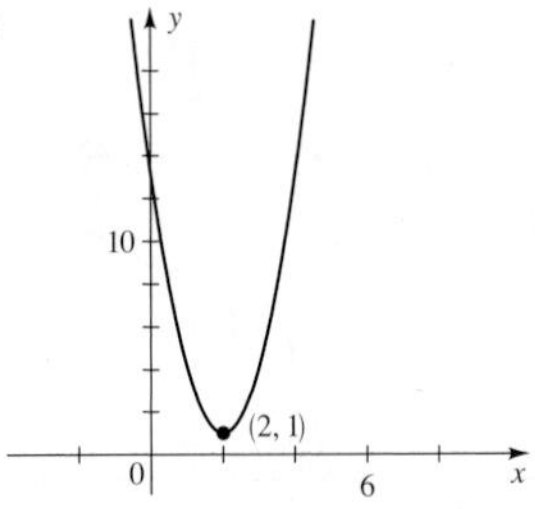

(c) Minimum $g(2) = 1$

31. (a) $h(x) = -\left(x + \frac{1}{2}\right)^2 + \frac{5}{4}$

(b)

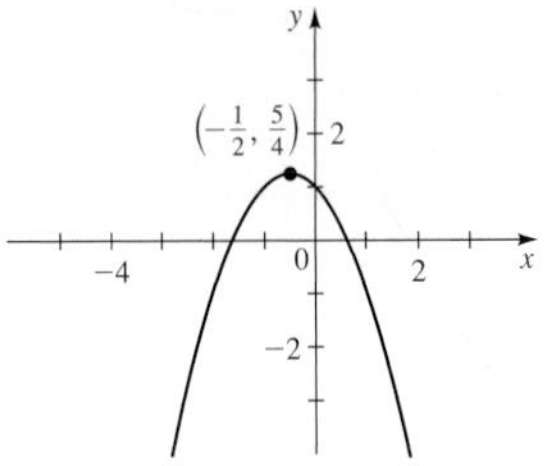

(c) Maximum $h\left(-\frac{1}{2}\right) = \frac{5}{4}$

33. Minimum $f\left(-\frac{1}{2}\right) = \frac{3}{4}$ **35.** Maximum $f(-3.5) = 185.75$

37. Minimum $f(0.6) = 15.64$ **39.** Minimum $h(-2) = -8$

41. Maximum $f(-1) = \frac{7}{2}$ **43.** $f(x) = 2x^2 - 4x$

45. $(-\infty, \infty)$, $(-\infty, 1]$ **47.** $(-\infty, \infty)$, $\left[-\frac{23}{2}, \infty\right)$

49. (a) -4.01 **(b)** -4.011025

51. Local maximum 2; local minima −1, 0

53. Local maxima 0, 1; local minima −2, −1

55. Local maximum ≈ 0.38 when $x \approx -0.58$; local minimum ≈ -0.38 when $x \approx 0.58$

57. Local maximum ≈ 0 when $x = 0$; local minimum ≈ -13.61 when $x \approx -1.71$; local minimum ≈ -73.32 when $x \approx 3.21$

59. Local maximum ≈ 5.66 when $x \approx 4.00$

61. Local maximum ≈ 0.38 when $x \approx -1.73$; local minimum ≈ -0.38 when $x \approx 1.73$ **63.** 25 ft

65. \$4000, 100 units **67.** 30 times **69.** 50 trees per acre

71. 600 ft by 1200 ft **73.** Width 8.40 ft, height of rectangular part 4.20 ft **75. (a)** $f(x) = x(1200 - x)$ **(b)** 600 ft by 600 ft **77. (a)** $R(x) = x(57{,}000 - 3000x)$ **(b)** \$9.50 **(c)** \$19.00

SECTION 3.2 ■ PAGE 243

1. II **2. (a)** (ii) **(b)** (iv) **3.** (a), (c) **4.** (a)

5. (a)

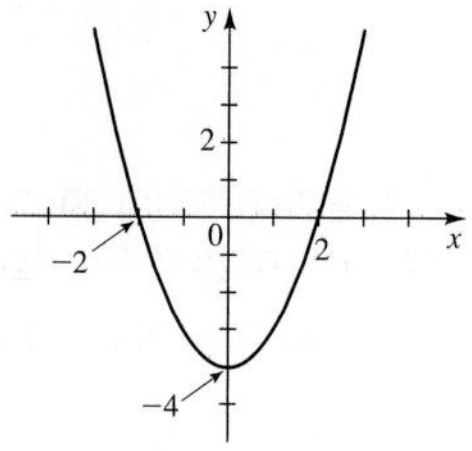

(b)

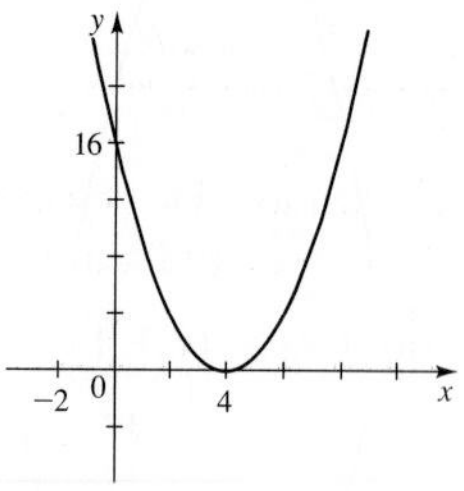

(c)

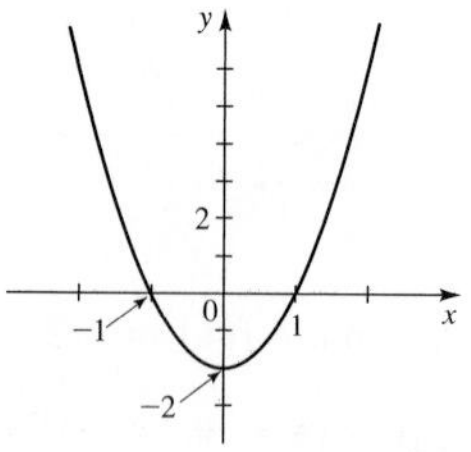

(d)

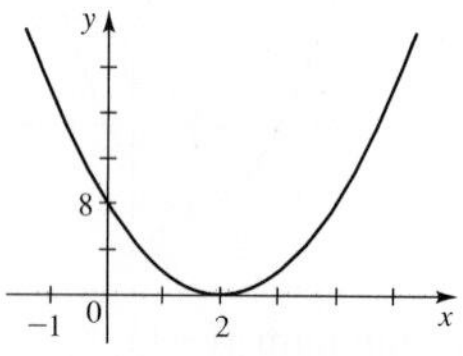

7. (a)

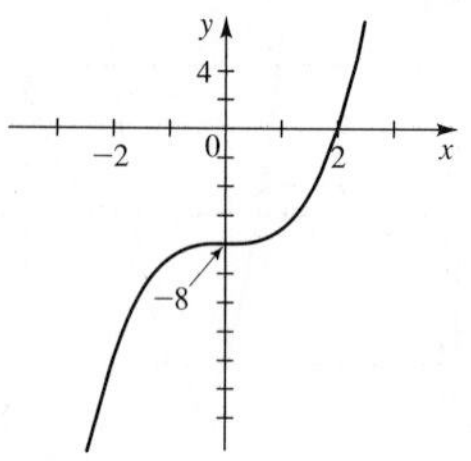

(b)

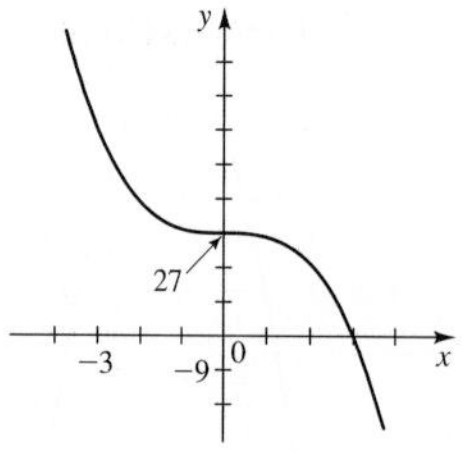

(c)

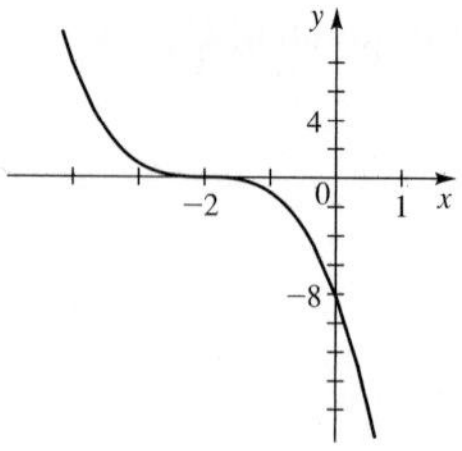

(d)

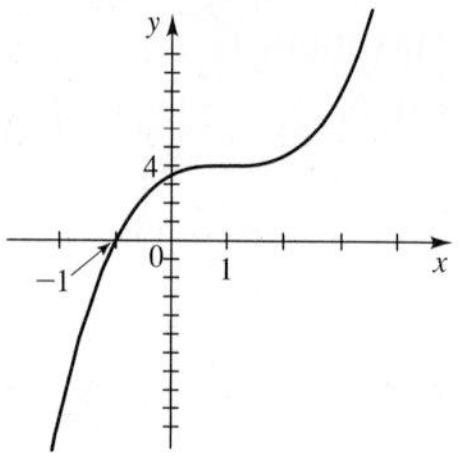

9. III **11.** V **13.** VI

15.

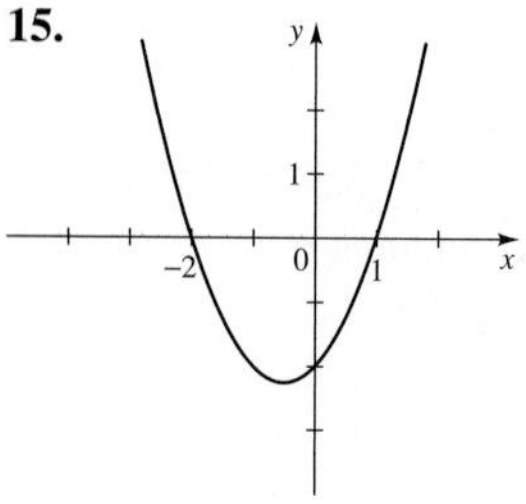

17.

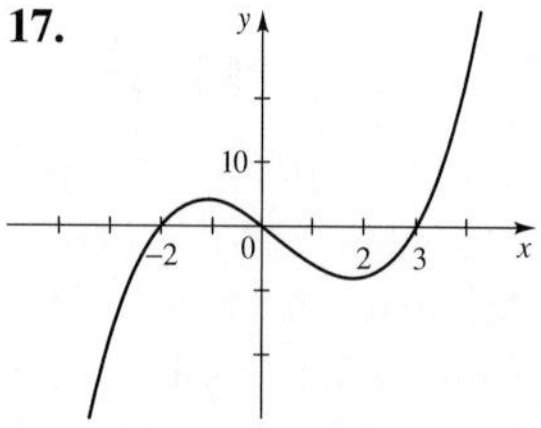

19.

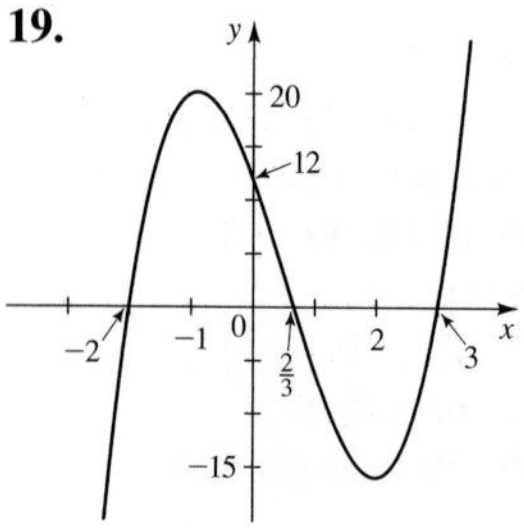

21.

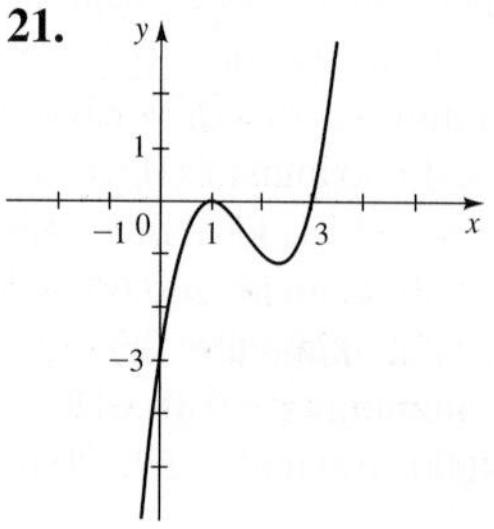

23.

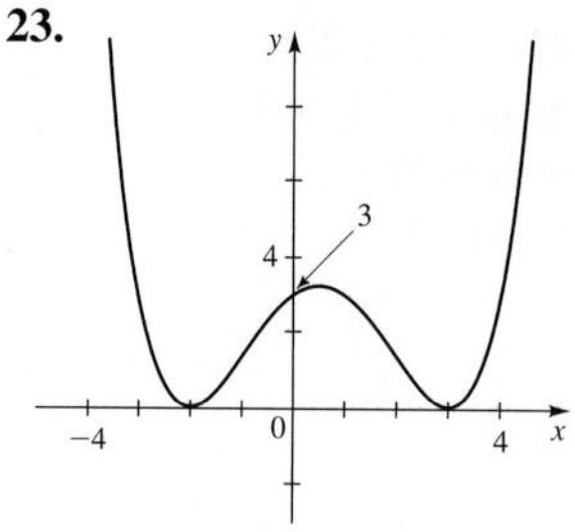

25.

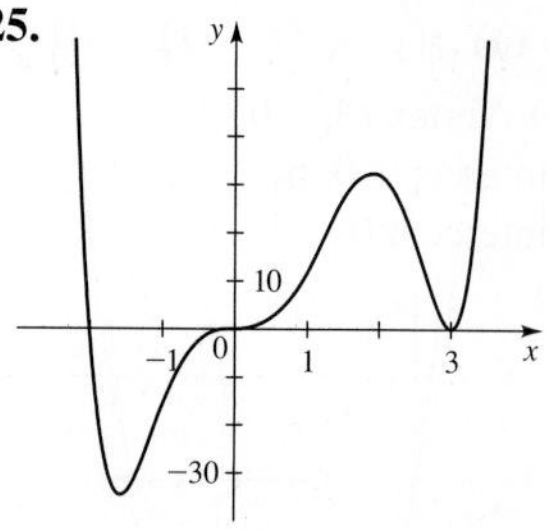

27. $P(x) = x(x + 2)(x - 3)$

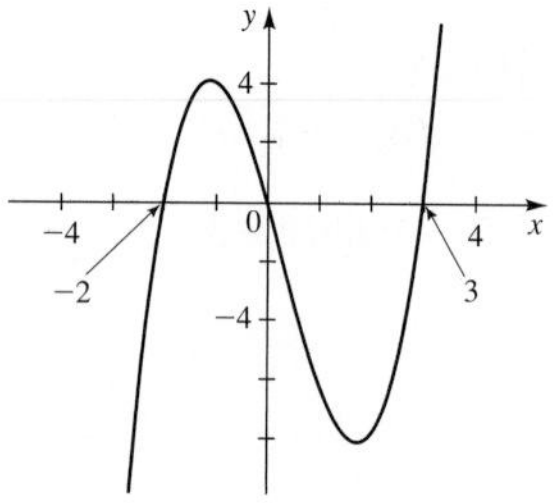

29. $P(x) = -x(x + 3)(x - 4)$

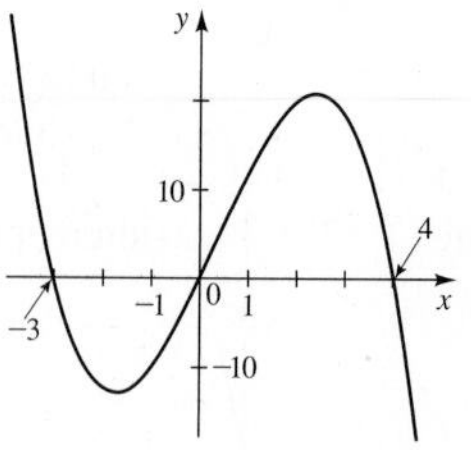

31. $P(x) = x^2(x - 1)(x - 2)$

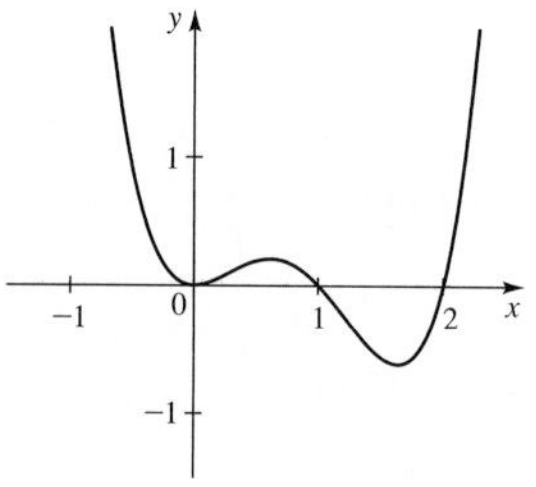

33. $P(x) = (x + 1)^2(x - 1)$

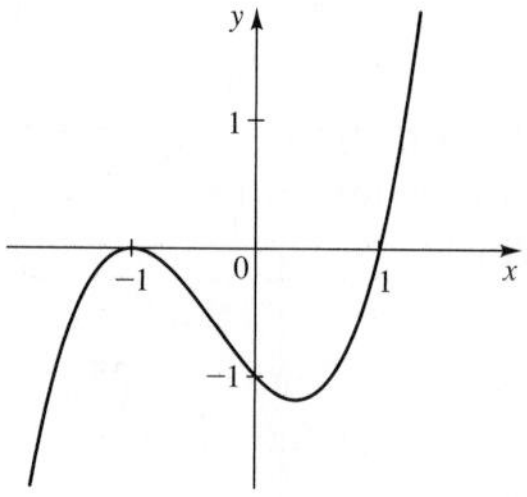

35. $P(x) = (2x - 1)(x + 3)(x - 3)$

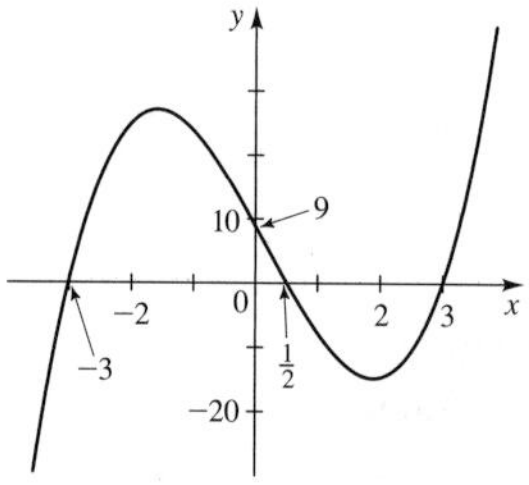

37. $P(x) = (x - 2)^2(x^2 + 2x + 4)$

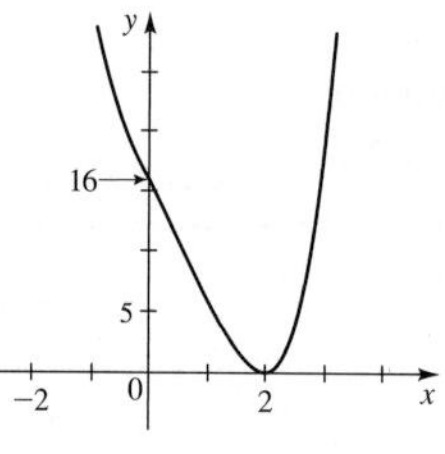

39. $P(x) = (x^2 + 1)(x + 2)(x - 2)$

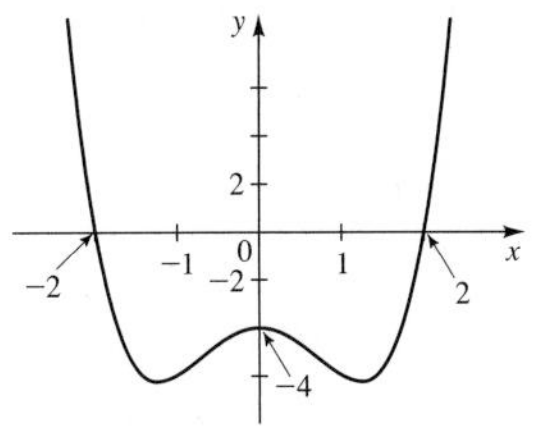

41. $y \to \infty$ as $x \to \infty, y \to -\infty$ as $x \to -\infty$
43. $y \to \infty$ as $x \to \pm\infty$
45. $y \to \infty$ as $x \to \infty, y \to -\infty$ as $x \to -\infty$
47. **(a)** x-intercepts 0, 4; y-intercept 0 **(b)** $(2, 4)$
49. **(a)** x-intercepts -2, 1; y-intercept -1 **(b)** $(-1, -2), (1, 0)$
51.

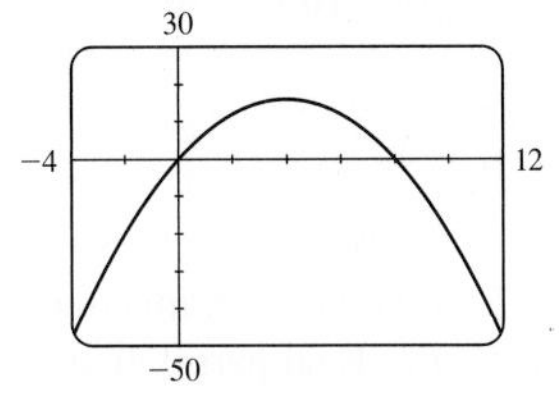

local maximum $(4, 16)$

53.

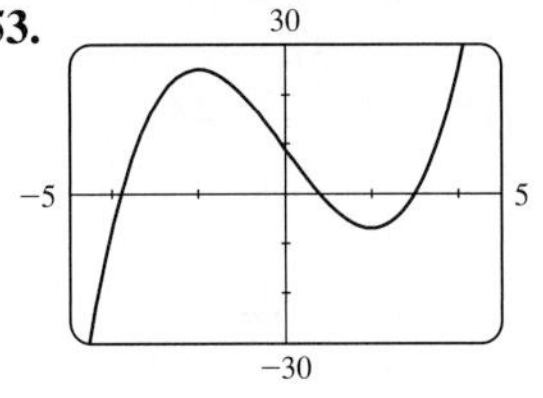

local maximum $(-2, 25)$,
local minimum $(2, -7)$

55.

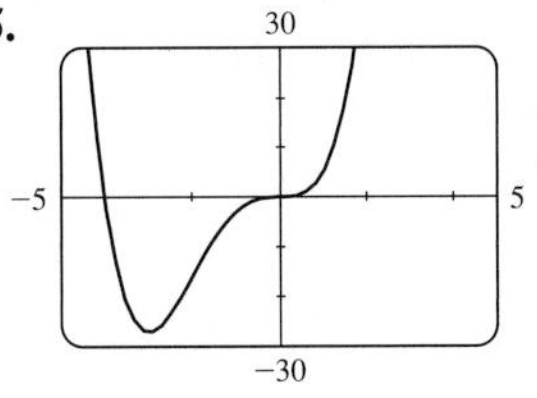

local minimum $(-3, -27)$

57.

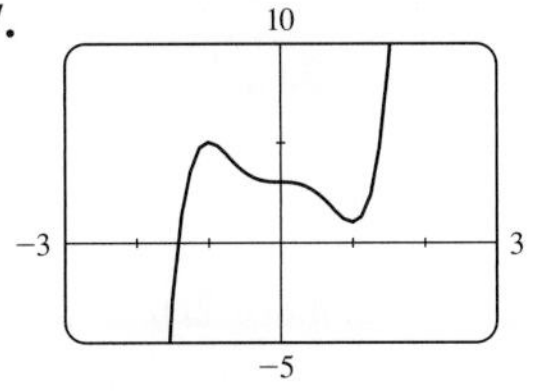

local maximum $(-1, 5)$,
local minimum $(1, 1)$
59. One local maximum, no local minimum
61. One local maximum, one local minimum
63. One local maximum, two local minima
65. No local extrema
67. One local maximum, two local minima
69.

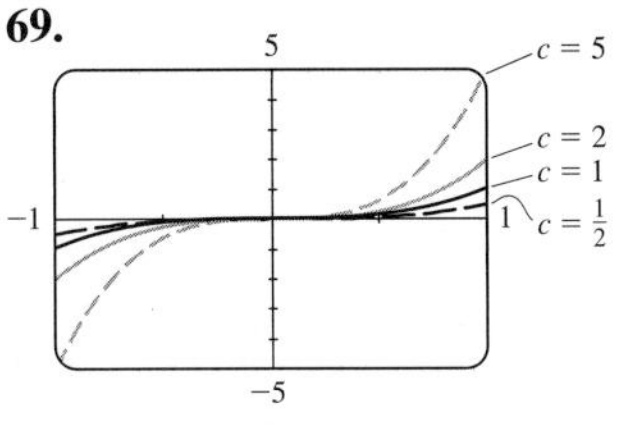

Increasing the value of c stretches the graph vertically.

71.

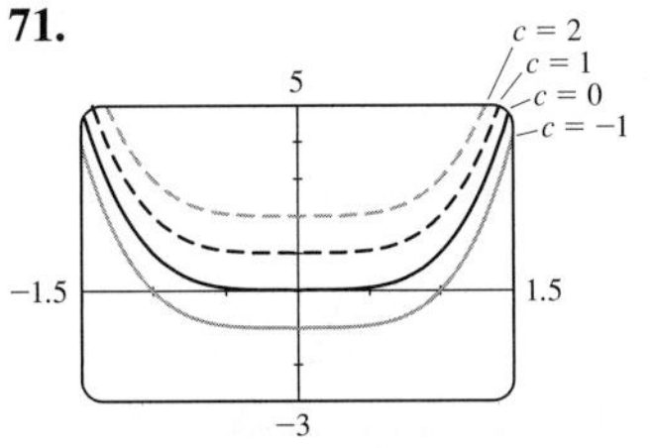

Increasing the value of c moves the graph up.

73.

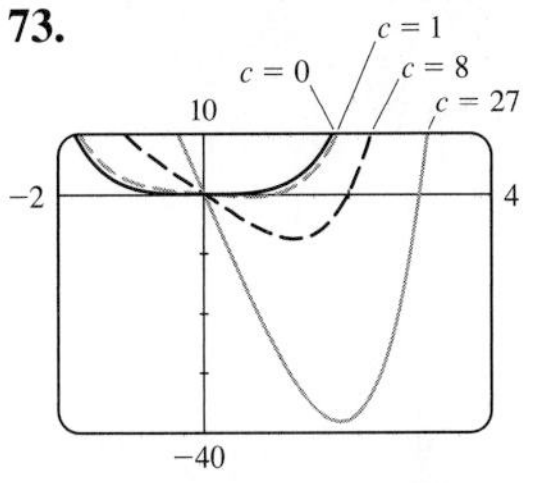

Increasing the value of c causes a deeper dip in the graph in the fourth quadrant and moves the positive x-intercept to the right.
75. **(a)**

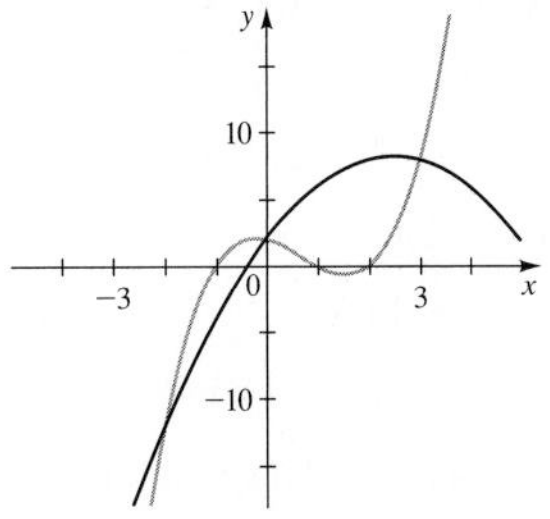

(b) Three **(c)** $(0, 2), (3, 8), (-2, -12)$
77. **(d)** $P(x) = P_O(x) + P_E(x)$, where $P_O(x) = x^5 + 6x^3 - 2x$ and $P_E(x) = -x^2 + 5$
79. **(a)** Two local extrema

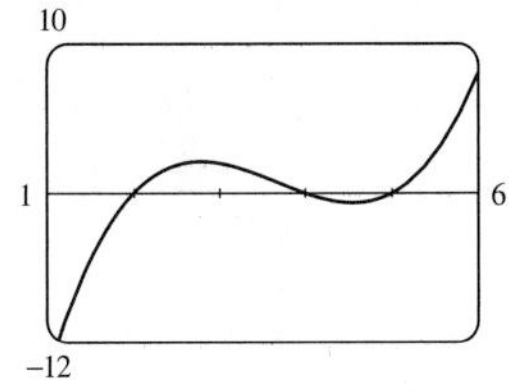

81. **(a)** 26 blenders **(b)** No; \$3276.22
83. **(a)** $V(x) = 4x^3 - 120x^2 + 800x$ **(b)** $0 < x < 10$
(c) Maximum volume $\approx$ 1539.6 cm^3

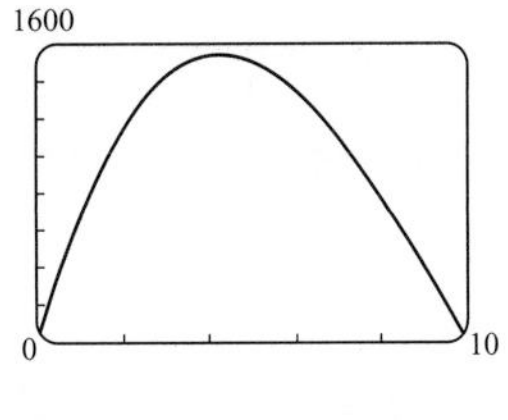

SECTION 3.3 ▪ PAGE 251

1. quotient, remainder **2.** **(a)** factor **(b)** k
3. $(x + 3)(3x - 4) + 8$ **5.** $(2x - 3)(x^2 - 1) - 3$
7. $(x^2 + 3)(x^2 - x - 3) + (7x + 11)$
9. $x + 1 + \dfrac{-11}{x + 3}$ **11.** $2x - \dfrac{1}{2} + \dfrac{-\frac{15}{2}}{2x - 1}$
13. $2x^2 - x + 1 + \dfrac{4x - 4}{x^2 + 4}$

In answers 15–37 the first polynomial given is the quotient, and the second is the remainder.
15. $x - 2, -16$ **17.** $2x^2 - 1, -2$ **19.** $x + 2, 8x - 1$
21. $3x + 1, 7x - 5$ **23.** $x^4 + 1, 0$ **25.** $x - 2, -2$
27. $3x + 23, 138$ **29.** $x^2 + 2, -3$ **31.** $x^2 - 3x + 1, -1$
33. $x^4 + x^3 + 4x^2 + 4x + 4, -2$ **35.** $2x^2 + 4x, 1$

37. $x^2 + 3x + 9, 0$ **39.** -3 **41.** 12 **43.** -7 **45.** -483 **47.** 2159 **49.** $\frac{7}{3}$ **51.** -8.279 **57.** $-1 \pm \sqrt{6}$ **59.** $x^3 - 3x^2 - x + 3$ **61.** $x^4 - 8x^3 + 14x^2 + 8x - 15$ **63.** $-\frac{3}{2}x^3 + 3x^2 + \frac{15}{2}x - 9$ **65.** $(x + 1)(x - 1)(x - 2)$ **67.** $(x + 2)^2(x - 1)^2$

SECTION 3.4 ■ PAGE 260

1. $a_0, a_n, \pm1, \pm\frac{1}{2}, \pm\frac{1}{3}, \pm\frac{1}{6}, \pm2, \pm\frac{2}{3}, \pm5, \pm\frac{5}{2}, \pm\frac{5}{3}, \pm\frac{5}{6}, \pm10, \pm\frac{10}{3}$ **2.** 1, 3, 5; 0 **3.** True **4.** False **5.** $\pm1, \pm3$ **7.** $\pm1, \pm2, \pm4, \pm8, \pm\frac{1}{2}$ **9.** $\pm1, \pm7, \pm\frac{1}{2}, \pm\frac{7}{2}, \pm\frac{1}{4}, \pm\frac{7}{4}$ **11. (a)** $\pm1, \pm\frac{1}{5}$ **(b)** $-1, 1, \frac{1}{5}$ **13. (a)** $\pm1, \pm3, \pm\frac{1}{2}, \pm\frac{3}{2}$ **(b)** $-\frac{1}{2}, 1, 3$
15. $-2, 1; P(x) = (x + 2)^2(x - 1)$
17. $-1, 2; P(x) = (x + 1)^2(x - 2)$ **19.** $2; P(x) = (x - 2)^3$
21. $-1, 2, 3; P(x) = (x + 1)(x - 2)(x - 3)$
23. $-3, -1, 1; P(x) = (x + 3)(x + 1)(x - 1)$
25. $\pm1, \pm 2; P(x) = (x - 2)(x + 2)(x - 1)(x + 1)$
27. $-4, -2, -1, 1; P(x) = (x + 4)(x + 2)(x - 1)(x + 1)$
29. $\pm2, \pm\frac{3}{2}; P(x) = (x - 2)(x + 2)(2x - 3)(2x + 3)$
31. $\pm2, \frac{1}{3}, 3; P(x) = (x - 2)(x + 2)(x - 3)(3x - 1)$
33. $-1, \pm\frac{1}{2}; P(x) = (x + 1)(2x - 1)(2x + 1)$
35. $-\frac{3}{2}, \frac{1}{2}, 1; P(x) = (x - 1)(2x + 3)(2x - 1)$
37. $-\frac{5}{2}, -1, \frac{3}{2}; P(x) = (x + 1)(2x + 5)(2x - 3)$
39. $-\frac{1}{2}, \frac{2}{5}, \frac{1}{2}; P(x) = (2x - 1)(5x - 2)(2x + 1)$
41. $-1, \frac{1}{2}, 2; P(x) = (x + 1)(x - 2)^2(2x - 1)$
43. $-3, -2, 1, 3; P(x) = (x + 3)(x + 2)^2(x - 1)(x - 3)$
45. $-1, -\frac{1}{3}, 2, 5; P(x) = (x + 1)^2(x - 2)(x - 5)(3x + 1)$
47. $-2, -1 \pm \sqrt{2}$ **49.** $-1, 4, \dfrac{3 \pm \sqrt{13}}{2}$ **51.** $3, \dfrac{1 \pm \sqrt{5}}{2}$
53. $\frac{1}{2}, \dfrac{1 \pm \sqrt{3}}{2}$ **55.** $-1, -\frac{1}{2}, -3 \pm \sqrt{10}$
57. (a) $-2, 2, 3$ **(b)**

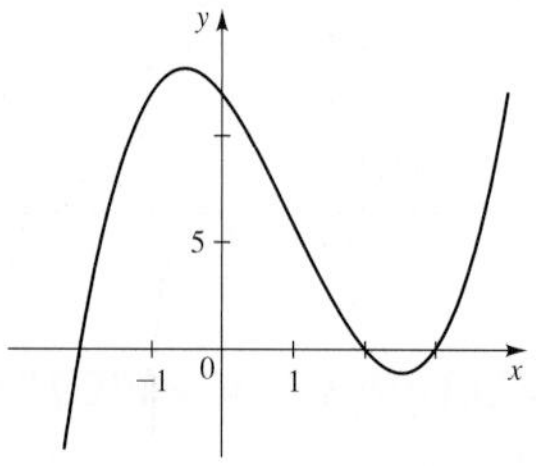

59. (a) $-\frac{1}{2}, 2$ **(b)**

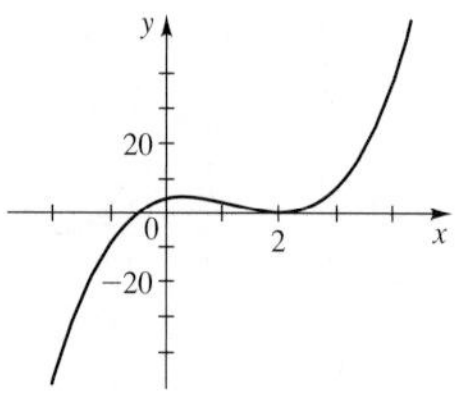

61. (a) $-1, 2$ **(b)**

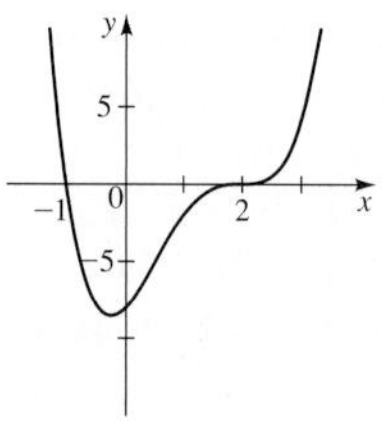

63. (a) $-1, 2$ **(b)**

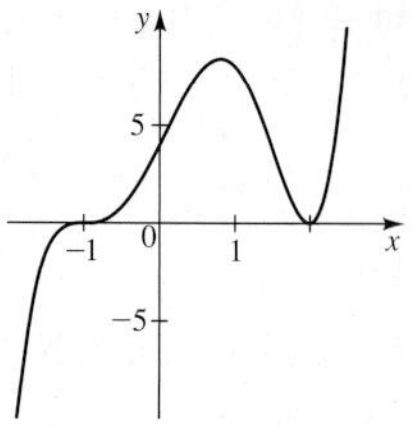

65. 1 positive, 2 or 0 negative; 3 or 1 real **67.** 1 positive, 1 negative; 2 real **69.** 2 or 0 positive, 0 negative; 3 or 1 real (since 0 is a zero but is neither positive nor negative) **75.** 3, -2 **77.** 3, -1 **79.** $-2, \frac{1}{2}, \pm1$ **81.** $\pm\frac{1}{2}, \pm\sqrt{5}$ **83.** $-2, 1, 3, 4$ **89.** $-2, 2, 3$ **91.** $-\frac{3}{2}, -1, 1, 4$ **93.** $-1.28, 1.53$ **95.** -1.50 **99.** 11.3 ft **101. (a)** It began to snow again. **(b)** No **(c)** Just before midnight on Saturday night **103.** 2.76 m **105.** 88 in. (or 3.21 in.)

SECTION 3.5 ■ PAGE 268

1. -1 **2.** 3, 4 **3. (a)** $3 - 4i$ **(b)** $9 + 16 = 25$ **4.** $3 - 4i$ **5.** Real part 5, imaginary part -7 **7.** Real part $-\frac{2}{3}$, imaginary part $-\frac{5}{3}$ **9.** Real part 3, imaginary part 0 **11.** Real part 0, imaginary part $-\frac{2}{3}$ **13.** Real part $\sqrt{3}$, imaginary part 2 **15.** $5 - i$ **17.** $3 + 5i$ **19.** $2 - 2i$ **21.** $-19 + 4i$ **23.** $-4 + 8i$ **25.** $30 + 10i$ **27.** $-33 - 56i$ **29.** $27 - 8i$ **31.** $-i$ **33.** 1 **35.** $-i$ **37.** $\frac{8}{5} + \frac{1}{5}i$ **39.** $-5 + 12i$ **41.** $-4 + 2i$ **43.** $2 - \frac{4}{3}i$ **45.** $-i$ **47.** $5i$ **49.** -6 **51.** $(3 + \sqrt{5}) + (3 - \sqrt{5})i$ **53.** 2 **55.** $-i\sqrt{2}$ **57.** $\pm7i$ **59.** $2 \pm i$ **61.** $-1 \pm 2i$ **63.** $-\dfrac{1}{2} \pm \dfrac{\sqrt{3}}{2}i$ **65.** $\frac{1}{2} \pm \frac{1}{2}i$ **67.** $-\dfrac{3}{2} \pm \dfrac{\sqrt{3}}{2}i$ **69.** $\dfrac{-6 \pm \sqrt{6}i}{6}$ **71.** $1 \pm 3i$

SECTION 3.6 ■ PAGE 276

1. 5, -2, 3, 1 **2. (a)** $x - a$ **(b)** $(x - a)^m$ **3.** n **4.** $a - bi$ **5. (a)** $0, \pm2i$ **(b)** $x^2(x - 2i)(x + 2i)$
7. (a) $0, 1 \pm i$ **(b)** $x(x - 1 - i)(x - 1 + i)$
9. (a) $\pm i$ **(b)** $(x - i)^2(x + i)^2$
11. (a) $\pm2, \pm2i$ **(b)** $(x - 2)(x + 2)(x - 2i)(x + 2i)$
13. (a) $-2, 1 \pm i\sqrt{3}$ **(b)** $(x + 2)(x - 1 - i\sqrt{3})(x - 1 + i\sqrt{3})$
15. (a) $\pm1, \frac{1}{2} \pm \frac{1}{2}i\sqrt{3}, -\frac{1}{2} \pm \frac{1}{2}i\sqrt{3}$
(b) $(x - 1)(x + 1)(x - \frac{1}{2} - \frac{1}{2}i\sqrt{3})(x - \frac{1}{2} + \frac{1}{2}i\sqrt{3}) \times (x + \frac{1}{2} - \frac{1}{2}i\sqrt{3})(x + \frac{1}{2} + \frac{1}{2}i\sqrt{3})$

In answers 17–33 the factored form is given first, then the zeros are listed with the multiplicity of each in parentheses.
17. $(x - 5i)(x + 5i); \pm5i\,(1)$
19. $[x - (-1 + i)][x - (-1 - i)]; -1 + i\,(1), -1 - i\,(1)$
21. $x(x - 2i)(x + 2i); 0\,(1), 2i\,(1), -2i\,(1)$
23. $(x - 1)(x + 1)(x - i)(x + i); 1\,(1), -1\,(1), i\,(1), -i\,(1)$
25. $16(x - \frac{3}{2})(x + \frac{3}{2})(x - \frac{3}{2}i)(x + \frac{3}{2}i); \frac{3}{2}\,(1), -\frac{3}{2}\,(1), \frac{3}{2}i\,(1), -\frac{3}{2}i\,(1)$
27. $(x + 1)(x - 3i)(x + 3i); -1\,(1), 3i\,(1), -3i\,(1)$
29. $(x - i)^2(x + i)^2; i\,(2), -i\,(2)$
31. $(x - 1)(x + 1)(x - 2i)(x + 2i); 1\,(1), -1\,(1), 2i\,(1), -2i\,(1)$
33. $x(x - i\sqrt{3})^2(x + i\sqrt{3})^2; 0\,(1), i\sqrt{3}\,(2), -i\sqrt{3}\,(2)$
35. $P(x) = x^2 - 2x + 2$ **37.** $Q(x) = x^3 - 3x^2 + 4x - 12$
39. $P(x) = x^3 - 2x^2 + x - 2$

41. $R(x) = x^4 - 4x^3 + 10x^2 - 12x + 5$

43. $T(x) = 6x^4 - 12x^3 + 18x^2 - 12x + 12$

45. $-2, \pm 2i$ **47.** $1, \dfrac{1 \pm i\sqrt{3}}{2}$ **49.** $2, \dfrac{1 \pm i\sqrt{3}}{2}$

51. $-\frac{3}{2}, -1 \pm i\sqrt{2}$ **53.** $-2, 1, \pm 3i$ **55.** $1, \pm 2i, \pm i\sqrt{3}$

57. 3 (multiplicity 2), $\pm 2i$ **59.** $-\frac{1}{2}$ (multiplicity 2), $\pm i$

61. 1 (multiplicity 3), $\pm 3i$ **63. (a)** $(x - 5)(x^2 + 4)$

(b) $(x - 5)(x - 2i)(x + 2i)$ **65. (a)** $(x - 1)(x + 1)(x^2 + 9)$

(b) $(x - 1)(x + 1)(x - 3i)(x + 3i)$

67. (a) $(x - 2)(x + 2)(x^2 - 2x + 4)(x^2 + 2x + 4)$

(b) $(x - 2)(x + 2)[x - (1 + i\sqrt{3})][x - (1 - i\sqrt{3})] \times$ $[x + (1 + i\sqrt{3})][x + (1 - i\sqrt{3})]$ **69. (a)** 4 real

(b) 2 real, 2 imaginary **(c)** 4 imaginary

SECTION 3.7 ▪ PAGE 289

1. $-\infty, \infty$ **2.** 2 **3.** $-1, 2$ **4.** $\frac{1}{3}$ **5.** $-2, 3$ **6.** 1

7. (a) $-3, -19, -199, -1999$; 5, 21, 201, 2001; 1.2500, 1.0417, 1.0204, 1.0020; 0.8333, 0.9615, 0.9804, 0.9980

(b) $r(x) \to -\infty$ as $x \to 2^-$; $r(x) \to \infty$ as $x \to 2^+$

(c) Horizontal asymptote $y = 1$

9. (a) $-22, -430, -40{,}300, -4{,}003{,}000$; $-10, -370, -39{,}700, -3{,}997{,}000$; 0.3125, 0.0608, 0.0302, 0.0030; $-0.2778, -0.0592, -0.0298, -0.0030$

(b) $r(x) \to -\infty$ as $x \to 2^-$; $r(x) \to -\infty$ as $x \to 2^+$

(c) Horizontal asymptote $y = 0$

11. x-intercept 1, y-intercept $-\frac{1}{4}$ **13.** x-intercepts $-1, 2$; y-intercept $\frac{1}{3}$ **15.** x-intercepts $-3, 3$; no y-intercept

17. x-intercept 3, y-intercept 3, vertical $x = 2$; horizontal $y = 2$

19. x-intercepts $-1, 1$; y-intercept $\frac{1}{4}$; vertical $x = -2, x = 2$; horizontal $y = 1$ **21.** Vertical $x = 2$; horizontal $y = 0$

23. Horizontal $y = 0$

25. Vertical $x = \frac{1}{2}, x = -1$; horizontal $y = 3$

27. Vertical $x = \frac{1}{3}, x = -2$; horizontal $y = \frac{5}{3}$

29. Vertical $x = 0$; horizontal $y = 3$

31. Vertical $x = 1$

33.

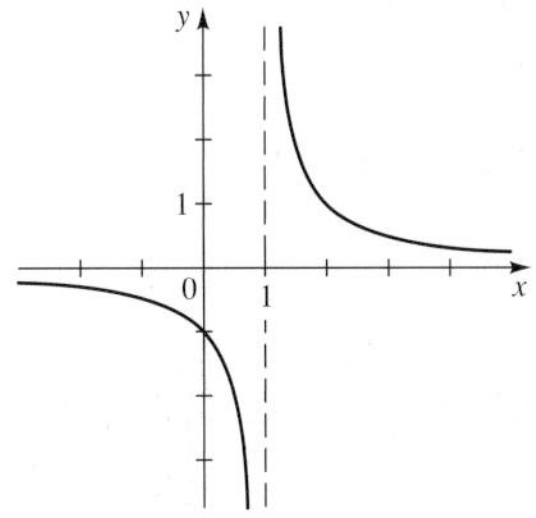

35.

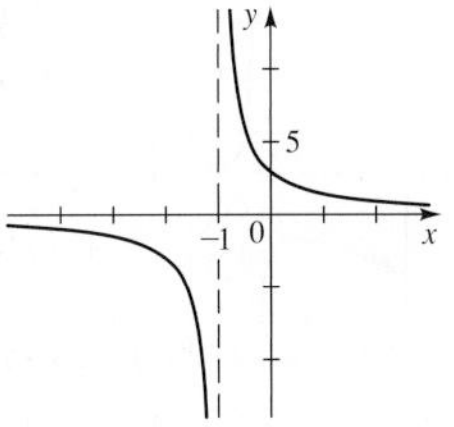

37.

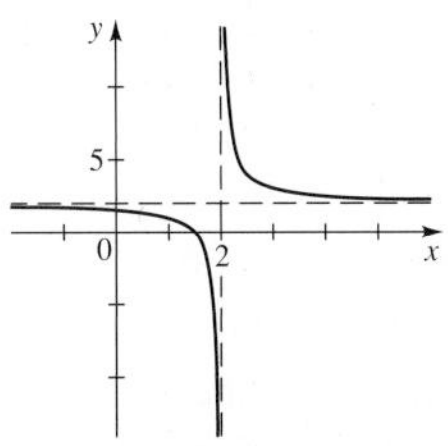

39.

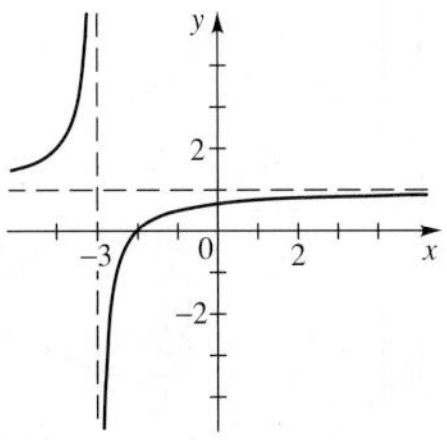

41.

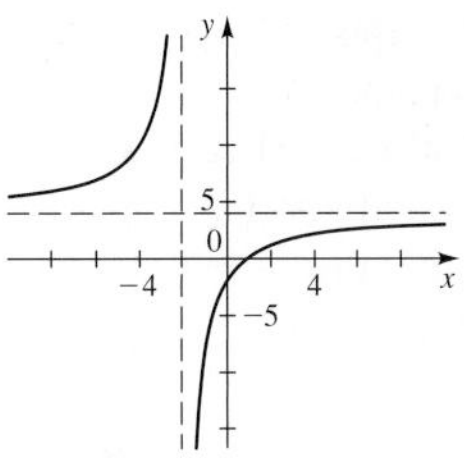

x-intercept 1
y-intercept -2
vertical $x = -2$
horizontal $y = 4$
domain $\{x \mid x \neq -2\}$
range $\{y \mid y \neq 4\}$

43.

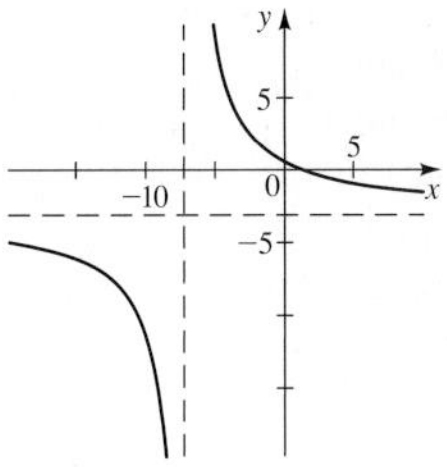

x-intercept $\frac{4}{3}$
y-intercept $\frac{4}{7}$
vertical $x = -7$
horizontal $y = -3$
domain $\{x \mid x \neq -7\}$
range $\{y \mid y \neq -3\}$

45.

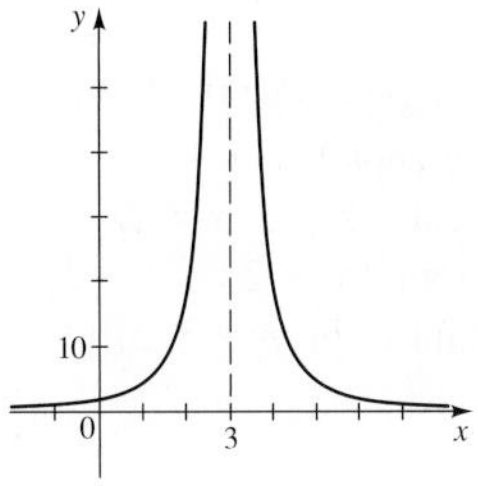

y-intercept 2
vertical $x = 3$
horizontal $y = 0$
domain $\{x \mid x \neq 3\}$
range $\{y \mid y > 0\}$

47.

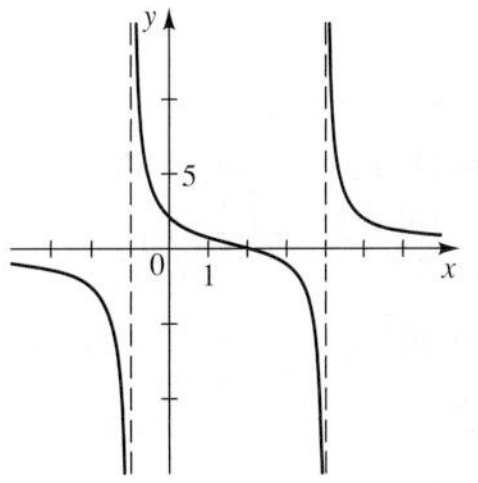

x-intercept 2
y-intercept 2
vertical $x = -1, x = 4$
horizontal $y = 0$
domain $\{x \mid x \neq -1, 4\}$
range $\mathbb{R}$

49.

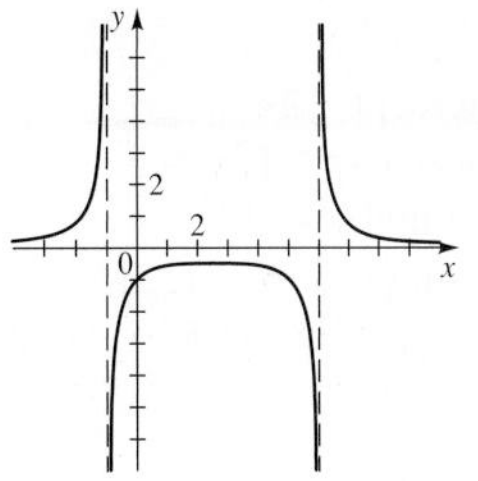

y-intercept -1
vertical $x = -1, x = 6$
horizontal $y = 0$
domain $\{x \mid x \neq -1, 6\}$
range $\{y \mid y \leq -0.5 \text{ or } y > 0\}$

51.

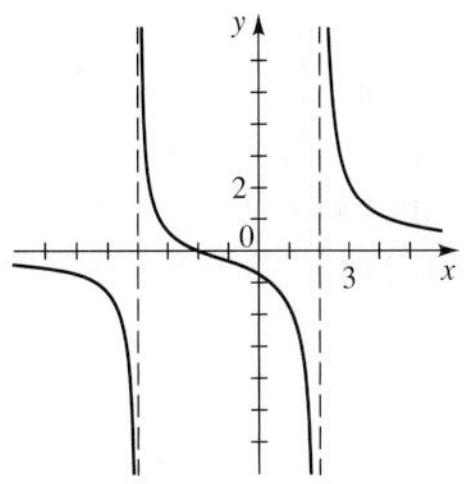

x-intercept -2
y-intercept $-\frac{3}{4}$
vertical $x = -4, x = 2$
horizontal $y = 0$
domain $\{x \mid x \neq -4, 2\}$
range $\mathbb{R}$

53.

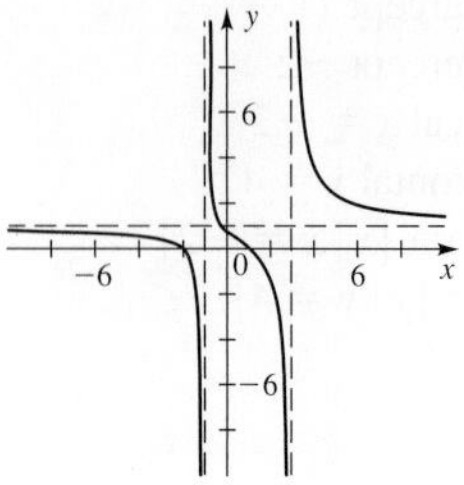

x-intercepts $-2, 1$
y-intercept $\frac{2}{3}$
vertical $x = -1, x = 3$
horizontal $y = 1$
domain $\{x \mid x \neq -1, 3\}$
range $\mathbb{R}$

55.

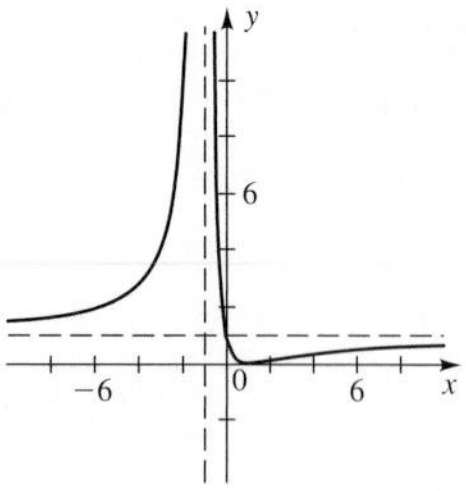

x-intercept 1
y-intercept 1
vertical $x = -1$
horizontal $y = 1$
domain $\{x \mid x \neq -1\}$
range $\{y \mid y \geq 0\}$

57.

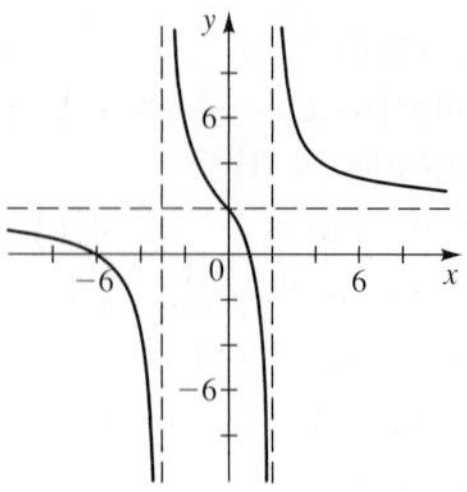

x-intercepts $-6, 1$
y-intercept 2
vertical $x = -3, x = 2$
horizontal $y = 2$
domain $\{x \mid x \neq -3, 2\}$
range $\mathbb{R}$

59.

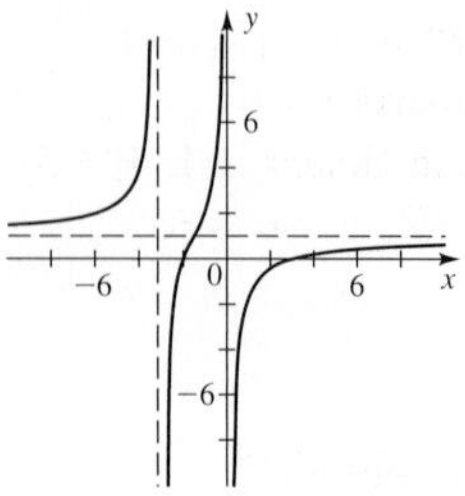

x-intercepts $-2, 3$
vertical $x = -3, x = 0$
horizontal $y = 1$
domain $\{x \mid x \neq -3, 0\}$
range $\mathbb{R}$

61.

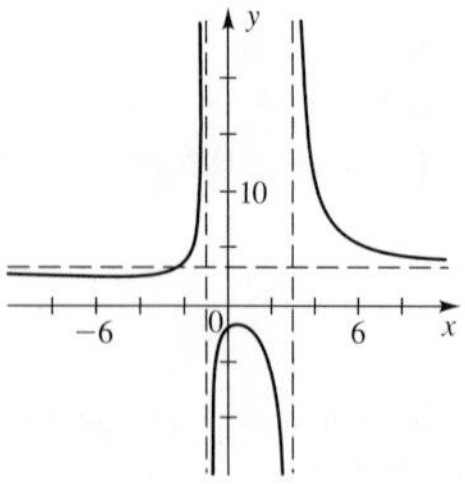

y-intercept -2
vertical $x = -1, x = 3$
horizontal $y = 3$
domain $\{x \mid x \neq -1, 3\}$
range $\{y \mid y \leq -1.5 \text{ or } y \geq 2.4\}$

63.

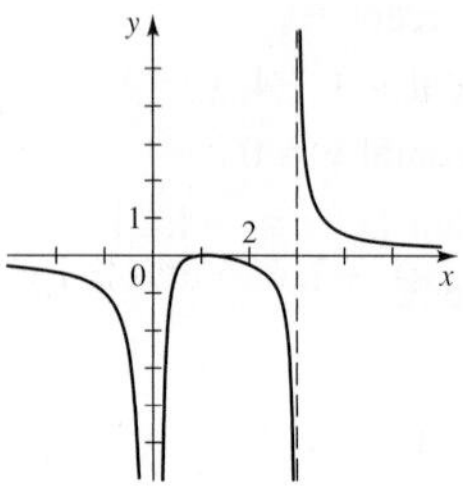

x-intercept 1
vertical $x = 0, x = 3$
horizontal $y = 0$
domain $\{x \mid x \neq 0, 3\}$
range $\mathbb{R}$

65.

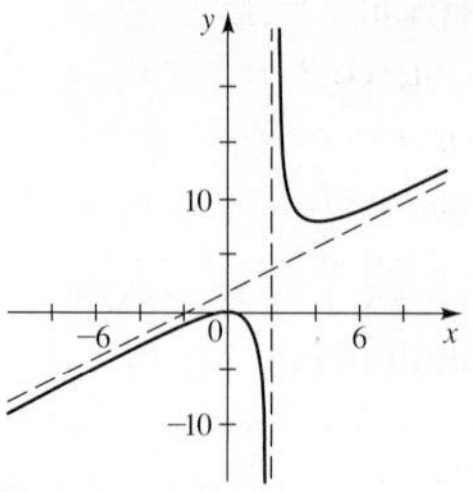

slant $y = x + 2$
vertical $x = 2$

67.

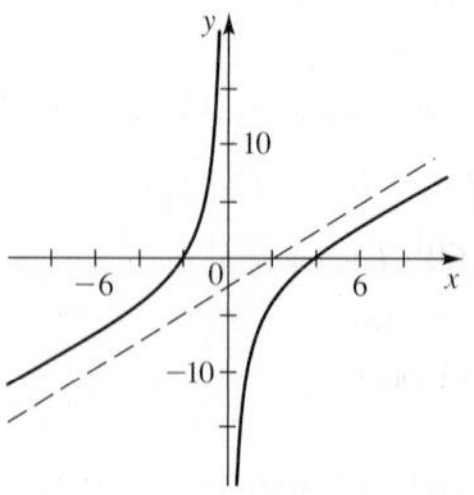

slant $y = x - 2$
vertical $x = 0$

69.

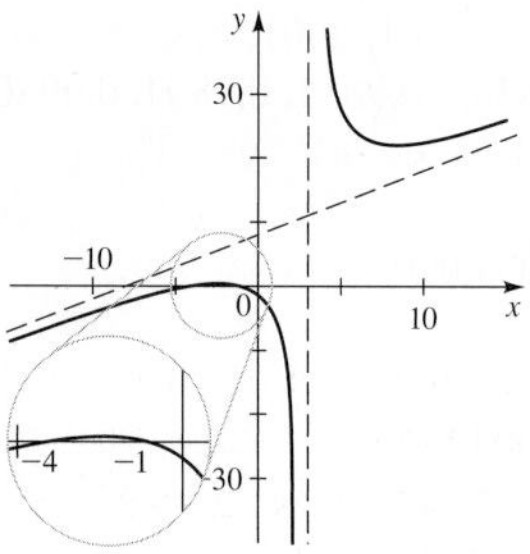

slant $y = x + 8$
vertical $x = 3$

71.

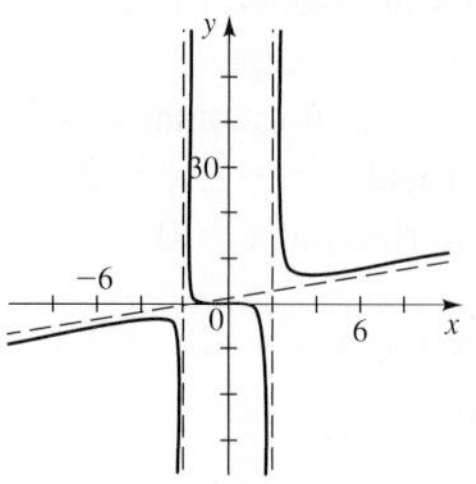

slant $y = x + 1$
vertical $x = 2, x = -2$

73.

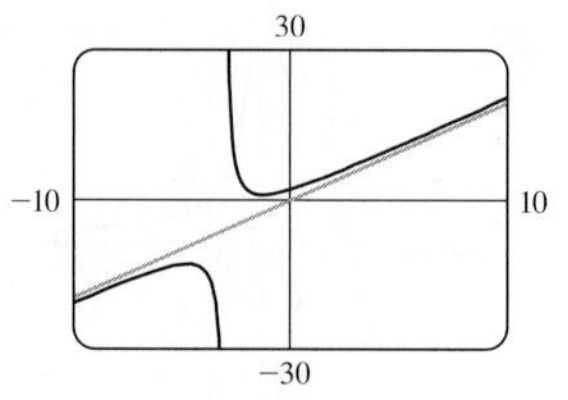

vertical $x = -3$

75.

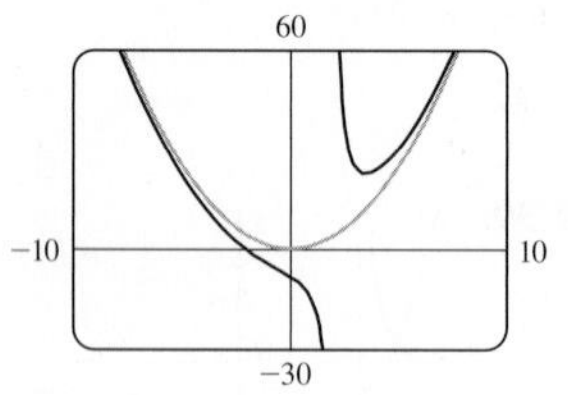

vertical $x = 2$

77.

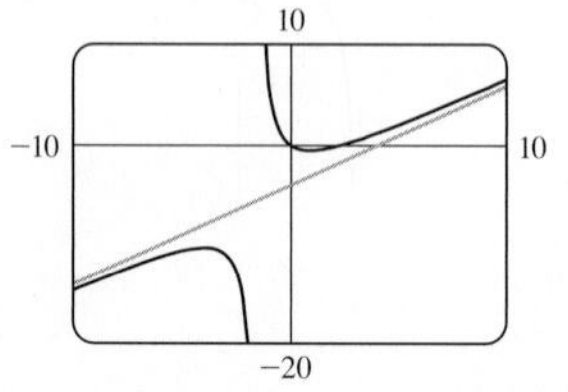

vertical $x = -1.5$
x-intercepts 0, 2.5
y-intercept 0, local maximum $(-3.9, -10.4)$
local minimum $(0.9, -0.6)$
end behavior: $y = x - 4$

79.

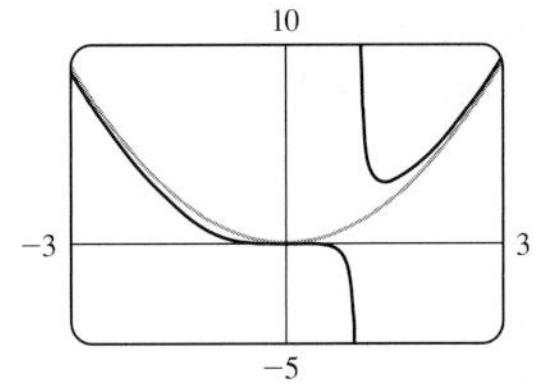

vertical $x = 1$
x-intercept 0
y-intercept 0
local minimum $(1.4, 3.1)$
end behavior: $y = x^2$

81.

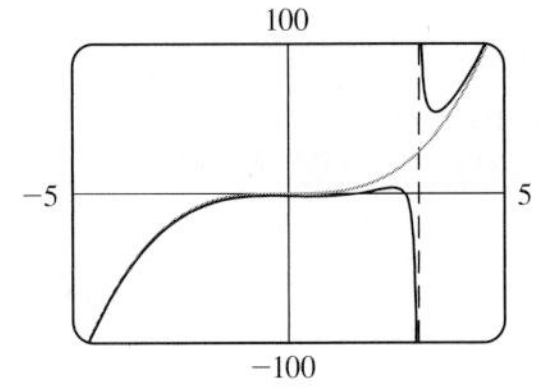

vertical $x = 3$
x-intercepts 1.6, 2.7
y-intercept -2
local maxima $(-0.4, -1.8)$, $(2.4, 3.8)$,
local minima $(0.6, -2.3)$, $(3.4, 54.3)$
end behavior $y = x^3$

83. (a)

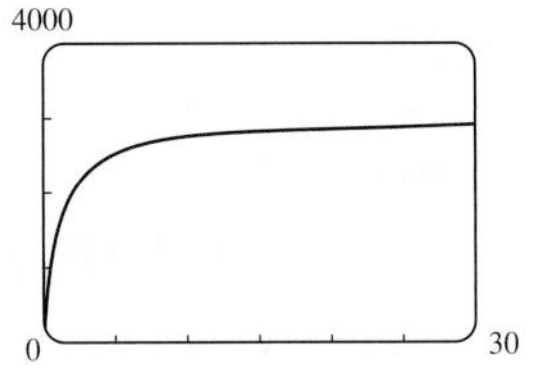

(b) It levels off at 3000.

85. (a) 2.50 mg/L **(b)** It decreases to 0. **(c)** 16.61 h

87.

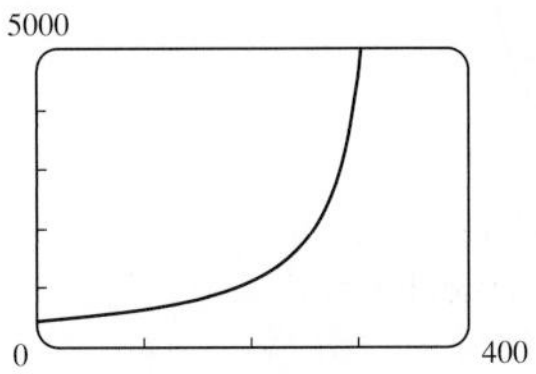

If the speed of the train approaches the speed of sound, then the pitch increases indefinitely (a sonic boom).

CHAPTER 3 REVIEW ■ PAGE 292

1. (a) $f(x) = (x + 2)^2 - 3$

(b)

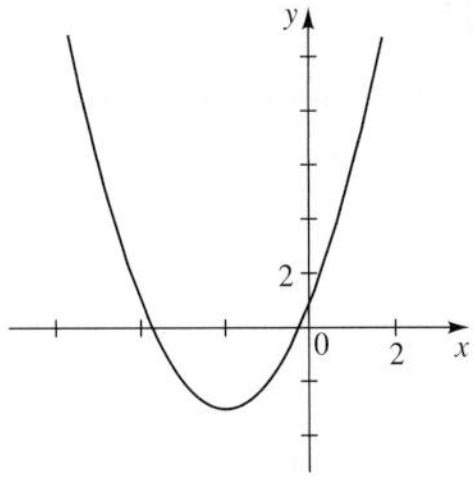

3. (a) $g(x) = -(x - 4)^2 + 17$

(b)

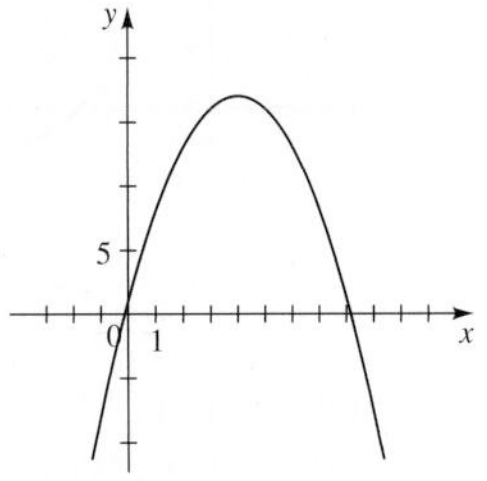

5. Minimum $f(-1) = -7$ **7.** 68 feet

9.

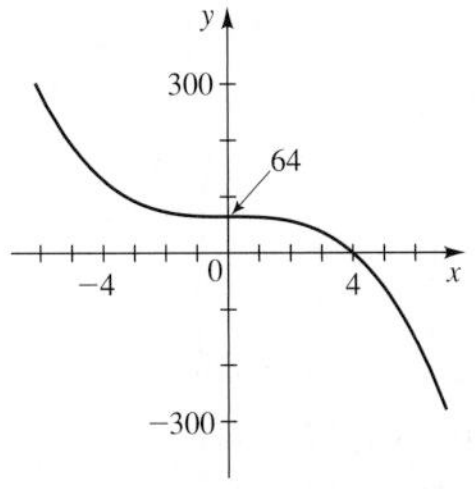

11.

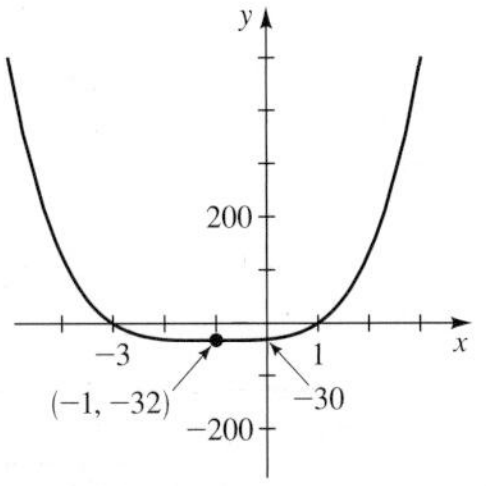

13.

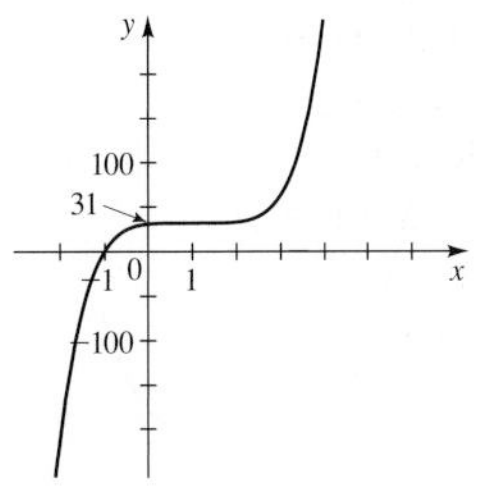

15. (a) 0 (multiplicity 3), 2 (multiplicity 2)

(b)

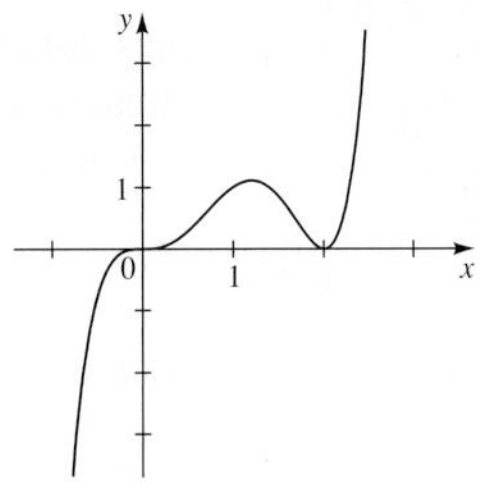

17.

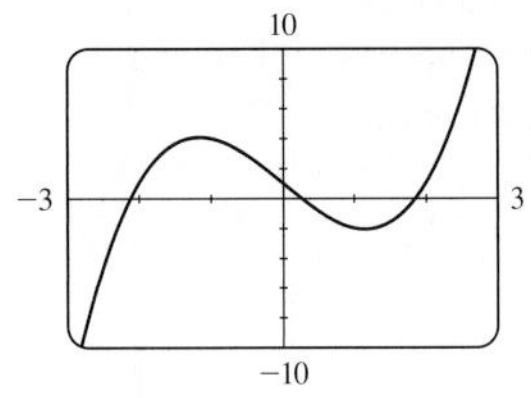

x-intercepts $-2.1, 0.3, 1.9$
y-intercept 1
local maximum $(-1.2, 4.1)$
local minimum $(1.2, -2.1)$
$y \to \infty$ as $x \to \infty$
$y \to -\infty$ as $x \to -\infty$

19.

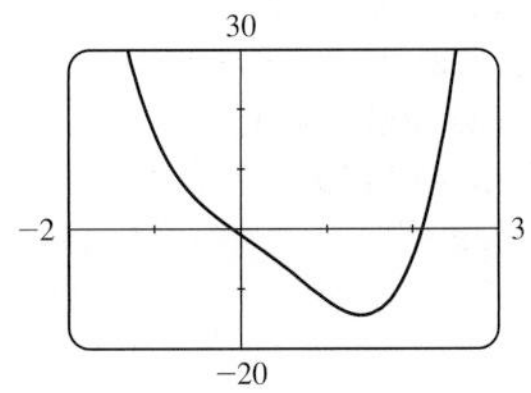

x-intercepts $-0.1, 2.1$
y-intercept -1
local minimum $(1.4, -14.5)$
$y \to \infty$ as $x \to \infty$
$y \to \infty$ as $x \to -\infty$

21. (a) $S = 13.8x(100 - x^2)$ **(b)** $0 \le x \le 10$

(c)

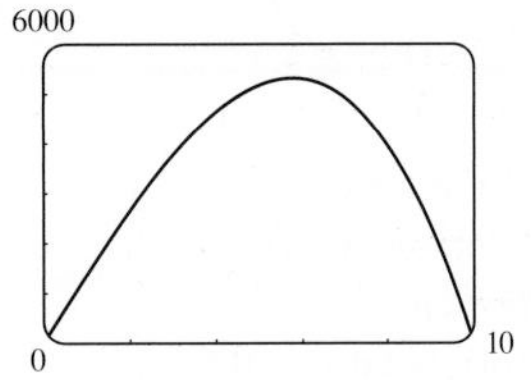

(d) 5.8 in.

In answers 23–29 the first polynomial given is the quotient, and the second is the remainder.

23. $x - 1, 3$ **25.** $x^2 + 3x + 23, 94$

27. $x^3 - 5x^2 + 17x - 83, 422$ **29.** $2x - 3, 12$

31. 3 **35.** 8 **37. (a)** $\pm1, \pm2, \pm3, \pm6, \pm9, \pm18$

(b) 2 or 0 positive, 3 or 1 negative

39. (a) $-4, 0, 4$

(b)

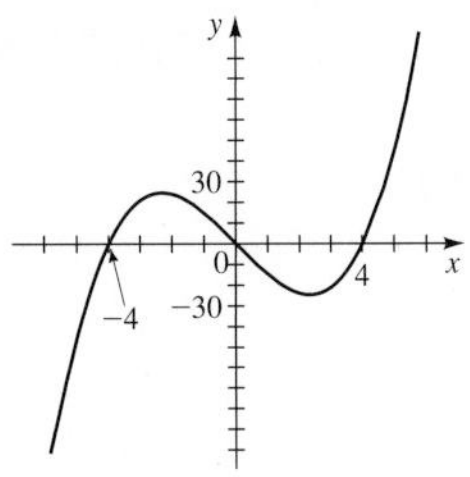

41. (a) $-2, 0$ (multiplicity 2), 1

(b)

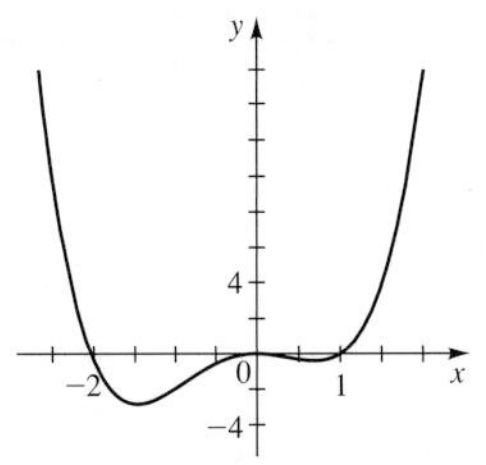

43. (a) −2, −1, 2, 3
(b)

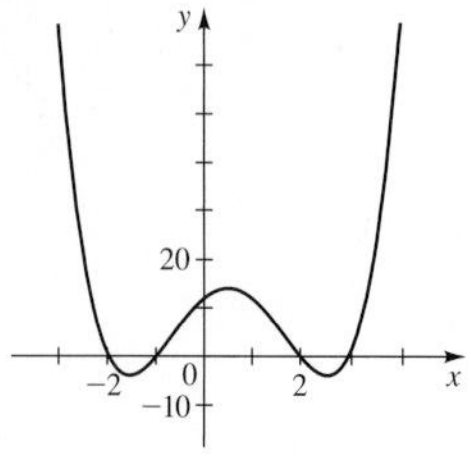

45. (a) $-\frac{1}{2}$, 1
(b)

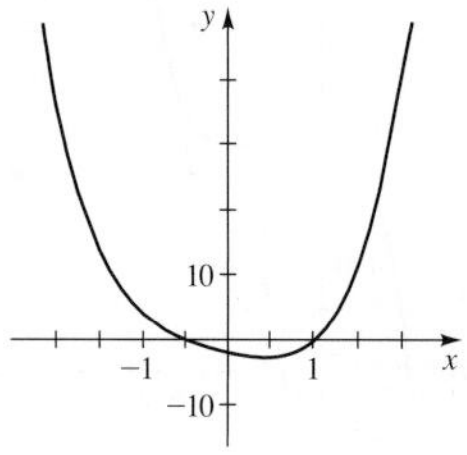

47. $3 + i$ **49.** $8 - i$ **51.** $\frac{6}{5} + \frac{8}{5}i$ **53.** i **55.** 2
57. $4x^3 - 18x^2 + 14x + 12$ **59.** No; since the complex conjugates of imaginary zeros will also be zeros, the polynomial would have 8 zeros, contradicting the requirement that it have degree 4.
61. −3, 1, 5 **63.** $-1 \pm 2i$, −2 (multiplicity 2)
65. ±2, 1 (multiplicity 3) **67.** $\pm 2, \pm 1 \pm i\sqrt{3}$
69. 1, 3, $\dfrac{-1 \pm i\sqrt{7}}{2}$ **71.** $x = -0.5, 3$ **73.** $x \approx -0.24, 4.24$
75. 2, $P(x) = (x - 2)(x^2 + 2x + 2)$

77.

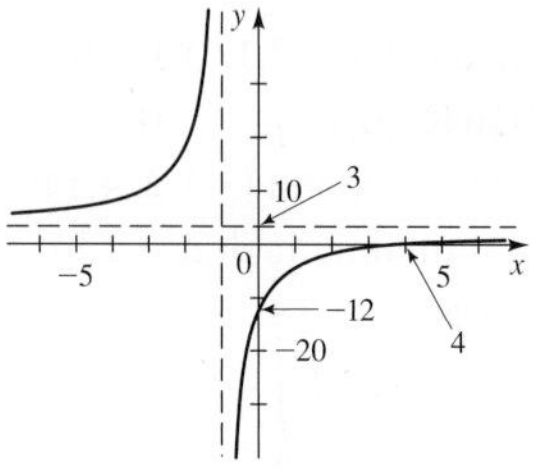

79.

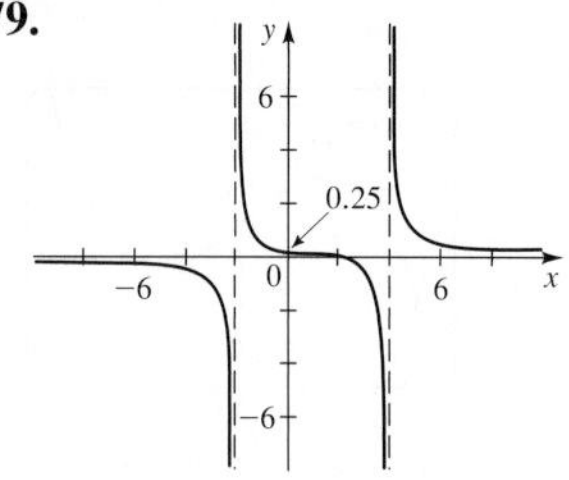

81.

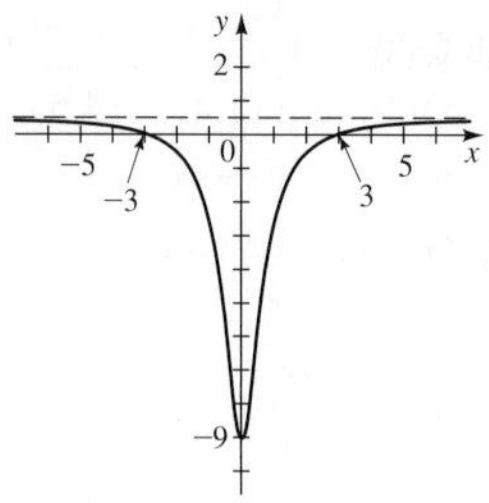

83.

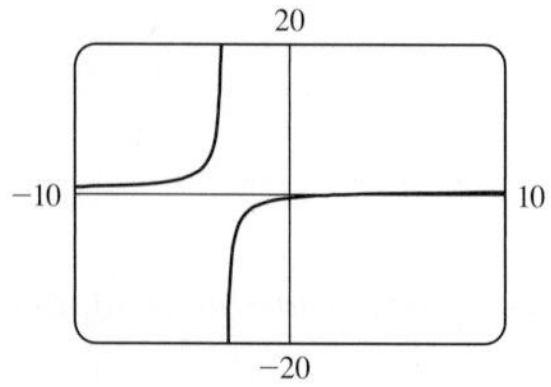

x-intercept 3
y-intercept −0.5
vertical $x = -3$
horizontal $y = 0.5$
no local extrema

85.

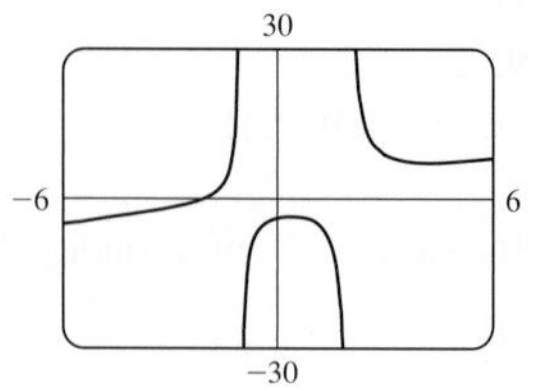

x-intercept −2
y-intercept −4
vertical $x = -1, x = 2$
slant $y = x + 1$
local maximum (0.425, −3.599)
local minimum (4.216, 7.175)

87. (−2, −28), (1, 26), (2, 68), (5, 770)

CHAPTER 3 TEST ■ PAGE 295

1. $f(x) = \left(x - \frac{1}{2}\right)^2 - \frac{25}{4}$

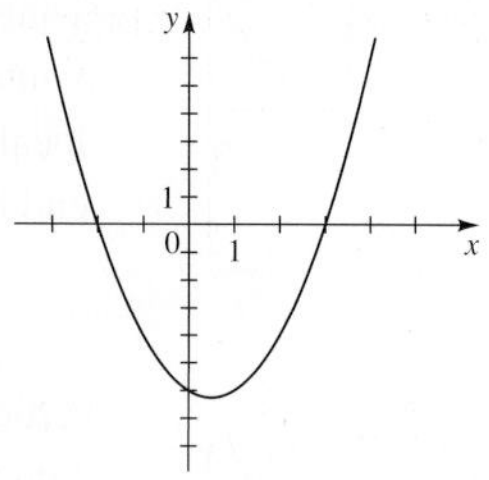

2. Minimum $f\left(-\frac{3}{2}\right) = -\frac{3}{2}$ **3. (a)** 2500 ft **(b)** 1000 ft
4.

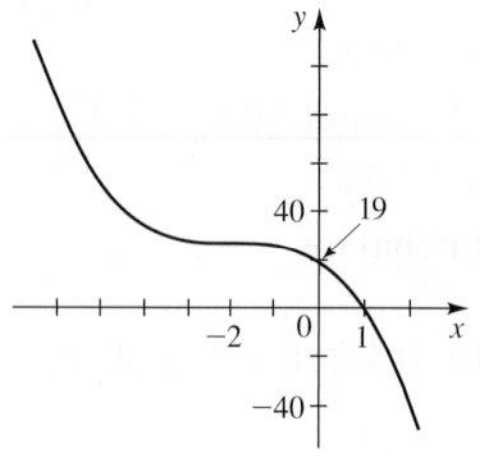

5. (a) $x^3 + 2x^2 + 2, 9$ **(b)** $x^3 + 2x^2 + \frac{1}{2}, \frac{15}{2}$
6. (a) $\pm 1, \pm 3, \pm \frac{1}{2}, \pm \frac{3}{2}$ **(b)** $2(x - 3)\left(x - \frac{1}{2}\right)(x + 1)$
(c) $-1, \frac{1}{2}, 3$ **(d)**

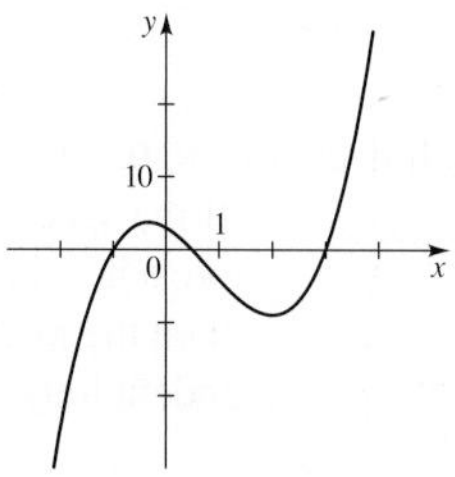

7. (a) $7 + i$ **(b)** $-1 - 5i$ **(c)** $18 + i$ **(d)** $\frac{6}{25} - \frac{17}{25}i$
(e) 1 **(f)** $6 - 2i$ **8.** 3, $-1 \pm i$ **9.** $(x - 1)^2(x - 2i)(x + 2i)$
10. $x^4 + 2x^3 + 10x^2 + 18x + 9$
11. (a) 4, 2, or 0 positive; 0 negative
(c) 0.17, 3.93

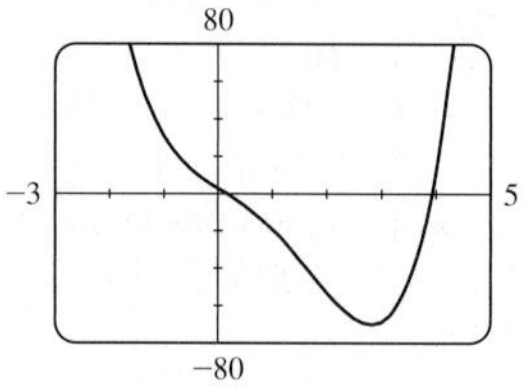

(d) Local minimum (2.8, −70.3)
12. (a) r, u **(b)** s **(c)** s **(d)**

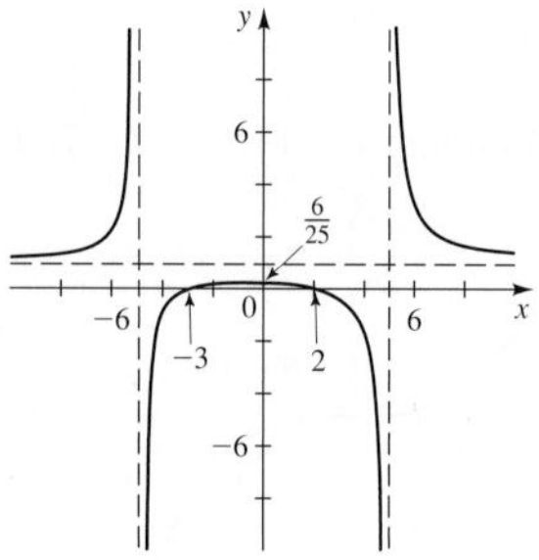

(e) $x^2 - 2x - 5$

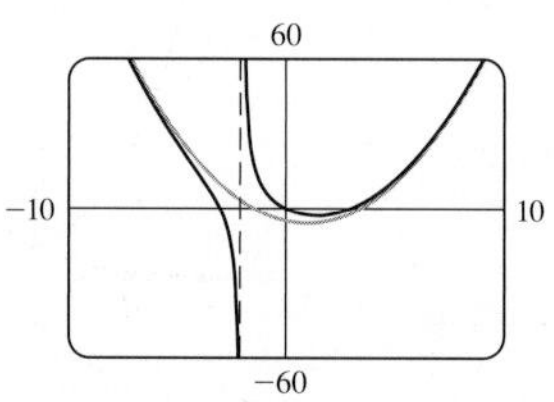

FOCUS ON MODELING ■ PAGE 298

1. (a) $y = -0.275428x^2 + 19.7485x - 273.5523$
(b)

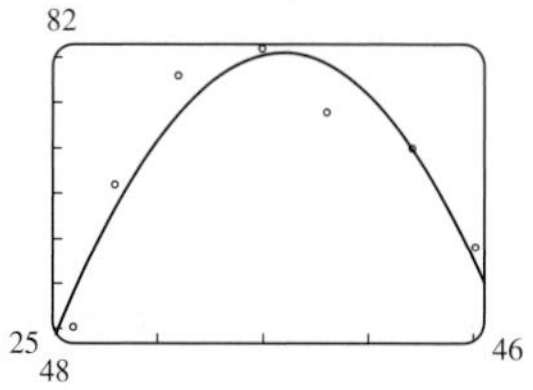

(c) 35.85 lb/in^2

3. (a) $y = 0.00203708x^3 - 0.104521x^2 + 1.966206x + 1.45576$
(b)

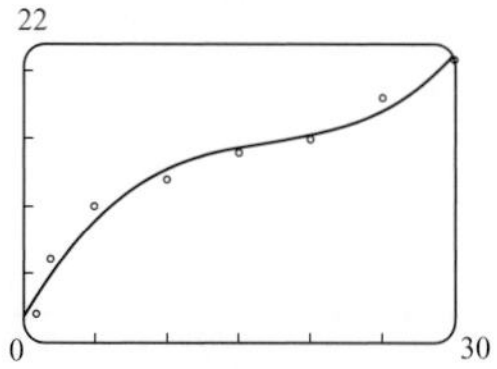

(c) 43 vegetables **(d)** 2.0 s

5. (a) Degree 2
(b) $y = -16.0x^2 + 51.8429x + 4.20714$

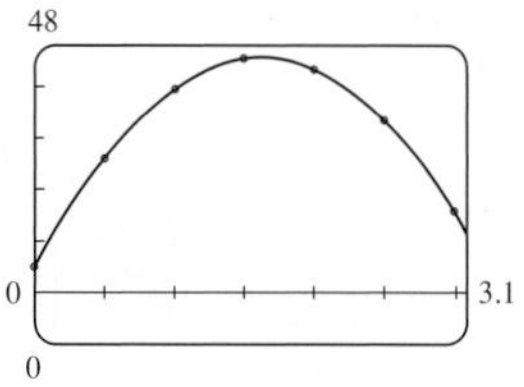

(c) 0.3 s and 2.9 s **(d)** 46.2 ft

CHAPTER 4

SECTION 4.1 ■ PAGE 307

1. 5; $\frac{1}{25}$, 1, 25, 15,625 **2. (a)** III **(b)** I **(c)** II **(d)** IV
3. (a) downward **(b)** right **4.** principal, interest rate per year, number of times interest is compounded per year, number of years, amount after t years; \$112.65 **5.** 2.000, 7.103, 77.880, 1.587
7. 0.885, 0.606, 0.117, 1.837

9.

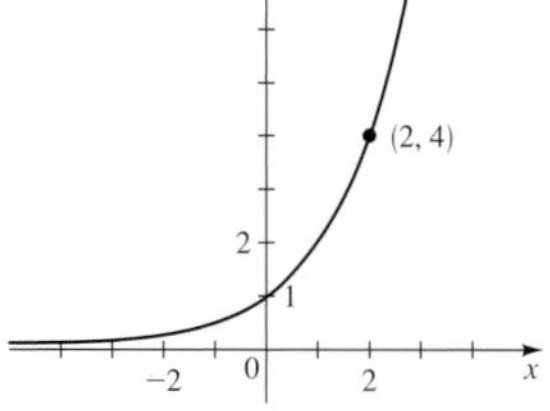

11.

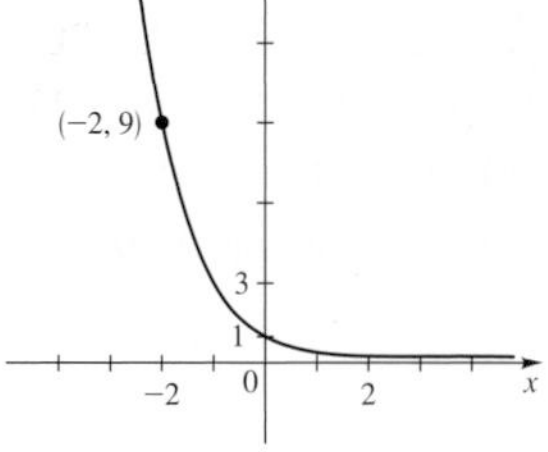

13.

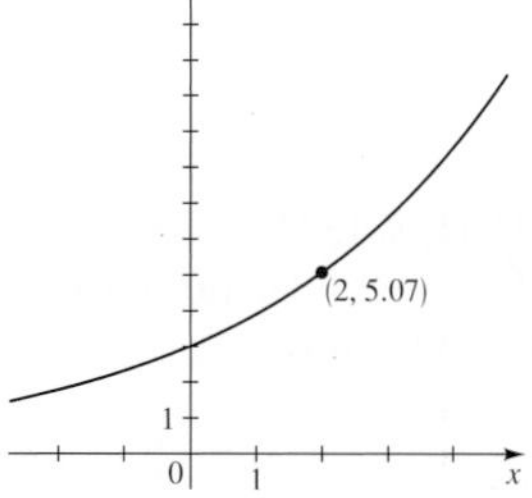

15.

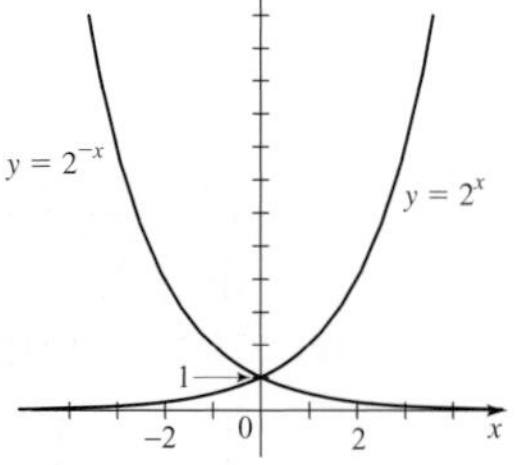

17.

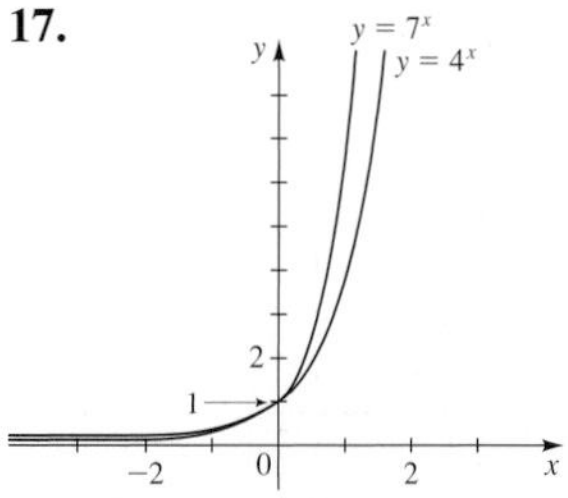

19. $f(x) = 3^x$ **21.** $f(x) = \left(\frac{1}{4}\right)^x$ **23.** II
25. $\mathbb{R}, (-\infty, 0), y = 0$

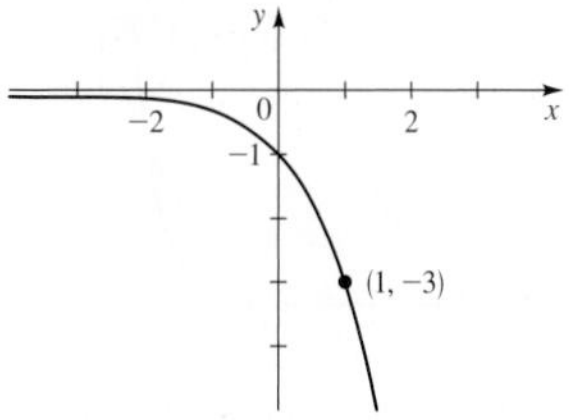

27. $\mathbb{R}, (-3, \infty), y = -3$

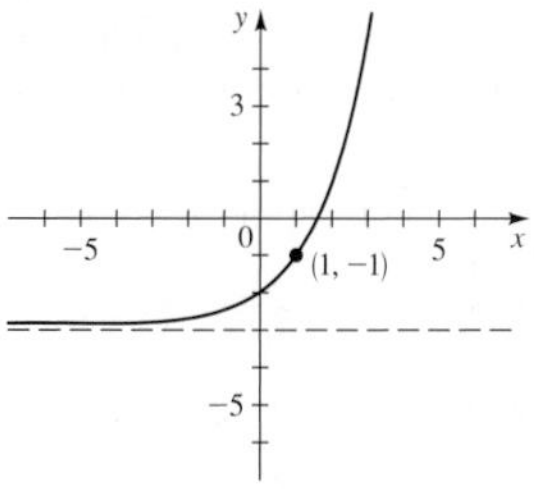

29. $\mathbb{R}, (4, \infty), y = 4$

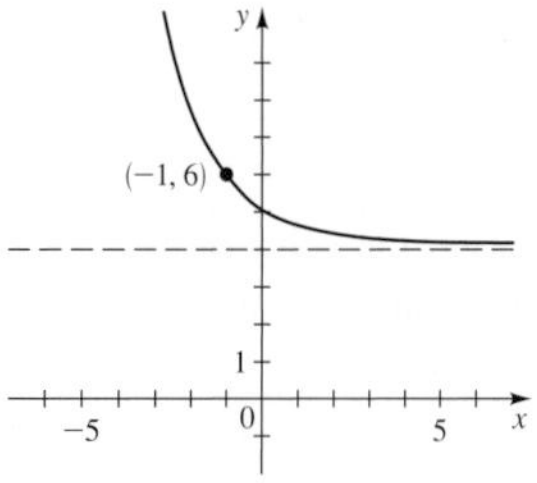

31. $\mathbb{R}, (0, \infty), y = 0$

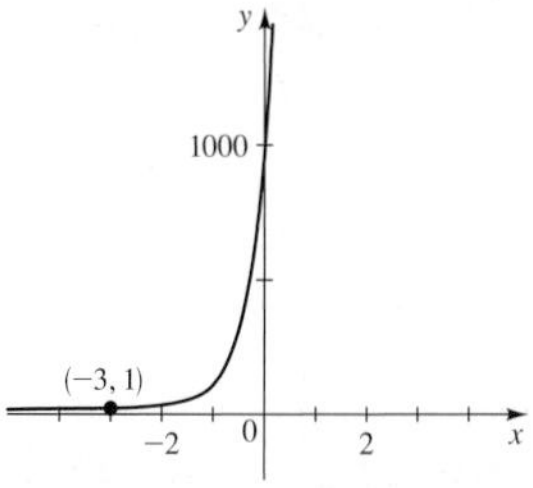

33. $\mathbb{R}, (1, \infty), y = 1$

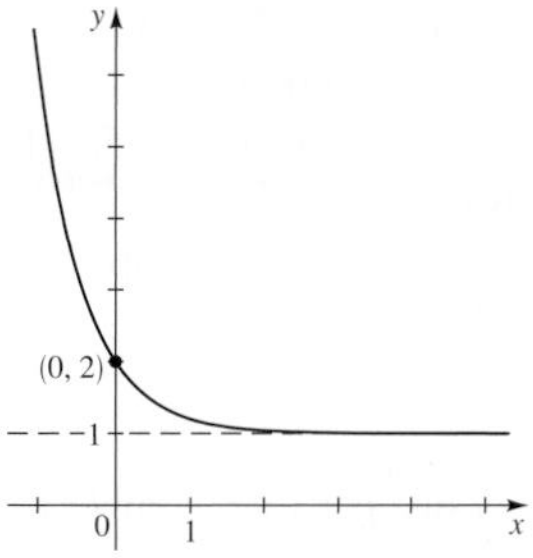

35. $\mathbb{R}, (-\infty, 3), y = 3$

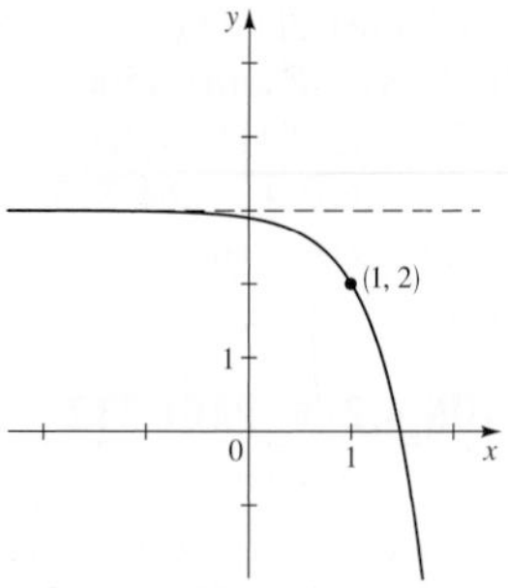

37. (a)

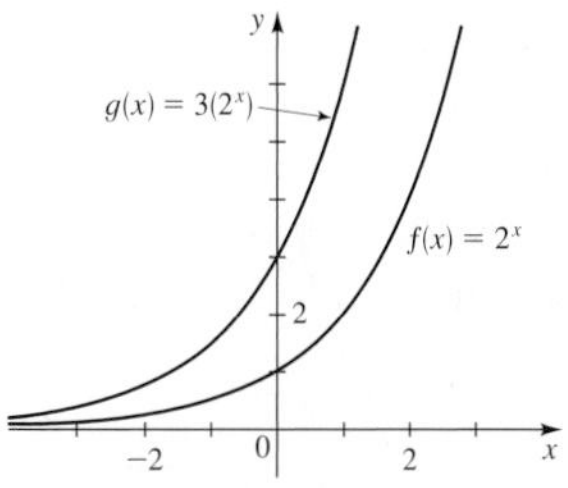

(b) The graph of g is steeper than that of f.

39.

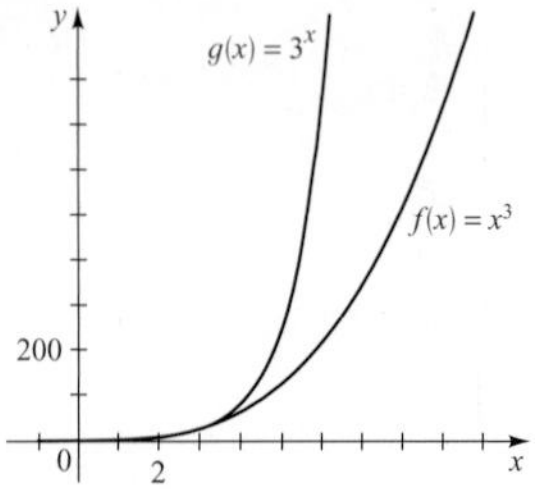

41. (a)

(i)

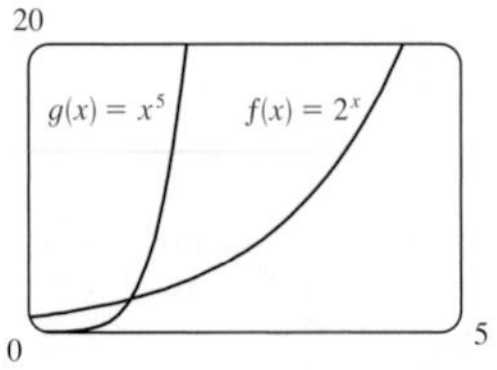

(ii)

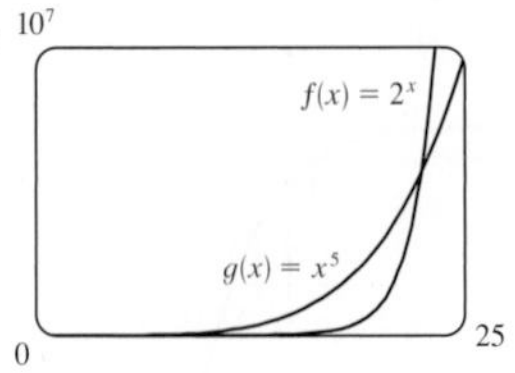

(iii)

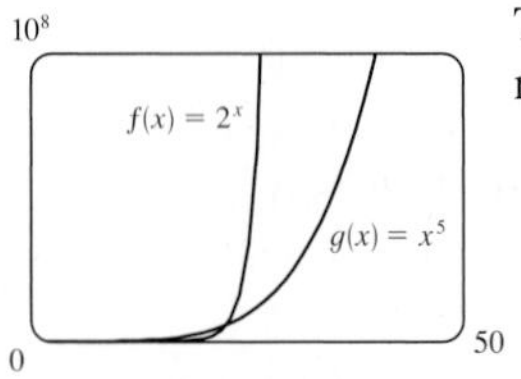

The graph of f ultimately increases much more quickly than that of g.

(b) 1.2, 22.4

43.

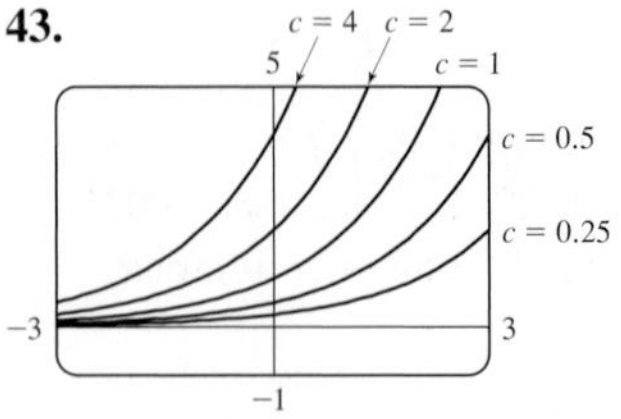

The larger the value of c, the more rapidly the graph increases.

45. (a) Increasing on $(-\infty, 0.50]$; decreasing on $[0.50, \infty)$ **(b)** $(0, 1.78]$ **47. (a)** $1500 \cdot 2^t$ **(b)** 25,165,824,000
49. \$5203.71, \$5415.71, \$5636.36, \$5865.99, \$6104.98, \$6353.71
51. (a) \$11,605.41 **(b)** \$13,468.55 **(c)** \$15,630.80
53. (a) \$519.02 **(b)** \$538.75 **(c)** \$726.23 **55.** \$7678.96
57. 8.30%

SECTION 4.2 ■ PAGE 312

1. natural; 2.71828 **2.** principal, interest rate per year, number of years; amount after t years; \$112.75
3. 20.085, 1.259, 2.718, 0.135

5.

x	y
−2	0.41
−1	1.10
−0.5	1.82
0	3
0.5	4.95
1	8.15
2	22.17

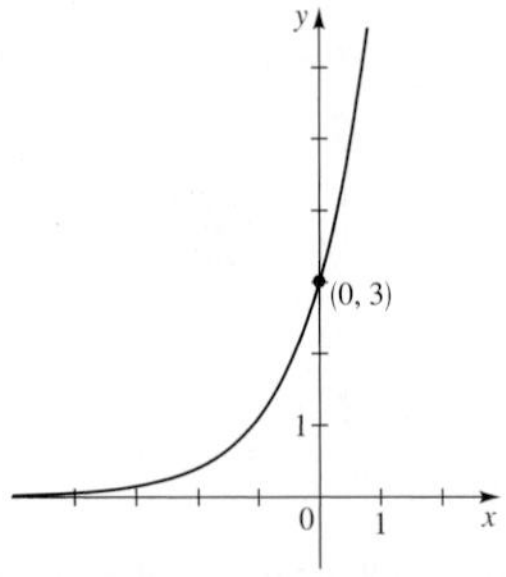

7. $\mathbb{R}, (-\infty, 0), y = 0$

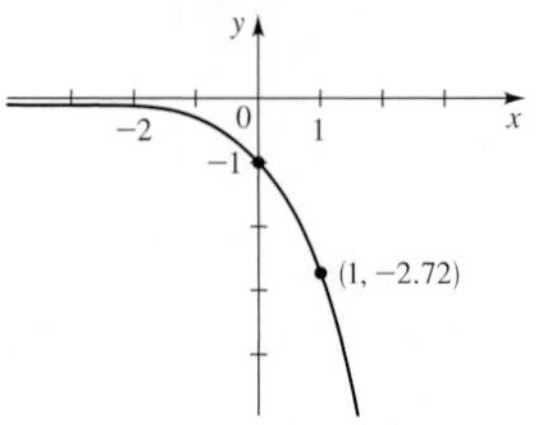

9. $\mathbb{R}, (-1, \infty), y = -1$

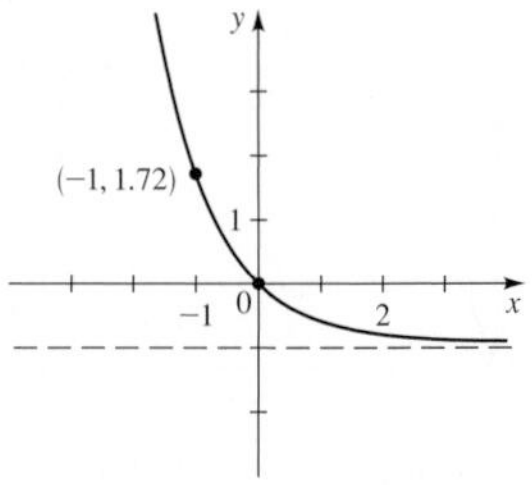

11. $\mathbb{R}, (0, \infty), y = 0$

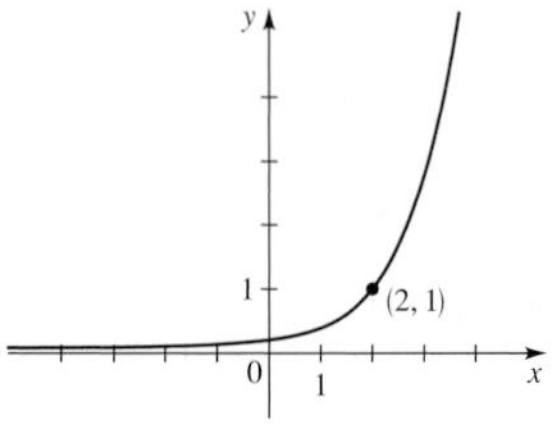

13. $\mathbb{R}, (-3, \infty), y = -3$

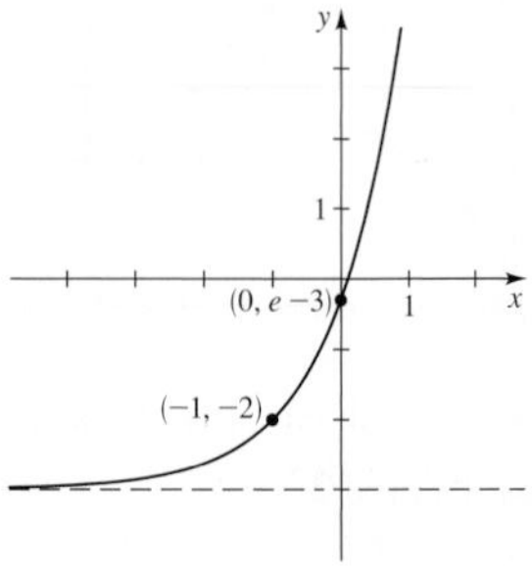

15. (a)

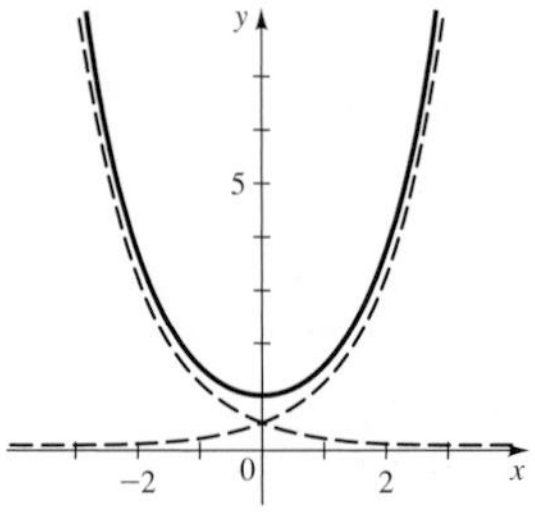

17. (a)

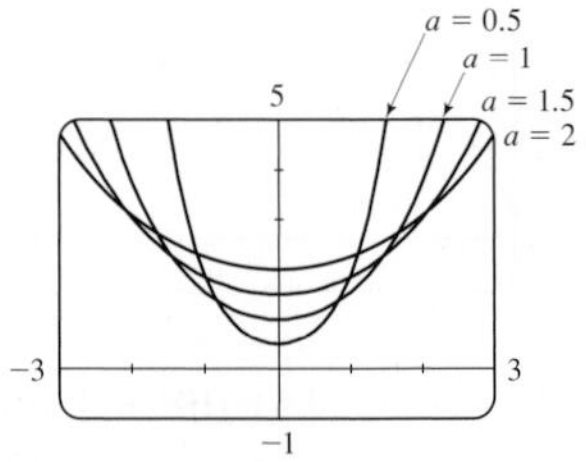

(b) The larger the value of a, the wider the graph.

19. Local minimum ≈ (0.27, 1.75)
21. (a) 13 kg **(b)** 6.6 kg
23. (a) 0 **(b)** 50.6 ft/s, 69.2 ft/s
(c)

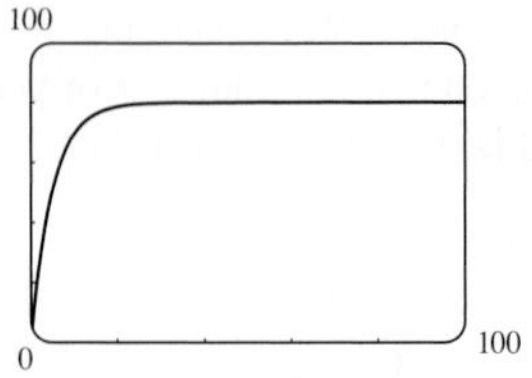

(d) 80 ft/s

25. (a) 100 **(b)** 482, 999, 1168 **(c)** 1200
27. (a) 11.79 billion, 11.97 billion
(b)

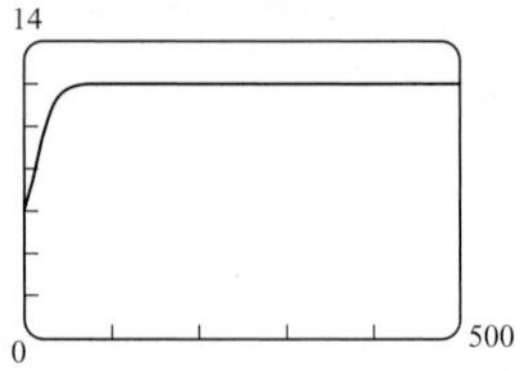

(c) 12 billion

29. \$7213.18, \$7432.86, \$7659.22, \$7892.48, \$8132.84, \$8380.52
31. (a) \$2145.02 **(b)** \$2300.55 **(c)** \$3043.92 **33. (a)** \$768.05 **(b)** \$769.22 **(c)** \$769.82 **(d)** \$770.42 **35.** (a) is best.

37. (a) $A(t) = 5000e^{0.09t}$ **(b)**

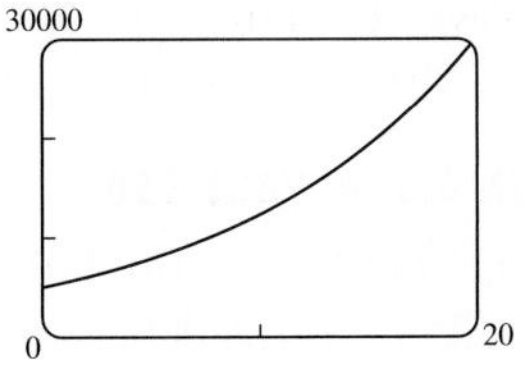

(c) After 17.88 yr

SECTION 4.3 ■ PAGE 322

1. 10^x

x	10^3	10^2	10^1	10^0	10^{-1}	10^{-2}	10^{-3}	$10^{1/2}$
$\log x$	3	2	1	0	-1	-2	-3	$\frac{1}{2}$

2. 9; 1, 0, -1, 2, $\frac{1}{2}$
3. (a) $\log_5 125 = 3$ **(b)** $5^2 = 25$ **4. (a)** III **(b)** II **(c)** I **(d)** IV

5.

Logarithmic form	Exponential form
$\log_8 8 = 1$	$8^1 = 8$
$\log_8 64 = 2$	$8^2 = 64$
$\log_8 4 = \frac{2}{3}$	$8^{2/3} = 4$
$\log_8 512 = 3$	$8^3 = 512$
$\log_8 \frac{1}{8} = -1$	$8^{-1} = \frac{1}{8}$
$\log_8 \frac{1}{64} = -2$	$8^{-2} = \frac{1}{64}$

7. (a) $5^2 = 25$ **(b)** $5^0 = 1$ **9. (a)** $8^{1/3} = 2$ **(b)** $2^{-3} = \frac{1}{8}$
11. (a) $e^x = 5$ **(b)** $e^5 = y$ **13. (a)** $\log_5 125 = 3$
(b) $\log_{10} 0.0001 = -4$ **15. (a)** $\log_8 \frac{1}{8} = -1$ **(b)** $\log_2 \frac{1}{8} = -3$
17. (a) $\ln 2 = x$ **(b)** $\ln y = 3$ **19. (a)** 1 **(b)** 0 **(c)** 2
21. (a) 2 **(b)** 2 **(c)** 10 **23. (a)** -3 **(b)** $\frac{1}{2}$ **(c)** -1
25. (a) 37 **(b)** 8 **(c)** $\sqrt{5}$ **27. (a)** $-\frac{2}{3}$ **(b)** 4 **(c)** -1
29. (a) 32 **(b)** 4 **31. (a)** 5 **(b)** 27 **33. (a)** 100 **(b)** 25
35. (a) 2 **(b)** 4 **37. (a)** 0.3010 **(b)** 1.5465 **(c)** -0.1761
39. (a) 1.6094 **(b)** 3.2308 **(c)** 1.0051

41.

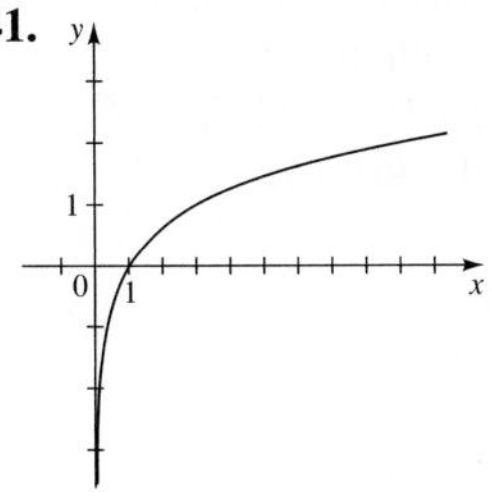

43.

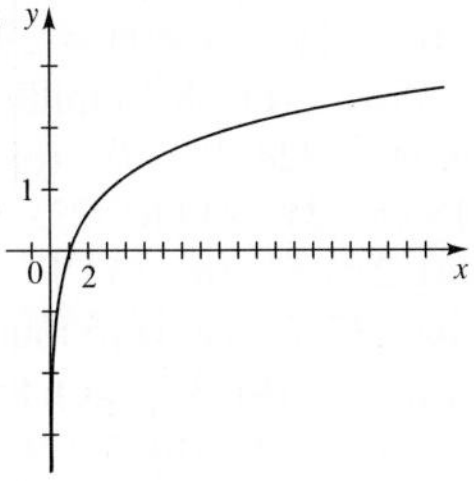

45. $y = \log_5 x$ **47.** $y = \log_9 x$ **49.** I

51.

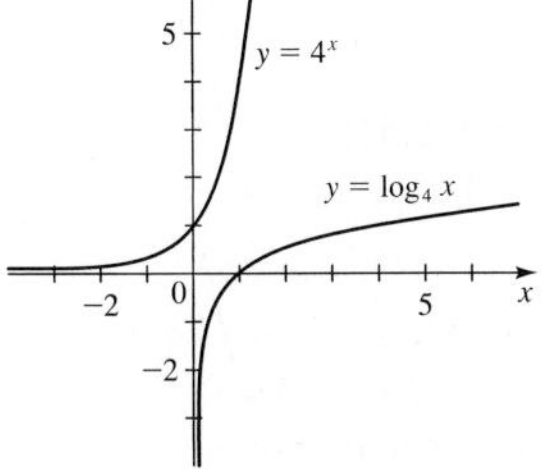

53. $(4, \infty), \mathbb{R}, x = 4$

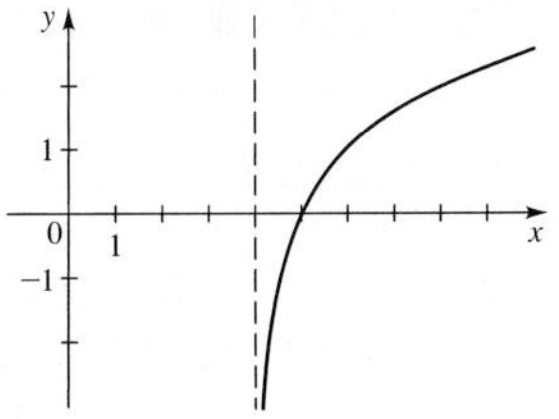

55. $(-\infty, 0), \mathbb{R}, x = 0$

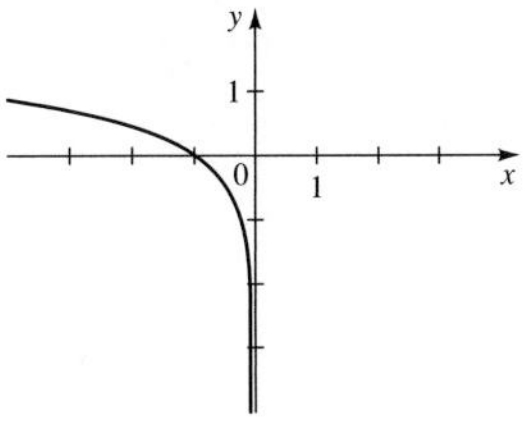

57. $(0, \infty), \mathbb{R}, x = 0$

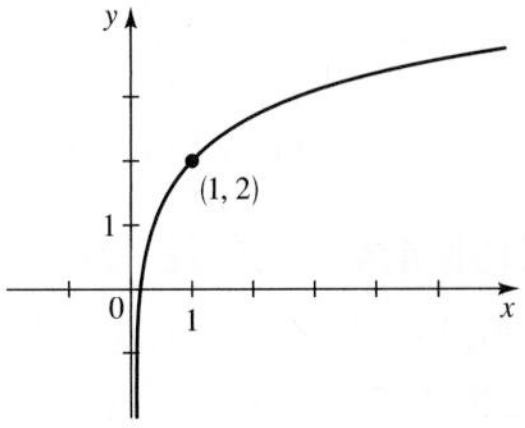

59. $(0, \infty), \mathbb{R}, x = 0$

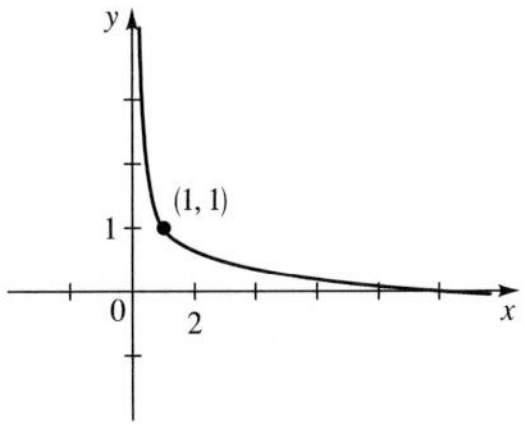

61. $(0, \infty), [0, \infty), x = 0$

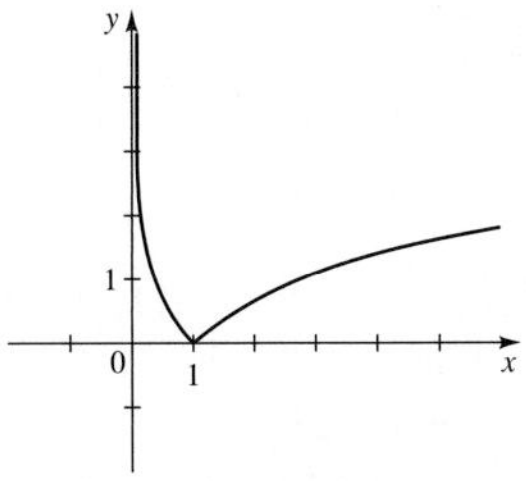

63. $(-3, \infty)$ **65.** $(-\infty, -1) \cup (1, \infty)$ **67.** $(0, 2)$

69.

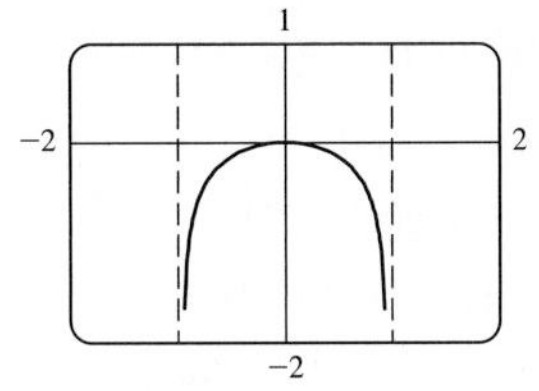

domain $(-1, 1)$
vertical asymptotes $x = 1$, $x = -1$
local maximum $(0, 0)$

71.

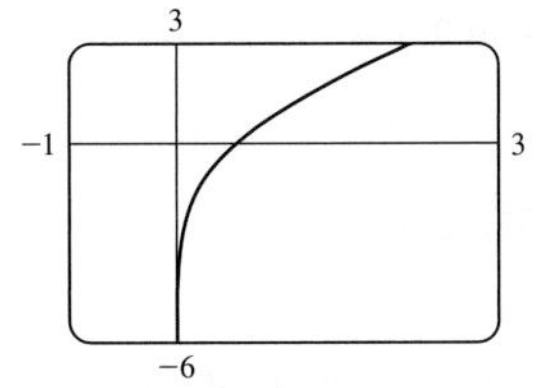

domain $(0, \infty)$
vertical asymptote $x = 0$
no maximum or minimum

73.

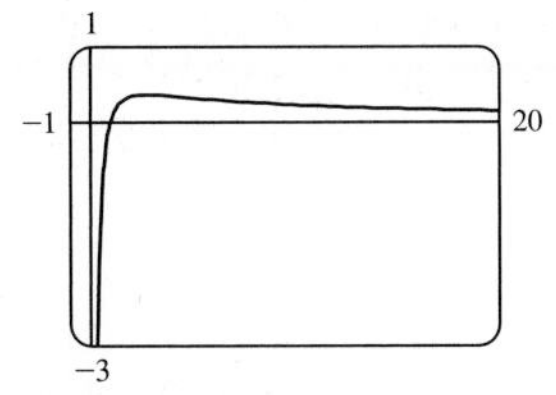

domain $(0, \infty)$
vertical asymptote $x = 0$
horizontal asymptote $y = 0$
local maximum $\approx (2.72, 0.37)$

75. $(f \circ g)(x) = 2^{x+1}, (-\infty, \infty)$; $(g \circ f)(x) = 2^x + 1, (-\infty, \infty)$
77. $(f \circ g)(x) = \log_2 (x - 2), (2, \infty)$;
$(g \circ f)(x) = \log_2 x - 2, (0, \infty)$
79. The graph of f grows more slowly than g.
81. (a)

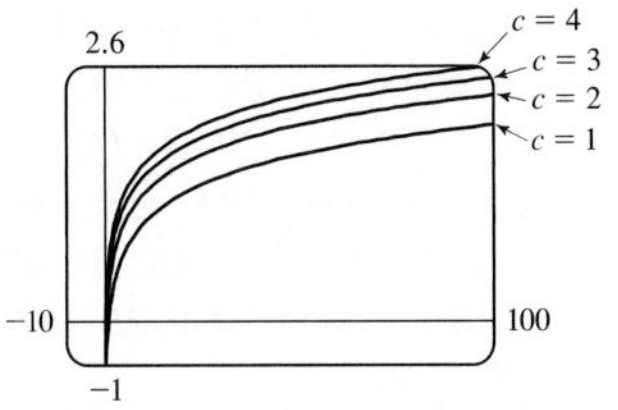

(b) The graph of $f(x) = \log(cx)$ is the graph of $f(x) = \log(x)$ shifted upward $\log c$ units.

83. **(a)** $(1, \infty)$ **(b)** $f^{-1}(x) = 10^{2^x}$

85. **(a)** $f^{-1}(x) = \log_2\left(\dfrac{x}{1 - x}\right)$ **(b)** $(0, 1)$ **87.** 2602 yr

89. 11.5 yr, 9.9 yr, 8.7 yr **91.** 5.32, 4.32

SECTION 4.4 ■ PAGE 329

1. sum; $\log_5 25 + \log_5 125 = 2 + 3$
2. difference; $\log_5 25 - \log_5 125 = 2 - 3$
3. times; $10 \cdot \log_5 25$
4. **(a)** $2 \log x + \log y - \log z$
(b) $\log\left(\dfrac{x^2 y}{z}\right)$

5. 10, e; Change of Base; $\log_7 12 = \dfrac{\log 12}{\log 7} = 1.277$

6. True **7.** $\frac{3}{2}$ **9.** 2 **11.** 3 **13.** 3 **15.** 200 **17.** 4
19. $1 + \log_2 x$ **21.** $\log_2 x + \log_2(x - 1)$
23. $10 \log 6$ **25.** $\log_2 A + 2 \log_2 B$ **27.** $\log_3 x + \frac{1}{2} \log_3 y$
29. $\frac{1}{3} \log_5(x^2 + 1)$ **31.** $\frac{1}{2}(\ln a + \ln b)$
33. $3 \log x + 4 \log y - 6 \log z$
35. $\log_2 x + \log_2(x^2 + 1) - \frac{1}{2} \log_2(x^2 - 1)$
37. $\ln x + \frac{1}{2}(\ln y - \ln z)$ **39.** $\frac{1}{4} \log(x^2 + y^2)$
41. $\frac{1}{2}[\log(x^2 + 4) - \log(x^2 + 1) - 2 \log(x^3 - 7)]$
43. $3 \ln x + \frac{1}{2} \ln(x - 1) - \ln(3x + 4)$ **45.** $\log_3 160$

47. $\log_2(AB/C^2)$ **49.** $\log\left(\dfrac{x^4(x - 1)^2}{\sqrt[3]{x^2 + 1}}\right)$

51. $\ln(5x^2(x^2 + 5)^3)$

53. $\log\left(\dfrac{x^2}{x - 3}\right)$ **55.** 2.321928 **57.** 2.523719

59. 0.493008 **61.** 3.482892
63.

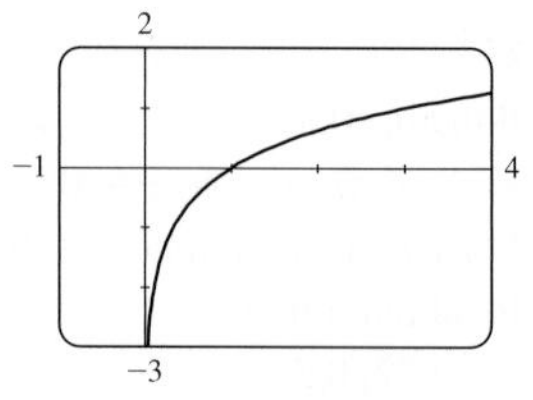

69. **(a)** $P = c/W^k$ **(b)** 1866, 64
71. **(a)** $M = -2.5 \log B + 2.5 \log B_0$

SECTION 4.5 ■ PAGE 338

1. **(a)** $e^x = 25$ **(b)** $x = \ln 25$ **(c)** 3.219
2. **(a)** $\log 3(x - 2) = \log x$ **(b)** $3(x - 2) = x$ **(c)** 3
3. 1.3979 **5.** −0.9730 **7.** −0.5850 **9.** 1.2040 **11.** 0.0767
13. 0.2524 **15.** 1.9349 **17.** −43.0677 **19.** 2.1492
21. 6.2126 **23.** −2.9469 **25.** −2.4423 **27.** 14.0055
29. $\ln 2 \approx 0.6931, 0$ **31.** $\frac{1}{2} \ln 3 \approx 0.5493$ **33.** ± 1 **35.** $0, \frac{4}{3}$
37. $e^{10} \approx 22026$ **39.** 0.01 **41.** $\frac{95}{3}$ **43.** −7 **45.** 5 **47.** 5
49. $\frac{13}{12}$ **51.** 4 **53.** 6 **55.** $\frac{3}{2}$ **57.** $1/\sqrt{5} \approx 0.4472$ **59.** 2.21
61. 0.00, 1.14 **63.** −0.57 **65.** 0.36
67. $2 < x < 4$ or $7 < x < 9$ **69.** $\log 2 < x < \log 5$

71. $f^{-1}(x) = \dfrac{\ln x}{2 \ln 2}$ **73.** $f^{-1}(x) = 2^x + 1$

75. **(a)** \$6435.09 **(b)** 8.24 yr **77.** 6.33 yr **79.** 8.15 yr

81. 13 days **83.** **(a)** 7337 **(b)** 1.73 yr **85.** **(a)** $P = P_0 e^{-h/k}$
(b) 56.47 kPa **87.** **(a)** $t = -\frac{5}{13} \ln(1 - \frac{13}{60} I)$ **(b)** 0.218 s

SECTION 4.6 ■ PAGE 350

1. **(a)** $n(t) = 10 \cdot 2^{2t/3}$ **(b)** 1.05×10^8 **(c)** After 14.9 h
3. **(a)** 3125 **(b)** 317,480
(c)

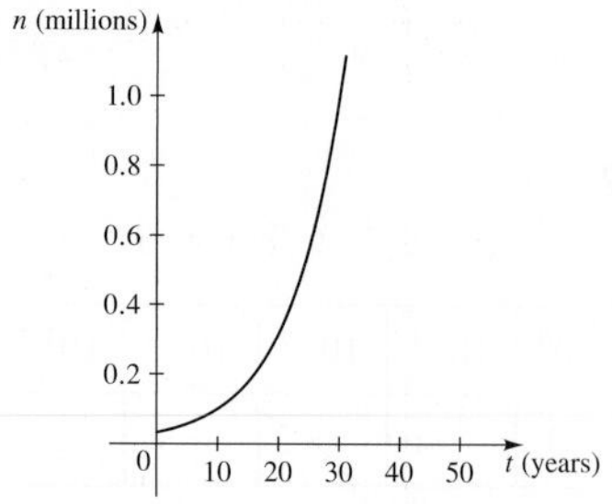

5. **(a)** $n(t) = 18{,}000 e^{0.08t}$ **(b)** 34,137
(c)

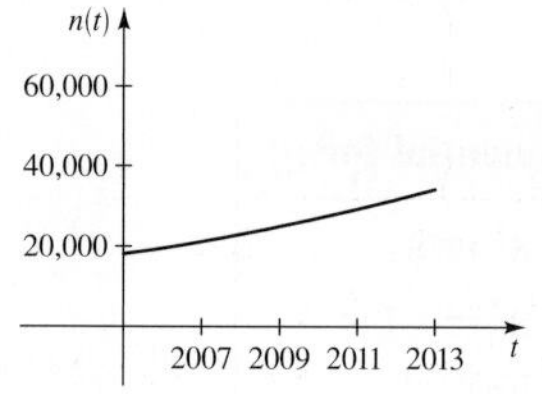

7. **(a)** 233 million **(b)** 181 million

9. **(a)** $n(t) = 112{,}000 \cdot 2^{t/18}$ **(b)** $n(t) = 112{,}000 e^{0.0385t}$
(c) **(d)** In the year 2045

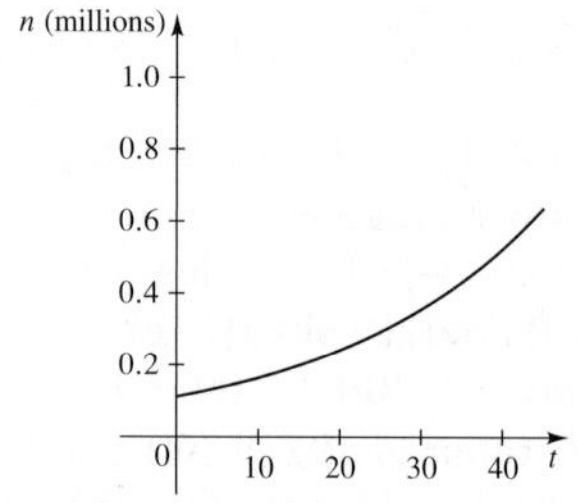

11. **(a)** 20,000 **(b)** $n(t) = 20{,}000 e^{0.1096t}$ **(c)** About 48,000
(d) 2017 **13.** **(a)** $n(t) = 8600 e^{0.1508t}$ **(b)** About 11,600
(c) 4.6 h **15.** **(a)** $n(t) = 29.76 e^{0.012936t}$ million
(b) 53.5 yr **(c)** 38.55 million **17.** **(a)** $m(t) = 22 \cdot 2^{-t/1600}$
(b) $m(t) = 22 e^{-0.000433t}$ **(c)** 3.9 mg **(d)** 463.4 yr
19. 18 yr **21.** 149 h **23.** 3560 yr
25. **(a)** 210°F **(b)** 153°F **(c)** 28 min
27. **(a)** 137°F **(b)** 116 min
29. **(a)** 2.3 **(b)** 3.5 **(c)** 8.3
31. **(a)** 10^{-3} M **(b)** 3.2×10^{-7} M
33. $4.8 \le \text{pH} \le 6.4$ **35.** $\log 20 \approx 1.3$ **37.** Twice as intense
39. 8.2 **41.** 73 dB **43.** **(b)** 106 dB

CHAPTER 4 REVIEW ■ PAGE 353

1. 0.089, 9.739, 55.902 **3.** 11.954, 2.989, 2.518
5. $\mathbb{R}$, $(0, \infty)$, $y = 0$ **7.** $\mathbb{R}$, $(3, \infty)$, $y = 3$

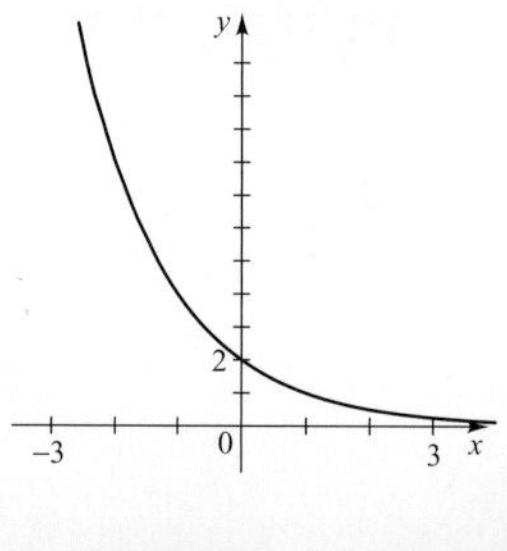

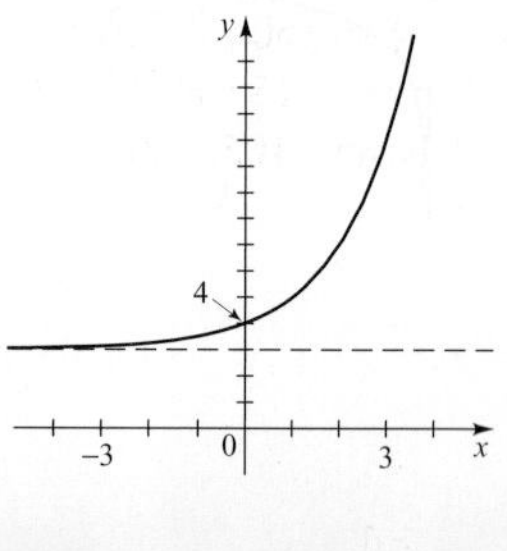

9. $(1, \infty), \mathbb{R}, x = 1$

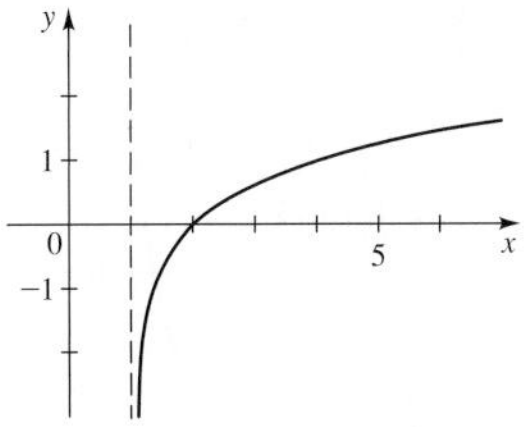

11. $(0, \infty), \mathbb{R}, x = 0$

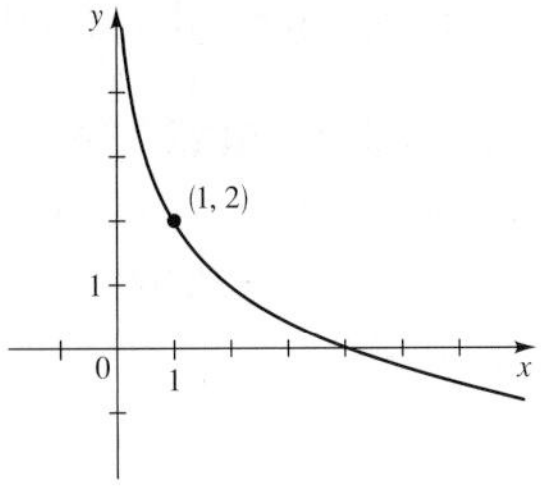

13. $\mathbb{R}, (-1, \infty), y = -1$

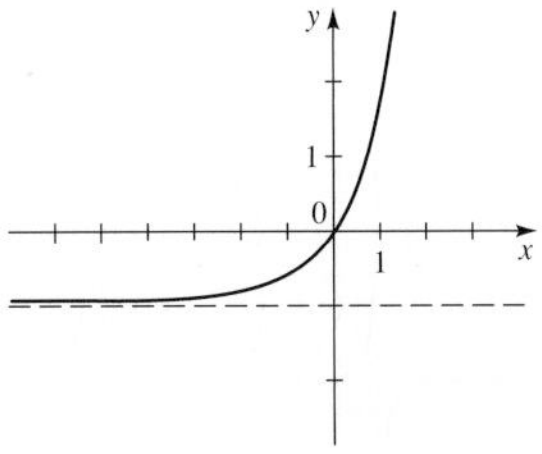

15. $(0, \infty), \mathbb{R}, x = 0$

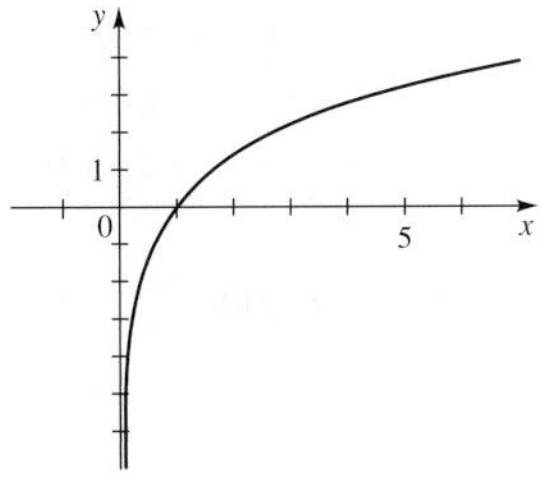

17. $\left(-\infty, \frac{1}{2}\right)$ **19.** $(-\infty, -2) \cup (2, \infty)$ **21.** $2^{10} = 1024$
23. $10^y = x$ **25.** $\log_2 64 = 6$ **27.** $\log 74 = x$ **29.** 7 **31.** 45
33. 6 **35.** -3 **37.** $\frac{1}{2}$ **39.** 2 **41.** 92 **43.** $\frac{2}{3}$
45. $\log A + 2 \log B + 3 \log C$ **47.** $\frac{1}{2}[\ln(x^2 - 1) - \ln(x^2 + 1)]$
49. $2 \log_5 x + \frac{3}{2} \log_5(1 - 5x) - \frac{1}{2} \log_5(x^3 - x)$
51. $\log 96$ **53.** $\log_2\left(\dfrac{(x - y)^{3/2}}{(x^2 + y^2)^2}\right)$ **55.** $\log\left(\dfrac{x^2 - 4}{\sqrt{x^2 + 4}}\right)$
57. 5 **59.** 2.60 **61.** -1.15 **63.** $-4, 2$
65. -15 **67.** 3 **69.** 0.430618
71. 2.303600

73.

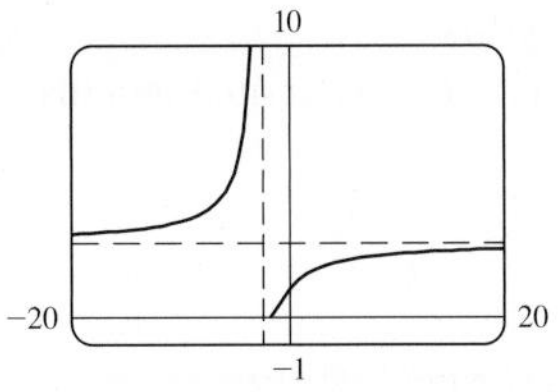

vertical asymptote
$x = -2$
horizontal asymptote
$y = 2.72$
no maximum or minimum

75.

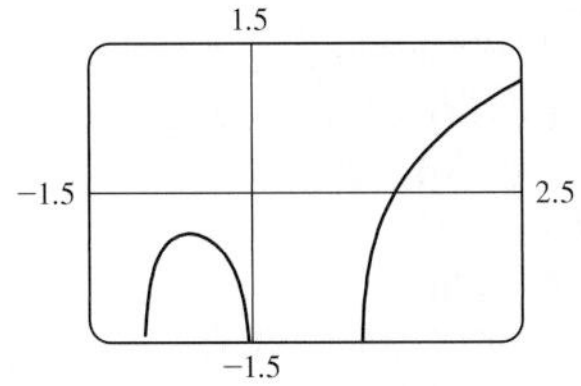

vertical asymptotes
$x = -1, x = 0, x = 1$
local maximum
$\approx (-0.58, -0.41)$

77. 2.42 **79.** $0.16 < x < 3.15$
81. Increasing on $(-\infty, 0]$ and $[1.10, \infty)$, decreasing on $[0, 1.10]$
83. 1.953445 **85.** -0.579352 **87.** $\log_4 258$
89. **(a)** \$16,081.15 **(b)** \$16,178.18 **(c)** \$16,197.64
(d) \$16,198.31 **91.** 1.83 yr **93.** 4.341%
95. **(a)** $n(t) = 30e^{0.15t}$ **(b)** 55 **(c)** 19 yr
97. **(a)** 9.97 mg **(b)** 1.39×10^5 yr
99. **(a)** $n(t) = 150e^{-0.0004359t}$ **(b)** 97.0 mg **(c)** 2520 yr
101. **(a)** $n(t) = 1500e^{0.1515t}$ **(b)** 7940
103. 7.9, basic **105.** 8.0

CHAPTER 4 TEST ■ PAGE 356

1. **(a)** $\mathbb{R}, (4, \infty), y = 4$ **(b)** $(-3, \infty), \mathbb{R}, x = -3$

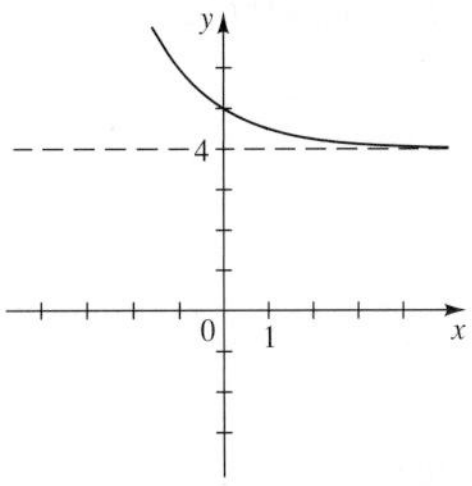

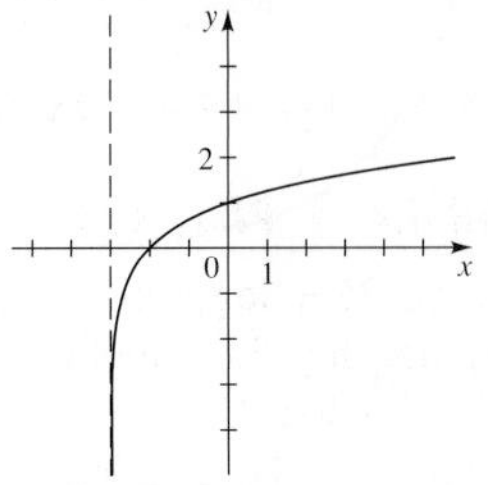

2. **(a)** $\log_6 25 = 2x$ **(b)** $e^3 = A$
3. **(a)** 36 **(b)** 3 **(c)** $\frac{3}{2}$ **(d)** 3 **(e)** $\frac{2}{3}$ **(f)** 2
4. $\frac{1}{3}[\log(x + 2) - 4 \log x - \log(x^2 + 4)]$
5. $\ln\left(\dfrac{x\sqrt{3 - x^4}}{(x^2 + 1)^2}\right)$ **6.** **(a)** 4.32 **(b)** 0.77 **(c)** 5.39 **(d)** 2
7. **(a)** $n(t) = 1000e^{2.07944t}$
(b) 22,627 **(c)** 1.3 h
(d)

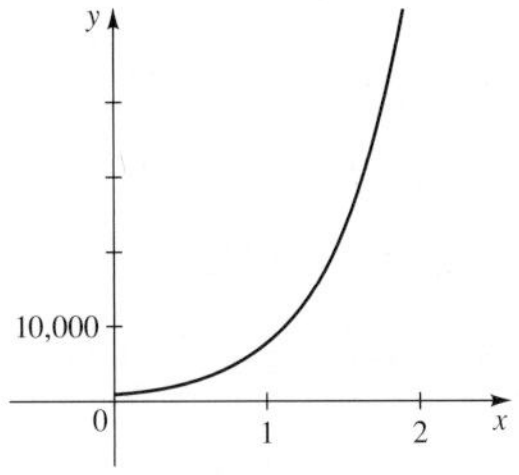

8. **(a)** $A(t) = 12{,}000\left(1 + \dfrac{0.056}{12}\right)^{12t}$
(b) \$14,195.06 **(c)** 9.249 yr
9. **(a)** $A(t) = 3e^{-0.069t}$ **(b)** 0.048 g **(c)** after 3.6 min
10. 1995 times more intense

FOCUS ON MODELING ■ PAGE 363

1. **(a)**

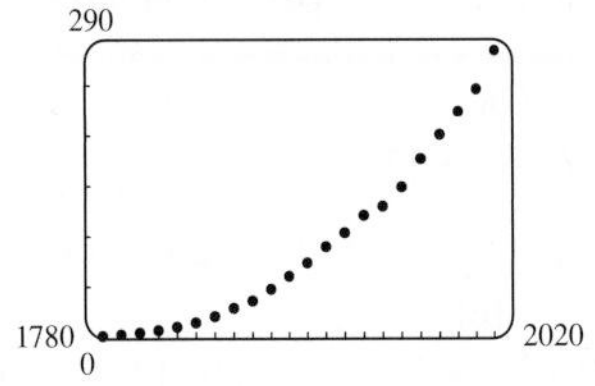

(b) $y = ab^t$, where $a = 1.180609 \times 10^{-15}$, $b = 1.0204139$, and y is the population in millions in the year t **(c)** 515.9 million
(d) 207.8 million **(e)** No
3. **(a)** Yes **(b)** Yes, the scatter plot appears linear.

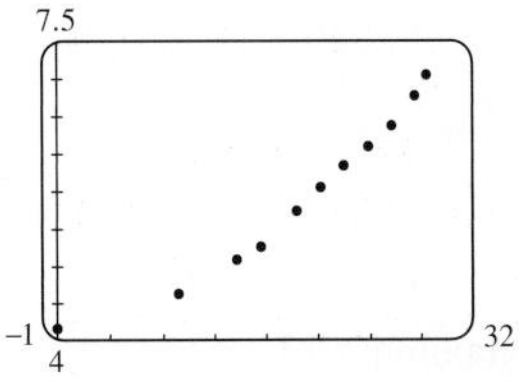

(c) $\ln E = 4.551436 + 0.092383t$, where t is years since 1970 and E is expenditure in billions of dollars
(d) $E = 94.76838139e^{at}$, where $a = 0.0923827621$
(e) 3478.5 billion dollars

5. (a) $I_0 = 22.7586444$, $k = 0.1062398$
(b)

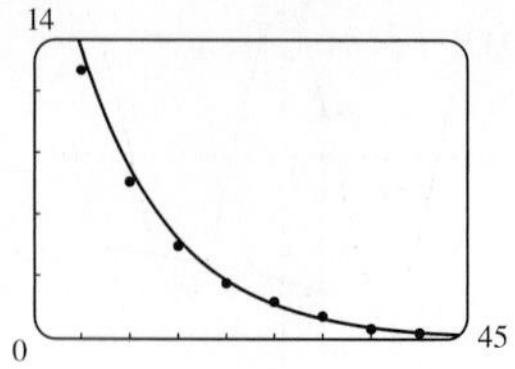

(c) 47.3 ft

7. (a) $S = 0.14A^{0.64}$
(b)

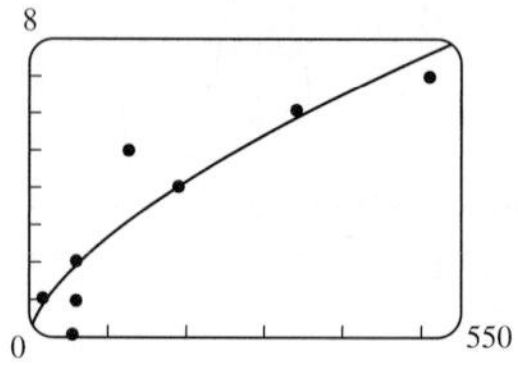

(c) 4 species

9. (a)

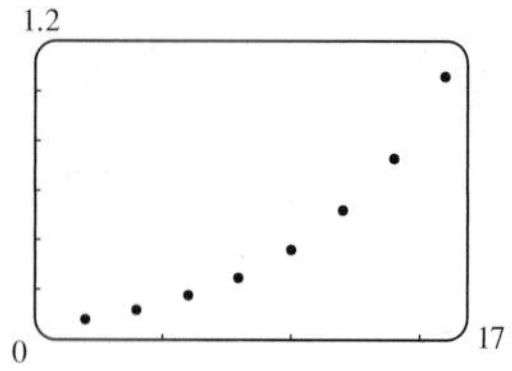

(b)

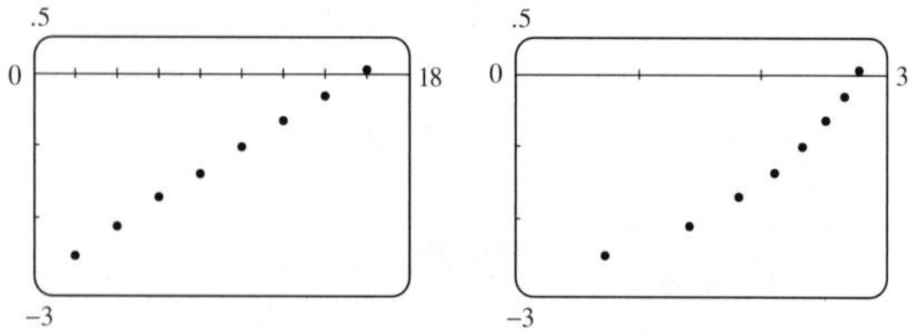

(c) Exponential function
(d) $y = ab^x$ where $a = 0.057697$ and $b = 1.200236$

11. (a) $y = \dfrac{c}{1 + ae^{-bx}}$, where $a = 49.10976596$, $b = 0.4981144989$, and $c = 500.855793$ **(b)** 10.58 days

CUMULATIVE REVIEW TEST FOR CHAPTERS 2, 3, AND 4 ■ PAGE 367

1. (a) $(-\infty, \infty)$ **(b)** $[-4, \infty)$ **(c)** $12, 0, 0, 2, 2\sqrt{3}$, undefined
(d) $x^2 - 4$, $\sqrt{x + 6}$, $-4 + h^2$ **(e)** $\frac{1}{8}$
(f) $f \circ g = x + 4 - \sqrt{x + 4}$, $g \circ f = |x - 2|$, $f(g(12)) = 0$, $g(f(12)) = 10$ **(g)** $g^{-1}(x) = x^2 - 4$, $x \ge 0$

2. (a) 4, 4, 4, 0, 1 **(b)**

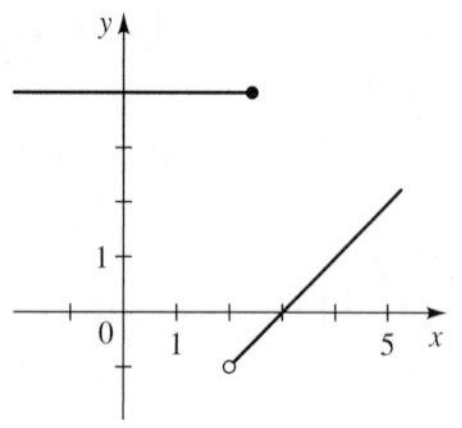

3. (a) $f(x) = -2(x - 2)^2 + 13$ **(b)** Maximum 13
(c)

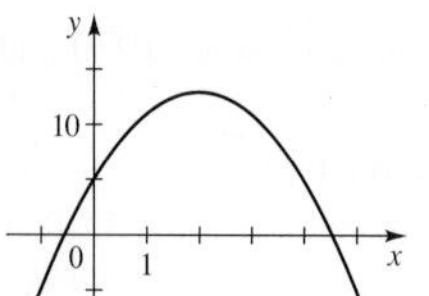

(d) Increasing on $(-\infty, 2]$; decreasing on $[2, \infty)$
(e) Shift upward 5 units
(f) Shift to the left 3 units

4. f, D; g, C; r, A; s, F; h, B; k, E
5. (a) $\pm 1, \pm 2, \pm 4, \pm 8, \pm \frac{1}{2}$ **(b)** $2, 4, -\frac{1}{2}$
(c) $P(x) = 2(x - 2)(x - 4)\left(x + \frac{1}{2}\right)$ **(d)**

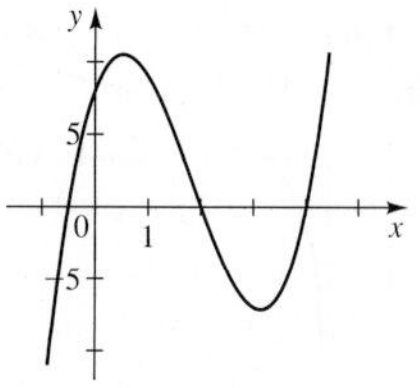

6. (a) 1 (multiplicity 2); -1, $1 + i$, $1 - i$ (multiplicity 1)
(b) $Q(x) = (x - 1)^2(x + 1)(x - 1 - i)(x - 1 + i)$
(c) $Q(x) = (x - 1)^2(x + 1)(x^2 - 2x + 2)$

7. x-intercepts $0, -2$; y-intercept 0; horizontal asymptote $y = 3$; vertical asymptotes $x = 2$ and $x = -1$

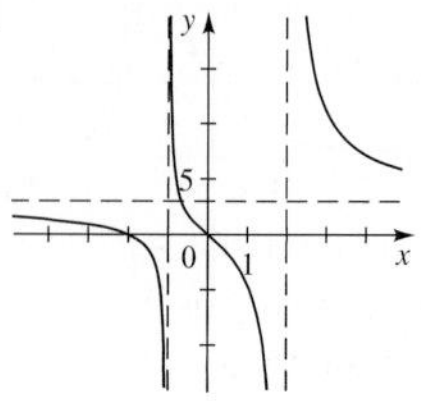

8.

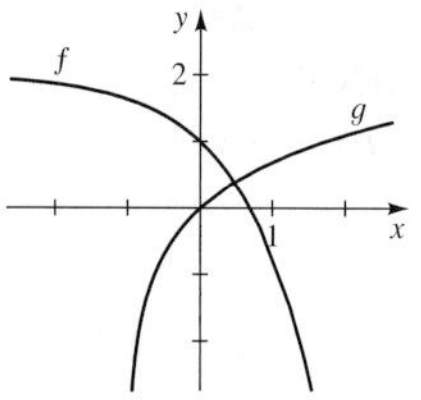

9. (a) -4 **(b)** $5 \log x + \frac{1}{2} \log(x - 1) - \log(2x - 3)$
10. (a) 4 **(b)** $\ln 2$, $\ln 4$ **11. (a)** \$29,396.15
(b) After 6.23 years **(c)** 12.837 years
12. (a) $P(t) = 120e^{0.0565t}$ **(b)** 917 **(c)** After 49.8 months

CHAPTER 5

SECTION 5.1 ■ PAGE 375

1. (a) $(0, 0)$, 1 **(b)** $x^2 + y^2 = 1$ **(c)** (i) 0 (ii) 0 (iii) 0 (iv) 0 **2. (a)** terminal **(b)** $(0, 1), (-1, 0), (0, -1), (1, 0)$
9. $-\frac{4}{5}$ **11.** $-2\sqrt{2}/3$ **13.** $3\sqrt{5}/7$ **15.** $P\left(\frac{4}{5}, \frac{3}{5}\right)$
17. $P\left(-\sqrt{5}/3, \frac{2}{3}\right)$ **19.** $P\left(-\sqrt{2}/3, -\sqrt{7}/3\right)$
21. $t = \pi/4, \left(\sqrt{2}/2, \sqrt{2}/2\right)$; $t = \pi/2, (0, 1)$;
$t = 3\pi/4, \left(-\sqrt{2}/2, \sqrt{2}/2\right)$; $t = \pi, (-1, 0)$;
$t = 5\pi/4, \left(-\sqrt{2}/2, -\sqrt{2}/2\right)$; $t = 3\pi/2, (0, -1)$;
$t = 7\pi/4, \left(\sqrt{2}/2, -\sqrt{2}/2\right)$; $t = 2\pi, (1, 0)$
23. $(0, 1)$ **25.** $\left(-\sqrt{3}/2, \frac{1}{2}\right)$ **27.** $\left(\frac{1}{2}, -\sqrt{3}/2\right)$
29. $\left(-\frac{1}{2}, \sqrt{3}/2\right)$ **31.** $\left(-\sqrt{2}/2, -\sqrt{2}/2\right)$
33. (a) $\left(-\frac{3}{5}, \frac{4}{5}\right)$ **(b)** $\left(\frac{3}{5}, -\frac{4}{5}\right)$ **(c)** $\left(-\frac{3}{5}, -\frac{4}{5}\right)$ **(d)** $\left(\frac{3}{5}, \frac{4}{5}\right)$
35. (a) $\pi/4$ **(b)** $\pi/3$ **(c)** $\pi/3$ **(d)** $\pi/6$
37. (a) $2\pi/7$ **(b)** $2\pi/9$ **(c)** $\pi - 3 \approx 0.14$ **(d)** $2\pi - 5 \approx 1.28$
39. (a) $\pi/3$ **(b)** $\left(-\frac{1}{2}, \sqrt{3}/2\right)$
41. (a) $\pi/4$ **(b)** $\left(-\sqrt{2}/2, \sqrt{2}/2\right)$
43. (a) $\pi/3$ **(b)** $\left(-\frac{1}{2}, -\sqrt{3}/2\right)$
45. (a) $\pi/4$ **(b)** $\left(-\sqrt{2}/2, -\sqrt{2}/2\right)$
47. (a) $\pi/6$ **(b)** $\left(-\sqrt{3}/2, -\frac{1}{2}\right)$
49. (a) $\pi/3$ **(b)** $\left(\frac{1}{2}, \sqrt{3}/2\right)$ **51. (a)** $\pi/3$ **(b)** $\left(-\frac{1}{2}, -\sqrt{3}/2\right)$
53. $(0.5, 0.8)$ **55.** $(0.5, -0.9)$

SECTION 5.2 ■ PAGE 384

1. $y, x, y/x$ **2.** 1, 1 **3.** $t = \pi/4$, $\sin t = \sqrt{2}/2$, $\cos t = \sqrt{2}/2$; $t = \pi/2$, $\sin t = 1$, $\cos t = 0$; $t = 3\pi/4$, $\sin t = \sqrt{2}/2$, $\cos t = -\sqrt{2}/2$; $t = \pi$, $\sin t = 0$, $\cos t = -1$; $t = 5\pi/4$, $\sin t = -\sqrt{2}/2$, $\cos t = -\sqrt{2}/2$; $t = 3\pi/2$, $\sin t = -1$, $\cos t = 0$; $t = 7\pi/4$, $\sin t = -\sqrt{2}/2$, $\cos t = \sqrt{2}/2$; $t = 2\pi$, $\sin t = 0$, $\cos t = 1$
5. **(a)** $\sqrt{3}/2$ **(b)** $-1/2$ **(c)** $-\sqrt{3}$
7. **(a)** $-1/2$ **(b)** $-1/2$ **(c)** $-1/2$
9. **(a)** $-\sqrt{2}/2$ **(b)** $-\sqrt{2}/2$ **(c)** $\sqrt{2}/2$
11. **(a)** $\sqrt{3}/2$ **(b)** $2\sqrt{3}/3$ **(c)** $\sqrt{3}/3$
13. **(a)** -1 **(b)** 0 **(c)** 0
15. **(a)** 2 **(b)** $-2\sqrt{3}/3$ **(c)** 2
17. **(a)** $-\sqrt{3}/3$ **(b)** $\sqrt{3}/3$ **(c)** $-\sqrt{3}/3$
19. **(a)** $\sqrt{2}/2$ **(b)** $-\sqrt{2}$ **(c)** -1
21. **(a)** -1 **(b)** 1 **(c)** -1 **23.** **(a)** 0 **(b)** 1 **(c)** 0
25. $\sin 0 = 0$, $\cos 0 = 1$, $\tan 0 = 0$, $\sec 0 = 1$, others undefined
27. $\sin \pi = 0$, $\cos \pi = -1$, $\tan \pi = 0$, $\sec \pi = -1$, others undefined
29. $\frac{4}{5}, \frac{3}{5}, \frac{4}{3}$ **31.** $-\sqrt{11}/4, \sqrt{5}/4, -\sqrt{55}/5$
33. $\sqrt{13}/7, -6/7, -\sqrt{13}/6$ **35.** $-\frac{12}{13}, -\frac{5}{13}, \frac{12}{5}$ **37.** $\frac{21}{29}, -\frac{20}{29}, -\frac{21}{20}$
39. **(a)** 0.8 **(b)** 0.84147 **41.** **(a)** 0.9 **(b)** 0.93204
43. **(a)** 1 **(b)** 1.02964 **45.** **(a)** -0.6 **(b)** -0.57482
47. Negative **49.** Negative **51.** II **53.** II
55. $\sin t = \sqrt{1 - \cos^2 t}$
57. $\tan t = (\sin t)/\sqrt{1 - \sin^2 t}$
59. $\sec t = -\sqrt{1 + \tan^2 t}$
61. $\tan t = \sqrt{\sec^2 t - 1}$
63. $\tan^2 t = (\sin^2 t)/(1 - \sin^2 t)$
65. $\cos t = -\frac{4}{5}$, $\tan t = -\frac{3}{4}$, $\csc t = \frac{5}{3}$, $\sec t = -\frac{5}{4}$, $\cot t = -\frac{4}{3}$
67. $\sin t = -2\sqrt{2}/3$, $\cos t = \frac{1}{3}$, $\tan t = -2\sqrt{2}$, $\csc t = -\frac{3}{4}\sqrt{2}$, $\cot t = -\sqrt{2}/4$
69. $\sin t = -\frac{3}{5}$, $\cos t = \frac{4}{5}$, $\csc t = -\frac{5}{3}$, $\sec t = \frac{5}{4}$, $\cot t = -\frac{4}{3}$
71. $\cos t = -\sqrt{15}/4$, $\tan t = \sqrt{15}/15$, $\csc t = -4$, $\sec t = -4\sqrt{15}/15$, $\cot t = \sqrt{15}$
73. Odd **75.** Odd **77.** Even **79.** Neither
81. $y(0) = 4$, $y(0.25) = -2.828$, $y(0.50) = 0$, $y(0.75) = 2.828$, $y(1.00) = -4$, $y(1.25) = 2.828$
83. **(a)** 0.49870 amp **(b)** -0.17117 amp

SECTION 5.3 ■ PAGE 396

1. 1, 2π

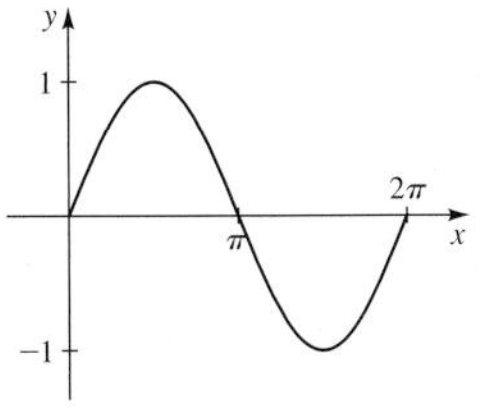

2. 3, π

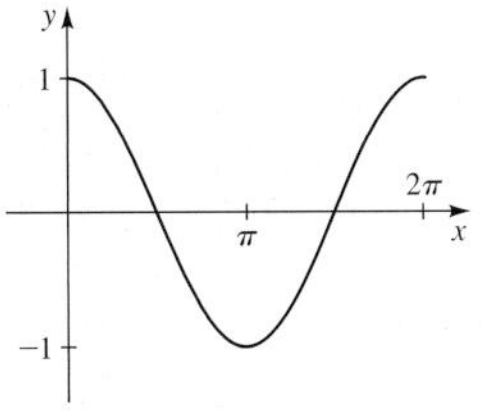

3.

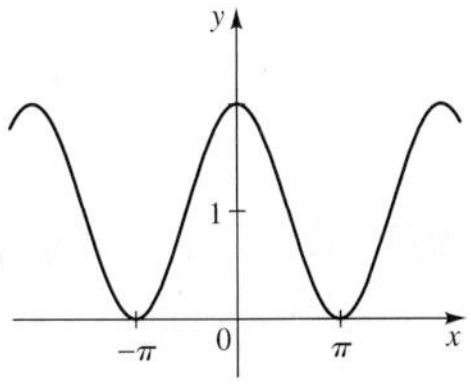

5.

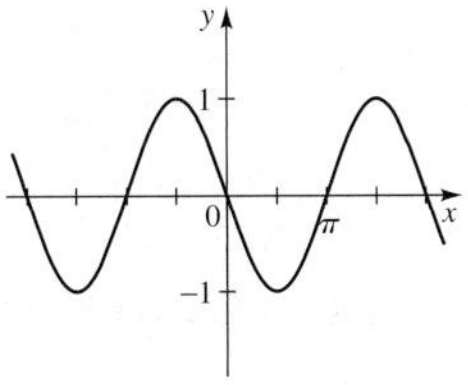

7.

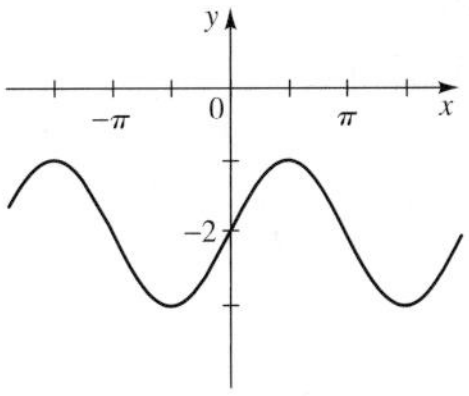

9.

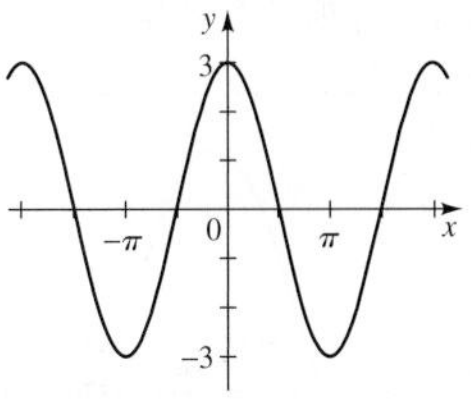

11.

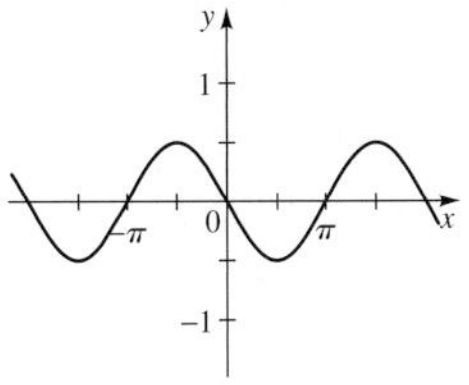

13.

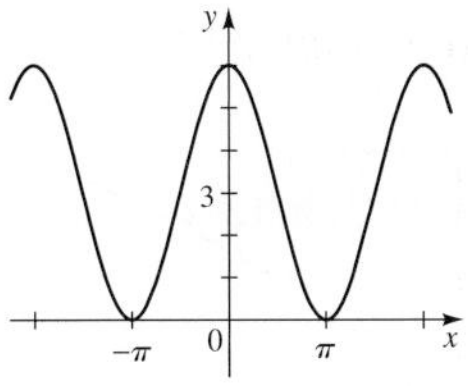

15.

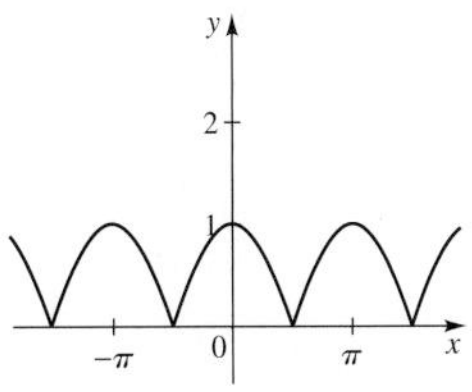

17. 1, π

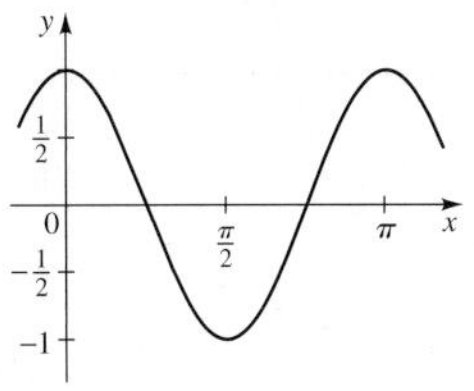

19. 3, $2\pi/3$

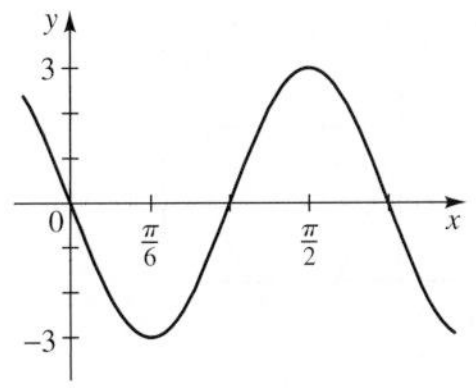

21. 10, 4π

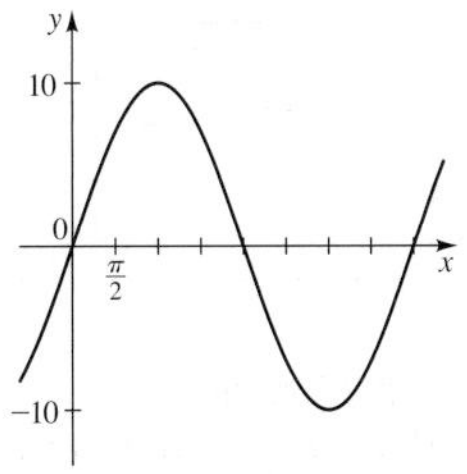

23. $\frac{1}{3}$, 6π

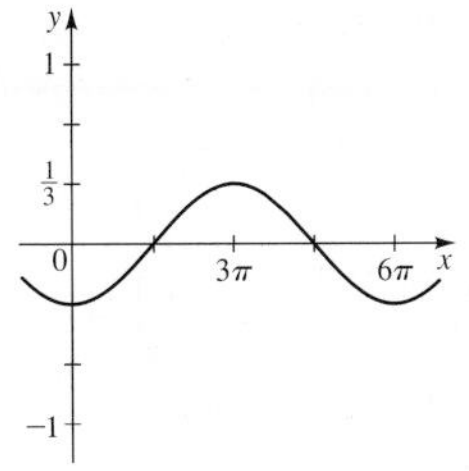

25. 2, 1

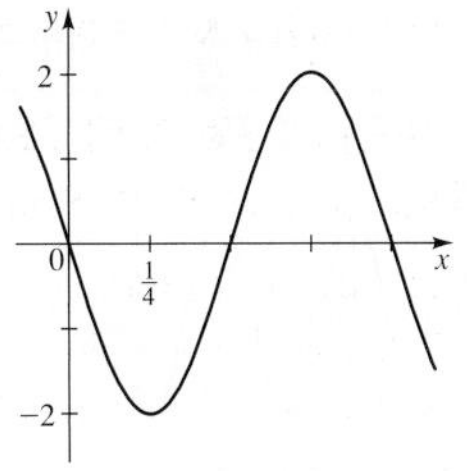

27. $\frac{1}{2}$, 2

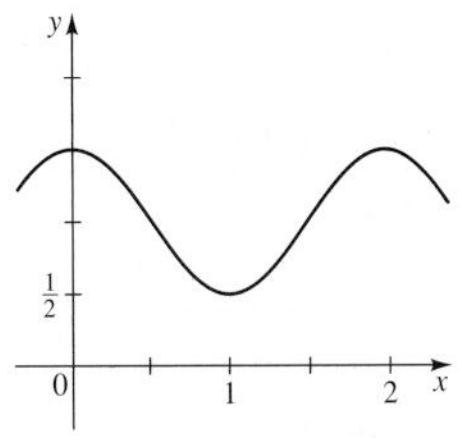

29. 1, 2π, $\pi/2$

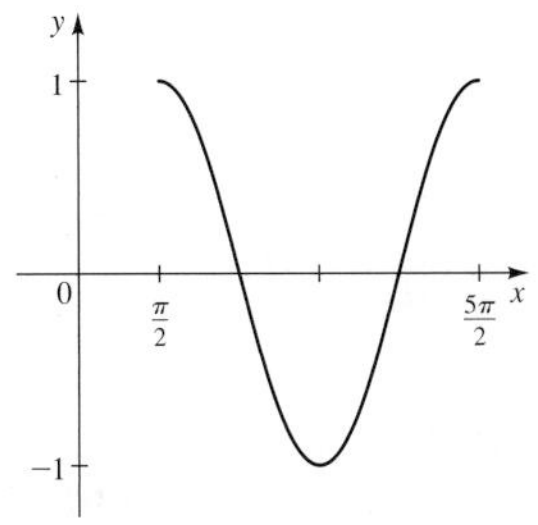

31. 2, 2π, $\pi/6$

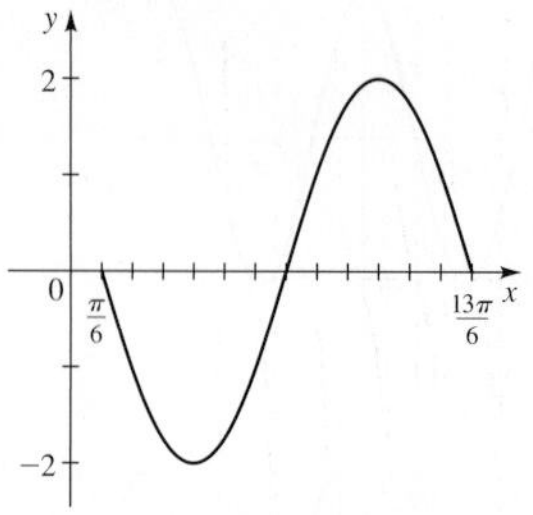

33. 4, π, $-\pi/2$

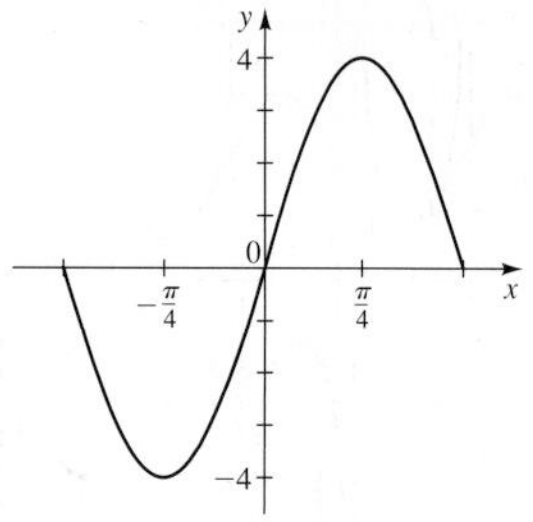

35. 5, $2\pi/3$, $\pi/12$

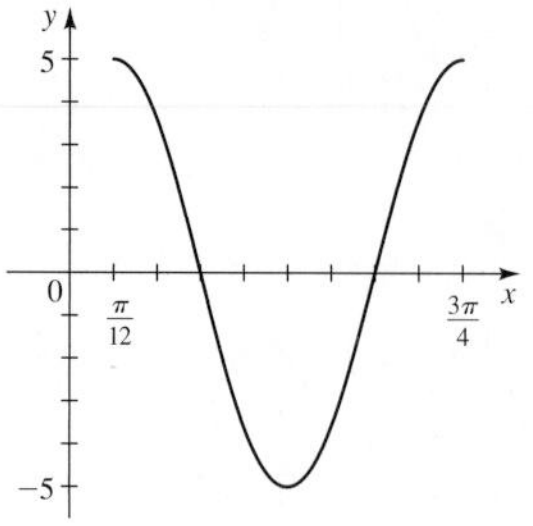

37. $\frac{1}{2}$, π, $\pi/6$

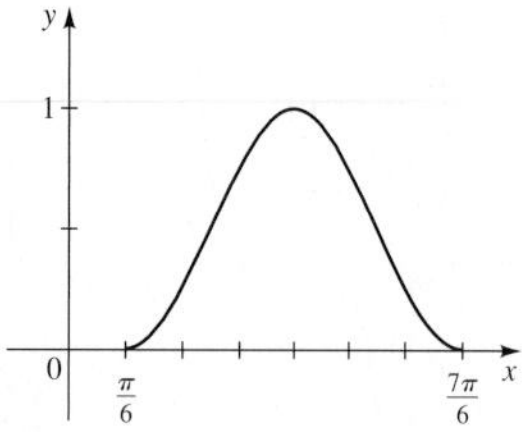

39. 3, 2, $-\frac{1}{2}$

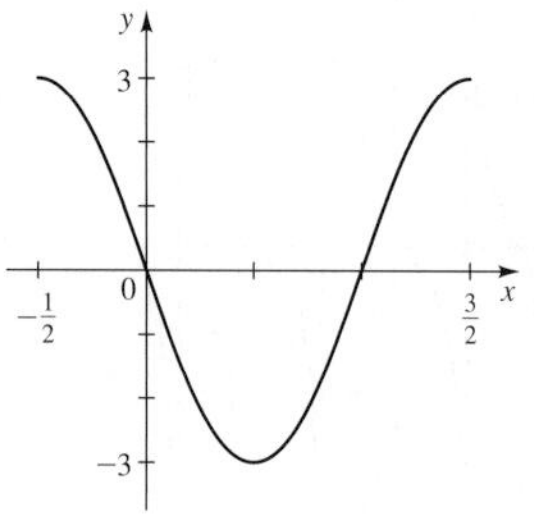

41. 1, $2\pi/3$, $-\pi/3$

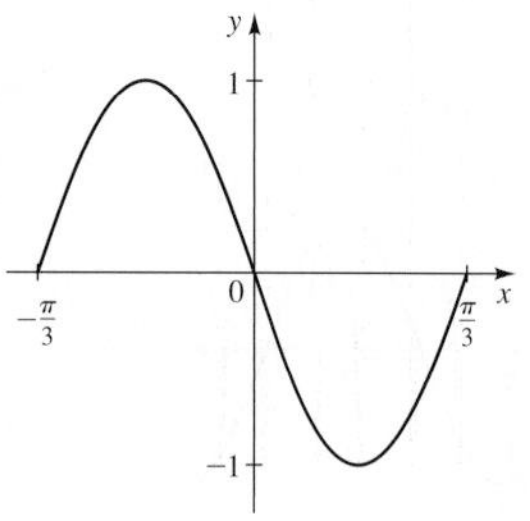

43. **(a)** 4, 2π, 0 **(b)** $y = 4 \sin x$
45. **(a)** $\frac{3}{2}, \frac{2\pi}{3}, 0$ **(b)** $y = \frac{3}{2} \cos 3x$
47. **(a)** $\frac{1}{2}, \pi, -\frac{\pi}{3}$ **(b)** $y = -\frac{1}{2} \cos 2(x + \pi/3)$
49. **(a)** $4, \frac{3}{2}, -\frac{1}{2}$ **(b)** $y = 4 \sin \frac{4\pi}{3}\left(x + \frac{1}{2}\right)$

51.

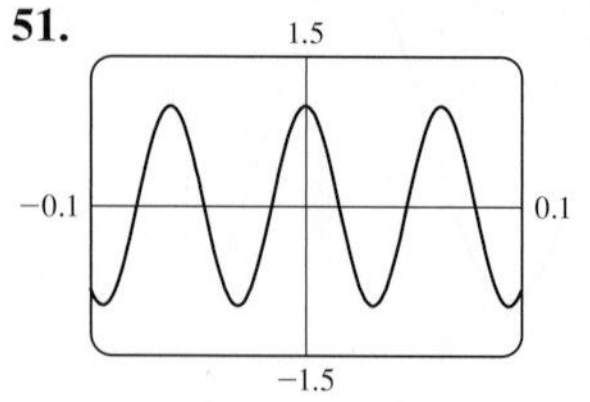

53.

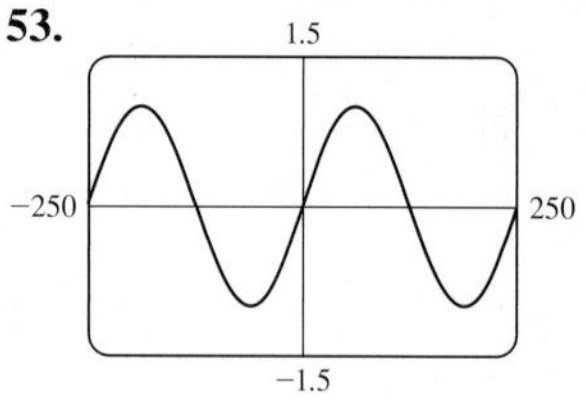

55.

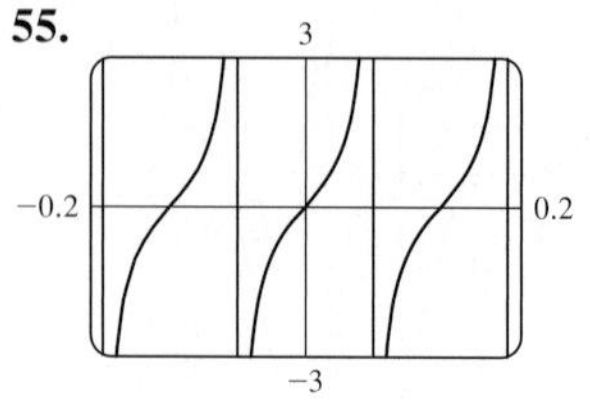

57.

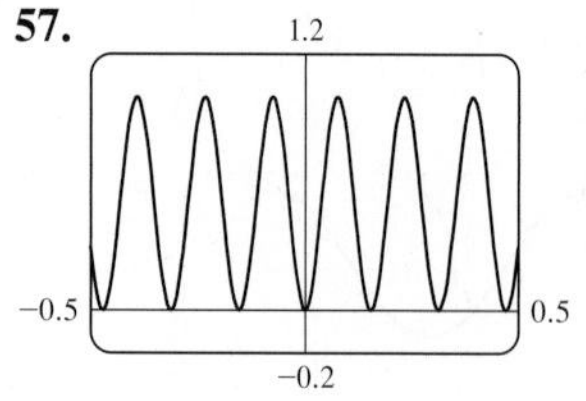

59.

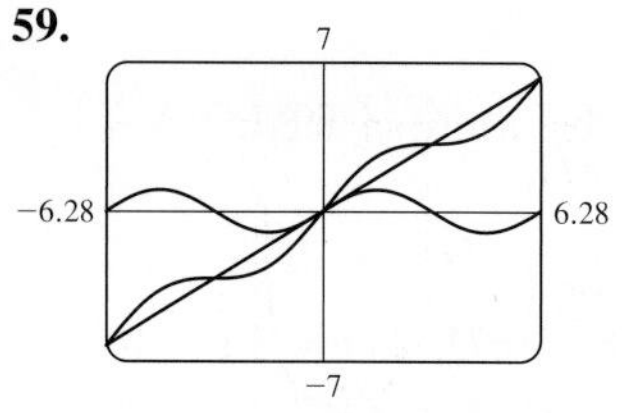

61.

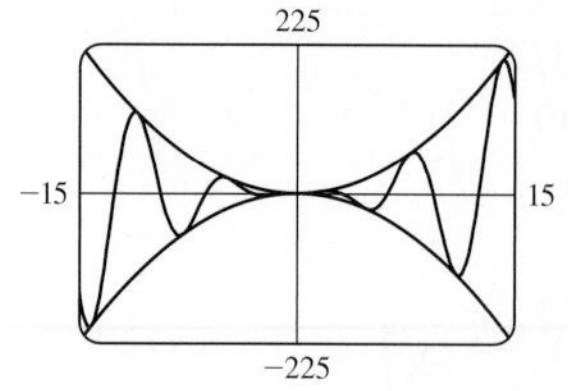

$y = x^2 \sin x$ is a sine curve that lies between the graphs of $y = x^2$ and $y = -x^2$

63.

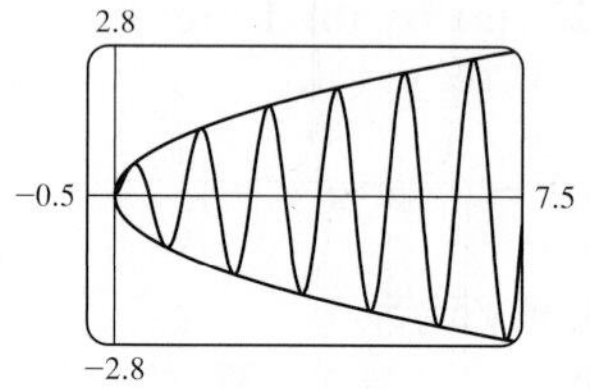

$y = \sqrt{x} \sin 5\pi x$ is a sine curve that lies between the graphs of $y = \sqrt{x}$ and $y = -\sqrt{x}$

65.

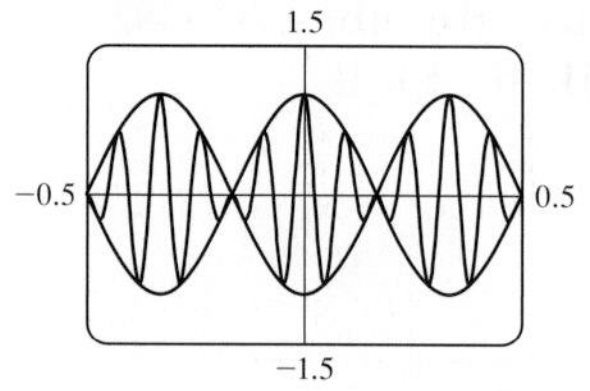

$y = \cos 3\pi x \cos 21\pi x$ is a cosine curve that lies between the graphs of $y = \cos 3\pi x$ and $y = -\cos 3\pi x$

67. Maximum value 1.76 when $x \approx 0.94$, minimum value -1.76 when $x \approx -0.94$ (The same maximum and minimum values occur at infinitely many other values of x.)
69. Maximum value 3.00 when $x \approx 1.57$, minimum value -1.00 when $x \approx -1.57$ (The same maximum and minimum values occur at infinitely many other values of x.)
71. 1.16 **73.** 0.34, 2.80
75. **(a)** Odd **(b)** $0, \pm 2\pi, \pm 4\pi, \pm 6\pi, \ldots$
(c)

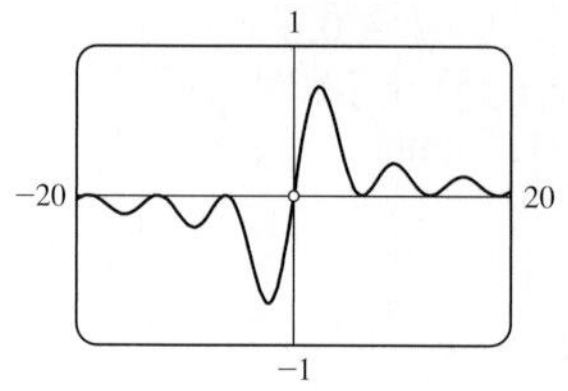

(d) $f(x)$ approaches 0
(e) $f(x)$ approaches 0

77. **(a)** 20 s **(b)** 6 ft
79. **(a)** $\frac{1}{80}$ min **(b)** 80
(c)

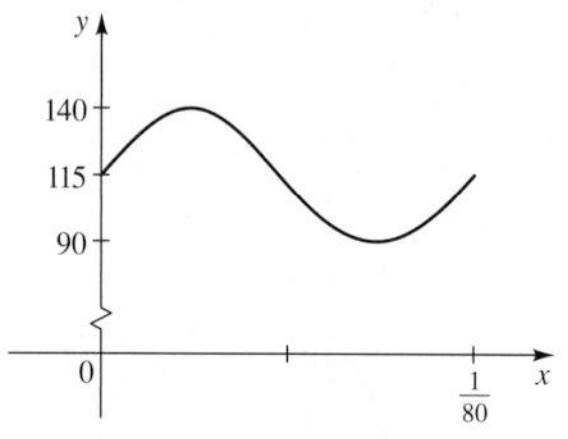

(d) $\frac{140}{90}$; it is higher than normal

SECTION 5.4 ■ PAGE 405

1. π; $\frac{\pi}{2} + n\pi$, n an integer

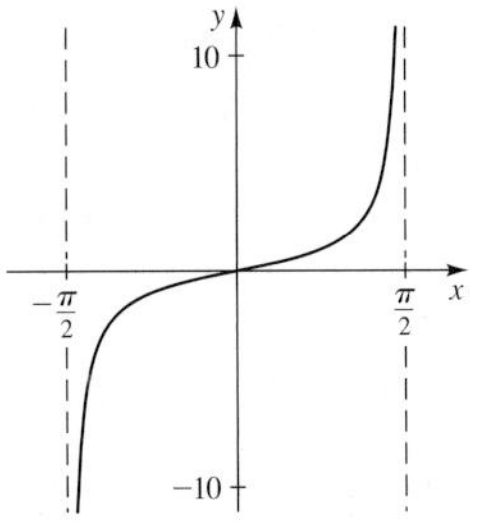

2. 2π; $n\pi$, n an integer

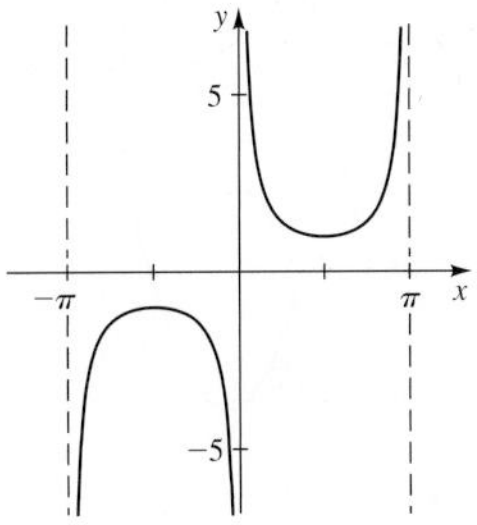

3. II **5.** VI **7.** IV

9. π

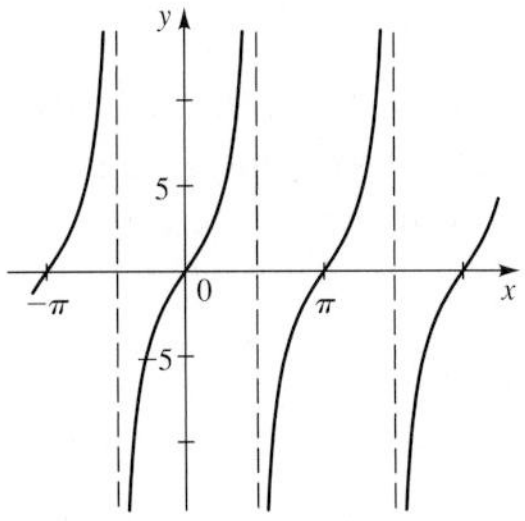

11. π

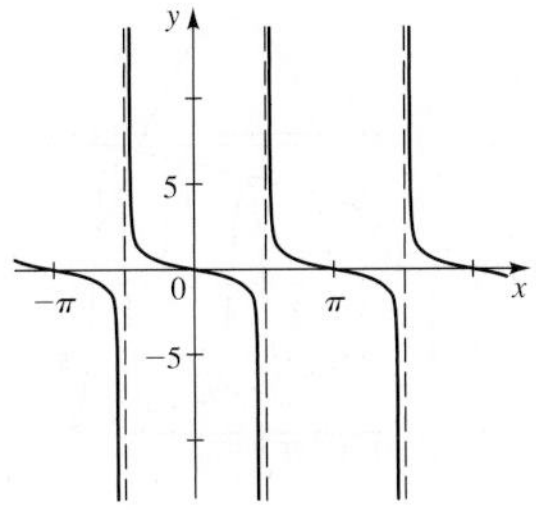

13. π

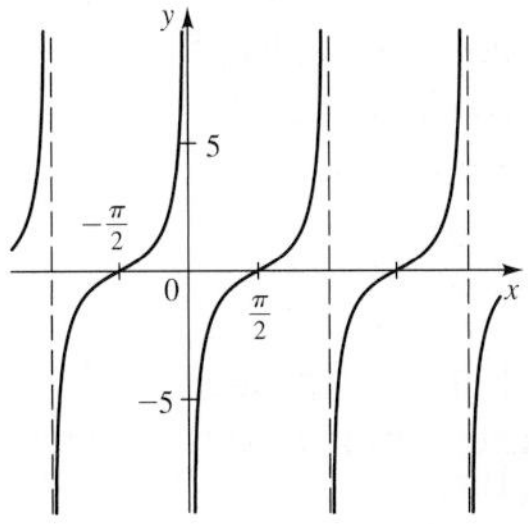

15. 2π

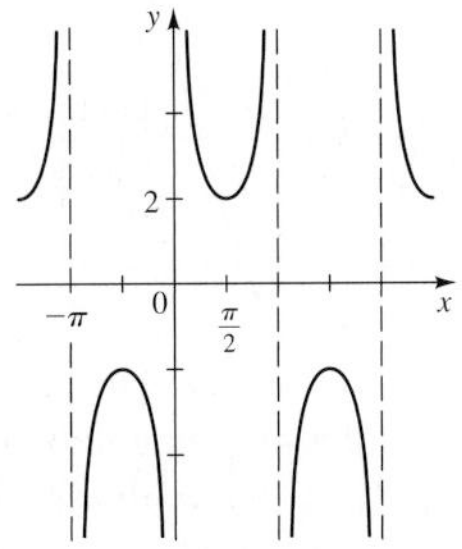

17. 2π

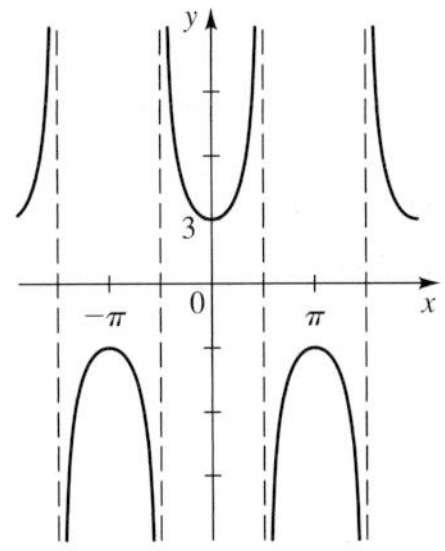

19. π

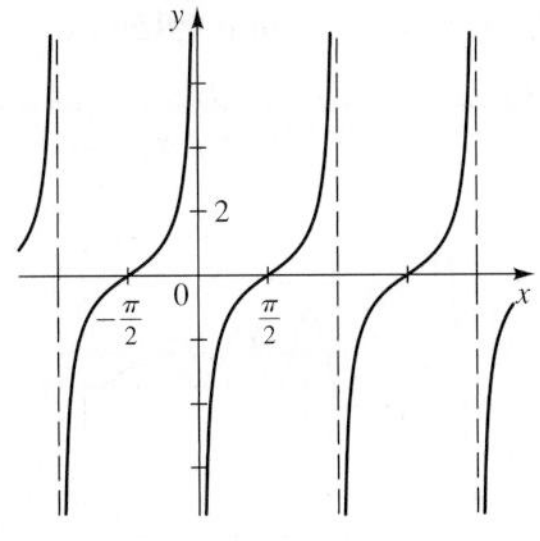

21. 2π

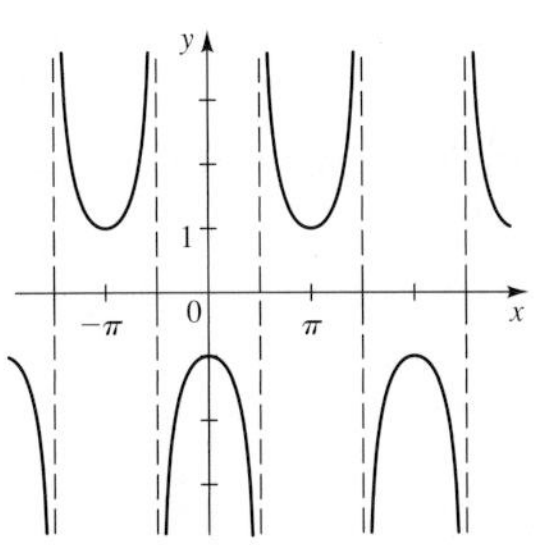

23. π

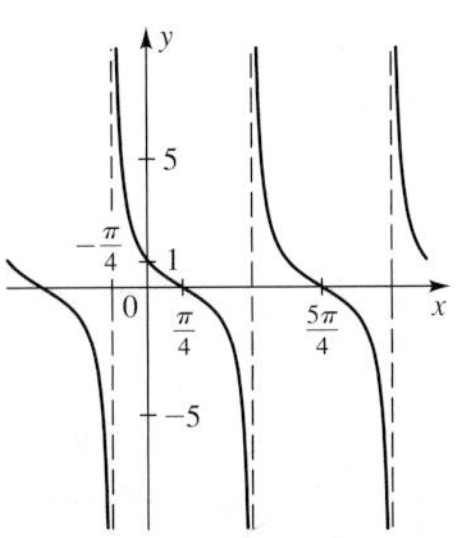

25. 2π

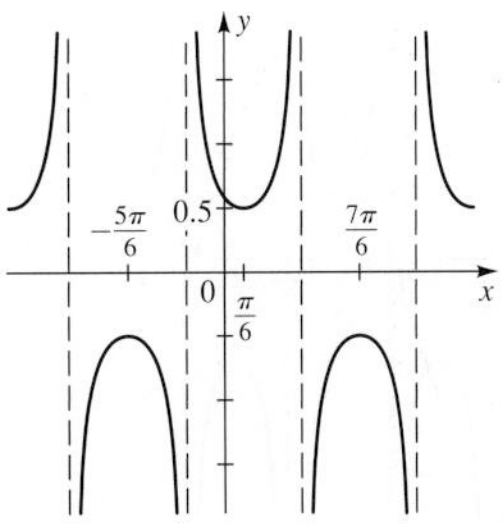

27. $\pi/4$

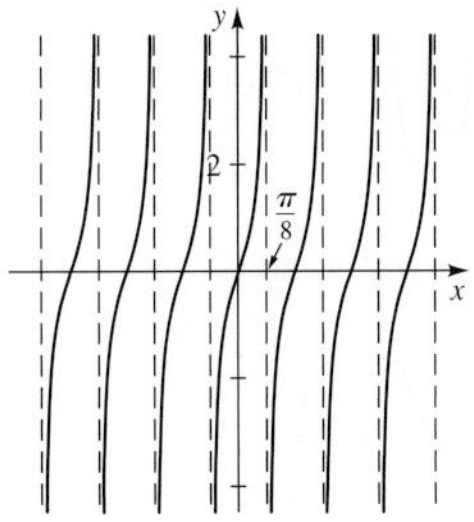

29. 4

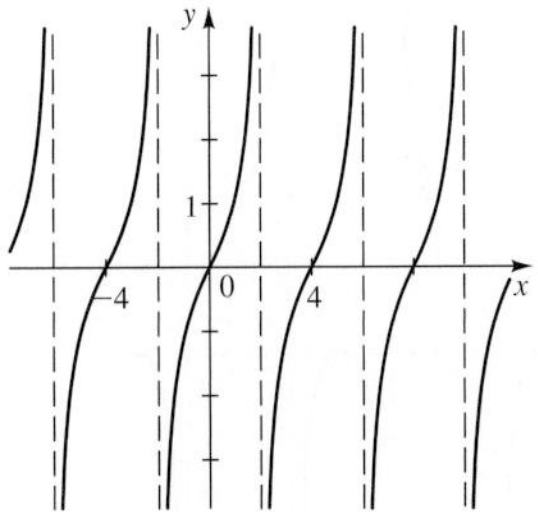

31. π

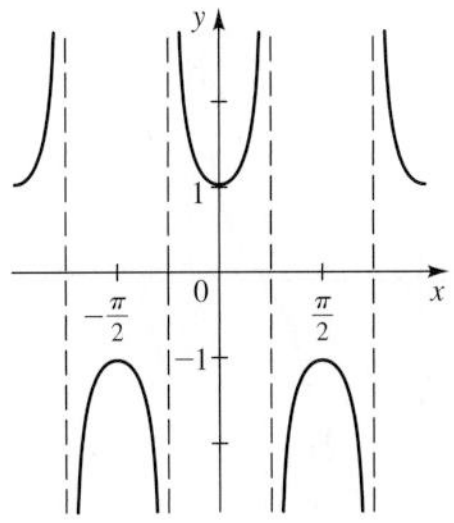

33. $\pi/2$

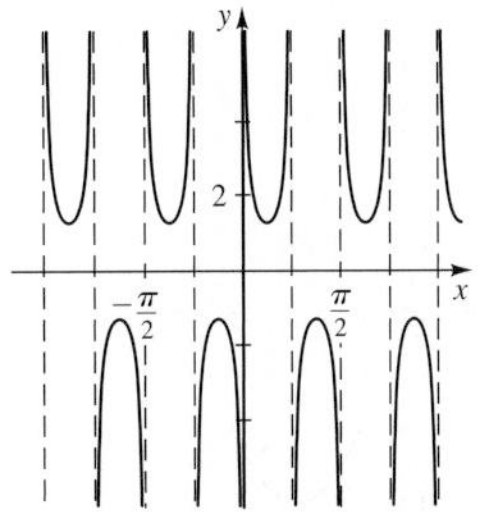

35. $\frac{1}{3}$

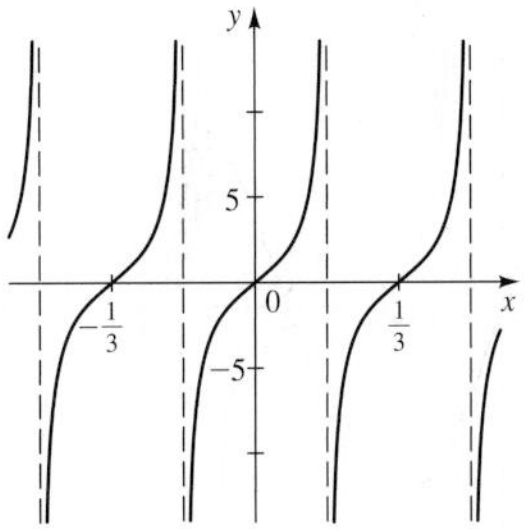

37. $\frac{4}{3}$

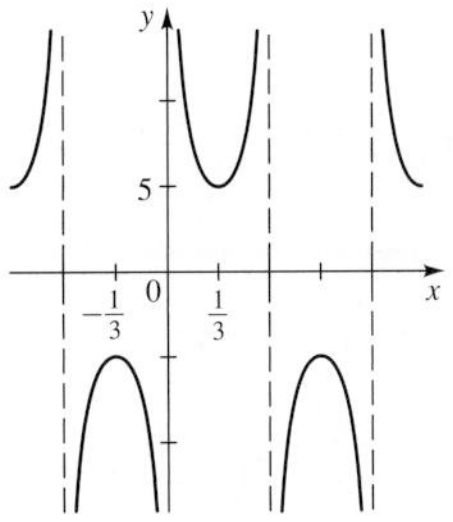

39. $\pi/2$

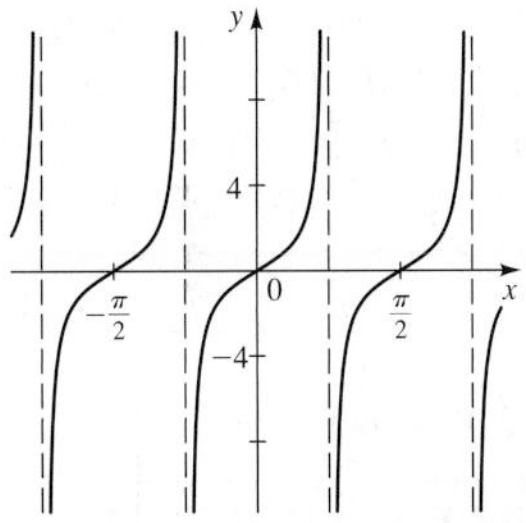

41. $\pi/2$

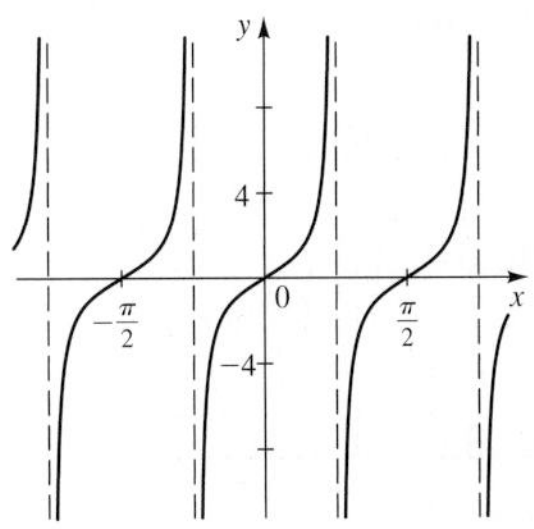

43. $\pi/2$

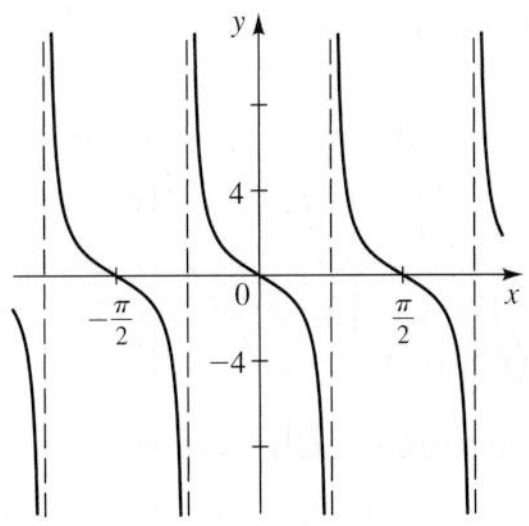

45. 2

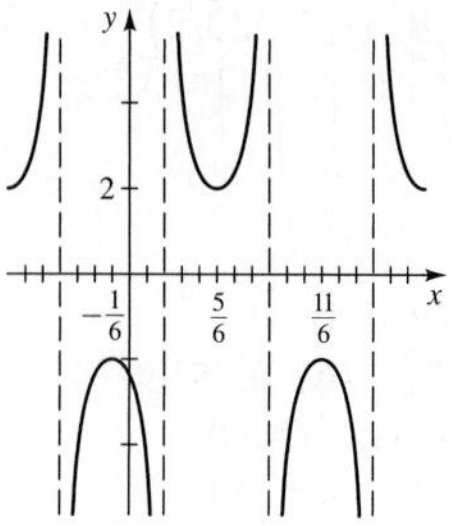

47. $2\pi/3$

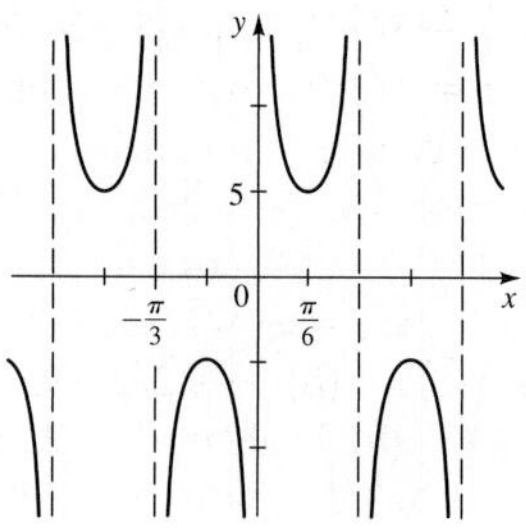

49. $3\pi/2$

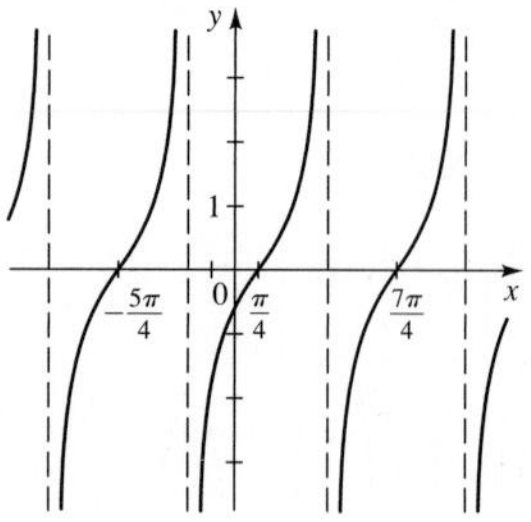

51. 2

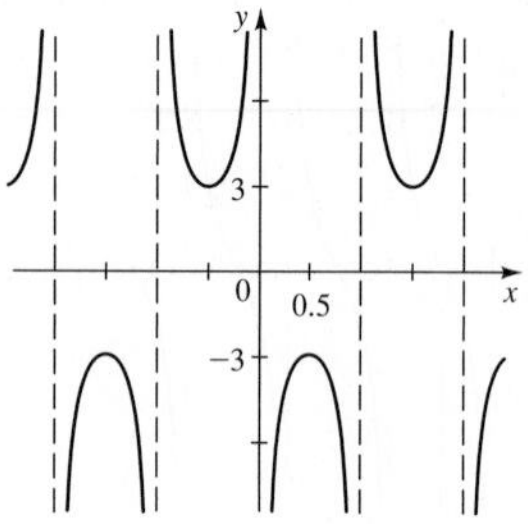

53. $\pi/2$

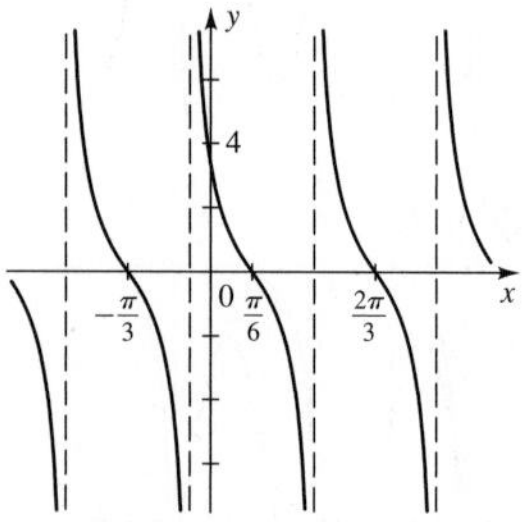

57. (a) 1.53 mi, 3.00 mi, 18.94 mi
(b)

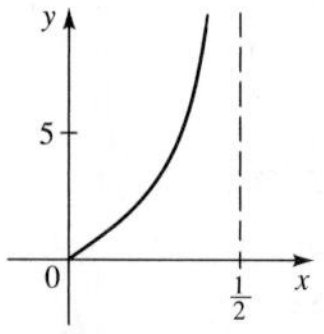

(c) $d(t)$ approaches ∞

SECTION 5.5 ■ PAGE 411

1. (a) $[-\pi/2, \pi/2]$, y, x, $\pi/6$, $\pi/6$, $\frac{1}{2}$
(b) $[0, \pi]$; y, x, $\pi/3$, $\pi/3$, $\frac{1}{2}$ **2.** $[-1, 1]$; (b)
3. (a) $\pi/2$ **(b)** $\pi/3$ **(c)** Undefined
5. (a) π **(b)** $\pi/3$ **(c)** $5\pi/6$
7. (a) $-\pi/4$ **(b)** $\pi/3$ **(c)** $\pi/6$
9. (a) $2\pi/3$ **(b)** $-\pi/4$ **(c)** $\pi/4$ **11.** 0.72973
13. 2.01371 **15.** 2.75876 **17.** 1.47113 **19.** 0.88998
21. -0.26005 **23.** $\frac{1}{4}$ **25.** 5
27. Undefined **29.** $5\pi/6$ **31.** $-\pi/6$ **33.** $\pi/6$ **35.** $\pi/6$
37. $-\pi/3$ **39.** $\sqrt{3}/3$ **41.** $\frac{1}{2}$ **43.** $-\sqrt{2}/2$

SECTION 5.6 ■ PAGE 420

1. (a) $a \sin \omega t$ **(b)** $a \cos \omega t$
2. (a) $ke^{-ct} \sin \omega t$ **(b)** $ke^{-ct} \cos \omega t$

3. (a) 2, $2\pi/3$, $3/(2\pi)$
(b)

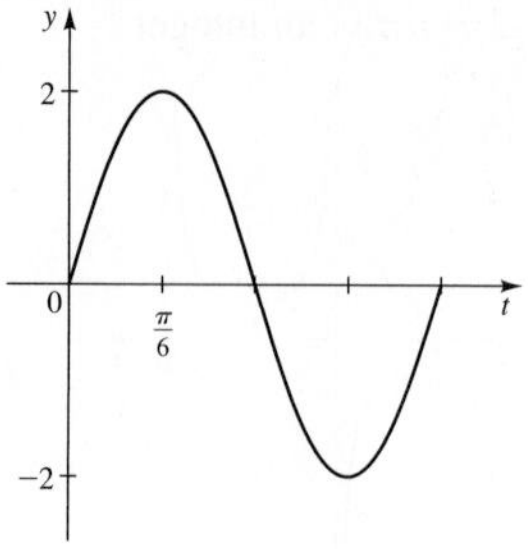

5. (a) 1, $20\pi/3$, $3/(20\pi)$
(b)

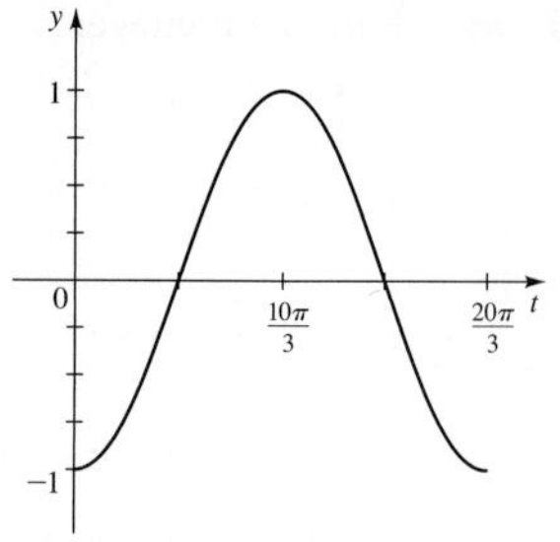

7. (a) $\frac{1}{4}$, $4\pi/3$, $3/(4\pi)$
(b)

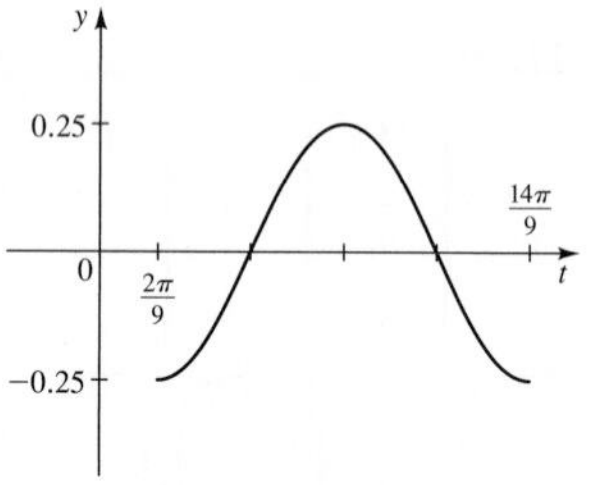

9. (a) 5, 3π, $1/(3\pi)$
(b)

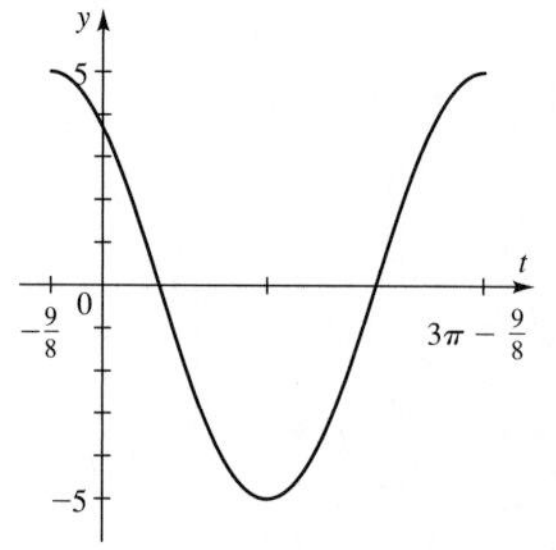

11. $y = 10 \sin\left(\frac{2\pi}{3}t\right)$ **13.** $y = 6 \sin(10t)$
15. $y = 60 \cos(4\pi t)$ **17.** $y = 2.4 \cos(1500\pi t)$

19. (a) $y = 2e^{-1.5t} \cos 6\pi t$
(b)

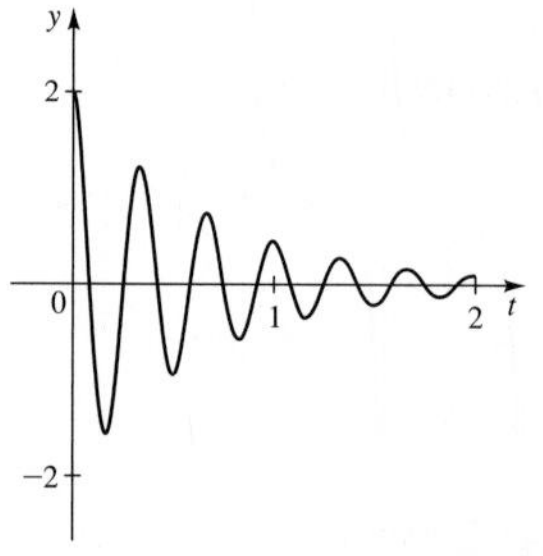

21. (a) $y = 100e^{-0.05t} \cos \frac{\pi}{2}t$

(b)

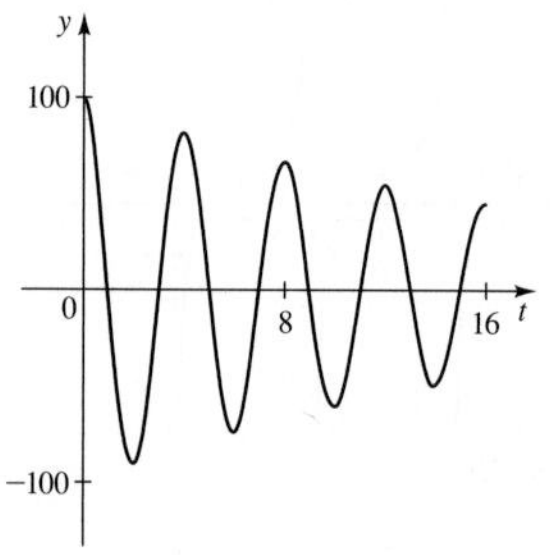

23. **(a)** $y = 7e^{-10t}\sin 12t$ **(b)**

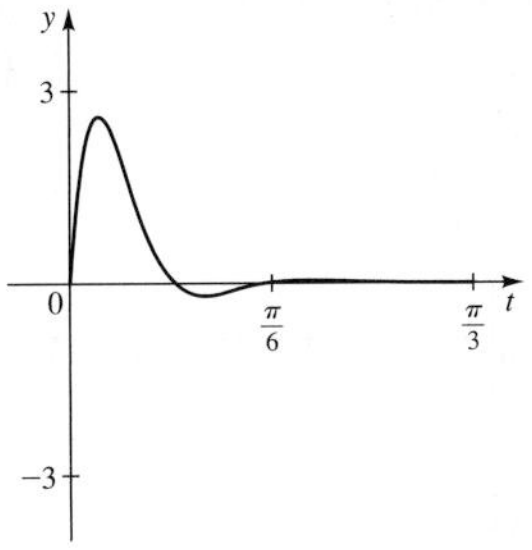

25. **(a)** $y = 0.3e^{-0.2t}\sin(40\pi t)$
(b)

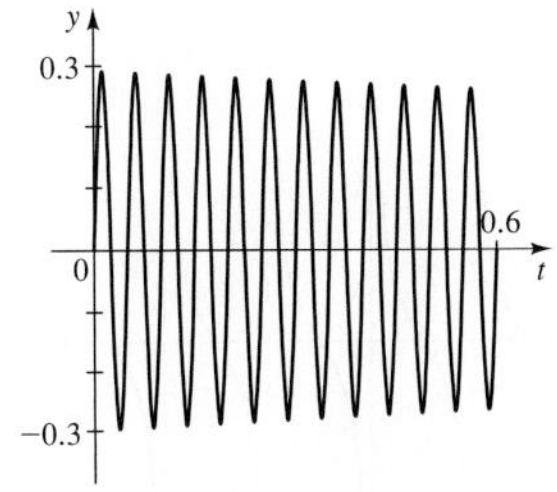

27. **(a)** 10 cycles per minute
(b)

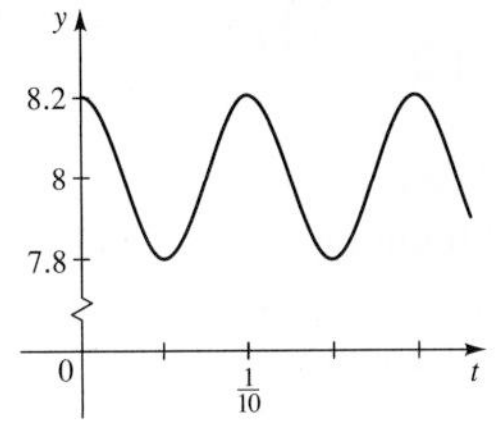

(c) 0.4 m

29. **(a)** 25, 0.0125, 80 **(b)**

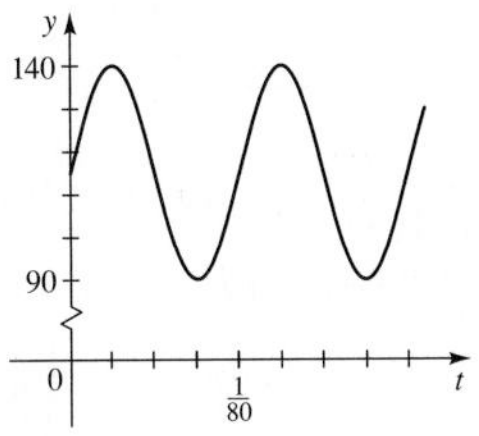

(c) The period decreases and the frequency increases.

31. $d(t) = 5\sin(5\pi t)$

33. $y = 21\sin\left(\dfrac{\pi}{6}t\right)$

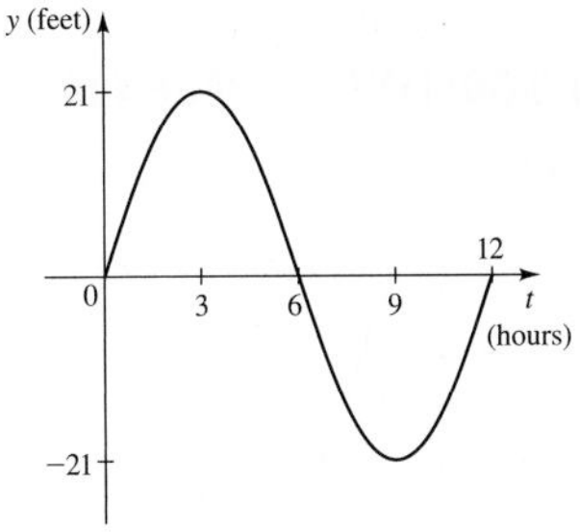

35. $y = 5\cos(2\pi t)$ **37.** $y = 11 + 10\sin\left(\dfrac{\pi t}{10}\right)$

39. $y = 3.8 + 0.2\sin\left(\dfrac{\pi}{5}t\right)$

41. $f(t) = 10\sin\left(\dfrac{\pi}{12}(t-8)\right) + 90$

43. **(a)** 45 V **(b)** 40 **(c)** 40 **(d)** $E(t) = 45\cos(80\pi t)$

45. $f(t) = e^{-0.9t}\sin \pi t$ **47.** $e = \dfrac{1}{3}\ln 4 \approx 0.46$

CHAPTER 5 REVIEW ■ PAGE 424

1. **(b)** $\frac{1}{2}, -\sqrt{3}/2, -\sqrt{3}/3$ **3.** **(a)** $\pi/3$ **(b)** $\left(-\frac{1}{2}, \sqrt{3}/2\right)$
(c) $\sin t = \sqrt{3}/2$, $\cos t = -\frac{1}{2}$, $\tan t = -\sqrt{3}$, $\csc t = 2\sqrt{3}/3$, $\sec t = -2$, $\cot t = -\sqrt{3}/3$
5. **(a)** $\pi/4$ **(b)** $(-\sqrt{2}/2, -\sqrt{2}/2)$
(c) $\sin t = -\sqrt{2}/2$, $\cos t = -\sqrt{2}/2$, $\tan t = 1$, $\csc t = -\sqrt{2}$, $\sec t = -\sqrt{2}$, $\cot t = 1$
7. **(a)** $\sqrt{2}/2$ **(b)** $-\sqrt{2}/2$ **9.** **(a)** 0.89121 **(b)** 0.45360
11. **(a)** 0 **(b)** Undefined **13.** **(a)** Undefined **(b)** 0
15. **(a)** $-\sqrt{3}/3$ **(b)** $-\sqrt{3}$ **17.** $(\sin t)/(1 - \sin^2 t)$
19. $(\sin t)/\sqrt{1 - \sin^2 t}$
21. $\tan t = -\frac{5}{12}$, $\csc t = \frac{13}{5}$, $\sec t = -\frac{13}{12}$, $\cot t = -\frac{12}{5}$
23. $\sin t = 2\sqrt{5}/5$, $\cos t = -\sqrt{5}/5$, $\tan t = -2$, $\sec t = -\sqrt{5}$
25. $(16 - \sqrt{17})/4$ **27.** 3

29. **(a)** $10, 4\pi, 0$
(b)

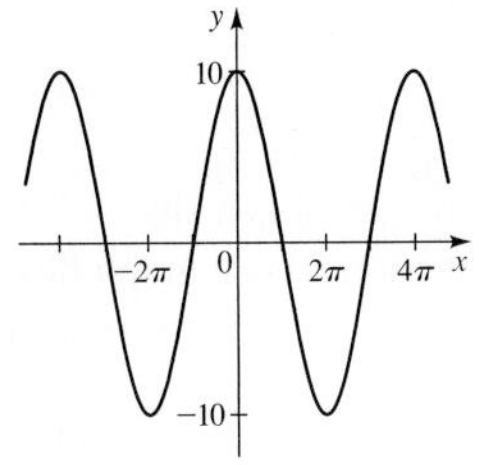

31. **(a)** $1, 4\pi, 0$
(b)

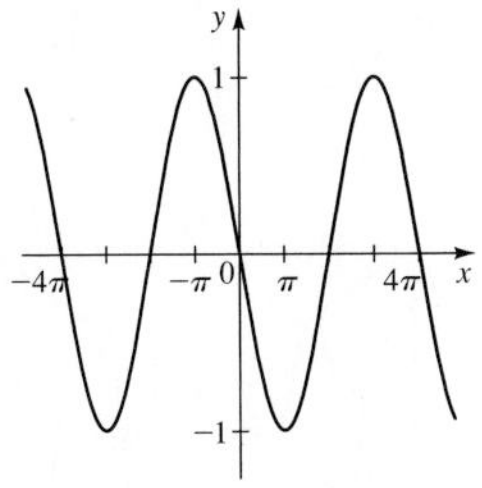

33. **(a)** $3, \pi, 1$
(b)

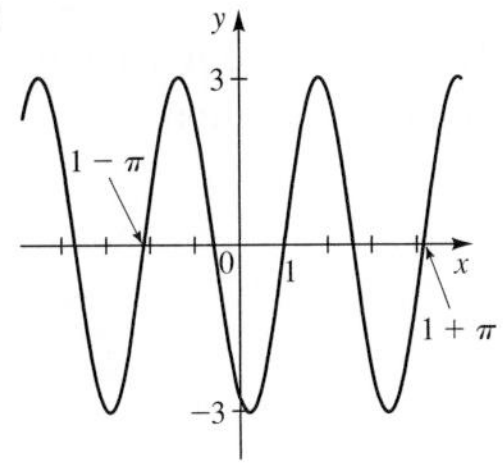

35. **(a)** $1, 4, -\frac{1}{3}$
(b)

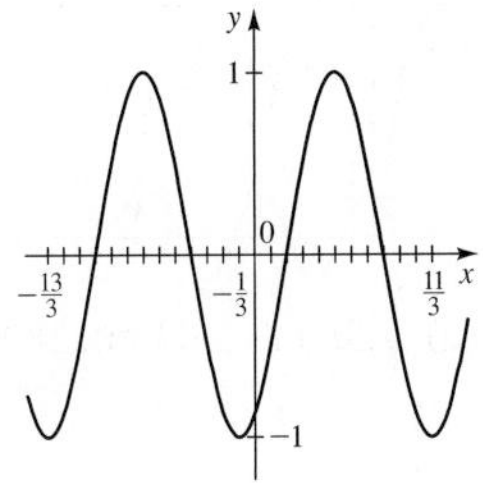

37. $y = 5\sin 4x$ **39.** $y = \frac{1}{2}\sin 2\pi\left(x + \frac{1}{3}\right)$

41. π

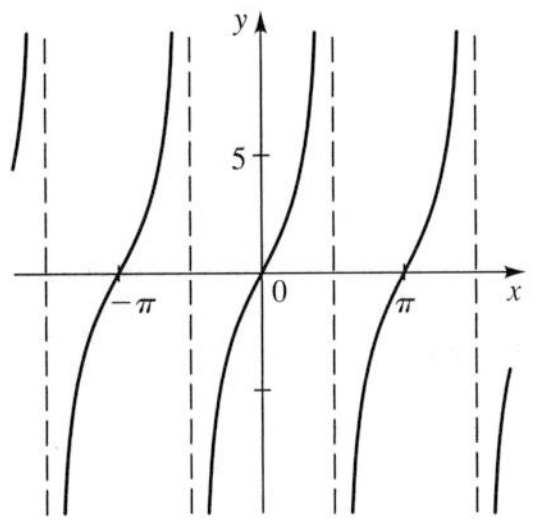

43. π

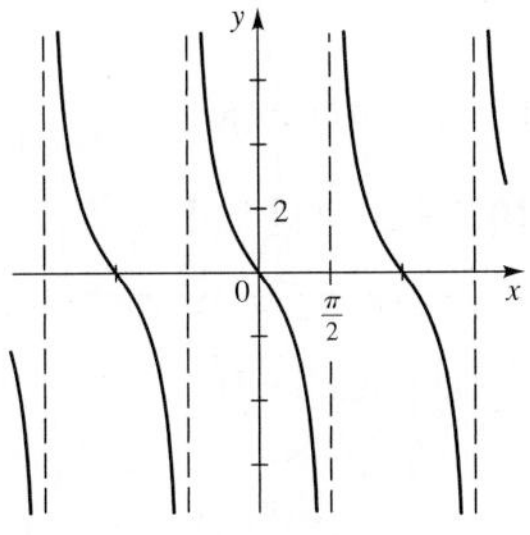

45. π

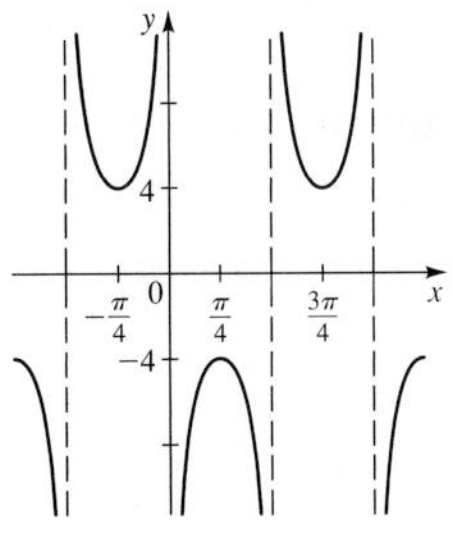

47. 2π

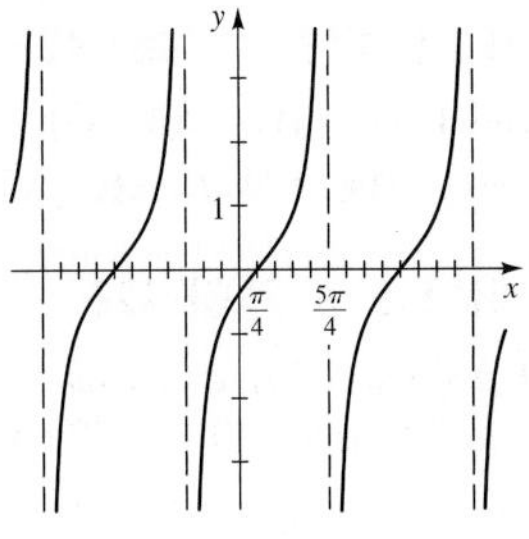

49. $\frac{\pi}{2}$ **51.** $\frac{\pi}{6}$

53. (a)

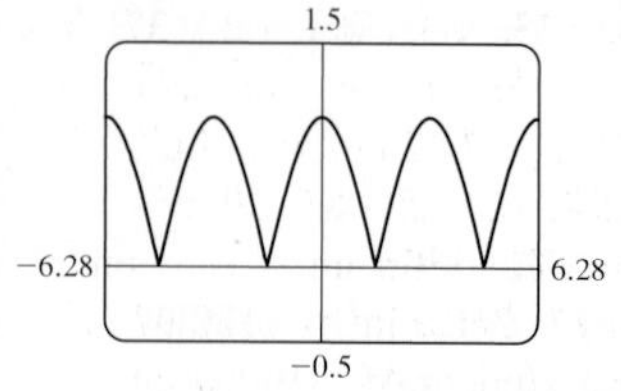

(b) Period π

(c) Even

55. (a)

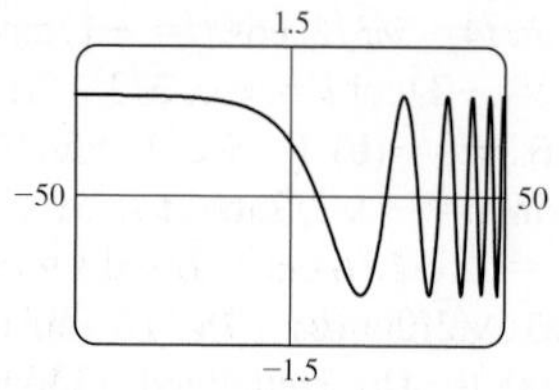

(b) Not periodic

(c) Neither

57. (a)

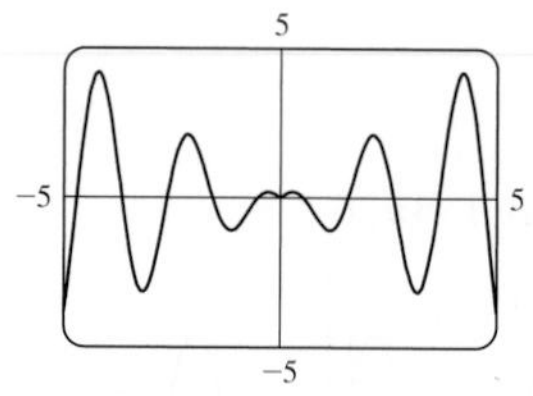

(b) Not periodic

(c) Even

59.

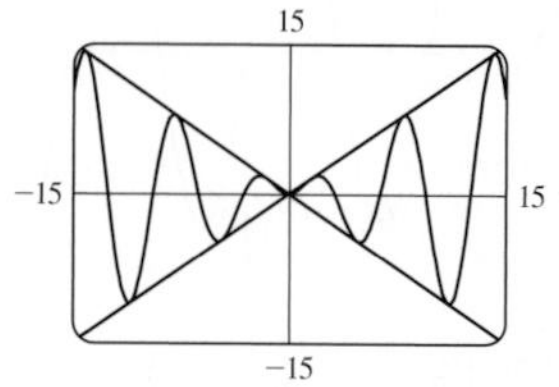

$y = x \sin x$ is a sine function whose graph lies between those of $y = x$ and $y = -x$

61.

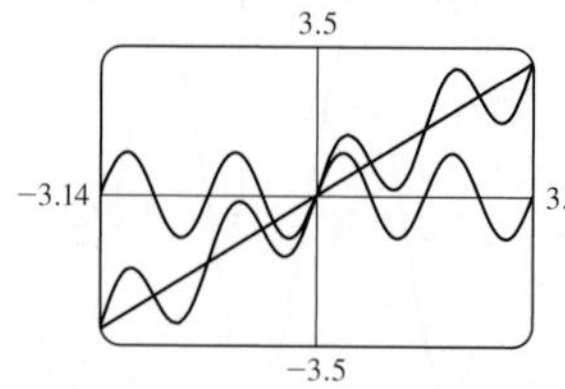

The graphs are related by graphical addition.

63. 1.76, −1.76 **65.** 0.30, 2.84

67. (a) Odd **(b)** $0, \pm\pi, \pm 2\pi, \ldots$

(c)

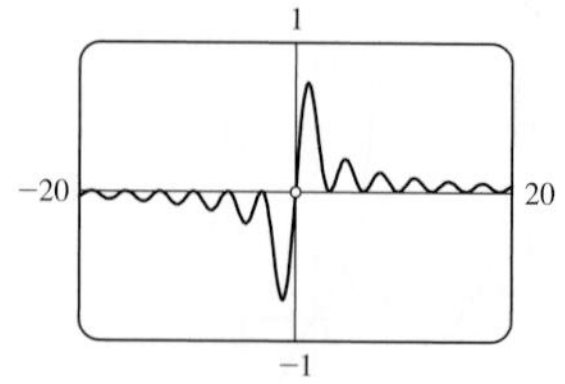

(d) $f(x)$ approaches 0

(e) $f(x)$ approaches 0

69. $y = 50 \cos(16\pi t)$

71. $y = 4 \cos(\frac{\pi}{6} t)$

CHAPTER 5 TEST ▪ PAGE 426

1. $y = -\frac{5}{6}$ **2. (a)** $\frac{4}{5}$ **(b)** $-\frac{3}{5}$ **(c)** $-\frac{4}{3}$ **(d)** $-\frac{5}{3}$

3. (a) $-\frac{1}{2}$ **(b)** $-\sqrt{2}/2$ **(c)** $\sqrt{3}$ **(d)** -1

4. $\tan t = -(\sin t)/\sqrt{1 - \sin^2 t}$ **5.** $-\frac{2}{15}$

6. (a) $5, \pi/2, 0$

(b)

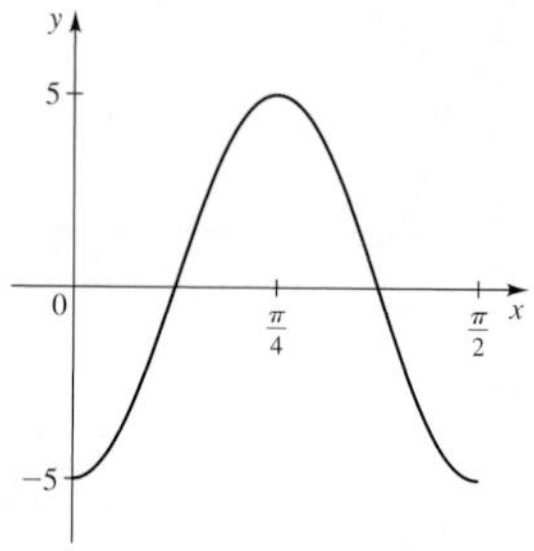

7. (a) $2, 4\pi, \pi/3$

(b)

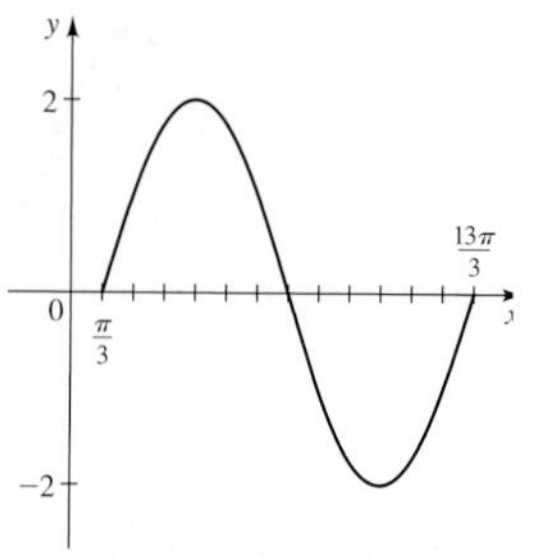

8. π

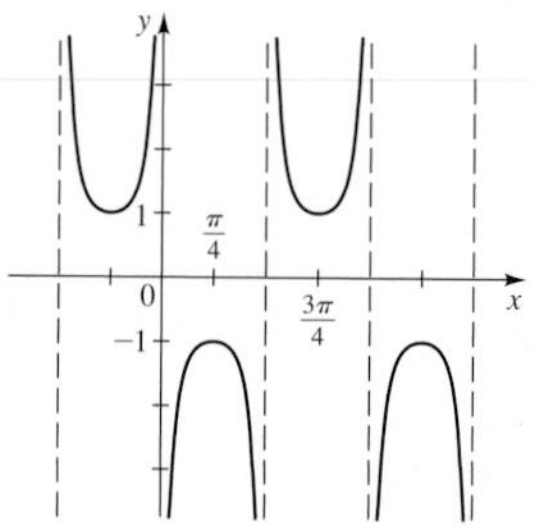

9. $\pi/2$

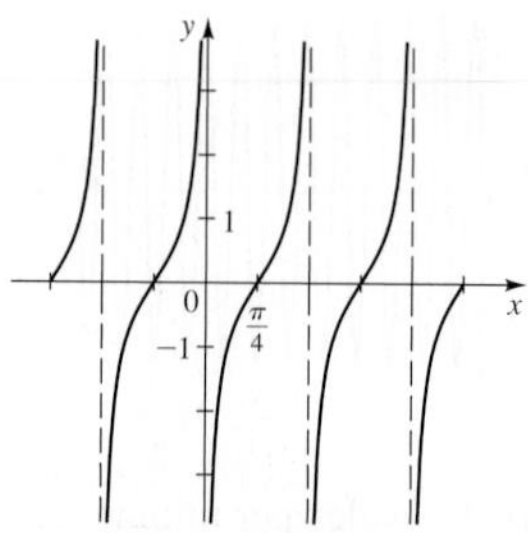

10. (a) $\pi/4$ **(b)** $5\pi/6$ **(c)** 0 **(d)** $1/2$

11. $y = 2 \sin 2(x + \pi/3)$

12. (a)

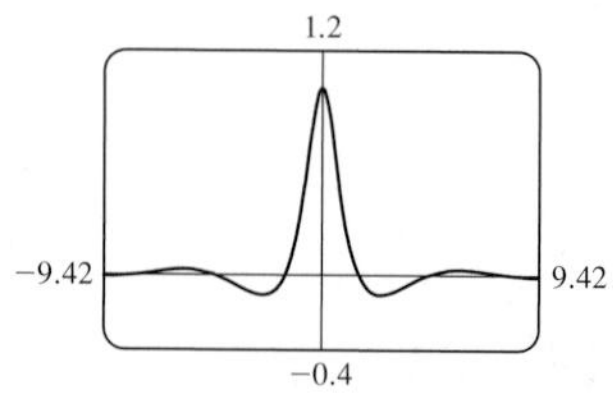

(b) Even

(c) Minimum value −0.11 when $x \approx \pm 2.54$, maximum value 1 when $x = 0$

13. $y = 5 \sin(4\pi t)$

14. $y = 16e^{-0.1t} \cos 24\pi t$

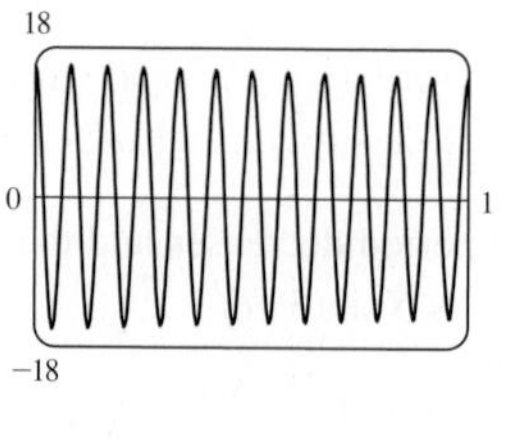

FOCUS ON MODELING ▪ PAGE 430

1. (a) and **(c)**

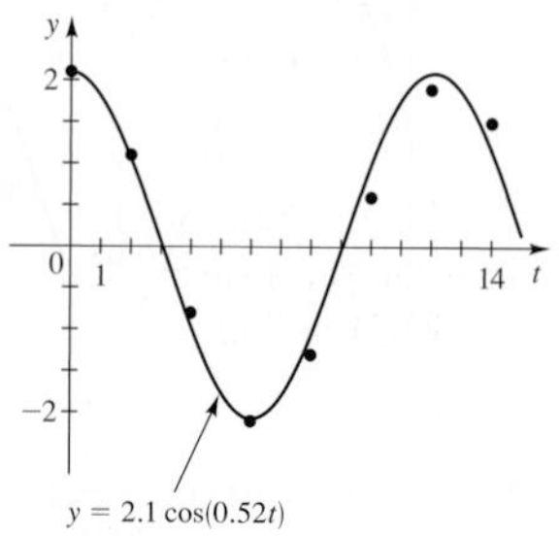

(b) $y = 2.1 \cos(0.52t)$

(d) $y = 2.05 \sin(0.50t + 1.55) - 0.01$ **(e)** The formula of (d) reduces to $y = 2.05 \cos(0.50t - 0.02) - 0.01$. Same as (b), rounded to one decimal.

3. **(a)** and **(c)**

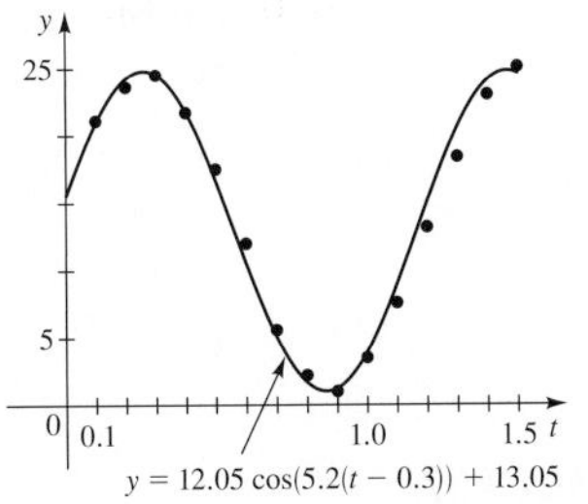

(b) $y = 12.05\cos(5.2(t - 0.3)) + 13.05$
(d) $y = 11.72\sin(5.05t + 0.24) + 12.96$ **(e)** The formula of (d) reduces to $y = 11.72\cos(5.05(t - 0.26)) + 12.96$. Close, but not identical, to (b).

5. **(a)** and **(c)**

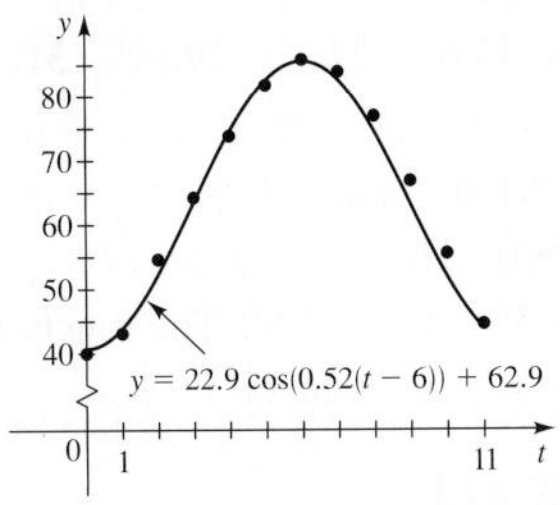

(b) $y = 22.9\cos(0.52(t - 6)) + 62.9$, where y is temperature (°F) and t is months (January = 0)
(d) $y = 23.4\sin(0.48t - 1.36) + 62.2$

7. **(a)** and **(c)**

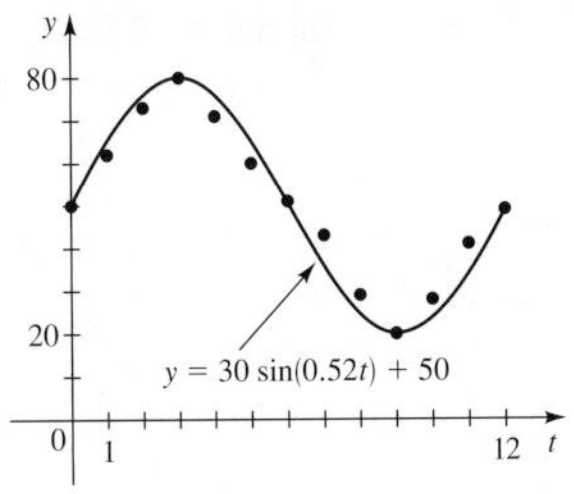

(b) $y = 30\sin(0.52t) + 50$ where y is the owl population in year t
(d) $y = 25.8\sin(0.52t - 0.02) + 50.6$

9. **(a)** and **(c)**

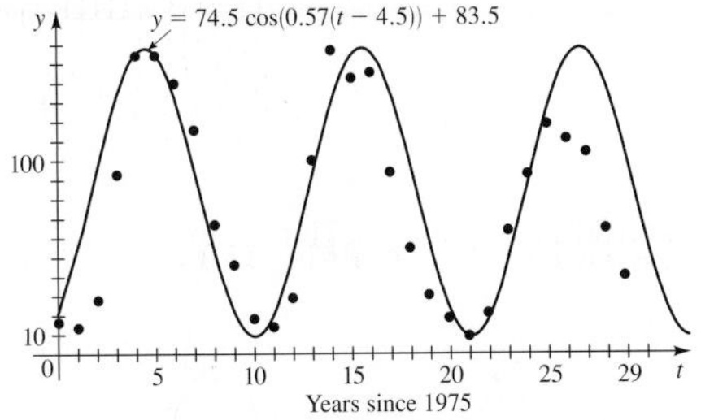

(b) $y = 74.5\cos(0.57(t - 4.5)) + 83.5$, where y is the average daily sunspot count, and t is the years since 1975
(d) $y = 67.65\sin(0.62t - 1.65) + 74.5$

Chapter 6

SECTION 6.1 ■ PAGE 440

1. **(a)** arc, 1 **(b)** $\pi/180$ **(c)** $180/\pi$ **2.** **(a)** $r\theta$ **(b)** $\frac{1}{2}r^2\theta$
3. $2\pi/5 \approx 1.257$ rad **5.** $-\pi/4 \approx -0.785$ rad
7. $-5\pi/12 \approx -1.309$ rad **9.** $6\pi \approx 18.850$ rad
11. $8\pi/15 \approx 1.676$ rad **13.** $\pi/24 \approx 0.131$ rad **15.** 210°
17. −225° **19.** $540/\pi \approx 171.9°$ **21.** $-216/\pi \approx -68.8°$
23. 18° **25.** −24° **27.** 410°, 770°, −310°, −670°
29. $11\pi/4$, $19\pi/4$, $-5\pi/4$, $-13\pi/4$
31. $7\pi/4$, $15\pi/4$, $-9\pi/4$, $-17\pi/4$ **33.** Yes **35.** Yes **37.** Yes
39. 13° **41.** 30° **43.** 280° **45.** $5\pi/6$ **47.** π **49.** $\pi/4$
51. $55\pi/9 \approx 19.2$ **53.** 4 **55.** 4 mi **57.** 2 rad ≈ 114.6°
59. $36/\pi \approx 11.459$ m **61.** **(a)** 35.45 **(b)** 25 **63.** 50 m^2
65. 4 m **67.** 6 cm^2 **69.** 13.9 mi **71.** 330π mi ≈ 1037 mi
73. 1.6 million mi **75.** 1.15 mi **77.** 360π in^2 ≈ 1130.97 in^2
79. **(a)** 90π rad/min **(b)** 1440π in./min ≈ 4523.9 in./min
81. $32\pi/15$ ft/s ≈ 6.7 ft/s **83.** 1039.6 mi/h **85.** 2.1 m/s
87. **(a)** 10π cm ≈ 31.4 cm **(b)** 5 cm **(c)** 3.32 cm
(d) 86.8 cm^3

SECTION 6.2 ■ PAGE 448

1. **(a)**

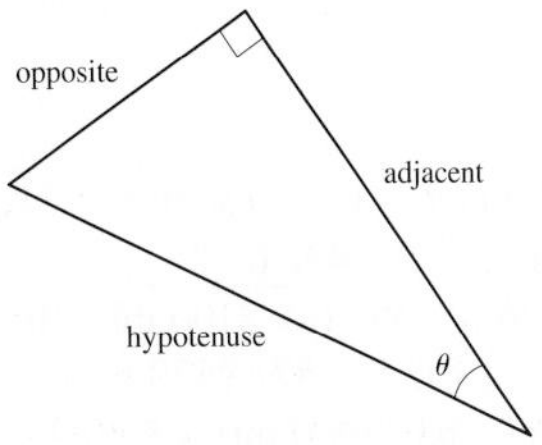

(b) $\frac{\text{opposite}}{\text{hypotenuse}}, \frac{\text{adjacent}}{\text{hypotenuse}}, \frac{\text{opposite}}{\text{adjacent}}$ **(c)** similar

2. $\sin\theta$, $\cos\theta$, $\tan\theta$
3. $\sin\theta = \frac{4}{5}$, $\cos\theta = \frac{3}{5}$, $\tan\theta = \frac{4}{3}$, $\csc\theta = \frac{5}{4}$, $\sec\theta = \frac{5}{3}$, $\cot\theta = \frac{3}{4}$
5. $\sin\theta = \frac{40}{41}$, $\cos\theta = \frac{9}{41}$, $\tan\theta = \frac{40}{9}$, $\csc\theta = \frac{41}{40}$, $\sec\theta = \frac{41}{9}$, $\cot\theta = \frac{9}{40}$
7. $\sin\theta = 2\sqrt{13}/13$, $\cos\theta = 3\sqrt{13}/13$, $\tan\theta = \frac{2}{3}$, $\csc\theta = \sqrt{13}/2$, $\sec\theta = \sqrt{13}/3$, $\cot\theta = \frac{3}{2}$
9. **(a)** $3\sqrt{34}/34$, $3\sqrt{34}/34$ **(b)** $\frac{3}{5}, \frac{3}{5}$ **(c)** $\sqrt{34}/5$, $\sqrt{34}/5$
11. $\frac{25}{2}$ **13.** $13\sqrt{3}/2$ **15.** 16.51658
17. $x = 28\cos\theta$, $y = 28\sin\theta$
19. $\cos\theta = \frac{4}{5}$, $\tan\theta = \frac{3}{4}$, $\csc\theta = \frac{5}{3}$, $\sec\theta = \frac{5}{4}$, $\cot\theta = \frac{4}{3}$

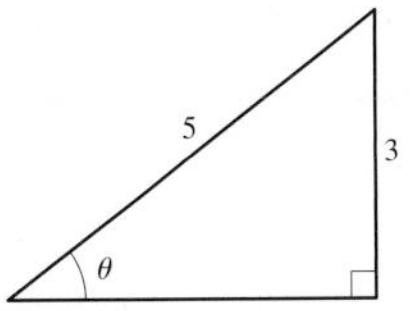

21. $\sin\theta = \sqrt{2}/2$, $\cos\theta = \sqrt{2}/2$, $\tan\theta = 1$, $\csc\theta = \sqrt{2}$, $\sec\theta = \sqrt{2}$

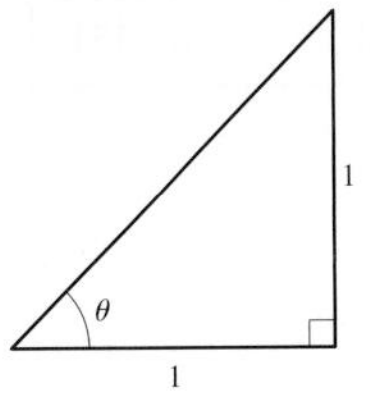

23. $\sin\theta = 3\sqrt{5}/7$, $\cos\theta = \frac{2}{7}$, $\tan\theta = 3\sqrt{5}/2$, $\csc\theta = 7\sqrt{5}/15$, $\cot\theta = 2\sqrt{5}/15$

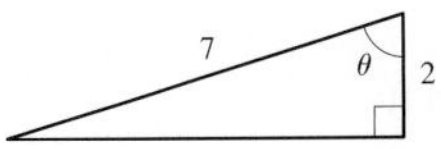

25. $(1 + \sqrt{3})/2$ **27.** 1 **29.** $\frac{1}{2}$

31.

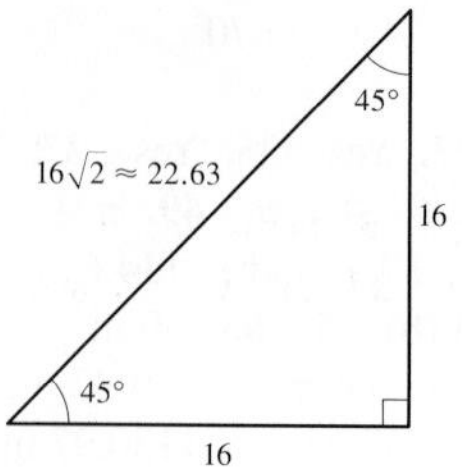

33.

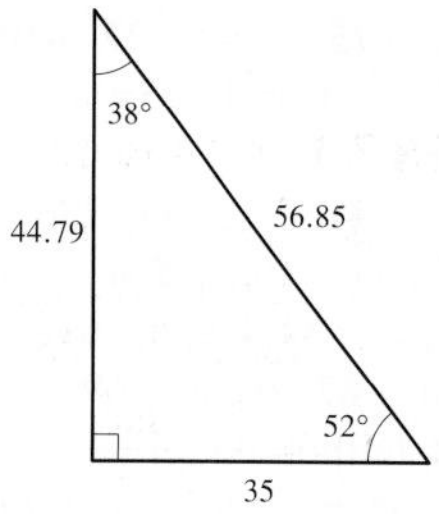

35.

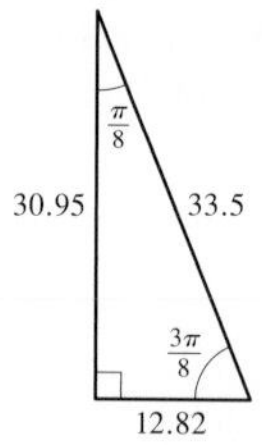

37.

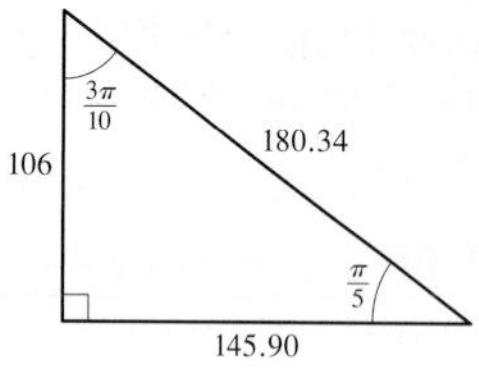

39. $\sin\theta \approx 0.45$, $\cos\theta \approx 0.89$, $\tan\theta = 0.50$, $\csc\theta \approx 2.24$, $\sec\theta \approx 1.12$, $\cot\theta = 2.00$ **41.** 230.9 **43.** 63.7
45. $x = 10\tan\theta\sin\theta$ **47.** 1026 ft **49. (a)** 2100 mi **(b)** No
51. 19 ft **53.** 345 ft **55.** 415 ft, 152 ft **57.** 2570 ft
59. 5808 ft **61.** 91.7 million mi **63.** 3960 mi **65.** 0.723 AU

SECTION 6.3 ▪ PAGE 459

1. y/r, x/r, y/x **2.** quadrant, positive, negative, negative
3. (a) 30° **(b)** 30° **(c)** 30° **5. (a)** 45° **(b)** 90° **(c)** 75°
7. (a) $\pi/4$ **(b)** $\pi/6$ **(c)** $\pi/3$ **9. (a)** $2\pi/7$ **(b)** 0.4π **(c)** 1.4
11. $\frac{1}{2}$ **13.** $-\sqrt{3}/2$ **15.** $-\sqrt{3}$ **17.** 1 **19.** $-\sqrt{3}/2$
21. $\sqrt{3}/3$ **23.** $\sqrt{3}/2$ **25.** -1 **27.** $\frac{1}{2}$ **29.** 2 **31.** -1
33. Undefined **35.** III **37.** IV
39. $\tan\theta = -\sqrt{1-\cos^2\theta}/\cos\theta$
41. $\cos\theta = \sqrt{1-\sin^2\theta}$
43. $\sec\theta = -\sqrt{1+\tan^2\theta}$
45. $\cos\theta = -\frac{4}{5}$, $\tan\theta = -\frac{3}{4}$, $\csc\theta = \frac{5}{3}$, $\sec\theta = -\frac{5}{4}$, $\cot\theta = -\frac{4}{3}$
47. $\sin\theta = -\frac{3}{5}$, $\cos\theta = \frac{4}{5}$, $\csc\theta = -\frac{5}{3}$, $\sec\theta = \frac{5}{4}$, $\cot\theta = -\frac{4}{3}$
49. $\sin\theta = \frac{1}{2}$, $\cos\theta = \sqrt{3}/2$, $\tan\theta = \sqrt{3}/3$, $\sec\theta = 2\sqrt{3}/3$, $\cot\theta = \sqrt{3}$
51. $\sin\theta = 3\sqrt{5}/7$, $\tan\theta = -3\sqrt{5}/2$, $\csc\theta = 7\sqrt{5}/15$, $\sec\theta = -\frac{7}{2}$, $\cot\theta = -2\sqrt{5}/15$
53. (a) $\sqrt{3}/2$, $\sqrt{3}$ **(b)** $\frac{1}{2}$, $\sqrt{3}/4$ **(c)** $\frac{3}{4}$, 0.88967 **55.** 19.1
57. 66.1° **59.** $(4\pi/3) - \sqrt{3} \approx 2.46$
63. (b)

θ	20°	60°	80°	85°
h	1922	9145	29,944	60,351

65. (a) $A(\theta) = 400\sin\theta\cos\theta$
(b)

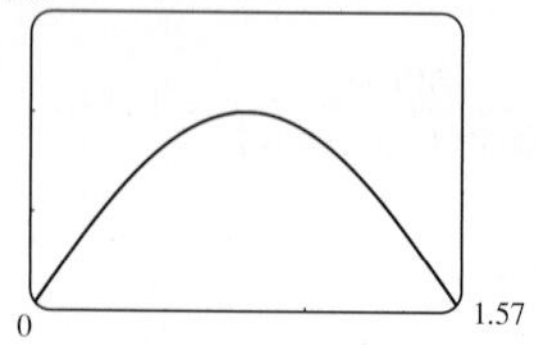

(c) width = depth ≈ 14.14 in.
67. (a) $9\sqrt{3}/4$ ft ≈ 3.897 ft, $\frac{9}{16}$ ft = 0.5625 ft
(b) 23.982 ft, 3.462 ft

69. (a)

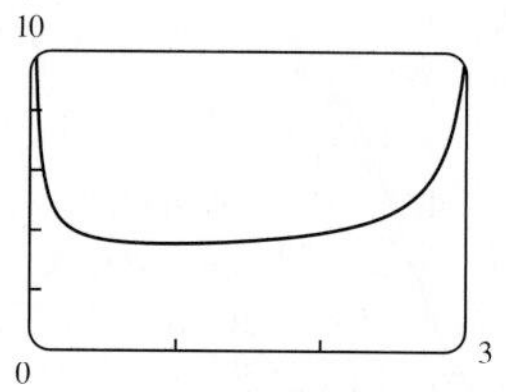

(b) 0.946 rad or 54°

SECTION 6.4 ▪ PAGE 467

1. (a) $[-1, 1]$, $[-\pi/2, \pi/2]$ **(b)** $[-1, 1]$, $[0, \pi]$ **(c)** $\mathbb{R}$, $(-\pi/2, \pi/2)$ **2. (a)** $\frac{8}{10}$ **(b)** $\frac{6}{10}$ **(c)** $\frac{8}{6}$ **3. (a)** $\pi/6$ **(b)** $5\pi/6$ **(c)** $-\pi/4$ **5. (a)** $-\pi/6$ **(b)** $\pi/3$ **(c)** $\pi/6$
7. 0.46677 **9.** 1.82348 **11.** 1.24905 **13.** Undefined
15. 36.9° **17.** 34.7° **19.** 34.9° **21.** 30°, 150°
23. 44.4°, 135.6° **25.** 45.6° **27.** $\frac{4}{5}$ **29.** $\frac{13}{5}$ **31.** $\frac{12}{5}$
33. $\sqrt{1-x^2}$ **35.** $x/\sqrt{1-x^2}$ **37.** 72.5°, 19 ft
39. (a) $h = 2\tan\theta$ **(b)** $\theta = \tan^{-1}(h/2)$
41. (a) $\theta = \sin^{-1}(h/680)$ **(b)** $\theta = 0.826$ rad
43. (a) 54.1° **(b)** 48.3°, 32.2°, 24.5°. The function $\sin^{-1}$ is undefined for values outside the interval $[-1, 1]$.

SECTION 6.5 ▪ PAGE 473

1. $\dfrac{\sin A}{a} = \dfrac{\sin B}{b} = \dfrac{\sin C}{c}$ **2.** ASA, SSA **3.** 318.8 **5.** 24.8
7. 44° **9.** $\angle C = 114°$, $a \approx 51$, $b \approx 24$ **11.** $\angle A = 44°$, $\angle B = 68°$, $a \approx 8.99$ **13.** $\angle C = 62°$, $a \approx 200$, $b \approx 242$

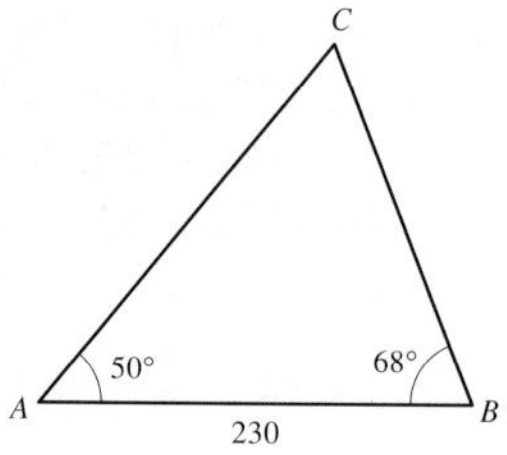

15. $\angle B = 85°$, $a \approx 5$, $c \approx 9$

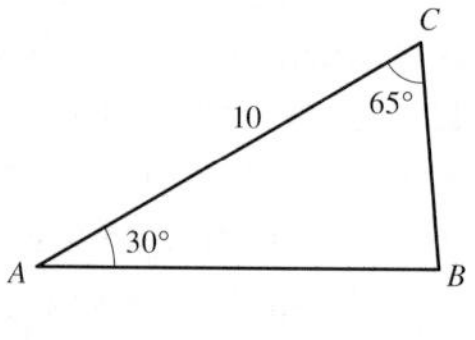

17. $\angle A = 100°$, $a \approx 89$, $c \approx 71$

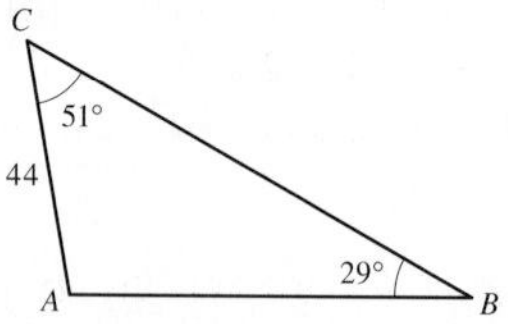

19. $\angle B \approx 30°$, $\angle C \approx 40°$, $c \approx 19$ **21.** No solution
23. $\angle A_1 \approx 125°$, $\angle C_1 \approx 30°$, $a_1 \approx 49$; $\angle A_2 \approx 5°$, $\angle C_2 \approx 150°$, $a_2 \approx 5.6$ **25.** No solution
27. $\angle A_1 \approx 57.2°$, $\angle B_1 \approx 93.8°$, $b_1 \approx 30.9$; $\angle A_2 \approx 122.8°$, $\angle B_2 \approx 28.2°$, $b_2 \approx 14.6$
29. (a) 91.146° **(b)** 14.427° **33. (a)** 1018 mi **(b)** 1017 mi
35. 219 ft **37.** 55.9 m **39.** 175 ft **41.** 192 m
43. 0.427 AU, 1.119 AU

SECTION 6.6 ▪ PAGE 480

1. $a^2 + b^2 - 2ab\cos C$ **2.** SSS, SAS **3.** 28.9 **5.** 47
7. 29.89° **9.** 15 **11.** $\angle A \approx 39.4°$, $\angle B \approx 20.6°$, $c \approx 24.6$

13. $\angle A \approx 48°, \angle B \approx 79°, c \approx 3.2$
15. $\angle A \approx 50°, \angle B \approx 73°, \angle C \approx 57°$
17. $\angle A_1 \approx 83.6°, \angle C_1 \approx 56.4°, a_1 \approx 193$;
$\angle A_2 \approx 16.4°, \angle C_2 \approx 123.6, a_2 \approx 54.9$ **19.** No such triangle
21. 2 **23.** 25.4 **25.** 89.2° **27.** 24.3 **29.** 54 **31.** 26.83
33. 5.33 **35.** 40.77 **37.** 3.85 cm^2 **39.** 2.30 mi **41.** 23.1 mi
43. 2179 mi **45. (a)** 62.6 mi **(b)** S 18.2° E **47.** 96°
49. 211 ft **51.** 3835 ft **53.** $165,554

CHAPTER 6 REVIEW ■ PAGE 483

1. (a) $\pi/3$ **(b)** $11\pi/6$ **(c)** $-3\pi/4$ **(d)** $-\pi/2$
3. (a) 450° **(b)** −30° **(c)** 405° **(d)** $(558/\pi)° \approx 177.6°$
5. 8 m **7.** 82 ft **9.** 0.619 rad ≈ 35.4° **11.** 18,151 ft^2
13. 300π rad/min ≈ 942.5 rad/min,
7539.8 in./min = 628.3 ft/min
15. $\sin\theta = 5/\sqrt{74}$, $\cos\theta = 7/\sqrt{74}$, $\tan\theta = \frac{5}{7}$,
$\csc\theta = \sqrt{74}/5$, $\sec\theta = \sqrt{74}/7$, $\cot\theta = \frac{7}{5}$
17. $x \approx 3.83, y \approx 3.21$ **19.** $x \approx 2.92, y \approx 3.11$
21. $A = 70°, a \approx 2.819, b \approx 1.026$
23. $A \approx 16.3°, C \approx 73.7°, c = 24$
25. $a = \cot\theta, b = \csc\theta$ **27.** 48 m **29.** 1076 mi **31.** $-\sqrt{2}/2$
33. 1 **35.** $-\sqrt{3}/3$ **37.** $-\sqrt{2}/2$ **39.** $2\sqrt{3}/3$ **41.** $-\sqrt{3}$
43. $\sin\theta = \frac{12}{13}$, $\cos\theta = -\frac{5}{13}$, $\tan\theta = -\frac{12}{5}$,
$\csc\theta = \frac{13}{12}$, $\sec\theta = -\frac{13}{5}$, $\cot\theta = -\frac{5}{12}$ **45.** 60°
47. $\tan\theta = -\sqrt{1 - \cos^2\theta}/\cos\theta$
49. $\tan^2\theta = \sin^2\theta/(1 - \sin^2\theta)$
51. $\sin\theta = \sqrt{7}/4$, $\cos\theta = \frac{3}{4}$, $\csc\theta = 4\sqrt{7}/7$, $\cot\theta = 3\sqrt{7}/7$
53. $\cos\theta = -\frac{4}{5}$, $\tan\theta = -\frac{3}{4}$, $\csc\theta = \frac{5}{3}$, $\sec\theta = -\frac{5}{4}$, $\cot\theta = -\frac{4}{3}$
55. $-\sqrt{5}/5$ **57.** 1 **59.** $\pi/3$ **61.** $2/\sqrt{21}$ **63.** $x/\sqrt{1 + x^2}$
65. $\theta = \cos^{-1}(x/3)$ **67.** 5.32 **69.** 148.07 **71.** 9.17
73. 54.1° or 125.9° **75.** 80.4° **77.** 77.3 mi **79.** 3.9 mi
81. 32.12

CHAPTER 6 TEST ■ PAGE 487

1. $11\pi/6, -3\pi/4$ **2.** 240°, −74.5°
3. (a) 240π rad/min ≈ 753.98 rad/min
(b) 12,063.7 ft/min = 137 mi/h **4. (a)** $\sqrt{2}/2$
(b) $\sqrt{3}/3$ **(c)** 2 **(d)** 1 **5.** $(26 + 6\sqrt{13})/39$
6. $a = 24\sin\theta, b = 24\cos\theta$ **7.** $(4 - 3\sqrt{2})/4$
8. $-\frac{13}{12}$ **9.** $\tan\theta = -\sqrt{\sec^2\theta - 1}$ **10.** 19.6 ft
11. (a) $\theta = \tan^{-1}(x/4)$ **(b)** $\theta = \cos^{-1}(3/x)$ **12.** $\frac{40}{41}$
13. 9.1 **14.** 250.5 **15.** 8.4 **16.** 19.5 **17.** 78.6° **18.** 40.2°
19. (a) 15.3 m^2 **(b)** 24.3 m **20. (a)** 129.9° **(b)** 44.9
21. 554 ft

FOCUS ON MODELING ■ PAGE 490

1. 1.41 mi **3.** 14.3 m **5. (c)** 2349.8 ft
7.

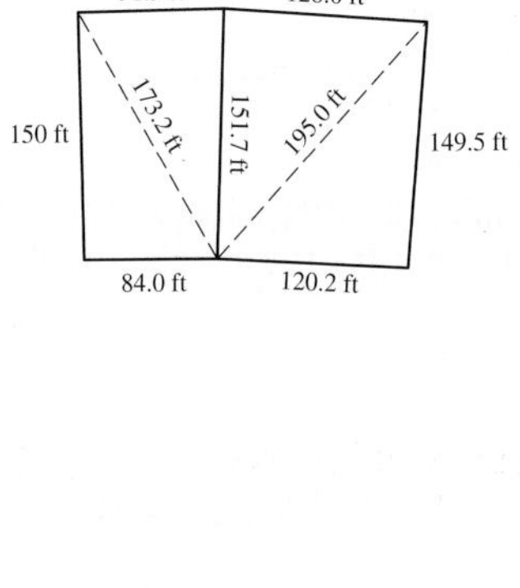

CHAPTER 7

SECTION 7.1 ■ PAGE 498

1. all; 1 **2.** $\cos(-x) = \cos x$ **3.** $\sin t$ **5.** $\tan\theta$ **7.** −1
9. $\csc u$ **11.** $\tan\theta$ **13.** 1 **15.** $\cos y$ **17.** $\sin^2 x$ **19.** $\sec x$
21. $2\sec u$ **23.** $\cos^2 x$ **25.** $\cos\theta$

27. (a) $\text{LHS} = \dfrac{1 - \sin^2 x}{\sin x} = \text{RHS}$

29. $\text{LHS} = \sin\theta\dfrac{\cos\theta}{\sin\theta} = \text{RHS}$

31. $\text{LHS} = \cos u\dfrac{1}{\cos u}\cot u = \text{RHS}$

33. $\text{LHS} = \sin B + \cos B\dfrac{\cos B}{\sin B}$
$= \dfrac{\sin^2 B + \cos^2 B}{\sin B} = \dfrac{1}{\sin B} = \text{RHS}$

35. $\text{LHS} = -\dfrac{\cos\alpha}{\sin\alpha}\cos\alpha - \sin\alpha = \dfrac{-\cos^2\alpha - \sin^2\alpha}{\sin\alpha}$
$= \dfrac{-1}{\sin\alpha} = \text{RHS}$

37. $\text{LHS} = \dfrac{\sin\theta}{\cos\theta} + \dfrac{\cos\theta}{\sin\theta} = \dfrac{\sin^2\theta + \cos^2\theta}{\cos\theta\sin\theta}$
$= \dfrac{1}{\cos\theta\sin\theta} = \text{RHS}$

39. $\text{LHS} = 1 - \cos^2\beta = \sin^2\beta = \text{RHS}$

41. $\text{LHS} = \dfrac{(\sin x + \cos x)^2}{(\sin x + \cos x)(\sin x - \cos x)} = \dfrac{\sin x + \cos x}{\sin x - \cos x}$
$= \dfrac{(\sin x + \cos x)(\sin x - \cos x)}{(\sin x - \cos x)(\sin x - \cos x)} = \text{RHS}$

43. $\text{LHS} = \dfrac{\frac{1}{\cos t} - \cos t}{\frac{1}{\cos t}} \cdot \dfrac{\cos t}{\cos t} = \dfrac{1 - \cos^2 t}{1} = \text{RHS}$

45. $\text{LHS} = \dfrac{1}{\cos^2 y} = \sec^2 y = \text{RHS}$

47. $\text{LHS} = \cot x\cos x + \cot x - \csc x\cos x - \csc x$
$= \dfrac{\cos^2 x}{\sin x} + \dfrac{\cos x}{\sin x} - \dfrac{\cos x}{\sin x} - \dfrac{1}{\sin x} = \dfrac{\cos^2 x - 1}{\sin x}$
$= \dfrac{-\sin^2 x}{\sin x} = \text{RHS}$

49. $\text{LHS} = \sin^2 x\left(1 + \dfrac{\cos^2 x}{\sin^2 x}\right) = \sin^2 x + \cos^2 x = \text{RHS}$

51. $\text{LHS} = 2(1 - \sin^2 x) - 1 = 2 - 2\sin^2 x - 1 = \text{RHS}$

53. $\text{LHS} = \dfrac{1 - \cos\alpha}{\sin\alpha} \cdot \dfrac{1 + \cos\alpha}{1 + \cos\alpha}$
$= \dfrac{1 - \cos^2\alpha}{\sin\alpha(1 + \cos\alpha)} = \dfrac{\sin^2\alpha}{\sin\alpha(1 + \cos\alpha)} = \text{RHS}$

55. $\text{LHS} = \dfrac{\sin^2\theta}{\cos^2\theta} - \dfrac{\sin^2\theta\cos^2\theta}{\cos^2\theta}$
$= \dfrac{\sin^2\theta(1 - \cos^2\theta)}{\cos^2\theta} = \dfrac{\sin^2\theta\sin^2\theta}{\cos^2\theta} = \text{RHS}$

57. $\text{LHS} = \dfrac{\sin x - 1}{\sin x + 1} \cdot \dfrac{\sin x + 1}{\sin x + 1} = \dfrac{\sin^2 x - 1}{(\sin x + 1)^2} = \text{RHS}$

59. $\text{LHS} = \dfrac{\sin^2 t + 2\sin t\cos t + \cos^2 t}{\sin t\cos t}$
$= \dfrac{\sin^2 t + \cos^2 t}{\sin t\cos t} + \dfrac{2\sin t\cos t}{\sin t\cos t} = \dfrac{1}{\sin t\cos t} + 2$
$= \text{RHS}$

61. $\text{LHS} = \dfrac{1 + \frac{\sin^2 u}{\cos^2 u}}{1 - \frac{\sin^2 u}{\cos^2 u}} \cdot \dfrac{\cos^2 u}{\cos^2 u} = \dfrac{\cos^2 u + \sin^2 u}{\cos^2 u - \sin^2 u} = \text{RHS}$

63. $\text{LHS} = \dfrac{\sec x}{\sec x - \tan x} \cdot \dfrac{\sec x + \tan x}{\sec x + \tan x}$
$= \dfrac{\sec x(\sec x + \tan x)}{\sec^2 x - \tan^2 x} = \text{RHS}$

65. $\text{LHS} = (\sec v - \tan v) \cdot \dfrac{\sec v + \tan v}{\sec v + \tan v}$
$= \dfrac{\sec^2 v - \tan^2 v}{\sec v + \tan v} = \text{RHS}$

67. $\text{LHS} = \dfrac{\sin x + \cos x}{\frac{1}{\cos x} + \frac{1}{\sin x}} = \dfrac{\sin x + \cos x}{\frac{\sin x + \cos x}{\cos x \sin x}}$
$= (\sin x + \cos x)\dfrac{\cos x \sin x}{\sin x + \cos x} = \text{RHS}$

69. $\text{LHS} = \dfrac{\frac{1}{\sin x} - \frac{\cos x}{\sin x}}{\frac{1}{\cos x} - 1} \cdot \dfrac{\sin x\cos x}{\sin x\cos x} = \dfrac{\cos x(1 - \cos x)}{\sin x(1 - \cos x)}$
$= \dfrac{\cos x}{\sin x} = \text{RHS}$

71. $\text{LHS} = \dfrac{\sin^2 u}{\cos^2 u} - \dfrac{\sin^2 u\cos^2 u}{\cos^2 u} = \dfrac{\sin^2 u}{\cos^2 u}(1 - \cos^2 u) = \text{RHS}$

73. $\text{LHS} = (\sec^2 x - \tan^2 x)(\sec^2 x + \tan^2 x) = \text{RHS}$

75. $\text{RHS} = \dfrac{\sin\theta - \frac{1}{\sin\theta}}{\cos\theta - \frac{\cos\theta}{\sin\theta}} = \dfrac{\frac{\sin^2\theta - 1}{\sin\theta}}{\frac{\cos\theta\sin\theta - \cos\theta}{\sin\theta}}$
$= \dfrac{\cos^2\theta}{\cos\theta(\sin\theta - 1)} = \text{LHS}$

77. $\text{LHS} = \dfrac{-\sin^2 t + \tan^2 t}{\sin^2 t} = -1 + \dfrac{\sin^2 t}{\cos^2 t} \cdot \dfrac{1}{\sin^2 t}$
$= -1 + \sec^2 t = \text{RHS}$

79. $\text{LHS} = \dfrac{\sec x - \tan x + \sec x + \tan x}{(\sec x + \tan x)(\sec x - \tan x)}$
$= \dfrac{2\sec x}{\sec^2 x - \tan^2 x} = \text{RHS}$

81. $\text{LHS} = \tan^2 x + 2\tan x\cot x + \cot^2 x = \tan^2 x + 2 + \cot^2 x$
$= (\tan^2 x + 1) + (\cot^2 x + 1) = \text{RHS}$

83. $\text{LHS} = \dfrac{\frac{1}{\cos u} - 1}{\frac{1}{\cos u} + 1} \cdot \dfrac{\cos u}{\cos u} = \text{RHS}$

85. $\text{LHS} = \dfrac{(\sin x + \cos x)(\sin^2 x - \sin x\cos x + \cos^2 x)}{\sin x + \cos x}$
$= \sin^2 x - \sin x\cos x + \cos^2 x = \text{RHS}$

87. $\text{LHS} = \dfrac{1 + \sin x}{1 - \sin x} \cdot \dfrac{1 + \sin x}{1 + \sin x} = \dfrac{(1 + \sin x)^2}{1 - \sin^2 x}$
$= \dfrac{(1 + \sin x)^2}{\cos^2 x} = \left(\dfrac{1 + \sin x}{\cos x}\right)^2 = \text{RHS}$

89. $\text{LHS} = \left(\dfrac{\sin x}{\cos x} + \dfrac{\cos x}{\sin x}\right)^4 = \left(\dfrac{\sin^2 x + \cos^2 x}{\sin x\cos x}\right)^4$
$= \left(\dfrac{1}{\sin x\cos x}\right)^4 = \text{RHS}$

91. $\tan\theta$ **93.** $\tan\theta$ **95.** $3\cos\theta$

97. Yes

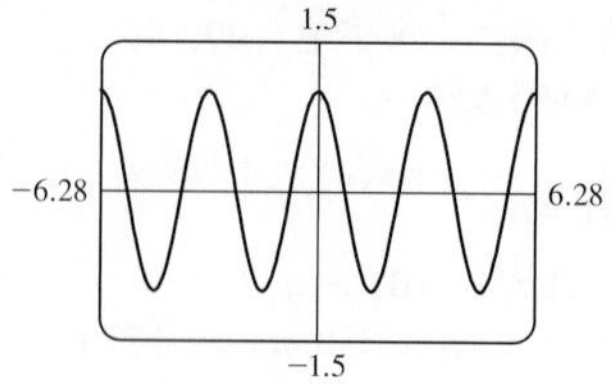

99. No

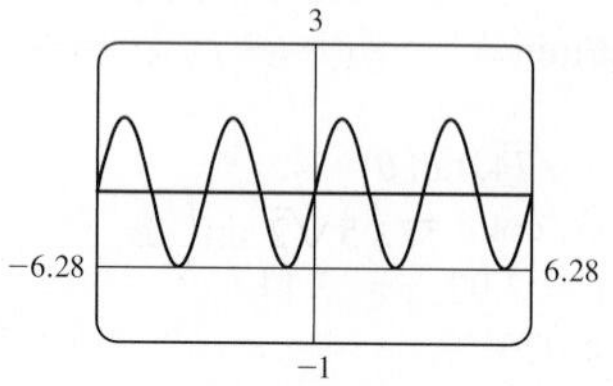

SECTION 7.2 ■ PAGE 505

1. addition; $\sin x\cos y + \cos x\sin y$
2. subtraction; $\cos x\cos y + \sin x\sin y$

3. $\dfrac{\sqrt{6} + \sqrt{2}}{4}$ **5.** $\dfrac{\sqrt{2} - \sqrt{6}}{4}$ **7.** $2 - \sqrt{3}$ **9.** $-\dfrac{\sqrt{6} + \sqrt{2}}{4}$

11. $\sqrt{3} - 2$ **13.** $-\dfrac{\sqrt{6} + \sqrt{2}}{4}$ **15.** $\sqrt{2}/2$ **17.** $\frac{1}{2}$ **19.** $\sqrt{3}$

21. $\text{LHS} = \dfrac{\sin(\frac{\pi}{2} - u)}{\cos(\frac{\pi}{2} - u)} = \dfrac{\sin\frac{\pi}{2}\cos u - \cos\frac{\pi}{2}\sin u}{\cos\frac{\pi}{2}\cos u + \sin\frac{\pi}{2}\sin u}$
$= \dfrac{\cos u}{\sin u} = \text{RHS}$

23. $\text{LHS} = \dfrac{1}{\cos(\frac{\pi}{2} - u)} = \dfrac{1}{\cos\frac{\pi}{2}\cos u + \sin\frac{\pi}{2}\sin u}$
$= \dfrac{1}{\sin u} = \text{RHS}$

25. $\text{LHS} = \sin x\cos\frac{\pi}{2} - \cos x\sin\frac{\pi}{2} = \text{RHS}$

27. $\text{LHS} = \sin x\cos\pi - \cos x\sin\pi = \text{RHS}$

29. $\text{LHS} = \dfrac{\tan x - \tan\pi}{1 + \tan x\tan\pi} = \text{RHS}$

31. $\text{LHS} = \cos x\cos\frac{\pi}{6} - \sin x\sin\frac{\pi}{6} + \sin x\cos\frac{\pi}{3} - \cos x\sin\frac{\pi}{3}$
$= \frac{\sqrt{3}}{2}\cos x - \frac{1}{2}\sin x + \frac{1}{2}\sin x - \frac{\sqrt{3}}{2}\cos x = \text{RHS}$

33. $\text{LHS} = \sin x\cos y + \cos x\sin y$
$-(\sin x\cos y - \cos x\sin y) = \text{RHS}$

35. $\text{LHS} = \dfrac{1}{\tan(x - y)} = \dfrac{1 + \tan x\tan y}{\tan x - \tan y}$
$= \dfrac{1 + \frac{1}{\cot x}\frac{1}{\cot y}}{\frac{1}{\cot x} - \frac{1}{\cot y}} \cdot \dfrac{\cot x\cot y}{\cot x\cot y} = \text{RHS}$

37. $\text{LHS} = \dfrac{\sin x}{\cos x} - \dfrac{\sin y}{\cos y} = \dfrac{\sin x\cos y - \cos x\sin y}{\cos x\cos y} = \text{RHS}$

39. LHS $= \dfrac{\sin x \cos y + \cos x \sin y - (\sin x \cos y - \cos x \sin y)}{\cos x \cos y - \sin x \sin y + \cos x \cos y + \sin x \sin y}$
$= \dfrac{2 \cos x \sin y}{2 \cos x \cos y} = \text{RHS}$

41. LHS $= \sin((x + y) + z)$
$= \sin(x + y) \cos z + \cos(x + y) \sin z$
$= \cos z[\sin x \cos y + \cos x \sin y]$
$+ \sin z[\cos x \cos y - \sin x \sin y] = \text{RHS}$

43. $\dfrac{\sqrt{1 - x^2} + xy}{\sqrt{1 + y^2}}$ **45.** $\dfrac{x - y}{\sqrt{1 + x^2}\sqrt{1 + y^2}}$

47. $\frac{1}{4}(\sqrt{6} + \sqrt{2})$ **49.** $\dfrac{3 - 2\sqrt{14}}{\sqrt{7} + 6\sqrt{2}}$ **51.** $-\frac{1}{10}(3 + 4\sqrt{3})$

53. $2\sqrt{5}/65$ **55.** $2 \sin\left(x + \dfrac{5\pi}{6}\right)$ **57.** $5\sqrt{2} \sin\left(2x + \dfrac{7\pi}{4}\right)$

59. (a) $g(x) = 2 \sin 2\left(x + \dfrac{\pi}{12}\right)$
(b)

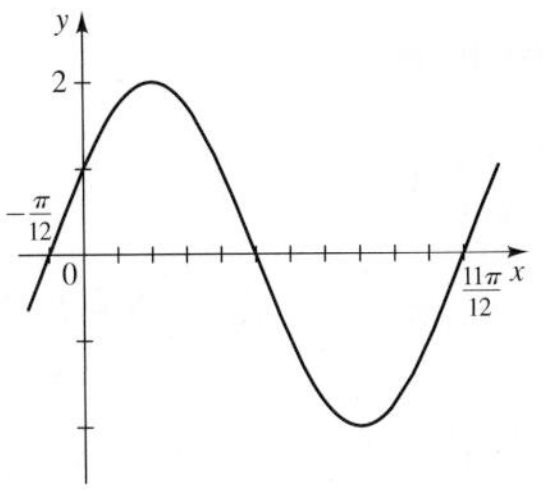

63. $\tan \gamma = \frac{17}{6}$

65. (a)

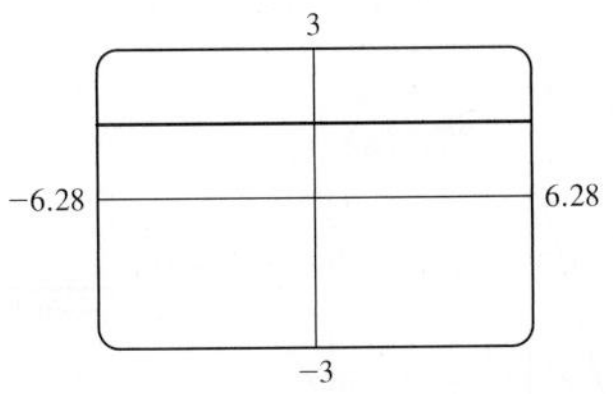

$\sin^2\left(x + \dfrac{\pi}{4}\right) + \sin^2\left(x - \dfrac{\pi}{4}\right) = 1$

67. $\pi/2$ **69. (b)** $k = 10\sqrt{3}, \phi = \pi/6$

SECTION 7.3 ■ PAGE 514

1. Double-Angle; $2 \sin x \cos x$

2. Half-Angle; $\pm\sqrt{(1 - \cos x)/2}$

3. $\frac{120}{169}, \frac{119}{169}, \frac{120}{119}$ **5.** $-\frac{24}{25}, \frac{7}{25}, -\frac{24}{7}$ **7.** $\frac{24}{25}, \frac{7}{25}, \frac{24}{7}$ **9.** $-\frac{3}{5}, \frac{4}{5}, -\frac{3}{4}$

11. $\frac{1}{2}(\frac{3}{4} - \cos 2x + \frac{1}{4} \cos 4x)$

13. $\frac{1}{16}(1 - \cos 2x - \cos 4x + \cos 2x \cos 4x)$

15. $\frac{1}{32}(\frac{3}{4} - \cos 4x + \frac{1}{4} \cos 8x)$

17. $\frac{1}{2}\sqrt{2 - \sqrt{3}}$ **19.** $\sqrt{2} - 1$ **21.** $-\frac{1}{2}\sqrt{2 + \sqrt{3}}$

23. $\sqrt{2} - 1$ **25.** $\frac{1}{2}\sqrt{2 + \sqrt{3}}$ **27.** $-\frac{1}{2}\sqrt{2 - \sqrt{2}}$

29. (a) $\sin 36°$ **(b)** $\sin 6\theta$ **31. (a)** $\cos 68°$ **(b)** $\cos 10\theta$

33. (a) $\tan 4°$ **(b)** $\tan 2\theta$ **37.** $\sqrt{10}/10, 3\sqrt{10}/10, \frac{1}{3}$

39. $\sqrt{(3 + 2\sqrt{2})/6}, \sqrt{(3 - 2\sqrt{2})/6}, 3 + 2\sqrt{2}$

41. $\sqrt{6}/6, -\sqrt{30}/6, -\sqrt{5}/5$

43. $\dfrac{2x}{1 + x^2}$ **45.** $\sqrt{\dfrac{1 - x}{2}}$ **47.** $\frac{336}{625}$ **49.** $\frac{8}{7}$ **51.** $\frac{7}{25}$

53. $-8\sqrt{3}/49$ **55.** $\frac{1}{2}(\sin 5x - \sin x)$ **57.** $\frac{1}{2}(\sin 5x + \sin 3x)$

59. $\frac{3}{2}(\cos 11x + \cos 3x)$ **61.** $2 \sin 4x \cos x$

63. $2 \sin 5x \sin x$ **65.** $-2 \cos \frac{9}{2}x \sin \frac{5}{2}x$ **67.** $(\sqrt{2} + \sqrt{3})/2$

69. $\frac{1}{4}(\sqrt{2} - 1)$ **71.** $\sqrt{2}/2$ **73.** LHS $= \cos(2 \cdot 5x) = \text{RHS}$

75. LHS $= \sin^2 x + 2 \sin x \cos x + \cos^2 x$
$= 1 + 2 \sin x \cos x = \text{RHS}$

77. LHS $= \dfrac{2 \sin 2x \cos 2x}{\sin x} = \dfrac{2(2 \sin x \cos x)(\cos 2x)}{\sin x} = \text{RHS}$

79. LHS $= \dfrac{2(\tan x - \cot x)}{(\tan x + \cot x)(\tan x - \cot x)} = \dfrac{2}{\tan x + \cot x}$
$= \dfrac{2}{\frac{\sin x}{\cos x} + \frac{\cos x}{\sin x}} \cdot \dfrac{\sin x \cos x}{\sin x \cos x} = \dfrac{2 \sin x \cos x}{\sin^2 x + \cos^2 x}$
$= 2 \sin x \cos x = \text{RHS}$

81. LHS $= \tan(2x + x) = \dfrac{\tan 2x + \tan x}{1 - \tan 2x \tan x}$
$= \dfrac{\frac{2 \tan x}{1 - \tan^2 x} + \tan x}{1 - \frac{2 \tan x}{1 - \tan^2 x} \tan x}$
$= \dfrac{2 \tan x + \tan x(1 - \tan^2 x)}{1 - \tan^2 x - 2 \tan x \tan x} = \text{RHS}$

83. LHS $= (\cos^2 x + \sin^2 x)(\cos^2 x - \sin^2 x)$
$= \cos^2 x - \sin^2 x = \text{RHS}$

85. LHS $= \dfrac{2 \sin 3x \cos 2x}{2 \cos 3x \cos 2x} = \dfrac{\sin 3x}{\cos 3x} = \text{RHS}$

87. LHS $= \dfrac{2 \sin 5x \cos 5x}{2 \sin 5x \cos 4x} = \text{RHS}$

89. LHS $= \dfrac{2 \sin\left(\frac{x + y}{2}\right) \cos\left(\frac{x - y}{2}\right)}{2 \cos\left(\frac{x + y}{2}\right) \cos\left(\frac{x - y}{2}\right)}$
$= \dfrac{\sin\left(\frac{x + y}{2}\right)}{\cos\left(\frac{x + y}{2}\right)} = \text{RHS}$

95. LHS $= \dfrac{(\sin x + \sin 5x) + (\sin 2x + \sin 4x) + \sin 3x}{(\cos x + \cos 5x) + (\cos 2x + \cos 4x) + \cos 3x}$
$= \dfrac{2 \sin 3x \cos 2x + 2 \sin 3x \cos x + \sin 3x}{2 \cos 3x \cos 2x + 2 \cos 3x \cos x + \cos 3x}$
$= \dfrac{\sin 3x(2 \cos 2x + 2 \cos x + 1)}{\cos 3x(2 \cos 2x + 2 \cos x + 1)} = \text{RHS}$

97. (a)

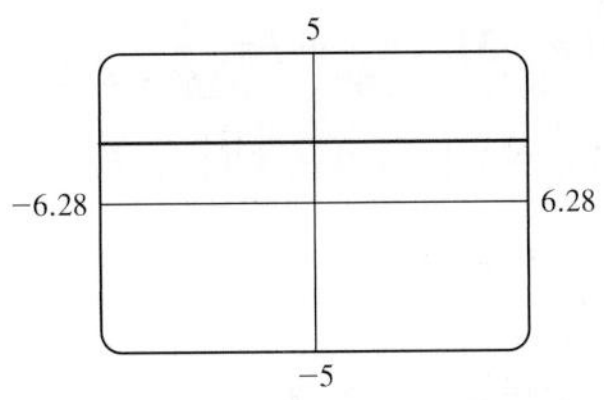

$\dfrac{\sin 3x}{\sin x} - \dfrac{\cos 3x}{\cos x} = 2$

99. (a)

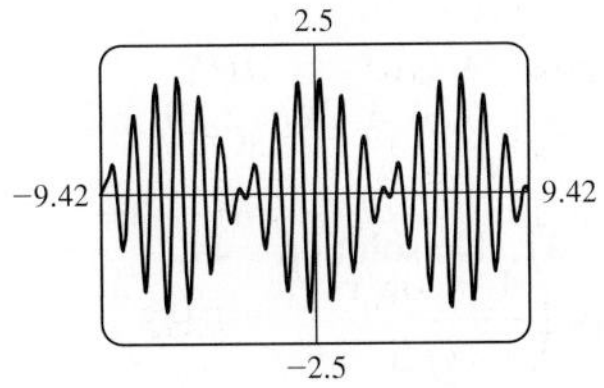

(c)

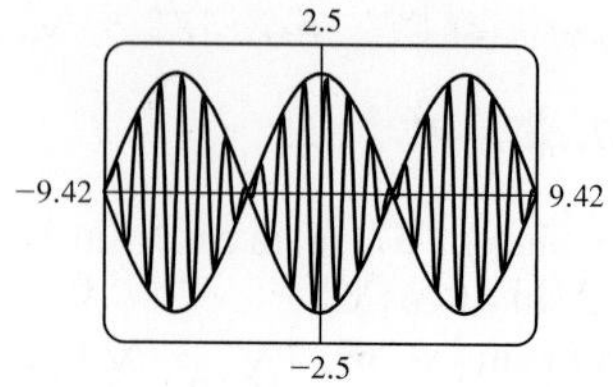

The graph of $y = f(x)$ lies between the two other graphs.

101. (a) $P(t) = 8t^4 - 8t^2 + 1$ **(b)** $Q(t) = 16t^5 - 20t^3 + 5t$

107. (a) and (c)

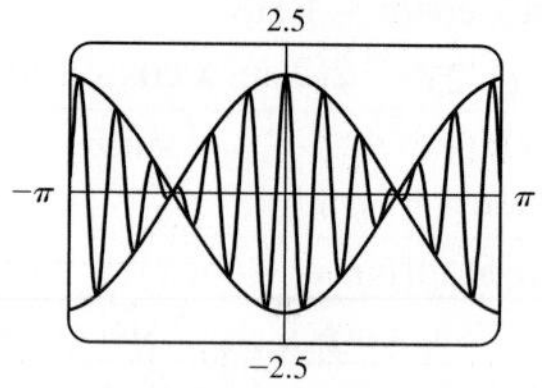

The graph of f lies between the graphs of $y = 2\cos t$ and $y = -2\cos t$. Thus, the loudness of the sound varies between $y = \pm 2\cos t$.

SECTION 7.4 ■ PAGE 522

1. infinitely many **2.** no, infinitely many
3. 0.3; $x \approx -9.7, -6.0, -3.4, 0.3, 2.8, 6.6, 9.1$
4. (a) 0.30, 2.84 **(b)** $2\pi, 0.30 + 2k\pi, 2.84 + 2k\pi$
5. $\frac{\pi}{3} + 2k\pi, \frac{2\pi}{3} + 2k\pi$
7. $(2k + 1)\pi$ **9.** $1.32 + 2k\pi, 4.97 + 2k\pi$
11. $-0.47 + 2k\pi, 3.61 + 2k\pi$
13. $-\frac{\pi}{3} + k\pi$ **15.** $1.37 + k\pi$
17. $\frac{5\pi}{6} + 2k\pi, \frac{7\pi}{6} + 2k\pi$; $-7\pi/6, -5\pi/6, 5\pi/6, 7\pi/6, 17\pi/6, 19\pi/6$
19. $\frac{\pi}{4} + 2k\pi, \frac{3\pi}{4} + 2k\pi$; $-7\pi/4, -5\pi/4, \pi/4, 3\pi/4, 9\pi/4, 11\pi/4$
21. $1.29 + 2k\pi, 5.00 + 2k\pi$; $-5.00, -1.29, 1.29, 5.00, 7.57, 11.28$
23. $-1.47 + k\pi$; $-7.75, -4.61, -1.47, 1.67, 4.81, 7.95$
25. $(2k + 1)\pi$ **27.** $-\frac{\pi}{4} + 2k\pi, \frac{5\pi}{4} + 2k\pi$
29. $0.20 + 2k\pi, 2.94 + 2k\pi$ **31.** $-\frac{\pi}{6} + k\pi, \frac{\pi}{6} + k\pi$
33. $\frac{\pi}{4} + k\pi, \frac{3\pi}{4} + k\pi$ **35.** $-1.11 + k\pi, 1.11 + k\pi$
37. $\frac{\pi}{4} + k\pi, \frac{3\pi}{4} + k\pi$
39. $-1.11 + k\pi, 1.11 + k\pi, \frac{2\pi}{3} + 2k\pi, \frac{4\pi}{3} + 2k\pi$
41. $\frac{\pi}{3} + 2k\pi, \frac{5\pi}{3} + 2k\pi$ **43.** $0.34 + 2k\pi, 2.80 + 2k\pi$
45. $\frac{\pi}{3} + 2k\pi, \frac{5\pi}{3} + 2k\pi$ **47.** No solution **49.** $\frac{3\pi}{2} + 2k\pi$
51. $\frac{\pi}{2} + k\pi, \frac{7\pi}{6} + 2k\pi, \frac{11\pi}{6} + 2k\pi$
53. $\frac{\pi}{2} + k\pi$ **55.** $k\pi, 0.73 + 2k\pi, 2.41 + 2k\pi$ **57.** 44.95°
59. (a) 0° **(b)** 60°, 120° **(c)** 90°, 270° **(d)** 180°

SECTION 7.5 ■ PAGE 528

1. $\sin x = 0, k\pi$ **2.** $\sin x + 2\sin x\cos x = 0$, $\sin x = 0, 1 + 2\cos x = 0$ **3.** $-\frac{\pi}{6} + 2k\pi, \frac{7\pi}{6} + 2k\pi, \frac{\pi}{2} + 2k\pi$
5. $(2k + 1)\pi, 1.23 + 2k\pi, 5.05 + 2k\pi$
7. $k\pi, 0.72 + 2k\pi, 5.56 + 2k\pi$ **9.** $\frac{\pi}{6} + 2k\pi, \frac{5\pi}{6} + 2k\pi$
11. $\frac{\pi}{3} + 2k\pi, \frac{5\pi}{3} + 2k\pi, (2k + 1)\pi$ **13.** $(2k + 1)\pi, \frac{\pi}{2} + 2k\pi$
15. $2k\pi$ **17. (a)** $\frac{\pi}{9} + \frac{2k\pi}{3}, \frac{5\pi}{9} + \frac{2k\pi}{3}$
(b) $\pi/9, 5\pi/9, 7\pi/9, 11\pi/9, 13\pi/9, 17\pi/9$
19. (a) $\frac{\pi}{3} + k\pi, \frac{2\pi}{3} + k\pi$ **(b)** $\pi/3, 2\pi/3, 4\pi/3, 5\pi/3$
21. (a) $\frac{5\pi}{18} + \frac{k\pi}{3}$ **(b)** $5\pi/18, 11\pi/18, 17\pi/18, 23\pi/18, 29\pi/18, 35\pi/18$ **23. (a)** $4k\pi$ **(b)** 0
25. (a) $4\pi + 6k\pi, 5\pi + 6k\pi$ **(b)** None
27. (a) $0.62 + \frac{k\pi}{2}$ **(b)** 0.62, 2.19, 3.76, 5.33
29. (a) $k\pi$ **(b)** $0, \pi$ **31. (a)** $\frac{\pi}{6} + k\pi, \frac{\pi}{4} + k\pi, \frac{5\pi}{6} + k\pi$
(b) $\pi/6, \pi/4, 5\pi/6, 7\pi/6, 5\pi/4, 11\pi/6$
33. (a) $\frac{\pi}{6} + 2k\pi, \frac{5\pi}{6} + 2k\pi, \frac{3\pi}{4} + k\pi$
(b) $\pi/6, 3\pi/4, 5\pi/6, 7\pi/4$

35. (a)

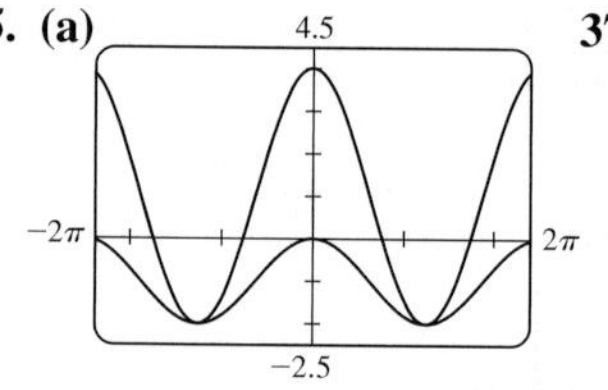

$(\pm 3.14, -2)$

(b) $((2k + 1)\pi, -2)$

37. (a)

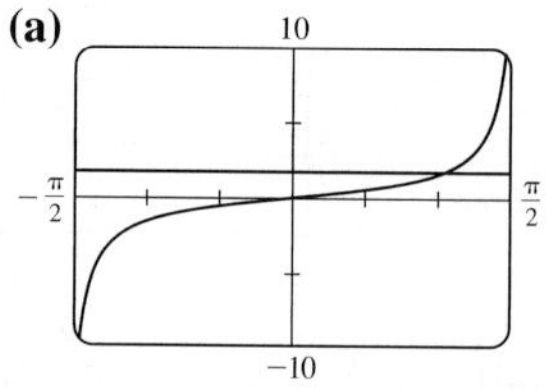

$(1.04, 1.73)$

(b) $\left(\frac{\pi}{3} + k\pi, \sqrt{3}\right)$

39. $\pi/8, 3\pi/8, 5\pi/8, 7\pi/8, 9\pi/8, 11\pi/8, 13\pi/8, 15\pi/8$
41. $\pi/3, 2\pi/3$ **43.** $\pi/2, 7\pi/6, 3\pi/2, 11\pi/6$ **45.** 0
47. $0, \pi$ **49.** $0, \pi/3, 2\pi/3, \pi, 4\pi/3, 5\pi/3$ **51.** $\pi/6, 3\pi/2$
53. $k\pi/2$ **55.** $\frac{\pi}{2} + k\pi, \frac{\pi}{9} + \frac{2k\pi}{3}, \frac{5\pi}{9} + \frac{2k\pi}{3}$
57. $0, \pm 0.95$ **59.** 1.92 **61.** ± 0.71
63. 0.94721° or 89.05279° **65. (a)** 34th day (February 3), 308th day (November 4) **(b)** 275 days

CHAPTER 7 REVIEW ■ PAGE 530

1. $\text{LHS} = \sin\theta\left(\frac{\cos\theta}{\sin\theta} + \frac{\sin\theta}{\cos\theta}\right) = \cos\theta + \frac{\sin^2\theta}{\cos\theta} = \frac{\cos^2\theta + \sin^2\theta}{\cos\theta} = \text{RHS}$

3. $\text{LHS} = (1 - \sin^2 x)\csc x - \csc x$
$= \csc x - \sin^2 x \csc x - \csc x$
$= -\sin^2 x \dfrac{1}{\sin x} = \text{RHS}$

5. $\text{LHS} = \dfrac{\cos^2 x}{\sin^2 x} - \dfrac{\tan^2 x}{\sin^2 x} = \cot^2 x - \dfrac{1}{\cos^2 x} = \text{RHS}$

7. $\text{LHS} = \dfrac{\cos x}{\frac{1}{\cos x}(1 - \sin x)} = \dfrac{\cos x}{\frac{1}{\cos x} - \frac{\sin x}{\cos x}} = \text{RHS}$

9. $\text{LHS} = \sin^2 x \dfrac{\cos^2 x}{\sin^2 x} + \cos^2 x \dfrac{\sin^2 x}{\cos^2 x} = \cos^2 x + \sin^2 x = \text{RHS}$

11. $\text{LHS} = \dfrac{2 \sin x \cos x}{1 + 2\cos^2 x - 1} = \dfrac{2 \sin x \cos x}{2\cos^2 x} = \dfrac{2 \sin x}{2 \cos x} = \text{RHS}$

13. $\text{LHS} = \dfrac{1 - \cos x}{\sin x} = \dfrac{1}{\sin x} - \dfrac{\cos x}{\sin x} = \text{RHS}$

15. $\text{LHS} = \frac{1}{2}[\cos((x + y) - (x - y))$
$- \cos((x + y) + (x - y))]$
$= \frac{1}{2}(\cos 2y - \cos 2x)$
$= \frac{1}{2}[1 - 2\sin^2 y - (1 - 2\sin^2 x)]$
$= \frac{1}{2}(2\sin^2 x - 2\sin^2 y) = \text{RHS}$

17. $\text{LHS} = 1 + \dfrac{\sin x}{\cos x} \cdot \dfrac{1 - \cos x}{\sin x} = 1 + \dfrac{1 - \cos x}{\cos x}$
$= 1 + \dfrac{1}{\cos x} - 1 = \text{RHS}$

19. $\text{LHS} = \cos^2 \frac{x}{2} - 2 \sin \frac{x}{2} \cos \frac{x}{2} + \sin^2 \frac{x}{2}$
$= 1 - \sin\left(2 \cdot \frac{x}{2}\right) = \text{RHS}$

21. $\text{LHS} = \dfrac{2 \sin x \cos x}{\sin x} - \dfrac{2\cos^2 x - 1}{\cos x}$
$= 2 \cos x - 2 \cos x + \dfrac{1}{\cos x} = \text{RHS}$

23. $\text{LHS} = \dfrac{\tan x + \tan \frac{\pi}{4}}{1 - \tan x \tan \frac{\pi}{4}} = \text{RHS}$

25. (a)

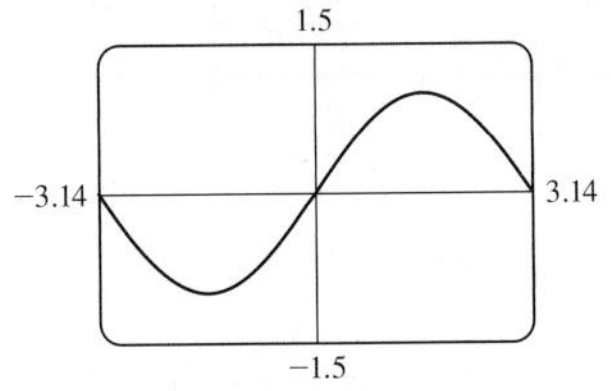

(b) Yes

27. (a)

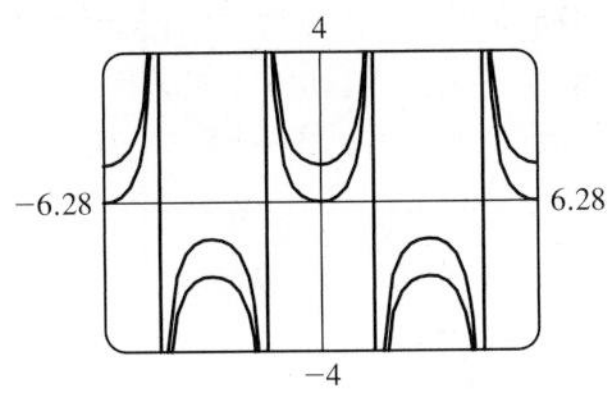

(b) No

29. (a)

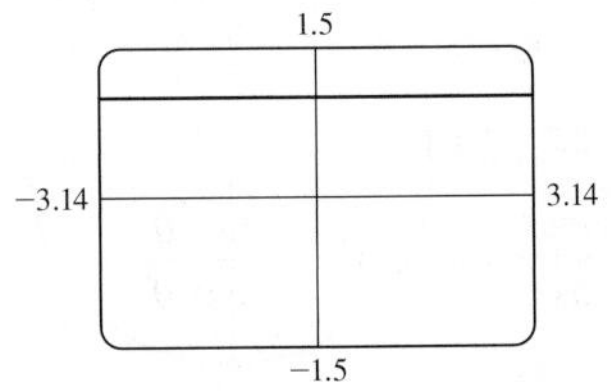

$2\sin^2 3x + \cos 6x = 1$

31. 0.85, 2.29 **33.** $0, \pi$ **35.** $\pi/6, 5\pi/6$ **37.** $\pi/3, 5\pi/3$
39. $2\pi/3, 4\pi/3$ **41.** $\pi/3, 2\pi/3, 3\pi/4, 4\pi/3, 5\pi/3, 7\pi/4$
43. $\pi/6, \pi/2, 5\pi/6, 7\pi/6, 3\pi/2, 11\pi/6$ **45.** $\pi/6$

47. 1.18 **49. (a)** 63.4° **(b)** No **(c)** 90° **51.** $\frac{1}{2}\sqrt{2 + \sqrt{3}}$
53. $\sqrt{2} - 1$ **55.** $\sqrt{2}/2$ **57.** $\sqrt{2}/2$ **59.** $\dfrac{\sqrt{2} + \sqrt{3}}{4}$

61. $2\dfrac{\sqrt{10} + 1}{9}$ **63.** $\frac{2}{3}(\sqrt{2} + \sqrt{5})$ **65.** $\sqrt{(3 + 2\sqrt{2})/6}$

67. $-\dfrac{12\sqrt{10}}{31}$ **69.** $\dfrac{2x}{1 - x^2}$ **71. (a)** $\theta = \tan^{-1}\left(\dfrac{10}{x}\right)$

(b) 286.4 ft

CHAPTER 7 TEST ■ PAGE 532

1. (a) $\text{LHS} = \dfrac{\sin\theta}{\cos\theta}\sin\theta + \cos\theta = \dfrac{\sin^2\theta + \cos^2\theta}{\cos\theta} = \text{RHS}$

(b) $\text{LHS} = \dfrac{\tan x}{1 - \cos x} \cdot \dfrac{1 + \cos x}{1 + \cos x} = \dfrac{\tan x(1 + \cos x)}{1 - \cos^2 x}$
$= \dfrac{\frac{\sin x}{\cos x}(1 + \cos x)}{\sin^2 x} = \dfrac{1}{\sin x} \cdot \dfrac{1 + \cos x}{\cos x} = \text{RHS}$

(c) $\text{LHS} = \dfrac{2 \tan x}{\sec^2 x} = \dfrac{2 \sin x}{\cos x} \cdot \cos^2 x = 2 \sin x \cos x = \text{RHS}$

2. $\tan\theta$ **3. (a)** $\frac{1}{2}$ **(b)** $\dfrac{\sqrt{2} + \sqrt{6}}{4}$ **(c)** $\frac{1}{2}\sqrt{2 - \sqrt{3}}$
4. $(10 - 2\sqrt{5})/15$
5. (a) $\frac{1}{2}(\sin 8x - \sin 2x)$ **(b)** $-2\cos \frac{7}{2}x \sin \frac{3}{2}x$ **6.** -2
7. (a) 0.34, 2.80 **(b)** $\pi/3, \pi/2, 5\pi/3$ **(c)** $2\pi/3, 4\pi/3$
(d) $\pi/6, \pi/2, 5\pi/6, 3\pi/2$ **8.** 0.58, 2.56, 3.72, 5.70 **9.** $\frac{1519}{1681}$
10. $\dfrac{\sqrt{1 - x^2} - xy}{\sqrt{1 + y^2}}$

FOCUS ON MODELING ■ PAGE 536

1. (a) $y = -5 \sin\left(\dfrac{\pi}{2}t\right)$

(b)

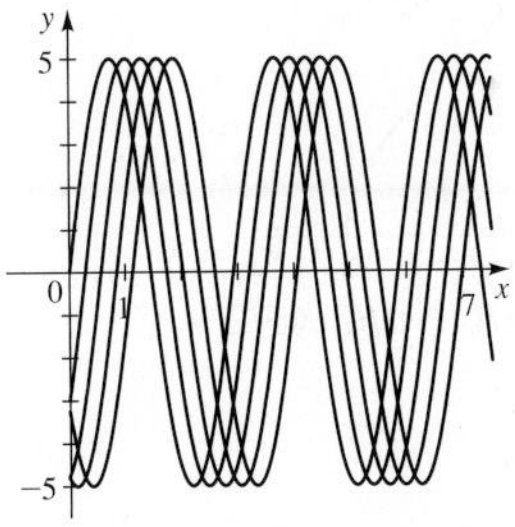

Yes, it is a traveling wave.
(c) $v = \pi/4$
3. $y(x, t) = 2.7 \sin(0.68x - 4.10t)$
5. $y(x, t) = 0.6 \sin(\pi x) \cos(40\pi t)$
7. (a) 1, 2, 3, 4
(b) 5:

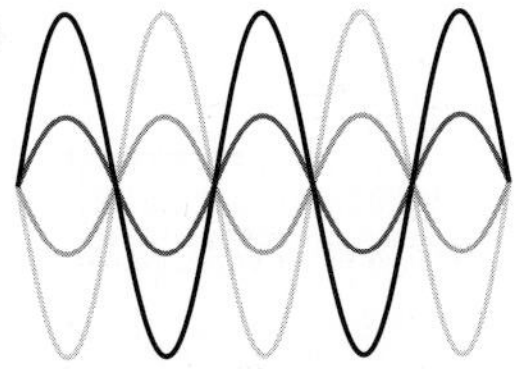

6:

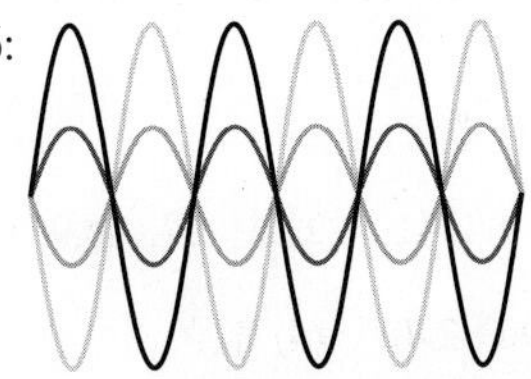

(c) 880π **(d)** $y(x, t) = \sin t \cos(880\pi t)$; $y(x, t) = \sin(2t)\cos(880\pi t)$; $y(x, t) = \sin(3t)\cos(880\pi t)$; $y(x, t) = \sin(4t)\cos(880\pi t)$

CUMULATIVE REVIEW TEST FOR CHAPTERS 5, 6 AND 7 ▪ PAGE 539

1. (a) $\sqrt{5}/3$ **(b)** $-2/3$ **(c)** $-\sqrt{5}/2$ **(d)** $3\sqrt{3}/5$
2. (a) $2\sqrt{10}/7$ **(b)** $7/3$ **(c)** $3\sqrt{10}/20$
3. (a) $-\sqrt{3}/2$ **(b)** -1 **(c)** $2\sqrt{3}/3$ **(d)** -1
4. $\sin t = -24/25$, $\tan t = -24/7$, $\cot t = -7/24$, $\sec t = 25/7$, $\csc t = -25/24$ **5. (a)** $2, \pi, \pi/4$ **(b)**

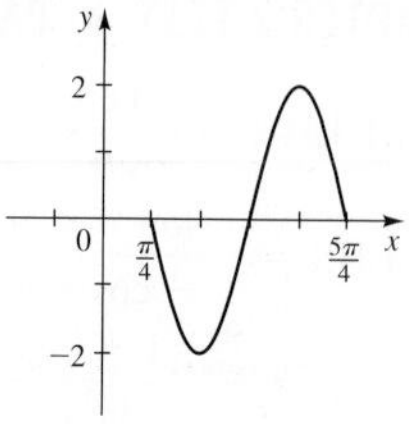

6. $y = 3\cos\frac{1}{2}(x - \frac{\pi}{3})$ **7. (a)** $h(t) = 45 - 40\cos 8\pi t$ **(b)** $2\sqrt{19} \approx 8.7$ cm **8. (a)** 7.2 **(b)** 92.9°

9. (a) $$\text{LHS} = \frac{(\sec\theta - 1)(\sec\theta + 1)}{\tan\theta\,(\sec\theta + 1)} = \frac{\sec^2\theta - 1}{\tan\theta\,(\sec\theta + 1)} = \frac{\tan^2\theta}{\tan\theta\,(\sec\theta + 1)} = \text{RHS}$$

(b) $\text{RHS} = 1 - (1 - 2\sin^2 2\theta) = 2\sin^2 2\theta = 2(2\sin\theta\cos\theta)^2 = \text{LHS}$

10. $2\cos\dfrac{7x}{2}\cos\dfrac{x}{2}$ **11. (a)** Domain $[-1, 1]$, range $[0, \pi]$

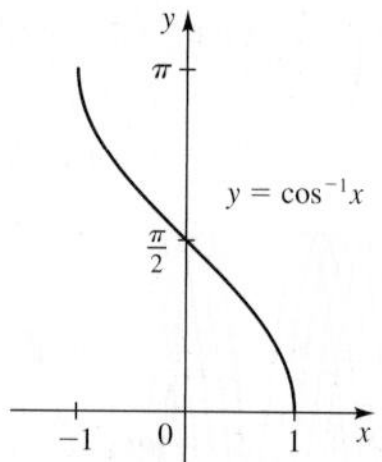

(b) $5\pi/6$ **(c)** $\sqrt{1 - x^2}/x$ **12.** $\pi/6, 5\pi/6, 3\pi/2$

CHAPTER 8

SECTION 8.1 ▪ PAGE 546

1. coordinate; $(1, 1), (\sqrt{2}, \pi/4)$ **2. (a)** $r\cos\theta, r\sin\theta$ **(b)** $x^2 + y^2, y/x$

3.

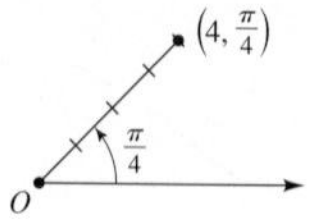

5.

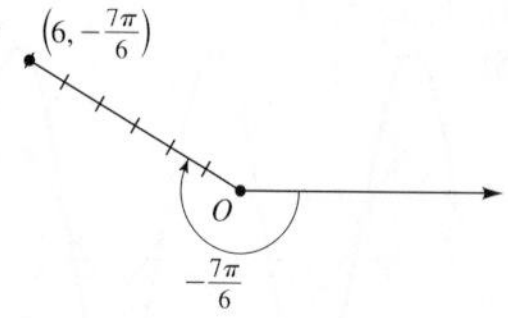

7.

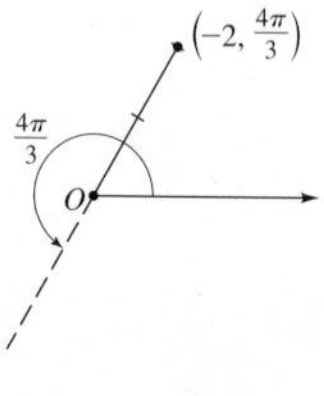

9.

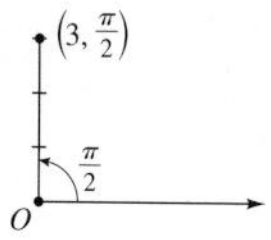

$\left(-3, \dfrac{3\pi}{2}\right), \left(3, \dfrac{5\pi}{2}\right)$

11.

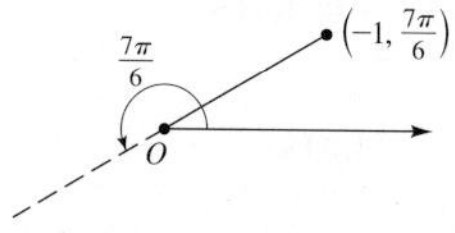

$\left(-1, -\dfrac{5\pi}{6}\right), \left(1, \dfrac{\pi}{6}\right)$

13.

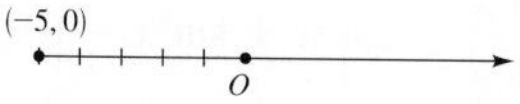

$(-5, 2\pi), (5, \pi)$

15. Q **17.** Q **19.** P **21.** P **23.** $(3\sqrt{2}, 3\pi/4)$ **25.** $\left(-\dfrac{5}{2}, -\dfrac{5\sqrt{3}}{2}\right)$ **27.** $(2\sqrt{3}, 2)$ **29.** $(1, -1)$ **31.** $(-5, 0)$
33. $(3\sqrt{6}, -3\sqrt{2})$ **35.** $(\sqrt{2}, 3\pi/4)$ **37.** $(4, \pi/4)$
39. $(5, \tan^{-1}\frac{4}{3})$ **41.** $(6, \pi)$ **43.** $\theta = \pi/4$ **45.** $r = \tan\theta\sec\theta$
47. $r = 4\sec\theta$ **49.** $x^2 + y^2 = 49$ **51.** $x = 0$ **53.** $x = 6$
55. $x^2 + (y - 2)^2 = 4$ **57.** $x^2 + y^2 = (x^2 + y^2 - x)^2$
59. $(x^2 + y^2 - 2y)^2 = x^2 + y^2$ **61.** $y - x = 1$
63. $x^2 - 3y^2 + 16y - 16 = 0$ **65.** $x^2 + y^2 = \dfrac{y}{x}$
67. $y = \pm\sqrt{3}x$

SECTION 8.2 ▪ PAGE 553

1. circles, rays **2. (a)** satisfy **(b)** circle, 3, pole; line, pole, 1
3. VI **5.** II **7.** I **9.** Symmetric about $\theta = \pi/2$
11. Symmetric about the polar axis
13. Symmetric about $\theta = \pi/2$
15. All three types of symmetry

17.

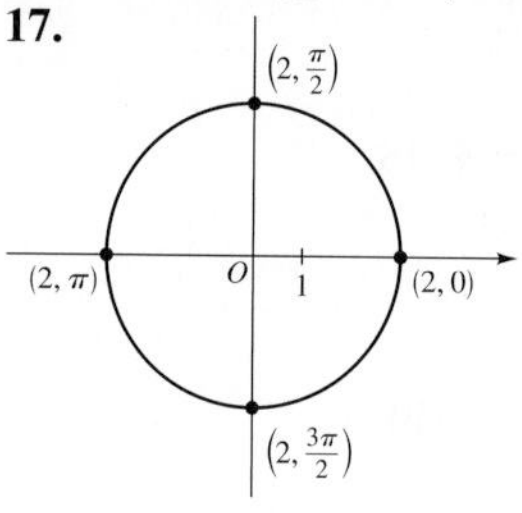

$x^2 + y^2 = 4$

19.

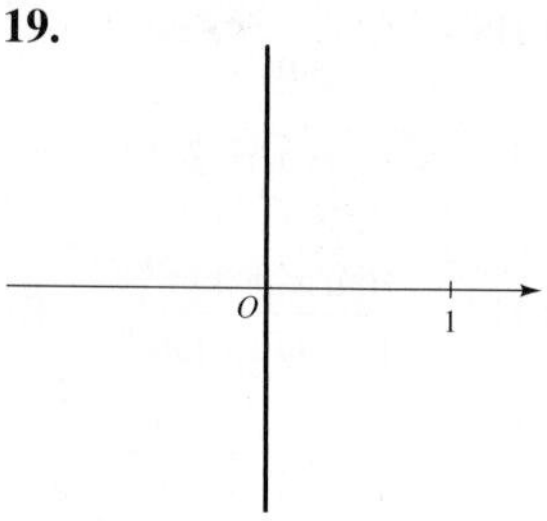

$x = 0$

21.

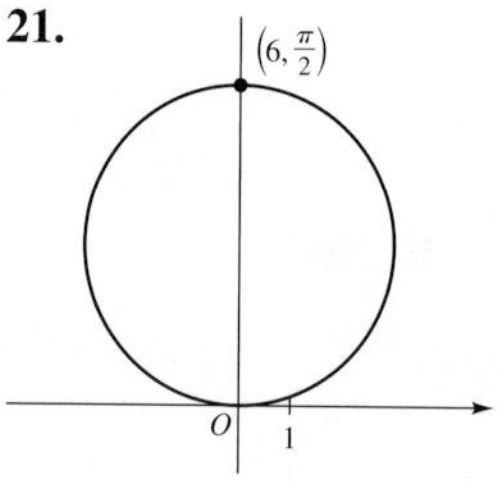

$x^2 + (y - 3)^2 = 9$

23.

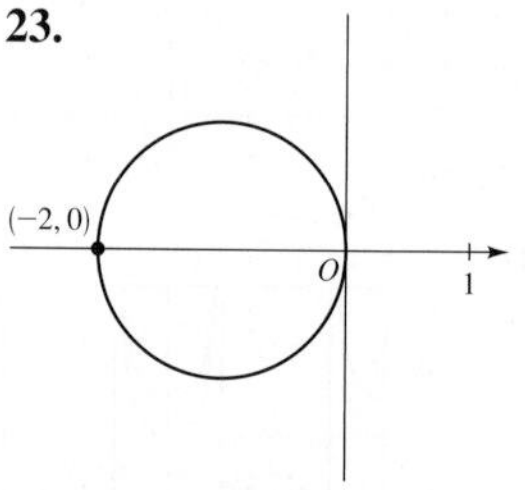

25.

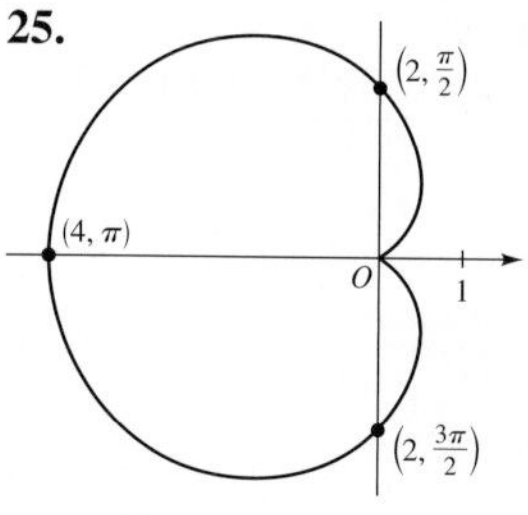

27.

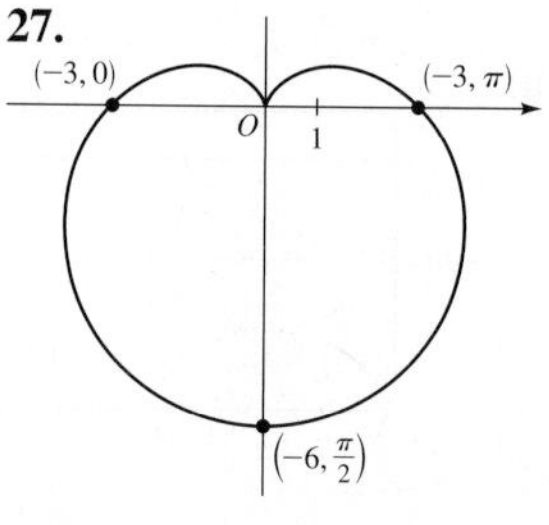

10. (a) $\langle 4, -3, 4 \rangle$ **(b)** $4x - 3y + 4z = 4$ **(c)** $\dfrac{\sqrt{41}}{2}$

11. $x = 2 - 2t, y = -4 + t, z = 7 - 2t$

CHAPTER 9 FOCUS ON MODELING ■ PAGE 626

1.

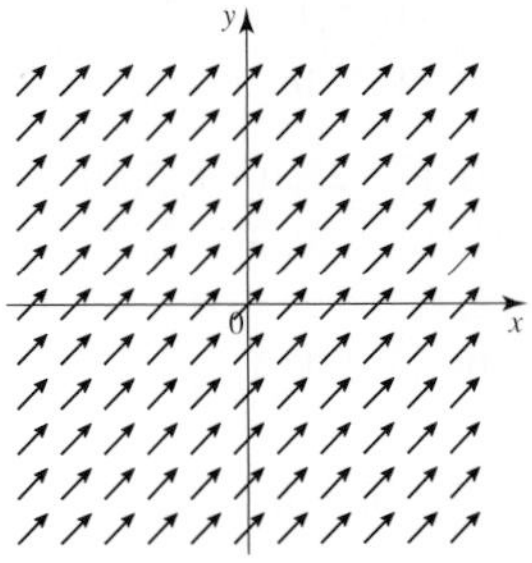

3.

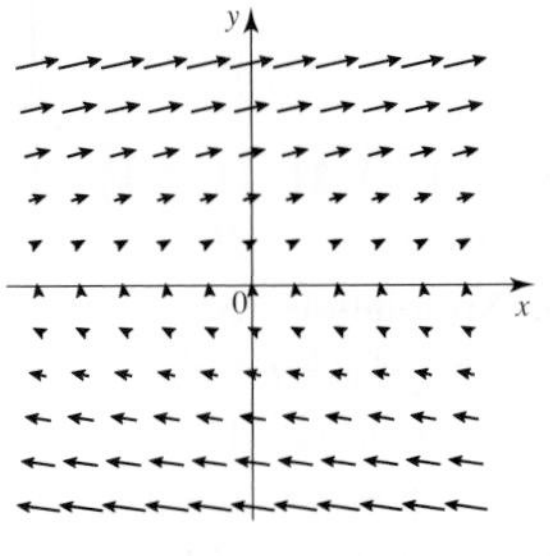

5.

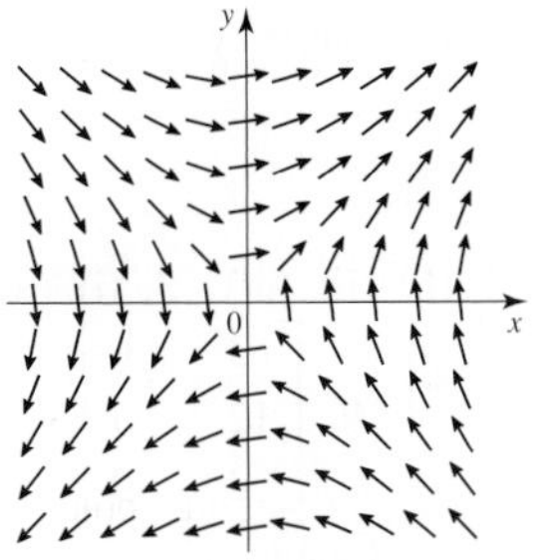

7.

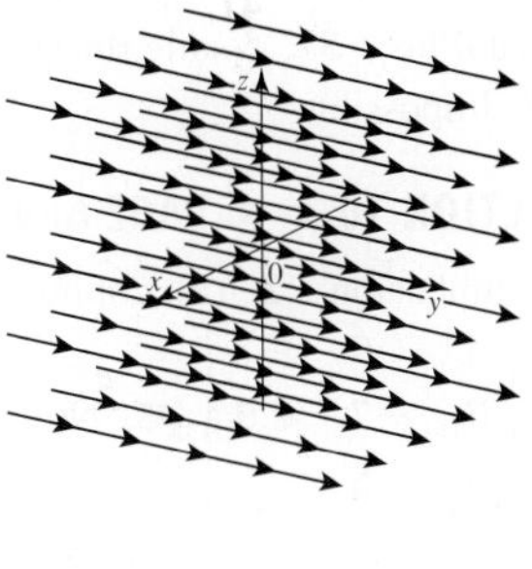

9.

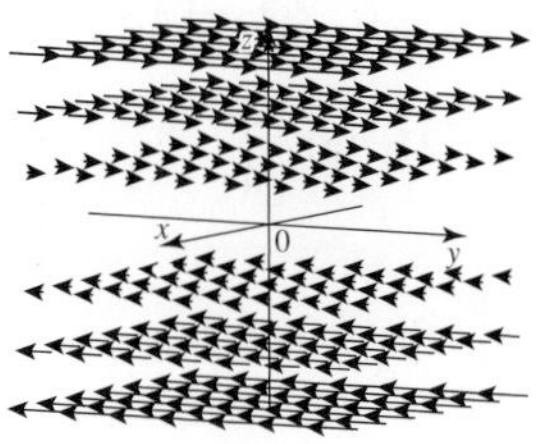

11. II **13.** I **15.** IV **17.** III

19.

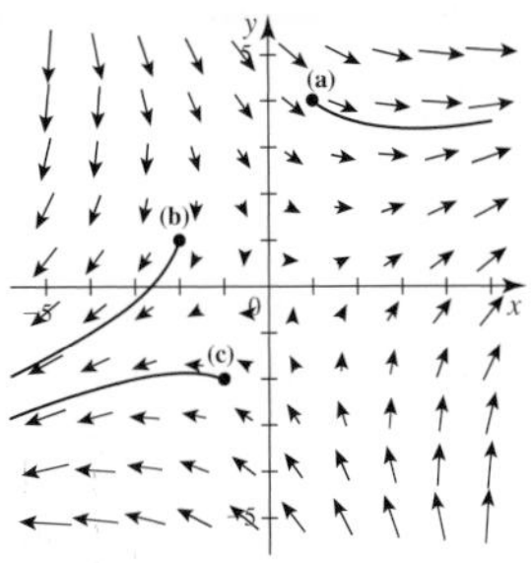

CUMULATIVE REVIEW TEST FOR CHAPTERS 8 AND 9 ■ PAGE 628

1. $(8\sqrt{2}, 7\pi/4), (-8\sqrt{2}, 3\pi/4)$

2. (a)

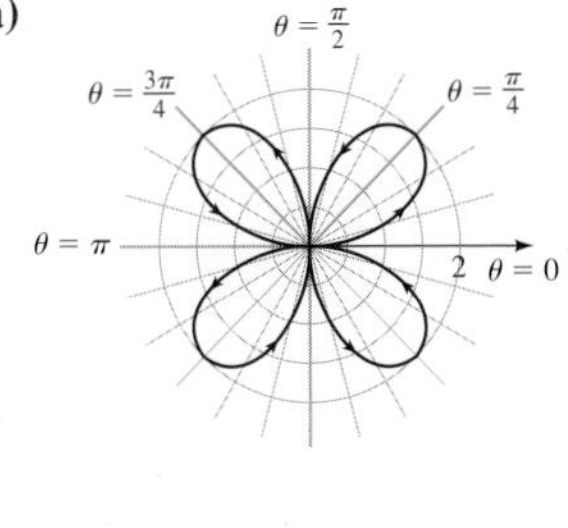

(b) $(x^2 + y^2)^{3/2} = 4xy$

3. (a) $z = 2\left(\cos\dfrac{11\pi}{6} + i\sin\dfrac{11\pi}{6}\right)$

(b) $zw = 12\left(\cos\dfrac{\pi}{4} + i\sin\dfrac{\pi}{4}\right) = 6\sqrt{2} + 6\sqrt{2}i$,

$z/w = \dfrac{1}{3}\left(\cos\dfrac{17\pi}{12} + i\sin\dfrac{17\pi}{12}\right)$

(c) $z^{10} = 1024\left(\cos\dfrac{\pi}{3} + i\sin\dfrac{\pi}{3}\right) = 512 + 512\sqrt{3}i$

(d) $\sqrt[3]{2}\left(\cos\dfrac{11\pi}{18} + i\sin\dfrac{11\pi}{18}\right)$, $\sqrt[3]{2}\left(\cos\dfrac{23\pi}{18} + i\sin\dfrac{23\pi}{18}\right)$, $\sqrt[3]{2}\left(\cos\dfrac{35\pi}{18} + i\sin\dfrac{35\pi}{18}\right)$

4. (a)

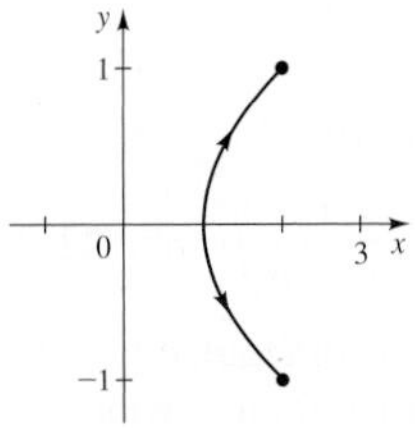

(b) $x = y^2 + 1$, parabola

5. (a)

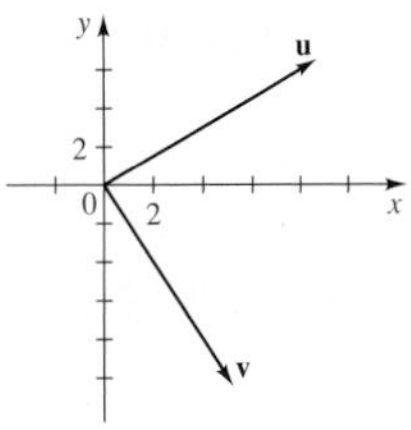

(b) $\mathbf{u} + \mathbf{v} = \langle 13, -4 \rangle$, $2\mathbf{u} - \mathbf{v} = \langle 11, 22 \rangle$, $\theta \approx 100.3°$, $\text{proj}_{\mathbf{v}}\,\mathbf{u} = \langle -\frac{4}{5}, \frac{8}{5} \rangle$ **(c)** 82

6. (a) 3 **(b)** $(x - 1)^2 + (y + 1)^2 + (z - 3)^2 = 9$
(c) $x = 1 + 2t, \quad y = -1 - t, \quad z = 3 - 2t$
7. (a) $\mathbf{a} \cdot \mathbf{b} = 0$, $\mathbf{a} \times \mathbf{b} = \langle 2, -13, -3 \rangle$, perpendicular
(b) $2x - 13y - 3z = 21$

CHAPTER 10

SECTION 10.1 ■ PAGE 638

1. x, y; equation; (2, 1) **2.** substitution, elimination, graphical **3.** no, infinitely many **4.** infinitely many; $1 - t$; (1, 0), (−3, 4), (5, −4) **5.** (3, 2) **7.** (3, 1) **9.** (2, 1) **11.** (1, 2) **13.** (−2, 3)
15. (2, −2) **17.** No solution

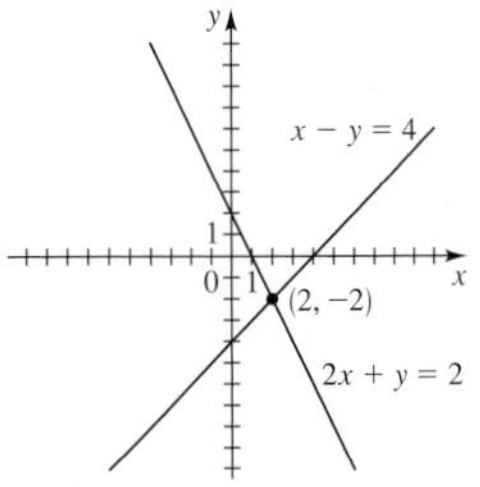

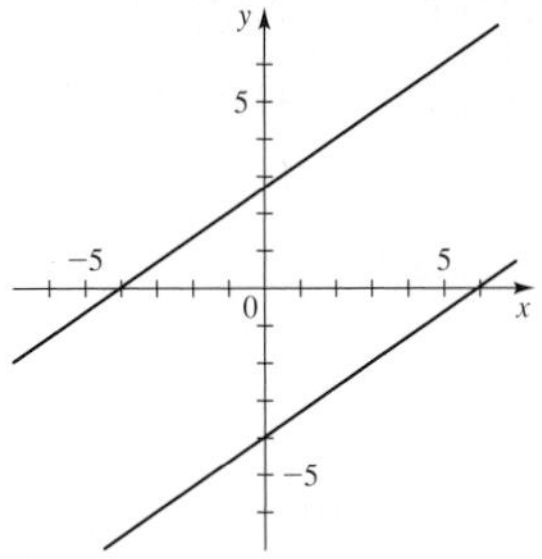

19. Infinitely many solutions

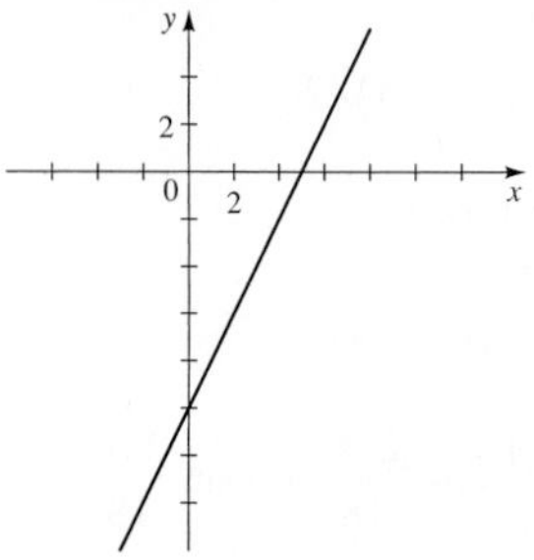

21. (2, 2) **23.** (3, −1) **25.** (2, 1) **27.** (3, 5) **29.** (1, 3) **31.** (10, −9) **33.** (2, 1) **35.** No solution **37.** $\left(x, \frac{1}{3}x - \frac{5}{3}\right)$ **39.** $\left(x, 3 - \frac{3}{2}x\right)$ **41.** (−3, −7) **43.** $\left(x, 5 - \frac{5}{6}x\right)$ **45.** (5, 10) **47.** No solution **49.** (3.87, 2.74) **51.** (61.00, 20.00)

53. $\left(-\dfrac{1}{a-1}, \dfrac{1}{a-1}\right)$ **55.** $\left(\dfrac{1}{a+b}, \dfrac{1}{a+b}\right)$ **57.** 22, 12

59. 5 dimes, 9 quarters **61.** 200 gallons of regular gas, 80 gallons of premium gas **63.** Plane's speed 120 mi/h, wind speed 30 mi/h **65.** 200 g of A, 40 g of B **67.** 25%, 10% **69.** $14,000 at 5%, $6,000 at 8% **71.** John $2\frac{1}{4}$ h, Mary $2\frac{1}{2}$ h **73.** 25

SECTION 10.2 ■ PAGE 646

1. $x + 3z = 1$ **2.** −3; $4y - 5z = -4$ **3.** Linear **5.** Nonlinear **7.** (1, 3, 2) **9.** (4, 0, 3) **11.** $\left(5, 2, -\frac{1}{2}\right)$

13. $\begin{cases} x - 2y - z = 4 \\ -y - 4z = 4 \\ 2x + y + z = 0 \end{cases}$ **15.** $\begin{cases} 2x - y + 3z = 2 \\ x + 2y - z = 4 \\ 3y + 7z = 14 \end{cases}$

17. (2, 1, −3) **19.** (1, 2, 1) **21.** (5, 0, 1) **23.** (0, 1, 2) **25.** $(1 - 3t, 2t, t)$ **27.** No solution **29.** No solution **31.** $(3 - t, -3 + 2t, t)$ **33.** $\left(2 - 2t, -\frac{2}{3} + \frac{4}{3}t, t\right)$ **35.** (1, −1, 1, 2) **37.** $30,000 in short-term bonds, $30,000 in intermediate-term bonds, $40,000 in long-term bonds **39.** 250 acres corn, 500 acres wheat, 450 acres soybeans **41.** Impossible **43.** 50 Midnight Mango, 60 Tropical Torrent, 30 Pineapple Power **45.** 1500 shares of A, 1200 shares of B, 1000 shares of C

SECTION 10.3 ■ PAGE 659

1. dependent, inconsistent

2. $\begin{bmatrix} 1 & 1 & -1 & 1 \\ 1 & 0 & 2 & -3 \\ 0 & 2 & -1 & 3 \end{bmatrix}$

3. **(a)** x and y **(b)** dependent **(c)** $x = 3 + t, y = 5 - 2t, z = t$ **4.** **(a)** $x = 2, y = 1, z = 3$ **(b)** $x = 2 - t, y = 1 - t, z = t$ **(c)** No solution **5.** 3 × 2 **7.** 2 × 1 **9.** 1 × 3

11. **(a)** Yes **(b)** Yes **(c)** $\begin{cases} x = -3 \\ y = 5 \end{cases}$

13. **(a)** Yes **(b)** No **(c)** $\begin{cases} x + 2y + 8z = 0 \\ y + 3z = 2 \\ 0 = 0 \end{cases}$

15. **(a)** No **(b)** No **(c)** $\begin{cases} x = 0 \\ 0 = 0 \\ y + 5z = 1 \end{cases}$

17. **(a)** Yes **(b)** Yes **(c)** $\begin{cases} x + 3y - w = 0 \\ z + 2w = 0 \\ 0 = 1 \\ 0 = 0 \end{cases}$

19. (1, 1, 2) **21.** (1, 0, 1) **23.** (−1, 0, 1) **25.** (−1, 5, 0) **27.** (10, 3, −2) **29.** No solution **31.** $(2 - 3t, 3 - 5t, t)$ **33.** No solution **35.** $(-2t + 5, t - 2, t)$ **37.** $x = -\frac{1}{2}s + t + 6, y = s, z = t$ **39.** (−2, 1, 3) **41.** No solution **43.** (−9, 2, 0) **45.** $x = 5 - t, y = -3 + 5t, z = t$ **47.** (0, −3, 0, −3) **49.** (−1, 0, 0, 1) **51.** $x = \frac{1}{3}s - \frac{2}{3}t, y = \frac{1}{3}s + \frac{1}{3}t, z = s, w = t$ **53.** $\left(\frac{7}{4} - \frac{7}{4}t, -\frac{7}{4} + \frac{3}{4}t, \frac{9}{4} + \frac{3}{4}t, t\right)$ **55.** 2 VitaMax, 1 Vitron, 2 VitaPlus **57.** 5-mile run, 2-mile swim, 30-mile cycle **59.** Impossible

SECTION 10.4 ■ PAGE 669

1. dimension **2.** **(a)** columns, rows **(b)** (ii), (iii) **3.** (i), (ii)

4. $\begin{bmatrix} 4 & 9 & -7 \\ 7 & -7 & 0 \\ 4 & -5 & -5 \end{bmatrix}$ **5.** No **7.** $\begin{bmatrix} 1 & 3 \\ 1 & 5 \end{bmatrix}$ **9.** $\begin{bmatrix} 3 & 6 \\ 12 & -3 \\ 3 & 0 \end{bmatrix}$

11. Impossible **13.** $\begin{bmatrix} 5 & 2 & 1 \\ 7 & 10 & -7 \end{bmatrix}$ **15.** $\begin{bmatrix} -1 & -\frac{1}{2} \\ 1 & 2 \end{bmatrix}$

17. No solution **19.** $\begin{bmatrix} 0 & -5 \\ -25 & -20 \\ -10 & 10 \end{bmatrix}$ **21.** **(a)** $\begin{bmatrix} 5 & -2 & 5 \\ 1 & 1 & 0 \end{bmatrix}$

(b) Impossible **23.** **(a)** $\begin{bmatrix} 10 & -25 \\ 0 & 35 \end{bmatrix}$ **(b)** Impossible

25. **(a)** Impossible **(b)** $\begin{bmatrix} 14 & -14 \end{bmatrix}$

27. **(a)** $\begin{bmatrix} -4 & 7 \\ 14 & -7 \end{bmatrix}$ **(b)** $\begin{bmatrix} 6 & -8 \\ 4 & -17 \end{bmatrix}$

29. **(a)** $\begin{bmatrix} 5 & -3 & 10 \\ 6 & 1 & 0 \\ -5 & 2 & 2 \end{bmatrix}$ **(b)** $\begin{bmatrix} -1 \\ 8 \\ -1 \end{bmatrix}$

31. **(a)** $\begin{bmatrix} 4 & -45 \\ 0 & 49 \end{bmatrix}$ **(b)** $\begin{bmatrix} 8 & -335 \\ 0 & 343 \end{bmatrix}$

33. **(a)** $\begin{bmatrix} 13 \\ -7 \end{bmatrix}$ **(b)** Impossible

35. $x = 2, y = -1$ **37.** $x = 1, y = -2$

39. $\begin{bmatrix} 2 & -5 \\ 3 & 2 \end{bmatrix}\begin{bmatrix} x \\ y \end{bmatrix} = \begin{bmatrix} 7 \\ 4 \end{bmatrix}$

41. $\begin{bmatrix} 3 & 2 & -1 & 1 \\ 1 & 0 & -1 & 0 \\ 0 & 3 & 1 & -1 \end{bmatrix}\begin{bmatrix} x_1 \\ x_2 \\ x_3 \\ x_4 \end{bmatrix} = \begin{bmatrix} 0 \\ 5 \\ 4 \end{bmatrix}$

43. Only ACB is defined. $ACB = \begin{bmatrix} -3 & -21 & 27 & -6 \\ -2 & -14 & 18 & -4 \end{bmatrix}$

45. (a) $\begin{bmatrix} 4{,}690 & 1{,}690 & 13{,}210 \end{bmatrix}$ **(b)** Total revenue in Santa Monica, Long Beach, and Anaheim, respectively.
47. (a) $\begin{bmatrix} 105{,}000 & 58{,}000 \end{bmatrix}$ **(b)** The first entry is the total amount (in ounces) of tomato sauce produced, and the second entry is the total amount (in ounces) of tomato paste produced.
49.

(a) $\begin{bmatrix} 1 & 0 & 1 & 0 & 1 & 1 \\ 0 & 3 & 0 & 1 & 2 & 1 \\ 1 & 2 & 0 & 0 & 3 & 0 \\ 1 & 3 & 2 & 3 & 2 & 0 \\ 0 & 3 & 0 & 0 & 2 & 1 \\ 1 & 2 & 0 & 1 & 3 & 1 \end{bmatrix}$ **(b)** $\begin{bmatrix} 2 & 1 & 2 & 1 & 2 & 2 \\ 1 & 3 & 1 & 2 & 3 & 2 \\ 2 & 3 & 1 & 1 & 3 & 1 \\ 2 & 3 & 3 & 3 & 3 & 1 \\ 1 & 3 & 1 & 1 & 3 & 2 \\ 2 & 3 & 1 & 2 & 3 & 2 \end{bmatrix}$

(c) $\begin{bmatrix} 2 & 3 & 2 & 3 & 2 & 2 \\ 3 & 0 & 3 & 2 & 1 & 2 \\ 2 & 1 & 3 & 3 & 0 & 3 \\ 2 & 0 & 1 & 0 & 1 & 3 \\ 3 & 0 & 3 & 3 & 1 & 2 \\ 2 & 1 & 3 & 2 & 0 & 2 \end{bmatrix}$

(d) $\begin{bmatrix} 3 & 3 & 3 & 3 & 3 & 3 \\ 3 & 0 & 3 & 3 & 0 & 3 \\ 3 & 0 & 3 & 3 & 0 & 3 \\ 3 & 0 & 0 & 0 & 0 & 3 \\ 3 & 0 & 3 & 3 & 0 & 3 \\ 3 & 0 & 3 & 3 & 0 & 3 \end{bmatrix}$ 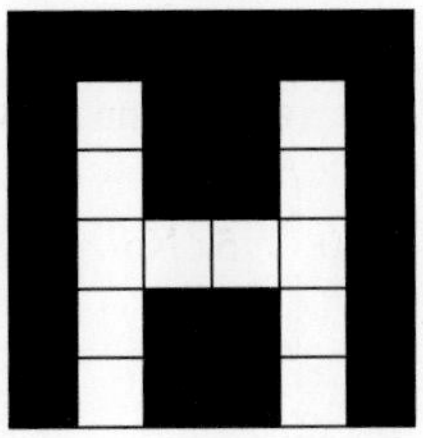

(e) The letter E

SECTION 10.5 ■ PAGE 680

1. (a) identity **(b)** A, A **(c)** inverse

2. (a) $\overset{A}{\begin{bmatrix} 5 & 3 \\ 3 & 2 \end{bmatrix}}\overset{X}{\begin{bmatrix} x \\ y \end{bmatrix}} = \overset{B}{\begin{bmatrix} 4 \\ 3 \end{bmatrix}}$ **(b)** $\begin{bmatrix} 2 & -3 \\ -3 & 5 \end{bmatrix}$

(c) $\overset{A^{-1}}{\begin{bmatrix} 2 & -3 \\ -3 & 5 \end{bmatrix}}\overset{B}{\begin{bmatrix} 4 \\ 3 \end{bmatrix}} = \overset{X}{\begin{bmatrix} x \\ y \end{bmatrix}}$ **(d)** $x = -1, y = 3$ **7.** $\begin{bmatrix} 1 & -2 \\ -\frac{3}{2} & \frac{7}{2} \end{bmatrix}$

9. $\begin{bmatrix} 3 & 5 \\ -2 & -3 \end{bmatrix}$ **11.** $\begin{bmatrix} 13 & 5 \\ -5 & -2 \end{bmatrix}$ **13.** No inverse

15. $\begin{bmatrix} 1 & 2 \\ -\frac{1}{2} & \frac{2}{3} \end{bmatrix}$ **17.** $\begin{bmatrix} -4 & -4 & 5 \\ 1 & 1 & -1 \\ 5 & 4 & -6 \end{bmatrix}$

19. No inverse **21.** $\begin{bmatrix} -\frac{9}{2} & -1 & 4 \\ 3 & 1 & -3 \\ \frac{7}{2} & 1 & -3 \end{bmatrix}$

23. $\begin{bmatrix} 0 & 0 & -2 & 1 \\ -1 & 0 & 1 & 1 \\ 0 & 1 & -1 & 0 \\ 1 & 0 & 0 & -1 \end{bmatrix}$ **25.** $x = 12, y = -8$

27. $x = 126, y = -50$ **29.** $x = -38, y = 9, z = 47$
31. $x = -20, y = 10, z = 16$ **33.** $x = 3, y = 2, z = 1$
35. $x = 3, y = -2, z = 2$ **37.** $x = 8, y = 1, z = 0, w = 3$

39. $\begin{bmatrix} 7 & 2 & 3 \\ 10 & 3 & 5 \end{bmatrix}$ **41.** $\frac{1}{2a}\begin{bmatrix} 1 & 1 \\ -1 & 1 \end{bmatrix}$

43. $\begin{bmatrix} 1 & -\dfrac{1}{x} \\ -\dfrac{1}{x} & \dfrac{2}{x^2} \end{bmatrix}$; inverse does not exist for $x = 0$

45. $\frac{1}{2}\begin{bmatrix} 1 & e^{-x} & 0 \\ e^{-x} & -e^{-2x} & 0 \\ 0 & 0 & 1 \end{bmatrix}$; inverse exists for all x

47. (a) $\begin{bmatrix} 0 & 1 & -1 \\ -2 & \frac{3}{2} & 0 \\ 1 & -\frac{3}{2} & 1 \end{bmatrix}$ **(b)** 1 oz A, 1 oz B, 2 oz C

(c) 2 oz A, 0 oz B, 1 oz C **(d)** No

49. (a) $\begin{cases} x + y + 2z = 675 \\ 2x + y + z = 600 \\ x + 2y + z = 625 \end{cases}$

(b) $\begin{bmatrix} 1 & 1 & 2 \\ 2 & 1 & 1 \\ 1 & 2 & 1 \end{bmatrix}\begin{bmatrix} x \\ y \\ z \end{bmatrix} = \begin{bmatrix} 675 \\ 600 \\ 625 \end{bmatrix}$ **(c)** $A^{-1} = \begin{bmatrix} -\frac{1}{4} & \frac{3}{4} & -\frac{1}{4} \\ -\frac{1}{4} & -\frac{1}{4} & \frac{3}{4} \\ \frac{3}{4} & -\frac{1}{4} & -\frac{1}{4} \end{bmatrix}$

She earns \$125 on a standard set, \$150 on a deluxe set, and \$200 on a leather-bound set.

SECTION 10.6 ■ PAGE 690

1. True **2.** True **3.** True **4. (a)** $2 \cdot 4 - (-3) \cdot 1 = 11$
(b)
$+1(2 \cdot 4 - (-3) \cdot 1) - 0(3 \cdot 4 - 0 \cdot 1) + 2(3 \cdot (-3) - 0 \cdot 2) = -7$
5. 6 **7.** -4 **9.** Does not exist **11.** $\frac{1}{8}$ **13.** 20, 20
15. $-12, 12$ **17.** 0, 0 **19.** 4, has an inverse
21. 5000, has an inverse **23.** 0, does not have an inverse
25. -4, has an inverse **27.** -18 **29.** 120 **31. (a)** -2
(b) -2 **(c)** Yes **33.** $(-2, 5)$ **35.** $(0.6, -0.4)$ **37.** $(4, -1)$
39. $(4, 2, -1)$ **41.** $(1, 3, 2)$ **43.** $(0, -1, 1)$ **45.** $\left(\frac{189}{29}, -\frac{108}{29}, \frac{88}{29}\right)$
47. $\left(\frac{1}{2}, \frac{1}{4}, \frac{1}{4}, -1\right)$ **49.** *abcde* **51.** 0, 1, 2 **53.** $1, -1$ **55.** 21
57. $\frac{63}{2}$

61. (a) $\begin{cases} 100a + 10b + c = 25 \\ 225a + 15b + c = 33\frac{3}{4} \\ 1600a + 40b + c = 40 \end{cases}$

(b) $y = -0.05x^2 + 3x$

SECTION 10.7 ■ PAGE 697

1. (iii) **2.** (ii) **3.** $\dfrac{A}{x-1} + \dfrac{B}{x+2}$

5. $\dfrac{A}{x-2} + \dfrac{B}{(x-2)^2} + \dfrac{C}{x+4}$ **7.** $\dfrac{A}{x-3} + \dfrac{Bx+C}{x^2+4}$

9. $\dfrac{Ax+B}{x^2+1} + \dfrac{Cx+D}{x^2+2}$

11. $\dfrac{A}{x} + \dfrac{B}{2x-5} + \dfrac{C}{(2x-5)^2} + \dfrac{D}{(2x-5)^3}$
$+ \dfrac{Ex+F}{x^2+2x+5} + \dfrac{Gx+H}{(x^2+2x+5)^2}$

13. $\dfrac{1}{x-1} - \dfrac{1}{x+1}$ **15.** $\dfrac{1}{x-1} - \dfrac{1}{x+4}$ **17.** $\dfrac{2}{x-3} - \dfrac{2}{x+3}$

19. $\dfrac{1}{x-2} - \dfrac{1}{x+2}$ **21.** $\dfrac{3}{x-4} - \dfrac{2}{x+2}$

23. $\dfrac{-\frac{1}{2}}{2x-1} + \dfrac{\frac{3}{2}}{4x-3}$ **25.** $\dfrac{2}{x-2} + \dfrac{3}{x+2} - \dfrac{1}{2x-1}$

27. $\dfrac{2}{x+1} - \dfrac{1}{x} + \dfrac{1}{x^2}$ **29.** $\dfrac{1}{2x+3} - \dfrac{3}{(2x+3)^2}$

31. $\dfrac{2}{x} - \dfrac{1}{x^3} - \dfrac{2}{x+2}$

33. $\dfrac{4}{x+2} - \dfrac{4}{x-1} + \dfrac{2}{(x-1)^2} + \dfrac{1}{(x-1)^3}$

35. $\dfrac{3}{x+2} - \dfrac{1}{(x+2)^2} - \dfrac{1}{(x+3)^2}$

37. $\dfrac{x+1}{x^2+3} - \dfrac{1}{x}$ **39.** $\dfrac{2x-5}{x^2+x+2} + \dfrac{5}{x^2+1}$

41. $\dfrac{1}{x^2+1} - \dfrac{x+2}{(x^2+1)^2} + \dfrac{1}{x}$ **43.** $x^2 + \dfrac{3}{x-2} - \dfrac{x+1}{x^2+1}$

45. $A = \dfrac{a+b}{2}, B = \dfrac{a-b}{2}$

SECTION 10.8 ■ PAGE 701

1. $(4, 8), (-2, 2)$ **3.** $(4, 16), (-3, 9)$
5. $(2, -2), (-2, 2)$ **7.** $(-25, 5), (-25, -5)$
9. $(-3, 4)\ (3, 4)$ **11.** $(-2, -1), (-2, 1), (2, -1), (2, 1)$
13. $(-1, \sqrt{2}), (-1, -\sqrt{2}), (\frac{1}{2}, \sqrt{\frac{7}{2}}), (\frac{1}{2}, -\sqrt{\frac{7}{2}})$
15. $(2, 4), (-\frac{5}{2}, \frac{7}{4})$ **17.** $(0, 0), (1, -1), (-2, -4)$
19. $(4, 0)$ **21.** $(-2, -2)$ **23.** $(6, 2), (-2, -6)$
25. No solution
27. $(\sqrt{5}, 2), (\sqrt{5}, -2), (-\sqrt{5}, 2), (-\sqrt{5}, -2)$
29. $(3, -\frac{1}{2}), (-3, -\frac{1}{2})$ **31.** $(\frac{1}{5}, \frac{1}{3})$
33. $(2.00, 20.00), (-8.00, 0)$ **35.** $(-4.51, 2.17), (4.91, -0.97)$
37. $(1.23, 3.87), (-0.35, -4.21)$
39. $(-2.30, -0.70), (0.48, -1.19)$ **41.** 12 cm by 15 cm
43. 15, 20 **45.** $(400.50, 200.25)$, 447.77 m **47.** $(12, 8)$

SECTION 10.9 ■ PAGE 708

1. equation; $y = x + 1$; test

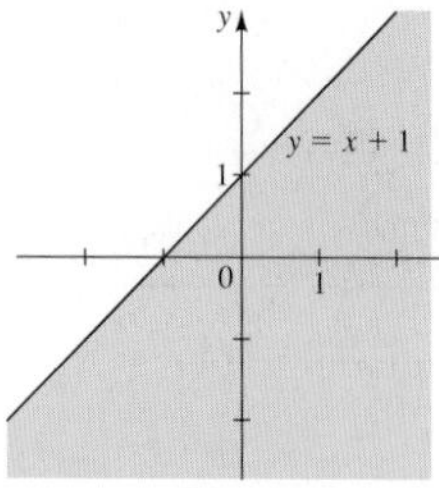

2. (a)

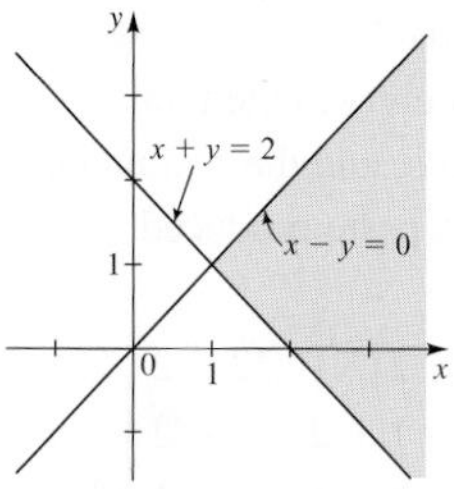

(b)

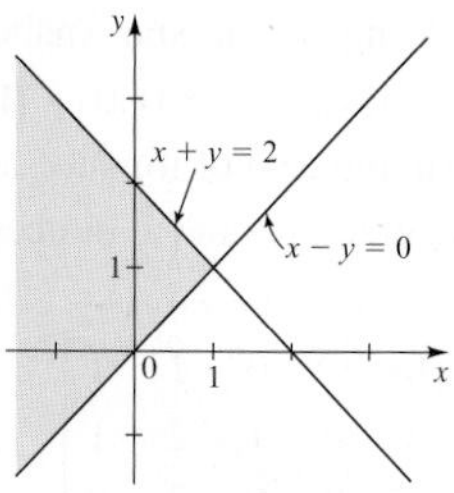

(c)

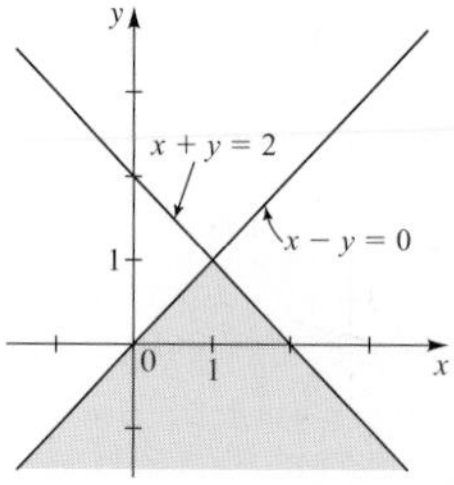

(d)

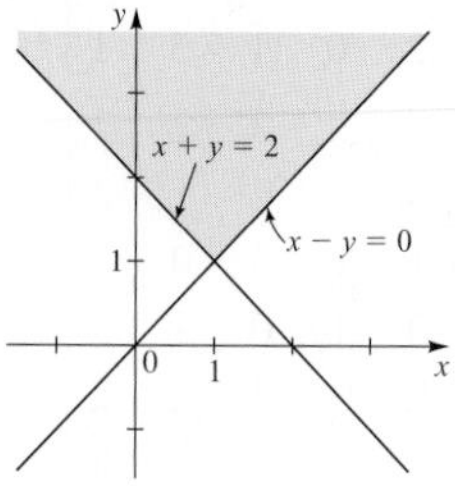

3.

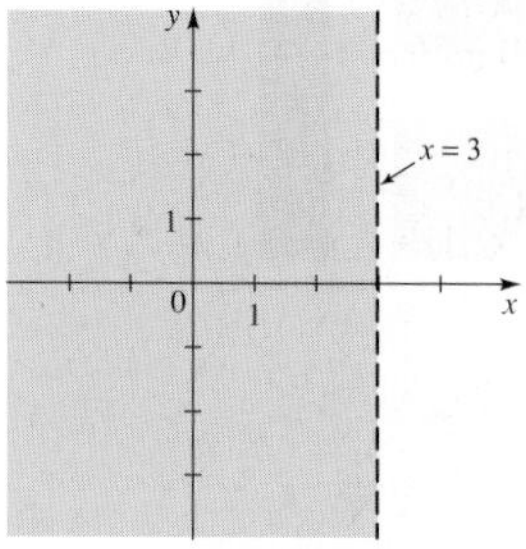

5.

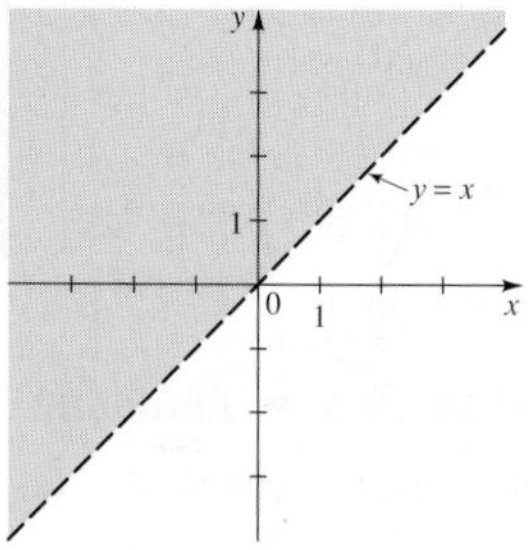

7.

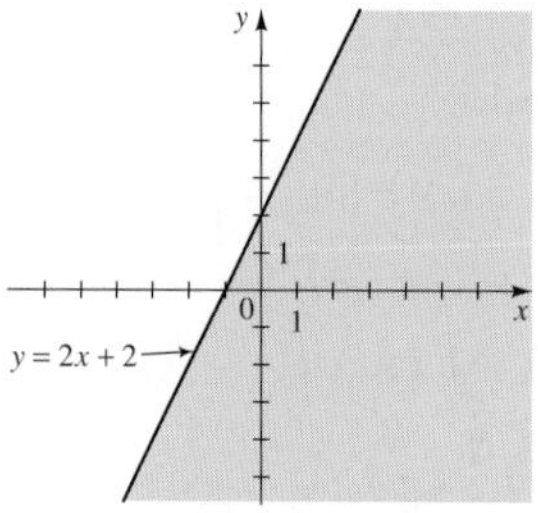

9.

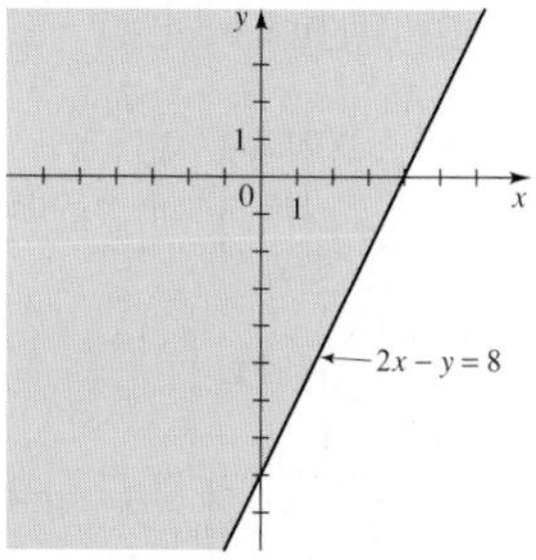

11.

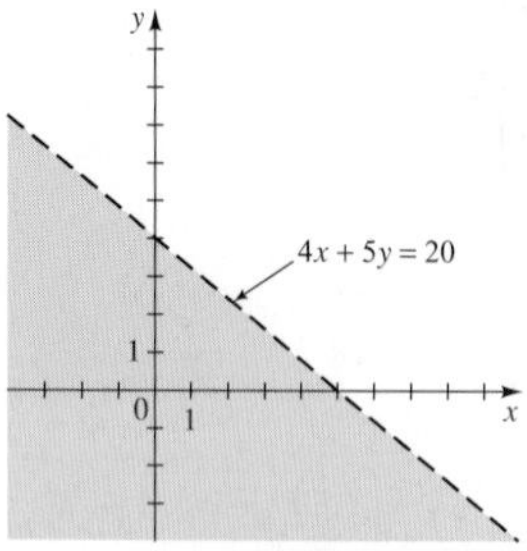

13.

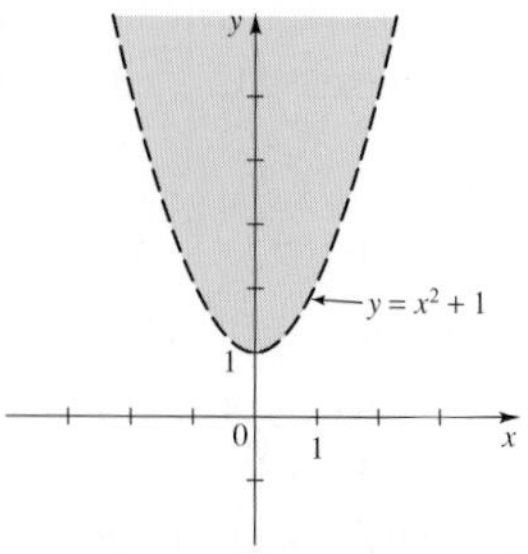

15.

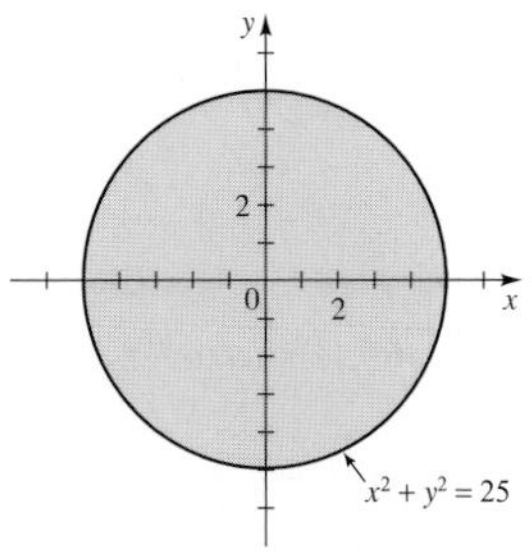

17. $y \leq \frac{1}{2}x - 1$ **19.** $x^2 + y^2 > 4$

21.

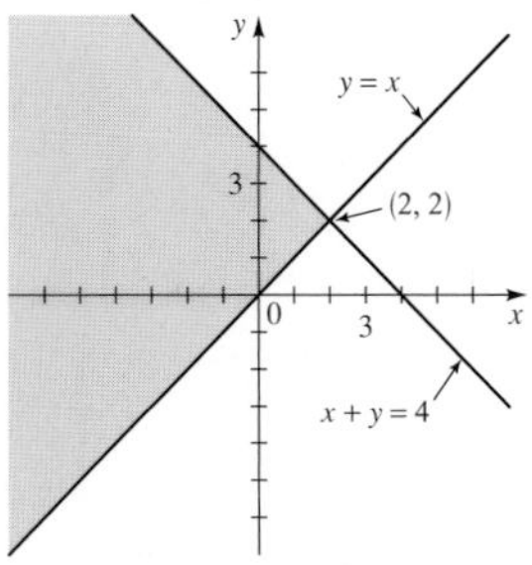

Not bounded

23.

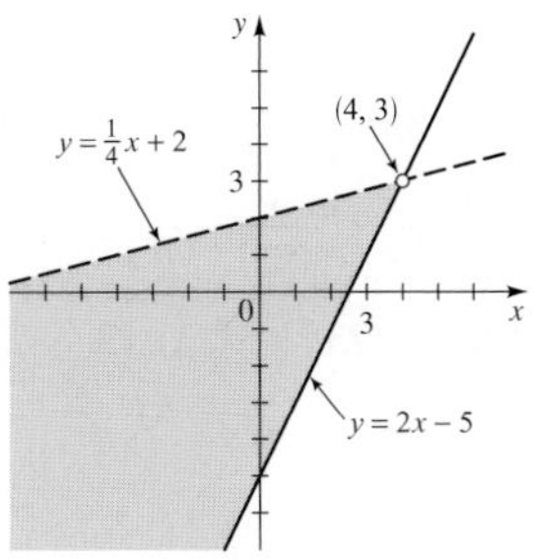

Not bounded

25.

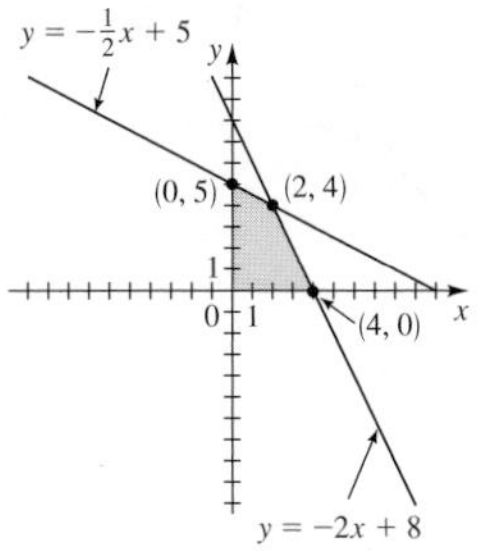

Bounded

27.

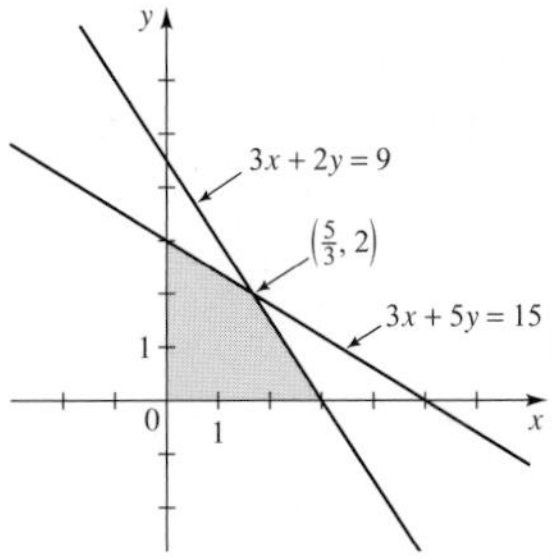

Bounded

29.

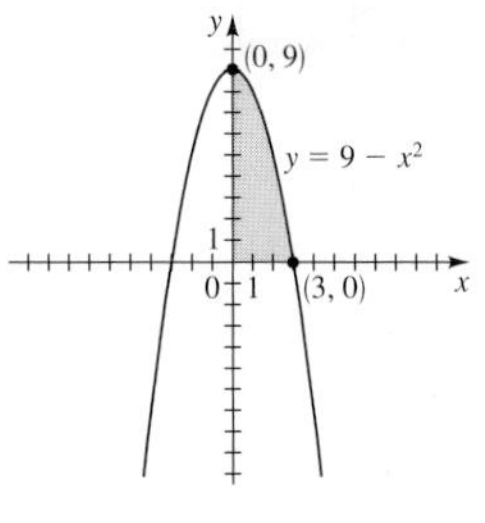

Bounded

31.

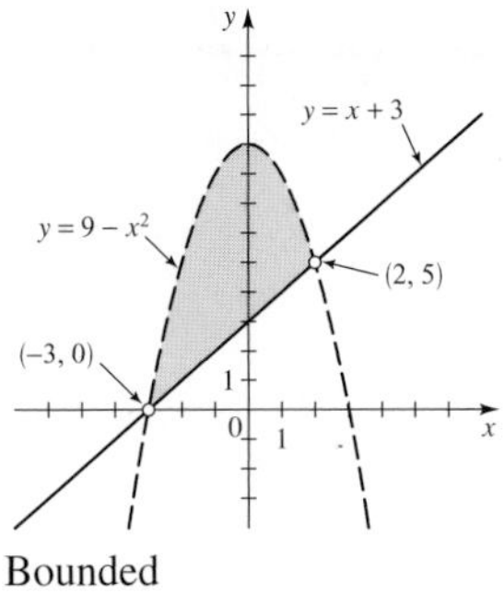

Bounded

33.

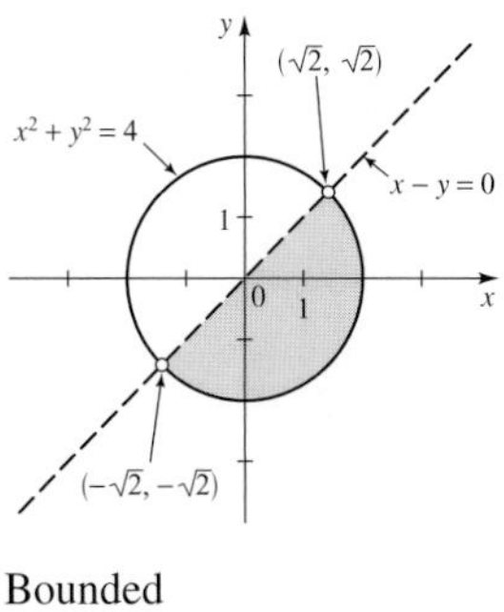

Bounded

35.

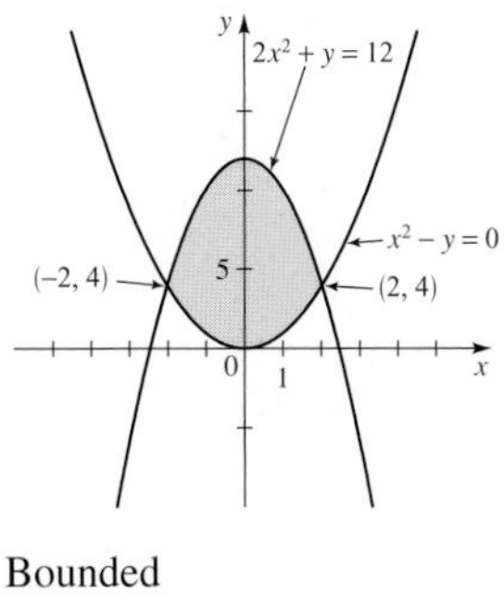

Bounded

37.

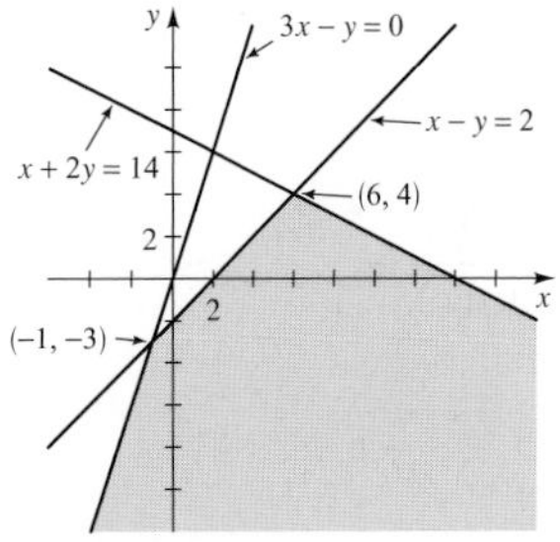

Not bounded

39.

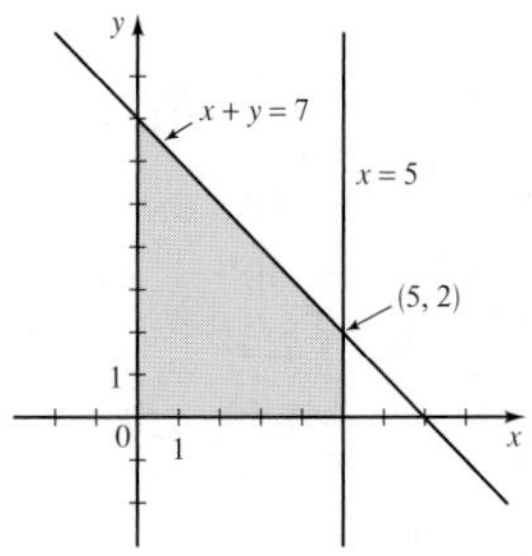

Bounded

41.

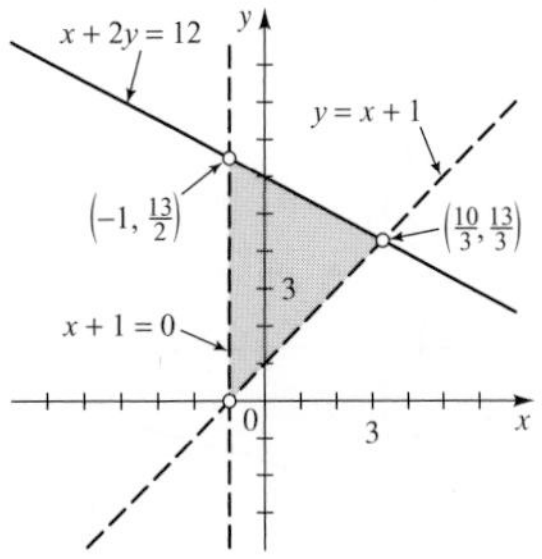

Bounded

43.

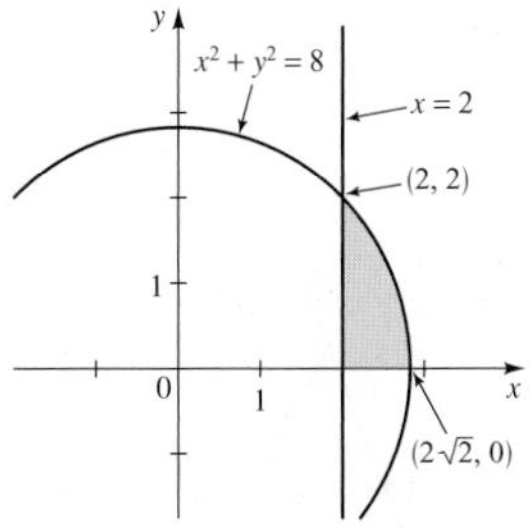

Bounded

45.

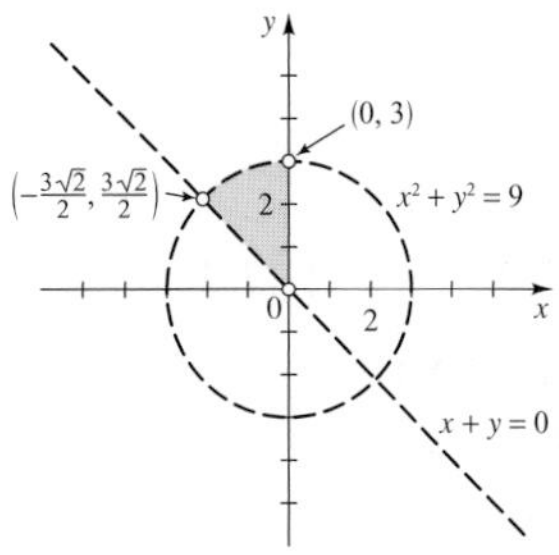

Bounded

47.

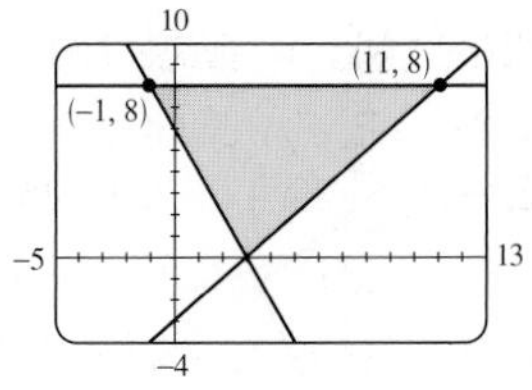

49.

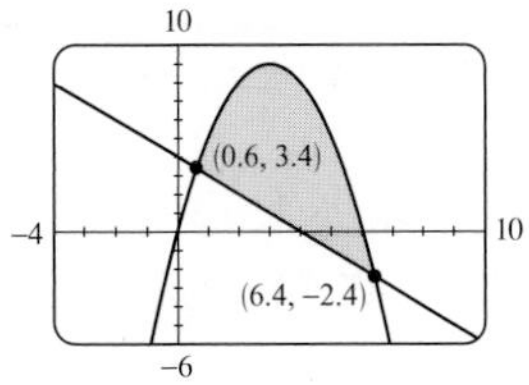

51. x = number of fiction books
y = number of nonfiction books

$$\begin{cases} x + y \leq 100 \\ 20 \leq y, \quad x \geq y \\ x \geq 0, \quad y \geq 0 \end{cases}$$

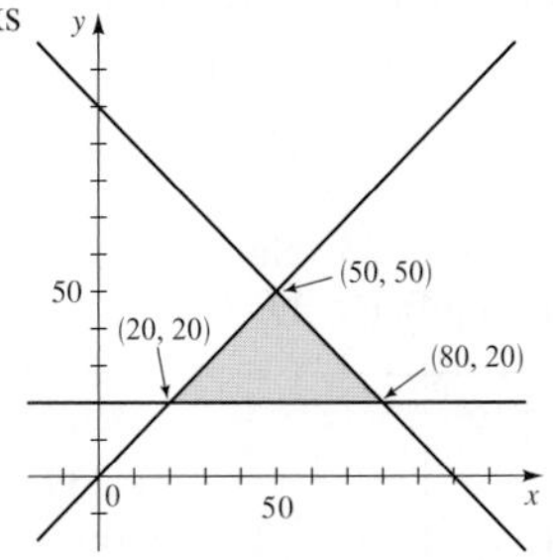

53. x = number of standard packages
y = number of deluxe packages

$$\begin{cases} \frac{1}{4}x + \frac{5}{8}y \le 80 \\ \frac{3}{4}x + \frac{3}{8}y \le 90 \\ x \ge 0, \quad y \ge 0 \end{cases}$$

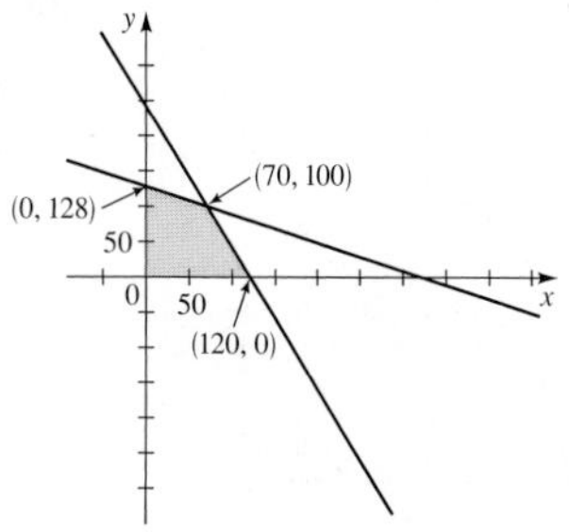

CHAPTER 10 REVIEW ■ PAGE 711

1. $(2, 1)$

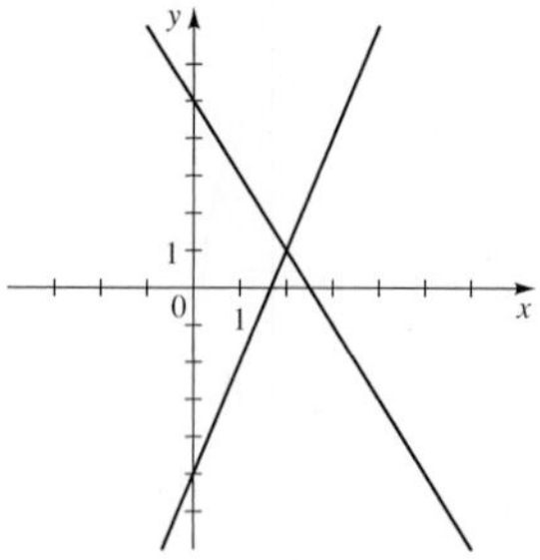

3. x = any number
$y = \frac{2}{7}x - 4$

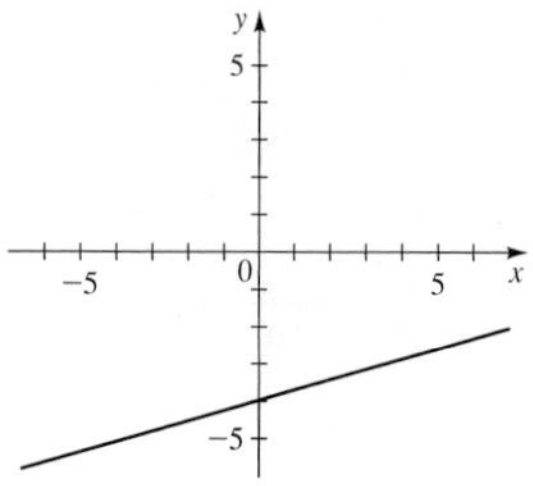

5. No solution

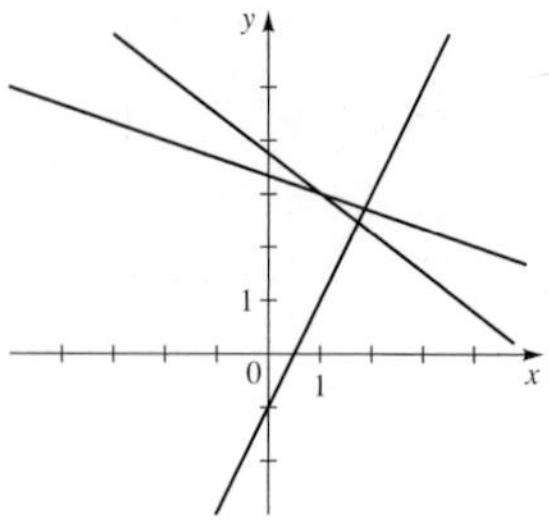

7. $(-3, 3), (2, 8)$ **9.** $\left(\frac{16}{7}, -\frac{14}{3}\right)$ **11.** $(21.41, -15.93)$
13. $(11.94, -1.39), (12.07, 1.44)$
15. **(a)** 2×3 **(b)** Yes **(c)** No
(d) $\begin{cases} x + 2y = -5 \\ \quad\;\; y = 3 \end{cases}$
17. **(a)** 3×4 **(b)** Yes **(c)** Yes
(d) $\begin{cases} x \quad + 8z = 0 \\ \quad y + 5z = -1 \\ \qquad\quad 0 = 0 \end{cases}$
19. **(a)** 3×4 **(b)** No **(c)** No
(d) $\begin{cases} \quad\; y - 3z = 4 \\ x + y \qquad = 7 \\ x + 2y + z = 2 \end{cases}$
21. $(1, 1, 2)$ **23.** No solution **25.** $(-8, -7, 10)$
27. No solution **29.** $(1, 0, 1, -2)$
31. $x = -4t + 1, y = -t - 1, z = t$
33. $x = 6 - 5t, y = \frac{1}{2}(7 - 3t), z = t$ **35.** $\left(-\frac{4}{3}t + \frac{4}{3}, \frac{5}{3}t - \frac{2}{3}, t\right)$
37. $(s + 1, 2s - t + 1, s, t)$ **39.** No solution
41. $(1, t + 1, t, 0)$ **43.** \$3000 at 6%, \$6000 at 7%
45. \$11,250 in bank A, \$22,500 in bank B, \$26,250 in bank C

47. Impossible **49.** $\begin{bmatrix} 4 & 18 \\ 4 & 0 \\ 2 & 2 \end{bmatrix}$

51. $\begin{bmatrix} 10 & 0 & -5 \end{bmatrix}$ **53.** $\begin{bmatrix} -\frac{7}{2} & 10 \\ 1 & -\frac{9}{2} \end{bmatrix}$ **55.** $\begin{bmatrix} 30 & 22 & 2 \\ -9 & 1 & -4 \end{bmatrix}$

57. $\begin{bmatrix} -\frac{1}{2} & \frac{11}{2} \\ \frac{15}{4} & -\frac{3}{2} \\ -\frac{1}{2} & 1 \end{bmatrix}$ **61.** $\frac{1}{3}\begin{bmatrix} -1 & -3 \\ -5 & 2 \end{bmatrix}$ **63.** $\begin{bmatrix} \frac{7}{2} & -2 \\ 0 & 8 \end{bmatrix}$

65. $\begin{bmatrix} 2 & -2 & 6 \\ -4 & 5 & -9 \end{bmatrix}$ **67.** $1, \begin{bmatrix} 9 & -4 \\ -2 & 1 \end{bmatrix}$ **69.** 0, no inverse

71. $-1, \begin{bmatrix} 3 & 2 & -3 \\ 2 & 1 & -2 \\ -8 & -6 & 9 \end{bmatrix}$ **73.** $24, \begin{bmatrix} 1 & 0 & 0 & -\frac{1}{4} \\ 0 & \frac{1}{2} & 0 & -\frac{1}{4} \\ 0 & 0 & \frac{1}{3} & -\frac{1}{4} \\ 0 & 0 & 0 & \frac{1}{4} \end{bmatrix}$

75. $(65, 154)$ **77.** $\left(-\frac{1}{12}, \frac{1}{12}, \frac{1}{12}\right)$ **79.** $\left(\frac{1}{5}, \frac{9}{5}\right)$ **81.** $\left(-\frac{87}{26}, \frac{21}{26}, \frac{3}{2}\right)$
83. 11 **85.** $\dfrac{2}{x - 5} + \dfrac{1}{x + 3}$ **87.** $\dfrac{-4}{x} + \dfrac{4}{x - 1} + \dfrac{-2}{(x - 1)^2}$
89. $\dfrac{-1}{x} + \dfrac{x + 2}{x^2 + 1}$
91. $(2, 1)$ **93.** $\left(-\frac{1}{2}, \frac{7}{4}\right), (2, -2)$ **95.** $x + y^2 \le 4$

97.

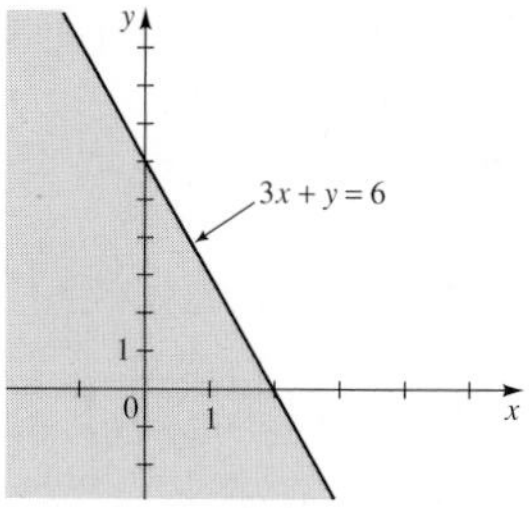

99.

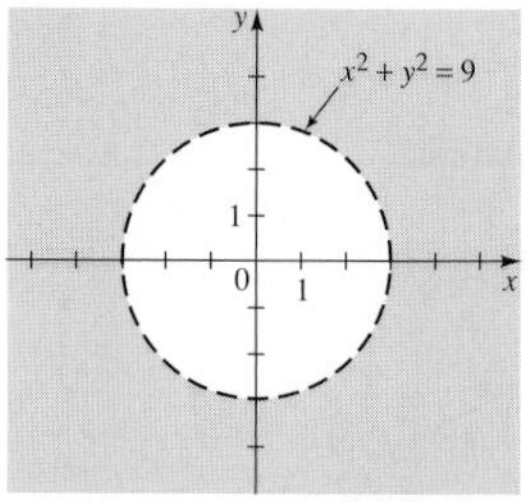

101.

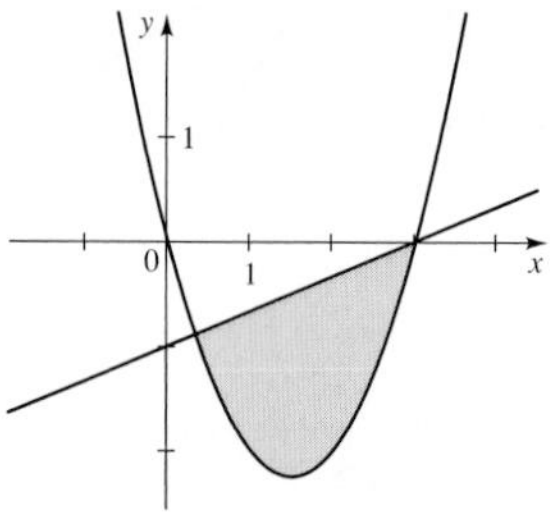

103.

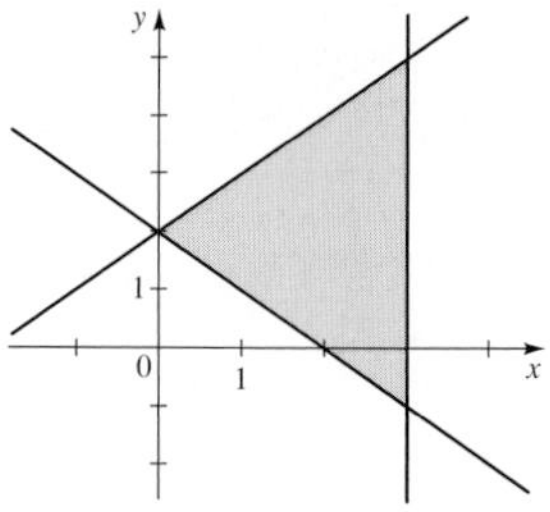

105.

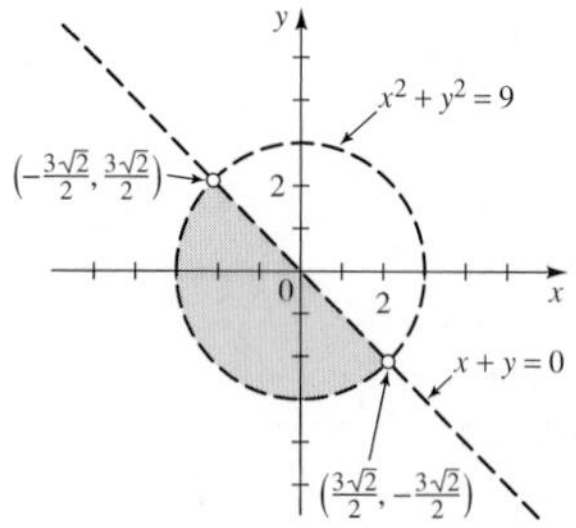

Bounded

107.

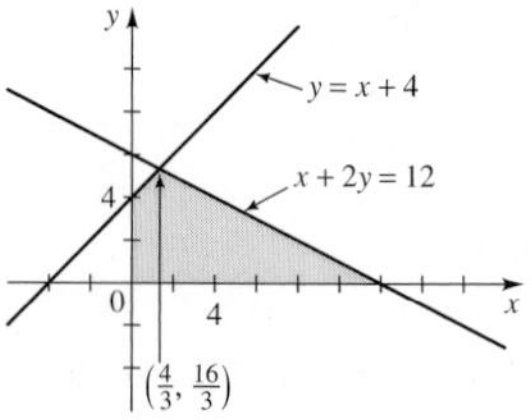

Bounded

109. $x = \dfrac{b + c}{2}, y = \dfrac{a + c}{2}, z = \dfrac{a + b}{2}$ **111.** 2, 3

CHAPTER 10 TEST ▪ PAGE 714

1. (a) Linear **(b)** $(-2, 3)$ **2. (a)** Nonlinear **(b)** $(1, -2), \left(\frac{5}{3}, 0\right)$
3. $(-0.55, -0.78), (0.43, -0.29), (2.12, 0.56)$
4. Wind 60 km/h, airplane 300 km/h
5. (a) Row-echelon form **(b)** Reduced row-echelon form **(c)** Neither **6. (a)** $\left(\frac{5}{2}, \frac{5}{2}, 0\right)$ **(b)** No solution
7. $\left(-\frac{3}{5} + \frac{2}{5}t, \frac{1}{5} + \frac{1}{5}t, t\right)$
8. Coffee \$1.50, juice \$1.75, donut \$0.75
9. (a) Incompatible dimensions **(b)** Incompatible dimensions
(c) $\begin{bmatrix} 6 & 10 \\ 3 & -2 \\ -3 & 9 \end{bmatrix}$ **(d)** $\begin{bmatrix} 36 & 58 \\ 0 & -3 \\ 18 & 28 \end{bmatrix}$ **(e)** $\begin{bmatrix} 2 & -\frac{3}{2} \\ -1 & 1 \end{bmatrix}$
(f) B is not square **(g)** B is not square **(h)** -3

10. (a) $\begin{bmatrix} 4 & -3 \\ 3 & -2 \end{bmatrix}\begin{bmatrix} x \\ y \end{bmatrix} = \begin{bmatrix} 10 \\ 30 \end{bmatrix}$ **(b)** $(70, 90)$

11. $|A| = 0, |B| = 2, B^{-1} = \begin{bmatrix} 1 & -2 & 0 \\ 0 & \frac{1}{2} & 0 \\ 3 & -6 & 1 \end{bmatrix}$

12. $(5, -5, -4)$

13. (a) $\dfrac{1}{x-1} + \dfrac{1}{(x-1)^2} - \dfrac{1}{x+2}$ **(b)** $-\dfrac{1}{x} + \dfrac{x+2}{x^2+3}$

14. (a)
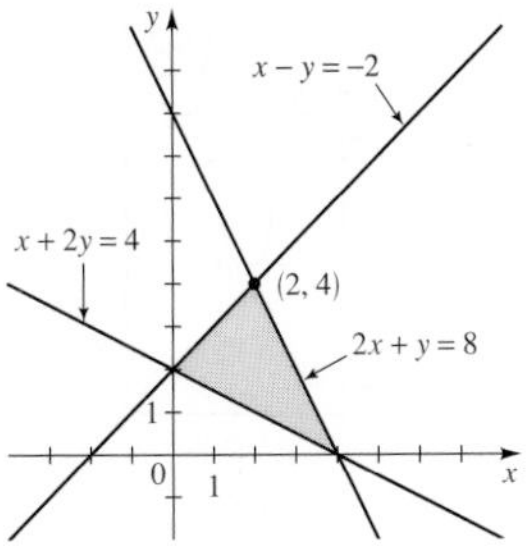

(b)
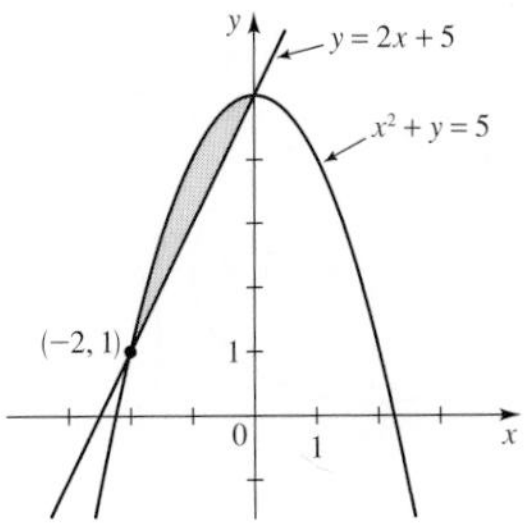

FOCUS ON MODELING ▪ PAGE 720

1. 198, 195
3.
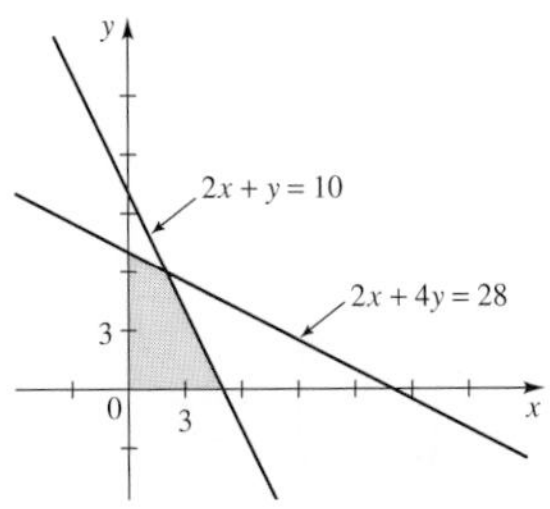

maximum 161
minimum 135

5. 3 tables, 34 chairs **7.** 30 grapefruit crates, 30 orange crates **9.** 15 Pasadena to Santa Monica, 3 Pasadena to El Toro, 0 Long Beach to Santa Monica, 16 Long Beach to El Toro
11. 90 standard, 40 deluxe **13.** \$7500 in municipal bonds, \$2500 in bank certificates, \$2000 in high-risk bonds
15. 4 games, 32 educational, 0 utility

CHAPTER 11

SECTION 11.1 ▪ PAGE 730

1. focus, directrix **2.** $F(0, p), y = -p, F(0, 3), y = -3$
3. $F(p, 0), x = -p, F(3, 0), x = -3$

4. (a)
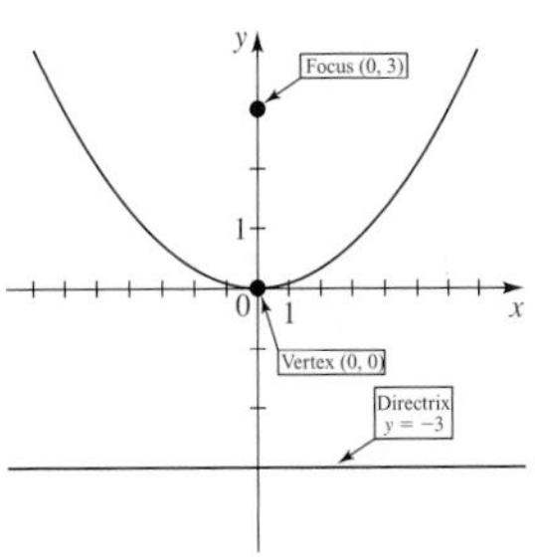

(b)
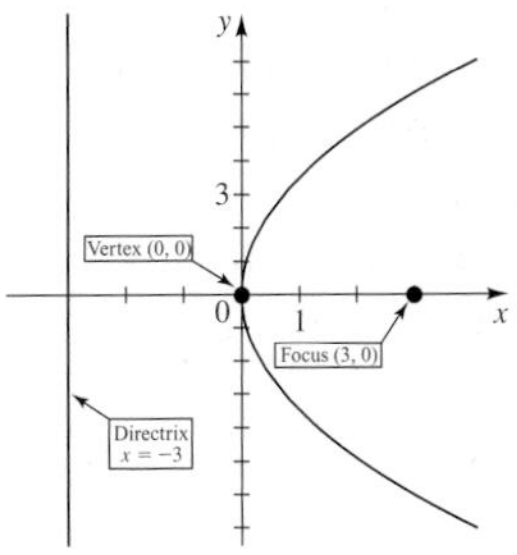

5. III **7.** II **9.** VI

Order of answers: focus; directrix; focal diameter

11. $F\left(0, \frac{9}{4}\right); y = -\frac{9}{4}; 9$
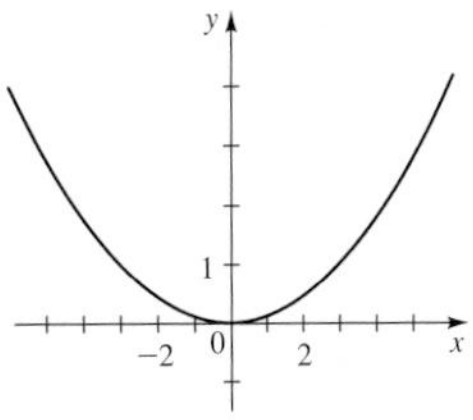

13. $F(1, 0); x = -1; 4$
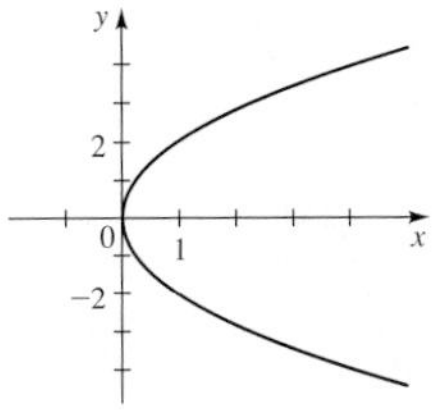

15. $F\left(0, \frac{1}{20}\right); y = -\frac{1}{20}; \frac{1}{5}$
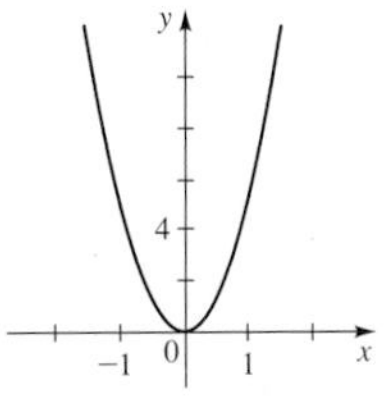

17. $F\left(-\frac{1}{32}, 0\right); x = \frac{1}{32}; \frac{1}{8}$
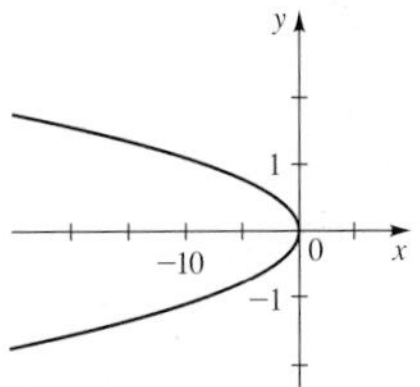

19. $F\left(0, -\frac{3}{2}\right); y = \frac{3}{2}; 6$
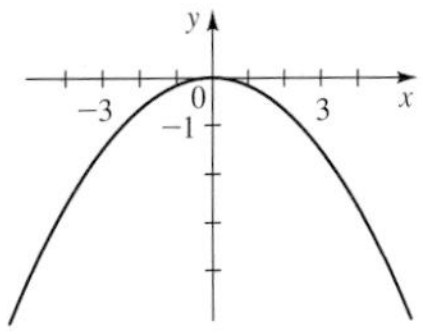

21. $F\left(-\frac{5}{12}, 0\right); x = \frac{5}{12}; \frac{5}{3}$
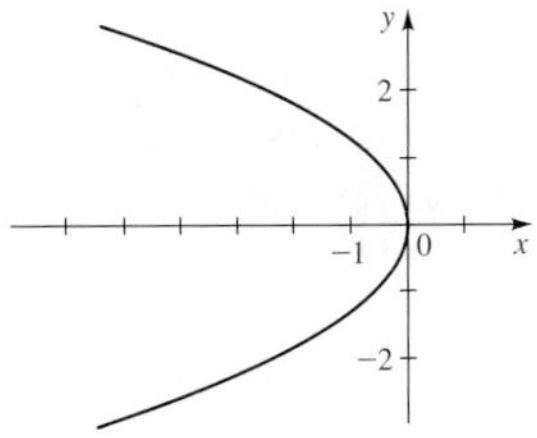

23.
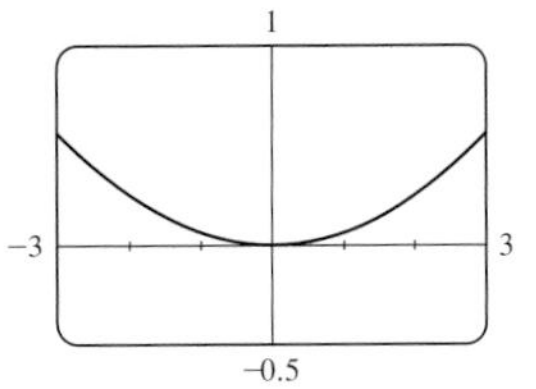

25.
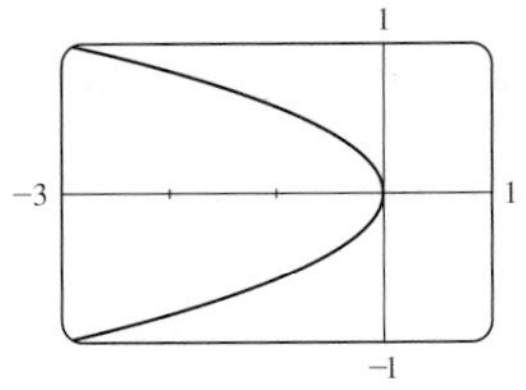

27.
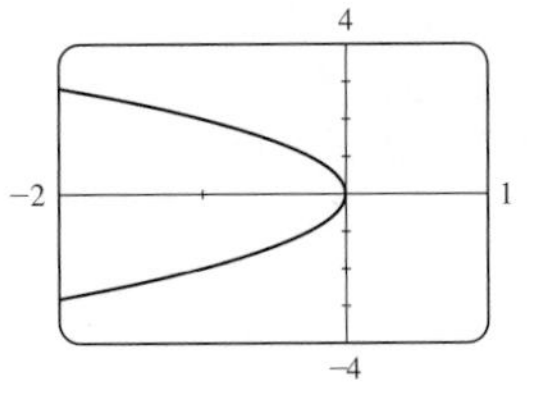

29. $x^2 = 8y$ **31.** $y^2 = -32x$ **33.** $y^2 = -8x$ **35.** $x^2 = 40y$ **37.** $y^2 = 4x$ **39.** $x^2 = 20y$ **41.** $x^2 = 8y$ **43.** $y^2 = -16x$ **45.** $y^2 = -3x$ **47.** $x = y^2$ **49.** $x^2 = -4\sqrt{2}y$

51. **(a)** $x^2 = -4py$, $p = \frac{1}{2}$, 1, 4, and 8

(b) The closer the directrix to the vertex, the steeper the parabola.

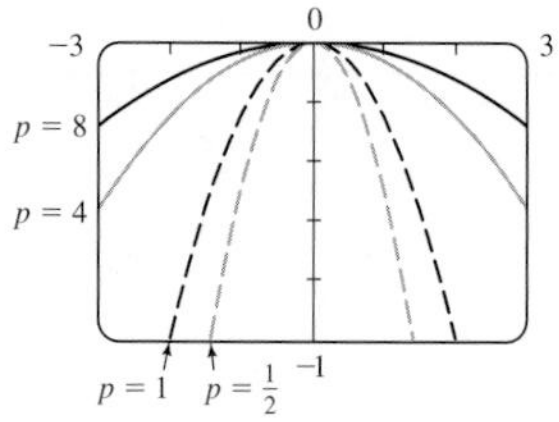

53. **(a)** $y^2 = 12x$ **(b)** $8\sqrt{15} \approx 31$ cm **55.** $x^2 = 600y$

SECTION 11.2 ■ PAGE 738

1. sum; foci

2. $(a, 0), (-a, 0)$; $c = \sqrt{a^2 - b^2}$; $(5, 0), (-5, 0), (3, 0), (-3, 0)$

3. $(0, a), (0, -a)$; $c = \sqrt{a^2 - b^2}$; $(0, 5), (0, -5), (0, 3), (0, -3)$

4. **(a)**

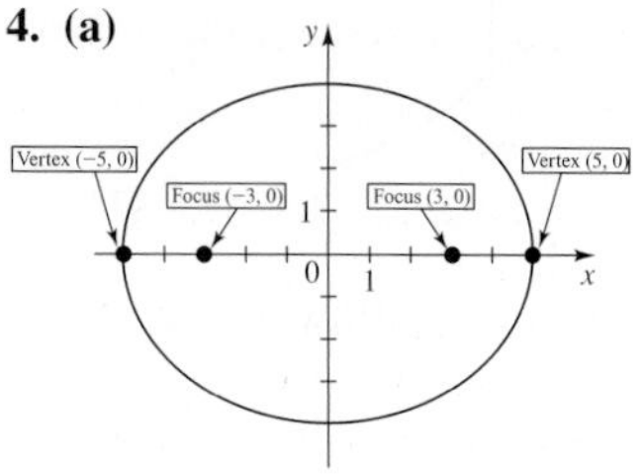

(b)

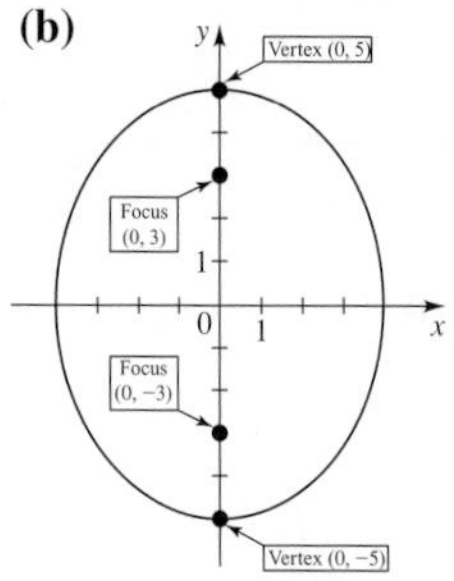

5. II **7.** I

Order of answers: vertices; foci; eccentricity; major axis and minor axis

9. $V(\pm 5, 0)$; $F(\pm 4, 0)$; $\frac{4}{5}$; 10, 6

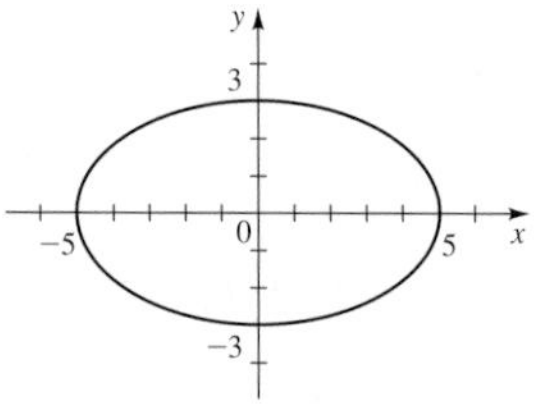

11. $V(0, \pm 3)$; $F(0, \pm\sqrt{5})$; $\sqrt{5}/3$; 6, 4

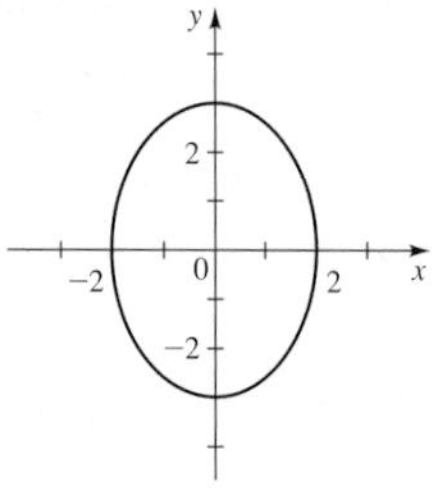

13. $V(\pm 4, 0)$; $F(\pm 2\sqrt{3}, 0)$; $\sqrt{3}/2$; 8, 4

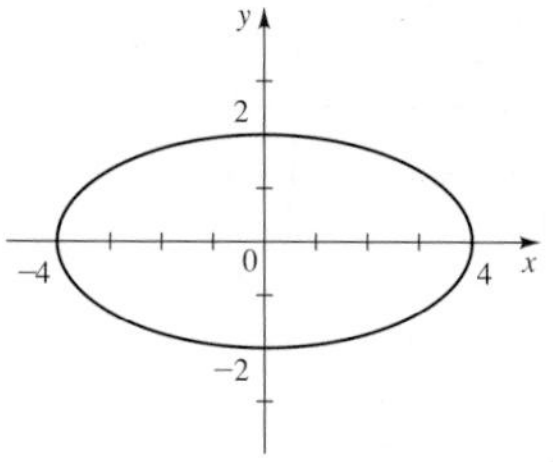

15. $V(0, \pm\sqrt{3})$; $F(0, \pm\sqrt{3/2})$; $1/\sqrt{2}$; $2\sqrt{3}$, $\sqrt{6}$

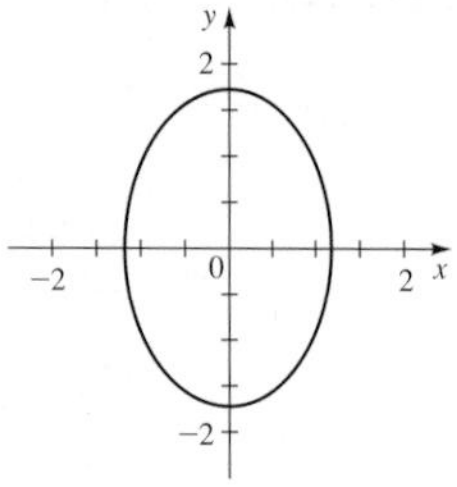

17. $V(\pm 1, 0)$; $F(\pm\sqrt{3}/2, 0)$; $\sqrt{3}/2$; 2, 1

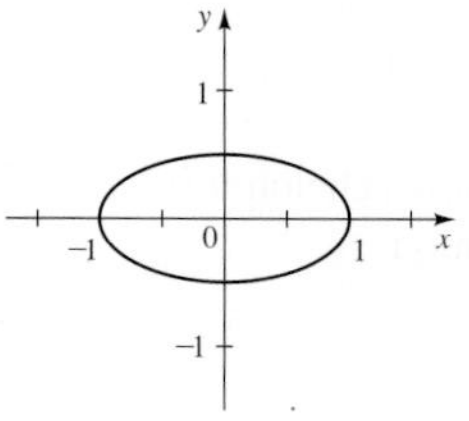

19. $V(0, \pm\sqrt{2})$; $F(0, \pm\sqrt{3/2})$; $\sqrt{3}/2$; $2\sqrt{2}$, $\sqrt{2}$

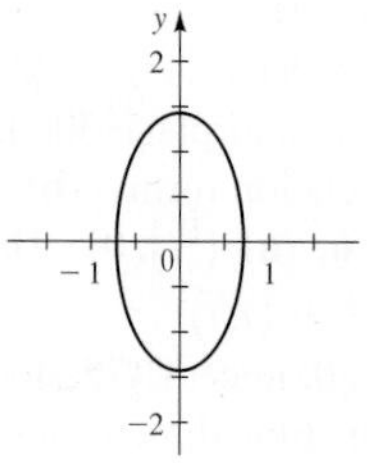

21. $V(0, \pm 1)$; $F(0, \pm 1/\sqrt{2})$; $1/\sqrt{2}$; 2, $\sqrt{2}$

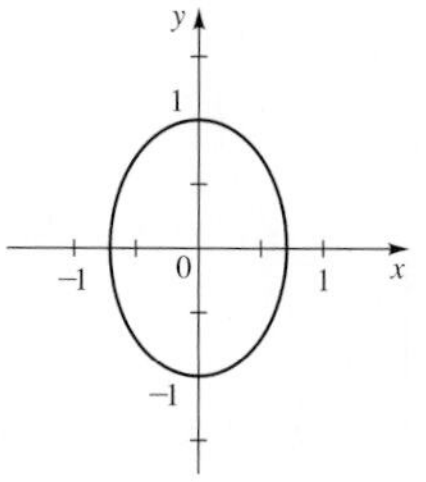

23. $\dfrac{x^2}{25} + \dfrac{y^2}{16} = 1$ **25.** $\dfrac{x^2}{4} + \dfrac{y^2}{8} = 1$ **27.** $\dfrac{x^2}{256} + \dfrac{y^2}{48} = 1$

29.

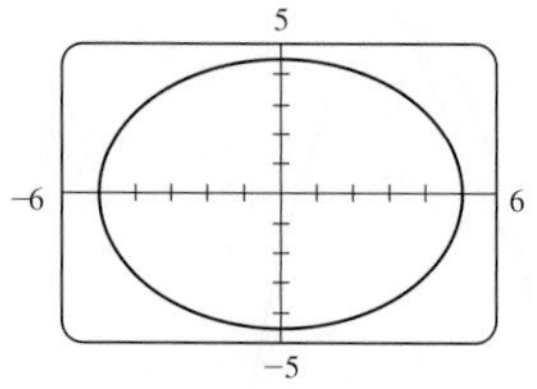

31.

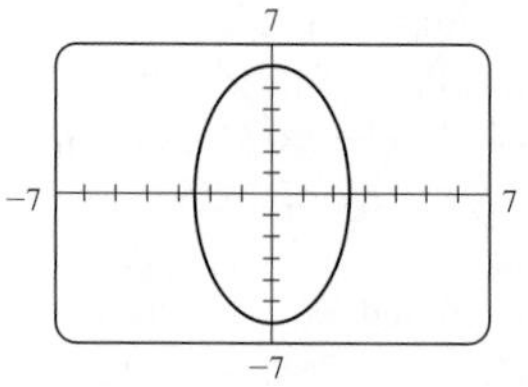

33. $\dfrac{x^2}{25} + \dfrac{y^2}{9} = 1$ **35.** $x^2 + \dfrac{y^2}{4} = 1$ **37.** $\dfrac{x^2}{9} + \dfrac{y^2}{13} = 1$

39. $\dfrac{x^2}{100} + \dfrac{y^2}{91} = 1$ **41.** $\dfrac{x^2}{25} + \dfrac{y^2}{5} = 1$ **43.** $\dfrac{64x^2}{225} + \dfrac{64y^2}{81} = 1$

45. $(0, \pm 2)$

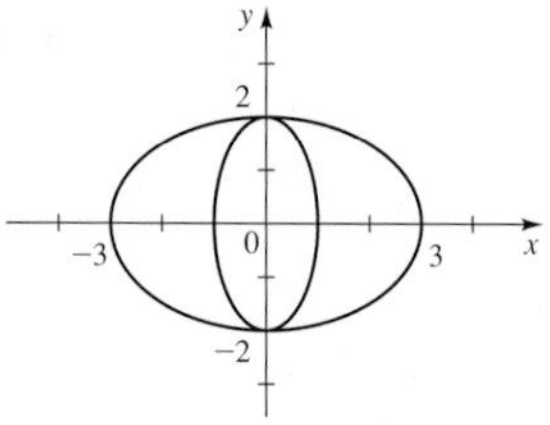

47. $(\pm 1, 0)$

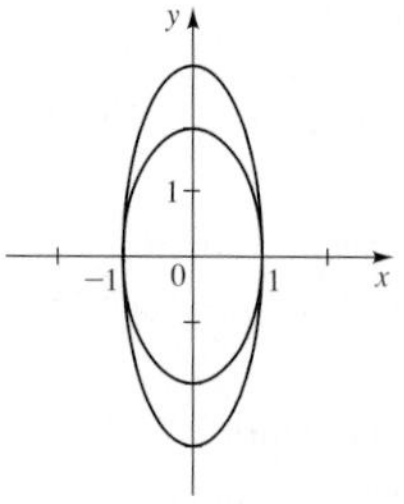

49. **(a)**

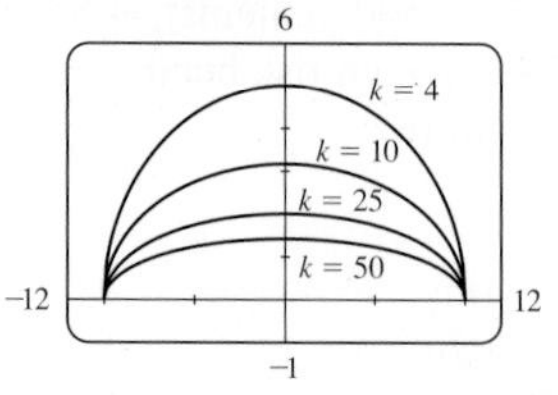

(b) Common major axes and vertices; eccentricity increases as k increases.

INDEX

SEQUENCES AND SERIES

Arithmetic

$a, a + d, a + 2d, a + 3d, a + 4d, \ldots$

$a_n = a + (n - 1)d$

$$S_n = \sum_{k=1}^{n} a_k = \frac{n}{2}[2a + (n - 1)d] = n\left(\frac{a + a_n}{2}\right)$$

Geometric

$a, ar, ar^2, ar^3, ar^4, \ldots$

$a_n = ar^{n-1}$

$$S_n = \sum_{k=1}^{n} a_k = a\frac{1 - r^n}{1 - r}$$

If $|r| < 1$, then the sum of an infinite geometric series is

$$S = \frac{a}{1 - r}$$

THE BINOMIAL THEOREM

$$(a + b)^n = \binom{n}{0}a^n + \binom{n}{1}a^{n-1}b + \cdots + \binom{n}{n-1}ab^{n-1} + \binom{n}{n}b^n$$

FINANCE

Compound interest

$$A = P\left(1 + \frac{r}{n}\right)^{nt}$$

where A is the amount after t years, P is the principal, r is the interest rate, and the interest is compounded n times per year.

Amount of an annuity

$$A_f = R\frac{(1 + i)^n - 1}{i}$$

where A_f is the final amount, i is the interest rate per time period, and there are n payments of size R.

Present value of an annuity

$$A_p = R\frac{1 - (1 + i)^{-n}}{i}$$

where A_p is the present value, i is the interest rate per time period, and there are n payments of size R.

Installment buying

$$R = \frac{iA_p}{1 - (1 + i)^{-n}}$$

where R is the size of each payment, i is the interest rate per time period, A_p is the amount of the loan, and n is the number of payments.

CONIC SECTIONS

Circles

$(x - h)^2 + (y - k)^2 = r^2$

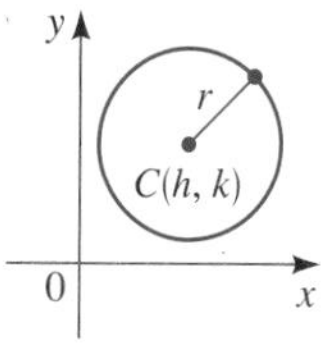

Parabolas

$x^2 = 4py$

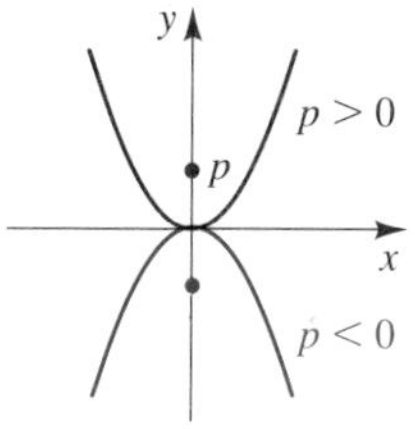

Focus $(0, p)$, directrix $y = -p$

$y^2 = 4px$

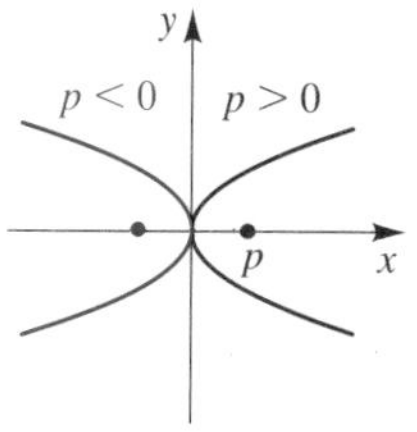

Focus $(p, 0)$, directrix $x = -p$

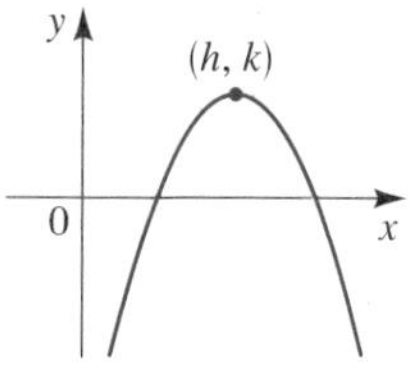

$y = a(x - h)^2 + k$,
$a < 0,\quad h > 0,\quad k > 0$

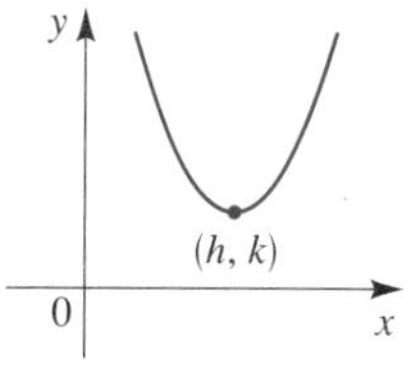

$y = a(x - h)^2 + k$,
$a > 0,\quad h > 0,\quad k > 0$

Ellipses

$$\frac{x^2}{a^2} + \frac{y^2}{b^2} = 1$$

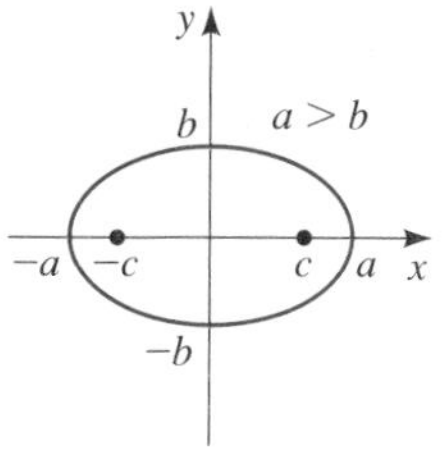

Foci $(\pm c, 0)$, $c^2 = a^2 - b^2$

$$\frac{x^2}{b^2} + \frac{y^2}{a^2} = 1$$

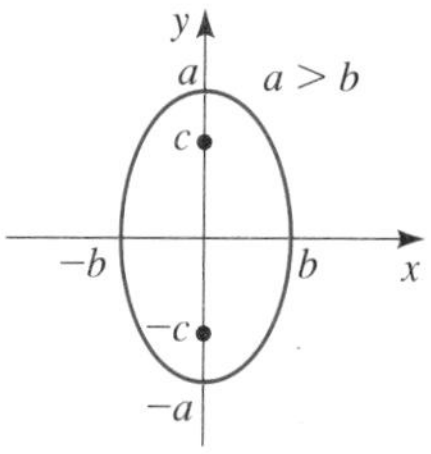

Foci $(0, \pm c)$, $c^2 = a^2 - b^2$

Hyperbolas

$$\frac{x^2}{a^2} - \frac{y^2}{b^2} = 1$$

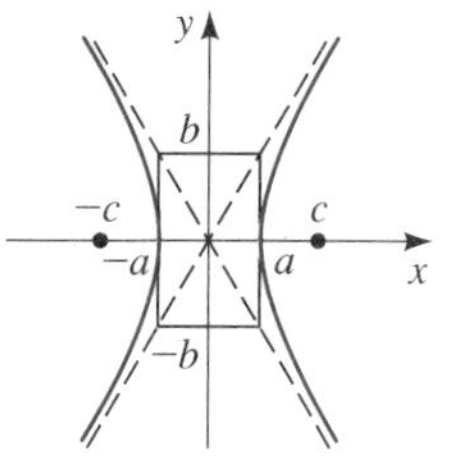

Foci $(\pm c, 0)$, $c^2 = a^2 + b^2$

$$-\frac{x^2}{b^2} + \frac{y^2}{a^2} = 1$$

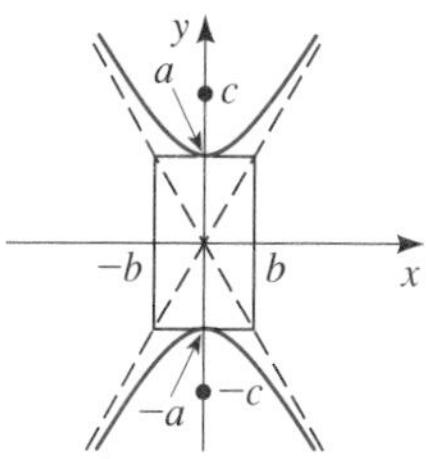

Foci $(0, \pm c)$, $c^2 = a^2 + b^2$

MENT

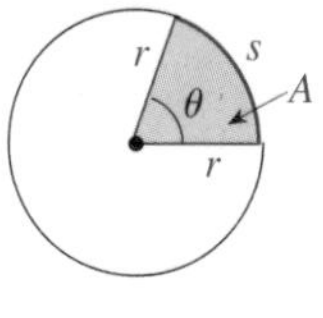

$$1 \text{ rad} = \frac{180°}{\pi}$$

$$A = \tfrac{1}{2}r^2\theta \quad (\theta \text{ in radians})$$

ert from degrees to radians, multiply by $\dfrac{\pi}{180}$.

convert from radians to degrees, multiply by $\dfrac{180}{\pi}$.

TRIGONOMETRIC FUNCTIONS OF REAL NUMBERS

$\sin t = y$ $\qquad \csc t = \dfrac{1}{y}$

$\cos t = x$ $\qquad \sec t = \dfrac{1}{x}$

$\tan t = \dfrac{y}{x}$ $\qquad \cot t = \dfrac{x}{y}$

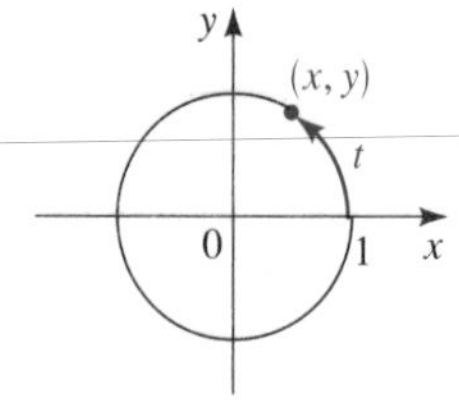

TRIGONOMETRIC FUNCTIONS OF ANGLES

$\sin \theta = \dfrac{y}{r}$ $\qquad \csc \theta = \dfrac{r}{y}$

$\cos \theta = \dfrac{x}{r}$ $\qquad \sec \theta = \dfrac{r}{x}$

$\tan \theta = \dfrac{y}{x}$ $\qquad \cot \theta = \dfrac{x}{y}$

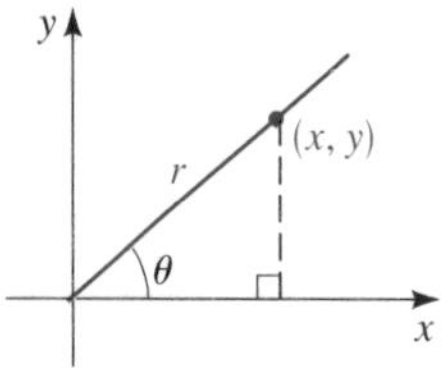

RIGHT ANGLE TRIGONOMETRY

$\sin \theta = \dfrac{\text{opp}}{\text{hyp}}$ $\qquad \csc \theta = \dfrac{\text{hyp}}{\text{opp}}$

$\cos \theta = \dfrac{\text{adj}}{\text{hyp}}$ $\qquad \sec \theta = \dfrac{\text{hyp}}{\text{adj}}$

$\tan \theta = \dfrac{\text{opp}}{\text{adj}}$ $\qquad \cot \theta = \dfrac{\text{adj}}{\text{opp}}$

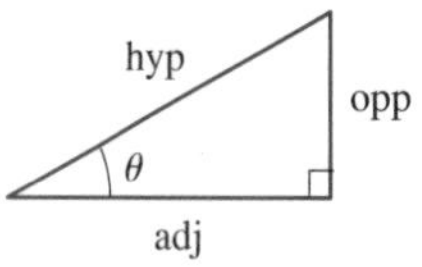

SPECIAL VALUES OF THE TRIGONOMETRIC FUNCTIONS

θ	radians	$\sin \theta$	$\cos \theta$	$\tan \theta$
0°	0	0	1	0
30°	$\pi/6$	$1/2$	$\sqrt{3}/2$	$\sqrt{3}/3$
45°	$\pi/4$	$\sqrt{2}/2$	$\sqrt{2}/2$	1
60°	$\pi/3$	$\sqrt{3}/2$	$1/2$	$\sqrt{3}$
90°	$\pi/2$	1	0	—
180°	π	0	-1	0
270°	$3\pi/2$	-1	0	—

SPECIAL TRIANGLES

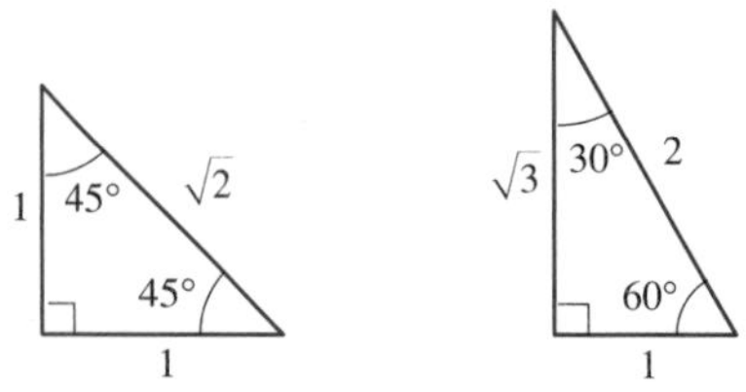

GRAPHS OF THE TRIGONOMETRIC FUNCTIONS

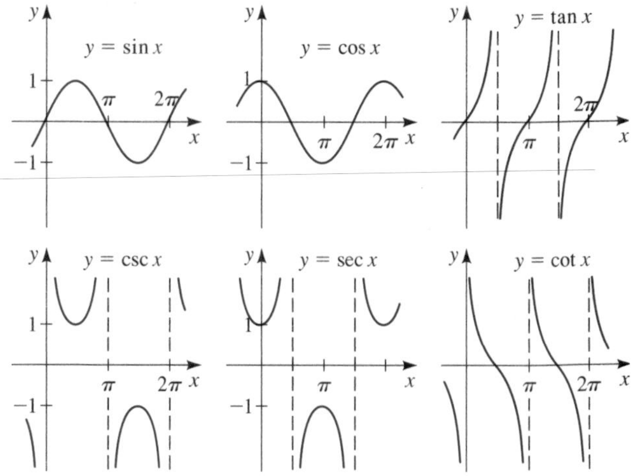

SINE AND COSINE CURVES

$y = a \sin k(x - b) \quad (k > 0)$ $\qquad y = a \cos k(x - b) \quad (k > 0)$

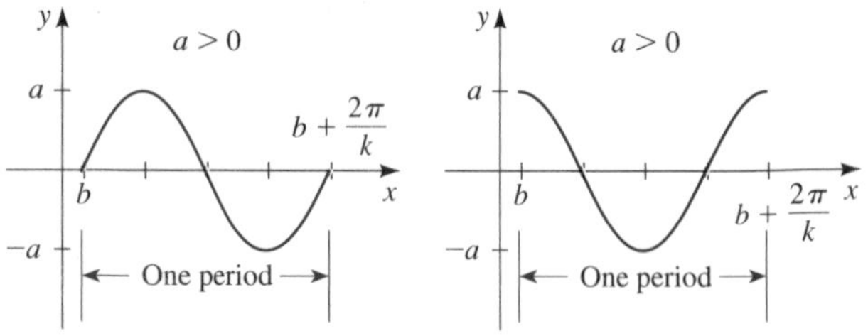

amplitude: $|a|$ $\qquad$ period: $2\pi/k$ $\qquad$ phase shift: b

GRAPHS OF THE INVERSE TRIGONOMETRIC FUNCTIONS

$y = \sin^{-1}x$ $\qquad y = \cos^{-1}x$ $\qquad y = \tan^{-1}x$

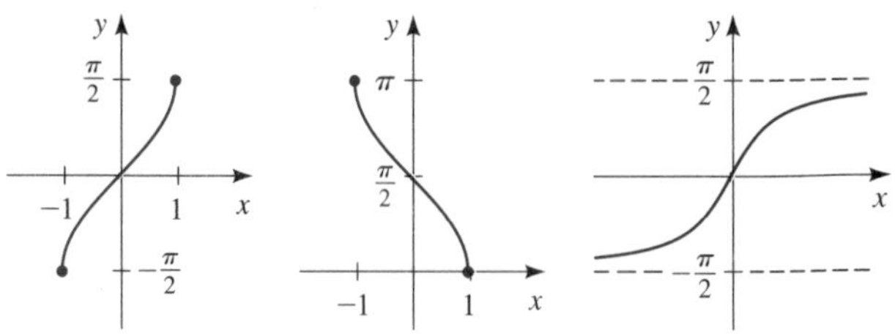